Satya Prakash's

MODERN INORGANIC CHEMISTRY

(For Students of B.Sc., M.Sc. and also Useful for Competitive Examinations)

Dr. R.D. Madan
M.Sc., Ph.D.
Former Reader
Department of Chemistry
N.R.E.C. College
Khurja, Uttar Pradesh

S. CHAND
PUBLISHING

S. Chand And Company Limited
(ISO 9001 Certified Company)

S. Chand And Company Limited
(ISO 9001 Certified Company)
Head Office: D-92, Sector–2, Noida – 201301, U.P. (India), Ph. 91-120-4682700
Registered Office: A-27, 2nd Floor, Mohan Co-operative Industrial Estate, New Delhi – 110 044, Phone: 011-49731800
www.**schandpublishing.com**; e-mail: **info@schandpublishing.com**

Marketing Offices:

Chennai	:	Ph: 23632120; chennai@schandpublishing.com
Guwahati	:	Ph: 2738811, 2735640; guwahati@schandpublishing.com
Hyderabad	:	Ph: 40186018; hyderabad@schandpublishing.com
Jalandhar	:	Ph: 4645630; jalandhar@schandpublishing.com
Kolkata	:	Ph: 23357458, 23353914; kolkata@schandpublishing.com
Lucknow	:	Ph: 4003633; lucknow@schandpublishing.com
Mumbai	:	Ph: 25000297; mumbai@schandpublishing.com
Patna	:	Ph: 2260011; patna@schandpublishing.com

First Edition 1987
Subsequent Editions and Reprints 1990, 92, 94, 96, 97, 98, 99, 2000, 2001, 2002, 2003, 2005, 2006, 2008, 2009, Third Revised Edition 2011, Reprints 2012, 2013, 2014, 2016 (Twice), 2017, 2018, 2019

Low Priced Students' Paperback Edition (LPSPE), 2019; Reprint 2022

Reprint 2023

ISBN: 978-93-528-3729-8 **Product Code:** H6MIC68CHEM10ENAC19L

PRINTED IN INDIA

By Vikas Publishing House Private Limited, Plot 20/4, Site-IV, Industrial Area Sahibabad, Ghaziabad – 201 010 and Published by S. Chand And Company Limited, A-27, 2nd Floor, Mohan Co-operative Industrial Estate, New Delhi – 110 044.

PREFACE TO THE THIRD EDITION

In the present edition of this book I have retained all the text matter that was judged good in the previous edition.

Main features of this new edition are :

(*i*) Emphasis has been put on the discussion of the correction between the structure and properties of elements/compounds.

(*ii*) At the end of each chapter some typical conceptual questions with their answers/solutions have been given. I strongly believe that these questions will be beneficial to the students during their examination period. These questions have been chosen with a view that they cover every important concept, reaction equation, mechanism and principle. Answers to these questions are clear, concise and correct. I hope that by doing these questions the students will develop confidence in their abilities in Inorganic Chemistry.

(*iii*) At the end of each chapter latest questions asked in various Indian Universities' examinations have been incorporated. By solving these questions the students can precisely know their own success level.

(*iv*) Subject index has been provided at the end of the book. This can be used for quick cross-reference and detailed contents.

Acknowledgments

I am extremely grateful to all my students and fellow-teachers all over the country for their support.

I wish to put on record my sincere thanks to Mrs. Nirmala Gupta (CMD), Mr. Amit Gupta (CEO), Mr. Navin Joshi, Vice President (Publishing) and Mr. Shishir Bhatnagar (Manager Editorial and Pre-Press) for the encouragement and guidance that I have received from them during the publication of the book.

Last but not the least I express my sincere thanks to Mr. Bhuvnshwar Kumar Tyagi, for all type of support in the revision of this book.

I would be glad to receive suggestions and comments for further improvement of the book.

Dr. R.D. MADAN
37, Bhagirathi Appartments,
Sector – 9, Rohini
Delhi - 110 085
Ph. 011-27860313(R)

PREFACE TO THE FIRST EDITION

Originally this book was written by famous scientist Professor Satya Prakash, D.Sc. He was Professor of Chemistry at Allahabad University, Allahabad. First edition of the book published in the year 1944. In the year 1986, S. Chand & Company Ltd. requested me to revise this book. For me it was a great challenge to revise the book by Prof. Satya Prakash ji. But due to confidence of publishers on me, I have done this great job with full sincerity. By revising this book, I have tried to explain the chemistry of elements on the basis of the latest theories of chemistry. Although the book is primarily written for B.Sc. students, yet it contains many topics useful for B.Sc. (Hons.) and M.Sc. (Part I) students of most of Indian Universities. First few chapters of the book are devoted to the study of fundamentals of chemistry like structure of the atom, periodic classification of the elements, chemical bonding, radioactivity etc. The others describe the extraction, properties and uses of the elements along with the preparation, properties, uses and structure of their important compounds. Every chapter of the book is followed by a large number of descriptive and objective type questions, asked in various years of B.Sc. examination of Indian Universities.

I wish to express my sincere thanks for Shri T.N. Goyal, Sales Manager, S. Chand & Company Ltd. for his valuable suggestions and able guidance received by me whenever required, during the period of the printing of the book. He has extended his full cooperation in the publication of the book. Finally a lot of thanks are also due to my family members, Smt. Madan, Km Sapna and Master Hero, for their full cooperation and patience.

A request to the fellow teachers and students. The real judges of a book are the teachers concerned and the students for whom the book has been written. I, therefore, request my fellow teachers as well as students to send me their comments and suggestions for the further improvement of the next editions of the book.

Dr. R.D. MADAN

Burj, Gali —
Opposite Oriental Bank of Commerce
Khurja
October 2, 1986

CONTENTS

Structure/Geometry of the Following Molecules/Ions are Given on Pages Shown Against Each Species

Group 1 (IA)

H_2 – 495, 559; H_2^+ – 560; Li_2 – 560; H_2^- – 569

Group 2 (II A)

Be_2 – 560; BeF_2 – 512; $BeCl_2$ – 506

Group 13 (III A)

B_2 – 560; BF_3 – 513, 564; BCl_3 – 508; BO_3^{3-} – 514; BH_4^- – 517; BF_4^- – 452; $B_3N_3H_6$ (Borazine or borazole) – 930; BN – 930

Group 14 (IV A)

(*i*) C_2 – 561; CH_4 – 510; CO – 453, 536, 563, 568; CO_2 – 538; C_2^{2-}(Acetylide ion) – 990; C_3^{4-} (Allylide ion) – 991; CO_3^{2-} – 539; $SnCl_2$ – 515

Group 15 (V A)

(*i*) N_2 –487, 499, 561; N_2^+ – 569; NH_3 – 497, 519; NH_4^+ – 453, 518; PF_5 – 523; NF_3 – 564; $NH_2 – NH_2$ or N_2H_4 – 1063; HN_3 – 1064, 1065; N_3^- – 567, 1064, 1086;

(*ii*) N_2O – 1073, 1089, 1090; NO – 563, 1073; NO^+ (Nitrosonium ion) – 563, 1097; N_2O_3 –1074; NO_2 – 1074, 1091, 1096; NO_2^+ – 566, 1096; N_2O_4 – 1074; N_2O_5 – 1075

(*iii*) $H_2N_2O_2$ and $N_2O_2^{2-}$ – 1083; HNO_2 – 1083; NO_2^- – 566, 1083, 1096; HNO_3 – 1084; NO_3^- – 564, 1084; HNO_4 – 1084;

(*iv*) $N(CH_3)_3$ (Trimethylamine) and $N(SiH_3)_3$ (Trisilylamine) – 1084; NH_2^- (Amide ion) – 1085; $N_2H_5^+$ (Hydrazinium ion) and $N_2H_6^{2+}$ (Hydrazonium ion) – 1088

(*v*) P_2O_3 or P_4O_6 – 1105; P_2O_5 or P_4O_{10} – 1107;

(*vi*) H_3PO_2 – 1117; H_3PO_3 – 1117; H_3PO_4 – 1117, 1125; $H_4P_2O_6$ – 1118; $H_4P_2O_7$ – 1118; HPO_3 – 1118; $[(PO_3)_3]^{3-}$ – 1124; $[(PO_3)_4]^{4-}$ – 1124; $[(PO_3)_n]^{n-}$ – 1124; H_3PO_5 – 1119; $H_4P_2O_8$ – 1119;

(*vii*) PCl_3 – 1120; PCl_5 – 1121; $[PCl_4]^+$ – 1121; $[PCl_6]^-$ – 1121; $[PBr_4]^+$ – 1127; $[PI_4]^+$ – 1127; $POCl_3$ – 1121

Group 16 (VI A)

(*i*) O_2 – 487, 499, 562; O_2^- (Superoxide ion) – 562; O_2^{2-} (Peroxide ion) – 569; O_2^+ – 569; O_3 – 452, 542; S_8 – 1131; H_2O – 496, 520; H_3O^+ – 453; H_2O_2 – 451.

(*ii*) SF_4 – 525, 564; SF_6 – 529; SCl_2 – 1136; S_2Cl_2 – 1137

(*iii*) SO_2 – 452, 544, 1138; SO_3 (Gaseous) – 452, 547, 1140; SO_3 (Solid) – 549; SO_3 (Solid α, β and r forms) – 1141; SeO_2 (Solid) – 1138; O_2F_2 – 1137

(*iv*) H_2SO_4 – 1161; SO_3^{2-} – 564; SO_4^{2-} – 452, 1161; $H_2S_2O_3$ – 1165; $S_2O_3^{2-}$ – 1165; $S_2O_7^{2-}$ – 1166; $H_2S_2O_6$ – 1168; $S_2O_6^{2-}$ – 1168; $H_2S_3O_6$, $S_3O_6^{2-}$, $H_2S_4O_6$, $S_4O_6^{2-}$, $H_2S_5O_6$, $S_5O_6^{2-}$, $H_2S_6O_6$, $S_6O_6^{2-}$ – 1171, 1172; H_2SO_5, $H_2S_2O_8$ – 1174; SO_5^{2-}, $S_2O_8^{2-}$ – 1175

(*v*) $SOCl_2$ – 1176; SO_2Cl_2 – 1177

Group 17 (VII A)

(*i*) F_2 – 497, 562; HF – 495; HF_2^- – 1198

(*ii*) ICl – 522; ClF_3 – 526; IF_5 – 530; IF_7 – 533

(*iii*) I_3^+ – 1255; ICl_2^+ – 521; IF_4^+ – 526; I_3^- –528; ICl_2^- – 528; IBr_2^- – 1282 ; ICl_4^- – 1282

(*iv*) O_2F_2 – 1137; Cl_2O – 1223; ClO_2 – 1224; Cl_2O_6 – 1238; Cl_2O_7 – 1225; I_2O_4 – 1237; I_2O_5 – 1238; I_4O_9 – 1237

(*v*) ClO^- – 1227; ClO_2^- – 1234; ClO_3^- – 1234; ClO_4^- – 1236; BrO_3^- –564, 1282; $CaOCl_2$ or Ca (OCl) Cl (Breaching powder) – 1231

(*vi*) $\bar{C}N$ (Cyanide ion) – 1255; $\bar{N}C$ (Isocynide ion) – 1255; $\bar{O}CN$ (Cyanate or oxycyanide ion) – 1255; $\bar{N}CO$ (Isocyanate ion) – 1255; SCN^- (Thiocyanate ion) – 1255; $NC\bar{S}$ (Isothiocyanate ion) – 1255; $\overset{-}{C}\overset{+}{N}\overset{-}{O}$ (Fulminate ion) – 1255

Group 18 (Zero)

He_2 – 560; Ne_2 – 562; XeO_3 – 564; XeO_4 – 1281; XeF_2 – 528; XeF_4 – 531, 564; XeF_6 – 534; $XeOF_4$ – 1276; $XeOF_2$ – 1282; XeO_2F_2 – 1282; XeO_3F_2 – 1282; XeO_2F_4 – 1282

CHAPTER 1

Structure of the Atom (i)
(Quantum Mechanical Approach : Dalton to Bohr-Sommerfeld)

Dalton's Atomic Theory

An English Chemist, Dalton in 1803 put forward atomic theory, known as *Dalton's Atomic Theory* according to which all matter is composed of extremely indivisible particles which he named *atoms* (*Greek* : atom = cannot be cut *i.e.*, indivisible.)

However this concept of indivisibility of atom did not hold long and it was proved by the experiments made by brilliant research workers like *J.J. Thomson* (1897), *Rutherford* (1911), *Neils Bohr* (1912), *Chadwick, Mosely* and others that the *atom itself has a complex structure, i.e.,* the atom consists of several particles, called *sub-atomic particles* like *electron, proton, neutron, positron, neutrino, meson* etc. Out of these, electrons, protons and neutrons are very important. The atom, of course, remains indivisible in all physical and chemical changes.

Cathode (or Negative) Rays—Discovery of Electron

Production of Cathode Rays.

The first indication of the complex nature of atom was obtained from a study of *conduction of electricity through gases at reduced pressure* by Julius Plucker (1859). He took a cylindrical glass tube, called *discharge tube* which was 30 cm long and 4 cm in diameter, closed at both ends, provided with metallic electrodes and a side tap meant for connecting the tube with a high vacuum pump and low pressure gauge. To obtain very low pressures, molecular pump and charcoal cooled by liquid air that has the property of absorbing most of the residual gas were used. The two electrodes were connected to a high voltage battery (voltage was of the order of 10,000 volts and more). The electrode connected to the negative terminal of the battery is known as *cathode* (–) while that connected to the positve terminal is called anode (+) or *anticathode*. Plucker pumped out most of the gas of the tube by means of high vacuum pump and low pressure gauge and thus reduced the pressure of the gas in the tube to about 0.001 mm Hg. Now he observed that when he applied a high voltage of the order of 10,000 volts across the metallic electrodes, the walls of the tube began to glow with a bluish or greenish light depending on the composition of the glass of which the tube is made. It thus appeared that the invisible rays of light were emitted from the cathode and moved towards anode at right angles and with large velocity (Fig. 1.1). These rays were called *cathode rays* by Goldstein (1876), since these originated from the cathode.

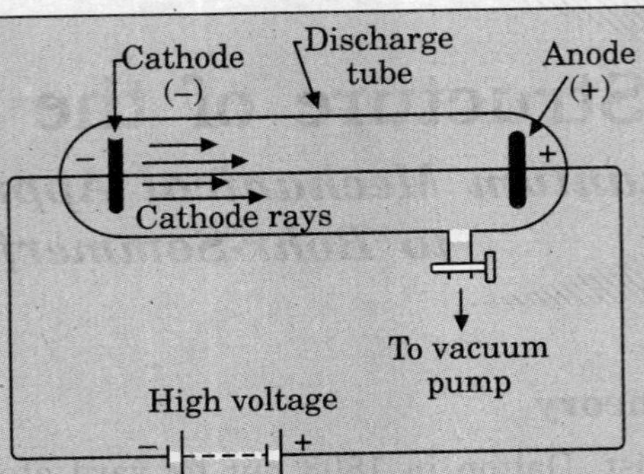

Fig. 1.1. Production of cathode rays in a discharge tube.

Properties of Cathode Rays.

Systematic investigations made by W. Crooks, J. J. Thomson, J. Perrin and others have established the following properties of cathode rays.

1. *The particles constituting the cathode rays are negatively charged.* This fact is confirmed by the following experiments.

(*a*) In 1897 J.J. Thomson showed that when cathode rays are passed through an electric field (*i.e,* when the rays are passed between two electrically charged plates), they are deflected towards the positively charged plate [Fig. 1.2 (*a*)], showing thereby that the particles constituting the cathode rays are negatively charged. These negatively charged particles were called **electrons** by Thomson.

(*b*) In 1895 J. Perrin showed that if a magnetic field is applied on cathode rays from top to bottom, perpendicular to the plane of the paper as shown in Fig. 1.2 (*b*), cathode rays are deflected upwards in the plane of the paper. This experiment also shows that the particles constituting the cathode rays are negatively charged.

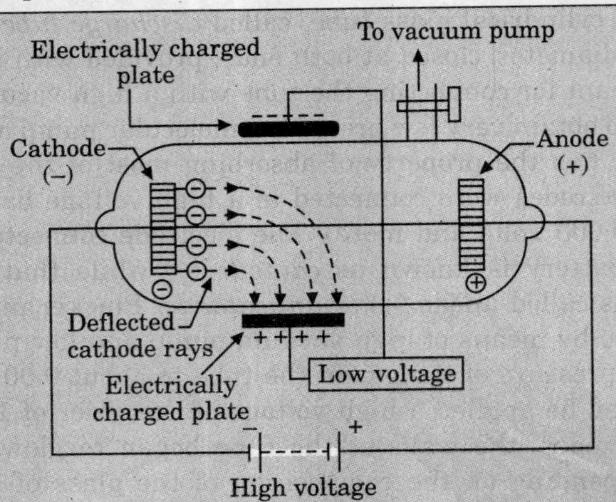

Fig. 1.2 (*a*) Cathode rays are deflected in an electric field, showing that the particles (*i.e.* electrons) constituting the cathode rays are negatively charged.

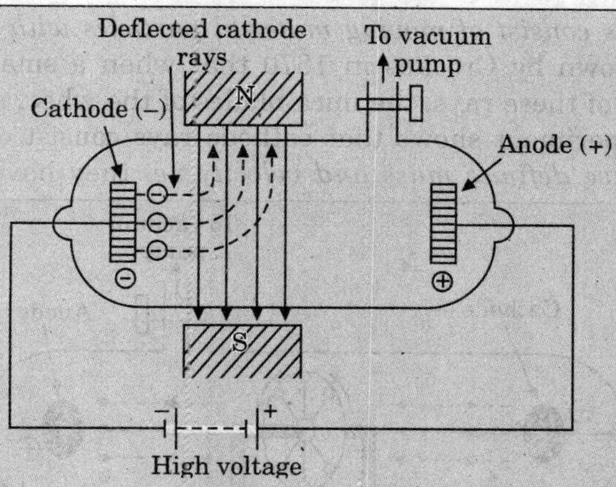

Fig. 1.2 (*b*) Cathode rays are deflected in a magnetic field, showing that the particles (*i.e.*, electrons) constituting the cathode rays are negatively charged.

Fig. 1.2. Experiment to show that the particles (*i.e.* electrons) constituting the cathode rays are negatively charged.

2. *They produce a greenish fluorescence on the walls of the glass tube.* They also produce flourescence in barium pentacyanide. rare earths (cerium etc.) natural minerals, sulphides and sulphates. When placed in path of these rays, *alumina glows red, tin stone yellow* and *willemite green.* The minerals which show fluorescence are called *phosphors.*

3. *The rays are shot out normally from the cathode.* The direction of the rays is not affected by the position of the anode.

4. *These rays travel in straight line perpendicular to the surface of the cathode.* This can be shown by the fact that if a metallic object like hinged aluminium cross is placed in the path of the rays, a sharp shadow of this object is cast on the walls opposite to the cathode (Fig. 1.3). This experiment shows that electrons have *particle properties.*

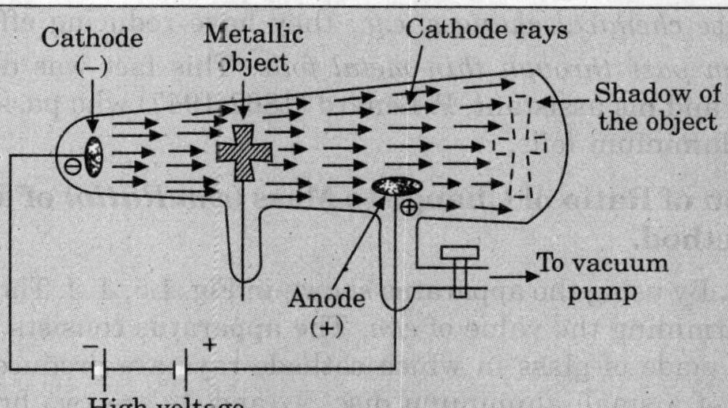

Fig. 1.3. Cathode ray cast shadow of the object placed in their path, showing thereby that these rays travel in straight line and hence electrons of which these rays are composed have particle properties.

5. *Cathode rays consist of moving material particles with definite mass and velocity.* It was shown by Crookes in 1870 that when a small paddle wheel is placed in the path of these rays, the mica blades of the wheel are set into motion (Fig. 1.4). This experiment shows that cathode rays consist of *moving material particles which have definite mass and velocity i.e.* they have *momentum*.

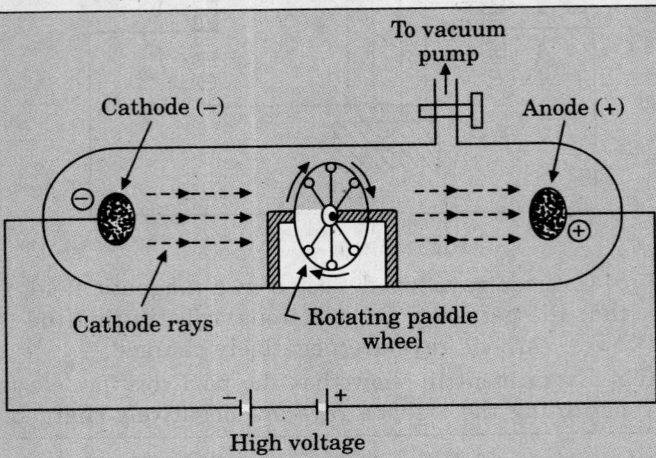

Fig. 1.4. Cathode rays rotate a small paddle wheel placed in their path, showing thereby that cathode rays consist of moving material particles with defenite mass and velocity.

6. *Cathode rays produce heat when they fall on matter.* When cathode rays from a concave cathode are focussed on a platinum foil, it begins to glow. The kinetic energy of the cathode rays is transformed into heat energy by their stoppage.

7. *Cathode rays produce X-rays.* When cathode rays strike a solid substance, particularly one with a large atomic weight, they give rise to penetrating radiations, called X-rays which have the same nature as light does.

8. *They ionise the gas through which they pass.*

9. *They couse chemical changes, e.g.,* they have reducing effect.

10. *They can pass through thin metal foils.* This fact was demonstrated by Heinrich Hertz and his assistant, P. Lenard (1862-1947) who passed cathode rays through thin aluminium foils.

Determination of Ratio of Change to Mass (*e/m Ratio*) of an Electron by Thomson's Method.

Apparatus. By using the apparatus shown in Fig. 1.5, J. J. Thomson succeeded in 1897 in determining the value of e/m. The apparatus consists of an evacuated discharge bulb made of glass in which cathode rays are produced from cathode which consists of a small aluminium disc. A_1 and A_2 are two brass discs which contain very narrow central holes. These holes convert the eletron beam into very fine beam beyond A_2. This beam strikes the fluorescent screen, SS' on which the deflections of the beam in electric and magnetic fields are noted. The surface of the screen is coated with some fluorescent material like ZnS. The plates P and P' between which an electric field in the plane of the paper is generated are placed on either side of the beam. Magnetic field is applied at the same place in a direction of the electric field, *i.e.* perpendicular to the plane of the paper.

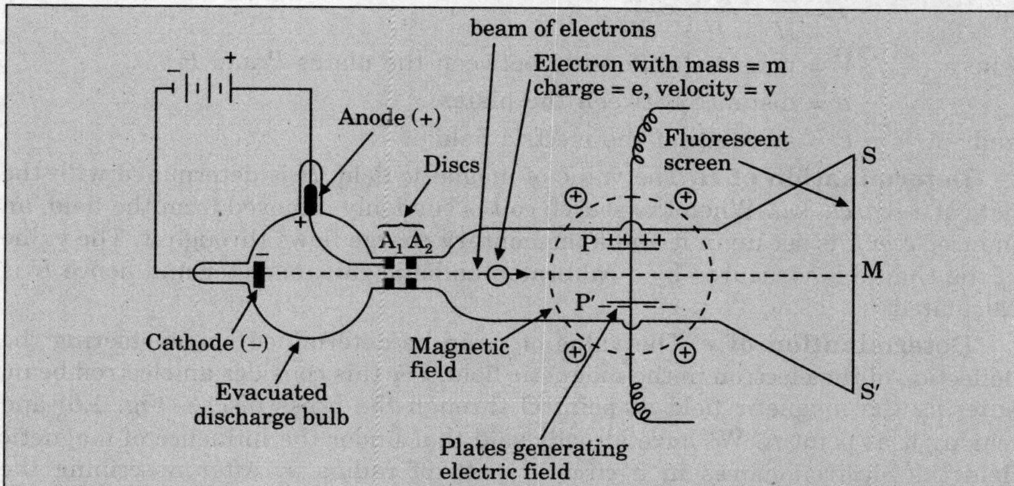

Fig. 1.5. Apparatus used in the determination of *e*/*m* value of the electron.

Working. A beam of electrons produced from cathode is first passed through a magnetic field of strength H which is being applied at right angles to the direction of the motion of electrons. Let the charge, mass and velocity of each electron be e, m and v respectively. Magnetic force acting on each of the electrons will then be equal to Hev. Under the influence of megnetic field the electron moves in a circular path. Let the radius of this circular path be r. After describing a circular path, the electron leaves the magnetic field and then moves in a straight line with the same velocity (= v) and ultimately strikes the screen at a point. In this case the centripetal (centrifugal) force (= mv^2/r) which is required for the orbital motion of the electron is supplied by the magnetic force (= Hev) acting on the eletron. Thus here both these forces are in equilibrium, *i.e.*

$$\frac{mv^2}{r} = Hev$$

or
$$\frac{e}{m} = \frac{v}{rH}$$

Knowing the values of v, H and r, the ratio e/m can be determined.

Determination of *v*. To find out the value of velocity v, apply the magnetic and electric fields on the electron beam *simultaneously at the right angles to each other.* Under this condition, according to Fleming's left-hand rule the electron will be deflected *downwards* by the magnetic field and *upwards* to the same extent by the electric field between the two plates and consequently the vertical deflection suffered by the electron would be zoo. Thus the electron would go straight (*i.e.* undeviated) and would strike the screen at point M. Since the deflection is zero, the downward magnetic force (= Hev) acting on the electron will be equal to the upward electric force (= eE) acting on the electron. Thus :

$$eE = Hev$$

or
$$v = \frac{E}{H} = \frac{V/d}{H} = \frac{V}{dH},$$

where V = potential difference between the plates P and P'

d = distance between the plates.

and E = strength of the electric field.

Determination of H. The value of magnetic field, H is determined with the help of a *search coil.* When the search coil is suddenly removed from the field, an induced *e.m.f.* is set up in it and a momentary charge flows through it. The value of the charge is measured by a calibrated ballistic galvanometer and hence H is calculated.

Determination of *r*. The value of r can be determined by considering the deflection of the electron in the magnetic field. For this consider an electron beam entering the magnetic field at point O through the brass disc A (Fig. 1.6) and leaving it at point R. We have already said that under the influence of magnetic field the electron moves in a circular path of radius, r. After describing the circular path, the electron leaves the magnetic field and then flies in a straight line with the same velocity and ultimately strikes the screen, MN at point N. Draw OC perpendicular on OM. Let RC and RN meet OM at S and L points respectively. Thus C is the centre of the circular path namely OR which is the length of the region in which magnetic field is applied.

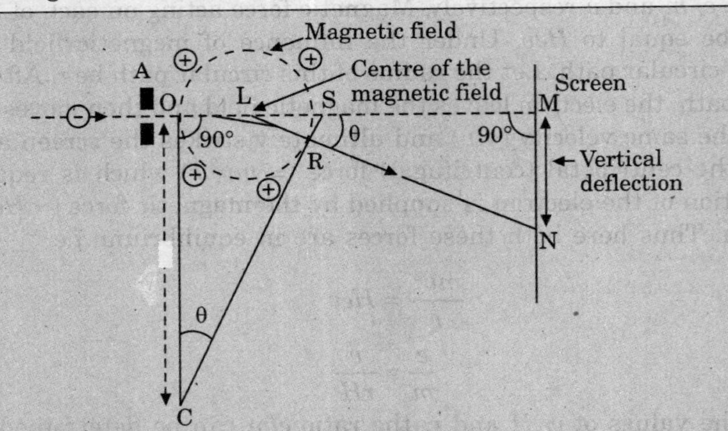

Fig. 1.6. Deflection of cathode rays in magnetic field and determination of the radius of the circular path described by the electron in the magnetic field.

Now $\angle OCR = \angle NLM = \theta$

$$\therefore \quad \tan\theta = \frac{OS}{OC} = \frac{MN}{ML}$$

or
$$\frac{OS}{OC} = \frac{MN}{ML}$$

or
$$\frac{OS}{r} = \frac{MN}{ML}$$

or
$$r = \frac{OS \times ML}{MN}$$

OS is very nearly equal to the length of the region, OR in which the magnetic field is applied. ML is equal to the distance between M and the centre of the magnetic field (*i.e.* L point) and MN is the vertical deflection suffered by the electron. The first two quantities *viz.* OS and ML are constants and are given by the makers of the apparatus and the third one can be easily measured.

Thomson carried out a number of experiments for the determination of *e*/*m* value of an electron in which he varied the metal forming the cathode as well as the gas in the evacuated discharge bulb. In every case he found the value of *e*/*m* to be equal to -1.758×10^8 coulomb/*g*.

Thomson's method suffers from the following two errors :

(*i*) Vertical deflection MN is very small.

(*ii*) There is lack of uniformity in the velocity of electrons constituting the cathode rays.

Determination of Charge on an Electron (*e value*) by Millikan's Oil Drop Method.

Apparatus. The apparatus used by Millikan (1909) for the determination of the charge on an electron consists of an earth connected metallic chamber having two window, *W* and *W′* one of which is used to illuminate the field of view and the other is used for passing X-rays to ionise the gas inside the chamber. The chamber is filled with air whose pressure can be adjusted by a vacuum pump. There are two electrodes, P and P' which are used to generate the known electric field in the space between the electrodes. The upper electrode has a hole O at its centre (Fig. 1.7).

Working. A fine spray of low pressure oil drops, as for example, of clock oil, is produced in the automiser, (a fine spray pump) and is then conveyed through the tube (T) to the chamber. A few drops pass through the hole (O) into the space between the two electrodes. As the drops pass through the hole O in P, the opening O is closed and velocity of the drop with which it is moving downwards under the action of gravity is determined by noting the time taken by the drop in moving down between the two points fiexed by cross hairs of a meter eye-piece having a scale. This eye piece is fitted in a microscope. An are lamp is used to illuminate the space between the electrodes through window, *W*. Evidently the velocity of the drop, v_1, with whcih it is moving downwards by the action of gravity is proportional to its weight, *mg* (*m* = mass of the drop, and *g* = acceleration due to gravity = 981 cm/sec^2), *i.e.*

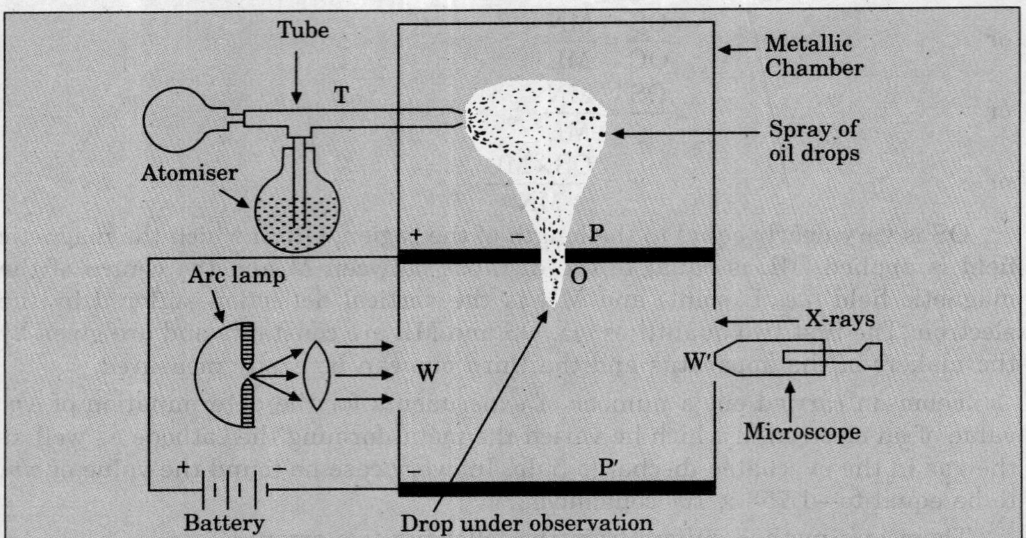

Fig. 1.7. Millikan's oil drop method for the dotermination of charge on an electron.

$$v_1 \propto mg \qquad ...(i)$$

Now the air space between the electrodes is ionised by X-rays passing into the chamber through the window, *W′*. The drop which is under observation, frequently takes up a gas ion and, thus gets charged. If now the electrodes, P and P′ are connected to a battery, electric field of strength, *E* comes into action and the drop moves upwards instead of downwards against the action of gravity with a uniform velocity, v_2. In this case the velocity, v_2 is proportional to *Ee—mg, i.e.*

$$v_2 \propto Ee - mg \qquad ...(ii)$$

Here *e* is the charge on the electron

From equations (*i*) and (*ii*), we get

$$\frac{v_1}{v_2} = \frac{mg}{Ee - mg} \qquad ...(iii)$$

The values of v_1 and v_2 are noted by means of the telescope and *E* is the strength of the applied electric fleld. Thus v_1, v_2 and *E* are the known quantities and, if the mass; *m* of the drop is determined, *e* can be calculated by equation (*iii*)

Determination of *m*. If oil drop is assumed to be a spherical particle, its mass, *m* is given by :

$$m = \frac{4}{3}\pi r^3 \rho$$

where r = radius of the spherical oil drop.

ρ = density of the oil of which the drop is composed of

The value of *r* to be used in the above equation can be known from the following relation, called *Stokes's relation,*

$$v_1 = \frac{2gr^2\rho}{9\eta}$$

or $$r = \left(\frac{9v_1\eta}{2g\rho}\right)^{1/2}$$

where v_1 = velocity of the oil drop which is moving down-wards under the influence of gravity alone.

η = coefficient of velocity of the medium.

With the help of the above two equations *m* is given by :

$$m = \frac{4}{3}\pi r^3\rho$$

$$= \frac{4}{3}\pi\left(\frac{9v_1\eta}{2g\rho}\right)^{3/2} . \rho$$

By varying the magnitude of the strength of the electric field, Millikan found that the charge on each drop was different. The smallest charge (*i.e.*, the highest common factor of all the values) is equal to -1.59×10^{-19} coulomb. More recent methods have shown that the value of $e = -1.602\times 10^{-19}$ coulomb. This quantity is taken to be equal to one unit of negative charge (–1). Thus the electron is said to carry one unit of negative charge on it. The charge taken by an electron is the smallest charge of electricity that has been measured so far. In *a.m.u.* and *e.s.u.* the charge carried by an electron is equal to 1.602×10^{-20} *a.m.u.* and 4.8×10^{-10} *e.s.u.* respectively.

Importance of the Determination of Charge on the Electron

The determination of the charge on an electron has enabled us to determine the following :

1. Determination of Avagadro's number. Avogadro's number is defined as the number of molecules per gram molecule of a substance or it is the number of atoms per gram atom of a substance. Now we know that the charge required to liberate one gm ion of H^+ ion is equal to 96,500 coulomb and the charge on one H^+ ion is equal (though of opposite sign) to that carried by an electron, *i.e.* 1.602×10^{-19} coulomb. Hence :

$$\text{Avogadro's number} = \frac{96,500 \text{ coulomb}}{1.602\times10^{-19} \text{ coulomb}}$$

$$= 6.023\times10^{23}$$

2. Mass of an electron in grams. The value of the electronic charge (*e*) and *e*/*m* enables us to determine the mass (*m*) of an electron as follows :

$$m = \frac{e}{e/m} = \frac{1.602\times10^{-19} \text{ coulomb}}{1.758\times10^{8} \text{ coulomb/g}}$$

$$= 9.108\times10^{-28}\, g$$

This is the mass of an electron moving at *low speed.* It is called its **rest mass** which is defined *as the mass of the electron when it is moving with a velocity much smaller than that of light.*

3. Mass of an electron relative to that of hydrogen atom. It has been found that the number of atoms in one mole of any element is equal to Avogadro's number, 6.023×10^{23}. Therefore :

The weight of 6.023×10^{23} atoms of hydrogen

$$= \text{At. wt of hydrogen} = 1.008 \text{ g}$$

$\therefore$ The weight of one atom of hydrogen

$$= \frac{1.008}{6.023\times10^{23}} \text{g}$$

$$= 1.67\times10^{-24} \text{g}$$

Now we know that the mass of an electron = 9.1×10^{-28}g

$$\therefore \frac{\text{Mass of hydrogen atom}}{\text{Mass of an electron}} = \frac{1.67\times10^{-24}\text{g}}{9.1\times10^{-28}\text{g}}$$

$$= 1.835\times10^{3} = 1835$$

or Mass of hydrogen atom = 1835 × Mass of an electron

or $\text{Mass of an electron} = \frac{1}{1835}\times\text{Mass of hydrogen atom}$ ...(*i*)

Thus we see that an atom of hydrogen is 1835 times heavier than an electron.

In other words. *the mass of an electron is* $\frac{1}{1835}$*th of the mass of a hydrogen atom.*

4. Mass of an electron in atomic mass units. On atomic weight scale, the mass of hydrogen atom has been found to be equal to 1.008 *amu*. Hence from equation (*i*), we have

On atomic weight scale, the mass of an electron

$$= \frac{\text{Mass of hydrogen atom in } amu}{1835}$$

$$= \frac{1.008 \; amu}{1835} = 0.0005493 \; amu$$

This mass is taken to be equal to zero. Thus an electron is said to be a massless particle.

5. Radius of an electron. Considering the whole mass of an electron to be electrical in nature it has been proved by J. J. Thomson that the mass, m (= 9.108×10^{-28} g) of an electron is related to its radius, r as follows :

$$m = \frac{2}{3}.\frac{\mu e^2}{r}$$

where μ = permeability of the medium = 1

e = charge on the electron = 1.602×10^{-20} *a.m.u.*

Thus :

$$r = \frac{2\mu e^2}{3m} = \frac{2\times 1\,(1.602\times 10^{-20})^2}{3\times 9.108\times 10^{-28}}$$

$$= 1.879\times 10^{-13}\ cm$$

Definition of Electron

Having known the mass and charge on an electron, it can be defined as *a subatomic particle which bears mass equal to 0.0005493 amu* (*which is taken to be equal to zero*) and negative charge equal to *–1 on it.* In other words *an electron is a massless particle which bears one unit of negative charge on it.* An electron is indicated as $_{-1}e^0$ where zero indicates that it has no mass and –1 stands to show the presence of one unit of negative charge on it.

Electrons are the Common Universal Constituents of all Atoms

It has been shown by Thomson that the electrons emitted in the cathode rays have the same *e/m* value (= -1.758×10^8 coulomb g.), irrespective of the nature of the material of the cathode and that of the gas taken in the discharge tube. Similarly it has been shown by Millikan that the value of *e* for the electrons is also the same (= -1.602×10^{-19} coulomb), irrespective of the gas used in the experiment. From these observations it can be concluded that *electrons are present in all forms of matter and are common or universal constituents of all atoms.*

Anode (or Positive) Rays–Discovery of Proton

Production of Positive Rays.

We have seen that since an atom, as a whole, is electrically neutral and it has negatively charged particles known as electrons, it should have some particles which should be positively charged. In order to show the presence of positively charged particles in an atom, attempts were made by many scientists. In 1886, German physicist, E. Goldstrein took a discharge tube provided with a cathode perforated with extremely fine holes and observed that when a large potential difference was applied between the electrodes, not only cathode rays of electrons were originated from the cathode, moving from cathode towards anode, but also new type of coloured rays were also simultaneously originated from the anode, moving from anode towards cathode through the fine holes or canals in the cathode (Fig. 1.8), and causing a glow on the wall opposite to anode.

Initially these rays were called canal rays, since they pass through the canals or holes of the cathode. These are also called *anode rays* as they seem to originate from the anode. J. J. Thomson has shown that these rays are positively-charged and hence he has called them *positive rays.*

The origin of these rays can easily be understood. In a discharge tube when gas atoms (say H atoms) lose negatively charged electrons, they acquire a positive charge. These positively charged particles then move from anode to cathode. (Fig. 1.8).

Properties of Positive Rays.

Some properties of these rays as investigated by J. J. Thomson are given below :

1. *The particles constituting the anode rays are positively charged.* This property can be shown by the following experiments :

(*a*) When these rays are passed through an electric field, they are deflected towards the negatively charged plate, showing thereby that these rays consist of positively-charged particles.

(*b*) In presence of magnetic field, these rays are deflected in a direction opposite to that of cathode. These anode rays consist of particles which are positively-charged.

2. The charge-to-mass ratio (*e/m*) for positive rays is smaller than that for an electron.

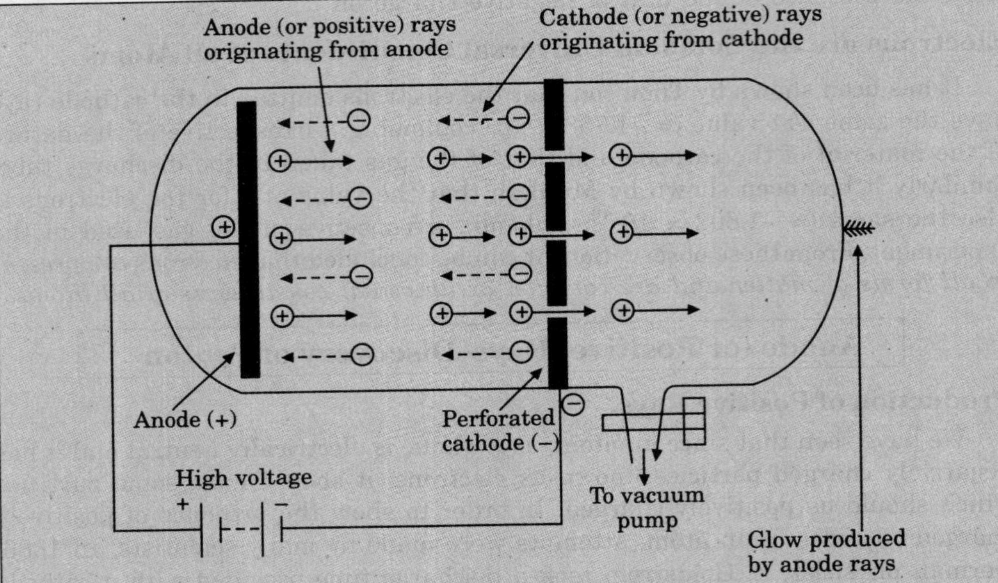

Fig. 1.8 Production of positive (anode) rays.

3. Positive rays produce flashes on ZnS plate.

4. These rays travel in a straight line in a direction opposite to the cathode.

5. *Charge-to-mass ratio* (*e/m*) *value.* For a number of gases charge-to-mass ratio (*e/m*) of positive particles constituting the positive rays has been determined and it has been found that *e/m* ratio for the positive rays is different for different gases. When hydrogen gas is used in the discharge tube, *e/m* value for the positive rays obtained from hydrogen gas has been found to be the maximum. Since the value of charge (*e*) on the positve particle obtained from all gases is the same, the value of *m* should be lowest for the positive particle obtained from hydrogen gas. Thus positive particle obtained from hydrogen gas is the lightest among all the positive particles obtained from different gases. This particle is called **proton**. Thus the positively-charged particle obtained from hydrogen gas is called a proton.

$$\underset{\text{Proton}}{H - e^- \rightarrow H^+}$$

This name was given by Rutherford.

6. *Charge on a proton.* Charge on a proton is equal in magnitude but opposite in sign to that of an electron. Thus the charge carried by a proton is equal to + 1.602 × 10^{-19} coulombs which is taken to be equal to one unit of positive charge (+1). Thus a proton is said to carry one unit of positive charge on it.

7. *Mass of a proton.* The mass of a proton has been found by Thomson and it is equal to 1.672 × 10^{-24}g. On the relative scale its mass is equal to 1.005757 *amu* which is taken to be equal to 1 *amu* Thus a proton is said to have one unit mass. The mass of a proton is about 1837 times the mass of an electron, *i.e.*

Mass of a proton = 1837 × Mass of an electron. Mass of a proton is $\frac{1835}{1837}$ times that of hydrogen atom.

Thus :

$$\text{Mass of a proton} = \frac{1835}{1837} \times \text{Mass of hydrogen atom.}$$
$$= 0.9989 \times \text{Mass of hydrogen atom.}$$

Definition of Proton

Having known the mass and charge on a proton, it can be defined as a *subatomic particle which bears mass equal to 1 amu and positive charge equal to + 1 on it.* In other words, a *proton is a subatomic particle which bears one unil mass and one unit of positive charge on it.* A proton is indicated as $^{1}_{+1}p$ or $^{1}_{+1}H$ where 1 indicates that it has mass equal to 1 *amu* and +1 stands to show the presence of one unit of positive charge on it.

Explanation of the Production of Positive Rays

When a high voltage is applied across the metallc electrodes of a discharge tube filled with a gas under reduced pressure, the molecules of the gas are dissociated into atoms. Now when high speed electrons (*i.e.,* cathode rays) originated from cathode strike these atoms, the atoms are converted into cations which pass through the holes of the cathode and constitute positive rays. The electrons obtained in the conversion of the gaseous atoms into cations move towards anode as negative rays (Fig. 1.9).

$$\underset{\text{Gaseous atoms}}{M} + e^{-}(\text{originated from cathode}) \rightarrow \underset{\text{Pass through the perforated cathode as positive rays}}{M^{+}} + \underset{\text{Move towards anode as negative rays}}{2e^{-}}$$

Now it is believed that *the protons like electrons are also the common universal constituents of all atoms, excepting hydrogen which has no neutron.*

Thomson's Atomic Model

On the basis of the properties of positive rays as discussed above, Thomson, in 1893, proposed that an atom can be regarded as a heavy positively charged sphere which consists of a positively charged substance distributed uniformly throughout the whole sphere. The electrons are embedded in the sphere so that the atom becomes electrically neutral. The electrons are embedded in the sphere as the seeds are embedded in a water melon. The mutual repulsion existing between different eletrons keeps them changed in the form of circular shells. (Fig. 1.10).

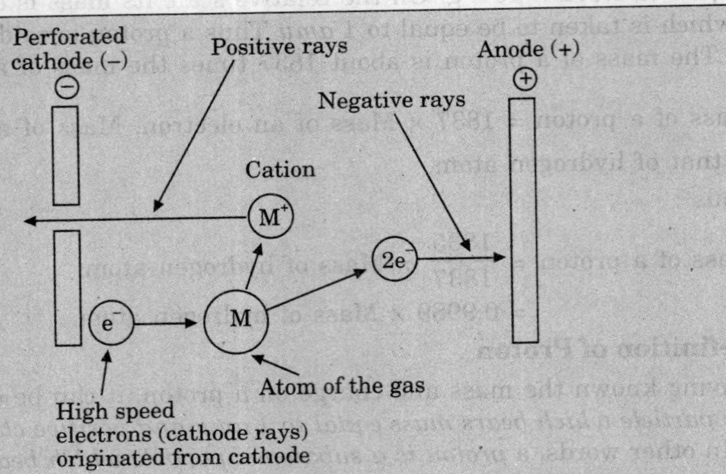

Fig. 1.9. Explanation of the production of positive rays (Illustration).

This atomic model was given up after some time, since it could not explain the large deflection suffered by α-particles in Rutherford's experiment which is described below.

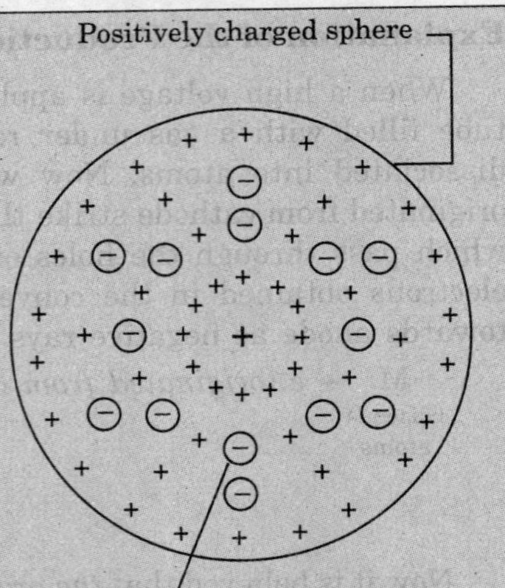

Fig. 1.10. Thomson's a atomic model.

Discovery of Neutron

Upto 1932 it was postulated than an atom is composed of only protons and electrons, and that the mass of an atom is due to the mass of protons present in the nucleus, since the mass of electrons is negligible. Mass of each proton is equal to 1.00375 atomic mass unit (*a.m.u*).

Rutherford noticed that the atomic masses of different atoms could not be explained if it is supposed that the atom is composed only of protons and electrons and, therefore, he predicted in 1920 that some kind of neutral particles having mass equal to protons must be present in an atom.

It was in 1932 that Chadwick discoved these neutral particles for which he was awarded the Nobel Prize in Physics in 1935. He obtained these neutral particles by bombarding light nuclei like Li and

Be with α-particles. He named these particles as **neutrons** on account of their neutral nature.

Experiment. In his experiment, Chadwick directed a stream of α-particles $\left({}^{4}_{2}\text{He}\right)$ produced from polonium source at a berrylium target $\left({}^{9}_{4}\text{Be isotope}\right)$ beyond which was placed a charge detector (Fig. 1.11). He found that penetrating radiations are produced. He named these radiations as neutrons on account of the fact that they have no charge on them as indicated by charge detector.

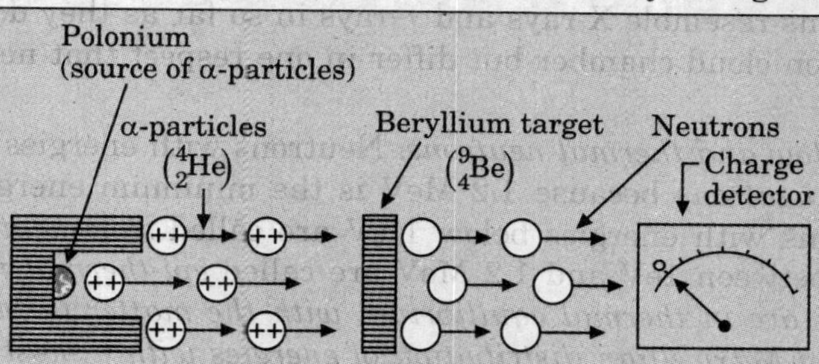

Fig. 1.11. Production of neutrons by the bombardment of α-particles on beryllium target.

The reaction which occurred in Chadwick's experiment is an example of artificial transmutation where an atom of beryllium $\left({}^{9}_{4}\text{B isotope}\right)$ is converted into a carbon atom through the nuclear reaction shown below :

$$\underset{\substack{\text{Alpha particle}\\ \text{(mass = 4 } amu\text{, at. no. = 2)}}}{{}^{4}_{2}\text{He}} + \underset{\substack{\text{Beryllium atom}\\ \text{(mass = 9 } amu\text{, at. no. = 4)}}}{{}^{9}_{4}\text{Be}} \rightarrow {}^{12}_{6}\text{C} + {}^{1}_{0}n$$

Mass and charge of a neutron

The mass of a neutron is equal to that of a proton. On the relative scale, since the mass of a proton is taken to be equal to 1 *a.m.u.,* although the exact mass of a neutron is 1.00898 *a.m.u.* In grams, the mass of a neutron is equal to 1.672×10^{-24} g. Neutron has no charge on it.

Definition of a Neutron

Having known the mass and charge of a neutron, it can be defined as a *sub-atomic particle which bears a mass equal to that of a proton (i.e.* equal to 1 *amu) and has no charge on it.* For this reason a neutron is represented as ${}^{1}_{0}n$.

Properties of Neutron

(*i*) *Decay of neutron.* A free neutron $({}_{0}n^{1})$ decays into a *proton* $\left({}^{1}_{+1}p\right)$ with the emission of an *electron* $\left({}^{0}_{-1}e\right)$ and a *neutrino* $\left({}^{0}_{0}v\right)$

$$\underset{\text{Neutron}}{{}_0n^1} \rightarrow \underset{\text{Proton}}{{}_{+1}p^1} + \underset{\text{Electron}}{{}_{-1}e^0} + \underset{\text{Neutrino}}{{}_0v^0}$$

(*ii*) *The neutrons are very penetrating.* Their intensity of penetration is reduced to one-half in passing through lead of thickness of about 1.5 cm. Neutrons are less penetrating than cosmic rays.

(*iii*) Neutrons have the property of expelling high speed protons from hydrogenous matter like paraffin, water, paper, cellophane etc. From the rang of protons knocked out from paraffin the energy of neutrons has been calculated.

(*iv*) Neutrons resemble X-rays and γ–rays in so far as they do not produce a track in a Wilson cloud chamber but differ in one respect that neutrons produce no ionisation.

(*v*) *Fast, slow and thermal neutrons.* Neutrons with energies above 1.2 MeV are called *fast neutrons* because 1.2 MeV is the minimum energy for fission of ${}_{92}U^{235}$. Neutrons with energies below 1 eV are called *slow neutrons.* Neutrons with energies between 1eV and 1.2 MeV are called *epi-thermal neutrons. Those neutrons which are in thermal equilibrium with the matter through which they pass and have a Maxwellian distribution of energies with a most probable value depending upon temperature are called thermal neutrons.* The most probable energy of thermal neutrons at 27^{0}C is 0.00260 eV.

Slow neutrons are usually more effective than fast ones, because high-speed neutrons may pass through a neucleus without bringing about transmutation.

Uses of Neutrons

(*i*) Almost all the elements can be transmulated by neutrons Some of the elements require fast neutrons whereas other require slow ones. Thus a fast neutron ejects an α-particle from a nitrogen nucleus, thereby converting it into a boron nucleus.

$${}_7N^{14} + {}_0n^{10} \rightarrow {}_5B^{11} + {}_2He^4$$

When slow moving neutrons fall on certain substances, they are captured with the emission of high energy γ-particle from a nitrogen nucleus, thereby converting it into a boron nucleus.

$${}_{29}Cu^{65} + {}_0n^1 \rightarrow \underset{\text{Radioactive}}{{}_{29}Cu^{66}} + hv$$

$${}_{29}Cu^{66} \rightarrow + {}_{30}Zn^{11} + {}_{-1}e^0.$$

(*ii*) The neutrons are now being widely used in biological and medical sciences. Because of their intense biological effects they are now being used in the treatment of cancer.

Summary of the Important Properties of Sub-atomic Particles : Electron, Proton and Neutron.

We have hitherto studied important properties of the three principal fundamental particles of the atom which are *electron, proton* and *neutron.* Their properties are summarised in table 1.1 given on page 17.

Nearly all of the ordinary properties of matter can be explained on the basis of the properties of these fundamental subatomic particles.

Table 1.1 Important properties of electron, proton and neutron

Particle	Mass			Charge		Charge units	Symbol	Discoverer	Location
	In g. (C.G.S. units)	In kg (S.I. units)	In a.m.u.	In e.s.u. (C.G.S. units)	In Coulomb (C) (S.I. units)				
Electron	9.108×10^{-28} g	9.108×10^{-31} kg	0.005493 a.m.u. ≈ 0.00 a.m.u.	4.8×10^{-10} e.s.u.	1.602×10^{-19} C	– 1	$_{-1}e^0$	J.J. Thomson	Outside the nucleus
Proton	1.672×10^{-27} g	1.672×10^{-27} kg	1.005757 a.m.u. ≈ 1.00 a.m.u.	—	1.602×10^{-19}C	+ 1	$_{+1}p^1$ or $_{+1}H^1$	J.J.Thomson	In the nucleus
Neutron	1.672×10^{-24} g	1.672×10^{-27} kg	1.00898 a.m.u. ≈ 1.00 a.m.u.	0	0	0	$_0n^1$	Chadwick	In the nucleus

Rutherford's Nuclear Atomic Model—Discovery of Nucleus

We have been that, according to Thomson's atomic model, an atom consists of electrons and protons. In order to test Thomson's model and to know as to how and where the electrons are located in the atom, Rutherford (1911), a British scientist, performed some experiments on the scattering of α-particles by a thin metal foil. Here it may be mentioned that α-particle is a helium nucleus which consists of two units of positive charge and its mass number is equal to 4. Hence it is represented by the symbol, $_{+2}\alpha^4$ or $_{+2}He^4$.

The pictorial representation, of Rutherford's experiment is given in Fig. 1.12. In his experiment he produced a narrow beam of α-particles from a radioactive substance like radium placed in a lead block and directed it at a thin sheet of gold foil of thickness of 0.00004 cm. A movable circular screen coated with ZnS was placed on the other side of the foil. This screen helped in detecting the direction in which the α-particles moved. When these particles struck the screen, a flash of light was observed.

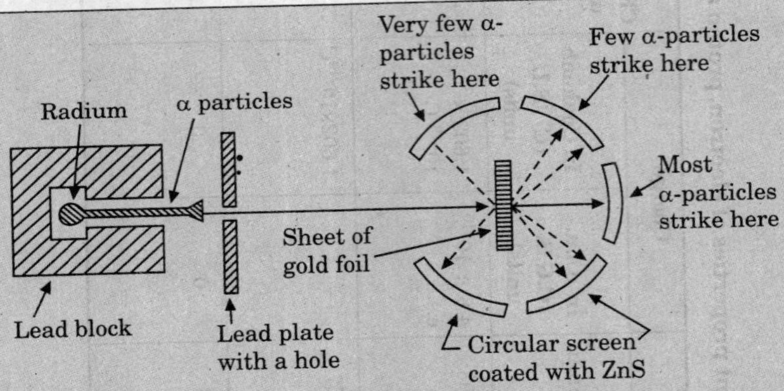

Fig. 1.12. Pictorial representation of Rutherford's experiment.

Rutherford observed that most of the α-*particles passed through the foil without any deflection from their path* and struck the screen at its centre, causing illuminations. *A few of them* were deflected at *some angles* (90^0 or wider angles) after passing through the foil. *Very few* (not more than one in 10,000) *turned back on their original path.* All the three types of deflections are shown in Fig. 1.13.

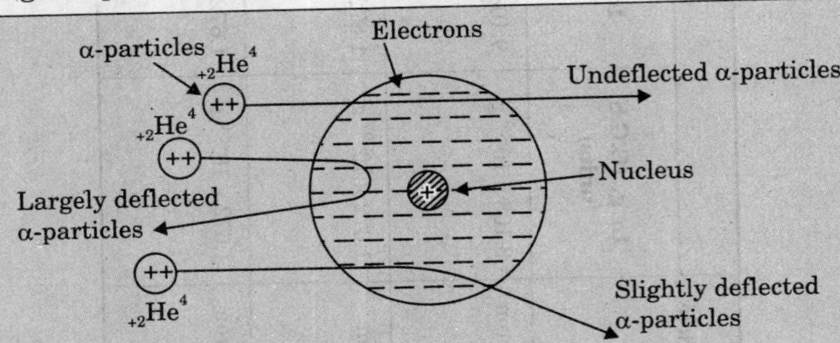

Fig. 1.13. Three types of deflection of α-particles by a thin metal foil shown in a simpler way.

On the basis of the observations made by Rutherford in his experiments, combined with the discovery of neutron made by Chadwick in 1932, he proposed a complete model of the atom which is named as *Rutherford's nuclear atomic model* after his name. This is also called *nuclear atom.* According to this model :

Postulates of Rutherford's Model

1. *Atom contains a massive (heavy) and positively charged part at its centre. This central part of the atom is called* **nucleus.** The presence of the nucleus at the central part of the atom is based on the observation that a few α-particles are deflected back. It is this positively charged nucleus which repells the positively charged α-particles to go back. All the protons and neutrons are present in the nucleus of the atom. Although the nucleus is small in size, it is heavy due to the presence of all protons and neutrons in it. Thus the entire mass of the atom resides in its nucleus and is equal to the sum of the masses of protons and neutrons, since the mass of the electrons is negligible. The remaining part of the atom round the nucleus is left empty. Thus the atom contains a lot of empty space round the nucleus. The nucleus has positive charge which is different in magnitude for different elements.

2. *The volume occupied by the nucleus is only a minute fraction of the total volume of the atom, i.e.* the size of the nucleus is very small as compared to that of the whole atom. This fact is based on the observation that a very small fraction of α-particles turned back on their original parts without undergoing any deflection.

3. Atom is not all solid, as was earlier suggested by Dalton, but is *extraordinarily hollow, since it consists of a lot of empty space round the nucleus.* Most of the space round the nucleus is empty, except for the presence of electrons which themselves are extremely minute. This conclusion has been drawn from the observation that most of α-particles passed straight through the foil undeflected. It is this empty space around the nucleus which allows the α-particles to pass undeflected through the foil.

4. *Electrons are revolving round the nucleus in closed orbits with a fast speed and hence almost all the space round the nucleus is occupied by the revolving electrons.* (Fig 1.14). The electrons keep revolving in orbits round the uncleus as the planets revolve round the sun. Thus Rutherford's model of atom bears a close resemblance with the solar system in which the massive sun plays the role of the massive nucleus and the planets play the role of the revolving electrons. It is because of the close resemblance of Rutherford's model with solar system that the electrons revolving round the nucleus are generally referred to as *planetary electrons.*

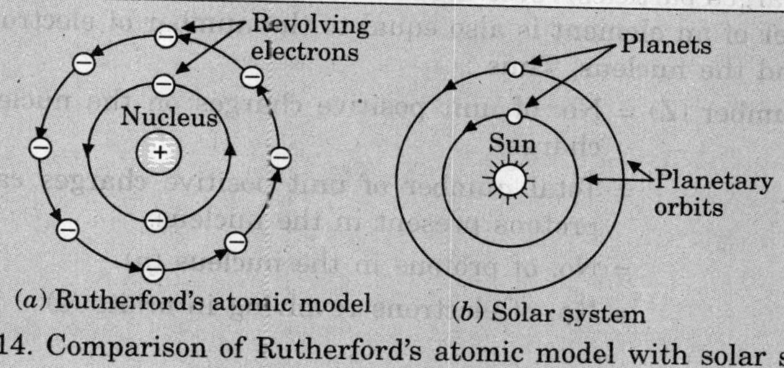

Fig. 1.14. Comparison of Rutherford's atomic model with solar system.

Drawbacks of Rutherford's Nuclear Atomic Model

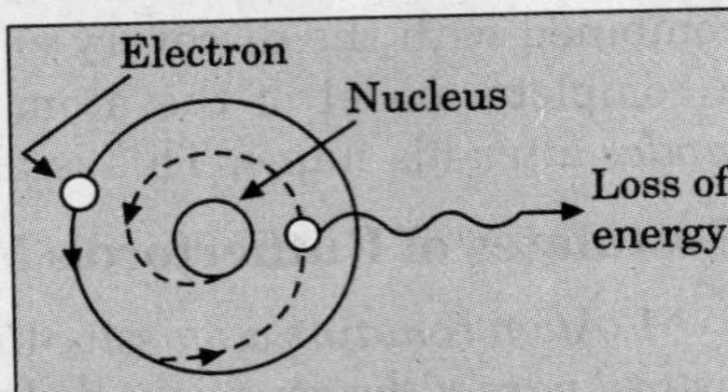

Fig. 1.15. An electron while revolving round the nucleus loses energy and its speed is decreased, and ultimately the electron falls into the nucleus by a spiral path.

We have seen that according to Rutherford's nuclear atomic model the electrons are revolving around the nucleus in orbits with a fast speed. The speed of orbiting electrons is so high that each of the electrons is accelerated. Now according to the *classical electromagnetic theory* (electrodynamics), if a charged particle (here this particle is electron) is accelerating around an oppositely charged particle (here this particle is nucleus), it (*i.e.* electron) would continuously lose or radiate energy and its speed will decrease and hence will gradually come closer to the nucleus by a spiral path. The ultimate result of this spiral path will be that the electron will fall into the nucleus as shown in Fig. 1.15, thereby making the atom unstable, *i.e.* according to Rutherford's model the atom should not exist. In actual practice this does not happen. Thus we see that Rutherford's model cannot explain the stability of the atom. This dilemma was solved in 1913 by Neils Bohr who proposed an improved form of Rutherford's atomic model.

What is Atomic Number (Z) ?

We have seen that, according to Rutherford, the atom has a nucleus which carries positive charge on it. The positive charge carried by the nucleus is called *nuclear charge* which is equal to the total number of positive charges on the nucleus of the atom of that element. *This total number of unit positive charges on the nucleus is called the atomic number of the element.* Now since each proton has +1 charge and neutron has no charge, the total number of unit positive charges carried by the nucleus is equal to the number of protons present in the nucleus of the atom. Thus atomic number of an element can also be defined *as the number of protons present in the nucleus. Thus* :

Atomic number of an element is defined as the number of unit positive charges on the nucleus (nuclear charge) of the atom of that element or as the number of protons present in the nucleus.

Now since the atom as a whole is electrically neutral, the number of protons (positively charged particles) in the nucleus is equal to the number of electrons (negatively charged particles) revolving in orbits around the nucleus. Consequently atomic number of an element is also equal to the number of electrons revolving in orbits round the nucleus. Thus :

Atomic number (Z) = No. of unit positive charges on the nucleus (Nucleus charge)

= Total number of unit positive charges carried by all protons present in the nucleus.

= No. of protons in the nucleus (p)

= No. of electrons revolving in orbits (e)

or $\boxed{Z = p = e}$

Atomic number of an element is also defined *as the serial number of its position in the periodic table, starting from hydrogen as the first element.*

Mosley's Law

Roentgen while making some experiments with a discharge tube in 1895, found that when cathode rays are allowed to fall on a metal target called *anti-cathode* or *target material* placed in their path, new type of radiations are produced. These radiations are called X-rays. These rays are also electromagnetic in character and are comparable to light waves. The common form of X-ray tube used for producing X-rays is shown in Fig. 1.16. The stream of cathode rays (*i.e.* electrons) produced from the cathode converges to a point through a concave surface on the anti-cathode and gives rise to X-rays.

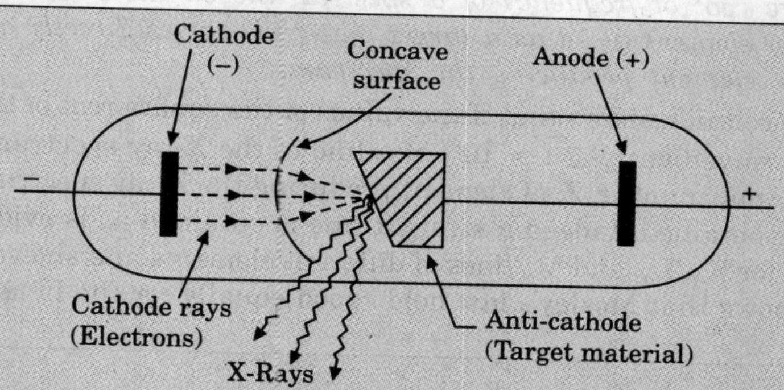

Fig. 1.16. Production of X-rays in X-ray tube

Mosley took different X-ray tubes provided with anti-cathode made from different elements. He found that when cathode rays are allowed to fall on the anti-cathode, X-rays are emitted the wave lengths of which are characteristic of the element of which the anti-cathode is made.

X-rays were passed through a slit in platinum plate and then emerged from X-ray apparatus through aluminium window. The emergent beam led on to the cleavage face of a crystal of potassium ferrocyanide (used as diffraction grating) which analysed the X-ray beam. The rays thus diffracted by the crystal were obtained in the form of *line X-ray spectrum* and were allowed to fall on a photographic plate. Line spectrum is characteristic of the target material used. In fact X-rays constituting the line spectrum are called *characteristic X-rays*. The characteristic X-ray spectrum consists of discrete spectral lines which can be grouped into K-series, L-series, M-series etc. Each of these series consists of various lines as shown below :

K-series $\rightarrow K_\alpha, K_\beta, K_\gamma$ *etc. lines (called K-lines).*

L-series $\rightarrow L_\alpha, L_\beta, L_\gamma$ *etc. lines (called L-lines).*

M-series $\rightarrow M_\alpha, M_\beta, M_\gamma$ *etc. lines (called M-lines).*

The number of the lines present in the spectrum depends both on the nature of the target material and the excitation voltage.

Mosley showed that the square root of the frequency $\sqrt{\upsilon}$ of a line of a given series of characteristic X-rays spectrum is related to the atomic number (Z) of the element used as anti-cathode (target material) by the following expression :

$$\sqrt{\upsilon} = a\,(Z-b) \quad ...(i)$$

where a = proportionality constant

and b = a constant whose value is the same for all the lines of a given series. For K_α and K_β lines of K-series, $b = 1$. b is called *screening constant* which takes into account the screening effect of electrons.

The values of a and b vary from one series to the other. *i.e.,* their values for the lines of K series are different from those for the lines of L-series, etc.

The relation represented by equation (*i*) is called *Mosley's law* which can be stated as :

The square root of frequency of a spectral line on the characteristic X-ray spectrum of an element (used as a target material) varies directly as the atomic number of the element producing the spectrum.

The above relation shows that if the values of the square root of the frequency, $\sqrt{\upsilon}$ (or more conveniently $\sqrt{\upsilon} \times 10^8$) of a line of the X-ray spectrum are plotted against the atomic number, Z, of element producing the X-ray spectrum, a straight line should be obtained. Indeed a straight line is obtained as is evident from the graph plotted for K_α, L_α, and M_α lines of different elements and shown in Fig. 1.17. This graph shows that Mosley's law holds good equally for the lines of K, L, and M series.

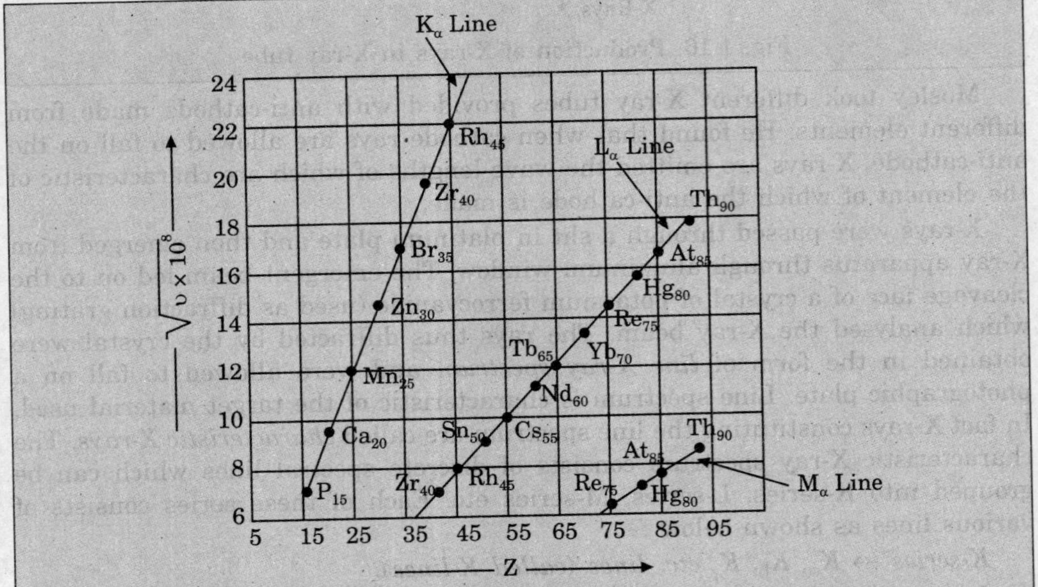

Fig. 1.17. Plot of $\sqrt{\upsilon}\times 10^8$ against Z of the element producing X-ray spectrum. υ is the frequency of the line and Z is the atomic number.

For the K_α and K_β lines of K-series for which $b = 1$, the above relation becomes :

$$\sqrt{\upsilon} = a(Z-1)$$

Thus for K_α and K_β lines Mosley law becomes *that the square root of the frequency is proportional to the atomic number of the element reduced by one.*

The integer, Z, which is called atomic number of the element, determines the exact position of the element in the periodic table. The atomic number of an element can be directly determined from the X-ray spectrum of the element and is equal to the number of unit positive charges on the nucleus of the atom concerned.

Importance of Mosley's Law

Mosley law has been found to be very useful. The improtance of this law is evident from the following points :

(*i*) We have seen that this law his established that the atomic number is always numerically equal to the number of electrons revolving round the nucleus. According to this law *it is the atomic number and not atomic weight on which the properties of the elements (both physical and chemical) depend.* It thus says that *the elements must be arranged in the periodic table in the increasing order of their atomic number and not atomic weight as originally enunciated by Mendeleef.* Accordingly, Mosley's law has been used to place elements in their proper order in the periodic table in certain questionable cases, *e.g.,* if the elements are placed according to their increasing atomic weight, K(Z = 19) should come before Ar (Z = 10); Ni (Z = 28) shold precede Co (Z = 27) etc. But Mosley law says that as per their atomic numbers their order should be reversed, *i.e.* K should be placed after Ar, Co should come before Ni etc. This fact is further supported by chemical properties.

(*ii*) Mosley law has led to the discovery of many new elements. The element like technitium (Z = 43), promethium (Z = 61), hafnium (Z = 72), rhenium (Z = 75) have been discovered on the basis of the gaps left by Mosley in Mosley's X-ray diagrams. These elements have been subsequently discovered.

(*iii*) This law has been helpful in determining the atomic number of rare-earths and has thus helped in fixing their position in the periodic table.

What is Mass Number (A) ?

We know that the nucleus of an atom contains neutrons and protons. *The sum of the number of neutrons (n) and protons (p) present in the nucleus of an atom is called the mass number (A) of that atom, i.e.*

Mass number (A) = No. of protons (p) + No. of neutrons (n)

or $$A = p + n$$

In cases where it is not necessary to differentiate between protons and neutrons, these particles are collectively called **nucleons.** Thus *the mass number (A) of an atom is equal to the tatal number of nucleons in the nucleus of an atom.* Obviously the mass number is a *whole number.*

Each different variety of atom as determined by the composition of its nucleus is called a **nucleide.**

Determination of the Composition of an Atom (Nucleide) in terms of Number of Electrons, Protons, and Neutrons

Using the relations namely :

Atomic number (Z) = No. of protons (p) = No. of electrons (e) ...(*i*)

Mass number (A) = No. of protons (p) = No. of neutrons (n) ...(*ii*),

we can find out the number of electrons, protons and neutrons in an atom. The composition of the atoms of first twenty elements (H to Ca) in terms of the number of electrons, protons and neutrons is shown in Table 1.2.

Table 1.2. Composition of the atoms of the first twenty elements in terms of the protons, neutrons and electrons.

Element	At. No. (Z)	Mass number (A) of the isotope having the maximum abundence	Protons (p) = Z	Nuetrons (n) = (A – Z) = (A – p)	Electrons (e) = Z = p
H	1	1	1	0	1
He	2	4	2	2	2
Li	3	7	3	4	3
Be	4	9	4	5	4
B	5	11	5	6	5
C	6	12	6	6	6
N	7	14	7	7	7
O	8	16	8	8	8
F	9	19	9	10	9
Ne	10	20	10	10	10
Na	11	23	11	12	11
Mg	12	24	12	12	12
Al	13	27	13	14	13
Si	14	28	14	14	14
P	15	31	15	16	15
S	16	32	16	16	16
Cl	17	35	17	18	17
Ar	18	40	18	22	18
K	19	39	19	20	19
Ca	20	40	20	20	20

Bohr's Atomic Model

We have seen that according to Rutherford's nuclear atomic model, an atom has a nucleus and negative electrons which are revolving round the nucleus in the same way as the planets revolve round the sun in the solar system. This model could not say anything as to how and where these electrons were arranged.

This model could also not explain how the spectral lines are produced by the simplest atoms like hydrogan atom when an electron jumps from one orbit to the other.

In order to explain why an electron revolving round the nucleus does not lose energy and consequently does not fall into the nucleus and how the spectral lines of the emission spectrum of H-atom are produced when an electron jumps from one energy level to the other, Niels Bohr, a Danish Physicist, in 1913, put forward a new atomic model which is based on Planck's quantum theory of radiation.

Before, we could deal with Bohr's atomic model we should have a clear understanding of the follwing topics :

(*i*) *Electromagnetic radiations (or waves) and nature of radiations emitted by electron.*

(*ii*) *Emission spectrum, continuous spectrum and line spectrum of hydrogen atom.*

(*iii*) *Plancks' quantum theory of radiation, and its application in explaining photoelectric effect and Compton effect.*

Electromagnetic Radiations or Electromagnetic Waves—Nature of Radiations Emitted by Electron

When an object moves up and down or vibrates continuously, energy, in the form of a wave, is sent or transmitted (or propagated) by the vibrating object to a distant place. The wave travels at right angle to the vibratory motion of the vibrating object and away from it (Fig. 1.18). Wave produced by a vibrating object can be represented by a wavy curve. One such wavy curve is shown in Fig. 1.18. The tops (*i.e.* maxima) of the wavy curve are called *crests* and the bottoms (*i.e.* minima) are called *troughs* (Fig. 1.18). When a stone is thrown in a pond of water, water molecules start to vibrate up and down and transmit their energy as waves on water surface which are seen travelling towards the bank of the pond.

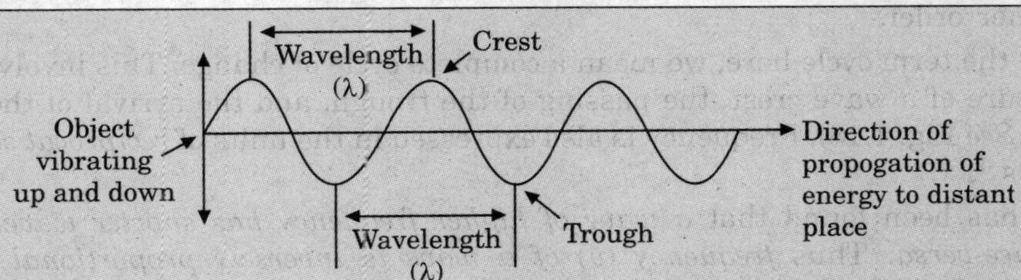

Fig. 1.18. Production of a wave by a vibrating object and definition of crests, troughs and wavelength in a wave.

In 1864 James Clark Maxwell showed that when an electric current is passed through a circuit (*e.g.* through a tungsten filament), it radiates energy in the form of waves. The energy radiated is called **radiate energy** and the waves produced are called **electromagnetic waves** or **electromagnetic radiations.** These waves are so called because similar sort of waves can also be obtained by moving a charged body as well as a magnet *to* and *fro* in a magnetic field. These waves do not require any medium for propagation or transmission; *e.g.,* light and radio waves (which are electromagnetic waves) from the sun and stars reach the earth by travelling through nonmaterial medium *i.e.,* through empty space. All these waves travel with the speed of light.

The radiations (or waves) emitted by the vibratory motion of an electron are also electromagnetic radiations. These radiations also travel with that speed of light.

Properties of Waves—Waves Parameters

Waves are characterised by the following properties :

(i) Wavelength (λ) *(ii) Frequency* (v) *(iii) Speed or velocity* (c) *(iv) Wave number* (v) These properties are called *wave parameters.*

Now let us discuss these wave parameters, one by one, as follows :

***(i)* Wave length (λ).** *The wave length of a wave is defined as the distance between two successive crests or two successive troughs of a wave* (Fig. 1.18). It is represented by *lambda* (λ) and is measured in terms of *centimetre (cm)* or *Angstrom* (A^0) *units (in C.G.S. units)* or *meters (m)* or *Angstrom units (in S.I. units)*. Wavelengths are also expressed in *micrometers* (μm) *and nanometers (nm)*. Various units are related to meter as follows :

$1\ A^0 = 10^{-10}\ \text{m} = 10^{-8}\ \text{cm}$

$1\ \mu m = 10^{-6}\ \text{m}$

$1\ nm = 10^{-9}\ \text{m}$

All the above mentioned units are related to each other as follows :

$1\ cm = 10^{-2}\ m = 10^{8} A^0 = 10^{4}\ \mu m = 10^{7}\ nm$

***(ii)* Frequency (v) and velocity (c).** *Frequency is defined as the number of waves or cycles which pass a given point on the wave in one second.* It is represented by *nu* (v) and is measured in terms of *cycles (or waves) per second (cps)* or *Hertz* (Hz) *units.* A frequency of 100 hertz means 100 cycles (or waves) per second. *Kilocycles per second* (ke/s), *megacycles per second* (Mc/s) and *kilomegacycles per second* (kMc/s) are the terms used to express the frequencies of higher order.

By the term cycle here, we mean a complete cycle of change. This involves the departure of a wave crest, the passing of the trough, and the arrival of the next crest. (*See* Fig. 1.18). Frequency is also expressed in the units of *reciprocal second,* 1/s or s^{-1}.

It has been found that *a wave of higher frequency has shorter wavelength and vice-versa.* Thus *frequency (v) of a wave is inversely proportional to its wavelength (λ), i.e.*

$$v \propto \frac{1}{\lambda}$$

or

$$v = \frac{c}{\lambda}, \qquad \text{...(1)}$$

where c is a constant which is called the *velocity of the wave. Velocity of a wave is defined as the distance travelled by the wave in one second.* It is, therefore, expressed in $cm.s^{-1}$ (in C.G.S. units) or in $m.s^{-1}$ (in S.I. units). Its value has been found to be constant for all types of electromagnetic radiations. Its experimental value has been found to be equal to 3×10^{10} $cm.s^{-1}$ or 3×10^{8} $m.s^{-1}$. As a matter of fact this value is the velocity of light in air. Thus c is equal to the velocity of light ($=3\times10^{-10}$ $cm.s^{-1}$ or 3×10^{8} $m.s^{-1}$) for all types of electromagnetic radiations.

Equation (1) clearly shows that :

	Velocity (c) = Wavelength (λ) × Frequency (v)
or	$c = \lambda \, v$

...(2)

The validity of equation (2) has been shown in Fig. 1.19, from which it is evident that the product of wavelength (λ) and frequency (*v*) for all the waves shown is constant (= 120), *i.e.* all types of radiations travel with the same speed or velocity.

(*a*)

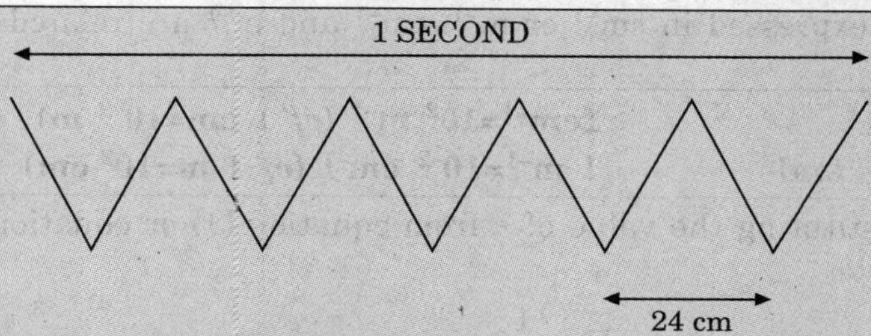

No. of waves = 5

∴ Frequency (v) = 5 waves/second

Wavelength (λ) = 24 cm/wave

∴ Velocity $(c) = v\lambda = 5 \times 24$ cm/s = 120 cm/s

(*b*)

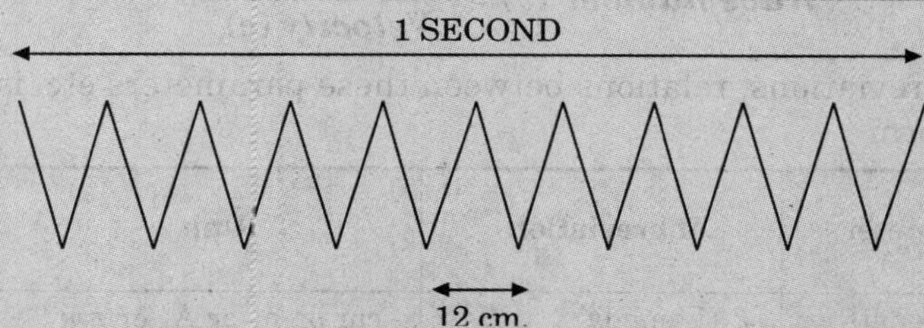

No. of waves = 10

∴ Frequency (v) = 10 waves/second

Wavelength (λ) = 12 cm/wave

∴ Velocity $(c) = v\lambda = 10 \times 12$ cm/s = 120 cm/s

(*c*)

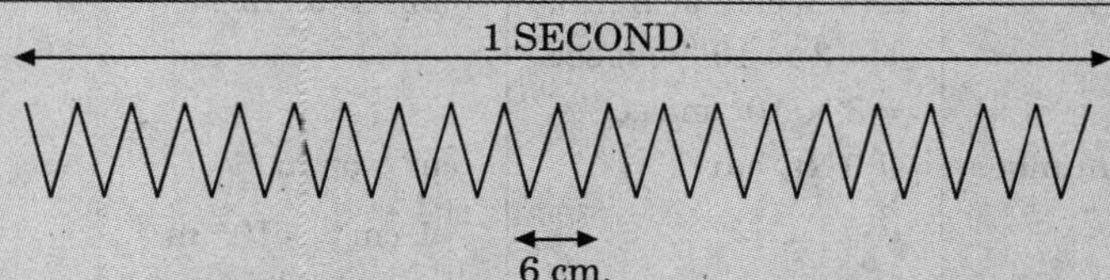

No. of waves = 20

∴ Frequency (v) = 20 waves/second

Wavelength (λ) = 6 cm/wave

∴ Velocity $(c) = v\lambda = 20 \times 6$ cm/s = 120 cm/s

Fig. 1.19. To show the validity of the relation, $c = \lambda v$. As is evident, the product of wavelength (λ) and frequency (*v*) is constant (= 120) for all the waves shown here.

***(iii)* Wave number** ($\bar{v}$), In spectroscopy, frequency of light is generally expressed in terms of wave number ($\bar{v}$) which is defined as the number of wavelengths per meter and is equal to the reciprocal of wavelength (λ). It is denoted by *nu bar* ($\bar{v}$). Thus :

$$\textbf{Waver number } (\bar{v}) = \frac{1}{\lambda} \qquad ...(3)$$

Since $\bar{v}$ is the reciprocal of wavelength and wavelength is expressed in cm or m, $\bar{v}$ is expressed in cm^{-1} or m^{-1}. cm^{-1} and m^{-1} are related with each other as follows :

$$1 cm^{-1} = 10^2 \; m^{-1} \; (cf\ 1 \; cm = 10^{-2} \; m)$$

and $$1 \; m^{-1} = 10^{-2} \; cm^{-1} \; (cf\ 1 \; m = 10^2 \; cm)$$

On substituting the value of λ from equation (1) or equation (2) in equation (3), we get :

$$\bar{v} = \frac{1}{c/v}$$

or $$\bar{v} = \frac{v}{c}$$

or $$\textit{Wave number } (\bar{v}) = \frac{\textit{Frequency } (v)}{\textit{Velocity } (c)} \qquad ...(4)$$

The abbreviations, relations between these parameters etc. is shown below in a tabular form :

Wave parameter	Abbreviation	Unit	How these parameters are related.
(i) Wavelength	λ (Lambda)	cm or m or A° or *nm* [1 cm = 10^{-2} m = 10^8 A° = 10^7 *nm*]	
(ii) Frequency	v (Nu)	cycles/sec or Hz or s^{-1}.	$\bar{v} = \frac{1}{\lambda} = \frac{v}{c}$
(iii) Velocity	c (Velocity of light) $c = 3 \times 10^{10}$ cm/sec $= 3 \times 10^8$ m/sec	cm/sec or m/sec	
(iv) Wave number	$\bar{v}$ (Nu bar)	cm^{-1} or m^{-1} $\begin{bmatrix} 1 \; cm^{-1} = 10^2 \; m^{-1} \\ 1 \; m^{-1} = 10^{-2} \; cm^{-1} \end{bmatrix}$	

Numerical Problems

Problem 1. (In S. I. units). *Calculate the wave number of the lines of frequencies 4×10^{14} Hz and 5×10^{16} Hz.*

Solution. We know that wave number, $\bar{v}$ is given by :

$$\bar{v} = \frac{1}{\lambda} = \frac{v}{c}$$

(*a*) Here $v = 4 \times 10^{14}$ Hz and $c = 3 \times 10^8$ ms^{-1}

$$\therefore \quad \bar{v} = \frac{4\times10^{14}\ \text{Hz}}{3\times10^8\ m.s^{-1}} = \frac{4\times10^{14}\ s^{-1}}{3\times10^8\ m.s^{-1}} \qquad (\because \text{Hz} = s^{-1})$$

$$= 1.33\times10^6\ \text{m}^{-1}$$

(*b*) Here $v = 5\times10^{16}$ Hz and $c = 3\times10^8\ m.s^{-1}$

$$\therefore \quad \bar{v} = \frac{5\times10^{16}\ \text{Hz}}{3\times10^8\ m.s^{-1}} = \frac{5\times10^{16}\ s^{-1}}{3\times10^8\ m.s^{-1}} \qquad (\therefore \text{Hz} = s^{-1})$$

$$= 1.66\times10^8\ m^{-1}.$$

Problem 2 (In S.I. units). *Calculate the wavelength of a wave of frequency 10^{11} Hz travelling with the speed of light = 3×10^8 m.s^{-1}.*

Solution. Here $\lambda = 10^{11}$ Hz and $c = 3\times10^8\ ms^{-1}$

$$\therefore \quad \lambda = \frac{c}{v} = \frac{3\times10^8\ ms^{-1}}{10^{11}\ \text{Hz}} = \frac{3\times10^{-3}\ ms^{-1}}{s^{-1}}$$

$$= 3\times10^{-3}\ \text{m}$$

Problem 3 (In S.I. units). *Calculate the frequency and wave number of yellow light having wavelength of 580 nm.*

Solution. Here $\lambda = 580\ nm = 580 \times 10^{-9}\ m$

and $c = 3 \times 10^8\ ms^{-1}$

$$\therefore \quad \lambda = \frac{c}{\lambda} = \frac{3\times10^8\ ms^{-1}}{580\times10^{-9}\ m} = \frac{3\times10^{17}}{580} s^{-1}$$

$$= \frac{3\times10^{17}}{580}\ \text{Hz} = 0.005172\times10^{17}\ \text{Hz}$$

$$= 5.172\times10^{14}\ \text{Hz}$$

$$v = \frac{1}{\lambda} = \frac{1}{580\times10^{-9}\ m} = \frac{1}{580}\times10^9\ m^{-1}$$

$$= 0.001724\times10^9\ m^{-1}$$

$$= 1.724\times10^6\ m^{-1}$$

Problem 4 (In S.I. units). *What is the frequency and wave number associated with light of wavelength equal to 6000 A° ?*

Solution. Here $\lambda = 6000$ A° $= 6000 \times 10^{-10}\ m$

and $c = 3 \times 10^8\ m.s^{-1}$

$$\therefore \text{Wave number, } \bar{v} = \frac{1}{\lambda}$$

$$= \frac{1}{6000\times10^{-10}\text{m}}$$

$$= \frac{1}{6\times10^3\times10^{-10}}\text{m}^{-1}$$

$$= \frac{10^7}{6} \text{ m}^{-1} = 0.1666 \times 10^7 \text{ m}^{-1}$$

$$= 1.666 \times 10^6 \text{ m}^{-1}$$

Frequency, $\upsilon = \frac{c}{\lambda} = \frac{3 \times 10^8 \text{ m.s}^{-1}}{6000 \times 10^{10} \text{ m}}$

$$= \frac{10^8}{2 \times 10^{-7}} s^{-1} = 0.5 \times 10^{15} \text{ s}^{-1}$$

$$= 5.0 \times 10^{14} \text{ s}^{-1}$$

Problem 5 (In S.I. units). *The wavelength of a certain wave is 6m and 30 waves pass a fixed point per minute. What is the speed of the waves ?*

Solution. Here λ = 6 m

and v = 30 cycles/minute

$$= \frac{30}{60} \text{ cycles .s}^{-1}$$

$$= \tfrac{1}{2} \text{ cycles .s}^{-1}$$

$\therefore$ Speed (c) of the wave

$$= v.\lambda$$

$$= \tfrac{1}{2} \times 6 \text{ m.s}^{-1}$$

$$= 3 \text{ m.s}^{-1}$$

Spectrum

What is Spectrum ?

A spectrum is an array of waves or particles which is spread out according to the increasing or decreasing of some property such as wavelength or frequency.

Types of Spectra

Depending on the nature of the source emitting the radiation, there are two types of spectra.

I. *Emission spectra*

II. *Absorption spectra.*

Emission spectra are further of two types :

(*a*) *Continuous spectrum*

(*b*) *Discontinuous spectrum* which may be a *band spectrum or line spectrum.* Line spectrum is also called *atomic spectrum.* Thus :

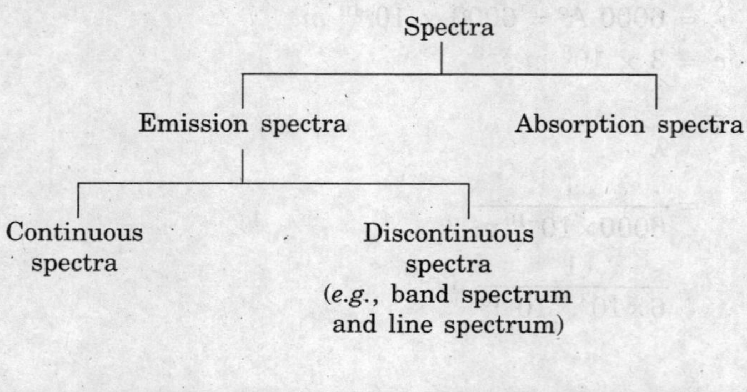

I. Emission Spectra

Emission spectra can be obtained from the substances which emit light on excitation. Their excitation can be done-as follows :

(*i*) By heating the liquid or solid substances in a flame at high temperature. These substances become incandescent at high temperature, *i.e.,* when these substances are heated at high temperature, they emit light. Electric light, lime light, arc light are the examples of such solid substances.

(*ii*) By passing an electric discharge through a gas at low pressure.

(*iii*) By passing electric current through thin filament of a high melting point metal like tungsten, when the filament becomes incandescent.

Types of Emission Spectra

Emission spectra are of the following two types :

1. **Continuous spectrum.** When a narrow beam of sun light or any white light (*e.g.* an arc lamp) is allowed to pass through a prism and then allowed to fall on a screen, it gets resolved into seven colours *viz, violet, indigo, blue, green, yellow, orange* and *red* (VIBGYOR) (Fig. 1.20). This phenomenon is called **dispersion** and the band of seven colours observed on the screen is called **visible spectrum.** Each of the seven colours of the spectrum has got a characteristic wavelength as shown in the figure. It may be seen that the wavelengths of different radiations of the visible spectrum lie in the range : 4000—7500Å.

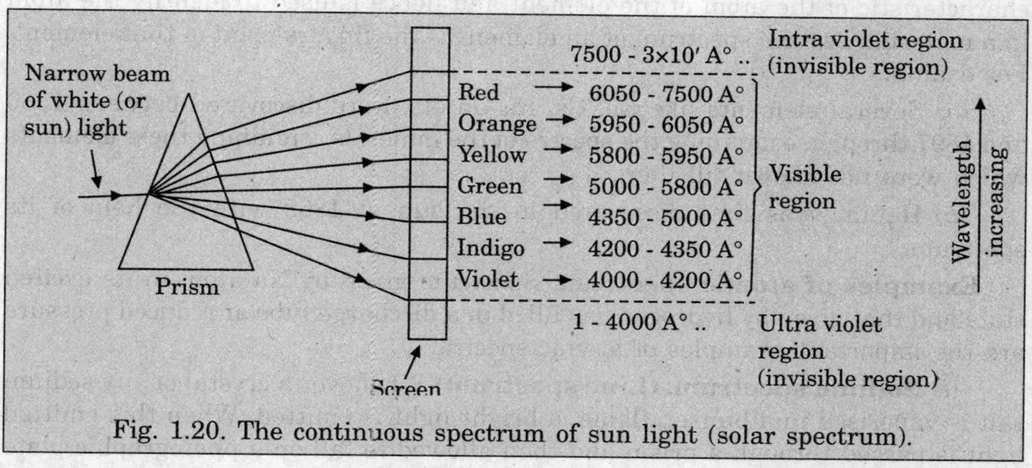

Fig. 1.20. The continuous spectrum of sun light (solar spectrum).

Since the wavelengths and frequencies of all the seven radiations of the spectrum can be detected by human eye, the region consisting of these colours is called *visible region or visible spectrum.* The visible spectrum is like the rainbow often seen in the sky after a rainfall or colour produced by an oil film on water. In the spectrum, since one colour merges into another without any break or discontinuity (*i.e.*, one colour changes gradually into another), the spectrum thus obtained is called *continuous spectrum.* The violet light of the spectrum has shortest wavelength (λ = 4000 Å) while the red light has longest wavelength (λ = 7500 Å). The region below the violet light is called *ultraviolet region.* The radiations of this region, have *lower wavelengths* (*i.e.* higher frequencies) than that of the violet rediaton (λ = 1—4000 Å). Similarly the region below the red radiation is called *infra-red region.* The radiations of this region have *higher wavelengths* (*i.e.* lower frequencies) than that of red radiation (λ = 7500—3 × 10^7 Å). Ultraviolet and infra-red regions are not visible to the human eye and hence are collectively called *invisible region or invisible spectra.* Our eyes are not sensitive to the radiations of invisible spectra.

2. **Discontinuous spectrum.** It may be band spectrum or line spectrum.

(*i*) **Band spectrum. (Molecular spectrum)** Band spectrum is the property of molecules and generally given by the compounds or gases like N_2 and O_2 molecules at low temperature and pressure. The band spectrum consists of a number of bright bands, separated by dark spaces. Each band is sharp at one end and fades gradually as we approach the other end. Due to its *fluted appearance,* band spectrum is sometimes referred to as *fluten spectrum.* Since this spectrum is characteristic of molecules, it is also called *molecular spectrum.*

(*ii*) **Atomic spectrum.** When an atom of an element, whether free or in combination, absorbs energy, it gets *excited.* The excited atom emits light radiations of a characteristic colour. When this emitted radiation is allowed to pass through a prism, it is resolved into individual components. The pattern of different components of the radiation is called *atomic spectrum.* It is so called because this spectrum is given by the atoms of the element This kind of spectrum is characteristic of the atom of the element and hence is used to identify the atom in a material. *i.e.,* the spectrum of an element is the fingers-print of that element. For example :

(*a*) Several elements like Rb, Cs, In, Ga etc. were discovered between 1860 and 1897 through, examining the spectra of the minerals containing these elements which were not known till then.

(*b*) Helium was first discovered in the sun in 1868 with the help of its spectrum.

Examples of atomic spectrum. Spectrum given by Na atom in its excited state and that given by hydrogen gas filled in a discharge tube at reduced pressure are the important examples of atomic spectra.

(*i*) **Sodium spectrum. (Line spectrum)** Whenever a crystal of any sodium salt is vaporised in a bunsen flame, a bright light is emitted. When this emitted light is passed through a prism and then allowed to fall on a photographic plate as shown in Fig. 1.21 (or when examined through a spectroscope), two bright isolated yellow lines, D_1 and D_2, separated by a dark space are found in the spectrum. The wavelength of these lines are found to be 5893 Å and 5896 Å. D_1 and D_2 lines appearing in the atomic spectrum constitute the *emission spectrum* of sodium. Line spectrum (or emission spectrum) given by sodium differs from that given be other elements.

Since the atomic spectrum consists of two isolated lines separated by a dark space, the atomic spectrum is also called *line spectrum.* Atomic spectrum is also known by other names like *line emission spectrum* or *atomic emission spectrum.*

(*ii*) **Hydrogen spectrum.** As mentioned above hydrogen spectrum is also an important example of atomic spectrum. This spectrum, like the atomic spectrum given by other elements, is also known by other names like *line emission spectrum* or *atomic emission spectrum of hydrogen.*

How is hydrogen spectrum obtained ?

When an electric discharge is passed through hydrogen gas filled in a discharge tube at a very low pressure, a bluish light is emitted from the discharge tube. When the ray of this bluish emitted light is passed through a prism (Fig. 1.21), a *discontinous line spectrum consisting* of several isolated sharp lines with dark area in between is obtained. Such a spectrum is called *line spectrum* of hydrogen. This spectrum gives a useful idea about the atomic structure.

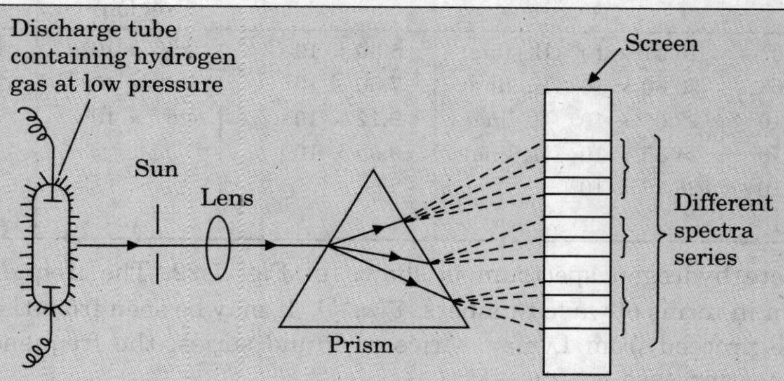

Fig. 1.21. Line spectrum of hydrogen atom observed on passing light emitted from a discharge tube (containing hydrogen) through a prism.

When an electric discharge is passed through hydrogen, the molecules of hydrogen break into atom. These absorb energy from the electric spark and get *excited.* The excited atoms give the extra energy in the form of emitted light (bluish light). When this emitted light is allowed to pass through a prism line spectrum of hydrogen is obtained. This spectrum is also called *line emission spectrum.* Here since it is the atom (of hydrogen) which is the source of the spectrum, hydrogen spectrum is also called *atomic spectrum of hydrogen.*

Classification of Spectral Lines in Hydrogen Spectrum.

As already said, hydrogen spectrum consists of several isolated sharp lines with dark area in between. The wavelengths of various spectral lines seen in hydrogen spectrum show that these lines lie in *visible, ultraviolet* and *infra-red regions* of the spectrum. All these spectral lines can be classified into five groups or series called **spectral series** which are named after the names of their discoverers as shown below :

Spectral series	***Discovered by***	***Appearing in***
(*i*) *Lyman series*	*Lyman*	*Ultraviolet region*
(*ii*) *Balmer series*	*Balmer*	*Visible region*

(iii) Paschen series	*Paschen*	*Infra-red region*
(iv) Brackett series	*Brackett*	*Infra-red region*
(v) Pfund series	*Pfund*	*Infra-red region*

The values of frequencies in terms of wave number, v (in m^{-1}) of various spectral lines, as determined experimentally by Ritz, are given in Table 1.3.

Table 1.3 Experimental values of frequencies in terms of wave numbers, $\bar{v}$ (in m^{-1}) of various spectral lines seen in hydrogen spectrum

Lyman series (Ultraviolet region)	Balmer series (Visible region)	Paschen series (Infra-red series)	Brackett series (Infra-red region)	Pfund series (Infra-red region)
82.20×10^5	15.21×10^5 (H_α line)	5.30×10^5	2.46×10^5	1.34×10^5
97.60×10^5	20.60×10^5 (H_β line)	7.80×10^5	3.80×10^5	2.14×10^5
102.70×10^5	23.05×10^5 (H_γ line)	9.12×10^5	4.61×10^5	
105.20×10^5	24.35×10^5 (H_δ line)	9.95×10^5		
106.20×10^5	25.18×10^5			
107.10×10^5				

Complete hydrogen spectrum is shown in Fig. 1.22. The frequencies have been shown in terms of wave numbers, $\bar{v}(m^{-1})$. It may be seen from the spectrum that as we proceed from Lyman series to Pfund series, the frequencies of the spectral lines increase.

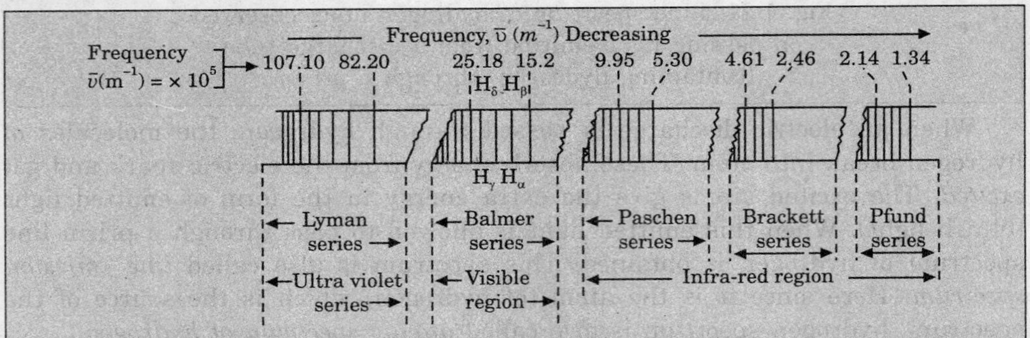

Fig. 1.22. Complete line spectrum of H-atom. The frequencies in wave number of different spectral lines have been shown in m^{-1}.

How to Calculate the Frequencies of Spectral Lines by Rydberg's Equation.

By measuring the wavelength of a large number of lines observed in different spectral series of hydrogen spectrum, Rydberg gave an empirical equation with the help of which the frequency in terms of wave number, $\bar{v}$ (reciprocal of wavelength, λ), of any line of a given series of hydrogen spectrum can be calculated. According to this equation the wave number of any line of a particular series appearing in the H-spectrum is represented as a difference of two terms. For a given series one of the two terms remains constant while the other varies from line to line in the same series. This equation can be written as :

$$\overline{\upsilon} = \frac{1}{\lambda} = \text{constant}\left(\frac{1}{x^2} - \frac{1}{y^2}\right) m^{-1}$$

$$= R\left(\frac{1}{x^2} - \frac{1}{y^2}\right) m^{-1},$$

where $\overline{\upsilon}$ = frequency in wave number (in m^{-1}) ; λ = wavelength ; R = a universal constant which is called **Rydberg's constant.** Its experimental value, as determined by Ritz, is $109.678 \times 10^5\ m^{-1}$, and x and y are simple whole numbers which are called Rydberg's integers. Equation given above is called **Rydberg's equation.**

For a given spectral series x remains constant while y varies from line to line in the same series. For example the frequency of H_α (= 15.21×10^5 m^{-1}), H_β (= 20.60×10^5 m^{-1}), H_γ (= 23.05×10^5 m^{-1}) and H_δ (= 24.35×10^5 m^{-1}) spectral lines appearing in the Balmer series can be obtained by putting x = 2 and y = 3, 4, 5 and 6 respectively in Rydberg's equation in which R = 109.687×10^5 m^{-1}. The values of x for various series and those of y for successive lines in the same series are given below in Table 1.4.

Table 1.4. Values of x and y for various spectral observed in hydrogen spectrum.

Spectral series	x	$y = (x + 1), (x + 2)... \infty$	Spectral region
Lyman	1	2, 3, 4, 5, 6.........∞	*Ultraviolet region*
Balmer	2	3, 4, 5, 6, ...∞	*Visible region*
Paschen	3	4, 5, 6 ....∞	*Infra-red region*
Brackett	4	5, 6, ...∞	*Infra-red region*
Pfund	5	6, ...∞	*Infra-red region*

This table clearly shows that the value of x remains fixed for all the lines in a particular series while y varies from line to line in the same series and is given by :

$$y = (x + 1), (x + 2)......\text{etc.}$$

Thus by assigning suitable values to x and y as indicated in the table, we can calculate the values of the wave number ($\overline{\upsilon}$) and wavelength (λ) of the spectral lines.

The table also makes it evident that Rydberg's equation for the five sectral series can be written as :

For Lyman series : x = 1 and y = 2, 3, 4, 5, 6, ... ∞

$$\overline{\upsilon} = \frac{1}{\lambda} = R\left(\frac{1}{1^2} - \frac{1}{y^2}\right) m^{-1} = R\left(1 - \frac{1}{y^2}\right) m^{-1}$$

For Balmer series : x = 2 and y = 3, 4, 5, 6 ... ∞

$$\overline{\upsilon} = \frac{1}{\lambda} = R\left(\frac{1}{2^2} - \frac{1}{y^2}\right) m^{-1} = R\left(\frac{1}{4} - \frac{1}{y^2}\right) m^{-1}$$

For Paschen series : $x = 3$ and $y = 4, 5, 6 ... \infty$

$$\bar{v} = \frac{1}{\lambda} = R\left(\frac{1}{3^2} - \frac{1}{y^2}\right)m^{-1} = R\left(\frac{1}{9} - \frac{1}{y^2}\right)m^{-1}$$

For Brackett series : $x = 4$ and $y = 5, 6 ... \infty$

$$\bar{v} = \frac{1}{\lambda} = R\left(\frac{1}{4^2} - \frac{1}{y^2}\right)m^{-1} = R\left(\frac{1}{16} - \frac{1}{y^2}\right)m^{-1}$$

For Pfund series : $x = 5$ and $y = 6, 7, ... \infty$

$$\bar{v} = \frac{1}{\lambda} = R\left(\frac{1}{5^2} - \frac{1}{y^2}\right)m^{-1} = R\left(\frac{1}{25} - \frac{1}{y^2}\right)m^{-1}$$

II. Absorption Spectra

If, in the arrangement to get a continuous spectrum of white light, a substance is placed between the white light source (*e. g.* sun) and the prism (or spectroscope), we get an **absorption spectrum.** This spectrum appears in the form of dark lines in a particular region of the continuous spectrum of white light. For example, if a solution of NaCl is placed between the sun (white light source) and prism, the continuous spectrum of white light is found to be crossed by two *dark lines* in the yellow region of the spectrum (Fig. 1.23). If the white light source is removed, the whole of the continuous spectrum disappears leaving behind only two, bright yellow lines. These lines are located at exactly the same place where two D_1 and D_2 lines of line spectrum of sodium were located. These dark lines are known as **absorption line spectrum.**

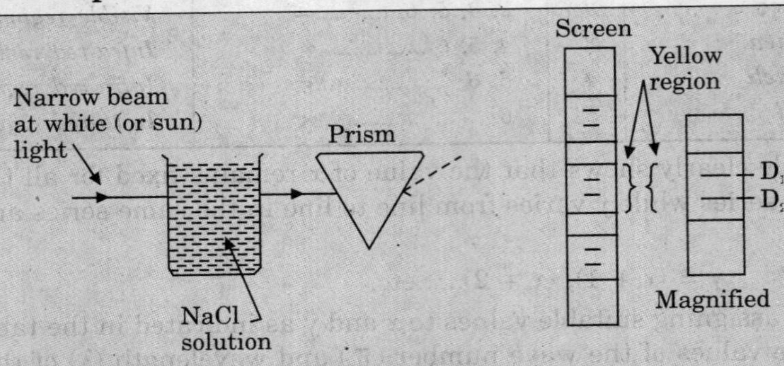

Fig. 1.23. Absorption spectrum.

Planck's Quantum Theory of Radiation : Quantisation of Energy.

In 1900 Max Planck studied the spectral lines obtained from thermal radiations emitted by a hot (black) body at different temperatures and put forward a theory which, after his name, is known as *Planck's quantum theory of radiation* (*quantum* is a Latin word which means *how much*).

Postulates of Plank's Quantum Theory of Radiation

Theory of Radiation. Various postulates of this theory are :

1. *A hot body emits radiant energy not continuously* (*i.e.* not as continous waves as suggested by the classical Max well's wave theory) *but discontinuously* (*discretely*). In other words this means that a *hot body emits radiant energy not as continuous waves but as small packets or bundles or discrete* (*separate*) *units of the waves.* Each of these units is called a **quantum** (plural is **quanta**) which can exist independently. Thus a quantum can be defined as follows :

A quantum is the smallest unit of waves in the form of which a hot body emits radiant energy and which can exist independently.

The emission of radiant energy in continuous waves and in discontinuous waves as individual quantum from a heated iron ball is shown in Fig. 1.24.

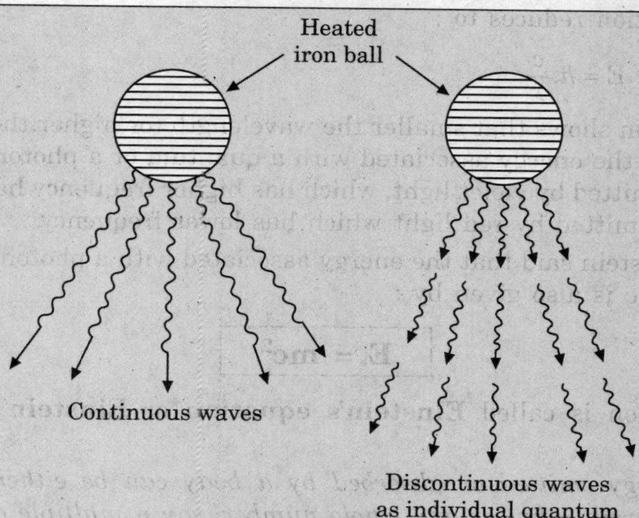

Fig. 1.24. Emission of radiant energy from a heated iron ball in (*a*) continuous waves (according to classical Maxwell's theory—old theory) and (*b*) discontinuous waves as individual quantum (according to Max Planck's quantum theory—modern theory).

In case of light which is also a form of radiation, light energy (*i.e.*, energy associated with light radiation) is emitted or absorbed in the form of packets or bundles each of which is called **photon** (instead of quantum). *Photon is not a material body.* It is considered to be a *massless bundle of energy.*

Thus we see that, according to this theory, *light is composed of massless particles which are called photons.* The presence of photons in light cannot be detected readily under normal conditions, since even light of low intensity consists of billions of photons. It was in 1905 when Einstein, while explaining the photoelectric effect, could prove the existence of photons.

2. *The energy* (E) *associated with each quantum or photon* (*in case of light*) *of a given radiation or light is proportional to the frequency* (v) *of the emitted radiation or light, i.e.*

$$E \propto v$$

or $$E = hv,$$

where h = a constant known as **Planck's constant** whose numerical value is 6.624×10^{-27} erg. s (in C.G.S. units) or 6.624×10^{-34} j-s (in S.I units).

E = energy associated with each quantum or photon of a given radiation. E is in ergs (in C.G.S. units) or in kJ (in S.I. units).

v = frequency of the emitted radiation or light.

This equation is applicable to all types of radiation and is called **planck's equation**. This equation shows that the energy associated with a quantum or a photon, E is equal to hv.

Now since $v = \frac{c}{\lambda}$, (c = velocity of radiation and λ = wave length of radiation), the above equation reduces to :

$$E = h.\frac{c}{\lambda}$$

This equation shows that smaller the wavelength (or higher the frequency) of radiation, large the energy associated with a quantum or a photon. For example a photon transmitted by violet light, which has higher frequency has more energy than that transmitted by red light which has lower frequency.

In 1905 Einstein said that the energy associated with a photon (E) with mass m and velocity c is also given by :

$$\boxed{\mathbf{E = mc^2}}$$

This equation is called **Einstein's equation** or **Einstein mass-energy relationship.**

3. *The energy emitted or absorbed by a body can be either equal to one quantum of energy (= hv) or any whole number, say n multiple of this unit i.e.*

Energy emitted or absorbed = $n \times hv$.

Thus the energy emitted or absorbed by a body can be equal to 1 hv, 2 hv, 3 hv, etc. but never equal to any fractional value of hv like 1.5 hv, 2.5 hv, 4.9 hv. Thus we find that the energy emitted or absorbed by a body is quantised and this is called the **concept of quantisation of energy.**

It is clear from the study of quantum theory of radiation as given above that the atoms would transfer energy in photon units. The absorption of a photon by an atom increases its energy by a definite quantity which is equal to hv. An atom which has absorbed energy in this way is said to be an *excited atom* or in the *excited state.* When an excited atom radiates energy, energy is given out in photon units.

The description given above makes it evident that this theory is also known as **photon theory**, *since according to the theory, light radiations are supposed to be composed of photons each of which is associated with energy equal to hv.*

Applications of Planck's Quantum Theory of Radiation

This theory has been used to explain many phenomena. For example :

(*i*) In 1905 Einsten applied this theory to explain the photoelectric effect.

(*ii*) In 1913 Bohr used this theory to explain the structure of the atom (See *Bohr's Model of Atomic Structure*) and the origin of spectral lines seen in hydrogen spectrum.

(*iii*) In 1922 Compton applied this theory to explain the phenomena of scattering of X-rays (*Compton Effect*).

Numerical Problems

Problem 1 (In S.I. units). *Two lines one voilet and the other red, in a spectrum, have wavelengths equal to 4000 and 6000 Å respectively. Calculate the energies of photons transmitted by them* ($h = 6.626 \times 10^{-39}$ *J-s*; $c = 3 \times 10^8\ m.s^{-1}$.)

Solution (*a*) *Energy of photon transmitted by violet light.*

Here $\lambda = 4000\ \text{Å} = 4000 \times 10^{-10}$ m

$$\therefore \quad E = h\nu = \frac{h.c}{\lambda}$$

$$= \frac{(6.626\times10^{-34}\ \text{J.s})\times(3\times10^{8}\ \text{ms}^{-1})}{4000\times10^{-10}\ \text{m}}$$

$$= \frac{6.626\times3\times10^{-16}}{4000}\ \text{J}$$

$$= \frac{6.626\times3\times10^{-19}}{4}\ \text{J}$$

$$= 4.9695\times10^{-19}\ \text{J}.$$

(*b*) *Energy of photon transmitted by red light.*

Here $\lambda = 6000\ \text{Å} = 6000 \times 10^{-10}$ m

$$\therefore \quad E = h\nu = \frac{h.c}{\lambda}$$

$$= \frac{(6.626\times10^{-34}\ \text{J.s})\times(3\times10^{8}\ \text{ms}^{-1})}{6000\times10^{-10}\ \text{m}}$$

$$= \frac{6.626\times3\times10^{-19}}{6}\ \text{J}$$

$$= 3.313\times10^{-19}\ \text{J}.$$

Problem 2. (In S.I. units) *Calculate the magnitude of the energy of the photon (or quantum) associated with light of wavelength* 6057.8 Å ($h = 6.6256 \times 10^{-44}$ *J-s*, $c = 3 \times 10^8\ ms^{-1}$)

Solution. Here $\lambda = 6057.8$ A $= 6057.8 \times 10^{-10}$ m. Now according to Planck's equation :

$$E = h\nu = \frac{h.c}{\lambda}$$

$$= \frac{(6.6256\times10^{-34}\ \text{J.s})\times(3\times10^{8}\ \text{ms}^{-1})}{6057.8\times10^{-16}\ \text{m}}$$

$$= 0.0032811\times10^{-16}\ \text{J}$$

$$= 3.2811\times10^{-19}\ \text{J}$$

Problem 3 (In S.I. units). *Electromagnetic radiation of wave length 242 nm is just sufficient to ionise the outermost electron of the sodium atom. Calculate the ionisation energy of sodium atom.* ($c = 3 \times 10^8$ *m.s*$^{-1}$, $h = 6.625 \times 10^{-34}$ *J.s*)

(Delhi B.Sc. Hons. 1975)

Solution. Here $c = 3 \times 10^8$ m.s^{-1} and $\lambda = 242$ nm $= 242 \times 10^{-9}$ m

$$v = \frac{c}{\lambda} = \frac{3\times10^8}{242\times10^{-9}} \text{ Hz}$$

$\therefore$ Energy (E) of this radiation $= hv$

$$= (6.625\times10^{-34})\times\frac{3\times10^8}{242\times10^{-9}} \text{ J/atom}$$

$= 8.21 \times 10^{-19}$ J

$= 8.21 \times 10^{-22}$ KJ

$= 8.21 \times 10^{-22} \times$ (Avogadro's number) KJ/mole

$= 8.21 \times 10^{-2} \times (6.023 \times 10^{23})$ KJ/mole

$= 494.48$ KJ/mole.

The radiation of energy equal to 494.48 KJ/mole is sufficient to ionise the outermost electron of Na atom. Thus ionisation energy of Na atom = 494.48 KJ/mole.

Problem 4 (In S.I. units). *Calculate the energy of a photon of light of frequency of* 3×10^{15} s^{-1}

Solution. Here $v = 3 \times 10^{-15}$s^{-1}

and $h = 6.624 \times 10$ J.s

$\therefore$ $E = hv = (6.624 \times 10^{-34}\text{Js}) \times (3 \times 10^{15}\text{s}^{-1})$

$= 6.624 \times 3 \times 10^{-19}$ J

$= 19.872 \times 10^{-19}$ J.

Problem 5 (In S.I. units). *Calculate the energy in KJ/mole associated with a radio wave of wavelength equal to* 3×10^{13} Å.

Solution. We know that :

$$E = hv = h.\frac{c}{\lambda}$$

Now since h is in Js, c is in ms^{-1} and λ is in Å, E cannot be obtained in joules. Thus in order to get E in joules, λ should be converted into m. Thus :

$\lambda = 3 \times 10^{13}$ Å $= 3 \times 10^{13} \times 10^{-10}$ m $= 3 \times 10^3$ m

$c = 3 \times 10^8$ ms^{-1}

$h = 6.626 \times 10^{-34}$ Js

$$\therefore \quad E = \frac{(6.626\times10^{-34}\text{ Js})\times(3\times10^8\text{ ms}^{-1})}{3\times10^3\text{ m}}$$

$= 6.626 \times 10^{-29}$ J/atom

Now since Avogadro's number $= 6.023 \times 10^{23}$

$\therefore$ $E = 6.626 \times 10^{-29} \times 6.023 \times 10^{23}$ J/mole

$$= \frac{6.626 \times 10^{-29} \times 6.023 \times 10^{23}}{10^{3}} \text{ KJ/mole}$$

$$= 6.626 \times 6.023 \times 10^{-9} \text{ KJ/mole}$$

$$= 0.6626 \times 6.023 \times 10^{-8} \text{ KJ/mole}$$

$$= 3.990 \times 10^{-8} \text{ KJ/mole}$$

Photoelectric Effect

What is Photoelectric Effect ?

In some of their experiments Sir J.J. Thomson and P. Lenard showed that *when a beam of light of suitable wavelength or frequency is allowed to fall on the surface of a metal, the electrons are emitted* (or *ejected*) *from the surface of the metal* (Fig. 1.25) This phenomenon of emission of electrons is known as **photoelectric effect** or **photoelectric emission** (*photo* means *light*). Although the electrons emitted from the surface are no different from all other electrons, it is customary to call there electrons as **photoelectrons.**

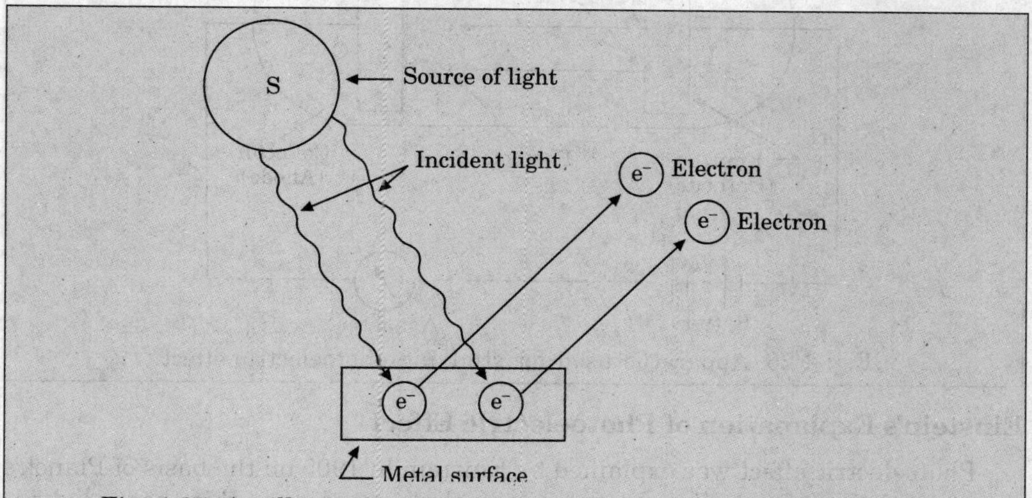

Fig. 1.25. An illustration of the phenomenon of photoelectric effect.

It may be noted that only a few metals can eject electrons when radiations from visible (which has less energy and hence is less energetic) fall on them, but many more metals are very much capable of ejecting electrons when radiations from ultraviolet light (which has more energy and hence is more energetic) fall on them.

Experimental Study of Photoelectric Effect

The phenomenon of photoelectric effect can be studied with the help of a simple apparatus shown in Fig. 1.26. The apparatus consists of two photosensitive plates, E (called *emitter*) and C (called *collector*) enclosed in an evacuated quartz tube which is called photoelectric cell. In the absence of light there is no flow the current in the circuit and the ammeter, A reads zero. When monochromatic light from source S falls on plate E connected to the negative terminal of the battery, current starts to flow as indicated by the ammeter. The explanation of the flow of the current lies in the fact that when plate E is irradiated with light, the incident photons colloid with the atoms of the plate and eject the electrons (called photoelectrons) from it. These photoelectrons are immediately attracted by the collector plate C and consequently current starts flowing as indicated by the ammeter. Since the effect is produced under the influence of light, it is called photoelectric effect (*photo* means light).

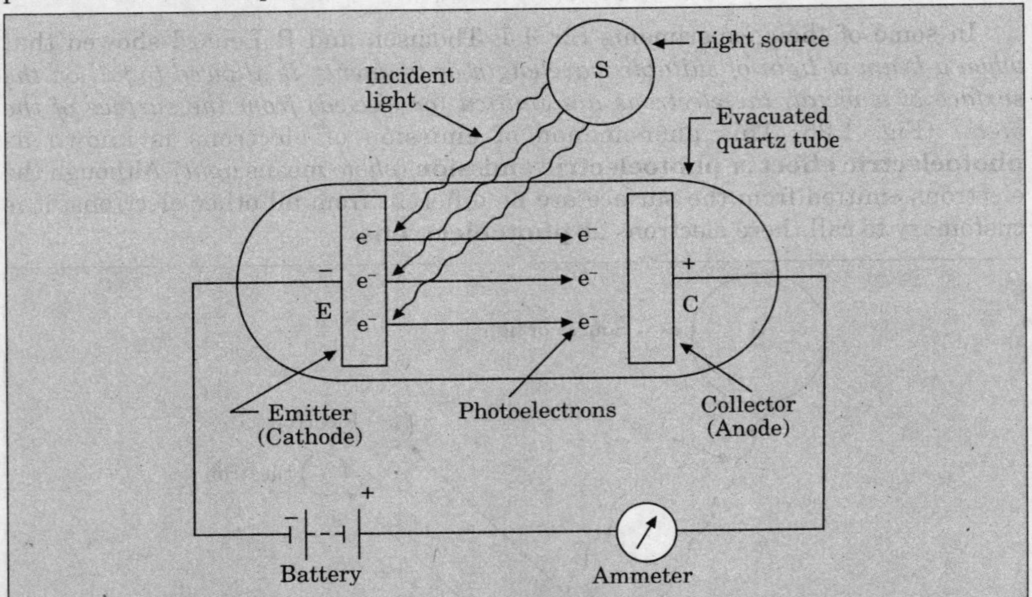

Fig. 1.26. Apparatus used for studying photoelectric effect.

Einstein's Explanation of Photoelectric Effect

Photoelectric effect was explained by Einstein in 1905 on the basis of Planck's quantum theory of radiation. We have seen that according to this theory light is composed of massless particles which are called photons. The energy associated with each photon, E_{photon} is given by Planck's equation :

$$E_{photon} = h\upsilon$$

When a photon of incident light which is associated with an energy equal to hv strikes a metal surface (Fig. 1.27), the photon is completely absorbed by the surface and transfers its energy (= $h\upsilon$) to an electron on the surface. The energy imparted by the photon is utilised for two purposes :

(*a*) Some portion of the photon energy (= $h\upsilon$) is used up in removing the electrons from the metal surface by overcoming the attraction between the electrons and the nucleus without giving them any velocity. Let the energy used up in this process be represented by W_0 which is called **photoelectric work function** or **threshold energy** of the metal and is defined as *the work to be*

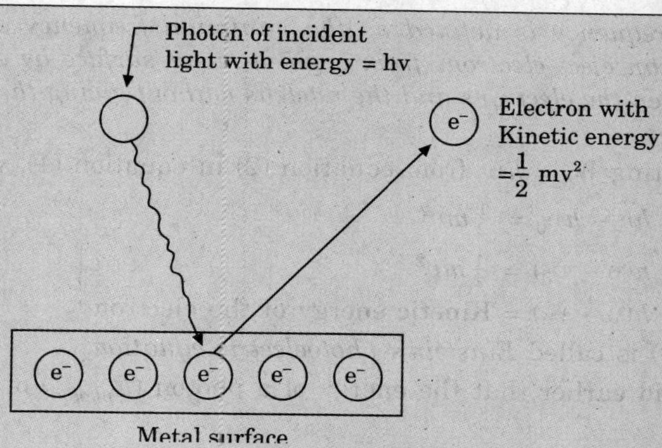

Fig. 1.27. When a photon of incident light with energy equal to $h\upsilon$ strikes a metal surface, an electron with kinetic energy equal to $\frac{1}{2}mv^2$ escapes from the metal surface.

done (or energy to be used up) to remove an electron from the metal surface with zero velocity and thus to convert it into photoelectron. For each metal surface there is a minimum value of W_0 to liberate a photoelectron.

(*b*) The remaining portion of photon energy (= $h\upsilon - W_0$) is converted into kinetic energy of the free electron (= $\frac{1}{2}mv^2$). Thus the kinetic energy of the free electron which is equal to $\frac{1}{2}mv^2$ (m = mass of the electron and v = velocity of the electron) is given by the difference of $h\upsilon$ and W_0, *i.e.*

$$h\upsilon - W_0 = \frac{1}{2}mv^2 \qquad ...(1)$$

The electron with kinetic energy equal to $\frac{1}{2}mv^2$ escapes from the metal surface.

Now if the energy of photon of incident light which is equal to $h\upsilon$ is just of such an amount that it can remove the electron only from the metal surface, then $h\upsilon$ will give no kinetic energy to the electron (*i.e.*, $\frac{1}{2}mv^2 = 0$) and will be equal to W_0, *i.e.*, in this case :

$$\frac{1}{2}mv^2 = 0$$

and hence from equation (1),

$$h\upsilon = W_0$$

Under this condition the incident light is of the minimum frequency which can emit the electrons from the metal surface. This minimum frequency is called **threshold frequency** or **cut-off-frequency** and is represented as υ_0. Hence :

$$h\upsilon_0 = W_0 \qquad ...(2)$$

Incident radiations having frequencies less than υ_0 will not eject the electrons from the metal surface while the incident radiations having frequencies greater than υ_0 can eject the electrons with a definite velocity and hence give kinetic energy to the free electron (= $\frac{1}{2}mv^2$). Quite obviously at the threshold frequency, υ_0 kinetic energy will be equal to zero. Thus υ_0 can be defined as follows :

Threshold frequency is defined as the minimum frequency of the incident radiation that can eject electrons from a given metal surface by overcoming the attraction between the electrons and the nucleus without giving them any velocity or kinetic energy.

On substituting $W_0 = h\upsilon_0$ from equation (2) in equation (1), we get :

$$hv - h\upsilon_0 = \tfrac{1}{2}mv^2$$

or $$h(\upsilon - \upsilon_0) = \tfrac{1}{2}mv^2$$

or $$h(\upsilon - \upsilon_0) = \text{Kinetic energy of the electron} \quad ...(3)$$

Equation (3) is called *Einstein's photoelectric equation.*

We have said earlier that the energy of a photon (E_{photon}) of frequency v is given by :

$$E_{photon} = h\upsilon$$

But the energy corresponding to the minimum frequency of photon (*i.e.* corresponding to thershold frequency, υ_0) is the minimum energy of the photon and is given by :

$$E_{photom} \text{ (minimum energy)} = h\upsilon_0$$

Here υ_0 is the minimum frequency of the photon.

Laws of Photoelectric Emission

From the experimental data collected by Richardson and Compton in 1921, photoelectric emission was found to be governed by the following laws.

1. *Relation between the intensity of incident light and the number of photo electrons.* If the frequency of the incident radiation is kept constant, then the number of electrons emitted from the metal surface in one second is directly proportional to the intensity or brightness of the incident light falling on the metal surface. Thus an increase in the intensity of incident light of a given frequency increases the number of electrons emitted per second from the metal surface, but the volocity and hence the kinetic energy of the emitted electron remains the same.

2. *Relation between the magnitude of frequency of incident light and the emission of electrons.* The emission of electrons from a metal surface depends on the magnitude of frequency of the radiations falling on the metal surface. Radiations of all frequencies can not make the metal surface to eject the electrons. Only those radiations which have a certain *critical* or *limiting minimum* value of frequency which is called *threshold frequency* or *cut-off frequency* (represented as υ_0) can eject the electrons. Below this value of υ_0 no electrons are emitted. From Einstein's photo-electric equation it is evident that if the frequency (υ) of the incident light is less than υ_0, $\frac{1}{2}mv^2$ (*i.e.* kinetic energy of the photoelectron) will be negative. Negative value of $\frac{1}{2}mv^2$ means that the incident radiation having υ less than υ_0 is not able to give kinetic energy to the electron and hence cannot eject electron from the metal surface. Thus an incident radiation having υ less than υ_0, no matter how long the radiation falls on the metal surface and how greater is its intensity, is not able to eject the electrons. On the other hand, since

the incident radiation having υ more than $υ_0$ is able to give velocity and hence kinetic energy (= $\frac{1}{2}mv^2$) to the ejected electron, it is able to produce the electron emission.

The magnitude of $υ_0$ depends on the nature of the metal, *e.g.*, $υ_0$ is quite low (visible light) for alkali metals like K, Na and Cs while it is quite high (ultraviolet light) for many other metals. $υ_0$ for Pt, Ag and K are 15.1×10^{11} s^{-1}, 11.3×10^{14} s^{-1} and 4.22×10^{14} s^{-1} respectively.

The wavlength corresponding to the value of $υ_0$ is called **threshold wavelength or long wavelength limit** and is represented as λ_0. No photoelectrons are emitted by the incident light having wavelength greater than λ_0.

3. *Relation between frequency of the incident light and kinetic energy of the photoelectron.* Einstein's photoelectric equation shows that since m (mass of the electron) remains constant, an increase in the magnitude of the frequency (υ) of the incident radiation will increase the velocity (v) with which the photoelectrons are ejected. An increase in the velocity will increase the kinetic energy of the photoelectrons and hence maximum kinetic energy of the emitted electron [(K. E.)$_{max}$] is directly proporticnal to the frequency of the incident radiation. Consequently the magnitude of kinetic energy of the photoelectrons will increase with an increase of the frequency of the incident radiation above the threshold frequency and hence a plot between the kinetic energy of the photoelectrons and the frequency of the incident radiation will be a straight line as shown in Fig. 1. 28.

The intercepts on the energy axis and frequency axis were found to be equal to $- W_0$ and W_0/h respectively. Now since W_0 is equal to $hυ_0$, W_0/h is equal to $hυ_0/h = υ_0$. Thus the intercept on the frequency axis is equal to the threshold

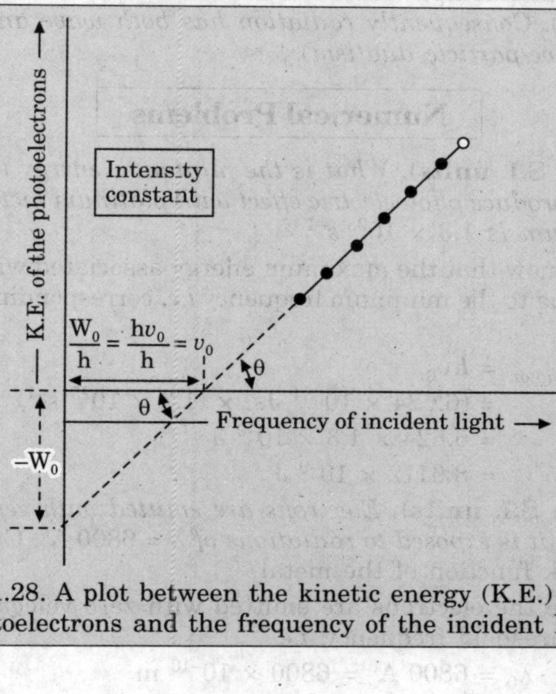

Fig. 1.28. A plot between the kinetic energy (K.E.) of the photoelectrons and the frequency of the incident light.

frequency, υ_0. The slope of the straight line was determined by Millikan who showed that the slope is given by :

$$\tan \theta = \frac{W_0}{W_0/h} = h$$

From a determination of this slope, Millikan found the value of h to be 6.570 × 10^{-34} J.s. This value is almost equal to that of Planck's constant, h (h = 6.624 × 10^{-34} Js) as determined experimentally by Planck himself from black-body measurements.

4. *Relation between the intensity of incident light and kinetic energy of photo electrons.* Since Eisten's photoelectric equation does not contain any term related to the intensity of radiation, kinetic energy remains independent of the intensity of radiation.

5. *The velocities of emitted electrons have values between zero and a definite maximum. The proportion of the electrons having a particular velocity is independent of the intensity of the radiation.*

6. *Electrons are emitted almost instantaneously even when the intensity of incident radiations is very low. The time log between the incidence of radiation and emission of first electrons is less than 10^{-8} seconds.*

The points given above could not be explained by the classical Maxwell's wave theory of radiation, since, if radiations are continuous as assumed by wave theory, any radiation should be able to accumulate enough energy to knock out the electrons from the metal surface. But this has not been found true in practice.

Thus we see that Einstein has explained photoelectric effect by assuming that radiations which were considered as waves (*Maxwell's wave theory of radiation*) are also considered as small particles which are called *photons.* (*Planck's quantum theory of radiation*). *Consequently radiation has both wave and particle nature simultaneously* (*wave-particle dualism*).

Numerical Problems

Problem 1 (In S.I. units). *What is the minimum energy that photons must possess in order to produce photoelectric effect with platinum metal ? The threshold frequency of platinum is* 1.3 × 10^{25} s^{-1}

Solution. We know that the maximum energy associated with a photon is the energy corresponding to the minimum frequency *i.e.* corresponding to the threshold frequency, υ_0.

$$\therefore \quad E_{photon} = h\upsilon_0$$
$$= (6.624 \times 10^{-34} \text{ Js}) \times (1.3 \times 10^{25} \text{ s}^{-1})$$
$$= 6.624 \times 1.3 \times 10^{-9} \text{ J}$$
$$= 8.6112 \times 10^{-9} \text{ J}$$

Problem 2 (In S.I. units). *Electrons are emitted with zero velocity from a metal surface when it is exposed to radiations of* λ = 6800 A^0. Calculate threshold frequency and work function of the metal.

Solution. Since the electrons are emitted with zero velocity from the metal surface, λ is the threshold frequency, *i.e.*

$$\lambda_0 = 6800 \text{ A}^0 = 6800 \times 10^{-10} \text{ m}$$

∴ Threshold frequency,

$$\upsilon_0 = \frac{c}{\lambda_0} = \frac{3\times10^8 \text{ m.s}^{-1}}{6800\times10^{-10} \text{ m}}$$

$$= \frac{3\times10^8}{68\times10^2\times10^{-10}} \text{ s}^{-1}$$

$$= \frac{3\times10^{16}}{68} \text{ s}^{-1} = 0.0441\times10^{16} \text{ s}^{-1}$$

$$= 0.441\times10^{15} \text{ s}^{-1}$$

Work function, $W_0 = h\upsilon_0$

$$= (6.624 \times 10^{-34} \text{ Js}) \times (0.441 \times 10^{15} \text{ s}^{-1})$$

$$= 6.624 \times 0.441 \times 10^{-19} \text{ J}$$

$$= 2.9145 \times 10^{-19} \text{ J}$$

Problem 3 (In C.G.S. units). *Calculate the kinetic energy of an electron emitted from a surface of potassium metal by light of wavelength* 5.5×10^{-8} cm (*work function* = 3.62×10^{-12} *erg*).

Solution. Here $\lambda = 5.5 \times 10^{-3}$ cm

$$\therefore \quad \upsilon = \frac{c}{\lambda} = \frac{3\times10^{10} \text{ cm.s}^{-1}}{5.5\times10^{-8} \text{ cm}} = 0.545\times10^{18} \text{ s}^{-1}$$

Now kinetic energy of the emitted electron

$$= \frac{1}{2}mv^2 = h\upsilon - h\upsilon_0$$

$$= h\upsilon - W_0$$

$$= (6.624\times10^{-27} \text{ erg.s})\times(0.545\times10^{18} \text{ s}^{-1}) - (3.62\times10^{-12} \text{ erg})$$

$$= (3.6100\times10^{-9} - 3.62\times10^{-12}) \text{ erg}$$

$$= \left[3.6100 - \frac{3.62}{10^3}\right]\times10^{-9} \text{ erg}$$

$$= (3.6100 - 0.00362)\times10^{-9} \text{ erg}$$

$$= 3.6063\times10^{-9} \text{ erg.}$$

Problem 4 (In S.I. units). *Photoelectric threshold wavelength of metallic copper is* λ = 3000 A^0. *Find out (i) the maximum kinetic energy of the photo-electrons ejected and (ii) the work function of the metal, when ultra violet light of* λ = 2536 A^0 *falls on the metal surface.*

Solution. Here $\lambda_0 = 3000 \text{ A}^0 = 3000 \times 10^{-10}$ m

$$\lambda = 2536 \text{ A}^0 = 2536 \times 10^{-10} \text{ m}$$

$$\therefore \quad \upsilon_0 = \frac{c}{\lambda_0} = \frac{3\times10^8 \text{ m.s}^{-1}}{3000\times10^{-10} \text{ m}} = 1\times10^{15} \text{ s}^{-1}$$

and $$\upsilon = \frac{3\times10^8 \text{ m.s}^{-1}}{2536\times10^{-10} \text{ m}} = 1.18\times10^{15} \text{ s}^{-1}$$

Maximum kinetic energy = $E_{max} = \frac{1}{2}mv^2_{max}$

$$= h\upsilon - h\upsilon_0$$
$$= h\ (\upsilon - \upsilon_0)$$
$$= 6.624 \times 10^{-34} \times (1.18 \times 10^{15} - 10^{15})\ \text{J.s.s.}^{-1}$$
$$= 6.624 \times 10^{-34} \times 10^{15} \times 0.18\ \text{J}$$
$$= 1.1923 \times 10^{-19}\ \text{J.}$$

Evidently, work function (W_0) is given by :

$$W_0 = h\upsilon_0$$
$$= (6.624 \times 10^{-34}\ \text{Js}) \times (10^{15}\ \text{s}^{-1})$$
$$= 6.624 \times 10^{-19}\ \text{J.}$$

Scattering of X-rays—Compton Effect

In 1923 during the course of his experiments on the scattering of X-rays by solid matter (called *scattering substance* or *scatterer*), A.H. Compton found that when X-rays of high energy are allowed to fall on solid matter like carbon block or some other light element, an electron is ejected from the carbon block and X-rays are scattered from their original path. In this process a photon from the incident X-rays collides with the loosely-bound electron in the carbon block (scatterer) and, after receiving an impact from the photon, the electron which was initially at rest gains some velocity and hence moves below the direction of the incident X-rays while the photon is deflected above. Thus we see that X-rays are scattered from their original path as a result of the collision between the photon and the electron. The collision between the photon and the electron can be regarded as a collision between two balls—one moving and the other at rest. It has been found that the scattered X-rays have longer wavelength (*i.e.* lower frequency or lower energy) than the incident X-rays. If the wavelengths of the incident X-rays and the scattered X-rays and λ and λ' respectively, then hc/λ and hc/λ' are the energies associated with the photon of the incident and scattered X-rays respectively. Thus the decrease in energy is equal to $\left(\frac{hc}{\lambda} - \frac{hc}{\lambda'}\right)$, since $hc/\lambda > hc/\lambda'$. This energy $\left(= \frac{hc}{\lambda} - \frac{hc}{\lambda'}\right)$ is transferred to the electron at rest and therefore, the electron gains some velocity and moves from its path in this process. *This phenomenon in which there is a change in wavelength of the scattered X-rays is called* **Compton effect.** The angle through which the incident X-rays are deflected from their original path is known as *angle of scattering* and is denoted by θ. Thus θ is the angle between the incident and the scattered X-rays (See Fig. 1.29).

By applying the laws of conservation of energy and of momentum, and by assuming that X-rays are composed of photons each of which has energy equal to $h\upsilon$, Compton showed that the wavelength, λ' of the scattered X-rays and the wavelength, λ of the incident X-rays are related to each other by the relation :

$$\lambda' - \lambda = \frac{h}{mc}(1 - \cos\theta)$$

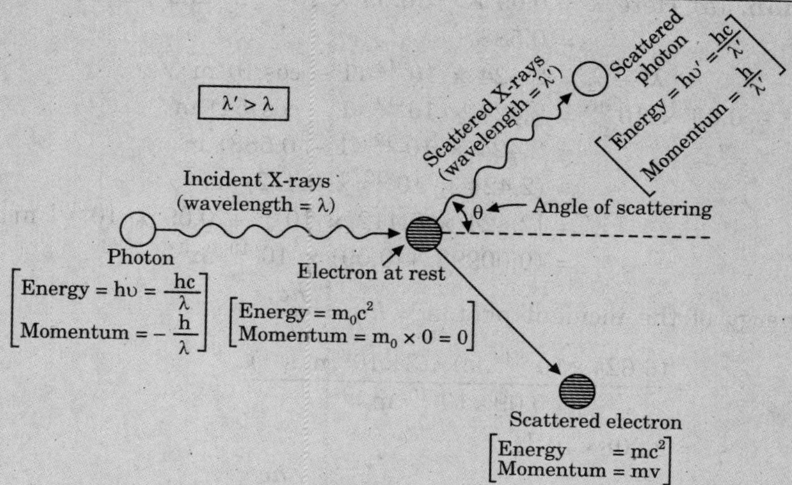

Fig. 1.29. Impact of photon from X-rays on an electron at rest—An illustration of Compton effect.

where h is the Planck's constant, m is the rest mass of the electron and c is the velocity of light. The difference in λ' and λ [*i.e.* $(\lambda' - \lambda)$] is called *Compton wavelength shift* and is denoted by $\Delta\lambda$. Thus :

$$\textit{Increase in wavelength} = \Delta\lambda = \lambda' - \lambda = \frac{h}{mc}(1 - \cos\theta).$$

This equation clearly shows that $\Delta\lambda$ depends on the magnitude of the angle of scattering, θ and is produced as a result of the collision between the photon of incident X-rays and electron at rest (photon-electron collision). This equation also shows that $\Delta\lambda$ is indepedent of the nature of the scattering substance and the wavelength of the incident X-rays. The term h/mc is called *Compton wavelength* whose value, in S.I. units, is equal to 2.424×10^{-12} metre as shown below :

$$\frac{h}{mc} = \frac{6.624\times10^{-34}\ \text{Js}}{(9.108\times10^{-31}\ \text{kg})\times(3\times10^{8}\ \text{ms}^{-1})}$$

$$= \frac{6.624\times10^{-11}\ \text{kg.m}^2.\text{s}^{-2}.\text{s}}{9.108\times3\ \text{kg.m.s}^{-1}} \quad (\because \text{J} = \text{kg.m}^2.\text{s}^{-2})$$

$$= 0.2424 - 10^{11}\ \text{m}$$

$$= 2.424\times10^{-12}\ \text{m}$$

Thus :

$$\Delta\lambda = \lambda' - \lambda = 2.424 \times 10^{-12}\,(1 - \cos\theta)\ \text{m}$$

$$= 2.424 \times 10^{-12} \times (2\sin^2\theta/2)\ \text{m}$$

$$= 4.848 \times 10^{-12}\ \sin^2\theta/2\ \text{m}$$

Numerical problem. *A beam of monochromatic X-rays with* λ = 0.09 A° *is incident on a carbon scatterer. The scattered X-rays are observed at an angle of* 54° *with the direction of the incident beam. Find* (*a*) *the wavelength of the scattered X-rays, and* (*b*) *the energies of the incident and scattered photons.*

Solution. (a) Here λ = 0.09 A° = 0.09×10^{-10} m and cos 54° = 0.588.

$$\therefore \quad \lambda' - \lambda = 2.424 \times 10^{-12}\,(1 - \cos\theta)\text{ m}$$

$$\lambda' - 0.09 \times 10^{-10} = 2.424 \times 10^{-12}\,(1 - \cos 54)\text{ m}$$

$$= 2.424 \times 10^{-12}\,(1 - 0.588)\text{ m}$$

$$= (2.424 \times 10^{-12} \times 0.412)\text{ m}$$

$$\therefore \quad \lambda' = [2.424 \times 0.412 \times 10^{-12} + 0.09 \times 10^{-10}]\text{ m}$$

$$= (0.00998 + 0.09) \times 10^{-10}\text{ m}$$

(b) Energy of the incident photon = $h\upsilon = \dfrac{hc}{\lambda}$

$$= \frac{(6.624\times10^{-34}\text{ Js})\times(3\times10^{8}\text{ m.s}^{-1})}{0.09\times10^{-10}\text{ m}}$$

$$= 2.208\times10^{-14}\text{ J}$$

(c) Energy of the scattered photon = $h\upsilon' = \dfrac{hc}{\lambda'}$

$$= \frac{(6.624\times10^{-34})\times(3\times10^{8})}{0.09998\times10^{-10}}\text{ J}$$

$$= 19.875\times10^{-16}\text{ J.}$$

Postulates of Bohr's Atomic Model

Bohr's atomic model retains the two essential features of Rutherford's atomic model which are :

(*i*) The atom has a very small positively charged nucleus at its centre. All the protons and neutrons are contained in the nucleus. Thus most of the mass of the atom is concentrated in the nucleus.

(*ii*) Negatively charged electrons are revolving round the nucleus in the same way as planets are revolving round the sun.

However, he applied Planck's quantum theory to the revolving electrons and thus explained why the electrons, while revolving round the nucleus, do not lose energy and consequently do not fall into the nucleus.

While describing his model of atomic structure, Bohr made the following postulates.

1. *Fixed circular orbits.* Bohr assumed that an electron is a material particle which is revolving round the nucles in concentric circular orbits situated at *definite* (*i.e.*, fixed) *distance* from the nucleus and with a *definite velocity.*

2. *Stationary energy levels.* As long as an electron remains in a particular orbit, it neither emits (*i.e.* radiates or loses) nor absorbs (*i.e.,* gains) energy. Thus in a particular orbit the energy of a revolving electron remains constant or stationary. Hence each of the fixed orbits is associated with a definite amount of energy, *i.e.*, with a definite whole number of quanta of energy. The orbits are, therefore, also called *stationary energy levels* or simply *energy levels or energy shells.* This concept of stationary energy levels explains the stability of the atom, since an electron cannot lose energy gradually and fall into the nucleus.

Different energy levels have been represented by n which can have integer (whole number) values like 1, 2, 3, ... ∞, starting from the nucleus. n has been called **principal quantum number** by Bohr. Different values of n are also represented by capital letters K, L, M, N, etc. Thus:

Principal quantum number $(n) \rightarrow$	*1*	*2*	*3......*
Letter designation of energy level $\rightarrow$	*K (1st shell)*	*L(2nd shell)*	*M(3rd shell)*

Thus for K shell (*1st* shell) $n = 1$, for L shell (*2nd* shell) $n = 2$ and so on. K, L, M......shells are also called $n = 1$ shell, $n = 2$ shell, $n = 3$ shell, etc. respectively. Representation of various orbits is depicted in Fig. 1.30.

The energy associated with a certain energy level increases with the increase of its distance from the nucleus. Thus, if E_1, E_2, E_3, ... , etc. are the energies associated with the energy levels numbered as 1 (*K*-shell), 2 (*L*-shell), 3 (*M*-shell), etc. the order of the energies is as :

$$E_1 < E_2 < E_3 < \text{.........}, \text{ etc.}$$

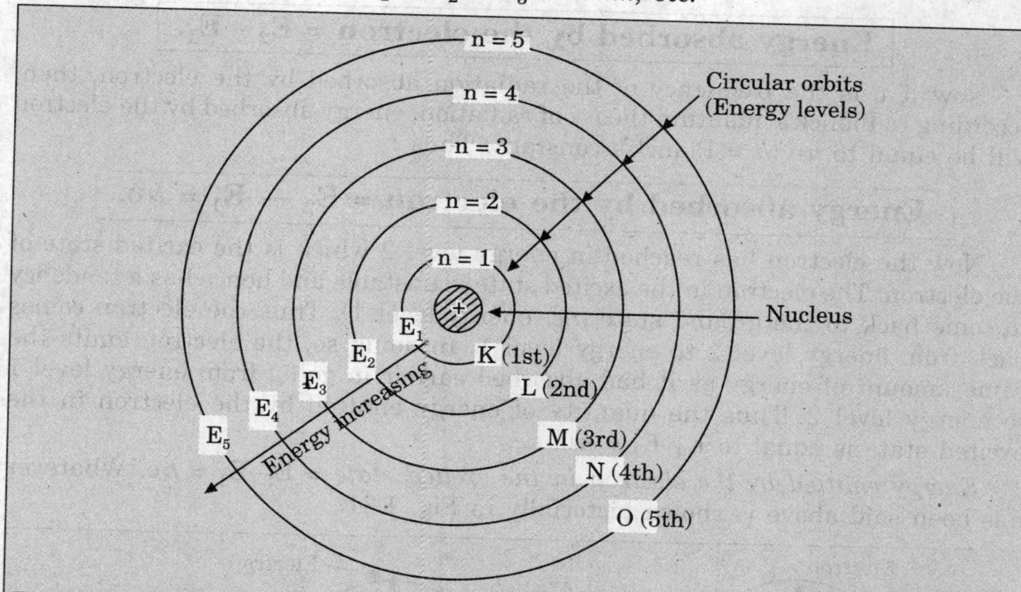

Fig. 1.30. Representation of various orbits (energy levels) round the nucleus. E_1, E_2, E_3......, etc. are the energies associated with orbits having $n = 1$ (K-shell), 2(L-shell)..., etc. Energies of different orbits are in the order : $E_1<E_2<E_3<E_4<E_5<...$

3. *Jumping of an electron from one energy level to the other—Ground state and excited state of an electron.* As long as an electron remains in a particular orbit, it neither emits (*i.e.*, radiates or loses) nor absorbs (*i.e.*, gains) energy. But when the electron is excited from a lower energy level to a higher energy level, it absorbs energy. On the other hand when it comes back from a higher energy level to a lower energy level, it emits energy. According to Planck's quantum theory of radiation, the absorption or emission of energy takes place not continuously but discontinuously (discretely), *i.e.*, the absorption or emission of energy takes place not as continuous waves but as small packets or bundles or discrete (separate) units of the waves each of which is called a quantum (plural is quanta) or photon. In short we say that the absorption or emission of energy takes place in the form of quanta or photons.

When an electron is present in energy level 1 which is the lowest energy level, the electron is said to be in the *ground state*. Thus energy level 1 is called the ground state of the electron or atom. Since energy level I (*i.e.*, ground state) has the minimum energy, ground state is the most stable state of the atom. Now if energy is supplied to the electron residing in energy level 1, it will absorb the supplied energy in the form of quanta or photons (Planck's quantum theory) and hence will move to one of the higher energy levels 2, 3 ......, etc. depending on the amount of energy absorbed by the electron. The electron present in these energy levels is said to be in the *excited state*.

Suppose an electron moves from energy level 1 to energy level 2. Let the energies associated with the electron present in these levels be E_1 and E_2 respectively. Then the amount of energy absorbed by the electron in coming from energy level 1 to energy level 2 is equal to $E_2–E_1$, *i.e.*

Energy absorbed by the electron = $E_2 - E_1$.

Now if υ is the frequency of the radiation absorbed by the electron, then, according to Planck's quantum theory of radiation, energy absorbed by the electron will be equal to $h\upsilon$ (h = Planck's constant) Thus :

Energy absorbed by the electron = $E_2 — E_1 = h\upsilon$.

Now the electron has reached in energy level 2 which is the excited state of the electron. The electron in the excited state is unstable and hence has a tendency to come back to the ground state (*i.e.* energy level 1). Thus the electron comes back from energy level 2 to energy level 1. In doing so, the electron emits the same amount of energy as it had absorbed earlier in going from energy level 1 to energy level 2. Thus the quantity of energy emitted by the electron in the excited state is equal to $E_2–E_1 = h\upsilon$, *i.e.*

Energy emitted by the electron in the excited state = $E_2–E_1 = h\upsilon$. Whatever has been said above is shown pictorially in Fig. 1.31.

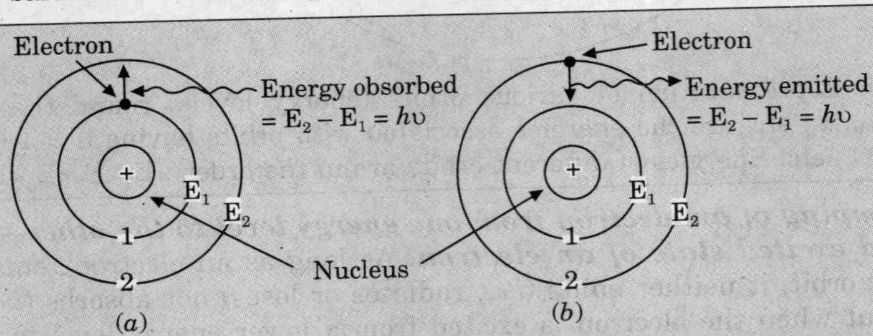

Fig. 1.31. (*a*) When an electron in energy level 1 (ground state) moves to energy level 2 (excited state), it absorbs energy equal to $E_2–E_1 = h\upsilon$. (*b*) When the electron in energy level 2 comes back to energy level 1, it emits the same amount of energy as it had absorbed earlier in going from energy level 1 to energy level 2, *i.e.*, it emits energy equal to $E_2 — E_1 = h\upsilon$.

4. *Principle of quantisation of angular momentum of the moving electron.* Bohr postulated that an electron cannot move in all the orbits. It can move only in that orbit in which the angular momentum of the electron (= mvr) moving round the nucleus is integral whole number multiple of $\frac{h}{2\pi}$ such as $1h/2\pi$, $2h/2\pi$, $3h/2\pi$, ..., $nh/2\pi$. Thus according to this postulate the angular momentum of a moving electron is given by :

$$mvr = n.\frac{h}{2\pi} \qquad ...(1)$$

Thus according to the postulate represented by equation (1), *angular momentum of a moving electron is quantinised.*

In equation (1) m = mass of the electron; v = velocity of the electron; h = Planck's constant = 6.624×10^{-27} erg-s; r = radius of the orbit in which the electron is moving; and n = an integral which has been called principal quantum number by Bohr. n gives the number of the orbit in which the electron is moving. It can have the values 1, 2, 3, ... for the orbit numbered as 1 (K orbit), 2 (L orbit), 3 (M orbit) from the nucleus. Thus for K, L, M, ... orbits n = 1, 2, 3, ... respectively as shown below :

Orbits →	K	L	M	N ...
Principal quantum number (n) →	1	2	3	4 ...

Bohr's Theory As Applied to Hydrogen Atom

By using the postulates of his theory and some classical laws of Physics, Bohr was able to calculate the radius of each orbit of H-atom, energy associated with the electron in each orbit of H-atom, velocity of electron in each orbit of H-atom, and the frequency and wavelength of the spectral lines observed in hydrogen spectrum.

1. Radii of Various Orbits in Hydrogen Atom

(*a*) **In C.G.S. units.** Let us consider a hydrogen atom. Quite obviously this atom contains only one electron revolving round the nucleus and only one proton present in the nucleus. Let the solitary electron have a charge equal to $-e$. Thus the proton will bear a charge equal to $1 \times (+e) = +1e$. Here 1 indicates the number of protons or the atomic number of hydrogen atom. Thus the charge carried by the nucleus of H-atom is equal to $+1e$. Let the radius of the circular orbit in which the electron is revolving round the nucleus be r. Let v be the tangential velocity and m, the mass of the revolving electron (See Fig. 1.32).

Now the revolving electron is being acted upon by the following two types of forces which act on the revolving electron simultaneously.

(*i*) *Electrostatic force of attraction*. This force acts between the nucleus and the revolving electron, and tends to attract the electron towards the nucleus. According to Coulomb's law this force is equal to :

$$\frac{e \times e}{r^2} = \frac{e^2}{r^2}$$

Here e is in electrostatic units (*e.s.u.*) and r is in centimetres (cm).

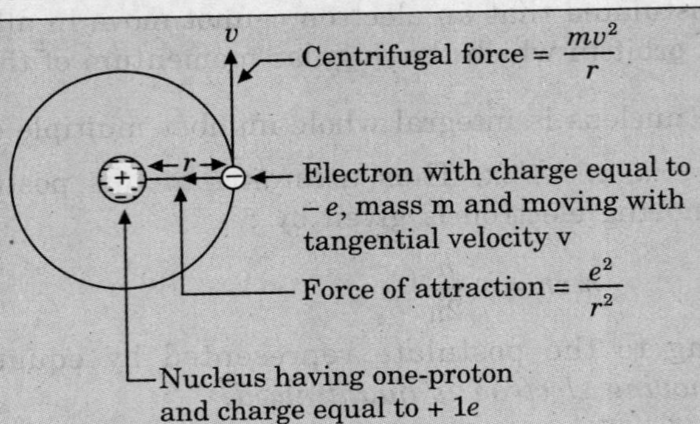

Fig. 1.32. Different types of forces acting on an electron revolving round the nucleus in hydrogen atom.

(*ii*) *Centrifugal or centripetal force*. This force tends to take the electron away from its orbit and is equal to:

$$\frac{mv^2}{r}$$

Here m is in grams.

Both the above forces act in opposite directions. Bohr assumed that in order to keep the electron revolving in its orbit, both these forces must be equal to each other, *i.e.*

$$\frac{e^2}{r^2} = \frac{mv^2}{r}$$

or $$v^2 = \frac{e^2}{r\,m} \qquad ...(1)$$

Now since the electron is revolving in a circular orbit, according to one of the postulates of Bohr's theory, the angular momentum of the revolving electron ($= mvr$) must be equal to $\frac{nh}{2\pi}$ where n is the number of the orbit in which the electron is revolving, *i.e.*

$$mvr = n.\frac{h}{2\pi}$$

or $$v^2 = \frac{n^2h^2}{4\pi^2\; m^2r^2} \qquad ...(2)$$

Here h is in erg. sec (erg-s)

Eliminating v^2 from equations (1) and (2), we get

$$\frac{e^2}{rm} = \frac{n^2h^2}{4\pi^2\; m^2r^2}$$

or $$r = \frac{n^2h^2}{4\pi^2\; me^2} \qquad ...(3)$$

If r_n is the radius of the nth orbit of hydrogen atom, then r_n will be given by :

$$\mathbf{r}_n = \frac{\mathbf{n^2h^2}}{4\pi^2 \ \mathbf{me^2}} \quad ...(4)$$

The value of r_n can be calculated by substituting the experimental values of h, m and e in equation (4) given as h (Planck's constant) = 6.624×10^{-27} *erg. s*; π (*pi*) = 3.14; m (mass of the revolving electron in grams) = 9.108×10^{-28} g; e (charge on electron in *e.s.u*) = 4.8×10^{-10} *e.s.u.* Thus :

$$r_n = \frac{n^2 \times (6.624 \times 10^{-27}\ erg.s)^2}{4 \times (3.14)^2 \times (9.108 \times 10^{-28}\ g) \times (4.8 \times 10^{-10}\ e.s.u.)^2}$$

Now since $erg = g.cm^2 \ . \ s^{-2}$

and $esu = g^{1/2} \ . \ cm^{3/2}. \ s^{-1}$,

$$r_n = \frac{n^2 \times (6.62 \times 10^{-27}\ g.cm^2.s^{-2}.s)^2}{4 \times (3.14)^2 \times (9.108 \times 10^{-28}\ g) \times (4.8 \times 10^{-10}\ g^{1/2}.cm^{3/2}.s^{-1})^2}$$

$$= n^2 \times 0.529 \times 10^{-8}\ cm$$

or $$\mathbf{r}_n = \mathbf{n^2 \times 0.529\ A^\circ} \quad (\because 10^{-8}\ cm = 1\ A^\circ) \quad ...(5)$$

On putting n = 1, 2, 3..., etc. in equation (5), the values of the radii of 1st (n = 1), 2nd (n = 2), 3rd (n = 3), etc. orbits of hydrogen can be calculated as shown below :

Radius of 1*st* orbit (n = 1), $r_1 = 1^2 \times 0.529$ Å = 0.529 Å.

Radius of 2*nd* orbit (n = 2), $r_2 = 2^2 \times 0.529$ Å = 2.11 Å.

Radius of 3*rd* orbit (n = 3), $r_3 = 3^2 \times 0.529$ Å = 4.75 Å.

Radius of 4*th* orbit (n = 4), $r_4 = 4^2 \times 0.529$ Å = 8.45 Å.

Radius of 5*th* orbit (n = 5), $r_5 = 5^2 \times 0.529$ Å = 13.22 Å.

Radius of the 1*st* orbit of hydrogen atom is also represented by other symbols like r_H, a_0 etc. Radius of the 1st orbit of hydrogen atom is called **Bohr radius.** Thus :

$$\mathbf{r_1 (or\ r_H\ or\ a_0)} = \frac{\mathbf{h^2}}{4\pi^2\mathbf{me^2}} \quad ...(6)$$

In order to find out the values of r_2, r_3, r_4 etc. in terms of r_1, we divide equation (4) by equation (6)

$$\frac{r_n}{r_1} = \frac{n^2h^2}{4\pi^2me^2} \times \frac{4\pi^2me^2}{h^2} = n^2$$

or $$\mathbf{r_n = r_1 \times n^2} \quad ...(7)$$

The values of the radii of the first five orbits of hydrogen atom, as calculated above, are represented in Fig. 1.33.

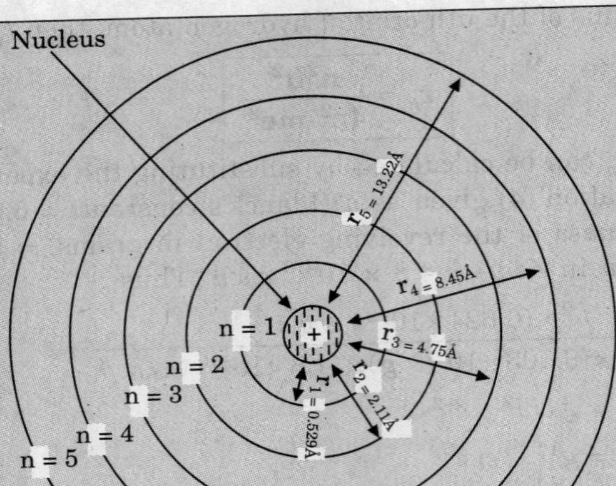

Fig. 1.33. Radii (in Å) of the first five orbits of hydrogen atom. The figure has not been drawn to scale.

(*b*) **In S.I. units.** In S.I. units the electrostatic force of attraction acting on the revolving electron is equal to :

$$\frac{1}{4\pi \in_0} \times \frac{e^2}{r^2}$$

Here e is the electronic charge in coulombs, r is the distance between the nucleus and the revolving electron, expressed in metres and $\in_0$ is a constant which is called absolute *permittivity* of the medium (*viz.* vacuum in this case). The value of $\in_0$ is equal to 8.85×10^{-12}.

Now in order to keep the electron revolving in its orbit, the electrostatic force of attraction $\left(= \frac{1}{4\pi \in_0} \times \frac{e^2}{r^2}\right)$ must be equal to the centrifugal force $\left[= \frac{mv^2}{r} \text{ in which } m \text{ is in kilogram (kg)}\right]$, *i.e.*,

$$\frac{1}{4\pi \in_0} \cdot \frac{e^2}{r^2} = \frac{mv^2}{r}$$

or $$v^2 = \frac{e^2}{4\pi \in_0 r\, m} \quad \text{...(8)}$$

Now, $$mvr = n.\frac{h}{2\pi}$$

or $$v^2 = \frac{n^2h^2}{4\pi^2m^2v^2} \quad \text{...(9)}$$

Here h is in joules-sec (J-s).

Eliminating v^2 from equations (8) and (9), we get

$$r = \frac{\in_0 n^2h^2}{\pi\ me^2}$$

i.e.

$$\boxed{\mathbf{r_n} = \frac{\in_0 \mathbf{n^2h^2}}{\pi\ \mathbf{me^2}}} \quad ...(10)$$

Substituting the experimental values of $\in_0$, h, m and e in equation (10) given as $\in_0 = 8.85 \times 10^{-12}$; $h = 6.624 \times 10^{-34}$ J-s; $\pi = 3.14$; $m = 9.108 \times 10^{-31}$ kg; and $e = 1.602 \times 10^{-19}$ C, we get

$$r_n = \frac{(8.85\times10^{-12})\times n^2\times(6.624\times10^{-34}\ \text{J-s})^2}{3.14\times(9.108\times10^{-31}\ \text{kg})\times(1.602\times10^{-19}\ \text{C})^2}$$

$$= \frac{8.85\times(6.62)^2\times10^{-11}\times n^2\times(\text{kg}.m^2.s^{-2}.s)^2}{3.14\times9.108\times(1.602)^2\times\text{kg}\times(\text{kg}^{1/2}.m^{3/2}.s^{-1})^2} \quad \left[\because \begin{array}{l}\text{J} = \text{kg}.m^2.s^{-2}\\ \text{C} = \text{kg}^{1/2}.m^{3/2}.s^{-1}\end{array}\right]$$

$$= \frac{0.885\times(6.62)^2\times10^{-10}\times n^2}{3.14\times9.108\times(1.602)^2}\,m$$

or

$$\boxed{\mathbf{r}_n = \mathbf{n^2}\times\mathbf{0.529}\times\mathbf{10^{-10}\ m}} \quad ...(11)$$

2. Radii of H-like Species

So far we have considered a hydrogen atom which contains one electron and one proton. Now let us consider hydrogen-like species such as He^+ (At. No., $Z = 2$), Li^{2+} ($Z = 3$), etc. in which the number of electrons is equal to 1 but the number of protons is more than 1. The number of protons in He^+ and Li^{3+} ions is 2 and 3 respectively. Thus the charge carried by the nucleus of these ions is equal to + 2*e* and + 3*e* respectively where 2 and 3 stand to indicate the atomic number of He and Li atoms. Consequently, in general, the charge carried by the nucleus of a hydrogen-like atom with atomic number, Z will be equal to + Ze. On the basis of the same arguments as used above, we can show that the electrostatic force of attraction acting between the nucleus and the revolving electron will be equal to

$$\frac{Ze\times e}{r^2} = \frac{Ze^2}{r^2}\ (\textit{in C.G.S. units})\ \text{or}\ \frac{Ze^2}{4\pi\in_0 e^2}\ (\textit{in S.I. units})$$

and hence the radius, r_n of the nth orbit of a hydrogen like atom with atomic number Z will be given by :

$$\left.\begin{array}{l} r_n = \dfrac{n^2h^2}{4\pi^2me^2Z} \\ \quad = \dfrac{n^2\times0.529\times10^{-8}}{Z}\ \text{cm} \\ \quad = \dfrac{n^2\times0.529}{Z}\ \text{A°} \end{array}\right\}\ (\textit{In C.G.S. units}) \quad ...11\ (a)$$

$$\left.\begin{aligned} r_n &= \frac{\epsilon_0\, n^2h^2}{\pi\, me^2Z} \\ &= \frac{n^2 \times 0.529 \times 10^{-10}}{Z}\ \text{m} \end{aligned}\right\} \text{(In S.I. units)} \qquad ...11\ (b)$$

3. Energy of an Electron in Various Orbits in Hydrogen Atom

(*a*) **In C.G.S. units.** The energy associated with a revolving electron at any instant is equal to the sum of its kinetic energy (K.E.) and potential energy (P.E.). Kinetic energy of a moving electron is due to its motion in the circular orbit while the potential energy arises because the electron lies in the electric field of the positive nucleus. Again we know that since K.E. $= \frac{1}{2}mv^2$ and P.E. $= -\frac{e^2}{r}$, total energy (E) of the electron is given by :

$$\text{E} = \text{K.E.} + \text{P.E.}$$

$$= \frac{1}{2}mv^2 + \left(-\frac{e^2}{r}\right)$$

or $$\text{E} = \frac{1}{2}mv^2 - \frac{e^2}{r} \qquad ...(12)$$

Putting $mv^2 = \frac{e^2}{r}$ from equation (1) in equation (12), we get :

$$\text{E} = \frac{1}{2}.\frac{e^2}{r} - \frac{e^2}{r}$$

or $$\text{E} = -\frac{e^2}{2r} \qquad ...(13)$$

Putting $r = \frac{n^2h^2}{4\pi^2me^2}$ from equation (3) in equation (13), we get :

$$\text{E} = -\frac{e^2}{2} \times \frac{4\pi^2me^2}{n^2h^2}$$

or $$\text{E} = -\frac{2\pi^2me^4}{n^2h^2}$$

If E_n is the energy associated with an electron revolving in the *n*th orbit, then E_n is given by :

$$\boxed{\mathbf{E}_n = -\frac{\mathbf{2\pi^2me^4}}{\mathbf{n^2h^2}}} \qquad ...(14)$$

The value of E_n can be calculated by substituting the experimental values of m (= 9.108 × 10^{-28} g); e (= 4.8 × 10^{-10} *esu*) and h (= 6.624 × 10^{-27} erg.s) in equation (14). Thus :

$$E_n = \frac{2 \times (3.14)^2 \times (9.108 \times 10^{-28}\ g) \times (4.8 \times 10^{-10}\ esu)^4}{n^2 \times (6.624 \times 10^{-27}\ \text{erg-s})^2}$$

$$= \frac{2\times(3.14)^2\times(9.108\times10^{-28})\times(4.8\times10^{-10})^4\ \text{g.}(esu)^4}{n^2\times(6.624\times10^{-27})^2.\text{erg}^2.\text{s}^2}$$

$$= \frac{2\times(3.14)^2\times(9.108)\times(4.8)^4\times10^{-68}\ \text{g.}(esu)^4}{n^2\times(6.624)^2\times10^{-54}.\text{erg}^2.\text{s}^2}$$

$$= \frac{2\times(3.14)^2\times0.009108\times(4.8)^4\times10^{-11}\ \text{g.}(esu)^4}{n^2\times(6.624)^2.\text{erg}^2.\text{s}^2}$$

$$= -\frac{2.179\times10^{-11}\ \text{g.}(\text{g}^{1/2}.\text{cm}^{3/2}.\text{s}^{-1})^4}{n^2\times(\text{g.cm}^2.\text{s}^{-2})^2.\text{s}^2}$$

$$= -\frac{2.179\times10^{-11}\ \text{g.g.}^2.\text{cm}^6.\text{s}^{-4}}{n^2.\text{g}^2.\text{cm}^4.\text{s}^{-4}.\text{s}^2}$$

$$= -\frac{2.179\times10^{-11}\ \text{g.cm.}^2}{n^2.\text{s}^2}$$

$$= -\frac{2.179\times10^{-11}}{n^2}(\text{g.cm}^2.\text{s}^{-2})$$

or

$$\boxed{\mathbf{E}_n = -\frac{\mathbf{2.179\times10^{-11}}}{\mathbf{n^2}}\ \textbf{erg/atom}} \quad ...(15)$$

Since E_n has a negative sign, it implies that energy increases directly with the square of n.

On putting n = 1, 2, 3, etc. in equation (15), we can get the energies associated with an electron revolving in *1st* (n = 1), *2nd* (n = 2), *3rd* (n = 3) etc. orbits of hydrogen atom as shown below :

Energy of electron in *1st* orbit (n = 1),

$$E_1 = -\frac{2.179\times10^{-11}}{1^2}\ \text{erg/atom} = -2.179\times10^{-11}\ \text{erg/atom}$$

Energy of electron in *2nd* orbit (n = 2),

$$E_2 = -\frac{2.179\times10^{-11}}{2^2} = -0.5448\times10^{-11}\ \text{erg/atom}$$

Energy of electron in *3rd* orbit (n = 3),

$$E_3 = -\frac{2.179\times10^{-11}}{3^2} = -0.2421\times10^{-11}\ \text{erg/atom}$$

Energy of electron in *4th* orbit (n = 4),

$$E_4 = -\frac{2.179\times10^{-11}}{4^2} = -0.1362\times10^{-11}\ \text{erg/atom}$$

Energy of electron in *5th* orbit (n = 5),

$$E_5 = -\frac{2.179\times10^{-11}}{5^2} = -0.08716\times10^{-11}\ \text{erg/atom}$$

The values of E_1, E_2, E_3, etc. calculated as above show that these go on increasing (*i.e.*, go on becoming less and less negative) as the electron moves away from the nucleus.

(*b*) **In S.I. units.** Kinetic energy of the electron, as before, is equal to $\frac{1}{2}mv^2$ while the potential energy of the electron moving at a distance of r from the nucleus is equal to $-\dfrac{e^2}{4\pi\epsilon_0 r}$. Here m, e and h are measured in S.I. units namely kg, coulomb and J.s. respectively. Thus total energy (E) of the electron is given by :

$$E = \frac{1}{2}mv^2 - \frac{e^4}{4\pi\epsilon_0 r} \qquad ...(16)$$

Putting $mv^2 = \dfrac{e^2}{4\pi\epsilon_0 r}$ from equation (8) in equation (16), we get :

$$E = \frac{1}{2}.\frac{e^2}{4\pi\epsilon_0 r} - \frac{e^2}{4\pi\epsilon_0 r}$$

$$E = -\frac{e^2}{8\pi\epsilon_0 r} \qquad ...(17)$$

Now put $r = \dfrac{\epsilon_0 n^2 h^2}{\pi m e^2}$ in equation (17) to get

$$E = -\frac{e^2}{8\pi\epsilon_0} \times \frac{\pi m e^2}{\epsilon_0 n^2 h^2}$$

or

$$\boxed{\mathbf{E_n = -\frac{me^4}{8\epsilon_0^2 n^2 h^2}}} \qquad ...(18)$$

The value of E_n can be calculated by putting

$m = 9.108 \times 10^{-31}$ kg;

$e = 1.602 \times 10^{-19}$ coulomb;

$\epsilon_0 = 8.85 \times 10^{-12}$;

and $h = 6.62 \times 10^{-34}$ J-s in equation (18). Thus :

$$E_n = -\frac{(9.108\times10^{-31}\text{ kg})\times(1.602\times10^{-19}\text{ C})^4}{8\times(8.85\times10^{-12})^2\times n^2\times(6.62\times10^{-34}\text{ J.s})^2}$$

$$= -\frac{9.108\times(1.602)^4\times10^{-15}\times\text{kg}\times(\text{kg}^{1/2}.\text{m}^{3/2}.\text{s}^{-1})^4}{8\times(8.85)^2\times n^2\times(6.62)^2\times(\text{kg.m}^2.\text{s}^{-2}.\text{s})^2}$$

$$(\text{C} = \text{kg}.^{1/2}\text{m}^{3/2}.\text{s}^{-1};\ \text{J} = \text{kg.m}^2.\text{s}^{-2})$$

$$= -\frac{59.9854\times10^{-15}\times\text{kg.kg}^2.\text{m}^6.\text{s}^{-4}}{27509.29\times n^2\times\text{kg}^2.\text{m}^4.\text{s}^{-2}}$$

$$= -\frac{0.0021805\times10^{-15}}{n^2}\text{kg.m}^2.\text{s}^{-2}$$

$$= -\frac{218.05\times10^{-20}}{n^2}\text{Joules/atom} \qquad [\because\ \text{J} = \text{kg.m}^2.\text{s}^{-2}]$$

$$= -\frac{218.05 \times 10^{-20} \times (\text{Avogadro's number})}{n^2} \text{ J/mole}$$

$$= -\frac{218.05 \times 10^{-20} \times 6.023 \times 10^{23}}{n^2} \text{ J/mole}$$

$$= -\frac{218.05 \times 10^{-20} \times 6.023 \times 10^{23}}{n^2 \times 10^3} \text{ KJ/mole}$$

or $$\boxed{\mathbf{E}_n = -\frac{\mathbf{1313.315}}{\boldsymbol{n^2}} \textbf{ KJ/mole}} \quad ...(19)$$

Substituting n = 1, 2, 3, 4, etc. in equation (19) we get the energies associated with an electron revolving in *1st* (n = 1), *2nd* (n = 2), *3rd* (n = 3), etc. orbits of hydrogen atom as shown below :

Energy of electron in *1st* orbit (n = 1),

$$E_1 = -\frac{1313.31}{1^2} = -1313.31 \text{ KJ/mole}$$

Energy of electron in *2nd* orbit (n = 2),

$$E_2 = -\frac{1313.31}{2^2} = -328.32 \text{ KJ/mole}$$

Energy of electron in *3rd* orbit (n = 3),

$$E_3 = -\frac{1313.31}{3^2} = -145.92 \text{ KJ/mole}$$

Energy of electron in *4th* orbit (n = 4),

$$E_4 = -\frac{1313.31}{4^2} = -82.08 \text{ KJ/mole}$$

Energy of electron in *5th* orbit (n = 5),

$$E_5 = -\frac{1313.31}{5^2} = -52.53 \text{ KJ/mole}$$

The energy values in KJ/mole associated with an electron in various orbits in hydrogen atom, calculated as above, are represented in Fig. 1.34.

On the same lines of argument it can be shown that total energy of an electron in *nth* orbit of a hydrogen-like atom with atomic number, Z is given by :

$$E_n = \text{K.E.} + \text{P.E.}$$

$$= \frac{1}{2}mv^2 + \left(-\frac{Ze^2}{r}\right)$$

$$= \frac{1}{2}mv^2 - \frac{Ze^2}{r} \text{ (in C.G.S. units)}$$

and $$E_n = \frac{1}{2}mv^2 + \left(-\frac{Ze^2}{8\pi\epsilon_0 r}\right)$$

$$= \frac{1}{2}mv^2 - \frac{Ze^2}{4\pi\epsilon_0 r} \text{ (in S.I. units)} \quad ...(A)$$

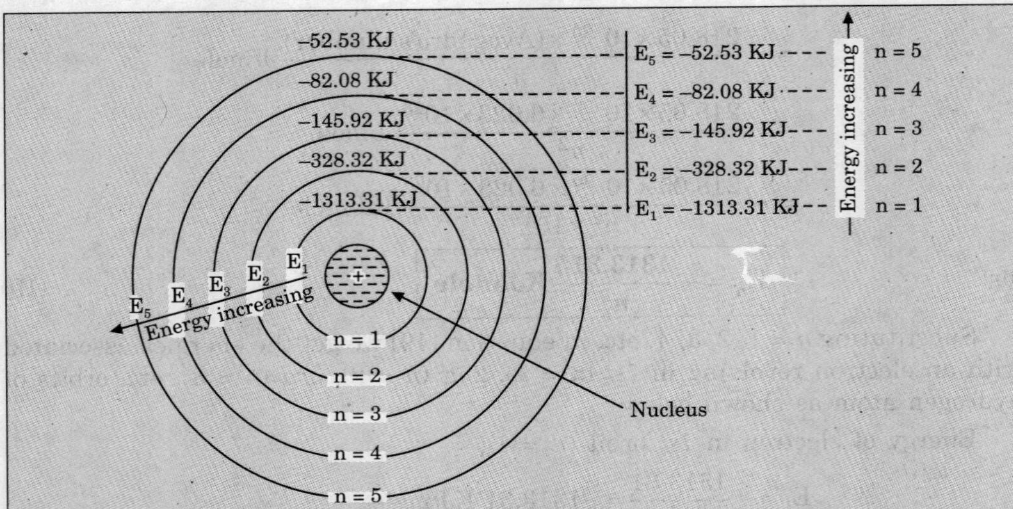

Fig. 1.34. Energy values (in KJ/mole) associated with an electron in various orbits in hydrogen atom.

On utilising the relations :

$$\frac{mv^2}{r} = \frac{Ze^2}{r^2}$$

or $$mv^2 = \frac{Ze^2}{r} \text{ (in C.G.S. units)} \quad ...(a)$$

$$\frac{mv^2}{r} = \frac{Ze^2}{4\pi \epsilon_0 r^2}$$

or $$mv^2 = \frac{Ze^2}{4\pi \epsilon_0 r} \text{ (in S.I. units)} \quad ...(b)$$

$$r = \frac{n^2h^2}{4\pi^2 me^2 Z} \text{ (in C.G.S. units)} \quad ...(c)$$

$$r = \frac{\epsilon_0 \, n^2h^2}{\pi me^2 Z} \text{ (in S.I. units)} \quad ...(d),$$

the equation (A) becomes :

$$\left.\begin{aligned} E_n &= -\frac{2\pi^2 me^4 Z^2}{n^2h^2} \\ &= -\frac{2.179 \times 10^{-11} \times Z^2}{n^2} \text{ erg/atom} \end{aligned}\right\} \text{(In C.G.S. units) ...(20)}$$

and $$\left.\begin{aligned} E_n &= -\frac{me^4 Z^2}{8 \epsilon_0^{\,2} n^2h^2} \\ &= -\frac{218.05 \times 10^{-20} \times Z^2}{n^2} \text{ J/atom} \\ &= -\frac{1313.315 Z^2}{n^2} \text{ KJ/mole} \end{aligned}\right\} \text{(In S.I. units) ...(21)}$$

4. Velocity of an Electron in Various Orbits of Hydrogen Atom.

We know that :

$$mvr = n.\frac{h}{2\pi}$$

Now since $r = n^2 \times 0.529 \times 10^{-8}$ cm, the above equation, reduces to :

$$mv \times n^2 \times 0.529 \times 10^{-8} = n.\frac{h}{2\pi}$$

$$\therefore \qquad v = \frac{h}{2\pi mn \times 0.529 \times 10^{-8}}$$

Substituting h = 6.624 × 10^{-27} erg-s, π = 3.14 and m = 9.108 × 10^{-28} g, we get :

$$v = \frac{6.624 \times 10^{-27} \text{ erg.s.}}{2 \times 3.14 \times (9.108 \times 10^{-28} \text{ g}) \times n \times (0.529 \times 10^{-8} \text{ cm})}$$

$$= \frac{6.624 \times 10^{-27} \text{ g.cm}^2.\text{s}^{-2}.\text{s}}{2 \times 3.14 \times (9.108 \times 10^{-28} \text{ g}) \times n \times (0.529 \times 10^{-8} \text{ cm})} \qquad (\because \text{erg} = \text{g.cm}^2.\text{s}^{-2})$$

$$= \frac{2.188 \times 10^8}{n} \text{ cm/sec}$$

Now if v_n is the velocity of the electron in the *nth* orbit, then v_n is given :

$$\boxed{\mathbf{v}_n = \frac{\mathbf{2.188 \times 10^8}}{\mathbf{n}} \textbf{ cm/s}}$$

Numerical Problems

Problem 1. (In C.G.S. units). *Calculate the excitation energy for the electronic transition from state n = 1 to state n = 2 in a hydrogen atom (h = 6.626 × 10^{-27} erg-s, e = 4.803 × 10^{-10} esu, m = 9.109 × 10^{-28} g).* *(Raj. B.Sc. I 1980)*

Solution. Let the energy associated with the orbit having n = 1 be E_1 and that associated wth the orbit with n = 2 be E_2.

$\therefore$ Excitation energy = $E_2 - E_1$

$$= -\frac{2\pi^2 me^4}{{n_2}^2 h^2} - \left(-\frac{2\pi^2 me^4}{{n_1}^2 h^2}\right)$$

$$= -\frac{2\pi^2 me^4}{h^2}\left(\frac{1}{{n_1}^2} - \frac{1}{{n_2}^2}\right)$$

Substituting π = 3.14; m = 9.109 × 10^{-28} g; e = 4.803 × 10^{-10} esu; h = 6.626 × 10^{-27} erg-s; n_1 = 1 and n_2 = 2, we get :

Excitation energy = 16.3526 × 10^{-22} erg/atom

Problem 2. (In C.G.S. units). *Calculate the radii of first and fourth orbit in hydrogen atom. Given that h = 6.6238 × 10^{-27} erg-s, m = 9.1072 × 10^{-28} g, π = 3.1416, e = 4.8022 × 10^{-10} esu.* *(Udaipur B.Sc. 1980)*

Solution. We know that :

$$r_n = \frac{n^2h^2}{4\pi^2me^2}$$

$$\therefore \quad r_1 = \frac{(1)^2 \times h^2}{4\pi^2me^2} \quad ...(1)$$

and $$r_4 = \frac{(4)^2 \times h^2}{4\pi^2me^2} \quad ...(2)$$

Substituting $h = 6.6238 \times 10^{-27}$ erg-s, $\pi = 3.1416$, $m = 9.1072 \times 10^{-28}$ g and $e = 4.8022 \times 10^{-10}$ esu, we get :

$$r_1 = (1)^2 \times 0.529 \times 10^{-8} \text{ cm}$$

or $$= 0.529 \times 10^{-8} \text{ cm}$$

and $$r_4 = (4)^2 \times 0.529 \times 10^{-8} \text{ cm}$$

$$= 8.45 \times 10^{-8} \text{ cm.}$$

Problem 3. (In C.G.S. units). *Calculate the velocity of an electron in the 3rd orbit of hydrogen atom ($\pi = 3.14$, $h = 6.6208 \times 10^{-27}$ erg-s, $e = 4.8029 \times 10^{-10}$ esu).*

(Raj. B.Sc. I 1983)

Solution. v_n is given by :

$$v_n = \frac{2.188 \times 10^8}{n} \text{ cm.s}^{-1}$$

$$\therefore \quad v_3 = \frac{2.188 \times 10^8}{3} \text{ cm.s}^{-1}$$

$$= 0.728 \times 10^8 \text{ cm.s}^{-1}$$

Problem 4. (In S.I. units). *The solitary electron in H-atom revolves in 2nd energy level. Calculate (i) The energy of the electron in this level (ii) Bohr's radius corresponding to this level. Given that $m = 9.109 \times 10^{-31}$ kg, $e = 1.602 \times 10^{-19}$ C, $\in_0 = 8.85 \times 10^{-12}$, $c = 3.0 \times 10^8$ m.s^{-1} and $h = 6.626 \times 10^{-34}$ J.s.*

Solution. We know that :

$$E_n = -\frac{me^4}{8\in_0^2 n^2h^2}$$

Substituting the values of different quantities

$$E_n = -\frac{9.109 \times 10^{-31} \times (1.602 \times 10^{-19})^4}{8 \times (8.85 \times 10^{-12})^2 \times 2^2 \times (6.626 \times 10^{-34})^2} \text{ J/atom}$$

$$= -5.45 \times 10^{-19} \text{ J/atom}$$

$$= -5.45 \times 10^{-19} \times 6.023 \times 10^{23} \text{ J/mole}$$

$$= -\frac{5.45 \times 10^{-19} \times 6.023 \times 10^{23}}{10^3} \text{ kJ/mole}$$

$$= -338.33 \text{ kJ/mole}$$

Radius of 2*nd* energy level is given by :

$$r_2 = \frac{\epsilon_0 \times (2)^2 \times h^2}{\pi m e^2}$$

$$= \frac{(8.85\times10^{-12})\times4\times(6.626\times10^{-34})^2}{3.14\times(9.109\times10^{-31})\times(1.602\times10^{-19})^2} \text{ m}$$

$$= 2.115\times10^{-10} \text{ m.}$$

Problem 5. (In S.I. units). *The first ionisation energy of hydrogen is* 2.179×10^{-18} *J. Calculate the second ioniastion energy of helium atom.*

(*Delhi. B.Sc. Hons.* 1977)

Solution. Since the first ionisation energy of hydrogen is 2.179×10^{-18} J, energy associated with an electron in the 1st orbit of H atom is equal to 2.179×10^{-18} J; *i.e.*

$$E_1 = \frac{me^4}{8\epsilon_0^2 n^2h^2} = 2.179\times10^{-18} \text{ J}$$

Now second ionisation energy of He atom corresponds to the removal of one electron from He^+ ion and this electron will be removed from 1st orbit of He atom, since the electronic configuration of He atom is 2.

∴ Second ionisation energy of He atom (Z = 2)

= Energy of the electron in the first orbit of He atom

$$= \frac{me^4Z^2}{8\epsilon_0^2 n^2h^2}$$

$$= 2.179 \times 10^{-18} \times Z^2 \text{ J}$$

$$= 2.179 \times 10^{-18} \times (2)^2 \text{ J}$$

$$= 8.716 \times 10^{-18} \text{ J.}$$

Origin of Hydrogen Spectrum on the Basis of Bohr's Theory

When an electronic discharge is passed through hydrogen gas filled in a discharge tube at a very low pressure, the molecules of hydrogen break into atoms. These atoms absorb energy from the electric spark and the solitary electron (present in 1st energy level) in different atoms of hydrogen gets *excited*, *i.e.*, the electron shifts from the energy level 1 (lowest energy level—*Ground state*) to different higher energy levels 2, 3, 4 etc. depending on the amount of energy absorbed by the atoms of hydrogen. The shifting of an electron from energy level 1 (*ground state*) to higher energy levels 2, 3, 4, 5, 6, 7, etc. (*Excited states*) has been shown in Fig. 1.35. These transitions are represented as 1→2, 1→3, 1→4, etc. respectively.

Now when the electron reaches in the higher energy levels, it is said to be in the excited state and hence has a tendency to come back to one of the lower energy levels or even to the lowest energy level (ground state) after a short interval of time, i.e. the electron which is unstable in the. excited state comes back to one of the lower energy levels or even to the lowest energy level as shown

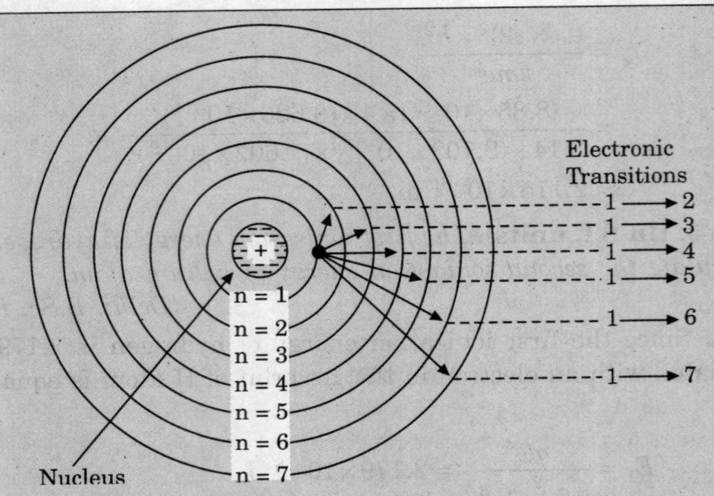

Fig: 1.35. Jump of hydrogen electron (shown by black circle) from the lowest energy level 1 (n = 1, ground state) to higher energy levels 2 (n = 2), 3 (n = 3), 4 (n = 4), 5 (n = 5), 6 (n = 6) and 7 (n = 7), etc.

in Fig. 1.36. When the electron in the excited state comes back to the lower energy levels, it *emits* energy it had absorbed earlier, in the form of photons of light of specific frequency and hence of specific wavelength. Thus spectral lines of different series observed in the emission spectrum of hydrogen are due to the emission of energy photons of different wavelengths when the excited electron in hydrogen atom comes back to lower energy levels. In simple words we say that

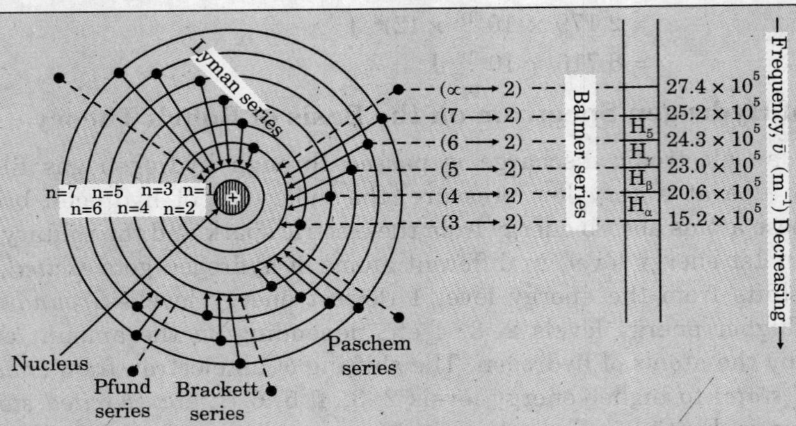

Fig. 1.36. Origin of hydrogen spectrum on the basis of Bohr's theory. This figure shows the jump of hydrogen electron (shown by black circle from higher energy levels to lower energy levels or even to the lowest energy level (n = 1) to produce spectral lines of various series. The jump of the electron from infinity to lower energy levels has been shown by dotted line and gives the limiting spectral line of each series. The frequency of the spectral lines of the Balmer series have also been shown.

the appearance of spectral lines in the line emission spectrum of H-atom is due to the jump of hydrogen electron from the higher energy levels to lower energy levels or even to the lowest energy level. For example the spectral lines of Lyman series are produced when the electron jumps from 2*nd*, 3*rd*, 4*th*, etc. levels (higher energy levels) to the 1*st* energy level (lowest energy level). The spectral lines of Balmer series originate when an electron jumps from 3*rd*, 4*th*, etc. levels (higher energy levels) to the 2*nd* energy level (lower energy level). In similar manner the origin of the spectral lines of other series can be explained.

Calculation of Frequency (υ and $\bar{\upsilon}$) and Wavelength (λ) of Spectral Lines of Hydrogen Spectrum by Bohr's Theory

While explaining the origin of the appearance of spectral lines of the line emission spectrum of hydrogen atom on the basis of Bohr's theory, we have said that when the electron in the higher energy levels jumps to the lower energy levels or to the lowest energy level, energy is liberated by the electron in the form of photon of light of specific frequency and wavelength which appears as spectral line in the spectrum. Frequency and wavelength of a spectral line depends on the quantity of energy liberated by the electron. Now suppose that an electron jumps from a higher energy level n_2 with energy E_{n_2} to a lower energy level n_1 whose energy is E_{n_1}. In this process of jumping the amount of energy liberated by the electron is equal to $E_{n_2} - E_{n_1}$. Now let us calculate the value of $E_{n_2} - E_{n_1}$ in terms of C.G.S. and S.I. units as follows :

In C.G.S. units, **In S.I. units**

$$E_{n_1} = -\frac{2\pi^2 me^4}{n_1^2 h^2} \qquad E_{n_1} = -\frac{me^4}{8\in_0^2 n_1^2 h^2}$$

$$E_{n_2} = -\frac{2\pi^2 me^4}{n_2^2 h^2} \qquad E_{n_2} = -\frac{me^4}{8\in_0^2 n_2^2 h^2}$$

Consequently the value of $\left(E_{n_2} - E_{n_1}\right)$ in terms of C.G.S. and S.I. units is given as :

In C.G.S. units **In S.I. units**

$$E_{n_2} - E_{n_1} = -\frac{2\pi^2 me^4}{n_2^2 h^2} - \left(-\frac{2\pi^2 me^4}{n_1^2 h^2}\right) \qquad E_{n_2} - E_{n_1} = -\frac{me^4}{8\in_0 n_2^2 h^2} - \left(-\frac{me^4}{8\in_0^2 n_1^2 h^2}\right)$$

$$= \frac{2\pi^2 me^4}{h^2}\left(\frac{1}{n_1^2} - \frac{1}{n_2^2}\right) \qquad = \frac{me^4}{8\in_0^2 h^2}\left(\frac{1}{n_1^2} - \frac{1}{n_2^2}\right)$$

***(a)* Calculation of frequency (υ).** We know that if υ is the frequency of the emitted spectral line in hydrogen spectrum, then

$$h\upsilon = E_{n_2} - E_{n_1}$$

(*a*) *In C.G.S. units,* $h\upsilon = \frac{2\pi^2 me^4}{h^2}\left(\frac{1}{n_1^2} - \frac{1}{n_2^2}\right)$

$$\therefore \qquad \upsilon = \frac{2\pi^2 me^4}{h^3}\left(\frac{1}{n_1^2} - \frac{1}{n_2^2}\right)$$

On substituting $\pi = 3.14$; $m = 9.108 \times 10^{-28}$ g; $e = 4.8 \times 10^{-10}$ *esu* and $h = 6.624 \times 10^{-27}$ erg-s in the above equation, we get the value of $2\pi^2\ me^4/h^3$ to be equal to 3.2906×10^{15} cycles/s. Thus the frequency (υ) of the spectral line emitted when the electron falls from energy level n_2 to energy level n_1 is given by :

$$\upsilon_{n_2 \to n_1} = \frac{2\pi^2 me^4}{h^3}\left(\frac{1}{n_1^2} - \frac{1}{n_2^2}\right) \text{ cycles / sec}$$

$$\text{or} \qquad \upsilon_{n_2 \to n_1} = 3.2906 \times 10^{15}\left(\frac{1}{n_1^2} - \frac{1}{n_2^2}\right) \text{ cycles / sec}$$

Any of the above equations can be used to calculate the value of υ in C.G.S. units.

(b) In S.I. units, $$h\upsilon = \frac{me^4}{8\epsilon_0^2 h^2}\left(\frac{1}{n_1^2} - \frac{1}{n_2^2}\right)$$

$$\therefore \qquad \upsilon = \frac{me^4}{8\epsilon_0^2 h^3}\left(\frac{1}{n_1^2} - \frac{1}{n_2^2}\right)$$

On substituting
$m = 9.108 \times 10^{-31}$ kg;
$e = 1.602 \times 10^{-19}$ *C*;
$\epsilon_0 = 8.85 \times 10^{-12}$;
$h = 6.624 \times 10^{-34}$ J-s.

we get the value of $me^4/8\epsilon_0^2\ h^3$ to be equal to 3.2906×10^{15} Hz. Thus :

$$\therefore \qquad \upsilon_{n_2 \to n_1} = \frac{me^4}{8\epsilon_0^2 h^3}\left(\frac{1}{n_1^1} - \frac{1}{n_2^2}\right) \text{ Hz}$$

$$\text{or} \qquad \upsilon_{n_2 \to n_1} = 3.2906 \times 10^{15}\left(\frac{1}{n_1^2} - \frac{1}{n_2^2}\right) \text{ Hz}$$

Any of the above equations can be used to calculate the value of υ in S.I. units. It may be noted that the values of υ obtained in both the units (C.G.S. and S.I. units) are the same.

***(b)* Calculation of frequency in wave number ($\bar{\upsilon}$).** We know that $\bar{\upsilon}$ is given by :

$$\bar{\upsilon} = \frac{\text{Frequency } (\upsilon)}{\text{Velocity of light } (c)}$$

$$\text{Now} \qquad \upsilon = \frac{2\pi^2 me^4}{h^3}\left(\frac{1}{n_1^2} - \frac{1}{n_2^2}\right) \textit{ (in C.G.S. units).}$$

Dividing both sides of the equation by c, we get :

$$\frac{\upsilon}{c} = \frac{2\pi^2 me^4}{h^3 c}\left(\frac{1}{n_1^{\ 2}} - \frac{1}{n_2^{\ 2}}\right)$$

Now since

$$\bar{\upsilon} = \frac{\upsilon}{c},$$

$$\therefore \qquad \bar{\upsilon} = \frac{2\pi^2 me^4}{h^3 c}\left(\frac{1}{n_1^{\ 2}} - \frac{1}{n_2^{\ 2}}\right) \text{ (in C.G.S. units).}$$

In this equation the expression $2\pi^2\ m\,e^4/h^3c$ is called *Rydberg's constant*, R and hence the equation reduces to :

$$\bar{\upsilon} = R\left(\frac{1}{n_1^{\ 2}} - \frac{1}{n_2^{\ 2}}\right) \text{ (in C.G.S. units)} \qquad ...(1)$$

Equation (1) gives the value of $\bar{\upsilon}$ in C.G.S. units. By putting π = 3.14; m = 9.108 × 10^{-28} g; e = 4.8 × 10^{-10} *e.s.u.*; h = 6.624 × 10^{-27} erg-s; c = 3 × 10^{10} cm-s^{-1}, the value of R in C.G.S. units has been obtained to be equal to 109.686 × 10^3 cm^{-1}. Thus :

$$\bar{\upsilon} = \frac{2\pi^2 me^4}{h^3 c}\left(\frac{1}{n_1^{\ 1}} - \frac{1}{n_2^{\ 2}}\right)\text{cm}^{-1}$$

$$\text{or} \quad \bar{\upsilon} = \text{R}\left(\frac{1}{n_1^{\ 2}} - \frac{1}{n_2^{\ 2}}\right)\text{cm}^{-1} \qquad ...(2)$$

$$\text{or} \quad \bar{\upsilon} = 109.686 \times 10^3\left(\frac{1}{n_1^{\ 2}} - \frac{1}{n_2^{\ 2}}\right)\text{cm}^{-1}$$

Any of the above equations can be used to calculate the value of $\bar{\upsilon}$ in C.G.S. units.

On similar grounds it can be shown that the value of $\bar{\upsilon}$ in S.I. units is given by :

$$\bar{\upsilon} = \frac{me^4}{8\epsilon_0^{\ 2}\ h^3 c}\left(\frac{1}{n_1^{\ 1}} - \frac{1}{n_2^{\ 2}}\right) \text{ (in S.I. units).}$$

By putting m = 9.108 × 10^{-3} kg; e = 1.602 × 10^{-19} C; ϵ_0 = 8.85 × 10^{-12}; h = 6.624 × 10^{-34} J-s and c = 3 × 10^8 m.s^{-1}, the value of R in S.I. units has been obtained to be equal to 109.686 × 10^5 m^{-1}. Thus :

$$\bar{\upsilon} = \frac{me^4}{8\epsilon_0^{\ 2}\ h^3 c}\left(\frac{1}{n_1^{\ 2}} - \frac{1}{n_2^{\ 2}}\right)\text{m}^{-1}$$

$$\text{or} \quad \bar{\upsilon} = \text{R}\left(\frac{1}{n_1^{\ 2}} - \frac{1}{n_2^{\ 2}}\right)\text{m}^{-1} \qquad ...(3)$$

$$\text{or} \quad \bar{\upsilon} = 109.686 \times 10^5\left(\frac{1}{n_1^{\ 2}} - \frac{1}{n_2^{\ 2}}\right)\text{m}^{-1}$$

Any of the above equations can be used to calculate the value of $\bar{\upsilon}$ in S.I. units.

(c) Calculation of wavelength (λ). We know that wavelength (λ) of a spectral line in hydrogen spectrum is given by any of the following relations :

$$\lambda = \frac{1}{\text{Frequency in wave number } (\bar{\upsilon})} = \frac{1}{\bar{\upsilon}} \quad ...(1)$$

$$\lambda = \frac{\text{Velocity of height } (c)}{\text{Frequency } (\upsilon)} = \frac{c}{\upsilon} \quad ...(2)$$

Any of the above relations can be used to calculate the value of λ. Considering equation (1) we have (in C.G.S. units)

$$\lambda = \frac{1}{\bar{\upsilon}}$$

$$= \frac{1}{\dfrac{2\pi^2 me^4}{h^3 c}\left[\dfrac{1}{n_1^2} - \dfrac{1}{n_2^2}\right] \text{cm}^{-1}}$$

Now we know that $\dfrac{2\pi^2 me^4}{h^3 c} = \text{R} = 109.686 \times 10^3 \text{ cm}^{-1}$

$$\therefore \quad \lambda = \frac{1}{\dfrac{2\pi^2 me^4}{h^3 c}\left[\dfrac{1}{n_1^2} - \dfrac{1}{n_2^2}\right]} \text{ cm}$$

$$= \frac{1}{\text{R}\left[\dfrac{1}{n_1^2} - \dfrac{1}{n_2^2}\right]} \text{ cm}$$

$$= \frac{1}{109.686 \times 10^3 \left[\dfrac{1}{n_1^2} - \dfrac{1}{n_2^2}\right]} \text{ cm}$$

Any of the above three equations can be used to calculate the value of λ in C.G.S. units. Similarly in S.I. units λ can be calculated by any of the following relations :

$$\lambda = \frac{1}{\dfrac{me^4}{8\,\varepsilon_0^2 h^3 c}\left[\dfrac{1}{n_1^2} - \dfrac{1}{n_2^2}\right]} \text{ m}$$

$$= \frac{1}{\text{R}\left[\dfrac{1}{n_1^2} - \dfrac{1}{n_2^2}\right]} \text{ m}$$

$$= \frac{1}{109.686 \times 10^5 \left[\dfrac{1}{n_1^2} - \dfrac{1}{n_2^2}\right]} \text{ m}$$

Calculation of Frequencies in Terms of Wave-number, $\overline{\upsilon}$ (in m^{-1}) of Different Spectral Lines of Different Series of Hydrogen Spectrum.

We have already shown that the frequency in terms of wave number, $\overline{\upsilon}$ (in m^{-1}) of a spectral line of a given series appearing in the hydrogen spectrum is given by :

$$\overline{\upsilon} = \frac{me^4}{8\,\varepsilon_0^{\,2} h^3 c}\left(\frac{1}{n_1^{\,2}} - \frac{1}{n_2^{\,2}}\right) \text{m}^{-1}$$

$$= R\left(\frac{1}{n_1^{\,2}} - \frac{1}{n_2^{\,2}}\right) \text{m}^{-1}$$

$$= 10.9686 \times 10^6 \left(\frac{1}{n_1^{\,2}} - \frac{1}{n_2^{\,2}}\right) \text{m}^{-1}$$

or

$$\overline{\upsilon} = 109.686 \times 10^5 \left(\frac{1}{n_1^{\,1}} - \frac{1}{n_2^{\,2}}\right) \text{m}^{-1} \qquad \text{...(1)}$$

Now let us use equation (1) for the calculation of frequencies of spectral lines of various series as follows :

1. Lyman series. The various spectral lines of this series are obtained when an electron jumps from the *2nd, 3rd, 4th* or any other *higher energy level* to the *1st energy level.* Thus the frequencies of various spectral lines of this series can be obtained by putting n_1 = 1 and n_2 = 2, 3, 4......∞ in equation (1) given above.

(*a*) *First line.* For this line n_1 = 1 and n_2 = 2, and hence :

$$\overline{\upsilon} = 109.686 \times 10^5 \left(\frac{1}{1^2} - \frac{1}{2^2}\right) \text{m}^{-1}$$

$$= 82.26 \times 10^5 \text{ m}^{-1}$$

(*b*) *Second line.* For this line n_1 = 1 and n_2 = 3, and hence :

$$\overline{\upsilon} = 109.686 \times 10^5 \left(\frac{1}{1^2} - \frac{1}{3^2}\right) \text{m}^{-1}$$

$$= 97.49 \times 10^5 \text{ m}^{-1}$$

(*c*) *Limiting line.* For this line n_1 = 1 and n_2 = ∞, and hence :

$$\overline{\upsilon} = 109.686 \times 10^5 \left(\frac{1}{1^2} - \frac{1}{\infty^2}\right) \text{m}^{-1}$$

$$= 109.686 \times 10^5 \text{ m}^{-1}$$

On similar lines of calculations the frequencies of other lines can also be calculated.

2. Balmer series. The various spectral lines of this series are obtained when an electron jumps from *3rd, 4th, 5th etc.*, energy levels to *2nd* energy level. Thus the frequencies of various spectral lines of this series can be obtained by putting n_1 = 2 and n_2 = 3, 4, 5......∞ in equation (1) given above.

(*a*) *First line* (*H_α line*). For this line n_1 = 2 and n_2 = 3, and hence :

$$\bar{\upsilon} = 109.686 \times 10^5 \left(\frac{1}{2^2} - \frac{1}{3^2}\right) \text{m}^{-1}$$

$$= 15.234 \times 10^5 \text{ m}^{-1}$$

(*b*) *Second line* (*H_β line*). For this line n_1 = 2 and n_2 = 4, and hence :

$$\bar{\upsilon} = 109.686 \times 10^5 \left(\frac{1}{2^2} - \frac{1}{4^2}\right) \text{m}^{-1}$$

$$= 20.566 \times 10^5 \text{ m}^{-1}$$

(*c*) *Limiting line.* For this line n_1 = 2 and n_2 = ∝, and hence :

$$\bar{\upsilon} = 109.686 \times 10^5 \left(\frac{1}{2^2} - \frac{1}{\propto^2}\right) \text{m}^{-1}$$

$$= \frac{109.686 \times 10^5}{4} \text{ m}^{-1}$$

$$= 27.421 \times 10^5 \text{ m}^{-1}$$

On similar lines of calculations the frequencies of other lines can also be calculated.

3. Paschen series. The various spectral lines of this series originate from the electronic jumps from the *4th, 5th, 6th* etc., energy levels to the *3rd* energy level. Thus the frequencies of various spectral lines of this series can be obtained by putting n_1 = 3 and n_2 = 4, 5, 6......∝ in equation (1) given above.

(*a*) *First line.* For this line n_1 = 3 and n_2 = 4, and hence :

$$\bar{\upsilon} = 109.686 \times 10^5 \left(\frac{1}{3^2} - \frac{1}{4^2}\right) \text{m}^{-1}$$

$$= 5.331 \times 10^5 \text{ m}^{-1}$$

(*b*) *Second line.* For this line n_1 = 3 and n_2 = 5, and hence :

$$\bar{\upsilon} = 109.686 \times 10^5 \left(\frac{1}{3^2} - \frac{1}{5^2}\right) \text{m}^{-1}$$

$$= 7.799 \times 10^5 \text{ m}^{-1}$$

(*c*) *Limiting line.* For this line n_1 = 3 and n_2 = ∝, and hence :

$$\bar{\upsilon} = 109.686 \times 10^5 \left(\frac{1}{3^2} - \frac{1}{\propto^2}\right) \text{m}^{-1}$$

$$= \frac{109.686 \times 10^5}{9} \text{ m}^{-1}$$

$$= 12.187 \times 10^5 \text{ m}^{-1}$$

On similar lines the frequencies of other lines can also be calculated.

4. Brackett series. The various spectral lines of this series originate when an electron jumps from *5th, 6th, 7th* etc. energy levels to *4th* energy level. Thus the frequency of various spectral lines of this series can be obtained by putting n_1 = 4 and n_2 = 5, 6, ...∞ in equation (1).

(a) *First line*. For this line n_1 = 4 and n_2 = 5, and hence :

$$\bar{\upsilon} = 109.686 \times 10^5 \left(\frac{1}{4^2} - \frac{1}{5^2} \right) \text{m}^{-1}$$

$$= 2.45 \times 10^5 \text{ m}^{-1}$$

(b) *Second line*. For this line n_1 = 4 and n_2 = 6, and hence :

$$\bar{\upsilon} = 109.686 \times 10^5 \left(\frac{1}{4^2} - \frac{1}{6^2} \right) \text{m}^{-1}$$

$$= 3.808 \times 10^5 \text{ m}^{-1}$$

(c) *Limiting line*. For this line n_1 = 4 and n_2 = ∝, and hence :

$$\bar{\upsilon} = 109.686 \times 10^5 \left(\frac{1}{4^2} - \frac{1}{\propto^2} \right) \text{m}^{-1}$$

$$= \frac{109.686 \times 10^5}{16} \text{m}^{-1}$$

$$= 6.855 \times 10^5 \text{ m}^{-1}$$

5. Pfund series. The various spectral lines of this series originate when an electron jumps from *6th, 7th, 8th* etc. energy levels to *5th* energy level. Thus the frequency of various spectral lines of this series can be obtained by putting n_1 = 5 and n_2 = 6, 7, 8...... ∝ in equation (1).

(a) *First line*. For this line n_1 = 5 and n_2 = 6, and hence :

$$\bar{\upsilon} = 109.686 \times 10^5 \left(\frac{1}{5^2} - \frac{1}{6^2} \right) \text{m}^{-1}$$

$$= 1.340 \times 10^5 \text{ m}^{-1}$$

(b) *Second line*. For this line n_1 = 5 and n_2 = 7, and hence :

$$\bar{\upsilon} = 109.686 \times 10^5 \left(\frac{1}{5^2} - \frac{1}{7^2} \right) \text{m}^{-1}$$

$$= 2.148 \times 10^5 \text{ m}^{-1}$$

(c) *Limiting line*. For this line n_1 = 5 and n_2 = ∝, and hence :

$$\bar{\upsilon} = 109.686 \times 10^5 \left(\frac{1}{5^2} - \frac{1}{\propto^2} \right) \text{m}^{-1}$$

$$= \frac{109.686 \times 10^5}{25} \text{m}^{-1}$$

$$= 4.387 \times 10^5 \text{ m}^{-1}.$$

The calculated values of the frequencies in terms of wave numbers, $\bar{\upsilon}$ (in m^{-1}) of the various spectral lines of all the five series of hydrogen spectrum and those determined experimentally have been given in Table 1.5. In the same table the values of n_1 and n_2, different types of electronic jumps due to which the various spectral lines are produced, etc. are also given. It may be seen from this table that there is an excellent agreement between the calculated and experimental values of frequency. *This excellent agreement offers another strong support to Bohr's theory of hydrogen spectrum.*

Table 1.5. Calculated and experimental values of frequency in terms of wave number, $\bar{\upsilon}$ (in m^{-1}) of various spectral lines of different series of hydrogen spectrum.

Spectral series	Frequency (m^{-1}) of various spectral lines		Values of n_1 and n_2 for various spectral lines the		Electronic jumps producing spectral lines
	Calculated (by Bohr)	Experimental	n_1	n_2	
1. Lyman series (Ultraviolet region)	82.26×10^5 97.49×10^5 102.80×10^5 105.20×10^5 106.30×10^5 107.20×10^5	82.20×10^5 97.60×10^5 102.70×10^5 105.20×10^5 106.20×10^5 107.10×10^5	1	2 3 4 5 6 7	$2 \rightarrow 1$ $3 \rightarrow 1$ $4 \rightarrow 1$ $5 \rightarrow 1$ $6 \rightarrow 1$ $7 \rightarrow 1$
2. Balmer series (Visible region)	15.23×10^5 (H_α line) 20.56×10^5 (H_β line) 23.00×10^5 (H_γ line) 24.30×10^5 (H_δ line) 25.20×10^5	15.21×10^5 (H_α line) 20.60×10^5 (H_β line) 23.05×10^5 (H_γ line) 24.35×10^5 (H_δ line) 25.18×10^5	2	3 4 5 6 7	$3 \rightarrow 2$ $4 \rightarrow 2$ $5 \rightarrow 2$ $6 \rightarrow 2$ $7 \rightarrow 2$
3. Paschen series (Infra-red region)	5.33×10^5 7.79×10^5 9.13×10^5 9.95×10^5	5.30×10^5 7.80×10^5 9.12×10^5 9.95×10^5	3	4 5 6 7	$4 \rightarrow 3$ $5 \rightarrow 3$ $6 \rightarrow 3$ $7 \rightarrow 3$
4. Brackett series (Infra-red region)	2.45×10^5 3.80×10^5 4.61×10^5	2.46×10^5 3.80×10^5 4.61×10^5	4	5 6 7	$5 \rightarrow 4$ $6 \rightarrow 4$ $7 \rightarrow 4$
5. Pfund series (Infra-red region)	1.34×10^5 2.14×10^5	1.34×10^5 2.14×10^5	5	6 7	$6 \rightarrow 5$ $7 \rightarrow 5$

Derivation of Rydberg's Equation from Bohr's Theory

We have already seen that, before Bohr's theory came to be known, Rydberg suggested that the wavelength (in terms of wave number, $\bar{\upsilon}$) of various lines in Lyman series, Balmer series etc. of hydrogen spectrum could be calculated with the help of the following empirical equation which is called Rydberg's equation.

$$\bar{\upsilon} = R\left(\frac{1}{x^2} - \frac{1}{y^2}\right) m^{-1} \text{ (Rydberg's equation)} \quad ...(1)$$

Here R is a constant which is called Rydberg's constant and its experimental value, as determined by Ritz, is 109.678×10^3 cm^{-1} or 109.678×10^5 m^{-1}, and x and y are two simple whole numbers. Later on Bohr obtained the above equation directly with the help of his theory. This is shown below :

We have seen that Bohr, with the help of his theory, deduced the following equations for the calculation of frequency in wave number, $\bar{\upsilon}$ (in cm^{-1} or m^{-1}) of a spectral line of a given series.

$$\bar{\upsilon} = \frac{2\pi^2 me^4}{h^3 c}\left(\frac{1}{n_1^{\,2}} - \frac{1}{n_2^{\,2}}\right) \text{cm}^{-1} \textit{ (in C.G.S. units)} \quad ...(2)$$

$$\bar{\upsilon} = \frac{me^4}{8\varepsilon_0^{\,2} h^3 c}\left(\frac{1}{n_1^{\,2}} - \frac{1}{n_2^{\,2}}\right) \text{m}^{-1} \textit{ (in S.I. units)} \quad ...(3)$$

(*Bohr's equations*)

It is obvious that Bohr's equations are equivalent to Ritz equation (1), if $x = n_1$, $y = n_2$ and $R = 2\pi^2 me^4/h^3c$ (in C.G.S. units) or $me^4/8\varepsilon_0^2h^3c$ (in S.I. units). Now let us calculate the value of R in both units.

(*a*) *In C.G.S. units*. The value of R is obtained by putting $\pi = 3.14$; $m = 9.108 \times 19^{-28}$ g; $e = 4.8 \times 10^{-19}$ esu; $h = 6.624 \times 10^{-27}$ erg-s; $c = 3 \times 10^{10}$ cm . s^{-1} in

$$R = \frac{2\pi^2 me^4}{h^3 c}$$

$$= \frac{2 \times (3.14)^2 \times (9.108 \times 10^{-28}) \times (4.8 \times 10^{-10} \text{ esu})^4}{(6.624 \times 10^{-27} \text{ erg-s})^3 \times (3 \times 10^{10} \text{ cm.s}^{-1})}$$

$$= 109.686 \times 10^3 \text{ cm}^{-1}.$$

(*b*) *In S.I. units*. The value of R is obtained by. putting m = 9.108×10^{-31} kg; $e = 1.602 \times 10^{-19}$ C; $\varepsilon_0 = 8.85 \times 10^{-12}$, $h = 6.624 \times 10^{-34}$ J-s; $c = 3 \times 10^8$ m.s^{-1} in

$$R = \frac{me^4}{8\varepsilon_0^{\,2} h^3 c}$$

$$= \frac{(9.108 \times 10^{-31} \text{ kg}) \times (1.602 \times 10^{-19} \text{ C})^4}{8 \times (8.85 \times 10^{-12})^2 \times (6.624 \times 10^{-34})^3 \times (3 \times 10^3 \text{ m.s}^{-1})}$$

$$= 109.686 \times 10^5 \text{ m}^{-1}.$$

Thus we find that the calculated value of R in both the units is the same and agrees closely with the experimental value (= 109.678×10^3 cm or 109.678×10^5 m^{-1}) obtained by Ritz. The agreement between the theoretical value of R as calculated by Bohr and that obtained experimentally by Ritz gives a further support to Bohr's theory.

Derivation of Mosley's Equation by Bohr's Theory

We have seen that Mosley, while working on X-rays, showed that the square root of the frequency ($\sqrt{\upsilon}$) of a line of a given series of characteristic X-ray spectrum is related to the atomic number (Z) of the element used as anticathode (target material) by the following equation which is called Mosley's equation :

$$\sqrt{\upsilon} = a(Z - b) \quad ...(1)$$

Here a is a proportionality constant and b is another constant which is called *screening constant* and takes into account the screening effect of electrons. The above equation can also be derived by Bohr's theory as follows :

We have seen that according to Bohr's theory, when an electron in an atom with atomic number, Z jumps from an orbit n_2 to the orbit n_1, the frequency (υ) of radiation given out is given by :

$$\upsilon = \frac{2\pi^2 Z^2 me^4}{h^3}\left(\frac{1}{n_1^{\,2}} - \frac{1}{n_2^{\,2}}\right) \text{ cycles/s}$$

$$= \frac{2\pi^2 me^4}{h^3}\left(\frac{1}{n_1^{\,2}} - \frac{1}{n_2^{\,2}}\right). Z^2 \text{ cycles/s}$$

Thus :

$$\upsilon \ \alpha \ Z^2$$

or
$$\sqrt{\upsilon} \ \alpha \ Z$$

Now if the screening constant, b which takes into account the screening effect of the electrons, is considered, the above relation becomes

$$\sqrt{\upsilon} \ \alpha \ (Z - b)$$

or
$$\sqrt{\upsilon} = a(Z - b)$$

This equation is the same, as equation (1) given above.

Numerical Problems

Problem 1 (In C.G.S. units). *A spectral line in the Lyman series of hydrogen atom has a frequency of 82, 200 cm*$^{-1}$*. What transition is responsible for this radiation? (R = 109,600 cm*$^{-1}$*)* (*Raj. B.Sc. I* 1982)

Solution. We know that the frequency of various spectral lines of Lyman series of hydrogen spectrum can be obtained by putting $n_1 = 1$ and $n_2 = 2, 3$, etc. in the following equation :

$$\bar{\upsilon} = \text{R}\left(\frac{1}{n_1^{\,2}} - \frac{1}{n_2^{\,2}}\right) \text{cm}^{-1}$$

Substituting $\bar{\upsilon}$ = 82,200, R = 109,600 and $n_1 = 1$ in the above equation we get :

$$82200 = 109600\left(\frac{1}{1^2} - \frac{1}{n_2^{\,2}}\right)$$

From this equation $n_2 = 2$ which means that the spectral line having $\bar{\upsilon}$ = 82200 cm^{-1} originates from the jump of the electron from 2*nd* orbit ($n = 2$) to 1*st* orbit ($n = 1$)

Problem 2 (In C.G.S. units). *Calculate the wavelength of light emitted from hydrogen for the electronic transmission for energy level n = 3 to n = 2 (R = 109,678 cm*$^{-1}$*)* (*Raj. P.M.T. 1981*)

Solution. We know that wavelength λ is given by :

$$\lambda = \frac{1}{\text{R}\left[\frac{1}{n_1^{\,2}} - \frac{1}{n_2^{\,2}}\right]} \text{ cm}$$

Substituting R = 109678, $n_1 = 2$ and $n_2 = 3$ in the above equation, we get :

$$\lambda = \frac{1}{109678\left[\frac{1}{2^2} - \frac{1}{3^2}\right]} \text{ cm}$$

$$= \frac{36}{109678 \times 5} = \frac{36}{109.678 \times 10^3 \times 5}$$

$$= \frac{36}{548.390 \times 10^3}$$

$$= 0.0656 \times 10^{-3} \text{ cm}$$

Problem 3. *Calculate the frequency of the line in the emission spectrum of hydrogen when the atoms of the gas contain electrons in the 3rd energy level.*

(Delhi B.Sc. Hons. Part I 1974)

Solution. (*a*) *When the electron falls from 3rd energy to 2nd energy level, i.e. $n_1 = 2$ and $n_2 = 3$.*

$$\therefore \quad \upsilon_{3\to2} = 3.290 \times 10^{15}\left(\frac{1}{n_1^2} - \frac{1}{n_2^2}\right) \text{Hz}$$

or $$= 3.290 \times 10^{15}\left(\frac{1}{2^2} - \frac{1}{3^2}\right) \text{Hz}$$

$$\therefore \quad = 0.4569 \times 10^{15} \text{ Hz}$$

(*b*) *When the electron falls from 3rd energy level to 1st energy level i.e. $n_1 = 1$ and $n_2 = 3$.*

$$\therefore \quad \upsilon_{3\to1} = 3.290 \times 10^{5}\left(\frac{1}{n_1^2} - \frac{1}{n_2^2}\right) \text{Hz}$$

$$= 3.290 \times 10^{15}\left(\frac{1}{1^2} - \frac{1}{3^2}\right) \text{Hz}$$

or $$= 2.924 \times 10^{15} \text{ Hz}$$

Problem 4 (In S.I. units). *Calculate the wave number and wavelength of the sepctral line observed in hydrogen spectrum with $n_1 = 2$ and $n_2 = 3$. The value of Rydberg's constant (R) = 109.686×10^6 m^{-1}.*

To which line does this line correspond in the spectral lines of Balmer series?

Solution. Substituting the values of n_1, n_2 and R in

$$\bar{\upsilon} = R\left(\frac{1}{n_1^2} - \frac{1}{n_2^2}\right) m^{-1}$$

$$= 109.686 \times 10^5 \left[\frac{1}{2^2} - \frac{1}{3^2}\right] \text{m}^{-1}$$

$$= \frac{109.686 \times 10^5 \times 5}{36}$$

$$= 15.234 \times 10^5 \text{ m}^{-1}$$

$$\lambda = \frac{1}{R\left[\frac{1}{n_1^2} - \frac{1}{n_2^2}\right]} \text{ m}$$

$$= \frac{1}{109.686\times10^5\left[\frac{1}{2^2} - \frac{1}{3^2}\right]}$$

$$= \frac{36}{109.686\times10^5}\times10^{-5} \text{ m}$$

$$= 0.0656\times10^{-5} \text{ m}$$

Quite obviously the spectral line with frequency, υ = 15.234 × 10^5 m^{-1} is Hα line in the Balmer series of hydrogen spectrum.

Problem 5. (In S.I. units). *Calculate the wave number and the energy of the spectral line in the hydrogen spectrum corresponding to $n_1 = 3$ and $n_2 = 5$. The value of Rydberg's constant (R) = 10.97 × 10^6 m^{-1}.*

Solution. Substituting the values of n_1, n_2 and R in

$$\overline{\upsilon} = R\left[\frac{1}{n_1^2} - \frac{1}{n_2^2}\right] \text{ m}^{-1}$$

$$= 10.97\times10^6\left[\frac{1}{3^2} - \frac{1}{5^2}\right] \text{ m}^{-1}$$

$$= \frac{10.97\times10^6\times16}{225} \text{ m}^{-1}$$

or

$$= 7.80\times10^5 \text{ m}^{-1}$$

Now the energy (E) of the spectral line is given by Planck's equation *viz.*

$$E = h\upsilon = h.\frac{c}{\lambda} \qquad \left(\because \frac{1}{\lambda} = \frac{\upsilon}{c}\right)$$

or

$$E = h.c\overline{\upsilon} \qquad \left(\because \overline{\upsilon} = \frac{1}{\lambda}\right)$$

Now on substituting h (Planck's constant) = 6.624 × 10^{-34} J-s, c (velocity of light) = 3 × 10^8 m/sec and $\overline{\upsilon}$ (wave number) = 7.8 × 10^5 m^{-1} (obtained as above) in the above equation, we get

$$E = hc\overline{\upsilon}$$

$$= (6.624 \times 10^{-34} \text{ J-s/} \times (3 \times 10^8 \text{ m.s}^{-1}) \times (7.8 \times 10^5 \text{ m}^{-1})$$

$$= 6.624 \times 3 \times 7.8 \times 10^{-21} \text{ J}$$

$$= 0.0624 \times 3 \times 7.8 \times 10^{-19} \text{ J}$$

$$= 1.55 \times 10^{-19} \text{ J}$$

$$= 1.55 \times 10^{-22} \text{ KJ}$$

Problem 6 (In S.I. units). *Energy in Bohr's orbit is given to be equal to $-A/n^2$, where A = 2.179 × 10^{-18} J. Calculate the frequency of radiation and also* *find out the wave number when the electron moves from the third orbit to the second orbit ($h = 6.62 \times 10^{-34}$ J-s).* *(Delhi B.Sc. Hons. 1979)*

Solution. In S.I. units energy of nth Bohr's orbit is given by :

$$E_n = -\frac{me^4}{8\varepsilon_0^2 n^2 h^2}$$

$$= -\frac{me^4}{8\varepsilon_0^2 h^2}\cdot\frac{1}{n^2}$$

According to question,

$$E_n = -\frac{A}{n^2} = -2.179\times10^{-18}\times\frac{1}{n^2}$$

$$\therefore \quad \frac{me^4}{8\varepsilon_0^2 h^2} = 2.179\times10^{-18}$$

or $$\frac{me^4}{8\varepsilon_0^2 h^2}\times\frac{1}{h} = \frac{2.179\times10^{-18}}{h} = \frac{2.179\times10^{-18}}{6.62\times10^{-34}}$$

$$= \frac{2.179\times10^{16}}{6.62}$$

$$= 3.29\times10^{15}$$

Now in S.I. units frequency, υ is given by :

$$\upsilon = \frac{me^4}{8\varepsilon_0^2 h^2}\left[\frac{1}{n_1^2}-\frac{1}{n_1^2}\right] \text{Hz}$$

$$= 3.29\times10^{15}\left[\frac{1}{2^2}-\frac{1}{n_2^2}\right] \text{Hz}$$

Substituting $n_1 = 2$ and $n_2 = 3$ in the above equation, we get :

$$\upsilon = 3.29\times10^{15}\left[\frac{1}{2^2}-\frac{1}{3^2}\right] \text{Hz}$$

$$\therefore \quad = \frac{3.29\times10^{15}\times5}{36} \text{Hz} = 0.45\times10^{15} \text{ Hz}$$

Now wave number, $\bar{\upsilon}$ is given by :

$$\bar{\upsilon} = \frac{me^4}{8\varepsilon_0^2 h^3 c}\left[\frac{1}{n_1^2}-\frac{1}{n_2^2}\right] \text{m}^{-1}$$

$$= \frac{3.29\times10^{15}}{3\times10^8}\left[\frac{1}{2^2}-\frac{1}{3^2}\right] \text{m}^{-1}$$

$$= \frac{3.29\times10^{15}\times5}{3\times36\times10^8} \text{m}^{-1}$$

$$= \frac{3.29\times5\times10^7}{3\times36} \text{m}^{-1} = 1.52\times10^6 \text{ m}^{-1}$$

Problem 7 (In S.I. units). *Calculate the frequencies of lines in the emission spectrum of hydrogen when the atoms of the gas contain the electrons in the 3rd energy level. Calculate the wavelength of any one of these. The energy of hydrogen electron in the ground state is* -2.18×10^{-18} *J/atom. The value of Planck's constant is* 6.625×10^{-34} *J-s. Velocity of light is* 3×10^{8} *m.s*$^{-1}$.

(Delhi B.Sc. Hons. Part I 1974; Mysore 1974)

Solution. Frequency (υ) of a spectral line is given by :

$$\upsilon = \frac{me^4}{8\varepsilon_0^2 h^3}\left(\frac{1}{n_1^2} - \frac{1}{n_2^2}\right) \text{ Hz}$$

and energy of the electron in the ground state (n = 1) is given by :

$$E_1 = \frac{me^4}{8\varepsilon_0^2 \times (1)^2 \times h^2} = -2.18 \times 10^{-18} \text{ J/atom}$$

$$\therefore \quad \frac{me^4}{8\varepsilon_0^2 h^3} = \frac{2.18 \times 10^{-18}}{h} = \frac{2.18 \times 10^{-18}}{6.625 \times 10^{-34}} = \frac{2.18 \times 10^{16}}{6.625}$$

$$= \frac{21.8 \times 10^{15}}{6.625} = 3.29 \times 10^{15}$$

$$\therefore \quad \upsilon = 3.29 \times 10^{15}\left[\frac{1}{n_1^2} - \frac{1}{n_2^2}\right] \text{ Hz}$$

Now the following cases are studied.

(*a*) *When the electron falls from 3rd energy level to 2nd energy level, i.e.* $n_2 = 3$ *and* $n_1 = 2$,

$$\therefore \quad \upsilon_{3\to 2} = 3.29 \times 10^{15}\left(\frac{1}{2^2} - \frac{1}{3^2}\right) \text{ Hz}$$

$$= \frac{3.29 \times 10^{15} \times 5}{36} \text{ Hz}$$

$$= 0.456 \times 10^{15} \text{ Hz}$$

(*b*) *When the electron falls from 2nd energy level to 1st energy level, i.e.* $n_2 = 2$ *and* $n_1 = 1$.

$$\therefore \quad \upsilon_{2\to 1} = 3.29 \times 10^{15}\left(\frac{1}{1^2} - \frac{1}{2^2}\right) \text{ Hz}$$

$$= 3.29 \times 10^{15} \times 0.75 \text{ Hz}$$

$$= 2.46 \times 10^{15} \text{ Hz}$$

(*c*) *The electron falls directly from 3rd energy level to 1st energy level, i.e.,* $n_2 = 3$ *and* $n_1 = 1$.

$$\therefore \quad \upsilon_{3\to 1} = 3.29 \times 10^{15}\left(\frac{1}{1^2} - \frac{1}{3^2}\right) \text{ Hz}$$

$$= \frac{3.29 \times 10^{15} \times 8}{9} \text{ Hz}$$

$$= 2.92 \times 10^{15} \text{ Hz}$$

Problem 8 (In C.G.S. units). *Find the wavelength in A° of the line in Balmer series that is associated with drop of the electron from the fourth orbit. The value of Rydberg's constant is 109, 676 cm*$^{-1}$.

Solution. Here $n_1 = 2$, $n_2 = 3$ and R = 109676 cm^{-1}. The wavelengths (λ) of the lines in Balmer series are given by :

$$\lambda = \frac{1}{R\left[\frac{1}{n_1^2} - \frac{1}{n_2^2}\right]} \text{ cm}$$

$$= \frac{1}{109676\left[\frac{1}{2^2} - \frac{1}{3^2}\right]} \text{ cm} = \frac{1 \times 36}{109676 \times 5} \text{ cm}$$

$$= 0.0000656 \text{ cm} = 6.56 \times 10^{-5} \text{ cm}$$

Defects of Bohr's Atomic Model

1. No explanation of fine structure—Existence of additional quantum numbers. We have seen that Bohr's theory had successfully explained the origin of hydrogen spectrum. But, when the hydrogen spectrum is observed by means of spectroscope of high resolving power, it is found that the individual spectral lines are not really single but consist of several very fine lines lying close together. For example H_α line in the Balmer series of hydrogen spectrum is found to consists of five component lines. Such a spectrum which consists of closely-spaced spectral lines is called **fine structure** or **multiple structure.** This fine structure of spectral lines in the hydrogen spectrum cannot be explained by Bohr's theory. Thus the appearance of the several lines in a single line suggests that only one quantum number *viz.*, principal quantum number (n) suggested by Bohr is not sufficient to explain the origin of various spectral lines. Rather, there are additional quantum numbers for a given principal quantum number. This, in other words, means that in a given orbit (or shell or energy level) there are several orbits (or shells or energy levels), called *sub-orbits* (or *sub-shells* or *sub-energy levels*), which are slightly different in energies.

2. No explanation for the spectra of multi-electron systems. Bohr's theory has successfully explained the origin of the spectrum of hydrogen atom and of hydrogen-like ions (*e.g.*, He^+, Li^{2+}, Be^{3+} etc.) which are *one-electron systems*, but it is not able to explain the origin of the spectrum of *many-electron* (or *multi-electron* or *poly-electron*) *systems*.

3. According to Bohr's theory the electrons move in circular orbits round the nucleus, but the modern researches have shown that the motion of the electron is not limited to a single plane but takes place in *three-dimensional space.* In other words, atomic model is not *flat* as suggested by Bohr, but has a *three-dimensional existence.*

4. No explanation for Zeeman and Stark effects. It has been observed that when an excited atom which gives a line emission spectrum is put in magnetic field, its spectral lines are further split up into still thinner (*i.e.*, closely-spaced) lines. This type of splitting of spectral lines into a number of closely-spaced lines is called **Zeeman effect.** For example when the source producing sodium spectrum is placed in a weak magnetic field, D_1 and D_2 lines of the spectrum are found to split up into components (Fig. 1.37).

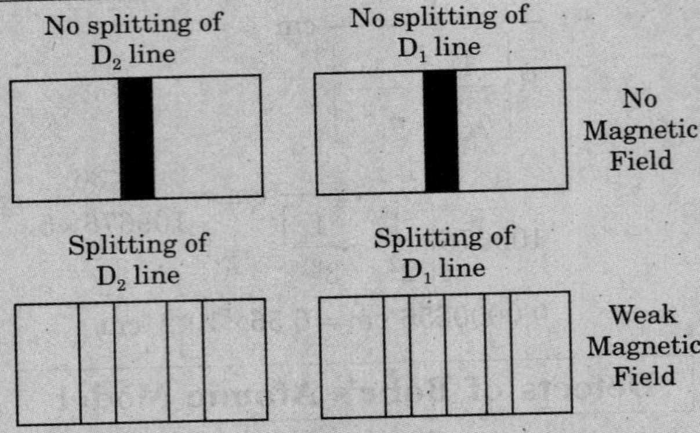

Fig. 1.37. Splitting of the D_2 and D_1 lines in the sodium spectrum by a weak magnetic field (*Illustration of Zeeman effect*).

Another similar splitting of the spectral lines into a number of thinner lines is also observed when the excited atom is placed in an electric field. This type of splitting is called **Stark effect.** Bohr's theory has no explanation for any of these effects.

5. Dual nature of electron and Heisenberg's uncertainty principle. We have seen that Bohr had assumed that an electron behaves as a *material particle* of small mass moving round the nucleus in fixed circular orbits. But de Broglie suggested that *an electron has a dual nature, i.e., an electron, like light, behaves both as a material particle and as a wave.*

We have also seen that Bohr had also assumed that an electron which is a material particle is revolving round the nucleus in circular orbits situated at *definite* (*i.e.*, fixed) distance from the nucleus and with a *definite velocity*. But any exact measurement of the position of an electron which also has a wave-like nature would create some uncertainty in the exact measurement of its velocity and vice-versa *i.e.*, for an electron it is impossible to measure simultaneously its exact position and exact velocity. This is called **Heisenberg's uncertainty principle** which can be stated as follows :

It is impossible to measure simultaneously the exact position and exact velocity of a sub-atomic particle like electron and neutron.

Thus we see that Bohr's assumption that an electron revolving in circular orbits situated at a fixed distance from nucleus and with fixed velocity is against the Heisenberg's uncertainty principle.

Bohr-Bury Scheme of Distribution of Electrons in Orbits

Having known that the number of planatory electrons in an atom is numerically equal to the atomic number of the element and they are revolving about the nucleus in closed orbits, the question arises as to how these electrons are arranged in these orbits.

In 1912 Bohr and Bury simultaneously put forward two identical schemes for the distribution of electrons in orbits. This scheme is now called **Bohr-Bury scheme.** According to this scheme :

1. *The maximum number of electrons in any orbit is given by the formula* $2n^2$ *where n is the number of the orbit. Thus :*

No. of. electrons in 1st orbit (n = 1, K-shell) = $2 \times 1^2 = 2$

" " in 2nd " (n = 2, L-shell) = $2 \times 2^2 = 8$

" " in 3rd " (n = 3, M-shell) = $2 \times 3^2 = 18$.

2. *The outermost orbit in an atom shall not have more than 8 electrons and next to the outermost not more than 18.*

3. *It is not necessary for an orbit to be complete before another commences to be formed. In fact a new orbit begins when the outermost orbit attains 8 electrons.*

4. *The outermost orbit cannot have more than 2 electrons and the orbit next to the outermost cannot have more than nine, so long as the next inner orbit in each case has not received the maximum number of electrons as required by rule (1) and (2) given above.*

Electronic Configuration of Inert Gases

On the basis of Bohr-Bury scheme the configuration of the atoms of inert gases is given in a tabular form as follows :

Inert gas	Atomic number	Distribution of electrons in various orbits					
		1st (K)	2nd (L)	3rd (M)	4th (N)	5th (O)	6th (P)
Helium (He)	2	2					
Neon (Ne)	10	2	8				
Argon (Ar)	18	2	8	8			
Krypton (Kr)	36	2	8	18	8		
Xenon (Xe)	54	2	8	18	18	8	
Radon (Rn)	86	2	8	18	32	18	8

Sommerfeld's Extension of Bohr's Atomic Model : Bohr-Sommerfeld's Atomic Model

To account for the *fine structure* in the spectrum of hydrogen atom Sommerfeld extended Bohr's theory in 1915 and suggested *that the moving electron might describe, in addition to the circular orbits, elliptical orbits as well, with the nucleus situated at one of the focii of the ellipse.*

It is well known that in a circle, the only coordinate that changes is the angle of revolution, ϕ while the radial coordinate, r (*i.e.*, the radius) remains constant. In an elliptical orbit like that shown in Fig. 1.38, however, both the polar coordinates *viz.*, radius vector, *r* (which is the distance of the electron from the nucleus situated at one of the focii, Z) and the vectorial angle, ϕ known as azimuthal angle (which is the angle between the radius vector and the major axis of the ellipse) change.

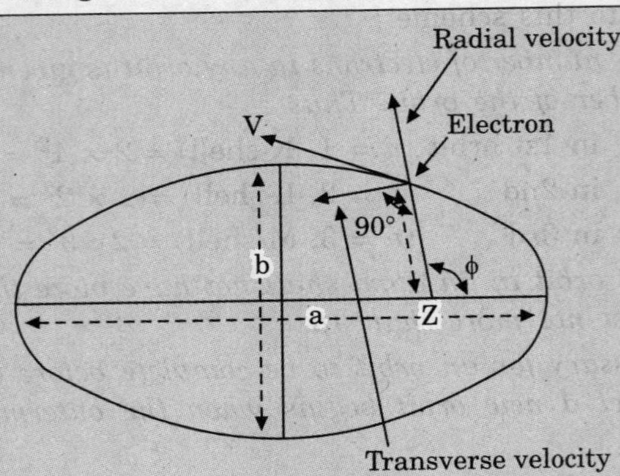

Fig. 1.38. Velocity of an electron in an elliptical orbit. *a* and *b* are the lengths of the ellipse. V represents the tangential velocity of the electron at any instant. Z is its the nucleus.

Now the tangential velocity, V of the electron at any instant can be resolved into two components one of which is along the radius vector and is called the *radial velocity* while the other is perpendicular to the radius vactor and is called *transverse* or *angular velocity*. These two velocities give rise to *radial momentum and angular* (or *azimuthal*) *momentum*. Sommerfeld assumed that these momentums must obey the quantum conditions, *i.e.*, these momentums must be integral multiple of $h/2\pi$. Thus :

$$\text{Radial momentum} = n_r \cdot \frac{h}{2\pi}$$

$$\text{Azimuthal momentum} = n_\phi \cdot \frac{h}{2\pi}$$

Here the integrals n_r and n_ϕ are two new quantum numbers which are called **radial quantum number** and **azimuthal quantum number** respectively. n_r and n_ϕ are related to Bohr's quantum number, *n* as follows :

$$n = n_r + n_\phi$$

n and n_ϕ are related to the geometry of the ellipse as :

$$\frac{n}{n_\varphi} = \frac{n_r + n_\varphi}{n_\varphi} = \frac{\textit{Length of the major axis (a)}}{\textit{Length of the minor axis (b)}} \quad ...(i)$$

Now let us consider the following possibilities :

1. *When* $n_\phi = 0$. In this case the length of the minor axis (*b*) of the ellipse will also become equal to zero. In other words the electronic path will be a straight line passing through the nucleus, Z which is impossible. Thus n_ϕ *cannot be equal to zero*.

2. *When* $n_\phi > n$. In this case the length of the minor axis (*b*) will be greater than the length of the major axis (*a*). Thus n_ϕ *cannot be greater than n*.

3. *When* $n_\phi = n$. In this case a becomes equal to b and the orbit becomes circular. Thus circular orbit is a special case of elliptical orbit.

4. *When* $n_\phi < n$. In this case the orbit will remain elliptical.

Thus when $n = 4$, $n_\phi = 1, 2, 3, 4$. Corresponding to these four values of n_ϕ for $n = 4$, we have four possible orbits which are designated as 4_1, 4_2, 4_3 and 4_4 respectively. The subscripts represent the values of n_ϕ corresponding to $n = 4$. Quite obviously one orbit (4_4 orbit) is circular while the other three orbits are elliptical as shown below :

n	n_ϕ	n_r	*Designation and shape of the orbit*
4	4	0	4_4 (Circular orbit)
4	3	1	4_3 (Elliptical orbit)
4	2	2	4_2 (Elliptical orbit)
4	1	3	4_1 (Elliptical orbit)

All the four possible orbits for $n = 4$ are shown in Fig. 1.39.

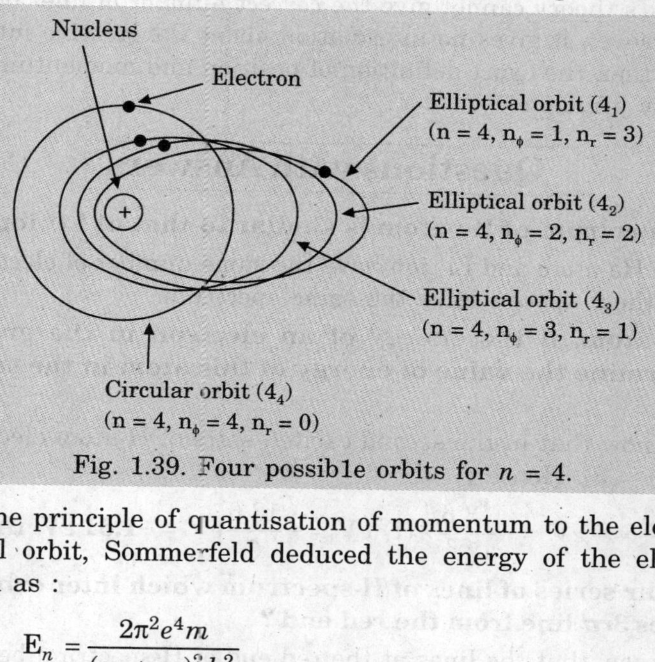

Fig. 1.39. Four possible orbits for $n = 4$.

Applying the principle of quantisation of momentum to the electron moving in an elliptical orbit, Sommerfeld deduced the energy of the electron in the hydrogen atom as :

$$E_n = \frac{2\pi^2 e^4 m}{(n_\phi + n_r)^2 h^2}$$

Thus we find that the energy depends not only on the principal quantum number, n but also to some extent on the azimuthal quantum number, n_ϕ. The frequency of the radiation emitted as a result of transition of an electron from a level of energy E_2 to a level of energy E_1 ($E_2 > E_1$) would, therefore, vary, depending on the various values of n_ϕ and as a result a group of closely-spaced lines would appear instead of a single line in the spectrum of hydrogen atom or hydrogen like atoms.

The concepts advanced by Sommerfeld evidently postulate existence of sub-energy level for electrons in each of the principal energy levels of the atom.

Weakness of Bohr-Sommerfeld's Atomic Model

Experimental observations and theoretical treatment based on wave mechanics show that the different values of azimuthal quantum number are : 0, 1, 2, 3, ..., $(n-2)$, $(n-1)$ (total n values) and not 1, 2, 3, ..., $(n-1)$, n (total n values). The new quantum number beginning with zero and ending at $(n-1)$ has been represented by l in order to avoid the confusion with n_φ. Thus l is related with n_ϕ and n by the following relations :

$$\left.\begin{aligned} l &= n_\phi - 1 \\ \text{or} \quad n_\phi &= l + 1 \end{aligned}\right\} \quad ...(ii)$$

$$l = 0, 1, 2, ..., (n-2), (n-1)$$

With the help of relation (*ii*), equation (*i*) becomes

$$\frac{n}{l+1} = \frac{a}{b}$$

Sommerfeld's theory cannot give the correct number of lines observed in fine structure. Moreover, it gives no information about the relative intensities of the "*fine line*". Further, the exact definition of position and momentum is contrary to the uncertainty principle.

Questions with Answers

Q.1 The spectrum of He atom is similar to that of Li^+ ion. Explain.

Ans. Since He-atom and Li^+ ion have the same number of electrons (He = $1s^2$, $Li^+ = 1s^2 2s^0$), these species have the same spectrum.

Q.2 In H-atom, if the energy of an electron in the ground state is 13.6 eV, determine the value of energy of this atom in the second excited state.

Ans. We know that in the second excited state of H-atom electron is present in 3rd orbit (n = 3). Hence :

$$E_3 = -\frac{13.6}{n^2} \times Z^2 eV = -\frac{13.6}{(3)^2} \times (1)^2 eV = -\frac{13.6}{9} eV = \mathbf{-1.51\ eV\ (Ans.)}$$

Q.3 In Bohr series of lines of H-spectrum which inter orbit jump of the electron gives *3rd* line from the red end ?

Ans. We know that the lines at the red end of H-spectrum belong to Balmer series. The lines of this series are obtained due to the following electronic jumps : $n_3 \to n_1$ (*1st* line), $n_4 \to n_1$ (*2nd* line), $n_5 \to n_1$ (*3rd* line), etc. Obviously *3rd* line is obtained due to $n_5 \to n_1$ or $5 \to 1$ electronic jump.

Q.4 If IP of H-atom is 13.6 eV, find out the IP of He^+ ion.

Ans. Since H-atom and He^+ ion both have one electron, IP of He^+ is given by :

$$(IP)_{He^+} = (IP)_H \times (\text{Atomic number of He})^2\ eV$$
$$= 13.6 \times 2^2\ eV = \mathbf{54.4\ eV\ (Ans.)}$$

Q.5 Find out the value of frequency (υ) of radiation emitted when an electron falls from n = 4 to n = 1 orbit in H-atom. Given : IE of H-atom = 2.18 × 10^{-18} J atom^{-1} and h = 6.625 × 10 Js.

Ans. ΔE for $n_4 \rightarrow n_1$ jump = $E_4 - E_1$

$$= -2.18\times10^{-18}\left(\frac{1}{n_4^{\ 2}}-\frac{1}{n_1^{\ 2}}\right)\text{J atom}^{-1}$$

$$= -2.18\times10^{-18}\left(\frac{1}{4^2}-\frac{1}{1^2}\right)\text{J atom}^{-1}$$

$$= +2.044\times10^{-18}\ \text{J atom}^{-1}$$

Now $\Delta E = h\upsilon$

or $$\upsilon = \frac{\Delta E}{h} = \frac{2.044\times10^{-18}\ \text{J atom}^{-1}}{6.625\times10^{-34}\ \text{Js}}$$

$$= \mathbf{3.085\times10^{15}\,s^{-1}}\ \textbf{(Ans.)}$$

Q.6 Which of the following ions has the same radius as that of the 1st orbit of H-atom ? He^+(n = 2), Li^{2+}(n = 2), Li^{2+}(n = 3) and Be^{3+}(n = 2).

Ans. We know that :

r_1 (n = 1) of H-atom = 0.529 Å.

Now let us calculate the radii of the given ions.

Radius of 2nd orbit (n = 2) of He^+(Z = 2)

$$= \frac{0.529\times n^2}{Z} = \frac{0.529\times 2^2}{2} = 0.529\times \text{Å}$$

Radius of 2nd orbit (n = 2) of Li^{2+}(Z = 3)

$$= \frac{0.529\times 2^2}{3} = 0.529\times\frac{4}{3}\ \text{Å}$$

Radius of 3rd orbit (n = 3) of Li^{2+} (Z = 3)

$$= \frac{0.529\times 3^2}{3} = 0.529\times 3\text{Å}$$

Radius of 2nd orbit (n = 2) of Be^{3+} (Z = 4)

$$= \frac{0.529\times 2^2}{4} = 0.529\text{Å}$$

Obviously 2nd orbit of Be^{3+} in has the same radius as the 1st orbit of H-atom.

Q.7 Calculate the wavelength of the electron emitted when in a H-atom electron falls from infinity to stationary state. Given R = 1.097 × 10^7 m^{-1}.

Ans. Frequency in terms of wave number ($\overline{\upsilon}$) is given by :

$$\overline{\upsilon} = R\left(\frac{1}{n_1^{\ 2}}-\frac{1}{n_2^{\ 2}}\right)$$

$$= 1.097\times10^7\left(\frac{1}{1^2}-\frac{1}{\infty^2}\right)\text{m}^{-1}$$

$$= 1.097\times10^7\,\text{m}^{-1}$$

$$\therefore \text{ Wavelength } (\lambda) = \frac{1}{\bar{\upsilon}} = \frac{1}{1.097 \times 10^7 \text{ m}^{-1}}$$

$$= \mathbf{0.91 \times 10^{-7} \text{ m (Ans.)}}$$

Q.8 Calculate the ratio of the area covered by 2nd orbit to the 1st orbit of H-atom.

Ans. We know that area covered by nth orbit of H-atom $= \pi r_n^2$

$$= \pi \times \left(\frac{0.529 \times n^2}{Z}\right)^2$$

$$= \frac{\pi \times (0.529)^2 \times n^4}{1^2}$$

$$\therefore \quad A_1 = \frac{\pi(0.529)^2 \times 1^4}{1^2}$$

and $$A_2 = \frac{\pi \times (0.529)^2 \times 2^4}{1^2}$$

$$\therefore \quad \frac{A_2}{A_1} = \frac{\pi \times (0.529)^2 \times 2^2}{1^2} \times \frac{1^2}{\pi \times (0.529)^2 \times 1^4}$$

$$= \frac{16}{1} \text{ (Ans.)}$$

Q.9 Which of the following one-electron ions has IE = 54.4 eV ? B^{4+}, Li^{2+}, Be^{2+} and He^{2+}

Ans. $(IE)_{B^{4+}}$ (Z = 5) = 13.6 × Z^2 eV = 13.6 × 5^2 eV = 340 eV

$(IE)_{Li^{2+}}$ (Z = 3) = 13.6 × Z^2 eV = 13.6 × 3^2 eV = 122.4 eV

$(IE)_{Be^{2+}}$ (Z = 4) = 13.6 Z^2 eV = 13.6 × 4^2 eV = 217.6 eV

$(IE)_{He^{2+}}$ (Z = 2) = 13.6 Z^2 eV = 13.6 × 2^2 eV = 54.4 eV

Obviously He^{2+} ion has IE = 54.4 eV **(Ans.)**

Q.10 Find out the value of most probable radius (in pm) for the electron in He^+ ion.

Ans. We know that the radius of 1st orbit (n = 1) of He^+ ion (Z = 2)

$$= \frac{0.529 \times n^2}{Z} \text{Å} = \frac{0.529 \times n^2}{Z} \times 10^2 \text{ pm} \qquad (\because 1 \text{ Å} = 10^2 \text{ pm})$$

$$= \frac{52.9 \times n^2}{Z} \text{ pm} = \frac{52.9 \times 1^2}{2} \text{ pm}$$

$$= \mathbf{26.5 \text{ pm Ans.}}$$

Q.11 Which of the following electronic transitions will have minimum wavelength ? $n_4 \rightarrow n_1$, $n_2 \rightarrow n_1$, $n_4 \rightarrow n_2$ and $n_3 \rightarrow n_1$.

Ans. We know that :

$$\Delta E = h\upsilon = \frac{hc}{\lambda} \qquad \left(\because \upsilon = \frac{c}{\lambda}\right)$$

or $$\lambda = \frac{hc}{\Delta E} \qquad ...(i)$$

In equation (*i*), since *hc* is constant, $\lambda \times \frac{1}{\Delta E}$. Now since $(\Delta E)_{n_4 \to n_1}$ corresponding to $n_4 \to n_1$ transition is maximum, λ for this transition is minimum.

University Questions

1. Describe Millikan's oil drop method for the determination of the charge on an electron. (*North Bengal 81; Poona 81*)

2. How is e/m ratio of the electron determined ? (*Tiruchirapalli 1985; Poona 86; Madras 85; Madurat 85*)

3. Describe the characteristics of proton. (*Venkateshwara B.Sc. 1979*)

4. Write a short note on, "Discovery of neutron". (*Garhwal 1988; Kanpur 1982*)

5. How and when was the neutron discovered ? Mention its use. (*Madras 86*)

6. Describe Rutherford's model of the atom. How was it improved by Bohr ? (*Banaras 80; Poona 81*)

7. Describe Mosley's work on X-ray spectra of elements with respect to instrument and equation. Bring out its theoretical importance. (*Poona 86; Madras 85*)

8. Write a note on photoelectric effect. (*Nagpur 81; Baroda 82; Madras 85*)

9. Write a note on "Compton Effect". (*Nagpur 81; Baroda 82*)

10. Write a note on "Planck's quantum theory of radiation". (*Kanpur 1982, 83, 84*)

11. Give a brief account of Bohr's model of hydrogen atom covering the following points: (*a*) postulates (*b*) energy and radius (*c*) short-comings. (*Allahabad 89*)

12. Explain how Bohr's theory accounts for the line spectrum of H-atom. (*Calicut 80; Osmania 81; Himachal 82; Raj 82, 86; Banaras 80*)

13. How did Sommerfeld modify the Bohr's model of atom? (*Raj B.Sc. 1987*)

14. K.E. of an electron is 5.76×10^{-15} J. Calculate the wavelength associated wit the electron. (Mass of electron = 9.1×10^{-31} kg and *h* 6.626×10^{-34} Js) (*Delhi 99*)

15. Deduce the following relation from Bohr's theory.

$$E_n = -\frac{2\pi^2 Z^2 me^4}{n^2 h^2}$$

[*Gauhati* (*General*) *2000*]

16. (*a*) Discuss Bohr's model of the atom. How does it account for the hydrogen spectra? What are its limitations?

(*b*) Calculate the frequency of spectral line emitted in case of hydrogen atom when the electron jumps from level 2 to level 1. (*Lucknow 2001*)

17. Write the demerits of Bohr's atomic model. (*Nagpur 2002*)

CHAPTER

2 Structure of the Atom (ii)

(*Wave Mechanical Approach*)

Dual Nature of Matter or Electron : Wave-Particle Nature of Matter—de Broglie's Concept of Matter Wave.

According to *Max Planck's quantum theory of radiation* some phenomena like photoelectric effect and black body radiation can be explained, if light radiation is supposed to behave *as a material particle* which is localised at a particular point in space, while according to *Maxwell's wave theory of radiation* certain other phenomena like diffraction, interference and polarisation can be explained, if light radiation is supposed to behave as a wave which is spread out over space. Thus we find that *light shows dual character, i.e. it behaves both as a material particle and as a wave.*

In 1924 a French physicist, Lonius de Broglie extended the idea of dual character of light to matter particles and suggested that *all matter particles in motion* (*e.g.* electrons, protons, neutrons, atoms or molecules etc.) *have a dual character, i.e. all matter particles possess characteristics of both a material particle and a wave.* The wave associated with matter particles is called **de Broglie's matter wave.** The revolutionary idea proposed by de Broglie that a material object possesses both the particle and the wave properties is known as the **wave-particle duality in matter.** The idea of wave -particle duality in matter led de Broglie to a very important conclusion that since an electron is also kind of matter particle, *it also, like matter particle, has a dual character i.e. an electron, like all other matter particles, behaves both as a material particle and as a wave.*

Difference Between de Broglie's Matter Waves and Electromagnetic Waves

Main differences are given below :

(*i*) de Broglie's matter wave are not radiated into empty space or emitted by oscillating particles, *i.e.*, they are never dissociated from the particle but remain attached to the particle. On the other hand the electromagnetic waves (*i.e.*, radiations) can be absorbed or emitted.

(*ii*) de Broglie's matter waves travel with speeds which depend on the nature of the matter particles while electromagnetic waves proceed with the speed of light.

(*iii*) de Broglie's matter waves are not electromagnetic in nature.

de Broglie's Equation

With the help of *Planck's equation,* $E = h\upsilon$ and *Einstein's mass-energy relationship,* $E = mc^2$ (E = energy of photon, h = Planck's constant, υ = frequency of radiation (photon), m = mass of photon and c = velocity of light), de Broglie derived a mathematical equation which gives the magnitude of wavelength, λ of a moving electron in terms of its mass, m and velocity, v. This equation is :

$$\boxed{\lambda = \frac{h}{mv}}$$

This equation is called **de Broglie's equation** and the wavelength, λ is called **de Broglie's wavelength.**

Since the product of m and v (*i.e.*, mv) is called the momentum of the moving electron, the above equation can also be written as :

$$\boxed{\textbf{de Broglie's wavelength } (\lambda) = \frac{\textbf{Planck's constant (h)}}{\textbf{Momentum of the electron (mv)}}}$$

The equation shows that, since h is constant of proportionality, *the momentum of a moving electrons* (= mv) *is inversely proportional to its wavelength* (= λ).

Derivation of de Broglie's Equation

We have seen that energy of a photon, E is related to its frequency, υ by *Planck's equation* as :

$$E = h\upsilon \text{ (Planck's equation)}$$

or $$E = h.\frac{c}{\lambda} \qquad \left(\because \upsilon = \frac{c}{\lambda}\right) \qquad ...(i)$$

Here λ is the wavelength of radiation (photon).

Again the energy of a photon, E with mass, m and velocity, c is also given by *Einstein's mass-energy relationship* as :

$$E = mc^2 \text{ (Einstein's mass-energy relationship)} \qquad ...(ii)$$

Equating equations (*i*) and (*ii*), we get

$$h.\frac{c}{\lambda} = mc^2$$

or $$\frac{h}{\lambda} = mc$$

or $$\lambda = \frac{h}{mc}$$

If the above equation is applied to a moving electron of mass m and moving with a velocity v, then the above equation reduces to :

$$\lambda = \frac{h}{mv} \quad \text{(de Broglie ? s equation)}$$

This is de Broglie's equation which gives the value of wavelength (λ) of a moving electron in terms of its mass (m) and velocity (v).

De Broglie's Equation as Applied to Large and Small Particles in Motion—Limitations of de Broglie's Equation.

Whether de Broglie's equation is applicable to large particles like a stone or small particles such as electrons, neutrons etc. can be known by calculating the magnitude of de Broglie's wavelengths for these particles and then comparing them.

(*a*) **Wavelength of large particles.** Let us consider a large particle like a stone in motion of mass of 1 g (= 1 × 10^{-3} kg) moving with a velocity equal to 1 m/sec. Then :

$$\lambda = \frac{h}{mv} = \frac{6.624 \times 10^{-34}}{(1 \times 10^{-3}) \times (1)} \text{ m}$$
$$= 6.624 \times 10^{-31} \text{ m.}$$

This value of λ is shorter than the wavelength of any electron-magnetic radiation (except cosmic rays) and hence cannot be measured by any conceivable method. Thus de Broglie's equation has no significance for large particles.

(*b*) **Wavelength of small particles.** Let us consider an electron in a hydrogen atom with mass = 9.108 × 10^{-31} kg and moving with a velocity = 2.188 × 10^{6} m/s. Then :

$$\lambda = \frac{h}{mv} = \frac{6.624 \times 10^{-34}}{(9.108 \times 10^{-31}) \times (2.188 \times 10^{6})} \text{ m}$$
$$= 0.3323 \times 10^{-9} \text{ m}$$

This value of λ is quite large and comparable to the wavelength of X-rays and hence can be measured.

As may be seen from the above calculations, de Broglie's wavelength of large particles in motion is very small and hence cannot be measured while the same for small particles is large enough and hence can be measured. Thus we find *that de Broglie's equation is applicable only for small particles* (*microscopic particles*) *like electrons, neutrons etc. and has no significant for large particles* (*macroscopic particles*).

Numerical Probelms

Problem 1. (In S.I. units). *Calculate the wavelength of a body of mass 1 mg moving with a velocity of 3 ms*$^{-1}$. *(Delhi B.Sc. Hons. 1975)*

Solution. We know that according to de Broglie's equation :

$$\lambda = \frac{h}{mv}$$

Substituting h = 6.624 × 10^{-34} Js, m = 1 mg = 1 × 10^{-6} kg, and v = 3 ms^{-1} in the above equation, we get

$$\lambda = \frac{6.624 \times 10^{-34} \text{ Js}}{(1 \times 10^{-6} \text{ kg}) \times (3 \text{ ms}^{-1})}$$
$$= \frac{6.624 \times 10^{-34} \text{ Js}}{3 \times 10^{-6} \text{ kg . ms}^{-1}}$$
$$= \frac{2.20 \times 10^{-28} \text{ Js}}{\text{kg.ms}^{-1}} = 2.20 \times 10^{-28}$$

$$= \frac{2.20 \times 10^{-28} \text{ kg.m}^2.\text{s}^{-2}.\text{s}}{\text{kg.m.s}^{-1}} \qquad (\because \text{J} = \text{kg.m}^2.\text{s}^{-2})$$

$$= 2.20 \times 10^{-28} \text{ m}$$

Problem 2. (In S.I. units). *Applying de Broglie's equation, calculate the wavelength associated with the motion of the earth, a stone and an electron. The masses and velocities of these objects are given below :*

	Mass	*Velocity*
Earth	*= 6 × 10^{24} kg*	*3 × 10^4 ms^{-1}*
Stone	*= 0.1 kg*	*1.0 ms^{-1}*
Electron	*= 10^{-30} kg*	*6 × 10^5 ms^{-1}*

Planck's constant = 6.6 × 10^{-34} Js. Which of these objects will have the measureable wavelength? (Delhi B.Sc. 1982)

Solution. (*a*) *For earth :*

$$\lambda = \frac{h}{mv} = \frac{6.6 \times 10^{-34}}{(6 \times 10^{24}) \times (3 \times 10^4)} = 3.68 \times 10^{-63} \text{ m}$$

(*b*) *For stone*

$$\lambda = \frac{h}{mv} = \frac{6.6 \times 10^{-34}}{(0.1) \times (1.0)} = 66 \times 10^{-34} \text{ m}$$

(c) *For electron :*

$$\lambda = \frac{h}{mv} = \frac{6.6 \times 10^{-34}}{(10^{-30}) \times (6 \times 10^5)} = 1.1 \times 10^{-9} \text{ m}$$

Quite obviously wavelength of electron is maximum and hence is a measurable quantity.

Problem 3. (In S.I. units). *Calculate the wavelength of an electron moving with a velocity of 2.5 × 10^7 ms^{-1} (h = 6.63 × 10^{-34} Js, mass of an electron = 9.11 × 10^{-31} kg).* (Bangalore B.Sc. 1981)

Solution. Applying de Broglie's equation

$$\lambda = \frac{h}{mv} = \frac{6.63 \times 10^{-34}}{(9.11 \times 10^{-31}) \times 2.5 \times 10^7} \text{ m}$$

$$= \frac{6.63 \times 10^{-10}}{9.11 \times 2.5} \text{ m}$$

$$= 0.2911 \times 10^{-10} \text{ m.}$$

Problem 4. (in C.G.S. units). *Calculate the wavelength of matter-wave associated with an electron moving with a velocity of 1.20 × 10^7 $cm.s^{-1}$. (Mass of electron = 9.1 × 10^{-28} g; Planck's constant h = 6.626 × 10^{-27} erg.s).*

Solution. Here v = 1.20 × 10^7 $cm.s^{-1}$, m = 9.1 × 10^{-28} g and

h = 6 626 × 10^{-27} erg.s.

$$\therefore \quad \lambda = \frac{h}{mv} = \frac{6.626 \times 10^{-27} \text{ erg.s}}{(9.1 \times 10^{-28} \text{ g}) \times (1.20 \times 10^7 \text{ cm.s}^{-1})}$$

$$= \frac{6.626 \times 10^{-6} \text{ g.cm}^2.\text{s}^{-2}.\text{s}}{9.1 \times 1.20 \text{ g.cm.s}^{-1}} = 0.6057 \times 10^{-6} \text{ cm}$$

Davisson and Germer's Experiment to Verify the Wave Nature of Moving Electrons

The first experimental proof of the fact that moving electrons have waves of definite wavelength associated with them was provided in 1927 by two American physicists, Davission and Germer. In their experiment they studied the scattering of slow moving electrons by reflecting them from the surface of nickel crystal.

Experimental procedure. In their experiment electrons were produced from a heated tungsten filament and were accelerated by passing them through charged plates kept at a potential difference of V volts. The accelerated electrons were passed through slits which collimate the accelerated electrons into a fine parallel pencil.

Due to the electrical force equal to Ve (where e = charge on the electron) acting on the electron, the electrons emerge out of the charged plates with a uniform velocity equal to v units. Thus in this case the kinetic energy acquired by an electron due to the electric field would be equal to $\frac{1}{2}mv^2$. This kinetic energy will be equal to the electrical Ve, *i.e.*

$$\frac{1}{2}mv^2 = eV$$

$$\because \quad v = \sqrt{\frac{2Ve}{m}}$$

Substituting this value of v in de Broglie's equation, we have :

$$\lambda = \frac{h}{mv} = \frac{h}{m} \times \sqrt{\frac{m}{2Ve}}$$

or $$\lambda = \frac{h}{\sqrt{2Vem}}$$

On putting h = 6.624 × 10^{-34} Js; e = 1.602 × 10^{-19} C; m = 9.108 × 10^{-31} kg; the above equation becomes :

$$\lambda = \frac{6.624 \times 10^{-34}}{\sqrt{2 \times 1.602 \times 10^{-19} \times 9.108 \times 10^{-31} \times \text{V volts}}}$$

$$= \frac{12.26 \times 10^{-10}}{\sqrt{\text{V volts}}} \text{ m}$$

or $$\lambda = \frac{12.26}{\sqrt{\text{V volts}}} \text{ Å} \qquad \textit{(de Broglie's equation)} \qquad ...(i)$$

If a potential difference of 10 volts is applied, the wavelength, λ of the electron-wave emerging out would be equal to 3.877 Å. Similarly, if the potential difference is varied between 10 and 10,000 volts, λ should vary between 3.877 Å. and 0.1226 Å. It is well known that X-rays have the wavelengths in this range. It is also known that the X-rays having the wavelength in the above range can undergo diffraction when they fall on the crystalline solids like nickel crystal, known as *target*.

When the accelerated electrons fall on the nickel crystal as shown in Fig. 2.1, the electrons of a definite wavelength are deflected from the crystal in different directions. The angle of diffraction (θ) can be measured with the help of an electron detector (called *collector*) which is connected with a sensitive galvanometer and can be moved along graduated circular scale. The electron detector has two walls C and D which are insulated from each other. The nickel crystal which is known to be of the face-centered cubic type is cut so as to have a smooth reflecting surface parallel to the lattice plane (1, 1, 1).

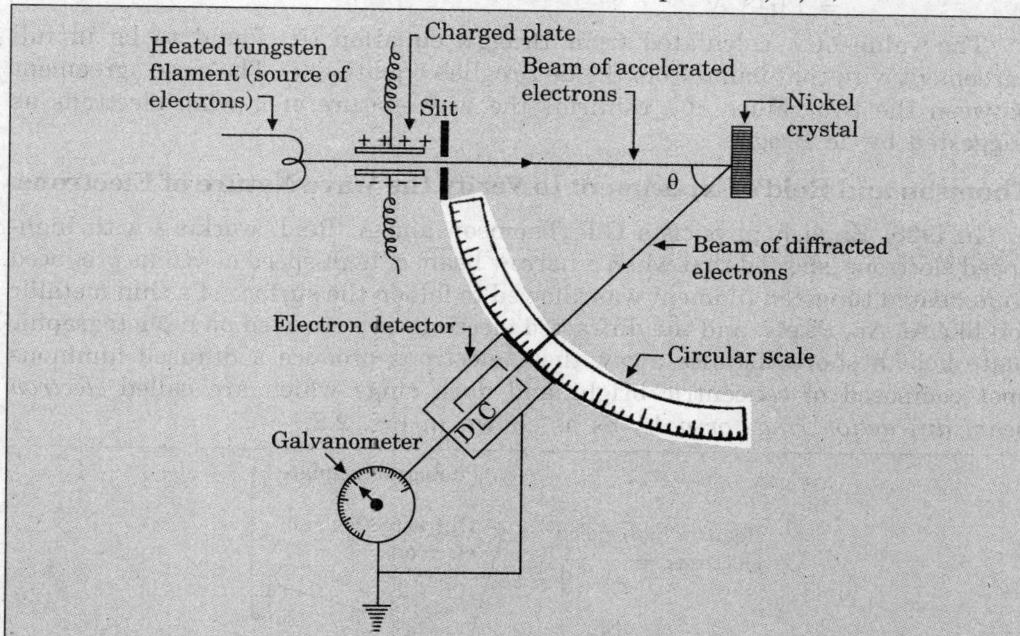

Fig. 2.1. Davisson and Germer's experiment to verify the wave-nature of moving electrons.

The intensities of electron waves scattered by the Ni crystal at different angles were measured and it was found that the most intense reflection took place when the angle of refraction was 50° and the electrons were accelerated through 54 volts. Substituting V = 54 volts in de Broglie's equation (*i*), we get the value of wavelength, λ of the electron-wave equal to 1.668 Å as shown below.

$$\lambda = \frac{12.26}{\sqrt{54 \text{ volts}}} \text{ Å} = \frac{12.26}{7.3484} \text{ Å}$$

or $\lambda = 1.688$ Å

This wavelength also lies in the X-rays range. Thus there is similarity between electron waves and X-rays. Since λ = 1.668 Å lies in the X-ray range, Bragg's equation viz.

$$n\lambda = 2d \sin \theta \text{ (Bragg's equation)} \qquad ...(ii)$$

which has been derived for X-ray diffraction, can be applied to electron diffraction as well and can be used to calculate the value of λ. In Bragg's equation n is a whole number (1, 2, 3 etc) of wavelength λ for a reflection of maximum intensity, d is the vertical distance (called *inter planar distance*) between two adjacent lattice planes of the crystal (nickel crystal in the present case) and θ is the angle of diffraction of the waves. By substituting $n = 1$ (first order Bragg's reflection occurs), $a = 50°$ and $d = 1.075$ Å, the value of λ is obtained as follows :

$$1 \times \lambda = 2 \times 1.075 \times \sin 50$$
$$= 2 \times 1.075 \times 0.7660$$
$$= 1.648 \text{ Å}$$

The value of λ calculated from Bragg's equation (*ii*) found to be in full agreement with that found from the de Broglie's equation (*i*). The close agreement between the two values of λ confirms the wave-nature of moving electrons as suggested by de Broglie.

Thomson and Reid's Experiment to Verify the Wave Nature of Electrons.

In 1928, English physicists G.P. Thomson and A. Reid, working with high-speed electrons, showed that when a narrow beam of high-speed electrons produced from heated tungsten filament was allowed to fall on the surface of a thin metallic foil like Al, Au, Pt etc. and the diffracted electrons are received on a photographic plate kept a short distance away, these electrons produce a diffused luminous spot composed of concentric bright and dark rings which are called *electron beam diffraction rings or patterns* as shown in Fig. 2.2.

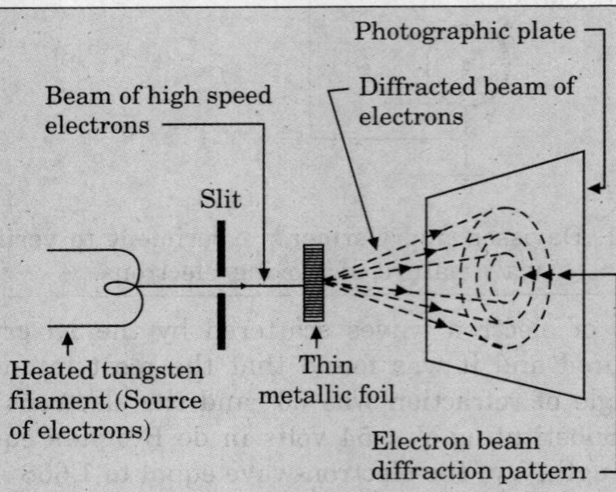

Fig. 2.2. Thomson and Reid's experiment to verify the wave-nature of electrons.

These diffraction patterns are obtained due to the fact that metals consist of many microscopic crystals arranged in random fashion so that some are always at the proper angles to give reflection in accordance with Bragg's equation. These diffraction rings given by electron beam are found to be similar to those given by X-rays (which are waves) when they are diffracted by the same metal foil. (See Fig 2.3). This similarity between the electron beam diffraction rings and those of X-rays rings confirms that electrons are also waves.

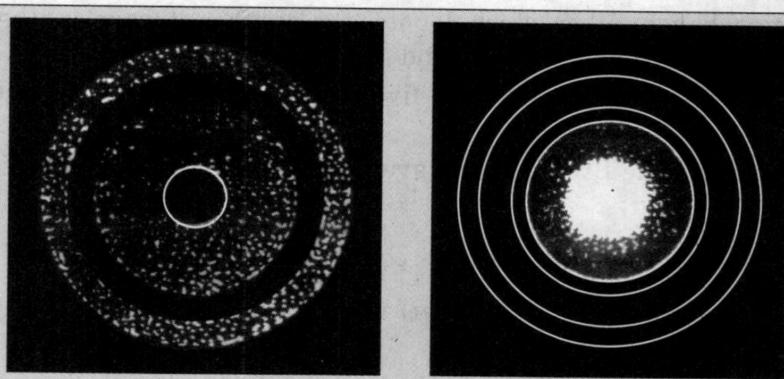

Fig. 2.3. (*a*) Electron beam diffraction rings (*b*) X-rays beam diffraction rings

Applications of Wave Nature of Electrons

(*i*) The wave nature of electrons has been utilised in the construction of electron microscope of high resolving power. This microscope is an important tool in research.

(*ii*) Wave nature of electrons has been used in the study of thin surface films.

Bohr's Theory vs de Broglie's Concept

Bohr's principle of quantisation of angular momentum of a moving electron can be derived from de Broglie's concept of wave nature of electron. This can be done as follows :

According to Bohr's principle of quantisation of angular momentum of a moving electron (Bohr had assumed an electron as a material particle), an electron can move only in that orbit in which the angular momentum (= mvr) is an integral whole number multiple of $h/2\pi$, *i.e.*, the principle can be represented as :

$$mvr = n.\frac{h}{2\pi} \text{ (Bohr's equation)},$$

where m = mass of the electron, v = velocity of the electron, r = radius of the orbit in which the electron is revolving, n = an integral whole number which can have values 1, 2 etc. and h = Planck's constant.

Now according to de Broglie an electron behaves as a standing or stationary wave which extends round the nucleus in a circular orbit (Fig. 2.4). If the two ends of the electron-wave meet to give a regular series of crests and troughs, the electron-wave is said to be in *phase*, *i.e.*, there is constructive interference of electron waves and the electron motion has a character of a standing wave or *non-energy radiating motion*. [Fig. 2.4 (*a*)]. Whatever be the path of the electron-wave round the nucleus, it is a necessary condition to get an electron-wave in phase that the circumference of the Bohr's orbit (= $2\pi r$) is equal to the whole number multiple of the wavelength (λ) of the electron wave, *i.e.*,

$$2\pi r = n.\lambda$$

or $$\lambda = \frac{2\pi r}{n} \quad ...(i)$$

Here n is a whole number which denotes the number of wavelengths associated with an electron-wave extending round the nucleus.

An electron-wave associated with five wavelengths (*i.e.*, $n = 5$) is shown in Fig. 2.4 (*a*).

Now according to de Broglie, the wave length (λ) of an electron which behaves as a wave is given by :

$$\lambda = \frac{h}{mv} \text{ (de Brogli's equation)} \qquad ...(ii)$$

On comparing equations (*i*) and (*ii*) we get

$$\frac{2\pi r}{n} = \frac{h}{mv}$$

or

$$mvr = n.\frac{h}{2\pi} \text{ (Bohr's equation)}$$

This equation which is Bohr's equation clearly shows that the angular momentum (= mvr) of an electron moving round the nucleus is a whole number multiple of $h/2\pi$. Thus we see that *Bohr's assumption namely the angular momentum (mvr) of an electron is a whole number multiple of h/2π is also confirmed if an electron is also supposed to behave as a wave as suggested by de Broglie*. In other words de Broglie's and Bohr's concepts are in agreement with each other.

If mvr is not a whole number multiple of $h/2\pi$ or if the circumference of the Bohr's orbit (= $2\pi r$) is bigger or smaller than $n\lambda$, the electron-wave is said to be *out of phase* [Fig. 2.4 (*b*)] and hence the electron-wave will not extend round the nucleus in circular orbit and ultimately will come out of the Bohr's orbit, *i.e.*, the destructive interference of waves will cause radiation of energy.

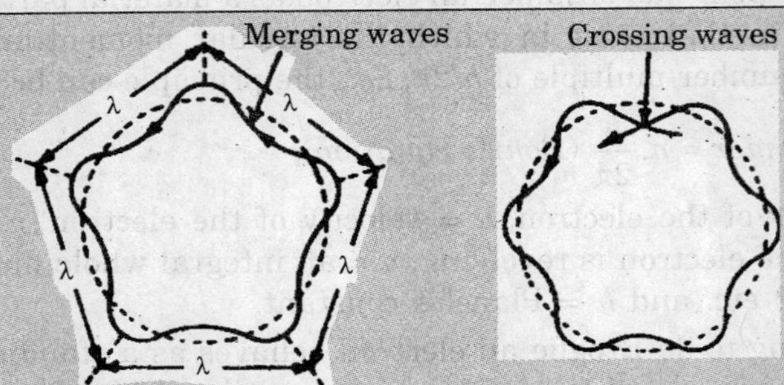

Fig. 2.4. de Broglie's electron wave extending round the nucleus and accomodated in Bohr's circular orbit (*a*) Electron wave in phase having $n = 5$ (n = No. of wavelengths)—Constructive-interference of electron waves (*b*) Electron wave out of phase—Destructive-interference of electron waves.

Heisenberg's Uncertainty Principle

We have seen that Bohr had assumed an electron to be a material particle of small mass revolving round the nucleus in circular orbits situated at definite (*i.e.*

fixed) distance from the nucleus and with a definite velocity (or momentum). Thus according to Bohr's theory which considers an electron as a material particle, the position and momentum of an electron can be determined with great accuracy. But, when an electron is considered as a wave as suggested by de Broglie, it is not possible for us to measure simultaneously its exact position in space at some given instant and its velocity, since the wave is extending throughout a region of space. This means that if an attempt is made to make the exact measurement of the position of an electron, the exact measurement of its velocity becomes uncertain and *vice-versa*. That the measurement of the position and momentum of an electron cannot be made simultaneously was suggested by Werner Heisenberg of Germany in 1927 in his principle which is known after his name as **Heisenberg's uncertainty principle.** This principle states as :

It is impossible to measure simultaneously the exact position and exact velocity (or momentum) of a sub-atomic particle like electron and neutron.

Illustration of the principle. The uncertainty principle can be better understood by considering the relations existing between a hostel suprintendent and a student living in the hostel. Suppose a student is very irresponsible and hence, for most of the time he prefers to keep away from his room. Now if, the superintendent becomes strict, the student stays in his room and hence the relations between the student and the superintendent become uncordial (*i.e.* uncertain). On the other hand if the superintendent becomes linient, the student does not stay in his room and hence the relations between them become cordial (*i.e.* certain). Thus, if the presence of the student in his room becomes certain; the relations between the two become uncertain.

Mathematical expression for the principle. If Δx represents the error (or uncertainty) in the measurement of the position and Δp represents the uncertainty in the measurement of the momentum of a sub-atomic particle like electron and neutron, then according to this principle, these two quantities are related as :

$$\boldsymbol{\Delta x . \Delta p \geq \frac{h}{2\pi}} \quad \textbf{(Uncertainty relation)}$$

where h is the Plank's constant and the sign $\geq$ indicates equal to or greater than.

The relation given above is called **uncertainty relation**. From this relation it is evident that, if Δx is small (*i.e.* if the position of a particle is measured accurately or with accuracy), Δp will be large (*i.e.* the momentum will be measured less accurately or with less accuracy) and *vice versa*. Thus, if one quantity is measured accurately, the other quantity is measured less accurately: This in other words means that certainty of the determination of one quantity introduces uncertainty for the determination of the other quantity. This fact can also be shown as follows :

If the position of a particle is measured accurately, Δx should be equal to zero and hence

$$\Delta p = \frac{h}{2\pi \times \Delta\pi} = \frac{h}{2\pi \times 0} = \infty$$

$\Delta p = \infty$ means that error in the determination of the momentum is infinite, *i.e.* the measurement of the momentum of the particle *cannot* be made. In other words the simultaneous measurement of position and momentum (or velocity) with absolute accuracy is impossible.

If m is the rest mass of the particle and is the uncertainty in the measurement of its velocity, then

$$\Delta p = \Delta v.m$$

On putting $\Delta p = \Delta v.m$ in the uncertainty relation, we get another form of the uncertainty relation *viz.*

$$\Delta x.\Delta v.m \geq \frac{h}{2\pi}$$

$$\text{or} \quad \Delta \mathbf{x}.\Delta \boldsymbol{v} \geq \frac{\boldsymbol{h}}{2\pi.\mathbf{m}}$$

Heisenberg's Uncertainty Principle as Applied to Large and Small Particles—Limitations of Heisenberg's Uncertainty Principle.

Whether Heisenberg's uncertainty principle is applicable to large particles like a ball or small particles such as electrons, neutrons etc. can be known be calculating the value of Δx or Δv of these particles from uncertainty relation, $\Delta x.\Delta v \geq \frac{h}{2\pi}$ and then comparing these values with the velocity or size (diameter) of the particle respectively. For example, if the calculated value of Δv for a given particle is very small compared to the velocity of that particle, uncertainty principle is meaningful for that particle. On the other hand if the value of Δv for a given particle is quite large compared to the velocity of that particle, this principle is meaningless for that particle. Similarly the value of Δx can be compared to the size (diameter) of the particle and the applicability of the principle to the particle can be predicted. Let us consider the following examples.

(a) Δv of large particles. Let us consider a large particle like a ball of mass equal to 1g (1×10^{-3} kg) whose uncertainty in position (Δx) is equal to 10^{-10} m. Then :

$$\Delta v = \frac{h}{2\pi.m.\Delta x} = \frac{6.624\times10^{-34}\ \text{J.s.}}{2\times3.14\times(10^{-10}\,\text{m})\times(10^{-3}\ \text{kg})}$$
$$= 1.0547\times10^{-21}\ \text{m.s.}^{-1}$$

This value of uncertainty in velocity (Δv) is very small and hence is negligible to the unusual velocity of the ball. Thus to talk about the trajectory of a large particle like a ball having such a small value of Δv is meaningful. In a similar way, if the value of uncertainty in position (Δx) is calculated for large particles, it is found to be very small and hence is negligible compared. to the size (diameter) of the particle. Consequently in case of large particles we can talk about the position and velocity simultaneously.

(b) Δv of small particles. Let us consider an electron in H-atom with mass $= 9.108 \times 10^{-31}$ kg whose uncertainty in position (Δx) is equal to 10^{-10} m. Then :

$$\Delta v = \frac{h}{2\pi.\Delta x.m} = \frac{6.614\times10^{-34}\ \text{J.s}}{2\times3.14\times(10^{-10}\text{m})\times(9.108\times10^{-31}\ \text{kg})}$$

$$= \frac{6.624\times10^{7}}{6.28\times9.108}\ \text{m.s}^{-1}$$

$$= 0.11580\times10^{7}\ \text{m.s}^{-1}$$

$$= 11.580\times10^{5}\ \text{m.s}^{-1}.$$

This value of uncertainty in velocity (Δv) is quite large compared to the velocity of the electron (velocity of an electron = 1.2×10^5 ms^{-1}). Thus to talk about the tragectory of a small particle like an electron having such a large value of Δv is meaningful. In a similar manner if the value of Δx is determined for small particles, it is found to be very large compared to the size (diameter) of the particle.

The above description clearly shows *that uncertainty principle is applicable only for small particles (microscopic particles) like electrons, neutrons etc and has no significance for large particles (macroscopic particles).*

Compton Effect and Uncertainty Principle

Compton effect provides a good illustration of the uncertainty principle. Let us attempt to determine the position and momentum of an electron moving in a Bohr's orbit. Suppose we wish to determine the position of the electron. We cannot obviously see it with visible light, since its wavelength is perhaps a million times as large as the diameter of the electron. So we try to use the radiation of shorter wavelength, say, X-rays of wavelength, λ. But in any case the uncertainty in the measurement of the position cannot be less than λ; at the best it can only be equal to λ. So the smaller the wave length used, the more accurate will be the measurement of the position.

We have seen in Compton effect that when a photon from X-rays strikes an electron at rest, the momentum of the electron changes from zero to mv where m = moving mass of the electron and v = velocity gained by the electron after an impact from the X-rays photon. Since this momentum change cannot be precisely measured, the measurement of the original momentum of the electron will be quite uncertain. Moreover, since the momentum of the X-rays photon, before impact, is equal to h/λ, the smaller the wavelength of X-rays, the greater is the photon momentum and, consequently, the greater the uncertainty in the measurement of the electron momentum. We thus see, the shorter wavelength we use in order to achieve greater accuracy in locating the position, the greater becomes the uncertainty in momentum and vice versa. It is impossible to determine both the original momentum as well as the position; and this uncertainty of either position or momentum is evidently not a matter of physical imperfections in the measuring apparatus, nor is it one that may be overcome by some improved technique. This uncertainty represents a fundamental limit of nature and sets a fundamental limit to our knowledge.

Bohr's Theory vs Heisenberg's Uncertainty Principle—Probability Concept of Electron

Heisenberg's uncertainty principle tells us that, due to the wave nature of electron, we cannot describe the exact path (*i.e.*, position) followed by an electron in an atom at any time. Thus Bohr's theory according to which the electrons are regarded as having fixed position (orbits) of definite energy is no longer correct. *Then how to describe the position of an electron in an atom*? In order to describe its position in an atom the maximum that we can do is that we can only predict the probability or relative chance of finding or locating an electron with a probable velocity in a particular region of space round the nucleus *i.e.*, we can only predict where an electron is most likely to be found. We have seen that Bohr's theory considers electrons as material particles revolving round the nucleus in circular orbits whereas the concept of probability visualises an electron in terms of probability of its being found in a particular region of space round the nucleus and thus does not define an exact path for the motion of the electron. Thus we see that the concept of probability rules out the possibility of strictly defined orbit for an electron as suggested by Bohr.

The probability of finding an electron at any position is given by the square of the wave function ψ *i.e.* the probability of finding an electron is given by ψ^2.

Numerical Problems

Problem 1 (In S.I. units). *Calculate the uncertainty in the position (Δx) of an electron, if Δv is 0.1 percent. Take velocity of electron = 2.2×10^6 ms^{-1}, mass of electron = 9.1×10^{-31} kg and h = 6.6×10^{-31} J-s. Comment on the result.*

Solution. Here Δv = 0.1% of the velocity of the electron.

$$= \frac{0.1 \times 2.2 \times 10^6}{100} \text{ ms}^{-1}$$

$$= 2.2 \times 10^3 \text{ m.s}^{-1}$$

Now $$\Delta x = \frac{h}{2\pi.m.\Delta v}$$

$$= \frac{6.6 \times 10^{-34} \text{ J.s}}{2 \times 3.14 \times (9.1 \times 10^{-31} \text{ kg}) \times (2.2 \times 10^3 \text{ m s}^{-1})}$$

$$= \frac{6.6 \times 10^{-6}}{2 \times 3.14 \times 9.1 \times 2.2} \text{ m}$$

$$= 0.0524953 \times 10^{-6} \text{ m}$$

$$= 524.953 \times 10^{-10} \text{ m}$$

Since the calculated value of Δx is much longer than the atomic diameter ($\approx 10^{-10}$ m), uncertainty principle is applicable to electron.

Problem 2. (In S.I. units). *Calculate the uncertainty in the position (Δx) of a dust particle, if 0.1% error in the measurement of velocity is allowed. Take velocity of the particle = 10^{-6} ms^{-1}, mass of the particle = 10^{-14} kg, h = 6.6×10^{-34} Js and diameter of the particle = 10^{-6} m. Comment on the result.*

Solution. Here $v = 10^{-6}$ m.s^{-1}, $m = 10^{-14}$ kg, $h = 6.6 \times 10^{-34}$, J-s, $\Delta v = 0.1\%$ of the velocity of the particle

$$= \frac{0.1 \times 10^{-6}}{100} \text{m.s}^{-1} = 10^{-9} \text{ m.s}^{-1}$$

$$\Delta x = \frac{h}{2\pi \times \Delta v \times m}$$

$$= \frac{6.6 \times 10^{-34} \text{ J.s}}{2 \times 3.14 \times (10^{-9} \text{ m.s}^{-1}) \times (10^{-14} \text{ kg})}$$

$$= \frac{6.6 \times 10^{-11}}{6.28} \text{m.}$$

$$= 1.05095 \times 10^{-11} \text{ m}$$

$$= \frac{1.05095 \times 10^{-6}}{10^5} \text{ m}$$

$$= 0.0000105 \times 10^{-6} \text{ m}$$

Since this value of Δx is much smaller as compared to the size (diameter) of the particle (= 10^{-6} m as given in the question), uncertainty principle is meaningful for this particle. Thus in case of this particle we can talk about the position and velocity simultaneously.

Problem 3. (In C.G.S. units). *An electron has a speed of 30000 cm/s accurate upto 0.001%. What is the uncertainty in locating its position? (Mass of an electron = 9.1×10^{-28} g, $h = 6.626 \times 10^{-27}$ erg-s).*

Solution. Here $h = 6.626 \times 10^{-27}$ erg-s, $m = 9.1 \times 10^{-28}$ g and velocity = 30000 cm/s.

$$\therefore \quad \Delta v = \frac{30000 \times 0.001}{100} \text{cm.s}^{-1} = 0.3 \text{ cm.s}^{-1}$$

Now according to Heisenberg's uncertainty principle :

$$\Delta x.m.\Delta v = \frac{h}{2\pi}$$

or

$$\Delta x = \frac{h}{2\pi \times m \times \Delta v}$$

$$= \frac{6.626 \times 10^{-27} \text{ erg.s}}{2 \times 3.14 \times (9.1 \times 10^{-28} \text{ g}) \times (0.3 \text{ cm.s}^{-1})}$$

$$= 386.48 \text{ cm.}$$

Problem 4. (In S.I. units). *A microscope using suitable photons is employed to locate an electron in an atom within a distance of 0.1 Å. What is the uncertainty involved in the measurement of its velocity. (Mass of electron = 9.1×10^{-31} kg, $h = 6.626 \times 10^{-34}$ J.s.)*

Solution. Here Δx = 0.1 Å = 0.1×10^{-10} m,
$m = 9.1 \times 10^{-31}$ kg and $h = 6.626 \times 10^{-34}$ J.s.

Now $\Delta x.m.\Delta v = \frac{h}{2\pi}$

$$\therefore \quad \Delta v = \frac{h}{2\pi \times \Delta x \times m}$$

$$= \frac{6.626 \times 10^{-34} \text{ J.s.}}{2 \times 3.14 \times (0.1 \times 10^{-10} \text{ m}) \times (9.1 \times 10^{-31} \text{ kg})}$$

$$= \frac{6.626 \times 10^{7}}{2 \times 3.14 \times 0.1 \times 9.1} \text{m.s}^{-1}$$

$$= 1.159 \times 10^{7} \text{ m.s}^{-1}.$$

Wave Mechanical Model of the Atom and Schrodinger's Wave Equation

Although Bohr's concept of strictly defined orbits round the nucleus was abondoned when de Broglie's concept of wave-nature of the electron and Heisenberg's uncertainty principle (idea of the uncertainty of the position of the electron in the atom) came in light, yet Bohr's concept of the existence of the stationary energy states (or energy levels) round the nucleus was retained in order to explain the atomic spectra. Now after the failure of Bohr's atomic model, the scientists were in search of a new model that could combine the following three concepts and explain the movement of an electron in an atom:

(i) Wave-nature of the electron (de Broglie's concept)

(ii) The idea of the uncertainty of the position of the electron in an atom (Heisenberg's uncertainty principle).

(iii) The idea of stationary energy levels (fixed energy levels) for the electrons round the nucleus (Bohr's postulate).

The atomic model that takes into account the wave and particle nature of the electron is refered to as **wave mechanical model of the atom.** In 1926 Erwin Schrodinger, an Austrian physicist developed the wave mechanical model of the atom. In his model he visualised the *atom as a positively charged nucleus surrounded by a standing or stationary electron wave which extends round the nucleus in Bohr's circular orbits.* Thus this model described the electron as a three dimensional wave in the electronic field of a positively charged nucleus.

Schrodinger proposed that if the electron has wave-like nature, it should obey the same equation of motion as all other known types of wave obey. On the basis of this simple idea, he derived an equation which describes the wave motion of an electron-wave along any of the three axes *viz*. x, y and z axes and is called Schrodinger's wave equation. This equation can be written in many forms one of which is given below :

$$\boldsymbol{\frac{d^2\psi}{dx^2} + \frac{d^2\psi}{dy^2} + \frac{d^2\psi}{dz^2} + \frac{8\pi^2 m}{h^2}(E - V)\,\psi = 0,}$$

where ψ [pronounced as sign (साइ) or shy (शाई)] is a mathematical function which is called **wave function.** ψ represents the amplitude of the electron-wave at various points surrounding the nucleus, Amplitude is expressed in terms of *x, y, z* space co-ordinates of the electron with respect to the nucleus (0, 0, 0); *m* is the mass of the electron; *h* is Planck's constant and E is the total energy of the electron, *i.e.* E is the sum of kinetic energy and potential energy (V) of the electron.

Schrodinger's wave equation is a differential equation of the second order. The first three terms on the left hand side of the above equation indicate the mathematical operations that we must do on ψ to solve the equation for E. These terms are represented by $\Delta^2\psi$ (pronounced as *del square sigh*). Δ^2 is known as **Laplacian operator.** Thus the above equation can also be written as :

$$\boldsymbol{\Delta^2\psi + \frac{8\pi^2 m}{h}(E - V)\,\psi = 0,}$$

where $$\Delta^2 = \frac{d^2}{dx^2} + \frac{d^2}{dy^2} + \frac{d^2}{dz^2}$$

Derivation of Schrodinger's Wave Equation

The wave motion of a wave of a vibrating string can be described by the equation :

$$\psi = A \sin \frac{2\pi x}{\lambda}, \qquad ...(1)$$

where ψ represents the amplitude of the wave and is called the *wave function*; A is a constant; *x* is the displacement of the wave in a given direction and λ is the wavelength.

On differentiating equation (1) with respect to *x*, we get :

$$\frac{d\psi}{dx} = \left(A \cos \frac{2\pi x}{\lambda}\right)\left(\frac{2\pi}{\lambda}\right)$$

or $$\frac{d\psi}{dx} = \frac{2\pi A}{\lambda} . \cos \frac{2\pi x}{\lambda} \qquad ...(2)$$

By differentiating again equation (2), we get :

$$\frac{d^2\psi}{dx^2} = \frac{2\pi A}{\lambda}\left(-\sin \frac{2\pi x}{\lambda}\right)\left(\frac{2\pi}{\lambda}\right)$$

or $$= -\frac{4\pi^2 A}{\lambda^2} . \sin \frac{2\pi x}{\lambda}$$

$$= -\frac{4\pi^2}{\lambda^2}\left(A . \sin \frac{2\pi x}{\lambda}\right)$$

$$= -\frac{4\pi^2}{\lambda^2} . \psi \qquad \left[\because \psi = A \sin \frac{2\pi x}{\lambda}\right]$$

or $$\boldsymbol{\frac{d^2\psi}{dx^2} + \frac{4\pi^2\psi}{\lambda^2} = 0} \qquad ...(3)$$

This is the *classical wave equation* which describes the wave motion of any particle vibrating along x-axis.

With the confirmation of the idea that an electron in an atom is described as a standing wave round the nucleus, the classical wave equation (3) which describes the wave motion of any particle vibrating along x-axis should be applicable to the standing wave of an electron.

We have seen in Bohr's theory that the total energy (E) of an electron is the sum of its kinetic energy (K.E.) and potential energy (V), *i.e.*

$$E = K.E. + V$$

or $$E = \frac{1}{2}mv^2 + V \quad ...(4)$$

In equation (4) m is the mass of the electron and v is the velocity with which the electron is moving.

Now from de Broglie's equation, we have

$$\lambda = \frac{h}{mv}$$

or $$mv = \frac{h}{\lambda}$$

or $$m^2v^2 = h^2/\lambda^2$$

or $$\frac{1}{2}m^2v^2 = \frac{h^2}{2\lambda^2}$$

or $$\frac{1}{2}mv^2 = \frac{h^2}{2\lambda^2 m} \quad ...(5)$$

Putting the value of $\frac{1}{2}mv^2$ from equation (5) in equation (4), we get :

$$E = \frac{h^2}{2\lambda^2 m} + V$$

or $$\frac{1}{\lambda^2} = \frac{2m}{h^2}(E - V) \quad ...(6)$$

Putting the value of $1/\lambda^2$ from equation (6) in equation (3), we get :

$$\frac{d^2\psi}{dx^2} + \frac{4\pi^2\psi \times 2m}{h^2}(E - V) = 0$$

or $$\boxed{\frac{d^2\psi}{dx^2} + \frac{8\pi^2 m}{h^2}(E - V)\,\psi = 0} \quad ...(7)$$

Equation (7) is the Schroding's wave equation of a wave moving in one dimension x, *i.e.* the wave is moving in one direction x. If an electron wave is moving in any of the three axes x, y, z, its wave motion can be described by the wave equation :

$$\boxed{\frac{d^2\psi}{dx^2} + \frac{d^2\psi}{dy^2} + \frac{d^2\psi}{dz^2} + \frac{8\pi^2 m}{h^2}(E - V)\,\psi = 0} \quad ...(8)$$

Equation (8) is called Schrodinger's wave equation.

Eigen-values and Eigen-functions

Being a differential equation of second order, Schrodinger's equation has several solutions for ψ but many of these are imaginary and hence are not valid. Only those values of ψ are valid which satisfy the following conditions :

(*i*) The wave function must be finite and continuous.

(*ii*) The solution must be single valued *i.e.* at a given point there can never be more than one value for the amplitude, ψ.

(*iii*) $\frac{d\psi}{dx}$, $\frac{d\psi}{dy}$, and $\frac{d\psi}{dz}$ must be continuous functions of *x, y*, and *z* respectively.

(*iv*) The solutions must be normalised *i.e.* they must satisfy the relation

$$\int_0^{+\infty} \psi^2 dr = 0,$$

where *dr* is a small volume element.

The above conditions clearly show that the wave function, ψ is always *finite, single-valued* and *continuous*. It has zero value at infinite distance.

Values of E obtained by substituting the values of ψ in Schrodinger's wave equation [equation (8)] satisfying the above conditions are called **eigen-values** while the values of ψ obtained by substituting the proper values of E in equation (8) are called **eigen-functions.** Thus E values are called eigen-values while ψ values are called eigen-functions. The eigen-function for an electron is called an atomic orbital. As the eigen-values (*i.e.* E values) correspond very nearly to the energy levels associated with different Bohr orbits, the occurrence of definite energy levels in an atom follows directly from the wave mechanical concepts.

Physical Significance of ψ and ψ²

In Schrodinger's wave equation the wave function, ψ represents the amplitude of the wave. From this equation the value of ψ for *1s* electron situated at a distance of *r* from the proton in the ground state of hydrogen atom is given by :

$$\psi_{1s} = C_1 . e^{-C_2 r}, \qquad ...(1)$$

where C_1 and C_2 are constants.

The energy for the same electron is given by

$$E_{1s} = 2.18 \times 10^{-11} \text{ ergs/atom} \qquad ...(2)$$

The wave function, ψ by itself has no physical meaning and all that we can extract from the wave mechanical approach to the internal structure of the atom is through its square (*i.e.* ψ^2) which measures the probability of finding an electron of given energy in an extremely small volume round the nucleus. The probability or chance of finding an electron in the space round the nucleus is called **electron probability function,** D whose value for an extremely small spherical shell of radius *r* and thickness, *dr* round the nucleus is given by :

Electron probability function, D

$= $ *Volume of the spherical shell of radius r and thickness dr* $\times \psi^2$

or $= 4\pi r^2.dr.\psi^2$

The above equation clearly shows that the magnitude of electron probability function, D depends on the value of *r*. In case of 1*s* electron of H-atom the following points may be noted :

(*a*) When $r = 0$ or ∞, D becomes equal to zero, *i.e.* there is no probability of finding an 1*s* electron of H-atom at the nucleus ($r = 0$) or at an infinite distance ($r = \infty$).

(*b*) When $r = 0.529$Å which is the radius of Bohr's first orbit, the probability of finding the electron is maximum (≈ 95%), *i.e.* the radius of maximum probability of finding the 1*s* electron of H-atom is 0.529 Å.

The facts mentioned at (*a*) and (*b*) above will be made more clear when we study *Radial Probability Distribution Curves* discussed in this chapter.

Applications of Schrodinger's Wave Equation

Important applications of Schrodinger's wave equation are given below :

1. To Calculate Energy and Wave Function of a Particle in One-Dimensional Box.

Let us consider a particle (say, an electron) of mass *m* bounding back and forth from one side of the box to the other as shown in Fig. 2.5. The length of the box is *a* and height is infinite. Suppose that the particle moves only in one direction, say, along the *x*-axis, and is confined between $x = 0$ and $x = a$. Inside the box, the potential energy, V of the particle is constant and may be taken, for convenience, equal to zero (V = 0) so that there is no restriction on the movement of the particle within the box. Thus *inside* the box V = 0. Let the value of V be infinite when $x < 0$ (*i.e.* negative value of *x*) and $x > a$, (*i.e.*, out side the box) the

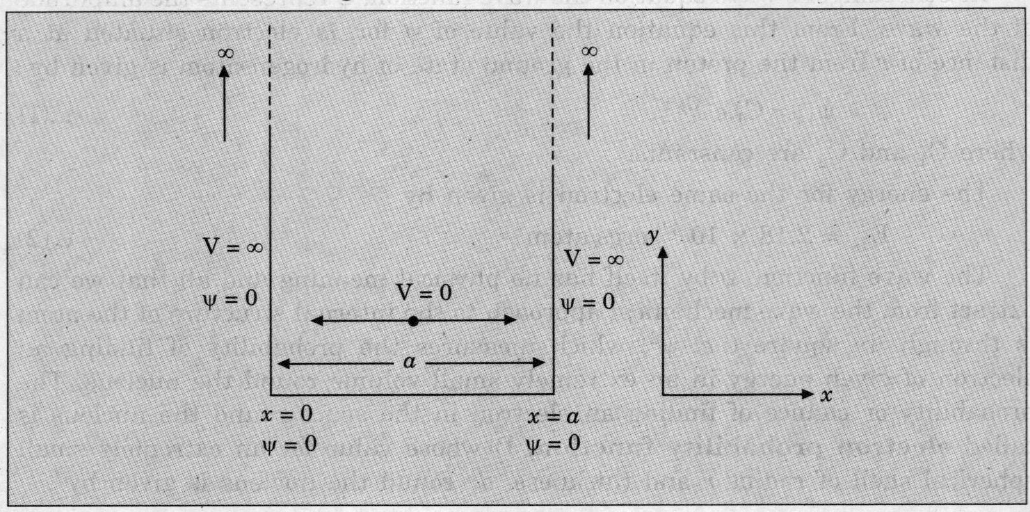

Fig. 2.5. Particle (shown by a dot) in one-dimensional box.

value of V is infinite. ($V = \infty$ at $x < 0$ and $x > a$). Obviously the wave function, ψ and hence the probability of finding the particle, ψ^2 must be zero when $x < 0$ (*i.e.* negative value of x), $x = 0$, $x = a$ and $x > a$. In other words this means that the particle is confined to the box and cannot escape the box, *i.e.* the particle does not exist outside the box.

Now our task is to find out the value of energy (E) and wave function (ψ) of the particle inside the box, *i.e.* between the region $x = 0$ and $x = a$.

We have seen that the equation given below is the Schrodinger's wave equation describing the motion of a wave in one dimension.

$$\frac{d^2\psi}{dx^2} + \frac{8\pi^2 m}{h^2}(E - V)\,\psi = 0 \quad ...(1)$$

Now since inside the box V = 0, the above equation becomes :

$$\frac{d^2\psi}{dx^2} + \frac{8\pi^2 m}{h^2}(E - 0)\,\psi = 0$$

or $$\boxed{\frac{d^2\psi}{dx^2} + \frac{8\pi^2 mE}{h^2}\psi = 0} \quad ...(2)$$

Equation (2) has the following solution for the wave function, ψ.

$$\psi = A.\sin bx + B.\cos bx \quad ...(3)$$

In equation (3) A and B are two parameters.

Differentiating equation (3) twice with respect to x, we get :

$$\frac{d^2\psi}{dx^2} = -\,b^2(A.\sin bx + B.\cos bx)$$

or $$\boxed{\frac{d^2\psi}{dx^2} = -\,b^2\psi} \quad \text{[From equation (3)]} \quad ...(4)$$

Comparing equations (2) and (4), we get

$$\boxed{b^2 = \frac{8\pi^2 mE}{h^2}} \quad ...(5)$$

Now applying the boundary condition that at $x = 0$, $\psi = 0$ to equation (3), we get :

$$0 = A.\sin b \times 0 + B.\cos b \times 0$$

$$= A.\sin 0 + B.\cos 0$$

$$= A \times 0 + B \times 1$$

or $$B = 0 \quad ...(6)$$

On putting B = 0 in equation (3) we get :

$$\psi = A.\sin bx + 0 \times \cos bx$$

or $$\boxed{\psi = A.\sin bx} \quad ...(7)$$

Applying the other boundary condition that at $x = a$, $\psi = 0$ to equation (3), we get :

$$0 = \text{A}.\sin ba + \text{B}.\cos ba$$

$$= \text{A}.\sin ba + 0 \times \cos ba \qquad (\because \text{B} = 0)$$

or $$0 = \text{A}.\sin ba \qquad ...(8)$$

Now A cannot be equal to zero ($\text{A} \neq 0$) because in that case the probability of finding the particle in the box will be zero which is not acceptable.

$\therefore$ $$\sin ba = 0$$

or $$\sin ba = \sin n\pi$$

or $$ba = n\pi$$

or $$\boxed{b = \frac{n\pi}{a}} \qquad ...(9)$$

or $$\boxed{b^2 = \frac{n^2\pi^2}{a^2}} \qquad ...(10)$$

Comparing equations (10) and (5), we get :

$$\frac{n^2\pi^2}{a^2} = \frac{8\pi^2 m\text{E}}{h^2},$$

which gives the value of the energy (E_n) of the particle satisfying equation (2) as follows :

$$\boxed{\mathbf{E}_n = \frac{\mathbf{n^2h^2}}{\mathbf{8ma^2}}} \qquad ...(11)$$

In equation (11) n is an integer which is called a quantum number and can have the values 1, 2, 3, ... This quantum number arises as a consequence of the boundary conditions imposed. n cannot be equal to zero, because then ψ and hence ψ^2 would be zero every where within the box. E_n values are usually called **eigen-values.** It is evident from equation (11) that the energy of the particle is quantised and so cannot vary continuously. Permissible values of E_n depend on the magnitude of n. E_n increases as n^2 increases and it cannot be equal to zero, since, as we have shown above, n cannot be equal to zero. E_n remains independent of x.

The values of energy that a particle in the box can possess are 1, 4, 9, 16...n^2 times the unit $h^2/8ma^2$. There are no energy values in between these values Thus an electron cannot have zero energy, since n is never zero. On the other hand, according to classical mechanics, all the energy values including E = 0 are permissible.

Substituting $b = n\pi/a$ from equation (9) in equation (7), we get the value of wave function, ψ_n given by :

$$\boxed{\psi_n = \mathbf{A}.\mathbf{sin}\left(\frac{\mathbf{n}\pi \boldsymbol{x}}{\mathbf{a}}\right)} \qquad ...(12)$$

ψ_n values are called **eigen-functions.**

Fig. 2.6 shows the plots of ψ and ψ^2 against x for the particle in the box. The following points may be noted from the figure :

(*i*) The probability of finding the particle (ψ^2) depends on x and energy, E

(*ii*) For the lowest or ground state one half-wave-length covers the box and hence there is no internal node, *i.e.* there is no point where the probability of finding the particle (ψ^2) is zero. In second energy level the wave function has one node and third has two nodes. Thus, in general, the number of nodes in the nth energy level is always equal to (n—1). Greater the number of nodes, more is the curvature in the wave function and greater is the total energy. This total energy is equal to kinetic energy, since in this case the potential energy has been assumed to be equal to zero inside the box.

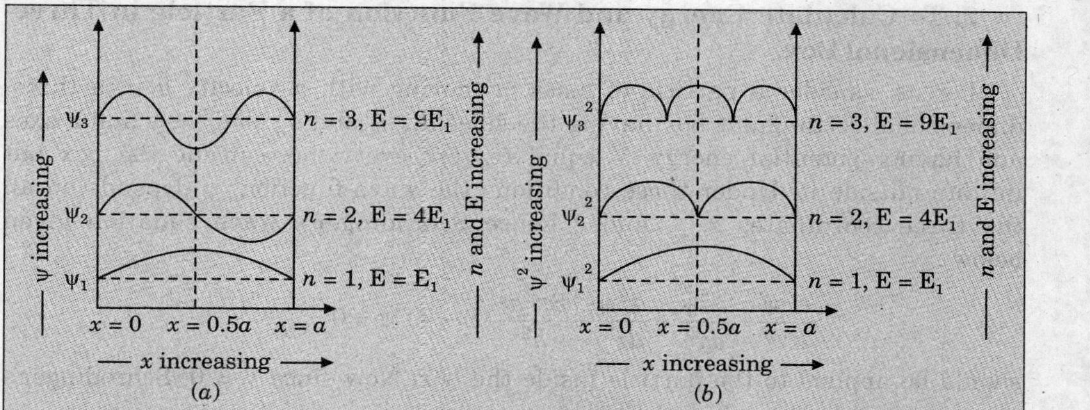

Fig. 2.6. Plot of wave function, ψ and its square, ψ^2 (called probability of finding the particle) of a particle in one-dimensional box against x (*a*) Plot of ψ against x (*b*) Plot of ψ^2 against x.

Determination of the value of A. The value of the constant A can be determined by *normalising* the wave function, ψ. Since the probability (expressed as ψ^2) of finding the particle within the box (*i.e.* between $x = 0$ and $x = a$) is unity, we have :

$$\int_0^a \psi^2 dx = 1 \qquad ...(13)$$

Substituting the value of ψ from equation (12) in this equation, we have :

$$\int_0^a A^2 \sin^2\left(\frac{n\pi x}{a}\right) dx = 1$$

or

$$\frac{1}{A^2} = \int_0^a \sin^2\left(\frac{n\pi x}{a}\right) dx$$

On solving the above equation we get :

$$A = \sqrt{\frac{2}{a}}$$

This mathematical operation is called **normalisation** and the constant value is called **normalisation factor.** Thus the normalised wave function (Ψ_n) is given by :

$$\boxed{\psi_n = \sqrt{\frac{2}{a}}.\sin\left(\frac{n\pi x}{a}\right)} \quad ...(14)$$

Equation (14) has been obtained by putting $A = \sqrt{\frac{2}{a}}$ in equation (12). In equation (14) the factor, $\sqrt{\frac{2}{a}}$ is called normalisation factor

2. To Calculate Energy and Wave Function of a Particle in Three-Dimensional Box.

Let us consider a particle of mass m moving with a velocity u in a three-dimensional rectangular box having the sides a_x, a_y and a_z along x, y and z axes and having potential energy, V equal to zero everywhere inside the box but infinite outside it. Under these conditions the wave function, ψ depends on all the three coordinates x, y, and z hence Schrodinger's wave equation given below :

$$\frac{d^2\psi}{dx^2} + \frac{d^2\psi}{dy^2} + \frac{d^2\psi}{dz^2} + \frac{8\pi^2 m}{h^2}(E - V)\psi = 0$$

should be applied to the particle inside the box. Now since V = 0, Schrodinger's wave equation given above becomes :

$$\frac{d^2\psi}{dx^2} + \frac{d^2\psi}{dy^2} + \frac{d^2\psi}{dz^2} + \frac{8\pi^2 m}{h^2} E\psi = 0 \quad ...(15)$$

Now, as already mentioned, since the wave function ψ depends on all the three coordinates *viz.* x, y and z, it should be equal to the product of three wave functions namely ψ_x, ψ_y, and ψ_z depending only on one axis *viz.* x, y or z respectively. Thus :

$$\psi = \psi_x.\psi_y.\psi_z \quad ...(16)$$

Now since ψ_x depends on x only,

$$\frac{d\psi}{dx} = \psi_y.\psi_z.\frac{d\psi_x}{dx}$$

or

$$\frac{d^2\psi}{dx^2} = \psi_y.\psi_z.\frac{d^2\psi_x}{dx^2} \quad ...(a)$$

Similarly the following relations can also be obtained :

$$\frac{d^2\psi}{dy^2} = \psi_x.\psi_z.\frac{d^2\psi_y}{dy^2} \quad ...(b)$$

$$\frac{d^2\psi}{dz^2} = \psi_x.\psi_y.\frac{d^2\psi_z}{dz^2} \quad ...(c)$$

Substituting the relations given at (16), (*a*), (*b*) and (*c*) in equation (15), we get :

$$\psi_y.\psi_z.\frac{d^2\psi_x}{dx^2} + \psi_x.\psi_z.\frac{d^2\psi_y}{dy^2} + \psi_x.\psi_y.\frac{d^2\psi_z}{dz^2} + \frac{8\pi^2 m}{h^2}.E\ \psi_x.\psi_y.\psi_z = 0$$

On dividing by $\psi_x\ \psi_y\ \psi_z$, we get :

$$\frac{1}{\psi_x}.\frac{d^2\psi_x}{dx^2} + \frac{1}{\psi_y}.\frac{d^2\psi_y}{dy^2} + \frac{1}{\psi_z}.\frac{d^2\psi_z}{dz^2} + \frac{8\pi^2 mE}{h^2} = 0$$

Thus we see that the above equation (Schrodinger's wave equation) can be broken into the following three differential equations :

$$\frac{1}{\psi_x}.\frac{d^2\psi_x}{dx^2} + \frac{8\pi^2 mE_x}{h^2} = 0 \qquad ...(d)$$

$$\frac{1}{\psi_y}.\frac{d^2\psi_y}{dy^2} + \frac{8\pi^2 mE_y}{h^2} = 0 \qquad ...(e)$$

$$\frac{1}{\psi_z}.\frac{d^2\psi_z}{dz^2} + \frac{8\pi^2 mE_z}{h^2} = 0 \qquad ...(f)$$

Here E_x, E_y and E_z are the three components of the total energy, E of the particle in the box along the three axes *x*, *y* and *z*.

The solutions of (*d*), (*e*) and (*f*) equations give the following expressions for E_x, E_y, and E_z

$$E_x = \frac{n_x^2 h^2}{8ma_x^2} = \frac{h^2}{8m}.\frac{n_x^2}{a_x^2}$$

$$E_y = \frac{n_y^2 h^2}{8ma_y^2} = \frac{h^2}{8m}.\frac{n_y^2}{a_y^2}$$

$$E_z = \frac{n_z^2 h^2}{8ma_z^2} = \frac{h^2}{8m}.\frac{n_z^2}{a_z^2}$$

In the above equations n_x, n_y and n_z are the quantum numbers in the direction of *x*, *y* and *z* axes respectively. Thus the total energy, E of the particle is given by :

$$E = E_x + E_y + E_z$$

or

$$\boxed{\mathbf{E = \frac{h^2}{8m}\left(\frac{n_x^2}{a_x^2} + \frac{n_y^2}{a_y^2} + \frac{n_z^2}{a_z^2}\right)}} \qquad ...(17)$$

It is evident from equation (17) that the energy of a particle in a three-dimensional rectangular box is *quantised*. Such a thing does not happen with the energy of the particle in one-dimensional box.

If the particle is considered to be present in a cubic box, then $a_x = a_y = a_z = a$ (say) and hence the energy, E of such a particle is given by :

$$\mathbf{E = \frac{h^2}{8ma^2}(n_x{}^2 + n_y{}^2 + n_z{}^2)} \quad ...(18)$$

From equation (18) it is evident that the energy of a particle in a cubic box of side 'a' depends on the *sum* of the squares of the three quantum numbers, n_x, n_y, and n_z. It is also evident that it is possible for an energy level to have independent states, each having specified quantum numbers. For example, if the energy of a level is equal to $6h^2/8ma^2$, then this energy level has three independent states having the following values of quantum numbers, n_x, n_y and n_z.

	Quantum numbers		
	n_x	n_y	n_z
Set No. 1	2	1	1
Set No. 2	1	2	1
Set No. 3	1	1	2

These three independent states have the same energy (= $6h^2/8ma^2$) and, since the number of states is three, these states are called *three-fold degenerate* or *triply degenerate states*. In other words the energy level having energy equal to $6h^2/8ma^2$ is triply degenerate. The degenerate states of an energy level having energy equal to $9h^2/8ma^2$, $11h^2/8ma^2$ and $14h^2/8ma^2$ are shown in Fig. 2.7. It may be noted from the figure that the states of energy levels with energy equal to $3h^2/8ma^2$ and $12h^2/8ma^2$ are *non-degenerate*, since there is only one set of quantum numbers for these states.

Since according to equation (14).

$$\psi_x = \sqrt{\frac{2}{a_x}}.\sin\left(\frac{n_x\pi x}{a_x}\right)$$

$$\psi_y = \sqrt{\frac{a}{a_y}}.\sin\left(\frac{n_y\pi y}{a_y}\right)$$

and $$\psi_z = \sqrt{\frac{2}{a_z}}.\sin\left(\frac{n_z\pi z}{a_z}\right),$$

$$\psi = \psi_x.\psi_y.\psi_z$$

$$= \left[\sqrt{\frac{2}{a_x}}.\sin\left(\frac{n_x\pi x}{a_x}\right)\right]\left[\sqrt{\frac{2}{a_y}}.\sin\left(\frac{n_y\pi y}{a_y}\right)\right]\left[\sqrt{\frac{2}{a_z}}.\sin\left(\frac{n_z\pi z}{a_z}\right)\right]$$

$$= \sqrt{\frac{8}{a_x.a_y.a_z}}.\sin\left(\frac{n_x\pi x}{a_x}\right).\sin\left(\frac{n_y\pi y}{a_y}\right).\sin\left(\frac{n_z\pi z}{a_z}\right)$$

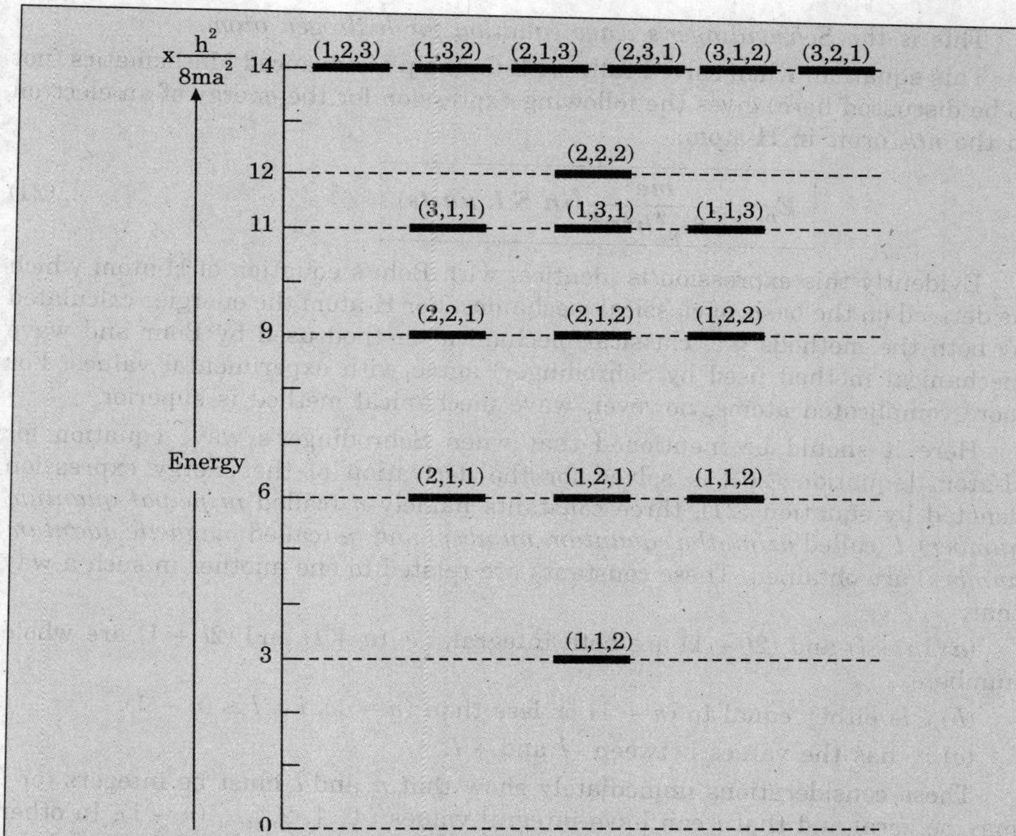

Fig. 2.7. Degenerate states of different energy levels having definite energy.

Now since the volume of the box, V = $a_x.a_y.a_z$, the above equation reduces to :

$$\psi = \sqrt{\frac{8}{V}}.\sin\left(\frac{n_x \pi x}{a_x}\right).\sin\left(\frac{n_y \pi y}{a_y}\right).\sin\left(\frac{n_z \pi z}{a_z}\right) \quad ...(19)$$

Equation (19) gives the value of normalised wave function, ψ. Here the factor, $\sqrt{8/V}$ is the normalisation factor.

3. To Derive an Expression for the Energy of an Electron in H-atom.

We know that H-atom has one proton in its nucleus and one electron revolving round the nucleus in an orbit of radius, say r. Let the charge on the proton be equal to $+e$. The charge on the electron would, therefore, be equal to $-e$. Consequently the potential energy, V of the electron would be equal to $(-e) \times (+e)/r = -e^2/r$. When Schrondinger's wave equation (equation given on page 10.4) is applied to H-atom, we get :

$$\frac{d^2\psi}{dx^2} + \frac{d^2\psi}{dy^2} + \frac{d^2\psi}{dz^2} + \frac{8\pi^2 m}{h^2}\left[E - \left(-\frac{e^2}{r}\right)\right] = 0$$

or

$$\frac{d^2\psi}{dx^2} + \frac{d^2\psi}{dy^2} + \frac{d^2\psi}{dz^2} + \frac{8\pi^2 m}{h^2}\left(E + \frac{e^2}{r}\right) = 0 \quad ...(20)$$

This is the *Schrondinger's wave equation for hydrogen atom.*

This equation, when solved fully with the help of advanced Mathematics (not to be discussed here) gives the following expression for the energy of an electron in the *nth* orbit in H-atom.

$$E_n = -\frac{me^4}{8\varepsilon_0{}^2 h^2 n^2} \text{(in S.I. units)} \quad ...(21)$$

Evidently this expression is identical with Bohr's equation of H-atom which he derived on the basis of classical mechanics. For H-atom the energies calculated by both the methods (*i.e.* classical mechanical method used by Bohr and wave mechanical method used by Schrodinger) agree with experimental values. For more complicated atoms, however, wave mechanical method is superior.

Here it should be mentioned that when Sohrodinger's wave equation for H-atom [equation (20)] is solved for the derivation of the energy expression denoted by equation (21), three constants namely n (called *principal quantum number*), l (called *azimuthal quantum number*) and m (called *magnetic quantum number*) are obtained. These constants are related to one another in such a way that;

(*a*) $(n + l)$ and $(2l + 1)$ are both integral, *i.e.* $(n + l)$ and $(2l + 1)$ are whole numbers.

(*b*) l is either equal to $(n - 1)$ or less than $(n - 1)$, *i.e.* $l \leq (n - 1)$.

(*c*) m has the values between $-l$ and $+ l$.

These considerations immediately show that n and l must be integers (or l may be zero) and that l can have integral values : 0, 1, 2, ... $(n - 1)$. In other words, the three constants n, l and m are the three quantum numbers and the relation between them arising from the solution of Schrodinger wave equation is also exactly the same which has been deduced by means of spectroscopic measurements.

Here it should be understood clearly that spin quantum number, s, does not appear in the solution of wave equation. It has, therefore, to be added.

4. Concept of Atomic Orbital.

If a line is drawn enclosing a region where an electron can be found, say 90 to 95% of time, such a region or space is called an **atomic orbital**. Thus :

The three-dimensional region or space (i.e., volume) round the nucleus of an atom where there is maximum probability of finding an electron having a certain energy is called atomic orbital.

OR

An atomic orbital represents the space round the nucleus in an atom where the electron spends most of its time while in constant motion.

OR

The thin, shell-like, region in space surrounding the nucleus for which electron probability has a high value is known as an atomic orbital.

The energy of the electron in an atomic orbital is always the same. The electron is free to be found anywhere in the atomic orbital at any given time. A spherical symmetrical orbital can be pictured as a tennis ball fluffed both inside and outside. For the **excited hydrogen** atom other orbitals of spherical as well as non- spherical shapes also exist and the probability of finding the electron may change not only with distance but also with direction from the nucleus.

Differences between Orbit and Orbital

The points of difference between an orbit and an orbital are given below in a tabular form.

Orbit	***Orbital***
(*i*) As postulated by Bohr, an orbit is a definite circular path at a definite distance from the nucleus in which the electron revolves round the nucleus. Orbits are numbered as 1, 2, 3, 4...... from the nucleus and are designated by the capital letters K, L, M, N......etc.	(*i*) As postulated by wave nature of an electron, an orbital is defined as a three-dimensional region or space around the nucleus within which the probability of finding an electron with a certain energy is maximum.
(*ii*) An orbit indicates an exact position or location of an electron in an atom. For example, according to Bohr's theory, an electron of hydrogen is found at a distance of 0.53Å from the nucleus.	(*ii*) An orbital does not specify the definite or exact position of an electron in an atom, since the electron, due to its wave-nature, can not be found at a fixed distance from the nucleus.
(*iii*) There is a certainty about the movement of an electron in an orbit.	(*iii*) Due to uncertainty principle, there is no certainty about the movement of an electron in an orbital.
(*iv*) It represents the planar motion of the electron.	(*iv*) It represents the three-dimensional motion of the electron round the nucleus.
(*v*) The maximum number of electrons in an orbit is equal to $2n^2$ where n is the number of the orbit.	(*v*) An orbital cannot accommodate more than two electrons. These two electrons must have opposite spins.
(*vi*) Orbits are circular in shape.	(*vi*) Orbitals have different shapes, *e.g.*, *s*-orbital is spherically symmetrical, *p* orbitals are dumb-bell shaped etc.

Quantum Numbers

The term quantum number is used to identify the various energy levels available with the atom in which the electron of the atom can reside. The word *quantum* is used to signify that all the energy levels which are available to an electron are governed by the laws of quantum mechanics.

Just as a college student is identified by (*i*) *his name* (*ii*) *the name of his college* (*iii*) *class in which he is studying and* (*iv*) *the locality in which his college is situated*, an electron in an atom is completely described by its four quantum numbers. Thus the quantum numbers serve as **identification numbers** or **labels** which completely define the electron in an atom, *i.e.* these numbers

(*i*) specify the position (or location) of an electron in an atom.

(*ii*) predict the direction of spin or self-rotation of the electron.

and, (*iii*) determine the energy and angular momentum of the electron.

Four quantum numbers which are required to completely specify the character of an electron are discussed below. Each quantum number refers to a particular character of an electron in an atom.

We have already seen that when Schrodinger's wave equation for H-atom is solved for the derivation of an expression giving the energy of an electron in H-atom, three constants called *principal quantum number* (n), *azimuthal quantum number* (l) and *magnetic quantum number* (m) are automatically obtained.

1. Principal Quantum Number (*n*).

We have seen that Bohr had assumed that electrons in an atom are revolving round the nucleus in circular orbits or shells which he had also called energy levels. Different energy levels in an atom have been represented by n which can have *non-zero positive integer* (whole number) *values upto infinity i.e.*

$$n = 1, 2, 3, ... \infty$$

Thus n can have infinite values and it has been called principal quantum number by Bohr.

Informations given by *n* values. (*i*) *The value of n represents the shells or energy levels* in which the electron revolves round the nucleus. These shells or energy levels are designated by capital letters K, L, M ... For example the shell for which $n = 1$ is called K shell (*1st* shell), that for which $n = 2$ is called L shell (*2nd* shell) and so on. Thus :

Value of n →	1	2	3	4	....
Designation of shells →	K (*1st shell*)	L (*2nd shell*)	M (*3rd shell*)	N (*4th shell*)	....

K, L, M etc shells are called $n = 1$ shell (*1st* shell), $n = 2$ shell (*2nd* shell) etc. Thus n values tell as to in which shell an electron is residing.

(*ii*) *n also determines the location of the electron in a shell, i.e. n* determines the distance of the electron from the nucleus. More is the value of n, more far is the electron from the nucleus, *e.g.*, the electron for which $n = 2$ is more far from the nucleus than that for which $n = 1$. Similarly the electron having $n = 3$ will be more far from the nucleus than an electron with $n = 2$. Thus :

	—*n values increasing*→
n-values →	$1 < 2 < 3 < 4 < \ldots\ldots$
r-values →	$r_1 < r_2 < r_3 < r_4 < \ldots\ldots$
	—*r values increasing* →

Thus we see that *n* represents the size of the electron orbits (*i.e.*, effective volume of the electron cloud). Higher is the value of *n*, higher is the size of the electron orbit or atom.

(*iii*) *n also determines the energy of the electron in a given orbit.* We know that the energy of an electron of hydrogen-like atoms or ions in an orbit with principal quantum number, *n* (*i.e.*, *nth* orbit) is given by :

$$E_n = -\frac{2\pi^2 me^4}{n^2h^2},$$

where E_n is the energy of the electron in *nth* orbit and π, *m, e* and *h* have their usual meanings. This equation shows that, due to the presence of negative sign, as the value of *n* increases, the value of energy becomes less negative *i.e.*, with the increase of the value of *n*, the magnitude of energy also increases. When *n* becomes equal to infinity, energy becomes zero. Thus the relation between *n* and E values of an electron in various orbits can be shown as :

	—*n* values increasing →
n-values →	$1 < 2 < 3 < 4 < \ldots\ldots$
E-values →	$E_1 < E_2 < E_3 < E_4 < \ldots\ldots$
	—E values increasing →

(*iv*) *n also gives the maximum number of electrons that a shell can hold.* A shell with principal quantum number, *n* can have $2n^2$ electrons. Thus :

Maximum number of electrons in K-shell (1st shell) with n = 1 } $= 2 \times 1^2 = 2$

Maximum number of) electrons in L-shell (2nd shell) with n = 2 } $= 2 \times 2^2 = 8$

Maximum number of) electrons in M-shell (3rd shell) with n = 3 } $= 2 \times 3^2 = 18$

Maximum number of) electrons in N-shell (4th shell) with n = 4 } $= 2 \times 4^2 = 32$

2. Azimuthal or Subsidiary Quantum Number (*l*)

We have seen that when hydrogen spectrum is observed by means of spectroscope of high resolving power, the individual spectral lines are found to consist of several very fine lines lying close together. For example H_α line in the

Balmer series of hydrogen spectrum is found to consist of five component lines. Such a spectrum which consists of closely-spaced spectral lines is called **fine structure**. The appearance of several lines in a single line has been explained by saying that each shell is divided into sub-shells which give closely-spaced lines in the spectrum. In order to represent these sub-shells only one quantum number viz., principal quantum number (n) is not sufficient. Rather, another quantum number which is called **azimuthal** or **subsidiary quantum number** (l) is used to represent the sub-shells of which the shells are composed.

The value of l depends on the value of n. For a given value of n, l can have the values from 0 to $(n - 1)$, *i.e.*,

$$l = 0, 1, 2, 3, \ldots\ldots\ldots (n - 2), (n - 1).$$

Thus l can have n values.

Informations given by *l* values. (*i*) Different values of l represent different sub-shells which are designated by small letters *s, p, d, f*... These letters stand *for sharp, principal, diffuse* and *fundamental* which are the spectral terms used to describe the series of lines observed in atomic spectrum. For example, the sub-shell for which $l = 0$ is called *s* sub-shell, that for which $l = 1$ is called *p* sub-shell and so on. Thus :

Value of l	→	*0*	*1*	*2*	*3* ....
Designation of sub-shell	→	*s*	*p*	*d*	*f* ....

Thus electrons with $l = 0, 1, 2$ and 3 are said to be residing in *s, p, d* and *f* sub-shells respectively. Consequently l value tells us as to in which sub-shell an electron is residing.

s, p, d, and *f* sub-shells are also called $l = 0$ sub-shell, $l = 1$ sub-shell, $l = 2$ sub-shell and $l = 3$ sub-shell respectively.

The energies of the sub-shells belonging to the same shell are in the following order :

$$s < p < d < f$$

Thus the sequence of filling of *s, p, d* and *f* sub-shells belonging to the same shell is as :

$$\xrightarrow[s,\ p,\ d,\ f]{}$$

(*ii*) *Values of l define the shape of the sub-shell (electron cloud) occupied by the electron, i.e.,* whether the sub-shell is *spherical, dumb-bell or with some other complicated shape.* Each value of l gives a particular shape to the sub-shell. For example the sub-shell (electron cloud) with $l = 0$ has a *spherical shape*, that with $l = 1$ has a *dumb-bell shape with distorted spheres of high probability of the electron density on each side of the nucleus* while that with $l = 2$ has a shape which approximates to four pears with their stems pointing towards the nucleus. Sub-shells with higher l values have *very complex shape.*

(*iii*) *l values enable us to calculate the total number of sub-shells which a main shell* (K, L, M etc., shells) *has. Total number of sub-shells in a main shell with principal quantum number, n is equal to the total number of l values for the given value of n.* For example :

(*a*) *K-shell (1st shell) for which n = 1.* When $n = 1, l = 0$. Since l has only one value, K-shell (1st shell) has only *one* sub-shell which is designated as *nl* or *1s*. Thus for 1s sub-shell $n = 1$ and $l = 0$. Thus :

$$n \rightarrow 1 \text{ (K-shell)}$$
$$l \rightarrow 0$$
Designation of sub-shell (nl) → 1s

(*b*) *L-shell (2nd shell) for which n = 2.* When $n = 2, l = 0, 1$. Since l has two values, L-shell (2nd shell) has *two* sub-shells which are designated as 2s and 2p. Thus :

$n \rightarrow$	2	2 (L-shell)
$l \rightarrow$	0	1
Designation of sub-shell (nl) →	2s	2p

(c) *M-shell (3rd shell) for which n = 3.* When $n = 3, l = 0, 1, 2$. Since l has three values, M-shell (3*rd* shell) has *three* sub-shells which are designated as 3*s*, 3*p* and 3*d*. Thus :

$n \rightarrow$	3	3	3 (M-shell)
$l \rightarrow$	0	1	2
Designation of sub-shell (nl) →	3s	3p	3d

(*d*) *N-shell (4th shell) for which n = 4.* When $n = 4, l = 0, 1, 2$ and 3. Since l has four values. N-shell (4*th* shell) has four sub-shells which are designated as *4s, 4p, 4d* and *4f*. Thus :

$n \rightarrow$	4	4	4	4 (N-shell)
$l \rightarrow$	0	1	2	3
Designation of sub-shell (nl) →	4s	4p	4d	4f

It is evident from the above discussion that the total number of l values for a given n value is equal to the value of n and hence *the number of sub-shells in a main shell with principal quantum number, n is equal to n.*

Whatever has been said so far can be summarised as follows :

(*a*) K-shell (*1st* energy level) having $n = 1$ contains only *one* sub-shell, called *s* sub-shell ($l = 0$) and designated as 1s ($n = 1, l = 0$).

(*b*) L-shell (*2nd* energy level) with $n = 2$ contains two sub-shells which are designated as 2*s* ($n = 2, l = 0$) and 2*p* ($n = 2, l = 1$)

(*c*) M-shell (*3rd* energy level) with $n = 3$ is composed of *three* sub-shells which are represented as 3*s* ($n = 3, l = 0$), 3*p* ($n = 3, l = 1$) and 3*d* ($n = 3, l = 2$).

(*d*) N-shell (*4th* energy level) with $n = 4$ consists of four sub-shells which are 4*s* ($n = 4, l = 0$), 4*p* ($n = 4, l = 1$), 4*d* ($n = 4, d = 2$) and 4*f* ($n = 4, l = 3$).

All the above points have been illustrated in Fig. 2.8.

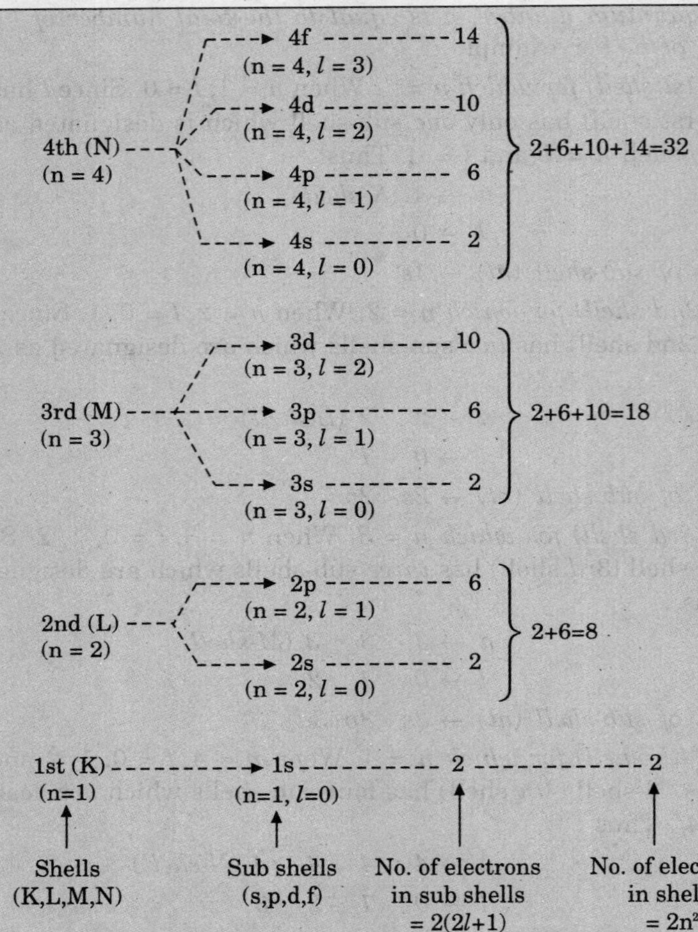

Fig. 2.8. K-shell (n = 1) contains one sub-shell (called s sub-shell), L-shell (n = 2) is composed of two sub-shells (*2s* and *2p*), M-shell (n = 3) consists of three sub-shells (*3s*, *3p* and *3d*) while N-shell (n = 4) contains four sub-shells (4*s*, 4*p*, 4*d* and 4*f*). Total number of electrons in shells (= $2n^2$) and sub-shells [$=2(2l + 1)$] have also been shown.

(iv) l values also enable us to calculate the total number of electrons in a given sub-shell (s, p, d and f sub-shells). Total number of electrons in a given sub-shell with a given l value is equal to $2(2l + 1)$. Thus *s, p, d* and *f* sub-shells have 2, 6, 10 and 14 electrons respectively as shown below : (Also see Fig. 2.8 and Table 2.1).

Sub-shell	→	*s*	*p*	*d*	*f*
Value of l	→	*0*	*1*	*2*	*3*
No. of electrons in a sub-shell = 2(2l + 1)	→	*2*	*6*	*10*	*14*

Table 2.1. Relation between principal quantum number (*n*) and azimuthal quantum number (*l*)

Value of *n* and the shell designation	**Value of *l* and designation of sub-shell (*nl*)**	**No. of sub-shells in main shell = *n***	**No. of electrons in sub-shells = 2(2*l*+1) and their designation [$nl^{2(2l+1)}$]**	**Total number of electrons in main shell = $2n^2$**
1 (K-shell or *1st* shell)	0 (*1s*)	1	$2(2 \times 0 + 1) = 2(1s^2)$	$2 \times 1^2 = 2$
2 (L-shell or *2nd* shell)	0 (*2s*) 1 (*2p*)	1+1 = 2	$2(2 \times 0 + 1) = 2(2s^2)$ $2(2 \times 1 + 1) = 6(2p^6)$	$2 \times 2^2 = 1$ *or* $2 + 6 = 8$
3 (M-shell or *3rd* shell)	0 (*3s*) 1 (*3p*) 2 (*3d*)	1+1+1 = 3	$2(2 \times 0 + 1) = 2(3s^2)$ $2(2 \times 1 + 1) = 6(3p^6)$ $2(2 \times 2 + 1) = 10(3d^{10})$	$2 \times 3^2 = 18$ *or* $2 + 6 + 10 = 18$
4 (N-shell or *4th* shell)	0 (*4s*) 1 (*4p*) 2 (*4d*) 3 (*4f*)	1+1+1+1 = 4	$2(2 \times 0 + 1) = 2(4s^2)$ $2(2 \times 1 + 1) = 6(4p^6)$ $2(2 \times 2 + 1) = 10(4d^{10})$ $2(2 \times 3 + 1) = 14(4f^{14})$	$2 \times 4^2 = 32$ *or* $2+6+10+14 = 32$

(*v*) *l* value is a measure of the *orbital angular momentum* of an electron during its orbital motion about the nucleus. It is for this reason that *l* is also called **angular momentum quantum number** or **orbital quantum number** and sometimes **subsidiary quantum number.**

(*vi*) Sub-shells having equal *l* values (but with different *n* values) have similar shapes but their sizes increase as the corresponding *n* values increase. Thus, although *1s* and *2s* sub-shells (for both the sub-shells $l = 0$) have the same shape, their sizes are different; *2s* sub-shell being greater in size than *1s* sub-shell; since *n* for *2s* sub-shell is 2 while *n* for *1s* sub-shell is equal to 1.

Similarly *2p, 3p, 4p* sub-shells for which $l = 1$ have similar shapes but their sizes increase in the order : *2p* < *3p* < *4p*, since *n* values for these sub-shells *viz.* 2, 3 and 4 respectively are also increasing.

3. Magnetic Quantum Number (*m*)

We have seen that when an excited atom which gives a line emission spectrum is placed in a strong magnetic field, its spectral lines are further resolved into still thinner (*i.e.* closely spaced) lines. This type of splitting of spectral lines into a number of closely spaced lines has been explained by Linde who assumes that under the influence of magnetic field each sub-shell is further sub-divided into sub-shells, called **orbitals** which give the thinner lines in the spectrum.

In order to represent these orbitals the first two quantum numbers *viz.* principal quantum number (*n*) and azimuthal quantum number (*l*) are not sufficient. Rather, a third quantum number which is called **magnetic quantum number** (*m*) is used to represent the orbitals of which a given sub-shell is composed.

Values of *m*. Different values of m depend on the value of l which itself depends on the value of n.

m can have integral values ranging from $-l$ through 0 to $+l$ *i.e.*

$$m = 0, \pm 1, \pm 2, \pm 3, \ldots\ldots, \pm l$$

Consequently :

When $l = 0$ (s sub-shell), $m = 0$ (only one value)

When $l = 1$ (p sub-shell), $m = 0, \pm 1$ (Three values)

When $l = 2$ (d sub-shell), $m = 0, \pm 1, \pm 2$ (Five values)

When $l = 3$ (f sub-shell), $m = 0, \pm 1, \pm 2, \pm 3$ (Seven values)

It may be seen from the above description that for a given value of l, total number of m values is $(2l + 1)$.

Information given by *m* values. Different values of m for a given value of l (*i.e.*, for a given sub-shell) give us the total number of different ways in which a given *s, p, d, f* sub-shell, in the presence of a magnetic field, can be arranged in space along *x, y* and *z* axes or the total number of orbitals into which a given *s, p, d* etc. sub-shell can be divided. Thus different values of m for a given value of l (a given sub-shell) give us the total number of different space orientations for the given *s, p, d* etc. sub-shells.

Examples (1) *s* sub-shell. We have seen that for *s* sub-shell $l = 0$. Now when $l = 0$, $m = 0$ (*only one value*). One value of m (= 0) for $l = 0$ (*s* sub-shell) implies that *s* sub-shell of any energy level has only one space orientation and hence can be arranged in space only in one way along *x, y* or *z*-axis (See Fig. 2.9). Consequently *s* sub-shell is *not* sub-divided into any other orbital or *s* sub-shell has no orbital and is present in all the energy levels. Thus :

$$l \rightarrow 0 \text{ (s sub-shell)}$$

$$m \rightarrow 0$$

Designation of orbital $\rightarrow s$

The shape of *s*-orbital has been depicted in Fig. 2.9 which clearly shows that if a straight line is drawn at a certain angle from the origin O (nucleus) to cut the surface at a point P, the length of the line remains the same in all the directions. This means that the probability of finding the electron (*i.e.* electron density) in all the directions from the nucleus is the same in *s*-orbital. Consequently *s*-orbital has a symmetrically spherical shape and hence is usually represented by a circle which, in turn, represents a cut of sphere. (Fig. 2.9).

As the value of the principal quantum number (n) for an energy level increases, the size of the *s*-orbital also increases. For example 2*s* orbital ($n = 2$) is bigger in size than 1*s* orbital ($n = 1$).

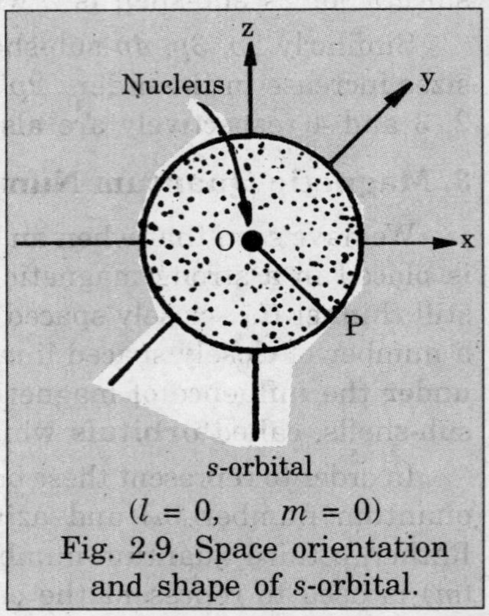

s-orbital
($l = 0$, $m = 0$)
Fig. 2.9. Space orientation and shape of *s*-orbital.

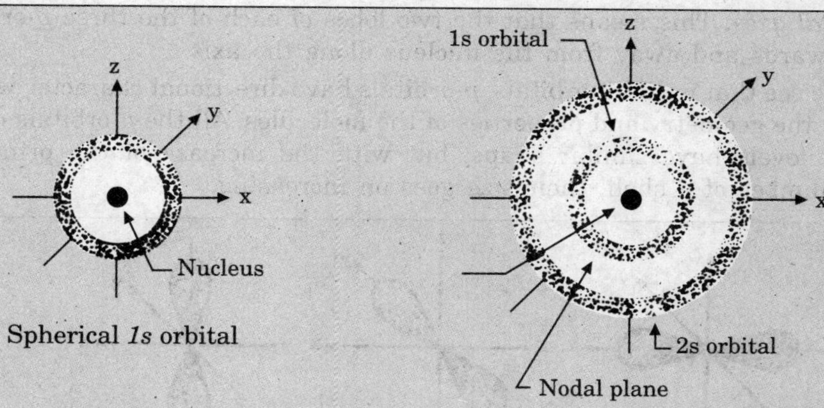

1s orbital lying completely within the *2s* orbital

Fig. 2.10. *2s* orbital is bigger in size than *1s* orbital, since the value of principal quantum number (n) for *2s* orbital (= 2) is higher than that for 1s orbital (= 1).

An electron in *2s* orbital is farther away from the nucleus and has greater energy than in *1s* orbital. *2s* orbital begins with 2nd energy level (n = 2) which, in addition to *1s* orbital, also contains *2s* orbital. Thus *1s* orbital lies completely within *2s* orbital which, in turn, is surrounded by *3s* orbital (which begins with *3rd* energy level with n = 3) and so on. Thus we see that 2*s* orbital has a spherical shell in between *1s* and *2s* orbitals where the probability of finding the electron (*i.e.*, electron density) is zero. This spherical shell is called *nodal plane* or *nodal surface*. Consequently there are two concentric dense shells in *2s* orbital and hence this orbital is *bigger in size, more diffused* and has *more energy* than *1s* orbital. In general the number of spherical nodal surfaces in *s*-orbital of a given main energy level with principal quantum number, n is equal to (n–1).

(2) *p* sub-shell (*a*) *Space orientation and shape*. For this sub-shell l = 1. Now when l = 1, m = 0, ±1 (Three values). Three values of m (= 0, ±1) for l = 1 (*p* sub-shell) imply that *p* sub-shell of any energy level (*p* sub-shell starts from *2nd* energy level with n = 2) has three space orientations and hence can be arranged in space in three different ways along *x*, *y* and *z* axes (Fig. 2.11). Consequently *p* sub-shell is sub-divided into three orbitals which, on the basis of the orientation of their lobes of maximum electron density with respect to *x*, *y* and *z* axes, are designated as p_x (m = 0), p_y (m = +1) and p_z (m = –l). Thus :

$$l \rightarrow 1 \ (p \ sub\text{-}shell)$$
$$m \rightarrow 0 \quad +1 \quad -1$$
$$Designation\ of\ orbitals \rightarrow p_x \quad p_y \quad p_z$$

Thus all the three *p*-orbitals have different space orientations. Since the three axes *viz.* *x*, *y* and *z* axes are perpendicular to each other, the three *p*-orbitals are also perpendicular to each other.

All the three *p*-orbitals namely p_x, p_y and p_z have *dumb-bell shape*. Each of the three *p*-orbitals has two *egg-shaped lobes* which touch each other at the origin (*i.e.* nucleus) and are *disposed symmetrically* along one of the three axes, called *orbital axes*. This means that the two lobes of each of the three *p*-orbital extend outwards and away from the nucleus along the axis.

Thus we see that unlike *s*-orbitals, *p*-orbitals have directional character which determines the geometry and properties of the molecules. All the *p*-orbitals of all the energy levels have similar shape, but with the increase of the principal quantum number of a shell, their size goes on increasing.

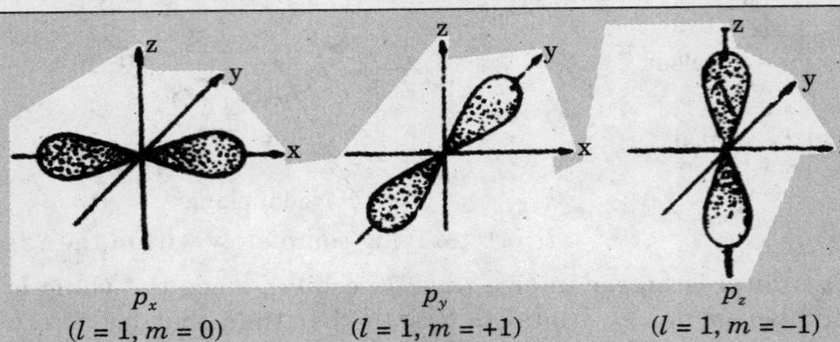

Fig. 2.11. Space orientation and shape of *p*-orbitals

(*b*) *Electron density and nodal plane*. The concept of electron density in *p*-orbitals can be understand by drawing a straight line OP at a certain angle from the origin O (nucleus) to cut the surface at a point P. The length of OP line is proportional to the probability of finding the electron anywhere in that direction. The probability decreases rapidly as the angle between OP and a given axis (*x*-axis, *y*-axis or *z*-axis) increases (Fig. 2.12). *x, y* and *z* axes are the directions of the maximum probability of electron density in case of p_x, p_y and p_z orbitals.

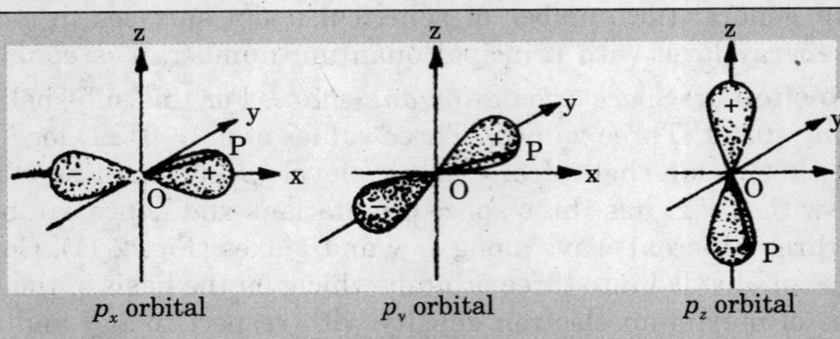

Fig. 2.12. Concept of electron density in *p*-oibitals.

The two lobes of a *p*-orbital can be separated by a plane which contains the nucleus and is perpendicular to the orbital axis. (See Fig. 2.13). The electron density (*i.e.* probability of finding the electron) at the nucleus as well as on the plane is zero. The nucleus at which the electron density is zero is called **node** and the plane passing through the orbital-axis (at this plane as well) the electron density is zero is called **nodal plane**.

It can be seen that each of the three *p*-orbitals has **one** such nodal plane. *e.g.*, for p_x orbital *yz* plane is the nodal plane.

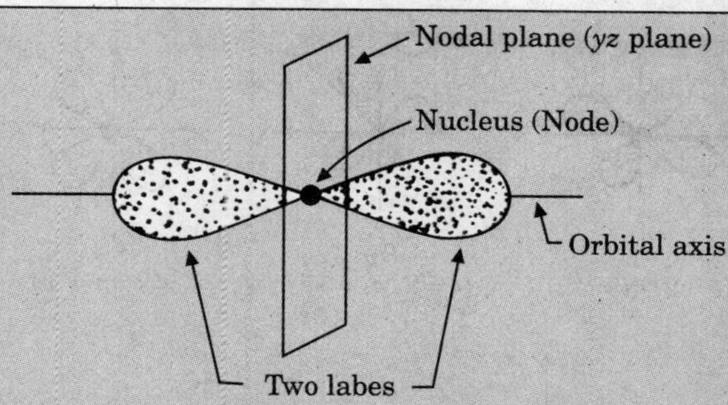

Fig 2.13. A *p*-orbital with two lobes and a nodal plane.

(*c*) *Energy*. In the absence of magnetic field, all the three *p*-orbitals have the same energy. Orbitals having the same energy are called *degenerate orbitals*. Since the number of *p*-orbitals having the same energy is three, these orbitals are said to be *three-fold degenerate or triply degenerate orbitals*.

(3) *d* sub-shell (*a*) *Space orientation, shape and electron density*. For this sub-shell $l = 2$. Now when $l = 2$, $m = 0, \pm 1, \pm 2$ (*Five values*). Five values of m (= 0, ±1, ±2) imply that *d* sub-shell of any energy level (*d* sub-shell starts from *3rd* energy level) has five space orientations and hence can be arranged in space in five different ways along *x, y* and *z* axes (Fig. 2.14). Consequently *d* sub-shell is sub-divided into five orbitals which, on the basis of the orientation of their lobes of maximum electron density with respect to *x, y* and *z* axes, are designated as $d_{xy}(m = -2)$, $d_{yz}(m = -1)$, $d_{zx}(m = +1)$, $d_{x^2-y^2}$ $(m = +2)$ and d_{z^2} $(m = 0)$. Thus :

$$l \rightarrow 2 \text{ (d sub-shell)}$$

$$m \rightarrow 0 \quad +1 \quad -1 \quad +2 \quad -2$$

$$\textit{Designation of orbitals} \rightarrow d_{z^2} \quad d_{xz} \quad d_{yz} \quad d_{x^2-y^2} \quad d_{xy}$$

All these five *d*-orbitals are not identical in shape. Four of the five *d*-orbitals namely d_{xy}, d_{yz}, d_{zx} and $d_{x^2-y^2}$ for a given energy level contain *four lobes* while the fifth orbital *viz.* d_{z^2} consists of only *two* lobes. On the basis of the orientation of the lobes of these orbitals with respect to *x, y* and *z* axes these *d*-orbitals have been divided into two sets

Set No. 1. d_{xy}, d_{yz} and d_{xz} orbitals are the members of this set and have their greatest electron density in the region *between the axes, i.e.*, the lobes of these orbitals are lying symmetrically between the axes. For example the lobes of d_{xy}, orbital are lying between *x* and *y* axes. Similar is the case with other orbitals. Group theory has called the three orbitals of this set as t_{2g} set where *t* refers to triply degenerate set.

Set No. 2. $d_{x^2-y^2}$ and d_{z^2} orbitals constitute this set and have their greatest electron density in the region *along the axes, i.e.*, the lobes of these orbitals are lying along the axes. For example the four lobes of $d_{x^2-y^2}$ orbital are lying along *x* and *y* axes while the two lobes of d_{z^2} orbital (which has dumb-bell shape) are lying along *z*-axis and contain a ring (also called *torus, collar* or *tyre*) of negative charge (electron density) surrounding the nucleus in *xy* plane. Group theory has called the two orbitals of this set as e_g set wherein *e* refers to doubly degenerate

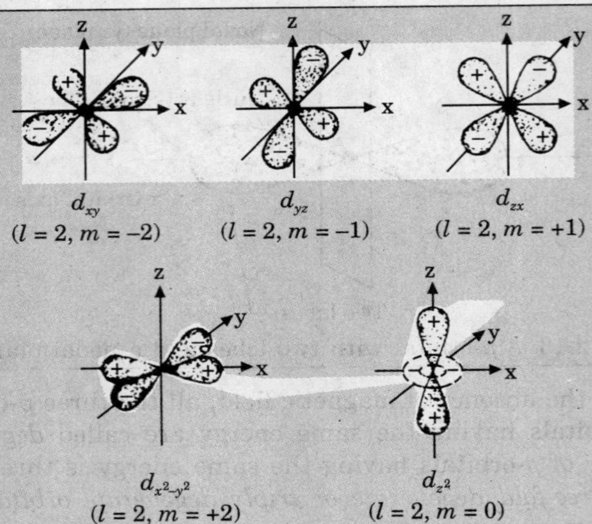

Fig. 2.14. Space orientation and shape of *d*-orbitals.

set. The torus of electron density of $d_z{}^2$ orbital in the *xy* plane is also referred to as a *doughnut* or as a *belly band*.

(*b*) *Energy*. In the absence of magnetic field all the five *d* orbitals have the same energy (degenerate orbitals). Since the number of *d*-orbitals having the same energy is five, these orbitals are said to be *five-fold degenerate orbitals*.

(4) *f* sub-shell. For this sub-shell l = 3. Now when l = 3, m = 0, ±1, ±2, ±3 (*seven values*). Since *m* has *seven* values (= 0, ± 1, +2, ±3) for l = 3 value (*f* sub-shell), *f* sub-shell has *seven* space orientations and hence can be arranged in space in seven different ways with respect to *x*, *y* and *z* axes. Consequently *f* sub-shell is sub-divided into *seven* orbitals which have complicated shape.

Since magnetic quantum number, as shown above, determines the orientation of the orbitals, it is also called **orbital orientation quantum number.**

Whatever has been said so far can be summarised as follows :

(*i*) *s* sub-shell which is present in all the energy levels contains only one orbital, called *s*-orbital.

(*ii*) *p* sub-shell which starts from *2nd* energy level (L-shell) with n = 2 contains *three* orbitals which are designated as p_x, p_y and p_z.

(*iii*) *d* sub-shell, starting from *3rd* energy level (M-shell) with n = 3, is composed of five orbitals which are represented as d_{xy}, d_{yz}, d_{zx}, $d_z{}^2$, $d_x{}^2{}_{-y}{}^2$.

(*iv*) *f* sub-shell which starts from 4th energy level (N-shell) with n = 4 is composed of seven orbitals which have complicated shape.

All the above points have been illustrated in Fig. 2.15.

Characteristics of orbitals. (*i*) In the absence of magnetic field all the orbitals of the same sub-shell of an isolated atom have the same energy, *i.e.*, they are degenerate. For example $2p_x$, $2p_y$ and $2p_z$ orbitals of 2*p* sub-shell have the same energy. Since the number of *p*-orbitals having the same energy is three, these orbitals are said to be *three-fold degenerate or triply degenerate orbitals.* Similarly five d orbitals are called *five-fold degenerate orbitals*. All the orbitals belonging to the same sub-shell have the same size and shape. For example p_x, p_y and p_z orbitals of the p-sub-shell have the same size and shape.

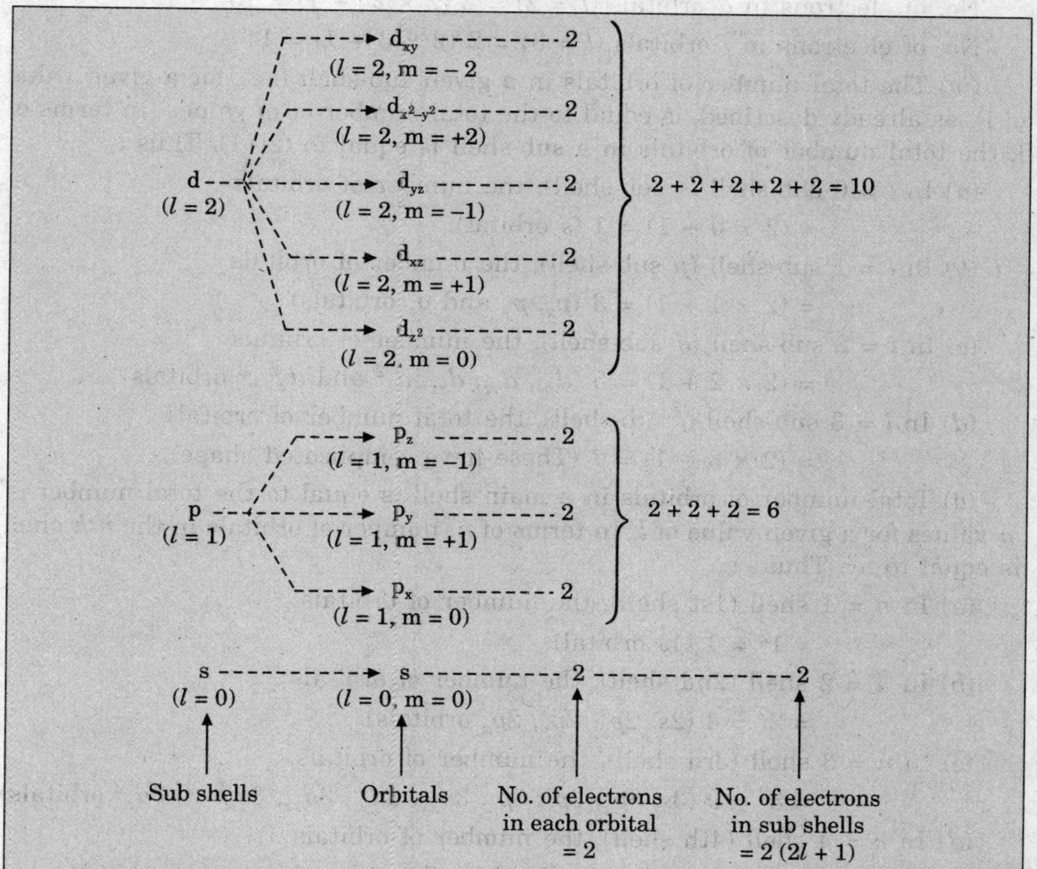

Fig. 2.15. s sub-shell (l = 0) contains one orbital (called s-orbital), p sub-shell (l = 0) is composed of three orbitals (p_x, p_y, and p_z) while d (l = 2) sub-shell contains five orbitals (d_{xy}, d_{yz}, d_{zx}, $d_z{}^2$, d_{x2-y2}).

(*ii*) All the orbitals of the same sub-sell differ in the direction of their space orientation or disposition. For example $2p_x$, $2p_y$ and $2p_z$ orbitals which belong to the same sub-shell namely $2p$ sub-shell (n = 2, l = 1) have different space orientations. Similarly $3d_{xy}$, $3d_{yz}$, $3d_{zx}$, $3d_z{}^2$ and $3d_{x-y}{}^2$ orbitals of $3d$ sub-shell (n = 3, l = 2) also have different space orientations.

(*iii*) Since the maximum capacity of an orbital to contain the electrons is two, total number of electrons in s, p, d and f sub-shells

$$= 2 \times \textit{No. of orbitals in a sub-shell}$$
$$= 2 \times (2l+1)$$
$$= 2\ (2l+1)$$

Thus :

No. of electrons in s orbital (l = 0) = 2 (2 × 0 + 1) = 2

No. of electrons in p orbitals (l = 1) = 2 (2 × 1 + 1) = 6

No. of electrons in d orbitals (l = 2) = 2 (2 × 2 + 1) = 10

No. of eleetrons in f orbitals (l = 3) = 2 (2 × 3 + 1) = 14

(*iv*) The total number of orbitals in a given sub-shell (i.e., for a given value of l), as already described, is equal to the total number of m values. In terms of l, the total number of orbitals in a sub-shell is equal to (2l+1). Thus :

(*a*) In l = 0 sub-shell (s sub-shell), the number of orbitals
= (2 × 0 + 1) = 1 (s orbital)

(*b*) In l = 1 sub-shell (p sub-shell), the number of orbitals
= (2 × 1 + 1) = 3 (p_x, p_y and p_z orbitals)

(*c*) In l = 2 sub-shell (d sub-shell), the number of orbitals
= (2 × 2 + 1) = 5 (d_{xy}, d_{yz}, d_{zx}, $d_z{}^2$ and $d_{x^2-y^2}$ orbitals)

(*d*) In l = 3 sub-shell (f sub-shell), the total number of orbitals
= (2 × 3 + 1) = 7 (These have complicated shape).

(*v*) Total number of orbitals in a main shell is equal to the total number of m values for a given value of l. In terms of n, number of orbitals in the nth shell is equal to n^2. Thus

(*a*) In n = 1 shell (1st shell), the number of orbitals
= 1^2 = 1 (1s orbital)

(*b*) In n = 2 shell (2nd shell), the number of orbitals
= 2^2 = 4 (2s, 2p_x, 2p_y, 2p_z orbitals)

(*c*) In n = 3 shell (3rd shell), the number of orbitals
= 3^2 = 9 (3s, 3p_x, 3p_y, 3p_z, 3d_{xy}, 3d_{yz}, 3d_{zx}, 3$d_{x^2-y^2}$, 3$d_z{}^2$ orbitals)

(*d*) In n = 4 shell (4th shell), the number of orbitals
= 4^2 = 16 (one 4s, three 4p, five 4d and seven 4f orbitals).

4. Spin Quantum Number (*s*).

When the spectral lines of alkali metals are observed by means of spectroscope of high resolving power, each line in the spectrum is found to consist of a pair of lines known as *double* or *doublet-line structure*. These lines are widely separated from each other. Here it should be understood clearly that this double or double-line structure is different from fine spectrum of hydrogen (*See* azimuthal quantum number), since the lines of the double-line structure are widely separated from each other while those of fine structure are closely spaced together.

In order to explain the double or doublet-line structure observed in the spectrum of alkali metals, Goudsmit and Uhlenbeck in 1925 suggested that *an electron, while moving in an orbit round the nucleus, also rotates or spins about its own axis (self-rotation) either in a clockwise direction or in an anti-clockwise direction* as shown in Fig. 2.16. Since a spinning charge is associated with a magnetic field, an electron which is spinning about its own axis also has a magnetic moment associated with it. The spinning of the electron in clock-wise and anticlock-wise directions produces opposite magnetic fields.

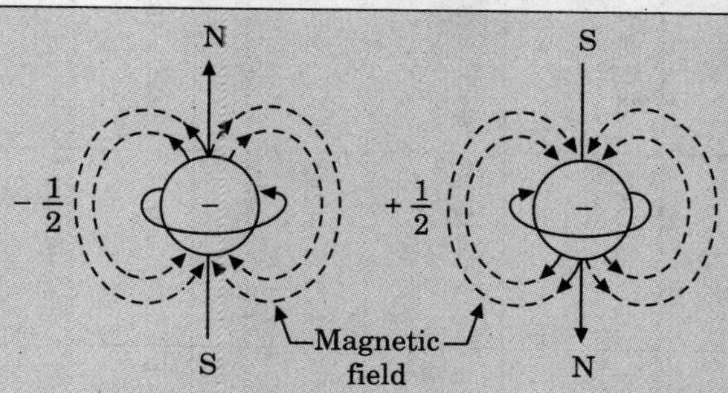

Fig. 2.16. Spinning of an electron about its own axis in clock-wise and anticlock-wise directions. Clock-wise and anticlock-wise spinning of the electron produces opposite magnetic fields.

In order to represent the direction of the spin or self-rotation of an electron about its own axis, Goudsmit and Uhlenbeck introduced another quantum number which they called spin quantum number (s), since it represents the direction of the spin of the electron. We know that since an electron can spin in any of the two directions *viz.* either in clockwise direction or in an anti-clock wise direction, spin quantum number can have two values namely $+\frac{1}{2}$ and $-\frac{1}{2}$ which differ from each other by unity as the values of other quantum numbers do. As a convention $s = +\frac{1}{2}$ is used to indicate the clockwise direction of the spin of the electron while $s = -\frac{1}{2}$ is used to indicate the anti-clockwise direction of the spin of the electron.

An electron spinning in a clockwise direction is represented by an arrow pointing upwards (↑) while that spinning in an anti- clockwise direction is represented by an arrow pointing downwards (↓). According to *Paulie's Exclusion Principle*, when there are two electrons in the same orbital, they have *opposite spins or antiparallel spins* $\left(+\frac{1}{2} \text{ and } -\frac{1}{2}\right)$ which oppose and cancel each other, since the magnetic field produced by one electron is cancelled by that produced by the other electron. Thus an electron pair in an orbital is represented as [↑↓]. Here the two electrons which have opposite spins are said to be *paired up electrons*.

Summary

The study of quantum numbers that we have made so far can be summarised as follows (Also see Table 2.2 and Fig. 2.17).

Table 2.2. Informations given by quantum numbers

n	l	Designation of sub-shell (nl)	m	Designation of orbital (nm)	No. of electrons in each orbital = 2	s for two electrons	No. of sub-shells in a main shell = n	No. of orbitals in a sub-shell = $(2l+1)$	No. of orbitals in a main shell = n^2	No. of electrons in a main shell = $2n^2$	No. of electrons in a sub-shell = $2(2l+1)$
1 (*1st* shell or K-shell)	0	*1s*	0	*1s*	2(↑↓)	+1/2, –1/2	1 (*1s*)	2×0+1 = 1	1^2 = 1 (*1s*)	2×1^2 = 2	2×1 = 2
2 (*2nd* shell or L-shell)	0	*2s*	0	*2s*	2(↑↓)	+1/2, –1/2	2 (*2s*, *2p*)	2×0+1 = 1	2^2 = 4 or	2×2^2 = 8	2×1 = 2
	1	*2p*	–1	$2p_z$	2(↑↓)	+1/2, –1/2			1+3 = 4(*2s*,		
			+1	$2p_y$	2(↑↓)	+1/2, –1/2			$2p_z$, $2p_y$, $2p_x$		2×3 = 6
			0	$2p_x$	2(↑↓)	+1/2, –1/2		2×1+1 = 3			
	0	*3s*	0	*3s*	2(↑↓)	+1/2, –1/2	3	2×0+1 = 1	3^2 = 9	2×3^2 = 18	2×1 = 2
3	1	*3p*	–1	$3p_z$	2(↑↓)	+1/2, –1/2	(*3s*,		or		
(*3rd* shell			+1	$3p_y$	2(↑↓)	+1/2, –1/2	*3p*,	2×1+1 = 3	1+3+5 = 9		2×3 = 6
or M-shell)			0	$3p_x$	2(↑↓)	+1/2, –1/2	*3d*)		(*3s*, $3p_z$, $3p_y$,		
	2	*3d*	0	$3d_{z^2}$	2(↑↓)	+1/2, –1/2			$3p_x$, $3d_{z^2}$,		
			+1	$3d_{xz}$	2(↑↓)	+1/2, –1/2			$3d_{xz}$, $3d_{yz}$,		
			–1	$3d_{yz}$	2(↑↓)	+1/2, –1/2		2×2+1 = 5	$3d_{x^2-y^2}$,		
			+2	$3d_{x^2-y^2}$	2(↑↓)	+1/2, –1/2			$3d_{xy}$)		2×5 = 10
			–2	$3d_{xy}$	2(↑↓)	+1/2, –1/2					

1. **Shells** (*a*) Electrons in an atom revolve round the nucleus in orbits or shells which are associated with some energy and hence are also called energy levels. These are represented as K, L, M etc. having principal quantum number, n = 1, 2, 3 etc. respectively.

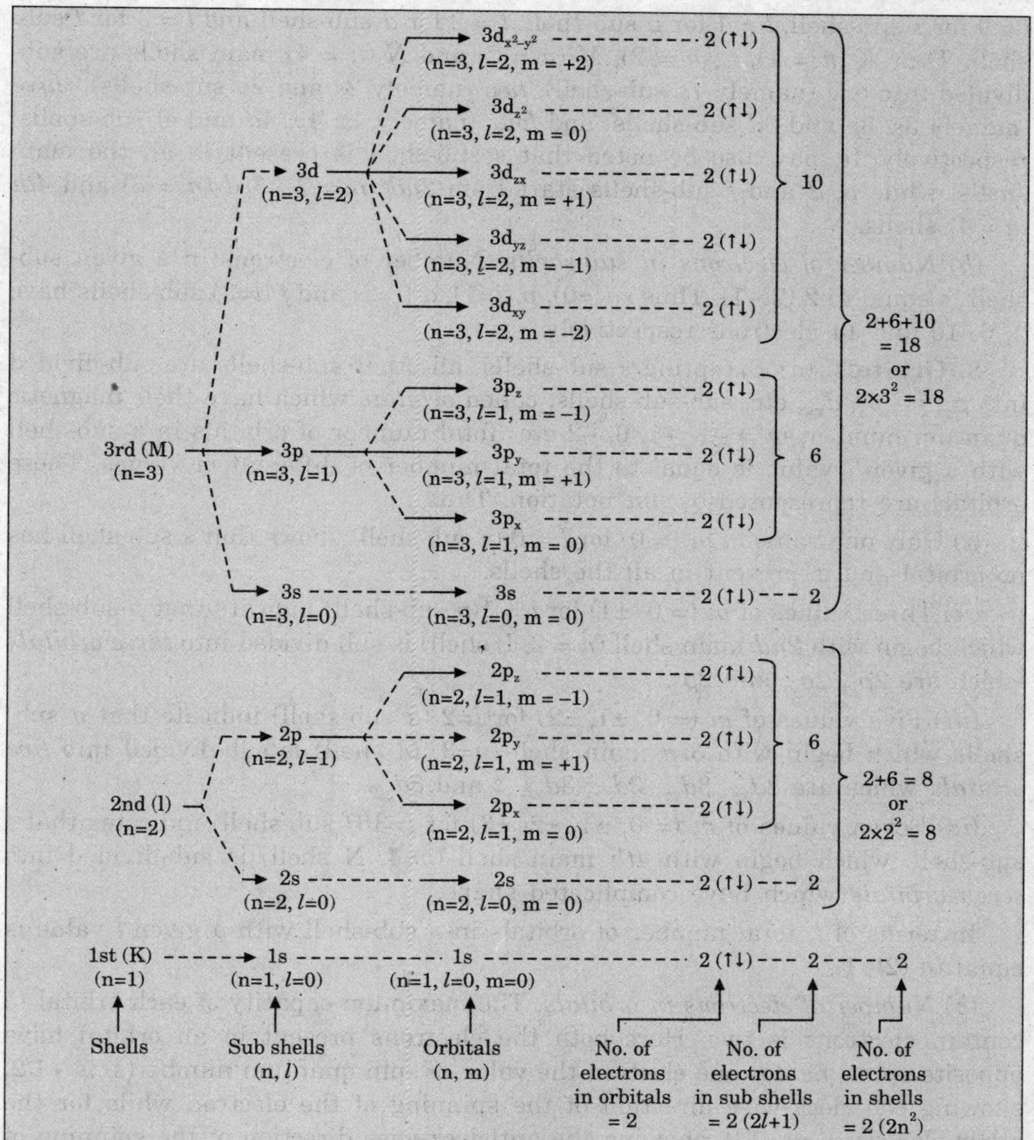

Fig. 2.17. Informations given by quantum numbers. Two electrons shown in each orbital [represented as (↑↓)] have opposite spins. For one electron the value of spin quantum number, s = + 1/2 while for the other s = –1/2.

(*b*) *No. of electrons in shells*. Number of electrons in a shell with a given n value is equal to $2n^2$.

2. **Sub-shells** (*a*) Each of the main shells is sub-divided into *s, p, d, f* sub-shells which have azimuthal quantum, $l = 0, 1, 2$ and 3 respectively. Total number of sub-shells in a main shell with given *n* value is equal to *n* value or total number of *l* values These sub-shells are represented by the notation *nl* where $l = 0$ for *s* sub-shell, $l = 1$ for *p* sub-shell, $l = 2$ for *d* sub-shell and $l = 3$ for *f* sub-shell. Thus *K* ($n = 1$), *L* ($n = 2$), *M* ($n = 3$) and *N* ($n = 4$) main shells are sub-divided into *one* (namely *1s* sub-shell), *two* (namely 2*s* and 2*p* sub-shells), *three* (namely 3*s*, 3*p* and 3*d* sub-shells) and *four* (namely 4*s*, 4*p*, 4*d* and 4*f* sub-shells) respectively. It may also be noted that s sub-shell is present in all the main shells, while *p*, *d* and *f* sub-shells start from *2nd* ($n = 2$), *3rd* ($n = 3$) and *4th* ($n = 4$) shells.

(*b*) *Number of electrons in sub-shells.* Number of electrons in a given sub-shell is equal to $2(2l+1)$. Thus *s* (l=0), *p* (l=1), *d* (l=2) and *f* (l=3) sub-shells have 2, 6, 10 and 14 electrons respectively.

3. **Orbitals.** (*a*) Excepting *s* sub-shells, all other sub-shells are sub-divided into p_x, p_y, p_z, d_{xy}, etc. sub-sub-shells, called *orbitals* which have their magnetic quantum number, $m = -1, +1, 0, -2$ etc. Total number of orbitals in a sub-shell with a given *l* value is equal to the total number of different *m* values. These orbitals are represented by *nm* notation. Thus :

(*i*) Only one value of *m* (= 0) for $l = 0$ (*s* sub-shell) shows that *s* sub-shell has no orbital and is present in all the shells.

(*ii*) Three values of *m* (= 0, ±1) for $l = 1$(*p* sub-shell) indicate that *p* sub-shell which begin with *2nd* main shell ($n = 2$, L shell) is sub-divided into *three orbitals* which are $2p_x$, $2p_y$, and $2p_z$.

(*iii*) Five values of *m* (= 0, ±1, ±2) for l=2 (*d* sub-shell) indicate that *d* sub-shells which begin with *3rd* main shell (n=3, M shell) is sub-divided into *five orbitals* which are $3d_{xy}$, $3d_{yz}$, $3d_{zx}$, $3d_{x^2-y^2}$ and $3d_{z^2}$.

(*iv*) Seven values of *m* (= 0, ±1, ±2, ±3) for l=3 (*f* sub-shell) indicates that *f* sub-shell, which begin with *4th* main shell (n=4, N shell) is sub-divided into *seven orbitals* which have complicated shape.

In terms of *l*, total number of orbitals in a sub-shell with a given *l* value is equal to $(2l+1)$.

(*b*) *Number of electrons in orbitals.* The maximum capacity of each orbital to contain electrons is two. Here both the electrons present in an orbital have opposite spins, *i.e.* for one electron the value of spin quantum number(s) is +1/2, showing the clock-wise direction of the spinning of the electron while for the other electron $s = -1/2$, showing the anticlock-wise direction of the spinning of the electron.

(*c*) *Number of orbitals in a main shell.* The number of orbitals in a main shell with a given *n* value is equal to n^2. Thus K ($n = 1$), L ($n = 2$) and M ($n = 3$) shells have *one* (namely *1s*), *four* (namely 2*s*, $2p_x$, $2p_y$, $2p_z$) and nine (namely 3*s*, $3p_x$, $3p_y$, $3p_z$, $3d_{xy}$ etc.) orbitals.

Problems Based On Quantum Numbers

Problem 1. *Which of the orbitals namely 1p, 2s, 2p, 3f are not possible ? Give reason for your answer.* *(Delhi B.Sc. 1976)*

Solution. The orbitals which are not possible are 1*p* and 3*f*. The reason for their non-existence is given as follows :

We know that when n = 1 (1st shell), l=0. l=0 value means that *1st* shell has only one sub-shell namely 1*s* for which l=0. Thus 1st shell does not contain 1*p* orbital.

It is known to us that when n=3, l=0, 1, 2. Three values of l mean that *3rd* shell has only three sub-shells namely *3s* (for which l=0), 3*p* (for which l=1) and 3*d* (for which l=2). Thus *3rd* shell does not contain 3*f* orbital.

Problem 2. *An electron is in 4f orbital. What possible values for the quantum number of n, 1, m and s it can have ?* *(Punjab B.Sc. 1981)*

Solution. We know that for 4*f* orbital n=4 and l=3. Now when l=3, m=0, ±1, ±2, ±3 (seven values). Seven values of m imply that the electron may be in any of the seven 4*f* orbitals and for this electron m may have any of the above seven values. For this electron $s = +\frac{1}{2}$ or $-\frac{1}{2}$.

Thus n, l, m and s values for an electron residing in 4*f* orbital are :

n= 4, l=3, m=any of the seven values namely 0, ±1, ±2, ±3 and $s = +\frac{1}{2}$ or $-\frac{1}{2}$.

Problem 3. *What is the maximum number of electrons in (i) all 4d orbitals (ii) all the orbitals with l = 1.* *(Kurukshetra B.Sc. I 1983)*

Solution. We know that the maximum number of electrons in an orbital with a given value of l is equal to 2 (2l+l). Thus :

(*i*) Maximum number of electrons in 4*d* orbitals (l=2)

$$= 2\ (2 \times 2 + 1) = 10$$

(*ii*) Maximum number of electrons in all the orbitals with l = 1

$$= 2\ (2 \times 1 + 1)$$
$$= 6$$

Problem 4. *Write the name of the orbital for which the quantum numbers are : n = 2 and l = 1.* *Jabalpur B.Sc. 1981; Himachal 1982)*

Solution. We know that n = 2 indicates *2nd* main shell. Now when n= 2, l= 0, 1. l= 0 stands for s sub-shell while l=1 indicates *p* sub-shell. Thus *p* (l=1) sub-shell belongs to 2nd shell and hence n=2 and l=1 stands for 2*p* sub-shell.

Problem 5. *Give the symbols for orbital types when the values of l are 0, 2 and 3.* *(Punjab B.Sc. 1972; Raj. 1972; Delhi 1974; Gorakhpur 1973)*

Solution. Obviously the orbitals having l=0, 2 and 3 are *s*, *d* and *f* respectively.

Problem 6. *Write correct orbital notation for each of the following set of quantum numbers :*

(*i*) $n = 1, l = 0, m = 0$

(*ii*) $n = 2, l = 1, m = 1$

(*iii*) $n = 3, l = 2, m = -1$ *(Meerut B.Sc. 1984)*

Solution. With reference to Table 2.2 given on page 131

(*i*) The orbital for which $n = 1$, $l = 0$ and $m = 0$ is 1*s*.

(*ii*) The orbital for which n=2, l=1 and m=1 is $2p_y$.

(*iii*) The orbital for which n=3, l=2 and m= –1 is $3d_{yz}$.

Problem 7. *What are n, l and m values for 3s, $3p_x$ and $3d_{xy}$ electrons ?*
(Meerut B.Sc. 1984)

Solution. With reference to Table 2.2 given on page 131

(*i*) For 3*s* electrons, n=3, l= 0, m = 0

(*ii*) For $3p_x$ electrons, n= 3, l=1, m = 0

(*iii*) For $3d_{xy}$ electrons, n= 3, l=2, m= –2

Problem 8. *What will be the values of other quantum numbers of an electron for which the principal quantum number is 3 ?*

Solution. The values of other quantum numbers namely *l*, *m* and *s* are as follows :

n	l	m	s
3	0	0	+1/2, –1/2
	1	0	+1/2, –1/2
		+1	+1/2, –1/2
		–1	+1/2, –1/2
	2	0	+1/2, –1/2
		+1	+1/2, –1/2
		–1	+1/2, –1/2
		+2	+1/2, –1/2
		–2	+1/2, –1/2

Problem 9. *What are the possible values of s for an electron with m = 0 ?*

Solution. *s* is always equal to +1/2 or –1/2, regardless of the values of other quantum numbers.

Problem 10. (*i*) *How are the electrons distributed in various sub-shells for n = 3 ? (ii) Give the quantum numbers for the electron in the first sub-shell appearing in n = 3 shell.*

Solution. (*i*) When n = 3, the shell is 3rd shell and l = 0 (3*s* sub-shell), 1 (3*p* sub-shell), 2 (3*d* sub-shell). Thus 3rd main shell contains three sub-shells which are 3*s*, 3*p* and 3*d*.

Now we know that the total number of electrons in 3rd main shell = 2×3^2 = 18. These 18 electrons in 3*s*, 3*p* and 3*d* sub-shells are distributed as follows :

No. of electrons in 3*s* sub-shell (l = 0) = 2(2l+1) = 2(2×0+1) = 2.

No. of electrons in 3*p* sub-shell (l = 1) = 2(2×1+1) = 6.

No. of electrons in 3*d* sub-shell (l = 2) = 2(2×2+1) = 10.

(*ii*) Quite evidently 1st sub-shell in 3rd main shell is 3*s* sub-shell which contains two electrons as shown above. The four quantum numbers for the two electrons in 3*s* sub-shell are as follows :

		Quantum numbers			
		n	*l*	*m*	*s*
Electron number	1st electron	3	0	0	+1/2
	2nd electron	3	0	0	—1/2

Problem 11. *Find the number of electrons in atoms which have the following levels in the ground state. (a) K shell, L shell, 3s sub-shell and half-filled 3p sub-shell. (b) K, L and M shells, and 4s, 4p and 4d sub-shells. What atoms are these?*

Solution. (*a*) No. of electrons in K-shell (n=1)

$$= 2n^2 = 2 \times 1^2 = 2$$

No. of electrons in L-shell (n=2) = $2 \times 2^2 = 8$

No. of electrons in 3*s* sub-shell (l=0) = $2(2l+1)$

$$= 2\ (2 \times 0 + 1) = 2$$

Half the number of electrons in 3*p* sub-shell (l=1)

$$= \frac{2(2\times1+1)}{2} = 3$$

Total number of electrons = 2 + 8 + 2 + 3 = 15

This number indicates the atomic number of the element. Thus the element is **phosphorus.**

(*b*) No. of electrons in K-shell (n = 1) = $2n^2 = 2 \times 1^2 = 2$

No. of electron in L-shell (n = 2) = $2 \times 2^2 = 8$

No. of electrons in M-shell (n = 3) = $2 \times 3^2 = 18$

No. of electrons in 4*s* sub-shell (l = 0) = $2\ (2l + 1)$

$$= 2\ (2 \times 0 + 1) = 2$$

No. of electrons in 4*p* sub-shell (l = 1) = $2\ (2 \times 1 + 1) = 6$

No. of electrons in 4*d* sub-shell (l = 2) = $2\ (2 \times 2 + 1) = 10$

∴ Total number of electrons = 2 + 8 + 18 + 2 + 6 + 10 = 46.

Thus the element is **palladium.**

Problem 12. *For n = 4 principal shell, find the following (i) name of the shell (ii) number of sub-shells in it (iii) total number of orbitals in each sub-shell (iv) numbers of orbitals in the principal shell.*

Solution. (*i*) n = 4 is the 4th shell (*i.e.*, N-shell).

(*ii*) No. of sub-shells in a main shell with a given value of n = 4 = n = 4.

(*iii*) We know that for n = 4 shell, l = 0 (4*s* sub-shell), 1(4*p* sub-shell), 2 (4*d* sub-shell), 3 (4*f* sub-shell). Now since total number of orbitals in a sub-shell with a given value of l = $(2l + 1)$,

No. of orbitals in 4*s* sub-shell (l = 0) = $(2 \times 0 + 1) = 1$

No. of orbitals in 4*p* sub-shell (l = 1) = $(2 \times 1 + 1) = 3$

No. of orbitals in $4d$ sub-shell ($l = 2$) = $(2 \times 2 + 1) = 5$

No. of orbitals in $4f$ sub-shell ($l = 3$) = $(2 \times 3 + 1) = 7$.

(*iv*) No. of orbitals in 4th shell ($n = 4$) = $4^2 = 16$

Pauli's Exclusion Principle

In 1925 Wolfgang Pauli put forward a principle which controls the assignment of values to the quantum numbers of an electron in an orbital. It imposes some restrictions on the assignment of the values of quantum numbers of electrons in the same orbital of an atom and hence the name *Exclusion Principle.* This principle can be stated in many ways some of which are given below :

(*i*) *It is imposible for two electrons residing in the same orbital of a given poly-electron atom (same atom) to have the same values of all the four quantum numbers.*

(*ii*) *In an atom no two electrons contained in the same orbital can have the same set of values of all the four quantum numbers.*

(*iii*) *No two electrons residing in the same orbital of an atom can be described by an identical set of four quantum numbers.*

(*iv*) *In the same orbital of an atom two electrons can have three quantum numbers identical at the most, but the fourth quantum number namely spin quantum number (s) will definitely be different for both the electrons, i.e.,* for one electron s will be +1/2 while for the other electron s will be –1/2. s = +1/2 indicates the clock-wise direction of the spinning of one electron while s = –1/2 stands to show the anticlock-wise direction of the other electron. This means that two electrons residing in the same orbital of an atom have opposite spins, since this state of the electrons having opposite spins (↑↓) gives lower energy to the orbital while the state with parallel spins (↑↑) gives higher energy to the orbital. According to thermodynamics a system with lower energy has greater stability. Thus an orbital having two electrons with opposite spins (↑↓) has lower energy and hence is more stable than that containing the two electrons with parallel or paired spins (↑↑).

(*v*) *The maximum number of identical quantum numbers which two electrons in an orbital can have is only three but never four.*

Illustration of the principle. In order to illustrate the principle let us consider helium atom which has two electrons (Atomic number = 2) in 1s orbital for which n = 1, l = 0 and m = 0. The four quantum numbers for the two electrons in 1s orbital are as follows :

Electron number	*Quantum numbers* n	l	m	s	*Set of values of quantum numbers* (n, l, m, s)
1st electron (↑)	1	0	0	+1/2	(1, 0, 0, +1/2)
2nd electron (↓)	1	0	0	–1/2	(1, 0, 0, –1/2)

Thus we see that for one electron shown as (↑) the set of values of quantum numbers is (1, 0, 0, +1/2) while for the electron shown as (↓) this set is (1, 0, 0, –1/2). As is evident, the two electrons have different sets of four quantum numbers which is in accordance with Pauli's exclusion principle.

Applications of Pauli's Exclusion Principle.

With the help of this principle, the maximum number of electrons that can be accommodated in an orbital, sub-shell and main shell can be calculated as shown below :

(i) **1st shell** (*K-shell with* $n = 1$). For this shell $n = 1$. The values of l and m for $n = 1$ are $l = 0$ and $m = 0$. Hence s can have a value either equal to +1/2 or —1/2. Thus these different values of n, l and m give the following **two** sets of the values of four quantum numbers, keeping in view Pauli's exclusion principle.

Set number	*n*	*l*	*m*	*s*	*Set*
1	1	0(1*s*)	0	+1/2	(1, 0, 0, + 1/2)
2	1	0(1*s*)	0	–1/2	(1, 0, 0, – 1/2)

Since the number of permitted sets of four quantum numbers is two for 1st shell ($n = 1$), this shell contains at the maximum two electrons which are contained in 1*s* orbital.

(ii) **2nd shell** (*L-shell with* $n = 2$). For this shell $n = 2$. For $n = 2$, the different values of l, m and s give the following **eight** sets of quantum numbers which indicate that 2*nd* shell contains eight electrons of which two are contained by 2*s* sub-shell while the remaining six are contained in 2p sub-shell. Six electrons in 2*p* sub-shell are distributed in $2p_x$, $2p_y$, and $2p_z$ orbitals as $2p_x^2$, $2p_y^2$, $2p_z^2$ in which the superscript indicates the number of electrons contained in the orbital.

Set number	*n*	*l*	*m*	*s*	*Set*
1	*2*	*0 (2s)*	*0 (2s)*	*+1/2*	*(2, 0, 0, + 1/2)*
2	*2*	*0 (2s)*	*0 (2s)*	*–1/2*	*(2, 0, 0, – 1/2)*
3	*2*	*1 (2p)*	*0* $(2p_x)$	*+1/2*	*(2, 1, 0, + 1/2)*
4	*2*	*1 (2p)*	*0* $(2p_x)$	*–1/2*	*(2, 1, 0, – 1/2)*
5	*2*	*1 (2p)*	*+1* $(2p_y)$	*+1/2*	*(2, 1, + 1, + 1/2)*
6	*2*	*1 (2p)*	*+1* $(2p_y)$	*–1/2*	*(2, 1, + 1, – 1/2)*
7	*2*	*1 (2p)*	*–1* $(2p_z)$	*+1/2*	*(2, 1, – 1, + 1/2)*
8	*2*	*1 (2p)*	*–1* $(2p_z)$	*–1/2*	*(2, 1, – 1, – 1/2)*

(iii) **3rd and 4th shells** (*M and N shells*). On the same lines of arguement it can be shown that 3*rd* shell ($n = 3$) can contain at the maximum 18 electrons (2 in 3*s* sub-shell, 6 in 3*p* sub-shell and 10 in 3*d* sub-shell) and 4*th* shell ($n = 4$) can have 32 electrons (2 in 4*s* sub-shell, 6 in 4*p* sub-shell, 10 in 4*d* sub-shell and 14 in 4*f* sub-shell).

Whatever has been said above has been shown in Table 2.3.

Table 2.3 Determination of the number of electrons present in orbitals, sub-shells and main shells by Pauli's exclusion principle.

n and shell designation	*l* and sub-shell designation (*nl*)	*m* and orbital designation	Value of *s*	No. of electrons in orbitals and sub-shells	No. of electrons present in main shells
1 (1*st* shell or K-shell)	0 (1*s*)	0 (1*s*)	+1/2, –1/2	2	2 or $2n^2 = 2\times1^2 = 2$
2 (2*nd* shell or L-shell)	0 (2*s*)	0 (2*s*)	+1/2, –1/2	2	2+6 = 8 or $2n^2 = 2\times2^2$ = 8
	1 (2*p*)	0 ($2p_x$)	+1/2, –1/2	2 } 6	
		+1 ($2p_y$)	+1/2, –1/2	2	
		–1 ($2p_z$)	+1/2, –1/2	2	
3 (3*rd* shell or M-shell)	0 (3*s*)	0 (3*s*)	+1/2, –1/2	2	2+6+10 = 18 or $2n^2 = 2\times3^2$ = 18
	1 (3*p*)	0 ($3p_x$)	+1/2, –1/2	2 } 6	
		+1 ($3p_y$)	+1/2, –1/2	2	
		–1 ($3p_z$)	+1/2, –1/2	2	
	2 (3*d*)	0 ($3d_{z^2}$)	+1/2, –1/2	2 } 10	
		+1 ($3d_{xz}$)	+1/2, –1/2	2	
		–1 ($3d_{yz}$)	+1/2, –1/2	2	
		+2 ($3d_{x^2-y^2}$)	+1/2, –1/2	2	
		–2 ($3d_{xy}$)	+1/2, –1/2	2	

Table 2.3 makes the following two rules evident :

(*a*) *An orbital like s, p_x, p_y, p_z, d_{xy}, d_{yz} etc., can have at the most two electrons and these two electrons must have opposite spins, i.e.,* one electron should have the value of spin quantum number (*s*) equal to +1/2 and the other equal to –1/2 so that both the electrons residing in the same orbital may have different sets of four quantum numbers in accordance with Pauli's exclusion principle.

(*b*) *Maximum number of electrons that can be accommodated by a shell with principal quantum number, n is equal to $2n^2$.*

These two rules have been used in the determination of the electronic configuration of the atoms of the elements.

Energy Level Diagrams

We have seen that the energy of an electron is determined by *n* and *l* quantum numbers. When we discuss the energy of orbitals, the following two cases may arise :

1. Energy Level Diagram of One-Electron Atoms (Hydrogen and Hydrogen Like-Atoms)

Hydrogen has only one electron which is present in the lowest energy level which is 1*st* energy level (*i.e.* 1*s* orbital) with $n = 1$. When hydrogen atom absorbs energy, its 1*s* electron may jump to higher energy levels with $n = 2$, (L-level), 3 (M-level) etc., or to higher sub-shells represented by their l values (*e.g.*, 2*s*, 2*p* etc., sub-shells). The spectral study of hydrogen atom has shown that its spectral lines correspond to the major energy levels only. Thus the energy of an electron present in a given principal energy level is the same irrespective of the orbitals of the same principal energy to which it belongs.

In other words this means that *the energies of all s, p, d, f orbitals of the same principal energy level are identical.* For example 3*s*, 3*p* and 3*d* orbitals which belong to 3*rd* energy level ($n = 3$) have the same energy (Fig. 2.18)

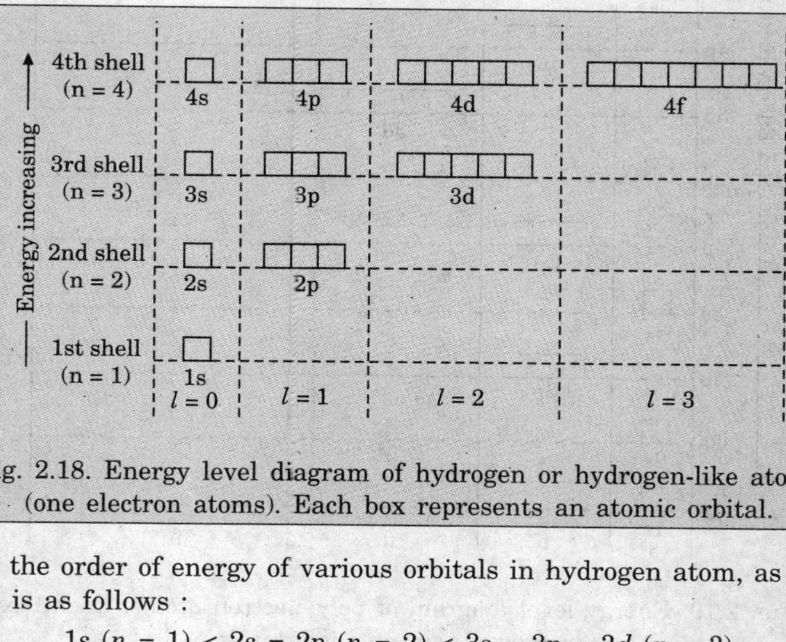

Fig. 2.18. Energy level diagram of hydrogen or hydrogen-like atoms (one electron atoms). Each box represents an atomic orbital.

Thus the order of energy of various orbitals in hydrogen atom, as shown in Fig. 2.18 is as follows :

$$1s\ (n = 1) < 2s = 2p\ (n = 2) < 3s = 3p = 3d\ (n = 3)$$
$$< 4s = 4p = 4d = 4f\ (n = 4)$$

Fig. 2.18, in which the above order has been shown is called *energy level diagram of hydrogen* or *hydrogen-like atoms.*

2. Energy Level Diagram of Poly-Electron Atoms.

For poly-electron atom the situation is, however, different. The relative order of energy values of different orbitals based on spectral data of these atoms is shown in Fig. 2.19 which is called *energy level diagram of poly-electron atoms.* From this diagram the following points may be noted :

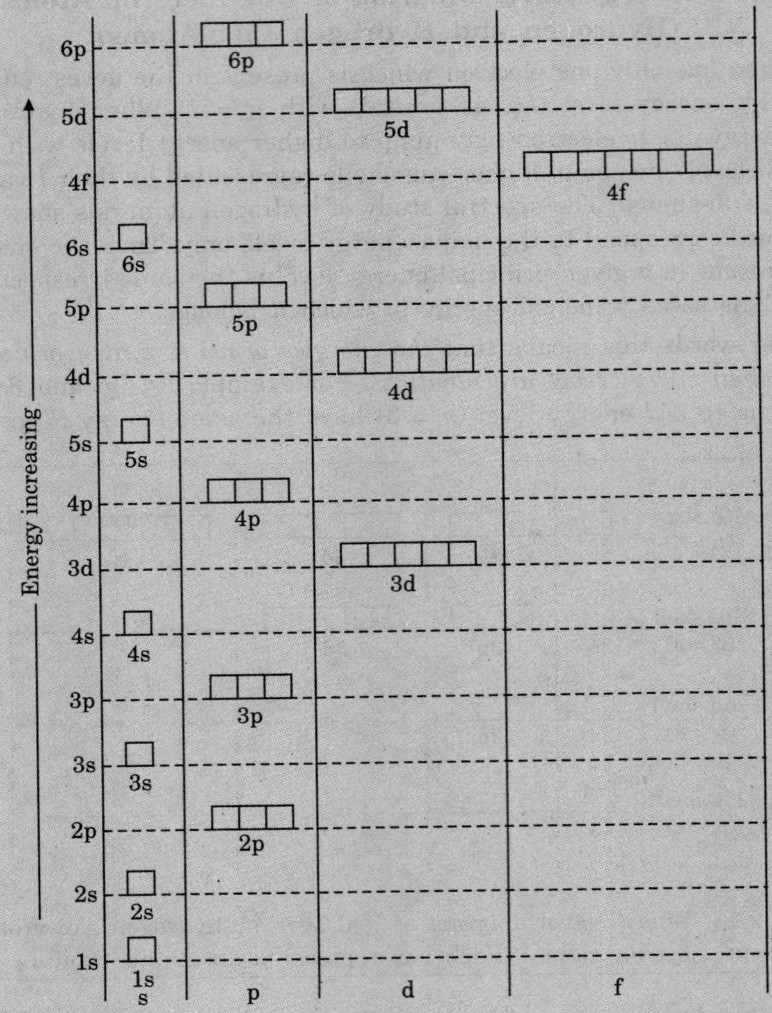

Fig. 2.19. Energy level diagram of poly-electron atoms. Each box represents an atomic orbital.

(*i*) The relative order of energy of various orbitals is as follows :

Energy Increasing →

1*s* < 2*s* < 2*p* < 3*s* < 3*p* < 4*s* < 3*d* < 4*p* < 5*s* < 4*d* < 5*p* < 6*d*
4*f* < 5*d* < 6*p* < ....

(*ii*) Orbitals of the same principal energy level have different energy values, *i.e.*, the orbitals with the same value of n but with different l values have different energy values. Thus the energy of *ns*, *np*, *nd* and *nf* orbitals which belong to the same energy level is in the order :

$$ns\ (l = 0) < np\ (l = 1) < nd\ (l = 2) < nf\ (l = 3)$$

The above order shows that with the increase of l value the energy of the orbital also increases.

Probability Distribution Curves

The probability of finding an electron in a given volume element can be represented by two types of probability distribution curves which are (1) *Radial probability distribution curves, and* (2) *Angular probability distribution curves.*

Now let us discuss these curves, one by one, as follows :

1. Radial Probability Distribution Curves.

These curves tell how the probability of finding an electron varies with its radial distance from the nucleus without any reference to its direction from the nucleus. The probability or chance of finding an electron in the space round the nucleus is called **electron probability function** which is denoted by D and gives an idea of the electron probability distribution in the space round the nucleus. Suppose that the space round the nucleus is divided into a number of thin concentric spherical shells of thickness *dr* and radius *r, i.e.*, the distance of this shell from the nucleus is *r*. Quite obviously the volume of such a spherical shell is equal to $4\pi r^2 dr$ and hence the value of electron probability function D is given by the product of $4\pi r^2 dr$ and ψ^2, *i.e.*,

***Electron probability function,* $\mathbf{D = 4\pi r^2 dr \psi^2}$**

The curves obtained by plotting the electron probability function; D $(= 4\pi r^2 dr \,.\, \psi^2)$ of the electron against its distance, *r* from the nucleus are called **radial probability distribution curves**. These curves for 1*s*, 2*s*, 2*p*, 3*s*, 3*p*, and 3*d* orbitals are given in Fig. 2.20.

From these curves the following points may be noted :

(*i*) In each case the probability of finding the electron at the origin (which is the nucleus) is zero. This means that the electron will never be found at the nucleus.

(*ii*) The distance for the maximum probability of finding an electron in an orbital increases with the value of its principal quantum number, *n*. For example this distance for 1*s*, 2*s* and 3*s* orbitals is in the order 1*s* < 2*s* < 3*s* as is evident from the curves of these orbitals.

(*iii*) The total number of peaks appearing in the curves for *s, p* and *d* orbitals is equal to *n*, (*n*–1) and (*n*–2) respectively (*n* the principal quantum number). For example the number of peaks for 1*s*, 2*s*, 3*s*, 2*p*, 3*p* and 3*d* orbitals is equal to 1, 2, 3, 1, 2 and 1 respectively as is evident from the curves.

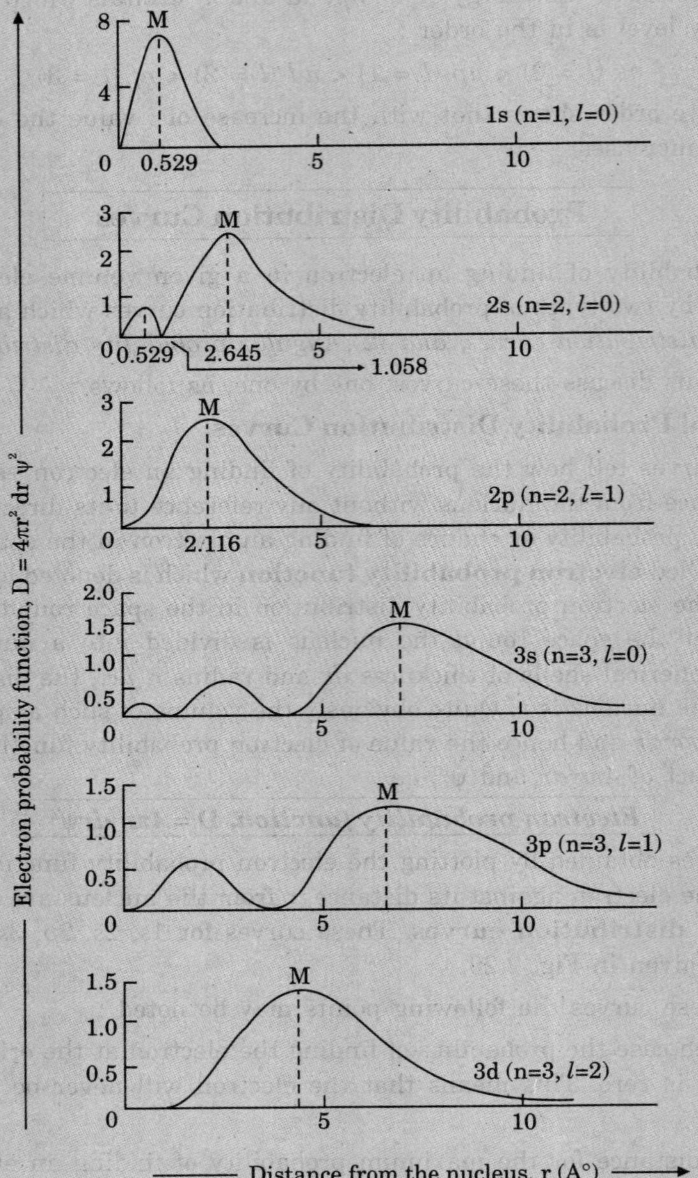

Fig. 2.20 Radial probability distribution curves for 1*s*, 2*s*, 2*p*, 3*s*, 3*p* and 3*d* orbitals. The point M in each curve represents the maximum electron density.

(*iv*) From the curve for 1*s*-electron it is evident that the value of D is zero at the nucleus where $r = 0$. D increases as r increases from zero, passes through a maximum (*i.e.* peak) at $r = 0.529$ A° which is equal to the radius of Bohr's first orbit, and finally falls to zero as r tends to infinity. Thus for 1*s*-electron D = 0 at $r = 0$ and ∞, and D = maximum at $r = 0.529$ A° (radius of Bohr's first orbit).

That D is maximum at r = 0.529 A° makes a difference between Bohr's theory and wave mechanical treatment of hydrogen atom. According to Bohr's theory, in hydrogen atom the electron under ordinary conditions (*i.e.* in ground state) always stays at a distance of 0.529 A° from the nucleus, while according to wave mechanical treatment, the *electron may exist at any distance from the nucleus but the maximum probability of locating it lies at a distance of 0.529 A° from the nucleus. In other words the radius of maximum probability* (≈ 95%) *of 1s electrons in H-atom is 0.529* A° as shown in Fig. 2.21 which depicts the probability of finding an electron round the nucleus in 1s orbital of H-atom.

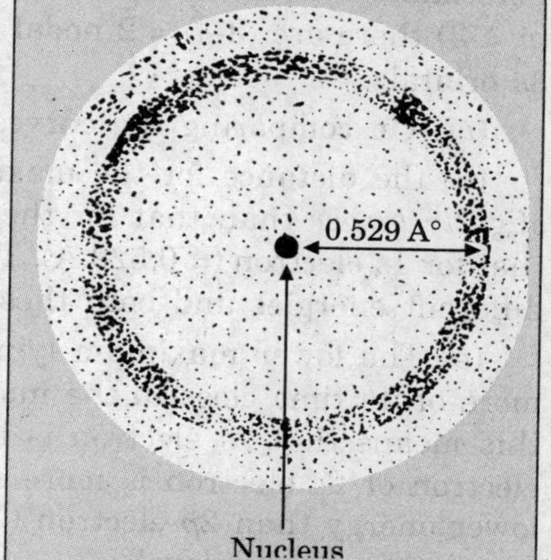

Fig. 2.21. Probability of finding the electron round the nucleus of 1s orbital of H-atom and the radius of maximum probability (= 0.529 A°).

In this figure the probability of finding the electron round the nucleus in 1s orbital has been shown by dots. The intensity of shading of dots is proportional to the probability of finding electron at that location. Such a representation of the probability of finding the electron (or electron density as it is called) is called **electron cloud** or **charge cloud representation of an orbital.** (Here the orbital is 1s orbital). It may be seen from the electron cloud that the electron density round the nucleus is denser at certain points and rarer at others. Denser is the cloud at a certain small volume, greater is the probability of finding the electron is that volume. The maximum blackening of the cloud shown in the figure at a distance of 0.529 A° indicates that 1s electron of H-atom spends most of its time at a distance of 0.529 A° from the nucleus *i.e.* the radius of maximum electron density in 1s electron is equal to 0.529 A°.

(v) For 2s electron the value of D is zero at r = 0. The value of D increases as the value of r = 0.529 Å increases, passes through a lower maximum at r = 0.529 Å and falls to zero at r = 1.058 Å. The value of D increases as r increases from 1.058 Å and passes through second higher maximum at $r = 5 \times 0.529$ Å and finally approaches zero as r tends to infinity. Thus for 2s-electron (a) D = 0 at r = 0, 1.058 Å, 0.529 Å and ∞ (b) lower maximum lies at r = 0.529 Å (c) higher maximum lies at r = 2.645 Å. Thus the distance for the maximum probability of 2s electron is equal to 2.645 Å (higher maximum) while the distance for the probability of 2s electron *very close* to the nucleus is equal to 0.529 Å (lower maximum). At an intermediate distance (= 1.058 Å) there is a surface at which the probability of finding the electron is zero. This implies that within the spherical orbital of a 2s electron, there is a spherical shell in which the probability of finding the electron is almost zero. This is called a **node** or **nodal plane** or **nodal surface.** Thus the surfaces at which the probability of finding the electron is zero are called **nodes**. *For any orbital having principal quantum number, n there are always (n–1) nodes or nodal planes* (neglecting the node of infinity). For example 1*s* orbital (n = 1) has *no* (1–1 = 0) nodal plane while in 3*s* orbital (n = 3) there are (3–1) = 2 nodal planes as is evident from the curves for 1*s* and 3*s* orbitals.

(*vi*) On comparing the curves for 2*s* and 2*p* electrons, we find that :

(*a*) the distance for the maximum probability for 2*s* electron (= 2.645 Å) is *slightly larger* than that for the 2*p* electron (= 2.116 Å) but much larger than that for 1*s* electron (= 0.529 Å). Thus 2*s* and 2*p* electrons or orbitals have slightly different energies and both these orbitals have greater energy than 1*s* orbital.

(*b*) The lower maximum lying at 0.529 Å indicates that 2*s* electron spends more of its time close to the nucleus than does the 2*p* electron. In other words this means that 2*s* electron is bound more tightly to the nucleus than the 2*p* electron or 2*s* electron is more stable than the 2*p* electron, *i.e.* 2*s* electron has lower energy than 2*p*-electron (2*s* < 2*p*), although both the electrons are of the same main energy level.

2. Angular Probability Distribution Curves—Space Orientations and Shapes or Boundary Surfaces of Orbitals.

These curves tell how the probability of finding the electron varies with its direction from the nucleus without any reference to its distance from the nucleus. These curves show how the various orbitals of a given sub-shell are disposed along *x, y* and *z* axes. The shapes, pace orientation, electron density etc. of orbitals have already been discussed under the study of *Magnetic Quantum Number*.

Questions with Answers

Q.1 Arrange the orbitals with the following quantum numbers in the increasing order of their energy.

(*a*) $n = 3, l = 0, m = 0, s = +\frac{1}{2}$

(*b*) $n = 3, l = 1, m = 1, s = +\frac{1}{2}$

(*c*) $n = 3, l = 2, m = 1, s = +\frac{1}{2}$

(*d*) $n = 4, l = 0, m = 0, s = +\frac{1}{2}$

Ans. Oribtals represented by the given quantum numbers are : (*a*) 3*s* (*b*) 3*p* (*c*) 3*d* (*d*) 4*s*. According to Aufban principle the energy of 3*s*, 3*p*, 3*d* and 4*s* orbitals increases as :

3*s* < 3*p* < 4*s* < 3*d* **(Ans.)**

Q.2 Write the expression representing the electron probability function (D).

Ans. Electron probability function (D) = $4\pi r^2 dr \psi^2$

Q.3 Determine the number of orbitals in the subshell having n = 4 and l = 3.

Ans. Sub-shell having $n = 4$ and $l = 3$ is $4f$ sub-shell. Obviously $4f$ sub-shell has seven orbitals, since when $l = 3$, $m = 0, \pm 1, \pm 2, \pm 3$ (seven values).

Q.4 Explain why the following sets of quantum numbers are not possible.

	n	*l*	*m*	*s*
(a)	**2**	**2**	**1**	**+ 1/2**
(b)	**1**	**0**	**–1**	**– 1/2**
(c)	**3**	**2**	**3**	**+ 1/2**

Ans. (*a*) $n = 2$ (2*nd* shell) and $l = 2$ (*d*-sub shell) represents $2d$ sub-shell. Since 2*nd* shell does not contain *d* sub-shell, $2d$ sub-shell does not exist. Consequently this set of quantum numbers is not possible.

(*b*) $n = 1$ and $l = 0$ represents $1s$ sub-shell. Since for this sub-shell *m* should have zero ($m = 0$) value and not –1 as given in question, the given set of quantum numbers is not possible.

(*c*) $n = 3$ and $l = 2$ represents $3d$ sub-shell. Since *d* sub-shell should have any of the five values of *m viz.* ± 2, ± 1, 0 and not 3 as given in the question, the given set of quantum numbers is not possible.

Q.5 The set of quantum numbers : $n = 3$, $l = 2$, $m \pm 2$ for an electron in the ground state of an atom (X) with Z = 19 is not possible. Explain.

Ans. $n = 3$ and $l = 2$ represents $3d$ sub shell. Complete E.C. of X atom with Z = 19 is $1s^2$, $2s^22p^6$, $3s^23p^6$, $4s^1$. Since this E.C. does not contain *d* sub-shell, the given set of quantum numbers is not possible.

Q.6 Find out the number of orbitals for which the following set of quantum numbers is possible. $n = 3$, $l = 2$ and $m = +2$.

Ans. $n = 3$ and $l = 2$ represents $3d$ sub shell. Since this sub shell has five values of *m* namely +2, –2, +1, –1 and 0, this sub shell has five $3d$ orbitals namely $3d_{xy}$, $3d_{yz}$, $3d_{zx}$, $3d_{x^2-y^2}$ and $3d_{z^2}$. $m = +2$ represents one of the five orbitals given above. Thus there is one orbital ($3d$ orbital) for which $n = 3$, $l = 2$ and $m = +2$.

Q.7 Find out the maximum number of electrons in an atom having quantum numbers as $n = 4$ and $m = +1$.

Ans. $n = 4$ represents 4*th* shell. This shell contains $4s$, $4p$, $4d$ and $4f$ sub shells. The values of *m* for these sub-shells are: for $4s$, $m = 0$ (one $4s$ orbital), for $4p$, $m = 0, +1, -1$ (three $4p$ orbitals), for $4d$, $m + 2, -2, +1, -1, 0$ (five $4d$ orbitals) and for $4f$, $m = +3, -3, +2, -2, +1, -1, 0$ (seven $4f$ orbitals). Thus the orbitals for which $m = +1$ may be $4p_x$, $4d_{xy}$, $4f_{z^3}$. Since each of these orbitals contains two electrons, maximum number of electrons in the atom = 2 + 2 + 2 = **6** **(Ans.)**

Q.8 There are three orbitals having quantum number values as $n = 5$, $l = 0$; $n = 4$, $l = 1$ and $n = 3$, $l = 2$. In which orbital will the electron enter first ?

Ans. The orbitals represented by the given values of quantum numbers are as : $5s$ ($n = 5$, $l = 0$), $4p$ ($n = 4$, $l = 1$) and $3d$ ($n = 3$, $l = 2$) respectively. The energy of these orbitals is in the order : $3d < 4p < 5s$. Since $3d$ orbital has the lowest energy, the electron will enter $3d$ orbital for which $n = 3$ and $l = 2$.

Q.9 Which of the orbitals having the following set of quantum numbers have the same energy? (*a*) $n = 2, l = 1, m = +1$ (*b*) $n = 2, l = 1, m = -1$ (*c*) $n = 2, l = 1, m = 0$ and (*d*) $n = 2, l = 0, m = 0$.

Ans. Orbitals represented by the given sets of quantum numbers are $2p_x$ ($n = 2, l = 1, m = +1$), $2p_y$ ($n = 2, l = 1, m = -1$), $2p_z$ ($n = 2, l = 1, m = 0$), and $2s$ ($n = 2, l = 0, m = 0$). All the three $2p$ orbitals are degenerate. Hence these orbitals have the same energy but $2s$ orbital has different energy.

Q.10 Find out the number of radial nodes in 3*s* and 2*p* orbitals.

Ans. Number of radial nodes in a given orbital $= n - l - 1$. Hence :

No. of radial nodes in $3s$ orbital $= 3 - 0 - 1 = 2$

No. of radial nodes in $2p$ orbital $= 2 - 1 - 1 = 0$

Q.11 Arrange *ns*, *np*, (*n*–1)*d* and (*n*–1)*f* orbitals in the increasing order of their energy.

Ans. The energy of orbitals increases with the increase of their $(n + l)$ value. If two orbitals have the same value of $(n + l)$, the orbital with higher value of n has higher energy. $(n + l)$ values for the given orbitals are given below :

Orbitals :	ns	np	$(n - 1)d$	$(n - 1)f$
$(n + l)$ values :	$n + 0$	$n + 1$	$n-1 + 2$	$n-1 + 3$
	$= n$	$= n + 1$	$= n + 1$	$= n + 2$

It may be seen that np and $(n - 1)d$ orbitals have the some value of $(n + l)$. Now since np orbital has higher value of $n (= n)$ than $(n - 1)d$ orbital ($= n - 1$), np orbital has higher energy than $(n - 1)d$ orbital. Thus the increasing order of energy is as $ns < (n - 1)d < np < (n - 1)f$.

Q.12 What is the atomic number (Z) of an atom having quantum numbers as $n = 3$, $l = 0$ and $m = 0$?

Ans. $n = 3$, $l = 0$ and $m = 0$ represents $3s$ orbital. E.C. of the atom having $3s$ orbital is $1s^2$, $2s^2 2p^6$, $3s^1$ (Z = 11) or $1s^2$, $2s^2 2p^6$, $3s^2$ (Z = 12). The atom having $n = 3$, $l = 0$ and $m = 0$ can have 11 or 12 as its atomic number.

University Questions

1. (*a*) Derive Schrodinger wave equation in three dimensions and what is the justification for this equation.

(*b*) Draw the shape of *d*-orbitals.

(*c*) Place the electrons identified by quantum numbers n and l (*i*) $n = 4, l = 1$ (*ii*) $n = 4, l = 0$ (*iii*) $n = 3, l = 2$ (*iv*) $n = 3, l = 1$ in the order of their decreasing energy. (*Himachal Pradesh 2000*)

2. (*a*) Derive Schrodinger wave equation. How does it help in determining position of electron in an atom?

(*b*) Write a note on Hamiltonian operator. (*Kanpur 2000*)

3. Write a note on "Magnetic quantum number". (*Kumaon 2000*)

4. (*a*) What are quantum numbers? Explain them clearly.

(b) Mention number of unpaired electrons in a subshell of the element whose atomic number is 29.

(c) Give all the four quantum numbers of 15th electron of the atom with atomic number 19.

(d) Designate the orbital for which : (i) $n = 3, l = 1, m = 0$ (ii) $n = 2, l = 1, m = 0$.

(e) Write down the general form of Schrodinger's wave equation and define each of its term.

(f) Explain Heisenberg uncertainty principle. How does its influence the concept of the electron?

(g) A moving cricket ball, weighing 200g, is to be located within 0.2 Å. What is the uncertainty in its velocity ? Comment on your result. Given : $h = 6.626 \times 10^{-34}$ Js. *(Lucknow 2001)*

5. (a) Write a note on significance of Ψ.

(b) What is the significance of four quantum numbers ? *(Delhi 2002)*

6. (a) Define orbital.

(b) What are n, l and m values for $2p_x$ and $3p_y$ electrons ?

(c) Explain briefly time independent Schrodinger wave equation. *(Delhi 2003)*

7. (a) Calculate screening constant for the elements with atomic numbers 11 and 17.

(b) State Paulies' exclusion principle. *(Nagpur 2003)*

8. *1st, 2nd, 3rd* and *4th* shells of an atom may certain maximum 2, 8, 18 and 32 electrons respectively. Explain this arrangement in terms of quantum numbers. *(Rohilkhand 2003)*

9. (a) Which of the following orbitals are not possible ? 1p, 2s, 2p, 2d and 3f.

(b) What are the values of question number of 10th electron of K-atom ?

(c) What are n, l and m values for $3p_x$ electron ? *(Meerut 2006)*

10. (a) Which of the following orbitals is nearest to the nucleus ? 4f, 5d, 6s and 6p ?

(b) Name the rule according to which the V.S.E.C. of C-atom is $2s^2\ 2p_x^1\ 2p_y^1\ 2p_z^0$ and not $2s^2\ 2p_x^2\ 2p_y^0\ 2p_z^0$. *(H.N. Bahuguna 2006)*

11. What we the values of n, m and l for $2p_x^1$ electron ? *(Meerut 2007)*

12. (a) Arrange the following orbitals in the increasing order of their energy : 5p, 3s, 4d, 6s.

(b) What are the values of n and l for 5d orbital ? *(Purvanchal 2007)*

13. (a) Write a note on Heisenberg's uncertainty principle.

(b) What are quantum numbers ? Explain their significance. *(H.N. Bahuguna 2007)*

14. (a) What is Schrödinger wave equation ? Give its derivation and significance of ψ and ψ^2.

(b) Give a brief account of wave nature of electron. Derive de Broglie equation.

(c) Write short notes on : (i) Heisenber's incerainty principle (ii) Paulie's exclusion principle. *(Meerut 2008)*

15. Discuss Heisenberg's uncertainty principle in detail alongwith its experimental verification. *(Agra 2008)*

16. (a) Explain Paulie's exclusion principle.

(b) Derive de-Broglie's equation for a particle of mass m and moving with a velocity of v. *(H.N. Bahuguna 2008)*

17. (a) Explain the concept of atomic orbital and explain the differences between orbit and orbital with suitable examples.

(b) What are quantum numbers ? Explain Hind's rule of maximum multiplicity. *(Gurukul Kangri 2008)*

18. (a) Explain quantum numbers and their significance in characterising an electron in an atom.

(b) What is Heisenberg's uncertainty principle ? *(Meerut 2009)*

CHAPTER 3

Modern Periodic Table and Electronic Configuration of Atoms

Periodic Table

Mendeleef's Periodic Law and Mendeleef's Periodic Table

Attempts were made to classify the elements in a number of ways. Most noteworthy of these is that made by Russian chemist, Mendeleef (*men-de-h-lay-eff*) who gave a law known as *Mendeleef's Periodic Law* which states as :

The properties of elements are a periodic function of their atomic weights, i.e., if the elements are arranged in the increasing order of their atomic weights, the properties of the elements (i.e., similar elements) are repeated after definite regular intervals or periods.

Working on this law, Mendeleef arranged the elements in the increasing order of their atomic weights in the form of a table which is known as *Mendeleef's Periodic Table* after his name.

In this table the elements are arranged in *groups* (*or columns*) and *periods* (*or rows*).

Defects of Mendeleef's Periodic Table

Mendeleef's periodic table suffers from the following defects :

1. Position of hydrogen. Hydrogen resembles both the alkali metals and the halogens. Its position in the periodic table is, therefore, anomalous.

2. Position of lanthanides and actinides. A group of 15 elements (At. No. 57 to 71) which is called rare earths or lanthanides does not find its proper place in the table and has been placed at one place in group III and period 6. Similarly, another group of 15 elements (At. No. 89 to 103) called actinides does not find its proper place and has been put at one place in group III and period 6.

3. Similar elements are separated while dissimilar elements are placed in the same group. Certain elements which possess similar properties are separated in the periodic table, as, for example, copper and mercury; silver and thalium; and barium and lead; while many dissimilar elements have been grouped together. For example Cu, Ag, and Au are grouped along with the alkali metals though there is little resemblance between them. Similarly manganese is grouped with the halogens.

4. Existence of four anomalous pairs of elements. The order of increasing atomic weight has been ignored in case of four pairs of elements in order to place them in a position justified by their properties. *Thus elements of higher atomic weights precede those of lower atomic weight at four places as shown below* :

(*a*) Ar (Z = 18, at. wt. = 40) precedes K (Z = 19, at. wt. = 39.0)

(*b*) Co (Z = 27, at. wt. = 59.9) precedes Ni (Z = 28, at. wt. = 58.6)

(*c*) Te (Z = 52, at. wt. = 127.6) precedes I (Z = 53, at. wt. = 126.9)

(*d*) Th (Z = 90, at. wt. = 232.12) precedes Pa (Z = 91, at. wt. = 231)

5. Position of isotopes. If the elements are arranged in the order of their increasing atomic weights, it is not possible to accommodate large number of *isotopes* in the periodic table.

6. Group does not represent valency. Excepting osmium, elements placed in group eight do not show a valency of 8. Also the elements lying in the middle of long periods show two or more valencies *e.g.* Cr, Mn etc.

Mosley's Modern Periodic Law

Mosley showed in 1911 that, since the physical and chemical properties of an element depend on the number of electrons and their arrangement in different orbitals of the atom, the classification of the elements should be based on the number of these electrons (*i.e.* atomic number) and their arrangement in different orbitals. This idea led Mosley to predict that most of the defects 0f Mendeleef's periodic table disappear, if the basis of classification of elements is changed to atomic number in place of atomic weight. Accordingly, Mosley put forward Modern Periodic Law which states as follows :

The properties of elements are a periodic function of their atomic numbers, i.e. if the elements are arranged in the increasing order of their atomic numbers, the properties of the elements (i.e. similar elements) are repeated after definite regular intervals or periods.

With the replacement of atomic weight by atomic number as the basis of classification of elements, many of the irregularities in the Mendeleef's table disappear as shown below

1. Position of hydrogen. The dual role of hydrogen is explained by the fact that it has one electron in its outer orbit. It has equal tendency of gaining or losing one electron for assuming a stable configuration. When it loses one electron to give H^+, it resembles alkali metals (which give Li^+, Na^+, K^+, etc. ions) while when it gains one electron to give H^-, it resembles halogens (which give Cl^-. Br^- etc.).

2. Anomalous pairs of elements. This anomally disappears altogether and the pairs Ar—K, Co—Ni, Te—I and Th—Pa are found arranged in the table in the order in increasing atomic numbers as shown below :

Pairs of elements →	*Ar*	*K*	*Co*	*Ni*	*Th*	*I*	*Ta*	*Pa*
Atomic numbers →	*18*	*19*	*27*	*28*	*52*	*53*	*90*	*91*
Atomic weights →	*40*	*39*	*59.9*	*58.6*	*127.6*	*126.9*	*232.12*	*231*

3. Position of rare earths. The arrangement of extranuclear electrons in all the rare earth elements can be represented as 2, 8, 18 (18 + x), 9, 2, where x

varies from 0 (for La) to 14 (for Lu). With this general arrangement of electrons, all of them possess the same valency and similar chemical properties. This justifies their grouping at the same place.

4. Position of isotopes. Since isotopes of the same element possess the same atomic number, all of them should occupy one and the same place in the periodic table.

5. Justification for dissimilar elements being placed together. The length of the periods is determined by arrangement of electrons in different orbits. The end of every period results from the completion of the last orbit (last number is always an inert gas). Different periods carry 2, 8, 18 and 32 elements.

When 18 elements are to be distributed among 8 groups; groups 1—7 get two elements each while group 0 gets only one. The three elements which cannot be arranged elsewhere are placed in a special group VIII. This lack of space is enough *justification for group VIII.*

Out of the two elements which every long period adds to a group, one resembles the typical element, the other does not. This gives rise to the formation of sub-groups. This explains *why dissimilar elements have been grouped together.*

Extended or Long Form of Periodic Table : Modern Periodic Table

In order to remove the defects of Mendeleef's periodic table, a number of tables have been suggested from time to time by various workers for classifying the elements in the increasing order of their atomic numbers. Out of the various tables the *Extended Long Form of Periodic Table* is the most simple and is widely accepted.

This table is shown on the next page and is also called *Bohr's Periodic Table,* since it is based on Bohr's scheme of the classification of the elements into four types depending on the number of incomplete shells of electrons in the atom. This table was proposed by Rang (1893) and then modified by Werner (1905) and extended by Bury (1921).

Groups

The vertical columns shown in the periodic table are called *groups* or *families* or simply *columns.*

(*a*) There are *nine groups* in all including VIII group consisting of three triads (Fe, Co, Ni; Ru, Rh, Pd; Os, Ir, Pt.) and zero group of inert gases. Groups I to VII are sub-divided into sub-groups A and B. Thus there are 18 vertical columns which are: IA, IIA, III A, IVA, VA, VIA, VII A, zero, IB, II B, III B, IV B, VB, VI B, VII B and three columns of group VIII.

(*b*) Elements of groups IA, IIA, IIIA, IVA, VA, VIA and VII A have their *outermost shells incomplete while each of their inner shells is complete.* These elements are called *normal or representative elements.* These elements consist of some metals, all non-metals and metalloids.

(*c*) Elements of groups I B, II B, III B (only Sc, Y, La and Ac) IVB, VB, VIB, VIIB and VIII have their *two outermost shells incomplete.* These are called *transition elements.* These elements are placed in the middle of the table. All these elements are *metals.*

Table: Extended or Long Form of Periodic Table (Modern or Bohr's Periodic Table). The number given below the symbol indicates the atomic number of the elements.

Metals (Active metals and Transition metals) | Non-metals Metalloids and Noble gases

Active metals: IA, IIA | Transition metals: IIIB–IIB

Periods and No of Elements	IA	IIA	IIIB	IVB	VB	VIB	VIIB	VIIB			IB	IIB	IIIA	IVA	VA	VIA	VIIA	Zero
1 (2) (H_1 — He_2)	H 1																	He 2
2 (8) (Li_3 — Ne_{10})	Li 3	Be 4											B 5	C 6	N 7	O 8	F 9	Ne 10
3 (8) (No_{11} — Ar_{18})	Na 11	Mg 12											Al 13	Si 14	P 15	S 16	Cl 17	Ar 18
4 (18) (K_{19} — Kr_{36})	K 19	Ca 20	Sc 21	Ti 22	V 23	Cr 24	Mn 25	Fe 26	Co 27	Ni 28	Cu 29	Zn 30	Ga 31	Ge 32	As 33	Se 34	Br 35	Kr 36
5 (18) (Rb_{37} — Xe_{54})	Rb 37	Sr 38	Y 39	Zr 40	Nb 41	Mo 42	Tc 43	Ru 44	Rh 45	Pd 46	Ag 47	Cd 48	In 49	Sn 50	Sb 51	Te 52	I 53	Xe 54
6 (32) (Cs_{55} — Rn_{86})	Cs 55	Sr 56	La 57	Hf 72	Ta 73	W 74	Re 75	Os 76	Ir 77	Pt 78	Au 79	Hg 80	Tl 81	Pb 82	Bi 83	Po 84	At 85	Rn 86
7 (19) (Fr_{87} — Lw_{103})	Fr 87	Ra 88	Ac 89	Ku 104	Ha 105													

Inner-transition Elements (Group III B, Period 6 and 7)

Ce 58	Pr 59	Nd 60	Pm 61	Sm 62	Eu 63	Gd 64	Tb 65	Dy 66	Ho 67	Er 68	Tm 69	Yb 70	Lu 71	Lanthanides
Th 90	Pa 91	U 92	Np 93	Pu 94	Am 95	Cm 96	Bk 97	Cf 98	Es 99	Fm 100	Md 101	No 102	Lw 103	Actinides

(*d*) Elements of group zero have *all their shells completely filled*. These elements are called *noble gases*. These are placed at the extreme right of the table.

(*e*) Two groups of 14 elements lying in group III B [Ce(Z = 58) to Lu (Z = 71) and Th (Z = 90) to Lw (Z = 103)] have their *three outermost shells incomplete*. These are called *lanthanides* and *actinides* respectively and have been placed at the bottom of the table.

Periods

The horizontal rows shown in the periodic table are called *periods* or simply *rows*. There are *seven periods* in the table.

(*a*) *1st* period consists of 2 elements which are H (Z = 1) and He (Z = 2).

(*b*) *2nd* and *3rd* periods have 8 elements each.

2nd period → Li (Z = 3) to Ne (Z = 10)

3rd period → Na (Z = 11) to Ar (Z = 18)

Both these periods are called *short periods*. The elements of *2nd* and *3rd* short periods were called *typical elements* by Mendeleef as these elements were considered to be the true representatives of their respective groups. But, now since it has been found that the elements of *2nd* period resemble more with the elements diagonally related to them than with the elements of their own groups, the elements of *2nd* period are not considered as true representatives of their respective groups. Now a days the elements of only *3rd period namely Na, Mg, Al, Si, P, S, Cl are called typical elements*. These elements have the following characteristics :

(*i*) They show their group characteristics as valency, chemical behaviour etc.

(*ii*) They form a connecting link between the first member of their respective groups and with the elements of B sub-groups. Hence these elements serve as a bridge between the two sub-groups. However, they show very little resemblance with elements of B sub-groups. For example Na and Cl which are the typical elements of I and VII groups respectively, resemble more closely with the elements of their own sub-groups and show a very little resemblance with the elements of B sub-groups.

(*c*) *4th* and *5th* periods have 18 elements each while *6th* period has 32 elements as shown below :

4th period → K (Z=19) to Kr (Z=36)

5th period → Rb (Z=37) to Xe(Z=54)

6th period → Cs (Z=55) to Rn (Z=86)

All these three periods are called *long periods*. *6th* period also includes 14 rare earths or lanthanides [Ce (Z=58) to Lu (Z=71)].

(*d*) *7th* period is an incomplete period and at present it consists of 19 elements which are Fr (Z=87) to Ha (Z= 105). All the elements of this period are radioactive. This period also includes 14 actinides [Th (Z=90) to Lw (Z = 103)]. The elements after U (Z=92) are called *transuranic elements*. These elements are the result of atomic research and hence are synthetic elements.

Thus we see that in this table the long periods have keen extended and short periods have been broken so as to accommodate the transitional elements in the long periods at their proper places between groups II A and III A.

General Characteristics of Groups

1. Number of valency electrons. On moving down a given group the number of valency electrons does not change, *i.e.* remains the same.

2. Valency. The valencies of all the elements of the same group are the same. The valency of an element with respect to oxygen is equal to its group number.

3. Properties of elements. All the elements of a given group possess very similar physical and chemical properties. There is a regular gradation in their properties when we move from top to bottom in a group. For example :

(*a*) The alkali metals (group IA) resemble each other and their base-forming tendency increases from Li to Cs.

(*b*) The reactivity of halogens (group VII A) decreases as we pass from F to I.

4. Size of atoms. *Size of atoms increases on descending a group.* For example in group IA, atomic size increases from Li to Cs. Thus :

$$Li < Na < K < Rb < Cs$$

5. Metallic character. *The metallic character of the elements increases in moving from top to bottom in a group.* This is particularly apparent in groups IVA, VA and VIA, which begin with non-metals (namely C, N and O respectively), and end with metals (namely Pb, Bi and Po respectively). For example, in group VA, N and P are non-metals, As and Sb are metalloids and Bi is a typical metal. Thus the metallic character of these elements increases from N to Bi as shown below :

Elements of group VA : $\underbrace{N, P,}_{\text{Non-metals}}$ $\underbrace{As, Sb,}_{\text{Metalloids}}$ $\underbrace{Bi}_{\text{Metal}}$

Metallic character : *—Metallic character increasing→*

It is because of a gradual increase of the metallic character of the elements from top to bottom that the *oxides of the elements become more and more basic in the same direction*. For example :

Oxides of the elements of group VA : $\underbrace{N_2O_3, P_2O_5,}_{\text{Acidic}}$ $\underbrace{As_2O_3, Sb_2O_3,}_{\text{Amphoteric}}$ $\underbrace{Bi_2O_3}_{\text{Basic}}$

Basic character : *—Basic character increasing→*

6. Number of electron shells. *In going down a group the number of electron shells increases by one at each step and ultimately becomes equal to the number of the period to which the element belongs* as shown below for the elements of Group IA.

Elements	*Electronic configuration*	*No. of shells*
Li	*2, 1*	*2*
Na	*2, 8, 1*	*3*

K	*2, 8, 8, 1*	4
Rb	*2, 8, 18, 8, 1*	5
Cs	*2, 8, 18, 18, 8, 1*	6
Fr	*2, 8, 18, 32, 18, 8, 1*	7

General Characteristics of Periods

1. Number of valency electrons. Number of valency electrons increases from 1 to 8 when we proceed from left to right in a period.

2. Valency. The valency of the elements with respect to (w.r.t.) hydrogen in each short period increases from 1 to 4 and then falls to one while the same with respect to oxygen increases from 1 to 7 as shown below for the elements of *2nd* and *3rd* period :

Elements of 2nd period	:	*Li*	*Be*	*B*	*C*	*N*	*O*	*F*
Hydrides of the elements	:	LiH	BeH_2	BH_3	CH_4	NH_3	H_2O	HF
Valency of the elements w.r.t hydrogen		1	2	3	4	3	2	1
Elements of 3rd period	:	*Na*	*Mg*	*Al*	*Si*	*P*	*S*	*Cl*
Oxides of the elements	:	Na_2O	MgO	Al_2O_3	SiO_2	P_2O_5	SO_3	Cl_2O_7
Valency of the elements w.r.t. oxygen	:	1	2	3	4	5	6	7

3. Size of atoms. *Size of atoms decreases from left to right in a period.* Thus alkali metals have the largest size while the halogens have the smallest size.

4. Properties of elements. The properties of the elements of a given period differ considerably but the elements in the two adjacent periods show marked similarity between them. For example, when we consider the elements of *2nd* and *3rd* periods, we find that Na resembles Li, Mg resembles Be, Al resembles B, Si resembles C, P resembles N, S resembles O, Cl resembles F and Ar resembles Ne.

5. Metallic character. On moving from left to right in a period the metallic character of the elements decreases. For example in *3rd* period, Na, Mg and Al are *metals* while Si, P, S and Cl are *non-metals* as shown below :

Elements of 3rd period :		Na, Mg, Al	Si, P, S, Cl
		Metals	*Non-metals*
Metallic character	:	*—Metallic character decreasing→*	

It is because of the gradual decrease of the metallic character from left to right that the oxides of the elements become less and less basic in the same direction. For example :

Oxides of the elements of 3rd period	:	Na_2O	MgO	Al_2O_3	SiO_2	P_2O_5	SO_3	Cl_2O_7
Basic character	:	*Strongly basic*	*Basic*	*Amphoteric*	*Feebly acidic*	*Acidic*	*More acidic*	*Most acidic*

—Basic character decreasing→

6. Number of shells. *In going from left to right in a period the number of electron shells remains the same and the number of a period corresponds to the number of the shells found in the elements of that period, e.g.* all the elements of *2nd* period have the electrons only in first *two shells* as shown below :

Elements of 2nd period	:	*Li*	*Be*	*B*	*C*	*N*	*O*	*F*	*Ne*
Atomic number	:	*3*	*4*	*5*	*6*	*7*	*8*	*9*	*10*
Electronic configuration	:	*2, 1*	*2, 2*	*2, 3*	*2, 4*	*2, 5*	*2, 6*	*2, 7*	*2, 8*
No. of shells	:	*2*	*2*	*2*	*2*	*2*	*2*	*2*	*2*

7. Diagonal relationship. It is one of the important features of the earlier elements of *2nd* period of the long form of the periodic table (*e.g.* Li, Be and B) that they resemble more with the elements lying at their right hand side in the *3rd* period (*i.e.* diagonally opposite elements) than with other members of their own sub-groups. Thus Li resembles more in its properties with Mg which lies at its lower right in the *3rd* period than with other members of its own sub-group IA. Similarly Be shows similarities with Al and B resembles Si. The two elements showing similarities in properties are called *diagonal neighbours* or *diagonal pair* and the *across-the-street* similarity in properties existing between the two elements of a diagonal pair is known as *diagonal relationship.* Thus :

Diagonal relationship is the resemblance of the properties of the elements of 2nd period with their diagonally opposite members lying in 3rd period.

Three diagonal pairs *viz.* Li—Mg, Be—Al and B—Si are shown below :

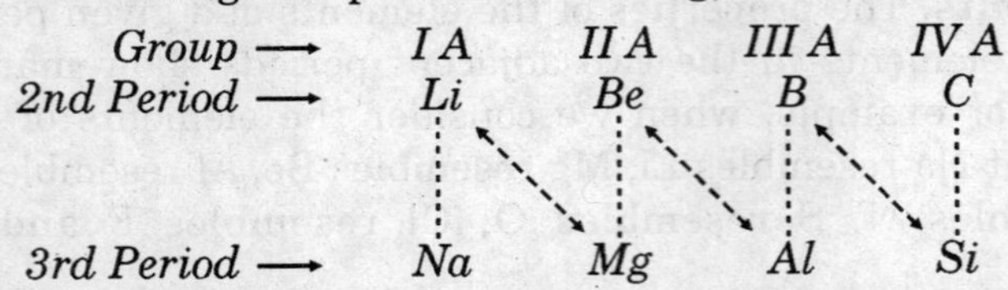

The two elements of a diagonal pair are also called *bridge elements*, since each diagonal pair forms a sort of bridge between the periods to which the members of a given diagonal pair belong.

The following are the important examples of diagonal relationship found in the periodic table.

(*i*) *Li—Mg diagonal relationship.* It will be discussed in Chapter No. 20

(*ii*) *Be—Al diagonal relationship.* It will be discussed in Chapter No. 22

(*iii*) *B—Si diagonal relationship*. It will be discussed in Chapter No. 25

Cause of diagonal relationship. Why the two members of a given diagonal pair show diagonal relationship (*i.e.* similarities in properties) has been explained on the basis of the following :

(*a*) *Applications of electronegativity.* For this see chapter 4 namely *"Periodic Properties"*

(*b*) *Application of polarisation of ions.* For this see chapter 7 namely *"Chemical Bonding (Lewis Theory)"*

Merits of Long Form of Periodic Table over Mendeleef's Periodic Table.

The long form of periodic table has a number of merits over the Mendeleef's periodic table in the following respects :

(1) The classification of the elements is based on a more fundamental property *viz.*, atomic number.

(2) It relates the position of an element to its electronic configuration. Thus each group contains elements with similar electronic configuration and hence similar properties. For example, all the alkali metals have similar valence-shell electronic configuration *viz.* ns^1 configuration and hence have similar properties.

Alkali metals	*Atomic number*	*Complete electronic configuration*	*Valence-shell configuration*
Li	*3*	*2, 1*	$2s^1$
Na	*11*	*2, 8, 1*	$3s^1$
K	*19*	*2, 8, 8, 1*	$4s^1$
Rb	*37*	*2, 8, 18, 8, 1*	$5s^1$
Cs	*55*	*2, 8, 18, 18, 8, 1*	$6s^1$
Fr	*87*	*2, 8, 18, 32, 18, 8, 1*	$7s^1$

(3) It explains the similarities and variations in the properties of the elements in terms of their electronic configurations and brings out clearly the trends in chemical properties across the long periods.

(4) The inert gases having completely filled electron shells have been placed at the end of each period. Such a location of the inert gases represents a logical completion of each period.

(5) In this form of the periodic table, the elements of the two sub-groups have been placed separately and thus dissimilar elements do not fall together.

(6) It provides a clear demarcation of different types of the elements like active metals, transition metals, non-metals, metalloids, inert gases, lanthanides and actinides. The elements of group IA and IIA are active metals and are located at the extreme left of the table. The transition metals are found in the middle of the table. The elements lying to the right of a dark line shown in the long form of periodic table in the form of ladder (i.e. steps) are noble gases, metalloids and non-metals while those lying to the left of the line are metals (*active metals and transition metals*).

(7) It is easier to remember, understand and reproduce.

Defects of Long Form of Periodic Table

Although the long form of the periodic table is superior to Mandeleef's periodic table in many respects, it retains some of the defects as such. For example :

(*i*) The problem of the position of hydrogen still remains unsolved.

(*ii*) Like the Mendeleef's table, it fails to accommodate the lathanides and actinides in the main body of the table.

(*iii*) The arrangement is unable to reflect the electronic configuration of many elements.

Electronic Configuration of Atoms

What is Electronic Configuration ?

The distribution of electrons in various shells, sub-shells and orbitals of an atom is called its *complete electronic configuration* or simply *electronic configuration*.

What is Valence-shell Configuration ?

The distribution of electrons in various sub-shells and orbitals of the valence-shell (or outermost shell) is called *valence-shell configuration*.

Representation of Electronic Configuration

The electronic configuration of an atom is written in terms of nl^x notation where l indicates the sub-shell as given below :

Value of l	:	0	1	2	3	....
Sub-shell	:	*s*	*p*	*d*	*f*	....

In the notation nl^x, super-script x written at the top of l indicates the number of electrons present in the sub-shell given by l, while n written to the left of l^x indicates the number of the shell to which the sub-shell denoted by l belongs. For example $3p^2$ in which $n = 3$, $x = 2$ and $l = 1$ (p sub-shell) indicates that two electrons ($x = 2$) are present in p sub-shell ($l = 1$) which belongs to *3rd* main shell ($n = 3$). (Fig. 3.1).

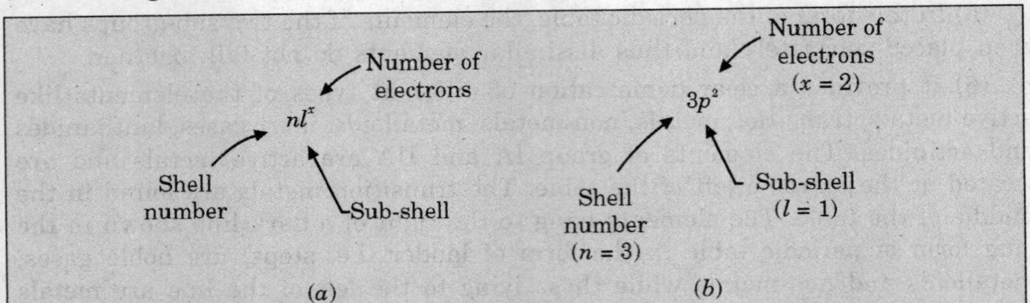

Fig. 3.1. (*a*) Representation of electronic configuration of an atom. (*b*) Representation of electronic configuration of an atom containing two electrons ($x = 2$) in p sub-sell ($l = 1$) belonging to *3rd* shell ($n = 3$).

Rules for Writing Electronic Configuration of a Given Atom

The rules used for writing the electronic configurations of the atoms of the elements in the ground state are :

1. **Maximum number of electrons in a shell.** According to *Bohr-Bury scheme* the maximum number of electrons that a shell can hold is equal to $2n^2$ where n is the number of the shell (*i.e.*, principal quantum number). Thus :

The maximum number of electrons in 1st shell (n = 1) = $2 \times 1^2 = 2$

The maximum number of electrons in 2nd shell (n = 2) = $2 \times 2^2 = 8$

The maximum number of electrons in 3rd shell (n = 3) = $2 \times 3^2 = 18$

The maximum number of electrons in 4th shell (n = 4) = $2 \times 4^2 = 32$

2. **Maximum number of electrons in a sub-shell.** We have seen during the study of azimuthal quantum number that the maximum number of electrons in a sub-shell is equal to $2(2l + 1)$ where l = 0, 1, 2 or 3 for *s, p, d* or *f* sub-shells respectively. Thus :

Value of l	:	0	1	2	3
Sub-shell	:	*s*	*p*	*d*	*f*
Max. number of electrons = $2(2l + 1)$	:	2	6	10	14

3. **Aufbau principle.** Aufbau is a German word which means *build up* or *construction*. It is for this reason that the principle is also often called *building up principle* or *construction principle*. It is pronounced as *of bow*. This principle gives us a sequence in which various orbitals are filled with electrons. This principle states as follows :

The orbitals are filled up with electrons in the increasing order of their energy.

The principle can also be restated in any of the following forms :

(*i*) *The orbitals of minimum energy are filled up first with electrons and then the orbitals of higher energy start to fill.*

(*ii*) *Electrons occupy the orbitals of minimum energy first and then they occupy the orbitals of higher enegry.*

(*iii*) *Orbitals with lowest energy are better seats for electrons and are, therefore, occupied first.*

The energy of various orbitals of an atom *a multi-electron atom* increases as :

$$1s < 2s < 2p < 3s < 3p < 4s < 3d < 4p < 5s < 4d < 5p < 6s$$
$$< 4f < 5d < 6p < 7s < 5f < 6d < 7p < 8s \ .....$$

Thus the sequence in which various orbitals are filled up with electrons is the same as the relative order of the energy of orbitals. Thus the orbitals will be filled up with electrons in the following sequence :

——— *Orbitals being filled* ———→
1*s*, 2*s*, 2*p*, 3*s*, 3*p*, 4*s*, 3*d*, 4*p*, 5*s*, 4*d*, 5*p*, 6*s*, 4*f*, 5*d*, 6*p*, 7*s*, 5*f*, 6*d*, ...

The sequence given above is represented diagramatically in Fig. 3.2.

Limitations (Deviations) of Aufbau principle. There are some elements in which the extra electron, while entering the orbital, does not obey Aufbau principle as shown by the following examples :

(i) Configuration of La (Z = 57) atom. We know that, according to Afbau principle, the configuration of Ba (Z = 56) atom is $[Xe]_{54}\ 6s^2$ where $[Xe]_{54}$ indicates the configuration of Xe (Z = 54) atom which is 2, 8, 18. 18, 8.

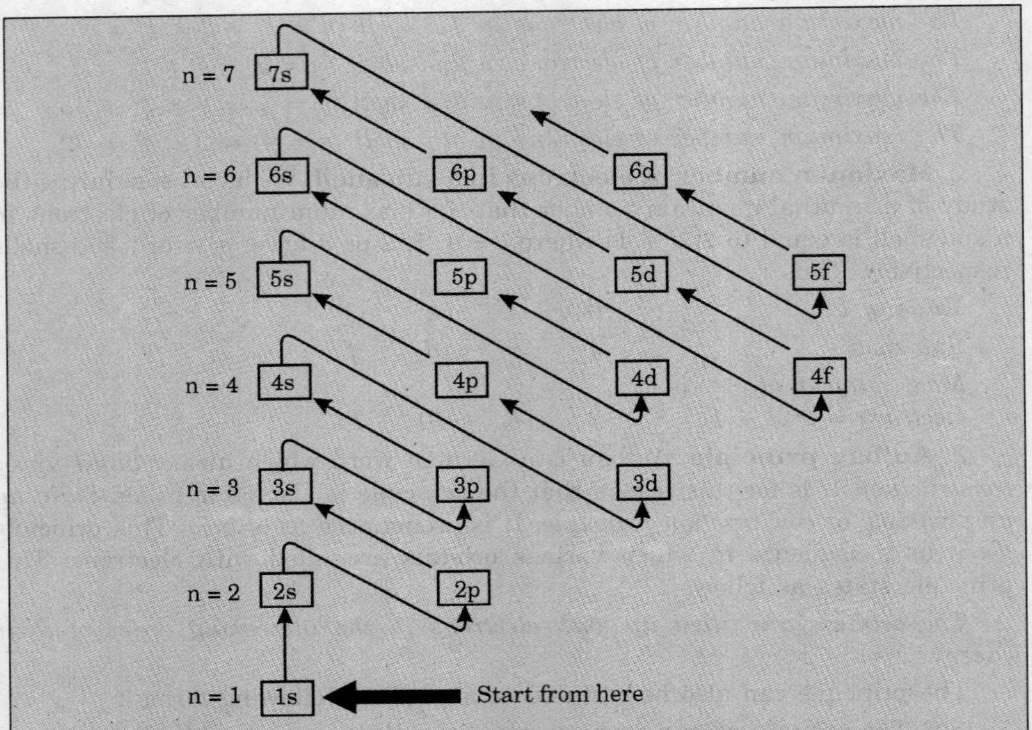

Fig. 3.2. The sequence in which various orbitals are filled up with electrons.

Now after 6s orbital has been completely filled at Ba, the next electron, according to Aufbau principle, must enter $4f$ orbital in La atom to give $[Xe]_{54}4f^1 5d^\circ 6s^2$ configuration to La atom; but it has been observed that the extra electron enters $5d$ orbital (instead of $4f$ orbital) to give $[Xe]_{54}4f^\circ 5d^1 6s^2$ configuration to La atom as shown below :

La (Z = 57) $\rightarrow$ $1s^2$, $2s^2$, $2p^6$, $3s^2$, $3p^6$, $4s^2$, $3d^{10}$, $4p^6$, $5s^2$, $4d^{10}$, $5p^6$, $6s^2$, $5d^1$.

$\rightarrow$ $1s^2$, $2s^2p^6$, $3s^2p^6d^{10}$, $4s^2p^6d^{10}$, $5s^2p^6d^1$, $6s^2$

$\rightarrow$ 2, 8, 18, 18, 9, 2

$\rightarrow$ $[Xe]_{54}$ $4f^\circ 5d^1 6s^2$

Why the extra electron goes to $5d$ orbital, instead of $4f$ orbital in La atom has been explained on the basis that at La, $5d$ and $4f$ orbitals are supposed to have almost the same energy and hence the extra electron is free to occupy any of these two orbitals.

(ii) Configuration of Ac (Z = 89) atom. We know that, according to Aufbau principle, the configuration of Ra (Z = 88) atom is $[Rn]_{86}$ $7s^2$ where $[Rn]_{86}$ is the configuration of Rn (Z = 86) atom which is 2, 8, 18, 32, 18, 8.

Now after $7s$ orbital has been completely filled at Ra, the next electron, according to Aufbau principle, must enter $5f$ orbital in Ac atom; but it has been observed that the extra electron enters 6d orbital (instead of $5f$ orbital) to give $[Rn]_{86}$ $5f^\circ.6d^1.7s^2$ configuration to Ac atom as shown below :

Ac (Z = 89) → $1s, 2s^2, 2p^6, 3s^2, 3p^6, 4s^2, 3d^{10}, 4p^6$
$5s^2, 4d^{10}, 5p^6, 6s^2, 4f^{14}, 5d^{10}, 6p^6, 7s^2, 6d^1$
→ $1s^2, 2s^2p^6, 3s^2p^6d^{10}, 4s^2p^6d^{10}f^{14}, 5s^2p^6d^{10}, 6s^2p^6d^1, 7s^2$
→ 2, 8, 18, 32, 18, 9, 2
→ $[Rn]_{86}\ 5f^{\circ}.6d^1.7s^2$

Why the extra electron goes to 6*d* orbital instead of 5*f* orbital in Ac atom has been explained on the basis that in Ac, 6*d* and 5*f* orbitals are supposed to be almost of the same energy and hence the extra electron is free to go to any of these two orbitals.

(n+l) Rule. The relative order of energy of different orbitals (or the sequence in which various orbitals are filled up with electrons) can also be determined with the help of $(n + l)$ value for a given orbital (n = principal quantum number, l = azimuthal quantum number). This rule is called $(n + l)$ rule. According to this rule :

The orbital having the lowest value of (n + l) has the lowest energy and hence is filled up first with electrons. When two or more orbitals have the same value of (n + l), the orbital with lower value of n is lower in energy and hence is filled up first with electrons.

Thus the order of energies of, for example, 3*d*, 4*p* and 5*s* orbitals for all of which $(n + l) = 5$ is as : 3*d* ($n = 3$) < 4p (n=4) < 5s (n=5), since n for these orbitals is 3, 4 and 5 respectively. Thus orbitals will be filled up with electrons in the following sequence :

Table 3.1. Determination of the relative order of energy of different orbitals on the basis of (n + l) rule.

Orbitals	*n*	l	(*n* + l)	Relative order of energy of different orbitals (Sequence of filling of orbitals)
1*s*	1	0	1	
2*s*	2	0	2	
2*p*	2	1	3	
3*s*	3	0	3	
3*p*	3	1	4	
3*d*	3	2	5	
4*s*	4	0	4	1*s* < 2*s* < 2*p* < 3*s* < 3*p*
4*p*	4	1	5	< 4*s* < 3*d* < 4*p* < 5*s* < 4*d*
4*d*	4	2	6	< 5*p* < 6*s* < 4*f* < 5*d*
4*f*	4	3	7	6*p* < 7*s* < 5*f* < 6*d* < 7*p*.
5*s*	5	0	5	
5*p*	5	1	6	
5*d*	5	2	7	
5*f*	5	3	8	
6*s*	6	0	6	
6*p*	6	1	7	
6*d*	6	2	8	
7*s*	7	0	7	
7*p*	7	1	8	

——— *Orbitals being filled* ———→

3*d* (n = 3), 4*p* (n = 4), 5*s* (n = 5)

The determination of relative order of energy of different orbitals, on the basis of ($n + l$) rule, has been shown in Table 3.1.

4. Pauli's exclusion principle. While filling up the various orbitals with electrons, we should obey Pauli's exclusion principle which imposes some restrictions on the assignment of the values of quantum numbers of electrons residing in an orbital of an atom. This principle, as we have already seen in the chapter: *Structure of the Atom Wave : Mechanical Approach* (Chapter 2), can be stated as :

In the same orbital of an atom, two electrons can have three quantum numbers identical at the most, but the fourth quantum number, namely spin quantum number (s), will definitely be different for both the electron, i.e., for one electron s will be + 1/2 while for the other electron it will be — 1/2. In other words, this means that two electrons residing in the same orbital of an atom have opposite spins, since this state with opposite spins (↑↓)gives lower energy to the system, while the state with parallel spins (↑↑) gives higher energy to the system. According to thermodynamics, a system with lower energy has greater stability. Thus an orbital having two electrons with opposite spins (↑↓) has lower energy and hence is more stable than that containing the two electrons with paired spins (↑↑). $s = +1/2$ indicates the clock-wise direction of the spinning of one electron while $s = -1/2$ stands to indicate the anticlock-wise direction of the spinning of other electron.

Pauli's exclusion principle, as mentioned above, clearly makes the following rules evident. These rules are used in the determination of the electronic configuration of the atoms of the elements.

(*i*) *An orbital like s, p_x, p_y, p_z, d_{xy}, d_{yz} etc can have at the most two electrons and these two electrons must have opposite spins i.e. one electron should have the value of s equal to +1/2 and the other equal to –1/2, so that both the electrons residing in the same orbital may have different sets of four quantum numbers, which is in accordance with Pauli's exclusion principle.*

(*ii*) *Maximum number of electrons that can be accommodated by a shell with principal quantum number, n is equal to $2n^2$.*

5. Hund's rule of maximum multiplicity. On the basis of magnetic measurements, Hund postulated an important empirical rule, popularly known after his name as Hund's rule of maximum multiplicity. This rule states as :

The distribution of electrons in a set of degenerate orbitals like p_x, p_y, p_z set, d_{xy}, d_{yz}, d_{zx}, d_{z^2}, $d_{x^2-y^2}$ set etc., of a given sub-shell takes place in such a way as to give the maximum number of unpaired electrons and these electrons must have the same direction of their spins.

This rule is based on the fact that electrons, being of the same charge, repel each other and hence try to keep farther apart from each other as much as possible. This they do by entering the degenerate orbitals as far as possible.

Their entering the orbitals in this manner minimises the inter-electronic repulsion energy of the system and hence we get a stable system. Thus we can also call this rule as a **rule of minimum energy.**

This rule can also be restated in any of the following forms :

(*i*) *When electrons enter a set of orbitals of a given sub-shell, electrons will pair up only when all the available orbitals have had at least one electron each.*

(*ii*) *The orbitals of a given sub-shell are first filled singly and then the pairing of electrons in each orbital begins.*

(*iii*) *When several orbitals of equal energy (degenerate orbitals) are available, electrons prefer to occupy separate orbitals, rather than getting paired in the same orbital.* Such electrons tend to have the same spins. This principle means that pairing of electrons begins with the introduction of *2nd* electron in *s*-orbital, the *4th* electron in *p*-orbitals, *6th* electron in *d*-orbitals and *8th* electron in *f*-orbitals.

(*iv*) *If the number of electrons is equal to or smaller than the number of available degenerate orbitals, the electrons prefer to occupy the orbitals singly. But when the number of electrons is greater than the number of available degenerate orbitals, the pairing of etectrons takes place.*

(*v*) *None of degenerate orbitals like* p_x, p_y, p_z *or* d_{xy}, d_{yz} *etc., will have two electrons as long as any other orbital is vacant or as long as each one of them has attained at least one electron.*

Illustration of the principle. In order to illustrate the principle, let us consider the electronic configuration of oxygen atom, whose atomic number is 8. Let the eight electrons in this atom be numbered as 1, 2, 3, ... 7, 8. Quite evidently the electron numbered as 1 will enter 1s-orbital (which is the lowest energy orbital), and the electron 2 will also enter 1s-orbital to fill it completely (See Fig. 3.3). These two electrons in 1*s*-orbital will have opposite spins, in accordance with Pauli's exclusion principle (For electron 1, n = 1, l = 0, m = 0

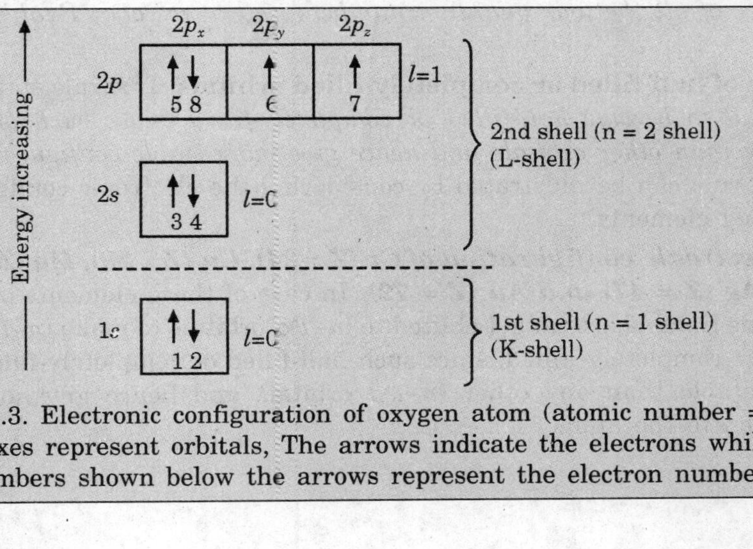

Fig. 3.3. Electronic configuration of oxygen atom (atomic number = 8). The boxes represent orbitals, The arrows indicate the electrons while the numbers shown below the arrows represent the electron number.

and s = +1/2 and for electron 2, n = 1, l = 0, m = 0 and s = –1/2). Electrons 3 and 4 will occupy 2s-orbital to make it completely filled. These electrons will also have opposite spins to obey Pauli's exclusion principle. Now, according to Hund's rule of maximum multiplicity, the degenerate $2p_x$, $2p_y$, and $2p_z$ orbitals will first be filled *singly* with electrons numbered as 5, 6 and 7 and all these three electrons will have the same spins. Electron number 8, which is the last electron will pair with the electron number 5 in $2p_x$ orbital. Thus the complete electronic configuration of oxygen atom, in its ground state, will be as given below :

$1s^{1,2}$, $2s^{3,4}$, $2p_x^{5,8}$ $2p_y^{6}$ $2p_z^{7}$ (Here the numbers used as super-scripts indicate the electron number).

or $1s^2$, $2s^2$, $2p_x^{2}$ $2p_y^{1}$ $2p_z^{1}$ (Here the numbers used as super-scripts indicate the number of electrons present in a given orbital).

The values of the four quantum numbers *viz.*, *n, l, m* and *s* for each electron are given below :

Shell	Orbital	Electron number	*n*	*l*	*m*	*s*
1st	*1s*	*1*	*1*	*0*	*0*	*+1/2*
		2	*1*	*0*	*0*	*–1/2*
2nd	*2s*	*3*	*2*	*0*	*0*	*+1/2*
		4	*2*	*0*	*0*	*–1/2*
	$2p_x$	*5*	*2*	*1*	*0*	*+1/2*
		8	*2*	*1*	*0*	*–1/2*
	$2p_y$	*6*	*2*	*1*	*+1*	*+1/2*
	$2p_z$	*7*	*2*	*1*	*–1*	*+1/2*

These values clearly show that *no two electrons in the same orbital have the same values of all the four quantum numbers and hence obey Pauli's exclusion principle.*

6. Rule of half-filled or completely-filled orbitals. This rule states *that the orbitals tend to become half-filled or completely-filled, since such orbitals are more stable than other orbitals and hence give more stable configuration to the atom.* This rule can be illustrated by considering the electronic configuration of the following elements.

(i) Electronic configuration of Cr (Z = 24), Cu (Z = 29), Mo (Z = 42), Pd (Z = 46), Ag (Z = 47) and Au (Z = 79). In case of these elements one or both the electrons from *ns* orbital get shifted to (*n*–1)*d* orbitals to make (*n*–1)*d* orbitals half-filled or completely- filled, since such half-filled or completely-filled orbitals are more stable than any other (*n*–1)*d* orbitals and hence give more stable configuration to the atom.

Element (Z = Atomic number)	Expected configuration	Actual (observed) configuration	Explanation of observed configuration
Cr (Z = 24)	$[Ar]_{18}\ 4s^2.3d^4$	$[Ar]_{18}\ 4s^1.3d^5$	One electron from 4*s* orbital gets shifted to 3*d* orbital to make 3*d* orbital half-filled ($3d^5$) which is more stable than $3d^4$ orbital.
Cu (Z = 29)	$[Ar]_{18}\ 4s^2.3d^9$	$[Ar]_{18}\ 4s^1.3d^{10}$	One electron from 4*s* orbital gets shifted to 3*d* orbital to make 3*d* orbital completely-filled ($3d^{10}$) which is more stable than $3d^9$ orbital.
Mo (Z = 42)	$[Kr]_{36}\ 5s^2.4d^4$	$[Kr]_{36}\ 5s^1.4d^5$	One electron from 5*s* orbital gets shifted to 4*d* orbital to make 4*d* orbital half-filled ($4d^5$) which is more stable than $4d^4$ orbital.
Pd (Z = 46)	$[Kr]_{36}\ 5s^2.4d^8$	$[Kr]_{36}\ 5s^0.4d^{10}$	Both the electrons from 5*s* orbital get shifted to 4*d* orbital to make 4*d* orbital completely-filled ($4d^{10}$) which is more stable than $4d^8$ orbital.
Ag (Z = 47)	$[Kr]_{36}\ 5s^2.4d^9$	$[Kr]_{36}\ 5s^1.4d^{10}$	One electron from 5*s* orbital gets shifted to 4*d* orbital to make 4*d* orbital completely-filled ($4d^{10}$) which is more stable than $4d^9$ orbital.
Au (Z = 79)	$[Xe]_{54}\ 6s^2.4f^{14}.5d^9$	$[Xe]_{54}\ 6s^1.4f^{14}.5d^{10}$	One electron from 6*s* orbital gets shifted to 5*d* orbital to make 5*d* orbital completely-filled ($5d^{10}$) which is more stable than $5d^9$ orbital.

(ii) Electronic configuration of Nb (Z = 41), Ru (Z = 44), Rh (Z = 45) and Pt (Z = 78). In case of these elements one of the two electrons from *ns* orbital gets shifted to (*n*–1)*d* orbital to make *ns* orbital half-filled, *since half-filled ns orbitals probably seem to be more stable than* ns^2 *orbitals in these cases.*

Element	Expected configuration	Actual (observed) configuration	Explanation of observed configuration
Nb (Z = 41)	$[Kr]_{36}\ 5s^2.4d^3$	$[Kr]_{36}\ 5s^1.4d^4$	One electron from 5*s* orbital gets shifted to 4*d* orbital so that 5*s* orbital becomes half-filled
Ru (Z = 44)	$[Kr]_{36}\ 5s^2.4d^6$	$[Kr]_{36}\ 5s^1.4d^7$	One electron from 5*s* orbital gets shifted to 4*d* orbital so that more stable $5s^1$ configuration is obtained.
Rh (Z = 45)	$[Kr]_{36}\ 5s^2.4d^7$	$[Kr]_{36}\ 5s^1.4d^8$	One electron from 5*s* orbital gets shifted to 4*d* orbital so that more stable $5s^1$ orbital is achieved.
Pt (Z = 78)	$[Xe]_{54}\ 6s^2.4f^{14}.5d^8$	$[Xe]_{54}\ 6s^1.4f^{14}.5d^9$	One electron from 6*s* orbital comes to 5*d* orbital to make 6*s* orbital half-filled ($6s^1$) and hence more stable

(iii) Electronic configuration of Gd (Z = 64) and Cm (Z = 96). In case of these elements one electron from *nf* orbital gets shifted to (*n* + *1*)*d* orbital to make *nf* orbital half-filled and hence more stable.

Element	Expected configuration	Actual (observed) configuration	Explanation of observed configuration
Gd (Z = 64)	$[Xe]_{54}$ $6s^2.4f^8.5d^0$	$[Xe]_{54}$ $6s^2.4f^7.5d^1$	One electron from 4*f* orbital gets shifted to 5*d* obrital so that 4*f* orbital becomes half-filled
Cm (Z = 96)	$[Rn]_{86}$ $7s^2.5f^8.6d^0$	$[Rn]_{86}$ $7s^2.5f^7.6d^1$	One electron from 5*f* orbital goes to 6*d* orbital to make 5*f* orbital half-filled.

Determination of the Electronic Configuration of the Elements According to their Position in a Period of the Periodic Table

Every element in a period has its atomic number (= No. of electrons) one unit more than its preceding element, *i.e.* an additional electron enters, as we move to the right in period. Thus the electronic configuration of the next higher element is obtained by adding one electron to the orbital, which is higher in energy than that which has already been filled up (*Aufbau principle*).

Total number of elements present in a period is equal the total number of electrons to be added in the whole of the period. Thus the orbitals that are filled up with electrons in each period can be shown as :

Period number	No. of elements (Z = atomic number)	Orbitals beings filled	Total number of electrons to be added in a period
1	2 [H(Z = 1) and He(Z = 2)]	1s	2
2	8 [Li(Z = 3) to Ne (Z = 10)]	2s, 2p	2 + 6 = 8
3	8 [Na(Z = 11) to Ar(Z = 18)]	3s, 3p	2 + 6 = 8
4	18[K(Z = 19) to Kr(Z = 36)]	4s, 3d, 4p	2 + 10 + 6 = 18
5	18[Rb(Z = 37) to Xe(Z = 54)]	5s, 4d, 5p	2 + 10 + 6 = 18
6	32[Cs(Z = 55) to Rn(Z = 86)]	6s, 4f, 5d, 6p	2 + 14 + 10 + 6 = 32
7	19[Fr(Z = 87) to Ha(Z = 105)]	7s, 5f, $6d^3$	2 + 14 + 3 = 19

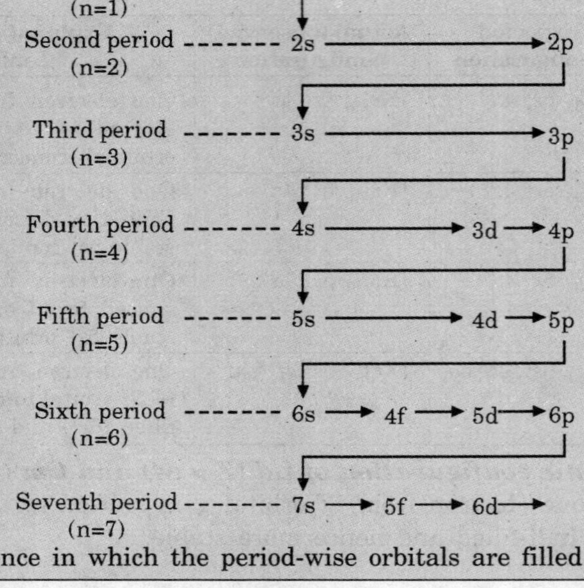

Fig. 3.4. Sequence in which the period-wise orbitals are filled with electrons.

It is obvious from the table given above that the sequence in which the period-wise orbitals are filled with electrons, can be shown as given in Fig. 3.4. This figure makes it evident that at the beginning of each period a new electron enters the *ns* orbital where *n* is the number of the period. For example at the beginning of *7th* period (*i.e.* at Fr with atomic number, Z = 87) a new electron enters 7*s* orbital, which belongs to *7th* energy level (n = 7). Consequently *each period in the periodic table begins with a new principal energy level.*

On the basis of the table given above and the sequence in which the period-wise orbitais are filled with electrons (Fig. 3.4), let us now proceed to determine the electronic configuration of each of the elements given in the periodic table. (Z is the atomic number of the element)

1st period (*1s*). Since this period has two elements *viz.* H(Z = 1) and He(Z = 2), only two electrons are added upto the end of this period and it is 1*s* orbital which is completely filled at the end of this period. One electron goes to 1*s* orbital giving 1*s* configuration to H atom while *2nd* electron also goes to 1*s* orbital giving $1s^2$ configuration to He atom. $1s^2$ configuration is also represented as 2 or $[He]_2$. Thus the configurations of H and He can be written as.

H (Z = 1) → $1s^1$
→ 1
He (Z = 2) → $1s^2$
→ 2
→ $[He]_2$

Here $[He]_2$ indicates the configuration of He atom (Z = 2) which is $1s^2$ or 2.

Thus we see that at the end of this period 1s orbital (or K-shell) is *completely filled.*

2nd period (2*s*, 2*p*). Since this period has 8 elements *viz.* Li (Z = 3) to Ne (Z = 10), only 8 electrons are added upto the end of this period and these electrons enter 2*s* and 2*p* orbitals which are filled completely at the end of the period.

At Li (Z = 3) 2*s* orbital starts to fill up and it is completely filled up at Be (Z = 4). Thus the configurations of Li and Be atoms are :

Li (Z = 3) → $1s^2, 2s^1$
→ 2, 1
→ $[He]_2\ 2s^1$
Be (Z = 4) → $1s^2, 2s^2$
→ 2, 2
→ $[He]_2\ 2s^2$

Now in the next six elements, namely B (Z = 5) to Ne (Z = 10) 2*p* orbital is being progressively filled. Thus at Ne, which is the last element of this period 2*p* orbital is completely filled as shown below :

B (Z = 5) → $1s^2\ 2s^2, 2p^1$
→ $1s^2, 2s^2p^1$
→ 2, 3
→ $[He]_2\ 2s^2p^1$

C (Z = 6) → $1s^2, 2s^2, 2p^2$
→ $1s^2, 2s^2p^2$
→ 2, 4
→ $[He]_2\ 2s^2p^2$

N (Z = 7) → $1s^2, 2s^2, 2p^3$
→ $1s^2, 2s^2p^3$
→ 2, 5
→ $[He]_2\ 2s^2p^3$

O (Z = 8) → $1s^2, 2s^2, 2p^4$
→ $1s^2, 2s^2p^4$
→ 2, 6
→ $[He]_2\ 2s^2p^4$

F (Z = 9) → $1s^2, 2s^2, 2p^5$
→ $1s^2, 2s^2p^5$
→ 2, 7
→ $[He]_2\ 2s^2p^5$

Ne (Z = 10) → $1s^2, 2s^2, 2p^6$
→ $1s^2, 2s^2p^6$
→ 2, 8
→ $[He]_2\ 2s^2p^6$
→ $[Ne]_{10}$

Here $[Ne]_{10}$ indicates the configuration of Ne atom (Z = 10) which, as shown above, is 2, 8. It may be seen from the configurations of the elements of *2nd* period that at the end of this period 1s (*i.e.* K- shell) and, 2s and 2p orbitals (*i.e.* L-shell) are completely filled.

3rd period (3*s*, 3*p*). Since this period has 8 elements *viz.* Na (Z = 11) to Ar (Z = 18), only 8 electrons are added upto the end of this period and these electrons enter 3*s* and 3*p* orbitals which are completely filled at the end of this period.

At Na (Z = 11) 3*s* orbital starts to fill up and it is completely filled at Mg (Z = 12). Thus the configurations of Na and Mg are as :

Na (Z = 11) → $1s^2, 2s^2, 2p^6, 3s^1$
→ $1s^2, 2s^2p^6, 3s^1$
→ 2, 8, 1
→ $[Ne]_{10}\ 3s^1$

Mg (Z = 12) → $1s^2, 2s^2, 2p^6, 3s^2$
→ $1s^2, 2s^2p^6, 3s^2$
→ 2, 8, 2
→ $[Ne]_{10}\ 3s^2$

Now in the next six elements, namely Al (Z = 13) to Ar (Z = 18) 3*p* orbital is being progressively filled. Thus at Ar winch is the last element of this period, 3*p* orbital is completely filled as shown below :

Al (Z = 13) → $1s^2, 2s^2, 2p^6, 3s^2, 3p^1$
→ $1s^2, 2s^2p^6, 3s^2p^1$
→ 2, 8, 3
→ $[Ne]_{10}\ 3s^2p^1$

Si (Z = 14) → $1s^2, 2s^2, 2p^6, 3s^2, 3p^2$
→ $1s^2, 2s^2p^6, 3s^2p^2$
→ 2, 8, 4
→ $[Ne]_{10}\ 3s^2p^2$

P (Z = 15) → $1s^2, 2s^2, 2p^6, 3s^2, 3p^3$
→ $1s^2, 2s^2p^6, 3s^2p^3$
→ 2, 8, 5
→ $[Ne]_{10}\ 3s^2p^3$

S (Z = 16) → $1s^2, 2s^2, 2p^6, 3s^2, 3p^4$
→ $1s^2, 2s^2p^6, 3s^2p^4$
→ 2, 8, 6
→ $[Ne]_{10}\ 3s^2p^4$

Cl (Z = 17) → $1s^2, 2s^2, 2p^6, 3s^2, 3p^5$
→ $1s^2, 2s^2p^6, 3s^2p^5$
→ 2, 8, 7
→ $[Ne]_{10}\ 3s^2p^5$

Ar (Z = 18) → $1s^2, 2s^2, 2p^6, 3s^2, 3p^6$
→ $1s^2, 2s^2p^6, 3s^2p^6$
→ 2, 8, 8
→ $[Ne]_{10}\ 3s^2p^6$
→ $[Ar]_{18}$

Here $[Ar]_{18}$ indicates the configuration of Ar atom (Z = 18) which is 2,8, 8.

We see that at the end of this period M-shell (3*rd* shell) whose maximum capacity to contain the electrons is 18 is *not completely filled*, since 3*d* orbitals of this shell are still vacant. K (1*st* shell) and L (2*nd* shell) shells are, however, completely filled at the end of this period, since 1s, 2s and 2p orbitals are completely filled.

4th period (4*s*, 3*d*, 4*p*). Since this period has 18 elements *viz.* K (Z = 19) to Kr (Z = 36), only 18 electrons are added upto the end of this period and these electrons enter 4*s*, 3*d* and 4*p* orbitals which are completely filled at the end of this period.

At K (Z = 19) 4*s* orbital starts to fill up and is completely filled at Ca (Z = 20). Thus the configurations of these two elements are as :

K (Z = 19) → $1s^2, 2s^2, 2p^6, 3s^2, 3p^6, 4s^1$
→ $1s^2, 2s^22p^6, 3s^2p^6, 4s^1$
→ 2, 8, 8, 1
→ $[Ar]_{18}\ 4s^1$

Ca (Z = 20) → $1s^2, 2s^2, 2p^6, 3s^2, 3p^6, 4s^2$
→ $1s^2, 2s^2p^6, 3s^2p^6, 4s^2$
→ 2, 8, 8, 2
→ $[Ar]_{18}\ 4s^2$

Now after the completion of 4*s* orbital at Ca [Ca → $[Ar]_{18}4s^2$], as shown above, the next electron, according to Aufbau principle, enters 3*d* orbital in the next ten elements namely Sc (Z = 21) to Zn (Z = 30) (*first transition series*). 3*d* orbital is thus completely filled at Zn, which is the last member of the first transition series. Thus the configurations of the members of first transition series can be given as shown below :

Sc (Z = 21) → $1s^2, 2s^2, 2p^6, 3s^2, 3p^6, 4s^2, 3d^1$
→ $1s^2, 2s^2p^6, 3s^2p^6d^1, 4s^2$
→ 2, 8, 9, 2
→ $[Ar]_{18}\ 3d^1.4s^2$

Ti (Z = 22) → $1s^2, 2s^2, 2p^6, 3s^2, 3p^6, 4s^2, 3d^2$
→ $1s^2, 2s^2p^6, 3s^2p^6d^2, 4s^2$
→ 2, 8, 10, 2
→ $[Ar]_{18}\ 3d^2.4s^2$

V (Z = 23) → $1s^2, 2s^2, 2p^6, 3s^2, 3p^6, 4s^2, 3d^3$
→ $1s^2, 2s^2p^6, 3s^2p^6d^3, 4s^2$
→ 2, 8, 11, 2
→ $[Ar]_{18}\ 3d^3.4s^2$

The configurations of the elements from Sc(Z = 21) to V(Z = 23), as given above, show that in all these elements 3*d* orbital is being progressively filled and hence the expected configuration of Cr (Z = 24) should be $[Ar]_{18}\ 3d^4.4s^2$, but it has been observed that the actual (or observed) configuration of Cr is $[Ar]_{18}3d^5.4s^1$ which is obtained by shifting one of the electrons from 4*s* orbital to 3*d* orbital. This type of electronic shifting makes 3*d* orbital half-filled ($3d^5$) and hence more stable than $3d^4$ configuration, as suggested by the rule of half-filled orbitals. Thus :

Cr (Z = 24) → $1s^2, 2s^2, 2p^6, 3s^2, 3p^6, 4s^1, 3d^5$
→ $1s^2, 2s^2p^6, 3s^2p^6d^5, 4s^1$
→ 2, 8, 13, 1
→ $[Ar]_{18}\ 3d^5.4s^1$

Mn(Z = 25) → $1s^2, 2s^2, 2p^6, 3s^2, 3p^6, 4s^2, 3d^5$
→ $1s^2, 2s^2p^6, 3s^2p^6d^5, 4s^2$
→ 2, 8, 13, 2
→ $[Ar]_{18}\ 3d^5.4s^2$

Fe(Z = 26) → $1s^2, 2s^2, 2p^6, 3s^2, 3p^6, 4s^2, 3d^6$
→ $1s^2, 2s^2p^6, 3s^2p^6d^6, 4s^2$

→ 2, 8, 14, 2
→ $[Ar]_{18}\ 3d^6.4s^2$
Co (Z = 27) → $1s^2, 2s^2, 2p^6, 3s^2, 3p^6, 4s^2, 3d^7$
→ $1s^2, 2s^2p^6, 3s^2p^6d^7, 4s^2$
→ 2, 8, 15, 2
→ $[Ar]_{18}\ 3d^7.4s^2$
Ni (Z = 28) → $1s^2, 2s^2, 2p^6, 3s^2, 3p^6, 4s^2, 3d^8$
→ $1s^2, 2s^2p^6, 3s^2p^6d^8, 4s^2$
→ 2, 8, 16, 2
→ $[Ar]_{18}\ 3d^8.4s^2$

In case of Cu (Z = 29) (as in case of Cr) one of the electrons from 4*s* orbital gets shifted to $3d^9$ orbital so that 3*d* orbital becomes completely-filled ($3d^{10}$) and hence more stable than $3d^9$ configuration, as suggested by the rule of completely-filled orbitals. Thus :

Cu(Z = 29) → $1s^2, 2s^2, 2p^6, 3s^2, 3p^6, 4s^1, 3d^{10}$
→ $1s^2, 3s^2p^6, 3s^2p^6d^{10}, 4s^1$
→ 2, 8, 18, 1
→ $[Ar]_{18}\ 3d^{10}.4s^1$
Zn(Z = 30) → $1s^2, 2s^2, 2p^6, 3s^2, 3p^6, 4s^2, 3d^{10}$
→ $1s^2, 2s^2p^6, 3s^2p^6d^{10}, 4s^2$
→ 2, 8, 18, 2
→ $[Ar]_{18}\ 3d^{10}.4s^2$

Now in the next six elements *viz.* Ga (Z = 31) to Kr (Z = 36) 4*p* orbital is being progressively filled. Thus at Kr, which is the last element of this period, 4*p* orbital is completely filled.

Ga (Z = 31) → $1s^2, 2s^2, 2p^6, 3s^2, 3p^6, 4s^2, 3d^{10}, 4p^1$
→ $1s^2, 2s^2p^6, 3s^2p^6d^{10}, 4s^2p^1$
→ 2, 8, 18, 3
→ $[Ar]_{18}\ 3d^{10}4s^2p^1$
Ge (Z = 32) → $1s^2, 2s^2, 2p^6, 3s^2, 3p^6, 4s^2, 3d^{10}, 4p^2$
→ $1s^2, 2s^2p^6, 3s^2p^6d^{10}, 4s^2p^2$
→ 2, 8, 18, 4
→ $[Ar]_{18}\ 3d^{10}4s^2p^2$
As (Z = 33) → $1s^2, 2s^2, 2p^6, 3s^2, 3p^6, 4s^2, 3d^{10}, 4p^3$
→ $1s^2, 2s^2p^6, 3s^2p^6d^{10}, 4s^2p^3$
→ 2, 8, 18, 5
→ $[Ar]_{18}\ 3d^{10}4s^2p^3$
Se (Z = 34) → $1s^2, 2s^2, 2p^6, 3s^2, 3p^6, 4s^2, 3d^{10}, 4p^4$
→ $1s^2, 2s^2p^6, 3s^2p^6d^{10}, 4s^2p^4$

$\rightarrow$ 2, 8, 18, 6

$\rightarrow$ $[Ar]_{18}\ 3d^{10}4s^2p^4$

Br (Z = 35) $\rightarrow$ $1s^2, 2s^2, 2p^6, 3s^2, 3p^6, 4s^2, 3d^{10}, 4p^5$

$\rightarrow$ $1s^2, 2s^2p^6, 3s^2p^6d^{10}, 4s^2p^5$

$\rightarrow$ 2, 8, 18, 7

$\rightarrow$ $[Ar]_{18}\ 3d^{10}4s^2p^5$

Kr (Z = 36) $\rightarrow$ $1s^2, 2s^2, 2p^6, 3s^2, 3p^6, 4s^2, 3d^{10}, 4p^6$

$\rightarrow$ $1s^2, 2s^2p^6, 3s^2p^6d^{10}, 4s^2p^6$

$\rightarrow$ 2, 8, 18, 8

$\rightarrow$ $[Ar]_{18}\ 3d^{10}4s^2p^6$

$\rightarrow$ $[Kr]_{36}$

Here $[Kr]_{36}$ indicates the configuration of Kr atom (Z = 36), which is 2, 8, 18, 8. It may be noted from the configuration of Kr that at the end of *4th* period, *1st*, *2nd* and *3rd* shells are completely filled, while the *4th* shell is not, since it has only 8 electrons. Note that the maximum capacity of *4th* shell to contain the electrons is 32.

5th period (5*s*, 4*d*, 5*p*). Since this period has 18 elements *viz.*, Rb (Z = 37) to Xe (Z = 54), only 18 electrons are added upto the end of this period and these electrons enter 5*s*, 4*d* and 5*p* orbitals which are completely filled at the end of this period.

At Rb (Z = 37), 5*s* orbital starts to fill up and is completely filled at Sr (Z = 38). Thus the configurations of these two elements are as given below :

Rb (Z = 37) $\rightarrow$ $1s^2, 2s^2, 2p^6, 3s^2, 3p^6, 4s^2, 3d^{10}, 4p^6, 5s^1$

$\rightarrow$ $1s^2, 2s^2p^6, 3s^2p^6d^{10}, 4s^2p^6, 5s^1$

$\rightarrow$ 2, 8, 18, 8, 1

$\rightarrow$ $[Kr]_{36}\ 5s^1$

Sr (Z = 38) $\rightarrow$ $1s^2, 2s^2, 2p^6, 3s^2, 3p^6, 4s^2, 3d^{10}, 4p^6, 5s^2$

$\rightarrow$ $1s^2, 2s^2p^6, 3s^2p^6d^{10}, 4s^2p^6, 5s^2$

$\rightarrow$ 2, 8, 18, 8, 2

$\rightarrow$ $[Kr]_{36}\ 5s^2$

Now, after the completion of 5*s* orbital at Sr [Sr $\rightarrow$ $[Kr]_{36}\ 5s^2$], as shown above, the next electron, according to Aufbau principle, enters 4*d* orbital in the next ten elements namely Y (Z = 39) to Cd (Z = 48) (*second transition series*). 4*d* orbital is thus completely filled at Cd which is the last member of second transition series. Thus the configurations of the members of the second transition series can be given as shown below :

Y (Z = 39) $\rightarrow$ $1s^2, 2s^2, 2p^6, 3s^2, 3p^6, 4s^2, 3d^{10}, 4p^6, 5s^2, 4d^1$

$\rightarrow$ $1s^2, 2s^2p^6, 3s^2p^6d^{10}, 4s^2p^6d^1, 5s^2$

$\rightarrow$ 2, 8, 18, 9, 2

$\rightarrow$ $[Kr]_{36}\ 4d^1.5s^2$

Zr (Z = 40) $\rightarrow$ $1s^2, 2s^2, 2p^6, 3s^2, 3p^6, 4s^2, 3d^{10}, 4p^6, 5s^2, 4d^2$
$\rightarrow$ $1s^2, 2s^2p^6, 3s^2p^6d^{10}, 4s^2p^6d^2, 5s^2$
$\rightarrow$ 2, 8, 18, 10, 2
$\rightarrow$ $[Kr]_{36}\ 4d^2.5s^2$

In case of Nb (Z = 41), Mo (Z = 42), Ru (Z = 44) and Rh (Z = 45) one of the electrons from 5*s* orbital gets shifted to 4*d* orbital to make 5*s* and 4*d* orbitals half-filled which, in accordance with the rule of half-filled orbitals are more stable. Thus :

Nb (Z = 41) $\rightarrow$ $1s^2, 2s^2, 2p^6, 3s^2, 3p^6, 4s^2, 3d^{10}, 4p^6, 5s^1, 4d^4$
$\rightarrow$ $1s^2, 2s^2p^6, 3s^2p^6d^{10}, 4s^2p^6d^4, 5s^1$
$\rightarrow$ 2, 8, 18, 12, 1
$\rightarrow$ $[Kr]_{36}\ 4d^4.5s^1$

Mo (Z = 42) $\rightarrow$ $1s^2, 2s^2, 2p^6, 3s^2, 3p^6, 4s^2, 3d^{10}, 4p^6, 5s^1, 4d^5$
$\rightarrow$ $1s^2, 2s^2p^6, 3s^2p^6d^{10}, 4s^2p^6d^5, 5s^1$
$\rightarrow$ 2, 8, 18, 13, 1
$\rightarrow$ $[Kr]_{36}\ 4d^5.5s^1$

Tc (Z = 43) $\rightarrow$ $1s^2, 2s^2, 2p^6, 3s^2, 3p^6, 4s^2, 3d^{10}, 4p^6, 5s^2, 4d^5$
$\rightarrow$ $1s^2, 2s^2p^6, 3s^2p^6d^{10}, 4s^2p^6d^5, 5s^2$
$\rightarrow$ 2, 8, 18, 13, 2
$\rightarrow$ $[Kr]_{36}\ 4d^5.5s^2$

Ru (Z = 44) $\rightarrow$ $1s^2, 2s^2, 2p^6, 3s^2, 3p^6, 4s^2, 3d^{10}, 4p^6, 5s^1, 4d^7$
$\rightarrow$ $1s^2, 2s^2p^6, 3s^2p^6d^{10}, 4s^2p^6d^7, 5s^1$
$\rightarrow$ 2, 8, 18, 15, 1
$\rightarrow$ $[Kr]_{36}\ 4d^7.5s^1$

Rh (Z = 45) $\rightarrow$ $1s^2, 2s^2, 2p^6, 3s^2, 3p^6, 4s^2, 3d^{10}, 4p^6, 5s^1, 4d^8$
$\rightarrow$ $1s^2, 2s^2p^6, 3s^2p^6d^{10}, 4s^2p^6d^8, 5s^1$
$\rightarrow$ 2, 8, 18, 16, 1
$\rightarrow$ $[Kr]_{36}\ 4d^8.5s^1$

In case of Pd (Z= 46) both the electrons from 5*s* orbital move to $4d^8$ orbital to make 4*d* orbital completely-filled ($4d^{10}$) and hence more stable than $4d^8$ configuration. Thus :

Pd (Z = 46) $\rightarrow$ $1s^2, 2s^2, 2p^6, 3s^2, 3p^6, 4s^2, 3d^{10}, 4p^6, 5s^0, 4d^{10}$
$\rightarrow$ $1s^2, 2s^2p^6, 3s^2p^6d^{10}, 4s^2p^6d^{10}, 5s^0$
$\rightarrow$ 2, 8, 18, 18, 0
$\rightarrow$ $[Kr]_{36}\ 4d^{10}.5s^0$

In case of Ag (Z = 47) one of the electrons from 5*s* orbital is shifted to $4d^9$ orbital so that 4*d* orbital becomes completely-filled ($4d^{10}$) and hence more stable. Thus :

Ag (Z = 47) $\rightarrow$ $1s^2, 2s^2, 2p^6, 3s^2, 3p^6, 4s^2, 3d^{10}, 4p^6, 5s^1, 4d^{10}$
$\rightarrow$ $1s^2, 2s^2p^6, 3s^2p^6d^{10}, 4s^2p^6d^{10}, 5s^1$

→ 2, 8, 18, 18, 1

→ $[Kr]_{36}\ 4d^{10}.5s^1$

Cd (Z = 48) → $1s^2, 2s^2, 2p^6, 3s^2, 3p^6, 4s^2, 3d^{10}, 4p^6, 5s^2, 4d^{10}$

→ $1s^2, 2s^2p^6, 3s^2p^6d^{10}, 4s^2p^6d^{10}, 5s^2$

→ 2, 8, 18, 18, 2

→ $[Kr]_{36}\ 4d^{10}.5s^2$

Now in the next six elements namely in (Z = 49) to Xe (Z = 54) 5*p* orbital is being progressively filled. Thus at Xe which is the last element of this period, 5*p* orbital is completely filled as shown below :

In (Z = 49) → $1s^2, 2s^2, 2p^6, 3s^2, 3p^6, 4s^2, 3d^{10}, 4p^6, 5s^2, 4d^{10}, 5p^1$

→ $1s^2, 2s^2p^6, 3s^2p^6d^{10}, 4s^2p^6d^{10}, 5s^2p^1$

→ 2, 8, 18, 18, 3

→ $[Kr]_{36}\ 4d^{10}5s^2p^1$

Sn (Z = 50) → $1s^2, 2s^2, 2p^6, 3s^2, 3p^6, 4s^2, 3d^{10}, 4p^6, 5s^2, 4d^{10}, 5p^2$

→ $1s^2, 2s^2p^6, 3s^2p^6d^{10}, 4s^2p^6d^{10}, 5s^2p^2$

→ 2, 8, 18, 18, 4

→ $[Kr]_{36}\ 4d^{10}5s^2p^2$

Sb (Z = 51) → $1s^2, 2s^2, 2p^6, 3s^2, 3p^6, 4s^2, 3d^{10}, 4p^6, 5s^2, 4d^{10}, 5p^3$

→ $1s^2, 2s^2p^6, 3s^2p^6d^{10}, 4s^2p^6d^{10}, 5s^2p^3$

→ 2, 8, 18, 18, 5

→ $[Kr]_{36}\ 4d^{10}5s^2p^3$

Te (Z = 52) → $1s^2, 2s^2, 2p^6, 3s^2, 3p^6, 4s^2, 3d^{10}, 4p^6, 5s^2, 4d^{10}, 5p^4$

→ $1s^2, 2s^2p^6, 3s^2p^6d^{10}, 4s^2p^6d^{10}, 5s^2p^4$

→ 2, 8, 18, 18, 6

→ $[Kr]_{36}\ 4d^{10}5s^2p^4$

I (Z = 53) → $1s^2, 2s^2, 2p^6, 3s^2, 3p^6, 4s^2, 3d^{10}, 4p^6, 5s^2, 4d^{10}, 5p^5$

→ $1s^2, 2s^2p^6, 3s^2p^6d^{10}, 4s^2p^6d^{10}, 5s^2p^5$

→ 2, 8, 18, 18, 7

→ $[Kr]_{36}\ 4d^{10}5s^2p^5$

Xe (Z = 54) → $1s^2, 2s^2, 2p^6, 3s^2, 3p^6, 4s^2, 3d^{10}, 4p^6, 5s^2, 4d^{10}, 5p^6$

→ $1s^2, 2s^2p^6, 3s^2p^6d^{10}, 4s^2p^6d^{10}, 5s^2p^6$

→ 2, 8, 18, 18, 8

→ $[Kr]_{36}\ 4d^{10}5s^2p^6$

→ $[Xe]_{54}$

Here $[Xe]_{54}$ indicates the configuration of Xe atom (Z = 54) which is 2, 8, 18, 18, 8.

6th period (6*s*, 4*f*, 5*d*, 6*p*). Since this period contains 32 elements viz Cs (Z = 55) to Rn (Z = 86), 32 electrons are added upto the end of this period and these electrons enter 6*s*, 4*f*, 5*d* and 6*p* orbitals which are completely filled at the end of this period.

At Cs (Z = 55), 6s orbital starts to fill up and is completely filled at Ba (Z = 56). Thus the configurations of these two elements are as given below :

Cs (Z = 55) → $1s^2, 2s^2, 2p^6, 3s^2, 3p^6, 4s^2, 3d^{10}, 4p^6, 5s^2, 4d^{10}, 5p^6, 6s^1$
→ $1s^2, 2s^2p^6, 3s^2p^6d^{10}, 4s^2p^6d^{10}, 5s^2p^6, 6s^1$
→ 2, 8, 18, 18, 8, 1
→ $[Xe]_{54}\ 6s^1$

Ba (Z = 56) → $1s^2, 2s^2, 2p^6, 3s^2, 3p^6, 4s^2, 3d^{10}, 4p^6, 5s^2, 4d^{10}, 5p^6, 6s^2$
→ $1s^2, 2s^2p^6, 3s^2p^6d^{10}, 4s^2p^6d^{10}, 5s^2p^6, 6s^2$
→ 2, 8, 18, 18, 8, 2
→ $[Xe]_{54}\ 6s^2$

Now after the completion of 6s orbital at Ba[Ba→$[Xe]_{54}\ 6s^2$], the next electron, according to Aufbau principle, must enter 4*f* orbital to give $[Xe]_{54}\ 4f^1.5d^0.6s^2$ configuration to La (Z = 57); but it has been observed that the extra electron enters 5*d* orbital (instead of 4*f* orbital) to give $[Xe]_{54}\ 4f^0.5d^1.6s^2$ configuration to La atom as shown below :

La (Z = 57) → $1s^2, 2s^2, 2p^6, 3s^2, 3p^6, 4s^2, 3d^{10}, 4p^6, 5s^2, 4d^{10}, 5p^6, 6s^2, 5d^1$
→ $1s^2, 2s^2p^6, 3s^2p^6d^{10}, 4s^2p^6d^{10}, 5s^2p^6d^1, 6s^2$
→ 2, 8, 18, 18, 9, 2
→ $[Xe]_{54}\ 4f^0\ .\ 5d^1\ .\ 6s^2$

Why the extra electron goes to 5*d* orbital, instead of 4*f* orbital, in La atom has been explained on the basis that at La, 5*d* and 4*f* orbitals are supposed to have almost the same energy and hence the extra electron is free to occupy any of these two orbitals.

In the next fourteen elements [Ce(Z = 58) to Lu (Z = 71)] called *lanthanides*, 4*f* orbital is being progressively filled. While filling 4*f* orbital with electrons we should keep the following rule in mind.

It is only in Ce (Z = 58), Gd (Z = 64) and Lu(Z = 71) that 5d orbital has one electron while in all the remaining lanthanides 5d orbital remains vacant.

Keeping in mind the above rule the configurations of the lanthanides can be written as shown below :

Ce (Z = 58) → $1s^2, 2s^2, 2p^6, 3s^2, 3p^6, 4s^2, 3d^{10}, 4p^6, 5s^2, 4d^{10}, 5p^6, 6s^2, 4f^1, 5d^1$
→ $1s^2, 2s^2p^6, 3s^2p^6d^{10}, 4s^2p^6d^{10}f^1, 5s^2p^6d^1, 6s^2$
→ 2, 8, 18, 19, 9, 2
→ $[Xe]_{54}\ 4f^1\ .\ 5d^1\ .\ 6s^2$

Some authors suggest that the configuration of Ce is $[Xe]_{54}\ 4f^2\ 5d^0\ .\ 6s^2$.

Pr (Z = 59) → $1s^2, 2s^2, 2p^6, 3s^2, 3p^6, 4s^2, 3d^{10}, 4p^6, 5s^2, 4d^{10}, 5p^6, 6s^2, 4f^3, 5d^0$
→ $1s^2, 2s^2p^6, 3s^2p^6d^{10}, 4s^2p^6d^{10}f^3, 5s^2p^6d^0, 6s^2$
→ 2, 8, 18, 21, 8, 2
→ $[Xe]_{54}\ 4f^3\ .\ 5d^0\ .\ 6s^2$

Nd (Z = 60) → $1s^2, 2s^2, 2p^6, 3s^2, 3p^6, 4s^2, 3d^{10}, 4p^6, 5s^2, 4d^{10}, 5p^6, 6s^2, 4f^4, 5d^0$
→ $1s^2, 2s^2p^6, 3s^2p^6d^{10}, 4s^2p^6d^{10}f^4, 5s^2p^6d^0, 6s^2$
→ 2, 8, 18, 22, 8, 2
→ $[Xe]_{54}\ 4f^4 . 5d^0 . 6s^2$

Pm (Z = 61) → $1s^2, 2s^2, 2p^6, 3s^2, 3p^6, 4s^2, 3d^{10}, 4p^6, 5s^2, 4d^{10}, 5p^6, 6s^2, 4f^5, 5d^0$
→ $1s^2, 2s^2p^6, 3s^2p^6d^{10}, 4s^2p^6d^{10}f^5, 5s^2p^6d^0, 6s^2$
→ 2, 8, 18, 23, 8, 2
→ $[Xe]_{54}\ 4f^5 . 5d^0 . 6s^2$

Sm (Z = 62) → $1s^2, 2s^2, 2p^6, 3s^2, 3p^6, 4s^2, 3d^{10}, 4p^6, 5s^2, 4d^{10}, 5p^6, 6s^2, 4f^6, 5d^0$
→ $1s^2, 2s^2p^6, 3s^2p^6d^{10}, 4s^2p^6d^{10}f^6, 5s^2p^6d^0, 6s^2$
→ 2, 8, 18, 24, 8, 2
→ $[Xe]_{54}\ 4f^6 . 5d^0 . 6s^2$

Eu (Z = 63) → $1s^2, 2s^2, 2p^6, 3s^2, 3p^6, 4s^2, 3d^{10}, 4p^6, 5s^2, 4d^{10}, 5p^6, 6s^2, 4f^7, 5d^0$
→ $1s^2, 2s^2p^6, 3s^2p^6d^{10}, 4s^2p^6d^{10}f^7, 5s^2p^6d^0, 6s^2$
→ 2, 8, 18, 25, 8, 2
→ $[Xe]_{54}\ 4f^7 . 5d^0 . 6s^2$

The configurations of the elements from Ce (Z = 58) to Eu (Z = 63) as given above, show that in all these elements 4*f* orbital is being progressively filled and hence the expected configuration of Gd (Z = 64) should be $[Xe]_{54}\ 4f^8 . 5d^0. 6s^2$, but it has been observed that the actual configuration of Gd is $[Xe]_{54}\ 4f^7 . 5d^1.6s^2$ which is obtained by shifting one of the electrons from 4*f* orbital to the vacant 5*d* orbital. This type of electronic shifting makes 4*f* orbital half-filled ($4f^7$) and hence more stable than $4f^8$ configuration as suggested by the rule of half-filled orbitals. Thus :

Gd (Z = 64) → $1s^2, 2s^2, 2p^6, 3s^2, 3p^6, 4s^2, 3d^{10}, 4p^6, 5s^2, 4d^{10}, 5p^6, 6s^2, 4f^7, 5d^1$
→ $1s^2, 2s^2p^6, 3s^2p^6d^{10}, 4s^2p^6d^{10}f^7, 5s^2p^6d^1, 6s^2$
→ 2, 8, 18, 25, 9, 2
→ $[Xe]_{54}\ 4f^7 . 5d^1 . 6s^2$

Tb (Z = 65) → $1s^2, 2s^2, 2p^6, 3s^2, 3p^6, 4s^2, 3d^{10}, 4p^6, 5s^2, 4d^{10}, 5p^6, 6s^2, 4f^9, 5d^0$
→ $1s^2, 2s^2p^6, 3s^2p^6d^{10}, 4s^2p^6d^{10}f^9, 5s^2p^6d^0, 6s^2$
→ 2, 8, 18, 27, 9, 2
→ $[Xe]_{54}\ 4f^9 . 5d^0 . 6s^2$

Dy (Z = 66) → $1s^2, 2s^2, 2p^6, 3s^2, 3p^6, 4s^2, 3d^{10}, 4p^6, 5s^2, 4d^{10}, 5p^6, 6s^2, 4f^{10}, 5d^0$
→ $1s^2, 2s^2p^6, 3s^2p^6d^{10}, 4s^2p^6d^{10}f^{10}, 5s^2p^6d^0, 6s^2$

→ 2, 8, 18, 28, 8, 2

→ $[Xe]_{54}\ 4f^{10}\ .\ 5d^0\ .\ 6s^2$

Ho (Z = 67) → $1s^2, 2s^2, 2p^6, 3s^2, 3p^6, 4s^2, 3d^{10}, 4p^6, 5s^2, 4d^{10}, 5p^6, 6s^2, 4f^{11}, 5d^0$

→ $1s^2, 2s^2p^6, 3s^2p^6d^{10}, 4s^2p^6d^{10}f^{11}, 5s^2p^6d^0, 6s^2$

→ 2, 8, 18, 29, 8, 2

→ $[Xe]_{54}\ 4f^{11}\ .\ 5d^0\ .\ 6s^2$

Er (Z = 68) → $1s^2, 2s^2, 2p^6, 3s^2, 3p^6, 4s^2, 3d^{10}, 4p^6, 5s^2, 4d^{10}, 5p^6, 6s^2, 4f^{12}, 5d^0$

→ $1s^2, 2s^2p^6, 3s^2p^6d^{10}, 4s^2p^6d^{10}f^{12}, 5s^2p^6d^0, 6s^2$

→ 2, 8, 18, 30, 8, 2

→ $[Xe]_{54}\ 4f^{12}\ .\ 5d^0\ .\ 6s^2$

Tm (Z = 69) → $1s^2, 2s^2, 2p^6, 3s^2, 3p^6, 4s^2, 3d^{10}, 4p^6, 5s^2, 4d^{10}, 5p^6, 6s^2, 4f^{13}, 5d^0$

→ $1s^2, 2s^2p^6, 3s^2p^6d^{10}, 4s^2p^6d^{10}f^{13}, 5s^2p^6d^0, 6s^2$

→ 2, 8, 18, 31, 8, 2

→ $[Xe]_{54}\ 4f^{13}\ .\ 5d^0\ .\ 6s^2$

Yb (Z = 70) → $1s^2, 2s^2, 2p^6, 3s^2, 3p^6, 4s^2, 3d^{10}, 4p^6, 5s^2, 4d^{10}, 5p^6, 6s^2, 4f^{14}, 5d^0$

→ $1s^2, 2s^2p^6, 3s^2p^6d^{10}, 4s^2p^6d^{10}f^{14}, 5s^2p^6d^0, 6s^2$

→ 2, 8, 18, 32, 8, 2

→ $[Xe]_{54}\ 4f^{14}\ .\ 5d^0\ .\ 6s^2$

Lu (Z = 71) → $1s^2, 2s^2, 2p^6, 3s^2, 3p^6, 4s^2, 3d^{10}, 4p^6, 5s^2, 4d^{10}, 5p^6, 6s^2, 4f^{14}, 5d^1$

→ $1s^2, 2s^2p^6, 3s^2p^6d^{10}, 4s^2p^6d^{10}f^{14}, 5s^2p^6d^1, 6s^2$

→ 2, 8, 18, 32, 9, 2

→ $[Xe]_{54}\ 4f^{14}\ .\ 5d^1\ .\ 6s^2$

Now the extra electron in the next *nine* elements namely Hf (Z = 72) to Hg (Z = 80) goes to 5*d* orbital which is already singly-filled at Lu, the last member of lanthanide series. 5*d* orbital is thus completely filled at Hg.

Hf (Z = 72) → $1s^2, 2s^2, 2p^6, 3s^2, 3p^6, 4s^2, 3d^{10}, 4p^6, 5s^2, 4d^{10}, 5p^6, 6s^2, 4f^{14}, 5d^2$

→ $1s^2, 2s^2p^6, 3s^2p^6d^{10}, 4s^2p^6d^{10}f^{14}, 5s^2p^6d^2, 6s^2$

→ 2, 8, 18, 32, 10, 2

→ $[Xe]_{54}\ 4f^{14}.5d^2.6s^2$

Ta (Z = 73) → $1s^2, 2s^2, 2p^6, 3s^2, 3p^6, 4s^2, 3d^{10}, 4p^6, 5s^2, 4d^{10}, 5p^6, 6s^2, 4f^{14}, 5d^3$

→ $1s^2, 2s^2p^6, 3s^2p^6d^{10}, 4s^2p^6d^{10}f^{14}, 5s^2p^6d^3, 6s^2$

→ 2, 8, 18, 32, 11, 2

→ $[Xe]_{54}\ 4f^{14}.5d^3.6s^2$

W (Z = 74) → $1s^2, 2s^2, 2p^6, 3s^2, 3p^6, 4s^2, 3d^{10}, 4p^6, 5s^2, 4d^{10}, 5p^6, 6s^2, 4f^{14}, 5d^4$

→ $1s^2, 2s^2p^6, 3s^2p^6d^{10}, 4s^2p^6d^{10}f^{14}, 5s^2p^6d^3, 6s^2$

→ 2, 8, 18, 32, 12, 2

→ $[Xe]_{54}\ 4f^{14}.5d^4.6s^2$

Re (Z = 75) → $1s^2, 2s^2, 2p^6, 3s^2, 3p^6, 4s^2, 3d^{10}, 4p^6, 5s^2, 4d^{10}, 5p^6, 6s^2, 4f^{14}, 5d^5$

→ $1s^2, 2s^2p^6, 3s^2p^6d^{10}, 4s^2p^6d^{10}f^{14}, 5s^2p^6d^5, 6s^2$

→ 2, 8, 18, 32, 13, 2

→ $[Xe]_{54}\ 4f^{14}.5d^5.6s^2$

Os (Z = 76) → $1s^2, 2s^2, 2p^6, 3s^2, 3p^6, 4s^2, 3d^{10}, 4p^6, 5s^2, 4d^{10}, 5p^6, 6s^2, 4f^{14}, 5d^6$

→ $1s^2, 2s^2p^6, 3s^2p^6d^{10}, 4s^2p^6d^{10}f^{14}, 5s^2p^6d^6, 6s^2$

→ 2, 8, 18, 32, 14, 2

→ $[Xe]_{54}\ 4f^{14}.5d^6.6s^2$

Ir (Z = 77) → $1s^2, 2s^2, 2p^6, 3s^2, 3p^6, 4s^2, 3d^{10}, 4p^6, 5s^2, 4d^{10}, 5p^6, 6s^2, 4f^{14}, 5d^7$

→ $1s^2, 2s^2p^6, 3s^2p^6d^{10}, 4s^2p^6d^{10}f^{14}, 5s^2p^6d^7, 6s^2$

→ 2, 8, 18, 32, 15, 2

→ $[Xe]_{54}\ 4f^{14}.5d^7.6s^2$

Now the expected configuration of Pt (Z = 78) should be $[Xe]_{54}\ 4f^{14}.5d^8.6s^2$, but the actual configuration is $[Xe]_{54}\ 4f^{14}.5d^9.6s^1$ which is obtained by shifting one of the two electrons from 6*s* orbital to $5d^8$ orbital. Thus :

Pt (Z = 78) → $1s^2, 2s^2, 2p^6, 3s^2, 3p^6, 4s^2, 3d^{10}, 4p^6, 5s^2, 4d^{10}, 5p^6, 6s^1, 4f^{14}, 5d^9$

→ $1s^2, 2s^2p^6, 3s^2p^6d^{10}, 4s^2p^6d^{10}f^{14}, 5s^2p^6d^9, 6s^1$

→ 2, 8, 18, 32, 17, 1

→ $[Xe]_{54}\ 4f^{14}.5d^9.6s^1$

Now the expected configuration of Au (Z = 79) is $[Xe]_{54}4f^{14}.5d^9.6s^2$, but the actual configuration is $[Xe]_{54}4f^{14}.5d^{10}.6s^1$ which is obtained by shifting one of the two electrons from 6*s* orbital to $5d^9$ orbital to make 5*d* orbital completely-filled ($5d^{10}$) which, according to the *rule of completely-filled orbitals*, is more stable than $5d^9$ orbital. Thus :

Au (Z = 79) → $1s^2, 2s^2, 2p^6, 3s^2, 3p^6, 4s^2, 3d^{10}, 4p^6, 5s^2, 4d^{10}, 5p^6, 6s^1, 4f^{14}, 5d^{10}$

→ $1s^2, 2s^2p^6, 3s^2p^6d^{10}, 4s^2p^6d^{10}f^{14}, 5s^2p^6d^{10}, 6s^1$

→ 2, 8, 18, 32, 18, 1

→ $[Xe]_{54}\ 4f^{14}.5d^{10}.6s^1$

Hg (Z = 80) → $1s^2, 2s^2, 2p^6, 3s^2, 3p^6, 4s^2, 3d^{10}, 4p^6, 5s^2, 4d^{10}, 5p^6, 6s^2, 4f^{14}, 5d^{10}$

→ $1s^2, 2s^2p^6, 3s^2p^6d^{10}, 4s^2p^6d^{10}f^{14}, 5s^2p^6d^{10}, 6s^2$
→ 2, 8, 18, 32, 18, 2
→ $[Xe]_{54}\ 4f^{14}.5d^{10}.6s^2$

Now in the next six elements namely Tl (Z = 81) to Rn (Z = 86) 6*p* orbital is being progressively filled. Thus at Rn which is the last element of this period 6*p* orbital is completely filled as shown below :

Tl (Z = 81) → $1s^2, 2s^2, 2p^6, 3s^2, 3p^6, 4s^2, 3d^{10}, 4p^6, 5s^2, 4d^{10}, 5p^6, 6s^2, 4f^{14}, 5d^{10}, 6p^1$
→ $1s^2, 2s^2p^6, 3s^2p^6d^{10}, 4s^2p^6d^{10}f^{14}, 5s^2p^6d^{10}, 6s^2p^1$
→ 2, 8, 18, 32, 18, 3
→ $[Xe]_{54}\ 4f^{14}.5d^{10}.6s^2p^1$

Pb (Z = 82) → $1s^2, 2s^2, 2p^6, 3s^2, 3p^6, 4s^2, 3d^{10}, 4p^6, 5s^2, 4d^{10}, 5p^6, 6s^2, 4f^{14}, 5d^{10}, 6p^2$
→ $1s^2, 2s^2p^6, 3s^2p^6d^{10}, 4s^2p^6d^{10}f^{14}, 5s^2p^6d^{10}, 6s^2p^2$
→ 2, 8, 18, 32, 18, 4
→ $[Xe]_{54}\ 4f^{14}.5d^{10}.6s^2p^2$

Bi (Z = 83) → $1s^2, 2s^2, 2p^6, 3s^2, 3p^6, 4s^2, 3d^{10}, 4p^6, 5s^2, 4d^{10}, 5p^6, 6s^2, 4f^{14}, 5d^{10}, 6p^3$
→ $1s^2, 2s^2p^6, 3s^2p^6d^{10}, 4s^2p^6d^{10}f^{14}, 5s^2p^6d^{10}, 6s^2p^3$
→ 2, 8, 18, 32, 18, 5
→ $[Xe]_{54}\ 4f^{14}.5d^{10}.6s^2p^3$

Po (Z = 84) → $1s^2, 2s^2, 2p^6, 3s^2, 3p^6, 4s^2, 3d^{10}, 4p^6, 5s^2, 4d^{10}, 5p^6, 6s^2, 4f^{14}, 5d^{10}, 6p^4$
→ $1s^2, 2s^2p^6, 3s^2p^6d^{10}, 4s^2p^6d^{10}f^{14}, 5s^2p^6d^{10}, 6s^2p^4$
→ 2, 8, 18, 32, 18, 6
→ $[Xe]_{54}\ 4f^{14}.5d^{10}.6s^2p^4$

At (Z = 85) → $1s^2, 2s^2, 2p^6, 3s^2, 3p^6, 4s^2, 3d^{10}, 4p^6, 5s^2, 4d^{10}, 5p^6, 6s^2, 4f^{14}, 5d^{10}, 6p^5$
→ $1s^2, 2s^2p^6, 3s^2p^6d^{10}, 4s^2p^6d^{10}f^{14}, 5s^2p^6d^{10}, 6s^2p^5$
→ 2, 8, 18, 32, 18, 7
→ $[Xe]_{54}\ 4f^{14}.5d^{10}.6s^2p^5$

Rn (Z = 86) → $1s^2, 2s^2, 2p^6, 3s^2, 3p^6, 4s^2, 3d^{10}, 4p^6, 5s^2, 4d^{10}, 5p^6, 6s^2, 4f^{14}, 5d^{10}, 6p^6$
→ $1s^2, 2s^2p^6, 3s^2p^6d^{10}, 4s^2p^6d^{10}f^{14}, 5s^2p^6d^{10}, 6s^2p^6$
→ 2, 8, 18, 32, 18, 8
→ $[Xe]_{54}\ 4f^{14}.5d^{10}.6s^2p^6$
→ $[Rn]_{86}$

Here $[Rn]_{86}$ indicates the configuration of Rn atom (Z = 86) which is 2, 8, 18, 32, 18, 8.

7th period (7*s*, 5*f*, 6*d*). At present, since this period has only 19 elements *viz.* Fr (Z = 87) to Ha (Z = 105), 19 electrons are added upto Ha and these electrons in 7*s*, 5*f* and 6*d* orbitals are distributed as : $7s^2$, $5f^{14}$, $6d^3$.

At Fr (Z = 87), 7*s* orbital starts to fill and is completely filled at Ra (Z = 88). Thus the configurations of these two elements can be written as :

Fr (Z = 87) → $1s^2, 2s^2, 2p^6, 3s^2, 3p^6, 4s^2, 3d^{10}, 4p^6, 5s^2, 4d^{10}, 5p^6, 6s^2, 4f^{14}, 5d^{10}, 6p^6, 7s^1$
→ $1s^2, 2s^2p^6, 3s^2p^6d^{10}, 4s^2p^6d^{10}f^{14}, 5s^2p^6d^{10}, 6s^2p^6, 7s^1$
→ 2, 8, 18, 32, 18, 8, 1
→ $[Rn]_{86}\ 7s^1$

Ra (Z = 88) → $1s^2, 2s^2, 2p^6, 3s^2, 3p^6, 4s^2, 3d^{10}, 4p^6, 5s^2, 4d^{10}, 5p^6, 6s^2, 4f^{14}, 5d^{10}, 6p^6, 7s^2$
→ $1s^2, 2s^2p^6, 3s^2p^6d^{10}, 4s^2p^6d^{10}f^{14}, 5s^2p^6d^{10}, 6s^2p^6, 7s^2$
→ 2, 8, 18, 32, 18, 8, 2
→ $[Rn]_{86}\ 7s^2$

Now after the completion of 7*s* orbital at Ra [Ra → $[Rn]_{86}\ 7s^2$], the next electron, according to Aufbau principle, must enter 5*f* orbital to give $[Rn]_{86}\ 5f^1.6d^0.7s^2$ configuration to Ac (Z = 89), but it has been observed that the extra electron enters 6*d* orbital (instead of 5*f* orbital) to give $[Rn]_{86}\ 5f^0.\ 6d^1\ .\ 7s^2$ configuration to Ac as shown below :

Ac (Z = 89) → $1s^2, 2s^2, 2p^6, 3s^2, 3p^6, 4s^2, 3d^{10}, 4p^6, 5s^2, 4d^{10}, 5p^6, 6s^2, 4f^{14}, 5d^{10}, 6p^6, 7s^2, 6d^1$
→ $1s^2, 2s^2p^6, 3s^2p^6d^{10}, 4s^2p^6d^{10}f^{14}, 5s^2p^6d^{10}, 6s^2p^6d^1, 7s^2$
→ 2, 8, 18, 32, 18, 9, 2
→ $[Rn]_{86}\ 5f^0\ .\ 6d^1\ .\ 7s^2$

Why the extra electron goes to 6*d* orbital instead of 5*f* orbital in Ac has been explained on the basis that in Ac, 6*d* and 5*f* orbitals are supposed to be almost of the same energy and hence the extra electron is free to go to any of these two orbitals.

In the next fourteen elements, [Th (Z = 90) to Lw (Z = 103)], called *actinides*, 5*f* orbital is being progressively filled. While filling 5*f* orbital with electrons we should keep the following rule in mind.

It is only in Th (Z = 90), Pa (Z = 91), U (Z = 92), Cm (Z = 96) and Lw (Z = 103) that 6d orbital has one electron while in all the remaining actinides 6d orbital remains vacant.

Keeping in mind the above rule the configurations of the actinides can be written as shown below :

Th (Z = 90) → $1s^2, 2s^2, 2p^6, 3s^2, 3p^6, 4s^2, 3d^{10}, 4p^6, 5s^2, 4d^{10}, 5p^6, 6s^2, 4f^{14}, 5d^{10}, 6p^6, 7s^2, 5f^1, 6d^1$
→ $1s^2, 2s^2p^6, 3s^2p^6d^{10}, 4s^2p^6d^{10}f^{14}, 5s^2p^6d^{10}f^1, 6s^2p^6d^1, 7s^2$

→ 2, 8, 18, 32, 19, 9, 2

→ $[Rn]_{86}\ 5f^1 . 6d^1 . 7s^2$

Some authors suggest that the configuration of Th is $[Rn]_{86}\ 5f^0 . 6d^2 . 7s^2$.

Pa (Z = 91) → $1s^2, 2s^2, 2p^6, 3s^2, 3p^6, 4s^2, 3d^{10}, 4p^6, 5s^2, 4d^{10}, 5p^6, 6s^2, 4f^{14}, 5d^{10}, 6p^6, 7s^2, 5f^2, 6d^1$

→ $1s^2, 2s^2p^6, 3s^2p^6d^{10}, 4s^2p^6d^{10}f^{14}, 5s^2p^6d^{10}f^2, 6s^2p^6d^1, 7s^2$

→ 2, 8, 18, 32, 20, 9, 2

→ $[Rn]_{86}\ 5f^2 . 6d^1 . 7s^2$

Some authors suggest that the configuration of Pa is $[Rn]\ 5f^1 . 6d^2 . 7s^2$.

U (Z = 92) → $1s^2, 2s^2, 2p^6, 3s^2, 3p^6, 4s^2, 3d^{10}, 4p^6, 5s^2, 4d^{10}, 5p^6, 6s^2, 4f^{14}, 5d^{10}, 6p^6, 7s^2, 5f^3, 6d^1$

→ $1s^2, 2s^2p^6, 3s^2p^6d^{10}, 4s^2p^6d^{10}f^{14}, 5s^2p^6d^{10}f^3, 6s^2p^6d^1, 7s^2$

→ 2, 8, 18, 32, 21, 9, 2

→ $[Rn]_{86}\ 5f^3 . 6d^1 . 7s^2$

Np (Z = 93) → $1s^2, 2s^2, 2p^6, 3s^2, 3p^6, 4s^2, 3d^{10}, 4p^6, 5s^2, 4d^{10}, 5p^6, 6s^2, 4f^{14}, 5d^{10}, 6p^6, 7s^2, 5f^5, 6d^0$

→ $1s^2, 2s^2p^6, 3s^2p^6d^{10}, 4s^2p^6d^{10}f^{14}, 5s^2p^6d^{10}f^5, 6s^2p^6d^0, 7s^2$

→ 2, 8, 18, 32, 23, 8, 2

→ $[Rn]_{86}\ 5f^5 . 6d^0 . 7s^2$

Some authors suggest that the configuration of Np is $[Rn]_{86}\ 5f^4 . 6d^1 . 7s^2$.

Pu (Z = 94) → $1s^2, 2s^2, 2p^6, 3s^2, 3p^6, 4s^2, 3d^{10}, 4p^6, 5s^2, 4d^{10}, 5p^6, 6s^2, 4f^{14}, 5d^{10}, 6p^6, 7s^2, 5f^6, 6d^0$

→ $1s^2, 2s^2p^6, 3s^2p^6d^{10}, 4s^2p^6d^{10}f^{14}, 5s^2p^6d^{10}f^6, 6s^2p^6d^0, 7s^2$

→ 2, 8, 18, 32, 24, 8, 2

→ $[Rn]_{86}\ 5f^6 . 6d^0 . 7s^2$

Some authors have suggested that the configuration of Pu is $[Rn]_{86}\ 5f^5 . 6d^1 . 7s^2$.

Am (Z = 95)→ $1s^2, 2s^2, 2p^6, 3s^2, 3p^6, 4s^2, 3d^{10}, 4p^6, 5s^2, 4d^{10}, 5p^6, 6s^2, 4f^{14}, 5d^{10}, 6p^6, 7s^2, 5f^7, 6d^0$

→ $1s^2, 2s^2p^6, 3s^2p^6d^{10}, 4s^2p^6d^{10}f^{14}, 5s^2p^6d^{10}f^7, 6s^2p^6d^0, 7s^2$

→ 2, 8, 18, 32, 25, 8, 2

→ $[Rn]_{86}\ 5f^7 . 6d^0 . 7s^2$

The configurations of the elements from Th to Am, as given above, show that in all these elements 5*f* orbital is being progressively filled and hence the expected configuration of Cm (Z = 96) should be $[Rn]_{86}\ 5f^8 . 6d^0 . 7s^2$, but it been observed that the actual configuration of Cm is $[Rn]_{86}\ 5f^7 . 6d^1 . 7s^2$ which is obtained by shifting one of the electrons from 5*f* orbital to the vacant 6*d* orbital. This type of electronic shifting makes 5*f* orbital half-filled and hence more stable than $5f^8$ configuration as suggested by the rule of half-filled orbitals. Thus :

Cm (Z = 96)→ $1s^2, 2s^2, 2p^6, 3s^2, 3p^6, 4s^2, 3d^{10}, 4p^6, 5s^2, 4d^{10}, 5p^6, 6s^2, 4f^{14}, 5d^{10}, 6p^6, 7s^2, 5f^7, 6d^1$.

→ $1s^2, 2s^2p^6, 3s^2p^6d^{10}, 4s^2p^6d^{10}f^{14}, 5s^2p^6d^{10}f^7, 6s^2p^6d^1, 7s^2$
→ 2, 8, 18, 32, 25, 9, 2
→ $[Rn]_{86}\ 5f^7 \cdot 6d^1 \cdot 7s^2$

Bk (Z = 97) → $1s^2, 2s^2, 2p^6, 3s^2, 3p^6, 4s^2, 3d^{10}, 4p^6, 5s^2, 4d^{10}, 5p^6, 6s^2, 4f^{14}, 5d^{10}, 6p^6, 7s^2, 5f^9, 6d^0$.
→ $1s^2, 2s^2p^6, 3s^2p^6d^{10}, 4s^2p^6d^{10}f^{14}, 5s^2p^6d^{10}f^9, 6s^2p^6d^0, 7s^2$
→ 2, 8, 18, 32, 27, 8, 2
→ $[Rn]_{86}\ 5f^9 \cdot 6d^0 \cdot 7s^2$

Some authors say that the configuration of Bk is $[Rn]_{86}\ 5f^7 \cdot 6d^2 \cdot 7s^2$.

Cf (Z = 98) → $1s^2, 2s^2, 2p^6, 3s^2, 3p^6, 4s^2, 3d^{10}, 4p^6, 5s^2, 4d^{10}, 5p^6, 6s^2, 4f^{14}, 5d^{10}, 6p^6, 7s^2, 5f^{10}, 6d^0$
→ $1s^2, 2s^2p^6, 3s^2p^6d^{10}, 4s^2p^6d^{10}f^{14}, 5s^2p^6d^{10}f^{10}, 6s^2p^6d^0, 7s^2$
→ 2, 8, 18, 32, 28, 8, 2
→ $[Rn]_{86}\ 5f^{10} \cdot 6d^0 \cdot 7s^2$

Some authors are of the opinion that the configuration of Cf is $[Rn]_{86}\ 5f^9 \cdot 6d^1 \cdot 7s^2$.

Es (Z = 99) → $1s^2, 2s^2, 2p^6, 3s^2, 3p^6, 4s^2, 3d^{10}, 4p^6, 5s^2, 4d^{10}, 5p^6, 6s^2, 4f^{14}, 5d^{10}, 6p^6, 7s^2, 5f^{11}, 6d^0$.
→ $1s^2, 2s^2p^6, 3s^2p^6d^{10}, 4s^2p^6d^{10}f^{14}, 5s^2p^6d^{10}f^{11}, 6s^2p^6d^0, 7s^2$
→ 2, 8, 18, 32, 29, 8, 2
→ $[Rn]_{86}\ 5f^{11} \cdot 6d^0 \cdot 7s^2$

Fm(Z = 100)→ $1s^2, 2s^2, 2p^6, 3s^2, 3p^6, 4s^2, 3d^{10}, 4p^6, 5s^2, 4d^{10}, 5p^6, 6s^2, 4f^{14}, 5d^{10}, 6p^6, 7s^2, 5f^{12}, 6d^0$.
→ $1s^2, 2s^2p^6, 3s^2p^6d^{10}, 4s^2p^6d^{10}f^{14}, 5s^2p^6d^{10}f^{12}, 6s^2p^6d^0, 7s^2$
→ 2, 8, 18, 32, 30, 8, 2
→ $[Rn]_{86}\ 5f^{12} \cdot 6d^0 \cdot 7s^2$

Md(Z = 101)→ $1s^2, 2s^2, 2p^6, 3s^2, 3p^6, 4s^2, 3d^{10}, 4p^6, 5s^2, 4d^{10}, 5p^6, 6s^2, 4f^{14}, 5d^{10}, 6p^6, 7s^2, 5f^{13}, 6d^0$.
→ $1s^2, 2s^2p^6, 3s^2p^6d^{10}, 4s^2p^6d^{10}f^{14}, 5s^2p^6d^{10}f^{13}, 6s^2p^6d^0, 7s^2$
→ 2, 8, 18, 32, 31, 8, 2
→ $[Rn]_{86}\ 5f^{13} \cdot 6d^0 \cdot 7s^2$

No(Z = 102) → $1s^2, 2s^2, 2p^6, 3s^2, 3p^6, 4s^2, 3d^{10}, 4p^6, 5s^2, 4d^{10}, 5p^6, 6s^2, 4f^{14}, 5d^{10}, 6p^6, 7s^2, 5f^{14}, 6d^0$
→ $1s^2, 2s^2p^6, 3s^2p^6d^{10}, 4s^2p^6d^{10}f^{14}, 5s^2p^6d^{10}f^{14}, 6s^2p^6d^0, 7s^2$
→ 2, 8, 18, 32, 32, 8, 2
→ $[Rn]_{86}\ 5f^{14} \cdot 6d^0 \cdot 7s^2$

Lw(Z = 103) → $1s^2, 2s^2, 2p^6, 3s^2, 3p^6, 4s^2, 3d^{10}, 4p^6, 5s^2, 4d^{10}, 5p^6, 6s^2, 4f^{14}, 5d^{10}, 6p^6, 7s^2, 5f^{14}, 6d^1$
→ $1s^2, 2s^2p^6, 3s^2p^6d^{10}, 4s^2p^6d^{10}f^{14}, 5s^2p^6d^{10}f^{14}, 6s^2p^6d^1, 7s^2$

$\rightarrow$ 2, 8, 18, 32, 32, 9, 2

$\rightarrow$ $[Rn]_{86}\ 5f^{14}\ .\ 6d^1\ .\ 7s^2$

Now the extra electron in the next two elements *viz.* Ku (Z = 104) and Ha (Z = 105) goes to 6*d* orbital which is already singly-filled at Lw, the last member of actinide series.

Ku(Z = 104) $\rightarrow$ $1s^2, 2s^2, 2p^6, 3s^2, 3p^6, 4s^2, 3d^{10}, 4p^6, 5s^2, 4d^{10}, 5p^6, 6s^2, 4f^{14}, 5d^{10}, 6p^6, 7s^2, 5f^{14}, 6d^2$

$\rightarrow$ $1s^2, 2s^2p^6, 3s^2p^6d^{10}, 4s^2p^6d^{10}f^{14}, 5s^2p^6d^{10}f^{14}, 6s^2p^6d^2, 7s^2$

$\rightarrow$ 2, 8, 18, 32, 32, 10, 2

$\rightarrow$ $[Rn]_{86}\ 5f^{14}\ .\ 6d^2\ .\ 7s^2$

Ha(Z = 105) $\rightarrow$ $1s^2, 2s^2, 2p^6, 3s^2, 3p^6, 4s^2, 3d^{10}, 4p^6, 5s^2, 4d^{10}, 5p^6, 6s^2, 4f^{14}, 5d^{10}, 6p^6, 7s^2, 5f^{14}, 6d^3$

$\rightarrow$ $1s^2, 2s^2p^6, 3s^2p^6d^{10}, 4s^2p^6d^{10}f^{14}, 5s^2p^6d^{10}f^{14}, 6s^2p^6d^3, 7s^2$

$\rightarrow$ 2, 8, 18, 32, 32, 11, 2

$\rightarrow$ $[Rn]_{86}\ 5f^{14}\ .\ 6d^3\ .\ 7s^2$

Problems Based on the Determination of the Electronic Configuration of a Given Element

Here we will consider some examples, showing how the rules given above can be used to determine the electronic configuration of a given element.

Problem 1. *Write the electronic configuration of the elements whose atomic numbers are 24, 29, 46, 56, 57, 58, 64 and 78.*

Solution. *Element with at no. = 24.* Using Aufbau principle and the rule of half-filled 3*d* orbital, one of the 4*s* electrons is shifted to 3*d* orbital and the configuration of the element with at. no. 24 becomes :

24 $\rightarrow$ $1s^2, 2s^2, 2p^6, 3s^2, 3p^6, 4s^1, 3d^5$

$\rightarrow$ $1s^2, 2s^2p^6, 3s^2p^6d^5, 4s^1$

$\rightarrow$ 2, 8, 13, 1

Element with at. no. = 29. Using Aufbau principle and the rule of completely-filled 3*d* and half-filled 4*s* orbital, one of the 4*s* electrons is shifted to 3*d* orbital and the configuration of the element with at. no. 29 becomes :

29 $\rightarrow$ $1s^2, 2s^2, 2p^6, 3s^2, 3p^6, 4s^1, 3d^{10}$

$\rightarrow$ $1s^2, 2s^2p^6, 3s^2p^6d^{10}, 4s^1$

$\rightarrow$ 2, 8, 18, 1

Element with at. no. = 46. Using Aufbau principle and the rule of completely-filled 4*d* orbitals, both the 5*s* electrons are shifted to 4*d* orbital and the configuration of the element with at. no. 46 becomes :

46 $\rightarrow$ $1s^2, 2s^2, 2p^6, 3s^2, 3p^6, 4s^2, 3d^{10}, 4p^6, 5s^0, 4d^{10}$

$\rightarrow$ $1s^2, 2s^2p^6, 3s^2p^6d^{10}, 4s^2p^6d^{10}, 5s^0$

$\rightarrow$ 2, 8, 18, 18, 0

Element with at. no. = 56. Using Aufbau principle the configuration of the element having at. no. 56 is :

56 → $1s^2, 2s^2, 2p^6, 3s^2, 3p^6, 4s^2, 3d^{10}, 4p^6, 5s^2, 4d^{10}, 5p^6, 6s^2$

→ $1s^2, 2s^2p^6, 3s^2p^6d^{10}, 4s^2p^6d^{10}, 5s^2p^6, 6s^2$

→ 2, 8, 18, 18, 8, 2

Element with at. no. = 57. Keeping in view the deviation of Aufbau principle the last electron in this element enters 5*d* orbital instead of 4*f* orbital and hence the configuration of this element becomes :

57 → $1s^2, 2s^2, 2p^6, 3s^2, 3p^6, 4s^2, 3d^{10}, 4p^6, 5s^2, 4d^{10}, 5p^6, 6s^2, 5d^1$

→ $1s^2, 2s^2p^6, 3s^2p^6d^{10}, 4s^2p^6d^{10}, 5s^2p^6d^1, 6s^2$

→ 2, 8, 18, 18, 9, 2

Element with at. no. = 58. The configuration of this element is built up in accordance with Aufbau principle as shown below :

58 → $1s^2, 2s^2, 2p^6, 3s^2, 3p^6, 4s^2, 3d^{10}, 4p^6, 5s^2, 4d^{10}, 5p^6, 6s^2, 4f^1, 5d^2$

→ $1s^2, 2s^2p^6, 3s^2p^6d^{10}, 4s^2p^6d^{10}f^1, 5s^2p^6d^1, 6s^2$

→ 2, 8, 18, 19, 9, 2

Element with at. no. = 64. Using Aufbau principle and the rule of half-filled 4*f* orbital, one of the 4*f* electrons is shifted to 5*d* orbitals and the configuration of the element with at. no. 64 becomes :

64 → $1s^2, 2s^2, 2p^6, 3s^2, 3p^6, 4s^2, 3d^{10}, 4p^6, 5s^2, 4d^{10}, 5p^6, 6s^2, 4f^7, 5d^1$

→ $1s^2, 2s^2p^6, 3s^2p^6d^{10}, 4s^2p^6d^{10}f^7, 5s^2p^6d^1, 6s^2$

→ 2, 8, 18, 25, 9, 2

Element with at. no. = 78. In the configuration of this element one of the 6*s* electrons gets shifted to 5*d* orbitals and the configuration of the element with at. no. 78 becomes :

78 → $1s^2, 2s^2, 2p^6, 3s^2, 3p^6, 4s^2, 3d^{10}, 4p^6, 5s^2, 4d^{10}, 5p^6, 6s^1, 4f^{14}, 5d^9$

→ $1s^2, 2s^2p^6, 3s^2p^6d^{10}, 4s^2p^6d^{10}f^{14}, 5s^2p^6d^9, 6s^1$

→ 2, 8, 18, 32, 17, 1

Problem 2. *A neutral atom of an element has two K, eight L and five M electron.. Give the following information : (i) Atomic number (ii) Total number of s-orbitals (iii) Total number of p-electrons (iv) Total number of d-electrons (v) Total number of f-electrons.* *(Raj. B. Sc. I. 1977)*

Solution. As given, the electronic configuration of the neutral atom is :

K	L	M
2	8	5
$1s^2$	$2s^2.2p^6$	$3s^2.3p^3.3d^0$

Thus : (*i*) Atomic number of the element = 2 + 8 + 5 = 15

(*ii*) Total number of *s*-orbitals = one 1*s* orbital + one 2*s* orbital + one 3*s* orbital = Three *s*-orbitals

(*iii*) Total number of *p*-electrons = 6 + 3 = 9

(*iv*) Total number of *d*-electrons = 0

(*v*) Total number of *f*-electrons = 0

Problem 3. *Write the electronic configuration (with electron spins) of the following :*

(*i*) *Elements with atomic number 4, 7, 21 and 42*

(*ii*) *Ions : V^{3+}, Fe^{2+}, Br^-, S^{2-}*

Solution. (*i*) Using Aufbau principle and Hund's rule of maximum multiplicity, the configuration of the elements with at no. 4, 7 and 21 are given below :

At No. = 4 → $1s^2$, $2s^2$ or $1s^2$ [↑↓] $2s^2$ [↑↓]

At No. = 7 → $1s^2$, $2s^2$, $2p^3$ or $1s^2$ [↑↓] $2s^2$ [↑↓] $2p^3$ [↑][↑][↑]

At No. = 21 → $1s^2$, $2s^2$, $2p^6$, $3s^2$, $3p^6$, $4s^2$, $3d^1$

→ $1s^2$, $2s^2p^6$, $3s^2p^6d^1$, $4s^2$

→ 2, 8, $3s^2$ [↑↓] $3p^3$ [↑↓][↑↓][↑↓] $3d^1$ [↑][ ][ ][ ][ ] $4s^2$ [↑↓]

At. No. 42. In case of this element one of the 5*s* electrons is shifted to 4*d* orbital to make it half-filled and hence the configuration of this element becomes :

$1s^2$, $2s^2$, $2p^6$, $3s^2$, $3p^6$, $4s^2$, $3d^{10}$, $4p^6$, $5s^1$, $4d^5$

or $1s^2$, $2s^2p^6$, $3s^2p^6d^{10}$, $4s^2p^6d^5$, $5s^1$

or 2, 8, 18, $4s^2$ [↑↓] $4p^6$ [↑↓][↑↓][↑↓] $4d^5$ [↑][↑][↑][↑][↑] $5s^1$ [↑]

(*ii*) The electronic configuration of the cations are obtained by removing the electrons from the configuration of their parent atom. The number of electrons removed is equal to the number of positive charges on the cation. On the other hand the electronic configuration of the anions are obtained by adding the electrons to the configuration of their parent atom. The number of electrons added is equal to the number of negative charges carried by the anion. Now :

∵ V (at. no. = 23) → $1s^2$, $2s^2$, $2p^6$, $3s^2$, $3p^6$, $4s^2$, $3d^3$

→ $1s^2$, $2s^2p^6$, $3s^2p^6d^3$, $4s^2$

∴ V^{3+} → $1s^2$, $2s^2p^6$, $3s^2p^6d^2$

→ 2, 8, $3s^2$ [↑↓] $3p^6$ [↑↓][↑↓][↑↓] $3d^2$ [↑][↑][ ][ ][ ]

∵ Fe (at. no. = 26) → $1s^2$, $2s^2$, $2p^6$, $3s^2$, $3p^6$, $4s^2$, $3d^6$

→ $1s^2$, $2s^2p^6$, $3s^2p^6d^6$, $4s^2$,

∴ Fe^{2+} → $1s^2$, $2s^2p^6$, $3s^2p^6d^6$.

$$\rightarrow 2, 8, \quad \underset{3s^2}{\boxed{\uparrow\downarrow}} \quad \underset{3p^6}{\boxed{\uparrow\downarrow}\boxed{\uparrow\downarrow}\boxed{\uparrow\downarrow}} \quad \underset{3d^6}{\boxed{\uparrow\downarrow}\boxed{\uparrow}\boxed{\uparrow}\boxed{\uparrow}\boxed{\uparrow}}$$

$\because$ Br (at. no. = 35) $\rightarrow 1s^2, 2s^2, 2p^6, 3s^2, 3p^6, 4s^2, 3d^{10}, 4p^5$

$\therefore$ Br^- $\rightarrow 1s^2, 2s^2, 2p^6, 3s^2, 3p^6, 4s^2, 3d^{10}, 4p^6$

$\rightarrow 1s^2, 2s^2p^6, 3s^2p^6d^{10}, 4s^2p^6$

$$\rightarrow 2, 8, 18, \quad \underset{4s^2}{\boxed{\uparrow\downarrow}} \quad \underset{4p^6}{\boxed{\uparrow\downarrow}\boxed{\uparrow\downarrow}\boxed{\uparrow\downarrow}}$$

$\because$ S (at. no. = 16) $\rightarrow 1s^2, 2s^2, 2p^6, 3s^2, 3p^4$

$\therefore$ S^{2-} $\rightarrow 1s^2, 2s^2, 2p^6, 3s^2, 3p^6$

$\rightarrow 1s^2, 2s^2p^6, 3s^2p^6$

$$\rightarrow 2, 8, \quad \underset{3s^2}{\boxed{\uparrow\downarrow}} \quad \underset{3p^6}{\boxed{\uparrow\downarrow}\boxed{\uparrow\downarrow}\boxed{\uparrow\downarrow}}$$

Problem 4. *How many orbits, orbitals and electrons are there in an atom having atomic weight 23 and atomic number 11?*

Solution. At wt. = No. of protons + No. of neutrons = 23

At. no. = No. of protons = No. of electrons = 11

$\therefore$ No. of neutrons = 23 – 11 = 12

Electronic configuration = $1s^2, 2s^2p^6, 3s^1$

No. of orbits = 3 which are K, L and M

No. of orbitals = 14 (one $1s$ + one $2s$ + three $2p$ + one $3s$ + three $3p$ + five $3d$)

No. of electrons = 11.

Problem 5. *There are two elements having atomic number 7 and 9.*

(*i*) *Write their electronic configuration*

(*ii*) *Which has maximum number of unpaired electrons?*

(*iii*) *What is the valency of each?*

Solution. (*i*) Electronic configuration of the element with at. no. 7 is :

$$1s^2, 2s^2 \,.\, 2p_x^1 \,.\, 2p_y^1 . 2p_z^1$$

Electronic configuration of the element with at. no. 9 is :

$$1s^2, 2s^2 \,.\, 2p_x^2 \,.\, 2p_y^2 \,.\, 2p_z^1$$

(*ii*) The element with at. no. 7 has 3 unpaired electrons while the element with at. no. 9 has 1 unpaired electron (See the electronic configurations given above).

(*iii*) Since the element having at. no. 7 has three unpaired electrons in its valence-shell, its valency is 3. Similarly the valency of the other element is 1.

Problem 6. *An atom has 2K, 8L and 2M electrons. Find the total number of:*

(*i*) *protons*

(*ii*) *principal quantum number*

(*iii*) *sub-levels*

(*iv*) *orbits*
(*v*) *p-electrons*
(*vi*) *paired electrons*
(*vii*) *unpaired electrons*
(*viii*) *valency of the atom.*

Solution. Electronic configuration of the atom is

$$\underbrace{1s^2}_{K} \qquad \underbrace{2s^2.2p^6}_{L} \qquad \underbrace{3s^2}_{M}$$

Total number of electrons = 2 + 8 + 2 = 12

(*i*) Total no. of protons = 12
(*ii*) Principal quantum number = 3
(*iii*) Sub-levels = 6 (one 1*s* + one 2*s* + three 2*p* + one 3*s*)
(*iv*) Orbits = 3(K, L, M)
(*v*) *p*-electrons = 6
(*vi*) Paired electrons = 12
(*vii*) Unpaired electrons = 0
(*viii*) Valency of the atom = 0

Problem 7. *The electronic configuration of carbon is* $1s^2$, $2s^2.2p_x^{\,1}.2p_y^{\,1}.2p_z^{\,0}$ *and not* $1s^2$, $2s^2.2p_x^{\,2}.2p_y^{\,0}.2p_z^{\,0}$. *Explain.* *(Lucknow B.Sc. 1981)*

Solution. According to Hund's rule of maximum multiplicity, when applied to the orbitals $2p_x$, $2p_y$, and $2p_z$ belonging to 2*p* sub-shell, these orbitals should first be filled singly and then, if necessary, the pairing of electrons in each orbital begins. Thus the correct configuration of carbon is $1s^2$, $2s^2$, $2p_x^{\,1}.2p_y^{\,1}.2p_z^{\,0}$ and not $1s^2$, $2s^2.2p_x^{\,2}.2p_y^{\,0}.2p_z^{\,0}$.

Division of the Elements into s-, p-, d- and f-Block Elements

A careful examination of the electronic configurations of the elements reveals that they may be divided into four *sections* or *blocks* in the long form of the periodic table on the basis whether the last electron (also called *differentiating electron or extra electron or additional electron*) goes to *s, p, d* or *f* orbital. Thus the four blocks into which the elements in the long form of the periodic table can be divided are *s-*, *p-*, *d-* and *f*-blocks (See Table 3.2 given on page 190). For the purpose of this type of division of elements we consider only the orbital of the highest energy into which the last electron goes. This orbital may not always belong to the outermost shell.

1. s-Block Elements. In the atoms of these elements *the last electron enters the s-orbital of the ultimate (outermost) shell, i.e., the last electron goes to ns-orbital* where *n* denotes the number of the outermost shell or the number of the period. In other words, in these elements *ns*-orbital is being progressively filled. Hence the name *s-block elements.* Consequently, the valence shell electronic configuration of these elements varies from ns^1 to ns^2.

Table 3.2. Division of the elements into s, p, d, and f block elements. Thick line shown in the form of ladder (i.e. steps) separates the metals from non-metals, metalloids and inert gases

← Metals →

← s-Block elements →

	IA	
1s	H 1	He 2 (← Inert gases)

	IA	IIA
2s	Li 3	Be 4
3s	Na 11	Mg 12
4s	K 19	Ca 20
5s	Rb 37	Sr 38
6s	Cs 55	Ba 56
7s	Fr 87	Ba 88

d-Block elements; ← III B →; ← f-Block elements (Inner transition elements) →

5d	La 57	4f	Ce 58	Pr 59	Nd 60	Pm 61	Sm 62	Eu 63	Gd 64	Tb 65	Dy 66	Ho 67	Er 68	Tm 69	Yb 70	Lu 71
6d	Ac 89	5f	Th 90	Pa 91	U 92	Np 93	Pu 94	Am 95	Cm 96	Bk 97	Cf 98	Es 99	Fm 100	Md 101	No 102	Lw 103

← d-Block elements (Transition elements) →

	III B	IV B	V B	VI B	VII B	VIII	VIII	VIII	I B	II B
3d	Sc 21	Ti 22	V 23	Cr 24	Mn 25	Fe 26	Co 27	Ni 28	Cu 29	Zn 30
4d	Y 39	Zr 40	Nb 41	Mo 42	Tc 43	Ru 44	Rh 45	Pd 46	Ag 47	Cd 48
5d		Hf 72	Ta 73	W 74	Re 75	Os 76	Ir 77	Pt 78	Au 79	Hg 80
6d		Ku 104	Ha 105	106	107	108	109	110	111	112

← Non-metals, Metalloids and Inert gases →

← p-Block elements →

	III A	IV A	V A	VI A	VII A	Zero
2p	B 5	C 6	N 7	O 8	F 9	Ne 10
3p	Al 13	Si 14	P 15	S 16	Cl 17	Ar 18
4p	Ga 31	Ge 32	As 33	Se 34	Br 35	Kr 36
5p	In 49	Sn 50	Sb 51	Te 52	I 53	Xe 54
6p	Tl 81	Pb 82	Bi 83	Po 84	At 85	Rn 86

The elements of group IA (H to Fr : *Alkali metals*), group II A (Be to Ra : *Alkaline earth metals*) and He (zero group) belong to this block, since their valence-shell configuration is ns^1 and ns^2 respectively where n is the number of the period or the principal quantum number. The elements of this block are collectively known as *normal elements or representative elements.*

The valence-shell electronic configurations of *s*-block elements are given in Table 3.3.

Table 3.3. Valence-shell electronic configuration of s-block elements. The number given above the symbol of each element indicates the atomic number of the element.

Period number (n) and the orbital being filled	**IA** ns^1	**IIA** ns^2	**Zero**
n = 1, 1s	1 H $1s^1$		2 He $1s^2$
n = 2, 2s	3 Li $2s^1$	4 Be $2s^2$	
n = 3, 3s	11 Na $3s^1$	12 Mg $3s^2$	
n = 4, 4s	19 K $4s^1$	20 Ca $4s^2$	
n = 5, 5s	37 Rb $5s^1$	38 Sr $5s^2$	
n = 6, 6s	55 Cs $6s^1$	56 Ba $6s^2$	
n = 7, 7s	87 Fr $7s^1$	88 Rn $7s^2$	

It may be seen from these configurations (Table 3.3) that the last electron goes into 1s orbital in H and He (*1st period, n = 1*) into 2s orbital in Li and Be (*2nd period, n = 2*), into 3s orbital in Na and Mg (*3rd period, n* = 3), into 4s orbital in K and Ca (*4th period, n* = 4), into 5s orbital in Rb and Sr (*5th period*, n = 5), into 6s orbital in Cs and Ba (*6th period, n* = 6) and into 7s orbital in Fr and Ra (*7th period, n* = 7). Thus we see that s-block elements are located at the extreme left of the periodic table and consist of active metals. It may also be seen from the electronic configurations of *s*-block elements that the principal quantum number (*n*) of the differentiating electron for these elements is equal to the period number in which these elements are present. The properties of *s*-block elements depend on the number of electrons present in *ns* orbital.

2. p-Block Elements. In the atoms of these elements *the last etectron enters the p-orbital of the ultimate (outermost) shell, i.e. the last electron goes to np orbital.* (*n* = number of the outermost shell or period). In other words, in these elements *np*-orbital is being progressively filled. Hence the name *p-block elements*. In the atoms of these elements the *ns*-orbital is completely filled and hence the valence-shell configuration of these elements varies from ns^2p^1 to ns^2p^6. The elements of groups III A (B to TI), IVA (C to Pb), VA (N to Bi), VIA (O to Po), VII A (F to At) and zero (Ne to Rn) belong to this block, since the valence-shell configurations are ns^2p^1, ns^2p^2, ns^2p^3, ns^2p^4, ns^2p^5 and ns^2p^6 respectively. The elements of this block, like *s*-block elements, are also called *normal elements or representative elements*.

The valence-shell electronic configurations of p-block elements are given in Table 3.4.

Table 3.4. Valence-shell electronic configuration of p-block elements. The number given above the symbol of each element indicates the atomic number of the element.

Period number (*n*) and the orbital being filled	III A ns^2p^1	IV A ns^2p^2	VA ns^2p^3	VI A ns^2p^4	VII A ns^2p^5	Zero ns^2p^6
n = 2, 2*p*	5 B $2s^2p^1$	6 C $2s^2p^2$	7 N $2s^2p^3$	8 O $2s^2p^4$	9 F $2s^2p^5$	10 Ne $2s^2p^6$
n = 3, 3*p*	13 Al $3s^2p^1$	14 Si $3s^2p^2$	15 P $3s^2p^3$	16 S $3s^2p^4$	17 Cl $3s^2p^5$	18 Ar $3s^2p^6$
n = 4, 4*p*	31 Ga $3d^{10}4s^2p^1$	32 Ge $3d^{10}4s^2p^2$	33 As $3d^{10}4s^2p^3$	34 Se $3d^{10}4s^2p^4$	35 Br $3d^{10}4s^2p^5$	36 Kr $3d^{10}4s^2p^6$
n = 5, 5*p*	49 In $4d^{10}5s^2p^1$	50 Sn $4d^{10}5s^2p^2$	51 Sb $4d^{10}5s^2p^3$	52 Te $4d^{10}5s^2p^4$	53 I $4d^{10}5s^2p^5$	54 Xe $4d^{10}5s^2p^6$
n = 6, 6*p*	81 TI $4f^{14}.5d^{10}6s^2p^1$	82 Pb $4f^{14}.5d^{10}6s^2p^2$	83 Bi $4f^{14}.5d^{10}6s^2p^3$	84 Po $4f^{14}.5d^{10}6s^2p^4$	85 At $4f^{14}.5d^{10}6s^2p^5$	86 Rn $4f^{14}.5d^{10}6s^2p^6$

It may be seen from these configurations that the last electron goes into 2*p* orbital in B to Ne (2*nd* period, *n* = 2), into 3*p* orbital in Al to Ar (3*rd* period,

$n = 3$), into $4p$ orbital in Ga to Kr (*4th* period, $n = 4$), into $5p$ orbital in In to Xe (*5th* period, $n = 5$) and into $6p$ orbital in TI to Rn (*6th* period, $n = 6$). Thus the principal quantum number for these elements is equal to the period number in which these elements are present.

It may also be seen that *p*-block elements are located at the extreme right of the periodic table and consist of *metals, non-metals, metalloids* and *inert gases*.

The properties of p-block elements are determined by the number of electrons present in *np* orbitals. These elements have a general electronic configuration of the outermost shell as $ns^2.np^{1...6}$ where n stands for the number of the period.

3. d-Block Elements (Transition Elements). In these elements, either in their atomic state or in any of their common oxidation state, *the last electron enters the d-orbital of the penultimate* (*inner to ultimate*) *shell, i.e. the last electron goes to* (*n–1*)*d orbital.* In other words, in these elements $(n - 1)d$ orbitals are being progressively filled. Hence the name *d-block elements*. With the exceptions of Cr, Cu, Nb, Mo, Ru, Rh, Pd, Ag, Pt and Au, in the atoms of these elements the *ns*-orbital is completely filled (ns^2 configuration). Consequently, the valence-shell configuration of most of these elements varies from $(n - 1)d^1.ns^2$ (Group III B) to $(n - 1)d^{10}.ns^2$ (Group II B) configuration. The elements of groups III B. IV B, VB, VI B. VII B, VIII, I B and II B belong to this block as is evident from their valence-shell configurations given in Table 3.5.

These elements are located in the middle of the periodic table [*i.e.* between *s* (electropositive elements) and p-block (electronegative elements) elements] and consist of metals only. *d*-block elements are also called *transition elements*, since their properties are intermediate between those of *s* and *p*-block elements.

These elements are classified into *four series viz.* $3d$, $4d$, $5d$, and $6d$ series corresponding to the filling of $3d$, $4d$, $5d$, and $6d$ orbitals of 3rd (M-shell, $n = 3$), 4th (N-shell, $n = 4$), 5th (O-shell, $n = 5$) and 6th (P-shell, $n = 6$) shells respectively.

(*a*) **3d-series** (*1st transition series*). In the elements of this series $3d$ orbitals are being progressively filled up with electrons. This series has *ten* elements, beginning with Sc (Z = 21) and ending at Zn (Z = 30). The elements of this series lie in 4th period ($n = 4$).

(*b*) **4d-series** (*2nd transition series*). In the elements of this series $4d$ orbitals are being progressively filled up with electrons. This series also has ten elements, beginning with Y (Z = 39) and ending at Cd (Z = 48). The elements of this series lie in 5th period ($n = 4$).

(*c*) **5d-series** (*3rd transition series*). In the elements of this series $5d$ orbitals are being filled up with electrons. This series also has ten elements *viz.* La (Z = 57), Hf (Z = 72) to Hg (Z = 80). The elements of this series lie in 6th period ($n = 6$).

(*d*) **6d-series** (*4th transition series*). In the elements of this series $5d$ orbitals are being filled up with electrons At present this series has three elements viz.

Table 3.5. Valence-shell electronic configuration of d-block elements (Transition elements). The number given above the symbol of each element indicates the atomic number of the element.

Period number (n), orbital being filled and the name of the series	III B	IV B	V B	VI B	VII B	VIII			I B	II B
n = 4, 3d, 3d series	21 Sc $3d^1.4s^2$	22 Ti $3d^2.4s^2$	23 V $3d^3.4s^2$	24 Cr $3d^5.4s^1$	25 Mn $3d^5.4s^2$	26 Fe $3d^6.4s^2$	27 Co $3d^7.4s^2$	28 Ni $3d^8.4s^2$	29 Cu $3d^{10}.4s^1$	30 Zn $3d^{10}.4s^2$
n = 5, 4d, 4d series	39 Y $4d^1.5s^2$	40 Zr $4d^2.5s^2$	41 Nb $4d^4.5s^1$	42 No $4d^5.5s^1$	43 Tc $4d^5.5s^2$	44 Ru $4d^7.5s^1$	45 Rh $4d^8.5s^1$	46 Pd $4d^{10}.5s^0$	47 Ag $4d^{10}.5s^1$	48 Cd $4d^{10}.5s^2$
n = 6, 5d, 5d series	57 La $4f^0.5d^1.6s^2$	72 Hf $4f^{14}.5d^2.6s^2$	73 Ta $4f^{14}.5d^3.6s^2$	74 W $4f^{14}.5d^4.6s^2$	75 Re $4f^{14}.5d^5.6s^2$	76 Os $4f^{14}.5d^6.6s^2$,	77 Ir $4f^{14}.5d^7.6s^2$,	78 Pt $4f^{14}.5d^9.6s^1$	79 Au $4f^{14}.5d^{10}.6s^1$	80 Hg $4f^{14}.5d^{10}.6s^2$
n = 7, 6d 6d series	89 Ac $5f^0.6d^1.7s^2$	104 Ku $5f^{14}.6d^2.7s^2$	105 Ha $5f^{14}.6d^3.7s^2$	106	107	108	109	110	111	112

Ac (Z = 89), Ku (Z = 104) and Ha (Z = 105). The elements of this series lie in 7th period (n = 7) which is an incomplete period. It is expected that the elements with atomic number 106, 107, 108, 109, 110, 111 and 112 which are still to be discovered will be the members of 6*d* series, *i.e.* in these unknown elements the additional electron will enter 6*d* orbital.

All the transition elements are metals, since the number of electrons in the outer-most shell is very small, being equal to 2. They are hard, malleable and ductile and possess high tensile strength. They exhibit all the three types of structures *viz. face centred cubic* (*fcc*), *hexagonal close packed* (*hcp*) and *body centrered cubic* (*bcc*).

4. *f*-Block Elements (Inner-transition Elements). In these elements, either in their atomic state or in any of their common oxidation state, *the last electron enters the f-orbital of ante-penultimate (inner to the penultimate) shell i.e. the last electron goes to (n–2) f-orbital.* In other words, in these elements (*n*–2)*f*-orbitals are being progressively filled. Hence the name *f-block elements*. The valence shell electronic configuration of the atoms of these elements is represented as :

$$(n-2)\ f^{1...7,\ 9...14}\ .\ (n-1)d^{0,\ 1}\ .\ ns^2$$ as shown in Table 3.6.

These elements are located in group III B and have been given a separate place at the bottom of the periodic table. *f*-block elements are also called *inner-transition elements* and are classified into two series *viz.* 4*f* and 5*f* corresponding to the filling of 4*f* and 5*f* orbitals of 4th (N-shell, *n* = 4) and 5th (O-shell, *n* = 5) shells respectively. (See Table 3.6).

(*a*) **4f-series.** (*1st inner transition series*). In the elements of this series, the extra electron enters 4*f* orbital, *i.e.* 4*f* orbitals are being progressively filled with electrons. This series has 14 elements beginning with Ce (Z = 58) and ending at Lu (Z = 71). The elements of this series lie in 6th period (*n* = 6) and are called *rare-earths or lanthanides or lanthanones.*

(*b*) **5f-series.** (*2nd inner transition series*). In the elements of this series, 5*f* orbitals are being progressively filled, *i.e.*, the extra electron goes to 5*f* orbital. This series, like 4*f* series, also has 14 elements beginning with Th (Z = 90) and ending at Lu (Z = 103). The elements of this series lie in 7*th* period (incomplete period with *n* = 7) and are called *actinides or actinones*. The elements after U(Z = 92) are called *trans-uranic elements* which have been obtained as a result of atomic research. Thus *trans-uranic elements* are synthetic (man-made) elements.

Classification of Elements on the Basis of their Complete and Incomplete Electron Shells

There are four types of elements based on the presence of the complete and incomplete electron shells in them.

1. Noble Gases. *In the atoms of these elements the outermost shell which has ns^2p^6 configuration* (He which has $1s^2$ configuration is an exception) *is completely filled* (See Table 3.4). ns^2p^6 configuration is stable and hence these elements generally do not enter into ordinary chemical reactions. These elements are present in zero group of the periodic table.

Table 3.6. Valence-shell electronic configuration of f-block elements (Lanthanides and Actinides). The number given above the symbol of each element indicates the atomic number of the element. All these elements belong to Group III B

Period number (n), orbital being filled and the name of the series	Elements with their atomic numbers and valence-shell electronic configurations.						
n = 6, 4f, 4f series (Lanthanides)	58 Ce $4f^1.5d^1.6s^2$ or $4f^2.5d^0.6s^2$	59 Pr $4f^3.5d^0.6s^2$	60 Nd $4f^4.5d^0.6s^2$	61 Pm $4f^5.5d^0.6s^2$	62 Sm $4f^6.5d^0.6s^2$	63 Eu $4f^7.5d^0.7s^2$	64 Gd $4f^7.5d^1.7s^2$
	65 Tb $4f^9.5d^0.6s^2$	66 Dy $4f^{10}.5d^0.6s^2$	67 Ho $4f^{11}.5d^0.6s^2$	68 Er $4f^{12}.5d^0.6s^2$	69 Tm $4f^{13}.5d^0.6s^2$	70 Yb $4f^{14}.5d^0.6s^2$	71 Lu $4f^{14}.5d^1.6s^2$
n = 7, 5f, 5f series (Actinides)	90 Th $5f^1.6d^1.7s^2$ or $5f^0.6d^2.7s^2$	91 Pa $5f^2.6d^1.7s^2$ or $5f^1.6d^2.7s^2$	92 U $5f^3.6d^1.7s^2$	93 Np $5f^5.6d^0.7s^2$ or $5f^4.6d^1.7s^2$	94 Pu $5f^6.6d^0.7s^2$ or $5f^5.6d^1.7s^2$	95 Am $5f^7.6d^0.7s^2$	96 Cm $5f^7.6d^1.7s^2$
	97 Bk $5f^9.6d^0.7s^2$ or $5f^7.6d^2.7s^2$	98 Cf $5f^{10}.6d^0.7s^2$ or $5f^9.6d^1.7s^2$	99 Es $5f^{11}.6d^0.7s^2$	100 Fm $5f^{12}.6d^0.7s^2$	101 Md $5f^{13}.6d^0.7s^2$	102 No $5f^{14}.6d^0.7s^2$	103 Lw $5f^{14}.6d^1.7s^2$

2. Representative or Normal Elements. *In the atoms of these elements the outermost shell (i.e. nth shell) is partly filled while the inner shells (i.e., 1st, 2nd,......(n–1)th shells) are completely filled* (Table 3.3 and 3.4). The electronic configuration of the outermost shell of the atoms of these elements varies from ns^1 to ns^2p^5. The elements of IA, IIA, IIIA, IVA, VA, VIA and VIIA groups belong to this type. The configurations of the outermost shell of these elements are given below :

Group IA (Li—Fr) → ns^1
Group IIA (Be—Ra) → ns^2
Group IIIA (B—TI) → ns^2p^1
Group IVA (C—Pb) → ns^2p^2
Group VA (N—Bi) → ns^2p^3
Group VIA (O—Po) → ns^2p^4
Group VIIA (F—At) → ns^2p^5

The configurations given above show that *s* and *p* block elements constitute representative or normal elements. The groups containing these elements are also known as main groups of the periodic table. Metals, non-metals and metalloids are present in these groups.

3. Transition Elements. *In the atoms of these elements two outermost shells [i.e. nth and (n–1)th] are partly filled while the remaining inner shells [i.e. 1st, 2nd), ... (n–2)th shells] are completely filled.* (Table 3.5). These elements lie in *4th, 5th, 6th* and *7th* periods of the long form of periodic table. The complete electronic configurations of these elements can be represented as follows :

Period number	Elements	Electronic configurations	Remarks
4	Sc (Z = 21) to Zn (Z = 30)	2, 8, $3s^2p^6d^x$, 2 (x = 1, 2, 3, ..., 10)	Last shell of Cr (Z = 24) and Zn (Z = 30) contains only one electron and hence in these elements x = 5 and 10 respectively.
5	Y (Z = 39) to Cd (Z = 48)	2, 8, 18, $4s^2p^6d^x$, 2 (x = 1, 2, ...10)	Last shell of Nb (Z = 41), Mo (Z = 42), Ru (Z = 44), Rh (Z = 45) and Ag (Z = 47) contains one electron while that of Pd (Z = 46) is empty. Thus for these elements x = 4, 5, 7, 8, 10 and 10 respectively.
6	La (Z = 57)	2, 8, 18, 18, $5s^2p^6d^x$, 2 (x = 1)	—
	Hf (Z = 72) to Hg (Z = 80)	2, 8, 18, 32, $5s^2p^6d^x$, 2 (x = 2, 3, 4, ..., 10)	Last shell of Pt (Z = 78) and Au (Z = 79) contains one electron only and hence for these elements x = 9 and 10 respectively.
7	Ac (Z = 89)	2, 8, 18, 32, 18, $6s^2p^6d^x$, 2 (x = 1)	—
	Ku (Z = 104) and Ha (Z = 105)	2, 8, 18, 32, 32, $6x^2p^6d^x$, 2 (x = 2, 3)	—

The configurations given above clearly show that these elements *generally* contain two electrons in their *ns* orbital while the differentiating electron enters an available $(n—1)d$ orbital and hence the electronic configuration of the outermost shell of these elements varies from $(n—1)d^1.\ ns^2$ to $(n—1)\ d^{10}.\ ns^2$. The elements of IIIB, IVB, VB, VIB, VIIB, VIII, IB and IIB groups belong to this type. Thus we see that these elements are located between electropositive elements of groups IA and IIA (*s*-block elements) and electronegative elements of groups IIIA to VIIA (*p*-block elements).

4. Inner-transition Elements. *In the atoms of these elements three outermost shells [i.e. nth, (n—l)th and (n—2)th shells] are partly filled while the remaining inner shells [i.e. 1st, 2nd,...(n—3)th shells] are completely filled.* (Table 3.6) Lanthanides [Ac (Z = 58) to Lu (Z = 71)—fourteen elements] and actinides [Th (Z = 90) to Lw (Z = 104)—fourteen elements] belong to this type. These elements lie in 6*th* and 7*th* periods respectively. The complete electronic configuration of these elements can be represented as follows :

Lanthanides : 2, 8, 18, $4s^2p^6d^{10}f^x$, $5s^2p^6d^y$, 2 (*Six shells*)

Actinides : 2, 8, 18, 32, $5s^2p^6d^{10}f^x$, $6s^2p^6d^y$, 2 (*Seven shells*)

The values of *x* and *y* for these elements are as follows :

Lanthanides	x	y	Actinides	x	y
Ce (Z = 58)	1 or 2	1 or 0	Th (Z = 90)	1 or 0	1 or 2
Pr (Z = 59)	3	0	Pa (Z = 91)	2 or 1	1 or 2
Nd (Z = 60)	4	0	U (Z = 92)	3	1
Pm (Z = 6l)	5	0	Np (Z = 93)	5 or 4	0 or 1
Sm (Z = 62)	6	0	Pu (Z = 94)	6 or 5	0 or 1
Eu (Z = 63)	7	0	Am (Z = 95)	7	0
Gd (Z = 64)	7	1	Cm (Z = 96)	7	1
Tb (Z = 65)	9	0	Bk (Z = 97)	9 or 7	0 or 2
Dy (Z = 66)	10	0	Cf (Z = 98)	10 or 9	0 or 1
Ho (Z = 67)	11	0	Es (Z = 99)	11	0
Er (Z = 68)	12	0	Fm (Z = 100)	12	0
Tm (Z = 69)	13	0	Md (Z = 101)	13	0
Yb (Z = 70)	14	0	No (Z = 102)	14	0
Lu (Z = 71)	14	1	Lw (Z = 103)	14	1

The configurations given above clearly show that these elements have not only partially filled $(n—1)d$ orbitals but also underlying partially filled $(n—2)f$ orbitals. *ns* orbital is completely filled (ns^2 configuration). Thus the ante-penultimate shell [*i.e.* $(n—2)th$ shell] is being expanded from 18 to 32 by the addition of electrons to $(n—2)\ f$ orbital. In case of lanthanides $(n—2)f$ orbitals are $4f$ orbitals while in case of actinides $(n—2)f$ orbitals are $5f$ orbitals. Thus in case of lanthanides $4f$ orbital while in case of actinides $5f$ orbital is being progressively filled. Elements within each of these series are very similar in chemical properties.

These elements are located in group IIIB and have been given a separate place at the bottom of the periodic table.

Characteristic of *s*-Block Elements

1. Size of atoms and ions. Size of an atom is expressed in terms of atomic radius, van der Waals radius or covalent radius. For comparative purposes any of the above radii can be used to express the size of an atom.

Size of an atom of *s*-block elements decreases on moving horizontally from group I A to group II A while the same increases on moving down the group as is evident from the values of covalent radii of these elements given in Table 3.7. The size of an ion is given by ionic radius. Ionic radii also show the same variation in the periodic table as covalent radii (See Table 3.8).

Table 3.7. Cavalent radii (in Å) of s-block elements

Group I A	Group II A
Li	Be
1.23	0.90
Na	Mg
1.54	1.36
K	Ca
2.03	1.74
Rb	Sr
2.16	1.91
Cs	Ba
2.35	1.98

Table 3.8. Ionic radii (in Å) of s-block elements

Group I A (M^+ ions)	Group II A (M^{2+} ions)
Li^+	Be^{2+}
0.60	0.31
Na^+	Mg^{2+}
0.95	0.65
K^+	Ca^{2+}
1.33	0.99
Rb^+	Sr^{2+}
1.48	1.13
Cs^+	Ba^{2+}
1.69	1.35

2. Atomic volumes. Since atomic radius decreases when we move from group I A to II A, atomic volume also decreases in the same direction. Similarly with the increase of atomic radius on moving down the group, the atomic volume also increases considerably in the same direction.

Group I A elements :	Li	Na	K	Rb	Cs
Atomic volume (*c.c.*) :	13.1	23.7	45.3	55.9	70.0
Group II A elements :	Be	Mg	Ca	Sr	Ba
Atomic volume (*c.c.*) :	5.0	14.0	29.9	33.7	39.0

3. Binding energies. The binding energies of *s*-block elements are low, since there are one (Group IA) or two (Group II A) *s*-electrons per atom available for bond formation in the crystal lattice of the metal. Binding energy decreases with the increase of the atomic radius. Thus binding energy decreases on going down a group while the same increases on moving horizontally from group I A to group II A.

The binding energies of these elements are lower than those of transitional elements lying to their right in the periodic table. This is due to the fact that the one or two s-electrons in the atoms of these elements are present outside a noble gas core and are shielded from direct attraction of the nucleus.

4. Melting and boiling points. Since the binding energies of s-block elements decrease on going down a group while the same increases on moving horizontally from group I A to group II A, the melting and boiling points of these elements invariably show the same trend. (See Table 3.9)

Table 3.9. Melting and boiling points (°C) of s-block elements.

Group IA elements	Group IIA elements
Li	Be
180.5; 1330	1277; 2770
Na	Mg
97.8; 892	650; 1107
K	Ca
63.7; 760	838; 1440
Rb	Sr
38.9; 688;	768; 1380
Cs	Ba
28.7; 690	714; 1640

5. Specific heats. Specific heats of s-block elements show the same pattern as the melting and boiling points of these elements.

6. Ionisation energies. Ionisation energy of an element is defined as the amount of energy required to remove the most loosely bound electron (*i.e.* outermost shell electron) from the isolated neutral gaseous atom in its lowest energy state (*i.e.* ground state) to convert it into a gaseous cation. Thus ionisation energy of an atom, M (g) can be defined by the following process.

$$\underset{\text{Isolated gaseous atom}}{M(g)} - e^-(g) + \textit{Energy supplied (called ionization energy)} \rightarrow \underset{\text{Gaseous cation}}{M^+(g)}$$

Ionisation energy defined as above is strictly called *first ionisation energy*, since it corresponds to the removal of *one* electron only.

Ionisation energy increases on moving horizontally from group I A to group II A and decreases on moving down a group as shown below.

Group I A elements :	Li	Na	K	Rb	Cs
First ionisation energy (*K.J./mole*) :	520.3	495.8	418.9	403.0	375.7
Group II A elements :	Be	Mg	Ca	Sr	Ba
First ionization energy (*KJ/mole*) :	899.5	737.7	589.8	549.5	502.9

The values of first ionisation energies of s-block elements as given above are comparatively lower than those of p-block elements. Their lower values are due to their lower binding energies. Due to low ionisation energies, these elements have a great tendency to lose their s-electrons easily in aqueous solution and can thus form M^+ (Group I A) and M^{2+} (Group II A) ions.

$$M\ (s) - e^- \rightarrow M^+\ (aq)\ \text{(Group I A)}$$
$$M\ (s) - 2e^- \rightarrow M^{2+}\ (aq)\ \text{(Group II A)}$$

7. Electropositive or metallic character. Metallic character of an element is defined as its tendency to lose one or more electrons to form a cation.

$$M - ne^- \rightarrow M^{n+}$$

Thus greater is the tendency of an element to lose the electron to form the cation, greater is its metallic or electropositive character. Electropositive character of an element depends on the magnitude of its ionisation energy. With the increase of ionisation energy, the metallic character of the element decreases. Thus since all the s-block elements have low values of ionisation energy, these elements have strongly electropositive character, *i.e.* all the s-block elements are metals. In other words these elements have a great tendency to lose one or two s-electrons to form M^+ or M^{2+} cations.

$$M - e^- \rightarrow M^+\ \text{(Group I A)}$$
$$M - 2e^- \rightarrow M^{2+}\ \text{(Group II A)}$$

Since the ionisation energy increases on moving horizontally from group IA to group IIA and decreases on moving down a group, the metallic or electropositive character of the s-block elements and hence the basic character of their oxides decreases as we move from group IA to group II A and increases on moving down a group. For example, Na_2O is more basic than MgO.

8. Electronegativity. When two *different* atoms in a molecule are bonded together by a covalent bond, the electron pair forming the covalent bond is not shared equally by both the atoms. The electron pair lies nearer to one atom than the other atom. *The relative tendency (or ability or power) of a bonded atom in a molecule to attract the shared electron pair towards itself is termed as its electronegativity.*

Since the electropositive character of s-block elements, as we have seen above, increases down the group and decreases on moving in a period, electronegativity which is opposite of electropositive character, will show opposite periodic variation, *i.e.*, the electronegativity of s-block elements decreases down the group and increases in a period as shown below :

Elements of group IA	:	H	Li	Na	K	Rb	Cs	Fr
Electronegativity	:	2.2	0.98	0.93	0.82	0.82	0.79	0.70
Elements group II A	:		Be	Mg	Ca	Sr	Ba	Ra
Electronegativity	:		1.57	1.31	1.00	0.95	0.89	0.90

The values of electronegativity as given above are Pauling's electronegativity values. The values show that s-block elements have low values of electronegativity.

9. Oxidation states. Since the first ionisation potential of the elements of group IA is very low, these elements can lose their single ns^1 electron easily to form M^+ cation and hence show an oxidation state of +1. On the other hand, since the *2nd* ionisation potential of these elements is very high, they cannot lose the second electron residing in the (n—1)*th* shell to form M^{2+} cation and hence cannot show +2 oxidation state.

Since the first and second ionisation potentials of the elements of group II A both are low, these elements can easily lose both ns^2 electrons to form M^+ and M^{2+} cations and hence show +2 oxidation state. Thus the elements of group I A are *monovalent* (+1 oxidation state) and those of II A are *divalent* (+2 oxidation state).

10. Density. In general, these metals have high density which is because of the fact that the atoms in metallic crystal are closely packed together. However, Li, Be and Mg have low densities which are due to low atomic weights of these elements.

Density increases on moving down the group as well as on moving from group I A to group II A , since the atomic weights of the atoms also increases in the same direction.

11. Electron affinity. *Electron affinity* (abbreviated as EA) *of an element is defined as the amount of energy released in adding an extra electron from outside to an isolated neutral gaseous atom in its lowest energy state (i.e. ground state) to convert it into gaseous anion.* Thus electron affinity of an atom, M(*g*) can be defined by the following process :

$$\underset{\text{Isolated gaseous atom}}{M(g)} + e^-(g) \rightarrow \underset{\text{Gaseous anion}}{M^-(g)} + \textit{Energy released} \ (= -\ EA_1)$$

Electron affinity defined above is strictly called *first electron affinity*, since it corresponds to the addition of one electron only. Hence it is represented as $-EA_1$ where negative (minus) sign represents the release of energy.

The above process which depicts the addition of an electron to a gaseous atom is an *exothermic process*, since in the addition of an electron energy is also released.

Electron affinity is also called *electron affinity energy*. It is measured in electron volts, kilocalories or kilojoules. The values of electron affinity are shown by a negative (minus) sign which is placed before them. Negative sign represents the release of energy in the addition of an electron to M(g) atom.

Values of first electron affinity (in KJ/mole) of *s*-block elements are given below. Here minus sign indicates the release of energy (exothermic reaction) when an electron is added to an atom to convert it into an anion.

Group I A	*Group II A*
$H \xrightarrow[-72.9]{+e^-} H^-$	
$Li \xrightarrow[-59.8]{+e^-} Li^-$	$Be \xrightarrow[\approx 0]{+e^-} Be^-$

$$\mathrm{Na} \xrightarrow[-52.9]{+e^-} \mathrm{Na^-} \qquad \mathrm{Mg} \xrightarrow[\approx 0]{+e^-} \mathrm{Mg^-}$$

$$\mathrm{K} \xrightarrow[-48.9]{+e^+} \mathrm{K^-} \qquad \mathrm{Ca} \xrightarrow[\approx 0]{+e^+} \mathrm{Ca^-}$$

$$\mathrm{Rb} \xrightarrow[-46.9]{+e^-} \mathrm{Rb^-} \qquad \mathrm{Sr} \xrightarrow[\approx 0]{+e^-} \mathrm{Sr^-}$$

It may be seen that EA values of alkali metals are decreasing when we move from Li to Rb. It may also be seen that the EA values of the elements of group II A are practically zero. This is because of the fact that *ns* orbitals of the valence shell of these elements are completely-filled and hence the addition of an extra electron from outside to these atoms is not possible.

12. Heats of atomisation. Heats of atomisation of the elements of group IA and group II A (in KJ/mole) are given below :

Group I A	*Group II A*
Li	Be
163.6	327
Na	Mg
104.2	147
K	Ca
83.2	178
Rb	Sr
79.5	164
Cs	Ba
75.7	177

In case of the elements of group IA the values fall when we pass from Li to Cs. Thus the metallic bond in Li has maximum strength while the same in Cs has minimum strength.

The heat of atomisation of Be is much higher than that of other elements of group II A which have almost similar values. This shows that metallic bond strength in Be atom is maximum.

It may also be seen that the heats of atomisation increase when we pass from group IA to group IIA. Thus the metallic bond strength in case of the elements of group IIA is greater than that of the elements of group IA. This also explains why the elements of group II A are harder than those of group IA.

13. Hardness. Elements of group IA (alkali metals) are soft and hence can be cut with a knife while those of group II A (alkaine earth metals) are harder. The greater hardness of akaline earth metals is due to the fact that the metallic bonding in alkaline earth metals is stronger than that in the alkali metals, because the alkali metals have one s-electron per atom while alkaline earth metals have two *s*-electrons per atom. The greater hardness of the elements of group II A can also be explained on the basis of the fact that the elements of group II A have higher values of heat of automisation than those of the elements of group IA.

14. Reducting property. The reducing property of an element depends on the magnitude of its ionisation energy. An element with lower value of ionisation energy can easily lose electron(s) and hence can act as a good reducing agent. Thus an element with lower value of ionisation energy will be a stronger reducing agent than the element with higher ionisation energy. Now since all the *s*-block elements have low ionisation energy values, all these elements are good reducing agents. Again since the ionisation energies of these elements decrease down the group and increase along the period, the reducing property of these elements increases on going down the group and decreases as we move along the period from group IA to group II A.

The reducing property of these elements in solution is measured by the magnitude of their standard reduction potential value (E° value) of the system:

$$M^+ + e^- \rightarrow M \text{ (Group IA)}$$

$$M^{2+} + 2e^- \rightarrow M \text{ (Group II A)}$$

If we make an observation of the electrochemical series, we find that the species having negative E° value, *i.e.*, the species lying below hydrogen in the electrochemical series, show a strong tendency to lose electron(s) to undergo oxdiation and hence these are strong reducing agents. E° values of s-block elements as given below :

Group IA (Alkali metals)	*Group IIA (Alkaline earth metals)*
$Li^+ + e^- \rightarrow Li$, E° = – 3.04 V	$Be^{2+} + 2e^- \rightarrow Be$, E° = – 1.85 V
$Na^+ + e^- \rightarrow Na$, E° = – 2.71 V	$Mg^{2+} + 2e^- \rightarrow Mg$, E° = – 2.37 V
$K^+ + e^- \rightarrow K$, E° = – 2.92 V	$Ca^{2+} + 2e^- \rightarrow Ca$, E° = – 2.87 V
$Rb^+ + e^- \rightarrow Rb$, E° = – 2.92 V	$Sr^{2+} + 2e^- \rightarrow Sr$, E° = – 2.89 V
$Cs^+ + e^- \rightarrow Cs$, E° = – 2.9 V	$Ba^{2+} + 2e^- \rightarrow Ba$, E° = – 2.90 V

show that these values are negative and quite high, and, therefore, *s*-block elements are strong reducing agents. Again it is known that with the decrease of E° values (*i.e.* as E° values becomes more negative), the tendency of species to lose electrons to undergo oxidation increases and hence the reducing power of the species also increases.

Now among the alkali metals since Li has the minimum E° value (*i.e.* Li has maximum negative E° value), it is the strongest reducing agent among all the alkali metals.

Among alkaline earth metals, since E° values are decreasing (*i.e.* E° values are becoming more and more negative) from Be to Ba, the reducing power of these metals is also increasing in the same direction, *i.e.*, Be with maximum E° value (*i.e.* with minimum negative E° value) is the weakest reducing agent, while Ba which has minimum E° value (*i.e.* maximum negative E° value) is the strongest reducing agent.

When we compare the E° values of alkali metals with those of alkaline earth metals, we find that E° values increase on moving from group IA to group IIA (*i.e.* E° values becomes less negative). Due to this increase in E° values, alkaline earth metals are weaker reducing agents than alkali metals.

15. Diagonal relationship. The *s*-block elements of *2nd* period show similarities in properties with the elements which are situated diagonally opposite to them in *3rd* period. For example Li (Group IA, Period = *2nd*) shows similar properties with Mg (Group IIA, Period = *3rd*) and Be (Group IIA, Period = *2nd*) shows similarities with Al (Group IIIA, Period = *3rd*)

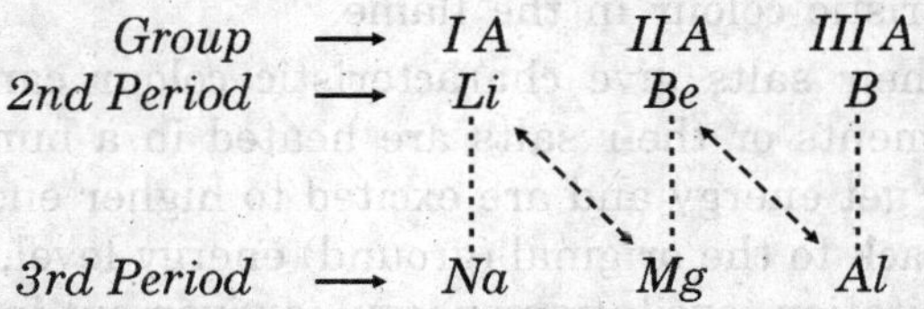

The two elements like Li and Mg or Be and Al showing similarities in properties are called diagonal neighbour or diagonal pair. The across-the-street similarity in properties existing between the two elements of a diagonal pair is known as diagonal relationship. Both the elements of each diagonal pair differ significantly in many properties from the other members of their own sub-groups. Rather they show many pronounced similarities with the elements situated diagonally opposite to them. Why the two elements of a given diagonal pair show diagonal relationship has been explained on the basis of the concept of eleotronegativity and polarisation of ions.

16. Hydration of ions. Ions of *s*-block elements are extensively hydrated in aqueous solutions. The degree of hydration of an ion depends on its size. The smaller the size of an ion, the greater is the degree of its hydration. Thus since the size of the ions increases from top to bottom in a group of *s*-block elements, the degree of hydration decreases down the group from Li^+ to Cs^+ (Group I A) and from Be^{2+} to Ba^{2+} (Group IIA) as shown below :

$$Li^+ > Na^+ > K^+ > Rb^+ > Cs^+ \text{ (Group IA)}$$

$$Be^{2+} > Mg^{2+} > Ca^{2+} > Sr^{2+} > Ba^{2+} \text{ (Group IIA)}$$

The decrease in the degree of hydration from Li^+ to Cs^+ and from Be^{2+} to Ba^{2+} as shown above is also confirmed by the fact that the heats of hydration ($-\Delta H_{hyd}$) of these ions also decrease in the same order :

M^+ ions :	Li^+	Na^+	K^+	Rb^+	Cs^+
Heat of hydration (ΔH_{hyd}) (*KJ/mole*) :	514.7	405.9	322.2	292.9	54.4
M^{2+} ions :	Be^{2+}	Mg^{2+}	Ca^{2+}	Sr^{2+}	Ba^{2+}
Heat of hydration (ΔH_{hyd}) (*KJ/mole*)	2406	1929	1632	1485	1276

It is because of the decrease of degree of hydration from Li^+ to Cs^+ and from Be^{2+} to Ba^{2+} that :

(*i*) The radii of these ions in water decrease in the order :

$Li^+(aq) > Na^+(aq) > K^+(aq) > Rb^+(aq) > Cs^+(aq)$ (Group IA)

$Be^{2+}(aq) > Mg^{2+}(aq) > Ca^{2+}(aq) > Sr^{2+}(aq) > Ba^{2+}(aq)$ (Group IIA)

(*ii*) The mobility of these ions in an electric field increases in the order :

$Li^+ < Na^+ < K^+ < Rb^+ < Cs^+$ (Group IA)

$Be^{2+} < Mg^{2+} < Ca^{2+} < Sr^{2+} < Ba^{2+}$ (Group IIA)

The above order of mobility explains why the electrical conductance of Cs^+ salts is the highest while that of Li^+ salts is the lowest.

17. Flame colouration. When *s*-block elements or their salts are heated in a bunsen flame, they give characteristic colour in the flame.

Why the *s*-block elements or their salts give characteristic colour can be explained as follows. When the elements or their salts are heated in a bunsen flame, the electrons of the elements get energy and are excited to higher energy levels. When these electrons come back to the original (ground) energy level, the energy absorbed by them during excitation (excitation energy) is given out in the form of light which appears in the visible region of the spectrum. Thus the following characteristic colours are given :

Li	Na	K	Rb Cs
Crimson	Yellow	Violet	Violet

Be	Mg	Ca	Sr	Ba	Ru
—	—	Brick red	Crimson	Apple green	Crimson

Since Be and Mg are smaller atoms, the electrons in these atoms are more strongly bound to the nucleus and hence are not excited by the energy of the flame to higher energy levels. Thus these elements do not give any colour in bunsen flame.

18. Action of air—Formation of oxides. These elements have a strong affinity for oxygen and tarnish quickly in air because they form a film of their oxide on their surface. The reactivity of the elements increases on moving down the group and decreases in a period from group IA to group II A. Thus the reactivity of alkaline earth metals (Group II A) towards O_2 is less than that of alkali metals (Group I A).

19. Action of water—Formation of hydroxides. These elements vigorously decompose H_2O with the formation of metallic hydroxide and liberation of H_2. The reaction with water increases on moving down the group and decreases as we move from group I A to group II A.

20. Action of hydrogen—Formation of hydrides. The elements combine with H_2 to form hydrides. The ease of formation of hydrides decreases on proceeding down the group as well as on moving from group I A to group II A.

21. Action of halogen—Formation of halides. All the metals combine directly with halogens at appropriate temperatures forming halides of MX (Group I A) and MX_2 (Group II A) type. Metals of group II A combine at higher temperatures.

Characteristics of p-Block Elements

1. Size of atoms or ions. Size of an atom of *p*-block elements decreases on moving from left to right in a period while the same increases on descending a group (see Table 3.10). As already said, the size of an ion is expressed as ionic radius. Ionic radii, like atomic radii, also decrease on moving from left to right in a period and increase on moving down the group (See Table 3.10).

Table 3.10. Covalent radii (in Å) of p-block elements. The values shown for the positive or negative oxidation states given in parentheses are ionic radii.

Group→ / Period ↓	III A	IV A	V A	VI A	VII A	Zero
2	B 0.82 0.20 (+3) —	C 0.77 0.15 (+4) 2.60 (−4)	N 0.75 0.11 (+5) 1.71 (−3)	O 0.73 0.09 (+6) 1.40 (−2)	F 0.72 0.07 (+7) 1.36 (−1)	Ne 0.71 — —
3	Al 1.18 0.50 (+3) —	Si 1.11 0.41 (+4) 2.71 (−1)	P 1.06 0.34 (+5) 2.12 (−3)	S 1.02 0.29 (+6) 1.84 (−2)	Cl 0.99 0.26 (+7) 1.81 (−1)	Ar 0.98 — —
4	Ga 1.26 0.62 (+3) 1.13 (+1)	Ge 1.22 0.53 (+4) 0.93 (+2)	As 1.20 0.47 (+5) 2.22 (−3)	Se 1.16 0.42 (+6) 1.98 (−2)	Br 1.14 0.39 (+7) 1.95 (−1)	Kr 1.12 — —
5	In 1.44 0.81 (+3) 1.32 (+1)	Sr 1.41 0.71 (+4) 1.12 (+2)	Sb 1.40 0.62 (+5) 2.45 (−3)	Te 1.36 0.56 (+6) 2.21 (−2)	I 1.33 0.50 (+7) 2.16 (−1)	Xe 1.31 — —
6	Tl 1.48 0.95 (+3) 1.40 (+1)	Pb 1.47 0.84 (+4) 1.20 (+2)	Bi 1.46 0.74 (+5) 1.20 (+3)	Po 1.46 — —	At 1.45 — —	Rn — — —

2. Ionisation energies. First ionisation energy of *p*-block elements are given in Table 3.11 from which it can be seen that these values generally increase on moving from left to right in a period and decrease on descending the group.

The increase in the period is not uniform *i.e.* there are certain elements which show irregular trends. For example, in 2*nd* period Be and N have higher ionisation energy values than their next neighbours namely B and O. Similarly in 3*rd* period Mg and P have higher values than Al and S A detailed explanation

Table 3.11. First ionisation energies (in KJ/mole) of p-block elements.

Group→ / Period↓	III A	IV A	V A	VI A	VII A	Zero
2	B 800.6	C 1086.4	N 1402.3	O 1314.0	F 1681.0	Ne 2080.7
3	Al 577.6	Si 786.5	P 1011.8	S 999.6	Cl 1251.1	Ar 1520.5
4	Ga 578.8	Ge 762.2	As 944.0	Se 940.9	Br 1139.9	Kr 1350.7
5	In 558.3	Sn 708.6	Sb 831.6	Te 869.3	I 1008.4	Xe 1170.4
6	Tl 589.3	Pb 715.5	Bi 703.3	Po 812.0	At —	Rn 1037.0

for this abnormal behaviour shown by Be, N, Mg and P has been given in the chapter on *Periodic Properties*. (Chapter No 4.)

3. Electropositive (or metallic or basic) character of the elements and the nature of their oxides. Electropositive character of an element depends on the magnitude of its ionisation energy. With the increase of ionisation energy, the metallic character of the elements decreases. Thus :

(*i*) Since the ionisation energy of *p*-block elements increases on moving from left to right in a period, the metallic character of the elements decreases. For example the elements lying at the extreme right of each period (*i.e.* halogens) show minimum electropositive character and hence are non-metals.

(*ii*) Since the ionisation energy of *p*-block elements decreases on descending a group, the metallic character of these elements increases in the same direction. The successive increase of metallic character of the elements on descending a group is particularly apparent in groups IV A, V A and VI A which begin with non-metals namely C, N and O and end with metals namely Pb, Bi and Po respectively.

The metallic elements of *p*-block give basic oxides (*e.g.* PbO_2, Bi_2O_3), metalloids give amphoteric oxides (*e.g.* GeO_2, As_2O_3, Sb_2O_3) while non metallic elements give acidic oxides (*e.g.* CO_2, P_4O_{10}, SO_3, Cl_2O_7). The variation of acidic/basic nature of these oxides in the periodic table depends on the metallic (electropositive) character of the element forming the oxide. With the decrease of the metallic character of the elements on moving across a period, the oxides of the elements of the same period become more and more acidic in the same direction. For example the oxides of the elements of 3*rd* period become more and more acidic when we proceed from Al_2O_3 to Cl_2O_7.

Group :	III A	IVA	V A	VI A	VII A
Oxides :	Al_2O_3 Amphoteric	SiO_2 Acidic	P_4O_{10} More acidic	SO_3 More acidic	Cl_2O_7 Most acidic

Asidic nature : —*Acidic nature increasing* →

With the increase of the metallic character of the elements, on descending a group, the oxides of the elements of the same group become more and more basic (*i.e.* less and less acidic) in the same direction. For example the oxides of MO_2 type of the elements of group IV A become more and more basic (*i.e.* less and less acidic) from CO_2 to PbO_2 as shown below :

$$\underbrace{CO_2,\ SiO_2,}_{\text{Acidic}}\quad \underbrace{GeO_2,\ SnO_2,}_{\text{Amphoteric}}\quad \underset{\text{Basic}}{PbO_2}$$

—*Basic character increasing* →

Similarly the oxides of M_2O_3 type of the elements of group VA become more and more basic (*i.e.* less and less acidic) from N_2O_3 to Bi_2O_3 as shown below :

$$\underbrace{N_2O_3,\ P_2O_3,}_{\text{Acidic}}\quad \underbrace{As_2O_3,\ Sb_2O_3,}_{\text{Amphoteric}}\quad \underset{\text{Basic}}{Bi_2O_3}$$

—*Basic character increasing* →

4. Electronegativity. Since the electropositive character of *p*-block elements, as we have seen above, decreases on moving in a period and increases down the group, electronegativity which is opposite of electropositive character will show opposite periodic variation, *i.e.* the electronegativity of *p*-block elements decreases down the group and increases in a period as is evident from the values given in Table 3.12.

Table 3.12. Electronegativity values (Pauling's values) of p-block elements.

Group→ / Period ↓	III A	IV A	V A	VI A	VII A	Zero
2	B 2.0	C 2.5	N 3.0	O 3.5	F 4.0	Ne –
3	Al 1.5	Si 1.8	P 2.1	S 2.5	Cl 3.0	Ar –
4	Ga 1.6	Ge 1.8	As 2.0	Se 2.4	Br 2.8	Kr 2.9
5	In 1.7	Sn 1.8	Sb 1.9	Te 2.1	I 2.5	Xe 2.6
6	Tl 1.8	Pb 1.8	Bi 1.9	Po 2.0	At 2.2	Rn –

5. Oxidation states and inert electron pair effect. The lighter elements of *p*-block like those of 2*nd* and 3*rd* period show a variety of oxidation states, both positive and negative, as shown below :

Group	III A	IV A	V A	VI A	VII A
Elements of 2nd period :	B	C	N	O	F
Positive oxidation states :	+3	+4	+5	–	–
Negative oxidation states :	–	–4	–3	–2	–1
Elements of 3rd period :	Al	Si	P	S	Cl
Positive oxidation states :	+3	+4	+5	+6	+7
Negative oxidation states :	–	–4	–3	–2	–1

It may be seen that on moving from left to right in each period the positive oxidation states increase while the negative oxidation states decrease.

Some of the heavier elements of *p*-block like those of groups III A (Ga, In, Tl), IV A (Ge, Sn, Pb), V A (As, Sb, Bi) and VI A (Te, Po) show two oxidation states. These oxidation states differ from each other by two units. The higher oxidation state for most of the elements is equal to their group number, G while the lower one is equal to (G–2) as shown below :

III A (ns^2p^1) *G = 3*	*IV A* (ns^2p^2) *G = 4*	*V A* (ns^2p^3) *G = 5*	*VI A* (ns^2p^4) *G = 6*
–	–		
–	–	–	
Ga	Ge	As	
+1, +3	+2, +4	+3, +5	
In	Sn	Sb	Te
+1, +3	+2, +4	+3, +5	+4. +6
Tl	Pb	Bi	Po
+1, + 3	+2, +4	+3, +5	+2, +4

Group number oxidation state, G (*i.e.* higher oxidation state) is obtained when all the *ns* and *np* electrons from ns^2p^x configuration of *p*-block elements (x = 1, 2, 3 and 4 for the elements of groups IIIA, IVA, VA and VIA respectively) are lost, while the lower oxidation state equal to (G—2) is obtained when only *np*-electrons are lost and the *ns*-electron pair, due to its extra stability, remains inert, *i.e.*, it is not lost. Such a pair of *ns*-electrons is called **inert electron pair** and the effect caused by it is known as **inert pair effect**.

How the concept of inert electron pair can be used to explain the two oxidation states (*i.e.* electrovalencies) of the elements given above can be shown by considering the following examples :

(a) Electrovalencies of Ga = + 1 and + 3. In chemical combinations Ga (Z = 31) atom which has $4s^2p^1$ as its valence-shell configuration initially loses its $4p^1$ electron, forming Ga^+ ion and thus shows an electrovalency of +1. In Ga^+ ion $4s^2$ electron pair remains inert due to its extra stability and hence does not participate in bonding. This $4s^2$ electron pair is called *inert electron pair*. If the $4s^2$ electron pair participates in bonding, both the electrons of $4s^2$ pair are lost by Ga^+ ion and Ga^{3+} ion is formed. Thus the variable electrovalencies of +1 and +3 of Ga atom are explained.

$$\underset{\substack{(2,\,8,\,18,\,4s^2.4p^1)\\(\text{O.S.}=0)}}{Ga^\circ} \xrightarrow[(-1e^-)]{4p^1 \text{ electron lost}} \underset{\substack{(2,\,8,\,18,\,4s^2)\\(\text{O.S.}=+1)}}{Ga^{1+}} \xrightarrow[(-2e^-)]{4s^2 \text{ electrons lost}} \underset{\substack{(2,\,8,\,18)\\(\text{O.S.}=+3)}}{Ga^{3+}}$$

Here O.S. indicates the oxidation state of the atom or ion as the case may be.

(b) *Electrovalencies of Tl = + 1 and +3.* These are explained as follows :

$$\underset{\substack{(2,\,8,\,18,\,32,\\18,\,6s^2.6p^1)\\(\text{O.S.}=0)}}{Tl^\circ} \xrightarrow[(-1e^-)]{6p^1 \text{ eletron lost}} \underset{\substack{(2,\,8,\,18,\,32,\,18,\,6s^2)\\(\text{O.S.}=+1)}}{Tl^{1+}} \xrightarrow[(-2e^-)]{6s^2 \text{ electrons lost}} \underset{\substack{(2,\,8,\,18,\,32,\,18)\\(\text{O.S.}=+3)}}{Tl^{3+}}$$

(c) *Electrovalencies of Sn = +2 and +4.* These are explained as follows :

$$\underset{\substack{(2,\,8,\,18,\,18,\,5s^2.5p^2)\\(\text{O.S.}=0)}}{Sn^\circ} \xrightarrow[(-2e^-)]{5p^2 \text{ electrons lost}} \underset{\substack{(2,\,8,\,18,\,18,\,5s^2)\\(\text{O.S.}=+2)}}{Sn^{2+}} \xrightarrow[(-2e^-)]{5s^2 \text{ electrons lost}} \underset{\substack{(2,\,8,\,18,\,18)\\(\text{O.S.}=+4)}}{Sn^{4+}}$$

(d) *Electrovalencies of As = +3 and +5.* These are explained as follows :

$$\underset{\substack{(2,\,8,\,18,\,4s^2.4p^3)\\(\text{O.S.}=0)}}{As^\circ} \xrightarrow[(-3e^-)]{4p^3 \text{ electrons lost}} \underset{\substack{(2,\,8,\,18,\,4s^2)\\(\text{O.S.}=+3)}}{As^{3+}} \xrightarrow[(-2e^-)]{4s^2 \text{ electrons lost}} \underset{\substack{(2,\,8,\,18)\\(\text{O.S.}=+5)}}{As^{5+}}$$

In general the variable electrovalency in these elements is due to the presence of ns^2 inert electron pair. The inertness of ns^2 electrons is found only (*i*) when the ns^2-electrons are in 4*th* or higher principal quantum number, and (*ii*) when their loss does not give a species with a noble gas configuration.

A partial explanation of the inert electron pair effect is given by the fact that the outer ns^2 electrons penetrate to some extent the 18-electron shell immediately below them [*i.e.* $(n—1)s^2p^6d^{10}$ electrons]. ns^2-electrons are, thus, less effectively screened from the attractive power of the nucleus and so are nearer to the nucleus and hence are more strongly held to the atom. As a consequence the energy required to remove these ns^2-electrons is very high and cannot be compensated by the energy released even on additional bond formation.

The inert electron pair effect increases as we move down a group. For example this effect increases as we move from Ge to Pb through Sn (Group IV A) (Ge < Sn < Pb), indicating that this effect is not much marked in Ge compounds. In order to show how this effect changes the properties of the elements we can consider the elements of group IV A. In this group the first two elements namely C and Si uniformly show +4 oxidation state while the remaining three elements *viz.* Ge, Sn and Pb show both +2 and +4 oxidation states. Now since the inert electron pair effect in Ge, Sn and Pb is in the order : Ge <Sn < Pb, stability of M^{2+} (divalent ions) and M^{4+} (tetravalent ions) ions is in the order :

$$\mathbf{Ge^{2+} < Sn^{2+} < Pb^{2+}}$$
$$\mathbf{Ge^{4+} > Sn^{4+} > Pb^{4},}$$

while that of M^{2+} and M^{4+} ions of the same element is in the order :

$$Ge^{2+} < Ge^{4+}$$
$$Sn^{2+} < Sn^{4+}$$
$$Pb^{2+} >> Pb^{4+}$$

The above order clearly shows that Ge^{2+} compounds are *unstable* and hence can easily be oxidised to Ge^{4+} compounds. In other words Ge^{2+} compounds are *strong reducing agents* while Ge^{4+} compounds are *stable*.

$$\underset{\text{Unstable (Reduced form, i.e. reducing agent)}}{Ge^{2+}} - 2e^- \xrightarrow{\text{Oxidation}} \underset{\text{Stable (Oxidised form, i.e. oxidising agent)}}{Ge^{4+}}$$

Similarly Sn^{2+} compounds are also *unstable* and hence can easily be oxidised to Sn^{4+} compounds. In other words Sn^{2+} compounds (*e.g.* SnO, $SnCl_2$) are *reducing agents* while Sn^{4+} compounds are *stable* and *covalent*.

$$\underset{\text{Unstable (Reduced form i.e. reducing agent)}}{Sn^{2+}} - 2e^- \xrightarrow{\text{Oxidation}} \underset{\text{Stable covalent (Oxidised form, i.e. oxidising agent)}}{Sn^{4+}}$$

Now since Pb shows the maximum inert electron pair effect, Pb^{2+} ion is much more stable than Pb^{4+} ion as shown above *i.e.* Pb^{4+} compounds are unstable and hence can easily be reduced to Pb^{2+} compounds. In other words Pb^{4+} compounds (*e.g.* PbO_2, $PbCl_4$) are *powerful oxidising agents* while Pb^{2+} compounds are *stable* and *ionic*.

$$\underset{\text{Unstable (Oxidised form i.e. oxidising agent)}}{Pb^{4+}} + 2e^- \xrightarrow{\text{Reduction}} \underset{\text{Stable and ionic (Reduced form i.e. reducing agent)}}{Pb^{2+}}$$

6. Electron affinity. Electron affinity values (EA values) of *p*-block elements are given in Table 3.13. Minus sign indicates the release of energy while plus sign indicates the absorption of energy.

From these values the following points may be noted :

(*i*) On moving from left to right in a period, EA values increase. However, there are certain elements in each period which have abnormal values. For example in *2nd* period N and Ne have abnormal values. Similarly in *3rd* period P and Ar have abnormal values. The abnormal values of N (*2nd* period) and P (*3rd* period) have been explained on the basis of the extra stability associated with their half-filled 2*p* (in N) and 3*p* (in P) orbitals. Being extraordinarily stable these orbitals are not able to accept the extra electron. Hence these elements have comparatively smaller EA values. On the basis of extra stability of ns^2p^6 configuration of Ne and Ar, their zero values can be explained.

Table 3.13. Electron affinity values (in KJ/mole) of p-block elements

Group→ / Period↓	III A	IV A	V A	VI A	VII A	Zero
1						$He \xrightarrow{+e^-} He^-$ ~ 0
2	$B \xrightarrow{+e^-} B^-$ –23	$C \xrightarrow{+e^-} C^-$ –122	$N \xrightarrow{+e^-} N^-$ –20.1	$O \xrightarrow{+e^-} O^-$ –141 $O \xrightarrow{+2e^-} O^{2-}$ +780	$F \xrightarrow{+e^-} F^-$ –322	
3	$Al \xrightarrow{+e^-} Al^-$ –44	$Si \xrightarrow{+e^-} Si^-$ –120	$P \xrightarrow{+e^-} P^-$ –74	$S \xrightarrow{+e^-} S^-$ –200.4 $S \xrightarrow{+2e^-} S^{2-}$ +590	$Cl \xrightarrow{+e^-} Cl^-$ –348.7	$Ar \xrightarrow{+e^-} Ar-$ ~ 0
4	$Ga \xrightarrow{+e^-} Ga^-$ –36	$Ge \xrightarrow{+e^-} Ge^-$ –116	$As \xrightarrow{+e^-} As^-$ –77	$Se \xrightarrow{+e^-} Se^-$ –195 $Se \xrightarrow{+2e^-} Se^{2-}$ +420	$Br \xrightarrow{+e^-} Br^-$ –324.5	$Kr \xrightarrow{+e^-} Kr-$ ~ 0
5	$In \xrightarrow{+e^-} In^-$ –34	$Sn \xrightarrow{+e^-} Sn^-$ –121	$Sb \xrightarrow{+e^-} Sb^-$ –101	$Te \xrightarrow{+e^-} Te^-$ –190.1	$I \xrightarrow{+e^-} I^-$ –295	$Xe \xrightarrow{+e^-} Xe-$ ~ 0

(*ii*) On moving down the group, EA values go on decreasing as is evident from these values for (*a*) C, Si and Ge (Group IV A) (*b*) S, Se and Te (Group VI A) (*c*) O, S and Se for two electrons (Group VI A) and (*d*) Cl, Br and I (Group VII A). However, EA values of some of the elements of 2*nd* period are lower than the EA values of the elements lying just below them in 3*rd* period. For example B < Al, N < P, < O < S and F < Cl. The lower values of the elements of 2*nd* period are explained by saying that, due to the smaller size of the atoms of the elements of 2*nd* period, the addition of an extra electron to these atoms produces high electron density round the resulted anions. This high electron density increases the repulsion between the electrons (called electron—electron repulsion) already present in the relatively compact 2*p* orbital of the 2*nd* shell of these atoms and the electron being added. Due to this electron—elctron repulsion, the atoms of the elements of 2*nd* period show lesser tendency to attract the extra electron from outside and hence lower value of electron affinity for these elements.

7. Oxidising and reducing property. The oxidising property of an element depends on the magnitude of its electron affinity. An element with higher value of electron affinity can easily accept electron(s) and hence can act as a good oxidising agent. Thus an element with higher value of electron affinity will be a stronger oxidising agent than the element with lower value of electron affinity. Now since electron affinity decreases on descending a group and increases along the period, the oxidising property of p-block elements decreases on descending the group and increases from left to right across the period. Thus the halogens which lie at the extreme right of the periodic table and have high values of electron affinity are strong oxidising agents.

The reducing property of an element depends on the magnitude of its ionisation energy. An element with lower value of ionisation energy can easily lose electron(s) and hence can act as a good reducing agent. Thus an element with lower value of ionisation energy will be a stronger reducing agent than the element with higher ionisation energy. Now since some of the *p*-block elements have low values of ionisation energy, they are good reducing agents. Again since the ionisation energies of these elements decrease down the group and increase along the period, the reducing property of *p*-block elements increases on going down the group and increases as we move along the period from group III A to group VII A.

The oxidising and reducing property of these elements in solution is measured by the magnitude of their standard reduction potential (E° values).

If we make an observation of electrochemical series, we find that some of the elements have positive E° values while others have negative E° values. The elements with positive E° values, *i.e.*, the elements lying above hydrogen in the series have a strong tendency to gain electrons to undergo reduction and hence they are strong oxidising agents. For example the non-metals like F_2, Cl_2, Br_2, I_2 and O_2 which have positive E° values are all strong oxidising agents. With the decrease of E value of a given element the tendency of that element to gain electrons to undergo the reduction decreases and hence the oxidising power of that element also decreases. For example among the non-metals mentioned above, F_2, which has the maximum E° value is the strongest oxidising agent while O_2 which has the lowest E° value is the weakest oxidising agent. Thus the oxidising power of these non-metals is in the order :

$$F_2 > Cl_2 > Br_2 > I_2 > O_2,$$

since E° values are also decreasing in the same order as shown below :

Reaction	E°	
$F_2 + 2e^- \rightarrow 2F^-$,	$E° = +2.85V$	E° values decreasing ↓ ; Oxidising power decreasing ↓
$Cl_2 + 2e^- \rightarrow 2Cl^-$,	$E° = +1.36V$	
$Br_2 + 2e^- \rightarrow 2Br^-$,	$E° = +1.06V$	
$I_2 + 2e^- \rightarrow 2I^-$,	$E° = +0.53V$	
$O_2 + 2H^+ + 2e^- \rightarrow 2OH^-$,	$E° = +0.40V$	

The elements with negative E° values *i.e.* the elements lying below hydrogen in the electrochemical series, have a strong tendency to lose electrons to undergo oxidation and hence they are strong reducing agents. For example both Al and Sn have negative E° values and hence are strong reducing agents.

$$Al^{3+} + 3e^- \rightarrow Al, \; E° = -1.67V$$

$$Sn^{2+} + 2e^- \rightarrow Sn, \; E° = -0.14V$$

With the decrease of E° value of a given element (*i.e.* as E° value becomes more negative), the tendency of that element to lose electrons to undergo oxidation increases and hence the reducing property of that element also increases. For example Al whose E° value is lower (*i.e.* E° value of Al is more negative) than that of Sn (*i.e.* E° value for Sn is less negative), Al is a stronger reducing agent as compared to Sn.

8. Diogonal relationship. Boron which is a *p*-block element and is present in the *2nd* period of the periodic table shows similarities in properties with silicon which is situated diagonally opposite to it in 3rd period. This type of across-the-street similarity in properties existing between B and Si is called diagonal relationship. This relationship can be explained on the basis of the concept of electronegativity and polarisation of ions.

9. Flame colouration. None of the *p*-block elements or its salt gives characteristic colour in the bunsen flame, when it is heated in it. The reason is that excitation energy given out by the electrons, when they come back to the original (ground) energy level, dces not appear in the vissible region of the spectrum.

10. Formation of hydrides. Many of the elements of *p*-block form hydrides. The stability of the hydrides of the elements of a given group decreases as we descend the group. For example the stability of the hydrides of group VA decreases from NH_3 to BiH_3.

$$NH_3 > PH_3 > AsH_3 > SbH_3 > BiH_3$$

—Stability decreases →

Again we know that, since the groups IV A, V A and VI A begin with non-metals (C, N and O) and end with metals (Pb, Bi and Po), the hydrides of non-metals are more stable than those of metals.

The aqueous solution of these hydrides show acidic character which increases from top to bottom in a group. For example the acidic nature of the hydrides *viz.*, HF, HCl, HBr and HI in aqueous solution increases from HF to HI.

$$HF < HCl < HBr < HI$$

—Acidic character increases →

11. Formation of halides. *p*-block elements also form halides. The covalent character of the halides containing different cations belonging to the *same period* of the periodic table but same halide ion increases from left to right in the period. For example in *2nd* period BCl_3 is partially ionic while CCl_4 is covalent. Similarly in *5th* period $SbCl_3$ is partly ionic whereas $TeCl_4$ is essentially covalent. The increase in covalent character from left to right in a period is due to the increase of the polarising power of the cation to polarise the anion due to the decrease of the size of the cations from left to right in a period (*One of the Fajans's rules*).

The covalent character of the halides containing the same cation but different anions belonging to the same group increases from top to bottom in a group. For example AlF_3 has essentially ionic character. $AlCl_3$ has intermediate character while $AlBr_3$ and AlI_3 have essentially covalent character.

AlF_3	$AlCl_3$	$AlBr_3$, AlI_3
Essentially ionic character	Intermediate character	Essentially covalent character

—Covalent character increasing →

The increase in covalent character from top to bottom is due to the increase of the polarisability of the anions or due to the increase of the size of the anions from top to bottom in a group (*One of the Fajans's rules*).

Characteristics of d-Block Elements

1. Metallic character. All the transition elements are metals, since the number of electrons in the outermost shell is very small; being equal to 2. They are hard, malleable and ductile and possess high tensile strength. They exhibit all the three types of structures : face centred cubic (*fcc*), hexagonal close packed (*hcp*), and body centered cubic (*bcc*). Metals of VIII and IB groups are softer and more ductile than other metals.

It appears that covalent and metallic bonding both exist in the atoms of transition metals. *The presence of unfilled d-orbitals favours covalent bonding.*

These metals are good conductors of heat and electricity.

2. Melting and boiling points. These are very heavy metals and have high melting and boiling points. The last member of each series, *i.e.*, Zn, Cd, and Hg have comparatively lower values. This is because these metals have *completely filled d-orbitals with no unpaired electrons* (*n*) as is evident from their valence-shell electronic configurations given below :

$$Zn \rightarrow 3d^{10}\ 4s^2 \text{ with } n = 0$$

$$Cd \rightarrow 4d^{10}\ 4s^2 \text{ with } n = 0$$

and

$$Hg \rightarrow 5d^{10}\ 6s^2 \text{ with } n = 0$$

3. Atomic volumes and densities. The atomic volumes of the metals are much lower than those of alkaline earth metals. As the inner orbitals get filled, the increased nuclear charge pulls the electronic cloud inward. The atomic volume, therefore, decreases. The decrease in atomic volume increases the density. Accordingly, the densities of transition elements are very high as compared to those of the metals of groups I A and II A.

4. Ionisation potentials. The ionisation potential values of most of the *d*-block elements lie between those of *s* and *p*-block elements. *Thus d-block elements are less electropositive than the s-block elements and more so than p-block elements.* Consequently, these elements *do not form ionic compounds so readily as s-block elements* (alkali and alkaline earth metals). In general, compounds with metals in lower valency state are ionic and others are covalent. The tendency to form ionic compounds decreases as the size of the atom increases.

5. Low reactivity. These elements have low reactivity and this tendency is most pronounced in Au and Pt which are, therefore, called noble metals. Their low reactivity is due to the high values of their melting and boiling points, heats of sublimation and ionisation potential.

6. Colour. The transition metals usually form coloured compounds. If an ion is coloured, it means that it absorbs light of a particular colour whose wavelength is in the visible region *viz.* λ = 400–7000 Å and reflects (*i.e.* transmits) the light which has the colour complementary to that of the absorbed light. This complementary colour which is actually the colour of the reflected light is the colour of the substance.

The colour is due to the promotion of unpaired *d*-electrons from one energy level to the other in *d* level. In general the atoms, ions or molecules which have unpaired electrons in *d*-orbitals (*i.e.* partially filled *d*-orbitals) are coloured because it is only in case of such orbitals that the promotion of unpaired *d*-electrons to higher energy level within the same sub-shell is possible. On the other hand, the atoms. ions or molecules having all the electrons paired in *d*-orbitals (*i.e.* completely filled *d*-orbitals) or no electrons (*i.e.* empty *d*-orbitals) are colourless, since in case of such *d*-orbitals there are no vacant *d*-orbitals to accept the electrons promoted to them. Thus, if n is the number of unpaired electrons, then the atoms, ions or molecules having $n = 0$ are *colourless* while those having $n = 1, 2, \ldots$ are *coloured* as shown below :

Ion	*Configuration*	n	*Colour*
Sc^{3+}	$3d^0$	0	Colourless
Ti^{4+}	$3d^0$	0	Colourless
Cu^+	$3d^{10}$	0	Colourless
Zn^{2+}	$3d^{10}$	0	Colourless
Ti^{3+}	$3d^1$	1	Purple
V^{3+}	$3d^2$	2	Green
Cr^{3+}	$3d^3$	3	Green
Fe^{2+}	$3d^6$	4	Pale green or faintly green
Fe^{3+}	$3d^5$	5	Yellow

7. Variable valency. Transition elements show variable valency. This is because in their case the core or kernel left after their atom loses valency electron is unstable and tends to lose one or more electrons further. For example, in case of iron we can represent the two valencies as below :

Fe 26 = 2, 8, 14, 2

Fe^{2+} 24 = 2, 8, 14, (Core)

The core being unstable loses one more electron to give Fe^{3+} ion

Fe^{3+} 23 = 2, 8, 13

8. Catalytic properties. Most of transition metals, their alloys and compounds are used as catalysts. Some common examples are: Pt, Ni, Fe, Cr, V_2O_5 etc. Catalytic power of these metals is due to either the use of their *d*-orbitals or due to the formation of interstitial compounds which absorb and activate the reacting substances.

9. Magnetic properties. Most of the transition metals are paramagnetic. *i.e.*, they tend to set themselves with their lengths parallel to the field, when they are placed between the magnetic poles. Thus these are attracted into magnetic field. This property is due to *the presence of unpaired electrons in d-orbitals*. Thus, the transition elements having all the electrons paired are diamagnetic. Fe and Co are ferromagnetic and can be magnetised.

10. Tendency to form complexes. The cations of transition elements have a great tendency to form complexes with several ligands. This tendency is due to the following reasons :

(*a*) The cations are relatively very small in size and hence have high positive density which makes it easy for the cations to accept the lone pairs of electrons from the ligands.

(*b*) The cations have *vacant* $(n—1)d$ orbitals which are of appropriate energy to accept lone pairs of electrons from the ligands for bonding with them.

11. Formation of interstitial compounds. Small non-metallic atoms (*e.g.* H, B, C, N, etc.) are able to fit into the interstitial sites of transition metal lattices to form the compounds which are called **interstitial compounds.** These have variable composition and cannot, therefore, be expressed by simple formulae, *i.e.* these are *non-stoichiometric*.

Characteristics of f-Block Elements

The characteristics of *f*-block elements have been discussed in chapter 50.

Questions with Answers

Q.1 Give different steps used for writing the valence-shell electronic configuration of a given element.

Ans. The technique used for writing the valence-shell configuration of an element consists of the following steps :

(*i*) Find out the difference between the atomic number of the given element and that of the noble gas present in the previous period. For example the difference between Pb_{82} (6*th* period) and Xe_{54} (5*th* period) = 82 – 54 = 28. Let this difference be denoted by *x*. This difference denotes the number of electrons present in the valence-shell of the given element.

(*ii*) Now find out the sub-shells which are to be filled up with *x* electrons.

(*iii*) Then fill up these sub-shells with *x* electrons. This gives the valence-shell configuration of the given element.

Example. *Find out the valence-shell configuration of the atom of the following elements :* Te_{52}, La_{57}, Pr_{59}, Pt_{78}, Pb_{82}, Ac_{89}, U_{92} *and* Ha_{105}.

Solution : (*i*) $\mathbf{Te_{52}}$: This element is present in 5*th* period and $x = Te_{52} - Kr_{36} = 16$.

Hence valence-shell configuration of Te_{52} (5*th* period) = $5s^2\ 4d^{10}\ 5p^4 = 4d^{10}\ 5s^2\ 5p^4$

(*ii*) $\mathbf{La_{57}}$: Here $x = La_{57} - Xe_{54} = 3$

Hence valence-shell configuration of Li_{57} (6*th* period) = $6s^2 5d^1 = 5d^1 6s^2$

(*iii*) $\mathbf{Pr_{59}}$: Here $x = Pr_{59} - Xe_{54} = 5$

∴ Expected valence-shell configuration of Pr_{59} (6*th* period) = $6s^2 5d^1 4f^2$

The actual valence-shell configuration is $6s^2 5d^0 4f^3 = 4f^3 5d^0 6s^2$

(*iv*) $\mathbf{Pt_{78}}$: Here $x = Pt_{78} - Xe_{54} = 24$

Expected valence-shell configuration of Pt_{78} (6*th* period) = $6s^2 5d^1 4f^{14} 5d^7 = 6s^2 5d^8 4f^{14}$

But the actual valence-shell configuration = $6s^1 5d^9 4f^{14} = 4f^{14} 5d^9 6s^1$

(*v*) $\mathbf{Pb_{82}}$ **:** Here $x = Pb_{82} - Xe_{54} = 28$

$\therefore$ Valence-shell configuration of Pb_{82} (6*th* period) $= 6s^25d^14f^{14}5d^96p^2 = 6s^25d^{10}4f^{14}6p^2 = 4f^{14}5d^{10}6s^26p^2$

(*vi*) $\mathbf{Ac_{89}}$ **:** Here $x = Ac_{89} - Rn_{86} = 3$

$\therefore$ Valence-shell configuration of Ac_{89} (7*th* period) $= 7s^26d^1 = 6d^17s^2$

(*vii*) $\mathbf{U_{92}}$ **:** Here $x = U_{92} - Rn_{86} = 6$

$\therefore$ Valence-shell configuration of U_{92} (7*th* period) $= 7s^26d^15f^3 = 5f^36d^17s^2$

(*viii*) $\mathbf{Ha_{105}}$ **:** Here $x = Ha_{105} - Rn_{86} = 19$

$\therefore$ Valence-shell configuration of Ha_{105} (7*th* period) $= 7s^26d^15f^{14}6d^2 = 7s^26d^35f^{14} = 5f^{14}6d^37s^2$

Q.2 What are different steps for writing the electronic configuration of an ion.

Ans. In order to write the electronic configuration of a given ion proceed as follows :

(*i*) Write the complete or valence-shell configuration of the element from which the ion has been derived.

(*ii*) (*a*) In case of a cation, remove the electron(*s*), equal to the number of positive charges on the cation from the outer-most shell of the parental atom. This gives the electronic configuration of the cation.

(*b*) In case of an anion, add electron(*s*), equal to the number of negative charges on the anion to the outer-most shell of the parental atom. This gives the configuration of the anion.

Examples. *(i) Write the complete and valence-shell configuration of Pb^{2+} ion (Z = 82)*

Solution : Configuration of Pb-atom (Z = 82) $= [Xe]_{54}\ 6s^25d^{10}4f^{14}6p^2 = [Xe]_{54}\ 4f^{14}5d^{10}6s^26p^2$

$\therefore$ Configuration of Pb^{2+} ion $= [Xe]_{54}\ 4f^{14}5d^{10}6s^2$

Valence-shell configuration of Pb-atom $= 4f^{14}5d^{10}6s^26p^2$

$\therefore$ Valence-shell configuration of Pb^{2+} ion $= 4f^{14}5d^{10}6s^2$

(ii) Write down complete and valence-shell configuration of O^{2-} ion (Z = 8)

Solution : Configuration of O-atom (Z = 8) $= [He]_2\ 2s^22p^4$

$\therefore$ Configuration of O^{2-} ion $= [He]_2 2s^22p^6$

Valence-shell configuration of O-atom (Z = 8) $= 2s^22p^4$

$\therefore$ Valence-shell configuration of O^{2-} ion $= 2s^22p^6$

Q.3 Give IUPAC nomenclature of the elements having Z > 100 (Superheavy elements).

Ans. Superheavy elements are trans-fermium elements, *i.e.*, superheavy elements are those which are placed after fermium, Fm (Z = 100) in the long form of periodic table. In other words these are the elements whose atomic number is greater than 100. For IUPAC nomenclature of these elements the following rules have been proposed.

(*i*) **Roots.** The names have three roots which indicate the three digits of the atomic number of the elements. In the last root-*ium* is added. The roots and symbols for the digit are given below :

Digit :	**0**	**1**	**2**	**3**	**4**	**5**	**6**	**7**	**8**	**9**
Root :	**nil**	**un**	**bi**	**tri**	**quad**	**pent**	**hex**	**sept**	**oct**	**enn**
Symbol :	**n**	**u**	**b**	**t**	**q**	**p**	**h**	**s**	**o**	**e**

(*ii*) **Symbol.** The symbol for the element has three letters derived from the roots.

(*iii*) **Name.** The name of the element starts with a small letter but the symbol of the element starts with a capital letter. The names are written as a complete word.

The above rules are clear from the following examples. In the IUPAC name column, a hyphen (-) has been put between each part of the names to make them more understandable. Present names and symbols of the elements with atomic number 101 to 109 have also been given. Present names of other elements have not been fixed as yet.

Atomic number	*IUPAC name*	*Symbol*	*Present name*	*Symbol*
101	un-nil-unium	Unu	Mendelevium	Md
102	un-nil-bium	Unb	Nobelium	No
103	un-nil-trium	Unt	Lawrencium	Lr
104	un-nil-quadium	Unq	Rutherfordium	Rf
105	un-nil-pentium	Unp	Dubnium	Db
106	un-nil-hexium	Unh	Seaborgium	Sg
107	un-nil-septium	Uns	Bohrium	Bh
108	un-nil-octium	Uno	Hassium	Hs
109	un-nil-ennium	Une	Meitnerium	Mt
110	un-un-nilium	Uun	—	—
111	un-un-unium	Uuu	—	—
112	un-un-bium	Uub	Eka mercury	—
113	un-un-trium	Uut	—	—
114	un-un-quadium	Uuq	—	—
115	un-un-pentium	Uup	—	—
116	un-un-hexium	Uuh	—	—
117	un-un-septium	Uus	—	—
118	un-un-octium	Uuo	—	—
119	un-un-ennium	Uue	—	—
120	un-bi-nilium	Ubn	—	—
130	un-tri-nilium	Utn	—	—
140	un-quad-nilium	Uqn	—	—
150	un-pent-nilium	Upn	—	—

Q.4 Give the electronic configuration of elements according to their position in different periods of long form of periodic table.

Ans. The electronic configurations of the elements based on their position in the long form of periodic table are given below :

1st period (n = 1)		
Orbitals being filled	**Atomic number and symbol**	**Electronic configuration**
↑ 1*s* ↓	1 (H) 2 (He)	$1s^1$ $1s^2$ or $[He]_2$
2nd period (n = 2)		
Orbitals being filled	**Atomic number and symbol**	**Electronic configuration**
↑ 2*s* ↓	3 (Li) 4 (Be)	$[He]_2\ 2s^1$ $[He]_2\ 2s^2$
↑ 2*p* ↓	5 (B) 6 (C) 7 (N) 8 (O) 9 (F) 10 (Ne)	$[He]_2\ 2s^2p^1$ $[He]_2\ 2s^2p^2$ $[He]_2\ 2s^2p^3$ $[He]_2\ 2s^2p^4$ $[He]_2\ 2s^2p^5$ $[He]_2\ 2s^2p^6$ or $1s^2$, $2s^2p^6$ or $[Ne]_{10}$
3rd period (n = 3)		
Orbitals being filled	**Atomic number and symbol**	**Electronic configuration**
↑ 3*s* ↓	11 (Na) 12 (Mg)	$[Ne]_{10}\ 3s^1$ $[Ne]_{10}\ 3s^2$
↑ 3*p* ↓	13 (Al) 14 (Si) 15 (P) 16 (S) 17 (Cl) 18 (Ar)	$[Ne]_{10}\ 3s^2p^1$ $[Ne]_{10}\ 3s^2p^2$ $[Ne]_{10}\ 3s^2p^3$ $[Ne]_{10}\ 3s^2p^4$ $[Ne]_{10}\ 3s^2p^5$ $[Ne]_{10}\ 3s^2p^6$ or $1s^2$, $2s^22p^6$, $3s^23p^6$ or $[Ar]_{18}$

4th period (n = 4)		
Orbitals being filled	**Atomic number and symbol**	**Electronic configuration**
↑ 4*s* ↓	19 (K)	$[Ar]_{18}\ 4s^1$
	20 (Ca)	$[Ar]_{18}\ 4s^2$
↑	21 (Sc)	$[Ar]_{18}\ 4s^2 3d^1$ or $[Ar]_{18}\ 3d^1 4s^2$
	22 (Ti)	$[Ar]_{18}\ 4s^2 3d^2$ or $[Ar]_{18}\ 3d^2 4s^2$
	23 (V)	$[Ar]_{18}\ 4s^2 3d^3$ or $[Ar]_{18}\ 3d^3 4s^2$
	24 (Cr)	$[Ar]_{18}\ 4s^1 3d^5$ or $[Ar]_{18}\ 3d^5 4s^1$
	25 (Mn)	$[Ar]_{18}\ 4s^2 3d^5$ or $[Ar]_{18}\ 3d^5 4s^2$
3*d*	26 (Fe)	$[Ar]_{18}\ 4s^2 3d^6$ or $[Ar]_{18}\ 3d^6 4s^2$
	27 (Co)	$[Ar]_{18}\ 4s^2 3d^7$ or $[Ar]_{18}\ 3d^7 4s^2$
	28 (Ni)	$[Ar]_{18}\ 4s^2 3d^8$ or $[Ar]_{18}\ 3d^8 4s^2$
	29 (Cu)	$[Ar]_{18}\ 4s^1 3d^{10}$ or $[Ar]_{18}\ 3d^{10} 4s^1$
↓	30 (Zn)	$[Ar]_{18}\ 4s^2 3d^{10}$ or $[Ar]_{18}\ 3d^{10} 4s^2$
↑	31 (Ga)	$[Ar]_{18}\ 4s^2 3d^{10} 4p^1$ or $[Ar]_{18}\ 3d^{10} 4s^2 4p^1$
	32 (Ge)	$[Ar]_{18}\ 4s^2 3d^{10} 4p^2$ or $[Ar]_{18}\ 3d^{10} 4s^2 4p^2$
	33 (As)	$[Ar]_{18}\ 4s^2 3d^{10} 4p^3$ or $[Ar]_{18}\ 3d^{10} 4s^2 4p^3$
4*p*	34 (Se)	$[Ar]_{18}\ 4s^2 3d^{10} 4p^4$ or $[Ar]_{18}\ 3d^{10} 4s^2 4p^4$
	35 (Br)	$[Ar]_{18}\ 4s^2 3d^{10} 4p^5$ or $[Ar]_{18}\ 3d^{10} 4s^2 4p^5$
↓	36 (Kr)	$[Ar]_{18}\ 4s^2 3d^{10} 4p^6$ or $[Ar]_{18}\ 3d^{10} 4s^2 4p^6$ or $1s^2$, $2s^2 2p^6$, $3s^2 3p^6 3d^{10}$, $4s^2 4p^6$ or $[Kr]_{36}$

5th period (n = 5)		
Orbitals being filled	**Atomic number and symbol**	**Electronic configuration**
↑ 5*s* ↓	37 (Rb)	$[Kr]_{36}\ 5s^1$
	38 (Sr)	$[Kr]_{36}\ 5s^2$
↑	39 (Y)	$[Kr]_{36}\ 5s^2 4d^1$ or $[Kr]_{36}\ 4d^1 5s^2$
	40 (Zr)	$[Kr]_{36}\ 5s^2 4d^2$ or $[Kr]_{36}\ 4d^2 5s^2$
	41 (Nb)	$[Kr]_{36}\ 5s^1 4d^4$ or $[Kr]_{36}\ 4d^4 5s^1$
	42 (Mo)	$[Kr]_{36}\ 5s^1 4d^5$ or $[Kr]_{36}\ 4d^5 5s^1$
	43 (Tc)	$[Kr]_{36}\ 5s^2 4d^5$ or $[Kr]_{36}\ 4d^5 5s^2$
4*d*	44 (Ru)	$[Kr]_{36}\ 5s^1 4d^7$ or $[Kr]_{36}\ 4d^7 5s^1$
	45 (Rh)	$[Kr]_{36}\ 5s^1 4d^8$ or $[Kr]_{36}\ 4d^8 5s^1$
	46 (Pd)	$[Kr]_{36}\ 5s^0 4d^{10}$ or $[Kr]_{36}\ 4d^{10} 5s^0$
	47 (Ag)	$[Kr]_{36}\ 5s^1 4d^{10}$ or $[Kr]_{36}\ 4d^{10} 5s^1$
↓	48 (Cd)	$[Kr]_{36}\ 5s^2 4d^{10}$ or $[Kr]_{36}\ 4d^{10} 5s^2$

5p	49 (In)	$[Kr]_{36}\ 5s^2 4d^{10} 5p^1$ or $[Kr]_{36}\ 4d^{10} 5s^2 5p^1$
	50 (Sn)	$[Kr]_{36}\ 5s^2 4d^{10} 5p^2$ or $[Kr]_{36}\ 4d^{10} 5s^2 5p^2$
	51 (Sb)	$[Kr]_{36}\ 5s^2 4d^{10} 5p^3$ or $[Kr]_{36}\ 4d^{10} 5s^2 5p^3$
	52 (Te)	$[Kr]_{36}\ 5s^2 4d^{10} 5p^4$ or $[Kr]_{36}\ 4d^{10} 5s^2 5p^4$
	53 (I)	$[Kr]_{36}\ 5s^2 4d^{10} 5p^5$ or $[Kr]_{36}\ 4d^{10} 5s^2 5p^5$
	54 (Xe)	$[Kr]_{36}\ 5s^2 4d^{10} 5p^6$ or $[Kr]_{36}\ 4d^{10} 5s^2 5p^6$ or $1s^2$, $2s^2 2p^6$, $3s^2 3p^6\ 3d^{10}$, $4s^2 4p^6 4d^{10}$, $5s^2\ 5p^6$ or $[Xe]_{54}$

6th period (n = 6)

Orbitals being filled	Atomic number and symbol	Electronic configuration
6s	55 (Cs)	$[Xe]_{54}\ 6s^1$
	56 (Ba)	$[Xe]_{54}\ 6s^2$
5d	57 (La)	$[Xe]_{54}\ 6s^2 5d^1$ or $[Xe]_{54}\ 5d^1 6s^2$
4f or 5d	58 (Ce)	$[Xe]_{54}\ 6s^2 5d^1 4f^1$ or $[Xe]_{54}\ 4f^1 5d^1 6s^2$
	59 (Pr)	$[Xe]_{54}\ 6s^2 4f^3$ or $[Xe]_{54}\ 4f^3 6s^2$
	60 (Nd)	$[Xe]_{54}\ 6s^2 4f^4$ or $[Xe]_{54}\ 4f^4 6s^2$
	61 (Pm)	$[Xe]_{54}\ 6s^2 4f^5$ or $[Xe]_{54}\ 4f^5 6s^2$
	62 (Sm)	$[Xe]_{54}\ 6s^2 4f^6$ or $[Xe]_{54}\ 4f^6 6s^2$
	63 (Eu)	$[Xe]_{54}\ 6s^2 4f^7$ or $[Xe]_{54}\ 4f^7 6s^2$
	64 (Gd)	$[Xe]_{54}\ 6s^2 4f^7 5d^1$ or $[Xe]_{54}\ 4f^7 5d^1 6s^2$
	65 (Tb)	$[Xe]_{54}\ 6s^2 4f^9$ or $[Xe]_{54}\ 4f^9 6s^2$
	66 (Dy)	$[Xe]_{54}\ 6s^2 4f^{10}$ or $[Xe]_{54}\ 4f^{10} 6s^2$
	67 (Ho)	$[Xe]_{54}\ 6s^2 4f^{11}$ or $[Xe]_{54}\ 4f^{11} 6s^2$
	68 (Er)	$[Xe]_{54}\ 6s^2 4f^{12}$ or $[Xe]_{54}\ 4f^{12} 6s^2$
	69 (Tm)	$[Xe]_{54}\ 6s^2 4f^{13}$ or $[Xe]_{54}\ 4f^{13} 6s^2$
	70 (Yb)	$[Xe]_{54}\ 6s^2 4f^{14}$ or $[Xe]_{54}\ 4f^{14} 6s^2$
	71 (Lu)	$[Xe]_{54}\ 6s^2 4f^{14} 5d^1$ or $[Xe]_{54}\ 4f^{14} 5d^1 6s^2$
5d	72 (Hf)	$[Xe]_{54}\ 6s^2 4f^{14} 5d^2$ or $[Xe]_{54}\ 4f^{14} 5d^2 6s^2$
	73 (Ta)	$[Xe]_{54}\ 6s^2 4f^{14} 5d^3$ or $[Xe]_{54}\ 4f^{14} 5d^3 6s^2$
	74 (W)	$[Xe]_{54}\ 6s^2 4f^{14} 5d^4$ or $[Xe]_{54}\ 4f^{14} 5d^4 6s^2$
	75 (Re)	$[Xe]_{54}\ 6s^2 4f^{14} 5d^5$ or $[Xe]_{54}\ 4f^{14} 5d^5 6s^2$
	76 (Os)	$[Xe]_{54}\ 6s^2 4f^{14} 5d^6$ or $[Xe]_{54}\ 4f^{14} 5d^6 6s^2$
	77 (Ir)	$[Xe]_{54}\ 6s^2 4f^{14} 5d^7$ or $[Xe]_{54}\ 4f^{14} 5d^7 6s^2$
	78 (Pt)	$[Xe]_{54}\ 6s^1 4f^{14} 5d^9$ or $[Xe]_{54}\ 4f^{14} 5d^9 6s^1$
	79 (Au)	$[Xe]_{54}\ 6s^1 4f^{14} 5d^{10}$ or $[Xe]_{54}\ 4f^{14} 5d^{10} 6s^1$
	80 (Hg)	$[Xe]_{54}\ 6s^2 4f^{14} 5d^{10}$ or $[Xe]_{54}\ 4f^{14} 5d^{10} 6s^2$

6p	81 (Tl) 82 (Pb) 83 (Bi) 84 (Po) 85 (At) 86 (Rn)	$[Xe]_{54}\ 6s^2 4f^{14} 5d^{10} 6p^1$ or $[Xe]_{54}\ 4f^{14} 5d^{10} 6s^2 6p^1$ $[Xe]_{54}\ 6s^2 4f^{14} 5d^{10} 6p^2$ or $[Xe]_{54}\ 4f^{14} 5d^{10} 6s^2 6p^2$ $[Xe]_{54}\ 6s^2 4f^{14} 5d^{10} 6p^3$ or $[Xe]_{54}\ 4f^{14} 5d^{10} 6s^2 6p^3$ $[Xe]_{54}\ 6s^2 4f^{14} 5d^{10} 6p^4$ or $[Xe]_{54}\ 4f^{14} 5d^{10} 6s^2 6p^4$ $[Xe]_{54}\ 6s^2 4f^{14} 5d^{10} 6p^5$ or $[Xe]_{54}\ 4f^{14} 5d^{10} 6s^2 6p^5$ $[Xe]_{54}\ 6s^2 4f^{14} 5d^{10} 6p^6$ or $[Xe]_{54}\ 4f^{14} 5d^{10} 6s^2 6p^6$ or $1s^2$, $2s^2 2p^6$, $3s^2 3p^6 3d^{10}$, $4s^2\ 4p^6\ 4d^{10}\ 4f^{14}$, $5s^2\ 5p^6\ 5d^{10}$, $6s^2 6p^6$ or $[Rn]_{86}$

7th period (n = 7)

Orbitals being filled	Atomic number and symbol	Electronic configuration
7s	87 (Fr) 88 (Ra)	$[Rn]_{86}\ 7s^1$ $[Rn]_{86}\ 7s^2$
5f or 6d	89 (Ac) 90 (Th) 91 (Pa) 92 (U) 93 (Np) 94 (Pu) 95 (Am) 96 (Cm) 97 (Bk) 98 (Cf) 99 (Es) 100 (Fm) 101 (Md) 102 (No) 103 (Lw)	$[Rn]_{86}\ 7s^2 6d^1$ or $[Rn]_{86}\ 6d^1 7s^2$ $[Rn]_{86}\ 7s^2 6d^1 5f^1$ or $[Rn]_{86}\ 5f^1 6d^1 7s^2$ $[Rn]_{86}\ 7s^2 6d^1 5f^2$ or $[Rn]_{86}\ 5f^2 6d^1 7s^2$ $[Rn]_{86}\ 7s^2 6d^1 5f^3$ or $[Rn]_{86}\ 5f^3 6d^1 7s^2$ $[Rn]_{86}\ 7s^2 6d^1 5f^4$ or $[Rn]_{86}\ 5f^4 6d^1 7s^2$ $[Rn]_{86}\ 7s^2 5f^6$ or $[Rn]_{86}\ 5f^6 7s^2$ $[Rn]_{86}\ 7s^2 5f^7$ or $[Rn]_{86}\ 5f^7 7s^2$ $[Rn]_{86}\ 7s^2 5f^7 6d^1$ or $[Rn]_{86}\ 5f^7 6d^1 7s^2$ $[Rn]_{86}\ 7s^2 5f^9$ or $[Rn]_{86}\ 5f^9 7s^2$ $[Rn]_{86}\ 7s^2 5f^{10}$ or $[Rn]_{86}\ 5f^{10} 7s^2$ $[Rn]_{86}\ 7s^2 5f^{11}$ or $[Rn]_{86}\ 5f^{11} 7s^2$ $[Rn]_{86}\ 7s^2 5f^{12}$ or $[Rn]_{86}\ 5f^{12} 7s^2$ $[Rn]_{86}\ 7s^2 5f^{13}$ or $[Rn]_{86}\ 5f^{13} 7s^2$ $[Rn]_{86}\ 7s^2 5f^{14}$ or $[Rn]_{86}\ 5f^{14} 7s^2$ $[Rn]_{86}\ 7s^2 5f^{14} 6d^1$ or $[Rn]_{86}\ 5f^{14} 6d^1 7s^2$
6d	104 [Rf or Ku or Unq Rutherfordium or Kurchatovium (Unnilquadium)]	$[Rn]_{86}\ 7s^2 5f^{14} 6d^2$ or $[Rn]_{86}\ 5f^{14} 6d^2 7s^2$
	105 [Ha or Db or Unp, Hahnium or Dubnium (Unnilpentium)]	$[Rn]_{86}\ 7s^2 5f^{14} 6d^3$ or $[Rn]_{86}\ 5f^{14} 6d^3 7s^2$
	106 [Sg or Unh Seaborgium (Unnihexium)]	$[Rn]_{86}\ 7s^2 5f^{14} 6d^4$ or $[Rn]_{86}\ 5f^{14} 6d^4 7s^2$

6d	107 [Ns or Bh or Uns Neilsbohrium or Bohrium (Unnilseptium)]	$[Rn]_{86}\ 7s^2 5f^{14} 6d^5$ or $[Rn]_{86}\ 5f^{14} 6d^5 7s^2$
	108 [Hs or Uno Hassnium (Unniloctium)]	$[Rn]_{86}\ 7s^2 5f^{14} 6d^6$ or $[Rn]_{86}\ 5f^{14} 6d^6 7s^2$
	109 [Mt or Une Meitnerium (Unnilennium)]	$[Rn]_{86}\ 7s^2 5f^{14} 6d^7$ or $[Rn]_{86}\ 5f^{14} 6d^7 7s^2$
	110 [Uun (Ununnilium) or Ekaplatinum]	$[Rn]_{86}\ 7s^2 5f^{14} 6d^8$ or $[Rn]_{86}\ 5f^{14} 6d^8 7s^2$
	111 [Uuu (Unununium) or Ekagold]	$[Rn]_{86}\ 7s^1 5f^{14} 6d^{10}$ or $[Rn]_{86}\ 5f^{14} 6d^{10} 7s^1$
	112 [Uub (Ununbium) or Ekamercury]	$[Rn]_{86}\ 7s^2 5f^{14} 6d^{10}$ or $[Rn]_{86}\ 5f^{14} 6d^{10} 7s^2$
7p	113 [Uut (Ununtrium)]	$[Rn]_{86}\ 7s^2 5f^{14} 6d^{10} 7p^1$ or $[Rn]_{86}\ 5f^{14} 6d^{10} 7s^2 7p^1$
	114 [Uuq (Ununquadium)]	$[Rn]_{86}\ 7s^2 5f^{14} 6d^{10} 7p^2$ or $[Rn]_{86}\ 5f^{14} 6d^{10} 7s^2 7p^2$
	115 [Uup (Ununpentium)]	$[Rn]_{86}\ 7s^2 5f^{14} 6d^{10} 7p^3$ or $[Rn]_{86}\ 5f^{14} 6d^{10} 7s^2 7p^3$

Q.5 Discuss the possibility of the atom for existing in the following electronic configurations :

(i) **$1s^2\ 2s^2\ 2p_x^1$** ***(ii)*** **$1s^2\ 2s^1\ 2p_x^2\ 2p_y^1\ 2p_z^1$**

(iii) **$1s^2\ 2s^2\ 2p_x^2\ 2p_y^1$** ***(iv)*** **$1s^2\ 2s^2\ 3s^2$**

Solution. *(i)* This electronic configuration is correct since it is in accordance with the rules for filling up of various orbitals.

(ii) This electronic configuration is wrong because it violates aufbau principle which states that an orbital with lower energy, *i.e.* 2*s* sub-shell in the present case should be completely filled before the electrons go to higher energy sub-shell, *i.e.* 2*p* in the present case.

(iii) This electronic configuration is not correct since it violates Hund's rule. According to this rule all the three 2*p* orbitals must have one electron each before the pairing occurs. But in the present case $2p_x$ orbital has two electrons while $2p_z$ orbital is empty.

(iv) This electronic configuration is wrong since after filling 2*s*-orbital, the electrons should go to 2*p*-orbital rather than 3*s*-orbital.

Q.6 Find the number of unpaired electrons present in phosphorus (atomic No. 15), chromium (atomic No. 24) and copper (atomic No. 29) after writing their orbital electronic configuration.

Solution.

Element	At. No.	Electronic configuration	No. of unpaired electrons
P	15	$1s^2, 2s^2\ 2p^6, 3s^2\ 3p_x^{\ 1}\ 3p_y^{\ 1}\ 3p_z^{\ 1}$	3
Cr	24	$1s^2, 2s^2\ 2p^6, 3s^2\ 3p^6\ 3d^5, 4s^1$	5 + 1 = 6
Cu	29	$1s^2, 2s^2\ 2p^6, 3s^2\ 3p^6\ 3d^{10}, 4s^1$	1

Q.7 Neutral atom of an element has 2K, 8L and 5M electrons. Find out the following from the data : (a) Atomic No. (b) Total No. of s electrons (c) Total No. of p-electrons (d) No. of protons in the nucleus and (e) Valency of the element.

Solution. The electronic configuration of the element with 2K, 8L and 5M electrons will be

$$\underbrace{1s^2}_{2K}\qquad \underbrace{2s^2\ 2p_x^{\ 2}\ 2p_y^{\ 2}\ 2p_z^{\ 2}}_{8L}\qquad \underbrace{3s^2\ 3p_x^{\ 1}\ 3p_y^{\ 1}\ 3p_z^{\ 1}}_{5M}$$

(*a*)Total No. of electrons = 2 + 8 + 5 = 15

∴ Atomic No. of the element = 15

(*b*) Total no. of s-electrons = 2 + 2 + 2 = 6

(*c*) Total no. of p-electrons = 6 + 3 = 9

(*d*) Since the atom is neutral,

∴ No. of protons = No. of electrons = Atomic No. = 15

(*e*) Since the element has only three half-filled atomic orbitals, therefore, valency of the element = 3.

Q.8 Write the electronic configurations of the elements with the following atomic numbers :

3, 8, 10, 14, 17, 25, 29, and 38.

Also mention the groups of the periodic table to which they belong.

Solution.

Atomic No.	Electronic configuration	Group of periodic table
3	$1s^2\ 2s^1$	1
8	$1s^2\ 2s^2\ 2p_x^{\ 2}\ 2p_y^{\ 1}\ 2p_z^{\ 1}$	16
10	$1s^2\ 2s^2\ 2p_x^{\ 2}\ 2p_y^{\ 2}\ 2p_z^{\ 2}$	18
14	$1s^2\ 2s^2\ 2p^6\ 3s^2\ 3p_x^{\ 1}\ 3p_y^{\ 1}$	14
17	$1s^2\ 2s^2\ 2p^6\ 3s^2\ 3p_x^{\ 2}\ 3p_y^{\ 2}\ 3p_z^{\ 1}$	17
25	$1s^2\ 2s^2\ 2p^6\ 3s^2\ 3p^6\ 3d^5\ 4s^2$	7
29	$1s^2\ 2s^2\ 2p^6\ 3s^2\ 3p^6\ 3d^{10}\ 4s^1$	11
38	$1s^2\ 2s^2\ 2p^6\ 3s^2\ 3p^6\ 3d^{10}\ 4s^2\ 4p^6\ 5s^2$	2

Q.9 An atom of an element contains 29 electrons and 35 neutrons. Deduce (*i*) the number of protons and (*ii*) the electronic configuration of the element.

Solution. (*i*) For neutral atom, number of protons = number of electrons = 29

(*ii*) Atomic number of the element = 29

Electronic configuration of element with Z = 29 will be $1s^2\ 2s^2\ 2p^6\ 3s^2\ 3p^6\ 3d^{10}\ 4s^1$ or $[Ar]_{18}\ 3d^{10}\ 4s^1$

Note. No. of neutrons are needed only to find mass number.

Mass number = No. of protons + No. of neutrons
= 29 + 35 = 64

Q.10 Give the electronic configuration of the following ions :

(*i*) Cu^{2+} (*ii*) Cr^{3+} (*iii*) Fe^{2+} and Fe^{3+} (*iv*) H^- (*v*) S^{2-}

Solution. During the formation of cations, electrons are lost while in the formation of anions, electrons are added to the valence shell. The number of electrons added or lost is equal to the numerical value of the charge present on the ion. Following this general concept, we can write the electronic configurations of all the ions given in the question.

(*i*) $Cu^{2+} = {}_{29}Cu - 2e^-$
$= 1s^2\ 2s^2\ 2p^6\ 3s^2\ 3p^6\ 3d^{10}\ 4s^1 - 2e^-$
$= 1s^2\ 2s^2\ 2p^6\ 3s^2\ 3p^6\ 3d^9$

(*ii*) $Cr^{3+} = {}_{24}Cr - 3e^-$
$= 1s^2\ 2s^2\ 2p^6\ 3s^2\ 3p^6\ 3d^5\ 4s^1 - 3e^-$
$= 1s^2\ 2s^2\ 2p^6\ 3s^2\ 3p^6\ 3d^3$

(*iii*) $Fe^{2+} = {}_{26}Fe - 2e^-$
$= 1s^2\ 2s^2\ 2p^6\ 3s^2\ 3p^6\ 3d^6\ 4s^2 - 2e^-$
$= 1s^2\ 2s^2\ 2p^6\ 3s^2\ 3p^6\ 3d^6$

$Fe^{3+} = {}_{26}Fe - 3e^-$
$= 1s^2\ 2s^2\ 2p^6\ 3s^2\ 3p^6\ 3d^5$

(*iv*) $H^- = {}_1H + 1e^- = 1s^1 + 1e^1 = 1s^2$

(*v*) $S^{2-} = {}_{16}S + 2e^-$
$= 1s^2\ 2s^2\ 2p^6\ 3s^2\ 3p_x{}^2\ 3p_y{}^1\ 3p_z{}^1 + 2e^-$
$= 1s^2\ 2s^2\ 2p^6\ 3s^2\ 3p_x{}^2\ 3p_y{}^2\ 3p_z{}^2$

Q.11 What atoms are indicated by the following electronic configurations ?

(*i*) $1s^2\ 2s^2\ 2p^1$ (*ii*) [Ar] $4s^2\ 3d^1$

Solution. (*i*) Total number of electrons in $1s^2\ 2s^2\ 2p^1$ = 2 + 2 + 1 = 5

∴ Atomic number of the element = 5

Hence the element is *Boron* (B)

(*ii*) Total number of electrons in [Ar] $4s^2\ 3d^1$ = 18 + 2 + 1 = 21

∴ Atomic number of the element = 21

Hence the element is *Scandium* (Sc)

Q.12 What are the rules for finding group number of a given element ?

Ans. Following are the rules for finding out the group number of an element whose electronic configuration is given.

(*i*) If the last shell contains 1 or 2 electrons, then the group number is 1 and 2 respectively.

(*ii*) If the last shell contains more than 2 electrons, then the group number is the total number of electrons in the last shell plus 10.

(*iii*) If electrons are present in $(n - 1)\ d$ orbitals in addition to those in the ns orbital, then the group number is equal to the total number of electrons present in the $(n - 1)\ d$ orbital and ns orbital.

Q.13 (*a*) Find out the E.C. of P-atom in H_3PO_4.

(*b*) Which of the following two ions have the same V.S.E.C. ? Cr^{3+}, Fe^{3+}, Mn^{2+} and Co^{3+}.

(*c*) Which of the following ions have $3d^2$ as their V.S.E.C. ? Ti^{2+}, Ti^{3+}, V^{3+}, V^{4+}, Cr^{4+}, Cr^{3+}, Mn^{5+} and Mn^{2+}.

Ans. (*a*) In H_3PO_4, P-atom is present as P^{5+}.

E.C. of P^{5+} ion = E.C. of P-atom $-5e^-$

$= 1s^2,\ 2s^22p^6,\ 3s^23p^3\ -5e^-$

$= 1s^2,\ 2s^22p^6$ **(Ans.)**

(*b*) V.S.E.Cs of the given ions are : $Cr^{3+} = 3d^3$, $Fe^{3+} = 3d^5$, $Mn^{2+} = 3d^5$ and $Co^{2+} = 3d^7$. Obviously Fe^{3+} and Mn^{2+} ions have the same V.S.E.C. (= $3d^5$)

(*c*) V.S.E.Cs of the given ions are $Ti^{2+} = 3d^2$, $Ti^{3+} = 3d^1$, $V^{3+} = 3d^2$, $V^{4+} = 3d^1$, $Cr^{4+} = 3d^2$, $Cr^{3+} = 3d^3$, $Mn^{5+} = 3d^2$ and $Mn^{2+} = 3d^5$. Obviously Ti^{2+}, V^{3+}, Cr^{4+} and Mn^{5+} ions have $3d^2$ configuration.

University Questions

1. Write a note on "Aufbau principle." (*Kanpur 2000*)
2. Explain the trends of the following properties of *p*-block elements of the periodic table : (*i*) Atomic radii (*ii*) ionisation potential (*iii*) Electronegativity (*iv*) Oxidation states. (*Kumaon 2000*)
3. (*a*) Give an account of the modern periodic table. Dicuss its advantages.
   (*b*) Write notes on (*i*) Heisenberg's uncertainty principle (*ii*) Pauli's exclusion principle. (*Kumaon 2000*)
4. Write a note on Heisenberg's uncertainty principle. [*Gauhati (General) 2000*]
5. Give in brief the general characteristics of *s*-block elements. How do the elements of ns^1 block differ from the elements of ns^2 block? (*Lucknow 2000*)
6. (*a*) Which has the lowest number of unpaired electrons? B, C, N.
   (*b*) Elements A, B and C have atomic number Z, Z +3 and Z + 6 respectively. The element A is an alkali metal. Answer the following :

(*i*) which group and block of P.T. the elements A, B and C are associated.

(*ii*) Decide the nature of the bond in the compound formed by combination of atoms of elements B and C.

(*iii*) Explain the shape of molecule formed by the combination of atoms of elements B and C.

(*iv*) Name the elements A, B and C when Z = 11, and write down their electronic configurations.

(*v*) Write down the reaction of water with compound formed by the combination of atoms of B and C. (*Avadh 2000*)

7. (*i*) Write a note on "Diagonal relationship".

(*ii*) What are *s*-block elements ? How do these elements differ *p*-block elements? Give their main characteristics. (*Lucknow 2001*)

8. Give an account of the modern periodic table. State how it is more useful than Mendeleef's periodic table ? (*Lucknow 2002*)

9. Write a note on Aufbau principle. (*Lucknow 2002*)

10. Give electronic configuration of group IA and IIA elements. (*Nagpur 2002*)

11. Write electronic configuration of Cr (24) and Cu (29). (*Nagpur 2002*)

12. (*a*) Write the electronic configuration of La(Z = 57) and Ce(Z = 58).

(*b*) What are the advantages of classifying the elements into *s, p, d* and *f* blocks in the periodic table? (*Delhi 2002*)

13. Correct the electronic configurations given below and give reasons for your answer :

(*a*) $1s^2, 2s^2\ 2p_x^2\ 2p_y^0\ 2p_z^0$ (*b*) $1s^2, 2s^2\ 2p^6, 3s^2\ 3p^4, 4p^1$

(*VBS Purvanchal 2003*)

14. On the basis of electronic configuration, which of the following ions has maximum number of unpaired electrons?

(*a*) Mg^{2+} (*b*) Ti^{3+} (*c*) V^{3+} (*d*) Fe^{2+} (*VBS Purvanchal 2003*)

15. (*a*) What is A U F B A U principle? Use this principle to explain the filling of electrons in 4s-orbital before 3d-orbitals.

(*b*) Write the most stable electronic configuration using *s, p, d* notation of atoms with atomic number 21 and 31 and decide the position of each in periodic table, indicating (*i*) The block (*ii*) The number of unpaired electrons for each. (*Rohilkhand 2003*)

16. (*a*) Elements A, B and C have atomic numbers 11, 29 and 35 respectively. Write their electronic configuration and classify them as *s, p* and *d* blocks.

(*b*) How many unpaired electrons are there in Cr^{2+} and Fe^{2+} ions?

(*Rohilkhand 2003*)

17. (*a*) What is Aufbau principle? Give the sequence in which energy levels are filled. Write the electronic configuration of elements having atomic numbers 24 and 29.

(*b*) State Hund's rule of maximum multiplicity. Give suitable example.

(*c*) Write the general valence shell electronic configuration of IA and IIA group elements.

18. (*a*) Write the electronic configuration of F^-, Al^{3+}, Mn, Cu
(Given : Z of F = 9, Al = 13, Mn = 25, Cu= 29).

(*b*) What are s-block elements? Write electronic configuration of IA group elements. (*Nagpur 2003*)

(*c*) Write the general electronic configuration of IA and IIA group elements. (*Nagpur 2003*)

19. Name the element whose electronic configuration is $[Xe]_{54}\ 4f^{14}5d^{1}6s^{2}$. (*H.N. Bahuguna 2004*)

20. (*a*) Name the orbital occupied by 19*th* electron of Cr atom (Z = 21).

(*b*) If the E.C. of V(Z = 23) is [Ar] $3d^{3}4s^{2}$, write the E.C. of the next element.

(*c*) Write the E.C. of the most electronegative element. (*H.N. Bahuguna 2004*)

21. Using Aufbau principle write down the E.C. of Cr(Z = 24) atom. (*H.N. Bahuguna 2005*)

22. Write a note on Aufbau principle. (*H.N. Bahuguna 2006*)

23. Find out the outer electronic configuration of Cu(Z = 29). (*Agra 2007*)

24. How many unpaired electrons are there in Cr atom (Z = 24)? (*Purvanchal 2007*)

25. (*a*) Write the number of electrons that can be accomodated by a *p*-orbital.

(*b*) Write the number of unpaired electrons in Fe^{3+} ion. (*H.N. Bahuguna 2007*)

26. Find out the number of unpaired electrons in Mn^{2+} ion. (*Agra 2008*)

CHAPTER

4 Periodic Properties

Periodicity of Properties and Magic Number

The term periodicity of properties indicates that the elements with similar properties reappear at certain regular intervals of atomic number in the periodic table. In order to understand the concept of periodicity of properties we may consider the properties of the elements of groups I A (*Alkali metals*), zero (*Inert gases*) and VII A (*Halogens*) given below :

Periodic number	Group I A	Group zero	Group VII A
1	*H(1)*	*H(2)*	
2	*Li (3)*	*Ne (10)*	*F (9)*
3	*Na (11)*	*Ar (18)*	*Cl (17)*
4	*K (19)*	*Kr (36)*	*Br (35)*
5	*Rb (37)*	*Xe (54)*	*I (53)*
6	*Cs (55)*	*Rn (86)*	*At (85)*
7	*Fr (87)*		

The examination of the properties of these elements will show that the elements belonging to the same group have similar properties, In other words we can say that the atomic number intervals at which the elements with similar properties reappear are 2, 8, 8, 18, 18, 32 and 32, *i.e.* we have to pass 2, 8, 8, 18, 18, 32, and 32 elements before we come across an element with similar properties. *The repetition of the elements with similar properties is the order of increasing atomic number as in the periodic table is called* **periodicity of properties** and the numbers 2, 8, 8 and 32 are called **magic numbers.**

Cause of Periodicity of Properties

What is the cause of periodicity of properties, *i.e.*, why do the elements placed in the same group have similar properties ?

In order to answer this question we may consider the valence- shell electronic configurations, for example, of the elements of groups I A, zero and VII A as given on the next page.

An examination of the valence-shell electronic configurations of the elements given in the table shows that the elements showing periodicity of properties *i.e.* the elements belonging to the same group have the same valence-shell electronic configuration. Thus we see that in any group similar valence-shell electronic configurations of the elements reoccur after certain regular intervals of atomic number and it is this reoccurrence of similar valance-shell electronic configurations of the elements at certain regular intervals of atomic number which becomes the cause of periodicity of properties.

Group I A	Group zero	Group VII A
H (1) $1s^1$	He (2) $1s^2$	
Li (3) $2s^1$	Ne (10) $2s^2p^6$	F (9) $2s^2p^5$
Na (11) $3s^1$	Ar (18) $3s^2p^6$	Cl (17) $3s^2p^5$
K (19) $4s^1$	Kr (36) $4s^2p^6$	Br (35) $4s^2p^5$
Rb (37) $5s^1$	Xe (54) $5s^2p^6$	I (53) $5s^2p^5$
Cs (55) $6s^1$	Rn (86) $6s^2p^6$	At (85) $6s^2p^5$
Fr (87) $7s^1$		

We have already said that, according to modern periodic law, the properties of the elements (i.e. atomic properties) like atomic volume, atomic and ionic radii, ionisation potential, electron affinity electronegativity etc. are periodic functions of their atomic number. Here the word *periodic* means *with the rise of atomic number of the elements in the same period or group, there is a gradual variation (i.e. increase or decrease) in a particular property of the element.* These properties depend on the electronic configuration (particularly valence-shell configuration) of the atoms of the elements. Some of such properties which are more useful to the students are: *(1) Atomic volume (2) Size of the atoms and ions (atomic and ionic radii) (3) Ionisation potential (4) Electron affinity (5) Electronegativity (6) Bond energy or bond strength (7) Polarisation of ions.*

Some Important Concepts

Before discussing the periodic properties we shall discuss some important concepts which are very useful in having a clear understanding of periodic properties. These concepts are (1) *Shielding or screening effect of the inner-shell electrons on the valence shell electron* and (2) *Effective nuclear charge.* These are discussed below.

I. Shielding or Screening Effect of Inner-shell Electrons on the Valence-shell Electron.

In a multi-electron atom the valence-shell electrons are attracted by the nucleus and also at the same time repelled by the electrons present between the nucleus and the valence-shell electrons (*i.e.* inner shell electrons) (See Fig. 4.1). Thus the nucleus exerts an attractive force on the valence-shell electrons while the inner-shell electrons exert a repulsive force on the valence-shell electron. The combined effect of these two forces is that the actual force exerted by the nucleus on the valence-shell electron is partially decreased or weakened by the presence of repulsive force exerted by the inner-shell electrons on the valence-shell electrons. Thus the valence-shell electrons experience less attraction or pull towards tile nucleus. *The decrease in the attractive force exerted by the nucleus on the valence-shell electron, which is obviously due to the presence of the electrons lying between the nucleus and valence-shell electrons,* (called intervening electrons) *is called* **shielding effect** or **screening effect.** In other words, the intervening electrons screen or shield the valence-shell electrons from the nucleus.

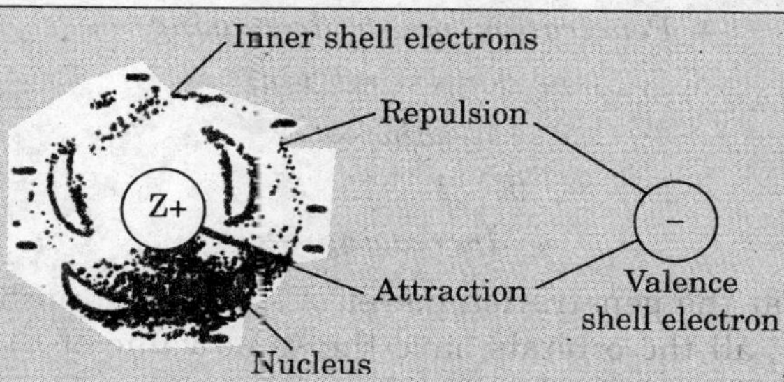

Fig. 4.1. The valence-shell electrons experience less attraction from the nucleus due to the presence of inner-shell electrons–called shielding or screening effect of inner shell electrons.

Factors Affecting the Magnitude of Shielding Effect

Following are the important factors on which the magnitude of shielding effect caused by the inner-shell electrons on the valence-shell electron depends.

***(i)* No. of inner-shell electrons or inner shells.** Greater is the number of inner-shell electrons or inner shells, greater is the magnitude of shielding effect caused by the inner electrons on the valence- shell electron. Thus as we move down a group, the number of inner-shells or inner shell electrons increases and hence the shielding effect also increases. For example in the elements of group IA, with the increase of the number of inner shells and electrons in them, the shielding effect caused by these electrons on the valence-shell electron also increases from Li to Cs as shown below :

Elements of group I A	Electronic configuration	No. of electrons in the inner shells	No. of inner shells	Order of shielding effect
Li	2, 1	2		
Na	2, 8, 1	2+8 = 10	2	Increasing
K	2, 8, 8, 1	2+8+8 = 18	3	
Rb	2, 8, 18, 8, 1	2+8+18+8 = 36	4	
Cs	2, 8, 18, 18, 8, 1	2+8+18+18+8 = 54	5	↓

***(ii)* Type of the orbital occupied by the electrons.** From the shapes of electron charge clouds of orbitals we find that s-electrons spend most of their time nearer the nucleus while *p*-electrons belonging to the same shell spend most of their time farther away from the nucleus. This, in other words means that an electron in an *s*-orbital of a given shell penetrates the nucleus more strongly than that in a *p*-orbital of the same shell. Thus for a given shell the penetration power (or penetration effect) of an *ns* electron is greater than that of an *np* electron (*ns* > *np*). The penetration power of an *np* electron is greater than that of an *nd* electron (*np* > *nd*) which has higher penetration power than an *nf* electron (*nd* > *nf*). Thus the penetration power of an electron in ns, np, nd and *nf* orbitals of a given *nth* shell is in the following order : (The values of *l* for all the orbitals are also given)

— Penetration power decreasing →

	ns > *np* > *nd* > *nf*
Value of n :	*—same—*
Value of l :	*0 1 2 3*
	— Increasing →

Thus we see that the penetration power of *s, p, d* and *f* orbitals belonging to the same shell (*i.e.*, all the orbitals have the same value of *n*) decreases, as the value of *l* increases. This order gives an idea of the extent of binding the electrons of various orbitals to the nucleus. The above order indicates that s-electrons are more tightly bound with the nucleus than *p*-electrons which, in turn, are bound more tightly than *d*-electrons and so on. Due to greater penetration power of s-electron, it screens the nucleus more effectively than a *p*-electron and a p-electron screens the nucleus more effectively than a d-electron and so on. Thus the order in which an electron in the intervening *ns, np, nd* and *nf* orbitals screens the nucleus (*i.e. nuclear charge*) is as :

— Screening power decreasing →

ns > *np* > *nd* > *nf*

This order shows that for the same shell, the *s*-electrons are more effective in shielding the nuclear charge than *p*-electrons, the *p* electrons are more effective than *d* electrons and *d* electrons are more effective than *f* electrons. This order also means that the electrons present in *d* or *f* orbitals are more effectively shielded than those in *s* or *p*-orbitals by the inner shell electrons. Thus an electron in *np* orbital is more effectively screened than the *ns* electron by the inner-shell electrons. This becomes evident when we consider Be and B atoms. 2*p* electron of B atom ($1s^2$, $2s^2$, $2p^1$) is more effectively shielded by 1*s* and 2*s* electrons while the 2*s* electron of Be atom ($1s^2$, $2s^2$) is not effectively shielded by the 1*s* electrons and the second 2*s* electron.

An electron present in *nth* orbital is more effectively screened by the electrons in the (*n*–1)*th* orbital while the same electron is *more effectively* shielded by the other electron present in the same orbital. For example in Li atom ($1s^2$, $2s^1$) $2s^1$ electron is *more effectively* screened by the two electrons present is 1*s* orbital while in He atom ($1s^2$) one of the two electrons in 1*s* orbital in not effectively screened by the other electron present in the same orbital *i.e.* 1*s* orbital.

Applications of Shielding Effect

(*i*) The concept of shielding effect has been used to explain why the ionisation potential values of the elements of a given group decrease on descending the group.

(*ii*) This concept has also been used to explain that when we proceed from an inert gas to an alkali metal, a large decrease in the value of ionisation potential is observed.

II. Effective Nuclear Charge (Z_{eff}) : Consequence of Shielding Effect

It is due to the shielding effect of the inner electrons on the outer-electrons that the valence electron experiences less attractive pull from the nucleus. The decrease in the attractive force reduces the nuclear charge, Z represented by the atomic number of the element. This decreased nuclear charge is called **effective nuclear charge** and is represented by Z_{eff}. It is given by the relation :

$$\boxed{Z_{eff} = Z - \sigma} \quad ...(i)$$

Here σ is a constant which is called **screening constant.** This constant is a measure of the extent to which the other electrons in an atom are able to screen the nucleus from the chosen electron. Thus :

Effective nuclear charge, Z_{eff} is defined as the actual nuclear charge, Z minus the screening effect caused by the electrons intervening between the nucleus and the outer electrons.

Greater is the number of electrons intervening between the nucleus and the outer electrons, more will be the magnitude of σ and hence the magnitude of effective nuclear charge, Z_{eff} will decrease.

By putting the value of σ in relation (*i*) the value of Z_{eff} for an element can be calculated. The value of σ can be calculated from *Slater's rules*. Thus we see that as a result of shielding effect, the effective nuclear charge, Z_{eff} that a valence shell electron experiences is not exactly the nuclear charge, Z.

Calculation of σ and Z_{eff} by Slater's Rules.

The value of σ and hence that of Z_{eff} can be calculated by using slater's rules. According to these rules the value of σ for a given electron is estimated as follows:

(1) Write down the complete electronic configuration of the element and divide the electrons into the following orbital groups starting from the inside of the atom. (1*s*); (2*s*, 2*p*); (3*s*, 3*p*); (3*d*); (4*s*, 4*p*); (4*d*); (4*f*); (5*s*, 5*p*); (6*s*, 6*p*) etc.

(2) Now select the electron for which the value of σ is to be calculated. For this calculation add up the contributions to σ for the other electrons according to the following rules :

Type of electron	**Contribution to σ for each electron of this type**
(*a*) All electrons in groups outside the electron chosen	0
(*b*) All other electrons in the same group as chosen one	0.35 (or 0.30 for 1*s* electron)

(*c*) All electrons in shell immediately inside 0.85

(*d*) All electrons further inside 1.00

In order to understand the above rules let us consider the following examples.

Example 1. *Calculate the effective nuclear charge experienced by the 4s electron in potassium atom.*

Solution. The electronic configuration of K atom (Z = 19) is $1s^2$, $2s^2p^6$, $3s^2p^6$, $4s^1$ and this atom has 4 shells in all. Evidently the magnitude of Z_{eff} experienced by 4*s* electron is given by :

$$Z_{eff} = Z - \sigma$$

$$= 19 - [(0.85 \times \text{No. of electrons in } 3rd \text{ shell}) + (1.00 \times \text{Total number of electrons in the inner shells})]$$

$$= 19 - [0.85 \times 8 + 1.00 \times (2 + 8)]$$

$$\text{or } Z_{eff} = 2.20$$

This is the effective nuclear charge experienced by 4*s* electron in potassium atom.

Example 2. *Calculate the effective nuclear charge of the last electron in an atom whose configuration is $1s^2$, $2s^2p^6$, $3s^2p^5$.*

Solution. Evidently the atomic number, Z of the element having this configuration is 17 and it has three shells in all

$$\therefore Z_{eff} = Z - \sigma$$

$$= 17 - [(0.35 \times \text{No. of electrons left in } 3rd \text{ shell}) + (0.85 \times \text{No. of electrons in } 2nd \text{ shell}) + (1.00 \times \text{Total number of electrons in the inner shell})]$$

$$= 17 - [(0.35 \times 6) + (0.85 \times 8) + (1 \times 2)$$

$$= 17 - 10.2 = 6.8$$

Variation of Screening Effect or σ in the Periodic Table.

The magnitude of screening constant is a measure of the extent to which the electrons intervening between the nucleus and the valance-shell shield valence-shell electron from the nucleus. Thus *greater is the magnitude of σ, greater is the amount of shielding effect caused by the intervening electrons on the valence-shell electron.*

Since σ increases on descending a group and on proceeding from left to right in a period of the periodic table, *the magnitude of screening effect also increases in the same direction on proceeding from top to bottom of a group and from left to right in a period.* This has been shown for the elements of group IA and 2nd period (Z = atomic number).

Elements of group IA :	Li	Na	K	Rb	Cs			
Value of Z :	3	11	19	37	55			
Value of σ :	1.7	8.8	16.80	34.80	52.80			
Elements of 2nd period :	Li	Be	B	C	N	O	F	Ne
Value of Z :	3	4	5	6	7	8	9	10
Value of σ :	1.70	2.05	2.40	2.75	3.10	3.45	3.80	4.15

Thus the variation of screening effect or σ in a period and a group can be shown as depicted in Fig. 4.2.

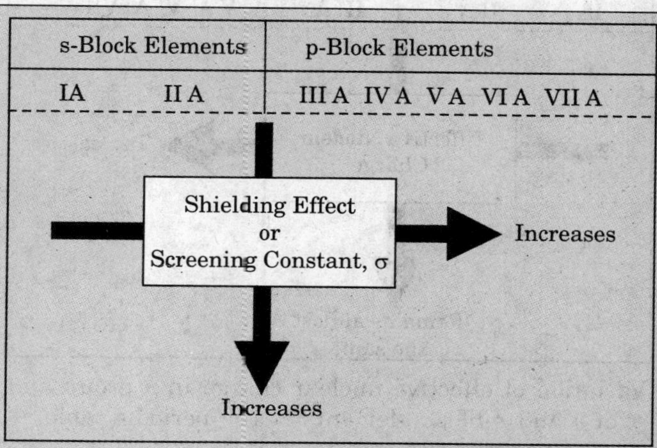

Fig. 4.2. Variation of shielding effect or screening constant, σ in the periodic table.

Variation of Effective Nuclear Charge in the Periodic Table.

(*a*) **In period.** The values of effective nuclear charge of different elements, a calculated by Slater's rules, show that *they increase when we proceed from left to right in a period of the periodic table.*

When we proceed from loft to right across a period, the atomic number (*i.e.* nuclear charge) increases by one at each next element and the next element has one more valence electron with its effective repulsion equal to 0.35 and therefore the effective nuclear charge increases by 1–0.35 = 0.65 from member to member. Consequently the effective nuclear charge of each next atom is greater by 0.65 than the previous atom. For example

Elements of 2nd period	:	Li	Be	B	C	N	O	F	Ne
Effective nuclear charge	:	1.30	1.95	2.60	3.25	3.90	4.55	5.20	5.85
Elements of 3rd period	:	Na	Mg	Al	Si	P	S	Cl	Ar
Effective nuclear charge	:	2.20	2.85	3.50	4.15	4.80	5.45	6.10	6.75

(*b*) **In group.** *When we descend a group, the values of effective nuclear charge remain almost the same* as is evident from these values for the elements of group IA and IIA. Li = 1.30, Na = 2.20, K = 2.20, Rb 2.20, Cs = 2.20, Fr = 2.20 (Group IA – Li is a typical case); Be = 1.95, Mg = 2.85, Ca = 2.85 and so on (Group IIA).

The variation of effective nuclear charge in the periodic table has been shown in Fig. 4.3.

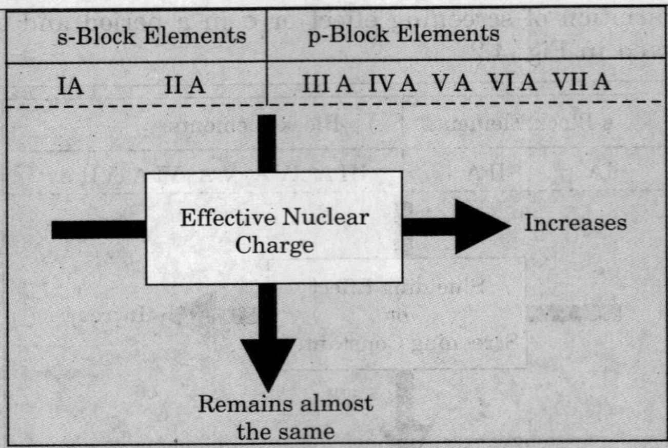

Fig. 4.3. Variation of effective nuclear charge in a group and a period of s and p-block elements of the periodic table.

Applications of Effective Nuclear Charge

A number of facts can be explained on the basis of the concept of effective nuclear charge. Some examples are given below :

***(i)* 4*s* orbital is filled before 3*d* orbitals.** We know that the configuration of argon (atomic number, Z = 18) is $1s^2$, $2s^2p^6$, $3s^2p^6$ while the configuration of potassium (Z = 19) is obtained by adding the extra electron to 4*s* orbital and not to 3*d* orbitals, giving $1s^2$, $2s^2p^6$, $3s^2p^6$, $4s^1$ configuration to this atom. Why 4*s* orbital is filled before 3*d* orbitals can be explained as follows on the basis of the concept of effective nuclear charge.

Suppose potassium can have any of the following two configurations :

(*a*) $1s^2$, $2s^2p^6$, $3s^2p^6d^0$, $4s^1$ (4*s* orbital is filled before 3*d* orbitals)

(*b*) $1s^2$, $2s^2p^6$, $3s^2p^6d^1$. $4s^0$ (3*d* orbitals are filled before 4*s* orbital)

Now let us calculate the effective nuclear charge acting on 4*s* electron [configuration (*a*)]

$[Z_{eff}]_{4s}$ = 19 –[(0.85 × No. of electrons in (n –1)th shell) + (1.00 × Total number of electrons in the inner shells)]

= 19–[(0.85 × 8) + (1.00 × 10)]

= 2.20

Now let us calculate the effective nuclear charge acting on 3*d* electron [configuration (*b*)]

$[Z_{eff}]_{3d}$ = 19 – 1.00 × Total number of inner electrons (*i.e.* $1s^2$, $2s^2p^6$, $3s^2p^6$ electrons)

= 19 – 1 × 18

= 1

Lower value of effective nuclear charge acting on 3d electron as compared to that acting on 4s electron makes it evident that in potassium atom 3d electron is less tightly bound to the nucleus than the 4*s* electron. Consequently the additional electron in potassium atom prefers to enter 4*s* orbital than 3*d* orbital.

***(ii)* 4s electrons are removed before 3d electrons in the conversion of 3d transition elements into cations.** Let us consider the configuration of vanadium atom (atomic number = 23) which is $1s^2$, $2s^2p^6$, $3s^2p^6d^3$, $4s^2$. Suppose this atom is to be converted into V^{2+} cation. Quite obviously this cation is formed by the removal of two electrons from 4*s* orbital and not from 3*d* orbital. Thus :

$$V(1s^2, 2s^2p^6, 3s^2p^6d^3, 4s^2) - 2e^- \rightarrow V^{2+}(1s^2, 2s^2p^6, 3s^2p^6d^3)$$

Why 4*s* electrons prefer to removed than 3*d* electrons can be explained by calculating the value of effective nuclear charge acting on one of the 4*s* or 3*d* electrons.

Effective nuclear charge acting on one of the 4*s* electrons is given by :

$[Z_{eff}]_{4s}$ = Z – σ

= 23 – [(0.35 × No. of electrons left in *nth* shell) + (0.85 × No. of electrons in (*n* – 1)*th* shell) + (1.00 × No. of electrons in the inner shells)]

= 23 – [0.35 × 1 + 0.85 × 11 + 1 × 10]

= 23 – 19.70

= 3.30

Effective nuclear charge acting on one of the 3*d* electrons is given by :

$[Z_{eff}]_{3d}$ = Z – σ

= 23 – [(0.35 × No. of electrons left in 3*d* orbitals) + (1.00 × Total number of electrons in the inner orbitals *i.e.* in 1*s*, 2*s*, 2*p*, 3*s* and 3*p* orbials)]

= 23 – (0.35 × 2 + 1 × 18)

= 23 – 18.70

= 4.30

Greater value of effective nuclear charge acting on one of the 3*d* electrons as compared to that acting on one of the 4*s* electrons in vanadium atom makes it evident that in this atom 3*d* electrons are more tightly bound to the nucleus than the 4*s* electrons. Consequently in the conversion of vanadium atom into V^{2+} cation the electrons to be removed are 4*s* electrons and not 3*d* electrons.

***(iii)* The successive ionisation potentials I_1, I_2, I_3, etc. of an atom are in the following order :**

$$I_1 < I_2 < I_3 < ......$$

This order can well be explained on the basis of effective nuclear charge concept.

***(iv)* In going from left to right in a period of s-and p-block elements atomic and ionic radii decrease with the increase of atomic number.** This can be explained on the basis of effective nuclear charge.

***(v)* A cation is smaller in size than its parent atom.** It has been explained on the basis of the concept of effective nuclear charge (see subsequent pages).

(vi) An anion is bigger in size than its parent atom. It has been explained on subsequent pages.

Atomic Volume

Atomic volume is defined as the volume in c.c. occupied by one gram atom of the element in the solid state and hence is commonly called gram atomic volume. It is obtained by dividing the atomic weight of the element by its density; *i.e.*

$$\text{Atomic volume} = \frac{\text{Atomic weight}}{\text{Density}}$$

In other words atomic volume is the volume in c.c. occupied by 6.023×10^{23} atoms of an element (Avogadro's number). Thus the atomic volumes of the elements should be the volumes occupied by the same number of the atoms. The volume occupied by a single atom may be obtained by dividing gram atomic volume with Avogadro's number of atoms. However, this conclusion is not very accurate due to certain variable factors which, in turn, depend on the manner in which the atoms are packed in various elements. It is a common practice to use the term gram atomic volume for volume per atom for assignment of sizes of atoms.

Lother Meyer plotted the atomic volumes (in c.c.) of different elements against their atomic numbers and obtained a graph shown n Fig. 4.4. An inspection of this graph shows that the curve has five complete well-defined loops and a fraction of the sixth, it may be seen that the maxima of the curve are occupied by the alkali metals (most electropositive elements) while the minima are occupied by C (At. number, Z = 6), Al (Z = 13), Ni (Z = 28), Rh (Z = 45) and Os (Z = 76). It may also be seen that when we proceed further from alkali metals, the atomic volumes on each occasion first decrease and then increase again till a sharp maximum is reached at the next alkali metal. Not only alkali metals but all the elements belonging to the same group lie at corresponding positions on the curve.

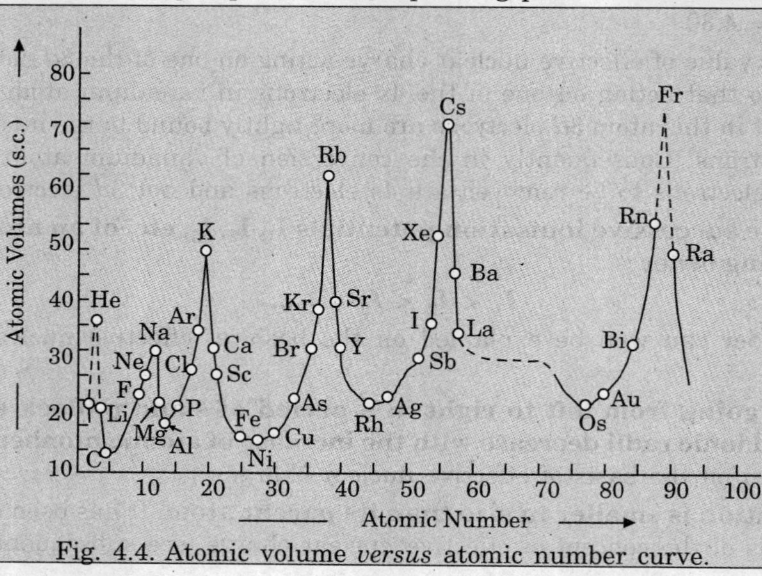

Fig. 4.4. Atomic volume *versus* atomic number curve.

The position of lanthanides [Ce (Z = 58) to Lu (Z = 78)] is to be noted. The expected increase in their atomic volumes is not observed in the curve because as the atomic number increases, the additional electrons enter into the inner $(n-2)th$ shell. This causes a contraction in the atomic volumes of lanthanides. The decrease in atomic volumes on proceeding from Ce to Lu is called *Lanthanide Contraction*.

Variation of Atomic Volume in a Period and a Group.

(*a*) **In a group.** *Atomic volume increases more or less regularly in going down a group* (*See* Table 4.1). The increase in atomic volume in going down a group is due to the increase in the number of shells. The larger the number of shells, the bigger is the volume of the atom.

Table 4.1. Atomic volumes (in c.c.) of *s*-and *p*-block elements. The numbers given in parentheses indicate the atomic numbers.

Group → / Period ↓	*s*-block elements		*p*-block elements					
	I A	*II A*	*III A*	*IV A*	*V A*	*VI A*	*VII A*	*Zero*
1	(1) H 14.1							(2) He 31.8
2	(3) Li 13.1	(4) Be 5.0	(5) B 4.6	(6) C 5.3	(7) N 17.3	(8) O 14.0	(9) F 17.1	(10) Ne 16.8
3	(11) Na 23.7	(12) Mg 14.0	(13) Al 10.0	(14) Si 12.1	(15) P 17.0	(16) S 15.5	(17) Cl 18.7	(18) Ar 24.4
4	(19) K 45.3	(20) Ca 29.9	(31) Ga 11.8	(32) Ge 13.6	(33) As 13.1	(34) Se 16.5	(35) Br 23.5	(36) Kr 32.2
5	(37) Rb 55.9	(38) Sr 33.7	(49) In 15.7	(50) Sn 16.3	(51) Sb 18.4	(52) Te 20.5	(53) I 25.7	(54) Xe 42.9
6	(55) Cs 70.0	(56) Ba 39.0	(81) Tl 17.2	(82) Pb 18.3	(83) Bi 21.3	(84) Po —	(85) At —	(86) Rn —

(*b*) **In a period.** *In going from left to right in a period, it varies cyclically, i.e., it decreases at first for some elements, becomes minimum in the middle and then increases* (*See* Table 4.1). The variation of atomic volume in going from left to right in a period is influenced by the following two factors :

(*i*) **Nuclear charge.** We know that the nuclear charge (*i.e.* atomic number) increases by one, as we move from left to right in a period. The increased nuclear charge attracts each electron more strongly towards the nucleus, resulting in a decrease in the volume of the atom.

***(ii)* No. of valence electrons.** Towards the close of a period, due to an increase in the number of valence-electrons (*i.e.* electrons in the valence shell) the volume of the atom increases, so that it may accommodate all the electrons.

These two factors, one causing an increase and the other causing a decrease, combine to result that in a period atomic volume decreases at first for some elements, becomes minimum in the middle and then increase.

Size of Atoms and Ions
(*Atomic and Ionic Radii*)

An atom is composed of a compact nucleus surrounded by electron cloud which does not have a sharply defined boundary because of its diffused nature (*Wave mechanical picture of atom*). The diffused nature of the electron cloud makes it difficult for us to give an exact definition of atomic size or atomic radius. Thus the atomic radius is an arbitrary concept and is influenced by the nature of neighbouring atoms. However, *it is usually considered as the distance from the centre of the nucleus to the point where the electron density is effectively zero* (*i.e. negligible*).

Types of Radii

Because of the vagueness of the concept of atomic radius, a number of radii have been defined for an atom which pertain to different types of bonding between the atoms. Some of these radii are : *Covalent radius, Crystal radius (Atomic or metallic radius), van der Waal's radius (Collision radius) and Ionic radius.*

Covalent (Atomic) Radius and Bond Length

The term covalent radius is used for non-metals. Before understanding the concept of covalent radius we should know the meaning of bond length. *The distance between the nuclei of two bonded atoms of a homonuclear diatomic molecule of A_2 type or heteronuclear diatomic molecule of AB type is called bond length (or bond distance or inter-nuclear distance).* In these molecules it is represented as *d* (A–A) and d (A–B) respectively.

Calculation of covalent radius.

In case of *homonuclear diatomic molecules* of A_2 type (*e.g.* F_2, Cl_2, Br_2 I, etc.), the bond length, *d*(A–A) is given by :

$$d\,(A–A) = r\,(A) + r\,(A)$$

or $$d(A–A) = 2 \times r\,(A)$$

or $$r\,(A) = \frac{d(A-A)}{2} \qquad ...(i)$$

Equation (*i*) shows that *in case of homonuclear diatomic molecule of A_2 type, the covalent radius of an atom A, r (A) is equal to one-half of the inter-nuclear distance, d(A–A) i.e. the covalent radius of an atom in a homonuclear diatomic molecule can be obtained by dividing the inter-nuclear distance by two.*

In case of *heteronuclear diatomic molecule* of AB type, bond length, *d*(A–B) is given by :

$$d(A–B) = r(A) + r(B) \quad ...(ii)$$

Here r (A) and r (B) are the covalent radii of A and B atoms. Equation (*ii*) shows that *in case of heteronuclear diatomic molecules of AB type, the bond length, d(A–B) is equal to the sum of the covalent radii of A and B atoms*. It is also obvious from equation (*ii*) that the magnitude of r(A) can be calculated if d(A–B) and r (B) are known.

The covalent radii obtained by equation (*i*) and (*ii*) are called **single bond covalent radii.** Equations (*i*) and (*ii*) reveal that the *single bond covalent radii are additive in nature, i.e.,* when the covalent radii of atoms in A_2 or AB molecules are added, we get the bond length, d (A–A) or d (A–B). This principle is called the ***principle of additivity of covalent radii.*** This principle has been used to calculate the covalent radii of the elements as shown by the following examples :

(i) Cl_2 molecule. The value of Cl–Cl bond distance as found experimentally is 1.98 A°. Thus :

$$r\,(Cl) = \frac{d\,(Cl-Cl)}{2} = \frac{1.98}{2} = 0.99\ A°$$

(ii) Diamond (C_2). The value of d(C – C) distance as found experimentally in a variety of saturated hydrocarbons is 1.54 A°. Thus :

$$r\,(C) = \frac{d\,(C-C)}{2} = \frac{1.54}{2} = 0.77\ A°$$

(iii) Carborundum, *SiC*. In case of this molecule the experimental value of d (Si–C) is 1.93. Thus :

$$d\,(Si–C) = r\,(Si) + r\,(C)$$

or

$$r\,(Si) = d\,(Si–C) - r\,(C)$$

$$= 1.93 - r\,(C)$$

Thus the covalent radius of silicon atom is obtained by substracting the covalent radius of carbon atom from Si–C bond distance, d (Si–C). Covalent radius of carbon atom, r (C) can also be calculated in a similar way, if that of Si atom, r (Si) is known.

(iv) CCl_4 molecule. In case of this molecule the experimental value of d (C–Cl) is 1.76 A°. Thus :

$$d\,(C–Cl) = r\,(C) + r(Cl)$$

or

$$r(C) = d\,(C - Cl) - r\,(Cl)$$

$$= 1.76 - r\,(Cl)$$

Thus the covalent radius of carbon atom can be obtained by substracting the covalent radius of Cl atom from d(C– Cl) bond length. The covalent radius of Cl atom can also be obtained, provided that the covalent radius of C atom is known.

The values of single bond covalent radii of *s*- and *p*-block elements, as determined experimentally, are given below in Table 4.2.

Table 4.2. Single bond covalent radii (in A°) of *s*- and *p*-block elements. The numbers given in parentheses indicate the atomic numbers.

Group → / Period ↓	*s*-block elements		*p*-block elements					
	IA	IIA	IIIA	IVA	VA	VIA	VIIA	Zero
1	(1) H 0.32							(2) He 0.93
2	(3) Li 1.23	(4) Be 0.90	(5) B 0.82	(6) C 0.77	(7) N 0.75	(8) O 0.73	(9) F 0.72	(10) Ne 0.71
3	(11) Na 1.54	(12) Mg 1.36	(13) Al 1.18	(14) Si 1.11	(15) P 1.06	(16) S 1.02	(17) Cl 0.99	(18) Ar 0.98
4	(19) K 2.03	(20) Ca 1.74	(31) Ga 1.26	(32) Ge 1.22	(33) As 1.20	(34) Be 1.16	(35) Br 1.14	(36) Kr 1.12
5	(37) Rb 2.16	(38) Sr 1.91	(49) In 1.44	(50) Sn 1.41	(51) Sb 1.40	(52) Te 1.30	(53) I 1.33	(54) Xe 1.31
6	(55) Cs 2.35	(56) Ba 1.98	(81) Ti 1.48	(82) Pb 1.47	(83) Bi 1.46	(84) Po 1.46	(85) At (1.45)	(86) Rn –

Characteristics of Covalent Radii

(*i*) *As already said, single bond covalent radii are additive in nature. The additive nature of covalent radii is called principle of additvity of covalent radii.*

(*ii*) *The values of covalent radii change with the multiplicity of the bonds between the atoms.* In case of atoms joined by double or triple covalent bonds, the covalent radius is called double bond covalent radius and triple bond covalent radius respectively. These radii are smaller than the single bond covalent radii due to greater overlapping of orbitals in the formation of double and triple covalent bonds.

Double and triple bond covalent radii (in A°) of some common elements are given on the next page.

(*iii*) *Covalent radius of an element is less than its van der Waals radius,* as the formation of a covalent bond involves the overlapping of the electron clouds of the two atoms.

(*iv*) *The magnitude of covalent radius undergoes a change depending on the nature of the space model of the atoms linked such as tetrahedral, square planar, octahedral etc.*

IIIA	IVA	VA	VIA
B = B 0.76 B≡B 0.68	C = C 0.67 C≡C 0.60	N = N 0.60 N≡N 0.55	O = O 0.57
	Si = Si 1.07 Si ≡ Si 1.00	P = P 1.00 P≡P 0.93	S = S 0.97 S≡S 0.94
	Ge = Ge 1.12	As = As 1.11	Se = Se 1.07
	Sn = Sn 1.30	Sb = Sb 1.30	Te = Te 1.27

More About Bond Length

As already said *bond length is defined as the distance between the nuclei of two bonded atoms*. For a homonuclear diatomic molecule of A_2 type it is calculated by :

$$\boxed{d(A\text{–}A) = 2 \times r\ (A)} \quad ...(i)$$

while in case of heterodiatomic molecule of AB type it is calculated by :

$$\boxed{d(A\text{–}B) = r(A) + r(B)} \quad ...(ii)$$

The experimental values of covalent bond lengths for some common homonuclear diatomic molecules are given below. These values are in full agreement with those calculated from equation (*i*).

Homonuclear diatomic molecule	***Bond***	***Bond length (A°)***
H_2	H–H	0.74
F_2	F–F	1.44
Cl_2	Cl–Cl	1.98
Br_2	Br–Br	2.28
$H_3C–CH_3$ (Ethane)	C–C	1.54

Although the values of *d* (A–B) calculated from equation (*ii*) for diatomic molecules of AB type have been found to the same as found experimentally, yet there are many AB molecules which show deviations, *i.e.* in these molecules *d* (A–B) is not equal to the sum of *r* (A) and *r* (B) as is evident from Table 4.3.

The deviation can also be seen when we consider the N–H bond length [calculated value of d(N–H) = r(N) = r(H) = 0.75 + 0.37 = 1.12 A°] in the following species.

d (N–H) in NH_3 = 1.008 Å

d (N–H) in $[NH_4]^+$ = 1.034 Å

d (N–H) in $[N_3H_6]^{2+}$ = 1.70 Å

d (N–H) in HN_3 = 1.020 Å

Factors Affecting the Magnitude of Bond Length

1. Electronegativity of the bonded atoms, A and B. The increase in the amount of ionic character in A–B bond decreases the bond length, d (A–B). Again we know that since the amount of ionic character in A–B bond depends on the difference in electronegativity of the elements A and B, the decrease in bond length ultimately is due to the difference in electronegativity of the bonded atoms A and B.

Taking into account the ionic character in A–B bond, Schomaker and Stevenson (1941) gave the following empirical equation by which the bond distance, d (A–B) can be calculated.

$$d\ (A\text{–}B) = r(A) + r(B) - 0.09\ (x_A - x_B) \quad ...(iii)$$

Here x_A and x_B are the electronegativities of the elements A and B.

The validity of the above equation can be shown by calculating the N–C distance in $(CH_3)_3N$ molecule by equations (*ii*) and (*iii*).

Table 4.3. Calculated and experimental values of some bond lengths (A°) of A–B bonds in AB molecules.

Bond	Calculated value of d(A–B) = r(A) + r(B)	Experimental value of d(A–B)
S—H	1.02 + 0.37 = 1.39	1.34
S—F	1.02 + 0.70 = 1.72	1.56
S—Cl	1.02 + 0.99 = 2.01	1.99
S—Br	1.02 + 1.14 = 2.16	2.27
B—F	0.82 + 0.70 = 1.52	1.30
B—Cl	0.82 + 0.99 = 1.81	1.72
B—Br	0.82 + 1.14 = 1.96	1.88
C—H	0.77 + 0.37 = 1.14	1.12
C—F	0.77 + 0.70 = 1.47	1.42
C—Br	0.77 + 1.14 = 1.91	1.94
C—I	0.77 + 1.33 = 2.10	2.14
Si—F	1.17 + 0.70 = 1.87	1.54 (in SiF_4)
Si—Cl	1.17 + 0.99 = 2.16	2.02
Si—Br	1.17 + 1.14 = 2.31	2.16
Si—I	1.17 + 1.33 = 2.50	2.44
N—H	0.75 + 0.37 = 1.12	1.00
O—H	0.73 + 0.37 = 1.10	0.96
B—F	0.90 + 0.72 = 1.62	1.42 (in $[BF_4]^-$ ion) 1.29 (in BF_3)

From equation (*ii*), d (N–C) $= r(N) + r(C)$

$= (0.75 + 0.77)$ Å

$= 1.52$ Å

This calculated value (= 1.52 A°) obtained as above is not in agreement with the experimental value (=1.47 Å) seen in $(CH_3)_3N$ molecule. But the value calculated from equation (*iii*) has been found in good agreement with the experimental value as shown below :

From equation (*iii*), $d(\text{N–C}) = r(\text{N}) + r(\text{C}) - 0.09\ (x_N - x_C)$

$$= 0.75 + 0.77 - 0.09\ (3.0 - 2.5)$$

$$= 1.475\ \text{Å}$$

2. **Bond order or bond multiplicity.** Bond order or bond multiplicity is defined as the number of covalent bonds between the two bonded atoms. *The bond length between the two atoms decreases with the increase of multiplicity of the bond between the atoms*. For example *d* (C–C) bond length found in ethane ($H_3C–CH_3$) is equal to 1.54 Å while *d* (C–C) bond length found in ethylene ($H_2C = CH_2$) is 1.34 Å and that found in acetylene (HC ≡ CH) is 1.20 Å. Thus we see that *d* (C–C) bond length decreases with the increase of the multiplicity of the bond between two carbon atoms. The decrease in bond length with the increase of the multiplicity of the bond is also evident from the following examples. Bond order is given in bracket.

B—B (1) 1.64 Å B = B (2) 1.52 Å B ≡ B (3) 1.36 Å	C—C (1) 1.54 Å C = C (2) 1.34 Å C ≡ C (3) 1.20 Å	N—N (1) 1.50 Å N = N (2) 1.20 Å N ≡ N (3) 1.12 Å	O—O (1) 1.46 Å O = O (2) 1.14 Å	C—N (1) 1.47 Å C = N (2) 1.29 Å C ≡ N (3) 1.16 Å
	Si—Si (1) 2.34 Å Si = Si (2) 2.14 Å Si ≡ Si (3) 2.00 Å	P—P (1) 2.12 Å P = P (2) 2.0 Å P ≡ P (3) 1.86 Å	S—S (1) 2.04 Å S = S (2) 1.94 Å S ≡ S (3) 1.88 Å	C—O (1) 1.43 Å C = O (2) 1.22 Å
	Ge—Ge (1) 2.44 Å Ge = Ge (2) 2.24 Å	As—As (1) 2.40 Å As = As (2) 2.22 Å	Se—Se (1) 2.32 Å Se = Se (2) 2.14 Å	
	Sn—Sn (1) 2.82 Å Sn = Sn (2) 2.60 Å	Sb—Sb (1) 2.80 Å Sb = Sb (2) 2.62 Å	Te—Te (1) 2.72 Å Te = Te (2) 3.54 Å	

Sometimes bond order increases due to resonance and bond shortening is observed. For example, in benzene molecule the bond distance between two carbon atoms is 1.39 A which is intermediate between the C–C and C=C bond lengths (C–C=154 Å, C=C= 1.34 Å). In other words the bond order between the two carbon atoms is more than 1 and less than 2.

3. Hybridisation. The change in hybridisation affects the shape of the molecule and hence the bond length is also changed in turn. *With the increase of the percentage of s-character in a hybrid orbital, the bond length decreases.* For example with the increase of the percentage of *s*-character from ethane to acetylene via ethylene, C–H and C–C bond lengths decrease in the same direction as shown below :

Molecule	Type of hybridisation	Percentage of s-character	C—H bond length (Å)	C—C bond length (Å)
Ethane ($H_3C—CH_3$)	sp^3	25	1.093	1.54
Ethylene ($H_2C = CH_2$)	sp^2	33.3	1.087	1.34
Acetylene ($HC \equiv CH$)	sp	50	1.057	1.20

Utility of Bond Length Determination

1. Stability of a molecule. The length of a given bond is a measure of its stability. A molecule with small bond length is more stable than that with large bond length.

2. Bond order. A molecule with smaller bond length has higher bond order. Thus the bond length can be used to infer the presence of π-bonding in a molecule. For example in $[BF_4]^-$ ion, the B–F bond length has been found to be equal to 1.42 Å, while B–F bond distance found in BF_3 molecule is 1.29 Å. We know that in $[BF_4]^-$ ion four F atoms are linked with B atom by four σ bonds. Thus the shortening of B–F bond distance in BF_3 molecule is due to the presence of π character in B–.F bonds. This character is produced by the donation of lone pair of electrons from the F atoms into the vacant $2p_z$ orbital of B atom ($p\pi$–$p\pi$ bonding).

Crystal Radius (*Atomic or Metallic Radius*)

The atomic radius as obtained from the metallic crystalline structure of a metal is referred to the crystal or metallic radius. Thus it is used for metal atoms which are assumed to be closely packed spheres in the metallic crystal. The metal atoms are supposed to touch one another in the crystals and *the crystal radius is defined as one-half of the distance between the nuclei of two adjacent metal atoms in the metallic close-packed crystal lattice in which metal exhibits a co-ordination number of 12.* For example the internuclear distance between two adjacent Na atoms in a crystal of sodium metal is 3.80 Å and hence the atomic radius of Na metal = 3.80/2 = 1.90 Å.

It may be noted that *the metallic radii are quite different from other types of radii viz. covalent, ionic or van der Walls radii.*

Since there are weak bonding forces between the metal atoms, the metallic radii are higher (10 to 15%) *than single bond covalent radii. The metallic radii are smaller than the van der Walls radii,* since the bonding forces in the metallic crystal lattice are much stronger than the van der Walls forces.

Metallic radii of *s*-and *p*-block elements are given in Table 4.4. These values are for coordination number of 12.

Table 4.4. Metallic (or atomic) radii of some common elements (in A°). These values are for co-ordination number of 12. The numbers given in parentheses indicate the atomic numbers

	***s*-block elements**		***p*-block elements**			
Group → / **Period ↓**	**IA**	**IIA**	**IIIA**	**IVA**	**VA**	**VIA**
2	*(3) Li 1.55*	*(4) Be 1.12*	*(5) B 0.98*	*(6) C 0.91*	*(7) N 0.92*	*(8) O —*
3	*(11) Na 1.90*	*(12) Mg 1.60*	*(13) Al 1.43*	*(14) Si 1.32*	*(15) P 1.28*	*(16) S 1.27*
4	*(19) K 2.35*	*(20) Ca 1.97*	*(31) Ga 1.41*	*(32) Ge 1.37*	*(33) As 1.39*	*(34) Se 1.40*
5	*(37) Rb 2.48*	*(38) Sr 2.15*	*(49) In 1.66*	*(50) Sn 1.62*	*(51) Sb 1.59*	*(52) Te 1.60*
6	*(55) Cs 2.67*	*(56) Ba 2.22*	*(81) Tl 1.71*	*(82) Pb 1.75*	*(83) Bi 1.70*	*(84) Po 1.76*

The values of the metallic radii depend to some extent on the crystal structure. For example the crystal radius of Ti in body-centred cubic lattice is 1.43 A° while in hexagonal close-packed lattice it is 1.49 A°.

van der Waals Radius (*Collision Radius*)

When two non-bonded isolated atoms or two non-bonded atoms belonging to two adjacent molecules of an element in the solid state approach each other without overlapping their electron clouds to form a bond between them, they cannot come closer than a minimum distance without forming a bond. This minimum distance is called *van der Waals distance* and half of this distance is called van der Waals radius. Thus :

The distance between the nuclei of two non-bonded isolated atoms or the distance between two non-bonded atoms belonging to two adjacent molecules of an element in the solid state is called **van der Waals distance** *while the half of this distance is called* **van der Waals radius.**

Illustration

(*i*) van der Waals radius of an atom in two non-bonded isolated atoms. In Fig. 4.5 two Cl atoms with their nuclei P and Q have been shown. When these two atoms are just in contact with each other (*No bond formation condition*), the distance between their nuclei P and Q is called van der Waals distance and half of this distance is called van der Waals radius (See Fig. 4.5).

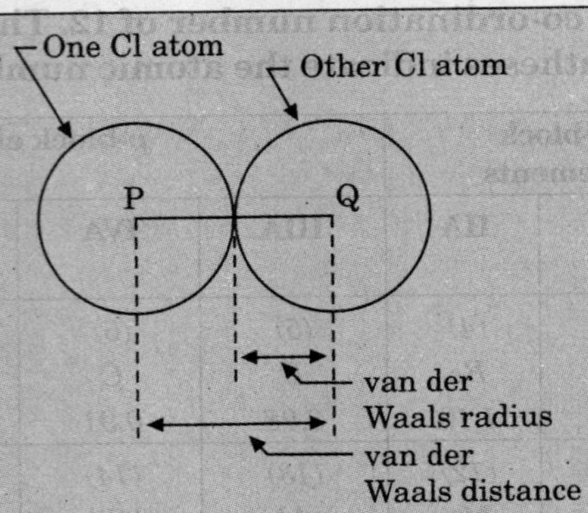

Fig. 4.5. Illustration of van der Waals radius of Cl atom in two non-bonded Cl atoms.

***(ii)* van der Waals radius of an atom in two non- bonded adjacent molecules.** In Fig. 4.6 two Cl_2 molecules have been shown. These molecules have resulted by the overlap of the electron clouds of two Cl atoms. Let these molecules be brought in close contact with each other (*No bond formation*). Let P and Q be the nuclei of one Cl_2 molecule and R and S of the other Cl_2 molecule. The distance between the nuclei Q and R is called van der Waals distance while half of this distance is called van der Waals radius. Thus :

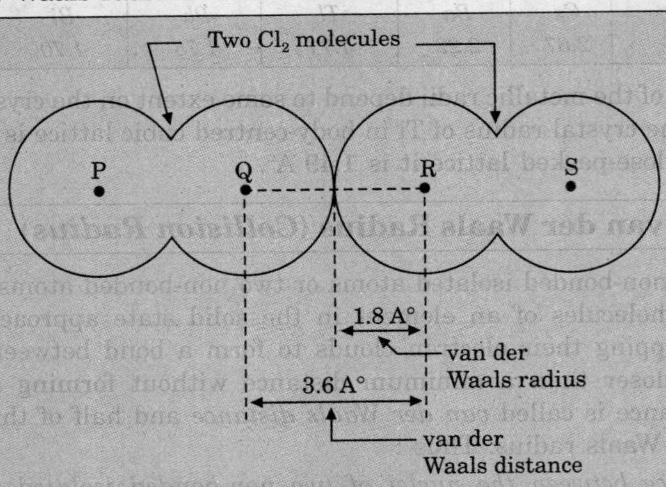

Fig 4.6. Illustration of van der Waals radius of Cl atom in two non-bonded adjacent Cl_2 molecules.

$\because$ van der Waals distance = QR = 3.6 A°

$\therefore$ van der Waals radius = QR/2 = 3.6/2 A° = 1.8 A°

The statement that the van der Waals radius of hydrogen is 1.2 A° means that nuclei of two H-atoms in two different H_2 molecules cannot approach each other nearer than 2 × 1.2 A° from each other without overlapping of their electron clouds to form a bond. Similarly since two Cl-atoms in two different Cl_2 molecules cannot approach each other nearer than 3.6 A°, the van der Waals radius of chlorine atom is 3.6/2 = 1.8 A°.

The name has been derived from the van der Waals forces which are operating between the neutral atoms of noble gases and between non-polar molecules of various substances in the solid state. van der Waals radius does not apply to metals. Its magnitude depends on the effective packing size of the atom.

Comparison of van der Waals Radius with Covalent Radius.

The values of van der Waals radii of some elements in the solid state as obtained from X-ray studies are given in Table 4.5 which also contains the covalent radii of the same elements for the purpose of comparison.

Table 4.5. van der Waals radii (A°) of some elements. The values in parentheses are covalent radii which are given for comparison.

Group→ Period ↓	VA	VIA	VIIA	Zero
1			H 1.2 (0.32)	He 1.20 (0.93)
2	N 1.5 (0.75)	O 1.40 (0.73)	F 1.36 (0.72)	Ne 1.60 (0.71)
3	P 1.9 (1.06)	S 1.85 (1.02)	Cl 1.80 (0.99)	Ar 1.91 (0.98)
4	As 2.0 (1.20)	Se 2.0 (1.16)	Br 1.95 (1.14)	Kr 2.0 (1.12)
5	Sb 2.2 (1.40)	Te 2.20 (1.36)	I 2.15 (1.33)	Xe 2.20 (1.31)

It may be seen from the table that *van der Waals radius of an element is higher than its covalent radius*. This is because of the fact that in the formation of covalent bond, atoms have to overlap each other while in van der Waals forces the atoms/molecules simply come closer to each other and do not enter into overlapping (*No bond formation*).

The comparison between van der Waals and covalent radii can be explained by considering the following examples :

(*i*) In Fig. 4.7 there are two Cl atoms with their nuclei P and Q. When these two atoms are just in contact with each other [Fig. 4.7 (*a*)] (*No bond formation condition*), the distance between their nuclei is called van der Waals distance and half of this distance is called van der Waals radius.

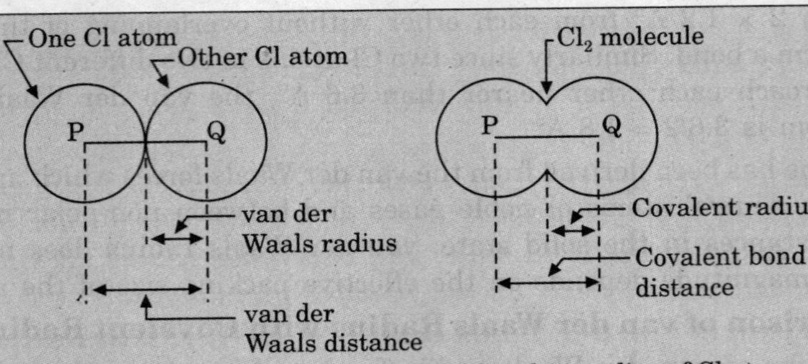

(*a*) van der Waals radius of Cl atom in two non-bonded isolated Cl atoms.

(*b*) Covalent radius of Cl atom in Cl_2 molecule.

Fig. 4.7. Comparison of van der Waals and covalent radii in two Cl atoms.

On the other hand when the electron clouds of the two Cl atoms merge with each other to form Cl_2 molecule [Fig. 4.7 (*b*)], the distance between the nuclei of Cl_2 molecule is called covalent bond distance and half of it is called covalent radius.

(*ii*) Suppose two Cl_2 molecules formed by the overlapping of electron clouds of two Cl atoms are brought in close contact with each other (*No bond formation*). Let P and Q be the nuclei of one Cl_2 molecule, and R and S of the other Cl_2 molecule (Fig. 4.8). It may be seen from the figure that half of the distance between the nuclei Q and R (called van der Waals distance) of the two non-bonded neighbouring Cl atoms belonging to two adjacent Cl_2 molecules is the van der Waals radius of Cl atom while half of the distance between the two nuclei P and Q (called covalent bond distance) of the same Cl_2 molecule is covalent radius of Cl atom. Thus :

van der Waals distance QR = 3.60 A°

∴ van der Waals radius = QR/2 = 3.60/2 = 1.80 A°

Covalent bond distance = PQ = 1.98 A°

∴ Covalent radius = PQ/2 = 1.98/2 = 0.99 A°

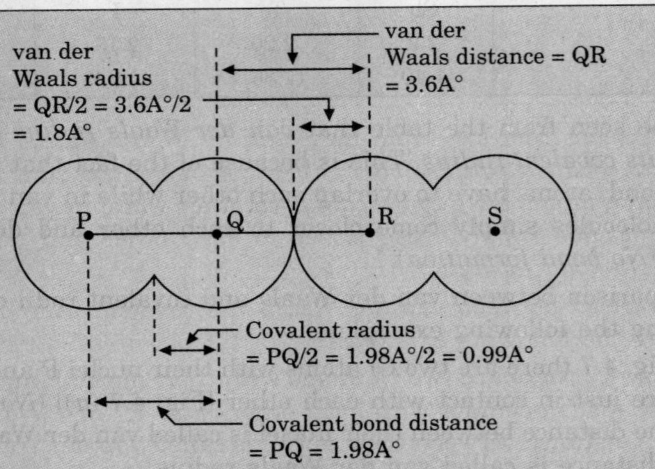

Fig. 4.8. Comparison of van der waals and covalent radii in two Cl_2 molecules.

For noble gases in the solid state the crystal radii (*i.e.* atomic radii) are actually van der Waals radii.

Ionic Radius

The ionic radius is defined as the distance between the nucleus of an ion and the point where the nucleus exerts its influence on the electron cloud.

Characteristics of Ionic Radii.

(*i*) The ionic radius of a particular ion is of more or less constant magnitude.

(*ii*) Ionic radii have an additive character. This means that if the ions of an ionic crystal are assumed to be spheres, the inter-nuclear distance between the two ions is equal to the sum of the radii of the ions of which the ionic crystal is composed. Thus in C^+A^- ionic crystal shown in Fig. 4.9.

$$r(C^+) + r(A^-) = d(C^+ - A^-)$$

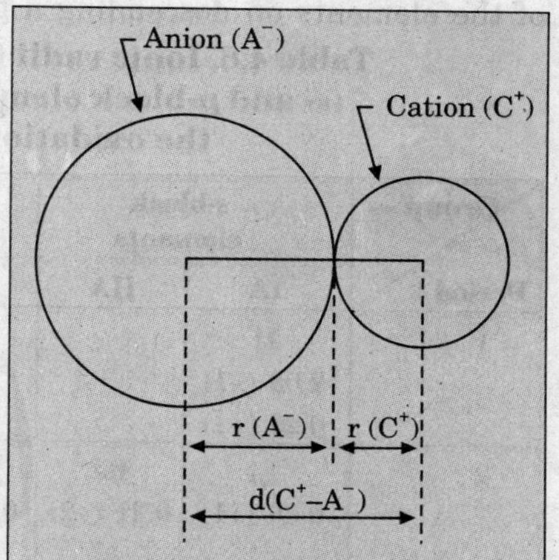

Fig. 4.9. Ionic radii have additive character.

Problem. *Ionic radius of Na^+ ion is 0.95 Å and the inter-nuclear distance between the ion-pairs in NaCl ionic crystal is equal to 2.76 Å. Calculate the radius of Cl^- ion.*

Solution. $d(Na^+ - Cl^-) = r(Na^+) + r(Cl^-)$

$\therefore$ $2.76 = 0.95 + r(Cl^-)$

$\therefore$ $r(Cl^-) = 2.76 - 0.95 = 1.81$ Å

Ionic radii of s- and p-block elements are given in Table 4.6.

While describing the sizes of various atoms or ions any of the radii described above can be used.

Factors Affecting the Magnitude of Atomic and Ionic Radii

Following are the factors which affect the magnitude of the size of atoms or ions.

1. Effective nuclear charge. We know that the magnitude of effective nuclear charge determines the magnitude of force of attraction exerted by the nucleus on the outermost electron cloud. Greater is the magnitude of effective nuclear charge, greater is the amount of the force which binds the outermost electrons to the nucleus and hence the electron cloud of the outermost shell is pulled inward nearer to the nucleus and consequently its distance from the nucleus (*i.e.* atomic and ionic radius) is reduced. Thus *with the increase of effective nuclear charge, the size of the atom or ion (i.e. atomic or ionic radii) decrease.*

This factor has been used in explaining the variation of atomic and ionic radius of the elements in a period of the periodic table.

2. Number of shells or principal quantum number (*n*). As the total number of shells or the magnitude of principal quantum number (*n*) increases, the outer electrons get farther and farther away from the nucleus and hence the atomic and ionic radii increase. *Thus with the increase of the number of shells or the principal quantum number, the atomic and ionic radii increase.*

This factor has been used to explain the increase of atomic and ionic radius of the elements on descending a group.

Table 4.6. Ionic radii (in Å) of representative elements (*s*- and *p*-block elements). In parentheses are given the oxidation states of the elements

Group → / Period ↓	*s*-block elements		*p*-block elements				
	IA	IIA	IIIA	IVA	VA	VIA	VIIA
1	H 2.08 (–1) 0.29 (+1)						
2	Li 0.60 (+1)	Be 0.31 (+2)	B 0.20 (+3)	C 2.60 (–4) 0.15 (+4)	N 1.71 (–3) 0.11 (+5)	O 1.40 (–2) 0.09 (+6)	F 1.36 (–1) 0.07 (+7)
3	Na 0.95 (+1)	Mg 0.65 (+2)	Al 0.50 (+3)	Si 2.71 (–1) 0.41 (+4)	P 2.12 (–3) 0.34 (+5)	S 1.84 (–2) 0.29 (+6)	Cl 1.81 (–1) 0.26 (+7)
4	K 1.33 (+1)	Ca 0.99 (+2)	Ga 1.13 (+1) 0.62 (+3)	Ge 0.93 (+2) 0.53 (+4)	As 2.22 (–3) 0.47 (+5)	Be 1.98 (–2) 0.42 (+6)	Br 1.95 (–1) 0.39 (+7)
5	Rb 1.48 (+1)	Sr 1.13 (+2)	In 1.32 (+1) 0.81 (+3)	Sn 1.12 (+2) 0.71 (+4)	Sb 0.45 (–3) 0.62 (+5)	Te 2.21 (–2) 0.56 (+6)	I 2.16 (–1) 0.50 (+7)
6	Cs 1.69 (+1)	Ba 1.35 (+2)	Ti 1.40 (+1) 0.95 (+3)	Pb 1.20 (+2) 0.84 (+4)	Bi 1.20 (+3) 0.74 (+5)	Po — —	At — —
7	Fr 1.76 (+1)	Ra 1.40 (+2)					

Variation of Atomic and Ionic Radii in a Period and a Group of Representative Elements. (*Periodic Trends of Atomic and ionic Radii*)

(*a*) **In a period.** We know that the number of shells in all the elements of a given period remains the same but the value of effective nuclear charge, as calculated by Slater's rules, *increases from left to right*. The increased effective nuclear charge pulls the electron cloud of the atom nearer to the nucleus and thus the size of the atoms and ions goes on decreasing from left to right. *Thus in going from left to right in a period of s-and p-block elements atomic and ionic radii decrease with the increase of atomic number.* This fact can be illustrated by considering the atomic (covalent) and ionic radii of the elements of 2*nd* period as shown below :

	Li	Be	B	C	N	O	F	Ne
Elements of 2nd period:	Li	Be	B	C	N	O	F	Ne
Effective nuclear charge:	1.30	1.95	2.60	3.25	3.90	4.55	5.20	5.85

—— *Increasing* ⟶

Atomic (covalent) radius (Å):	1.23	0.90	0.82	0.77	0.75	0.73	0.72	0.71

—— *Decreasing* ⟶

Ionic radius (Å):	0.60 (Li^+)	0.31 (Be^{2+})	0.20 (B^{3+})	0.15 (C^{4+})	0.11 (N^{5+})	0.09 (O^{6+})	0.07 (F^{7+})	–

—— *Decreasing* ⟶

Thus in any period the alkali metals (that are present at the extreme left of the periodic table) have the largest size while the halogens (that are present at the extreme right, excluding zero group elements) have the smallest size. However, the size of the atoms of inert gases is larger than that of the preceeding halogen.

(*b*) **In a group.** *In going down a group of s-and p-block elements the atomic and ionic radii both increase with the increase of atomic number.* For example the atomic (covalent) and ionic radii of alkali metals increase on proceeding from Li to Cs as shown below :

Alkali metal (atomic numbers given in parentheses)	Effective nuclear charge	Electronic configuration	No. of shells or principal quantum number (n)	Atomic (covalent) radius (Å)	Ionic radius (M^+ ions) (Å)
Li (3)	1.30	2, 1	2	1.23	0.60
Na (11)	2.20	2, 8, 1	3	1.54	0.95
K (19)	2.20	2, 8, 8, 1	4	2.03	1.33
Rb (37)	2.20	2, 8, 18, 8, 1	5	2.16	1.48
Cs (55)	2.20	2, 8, 18, 18, 8, 1	6	2.35	1.60

We have seen that on descending a group the magnitude of effective nuclear charge acting on the valence-shell electron of the elements remains the same (The first element is a typical case). Thus the concept of effective nuclear charge cannot be used to explain the successive increase in the atomic or ionic radii of the elements of a given group. However, the other factor namely *the number of shells or principal quantum number* (*n*) can be used to explain the increase in radii. As the number of shells or principal quantum number (*n*) increases from 2 (in case of Li) to 6 (in case of Cs), the outer-most shell electrons get farther and farther away from the nucleus and hence atomic and ionic radii increase. Thus it is due to the progressive addition of a new shell (or the increase in the number of shells) that the atomic or ionic radii increase when we proceed from top to bottom in a group.

The variation of atomic and ionic radii of representative elements (*s*-and *p*-block elements) in a period and a group of the periodic table can be shown as given in Fig. 4.10.

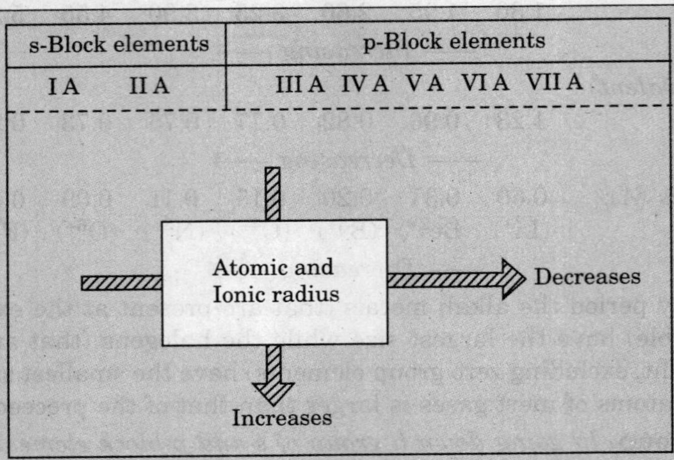

Fig. 4.10. Variation of atomic and ionic radii in a period and a group of representative elements (*s*- and *p*-block elements) of the long form of periodic table.

A Cation is Smaller in Size than its Parent Atom

A cation is smaller in size than the atom from which it has been derived, *i.e.* a cation is smaller in size than its parent atom.

The decrease in size of the cation as compared to its parent atom can be explained as follows on the basis of the concept of effective nuclear charge.

We know that a cation is formed by the removal of one or more electrons from the atom. Thus a cation has lesser number of electrons than its parent atom. With the decrease of the number of electrons the magnitude of screening constant, σ also decreases. The decrease in the magnitude of σ increases the magnitude of effective nuclear charge. The increased effective nuclear charge pulls the electron cloud of the cation inward nearer to the nucleus and thus makes the cation smaller in size than its parent neutral atom.

Thus alkali metal cations are smaller in size than the alkali metal atoms from which they have been derived as shown below :

Alkali metal (Atomic number, Z given in bracket)	σ	$Z_{eff} = Z-\sigma$	Atomic (covalent) radius (Å)	Cation (No. of electrons given in bracket)	σ	$Z_{eff} = Z-\sigma$	Radius of cation (Å)
Li (3)	1.70	1.30	1.22	Li^+ (2)	0.35	1.65	0.60
Na (11)	8.80	2.20	1.54	Na^+ (10)	4.15	6.85	0.95
K (19)	16.80	2.20	2.03	K^+ (18)	11.25	7.75	1.33
Rb (37)	34.80	2.20	2.16	Rb^+ (36)	27.75	9.25	1.48
Ce (55)	52.80	2.20	2.35	Cs^+ (54)	45.75	9.25	1.69

On similar lines of argument it can also be explained that *di, tri, tetra, valent* positive ions would possess successively smaller sizes. For example:

$$Fe^{2+}\ (0.76\ \text{Å}) > Fe^{3+}\ (0.64\ \text{Å})$$

$$Pb^{2+}\ (1.20\ \text{Å}) > Pb^{4+}\ (0.84\ \text{Å})$$

The numbers given in the brackets indicate the numerical values of the ionic radii of the cations. From this we conclude that *the size of various cations in different oxidation states derived from the same element decreases with the increase in their positive oxidation state.*

An Anion is Bigger in Size than its Parent Atom

An anion is bigger in size than the atom from which it has been derived, *i.e.* an anion is bigger in size than its parent atom.

The increase in size of the anion as compared to its parent atom can be explained as follows on the basis of the concept of effective nuclear charge.

We know that an anion is formed by adding one or more electrons to the neutral atom. Thus an anion has more electrons than its parent atom. With the increase of the number of electrons the magnitude of screening constant, σ also increases. The increase in the magnitude of σ decreases the magnitude of effective nuclear charge. The decreased effective nuclear charge pulls the electron cloud of the anion outward away from the nucleus and thus makes the anion larger in size than its parent atom.

Thus halide ions are bigger in size than the halogen atoms from which they have been derived as shown below :

Halogens (Atomic number, Z given in bracket)	σ	Z_{eff} = Z–σ	Atomic (covalent) radius (Å)	Halide ions (No. of electrons given in bracket)	σ	Z_{eff} = Z–σ	Radius of halide ion (Å)
F (9)	3.80	5.20	0.72	F^- (10)	4.15	4.85	1.36
Cl (17)	10.90	6.10	0.99	Cl^- (18)	11.25	5.75	1.81
Br (35)	27.40	7.60	1.14	Br^- (36)	27.75	7.25	1.95
I (53)	45.40	7.60	1.33	I^- (54)	45.75	7.25	2.16

Radii of Iso-Electronic Species

Iso-electronic species are the atoms or ions which have the same number of electrons and same electronic configuration but different magnitude of nuclear charge. For example C^{4-}, N^{3-}, O^{2-}. F^-, Ne^0, Na^+ etc. are iso-electronic species, since these have the same number of electrons (namely 10 electrons) and same configuration *viz.* Ne-type configuration ($1s^2$, $2s^2p^6$). *The size of iso-electronic species decreases with the increase of their atomic number* (*Z*) as shown below for the iso-electronic species given above.

Iso-electronic species (Atomic number, Z given in bracket)	$Z_{eff} = Z - \sigma$ $= Z - 4.15$	Atomic or ionic radius (Å)
C^{4-} (6)	6 − 4.15 = 1.85	2.60
N^{3-} (7)	7 − 4.15 = 2.85	1.71
O^{2-} (8)	8 − 4.15 = 3.85	1.40
F^- (9)	9 − 4.15 = 4.85	1.36
Ne^0 (10)	10 − 4.15 = 5.85	1.12
Na^+ (11)	11 − 4.15 = 6.85	0.95
Mg^{2+} (12)	12 − 4.15 = 7.85	0.65
Al^{3+} (13)	13 − 4.15 = 8.85	0.50
Si^{4+} (14)	14 − 4.15 = 9.85	0.41
P^{5+} (15)	15 − 4.15 = 10.85	0.34
S^{6+} (16)	16 − 4.15 = 11.85	0.29
Cl^{7+} (17)	17 − 4.15 = 12.85	0.26

The decrease in size with the increase of atomic number can be explained as follows on the basis of effective nuclear charge.

All the species shown above have $1s^2, 2s^2p^6$ configuration and hence the value of screening constant, σ in all these cases is equal to 4.15 as shown below :

$\sigma = 0.35 \times$ No. of electrons left in *2nd* shell + 0.85 × No. of electrons in *1st* shell

$= 035 \times 7 + 0.85 \times 2$

$= 2.45 + 1.70$

$= 4.15$

With the increase of atomic number from C^{4-} to Cl^{7+}, the magnitude of effective nuclear charge, $Z_{eff} = Z - \sigma = Z - 4.15$ also increases. The increased effective nuclear charge pulls the electron cloud of the isoelectronic species inward nearer to the nucleus and thus makes their size decrease successively.

Evaluation of Ionic Radii

The following are the important methods for the evaluation of ionic radii.

1. Lande's Method. Lande (1920) determined the radius of anions by assuming that in an ionic crystal having very small cation and very large anion, the anions touch one another. For example in Li^+I^- ionic crystal which has very small Li^+ cation and very large I^- ion (the radius of I^- ion is 3.6 times that of Li^+ ion), large-sized ions touch each other (Fig. 4.11). Li^+ ions which are very small as compared to I^- ions are accommodated in octahedral holes without disturbing the close packing.

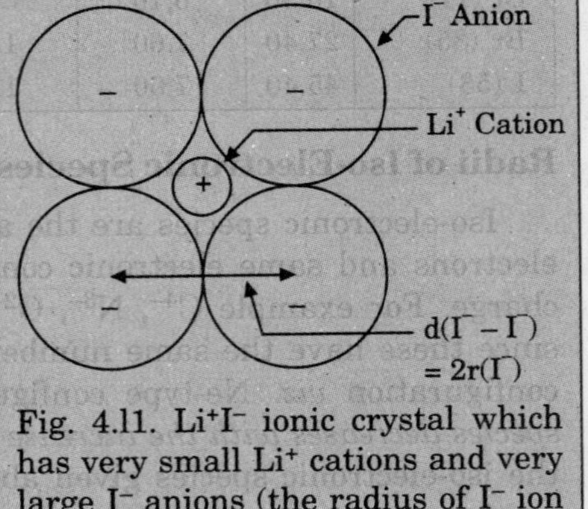

Fig. 4.11. Li^+I^- ionic crystal which has very small Li^+ cations and very large I^- anions (the radius of I^- ion is 3.6 times that of Li^+ ion) and hence I^- ions touch each other, but Li^+ cation does not touch I^- anions.

Since I^- ions touch each other, the anion–anion distance [*i.e.* $I^- - I^-$ inter-ionic distance), $d(I^- - I^-)$ in $Li^+ I^-$ ionic crystal can be assumed to be equal to double the radius of I^- ions, $r(I^-)$, *i.e.*

$$d(I^- - I^-) = 2r\,(I^-)$$

or $$r(I^-) = \frac{d(I^- - I^-)}{2} \quad ...(i)$$

The value of $d(I^- - I^-)$ has been found to be equal to 4.32 Å by diffraction technique. Thus from equation (*i*), $r(I^-)$ is given by :

$$r(I^-) = \frac{4.32}{2} = 2.16 \text{ Å}$$

Thus we see that half of the anion–anion distance is equal to the radius of I^- ion. But in case of ionic crystals like K^+I^- having the cations and anions of comparable size, the anions do not touch each other and hence half of the anion–anion distance is not equal to the radius of the anion (Fig. 4.12). In such crystals the cation touches the anions and hence the cation–anion distance is equal to the sum of the radii of the cation and anion. Thus for K^+I^- ionic crystal,

$$d(K^+ - I^-) = r(K^+) + r(I^-)$$

or $$r(K^+) = d\,(K^+ - I^-) - r(I^-) \quad (ii)$$

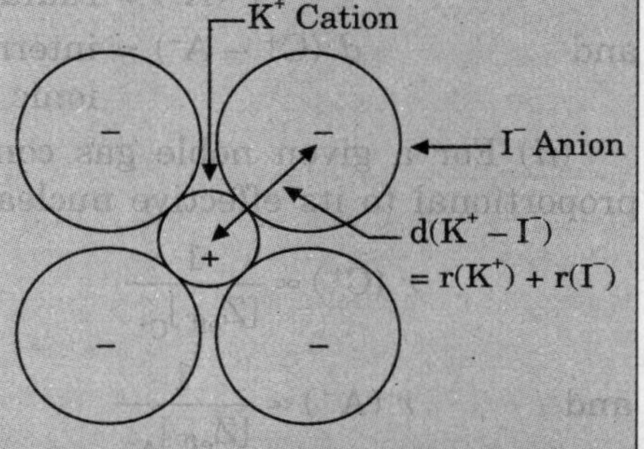

Fig. 4.12. K^+I^- Ionic crystal which has K^+ and I^- ions of comparable size and hence K^+ cation touches I^- anions, but I^- anions do not touch each other.

Thus the ionic radius of K^+ ion may be obtained by subtracting the known radius of I^- ion (as calculated from Li^+I^- crystal) from the cation–anion distance. For example in KI ionic crystal $d(K^+ - I^-)$ has been found to be equal to 3.49 Å and hence

$$r(K^+) = d(K^+ - I^-) - r(I^-)$$
$$= 3.49 - 2.16$$
$$= 1.33 \text{ Å}.$$

Bragg and West (1926) used this method to obtain the radius for the oxide ion from the silicates and then calculated the ionic radii of over 80 ions.

2. Pauling's Method. Pauling has calculated the radii of the ions on the basis of the observed internuclear distances in four crystals namely NaF, KCl, RbBr and CsI. In each ionic crystal the cations and anions are isoelectronic (*i.e.* have the same number of electrons) with inert gas configuration as shown below :

NaF crystal : $Na^+ \rightarrow 2, 8$; $F^- \rightarrow 2, 8$ } Ne (Z = 10) type configuration

KCl crystal : $K^+ \rightarrow 2, 8, 8$; $Cl^- \rightarrow 2, 8, 8$ } Ar (Z = 18) type configuration

RbBr crystal : $Rb^+ \rightarrow 2, 8, 18, 8$; $Br^- \rightarrow 2, 8, 18, 8$ } Kr (Z = 36) type configuration

CsF crystal : $Cs^+ \rightarrow 2, 8, 18, 18, 8$
$F^- \rightarrow 2, 8, 18, 18, 8$ } Xe (Z = 54) type configuration

Further the following two assumptions are made to assign the ionic radii :

(*i*) The cations and anions of an ionic crystal are assumed to be in contact with each other and hence the sum of their radii will be equal to the inter-nuclear (inter-ionic) distance between them, *i.e.*

$$r\,(C^+) + r\,(A^-) = d\,(C^+ - A^-) \quad ...(i)$$

where $r\,(C^+)$ = radius of cation, C^+

$r\,(A^-)$ = radius of anion, A^-

and $d\,(C^+ - A^-)$ = internuclear distance between C^+ and A^- ions in C^+A^- ionic crystal

(*ii*) For a given noble gas configuration, the radius of an ion is inversely proportional to its effective nuclear charge, *i.e.*

$$r\,(C^+) \propto \frac{1}{[Z_{eff}]_{C^+}} \quad ...(ii)$$

and
$$r\,(A^-) \propto \frac{1}{[Z_{eff}]_{A^-}} \quad ...(iii)$$

Here $[Z_{eff}]_{C^+}$ and $[Z_{eff}]_{A^-}$ are the effective nuclear charges on cation (C^+) and anion (A^-) respectively.

On combining relations (*ii*) and (*iii*) we get.

$$\frac{r\,(C^+)}{r\,(A^-)} = \frac{[Z_{eff}]_{A^-}}{[Z_{eff}]_{C^+}} \quad ...(iv)$$

Equations (*i*) and (*iv*) can be used to evaluate the values of $r\,(C^+)$ and $r\,(A^-)$ provided that the values of $d(C^+ - A^-)$, $[Z_{eff}]_{C^+}$ and $[Z_{eff}]_{A^-}$ are known.

Illustration. The method can be illustrated by calculating the radii of K^+ and Cl^- ions in KCl ionic crystal. The internuclear distance between K^+ and Cl^- ions in KCl ionic crystal has been found to be equal to 3.14 Å, *i.e.*

$$r\,(K^+) + r\,(Cl^-) = d\,(K^+ - Cl^-)\ 3.14\ \text{Å} \quad ...(v)$$

Now since K^+ and Cl^- ions have Ar(Z = 18) type configuration ($1s^2, 2s^2p^6, 3s^2p^6$), $[Z_{eff}]_{K^+}$ and $[Z_{eff}]_{Cl^-}$ can be calculated as follows :

$$\begin{aligned}
[Z_{eff}]_{K^+} &= Z - \sigma \\
&= 19 - [(0.35 \times \text{No. of electrons left in } nth \text{ shell}) \\
&\quad + (0.85 \times \text{No. of electrons in } (n-1)th \text{ shell}) \\
&\quad + (1.00 \times \text{No. of electrons in the inner shell})] \\
&= 19 - [0.35 \times 7 + 0.85 \times 8 + 1 \times 2] \\
&= 19 - 11.25 \\
&= 7.75 \\
[Z_{eff}]_{Cl^-} &= 17 - 11.25 \\
&= 5.75
\end{aligned}$$

$$\therefore \frac{r(K^+)}{r(Cl^-)} = \frac{[Z_{eff}]_{Cl^-}}{[Z_{eff}]_{K^+}} = \frac{5.75}{7.75} \quad ...(vi)$$

With the help of equations (*v*) and (*vi*) we find that

$r\,(K^+) = 1.33$ Å

and $r\,(Cl^-) = 18.1$ Å.

The values of ionic radii for univalent ions as found by Pauling have been found to be in good agreement with the observed inter-ionic distances.

Bond Energy or Bond Strength

In case of a diatomic molecule bond energy is defined as *the energy released when two neutral atoms A and B get bonded together to form A–B molecule. Thus* :

$$\underbrace{A + B}_{\text{Neutral atoms}} \longrightarrow \underset{\text{AB molecule}}{A - B} + \textit{Energy released (called bond energy)}$$

When a bond is formed, energy is evolved and hence energy is written with a negative sign.

Bond energy can also be defined as the energy required to break A–B molecule into its neutral atoms, A and B. Thus :

$$\underset{\text{AB molecule}}{A–B} + \textit{Energy required (called bond energy)} \longrightarrow \underbrace{A + B}_{\text{Neutral atoms}}$$

When a bond is broken, energy is required and hence bond energy is written with a positive sign. When we consider the dissociation of a molecule into atoms, bond energy is called **dissociation energy** *of the molecule.*

For a poly-atomic molecule two kinds of bond energies are considered. These are (*i*) *Bond dissociation energy*, D and (*ii*) *Average bond energy*, E.

Bond dissociation energy is defined as the energy required to break a particular bond in a polyatomic molecule, while the average bond energy is equal to the energy required to break the molecule into atoms divided by the number of bonds present in the polyatomic molecule. Thus :

$$\textbf{Average bond energy} = \frac{\textbf{Energy required to break the molecule into its atoms}}{\textbf{No. of bonds}}$$

Since the energy required to break the molecule into its atoms is equal to the energy of its formation,

$$\textbf{Average bond energy} = \frac{\textbf{Heat of formation of the molecule}}{\textbf{No. of bonds}}$$

For a polyatomic molecule AB_n having all identical A–B bonds, there are *n* dissociation energies but this molecule has only one average bond energy. For example there are two dissociation energies for H_2O, three for NH_3 and four for

CH_4 molecules, but there is only one average bond energy in each case. Two successive bond dissociation energies ΔH_1 and ΔH_2 for H_2O molecule are shown below :

$$\text{H–O–H} \rightarrow \text{H + O–H}; \quad \Delta H_1 = 501.7 \text{ KJ/mole}$$
$$\text{O–H} \rightarrow \text{H + O}; \quad \Delta H_2 = 423.5 \text{ KJ/mole}$$

Thus the energy required to break both O–H bonds simultaneously in one molecule of water

$$= 501.7 + 423.5 = 925.2 \text{ KJ/mole}$$

or

$$\begin{matrix}\text{H} \\ \text{H}\end{matrix}\!\!>\!\text{O} \longrightarrow \text{O + H + H}; \ \Delta H = 925.2 \text{ KJ/mole}$$

and hence the average O—H bond energy $= \dfrac{925.2}{2}$ KJ/mole

$$= 462.2 \text{ KJ/mole}$$

Here we see that the two successive bond dissociation energies for H_2O molecule are different from each other and from the average O–H bond energy. It is also clear that the bond energy of O–H bond is one-half of the energy required to break H_2O molecule (H–O–H) into elements. For NH_3 molecule the three successive bond dissociation energies ΔH_1, ΔH_3 and ΔH_3 are given below :

$$\begin{matrix}\text{H} \\ | \\ \text{N–H} \\ | \\ \text{H}\end{matrix} \rightarrow \text{H} + \begin{matrix}\text{H} \\ | \\ \text{N–H}\end{matrix}; \quad \Delta H_1 = 435.1 \text{ KJ/mole}$$

$$\begin{matrix}\text{H} \\ | \\ \text{N–H}\end{matrix} \rightarrow \text{H + N–H}; \quad \Delta H_2 = 397.5 \text{ KJ/mole}$$

$$\text{N–H} \rightarrow \text{H + N}; \quad \Delta H_3 = 338.9 \text{ KJ/mole}$$

Thus the energy required to break all the, three N–H bonds simultaneously in one molecule of ammonia

$$= 435.1 + 397.5 + 338.9 = 1171.5 \text{ KJ/mole}$$

and hence the average N–H bond energy $= \dfrac{1171.5}{3}$

$$= 390.5 \text{ KJ/mole.}$$

Thus the bond energy of N–H bond is one-third of the energy required to break NH_3 $\left(\begin{matrix}\text{H} \\ | \\ \text{H–N–H}\end{matrix}\right)$ molecule into elements.

Bond dissociation energies can be determined by calorimetric method, electronic absorption or absorption spectroscopic analysis, mass spectroscopic analysis, magnetic susceptibility measurements, magnetic resonance spectroscopy, molar refraction etc. The bond dissociation energies of some common bonds are given in Table 4.7.

Table 4.7. Bond dissociation energies of some common bonds (KJ/mole)

Bond	Bond dissociation energy	Bond	Bond dissociation energy
H–H	431.0	H–F	560.7
C–C	347.3	H–Cl	426.8
C=C	606.7	H–Br	364.0
C≡C	803.3	H–I	297.1
N–N	163.2	C–H (average in CH_4)	414.2
N=N	418.4	C–N	284.5
N≡N	945.6	C=N	615.1
O–O	146.4	C≡N	866.1
O=O	497.9	C–O	334.7
S–S	225.9	C=O	694.5
F–F	150.6	C≡O	1071.3
Cl–Cl	242.7	C–S	272.0
Br–Br	188.3	C–F	447.7
I–I	150.6	C–Cl	326.4
H–N (average in NH_3)	390.5	C–Br	284.5
H–O (average in H_2O)	462.6	C–I	213.4

Standard ΔH° Values for Atoms

Bond energy, ΔH for H_2 molecule has been experimentally found to be equal to 435.4 KJ/mole. Half of this (217.7 KJ/mole) is called standard ΔH° value for H atom. Similarly, since ΔH for I_2 molecule has been found equal to 150.6 KJ/mole, standard ΔH° value for I atom is half of 150.6 KJ/mole (= 75.3 KJ/mole). Standard ΔH° values for some common atoms (in KJ/mole) in gaseous state are given below :

H(*g*) = 217.7	C(*g*) = 718.3
O(*g*) = 247.5	Cl(*g*) = 121.3
N(*g*) = 473.3	Br(*g*) = 96.42

Calculation of Bond Energy

Standard ΔH° values of atoms can be used to calculate the bond energy of a given bond. For example, since, as we have seen, the standard ΔH° values of H and I atoms are 217.7 KJ/mole and 75.3 KJ/mole, respectively bond energy of H–I bond

= ΔH° value for H atom + ΔH° value for I atom

= 217.7 KJ/mole + 75.3 KJ/mole

= 293.0 KJ/mole

This calculated value (= 293.0 KJ/mole) has been found in close agreement with the experimental value (= 297.1 KJ/mole)

Problem. *From ΔH° values of the elements as given above, calculate the bond energies of (a) O–H (b) H–N (c) H–Cl (d) H–Br (e) C–Cl (f) C–Br (g) C–O, and (h) C–N bonds.*

Solution. (*a*) Bond energy of O–H bond
= ΔH° value for O atom + ΔH° value for H atom
= 247.5 KJ/mole + 217.7 KJ/mole
= 465.2 KJ/mole

On similar lines other bond energies can be calculated.

The average bond energy can also be calculated with the help of Hesse's law. For example average bond energy of C–H bond in CH_4 molecule can be calculated as follows :

For this purpose we need the heat change in the reaction :

$$C(g) + 4H(g) \rightarrow CH_4(g)$$

The magnitude of heat change for the above reaction can be obtained by the following reactions :

$C(s) + 2H_2(g) \rightarrow CH_4(g)$; $\Delta H_1 = -\ 74.4$ KJ/mole

$C(g) \rightarrow C(s)$; $\Delta H_2 = -\ 713.1$ KJ/mole

$4H(g) \rightarrow 2H_2(g)$; $\Delta H_3 = -\ 872.1$ KJ/mole

On adding all these equations we get

$C(g) + 4H(g) \rightarrow CH_4(g)$; $\Delta H = \Delta H_1 + \Delta H_2 + \Delta H_3$

$= -74.4 + (-713.1) + (-872.1)$

$= -1659.6$ KJ/mole

Here minus sign indicates the release of energy and 1659.6 KJ/mole represents the bond energy of four C–H bonds in CH_4 molecule. Now since all the four C–H bonds in CH_4 molecule are identical, the average bond energy of C–H bond in this molecule is equal to $\frac{1}{4} \times 1659.6 = 414.9$ KJ/mole.

Bond energy of A–B bond can also be calculated by using Pauling's relationship which is given below :

$$E(A-B) = \frac{D(A-A) + D(B-B)}{2} + 23(x_A - x_B)^2,$$

where E(A–B) = Bond energy of A–B bond

D(A–A) and D(B–B) = Dissociation energies of A–A and B–B bonds

x_A and x_B = Electronegativities of A and B atoms.

Multiple Bond Energies

A double bond in a diatomic molecule has a higher bond energy than a single bond, and a triple bond has a higher bond energy than a double bond. Thus the bond energies of various types of bonds are in the order :

Bond Energy of a triple bond > Bond energy of a double bond > Bond energy of a single bond

Thus we see that *for diatomic molecules bond dissociation energy is directly proportional to the bond order or bond multiplicity*. (Bond order is the number of covalent bonds by which the two atoms in a diatomic molecule are joined together). This fact is evident from the following values of bond energies (KJ/mole) of some single, double and triple bonds.

Bond	Bond order	Bond energy (KJ/mole)	Bond	Bond order	Bond energy (KJ/mole)
C–C	1	347.3	C–O	1	334.7
C=C	2	606.7	C=O	2	694.5
C≡C	3	803.3	C≡O	3	1071.3
N–N	1	163.2	N–O	1	200.8
N=N	2	418.5	N=O	2	611.0
N≡N	3	945.6			
C–N	1	284.5	O–O	1	146.4
C=N	2	615.1	O=O	2	497.9
C≡N	3	866.1			

The strength of a given bond depends on its bond energy. The strength of a bond is proportional to its bond energy value. Thus, the strength of various types of bonds will be in the following order :

Strength of a triple bond > Strength of a double bond > Strength of a single bond

Factors Affecting the Magnitude of Bond Energy

The factors that affect the magnitude of bond energy are :

(*i*) The extent to which the orbitals overlap with each other to form the bond.

(*ii*) The type of hybrid orbital formed.

(*iii*) The resonance energy of the molecule, if resonance is involved.

(*iv*) The type of bonding orbitals involved.

(*v*) Size of the atoms forming the bond.

(*vi*) The repulsion due to the presence of non-bonding electron pairs.

(*vii*) The amount of ionic character in the bond i.e., polarity of the bond.

Ionisation Energy or Ionisation Potential (IE or I or $\Delta_i H$)

Ionisation energy or ionisation potential is an atomic property which is shown by the atoms of all the elements. This property gives us an idea of the tendency of the elements to lose the outer-most shell electron to form a cation. With the help of this property a number of other properties of the elements like reducing power, metallic (or electropositive) character etc. can be explained.

What is Ionisation Potential ?

Ionisation energy of an element is defined as the amount of energy required to remove the most loosely bound electron (i.e. outermost shell electron) from isolated neutral gaseous atom in its lowest energy state (i.e. ground state) to convert it into a unipositive gaseous cation. Thus the ionisation energy of an atom, M(g) can be defined by a process represented by the following equation :

$$\underset{\text{Isolated gaseous atom}}{M(g)} + \textit{Energy} \text{ supplied } (+ I_1) \xrightarrow{-e^-} \underset{\text{Gaseous cation}}{M^+(g)} + e^-$$

Ionisation energy defined as above is strictly called **first ionisation energy,** since it corresponds to the removal of *one* electron only. Hence it is represented as I_1 or $+I_1$, where positive (plus) sign represents the absorption of energy by atom, M(*g*). It is also represented as $\Delta_i H$

The process represented by the above equation which depicts the removal of an electron from a gaseous atom is an *endothermic process*, since the removal of the electron requires energy.

Ionisation energy is also called **ionisation potential,** since it represents the amount of potential (or voltage) required to remove the most loosely bound electron from the gaseous atom of an element to get a gaseous cation.

Ionisation energy is measured in electron volts per atom (eV/atom), kilo calories per mole (Kcal/mole) or kilo joules per mole (KJ/mole). These quantities are related as :

1 *e*V/atom = 96.4705 KJ/mole = 23.06 Kcal/mole
1 Kcal/mole = 4.185 KJ/mole.

The values of ionisation potential are shown by a positive (plus) sign which is placed before them. *Positive sign represents the absorption of energy by the atom, M(g).*

Values of first ionisation potentials (in KJ/mole) of representative element (*s*-and *p*-block elements) are given Table 4.8. Positive sign has not been shown before the values.

Successive (Higher) Ionisation Potentials (I_1, I_2, I_3 ...)

In addition to first ionisation potential (I_1) defined above, second, third. etc ionisation potentials are also known. **Second ionisation potential (I_2)** *is the energy required to remove one more electron from the gaseous cation, $M^+(g)$ to get the doubly positively charged gaseous cation, $M^{2+}(g)$, i.e,*

$$M^+(g) + I_2 \xrightarrow{-e^-} M^{2+}(g) + e^-$$

Similarly, **third ionisation potential (I_3)** is the energy required to remove still one more electron from $M^{2+}(g)$ cation to get $M^{3+}(g)$ cation, *i.e.*

$$M^{2+}(g) + I_3 \xrightarrow{-e^-} M^{3+}(g) + e^-$$

Similarly ionisation potentials of higher and higher grades are also known. These are obtained by removing the electrons one by one. The values of eight successive ionisation potentials of the first ten elements are given in Table 4.9

Table 4.8. First ionization potentials (in KJ/mole) of representative elements (s-and p-block elements) (Positive sign has not been shown before the values).

Group → / Period ↓	*s*-block elements		*p*-block elements					
	IA	IIA	IIIA	IVA	VA	VIA	VIIA	Zero
1	1 H 1312.0							2 He 2372.3
2	3 Li 520.2	4 Be 899.5	5 B 800.6	6 C 1086.4	7 N 1402.3	8 O 1314.0	9 F 1681.0	10 Ne 2080.7
3	11 Na 495.8	12 Mg 737.7	13 Al 577.6	14 Si 786.5	15 P 1011.8	16 S 999.6	17 Cl 1251.1	18 Ar 1520.5
4	19 K 418.9	20 Ca 589.8	31 Ga 578.8	32 Ge 762.2	33 As 944.0	34 Se 940.9	35 Br 1139.9	36 Kr 1350.7
5	37 Rb 403.0	38 Sr 549.5	49 In 558.3	50 Sn 708.6	51 Sb 831.6	52 Te 869.3	53 I 1008.4	54 Xe 1170.4
6	55 Cs 375.7	56 Ba 502.9	81 Tl 589.3	82 Pb 715.5	83 Bi 703.3	84 Po 812.0	85 At –	86 Rn 1037.0

It may be seen from Table 4.9 that for a given element the magnitudes of its successive ionisation potentials are in the order :

$$I_1 < I_2 < I_3 < I_4 \text{... and so on.}$$

The successive increase in the values of successive ionisation potentials can be explained as follows on the basis of the concept of effective nuclear charge.

We know that the cations are formed by removing the electrons from the atom/cation one by one.

$$M(g) \xrightarrow{-e^-} M^+(g) \xrightarrow{-e^-} M^{2+}(g) \ldots \xrightarrow{-e^-} M^{n+}(g)$$

Thus a given cation has lesser number of electrons than its parent atom or cation. With the decrease of the number of electrons, the magnitude of the effective nuclear charge increases from $M(g)$ to $M^{n+}(g)$. With the increase of effective nuclear charge, the magnitude of the force of attraction between the nucleus and the outer-most shell electron cloud also increases as shown below :

Atom / cation : $M(g)$ $M^+(g)$ $M^{2+}(g)......M^{n+}(g)$

Effective nuclear charge : — *Increasing* →

Force of attraction : — *Increasing* →

With the increase of force of attraction, the magnitude of energy to remove the outer-most shell electron (*i.e.* ionisation potential) also goes on increasing from M(*g*) to $M^{n+}(g)$ and hence :

$$I_1 < I_2 < I_3 < \ldots\ldots \text{ and so on.sss}$$

Table 4.9. Values of I_1 ... I_2 I_s (in kJ/mole) of the ten elements

Elements with atomic number and electronic configuration	Values of successive ionisation potentials							
	I_1	I_2	I_3	I_4	I_5	I_6	I_7	I_8
H (1) ($1s^1$)	1312.0							
He (2) ($1s^2$)	2372.3	5250.4						
Li (3) ($1s^2$, $2s^1$)	520.3	7298.1	11814.9					
Be (4) ($1s^2$, $2s^2$)	899.5	1757.1	14848.7	21006.5				
B (5) ($1s^2$, $2s^2p^1$)	800.6	2427.0	3659.8	25025.7	32826.6			
C (6) ($1s^2$, $2s^2p^2$)	1086.4	2352.6	4620.5	6222.6	37830.4	47276.9		
N (7) ($1s^2$, $2s^2p^3$)	1402.3	2856.1	4578.1	7475.1	9444.9	53266.4	64359.8	
O (8) ($1s^2$, $2s^2p^4$)	1314.0	3388.2	5300.4	7469.3	10989.5	13326.4	71334.5	84077.4
F (9) ($1s^2$, $2s^2p^5$)	1681.0	3374.2	6050.4	8407.7	11022.7	15164.0	17867.7	92037.8
Ne (10) ($1s^2$, $2s^2p^6$)	2080.7	3952.3	6122.0	9370.0	12178.0	15238.0	19999.0	23069.0

When the term ionisation potential is not qualified, it is taken as first ionisation potential.

The total energy required to remove two electrons from a gaseous M(*g*) atom to get $M^{2+}(g)$ cation is the sum of first (I_1) and second (I_2) ionisation potentials, *i.e.*

$$M(g) + (I_1 + I_2) \xrightarrow{-2e^-} M^{2+}(g) + 2\,e^-$$

Factors Affecting the Magnitude of Ionisation Potential

Following are the factors which influence the magnitude of ionisation potential of atoms of elements.

1. Effective nuclear charge. Greater is the magnitude of effective nuclear charge, greater is the electrostatic force of attraction exerted by the nucleus on the outer electrons. Thus it would be more difficult to remove the outermost shell electron from an atom with higher effective nuclear charge. This means that *greater the magnitude of effective nuclear charge, higher is the amount of energy* (*i.e.* ionisation potential) *needed to remove the outermost shell electron.* Thus with the *increase of the magnitude of effective nuclear charge, the magnitude of ionisation potential also increases.*

This factor has been used to explain the variation of ionisation potential values of the elements of a period of the representative elements of the periodic table as shown below :

We know that on passing from left to right in a period, the effective nuclear charge increases by 0.65 at each next element. The increase in effective nuclear charge increases the magnitude of ionisation potential as shown below for the elements of *2nd* period (Note the abnormal values of Be and N as compared with those of B and O).

Elements of 2nd period :

Li	Be	B	C	N	O	F	Ne
Effective nuclear charge :							
1.30	1.95	2.60	3.25	3.55	4.55	5.20	5.85
Increasing →							
Ionisation potential (KJ/mole) :							
520.3	899.5	800.6	1086.4	1402.3	1314.0	1681.0	2080.7

The abnormal values of Be and N as compared to those of B and 0 have been explained on subsequent pages.

2. Size of the atom (*r*). Greater is the magnitude of r of an atom, more far is the outer-most shell electron from the nucleus and hence lesser will be the force of attraction exerted by the nucleus on the outer most shell electron. Thus with the increase of *r*, it would be more and more easy to remove an electron from the outermost shell, *i.e. higher is the value of r of an atom, lower will be its ionisation energy.*

This fact has been used to explain the variation of ionisation potential values of the elements of a group and a period of the periodic table as shown below :

(a) Variation of ionisation potential of the elements in a group. We know that on proceeding from top to bottom in a group, *the value of r increases and hence the magnitude of ionisation potential decreases* as shown below for the elements of group IA.

Elements of group IA	Atomic (covalent) radius (A)	Ionisation potential (KJ/mole)
H (1)	0.32	1312.0
Li (3)	1.23	520.3
Na (11)	1.54	495.8
K (19)	2.03	418.9
Rb (37)	2.16	403.4
Cs (55)	2.35	375.7
	Increasing ↓	*Decreasing* ↓

(b) Variation of ionisation potential of the elements in a period. We know that on proceeding from left to right in a period *the value of r decreases and hence the magnitude of ionisation potential increases* as shown below for the elements of 3*rd* period. The abnormal values of Mg and P can be noted. These have been explained in subsequent pages.

Elements of 3rd period :

Na	*Mg*	*Al*	*Si*	*P*	*S*	*Cl*	*Ar*
Atomic (covalent) radius (Å) :							
1.54	*1.36*	*1.18*	*1.11*	*1.06*	*1.02*	*0.99*	*0.98*
Ionisation potential (KJ/mole) :							
495.8	*737.7*	*577.6*	*786.5*	*1011.8*	*999.6*	*1251.1*	*1520.5*

3. Principal quantum number (*n*). Greater is the value of *n* for the valence-shell electron of an atom, farther away this electron will be from the nucleus and hence lesser will be the force of attraction exerted by the nucleus on it. This means that lesser amount of energy will be required to remove the valence-shell electron. Thus with the increase of the value of *n* of the valence-shell electron, ionisation potential decreases and vice versa. In simple words the energy required (*i.e.* ionisation potential) to remove an electron from an orbital with higher *n* value (*i.e.* belonging to a higher shell) will be less than that required to remove an electron from an orbital with a lower *n* value (*i.e.* belonging to a lower shell.) *Thus with the increase of the principal quantum number (n) of the orbital from which the electron is to be removed, the magnitude of ionisation potential decreases.*

This factor has been used to explain the following :

(a) Variation of ionisation potential of elements in a group. On moving down a group the value of the principal quantum number (*n*) of the orbital from which the electron is to be removed increases and hence the magnitude of the ionisation potential decreases as shown below for the elements of group II A.

Elements of group II A	Valence shell electronic configuration	*n*	Ionisation potential (KJ/mole)
Be	$2s^2$	2	899.5
Mg	$3s^2$	3	737.7
Ca	$4s^2$	4	589.8
Sr	$5s^2$	5	549.5
Ba	$6s^2$	6	502.9

(b) Successive ionization potentials of an element. (*i*) Values of I_1 and I_2 of Li atom ($1s^2, 2s^1$) which are 520.3 KJ/mole and 7298.1 KJ/mole respectively can be represented by the following equations :

$$\text{Li}\ (1s^2, 2s^1) \xrightarrow[I_1\ =\ 520.3\ \text{KJ/mole}]{-e^-\ (\text{From } 2s \text{ orbital with } n\ =\ 2)} \text{Li}^+(1s^2)$$

$$\xrightarrow[I_2\ =\ 7298.1\ \text{KJ/mole}]{-e^-\ (\text{From } 1s \text{ orbital with } n\ =\ 1)} \text{Li}^{2+}(1s^1)$$

Why I_2 is greater than I_1 can be explained by saying that in case of I_1 the electron is removed from $2s$ orbital which has higher value of $n(=2)$ while in case of I_2 the electron is removed from $1s$ orbital which has lower value of $n(=1)$.

(*ii*) I_3 of Mg is very high as compared to I_1 and I_2 (I_1 = 737.7 KJ/mole, I_2 = 1450.7 KJ/mole, I_3 = 7732.8 KJ/mole). This can be explained by considering I_1, I_2 and I_3 as represented by the following equations :

$$\text{Mg}\,(1s^2, 2s^2p^6, 3s^2) \xrightarrow[I_1 = 737.7 \text{ KJ/mole}]{-e^- \text{ (From } 3s \text{ orbital with } n = 3)} \text{Mg}^+(1s^2, 2s^2p^6, 3s^1)$$

$$\xrightarrow[I_2 = 1450.7 \text{ KJ/mole}]{-e^- \text{ (From } 3s \text{ orbital with } n = 3)} \text{Mg}^{2+}(1s^2, 2s^2p^6)$$

$$\xrightarrow[I_3 = 7732.8 \text{ KJ/mole}]{-e^- \text{ (From } 2p \text{ orbital with } n = 2)} \text{Mg}^{3+}(1s^2, 2s^2p^5)$$

In case of I_1 and I_2 the electron is removed from an orbital with n = 3 while in case of I_3 the electron is removed from an orbital having a lower value of n (=2). Hence I_3 is very high as compared to I_1 and I_2.

(*iii*) The values of I_1, I_2 and I_3 (in KJ/mole) of Be and B are given below :

	I_1	I_2	I_3
Be	899.5	1757.1	14848.7
B	800.6	2427.0	3659.8

It may be seen from these values that there is a big jump in ionisation potentials of Be from I_2 (= 1757.1 KJ/mole) to I_3 (= 14848.7 KJ/mole) as compared with I_2 (= 2427.0 KJ/mole) to I_3 (= 3659.8 KJ/mole) of B. How to explain this big jump? It has been explained as follows :

$$\text{Be}\,(1s^2, 2s^2) \xrightarrow[I_1 = 899.5 \text{ KJ/mole}]{-e^- \text{ (From } 2s \text{ orbital with } n = 2)} \text{Be}^+(1s^2, 2s^1)$$

$$\xrightarrow[I_2 = 1757.1 \text{ KJ/mole}]{-e^- \text{ (From } 2s \text{ orbital with } n = 2)} \text{Be}^{2+}(1s^2)$$

$$\xrightarrow[I_3 = 14848.7 \text{ KJ/mole}]{-e^- \text{ (From } 1s \text{ orbital with } n = 1)} \text{Be}^{3+}(1s^1)$$

$$\text{B}\,(1s^2, 2s^2\, 2p^1) \xrightarrow[I_1 = 800.6 \text{ KJ/mole}]{-e^- \text{ (From } 2p \text{ orbital with } n = 2)} \text{B}^+(1s^2, 2s^2)$$

$$\xrightarrow[I_2 = 2427.0 \text{ KJ/mole}]{-e^- \text{ (From } 2s \text{ orbital with } n = 2)} \text{B}^{2+}(1s^2, 2s^1)$$

$$\xrightarrow[I_3 = 3659.7 \text{ KJ/mole}]{-e^- \text{ (From } 2s \text{ orbital with } n = 2)} \text{B}^{3+}(1s^2)$$

It may be seen that in case of I_3 of Be atom the electron is removed from $1s$ orbital having lower value of n(= 1) while in case of I_3 of B atom the electron is removed from $2s$ orbital having higher value of n (= 2).

(c) Ionisation potentials of inert gases as compared to those of alkali metals. When we proceed from an inert gas (*zero group*) to an alkali metal (*IA group*), there is a large decrease in the value of ionisation potential as shown below :

Inert gas (Zero group)	Alkali metal (IA group)
He (Z = 2) = 2372.3 KJ/mole $1s^2$	Li (Z = 3) = 520.3 KJ/mole $1s^2$, $2s^1$
Ne(Z = 10) = 2080.7 KJ/mole $1s^2$, $2s^2p^6$	Na(Z = 11) = 495.8 KJ/mole $1s^2$, $2s^2p^6$, $3s^1$
Ar(Z = 18) = 1520.7 KJ/mole $1s^2$, $2s^2p^6$, $3s^2p^6$	K(Z = 19) = 418.9 KJ/mole $1s^2$, $2s^2p^6$, $3s^2p^6$, $4s^1$
Kr(Z = 36) = 1350.7 KJ/mole $1s^2$, $2s^2p^6$, $3s^2p^6d^{10}$, $4s^2p^6$	Rb(Z = 37) = 403.0 KJ/mole $1s^2$, $2s^2p^6$, $3s^2p^6d^{10}$, $4s^2p^6$, $5s^1$
Xe(Z = 54) = 1170 KJ/mole $1s^2$, $2s^2p^6$, $3s^2p^6d^{10}$, $4s^2p^6d^{10}$, $5s^2p^6$	Cs(Z = 35) = 375.7 KJ/mole $1s^2$, $2s^2p^6$, $3s^2p^6d^{10}$, $4s^2p^6d^{10}$, $5s^2p^6$, $6s^1$

The large decrease is explained on the basis of the fact that in case of an alkali metal it is easier to remove an electron from an orbital belonging to a higher shell while in case of an inert gas it is more difficult to remove the same from the orbital belonging to the lower shell.

4. Shielding effect. We have already said that in a multi- electron atom the electrons lying between the nucleus and the valence-shell (called intervening electrons) shield the outermost shell electron from the nucleus. The effect of shielding the valence-shell electron from the nucleus which is caused by the intervening electrons is called *shielding* or *screening effect*. The magnitude of shielding effect determines the magnitude of the force of attraction between the nucleus and the valence-shell electron. Greater is the magnitude of shielding effect working on the valence-shell electron, smaller is the magnitude of the force of attraction between the nucleus and the valence-shell electron and hence lower is the amount of energy (*i.e.* ionisation potential) to remove the valence-shell electron. *Thus with the increase of shielding effect, ionisation potential decreases.*

The magnitude of shielding effect is determined by the magnitude of shelding constant, σ. Greater the value of σ, greater is the amount of shielding effect caused by the intervening electrons on the valence-shell electron and hence lower is the value of ionisation potential. Thus *with the increase of σ, the magnitude of ionisation potential decreases.*

This factor can be used to explain the variation of ionisation potential of the elements of a given group of the periodic table. As we move down a group, the number of inner-shell electrons (*i.e.* intervening electrons) increases and hence the magnitude of shielding effect (or the magnitude of σ) caused by these electrons on the valence-shell electron increases. With the increase of the magnitude of shielding effect or σ, the ionisation potential decreases. This can be shown by considering the ionisation potential values of the elements of group IA as given below :

Elements of group IA	Electronic configuration	No. of inner shell electrons (↓ Increasing)	Value of σ (↓ Increasing)	Ionisation potential (KJ/mole) (↓ Decreasing)
Li (3)	2, 1	2	1.70	520.3
Na (11)	2, 8, 1	10	8.80	495.8
K (19)	2, 8, 8, 1	18	16.80	418.9
Rb (37)	2, 8, 18, 8, 1	36	34.80	403.0
Cs (55)	2, 8, 18, 18, 8, 1	54	52.80	375.7

This factor has also been used to explain the large decrease in the value of ionisation potential when we pass from an inert gas (*zero group*) to an alkali metal (*IA group*). For example when we pass from He to Li, there is large decrease in the value of ionisation potential (He = 2372.3 KJ/mole, Li 520.3 KJ/mole). This large decrease has been explained as follows :

In Li atom ($1s^2$, $2s^1$) the electron to be removed from $2s$-orbital is *effectively screened* or *shielded* by the two electrons in $1s$-orbital while in He atom ($1s^2$) the electron to be removed from the $1s$-orbital is *not effectively* screened by the other electron in $1s$-orbital, since both the electrons in He atom reside in the same orbital *viz* $1s$ orbital.

On similar grounds a large decrease in the ionisation potential on proceeding from Ne (Z = 10) to Na (Z = 11), from Ar(Z = 18) to K(Z = 19), from Kr(Z = 36) to Rb(Z = 37) and from Xe(Z = 54) to Cs(Z = 55) can also be explained.

5. Half-filled and completely filled orbitals. According to Hund's rule, *half-filled* (ns^1, np^3, nd^5) or *completely-filled* (ns^2, np^6, nd^{10}) *orbitals* are comparatively more stable and hence more energy is needed to remove an electron from such orbitals. This means that *the ionisation potential of an atom having half-filled or completely-filled orbitals in its electronic configuration is relatively higher than that expected normally from its position in the periodic table.*

This factor has been used to explain the higher value of ionisation potential of some of the elements as compared to that of their left-hand neighbour in a period of the periodic table. For example as we proceed from left to right in a period of the long form of periodic table, usually the ionisation potential increases due to the successive increase in nuclear charge (*i.e.* atomic number) and decrease in atomic size. However, there are certain elements which show irregular trends. For example :

(*i*) In 2*nd* period Be (Z = 4) has higher ionisation potential than B (Z = 5). Similarly N (Z = 7) has higher value of ionisation potential than O (Z = 8).

(*ii*) In 3*rd* period Mg (Z = 12) has higher value of ionisation potential than Al (Z = 13). Similarly P (Z = 15) has higher ionisation potential than S (Z = 16).

These irregularities can be explained as follows on the basis of the concept of extra stability associated with completely-filled and half-filled orbitals.

(a) Ionisation potential of Be compared to that of B, and of Mg compared to that of Al. The equations representing the ionisation potential of Be and B can be written as :

$$Be\ (1s^2, 2s^2) \xrightarrow[I_1 = 899.5\ KJ/mole^{-1}]{-e^-\ \text{(From completely-filled 2s orbital)}} Be^+(1s^2, 2s^1)$$

$$B\ (1s^2, 2s^2, 2p^1) \xrightarrow[I_1 = 800.6\ KJ/mole^{-1}]{-e^-\ \text{(From partially filled 2p orbital)}} B^+(1s^2, 2s^2)$$

In case of Be it is more difficult to remove an electron from the completely-filled 2*s*-orbital while in case of B it is easier to remove the same from a *partially-filled* 2*p*-orbital. Thus the ionisation potential of Be is higher than that of B. Note that Be and B both have the same value of principal quantum number ($n = 2$).

Similarly the higher value of ionisation potential of Mg as compared to that of Al is because of the fact that in case of Mg the electron is to be removed from the *completely-filled* 3*s*-orbital while in case of Al the same is removed from the *partially-filled* 3*p*-orbital.

$$Mg\ (1s^2, 2s^2p^6, 3s^2) \xrightarrow[I_1 = 737.7\ KJ/mole^{-1}]{-e^-\ \text{(From completely-filled 3s orbital)}} Mg^+(1s^2, 2s^2p^6, 3s^1)$$

$$Al\ (1s^2, 2s^2p^6, 3s^2p^1) \xrightarrow[I_1 = 577.6\ KJ/mole^{-1}]{-e^-\ \text{(From partially-filled 3p orbital)}} Al^+(1s^2, 2s^2p^6, 3s^2)$$

(b) Ionisation potential of N compared to that of O, and of P compared to that of S. The equations representing the ionisation potential of N and O can be written as :

$$N(1s^2, 2s^2 2p^3) \xrightarrow[I_1 = 1402.3\ KJ/mole]{-e^-\ \text{(From half-filled 2p orbital)}} N^+(1s^2, 2s^2.2p^2)$$

$$O(1s^2, 2s^2 2p^4) \xrightarrow[I_1 = 1314.0\ KJ/mole]{-e^-\ \text{(From partially-filled 2p orbital)}} O^+(1s^2, 2s^2.2p^3)$$

In case of N it is more difficult to remove an electron from the *half -filled 2p-orbital* while in case of O it is easier to remove the same from a *partially filled* 2*p*-orbital. Thus the ionisation potential of N is higher than that of O.

On similar grounds. the higher ionisation potential of P as compared to S can be explained.

$$P\ (2, 8, 3s^2 3p^3) \xrightarrow[I_1 = 1011.8\ KJ/mole^{-1}]{-e^-\ \text{(From half-filled 3p orbital)}} P^+(2, 8, 3s^2.3p^2)$$

$$S\ (2, 8, 3s^2 3p^4) \xrightarrow[I_1 = 999.6\ KJ/mole]{-e^-\ \text{(From partially-filled 3p orbital)}} S^+(2, 8, 3s^2.3p^3)$$

Noble gases (ns^2p^6), Zn ($3d^{10}.4s^2$), Cd ($4d^{10}.5s^2$) and Hg ($4f^{10}.5d^{10}.6s^2$) which have completely-filled orbitals possess very high ionisation potentials.

6. Nature of orbitals. The nature of the orbitais of the valence-shell from which the electron is to be removed also influences the magnitude of ionisation potential. The relative order of energy of *s*, *p*, *d* and *f* orbitals of a given nth shell is as :

$$ns < np < nd < nf$$

The above order means that the relative order of ease with which the electron residing in these orbitals can be removed will be as :

—Order of ease of removing electron increasing →

$$ns < np < nd < nf$$

This order clearly indicates that to remove an electron from *f*-orbital will be the easiest while to remove the same from s-orbital will be the most difficult, *i.e.* the amount of energy required to remove an electron from *f*-orbital is minimum while that required to remove an electron from *s*-orbital is maximum. In other words, other factors remaining the same, the ionisation potentials of an electron in *s*, *p*, *d* and *f*-orbitals are as the following order :

$$ns > np > nd > nf$$

This factor can be used to explain the successive ionisation potentials of an element and higher values of ionisation potential of Be and Mg as compared to that of B and Al respectively.

(a) Successive ionisation potentials of an element. 1*st* (I_1) and 2*nd* (I_2) ionisation potentials of B atom ($1s^2$, $2s^2\ 2p^1$) are : I_1 = 800.6 KJ/mole^{-1}, I_2 = 2427.0 KJ/mole^{-1} and can be represented by the following equations :

$$\text{B}\ (1s^2, 2s^2 2p^1) \xrightarrow[\text{I}_1\ =\ 800.6\ \text{KJ.mole}^{-1}]{-e^-\ (2p^1)} \text{B}^+(1s^2, 2s^2)$$

$$\text{B}^+(1s^2, 2s^2) \xrightarrow[\text{I}_2\ =\ 2427.0\ \text{KJ.mole}^{-1}]{-e^-\ (2s^1)} \text{B}^{2+}(1s^2, 2s^1)$$

Why I_1 is lower than I_2 is explained by saying that since it is easier to remove an electron from 2*p* orbital than to remove the same from 2*s* orbital, I_1 which corresponds to the removal of a 2*p* electron is lower than I_2 which corresponds to the removal of 2*s* electron. Here it may be noted that both the orbitals *viz.* 2*p* and 2*s* belong to the same shell i.e. the value of principal quantum number (*n*) of 2*p* and 2*s* orbitals is the same *viz.* $n = 2$.

(b) 1st ionisation potentials of two different elements of the same period. In general when we proceed from left to right in a period, ionisation potentials increase. But some of the elements show irregular trends. For example first ionisation potential of Be = 899.5 KJ. mole^{-1} is higher (instead of lower) than that of B = 800.6 KJ. mole^{-1}. These ionisation potentials can be represented as follows :

$$\text{Be}\ (1s^2, 2s^2) \xrightarrow[\text{I}_1\ =\ 899.5\ \text{KJ.mole}^{-1}]{-e^-\ (2s^1)} \text{Be}^+(1s^2, 2s^1)$$

$$\text{B}\ (1s^2, 2s^2 2p^1) \xrightarrow[\text{I}_2\ =\ 800.6\ \text{KJ.mole}^{-1}]{-e^-\ (2p^1)} \text{B}^+(1s^2, 2s^2)$$

Ionisation potential of Be is higher than that of B, since to remove an electron from a 2*s*-orbital of Be requires more energy than to remove the same from a 2*p*-orbital of B.

On similar lines of reasoning it can also be explained why ionisation potential of Mg is higher than that of Al (Mg = 737.7 KJ. mole^{-1}, Al = 577.6 KJ. mole^{-1}).

$$\text{Mg}\ (2, 8, 3s^2) \xrightarrow[\text{I}_1 = 737.7\ \text{KJ.mole}^{-1}]{-e^-\ (3s^1)} \text{Mg}^+(2, 8, 3s^1)$$

$$\text{Al}\ (2, 8, 3s^2 3p^1) \xrightarrow[\text{I}_2 = 577.6\ \text{KJ.mole}^{-1}]{-e^-\ (3p^1)} \text{Al}^+(2, 8, 3s^2)$$

All the factors mentioned above are independent of each other and hence it becomes impossible to predict the exact extent of each to which they can affect the magnitude of ionisation potential.

Variation of Ionisation Potential in a Period and a Group of Representative Elements :

(*a*) **In a period.** *In general, as we move from left to right in a period, the ionisation. potential of the elements increases due to the successive increase in the nuclear charge (i.e. atomic number) and decrease in atomic size.* However there are certain elements which show irregular trends.

1st period. When we pass from H to He, there is a large increase in the value of ionisation potential (H = 1320.0 KJ.mole^{-1}, He = 2372.3 KJ. mole^{-1}). This large increase in case of He atom is explained on the basis of the fact that, since in case of He both the electrons are present in the same orbital namely 1*s* orbital, the first electron is not able to shield the second. This increases the ionisation potential value of He atom.

$$\text{H}\ (1s^1) \xrightarrow[\text{I}_1 = 1320.0\ \text{KJ.mole}^{-1}]{-e^-\ (1s^1)} \text{H}^+(1s^0)$$

$$\text{He}\ (1s^2) \xrightarrow[\text{I}_1 = 2372.3\ \text{KJ.mole}^{-1}]{-e^-\ (1s^1)} \text{He}^+(1s^1)$$

2nd period. Now when we proceed from He to Li, there is a large decrease in the value of ionisation potential (He = 2372.3 KJ.mole^{-1}, Li = 520.3 KJ.mole^{-1}). This large decrease has been explained by the following two arguments.

(*i*) In case of Li atom ($1s^2$, $2s^1$) it is easier to remove an electron from the 2nd shell (n = 2) while in case of He atom ($1s^2$) it is comparatively more difficult to remove the same from the 1st shell (n = 1). This means that the removal of an electron from 2*s* orbital of Li atom requires less energy while the removal of the same from 1*s* orbital of He atom requires more energy. Hence the ionisation potential of He is much higher than that of Li.

$$\text{He}\ (1s^2) \xrightarrow[\text{I}_1 = 2372.0\ \text{KJ.mole}^{-1}]{-e^-\ (1s^1)} \text{He}^+(1s^1)$$

$$\text{Li}\ (1s^2, 2s^1) \xrightarrow[\text{I}_1 = 520.3\ \text{KJ.mole}^{-1}]{-e^-\ (2s^1)} \text{Li}^+(1s^2)$$

(*ii*) In Li atom ($1s^2$, $2s^1$) the electron to be removed from 2*s*-orbital is *effectively screened or shielded* by the two electrons in 1*s* orbital while in He atom ($1s^2$) the electron to be removed from the 1*s*-orbital is not effectively screened by the other electron in 1*s*-orbital, since both the electrons in He atom reside in the same orbital *viz.* 1*s*-orbital.

On passing from Li to Ne via Be, B etc. generally the ionisation potential increases with the increase of effective nuclear charge and decrease of atomic size. However Be and N show irregular trends as is evident from the fact that, although Be and N have smaller effective nuclear charges than their next elements namely B and O respectively (Be = 1.95. B = 2.60, N = 3.90, O = 4.55), these elements have higher ionization potential values than B and O respectively. These higher values can be explained as follows.

In case of Be ($1s^2$, $2s^2$) it is more difficult to remove an electron from the *completely filled 2s-orbital* while in case of B ($1s^2$, $2s^22p^1$) it easier to remove the same from a *partially-filled* 2*p*-orbital. Since to remove an electron from a 2*s*-orbital of Be atom requires more energy than to remove the same from a 2*p*-orbital of B atom, ionisation potential of Be is higher than that of B.

In case of N ($1s^2$, $2s^2$, $2p^3$) it is more difficult to remove an electron from the *half-filled 2p.orbital* while in case of O ($1s^2$. $2s^22p^4$) it is easier to remove the same from a *partially-filled 2p-orbital* Thus the ionization potential of N is higher than that O.

Now when pass from Ne (Z = 10) to Na (Z = 11), there is a large decrease in the value of ionization potential of Na as compared to that of Ne (Ne = 2080.7 KJ.mole^{-1}, Na = 495.8 KJ.mole^{-1}). This large decrease has been explained by the following two arguments :

(*i*) In case of Na atom ($1s^2$, $2s^2p^6$, $3s^1$) it is easier to remove an electron from 3*rd* shell (n = 3) while in case of Ne atom ($1s^2$, $2s^2p^6$) it is comparatively more difficult to remove the same from the 2*nd* shell (n = 2). This means that the removal of an electron from 3*s*-orbital of Na atom requires less energy while the removal of the same from 2*p*-orbital of Ne atom requires more energy. Hence the ionization potential of Ne is much higher than that of Na.

$$\text{Ne}\,(1s^2,\,2s^2p^6) \xrightarrow[\text{I}_1\,=\,2080.7\text{ KJ.mole}^{-1}]{-e^-\,(2p^1)} \text{Ne}^+(1s^2,\,2s^2p^5)$$

$$\text{Na}\,(1s^2,\,2s^2p^6,\,3s^1) \xrightarrow[\text{I}_1\,=\,495.8\text{ KJ.mole}^{-1}]{-e^-\,(3s^1)} \text{Na}^+(1s^2,\,2s^2p^6)$$

(*ii*) In case of Na atom the electron to be removed from 3*s*-orbital is *effectively* screened by the two electrons in 1*st* shell and eight electrons in 2*nd* shell while in case of Ne atom the electron to be removed from 2*p*-oribital is not *effectively* screened by the other five electrons of 2*p*-orbital.

3rd period. The elements of this period also show a general increase in their ionisation potential value with a few exceptions. For example ionisation potentials of Mg and P are higher than those of Al and S.

Higher value of ionisation potential of Mg as compared to Al is explained on the same lines of argument as used in explaining the higher value of Be compared to B in 2*nd* period. Similarly higher value of P as compared to S has been explained on the same lines of argument as used in explaining the higher value of N compared to O in 2*nd* period.

(*b*) **In a group.** *When we proceed from top to bottom in a group of representative elements, the ionisation potential values of the elements go on decreasing.* This decrease in the ionisation potential values on descending a group can be accounted for as follows.

We have already said that on descending a group, the magnitude of effective nuclear charge acting on the valence-shell electron remains the same (First element is a typical case). Thus the concept of effective nuclear charge cannot be used to explain the decrease in ionisation potential values. However, the decrease has been explained on the basis of the concept of atomic size, the value of principal quantum number (n) of the orbital from which the electron is to be removed and shielding effect (or the magnitude of screening constant, σ) caused by the intervening electrons on the valence-shell electron. It is already known that all these quantities increase on descending a group and *their increase on proceeding from top to botton in a group decreases* the ionisation potential values as shown below for the elements of group IA.

Etements of group IA (Atomic number) given in bracket)	Electronic configuration	Atomic (covalent radius (Å)	(n)	σ	Ionisation potential KJ/mole
Li (3)	2, 1	1.23	2	1.70	520.3
N (11)	2, 8, 1	1.54	3	8.80	495.8
K (19)	2, 8, 8, 1	2.03	4	16.80	418.9
Rb (37)	2, 8, 18, 8, 1	2.16	5	34.80	403.4
Cs (55)	2, 8, 18, 18, 8, 1	2.35	6	52.80	375.7

The variation of ionisation potential in a period and a group of representative elements (s and p block elements) of the long form of periodic table can be shown as given in Fig. 4.13.

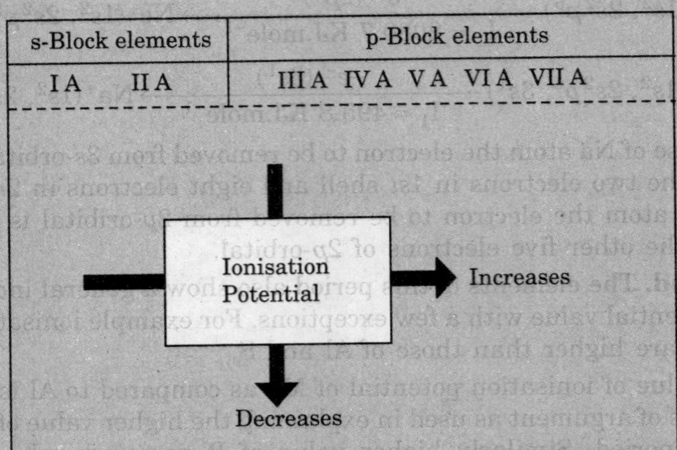

Fig. 4.13. Variation of ionisation potential in a period and a group of representative elements (s- and p-block elements) of the long form of periodic table.

Electron Affinity (EA or $\Delta_{eg}H$)

Electron affinity (abbreviated as EA) is an atomic property which gives us an idea of the tendency of the element to accept the electron to form an anion. With the help of this property a number of properties of the elements like oxidising power and non-metallic (or electronegative) character can well be explained.

What is Electron Affinity ?

Electron affinity of an element is defined as the amount of energy released in adding an extra electron from outside to an isolated neutral gaseous atom in its lowest energy state (i.e. ground state) to convert it into a gaseous anion. Thus electron affinity of an atom, M(g) can be defined by the following process:

$$\underset{\text{Isolated gaseous atom}}{M(g)} + e^-(g) \rightarrow \underset{\text{Gaseous anion}}{M^-(g)} + \textit{Energy released} \ (= -\ EA_1)$$

Electron affinity defined above is strictly called **first electron affinity,** since it corresponds to the addition of *one* electron only. Hence it is represented as $-EA_1$ where negative (minus) sign represents the release of energy. It is also represented as E or Δ_{eg} H.

The above process which depicts the addition of an electron to a gaseous atom is an *exothermic process*, since in the addition of an electron energy is released.

Electron affinity is also called **electron affinity energy.** It is measured in electron volts, kilocalories or kilojoules. The values of electron affinity are shown by a negative (minus) sign which is placed before them. Negative sign represents the release of energy in the addition of an electron to M(*g*) atom.

Values of first electron affinity (in KJ/mole) of *s*-and *p*-block elements are given in Table 4.10.

Second Electron Affinity

In addition to first electron affinity (E_1) defined above, second electron affinity (E_2) of some elements like O, S and Se is also known. *Second electron affinity of an element, M(g) is defined as the amount of energy required to add one more electron to its mono-negative anion, $M^-(g)$ to convert it into di-negative anion, $M^{2-}(g)$. Thus.*

$$M^-(g) + e\ (g) + \text{Energy supplied } (= +\ E_2) \rightarrow M^{2-}(g)$$

In the process of adding one more electron to $M^-(g)$ anion against the electrostatic repulsion between the extra electron being added to $M^-(g)$ and the negative charge on M^- (*g*) anion, energy, instead of being released, (as in case of first electron affinity) is supplied to $M^-(g)$ ion to convert it into $M^{2-}(g)$ ion. Thus the process of adding a second electron to $M^-(g)$ anion is an *endothermic process*. Second electron affinity is represented by a positive sign, since it represents the energy absorbed (*i.e.* supplied) and not released as in case of first electron affinity.

In the formation of a dinegative ion, $M^{2-}(g)$ from the neutral M(*g*) atom, electrons are added one by one. Energy equal to first electron affinity (E_1) is

Table 4.10. Electron affinities of *s*- and *p*-block elements in KJ/mole. Negative sign indicates the release of energy (exothermic reaction) when an electron is added to an atom to convert it into an anion while the positive sign indicates the absorption of energy (endothermic reaction) for the same process

Group → / Period ↓	IA	IIA	IIIA	IVA	VA	VIA	VIIA	Zero
1	$H \xrightarrow{+e^-} H^-$ –72.9							$He \xrightarrow{+e^-} He^-$ ~ 0
2	$Li \xrightarrow{+e^-} Li^-$ –59.8	$Be \xrightarrow{+e^-} Be^-$ ~ 0	$B \xrightarrow{+e^-} B^-$ –23	$C \xrightarrow{+e^-} C^-$ –122	$N \xrightarrow{+e^-} N^-$ +20.1	$O \xrightarrow{+e^-} O^-$ –141 $O \xrightarrow{+2e^-} O^{2-}$ + 780	$F \xrightarrow{+e^-} F^-$ –322	$Ne \xrightarrow{+e^-} Ne^-$ ~ 0
3	$Na \xrightarrow{+e^-} Na^-$ –52.9	$Mg \xrightarrow{+e^-} Mg^-$ ~ 0	$Al \xrightarrow{+e^-} Al^-$ –44	$Si \xrightarrow{+e^-} Si^-$ –120	$P \xrightarrow{+e^-} P^-$ –74	$S \xrightarrow{+e^-} S^-$ –200.4 $S \xrightarrow{+2e^-} S^{2-}$ + 590	$Cl \xrightarrow{+e^-} Cl^-$ –348.7	$Ar \xrightarrow{+e^-} Ar^-$ ~ 0
4	$K \xrightarrow{+e^-} K^-$ –48.9	$Ca \xrightarrow{+e^-} Ca^-$ ~ 0	$Ga \xrightarrow{+e^-} Ga^-$ –36	$Ge \xrightarrow{+e^-} Ge^-$ –116	$As \xrightarrow{+e^-} As^-$ –77	$Se \xrightarrow{+e^-} Se^-$ –195 $Se \xrightarrow{+2e^-} Se^{-2}$ + 420	$Br \xrightarrow{+e^-} Br^-$ –324.5	$Kr \xrightarrow{+e^-} Kr^-$ ~ 0
5	$Rb \xrightarrow{+e^-} Rb^-$ –46.9	$Sr \xrightarrow{+e^-} Sr^-$ ~ 0	$In \xrightarrow{+e^-} In^-$ –34	$Sn \xrightarrow{+e^-} Sn^-$ –121	$Sb \xrightarrow{+e^-} Sb^-$ –101	$Te \xrightarrow{+e^-} Te^-$ –190.1	$I \xrightarrow{+e^-} I^-$ –295	$Xe \xrightarrow{+e^-} Xe^-$ ~ 0

liberated in the addition of one electron to M(*g*) atom to convert it into $M^-(g)$ anion and the process is *exothermic*

$M(g) + e^-(g) \rightarrow M^-(g)$ + *Energy liberated* $(= -E_1)$.....*Exothermic process* ...(*i*)

Negative sign placed before E_1 indicates the *liberation* of energy.

Energy equal to second electron affinity (E_2) is to be supplied in the addition of one more electron to $M^-(g)$ to convert it into $M^{2-}(g)$and the process is *endothermic.*

$M^-(g) + e^-$ + *Energy supplied* $(= + E_2) \rightarrow M^{2-}(g)$...*Endothermic process* ...(*ii*)

Plus sign placed before E indicates the *absorption* of energy.

Now since $E_2 > E_1$, there is net absorption of energy in the formation of a dinegative ion, $M^{2-}(g)$ from the neutral M(*g*) atom and the amount of energy absorbed in this process is equal $E_2–E_1$ which is obtained by adding equations (*i*) and (*ii*) given above. Thus :

$M(g) + 2e^-(g)$ + *Energy absorbed* $(= E_2 - E_1) \rightarrow M^{2-}(g)$

Examples to Illustrate the Concept of Second Electron Affinity

The following examples may be used for the illustration of the concept of second electron affinity.

(*i*) Formation of O^{2-}(g) from O(g). The formation of $O^{2-}(g)$ from O(*g*) can be shown by the following equations :

$O(g) + e^-(g) \rightarrow O^-(g) + E_1(= -141$ KJ/mole) ...*Exothermic process*

$O^-(g) + e^-(g) + E_2(= +921$ KJ/mole) $\rightarrow O^{2-}(g)$ ...*Endothermic process*

On adding :

$O(g) + 2e^-(g)$ + Energy (= + 921 – 141 = + 780 KJ/mole) $\rightarrow O^{2-}(g)$

Thus the total electron affinity for two electrons for oxygen atom = + 780 KJ/mole.

(*ii*) Formation of S^{2-}(g) from S(g).

$S(g) + e^-(g) \rightarrow S^-(g) + E_1(= -200.4$ KJ/mole) ...*Exothermic process*

$S^-(g) + e^-(g) + E_2(= +790.4$ KJ/mole) $\rightarrow S^{2-}(g)$ ...*Endothermic process*

On adding :

$S(g) + 2e^-(g)$ + Energy (= + 790.4 – 200.4 = + 590 KJ/mole) $\rightarrow S^{2-}(g)$

Thus the total electron affinity for two electrons for sulphur atom = + 590 KJ/mole.

(*iii*) Formation of Se^{2-}(g) from Se(g)

$Se^-(g) + e^-(g) \rightarrow Se^-(g) + E_1(= -195$ KJ/mole) ...*Exothermic process*

$Se^-(g) + e^-(g) + E_2(= +615$ KJ/mole) $\rightarrow Se^{2-}(g)$ ...*Endothermic process*

On adding :

$Se(g) + 2e^-(g)$ + Energy (= + 615 – 195 = + 420 KJ/mole) $\rightarrow Se^{2-}(g)$

Thus the total electron affinity for two electrons for Se atom = + 420 KJ/mole.

Factors Affecting the Magnitude of Electron Affinity

The factors on which the magnitude and sign of electron affinities depend are, in general, similar to those which determine the ionisation energies but in the reverse order. These factors are :

1. Atomic radius. Larger the atomic radius, lesser the tendency of the atom to attract the additional electron towards itself, lesser the force of attraction exerted by the nucleus on the extra electron being added to the outer-shell of the atom and hence lesser the amount of energy released when the extra electron is added to the atom to form an anion. Thus *smaller atoms have higher electron affinities.*

2. Effective nuclear charge. Higher the effective nuclear charge, greater the tendency of the atom to attract the additional electron towards itself, greater the force of attraction exerted by the nucleus on the extra electron being added to the outer-shell of the atom and hence higher the amount of energy released when the extra electron is added to the atom. *Thus the atoms with higher effective nuclear charge have higher electron affinities.*

3. Electronic configuration. The effect of this factor on the magnitude of electron affinity of an element can be explained by the following examples :

(*a*) Electron affinity values of the elements of group IIA. *ns* orbital of the valence-shell of the atoms of these elements is completely filled and hence the addition of any extra electron from outside to this *ns* orbital is not possible. *Consequently the elements of group IIA have practically zero electron affinity.*

(*b*) Electron affinity values of N and P. The valence-shell electronic configuration of N and P are :

$$N \rightarrow 2s^2p^3 \; ; \quad P \rightarrow 3s^2p^3$$

$2p$ and $3p$ orbitals in N and P respectively are half-filled and hence are extra ordinarily stable. Thus the addition of any extra electron from outside to these orbitals is not possible. *Consequently N and P have very low electron affinity values* (N = + 20.1 KJ/mole, P = –74 KJ/mole).

(*c*) Electron affinity values of halogens. The valence-shell configuration of halogen atoms (ns^2p^5) has an appetite for one electron to stabilise its configuration by attaining the noble gas configuration (ns^2p^6) which is very stable. Thus halogen atoms have a strong tendency to accept an extra electron from outside and the addition an extra electron makes the configuration of the resulted halide ion similar to that of the noble gas and hence very stable. *Consequently the halogen atoms have very high values of electron affinity.*

(*d*) Electron affinity values of noble gases. The noble gases which have extra-ordinarily stable configuration namely ns^2p^6 show no tendency to accept the additional electron from outside. *Hence the electron affinity values of inert gases are practically zero.*

Variation of Electron Affinity in a Period and a Group : *Periodic Trends of Electron Affinity*

It is difficult to make general views about the periodic variation of electron affinity of various elements for the following reasons

(*i*) The electron affinities of only a very few atoms are known with certainty.

(*ii*) The elements like Be, Mg and noble gases have practically zero electron affinity values, and N and P have very small values.

(*iii*) The elements of 2nd period of the periodic table do not follow the general trend of decrease of electron affinity values on moving from top to bottom in a group.

However, the general trend of variations of electron affinity of various elements can be discussed as follows :

(*a*) **In a period.** On moving from left to right in a period, the size of the atoms decreases and the effective nuclear charge increases. Both these factors favour an increase in the force of attraction exerted by the nucleus on the electron. Consequently the atom has a greater tendency to attract an extra electron from outside towards itself and hence its *electron affinity increases from left to right*. Thus the metals (*e.g.* alkali metals) which lie at the left hand portion of the periodic table have low values of electron affinity while the non-metals (*e.g.* halogens) which lie at the right hand portion of the periodic table have high values of electron affinity.

Exceptions. There are certain elements in each period which have abnormal values. For example in *2nd* period Li, Be, N and Ne have abnormal values. Similarly in *3rd* period Na, Mg, P and Ar have abnormal values. The abnormal values of Li and Na cannot be explained by simple mechanism while those of Be, N, Mg, P, Ne and Ar have already been explained.

(*b*) **In a group.** On moving down a group both the size of the atom and the effective nuclear charge increase. The increase in atomic size tends to decrease the electron affinity values while the increase in nuclear charge tends to increase the electron affinity values. The net result is that the effect produced by the progressive increase of the atomic size outweighs the effect produced by the progressive increase in nuclear charge and *consequently the electron affinity goes on decreasing as we move from top to bottom in a group* as is evident from EA values of (*i*) alkali metals (Group IA) (*ii*) C, Si and Ge (Group IVA) (*iii*) S, Se and Te (Group VIA) (*iv*) O, S and Se for two electrons (Group VI A) and (*v*) Cl, Br and I (Group VII A).

Exceptions. Although the elements of *2nd* period of the periodic table have smaller size than the elements of *3rd* period, the EA values of some of the elements of *2nd* period are lower than the EA values of the elements lying just below them in *3rd* period. For example B<Al, N<P, O<S and F<Cl. The lower values for the elements of *2nd* period are explained by saying that, due to the smaller size of the atoms of the elements of *2nd* period, the addition of an extra electron to these atoms produces high electron density round the resulted anions. This high electron density increases the repulsion between the electrons (called electron–electron repulsion) already present in the relatively compact 2*p* orbital of the *2nd* shell of these atoms and the electron being added. Due to this electron–electron repulsion, the atoms of the elements of *2nd* period show lesser tendency to attract the extra electron from outside and hence lower values of electron affinity for these elements.

The variation of electron affinity of *s*- and *p*- block elements in a group and a period is shown in Fig. 4.14.

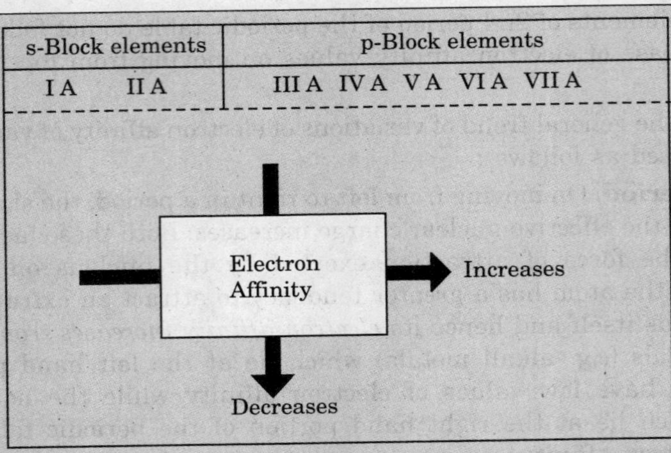

Fig. 4.14. Variation of electron affinity of *s*- and *p*-block elements in the periodic table.

Applications of Ionisation Potential and Electron Affinity

1. Tendency of elements to form cations and anions. We have seen that the magnitude of ionisation potential of an element gives us an idea of the tendency of the element to lose an electron and hence to form a cation. Greater the value of ionisation potential of an element, lower its tendency to lose the electron and hence to form cation. Thus the elements with low values of ionisation potential (*e.g.* alkali metals and alkaline earth metals) have a strong tendency to lose their outermost shell electron and hence form cations quite easily while those having high values (*e.g.* halogens) are not able to lose the electron and hence have no tendency to form cations. Rather, such elements, due to their high electron affinity values, show a strong tendency to accept the electrons and thus form anions.

Again we have seen that the magnitude of electron affinity of an element gives us an idea of the tendency of the elements to accept an electron and hence to form an anion. Greater the value of electron affinity of an element, greater its tendency to accept the electrons and hence to form anions. Thus the elements with high electron affinity values (*e.g.* halogens) have a strong tendency to accept an electron and hence form an anion quite easily while the elements with low electron affinity values (*e.g.* alkali metals and alkaline earth metals) are not able to accept an electron and hence have no tendency to form an anion. Rather, such elements, due to low ionisation potential values, show a strong tendency to lose the electrons and thus form cations.

The above description can be summarised by saying that *with the increase of ionisation potential values of the elements from left to right in a period, the tendency of the elements to lose the electrons and thus to form the cations decreases while with the decrease of ionisation potential values from top to bottom in a group, this tendency increases.* (Fig. 4.15).

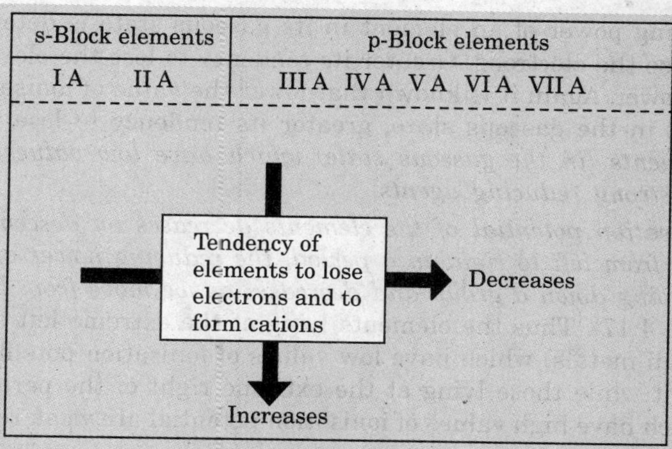

Fig. 4.15. Variation of the tendency of *s*- and *p*-block elements to lose electrons and to form cations in the periodic table.

Similary *with the increase of electron affinity values of the elements from left to right in a period, the tendency of the elements to accept the electrons and thus to form the anions increases while with the decrease of electron affinity values from top to bottom in a group, this tendency decreases* (Fig. 4.16).

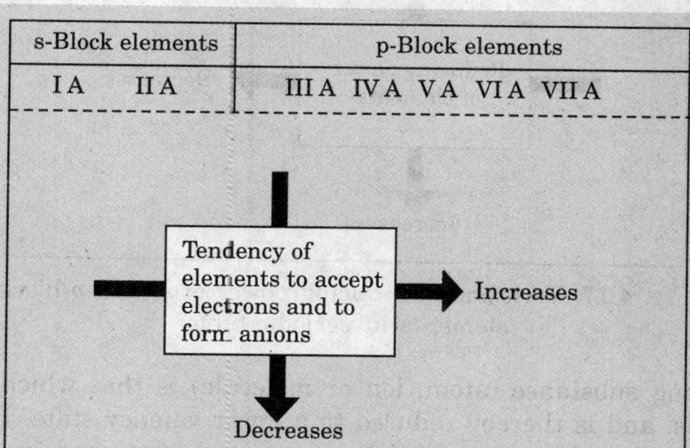

Fig. 4.16. Variation of the tendency of *s*- and *p*-block elements to accept electrons and to form anions in the periodic table.

2. Reducing and oxidising power of elements. A reducing substance (atom, ion or molecule) is that which loses one or more electrons and is thereby oxidised to a higher valency state. For example if an element in its gaseous state, $M(g)$ loses one electron and is converted into $M(g)^+$ cation, the element $M(g)$ is said to be acting as a reducing agent.

$$\underset{\substack{\text{Reducing agent} \\ \text{(Lower valency} \\ \text{state = 0)}}}{M(g) - e^-} \longrightarrow \underset{\substack{\text{(Higher valency} \\ \text{state = + 1)}}}{M^+(g)}$$

The reducing power of an element in its gaseous state is determined by its tendency to lose the electrons. Greater its tendency to lose the electrons, greater its reducing power. Again it is known that lower the value of ionisation potential of an element in the gaseous state, greater its tendency to lose the electrons. *Thus the elements (in the gaseous state) which have low values of ionisation potential are strong reducing agents.*

Since ionisation potential of the elements decreases on descending a group and increases from left to right in a period, the reducing power of the elements increases in going down a group and decreases as we move from left to right in a period. (Fig. 4.17). Thus the elements lying at the extreme left of the periodic table (*i.e.* alkali metals) which have low values of ionisation potential are strong reducing agent while those lying at the extreme right of the periodic table (*i.e.* halogens) which have high values of ionisation potential are weak reducing agents (Fig. 4.17).

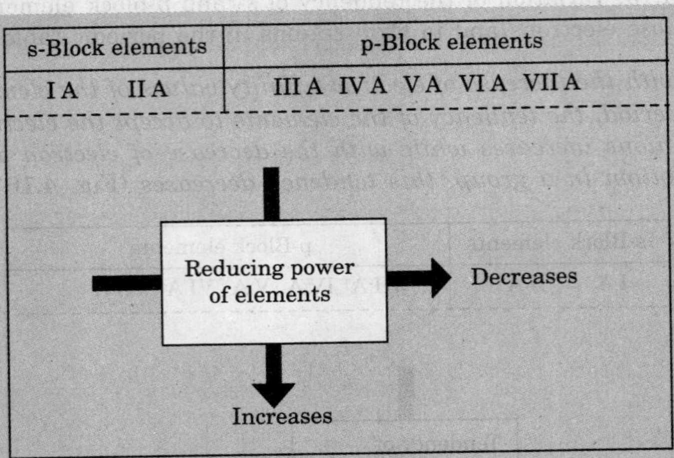

Fig. 4.17. Variation of reducing power of *s*- and *p*-block elements in periodic table.

An oxidising substance (atom, ion or molecule) is that which gains one or more electrons, and is thereby reduced to a lower valency state. For example if an element in its gaseous state, M(*g*) gains one electron and is converted into $M^-(g)$ anion, the element M(*g*) is said to be acting as an oxidising agent.

$$\underset{\substack{\text{Oxidising agent}\\\text{(Higher valency}\\\text{state = 0)}}}{M(g)} + e^- \longrightarrow \underset{\substack{\text{(Lower valency}\\\text{state = -1)}}}{M^-(g)}$$

The oxidising power of an element in its gaseous state is determined by its tendency to accept the electrons. Greater its tendency to accept the electrons, greater its oxidising power. Again it is known that higher the value of electron affinity of an element in the gaseous state, greater its tendency to accept the electrons. *Thus the elements (in the gaseous state) which have high electron affinity values are strong oxidising agents.*

Since electron affinity of the elements decreases on descending a group and increases from Left to right in a period, the oxidising power of the elements decreases on descending a group and increases from left to right in a period. (Fig. 4.18). Thus the elements lying at the extreme left of the periodic table (*i.e.* alkali metals) which have low values of electron affinity are poor oxidising agents while those lying at the extreme right of the pariodic table (*i.e.* halogens) which have high values of electron affinity are strong oxidising agents.

Here it may be noted that we have considered the reducing and oxidising powers of the elements *only in their gaseous state*. If the chemical reaction is carried out in solution using water as the main solvent, reducing and oxidising powers are better understood in terms of standard oxidation potential value of the elements.

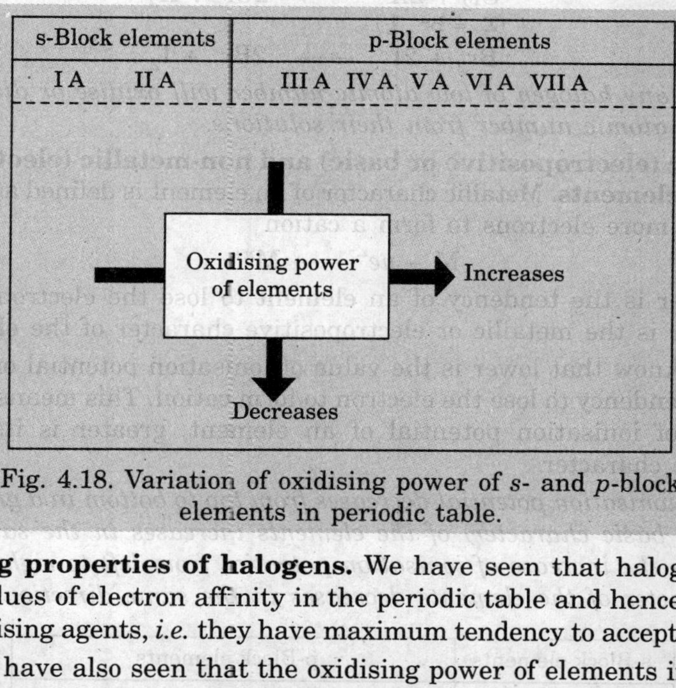

Fig. 4.18. Variation of oxidising power of *s*- and *p*-block elements in periodic table.

Oxidising properties of halogens. We have seen that halogens have the maximum values of electron affinity in the periodic table and hence are the most powerful oxidising agents, *i.e.* they have maximum tendency to accept the electrons.

Again we have also seen that the oxidising power of elements increases with the increase of their electron affinity values. Thus the oxidising power of halogens must be in the following order : (Electron affinity values in KJ. mole^{-1} are also given in parenthesis)

$$Cl(-348.7) > Br(-324.5) > F(322.0) > I(-295.0)$$

In fact, F has been found to be the most powerful oxidising agent and the oxidising power of halogens decreases from F to I as shown below:

$$F > Cl > Br > I$$

The unexpected order of oxidising power of halogens shown above is due to the fact their electron affinity alone does not control the oxidising power of elements but there are many other energy terms like *heat of fusion, heat of evaporation, heat of dissociation and heat of hydration* and their sum on which the oxidising power of a given halogen depends. When all these energy terms for different halogens are summed up with the help of Born-Haber cycle, the sum is found in the following order :

$$F > Cl > Br > I$$

The order given above is actually the order of the oxidising power of halogens.

That F is the most powerful oxidising agent among all the halogens is evident from the fact that F_2 will *oxidise* or *displace* any of the other halide ions (X^-) *both in solution and dry state*

$$F_2 + 2X^- \longrightarrow 2F^- + X_2$$
$$(X^- = Cl^-, Br^-, I^-)$$

Similarly Cl_2 gas will displace Br^- and I^- ions from their solutions and Br_2 will displace only I^- ions.

$$Cl_2 + 2X^- \longrightarrow 2Cl^- + X_2$$
$$(X^- = Br^-, I^-)$$
$$Br_2 + 2I^- \longrightarrow 2Br^- + I_2$$

In general *any halogen of low atomic number will oxidise or displace halide ions of higher atomic number from their solutions.*

3. Metallic (electropositive or basic) and non-metallic (electronegative) character of elements. Metallic character of an element is defined as its tendency to lose one or more electrons to form a cation

$$M - ne^- \longrightarrow M^{n+}$$

The greater is the tendency of an element to lose the electron to form the cation, greater is the metallic or electropositive character of the element.

Again we know that lower is the value of ionisation potential of an element, greater is its tendency to lose the electron to form cation. This means that smaller is the value of ionisation potential of an element, greater is its metallic or electropositive character.

Now since ionisation potential decreases from top to bottom in a group, metallic character (i.e. basic character) of the elements increases in the same direction. Similarly *with the increase of ionisation potential from left to right in a period, metallic character of the elements decreases in the same direction.* Fig. (4.19.)

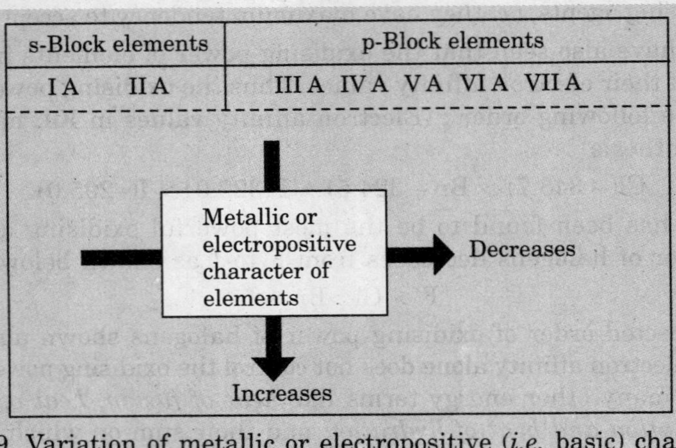

Fig. 4.19. Variation of metallic or electropositive (*i.e.* basic) character of representative elements (*s*- and *p*-block elements) of the periodic table.

The successive increase of metallic character of the elements on descending a group is particularly apparent in groups IVA, VA and VIA which begin with non-metals namely C, N and O and end with metals namely Pb, Bi and Po respectively. The general increase of metallic character in going down a group is also confirmed by the fact that the oxides of the elements of a group become more and more basic when we proceed from top to bottom in a group. For example the oxides of M_2O_3 type of the elements of group VA become more and more basic (*i.e.* less and less acidic) from N_2O_3 to Bi_2O_3 as shown below :

N_2O_3, P_2O_3,	As_2O_3, Sb_2O_3,	Bi_2O_3
Acidic	Amphoteric	Basic

—— *Basic character increasing* ⟶

The general decrease of metallic character of the elements in going from left to right in a period is also confirmed by the fact that the oxides of the elements of a given period become *less and less basic* or *more and more acidic* as we proceed from left to right. For example the oxides of the elements of 3rd period show this type of behaviour as shown below :

Na_2O,	MgO,	Al_2O_3,	SiO_2,	P_2O_5,	SO_3,	Cl_2O_7
Strongly basic	Basic	Amphoteric	Feebly acidic	Acidic	More acidic	Most acidic

—— *Basic character decreasing* ⟶

Non-metallic character of an element is defined as its tendency to accept one or more electrons to form an anion.

$$M + ne^- \rightarrow M^{n-}$$

The greater is the tendency of an element to accept the electron to form the anion, greater is its non-metallic character.

Again we know that greater is the value of electron affinity of an element, greater is its tendency to accept the electron to give the anion. This means *that greater is the value of electron affinity of an element, greater is its non metallic or electronegative character.*

Now since electron affinity values decrease from top to bottom in a group, non-metallic character of the elements also decreases in the same direction. Similarly with the increase of electron affinity values from left to right in a period non-metallic character of the elements also increases. (Fig. 4.20).

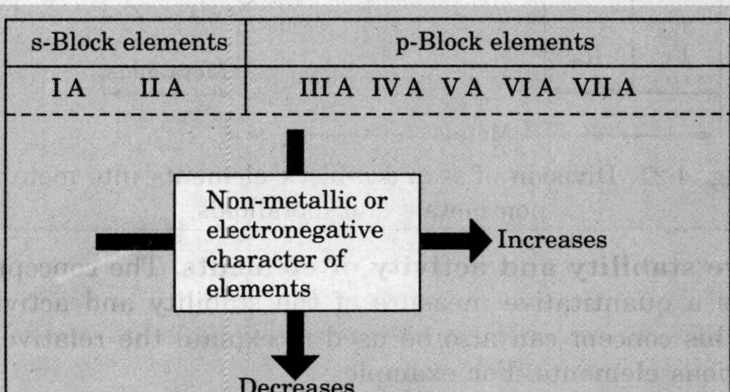

Fig. 4.20. Variation of non-metallic or electronegative character of *s*- and *p*-block elements in the periodic table.

The variation of non-metallic or electronegative character of *s*- and *p*-block elements as described above makes it clear that the elements lying at the extreme right of the periodic table (*i.e.* halogens) show maximum electronegative character and hence are non-metals, while the elements located at the extreme left of the periodic table (*i.e.* alkali metals) show minimum electronegative character and hence are metals.

Metals, non-metals and metalloids. The trend of variation in metallic and non-metallic character of the elements as described above has been used to classify *s*- and *p*-block elements into metals, non-metals and metalloids. The trend of variation shows that *s*- and *p*-block elements lying at the lower left hand side of the periodic table show maximum electropositive character and hence are **metals** while the elements lying at the upper right hand side show minimum electropositive character and hence are **non-metals.** The elements lying in between metals and non-metals (*i.e.* the elements lying in a diagonal strip) show (characteristics of both metals and non-metals and hence are called **metalloids.**

The divisions of *s*- and *p*-block elements into metals, non- metals and metalloids is shown in Fig. 4.21.

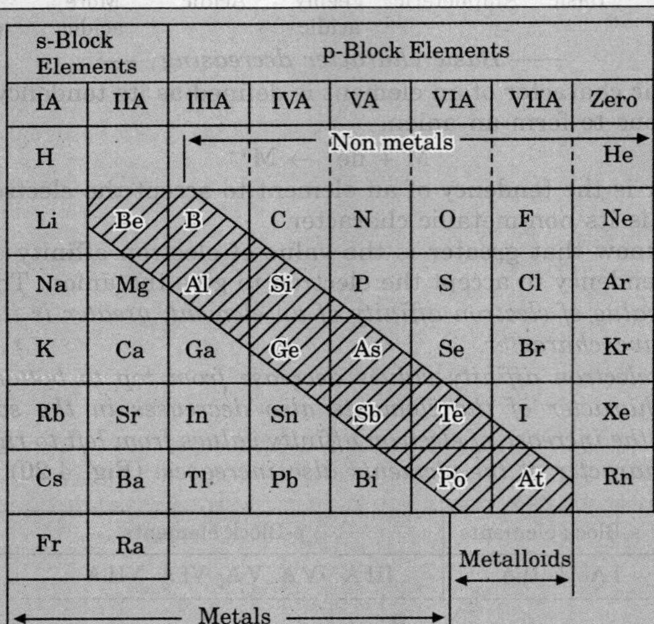

Fig. 4.21. Division of *s*- and *p*-block elements into metals, non-metals and metalloids.

4. Relative stability and activity of elements. The concept of ionisation potential gives a quantitative measure of the stability and activity of various elements *i.e.* this concept can also be used to explain the relative stability and activity of various elements. For example :

(*a*) We know that in a period the ionisation potential increases from left to right, *i.e.* as we move from left to right in a period, more and more energy is required to remove an electron which means that *stability of the elements goes on increasing or their activity goes on decreasing*. Thus in the 3rd period the relative activity of Na, Mg and Al are in the following order :

$$Na > Mg > Al$$

(*b*) In going down a group the ionisation energy decreases from top to bottom, *i.e.* as the group is descended less and less energy is required to remove an electron which means that *the stability of the elements goes on decreasing or their activity goes on increasing*. Thus in the 1*st* group the relative activity of Na, K and Rb is in the following order :

$$Na < K < Rb.$$

The relation between the ionisation potential and the activity of the elements studied as above shows that the elements having high values of ionisation potential are less reactive (*e.g.* inert gases), while those having low values are very reactive (*e.g.* alkali metals and alkaline earth metals).

Exceptions. This rule is applicable only to those elements which have dominant electro-positive character (*i.e.* alkali metals and alkaline earth metals) and not to those which are highly electro-negative, *e.g.*, ionisation potential of F is 1681.0 KJ/mole, which is very high and, therefore, shows that F should be inert, but F is actually very reactive.

Electronegativity (EN)

What is Electronegativity ?

When two different atoms in a molecule are bonded together by a covalent bond, the electron pair forming the covalent bond is not shared equally by both the atoms. Rather, the electron pair lies nearer to one atom than the other. *The relative tendency (or ability or power) of a bonded atom in a molecule to attract the shared electron pair towards itself is termed as its electronegativity.*

This definition was given by Pauling in 1932. Electronegativity is an inherently fundamental property of the atom and is fundamentally different from electron affinity, since electron affinity represents the tendency of an *isolated gaseous atom* to attract the electrons while electronegativity represents the tendency of a bonded atom (*i.e.* of an atom in a molecule) to attract the shared electron pair.

Scales of Electronegativity : Measurement of Electronegativity.

A number of scales have been devised to measure the electronegativity of the atoms. These scales are arbitrary and are based on various types of experimental data like bond energy, dipole moment, ionisation potential and electron affinity. More commonly used scales are described below :

1. Pauling's Scale (1932). This scale is based on an empirical relation between the energy of a bond (called *bond energy*) and the electronegativities of the bonded atoms. Let us consider a A–B bond between two dissimilar atoms, A and B of a molecule AB. Let the bond energies of A–A, B–B and A–B bonds be represented as E_{A-A}, E_{B-B} and E_{A-B} respectively.

It has been seen that the energy of A–B bond (E_{A-B}) is almost always greater than the geometric mean of the energies of A–A (E_{A-A}) and B–B (E_{B-B}) bonds, *i.e.*

$$E_{A-B} > \sqrt{E_{A-A} \times E_{B-B}}$$

The difference between E_{A-B} and $\sqrt{E_{A-A} \times E_{B-B}}$ is called **ionic-resonance energy** of A–B bond and is denoted by Δ_{A-B}. It is thus given by :

$$\boxed{\Delta_{A\text{-}B} = E_{A\text{-}B} - \sqrt{E_{A\text{-}A} \times E_{B\text{-}B}}} \quad ...(i)$$

Δ_{A-B} values do not possess additive property, *i.e.* if we consider three covalent bonds *viz.* A–B, B–C and C–A, then the sum of Δ_{A-B} and Δ_{B-C} is not equal to Δ_{C-A} which means that :

$$\boxed{\Delta_{A\text{-}B} + \Delta_{B\text{-}C} \neq \Delta_{C\text{-}A}} \quad ...(ii)$$

The relation (*i*) is true when the electronegativities of A, B and C are in the order : $x_A > x_B > x_C$.

The validity of relation (*i*) given above can be verified by calculating the values of Δ_{F-O}, Δ_{O-Si} and Δ_{F-Si} for F–O, O–Si and F–Si covalent bonds (electronegativities of F, O and Si are in the order : $x_F > x_O > x_{Si}$) by using the given bond energy values of F–O, O–Si, F–Si, F–F, O–O and Si–Si bonds as :

E_{F-O} = 184.14 KJ. mole^{-1}; E_{O-Si} = 372.46 KJ. mole^{-1}

E_{F-Si} = 535.58 KJ. mole^{-1}; E_{F-F} = 154.84 KJ. mole^{-1}

E_{O-O} = 138.10 KJ. mole^{-1}; E_{Si-Si} = 179.95 KJ. mole^{-1}

Thus by using relation (*i*), we can calculate the values of Δ_{F-O}, Δ_{O-Si} and Δ_{F-Si} for F–O, O–Si and F–Si bonds as shown below :

$$\Delta_{F-O} = E_{F-O} - \sqrt{E_{F-F} \times E_{O-O}}$$
$$= 184.14 - \sqrt{154.84 \times 138.10}$$
$$= 184.14 - 146.23 = 37.91 \text{ kJ.mole}^{-1}$$

$$\Delta_{O-Si} = E_{O-Si} - \sqrt{E_{O-O} \times E_{Si-Si}}$$
$$= 372.46 - \sqrt{138.10 \times 179.95}$$
$$= 372.46 - 157.64 = 214.82 \text{ kJ.mole}^{-1}$$

$$\Delta_{F-Si} = E_{F-Si} - \sqrt{E_{F-F} \times E_{Si-Si}}$$
$$= 535.58 - \sqrt{154.84 \times 179.95}$$
$$= 535.58 - 166.92 = 368.66 \text{ kJ.mole}^{-1}.$$

With the help of the values of Δ_{F-O}, Δ_{O-Si} and Δ_{F-Si} as calculated above it can be shown that

$$\Delta_{F-O} + \Delta_{O-Si} \neq \Delta_{F-Si}$$

or $$37.91 + 214.82 \neq 368.66$$

or $$252.73 \neq 368.66$$

We have already mentioned that the values of ionic-resonance energy (*i.e.* Δ_{A-B} values) do not show additive property. However it has been found that the square roots of these values do show nearly the additive property, *i.e.* if we again consider the bonds *viz.* A–B, B–C and C–A, then the sum of $\sqrt{\Delta_{A-B}}$ and $\sqrt{\Delta_{B-C}}$ is nearly equal to $\sqrt{\Delta_{C-A}}$, *i.e.*

$$\boxed{\sqrt{\Delta_{\mathbf{A\text{-}B}}} + \sqrt{\Delta_{\mathbf{B\text{-}C}}} \approx \sqrt{\Delta_{\mathbf{C\text{-}A}}}} \qquad ...(iii)$$

The validity of relation (*iii*) can be verified by the fact that for F–O, O–Si and F–Si bonds the sum of $\sqrt{\Delta_{F-O}}$ and $\sqrt{\Delta_{O-Si}}$ is almost equal $\sqrt{\Delta_{F-Si}}$ to as shown below :

$$\sqrt{\Delta_{F-O}} + \sqrt{\Delta_{O-Si}} \approx \sqrt{\Delta_{F-Si}}$$

or $$\sqrt{37.91} + \sqrt{214.82} \approx \sqrt{368.66}$$

or $$6.15 + 14.65 \approx 19.20$$

or $$20.80 \approx 19.20$$

The square root of ionic-resonance energy (*i.e.* $\sqrt{\Delta_{A-B}}$) is a measure of the partial ionic character of an A–B covalent bond. With the increase of ionic character of A–B covalent bond, the magnitude of $\sqrt{\Delta_{A-B}}$ also increases. The difference in electronegativity between the bonded atoms (A and B), $x_A – x_B$ ($x_A > x_B$) also increases with the increase of partial ionic character of A–B bond. Thus :

$\sqrt{\Delta_{A-B}} \propto$ *Amount of ionic character in A–B bond*

and

$x_A – x_B \propto$ *Amount of ionic character in A–B bond*

Consequently,
$$\boxed{\boldsymbol{x}_{\mathbf{A}} - \boldsymbol{x}_{\mathbf{B}} \propto \sqrt{\Delta_{\mathbf{A\text{-}B}}}}$$

or
$$\boxed{\boldsymbol{x}_{\mathbf{A}} - \boldsymbol{x}_{\mathbf{B}} = \mathrm{K}\sqrt{\Delta_{\mathbf{A\text{-}B}}}}$$

Now on putting the value of Δ_{A-B} from equation (*i*) we get

$$\boxed{\boldsymbol{x}_{\mathbf{A}} - \boldsymbol{x}_{\mathbf{B}} = \mathrm{K}\left[\mathbf{E}_{\mathbf{A\text{-}B}} - \sqrt{\mathbf{E}_{\mathbf{A\text{-}A}} \times \mathbf{E}_{\mathbf{B\text{-}B}}}\right]^{1/2}}$$

The value of the constant, K has been found to be equal to 0.208 which comes from the conversion of experimental values of E_{A-B} measured in Kcals/mole into e*V*. Thus :

$$\boxed{\boldsymbol{x}_{\mathbf{A}} - \boldsymbol{x}_{B} = \mathbf{0.208}\left[\mathbf{E}_{\mathbf{A-B}} - \sqrt{\mathbf{E}_{\mathbf{A-A}} \times \mathbf{E}_{\mathbf{B-B}}}\right]^{1/2}}$$

This equation gives only the differences in electronegativity values. This equation can be used to calculate x value of an atom provided that x value of the other element is known. Pauling assigned an arbitrary value of 4.0 for the electronegativity of fluorine and calculated that of other elements with the help of this equation. It must be remembered that in the calculation of electronegativity values Pauling has expressed ionic-resonance energies not in kcal, but in electron-volts (1 *eV* per bond = 23 kcal per gram bond).

The values of electronegativities of the atoms of elements which Pauling has calculated are given in Table 4.11. These values are arbitrary and only relative.

The disadvantage of Pauling's scale is that bond energies are not known with any degree of accuracy for many solid elements.

Table 4.11. Electronegativity values of elements as estimated by Pauling (Scale used : F = 4.0)

Group→ / Period↓	IA	IIA	IIIB	IVB	VB	VIB	VIIB	←VIII→			IB	IIB	IIIA	IVA	VA	VIA	VIIA	Zero
1	1 H 2.1																	2 He –
2	3 Li 1.0	4 Be 1.5											5 B 2.0	6 C 2.5	7 N 3.0	8 O 3.5	9 F 4.0	10 Ne –
3	11 Na 0.9	12 Mg 1.2											13 Al 1.5	14 Si 1.8	16 P 2.1	16 S 2.5	17 Cl 3.0	18 Ar –
4	19 K 0.8	20 Ca 1.0	21 Sc 1.3	22 Ti 1.5	23 V 1.6	24 Cr 1.6	25 Mn 1.5	26 Fe 1.	27 Co 1.8	28 Ni 1.8	29 Cu 1.9	30 Zn 1.6	31 Ga 1.6	32 Ge 1.8	33 As 2.0	34 Se 2.4	35 Br 2.8	36 Kr –
5	37 Rb 0.8	38 Sr 1.0	39 Y 1.2	40 Zr 1.4	41 Nb 1.6	42 Mo 1.8	43 Tc 1.9	44 Ru 2.2	45 Rh 2.2	46 Pd 2.2	47 Ag 1.9	48 Cd 1.7	49 In 1.7	50 Sn 1.8	51 Sb 1.9	52 Te 2.1	53 I 2.5	54 Xe –
6	55 Cs 0.7	56 Ba 0.9	57 71 La–Lu 1.1 1.2	72 Hf 1.3	73 Ta 1.5	74 W 1.7	75 Re 1.6	76 Os 2.2	77 Ir 2.2	78 Pt 2.2	79 Au 2.4	80 Hg 1.9	81 Tl 1.8	82 Pb 1.8	83 Bi 1.9	84 Po 2.0	85 At 2.2	86 Rn –
7	87 Fr 0.7	88 Ra 0.9	89 – 95 Ac–Am 1.1 1.3															

2. Mulliken's Scale (1934), R.S. Mulliken has suggested that the electronegativity is related to its ionisation potential and its electron affinity, since these properties of an element measure its tendency to lose its valence shell electron and to attract an extra electron respectively. Therefore an atom having a high value of ionisation potential (*i.e.* small tendency to lose its valence-shell electron) and a high value of electron affinity (*i.e.* having a great tendency to attract an electron) has a high value of electronegativity.

Mulliken considers the formation of a covalent bond between two atoms A and B as follows. Suppose :

x_A and x_B = Electronegativities of atoms A and B

$(IP)_A$ and $(IP)_B$ = Ionisation potentials of atoms A and B

and $(EA)_A$ and $(EA)_B$ = Electron affinities of atoms A and B

According to Mulliken :

(*i*) If $x_A > x_B$, then A^-B^+ structure will be obtained and its formation can be regarded as taking place through the following steps;

Step (*i*) $\quad B + \textit{Energy supplied } [= (IP)_B] \xrightarrow{-e^-} B^+$

Step (*ii*) $\quad A \xrightarrow{+e^-} A^- + \textit{Energy released } [= (EA)_A]$

On adding :

$$B + A + (IP)_B \rightarrow B^+A^- + (EA)_A$$

or

$$B + A + (IP)_B - (EA)_A \rightarrow B^+A^-$$

∴ *Energy change in the formation of B^+A^- structure*

$$= (IP)_B - (EA)_A$$

(*ii*) If $x_B > x_A$, the A^+B^- structure will be obtained and the change in energy in its formation = $(IP)_A$–$(EA)_B$.

Quite obviously in the formation of a covalent bond between A and B (*i.e.* in the formation of A–B structure) x_A will be equal to x_B and the energy changes shown above for B^+A^- and A^+B^- structures must be equal, *i.e.*,

$$(IP)_B - (EA)_A = (IP)_A - (EA)_B$$

or

$$(IP)_A + (EA)_A = (IP)_B + (EA)_B \quad ...(i)$$

On the basis of relation (*i*) Mulliken proposed that *the electronegativity of an atom can be taken to be equal to the average of its IP and EA values, i.e.,*

$$\boxed{x_A = \frac{(IP)_A + (EA)_A}{2}} \quad ...(ii)$$

Relation (*ii*) holds good when IP and EA of the element are given in electron volts. If these quantities are given in kilo calories, then the value of x_A is obtained by dividing the right hand side of equation (*ii*) by 62.5. Thus :

$$\boxed{x_A = \frac{(IP)_A + (EA)_A}{2 \times 62.5}}$$

Mulliken's values of electronegativity are about 2.8 times as large as Pauling's values. Hence to make the Mulliken's values approximately equal to the Pauling's values, right hand side of equation (*ii*) is divided by 2.8. Thus :

$$\boldsymbol{x_A = \frac{(IP)_A + (EA)_A}{2 \times 2.8} = \frac{(IP)_A + (EA)_A}{5.6}}$$

The constant 1/5.6 is called **scale adjustment factor.** This factor is used when IP and EA are expressed in electron volts.

Example. *Calculate the electronegativity values of fluorine aid chlorine on Mulliken's scale, given that $(IP)_F$ = 17.4 eV/atom, $(EA)_F$ =3.62 eV/atom, $(IP)_{Cl}$ = 13.0 eV/atom and $(EA)_{Cl}$ = 4.0 eV*

Solution. The electronegativities of F and Cl denoted by x_F and x_{Cl} are given by :

$$x_F = \frac{(IP)_F + (EA)_F}{5.6} = \frac{17.4 + 3.62}{5.6} = \frac{21.02}{5.6} = 3.75$$

and $$x_{Cl} = \frac{(IP)_{Cl} + (EA)_{Cl}}{5.6} = \frac{13.0 + 4.0}{5.6} = \frac{17}{5.6} = 3.03$$

Disadvantages of Mulliken's Scale. (*i*) Although Mullikan's scale is less empirical than Pauling's scale, yet it suffers from a serious disadvantage that electron affinities, with the exception of a few elements, are not reliably known.

(*ii*) The values of ionisation potential and electron affinity refer to the transfer of electrons between the atomic orbitals. However, the exact constitution of such atomic orbitals is not always known. Further ionisation potential and electron affinity values will vary with the nature of valence state. Thus, if the valence electrons in nitrogen atom in its valence state are s^2p^3 (two electrons in *s*-orbital and three electrons in *p* orbitals of valency energy level), the electronegatvity value comes to 2.33 while if the electrons are in sp^4 state which represents the excited state of nitrogen atom one electron in *s*-orbital and four in *p*-orbitals), the electronegativity value becomes 2.55.

3. Sanderson's Scale (1955). Sanderson's electronegativity scale is based on a new quantity called *stability ratio*, SR for an atom which is defined *as the ratio of the average electron density, ED* (electron density differs from point to point as electrons are not evenly spaced round the nucleus) *round the nucleus and its hypothetical electron density, ED_h* which the atom would have, if it were an inert gas atom. Thus :

$$SR = \frac{ED}{ED_h}$$

According to Sanderson the *stability ratio (SR) of an atom, A is a measure of its electronegativity, x_A, i.e.*

$$\boldsymbol{(x_A)_{Sanderson} = SR = \frac{ED}{ED_h}} \quad ...(i)$$

Equation (*i*) is called ***Sanderson's equation***

According to Sanderson ED is a measure of comparative compactness of the atom and it is obtained by dividing the number of electrons (*i.e.* nuclear charge, Z) round the nucleus by $\frac{4}{3}\pi r^3$ where *r* is the non-polar covalent radius of the atom. Thus :

$$ED = \frac{Z}{\frac{4}{3}pr^3} = \frac{3Z}{4pr^3} = \frac{Z}{4.19\,r^3}$$

The value of ED_i for a particular atomic number is found from interpolation obtained by plotting electron densities of inert gas atoms against atomic numbers.

Sanderson's electronegativity values obtained by the above equation can be converted empirically to Pauling's values by using the relation :

$$x_{Pauling} = 0.21\ x_{Sanderson} + 0.77$$

4. Allred and Rochow's Scale (1958). In 1958 Allred and Rochow suggested another scale of electronegativity which is based on covalent radii. According to this scale *electronegativity of an atom is the force of attraction between the nucleus of one atom and an electron of an adjacent atom bonded to it and separated from the nucleus by the covalent radius.*

Thus, if Z_{eff} is the effective nuclear charge, the effective nuclear charge felt by the electron under the influence of the nucleus is Z_{eff} e and the force of attraction between the nucleus and the electron is, therefore, equal to $(Z_{eff}\,e) \times e/r^2$ where r is the distance (in cms) between the nucleus and the electron. This force of attraction is equal to the electronegativity, x_A of an atom A. Thus :

$$x_A = \frac{(Z_{eff}e)\times e}{r^2} = \frac{Z_{eff}e^2}{r^2}$$

In order to correlate Allred and Rochow's electronegativity values (which are absolute values) given by the above equation with Pauling's values (which are arbitrary and only relative), Allred and Rochow's values are plotted against Pauling's values and the best straight line is drawn through the points. From the slope and the intercept of this line they obtained the following relation for the electronegativity values [Allred and Rochow's values denoted by $(x_A)_{AR}$]

$$(x_A)_{AR} = 0.359\frac{Z_{eff}}{r^2} + 0.744$$

In this equation r is in angstrom units (Å units). The value of Z_{eff} can be determined by *Slatter's rules* according to which $Z_{eff} = Z - \sigma$ where σ is the screening constant. Thus the above equation becomes :

$$(x_A)_{AR} = 0.359\frac{Z-\sigma}{r^2} + 0.744$$

Example. *Find out the electronegativity of lead with the help of the given values : screening constant (σ) of Pb = 76.70, atomic number of Pb(Z) = 82 and covalent radius of Pb(r) = 5.3 Å.*

Solution. On substituting the values of σ, Z and r in the equation :

$$(x_{Pb})_{AR} = 0.359\frac{Z-\sigma}{r^2} + 0.744,$$

we get :

$$(x_{Pb})_{AR} = 0.359\frac{82-76.70}{(5.3)^2} + 0.744 = 1.55$$

This scale has gained wide acceptance, as it has the advantage that Z_{eff} can be estimated by Slatter's rules and r can be determined experimentally for most elements. It does not depend on the quantities like electron affinities and bond dissociation energies which are known for comparatively few elements.

The most serious disadvantage of this scale is that the values of covalent radii are unreliable.

None of the electronegativity scales discussed above has a distinct advantage over the others. Drago (1960), who has critically examined various proposed scales, is of the view that Pauling's scale with all its limitations is the best one.

Factors Affecting the Magnitude of Electronegativity

Following are the important factors on which the magnitude of electronegativity depends

1. Size of the atom. The smaller the size of an atom, greater is its tendency to attract towards itself the shared pair electrons.

Thus the smaller atoms have greater electronegativity values than the larger atoms. For example :

(*a*) Electronegativity values of the elements of group IA decrease from H(Z = 1) to Cs(Z = 55), since the atomic (covalent) radii of these elements increase in the same order as shown below :

Elements of group IA	Atomic (covalent) radius (Å)	Electronegativity
H (1)	0.32	2.1
Li(3)	1.23	1.0
Na (11)	1.54	0.9
K (19)	2.03	0.8
Rb (37)	2.16	0.8
Cs (55)	2.35 ↓ *Increasing*	0.7 ↓ *Decreasing*

(*b*) Electronegativity values of the elements of 2*nd* period increase from Li (Z = 3) to F(Z = 9), since the atomic (covalent) radii of these elements decrease in the same order as shown below :

Elements of 2nd period	*Li*	*Be*	*B*	*C*	*N*	*O*	*F*
Atomic (covalent) radius (Å)	*1.23*	*0.90*	*0.82*	*0.77*	*0.75*	*0.73*	*0.72*
	Decreasing →						
Electronegativity	*1.0*	*1.5*	*2.0*	*2.5*	*3.0*	*3.5*	*4.0*
	Increasing →						

2. Number of inner shells. *The atom with greater number of inner shells (i.e. the shells between the nucleus and the outer-most shell) has less value of electronegativity than the atom with smaller number of inner shells.* For example the electronegativity values of halogens decrease from F(Z = 9) to At(Z = 85), since the number of inner shells increases in the same order as shown below :

Halogens (Elements of VII A group)	No. of inner shells (Complete electronic configurations are given in parenthesis)	Electronegativity
F(9)	1(2, 7)	4.0
Cl (17)	2 (2. 8, 7)	3.0
Br (35)	3(2, 8, 18. 7)	2.8
I (53)	4(2, 8, 18, 18, 7)	2.5
At (85)	5(2, 8, 18, 32, 18, 7) ←Increasing—	2.2 ←Decreasing—

3. Charge on the ion (*Type of the ion*). A cation attracts the electron pair more readily towards itself than the atom from which it has been derived. This is due to the smaller size of the cation as compared to its parent atom ($M^+ < M$). *Thus a cation, M^+ has higher electronegativity than its parent atom, M ($M^+ > M$). (See Table 4.12).*

If an element shows many positive oxidation states, *the element in higher oxidation state has more value of electronegativity than that in the lower oxidation state* (Table 4.12). This is due to the fact that the atom in a higher positive oxidation state is smaller in size and hence has greater attraction for electrons. Thus we can say that an increase in the positive oxidation state increases the tendency of that element to attract electron pair towards itself and hence *electronegativity increases with the increase in positive oxidation state* as shown in Table 4.12 (Atoms from which cation has been derived have also been included in the table).

Table 4.12. Electronegativity values of some cations as compared to their parent atoms.

Atom/cation	Atomic (covalent) radius (Å)	Oxidation state of atom/cation	Electro-negativity
Li	1.23	0	1.0
Li^+	0.60	+1	2.5
Mo	1.30	0	1.8
Mo^{4+}	0.68	+4	2.24
Mo^{6+}	0.62	+6	2.35
Fe	1.17	0	1.8
Fe^{2+}	0.76	+2	1.83
Fe^{3+}	0.64	+3	1.96
Sn	1.41	0	1.80
Sn^{2+}	1.12	+2	1.81
Sn^{4+}	0.71	+4	1.96
Pb	1.47	0	1.80
Pb^{2+}	1.20	+2	1.87
Pb^{4+}	0.84	+4	2.33

Table 4.12 makes it evident that, in general, the electronegativity values of M, M^+ and M^{2+} species are in the order :

$$M^{2+} > M^+ > M,$$

while the size of these species is in the reverse order as shown below :

$$M^{2+} < M^+ < M$$

An anion attracts the electron pair less readily than its parent atom. This is also due to the fact that an anion is larger in size than its parent atom ($X^- > X$). *Thus an anion, X^- has less electronegativity than its parent atom, X, i.e.*

Order of electronegativity : $X^- < X$

Order of size : $X^- > X$

The above point is clear from the fact that F^- ion has less electronegativity value than F atom ($F^- = 0.8$. $F = 4.0$).

4. Number and nature of atoms to which the atom is bonded. We have seen that, since electronegativity of an atom is not the property of this atom in its isolated state, it depends on the number and nature of the atoms to which the atom is bonded. For this reason the electronegativity value of an atom is not constant. For example electronegativity value of P atom in PCl_3 molecule is different from that in PF_5 molecule in which the number and nature of the atoms both to which P atom is bonded change.

5. Ionisation energy and electron affinity. The numerical value of electronegativity of the atom of an element also depends both on the magnitude of ionisation energy and electron affinity. Higher ionisation energy of an atom means that it is difficult to remove the most loosely bonded electron from the atom which leads us to expect that the electron affinity of that atom will also be greater. *Thus the atoms of the elements which have higher values of ionisation energy and electron affinity also have higher values of electronegativity.* For example, the elements of group VII A (halogens) which have the highest ionisation energies and electron affinities also have the highest values of electronegativity. Similarly, the elements of group IA (alkali metals) which have the lowest ionisation energies and electron affinities have the lowest values of electronegativity.

6. Type of hybridisation. The magnitude of electronegativity of an atom also depends on the type of hybridisation which the atom undergoes in the formation of different bonds in the molecule. *The magnitude of electronegativity increases as the s-character in hybrid orbitals increases.*

Example : The electronegativity of carbon atom in methane (sp^3 hybridisation with 25% *s*-character), ethylene (sp^2 hybridisation with 33% *s*-character) and acetylene (*sp* hybridisation with 50% *s*-character) is in the increasing order as shown below :

Hydrocarbon	:	*Methane* (CH_4)	*Ethylene* ($H_2C = CH_2$)	*Acetylene* ($CH \equiv CH$)
Type of hybridisation	:	sp^3	sp^2	sp
Amount of s-character in hybrid orbitals	:	25%	33%	50%
Order of electronegativity of carbon atom	:	———	*Increasing*	———→

The increasing order of the magnitude of the electronegativity of carbon atom in the above hydrocarbons shows that methane is unreactive, ethylene is *more reactive* while acetylene is *highly reactive*. The highest value of electronegativity of carbon atom in acetylene also accounts for its highly acidic hydrogen atom (*i.e.* acidic properties of acetylene).

Variation of Electronegativity in a Period and a Group of Representative Elements : *Periodic Trends of Electronegativity*

(*a*) **In a period.** *In going from left to right in a period of s- and p-block elements, the electronegativity values increase.* This increase can be explained on the basis of any of the following facts.

(*i*) On moving from left to right in a period, there is a decrease in the size of the atoms. Smaller atoms have greater tendency to attract the electrons towards themselves *i.e. smaller atoms have higher electronegativity values.*

(*ii*) On moving from left to right in a period there is an increase of ionisation energy and electron affinity of the elements. The atoms of the elements which have higher value of ionisation energies and electron affinities also have higher electronegativities.

(*b*) **In a group.** *In going down a group of s- and p-block elements, the electronegativity values decrease.* This decrease can also be explained on the basis of any of the following facts.

(*i*) As we move down a group, there is an increase in the size of the atoms. With the increase in size of the atoms, their electronegativity values decrease.

(*ii*) Ionisation energy and electron affinity on which electronegativity depends decrease as the group is descended. With the decrease of these quantities the electronegativity values also decrease.

The heavier elements of group III A (*i.e.* Ga, In and Tl) show *reverse trend* due to the intervening transition series.

The variation of electronegativity values discussed above reveals that the *halogens* (VII A group elements) which lie on the extreme right of the periodic table are the *most electronegative* (*i.e. least electropositive*) *elements and the alkali metals* (IA group elements) which lie on the extreme left of the periodic table are the *least electronegative* (*i.e. most electropositive*) *elements*. Thus we see that the *most electronegative element is flourine* which occurs at the top right hand corner and the *least electronegative* element is *cesium* which occurs at the bottom left hand corner of the periodic table. Being the most electronegative, F does not show any basic character, *i.e.*, it has no tendency to form positive ions in any of its known compounds. On the other hand, there is, however, evidence to show that Cl, Br and I have a tendency to form positive ions.

The variation of electronegativity in a period and a group of (*s*- and *p*-block elements) is shown in Fig. 4.22.

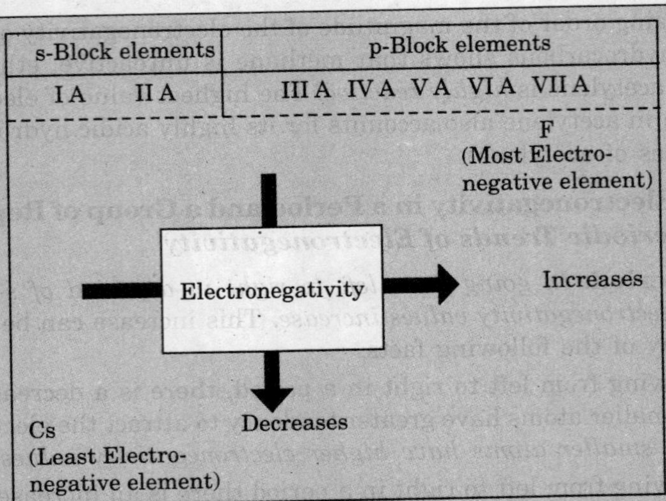

Fig. 4.22 Periodic variation of electronegativity in *s*- and *p*-block elements.

Applications of Electronegativity

1. Nature of bond. The concept of electronegativity can be used to predict the nature of bond formed between two similar or dissimilar atoms, i.e. the concept of electronegativity can predict whether the bond between two similar or dissimilar atoms is *non-polar covalent bond, polar covalent bond* or *ionic bond* as discussed below :

(*i*) **Non-polar covalent bond.** The bonding electron pair constituting the covalent bond between two similar atoms which have the same electronegativity or zero electronegativity difference *is equally shared by both the linked atoms i.e.* the shared electron pair is placed at the centre of the distance between the nuclei of the linked atoms. The equal sharing of the electron pair is because of the fact that both the linked atoms, due to their same electronegativity or zero electronegativity difference, have the same tendency to attract the shared electron pair towards them. It is because of the equal sharing of the electron pair that the resulted covalent bond has no polarity or ionic character, *i.e.* the bond is *non-polar* or *non-ionised*. This bond is therefore, called **non-polar covalent bond** or simply **covalent bond.**

(*ii*) **Polar covalent bond.** The bonding electron pair constituting the covalent bond between two dissimilar atoms like H and Cl which have different electronegativity values (H = 2.1, Cl = 3.0) *is not equally shared by the linked atoms i.e.*, the electron pair is not placed at the centre of the distance between the nuclei of the two atoms *viz.* H and Cl. Rather it is partially displaced (*i.e.* attracted) towards the more electronegative Cl atom owing to its greater affinity for electron pair. This type of partial displacement of the shared electron pair towards the more electronegative Cl atom develops a small (fractional) negative charge (represented as $\delta-$) on more electronegative Cl atom and an equal amount of small positive charge (represented as δ+) on less electronegative H atom and HCl molecule appears to be containing two oppositely charged poles (called *charge-centres* or *electrical poles*) namely $H^{\delta+}$ and $Cl^{\delta-}$ at the end of the bond and the molecule is depicted as $H^{\delta+} - Cl^{\delta-}$. The covalent bond between H and Cl atoms develops some polarity or partial ionic character and hence is called **polar covalent bond**.

(*iii*) **Ionic or polar bond.** *When the two linked atoms have very different electronegativity or have large electronegativity difference, the bond established between the atoms is called* **ionic bond.** Consequently this bond is established between the two atoms one of which is electropositive (*e.g.* alkali metals or alkaline earth metals) and the other is electronegative (*e.g.* halogens). In the formation of this bond between, for instance Li and F which have very different electronegatives (Li = 1.1, F = 4.01), the electron pair being shared by the two atoms is completely displaced or transferred towards the more electronegative F atom and consequently less electronegative Li atom gains positive (+) charge while the more electronegative F atom aquires equal negative (–) charge and the bond is represented as $Li^+ F^-$. Ionic bond is totally free from covalent character. Thus it is a purely ionic bond.

Whatever has been said above in (*i*), (*ii*) and (*iii*) can be summarised as follows (x_A and x_B are the electronegativity values of atoms A and B in AB molecule) :

(*a*) When $x_A = x_B$, *i.e.* $x_A - x_B = 0$, then A–B bond is non-polar covalent bond or simply *covalent bond* and is represented as A–B. For example H–H bond in H_2 molecule is a covalent bond and is represented as H–H bond.

(*b*) When $x_A > x_B$, *i.e.* $x_A - x_B$ is *small*, then A–B bond is *polar covalent bond* and is represented as $A^{\delta-} + B^{\delta+}$, since $x_A > x_B$. For example both O–H covalent bonds in H_2O molecule are polar covalent bonds and are represented as $O^{\delta-}-H^{\delta+}$, since $x_O > x_H$ and $x_O - x_H$ is small.

(*c*) When $x_A \gg x_B$, *i.e.*, $x_A - x_B$ is *very large*, A–B bond is an *ionic or polar bond* and is represented as $A^- - B^+$, since $x_A \gg x_B$. For example Na–Cl bond in NaCl molecule is an ionic bond and is represented as Na^+Cl^- (Here Cl = A and Na = B).

2. Percentage of ionic character in a polar covalent bond. The concept of electronegativity can also be used to calculate the percentage of ionic character in a polar covalent bond. If two atoms A and B are linked by a polar covalent bond ($A^{\delta-} - B^{\delta+}$ or simply A–B), the amount of ionic character in this bond depends on the difference of electronegativity values of A and B. *Greater is the difference* $(x_A - x_B)$, *greater is the percentage of ionic character in A–B bond.* Here the electronegativity of atom A (x_A) has been assumed to be higher than that of B (x_B). Pauling has estimated the percentage of ionic character in various $A^{\delta-} - B^{\delta+}$ polar covalent bonds from known $(x_A - x_B)$ values and has derived the following conclusions :

(*i*) When $(x_A - x_B) = 1.9$, the amount of ionic character in $A^{\delta-} - B^{\delta+}$ bond is 50% and that of covalent character is also 50%. Thus A–B bond is 50% ionic and 50% covalent.

(*ii*) When $(x_A - x_B) < 1.9$, the amount of ionic character in $A^{\delta-} - B^{\delta+}$ bond is less than 50% and that of covalent character is more than 50%. Thus $A^{\delta-} - B^{\delta+}$ bond is *predominantly* covalent and hence is represented as A–B.

(*iii*) When $(x_A - x_B) > 1.9$, the amount of ionic character in $A^{\delta-} - B^{\delta+}$ bond is more than 50% and that of covalent character is less than 50%. Hence $A^{\delta-} - B^{\delta+}$ bond is predominantly ionic and hence is represented as A^-B^+.

Above discussion shows that if $(x_A - x_B)$ increases from 1.9, percent of ionic character in A – B bond also increases as shown below :

$(x_A - x_B)$ value :	0.6	0.8	1.2	1.9	2.2	2.8	3.2
	Increases →						
Percent ionic character in A – B bond	10	15	30	50	70	86	92
	Increases →						

3. Acidic and basic character of XOH molecule in aqueous solution. With the help of the concept of electronegativity we can predict whether a given compound of XOH type (X = Metal or non-metal) will behave as an acid or a base in aqueous solution. This prediction can be made by considering the ionisation of XOH molecule in aqueous solution as follows :

(*a*) If $x_O - x_H > x_O - x_X$, O–H bond will be more polar than O–X bond and hence the ionisation of XOH molecule in aqueous solution will take place at O–H bond and H_3O^+ ions will be produced as shown below :

$$\text{X–O–H} \longrightarrow \text{XO}^- + \text{H}^+$$
$$\text{H}^+ + \text{H}_2\text{O} \longrightarrow \text{H}_3\text{O}^+$$
$$\overline{\text{X–O–H} + \text{H}_2\text{O} \longrightarrow \text{XO}^- + \text{H}_3\text{O}^+}$$

The production of H_3O^+ ions confirms that XOH molecule will behave as an acid in its aqueous solution.

(*b*) If $x_O - x_H < x_O - x_X$, X–O bond will be more polar than O–H bond and hence the ionisation of XOH molcule in aqueous solution will take place at X–O bond and OH^- ions will be produced as shown below :

$$\text{X–O–H} \longrightarrow \text{X}^+ + \text{OH}^-$$
$$\text{X}^+ + \text{H}_2\text{O} \longrightarrow [\text{X} \leftarrow \text{OH}_2]^+$$
$$\overline{\text{X–O–H} + \text{H}_2\text{O} \longrightarrow [\text{X} \leftarrow \text{OH}_2]^+ + \text{OH}^-}$$

The production of OH^- ions confirms the fact that XOH molecule will behave as a base in its aqueous solution.

Example. *Predict which of the following hydroxy compounds is acidic and which is basic in aqueous solution*? CsOH, IOH

Solution. The electronegativity values of different elements present in CsOH and IOH compounds and those of $x_O - x_H$ and $x_O - x_X$ (X = Cs or I) as given below :

$$\underset{0.7}{\text{Cs}}\overset{2.8}{—}\underset{3.5}{\text{O}}\overset{1.4}{—}\underset{2.1}{\text{H}} \quad ; \quad \underset{2.5}{\text{I}}\overset{1.0}{—}\underset{3.5}{\text{O}}\overset{1.4}{—}\underset{2.1}{\text{H}}$$

show that CsOH will give OH^- ions while IOH will give H^+ ions (or $H + H_2O \rightarrow H_3O^+$). Consequently CsOH behaves as a base while IOH behaves as an acid.

The above description clearly shows that the *hydroxy compounds of metals like alkali metals and alkaline earth metals behave as bases in aqueous solution while those of non-metals like halogens behave as acids.*

4. Cause of diagonal relationship. It is one of the important properties of the earlier elements of 2*nd* period of the long form of periodic table (*e.g.* Li, Be and B) that they resemble more with the elements lying at their right hand side in the 3*rd* period (*i.e.* diagonally opposite elements) than with the other members of their own sub-groups. Thus Li resembles in its properties more with Mg which lies to its lower right in the 3*rd* period than with other members of its own sub-group IA. Similarly Be resembles Al, and B resembles Si. This type of similarity in properties in the two diagonally opposite elements is called diagonal relationship. Why the two elements show diagonal relationship has been explained on the basis of the concept of electronegativity as below

We know that the electronegativity of elements increases from left to right in a period and decreases on descending a group. Thus, Be is more electronegative than Li (Be>Li) and Mg is slightly less electronegative than Be (Be>Mg). *Li and Mg, being slightly less electronegative than Be, can be regarded as having almost the same value of electronegativity* and hence have many similar properties, *i.e.* they show diagonal relationship with each other.

4. Bond angle. The concept of electronegativity has been found to be very useful to explain the variation of bond angle. Here we shall consider the following two cases :

(*i*) First of all we shall see how the B–A–B bond angle in the molecule of AB_x type (A is the central atom, B are the atoms attached with atom A and x is the number of B atoms) is changed with the change of eleotronegativity of the central atom. It has been observed that *with the decrease of the electronegativity of the central atom, the B–A –B bond angle also decreases.* For example the bond angle H–A–H (A = N, P, As, Sb) in NH_3, PH_3, AsH_3 and SbH_3 hydrides (hydrides of VA group) decreases from ∠H–N–H to ∠H–Sb–H as the electronegativity of the central atom (*i.e.* N, P, As or Sb) decreases from N to Sb.

Hydrides, AH_3 (A = N, P, As, Sb)	***Electronegativity of the central atom, A***	***H–A–H bond angle***
NH_3	3.0	107.5°
PH_3	2.1	93.3°
AsH_3	2.0	91.8°
SbH_3	1.9	91.3°

With the decrease of electronegativity from N (3.0) to Sb (1.9), the A–H bonding electron pairs in AH_3 molecules are drawn farther away from the central atom (A) but nearer to H-atom. The effect of this shifting of the bonding electron pairs towards H-atom is that the repulsion between the bonding electron pair (*bp*) and lone pair (*lp*) on the central atom [(*bp–lp*) repulsion] *increases*. The increase in (*bp–lp*) repulsion results in that bonding electron pairs come still close to each other and hence the bond angle progressively decreases from NH_3 to SbH_3.

On similar grounds the decrease in the bond angle in the hydrides H_2O, H_2S, H_2Se and H_2Te can also be explained.

Hydrides, AH_2 (A = O, S, Se, Te)	***Electronegativity of the central atom, A***	***H–A–H bond angle***
H_2O	3.5	104.5°
H_2S	2.5	92.3°
H_2Se	2.4	91°
H_2Te	2.1	90°

(*ii*) Now let us see how the B–A–B bond angle in AB_x type molecules is changed with the change of the electronegativity of the atom B. It has been observed that *B–A–B bond angle increases with the decrease of the electronegativity of atom B*. This is evident from the following examples :

AB_x molecules	**Electronegativity of B atom**	**B–A–B bond angle**
PCl_3	3.0	100°
PBr_3	2.8	101.5°
PI_3	2.5	102
$AsCl_3$	3.0	98.4°
$AsBr_3$	2.8	100.5°
AsI_3	2.5	101.5°
NF_3	4.0	102°
NH_3	2.1	107°
OF_2	4.0	103°
OH_2	2.1	104.5°

The increase in bond angle with the decrease of electronegativity of atom B is due to the fact that with the decrease of the electronegativity of atom B, the bonding electron pair forming A–B bond is drawn farther away from B atom but nearer to the central atom, A. The net effect of this type of shifting of A–B bonding electron pair towards the central atom A is that the repulsion between the bonding electron pair and the lone pair on the central atom decreases. The decrease in (*bp–lp*) repulsion results in that the bonding electron pairs get farther away from each other, *i.e.* the B–A–B bond angle increases.

6. Mechanism of hydrolysis. Consider the hydrolysis of Cl_2O and BCl_3 molecules. In Cl_2O molecule O atom has a small negative charge (–δ) on it and each of the two Cl atoms has a small positive charge (+δ), because O atom has higher electronegativity than Cl. When H_2O attacks Cl_2O, the positively charged H atom ($H^{\delta+}$) of H_2O attacks at $O^{\delta-}$ atom of Cl_2O and the product of hydrolysis is hypochlorous acid (HOCl)

In BCl_3, B atom has a small positive charge on it while each of the three Cl atoms bears a negative charge because B atom has less electronegativity than Cl atom. When H_2O attacks BCl_3, the negatively charged O atom ($O^{\delta-}$) of H_2O attacks at $B^{\delta+}$ atom of BCl_3 and the products of hydrolysis are boric acid, $B(OH)_3$ or H_3BO_3 and HCl.

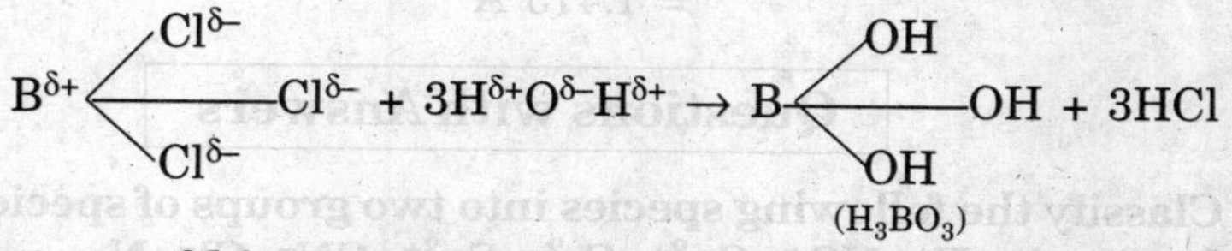

7. Nomenclature of binary compounds. Generally binary compounds are regarded as derivatives of more electronegative elements. For example :

(*i*) The binary compound of iodine and chlorine is called iodine chloride and not chlorine iodide and hence is represented as ICl and not ClI, since electronegativity of Cl is higher than that of I (Cl = 3.2, I = 2.7).

(*ii*) The binary compound of oxygen and fluorine is called oxygen difluoride rather than fluorine monoxide and hence is represented as OF_2, and not as F_2O, since electronegativity of F is higher than that of O (F = 4.0, O = 3.5). Note that Cl_2O is called chlorine monoxide and not oxygen dichloride since O has higher electronegativity than Cl (O = 3.5, Cl = 3.0).

Exceptions. Names of some binary compounds have been retained for historical reasons. For example ammonia is written as NH_3 and not as H_3N, although the electronegativity of first element namely N is higher than that of H (N = 3.0, H = 2.1).

8. Calculation of bond length. Generally in case of heterodiatomic molecules of AB type, bond length, $d(A - B)$ is equal to the sum of covalent radii of A and B atoms, *i.e.*

$$\boldsymbol{d\ (A–B) = r\ (A) + r\ (B)} \quad ...(i)$$

Although equation (*i*) holds good in most of the heterodiatomic molecules, yet the experimental value of d (A–B) in case of many AB type molecules is slightly less than the sum of r (A) and r (B). The decrease in the bond length is due to the ionic character in A–B bond. The increase in the amount of ionic character in A–B bond decreases the bond length, d (A–B).

Taking the electronegativity difference, $(x_A–x_B)$ into consideration Schomakar and Stevenson (1941) proposed the following empirical equation for the calculation of bond length.

$$\boldsymbol{d\ (A–B) = r\ (A) + r\ (B) – 0.09\ (x_A–x_B)} \quad ...(ii)$$

Here x_A and x_B are the electronegativities A and B elements.

The validity of equation (*ii*) can be shown by calculating the N–C distance in $(CH_3)_3N$ molecule by equations (*i*) and (*ii*) as shown below :

From equation (*i*), d (N–C) $= r$ (N) $+ r$ (C)

$$= (0.75 + 0.77)\ \text{Å}$$
$$= 1.52\ \text{Å}$$

The calculated value (= 1.52 Å) obtained as above is not in agreement with the experimental value (= 1.47 Å) seen in $(CH_3)_3N$ molecule. But the value calculated from equation (*ii*) has been found in good agreement with the experimental value as shown below :

From equation (*ii*), d (N–C) = r (N) + r (C) – 0.09 (x_N–x_C)

= 0.75 + 0.77 – 0.09 (3.0 – 2.5)

= 1.475 Å

Questions with Answers

Q.1 Classify the following species into two groups of species which should be isoelectronic. K^+, NO^+, Ca^{2+}, C_2^{2-}, Sc^{3+}, CN^-, Cl^-, N_2.

Ans. Find out the number of electrons in the given species. K^+ = 19 – 1 = 18, NO^+ = 7 + 8 – 1 = 14, Ca^{2+} = 20 – 2 = 18, C_2^{2-} = 6 × 2 + 2 = 14, Sc^{3+} = 21 – 3 = 18, CN^- = 6 + 7 + 1 = 14, Cl^- = 17 + 1 = 18, N_2 = 2 × 7 = 14. The given species can be classified into the following two groups :

(*i*) K^+, Ca^{2+}, Sc^{3+}, Cl^- (*All contain 18 electrons*)

(*ii*) NO^+, C_2^{2-}, CN^-, N_2 (*All contain 14 electrons*)

Q.2 Arrange the following atoms/ions in the increasing order their size.

(*i*) F^-, Al, Na^+, Mg^{2+}

(*ii*) Ca^{2+}, Ar, K^+, Cl^-, S^{2-}

Ans. (*i*) Since F^-, Na^+ and Mg^{2+} ions are isoelectronic (F^- = 9 + 1 = 10, Na^+ = 11 – 1 = 10, Mg^{2+} = 12 – 2 = 10), their size increases with the decrease of their atomic number. Thus :

$$Mg^{2+} (Z = 12) < Na^+ (Z = 11) < F^- (Z = 9)$$

Now since F^- < Al, the overall increasing order is :

$$\mathbf{Mg^{2+} < Na^+ < F^- < Al \text{ (Ans.)}}$$

(*ii*) Since all the given species are isoelectronic (Ca^{2+} = 20 – 2 = 18, Ar = 18, K^+ = 19 – 1 = 18, Cl^- = 17 + 1 = 18, S^{2-} = 16 + 2 = 18), their size increases with the decrease of their atomic number. Thus :

$$\left.\begin{array}{lccccc} & \mathbf{Ca^{2+}} & \mathbf{< K^+} & \mathbf{< Ar} & \mathbf{< Cl^-} & \mathbf{< S^{2-}} \\ \mathbf{Z =} & \mathbf{20} & \mathbf{19} & \mathbf{18} & \mathbf{17} & \mathbf{16} \end{array}\right\} \mathbf{(Ans.)}$$

Q.3 Arrange the elements with E.Cs [He] $2s^1$, [Ne] $3s^1$ and [Ar] $4s^1$ in the decreasing order of their IE_1 values.

Ans. E.Cs show that the given elements belong to group 1. Since IE_1 values decrease on moving down the group, IE_1 values of the given elements decrease as :

[He] $2s^1$ > [Ne] $3s^1$ > [Ar] $4s^1$ (Ans.)

Q.4 Find out the number of valence electrons in an atom whose successive IE values (in $KJmol^{-1}$) are as : IE_1 = 410, IE_2 = 820, IE_3 = 1100, IE_4 = 1500 and IE_5 = 3200.

Ans. IE values show that since the magnitude of energy required for removing 5*th* electron from the atom is more than the double the magnitude of energy needed for removing 4*th* electron, the number of valence electrons in the atom is 4.

Q.5 Arrange F, Cl, O and S in the increasing order of their EA values.

Ans. EA values of the given elements are in the following increasing order :

O < S < F < Cl

Explanation. On the basis of the position of the given elements in the long form of periodic table the variation of their EA values can be shown as :

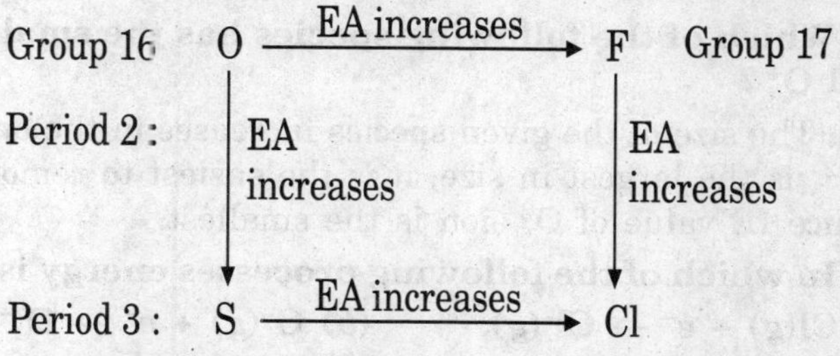

Hence the increasing order of EA values is O < S < F < Cl.

Q.6 The electronic configurations for four elements A, B, C and D are as : A = $1s^2, 2s^22p^6, 3s^1$; B = $1s^2, 2s^22p^6, 3s^2$; C = $1s^2, 2s^22p^1$; D = $1s^2, 2s^22p^6, 3s^23p^1$. Predict which of these elements has maximum value of ($IE_2 - IE_1$).

Ans. IE_1 and IE_2 values for the given elements correspond to the following ionisation processes.

$$\underset{(A)}{1s^2, 2s^22p^6, 3s^1} \xrightarrow[IE_1]{-3s^1(n=3)} 1s^2, 2s^22p^6 \xrightarrow[IE_2]{-2p^1(n=2)} 1s^2, 2s^22p^5$$

$$\underset{(B)}{1s^2, 2s^22p^6, 3s^2} \xrightarrow[IE_1]{-3s^1(n=3)} 1s^2, 2s^22p^6, 3s^1 \xrightarrow[IE_2]{-3s^1(n=3)} 1s^2, 2s^22p^6$$

$$\underset{(C)}{1s^2, 2s^22p^1} \xrightarrow[IE_1]{-2p^1(n=2)} 1s^2, 2s^2 \xrightarrow[IE_2]{-2s^1(n=2)} 1s^2, 2s^1$$

$$\underset{(D)}{1s^2, 2s^22p^6, 3s^23p^1} \xrightarrow[IE_1]{-3p^1(n=3)} 1s^2, 2s^22p^6, 3s^2$$

$$\xrightarrow[IE_2]{-3s^1(n=3)} 1s^2, 2s^22p^6, 3s^1$$

In case of A 1*st* electron is removed from 3*s* orbital (*n* = 3) and 2*nd* electron is removed from 2*p* orbital (*n* = 2). Since 3*s* orbital belongs to outer orbit (*n* = 3) and 2*p* orbital belongs to inner orbit (*n* = 2), ($IE_2 - IE_1$) for this element is maximum. In case of other elements the orbitals from which electron(s) is (are) removed belong to the same shell.

Q.7 Arrange C, N, O and F in the decreasing order of their IE_2 values.

Ans. The decreasing order of IE_2 values is : O > F > N > C.

Explanatin. IE_2 value of an element A is the energy required to remove an electron from A^+ cation. V.S.E.Cs of $\overset{+}{C}$, $\overset{+}{N}$, $\overset{+}{O}$ and $\overset{+}{F}$ cations are as : $\overset{+}{C} = 2s^22p^1$,

$\overset{+}{N} = 2s^2 2p^2$, $\overset{+}{O} = 2s^2 2p^3$ and $\overset{+}{F} = 2s^2 2p^4$. Since $\overset{+}{O}$ cation has half-filled $2p$ orbitals, $2s^2 2p^3$ configuration is stable. Hence IE_2 of O-atom would be higher than that of F(O > F). Now since C-atom is larger in size than N-atom, IE_2 of C would be lower than that of N(N > C). Thus the overall decreasing order is : O > F > N > C.

Q.8 Which of the following species has the smallest value of IE. O, O^-, O^{2-} and O^+?

Ans. The size of the given species increases in the order : $O^+ < O < O^- < O^{2-}$. Since O^{2-} is the largest in size, it is the easiest to remove an electron from this ion. Hence IE value of O^{2-} ion is the smallest.

Q.9 In which of the following processes energy is absorbed ?

(a) $Cl(g) + e^- \rightarrow Cl^-(g)$, **(b)** $O^-(g) + e^- \rightarrow O^{2-}(g)$

(c) $O(g) + e^- \rightarrow O^-(g)$, **(d)** $S(g) + e^- \rightarrow S^-(g)$

Ans. When an electron is added to a neutral gaseous atom (A), energy is released. Thus in the processes (*a*), (*c*) and (*d*), energy is released. On the other hand, when an electron is added to a given A^- ion, energy is absorbed (required). Thus in process (*b*) energy will be absorbed. In process (*b*) energy is absorbed to overcome the repulsions between the incoming electron and negative charge on $O^-(g)$ ion.

University Questions

1. (*a*) Give Slater's rules for calculating shielding constant or Z_{eff}.
   (*b*) Give Pauling's method to determine the electronegativity of an atom.
   (*c*) On the basis of Slater's rule, calculate the value of Z_{eff} for (*i*) a 4*s* electron (*ii*) a 3*d* electron. in chromium (*Himachal Pradesh 2000*)
2. Explain why metallic character increases as we move down the group. (*Himachal Pradesh, 2000*)
3. (*a*) Define ionisation potential of an atom. Which factors influence the ionisation potential ? What is effective nuclear charge?
   (*b*) I.P. of H is 13.6 eV. Calculate the energy required to produce H hoot 0.5 g of hydrogen. (*Kanpur, 2000*)
4. Write notes on : (*i*) Ionisation potential (*ii*) Electron affinity (*iii*) Electronegativity. (*Kumaon, 2000*)
5. Explain the following :
   (*i*) Electron affinities of Be, Mg and noble gases are zero while those nitrogen (= 0.20 eV) and phosphorus (= 0.80 eV) are very low.
   (*ii*) The formation of F^- (*g*) ion from F(*g*) atom is exothermic whereas that of O^2 (*g*) ion from O(*g*) is endothermic.
   (*iii*) In a period ionisation energy increases from left to right.
   (*iv*) In a group electronegativity decreases from top to bottom.
   (*v*) Group IA alkali metals form ionic compounds with group VII A halogens. [*Gauhati, 2000* (*General*)]

6. What do you understand by the term, "Periodic Properties"? Distinguish between electron affinity and electronegativity. Flourine is more electronegative than chlorine, although its electron affinity is less than that of chlorine. Explain. *(Lucknow, 2000)*

7. (*i*) Arrange the following in the order of their increasing radii : O, F, Ne.

(*ii*) First ionisation potential of nitrogen is more than that of oxygen but 2*nd* ionisation potential of oxygen is more than that of nitrogen. Explain.

(*iii*) Which of the following has the highest electron affinity? F, Cl, Br

(*iv*) The electronic configuration of some elements are given as : (*a*) $1s^2$, $2s^2$ $2p^6$, $3s^2$ (*b*) $1s^2$, $2s^2$ $2p^6$ (*c*) $1s^2$, $2s^2$ $2p^4$ (*d*) $1s^2$, $2s^2$, $2p^6$, $3s^2$ $3p^6$. Which of these elements has the lowest ionisation potential?

(*v*) Arrange the bond angle in the decreasing order of the following hydrides: NH_3, AsH_3 and PH_3.

(*vi*) Arrange the following in the order of increasing covalent character : C – Cl, B –Cl, Be – Cl, Be – F, Cl – Cl.

(*vii*) What is meant by electronegativity ? How does it depend on I.P. and E.A.?

(*viii*) How is electronegativity related with the polarity of a covalent bond? *(Avadh 2000)*

8. (*a*) What is ionization potential ? Discuss the factors on which ionization potential depends. Arrange the following according to decreasing ionization potential: Mg, Ra, Ba, Ca, Be.

(*b*) Atomic radii of the element in a series in the periodic table decrease with increasing atomic number upto VIIth group, but increase in inert gases. Explain giving reasons. *(Avadh 2000)*

9. Explain ionisation potential and electronegativity. *(Lucknow 2001)*

10. What is ionisation potential of an element. Explain why IE of N is higher than that of O ? *(Nagpur 2002)*

11. Write a note on Electron Affinity. *(Lucknow 2002)*

12. Explain why alkali metal cations are generally smaller in size than their corresponding atoms ? *(Delhi 2003)*

13. (*a*) Define electronegativity.

(*b*) Give reasons for the following :

(*i*) The second ionisation energy of sodium is very high as compared to its first ionisation energy.

(*ii*) First ionisation energy of nitrogen is higher than tht of oxygen.

(*c*) Define the followng and explain their trends across the period and down the group in periodic table : (*i*) Atomic radius and (*ii*) Electron affinity. *(Nagpur 2003)*

14. (*a*) Differentiate between covalent reactions and Van der Waal's radius.

(*b*) What are iso-electronic ions ? Arrange the given ions in order of their decreasing size : F^-, O^{2-}, Mg^{2+} and Na^+.

(c) What is screening effect ? How does it govern the ionisation energy of an atom ? (*GND Amritsar 2004*)

15. Explain why EA values of N and P atoms are low. (*Meerut 2005, 2008*)

16. (a) Which of the following ions has the smallest size ? O^{2-}, F^-, Na^+ and Mg^{2+}.

(b) Define EA and state how if changes in a period. (*H.N. Bahuguna 2005*)

17. Explain the following :

(i) $(IP)_3$ of Mg is very high.

(ii) EA values of N and P atoms are low. (*Meerut 2006*)

18. (a) Arrange N, O, Cl and F in the increasing order of their EA values.

(b) In a covalent molecule AB, EN values of A and B are 2.1 and 3.5 respectively. Calculate % ionic character in A – B bond. (*H.N. Bahuguna 2006*)

19. Explain why Cl-atom is smaller in size than Cl^- ion. (*Meerut 2007*)

20. (a) First IP of N is higher than that of O. Explain.

(b) What is the trend in size of N^{3-}, O^{2-} and F^- ?

(c) EA of Cl is higher than that of F. Explain. (*Purvanchal 2007*)

21. Define and discuss electronegativity in detail along with the factors affecting it ? (*Meerut 2008*)

22. (a) Which of the following elements has the highest IP value ? Li, Na, K and Rb.

(b) Which of the following elements is the most electronegative ? F, Cl, Br and I. (*Agra 2008*)

CHAPTER 5

Radioactivity, Isotopes, Isobars and Isotones

Radioactivity

What is Radioactivity? (Discovery of Radioactivity)

Shortly after the discovery of X-rays, French physicist, *Henri Bacquerel*, in 1895, while trying to establish a possible relationship between the fluorescence of the glass walls of an X-ray tube and the phosphorescence of certain fluorescent salts caused by sunlight, left *accidently* a little pile of a uranium mineral on a photographic plate wrapped in a thick black paper lying in the dark room. On developing the plate he found a dark spot on it right below the position of the pile of the mineral. The black spot was a silhourette of the pile of the mineral. Bacquerel concluded that uranium emitted certain rays which penetrated paper wrapper and affected the photographic plate. These rays are invisible to naked eye and are given off by metallic uranium and also by uranium salts. These rays pass through thin sheets of glass or metal, produce ionisation in the gases through which they pass and rapidly discharge the charged leaves of a well-insulated electroscope. Bacquered called these rays as *radioactive rays* (*or radioactive emanations or radioactive radiations*) and the property or phenomenon of giving out these rays is called *radioactivity*. The substances which emit these radioactive rays are called *radioactive substances*. The term radioactivity was first used by *Marie Curie*.

Radioactivity is essentially a nuclear phenomenon and is a drastic process because the element changes in kind. It is spontaneous and an irreversible self-disintegrating activity because the element breaks itself up for good.

Discovery of Polonium and Radium

A year later, in 1896, *Marie Curie* undertook a systemetic study of the substances emitting radioactive rays for his doctorate thesis. During the study she found that pitchblende, a mineral of uranium, was far more radioactive than it could be due to its uranium content. She suggested that the greater activity shown by the mineral might be due to the presence of some new element in it. She and her husband, *Pierre Curie* worked together day and night, hardly stopping to eat or sleep, for the separation of this new hypothetical element. They took a ton of pitchblende and separated it into its constituents and found the radioactivity of each constituent. From the fraction containing bismuth they isolated a new radioactive element which they named **polonium** in honour of Marie Curies' native land of Poland. After a long and tedious process of extraction and fractional crystallisation of barium fraction, wife-husband team obtained, in 1898, a few milligrams of another radioactive element which they named **radium** after the Latin word for ray. This new element was found to be three million times more radioactive than uranium. As the news of the discovery of the new radioactive elements spread, scientists in different parts of the world discovered many other radioactive elements, *e.g.*, in 1900 *Debierne* reported the discovery of another radioactive element which he called **actinium.**

Types of Radioactive Rays

Rutherford and his co-workers, in 1904, passed the radiations emitted by radioactive elements through a strong electric field between parallel plates and thus separated them into *three* quite different kinds which are : (*See* Fig. 5.1).

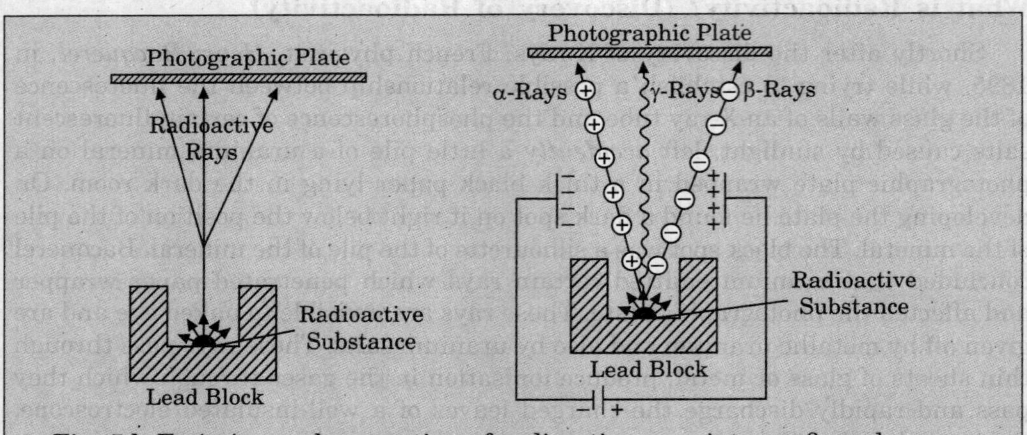

Fig. 5.l. Emission and separation of radioactive rays into α-, β- and γ-rays.

***(i)* Alpha (α) rays.** The rays which are deflected towards the negative plate are *positively charged* and are called *alpha* (α) *rays or alpha particles*. These rays consist merely of *helium nuclei*.

***(ii)* Beta (β) rays.** The rays which are deflected towards the positive plate are *negatively charged* and are called *beta* (β) *rays* or *beta particles*. These rays merely consist of electrons.

***(iii)* Gamma (γ) rays.** These rays which are *not* deflected at all are neutral and are called *gamma* (γ) rays.

General Properties of Radioactive Radiations

(*i*) These radiations are highly penetrating, they affect photographic plates, ionize gases, cause scientillations on fluorescent screen, develop heat and produce chemical changes.

(*ii*) As radiations are given out, new elements are formed in an irreversible process—the new elements themselves being usually radioactive.

(*iii*) The emission of radiations is spontaneous and is not affected by external agents.

(*iv*) The emission is not instantaneous but is prolonged i.e., it is extended over a period of time, otherwise it would not have been discovered at all.

(*v*) Except for radioactivity, there is nothing abnormal about the radioactive elements as regards their physical and chemical properties.

(*vi*) *Effect of electric and magnetic fields—Electrical nature.* (*a*) *α-rays*. If a narrow beam of radioactive radiations proceeding from a deep cavity in a block of lead is subjected to an electric field between parallel plates, the α-particles are deflected towards the negative plate. This shows that α-rays consist of *positively charged particles* which have two units of positive charge and a mass four times the mass of a hydrogen atom. In other words, α-particles are *double charged helium ions*, each consisting of two neutrons and two protons but having no electrons.

If a magnetic field is applied perpendicular to the plane of the paper and directed from top to bottom, the α-rays are deflected towards the left, as shown in Fig. 5.2. Application of Fleming's left hand rule indicates that α-rays consist of *positively charged particles*. The value of *e/m* has been determined on the same lines as for positive rays and this determination indicates that α-particles are *nuclei of helium atoms*.

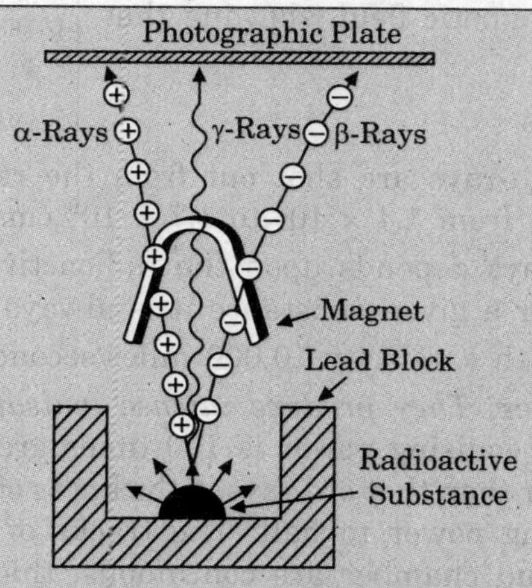

Fig. 5.2. Deflection of α, β and γ-rays in a magnetic field.

It was conclusively proved for the first time by Rutherford and Royds in England in 1909 that α-particles are nothing else *but doubly ionized helium nuclei*. A special discharge tube made of glass (Fig. 5.3) was used for this purpose. A thin-walled glass tube A containing radon gas was sealed inside a thick-walled tube B containing mercury. At the top was a glass capillary tube C with two metal electrodes sealed through glass as shown. Tubes B and C were evacuated through the side tube D after which mercury was raised to the level shown in the figure.

After some time, a few α-particles from radon gas pass through thin walls of A and enter B where they acquire electrons and become neutral atoms of helium. After standing for a few days, mercury level is raised up to L, thereby forcing the helium gas into the discharge tube C. A high-voltage is applied to the electrodes of C and the light from the discharge is observed with a spectroscope. The line spectrum so obtained is found to be identical with that obtained by passing a discharge through another tube containing regular helium gas.

(*b*) β-*rays*. β-rays are also deflected by electric and magnetic fields. Their direction of deflection has shown that they are *negatively charged particles*. The *e/m* value has been found on the same lines as for cathode rays and this determination shows that β-rays are identical with cathode rays, *i.e.*, these are fast-moving *electrons*.

(*c*) γ-*rays*. γ-rays are similar to X-rays and are short-wave electromagnetic radiations similar to light. They have extremely small wavelengths of the order of 10^{-8} to 10^{-11} cm. They have no mass, hence they cannot be thought to be material particles. They *cannot* be deflected even in the strongest electric or magnetic field, showing that they *carry no charge*.

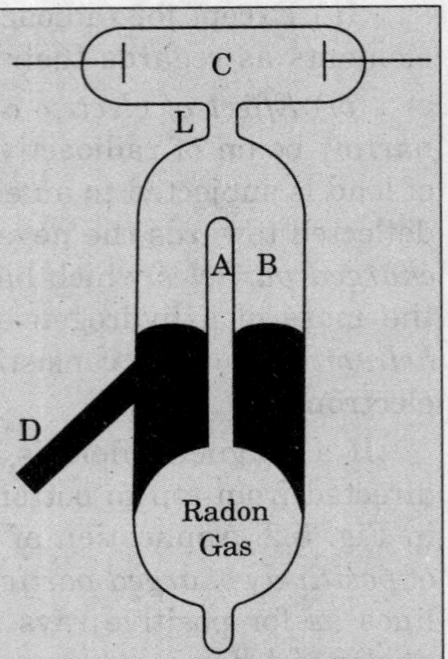

Fig. 5.3. An experimental proof to show that α-particles are doubly charged helium nuclei.

Properties of α-rays

***(i)* Velocity.** The α-rays are shot out from the radioactive material with large velocities ranging from 1.4×10^9 to 1.7×10^9 cms. per second.

The velocity of α-rays depends upon the radioactive substance from which they are ejected and for a given substance it is always the same.

α-particles move with a velocity 10,000 miles/second.

***(ii)* Ionising power.** *They produce intense ionisation* in the gas through which they pass. Their ionising power is 100 times greater than that of β-rays and 10,000 times greater than that of γ-rays. A thickness of 0.0005 cm. of aluminium foil reduces the ionising power to half. The tracks of α-particles observed by means of a Wilson cloud chamber are continuous, thick, straight lines with a slight bend in the end in certain cases.

***(iii)* They affect a photographic plate.** The effect is, of course, very feeble.

***(iv)* Penetrating power.** Out of the three radioactive radiations, they have the shortest penetrating power. It is so because α-particles produce the greatest number of ions in a given path, hence they penetrate the shortest distance. In *fact, their penetrating power is inversely proportional to their ionizing power.*

***(v)* Fluorescence.** α-rays emitted by a radioactive material (*e.g.* radium salt) produce fluorescence (*i.e.* lumiscence) when they strike a fluorescent screen like zinc sulphide or barium platinocyanide. In other words, when α-particle strikes a fluorescent screen, a tiny flash of light which is called **scintillation** is produced. Each individual scintillation can be observed with the help of an apparatus called **spin thariscope** which is a low-power microscope. On observing the fluorescence through spin thariscope, it is found to consist of successive *scintillations*, produced by the impact of an individual α-particle. The impact of a single α-particle can thus be observed and the number of such particles given out per second by a radioactive substance can be found out. The fact that these scientillations are due to α-particles and not due to β or γ-rays can be proved by interposing a thin sheet of mica between the radium salt and the fluorescent screen. Since α-rays are absorbed by mica sheet, the scientillations disappear.

***(vi)* Range of α-particles.** Scintillations from α-particle cease abruptly after it has traversed a certain distance through matter. *The distance that an α-particle can travel in air, at atmospheric pressure, is called its range in air*. It varies from 2.70 cms. for α-particles given by uranium to 8.62 cms. for those of polonium. The range of an α-particle depends upon

(*a*) the radioactive substance from which the rays are given out.

(*b*) the nature of the medium through which the rays travel.

(*c*) velocity of emission. The range is proportional to V^3.

The number of α-particles in a beam remains constant upto the end of the range, but their energy goes on decreasing. Ultimately the α-particles do not have sufficient energy to produce either scientillations or ionisation.

The ionisation first increases slightly, reaches a maximum and then suddenly drops almost to zero value. The distance at which the ionisation drops almost to zero gives the range of α-particles.

When the radioactive sample, say, polonium (P) is moved farther and farther away by pulling the rod (R) back, a point is reached when scientillations are no longer observed through the low-power microscope (M). The distance *d* where the α-particles just fail to reach the screen and hence produce no scientillation gives the range of the α-particles. (*See* Fig. 5.4).

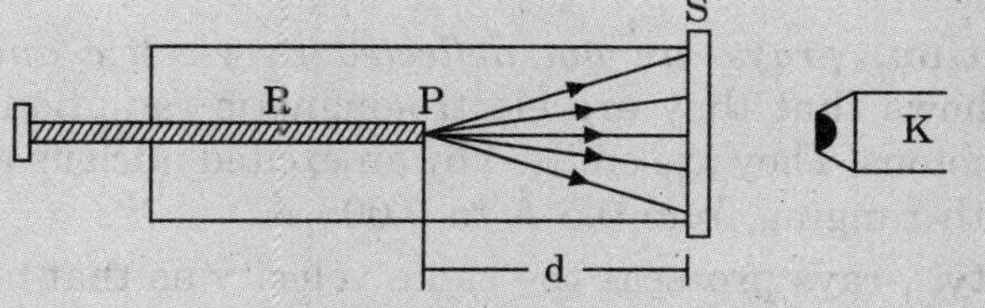

Fig. 5.4. Measurement of the range of an α-particle by scientillations produced on a fluorescent screen. R = Rod; P = Radioactive sample (polonium); S = fluorescent screen; M = Low-power microscope; *d* = Range of α-particles.

***(vii)* The α-rays are scattered when they pass through thin sheets of mica, gold foil, etc.** The divergence of α-particle from its straight line path is 2 to 3 degrees. Geiger and Marsden found that a few particles, sometimes, were deflected by more than 90°. This was explained by Rutherford to be due to the repulsion between the α-particle and the nucleus of the atom scattering it.

***(viii)* The rays are deflected by electric and magnetic fields showing that they are charged particles.**

***(ix)* They produce a heating effect.** A quantity of radium always maintains itself at a temperature higher than that of the surroundings. The evolution of heat is due to the stoppage of α, β and γ-rays by the radioactive substance.

(x) **When exposed to α-rays the body suffers incurable burns.**

Properties of β-rays

***(i)* Velocity and range of β-rays.** β-particles do not emerge out of the nucleus with the same range and velocity; rather they have a wide band of velocity ranging from 1 per cent to 99 per cent of the velocity of light.

***(ii)* Ionisation in air.** β-rays produce ionisation in air but the number of ions produced is hardly $\frac{1}{1000}$th of those produced by α-rays. Although their velocity is very large, they possess a comparatively small mass and hence have small kinetic energy. As β-particles change their path, their tracks in a Wilson cloud chamber are scattered not continuous as are for the α-particles.

***(iii)* Penetration.** β-rays can penetrate through large thickness of matter, *e.g.* they can easily pass through 1 cm. thickness of aluminium sheet.

***(iv)* Deflection.** *β-rays are deflected, by electric and magnetic fields*. Their direction of deflection shows that they are *negatively charged* particles. The *e/m* value has been found on the same lines as for cathode rays and this determination shows that β-rays are fast-moving electrons which are of *nuclear* origin although they do not revolve in orbits as do the extra-nuclear electrons.

***(v)* They are readily scattered when passing through matter** because of their extremely small mass. Generally, they undergo multiple scattering although single scattering also takes place by impact both with electrons and with nuclei.

***(vi)* They affect photographic plates more strongly than a-particles.**

***(vii)* They produce fluorescence** in barium platinocyanide, calcium tungstate, Willemite etc.

Properties of γ-rays

***(i)* Deflection.** *γ-rays are not deflected by electric and magnetic fields*. This property shows that they are electromagnetic radiations which consist of very energetic photons. They are emitted by an excited nucleus and have extremely short wave-length ranging from 0.5 Å to 0.005 Å.

***(ii)* Velocity.** γ-rays prossess the same velocity as that of light, *i.e.* 3×10^{10} cms/sec.

***(iii)* Diffraction.** They are diffracted by crystals in the same way as X-rays.

***(iv)* Ionisation in gas and genetrating power.** They ionize the gas through which they pass, although the ionization produced by them is comparatively small. Consequently, their penetrating power is very high as compared to that of β-rays, *e.g.* they can pass easily through 30 cms. thickness of iron.

(v) They are capable of producing fluorescence and affecting photographic plates more intensely than β-rays.

Applications of Radioactive Radiations

(i) Many radiation sources such as cobalt—60 (γ-ray emitter) have been used for industrial radiography *i.e.* for investigating the interiors of metallic castings for detecting any flaws or defects.

(*ii*) Radioactive nuclides emitting α- or β-particles have been used for the production of electric power by thermo-electric conversion. When α- or β-particles are absorbed in matter, the energy of the radiation is converted into heat which can be utilized for production of electric power. Such radio nuclide power generators are particularly suitable for space vehicles because of their light weight, long life and absolute reliability.

(*iii*) Radio nuclides have been used as compact sources of heat energy because of which they find many space-related applications. Radio nuclides of promethium—147, polonium—210 and plutonium—238 have been used to heat the propellant gas (hydrogen) in low-thrust rockets.

(*iv*) Chemical changes initiated by radiations from radioactive substances is the basis of a newly developed subject called *radiation chemistry* which has been applied in the production and modification of plastics and in the production of irradiated wood-plastic combinations.

(*v*) Nuclear radiations like γ-rays have been utilized for the preservation of food. The irradiation of food-stuff mainly meat, poultry, fish and fruits is achieved by exposure to γ-rays from cobalt—60 or calcium—137. A dose of 2 to 5 million rads, is sufficient to destroy almost all bacteria in food. This increases the shelf-life of these articles without refrigeration. Radiations have also been employed for insect disinfestation of wheat and flour.

(*vi*) Gamma radiations from cobalt—60 are used in hospitals for sterilization of materials like dressings, hypodermic syringes and surgical sutures.

(*vii*) Population of insects which causes considerable damage to both plant crops and livestock can be controlled by irradiating the male members of these insects so that they become sterile.

(*viii*) Radiation mutations in plants have been practised to produce new varieties of these plants.

(*ix*) Self-luminous paints for use on instruments and watch dails have been made by adding a natural α-emitting radioactive substance to phosphur.

(*x*) The ionization produced by β-particles has been widely used for the elimination of static electricity which constitutes a serious fire and explosion hazard in the paper, textile, rubber and plastic industries.

Types of Radioactivity

Radioactivity is of two types :

1. Natural radioactivity. We have seen that the naturally occurring elements like uranium, polonium, radium etc. are constantly undergoing a spontaneous change (*i.e.* change by itself) and as a result of this they are emitting α-, β- and γ-rays and thus change into other elements. *This spontaneous change is called natural radioactivity.* In natural radioactivity only a single nucleus is involved It is always found in heavier elements in the periodic table.

2. Artificial or induced radioactivity. Artificial or induced radioactivity *is a process by which an element is converted into a new radioactive isotope of a known element by artificial means.*

Detection and Measurement of Radioactivity

The detection and measurement of radioactivity of a substance can be made by a number of methods some of which are described below :

1. Electroscope Method. The electroscope used for the purpose consists of an ionisation chamber and electroscope box (Fig. 5.5). The ionisation chamber is made of lead and consists of two metallic plates P and P′ separated by the gas which is exposed to the rays. The lower plate is connected with the wall of the chamber and earthed. The upper plate which is insulated by a block of sulphur and connected with the gold leaf system, G is initially charged to a high potential. The radioactive material is then placed in the cavity of the plate P′. The air between the two plates gets ionised and the charge leakes away between P and P′. The gold leaf falls and its rate of fall is proportional to the rate of leakge and thus to the intensity of the rays. The rate of leakage so found is compared with that of a substance of known radioactivity under exactly similar conditions and thus the total radioactivity of the substance under examination is determined.

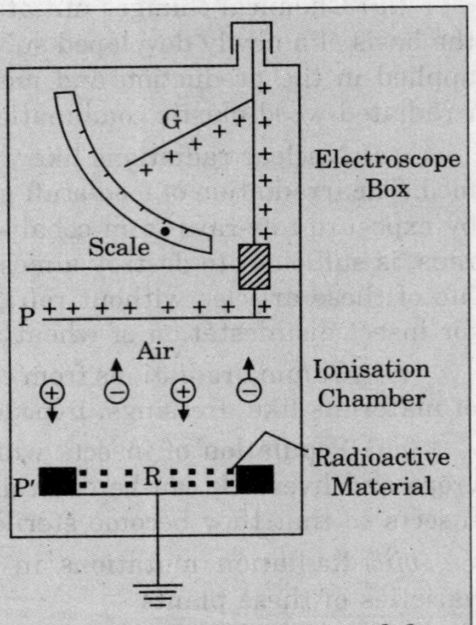

Fig. 5.5. Electroscope used for the measurement of radioactivity of a substance.

2. Wilson's Cloud Chamber Method. The apparatus used in this method was devised by Wilson (1912) and is known as *Wilson's Cloud Chamber* (Fig. 5.6). This is a new technique which is based on the production of ions by radioactive emissions. Such ions when produced in a chamber supersaturated with vapours (say of water) act as nuclei for the condensation of vapours. Since ions are produced along the path of radioactive emission, the track of fog formed indicates the path followed by particle. This path can be photographed.

In actual practice, a gas or dust-free air saturated with water (or alcohol) vapour is enclosed in the chamber. The piston is rapidly moved up to bring about cooling by sudden expansion. Thus the air contained in the chamber becomes supersaturated with water vapour. At this stage the radioactive emissions are led into chamber. A fog track is at once formed along the path transversed by the radioactive emission. This is immediately

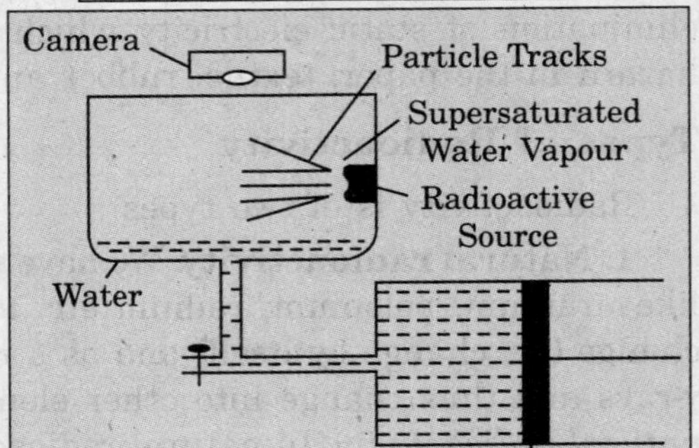

Fig. 5.6. Wilson cloud chamber used for the measurement of radioactivity of a substance.

photographed by the camera. In more recent sophisticated devices, the cloud chamber operates automatically.

3. Geiger-Muller Counter Method. The apparatus used in this method has been shown in Fig. 5.7. This apparatus consists of a hollow cylinderical tube fitted with a thin metal rod along its axis and insulated from the tube. The tube is filled with a gas like oxygen or argon at a low pressure (say, 0.1 atm). A high potential upto 1000 volts is applied at the rod (acting as anode) and the outer tube (acting as cathode). When α and β particles of high energy are allowed to enter the metallic tube through the quartz window, they produce enough ions which cause a pulse of a current to flow between the rod and the tube. This pulse can be amplified and recorded on a loud-speaker as a sharp click. In the more sophisticated instruments, the loud speaker has been replaced by an automatic recorder which counts the number of pulses.

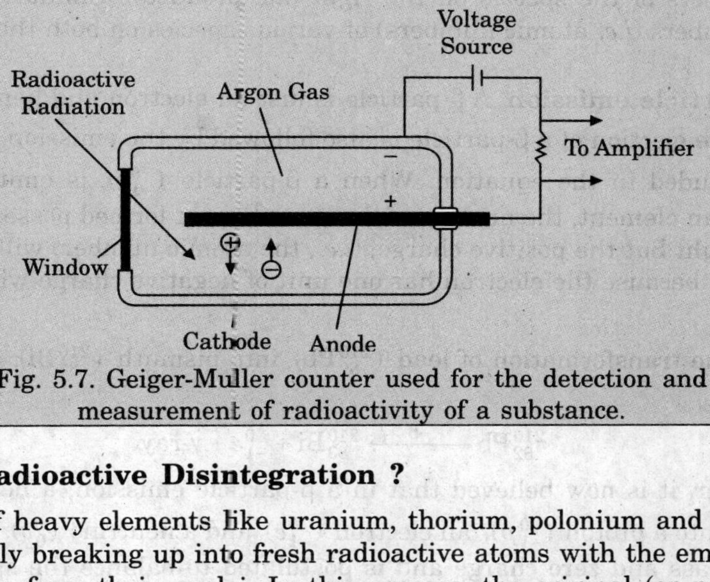

Fig. 5.7. Geiger-Muller counter used for the detection and measurement of radioactivity of a substance.

What is Radioactive Disintegration ?

Atoms of heavy elements like uranium, thorium, polonium and radium etc. are constantly breaking up into fresh radioactive atoms with the emission of α-, β- and γ-rays from their nuclei. In this process, the original (or parent) atom disappears and gives rise to new (daughter) atom. These new atoms are also, in general, radioactive and hence spontaneously break up in their turn, thereby leading to a long chain of different radioactive elements in the form of a series until an inactive element (usually lead) is reached. This *spontaneous breaking up of the nucleus is known as radioactive disintegration or radioactive decay.*

Examples of Modes of Decay

The disintegration process may proceed in one of the following first two ways :

1. α-particle emission. When an α-particle (^4_2He) is emitted from the nucleus of an atom (*called parent element*), the nucleus of the new element formed (*called daughter element*) possesses atomic weight or atomic mass number less by four units and nuclear charge (*i.e.* atomic number) less by two units because α-particle has an atomic mass of four units and nuclear charge of two units. Thus an atom of radium ($^{226}_{88}Ra$) changes on emission of an α-particle into an atom of radon ($^{222}_{86}Rn$). Here the subscript to the bottom left gives the *atomic number* (*i.e.* the number of protons or positive charges on the nucleus) and the superscript on the top right gives the *atomic weight* (*i.e.* weight of the protons and neutrons put together) or *atomic mass number*.

Thus the disintegration of $^{226}_{88}Ra$ into $^{222}_{86}Rn$ can be represented by the following equation which is called *nuclear equation* (*or transmutation equation*).

$$^{226}_{88}Ra \xrightarrow{-\alpha} {}^{222}_{86}Rn + \alpha$$

or $$^{226}_{88}Ra \longrightarrow {}^{222}_{86}Rn + {}^{4}_{2}He$$

In a balanced nuclear equation the sum of the mass numbers (*i.e.* atomic mass numbers) of various species on the left (*i.e.* reactants) is equal to the sum of the mass numbers of the species on the right (*i.e.* products). Similarly, the sum of charge numbers (*i.e.* atomic numbers) of various species on both the sides should be equal.

2. β-particle emission. A β-particle is just an electron and hence its symbol is $_{-1}^{0}e$. Since ejection of a β-particle is also followed by the emission of γ-ray, γ-ray is also included in the equation. When a β-particle ($_{-1}^{0}e$) is emitted from the nucleus of an element, the nucleus of the new element formed possesses the same atomic weight but the positive charge, (*i.e.*, the atomic number) will be increased by one unit because the electron has one unit of negative charge with practically no mass.

Thus the transformation of lead ($^{210}_{82}Pb$) into bismuth ($^{210}_{83}Bi$) can be shown as :

$$^{210}_{82}Pb \xrightarrow{-\beta} {}^{210}_{83}Bi + {}_{-1}^{0}e + \gamma\text{-ray}$$

However, it is now believed that in a β-particle emission, a neutron ($^{1}_{0}n$) is converted into a proton ($_{+1}^{0}p$), an electron ($_{-1}^{0}e$) and a neutrino ($^{0}_{0}v$). The neutrino has zero mass and zero charge and is postulated to balance the spins. Thus :

$$\textit{Neutron} \rightarrow \textit{Proton} + \textit{Electron} + \textit{Neutrino}$$

or $$^{1}_{0}n \rightarrow {}_{+1}^{1}p + {}_{-1}^{0}e + {}^{0}_{0}v$$

The electron produced escapes as β-particle leaving the proton behind in the nucleus. The increase in the atomic number of the necleus by one unit is due to this proton left in the nucleus.

3. γ-ray emission. We have already seen that γ-rays are short wave electro magnetic radiations with *no charge* and *no mass*. Thus the emission of γ-rays from a radio-active element will not produce a new element. The emission of γ-rays only results in decrease in energy of the nucleus. When α or β-particle is ejected from the nucleus, it is left in an excited state and emits γ-rays.

Laws of Radioactive Disintegration

Radioactive disintegration is found to obey the following two laws :

1. *Atoms of all radioactive elements are undergoing spontaneous disintegration i.e., are constantly breaking to form fresh rodioactive products with the emission of α, β- and γ-rays.*

2. *The rate of breaking is not affected by external factors like temperature, pressure, chemical combination, etc. but depends entirely on the law of chance, i.e., the number of atoms breaking per second at any instant is proportional to the number present. In other words, the quantity of a radioactive element which disintegrates in unit time is directly proportional to the amount of the radioactive element present.*

Rate of Disintegration

Suppose an element A disintegrates into another element B. Let the number of atoms of A in the beginning (*i.e.*, at $t = 0$) be N_0. Now as the time passes, the element A undergoes continuous disintegration into B and hence the number of atoms of A goes on decreasing. Let the number of atoms of A left undisintegrated in time t be N. Thus :

$$\begin{array}{ll} & A \rightarrow B \\ \text{At } t = 0 & N_0 \\ \text{At } t = t & N \end{array}$$

If dN atoms of A disintegrates in a small time dt, then the *rate of disintegration* (*i.e., rate of decrease*) of A into B which is also called *activity* is equal to $-\dfrac{dN}{dt}$ which is proportional to N, *i.e.*

$$-dN/dt \propto N$$

or

$$\frac{-dN}{dt} = \lambda N \quad ...(i)$$

Here λ is a constant of proportionality which has been called by Rutherford and Soddy as *radioactive* (or *decay* or *disintegration*) *constant*.

Since the intensity and the number of atoms of A decrease with time, the rate of disintegration has been indicated by negative sign.

The differential equation (*i*) can be put in the form

$$\frac{dN}{N} = -\lambda dt$$

Integrating we have

$$\int \frac{dN}{N} = -\int \lambda dt + \textit{constant of integration, C}$$

$$\log_e N = -\lambda t + C \quad ...(ii)$$

Now when $t = 0, N = N_0$

$\because \quad \log_e N_0 = C$

Substituting in (*ii*), we get

$$\log_e \frac{N}{N_0} = -\lambda t$$

or $$\frac{N}{N_0} = e^{-\lambda t}$$

$\therefore$ $$\boxed{N = N_0 e^{-\lambda t}} \quad ...(iii)$$

This decay equation may be used to find the number of the atoms of the original kind present at any time t.

If M_0 is the original mass of the material and M its present mass, then

$$\boxed{M = M_0 e^{-\lambda t}}$$

Definition of Radioactive Constant (λ).

(*a*) Eq. (*i*) given above may be put as

$$\lambda = \frac{-dN/dt}{N}$$

Hence, radioactive constant may be defined as *the ratio of the amount of the substance which disintegrates in a unit time to the amount of the substance present.*

(*b*) Suppose, we put $t = 1/\lambda$ in Eq. (*iii*) above. Then the number of the original atoms present after this tune is

$$N = N_0 e^{-\lambda \times 1/\lambda} = N_0 e^{-1}$$

$$= \frac{N_0}{e} = \frac{N_0}{2.718}$$

$$\boxed{N = 0.368\ N_0 \approx 0.37\ N_0}$$

Hence, radioactive constant may also be defined as the *reciprocal of the time during which the number of atoms of a radioactive substance falls to 37 per cent of its original value.*

Definition of Half-life Period ($t_{1/2}$ or T).

Suppose after time T (or $t_{1/2}$) half of the atoms of the radioactive element has disintegrated, *i.e.*

when $t = T$, $N = N_0/2$

Thus on putting $t = T$ and $N = N_0/2$ in equation (*iii*), we get

$$\frac{N_0}{2} = N_0 . e^{-\lambda T}$$

or $$\frac{1}{2} = e^{-\lambda T}$$

or $$2 = e^{+\lambda T}$$

or $$\lambda T = \log_e 2$$

or $$T = \frac{\log_e 2}{\lambda}$$

Hence $$\boxed{t_{1/2} \text{ or } T = \frac{0.6931}{\lambda}} \quad ...(iv)$$

As is evident from equation (*iv*), *half-life period* (T or $t_{1/2}$) *of a radioactive substance is the time required to disintegrate one-half of the original amount of the substance.* In other words, *the half-life period of a radioactive substance is the time in which half of the radioactive substance will disintegrate.*

As seen from eqn. (*iv*), T is inversely proportional to the radioactive constant and hence is a characteristic constant which can be employed to differentiate radio elements. For example, it takes nearly 1620 years for one-half of a given quantity of radium to change into radon. In another 1620 years, 1/2 of the remainder would have disintegrated leaving 1/4 of the original amount behind. Accordingly, we say that half-life period of radium is 1620 years. However, radon has a half-life of only 38 days.

In general, it may be said that at the end of time T, only 50% of the radioactive atoms remain unchanged, at the end of $2T$ only 25%, after $3T$ only 12.5%, after $4T$ only 6.25% and at the end of $10T$ only 0.1%. Hence, the life time of a radioactive element, though theoretically infinite, is finite for all practical purposes.

As already seen, λN represents the rate at which particles are emitted by N radioactive atoms.

$$\boxed{\therefore \textbf{ Rate of emission of particles or activity } = \lambda N = \frac{0.693N}{T}}$$

If T is expressed in seconds, then the above equation gives the emission rate in terms of the number of particles expelled per second from N atoms of the given nuclide.

Evidently, half-life period of a given substance is independent of the amount of the substance present initially. It depends only on the disintegration constant (λ) of the element.

Half-life period of a radioactive substance is a measure of the radioactivity of the element, since shorter the half-life period of a radio-active element, larger the number of its atoms that are disintegrating in a unit time.

Determination of half-life period.

To find the half-life period, of a radioactive substance the value of radioactive constant λ is determined and the half-life period is calculated from relation (*iv*).

The radioactive constant λ can be determined experimentally by measuring the activity of a radioactive substance at two instances having an interval of time t. The apparatus used is the same as for measuring the ionisation current with a quadrant electrometer. The radioactive substance is placed in a thin layer on the lower plate and the ionisation current is measured at various intervals of time. A graph is then plotted between log I_t taken along the Y-axis and the time t along the X-axis. Now

$$I_t = I_0\, e^{-\lambda t}$$

$$\because \quad \log_e I_t = \log_e I_0 - \lambda t$$

or $$y = a - \lambda t$$

where $\log_e I_t = y$ and $\log I_0 = a$ (constant) ...(*v*)

The graph is a straight line. Differentiating equation (*v*), we have

$$\frac{dy}{dt} = -\lambda$$

Hence the slope of the graph gives the radioactive constant.

Average Life Period (*l*)

Since atoms of a radioactive element are constantly disintegrating one after another, the life of every atom is different. The atoms which disintegrate earlier have very short life whereas others which disintegrate at the end have a long life. Hence, the process of disintegration will continue indefinitely. Total decay period of any radioactive element is, therefore, infinity. Use of total decay periods or total life periods for radio-elements are thus meaningless.

Instead of total decay period, another term which is generally used is **average life period.** This is also called **mean life period** or **life expectancy** and is defined as follows :

Average life period of a radioactive element is the reciprocal of the disintegration constant (λ). Thus :

$$\boxed{l = \frac{1}{\lambda}}$$

We have already seen that

$$t_{1/2} = \frac{0.693}{\lambda}$$

or $$\frac{1}{\lambda} = \frac{t_{1/2}}{0.693} \quad ...(a)$$

and $$l = \frac{1}{\lambda} \quad ...(b)$$

Hence, from equation, (*a*) and (*b*), we get

$$l = \frac{t_{1/2}}{0.693} = 1.34 \times t_{1/2}$$

Thus :

$$\boxed{\textit{Average life period } (l) = 1.44 \times \textit{Half-life period } (t_{1/2})}$$

Radioactive Equilibrium : Law of Successive Disintegration

Consider a long-lived radioactive element *A*. This element, on standing, disintegrates successively into a decay series of radioactive atoms B, C, D etc.

$$A \rightarrow B \rightarrow C \rightarrow D \rightarrow ......$$

In the successive disintegration of radioactive substances *A, B, C, D* etc., any two adjacent elements may be considered as *parent* and *daughter*, the former being that which by its own decay produces the later, *e.g.*, A which produces B, is the parent with respect to *A*, the daughter. Evidently in a series, the parent of the following element will be daughter of the preceding one. The law governing successive disintegrations deals with the quantity of the daughter present at any instant and determines the condition of equilibrium between the daughter and the parent.

After some time the rate at which the element *B* is formed from *A* becomes equal to the rate of its decay into *C*. *Thus a stage will be reached when the number of atoms of B formed (from A) per second will become equal to the number of atoms which disintegrate into C per second.* The element B is then said to be in a **radioactive equilibrium** with the given amount of *A*. In the same way, the element *C* and the element *D* will also reach the state of radioactive equilibrium with the preceding number of the decay series. Some authors prefer to call the equilibrium state as **steady state** because equilibrium, as used in a chemical reaction, implies existence of a reversible reaction whereas radioactive disintegrations are *irreversible*.

If N_1, N_2, N_3 etc. are the number of atoms of various members of the decay series at the steady state and λ_1, λ_2, λ_3 etc. are their disintegration constants,

Decay series : $A \rightarrow B \rightarrow C \rightarrow D \rightarrow ......$

No. of atoms : $N_1 \quad N_2 \quad N_3$

Disintegration constants : $\lambda_1 \quad \lambda_2 \quad \lambda_3$

then $\lambda_1 N_1 = \lambda_2 N_2 = \lambda_3 N_3$ etc.

Now if we consider the elements *A* and *B*, then

$$\frac{N_1}{N_2} = \frac{\lambda_2}{\lambda_1} \qquad ...(vi)$$

Now since average life periods are reciprocal of disintegration constants,

$$\frac{\lambda_2}{\lambda_1} = \frac{l_1}{l_2} \qquad ...(vii)$$

With the help of equation (*vii*), equation (*vi*) reduces to :

$$\frac{N_1}{N_2} = \frac{\lambda_2}{\lambda_1} = \frac{l_1}{l_2} \qquad ...(viii)$$

where l_1 and l_2 are the average life periods of the elements *A* and *B*. Similar equations can be found by considering the elements *B* and *C*, *C* and *D* etc.

Relation (*viii*) shows that *at the steady state, the radioactive elements, A, B, C etc. are present in the direct ratio of their average-life periods*. Thus, at the steady state, the element having the highest average life period is present in the largest amount.

Radioactive equilibrium can be had only when the average life-period of the parent element is greater than that of the daughter element.

Numerical Problems

Example 1. *1 gram of radium is reduced by 2.1 mg in 5 years by α decay. Calculate the half-life period of radium.*

Solution. Initial mass of radium (N_0) = 1 gram

Mass of radium left behind (N) = 1 – 0.0021

= 0.9979 gram

$$\frac{N}{N_0} = e^{-\lambda t}$$

Here $$\frac{N}{N_0} = \frac{0.9979}{1}$$

and $$t = 5 \text{ years}$$

Hence $$0.9979 = e^{-5\lambda}$$

or $$e^{5\lambda} = \frac{1}{0.9979}$$

$$5\lambda = 2.3026 \log_{10}\left[\frac{1}{0.9979}\right]$$

$$\lambda = \frac{2.3026 \times 0.0009}{5}$$

or $$\lambda = 41.4468 \times 10^{-45} \text{ per year}$$

$$T = \frac{0.693}{\lambda} = \frac{0.693}{41.4468 \times 10^{-5}} = \mathbf{1672\ years}$$

Example 2. *A radioactive substance has a half-life period of 30 days. Calculate (i) the radioactive disintegration constant (ii) the average life period (iii) the time taken by $\frac{3}{4}$ of the original number of atoms to disintegrate and (iv) the time taken by $\frac{1}{8}$ of the original number of atoms to remain unchanged.*

Solution. Here T = 30 days

(*i*) $$\lambda = \frac{0.693}{T}$$

or $$\lambda = \frac{0.693}{30} = \mathbf{0.0231} \text{ per day}$$

(*ii*) Average life-period, $l = \frac{1}{\lambda}$

$$l = \frac{1}{0.0231} = \mathbf{43.29\ days}$$

(*iii*) Number of atoms disintegrated = $\frac{3}{4} N_0$

Number of atoms left behind $(N) = N_0 - \frac{3}{4}N_0 = \frac{1}{4}N_0$

$$N = N_0\, e^{-\lambda t}$$

$$\frac{N}{N_0} = e^{-\lambda t}$$

Here $$\frac{N}{N_0} = \frac{N_0/4}{N_0} = \frac{1}{4}$$

$\therefore$ $$\frac{1}{4} = e^{-\lambda t}$$

$$\lambda t = \log_e 4$$

$$t = \frac{2.3026 \log_{10} 4}{0.0231}$$

$$= \frac{23.026 \times 0.6021}{0.0231}$$

$$= \mathbf{60\ days}$$

(*iv*) Here $$\frac{N}{N_0} = \frac{1}{8}$$

$$\frac{N}{N_0} = e^{-\lambda t}$$

$$\frac{1}{8} = e^{-\lambda t}$$

$$\lambda t = \log_e 8 = 2.3026 \log_{10} 8$$

or $$t = \frac{2.3026 \times 0.9031}{0.0231} = \mathbf{90\ days}$$

Example 3. *Calculate the time in which the activity of a sample of thorium reduces to 90% of its original value. Assume the half-life of thorium to be 1.4 × 10^10 years.*

Solution. Here T = 1.4 × 10^{10} years

$$T = \frac{0.693}{\lambda}$$

$\therefore$ $$\lambda = \frac{0.693}{T}$$

$$\lambda = \frac{0.693}{1.4 \times 10^{10}} \text{ per year}$$

$$N = N_0 e^{-\lambda t}$$

$$\frac{N}{N_0} = e^{-\lambda t}$$

$$\frac{9}{10} = e^{-\lambda t}$$

$$\lambda t = \log_e 1.11$$

$$t = \frac{2.3026\ [\log_{10} 1.11] \times 1.4 \times 10^{10}}{0.693}$$

$$t = \mathbf{2.108 \times 10^9\ years}$$

Theory of Radioactive Disintegration : Stability of the Nucleus and Ratio of Neutrons and Protons.

The stability of a nucleus depends qualitatively on the ratio of the number of neutrons (N) and protons (P) present in it. In case of elements with atomic numbers equal to 20, the ratio N/P is quite close to *unity*, *i.e.* for such elements the number of neutrons is equal to that of protons. For the elements with higher atomic number ratio N/P increases progressively to about 1.6. In such elements the repulsion between them increases.

By plotting a graph between N and P for the nuclei of various elements it has been found that most of the stable nuclei (*i.e. non-radioactive nuclei*) lie in a well-defined belt which is known as the **zone** or **belt of stability** because it contains the stable nuclei. (See Fig. 5.8). *Nuclei whose N/P ratio lies above or below this belt are unstable and, therefore, undergo spontaneous radioactive disintegration from one atom to another, giving α– or β–particles. This process of disintegration continues until a stable nucleus is obtained.* For example, a radium nucleus changes successively into nine other nuclei emitting in all 5α and 4β particles and a stable nucleus of lead, Pb^{206} is obtained.

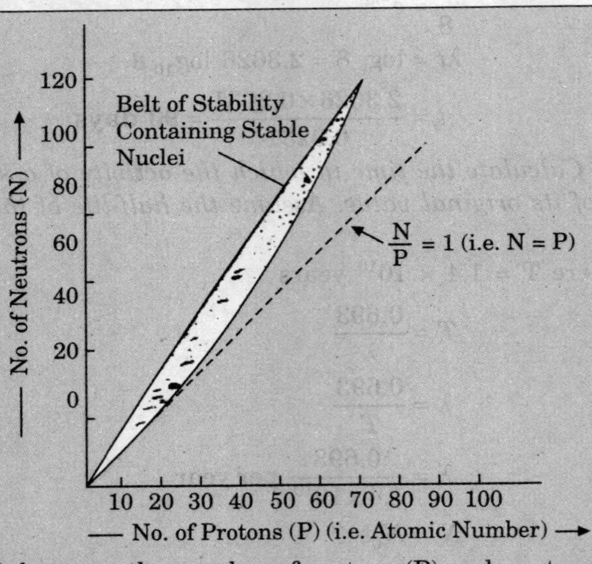

Fig. 5.8. A graph between the number of protons (P) and neutrons (N) having the belt of stability which contains stable nuclei. Nuclei whose N/P ratio lies above or below this belt are unstable and hence undergo spontaneous disintegration.

Group Displacement Law

The changes that occur when an α-particle or a β-particle is emitted from a radioactive element have been stated by *Fajan, Russel* and *Soddy* in 1913 in the form of a law which is known as **Group Displacement Law.** This law states as follows :

When an α-particle is emitted from a radioactive element (i.e. parent element), the new element (i.e. daughter element) formed has atomic number two units less and mass number four units less than the parent element. The element is consequently displaced two places (i.e. two groups) to the left of the periodic table. Similarly, when a β-particle is emitted, the new element has atomic number one unit more but its mass number remains the same as that of the parent element. Consequently the new element is displaced one place (i.e. one group) to the right of the periodic table.

Explanation of the law. Since α-particle is a helium nucleus of mass number 4 (2 protons and 2 neutrons), the emission of an α-particle from a radioactive element causes the *mass number to decrease by four units*. Further, since two protons have been lost, *the atomic number falls by two units*. Consequently, the new element formed moves two places (*i.e. two groups*) to the left of the periodic table.

The β-particle is an electron and is emitted from the nucleus. But, as there are no electrons in the nucleus, its origin is uncertain. It is, therefore, postulated that a neutron changes into a proton and an electron. The electron is emitted as a β-particle leaving a proton behind. The number of protons in the nucleus increases by one unit, *i.e.* the positive charge (*i.e.* atomic number) increases by one unit. Consequently the daughter element moves one place (*i.e. one group*) to the right of the periodic table.

Examples to illustrate the law. $^{215}_{84}Po$ which is in group VIA of the periodic table, on losing an α-particle ($^{4}_{2}He$) is transformed into radioactive lead, $^{211}_{82}Pb$ which belongs to group IVA. Thus we see that the daughter element *viz.* $^{211}_{82}Pb$ is displaced two places to the left of the periodic table.

$$\underset{\text{VIA}}{^{215}_{84}Po} \xrightarrow{-\alpha} \underset{\text{IVA}}{^{211}_{82}Pb} + {^{4}_{2}He}$$

Radioactive $^{211}_{82}Pb$ emits a β-particle ($^{0}_{-1}e$) to give radioactive $^{211}_{83}Bi$ which belongs to group VA, *i.e.* one place to the right of the parent element *viz.* $^{211}_{82}Pb$.

$$\underset{\text{IVA}}{^{211}_{82}Pb} \xrightarrow{-\beta} \underset{\text{VA}}{^{211}_{83}Bi} + {^{0}_{-1}e}$$

Radioactive bismuth, $^{211}_{83}Bi$ loses a β-particle resulting in the formation of $^{211}_{84}Po$ which belongs to group VIA, *i.e.*, one place to right of the parent element, $^{211}_{83}Bi$.

$$\underset{\text{VA}}{^{211}_{83}Bi} \xrightarrow{-\beta} \underset{\text{VIA}}{^{211}_{84}Po} + {^{0}_{-1}e}$$

The transformations shown above can be summarised as follows :

$$\underset{\text{(VIA)}}{^{215}_{84}Po} \xrightarrow{-\,^{4}_{2}He} \underset{\text{(IVA)}}{^{211}_{82}Pb} \xrightarrow{-\,^{0}_{-1}e} \underset{\text{(VA)}}{^{211}_{83}Bi} \xrightarrow{-\,^{0}_{-1}e} \underset{\text{(VIA)}}{^{211}_{84}Po}$$

In the above example $^{215}_{84}Po$ and $^{211}_{84}Po$ are isotopes while $^{211}_{82}Pb$, $^{211}_{83}Bi$ and $^{211}_{84}Po$ are isobars. Thus we conclude that a β-particle emission gives isobar while the combined emission of one α- and two β-particles results in the formation of an isotope.

Numerical problem. *Calculate the number of α- and β-particle, emitted when $^{238}_{92}U$ changes into $^{206}_{82}Pb$.*

Solution. Let the element formed between $^{238}_{92}U$ and $^{206}_{82}Pb$ be X. The mass number of X should be the same as that of Pb because there is no loss in the mass number of the daughter element formed as a result of β-emission. Now let the number of α- and β-particles emitted be x and y respectively. The disintegration sequence will be as follows :

$$^{238}_{92}U \xrightarrow{-x\alpha} {}^{206}X \xrightarrow{-y\beta} {}^{206}_{82}Pb$$

Thus :

$$238 - 4x = 206$$

$$\therefore \quad x = 32/4 = 8$$

The atomic number of

$$X = 92 - 2 \times 8 = 92 - 16 = 76$$

Consequently

$$y = 82 - 76 = 6$$

Thus the number of various particles emitted is :

α-particles = 8

β-particles = 6

Radioactive Series

Atoms of heavy elements like uranium, thorium, polonium and radium etc. are constantly breaking up into fresh radioactive atoms with the emission of α-, β- and γ-rays from their nuclei. In the process, the original (or parent) atom disappears and gives rise to new (daughter) atom. These new atoms are also, in general, radioactive and hence spontaneously break up in their turn, thereby leading to a long chain of different radioactive elements in the form of a series until an inactive element (usually lead) is reached. *The series of elements thus obtained by the successive disintegration of the new atoms is known as* **radioactive disintegration series** and the *spontaneous breaking up of the nucleus is known as* **radioactive disintegration.**

All the *naturally occurring radioactive elements* belong to one of the following three series : (*1*) *Uranium series* [(*4n* + *2*) *series*]. (*2*) *Thorium series* [*4n series*], (*3*) *Actinium series* [(*4n* + *3*) *series*].

These series have been named after the name of the element at or near the head of the respective series. All these three series which are also called **natural radioactive series** end with a stable isotope of lead.

Some elements of each series emit α-particles whereas some other elements emit β-particles. Although no one atom can go both ways, some atoms go either of the two ways and cause branches in the series. No matter which way the parent goes, the daughter the goes other way so that even though the series branch, they always come together again.

1. Uranium series [(4n + 2) *series*]. This series is also called (4n + 2) series because the mass number of the elements of this series are given by this expression in which n is an integer whose value necreases by unity when we go from one radioactive element to the next one below it. The mass numbers of the members of this series give a remainder of 2 when divided by 4. In this series $^{238}_{92}U$ is the parent element and through the successive disintegrations it is finally transformed into a stable isotope of lead, $^{206}_{82}Pb$ (*See* Fig. 5.9).

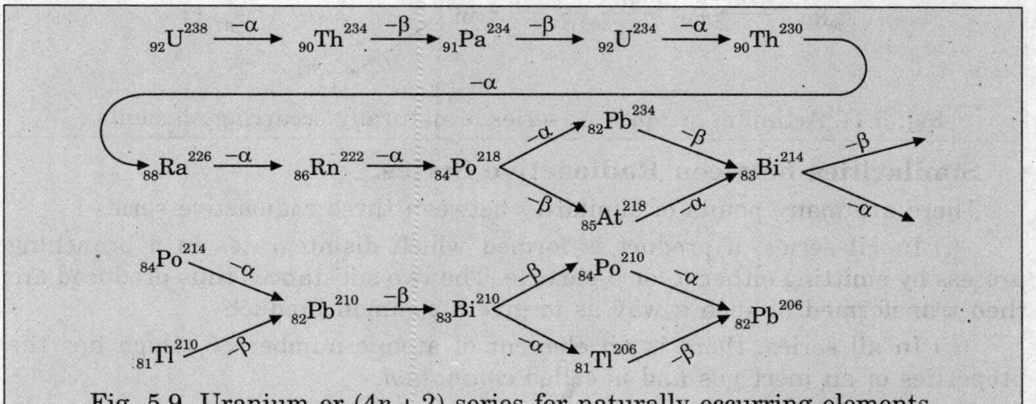

Fig. 5.9. Uranium or (4n + 2) series for naturally occurring elements.

2. Thorium series (4n *series*). This series is also known as 4n series because mass numbers of the members of this series are divisible by 4. This series starts with $^{232}_{90}Th$. As shown in Fig. 5.10 by a successive transmutation or disintegration it ends up in a stable isotope of lead, $^{208}_{82}Pb$.

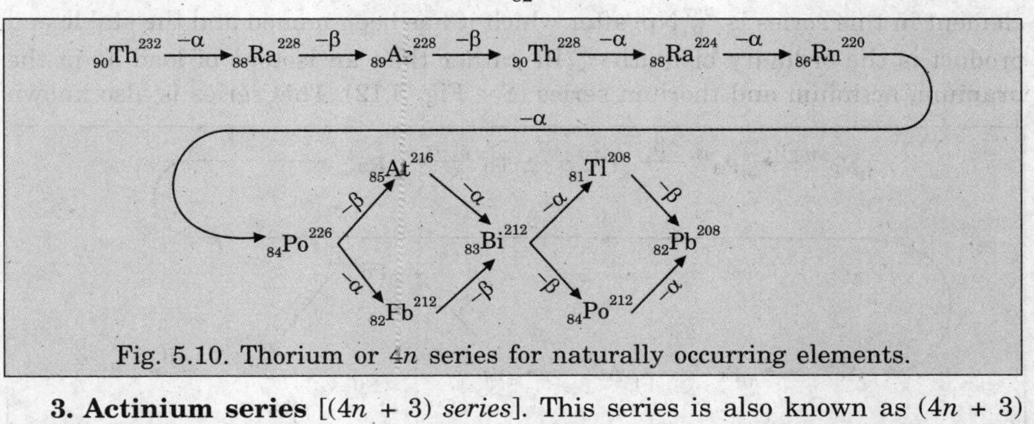

Fig. 5.10. Thorium or 4n series for naturally occurring elements.

3. Actinium series [(4n + 3) *series*]. This series is also known as (4n + 3) series because the mass numbers of the members of this series give a remainder of 3 when divided by 4. Actinium was one time thought to be the starting element of this series, but now it is known that the true starting element is $^{235}_{92}U$ which by successive transformations ends up in a stable isotope of lead, $^{207}_{82}Pb$. The whole chain of elements is shown in Fig. 5.11.

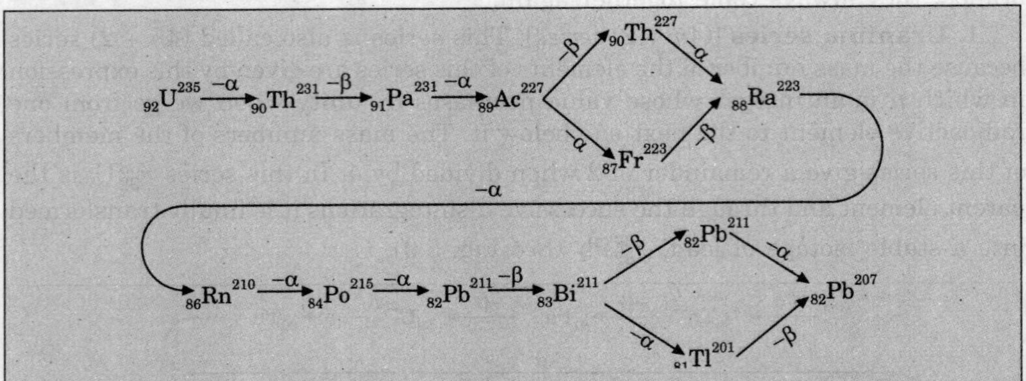

Fig. 5.11. Actinium or (4*n* + 3) series of naturally occurring elements.

Similarities between Radioactive Series.

There are many points of similarity between three radioactive series :

(*i*) In all series, a product is formed which disintegrates in a branching process by emitting either α- or β-particle. The two substances thus produced are then transformed in such a way as to give a common product.

(*ii*) In all series, there is an element of atomic number 86 which has the properties of an inert gas and is called *emanation*.

(*iii*) The stable end-product in all the three series having an atomic number of 82, is an isotope of lead *i.e.* Pb^{206}, Pb^{207} and Pb^{208}.

Neptunium Series [(4*n* + 1) *series*).

In addition to the three series described above, there is a fourth series which has been obtained from an *artificially-produced* radioactive material. The first element in this series is $^{237}_{93}Np$ after which it has been named and the stable end product is the ordinary bismuth $^{209}_{83}Bi$ rather than an isotope of lead as in the uranium, actinium and thorium series (*See* Fig. 5.12). This series is also known

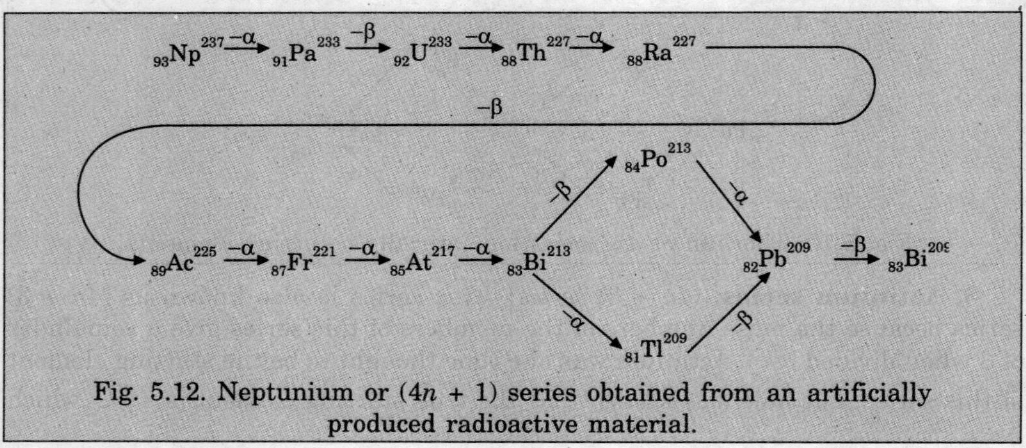

Fig. 5.12. Neptunium or (4*n* + 1) series obtained from an artificially produced radioactive material.

as $(4n + 1)$ series. All the members of this series are either unknown or extremely rare in nature.

It is quite obvious from the study of these series that all the members of a particular series are characterised by having $(4n + q)$ nucleons, where q is one of the integers 0, 1, 2, 3 characteristic of a given series and n may be any integer. Evidently, $(4n + q)$ gives the mass number A of the members of the series. Thus :

For thorium series $A = (4n + 0), q = 0$

For neptunium series $A = (4n + 1), q = 1$

For uranium series $A = (4n + 2), q = 2$

For actinium series $A = (4n + 3), q = 3$

Contributions of Radioactivity

1. Discovery of new fundamental particles. The study of radioactivity has led to the discovery of a number of new fundamental particles such as *neutrons, positrons, deuterons,* α-, β-, and *γ-particles etc.* These particles have been used as bombarding particles (*projectiles*) in the artificial transmutation of elements and have added to our knowledge of atomic structure. For example :

(*a*) The scattering of α-particles proved that the *nucleus is positively charged.*

(*b*) The extent of the deflection of α-particles indicates that the *atomic number is numerically equal to nuclear charge.*

(*c*) The production of protons (^1_1H) in the bombardment of lighter elements like ^6_3Li by deutrons (^2_1D) shows *the presence of protons in the nucleus.*

$$^6_3Li - {}^2_1D \rightarrow {}^7_3Li + {}^1_1H$$

(*d*) The bombardment of $^{24}_{12}Mg$ by ^4_2He gives $^{27}_{14}Si$ which is radioactive and hence disintegrates spontaneously to give stable isotope $^{27}_{13}Al$ and positrons ($_{+1}^{0}e$).

$$^{24}_{12}Mg + {}^4_2He \rightarrow \underset{\text{Radioactive}}{^{27}_{14}Si} + {}^1_0n$$

$$^{27}_{14}Si \rightarrow \underset{\substack{\text{Non-radio-}\\\text{active isotope}\\(i.e.\text{ stable})}}{^{27}_{17}Al} + \underset{\substack{\text{Posi-}\\\text{tron}}}{^{0}_{+1}e}$$

The production of positron during the disintegration of $^{27}_{14}Si$ shows the *presence of positron.*

2. Discovery of isotopes and isobars. The existence of isotopes and isobars was first recognised during the study of radioactive series.

3. Discovery of new elements. The investigations made in the field of artificial radioactivity have helped us in the discovery of transuranium elements (*i.e.* elements having atomic numbers higher than 92) which do not exist in nature. Also, a large number of radioisotopes of known elements have been discovered by transmutation reactions.

4. Release of nuclear energy. The processes involving nuclear fission and nuclear fussion release a huge amount of energy which can be used both in peace and war.

5. Production of radioisotopes. Artificial radioactivity has been used in producing radioisotopes of many elements which are used for various purposes such as :

(*i*) Use of radioisotopes as radioactive tracers.

(*ii*) Use of radioisotopes in the treatment of diseases (*Radiotherapy*).

(*iii*) Determination of the age of the earth by rock dating method.

(*iv*) Determination of the age of the recent objects by radiocarbon dating method.

All the above uses of radioisotopes are described under "*Uses of Radioisotopes*".

Isotopes

What are Isotopes ?

Different kinds of atoms of the same element which have the same atomic number but different mass numbers or atomic masses (or atomic weights) are called isotopes of that element. Since isotopes have the same atomic number, these can also be defined as *different atoms of the same element which have the same number of protons but different number of neutrons in their respective nuclei.*

The name isotope was first introduced by Soddy. The word *isotopes* has been derived form the Greek word meaning *same position* (*isos* = same; *topes* = position), since the isotopes occupy the same position in the periodic table. Isotopes have identical chemical properties and differ slightly only in respect of such properties as depend on their mass numbers.

Symbolic Representation of Isotopes

It is a common practice to represent each isotope of a given element by an appropriate symbol which is internationally adopted. In the representation of an isotope of a given element, the mass number (A) is shown as superscript at the head of the symbol of the element while the atomic number (Z) is shown as subscript at its bottom. The superscripts and subscripts may be shown at the left or right of the symbol of the element. The superscript (*i.e.* mass number) gives the sum of the number of protons and neutrons present in the nucleus while the subscript (*i.e.* atomic number) gives the number of protons or electrons. The symbolic representation of an isotope of an element, X has been shown in Fig. 5.13.

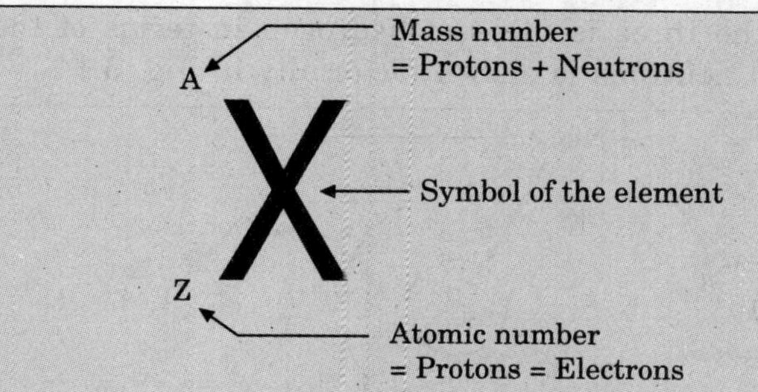

Fig. 5.13. Symbolic representation of an isotope of an element, X

As an example three isotopes of carbon (at no = 6) having mass numbers equal to 12, 13 and 14 can be shown as :

(*i*) ${}^{12}_{6}C$ or ${}^{12}C$ (*ii*) ${}^{13}_{6}C$ or ${}^{13}C$ (*iii*) ${}^{14}_{6}C$ or ${}^{14}C$.

Alternatively the above three isotopes of carbon can also be shown as C–12, C–13 and C–14 which are read as *carbon-twelve*, *carbon-thirteen* and *carbon-fourteen* respectively.

Structure of Isotopes

Since isotopes of an element have the same atomic number, each of these isotopes contains equal numbers of protons in the nucleus and equal number of electrons revolving in different orbits around the nucleus. Now since they have different mass numbers, they have different number of neutrons in their respective nuclei. Thus the number of protons (p), neutrons (n) and electrons (e) in an isotope with atomic number, Z and mass number, A $\left({}^{A}_{Z}X \text{ isotope}\right)$ is given by the following relations :

(*i*) No. of protons, $p = Z$

(*ii*) No. of neutrons, $n = A–Z$ $(\because A = n + p = n + Z)$

(*iii*) No. of electrons, $e = Z$

Now let us consider the structure of the isotopes of some elements in terms of the number of electrons, protons and neutrons in them.

1. Isotopes of hydrogen. There are three isotopes of hydrogen (At No. Z = 1) with mass number (A) 1, 2 and 3. These are : (*i*) *Ordinary hydrogen or protium*, ${}^{1}_{1}H$ (*ii*) *Heavy hydrogen or deuterium*, ${}^{2}_{1}H$ or D, and (*iii*) *Tritium*, ${}^{3}_{1}H$ or T. As is evident, each of these isotopes has one electron. The nucleus of protium has one proton only; that of deuterium has one proton and one neutron while that of tritium has one proton and two neutrons. The number of electrons, protons and neutrons in each of these three isotopes in given below :

Isotope	**Atomic number (Z)**	**Mass number (A)**	**Electrons (e) = Z**	**Protons (p) = Z**	**Neutrons (n) = (A–Z)**
${}^{1}_{1}H$	1	1	1	1	1–1 = 0
${}^{2}_{1}H$ or D	1	2	1	1	2–1 = 1
${}^{3}_{1}H$ or T	1	3	1	1	3–1 = 2

The structure of the three isotopes of hydrogen in terms of the number of electrons, protons and neutrons is shown pictorially in Fig. 5.14.

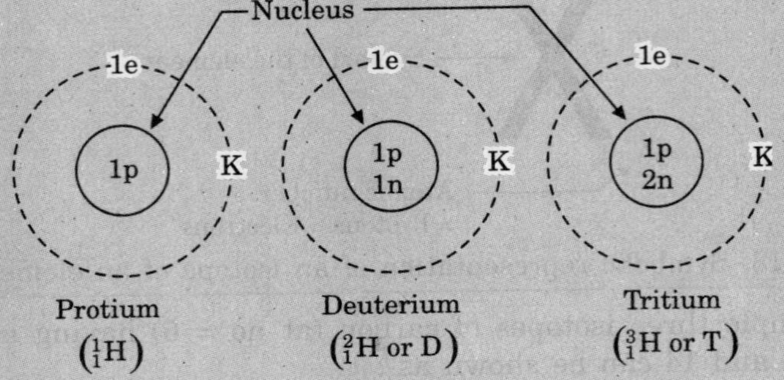

Fig. 5.14. Structure of the three isotopes of hydrogen in terms of the number of electrons, protons and neutrons.

2. Isotopes of oxygen. Oxygen has three isotopes with atomic number, Z = 8 and mass number, A equal to 16, 17 and 18. These are thus represented as $^{16}_{8}O$, $^{17}_{8}O$ and $^{18}_{8}O$ respectively. The number of electrons, protons and neutrons in each of them is given below :

Isotope	**Atomic number (Z)**	**Mass number (A)**	**Electrons (e) = Z**	**Protons (p) = Z**	**Neutrons (n) = (A–Z)**
$^{16}_{8}O$	8	16	8(2, 6)	8	16 – 8 = 8
$^{17}_{8}O$	8	17	8(2, 6)	8	17 – 8 = 9
$^{18}_{8}O$	8	18	8(2, 6)	8	18 – 8 = 10

The structure of the three isotopes is shown in Fig. 5.15.

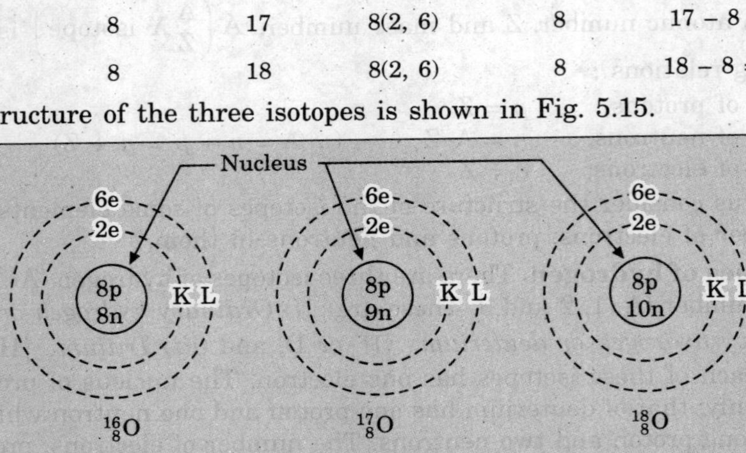

Fig. 5.15. Structure of the three isotopes of oxygen in terms of the number of electrons, protons and neutrons.

3. Isotopes of neon. Neon has three isotopes with atomic number, Z = 10, and mass number, A equal to 20, 21 and 22. These are thus represented as : $^{20}_{10}Ne$, $^{21}_{10}Ne$, $^{22}_{10}Ne$ respectively. The number of electrons, protons and neutrons in each of them is given below :

Isotope	Atomic number (Z)	Mass number (A)	Electrons (e) = Z	Protons (p) = Z	Neutrons (n) = (A–Z)
$^{20}_{10}Ne$	10	20	10 (2, 8)	10	20 – 10 = 10
$^{21}_{10}Ne$	10	21	10 (2, 8)	10	21 – 10 = 11
$^{22}_{10}Ne$	10	22	10 (2, 8)	10	22 – 10 = 12

The complete structure of these isotopes is shown below in Fig. 5.16.

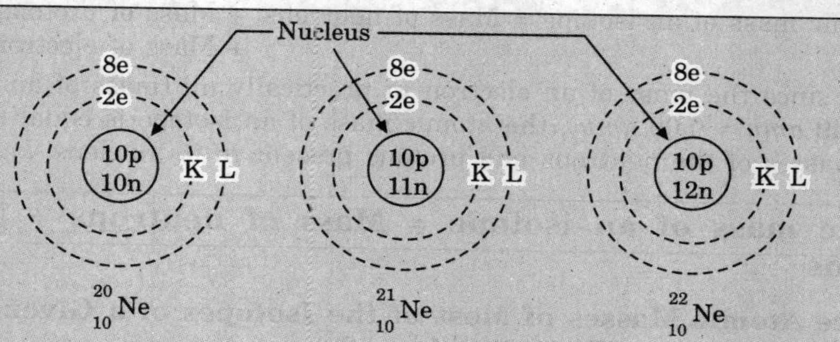

Fig. 5.16. Structure of the three isotopes of neon in terms of the number of electrons, protons and neutrons.

4. Isotopes of chlorine. Chlorine is a mixture of two isotopes with atomic number, Z = 17 and mass number A equal to 35 and 37. Thus these are represented as $^{35}_{17}Cl$ and $^{37}_{17}Cl$ respectively. The number of electrons, protons and neutrons in each of these isotopes is as follows :

Isotope	Atomic number (Z)	Mass number (A)	Electrons (e) = Z	Protons (p) = Z	Neutrons (n) = (A–Z)
$^{35}_{17}Cl$	17	35	17 (2, 8, 7)	17	35 – 17 = 18
$^{37}_{17}Cl$	17	37	17 (2, 8, 7)	17	37 – 17 = 20

The complete structure of these isotopes is given below in Fig. 5.17 :

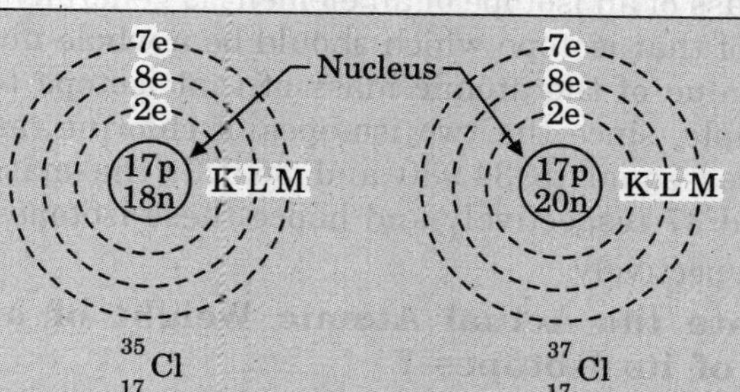

Fig. 5.17. Structure of the two isotopes of chlorine in terms of the number of electrons, protons and neutrons.

Almost all the elements are found in nature as a mixture of isotopes. The isotopes of some elements and their structure in terms of the number of electrons, protons and neutrons is given in Table 5.1.

It may be seen from the table that some elements like beryllium and fluorine are found in nature as a single isotope. Such elements are called **monoisotopic.** Only about 20 elements have been found to be monoisotopic.

What is Atomic Mass of an Isotope ?

We know that an atom is made up of neutrons, protons and electrons. The sum of the masses of these constituent particles in a given isotope of an element is known as its atomic mass. Thus :

Atomic mass of an isotope = Mass of neutrons + Mass of protons + Mass of electrons

Now since the mass of an electron in practically nil (mass of an electron = 0.0005449 *amu* ≈ 0.00 *amu*), the atomic mass of an isotope is equal to the sum of the masses of the neutrons and protons present in its nucleus, *i.e.*

Atomic mass of an isotope = Mass of neutrons + Mass of protons

Why are Atomic Masses of Most of the Isotopes of a Given Element Fractional and not Whole Numbers ?

We have seen that, since the mass of an electron is practically equal to zero, the atomic mass of an isotope of a given element is equal to the sum of the masses of neutrons and protons present in the nucleus of the isotope under consideration. Now since the mass of each of the protons and neutrons is not a whole number on the atomic weight scale (mass of a proton = 1.003757 *amu*, mass of a neutron = 1.00898 *amu*), the atomic mass of an isotope which is equal to the sum of the masses of the protons and neutrons is *not a whole number, i.e.* it is a fractional number. For example the atomic masses of the two isotopes of chlorine are 34.980 *amu* and 36.977 *amu* which are fractional numbers.

How to Calculate the Mass Number of an Isotope From its Atomic Masses ?

We have seen that mass number (*A*) of an isotope is a whole number. Now since the atomic mass of an isotope of an element is generally a fractional number, the mass number of that isotope which should be a whole number is obtained by rounding off the value of the atomic mass of that isotope to the nearest whole number. For example, since the two isotopes of chlorine (atomic number = 17) have atomic masses equal to 34.980 and 36.977, the mass numbers of these isotopes are 35 and 37 respectively and hence these isotopes are represented as $^{35}_{17}Cl$ and $^{37}_{17}Cl$ respectively.

How to Calculate the Actual Atomic Weight of an Element From Atomic Masses of its Isotopes ?

The atomic weight of an element is a statistical mean of the atomic masses of different isotopes of the elements. The value of atomic weight depends on the relative proportions of the various isotopes. For example :

(*i*) The actual atomic weight of neon which is found to consist of $^{20}_{10}Ne$, $^{21}_{10}Ne$ and $^{22}_{10}Ne$ isotopes in the percentage : ^{20}Ne = 90.92%, ^{21}Ne = 0.26% and ^{22}Ne = 8.82% is :

Table 5.1 Isotopes of some elements and their structure

Element and atomic number	Isotopes	Mass number (A)	Electrons, e = Z The distribution of electrons in K, L and M shells is given in bracket	Protons, p = Z	Neutrons, n = A–Z
Hydrogen (Z = 1)	$^{1}_{1}H$	1	1(1)	1	1–1 = 0
	$^{2}_{1}H$ or D	2	1(1)	1	2–1 = 1
	$^{3}_{1}H$ or T	3	1(1)	1	3–1 = 2
Helium (Z = 2)	$^{3}_{2}He$	3	2(2)	2	3–2 = 1
	$^{4}_{2}He$	4	2(2)	2	4–2 = 2
Lithium (Z = 3)	$^{6}_{3}Li$	6	3(2, 1)	3	6–3 = 3
	$^{7}_{3}Li$	7	3(2, 1)	3	7–3 = 4
Beryllium (Z = 4)	$^{9}_{4}Be$	9	4(2, 2)	4	9–4 = 5
Boron (Z = 5)	$^{10}_{6}B$	10	5(2, 3)	5	10–5 = 5
	$^{11}_{5}B$	11	5(2, 3)	5	11–5 = 6
Carbon (Z = 6)	$^{12}_{6}C$	12	6(2, 4)	6	12–6 = 6
	$^{13}_{6}C$	13	6(2, 4)	6	13–6 = 7
Nitrogen (Z = 7)	$^{14}_{7}N$	14	7(2, 5)	7	14–7 = 7
	$^{15}_{7}N$	15	7(2, 5)	7	15–7 = 8
Oxygen (Z = 8)	$^{16}_{8}O$	16	8(2, 6)	8	16–8 = 8
	$^{17}_{8}O$	17	8(2, 6)	8	17–8 = 9
	$^{18}_{8}O$	18	8(2, 6)	8	18–8 = 10
Fluorine (Z = 9)	$^{19}_{9}F$	19	9(2, 7)	9	19–9 = 10

Element and atomic number	Isotopes	Mass number (A)	Electrons, e = Z The distribution of electrons in K, L and M shells is given in bracket	Protons, p = Z	Neutrons, n = A–Z
Neon (Z = 10)	$^{20}_{10}Ne$	20	10(2, 8)	10	20–10 = 10
	$^{21}_{10}Ne$	21	10(2, 8)	10	21–10 = 11
	$^{22}_{10}Ne$	22	10(2, 8)	10	22–10 = 12
Chlorine (Z = 17)	$^{35}_{17}Cl$	35	17(2, 8, 7)	17	35–17 = 18
	$^{37}_{17}Cl$	37	17(2, 8, 7)	17	37–17 = 20
Argon (Z = 18)	$^{36}_{18}Ar$	36	18(2, 8, 8)	18	36–18 = 18
	$^{40}_{18}Ar$	40	18(2, 8, 8)	18	40–18 = 22

$$= \frac{90.92 \times 20 + 0.26 \times 21 + 8.82 \times 22}{(90.92 + 0.26 + 8.82)}$$

$$= \frac{1818.40 + 5.46 + 194.04}{100} = \frac{2017.90}{100}$$

$$= 20.1790 \approx 20.18$$

(*ii*) Ordinary chlorine consists of 75.4% of ^{35}Cl isotope and 24.6% of ^{37}Cl isotope. Hence its atomic weight as :

$$= \frac{75.4 \times 35 + 24.6 \times 37}{75.4 + 24.6} = 35.492$$

(*iii*) Two isotopes of ordinary hydrogen with mass numbers 1 and 2 have atomic masses equal to 1.007825 and 2.01473 and are in the ratio of 6400 : 1. Therefore, the atomic weight of hydrogen is :

$$= \frac{1.007825 \times 6400 + 2.01473 \times 1}{6400 + 1} = 1.0080$$

(*iv*) Two isotopes of Li have atomic masses equal to 6.017 *amu* and 7.018 *amu* and are present in the ratio 1 : 11.7. Thus the atomic weight of lithium is

$$= \frac{6.017 \times 1 + 7.018 \times 11.7}{1 + 11.7} = 6.93$$

List of Isotopes of Some Elements

The isotopes of some elements along with their atomic masses (in *amu*), mass numbers etc. are given in Table 5.2.

Table 5.2. Isotopes of some elements

Element	No. of Isotopes	Atomic Mass (amu)	Mass Number	Relative abundance of isotopes (%)	Actual atomic weight of the element
Boron (Z = 5)	2	10.016 11.012	10 11	18.5 81.5	10.811
Carbon (Z = 6)	2	12.003 13.007	12 13	98.8 1.2	12.0111
Nitrogen (Z = 7)	2	14.007 15.004	14 15	99.63 0.37	14.0067
Oxygen (Z = 8)	3	16.000 17.004 18.004	16 17 18	99.76 0.04 0.20	15.9994
Fluorine (Z = 9)	1	18.9984	19	100.00	18.9984
Neon (Z = 10)	3	19.993 21.000 21.998	20 21 22	90.92 0.26 8.82	20.183
Sodium (Z = 11)	1	22.9898	23	100.00	22.9898
Silicon (Z = 14)	3	27.985 28.985 29.983	28 29 30	92.3 4.6 3.1	28.086
Sulphur (Z = 16)	4	31.982 32.981 33.978 35.978	32 33 34 36	95.1 0.74 4.20 0.016	32.064
Chlorine (Z = 17)	2	34.980 36.977	35 37	75.4 24.6	35.453

Identification of Isotopes

Different isotopes of an element can be identified by any of the following mass spectrographs.

1. Aston's mass spectrograph. In 1919 Aston devised a mass spectrograph (Fig. 5.18) which can accurately sort out the positive ions of different isotopes of an element and determine their masses. In this spectrograph, a beam of positive rays obtained from a gaseous element in a discharge tube is passed through two slits S_1 and S_2 which are so arranged that only a thin ribbon of rays passes through an electric field. The thin rays consisting of positive ions of various isotopes of the element are sent between electrically charged plates P_1 and P_2. Here the beam of positive ions is deflected towards the negatively-charged plate, P_2.

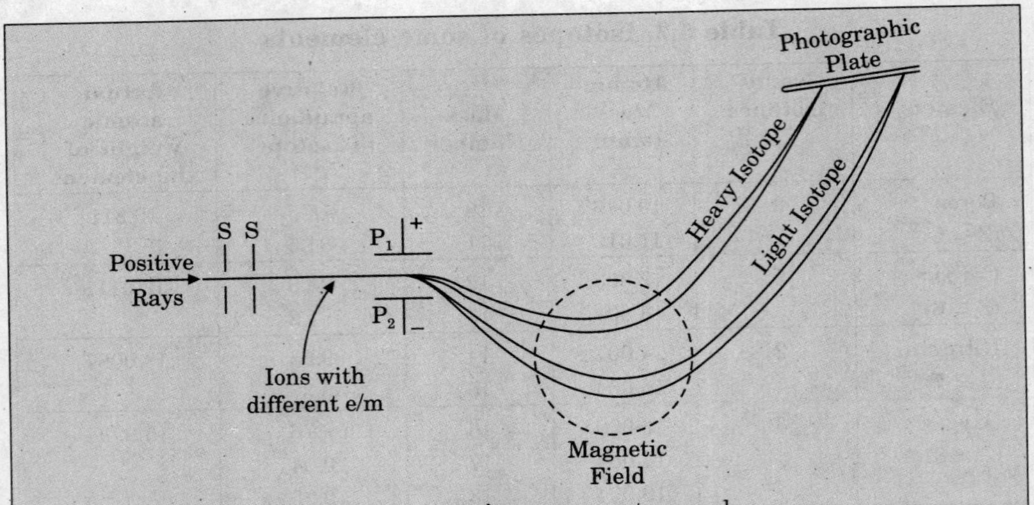

Fig. 5.18. Aston's mass spectrograph.

The angle of deflection depends on the mass, charge and velocity of particles constituting the positive rays. The slow moving ions of the same isotope are deflected more than the fast moving ions which cause a broadening of the beam. The broadened beam of ions is then allowed to pass through a magnetic field (shown by a dotted circle in the Figure) at right angles to the plane of the charged plates and is thus sent in a direction opposite to that caused by the electric field. Here also the slow moving particles (heavier particles) undergo maximum deflection. The strengths of electric and magnetic fields are so adjusted that all the ions of the same mass are focussed at the same point on the photographic plate where a sharp line or spot is obtained. The developed plate is known as **mass spectrogram.** Each line or spot seen on the photographic plate shows the existence of separate isotope. Further, the intensity of the line or spot in comparison with the lines of other isotopes gives the relative abundence of this particular isotope. This point is evident from the fact that an examination of the Aston's mass spectrogram of a sample of ordinary neon shows that the sample consists of two lines (see Fig. 5.19) which indicate that neon in a mixture of ^{20}Ne and ^{22}Ne isotopes. The relative intensities of these lines show that the relative abundance of these isotopes is as ^{20}Ne : ^{22}Ne = 9 : 1. Similarly the mass spectrogram of a sample of chlorine has two lines, indicating the presence of ^{35}Cl and ^{37}Cl isotopes (Fig. 5.19). The relative intensities of these lines show that the relative abundance of these isotopes is as : ^{35}Cl : ^{37}Cl = 3 : 1.

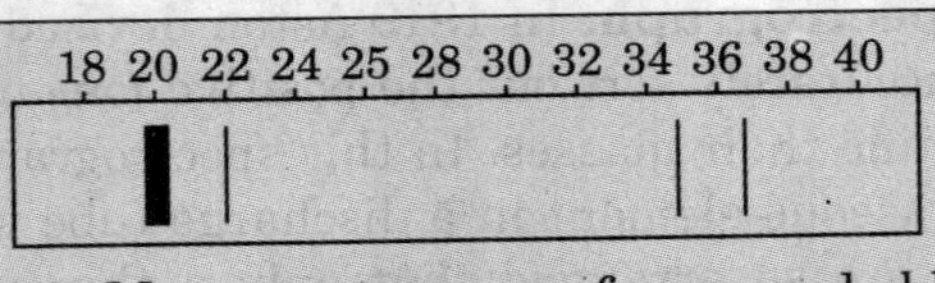

Fig. 5.19. Mass spectrogram of neon and chlorine.

Aston's mass spectrograph not only helps in identifying the isotopes present in an element but also helps in determining the actual atomic weight of a given element as is clear from the following example.

Example. *A sample of ordinary neon is found to consist of ^{20}Ne, ^{21}Ne and ^{22}Ne isotopes in the percentage ^{20}Ne = 90.92%, ^{21}Ne = 0.26% and ^{22}Ne = 8.82%. Calculate the actual atomic weight of neon.*

Solution. Actual atomic weight of neon

$$= \frac{90.92 \times 20 \times 0.26 \times 21 + 8.82 \times 22}{90.92 + 0.26 + 8.82}$$

$$= \frac{1818.40 + 5.46 + 194.04}{100} = \frac{2017.90}{100}$$

$$= 20.1790 \approx 20.18$$

2. Dempster's mass spectrograph. This spectrograph is shown in Fig. 5.20. In this apparatus a slow stream of gas or vapour of the sample is allowed to pass in between two perforated plates. Between these plates the stream is bombarded with the help of high-energy electrons which are produced by an electric gun. The gas atoms (A) are thus ionised and are converted into monopositive ions (A^+) ($A \xrightarrow{-e^-} A^+$). Now when a potential of 500 – 2000 volts is applied between the perforated accelerating plates, the positively-charged ions are strongly attracted towards the negatively charged plate. The positively-charged ions moving with accelerated speed then enter the magnetic field at right angles to their path and are thus made to move in a circular path.

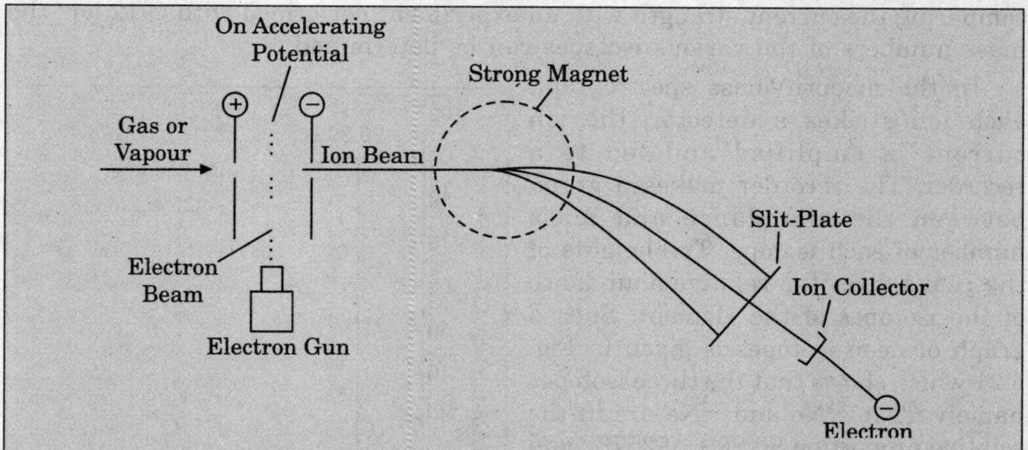

Fig. 5.20. Dempster's mass spectrograph shown in a simplified form.

Principle. If V is the potential applied between the accelerating plates and *e* is the charge on each positive particle, then the electrical energy is equal to Ve. This energy is imparted to the positive particles as kinetic energy, 1/2 mv^2, *i.e.*

$$Ve = \frac{1}{2} mv^2 \qquad ...(1)$$

Here *m* is the mass and *v* is the velocity of the particle.

If the strength of the magnetic field is H, then the magnetic force acting on the particle is equal to HeV which is again equal to centrifugal force, mv^2/r (r = radius of the circular path). Thus :

$$\text{H}ev = \frac{mv^2}{r} \quad \text{...(2)}$$

or
$$\text{H}^2e^2 = \frac{m^2v^2}{r^2}$$

Since
$$Ve = \tfrac{1}{2}mv^2 \quad \text{[From eqn. (1)]}$$

$$\frac{\text{H}^2e^2}{\text{Ve}} = \frac{m^2v^2}{r^2} \times \frac{2}{mv^2} = \frac{2m}{r^2}$$

or
$$e/m = \frac{2\text{V}}{\text{H}^2r^2} \quad \text{...(3)}$$

In the above equation [eqn. (3)], e, being the unit electrical charge, is constant. The radius (r) of the circular path is constant for a given spectrograph. Now if the strength of the magnetic field (H) is kept constant, than m is proportional to 1/V, *i.e.*

$$m \propto \frac{1}{\text{V}}$$

Thus by adjusting the accelerating potential (V), positively charged particles of mass m are made to fall on the collector plate. Each particle sets up a minute electric current which passes to the electrometer. The strength of the current thus gives the relative abundance of the individual particles (*i.e.* isotopes). By comparing the current strength with an experiment performed with C-12 ion, the mass numbers of the various isotopes can be determined.

In the modern mass spectrograph each ion strikes a detector, the ion current is amplified and fed to a recorder. The recorder makes a graph between the abundance and mass number of each isotope. The heights of the peaks gives the relative abundance of the isotopes of the element. Such a graph of neon isotopes is given in Fig. 5.21 which shows that the three isotopes namely ^{20}Ne, ^{21}Ne and ^{22}Ne are in the relative proportion 90.20%, 0.257%, and 8.82% respectively.

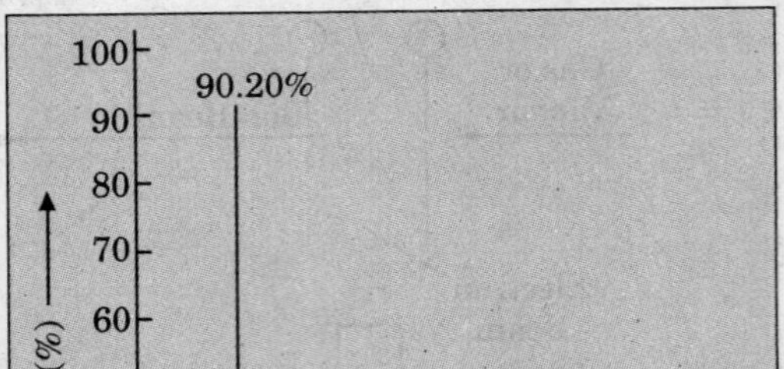

Fig. 5.21. Graph between the relative abundance (%) of the three isotopes of neon against their mass numbers.

Separation of Isotopes

Since all the isotopes of a given element have similar properties, it is almost impossible to separate them by chemical methods. However, some physical methods have been used for this purpose. In these methods the difference in physical properties of the isotopes which depend on the mass of the atom has been utilised. Some of the physical methods are discussed below :

1. Gaseous diffusion method. According to Grahm's law of diffusion, the rate of diffusion (r) of a gas is inversely proportional to the square root of its molecular weight (M), *i.e.*

$$r \propto \sqrt{\frac{1}{M}}$$

Thus if a mixture consisting of two gaseous isotopes is allowed to diffuse through a porous partition, the lighter isotope passes through the porous partition more rapidly than the heavier one. Gaseous ^{20}Ne and ^{22}Ne isotopes of neon, and ^{16}O and ^{18}O isotopes of oxygen have been separated by this method. The separation of gases ^{16}O and ^{18}O isotopes has been shown in Fig. 5.22. The mixture of gaseous isotopes is passed through a porous tube sealed in an outer jacket. The lighter isotope (^{16}O isotope) passes into the jacket while the heavier isotope (^{18}O isotope) goes into the residual gas.

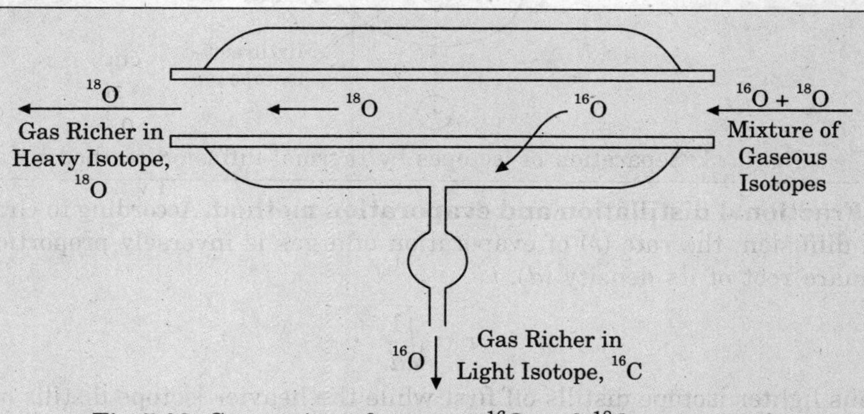

Fig 5.22. Separation of gaseous ^{16}O and ^{18}O isotopes of oxygen by gaseous diffusion method.

2. Thermal diffusion method. The apparatus used in this method consists of a long vertical cylindrical tube which is heated electrically to 600°C or above from inside by means of a thin platinum wire running down the axis of the tube (Fig. 5.23). This method is based on the tendency of heavier molecules of a gas to concentrate in the cooler region of the vertical tube. Thus when a mixture of gaseous isotopes is introduced into the tube, the molecules of the lighter isotope diffuse more rapidly to the central hot region of the tube. Here they are carried upwards by convection currents. The molecules of heavier isotope travel to the cooler inner surface of the tube and thus sink to the bottom. Thus lighter isotope collects at the top while the heavier isotope collects at the bottom. The process is repeated several times. This method has been used by Clausius for the separation of ^{35}Cl and ^{37}Cl isotopes of chlorine.

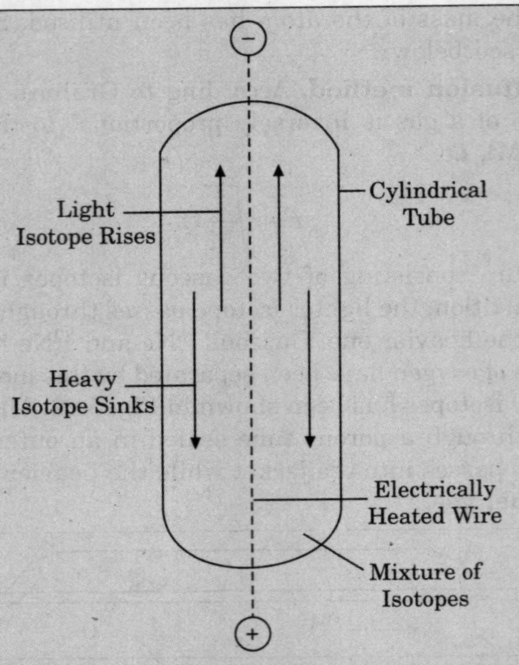

Fig. 5.23. Separation of isotopes by thermal diffusion method.

3. Fractional distillation and evaporation method. According to Graham's law of diffusion, the rate (r) of evaporation of a gas is inversely proportional to the square root of its density (d), *i.e.*

$$r \propto \sqrt{\frac{1}{d}}$$

Thus lighter isotope distills off first while the heavier isotope distills off last. For example, when liquid oxygen is distilled, the lighter isotope (^{16}O) comes off first and the heavier isotope (^{18}O) comes off in the end. The isotopes of neon have been separated by this method.

The separation of the two isotopes of Hg has been achieved by fractional evaporation at 40°C to 50°C in vacuum as shown in Fig. 5.24. The vapours of Hg are condensed on a surface cooled in liquid air which is placed a little above the evaporating Hg. After some time the liquid Hg left behind is found to consist of mostly heavier isotope which is taken out by opening stop-cock at the bottom. The frozen Hg collected at the cooler surface and consists of lighter isotope. It is removed, melted and evaporated under vacuum again. The whole process is repeated several times.

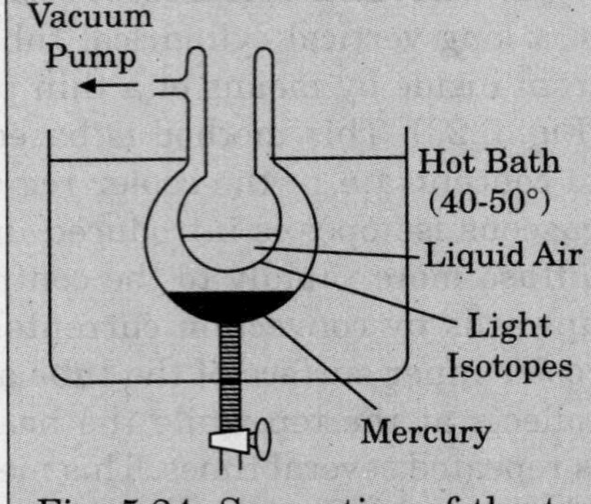

Fig. 5.24. Separation of the two isotopes of mercury by fractional evaporation

4. Electromagnetic method. This method is based on the principle of Dempster's mass spectrograph. This method has been used for the separation of the three isotopes of neon. The beam of ions of the three isotopes of neon namely ^{20}Ne, ^{21}Ne and ^{22}Ne as obtained in Dempster's mass spectrograph is passed between the poles of a magnet (Fig. 5.25). The different isotopes get difflected at different angles and are collected in cooled chambers which are placed in appropriate positions. The separation achieved by this method is complete, but the quantities of different isotopes obtained are very small.

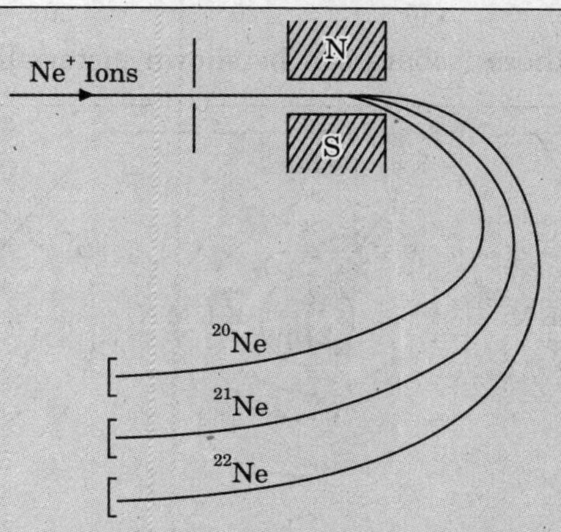

Fig. 5.25. Separation of the three isotopes of neon (^{20}Ne, ^{21}Ne and ^{22}Ne) by electromagnetic method making the use of Dempster's mass spectrograph.

Isobars

What are Isobars ?

The atoms of different elements which have the same mass number but different atomic number are called **isobars.** Since isobars have the same mass number, these can also be defined as atoms of different elements which have the same sum of protons and neutrons in the necleus of each of these atoms but different atomic number. The word *isobar* has been derived from the Greek word meaning *equally heavy* (*isos* = equal; *barys* = heavy). $^{40}_{18}Ar$, $^{40}_{19}K$ and $^{40}_{20}Ca$ are the examples of isobars, since each of them has the same mass number *i.e.* the sum of protons and neutrons is the same.

Structure of Isobars

Since isobars have the same mass number (A), the sum of protons and neutrons in the nucleus of each of these isobars is equal. Quite obviously the number of protons (p) and the number of electrons (e) is equal to atomic number (Z) while the number of neutrons (n) will be equal to (A–Z). Thus :

$$p = Z = e \qquad ...(i)$$

$$n = (A - Z) \qquad ...(ii)$$

On the basis of equations (*i*) and (*ii*), the number of electrons, protons and neutrons in each of the three isobars namely $^{40}_{18}Ar$, $^{40}_{19}K$ and $^{40}_{20}Ca$ is shown below :

Isobars	Atomic number (Z)	Mass number (A)	Electrons (e) = Z	Protons (p) = Z	Neutrons (n) = (A–Z)
$^{40}_{18}Ar$	18	40	18	18	40 – 18 = 22
$^{40}_{19}K$	19	40	19	19	40 – 19 = 21
$^{40}_{20}Ca$	20	40	20	20	40 – 20 = 20

The structure of these isobars can be shown pictorially as follows :

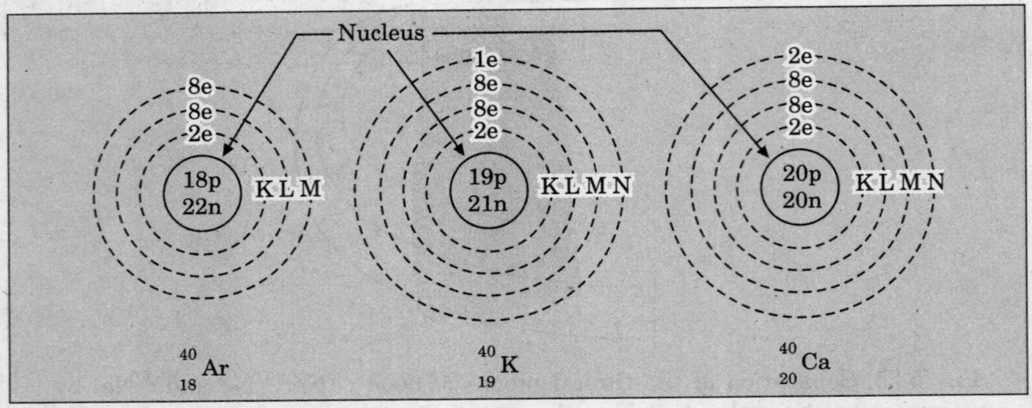

As can be seen from the structures given, the isobars have the same sum of protons and neutrons in the nucleus.

Isotones

What are Isotones ?

The atoms of different elements which have the same number of neutrons but different atomic number are called **isotones.** Since isotones have different atomic number, these can also be defined as the atoms of different elements which have the same number of neutrons but different number of protons and electrons. $^{14}_{6}C$, $^{15}_{7}N$ and $^{16}_{8}O$ are isotones, since each has the same number of neutrons as shown below :

Isotones	Atomic number (Z)	Mass number (A)	Electrons (e) = Z	Protons (p) = Z	Neutrons (n) = (A–Z)
$^{14}_{6}C$	6	14	6	6	14 – 6 = 8
$^{15}_{7}N$	7	15	7	7	15 – 7 = 8
$^{16}_{8}O$	8	16	8	8	16 – 8 = 8

The structure of these isotones can be shown pictorially as follows :

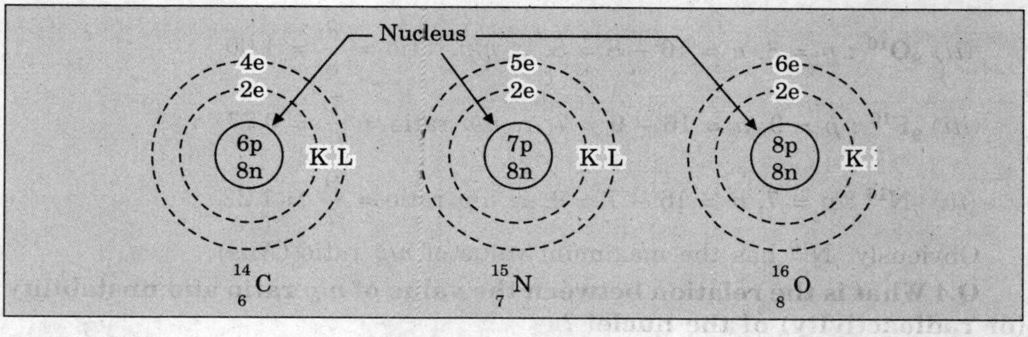

Questions with Answers

Q.1 What are magic numbers ? How is the stability of nuclei related with magic number ?

Ans : The nuclei which contain 2, 8, 20, 50, 82 or 126 protons (p) or neutrons (n) have been found to be extra stable, have large number of isotopes and have their nuclear shells completely-filled (*i.e* closed shells). The numbers 2, 8, 20, 50, 82 and 126 are called magic numbers for the nuclear shells.

Examples of stable nuclei (*i*) The nuclei $_{18}Ar^{38}$ (n = 38 – 18 = 20), $_{40}Zr^{90}$ (n = 90 – 40 = 50) and $_{56}Ba^{138}$ (n = 138 – 56 = 82) are stable, since the number of neutrons in these nuclei are magic numbers.

(*ii*) $_2He^4$ (p = 2, n = 4 – 2 = 2), $_8O^{16}$ (p = 8, n = 16 – 8 = 8), $_{20}Ca^{40}$ (p = 20, n = 40 – 20 = 20) and $_{82}Pb^{208}$ (p = 82, n = 208 – 82 = 126) nuclei are extra stable, since both the numbers of protons (p) and netrons (n) in these nuclei are magic numbers.

(*iii*) The end products in thorium (4n), uranium [(4n + 2)] and actinium [(4n + 3)] series are $_{82}Pb^{208}$, $_{82}Pb^{206}$ and $_{82}Pb^{207}$ respectively. All these isotopes are stable nuclei of $_{82}Pb$, since the number of protons (p = 82) in each isotope is a magic number.

Q.2 Which of the following nuclei would you expect to be radioactive and why ?

$$^{4}_{2}\mathbf{He},\ ^{39}_{20}\mathbf{Ca},\ ^{210}_{85}\mathbf{At}.$$

Ans : (*i*) ^4_2He has protons = 2, neutrons = 2, *i.e.*, both are magic numbers. Hence, it is expected to be *stable* (*non-radioactive*).

(*ii*) $^{39}_{20}Ca$ has protons = 20 (even number) and neutrons = 19 (odd number). 20 is one of the magic numbers. Nevertheless, n/p ratio is less than 1. Hence, it would lie below the stability belt. Hence, it is expected to be *radioactive*.

(*iii*) $^{210}_{85}At$ is *radioactive* as there is no stable nuclei with atomic number > 83.

Q.3 which of the following species has the maximum value of n/p ratio ? (*i*) Ne^{16} (*ii*) O^{16} (*iii*) F^{16} (*iv*) N^{16}

Ans : (*i*) $_{10}Ne^{16}$: p = 10, n = 16 – 10 = 6, $\therefore$ n/p ratio = $\frac{6}{10}$ = 0.60

(*ii*) $_8O^{16}$: $p = 8$, $n = 16 - 8 = 8$, $\therefore$ n/p ratio $= \dfrac{8}{8} = 1.00$

(*iii*) $_9F^{16}$: $p = 9$, $n = 16 - 9 = 7$, $\therefore$ n/p ratio $= \dfrac{7}{9} = 0.77$

(*iv*) $_7N^{16}$: $p = 7$, $n = 16 - 7 = 9$, $\therefore$ n/p ratio $= \dfrac{9}{7} = 1.22$

Obviously $_7N^{16}$ has the maximum value of n/p ratio (**Ans**).

Q.4 What is the relation between the value of n/p ratio and unstability (or radioactivity) of the nuclei ?

Ans : The nuclei having higher value of n/p ratio (or lower value of p/n ratio) are more unstable and hence are radioactive.

Q.5 In the emission of a $\overline{\beta}$-particle the atomic number of the daughter element increases by 1. Explain why ?

Ans : Let us consider the following nuclear reaction which involves $\overline{\beta}$-emmission.

$_6C^{14}$	$\longrightarrow$ $_7N^{14}$	+ $_{-1}e^0$	+ $_0\upsilon^0$
$p = 6$	$p = 7$	$\overline{\beta}$-particle	Neutrino
$n = A - p$	$n = A - p$		
$= 14 - 6$	$= 14 - 7$		
$= 8$	$= 7$		

This reaction shows that the number of protons increases by 1 ($6 \rightarrow 7$) and that of neutrons decreases by 1 ($8 \rightarrow 7$). The increase in the number of protons is due to the fact thus one neutron is converted into proton.

$$_0n^1 \rightarrow {}_{+1}p^1 + {}_{-1}e^0$$

Due to the increase in the number of protons the atomic number of the daughter element increases by 1.

Q.6 Calculate the amount of a radioactive element ($t_{1/2}$ = 140 days) to which 1 g of the element will reduce in 560 days.

Ans : $N_o = 1$ g, $t = 560$ days, $t_{1/2} = 140$ days.

$\therefore$ No. of half-lives in 560 days $(n) = \dfrac{t}{t_{1/2}}$

$$= \frac{560}{140} = 4$$

$\therefore$ Amount of the element left after 560 days or 4 half-periods

$$= \frac{N_0}{2^n} = \frac{1}{2^4}\text{ g} = \mathbf{\frac{1}{16}\text{ g}}\ \textbf{(Ans)}$$

Q.7 I^{131} has half-life period of 13.3 hours. After 79.8 hours, what fraction of I^{131} will remain ?

Ans : $t_{1/2}$ = 13.3 hours

$\therefore$ No. of half-lives in 79.8 hours $= \dfrac{79.8}{13.3} = 6$ half-lives

Amount left after n half-lives (N) $= \dfrac{N_0}{2^n}$

or Fraction left $\left(\dfrac{N}{N_0}\right) = \dfrac{1}{2^n} = \dfrac{1}{2^6} = \dfrac{1}{64}$.

Q.8 $^{234}_{90}Th \rightarrow 7\alpha + 6\beta + {}^{A}_{Z}X$**. Identify** $^{A}_{Z}X$**.**

Ans : $^{234}_{90}Th \rightarrow 7\,{}^{4}_{2}\alpha + 6\,{}^{0}_{-1}\beta + {}^{A}_{Z}X$

Balancing the atomic number and mass number on both sides, we have

$$234 = 7 \times 4 + 6 \times 0 + A$$

or $$A = 234 - 28 = 206$$

$$90 = 7 \times 2 + 6\,(-1) + Z$$

or $$Z = 90 - 14 + 6 = 82$$

Hence, $^{A}_{Z}X$ is $^{206}_{82}Pb$.

Q.9 ^{227}Ac has a half-life of 22.0 years with respect to radioactive decay. The decay follows two parallel paths, one leading to ^{227}Th and the other to ^{223}Fr. The percentage yields of these two daughter nuclides are 2.0 and 98.0 respectively. What are the decay constants (λ) for each of the separate paths ?

Ans : The decay of ^{227}Ac into ^{227}Th and ^{223}Fr can be shows as :

$$^{227}Ac \xrightarrow{\lambda_1} {}^{227}Th\left(2\% = \frac{2}{100}\right)$$

$$^{227}Ac \xrightarrow{\lambda_2} {}^{223}Fr\left(98\% = \frac{98}{100}\right)$$

Here λ_1 and λ_2 are the disintegration constants for the two paths and over all disintegration constant is λ.

$$\lambda = \frac{0.693}{t_{1/2} \text{ of } {}^{227}Ac} = \lambda_1 + \lambda_2$$

or $$\frac{0.693}{22 \text{ yrs}} = \lambda_1 + \lambda_2$$

or $$\lambda_1 + \lambda_2 = 0.0315 \text{ yrs}^{-1}$$

Fractional yield of ^{227}Th $= \dfrac{2}{100} = \dfrac{\lambda_1}{\lambda_1 + \lambda_2}$

or $$\frac{2}{100} = \frac{\lambda_1}{0.0315}$$

This equatian gives λ_1 = **6.3×10^{-4} yrs^{-1} (Ans)**

Similarly,

$$\text{Fractional yield of } ^{223}\text{Fr} = \frac{98}{100} = \frac{\lambda_2}{\lambda_1 + \lambda_2}$$

or $$\frac{98}{100} = \frac{\lambda_2}{0.0315}$$

This equantian gives λ_2 = **308.7×10^{-4} yrs^{-1} (Ans)**

Q.10 ^{64}Cu (half-life = 12.8 *h*) decays by β^- emission (38%), β^+ emission (19%) and electrons capture (43%). Write the decay products and calculate partial half-lives for each of the decay processes.

Ans.

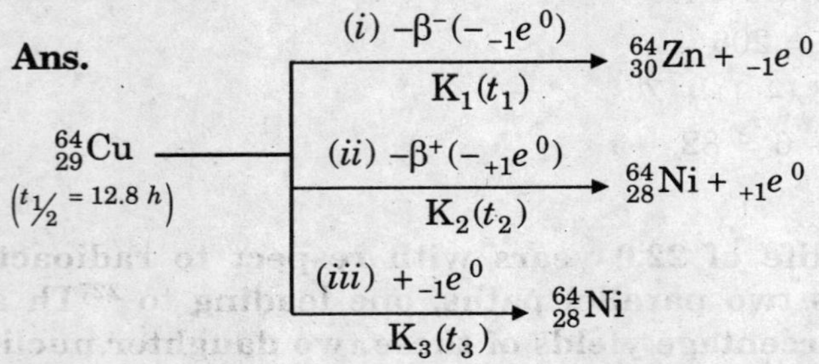

If t_1, t_2 and t_3 are the half-life for the partial processes (*i*), (*ii*) and (*iii*) respectively, then :

$$t_1 = \frac{0.693}{K_1} = \frac{0.693}{0.38\ K}$$

(Here K_1 is the disintegration constant for process (*i*) and K is the rate constant for the over all process)

$$= \frac{0.693}{0.38 \times 0.693 / t_{1/2} \text{ of Cu}}$$

$$= \frac{0.693 \times t_{1/2} \text{ of Cu}}{0.38 \times 0.693} = \frac{12.8}{0.38} = \mathbf{33.68\ hr\ (Ans.)}$$

Similarly :

$$t_2 = \frac{12.8}{0.19} = \mathbf{67.37\ hr\ (Ans.)}$$

and, $$t_3 = \frac{12.8}{0.43} = \mathbf{29.77\ hr\ (Ans.)}$$

University Questions

1. Describe the nature and properties of the rays emitted by radioactive materials. (*Meerut 1981*)
2. Describe the properties of α, β and γ-rays. (*North Bengal 81; Kerala 80; Saugar 82*)

3. Select any two members of the same radioactive series : $_{92}U^{238}$, $_{88}Ra^{224}$, $_{82}Pb^{206}$, $_{92}U^{235}$, $_{82}Pb^{208}$, $_{86}Rn^{22}$, $_{83}Bi^{214}$. *(Lucknow 85)*
4. Name the different radioactive series you know. How can you find that a particular isotope belongs to which radioactive series? To which radioactive series respectively $_{90}Th^{231}$, $_{82}Pb^{266}$ and $_{89}Ac^{228}$ belong? *(Kanpur 86)*
5. Calculate the value of disintegration constant for $_{86}Rn^{222}$, if its half-life period is 3.82 days. *(Kanpur 85)*
6. If the activity of a radioactive element drops to 1/6 of its initial value in one hour and twenty minutes, what are its disintegration constant and half-life period. *(Raj. 84)*
7. Explain the theory of radioactive disintegration. *(Meerut 1981)*
8. Write short note on "Isotopes". *(Calicut 80; Banaras 80; Saugar 82; Raj. 81, 84; MD Rohtak 84; Garhwal 88)*
9. Explain the principle of Aston's mess spectrograph. *(Raj. 82; Bundelkhand 80)*
10. Why can the isotopes of an element not be separated by chemical methods ? *(Himachal 82; Kashmir 83; GND 82)*
11. Discuss Dempster's mass spectrographic method for the identification of isotopes. *(GND 82)*
12. Write a note on "Separation of isotopes". *(Jiwaji 84)*
13. Write a note on "Isobars". *(Calicut 80; Banaras 80; Saugar 82; Raj. 81, 84; MD Rohtak 84; Garhwal 88; Madras 87; Andhra 87)*
14. What is the difference between isotope and isotone ? Explain with examples. *(Meerut M.Sc. 89)*
15. Discuss the applications of radioactive isotopes. *(Kanpur 99)*
16. Discuss the uses of isotopes as tracers. *(Gujrat 2003)*
17. Name the isotopes of hydrogen. *(Nagpur 2003)*
18. (*i*) What do you mean by half-hfe period of a radioactive element ? Deduce the formula for it.

    (*ii*) 1 gram of $_{79}Au^{198}$ ($T_{1/2}$ = 65 hrs) emits a β-particle and gives a stable isotope of Hg. Calculate the amount of Hg after 260 hrs.

    (*iii*) Write a note on "Disintegration Series" or "Radioactive Equilibrium".

    (*iv*) Write the differences between nuclear changes and chemical changes.

    (*iv*) Write a note on "Group Displacement Law". How many α and β particles should be emitted to change $_{90}Th^{232}$ to $_{82}Pb^{208}$ (stable). *(Meerut 2009)*

CHAPTER

6 Nuclear Transmutations and Artificial Radioactivity

Nuclear Transmutation

What is Nuclear Transmutation ?

Nuclear transmutation, in general, may by defined as the conversion of one element into another, *i.e.*, one type of atom into another. Nuclear transmutation *is such a conversion achieved by artificial means, as distinct from the spontaneous transmutation witnessed in natural radioactive substances.* Nuclear transmutation is also called **nuclear transformation** or **disintegration** or **nuclear reaction,** since the transmuted element disintegrates forming a different new element.

The study of the natural radioactive phenomena paved the way for artificial transmutation. For, the occurrence of spontaneous disintegration of the nucleus in the heavy radioactive elements readily suggested the idea whether it might not be possible to break up nuclei of other ordinarily stable elements by bombarding them artificially with high-speed particles Rutherford took up the hint and performed a series of experiments in which the atoms of different stable elements such as nitrogen, aluminium, phosphorus, etc., were subjected to an intense bombardment by the high velocity α-particles (as *projectiles*) obtained from natural radioactive substances and finally, in 1919, succeeded in transmuting these elements. Now a days other bombarding particles (*i.e.*, projectiles) like protons $\left({}_{1}^{1}\text{H}\right)$, deuterons $\left({}_{1}^{2}\text{H}\right)$ etc., are also used for the same purpose.

Rutherford, in his first experiment which he performed to demonstrate the first artificial transmutation, directed a beam of α-particles obtained from radium against a thin layer of nitrogen. The outcoming α-particles were then absorbed by placing some material (*e.g.* 1 *mm* thick aluminium foil) in between the nitrogen and ZnS screen. This indicated the presence of some new particles, which were identified from their deflection in the magnetic field as high energy protons $\left({}_{1}^{1}\text{H}\right)$. The capture of an α-particle followed by the ejection of a proton led Rutherford to believe that ${}_{7}^{14}\text{N}$ is transformed into a rare isotope of oxygen, ${}_{8}^{17}\text{O}$. This transformation (or *disintegration* or *nuclear reaction* as it is called) may be represented by an equation similar to that used for chemical reactions :

$$ {}_{7}^{14}\text{N} + \underset{\alpha\text{-particles}}{{}_{2}^{4}\text{He}} \rightarrow {}_{8}^{17}\text{O} + \underset{\text{proton}}{{}_{1}^{1}\text{H}} $$

Since, like chemical equations, such equations must balance the charge and mass numbers have to be the same on the two sides of these equations.

The above reaction may also be represented by the abbreviated notation : ^{14}N (α, p) ^{17}O which would be interpreted as follows : when a nitrogen (^{14}N) nucleus, called *target nucleus*, interacts with an alpha particle (α), referred to as *incident* or *projectile* (*bombarding particle* or *bullet*), a proton (p) is ejected and an oxygen (^{17}O) nucleus called *recoil* nucleus remains behind.

The exact nature of the nuclear reaction which takes place when an α-particle makes head-on collision with a nitrogen nucleus was not clearly understood at first. However, soon a satisfactory explanation backed by experimental verification was achieved. It was proved that α-particle is captured by nitrogen nucleus thus forming an *unstable compound nucleus* (*fluorine atom, not found in nature*) which immediately disintegrates by ejecting a proton (^1_1H). The intermediate unstable nucleus is shown in bracket.

$$^{14}_{7}N + ^{4}_{2}He \rightarrow \underset{\substack{\text{Unstable}\\\text{intermediate}\\\text{compound}\\\text{nucleus}}}{[^{18}_{9}F]} \rightarrow ^{17}_{8}O + ^{1}_{1}H$$

Hence, it is seen that starting *with two stable nuclei of helium and nitrogen, we get two stable nuclei of oxygen and hydrogen.* In other words, an atom of nitrogen has been transformed into an atom of oxygen. *This process is called* **transmutation of elements.** Since the intermediate compound molecule has temporary existence, its is often omitted from any discussion of the nuclear reaction.

The $^{17}_{8}O$ nucleus formed in this nuclear reaction is stable and hence this nuclear reaction does not lead to the production of artificial radioactivity.

Rutherford, encouraged by this first success, continued his researches and in collaboration with Chadwick (1921 – 24) showed that many other light elements, from boron (Atomic No. 5) to potassium (Atomic No. 19) with the exception of carbon and oxygen, could be transmuted by bombardment with α-particles, resulting in the emission of protons (^1_1H). Some of these nuclear reactions are given below :

Light elements		***α-particles***	***Intermediate unstable compound nuclie***		***Stable elements***		***Protons*** (^1_1H)
$^{10}_{5}B$	+	$^{4}_{2}He$	$\rightarrow [^{14}_{7}N]$	$\rightarrow$	$^{13}_{6}C$	+	$^{1}_{1}H$
$^{23}_{11}Na$	+	$^{4}_{2}He$	$\rightarrow [^{27}_{13}Al]$	$\rightarrow$	$^{26}_{12}Mg$	+	$^{1}_{1}H$
$^{27}_{13}Al$	+	$^{4}_{2}He$	$\rightarrow [^{31}_{15}P]$	$\rightarrow$	$^{30}_{14}Si$	+	$^{1}_{1}H$
$^{32}_{16}S$	+	$^{4}_{2}He$	$\rightarrow [^{36}_{18}Ar]$	$\rightarrow$	$^{35}_{17}Cl$	+	$^{1}_{1}H$
$^{39}_{19}K$	+	$^{4}_{2}He$	$\rightarrow [^{43}_{21}Sc]$	$\rightarrow$	$^{42}_{20}Ca$	+	$^{1}_{1}H$

It will be seen that in each case, charge of the product nucleus is increased by one unit and the mass is increased by three units. Such reactions are also referred to as (α, p) reactions.

A serious difficulty that arose during these transmutation reactions was that the yield of protons produced rapidly decreased with increasing atomic number of the target nuclei. This was due to that fact the α-particles, being positively charged, are repelled by the positive charge on the target nuclei. Thus α-particles are deflected away or scattered as they approach the positive field of the target nucleus. Consequently, α-particles are not focused properly on to the target nuclei. This indicated that α-particles did not have sufficient energy to penetrate the target nucleus. Hence, attention was directed towards exploring the means to obtain high-velocity charged particles from suitable high-voltage accelerators. In 1932, Cockcroft and Walton, working in Rutherford's laboratory, constructed the first high tension machine (atom smascher) which could accelerate protons to one million electron volts. Using the artificially accelerated protons of energy 0.15 MeV they brought about the transmutation of ${}^{7}_{3}\text{Li}$. This was the first case of a nuclear disintegration brought about by purely artificial means.

$${}^{7}_{3}\text{Li} + {}^{1}_{1}\text{H} \rightarrow \underset{\text{Unstable nucleus}}{[{}^{8}_{4}\text{Be}]} \rightarrow {}^{4}_{2}\text{He} + {}^{4}_{2}\text{He}(p, 2\alpha)$$

As shown in the nuclear equation given above, a portion of energy 0.15 MeV enters a lithium nucleus to form a new but unstable beryllium nucleus. This nucleus splits up into two α-particles which are driven apart with great violence. The production of α-particles in the above nuclear reaction was confirmed by following their track in cloud chambers. Meanwhile, E.O. Lawrence and co-workers, at the University of California, devised an instrument called **cyclotron** for the purpose. A modified instrument called **synchroton** has been built and designed to produce the protons of ten billion electron volts energy.

Bohr's Theory of Nuclear Reactions

According to Bohr's theory, the bombarding particle combines with the target nucleus to form an intermediate compound nucleus which is in the excited state and hence is unstable. This unstable intermediate compound nucleus has small life period (10^{-12} to 10^{-14} seconds) and hence breaks up to give the final product, called recoil nucleus. The nature of the final product depends on the amount of energy associated with the unstable intermediate compound nucleus. For example when ${}^{27}_{13}\text{Al}$ is bombarded with protons (${}^{1}_{1}\text{H}$), an unstable intermediate compound nucleus, $[{}^{28}_{14}\text{Si}]$ is formed which breaks up in any of the following ways depending on the amount of energy associated with the intermediate compound nucleus.

$$\underset{\substack{\text{Target}\\\text{nucleus}}}{{}^{27}_{13}\text{Al}} + \underset{\substack{\text{Bombarding}\\\text{particle}}}{{}^{1}_{1}\text{H}} \rightarrow \underset{\substack{\text{Unstable}\\\text{intermediate}\\\text{compound}\\\text{nucleus}}}{[{}^{28}_{14}\text{Si}]} \begin{cases} \rightarrow {}^{27}_{14}\text{Si} + {}^{1}_{0}n \\ \rightarrow {}^{24}_{12}\text{Mg} + {}^{4}_{2}\text{He} \\ \rightarrow \underset{\substack{\text{Recoil}\\\text{nucleus}\\\text{(Final}\\\text{product)}}}{{}^{24}_{11}\text{Na}} + {}^{1}_{0}n + 3{}^{1}_{1}\text{H} \end{cases}$$

Classification of Nuclear Reactions

The classification of nuclear reactions can be made (1) on the basis of overall energy transformation that occurs during the nuclear reaction, and (2) on the basis of the nature of the bombarding particle used to bombard the target nucleus. Thus the following two classifications are obtained :

I. Classification based on overall energy transformation. This classification gives the following types of nuclear reactions :

(1) Capture reactions. In these reactions the bombarding particle is captured or absorbed by the target nucleus with the emission of γ-rays. For example :

$$^{85}_{37}\text{Rb} + ^{1}_{0}n \rightarrow ^{86}_{37}\text{Rb} + ^{0}_{0}\gamma$$

$$^{12}_{6}\text{C} + ^{1}_{1}\text{H} \rightarrow ^{13}_{7}\text{N} + ^{0}_{0}\gamma$$

$$^{238}_{92}\text{U} + ^{1}_{0}n \rightarrow ^{239}_{92}\text{U} + ^{0}_{0}\gamma$$

(2) Particle-particle reactions. In these reactions the bombarding particle is absorbed by the target nucleus to form the compound nucleus which breaks down to give the final product. For example :

$$\begin{array}{lll} ^{24}_{11}\text{Na} + ^{4}_{2}\text{He} \rightarrow & & \rightarrow ^{26}_{13}\text{Al} + 2\,^{1}_{0}n \\ ^{27}_{13}\text{Al} + ^{1}_{0}n \rightarrow & \left(\begin{array}{c} \\ ^{28}_{13}\text{Al} \\ \\ \end{array}\right) & \rightarrow ^{28}_{13}\text{Al} + ^{0}_{0}\gamma \\ ^{26}_{12}\text{Mg} + ^{2}_{1}\text{H} \rightarrow & & \rightarrow ^{27}_{12}\text{Mg} + ^{1}_{1}\text{H} \\ ^{27}_{12}\text{Mg} + ^{1}_{1}\text{H} \rightarrow & \text{Compound nucleus} & \rightarrow ^{24}_{11}\text{Na} + ^{4}_{1}\text{He} \end{array}$$

(3) Fission reactions. These are the reactions in which a heavier nucleus breaks down into two or more lighter nuclei of almost equal size and a large amount of energy is also liberated. For example $^{235}_{92}\text{U}$ breaks down into $^{144}_{56}\text{Ba}$ and $^{90}_{36}\text{Kr}$ nuclei by slow neutrons.

$$^{235}_{92}\text{U} + ^{1}_{0}n\ (\text{slow}) \rightarrow ^{144}_{56}\text{Ba} + ^{90}_{36}\text{Kr} + 2\,^{1}_{0}n + \text{Energy}$$

(4) Fusion reactions. These are the reactions in which two lighter nuclei are combined or fused together to give a stable and heavier nucleus, and a large amount of energy is liberated. For example the fusion or combination of deuterium ($^{2}_{1}\text{H}$) and tritium ($^{3}_{1}\text{H}$) gives a stable nucleus of helium ($^{4}_{2}\text{He}$) and energy equal to 17.6 MeV is also liberated.

$$^{2}_{1}\text{H} + ^{3}_{1}\text{H} \rightarrow ^{4}_{2}\text{He} + ^{1}_{0}n + 17.6\ \text{MeV} \rightarrow$$

(5) Spallation reactions. These reactions were discovered by Seaborg and Perimen in 1947. In these reactions the high-speed bombarding particle is absorbed to form a normal nucleus and a large number of light particles. For example, the bombardment of $^{63}_{29}\text{Cu}$ nucleus by high-speed α-particle having energy = 400 MeV gives $^{37}_{17}\text{Cl}$ nucleus 14 protons and 16 neutrons.

$$^{63}_{29}\text{Cu} + ^{4}_{2}\text{He} \rightarrow ^{37}_{17}\text{Cl} + 14\ ^{1}_{1}\text{H} + 16\ ^{1}_{0}n.$$

II. Classification based on the nature of the bombarding particle. This classification gives the following types of nuclear reactions :

(1) Nuclear reactions induced by alpha particles ($^4_2$He). These reactions may be of following types :

(*a*) (α, *p*) *reactions*. These reactions are common with the elements of low atomic number. Examples are :

$$^{14}_{7}\text{N} + ^{4}_{2}\text{He} \rightarrow ^{17}_{8}\text{O} + ^{1}_{1}\text{H}$$
$$^{10}_{5}\text{B} + ^{4}_{2}\text{He} \rightarrow ^{13}_{6}\text{C} + ^{1}_{1}\text{H}$$
$$^{23}_{11}\text{Na} + ^{4}_{2}\text{He} \rightarrow ^{26}_{12}\text{Mg} + ^{1}_{1}\text{H}$$
$$^{37}_{13}\text{Al} + ^{4}_{2}\text{He} \rightarrow ^{30}_{14}\text{Si} + ^{1}_{1}\text{H}$$

(*b*) (α, *n*) *reactions*. By using more energetic α-paticles these reactions may be obtained. Examples are :

$$^{9}_{4}\text{Be} + ^{4}_{2}\text{He} \rightarrow ^{12}_{6}\text{C} + ^{1}_{0}n$$
$$^{40}_{18}\text{Ar} + ^{4}_{2}\text{He} \rightarrow ^{43}_{20}\text{Ca} + ^{1}_{0}n$$
$$^{7}_{3}\text{Li} + ^{4}_{2}\text{He} \rightarrow ^{10}_{5}\text{B} + ^{1}_{0}n$$

(*c*) (α, γ) *reactions*. Following is a typical example of this type of reaction.

$$^{7}_{3}\text{Li} + ^{4}_{2}\text{He} \rightarrow ^{11}_{5}\text{B} + ^{0}_{0}\gamma$$

(2) Nuclear reactions induced by protons ($^1_1$H). These reactions are of the following types.

(*a*) (*p*, α) *reactions*. Examples are :

$$^{7}_{3}\text{Li} + ^{1}_{1}\text{H} \rightarrow ^{4}_{2}\text{He} + ^{4}_{2}\text{He}$$
$$^{19}_{9}\text{F} + ^{1}_{1}\text{H} \rightarrow ^{16}_{8}\text{O} + ^{4}_{2}\text{He}$$
$$^{14}_{7}\text{N} + ^{1}_{1}\text{H} \rightarrow ^{11}_{6}\text{C} + ^{4}_{2}\text{He}$$
$$^{9}_{4}\text{Be} + ^{1}_{1}\text{H} \rightarrow ^{6}_{3}\text{Li} + ^{4}_{2}\text{He}$$

(*b*) (*p*, *n*) *reactions*. If proton has sufficient energy, probability of (*p*, *n*) type reactions becomes high. Some examples are :

$$^{18}_{8}\text{O} + ^{1}_{1}\text{H} \rightarrow ^{18}_{9}\text{F} + ^{1}_{0}n$$
$$^{58}_{28}\text{Ni} + ^{1}_{1}\text{H} \rightarrow ^{58}_{29}\text{Cu} + ^{1}_{0}n$$
$$^{65}_{29}\text{Cu} + ^{1}_{1}\text{H} \rightarrow ^{65}_{30}\text{Zn} + ^{1}_{0}n$$
$$^{23}_{11}\text{Na} + ^{1}_{1}\text{H} \rightarrow ^{23}_{12}\text{Mg} + ^{1}_{0}n$$

(*c*) (*p*, γ) *reactions*. With proton as the projectile, *radiative capture* processes of (*p*, γ) type have been observed for a number of lighter elements.

$$^{7}_{3}\text{Li} + ^{1}_{1}\text{H} \rightarrow ^{8}_{4}\text{Be} + ^{0}_{0}\gamma$$
$$^{27}_{13}\text{Al} + ^{1}_{1}\text{H} \rightarrow ^{28}_{14}\text{Si} + ^{0}_{0}\gamma$$
$$^{12}_{6}\text{C} + ^{1}_{1}\text{H} \rightarrow ^{13}_{7}\text{N} + ^{0}_{0}\gamma$$

$$^{14}_{7}N + ^{1}_{1}H \rightarrow ^{15}_{8}O + \gamma$$

$$^{50}_{24}Cr + ^{1}_{1}H \rightarrow ^{51}_{25}Mn + \gamma$$

(*d*) (*p*, *D*) *reactions.* These reactions are more common with heavier elements and protons are of fairly high energy.

$$^{9}_{4}Be + ^{1}_{1}H \rightarrow ^{8}_{4}Be + ^{2}_{1}H \text{ or D}$$

$$^{7}_{3}Li + ^{1}_{1}H \rightarrow ^{6}_{3}Li + ^{2}_{1}H \text{ or D}$$

(3) Nuclear reactions induced by deuterons ($^{2}_{1}$H or D). Some of the nuclear reactions induced by deuterons are given below :

(*a*) (*D*, α) *reactions.*

$$^{6}_{3}Li + ^{2}_{1}H \rightarrow ^{4}_{1}He + ^{4}_{2}He$$

$$^{16}_{8}O + ^{2}_{1}H \rightarrow ^{14}_{7}N + ^{4}_{2}He$$

$$^{40}_{20}Ca + ^{2}_{1}H \rightarrow ^{38}_{19}K + ^{4}_{2}He$$

(*b*) (*D*, *p*) *reactions.*

$$^{31}_{15}P + ^{2}_{1}H \rightarrow ^{32}_{15}P + ^{1}_{1}H$$

$$^{23}_{11}Na + ^{2}_{1}H \rightarrow ^{24}_{11}Na + ^{1}_{1}H$$

(*c*) (*D*, *n*) *reactions.*

$$^{12}_{6}C + ^{2}_{1}H \rightarrow ^{13}_{7}N + ^{1}_{0}n$$

$$^{30}_{52}Te + ^{2}_{1}H \rightarrow ^{131}_{53}I + ^{1}_{0}n$$

$$^{209}_{83}Bi + ^{2}_{1}H \rightarrow ^{210}_{84}Po + ^{1}_{0}n$$

$$^{2}_{1}H + ^{2}_{1}H \rightarrow ^{4}_{2}H + ^{1}_{0}n$$

(*d*) (*D*, *T*) *reactions.*

$$^{7}_{3}Li + ^{2}_{1}H \rightarrow ^{6}_{3}Li + ^{3}_{1}T \text{ (or } ^{3}_{1}H)$$

$$^{109}_{47}Ag + ^{2}_{1}H \rightarrow ^{108}_{47}Ag + ^{3}_{1}T$$

$$^{31}_{15}P + ^{2}_{1}H \rightarrow ^{30}_{15}P + ^{3}_{1}T$$

(4) Nuclear reactions induced by neutrons ($^{1}_{0}$n). In these reactions the neutrons are captured resulting in the emission of any of the following : (*a*) γ-rays (*b*) α-particle (*c*) proton (*d*) neutron. Thus these reactions may be of the following types :

(*a*) (*n*, γ) *reactions.* Examples are given below :

$$^{27}_{13}Al + ^{1}_{0}n \rightarrow ^{28}_{13}Al + ^{0}_{0}\gamma$$

$$^{115}_{49}In + ^{1}_{0}n \rightarrow ^{116}_{49}In + ^{0}_{0}\gamma$$

$$^{238}_{92}U + ^{1}_{0}n \rightarrow ^{239}_{92}U + ^{0}_{0}\gamma$$

(*b*) (*n*, α) *reactions.* Examples are :

$$^{6}_{3}Li + ^{1}_{0}n \rightarrow ^{3}_{1}H + ^{4}_{2}He$$

$$^{10}_{5}B + ^{1}_{0}n \rightarrow ^{7}_{3}Li + ^{4}_{2}He$$

$$^{27}_{13}Al + ^{1}_{0}n \rightarrow ^{24}_{11}Na + ^{4}_{2}He$$

(*c*) (*n*, *p*) *reactions.* In these reactions the final product is a new element whose atomic number is one unit less than the target nucleus and atomic weight is the same. Examples are :

$$^{24}_{12}Mg + ^{1}_{0}n \rightarrow ^{24}_{12}Mg + ^{1}_{1}H + ^{0}_{-1}e$$

$$^{14}_{7}N + ^{1}_{0}n \rightarrow ^{14}_{7}N + ^{1}_{1}H + ^{0}_{-1}e$$

$$^{64}_{30}Zn + ^{1}_{0}n \rightarrow ^{64}_{30}Zn + ^{1}_{1}H + ^{0}_{-1}e$$

Thus we find that the final product is the same as the target nucleus and hence the overall effect is that the bombarding neutron splits up into a proton and an electron.

(*d*) (*n*, 2*n*) *reactions.* When the kinetic energy of the bombarding neutron is about 10 MeV, it is able to eject two neutrons from nucleus. For higher energies, more than two neutrons may be ejected. Examples are :

$$^{39}_{19}K + ^{1}_{0}n \rightarrow ^{38}_{19}K + 2^{1}_{0}n$$

$$^{12}_{6}C + ^{1}_{0}n \rightarrow ^{11}_{6}C + 2^{1}_{0}n$$

$$^{238}_{92}U + ^{1}_{0}n \rightarrow ^{237}_{92}U + 2^{1}_{0}n$$

In these cases the product nucleus invariably decays with the emission of a positron. For example :

$$^{38}_{19}K \rightarrow ^{38}_{18}Ar + \underset{\text{Positron}}{^{0}_{+1}e}$$

(5) Nuclear reactions induced by tritons ($^{3}_{1}$H or $^{3}_{1}$T). The nuclear reactions induced by tritons are of the following types :

(*a*) (*T*, *p*) *reactions*

$$^{59}_{27}Co + ^{3}_{1}H \rightarrow ^{61}_{27}Co + ^{1}_{1}H$$

(*b*) (*T*, *d*) *reactions*

$$^{6}_{3}Li + ^{3}_{1}H \rightarrow ^{7}_{3}Li + ^{2}_{1}H$$

$$^{63}_{29}Cu + ^{3}_{1}H \rightarrow ^{61}_{29}Cu + ^{2}_{1}H$$

(*c*) (*T*, *n*) *reactions*

$$^{32}_{16}S + ^{3}_{1}H \rightarrow ^{34}_{17}Cl + ^{1}_{0}n$$

(*d*) (*T*, α) *reactions*

$$^{7}_{3}Li + ^{3}_{1}H \rightarrow ^{6}_{2}He + ^{4}_{2}He$$

(6) Nuclear reactions induced by photons or γ-rays ($^{0}_{0}\gamma$). The nuclear reactions induced by photons are called *photo disintegration*. The type of nuclear reactions induced by photons depends on the energy of the incident photon. Various types of reactions produced by photons are given below :

(*a*) (γ, *n*) *reactions*

$$^{2}_{1}H + ^{0}_{0}\gamma \rightarrow ^{1}_{1}H + ^{1}_{0}n$$

$$^{9}_{4}Be + ^{0}_{0}\gamma \rightarrow ^{8}_{4}Be + ^{1}_{0}n$$

$$^{31}_{15}P + ^{0}_{0}\gamma \rightarrow ^{30}_{15}P + ^{1}_{0}n$$

(*b*) (γ, *p*) *reactions*

$$^{25}_{12}\text{Mg} + ^{0}_{0}\gamma \rightarrow ^{24}_{11}\text{Na} + ^{1}_{1}\text{H}$$

(*c*) (γ, 2*p*, *n*) *reactions*

$$^{27}_{13}\text{Al} + ^{0}_{0}\gamma \rightarrow ^{24}_{11}\text{Na} + 2^{1}_{1}\text{H} + ^{1}_{0}\text{n}$$

Nuclear Reactions Vs Chemical Reactions

Nuclear reactions differ from chemical reactions in the following respects :

(*i*) In chemical reactions rearrangement of outermost electrons of the atom takes place while the nuclei remain uneffected. Thus no change occurs in the atomic species taking part in the chemical reaction. In other words, no new atomic species are formed in chemical reaction. On the other hand, in nuclear reactions, changes in the number of nucleons present in the nucleus take place and consequently there is formation of new atomic species. *Thus nuclear reactions bring about the atomic transmutation or transformation.*

(*ii*) *The amount of energy released in nuclear reactions is million times greater than that produced in chemical reactions.* This is because of the fact that, in nuclear reaction, unlike in a chemical reaction, a small but measurable *loss in mass* takes place. This loss in mass is converted into energy, called *nuclear reaction energy*, in accordance with *Eienstein's mass-energy relationship,* $E = mc^2$ where m = loss in mass, E = energy and c = velocity of light.

Mass Defect and Binding Energy of the Nucleus

It has been observed that the experimental value of atomic mass (A) of an isotope of an element is invariably less than the sum (M) of the masses of protons, neutrons (called nucleons) and electrons present in the atom of the element. The difference between the two masses is called **mass defect** and is denoted by Δm. Thus :

Mass defect **(Δm) = *Sum of the masses of protons, neutrons and electrons (M)—Experimental value of atomic mass of the isotope (A)***

or $\Delta m = \text{M} - \text{A}$

When the nucleons are allowed to combine together to form the nucleus, some mass disappears in the form of energy, *i.e.* in the combination of nucleons to form the nucleus, some amount of energy is released. *The energy released in the formation of a nucleus from its constituent nucleons is called* **binding energy of the nucleus.** Conversely, the same amount of energy will naturally be required to break the nucleus into its component nucleons.

If Δm is the mass defect in *amu*, then binding energy of the nucleus in MeV is equal to $\Delta m \times 931$ MeV, since 1 *amu* = 931 MeV of energy. The greater the mass defect, the greater is the binding energy of the nucleus. The binding energy of a nucleus divided by the total number of nucleons (*i.e.* neutrons + protons) is called the binding energy per nucleon in the atom, *i.e.*

$$\textbf{Binding energy per nucleon} = \frac{\textbf{Binding energy}}{\textbf{Neutrons + Protons}} = \frac{\Delta m \times 931}{\text{A}} \textbf{ MeV}$$

The binding energy per nucleon is a measure of the stability of the nucleus. The greater the binding energy per nucleon, greater is the stability associated with the nucleus, *i.e.* the greater the amount of binding energy per nucleon, greater is the force holding the nucleons in the nucleus together.

Calculation of Binding Energy

How the binding energy of a given nucleus can be calculated can be learnt from the following solved examples :

Example 1. *Calculate the binding energy of lithium nucleus ($_3Li^7$), given that :*

mass of proton = 1.00814 amu

mass of neutron = 1.00893 amu

mass of lithium nucleus = 7.01822 amu (*A.M.I.E. Summer 1978*)

Solution. Obviously, the lithium isotope considered is $_3\text{Li}^7$. It contains 3 protons and (7 – 3) = 4 neutrons.

Mass of 3 protons = 3 × 1.00814 = 3.02442 amu

Mass of 4 neutrons = 4 × 1.00893 = 4.03572 amu

Total mass of 7 nucleons, M = 7.06014 amu

Mass of the lithium nucleus, A = 7.01822 amu

Mass defect, Δm = M – A = 7.06014 – 7.01822

= 0.04192 amu

∴ Binding energy = 931 × 0.04192 = **39 MeV**

Example 2. *The mass of $_{17}Cl^{35}$ is 34.9800 amu. Calculate its binding energy. What is the binding energy per nucleon* ?

Mass of $_0n^1$ = 1.008665 amu and $_1H^1$ = 1.007825 amu.

Solution. The given chlorine atom has 17 protons and (35 – 17) = 18 neutrons.

Mass of 17 protons = 17 × 1.007825 = 17.133025 amu

Mass of 18 neutrons = 18 × 1.008665 = 18.155970 amu

Total mass of separate constituents = 35.288995 amu

Mass defect, Δm = 35.288995 – 34.9800

= 0.308995 amu

Now, 1 amu = 931 MeV

∴ Binding energy of the nucleus = 931 × 0.308995 = **288 MeV**

No. of nucleons (*i.e.* protons and neutrons) = 35

∴ Binding energy per nucleon = 288/35 = **8.22 MeV**

Example 3. *Calculate the binding energy of a deuteron nucleus, given that mass of neutron = 1.008665 amu, mass of proton = 1.007825 amu and mass of deuteron nucleus = 2.014103 amu.*

Solution. Neutron mass, $_0n^1$ = 1008665 amu

Proton mass, $_1\text{H}^1$ = 1.007825 amu

Sum = 2.016490 amu

Deuteron mass, $_1H^2 = 2.014103$ amu

Hence $M = 2.014103$ amu

and $A = 2.016490$ amu

$\therefore$ Mass defect, $\Delta m = M - A = (2.014103 - 2.016490)$

$= -0.002387$

Binding energy $= (\Delta m \times 931)$ MeV

$= -(0.002387 \times 931)$ MeV

$= -2.22$ MeV

Example 4. *Calculate the binding energy per nucleon of oxygen atom, $^{16}_{8}O$, given that mass of $^{16}_{8}O$ = 15.994910 amu, mass of neutron = 1.008665 amu, and mass of proton = 1.007277 amu and mass of electron = 0.0005486 amu.*

Solution. We know that $^{16}_{8}O$ contains 8 protons, (16 – 8) = 8 neutrons and 8 electrons. Now :

Mass of 8 protons = 8 × 1.007277 = 8.058216 amu

Mass of 8 neutrons = 8 × 1.008665 = 8.069320 amu

Mass of 8 electrons = 8 × 0.0005486 = 0.0043888 amu

Sum, M = 16.1319248 amu

Actual mass of oxygen atom, A = 15.994910 amu

$\therefore$ Mass defect, $\Delta m = M - A$

= (16.1319248 – 15.994910) amu

= 0.1370148 amu

Now, since 1 amu = 531 MeV of energy,

Binding energy of the nucleus $= (\Delta m \times 931)$ MeV

$= 0.1370148 \times 931$ MeV

$= 127.56077$ MeV

Now, since total number of nucleons = 8 + 8 = 16,

$$\text{Binding energy per nucleon} = \frac{127.56077}{16} \text{ MeV} = 7.9725 \text{ MeV}$$

Binding Energies and Stability of Nuclei

If the binding energies per nucleon (in MeV) of a number of isotopes are plotted against their mass numbers (A), a graph of the type shown in Fig. 6.1 is obtained. This graph is called *binding energy curve*. The following conclusions may be drawn from this curve.

(*i*) Lighter nuclei have very small binding energy and hence are unstable. Therefore such nuclei (shown in region A) have tendency to combine or fuse together (*Fusion*) to give the nuclei of intermediate mass number (atomic weight) and higher binding energy or stability. Similarly heavier nuclei are seen to have low binding energy and hence are very unstable. Therefore such nuclei (shown in region B) may break down (*Fission*) to form the nuclei of intermediate atomic weight that possess higher binding energies and greater stability.

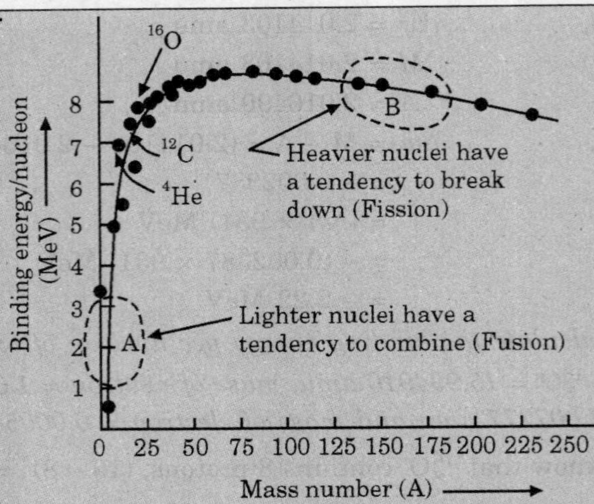

Fig. 6.1. Binding energy curve. It is obtained by plotting binding energies per nucleon (in MeV) of a number of isotopes against their mass numbers (A). The lighter nuclei (shown in region A) have a tendency to combine (Fusion) while the heavier nuclei (shown in region B) have a tendency to break down (Fission).

(*ii*) The binding energies of some of the light nuclei are very small. The binding energy per nucleon rises sharply and reaches a maximum value of 8.7 MeV in the neighbourhood of a nucleus with mass number, A = 56 (Iron). Thus nucleus of iron is thermodynamically most stable and hence is found in abundance in earth. The graph for the nuclei having mass numbers in the range of 50 to 100 is almost flat. Thus the addition of a single nucleon to any nucleus in the range of A = 50 – 100 increases the binding energy roughly by the same amount. For the nuclei having high mass numbers the value decreases to 7.6 MeV for uranium.

(*iii*) The points for helium (A = 4), carbon (A = 12) and oxygen (A = 16) lie quite high in the graph. Thus the nuclei of these elements are *exceptionally* stable.

(*iv*) Over a considerable range of mass numbers, the binding energy per nucleon is 8 MeV which is usually taken as the average value of binding energy for all nucleons.

Meson Theory of Nuclear Forces

We have already said that the neutrons and protons (*i.e.*, nucleons) are packed together in the nucleus in a very small space, since the nucleus is about $\frac{1}{10^5}$ in volume as compared to the entire atom. Moreover the protons (positively-charged particles) should ordinarily repell each other, but they are present together in the nucleus. Therefore, a question arises here which is : *What is the nature of forces that hold the protons and neutrons together in such a small space* ? The forces which hold protons and neutrons together in the nucleus are attractive forces and are called **nuclear forces.** These forces are different from electrostatic forces in that the former forces operate only within distances of approximately 0.10 fermi

(= 10^{-13} cm, 1 *fermi* = 10^{-12} cm) while the latter operate over long range also. Nuclear forces fall rapidly to zero at distances greater than 14×10^{-13} cm and are referred to as *short range forces*. These forces do not obey the inverse square law.

The exact nature of the forces was explained by Hideki Yukawa, a Japanese theoretical physicist, who suggested in 1935 that the nucleons are surrounded by a cloud of negatively-charged and positively-charged π-mesons ($_{-1}\pi$ and $_{+1}\pi$ mesons) as shown in Fig. 6.2.

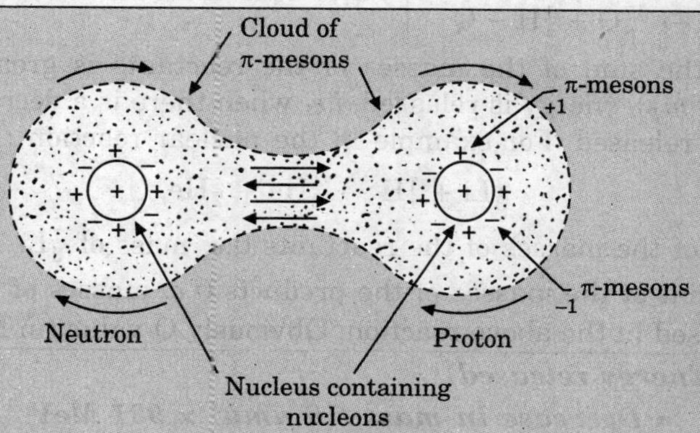

Fig. 6.2. Nature of nuclear forces holding nucleons in the nucleus.

These mesons rapidly jump back and forth between the neighbouring nucleons with a velocity approximately equal to that of light. This sort of jumping of mesons keeps the nucleons bound together in the nucleus. Negatively-charged mesons ($_{-1}\pi$ mesons) interact with protons ($_{+1}p$) while positively-charged mesons ($_{+1}\pi$ mesons) interact with neutrons ($_0n$) and by these interactions a proton is changed into a neutron and a neutron is changed in a proton. This change takes place for a part of time only.

$$_{+1}p + {}_{-1}\pi \rightleftarrows {}_0n$$

$$_0n + {}_{+1}\pi \rightleftarrows {}_{+1}p$$

Q-Value of a Nuclear Reaction

All types of nuclear reactions are accompained by either the absorption of energy or release of energy. The amount of energy absorbed or released in a nuclear reaction is generally called Q-value of the nuclear reaction. Thus the following two cases arise :

(*i*) When the sum of the masses of the reactants (m_R) is less than that of the products (m_P) ($m_R < m_P$), energy is absorbed, *i.e.*, when there is an increase in the mass, energy is absorbed. For example, in the nuclear reaction :

$$^{14}_{7}N + {}^{4}_{2}He \rightarrow {}^{17}_{8}O + {}^{1}_{1}H,$$

since the sum of the masses of the reactants (*i.e.* mass of $^{14}_{7}N$ + mass of $^{4}_{2}He$) is less than that of the masses of the products (*i.e.* mass of $^{17}_{8}O$ + mass of $^{1}_{1}H$), energy is absorbed in this reaction. Obviously Q-value (in MeV) is given by :

Q-value (Energy absorbed)
= Increase in mass (in amu) × 931 MeV
= $(m_P - m_R)$ × 931 MeV

Q-value which is absorbed is generally represented by a negative sign. Now if Q-value is incorporated in the above reaction, the reaction can be written as :

$$\left.\begin{array}{l} {}^{14}_{7}\mathrm{N} + {}^{4}_{2}\mathrm{He} + \mathrm{Q} \rightarrow {}^{17}_{8}\mathrm{O} + {}^{1}_{1}\mathrm{H} \\ \text{or } {}^{14}_{7}\mathrm{N} + {}^{4}_{2}\mathrm{He} \rightarrow {}^{17}_{8}\mathrm{O} + {}^{1}_{1}\mathrm{H} - \mathrm{Q} \end{array}\right\} \text{ Q is absorbed and hence is negative}$$

(*ii*) When the sum of the masses of the reactants is greater than that of products ($m_R > m_P$), energy is released, *i.e.* when there is a decrease in the mass, energy will be released. For example in the nuclear reaction :

$${}^{7}_{3}\mathrm{Li} + {}^{1}_{1}\mathrm{H} \rightarrow {}^{4}_{2}\mathrm{He} + {}^{4}_{2}\mathrm{He},$$

since the sum of the masses of the reactants (*i.e.* mass of ${}^{7}_{3}\mathrm{Li}$ + mass of ${}^{1}_{1}\mathrm{H}$) is greater than that of the masses of the products (*i.e.* masses of two ${}^{4}_{2}\mathrm{He}$ nuclei), energy is released in the above reaction. Obviously Q-value (in MeV) is given by :

Q-value (Energy released)
= Decrease in mass (in amu) × 931 MeV
= $(m_R - m_P)$ × 931 MeV

Q-value which is released is generally represented by a *positive* sign. Now if Q-value is incorporated in the above reaction, the reaction can be written as :

$$\left.\begin{array}{l} {}^{7}_{3}\mathrm{Li} + {}^{1}_{1}\mathrm{H} \rightarrow 2{}^{4}_{2}\mathrm{He} + \mathrm{Q} \\ \text{or } {}^{7}_{3}\mathrm{Li} + {}^{1}_{1}\mathrm{H} - \mathrm{Q} \rightarrow 2{}^{4}_{2}\mathrm{He} \end{array}\right\} \text{ Q is released and hence is positive}$$

Solved example. *Calculate Q-value of the following nuclear reactions*

(*i*) ${}^{14}_{7}N + {}^{4}_{2}He \rightarrow {}^{17}_{8}O + {}^{1}_{1}H$

(*ii*) ${}^{7}_{3}Li + {}^{1}_{1}H \rightarrow 2{}^{4}_{2}He$

Solution. (*i*) ${}^{14}_{7}\mathrm{N} + {}^{4}_{2}\mathrm{He} \rightarrow {}^{17}_{8}\mathrm{O} + {}^{1}_{1}\mathrm{H}$

The change in mass in atomic-mass-units (amu) can be found as follows :

Interacting particles	***Product particles***
${}^{4}_{2}\mathrm{He}$ = 4.00387	${}^{17}_{8}\mathrm{O}$ = 17.00450
${}^{14}_{7}\mathrm{N}$ = 14.00753	${}^{1}_{1}\mathrm{H}$ = 1.00814
Total = 18.01140 = m_R	Total = 18.01264 = m_P

Increase in mass = $(m_P - m_R)$ amu
= 18.01264 – 18.01140 = 0.00124 amu

∴ Q-value (*Energy absorbed*) = 0.00124 × 931 = 1.15 MeV

Since $m_R < m_P$, energy is absorbed in this reaction and hence Q-value is negative, *i.e.* Q-value = – 1.15 MeV.

(*ii*) ${}^{7}_{3}\mathrm{Li} + {}^{1}_{1}\mathrm{H} \rightarrow {}^{4}_{2}\mathrm{He} + {}^{4}_{2}\mathrm{He}$

Here, the mass values are as under :

Interacting particles	***Product particles***
$^1_1He = 1.00814$	$^4_2He = 4.00387$
$^7_3Li = 7.01822$	$^4_2He = 4.00387$
Total = $8.02636 = m_R$	Total = $8.00774 = m_P$

Decrease in mass = $(m_R - m_P)$ amu

$$= 8.02636 - 8.00774 = 0.01862 \text{ amu}$$

$\therefore$ Q-value (*Energy released*) = 0.01862 × 931 MeV = 17.3 MeV

Since $m_R > m_P$, energy is released in this reaction and hence Q-value is positive, *i.e.* Q-value = + 1.73 MeV.

Nuclear Fission

What is Nuclear Fission ?

The binding energy curve given in Fig. 6.1 indicates that the heavier nuclei, due to their low binding energies, are unstable and hence break down to produce the nuclei of intermediate atomic weights (*i.e.* mass numbers) that possess higher binding energies and hence greater stability. They do so when they are bombarded with energetic particles such as neutrons. Thus :

The nuclear reaction in which one heavier nucleus is split up into two lighter nuclei of almost equal size with the release of a huge amount of energy is called **nuclear fission or atomic fission.**

In 1939, German radio chemists, *Otto Hahn* and his associates, *Meitner* and *Strassman* bombarded $^{235}_{92}U$ with *slow moving* neutrons and observed that ^{235}U nucleus, after capturing neutron, gave ^{236}U nucleus which was *radioactive* (called *compound nucleus*) and hence had broken up to form ultimately a pair of stable products (called *fission products*) consisting of $^{141}_{56}Ba$ and $^{92}_{36}Kr$ and liberated *three* neutrons (called *secondary neutrons*) in addition to a huge amount of energy, called *fission energy*.

$$^{235}_{92}U + ^1_0n \rightarrow \underset{\text{Compound nucleus}}{[^{236}_{92}U]} \rightarrow \underbrace{^{141}_{56}Ba + ^{92}_{36}Kr}_{\text{Pair of stable products (Fission products)}} + 3\,^1_0n + \text{Energy } (\textit{Fission Energy})$$

Although originally Hahn and Strassman identified by analysis $^{141}_{56}Ba$ and $^{92}_{36}Kr$ as the fission products of $^{235}_{92}U$, now it is known to split at least in 50 different ways. What particular pair of stable products will be formed *actually depends on chance*. The number of neutrons released depends on the mode of fission and on the energy of the neutrons which have been used to induce the fission. The neutrons are emitted by the fission products and not by the compound nucleus.

In the nuclear fission reaction a huge amount of energy is also liberated. How this energy is produced in a nuclear fission can be explained as follows : Suppose we initiate fission of one ^{235}U nucleus by one neutron obtained either by cosmic rays or by some radioactive source. In this fission process each ^{235}U nuclear suffering fission will liberate three neutrons. These three neutrons will in turn cause the fission of other three ^{235}U nuclei and will liberate three more neutrons in the fission of each ^{235}U nucleus. Thus nine neutrons will be obtained and so on. (*See* Fig. 6.3). Consequently, a self-propagating or self sustaining *chain reaction* (also called *autocatalytic reaction*) starts in which the number of neutrons keeps on multiplying rapidly (in geometrical progression). This chain reaction continues till the whole of U^{235} has undergone fission. This chain reaction is also accompanied by the liberation of a huge amount of energy which is called **Fission Energy** or **Atomic Energy.** This energy can be either very fast resulting in a violent explosion (as in an atom bomb) or it can be controlled and maintained at a steady rate (as in chain reacting nuclear piles or nuclear reactors).

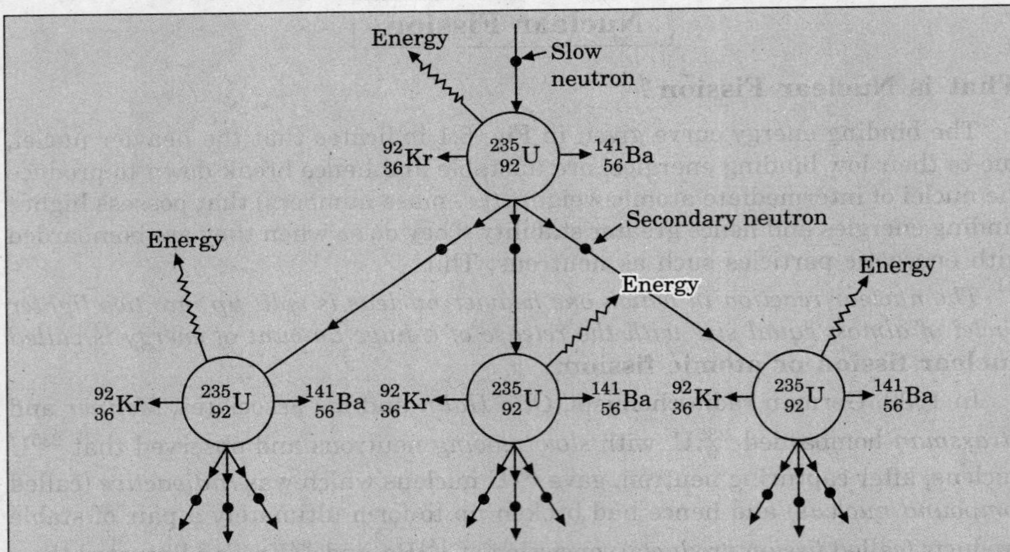

Fig. 6.3. Schematic representation of chain reaction taking place in the fission of one nucleus of $^{235}_{92}U$.

The amount of energy released in a fission reaction varies from one reaction to another. But, on the average, approximately 200 MeV of energy is released by the fission of one U^{235} nucleus. This energy (= 200 MeV) is about 500 million times as great as that released per atom in the combination of hydrogen which is the primary heat-producing constituent of coal or oil.

Calculation of Energy Released in Nuclear Fission : Q-Value of Nuclear Fission

The amount of energy released in a nuclear fission can be calculated by the *mass-defect method* which is based on the fact that *in a nuclear fission reaction, the total mass of the products is always less than that of the reacting nuclei* (*i.e. parent nuclei*). This decrease in mass is called *mass defect* which is converted into energy according to mass-energy relationship (Einstein's equation) *viz.* $E = mc^2$. Thus for the nuclear fission of the type :

$$^{235}_{92}U + ^{1}_{0}n \rightarrow ^{141}_{56}Ba + ^{92}_{36}Kr + 3^{1}_{0}n + \text{Energy},$$

energy has been calculated to be equal to 250 MeV as shown below :

Interacting particles	*Product particles*
$^{235}_{92}U = 235.11750$ *a.m.u.*	$^{141}_{56}Ba = 140.95770$ *a.m.u.*
$^{1}_{0}n = 1.00898$ *a.m.u.*	$^{92}_{36}Kr = 91.92640$ *a.m.u.*
	$3^{1}_{0}n = 3 \times 1.00898 = 3.02694$ *a.m.u.*
Total = 236.12648 *amu*	Total = 235.91104 *amu*

$\therefore$ Loss in mass (*Mass defect*) = 236.12648 − 235.91104
= 0.21544 *amu*

Therefore, energy released per fission of ^{235}U nucleus
= 0.21544 × 931 MeV
= 200.5 MeV.

Out of this energy 170.5 MeV is carried by the fission fragments as kinetic energy, 5 MeV by fission neutrons, 15 MeV by β and γ-rays which are produced due to the radioactivity of the fission fragments and 10 MeV by the neutrons associated with β-decay.

Mechanism of Nuclear Fission—Liquid Drop Model of Nuclear Fission

The phenomena of nuclear fission has been explained by liquid drop model suggested by Bohr-Wheeler, since an atomic nucleus has many similarities with a liquid drop. A liquid drop has a spherical shape due to its surface tension. If sufficient energy from outside is applied on the drop to overcome the forces of surface tension, the drop may change it spherical shape to an elliptical shape. If the external energy is sufficiently large, the elliptical shape may change to a dumb-bell shape and ultimately may break into two portions of spherical shape.

As said above, an atomic nucleus has many similarities with a liquid drop and hence behaves as a liquid drop. When sufficient energy is applied on a uranium nucleus through the neutron bombardment, the compound nucleus formed (behaving as a liquid drop) changes its spherical shape to elliptical shape (Fig. 6.4). If the energy applied on the uranium nucleus through the neutron bombardment is sufficiently large, the elliptical shape of the nucleus may change to a dumb-bell shape. Both the parts of the dumb-bell carry positive charge (critical shape). The energy absorbed by the uranium nucleus to change from spherical shape to critical shape is called **threshold energy** required by the fission reaction to occur. Since the two parts of the critical shape have similar (positive) charges, they repell each other and ultimately are separated from each other into two nuclei having the spherical shape and almost the same size.

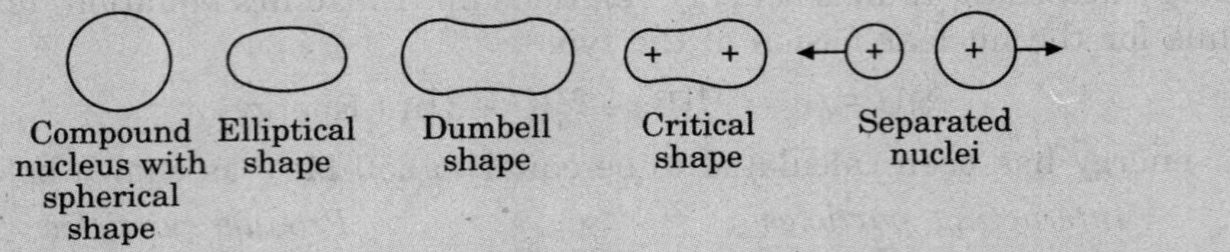

Fig. 6.4. Mechanism of nuclear fission by liquid drop model.

Uses of Nuclear Fission

The enormous energy liberated in nuclear fission and the occurrence of chain reaction have been used in *Atomic Bomb (Fission Bomb) and in Atomic (or Nuclear) Pile or Nuclear Reactor.*

Atomic Bomb (Fission Bomb)

We have already seen that a chain reaction takes place in a nuclear fission reaction and a huge amount of energy is produced. Each ^{235}U nucleus liberates three neutrons (called *secondary neutrons*). Each of these secondary neutrons strikes a fresh ^{235}U nucleus and causes its fission with the liberation of three more secondary neutrons. In reality it has been observed that if the size of ^{235}U nucleus being used for fission is smaller than a minimum size, called **critical size,** some of the secondary neutrons escape into the air and hence no chain reaction takes place to cause explosion and also no vast amount of energy is liberated in an explosive fashion. This size of ^{235}U nucleus which is smaller than the critical size is called **sub-critical size** and is quite safe [See Fig. 6.5 (*a*)] Thus we see that in order that the secondary neutrons do not escape into the air, the size of ^{235}U nucleus should be bigger than the critical size. This size of ^{235}U which is bigger than the critical size is called **super-critical size.** In this condition a large number of neutrons are captured by ^{235}U nucleus and hence a chain reaction is set up to cause the explosion.

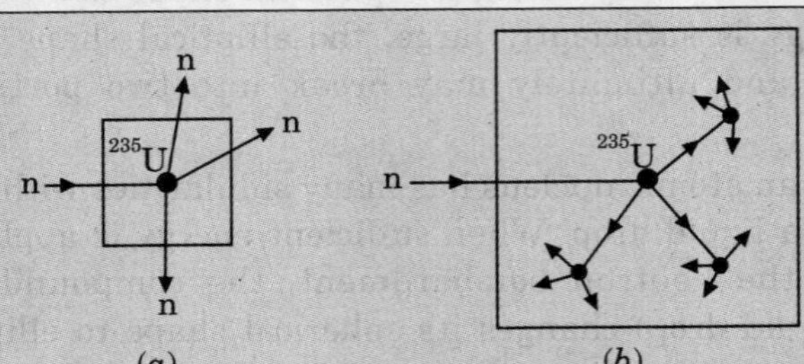

Fig. 6.5. Principle of atomic bomb. (*a*) ^{235}U nucleus is in sub-critical size and hence no chain reaction takes place to cause the explosion (*b*) ^{235}U nucleus is in super-critical size and hence chain reaction takes place to cause explosion.

In its safety position an atomic bomb consists of several thousands of pieces of ^{235}U all of which are of sub-critical size. At the time of explosion these are driven together rapidly by firing high explosives like TNT (trinitrotoluene) lying behind each of ^{235}U pieces so that these ^{235}U pieces form one large piece of super-critical size. Now neutrons from Ra–Be source S (Fig. 6.6) strike the large piece of ^{235}U and a rapid chain reaction is set up. This chain reaction results in a violent explosion with the release of vast amount of energy.

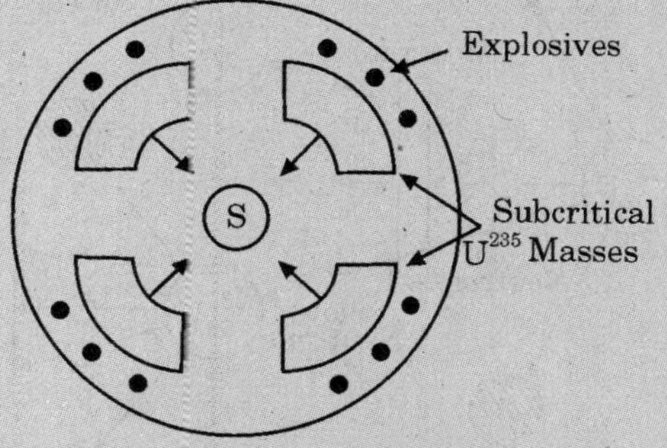

Fig 6.6. Atomic bomb.

The amount of energy liberated in an atom bomb is equal to that produced by the detonation of about 20,000 tons of TNT. This energy rises the temperature to the order of 10,000,000°C (10^7 °C) and more and produces a pressure of several millions atmosphere. The explosion also produces a violent and intense blast of visible, ultraviolet, X-ray and γ-ray radiations which cause blinding flash.

When fission occurs, the fragments fly apart with tremendous speeds. They collide with each other and then kinetic energy is changed into heat energy. The radioactive fragments formed scatter over wide areas and get deposited on the surrounding objects. Thus, we see that in atomic bomb the chain reaction taking place in nuclear fission reaction is *uncontrolled*.

Atomic Pile or Nuclear Reactor

The study of nuclear reactor can be made under the following headings.

What is a nuclear reactor ? A nuclear reactor is *an apparatus in which nuclear fission is produced in the form of a controlled self-sustaining chain reaction.* In other words, it is a controlled chain-reacting system supplying nuclear energy. Thus we see that in a nuclear reactor the fission chain reaction is *controlled i.e.* the explosive effect of the nuclear chain reaction taking place in a nuclear reactor is considerably reduced. This is achieved by using natural uranium in place of ^{235}U, since the natural uranium contains a very large amount of ^{238}U isotope which absorbs slow neutrons and thus reduces the explosive effect of the nuclear fission chain reaction seen in case of ^{235}U to a great extent.

The explosive effect of the nuclear fission chain reaction can also be reduced by using some substances, called **moderators** which slow down the speed of secondary neutrons which are fast neutrons. The secondary neutrons pass through the moderator and lose some of their kinetic energy and their speed is reduced. Thus the use of a moderator controls the explosive effect of the nuclear fission chain reaction (Fig. 6.7).

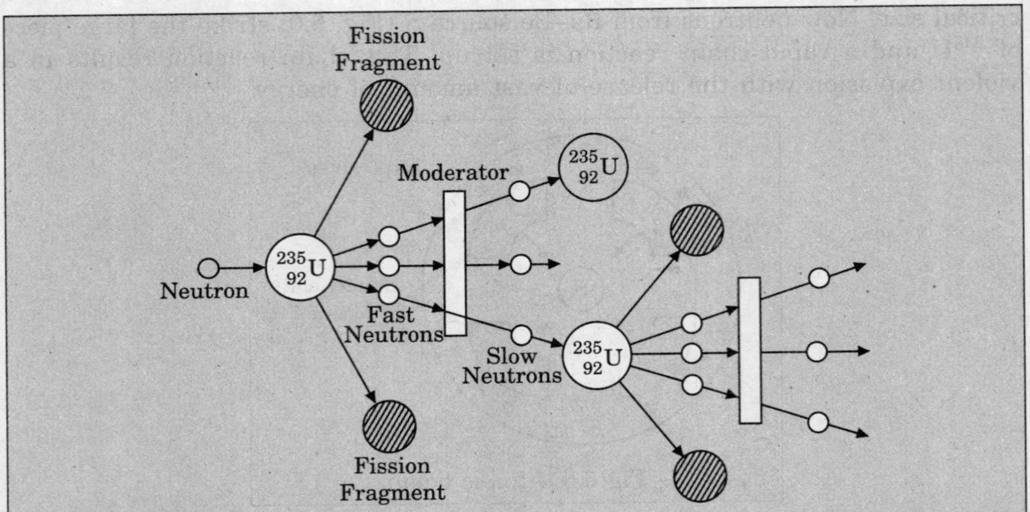

Fig 6.7. The use of a moderator reduces the speed of secondary neutrons (fast neutrons) and thus the explosive effect of the nuclear fission chain reaction occurring in a nuclear reactor is controlled.

The materials like D_2O, graphite etc. have been used as moderators. When graphite is used as a moderator, the reactor is called *Atomic Pile* and when D_2O is used, it is called *Swimming Pool Reactor*. In atomic pile a huge block of specially purified graphite containing holes in which uranium rods are inserted, acts as moderator. In the swimming pool reactor D_2O is used in a tank in which the rods of uranium are suspended.

Components of a nuclear reactor. The essential components of a nuclear reactor are as follows (Also see Fig. 6.8).

(1) Core. It consists of aluminium plated uranium rods usually 1.5" thick and 4" long in a number of holes. It is this part of a nuclear power plant where fission chain reaction is made to occur and where fission energy is liberated in the form of heat for operating power conversion equipment.

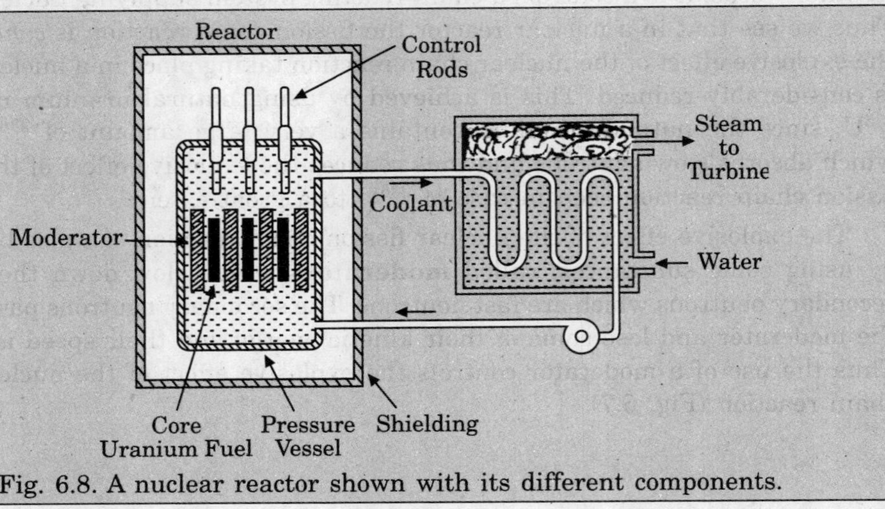

Fig. 6.8. A nuclear reactor shown with its different components.

(2) Moderator. It has already been discussed.

(3) Control rods. These are the rods of cadmium or boron steel. These have the ability to absorb neutrons and thus regulate the number of neutrons that can cause fission and in this way the rate of fission and hence the amount of energy generated per unit time is controlled. When the process of nuclear fission is to be slowed, these rods are pushed in. As soon as these rods are raised, the neutrons begin to fly about and the splitting of ^{235}U nucleus grows in intensity.

(4) Coolant. The heat produced in the controlled nuclear fission reaction is removed from the reactor by a liquid called *Coolant*. Since water absorbs neutrons, it is not fit to be used as coolant. In actual practice a liquid alloy of sodium and potassium is used as a coolant. The alloy takes heat to the heat exchanger for raising steam that runs a turbine generator to produce electricity. Heavy water, organic compounds (*e.g.* polyphenyls), air and CO_2 have also been used as coolants.

(5) Shielding or protective screen. It is an important component of a reactor, although it plays no role in its operation. It stops the heat, γ-rays and neutrons from escaping form the reactor so that they should not be able to harm the persons living in the immediate vicinity. Protective screen is made of special concrete about 8 feet think.

Nuclear Fusion

What is Nuclear Fusion ?

The binding energy curve given in Fig. 6.1 on page 396 indicates that the lighter nuclei have very small binding energies and hence are unstable. Therefore two or more such nuclei have a tendency to combine or fuse together to give the nucleus of higher mass number (atomic weight) and higher binding energy or stability. *Such a nuclear reaction in which two lighter nuclei are combined or fused together to form a heavier (and hence stabler) nucleus is called* **nuclear fusion.** In this respect, nuclear fusion is the opposite of nuclear fission in which one heavier nucleus is split up into two lighter nuclei. In nuclear fusion also, large amount of energy is released because the mass of the product nucleus is *less* than the masses of the two nuclei which are fused together. Once fusion reaction is initiated, the energy released is sufficient to maintain the temperature and keep the reaction going on. Nuclear fusion occurs at reasonable rates only at *very high temperatures*. It is because of this that nuclear fusion reactions are also called *thermonuclear reactions i.e. temperature-dependent reactions.*

Uses of Nuclear Fusion

Nuclear fusion adequately explains the energy of the solar system, called *stellar energy* and *the basic principle of hydrogen bomb* (*Fusion bomb*).

Stellar Energy

Stellar energy is the energy of the sun and other stars of the main sequence. Van Weizsacher and Hans Beth (1932) proposed that stellar energy is due to a series of nuclear reactions involving carbon-hydrogen-nitrogen cycle which operates through the following six steps :

$$^{12}_{6}C + ^{1}_{1}H \rightarrow ^{13}_{7}N + ^{0}_{0}\gamma + \text{Energy} \quad ...(i)$$

$$^{13}_{7}N \rightarrow ^{13}_{6}C + \underset{\text{Positron}}{^{0}_{1}e} + \text{Energy} \quad ...(ii)$$

$$^{13}_{6}C + ^{1}_{1}H \rightarrow ^{14}_{7}N + ^{0}_{0}\gamma + \text{Energy} \quad ...(iii)$$

$$^{14}_{7}N + ^{1}_{1}H \rightarrow ^{15}_{8}O + ^{0}_{0}\gamma + \text{Energy} \quad ...(iv)$$

$$^{15}_{8}O \rightarrow ^{15}_{7}N + ^{0}_{1}e + \text{Energy} \quad ...(v)$$

$$^{15}_{7}N + ^{1}_{1}H \rightarrow ^{12}_{6}C + ^{4}_{2}He + \text{Energy} \quad ...(vi)$$

On adding the nuclear reactions from (*i*) to (*vi*) we have :

$$\underset{\text{Proton}}{4^{1}_{1}H} \rightarrow \underset{\text{Helium}}{^{4}_{2}He} + \underset{\text{Positron}}{2^{0}_{1}e} + \text{Energy} \quad ...(1)$$

Nuclear reaction (1) indicates that the stellar energy is produced due to the fusion reaction in which four H-atoms (or protons) are fused together to produce one helium nucleus and two positrons. On the sun the fusion of four H-nuclei into helium nucleus does not take place through a single step but takes place through deuterium ($^{2}_{1}H$) as shown below :

$$\left[2^{1}_{1}H \rightarrow ^{2}_{1}H + ^{0}_{1}e + \text{Energy}\right] \times 2 \quad ...(i)$$

$$\left[^{2}_{1}H + ^{1}_{1}H \rightarrow ^{3}_{2}He + \text{Energy}\right] \times 2 \quad ...(ii)$$

$$2^{3}_{2}He \rightarrow ^{4}_{2}He + 2^{1}_{1}H + \text{Energy} \quad ...(iii)$$

On adding : $4^{1}_{1}H \rightarrow ^{4}_{2}He + 2^{0}_{1}e + \text{Energy}$

Energy Released in the Nuclear Fusion of $4^{1}_{1}H$ Nuclei into one $^{4}_{2}He$ and two Positrons ($^{0}_{1}e$).

The amount of energy, Q released in the nuclear fusion reaction,

$$4^{1}_{1}H \rightarrow ^{4}_{2}He + 2^{0}_{1}e$$

can be calculated by the *mass-defect method* as shown below (all the masses are given in *amu*)

$$4^{1}_{1}H \rightarrow ^{4}_{2}He + 2^{0}_{1}e + Q$$

$$4 \times 1.0084167 = 4.003873 + 2 \times 0.000558 + Q$$

or $4.0336668 = 4.003873 + 0.001116 + Q$

or $4.0336668 = 4.004989 + Q$

$\therefore$ $Q = 4.0336668 - 4.004989 = 0.0286778 \; a.m.u$

$= 0.0286778 \times 931 \text{ MeV} = 26.699032 = 26.7 \text{ MeV}$

It is worth noting that the above energy release is less than that in a nuclear fission. However, its value is 26.7/4 = 6.7 MeV per nucleon as compared to less than 1 MeV per nucleon in fission process.

Hydrogen Bomb (Fusion Bomb).

We have already seen that the nuclear fusion reaction is a temperature-dependent reaction, *i.e.* the nuclear fusion reaction takes place only at drastically high temperature. Such a high temperature which is necessary for bringing about nuclear fusion reaction can be obtained only by the fission of ^{235}U nucleus which is an explosive chain reaction. This forms the principle of hydrogen bomb. In the preparation of hydrogen bomb a suitable quantity of all the isotopes of hydrogen ($^1_1$H, $^2_1$H and $^3_1$H) is combined with an atomic bomb. The function of atomic bomb, as said above, is to provide the high temperature which is necessary for bringing about the nuclear fusion reactions in hydrogen bomb.

In hydrogen bomb the following nuclear fusion reactions take place :

$$^1_1\text{H} + \underset{\text{Deuterium}}{^2_1\text{H}} \rightarrow {^3_2\text{He}} + 5 \text{ MeV}$$

$$^2_1\text{H} + {^2_1\text{H}} \rightarrow {^3_2\text{He}} + {^1_0\text{n}} + 3.2 \text{ MeV}$$

$$\underset{\text{Tritium}}{^3_1\text{H}} + {^2_1\text{H}} \rightarrow {^4_2\text{He}} + {^1_0\text{n}} + 17.6 \text{ MeV}$$

We see that in each of the above nuclear fusion reactions energy is released. Thus in hydrogen bomb isotopes of hydrogen are combined (fused) with deuterium ($^2_1$H) and form heavier nuclei with the release of enormous energy.

Hydrogen bomb is 1000 times more powerful than the atom bomb which is based on *nuclear fission*. Its more power is due to the rapid release of a tremendous amount of energy, in the fusion reactions taking place in hydrogen bomb.

Why is H-bomb (fusion bomb) superior to atom bomb (fission bomb) ?

A fusion bomb is superior to a fission bomb because of the following reasons :

(*i*) The energy release in a hydrogen bomb is *open-ended i.e.* it has no upper limit. It depends on how much fusible material is present in the bomb.

(*ii*) It has no limitation of a critical size of the fusible material unlike an atomic bomb. If the active material in an atomic bomb exceeds the critical size, *spontaneous* explosion results. Hydrogen bomb cannot explode unless 'ignited' *i.e.*, heated to critical ignition temperature and any amount of fusible material is safe until ignited. Thus the amount of fusible material in a hydrogen bomb is not limited.

Comparison of Fission and Fusion

(*a*) **Point of similarity.** One thing common between the two nuclear processes is that they release very large amount of energy.

(*b*) **Points of dissimilarities.** There are many differences in the mechanisms of the two processes.

(*i*) Fission involves *breaking* up of a heavy nucleus into lighter nuclei. Fusion, on the other hand, involves *combining* of two lighter nuclei into one heavy nucleus.

(*ii*) The links of the fission process are neutrons while the links of a fusion process are protons.

(*iii*) Fission proceeds best with *thermal neutrons* where thermal means room temperature. Fusion proceeds best with *thermal particles* where thermal means temperatures of millions of °K.

Artificial Or Induced Radioactivity

What is Artificial Radioactivity ?

There are many artificial transmutations (*i.e. nuclear reactions*) in which the product nucleus is *unstable* (*i.e. radioactive isotope*) and hence disintegrates *spontaneously* (*i.e.* by itself) like naturally occurring radioactive elements such as uranium, radium etc. to give a stable isotope of some other element. The radioactive isotope while giving the stable isotope during its disintegration, emits usually either electrons ($_{-1}^{0}e$) or positrons ($_{+1}^{0}e$). The spontaneous disintegration of the radioactive isotope takes place according to the same laws that apply to the disintegration of naturally-occurring radioactive elements. Thus :

The process in which an element is converted into a new radioactive isotope (i.e. unstable isotope) by artificial means (i.e. by bombarding the element with fast moving particles like proton, deutron, helium etc) is called **artificial** *or* **induced radioactivity.**

Discovery of Artificial Radioactivity

Mme. Iren Curie Joliot (daugther of Marie Curie) and her husband, *M. Frederick Joliot* found that when $_{13}^{27}Al$ was bombarded with $_{2}^{4}He$ particles obtained from polonium and the energy of the ejected neutrons was measured, the detector continued to respond to some penetrating radiations, even after the polonium source was taken away. These penetrating radiations were later on found to be *positrons* ($_{+1}^{0}e$). These observations made Curie-Joliots pair draw the conclusion that the product nucleus formed during the bombardment of $_{13}^{27}Al$ by $_{2}^{4}He$ particles was unstable (*i.e. radioactive*) and hence disintegrated spontaneously emitting positron and finally changing into a stable $_{14}^{30}Si$ isotope.

The whole process can be shown by the following nuclear equations :

$$\underset{\text{stable isotope}}{_{13}^{27}Al} + {_{2}^{4}He} \rightarrow \underset{\substack{\text{unstable isotope} \\ (i.e.\text{ Radioactive})}}{_{15}^{30}P} + {_{0}^{1}n}$$

$$_{15}^{30}P \rightarrow \underset{\text{stable isotope}}{_{14}^{30}Si} + \underset{\text{positron}}{_{+1}^{0}e} \quad (t_{1/2} = 3.25 \text{ minutes})$$

This was how the positron which is a counterpart of electron and has a unit positive charge and no mass was first discovered.

The reactions in case of $_{12}^{24}Mg$ and $_{5}^{10}B$ can be shown as follows :

(*i*)
$$_{12}^{24}Mg + {_{2}^{4}He} \rightarrow \underset{\text{unstable isotope}}{_{14}^{27}Si} + {_{0}^{1}n}$$

$$_{14}^{27}Si \rightarrow \underset{\text{stable isotope}}{_{13}^{27}Al} + {_{+1}^{0}e}$$

(*ii*)
$$_{5}^{10}B + {_{2}^{4}He} \rightarrow \underset{\text{unstable isotope}}{_{7}^{13}N} + {_{0}^{1}n}$$

$$_{7}^{13}N \rightarrow \underset{\text{stable isotope}}{_{6}^{13}C} + {_{+1}^{0}e}$$

Researches on Induced Radioactivity

As soon as Curie and Joliot announced their discovery, the study of artificial radioactivity was taken up by many other workers and new cases of it were announced at a rapid rate. It was soon found that the phenomenon occurred not only under α-particle bombardment, but also with other projectiles, as protons, deuterons and neutrons and that the artificially produced radioelements not only emitted positrons but also electrons in several cases. Some examples are given below :

(i) Reactions emitting electrons ($_{-1}^{0}e$). Electrons are emitted in (*n*, γ), (*n*, *p*), (*n*, α) and (D, *p*) reactions.

(*a*) (*n*, γ) $\quad {}_{49}^{115}\text{In} + {}_{0}^{1}n \rightarrow \underset{\text{unstable}}{{}_{49}^{116}\text{In}} + {}_{0}^{0}\gamma$

$${}_{49}^{116}\text{In} \rightarrow \underset{\text{stable}}{{}_{50}^{116}\text{Sn}} + {}_{-1}^{0}e \; (t_{1/2} = 13 \text{ seconds})$$

(*b*) (*n*, *p*) $\quad {}_{7}^{14}\text{N} + {}_{0}^{1}n \rightarrow \underset{\text{unstable}}{{}_{6}^{14}\text{C}} + {}_{1}^{1}\text{H}$

$${}_{6}^{14}\text{C} \rightarrow \underset{\text{stable}}{{}_{7}^{14}\text{N}} + {}_{-1}^{0}e \; (t_{1/2} = 5568 \text{ years})$$

(*c*) (*n*, α) $\quad {}_{13}^{27}\text{Al} + {}_{0}^{1}n \rightarrow \underset{\text{unstable}}{{}_{11}^{24}\text{Na}} + {}_{2}^{4}\text{He}$

$${}_{11}^{24}\text{Na} \rightarrow \underset{\text{stable}}{{}_{12}^{24}\text{Mg}} + {}_{-1}^{0}e \; (t_{1/2} = 15 \text{ hours})$$

(*d*) (*D*, *p*) $\quad {}_{15}^{31}\text{P} + {}_{1}^{2}\text{H} \rightarrow \underset{\text{unstable}}{{}_{15}^{32}\text{P}} + {}_{1}^{1}\text{H}$

$${}_{15}^{32}\text{P} \rightarrow \underset{\text{stable}}{{}_{16}^{32}\text{S}} + {}_{-1}^{0}e \; (t_{1/2} = 14.3 \text{ days})$$

(ii) Reactions emitting positrons ($_{+1}^{0}e$). Positrons are emitted in (*p*, γ), (*p*, *n*), (α, *n*), (D, *n*) and (γ, *n*) reactions.

(*a*) (*p*, γ) $\quad {}_{6}^{12}\text{C} + {}_{1}^{1}\text{H} \rightarrow \underset{\text{unstable}}{{}_{7}^{13}\text{N}} + {}_{0}^{0}\gamma$

$${}_{7}^{13}\text{N} \rightarrow \underset{\text{stable}}{{}_{6}^{13}\text{C}} + {}_{+1}^{0}e \; (\text{T} = 10 \text{ minutes})$$

(*b*) (*p*, *n*) $\quad {}_{28}^{58}\text{Ni} + {}_{1}^{1}\text{H} \rightarrow \underset{\text{unstable}}{{}_{29}^{58}\text{Cu}} + {}_{0}^{1}n$

$${}_{29}^{58}\text{Cu} \rightarrow \underset{\text{stable}}{{}_{28}^{58}\text{Ni}} + {}_{+1}^{0}e \; (\text{T} = 2.6 \text{ seconds})$$

(*c*) (α, *n*) $\quad {}_{7}^{14}\text{N} + {}_{2}^{4}\text{He} \rightarrow \underset{\text{unstable}}{{}_{9}^{17}\text{F}} + {}_{0}^{1}n$

$${}_{9}^{17}\text{F} \rightarrow \underset{\text{stable}}{{}_{8}^{17}\text{O}} + {}_{+1}^{0}e \; (\text{T} = 70 \text{ seconds})$$

(*d*) (*D*, *n*) $\quad {}_{7}^{14}\text{N} + {}_{1}^{2}\text{H} \rightarrow \underset{\text{unstable}}{{}_{8}^{15}\text{O}} + {}_{0}^{1}n$

$${}_{8}^{15}\text{O} \rightarrow \underset{\text{stable}}{{}_{7}^{15}\text{N}} + {}_{+1}^{0}e \; (\text{T} = 2.1 \text{ minutes})$$

(*e*) (γ, *n*) $$^{31}_{15}P + ^{0}_{0}\gamma \rightarrow \underset{\text{unstable}}{^{30}_{15}P} + ^{1}_{0}n$$

$$^{30}_{15}P \rightarrow \underset{\text{stable}}{^{30}_{14}Si} + ^{0}_{+1}e \text{ (T = 2.5 minutes)}$$

Preparation of Transuranium Elements

Until recently, it was believed that there are only 92 elements (Atomic number 1 to 92). As a result of study of radioactive processes, however, it was observed that atoms with atomic number higher than 92 could be produced. At present 13 elements with atomic weight higher than that of uranium (At. No. = 92) and with atomic numbers 93 to 105 are called **transuranium elements or transuranic elements** because they lie beyond uranium in the periodic table.

It was only in 1940 that the existence of these elements was definitely established.

The *names, symbols* and *atomic numbers* of these elements are given below. The numbers given on the bottom left of the symbols indicate their atomic numbers.

Neptunium ($_{93}$Np), *Plutonium* ($_{94}$Pu), *Amercium* ($_{95}$Am), *Curium* ($_{96}$Cm), *Berkelium* ($_{97}$Bk), *Californium* ($_{98}$Cf), *Einsteinium* ($_{99}$Es), *Fermium* ($_{100}$Fm), *Mendelevium* ($_{101}$Md), *Nobelium* ($_{102}$No), *Lawrencium* ($_{103}$Lw), *Kurchatovium* ($_{104}$Ku), and *Hahnium* ($_{105}$Ha).

$_{93}$Np. In 1934 *Fermi* suggested that the bombardment of uranium with neutrons might give transuranium elements. In May 1940 *Mc Millan* and *Abelson* bombarded $^{238}_{92}U$ with slow neutrons and thus got a new isotope, $^{239}_{92}U$ which is radioactive with a half-life period of 23 minutes and hence disintegrates emitting β-particles (*i.e.* $^{0}_{-1}e$ or $^{0}_{-1}\beta$) to produce a new element called **Neptunium**, Np with atomic number, Z = 93.

$$^{238}_{92}U + ^{1}_{0}n \rightarrow \underset{\text{unstable}}{^{239}_{92}U} + ^{0}_{0}\gamma \; (t_{1/2} = 23.5 \text{ minutes})$$

$$^{239}_{92}U \rightarrow ^{0}_{-1}e + ^{239}_{93}Np \; (\textit{Neptunium}) \; (t_{1/2} = 23 \text{ minutes})$$

In all, 20 isotopes of neptunium have been produced with mass numbers ranging from 231 to 240 by bombarding uranium with (*i*) fast deuterons (up to 100 MeV energy) from a cyclotron (*ii*) fast neutrons and (*iii*) accelerated α-particles.

$_{94}$Pu. Later on it was found that $^{239}_{93}Np$ was also radioactive with a half-life period of 2.3 days and hence disintegrated emitting β-particles into another transuranium element now called **Plutonium,** Pu with Z = 94.

$$^{239}_{93}Np \rightarrow ^{0}_{-1}e + ^{239}_{94}Pu \; (\textit{Plutonium}) \; (t_{1/2} = 2.3 \text{ days})$$

The isotope of plutonium is itself an α-emitter and has a half-life of 24,400 years. It is manufactured in large amounts in nuclear reactors and can be used for the production of nuclear power or atom bomb.

$_{95}$Am and $_{96}$Cm. In 1945, *Hamilton, Seaberg* and collaborators prepared two more trans-uranium elements *viz.* **Americium,** Am with Z = 95 and **Curium,** Cm with Z = 96 by the bombardment of $^{238}_{92}U$ and $^{239}_{94}Pu$ respectively with 40 MeV helium produced in the 60 inch Berkeley cyclotron.

$$^{238}_{92}U + ^{4}_{2}He \rightarrow ^{241}_{94}Pu + ^{1}_{0}n$$

$$^{241}_{94}Pu \rightarrow ^{0}_{-1}e + ^{241}_{95}Am \text{ (Americium) } (t_{1/2} = 14 \text{ years})$$

$$^{239}_{94}Pu + ^{4}_{2}He \rightarrow ^{1}_{0}n + ^{242}_{96}Cm \text{ (Curium)}$$

$_{97}$Bk and $_{98}$Cf. In 1950, *Thomson, Ghiorso* and *Seaborg* obtained **Berkelium.** Bk with Z = 97 and **Californium,** Cf with Z = 98 by bombarding $^{241}_{95}Am$ and $^{242}_{96}Cm$ respectively with 35 MeV helium produced in the same 60 inch cyclotron.

$$^{241}_{95}Am + ^{4}_{2}He \rightarrow 2\,^{1}_{0}n + ^{243}_{97}Bk \text{ (Berkelium)}$$

$$^{242}_{96}Cm + ^{4}_{2}He \rightarrow 2\,^{1}_{0}n + ^{244}_{98}Cf \text{ (Californium)}$$

$_{99}$Es. *Ghiorso, Rossi, Harvey* and *Thompson* under the leadership of *Seaborg*, bombarded $^{239}_{92}U$ with accelerated nucleus of $^{14}_{7}N$ to get an isotope of **Einsteinium,** Es with Z = 99.

$$^{239}_{92}U + ^{14}_{7}N \rightarrow 6\,^{1}_{0}n + ^{247}_{99}Es \text{ (Einsteinium)}$$

$_{100}$Fm. *Ghiorso, Harvey, Thompson* and *Choppin*, under the direction of *Seaborg*, discovered the element, **Fermium,** Fm with Z = 100 by bombarding $^{239}_{94}Pu$ with neutrons from an atomic reactor.

$$^{239}_{94}Pu + 15\,^{1}_{0}n \rightarrow 6\,^{0}_{-1}e + ^{254}_{100}Fm \text{ (Fermium)}$$

$^{250}_{100}Fm$ isotope can be prepared by bombarding $^{238}_{92}U$ with accelerated nucleus of $^{16}_{8}O$.

$$^{238}_{92}U + ^{16}_{8}O \rightarrow 4\,^{1}_{0}n + ^{250}_{100}Fm \text{ (Fermium)}$$

$_{101}$Md. In April 1955, *Seaborg, Ghiorso* and coworkers discovered **Mendelevium,** Md with Z = 101 by bombarding $^{253}_{99}Es$ with 48 MeV helium ions.

$$^{253}_{99}Es + ^{4}_{2}He \rightarrow ^{1}_{0}n + ^{256}_{101}Md \text{ (Mendelevium)}$$

$_{102}$No. This element is an α-emitter. Very recently (in 1957) the *Noble Institute of Physics Stockholm* obtained the element **Nobelium,** No with Z = 102 by bombarding $^{246}_{96}Cm$ with accelerated $^{12}_{6}C$ ions. Russian scientists have proposed the name *Joliocium* for nobelium.

$$^{246}_{96}Cm + ^{12}_{6}C \rightarrow 6\,^{1}_{0}n + ^{252}_{102}No \text{ (Nobelium)}$$

This element is an α-emitter and has a half-life period on 10 to 12 minutes.

$_{103}$Lw. The element, **Lawrencium,** Lw with Z = 103 has been prepared by bombarding $^{251}_{98}Cf$ with accelerated $^{10}_{5}B$ ions.

$$^{251}_{98}Cf + ^{10}_{5}B \rightarrow 4\,^{1}_{0}n + ^{257}_{103}Lw \text{ (Lawrencium)}$$

$_{104}$Ku and $_{105}$Ha. Kurchatovium. Ku with Z = 104 and **Hahnium,** Ha with Z = 105 are the latest addition to the list of transuranium elements. American scientists have suggested the name *Rutherfordium* for the element with Z = 104.

Uses of Radioisotopes

Some important uses of radioisotopes are given below :

1. Uses of Radioisotopes as Radioactive Tracers.

First of all we should understand: "*What are radioactive tracers* ?" Radioactive isotopes have a property due to which they can easily be detected and estimated quantitatively. Further, they reach almost exactly as the corresponding non-radioactive isotopes do. As such, if we want to trace the course of an atomic species in a chemical or a physical change, we mix a small quantity of the radioactive isotope with non-radioactive isotope. Here we say that the radioactive istope is acting as a **Radioactive tracer** or **Radioactive indicator** or simply as **Labelled** *or* **Tagged element.** The presence of radioactive tracer is later on detected in the products with the help of sensitive instruments. Some important uses of radioactive tracers are given below :

(*i*) **Uses of tracers in studying reaction mechanisms.** **(*a*) *Photosynthesis.*** The mechanism of the process of photosynthesis taking place in plants has been studied by reacting carbon dioxide containing ^{18}O (CO_2^{18}) with H_2O. In this process the green plants take up CO_2 and H_2O in presence of sunlight and chlorophyll to form sugar and starch, giving out oxygen gas.

$$6CO_2 + 6H_2O \rightarrow C_6H_{12}O_6 + 6O_2$$

Here a very important question arises—whether the oxygen evolved in this process comes from CO_2 or from H_2O or from both. This question has been answered by the use of radioactive ^{18}O isotope. What we do is that H_2O is made to react with carbon dioxide containing ^{18}O (*i.e.* $CO_2{}^{18}$) and it is found that the oxygen evolved does not contain ^{18}O, *i.e.*, O^{18} present in $CO_2{}^{18}$ goes to sugar and water while that of water is evolved as free oxygen. Thus oxygen evolved in the process of photosynthesis comes from H_2O and not from carbon dioxide and consequently, the reaction shown above should be shown as :

$$6CO_2{}^{18} + 12H_2O \rightarrow C_6H_{12}O^{18} + 6H_2O^{18} + 6O_2$$

(*b*) *Hydrolysis of ester.* Mechanism of ester hydrolysis has been studied by using O^{18} isotope. It is observed that when an ester is hydrolysed by water containing O^{18} (*i.e.* H_2O^{18}), the alcohol formed does not have any O^{18} while the acid has the whole of it. The presence of O^{18} in the acid proves that the—OR′ group in the ester is wholly replaced by —O^{18}H group in H_2O^{18}. Thus the hydrolysis should correctly be shown as :

$$\underset{\text{Ester}}{\text{K?}\overset{\overset{\displaystyle O}{\|}}{\text{C}}\text{?}\boxed{\text{OR}' + \text{H}}}\underset{\substack{\text{Water}\\ \text{containing } O^{18}}}{\text{O18 H}} \rightarrow \underset{\text{Acid}}{\text{R?}\overset{\overset{\displaystyle O}{\|}}{\text{C}}\text{?O18H}} + \underset{\text{Alcohol}}{\text{R}'\text{OH}}$$

(*ii*) **Use of tracers in the diagnosis of diseases.** Radioisotopes have proved very useful in the diagnosis and understanding many of the diseases. Some examples are given below :

(*a*) *Location and detection of brain tumour.* Labelled iodine has been used to locate and detect the presence of tumour. To locate it a small quantity of radioisotope of iodine mixed with an organic dye is injected into the body of the patient. Dye with radioisotope is strongly absorbed by the tumour and radiations are produced. Detection of these radiations gives valuable information regarding the limits of the growth and location of the tumour.

(b) Circulation of blood. Radiosodium, Na^{24} ($t_{1/2}$ = 15 hours, β-particle emitter) has been used to study cases of restricted circulation of blood. A small quantity of NaCl solution which contains labelled-sodium (Na^{24}) is injected into, the vein of a patient's forearm and a γ-ray detector is then placed in contact with one of his feet. If blood circulation is normal, the presence of rodioactivity is soon discovered in the foot. It increases rapidly and becomes maximum within an hour. However, if there is any circulatory impairment, radioactivity will increase slowly showing that blood has some difficulty in reaching the foot. By moving the detector to different parts of the body, the position of restriction can be located and the necessary treatment applied.

(c) Pumping action of heart. Pumping action of heart has been studied by using radiosodium or radioiodine. A small quantity of labelled NaCl solution is injected into the blood stream and a detector, attached to an automatic pen recorder, is placed on the heart. As radiosodium enters the right side of the heart (i.e. atrium), the pen of the recorder rises and then drops as the venous blood enters the lungs for oxygenation. After some time, the radiosodium appears with the arterial blood on left side of the heart (*i.e.* ventricle) and there is another rise and fall of the recording pen. By examining the resulting curve, the pumping action of the two sides of the heart can be compared and abnormalities, if any, can be discovered.

(d) Functioning of thyroid gland. Radioisotope of iodine, I^{131} with $t_{1/2}$ = 8 days is given to a patient with thyroid disorders. It decays β-rays. Detection of these rays gives an information regarding the functioning of the thyroid gland.

(e) Disorders associated with pregnancy. Tracer iron has been used in studying the disorders associated with pregnancy.

(f) Test of cancer growth. ^{50}Co emits high energy γ-rays which are used for testing deeply seated cancer growth.

(*iii*) **Uses of tracers in industry.** Radioactive isotopes have found many applications in industry both in research and in process control. Some of the examples are given below :

(a) Determination of the wear of piston rings. The wear of piston rings and of gears in engines and its prevention by means of suitable lubricants have been studied by radioactivity. A steel piston ring is made radioactive by exposing it to neutrons in a nuclear reactor. The piston ring is then fitted into the cylinder of an internal combustion engine which is operated normally with a particular lubricating oil. By determining the rodioactivity removed by the oil, the amount of piston wear can be determined.

(b) Self-diffusion of metals. The phenomenon of self-diffusion in metals *i.e.* the movement of the atoms of a metal within the crystal lattice has been studied with the help of isotopes.

(c) Uniform mixing. Uniformity of mixing during the blending of petrols, lubricating oils and greases etc. has been achieved by labelling one of the constituents with a radioactive tracer.

(d) As catalyst. γ-rays obtained from Co^{60} have been used as catalyst in the manufacture of C_2H_5 Br from ethylene and HBr.

(e) Treatment of a metal. By incorporating radioactive isotope in the metal it is possible to know as to what happens when a metal is subjected to a particular treatment like age-hardening, annealing, quenching, coldroling etc.

(f) Mechanism of lubricants. Radioactive isotopes have been used in studying the mechanism of the effectiveness of various lubricants.

(g) Mechanisms of reactions. Radioactive carbon, C^{15} has been used as a tracer in studying mechanisms involved in many reactions of industrial importance such as alkylation, polymerisation, catalytic synthesis, etc.

(*iv*) **Uses of tracers in agriculture.** ***(a) Uptake of phosphorus by plants.*** When a plant grows, it absorbs phosphorus both from the soil and from the added phosphatic fertilizer. In order to check the usefulness of the fertilizer, it is important to know what proportion of phosphorus comes from each source. Ordinary analytical methods cannot answer this question because phosphorus from both sources is exactly the same. However, if radiophosphorus (P^{32}) is added in the phosphatic fertilizer, the exact proportion taken up by the plant can be determined accurately.

The efficiency of different phosphatic fertilizers can be compared with the help of radioactive phosphorus being used as a tracer element. Fertilizer which is absorbed quickly is more efficient.

(b) Transportation of mineral in plants. The transportation of mineral from roots to leaves and its subsequent redistribution within the plants can be followed through the use of radioactive isotopes by means of radioautography. By this technique the distribution of S^{35}, supplied in the form of sulphate, in sugar beet plants has been followed.

(c) Irradiation of maize with radioactive cobalt produces 15% increase in the quantity of green part over those plants growing in the same field, but not irradiated. In some cases, solutions of radioactive substances are used in place of irradiation. The seeds are dipped in these solutions before planting. This also raises the yield of cane sugar, wheat, maize etc.

(*v*) **Uses of tracers in analytical chemistry.** Suppose we want to determine the solubility of $PbSO_4$ in water. A known quantity of *radioactive* lead (Pb*) is mixed with ordinary lead (Pb) in a known proportion. The entire amount consisting of radioactive and ordinary lead is dissolved in dil. HNO_3 and then precipitated as lead sulphate on the addition of dil. H_2SO_4.

$$Pb^* + Pb \xrightarrow{HNO_3} Pb^*(NO_3)_2 + Pb(NO_3)_2 \xrightarrow{H_2SO_4} Pb^*SO_4\downarrow + PbSO_4\downarrow$$

After filtration, the radioactivity of the solution is measured and thus the amount of lead sulphate present in solution is calculated. This, of course, gives the amount of radioactive lead present in solution. If the ratio of the two varieties of lead originally taken is known, the amount of ordinary lead present in solution can be calculated.

2. Uses of Radioisotopes in the Treatment of Diseases (*Radiotherapy*).

The treatment of diseases by the use of radioisotopes is called *Radiotherapy*. Some examples of radiotherapy are given below :

(a) Treatment of thyroid complaints. Radioactive iodine I^{131} has been used for the treatment of certain thyroid complaints. Stable iodine and radioactive iodine are preferentially absorbed by the thyroid gland. This fact is utilised in the treatment of *hyperthyroidism* and *cancer* of the thyroid. Radioactive iodine destroys abnormal tissues in the thyroid gland more rapidly and in this way checks the abnormal growth.

(b) Treatment of blood disorders. Radioactive phosphorus in the from of phosphate is widely used in the treatment of certain blood disorders.

(c) Treatment of cancer growth. Radiocobalt (Co^{60}) produced by the action of neutrons from a nuclear pile, $t_{1/2}$ = 5.3 days, γ-rays emittter and *radiogold* (Au^{198}), T = 3 days, γ-rays emitter are used in the treatment of some forms of cancer. Co^{60} has been called *Poor Man's Radium.*

(d) Treatment of leukemia. Radioisotope of phosphorus is being used for the treatment of leukemia.

(e) Treatment of skin diseases. Radiophosphorus (P^{30}), $t_{1/2}$ = 14.3 days, β-emitter is found good for skin diseases.

3. Determination of the Age of the Earth by Rock Dating Method

The age of a rock (*i.e.* the time that elapsed since the rock crystallied from the molten state) gives an approximate idea of the age of the earth provided that it is assumed that the rock or the mineral under examination is as old as the earth. Let us consider a rock containing U-238 isotope formed many years ago. The age of this rock can be determined by considering its radioactive disintegration which, as we have already seen, is governed by the relation

$$N = N_0 . e^{-\lambda t}, \qquad ...(i)$$

where N_0 = amount of U-238 isotope originally present in a small quantity of the rock or mineral at the time the rock was formed, N = Amount of U-238 still left undecayed after the lapse of time, t which represents the age of the rock, and λ = disintegration constant of U-238.

U-238 started to decay giving rise to U-238—Pb-206 series. Some of its quantity decayed into Pb-206 in time, *t*, while the remaining quantity of U-238 remained undisintegrated. If the quantity of U-238 remained undisintegrated is U^{238} (*i.e.* U^{238} represents the quantity of U-238 existing at the present time in the rock) and that of Pb-206 formed as a result of disintegration of U-238 is Pb^{206} (*i.e.* Pb^{206} represents the quantity of Pb-206 existing at present in the rock), then the quantity of U-238 present at zero time (*i.e.* before disintegration started) is equal to (U^{238} + Pb^{206}). Thus :

$$N = U^{238} \text{ and } N_0 = (U^{238} + Pb^{206})$$

and hence eqn. (*i*) given above becomes :

$$U^{238} = (U^{238} + Pb^{206})\, e^{-\lambda t} \qquad ...(ii)$$

or $$\frac{U^{238}}{(U^{238} + Pb^{206})} = e^{-\lambda t}$$

or $$e^{\lambda t} = \frac{U^{238} + Pb^{206}}{U^{238}}$$

$$= \left(1 + \frac{Pb^{206}}{U^{238}}\right)$$

or $$\lambda t = 2.303 \log_{10}\left(1 + \frac{Pb^{206}}{U^{238}}\right)$$

Now we know that the value of decay constant (λ) of U-238 is given by :

$$\lambda = \frac{0.693}{t_{1/2}}$$

where $t_{1/2}$ is the half-life period of U-238.

Thus

$$t = \frac{2.303\, t_{1/2}}{0.693} \log_{10}\left(1 + \frac{Pb^{206}}{U^{238}}\right)$$

Since the values of Pb^{206} and U^{238} can be determined experimentally, the value of t which is the age of the rock (or of earth) can be found out.

If, in place of rock containing U-235, we consider the rock containing U-235, then the isotope of lead obtained from the disintegration of U-235 is Pb-207 and the relation (*ii*) reduces to (*iii*) given below :

$$U^{235} = (U^{235} + Pb^{207})\, e^{-\lambda' t} \qquad ...(iii)$$

where λ' is the disintegration constant of U-235 isotope.

Now equation (ii) can also be written as :

$$\frac{U^{238}}{U^{238}} = \left(\frac{U^{238}}{U^{238}} + \frac{Pb^{206}}{U^{238}}\right) e^{-\lambda t}$$

or $$1 = \left(1 + \frac{Pb^{206}}{U^{238}}\right) e^{-\lambda t}$$

or $$e^{\lambda t} = 1 + \frac{Pb^{206}}{U^{238}}$$

or $$(e^{\lambda t} - 1)\, U^{238} = Pb^{206} \qquad ...(iv)$$

Likewise equation (*iii*) reduces to :

$$(e^{\lambda' t} - 1)\, U^{235} = Pb^{207} \qquad ...(v)$$

On dividing equations (*iv*) and (*v*) by each other, we get

$$\frac{Pb^{206}}{Pb^{207}} = \frac{U^{238}(e^{\lambda t} - 1)}{U^{235}(e^{\lambda' t} - 1)}$$

The present ratio, U^{238}/U^{235} is known to be 139 and hence

$$\frac{Pb^{206}}{Pb^{207}} = \frac{139\,(e^{\lambda t} - 1)}{(e^{\lambda' t} - 1)}$$

When the values of λ and λ′ as determined from the half-life periods of the isotopes *viz.* U-238 and U-235 and the ratio Pb^{206}/Pb^{207} as determined by a mass spectrometer are put in the above equation, we get the value of t which is the age of the rock or earth.

The above method is useful for dating minerals and rocks of considerable age.

Example. *The ratio of mass of Pb^{206} to the mass of U^{238} in a certain rock specimen is found to be 0.5. Assuming that the rock originally contained no lead, estimate its age. Half-life of uranium is 4.5×10^9 years.*

Solution. We have already seen that the relation between the existing masses of U^{238} and Pb^{206} in any rock specimen is given by :

$$U^{238} = (U^{238} + Pb^{206})e^{-\lambda t}$$

or
$$1 = \left(1 + \frac{Pb^{206}}{U^{238}}\right)e^{-\lambda t}$$

or
$$1 = (1 + 0.5)e^{-\lambda t} = 1.5e^{-\lambda t}$$

∴
$$e^{\lambda t} = 1.5$$

or
$$\lambda t = 2.3 \log_{10} 1.5 = 2.3 \times 0.1761$$

∴
$$\frac{0.693}{4.5 \times 10^9} \times t = 0.23 \times 0.1761$$

$$t = \mathbf{2.63 \times 10^9}\ \textbf{years}$$

4. Determination of the Age of Recent Objects by Radiocarbon Dating Method.

The age of recent objects such as those of animal or vegetable origin like a piece of wood or animal fossil can be determined by radio-carbon dating method, since rock dating method is useful only for the minerals or rocks of considerable age. This method was developed by Willard Libby who was awarded Nobel Prize for this brilliant work.

The determination of the age of a sample of wood (*i.e.* the time which elapsed after the death of the living plant) consists of determining the ratio of the amount of ${}_6C^{14}$ to that of ${}_6C^{12}$ in both the pieces of wood, i.e. in fresh (living) piece and dead (cut) piece. We know that the atmospheric carbon dioxide is a mixture of $C^{14}O_2$ and $C^{12}O_2$ which are present in a fixed ratio. Plants absorb carbon dioxide from the atmosphere and prepare cellulose (wood). As long as the plant is alive, the ratio of ${}_6C^{14}$ to ${}_6C^{12}$ atoms in the wood of the plant is the same as in the atmosphere but when the tree is cut (*i.e.* when the plant dies), the ratio of ${}_6C^{14}$ to ${}_6C^{12}$ begins to decrease continuously due to the continuous decrease in the amount of ${}_6C^{14}$ in the plant. This decrease in the amount of ${}_6C^{14}$ is due to its continuous disintegration emitting β-radiations as shown :

$${}_6C^{14} \rightarrow {}_7N^{14} + \beta\text{-particle } (t_{1/2} = 5568 \text{ years})$$

Now we know that :

$$N = N_0.e^{-\lambda t}$$

or $$t = \frac{2.303\ T}{0.693} \log_{10} \frac{N_0}{N}$$

On putting $N_0 =$ C^{14}/C^{12} ratio in the living (*i.e.* fresh) plant,

and $N =$ C^{14}/C^{12} ratio in the dead (*i.e.* cut) plant, in the above equation, the age of the wood, t is given by :

$$t = \frac{2.203\ t_{1/2}}{0.693} \log \left[\frac{C^{14}/C^{12} \text{ ratio in the living fresh plant}}{C^{14}/C^{12} \text{ ratio in the dead plant}} \right] \quad ...(i)$$

Here $t_{1/2}$ is half-life period of ${}_6C^{14}$

If the ratio of the amounts of C^{14} isotope in the fresh and dead wood is known, then

$$t = \frac{2.303\ t_{1/2}}{0.693} \log \left[\frac{\text{Amount of } C^{14} \text{ in living (fresh) wood}}{\text{Amount of } C^{14} \text{ in dead wood}} \right] \quad ...(ii)$$

Example 1. *The amount of ${}_5C^{14}$ isotope in a piece of wood is found to be one-sixth present in a fresh piece of wood. Calculate the age of wood. Half-life of ${}_6C^{14}$ = 5577 years.*

Solution. With the help of equation (*ii*) given above we get

$$t = \frac{2.303 \times 5577}{0.693} \log \frac{1}{1/6}$$

$$= \frac{2.303 \times 5577}{0.693} \log 6$$

$$= \frac{2.303 \times 5577 \times 0.7782}{0.693}$$

$$t = \mathbf{20170\ years}$$

Example 2. *A piece of wood was found to have C^{14}/C^{12} ratio 0.7 times that in a living plant. Calculate the period when the plant died. Half-life of C^{14} = 5760 years.*

Solution. $$t = \frac{2.303\ T}{0.693} \log \frac{1.00}{0.70}$$

$$= \frac{2.303 \times 5760 \times 0.155}{0.693}$$

$$\therefore \quad t = \mathbf{2970\ years}$$

Example 3. *The C^{14} to C^{12} ratio in a certain piece of wood is 13% that of the atmosphere. Calculate the age of wood. given that half-life of C^{14} = 5580 years.*

Solution. We know that

$$t = \frac{2.303 \times t_{1/2}}{0.693} \log \frac{100}{13}$$

Therefore, in this case

$$t = \frac{2.303 \times 5580 \times 0.8161}{0.693}$$

$$t = \mathbf{16430\ years}$$

Questions with Answers

Q.1 If the enrgy released in the reaction, $4{}_1^1H \rightarrow {}_2^4He + 2{}_1^0e$ is 26×10^8 KJ mol^{-1}of He, calculate the magnitude of the change in mass (m) in g mol^{-1} of He.

Ans : Energy released of He

$$= 26 \times 10^8 \text{ kJ mol}^{-1}$$
$$= 26 \times 10^8 \times 10^3 \text{ J mol}^{-1}$$
$$= 26 \times 10^{11} \text{ J mol}^{-1}$$

Now we know that change in mass (m) is given by :

$$E = mc^2$$

or $$m = \frac{E}{c^2} \quad ...(i)$$

Now put $E = 26 \times 10^{11}$ J mol^{-1} and $c = 3 \times 10^8$ ms^{-1} in equation (i) to get :

$$m = \frac{26 \times 10^{11} \text{ J mol}^{-1}}{(3 \times 10^8)^2 \ m^2 s^{-2}}$$
$$= 2.9 \times 10^{-5} \frac{J}{m^2 s^{-2}} \text{ mol}^{-1}$$
$$= 2.9 \times 10^{-5} \text{ kg mol}^{-1} \quad (\because 1J = \text{kg } m^2 s^{-2})$$
$$= 2.9 \times 10^{-5} \times 10^3 \text{ g mol}^{-1}$$
$$= 0.029 \text{ g mol}^{-1}$$

Now since energy is released in the given reaction, mass of products < mass of reactants. Hence the value of m should be negative, *i.e.* change in mass (m) = **– 0.029 g mol^{-1} (Ans)**

Q.2 Find out the value of binding energy per nucleon of ${}_8O^{16}$, if the binding energy of ${}_8O^{16}$ is 127 Mev.

Ans : The symbol, ${}_8O^{16}$ shows that the number of nucleons in ${}_8O^{16}$ = 16 Thus :

Binding energy of ${}_8O^{16}$ per nucleon

$$= \frac{\text{Binding energy of } {}_8O^{16}}{\text{No. of nucleons in } {}_8O^{16}}$$
$$= \frac{\text{Binding energy of } {}_8O^{16}}{\text{Mass number of } {}_8O^{16}}$$
$$= \frac{127}{16} \text{ MeV} = \mathbf{7.94 \text{ MeV (Ans.)}}$$

Q.3 Find out the age of artifact, if the artifact and a freshly cut down tree give 7.6 and 15.2 counts min^{-1} g^{-1} of carbon ($t_{1/2}$ = 5760 yrs).

Ans : The age of the artifact (t) is given by :

$$t = \frac{2.303}{\lambda} \log \frac{\text{No. of counts of freshly cut tree}}{\text{No. of counts of artifact}} \quad ...(i)$$

λ is given by :

$$t = \frac{0.693}{t_{1/2} \text{ of carbon}} = \frac{0.693}{5760 \text{ yrs}} = \frac{0.693}{5760} \text{yrs}^{-1}$$

Put the value of λ in equation (*i*)

$$t = \frac{2.303}{0.693/5760 \text{ yrs}^{-1}} \log \frac{15.2}{7.6}$$

$$= \frac{2.303 \times 5760}{0.693} \log 2 \text{ yrs}$$

$$= \mathbf{5761 \text{ yrs (Ans.)}}$$

Q.4 Find out the value of binding energy of $^{9}_{4}X$ per nucleon, if its mass defect is 0.090 amu. (1 amu = 931.5 Mev).

Ans : Binding energy of $^{9}_{4}X$ per nucleon

$$= \frac{\text{Mass defect in amu} \times 931.5 \text{ MeV}}{\text{No. of nucleons in } ^{9}_{4}X}$$

$$= \frac{0.090 \times 931.5}{9} \text{ MeV} = \mathbf{9.315 \text{ MeV. (Ans.)}}$$

Q.5 Mass of $^{7}_{3}Li$ is 7.016005 amu. Mass of a proton = 1.007277 amu and mass of a neutron = 1.008665 amu. Find out the mass defect of $^{7}_{3}Li$ nucleus.

Ans : No. of protons in $^{7}_{3}Li$ = 3

No. of neutrons in $^{7}_{3}Li$ = 7 – 3 = 4

∴ Mass defect of $^{7}_{3}Li$ nucleus

= (Mass of 3 protons + Mass of 4 neutrons) – Mass of $^{7}_{3}Li$ nucleus

= (3 × 1.007277 + 4 × 1.008665) – 7.016005

= 3.021831 + 4.034660 – 7.016005

= 7. 056491 – 7.016005 = **0.040486 amu (Ans.)**

Q.6 Calculate the magnitude of energy released per atom of He in the nuclear fusion : $^{2}_{1}H + ^{2}_{1}H \rightarrow ^{4}_{2}He$ (given that mass of $^{2}_{1}H$ = 2.014 amu and mass of $^{4}_{2}He$ = 4.003 amu).

Ans :

$$\underbrace{\underset{2.014 \text{ amu}}{^{2}_{1}H} + \underset{2.014 \text{ amu}}{^{2}_{1}H}}_{= 4.028 \text{ amu}} \rightarrow \underset{4.003 \text{ amu}}{^{4}_{2}He} \quad 4.003 \text{ amu}$$

∴ Mass lost = Mass of reactants – Mass of products

= (4.028 – 4.003) amu

= 0.025 amu

∴ Energy released per atom of He = [Mass lost (in amu) × 931] Mev

= 0.025 × 931 Mev

= **23.275 Mev (Ans.)**

Q.7 Calculate binding energy of ^{16}O (in J mol^{-1}), if its mass defect is 0.210 amu.

Ans : Binding energy = Mass defect × 931 MeV

$$= 0.210 \times 931 \text{ MeV}$$
$$= 195.5 \text{ MeV}$$
$$= 195.5 \times 10^6 \times \text{eV}$$
$$= 195.5 \times 10^6 \times 1.602 \times 10^{-19} \times 6.02 \times 10^{23} \text{ Jmol}^{-1}$$
$$= \mathbf{1.89 \times 10^{13} \ Jmol^{-1}}$$

Q.8 If the number of atoms of parent element and daughter element after time *t* in a transformation process is P and D respectively, then *t* is given by :

$$t = \frac{1}{\lambda} \ln \left(1 + \frac{D}{P}\right)$$

Prove this relation (λ = Diseintegration constant).

Ans : Suppose the number of atoms of parent element in the beginning (*i.e* at $t = 0$) is N_0 and after time t the number of atoms of parent element is N. Thus the number of atoms of daughter element at time $t = 0$ is zero and after time t the number of atoms of daughter element is $N_0 - N$. Hence :

	Parent element →	Daughter element
No. of atoms at $t = 0$	N_0	zero
No. of atoms after time, t	N = P	$(N_0 - N) = D$

Above transformation process shows that :

$$P = N \quad ...(i)$$

and $D = N_0 - N$

or $D = N_0 - P$

or $N_0 = D + P \quad ...(ii)$

Now we know that t, in terms of N_0 and N, is given by :

$$t = \frac{1}{\lambda} \ln \frac{N_0}{N} \quad ...(iii)$$

Put N = P and $N_0 = D + P$ from equations (*i*) and (*ii*) respectively in equation (*iii*) to get :

$$t = \frac{1}{\lambda} \ln \frac{D+P}{P}$$
$$= \frac{1}{\lambda} \ln \left(1 + \frac{D}{P}\right)$$

Q.9 What is the binding energy of $^{11}_{5}B$ nucleus if its mass defect is 0.08181 amu.

Ans : Binding energy of $^{11}_{5}B$ nucleus = (Mass defect in amu × 931.5) Mev

$$= 0.08181 \times 931.5 \text{ MeV}$$
$$= \mathbf{76.20 \ MeV \ (Ans)}$$

Q.10 Find out the packing fraction of $^{238}_{92}U$, if its isotopic mass is 238.125 amu.

Ans : Packing fraction of $^{238}_{92}U$

$$= \frac{\text{Isotopic mass of } ^{238}_{92}U - \text{Mass number of } ^{238}_{92}U}{\text{Mass number of } ^{238}_{92}U} \times 10^4$$

$$= \frac{238.125 - 238}{238} \times 10^4 = \mathbf{5.25\ (Ans.)}$$

University Questions

1. Calculate the mass defect and binding energy in MeV of oxygen isotope, $^{16}_{8}O$ whose mass number is 15.99491 amu. Given that mass of electron = 0.0005486 amu, mass of a proton = 1.007277 amu, mass of a neutron = 1.008665 amu. *(Delhi 99)*

2. Write a note on "Nuclear fusion". *(Kanpur 99)*

3. Name the elements which are used as nuclear fuels. Explain their action. *(Himachal Pradesh 99)*

4. (*a*) What is role of meson in the nucleus of an atom?

(*b*) Write a note on "Packing Fraction". *(Allahabad 99)*

5. Write short notes on (*i*) Transuranic elements (*ii*) Uranium as nuclear fuel. *(Himachal Pradesh 2000)*

6. What are nuclear reactions ? Describe nuclear reactions with deutrons, neutrons, tritones and photons with suitable examples. *(Kumaon 2000)*

7. Discuss the stability of nucleus on the basis of : (*i*) n/p ratio (*ii*) odd-even rule. *(Lucknow 2000)*

8. (*a*) Explain "Nuclear Reactor".

(*b*) Derive an expression for the disintegration constant of a radioactive element and show that decay of the element is exponential.

(*c*) The amount of C^{14} in a sample of wood is found to be one-fourth of its amount present in a fresh piece of wood. Calculate the age of the wood. (Half life = 5580 yrs.)

(*d*) How much energy in MeV is released in the formation of He nucleus (mass = 4.0017 amu) ? Given m_n = 1.0087, amu m_p = 1.0072 amu

(*e*) Write a note on carbon dating. *(Lucknow 2001)*

9. Describe fundamental particles of nucleons. or what are nucleons ? *(Kurushetra 2002)*

10. Describe the Meson theory of stability of nucleus. *(Kurushetra 2002)*

11. Describe different methods of detection and measurement of radioactivity. *(Guwahati 2002)*

12. What are diffirent types of nuclear reactions depending on transformation of nucleus. *(Guwahati 2002)*

13. Discuss theory of radioactive disintegration with suitable examples. (*Culcutta 2005*)

14. Define artificial radioactivily with on example. (*Meerut 2006*)

15. (*a*) What are nuclear reactions ?

(*b*) Explain Bohr's mechanism of nuclear reactions.

(*c*) Classify nuclear reactions based on energy.

16. What do you mean by artificial transformation and artificial radioactivity ? Give two reactions of each type. (*Meerut 2008*)

17. (*a*) Discuss nuclear reactions in detail.

(*b*) Define nuclear fission. (*Meerut 2009*)

CHAPTER 7

Chemical Bonding : Lewis Theory

What is Chemical Bond?

Atoms rarely occur in the free state at ordinary temperature. Except noble gases, which are chemically unreactive, atoms of all other elements have a tendency to combine either with each other or with the atoms of other elements to form cluster or aggregates of atoms with definite composition. A cluster formed may be either a molecule or an ion. The attraction between atoms forming the cluster is called a *chemical bond* or *valence bond*. Thus :

A chemical bond is defined as the attractive force that holds two or more atoms together in a molecule or an ion.

Our main concern in this chapter is to consider the interactions between atoms in the light of the structure of atom and find out the answers of the following questions

(*a*) *Why do atoms combine*?

(*b*) *How do atoms combine together*?

(*c*) *How can the properties of compounds be understood and determined in terms of the forces that keep the atoms together in molecules.*

Why Do Atoms Combine? (*Cause of Chemical Bonding*)

By a close study of atoms and molecules it has been found that atoms combine chemically for the following reasons :

1. Net attractive force between atoms. Atoms consist of strongly positive nucleus and negative electrons. When two atoms come closer to combine with each other to form a bond between them, the attractive and repulsive forces begin to operate between them. The attractive forces are between the electrons of one atom and the nucleus of the other atom while the repulsive forces are between the electrons or the nuclei of the two atoms (Fig. 7.1).

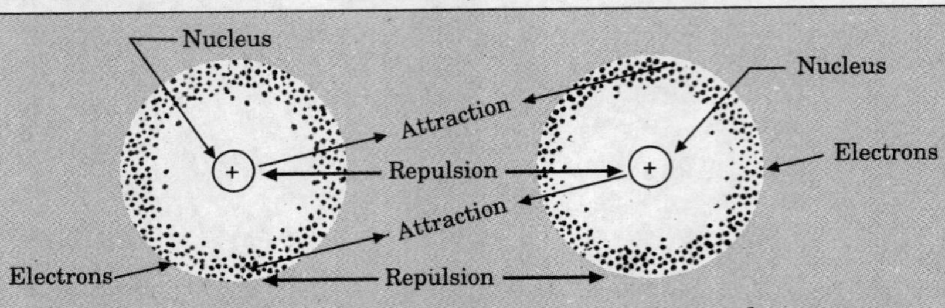

Fig. 7.1. Attractive and repulsive forces operating between two atoms when they come closer to each other.

When the two atoms approach closer to each other, these forces counteract each other. The net result of these forces may be either *attraction* or *repulsion* between the atoms. If the attractive forces become dominant over the repulsive forces, the net result is the attraction between the atoms and hence they combine together to form a chemical bond between them. On the other hand if the repulsive forces become dominant over the attractive forces, the atoms do not combine and hence no chemical bond is established between them. For example in case of hydrogen atoms, the net result is *attraction and hence two H-atoms combine together to form* H_2 *molecule.* On the other hand in case of helium atoms, the net result is *repulsion and hence two He-atoms do not combine together to form* He_2 *molecule.*

2. Octet rule or rule of eight. (*Electronic Theory of valency or octet theory of valency*). Lewis, Kossel and Longmuir (1916) tried, for the first time, to explain *why atoms combine together* on the basis of the electronic configuration of noble gases as given below.

Noble gas	At. No.	Electronic configuration
He	2	2
Ne	10	2, 8
Ar	18	2, 8, 8
Kr	36	2, 8, 18, 8
Xe	54	2, 8, 18, 18, 8
Rn	86	2, 8, 18, 32, 18, 8

They assumed that since the atoms of noble gases do not normally react with other atoms to form compounds, it is reasonable to assume that the outermost shell configuration of the atoms of noble gases is a stable configuration of 8 electrons which they called an *octet*. They also concluded that the two electrons in case of helium (called *duplet*) is also as stable as an octet present in other noble gases. *Since the octet of electrons is so stable in the gases, one can reasonably assume that when atoms of other elements combine to form a molecule, the electrons in their outermost orbits are arranged between themselves in such a way that they achieve an octet of electrons which is stable and thus a chemical bond is established between the atoms.*

The tendency of the atoms to have eight electrons in their outermost shell is known as **octet rule** or **rule of eight.** Since helium atom has only two electrons, this rule is called **doublet rule** or **rule of two** in case of helium. Octet rule was given in the form of a theory which is known as *octet theory of valency or electronic theory of valency* which states that :

In the formation of a chemical bond atoms interact with each other by losing, gaining or sharing of electrons so as to acquire a stable outer-shell of eight electrons.

The main points of this theory can be summarised as follows :

(*i*) Atom with 8 electrons in the outer most shell (2 in case of helium) are chemically stable and hence are incapable of chemical combination.

(*ii*) An atom having less than 8 electrons it its outer-most shell is chemically active and hence has a tendency to combine with other atoms. The atoms possessing less than 4 electrons in their ultimate shell usually tend to lose them, while those having more the 4 electrons in the outermost shell tend to gain the electrons during the chemical combination or bond formation to attain stable configuration of the nearest inert gas.

(*iii*) Atoms combine chemically as a result of transferring of electrons from the outer-most shell of one atom to that of the other or by sharing one, two or three electron pairs between the valence-shell of both the combining atoms. The transfer of electrons or sharing of electron pairs gives a stable configuration of 8 electrons to the valence-shell of both the atoms.

(*iv*) The tendency of an atom for transference or sharing its electron pairs is a measure of its chemical activity.

3. Lowering of energy of combining atoms. When two atoms combine together to form a bond, there is an over all decrease in the potential energy of the combining atoms, *i.e.*, a system having bonded atoms has lower energy than that having the unbonded atoms. This implies that the system of bonded atoms having lower energy is more stable than that of unbonded atoms having higher energy. *From this it, therefore, follows that the process of chemical bonding between the atoms decreases the energy of the combining atoms and gives rise to the formation of a system which has lower energy and hence has greater stability.*

Potential Energy Curve

The curve given in Fig. 7.2 shows the variation of potential energy with the distance between the nuclei of two atoms A and B which are approaching closer to each other to form a bond between them. The trend of the curve from right to left should be observed. When the two atoms A and B are far away, say at an infinite distance, from each other, the attraction between them is zero and hence, by convention, the energy of each of the atoms in arbitrarily taken to be zero and for stable system it is negative. It is for this

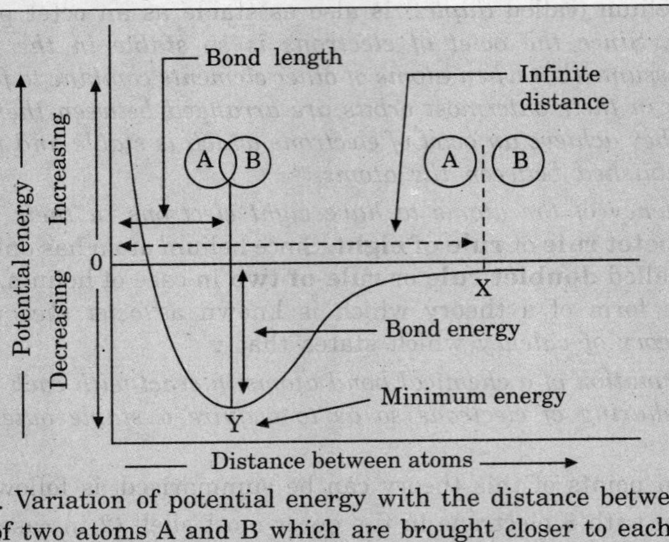

Fig. 7.2. Variation of potential energy with the distance between the nuclei of two atoms A and B which are brought closer to each other to form a bond between them (Potential Energy Curve)

reason that there, is no possibility of the formation of any bond between them. This situation has been represented by point X shown at the far end towards the right of the curve.

As the two atoms are brought closer to each other, *i.e.* as the distance between the atoms is decreased, the attractive forces between the electrons and the nucleus become more dominant than the repulsive forces operating between the electrons of the two atoms and hence the energy of the system goes on decreasing as shown by the downward trend of the curve. This decrease of energy continues till a certain minimum value shown by point Y in the curve is obtained. Now if the atoms are brought still closer together, the repulsive forces between the two nuclei at such small inter-nuclear distance (inter-nuclear distance between the nuclei of the two atoms) becomes dominant and hence the energy of the system starts increasing as shown by the upward trend of the curve.

Point Y which is the minimum of the curve has the following characteristics :

(*a*) At this point the attractive and repulsive forces are in equilibrium and hence a bond is said to be established between the atoms at this point.

(*b*) The overlapping between the orbitals of the two atoms is maximum at this point.

(*c*) The energy of the system at this point is minimum and hence AB molecule is in the stablest position. The inter-nuclear distance is also minimum at this point. The minimum energy value and inter nuclear distance corresponding to this point are called respectively *bond energy* and *bond length* of AB molecule. For H_2 molecule the bond energy is 103.2 kcal/mole and bond length is 0.74Å.

The above description makes it evident that a chemical bond is formed between the two atoms when the potential energy of the combining atoms is minimum and when they are at the minimum distance from each other. In order to break the bond between the atoms as much energy has to be put in as was earlier lost in the formation of the bond. The amount of energy lost in the process of bond formation or needed to break the bond is a measure of the strength of the bond. *More the energy decreased during bond formation (or combination of atoms in a molecule), the stronger is the bond between the atoms.*

How Do Atoms Combine ?

The process by which the atoms of the elements rearrange their outer-most shell electrons to get eight-electron outer-most shell configuration which is a stable configuration takes place in the following ways :

(*i*) In this way the transfer of one or more electrons from the valence-shell of an atom to the valence-shell of another atom takes place (*Formation of ionic bond*).

(*ii*) In this way one, two or three electron pairs of the valence- shell of both the combining atoms are shared between them. The shared electron pairs may be contributed equally by both the atoms (*Formation of covalent bond*) or may be contributed by only one of the combining atoms (*Formation of co-ordinate bond*).

Types of Bonds

Corresponding to the above two ways by which the two atoms rearrange their outer-most shell electrons to get an eight-electron outer-most shell configuration, we have the following four types of bonds (or linkages) which hold the atoms together in a molecule.

(*i*) *Ionic or Electrovalent Bond*. This type of bond is established by the transfer of one or more valence electrons from one atom to the other.

(*ii*) *Covalent Bond*. This type of bond is established by the sharing of one, two or three electron pairs between the combining atoms. Each of the two bonded atoms contributes one electron to make the shared electron pair and has equal claim on the shared electron pair, *i.e.*, the resulting electron pair fills the outer-shell of both the atoms and thus both the atoms attain the stable configuration of the nearest inert gas.

(*iii*) *Co-ordinate Bond*. A covalent bond in which both electrons of the shared electron pair come from one of the two atoms is called a coordinate bond.

(*iv*) *Metallic Bond*. It is a typical bond and is formed in metals. In this bond a variable number of electrons are shared simultaneously by a variable number of atoms of the metal. This bond is altogether different from the three bonds mentioned above. This bond will be described in detail in subsequent pages of this chapter.

All these four bonds are called **strong bonds.** There are, however, attractive interactions between the atoms which are comparatively weaker than the bonds mentioned above. These weaker interactions are called **weaker bonds.** In these bonds the bonding atoms do not lose their identity. These weaker bonds are of the following types.

(*i*) *Hydrogen Bond*. This bond involves the bonding of a hydrogen atom with two strongly electronegative atoms (*e.g.* N, O and F) simultaneously.

(*ii*) *van der Walls Interaction*. This bond involves the interaction between atoms or molecules having inert gas configuration.

Thus the classification of bonds can be illustrated as follows :

Chemical Bonds

- Strong Bonds
  - Ionic or Electrovalent Bond
  - Covalent Bond
  - Coordinate or Dative Bond
  - Metallic Bond
- Weak Bonds
  - Hydrogen Bond
  - van der Walls Interaction

Ionic or Electrovalent Bond

(*Bonding by Transference of Electrons*)

What is Ionic Bond ?

The chemical bond formed between two atoms by the transfer of one or more valence electrons from one atom to the other is called ionic bond. This bond is also called **electrovalent** or **polar bond.**

One of the combining atoms has excess of electrons than the stable number (2 or 8) in its valence-shell while the other atom is short of electrons and hence needs electrons to complete its octet. When they combine, the former surrenders surplus electrons to the latter and as a result of this transfer of electrons each of the atoms attains the stable configuration of the nearest inert gas (*i.e.* ns^2p^6 configuration). The compounds which contain electrovalent bonds are called **electrovalent** or **ionic compounds.**

Illustration of the formation of ionic bond. How an ionic bond is established between two atoms can be explained by considering a general case in which an atom A forms an ionic bond with another atom B. Atom A has one electron in its valence-shell while atom B has seven electrons. Thus A has one electron in excess and B has one electron short than the stable octet. Therefore A transfers an electron to B and in this transaction both the atoms acquire a stable electron-octet. The resulting positive ion (cation), A^+ and negative ion (anion), B^- are held together by electrostatic force of attraction which is called ionic bond or electrovalent bond. (See Fig. 7.3).

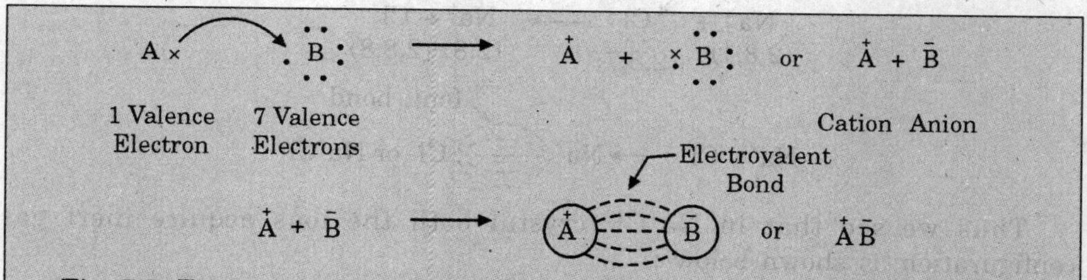

Fig. 7.3. Formation of an ionic bond between A and B atoms (Illustration)

Thus an ionic bond can also be defined as the electrostatic force of attraction between cation and anion which are produced by the transfer of electrons.

Examples of Ionic Compounds ($NaCl$, MgO, CaF_2, Al_2O_3).

The symbol of an element surrounded by a number of dots (.) or crosses (×) equal to the number of valence electrons is called the *Lewis symbol* of the element. The structural formulae of molecules built by the union of Lewis symbols of the component atoms are called *electron-dot formulas* or *electron dot structures* or *Lewis structures*.

Now let us write the Lewis structure of some ionic compounds.

1. NaCl molecule. Here Na (2, 8, 1) transfers its excess one electron to Cl atom (2, 8, 7) and thus Na atom acquires the configuration of Ne (2, 8) and Cl acquires the configuration of Ar (2, 8, 8).

The electron lost by Na atom is accepted by Cl atom and consequently Na atom is converted into a *positively charged ion* (*i.e. cation*) and Cl atom is converted into a *negatively charged ion* (*i.e. anion*). *The two ions thus formed attract each other by electrostatic force of attraction which leads to the formation of an ionic or electro-valent bond between* Na^+ *and* Cl^- *ions.* Different steps are shown below in Fig 7.4

(a) $\underset{(2,8,1)}{Na^{\times}} \xrightarrow[(-e)]{\text{Loss of electron}} \underset{(2,8)}{\left[\overset{\times\times}{\underset{\times\times}{\times Na \times}}\right]^{+}}$ or Na^{+}

(b) $\underset{(2,8,7)}{:\ddot{\underset{..}{Cl}}\cdot} \xrightarrow[\text{by Na atom } (+e^{-})]{\text{Addition of electron lost}} \underset{(2,8,8)}{\left[:\ddot{\underset{..}{Cl}}\overset{\times}{:}\right]^{-}}$ or Cl^{-}

(c) $\underset{(2,8)}{Na^{+}} + \underset{(2,8,8)}{Cl^{-}} \xrightarrow{\text{Combination of ions}} Na^{+}\,Cl^{-}$ Crystal

Fig. 7.4. Simple way to show the formation of NaCl ionic crystal.

The formation of $Na^{+}Cl^{-}$ ionic crystal shown by equations (*a*), (*b*), and (*c*) in Fig. 7.4 can also be depicted as :

Electron transfer

$$\underset{(2,8,1)}{Na^{\times}} + \underset{(2,8,7)}{\cdot\ddot{\underset{..}{Cl}}:} \longrightarrow \underset{(2,8)}{Na^{+}} + \underset{(2,8,8)}{Cl^{-}}$$

Ionic bond

$$Na^{+} + Cl^{-} \longrightarrow Na^{+} \Leftarrow\Rightarrow Cl^{-} \text{ or } Na^{+}Cl^{-}$$

Thus we see that in $Na^{+}Cl^{-}$ crystal both the ions acquire inert gas configuration as shown below :

$Na^{+} \rightarrow 2, 8$ (*Neon configuration*)

$Cl^{-} \rightarrow 2, 8, 8$ (*Argon configuration*)

The dots (•) and crosses (×) indicate the electrons in the outer-most shells of Cl and Na atoms.

2. MgO molecule. In the formation of this molecule Mg atom loses its two electrons from its outer-most shell (Mg → 2, 8, 2) which are taken up by oxygen atom (O → 2, 6). This converts Mg atom into Mg^{+2} ion and O atom into O^{2-} ion. The ions thus formed attract each other and form MgO molecule.

$$\underset{(2,\,8,\,2)}{Mg:} \xrightarrow{-2e^{-}} \underset{(2,\,8)}{\left[:\overset{..}{\underset{..}{Mg}}:\right]^{2+}} \textit{ or } Mg^{2+} + 2e^{-}$$

$$\underset{(2,\,6)}{:\ddot{O}:} + 2e^{-} \text{ (From Mg atom)} \longrightarrow \underset{(2,\,8)}{\left[:\ddot{\underset{..}{O}}:\right]^{2-}} \textit{ or } O^{2-}$$

$$\underset{(2,\,8)}{Mg^{2+}} + \underset{(2,\,8)}{O^{2-}} \longrightarrow Mg^{2+}\,O^{2-} \textit{ or } Mg\,O$$

$$\textit{or } \underset{(2,\,8,\,2)}{Mg\binom{\times}{\times}} + \underset{(2,\,6)}{\cdot\ddot{\underset{..}{O}}:} \longrightarrow \underset{(2,\,8)}{Mg^{2+}} + \underset{(2,\,8)}{O^{2-}} \text{ or } Mg^{2+}\,O^{2-}$$

3. CaF_2 molecule. In the formation of this molecule each Ca atoms loses two electrons and is converted into Ca^{2+} ion. Each F atom accepts one electron and is converted into F^- ion. One Ca^{2+} ion attracts two F^- ions and forms CaF_2 molecule.

$$\overset{\times\times}{Ca} \xrightarrow{-2e^-} \left[\overset{\times\times}{\underset{\times\times}{\times Ca \times}} \right]^{2+} \text{ or } Ca^{2+} + 2e^-$$

(2, 8, 8, 2) (2, 8, 8)

$$2 : \ddot{\underset{\cdot\cdot}{F}} : + 2e^- \longrightarrow 2 \left[: \ddot{\underset{\cdot\cdot}{F}} : \right]^- \text{ or } 2F^-$$

(2, 7) (2, 8)

$$Ca^{2+} + 2F^- \longrightarrow F^- Ca^{2+} F^- \text{ or } Ca^{2+}\, 2F^- \text{ or } Ca^{2+} F_2^-$$

$$\textit{or } Ca\,\overset{\times}{\underset{\times}{}} + \cdot\ddot{\underset{\cdot\cdot}{F}} : + \cdot\ddot{\underset{\cdot\cdot}{F}} : \longrightarrow Ca^{2+} + F^- + F^- \text{ or } Ca^{2+}F_2^-$$

(2,8,8,2) (2,7) (2,7) (2,8,8) (2,8) 2,8)

4. Al_2O_3 molecule. Here Al atom has three electrons in its valence shell (2, 8, 3) while oxygen has six (2, 6). Two atoms of Al transfer their six electrons to three oxygen atoms. Thus the electron-octets of the two Al and three O atoms are achieved. The two Al atoms deprived of three electrons each, give two Al^{3+} ions while the three O atoms having gained two electrons each, give three O^{2-} ions. In this way we get $Al_2^{3+}\ O_3^{2-}$ or Al_2O_3 molecule (See Fig. 7.5).

$$\text{or } 2\,Al + 3\,O \longrightarrow 2\,Al^{3+} + 3\,O^{2-} \text{ or } (Al^{3+})_2(O^{2-})_3$$

Fig. 7.5. Formation of Al_2O_3 molecule.

Nature of Ionic Bond

When two oppositely charged ions combine together, they attract each other by electrostatic force of attraction and are thus held together by the lines of force of attraction.

Since an ion can attract other ions having opposite charge on them from any direction, ionic bond is *non-directional in nature and extends equally in all directions.* If the ions are regarded as charged spheres, the attraction between the ions having opposite charge can be shown as follows :

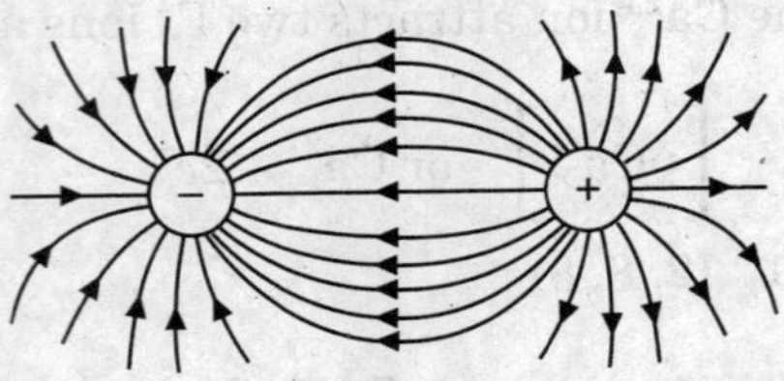

When an ionic compound is dissolved in water, the lines of force of attraction holding the ions together are broken by the high dielectric constant value of water and thus the ions constituting the ionic compound are separated from each other as shown below for Na^+Cl^- ionic crystal.

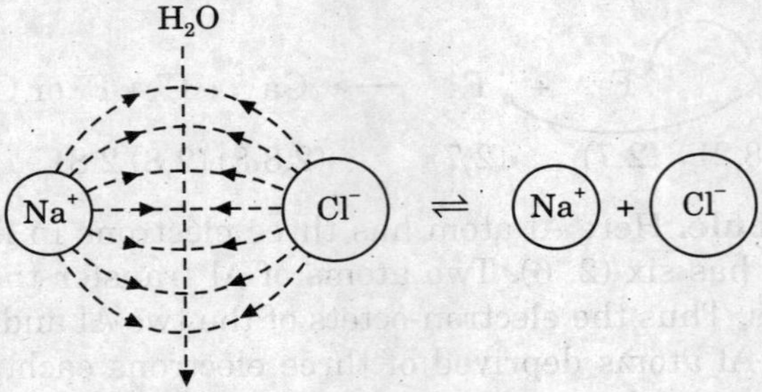

Factors Favouring the Formation of Ionic Compounds : Conditions for the Formation of Ionic Compounds

The formation and the stability of the ionic compounds depend on how easily the ions constituting the ionic compound are formed. Following are the important factors which favour the formation of ions (*i.e.* of ionic compounds) :

1. *Number of valence electrons.* The atom A which is converted into cation, A^+ should possess 1, 2 or 3 valence electrons while the atom B which is converted into anion, B^- should have 5, 6 or 7 valence electrons. The element of groups IA, IIA and IIIA satisfy this condition for atom A and those of groups VA, VIA and VIIA satisfy this condition for atom B.

2. *The ionisation energy of the metal atom should be low.* Ionisation energy is the minimum amount of energy required to remove the most loosely bound electron from a neutral isolated gaseous atom. Thus :

$$\underset{\text{Gaseous neutral atom}}{A\,(g)} + \textit{Energy required (Ionisation energy)} \rightarrow \underset{\text{Cation}}{A^+\,(g)} + e^-$$

Quite obviously *if an atom has low ionisation energy, it will be easy for it to lose the electron and hence get converted into a cation. Hence low ionisation energy of the metal will favour the formation of cation.*

3. *Electron affinity of the non-metal should be high.* Electron affinity is the amount of energy released when an electron is added to a neutral isolated gaseous atom. Thus :

$$\underset{\text{Gaseous neutral atom}}{B\ (g)} + e^- \rightarrow \underset{\text{Anion}}{B^-\ (g)} + \textit{Energy released (Electron affinity)}$$

Now, since the release of energy is the sign of stability of a system, *the atoms with high electron affinity will form the anions quite easily.*

4. *The lattice energy of the ionic compound formed should be high.* The energy released when one gram mole of a crystal is formed from its gaseous ions is called the lattice energy of the crystal. Thus :

$$\underset{\text{1 mole}}{A^+(g)} + \underset{\text{1 mole}}{B^-\ (g)} \rightarrow \underset{\text{1 mole}}{A^+\ B^-\ (crystal)} + \textit{Energy released (Lattice energy)}$$

Higher the value of lattice energy of a crystal, the greater is the ease of its formation, i.e., greater will be the stability of the ionic crystal.

The energy released in the formation of ionic molecule, $A^+\ B^-$ will lower the energy of the system, *i.e., for the formation of stable ionic compound there must be a net lowering of the energy.*

For lattice energy to have a high value, the electrostatic force of attraction between the constituent ions of the ionic compound should be high. According to Coulomb's law the force of attraction (F) between two oppositely charged ions in air with charges equal to q_1 and q_2 and separated by a distance equal to d is given by :

$$F = \frac{q_1 q_2}{d^2} = \frac{q_1 q_2}{(r_A{}^+ + r_B{}^-)^2}$$

d is equal to the sum of the radii of the positive and negative ions. From this relation it is clear that to have a large value of force of attraction (or lattice energy) the following two condition should be satisfied.

(a) The electric charges on the ions (i.e. q_1 and q_2) should be large. For example the force of attraction between the doubly-charged Mg^{2+} ion and the doubly-charged O^{2-} ion in $Mg^{2+}\ O^{2-}$ ionic crystal is 4 times greater than that between monovalent Na^+ and Cl^- ions in $Na^+\ Cl^-$ crystal. Here it has been assumed that the distance between the ions in both the ionic compounds is the same. Consequently MgO is more stable than NaCl.

(b) d or ($r_A{}^+ + r_B{}^-$) should be small, i.e., the cation and the anion constituting the ionic compound should be small in size. Thus for a given anion, the smaller the radius of the cation, the greater is the strength of the ionic bond, *i.e.* greater is the stability of the ionic crystal. For example, since Na^+ ion is smaller than Cs^+ ion, the force of attraction between Na^+ and Cl^- ions in $Na^+\ Cl^-$ ionic crystal is more than that between Cs^+ and Cl^- ions in $Cs^+\ Cl^-$ ionic crystal. Thus the lattice energy of $Na^+\ Cl^-$ is higher than that of Cs^+Cl^-. Consequently Na^+Cl^- compound will be formed more easily than $Cs^+\ Cl^-$ compound. In other words Na^+Cl^- is more stable than $Cs^+\ Cl^-$.

Similarly, for a given cation, the smaller the radius of the anion, the greater is the strength of the ionic bond, *i.e.*, the greater is the stability of the ionic compound.

5. *Electronegativity difference of A and B should be high.* From the line of arguments used in *2nd*, *3rd*, and *4th* conditions discussed above, we can say that atoms A and B will form an ionic bond, if they have greatly different electronegativities. In fact, a difference of 2 or more is necessary for the formation of an ionic bond between atoms A and B. For example, since the electronegativity difference between Na and Cl is 2.1 (Na = 0.9, Cl = 3.0), Na and Cl will form an ionic bond in NaCl molecule.

Formation of Ionic Bonding and Periodic Table

The relative tendency of the atoms to turn into cations and anions depends on their position in the periodic table. The elements having *low ionisation energies, i.e., elements of groups IA and IIA have a strong tendency to turn into cations.* The elements having *high electron affinities, i.e., the higher elements of groups VA and VIA and those of group VIIA, have a strong tendency to turn into anions.*

Thus the binary compounds having ionic bonding are most readily formed by the elements of groups IA and IIA with oxygen (group VIA) and halogens (VIIA group) elements.

Electrovalency

Electrovalency of an element is its combining capacity in an ionic compound, *i.e.*, when an element forms electrovalent bond, its valence is known as its electrovalency. *Electrovalency of an element is equal to the number of electrons lost by an atom of that element in forming a positive ion or gained by it in forming a negative ion, both having the noble gas configuration (s^2p^6 configuration) in their outermost shell.* The elements which lose electrons show positive electrovalency while the elements which gain electrons show negative electrovalency. For example in the formation of NaCl the electrovalency of Na is equal to +1 while that of Cl is equal to –1. An element which gains or loses *one, two, three, four* ... etc. electrons is called *mono (or uni) valent, di (or bi) valent, trivalent, tetravalent* etc. ... element. Examples are :

Na, Cl, F ...	*Monovalent elements*
Mg, Ca, O ...	*Divalent elements*
Al, B, ...	*Trivalent elements*
C, Si ...	*Tetravalent elements*

Variable Electrovalency

There are many elements which show different values of electrovalency in different electrovalent compounds. This phenomenon is called *variable electrovalency* which is due to the following two reasons (*i*) *Unstable configuration of the core or kernel,* and (*ii*) *Inert electron pair effect.*

1. Unstable configuration of the core. The outermost levels of the elements like Fe, Co, Ni, Cu etc. contain only one or two electrons. When these electrons are lost, the remaining part which is called the *core* or *kernel*, is unstable and hence some more electrons from the core can be further lost, giving variable electrovalency to the ion. For example :

(*a*) The variable electrovalency of iron equal to + 2 in ferrous compounds and equal to +3 in ferric compounds can be explained as follows :

$Fe\ (26) \rightarrow 2, 8, 14, 2$ or $2, 8, 3s^2p^6d^6, 4s^2$

$Fe^{2+}\ (24) \rightarrow 2, 8, 14,$ or $2, 8, 3s^2p^6d^6$

$3s^2p^6d^6$ configuration is unstable and hence loses one more electron and thus forms Fe^{3+} ion

$Fe^{3+}\ (23) \rightarrow 2, 8, 13,$ or $2, 8, 3s^2p^6d^5$

(*b*) An increase in the nuclear charge tends to prevent the removal of electrons from a lower energy level. For example Co (Z = 27) atom forms Co^{3+} ion with great difficulty because 3*d* electrons are more firmly retained by cobalt nucleus with positive charge equal to +27 which is one unit higher than that on iron nucleus (= + 26)

$Co\ (27) \rightarrow 2, 8, 15, 2$ or $2, 8, 3s^2p^6d^7, 4s^2$

$Co^{2+}\ (25) \rightarrow 2, 8, 15$ or $2, 8, 3s^2p^6d^7$

$Co^{3+}\ (24) \rightarrow 2, 8, 14$ or $4, 8, 3s^2p^6d^6$

(*c*) The variable electrovalency of Cu equal to +1 (cuprous ion) and +2 (cupric ion) is easily understood from the electronic configurations given below :

$Cu\ (29) \rightarrow 2, 8, 18, 1$ or $2, 8, 3s^2p^6d^{10}, 4s^1$

$Cu^{+}\ (28) \rightarrow 8, 8, 18$ or $2, 8, 3s^2p^6d^{10}$

$Cu^{2+}\ (27) \rightarrow 2, 8, 17$ or $2, 8, 3s^2p^6d^9$

2. Inert electron pair effect. Some of the heavier elements of *p*-block of the long form of periodic table like those of groups IIIA (Ga, In, Tl), IVA (Ge, Sn, Pb), VA (As, Sb, Bi) and VI A (Te Po) show two oxidation states (*i.e.* electrovalencies) as shown below :

Group IIIA (G = 3) → Ga (+1, +3), In (+1, +3), Ti (+1, + 3)

Group IVA (G = 4) → Ge (+2, +4), Sn (+2, +4), Pb (+2, +4)

Group VA (G = 5) → As (+3, +5) Sb (+3, +5), Bi (+3, +5)

Group VIA (G = 6) → Te (+4, +6), Po (+2, +4)

These two oxidation states differ from each other by two units. The higher oxidation state for most of the elements is equal to their group number, G while the lower ones are equal to (G–2) as shown above. Group number oxidation state (= G) is obtained when all the *ns* and *np* electrons from ns^2p^x configuration of *p*-block elements (x = 1, 2, 3 and 4 for the elements of groups IIIA, IVA, VA and VIA respectively) are lost while the lower oxidation state is obtained when only *np* electrons are lost and the *ns* electron pair, due to its extra stability, remains inert, *i.e.*, it is not lost. Such a pair of *ns* electrons is called **inert electron pair** and the effect caused by it is called **inert electron pair effect.**

The inert electron pair effect has been used to explain the electrovalencies of the elements given above.

Formation of an Ionic Bond also Involves Decrease in Energy

As a general rule, *the formation of an ionic bond also involves a decrease in energy.* This fact can be illustrated by considering the formation of one gm molecule of solid sodium chloride, NaCl (*s*) through the following steps :

1. *Conversion of Na atom from the solid state to the gaseous state* [Na (*s*) → Na (*g*)]. This type of conversion is called *sublimation of solid sodium into gaseous sodium.* Since in the solid state Na atoms are fairly close to each other while in the gaseous state they are relatively far apart, the conversion of Na(*s*) into Na(*g*) requires the *absorption* of energy, *i.e., this conversion is an endothermic reaction.* The energy required to convert one mole of Na(*s*) to Na(*g*) is called *sublimation energy* which has been found to be equal to 108.81 kJ/mole. *Energy absorbed in the reaction is shown at the left hand side of the equation.* Thus :

Na(*s*) + *Sublimation energy* (= 108.81 kJ. mole^{-1}) → Na (*g*)

(*Endothermic reaction*)

2. *Dissociation of $Cl_2(g)$ molecules into Cl(g) atoms* [$\frac{1}{2}$ Cl_2 (*g*) → Cl(*g*)]. Here, again, energy is to be *supplied* in order to break the bond between two Cl atoms in Cl_2 molecule (Cl–Cl bond). In this step half a molecule of Cl_2 (*g*) absorbs energy equal to half of the *dissociation energy* of Cl_2(*g*) and is converted into Cl(*g*). Dissociation energy of Cl_2(*g*) has been found equal to 242.7 kJ/mole. *Absorption of energy means that the process is an endothermic reaction.* Thus :

$\frac{1}{2}$ Cl_2(*g*) + *Dissociation energy* [= 242.7 kJ/mole of Cl_2(*g*) = 121.35 kJ/mole of Cl(*g*)] → Cl(*g*)

(*Endothermic reaction*).

3. *Removal of an electron from gaseous Na atom* [Na(*g*)– *e*$^-$ → Na^+(*g*)]. The removal of an electron from gaseous Na atom requires energy equal to the *ionisation energy* of Na. Thus the removal of an electron from Na (*g*) is an *endothermic reaction.* Ionisation energy of sodium is equal to 495.8 kJ/mole. Thus :

Na (*g*) + *Ionisation energy* (495.8 kJ/mole) → Na^+(*g*) + *e*$^-$

(*Endothermic reaction*).

4. *Conversion of Cl(g) atom into Cl^-(g) ion* [(*Cl*(*g*) + e$^-$ → *Cl^-*(*g*)]. Here Cl(*g*) atom adds the electron given by Na(*g*) atom in step (3) to form Cl^-(*g*). In this process Cl(*g*) *releases* energy equal to its *electron affinity* (*or affinity energy*) which is equal to 348.7 kJ/mole. *The energy released is shown at the right hand side of the equation.* The release of energy means that the conversion of Cl(*g*) to Cl^-(*g*) is an *exothermic reaction*. Thus :

Cl (*g*) + *e*$^-$ → Cl^- (*g*) + *Affinity energy* (= 348.7 kJ/mole)

(*Exothermic reaction*).

5. *Combination of Na^+(g) and Cl^-(g) ions to form NaCl (s)* [Na^+(*g*) + Cl^-(*g*) → NaCl (*s*)]. This is the final step in which Na^+(*g*) and Cl^-(*g*) ions formed in steps (3) and (4) respectively combine together to form one mole of solid sodium chloride crystal, NaCl (*s*). Here attractive forces between Na^+(*g*) and Cl^-(*g*) operate and thereby decrease the energy of the system. *Thus the formation of NaCl (s) from the combination of Na^+(g) and Cl^-(g) is accompanied by the release of energy.* The energy released is called *lattice energy* of NaCl which has been found (for NaCl) to be equal to 769.2 kJ/mole. Thus :

Na^+(*g*) + Cl^-(*g*) → NaCl(*s*) + *Lattice energy* (= 769.2 kJ/mole)

(*Exothermic reaction*).

Now,

Total energy absorbed = 108.81 + 121.35 + 495.80 = 725.96 kJ.

and, *Total energy released* = 348.7 + 769.2 = 1117.9 kJ.

Since total energy released is larger than the total energy absorbed, in the formation of NaCl ionic crystal energy is released and its magnitude is equal to 1117.90 –725.96 = 391.94 kJ. Thus we see that the net process of the formation of NaCl crystal by the transfer of an electron from Na atom to Cl atom involves the evolution of energy. Hence the formation of ionic bond between Na^+ and Cl^- ions of NaCl(*s*) ionic crystal involves decrease in energy. The release of energy in the formation of an ionic bond in NaCl solid also shows that *the ionic bond is a very strong bond.*

Properties of Ionic Compounds

1. *Physical state.* Ionic compounds consist of three dimensional solid aggregates of cations and anions which are arranged in a well-defined geometrical pattern. Thus ionic compounds are *crystalline solids* at room temperature. They are never liquids or gases under the ordinary conditions of temperature and pressure, since their ions lack the freedom of movement characteristic of the liquid or gaseous state.

2. *Electrical conductivity.* Ionic compounds do not conduct electricity when they are in the solid state. The reason is that the cations and anions, on account of electric force of attraction existing between them, remain tightly held together with each other in the ionic compounds and hence occupy their fixed positions in the crystal lattice. The ions, therefore, cannot move freely to any large extent when an electric current is passed through the ionic solids.

However, *the ionic solids conduct electricity when they are in water solution or in the fused* (*molten*) *state*. The reason for this can be explained as follows : As the temperature is raised, the kinetic energy of the ions also increases. Ultimately when the ionic compound goes into the molten state, the kinetic energy of the ions becomes so large that the attractive forces acting between the ions are overcome and the well-ordered arrangement of the ions in the ionic crystal is destroyed and consequently the ions become free to move about in the liquid medium under the influence of the applied electric field. *Thus the ionic compounds are good conductors of electricity in the molten state, i.e., molten ionic compounds or their aqueous solutions conduct a current when placed in an electrolytic cell.*

3. *They are quite hard, have low volatility and high melting and boiling points.* Since in ionic solids, the cations and anions are held together very tightly in their allotted positions by very strong electrostatic forces of attraction, very high amount of energy (in the form of heat) is required to separate the cations and anions from one another against the force of attraction and thus to make them free to move, as in a liquid. Consequently the ionic solids *are quite hard* (though brittle), *have low volatility (i.e., have low vapour pressure) and have high melting and boiling points.*

4. *Solubility in polar and non-polar solvents.* Ionic solids are freely soluble in *polar solvents* like H_2O, liq. NH_3 etc., because the electrostatic force of attraction holding the cations and anions together in the ionic solids *is reduced* by the high value of dielectric constant of the polar solvent. The reduction in the electrostatic force of attraction by the high dielectric constant value of polar solvent makes the ions move freely and interact with solvent molecules to form the solvated ions. On the other hand ionic solids are *insoluble* or *slightly soluble in non-polar solvents* (organic solvents) such as C_6H_6, CCl_4 etc. Such solvents, due to their low value of dielectric constant, do not allow the ions to move freely and interact with them to form the solvated ions.

The solubility of ionic solids in a polar solvent like water can also be explained by saying that a water molecule is a dipole and hence the positive end of water dipole interacts with the negative ion of the ionic solids and the negative end of the dipole interacts with the positive ion of the same ionic crystal as shown in Fig. 7.6 in which the dissolution of NaCl ionic crystal in water has been shown. The interaction between the water dipoles and the ions of the crystal lowers down the energy of the system and thus the force of attraction between the cations and anions of the ionic solid is weakened. Consequently the water molecules tear off the ions from the crystal lattice and make them float in the bulk of water. Further, they are surrounded by water dipoles with the oppositely charged ends directed towards them (See Fig. 7.6).

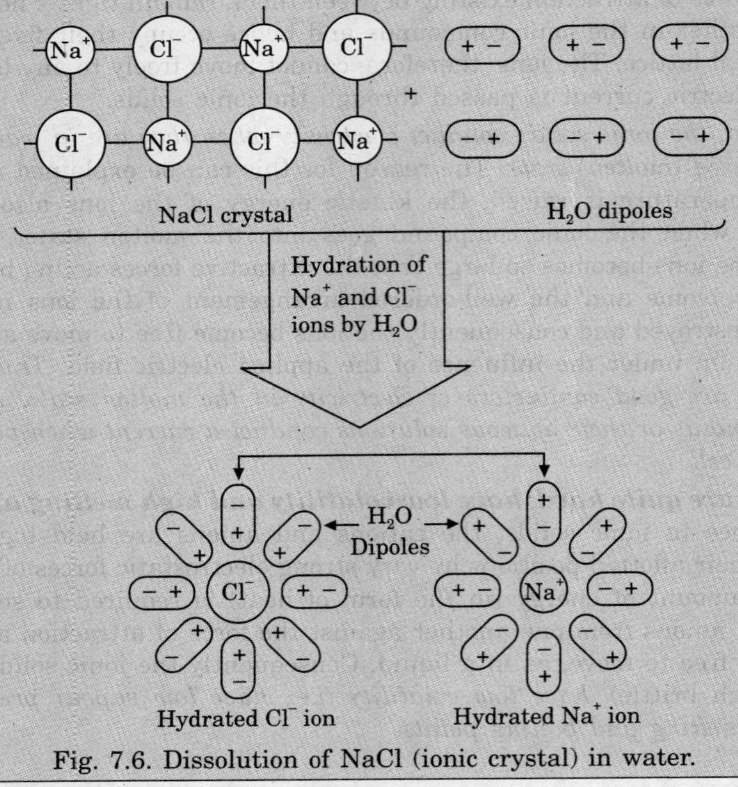

Fig. 7.6. Dissolution of NaCl (ionic crystal) in water.

Explanation of dissolution on the basis of relative magnitutes of lattice and hydration energies. The dissolution of an ionic compound in a polar solvent can also be explained on the basis of relative values of lattice and hydration energies. The dissolution of an ionic compound in a polar solvent takes place through the following two steps :

Step (a). When an ionic solid, MX(*s*) is dissolved in a polar solvent, the ionic solid is broken into its isolated gaseous ions *viz.* $M^+(g)$ and $X^-(g)$. In this process of breaking, some amount of energy is required. This is called **lattice energy** of MX(*s*) ionic crystal. Thus :

$$\underset{\text{Ionic crystal}}{MX(s)} + \textit{Energy required (called lattice energy)} \rightarrow \underbrace{M^+(g) + X^-(g)}_{\text{Isolated gaseous ions}}$$

Step (b). In this step the isolated gaseous ions formed in step (*a*) go into the solvent and interact with their molecules so that they get surrounded by a definite, but unknown, number of solvent molecules and are consequently converted into **solvated ions** which are represented as $[M(solv)_x]^+$ and $[X(solv)_y]^-$. Here *solv* stands to show solvent molecules. The formation of these solvated ions is due to the *ion-solvent interaction.* If water is used as a solvent, the gaseous ions are said to be converted into **hydrated ions** which are represented as $[M(H_2O)_x]^+$ or $[M(aq)]^+$ and $[(X(H_2O)_y]^-$ or $[X(aq)]^-$. This representation of hydrated ions indicates that the ions are in aqueous phase. *The process by which the solvated (or hydrated) ions are produced is called solvation (or hydration) process.* The process of solvation (or hydration) is accompanied by the release of a certain amount of energy which is called **salvation** (or **hydration**) **energy.**

$$M^+(g) + x(solv) \rightarrow \underset{\text{Solvated cation}}{[M\ (solv)_x]^+} + \textit{Energy released} \text{ (called } \textit{solvation energy}\text{)}$$

$$X^-(g) + y(solv) \rightarrow \underset{\text{Solvated anion}}{[X(solv)]^-} + \textit{Energy released} \text{ (called } \textit{solvation energy}\text{)}$$

Thus *the amount of energy released when one gm mole of a gaseous ion is dissolved in a solvent (or water) is called solvation (or hydration) energy of the solvent.* Solvation (or hydration) energy is also called *heat of solvation* (or *heat of hydration*). Obviously the process of solvation (or hydration) is an *exothermic process*.

Now for an ionic solid to dissolve in a polar solvent, the magnitude of hydration energy of the solvent should be such that it should be able to overcome (*i.e.* offset, counteract or break down) the lattice energy of the ionic solid which is responsible for holding the ions together in the ionic crystal. Thus *for an ionic solid to dissolve in a polar solvent, the solvation energy of the solvent must be greater than the lattice energy of the ionic solid so that the solvation energy may overcome the lattice energy.*

When we consider the possibility of the solubility of ionic solids in non-polar solvents, we find that the non-polar solvents are not able to interact with the gaseous ions to form the solvated ions and hence no energy is released in the form of solvation energy which may break down the lattice energy. Thus ionic solids are not soluble in non-polar solvents.

5. *Stability*. Ionic crystals are very stable compounds and for maximum stability of these crystals *the oppositively charged ions are close to one another and the similarly charged ions are as away from one another as possible.*

6. *Crystal structure* (*Packing of ions in ionic crystals*). Single ionic molecules (*e.g.* Na^+Cl^-) do not exist as such in the solid form of ionic compounds, *i.e.*, ionic solids do not exist as individual neutral independent molecules. Rather, a number of cations and anions attract each other due to electrostatic force of attraction which is nondirectional and hence extends in all directions. In order to occupy minimum space the ions arrange themselves systematically in an alternating cation-anion pattern called **crystal** or **lattice**. Thus we see that *ionic solids consist of three dimensional solid aggregates.*

The structure (*i.e.* geometry) of an ionic crystal depends on radius ratio of the cation and anion (r_c^+/r_a^- ratio). For example CsF has the same structure as NaCl, since both the ionic molecules have the same radius ratio.

Ionic crystals possess typical structure of their own. In each type of the structure of an ionic crystal each ion of one type is surrounded by a certain number of ions of opposite charge equidistant form it. The structure of NaCl ionic crystal can be considered as an example. It has *cubical* shape (Fig. 7.7): X-ray study of the structure of this ionic crystal has shown that :

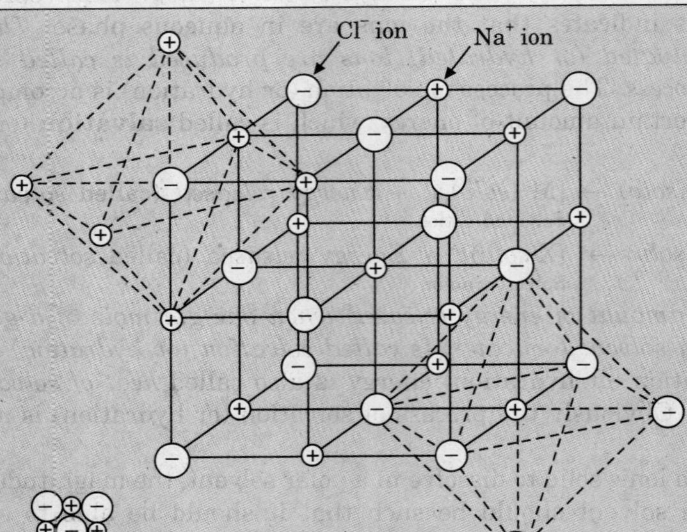

Fig. 7.7. Cubical structure of NaCl ionic crystal. The bottom figure represents a cross section of NaCl crystal.

(*a*) The distance between two adjacent ions of the same kind like, say Cl^- ions (*i.e.*, length of the edge), is equal to 5.63 Å while the distance between the two adjacent ions of different kind like, say Na^+ and Cl^- ions, is equal to 5.63/ 2 = 2.815 Å.

(*b*) Na^+ and Cl^- ions are so located that each Na^+ ion is surrounded by equally spaced *six* Cl^- ions placed at the corners of a regular octahedron and similarly each Cl^- ion is surrounded by equally spaced *six* Na^+ ions placed at the corners of an octahedron (See Fig. 7.7). Thus the coordination number of each ion is *six* and NaCl crystal has *octahedral structure*. In other words, NaCl structure has 6 : 6 coordination.

It is worth noting that Na^+ and Cl^- ions are not connected to one another by pairs, since all the six Cl^- ions are the same distance away from any one Na^+ ion. Thus for maximum stability of NaCl crystal the oppositely charged ions are close to one another and the similarly charged ions are as away from one another as possible.

Although discrete molecules (Na^+Cl^-) do not exist in the solid from of ionic crystals, independent molecules do exist in the vapour form of ionic compounds.

7. *Highly brittle.* Ionic solids are highly brittle, *i.e.,* if a little external force is applied on ionic crystals, they are generally easily broken. This property is called brittleness and can be explained as follows :

We know that ionic solids are composed of parallel layers which contain cations and anions in alternate positions so that the opposite ions in the various parallel layers lie over each other. When a little external force is applied on an ionic crystal, one layer of ions slides a bit over the other layer along a plane. The sliding of a layer over the other results in that the like ions come in front of each other and hence begin to repell each other (Fig. 7.8). Consequently the application of a little external force brings about repulsion between the two layers and the ionic solid breaks. Hence ionic solids are highly brittle.

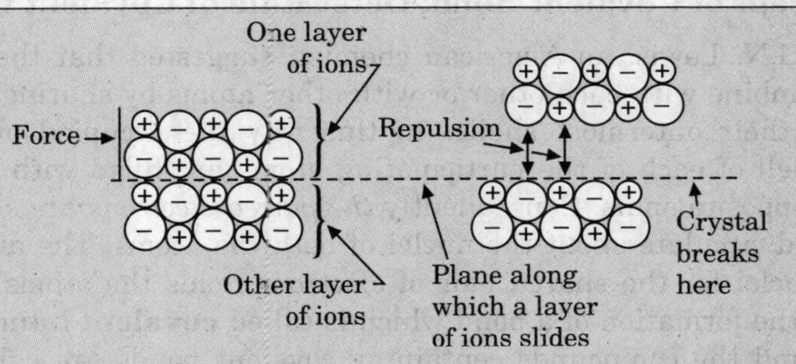

Fig. 7.8. Explanation of high brittleness of ionic solids.

8. *High density.* The electrostatic force of attraction existing between the cations and anions in an ionic crystal brings these ions very close to one another. This decreases the volume of the crystal and consequently the ionic crystals have *high density*.

9. *They undergo ionic reactions.* The reactions of an ionic compound are precisely the reactions of its constituent ions. Such reactions which are called *ionic reactions* take place as a result of collision between cations and anions, giving rise to the formation of new ionic compounds. Ionic reactions are practically *rapid* and *instaneous* (*i.e.* precipitation). For example Cl^- ions present in NaCl give white precipitate of AgCl as soon as $AgNO_3$ solution is added to that of NaCl.

$$NaCl \rightarrow Ni^+ + Cl^-$$

$$AgNO_3 \rightarrow Ag^+ + NO_3^-$$

$$Cl^- \text{ (From NaCl)} + Ag^+ \text{ (From } AgNO_3) \rightarrow \underset{\text{white ppt.}}{AgCl}$$

10. *They show isomorphism.* Ionic solids made up of ions with identical electronic configurations show an identity of crystalline form which is called *isomorphism*. Two such pairs of isomorphous compounds are (*a*) NaF and MgO, and (*b*) $CaCl_2$ and K_2S.

	Sodium	*fluoride*	and	*Magnessium*	*oxide*
(*a*)	Na^+	F^-		Mg^{2+}	O^{2-}
	(2, 8)	(2, 8)		(2, 8)	(2, 8)

(*b*) *Calcium*	*chloride*		and	*Potassium*	*sulphide*
Cl^-	Ca^{2+}	Cl^-	K^+	S^{2-}	K^+
(2, 8, 8)	(2, 8, 8)	(2, 8, 8)	(2, 8, 8)	(2, 8, 8)	(2, 8, 8)

11. *They do not exhibit isomerism.* Ionic bond involving electrostatic lines of force between opposite ions is non-rigid and non-directional. The ionic compounds are, therefore, incapable of exhibiting stereoisomerism.

Covalent Bond

(*Bonding by Mutual Sharing of Electrons*)

Lewis Concept of Covalent Bond: Octet Rule of Covalent Bond

In 1916 G.N. Lewis, an American chemist, suggested that there are atoms which can combine with each other or with other atoms by sharing the unpaired electrons in their outermost shells. In this way the occupied orbitals of the outermost shell of each of the participating atoms are filled with two electrons which have opposite spins. Consequently *the paired electrons are shared by both the atoms* and circulate about the nuclei of both the atoms. The attractive force of the two nuclei for the shared pair of electrons holds the atoms together and gives rise to the formation of a bond which is called **covalent bond** or **electron pair bond** and the compounds containing covalent bonds are called **covalent compounds.** Thus :

The chemical bond between two atoms in which the electrons (*in pairs*) *are shared by both the participating atoms is called covalent bond.*

Each of the two combining atoms contributes one electron to the electron pair and has equal claim on the shared electron pair.

According to Lewis concept when two atoms form a covalent bond between them, each of the atoms attains the stable configuration of the nearest inert gas, by completing its *octet* (*i.e.* 8 electrons in the outermost shell) or *duplet* (*i.e.* 2 electrons in case of hydrogen). The covalent bond is established between the atoms of the same or different elements. Since a covalent bond between atoms results by the interaction of their electrons which become common to both the atoms, it is also called an **atomic bond.**

Types of Covalent Bond (*Single, double and triple covalent bonds*).

The covalent bond formed by the sharing of *one, two* or *three electron pairs* between the participating atoms is called *single, double* and *triple covalent bond* respectively. These bonds are represented as follows :

Type of bond	Represented by	No. of electron pairs involved	Examples
Single covalent bond	*Single dash* (–)	*1 pair = 1 × 2 = 2 electrons*	*H–H, Cl–Cl, H–Cl, H–O–H*
Double covalent bond	*Double dash* (=)	*2 pairs = 2 × 2 = 4 electrons*	*O = O*
Triple covalent bond	*Triple dash* (≡)	*3 pairs = 3 × 2 = 6 electrons*	*N ≡ N*

Double and triple covalent bonds are called **multiple covalent bonds.**

Illustration of the formation of covalent bond. How a covalent bond is formed between two atoms can be understood by considering a general case in which an atom A has one valence electron and another atom B has seven valence electrons. As these atoms approach nearer to each other, each atom contributes one electron and the resulting electron pair fills the outer shell of both the atoms. Thus atom A acquires 2 electrons and B, 8 electrons in their respective outer shells and the shared electron pair constitutes a covalent bond between A and B.

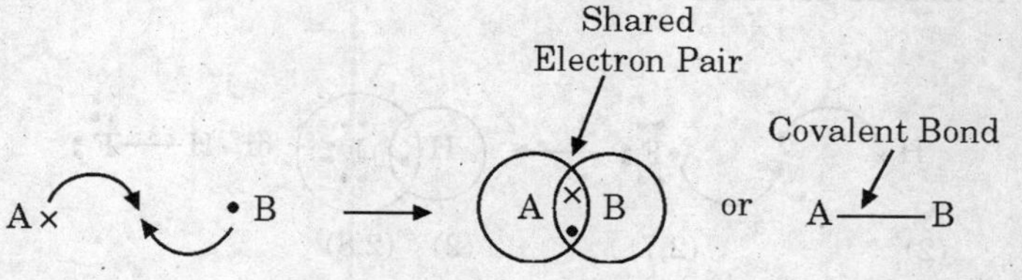

Writing the Electronic Formulae (Lewis Formulae) for Covalent Compounds (H_2, F_2, HF, H_2O, H_2S, NH_3, BF_3, CH_4, O_2, N_2, CO_2, C_2H_4, C_2H_2).

The method used for writing the electronic formulae of covalent compounds consists of the following steps :

(*a*) Around the symbol of each atom in the formula, we place dots, crosses etc. equal to the number of electrons present in the valence-shell of the atom.

(*b*) The shared pair of electrons are indicated by electron signs like dots (•), crosses (×) etc which are placed between the symbols of the atoms.

(*c*) Symbol of each atom and its octet is enclosed in a circle so that the two circles overlap the shared electrons. Thus each circle now has *eight* electrons (except *two* in case of H atom).

This method of writing the structure of covalent compounds can be illustrated by the following examples. For the differentiation of electrons, they have been shown by dots (•) and crosses (×). The structures obtained are called **Lewis structures** or **electron dot structures.**

1. *H_2, F_2 and HF molecules.* (*a*) H_2 molecule is composed of two H atoms, each having one valence electron. Each contributes an electron to the shared pair and both atoms acquire stable helium configuration. Thus stable H_2 molecule results.

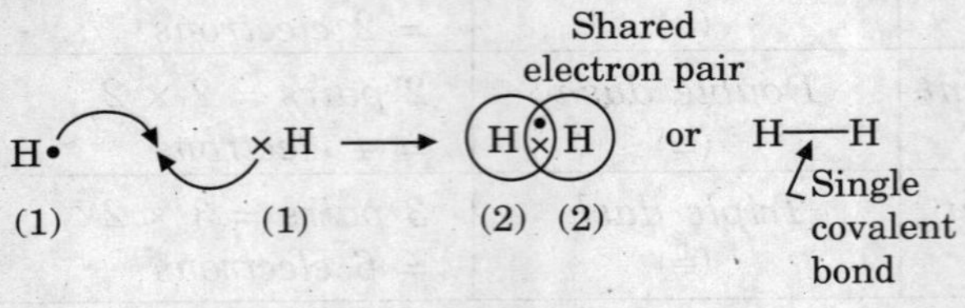

(*b*) In F_2 molecule each F atom (2, 7) has seven valence electrons. The two F atoms achieve a stable electron octet by sharing a pair of electrons.

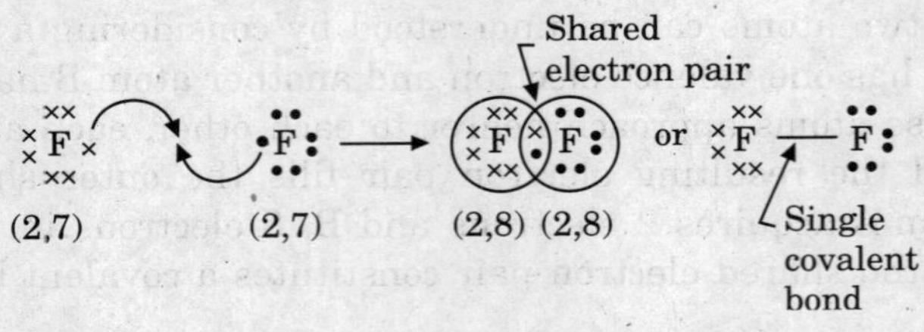

(*c*) In HF molecule H atom attains a doublet while F atom achieves an octet of electrons.

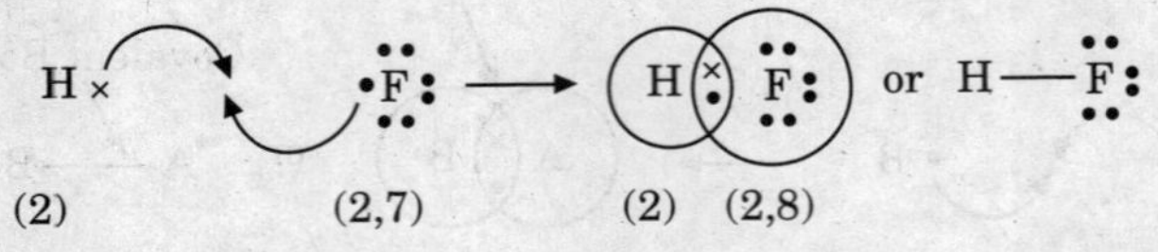

2. *H_2O and H_2S molecules.* (*a*) Oxygen atom (2, 6) has six valence electrons and achieves the stable octet by sharing two electrons, one with each H atom. Thus Lewis structure of water can be written as :

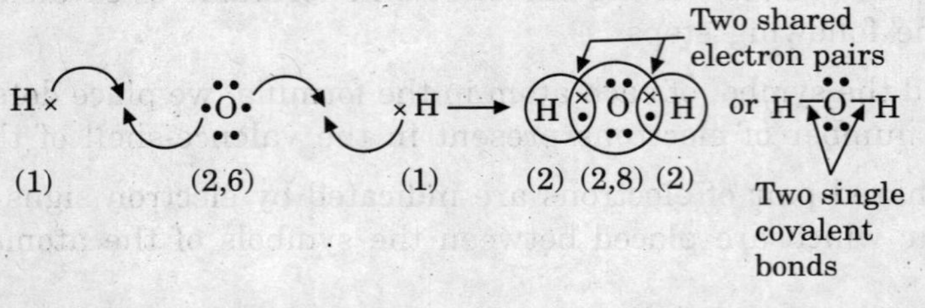

(*b*) On the same lines of argument Lewis structure of H_2S can also be written. (S→2, 8, 6).

3. *NH_3 and BF_3 molecules.* (*a*) N atom (2, 5) has five valence electrons and can achieve the octet by sharing three electrons, one each with three H atoms. This gives the following Lewis structure to NH_3 molecule.

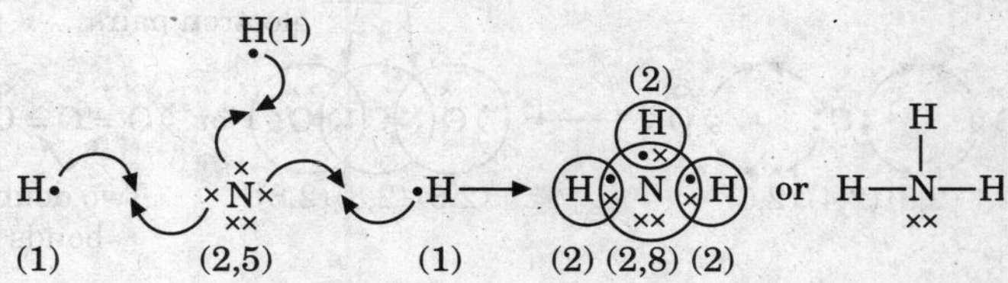

(*b*) On the same lines of agrument Lewis structure of BF_3 can also be written (B→2, 3 ; F→2, 7).

4. *Methane molecule (CH_4).* Carbon atom (2, 4) has four electrons in the valence shell. It can achieve the stable octet by sharing these electrons with H atoms, one with each H atom. Thus the Lewis structure of CH_4 can be written as :

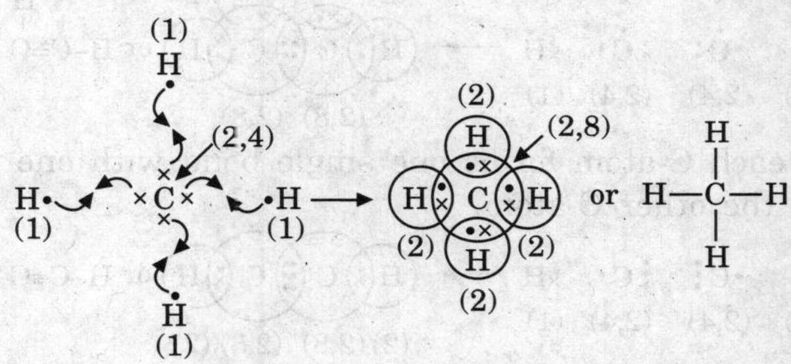

5. *O_2 molecule.* The conventional Lewis structure of O_2 molecules is written by sharing two pairs of electrons between two O-atoms (2, 6). In this way both O-atoms achieve the octet and both O-atoms are linked by a double bond.

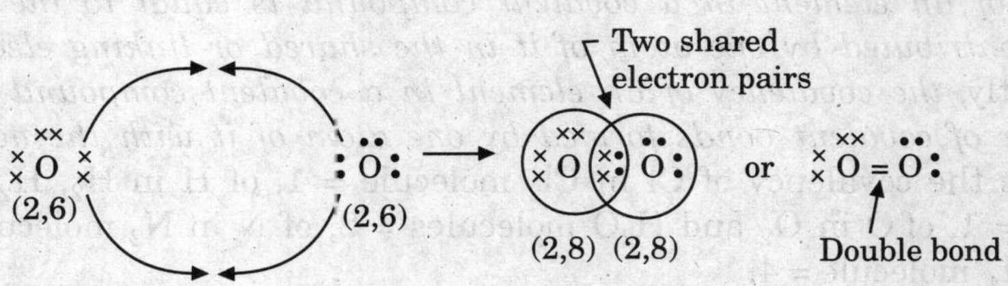

6. *N_2 molecule.* The two N atoms (2, 5), each having five electrons in the valence shell, achieve the octet by sharing three electron pairs between them.

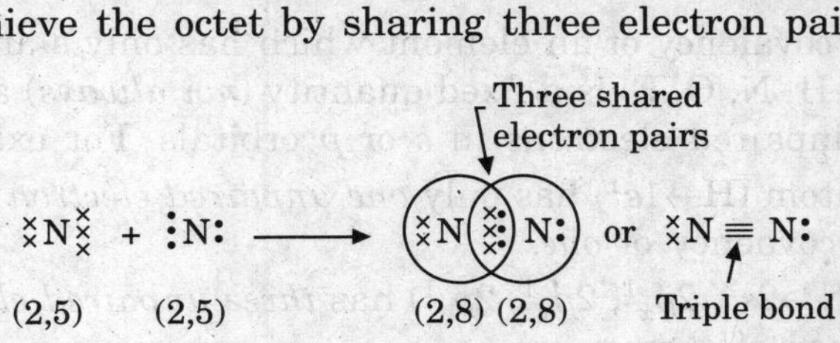

Thus we see that both N-atoms in N_2 molecule are linked by a triple bond.

7. *CO_2 molecule.* Carbon (2, 4) has four electrons in its valence-shell. It shares two electrons with each oxygen atom (2, 6). Thus C-atom and O-atom both achieve their octet and carbon atom is linked with both O-atoms by a double bond.

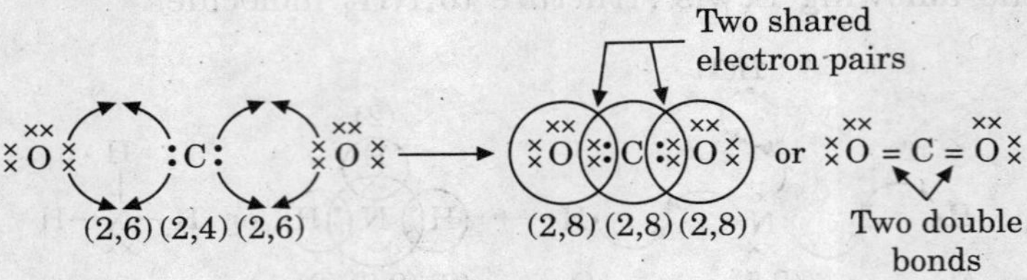

8. *Ethylene (C_2H_4) and acetylene (C_2H_4) molecules.* (*a*) In C_2H_4 each C-atom achieves the stable octet by sharing its four electrons with two H-atoms and the other C atom. In this way each C-atom forms two single bonds with two H-atoms and one double bond with the other C-atom.

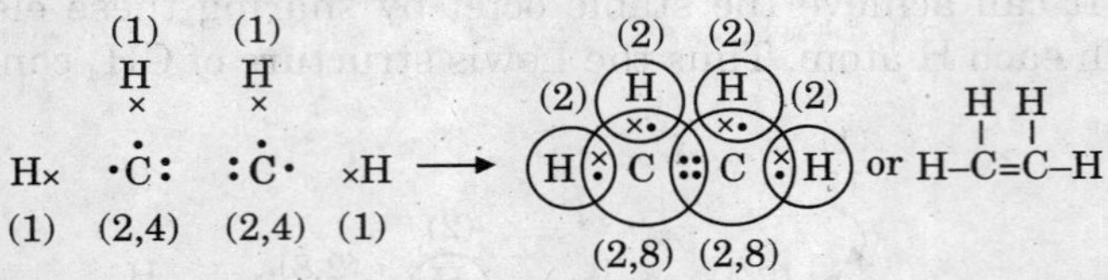

(*b*) In C_2H_2, each C-atom forms one single bond with one H-atom and one triple bond with the other C-atom.

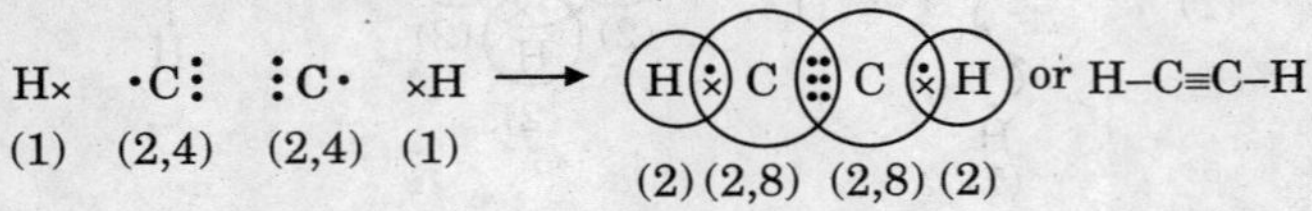

Covalency

The valency of an element in a covalent compound is called its covalency. *Covalency of an element in a covalent compound is equal to the number of electrons contributed by one atom of it in the shared or linking electron pairs.* Consequently, *the covalency of an element in a covalent compound is equal to the number of covalent bonds formed by one atom of it with the neighbouring atom.* Thus the covalency of Cl in Cl_2 molecule = 1, of H in H_2, H_2O and HCl molecules = 1, of O in O_2 and H_2O molecules = 2, of N in N_2 molecule = 3, and of C in CH_4 molecule = 4.

Variable Covalency

Generally the covalency of an element which has only s and p orbitals in its valency-shell (*e.g.* H, N, O, F) is a fixed quantity (*not always*) and is equal to the total number of unpaired electrons in *s* or *p* orbitals. For example :

(*i*) Since H atom ($H \to 1s^1$) has only *one unpaired electron* in its orbital, this element shows a covaency of *one.*

(*ii*) N atom ($N \to 2s^2, 2p_x^{\ 1}, 2p_y^{\ 1}, 2p_z^{\ 1}$) has *three unpaired electrons* and hence exhibits a covalency of *three.*

(*iii*) O atom ($O \to 2s^2, 2p_x^{\ 2}, 2p_y^{\ 1}, 2p_z^{\ 1}$) which has *two unpaired electrons* shows a covalency of two.

(*iv*) F atom ($2s^2$, $2p_x^2$, $2p_y^2$, $2p_z^1$) has *one unpaired electron* and, therefore, shows a covalency of *one*.

Note that the elements mentioned above have only *s* and *p*-orbitals in their valence-shells, *i.e.*, these elements do not contain vacant *d*-orbitals. However, the elements containing vacant *d*-orbitals (*e.g.* P, S, Cl etc.) show different values of covalency in different covalent compounds. This phenomenon is called *variable covalency* and is due to the presence of vacant *d*-orbitals in the valency-shell of these elements. Examples of elements, different covalencies shown by them and the compounds in which these covalencies are shown are given below :

Elements	**Different covalencies**	**Examples of compounds**
Phosphorus	3	PCl_3
	5	PCl_5
Sulphur	2	SCl_2
	4	SF_4
	6	SF_6
Halogen (other than flourine)	1	ICl
	3	ICl_3
	5	IF_5
	7	IF_7

Explanation of Variable Covalency

Variable covalency of an element arises due to the increase in the number of unpaired electrons in the different excited states of the atom. For example the covalencies of P, S and Cl atoms can be explained as follows :

1. Variable covalency of P atom (3, 5). Covalencies of P atom equal to 3 and 5 can be explained as follows :

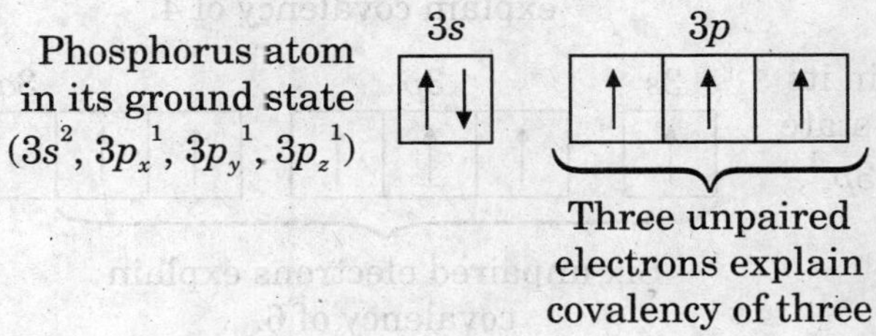

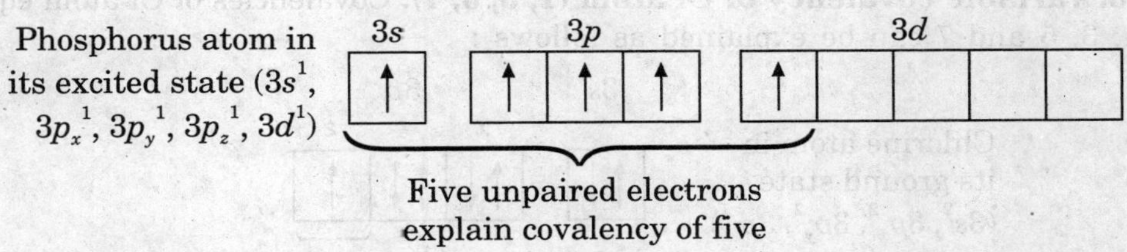

2. Variable covalency of S atom (2, 4, 6). ***(a) Covalency of 2.*** We know that the normal configuration of the valence shell of sulphur atom is $3s^2$, $3p_x^{\,2}$, $3p_y^{\,1}$, $3p_z^{\,1}$. Sulphur atom having this configuration is said to be in the *ground state* and the configuration in this state is called *ground state configuration.* Thus normal configuration and ground state configuration is the same. The presence of two unpaired electrons explains a covalency of two for sulphur atom.

Sulphur atom in its ground state ($3s^2$, $3p_x^{\,2}$, $3p_y^{\,1}$, $3p_z^{\,1}$)

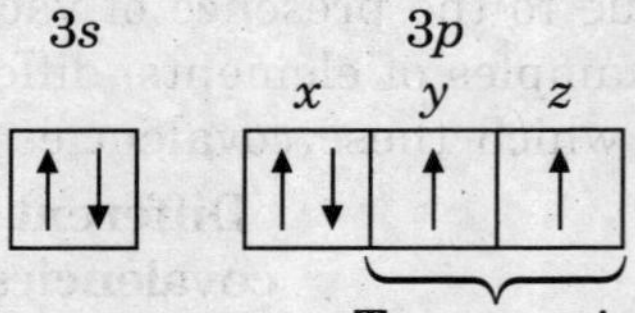

Two unpaired electrons explain a covalency of two

(b) Covalencies of 4 and 6. When sulphur atom forms covalent bonds with other atoms, energy is set free during the formation of these bonds and this energy makes one 3*p* electron or both 3*s* and 3*p* electrons jump to the vacant 3*d*-orbital. Now sulphur atom is said to be in the *excited state*. When only 3*p* electron jumps, we get *four* unpaired electrons which explain the covalency of 4 for sulphur atom. On the other hand when both 3*s* and 3*p* electrons jump, we get *six* unpaired electrons which account for the covalency of 6 for sulphur atom. How 4 and 6 covalencies of sulphur arise is shown below :

Sulphur atom in its first excited state ($3s^2$, $3p_x^{\,1}$, $3p_y^{\,1}$, $3p_z^{\,1}$, $3d^1$)

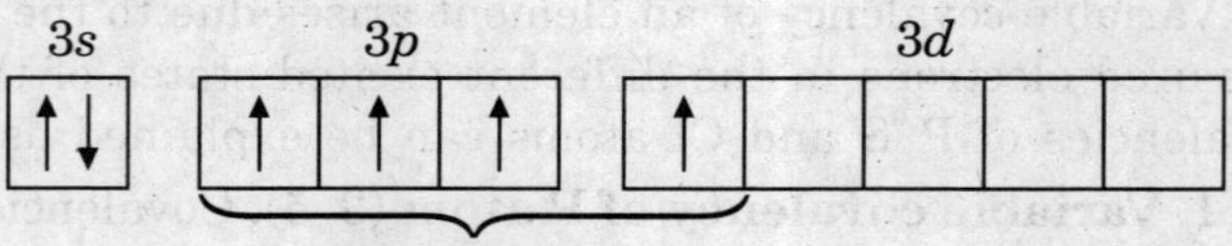

Four unpaired electrons explain covalency of 4.

Sulphur atom in its second excited state ($3s^1$, $3p_x^{\,1}$, $3p_y^{\,1}$, $3p_z^{\,1}$, $3d^1$, $3d^1$)

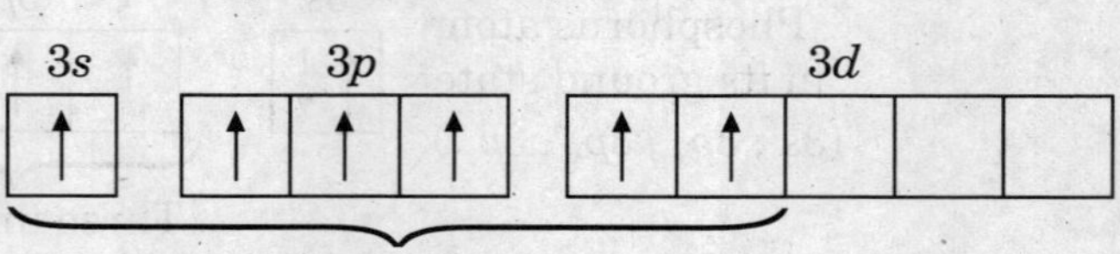

Six unpaired electrons explain covalency of 6.

3. Variable covalency of Cl atom (1, 3, 5, 7). Covalencies of Cl atom equal to 1, 3, 5 and 7 can be explained as follows :

Chlorine atom in its ground state ($3s^2$, $3p_x^{\,2}$, $3p_y^{\,2}$, $3p_z^{\,1}$)

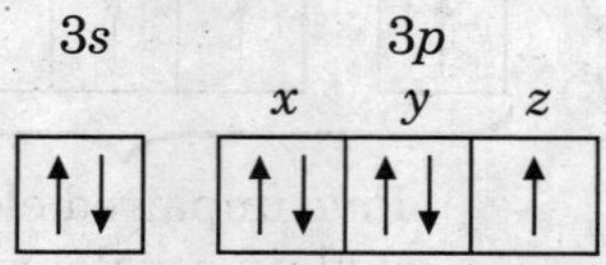

One unpaired electron explains covalency of one.

Chlorine atom in its first excited state ($3s^2$, $3p_x^2$, $3p_y^1$, $3p_z^1$, $3d^1$)

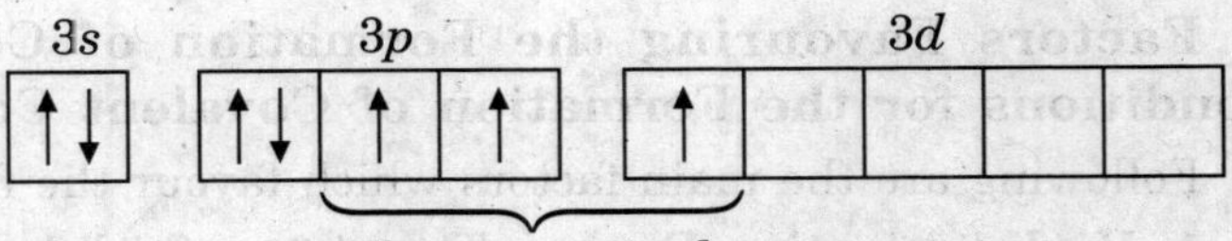

Three unpaired electrons explain covalency of three.

Chlorine atom in its second excited state ($3s^2$, $3p_x^1$, $3p_y^1$, $3p_z^1$, $3d^1$, d^1)

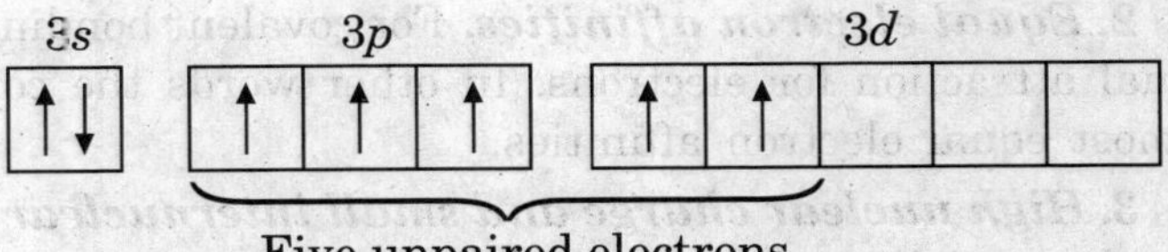

Five unpaired electrons explain covalency of five.

Chlorine atom in its third excited state ($3s^1$, $3p_x^1$, $3p_y^1$, $3p_z^1$, $3d^1$, $3d^1$)

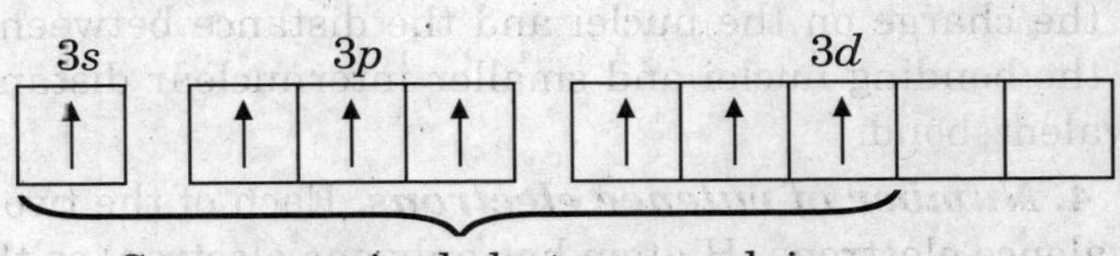

Seven unpaired electrons explain covalency of seven.

It may be seen from the above discussion that variable covalency is shown only by those elements whose atoms have vacant d-orbitals in their valence shell so that the unpairing of s- and p-electrons by promoting them to vacant d-orbitals may be possible. The elements having no d-orbitals (i.e. elements of 2nd period) do not exhibit variable covalency. Thus we can explain why molecules like NCl_5 and OF_6 do not exist.

Maximum Covalency

Maximum covalency of an element may be defined as the maximum number of covalent bonds formed by that element. For example the maximum covalency of P, S, and Cl atoms is 5, 6, and 7 respectively. It may be seen form the discussion on *Variable Covalency* that the maximum covalency exhibited by an element is equal to the number of unpaired electrons obtained after promoting all the *s*- and *p*-electrons to *d*-orbitals. This type of promotion makes all the electrons unpaired and the number of unpaired electrons obtained is equal to the maximum covalency of the atom under consideration.

Formation of Covalent Bonding and Periodic Table

Covalent bonds are formed between the same or different kinds of atoms which should have high electronegativity, since the elements with high electronegativity do not ionise and have an equal attraction for electrons to complete their octets. Such elements are mostly located in right-hand portion of the periodic table.

Factors Favouring the Formation of Covalent Compounds : Conditions for the Formation of Covalent Compounds

Following are the main factors which favour the formation of covalent bond.

1. ***High ionisation Energy.*** The atoms which have high value of ionisation energy are incapable of losing electrons to form cations. Thus these elements cannot form ionic bonds. Rather, they, due to their high ionisation energy, can form covalent bonds between them.

2. ***Equal electron affinities.*** For covalent bonding the two atoms must have equal attraction for electrons. In other words the combining atoms must have almost equal electron affinities.

3. ***High nuclear charge and small internuclear distance.*** In covalent bond the electron charge becomes highly concentrated in the region between the atomic nuclei. This attracts both the nuclei towards itself. This force of attraction depends on the charge on the nuclei and the distance between them. Hence high charge on the bonding nuclei and smaller internuclear distance favour the formation of covalent bond.

4. ***Number of valence electrons.*** Each of the two atoms should have 5, 6 or 7 valence electrons (H atom has only one electron) so that both the atoms achieve the stable octet by sharing 3, 2 or 1 electron pair. The non-metals of VA, VIA and VIIA groups respectively satisfy this condition.

5. ***Equal electronegativity.*** Both atoms should have equal electronegativity so that the transfer of electron (*s*) from one atom to the other may not take place, *i.e.,* ionic bond may not be formed. When the electronegativity of both the atoms is equal, sharing of electron pair(s) occurs and covalent bond is established.

Properties of Covalent Compounds

1. ***Physical state.*** Covalent compounds usually consist of *discrete molecules* and the force of attraction between adjacent covalent molecules is *weak*. It is due to these weak forces that most of the covalent compounds exist as *gases or liquids of low boiling points under the normal conditions of temperature and pressure. However, they may exist* as soft solids only when their molecular weights are high; *e.g.* chlorine (mole. wt. = 71) is a *gas*, bromine (mol. wt. = 160) is a *liquid*, while iodine (mol. wt. = 254) is a *solid*.

2. ***Crystal structure.*** The crystals of covalent solids are of the *two* types :

(*a*) Those in which every atom is bonded with other atoms by covalent bonds resulting in the formation of *giant molecules*. Examples of giant molecules are *diamond, silicon carbide (SiC), aluminium nitride* (AIN) etc.

(*b*) Those which consist of *separate layers*. The covalent compounds containing separate layers are said to have *layer lattice structure.* Examples of covalent compounds having layer structure are *graphite*, CdI_2, $CdCl_2$, *BN* etc.

3. ***Melting and boiling points.*** With the exception of covalent solids consisting of *giant molecules* (*e.g.* diamond, SiC, AIN etc), other covalent solids have relatively *lower melting and boiling, points than the ionic solids, e.g.* boiling

points of $SiCl_4$ (covalent compound) and NaCl (ionic compound) are 58°C and 1440°C respectively. Low boiling points are due to the fact that the attractive forces between covalent molecules are *weak van der Waal's forces*.

4. *Electrical conductivity.* Covalent solids consisting of giant molecules are *bad conductors of electricity*, since they do not contain charged particles (*i.e.* ions) or electrons to carry the current. However the covalent solids having the layer lattices (*e.g.* graphite) are good conductors of electricity, since in such solids electrons can pass from one layer to the other and thus current can be carried. Certain covalent substances like HCl ionise in solution and their solutions conduct electricity.

5. *Solubility in polar and non-polar solvents.* With the exception of covalent solids consisting of giant molecules, all other covalent solids are *insoluble in polar solvents* like H_2O but are *readily soluble in non-polar solvents like* C_6H_6, CCl_4 etc. Their solubility in non-polar solvents is due to the similarity in covalent nature of the molecules of the solute and solvent, *i.e.*, their solubility is based on the principle : "*Like dissolves like*". Covalent solids having the giant molecules are insoluble in all the solvents. This is because of the fact that these solids, due to their big size, are not able to interact with the solvent molecules. *Some of the covalent compounds like alcohol, amines etc. are soluble in H_2O due to hydrogen bonding.*

6. *Neither hard nor brittle.* While ionic compounds are hard and brittle, covalent compounds are neither hard nor brittle. Rather they are *soft* and *waxy*, since they usually consist of separate molecules. There are weak forces holding the molecules in the solid crystal lattice. A molecular layer in the crystal easily slips relative to other adjacent layers and there are no forces of repulsion between the layers like those in ionic compounds. Thus covalent crystals are easily broken and there is no sharp cleavage between the layers on application of external force.

7. *Molecular reactions.* Covalent compounds undergo molecular reactions in solutions, *i.e.*, in solution covalent compounds give reactions where the molecule as a whole undergoes a change. Since there are no strong electrical forces to speed up the reaction between molecules, these reactions are slow and hence need a control of temperature and pressure.

8. *Isomerism.* Since covalent bonds are *rigid* and *directional*, they can give rise to different arrangements of atoms in space. So, a single molecular formula of a covalent compound may represent a number of different compounds with different properties. This means that covalent compounds can show isomerism. This phenomenon of isomerism is shown mostly by organic compounds.

Comparison of the Properties of Ionic and Covalent Compounds

The comparison of the properties of ionic and covalent compounds is shown in Table 7.1.

Table 7.1. Comparison of the properties of ionic and covalent compounds

Property	Ionic compounds (Electrovalent compounds)	Covalent compounds
(*i*) Physical state	These are crystalline solids at room temperature. They are never liquids or gases under the ordinary conditions of temperature and pressures.	Most of the covalent compounds exist as gases or liquids. However, the covalent compounds having high molecular weight can exist as solids.
(*ii*) Crystal structure	They consist of three dimensional solid aggregates (ionic solids).	The crystals of covalent compounds are of two types : (*a*) those consisting of giant molecules (*e.g.* diamond etc.) (*b*) those consisting of separate layers (*e.g.* graphite, CdI_2 etc.).
(*iii*) Hardness and brittleness	These are hard and brittle, since the cations and anions are held together very tightly by very strong electrostatic forces of attraction.	Solid covalent compounds are soft and waxy, since they usually consist of separate molecules. They are much readily broken.
(*iv*) Nature of reactions	They undergo ionic reactions (in solution) which are fast and instantaneous.	They undergo molecular reactions (in solution) which are slow.
(*v*) Melting and boiling points	They have usually high melting and boiling points, since very high amount of energy (in the form of heat) is required to separate the cations and anions against the forces holding them together in the ionic solids.	With the exception of giant molecules, other covalent solids have low melting and boiling points. This is because of the reason that the attractive forces between covalent molecules are weak van der Waal's forces.
(*vi*) Solubility in polar and non-polar solvents	They are freely soluble in polar solvents and insoluble or slightly soluble in nonpolar solvents. Their solubility in polar solvents is because of the reason that the solvation energy of the solvent is greater than the lattice energy of the ionic solid.	They are insoluble in polar solvents and readily soluble in non-polar solvents. Their solubility in non-polar solvents is based on the principle : "*Like dissolves like*".
(*vii*) Electrical conductivity	These are bad conductors when they are in the solid state. However, they conduct electricity in fused state or in solution in which their ions are free to migrate.	Covalent solids consisting of giant molecules are poor conductors because they do not contain charged particles or electrons to carry the current.
(*viii*) Nature of bonds	(*a*) Formed by the transfer of electrons from a metal to a non-metal. (*b*) Consist of electrostatic force between cations and anions. (*c*) Non-rigid and non-directional. (*d*) Cannot cause isomerism	(*a*) Formed by the sharing of electron pair(*s*) between non-metal atoms (*b*) Consist of shared pair(*s*) or electrons between atoms (*c*) Rigid and directional (*d*) Cause streoisomerism.

Failure of Octet Rule (Lewis Concept) in Covalent Compounds

In the formation of a covalent bond, the atoms attain an inert gas configuration with an octet of electrons (*i.e.*, ns^2p^6 configuration). This is known as **octet rule or rule of eight.** In H_2 molecule each of the two H.atoms. however, attains $1s^2$ configuration (*doublet*).

$$H\cdot + \cdot H \rightarrow H : H \text{ or } H\text{–}H$$

But there are many covalent molecules in which the central atom which is covalently bonded with other atoms has electrons either *less* than eight (*incomplete octet*) or *more* than eight (*expansion of octet*) in its outermost shell, *i.e.* in such molecules the central atom is surrounded by electrons either less or more than eight. In this way we come across many molecules which have *non-octet structure*. Thus there are two types of deviations from the octet rule in covalent compounds.

1. Molecules in which the octet remains incomplete : *Incomplete octet.* Consider the structures of $BeCl_2$, BF_3, $(CH_3)_3Ga$ and NO molecules as given below :

(*a*) *BeCl$_2$ molecule.* In this molecule Be atom (central atom) has only *four* electrons in its outermost shell–two of its own (shown by crosses) and the other two (shown by dots) from the two covalently bonded Cl atoms.

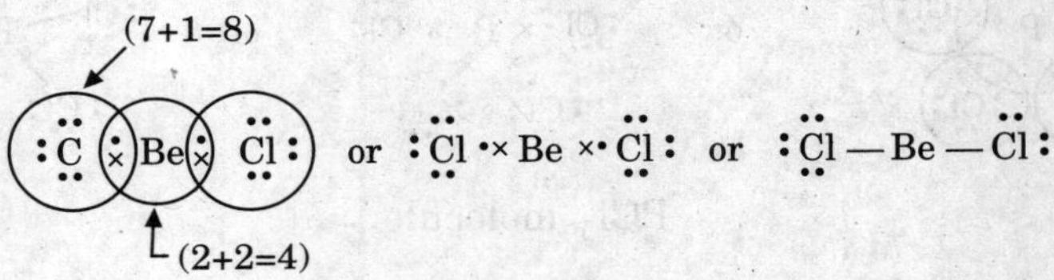

(*b*) *BF$_3$ molecule.* In this molecule the boron atom (central atom) has only *six* electrons in its outermost shell–three of its own and the other three from the three covalently bonded F atoms.

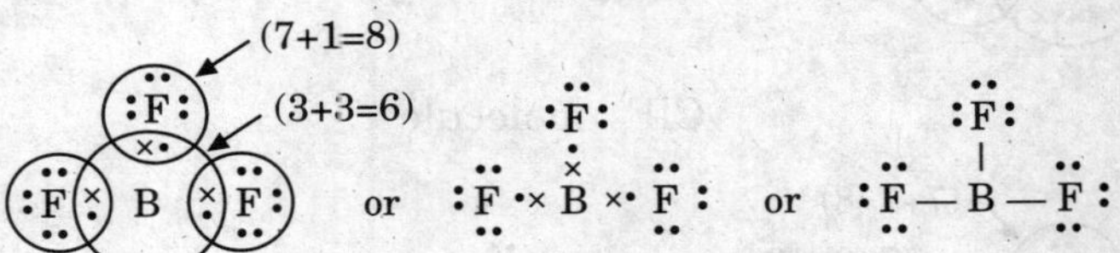

(*c*) *(CH$_3$)$_3$Ga molecule.* Here Ga atom has *six* electrons in its outer-most shell.

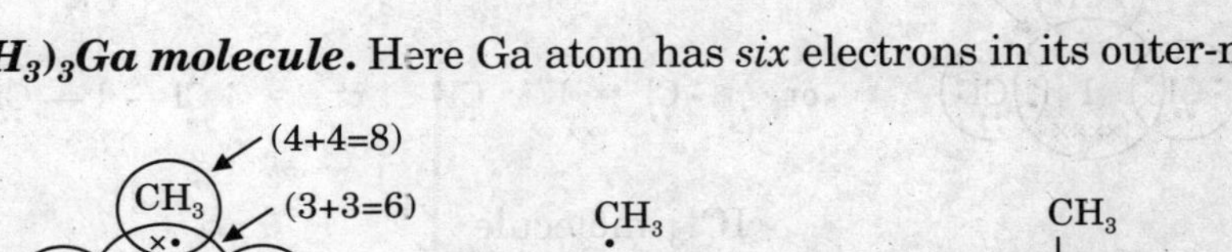

(*d*) *NO molecule.* Here N atom has only *seven* electrons (*septet*) in its outermost shell–five of its own and the other two from oxygen atom with which it is linked by a double bond. Note that oxygen atom has *eight* electrons in its outermost shell.

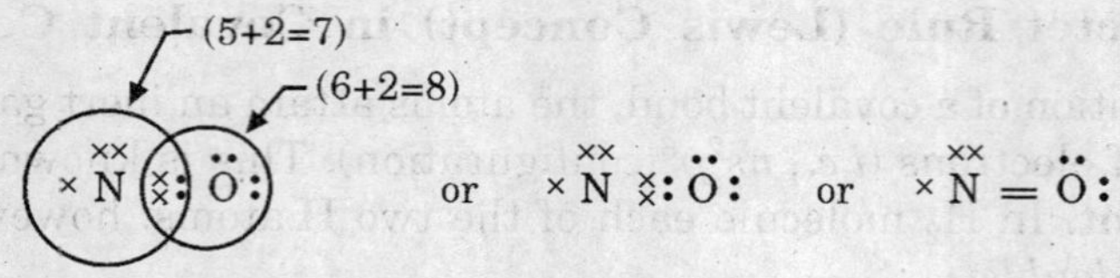

2. Molecules in which the octet is expanded : *Expansion of octet.* Consider the structures of PCl_5, ClF_3, ICl_3, SF_6, IF_7, OsF_8 and OsO_4 molecules as given below :

(a) PCl_5, ClF_3* and *ICl_3 molecules. In PCl_5 molecule the total number of electrons in the outermost shell of phosphorus atom (central atom) is *ten*–five electrons of its own ($P \rightarrow 3s^2p^3$) and five electrons gained by it in forming five covalent bonds with five chlorine atoms. Similarly in ClF_3 and ICl_3 molecules the number of electrons in the outermost shell of Cl and I atoms (central atoms) is also *ten*–seven electrons of their own ($Cl \rightarrow 3s^2p^5$, $I \rightarrow 5s^2p^5$) and three electrons gained by Cl and I atoms in forming three covalent bonds with three halogen atoms.

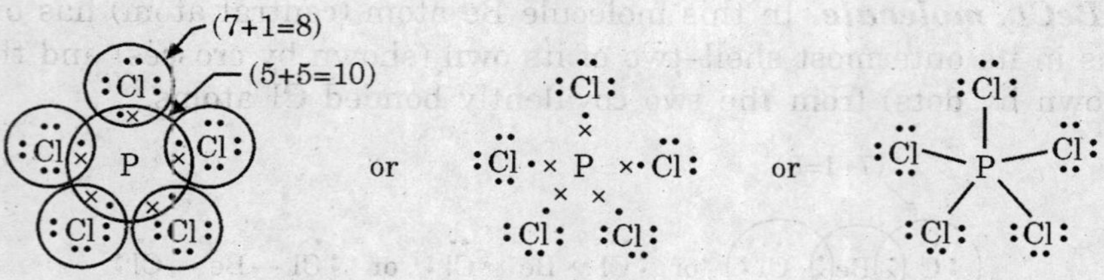

PCl_5 molecule

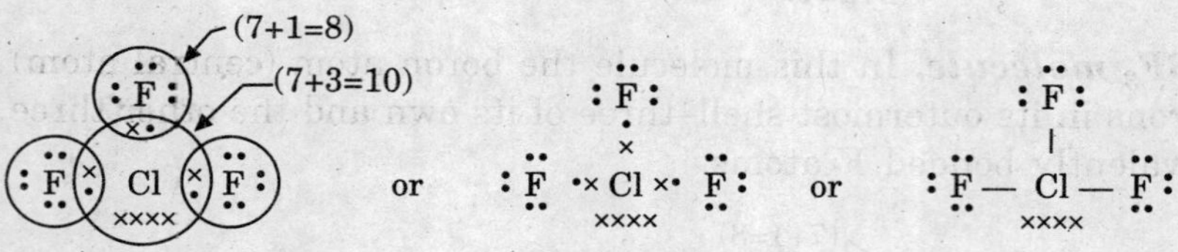

ClF_3 molecule

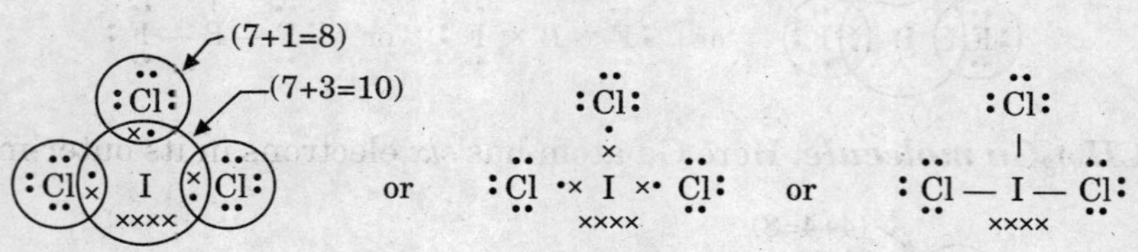

ICl_3 molecule

(b) SF_6 molecule. In SF_6 molecule the number of electrons in the outermost shell of sulphur atom (central atom) is *twelve*—six electrons of its own ($S \rightarrow 3s^2p^4$) and six electrons gained by S in forming six covalent bonds with six F atoms.

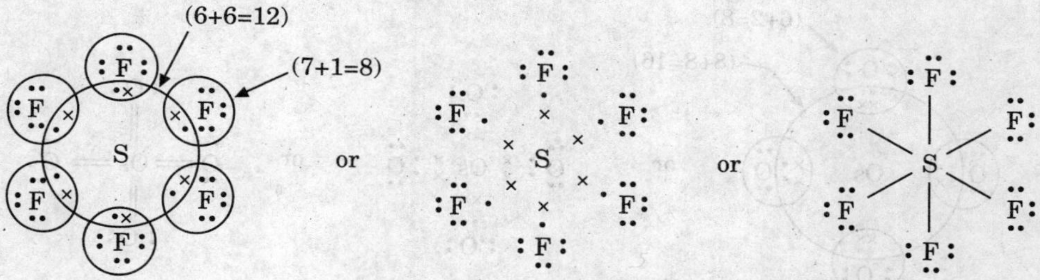

SF_6 molecule

(c) *IF_7 molecule.* In IF_7 molecule the number of electrons in the outermost shell of I atom (central atom) is *fourteen*—seven electrons of its own ($I \rightarrow 5s^2p_5$) and seven electrons gained by I atom in forming seven covalent bonds with seven F atoms.

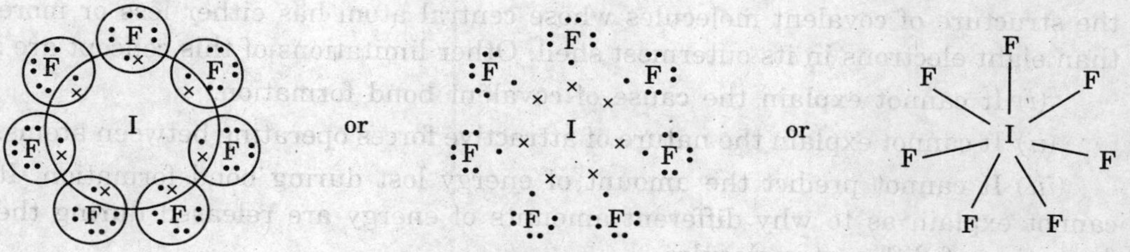

IF_7 molecule

(d) *OsF_8 and OsO_4 molecules.* In OsF_8 molecule the number of electrons in the outermost shell of Os atom (central atom) is *sixteen*—eight electrons of its own ($Os \rightarrow 5d^6\ 6s^2$) and eight electrons gained by Os in forming eight covalent bonds with eight F atoms. Similary in OsO_4 molecule the number of electrons in the outermost shell of Os atom (central atom) is *sixteen*–eight electrons of its own and eight electrons gained by Os in forming four double covalent bonds with four oxygen atoms.

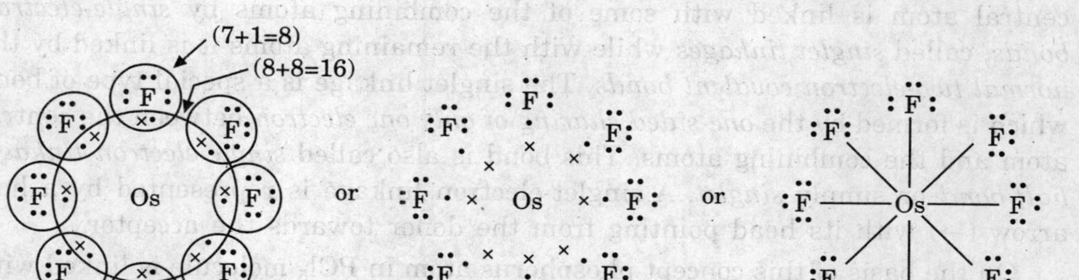

OsF_8 molecule

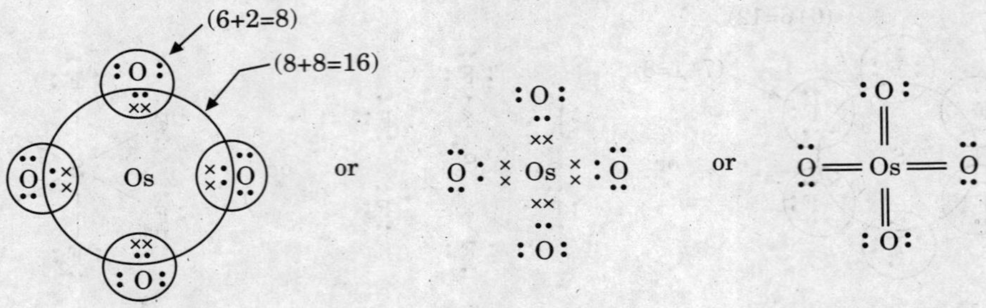

OsO_4 molecule

Thus we see that in all the above molecules the octet of the central atom is expanded. The molecules containing central atoms with more than an octet of electrons are called **super octet structures.**

The above study makes it quite clear that Lewis concept is not able to explain the structure of covalent molecules whose central atom has either less or more than eight electrons in its outermost shell. Other limitations of this concept are :

(*i*) It cannot explain the cause of covalent bond formation.

(*ii*) It cannot explain the nature of attractive forces operating between atoms.

(*iii*) It cannot predict the amount of energy lost during bond formation. It cannot explain as to why different amounts of energy are released during the formation of different molecules.

(*iv*) It cannot predict the geometry of a molecule (*i.e.* arrangement of atoms in space). For example, it can not explain as to why BF_3 is planar and NH_3 is pyramidal.

Explanation of the Failure of Octet Rule.

The failure of octet rule in the formation of covalent molecules like those mentioned above has been explained with the help of the following concepts :

1. Sugden's concept of single linkages. Sugden postulated that the central atom of the molecules like PCl_5. SF_6 etc. maintains its octet and in doing so the central atom is linked with some of the combining atoms by *single-electron bonds*, called *singlet linkages* while with the remaining atoms it is linked by the *normal two-electron covalent bonds*. The singlet linkage is a special type of bond which is formed by the *one-sided sharing of only one electron* between the central atom and the combining atoms. This bond is also called *single-electron linkage, half-bond* or simply *singlet*. A singlet-electron linkage is represented by a half arrow ($\rightharpoonup$) with its head pointing from the donor towards the acceptor.

On the basis of this concept phosphorus atom in PCl_5 molecule is linked with three Cl atoms by three covalent bonds (*i.e.* three electron pairs are being shared between phosphorus and three chlorine atoms) and with each of the remaining two chlorine atoms it is linked by a singlet linkage. Thus the structure of PCl_5 can be shown as follows :

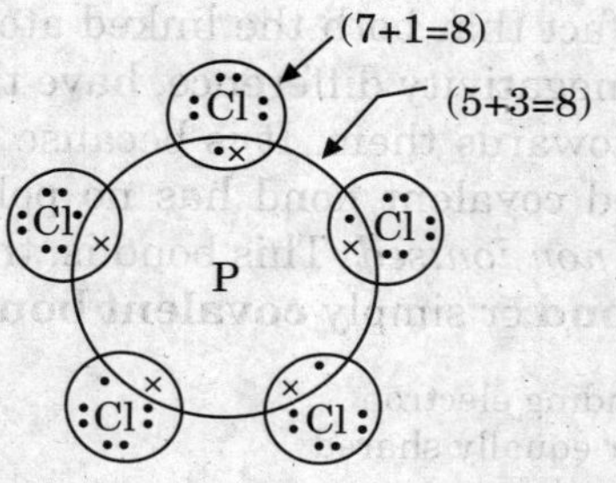

or

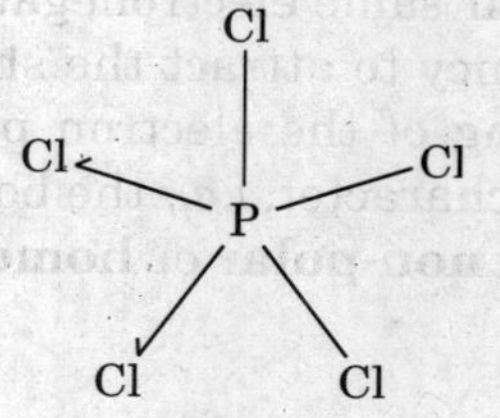

Singlet linkage is weaker than a covalent bond and this explains why PCl_5 readily dissociates into PCl_3 and Cl_2.

In SF_6 molecule sulphur atom is linked with two F atoms by two covalent bonds and with four F atoms by four singlet linkages as shown below :

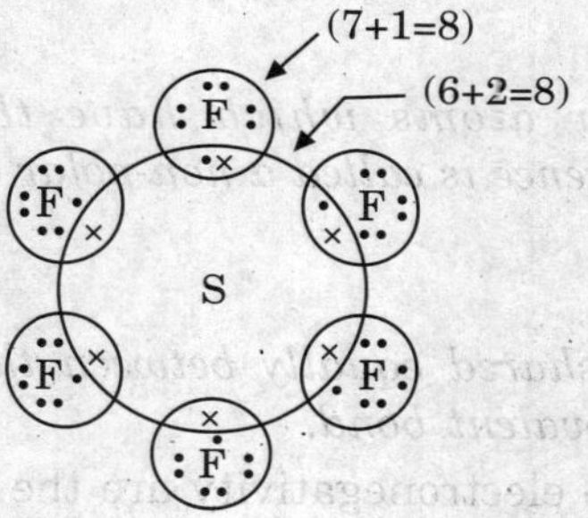

or

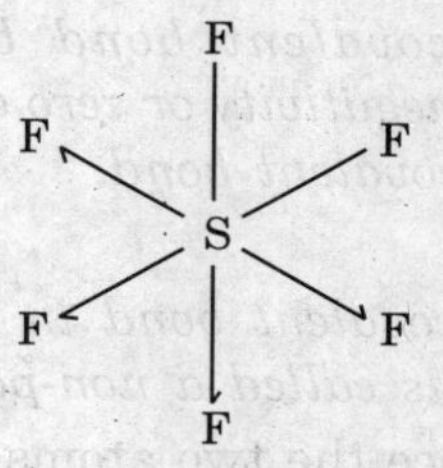

2. Sidgwick's concept of maximum covalency. According to this concept it is not necessary for an element to have a maximum covalency of four, *i.e.* it is not essential that an element be surrounded by 4 × 2 = 8 electrons for its stability. The covalency of an element may exceed four and the octet can be expanded. The maximum covalency of an element actually depends on the period in which the element concerned is present. For hydrogen (first period with n = 1) it is 2, for the elements of 2nd period with n = 2 (Li to F) it is 4, for the elements of 3rd (n = 3) and 4th (n = 4) periods it is equal to 6 and for those in higher periods (n > 4) it is equal to 8. Consequently maximum capacity of the valence shell of an atom to have electrons (or maximum number of electrons being shared) for the elements mentioned above is equal to 2 × 2 = 4, 4 × 2 = 8, 6 × 2 = 12 and 8 × 2 = 16 respectively. Thus we see that there is no anomaly in compounds like PCl_5, SF_6, OsF_8 etc. having central atoms making five, six and eight covalent bonds with the combining atoms respectively.

Covalent Bonds Having Partial Ionic Character : Non-polar and Polar Covalent Bonds

1. *Covalent bond between two similar atoms–Non-polar covalent bond.* The bonding electron pair constituting the covalent bond between two similar atoms which have the same electronegativity or zero electronegativity difference is *equally shared by both the linked atoms, i.e.* the shared electron pair is placed at the centre of the distance between the nuclei of the linked atoms. The equal sharing of the electron pair is because of the fact that both the linked atoms, due to their same electronegativity or zero electronegativity difference, have the same tendency to attract the shared electron pair towards them. It is because of equal sharing of the electron pair that the resulted covalent bond has no polarity or ionic character, *i.e.,* the bond is *non-polar* and *non-ionised*. This bond is, therefore, called **non-polar** or **homo-polar covalent bond** or simply **covalent bond.**

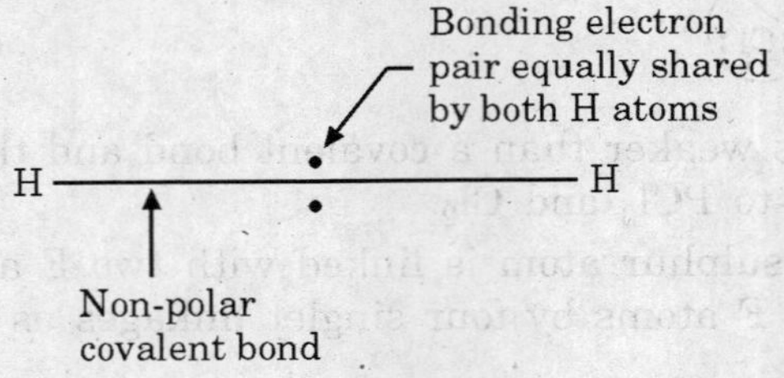

Thus :

A covalent bond between two similar atoms which have the same electronegativity or zero electronegativity difference is called a non-polar or homo-polar covalent bond.

OR

A covalent bond in which electrons are shared equally between the linked atoms is called a non-polar or homo-polar covalent bond.

Since the two atoms which have the same electronegativity are the atoms of the same element, *all covalent bonds between two similar atoms are non-polar or homo-polar covalent bonds.*

A non-polar covalent bond is a true or purely covalent bond, since it has no or negligible polarity or ionic character. H–H, F–F etc are the examples of non-polar covalent bonds. The molecules like H_2, F_2 etc which have non-polar covalent bonds are called **non-polar molecules.** Such molecules are the true covalent compounds.

2. *Covalent bond between two dissimilar atoms–Polar covalent bond.* The bonding electron pair constituting the covalent bond between two dissimilar atoms like H and Cl which have different electronegativity values (H = 2.1, Cl = 3.3) *is not equally shared by the linked atoms, i.e.* the electron pair is not placed at the centre of the distance between the nuclei of the two atoms *viz.* H and Cl. Rather, it is *partially displaced* (*i.e. attracted*) towards the more electronegative Cl atom owing to its greater affinity for electron pair. This type of partial displacement of the shared electron pair towards the more electronegative Cl atom develops a small (fractional) negative charge (represented as δ–) on more electronegative Cl atom and an equal amount of small positive charge (represented as δ+) on less electronegative H atom, and HCl molecule appears to be containing two oppositely charged poles (called *charge-centres* or *electrical poles*) namely $H^{\delta+}$ and $Cl^{\delta-}$ at the end of the bond and the molecule is depicted as $H^{\delta+}–Cl^{\delta-}$. The covalent bond between H and Cl atoms develops some polarity or partial ionic character and hence is called *polar covalent bond*.

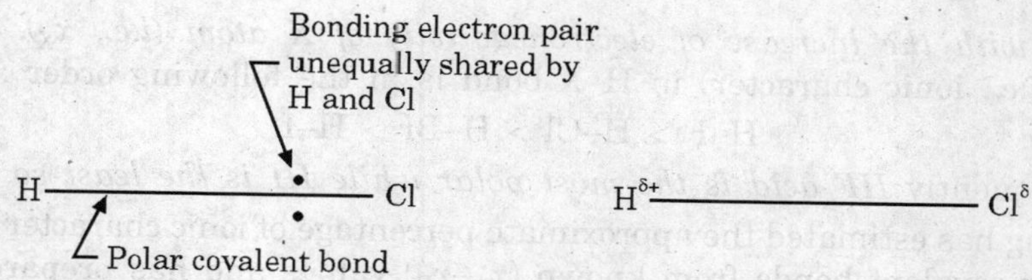

Thus polar covalent bond can be defined as follows :

A covalent bond between two dissimilar atoms which have different electronegativity values is called a polar covalent bond.

OR

A covalent bond in which electrons are shared unequally between the linked atoms and the linked atoms acquire fractional positive and negative charge is called a polar covalent bond.

Polar covalent bond is not a true or purely covalent bond, since it has some polarity or ionic character. Thus it is neither wholly covalent nor wholly ionic but has an intermediate character. The molecules like HCl which have polar covalent bonds are called **polar molecules.**

Since the linked atoms having different electronegativity may not be the atoms of the same elements, *all covalent bonds between two dissimilar atoms are polar covalent bonds.* For example :

(*a*) H_2O molecule contains two O–H covalent bonds. Since H and O atoms have different electronegativity (H = 2.1, O = 3.5), both O–H covalent bonds are polar covalent bonds and H_2O is a polar molecule.

(*b*) NH_3 molecule has three N–H covalent bonds. Since H and N have different electronegativity (H = 2.1, N = 3.0), all the three N–H covalent bonds are polar covalent bonds and NH_3 is a polar molecule.

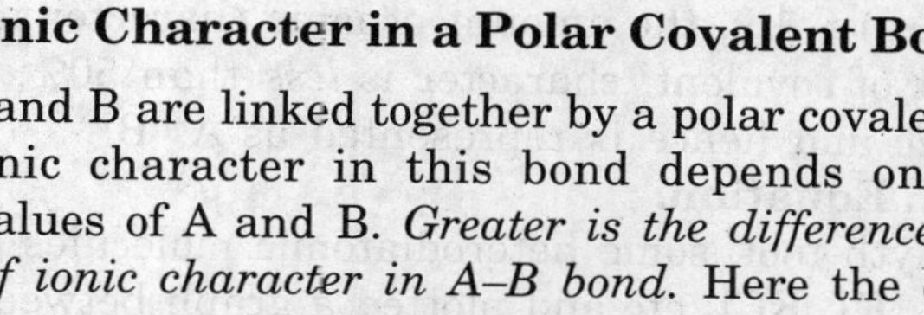

Percentage of Ionic Character in a Polar Covalent Bond

If two atoms A and B are linked together by a polar covalent bond ($A^{\delta-} - B^{\delta+}$), the amount of ionic character in this bond depends on the difference of electronegativity values of A and B. *Greater is the difference, ($x_A - x_B$), greater is the percentage of ionic character in A–B bond.* Here the electronegativity of atom A (x_A) has been assumed to be higher than that of B (x_B). This point can be illustrated by considering the nature of X–H bond in halogen acids of HX type (*e.g.* HF, HCl, HBr and HI). *The polarity of H–X bond in HX acids increases with the increase of electronegativity difference, (x_X–x_H). Since electronegativity of H atom (i.e. x_H) remains the same throughout the series, the polarity of H–X bond* *increases with the increase of electronegativity of X atom (i.e., x_X).* Thus the polarity (*i.e.*, ionic character) in H–X bond is in the following order :

H–F > H–Cl > H–Br > H–I.

Consequently *HF acid is the most polar while HI is the least so.*

Pauling has estimated the approximate percentage of ionic character in various A–B polar covalent bonds from known (x_A–x_B) values and has prepared a table (given below) which gives a relation between (x_A–x_B) values, percentage ionic character in A–B bond and the nature of this bond.

Table. Relation between (x_A–x_B) values, percentage ionic character in A–B bond and the nature of A–B bond ($x_A > x_B$).

($x_A - x_B$)	% ionic character in A–B bond	Nature of A–B bond and its representation	Examples of bonds
0	0	Purely covalent (A–B)	N≡N, O=O, Cl–Cl, H–H.
0.1–0.8	0.5–15	Covalent (A–B)	C–S, C–I, N–Cl, P–H
0.9–1.6	19–47	Polar covalent ($A^{\delta-} - B^{\delta+}$)	Both O–H bonds in H_2O molecule are polar covalent bonds and are represented as $O^{\delta-}$–$H^{\delta+}$
1.9	50	50% ionic and 50% covalent	—
1.9–2.0	55–63	Ionic (A^-–B^+)	—
2.0–2.3	67–74	Ionic (A^-–B^+)	Na–Cl (Na^+Cl^-)
2.3–3.3	76–93	Ionic (A^-–B^+)	Cs–F (Cs^+F^-)

The following points may be noted from the above table :

(*i*) When (x_A–x_B) = 1.9, the amount of ionic character in A–B bond is 50% and that of covalent character is also 50%. Thus A–B bond is 50% ionic and 50% covalent.

(*ii*) When (x_A–x_B) < 1.9, the amount of ionic character in A–B bond is *less* than 50% and that of covalent character is *more* than 50%. Thus A–B bond is *predominantly covalent* and hence is represented as A–B.

(*iii*) When (x_A–x_B) > 1.9, the amount of ionic character in A–B bond is *more* than 50% and that of covalent ,character is *less* than 50%. Hence A–B bond is *predominantly ionic* and hence is represented as A^- B^+.

Hanny and Smyth Equation.

Hanny and Smyth took some heterodiatomic molecules of AB type like HF, HCl, HBr, HI, NaI, KI, KCl, etc and plotted a graph between the percentage of ionic character in A–B bond and electronegativity difference, (x_A–x_B). With the help of this graph they obtained the following equation which gives the percentage of ionic character in a given A–B bond in AB molecule.

$$\textbf{\% ionic character in A–B bond} = [0.16(x_A-x_B) + 0.035(x_A-x_B)^2] \times 100\%$$
$$= [16(x_A-x_B) + 3.5(x_A-x_B)^2]\% \quad ...(i)$$

Solved Examples

Example 1. *Calculate the percentage of ionic character in Cs–F bond in CsF molecule. The electronegativity values of Cs and F are 0.7 and 4.0 respectively. Predict the nature of CsF molecule.*

Solution. $x_A - x_B = x_F - x_{Cs} = 4.0 - 0.7 = 3.3$

$\therefore$ % ionic character in Cs–F bond $= [16 \times 33 + 3.5 \times (3.3)^2]\%$

$= 90.9\%$

Since the ionic character is more than 50%, CsF is a purely ionic molecule.

Problem 2. *Calculate the percentage of ionic character in K–Cl bond in KCl molecule. The electronegativity values of K and Cl are 0.8 and 3.0 respectively. Also predict the nature of KCl molecule.*

Solution. $x_A - x_B = x_{Cl} - x_K = 3.0 - 0.8 = 2.2$

$\therefore$ % ionic character in K–Cl bond $= [16 \times 2.2 + 3.5 \times (2.2)^2]\%$

$= 52.1\%$

Since the ionic character is more than 50%, KCl is an ionic molecule.

Problem 3. *Calculate the percentage of ionic character in Na–Cl bond in NaCl molecule. Also predict the nature of NaCl molecule* ($x_{Na} = 0.9$ and $x_{Cl} =$ 3.0).

Solution. $x_A - x_B = x_{Cl} - x_{Na} = 3.0 - 0.9 = 2.1$

% ionic character in Na–Cl bond $= [16 \times 2.1 + 3.5 \times (2.1)^2]\%$

$= 49\%$

Since the ionic character is less than 50%, NaCl is a covalent molecule. (According to Hanny and Smyth's equation)

Example 4. *Arrange the molecules HF, HCl, HBr and HI in the decreasing order of the percentage of ionic character in H–X bond (X = F, Cl, Br or I). Electronegativity values are as : F = 4.0, Cl = 3.0, Br = 2.8, I = 2.2, and H = 2.1.*

Solution. If we substitute the electronegativity values of F, Cl, Br I and H atoms in Hanny and Smyth's equation [equation (*i*)], the percentage of ionic character in H–X bonds in HF, HCl, HBr and HI molecules is found as given below :

% ionic character in H–F bond and HF molecule = 43%
" " " " " H–Cl " " " " HCl " = 17%
" " " " " H–Br " " " " HBr " = 13%
" " " " " H–I " " " " HI " = 7%

Thus we see that the percentage of ionic character in H–X bonds in the above molecules is in the following decreasing order :

H–F > H–Cl > H–Br > H–I

Properties of Polar Molecules

Polar molecules have the following properties :

(*i*) The properties of polar compounds are intermediate between those of purely covalent and purely ionic compounds.

(*ii*) A polar molecule like HCl ($H^{\delta+} - Cl^{\delta-}$) which has positive and negative *charge centres* or *electrical poles*, as they are called, at the ends of the covalent bond, becomes *dipolar* and hence is called a **dipole** (*two poles*). Sidgwick (1930) proposed that a dipole of a bond may be shown by an arrow with a crossed tail. The arrow is placed parallel to the line joining the points of charge and should be from positive to negative end of the dipole. Thus a dipole of HCl molecule may be depicted as shown in Fig. 7.9. The dipole behaves as a small magent.

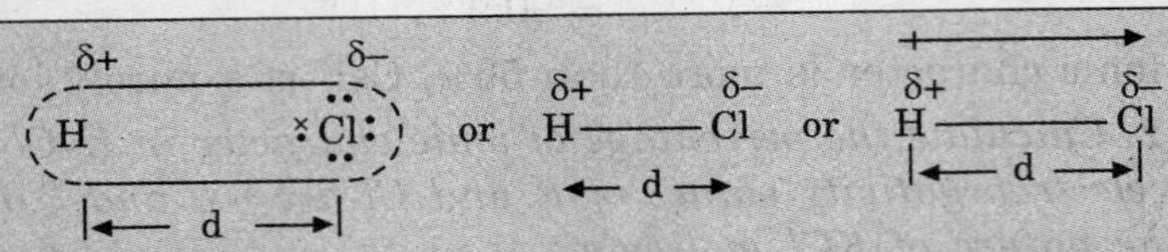

Fig. 7.9. A dipole of HCl molecule, Here *d* is the distance between the positive and negative centres of the dipole and is called bond length.

(*iii*) *Dissolution in polar solvents.* When a polar molecule is dissolved in a polar solvent like H_2O, the polar molecules of water surround the positive and negative ends of the polar covalent compound (solute) and split it into cations and anions which are surrounded by a definite, but unknown, number of water molecules. For example when HCl molecule, which is a polar covalent molecule, is dissolved in water, the polar molecules of H_2O break HCl ($H^{\delta+} - Cl^{\delta-}$) molecule into $H^{\delta+}$ and $Cl^{\delta-}$ ions which are surrounded by water dipoles as shown in Fig. 7.10.

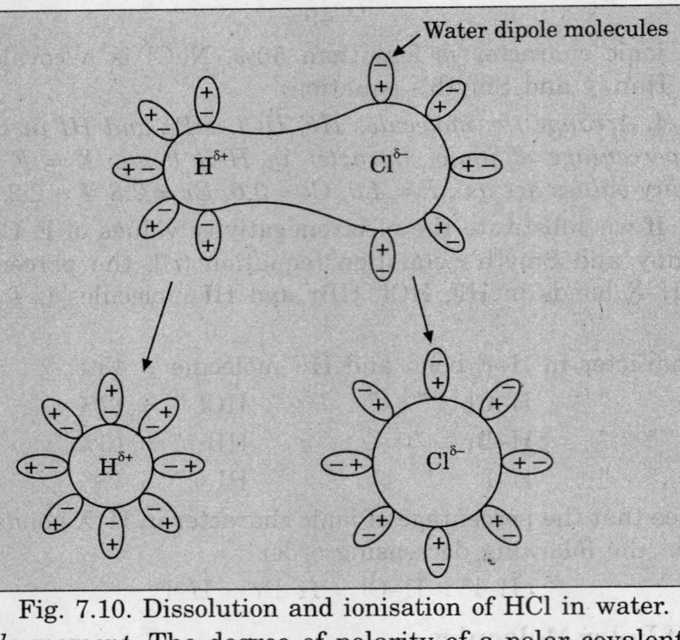

Fig. 7.10. Dissolution and ionisation of HCl in water.

(*iv*) *Dipole moment.* The degree of polarity of a polar covalent bond or of a polar molecule (*i.e.* a molecule having a polar covalent bond) is expressed in terms of its *dipole moment* (μ) which is equal to the product of the magnitude of electric charge (*e*) in *esu* and distance (*d*) in A° between the positive and negative centres (*i.e.* bond length).

Thus :

$$\mu = e \times d$$

As e is of the order of 10^{-10} *esu* and d of the order of 10^{-8} *cm*, μ is of the order of $10^{-10} \times 10^{-8}$ *esu. cm* or 10^{-18} *esu. cm* and this unit of μ is known as *Debye* (D). Thus :

$$1 \text{ D} = 10^{-18} \textit{ esu. cm}$$
$$= 10^{-10} \textit{ esu. } \text{A}^\circ$$
$$= 3.33 \times 10^{-28} \textit{ coulomb. cm}$$

μ is a vector quantity.

Magnitude of dipole moment of a polar molecule depends on the difference in electronegativities of the bonded atoms. *The greater the difference in electronegativities, the greater is the dipole moment. Consequently greater is the value of dipole moment of a polar molecule, greater is the degree of polarity of the polar covalent bond between the linked atoms* (*See Table 7.1*).

Table 7.1. Relation between the dipole moment value and the degree of polarity (i.e. percent of ionic character) of H–X covalent bond.

Molecule	Dipole moment value (in D)	% ionic character of H–X bond (X = I, Br, Cl, F)
H—I	0.38	5
H—Br	0.87	12
H—Cl	1.03	17
H—F	1.92	43

(*v*) Polar molecules have great attraction for each other and arrange themselves into an endless chain and thus get associated with each other as shown below in Fig. 7.11 for $H^{\delta+} - F^{\delta-}$ polar molecules.

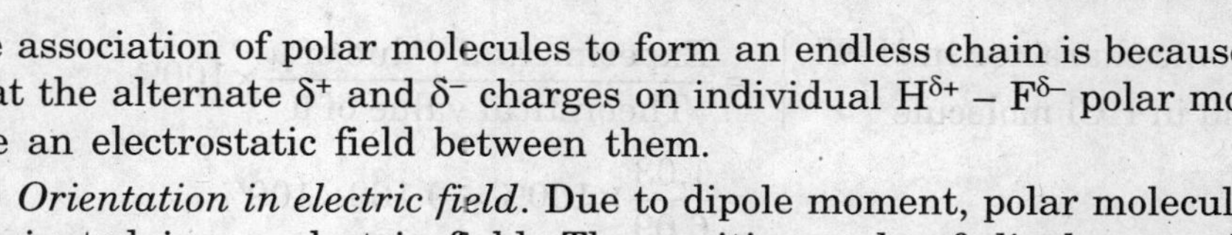

Fig. 7.11. Association of numberless $H^{\delta+} - F^{\delta-}$ polar molecules.

The association of polar molecules to form an endless chain is because of the fact that the alternate δ^+ and δ^- charges on individual $H^{\delta+} - F^{\delta-}$ polar molecules produce an electrostatic field between them.

(*vi*) *Orientation in electric field*. Due to dipole moment, polar molecules tend to be oriented in an electric field. The positive ends of dipoles are directed towards the negative electric plate while the negative ends towards the positive plate (Fig. 7.12). Measurements based on the orientation of polar molecules in electric field have been used to calculate the dipole moments.

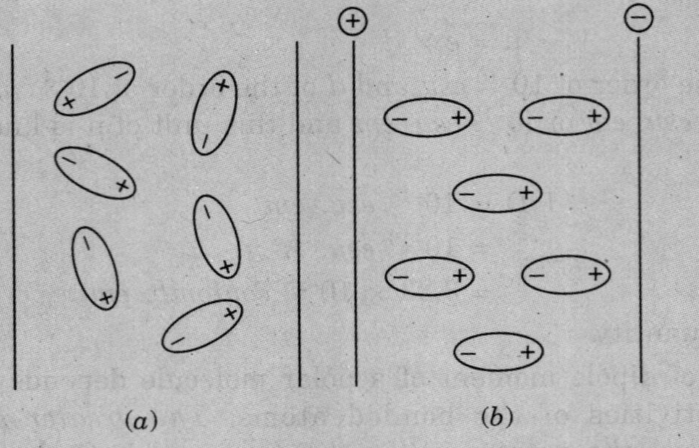

Fig. 7.12. Orientation of polar molecules in an electric field. (*a*) Field off (*b*) Field on.

Applications of Dipole Moment

1. Amount of ionic character in a bond. The value of dipole moment can be used for determining the amount of ionic character in a bond. The dipole moments of diatomic non-polar molecules like H_2, Cl_2, F_2 etc are *zero*. Greater is the degree of polarity of a diatomic molecule, the larger will be its dipole moment.

Let us find out the percent of ionic character of H–Cl bond in HCl molecule with the help of dipole moment value of this molecule. Suppose that HCl molecule is *completely* ionic. Under this condition H^+ and Cl^- ions would bear a unit charge equal to 4.80×10^{-10} *e.s.u.* (= *e*) and the bond distance between H and Cl atoms = 1.27 A° (= *d*).

Thus :

$$\begin{aligned}\mu &= e \times d \\ &= (4.80 \times 10^{-10}\ e.s.u.) \times (1.27\ \text{A}^\circ) \\ &= 4.8 \times 1.27 \times 10^{-10}\ e.s.u.\ \text{A}^\circ \\ &= 6.19 \times 10^{-10}\ e.s.u.\ \text{A}^\circ \\ &= 6.09\ \text{D} \qquad (\because 1\ \text{D} = 10^{-10}\ e.s.u.\ \text{A}^\circ)\end{aligned}$$

This is the theoretical value of μ. But, since the experimental value of μ for HCl is 1.03,

$$\left.\begin{array}{l}\%\ \text{ionic character in H–Cl} \\ \text{bond in HCl molecule}\end{array}\right\} = \frac{\text{Experimental value of }\mu}{\text{Theoretical value of }\mu} \times 100\%$$

$$= \frac{1.03}{6.09} \times 100\% = 0.169 \times 100\%$$

$$= 16.9\% \approx 17\%$$

This means that H–Cl bond in HCl molecule has 17% ionic character and hence 83% covalent character.

2. Geometry and shape of molecules. Dipole moment value of a molecule also helps in determining its geometry and shape.

The diatomic molecules composed of *similar atoms* (*e.g.* H_2, Cl_2 etc.) have *no dipole moment*, but those having *different atoms* (*e.g.* HF, HCl, HBr, HI etc.) do have *some value of dipole moment. Symmetrical linear polyatomic molecules* like CCl_4, CO_2, CS_2, HgX_2 (X = Cl, Br, I) etc. have *zero resultant dipole moment* and hence are *non-planar*, since dipole moments for all the bonds present in the molecule cancel each other in opposite directions. On the other hand *unsymmetrical non-linear polyatomic molecules* like H_2O, H_2S, SO_2, NO_2 etc. have some value of dipole moment and hence are *polar molecules*. Thus the value of the dipole moment of a molecule gives an information about its geometry. Following examples will illustrate this point.

(a) ***CO_2, CS_2, and HgX_2 molecules.*** These are triatomic symmetrical molecules. CO_2 molecule (O=C=O) is non-polar, although C=O bonds are polar as the shared electron pair lies closer to O-atom than to C-atom on account of higher electronegativity of O-atom. This molecule has zero dipole moment. Zero dipole moment value and non-polar nature of the molecule can be explained only when it is assumed that the molecule is *linear* so that the dipole moment of one $C^{\delta+}-O^{\delta-}$ bond on one side cancels that of the same bond on the other side.

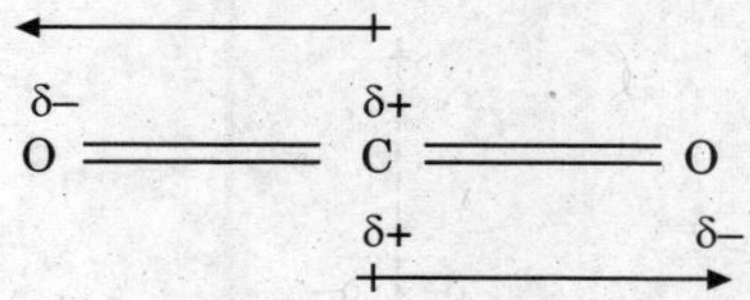

Similar arguments can be made to explain that CS_2 and HgX_2 molecules are non-polar, have zero dipole moment and hence are linear.

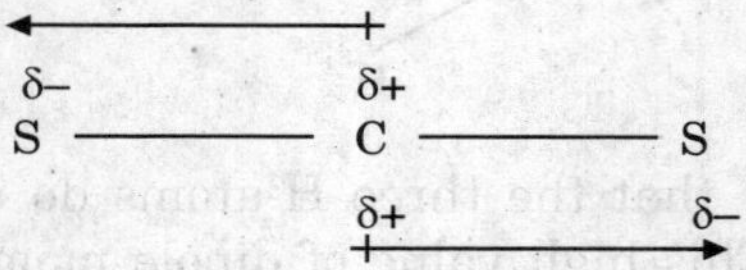

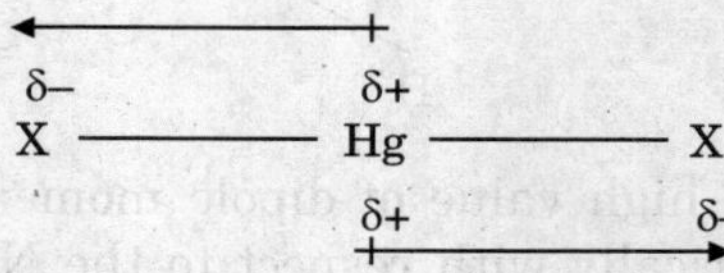

(b) ***H_2O, H_2S, SO_2 and NO_2 molecules.*** Although these are also triatomic molecules like CO_2, CS_2 and HgX_2 molecules, yet these have some value of dipole moment (*e.g.* H_2O = 1.85 D, SO_2 = 1.60 D, etc). This clearly shows that there is no possibility for these molecules to have a linear shape as CO_2, CS_2 and HgX_2 have. Rather these molecules have *bent structure* (Fig. 7.13) to give rise to the resultant dipole moment. In these molecules there are two dipolar bonds and the resultant of the dipole moment of these individual dipolar bonds gives the dipole moment of the molecule, called *molecular dipole moment*. The molecular dipole moment of these molecules has been shown by a dotted arrow.

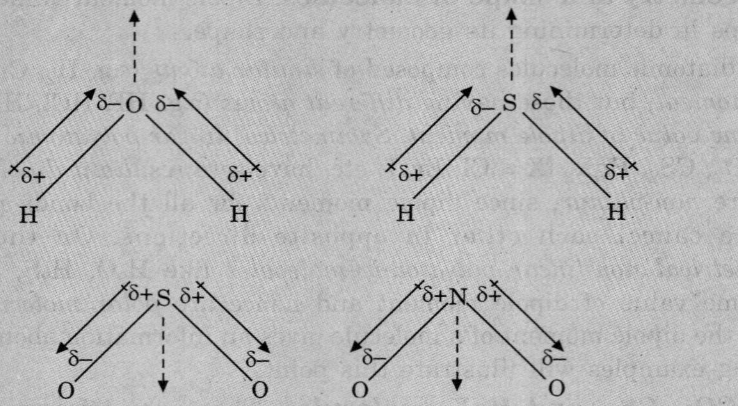

Fig. 7.13. Bent structure of H_2O, H_2S, SO_2, and NO_2 molecules. The resultant dipole moment (called molecular dipole moment) has been shown by a dotted arrow.

(c) NH_3 molecule. It is a tetratomic molecule and contains three dipolar $N^{\delta-}$–$H^{\delta+}$ bonds and their resultant gives the molecular dipole moment which has been found experimentally equal to 1.46 D.

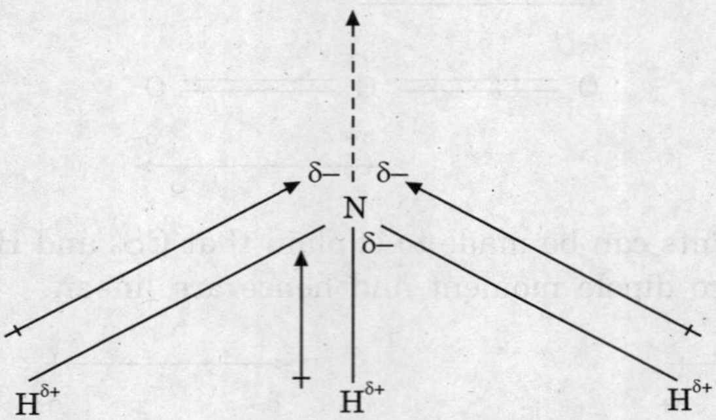

This high value of dipole moment shows that the three H-atoms do not lie symmetrically with respect to the N-atom. This high value of dipole moment is explained by giving a *triangular pyramidal structure* to this molecule, in which N-atom is situated at the apex and three H-atoms lie at the other three corners (Fig. 7.14).

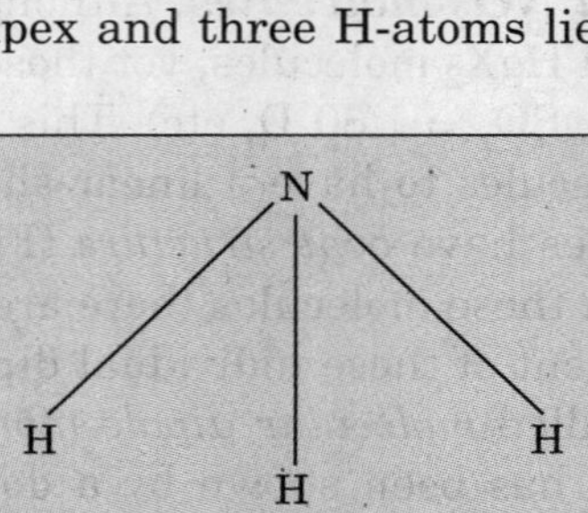

Fig. 7.14. Triangular pyramidal structure of NH_3 molecule.

Polarisation of Ions and Fajans's Rules

What is Polarisation of Ions?

When a cation, C^+ of an ionic molecule, C^+A^- approaches closely the anion, A^-, it withdraws the electrons (*i.e.* electron cloud) of the anion towards itself and the symmetrical shape of the electron cloud of the anion gets distorted. Thus the electron cloud of the anion no larger remains symmetrical but is elongated towards the cation. In other words we say that the A^- anion is *distorted* (or *deformed* or *polarised*) by C^+ cation and this phenomenon is called **distortion** (or **deformation** or **polarisation**) of A^- anion by C^+ cation (Fig. 7.15).

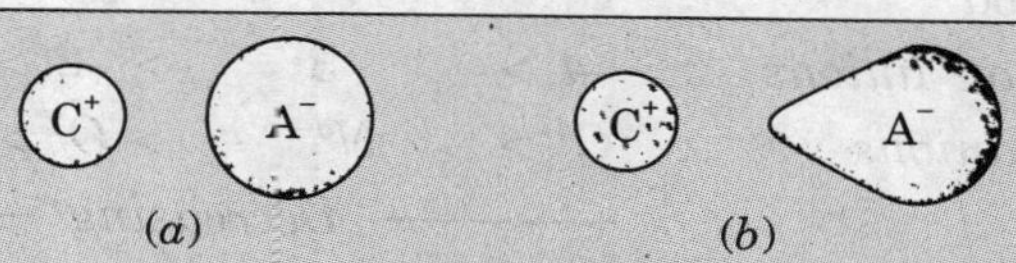

Fig. 7.15. Schematic representation of polarisation of an anion, A^- by a cation, C^+ in C^+A^- ionic molecule. (*a*) No polarisation (*b*) Polarised anion.

The cation, C^+ is also deformed by the anion, A^-, but due to smaller size of the cation, its electrons are strongly held to the nucleus and hence the cation is not polarised to an appreciable extent by a nearby anion, *i.e.*, the polarisation of a cation by an anion is regarded as negligible and hence we generally do not consider the polarisation of a cation by anion.

The ability of a cation to polarise a nearby anion is called its **polarising power** or **polarising ability** and *the tendency of an anion to get distorted or polarised by a cation is called its* **polarisability.**

Factors Affecting the Magnitude of Polarising Power of a Cation and Polarisability of an Anion–Fajans's Rules

Following are the factors on which the magnitude of polarising power of a cation to polarise a nearby anion and the polarisability of an anion depends. These factors have been suggested by Fajans and hence are called **Fajans's Rules.**

1. Charge on cation or anion. The cation with higher positive charge attracts the electron cloud of the anion more strongly towards itself than the cation with smaller positive charge and hence polarises the anion more strongly. Consequently *higher is the positive charge on the cation, greater is its polarising power to polarise a given nearby anion.* For example in LiCl, $BeCl_2$, BCl_3 and CCl_4 molecules, the polarising power of Li^+, Be^{2+}, B^{3+} and C^{4+} cations (*cations of the same period*) to polarise Cl^- anion *increases with the increase of positive charge on them* as shown below :

Molecules :	*LiCl*	*$BeCl_2$*	*BCl_3*	*CCl_4*
Cations (2nd period) :	*Li^+*	*Be^{2+}*	*B^{3+}*	*C^{4+}*
Positive charge on cations :	$+1$ <	$+2$ <	$+3$ <	$+4$
Polarising power of cations :	*Li^+* <	*Be^{2+}* <	*B^{3+}* <	*C^{4+}*

——— *Increasing* ———→

An anion with higher negative charge repells the electron cloud of its outer-most shell towards the cation more effectively and hence is polarised by the cation more strongly. Consequently *higher is the negative charge on the anion, more strongly it will be polarised by a given cation i.e. more will be its polariability.* For example in CH_4, NH_3, H_2O and HF molecules, the polarisability of C^{4-}, N^{3-}, O^{2-} and F^- anions (*anions of the same period*) to be polarised by H^+ cation decreases from C^{4-} to F^-, since the magnitude of negative charge on these anions also decreases in the same direction as shown below :

Molecules	:	CH_4	NH_3	H_2O	HF
Anions (2nd period)	:	C^{4-}	N^{3-}	O^{2-}	F^-
Negative charge on anions	:	–4 >	–3	> –2 >	–1
Polarisability of anions	:	C^{4-}>	N^{3-}	> O^2 >	F^-

——— *Decreasing* ———→

Thus in order to make a cation to polarise an anion *effectively,* both the positive charge on the cation and the negative charge on the anion should be *high.*

2. Size of the cation. The nucleus of a smaller cation is shielded to a lesser extent by the electron shells and, therefore, such a cation attracts the electron cloud of the anion more strongly than a larger cation. This means that *smaller is the size of the cation, higher is its polarising power to polarise a given nearby anion.* For example in $BeCl_2$, $MgCl_2$, $CaCl_2$, $SrCl_2$ and $BaCl_2$ molecules the polarising power of Be^{2+}, Mg^{2+}, Ca^{2+}, Sr^{2+} and Ba^{2+} cations (*cations of the same group*) to polarise Cl^- ion decreases with the increase of their size as shown below :

Molecules	:	$BeCl_2$	$MgCl_2$	$CaCl_2$	$SrCl_2$	$BaCl_2$
Cations (Group IIA)	:	Be^{2+}	Mg^{2+}	Ca^{2+}	Sr^{2+}	Ba^{2+}
Size of cations (Å)	:	0.31	< 0.65	< 0.99	< 1.13 <	1.35
Polarising power of cations	:	Be^{2+}	> Mg^{2+}	> Ca^{2+}	> Sr^{2+} >	Ba^{2+}

——— *Decreasing* ———→

3. Size of the anion. The electrons of the outer-most shell of a larger anion are less firmly bound to its nucleus and hence can be withdrawn by the cation more easily. Consequently a larger anion will be polarised by the cation more easily than a smaller anion. This means that *larger is the size of the anion, more strongly or easily it will be polarised by a given cation, i.e. more will be its polarisability.* For example in halides of a given cation (*e.g.* MX halides) the polarisability of halide ions (*anions of the same group*) will increase from F^- to I^-, since the size of these ions also increases in the same direction.

Halides	:	*MF*	*MCl*	*MBr*	*MI*
Halide ions (Group VII A)	:	F^-	Cl^-	Br^-	I^-
Size of halide ions (Å)	:	1.36	< 1.81	< 1.95 <	2.16
Polarisability of halide ions	:	F^-	< Cl^-	< Br^- <	I^-

——— *Increasing* ———→

4. Electronic configuration of the cation. Consider two different cations like Cu^+ and Na^+ which have nearly the same size (Cu^+ = 0.96 Å and Na^+ = 0.95 Å) and same charge (= +1) but have different valence-shell electronic configurations—Cu^+ has 18-electron valence-shell configuration ($Cu^+ \rightarrow 3s^2p^6d^{10}$) while Na^+ has 8-electron valence-shell configuration ($Na^+ \rightarrow 2s^2p^6$). Out of these cations, Cu^+ cation polarises the anions more strongly than Na^+ cation. (The anion to be polarised must be the same). This means that *a cation with 18-electron valence shell configuration ($ns^2p^6d^{10}$ configuration) has greater polarising power than that with 8-electron valence-shell configuration (ns^2p^6 configuration).* Greater polarising power of the cation with $ns^2p^6d^{10}$ valence-shell configuration is because of the fact that the *d*-electrons of the 18-electron shell shield the nuclear charge of the cation less effectively than the *s*- and *p*-electrons of 8-electron shell. Hence the cation with 18-electron valence shell configuration behaves, as if it has greater nuclear charge on it. Greater nuclear charge increases the polarising power of the cation having 18-electron valence-shell configuration.

On similar grounds it can be explained that Ag^+ cation ($4s^2p^6d^{10}$) has greater polarising power than K^+ cation ($3s^2p^6$).

Variation of Polarising Power of Cations in a Period and a Group of the Periodic Table (*Periodic Tends of Polarising Power of Cations*)

(*a*) **In a period.** In moving from left to right in a period, the size of the cations decreases and the positive charge increases. *Both these factors increase the polarising power of the cations from left to right in a period.* For example the polarising power of the cations of 2*nd* period *viz.* Li^+, Be^{2+}, B^{3+} and C^{4+} to polarise Cl^- anion in LiCl, $BeCl_2$, BCl_3 and CCl_4 *molecules increases from Li^+ to C^{4+} with the decrease of their size and increase in their positive charge from +1 to + 4 as shown below :*

Molecules	:	*LiCl*	*$BeCl_2$*	*BCl_3*	*CCl_4*
Cations (2nd period)	:	Li^+	Be^{2+}	B^{3+}	C^{4+}
Size of cations (Å)	:	*0.60*	*> 0.31*	*> 0.20 >*	*0.15*
Positive charges	:	*+ 1*	*< + 2*	*< + 3 <*	*+ 4*
Polarising power of cations	:	Li^+	$< Be^{2+}$	$< B^{3+} <$	C^{4+}

Increasing ⟶

Similarly the polarising power of Na^+, Mg^{2+} and Al^{3+} cations (cations of 3*rd* period) to polarise Cl^- anion in NaCl, $MgCl_2$ and $AlCl_3$ molecules increases from Na^+ to Al^{3+} with the decrease of their size and increase in their positive charge from + 1 to +3 as shown below :

Molecules	:	*NaCl*	*$MgCl_2$*	*$AlCl_3$*
Cations (3rd period)	:	Na^+	Mg^{2+}	Al^{3+}
Size of cations (Å)	:	*0.95 >*	*0.65 >*	*0.50*
Positive charges	:	*+ 1 <*	*+ 2 <*	*+ 3*
Polarising power of cations	:	$Na^+ <$	$Mg^{2+} <$	Al^{3+}

Increasing ⟶

(*b*) **In a group.** On descending a group the magnitude of positive charge on the cations remains the same and hence it is the size of the cation alone which affects the magnitude of the polarising power of the given cation. *On moving down a group the size of the cations increases and hence the polarising power of the cations goes on decreasing*. For example the polarising power of the cations of the elements of Group IIA *viz*. Be^{2+}, Mg^{2+}, Ca^{2+}, Sr^{2+} and Ba^{2+} to polarise Cl^- ion in $BeCl_2$, $MgCl_2$, $CaCl_2$, $SrCl_2$ and $BaCl_2$ molecules *decreases* with the *increase* of their size as shown below :

Molecules	:	$BeCl_2$	$MgCl_2$	$CaCl_2$	$SrCl_2$	$BaCl_2$
Cations (Group IIA)	:	Be^{2+}	Mg^{2+}	Ca^{2+}	Sr^{2+}	Ba^{2+}
Size of the cations (Å)	:	0.31	< 0.65 <	0.99 <	1.13 <	1.35
Polarising power of cations	:	Be^{2+}	> Mg^{2+} >	Ca^{2+} >	Sr^{2+} >	Ba^{2+}

Decreasing ⟶

The variation of polarising power of cations in the periodic table has been shown in Fig. 7.16.

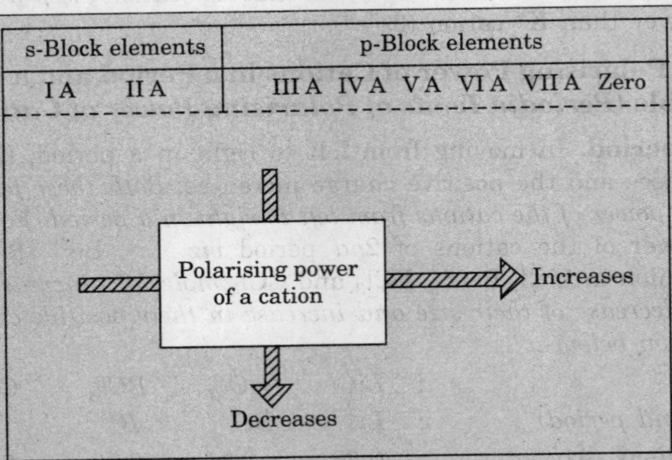

Fig. 7.16. Variation of polarising power of cations in the periodic table.

Variation of Polariability of Anions in a Period and a Group of the Periodic Table. (*Periodic Trends of Polarisability of Anions*)

(*a*) **In a period.** We know that on proceeding from left to right in a period, both negative charge on the anions and their size decrease. *Both these factors decreases the polarisability of the anions by a given cation from left to right in a period of the long form of periodic table.* For example the polarisability of C^{4-}, N^{3-}, O^{2-} and F^- anions (anions of 2nd period) to be polarised by M^+ cations in M_4C, M_3N, M_2O and MF molecules decreases from C^{4-} to F^-, since both negative charge on these anions and their size decrease in the same direction as shown below :

Molecules	:	M_4C	M_3N	M_2O	MF
Anions (2nd period)	:	C^{4-}	N^{3-}	O^{2-}	F^-
Negative charge on anions	:	– 4	> – 3	> – 2	> – 1
Size of anions (Å)	:	2.60	> 1.71	> 1.40	> 1.36
Polarisability of anions	:	C^{4-}	$< N^{3-}$	$< O^{2-}$	$< F^-$
		——— *Decreasing* ———→			

(*b*) **In a group.** On descending a group, the magnitude of negative charge on the anions remains the same and hence it is the size of the anion alone which affects the magnitude of polarisability of a given anion. *On moving down a group the size of the anions increases and hence the polarisability of the anions also goes on increasing.* For example in the halides of a given cation (*e.g.* MX type halides) the polarisability of halide ions (*anions of the same group*) will increase from F^- to I^-, since the size of these anions also increases in the same direction as shown below :

Halides	:	MF	MCl	MBr	MI
Halide ions (Group VII A)	:	F^-	Cl^-	Br^-	I^-
Size of halide ions (Å)	:	1.36 <	1.81 <	1.95 <	2.16
Charge on halide ions	:	– 1	– 1	– 1	– 1
Polarisability of halide ions	:	F^- <	Cl^- <	Br^- <	I^-
		——— *Increasing* ———→			

The variation of polarisability of anions by a given cation in the periodic table has been shown in Fig. 7.17.

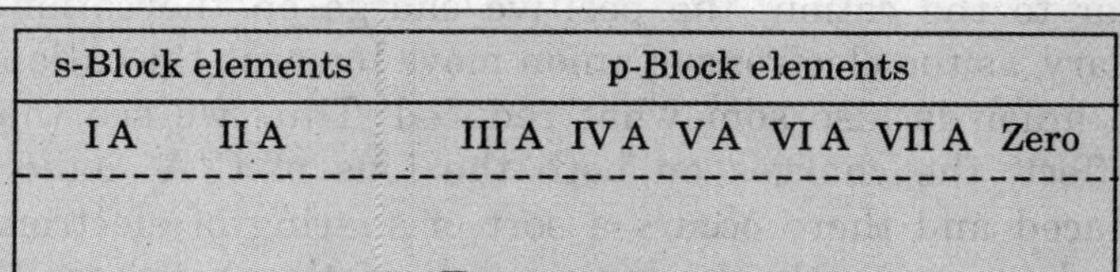

Fig. 7.17. Variation of polarisability of anions by a given cation in the periodic table.

Applications of Polarisation of Ions : Fajans's Rules

A. Degree of covalent character in ionic bond : Tendency of ions to form covalent bond : Fajans's rules. When the ions C^+ and A^- of an ionic molecule, C^+A^- whose both the ions are symmetrical [Fig. 7.18 (*a*)] come closer to each other, each of the ions gets distorted or polarised by the other ion. But the polarisation produced by C^+ cation in A^-anion is *appreciable* while that produced by A^- in C^+ cation is negligible, *i.e.* C^+ cation is distorted by A^- anion to a smaller extent while A^- anion is distorted by C^+ cation to a larger extent, Now C^+A^- molecule has distorted ions [Fig. 7.18 (*b*)].

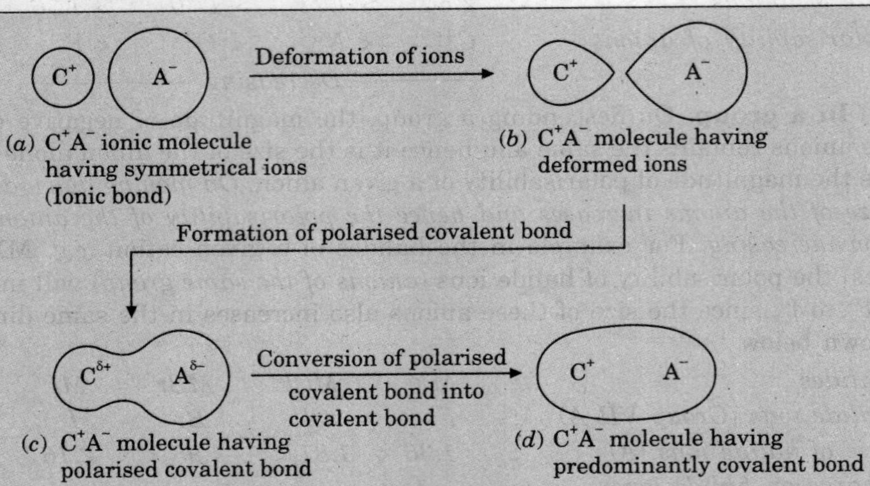

Fig. 7.18. Conversion of ionic bond into covalent bond as a result of polarisation.

In the polarisation process the electrons of the anion are withdrawn by the cation towards itself. As the electrons withdrawn by the cation from the anion approach closer to the cation, the positive charge on the cation is somewhat reduced. Similary, as the electrons of anion move towards the cation, the negative charge on the anion is also somewhat reduced. Thus we see that, due to the polarisation effect, the charges on both the ions of C^+A^- ionic molecule are somewhat reduced and there occurs a sort of sharing of electrons between C^+ and A^- ions and consequently the ionic bond existing between C^+ and A^- ions develops partial covalent character, *i.e.* at this stage C^+A^- molecule has partial ionic bond and partial covalent bond character or in other words C^+A^- molecule has *polarisd covalent bond* [Fig. 7.18 (*c*)]. Now, if the polarisation produced by C^+ in A^- is very high in magnitude, the polarised covalent bond is converted into *predominantly covalent bond* [Fig. 7.18 (*d*).

Thus we see that the process of polarisation produces some amount of covalent character in the ionic bond. The magnitude of covalent character produced due to polarisation effect depends on the amount of polarisation produced in the anion by the cation (or on the magnitude of polarising power of the cation) or polarisability of the anion. *Greater the magnitude of polarising power of a cation or polarisability of an anion, greater the amount of covalent character produced in the ionic bond.* Consequently the conditions which bring about the maximum polarisation of anions by given cations are, in fact, the conditions to make the bond covalent. These conditions are called **Fajans's rules** and are given below :

1. Charge either on cation or on anion should be large. Higher the positive charge on the cation, greater its polarising power to polarise a nearby anion. The increase in polarising power of the cation increases the covalent character of the bond. Similarly higher the negative charge on the anion, higher will be its polarisability. The increase in polarisability is accompanied by an increase in the covalent character of the bond.

Examples. In order to understand how the large charge either on cation or on anion increases the covalent character of the bond, the following examples may be considered.

(*a*) The covalent character of $Li^+–Cl^-$, $Be^{2+}–Cl^-$, $B^{3+}–Cl^-$ and $C^{4+}–Cl^-$ bonds in $LiCl$, $BeCl_2$, BCl_3 and CCl_4 molecules increases in the following order :

$$Li^+–Cl^- < Be^{2+}–Cl^- < B^{3+}–Cl < C^{4+}–Cl^-,$$

since the polarising power (or positive charge) of Li^+, Be^{2+}, B^{3+} and C^{4+} cations (cations of the same period) to polarise Cl^- anion also increases in the same direction as shown below : (charges are given in parentheses)

$$Li^+(+1) < Be^{+2}(+2) < B^{+3}(+3) < C^{+4}\ (+4)$$

(*b*) Similarly the covalent character of $Na^+–Cl^-$, $Mg^{+2}–Cl^-$ and $Al^{+3}–Cl^-$ bonds in $NaCl$, $MgCl_2$ and $AlCl_3$ molecules increases in the following order :

$$Na^+–Cl^- < Mg^{+2}–Cl^- < Al^{+3}–Cl,$$

since the polarising power (or positive charge) of Na^+, Mg^{+2} and Al^{+3} cations (cations of the same period) to polarise Cl^- anion also increases in the same direction as shown below :

$$Na^+(+1) < Mg^{+2}(+2) < Al^{+3}(+3)$$

(*c*) The covalent character (*i*) of $Ge^{2+}–X^-$ and $Ge^{4+}–X^-$ bonds in GeX_2 and GeX_4 molecules (*ii*) of $Sn^{2+}–X^-$ and $Sn^{4+}–X^-$ bonds in SnX_2 and SnX_4 molecules, and (*iii*) of $Pb^{2+}–X^-$ and $Pb^{4+}–X^-$ bonds in PbX_2 and PbX_4 molecules increases in the order :

$$Ge^{2+}–X^- < Ge^{4+}–X^-$$

$$Sn^{2+}–X^- < Sn^{4+}–X^-$$

$$Pb^{2+}–X^- < Pb^{4+}–X^-,$$

since the polarising power (or the positive charge) of different cations (*same group*) to polarise X^- anion also increases in the same direction as shown below (positive charges carried by the cations are given in parentheses) :

$$Ge^{2+}(+2) < Ge^{4+}(+4)$$

$$Sn^{2+}(+2) < Sn^{4+}(+4)$$

$$Pb^{2+}(+2) < Pb^{4+}(+4)$$

Thus the halides of Ge^{2+}, Sn^{2+} and Pb^{2+} are largely ionic while those of Ge^{4+}, Sn^{4+} and Pb^{4+} are largely covalent. Note that all the cations are of the same group (Group IVA).

In general we can say that *if a metal forms halides in different oxidation states, the halides having the metal ion in lower oxidation state are largely ionic while those having the metal in higher oxidation state are largely covalent.*

(*d*) The covalent character of $H^+–C^{4-}$, $H^+–N^{3-}$, $H^+–O^{2-}$ and $H^+–F^-$ bonds in CH_4, NH_3, H_2O and HF molecules decreases in the order :

$$H^+–C^{4–} > H^+–N^{3–} > H^+–O^{2–} > H^+–F^–,$$

since the polarisability (or negative charge) of $C^{4–}$, $N^{3–}$, $O^{2–}$ and $F^–$ anions (*same period*) to be polarised by H^+ cation also decreases in the same direction as shown below (negative charges carried by the anions are given in parentheses):

$$C^{4–}(–4) > N^{3–}(–3) > O^{2–}(–2) > F^–(–1)$$

2. Size of the cation should be small. Smaller the size of the cation, greater its polarising power and hence greater its tendency to form covalent compounds.

Example. The covalent character of $Be^{2+}–Cl^–$, $Mg^{2+}–Cl^–$, $Ca^{2+}–Cl^–$, $Sr^{2+}–Cl^–$ and $Ba^{2+}–Cl^–$ bonds in $BeCl_2$, $MgCl_2$, $CaCl_2$, $SrCl_2$ and $BaCl_2$ molecules *decreases* in the following order :

$$Be^{2+}–Cl^– > Mg^{2+}–Cl^– > Ca^{2+}–Cl^– > Sr^{2+}–Cl^– > Ba^{2+}–Cl^–,$$

since the polarising power (or size) of Be^{2+}, Mg^{2+}, $Ca^{2–}$, Sr^{2+} and Ba^{2+} cations (*cations of the same group*) to polarise $Cl^–$ anion *increases* in the same direction as shown below :

$$Be^{2+}(0.31 \text{ Å}) < Mg^{2+}(0.65 \text{ Å}) < Ca^{2+}(0.99 \text{ Å}) < Sr^{2+}(1.13 \text{ Å}) < Ba^{2+}(1.35 \text{ Å})$$

Due to maximum polarising power (or smallest size) of Be^{2+} ion, it forms largely covalent bond with $Cl^–$ ion or with any other anion while other cations of IIA group form largely ionic bonds, *i.e.* *$BeCl_2$ is largely covalent while other chlorides are largely ionic.*

3. Size of the anion should be large. Larger the size of the anion, more strongly or easily it will be polarised by a given cation, *i.e.* more will be its polarisability and hence greater will be the tendency of the anion to form covalent compounds.

Examples (*a*) The covalent character of $Ca^{2+}–F^–$, $Ca^{2+}–Cl^–$, $Ca^{2+}–Br^–$ and $Ca^{2+}–I^–$ bonds in CaF_2, $CaCl_2$, $CaBr_2$ and CaI_2 molecules respectively increases in the following order :

$$Ca^{2+}–F^– < Ca^{2+}–Cl^– < Ca^{2+}–Br^– < Ca^{2+}–I^–,$$

since the polarisability or size of $F^–$, $Cl^–$, $Br^–$ and I anions (*anions of the same group*) to be polarised by Ca^{2+} cation also *increases* in the same direction as shown below :

$$F^– (1.36 \text{ Å}) < Cl^– (1.81 \text{ Å}) < Br^– (1.95 \text{ Å}) < I^– (2.16 \text{ Å})$$

$F^–$ anion has so small polarisability or size that CaF_2 is *largely ionic* in nature while other halides of Ca^{2+} are largely *covalent in character*.

(*b*) Similarly the covalent character of $Na^+–F^–$, $Na^+–Cl^–$, $Na^+–Br^–$ and $Na^+–I^–$ bonds in NaF, NaCl, NaBr and NaI molecules respectively increases in the following order :

$$Na^+–F^– < Na^-–Cl^– < Na^+–Br^– < Na^+–I^–$$

(*c*) On similar grounds it can be shown that the covalent character of halogen acids *viz.* HF, HCl, HBr and HI containing the anions of the same group is in the order :

$$HF < HCl < HBr < HI$$

(*d*) Due to the increase in polarisation of X^- anion by Al^{3+} cation from F^- to I^- or due to the increase of the size of the anions from F^- to I^-, the covalent character of the trihalides of Al is in the order :

$$AlF_3 < AlCl_3 < AlBr_3 < AlI_3$$

Thus AlI_3 and $AlBr_3$ have largely covalent character, $AlCl_3$ has intermediate character while AlF_3 has largely ionic character.

4. Cation should have 18 electron valence-shell configuration ($ns^2p^6d^{10}$ configuration). The cation with $ns^2p^6d^{10}$ configuration has greater polarising power than the cation with 8-electron valence-shell configuration (ns^2p^6 configuration) and hence has greater tendency to form covalent compound.

Examples (*a*) Cu^+–Cl^- bond is CuCl is more covalent than Na^+–Cl^- bond in NaCl, since Cu^+ has $3s^2p^6d^{10}$ configuration while Na^+ has $2s^2p^6$ configuration. Due to 18-electron valence-shell configuration, Cu^+ ion has greater polarising power to polarise Cl^- ion than Na^+ ion having 8-electron valence-shell configuration.

(*b*) Due to the 18-electron valence-shell configuration ($4s^2p^6d^{10}$), the polarising power of Ag^+ ion to polarise I^- ion in AgI molecule is so great that it makes AgI *covalent* while, due to 8-electron configuration ($2s^2p^6$), the polarising power of Na^+ ion to polarise I^- ion in NaI molecule is so small that it makes NaI *ionic*. (See Fig. 7.19).

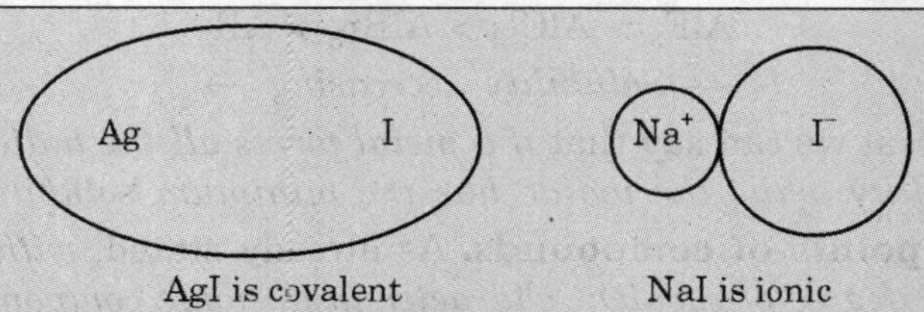

Fig. 7.19. Due to strong polarising power of Ag^+ ion to polarise I^- anion, AgI is covalent, while, due to weak polarising power of Na^+ ion to polarise the same anion, NaI is ionic.

B. Solubility of compounds in non-polar (organic) and polar solvents. *As already stated, with the increase of polarisation, the degree of covalent character in an ionic molecule increases. This increase in covalent character is accompanied by an increase of the solubility of the compound in non-polar (organic) solvents like alcohol, but its solubility in polar solvents like H_2O decreases with the increase in covalent character.* For example :

(*i*) Since the polarisability of S^{2-} anion to be polarised by a metallic cation, say, M^+ is greater than that of O^{2-} anion, M^+–S^{2-} bond has more covalent character than M^+–O^{2-} bond in M_2S and M_2O molecules respectively. With the increase of covalent character in M_2S, M_2S is *more soluble* than M_2O in non-polar (organic) solvents like alcohol etc. but *less so* in polar solvents like H_2O. *Thus in general sulphides are less soluble in water than the oxides of the same metal.*

(*ii*) Li^+ cation has the maximum polarising power of all the alkali metal cations and hence gives *largely covalent compounds* while other alkali metal cations give *largely ionic compounds*. With the increase of covalent character in Li^+ compounds these are readily soluble in non-polar (organic) solvents like alcohol but insoluble in polar solvents like H_2O, while the compounds of other alkali metal cations are insoluble in organic solvents but readily soluble in polar solvents like H_2O.

(*iii*) On similar grounds, it can explained why the compounds of Be^{2+} are insoluble in H_2O while the compounds of other alkaline earth metal cations are readily soluble in H_2O.

(*iv*) We know that CuX, AgX, and AuX halides (X = Cl, Br and I) are much more covalent in nature than the halides of alkali metal cations having similar size and charge as Cu^+, Ag^+ and Au^+ ions. Due to the increased covalent character, CuX, AgX and AuX are insoluble in water while the alkali metal halides are freely soluble.

(*v*) We have already seen that the covalent character of the halides of Al^{3+} increases from AlF_3 to AlI_3.

$$AlF_3 < AlCl_3 < AlBr_3 < AlI_3$$

— *Covalent character increasing* →

With the increase of covalent character from AlF_3 to AlI_3, solubility of these halides in polar solvents like H_2O decreases in the same direction as shown below :

$$AlF_3 > AlCl_3 > AlBr_3 > AlI_3$$

— *Solubility decreasing* →

Thus in general we can say that *if a metal forms all the halides, fluoride has maximum solubility while the iodide has the minimum solubility.*

C. Melting points of compounds. As already stated, *with the increase of polarisation, the degree of covalent character in an ionic compond also increases and this increased covalent character decreases the melting point of the compound.* For example :

(*i*) The increase in covalent character in $Na^+–X^-$ bonds of NaX halides (X = F, Cl, Br and I) from NaF to NaI decreases the melting points of these halides in the same direction as shown below :

Halides	:	*NaF*		*NaCl*		*NaBr*		*NaI*
Covalent character of $Na^+ - X^-$ bond	:	$Na^+ - F^-$	<	$Na^+ - Cl^-$	<	$Na^+ - Br^-$	<	$Na^+ - I^-$
Melting point (°C)	:	988	>	801	>	755	>	651

Similarly the increase in covalent character in $Ca^{2+}–X^-$ bonds of CaX_2 halides (X = F, Cl, Br and I) from CaF_2 to CaI_2 decreases the melting points of these halides in the same direction as shown below :

Halides	:	CaF_2		$CaCl_2$		$CaBr_2$		CaI_2
Covalent character of $Ca^{2+} - X^-$ bond	:	$Ca^{2+} - X^-$	<	$Ca^{2+} - Cl^-$	<	$Ca^{2+} - Br^-$	<	$Ca^{2+} - I^-$
Melting point (°C)	:	1392	>	772	>	730	>	575

(*ii*) The decrease in covalent character in M^{2+}–Cl^- bond (M^{2+} = Be^{2+}, Mg^2, Ca^{2+}, Sr^{2+}, Ba^{2+}) of MCl_2 chlorides from $BeCl_2$ to $BaCl_2$ increases the melting points of these chlorides in the same direction.

Chlorides :	$BeCl_2$	$MgCl_2$	$CaCl_2$	$SrCl_2$	$BrCl_2$
Covalent character of $M^{2+}-Cl^-$ bond :	$Be^{2+}-Cl^-$ >	$Mg^{2+}-Cl^-$ >	$Ca^{2+}-Cl^-$ >	$Sr^{2+}-Cl^-$ >	$Ba^{2+}-Cl^-$
Melting point (°C) :	405 <	712 <	772 <	872 <	960

(*iii*) We have seen that in NaCl, $MgCl_2$ and $AlCl_3$ molecules the covalent character of the bond increases from Na^+–Cl^- to Mg^{2+}–Cl^- to Al^{3+}–Cl^-. This successive increase in the covalent character of the bond decreases the melting points of these chlorides in the same direction as shown below :

Chlorides :	*NaCl*	$MgCl_2$	$AlCl_3$
Covalent character of M – Cl (M = Na, Mg, Al bond :	Na^+-Cl^- <	$Mg^{2+}-Cl^-$ <	$Al^{3+}-Cl^-$
Melting point (°C) :	800 >	712 >	*Sublimes*

(*iv*) We know that since Sn^{4+}–Cl^- bond in $SnCl_4$ has more covalent character than Sn^{2+}–Cl^- bond in $SnCl_2$, $SnCl_4$ has less melting point than $SnCl_2$, ($SnCl_4$ = –33°C, $SnCl_2$ = 246°C).

Similarly $PbCl_4$ which has greater covalent character than $PbCl_2$ has less melting point ($PbCl_4$ = –15°C, $PbCl_2$ = 501°C). $TlCl_3$ has more covalent character than TlCl and hence has less melting point ($TiCl_3$ = 25°C, TlCl = 430°C).

(*v*) Due to the presence of Cu^+ cation with $ns^2p^6d^{10}$ configuration, CuCl has more covalent character than NaCl which has Na^+ cation with ns^2p^6 configuration. CuCl has, therefore, less melting point than NaCl (CuCl = 422°C, NaCl = 801°). Other such examples with their melting points are :

AgCl ($Ag^+ \rightarrow 4s^2p^6d^{10}$) = 455°C | AuCl ($Au^+ \rightarrow 5s^2p^6d^{10}$) = 170°C

KCl ($K^+ \rightarrow 3s^2p^6$) = 776°C | RbCl ($Rb^+ \rightarrow 4s^2p^6$) = 715°C

CdCl ($Cd^{2+} \rightarrow 4s^2p^6d^{10}$) = 568°C

CaCl ($Ca^{2+} \rightarrow 3s^2p^6$) = 772°C

D. Cause of diagonal relationship. The cause of diagonal relationship can also be explained on the basis of the concept of polarisation of ions as follows :

We have seen that on moving from left to right in a period the positive charge on cations increases and their size decreases. Both these factors increase the polarising power of the cations in a period. However, on moving down a group, the positive charge on the cations remains the same but the size of the cations increases. The increase in the size decrease the polarising power of the cations. On moving diagonally in the periodic table, these two effects partly cancel each other and a good similarity in properties of the two members of a diagonal pair is observed.

Summary of Fajans's Rules

Whatever has been studied so for reveals that Fajans's rules can be summarised by saying that the conditions which bring about maximum polarisation of the anion, A^- by the cation, C^+ in C^+A^- molecule or which create covalent character in $C^+–A^-$ ionic bond in C^+A^- molecule are :

1. The charge either on cation or on anion should be large. The following examples illustrate this point.

(*a*) The covalent character of LiCl, $BeCl_2$, BCl_3 and CCl_4 molecules containing the cations of the same period (2*nd* period) is in the order :

$$LiCl < BeCl_2 < BCl_3 < CCl_4$$

(*b*) The covalent character of $GeCl_2$, $GeCl_4$; $SnCl_2$, $SnCl_4$ and $PbCl_2$, $PbCl_4$ pairs is in the order :

$$GeCl_4 > GeCl_2$$
$$SnCl_4 > SnCl_2$$
$$PbCl_4 > PbCl_2$$

(*c*) The covalent character of CH_4, NH_3, H_2O and HF molecules containing the anions of the same period (2*nd* period) is in the order :

$$CH_4 > NH_3 > H_2O > HF$$

2. The size of the cation should be small. The covalent character of $BeCl_2$, $MgCl_2$, $CaCl_2$, $SrCl_2$ and $BaCl_2$ molecules containing the cations of the same group (IIA group) is in the order :

$$BeCl_2 > MgCl_2 > CaCl_2 > SrCl_2 > BaCl_2$$

3. The size of the anion should be large. The following examples may be studied.

(*a*) The covalent character of CaF_2, $CaCl_2$, $CaBr_2$ and CaI_2 containing the anions of the same group (VIIA group) is in the order :

$$CaF_2 < CaCl_2 < CaBr_2 < CaI_2$$

(*b*) The covalent character of halogen acids namely HF, HCl, HBr and HI containing the anions of the same group (VIIA group) is in the order :

$$HF < HCl < HBr < HI$$

4. The cation should have 18-electron valence-shell configuration. The following examples are worth noting.

(*a*) CuCl is more covalent than NaCl.

(*b*) AgI is covalent while NaI is ionic.

Variation of the Tendency of Cations and Anions to Form Covalent Compounds in the Periodic Table

The tendency of a given cation to form covalent compounds depends on the magnitude of its polarising power. Greater is its polarising power, greater is its tendency to give covalent molecule. Further, we have already seen that the polarising power of a cation or a given element increases when we proceed from left to right in a period of the long form of periodic table while the same decreases on descending a group. *Thus with the increase of the polarising power of a cation from left to right in a period, its tendency to form covalent compounds also increases in the same direction in a period. Similarly with the decrease of the polarising power of a cation from top to bottom in a group, its tendency to form covalent compounds also decreases in the same direction in a group.* This variation has been shown pictorially in Fig. 7.20.

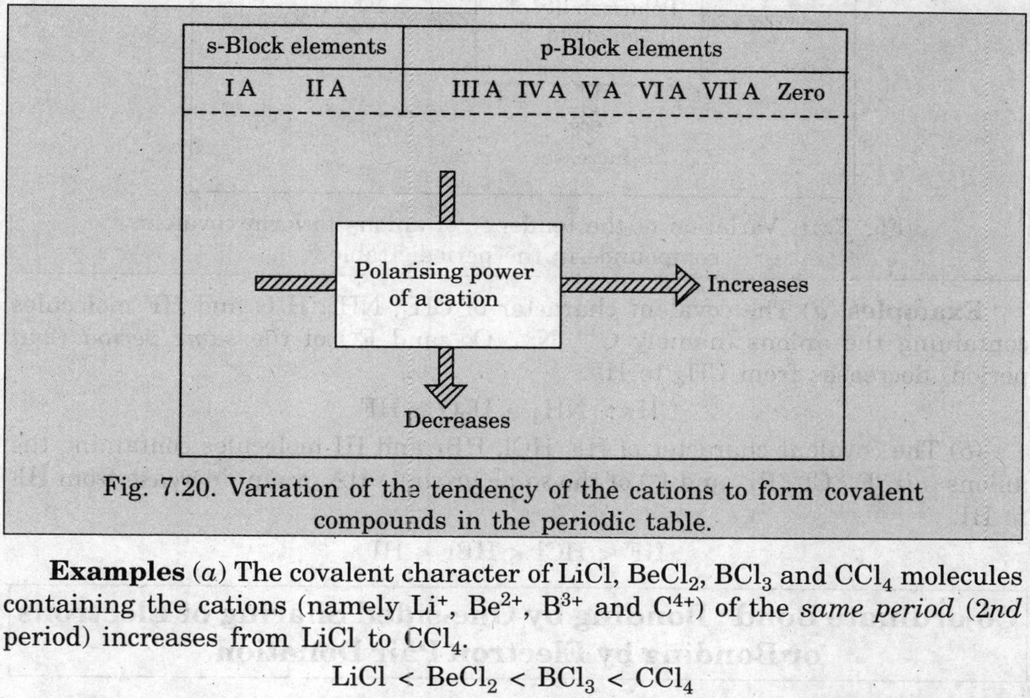

Fig. 7.20. Variation of the tendency of the cations to form covalent compounds in the periodic table.

Examples (*a*) The covalent character of $LiCl$, $BeCl_2$, BCl_3 and CCl_4 molecules containing the cations (namely Li^+, Be^{2+}, B^{3+} and C^{4+}) of the *same period* (2nd period) increases from $LiCl$ to CCl_4.

$$LiCl < BeCl_2 < BCl_3 < CCl_4$$

(*b*) The covalent character of $BeCl_2$, $MgCl_2$, $CaCl_2$, $SrCl_2$ and $BaCl_2$ molecules containing the cations (namely Be^{2+}, Mg^{2+}, Ca^{2+}, Sr^{2+} and Ba^{2+}) of the *same group* (IIA group) decreases from $BeCl_2$ to $BaCl_2$.

$$BeCl_2 > MgCl_2 > CaCl_2 > SrCl_2 > BaCl_2$$

The tendency of an anion to form covalent compounds depends on the magnitude of its tendency to be polarised by a given cation, *i.e.* on its polarisability. Greater is its polarisability, greater is its tendency to give covalent molecule. Further, we have already seen that the polarisability of an anion of a given element decreases when we proceed from left to right in a period of the long from of periodic table while the same increases on descending a group. *Thus with the decrease of the polarisability of an anion from left to right in a period, its tendency to form covalent compounds also decreases in the same direction in a period. Similarly with the increase of the polarisability of an anion from top to bottom in a group, its tendency to form covalent compounds also increases in the same direction in a group.* This variation has been shown pictorially in Fig. 7.21.

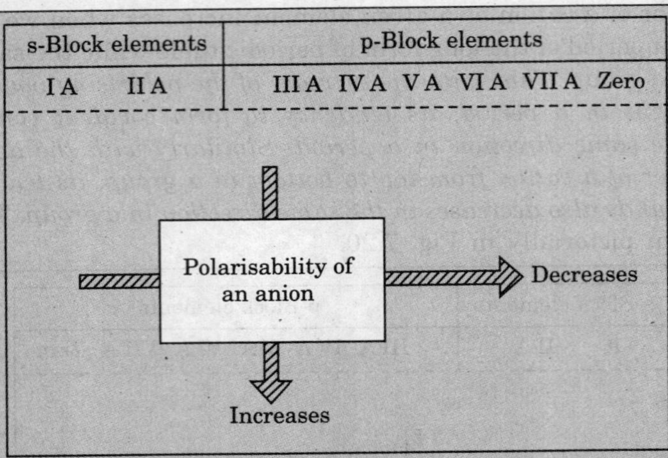

Fig. 7.21. Variation of the tendency of anions to form covalent compounds in the periodic table.

Examples (*a*) The covalent character of CH_4, NH_3, H_2O and HF molecules containing the anions (namely C^{4-}, N^{3-}, O^{2-} and F^-) of the *same period* (2nd period) decreases from CH_4 to HF.

$$CH_4 > NH_3 > H_2O > HF$$

(*b*) The covalent character of HF, HCl, HBr and HI molecules containing the anions (*viz.* F^-, Cl^-, Br^- and I^-) of the *same group* (VIIA group) increase from HF to HI.

$$HF < HCl < HBr < HI$$

Co-ordinate Bond : Bonding by One-sided Sharing of Electrons or Bonding by Electron Pair Donation

What is Co-ordinate Bond ?

We have seen that in the formation of a normal covalent bond each of the two bonded atoms contributes one electron for the shared pair of electrons which form a covalent bond between the two atoms. However in some compounds the formation of a covalent bond between two atoms takes place by the mutual sharing of two electrons both of which are provided entirely by one of the bonded atoms. This bond is called coordinate bond. Thus :

A covalent bond which is formed by the mutual sharing of two electrons both of which are provided entirely by one of the linked atoms (or ions) is called a co-ordinate bond.

Co-ordinate bond is also sometimes referred to as *co-ordinate covalent bond or dative bond*. The pair of shared electrons is called *lone pair*. The atom which furnishes the electron pair is called *donor* or *ligand* while the other atom which accepts the electron pair is called *acceptor*. A coordinate bond is represented by an arrow which points away from the donor to the acceptor.

Illustration of the Formation of a Coordinate Bond.

The formation of a co-ordinate bond between two atoms, say A and B, can be illustrated as :

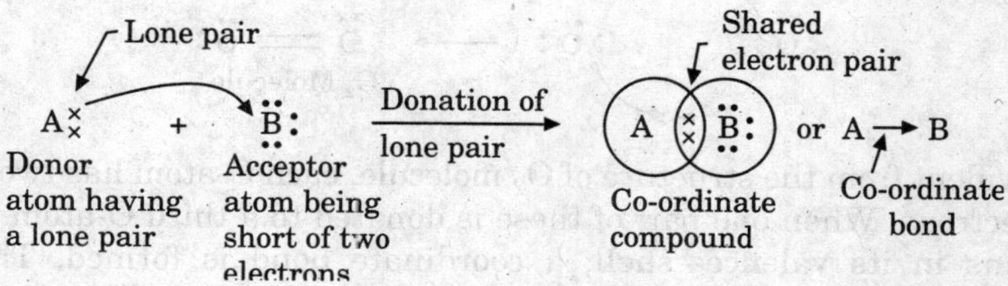

The donor atom A has a spare lone pair of electrons on it while the acceptor atom B is short of two electrons than the octet in its valence-shell. A donates its lone pair to B which accepts it. Thus the two electrons of the lone pair which originally belonged to A atom are now shared by both the atoms and this mutual sharing of electron pair results in the formation of a co-ordinate bond between A and B (A $\rightarrow$ B). Although the arrow head indicates the origin of electrons, once a co-ordinate bond has been established, it becomes quite identical with a normal covalent bond. Both are established by a shared electron pair and have similar properties.

Conditions for the Formation of a Co-ordinate Bond.

Conditions which are necessary for the formation of a coordinate bond are :

(*i*) The atom acting as a donor should have a lone pair of electrons.

(*ii*) The atom acting as an acceptor should have a vacant orbital to accept the electron pair donated by the donor.

Writing the Electronic Formulae (Lewis Formulae) of the Compounds or Ions Having Co-ordinate Bonds

The compounds which contain coordinate bonds are called co-ordinate compounds. Some of them are given below :

1. H_2O_2 molecule. This molecule can be regarded as being formed by the combination of H_2O molecule and oxygen atom. Oxygen atom of H_2O molecule has two lone pairs of electrons on it.

In the formation of H_2O_2 molecule by the combination of H_2O molecule and oxygen atom, one of the two lone pairs on oxygen atom of H_2O molecule is donated to the new oxygen atom and thus a coordinate bond is established between oxygen atom of H_2O molecule and the new oxygen atom.

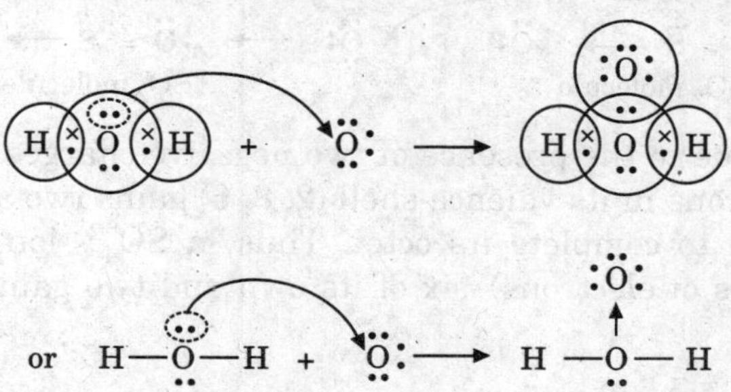

2. O_3 molecule. We know that O_2 molecule is made up of two O-atoms which are joined together by covalent bonds.

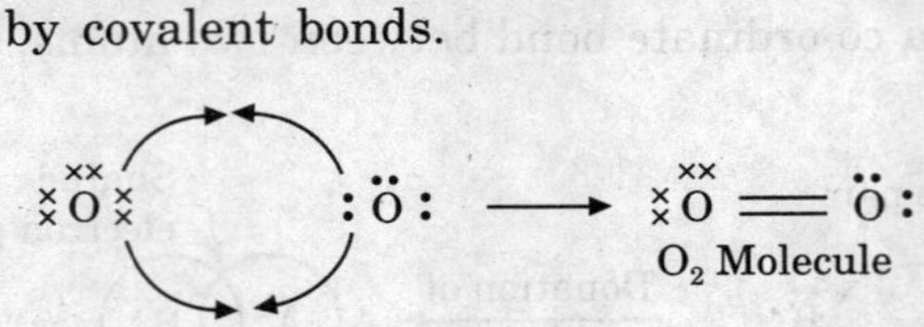

As is evident from the structure of O_2 molecule, each O-atom has two unshared pairs of electrons. When one pair of these is donated to a third O-atom which has six electrons in its valence- shell, a coordinate bond is formed. Thus Lewis structure of O_3 may be shown as :

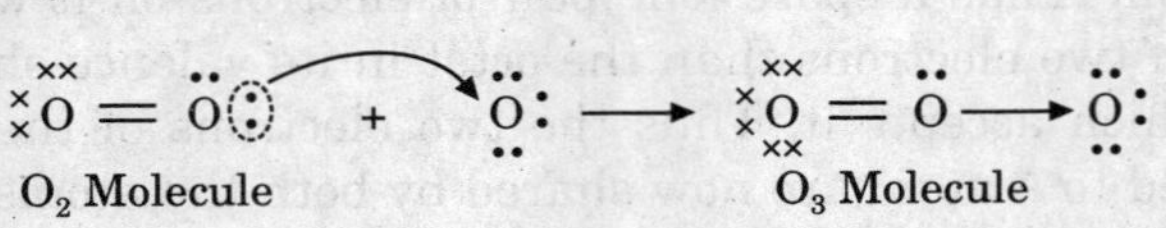

3. Fluoroborate ion, BF_4^-. It is formed when BF_3 molecule shares a pair of electrons supplied by F^- ion.

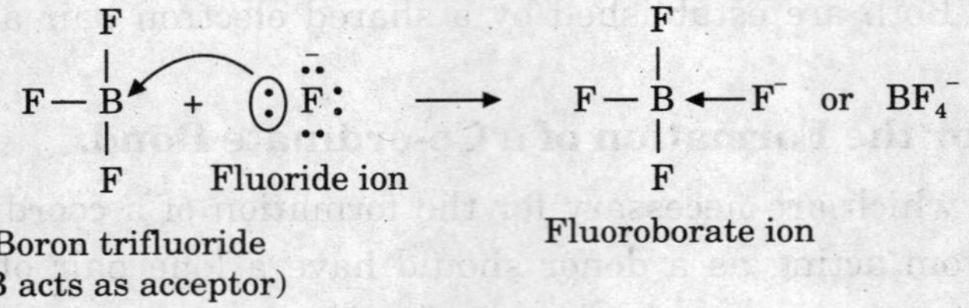

4. SO_2 and SO_3 molecules. Sulphur achieves its octet by forming two covalent bonds with one O-atom, giving SO species. The S-atom in SO has two lone pairs, one of which is shared with a second O-atom to form SO_2.

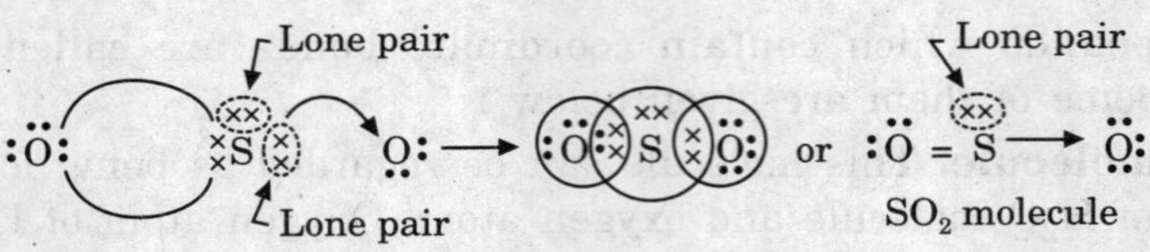

S-atom in SO_2 still has one lone pair which it donates to a third O-atom forming SO_3 molecule.

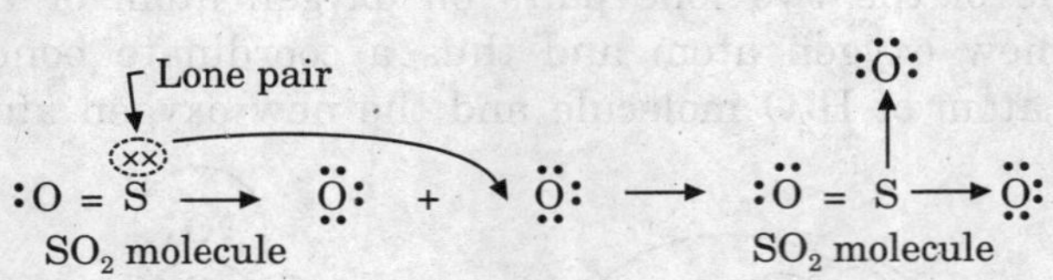

5. SO_4^{2-} ion. Due to the presence of two negative charges on the ion, S-atom which has six electrons in its valence-shell (2, 8, 6) gains two additional electrons from the metal ion to complete its octet. Thus in SO_4^{2-} ion, S-atom has eight electrons (four pairs of electrons)–six of its own and two gained from metal ion.

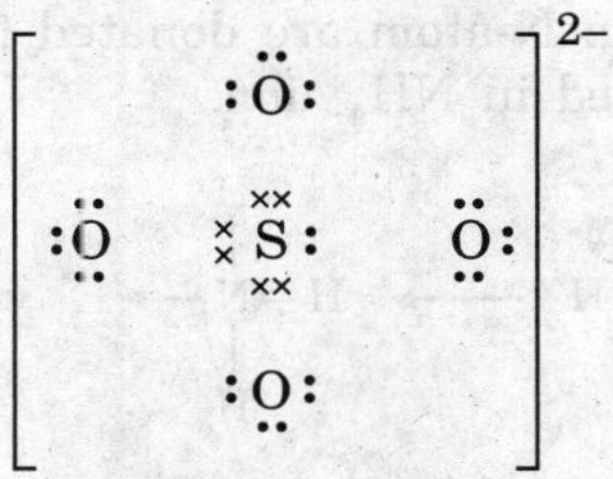

Now the four electron pairs round S-atom in SO_4^{2-} ion are donated to four O-atoms each of which thus achieves its octet. Thus we see that S-atom is linked with four O-atoms by four co-ordinate bonds.

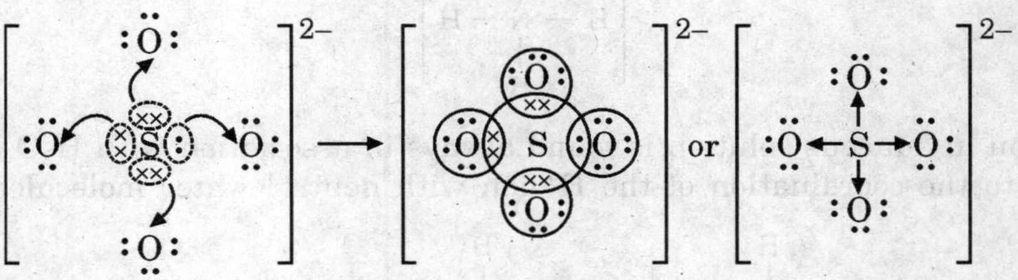

Here it may be noted that Lewis structure of H_2SO_4 is different from that of SO_4^{2-} ion as shown below.

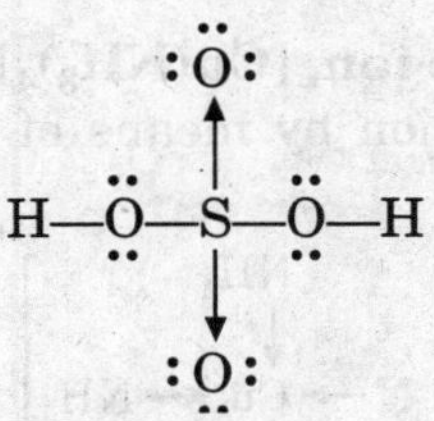

Thus we see that in H_2SO_4 molecule S-atom is linked with two O-atoms by two covalent bonds and with two O-atoms by two co-ordinate bonds.

6. CO molecule. C-atom has four valence-electrons (2, 4) while O-atom has six (2, 6). By forming two covalent bonds between them, O-atom achieves octet but C-atom has only six electrons. Therefore O-atom donates an unshared electron pair to C-atom and a coordinate covalent bond is established between the two atoms. Thus CO has two covalent bonds and one coordinate bond (O→C bond).

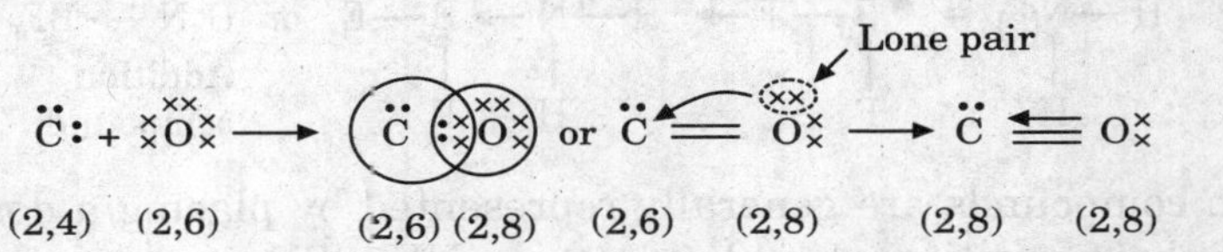

7. Ammonium (NH_4^+) and hydroxonium (H_3O^+) ions. NH_4^+ ion may be regarded as being obtained by the combination of NH_3 molecule and H^+ ion. In NH_3 molecule each of the three H-atoms is linked to N-atom by a covalent bond. Thus N-atom in this molecule is left with a lone pair of electrons after completing its octet by sharing three of its five valence-shell electrons with three H-atoms.

The electrons of lone pair on N-atom are donated to H^+ ion and thus a N→H coordinate bond is established in NH_4^+ ion.

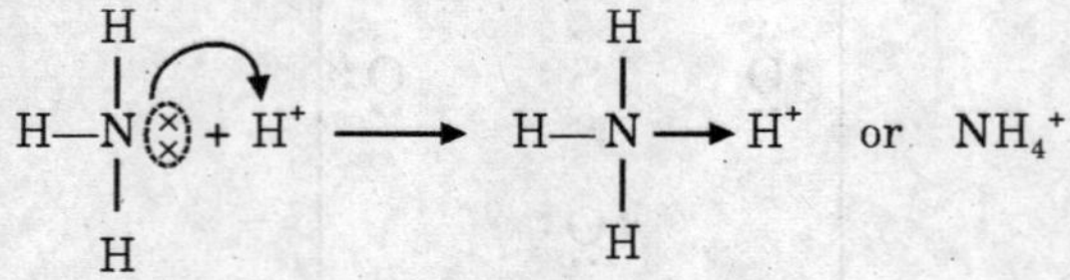

However, after the formation of NH_4^+ ion all the four N–H bonds become identical and hence NH_4^+ ion is represented as shown in the margin.

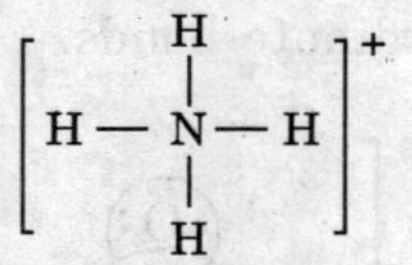

H^+ ion in aqueous solution is found always in association with H_2O as H_3O^+ ion due to the coordination of the H^+ ion with neutral water molecule.

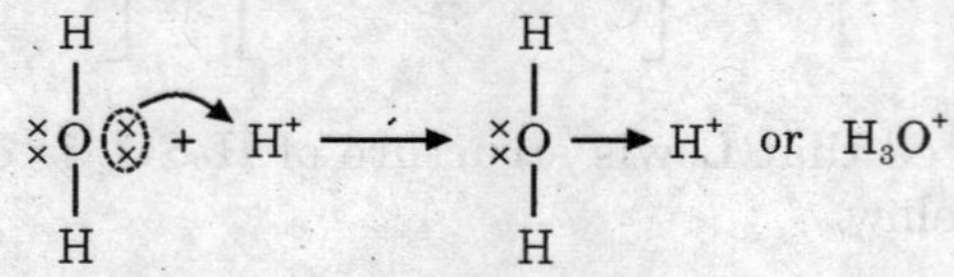

8. Copper (II) tetrammine ion, $[Cu(NH_3)_4]^{2+}$. This ion is formed by linking four NH_3 molecules with Cu^{2+} ion by means of four coordinate bonds.

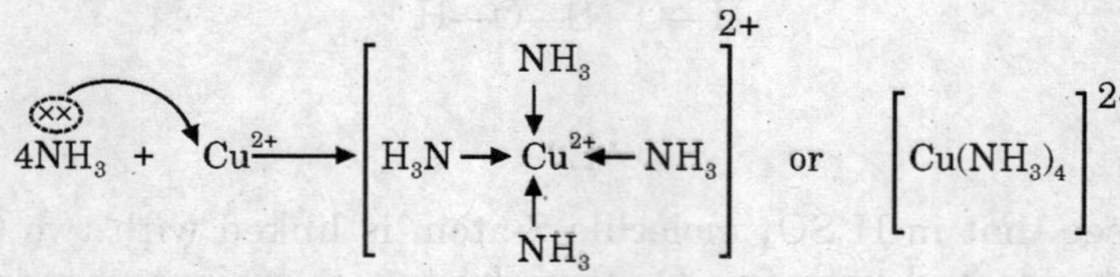

9. Addition compounds. Coordinate bond is also formed between two atoms both of which are the parts of a stable molecule. For example this type of bond is found in boron *fluoride–ammonia complex,* $H_3N \rightarrow BF_3$.

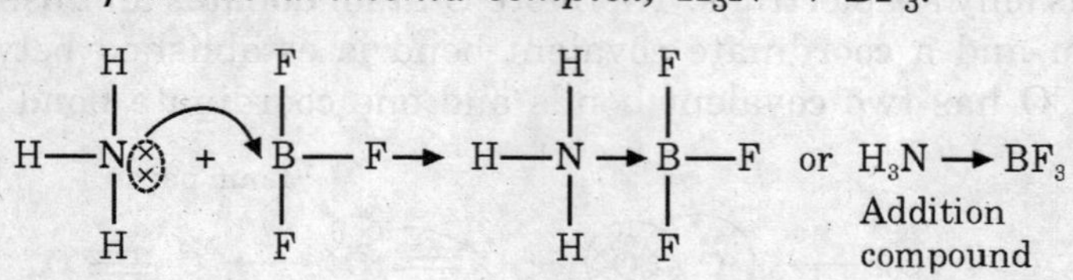

Addition compounds are generally represented by placing a dot between the formulae of their constituent molecules, *e.g.* NH_3. BF_3.

10. Aluminium chloride, Al_2Cl_6. Al atom has three valence-electrons which it shares with three Cl-atoms, forming three covalent bonds. Thus in $AlCl_3$, Al atom acquires six electrons in its outershell. Now each of the three Cl-atoms in $AlCl_3$ molecule has three lone pairs. One of these lone pairs is donated to Al atom of another $AlCl_3$ molecule. Thus both Al atoms achieve an octet and a stable Al_2Cl_6 molecule results.

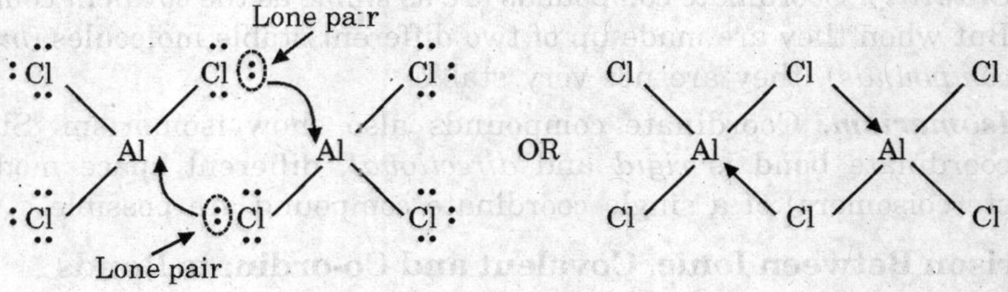

How is a Co-ordinate Bond Formed?

The formation of a coordinate bond (A → B) between two atoms *viz.* A and B may be regarded to occur in the following two steps :

1st step **:** In this step the donor atom A transfers one electron of its lone pair to the acceptor atom B. This results in that atom A develops *unit positive charge* (+) and atom B develops a *unit negative charge* (–). This charge is known as *formal charge*. This step is similar to the formation of an ionic bond.

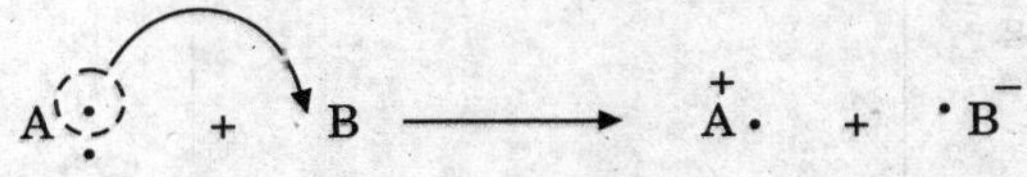

2nd step **:** In this step the two electrons, one each with A^+ and B^- are shared by both the ions. This step is similar to the formation of a covalent bond.

$$A\bullet^{+} + \bullet B^{-} \rightarrow A : B \quad or \quad A \rightarrow B$$

Thus we see that a co-ordinate bond is equivalent to a combination of an electrovalent bond (polar bond) and a covalent bond (non-polar bond). It is for this reason that a coordinate bond is also sometimes called a *semi-polar bond.*

Properties of Co-ordinate Compounds.

The main characteristics are given below :

1. ***Melting / boiling points and viscosity.*** We have seen that a co-ordinate bond is, in a way, a combination of an ionic bond and a covalent bond. It is for this reason that coordinate compounds have melting and boiling points, and viscosities which are *higher* than those of purely covalent compounds but *lower* than those of purely ionic compounds.
2. ***Semi-polar character.*** These compounds are semi-polar in character (A^+–B^-), *i.e.* they are more polar than covalent compounds and less polar than the ionic compounds.
3. ***Physical state.*** These compounds are gases, liquids or solids.
4. ***Solubility.*** These are usually insoluble in polar solvents like water but are soluble in non-polar (*i.e.* organic) solvents.
5. ***Conductivity.*** Like covalent compounds, coordinate compounds are also *non-ionic, i.e.* they do not conduct electric current through their *aqueous solutions or fused mass.*

6. ***Molecular reactions.*** Coordinate compounds are molecular and hence undergo molecular reactions which are slow.
7. ***Stability.*** Coordinate compounds are as stable as the covalent compounds. But when they are made up of two different stable molecules (*molecular compounds*), they are not very stable.
8. ***Isomerism.*** Coordinate compounds also show isomerism. Since the coordinate bond is *rigid* and *directional*, different space models (*i.e.* stereoisomers) of a single coordinate compound are possible.

Comparison Between Ionic, Covalent and Co-ordinate Bonds

The comparison between the three bonds is given in Table 7.2.

Table 7.2. Comparison between ionic, covalent and co-ordinate bonds

Ionic Bond	Covalent Bond	Coordinate Bond
(*i*) Ionic bond is formed by the transfer of electrons from a metal atom (A) which has 1, 2 or 3 valence-electrons to a non-metal (B) having 5, 6 or 7 valence electrons. $A^{\times} + \cdot\overset{..}{\underset{..}{B}}: \rightarrow [A]^{+}\left[\overset{..}{\underset{..}{\times B}}:\right]^{-}$ or $A^{+}B^{-}$	(*i*) Covalent bond is formed by sharing two electrons between non-metal atoms having 1, 4; 5, 6, or 7 valence electrons. $\underset{\times\times}{\overset{\times\times}{\times A}} + :\overset{..}{\underset{..}{B}}: \rightarrow \underset{\times\times}{\overset{\times\times}{\times A}}\;:\overset{..}{\underset{..}{B}}:$ or A–B	(*i*) It is formed by the sharing of two electrons between two atoms, both electrons coming from one atom. $\underset{\times\times}{\overset{\times\times}{\times A \times}} + \overset{..}{\underset{..}{B}}: \rightarrow \underset{\times\times}{\overset{\times\times}{\times A \times}}\,\overset{..}{\underset{..}{B}}:$ or $A \rightarrow B$
(*ii*) Ionic bond consists of electrostatic force between cations and anions.	(*ii*) Covalent bond consists of two electrons that hold the atoms together.	(*ii*) It consists of an electron pair between the linked atoms.
(*iii*) It is a weak bond, since the electro-static force can be broken easily.	(*iii*) It is a strong bond, since the paired electrons cannot be separated easily.	(*iii*) It is also a strong bond, since the paired electrons cannot be separated easily.
(*iv*) It is a polar bond (A^{+}–B^{-})	(*iv*) It is a non-polar bond (A–B)	(*iv*) It is a semi-polar bond ($A^{+} - B^{-}$)

Metallic Bond

What is Metallic Bond ?

Since all the atoms in a metal crystal are identical, these cannot be linked by ionic bonds as cations and anions are bonded together in ionic crystals, Moreover, ionic bond cannot be formed between the metal atoms, since the metal atoms have the same electronegativities.

Also, each atom in a metal crystal cannot be bonded to other 8 or 12 atoms (X-ray study of metal crystals has shown that each metal atom is surrounded by

8 or 12 other metal atoms) by the covalent bond (sharing of electrons), since in metals the number of valence electrons is insufficient for the formation of covalent bonds with all its 8 or 12 neighbouring metal atoms. For example Na atom which has only one valence electron cannot be expected to form covalent bonds with 8 nearest neighbouring Na atoms in its crystal.

Obviously, the metal atoms in a metal crystal are bonded together by a bond which is neither an ionic bond nor a covalent bond, but it is a special type of bond which is called *metallic bond*. This bond binds the metal atoms together in a metal crystal and is of its own type.

The peculiar type of bonding which holds the metal atoms together in a metal crystal is called metallic bonding.

To Explain the Nature of Metallic Bond by Electron-Gas Theory

In order to explain how the metal atoms in a metal crystal are bonded together, a simple theory, first proposed by Drude (1900) and later on developed by Lorentz (1916), is known as *electron-cloud* or *electron-pool* or *electron-gas theory*. This theory can well explain the nature of forces holding the metal atoms together in a metal crystal (Metallic bond).

Since the ionisation energies of metals are low, they readily give up their valence electrons to form the metal ions (called *positive cores or kernels*). The electrons given by the metal atoms move from place to place through the empty valence orbitals of the closed packed metal ions and are shared simultaneously by all the atoms together by a characteristic type of bond which is called *metallic bond.*

The electrons now no longer belong to individual metal atoms but belong to the crystal as a whole. As a result of delocalisation of electrons, the positive metal ions that are produced, remain fixed in the crystal lattice while the delocalised electrons move freely in the vacant valence orbitals. Thus on the basis of this theory :

A metal is regarded as a group of positive metal ions packed together as closely as possible in a regular geometric pattern and immersed in a sea of electrons (called electron-pool or electron-gas or electron-cloud) which move about freely (mobile or delocalised electrons) in the vacant valence orbitals. The *attractive force that binds the metal ions to the mobile electrons is called* **metallic bond** *and the force of attraction between the metal ions and the mobile electrons holds the atoms together.*

The electrons which are moving from place to place through the space between the metal ions are called *electron-gas* because of their mobile nature.

Diagrammatic representation of metallic bond in sodium metal is shown in Fig. 7.22. Circles with + sign at the centre represent Na^+ ions (positive metal cores) and – sign represents an electron. Attraction between the positive metal cores and mobile electrons has been shown by arrows and this constitutes the metallic bond.

The bonding which holds the metal atoms firmly together as a result of the attraction between the positive metal ions and surrounding freely mobile electrons, is called the **metallic bonding.**

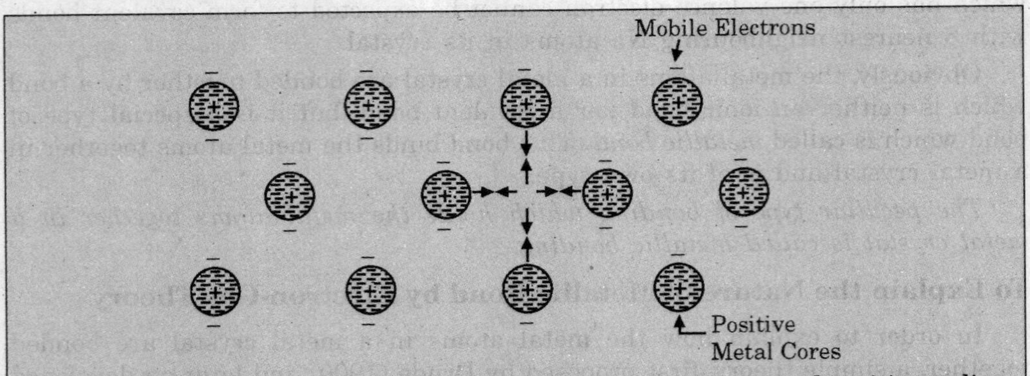

Fig. 7.22. A schematic representation of sea of mobile electrons and positive sodium metal cores. The attraction between the positive metal sodium cores and mobile electrons constitute the metallic bond. (Attraction has been shown by arrows).

The electron-gas model of metals can be compared with a locality consisting of a number of houses with lanes around them (Fig. 7.23). Children (*electrons*) from different houses (*metal atoms*) are free to play or move anywhere in the lanes (*vacant valence orbitals*) within the locality (*structure of the metal*) itself, but are not allowed by their parents (*positive metal cores*) to go away from the locality.

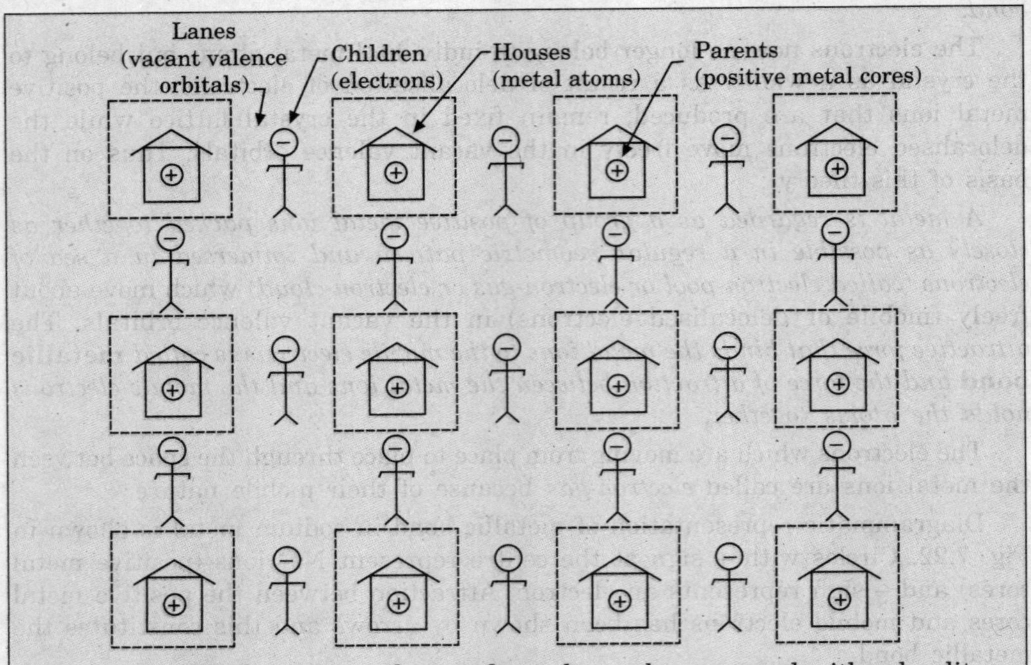

Fig. 7.23. The electron-gas theory of metals can be compared with a locality consisting of a number of houses with lanes around them.

Explanation of Physical Properties of Metals

The electron gas theory of metals can well explain various physical properties of metals. This has been shown as follows :

1. ***Metals are good conductors of electricity*** **(*Electrical conductivity*).** When an electric field is applied between the two ends of a metal, the mobile electrons begin to move towards the positive pole and the new electrons from negative pole take their position. Thus with the movement of electrons, the electric current starts to flow in the metal and the metal, therefore, acts as a good conductor of electricity [See Fig. 7.24 (*a*)].

It has been observed that the electrical conductivity of most of the metals decreases with the increase of temperature. This is because of the fact that with increase of temperature the positive metal ions also begin to vibrate and their motion hinders the free movement of the mobile electrons between the positive metal ions [Fig. 7.24 (*b*)]. Thus the vibrating positive metal cores create resistance to the flow of the electrons towards the positive pole and hence the movement of the electrons towards the positive pole falls and so does the electrical conductivity of the metals at higher temperature.

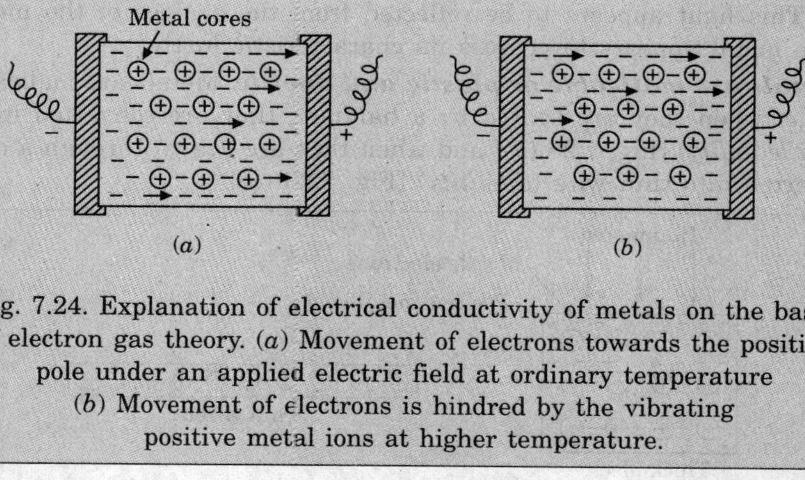

Fig. 7.24. Explanation of electrical conductivity of metals on the basis of electron gas theory. (*a*) Movement of electrons towards the positive pole under an applied electric field at ordinary temperature (*b*) Movement of electrons is hindred by the vibrating positive metal ions at higher temperature.

2. ***Metals are good conductors of heat*** **(*Thermal conductivity*).** It has been observed that the metals which are good conductors of electricity are also good conductors of heat. Thermal conductivity of metals can also be explained on the basis of mobile electrons present is the metallic lattice. When a piece of metal is heated at one end, the mobile electrons at this end absorb heat energy from this end and move very rapidly through the metallic lattice towards the cooler end. During this process they collide with adjacent electrons and thus transfer their heat energy to them (Fig. 7.25). Thus the mobility of electrons allows heat transfer to the other end.

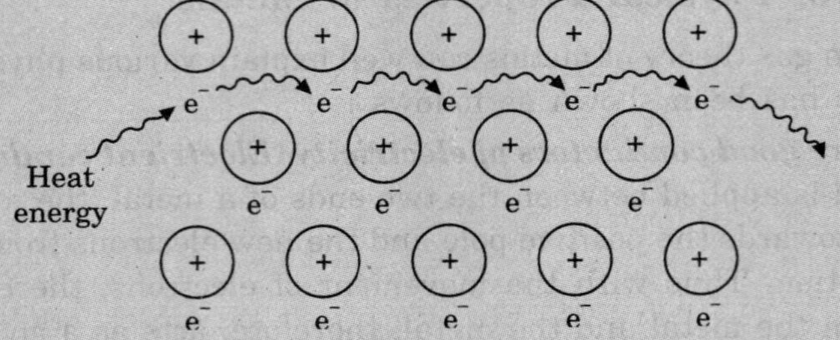

Fig. 7.25. Explanation of thermal conductivity of a metal on the basis of electron-gas theory.

3. *Metals are opaque and have lustre or colour.* The light that falls on metals is absorbed completely by the electronic transitions of the sea of electrons. Since no light is permitted to pass through the metal, it is opaque.

When freshly cut or scratched, most of the metals possess metallic lustre, *i.e. they have a shining surface*. The lustre of metals can also be explained on the basis of mobile electrons. When light falls on the surface of the metal, the incident light collides with the mobile electrons which are excited. When these excited electrons revert to their original position, they give off energy in the form of light. This light appears to be reflected from the surface of the metal and hence the metal appears to possess its characteristic lustre.

4. *Metals are malleable or plastic and ductile.* Metals are malleable and ductile, *i.e.* when they are beated by a hammer, they are converted into their seets (*malleability*) [Fig. 7.26 (*a*)] and when they are drawn through a die, they are converted into thin wire (*ductility*) [Fig. 7.26 (*b*)].

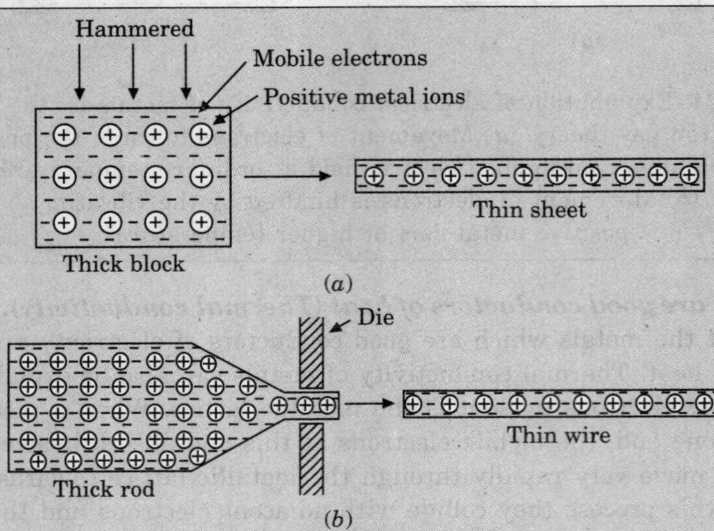

Fig. 7.26. (*a*) Metals are malleable (*b*) Metals are ductile.

When a metal is beaten, say by a hammer, one layer of positive metal ions moves along another layer through a distance and occupies new positions. Since the sea of electrons (*i.e.* valence electrons) are mobile, they also move along with the positive metal ions and thus the structure of the metal does not alter but remains the same and the crystal structure is restored (Fig. 7.27). This results in that the thickness of the metal is reduced and the metal is converted into thin sheet or thin wire.

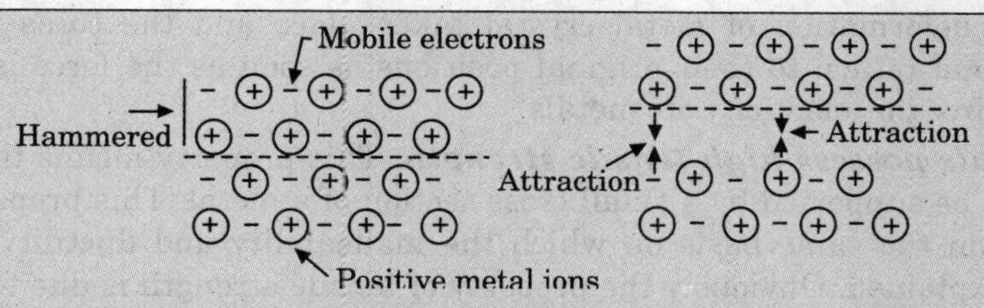

Fig. 7.27. Explanation of malleability and ductility of metals.

However, if an alloying metal is added to a metal, the structural homogeneity of the principal metal is disturbed and hence the resulting alloy is hard and brittle.

Brittleness of ionic crystals. When a little force is applied on an ionic crystal, the ions in a layer slide a bit over another layer and the ions having similar charges in the two layers come in contact with each other and hence begin to repell each other (Fig. 7.28). The force of repulsion created by this situation shatters the ionic crystal, *i.e. ionic solids are highly brittle*. In case of metals, as we have seen above, this does not happen and hence in this respect the metals differ from the ionic crystals.

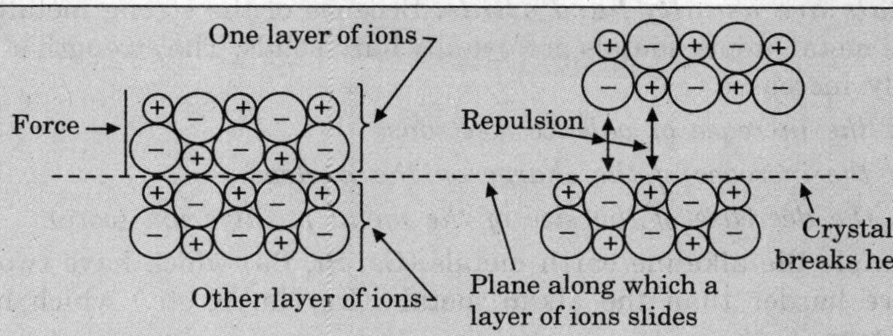

Fig. 7.28. The sliding of a layer of ions in an ionic crystal shatters it and thus the brittleness of ionic crystals is explained.

5. *Metals form alloys.* Metallic mixtures are called alloys. Metals form alloys easily. In these alloys, the spherical ions of different metals share the same sea of electrons. (Fig. 7.29).

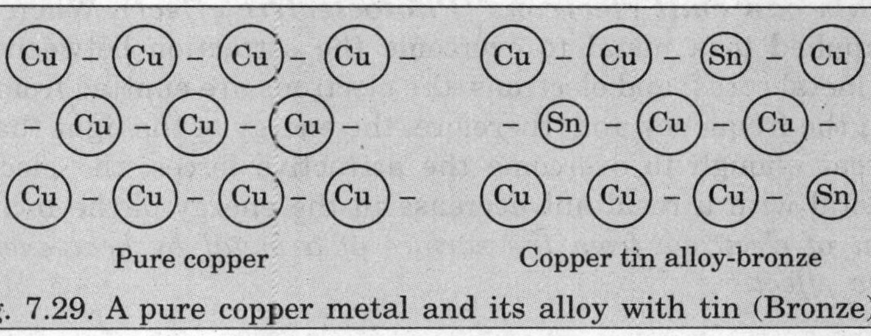

Fig. 7.29. A pure copper metal and its alloy with tin (Bronze).

Only such metals can form the alloys the atoms of which when mixed can fit into the crystal lattice of the other. This is only possible, if the geometry of the mixing atoms allows this fitting in, *i.e.* if the diameter of the atom of one metal is much greater than that of the other, no alloy can be formed by the two metals.

6. *Metals have elasticity.* Under the influence of comparatively less force, temporary deformation of metal crystal takes place and the cores with their electron-cloud return to their original positions as soon as the force is removed. This explains the elasticity in metals.

7. *Metals possess high tensile strength.* This property means that a large weight can be supported by a small cross section of a metal. This property is also explained on the same basis on which the malleability and ductility of metals has been explained. Obviously the property of tensile strength is due to the great attraction between the positive metal ions and the mobile electrons.

8. *Metals are solids and have high density.* Metals are solids (mercury being the only exception) and have high density. The high density is due to the fact that the metallic bond keeps the metal atoms closely packed in the metallic crystal.

9. *Melting and boiling points.* In general, metals have intermediate melting and boiling points as compared to those of covalent and ionic compounds. This is due to the reason that the attractive forces (*i.e.* metallic bonds) that hold the metal atoms together in metallic crystal are intermediate between those in covalent and ionic compounds.

10. *Metals are usually hard solids.* Because of the strong metallic bonds between the metal atoms, metals are usually hard solids. The strength of metallic bond usually increases

(*a*) *with the increase of valence electrons.*

(*b*) *with the increase of the charge on the nucleus.*

(*c*) *with the decrease of the size of the metal positive ion (core).*

For example the alkaline earth metals (Ca, Sr, Ba) which have two valence electrons are harder than the alkali metals (Li, Na, K etc.) which have one valence-electron.

Similarly the hardness of Li, Na and K is in the order

$$Li > Na > K,$$

because the size of Li^+, Na^+ and K^+ ions is in the order :

$$Li^+ < Na^+ < K^+$$

11. *Metals can emit electrons (Photoelectric effect).* When enough heat energy is applied to a metal to overcome the attraction between the positive metal ions (metal cores) and electrons, the electrons are emitted from the metallic atom. When the frequency and, therefore, the energy of the light that strikes the metal is great enough to overcome the attractive forces, the electrons escape from the metal with a resultant decrease in the energy of the incident photon. *The removal of electrons from the surface of a metal by heat energy is called photoelectric effect.*

Conditions for the Formation of Metallic Bond

The above discussion clearly shows that the metals forming metallic bond should obey the following conditions :

(*i*) The metals should have *low ionisation energies* so that the metals should readily give up their valence-electrons to form the positive metal ions (cores).

(*ii*) The metals should have *vacant valence orbitals* so that the positive metal ions may move freely in these vacant orbitals.

Differences between Metallic and Ionic Bond

(*i*) Since all the atoms in a metal crystal are similar, there is no possibility of the linking of the metals atoms by ionic bond.

(*ii*) Further a metal conducts electricity by the motion of electrons from one end to the other. This is quite different from the way in which electrolytes conduct electricity by the migration of their ions. This too confirms that metallic bond differs from ionic bond.

Differences between Metallic and Covalent Bond

(*i*) Most of the metals have less than four electrons in their valence shell (many metals have only one or two electrons). With so few valence electrons in valence shell, it is not possible for a metal atom to be linked to other eight or twelve atoms (in metal crystal each metal atom is surrounded by 8 or 12 other metal atoms) by sharing the electrons with them and thus attrain the configuration of the nearest inert gas with 8 electrons in its valence shell (*octet rule*). Thus the metal atoms, due to insufficient number of valence electrons, cannot be linked to other metals atoms by covalent bond.

(*ii*) Since the valence electrons in a covalent bond are *localised*, a covalent bond has a *directional character*. On the other hand, since the valence electrons involved in metallic bond are spead all over the metallic crystal, the metallic bond has *non-directional character.*

(*iii*) Further, in covalent bond, due to the presence of localised electrons, there is much more attraction between the valence electrons and the nuclei of the atoms, *i.e.* valence electrons in covalent bond experience strong attractive pull by the nuclei of the atoms. Thus covalent bond is quite strong. On the other hand, in metallic bond, due to the presence of delocalised electrons, (mobile electron) the force of attraction existing between the valence electrons and the nuclei of the atoms is weak, *i.e.* valence electrons in metallic bond are weakly attracted by the nuclei. It is because of this that *metallic bond is weaker than a covalent bond.* For example the energy required to vaporise a mole of metal like copper to the vapor state is smaller than the energy required to vaporise a mole of a covalent material like graphite.

Hydrogen Bond

What is Hydrogen Bond ?

In order to understand the concept of hydrogen bond let us consider a molecule, say, AH in which H atom is linked with a *strongly electronegative but very small atom* A (A may be N, O or F) by a normal covalent bond. The electron pair being shared between H and the strongly electronegative atom A will evidently lie far away from H and thus partial positive and negative charges will be developed on H and A atoms respectively. Consequently AH molecule will behave as a dipole which is represented as :

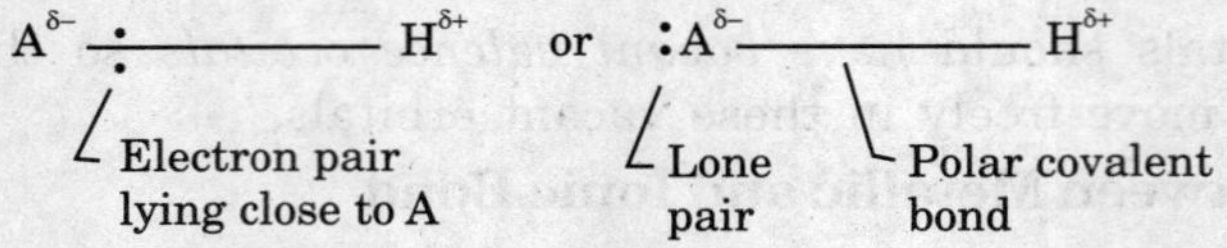

Evidently the dipole has A as its negative end and H as its positive end. Since the electronegative atom, A attracts the electron pair constituting the covalent bond between A and H almost completely towards itself, the atom A takes almost full possession of the electron pair, *i.e.* the atom A will have a lone pair of electrons. This leaves H atom with a large partial positive charge and AH molecule, therefore, becomes highly polar and hydrogen end of it becomes nearly bare hydrogen nucleus or proton (H^+), *i.e.* H-atom is reduced to a proton which is almost devoid of electrons.

Now if another molecule like AH (same molecule) or BH (different molecule) (A and B are strongly electronegative atoms) which also forms a dipole $A^{\delta-} - H^{\delta+}$ or $B^{\delta-} - H^{\delta+}$ respectively is brought near $A^{\delta-} - H^{\delta+}$ dipole, these two dipoles will be attracted towards each other by electrostatic force of attraction which is represented by a dotted or dashed line and is called hydrogen bond or hydrogen bonding.

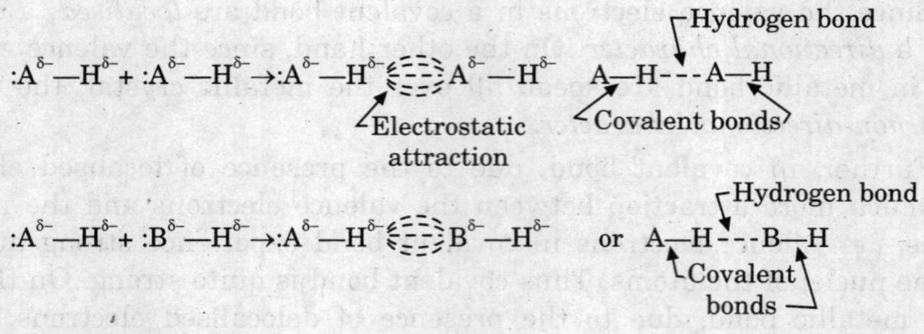

Thus we see that in H-bond H-atom forms a bridge between two electronegative atoms and for this reason H-bond is also called a H-*bridge*. Thus H-bond can be defined as :

The attractive electrostatic force between a hydrogen atom which is already covalently attached with a strongly electronegative atom of a molecule and another electronegative atom of some other molecule (same molecule or different molecule) is known as hydrogen bond.

OR

Under appropriate conditions a hydrogen atom may be linked to two similar or different electronegative atoms. It is bonded to one of the two atoms by a covalent bond while to the other atom it is attached by a special type of bond which is much weaker than the covalent bond and is called hydrogen bond.

Nature of Hydrogen Bond

It is difficult to ascertain the exact nature of hydrogen bond. We know that the covalency of H-atom is limited to one, *i.e.*, H- atom can be linked to only one atom, say A. This is because of the fact that 1*s* atomic orbital of H-atom becomes completely filled after it has formed a covalent bond, A–H. Thus H-atom already covalently bonded to atom A in A–H bond cannot form a second covalent bond with another atom, say B. This shows that A–H–B in which H-atom shows bivalency is unlikely, because, if it is likely at all, it would require the use of 2*s*- or 2*p* -orbitals of H-atom which are of much higher energy and consequently are useless for bonding. Further if A–H–B is supposed to be correct, we should expect H-atom to be equidistant from A and B atoms, if the elecnegativities of A and B are the same. However, it has been found by experiments that H-atom in hydrogen bond is close to that atom with which it forms a covalent bond. This shows *that hydrogen bond is electrostatic in nature, i.e. hydrogen bond is merely an electrostatic force rather than a chemical bond.*

A question may now arise as to why H-atom alone is capable of forming hydrogen bond. The reason, probably, is that H-atom has small size with only one electron in its only one energy shell. When this electron is taken away, the proton (H^+) left behind can easily manage to stap in between two electronegative atoms bringing them closer together by creating an *electrostatic field*, though weak. That is why the proton is in a state of oscillation between the electronegative atoms.

Atoms of sulphur, phosphorus, chlorine, bromine and iodine are also highly electronegative, but they cannot form hydrogen bond. Their incapability is due to their larger atomic size, which results in extremely weak electrostatic field about them.

Properties of Hydrogen Bond

(*i*) A H-bond is a bond of hydrogen between two electronegative atoms only. It never involves more than two atoms (excluding H-atom).

(*ii*) Bond energy of a H-bond is in the range of 3–10 kcal/ mole while that of a normal covalent bond is in the range of 50–100 kcal/mole. Thus a H-bond (*i.e.*, H...B bond) is much weaker and longer than a covalent bond, A–H and hence it can be readily ruptured. The difference in energy between A–H and H...B bonds indicates that these have different bond lengths which, in turn, show that H-atom in A–H...B is never midway between the two atoms, A and B ; it is rather always nearer to atom A which is covalently bonded to H-atom. H-bond has more energy (= 3–10 kcal/mole) than van der Waals forces (= 1 kcal/ mole).

With the increase of electronegativity of the atom to which H-atom is covalently linked, the strength of H-bond also increases. Thus the strength of H-bonds in N–H...N, O–H...O and F–H...F is in the following order :

Order of strength : $N-H...N < O-H...O < F-H...F$

Order of electronegativity values : $N (= 3.0) < O (= 3.5) < F (= 4.0)$

The numbers given in brackets indicate the electronegativity values of the elements concerned.

The strength of H-bond also depends on the size of the atom to which H-atom is linked. For example the larger Cl and S atoms whose electronegativities are almost the same as that of N form H-bond to a lesser degree.

(*iii*) The formation of a H-bond does not involve any sharing of electron pairs. It is, therefore, quite different from a covalent bond.

(*iv*) Only O, N and F which have high electronegativity and small atomic size, are capable of forming H-bonds.

(*v*) Hydrogen bonding results in the formation of long chains or clusters of a large number of associated molecules like as many tiny magnets.

(*vi*) Like a covalent bond, H-bond has a preferred bonding direction. This is attributed to the fact that H-bonding occurs through *p* orbitals which contain the lone pair of electrons on A atom. This implies that all the three atoms in A–H...A will be in a straight line.

Types of Hydrogen Bond

Hydrogen bond is two types :

1. Inter-molecular hydrogen bond (Association). *This type of H-bond occurs between two or more molecules of the same or different compound. Thus* NH_3, H_2O *and HF molecules are associated by inter-molecular H-bond.*

2. Intra-molecular hydrogen bond (Chelation). This type of H-bond is formed between a H-atom and an electronegative atom present in the same molecule (*intra* means within). In intramolecular H-bonding, the H-atom is bonded to two atoms of the same molecule. This type of H-bonding may lead to the linking of two groups to form a ring structure and such an effect is one kind of *chelation*. The occurrence of this type of H-bond does not disturb the normal bond angles.

Examples of molecules showing intra-molecular hydrogen bonding are provided by *o*-nitropenol, *o*-hydroxy benzaldehyde, *o*-chlorophenol, *o*-hydroxy benzonic acid (salicylic acid), *o*-nitro benzoic acid, maleic acid etc.

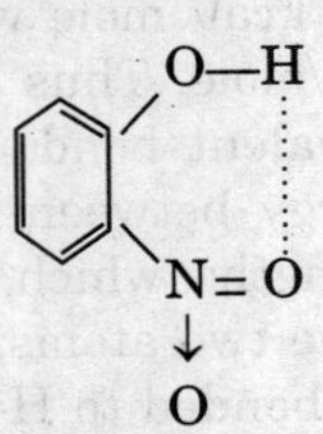

o-nitro phenol

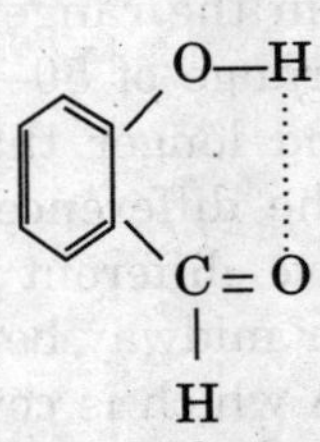

o-hydroxy benzaldehyde

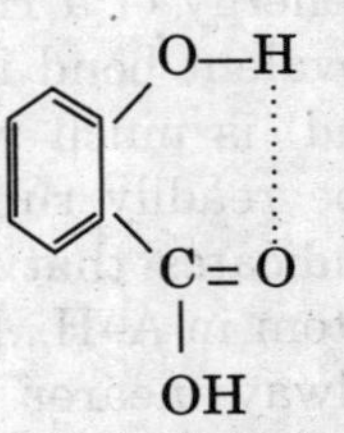

o-hydroxy benzoic acid (salicylic acid)

Thus we see that intra-molecular H-bonding in these molecules leads to the formation of a six-membered ring (*chelation*). The occurrence of H-bond of this type is not possible in *m*-nitrophenol and *p*-nitrophenol because of the size of the ring that would result. It is because of the non-existence of H-bond in *m*- and *p*-isomers that *o*-nitrophenol has different properties from those of other isomers. For example :

(*i*) Melting point of *o*-nitrophenol is 214° while those of its *m*- and *p*-isomers are 290° and 279° respectively.

(*ii*) *o*-nitrophenol is volatile in steam and less soluble in water than the other two isomers.

p-nitrophenol shows inter-molecular H-bonding, resulting in the association of two molecules as shown below.

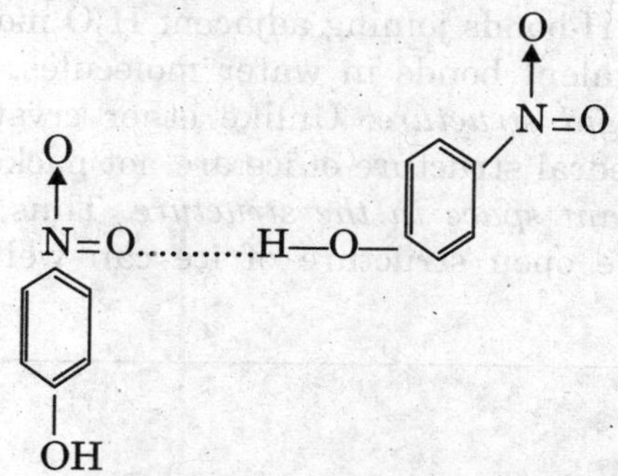

Physical Properties Explained by Hydrogen Bond — Consequences of Hydrogen Bond.

Some of the unusual properties of the compounds containing H-bond can be explained on the basis of the concept of H-bonding. For example :

A. Association of molecules. Because of the fact that N,O and F atoms are of small size and high electronegativity, their hydrides *viz.* NH_3, H_2O and HF associate by H-bonds and form the bigger aggregates (polymerised molecules) of various sizes. The formation of aggregates (bigger units) is called *association* or *polymerisation*. The bigger aggregates are *polymerised molecules*. Within the associated unit, the molecules are bonded together by H-bonding. The associated unit can be broken to smaller unit or to single molecules on heating, *i.e.* by suppling enough energy which overcomes H-bonding present in the associated unit. Now let us discuss the association in NH_3, H_2O and HF molecules.

(1) **Association of liq. NH_3 molecules.** In NH_3 molecules there are three H-atoms which are covalently bonded to the highly electronegative N-atom which has a lone pair of electrons on it. This N-atom forms a H-bond with H-atom of the neighbouring NH_3 molecules through its lone pair and gives rise to the formation of an associated molecule, $(NH)_x$ in the liquid state as shown below :

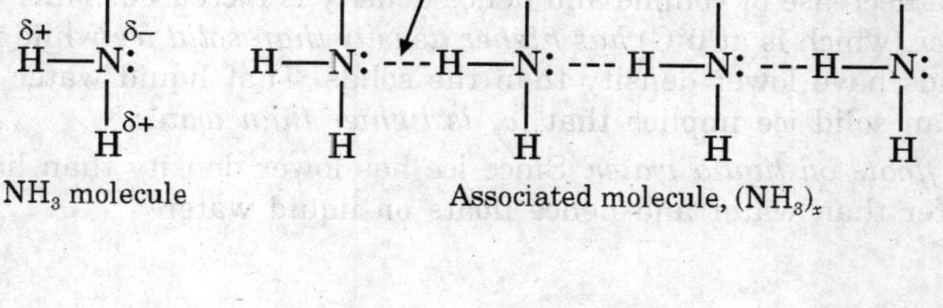

NH_3 molecule   Associated molecule, $(NH_3)_x$

(2) **Association of H_2O molecules.** ***(i) Solid state of water (i.e. ice).*** Here each O-atom of each H_2O molecule is linked with two H-atoms of its own molecule by covalent bonds and with two H- atoms of the neighbouring H_2O molecule by H-bonds. Two H-bonds are formed between the two lone pairs of electrons present on O-atom and two H-atoms of the neighbouring H_2O molecules as shown in Fig. 7.30. This type of bonding between O-atom and four H-atoms gives a *tetrahedral structure to ice*. In this structure two H-atoms are linked covalently and lie close to O-atom at a distance of 1.0A° whereas the other two H-atoms (of other H_2O molecule) are linked by H-bonding and are at a distance of 1.76A° from O-atom. Since the H-bonds joining adjacent H_2O molecules are weaker and longer than normal covalent bonds in water molecules, the structure of ice so obtained is *cage-like rigid structure*. Unlike other crystalline structures, H_2O molecules in the tetrahedral structure of ice are not packed very closely together and hence there is *vacant space in the structure*. Thus, ice has *cage-like open structure*. This cage-like open structure of ice can well explain the following properties of water :

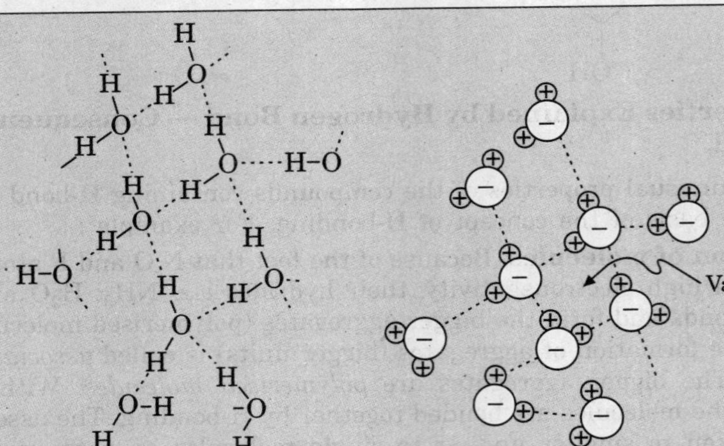

Fig. 7.30. Tetrahedral structure of ice (Cage-like open structure). The oxygen atoms are situated tetrahedrally to one another.

(*a*) *Liquid water has higher density than solid ice.* We have seen that the cage-like open structure of ice has vacant space in it. The presence of vacant space gives a large volume to a given mass of ice. When ice is allowed to melt to come in the liquid state, some of the H-bonds in the cage-like structure of ice are broken to some extent and water molecules come closer to each other. This results in a decrease of volume and hence density is increased. Thus we see that *liquid water* (which is at 0°C) *has higher density than solid ice* while most of the other liquids have lower density than the solids. That liquid water has higher density than solid ice implies that *ice is lighter than water.*

(*b*) *Ice floats on liquid water.* Since ice has lower density than liquid water, ice is lighter than water and hence floats on liquid water.

(c) *Density of water is maximum at 4°C.* As the temperature of liquid water (= 0°C) is increased above 0°C, density of water also increases. The increase in density continues only upto 4°C. Above 4°C, kinetic energy of H_2O molecules increases and hence H_2O molecules move away from each other, *i.e.* the volume of water molecules is increased. The increase in volume decreases the density of water. Therefore, above 4°C, the density of water starts decreasing again. Thus we see that *the density of water is maximum at 4°C.*

(ii) Liquid state of water. Here each O atom of each H_2O molecule is covalently linked with two H-atoms of its own molecule and with another H-atom of the adjacent H_2O molecule by H-bonding. The formation of H-bond occurs through one of the two lone pairs of electrons present on O-atom in H_2O. molecule. The association of different H_2O molecules by H-bonding gives an associated unit or cluster, $(H_2O)_x$ which consists of a large number of H_2O molecules as shown below :

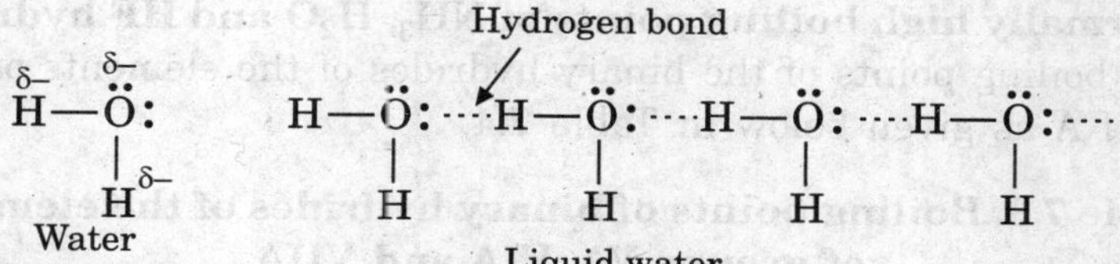

Water molecule Liquid water

However the most stable cluster is the double molecule, $(H_2O)_2$ which consists of two H-bonds as shown below :

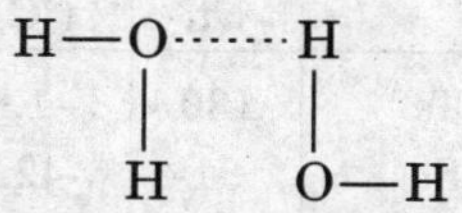

The formation of polymerised molecule, $(H_2O)_x$ by H-bonding can well explain the fact that water exists as a liquid under the ordinary conditions while the hydrides of other elements of the same group (VI A group) namely H_2S, H_2Se and H_2Te exist as gases under the same conditions. This is explained as follows :

Due to the fact that the electronegativity of O-atom is higher than that of S, Se and Te atoms, there is considerable H-bonding in liquid H_2O molecules while the same is absent in H_2S, H_2Se and H_2Te molecules. Thus H_2O molecules can associate to form a *polymerised molecule.* $(H_2O)_x$ while H_2S, H_2Se and H_2Te cannot do so. It is due to the formation of polymerised molecule, $(H_2O)_x$ that *water exists as a liquid under the ordinary conditions while H_2S, H_2Se and H_2Te exist as gases under the same conditions.*

(3) **Association of HF molecules** (*i*) HF *crystals* contain infinitely long zig-zag chains of HF molecules which are associated to $(HF)_x$ unit by H-bonding. F atom is covalently bonded to the H-atom of its own molecule while the same atom is bonded to the H- atom of the neighbouring HF molecule by a H-bond. F atom forms H-bond with H-atom of the other HF molecule by its lone pair present in *p*-orbitals. H–F and H...F bond lengths are 1.0Å and 1.55Å respectively.

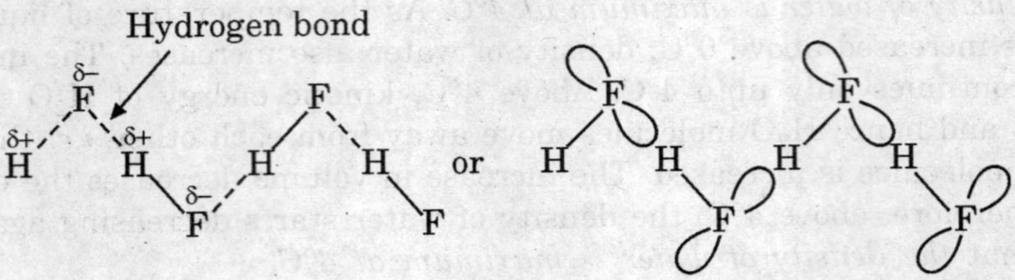

On heating the length of the chains shortens progressively till the unit, $(HF)_x$ becomes fairly small and x assumes any value upto six.

(*ii*) *Gaseous HF molecule* contains molecular species, $(HF)_2$, $(HF)_3$ and so on upto $(HF)_6$ as well as some single HF molecules.

(*iii*) *In aqueous solution* HF ionises to give HF_2^- ion (F^-...H–F) rather than F^- ion.

B. Abnormally high boiling points of NH_3, H_2O and HF hydrides. Let us consider the boiling points of the binary hydrides of the elements of groups VA, VI A and VII A as given below in Table 7.4.

Table 7.3. Boiling points of binary hydrides of the element of groups VA, VIA and VIIA

Group VA			Group VIA			Group VIIA		
Hydride	**Mol. wt.**	**B. pt. (°C)**	**Hydride**	**Mol. wt.**	**B. pt. (°C)**	**Hydride**	**Mol. wt.**	**B. pt. (°C)**
SbH_3	125	−17	H_2Te	130	−1.8	HI	128	−35.5
AsH_3	78	−55	H_2Se	81	−42.0	HBr	81	−67.1
PH_3	34	−85	H_2S	34	−59.6	HCl	365	−85.0
NH_3	17	−33	H_2O	18	100	HF	20	19.4

It may be seen from the table that on proceeding from SbH_3 to PH_3 (Group VA), from H_2Te to H_2S (Group VIA) and from HI to HCl (Group VIIA), the boiling points of the hydrides decrease with the decrease of their molecular weights while those of NH_3, H_2O and HF show a sudden increase, although they have the minimum molecular weights in their respective groups. The abnormal high boiling points of NH_3, H_2O and HF are due to the fact that these molecules are capable of forming the associated molecules, $(NH_3)_x$, $(H_2O)_x$ and $(HF)_x$ by H-bonding between them, while other molecules exist as single molecules as they are incapable of forming associated molecules (clusters) by H-bonding. In order to break H-bonds existing in the associated molecules and thus to separate molecules of the clusters as they enter the gaseous state or the liquid state, relatively more energy is required and hence NH_3, H_2O and HF molecules have abnormally higher boiling points compared to other hydrides of the elements of the same group which do not form any H-bonds.

C. **High solubilities of some covalent compounds.** Hydrogen bonding also accounts for the abnormally high solubilities of some covalent compounds containing N, O and F such as NH_3 and CH_3OH in certain hydrogen containing solvents noteably water. For example NH_3 and CH_3OH are highly soluble in H_2O because their molecules can associate with H_2O (solvent) molecules through H-bonds as shown below :

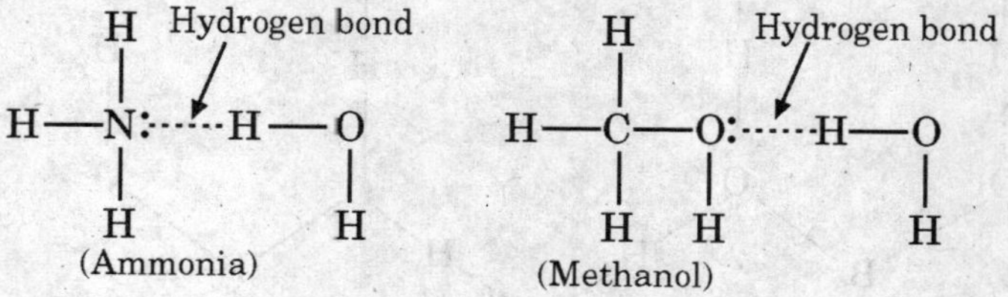

In addition, compounds containing oxygen such as sulphates freely dissolve in water through H-bond formation.

D. High viscosity, high heat of vaporisation and high dielectric constant. Glycerol ($CH_2OH–CHOH–CH_2OH$) is far more viscous than ethyl alcohol (C_2H_5OH) because of the fact that glycerol with three OH groups in its molecule can form many more H-bonds per molecule. This results in greater attraction among the molecules of glycerol and hence its resistance to flow is increased. Thus glycerol is more viscous.

The molecules which are capable of forming associated molecules through H-bonding have naturally high heat of vaporisation and high dielectric constant.

Significance of Hydrogen Bonding

Apart from providing explanation for a number of abnormal properties of certain compounds containing H-bonding, it is also useful in explaining physiological and geological processess occurring around us. For example :

1. *Physical state of water*. Without H-bonding, H_2O would have existed as a gas like H_2S. In that case no life would have been possible without liquid H_2O.

2. *Structure of compounds*. H-bonding is directional and on this account it helps in studying and establishing the structure of many compounds like ice, solid hydrogen fluoride, hydrates, solid acids, acid salts etc. For example H-bonding accounts for the tetrahedral structure of ice, zig-zag arrangement of HF molecules in solid hydrogen fluoride and sheet structure of boric acid (Fig. 7.31) which is obtained by joining molecules of boric acid (H_3BO_3) by H-bonding, resulting in the formation of infinite parallel sheet.

3. *Explanation of polymerisation*. H-bonding gives a satisfactory explanation for the mechanism of polymerisation and hence of the formation of bigger aggregates of organic as well as inorganic molecules.

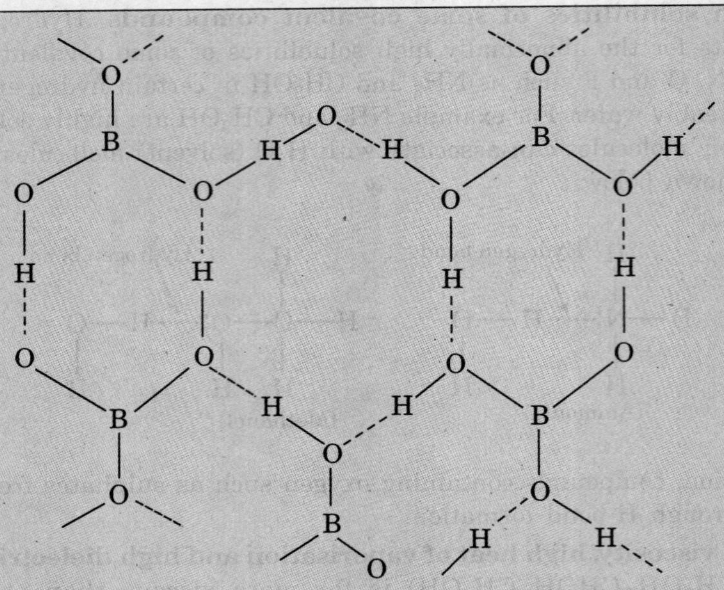

Fig. 7.31. Sheet structure of boric acid.

4. *Applications in biological investigations.* H-bonding also exists in molecules of living systems proteins like various tissues, organs, blood, skin and bones in animals. Fibrous like those found in hair, silk and muscles consist of long chains of a large number (sometimes 1000 or so) of amino acids as shown below. Here R = methyl isopropyl or butyl group.

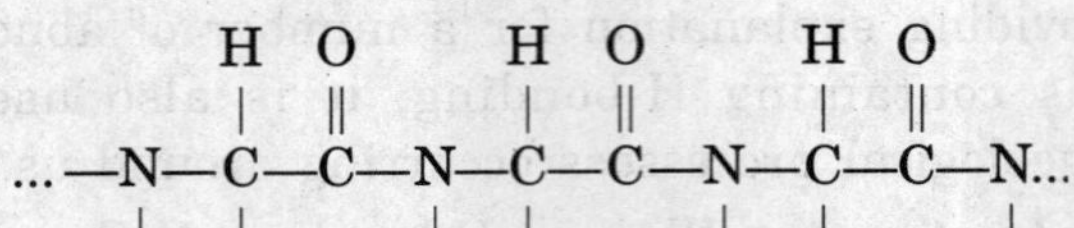

The long chains of amino acids are coiled about one another into a spiral celled *helix.* Such a helix may be either right-handed (called α-helix) or left-handed (celled β-helix) as in the case screws.

It has been found that α-helix constitutes the more stable arrangement. N–H and C=O groups which are vertically adjacent to one another in the helix are linked together by H-bonds. The H-bonds link one spiral to the other (Fig. 7.32). X-ray studies have shown that on an average there are 2.7 amino acid units for each turn of the helix (coil).

Similarly *deoxy ribonucleic acid* (DNA) has two spiral chains which are coiled about each other on a common axis to form a double helix, 18–20 Å in diameter. They are linked together by H- bonding between their sub-units. (Fig. 7.33).

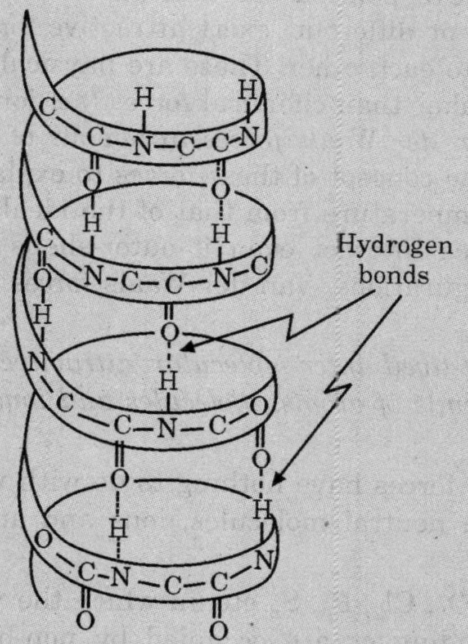

Fig. 7.32. Representation of an α-helix. Only the amino acids of the front portion have been shown. H and R groups attached with C-atom have not been shown for the sake of simplicity.

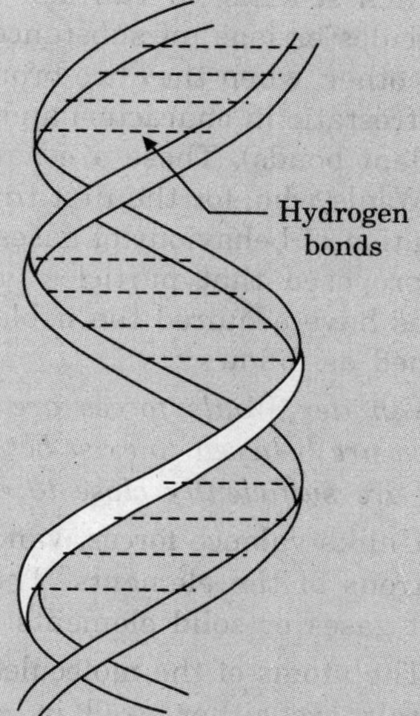

Fig. 7.33. Hydrogen bonding in DNA double helix.

5. ***Applications in geology.*** Many geological products such as organic compounds contained in petroleum, coal products and complex hydrates of minerals are formed and established by means of H-bonding.

6. ***In paints and dyes.*** The adhesive action (*i.e.* stickness) of glue, honey, dyes and paints is also due to the presence of H-bonding in them.

7. ***In clothing.*** H-bond is of vital importance for our clothing. For example the rigidity and tensile strength of cotton, silk or synthetic fibres is due to the H-bonding in them.

8. ***Food materials.*** Molecules of most of our food materials like sugars, carbohydrates etc. also consist of H-bonding. For example O-atom of one – OH group of one molecule of sugar and carbohydrate is bonded with –OH group of another molecule through H-bonding.

9. ***Wood fibres.*** H-bonding also plays a significant role in making wood fibres more rigid so that they may be used to meet the requirements of housing, furnitures etc.

van der Waals Forces : Intermolecular Forces

What are van der Waals Forces ?

Till now we had been dealing with the forces or bonds between atoms leading to the formation of molecules. These are called *inter-atomic* or *chemical forces.*

A Dutch scientist, J van der Waals in 1873 pointed out that particles (atoms, molecules or ions) of substances (similar or different) exert attractive forces on each other, when they are brought near to each other. These are physical forces (electrostatic in character) and much weaker than chemical forces (*i.e.* ionic and covalent bonds). These are named as *van der Waals forces* in honour of J. van der Waals who, for the first time, used the concept of these forces to explain the deviation of behaviour of gases at low temperature from that of the ideal gases. He proposed that particles of substances interact even if outer-shells of the atoms have acquired the noble gas configurations. Van der Waals forces can be defined as follows :

Van der Waals forces are very short-lived inter-molecular attractive forces which are believed to exist between all kinds of atoms, molecules and ions when they are sufficiently close to each other.

Unlike valence forces, van der Waals forces have nothing to do with valence electrons of the elements. They exist in neutral molecules, ions and atoms of inert gases or solid elements alike.

The atoms of the molecules like N_2, O_2, Cl_2, P_4, S_8 etc. in which the valence orbitals are either used in normal bonding or are occupied by non-bonding electrons, are rather firmly held together in the liquid or solid states by van der Waals forces or bonds. These forces are almost absent when the molecules of a gas are far apart and are in rapid kinetic motion. But as the atoms or molecules of the gas are brought nearer to each other by increasing the pressure and kinetic energy is withdrawn by cooling, van der Waals forces are capable of holding the molecules together to form the liquid or solid state. Thus a gas can be liquified by allowing it to expand suddenly from high pressure. During expansion the gas does work in overcoming intermolecular forces of attraction. The energy required for this purpose is obtained from the gas itself, thereby lowering the temperature and causing the liquifaction.

Types of van der Waals Forces

There are four types of van der Waals forces. The origin of each one of these has been explained by a different scientist. The name of scientist has been associated with that type to commemorate his contribution. The four types are :

1. Dipole-dipole interactions. These forces are found in polar molecules having permanent polarity in them. We know that a polar molecule has separate centres of positive and negative charge and possess permanent dipole moment. When polar molecules are brought nearer to each other, they orient themselves in such a way that the positive end of one dipole (polar molecule) attracts the negative end of another dipole and *vice-versa* as shown in Fig. 7.34. Due to this dipole-dipole interaction many molecules are held together. Dipole-dipole interactions between polar molecules are also called *Keesom Forces* and are the strongest of all other types of van der Waals forces. The intensity of these forces is generally hampered by an increase in temperature.

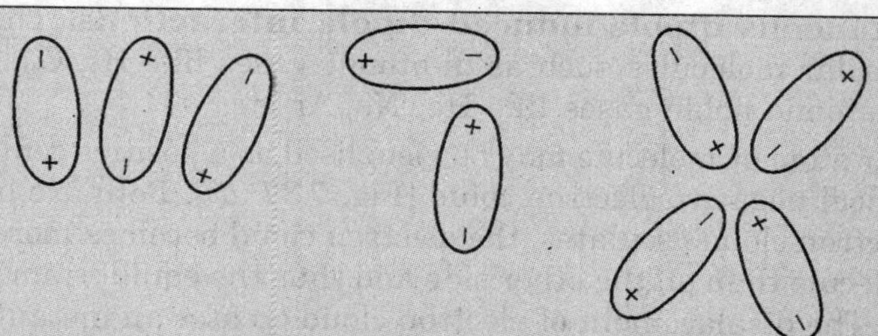

Fig. 7.34. Illustration of dipole-dipole interactions

2. Ion–dipole interactions. Polar molecules are attracted towards ions. The negative end of dipoles is attracted towards the cation while the positive end towards the anion (Fig. 7.35).

This type of interaction is called *ion-dipole interaction.* Ion- dipole interactions have been used to explain the dissolution of NaCl in H_2O. When NaCl is put in H_2O, it dissolves in it since the negative ends of water molecule dipoles aggregate around Na^+ ions and the positive ends around Cl^- ions.

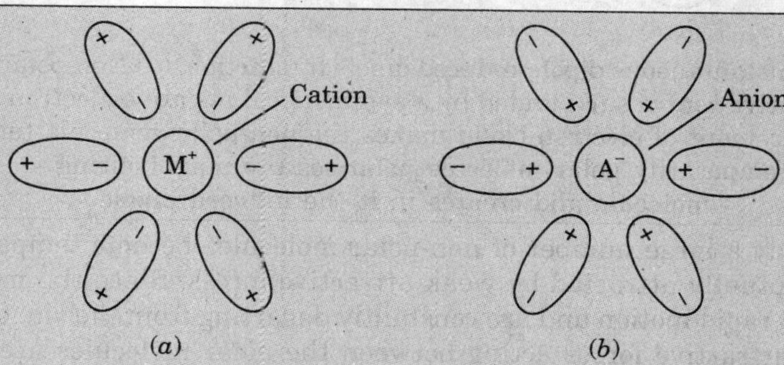

Fig. 7.35. Ion-dipole interaction. (*a*) Cation (M^+) attracts the negative end of the dipole. (*b*) Anion (A^-) attracts the positive end.

3. Dipole-induced dipole interactions. This type of force is found in a mixture containing polar and non polar molecules. When a non-polar molecule is brought near to a polar molecule, the positive end of the polar molecule attracts the mobile electrons of non-polar molecule and thus polarity is induced in non-polar molecule as shown in Fig. 7.36. Now both the molecules become dipoles and hence the positive end of the polar molecule attracts the displaced electron cloud of non-polar molecule. Thus the two types of molecules are held together by forces which are also called *Debye Forces*.

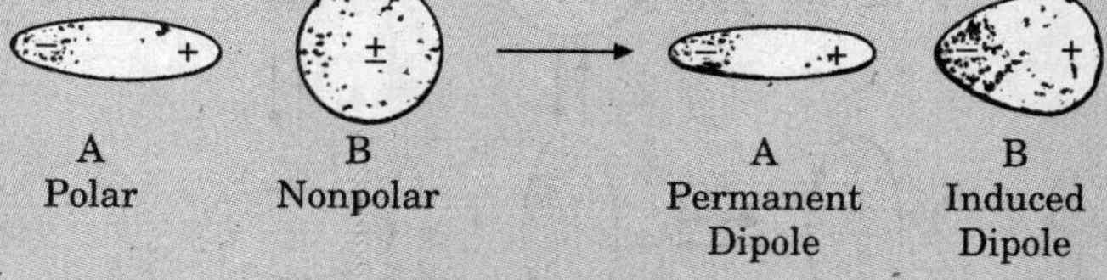

Fig. 7.36. Dipole-induced dipole interaction.

4. Instantaneous dipole-induced dipole interactions. These forces are found in non-polar molecules such as di-atomic gases like H_2, O_2, Cl_2 N_2 etc. as well as mono-atomic noble gases like He, Ne, Ar etc.

A non-polar atom or molecule may be visualised as a positive centre surrounded by assymmetrical negative electron colud [Fig. 7.37 (*a*)]. Both are in equilibrium. But as the electron cloud oscillates, the electron cloud becomes more dense on one side of the molecule than on the other side and thus the equilibrium gets disturbed for a moment. The displacement of electron cloud creates an instantaneous dipole temporarily. Thus the non-polar molecule is momentarily self-polarised and becomes temporarily polar [Fig. 7.37. (*b*)].

Now when the self-polarised (*i.e.* temporarily polar) molecule is brought near to a non-polar molecule, it polarises the neighbouring molecule by disturbing its electronic distribution and an induced dipole is created in it, *i.e.* the non-polar molecule momentarily becomes polar [Fig. 7.37 (*c*)].

(*a*) (*b*) (*c*)

Fig. 7.37. Instantaneous dipole-induced dipole interaction. (*a*) Non-polar molecule with a positive centre surrounded by a symmetrical negative electron colud. (*b*) The displacement of electron cloud makes the non-polar molecule temporarily polar. (*c*) Temporarily polar molecule polarises the neighbouring molecule and creates in it the induced dipole.

In this way a large number of non-polar molecules become temporarily polar which are mutually attracted by weak attractive forces. Since the molecules are in ceaselessly rapid motion and are constantly departing from the site of temporary dipoles, the attractive forces acting between the polar molecules are very weak. Since the electron cloud densities are continuously fluctuating, these attractive forces may be formed and broken quickly. According to Frintz London (1930) the instantaneous formation and subsequent decay of dipoles in a gas due to temporary distortion of electron cloud may be depicted as in Fig. 7.38.

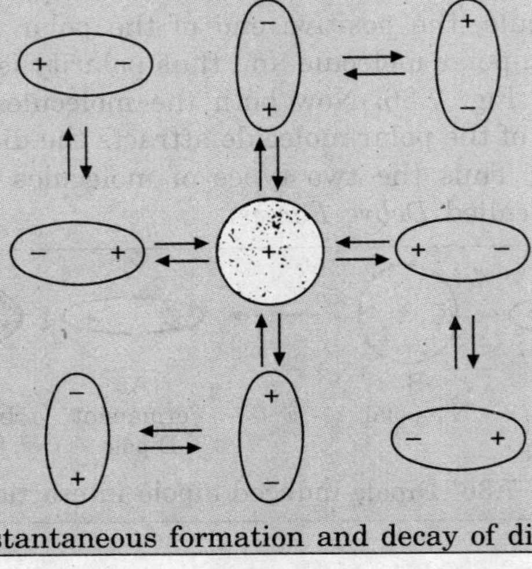

Fig. 7.38. Instantaneous formation and decay of dipoles in a gas

The weak inter molecular forces operating in gases due to instantaneous polarisation of non-polar molecule are also called *London Forces*. These forces are very weak and are known to operate in all types of molecules. These forces are weaker in most cases than dipole-dipole or ion-dipole forces, but are responsible for bringing about condensation of gases, even of noble gases. These forces become stronger with the increase of number of electrons in molecules/atoms. The forces become stronger with the increase in boiling points of the molecules as shown below :

Molecules/Atoms	:	*He*	H_2	*Ne*	N_2	O_2	*Ar*	F_2
No. of electrons	:	*2*	*2*	*10*	*14*	*16*	*18*	*18*
Boiling points (°C)	:	*–269*	*–253*	*–246*	*–196*	*–183*	*–186*	*–186*
Strength of London forces	:	——— *Strength increasing* ———→						

Nature of van der Waals Forces

van der Waals forces are much weaker than both ionic and covalent bonds, *e.g.* the energy needed to dissociate a Cl_2 molecule (containing atoms bonded together by a covalent bond, Cl–Cl) into its atoms is 58.0 Kcal while that needed for sublimating crystalline chlorine (containing atoms bonded together by van der Waals forces) is only 6.02 Kcal.

These forces may also be present, to a lesser extent, in molecules which are partly polar and partly non-polar and even in those which are completely non-polar.

Origin of van der Waals Forces

When two atoms or molecules approach each other, both attractive and repulsive forces operate between their negative electrons and positive protons (nucleus) as shown in Fig. 7.39. Protons have been shown by + sign.

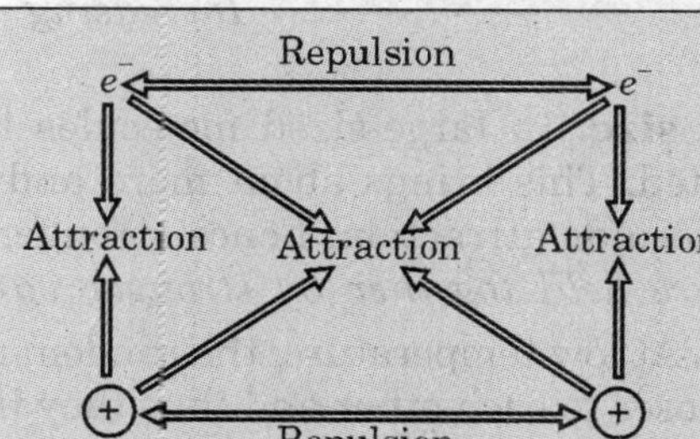

Fig. 7.39. Attractive and repulsive forces operating between electrons and protons (shown by + sign) when the atoms/molecules approach each other.

These forces are :

(i) Attractive forces. These forces act between the nucleus of a molecule and its own electrons, and between the nucleus of one molecule and electrons of the other molecule.

(ii) Repulsive forces. These forces act between the electrons of two different molecules, and between the nuclei of the two different molecules.

The attractive and repulsive forces are in equilibrium at an inter-molecular distance which is, approximately equal to 4 Å (1 Å = 10^{-10} *m*). However, at inter-molecular distance between 4 Å to 10 Å, attractive forces predominate. It is these attractive forces which are acting between the molecules and are called van der Waals forces.

Factors Affecting the Strength and Magnitude of van der Waals Forces

Following are the important factors that determine the strength and magnitude of van der Waals forces.

1. Large number of electrons in molecules. *With the increase in the number of electrons in a molecule, the magnitude of van der Waals forces between the molecules also increases.* On account of larger number of electrons in a molecule, there is a greater diffusion of electron clouds. This results in a greater polarisation and hence inter-molecular interactions (*i.e.* van der Waals forces) also increase. Since the molecular weight is roughly proportional to the number of electrons in the molecule, the van der Waals forces increase with the increase in the molecular weight. The boiling point or melting point of a substance is taken as a measure of the magnitude of van der Waals forces.

With the increase of van der Waals forces the boiling point or melting point of a substance also increases. For inert gases the relationship between the number of electrons, molecular weight and boiling point, and magnitude of van der Waals forces is shown below :

Inert gases	:	*He*	*Ne*	*Ar*	*Kr*	*Xe*	*Rn*
No. of electrons	:	*2*	*10*	*18*	*36*	*54*	*86*
Mol. wt.	:	*4*	*20*	*40*	*83.7*	*131*	*222*
Boiling point (°C)	:	*–269*	*–246*	*–186*	*–152*	*–107*	*–62*
Magnitude of van der Waals forces	:	——— *Increasing* ———→					

2. Large molecular size. In large-sized molecules the electron clouds are more diffused and distorted. This brings about more and more polarisation and more and more inter-molecular attraction, hence stronger van der Waals forces. Thus, *larger molecules are held together by stronger van der Waals forces.*

3. Low temperature. At low temperature, the random movements of molecules decrease and they come closer to each other and thus they interact more effectively. Hence, *low temperature increases the strength and magnitude of van der Waals forces.*

4. High pressure. At high pressures the molecules come closer and it results into strong inter-molecular interactions or van der Waals forces. Thus, *higher the pressure, stronger will be the van der Waals forces.*

Applications of van der Waals Forces

Although van der Waals forces are weak, yet they have been used to explain many important phenomena. For example :

(*i*) It is the van der Waals forces between the molecules that cause substances like inert gases, halogens etc. to condense to liquids and to freeze into solids, when the temperature is considerably lowered. Thus these forces are responsible for bringing about condensation and crystallisation (at low temperatures) in the noble gases and halogens.

(*ii*) These forces account for varied hardness of certain solids. If the atoms or molecules in any crystal are united only by van der Waals forces, the crystals can be broken down easily. Such crystals will be soft and possess low melting points.

(*iii*) These forces, though extremely weak, can explain the deviation of the behaviour of gases at low temperature from that of the ideal gases. The details of this application may be found in the chapter on *Kinetic Theory of Gases* which may be studied from some *Physical Chemistry Book.*

Questions with Answers

Q.1 Arrange the following compounds in the order of their increasing melting points :

(*i*) LiF, LiCl, LiBr, LiI **(*ii*) LiCl, NaCl, KCl, RbCl, CsCl.**

Ans : The answer to this question can be given either on the basis of Fajans'rules or on the basis of the relation between the electronegativity diffirence between two concerned atoms and ionic character in the bond between two atoms.

(*i*) Greater the electronegativity difference between the two concerned atoms, greater would be the ionic character and higher would be the melting point.

LiF > LiCl > LiBr > LiI

(*ii*) Due to covalent character LiCl has lower melting point.

∴ LiCl < NaCl > KCl > RbCl > CsCl.

Q.2 The ionic character in certain A–B bond is 76.81% and the bond length A–B is 159.0 pm. Calculate dipole moment of AB molecule.

Ans : Ionic character in A–B bond = 76.81% (Given)

Bond length (d) = 159.0 pm (Given)

$= 159.0 \times 10^{-10}$cm (∵ 1pm = 10^{-10}cm)

We know that $q = 4.8 \times 10^{-10}$esu

$\mu_{exp} = ?$

$\mu_{cal} = q \times d = (4.8 \times 10^{-10} \text{ esu}) \times (159 \times 10^{-10} \text{ cm})$

$= 4.8 \times 159 \times 10^{-20}$ esu.cm

$$\% \text{ ionic character} = \frac{\mu_{exp}}{\mu_{cal}} \times 100$$

or
$$76.81 = \frac{\mu_{exp} \times 100}{4.8 \times 159 \times 10^{-20} \text{ esu.cm}}$$

∴
$$\mu_{exp} = \frac{76.81 \times 4.8 \times 159 \times 10^{-20}}{100} \text{esu.cm}$$

$$= \frac{76.81 \times 4.8 \times 159 \times 10^{-2}}{10^2} \times 10^{-18}\text{esu.cm}$$
$$= 76.81 \times 4.8 \times 159 \times 10^{-4}\text{D} \quad (\because \quad 10^{-18}\text{esu.cm} = 1\text{D})$$
$$= 58621.392 \times 10^{-4}\text{D}$$
$$= \mathbf{5.86\ D\ (Ans)}$$

Q.3 The observed value of dipole moment of H_2O molecule is found to be 1.84 D. Calculate the H–O–H bond angle in H_2O molecule, given that the bond moment of O–H bond is 1.5 D.

Ans : μ_{H_2O} = 1.84 D (*Given*)

μ_{OH} = 1.5 D (*Given*)

If H–O–H bond angle is $\propto$, then we have :

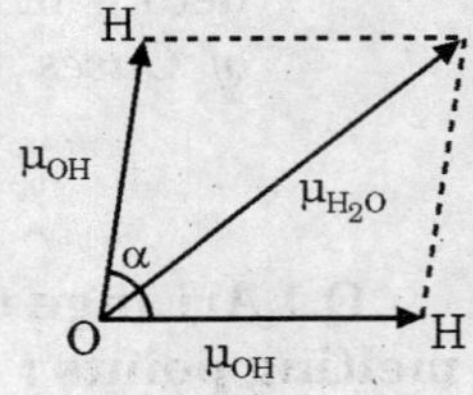

$$\mu^2_{H_2O} = \mu^2_{OH} + \mu^2_{OH} + 2 \times \mu_{OH} \times \mu_{OH} \times \text{Cos} \propto$$
$$= (1.5)^2 + (1.5)^2 + 2 \times 1.5 \times 1.5 \times \text{Cos} \propto$$
$$= 2.25 + 2.25 + 4.5 \times \text{Cos} \propto$$

or $(1.84)^2 = 4.5 + 4.5 \text{ Cos} \propto$

$$\therefore \quad \text{Cos} \propto = \frac{(1.84)^2 - 4.5}{4.5} = \frac{3.3856 - 4.5}{4.5}$$

or $\text{Cos} \propto = -\ 0.2476$

or $\propto = \mathbf{104° \ 28' \ (Ans)}$

Q.4 Arrange the following molecules in the decreasing order of their dipole moment : CH_4, CH_3Cl, CH_2Cl_2 (cis form) and $CHCl_3$.

Ans : The structeres of the given molecules are given below. The direction of bond moment has also been shown

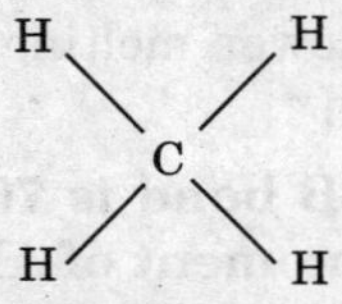

CH_4 (tetrahedral-symmetrical shape), μ = 0

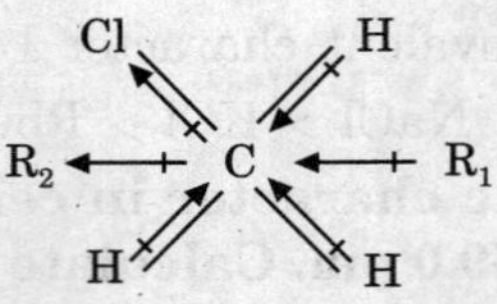

CH_3Cl (μ = 1.86 D)

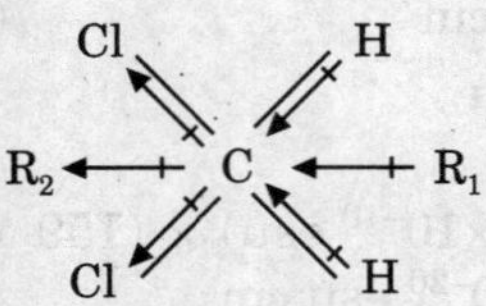

CH_2Cl_2 (μ = 1.62 D)

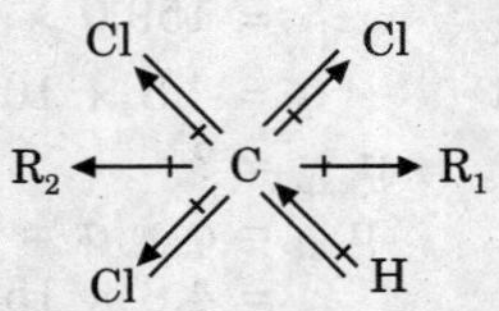

$CHCl_3$ (μ = 1.03 D)

CH_4 molecule (μ = 0) This molecule has tetrahedral shape (symmetrical shape). Hence μ for this molecule = 0

CH_3Cl molecule (μ = 1.86 D). In this molecule, the resltant (R_1) of two C–H bond moments adds to the resultant (R_2) of C–H and C–Cl bond moments.

CH_2Cl_2 molecule (μ = 1.62 D). In this molecule, the resultant (R_1) of two C–H bond moments adds to the resultant (R_2) of two C–Cl bond moments.

$CHCl_3$ molecule (μ = 1.03 D). In this molecule the resltant (R_1) of C–H and C–Cl bond moments oppose the resultant (R_2) of two C–Cl bond moments.

In CH_3Cl and CH_2Cl_2 molecules, the resultants R_1 and R_2 are equal but R_2 in CH_3Cl is more than R_2 in CH_2Cl_2. Hence the sum, ($R_1 + R_2$) for CH_3Cl is greater than the sum, ($R_1 + R_2$) for CH_2Cl_2. In other words μ of CH_3Cl is greater than that of CH_2Cl_2. In $CHCl_3$, since R_1 and R_2 oppose each other, μ for $CHCl_3$ is less than CH_3Cl and CH_2Cl_2 both.

Above discussion shows that the decreasing order of dipole moments values are :

CH_3Cl	>	CH_2Cl_2	>	$CHCl_3$	>	CH_4
1.86 D		1.62 D		1.03 D		0.0 D

Q.5 Which of the compounds *viz* $SnCl_2$, $SnCl_4$, SnF_2 and SnF_4 is the least covalent?

Ans : Out of given halides, SnF_2 is the least covalent. This can be explained as follows with the help of Fajan's rules.

Due to smaller size of Sn^{4+} ion as compared to that of Sn^{2+} ion, $SnCl_4$ is more covalent than $SnCl_2$ ($SnCl_4 > SnCl_2$). Simiarly SnF_4 is more covalent than SnF_2 ($SnF_4 > SnF_2$). Out of SnF_2 and $SnCl_2$, due to large size of Cl^- ion as compared to that of F^- ion, $SnCl_2$ is more covalent than SnF_2. Thus SnF_2 is the least covalent.

Q.6 In each of the following pairs of compounds, which one is more covalent and why?

***(i)* AgCl, AgI** ***(ii)* $BeCl_2$, $MgCl_2$** ***(iii)* $SnCl_2$, $SnCl_4$** ***(iv)* CuO, CuS**

Ans : Applying Fajan's rules, the result can be obtained in each case, as follows :

(i) AgI is more covalent than AgCl. This is because I^- ion is larger in size than Cl^- ion and hence is more polarized than Cl^- ion.

(ii) $BeCl_2$ is more covalent than $MgCl_2$. This is because Be^{2+} ion is smaller in size than Mg^{2+} ion and hence has greater polarizing power.

(iii) $SnCl_4$ is more covalent than $SnCl_2$. This is because Sn^{4+} ion has greater charge and smaller size than Sn^{2+} ion and hence has greater polarizing power.

(iv) CuS is more covalent than CuO. This is because S^{2-} ion has larger size than O^{2-} ion and hence is more polarized than O^{2-} ion.

Q.7 $BeCO_3$ is less stable than $MgCO_3$. Explain.

Ans : Lower stability of $BeCO_3$ as compared to that of $MgCO_3$ can be explained on the basis of Fajan's rules. Since Be^{2+} ion is smaller in size than Mg^{2+} ion, $BeCO_3$ is covalent (lesser ionic) while $MgCO_3$ is ionic. Being covalent, $BeCO_3$ is less stable than $MgCO_3$ which is ionic. It in due to lesser stability of $BeCO_3$ that it gets decomposed into BeO and CO_2 on heating

$$BeCO_3 \xrightarrow{\Delta} BeO + CO_2$$

Q.8 How many H-bonds are formed by each H_2O molecule in the structure of liquid water and how many water molecules are attached to each H_2O molecule and in what direction ?

Ans : If we examine the structure of liquid water, we find that :

(*i*) Each H_2O molecule forms **four H-bonds :** two H-bonds are formed by O-atom of H_2O molecule and two H-bonds are formed by two H-atoms of the same H_2O molecule.

(*ii*) Each H_2O molecule is linked **tetrahedrally** with four H_2O molecules through H-bonds.

Q.9 Discuss the structuer of ice.

or **Explain why ice has lower density than water.**

or **Explain why ice floats on water.**

or **Explain why ice is lighter than water.**

Ans : When liquid water is allowed to freez, it is converted into solid, called ice. Thus ice is the solid form of liquid water. Ice has larger volume than liquid water.

$$\underset{\text{(Less volume)}}{\text{Liquid water}} \xrightarrow{\text{Freezing}} \underset{\text{(Larger volume)}}{\text{Solid water (Ice)}}$$

X-ray studies of the structure of hexagonal ice have shown that the crystal of ice contains H_2O molecules. In this structuer each H_2O molecule is tetrahedrally linked will four other H_2O molecules, *i.e.* one H_2O molecule may be supposed to be placed at the centre of the tetrahedron and this central H_2O molecule is linked with other four H_2O molecules placed at the four corners of the tetrahedron as shown in Fig. 7.40.

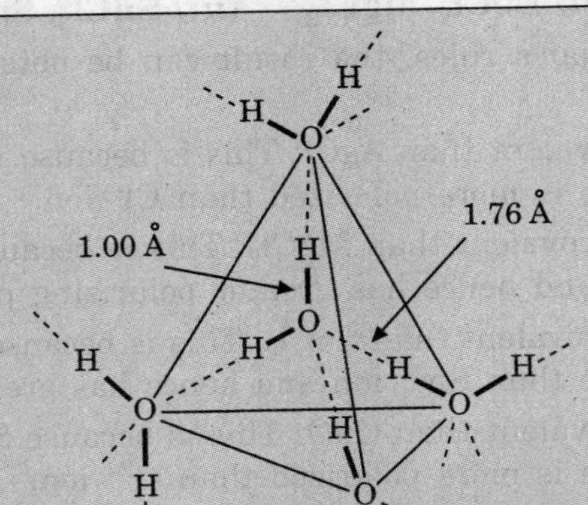

Fig. 7.40. Tetrahedral arrangement of four H_2O molecules round each H_2O molecule in the structure of ice.

It may be seen from Fig. 7.40. that each O-atom of each H_2O molecule is linked with four H-atoms. O-atom is linked with two H-atoms of its own H_2O molecule by strong covalent bonds (shown by solid lines) and with two H-atoms of the neighbouring two H_2O molecules by two H-bonds (shown by dotted lines).

This structure shows that O-atom of each H_2O molecule can form maximum two H-bonds with other two H_2O molecules and each H-atom of each H_2O molecule can form only one H-bond with O-atom of the neighbouring H_2O molecule. Thus each H_2O molecule, as a whole, can form maximum four H-bonds with other four H_2O molecules, ***i.e.* each H_2O molecule is linked tetrahedrally with four H_2O molecules by H-bonds.**

Since O...H hydrogen bonds are longer (= 1.76 Å) than O–H covalent bonds (= 1.00 Å), H_2O molecules in the crystal structure of ice are not closely packed together and hence there exist a number of vacant spaces between H_2O molecules. The presence of vacant spaces gives **open cage-like structure** to ice (see Fig. 7.41).

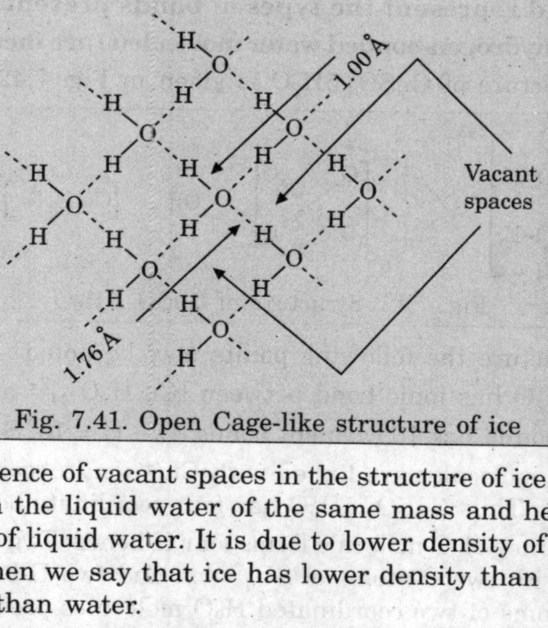

Fig. 7.41. Open Cage-like structure of ice

Due to the presence of vacant spaces in the structure of ice, ice has relatively larger volume than the liquid water of the same mass and hence density of ice is lower than that of liquid water. It is due to lower density of ice that it (*i.e.*ice) floats on water. When we say that ice has lower density than water, this means that ice is lighter than water.

Q.10 Find out the number of bonds formed by O-atom of each H_2O molecule in ice.

Ans : If we examine the structure of ice, we find that O-atom of each H_2O molecule is linked with its two H-atoms by two O — H covalent bonds and with two H-atoms of other two H_2O molecules by two O...H hydrogen bonds. Thus total number of bonds formed by O-atom is 4.

Q.11 Discuss the action of heat on ice.

or Explain that water has maximum density at 4° C.

Ans : We know that ice has open cage-link structure which contains a number of H_2O molecules which are linked with one another by H-bonds. This structure also contains vacant spaces between H_2O molecules. When ice is heated, it melts and is converted into liquid water. During melting H-bonds existing between different H_2O molecules in the open cage-link structure of ice start breaking and water molecules are separted from ice. These separated water molecules occupy vacant spaces exiting in the open-cage structure of ice. This results in that water molecules come closer to each other and hence volume decreases. With the decrease in volume, the density of liquid water increases. The increase in density of water takes place only in the temperatuer range 0°C to 4°C. After 4°C, the kinetic energy of water molecules increase and hence they once again start moving away from each other. This causes expansion in volume of liquid water. Thus the volume of liquid water increases. The increases in volume decreases the density of liquid water. Thus above 4°C the density of water starts decreasing. The above discussion shows that when ice is melted, it is converted into liquid water. The density of liquid water increases upto 4°C but beyond 4°C the density starts decreasing. Thus water has maximum density at 4°C.

Q.12 Name and represent the types of bonds present in $CuSO_4.5H_2O$

***or* How many hydrogen-bonded water molecule(s) are there in $CuSO_4.5H_2O$?**

Ans : The structure of $CuSO_4.5H_2O$ is given in Fig. 7.42.

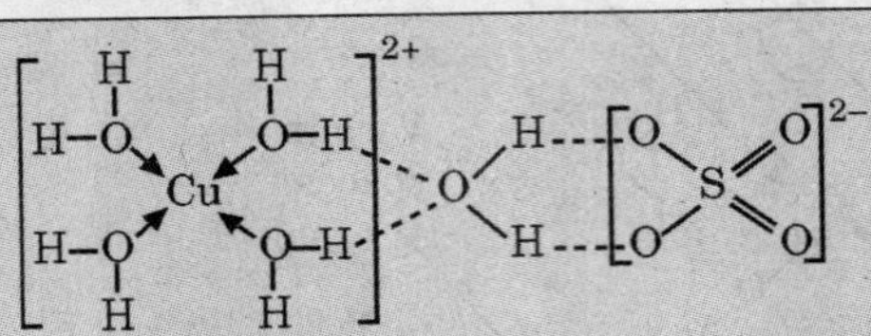

OR $[Cu(H_2O)_4]^{2+}\ SO_4^{2-}.H_2O$

Fig. 7.42. Structure of $CuSO_4.5H_2O$

From this structure the following points may be noted :

(*i*) $CuSO_4.5H_2O$ has ionic bond between $[Cu(H_2O)_4]^{2+}$ and SO_4^{2-} ions,

(*ii*) This compound has 16 covalent bonds (O – H = 10, S–$\bar{O}$ = 2, S = O = 4).

(*iii*) Four H_2O molecules are linked with Cu^{2+} ion in coordination sphere by 4 coordinate bonds. Thus $CuSO_4.5H_2O$ has four coordinate bonds ($H_2O \rightarrow Cu^{2+}$).

(*iv*) The remaining H_2O molecule forms four H-bonds. Two H-bonds are formed by two H-atoms with two O-atoms of SO_4^{2-} ion and two H-bonds are formed by O-atom with H-atoms of two coordinated H_2O molecules present in coordination sphere.

Q.13 Explain the mechanism for the formation of KHF_2 by the combination of KF and HF in aqueous solution.

***or* KF combines with HF to form KHF_2 but KCl, KBr and KI donot combine with HCl, HBr and HI to form $KHCl_2$, $KHBr_2$ and KHI_2 respectively. Explain why ?**

***or* KHF_2 exists but $KHCl_2$, $KHBr_2$ and KHI_2 do not exist. Explain why?**

or KHF_2 contains K^+ and HF_2^- ions but does not contain K^+, H^+ and $2F^-$ ions. Explain.

Ans : When KF combines with HF in aqueous solution, KHF_2 is formed.

$$KF \rightarrow K^+ + F^-$$

$$F^- + HF \rightarrow F^- ... H - F \text{ or } [HF_2]^-$$

On adding : $KF + HF \rightarrow K^+ [HF_2]^-$

The formation of HF_2^- ion takes place due to the attachment of F^- ion to H-atom of HF molecule through H-bonding. Due to smaller electronegative of Cl^-, Br^- and I^- these ions cannot combine with HCl, HBr and HI to form $[HCl_2]^-$, $[HBr_2]^-$ and $[HI_2]^-$ ions respectively. Due to non-formation of these ions, $K[HCl_2]$, $K[HBr_2]$ and $K[HI_2]$ donot exits. In other words we can say that KCl, KBr and KI cannot combine with HCl, HBr and HI and cannot form $K[HCl_2]$, $K[HBr_2]$ and $K[HI_2]$ respectively. Above discussion shows KHF_2 contains K^+ and $[HF_2]^-$ and not K^+, H^+ and F^- ions.

University Questions

1. Discuss any two rules proposed by Fajan. *(Himachal Pradesh 2000)*

2. (*a*) What is inert pair effect? List the elements of p-block for which inert pair effect is predominant.

(*b*) Explain (*i*) Boron forms no compounds in unipositive oxidation but thallium in unipositive state is quite stable. (*ii*) Al forms the ion, $[AlF_6]^{3-}$ but boron does not form the ion, $[BF_6]^{3-}$. (*iii*) Bond angle in oxides decreases as $Br_2O > Cl_2O > OF_2$. *(Himachal Pradesh 2000)*

3. (*a*) Nitrogen forms only NCl_3 but phosphorus forms both PCl_3 as well as PCl_5. Explain why ?

(*b*) Write a note on (*i*) Sidgwick's maximum covalency rule. (*ii*) Variable valency. *(Kanpur 2000)*

4. (*a*) Write a note on "Free electron theory of metallic bonding".

(*b*) Give an account of odd electron bond. *(Kumaon 2000)*

5. (*a*) Explain the following :

(*i*) NaCl is soluble in water but not in chloroform.

(*ii*) Though ethanol is a covalent compound it is soluble in water.

(*b*) Write a note on "Resonance". *[Gauhati (General) – 2001]*

6. Explain the following :

(*i*) NCl_5 is not formed while PCl_5 is formed.

(*ii*) Bond angle of PH_3 is less than NH_3.

(*iii*) Al forms $[AlF_6]^{3-}$ ion but boron does not form $[BF_6]^{-3}$.

(*iv*) NH_3 has a dipole moment but BF_3 is non polar. *(Lucknow 2000)*

7. (*i*) Explain the following :

(*a*) AlF_3 is ionic while $AlCl_3$ is covalent.

(*b*) AgI is covalent whereas Nails ionic.

(*c*) Phosphorus forms PCl_3 and PCl_5 both while nitrogen forms only NCl_3.

(*ii*) Write short note on : (*a*) Diagonal relationship (*b*) inert pair effect (*c*) Hydrogen bond (*d*) Electropositivity. *(Avadh 2000)*

8. (*i*) Explain why LiCl is insoluble in water.

(*ii*) Write a note on "Inert pair effect".

(*iii*) Explain σ and π bonds. *(Lucknow 2001)*

9. Which of the following compounds does not contain ionic bond ? NaOH, K_2S, HCl and LiH. *(H.N. Bahuguna 2006)*

10. Which of the following is the most covalent ? $BeCl_2$, $MgCl_2$, $CaCl_2$ and $SrCl_2$. *(Agra 2008)*

11. (i) Write a note an Fajans rules *(Meerut 2008)*

(ii) Explain intermolecular H-bonding

12. Cu_2Cl_2 is more covalent than NaCl. Why ? *(Meerut 2009)*

CHAPTER

8

Chemical Bonding : Orbital Concept

Linnett's Double-Quartet (D-Q) Theory

The octet theory given by Lewis and Langmuir has been found unsuccessful in explaining the paramagnetic behaviour of some molecules.

In 1961 Linnett modified octet theory and gave a satisfactory explanation of the paramagnetism of the molecules. According to Linnett's D-Q theory, eight valence-shell electrons (*i.e.* an octet) of an atom can be considered to be made up of two groups of four electrons instead of a group of four localised pairs of electrons. One group has 4 electrons of one spin while the other group also has 4 electrons of the other spin. Thus according to Linnett, the octet of valence shell electrons is treated as a *double-quartet* (*D-Q*) of electrons. The distribution of electrons round a given nucleus is governed by two factors.

1. Spin correlation. It is because of spin correlation that the electrons of the like spin tend to keep apart and those of unlike spin tend to come together (*Pauli's exclusion principle*).

2. Charge correlation. This arises because of the repulsion between the particles of like charge and it also keeps the electrons of the like spin apart.

Thus we see that for paired electrons the spin correlation is opposed to the charge correlation. In order to make the inter-electronic repulsion minimum four electrons of each quartet set tend to arrange themselves at the vertices of two inter-connected tetrahedra. The state of minimum inter-electronic repulsion is achieved when the vertices of one tetrahedron lie at the centre of the faces of the other tetrahedron (Fig. 8.1).

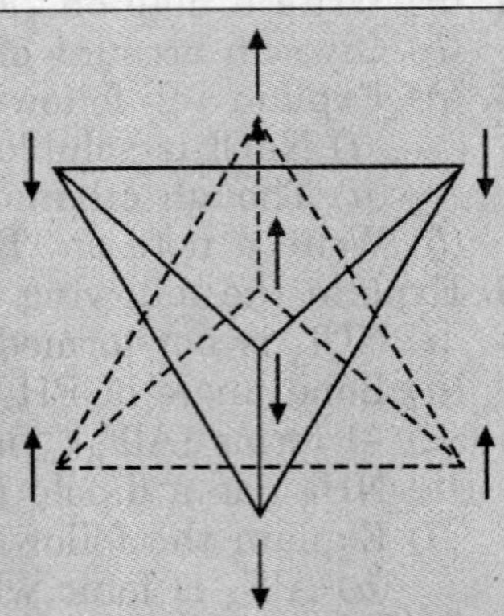

Fig. 8.1. Two inter-connected tetrahedra having two sets of 4 electrons. Four electrons of one set having like spin have been shown as ↑ while the four electrons of the other set have been shown as ↓. The vertices of one tetrahedron lie at the centre of the faces of the other tetrahedron. This arrangement gives minimum inter-electronic repulsion.

To Explain the Magnetic Behaviour of O_2 and N_2 Molecules on the Basis of D-Q Theory

Now let us take the examples of O_2 and N_2 molecules to illustrate the use of Linnett's D-Q theory in explaining their structure and magnetic behaviour.

1. O_2 molecule. On the basis of D-Q theory, 12 valence- shell electrons of O_2 molecule can be considered to be made up of two groups—one group has 5 electrons of one spin while the other group has 7 electrons of the other spin. Five electrons of one group having like spin arrange themselves in two tetrahedra sharing a face (Fig. 8.2) while the seven electrons of the other group having like spin arrange themselves in two tetrahedra sharing one apex (Fig 8.3). Both the arrangements have minimum energy.

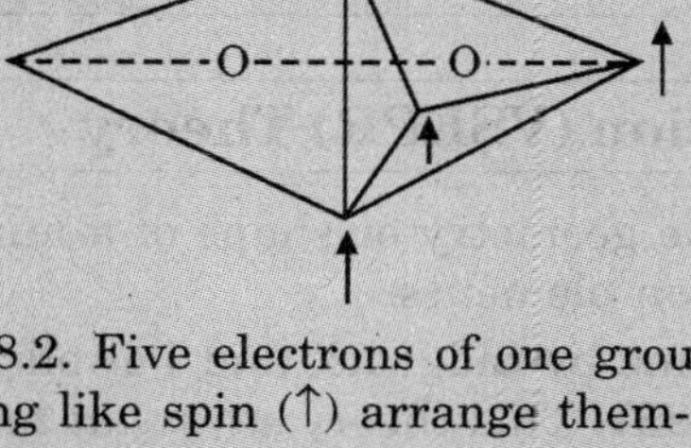

Fig. 8.2. Five electrons of one group having like spin (↑) arrange themselves in two tetrahedra sharing a face.

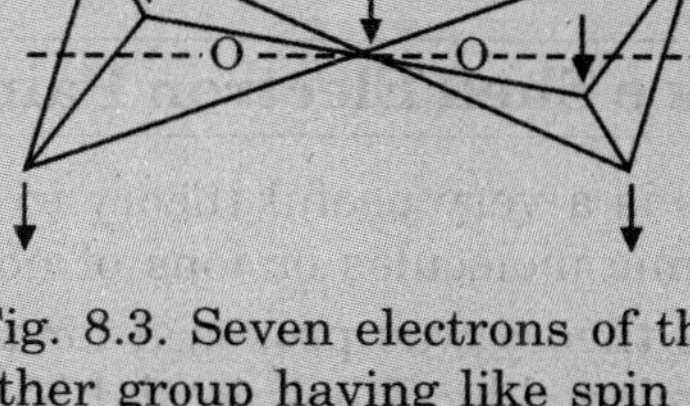

Fig. 8.3. Seven electrons of the other group having like spin (↓) arrange themselves in two tetrahedra sharing one apex.

On superimposing the two arrangements shown in Fig. 8.2 and 8.3, we obtain the structure of O_2 molecule in the ground state as :

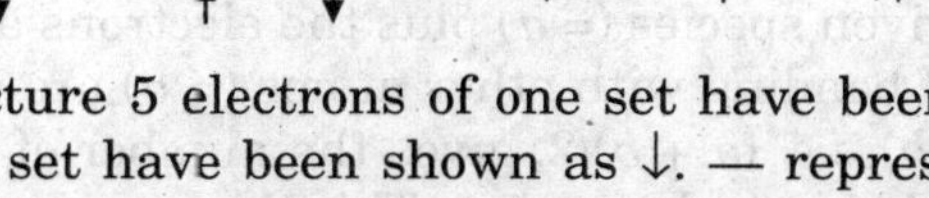

In the above structure 5 electrons of one set have been shown as ↑ while 7 electrons of the other set have been shown as ↓. — represents an electron pair of opposite spins. This structure of O_2 molecule in its ground state satisfies all the requirements *viz.* 2 unpaired electrons (*paramagnetic behaviour*), 8 electrons around each O-atom and 4 electrons in the inter-nuclear region.

Twelve electrons of O_2 molecule in the excited state (159 KJ. mole^{-1} above the ground state) can be consider to be made up of two sets of 6 electrons which are arranged as 6 pairs at the corners of two tetrahedra sharing one edge between them (Fig. 8.4). This structure does not contain any unpaired electrons and hence well explains the diamagnetic nature of O_2 molecule in the excited state.

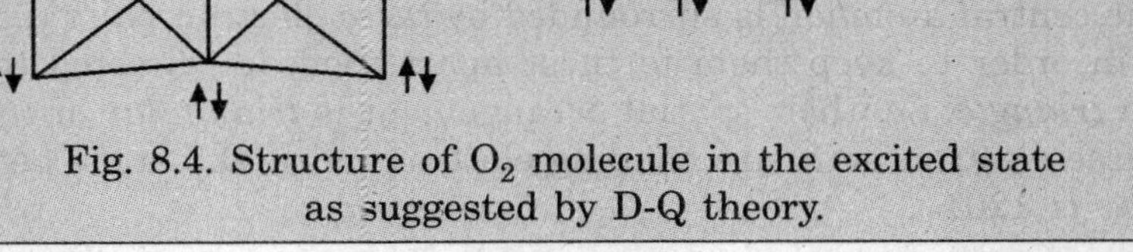

Fig. 8.4. Structure of O_2 molecule in the excited state as suggested by D-Q theory.

2. N_2 molecule. According to D-Q theory 10 valence-shell electrons of N_2 molecule are made up of two sets each of which has 5 electrons of the same spin

(See Fig. 8.5). Thus we find that both octet theory and D-Q theory suggest that N_2 molecule has bond order of three (six electrons between the nuclei) and is *diamagnetic* because of spin cancellation.

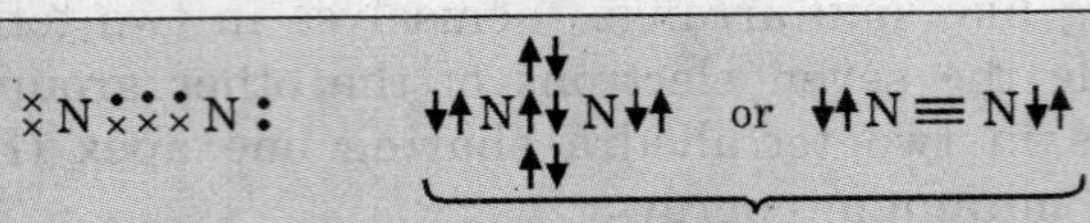

(*a*) (*b*)

Fig. 8.5. Structures of N_2 molecule. Structure (*a*) is based on octet theory while structure (b) is based on D-Q theory.

Valence Shell Electron Pair Repulsion (VSEPR) Theory

This is a very useful theory to predict the geometry or shape of a number of polyatomic molecules or ions of non-transition elements.

This theory was proposed for the first time by Sidgwick and Powell in 1940 and developed by Gillespie and Nyholm in 1957. This theory says *that the shape of a given species (molecule -or ion) depends on the number and nature of electron pairs surrounding the central atom/ion of the species.*

This theory makes the following assumptions :

Assumptions of VSEPR Theory

1. Spatial arrangement of electron pairs round the central atom/ion of a given molecule/ion. The electrons already present in the valence-shell of the central atom/ion of a given species (= *a*) plus the electrons acquired by the central atom/ion as a result of bonding with other atoms (= *b*) are called its *valence-shell electrons.* Half of (*a* + *b*), *i.e.* (*a* + *b*)/2, gives the number of electron pairs present in the valence-shell of the central atom/ion. This theory assumes that these electron pairs occupy localised orbitals which arrange themselves in space in such a way that they keep apart from one another as far as possible so that they may experience minimum electrostatic repulsion between them and hence may give minimum energy and maximum stability to the species.

According to this assumption :

(*a*) If the central atom/ion is surrounded *by two* electron pairs, *i.e.* if the sum of bonding electron pairs (*bps*) and non-bonding electron pairs, also called lone pairs of electrons (*lps*) is *two*, then in order to keep them farthest apart these are directed in opposite directions along the same axis, *i.e.* their spatial arrangement is *linear* (*diagonal*) and the angle between them (*i.e.* bond angle) is 180°.

(*b*) If the central atom/ion is surrounded by *three* electron pairs (*i.e. bps* + *lps* = 3), then in order to keep them farthest apart they are directed towards the corners of a *triangle*, *i.e.* their spatial arrangement is *triangular* or *trigonal* and if all the three electron pairs are bonding electron pairs, the angle between each electron pair is 120°.

(*c*) If the central atom/ion is surrounded by four electron pairs (*i.e. bps* + *lps* = 4), then these electron pairs are directed towards the corners of a *regular tetrahedron i.e.* their spatial arrangement is *tetrahedral* and if all the four electron pairs are *bps*, the bond angle is 109.5°.

(*d*) If the central atom/ion is surrounded by *five* electron pairs (*i.e. bps* + *lps* = 5), then these electron pairs are directed towards the corners of a *trigonal bipyramid, i.e.* their spatial arrangement is trigonal bipyramidal and if all the five electron pairs are bonding pairs, the bond angle between the two equatorial (*i.e.* basal) electron pairs is 120° while that between axial and equatorial electron pairs is 90°. The angle between the two axial electron pairs is 180°.

(*e*) If the central atom/ion is surrounded by *six* electron pairs (*i.e. bps* + *lps* = 6), then these electron pairs are directed towards the corners of a *octahedron, i.e.,* their spatial arrangement is octahedral and the bond angle between the two adjacent (*cis*) electron pairs is 90° while the angle between the axial (*trans*) electron pairs is 180°.

(*f*) If the central atom/ion is surrounded, by *seven* electron pairs (*i.e. bps* + *lps* = 7), then these electron pairs are directed towards the corners of a *pentragonal bipyramid* and hence their spatial arrangement is pentagonal bipyramidal. The angle between the two equatorial electron pairs is 72° while that between the equatorial and axial electron pairs is 90°. The angle between the two axial electron pairs is 180°.

Relationship between the number of electron pairs (*i.e. bps* + *lps*) round the central atom/ion of a given species, the spatial arrangement of the electron pairs and the bond angle (provided that all the electron pairs are bonding electron pairs. *i.e.* there are no lone pairs of electrons) is shown as follows :

No. of electron pairs (bps + lps) :	2	3	4	5	6	7
Spatial arrangement of electron pairs round the central atom / ion :	Linear	Trigonal planar	Tetrahedral	Trigonal bipyramidal	Octahedral	Pentagonal biyramidal
Bond angles provided that all the electron pairs are bps :	180°	120°	109.5	120°, 90°	90°	72°, 90°

2. Regular and irregular geometry of the molecule/Ion. The electron pairs surrounding the central atom/ion are either *bps* or some of them are *bps* and the remaining are *lps*. If the central atom/ion is surrounded *only by the bps*, then the species has a *regular geometry, i.e.* there is no distortion in the shape of the species. If, however, the central atom/ion is surrounded by the *bps* as well as by *lps*, the bond angle gets altered from the value expected for a particular geometry of the molecule or ion. Now since the magnitude of bond angle gives an idea of the geometry of the molecule or ion, the change in the magnitude of the bond angle also changes the geometry of the molecule or ion, *i.e.* with the change of the magnitude of the bond angle, the shape of the molecule or ion gets distorted. In short we can say that *the presence of one or more lone pairs of electrons on* *the central atom changes the magnitude of bond angle which, in turn, changes the geometry of the molecule or ion i.e.* the presence of one or more lone pairs of electrons in the valence shell of the central atom or ion gives *irregular* or *distorted geometry* to the molecule or ion.

Cause of change in bond angle. The change in the magnitude of bond angle is due to the fact that (*lp*—*lp*) repulsion is greater than (*lp*—*bp*) repulsion which, in turn, is greater than (*bp*—*bp*) repulsion, *i.e.*

(lp—lp) repulsion > (lp—bp) repulsion > (bp—bp) repulsion

Here *lp* and *bp* have been used to represent the lone pairs and bonding pairs of electrons surrounding the central atom. Since a *bp* experiences less repulsion from another *bp* than from a *lp*, it (*i.e. bp*) becomes closer to the *bp* and a contraction in the bond angle (*i.e.*, the angle between the two *bps*) occurs.

More the number of *lps* on a central atom, the greater is the contraction caused in the angle between the *bps*. This fact is clear when we compare the bond angles in CH_4, NH_3 and H_2O molecules.

Molecules	:	CH_4	NH_3	H_2O
No. of lps on the central atom	:	0	1	2
Bond angle (Angle between two bps)	:	109.5°	107.5°	105.5°
Contraction in bond angle relative to that of CH_4	:	0	109.5°—107.5° = 2°	109.5° –105.5° = 4°

The comparison of bond angles of NH_3 and H_2O with that of CH_4 shows that each of H—O—H bond angles in H_2O is decreased from a tetrahedral angle (= 109.5°) to a greater extent than each of H—N—H bond angles in NH_3. The greater decrease in case of H_2O is explained as follows :

The valence-shell of O-atom in H_2O molecule has four electron pairs as N-atom in NH_3 molecule has. Two of these electron pairs are *bps* each of which is attracted by two nuclei (of H and O atoms) while the remaining electron pairs are *lps* each of which is attracted only by one nucleus (of O-atom), since these *lps* orginate from O-atom only. Thus we see that O-atom in H_2O molecule has two *lps* while N-atom in NH_3 molecule has only one *lp* on it. Consequently in H_2O molecule there are three types of electron pair—electron pair repulsions which are (*i*) (*lp*—*lp*) *repulsion* (*ii*) (*lp*—*bp*) *repulsion and* (*iii*) (*bp*—*bp*) repulsion while in NH_3 molecule there are only two types of repulsions which are (*i*) (*lp* —*bp*) repulsion and (*ii*) (*bp*—*bp*) repulsion (See Fig. 8.6.).

The magnitude of these repulsions, as shows above, is the following order :

(lp—lp) repulsion > (lp—bp) repulsion > (bp—bp) repulsion

The net effect of this difference in electron pair—electron pair repulsions is that the two *lps* on O-atom force the two (O—H) *bps* to come more closer to each other than one *lp* on N-atom in NH_3 molecule forces three (N—H) *bps* to come closer to each other. Thus each of H—O—H bond angles in H_2O is decreased from a tetrahedral angle to a *greater extent* than each of H—N—H bond angles in NH_3

(H—N—H bond angle in NH_3 = 107.5°, H—O—H bond angle in H_2O = 105.5°). Due to the decrease in the expected tetrahedral angle, shape of H_2O molecule gets distorted and hence H_2O molecule assumes *planar V-shape* (*angular or bent shape*).

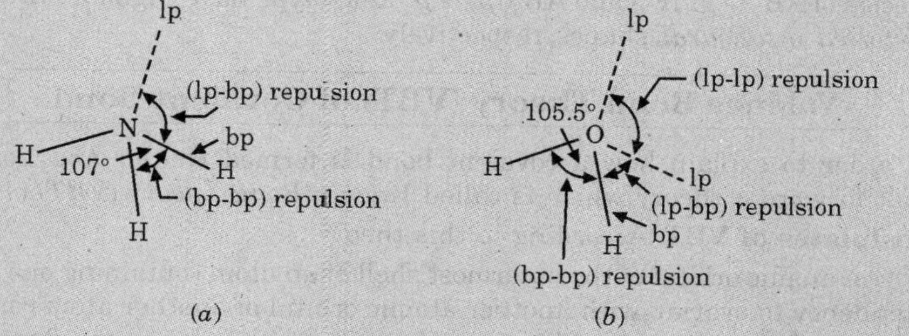

Fig. 8.6. Different types of electron pair—electron pair repulsions in (*a*) NH_3 and (*b*) H_2O molecules.

Shape of Molecules/Ions on the basis of VSEPR Theory

1. Shape of molecules/ions whose central atom/ion has two electron pairs. Such molecules may be of AB_2 (*e.g.* BeF_2) type. These have two σ-bonding electron pairs and no lone pair. (*bps* = 2, *lp* = 0) These have *linear shape*. Here A is the central atom/ion and B are the atoms bonded to atom A by σ-bonds only. π-bonds are not present.

2. Shape of molecules/ions whose central atom/ion has three electron pairs. Such species may be of AB_3 (*bps* = 3, *lp* = 0) and $AB_2(lp)$ (*bps* = 2, *lp* = 1) type. The species of AB_3 (*e.g.* BF_3, $GaCl_3$) and AB_2 (*lp*) [*e.g.* $SnCl_2$ (*gaseous*), $PbCl_2$] type have *trigonal planar* and *angular* or *V-shape* respectively. Here *lp* represents the lone pair of electrons in the valence-shell of the central atom/ion, A.

3. Shape of molecules/ions whose central atom/ion has four electron pairs. Such species may be of AB_4 (*bps* = 4, *lp* = 0), $AB_3(lp)$ (*bps* = 3, *lp* = 1), $AB_2(lp)_2$ (*bps* = 2, *lps* = 2) and $AB(lp)_3$ (*bp* = 1, *lps* = 3) type. The species of AB_4 (*e.g.* CH_4, PCl_4^+, $SiCl_4$, BH_4^-, NH_4^+), $AB_3(lp)$ (*e.g.* NH_3, PH_3), $AB_2(lp)_2$ (H_2O, ICl_2^+, SCl_2) and $AB(lp)_3$ (*e.g.* ICl) type have *regular tetrahedral, trigonal pyramidal, angular* or *V-shape* and *linear shape* respectively.

4. Shape of molecules/ions whose central atom/ion has five electron pairs. Such species may be of AB_5 (*bps* = 5, *lp* = 0), $AB_4(lp)$ (*bps* = 4, *lp* = 1). $AB_3(lp)_2$ (*bps* = 3, *lps* =) and AB_2 $(lp)_3$ (*bps* = 2, *lps* = 3) type. The species of AB_5 (*e.g.* PF_5, PCl_5 (*g*), $AB_4(lp)$ (*e g.* SF_4, IF_4^+), AB_3 $(lp)_2$ (*e.g.* ClF_3 BrF_3), AB_2 $(lp)_3$ (*e.g.* XeF_2 I_3^-, ICl_2^-) type have *trigonal bipyramidal, distorted trigonal bipyramidal, T-shape* and *linear* structures respectively. The lone pairs occupy the *equatorial positions.*

5. Shape of molecules/ions whose central atom/ion has six electrons pairs. Such species may be of AB_6 (*bps* = 6, *lp* = 0), $AB_5(lp)$ (*bps* = 5, *lp* = 1) and $AB_4(lp)_2$ (*bps* = 4, *lps* = 2) type. The species of AB_6 (*e.g.* SF_6, PCl_6^-), $AB_5(lp)$ (*e.g.* SbX_5^{-2}, BrF_5, IF_5) and $AB_4(lp)_2$ (*e.g.* XeF_4, ICl_4^-) type have *regular octahedral, square* *pyramidal* and *square planar* shapes respectively. The lone pairs occupy the *axial positions*.

6. Shape of molecules/ions whose central atom/ion has seven electron pairs. Such species may be of AB_7 (*bps* = 7, *lp* = 0) and $AB_6(lp)$ (*bps* = 6, *lp* = 1). The species of AB_7 (*e.g.* IF_7) and $AB_6(lp)$ (*e.g.* XeF_6) type have *trigonal bipyramidal* and *distorted octahedral* shapes respectively.

Valence Bond Theory (VBT) of Covalent Bond

In order to explain how a covalent bond is formed Heitler and London in 1927 put forward a theory which is called *Valence Bond Theory (VBT) or Atomic*

Poslulates of VBT. According to this theory :

(*i*) An atomic orbital of the outermost shell of an atom containing one electron has a tendency to overlap with another atomic orbital of another atom containing one electron of opposite spin. This type of overlap gives rise to the formation of a bond which is called *covalent bond.* Thus :

According to atomic orbital theory a covalent bond between two atoms results by the overlap of half-filled atomic orbitals belonging to the outermost shell of these two atoms.

OR

The bond formed by the overlap of two atomic orbitals belonging to the outermost shell of two atoms and containing two electrons of opposite spin is called a covalent bond.

(*ii*) The overlap of two atomic orbitals gives rise to a single *bond orbital which is a localised orbital* and is occupied by both the electrons.

(*iii*) The two electrons that occupy the bond orbital have opposite spins.

(*iv*) Each electron occupies the entire bond orbital and hence may be considered to belong to both the atomic orbitals, *i.e.*, the electron pair present in the bond orbital now belongs to each of the two atomic orbitals.

(*v*) As a result of overlapping there is maximum electron density some where between the two atoms. *A large part of the bonding force of covalent bond results from the electrostatic attraction between the nuclei and the accumulated electron clouds between them.*

Extension of VB Theory

The above theory was extended by Pauling and Slater in 1931. According to the extension of this theory :

(*i*) The overlapping of orbitals involves a release of energy and hence the molecule formed has less energy and consequently is more stable than the isolated atoms. This implies that the formation of a covalent bond is always accompanied by the evolution of energy. The energy released as a result of overlapping of orbitals stabilises the system. The amount of energy released per mole at the time of overlapping of orbitals to form the covalent bond is termed as **bond energy** or **stabilisation energy.** This means that this much energy is required to separate the atoms in a molecule, *i.e.* to break the bond. The energy required to break the bond in the molecule is called **bond dissociation energy.** The overlapping of atomic orbitals stops at a point where the atomic nuclei of both the atoms have come close enough to exercise a repulsive force on each other. Here the force of repulsion exactly balances the force of merger or attraction. *The equilibirium distance at which the two atomic nuclei are now held is called* **bond length.**

(*ii*) Overlapping takes place only between those valence- shell orbitals which have unpaired electrons. The orbitals containing paired electrons do not participate in overlapping process and hence do not contribute to bond formation.

(*iii*) Between the two orbitals of the same energy or stability, the orbital which is non-spherical (*e.g.* dumb-bell shaped *p*-orbital) forms stronger bonds than that which is spherically symmetrical (*e.g. s*-orbital).

(*iv*) A spherically symmetrical orbital does not show any preference in direction whereas the non-spherical orbitals tend to form a bond in the direction of maximum electron density within the orbital, *i.e.* along their axis.

Types of Overlap of Atomic Orbitals

Covalent bond between two atoms results by the following three types of overlap of atomic orbitals :

(*i*) s-s overlap. In this type of overlap half-filled *s*-orbital of one atom overlaps with the half filled s-orbital of the other atom [Fig. 8.7 (*a*)]

(*ii*) s-p overlap. Here half-filled *s*-orbital of one atom overlaps with the half-filled *p*-orbital of the other atcm [Fig. 8.7 (*b*)].

(*iii*) p-p overlap. Here singly-filled *p*-orbital of one atom overlaps with the half-filled *p*-orbital of the other atom. Due to different spatial disposition of the *p*-orbital along the three cartesian coordinate axis the two *p*-orbitals may overlap in any of the following two ways :

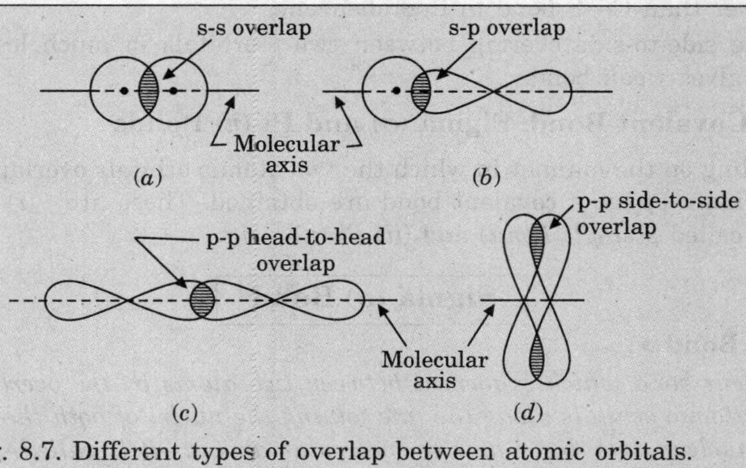

Fig. 8.7. Different types of overlap between atomic orbitals.
(*a*) *s-s* overlap
(*b*) *s-p* overlap
(*c*) *p-p* head-to-head overlap
(*d*) *p-p* side-to-side overlap. Shaded portion contains an electron pair whose electrons have opposite spin.

(*a*) p-p head-to-head overlap. This is also called *head-on* or *end-on* or *end-to-end* or *linear overlap.* Here *the overlap of the two p-orbitals takes place along the line joining the nuclei of two atoms* (called *nuclear axis, bond axis* or *molecular axis*) [Fig. 8.7 (*c*)]. Hence the name linear overlap.

(*b*) p-p side-to-side overlap. This is also called *side-wise* or *side-way* or *lateral overlap. Here the overlap of two p-orbitals takes place along a line perpendicular to the molecular axis.* [Fig. 8.7 (*d*)]. Hence the name lateral overlap. This means that for this type of overlap the two *p*-orbitals must be held parallel to each other, *i.e.* the orbital axis should be parallel.

Strength of Covalent Bond—Effectiveness of Overlap

The strength of a covalent bond can be predicted from the following general conclusions.

(*i*) In the covalent bond formation, greater the extent of overlapping of orbitals, greater is the energy released, *i.e.* the higher will be the strength of the covalent bond. *This implies that the strength of a covalent bond is proportional to the extent of overlapping between the atomic orbitals* and that for a bond to be stronger, greater should be the overlap which in turn shortens the distance between the nuclei. *Thus a stronger bond has a shorter bond length.*

(*ii*) Bonding between two *s*-orbitals is weak particularly when two orbitals are of different energy (*e.g.* Li and K). Strongest bond between s-orbitals is between two 1s-orbitals of two H-atoms.

(*iii*) A strong bond is given by the head-to-head overlap of two *p*-orbitals which are of the same energy. The bond is much weaker, if the orbitals have different principal quantum numbers and bond energy becomes zero as the distance between the two levels increases. For example, C—Br bond in bromobenzene is much weaker than C—F bond in fluorobenzene.

(*iv*) The side-to-side overlap between two *p*-orbitals in much less effective and hence gives weak bond.

Types of Covalent Bond: Sigma (σ) and Pi (π) Bonds

Depending on the manner in which the two atomic orbitals overlap with each other, two main types of covalent bond are obtained. These are : (*i*) **Sigma** (σ) bond (also called a *single bond*) and (*ii*) *Pi* (π) *bond.*

Sigma (σ) Bond

What is σ Bond >

A covalent bond which is formed between two atoms by the overlap of their half filled atomic orbitals along the line joining the nuclei of both the atoms (i.e. along the nuclear axis, bond axis or molecular axis, as it is called) is called a σ*-bond.* In other words σ*-bond is produced by the head-to-head overlap of the half filled atomic orbitals of the two atoms.*

Characteristics of σ-Bond.

σ-bond formed as above has the following characteristics :

(*i*) *The boundary surface of σ-bond takes up an avoid shape.*

(*ii*) *The electron choud of this bond is symmetrical about the bond axis, i.e. the electron density in this bond is distributed symmetrically about the bond axis.*

(*iii*) *This bond has two electrons which have opposite spins. Although these two electrons may be present any where in the bond, the probability of their finding is maximum in the region between the two nuclei on the bond axis. Therefore they are attracted equally by the nuclei of both the atoms.*

How this bond is formed between two atoms can be understood by considering the formation of covalent bonds in the molecules like H_2, HF, H_2O, NH_3 and F_2 on the basis of atomic orbital overlap theory of covalent bond described above.

Formation of σ Bonds in Some Molecules

1. H_2 molecule (*s—s overlap*). The two H-atoms of which H_2 molecule is composed have 1*s* orbital each which contains only one electron (H→1s^1). These 1*s* orbitals of both H-atoms approach each other closely and when they reach a point where the attractive forces (between electrons and nuclei) are balanced by the repulsive forces (between electrons and electrons, between the two nuclei), the potential energy becomes minimum, and hence they overlap and give rise to the formation of H—H σ bond. Since σ bond is produced by the overlap of two *s*-orbitals, it is also called *s—s* σ-bond. Thus a σ bond can be defined as

Various steps for the formation of a σ-bond in H_2 molecule are shown in Fig. 8.8.

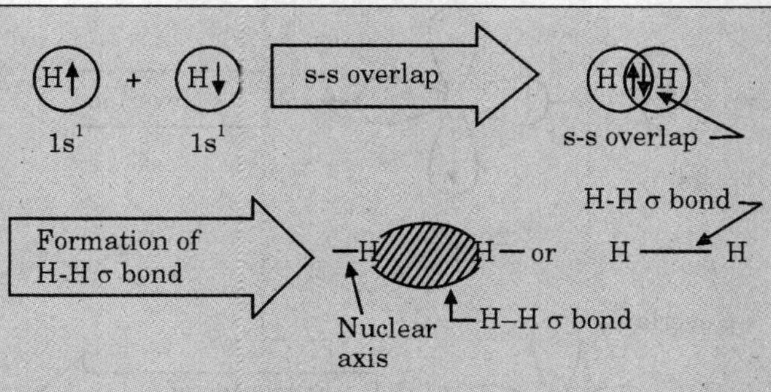

Fig. 8.8. Pictorial representation of various steps for the formation of H—H σ bond in H_2 molecule.

For reasons of space and convenience, a pictorial representation of the formation of σ-bond is not always possible. Therefore in actual practice a single horizontal line (—) between the two linked atoms is used to depict a σ-bond.

2. HF molecule (*s—p overlap*). Here the *singly-filled* 1*s*-orbital of H-atom overlaps with *half-filled* $2p_x$ orbital on F atom (F→$2s^2\ 2p_x^{\ 1}\ 2p_y^{\ 2}\ 2p_z^{\ 2}$) to form H—F σ bond. This bond has one large lobe and one small lobe (Fig. 8.9). The other orbitals on F atom namely 2*s*, $2p_y$, and $2p_z$, being completely filled, do not participate in the overlapping or bond formation process and hence have not been shown in Fig. 8.9. The resulting σ bond is also symmetrical about the nuclear axis and has a nodal plane that stands perpendicular at the right angles to the nuclear axis but does not contain nuclear axis.

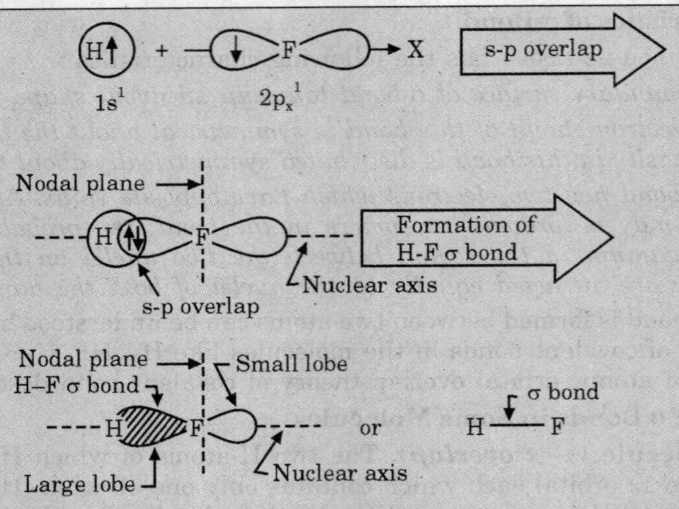

Fig. 8.9. Pictorial representation of various steps for the formation of H—F σ bond in HF molecule.

3. H_2O molecule (*s-p overlap*). The valence shell configuration of O-atom *viz.* $2s^2\ 2p_x^{\ 1}\ 2p_y^{\ 2}\ 2p_z^{\ 1}$ shows that it has two unpaired electrons. In the formation of H_2O molecule the half-filled $2p_x$ and $2p_z$ orbitals overlap with half-filled 2*s*-orbitals of two H-atoms and thus form two O—H σ-bonds as shown in Figure 8.10. Since these

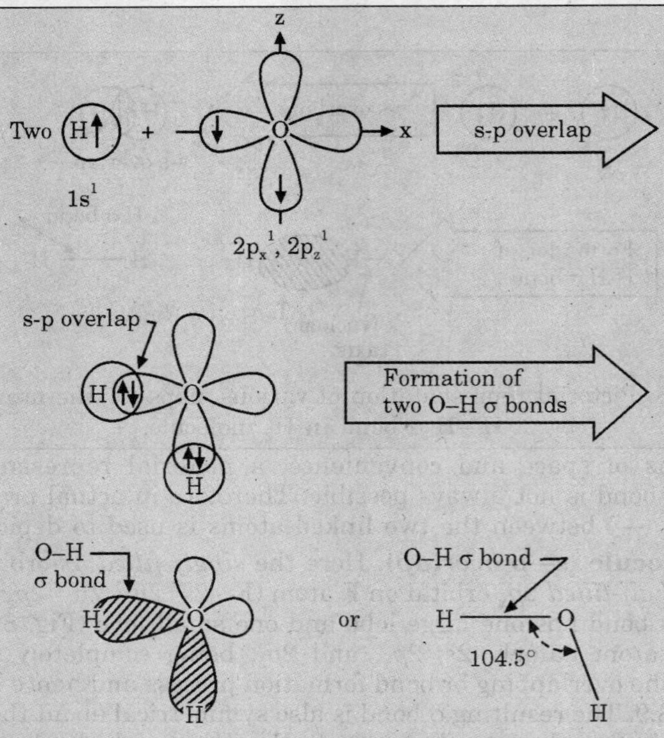

Fig. 8.10. Pictorial representation of various steps for the formation of two O—H σ bonds in H_2O molecule.

O—H σ-bonds are produced by *s-p* overlap, these bonds are also called s–p_x and s–p_z bonds. Now since $2p_x$ and $2p_z$ orbitals are at right angles to each other, an angle of 90° between two O—H σ-bonds (*i.e.* H—O—H angle) is expected, but the actual angle is 104.5° instead of 90°. The increase in angle is due to the mutual electrostatic repulsion between the electron pairs constituting the two O—H bonds.

2*s* and $2p_y$ orbitals of O-atom, being completely filled, do not participate in the overlapping or bond formation process and hence have not been shown in Fig. 8.10.

4. NH_3 molecule (*s-p overlap*). The valence shell configuration of N-atom *viz.* $2s^2\ 2p_x^1\ 2p_y^1\ 2p_z^1$ shows that all its *p*- orbitals have unpaired electrons and hence overlap with singly-filled 1*s*-orbitals of three H- atoms to form three *s-p* σ bonds (Fig. 8.11). 2*s* orbital of N atom, being completely filled, has not been shown in the Figure. Due to mutual repulsion of the electron pairs constituting three N—H σ bonds, the H—N—H angles are 107° instead of 90°.

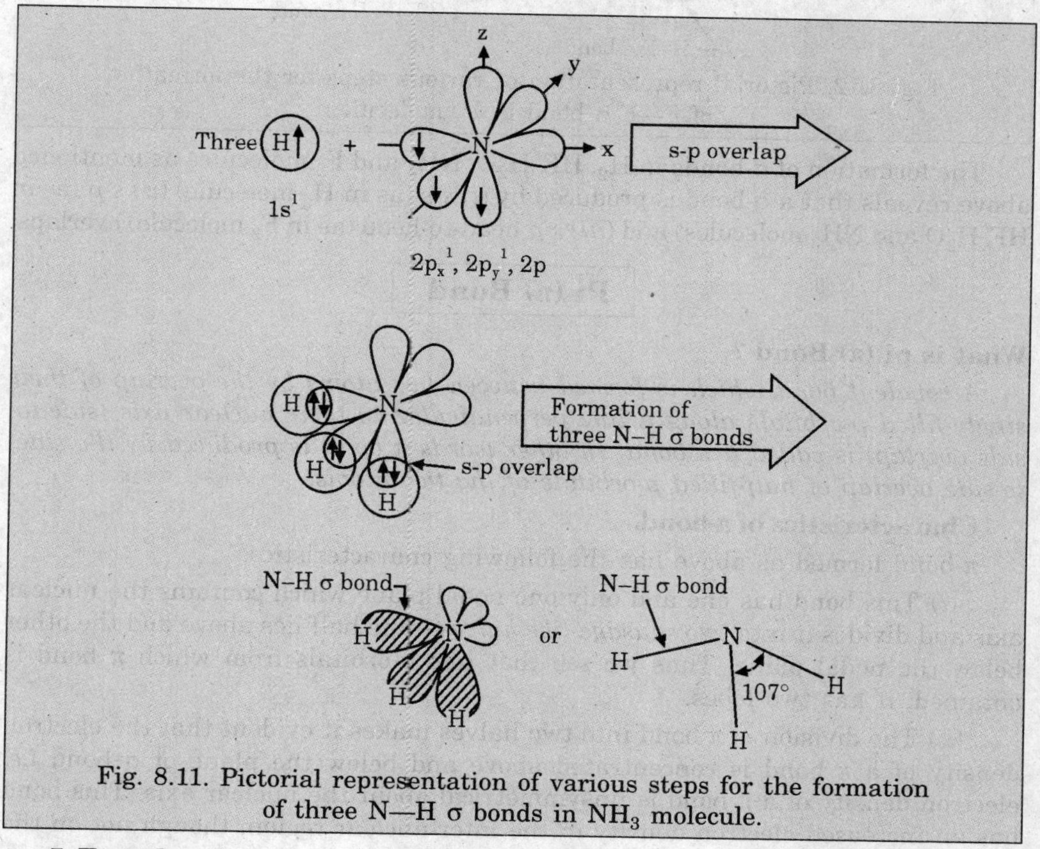

Fig. 8.11. Pictorial representation of various steps for the formation of three N—H σ bonds in NH_3 molecule.

5. F_2 molecule (*p-p head-to-head overlap*). Here *half-filled* $2p_x$ orbital on one F atom (F → $2s^2\ 2p_x^1\ 2p_y^2\ 2p_z^2$) overlaps with similar *half-filled* $2p_x$ orbital on the other F atom in a head-to-head fashion and forms F—F σ bond (Fig. 8.12), since the bond thus formed is symmetrical about the axis. 2*s*, $2p_y$ and $2p_z$ orbitals of both F atoms, being completely filled, do not participate in the bond formation process and hence have not been shown in the Figure. *σ-bond formed has two nodal planes which are perpendicular to the nuclear axis but do not contain the nuclear axis.*

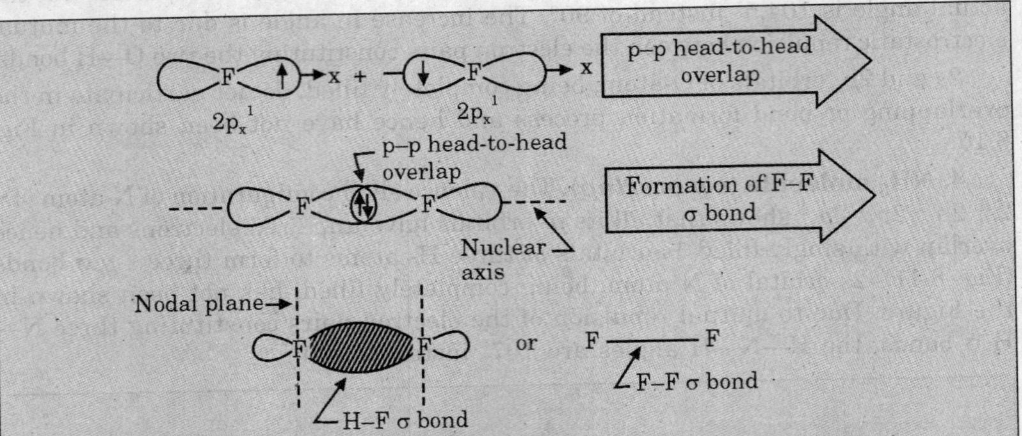

Fig. 8.12. Pictorial representation of various steps for the formation of F—F σ bond in F_2 molecule.

The formation of σ-bonds in H_2, HF, H_2O, NH_3 and F_2 molecules as mentioned above reveals that a σ-bond is produced by (*i*) *s-s* (as in H_2 molecule) (*ii*) *s-p* (as in HF, H_2O and NH_3 molecules) and (*iii*) *s-p* head-to-head (as in F_2 molecule) overlaps.

Pi (π) Bond

What is pi (π) Bond ?

A covalent bond which is formed between two atoms by the overlap of their singly-filled p-orbitals along a line perpendicular to their nuclear axis (side-to-side overlap) is called a π bond. In other words π bond is produced by the side-to-side overlap of half-filled p-orbitals of the two atoms.

Characteristics of π-bond.

π-bond formed as above has the following characteristics.

(*i*) This bond has one and only one nodal plane which contains the nuclear axis and divides it into two *sausage-like halves*—one half lies above and the other below the nodal plane. Thus we see that like *p*-orbitals from which π bond is obtained, it has two lobes.

(*ii*) The division of π bond into two halves makes it evident that the electron density of a π bond is concentrated above and below the plane of σ bond *i.e.* electron density of a π bond is unsymmetrical about the nuclear axis. This bond has an increased electron density in the inter-nuclear region, though not on the bond axis.

Formation of π Bonds in O_2 and N_2 Molecules

How this bond is formed can be understood by considering the formation of some simple molecules like O_2 and N_2 molecules.

1. O_2 molecule (*p-p head-to-head and p-p side-to-side overlaps*). Valence shell configuration of O-atom *viz.* $2s^2\ 2p_x 1\ 2p_y^1\ 2p_z^2$ shows that each of the O-atoms of O_2 molecule has two singly filled *p*-orbitals namely $2p_x$ and $2p_y$ which are perpendicular to each other. Singly-filled $2p_x$ orbitals on both O-atoms overlap in a *head-to- head manner* and form O—O σ bond while the remaining singly-filled $2p_y$ orbitals on both O-atoms are parallel to each other and hence overlap in a *side-to-side manner* to form a new type of bond around O—O σ bond which is called *pi* (π) *bond*. The upper and lower overlaps of the two $2p_y$ orbitals taken together constitute the π bond (Fig. 8.13).

Here it may be noted that 2*s* and $2p_z$ orbitals of both O atoms, being completely filled, do not participate in bond formation or overlapping process and hence have not been shown in Figure 8.13.

We see from Fig 8.13. that O_2 molecule has two bonds one of which is a σ-bond (p_x–p_x σ bond) and the other is a π-bond (p_y–p_y π bond). Both the bonds taken together are referred to as a *double bond*. Although both bonds (σ and π) are different from each other, for convenience both are shown by two equal parallel lines.

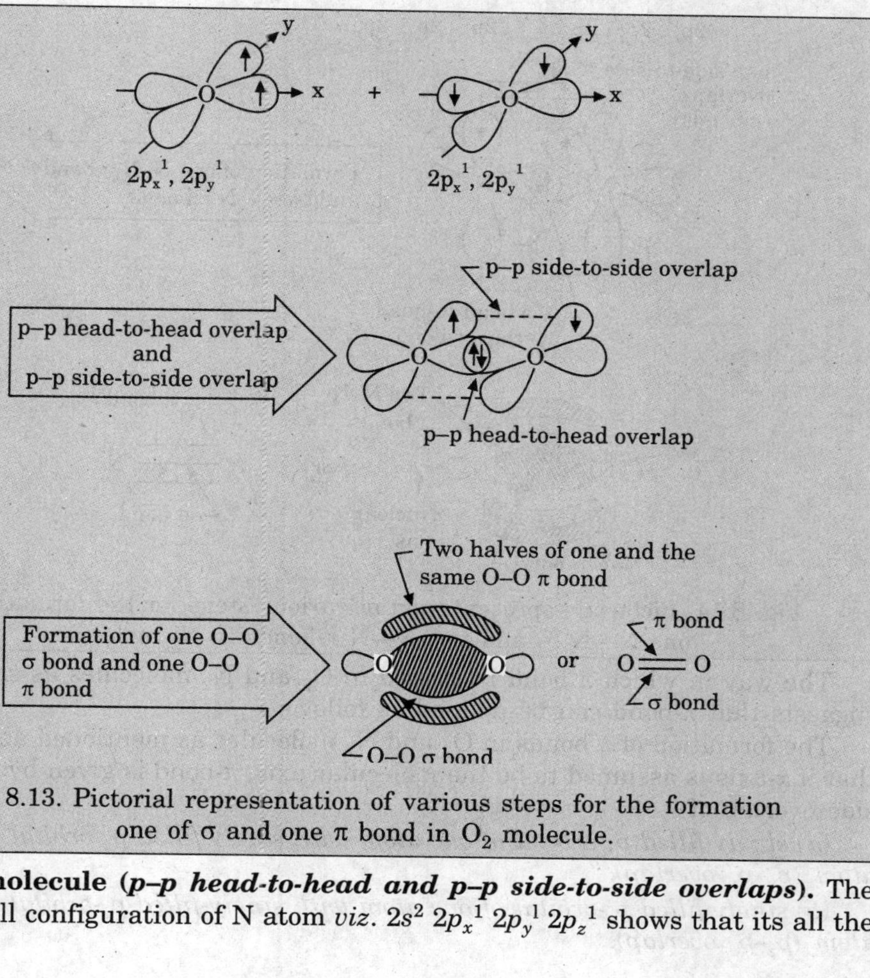

Fig. 8.13. Pictorial representation of various steps for the formation one of σ and one π bond in O_2 molecule.

2. N_2 molecule (*p–p head-to-head and p–p side-to-side overlaps*). The valence-shell configuration of N atom *viz.* $2s^2\ 2p_x^1\ 2p_y^1\ 2p_z^1$ shows that its all the three *p*-orbitals are singly-filled and hence will participate in the bond formation with other N-atom. Singly filled $2p_x$ orbitals of both N atoms overlap in a *head-to-head manner* and form N—N σ-bond (p_x—p_x σ-bond) while the remaining singly-filled $2p_y$ and $2p_z$ orbitals of this N atom overlap with the similar *p*-orbitals of the other N atom in a *side-to-side manner* and form two N—N π-bonds (p_y—p_y π-bond and p_z–p_z π-bond) as shown in Fig. 8.14. Note that 2*s* orbital, being completely filled does not participate in bond formation and hence has not been shown in the Figure.

We see from Fig. 8.14. that N_2 molecule has three bonds one of which is a σ-bond (p_x–p_x σ bond) and the other two are π-bonds (p_y–p_y π-bond and p_z–p_z π-bond). All the three bonds taken together are referred to as a *triple bond.*

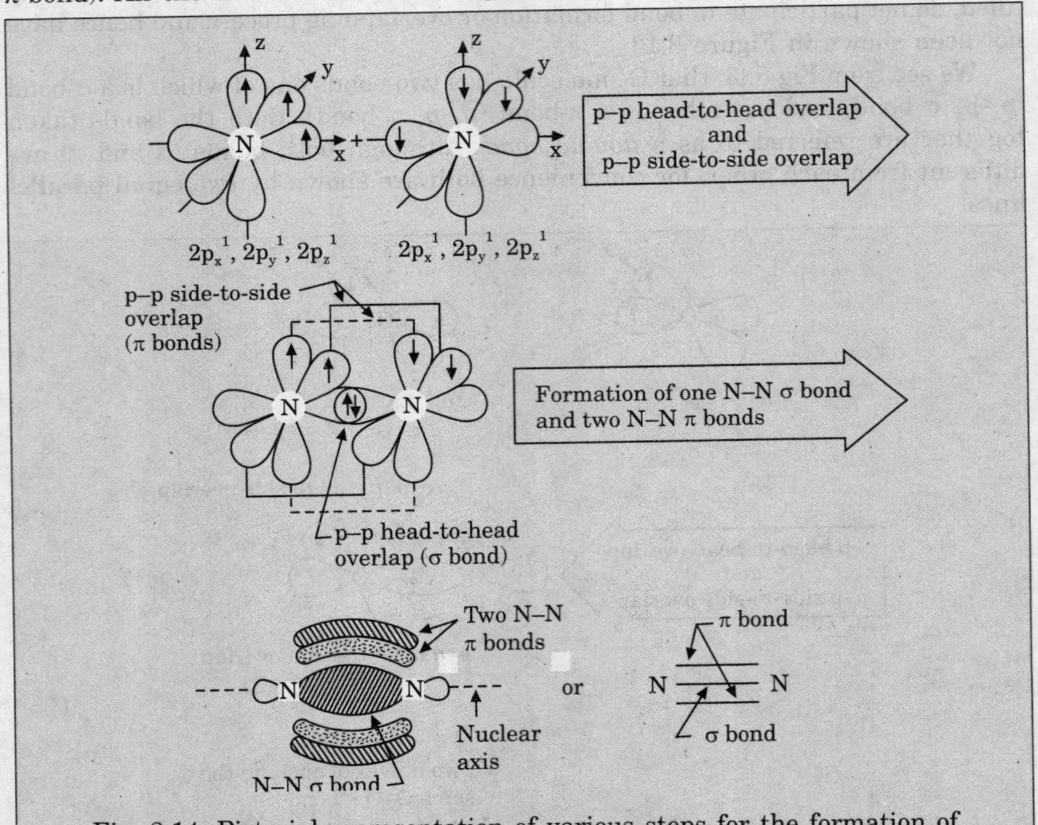

Fig. 8.14. Pictorial representation of various steps for the formation of one N—N σ- and two N—N π-bonds in N_2 molecule.

The way in which π bond is formed in O_2 and N_2 molecules as shown above suggests that π-bond can be defined as follows :

The formation of π bonds in O_2 and N_2 molecules as mentioned above reveals that if *x*-axis is assumed to be the molecular axis, π-bond is given by the side-to-side overlap of :

(*a*) *singly-filled* p_y*-orbital of one atom with singly-filled* p_y*-orbital of the other atom* (p_y–p_y *overlap*).

(*b*) *singly-filled* p_z*-orbital of one atom with singly-filled* p_z*-orbital of the other atom* (p_z–p_z *overlap*).

Comparison of σ- and π-Bonds

The distinguishing features of a σ-bond from a π-bond are given in the following Table :

σ-Bond	π-Bond
(*i*) A σ-bond obtained by the *head-to-head* overlap *i.e.* (overlap of orbitals along the nuclear axis) of : (*a*) two *s*-orbitals on two different atoms (*s–s* overlap). (*b*) one *s*-orbital of one atom with *p*-orbital of the other atom (*s–p* overlap). (*c*) two *p*-orbitals on two different atoms (*p–p* overlap).	(*i*) If *x*-axis is assumed to be the nuclear axis, π-bond is given by the *side-to-side* overlap (*i.e.* overlap of orbitals along a line perpendicular to the nuclear axis) of : (*a*) p_y orbital of one atom with p_y orbital of the other atom (*p_y–p_y overlap*). (*b*) p_z orbital of one atom with p_z orbital of the other atom (p_z–p_z overlap).
(*ii*) Since the extent of overlapping of orbitals along the nuclear axis is relatively greater than the extent of overlapping of orbitals along a line perpendicular to the nuclear axis, there is *more* decrease in energy of the molecule formed and hence σ-bond is a stronger bond.	(*ii*) Since the side-wise overlap of two *p*-orbitals which gives a π-bond is partial, there is lesser *decrease* in energy of the molecule formed and hence π-bond is a weaker bond.
(*iii*) In the formation of a σ-bond by the overlap of two *p*-orbitals, only one lobe of one *p*-orbital of one atom overlaps with only one lobe of *p*-orbital of the other atom in a head-to-head manner.	(*iii*) In the formation of a π-bond by the overlap of two *p*-orbitals, both lobes of both *p*-orbitals overlap with each other in a side-to-side manner.
(*iv*) The probability of finding the electrons between the two nuclei (*i.e.* on the bond axis) is *maximum*, since they are attracted equally by both the nuclei. It is for this reason that σ-bond is a very strong bond.	(*iv*) The probability of finding the electrons between the two nuclei (*i.e.* on the bond axis) is *poor* and hence π-bond is a poor bond.
(*v*) The electron density in this bond is distributed *symmetrically* about the nuclear axis.	(*v*) The distribution of electron density in this bond is *unsymmetrical* about the nuclear axis.
(*vi*) There can be *free rotation* of the atoms relative to one another round the bond axis of σ-bond (*i.e.* nuclear axis).	(*vi*) There is no possibility of free rotation of the atoms relative to one another round the bond axis of π-bond, since the electron clouds of *p*-orbitals of the two atoms overlap above and below the plane of the atoms. If an attempt to rotate the atoms relative to one another along the bond axis of π-bond is made, the lobes of *p*-orbitals will no longer remain co-planar and the extent of overlapping of *p*-orbitals decreases, thus weakening the π-bond still further. This property of *restricted rotation* in case of π-bond is very useful in explaining the phenomena of *cis-trans isomerism* (*geometrical isomerism*).

Concept of Resonance

What is Resonance ?

Sometimes it is not possible to represent a molecule by a single valence bond structure which can correctly account for the properties of the molecule. In such cases valence bond method introduces the **resonance concept** *according to which if two (or more) alternate valence bond structures can be written for a molecule, the actual structure of the molecule is said to be a* **resonance** or **mesomeric hybrid** *of all these alternate structures.* For example, for H_2 molecule this concept implies that neither structure (*i*) nor structure (*ii*) nor (*iii*) by itself provides a correct description of this molecule, but only a combination (*i.e.* hybrid) of the three can do so. It is expressed by saying that H_2 molecule resonates between the three structures (*i*), (*ii*) and (*iii*) or H_2 molecule is a **resonance hybrid** of these three structures :

$$\overset{+}{H}_A\ \overset{-}{H}_B \leftrightarrow H_A - H_B \leftrightarrow \overset{-}{H}_A\ \overset{+}{H}_B$$

Ionic structure (*i*)	Covalent structure (*ii*)	Ionic structure (*iii*)

This type of expression should not be taken to mean that H_2 molecule sometimes has structure (*i*) and sometimes structures (*ii*) or (*iii*) and that the molecule oscillates between these three structures. The actual structure of the molecule is not a mixture of these three structures. Rather it is a single structure of its own, but as it cannot be written on paper, it is convenient to think of it in terms of the structures which can, of course, be written down on paper. The various structures of which the molecule is said to be a resonance hybrid are known as **contributing resonance** (or **canonical**) **forms** (or **extremes**) and have no physical reality *i.e.* these cannot be isolated. They are merely mental aids to rationalise the experimental behaviour of the molecules. Ingold has called the canonical forms as **mesomeric forms** or **mesomeric structures** and the conception is consequently known as **mesomerism** (between the parts).

To sum up :

Resonance is the description of the electronic structures of a molecule or an ion by means of several schemes of pairing of electrons, with the features of each scheme contributing in the description.

A double-headed arrow (↔) is placed between the different contributing forms to indicate that there is resonance between the various contributing structures.

Various structures of H_2 molecule given above provide an example of *covalent-ionic resonance* in one particular bond.

Resonance Energy

In a molecule which exhibits resonance it has been found *that in every case the observed (i.e. experimental) heat of formation of the molecule is greater than the calculated heat of formation, i.e.* the actual molecule is more stable than the resonating structure considered for calculating the heats of formation. The resonating structure being used as the basis for calculating heats of formation must be the most stable of all the resonating structures among which resonance occurs. The difference between observed and calculated heat of formation is called **resonance energy,** usually represented as ΔE. Thus ΔE *is the difference between the bond energy of the actual structure* (E_{actual}) *and that of the most stable resonating structure.*

Calculation of Resonance Energy

Let us calculate the resonance energy of CO_2 molecule which is supposed to be resonance hybrid of the following three structures.

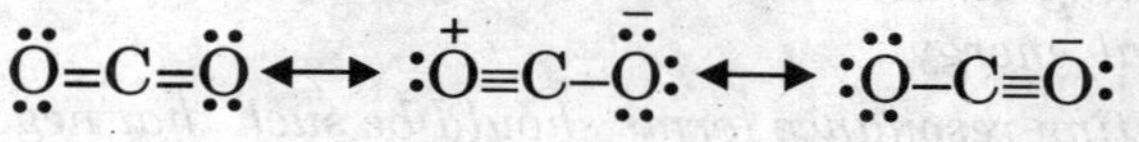

(I) (II) (III)

Symmetrical structure (Covalent (structures) — (I)

Unsymmetrical structures (Ionic structures) — (II), (III)

Of these three resonating structures, structure (I) has the *maximum bond energy* and hence is the *most stable*. Thus resonance energy, ΔE of CO_2 molecule relative to resonating structure (I) is given by :

$$\Delta E = E_{actual} - E_1$$

$$= \begin{bmatrix} \text{Experimental value of} \\ \text{heat of formation of } CO_2 \end{bmatrix} - \begin{bmatrix} \text{Calculated value of heat of} \\ \text{formation of structure (I)} \end{bmatrix}$$

$$= 1602.8 - 2 \times \text{Energy of C} = O \text{ bond (in KJ/mole)}$$

$$= 1602.8 - 2 \times 732.3$$

$$= 1602.8 - 1464.6$$

$$= 138.2 \text{ kJ/mole.}$$

ΔE (= 138.2 KJ/mole) is the energy lost when one molecule of CO_2 takes the resonance form (I). This also represents the additional energy which has to be supplied to disrupt the bonds due to their increased strength because of resonance. Resonance energy represents the extent to which the actual structure of CO_2 is more stable than the most stable resonating structure (I). ΔE (= 138.2 KJ/molecule) stabilises the CO_2 molecule relative to the resonance form (I).

The calculated O—O bond distance in structure (I) is 2.44 Å while the experimental value is 2.30 Å. Thus we find that *in the actual molecule (i.e., resonance hybrid) there is an increase in the value of heat of formation and shortening of bond lengths.*

Requirements for Contributing Resonance Forms

For the resonance to be possible the different canonical structures must satisfy the following conditions :

(*i*) *The relative position of all the atoms in each of the resonance forms must be the same; the arrangement of electrons may differ.*

(*ii*) *The number of unpaired and paired electrons in each of resonance forms must be the same* so that a continuous change from one bond type to another may occur. For example,

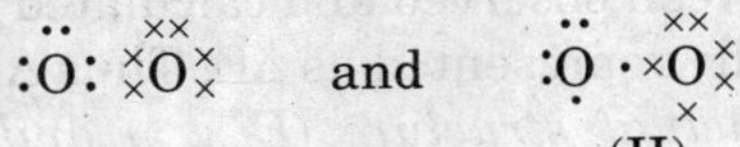

(I) (II)

cannot be the contributing structures of O_2 molecule, since the number of paired and unpaired electrons in both the forms is different. Structure (II) contains two unpaired electrons and structure (I) does not contain any unpaired electron.

(*iii*) *Contributing structures should be of comparable energy i.e. they should have almost equal energy.*

(*iv*) *Contributing resonance forms should be such that negative charge resides on an electronegative element and positive charge resides on an electropositive element.* For example, of the following three possible structures (I), (II) and (III) of HF molecule,

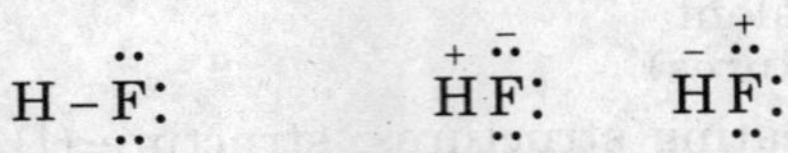

(I) (II) (III)

Covalent structure — Ionic structures

structure (III) is ruled out, since in this structure negative charge resides on an electropositive element (H) and positive charge resides on an electronegative element (F). The actual structure of HF molecule is, therefore, considered to be a resonance hybrid of the structures (I) and (II) as shown below :

$$\underset{\text{(I)}}{H-F} \quad \leftrightarrow \quad \underset{\text{(II)}}{\overset{+}{H}\overset{-}{F}}$$

Structure (II) provides an example of *covalent-ionic resonance* in one particular bond.

(*v*) *The greater the number of covalent bonds in a contributing structure, greater is the importance of that structure.* For example since the structures of BF_3 molecule shown below have double bond (B = F double bond), these are important.

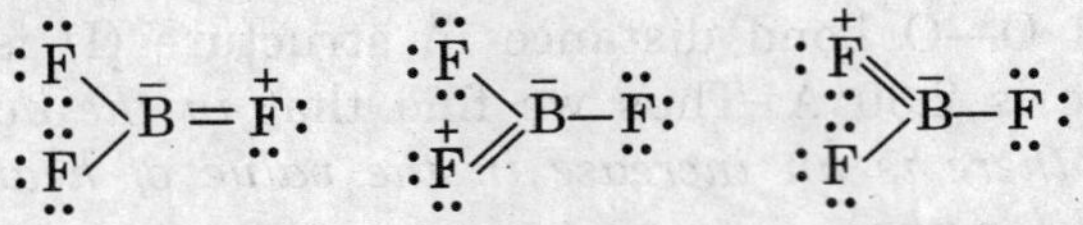

In these structures it may be noted that F atom which is an electronegative atom acquires positive charge. This is contrary to rule (*iv*) given above. F atom acquires positive charge so that the electron deficiency of B atom is removed. The formation of an additional B = F double bond has a couter balancing effect.

Resonance Structures of Some Species

1. **CO molecule.** See *"Structure of CO molecule"* given in this chapter.
2. **CO_2 molecule.** See *"Structure of CO_2 molecule"* given in this chapter.
3. **CO_3^{2-} ion.** See *"Structure of CO_3^{2-} ion"* given in this chapter.
4. **O_3 molecule.** See *"Structure of O_3 molecule"* given in this chapter.
5. **SO_2 molecule.** See *"Structure of SO_2 molecule"* given in this chapter.
6. **SO_3 molecule.** See *"Structure of SO_3 molecule"* given in this chapter.

Concept of Promotion of Electrons

Although the atomic orbital theory of covalent bond has satisfactorily explained the formation of covalent bonds in simple molecules like H_2, HF, H_2O, NH_3, F_2, O_2 and N_2, yet it is not able tc explain the formation of covalent bonds in some molecules like BeF_2, BF_3, CH_4 etc. as shown below.

We have seen that according to atomic orbital theory, for the formation of a covalent bond between two atoms, these atoms must have unpaired electrons in their valence-shell. Thus according to this theory Be ($2s^2$), B ($2s^2p^1$) and C ($2s^2p^2$) which have 0, 1 and 2 unpaired electrons are expected to form *no covalent bond, one covalent bond and two single covalent bonds* with monovalent atoms like H, F etc., *i.e.* these elements shoɹld show a covalency of zero (zero covalency), a covalency of one (monocovalency) and a covalency of two (discovalency) as shown below :

Elements	Valence-shell ground state configuration	No. of unpaired electrons	Expected covalency or number of expected single covalent bonds
Be ($2s^2$)	2s: [↑↓] 2p: [ ] [ ] [ ]	0	0
B ($2s^2p^1$)	2s: [↑↓] 2p: [↑] [ ] [ ]	1	1
C ($2s^2p^2$)	2s: [↑↓] 2p: [↑] [↑] [ ]	2	2

However, the compounds of these elements show that these elements show a covalency of 2, 3 and 4 respectively. In other words these elements form the compounds like $BeCl_2$ (covalency of Be = 2), BCl_3 (covalency of B = 3) and CH_4 (covalency of C = 4).

In order to account for the covalency of 2, 3 and 4 for Be, B and C atoms, these elements should have 2, 3 and 4 unpaired electrons in their valence-shell respectively. This can be possible only when it is supposed that one of the 2*s* electrons is promoted to the next vacant higher energy 2*p* orbital as shown below :

Elements	Valence-shell ground state configuration	Valence-shell excited state configuration
Be ($2s^2 2p^0$)	2s: [↑↓] 2p (x y z): [ ][ ][ ] ($2s^2$) No. of unpaired electrons or covalency = 0	2s: [↑] 2p (x y z): [↑][ ][ ] ($2s^1p^1$) No. of unpaired electrons or covalency = 2
B ($2s^2 2p^1$)	[↑↓] [↑][ ][ ] ($2s^2p^1$) No. of unpaired electrons or covalency = 1.	[↑] [↑][↑][ ] ($2s^1p^2$) No. of unpaired electrons or covalency = 3
C ($2s^2 2p^2$)	[↑↓] [↑][↑][ ] ($2s^2p^2$) No. of unpaired electrons or covalency = 2	[↑] [↑][↑][↑] ($2s^1p^3$) No. of unpaired electrons or covalency = 4

The above description which is based on the promotion of one of the two $2s$ electrons to the vacant higher energy $2p$ orbital well explains the covalency of 2, 3 and 4 of Be. B and C atoms.

Hybridisation

What is Hybridisation ?

In order to understand the concept of hybridisation let us consider the formation of two Be—Cl bonds in $BeCl_2$ molecule, three B—Cl bonds in BCl_3 molecule and four C—H bonds in CH_4 molecule on the basis of atomic orbital theory as given below :

1. Formation of two Be—Cl bonds in $BeCl_2$ molecule—sp hybridisation. On the basis of atomic orbital theory the two Be—Cl bonds in $BeCl_2$ molecule are of two types : (*i*) 2*s* (Be) —$3p_x$(Cl) bond and (*ii*) $2p_x$(Be)—$3p_x$(Cl) bond as shown in Fig. 8.15. Both these bonds are σ-bonds. These are different in energy and hence different in length from each other. But experiments have shown that in $BeCl_2$ molecule both Be—Cl bonds have the same energy and same length and hence are identical to each other. These are directed in opposite direction along the axis of $2p_x$ orbital, *i.e.* their arrangement round the Be atom

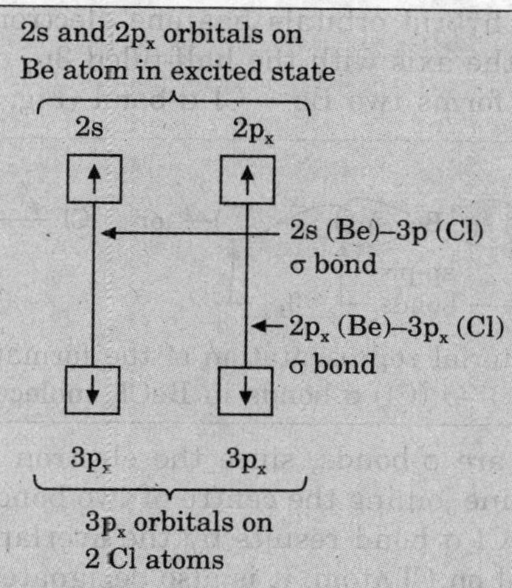

Fig. 8.15. Formation of $2s$(Be)–$3p_x$(Cl) and $2p_x$(Be)–$3p_x$(Cl) bonds in $BeCl_2$ molecule on the basis of atomic orbital theory. Both these bonds are σ-bonds.

is linear and they make an angle of 180° with each other. Then how to account for the identical nature of Be—Cl bonds ? For this it is assumed that $2s$ and $2p_x$ orbitals of Be atom in the excited state *mix* or *hybridise* or *merge* together and redistribute their energy to give a set of two new equivalent (*i.e.* of the same energy and shape) orbitals which are known as *sp hybrid orbitals* because they are formed by the interaction of one *s* and one *p* orbital. These hybrid orbitals are also called *sp orbitals* and Be atom is said to be in *sp hybridised state* and *the process of mixing one s and one p orbital to produce two equivalent (identical) sp hybrid orbitals is called* **sp hybridisation.** The formation of two *sp* hybrid orbitals in case of Be atom is shown in Fig. 8.16.

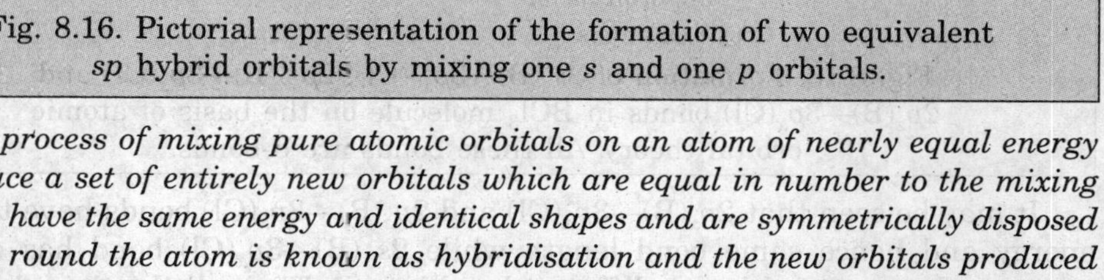

Fig. 8.16. Pictorial representation of the formation of two equivalent *sp* hybrid orbitals by mixing one *s* and one *p* orbitals.

The process of mixing pure atomic orbitals on an atom of nearly equal energy to produce a set of entirely new orbitals which are equal in number to the mixing orbitals, have the same energy and identical shapes and are symmetrically disposed in space round the atom is known as hybridisation and the new orbitals produced are called hybrid orbitals or simply hybrids. The process of hybridisation is a hypothetical process.

Each of the two *sp* hybrid orbitals has one electron. Each of these hybrid orbitals overlaps along the axis with the half-filled $3p_x$ orbital of Cl atom (Cl → $3s^2$, $3p_x^1\ 3p_y^2\ 3p_z^2$) and forms two Be— Cl σ-bond (Fig. 8.16).

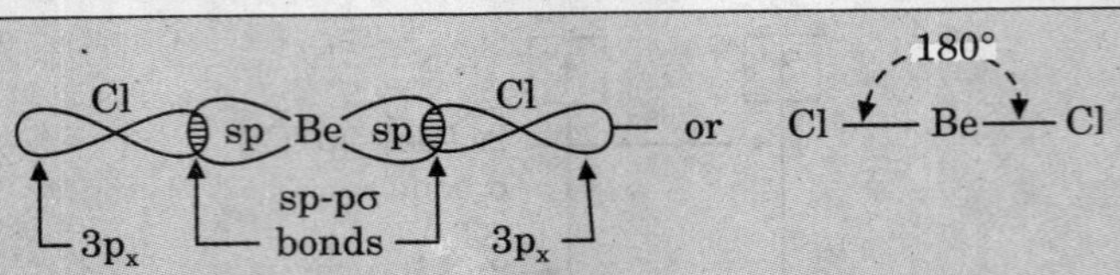

Fig. 8.17. Pictorial representation of the formation of two *sp* (Be)—*p* (Cl) σ-bonds in $BeCl_2$ molecule.

Both Be—Cl bonds are σ-bonds, since the electron density in each bond is symmetrical about the line joining the centre of two bonded atoms *viz.* Be and Cl atoms. Since each Be—Cl σ bond results by the overlap of *sp* hybrid orbital on Be atom with $3p_x$ orbital on Cl atom, it is also designated as *sp* (Be)—$3p_x$ (Cl) σ-bond or *sp*—*p* σ-bond. Both the bonds are directed in the opposite directions from the centre of Be atom. Thus Cl—Be—Cl bond angle is of 180° and both the bonds have the same energy and bond length. Thus $BeCl_2$ molecule has a linear shape with both Be—Cl bonds *identical* in nature.

2. Formation of three B—Cl bonds in BCl_3 molecule—sp^2 hybridization. On the basis of atomic orbital theory the three B—Cl bonds in BCl_3 molecule are of three types *viz.* (*i*) 2*s* (B)—$3p_x$(Cl) σ bond (*ii*) $2p_x$(B)—$3p_x$(Cl) σ-bond and (*iii*) $2p_z$ (B)—$3p_x$ (Cl) σ bond as shown in Fig. 8.18.

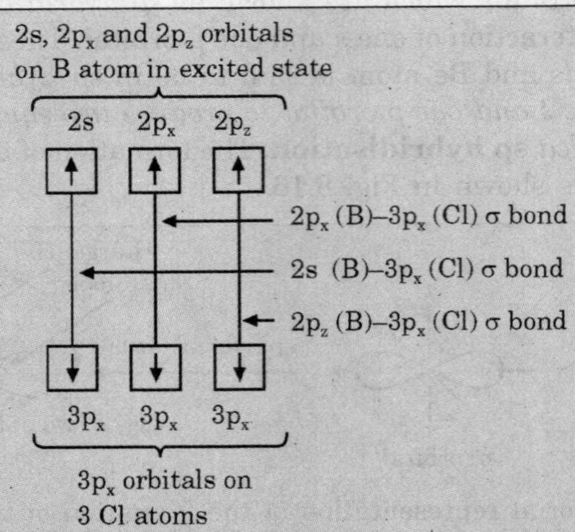

Fig. 8.18. Formation of 2*s* (B)—$3p_x$ (Cl), $2p_x$ (B)—$3p_x$(Cl) and $2p_z$(B)—$3p_x$(Cl) bonds in BCl_3 molecule on the basis of atomic orbital theory. All these bonds are σ-bonds.

It may be seen that $2p_x$(B)—$3p_x$(Cl) and $2p_z$(B)—$3p_x$(Cl) bonds have the same energy and hence same bond length while 2*s* (B)—$3p_x$(Cl) bond has different amount of energy and hence different bond length. Thus all the three bonds are not identical. But experiments have shown that all the three B—Cl bonds in BCl_3 molecule are identical and are inclined at an angle of 120° to each other, *i.e.* their arrangement round B atom is trigonal planar. Then how to explain the identical nature of B—Cl bonds ? For this it is assumed that $2s$, $2p_x$ and $2p_z$ orbitals of B atom in the excited state hybridise together to form a set of three new equivalent (*i.e.* of the same energy and shape) orbitals which are known as sp^2 (read as *ess-pee-two*) *hybrid orbitals*, because they are formed by the interaction of one s and two p orbitals. These are also called *sp^2 orbitals and B atom is said to be in sp^2 hybridised state and the process of mixing one s and two p orbitals to produce three equivalent (i.e. identical) sp^2 hybrid orbitals is called sp^2 hybridisation.* The formation of three sp^2 orbitals in case of B atom is shown in Fig. 8.19.

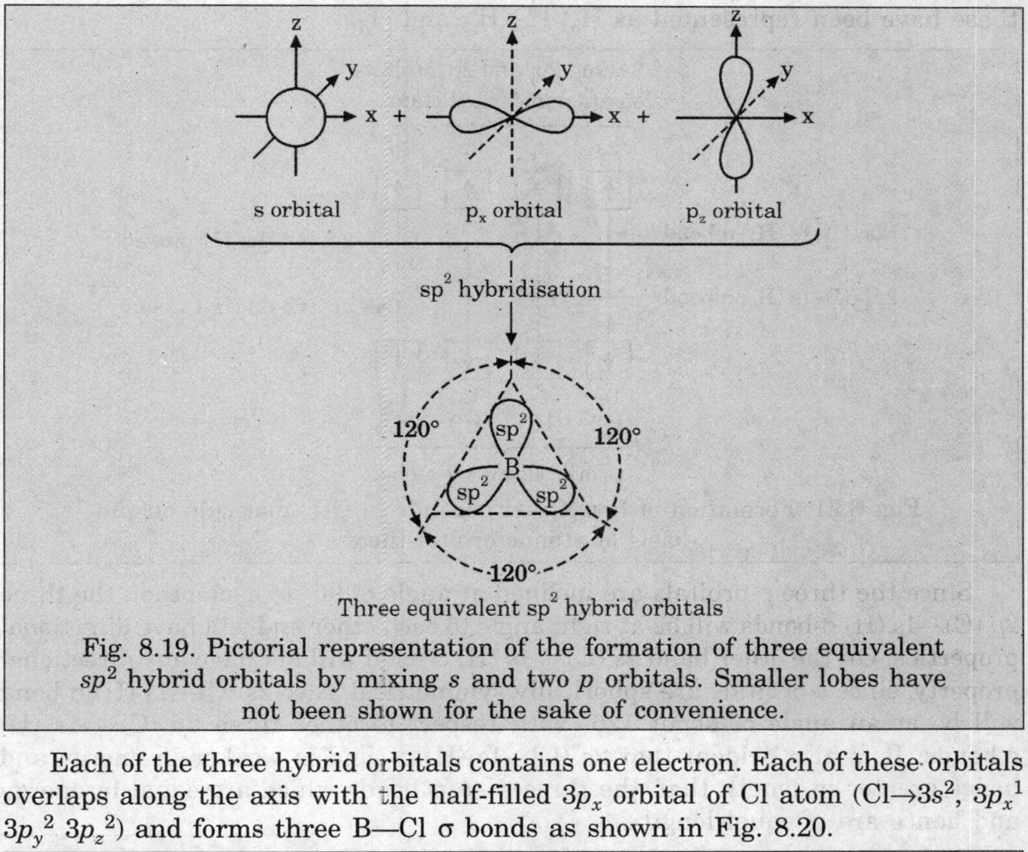

Fig. 8.19. Pictorial representation of the formation of three equivalent sp^2 hybrid orbitals by mixing s and two p orbitals. Smaller lobes have not been shown for the sake of convenience.

Each of the three hybrid orbitals contains one electron. Each of these orbitals overlaps along the axis with the half-filled $3p_x$ orbital of Cl atom (Cl→$3s^2$, $3p_x^{\,1}$ $3p_y^{\,2}$ $3p_z^{\,2}$) and forms three B—Cl σ bonds as shown in Fig. 8.20.

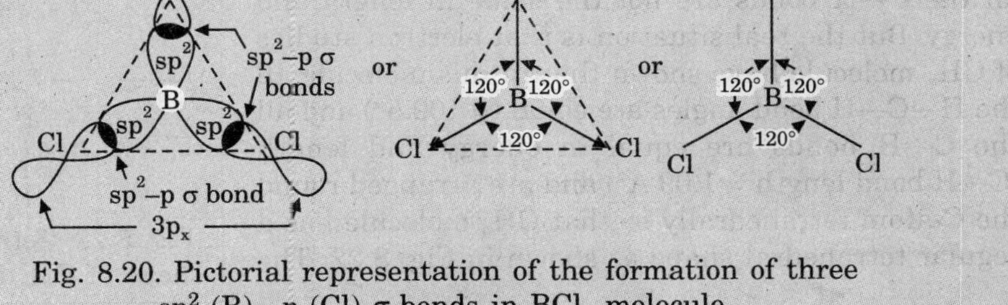

Fig. 8.20. Pictorial representation of the formation of three sp^2 (B)—p (Cl) σ-bonds in BCl_3 molecule.

All the three B—Cl bonds are σ bonds. All the bonds are directed towards the corners of a triangular plane and hence each Cl—B—Cl bond angle is of 120° and all the bonds have the same bond length and energy. Thus BCl_3 molecule has a triangular shape with all the three B—Cl bonds *identical* in nature.

3. Formation of four C—H bonds in CH_4 molecule—sp^3 hybridisation. On the basis of atomic orbital theory the four C—H bonds in CH_4 molecule are of four types namely *(i)* 2*s* (C) —1*s* (H_a) σ-bond *(ii)* $2p_x$ (C) —1*s* (H_b) σ-bond *(iii)* $2p_y$ (C) —1*s* (H_c) σ bond and *(iv)* $2p_z$ (C)—1*s*(H_d) σ bond. The formation of these bonds has been shown in Fig 8.21. For the sake of differentiating the four H-atoms these have been represented as H_a, H_b, H_c, and H_d.

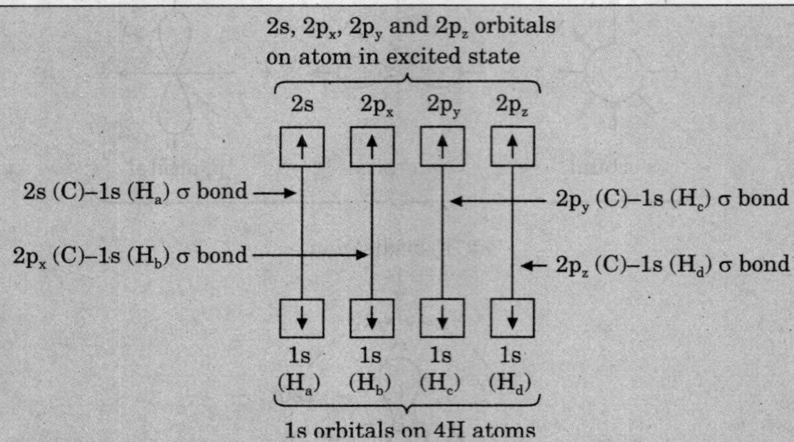

Fig. 8.21. Formation of four C—H σ bonds in CH_4 molecule on the basis of atomic orbital theory

Since the three *p*-orbitals are inclined at angle of 90° to each other, the three 2*p* (C)—1*s* (H) σ-bonds will be at right angle to each other and will have directional properties. On the other hand 2*s* (C) —1*s* (H) σ bond will not have any directional property, since *s*-orbitals are spherically symmetrical. Also 2*s* (C)—1*s* (H) σ bond will be at an angle of about 125° with respect to other three 2*p* (C) —1*s* (H) σ bonds. It is also evident that 2*s* (C)—1*s* (H) σ-bond is weaker in energy and hence greater in length than the three *s*—*p* σ bonds which are equal in energy and hence are of equal length.

With this picture of carbon atom we may expect that all the H—C—H bond angles are not equal and all the C—H bonds are not the same in length and energy. But the real situation is that electron studies of CH_4 molecule have shown that in this molecule all the H—C—H bond angles are equal (= 109.5°) and all the C—H bonds are equal in energy and length (C—H bond length = 1.09 A°) and are arranged round the C-atom tetrahedrally so that CH_4 molecule has a regular tetrahedral shape as shown in Fig. 8.22. The

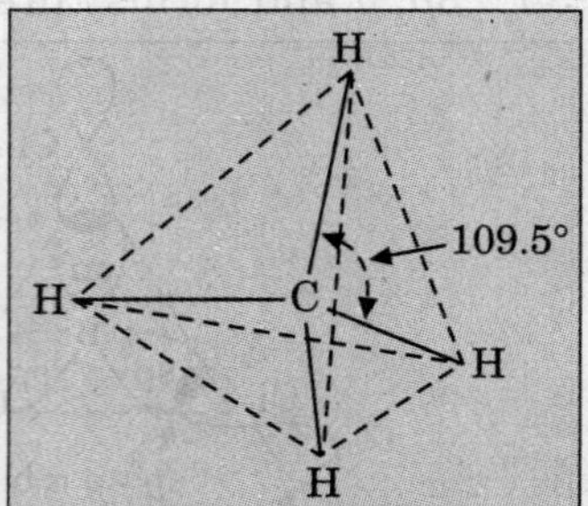

Fig. 8.22. Tetrahedral shape of CH_4 molecule

energy required to break any of the four bonds is the same (= 426.87 KJ mole^{-1}). Thus all the C—H bonds in CH_4 molecule are identical.

Then how account for the formation of four identical C—H bonds in CH_4 molecule. For this it is assumed that the 2*s* and three 2*p* orbitals of carbon atom in the excited state *mix* or *hybridise* or *merge* together and redistribute their energy to give a set of four new equivalent (*i.e.* of the same energy and shape) orbitals which are known as sp^3 (pronounced as *ess-pee-three*) *hybrid orbitals* because they are formed by interaction of one *s* and three *p* orbitals. These sp^3 hybrid orbitals are also known as sp^3 orbitals. Here carbon atom is said to be in *sp^3 hybridised state and the process of mixing one s and three p orbitals to produce four sp^3 hybrid orbitals is called* **sp^3 hybridisation.** The formation of four sp^3 hybrid orbitals in case of carbon atom is shown in Fig. 8.23.

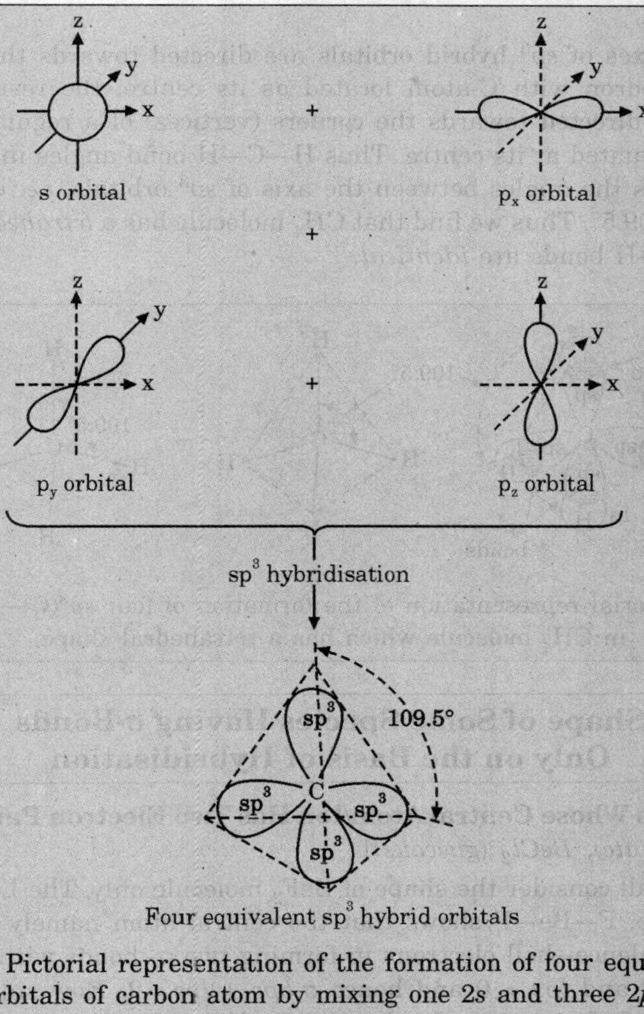

Fig. 8.23. Pictorial representation of the formation of four equivalent sp^3 hybrid orbitals of carbon atom by mixing one 2*s* and three 2*p* orbitals. Smaller lobes have not been shown for the sake of convenience.

All the four sp^3 hybrid orbitals are identical, *i.e.*, they have the same energy and shape. These are directed towards the corners of a regular tetrahedron at the centre of which is placed the atom that has undergone sp^3 hybridisation. Because of the tetrahedral disposition of sp^3 hybrids, sp^3 hybridisation is also called *tetrahedral hybridisation* and the four sp^3 hybrid orbitals are also called *tetrahedral hybrid orbitals*. Each of the four sp^3 hybrid orbitals on carbon atom is singly-filled. Each of these hybrid orbitals overlaps along the axis with the half-filled 1s orbital on H-atom and forms four C—H σ bonds (See Fig. 8.24). All the four C—H bonds are σ bonds, since the electron density in each bond is symmetrical about the line joining the centre of two bonded atoms *viz.* C and H atoms. Since each C—H σ bond results by overlap of sp^3 hybrid orbital on C-atom with 1s orbital on H-atom, it is also designated as sp^3(C)—s(H) σ-bond or sp^3 — s σ-bond.

Since the axes of sp^3 hybrid orbitals are directed towards the corners of a regular tetrahedron with C-atom located as its centre, the resulting C—H σ bonds are also directed towards the corners (vertices) of a regular tetrahedron with C-atom situated at its centre. Thus H—C—H bond angles in CH_4 molecule are the same as the angles between the axis of sp^3 orbitals, *i.e.* each H—C—H bond angle is 109.5°. Thus we find that CH_4 molecule has a *tetrahedral shape* and all the four C—H bonds are *identical*.

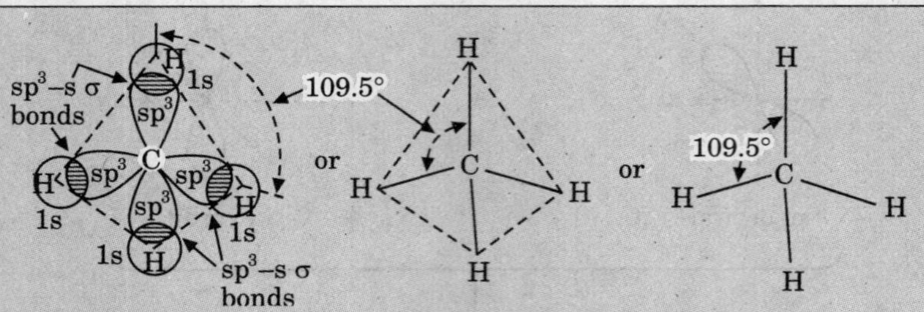

Fig 8.24. Pictorial representation of the formation of four sp^3(C)—s(H)σ bonds in CH_4 molecule which has a tetrahedral shape.

Shape of Some Species Having σ-Bonds Only on the Basis of Hybridisation

[A]. Species Whose Central Atom/Ion Has Two Electron Pairs [*Examples* : *BeF_2 (molten state), $BeCl_2$ (gaseous)*].

Here we shall consider the shape of BeF_2 molecule only. The Lewis structure of this molecule, F—Be—F shows that the central atom namely Be atom ($2s^2$) uses both its valence-shell electrons in forming two σ- bonds with two F atoms. Thus σ-*bps* = 2 and *lps* = 0 and hence σ-*bps* + *lps* = 2. Now since the sum of σ-*bps* and *lps* = 2, Be atom (central atom) in this molecule is supposed to be *sp* hybridised with bonding electrons = 2 and *lps* = 0 as shown in Fig. 8.25.

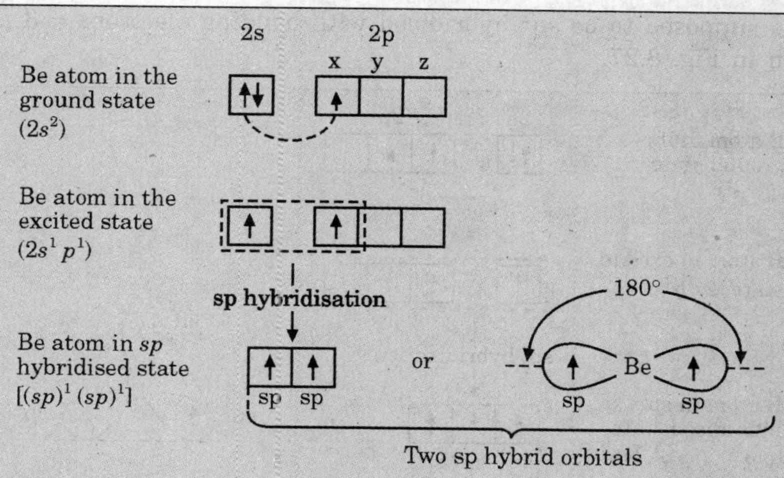

Fig. 8.25. *sp* hybridisation undergone by Be atom in its excited state.

sp hybridisation scheme shown in Fig. 8.25 makes it evident that both the *sp* hybrid orbitals are singly-filled. These orbitals make a *head-to-head (linear)* overlap with the singly-filled $2p_x$ orbitals on each of the two F atoms ($F \rightarrow 2s^2$, $2p_x^{\,1}\ 2p_y^{\,2}\ 2p_z^{\,2}$) and form two *sp* (Be)—$2p_x$(F)σ bonds as shown in Fig. 8.26. *sp* hybridisation undergone by Be atom gives *linear shape* to BeF_2 molecule.

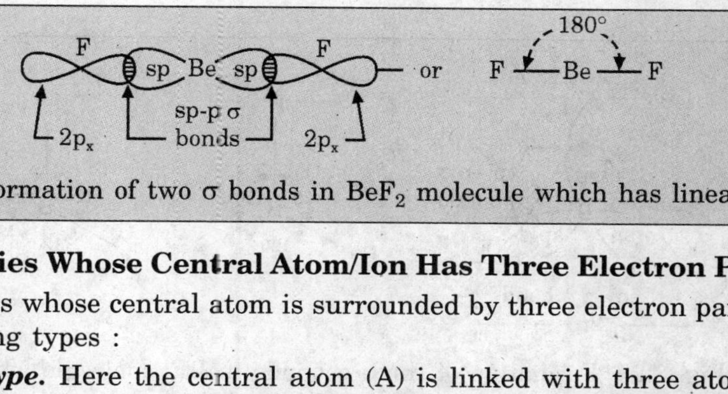

Fig. 8.26. Formation of two σ bonds in BeF_2 molecule which has linear shape.

[B]. Species Whose Central Atom/Ion Has Three Electron Pairs.

The species whose central atom is surrounded by three electron pairs may be of the following types :

(*a*) *AB_3* *type*. Here the central atom (A) is linked with three atoms (B) by σ-bonds and is surrounded by three σ-*bps*. Example are : BX_3(X = F, Cl, Br, I), BO_3^{3-} etc.

(*b*) *$AB_2(lp)$* *type*. The central atom is surrounded by two σ-*bps* and one *lp*. Examples are : gaseous $SnCl_2$, $PbCl_2$ etc.

Here shall consider the shapes of BF_3, BO_3^{3-}, and $SnCl_2$ species.

1. BF_3 molecule. The Lewis structure of this molecule, $F-\overset{\overset{F}{|}}{B}-F$ shows that the central atom namely B atom ($2s^2p^1$) uses all its valence-shell three electrons in forming three σ-bonds with three F atoms. Thus σ-*bps* = 3 and *lps* = 0 and hence σ-*bps* + *lps* = 3. Now since σ-*bps* + *lps* = 3, B atom (central atom) in this molecule is supposed to be sp^2 hybridised with bonding electrons = 3 and *lps* = 0 as shown in Fig. 8.27.

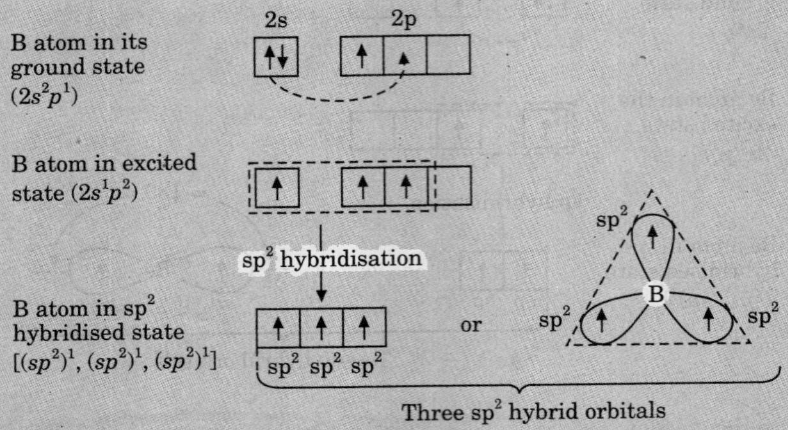

Fig. 8.27. sp^3 hybridisation undergone by B-atom in its excited state.

sp^2 hybridisation scheme shown in Fig. 8.27 makes it evident that all the three sp^2 hybrid orbitals are singly-filled. These orbitals make a *head-to-head* (*linear*) overlap with the singly-filled $2p_x$ orbitals on each of the three F atoms ($F \rightarrow 2s^2, 2p_x{}^1\, 2p_y{}^2\, 2p_z{}^2$) and form three $sp^2(B)$—$2p_x(F)$ σ-bonds as shown in Fig. 8.28. sp^2 hybridisation undergone by B atom gives a *triangular planar shape* to BF_3 molecule.

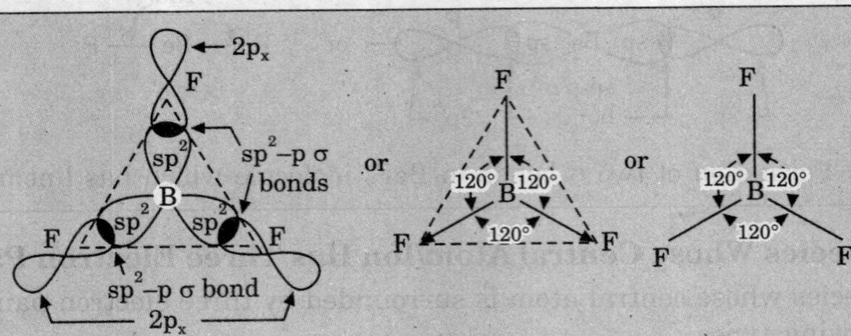

Fig. 8.28. Formation of three σ bonds in BF_3 molecule which has triangular planar shapes.

2. BO_3^{3-} ion. The Lewis structure of this ion is 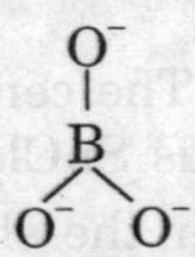 which shows that the central atom namely B atom uses all its valence-shell three electrons ($B \rightarrow 2s^2p^1$) in forming three σ bonds with three O^- ions. Thus σ-*bps* = 3 and *lps* = 0 and hence σ-*bps* + *lps* = 3. Now since the sum, *bps* + *lps* = 3, B atom (central atom) in BO_3^{3-} ion is supposed to be sp^2 hybridised with bonding electrons = 3 and *lps* = 0 as shown below in Fig. 8.29.

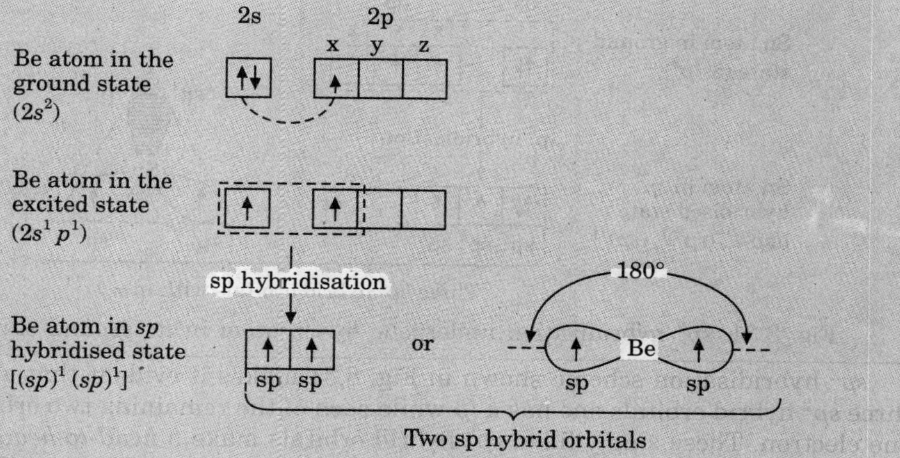

Fig. 8.29. sp^2 hybridisation undergone by B atom in its excited state.

sp^2 hybridisation scheme shown in Fig. 8.29 makes it evident that all the three sp^2 hybrid orbitals are singly-filled. These orbitals make a *head-to-head* (*linear*) overlap with the singly-filled $2p_x$ orbitals on each of the three O^- ions $(2s^2,\ 2p_x^1.\ 2p_y^1.\ 2p_z^2) + e^- \rightarrow O^-\ (2s^2,\ 2p_x^1.\ 2p_y^2.\ 2p_z^2)]$ and form three sp^2 (B) – $2p_x$ (O^-) σ bonds as shown in Fig. 8.30. sp^2 hybridisation undergone by B atom gives a *trigonal planar* shape to BO_3^{3-} ion.

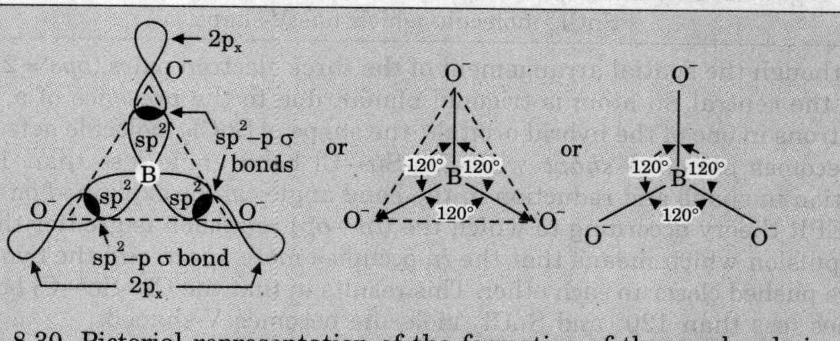

Fig. 8.30. Pictorial representation of the formation of three σ bonds in BO_3^{3-} ion which has trigonal planar shape.

3. $SnCl_2$ molecule. The Lewis structure of this molecule, Cl–$\ddot{\text{Sn}}$–Cl shows that the central atom (*i.e.* Sn atom) uses only two electrons out of its four valence-shell electrons. (Sn → $5s^2p^2$) in forming two σ bonds with 2Cl atoms while the remaining two electrons (*i.e.* one electron pair) remain as non-bonding electrons. Thus σ-*ps* = 2 and *lps* = 1 and σ-*bps* + *lps* = 3. Now since the sum of σ-*bps* and *lps* is 3, Sn atom (central atom) in $SnCl_2$ is supposed to be sp^2 hybridised with *bps* = 2 and *lps* = 1 as shown in Fig. 8.31.

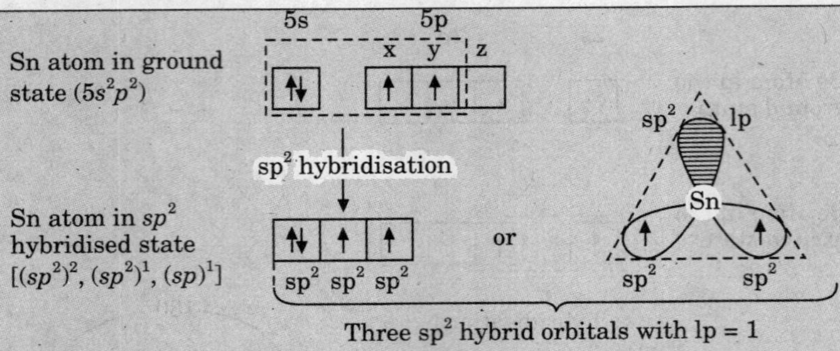

Fig. 8.31. sp^2 hybridisation undergone by Sn atom in its ground state.

sp^2 hybridisation scheme shown in Fig. 8.31 makes it evident that out of the three sp^2 hybrid orbitals one has a *lp* while each of the remaining two orbitals has one electron. These singly-filled sp^2 hybrid orbitals make a *head-to-head* (linear) overlap with the singly-filled $3p_x$ orbitals on each of the two Cl atoms (Cl $\rightarrow$ $3s^2$, $3p_x^{\ 1}$ $3p_y^{\ 2}$, $3p_z^{\ 2}$) and form two sp^2(Sn)—$3p_x$(Cl) σ—bonds as shown in Fig. 8.32.

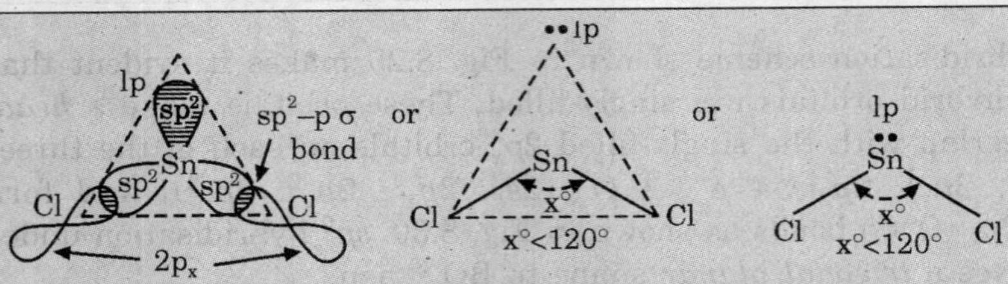

Fig. 8.32. Pictorial representation of the formation of two σ bonds in $SnCl_2$ molecule which has V-shape.

Although the spatial arrangement of the three electron pairs (*bps* = 2, *lps* = 1), round the central Sn atom is trigonal planar, due to the presence of a lone pair of electrons in one of the hybrid orbitals, the shape of $SnCl_2$ molecule gets distorted and becomes *planar V-shape* with Cl—Sn—Cl bond angle less than 120°. The distortion in shape and reduction in the bond angle can be explained on the basis of VSEPR theory according to which the (*lp*—*bp*) repulsion is greater than (*bp*—*bp*) repulsion which means that the *lp* occupies more space and the two (Sn—Cl) *bps* are pushed closer to each other. This results in that the Cl—Sn—Cl bond angle becomes less than 120° and $SnCl_2$ molecule becomes V-shaped.

[C]. Species Whose Central Atom/Ion Has Four Electron Pairs.

The molecules whose central atom is surrounded by four electron pairs may be of the following *four* types :

(a) AB_4 type. Here the central atom (A) is surrounded by four σ-*bps*. Examples are : CH_4, CX_4 (X = F, Cl, Br, I), $SiCl_4$, BH_4^-, AlH_4^-, NH_4^+, PH_4^+.

(b) AB_3 (lp) type. The central atom is surrounded by three σ-*bps* and one *lp*. Examples are : NH_3, NF_3, PX_3 (X = F, Cl, Br, I), PH_3.

(c) AB_2 $(lp)_2$ type. The central atom is surrounded by two σ-*bps* and two *lps*. Examples are : H_2O, H_2S, OX_2 (X = F, Cl, Br), SCl_2, ICl_2, NH^-_2.

(d) AB $(lp)_3$ type. The central atom is surrounded by one σ-*bp* and three *lps*. ICl molecule is an example of this type of molecules.

Here we shall consider the shapes of CH_4, BH_4^-, NH_4^+, NH_3, H_2O, ICl_2^+ and ICl species.

1. CH_4 molecule. The structure of CH_4 molecule has already been discussed.

2. Borohydride ion (BH_4^-). Due to the presence one unit of negative charge on BH_4^- ion, B atom (central atom) can be regarded as having 4 electrons instead of 3 in its valence-shell. [B($2s^2p^1$)+e^- → B^- ($2s^2p^2$)]. The Lewis structure of this ion

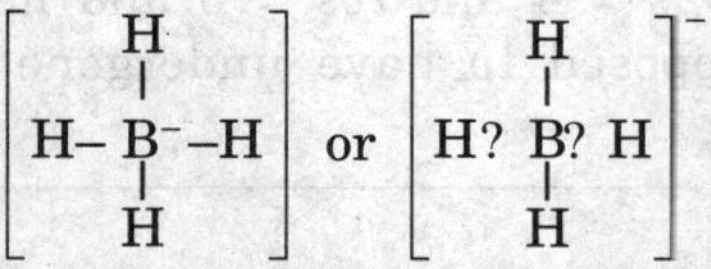

, shows that B atom uses all its 4 electrons in forming 4σ bonds with four H atoms. Thus σ-*bps* = 4 and *lps* = 0 and hence σ-*lps* + *lps* = 4. Consequently B atom is supposed to have undergone sp^3 hybridisation in the given ion as shown in Fig. 8.33.

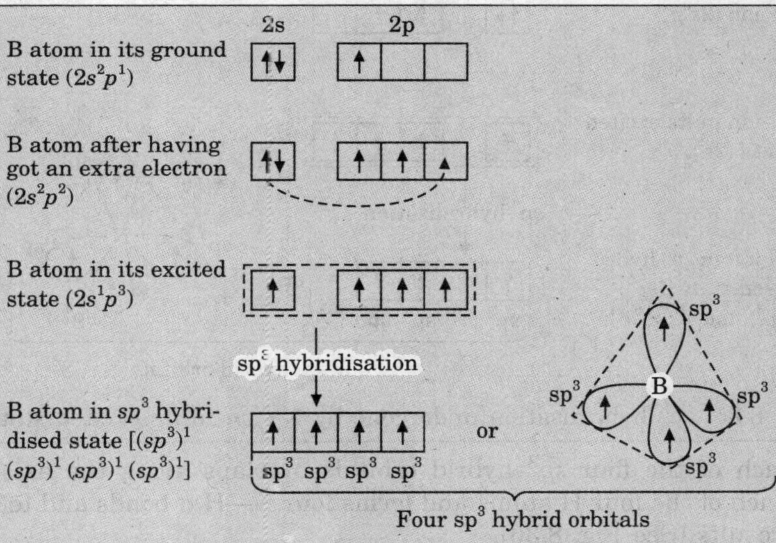

Fig. 8.33. sp^3 hybridisation undergone by B atom in its excited state.

Each of the four sp^3 hybrid orbitals on B atom is singly filled. Each of these hybrid orbitals overlaps along the axis with the half-filled 1s orbital on H-atom and forms four B—H σ bonds and consequently BH_4^- ion has a tetrahedral shape (See Fig. 8.34).

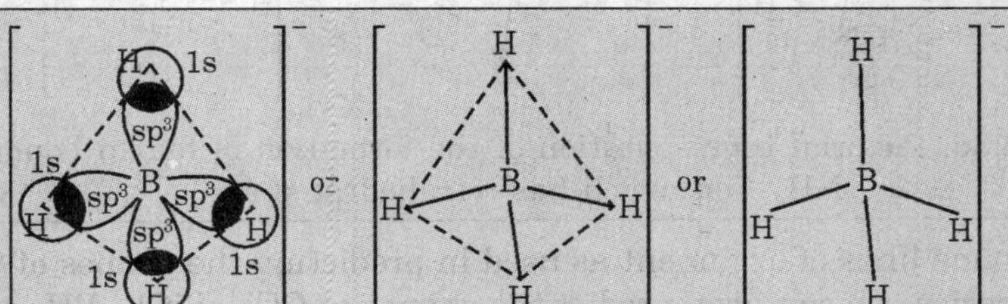

Fig. 8.34. Pictorial representation of the formation of four σ bonds in BH_4^- ion which has tetrahedral shape.

3. Ammonium ion (NH_4^+). Due to the presence of one unit of positive charge on NH_4^+ ion, N atom (central atom) can be regarded as having 4 electrons instead of 5 in its valence-shell [$N(2s^2p^3) - e^- \rightarrow N^+(2s^2p^2)$]. The Lewis structure of this ion,

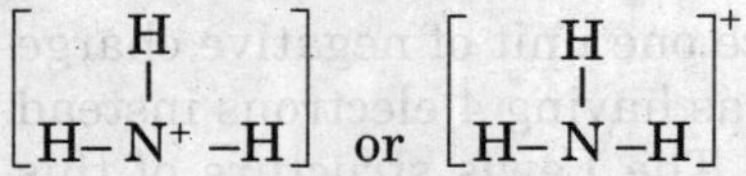

shows that N atom uses all its four electrons in forming four σ bonds with 4H atoms. Thus σ-*bps* = 4 and *lps* = 0 and hence σ-*bps* + *lps* = 4. Consequently N atom is supposed to have undergone sp^3 hybridisation in NH_4^+ ion as shown in Fig. 8.35.

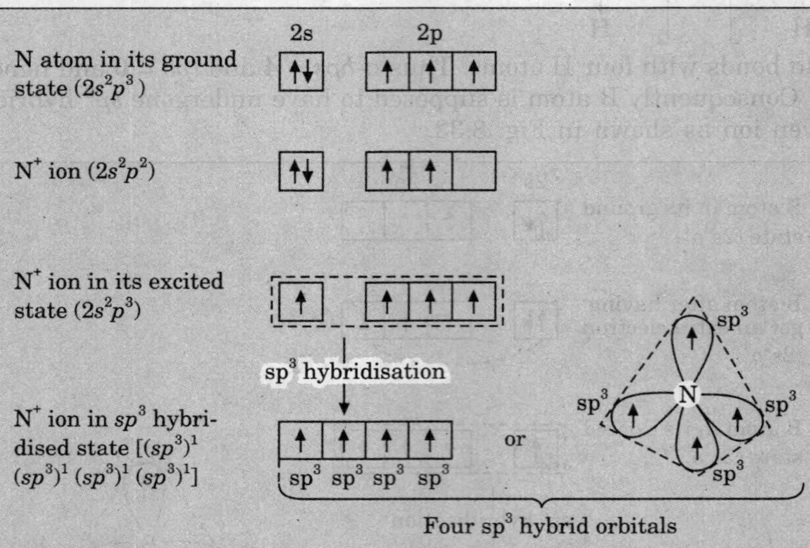

Fig. 8.35. sp^3 hybridisation undergone by N^+ ion in its excited state.

Now each of the four sp^3 hybrid orbitals overlaps along the axis with 1*s* orbital of each of the four H atoms and forms four N—H σ bonds and tetrahedral NH_4^+ ion results (See Fig. 8.36).

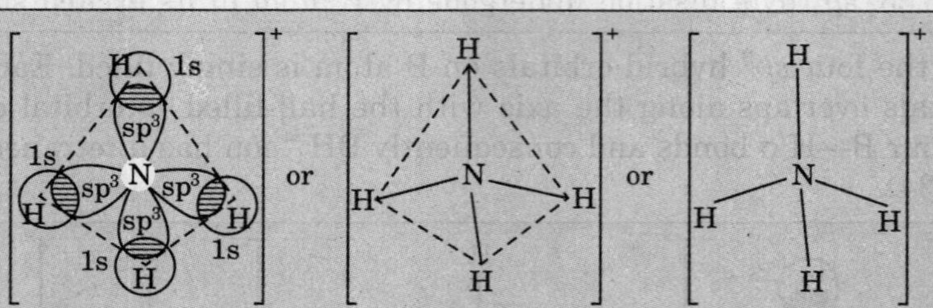

Fig. 8.36. Pictorial representation of the formation of four σ bonds in NH_4^+ ion which has tetrahedral shape.

On the same lines of argument as used in predicting the shapes of CH_4, BH_4^- and NH_4^+ species, we can also, predict the shapes of CCl_4, $SiCl_4$, PH_4 etc. species whose central atoms are sp^3 hybridised and hence all the species have a tetrahedral shape.

4. NH_3 molecule. The Lewis structure of this molecule, H–$\ddot{\text{N}}$–H (with H bonded below N) shows that the central atom (*i.e.* N atom) uses only three electrons. Out of its five valence-shell electrons ($N \rightarrow 2s^2p^3$) three are used up in forming 3σ bonds with 3H atoms while the remaining two electrons (*i.e.* one electron pair) remain as non-bonding electrons on N-atom. Thus σ-*bps* = 3 and *lps* = 1 and σ-*bps* + *lps* = 4. Now since the sum of σ-*bps* and *lps* is 4, N atom in NH_3 molecule is supposed to be sp^3 hybridised as shown in Fig. 8.37.

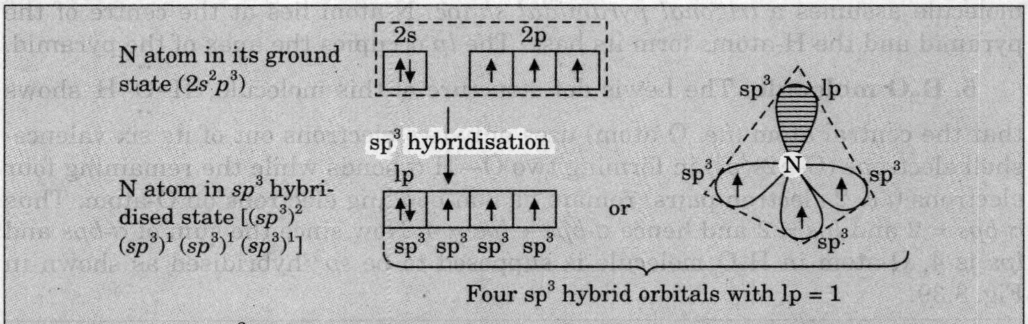

Fig. 8.37. sp^3 hybridisation undergone by N atom in its ground state.

sp^3 hybridisation scheme shown in Fig. 8.37 makes it evident that out of the four sp^3 hybrid orbitals one contains a lone pair while each of the remaining three orbitals has one electron. The hybrid orbital containing the lone pair does not participate in the overlapping or N—H bond formation process while the singly-filled sp^3 hybrid orbitals make a *head-to-head* (*linear*) overlap with the singly-filled 1s orbitals on each of the three H atoms ($H \rightarrow 1s^1$) and form three sp^3(N)—1*s* (H) σ bonds as shown in Fig. 8.38.

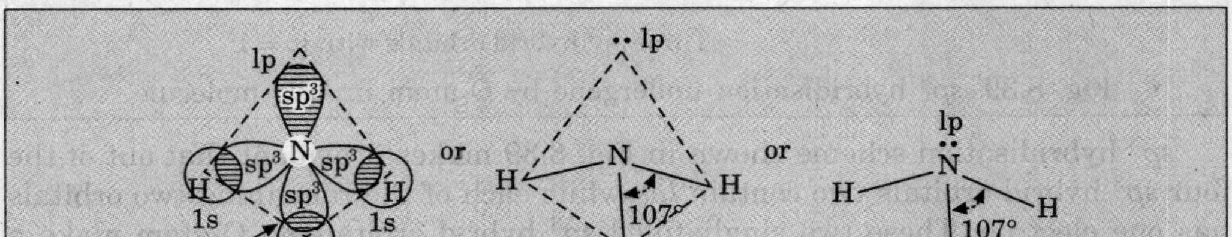

Fig. 8.38. Pictorial representation of the formation of three σ bonds in NH_3 molecule which has trigonal pyramidal shape.

Geometry of NH_3 molecule and H—N—H bond angle. Although the spatial arrangement of the four electron pairs (*bps* = 3, *lps* = 1) round the central N atom is tetrahedral, due to the presence of a lone pair of electrons in one of the hybrid orbitals, the shape of NH_3 molecule gets distorted and becomes *trigonal pyramidal* and each of three H—N—H bond angles decreases from the expected tetrahedral angle (= 109.5°) to 107°. The distortion in shape and hence reduction in the bond angle can be explained on the basis of VSEPR theory according to which, since *lp* belongs to N-atom only, the hybrid orbital containing the *lp* occupies more space than the other three hybrid orbitals each of which contains a *lp* and belongs to three H-atoms and one N-atom jointly. The greater spread in space of the hybrid orbital containing the *lp* results in that a *lp* repells a *bp* more strongly than a *bp* repells another *bp*, *i.e.* (*lp*—*bp*) repulsion becomes greater than (*bp*—*bp*) repulsion. The net effect of this difference in the electron pair—electron pair repulsions is that the lone pair on N-atom forces the three (N—H) *bps* to come closer to each other and hence each of H—N—H bond angle decreases from tetrahedral angle (= 109.57) to 107°. Due to the decrease in the bond angle, the expected tetrahedral shape of NH_3 molecule gets a little distorted and NH_3 molecule assumes a *trigonal pyramidal shape*. N-atom lies at the centre of the pyramid and the H-atoms form its base. The *lp* occupies the apex of the pyramid.

5. H_2O molecule. The Lewis dot structure of this molecule, $H-\ddot{\underset{\cdot\cdot}{O}}-H$ shows that the central atom (*i.e.* O atom) uses only two electrons out of its six valence-shell electrons ($O \rightarrow 2s^2p^4$) in forming two O—H σ bonds while the remaining four electrons (*i.e.* 2 electron pairs) remain as non-bonding electrons on O-atom. Thus σ-*bps* = 2 and *lps* = 2 and hence σ-*bps* + *lps* = 4. Now since the sum of σ-*bps* and *lps* is 4, O-atom in H_2O molecule is supposed to be sp^3 hybridised as shown in Fig. 8.39.

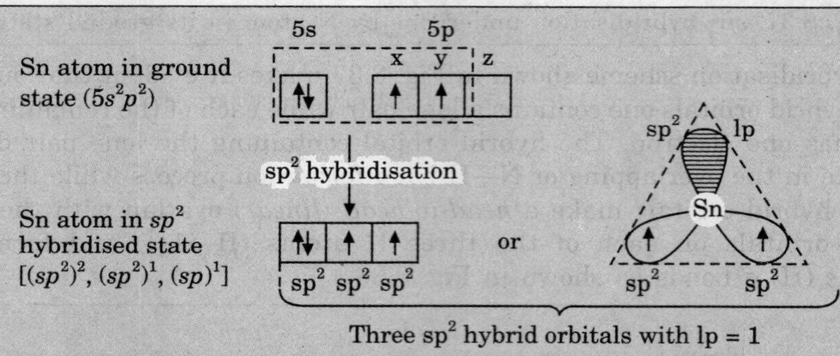

Fig. 8.39. sp^3 hybridisation undergone by O-atom in H_2O molecule.

sp^3 hybridisation scheme shown in Fig. 8.39 makes it evident that out of the four sp^3 hybrid orbitals two contain *lps* while each of the remaining two orbitals has one electron. These two singly-filled sp^3 hybrid orbitals on O-atom make a *head-to-head* (linear) overlap with the singly-filled 1*s* orbitals on each of the two H-atoms ($H \rightarrow 1s^1$) and form two sp^3 (O)—1*s* (H) σ bonds as shown in Fig. 8.40.

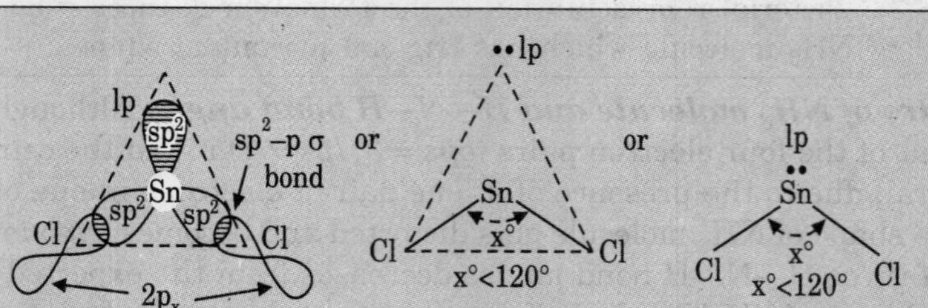

Fig. 8.40. Pictorial representation of the formation of two σ bonds in H_2O molecule (*bps* = 2, *lps* = 2) which has V-shape.

Geometry of H_2O molecule and H—O—H bond angle. Although the spatial arrangement of the four electron pairs (*bps* = 2, *lps* = 2) round the central O-atom is tetrahedral, due to the presence of two lone pairs of electrons in two hybrid orbitals, the shape of H_2O molecule gets distorted and becomes *V-shape* (*angular or bent shape*), and H—O—H bond angle decreases form the expected tetrahedral angle (= 109.5°) to 104.5°. The distortion in shape and decrease in bond angle can be explained on the basis of VSEPR theory according to which the distortion in shape and decrease in H—O—H bond angle is due to the presence of two *lps* on O-atom and the difference in the magnitude of three types of electron pair—electron pair repulsions shown below :

(lp—lp) repulsion > (lp—bp) repulsion > (bp—bp) repulsion

The net effect of the difference in the magnitude of electron pair—electron pair repulsions shown above is that the two *lps* on O-atom force the two (O—H) *bps* to come closer to each other and hence each of the H—O—H bond angles in H_2O becomes less than the tetrahedral angle (= 109.5°) and H_2O molecule assumes V-shape which is a distorted shape of a tetrahedron.

6. $[ICl_2]^+$ ion. Due to the presence of one unit of positive charge on the ion, I atom (central atom) can be regarded as having six electrons instead of seven in its valence-shell [I $(5s^2p^5)$—e^- → I $(3s^2p^4)$]. The Lewis structure of this ion, $\left[\text{Cl–}\overset{+}{\underset{\cdot\cdot}{\ddot{\text{I}}}}\text{–Cl}\right]$ or $\left[\text{Cl–}\underset{\cdot\cdot}{\ddot{\text{I}}}\text{–Cl}\right]^+$ shows that I atom uses two of its six electrons in forming 2σ bonds with two Cl atoms. Thus σ-*bps* = 2 and *lps* = 2 and hence σ-*bps* + *lps* = 4. Consequently I atom is supposed to be sp^3 hybridised in ICl_2^+ ion as shown below in Fig. 8.41.

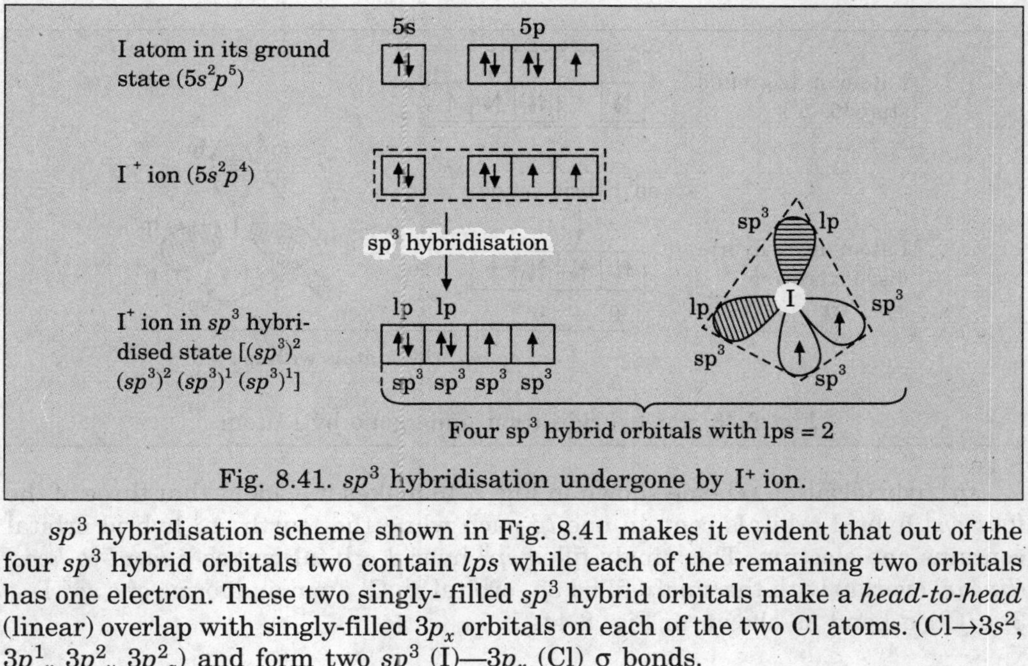

Fig. 8.41. sp^3 hybridisation undergone by I^+ ion.

sp^3 hybridisation scheme shown in Fig. 8.41 makes it evident that out of the four sp^3 hybrid orbitals two contain *lps* while each of the remaining two orbitals has one electron. These two singly- filled sp^3 hybrid orbitals make a *head-to-head* (linear) overlap with singly-filled $3p_x$ orbitals on each of the two Cl atoms. (Cl→$3s^2$, $3p^1_x$ $3p^2_y$ $3p^2_z$) and form two sp^3 (I)—$3p_x$ (Cl) σ bonds.

Although the spatial arrangement of four electron pairs (*bps* = 2, *lps* = 2) round the central I atom in space is tetrahedral, due to the presence of two lone pairs, the shape of ICl_2^+ ion gets distorted and becomes angular or V-shape as shown in Fig. 8.42.

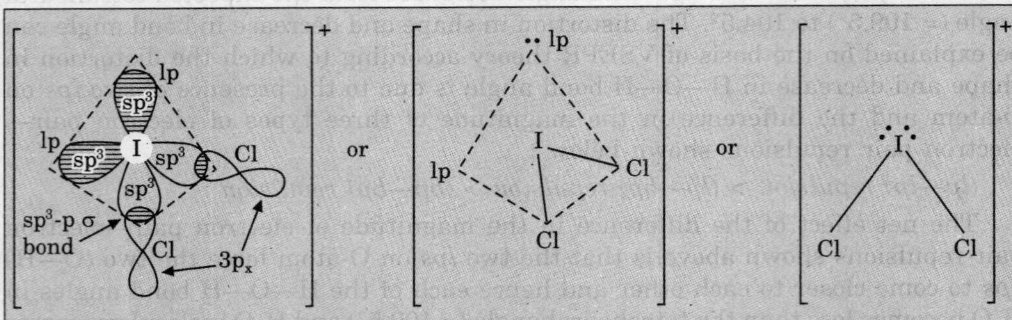

Fig. 8.42. Pictorial representation of the formation of 2σ bonds in $[ICl_2]^+$ ion (*bps* = 2, *lps* = 2) which has V-shape.

7. ICl molecule. The Lewis structure of this molecule, $:\ddot{\underset{\cdot\cdot}{I}}-Cl$ shows that the central atom (*i.e.* I atom) uses only one of its seven valence-shell electrons (I → $5s^2p^5$) in forming 1σ bond with one Cl atom while the remaining six electrons (*i.e.*, 3 electron pair) remain as non-bonding electrons on I-atom. Thus σ-*bps* = 1 and *lps* = 3 and σ-*bps* + *lps* = 4. Now since the sum of σ-*bps* and *lps* is 4, I atom in ICl molecule is supposed to be sp^3 hybridised as shown in Fig. 8.43.

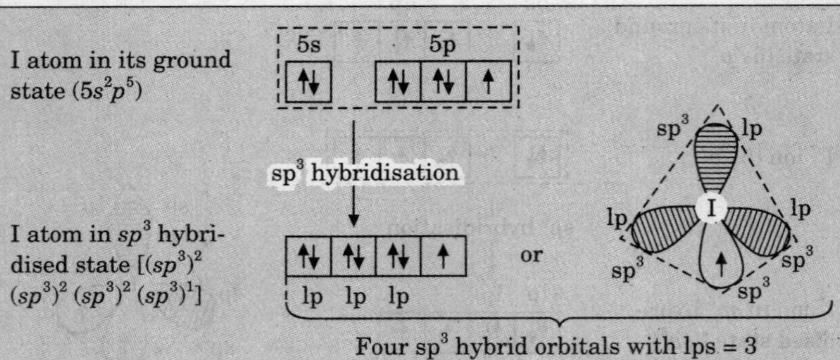

Fig. 8.43. sp^3 hybridisation undergone by I atom.

sp^3 hybridisation scheme shown in Fig. 8.43 makes it evident that three of the four sp^3 hybrid orbitals contain one *lp* each while the fourth sp^3 hybrid orbital contains one electron. This singly-filled sp^3 hybrid orbital makes a *head-to-head (linear)* overlap with the singly-filled $3p_x$ orbital on Cl atom and forms one sp^3(I)—$3p_x$ (Cl) σ bond as shown in Fig. 8.44.

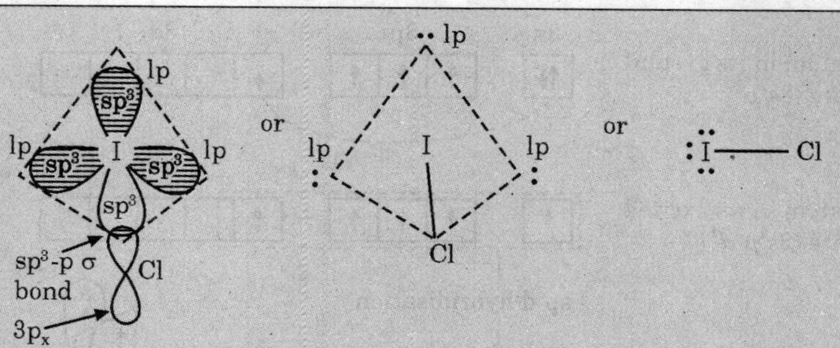

Fig. 8.44. Pictorial representation of the formation of I—Cl σ-bond in ICl molecule (bp = 1, lps = 3) which has linear shape.

Although the spatial arrangement of the four electron pairs (bp = 1 and lps = 3) round the central I atom is tetrahedral, due to the presence of three lone pairs of electrons in three hybrid orbitals, the shape of ICl molecule gets distorted and becomes linear.

[D] Species Whose Central Atom/Ion Has Five Electron Pairs.

The molecules whose central atom is surrounded by five electron pairs may be of the following *four* types :

(a) AB_5 type. Here the central atom (A) is surrounded by five σ-*bps*. Example are : PX_5 (X = F, Cl, Br) and SbF_5 (all in gaseous phase).

(b) AB_4 (lp) type. Here the central atom is surrounded by four σ-*bps* and one *lp*. Examples are : SF_4, SeF_4, TeF_4 (all in gaseous phase) and IF_4^+.

(c) AB_3 $(lp)_2$ type. The central atom is surrounded by three σ-*bps* and two *lps*. Examples are ClF_3, BrF_3, IF_3.

(d) AB_2 $(lp)_3$ type. The central atom is surrounded by two σ-*bps* and three *lps*. Examples are : XeF_2, I_3^-, ICl_2^-.

Here we shall consider the shape of PF_5, SF_4, IF_4^+, ClF_3, XeF_2, I_3^- and ICl_2^- species.

1. PF_5 molecule in liquid and gaseous phase. The Lewis structure of this molecule, 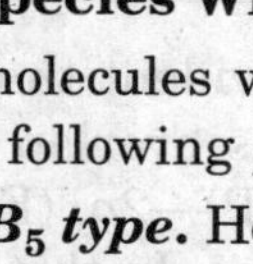

shows that the central atom (*i.e.* P atom) uses all its five electrons from its valence-shell (P → $3s^2p^3$) in forming 5σ bonds with 5F atoms. Thus σ-*bps* = 5 and *lps* = 0 and hence σ-*bps* + *lps* = 5, showing that P-atom is sp^3d hybridised in this molecule as shown in Fig. 8.45. The axial and equatorial (basal) hybrid orbitals have been shown by a and e respectively.

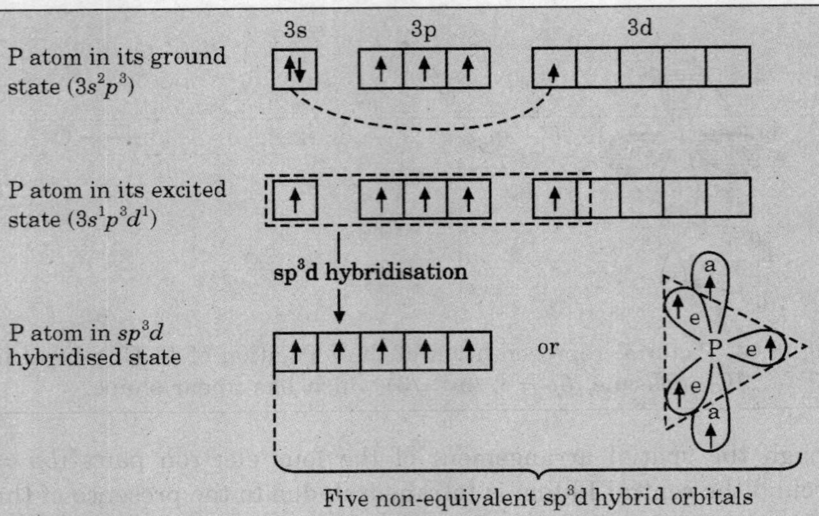

Fig. 8.45. sp^3d hybridisation undergone by P atom in its excited state. *a* and *e* represent the axial and equatorial (basal) hybrid orbitals respectively.

sp^3d hybridisation scheme shown in Fig. 8.45 makes it evident that each of the five hybrid orbitals is singly-filled. These hybrid orbitals overlap with $2p_x$ orbital of 5F atoms (F→$2s^2$, $2p_x{}^1.2p_y{}^2.2p_z{}^2$) and form five sp^3d(P)—$2p_x$ (F) σ bonds as shown in Fig. 8.46 and PF_5 has a trigonal bipyramidal shape. In this molecule the three equatorial bonds (P—F_e bonds) are equal in length but shorter than the two axial bonds (P—F_a bonds). This is because of the fact that the axial P—F_a bonds are less strong than the equatorial P—F_e bonds. This confirms that the five sp^3d hybrid orbitals are non-equivalent.

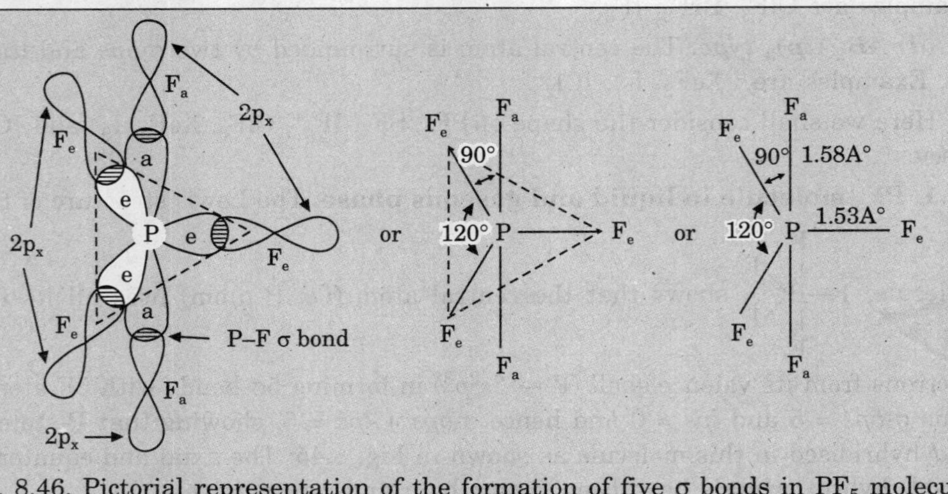

Fig. 8.46. Pictorial representation of the formation of five σ bonds in PF_5 molecule (gaseous) which has trigonal bipyramidal shape. F_a and F_e represent the axial and equatorial (or basal) F atoms.

On similar lines of argument it can be shown that PCl_5, PBr_5, SbF_5 (all in gaseous phase) molecules also have a trigonal bipyramidal shape.

2. SF_4 molecule in gaseous phase. The Lewis structure of this molecule

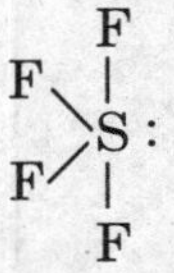

shows that the central atom (*i.e.* S atom) uses only four out of its six valence-shell electrons ($S \rightarrow 3s^2p^4$) in forming 4σ bonds with 4F atoms and remaining one electron pair remains as a *lp* on S atom. Thus σ-*bps* = 4 and *lp* = 1 and hence σ-*bps* + *lps* = 5, showing that S atom in SF_4 molecule is sp^3d hybridised as shown in Fig. 8.47.

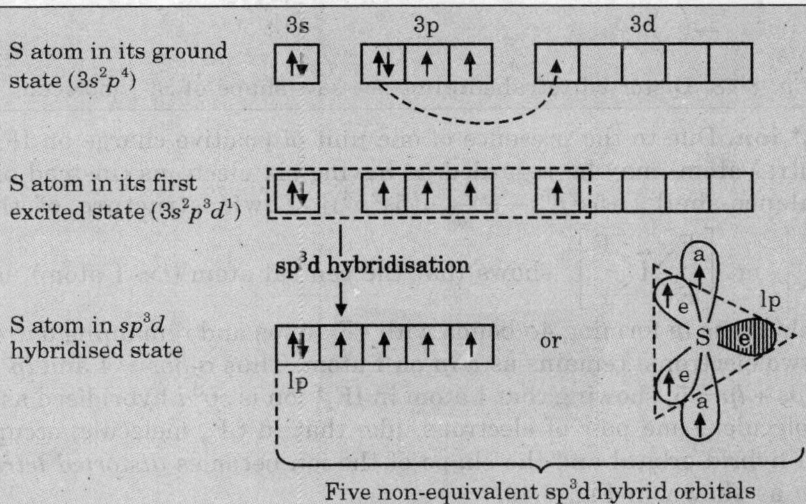

Fig. 8.47. sp^3d hybridisation undergone by S-atom in its first excited state. Note that the lone pair (*lp*) occupies the basal (equatorial) hybrid orbital of the pentagonal bipyramid.

sp^3d hybridisation scheme shown in Fig. 8.47 makes it evident that out of five sp^3d hybrid orbitals one equatorial hybrid orbital contains a *lp*, because in this case (*lp*—*bp*) repulsion is minimum. Other four hybrid orbitals which are singly-filled overlap with the singly-filled $2p_x$ orbitals of 4F atoms and form four sp^3d (S)—$2p_x$(F) σ bonds. Although the spatial arrangement of five electron pairs (*bps* = 4 and *lp* = 1) round the central S atom is trigonal bipyramidal, due to the presence of one lone pair of electrons in one of the equatorial hybrid orbitals, the shape of SF_4 molecule gets distorted and becomes *distorted tetrahedral* or *see-saw*. In this molecule the two axial S—F bonds (S—F_a bonds) are slightly bent towards the equatorial S—F bonds, (S—F_e bonds), showing the repulsion effect of lone pair. The S—F_a bonds (= 1.64 Å) are appreciably longer than the S—F_e bonds (= 1.55 Å).

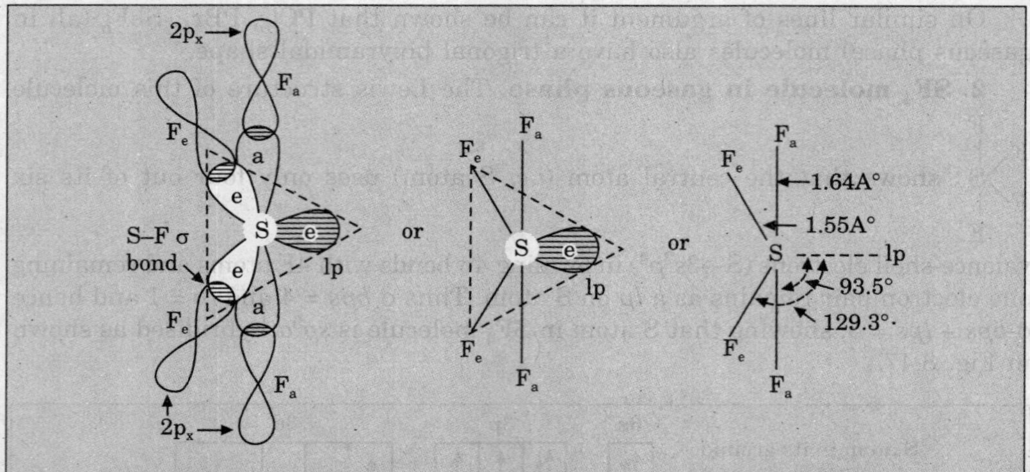

Fig. 8.48. Distorted tetrahedral or see-saw shape of SF_4 molecule.

3. IF_4^+ ion. Due to the presence of one unit of positive charge on IF_4^+ ion, I atom (central atom) may be regarded as having six electrons (instead of seven) in its valence-shell [I($5s^2p^5$)—$e^-\rightarrow$ I($5s^2p^4$)]. Lewis structure of this ion,

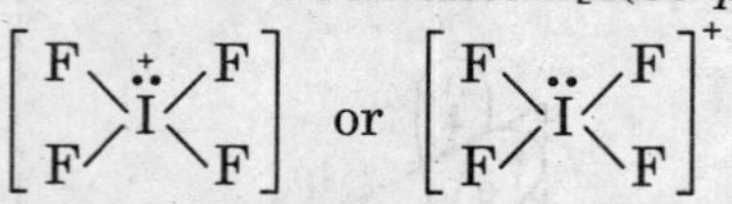

shows that the central atom (*i.e.* I atom) uses four of its six electrons in forming 4σ bonds with 4F atoms and remaining one electron pair (*i.e.* two electrons) remains as a *lp* on I atom. Thus σ-*bps* = 4 and *lp* = 1 and hence σ-*bps* + *lp* = 5, showing that I atom in IF_4^+ ion is sp^3d hybridised as S atom in SF_4 molecule. Lone pair of electrons, like that in SF_4 molecule, occupies the equatorial hybrid orbital and the shape of the ion becomes *distorted tetrahedral or see-saw* as shown in Fig. 8.49.

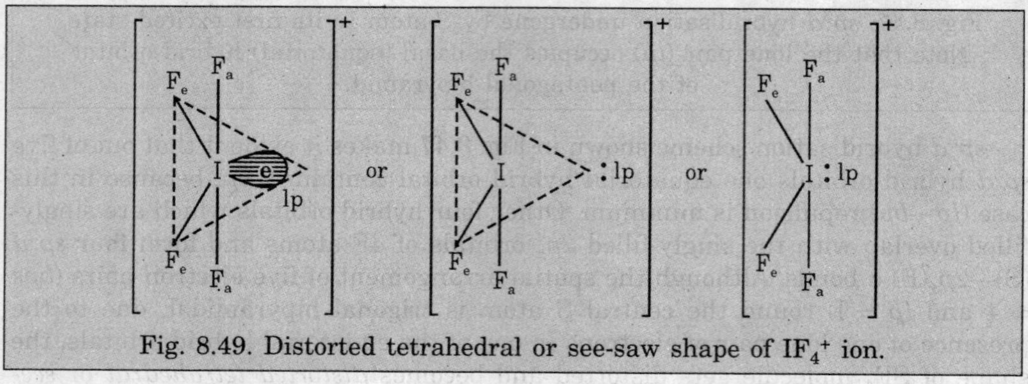

Fig. 8.49. Distorted tetrahedral or see-saw shape of IF_4^+ ion.

4. ClF_3 molecule. The electronic structure of this molecule, 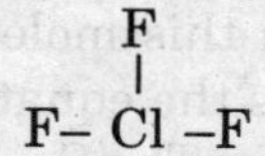 shows that the central atom namely Cl atom uses three out of its seven valence-shell electrons in forming 3σ bonds with 3F atoms while the remaining four electrons (*i.e.* two electron pairs) are left as *lps* on Cl atom. Thus σ-*bps* + *lps* = 3 + 2 = 5 and hence Cl atom in ClF_3 molecule can be supposed to be sp^3d hybridised as shown in Fig. 8.50. The *lps* occupy the equatorial hybrid orbitals.

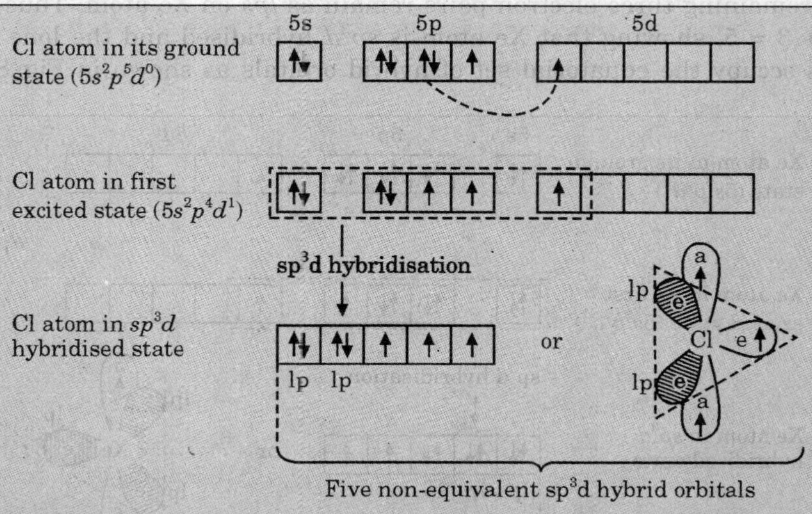

Fig. 8.50. sp^3d hybridisation undergone by Cl atom in its first excited state. Lone pairs (*lps*) occupy the basal (equatorial) hybrid orbitals.

sp^3d hybridisation scheme shown in Fig. 8.50 makes it evident that out of five hybrid orbitals two (equatorial) contain *lps* while the remaining three contain one electron each. The singly-filled hybrid orbitals overlap with the singly filled $2p_x$ orbitals on 3F atoms and form three Cl—F σ bonds. Although the spatial arrangement of five electron pairs (*bps* = 3, *lps* = 2) round the central Cl atom in space is trigonal bipyramidal, due to the presence of two lone pairs, the shape of ClF_3 molecule gets distorted and becomes *slightly bent T-shape* with bond angles equal to 87.5°. The lone pairs occupy the hybrid orbitals directed towards the equatorial positions of the trigonal bipyramid (See Fig. 8.51).

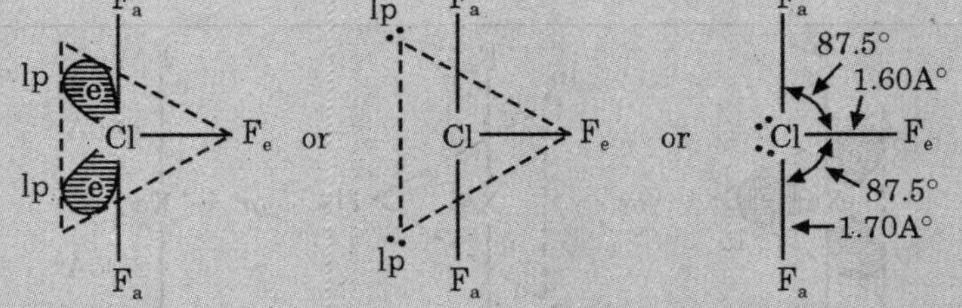

Fig. 8.51. Bent T-shape of ClF_3 molecule with *bps* = 3 and *lps* = 2. F_a are the F atoms occupying the axial positions of the trigonal bipyramid while F_e is the F atom occupying the basal position.

On similar grounds if can be shown that BrF_3 and IF_3 both have T-shaped structure. In BrF_3 the axial bonds (Br—F_a = 1.81 Å) are appreciably larger than the equatorial bonds (Br—F_b = 1.72 Å). Such parameters for IF_3 are not known.

5. XeF_2 molecule. The Lewis structure of this molecule, F–Xe–F shows that the central atom (*i.e.* Xe atom) uses two electrons in forming two Xe—F σ bonds and the remaining three electron pairs remain as *lps* on Xe atom. Thus σ-*bps* + *lps* = 2 + 3 = 5, showing that Xe atom is sp^3d hybridised and the lone pairs of electrons occupy the equatorial set of hybrid orbitals as shown in Fig. 8.52.

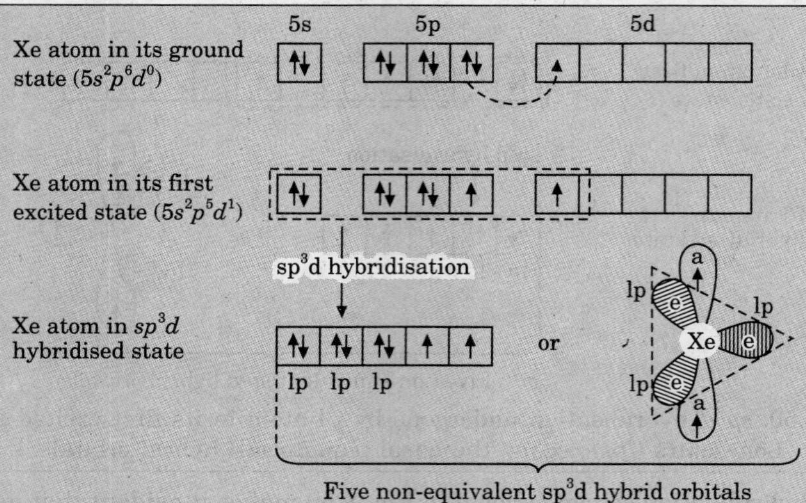

Fig. 8.52. sp^3d hybridisation undergone by Xe atom in its first excited state. Lone pairs (*lp*) occupy the basal (equatorial) set of hybrid orbitals.

sp^3d hybridisation scheme given in Fig. 8.52 shows that three equatorial hybrid orbitals (*e*) have lone pairs and the remaining axial hybrid orbitals (*a*) have one electron each. These orbitals overlap with $2p_x$ orbitals of two F atoms ($F \rightarrow 2s^2, 2p_x^1\ 2p_y^2\ 2p_z^2$) and form two Xe—F σ bonds. Due to the presence of *lps*, XeF_2 molecule assumes a *linear shape* as shown in Fig. 8.53 with Xe—F bond lengths equal to 2.00 A°.

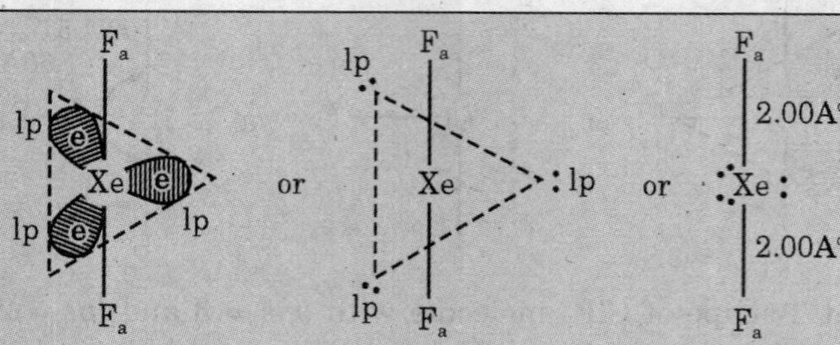

Fig. 8.53. Linear shape of XeF_2 molecule with *bps* = 2 and *lps* = 3.

6. ICl_2^- and I_3^- ions. On the same lines of argument as used in case of XeF_2, it can be shown that both ICl_2^- and I_3^- ions are also linear in shape. In both the ions, due to the presence of one unit of negative charge, the central atom namely

I atom is regarded as having eight electrons [$I(5s^2p^5) + e^- \rightarrow I^- (5s^2p^6)$] (instead of seven). The linear shapes of ICl_2^- = and I_3^- ions are given in Fig. 8.54.

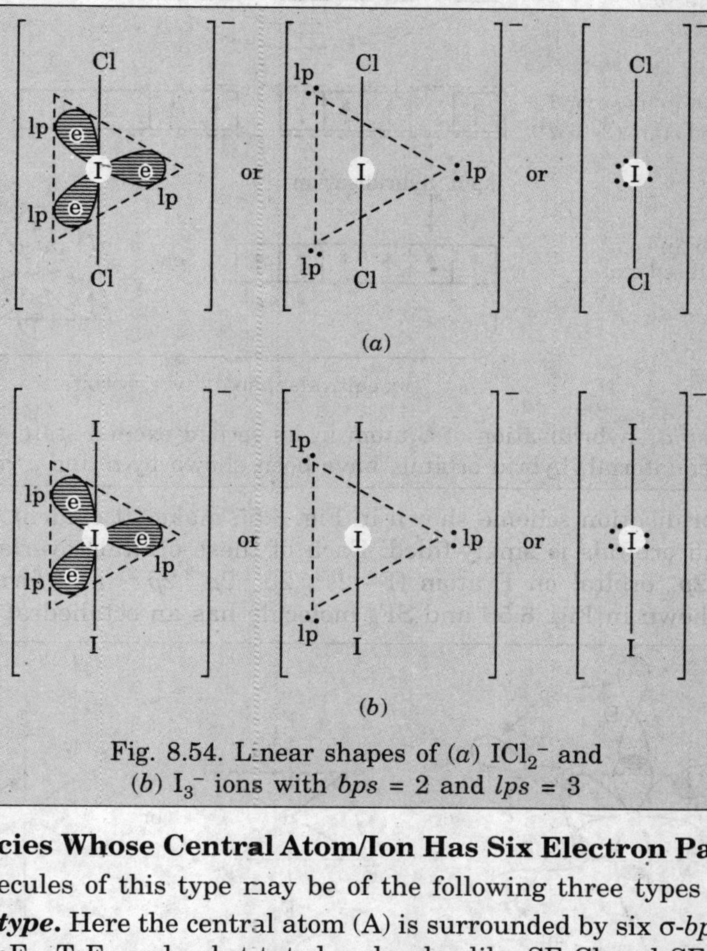

Fig. 8.54. Linear shapes of (*a*) ICl_2^- and (*b*) I_3^- ions with *bps* = 2 and *lps* = 3

[E] Species Whose Central Atom/Ion Has Six Electron Pairs

The molecules of this type may be of the following three types :

(a)* AB_6 *type. Here the central atom (A) is surrounded by six σ-*bps*. Examples are : SF_6, SeF_6, TeF_6 and substituted molecules like SF_5Cl and SF_5Br.

(b)* $AB_5(lp)$ *type. Here the central atom is surrounded by five σ-*bps* and one *lp*. Examples are : IF_5, BrF_5 etc.

(c)* $AB_4(lp)_2$ *type. The central atom is surrounded by four σ-*bps* and two *lps*. Examples are : XeF_4, ICl_4^- etc.

Here shall consider the shape of SF_6, IF_5, XeF_4 molecules.

1. SF_6 molecule. Lewis structure of this molecule, 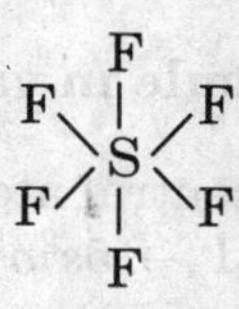 shows that the central atom namely S atom uses all its six valence-shell electrons in forming six S—F σ-bonds. Thus σ-*bps* = 6 and *lps* = 0 and hence σ-*bps* + *lps* = 6, showing that S atom is sp^3d^2 hybridised in this molecule as shown in Fig. 8.55.

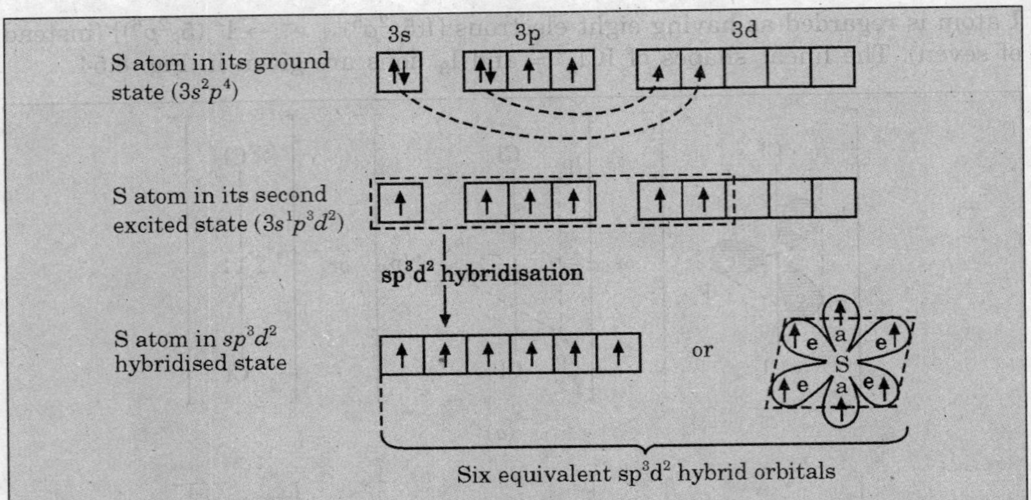

Fig. 8.55. sp^3d^2 hybridisation of S atom in its second excited state. The axial and basal (equatorial) hybrid orbitals have been shown by *a* and *e* respectively.

sp^3d^2 hybridisation scheme shown in Fig. 8.55 makes it evident that each of the six hybrid orbitals is singly-filled. Each of these orbitals overlaps with the singly-filled $2p_x$ orbital on F atom ($F \rightarrow 2s^2, 2p_x^1.2p_y^2.2p_z^2$) and forms six S—F σ-bonds as shown in Fig. 8.56 and SF_6 molecule has an octahedral shape.

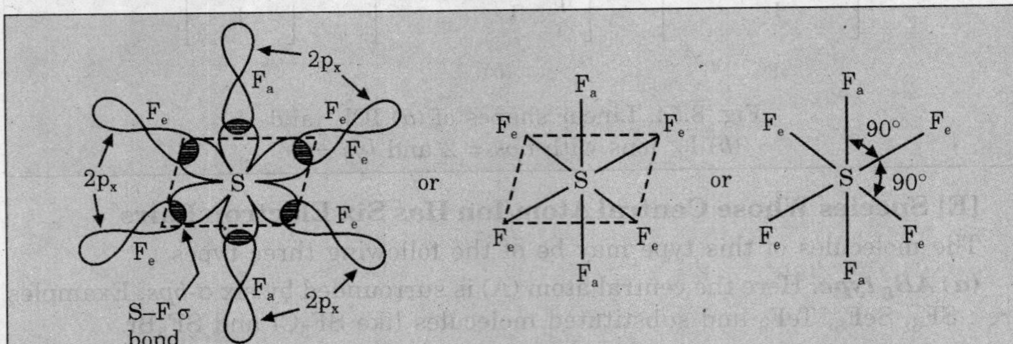

Fig. 8.56. Octahedral shape of SF_6 molecule.

On the similar lines of arguement it can be shown that SeF_6, TeF_6, SF_5CI and SF_5Br molecules also have octahedral shape.

2. IF_5 molecule in vapour state. Lewis structure of this molecule,  shows that the central atom namely I atom uses only five out of its seven valence-shell electrons ($I \rightarrow 5s^2p^5$) in forming five I—F σ-bonds and the remaining one electron pair remains as a *lp* on I atom. Thus σ-*bps* = 5 and *lp* = 1 and hence σ-*bps* + *lps* = 6, showing that I atom in IF_5 molecule is sp^3d^2 hybridised as shown in Fig. 8.57 The *lp* occupies the axial hybrid orbital.

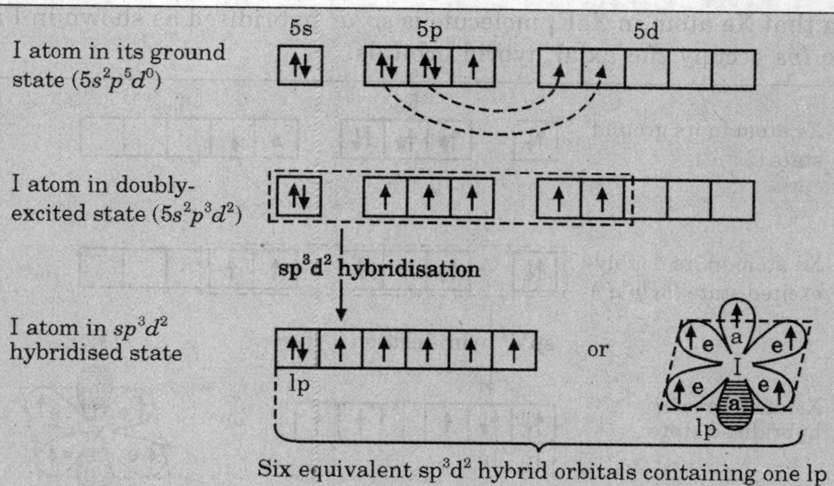

Fig. 8.57. sp^3d^2 hybridisation undergone by I atom in its doubly-excited state. Note that the lone pair (*lp*) occupies the axial hybrid orbital.

sp^3d^2 hybridisation shown in Fig. 8.57 makes it evident that out of six sp^3d^2 hybrid orbitals one axial hybrid orbital contains a *lp*, because in this case the (*lp—bp*) repulsion is minimum. Other five hybrid orbitals which are singly-filled overlap with singly-filled $2p_x$ orbitals of 5F atoms and form five I—F σ-bonds. Although the spatial arrangement of six electron pairs (*bps* = 5 and *lp* = 1) round the central I atom is octahedral, due to the presence of one lone pair of electrons in the axial hybrid orbital, the shape of IF_5 molecule gets distorted and becomes *square pyramidal*. A recent single crystal structure study of this molecule made by Burbank and Jones has shown that the basal F atoms (F_e atoms) are slightly displaced upwards from the base of the square pyramid and hence IF_5 molecule assumes the structure as shown in Fig. 8.58. F_aIF_e bond angle has been found to be equal to 81.9°. I—F_a bond distance (= 1.73 A°) is appreciably shorter than the I—F_e bond distance (= 1.87 A°).

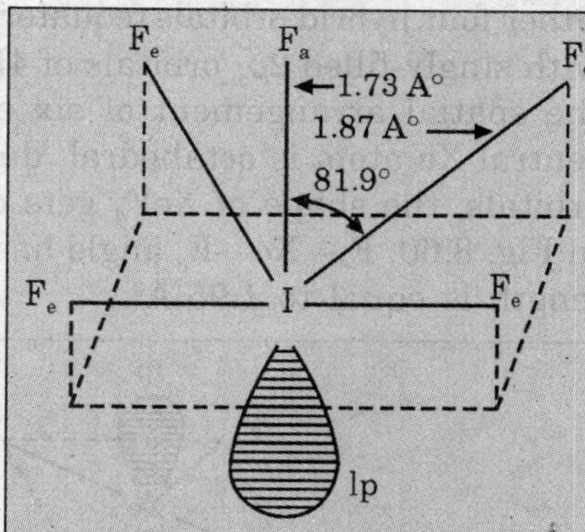

Fig. 8.58. Square pyramidal structure of IF_5 molecule.

On the basis of similar arguements it can be shown that ClF_5 and BrF_5 molecules also have square pyramidal shape.

3. XeF_4 molecule. Lewis structure of this molecule, F F :Xe: F F (four F atoms bonded to Xe) shows that the central atom namely Xe atom uses only four of its eight valence-shell electrons (Xe → $5s^2p^6$) in forming four Xe—F σ-bonds and the remaining two electron pairs remain as *lps* on Xe atom. Thus σ-bps = 4 and *lps* = 2 and hence σ-*bps* + *lps* = 6, showing that Xe atom in XeF_4 molecule is sp^3d^2 hybridised as shown in Fig. 8.59. The two *lps* occupy the axial hybrid orbitals.

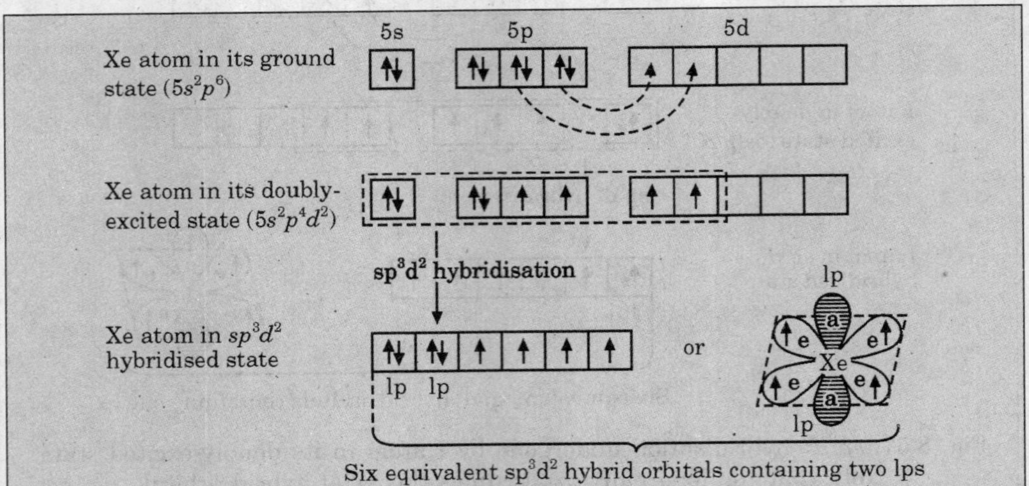

Fig. 8.59. sp^3d^2 hybridisation undergone by Xe atom in its doubly-excited state. Note that the two lone pairs (*lps*) occupy the axial hybrids.

sp^3d^2 hybridisation shown in Fig. 8.59. makes it evident that out of six sp^3d^2 hybrid orbitals two axial hybrid orbitals contain two *lps*, because, in order to have minimum lone pair repulsion, the lone pairs must be as far from each as possible. Other four hybrid orbitals (equatorial hybrid orbitals) which are singly-filled overlap with singly-filled $2p_x$ orbitals of 4F atoms and form four Xe—F σ-bonds. Although the spatial arrangement of six electron pairs (*bps* = 4 and *lps* = 2) round the central Xe atom is octahedral, due to the presence of two *lps* in the axial hybrid orbitals, the shape of XeF_4 gets distorted and becomes *square planar* as shown in Fig. 8.60. F_e—Xe—F_e angle has been found to be equal to 90° and Xe—F_e bond length is equal to 1.95 A°.

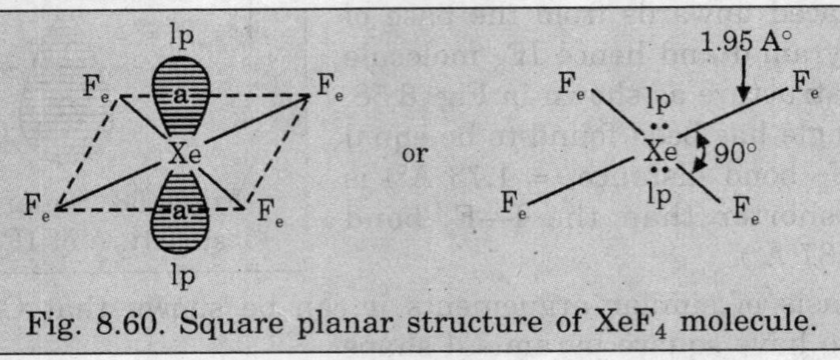

Fig. 8.60. Square planar structure of XeF_4 molecule.

On the basis of similar arguments it can be shown that ICl_4^- ion (I is the central atom) also has a square planar shape as depicted in Fig. 8.61. Due to the presence one unit of negative charge on the ion, I atom (central atom) is supposed to contain eight electrons (instead of seven) in its valence-shell [I($5s^2p^5$) + e^- → I^- ($5s^2p^6$)]. Four electrons are used to from four I—Cl σ-bonds and four electrons (*i.e.* two electron pairs) remain as *lps* on I atom.

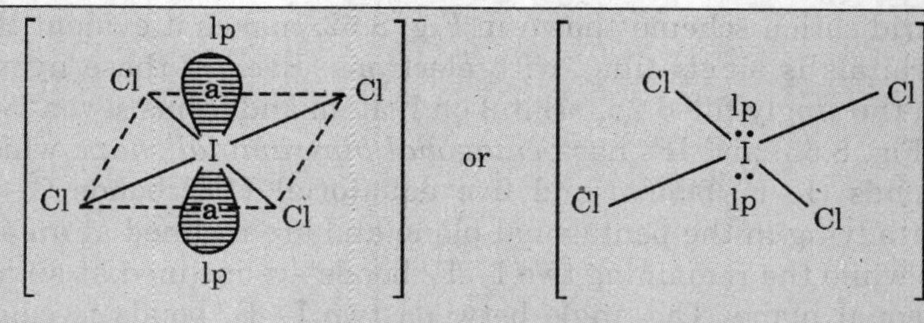

Fig. 8.61. Square planar shape of ICl^-_4 ion

BrF_4^- (central atom is Br) and IF_4^- (central atom is I) ions have analogous shapes.

[F] Species Whose Central Atom/Ion Has Seven Electron Pairs

The molecules of this type may be of the following *two* types :

(a) AB_7 type. Here the central atom (A) is surrounded by seven σ-*bps*. IF_7 is an important example of this type of molecules.

(b) $AB_6(lp)$ type. Here the central atom is surrounded by six σ-*bps* and one *lp*. XeF_6 is an example of this type of molecules.

1. IF_7 molecule. Lewis structure of this molecule, 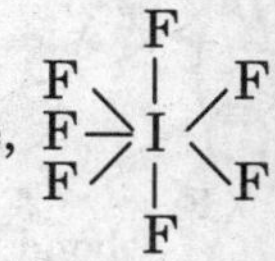 shows that the central atom namely I atom uses all its seven valence-shell electrons in forming seven I—F σ-bonds. Thus σ-*bps* = 7 and *lps* = 0 and hence σ-*bps* + *lps* = 7, showing that I atom is sp^3d^3 hybridised in this molecule as shown in Fig. 8.62.

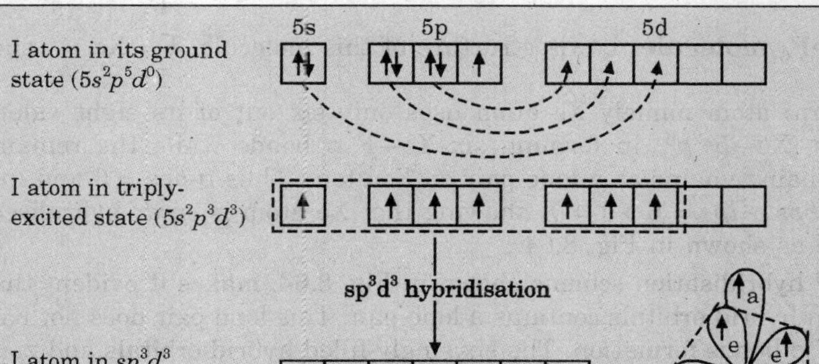

Fig. 8.62. sp^3d^3 hybridisation of I atom in its triply excited state. The axial and basal (equational) hybrid orbitals have been shown by *a* and *e* respectively.

sp^3d^3 hybridisation scheme shown in Fig. 8.62. makes it evident that each of the hybrid orbitals is singly filled with electrons. Each of these hybrid orbitals overlaps with the singly-filled $2p_x$ orbital on F atom and forms seven I—F σ bonds as shown in Fig. 8.63. and IF_7 has *pentagonal bipyramidal shape* which has two axial I—F bonds (I—F_a bonds) and five equatorial I—F bonds (I—F_e bonds). I—F_e bonds are lying in the pentagonal plane and are inclined at an angle of 72° to each other while the remaining two I—F_a bonds are inclined at an angle of 90° to the pentagonal plane. The angle between two I—F_a bonds is equal to 180°. Thus F_a—I—F_e bond angle = 90°, F_e—I—F_e bond angle = 72° and F_a—I—F_a bond angle = 180°. Both the I—F_a bond distances which are equal to each other are greater than both I—F_e bond distances which are again equal to each other.

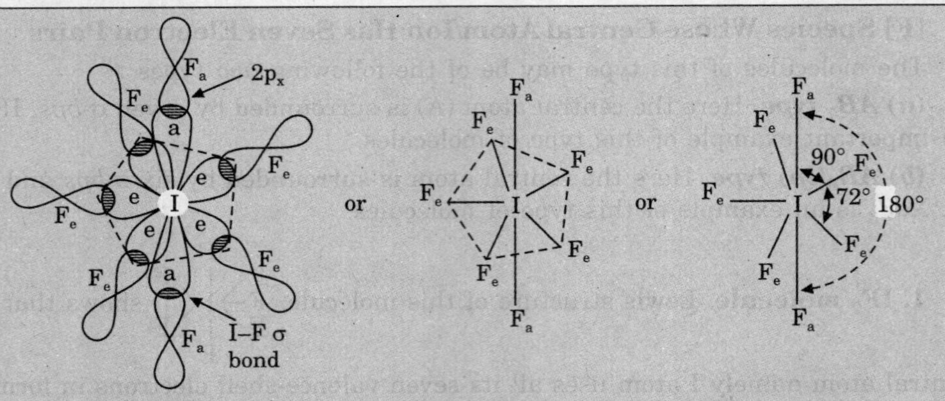

Fig. 8.63. Pentagonal bipyramidal shape of IF_7 molecule.

2. XeF_6 molecule. Lewis structure of this molecule, 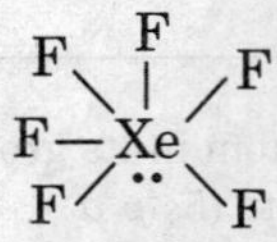 shows that the central atom namely Xe atom uses only six out of its eight valence shell electrons (Xe→$5s^2p^6$) in forming six Xe—F σ bonds while the remaining one electron pair remains as a lone pair on Xe atom. Thus σ-*bps* = 6 and *lp* = 1 and hence σ-*bps* + *lps* = 6 + 1 = 7, showing that Xe atom is sp^3d^3 hybridised in this molecule as shown in Fig. 8.64.

sp^3d^3 hybridisation scheme shown in Fig. 8.64. makes it evident that one of the seven hybrid orbitals contains a lone pair. This lone pair does not participate in Xe—F σ bonds formation. The six singly-filled hybrid orbitals and $2p_x$ orbitals on six F atoms overlap together and give rise to the formation of six Xe—F σ bonds. Thus XeF_6 should have a distorted pentagonal bipyramidal shape having one lone pair at its axial position. Monomeric XeF_6 in the liquid and gaseous state has been shown to contain a *slightly distorted octahedral shape* with one lone pair which appears to occupy less space than the bonding electron pair. The lone pair is trying to emerge either through a face of the octahedron or through an edge of the octahedron (See Fig. 8.65). In the latter case the molecule would have some resemblance with a distorted pentagonal bipyramid (*cf* IF_7). In either

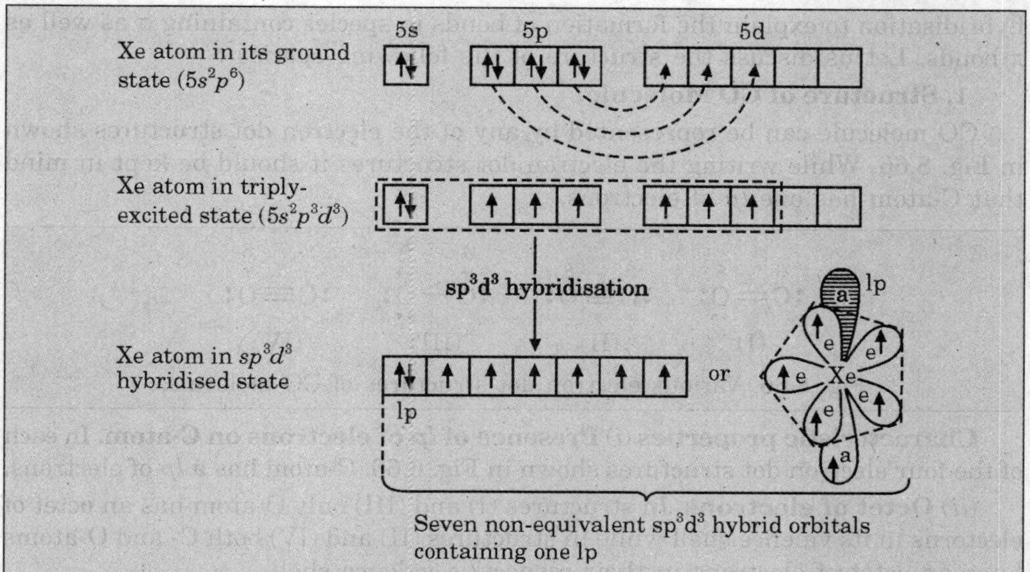

Fig. 8.64. sp^3d hybridisation of Xe atom in its triply excited state. The axial and basal (equatorial, hybrid orbitals have been shown by a and e respectively.

structure, the Xe—F bonds near the lone pair appear to be somewhat lengthened and distorted away from the lone pair. The fact that XeF_6 forms adducts like $XeF_6.SbF_5$ confirms the presence of a lone pair in the valence-shell of Xe atom.

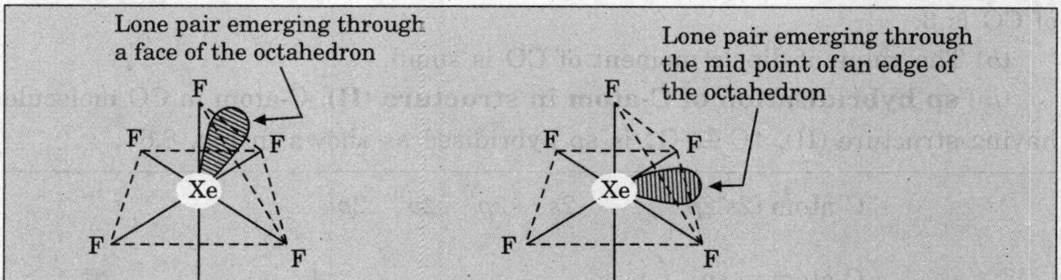

Fig. 8.65. Distorted octahedral shape of XeF_6 molecule containing six Xe—F σ-bonds and one pair of electrons.

A cubic phase of XeF_6 studied at —80°C contains XeF_5^+ and F^- ions that are associated in tetrameric and hexameric rings.

Shape of Species Having σ As Well As π Bonds On the Basis of Hybridisation.

So far we have used the concept of hybridisation to explain the shape of the species containing only σ bonds. In this section we shall use the concept of hybridisation to explain the formation of bonds in species containing σ as well as π bonds. Let us discuss the structure of the following species :

1. Structure of CO Molecule.

CO molecule can be represented by any of the electron dot structures shown in Fig. 8.66. While writing the electron dot structures it should be kept in mind that C-atom has one *lp* of electrons.

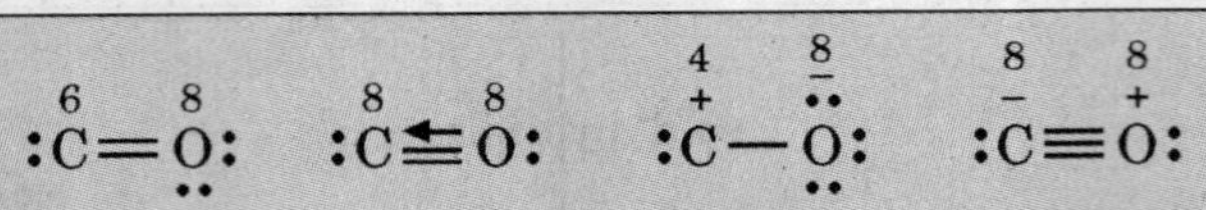

(I) (II) (III) (IV)

Fig. 8.66. Various electron dot structures of CO molecule.

Characteristic properties (*i*) **Presence of *lp* of electrons on C-atom.** In each of the four electron dot structures shown in Fig. 6.66. C-atom has a *lp* of electrons.

(*ii*) **Octet of electrons**. In structures (I) and (III) only O-atom has an octet of electorns in its valence shell while in structures (II) and (IV) both C- and O-atoms have an octet of electrons in their respective valence shell.

(*iii*) **Presence of a triple bond**. Both structres (II) and (IV) contain a triple bond between C and O atoms. One bond is a σ-bond and the two bonds are π-bonds. The presence of a triple bond between C and O-atoms in CO molecule is supported by the following facts.

(*a*) The bond length between C and O-atoms in CO molecule is 1.128 A° which is very close to C ≡ O triple bond length (= 1.10 A°). This shows that bond order of CO is 3.

(*b*) The value of dipole moment of CO is small.

(*iv*) **sp hybridisation of C-atom in structure (II).** C-atom in CO molecule having structure (II), :C ⇇ O: is sp hybridised as shown in Fig. 8.67.

C-atom ($2s^2 2p^2$) : $2s^2$ $2p^1$ $2p^1$ $2p^0$

C-atom in sp hybridised state } : $(sp)^2$ $(sp)^1$ — Two *sp* hybrid orbitals; $2p^1$ $2p^0$ — Two unhybridised 2*p* orbitals

Fig. 8.67. *sp* hybridisation of C-atom CO_2 molecule having the structure, :C ⇇ O:

It may be seen that two 2*p* orbitals remain unhybridised.

(*v*) **Formation of different bonds in CO molecule.** Formation of different bonds in CO molecule has been shown in Fig. 8.68. (C – O) σ bond is obtained by the overlap of *sp* hybrid orbital of C-atom with 2*p* orbital of O-atom. (C – O) π bond is obtained by the overlap between 2*p* orbital on C and O atoms. Thus (C – O) π bond is a (*p*π – *p*π) π bond. O → C coordinate bond is a [$2p^2$(O) → $2p^0$(C)] bond.

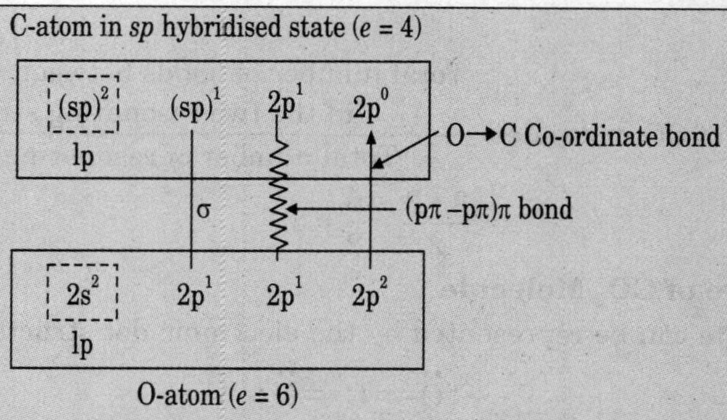

Fig. 8.68. Formation of different bonds in CO molecule having :C$\stackrel{\leftarrow}{=}$O: structure.

(*vi*) **Resonance in CO molecule.** X-ray diffraction study of CO molecule has shown that carbon-oxygen bond length in this molecule is equal to 1.13 A°. This exprimentally determined value is niether equal to the calculated value of C ≡ O triple bond length (= 1.10 A°) nor equal to C = O double bond length (= 1.22 A°) or C – O single bond length (= 1.43 A°). Thus the true structuer of CO molecule is the resonance hybrid of the following resonating structures.

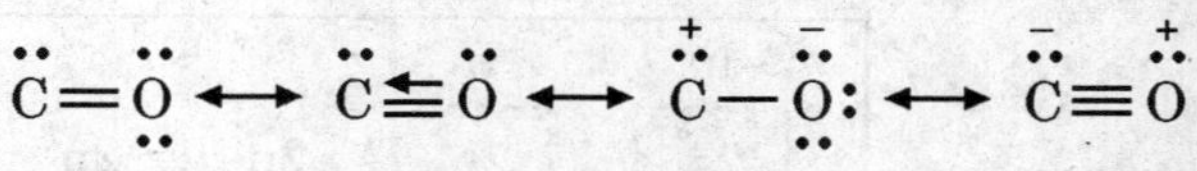

(I) (II) (III) (IV)

Dipole moment of CO is very small (= 0.12 D)

Since the exprerimental value of carbon-oxygen bond length in CO (= 1.13 A°) is very close to the calculated value of C ≡ O triple bond length (= 1.10 A°), the true structure of CO is generally regarded as a resonance hybrid of the following resonating structures.

(II) (IV)

(*vii*) **Acting as Lewis base or ligand.** Due to the presence of *lp* of electrons on C-atom, CO molecule acts as a Lewis base or a Ligand in metal carbonyls. The *lp* of electrons is used in forming metal ⟵ carbon coordinate bond in carbonyls

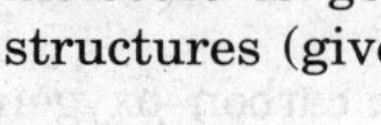

(*ix*) **Bond order (B.O).** We have already said that the true structure of CO molecule is generally regarded as a resonance hybrid of the two resonating structures (given below) containg triple bond between C and O atoms.

(II) (IV)

Thus :

$$\text{Bond order of CO molecule} = \frac{\text{Total number of bonds between C and O atoms in the two resonating structures}}{\text{Total number of resonating structures}}$$

$$= \frac{3+3}{2} = \frac{6}{2} = 3$$

2. Structure of CO_2 Molecule

CO_2 molecule can be represented by the electronic dot structure given below.

$$:\ddot{O}=C=\ddot{O}:$$

All the three atoms in CO_2 molecule obey octet rule.

Characteristic properties (*i*) ***sp* hybridisation of C-atom**. Since the central C-atom is surrounded by two σ-bonds and no *lp* of electrons, this atom in its excited state (C in excited state = $2s^1\,2p^3$) is *sp* hybridised. In *sp* hybridisation, two $2p$ orbitals of C-atom remain unhybridised and form two (C – O) π bonds with two O-atoms. Both π bonds are (pπ – pπ) π bonds. The formation of two (C – O) σ bonds and two (C – O) π bonds in CO_2 molecule by *sp* hybridisation is shown in Fig. 8.69

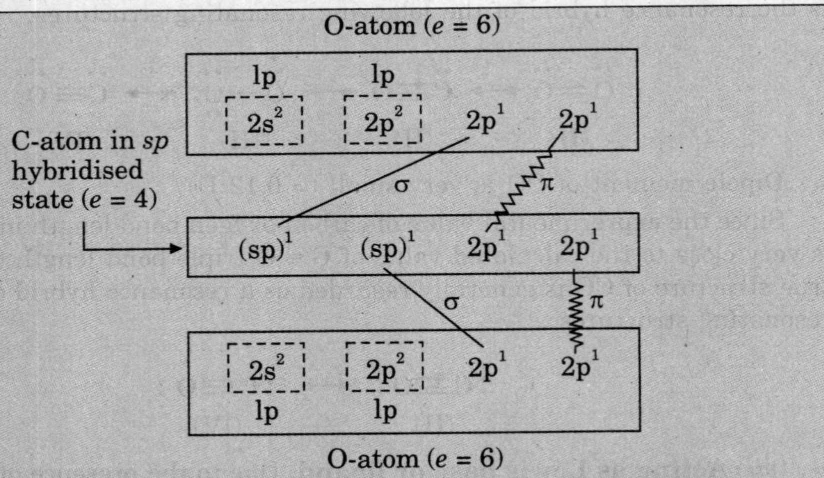

Fig. 8.69. Formation of different bonds in CO_2 molecule having the electron dot structure, $\ddot{O}=C=\ddot{O}:$ by the use of *sp* hybridisation.

(*ii*) **Linear shape of CO_2 molecule.** Due to *sp* hybridisation of C-atom and absence of *lp* of electrons on C-atom, CO_2 molecule has linear shape. Thus CO_2 is *linear monomeric covalent molecule*. Both carbon–oxygen bond lengths are equal to 1.15 A°.

(*iii*) **Resonance in CO_2 molecule.** We have said above that both carbon–oxygen bond lengths in CO_2 molecule are equal to 1.15 A° (experimental value).

This value of bond length is shorter than the calculated C = O double bond length (= 1.22 A°) and larger than the calculated C ≡ O triple bond length (= 1.10 A°). In order to account for the discerpancy between the experimental and calculated values of bond length, it has been suggested that the actual structure of CO_2 is a resonance hybrid of the following contributing structures :

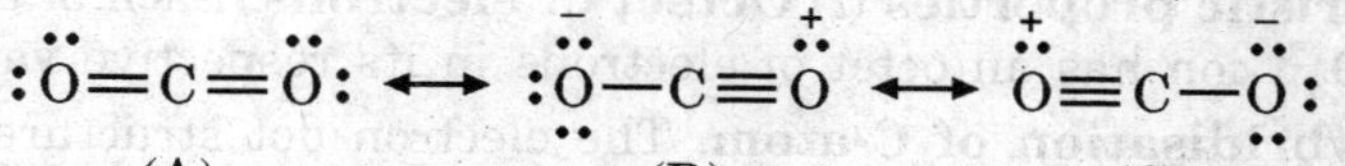

(A) (B) (C)

Structure (A) is covalent while structures (B) and (C) are ionic structures. It is due to resonance that the experimentally determined value of carbon-oxygen bond lengths (= 1.15 A°) is in between that of C = O double bond (= 1.22 A°) and C ≡ O triple bond (= 1.10 A°). The dipoles of structure (B) and (C) are equal and act in opposite directions so that the resultant dipole moment of the actual structure become zero. Observed dipole moment of CO_2 molecule is also equal to zero. Thus it can be concluded that structure (B) and (C) make equal and opposite contributions to the overall structure of this molecule. The shortening of the carbon-oxygen bond lengths mentioned above is due to the stabilisation acquired through resonance.

(*iv*) **To calculate bond order (B.O.) of CO_2 molecule.** CO_2 shows resonance and it has three resonating structure given below :

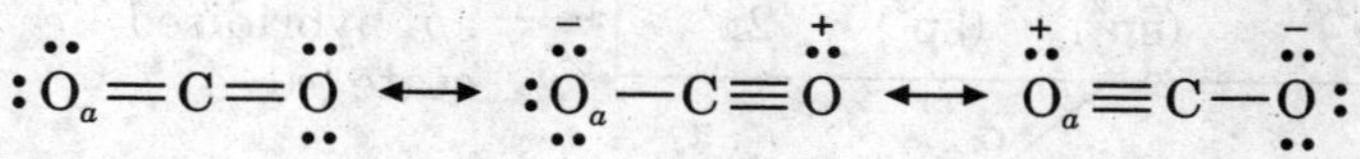

(A) (B) (C)

Thus :

$$\text{B.O. of } CO_2 = \frac{\text{Total number of bonds between C and } O_a \text{ atoms in all the three resonating structures}}{\text{Total number of resonating structures}}$$

$$= \frac{2+1+3}{3} = \frac{6}{3} = 2$$

3. Structure of Carbonate Ion, CO_3^{2-} ion.

Since O-atom is more electronegative than C-atom, the two negative (–) charges on CO_3^{2-} ion are present on two O-atoms and not on C-atoms. Thus CO_3^{2-} ion can be represented by the electron dot structure given below.

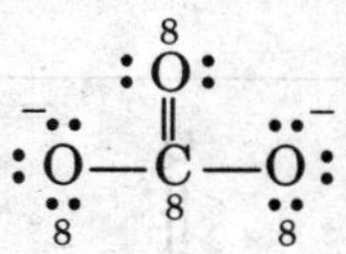

(I)

Since carbonic acid (H_2CO_3) is the parental acid of CO_3^{2-} ion, the structure of CO_3^{2-} ion can also be obtained by removing two H-atoms from two O–H groups attached directly with the central C-atoms in H_2CO_3 molecule (H_2CO_3 is a dibasic acid).

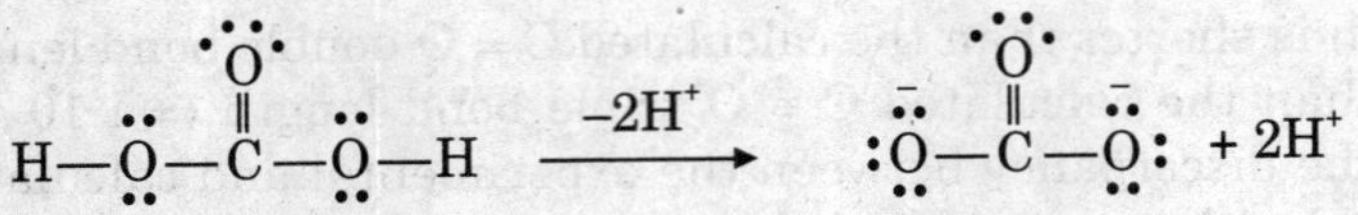

H_2CO_3 molecule $\qquad\qquad$ CO_3^{2-} ion (I)

Characteristic properties (*i*) **Octect of electrons**. Each of the four atoms present in CO_3^{2-} ion has an octet of electrons in its respective valence-shell.

(*ii*) ***sp*²** **hybridisation of C-atom**. The electron dot structure of CO_3^{2-} ion given above shows that since the central C-atom is surrounded by *three σ-bps and no lp* of electrons, this atom in its excited state is sp^2 hybridised as shown in Fig. 8.70 The formation of different bonds in CO_3^{2-} ion is shown in the same Figure. It is evident from sp^2 hybridisation scheme shown in Fig. 8.70 that one *2p* orbital of C-atom remains unhybridised and contains one electron. This orbital is used to form C–O π bond. Obviously this π-bond is a (*p*π–*p*π) bond.

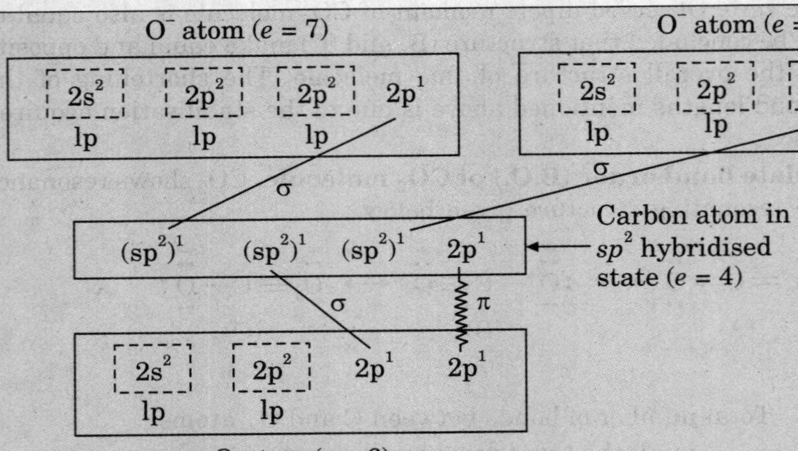

Fig. 8.70. Formation of different bonds in CO_3^{2-} ion having

electron dot structure, 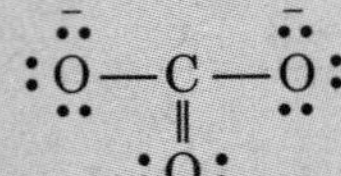

(*iii*) **Coplanar trigonal geometry.** Due to sp^2 hybridisation and absence of *lp* of electrons on C-atom, CO_3^{2-} ion has *coplanar tringonal geometry* with OCO bond angle equal to 120°. Trignale planar geometry of CO_3^{2-} ion is shown in Fig. 8.71.

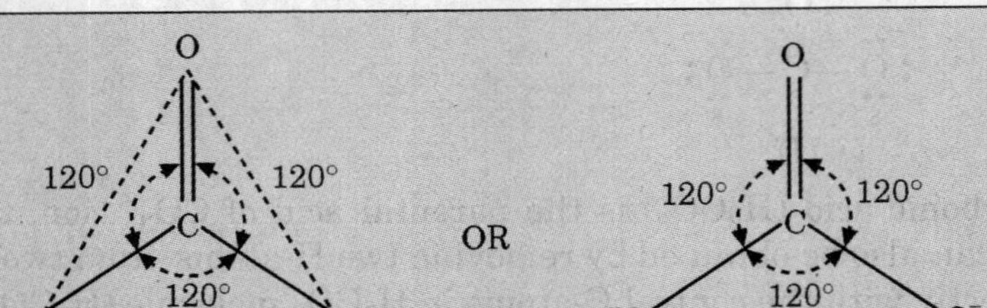

Fig. 8.71. Tringonal palanr geomatry of CO_3^{2-} ion

(*iv*) **Resonance in CO_3^{2-} ion**. X-ray examination of CO_3^{2-} ion has indicated that all the three carbon-oxygen bonds in CO_3^{2-} ion are of the same length (= 1.31 A°). This value is neither equal to the calculated value of C — O single bond length (= 1.43 A°) nor equal to C = O double bond length (= 1.22 A°). This value is inbetween the C — O single bond and C = O double bond lengths. In order to account for the discrepency between exprimental value (= 1.31 A°) and calculated values (= 1.43 A° and 1.22 A°) of carbon–oxygen bond lengths, it has been suggested that the true structure of CO_3^{2-} ion is not that shown in Fig. 6.71, but it is a resonance hybrid of the resonating structure shown in Fig. 8.72.

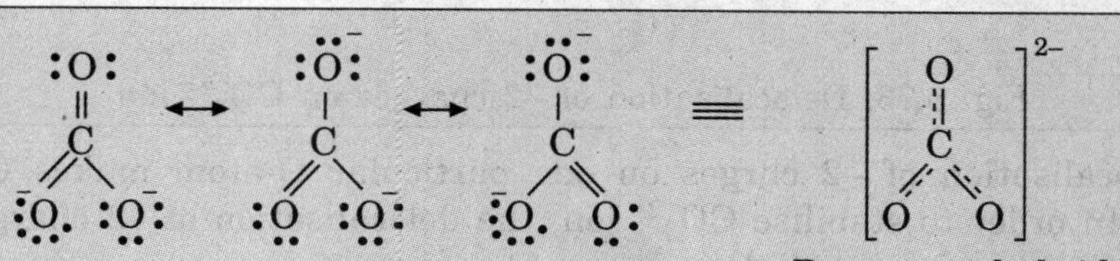

Fig. 8.72. Resonating structures of CO_3^{2-} ion.

The dotted lines shown in the resonance hybrid indicates that π-electron cloud in the ion is not localised between C-atom and any particular O-atom. Instead, it is spread over all the three carbon-oxygen bonds. It is due to the delocalisation of π-electron cloud that all the three carbon-oxygen bond lengths in CO_3^{2-} ion are equal.

(*v*) **To Calculate bond order (B.O.) of CO_3^{2-} ion.** We know that B.O of an ion showing resonance is given by :

$$\text{B.O.} = \frac{\text{Total number of bonds between any two bonded atoms in all the resonating structures}}{\text{Total number of resonating structures}}$$

Now CO_3^{2-} ion has the following three resonanting structures.

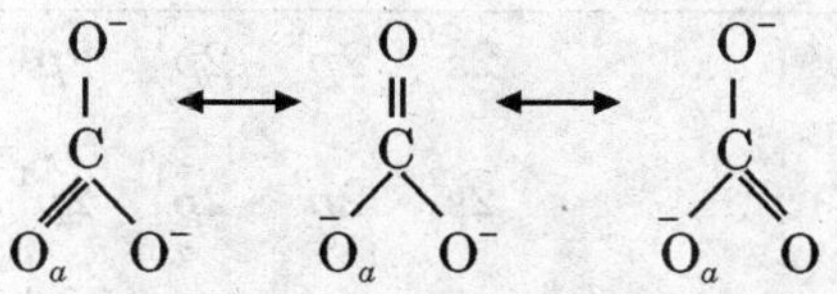

$$\therefore \text{B.O. of } CO_3^{2-} \text{ ion} = \frac{\text{Total number of bonds between C and } O_a \text{ atoms in all the three resonating structures}}{\text{Total number of resonating structures}}$$

$$= \frac{2+1+1}{3} = \frac{4}{3} = \mathbf{1.33}$$

(*vi*) **Delocalisation of – 2 charges.** The two negative charges on CO_3^{2-} ion are not localised on any particular two O-atoms. Instead, these are spread over all the three O-atoms. Thus each O-atom can be regarted as having a charge equal to – 2/3 as shown below. (Also see Fig. 6.73).

$$\text{No. of charges on one O-atom} = \frac{\text{Total charges on } CO_3^{2-} \text{ ion}}{\text{No. of O-atoms}} = \frac{-2}{3}$$

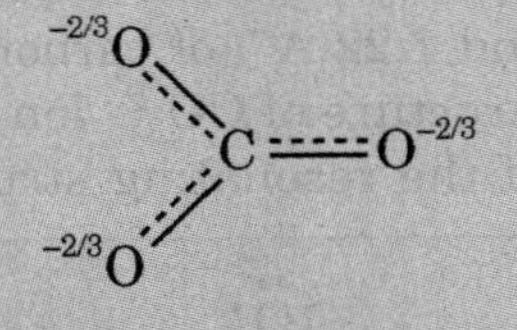

Fig. 8.73. Delocalisation of –2 charges on CO_3^{2-} ion.

The localisation of –2 chrges on one particular O-atom makes CO_3^{2-} ion unstable. In order to stabilise CO_3^{2-} ion, the delocalisation of –2 charges on all the three O-atoms is essential.

4. Stucture of O_3 Molecule.

O_3 molecule can be represented by any of the electronic sturucers given in Fig. 8.74.

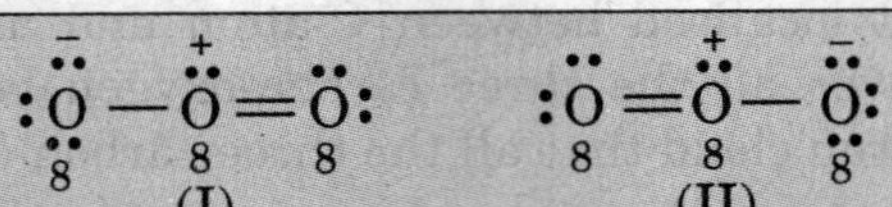

(I) (II)

Fig. 8.74. Electronic structures of O_3 molecule.

(*i*) **Octet rule.** Each of the three O-atoms in O_3 molecule is surrounded by an octet of electrons. Thus O_3 molecule obeys octet rule.

(*ii*) ***sp*^{2} hybridisation of O$^+$ ion and formation of different bonds.** Since the central O^+ ion is surrounded by two σ-*bps* and one *lp* of electrons, O^+ ion is sp^2 hybridised as shown in Fig. 8.75.

O-atom ($2s^2 2p^4$) ...... $2s^2$ $2p^2$ $2p^1$ $2p^1$

O^+ ion ($2s^2 2p^3$) ...... $2s^2$ $2p^1$ $2p^1$ $2p^1$

O^+ ion in sp^2 hybridised state .... $(sp^2)^2$ $(sp^2)^1$ $(sp^2)^1$ (Three sp^2 hybrid orbitals) $2p^1$ (Unhybridised 2*p* orbital)

Fig. 8.75 sp^2 hybridisation of O^+ ion in O_3 molecule

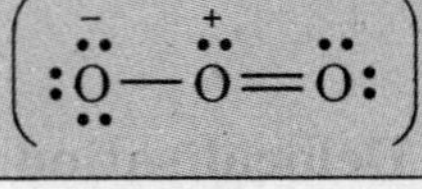

Formation of (O^+ – O) and (O^+ – O^-) σ-bonds and (*p*π–*p*π) π bond in O_3 molecule has been shown in Fig. 8.76.

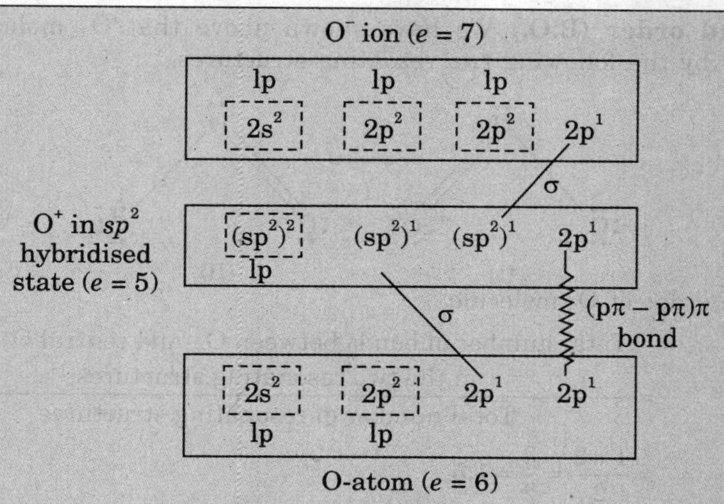

Fig. 8.76. Formation of different bond in O_3 molecule, $:\ddot{\bar{O}}-\ddot{\overset{+}{O}}=\ddot{O}:$

(*iii*) **Angular geometry of O_3 molecule.** Due to sp^2 hybridisation and presence of one *lp* of electrons on O^+ atom, O_3 molecule has angular geometry with bond lengths equal to 128pm. (See Fig.10.7)

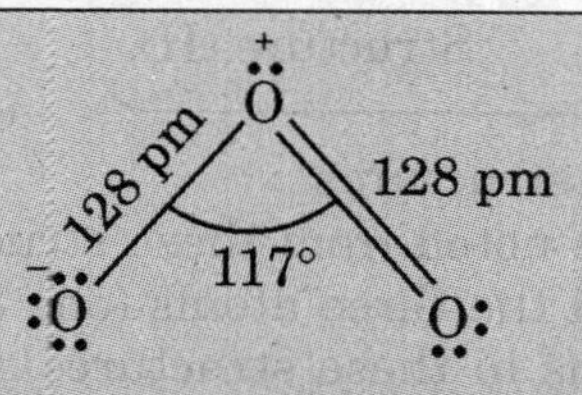

Fig. 8.77. Angular geometry of O_3 molecule

(*iv*) **Resonance.** In O_3 molecule, since both oxygen–oxygen bond lengthes are the same (= 128 pm) and this distance lies in between single O – O bond distance (= 148 pm) and double O = O bond distance (= 110 pm),the true sturcture of O_3 molecule is a resonance hybride of the two resonating structures given in Fig.8.78.

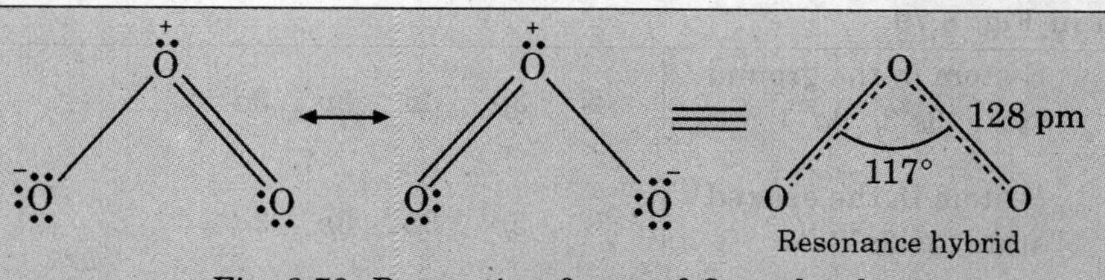

Fig. 8.78. Resonating forms of O_3 molecule

The following resonating structures can also be written for O_3 molecule.

$$:\ddot{\underset{\cdot\cdot}{O}}\leftarrow\ddot{O}=\ddot{\underset{\cdot\cdot}{O}}: \quad \longleftrightarrow \quad :\ddot{O}=\ddot{O}\rightarrow\ddot{\underset{\cdot\cdot}{O}}:$$

(*v*) **Bond order (B.O.)** We have shown above that O_3 molecule can be represented by the following two resonting structures.

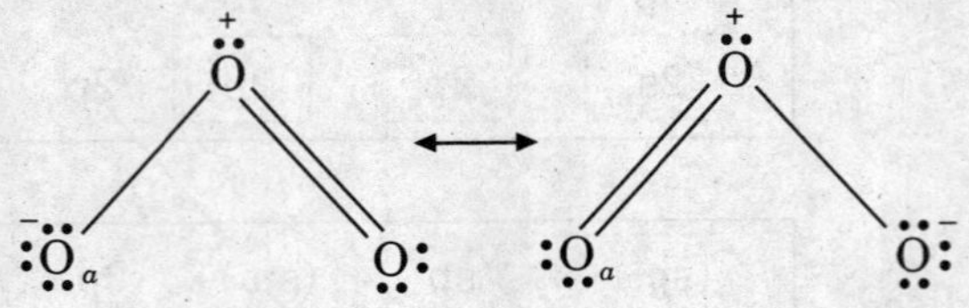

(I) (II)

$\therefore$ Bond order of O_3 molecule

$$= \frac{\text{Total number of bonds between } O_a \text{ and central } O^{+}\text{atom in the two resonating structures}}{\text{Total number of resonating structures}}$$

$$= \frac{1+2}{2} = \frac{3}{2} = \mathbf{1.5}$$

5. Structre and Geometry of SO_2 Molecule.

In the gaseous as well as in the soild state, SO_2 exists as discrete molecule which can be represented by any of the following three structures. Structures (I) and (II) are *covalent structures* while structure (III) is an *ionic structure.*

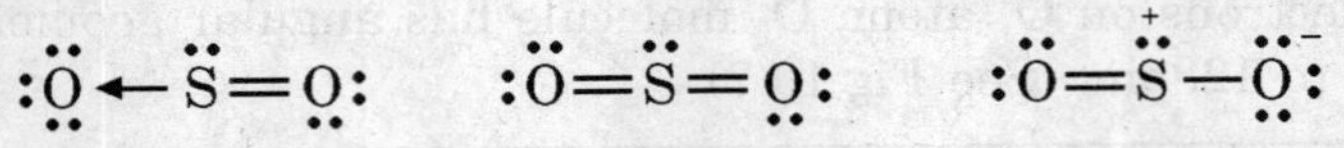

Structure (I) Structure (II) — Covalent structures

Structure (III) (Ionic structure)

(*i*) **Octet rule.** It may be noted from these structures that structures (I) and (III) obey octet rule, since all the three atoms (one S-atom and two O-atoms) are surrounded by eight electrons in these structures. On the other hand, structure (II) does not obey octet rule bacause S-atom is surrounded by ten electrons while each of the two O-atoms is surrounded by eight electrons.

(*ii*) ***sp*^{2} hybridisation of S-atom and formation of (S – O)σ and (S – O)π bonds.** Let us consider structure (II) only for the formation of different bonds in it. Since S-atom (central atom) in this structured is surrounded by two (S – O)σ bonds and one *lp* of electrons, this atom in its excited state is sp^2 hybridised as shown in Fig. 8.79.

S-atom in the ground state ($3s^2 3p^4 3d^0$)	$3s^2$ $3p^2$ $3p^1$ $3p^1$ $3d^0$	
S-atom in the excited state ($3s^2 3p^3 3d^1$)	$3s^2$ $3p^1$ $3p^1$ $3p^1$ $3d^1$	
S-atom in sp^2 hybridised state	$(sp^2)^2$ $(sp^2)^1$ $(sp^2)^1$ — Three sp^2 hybrid orbitals	$3p^1$ $3d^1$ — Two unhybridised atomic orbitals

Fig. 8.79. sp^2 hybridisation of S-atom in SO_2 molecule having $(:\ddot{O}=\ddot{S}=\ddot{O}:)$ structuer

Each of the two (S – O)σ bonds is obtained by end-to-end overlap (*axial* or *liner overlop*) between half-filled sp^2 hybrid orbital on S-atom and half-filled $2p$ orbital on O-atom (O = $2s^2\ 2p^2\ 2p^1\ 2p^1$). Thus each (S – O)σ bond is [sp^2(S) – $2p$ (O)]σ bond. One of the two (S – O)π bonds results by the *side-to- side* overlap (*lateral* overlap) between half-filled $3p$ orbital on S-atom and half-filled $2p$ orbital on O-atom as shown in Fig. 6.80(*a*). Thus this π bond is a [$3p$ (S) – $2p$ (O)]π bond. which is called (pπ – pπ)π bond. Other (S – O)π bond is obtained by the *side-to-side overlap (lateral overlap)* between half-filled $3d$ orbital on *S*-atom and half-filled $2p$ orbital on O-atom as shown in Fig. 8.80(*b*). Thus this (S – O)π bond is a [$3d$ (S) – $2p$(O)]π bond which is called (dπ – pπ)π.

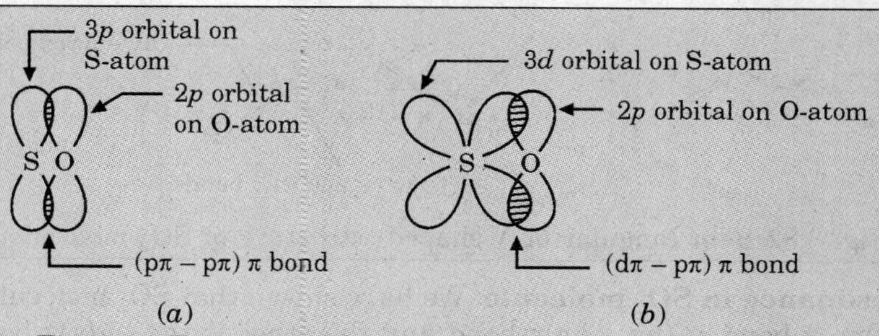

Fig. 24.2 Formation of (pπ – pπ) and (dπ – dπ) π bonds in SO_2 molecule ($:\ddot{O}=\ddot{S}=\ddot{O}:$)

Formation of both (S – O)σ and both (S – O)π bonds in structure (II) of SO_2 molecule has been shown in simplified manner in Fig. 8.81.

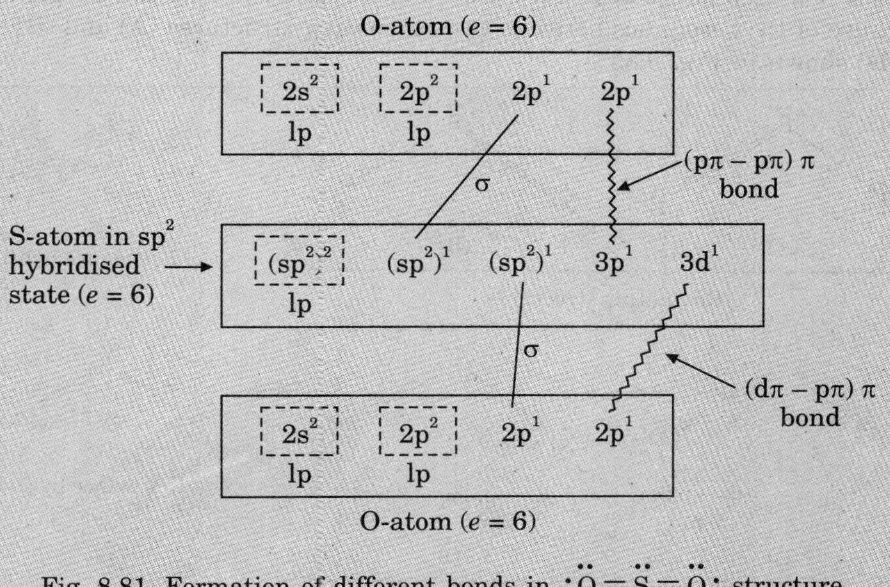

Fig. 8.81. Formation of different bonds in $:\ddot{O}=\ddot{S}=\ddot{O}:$ structure of SO_2 molecule. *e* indicates the number of electrons present in the valence shell of the atom.

(*iii*) **Angular shape of SO_2 molecule and OSO bond angle.** It is due to the presence of one *lp* of electrons in one of the three sp^2 hybrid orbitals on S-atom that SO_2 molecule does not have the expected trigonal planer shape corresponding to sp^2 hybridisation of S-atom and OSO bond angle is also not equal to the expected angle of 120° for trigonal planar geometry. Rather, its expected trigonal shape gets distorted and becomes *bent* or *angular* (V-shape). Since (*lp* – *bp*) repulsion is greater than (*bp* – *bp*) replusion, OSO bond angle gets sligthly reduced from 120° to 119.5° [See Fig. 8.82]

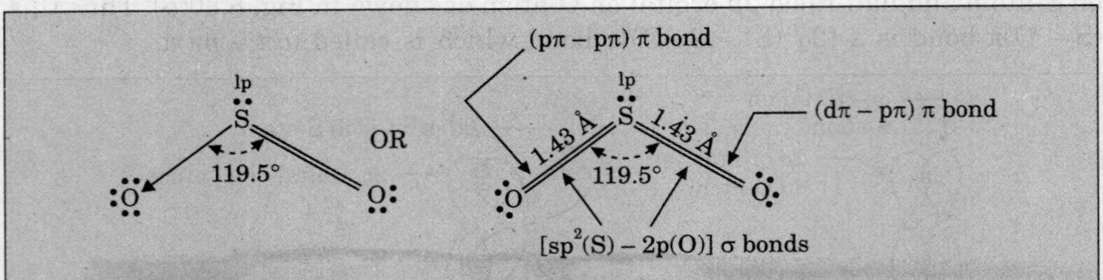

Fig. 8.82 Bent (angular or V-shaped) structure of SO_2 molecule.

(*iv*) **Resonance in SO_2 molecule.** We have shown that SO_2 molecule has two π bonds. One π bond is $(p\pi - p\pi)\pi$ bond and the other is $(p\pi - d\pi)\pi$ bond. Since these bonds are formed by the overlap of different orbitals, sulphur—oxygen bonds should have different bond lengths. However, expriments have shown that both the bonds are of equal length (= 1.43 A°). Each bond length is intermediate between the calculated value of S – O single bond length and S = O double bond length. In order to explain the equivalence of both sulphur—oxygen bond lengthe in SO_2 it has been suggested that equal length of the two sulphur–oxygen bonds is because of the resonance between two resonating structures (A) and (B) or (C) and (D) shown in Fig. 8.83.

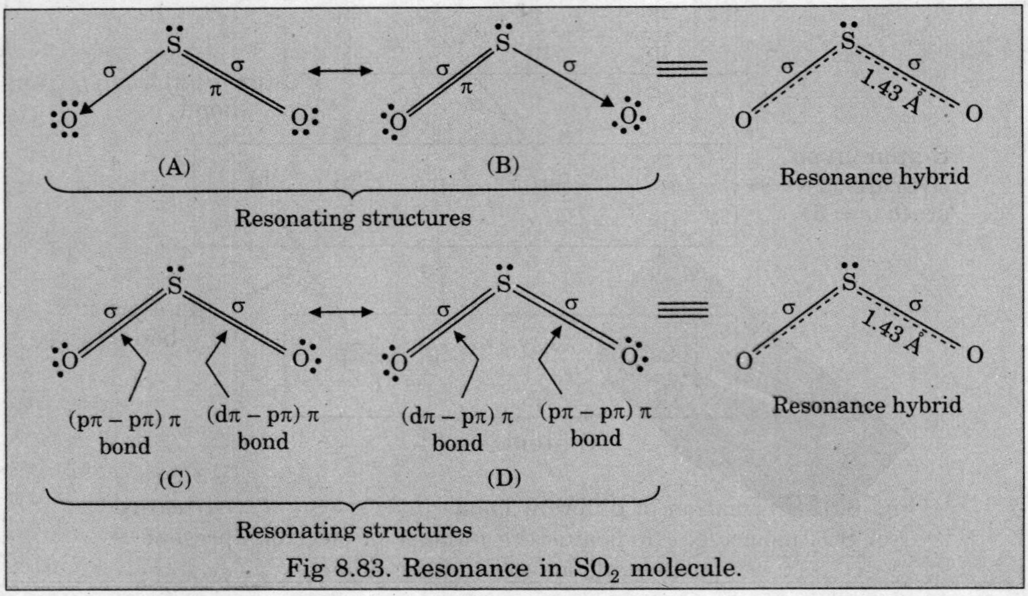

Fig 8.83. Resonance in SO_2 molecule.

6. Structure of SO_3 molecule in the Gaseous State.

In the gaseous state SO_3 molecule exists as a monomeric molecule, SO_3. This molecule can have any of the following two structures :

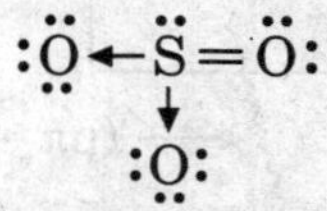

Structure (I)

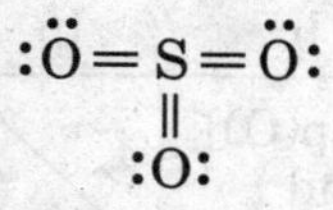

Structure (II)

(*i*) **Octet rule.** It may be noted from these structures that it is only structure (I) which obeys octet rule since each atom in this structure is surrounded by eight electrons. Structure (II) does not obey octet rule, since the central S-atom is surrounded by 12 electrons but each of the three O-atoms is surrounded by an octet of electrons

(*ii*) ***sp*² hybridisation of S-atom and formation of (S – O)σ and (S – O)π bonds.** Let us consider structure (II) only for the formation of different bonds in it. Since S-atom (central atom) in this structure is surrounded by three (S – O)σ bonds and no *lp* of electrons, this atom in its excited state is sp^2 hybridised as shown in Fig. 8.84

S-atom in the ground state ($3s^2 3p^4 3d^0$)	$3s^2$ $3p^2$ $3p^1$ $3p^1$ $3d^0$ $3d^0$
S-atom in the 2nd excited state	$3s^1$ $3p^1$ $3p^1$ $3p^1$ $3d^1$ $3d^1$
S-atom in sp^2 hybridised state	$(sp^2)^1$ $(sp^2)^1$ $(sp^2)^1$ — Three sp^2 hybrid orbitals; $3p^1$ $3d^1$ $3d^1$ — Three unhybridised atomic orbitals

Fig. 8.84. sp^2 hybridisation of S-atom in SO_3 molecule having 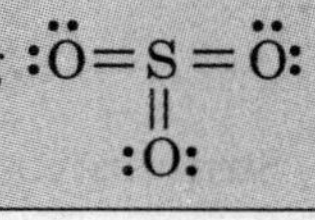 structure.

Each of the three (S – O)σ bonds is obtained by the overlap between singly-filled sp^2 hybrid orbital on S-atom and half-filled 2*p* orbital on O-atom.. Thus each of the three (S – O)σ bonds is a [sp^2 (S) – 2*p* (O)]σ bond. One (S – O)π bond is obtained by the overlap between half-filled 3*p* orbital on S-atom and half-filled 2*p* orbital on O-atom. Thus this (S – O)π bond is a (*p*π – *p*π)π bond. Each of the remaining two (S – O)π bonds is obtained by the overlap between half-filled 3*d* orbital on *S*-atom and half-filled 2*p* orbital on two O-atom. Thus each of the two (S – O)π bonds is a (*d*π – *p*π)π bond. The formation of [sp^2 (S) – 2*p* (O)]σ bonds, (*p*π – *p*π)π and (*d*π – *p*π)π bonds has been shown in Fig. 8.85.

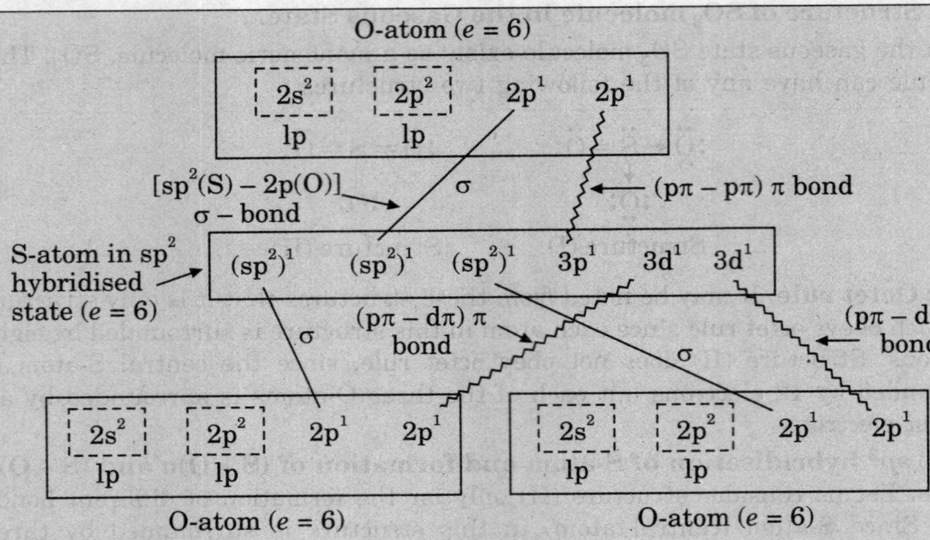

Fig. 8.85. Formation of different bonds in SO_3 molecule (Ö=S=Ö, with :O: double-bonded to S)

(*iii*) **Shape of SO_3 molecule and OSO bond angle.** It is due to sp^2 hybridisation of the central S-atom and absence of any lone pair of electrons on the central S-atom that SO_3 molecule has the expected *trigonal planar geometry* (See Fig. 8.86. Each of the three OSO bond angles is equal to 120° which is the expected angle for a *trigonal planar geometry*.

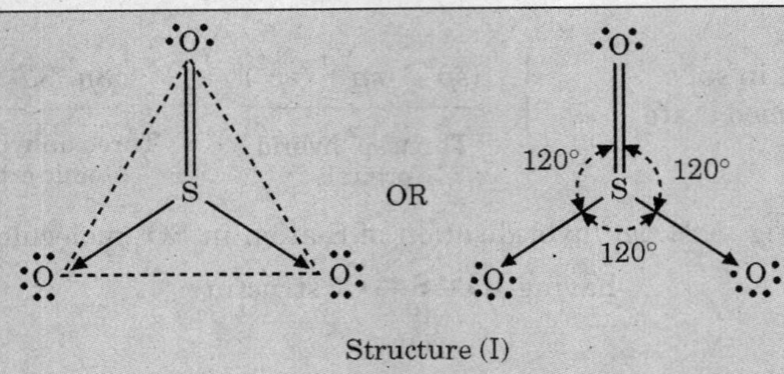

Fig.8.86 Trigonal planar shape of SO_3 molecule.

(*iv*) **Resonance in SO_3.** We have already shown that SO_3 molecule has three π bonds. One π bond is $(p\pi - p\pi)\pi$ bond and the other two π-bonds are $(p\pi - d\pi)\pi$ bonds. Length of the two $(p\pi - d\pi)\pi$ bonds should be the same but different from that of $(p\pi - p\pi)\pi$ bond. Expriments have, however, shown that all the three sulphur-oxygen bonds in SO_3 are of equal length (= 1.42 A°). Each bond length is intermediate between the calculated value of S – O single bond length and S = O double bond length. In order to explain the equivalence of all the three sulphur-oxygen bond lengths in SO_3,it has been suggested that equal lengths of the bonds is because of resonance between the three resonating structures (A), (B) and (C) shown in Fig. 8.87.

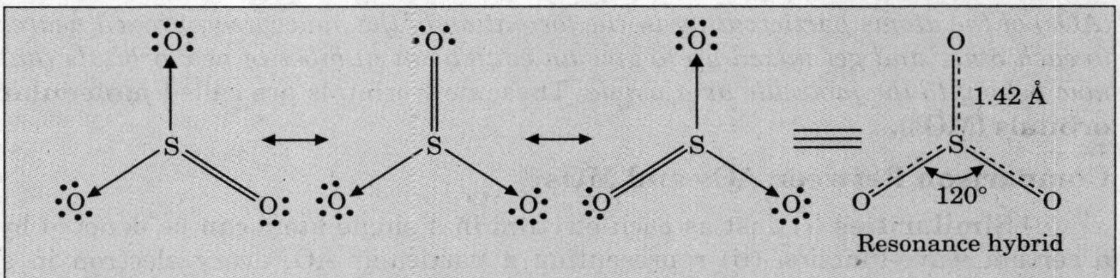

Fig.8.87. Resonance in SO_3 molecule.

7. Structure of SO_3 Molecule in the Solid State.

α–form of SO_3 solid molecule gets polymerised and gives a trimer, $(SiO_3)_3$. Which has a ring (cyclic) structure shown at (a) of Fig. 6.88. On the other hand, β and γ–forms of solid SO_3 have linear layer (or chain) structure shown at (*b*) of Fig. 8.88. This structure consists of a number of SO_3 chains linked together. These chains are produced when many SO_4 tetrahedrons get linked together by the sharing of O-atom.

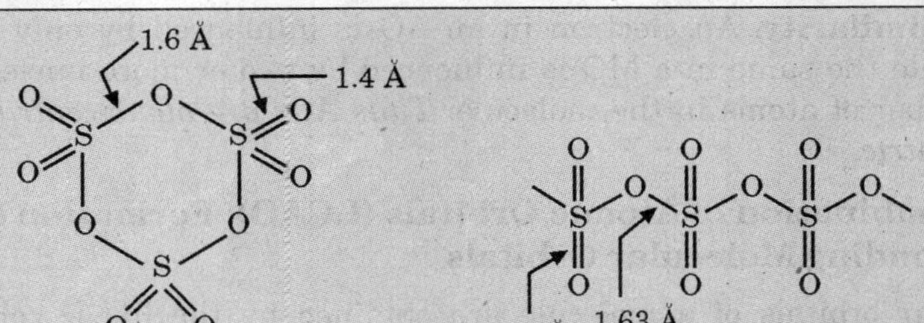

Fig. 8.88 : (*a*) Ring (cyclic) trimeric structure of α-form of SO_3 molecule in the solid state (*b*) Chain (or layer) structure of β and γ-form of SO_3 molecule in the solid state.

Molecular Orbital Theory (MOT) of Covalent Bond: Hund-Mullikan Theory

We have seen that the atomic orbital theory (Valence Bond Theory, VBT) of covalent bond is based on the assumption that the formation of a bond between the atoms of a given molecule involves the overlapping of the atomic orbitals which are half-filled. These half-filled orbitals (called bonding orbitals) merge (*i.e.* overlap or mix) into one another and give a bond (or an orbital) which is responsible for the stability of the resulted molecule. All other orbitals which are completely-filled do not participate in the process of overlapping and hence do not contribute to bond formation. In other words all the completely-filled orbitals maintain their individual indentity. However, this seems to be an exaggeration because the nucleus of an approaching atom is bound to affect the electron- waves of nearly all orbitals. Apart from this, VBT fails to explain some of the properties of some molecules like O_2.

According to MOT, put forward by Hund and Mullikan *all the atomic orbitals (AOs) of the atoms participating in the formation of the molecule approach nearer to each other and get mixed up to give an equivalent number of new orbitals that now belong to the molecule as a whole.* These new orbitals are called **molecular orbitals (MOs).**

Comparison Between AOs and MOs

(*a*) **Similarities** (*i*) Just as each electron in a single atom can be denoted by a certain wave-function (ψ) representing a particular AO, every electron in a molecule can like wise be denoted by a similar wave function representing a particular MO.

(*ii*) Like an AO, a MO can accomodate a maximum of two electrons which should have opposite spins. In other words, MOs, like AOs, obey Paulie's exclusion principle and Hund's rule.

(*iii*) Like AOs, various MOs of a molecule have different energies and different shapes.

(*iv*) Electrons in the molecule (except those residing in the inner-shells of the participating atoms) enter MOs in the order of increasing energy (*Aufbau principle*).

(*b*) **Dissimilarity.** An electron in an AO is influenced by only one positive nucleus while the same in a MO is influenced by two or more nuclei, depending on the number of atoms in the molecule. *Thus AOs are mono-centric while MOs are poly-centric.*

Linear Combination of Atomic Orbitals (LCAO): Formation of Bonding and Antibonding Molecular Orbitals

Molecular orbitals of a molecule are obtained by the linear combination of atomic orbitals of the bonded atoms. The linear combination of two AO wave functions is brought about either by adding or by subtracting the two wave functions of two AOs. Thus if ψ is the wave function of the MO obtained from two AOs of two atoms A and B having ψ_A and ψ_B respectively as their wave functions, then :

$$\boxed{\Psi = \Psi_A \pm \Psi_B} \quad ...(i)$$

Equation (*i*) clearly shows that the linear combination of ψ_A and ψ_B gives two MOs. One of these is represented as ψ^b and the other as ψ^a. These are thus given by

$$\boxed{\Psi^b = \psi_A + \Psi_B} \quad ...(ii)$$

and

$$\boxed{\Psi^a = \psi_A - \Psi_B} \quad ...(iii)$$

Ψ^b has lower energy than that of either of the AOs from which it is formed. This MO thus leads to the formation of a stable molecule, AB and is, therefore, called a **bonding molecular orbital.** The other MO namely ψ^a has higher energy than that of Ψ_A and Ψ_B and is, therefore, called an ***antibonding molecular orbital*** (Fig. 6.89). The superscript *b* used in Ψ^b indicates that the molecular orbital is bonding molecular orbital while *a* used in ψ^a indicates that the molecular orbital is antibonding. The number of molecular orbitals is always equal to the number of atomic orbitals involved in the formation of molecular orbitals.

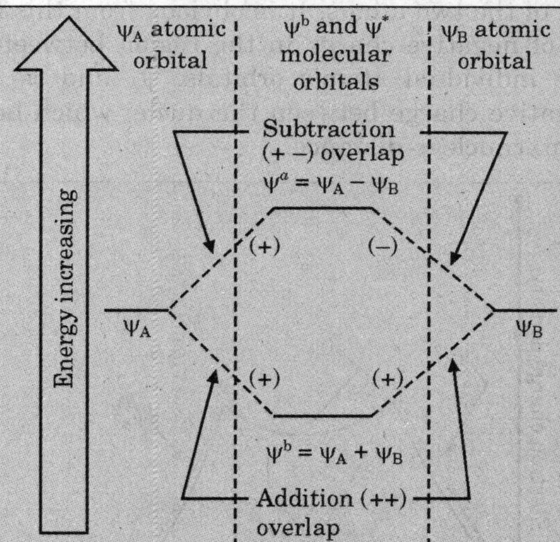

Fig. 8.89. Formation of bonding (ψ^b) and antibonding (ψ^a) molecular orbitals by the combination of ψ_A and ψ_B atomic orbital.

The plus sign in equation (*ii*) indicates that the Ψ^b MO is obtained when the regions of AOs marked with positive sign overlap with each other. In this case it is said that the two AOs have been *added*. On the other hand in equation (*iii*) negative sign indicates that ψ^a is obtained when the region of positive sign of one AO (say of Ψ_A) overlaps with the region of negative sign of the other AO (say of Ψ_B). In this case we say that the two AOs have been *subtracted*.

Electronic Charge Density of Bonding (Ψ^b) and Anti-bonding (Ψ^a) Molecular Orbitals

In order to understand the significance of bonding and anti-bonding molecular orbitals in terms of wave functions it is necessary to know the electronic charge distribution of these orbitals.

Squaring the equation (*ii*) we get :

$$(\psi^b)^2 = \psi_A{}^2 + \psi_B{}^2 + 2\psi_A\psi_B \qquad ...(iv)$$

The terms $\psi_A{}^2$ and $\psi_B{}^2$ indicate the electronic charge densities of the wave functions ψ_A and ψ_B of the isolated atoms A and B while the term $(\psi^b)^2$ indicates that of the wave function ψ^b of the bonding molecular orbital. Obviously greater the value of $(\psi^b)^2$, greater is the charge density between the two atomic nuclei and stronger is the bond between them. It may be seen from equation (*iv*) that $(\psi^b)^2 > (\psi_A^2 + \psi_B^2)$ by a term $2\psi_A\psi_B$. This term results from the *interaction* or *overlap* of atomic orbitals. Thus larger is the overlap of atomic orbitals, greater is the charge density between the nuclei and more stable is the bond formed between A and B.

Fig. 8.90. contains the plot of electron charge density for bonding molecular orbital, $(\psi^b)^2$ and for individual atomic orbitals (ψ_A^2 and ψ_B^2) against the internuclear distance (*i.e.* the distance between the two nuclei A and B). $(\psi^b)^2$ has been shown by solid lines and ψ_A^2 and ψ_B^2 by dotted lines. The points A and B represent the nuclei of the two atoms. It is obvious from this figure that there is more concentration of negative charge in the region between the two nuclei, A and B than for the individual atomic orbitals, ψ_A and ψ_B. Thus there is an accumulation of negative charge between the nuclei which holds these together at an equilibrium internuclear distance.

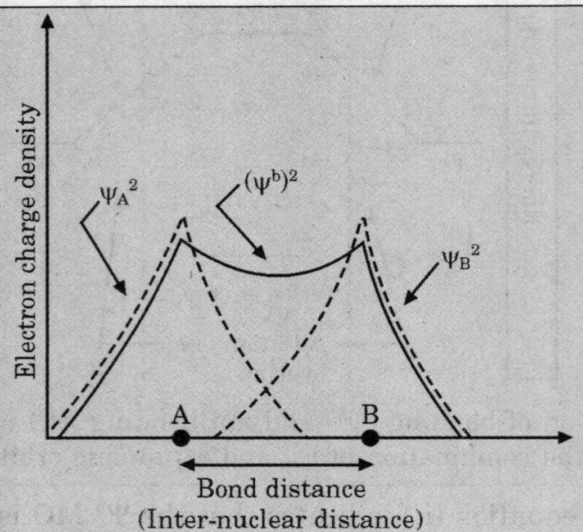

Fig. 8.90. Plots of electron charge density (square of wave function) for individual AOs (dotted lines) and ψ^b MO (solid lines) against internuclear distance between A and B.

Squaring equation. (*iii*), we get :

$$(\psi^a)^2 = \Psi_A^2 + \Psi_B^2 - 2\Psi_A\Psi_B \quad ...(v)$$

It may be seen from this equation that $(\Psi^a)^2 < (\psi_A^2 + \psi_B^2)$ by a term $2\Psi_A\Psi_B$. This means that in antibonding molecular orbital there is less charge density between the interacting atoms than there would be between the non-interacting atoms. Thus the energy of this antibonding molecular orbital is greater than the sum of the energies of the two interacting atoms A and B and the resulting bond formed between A and B is less stable.

In Fig. 8.91 $(\Psi^a)^2$ is shown by the solid line while Ψ_A^2 and Ψ_B^2 by dotted line. Plot of $(\Psi^a)^2$ touches the axis at midpoint N between the nuclei A and B. This point is called a **node point** and at this point the electron density is zero. Thus the formation of Ψ^a molecular orbital decreases the electron charge density in the region between the two nuclei A and B over that of the individual atoms. Negative charge is withdrawn from the region between the nuclei and is concentrated round the nucleus. This makes the electrons repel each other in the region between A and B. Thus we can say that in Ψ^a molecular orbital an electron spends half of its time on each of the individual atoms and consequently is never found in the region between A and B *i.e.* the probability of finding the electron in the region between A and B is decreased.

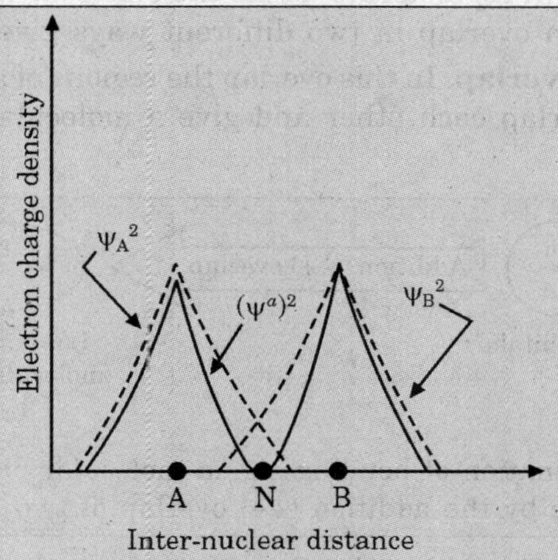

Fig. 8.91. Plots of electron charge density for individual AOs (dotted line) and ψ^a MO (solid lines) against internuclear distance between A and B.

We can say, therefore, that *a build-up of electronic charge between A and B is characteristic of the bonding or attractive state and a depletion of charge density between A and B is characteristic of antibonding or repulsive state.*

Differences Between Bonding and Anti-bonding Molecular Orbitals.

The differences between bonding and anti-bonding molecular orbitals can be summarised as follows :

Bonding molecular orbital	Anti-bonding molecular orbital
(*i*) A bonding molecular orbital is formed by the combination of electron-waves of the *same* sign, *i.e.* by the *addition-overlap* of atomic orbitals.	(*i*) An anti-bonding molecular orbital is formed by the combination of electron waves of the *opposite* sign, *i.e.* by the *subtraction-overlap* of atomic orbitals.
(*ii*) Bonding molecular orbital possesses *lower* energy than the atomic orbitals from which it is formed.	(*ii*) Anti-bonding molecular orbital possesses higher energy than the atomic orbitals from which it is formed.
(*iii*) The electron density in between the nuclei is *high* and hence the attraction between the nuclei is high. This explains the formation of a bond between the atoms.	(*iii*) The electron density in between the nuclei is low and hence the repulsion between the nuclei is high. This explains the non-formation of the bond between the atoms.

Formation of Molecular Orbitals from s-Atomic Orbitals

We may consider the formation of molecular orbitals obtained by the combination of two 1s orbitals on two H-atoms in H_2 molecule. In this combination 1s orbital of one H-atom overlaps with the 1*s* orbital of the other H-atom to give *two* molecular orbitals in H_2 molecule. Since the electrons have wave-like-nature, the two 1s orbitals can overlap in two different ways given below :

1. Addition (+ +) overlap. In this overlap the regions of atomic orbitals marked with positive sign overlap each other and give a molecular orbital as shown in Fig. 8.92.

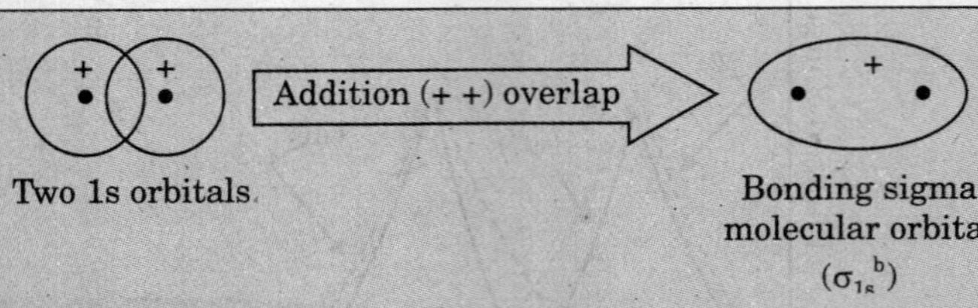

Fig. 8.92. Formation of bonding sigma molecular orbital ($\sigma_{1s}{}^{b}$) in H_2 molecule by the addition (++) overlap of two 1*s* orbitals.

It is clear from this figure that the electron in the molecular orbital will spend most of its time in the region between the two nuclei (shown by dots). The electron is now stabilished by two nuclear charges instead of one unlike an atomic orbital. *The resulting molecular orbital has lower energy than either of the two 1s orbitals and hence is called bonding molecular orbital.* Since the molecular orbital is symmetric to rotation about the inter-nuclear axis, *i.e.* it does not show any change of sign on rotation through 180° through the inter-nuclear axis, it is a sigma (σ) molecular orbital. Thus this molecular orbital is a **bonding sigma molecular orbital** and hence is represented as $\sigma_{1s}{}^{b}$.

2. Subtraction (+ –) overlap. In this overlap the region of positive sign of one atomic orbital overlaps with the region of negative sign of the other atomic orbital and gives a molecular orbital as shown in Fig. 8.93.

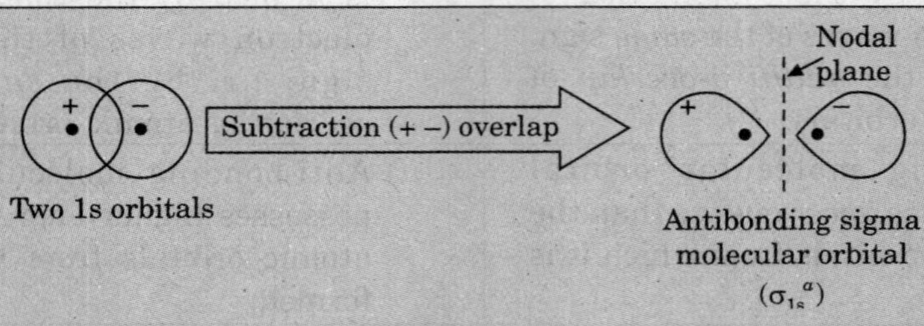

Fig. 8.93. Formation of antibonding sigma molecular orbital ($\sigma_{1s}{}^{a}$) in H_2 molecule by the subtraction (+ –) overlap of two 1*s* orbitals.

The resulting molecular orbital has zero electron density (*nodal plane*) in the region between the two nuclei (shown by dots). The electron in this molecular orbital is concentrated in region away from each of the nuclei. *The resulting molecular orbital is of higher energy than either of the two 1s-orbitals and hence is called antibonding molecular orbital.* Since this molecular orbital is also symmetric to rotation about the inter-nuclear axis, it is also a sigma (σ) molecular orbital. Thus this molecular orbital is an **antibonding sigma molecular orbital** and is, therefore, represented as $\sigma_{1s}{}^{a}$.

Similarly the combination of two 2*s* orbitals gives $\sigma_{2s}{}^{b}$ and $\sigma_{2s}{}^{a}$ molecular orbitals. Their formation can be shown in the same way as that of $\sigma_{1s}{}^{b}$ and $\sigma_{1s}{}^{a}$ molecular orbitals has been shown in Figures 8.92. and 8.93.

Order of Energy of $\sigma_{1s}{}^{b}$ and $\sigma_{1s}{}^{a}$ Molecular Orbitals : Energy Level Diagram

The above discussion makes it evident that the combination of two 1*s* orbitals on two H-atoms of H_2 molecule gives rise to the formation of two molecular orbitals which are $\sigma_{1s}{}^{b}$ and $\sigma_{1s}{}^{a}$. $\sigma_{1s}{}^{b}$ molecular orbital has lower energy than either of the two 1*s* orbitals of two H-atoms from which it is formed while $\sigma_{1s}{}^{a}$ molecular orbital has higher energy than each of the 1*s* atomic orbitals. Thus the order of energy of various orbitals is $\sigma_{1s}{}^{b} < 1s < \sigma_{1s}{}^{a}$ as shown in Fig. 8.94. which is called energy level diagram of $\sigma_{1s}{}^{b}$ and $\sigma_{1s}{}^{a}$ molecular orbitals.

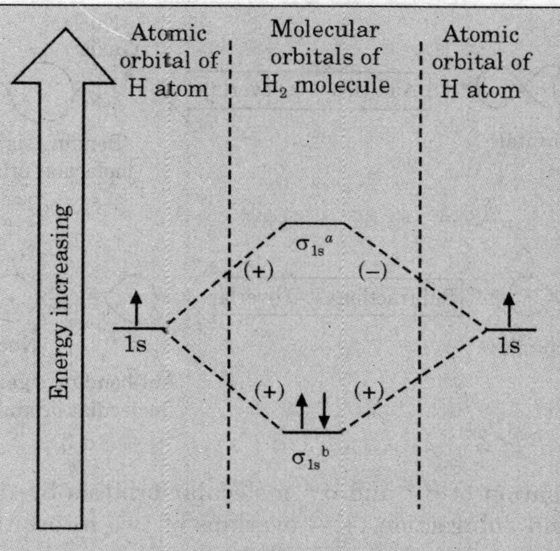

Fig. 8.94. Energy level diagram for $\sigma_{1s}{}^{b}$ and $\sigma_{1s}{}^{a}$ molecular orbitals formed by the combination of 1*s* orbitals on two H-atoms in H_2 molecule.

It is the lower energy $\sigma_{1s}{}^{b}$ molecular orbital which makes the bonding between two H-atoms in H_2 molecule possible. In H_2 molecule both the electrons occupy $\sigma_{1s}{}^{b}$ molecular orbital, since this molecular orbital is of lower energy (Aufbau principle). Thus $\sigma_{1s}{}^{a}$ molecular orbital which has higher energy remains vacant.

Formation of Molecular Orbitals from p-Atomic Orbitals

Since there are three *p* orbitals namely p_x, p_y and p_z, their combination gives rise to the formation of two types of molecular orbitals, depending on the way in which they overlap with each other. The following two cases may be studied.

1. Combination of two p_x orbitals. If *x*-axis is assumed to be the axis joining the two nuclei in a molecule (*i.e.* nuclear or molecular axis, as it is called), then the combination of two p_x orbitals will take place along the molecular axis (head-to-head overlap). This type of overlap gives rise to the formation of two molecular orbitals that are symmetrical about the molecular axis and hence are σ molecular orbitals. + + (addition) overlap of two p_x orbitals gives $\sigma_x{}^b$ molecular orbital while + – (subtraction) overlap of the same orbitals gives $\sigma_x{}^a$ molecular orbital (Fig. 8.95).

In case of $\sigma_x{}^b$ molecular orbital since the repulsion between the nuclei is minimum, the energy of this molecular orbital is lower than that of any of the two p_x orbitals. In case of $\sigma_x{}^a$ molecular orbital since the electron waves cancel each other, the electron density between the nuclei is very low and hence the nuclei *fly apart*. The energy of $\sigma_x{}^a$ molecular orbital is higher than that of $\sigma_x{}^b$ molecular orbital. $\sigma_x{}^b$ molecular orbital has no node while $\sigma_x{}^a$ molecular orbital has one node.

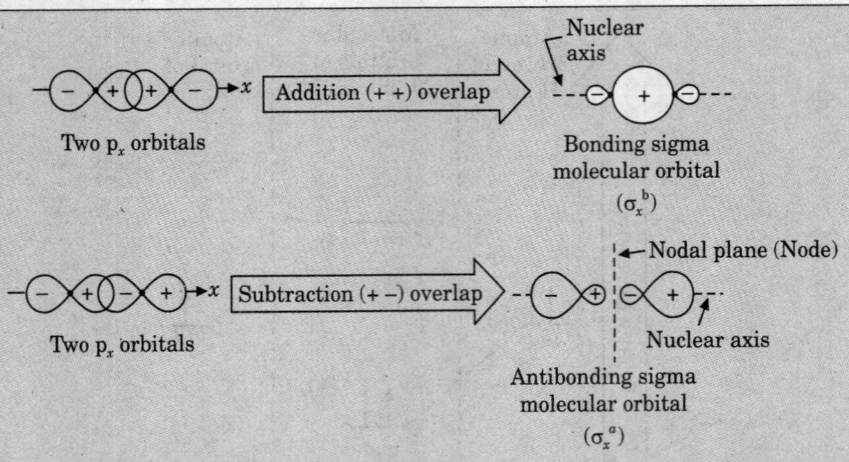

Fig. 8.95. Formation of $\sigma_x{}^b$ and $\sigma_x{}^a$ molecular orbitals by the addition (+ +) and subtraction (+ –) overlaps of two p_x orbitals.

2. Combination of two p_y or two p_z orbitals. Since *x*-axis has been assumed to be the molecular axis, two p_y (or two p_z) orbitals will overlap with each other along a line perpendicular to the molecular axis. This type of overlap gives rise to the formation of two molecular orbitals that are not symmetrical about the molecular axis and hence *are not sigma molecular orbitals*. Rather such orbitals are called *pi* (π) molecular orbitals. + + (addition) overlap of, say, two p_z orbitals gives $\pi_z{}^b$ molecular orbital while + – (subtraction) overlap of the same orbitals gives $\pi_z{}^a$ molecular orbital (Sec Fig. 8.96).

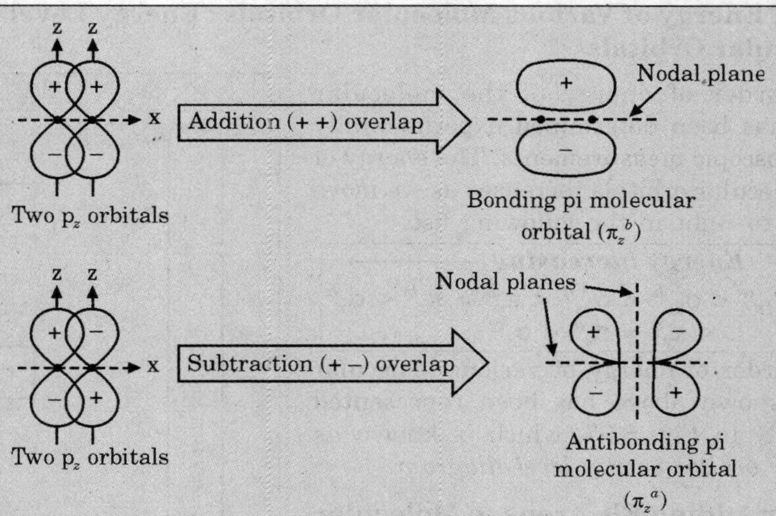

Fig. 8.96. Formation of π_z^b and π_z^a molecular orbitals by the addition (+ +) and subtraction (+ –) overlaps of two p_z orbitals.

Similarly the overlap of two p_y orbitals gives π_y^b and π_y^a molecular orbitals. The electron density in π_z^b or π_y^b molecular orbital is zero on a plane that contains the nuclear axis (nodal plane). The electron density is concentrated in the region above and below this nodal plane. π_z^a or π_y^a molecular orbital has the least energy in the region between the nuclei of the two atoms.

Since p_y and p_z atomic orbitals on an atom are degenerate (are of equal energy), the π molecular orbitals formed by their combination are also degenerate. Thus π_y^b and π_z^b molecular orbitals have equal energy and the same situation prevails with π_y^a and π_z^a molecular orbitals.

The above discussion makes it evident that the combination of two sets of three *p* orbitals on two isolated atoms gives rise to the formation of six molecular orbitals which are σ_x^b, σ_x^a, π_y^b, π_y^a, π_z^b and π_z^a.

Molecular Orbitals of a Homo-nuclear Diatomic Molecule

Now let us see how many molecular orbitals are formed in a homo-nuclear diatomic molecule whose both the atoms are of the elements belonging to the 2*nd* period of the long form of periodic table. Obviously these molecules may be Li_2, Be_2, B_2, C_2, N_2, O_2, F_2 and Ne_2. Since the isolated atoms of Li_2 and Be_2 molecules contain 1s and 2*s* orbitals only, these orbitals, on combination, give rise to the formation of *four* molecular orbitals which are σ_{1s}^b, σ_{1s}^a, σ_{2s}^b, and σ_{2s}^a. In case of B_2, C_2, N_2, O_2, F_2 and Ne_2 molecules, since the isolated atoms contain 1*s*, 2*s* and 2*p* orbitals, these orbitals, on combination, give rise to the formation of the following *ten* molecular orbitals :

$$\boldsymbol{\sigma_{1s}^b, \sigma_{1s}^a, \sigma_{2s}^b, \sigma_{2s}^a, \sigma_x^b, \sigma_x^a, \pi^b, \pi_y^a, \pi_z^b, \pi_z^a}$$

Order of Energy of Various Molecular Orbitals : Energy Level Diagram of Molecular Orbitals

The order of energy of the molecular orbitals has been determined experimentally by spectroscopic measurements. The energy of these molecular orbitals increases as we move from left to right in the following list.

——— ***Energy increasing*** ———→
$\sigma_{1s}{}^b < \sigma_{1s}{}^a < \sigma_{2s}{}^b < \sigma_{2s}{}^a < \pi_y{}^b = \pi_z{}^b < \sigma_x{}^b$ $< \pi_y{}^a = \pi_z{}^a < \sigma_x{}^a$

The order of energy of various molecular orbitals shown above has been represented graphically in Fig. 8.97. which is known *as molecular orbital energy level diagram.*

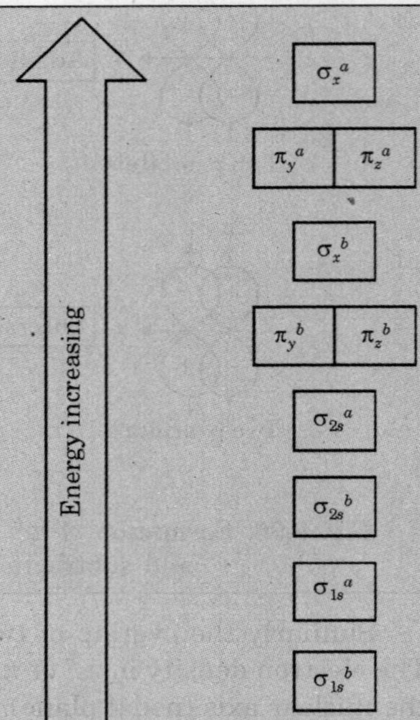

Fig. 8.97. Energy level diagram of various molecular orbitals.

Rules for Adding Electrons to Molecular Orbitals

Electrons are added to the molecular orbitals in the same way as they are added to the atomic orbitals. The rules are given below :

(*i*) The molecular orbitals are filled with electrons in the increasing order of their energy, *i.e.*, the molecular orbital having the lowest energy is filled up first while that with the highest energy is filled up in the end.

(*ii*) Hund's rule is also obeyed by the degenerate molecular orbitals. According to this rule the degenerate orbitals are first singly filled and pairing starts when more electrons are to be accommodated.

(*iii*) The maximum number of electrons that a molecular orbital can contain is two. These two electrons must have opposite spins.

Bond Order (B.O) or Bond Multiplicity

The electrons in the bonding molecular orbitals help in the formation of bonds while those in the antibonding molecular orbitals oppose it. Bond order (B.O) is defined as the number of covalent bonds between the two combining atoms of a molecule and *is equal to one-half of the difference between the number of electrons in bonding molecular orbitals* (n_b) *and that in the antibonding molecular orbitals* (n_a).

Thus :

$$\textbf{Bond Order} = \frac{1}{2}(n_b - n_a)$$

In common practice, the electrons residing in the molecular orbitals formed only from the orbitals of the valence-shell are considered for determining the bond order.

Conditions for Effective Combination of Atomic Orbitals

There are certain conditions which are necessary for the AOs to be able to combine effectively. These conditions are given below :

(*i*) The energies of the AOs combining together must be similar in magnitude or the AOs should have comparable energies. Thus in case of the formation of a homo-nuclear diatomic molecule of A_2 type, the 1*s* AO of one atom A will not combine with the 2*s* AO of the other atom since their energies are not equal. Similarly since the energy difference between 2*s*- and 2*p*- AOs is too great, they will also not combine.

But in case of the formation of hetero-nuclear diatomic molecule of AB type, such combinations may be expected.

(*ii*) The charge clouds of AOs must overlap one another as much as possible, if they are going to combine together to form the MOs. This condition is often referred to as the *principal of maximum overlap.*

(*iii*) The AOs should have the same symmetry about the molecular axis. This condition is known as *symmetry condition for the combination of AOs*.

On the basis of this symmetry condition it is noted that some of the AOs which have comparable energies do overlap, but cannot combine to give MOs. Thus MOs cannot be formed by the overlap of an *s*-atomic orbital of atom A and one *p*-atomic orbital of atom B perpendicular to the molecular axis. The cause of non-formation of MOs is that the symmetry of *s*-orbital is not the same as that of *p*-orbital. Alternatively it can be said that + + overlap is neutralised by the + – overlap or in other words the + – overlap cancels the bonding contribution from the + + overlap.

Molecular Orbital Configuration of Homo-nuclear Diatomic Species

For diatomic species (homo-nuclear and hetero-nuclear both) obtained from atoms having more than one electronic shells, the atomic orbitals of the inner-shells are deeply seated and the molecular orbitals formed by the overlap of these orbitals do not affect the bonding and hence can be omitted. The electrons in these molecular orbitals are called *non-bonding electrons* and the molecular orbitals containing these electrons are known as *non-bonding molecular orbitals*. Thus when we are considing a diatomic molecule, we consider only those molecular orbitals that are formed by the overlap of atomic orbitals belonging to the valence-shell only.

1. H_2 molecule ($H \rightarrow 1s^1$, $e = 1$). We have already said that the combination of two 1*s* orbitals on two H-atoms in H_2 molecule gives rise to the formation of $\sigma_{1s}{}^b$ and $\sigma_{1s}{}^a$ molecular orbitals whose energies are in the order : $\sigma_{1s}{}^b < \sigma_{1s}{}^a$. Both the electrons go to the lower energy $\sigma_{1s}{}^b$ molecular orbital and hence $\sigma_{1s}{}^a$ molecular remains vaccant. Thus the molecular orbital configuration of H_2 molecule is $(\sigma_{1s}{}^b)^2$ $(\sigma_{1s}{}^a)^0$, showing that $n_a = 2$, $n_b = 0$, and hence :

$$\text{Bond order} = \frac{2}{2} = 1$$

Bond order equal to 1 implies that two H-atoms in H_2 molecule are bonded together by *one* σ bond (H—H).

2. Hydrogen molecule ion, H_2^+ ($H \rightarrow 1s^1$, $H^+ \rightarrow 1s^0$, e = 1). This ion has one hydrogen atom and one H^+ ion linked together. Each of these has 1s-atomic orbital. Using the LCAO method the two 1s-orbitals will combine together to give $\sigma_{1s}{}^b$ and $\sigma_{1s}{}^a$ molecular orbitals. H_2^+ ion has only one electron which occupies lower energy $\sigma_{1s}{}^b$ molecular orbital and hence $\sigma_{1s}{}^a$ molecular orbital remains vaccant. Thus the molecular orbital configuration of H_2^+ ion is $(\sigma_{1s}{}^b)^1$ $(\sigma_{1s}{}^a)^0$, showing that $n_b = 1$, $n_a = 0$, and hence bond order of H^+ ion is one-half $\left(\frac{1}{2}\sigma\right)$ of a normal covalent bond. Low bond order of H_2^+ ion explains its low bond dissociation energy (= 255.28 KJ. mole^{-1}) and large bond length (= 1.06 A°)

3. He_2 molecule ($He \rightarrow 1s^2$, e = 4). He atom has the configuration 1s^2. For the formation of He_2 molecule $\sigma_{1s}{}^b$ and $\sigma_{1s}{}^a$ molecular orbitals are required. Four electrons (two contributed by each He atom) are to be accommodated in these molecular orbitals. Thus the molecular orbital configuration of He_2 molecule is $(\sigma_{1s}{}^b)^2$ $(\sigma_{1s}{}^a)^2$, showing that $n_b = 2$, $n_a = 2$, and hence bond order = (2–2)/2 = 0, Bond order equal to zero implies that He_2 molecule does not exist.

4. Li_2 molecule (*Li*$\rightarrow 2s^1$, e = 2). Each of the two Li atoms of Li_2 molecule has 1s and 2s atomic orbitals. Since 1s-atomic orbitals of both the atoms are deeply seated, the molecular orbitals formed by the overlap of these orbitals are considered to be non-bonding. Thus we are left only with the orbitals of the valence-shell namely 2s orbitals whose overlap should be considered. The overlap of two 2s orbitals on two Li atoms gives rise to the formation of $\sigma_{2s}{}^b$ and $\sigma_{2s}{}^a$ molecular orbitals whose energies are in the order : $\sigma_{2s}{}^b < \sigma_{2s}{}^a$. Both the electrons go to the lower energy $\sigma_{2s}{}^b$ molecular orbital and hence $\sigma_{2s}{}^a$ molecular orbital remains vaccant. Thus the molecular orbital configuration of Li_2 molecule is $(\sigma_{2s}{}^b)^2$ $(\sigma_{2s}{}^a)^0$, showing that $n_b = 2$, $n_a = 0$ and hence bond order = 2/2 = 1, *i.e.* two Li atoms in Li_2 molecule are linked together by a single covalent bond.

5. Be_2 molecule ($Be \rightarrow 2s^2$, e =4). Valence-shell of Be atom has 2s orbital. Two 2s orbitals on two Be atoms in Be_2 molecule overlap each other and form $\sigma_{2s}{}^b$ and $\sigma_{2s}{}^a$ molecular orbitals. Both the molecular orbitals are filled with four electrons. Thus the molecular orbital configuration of Be_2 molecule is $(\sigma_{2s}{}^b)^2$ $(\sigma_{2s}{}^a)^2$, showing that $n_b = 2$, $n_a = 2$ and hence bond order = (2 – 2)/2 = 0, *i.e.* Be_2 molecule does not exist.

6. B_2 molecule ($B \rightarrow 2s^2$, $2p_x{}^1$, $2p_y{}^0$, $2p_z{}^0$, e = 6). It is the first element in which p-orbital starts. Its valence-shell configuration is $2s^2$, $2p_x{}^1$, $2p_y{}^0$ $2p_z{}^0$, with three electrons. Thus B_2 molecule has in all 3 + 3 = 6 electrons and its molecular orbital configuration is :

$$(\sigma_{2s}{}^b)^2, (\sigma_{2s}{}^a)^2, (\pi_y{}^b)^1 = (\pi_z{}^b)^1 \quad \text{(See Fig. 8.98)}$$

Bond order of B_2 molecule in equal to $\frac{1}{2}$ (4–2) = 1 which shows that B_2 molecule is formed by a weak π-bond and has two unpaired electrons in π-molecular orbitals. This has been verified by experiments. Due to the presence of two unpaired electrons, B_2 molecule is *paramagnetic*.

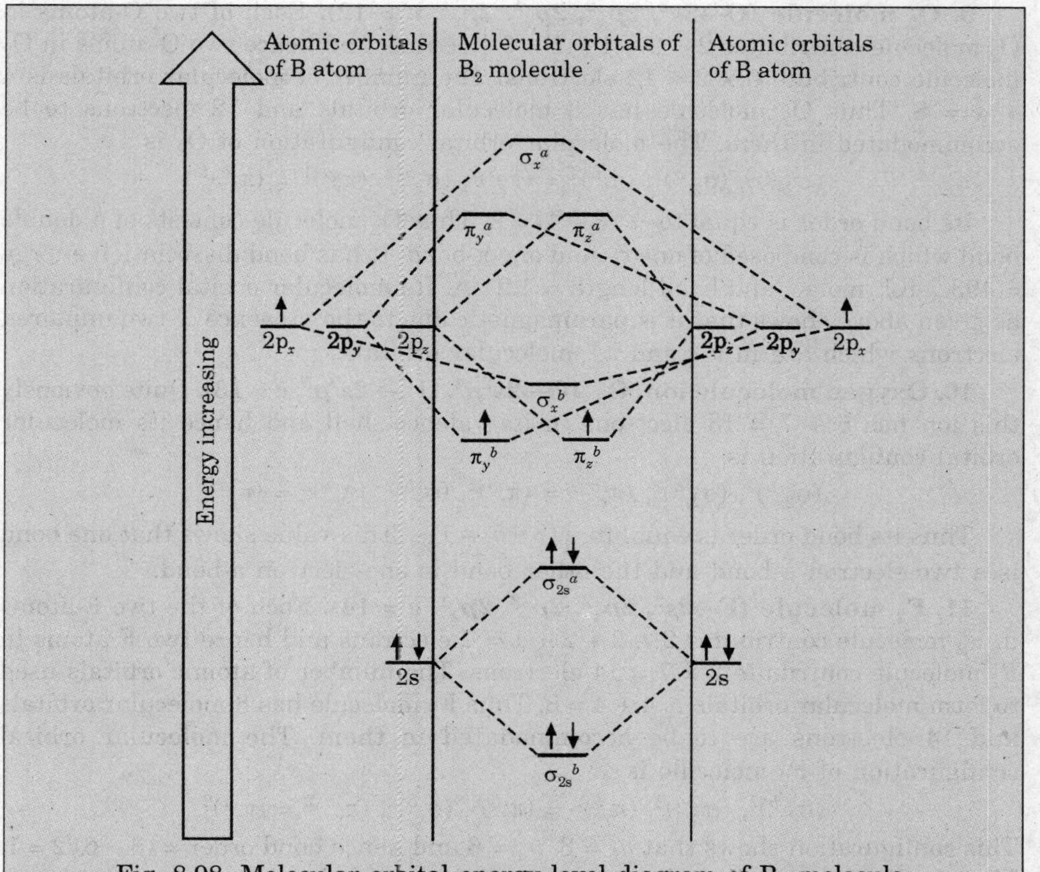

Fig. 8.98. Molecular orbital energy level diagram of B_2 molecule.

7. C_2 molecule ($C \rightarrow 2s^2, 2p_x^1. 2p_y^1. 2p_z^0, e = 8$). C_2 molecule has in all 4 + 4 = 8 electrons and its molecular orbital configuration is $(\sigma_{2s}{}^b)^2$, $(\sigma_{2s}{}^a)^2$, $(\pi_y{}^b)^2 = (\pi_z{}^b)^2$, showing that $n_b = 6$, $n_a = 2$ and hence bond order = (6 – 2)/2 = 2, *i.e.* the two C-atoms in C_2 molecule are joined together by two covalent bonds one of which is a σ bond and the other is a π bond. Since the molecular orbital configuration of C_2 molecule does not have unpaired electron, this molecule is expected to be diamagnetic. Experimentally it has been found diamagnetic.

8. N_2 molecule ($N \rightarrow 2s^2, 2p_x^1. 2p_y^1. 2p_z^1, e = 10$). Each of the two N-atoms in N_2 molecule contributes 2 + 1 + 1 + 1 = 5 electrons and hence two N-atoms in N_2 molecule contribute 5 × 2 = 10 electrons. The number of atomic orbitals used to form molecular orbitals is 4 + 4 = 8. Thus N_2 molecule has 8 molecular orbitals and 10 electrons to be accommodated in them. The molecular orbital configuration of N_2 molecule is :

$$(\sigma_{2s}{}^b)^2, (\sigma_{2s}{}^a)^2, (\pi_y{}^b)^2 = (\pi_z{}^b)^2, (\sigma_x{}^b)^2$$

and its bond order is equal to $\frac{1}{2}$ (8–2) = 3 which indicates that the molecule has one σ and two π-bonds. Bond order of three is in accordance with the very high bond energy (= 945.6 KJ. mole^{-1}).

9. O_2 molecule ($O \rightarrow 2s^2, 2p_x^2. 2p_y^1. 2p_z^1$, e = 12). Each of two O-atoms in O_2 molecule contributes 2 + 2 + 1 + 1 = 6 electrons and hence two O-atoms in O_2 molecule contribute 6 × 2 = 12 electrons. The number of molecular orbitals is 4 + 4 = 8. Thus O_2 molecule has 8 molecular orbitals and 12 electrons to be accommodated in them. The molecular orbital configuration of O_2 is :

$$(\sigma_{2s}{}^b)^2, (\sigma_{2s}{}^a)^2, (\pi_y{}^b)^2 = (\pi_z{}^b)^2, (\sigma_x{}^b)^2, (\pi_y{}^a)^1 = (\pi_z{}^a)^1$$

Its bond order is equal to $\frac{1}{2}$ (8 – 4) = 2. Thus O_2 molecule consists of a double bond which is composed of one σ- and one π-bond. It has bond dissociation energy = 493.7 KJ. mole^{-1} and bond length = 1.21 Å. Its molecular orbital configuration as given above shows that it is paramagnetic due to the presence of two unpaired electrons which are in $\pi_y{}^a$ and $\pi_z{}^a$ molecular orbitals.

10. Oxygen molecule ion, O_2^- ($O \rightarrow 2s^2p^4$, $O^- \rightarrow 2s^2p^5$, e = 13). Quite obviously this ion has 6 + 7 = 13 electrons in its valence-shell and hence its molecular orbital configuration is :

$$(\sigma_{2s}{}^b)^2, (\sigma_{2s}{}^a)^2, (\pi_y{}^b)^2 = (\pi_z{}^b)^2, (\sigma_x{}^b)^2, (\pi_y{}^a)^2 = (\pi_z{}^a)^1$$

Thus its bond order is equal to $\frac{1}{2}(8-5) = 1\frac{1}{2}$. This value shows that one bond is a two-electron σ-bond and the other bond is one-electron π-bond.

11. F_2 molecule ($F \rightarrow 2s^2, 2p_x^2. 2p_y^2. 2p_z^1$, e = 14). Each of the two F-atoms in F_2 molecule contributes 2 + 2 + 2 + 1 = 7 electrons and hence two F atoms in F_2 molecule contribute 2 × 7 = 14 electrons. The number of atomic orbitals used to form molecular orbitals is 4 + 4 = 8. Thus F_2 molecule has 8 molecular orbitals and 14 electrons are to be accommodated in them. The molecular orbital configuration of F_2 molecule is :

$$(\sigma_{2s}{}^b)^2, (\sigma_{2s}{}^a)^2, (\pi_y{}^b)^2 = (\pi_z{}^b)^2, (\sigma_x{}^b)^2, (\pi_y{}^a)^2 = (\pi_z{}^a)^2.$$

This configuration shows that n_b = 8, n_a = 6 and hence bond order = (8 – 6)/2 = 1. Thus in F_2 molecule both F_2 atoms are linked to each other by a single bond. Since there are no unpaired electrons, F_2 molecule is *diamagnetic*.

12. Ne_2 molecule ($Ne \rightarrow 2s^2, 2p_x^2. 2p_y^2. 2p_z^2$, e = 16). The molecular orbital configuration of Ne_2 molecule is :

$$(\sigma_{2s}{}^b)^2, (\sigma_{2s}{}^a)^2, (\pi_y{}^b)^2 = (\pi_z{}^b)^2, (\sigma_x{}^b)^2, (\pi_y{}^a)^2 = (\pi_z{}^a)^2, (\sigma_x{}^a)^2.$$

This configuration shows that n_b = 8, n_a = 8 and hence bond order = (8 – 8)/2 = 0 which means that Ne_2 molecule does not exist.

Molecular Orbital Configuration of Some Hetero-nuclear Diatomic Species

Hetero-nuclear diatomic species have two different atoms linked together. Thus these are AB type species. Here we shall consider those AB type species in which B has higher electronegativity than A and both the atoms A and B have 2*s* and 2*p* atomic orbitals, *i.e.* A and B are *p* block elements of second period of the periodic table.

The MO treatment of such types of hetero-nuclear diatomic molecules and molecule ions by LCAO approximation method is not fundamentally different from that of homo-nuclear diatomic molecules of A_2 type, except that MOs are not symmetric with respect to a plane perpendicular to and bisecting the internuclear axis.

Let us consider the following hetero-nuclear diatomic species.

1. CO molecule (C→2s^2p^2, O→2s^2p^4, e = 10). Evidently CO molecule has 10 valence electrons : four from carbon atom ($2s^2p^2$) and six from oxygen atom ($2s^2p^4$). Thus its molecular orbital configuration is :

$$(\sigma_{2s}{}^{b})^2, (\sigma_{2s}{}^{a})^2, (\pi_y{}^{b})^2 = (\pi_z{}^{b})^2, (\sigma_x{}^{b})^2$$

Thus its bond order is equal to $\frac{1}{2}(8-2) = 3$, showing that carbon and oxygen atoms in CO molecule are linked together by one σ-and two π-bonds. As is evident from the molecular orbital configuration, the molecule has *no unpaired electron* and hence it is *diamagnetic*.

2. NO molecule (N→2s^2p^3, O→2s^2p^4, e = 11). This molecule has 11 electrons in its valence-shell : five from nitrogen atom ($2s^2p^3$) and six from oxygen atom ($2s^2p^4$). Thus it has the following molecular orbital configuration :

$$(\sigma_{2s}{}^{b})^2, (\sigma_{2s}{}^{a})^2, (\pi_y{}^{b})^2 = (\pi_z{}^{b})^2, (\sigma_x{}^{b})^2, (\pi_y{}^{a})^1 = (\pi_z{}^{a})^0$$

The configuration given above shows that $n_b = 8$, $n_a = 3$ and hence bond order $= (8-3)/2 = 2\frac{1}{2}$ Bond order equal to $2\frac{1}{2}$ means that N and O atoms in NO molecule are linked together by one σ bond (due to the presence of electron pair in $\sigma_x{}^b$ molecular orbital) and two π bonds (due to the presence of two electron pairs in $\pi_y{}^b$ and $\pi_z{}^b$ molecular orbitals) *less* the effect of one unpaired electron present in $\pi_y{}^a$ molecular orbital. In other words we can say that NO molecule has two 2-electron bonds and one 3-electron bond and can, therefore, be represented as : $\mathrm{N} \overset{\bullet\times\times}{=} \mathrm{O}^{\times}_{\times}$. The presence of one unpaired electron in $\pi_y{}^a$ molecular orbital makes NO molecule *less* stable than N_2 molecule. This is confirmed by the fact that bond dissociation energy of NO molecule (= 667.8 KJ/mole) is less than that of N_2 molecule (= 945.6 KJ/mole). The presence of one unpaired electron also accounts for the paramagnetic nature of NO molecule. Molecular orbital configuration of NO molecule as given above also explains that it should be easy for this molecule to form nitrosonium ion (NO^+) by losing one electron residing in $\pi_y{}^a$ molecular orbital and NO^+ ion so produced should have shorter and hence stronger nitrogen–oxygen bond. The experimental values of nitrogen–oxygen bond length and bond energy for these species as given below confirm this fact.

$:\mathrm{N} \overset{\bullet\times\times}{=} \mathrm{O}^{\times}_{\times}$	$[:\mathrm{N} \equiv \mathrm{O}^{\times}_{\times}]^+$
1.14 A°	1.06 A°
667.8 KJ/mole	1019.9 KJ/mole

3. Nitrosyl ion, NO^+ (N→2s^2p^3, O^+→2s^2p^3, e = 10). As is evident, this ion has 5 + 5 = 10 electrons and its molecular orbital configuration is : $(\sigma_{2s}{}^{b})^2$, $(\sigma_{2s}{}^{a})^2$, $(\pi_y{}^{b})^2 = (\pi_z{}^{b})^2$, $(\sigma_x{}^{b})^2$, showing that its bond order = (8 – 2)/2 = 3, *i.e.* this ion has one σ and two π bonds, $:\mathrm{N} \equiv \overset{+}{\mathrm{O}}:$ or $[:\mathrm{N} \equiv \mathrm{O}:]^+$. Since the ion has no unpaired electron, it has diamagnetic character.

Valency Bond Theory *vs* Molecular Orbital Theory

(A) Similarities

(*i*) Both the theories are the interpretation of covalent bond.

(*ii*) According to both the theories the AOs of the two atoms must overlap to form a covalent bond. The AOs overlapping each other must have the same symmetry and must be of the same energy.

(*iii*) According to both, the electronic charge resides in the region between the atomic nuclei.

(*iv*) Both the theories postulate that a covalent bond possesses directional properties.

(B) Differences

MOT	VBT
(*i*) MOs are formed by LCAO approximation method. The AOs involved in the formation of these MOs are from valence shell of two atoms.	(*i*) Two AOs give an interatomic orbital obtained by the combination of the space functions of two unpaired electrons; one being in each of the two AOs.
(*ii*) AOs of the resulting molecule lose their individual identity.	(*ii*) The resulting molecule consists of atoms which, although, interacting, retain a large measure of their individual character.
(*iii*) MOT starts with the nuclei of the constituent atoms.	(*iii*) VBT starts with the individual atoms and considers the interaction between them.
(*iv*) MOT offers a satisfactory explanation of the paramagnetic character of O_2 molecule.	(*iv*) VBT fails to account for the paramagnetic character of O_2 molecule.
(*v*) Resonance plays no role in MOT.	(*v*) Resonance plays an important role in VBT.

Questions with Answers

Q.1 We are given some molecules / ions that have been grouped into four pairs :

(*i*) BrO_3^-, XeO_3 (*ii*) SF_4, XeF_4 (*iii*) SO_3^{2-}, NO_3^- (*iv*) BF_3, NF_3. Out of these pairs which pair is isostructural ?

Ans : The structure of the given pairs are given below :

(*i*)

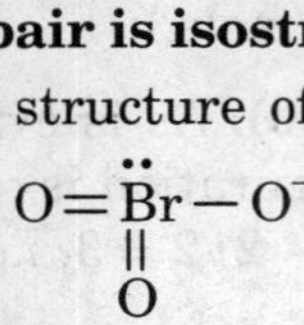

BrO_3^-

σ-bps + lps = 3 + 1 = 4

(Trigonal pyramidal)

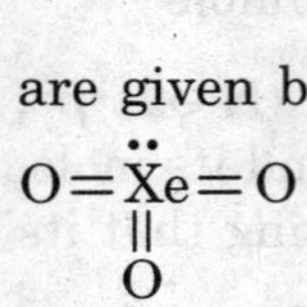

XeO_3

σ-bps + lps = 3 + 1 = 4

(Trigonal pyramidal)

(*ii*)

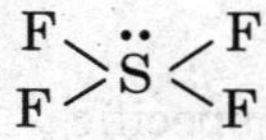

SF_4

σ-bps + lps = 4 + 1 = 5

(See-saw geometry)

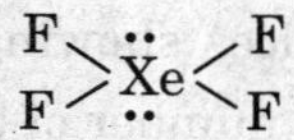

XeF_4

σ-bps + lps = 4 + 2 = 6

(Square planar)

(*iii*)

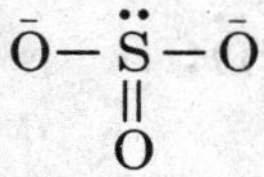

SO_3^{2-}

σ-bps + lps = 3 + 1 = 4

(Trigonal pyramidal)

NO_3^-

σ-bps + lps = 3 + 0 = 3

(Trigonal planar)

(*iv*)

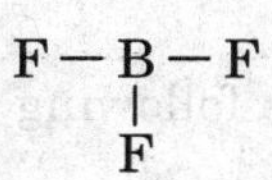

BF_3

σ-bps + lps = 3 + 0 = 3

(Trigonal planar)

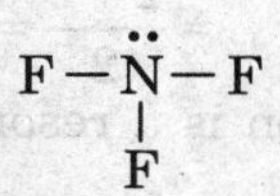

NF_3

σ-bps + lps = 3 + 1 = 4

(Trigonal pyramidal)

Since BrO_3^- – XeO_3 pair has the same geometry (trigonal pyramidal), this pair is isostructural.

Q.2 How many sigma (σ) and pi (π) bonds are there in allyl isocyanide ?

Ans : The radical obtained by removing H-atom from propence (also called propylence), $CH_2 = CH - CH_3$ is called allyl radical.This radical has positive charge.

$$CH_2 = CH - CH_3 \xrightarrow{-H} CH_2 = CH - \overset{+}{C}H_2$$

Propene or propylene   Allyl radical

Obviously the structure of allyl isocyanide molecule can be written as :

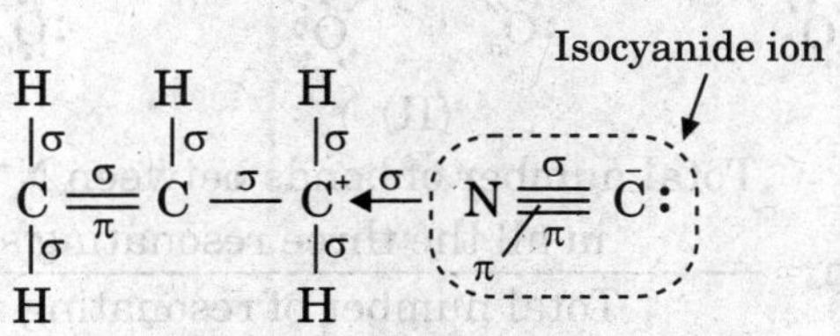

Obviously ally isocyanide has **9σ** and **3π** bonds.

Q.3 (a) Calculate bond order (B.O.) in NO_2^+, NO_2^-, NO_3^-, CO_3^{2-} and N_3^- ions.

(b) Compare the nitrogen–oxygen bond lengths in NO_2^+, NO_2^- and NO_3^- ions.

Ans : (*a*) We know that :

B.O. of a given species showing resonance

$$= \frac{\text{Total number of bonds between two specific atoms in all the resonating structures}}{\text{Total number of resonating structures}}$$

(*i*) **NO_2^+ ion.** This ion shows resonance. It is aresonance hybrid of the following two resonating structures.

(I) (II)

$$\therefore \text{Bond order of } NO_2^+ \text{ ion} = \frac{\text{Total number of bonds between N and } O_a \text{ atoms in the two resonating structures shown above}}{\text{Total number of resonating structures}}$$

$$= \frac{2+2}{2} = \frac{4}{2} = \mathbf{2\ (Ans.)}$$

(*ii*) **NO_2^- ion.** This ion is a resonance hybrid of the following two resonating structures.

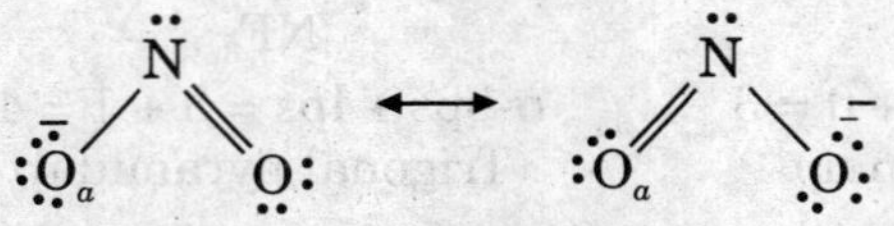

$$\therefore \text{B.O. of } NO_2^- \text{ ion} = \frac{\text{Total number of bonds between N and } O_a \text{ atoms in the two resonating structures}}{\text{Total number of resonating structures}}$$

$$= \frac{1+2}{2} = \frac{3}{2} = \mathbf{1.5\ (Ans.)}$$

(*iii*) **NO_3^- ion.** This ion has the following three resonating structures.

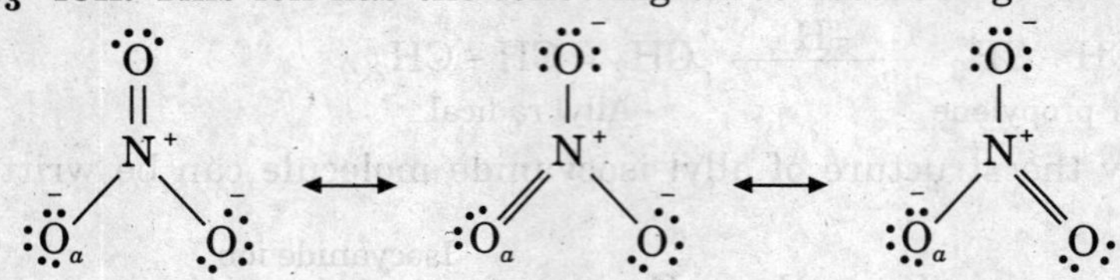

(I) (II) (III)

$$\therefore \text{B.O. of } NO_3^- \text{ ion} = \frac{\text{Total number of bonds between } N^+ \text{ and } O_a \text{ atoms in all the three resonating structures}}{\text{Total number of resonating structures}}$$

$$= \frac{1+2+1}{3} = \frac{4}{3} = \mathbf{1.33\ (Ans.)}$$

(*iv*) **CO_3^{2-} ion.** CO_3^{2-} ion can have the following three resonating structures

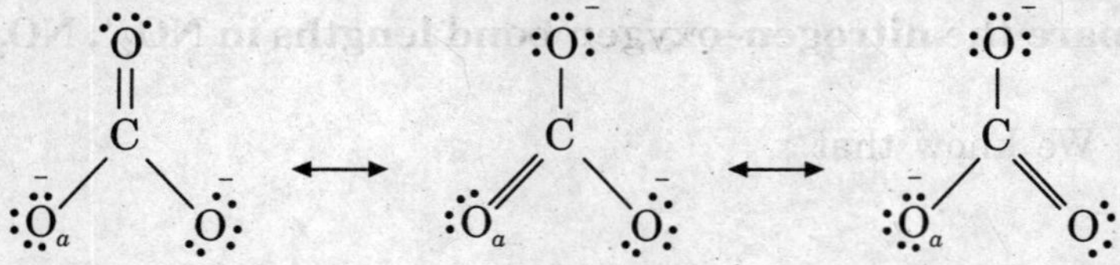

(I) (II) (III)

$$\therefore \text{ B.O. of } CO_3^{2-} \text{ ion} = \frac{\text{Total number of bonds between C and } O_a \text{ atoms in all the three resonating structures}}{\text{Total number of resonating structures}}$$

$$= \frac{1+2+1}{3} = \frac{4}{3} = \mathbf{1.33\ (Ans.)}$$

(*v*) **N_3^- ion. (Azide ion).** We know that this ion is a resonance hybrid of the following three structures

$$:\overset{-}{\ddot{N}}_a = \overset{+}{N} = \overset{-}{\ddot{N}}: \longleftrightarrow :\overset{-2}{\ddot{\underset{\cdot\cdot}{N}}}_a - \overset{+}{N} \equiv \ddot{N} \longleftrightarrow :N_a \equiv \overset{+}{N} - \overset{-2}{\ddot{\underset{\cdot\cdot}{N}}}:$$

(I) (II) (III)

$$\therefore \text{ B.O. of } N_3^- \text{ ion} = \frac{\text{Total number of bonds between } N_a \text{ atom and the central } \overset{+}{N}\text{-atom in all the three resonating structures}}{\text{Total number of resonating structures}}$$

$$= \frac{2+1+3}{3} = \frac{6}{3} = \mathbf{2\ (Ans.)}$$

(*b*) Since B.O. of NO_2^+, NO_2^- and NO_3^- ions is in the order : NO_2^+ (B.O. = 2.00) > NO_2^- (B.O. = 1.50) > NO_3^- (B.O. = 1.33), nitrogen-oxygen bond lengths are in the reverse order, *i.e.*nitrogen-oxygen bond lengths are in the order : NO_2^+ (= 1.55 A°) < NO_2^- (= 1.22 A°) < NO_3^- (= 2.18 A°)

Q.4 (*i*) Calculate (*a*) Angle between two *sp* hybrid obitals (*sp* hybridisation) and three sp^2 hybrid orbitals (sp^2 hybridsation) (*b*) % s-character in *sp*, sp^2 and sp^3 hybrid orbitals (*ii*) Arrange the given hybrid orbitals in the decreasing order of percentage of s-character in them.

Ans : (*i*) (*a*) Angle (∝) between two hybrid orbitals is given by :

$$\text{Cos} \propto = -\frac{1}{m}$$

Here *m* is the number of *p*-orbitals that take part in hybridisation. For *sp*, sp^2 and sp^3 hybridisation m = 1, 2 and 3 respectively. *m* is called **hybridisation index**. Thus :

For *sp* hybridisation $\cos \propto = -\frac{1}{m} = -\frac{1}{1}$

$\therefore \propto =$ **180° (Ans)**

For sp^2 hybridisation $\cos \propto = -\frac{1}{m} = -\frac{1}{2}$

$\therefore \propto =$ **120° (Ans)**

(*b*) % s-character is given by :

$$\% \text{ } s\text{-character} = \frac{1}{1+m} \times 100$$

Thus :

% s-character in *sp* hybrid orbitals

$$= \frac{1}{1+1} \times 100 = \mathbf{50\%\ (Ans)}$$

% s-character in sp^2 hybrid orbitals

$$= \frac{1}{1+2} \times 100 = \mathbf{33.3\%\ (Ans)}$$

% s-character in sp^3 hybrid orbitals

$$= \frac{1}{1+3} \times 100 = \mathbf{25\%\ (Ans)}$$

(*ii*) Obviously the decreasing order of the percentage of *s*-character is :

$$sp\ (= 50\%) > sp^2\ (= 33.3\%) > sp^3\ (= 25\%)$$

Q.5 Bond order of CO⁺ ion is higher than that of CO molecule. Explain

OR

Carbon-oxygen bond length in CO^+ ion is lower than that in CO molecule. Explain.

Ans : We know that the MOEC of CO molecule (e = 4 + 6 = 10) is :

$$(\sigma_{2s}{}^b)^2\ (\sigma_{2s}{}^a)^2\ (\pi_y{}^b)^2 = (\pi_z{}^b)^2\ (\sigma_x{}^b)^2$$

$$\therefore \text{B.O. of CO} = \frac{n_b - n_a}{2} = \frac{8-2}{2} = 3$$

In case of CO^+ ion it has been suggested that 2*s* orbital of C-atom has higher energy than 2*s* orbital of O-atom. When these 2*s* orbitals mix together to form $\sigma_{2s}{}^b$ and $\sigma_{2s}{}^a$ MOs, $\sigma_{2s}{}^a$ MO has so high energy that it occupies a position in the energy level diagram above $\pi_y{}^b$, $\pi_z{}^b$ and $\sigma_x{}^b$ MOs. Thus MOEC of CO^+ ion (e = 4 + 6 − 1 = 9) is :

$$(\sigma_{2s}{}^b)^2\ (\pi_y{}^b)^2 = (\pi_z{}^b)^2\ (\sigma_x{}^b)^2\ (\sigma_{2s}{}^a)^1$$

$$\therefore \text{B.O. of CO}^+ = \frac{n_b - n_a}{2} = \frac{8-1}{2} = 3.5$$

Obviously B.O. of CO^+ ion (= 3.5) is higher than that of CO molecule (= 3). It is because of higher B.O. of CO^+ ion that the carbon-oxygen bond length in CO^+ ion (= 1.115 A°) is lower than carbon-oxygen bond length in CO molecule (= 1.128 A°).

Q.6 Explain the following, giving appropriate reasons.

(*i*) H_2^+ and H_2^- ions have the same bond order (= 0.5), yet H_2^- ion is slightly less stable than H_2^+ ion.

(*ii*) N_2 has greater bond dissociation energy than N_2^+ whereas O_2^+ has greater bond dissociation energy than O_2.

(*iii*) Bond in H_2^+ is larger than that in H_2.

(*iv*) Superoxide ion (O_2^-) is paramagnetic but peroxide ion (O_2^{2-}) is not.

(*v*) Sigma bond in H_2 molecule (H – H) is stronger than that in Li_2 molecule (Li – Li).

(vi) Out of H and H_2, the latter has higher IE while out of O and O_2, the former has higher IE.

Ans : (*i*) MOECs of H_2^+ and H_2^- ions are as :

H_2^+ ion $(e = 1 + 1 - 1 = 1) = (\sigma_{1s}{}^b)^1 (\sigma_{1s}{}^a)^0 \ (n = 1)$

H_2^- ion $(e = 1 + 1 + 1 = 3) = (\sigma_{1s}{}^b)^2 (\sigma_{1s}{}^a)^1 \ (n = 1)$

These configurations show that H_2^- ion contains one unpaired electron in antibonding MO ($\sigma_{1s}{}^a$) which produces destablising effect and hence decreases the stability of H_2^- ion. On the other hand H_2^+ ion does not contain any electron in antibonding MO.

(*ii*) MOECs and BOs of the given species are given below :

Species	MOECs	B.O.
N_2 $(e = 10)$	$(\sigma_{2s}{}^b)^2 (\sigma_{2s}{}^a)^2 (\pi_y{}^b)^2 = (\pi_z{}^b)^2 (\sigma_x{}^b)^2$	$\frac{8-2}{2} = 3$
N_2^+ $(e = 9)$	$(\sigma_{2s}{}^b)^2 (\sigma_{2s}{}^a)^2 (\pi_y{}^b)^2 = (\pi_z{}^b)^2 (\sigma_x{}^b)^1$	$\frac{7-2}{2} = 2.5$
O_2 $(e = 12)$	$(\sigma_{2s}{}^b)^2 (\sigma_{2s}{}^a)^2 (\pi_y{}^b)^2 = (\pi_z{}^b)^2 (\sigma_x{}^b)^2 (\pi_y{}^a)^1 = (\pi_z{}^a)^1$	$\frac{8-4}{2} = 2$
O_2^+ $(e = 11)$	$(\sigma_{2s}{}^b)^2 (\sigma_{2s}{}^a)^2 (\pi_y{}^b)^2 = (\pi_z{}^b)^2 (\sigma_x{}^b)^2 (\pi_y{}^a)^1 = (\pi_z{}^a)^0$	$\frac{8-3}{2} = 2.5$

Since $(B.O.)_{N_2}$ (= 3.0) > $(B.O.)_{N_2^+}$ (= 2.5), N_2 has greater bond dissociation energy. On the other bond, since $(B.O.)_{O_2^+}$ (= 2.5) > $(B.O.)_{O_2}$ (= 2.0), O_2^+ has greater bond dissociation energy.

(*iii*) MOECs and BOs of H_2^+ and H_2 are :

H_2^+ $(e = 2 - 1 = 1) = (\sigma_{1s}{}^b)^1$, $(B.O.)_{H_2^+} = \frac{1}{2} = 0.5$

H_2 $(e = 2) = (\sigma_{1s}{}^b)^2$, $(B.O.)_{H_2} = \frac{2}{2} = 1.0$

Now since $(B.O.)_{H_2^+}$ (= 0.5) < $(B.O.)_{H_2}$ (= 1.0), bond in H_2^+ is larger than that in H_2.

(*iv*) MOECs of O_2^- and O_2^{2-} ion are given below :

O_2^- $(e = 6 \times 2 + 1 = 13) = (\sigma_{2s}{}^b)^2 (\sigma_{2s}{}^a)^2 (\pi_y{}^b)^2 = (\pi_z{}^b)^2 (\sigma_x{}^b)^2$
$(\pi_y{}^a)^2 = (\pi_z{}^a)^1 \ (n = 1)$

O_2^{2-} $(e = 6 \times 2 + 2 = 14) = (\sigma_{2s}{}^b)^2 (\sigma_{2s}{}^a)^2 \quad (\pi_y{}^b)^2$
$= (\pi_z{}^b)^2 (\sigma_x{}^b)^2 (\pi_y{}^a)^2 = (\pi_z{}^a)^2 \ (n = 0)$

Now since O_2^- ion has one unpaired electron ($n = 1$), this ion is paramagnetic. On the other hand since O_2^{2-} ion does not have any paired electron, this ion not paramagnetic. Rather, O_2^{2-} ion is diamagnetic.

(*v*) On the basis of MO theory, MO configuration of H_2 and Li_2 molecules can be written as :

$$H_2\ (e = 2) = (\sigma_{1s}{}^{b})^2,\ (\text{B.O.})_{H_2} = \frac{2-0}{2} = 1$$

$$Li_2\ (e = 2) = (\sigma_{2s}{}^{b})^2,\ (\text{B.O.})_{Li_2} = \frac{2-0}{2} = 1$$

Bond order equal to 1 indicates that the two atoms in both the molecules are bonded together by a single sigma bond. Thus these molecules can be represented as H–H and Li–Li respectively. It has been observed that in H_2 molecule H–H bond length = 0.74 A° and bond dissociation energy = 458 KJ mol^{-1}. Similary, in Li_2 molecule, Li – Li bond length = 2.67 A° and bond dissociation energy = 105 KJ mol^{-1}. These values show that H–H sigma bond in H_2 molecule is stronger than Li – Li bond in Li_2 molecule. This has been explained as follows.

Explanation : (*a*) Since 2*s* orbital of Li-atom is larger in size than 1*s* orbital of H-atom, 2*s* – 2*s* overlap in Li_2 molecule is less effective than 1*s* – 1*s* overlap in H_2 molecule. These overlaps give $\sigma_{2s}{}^{b}$ and $\sigma_{1s}{}^{b}$ molecular orbitals in Li_2 and H_2 molecules respectively.

(*b*) In Li_2 molecule, the outer two electrons present in $\sigma_{2s}{}^{b}$ MO are shielded by the inner four electrons present in two 1*s* orbitals on two Li atoms. Therefore, the attraction between the nuclei and the two electrons in $\sigma_{2s}{}^{b}$ MO is less.

(*vi*) The configurations of given species are given below :

H $(e = 1) = 1s^1$, H_2 $(e = 1 \times 2 = 2) = (\sigma_{1s}{}^{b})^2$

O $(e = 6) = 2s^2 2p^4$, O_2 $(e = 2 \times 6 = 12) = (\sigma_{2s}{}^{b})^2\ (\sigma_{2s}{}^{a})^2\ (\pi_y{}^{b})^2\ = (\pi_z{}^{b})^2\ (\sigma_x{}^{b})^2\ (\pi_y{}^{a})^1 = (\pi_z{}^{a})^1$

In H-atom electron is to be removed from 1*s* AO while in H_2 molecule electron is to be removed from $\sigma_{1s}{}^{b}$ MO. Now since $\sigma_{1s}{}^{b}$ MO has lower energy (more stable) than 1*s* AO, H_2 molecule has higher IE than H-atom.

In O-atom electron has to be removed from 2*p* AO while in O_2 molecule electron has to be removed from $\pi_y{}^{a}$ or $\pi_z{}^{a}$ MO. Now since 2*p* AO of O-atom has lower energy (more stable) than $\pi_y{}^{a}$ or $\pi_z{}^{a}$ MO in O_2 molecule, O-atom has higher IE than O_2 molecule.

University Questions

1. Mention any two exceptions of VSEPR theory. (*Punjab B.Sc. 1986*)
2. Distinguish two aspects of the following : (*i*) Atomic and molecular orbital (*ii*) Bonding and anti bonding MO. (*iii*) Sigma and pi MO. (*Himachal Pradesh 2000*)
3. In which of the following species sulphur does not undergo sp^3 hybridisation ? $SO_4{}^{2-}$, H_2S, SF_4 and SF_2. (*Himachal Pradesh 2000*)
4. Discuss the structures of $ICl_2{}^-$, $SO_4{}^{2-}$ and Cl_2O. (*Himachal Pradesh 2000*)
5. What is hybridisation? On the basis of hybridisation of orbitals deduce the shape of PCl_5, SF_6 and NH_3 molecules. (*Kanpur 2000*)

6. Give an account of *sp*, sp^2 and sp^3 hybridisation. (*Kumaon 2000*)

7. Give the structure of the following with the help of orbital diagram : PCl_5, ICl_4^-, $XeOF_4$ and $SOCl_2$ (*Lucknow 2000*)

8. (*a*) What do you understand by hybridisation ? Describe its important features.

(*b*) Deduce the shapes of following molecules and predict the bond angle in each case. (*i*) BF_3 (*ii*) NH_3 (*iii*) H_2O. (*Avadh 2000*)

9. Draw and explain the structures of BF_3, PCl_5, SF_6 and XeF_6 on the basis of VSEPR theory. (*Bharathiar 2000*)

10. (*i*) Explain why hydrogen forms diatomic molecule while helium remains monoatomic.

(*ii*) Give the salient features of VSEPR theory and discuss the shape of ICl_3, XeF_4, PCl_5 and $TeCl_4$ molecules with the help of this theory.

(*iii*) Use MO theory to predict the bond order and the number of unpaired electrons in O_2^{2-}, O_2, O_2^+, NO and CO. (*Lucknow 2001*)

11. (*a*) What is the basic difference in valence bond theory and molecular orbital theory? Draw MO. diagram and discuss stability, electronic configuration and characteristics of the following :

Be_2, He_2^+, N_2 and O_2.

(*b*) Explain the term hybridisation. Discuss the structure, shape and bond angles of the following molecules: BF_3, SiF_4, NH_3, H_2O and PCl_5. (*Lucknow 2001*)

12. (*a*) What is hybridisation ? On the basis of hybridisation of orbitals, leduce the shape of the following molecuels / ions : H_3O^+, BF_3, ClF_3 and SF_4.

(*b*) On the basis MOT calculate B.O. of : He_2^+, O_2^-, N_2^+, F_2^-, B_2. (*Purvanchal 2007*)

13. Which of the following compounds shows sp^2 hybridisation ? $BeCl_2$, BF_3, PCl_5 and FF_7. (*Agra 2008*)

14. (*a*) In sp^2 hybridisation, the value of bond angle is : 120°, 109°28', 180° or 108°.

(*b*) The state of hybridisation of S-atom of SO_2 molecule is : sp^2, *sp*, dsp^2 or sp^3. (*H.N. Bhauguna 2008*)

15. (*a*) There is sp^3 hybridisation in methane, ammonia and water but their bond angles are different Explain.

(*b*) O_2 is paramagnetic but N_2 is diamagnetic. Explain. (*Meerut 2008*)

16. Discuss MOT of chemical bonding. Describe the MOC of O_2, N_2 and CD molecules. (*Meerut 2009*)

CHAPTER 9

Structure of Solids

Crystalline and Amorphous Solids : Classification of Solids on the Basis of Regular or Haphazard Arrangement of the Building Constituents.

1. Crystalline solids. In crystalline solids the building constituents (atoms, ions or molecules) arrange themselves in a *three dimensional recurring regular geometric pattern* which extends to very large distance compared to the inter-particle distances. Thus the crystalline solids have *long-range order*.

Crystalline solids exist as crystals and hence these are also called simply **crystals.** These crystals may be big or small. Some crystals are so tiny that their crystalline shape can be seen with a microscope only. A crystalline solid has a sharp melting point.

2. Amorphous solids. The word amorphous is a Greek word which means *shapeless*. Thus in these solids the arrangement of different building constituents is *not regular* but *haphazard i.e.* the constituents are arranged at random in the same disorderly way as in liquids. It is for this reason that amorphous substances like glass, pitch and resins are regarded as *super-cooled liquids* or *as intermediate between solids and liquids.* An amorphous substance does not have a sharp melting point.

Differences between Crystalline and Amorphous Solids

These two classes of solids differ from one another in the following properties.

1. Geometrical shape. The crystals of every crystalline solid have a *define geometrical shape* due to definite and orderly arrangement of particles in three-dimensional shape, *e.g.* crystals of NaCl have a *cubical shape*, those of *calcite* ($CaCO_3$) have *rhombohedral shape* while those of *copper* metal have an *octahedral shape*. (Fig. 9.1).

For a given crystal the angles at which the surfaces or faces intersect are always the same.

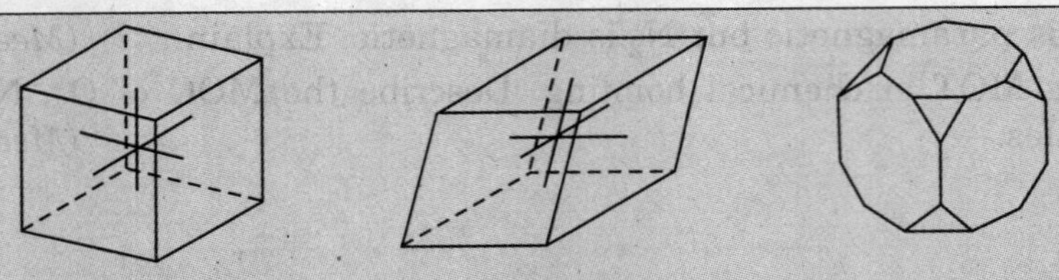

Cubical shape of NaCl | Rhombohedral shape of calcite ($CaCO_3$) | Octahedral shape of copper metal

Fig. 9.1. Shapes of some crystalline solids.

Most of the crystalline solids are *poly crystals. i.e.* they contain aggegate of many inter-locking small crystals. Such solids have irregular shape but their internal structure is regular. *Mono crystals* (*i.e.* single crystals) are sometimes found in nature. They can also be prepared by artificial means. Sugar is an example of a crystal composed of many single crystals.

An amorphous solid does not have any orderly pattern of arrangement of particles and, therefore, does not have any definite geometrical shape.

2. Melting point. There are many crystalline solids which do not change directly to the liquid state and also there are many crystalline solids which decompose before going into the liquid state. The crystalline solids which directly change into liquid state do so at a definite temperature *i.e.* the melting point of such crystalline solids is *definite*.

Amorphous substances like glass do not have definite melting point.

3. Cleavage planes. When a crystal of a crystalline solid is hammered, it readily breaks up into smaller crystals along particular planes which are called *cleavage planes*. These planes are inclined to one another at a particular angle for a given crystalline solid. Thus the magnitude of this angle varies from substance to substance.

Amorphous substances do not have such well-defined cleavage planes.

4. Anisotropic and isotropic properties. Magnitude of some of the physical properties of crystalline solids like refractive index, coefficient of thermal expansion, electrical and thermal conductivities etc. is different in different directions within the crystal, *e.g.* the coefficient of thermal expansion of a crystal of AgI is positive in one direction and negative in the other direction. Such properties are called *anisotropic properties* and the phenomenon is referred to as *anisotropy*. These properties are due to the fact that the orderly arrangement of particles in crystalline solids is different in different directions.

The above said properties of isomorphous substances are the same in all directions as those of liquids and gases, *i.e. amorphous substances have isotropic properties*. This is because of the fact that in isomorphous solids, as in liquids and gases, the arrangement of particles is random and, therefore their isotropic properties are the same in all the directions.

5. Symmetry. Crystalline solids have crystal symmetry, *i.e.* when a crystalline solid is rotated about an axis, its appearance does not change (*i.e.* remains the same).

Amorphous substances do not have symmetry.

Various Types of Symmetry Found in Crystals

Symmetry is a very important property of crystals. Only three of these are described here. (1) *Centre of symmetry* (2) *Axis of symmetry,* and (3) *Plane of symmetry.*

1. Centre of symmetry. *It is such an imaginary point within the crystal that any straight line drawn through it will intersect the faces, edges or solid angles of the crystal at equal distances on opposite sides.* A crystal may have either one or no centre of symmetry. It can never have more than one centre of symmetry. A cubical crystal like NaCl crystal has one centre of symmetry as shown in Fig. 9.2.

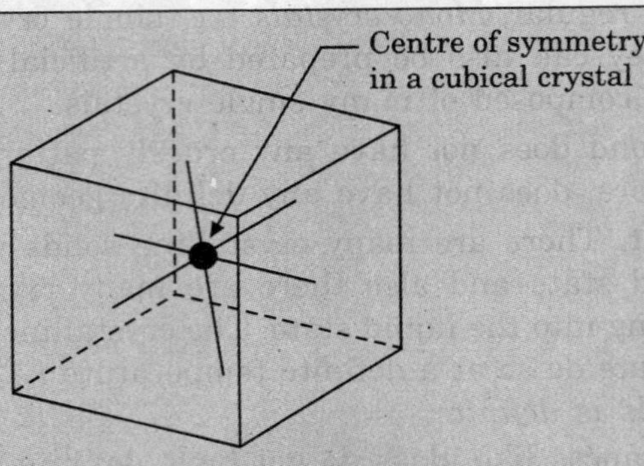

Fig. 9.2. Centre of symmetry in a cubical crystal.

2. Axis of symmetry. *It is an imaginary straight line about which, if the crystal is rotated, it will present the same appearance more than once during its complete rotation.*

In all there are thirteen axes of symmetry possessed by a cubical crystal like NaCl as shown below :

(*a*) *Six axes of two-fold symmetry.* Each of these six axes is called *diad axis.* When the crystal is rotated about any of these axes, it presents the same appearance two times in a complete rotation. One of such axes is shown in Fig. 9.3(*a*). Each of these three axes intersects at the middle points of the two opposite edges of the cube.

(*b*) *Four axes of three-fold symmetry.* Each of these axes is called *triad axis.* When the crystal is rotated about any of these axes, it presents the same appearance *three times* in a complete rotation. One of such axes is shown in Fig. 9.3(*b*). Each of these four axes intersects at the two opposite solid angles (*i.e.* corners) of the cube.

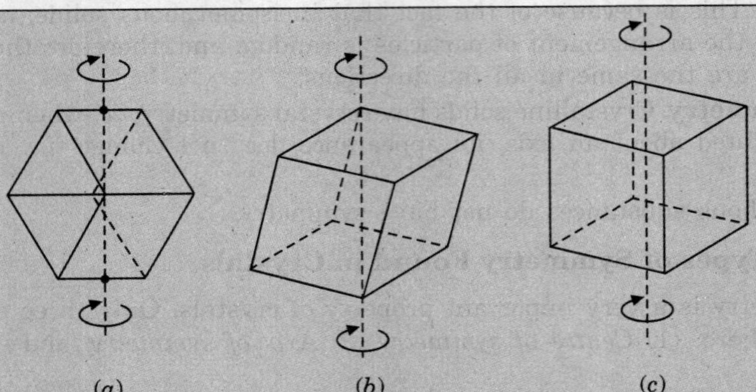

(*a*) Axis of two-fold symmetry (Six) (*b*) Axis of three-fold symmetry (Four) (*c*) Axis of four-fold symmetry (Three)

Fig. 9.3. Thirteen axes of symmetry in a cubical crystal.

(*c*) *Three axes of four-fold symmetry.* Each of these three axes is called *tetrad axis*. When the crystal is rotated about any of these axes, it presents the same appearance *four times* in a complete rotation. One of such axes is shown in Fig. 9.3(*c*). Each of these three axes intersects at the middle points of the two opposite faces of a cube. All the three axes are at right angles to one another.

3. Plane of symmetry. *It is an imaginary plane which divides a crystal into two such parts that one is the exact mirror image of the other.* In other words, a plane of symmetry divides the crystal into two identical and similarly placed halves.

In all, there are nine planes of symmetry possessed by a cubical crystal like NaCl as shown below :

(*a*) *Three rectangular planes of symmetry*. One such rectangular plane of symmetry is shown in Fig. 9.4(*a*). There are in all three such rectangular planes of symmetry which are at right angles to each other.

(*b*) *Six diagonal planes of symmetry*. One such diagonal plane of symmetry is shown in Fig. 9.4 (*b*). There are in all *six* such diagonal planes of symmetry which pass diagonally through the cube.

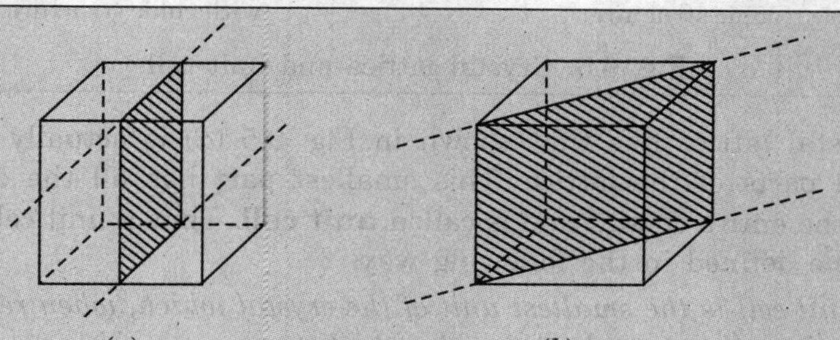

(*a*) Rectangular planes of symmetry (Three)

(*b*) Diagonal planes of symmetry (Six)

Fig. 9.4. Nine planes of symmetry for a cubical crystal like NaCl.

Crystal Lattice and Unit Cell

The internal structure of crystals (*i.e.* the arrangement of the particles *viz.* atoms, molecules or ions of which the crystal is composed) has been determined by *X-ray diffraction, electron diffraction, neutron diffraction* etc. methods. These methods have shown that the constituent particles are situated at strictly definite positions in space. In crystals, the constituent particles are represented by points. These points are called **lattice points** or **lattice sites** and the arrangement of points in the crystal is called **crystal lattice** or simply **crystal**. Crystal lattice is also called by other names like **space lattice** or **lattice array.** Thus a crystal lattice can be defined as follows :

A crystal lattice is an array of points which show how the constituent particles (atoms, ions or molecules) of a crystal are arranged at different sites in three-dimensional space.

In other words *a crystal lattice is an infinitely extended regular arrangement of different constituent particles of a crystal.* Each point in a crystal lattice has the same environment as any other similar point. A crystal lattice with cubic structure with three-dimensional net work is shown in Fig. 9.5 (*a*).

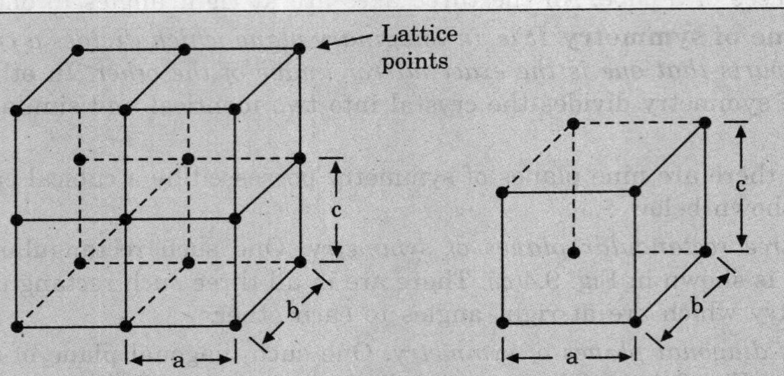

(*a*)
A crystal lattice with cubic structure

(*b*)
Unit cell of crystal lattice with cubic structure

Fig. 9.5. Crystal lattice and unit cell.

The crystal lattice like that shown in Fig. 9.5 (*a*) is actually composed of many small parts of the lattice. This smallest part has all the characteristic features of the entire crystal and is called **unit cell.** Thus a unit cell of a crystal lattice can be defined in the following ways :

(*i*) *A unit cell is the smallest unit of the crystal which, when repeated again and again, gives the crystal of the given substance.*

(*ii*) *A unit cell is the smallest sample that represents the picture or definite pattern of the entire crystal.*

(*iii*) *A unit cell of a crystal lattice is the smallest block or geometrical figure from which the entire crystal can be built up by its translational repetition in three dimensions.*

Thus the entire crystal consists of a large number of unit cells adjacent to one another in all three dimensions. The unit cell of the crystal lattice shown in Fig. 9.5 (*a*) is shown in the same figure at (*b*).

A unit cell of a crystal possesses all the structural properties of the given crystal. A large number of unit cells of the crystal combine together to arrange themselves in a regular order and thus form the crystal lattice.

It may be noted that the properties of crystalline solids depend not only on their composition and external conditions, but also on their internal structure. For example, although diamond and graphite are composed of the same substance *viz.*, carbon, they have different properties. The difference in their properties is due to the difference in their internal structure.

Seven Crystal Systems

Depending on the nature of the constituents (which may be atoms, molecules or ions) of which a given crystal is composed, we have different types of crystal. The shape of a given crystal can be described by the lengths of the three sides or edges (*a, b* and *c*) of its unit cell and the three angles (α, β and γ) between the three axes of the unit cell. The lengths of the sides of the unit cell are called *primitives* or *crystal axes* and the angles between the three axes are called *interfacial angles* (See Fig. 9.6).

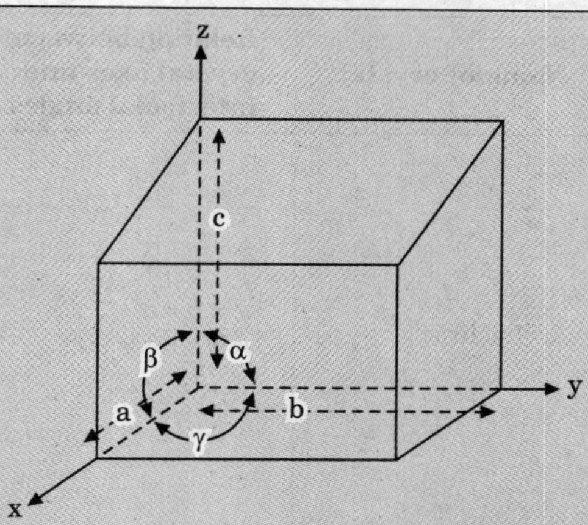

Fig. 9.6. A unit cell with the lengths of its three sides *a, b* and *c* and three interfacial angles α, β and γ

There are seven crystal systems corresponding to the seven distinct types of unit cells (sub-units). All these unit cells are parallelopipeds and their shapes are determined by the lengths of three crystal axes, *a, b* and *c* and the magnitude of the three interfacial angles α, β and γ. The seven crystal systems are given in Table 9.1.

Table 9.1. Seven systems of crystals

Name of crystal	Relation between crystal axes and interfacial angles	Shape of unit cell of the system and examples of crystals
1. Cubic	$a = b = c$ $\alpha = \beta = \gamma = 90°$	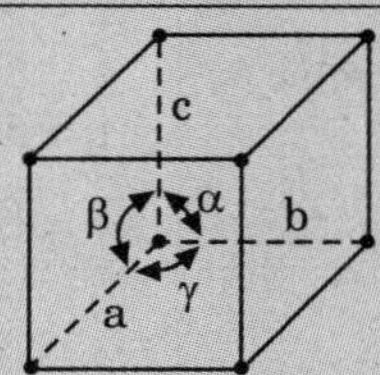 Examples : NaCl, KCl, CaF_2, $NaClO_2$ alums, diamond
2. Monoclinic	$a \neq b \neq c$ $\alpha = \beta = 90°; \gamma \neq 90°$	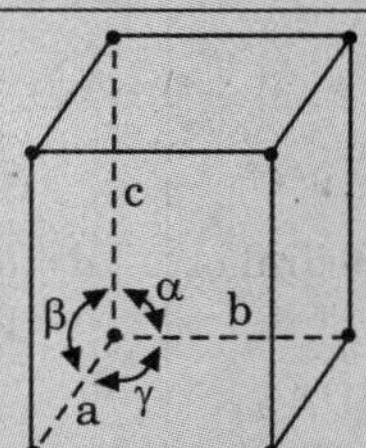 Examples : $Na_2SO_4.10H_2O$, $NaHCO_3$, $FeSO_4$, $7H_2O$, monoclinic sulphur (SM)

Name of crystal	Relation between crystal axes and interfacial angles	Shape of unit cell of the system and examples of crystals
3. Triclinic	$a \neq b \neq c$ $\alpha \neq \beta \neq \gamma \neq 90°$	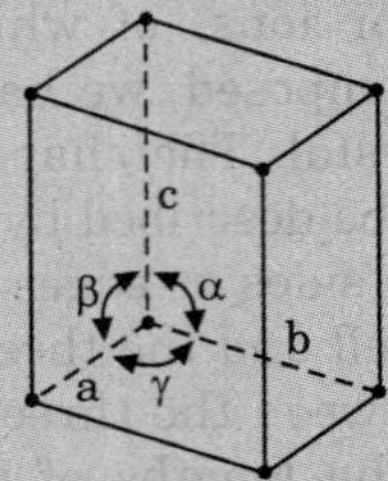 Examples : $CuSO_4.5H_2O$, $K_2Cr_2O_7$, H_3BO_3
4. Tetragonal	$a = b \neq c$ $\alpha = \beta \neq \gamma = 90°$	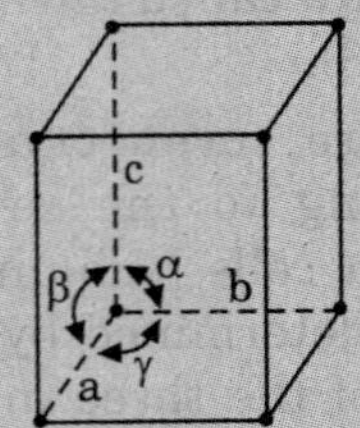 Examples : $NiSO_4$, white tin, $SnCl_2$, TiO_2 etc.
5. Orthorhombic or Rhombic	$a = b \neq c$ $\alpha = \beta = \gamma = 90°$	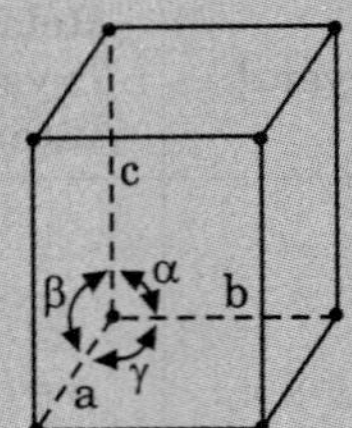 Examples : KNO_3, $BaSO_4$, $MgSO_4.7H_2O$, Rhombic sulphur (SR)
6. Rhombohedral or Trigonal	$a = b = c$ $\alpha = \beta = \gamma \neq 90°$	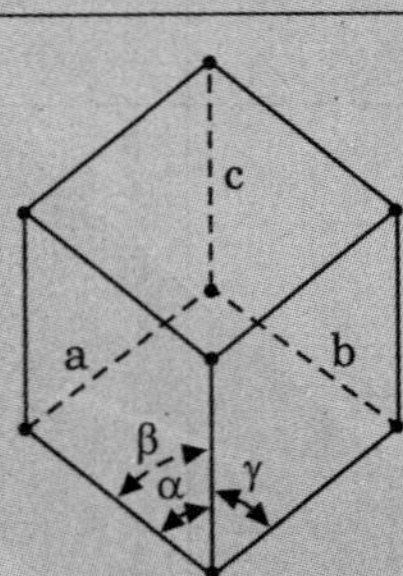 Examples : $CaSO_4$, $NaNO_3$, $KMnO_4$

Name of crystal	Relation between crystal axes and interfacial angles	Shape of unit cell of the system and examples of crystals
7. Hexagonal	$a = b \neq c$ $\alpha = \beta = 90°; \gamma = 120°$	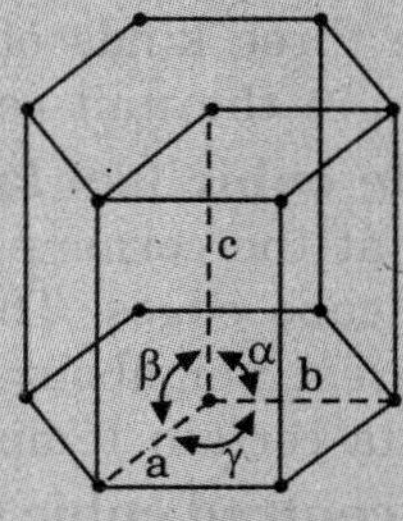 Examples : Graphite, SiO_2, AgI, ice, PbI_2, ZnO

From the above table the following points may be noted :

(*i*) For cubic and trigonal systems, the three edges are of equal lengths while for the remaining five systems the edges are not equal.

(*ii*) All the three interfacial angles are of 90° for cubic, tetragonal and orthorhombic systems while these angles for other systems are not of 90°.

Fourteen Bravais Lattices

It has been shown methametically by Bravais (1848) that there are 14 independent ways of arranging the similar lattice points in a three-dimensional space, *i.e.* there are 14 space lattices in the seven systems of crystals mentioned above. These are : *three cubic, two monoclinic, one triclinic, two tetragonal, four orthorhombic, one rhombohedral* (*trigonal*) and *one hexagonal* as shown in Table 9.2.

Table 9.2. Fourteen Bravais lattices in seven systems of crystals

Crystal system	No. of Bravais lattices	Bravais lattices
1. Cubic	3	(*i*) Simple cubic (sc) lattice (*ii*) Body-centred cubic (bcc) lattice (*iii*) Face-centred cubis (fcc) lattice
2. Monoclinic	2	(*i*) Simple monoclinic lattice (*ii*) End-centred monoclinic lattice
3. Triclinic	1	Simple triclinic lattice
4. Tetragonal	2	(*i*) Simple tetragonal lattice (*ii*) Body-centred tetragonal lattice
5. Orthorhombic or Rhombic	4	(*i*) Simple orthorhombic lattice (*ii*) Body-centred orthorhombic lattice (*iii*) Face-centred orthorhombic lattice (*iv*) End-centred orthorhombic lattice
6. Rhombohedral or Trigonal	1	Simple rhombohedral lattice
7. Hexagonal	1	Simple hexagonal lattice

Bravais Lattices of Cubic System

Bravais lattices of the cubic system which is the simplest and also most common are considered in detail as follows. The crystals belonging to cubic system have three kinds of Bravais lattices depending on the positions of the lattice points in the unit cell of cubic system.

1. Simple cubic (sc) lattice. The unit cell of this lattice has atoms (lattice points) only at the corners of the cube [Fig. 9.7(*a*)]. Consequently, these atoms touch along cube edges. This structure is *loosely-packed*, since each atom has only six nearest neighbours. Only one element namely polonium exhibits this structure in a certain temperature region.

2. Body-centred cubic (bcc) lattice. The unit cell of this lattice has one atom at each of the eight corners and one atom at the centre of the body of the cube. [Fig. 9.7(*b*)]. The atom at the centre of the cube belongs entirely to the unit cell. The atoms are in contact along body diagonals. The elements which passess *bcc* lattice are listed below and the value of primitive, *a* (in A°) is also given in parentheses : Ba (= 5.025), α-Fe (= 2.867), Rb (= 5.630), Na (= 4.291), Ti (= 3.306), W (= 3.115), U (= 3.474), Zr (= 3.620).

3. Face-centred cubic (fcc) lattice. The unit cell of this lattice has atoms at the six corners as well as at the centre of each of the six faces of the cube [Fig. 9.7(*c*)]. This structure is close-packed because each atom has 12 nearest neighbours. The atoms are in contact with the diagonal atoms. The elements which crystallise into *fcc* lattice are given below and the value of primitive a (in A°) is also given in parentheses : Al (= 4.049), Cu (= 3.615), Au (= 4.090), γ-Fe (= 3.591), Ni (= 3.524), Pt (= 4.086), Ag (= 4.086).

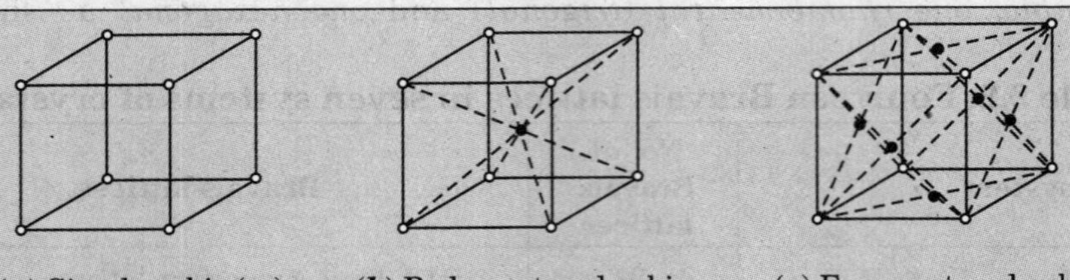

(***a***) Simple cubic (sc) lattice (***b***) Body-centered cubic (bcc) lattice (***c***) Face-centered cubic (fcc) lattice

Fig. 9.7. Three Bravais lattices of cubic system

Classification of Crystalline Solids or Crystals : Types of Crystals

There are three ways to classify the crystals. But in chemistry the crystals are classified on the basis of the nature of the building particles occupying the lattice points (or lattice sites) in the crystal lattice and the nature of the bonds existing between the building particles. This basis of classification gives the following four types of crystals :

1. Ionic crystals. Ionic crystals are those crystals in which the particles forming the crystal are positively and negatively charged ions (*i.e.* cations and anions) which are held together by strong electrostatic force of attraction (*i.e.* by ionic bonds). Hence the name ionic crystals. Crystals of NaCl, $CuSO_4 . 5H_2O$ etc belong to this class.

2. Metallic crystals. Metallic crystals are those in which the particles forming the crystals are metallic positive ions (called *positive cores* or *Kernels*) which are surrounded by a *sea of electrons* (also called *electron-gas* or *electron-cloud*) and are held together by metallic bond (Fig. 9.8). The positive metal ions are obtained when the metal atoms lose their valence-shell electrons. The electrons obtained form a *sea of* electrons. These electrons are not bond up with any particular metal ion and hence are free to move throughout the metallic crystal. Due to their mobile nature these electrons are called *mobile* or *delocalised electrons*. It is these mobile electrons which keep the metal positive ions held together.

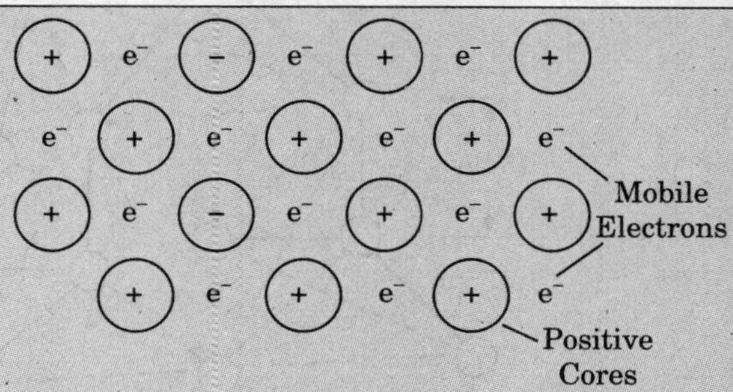

Fig. 9.8. Positive cores surrounded by mobile electrons

3. Atomic or covalent crystals. Atomic crystals are those in which the particles forming the crystal are neutral atoms of the same element (as in diamond) or of different elements (as in SiC) which are held together by covalent bonds. Hence atomic crystals are also called *covalent crystals*. Atomic crystals are of two types :

(*i*) Those in which atoms are bonded with each other by covalent bonds resulting in the formation of *giant molecules*. Examples of giant molecules are *diamond, silicon carbide* (SiC). *aluminium nitride* (AlN) etc.

(*ii*) Those which consist of *separate layers*. Examples of covalent crystals containing separate layers are *graphite*, CdI_2, $CdCl_2$, *BN* etc.

Structure of diamond crystal. The unit cell of the diamond crystal is a *regular tetrahedron* having one C-atom at each of its four corners and one C-atom lying at its centre [Fig. 9.9 (*a*)]. In this unit cell, central carbon atom is linked to the four corner C-atoms by covalent bonds.

The structure of diamond is obtained by uniting a number of such unit cells in such a way that each of the four corner C-atoms of each unit cell is linked with the central C-atom of its own unit cell and with the other three corner C-atoms belonging to three different unit cells [Fig. 9.9 (*b*)]. Thus we find that in the structure of diamond each of the C-atoms is linked with four carbon atoms. The bonds between carbon atoms are covalent bonds which run through the crystal in three dimensions. Each of the corbons in the diamond crystal is sp^3 hybridised and hence each of the four covalent bonds by which each carbon is linked with the three carbons is $sp^3(C)$—sp^3 (C) σ bond. The tetrahedral arrangement of four C-atoms round the central C-atom gives rise to rigid three *dimensonal network* which runs throughout the crystal. The whole lattice is, therefore, continuous and because of the continuity of C—C covalent bonding, the entire diamond crystal behaves as a single huge or giant three dimensional carbon molecule which is called **macro molecule.**

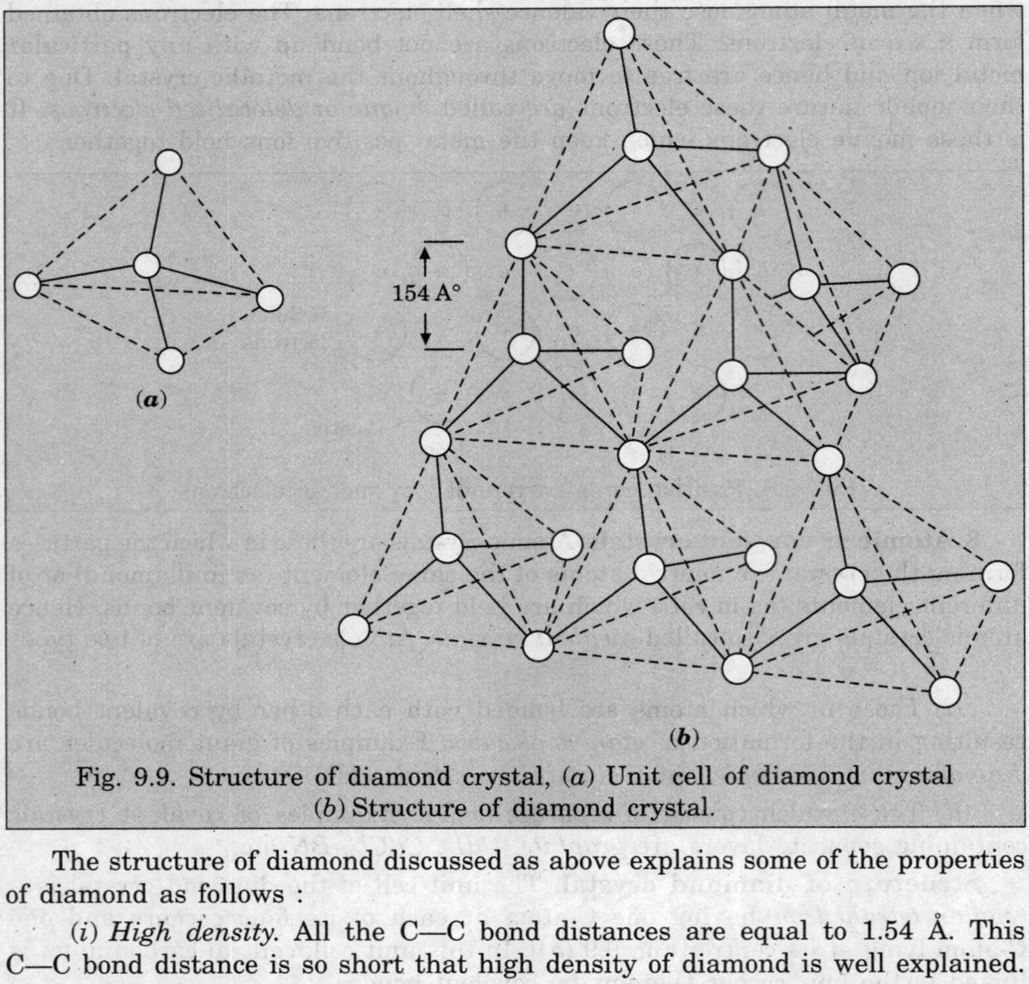

Fig. 9.9. Structure of diamond crystal, (*a*) Unit cell of diamond crystal (*b*) Structure of diamond crystal.

The structure of diamond discussed as above explains some of the properties of diamond as follows :

(*i*) *High density.* All the C—C bond distances are equal to 1.54 Å. This C—C bond distance is so short that high density of diamond is well explained.

(*ii*) *Non-conductor of electricity*. We have seen that, since all the four electrons present in the valence-shell of carbon atom are used up in forming four C—C σ bonds, no mobile electrons are left in the diamond crystal to allow the conduction of electricity. Thus diamond crystal is a non-conductor of electricity.

(*iii*) *Extremely hard, high melting point and high boiling point.* We have seen that, since the structure of diamond crystal consists of a large number of C—C covalent bonds, the force acting between carbon atoms is very strong and hence it is very difficult, to break these bonds. This explains the extreme hardness of diamond crystal. Also at the same time, in order to break this large number of C—C bonds, a large amount of energy is needed. This means that diamond crystal has high melting and boiling point.

Structure of graphite. Graphite has a layer structure, *i.e.* it has a number of flat parallel layers (or planes or sheets) of carbon atoms. In each layer carbon atoms are arranged in a regular flat hexagon as in hexane, napthalene etc. C—C distance in each layer is 1.42Å which is intermediate between the single C—C bond distance (= 1.54Å) and double C=C bond distance (= 1.33Å). Note that the C=C distance in benzene compounds is 1.39Å. Each of the carbon atoms of a given layer is sp^2 hybridised and is thus bonded to the adjacent three carbon atoms of the same layer by three sp^2(C) —sp^2(C) σ-bonds. Each of these three C—C σ-bonds has *two-third single bond character* and *one-third double bond character* which is due to the resonance existing between single and double bonds as shown below :

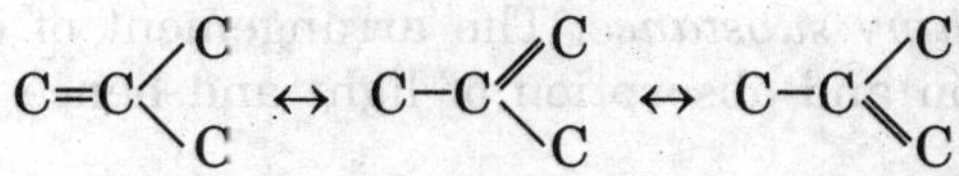

Since there are four valence electrons in each C-atom, after forming three C—C σ-bonds, each C-atom is left with one spare electron in its $2p_z$ orbital. These $2p_z$ orbitals which are single-filled overlap together to form a delocalised π-system which extends above and below each layer. This π-system gives an aromatic quinone type (*i.e.* hexagonal) structure to graphite as shown in Fig. 9.10.

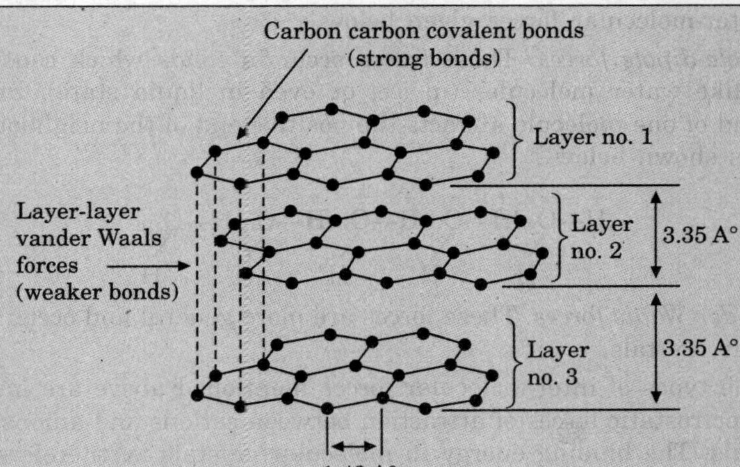

Fig. 9.10. Layer structure of graphite. Black circles indicate carbon atoms.

This hexagonal structure is supported by the fact that when graphite is completely oxidised, it gives a benzene derivative *viz. mellitic acid*, $C_6(COOH)_6$ which is benzene hexacarboxylic acid. The structure of this acid is shown in the margin. The layer structure of graphite as given above explains some of the properties of graphite as follows :

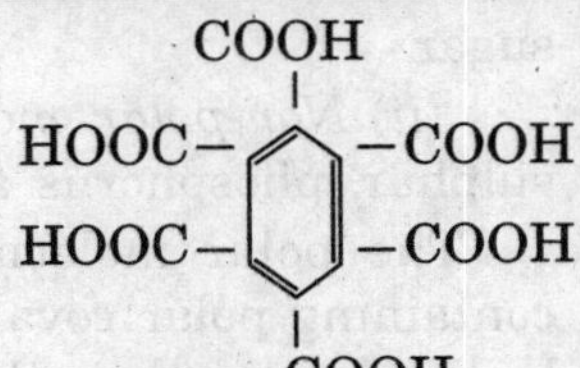

(*i*) *Good conductor of electricity.* In each delocalised π-system the electrons are free to move within the layer and this accounts for the fact that graphite conducts electricity in the plane of the layer but not in the direction perpendicular to its laminal crystals. *Thus graphite is a good conductor of electricity.*

(*ii*) *Soft, flaky and slippery substance.* Every two adjacent layers in graphite are at a distance of 3.35Å. This distance is so large that the possibility of the formation of covalent bonds between the atoms of the adjacent layers is ruled out. These layers are, therefore, loosely linked together by weak *Van der Waals forces*. These van der Waals forces which hold the graphite crystal layers together are so weak that even a slight pressure causes the layers to slide over one another. This explains why (*a*) graphite is a *soft substance* and hence is used as a *solid lubricating agent* (*b*) graphite is *flaky* substance and *feels slippery* between the figures.

(*iii*) *Black and shiny substance*. The arrangement of carbon atoms in each layer causes reflection and absorption of light and hence graphite is *black* and *shiny*.

(*iv*) *Less dense than diamond.* The comparatively more open structure of graphite makes it less dense than diamond (sp. gra. of diamond = 3.52, sp. gra. of graphite = 2.25) but more reactive chemically.

4. Molecular crystals. Molecular crystals are those in which the particles forming the crystal are polar or non-polar whole molecules, except in solidified noble gases in which the particles are *atoms*, which are held together by two types of inter-molecular forces given below :

(*a*) *Dipole-dipole forces*. These forces occur in solids which consist of polar molecules like water molecules (in ice, or even in liquid state). In water the negative end of one molecule attracts the positive end of the neighbouring water molecule as shown below :

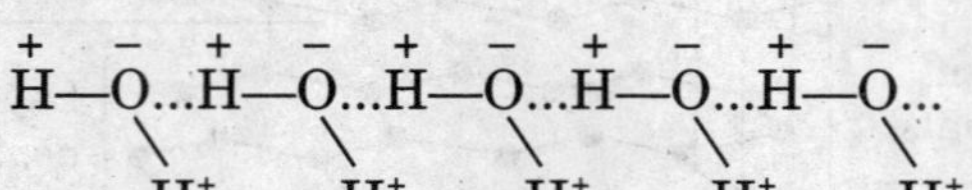

(*b*) *van der Waals forces*. These forces are more general and occur in all kinds of molecular crystals.

Both the types of inter-molecular forces mentioned above are much weaker than the electrostatic forces of attraction between cations and anions existing in ionic crystals. The binding energy in molecular crystals is, therefore, less than that in ionic crystals. The molecular crystals are of two types :

(*i*) *Polar molecular crystals* [*See Fig. 9.11*(*a*)]. Examples are water (ice) and sugar.

(*ii*) *Non-polar molecular crystals* [*See Fig. 9.11*(*b*)]. Examples are iodine, sulphur, phosphorus and solid carbon dioxide.

The polar molecular crystals are formed from unsymmetrical molecules containing polar covalent linkages and have comparatively higher melting and boiling points than the non-polar molecular crystals. This is so because in the former there are stronger dipole-dipole forces.

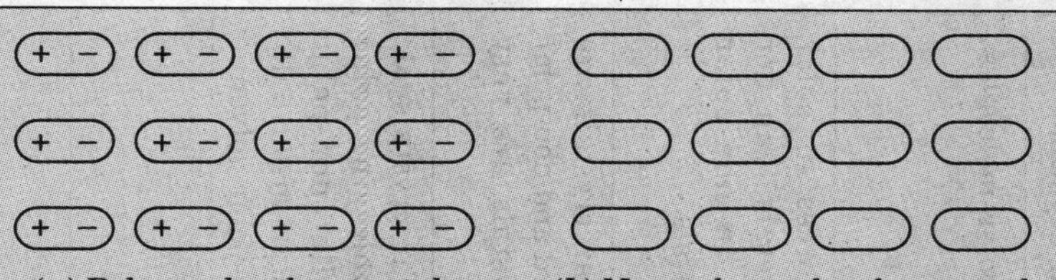

Fig. 9.11. Polar and non-polar molecular crystals.

Comparison Between Different Types of Crystals

The comparison of properties of different types of crystals is given in Table 9.3.

Polymorphism, Allotropy Transition Temperature, Enantiotropy and Monotropy

The phenomenon by which a substance (compound or element) exists in two or more different crystalline forms is called **polymorphism** and the different crystalline forms are said to be **polymorphous** with each other. For example As_2O_3 exists in two polymorphic forms *viz. rhombic and octahedral*. These two forms are said to be polymorphous with each other. Different crystalline forms of the same element are called its **allotropes** or **allotropic forms** and the phenomenon is called **allotropy.** For example different crystalline forms of sulphur are *rhombic sulphur* or *alpha sulphur* (represented as S_R or S_α) and *monocilic sulphur* or *beta sulphur* (represented as S_M or S_β).

In many cases one of the allotropic forms of a given element is stable at all the temperatures below a certain temperature which is called **transition temperature** while the other form is stable above this temperature. This phenomenon is called **enantiotropy** and the allotropic forms showing this phenomenon are called **enantiotropes** or **enantiotropic substances.** For example rhombic form of sulphur is stable at all temperatures below 96°C and above this temperature it is monoclinic form which is stable. At 96°C which is transition temperature in this ease both the forms can co-exist. The transformation of one form to the other takes place at this transition temperature and can occur in either direction. This change can be shown as :

$$S_R \overset{96°C}{\rightleftarrows} S_M$$

It may be noted that the transition temperature of the enantiotropes of an element is always less than the melting paint of that element. There are certain solid substances (*e.g.* bonzophenone and iodine chloride) whose one solid crystalline form is stable below the melting point of that substance and the other form has a metastable existence. This phenomenon is called **monotropy.** The transition temperature of the substances showing monotropy is always above their melting point.

Table 9.3. Comparison of the properties of different types of crystals

Property	Different types of crystals			
	Ionic crystals	**Metallic crystals**	**Atomic or covalent crystals**	**Molecular crystals**
(*i*) **Constituent particles forming the crystal.**	Positive and negative ions *i.e.* etions and anions.	Metallic positive ions surrounded by sea of electrons.	Neutral atoms	Polar or non-polar molecules
(*ii*) **Forces holding the constituent particles.**	Electrostatic force of attraction acting between the cations and anions.	Electrostatic force of attraction acting between metallic positive ions and sea of electrons.	Covalent bonds (Shared pairs of electrons).	Weak polar forces in polar molecular crystals, van der Waals forces in non-polar molecular crystals.
(*iii*) **Hardness**	Hard and brittle.	Usually hard.	Usually very hard.	Polar molecular crystals are moderately hard and non-polar molecular crystals are very soft.
(*iv*) **Solubility**	Soluble in polar solvents.	Dissolve in solvents by chemical action.	Insoluble in most solvents.	Polar molecular crystals ale usually soluble in polar solvents while non-polar molecular crystals are usually soluble in non-polar solvents.
(*v*) **Melting and boiling points.**	High	High	Very high	Low
(*vi*) **Electrical conductivity.**	High in fused state.	High	Usually low	Extremely low.
(*vii*) **Examples**	NaCl	Cu metal	Diamond, graphite	Ice (polar molecular crystal), solid carbon dioxide (non-polar molecular crystal).

Allotropic Forms of Sulphur

Sulphur exists in two crystalline forms : (*i*) *Rhombic or octohedral or alpha sulphur* (*ii*) *Monoclinic or prismatic or beta sulphur.* Both the forms are found to exist in *octaatomic crystals* in the form of a *puckered ring*, *i.e.* in both the forms sulphur exists in the form of octagonal ring molecules (S_8 molecules) in which the eight sulphur atoms are joined by covalent bonds (Fig. 9.12).

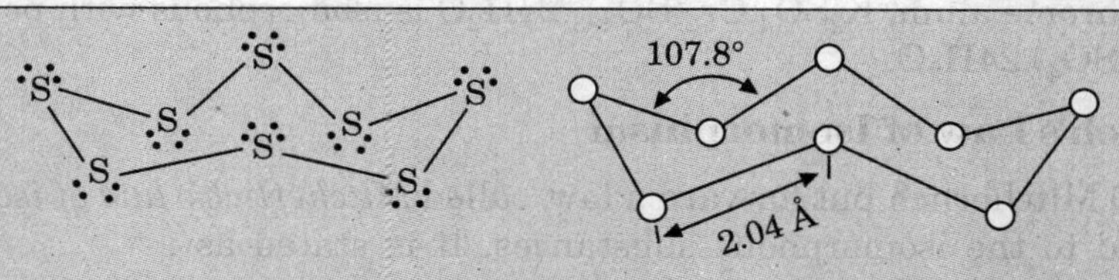

Fig. 9.12. Structure of sulphur S_8 molecule.

Such rings give molecular crystal in which the rings are held together by relatively weak polar forces. Consequently both the forms have low melting points.

Rhombic sulphur has *octahedral shape* while monoclinic sulphur has long *needle like shape* (Fig. 9.13).

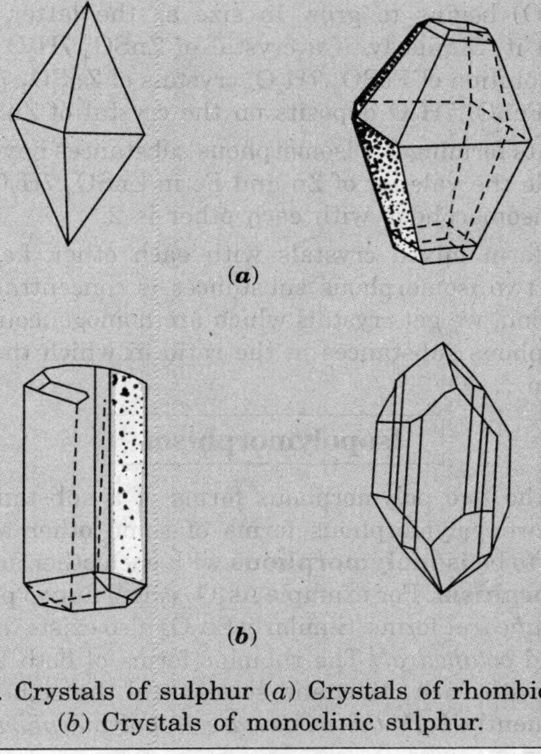

Fig. 9.13. Crystals of sulphur (*a*) Crystals of rhombic sulphur (*b*) Crystals of monoclinic sulphur.

Isomorphism

When two or more crystalline substances have the same number of atoms and the same shape (*i.e.* the atoms in them are united in a similar manner), the phenomenon is called **isomorphism** and such substances are said to be **isomorphous** substances with each other. (*Isomorphous* = having the same shape). For example $ZnSO_4$. $7H_2O$ and $FeSO_4$. $7H_2O$ are isomorphous with each other. Similarly chrome alum, $K_2SO_4.Cr_2(SO_4)_3.24H_2O$ is isomorphous with potash alum, $K_2SO_4.Al_2(SO_4).24H_2O$.

Mitcherlich's Law of Isomorphism

In 1820 Mitcherlich put forward a law, called *Mitcherlich's law of isomorphism* with regard to the isomorphous substances. It is stated as :

Isomorphous substances have an equal number of atoms which are united in a similar way, i.e. they have similar chemical formulae or constitution.

Properties of Isomorphous Substances

The chief characteristics of isomorphous substances are :

(*i*) The crystals of these substances have the same shape.

(*ii*) The crystals of one substance can be grown in a saturated solution of the other isomorphous substance. For example, if a crystal of $FeSO_4.7H_2O$ is suspended in a saturated solution of $ZnSO_4.7H_2O$, the former (*i.e.* $FeSO_4.7H_2O$) begins to grow in size as the latter (*i.e.* $ZnSO_4.7H_2O$) deposits on it. Similarly, if a crystal of $ZnSO_4.7H_2O$ is suspended in a saturated solution of $FeSO_4.7H_2O$, crystals of $ZnSO_4.7H_2O$ begin to grow in size as $FeSO_4.7H_2O$ deposits on the crystal of $ZnSO_4.7H_2O$.

(*iii*) The elements forming the isomorphous substances have the same valency. For example the valency of Zn and Fe in $ZnSO_4.7H_2O$ and $FeSO_4.7H_2O$ which are isomorphous with each other is 2.

(*iv*) They can form mixed crystals with each other, *i.e.* when a solution containing two isomorphous substances is concentrated to the point of crystallisation, we get crystals which are homogeneous and contain both the isomorphous substances in the ratio in which they were present in the solution.

Isopolymorphism

When each of the two polymorphous forms of a substance is isomorphous with each of the two polymorphous forms of some other substance, the two substances are said to be **isopolymorphous** with each other and the phenomenon is called **isopolymorphism.** For example As_2O_3 exists in two polymorphous forms *viz. rhombic* and *octahedral* forms. Similarly Sb_2O_3 also exists in two polymorphous forms : *rhombic* and *octahedral*. The rhombic forms of both the oxides are said to be isomorphous with each other and so are said the octhedral forms of both the oxides. Consequently *both the oxides are isopolymorphous with each other.* This is shown below :

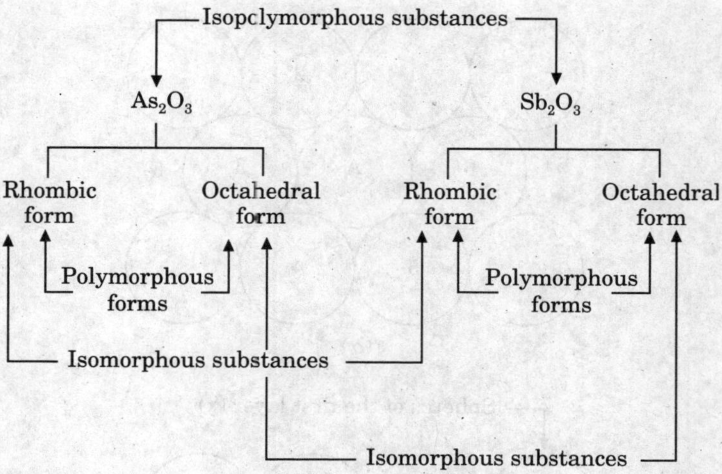

Close Packing of Spheres

It has been observed that the structures of most of the crystals studied in inorganic chemistry is obtained by the most efficient packing of the units of which these crystals are composed. These units may be either atoms, ions or molecules which are of approximately spherical shape, having equal size. It is, therefore, essential to study the ways in which these units are packed together in three dimensions.

Place some solid spheres of equal size in one row on the table. Place another row of the spheres in such a manner that the spheres of the second row rest between the depressions of every two spheres of the first now. In this way place some rows of the spheres as shown in Fig. 9.14(*a*) in which three rows of spheres have been shown. The spheres of this layer have been represented by full line circles.

The following points may be noted from Fig. 9.14(*a*).

(*i*) Every sphere is surrounded by six other similar spheres. For example the sphere marked with A is surrounded by six other similar spheres marked as 1, 2, 3, 4, 5 and 6. Thus we see that every sphere has 6 nearest neighbours. The number of nearest neighbours is called the **coordination number (C. N.).** Thus the C. N. of each of the spheres arranged in this manner is six. It has been shown by calculation that in this arrangement only 60.4% of the space is occupied by the spheres while the remaining 39.6% of the space remains unoccupied (vacant). This vacant space is called **void volume.**

(*ii*) There is an opening between each set of three touching spheres. These openings have been represented as • and ×.

Now place a second layer of spheres (shown by dotted line spheres) on the spheres of the first layer so that the openings shown as • are covered by them while those shown as × remain uncovered [Fig. 9.14 (*b*)].

Now suppose we have to build up a third layer of spheres over the second layer. It can be done in any of the following two ways :

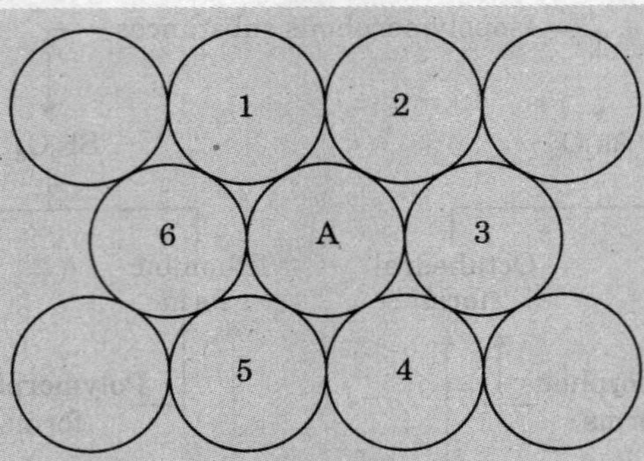

(*a*)

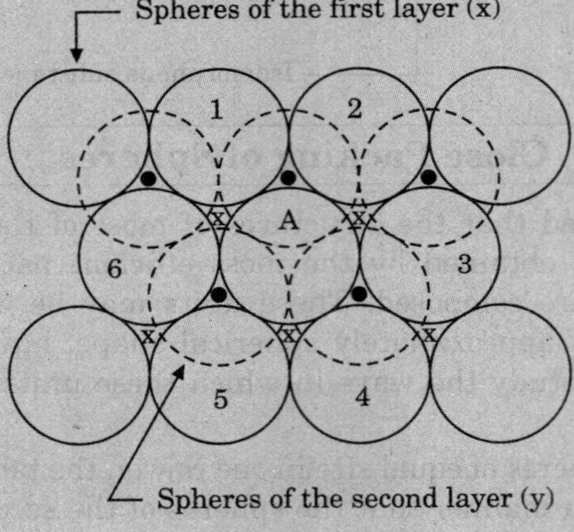

(*b*)

Fig. 9.14. Close packing of spheres. (*a*) Close packing of spheres in one layer. (*b*) Close packing of spheres in two layers.

First way. In this way, the spheres of *3rd* layer (not shown in the Figure) are placed so that the centres of the spheres of the 3rd layer are directly above the centres of the spheres of the first layer. Thus in this arrangement the spheres of the first layer and those of the third layer are exactly identical. This arrangement of close-packing of spheres is referred to as X Y X arrangement of spheres where X, Y and X represent the 1*st*, 2*nd* and 3*rd* layers of spheres. Quite obviously 1*st* and 3*rd* layers are identical.

When X Y X Y X Y ... arrangement of packing of spheres is continued indefinitely, the system obtained is found to possess *hexagonal symmetry* (This system should not be confused with the hexagonal pattern of spheres), *i.e.* this system possesses *one 6-fold axis of symmetry* which is perpendicular to the planes of the close- packed spheres. Such as arrangement of three dimensional packing of spheres is shown in Fig. 9.15. Because of its hexagonal symmetry this arrangement is called **hexagonal close-packing** of spheres and is abbreviated as **hcp.**

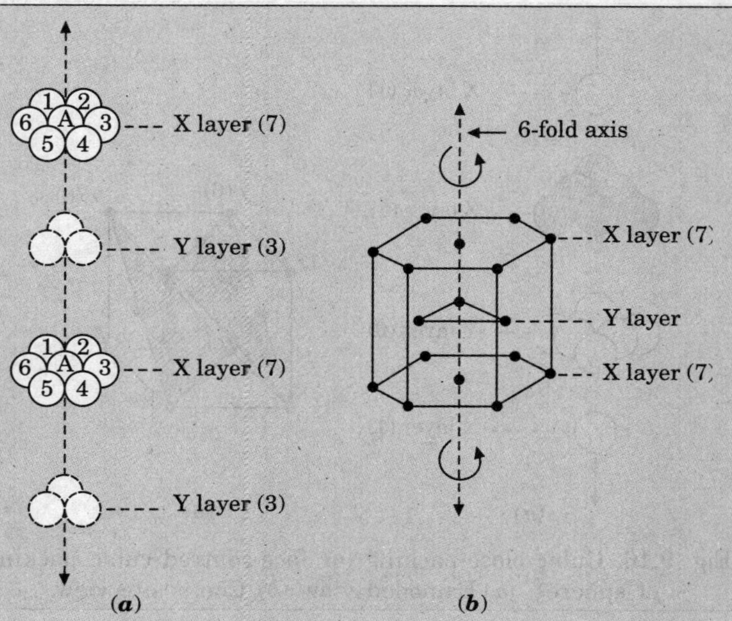

Fig. 9.15. Hexagonal close-packing of spheres (*a*) Exploded view (*b*) Composite view.

In the hexagonal close-packing of spheres the spheres in the two X-layers occupy the sites of the unit cell of the hexagonal lattice while the three spheres of Y-layer are situated within the unit cell and do not occupy the lattice sites. It can be seen from the figure that when the hexagonal structure is rotated about the axis perpendicular to the layers and passing through one sphere, it presents identical view thrice in one rotation.

Second way. In this way the spheres of the third layer are placed on the spheres of the 2*nd* layer so that the openings marked as × in the first layer are covered. Note that these openings were left uncovered in arranging the second layer of spheres. It may also be noted that in this arrangement the spheres of the third layer do not come over those of the first layer. This arrangement of close- packing of spheres is referred to as X Y Z arrangement of spheres where X, Y and Z represent the three layers. When X Y Z X Y Z...arrangement of packing of spheres is continued indefinitely, the system obtained is found to possess *cubic symmetry*, *i.e.* this system has *four 3-fold axes of symmetry* which pass through the diagonals of the cube. Such an arrangement of three dimensional packing of spheres is shown in Fig. 9.16. Because of its cubic symmetry this arrangement is called *cubic close-packing* of spheres and is abbreviated as **ccp.** The unit cell of this structure is a cube which contains 14 spheres. An examination of the arrangement of 14 spheres in the unit cell of ccp structure reveals that this structure has one sphere at the centre of each of the six faces of the unit cell. It is for this reason that ccp structure is also sometimes called **face-centred cubic (fcc) structure.**

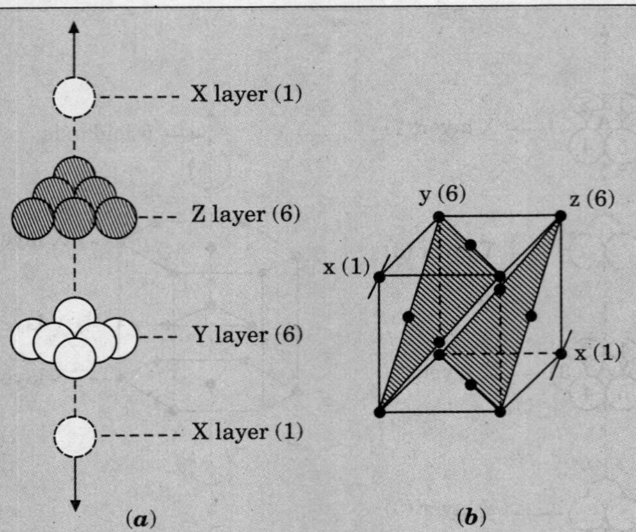

Fig. 9.16. Cubic close-packing (or face-centred cubic packing) of spheres. (*a*) Exploded view (*b*) Composite view.

Although the two closest packing of spheres (*hcp* and *ccp*) described above differ in the mode of arranging the spheres, yet they have the following common characteristics :

(*i*) In both the packings of spheres, each sphere is surrounded by 12 other equidistant spheres, *i.e.* coordination number of each sphere in both the packings of spheres is 12. This fact has been illustrated in Fig. 9.17 (*a*) and (*b*).

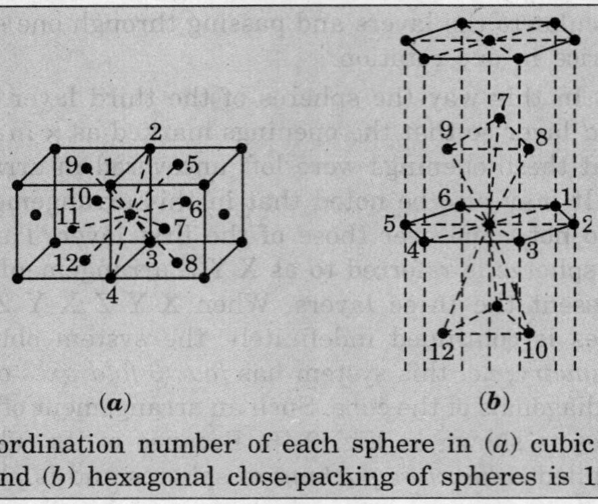

Fig. 9.17. Coordination number of each sphere in (*a*) cubic close-packing and (*b*) hexagonal close-packing of spheres is 12.

(*ii*) In cubic close-packing of spheres [Fig. 9.17 (*a*)] the sphere lying at the centre of the middle face has four nearest neighbours marked as 1, 2, 3 and 4 lying at the corners of that face and eight more marked as 5, 6, 7, 8, 9, 10, 11 and 12 at the same distance at the centres of four faces of adjoining cubes.

(*iii*) In hexagonal close-packing of spheres [Fig. 9.17 (*b*)] the sphere lying at the centre of the hexagon has six nearest neighbours marked as 1, 2, 3, 4, 5 and 6 lying at the corners of the hexagon and six more marked as 7, 8, 9, 10, 11 and 12 at the same distance at the corners of the two adjascent triangular planes, one above and the other below the hexagonal plane. All the three planes are parallel to each other.

Structure of Metallic Crystals

Most of the metals belonging to *s*- and *d*-block elements of the long form of periodic table have any of the following structures.

1. Hexagonal close-packed (hcp) structure. In this structure the atoms are located at the corners and centres of two hexagons which are placed parallel to each other and three more atoms are placed in a parallel triangular plane lying mid way in between the two hexagonal planes [Fig. 9.18 (*a*)]. This structure is obtained by A B A B A... type of close-packing of atoms. Each atom in this structure has 12 nearest neighbours, *i.e.* the coordination number of each of the atoms in *hcp* structure is 12. Examples of metals having *hcp* structure are Cr, Mo, V, Mg, Zn etc.

2. Cubic close-packed (ccp) or face-centred cubic (fcc) structure. In this structure the atoms are located at the corners and at the centres of all the six faces of a cube (Fig. 9.18 (*b*)). This structure is obtained by ABC ABC type of close-packing of atoms. Each atom in this structure has 12 nearest neighbours, *i.e.* coordination number of each atom in this structure is 12 as already discussed. Examples of metals having *ccp* or *fcc* structure are Cu, Ag, Au, Ni, Pt etc.

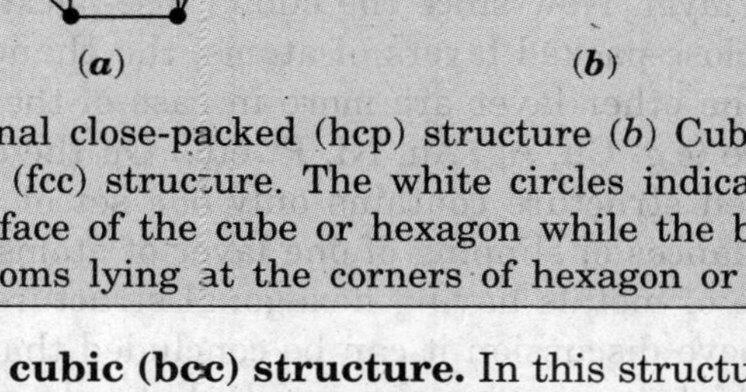

(*a*) (*b*)

Fig. 9.18 (*a*) Hexagonal close-packed (hcp) structure (*b*) Cubic close-packed (ccp) or face-centred cubic (fcc) structure. The white circles indicate the atoms placed at the centre of the face of the cube or hexagon while the black circles indicate the atoms lying at the corners of hexagon or cube.

3. Body-centred cubic (bcc) structure. In this structure the atoms are not closely packed as in the above two structures. This structure is obtained by the following two steps :

(*i*) Place some spheres (represented by full line circles) side by side so that *they are slightly opened up*. By doing so none of the spheres touches each other [See Fig. 9.19 (*a*)].

(*ii*) Place the second layer of spheres (represented by clotted line) on the openings between the spheres of the first layer so that each sphere of the 2*nd* layer is in contact with four spheres of the first layer.

(*iii*) Now place a third layer of spheres so that the openings in the second layer (marked as ×) are covered by them. Thus we see that third layer is exactly identical to first layer. This type of arrangement of spheres gives *body-centred cubic* (*bcc*) *structure* in which each of the spheres (or atoms) is attached with eight nearest spheres (four spheres in the layer just above and four spheres in the layer just below) and so the *coordination number in body-centred cubic structure is only eight*. It is for this reason that the atoms in this structure are *not closely packed*, and only 68% of the total volume is actually occupied by the atoms. However, the atoms are in *close contact along body diagonals*. The unit cell of this structure has one atom at each corner and one atom at the centre of the cube [Fig. 9.19 (*b*)]. Alkali metals have body centred cubic structure.

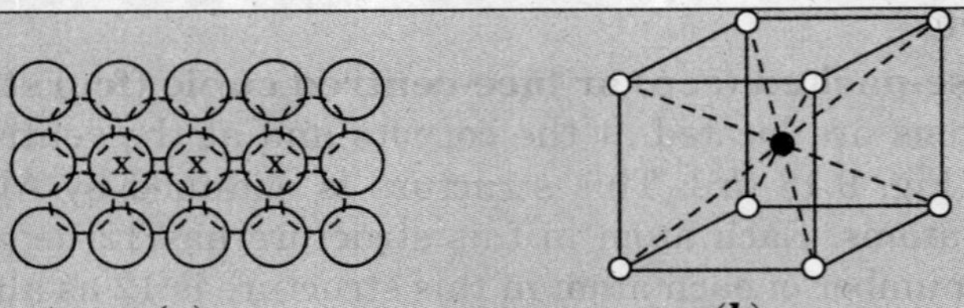

Fig. 9.19. Body-centred cubic structure (*a*) Arrangement of spheres in three layers to get body-centred cubic structure (*b*) Unit cell of body-centred cubic structure.

Malleability and Ductility of Metals

Malleability and ductility of metals are related with the deformation in the crystals of the metals. Deformation in crystals means that one layer of atoms slides over the other layer. Now since the cubic close-packed structure contains *four* sets of parallel close-packed layers of atoms, the chances of slipping of one layer of atoms over the other layer are more in case of the metals having cubic close-packed structure (*e.g.* Cu, Ag, Au, Ni, Pt etc.) On the other hand, since the hexagonal close-packed structure contains only *one* set of parallel close-packed layer of atoms, the chances of slipping of one layer of atoms over the other layer are very little in case of metals having hexagonal structure (*e.g.* Cr, Mo, V, Mg, Zn etc.). From the above discussion it can be concluded that the metals like Cu, Ag, Au, Ni, Pt etc which have cubic close-packed structure are more malleable (*i.e.* can be beaten into sheets) and ductila (*i.e.* can be drawn into wires) than the metals like Cr, Mo, V, Mg, Zn etc which have hexagonal close-packed structure.

Since Fe can adopt both these structures depending on temperature, this metal shows a wide variety of properties.

Characteristics of hcp, ccp and bcc Structures

The characteristics of *hcp*, *ccp* and *bcc* structures mentioned above have been summarised in Table 9.4.

Table 9.4. Characteristics of hexagonal close-packed (hcp), cubic close-packed (ccp) and body centred-cubic (bcc) structures.

Characteristics	Structure		
	Hexagonal close-packed (hcp)	**Cubic close packed (ccp)**	**Body centred cubic (bcc)**
1. Nature of packing of atoms	Close-packed	Close-packed	Not close-packed
2. Volume occupied by atoms	74% of the total volume	74% of the total volume	68% of the total volume
3. Type of packing of atoms	ABABABA.........	ABCABC.........	...
4. Coordination number	12	12	8
5. Malleability and ductility	Less malleable and ductile	More malleable and ductile	Malleable and ductile
6. Examples of metals	Cr, Mo, V, Mg, Zn etc.	Cu, Ag, Au, Ni, Pt etc.	Alkali metals.

Interstitial Sites in Closely-packed Arrangement of Atoms

The interstitial sites formed in between the closely-packed atoms of a given crystal are of the following types :

1. Triangular or trigonal site. When the spheres are placed at the vertices of an equilateral triangle, a triangular arrangement of close-packed spheres is obtained and some empty space (hole or void) is left between the three touching spheres. This empty space is called *triangular or trigonal site*. The corners of the triangle are occupied by the spheres [Fig. 9.20 (*a*)]. It has been shown by calculations that the radius of the sphere which can fit into the trigonal site without disturbing the close-packed arrangement of the surrounding large spheres should not be greater than 0.155 times that of the surrounding large close-packed spheres.

2. Tetrahedral site. When a sphere is placed on the three spheres placed at the vertices of an equilateral triangle and touching each other, a tetrahedral arrangement of close-packed spheres is obtained. In this arragement the centres of the four spheres lie at the apices of a regular tetrahedron [Fig. 9.20 (*b*)]. The empty space left at the centre of this tetrahedron is called *tetrahedral site*. It may be mentioned here that it is not the shape of the void which is tetrahedral but it is the arrangement of the spheres which is tetrahedral. In a close-packed arrangement of spheres, each sphere is in contact with three spheres in the layer above it and other spheres in the layer below it. Thus there are two tetrahedral sites associated with each sphere. The size of the tetrahedral site is much smaller than the size of the sphere. With the increase of the size of the spheres, the size of the tetrahedral site also increases. Calculations have shown that a tetrahedral site can accomodate a sphere whose radius should not be greater than 0.225 times that of the spheres forming the tetrahedral site.

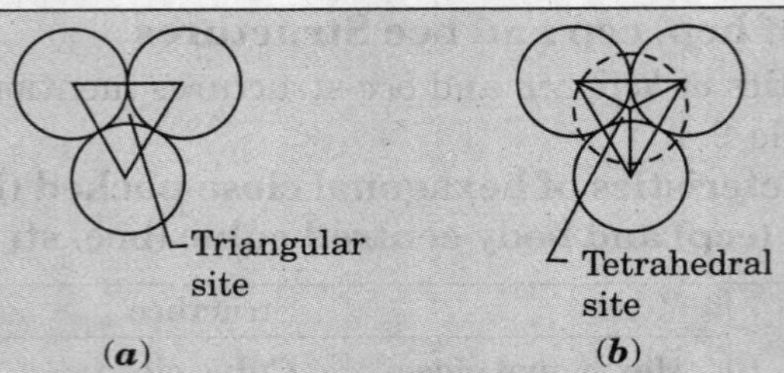

Fig. 9.20. Formation of (*a*) triangular or trigonal site by close-packed arrangement of three spheres (*b*) tetrahedral site by close-packed of arrangement four spheres.

3. Octahedral site. This interstitial site is formed by joining six spheres whose centres lie at the apices of a regular octahedron as shown in Fig. 9.21. This site is found in both *hcp* and *ccp* systems. Fig. 9.21 shows that an octahedral site is generated by two equilateral triangles whose apices point in opposite directions. One of these triangles joins the centres of three spheres in one plane while the other triangle (shown by dotted lines) joins the centres of the three spheres (shown by dotted lines) in the second plane. The octahedral site marked as × is formed where two triangles of spheres belonging to different layers are super-imposed on each other. It may also be noted that there is only one octahedral site for each sphere. Thus the number of octahedral sites is half the number of tetrahedral sites. The sizes of different sites studied so far are related as :

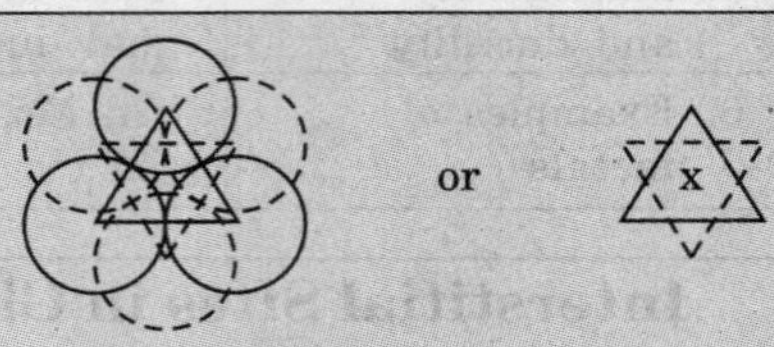

Fig. 9.21. Formation of octahedral site by close-packing of six spheres.

octahedral site > tetrahedral site < trigonal site.

Calculations have shown that an octahedral site can accomodate a sphere whose radius should not be greater than 0.414 times that of the spheres forming octahedral site.

Radius Ratio Rule and its Effect on the Shape of Ionic Crystals.

The ratio of the radii of the positive and negative ions is called the radius ratio. Thus :

$$\textit{Radius ratio} = \frac{\textit{Radius of positive ion (cation)}}{\textit{Radius of negative ion (anion)}}$$

The effect of radius ratio in determining the coordination number and the shape of the ionic crystals is known as *radius ratio effect*. The radius ratio can be used in determining the most stable form of arrangement of ions in the crystal lattice of an ionic solid.

In order to understand the application of radius ratio concept, let us suppose that the coordination number of a cation C^+, in an ionic crystal, C^+A^- is 3. Thus three A^- anions are in contact with one C^+ ion [Fig. 9.22 (*a*)].

1. NaCl crystal (6 : 6 co-ordination crystal). A unit cell representation of NaCl ionic crystal is shown in Fig. 9.23. The salient features of this structure are summarised below :

(*i*) Cl^- ions have cubic close packed arrangement in which all the octahedral holes are filled by Na^+ ions.

(*ii*) In this ionic crystal, the radius ratio, $r(Na^+)/r(Cl^-) = 0.95/1.81 = 0.524$. Since this value lies between 0.414 and 0.732, the co-ordination number of each of the ions is *six*, *i.e.* each Na^+ ion is surrounded by *six* Cl^- ions and similarly each Cl^- ion is surrounded by *six* Na^+ ions which are placed at the six corners of a regular octahedron. Thus we may say that this structure has 6 : 6 co-ordination. The octahedral arrangement of six Cl^- ions marked 1, 2, 3, 4, 5, 6 round one Na^+ ion marked as 7, 8, 9, 10, 11, 12 round one Cl^- ion marked as B has been shown in Fig. 9.23.

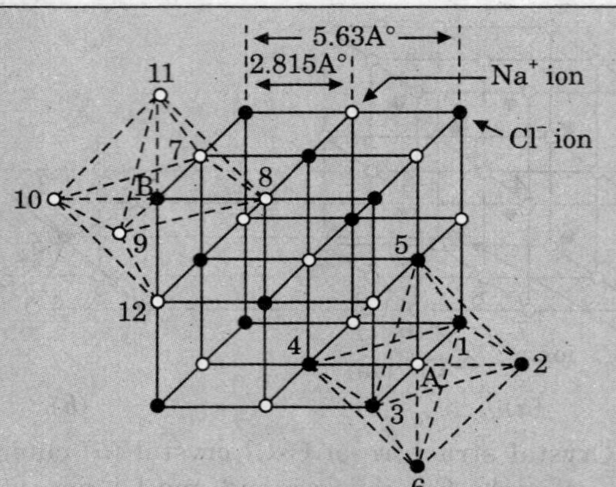

Fig. 9.23. Crystal structure of NaCl crystal.

An examination of the cubical structure of NaCl crystal as shown in Fig. 9.23 makes it clear that the distance between two adjascent ions of the same kind like, say Cl^- ions (*i.e.* length of the edge) is equal to 5.63 Å while the distance between two adjacent ions of different kind like, say Na^+ and Cl^- ions, is equal to 5.63/2 = 2.815 Å.

It is worth noting that Na^+ and Cl^- ions are not connected to one another by pairs, since all the six Cl^- ions are the same distance away from any one Na^+ ion. Thus for the maximum stability of NaCl crystal the oppositely charged ions are close to one another and the similarly charged ions are as away from one another as possible.

Although discrete molecules (Na^+Cl^-) do not exist in the solid form of ionic crystals, independent molecules do exit in the vapour form of ionic compounds.

Examples of some other compounds which have NaCl structures are : NaI, KCl, RbI and RbF.

2. CsCl crystal (8 : 8 co-ordination crystal). The structure of CsCl is shown in Fig. 9.24. The salient features of this structure are summarised below :

(*i*) In this structure, the radius ratio, $r(Cs^+)/r(Cl^-) = 1.69/1.81 = 9.33$ Since this value lies between 0.732 and 1.000, the co-ordination number of the ions is 8, *i.e.* each Cs^+ ion is surrounded by eight Cl^- ions and similarly each Cl^- ion is surrounded by eight Na^+ ions. Thus we can say that this structure has 8 : 8 co-ordination.

(*ii*) The arrangement of the ions in this crystal is body-centred cubic, *i.e.* the unit cell of its structure has one ion at the centre and the oppositely-charged ions at the corner of the cube. The cubic arrangement of eight Cs^+ ions round one Cl^- ion marked as A has been shown in Fig. 9.24 (*b*). Here it should be noted that the arrangement of the ions in this crystal is not strictly body-centred cubic, since in a body-centred cubic arrangement the atom/ion at the centre and those at the corners of the cube are identical whereas in CsCl crystal these atoms or ions are different as mentioned above.

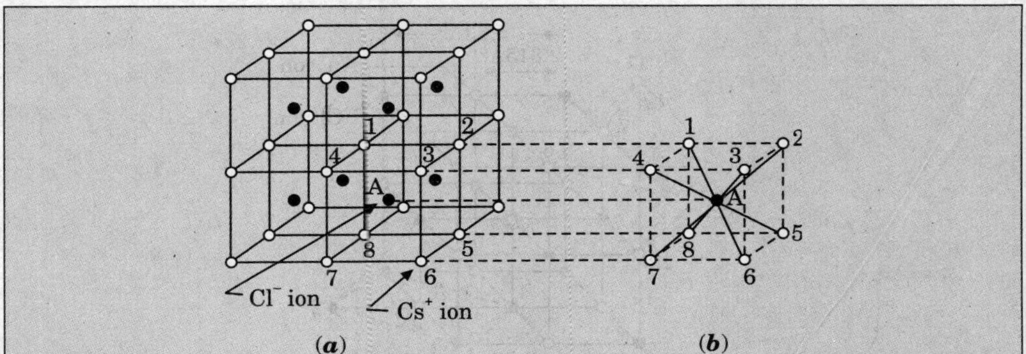

Fig. 9.24. (*a*) Crystal structure of CsCl crystal (*b*) cubic arrangement of eight Cs^+ ions around one Cl^- ion.

CsCl type structure is found in those crystals in which the cations are comparable in size to anions.

Examples of some other compounds which have CsCl structure are CsBr, CaI, TlCl, TlBr and TlI.

3. Zinc blende (ZnS) crystal (4 : 4 co-ordination crystal). A unit cell representation of the structure of this crystal is shown in Fig. 9.25. The important characteristics of this structure are :

(*i*) S^{2-} ions are in the cubic close-packed arrangement, *i.e.* S^{2-} ions are placed at the corners and at the centre of each face of the cube.

(*ii*) The co-ordination number of each ion is 4, *i.e.* each Zn^{2+} ion is tetrahedrally surrounded by four S^{2-} ions and similarly each S^{2-} ion is tetrahedrally surrounded by four Zn^{2+} ions. The tetrahedral arrangement of four

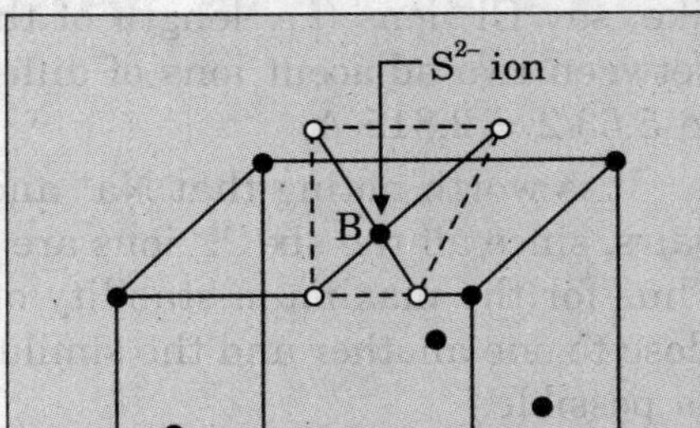

Fig. 9.25. Unit cell representation of zinc blede (ZnS) structure.

S^{2-} ions round one Zn^{2+} ion marked as A and the tetrahedral arrangement of four Zn^{2+} ions round one S^{2-} ion marked as B has been shown in Fig. 9.25. Thus zinc blende is 4 : 4 co-ordination compound.

Examples of some other compounds which have zinc blende structure are : CuCl, CuBr, CuI, AgI and BeS.

4. Wurtzite (ZnS) cystal (*4 : 4 coordination crystal*). A unit cell representation of the structure of this crystal is shown in Fig. 9.26. The important characteristics of this structure are :

(*i*) S^{2-} ions are in the hexagonal close-packed arrangment.

(*ii*) The coordination number of each ion is 4, *i.e.* each Zn^{2+} ion is tetrahedrally surrounded by four S^{2-} ions and similarly each S^{2-} ion is tetrahedrally surrounded by four Zn^{2+} ions. The tetrahedral arrangement of four S^{2-} ions around one Zn^{2+} ion marked as A and the tetrahedral atrangement of four Zn^{2+} ions round one S^{2-} ion marked as B has been shown in Fig. 9.26. Thus wurtzite is 4 : 4 coordination compound.

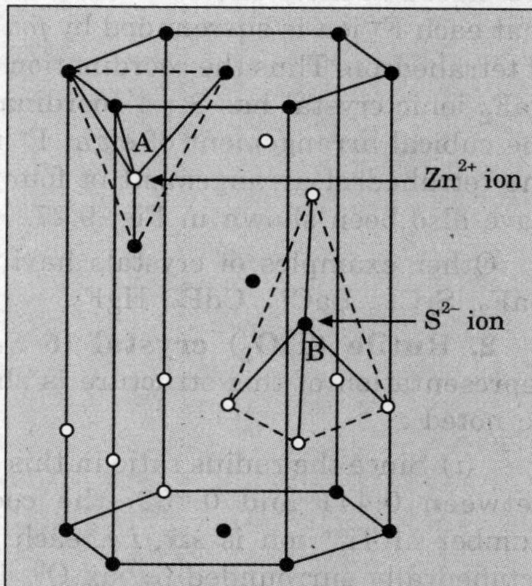

Fig. 9.26. Unit cell representation of wurtzite (ZnS) structure.

Other compounds having wurtzite structure are : ZnO, CdS and BeO.

The structures of zinc blende and wurtzite as described above show that they are very similar to each other. The only difference is in the arrangement of close-packed layers of S^{2-} ions. In zinc blends S^{2-} ions have cubic close-packed arrangement while in wurtzite the same ions have hexagonal close-packed arrangement.

Structures of Ionic Crystals of AB_2 Type

Here we shall discuss the structures of fluorite (CaF_2) and rutile (TiO_2) which are the representatives of the structural arrangements by which similar compounds are expressed.

1. Fluorite, (CaF_2) crystal (*8 : 4 coordination crystal*). The unit cell representation of the structure of CaF_2 crystal is shown in Fig. 9.27. The important points are :

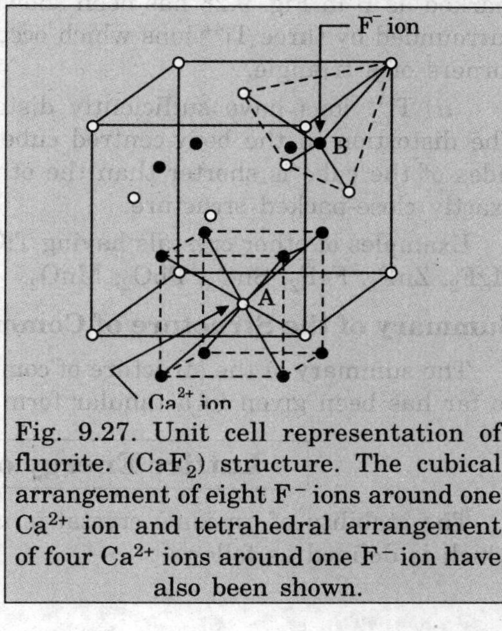

Fig. 9.27. Unit cell representation of fluorite. (CaF_2) structure. The cubical arrangement of eight F^- ions around one Ca^{2+} ion and tetrahedral arrangement of four Ca^{2+} ions around one F^- ion have also been shown.

(*i*) Ca^{2+} ions are in cubic close-packed arrangement, *i.e.* Ca^{2+} ions occupy all the corners and central positions of each face of the cube.

(*ii*) Since the radius ratio value of this crystal is 0.73, the coordination number of Ca^{2+} ion is 8, *i.e.* each Ca^{2+} ion is surrounded by eight F^- ions which are placed at the corners of a cube. F^- ions are so placed at the corners of a cube that each F^- ion is surrounded by *four* Ca^{2+} ions which are placed at the vertices of tetrahedron. Thus the coordination number of each F^- ion is 4 and consequently CaF_2 ionic crystal has 8 : 4 coordination. For the sake of better understanding the cubical arrangement of eight F^- ions around one Ca^{2+} ion marked as A and the tetrahedral arrangement of four Ca^{2-} ions around one F^- ion marked as B have also been shown in Fig. 9.27.

Other examples of crystals having fluorite type crystal structure are SrF_2, BaF_2, $SrCl_2$, $BaCl_2$, CdF_2, HgF_2.

2. Rutile (TiO_2) crystal (*6 : 3 coordination crystal*). The unit cell representation of this structure is shown in Fig 9.28. The following points may be noted :

(*i*) Since the radius ratio in this crystal lies between 0.441 and 0.732, the coordination number of Ti^{4+} ion is *six*, *i.e.* each Ti^{4+} ion is octahedrally surrounded by six O^{2-} ions placed at the corners of an octahedron. For example Ti^{4+} ion marked as A in Fig. 9.28 has been shown to be surrounded by six O^{2-} ions which are placed at the corners of an octahedron. Each O^{2-} ion is surrounded by three Ti^{4+} ions placed at the corners of a triangle. For example O^{2-} ion marked as B in Fig. 9.28 has been shown to be surrounded by three Ti^{4+} ions which occupy the corners of a triangle.

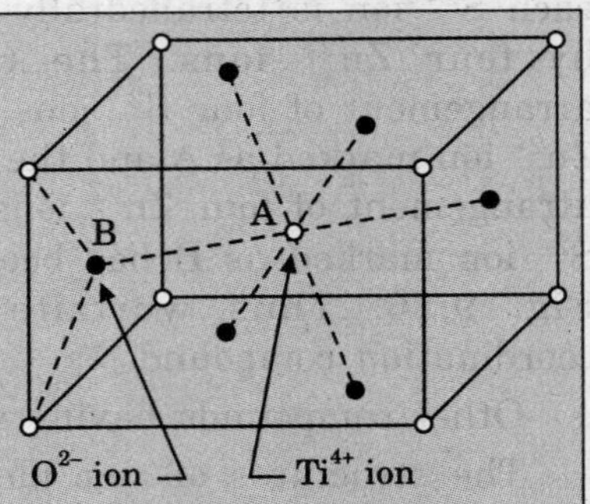

Fig. 9.28. Unit cell representation of rutile (TiO_2) structure.

(*ii*) Ti^{4+} ions have sufficiently distorted body-centred cubic arrangement. The distortion in the body-centred cube is because of the fact that one of the sides of the cube is shorter than the other sides. Thus we see that TiO_2 is not exactly close-packed structure.

Examples of other crystals having TiO_2 type structure are NiF_2, CoF_2, MnF_2, MgF_2, ZnF_2, FeF_2, SnO_2, PbO_2, MnO_2.

Summary of the Structure of Common Ionic Crystals

The summary of the structure of common ionic crystals that we have studied so far has been given in a tabular form in Table 9.6.

Lattice Energy of Ionic Crystals

The stability of an ionic crystal is measured in terms of its lattice energy which is defined as follows :

Lattice energy of an ionic crystal is defined as the energy released when a mole of gaseous cations (i.e. N gaseous cations where N is Avogadro's number) and a mole of gaseous anions (i.e. N gaseous anions) separated from each other by an infinite distance are brought to their equilibrium distance to form one mole of the ionic crystal (See Fig. 9.29).

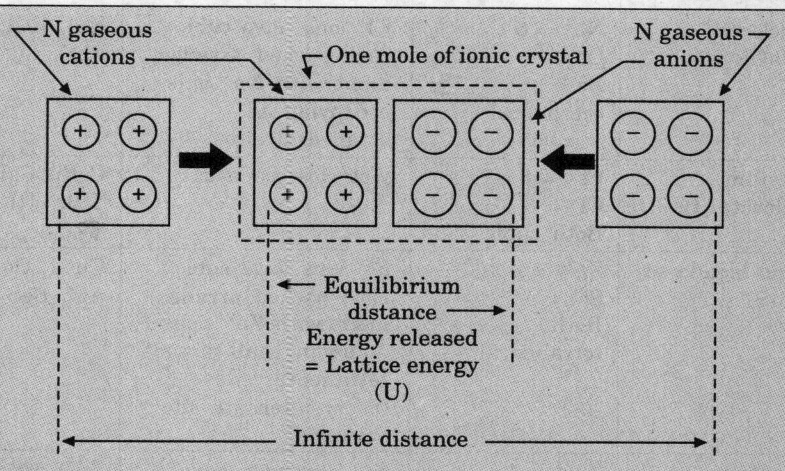

Fig. 9.29. Definition of lattice energy. N is the Avogadro's number.

The formation of a mole of an ionic solid, $A^+B^-(s)$ from its constituent gaseous ions, $A^+(g)$ and $B^-(g)$ may be represented by the equation :

$$A^+(g) + B^-(g) \rightarrow \underset{\text{one mole of ionic solid}}{A^+B^-(s)} + \textit{Energy released (called lattice energy)}.$$

At equilibrium distance the ionic crystal has the *lowest energy* and hence is in the *most stable state* at this distance.

Lattice energy can also be defined as :

Lattice energy is the energy required to move the ions of one mole of an ionic crystal separated from each other by an equilibrium distance to an infinite distance. In simple words lattice energy of an ionic crystal, $A^+B^-(s)$ is numerically equal to that which is required to convert one mole of the same ionic solid into its constituent gaseous ions, *i.e.*

$$\underset{\text{one mole of ionic solid}}{A^+B^-(s)} + \underset{\text{(equal to lattice energy)}}{\text{Energy absorbed}} \rightarrow A^+(g) + B^-(g)$$

Lattice energy is presented by U. It is generally represented by placing a negative sign before its value. Negative sign indicates that the formation of an ionic solid from its gaseous cations and anions is an *exothermic process, i.e.*, in the formation of the ionic solid there is a release of energy and hence the energy of the system is decreased. This is because of the decrease in energy that the ionic solid is more stable than the individual ions of which the ionic solid is composed.

Greater is the value of lattice energy of an ionic solid, more stable is the ionic solid.

Table 9.6. Summary of the structure of common ionic crystals

Crystal	Coordination number of each ion	Nature of arrangement of ions	Examples
AB type Crystal			
(1) Rock salt (NaCl)	Na^+ = 6 Cl^- = 6 Both octahedral	Cl^- ions have cubic close-packed arrangement with Na^+ ions occupying all octahedral sites.	NaI, KCl, RbI, RbF
(2) Cesium chloride (CsCl)	Cs^+ = 8 Cl^- = 8 Both cubic	Not close-packed	CsBr, CsI, TlCl, TlBr, TlI
(3) Zinc blende (ZnS)	Zn^{2+} = 4 S^{2-} = 4 Both tetrahedral	S^{2-} ions have cubic close-packed arrangement with Zn^{2+} ions occupying half of the tetrahedral sites (every alternate site occupied)	CuCI, CuBr, CuI, AgI, BeS
(4) Wurtzite (ZnS)	Zn^{2+} = 4 S^{2-} = 4 Both tetrahedral	S^{2-} ions have hexagonal close-packed arrangement with Zn^{2+} ions occupying half of the tetrahedral sites.	ZnO, CdS, CdSe, BeO
AB_2 type Crystals			
(1) Fluorite (CaF_2)	Ca^{2+} = 8 F^- = 4 Cubic and tetrahedral	Ca^{2+} ions have cubic closed-packed arrangement with F^- ions occupying all the tetrahedral sites.	BaF_2, SrF_2, CaF_2, $SrCl_2$, $SrCl_2$, $BaCl_2$, CdF_2, HgF_2
(2) Rutile (TiO_2)	Ti^{4+} = 6 O^{2-} = 3 Octahedral and triangular	Not close-packed. Ti^{4+} ions lie in a sufficiently distorted body-centred cube.	NiF_2, CoF_2, FeF_2, MnF_2, MgF_2, ZnF_2, SnO_2, PbO_2, MnO_2,

Calculation of Lattice Energy

Let us calculate the lattice energy of an ionic crystal, $M^{z+}X^{z-}$ having Na^+Cl^- type crystal structure.

At least two forces are operating between the constituent ions of the crystal under consideration. These are :

(i) Electrostatic forces of attraction which arise due to the opposite charges on the ions.

(ii) Repulsive forces which arise due to the overlap of the electron clouds of the ions.

Now let us calculate the magnitude of attractive potential energy and repulsive potential energy of $M^{z+}X^{z-}$ crystal. The former type of energy is due to the electrostatic forces between the oppositely-charged ions while the latter is due to the repulsive forces arising due to the overlap of the electron clouds of the ions when they are brought nearer to each other from an infinite distance.

Attractive Potential Energy. According to Coulomb's law of inverse square, the electrostatic force of attraction between the oppositely-charged ions varies directly with the product of the charges on the ions and inversely with the square of the distance between them. Thus the force of attraction (F) between the ions M^{z+} and X^{z-} of the ionic crystal, $M^{z+}X^{z-}$ in air is given by :

$$F = \frac{(z^+e)\,(-z^-e)}{r^2}$$

where z^+e and z^-e = number of charges of the cation and anion respectively

z^+ and z^- = positive integers (*i.e.*, ionic charge numbers)

e = charge on an electron which is equal to 4.8×10^{-10} *e.s.u.*

r = inter-ionic distance between the ions.

Now, if one of the ions, under this force of attraction is moved from infinity through a small distance dr, work done in this process will be given by :

$$\text{Work done} = \frac{(z^+e)\,(-z^-e)}{r^2}\,dr \qquad \text{...(A)}$$

The work done in moving one of the ions from infinity through a distance, r will be given by integrating equation (A) between the limits r and ∞ (infinity).

$$\therefore \text{ Work done} = \int_{r=\infty}^{r=r} \frac{(z^+e)\,(-z^-e)}{r^2}\,dr = \frac{-z^+z^-e^2}{r}$$

$$= \frac{(z^+e)\,(-z^-e)}{r} \qquad \text{...(B)}$$

By definition work done given by equation (B) is equal to the electrostatic (or attractive) potential energy between the two ions (*i.e.* potential energy of the ion-pair) *viz.* M^{z+} and X^{z-} separated from each by a distance r.

$$\therefore \left.\begin{array}{l}\textit{Attractive potential energy between } M^{z+} \textit{ and} \\ X^{z-} \textit{ ions placed at a distance of } r \textit{ from each other}\end{array}\right\} = \frac{(z^+e)\,(-z^-e)}{r} \qquad \text{...(C)}$$

Thus, according to Coulomb's law, the potential energy between two ions is equal to the product of their respective charges divided by the distance between the ions [*See* equation (C)].

Now we know that our ionic crystal contains not only two ions but consists of numberless ions which are arranged in an ordered pattern. Thus in an ionic crystal each ion is surrounded not only by a single ion of opposite charge but also by a number of other ions (positive and negative both). Consequently in order to calculate the attractive (electrostatic) potential energy of the crystal we must consider all the interactions given below :

(*i*) Interaction of a cation with all the anions surrounding it. This interaction gives the *attractive* (*i.e.* electrostatic) *energy.*

(*ii*) Interaction of the same cation with all the cations surrounding it. This interaction gives the *repulsive energy*.

(*iii*) Interaction of an anion with all the cations surrounding it.

(*iv*) Interaction of the same anion with all the anions surrounding it.

Considering all the interections mentioned above, we can calculate the potential energy of interaction between different ions of $M^{z+}X^{z-}$ ionic crystal as follows :

Since, as we have already said, the crystal structure of $M^{z+}X^{z-}$ crystal is the same as that of NaCl, each M^{z+} cation is surrounded :

(*a*) *by six X^{z-} anions* at a distance of $\sqrt{1}r$ or r where r is the distance between M^{z+} and the nearest X^{z-}.

(*b*) *by twelve M^{z+} cations* at a distance of $\sqrt{2}r$.

(*c*) *by eight X^{z-} anions* at a distance of $\sqrt{3}r$.

(*d*) *by six M^{z+} cations* at a distance of $\sqrt{4}r$ or $2r$ etc. as is evident from Fig. 9.30

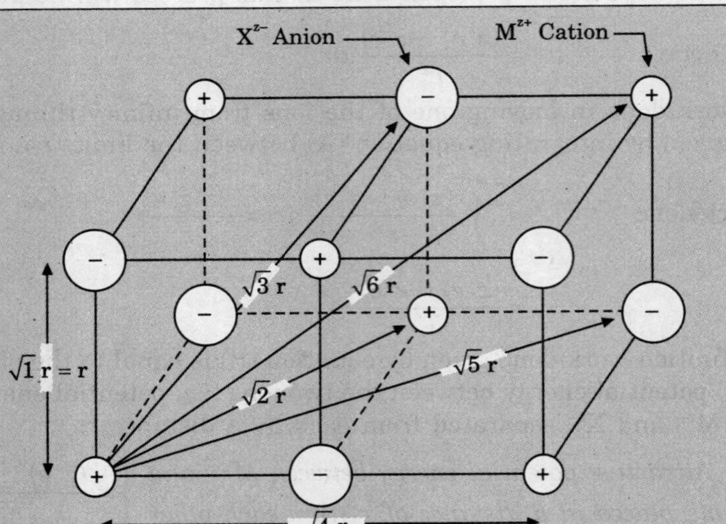

Fig. 9.30. Distances between the neighbouring ions in $M^{z+}X^{z-}$ ionic crystal having NaCl crystal structure. r is the distance between M^{z+} cation and the nearest X^{z-} anion.

From equation (C) :

(*i*) The attractive potential energy between M^{z+} cation and six X^{z-} anions which are placed at a distance of $\sqrt{1}r$ from M^{z+} cation.

$$= \frac{(z^+e) \times 6(-z^-e)}{\sqrt{1}r} = \frac{-6\, z^+z^-e^2}{\sqrt{1}r}$$

(*ii*) The repulsive potential energy between M^{z+} cation and twelve M^{z+} cations which are placed at a distance of $\sqrt{2}r$ from M^{z+} cation.

$$= \frac{(z^+e) \times 12(z^+e)}{\sqrt{2}r} = \frac{12(z^+)^2 e^2}{\sqrt{2}r}$$

(*iii*) The attractive potential energy between M^{z+} cation and eight X^{z-} anions which are placed at a distance of $\sqrt{3}r$ from M^{z+} cation

$$= \frac{(z^+e) \times 8(-z^-e)}{\sqrt{3}r} = \frac{-8\, z^+z^-e^2}{\sqrt{3}r}$$

(*iv*) The repulsive potential energy between M^{z+} and six M^{z+} cations which are placed at a distance of $\sqrt{4}r$ from M^{z+} cation.

$$= \frac{(z^+e) \times 6(z^+e)}{\sqrt{4}r} = \frac{6(z^+)^2 e^2}{\sqrt{4}r}$$

Thus the value of attractive (or electrostatic) potential energy of M^{z+} cation, $(PE)_{M^{z+}}$ when we consider the interaction of this cation with all other ions (cations and anions both) surrounding it like six X^{z-} anions placed at a distance of $\sqrt{1}r$, twelve M^{z+} cations placed at a distance of $\sqrt{2}r$, eight X^{z-} anions placed at a distance of $\sqrt{3}r$ and so on in $M^{z+}X^{z-}$ ionic crystal is obtained by summing up all the energy terms calculated as above. Thus :

$$(PE)_{M^{z+}} = \frac{-6\, z^+z^-e^2}{\sqrt{1}r} + \frac{12(z^-)^2e^2}{\sqrt{2}r} - \frac{8\, z^+z^-e^2}{\sqrt{3}r} + \frac{6(z^+)^2e^2}{\sqrt{4}r} + \ldots\ldots$$

$$= -\frac{z^+z^-e^2}{r}\left(\frac{6}{\sqrt{1}} - \frac{12\, z^+}{\sqrt{2}\, z^-} + \frac{8}{\sqrt{3}} - \frac{6\, z^+}{\sqrt{4}\, z^-} + \ldots\right) \qquad \ldots(i)$$

Since the ratio of charges on the ions (z^+/z^-) is constant for a particular type of structure (*e.g.* z^+/z^- ratio for NaCl = 1, z^+/z^- ratio for CaF_2 = 2 etc). the sum of all the terms of the infinite series given in bracket is constant for an ionic crystal of given geometry. The value of this constant depends on the geometry of the crystal and remains indepedent of the value of *r*. This constant is called **Mandelung constant** and is generally represented by A. The value of A for NaCl crystal may be obtained by putting z^+/z^- = 1 in the infinite series given in the bracket. Thus A for NaCl is given by :

$$A_{NaCl} = \left(\frac{6}{\sqrt{1}} - \frac{12}{\sqrt{2}} + \frac{8}{\sqrt{3}} - \frac{6}{\sqrt{4}} + \ldots\right)$$

$$= 1.7476.$$

As a matter of fact the value 1.7476 is the value of the Mandelung constant for all those ionic crystals which have NaCl type structure. Values of Mandelung constants for other ionic crystals can be derived by a similar method. These values for some ionic crystals are given below in Table 9.7.

Table 9.7. Values of Mandelung constant (A values) for different crystal structure.

Crystal structure	Values of Mandelung constant (A)
Sodium chloride (NaCl)	1.7476
Caesium chloride (CsCl)	1.7627
Zinc blende or sphalerite (ZnS)	1.6381
Wurtzite (ZnS)	1.6413
Fluorite (CaF_2)	2.5193
Rutile (TiO_2)	2.4080
Corundum (Al_2O_3)	4.1718
Anastase (TiO_2)	4.800
Cadmium iodide (CdI_2)	4.71
β-quartz (SiO_2)	4.4394

Thus writting A (Mendelung constant) for the sum of all the terms of the infinite series given in bracket in equation (*i*) we get :

$$(PE)_{M^{z+}} = -\frac{z^+z^-e^2\ A}{r} \qquad ...(ii)$$

Now when we consider the interaction of X^{z-} anion with all other ions (cations and anions both) surrounding it, the value of electrostatic potential energy of this anion, $(PE)_{X^{z-}}$ can be calculated by using the same procedure as employed for the calculation of $(PE)_{M^{z+}}$. The value of $(PE)_{M^{z-}}$ comes out to be the same as that of $(PE)_{M^{z+}}$ given by equation (*ii*), since cations and anions have the same arrangement of ions surrounding them in NaCl structure. Thus :

$$(PE)_{X^{z-}} = (PE)_{M^{z+}} = -\frac{z^+z^-e^2\ A}{r} \qquad ...(iii)$$

The total attractive or electrostatic interaction potential energy. $(PE)_{attraction}$ is then given by half the sum of $(PE)_{M^{z+}}$ and $(PE)_{X^{z-}}$, *i.e.*

$$\begin{aligned}(PE)_{attraction} &= \tfrac{1}{2}[(PE)_{M^{z+}} + (PE)_{X^{z-}}] \\ &= \frac{1}{2}\left[-\frac{z^+z^-e^2\ A}{r} - \frac{z^+z^-e^2\ A}{r}\right] \\ &= -\frac{z^+z^-e^2\ A}{r} \qquad ...(iv)\end{aligned}$$

The factor $\frac{1}{2}$ is needed since the addition of $(PE)_{M^{z+}}$ and $(PE)_{X^{z-}}$ takes the interaction in account twice.

Repulsive Potential Energy. So far we have considered only the attraction between the ions but when the ions are brought near to each other from an infinite distance, they also repell each other due to the overlap of their electron clouds. Thus the attraction between the ions is also opposed by the repulsion between them. In other words, the total attractive interaction potential energy, $(PE)_{attraction}$ which represents a net attraction between ions, is opposed by repulsive potential energy, $(PE)_{repulsion}$. A number of expressions have been suggested for the repulsive potential energy. According to one expression given by Born and Lande, $(PE)_{repulsion}$ is inversely proportional to the *nth* power of the distance, *r* between the ions. Thus according to them, $(PE)_{repulsion}$, is given by the expression :

$$(PE)_{repulsion} = \frac{B}{r^n} \qquad ...(v)$$

Here B is a constant which is called **Born coefficient** or **repulsion coefficient** and is approximately proportional to the number of nearest neighbours (*i.e.* co-ordination number). This constant expresses the repulsion between the ions. *n* is another constant which is called **Born exponent** and can be estimated by compressibility measurements. Average value of Born exponent (*n* values) for different electronic configurations of the ions are given in Table 9.8.

Table 9.8. Average values of Born exponent (*n*-values) for different electronic configurations of ions.

Electronic configuration	*n* values	Examples
He	5	Li^+, Be^{2+}
Ne	7	Na^+, Mg^{2+}, O^{2-}, F^-
Ar	9	K^+, Ca^{2+}, S^{2-}, Cl^-, Cu^+
Kr	10	Rb^+, Br^-, Ag^+
Xe	12	Cs^+, I^-, Au^+
Rn	14	

n-values of an ionic crystal composed of ions having different configurations (*i.e.* non-isoeleotronic ions) is taken as an average of the *n*-values of the two constituent ions, *e.g.* *n*-value of NaCl in which Na^+ and Cl^- ions are not isoelectronic is given by :

$$n\text{-value of } NaCl = \tfrac{1}{2}(n\text{-value of } Na^+ + n\text{-value of } Cl^-)$$
$$= \tfrac{1}{2}(7+9) = 8$$

Thus the total potential energy, $(PE)_{total}$ of a pair of ions separated by a distance of *r* within the ionic crystal, $M^{z+}X^{z-}$ which has NaCl type structure and for which all the types of forces due to the neighbouring ions have been considered, is given by the sum of $(PE)_{attraction}$ and $(PE)_{repulsion}$. Thus :

$$(PE)_{total} = (PE)_{attraction} + (PE)_{repulsion}$$
$$= \left[-\frac{z^+z^-e^2\,A}{r} + \frac{B}{r^n}\right] \text{ per ion-pair} \qquad ...(vi)$$

The energy given by equation (*vi*) is the energy released when $M^{z+}X^{z-}$ crystal is formed at a distance of *r* between M^{z+} cation and X^{z-} anion. In other words $(PE)_{total}$ is the energy released when a cation (M^{z+}) and an anion (X^{z-}) separated by an infinite distance are brought nearer to each other until they are separated by a distance *r*. $(PE)_{total}$ is the potential energy of an ion-pair separated by a distance *r*.

Estimation of the value of B. The value of B can be estimated by utilising the fact that at equilibrium distance between the ions in the crystal (r_0), the attractive and repulsive forces in the crystal are exactly balanced and hence the value of $(PE)_{total}$ becomes minimum, *i.e.* when $r = r_0$, $d\,(PE)/dr = 0$. Thus :

$$\left(\frac{d\,(PE)}{dr}\right)_{r=r_0} = \frac{d}{dr}\left(-\frac{z^+z^-e^2\,A}{r}+\frac{B}{r^n}\right) = 0$$

$$= \frac{z^+z^-e^2\,A}{r_0^{\,2}} - \frac{nB}{r_0^{\,n+1}} = 0$$

or
$$\frac{z^+z^-e^2\,A}{r_0^{\,2}} = \frac{nB}{r_0^{\,n+1}}$$

$\therefore$
$$\boxed{\mathbf{B} = \frac{z^+z^-e^2\,A\,r_0^{\,n-1}}{n}}$$

Now put the above value of B and $r = r_0$ in equation (*vi*) to get :

$$(PE)_{r=r_0} = \frac{z^+z^-e^2\,A}{r_0} + \frac{z^+z^-e^2\,A\,r_0^{\,n-1}}{n\,r_0^{\,n}}$$

$$= -\frac{z^+z^-e^2\,A}{r_0}\left(1-\frac{1}{n}\right) \text{ per ion-pair in } cgs \text{ units} \qquad ...(vii)$$

Potential energy, $(PE)_{r=r_0}$ given by equation (*vii*) is the energy released when a gaseous cation (M^{z+}) and a gaseous anion (X^{z-}) separated from each other by an infinite distance are brought to an equilibirium distance (r_0) to form the ionic crystal, $M^{z+}X^{z-}$. Consequently potential energy released, when N gaseous cations and N gaseous anions (N is the Avogadro's number which is defined as the number of molecules in a gram molecule and has the value 6.0238×10^{23} atoms/g. atom) separated from each other by an infinite distance are brought to their equilibrium distance (r_0) to form a mole of the ionic crystal, is obtained by multiplying $(PE)_{r=r_0}$ by N. The product, $(PE)_{r=r_0} \times N$, by definition, gives the lattice energy (U) of the crystal, *i.e.*,

$$(PE)_{r=r_0} \times N = \text{Lattice energy (U)}$$

$$= -\frac{z^+z^-e^2\,AN}{r_0}\left(1-\frac{1}{n}\right) \text{ per mole (in } cgs \text{ units)} \qquad ...(viii)$$

or
$$\boxed{\mathbf{U} = -\frac{z^+z^-e^2\,AN}{r_0}\left(1-\frac{1}{n}\right) \textbf{ per mole}} \qquad ...(ix)$$

Equation (*ix*) gives the potential energy of one mole of the crystal and is called ***Born-Lande equation.*** The derivation of the above equation is based on the assumption that both the cations and anions are hard spheres and the solid has an essentially ionic structure, *i.e.* the ions are *not distorted* (*i.e. polarised*) by the neighouring ions. If the ions are distorted, M—X bonds achieve some covalent character and hence the lattice energy calculated from Born-Lande equation differs considerably form that found experimentally (*e.g.* by Born-Haber cycle) as is evident from Table 9.9. Consequently the difference in the calculated and experimental values of lattice energy of an ionic solid is due to some covalent character in the solid. This covalent character is developed due to the polarisation of ions.

Table 9.9. Calculated and experimental values of lattice energies of alkali metal halides (Kcal/mole)

Alkali metal halides	Calculated values from Born-Lende equation [Equ. (*ix*)]	Experimental values from Born-Haber cycle
LiF	240.8	247.2
LiCl	193.9	200.8
LiBr	183.1	186.7
LiI	169.3	171.7
NaF	215.6	218.5
NaCl	182.0	184.0
NaBr	171.8	174.1
NaI	158.5	162.7
KF	190.6	194.1
KCl	164.3	167.6
KBr	157.6	160.4
KI	148.9	151.1
RbF	181.9	186.5
RbCl	158.9	163.1
RbBr	152.1	156.3
RbI	144.0	147.4
CsF	172.8	177.8
CsCl	148.8	150.5
CsBr	143.3	146.4
CsI	135.8	139.7

Theoretical values calculated by Born-Lande equation can be improved by taking into account the following :

(*i*) Van der Waals forces between the ions

(*ii*) Zero point vibrational energy of the ions

(*iii*) Resonance energy which arises because all ionic bonds have some covalent character.

Example. *Calculate the lattice energy of NaCl crystal with the help of Born-Lande equation.*

Solution. The Born-Lande equation can be used as follows for the calculation of lattice energy of Na*Cl* crystal by using the following values of various constant $z^+ = z^- = 1$; e = 4.8×10^{-10} *esu*; A = 1.7476; N = 6.0238×10^{23} atoms/g. atom; $r_0 = (r_{Na^+} + r_{Cl^-})$A° = (0.95 + 1.81) Å = 2.76A° = 2.76×10^8 cm; n = $\frac{1}{2}(7 + 9) = 8$

$$\text{Now,} \quad U = -\frac{z^+z^-e^2AN}{r_0}\left(1-\frac{1}{n}\right)$$

$$\therefore \quad U_{NaCl} = -\frac{1\times1\times(4.8\times10^{-10}\ esu)^2\times1.7476\times6.0238\times10^{23}}{2.76\times10^8\ cm}\left(1-\frac{1}{8}\right)$$

$$= -\ 7.70\times10^{12}\ \text{ergs/mole} \qquad \left[\because \frac{(esu)^2}{cm} = \text{erg.}\right]$$

$$= -\ 184.03\ \text{Kcal/mole} \qquad [\because 1\ \text{erg} = 2.390\times10^{-11}\ \text{Kcal}].$$

This value agrees closely with the experimentally determined value of 180 Kcal/mole.

It may be seen from Table 9.9 that the calculated and experimetal values agree to some extent. But in case of salts which contain ions of high charge or pseudo-intert gas type cations, the experimental values are much higher than the calculated values and hence the agreement between calculated and experimental values is very poor as is evident from the values of some salts given below :

Salt	Calculated value of lattice energy (kcal/mole)	Experimental value of lattice energy (kcal/mole)
MnI_2	481	563
FeI_2	493	589
CoI_2	502	605
PbI_2	460	510
Ag_2O	585	714
CuS	612	673
HgS	774	854

The discrepencies between the calculated and experimental values are due to the fact that these salts, although predominantly ionic, also contain substantial amount of covalent character.

Experimental Determination of Lattice Energy by Born-Haber Cycle.

The lattice energy of an ionic solid can be determined experimentally by a process known as **Born-Haber cycle** devised by Born-Haber in 1919. Let us illustrate this cycle by considering NaCl ionic crystal. The determination of lattice energy of NaCl by this cycle is based on the assumption that the formation of one mole of crystalline NaCl can occur by any of the following two processes.

1. Process consisting of direct combination of Na(s) and $\frac{1}{2}$Cl(g) to form NaCl(s). In this process one gram atom of sodium, Na(*s*) and half mole of chlorine gas, $\frac{1}{2}Cl_2(g)$ are combined together to form one mole of crystalline sodium chloride, NaCl(*s*). In this reaction heat, called *heat of formation* of NaCl(*s*) (represented by Q) is also *produced*. Thus it is an *exothermic reaction*. Heat produced is shown with a *negative sign*. Thus :

$Na(s) + \frac{1}{2}Cl_2(g) \rightarrow NaCl(s)$; *Heat produced* = *Heat of formation of* $NaCl(s) = -Q$

2. Process consisting of the following five steps (*a*) *Conversion of Na atoms from solid state to gaseous state* [$Na(s) \rightarrow Na(g)$]. Here solid sodium, $Na(s)$ is converted into gaseous state, $Na(g)$. This type of conversion is called *sublimation*. In this reaction heat, called *heat of sublimation* of sodium (represented by S) is *required*, *i.e.* it is an *endothermic reaction*. Heat required (or absorbed) in a reaction is shown with a *positive sign*. Thus :

$Na(s) \rightarrow Na(g)$; *Heat required* = *Heat of sublimation of Na* = + S

(*b*) *Conversion of Na(g) into Na⁺(g) ion* [$Na(g) \rightarrow Na^+(g) + e^-$]. Here sodium atom in the gaseous state, $Na(g)$ formed in step (*a*) gives up an electron and is converted into gaseous ion, $Na^+(g)$. This type of conversion is called *ionisation*. For this reaction, energy, called ionisation energy of sodium atom (represented as I) is required, *i.e.* this reaction is an *endothermic reaction*. Thus :

$Na(g) \rightarrow Na^+(g) + e^-$; *Energy required* = *Ionisation energy of Na atom* = + I

(*c*) *Conversion of* $\frac{1}{2}Cl_2(g)$ *molecule into Cl(g) atom* [$\frac{1}{2}Cl_2(g) \rightarrow Cl(g)$]. Here chlorine molecule in the gaseous state, $Cl_2(g)$ is converted into gaseous chlorine atom, $Cl(g)$. This type of conversion is called *dissociation*. For this reaction heat, called *heat of dissociation of chlorine molecule* (represented as D) is *required*, *i.e.* this conversion is an *endothermic reaction*. Thus :

$\frac{1}{2}Cl_2(g) \rightarrow Cl(g)$; *Heat required* = $\frac{1}{2} \times$ *Heat of dissociation of* $Cl_2(g)$ *molecule* $= +\frac{1}{2}D$

(*d*) *Conversion of Cl(g) into Cl⁻(g)* [$Cl(g) + e^- \rightarrow Cl^-(g)$]. Here chlorine atom in the gaseous state, $Cl^-(g)$ formed in step (*c*) adds up an electron given up by $Na(g)$ in step (*b*) and is converted into gaseous ion, $Cl^-(g)$. In this reaction energy, called *electron affinity energy* of Cl atom (represented as E) is *produced*, *i.e.* this conversion is an *exothermic reaction*. Thus :

$Cl(g) + e^-$[Lost by $Na(g)$ in step (*b*)] $\rightarrow Cl^-(g)$; *Heat produced* = *Electron affinity of Cl atom* $= -E$

(*e*) *Combination of Na⁺(g) and Cl⁻(g) ions to form NaCl(s)* [$Na^+(g) + Cl^-(g) \rightarrow NaCl(s)$]. $Na^+(g)$ and $Cl^-(g)$ ions formed in steps (*b*) and (*d*) respectively combine together and form one mole of solid sodium chloride, $NaCl(s)$. Here attractive forces between $Na^+(g)$ and $Cl^-(g)$ operate and hence energy of the system is decreased, *i.e.* this reaction is accompanied by the release of energy (*exothermic reaction*) which is obviously called *lattice energy* (U) of $NaCl(s)$. Thus :

$Na^+(g) + Cl^-(g) \rightarrow NaCl(s)$; *Energy released* = *Lattice energy of NaCl(s)* = – U

Diagramatically the formation of one mole of crystalline sodium chloride $NaCl(s)$ by the two processes mentioned above has been shown in Fig. 9.31 which is called **Born-Haber cycle.**

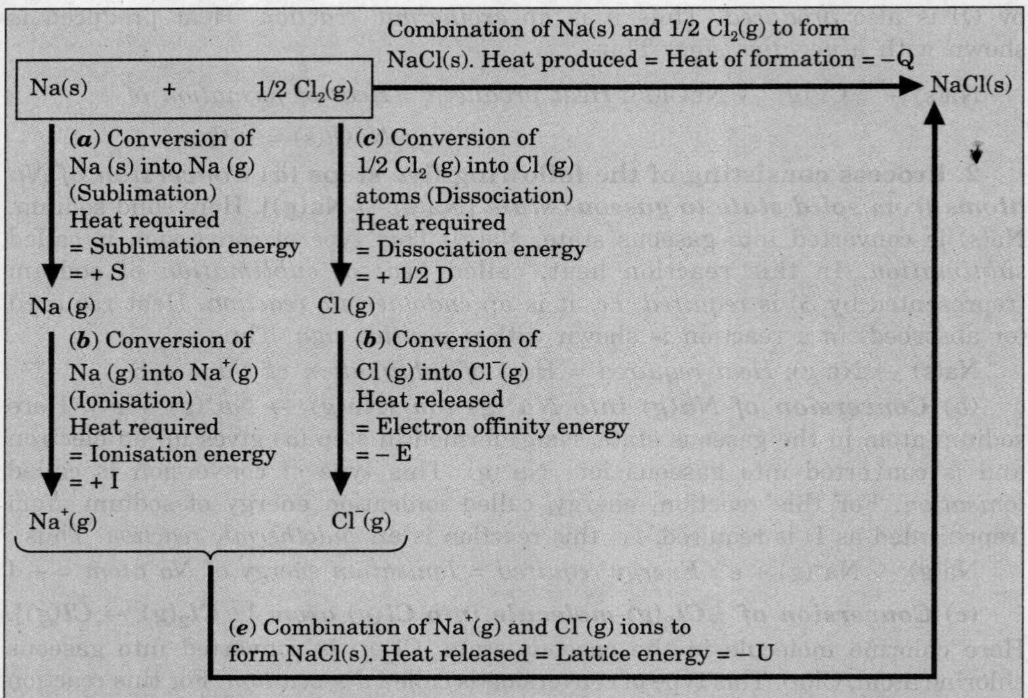

Fig. 9.31. Born-Haber cycle illustrating the formation of one mole of crystalline sodium chlorine, NaCl(*s*). Positive sign associated with energy symbols indicates the absorption of energy (energy required) while the negative sign indicates the release of energy (energy released or produced).

Since the two processes are independent of each other, by Hess's law the heat of formation of NaCl(*s*), Q must be equal to the sum of all other energy terms. Thus :

$$\mathbf{-\,Q = +\,S + I + \tfrac{1}{2}D - E - U} \quad \text{...(1)}$$

With the help of this equation we determine the value of U, if the quantitative values of other energy terms are known. *The quantitative value of each term is substituted in the above equation along with its sign.*

Solved Example. *When a mole of crystalline NaCl is prepared from 1 gram atom of Na and 0.5 mole of chlorine gas, 410 KJ of heat is produced. The heat of subilmation of Na metal 108.8 KJ. The heat of dissociation of chlorine gas into atoms is 242.7 KJ, the ionisation energy of Na is 493.7 KJ and the electron affinity of Cl is 368.2 KJ. Calculate the lattice energy of NaCl.*

(*Delhi B.Sc. Hons. 1972*)

Solution. Here Q = 410 KJ, S = 108.8 KJ, I = 493.7 KJ, D = 242.7 KJ, E = 368.2 KJ and U is to be calculated. Substituting these values, along with their signs, in equation (1) given above, we get :

$$-\,410 = 108.8 + 493.7 + \tfrac{1}{2} \times 242.7 \times 368.2 + U$$

or $\quad -\,410 = 355.65 + U$

or $\quad U = -\,410 - 355.65 = -\,765.65$ KJ/mole.

Defects in Crystals

An ideal crystal is that which has the same unit cell containing the same lattice points throughout the whole crystal.

At absolute zero most of the ionic crystals contain well-ordered arrangement of constituent ions and hence the crystals have no imperfections or defects. With the rise of temperature, one or more the lattice sites become unoccupied or vacant and this gives rise to a defect in the regular arrangement of the constituent ions of the crystal. The number (n) of such defects per cm^3 at a given temperature (T) is given by :

$$n = N.e^{-w/2RT}$$

Here N = total number of sites per cm^3; w = work (energy) required to produce a defect; R = gas constant and e = base of natural logrithm.

Above discussion makes it evident that above absolute zero all crystalline solids have defects in the regular arrangement of the constituent ions of which the ionic crystals are composed. These defects are of two types : (1) *Stoichiometric defects*, and (2) *Non-stoichiometric defects*.

1. Stoichiometric Defects

Stoichiometric compounds are those in which the ratio of the number of A^+ ions to the number of B^- ions (in an ionic crystal, A^+B^-) is equal to one, *i.e.* the number of A^+ ions is equal to that of B^- ions. These defects are of two types : (*i*) *Schottky defect*, and (*ii*) *Frankel defect*.

(*i*) **Schottky Defect.** *This defect arises when some cations and equal number of anions are missing from their own position and thus leave lattice vacancies or holes in the crystal lattice.* Thus some holes are created due to the missing of cations and equal number of holes are created due to the missing of anions. This defect created due to the missing of one cation and one anion is illustrated in Fig. 9.32. Since the number of missing cations is equal to the number of missing anions, the crystal as a whole remains neutral. In this defect the missing cations and missing anions leave their original position and move to the surface. This defect is mostly found in compounds (*i*) *which have high coordination numbers* (*ii*) *in which the ions* (*both cations and anions*) *are almost of the same size, i.e.* the radius ratio. r_+/r_- is not far below unity. Thus alkali metal halides like NaCl (C.N. = 6) and CsCl (C.N. = 8) have Schottky defect.

(*ii*) **Frenkel Defect.** *This defect arises when one of the ions (mostly cation) leaves its own position and occupies another one (hole) in between other ions in the lattice.* Cations, being smaller in size, are more likely than the anions to leave their own position and occupy the other one in between the ions in the lattice. It may be noted that here again, since the number of cations is equal to that of anions, the crystal as a whole remains neutral. In this defect the cation moves to another position between the two layers and is thus surrounded by a greater number of anions. (See Fig. 9.33). This defect is mostly found in compounds (*i*) *which have low coordination number* (*ii*) *in which the negative ions are much larger in size than the positive ions, i.e. the radius ratio* r_+/r_- *is low, and* (*iii*) *which have a highly polarising cation and an easily polarisable anion.* ZnS (both zinc blonde and wurtzite), AgBr etc. are the examples of compounds having Frenkel defect. In ZnS, some of the Zn^{2+} ions are missing from their regular positions and occupy positions in between other ions. Similarly in AgBr, some of Ag^+ ions are entrapped in the interstitial space and thus holes are created in the lattice.

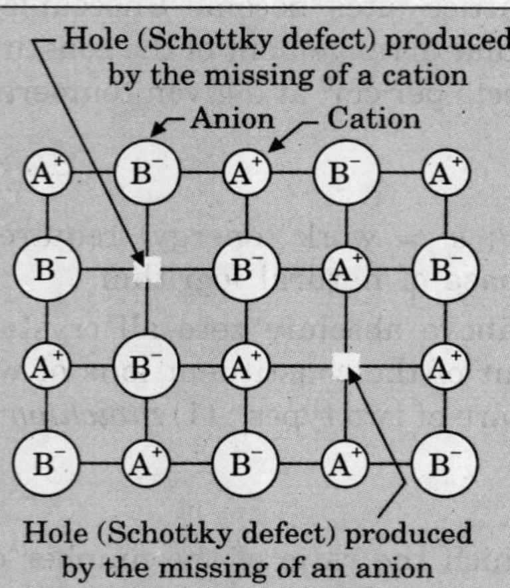

Fig. 9.32. Illustration of Schottky defect produced by the missing of one cation and one anion from their own position.

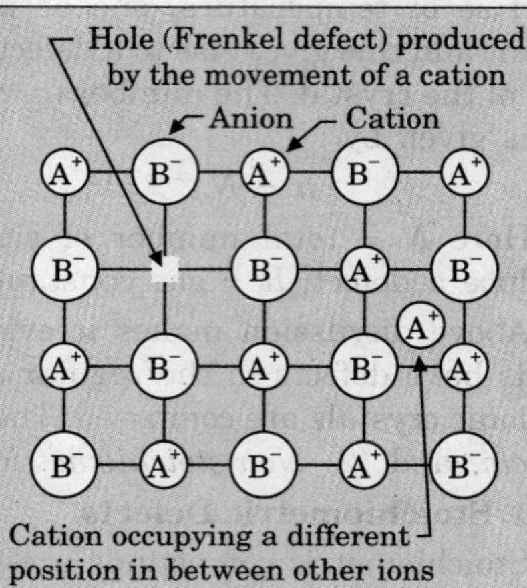

Fig. 9.33. Illustration of Frenkel defect produced by the movement of a cation and occupying a different position in between other ions.

Solids generally contain both the types of defects mentioned above, but one is more prominent than the other. The number of defects in a crystal increases generally with the increase of temperature. For example at room temperature NaCl ionic crystal has *one defect* for 10^{15} lattice sites while at 500°C it has 10^9 defects for the same number of lattice sites. Normally, Schottky defects are easier to form than Frenkel defects, since the former require less energy for their formation.

Consequences of Schottky and Frenkel Defects. (*a*) We have seen that in Frenkel defect similar charges come closer to each other. This closeness leads to an increase in the dielectric constant of the crystal.

(*b*) We have seen that is Schottky defect, the number of ions in the lattice of the crystal is decreased and this decrease in the number of ions decreases the density of the ionic solids.

(*c*) The presence of holes in both types of defects lowers the lattice energy or stability of the crystal.

(*d*) As a result of both the types of defects, the crystal is able to conduct electric current when electric field is applied between the two ends of a crystal. The conduction of electric current takes place by an ionic mechanism. In this mechanism, when an electric field is applied, a near-by ion leaves its original position and occupies another one (hole). This results in creating a new hole and another near-by ion leaves its original position and occupies this new hole and so on. This process continues and a hole migrates from one end to the other end and the conduction of electric current takes place in the whole crystal.

2. Non-stoichiometric Defects

Non-stoichiometric compounds are those in which there is a departure from stoichiometry based on the law of constant proportions, *i.e.* these compounds do not obey the law of constant proportion and hence the ratio of the number of A^+ ions to the number of B^- ions does not correspond to a simple whole number as suggested by the formula. In other words there is an excess of either positive charges (cations) or negative charges (anions) in the non-stoichiometric crystals. The excess of either positive ions or negative ions makes the crystal structure irregular and hence defects, which are called *non-stoichiometric defects*, are produced in the crystal lattice. These defects may exist in a crystal in addition to Schottky and Frenkel defects. The above discussion makes it evident that non-stoichiometric defects may be of two types, depending on the nature of the ions present in excess. These two types are evidently : (*i*) *Metal (positive ions) excess defects, and (ii) Non-metal (negative ions) excess defects or metal deficiency defects.*

(*i*) **Metal (positive ions) excess defects.** As the name suggests, this defect is produced due to the presence of excess of positive ions (metal) ions in the crystal lattice. The excess of positive ions can be achieved by any of the following two ways :

(*a*) In the first way, *one of the negative ions is missing from its real position, leaving a hole and this hole is occupied by an extra free electron so that the neutral nature of the crystal lattice may be maintained* (Fig. 9.34). The negative ion leaves the surface in the form of gas. For example, when excess of Na is burnt in Cl_2, we get yellow NaCl. The yellow colour is produced because the Cl^- ions leave the lattice and electrons occupy their position.

$$\begin{array}{c} Na^+\ Cl^-\ Na^+\ Cl^- \\ Cl^-\ Na^+\ e^-\ Na^+ \\ Na^+\ Cl^-\ Na^+\ Cl^- \end{array}$$

The electrons occupying the position vacated by the missing anions (Cl^- ions) are called *F centres*. (F = Farbe which is a German word and means colour). This defect is found in those crystals which have Schottky defects. Note that this defect contains only one hole which is occupied by an extra electron while the Schottky defect has two holes.

(*b*) In the second way, *an extra positive ion occupies a position in the lattice and in order to maintain the electrical neutrality of the crystal an electron also occupies a position in the lattice* (Fig. 9.35). This defect is found in those crystals which have Frenkel defects. This defect has no hole while Frenkel defect contains one hole in the lattice.

Consequences of Metal Excess Defects (*a*) The crystals having metal excess defects produced by any of the above two ways contain *free electrons*. Such crystals can, therefore, conduct electricity to some extent with the help of these free electrons. Since the number of free electrons in such crystals is very small, such crystals can conduct small amount of electricity. The amount of electricity conducted by such crystals is smaller even than that conducted by metals, fused salts and water solution of salts. It is for this reason that the crystals having metal excess defects are called **semiconductors.**

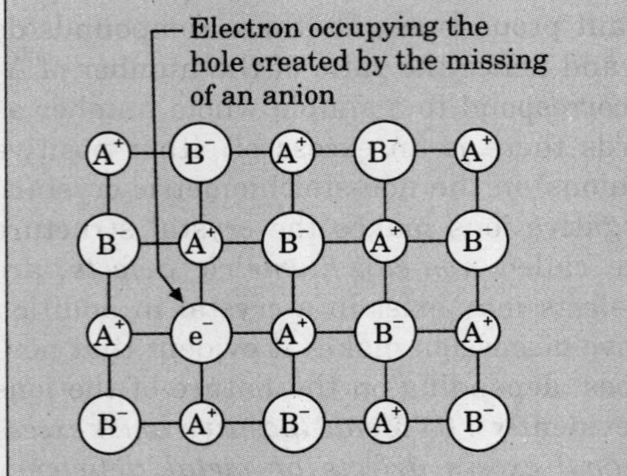

Fig. 9.34. Metal (positive ions) excess defect produced by the presence of an electron occupying the hole created by the missing of an anion.

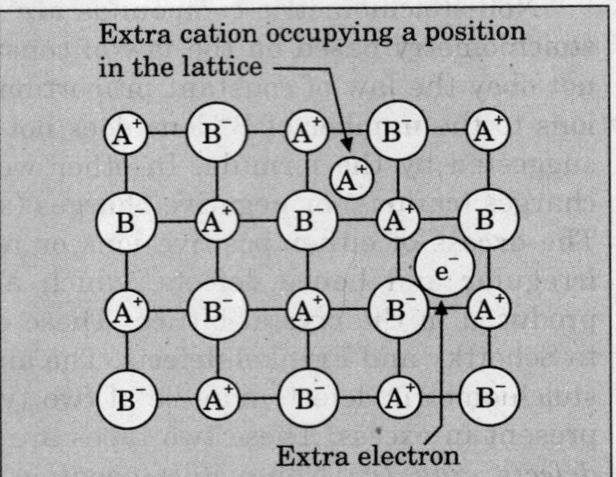

Fig. 9.35. Metal (positive ions) excess defect produced by the presence of an extra cation occupying a position in the crystal lattice.

(*b*) Due to the presence of free electrons, the crystals having metal excess defects are generally coloured. The free electrons get excited easily to higher energy levels by the absorption of certain wavelength from the visible white light and, therefore, the compounds appear coloured. For example non-stoichiometric sodium chloride is *yellow* and zinc oxide which is white in cold, *appears yellow when hot.*

(*ii*) **Non-metal (negative ions) excess defects or metal deficiency defects.** As the name suggests, this defect is produced by the presence of excess of negative ions in the crystal lattice. The excess of negative ions can be achieved by any of the following ways :

(*a*) In the first way, one of the positive ions is missing from its position and in order to keep the electrical neutrality of the crystal, one of the adjascent cations acquires one extra positive charge on it, becoming A^{2+} (Fig. 9.36). The cation A^+ acquiring are extra positive charge on it should be able to exhibit variable valency, *i.e.* such a cation should be of transition metals.

(*b*) In the second way, an extra negative ion occupies a position in the lattice and in order to maintain the electrical neutrality of the crystal, one of the adjascent cations acquires an extra positive charge on its becoming A^{2+} (Fig. 9.37).

Here it is essential to understand that since the size of the anions is usually large, it is difficult for the anions to fit into the interstitial sites. As a matter of fact, no crystals containing extra anions occupying the interstitial positions in the crystal lattice are known and hence the metal deficiency defect shown in Fig. 9.37 represents only a *theoretical prediction or possibility.*

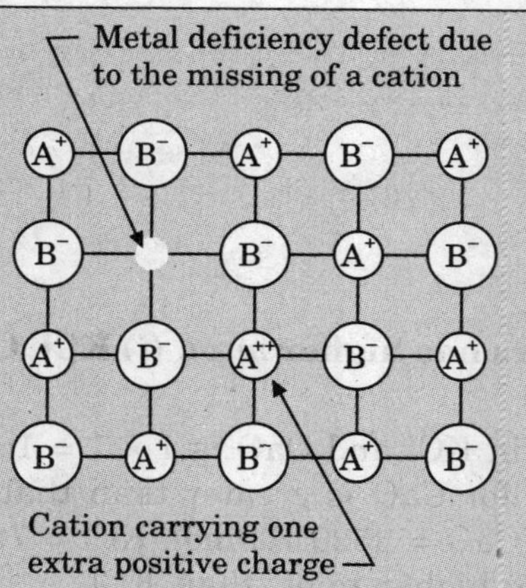

Fig. 9.36. Metal deficiency defect produced by the missing of one cation.

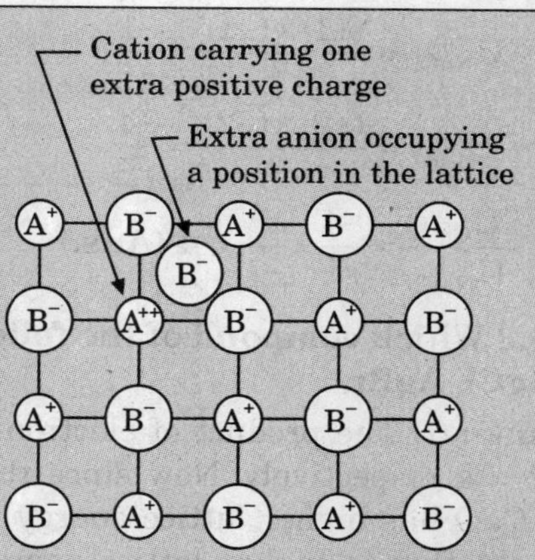

Fig. 9.37. Metal deficiency defect produced by the presence of an anion occupying a position in the lattice.

Consequences of Metal Deficiency Defects. We have seen that in metal deficiency defects one of the cation, say A^+, is changed into A^{2+} cation. This type of conversion takes place through the movement of an electron from A^+ to A^{2+}. Such type of movement is called *movement of positive hole* and makes the crystals showing metal deficiency defects behave as semi-conductors. Consequently the crystals which contain metal deficiency defects also act as semi-conductors.

Questions with Answers

Q.1 Which of the following alkali metal halides has the lowest lattice energy.

(*a*) LiF (*b*) NaCl (*c*) KBr (*d*) CsI.

Ans : (*d*).

Explanation. We know that lattice energy of an ionic salt decreases with the increase of the size of its cation and with the increase of the size of its anion. Thus an ionic salt having the largest cation and largest anion has the lowest lattice energy. Thus since Cs^+ ion is the largest cation among Li^+, Na^+ and K^+ ions and I^- ion is the largest anion among F^-, Cl^- and Br^- ions, CsI has the lowest value for its lattice energy.

Q.2 CaO and NaCl have same crystal structure and approximately the same inter-ionic distance. Calculate the ratio between U_{CaO} and U_{NaCl}.

Ans : We know that :

$$\text{Lattice energy of an ionic crystal} = \frac{q_1 \times q_2}{r^2},$$

where q_1 and q_2 are the charges on cation and anion and r is the inter-ionic distance. Since r is the same for NaCl and CaO,

$$U_{NaCl} = \frac{(+1)\times(-1)}{r^2} = \frac{-1}{r^2}$$

$$U_{CaO} = \frac{(+2)\times(-2)}{r^2} = \frac{-4}{r^2}$$

$$\therefore \frac{U_{CaO}}{U_{NaCl}} = \frac{-4}{r^2}\times\frac{r^2}{-1} = \mathbf{4\ (Ans.)}$$

Q.3 Which compond of the following pairs has higher *m.pt.* (*i*) KCl, CaO (*ii*) AgCl, AgBr.

Ans : (*i*) The product of electronic charges in KCl and CaO is $1 \times 1 = 1$ and $2 \times 2 = 4$ respectively. Now since this product for CaO is greater than that for KCl, CaO has higher lattice energy than KCl (CaO = 3520 KJ/mol, KCl = 701.2 KJ/mol). Due to higher lattice energy CaO has higher m.pt. than KCl.

(*ii*) Since both the halides contain the same cation *viz* Ag^+ but different halide ions, their lattice energy depends on the relative size of halide ions. Now since Br^- ion is larger in size than Cl^- ion, lattice energy of AgBr is lower than that of AgCl (AgBr = 883KJ/mol, AgCl = 895 KJ/mol) and hence AgBr has lower m.pt. than AgCl (AgBr = 434°C, AgCl = 455°C), *i.e.* AgCl has higher m.pt. than AgBr.

Q.4 Which of the following species has the highest lattice energy ? LiF, LiCl, NaCl and MgO.

Ans : The given species can be represented as Li^+F^-, $Li^+\ Cl^-$, Na^+Cl^- and $Mg^{2+}\ O^{2-}$. We know that lattice energy increases with the increase of charges on the ions. Now since MgO has +2 charges on Mg and –2 charges on O-atom, it has the highest lattice energy.

Q.5 If the close packed cations in an AB type solid have a radius of 75 pm, what would be the maximum and minimum sizes of the anions filling the voids ?

Ans : Since the cations in AB solid have cubic (fcc) unit cell or NaCl type structure and hence C.N. of each ion = 6. Thus the value of r_{A^+}/r_{B^-} should lie in the range 0.414 – 0.732, *i.e.*

$$\frac{r_{A^+}}{r_{B^-}} = 0.414 - 0.732$$

or $$\frac{75}{r_{B^-}} = 0.414 - 0.732$$

(*a*) Maximum value of r_{B^-}, $(r_{B^-})_{max}$ is obtained, if

$$\frac{75}{(r_{B^-})_{max}} = 0.414$$

or $$(r_{B^-})_{max} = \frac{75}{0.414}\ \text{pm} = \mathbf{181.2\ pm\ (Ans.)}$$

(*b*) Minimum value of r_{B^-}, $(r_{B^-})_{mini}$ is obtained, if

$$\frac{75}{(r_{B^-})_{mini}} = 0.732$$

or $$(r_{B^-})_{mini} = \frac{75}{0.732} \text{ pm} = \mathbf{102.5\ pm\ (Ans.)}$$

Q.6 (*i*) Br^- ions form a close packed structure. If the radius of Br^- ion is 195 pm, calculate the radius of the cation that just fits into the tetrahedral hole.

(*ii*) Can a cation A^+ having a radius of 82 pm be slipped into the octahedral hole of the crystal $A^+ Br^-$?

Ans : (*i*) Radius of the cation that can just fit into the tetrahedral hole should be equal to the radius of tetrahedral hole. Thus :

Radius of cation = Radius of tetrahedral void (hole)

= 0.225 × Radius of Br^- ions having close-packed structure

= 0.225 × 195 pm = **43.87 pm (Ans)**

(*ii*) Radius ratio $= \frac{r_{A^+}}{r_{Br^-}} = \frac{82}{195} = 0.4205$

Since the value of r_{A^+}/r_{Br^-} lies in the range 0.414 – 0.732, cation A^+ can fit into the octahedral hole.

Q.7 In a mixed oxide, oxide ions have *ccp* arrangement. Cations A occupy 12.5% of tetrahedral holes and cations B are present in 50% of octahedral holes. What is the simplest formula of the mixed oxide ?

Ans : Since oxide ions (O^{2-}) have *ccp* arrangement, each O^{2-} ion has two tetrahedral holes and one octrahedral hole.

12.5% of two tetrahedral holes $= \frac{2 \times 12.5}{100} = 0.25 = \frac{1}{4}$

50% of one octahedral hole $= \frac{1 \times 50}{100} = \frac{1}{2}$

∴ No. of A cations per O^{2-} ion = $A_{1/4}$

No. of B cations per O^{2-} = $B_{1/2}$

∴ Formula of the mixed oxide = $A_{1/4}\ B_{1/2}\ O$ = $\mathbf{AB_2O_4}$ **(Ans)**

Q.8 If AgI crystallises in a zinc blend structure with I^- ions forming the lattice, then calculate the fraction of tetrahedral voids occupied by Ag^+ ions.

Ans : Since zinc blende crystal has face-centerd cubic (*fcc*) unit cell, AgI is also a *fcc* crystal. Thus in AgCl crystal, I^- ions have *ccp* arrangement and Ag^+ ions occup altternate out of 8 available tetrahedral voids. Thus :

Fraction of tetrahedral voids occupied by Ag^+ ions $= \frac{4}{8} = \frac{1}{2}$ = **50% (Ans.)**

Q.9 The composition of a sample of wustile is $Fe_{0.93}O$. What percentage of iron is present in the form of Fe (III) ?

Ans : As a matter of fact the given compound, $Fe_{0.93}O$ contains Fe as Fe^{2+} and Fe^{3+} ions. If the formula contains x atoms of Fe^{3+} ions, then the number of Fe^{2+} ions present in the formula would be equal to $(0.93 - x)$. Hence the formula would be :

$$Fe^{2+}_{(0.93-x)}\ Fe^{3+}_{x}\ O^{2-}$$

Since the compound is neutral, the sum of charges should be equal to zero. Thus :

$$2(0.93 - x) + 3x - 2 = 0$$

or $\quad 1.86 - 2x + 3x - 2 = 0$

or $\quad x = 0.14$

Thus the actual formula of the compound is :

$$Fe^{2+}_{(0.93-0.14)}\ Fe^{3+}_{0.14}\ O^{2-} \text{ or } Fe^{2+}_{0.79}\ Fe^{3+}_{0.14}\ O$$

Molar mass of the compound $= 0.79 \times 56 + 0.14 \times 56 + 1 \times 16$

$= 56\ (0.79 + 0.14) + 16$

$= 56 \times 0.93 + 16$

$= 68.08$ g

Weigth of Fe^{3+} present in the compound $= 0.14 \times 56 = 7.84$ g

$\therefore$ % of Fe present as Fe (III) $= \dfrac{7.84}{68.08} \times 100\% =$ **11.5% (Ans.)**

Q.10 Why is diamond a bad conductor of electricity but a good conductor of heat.

Ans : Conduction of electricity essentially requires movement of electrons while conduction of heat does not necessarliy require movement of electrons.

Q.11 Make a comparison between hexagonal close packing (*hcp*) and cubic close packing (*ccp*) of spheres.

Ans : (A) Similarities. (*i*) In both the packings, each sphere (atom) is in direct contact with 6 other spheres in its own layer, 3 other spheres in the upper layer and 3 other spheres in the lower layer.. Thus in these structures each sphere is in direct contact with 6 + 3 + 3 = 12 spheres. Hence C.N of each atom in these structures is equal to 12.

(*ii*) In both the structures, 74% of the available space is occupied by spheres. The remaining 26% of the space remains vacant, *i.e.* 26% of the space has voids or holes.

(B) Dissimilarities. (*i*) In *hcp* structure the different layers of atoms are packed in AB AB AB... patter. A layer has 7 spheres (atoms) and B layer contains 3 atoms. As a result, AB AB... pattern can also be represented as 73, 73... pattern. Here 7 and 3 represent the number of atoms present in A and B layers respectively. In *ccp* structure, different layers of atoms are packed in ABC ABC ABC... pattern. A layer has 1 atom, B and C both the layers have 6 atoms each. As a result ABC ABC... pattern can also be respresent as 166, 166... pattern. Here 1, 6 and 6 represent the number of atoms in A, B and C layers respectively.

(*ii*) In *hcp* structure, third layer of atoms covers the tetrahedral voids of the second layer.In *ccp* structure, third layer of atoms covers the octahedral voids of the second layer.

(*iii*) In *hcp* structure, third layer is similar to the first layer. Hence this structure has AB AB... arrangement. In *ccp* structure, every fourth layer is similar to the first layer and hence this structure has ABC ABC... arrangement.

Q.12 Determine the following :

(*i*) Number of ZnS units in a unit cell of zinc blende.

(*ii*) Number of CaF_2 units in a unit cell of fluorite.

(*iii*) Number of NaCl units in a unit cell of rock salt.

(*iv*) Number of CsCl units in a unit cell of CsCl crystal.

Ans : (*i*) The unit cell of zinc blende has *ccp* (or *fcc*) arrangement (or packing) of S^{2-} ions, *i.e.* in the unit cell of zinc blende S^{2-} ions are present on each of the 8 corners and on each of the six face centres. Unit cell of zinc blende has four Zn^{2+} ions within the body of the unit cell. Each Zn^{2+} ion is surrounded tetrahedrally by S^{2-} ions. Thus :

No. of Zn^{2+} ions in the unit cell of zinc blende

= 4 (within the body of the unit cell).

and, No. of S^{2-} ions in the unit cell of zinc blende

$$= 8 \times \frac{1}{8} \text{ (corner } S^{2-} \text{ ions)} + 6 \times \frac{1}{2} \text{ (face-centred } S^{2-} \text{ ions)}$$

$$= 1 + 3 = 4$$

$\therefore$ Formoula : $Zn_4S_4 \equiv 4\ ZnS$

Obviously the number of ZnS units in the unit cell of zinc blende = 4.

(*ii*) The unit cell of fluorite has *ccp* (or *fcc*) arrangement of Ca^{2+} ions, *i.e.* in the unit cell of fluorite, Ca^{2+} ions are present on each of the 8 corners and one on each of the six face centres. Unit cell of fluorite also has eigth F^- ions within the body of the unit cell. Thus :

No. of Ca^{2+} ions in the unit cell of fluorite (CaF_2)

$$= 8 \times \frac{1}{8} \text{ (corner } Ca^{2+} \text{ ions)} + 6 \times \frac{1}{2} \text{ (face-centered } Ca^{2+} \text{ ions)}$$

$$= 1 + 3 = 4$$

No. of F^- ions = 8 (within the body of the unit cell)

$\therefore$ Formula : $Ca_4F_8 \equiv 4CaF_2$

Obviously the unit cell of fluorite has four CaF_2 units.

(*iii*) The unit cell of rock salt has fcc arrangement (or packing) of Cl^- ions, *i.e.* in the unit cell of rock salt, Cl^- ions are present on each of 8 corners and one on each of the six face centres. Unit cell of rock salt also contains one Na^+ ion at its body centre and 12 Na^+ ions on 12 edge centres. Thus :

No. of Na^+ ions in the unit cell of rock salt

$= 1$ (body-centred Na^+ ion) + $12 \times \frac{1}{4}$ (edge-centerd Na^+ ions)

$= 1 + 3 = 4$

No. of Cl^- ions in the cell of rock salt

$= 8 \times \frac{1}{8}$ (corner Na^+ ions) + $6 \times \frac{1}{2}$ (face-centerd Na^+ ions)

$= 1 + 3 = 4$

$\therefore$ Formula : $Na_4Cl_4 \equiv 4NaCl$

Obviously unit cell of rock salt has four NaCl units.

(*iv*) The unit cell of CsCl crystal has $8Cl^-$ ions on eight corners of the unit cell and one Cs^+ ion of the body centre. Hence :

No. of Cl^- ions present in unit cell $= 8 \times \frac{1}{8} = 1$

No. of Cs^+ ions present in unit cell = 1 (in the body of the unit cell)

$\therefore$ Formula : CsCl

Obviously the number of CsCl units in the unit cell of CsCl is 1.

Q.13 What are the diffirences between Schottky and Frenkel defects.

Ans : The main points showing the diffirences between these two types of defects are given in the following Table.

Table. Differences between sclottky and Frenkel defects.

Schottky defect	Frenkel defect
(*i*) This defect is created when equal number of cations and anions are completely missing from their lattice sites.	(*i*) This defect is created when ions (usually cations) leave their original sites and occupy the vacant interstial sites else where in ionic crystal.
(*ii*) This defect decreases the density of ionic crystal.	(*ii*) This defect does not change the density of the crystal.
(*iii*) This defect does not change the dielectric constant value of ionic crystal.	(*iii*) This defect increases the value of dielectric constant value of the crystal.
(*iv*) This defect is shown by those ionic crystals in which C.N. of cations and anions is high and these ions have small diffirence in their size.	(*iv*) This defect is shown by those ionic crystals in which C.N. of cations and anions is low and these ions have large difference in their size.
(*v*) This defect is shown by alkali metal halides.	(*v*) Alkali metal halides do not show this defect.
(*vi*) Examples : NaCl, KCl, AgBr, CsCl.	(*vi*) Examples : AgCl, AgBr, AgI, Zns etc.

Q.14 ZnO is white but it turms yellow and shows enhanced electrical conductivity on heating. Explain.

Ans : When ZnO is heated, it loses O_2 gas and Zn^{2+} ions are produced.

$$ZnO \xrightarrow{\Delta} Zn^{2+} + \frac{1}{2}O_2 + 2e^-$$

Zn^{2+} ions thus produced occupy the vacant interstial sites and liberated electrons occupy the neighbouring vacant intesstial sites as shown in Fig. 9.38 so that the neutrality of ionic crystal (ZnO) may be maintained. It is due to the presence of electrons that white ZnO becomes yellow on heating and its electrical conductivity also increases.

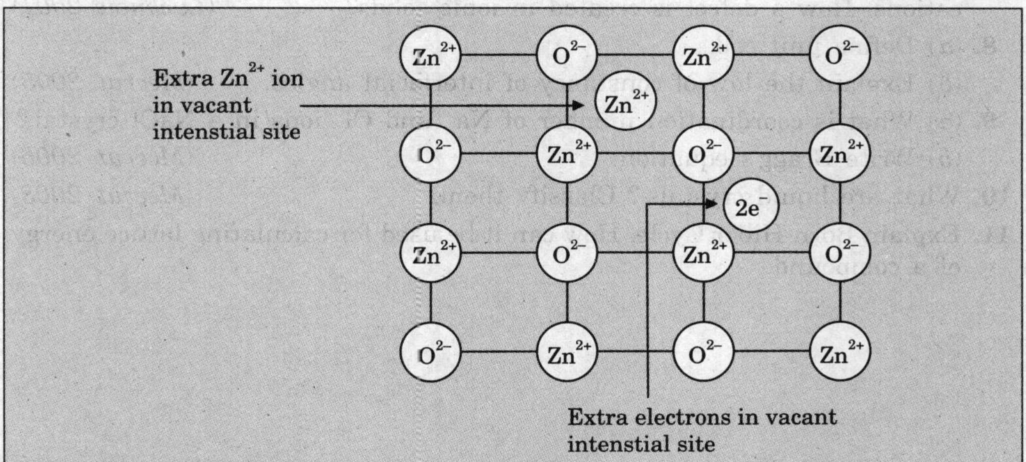

Fig. 9.38. To explain why ZnO (white) becomes yellow and shows enhanced conductivity on heating.

University Questions

1. (*a*) What is Born-Haber cycle ? Explain how it is used for calculating lattice energy of NaCl ?

(*b*) What is the total number of ions in a unit cell of NaCl?

(*Himachal Pradesh 2000*)

2. Name important defects found in crystals. How are they caused ? What is the effect of these defects on the solubility of the crystals?

(*Kanpur 2000*)

3. Write a note on : (*i*) Radius ratio and its limitations (*ii*) Metallic bond (*iii*) Lattice energy. (*Gauhati* (*General*) *2000*)

4. (*a*) Explain why metal excess non-stoichiometric compounds behave as semiconductors.

(*b*) Distinguish between octahedral and tetrahedral sites (holes) formed in closed-packed ionic solids. Derive the limiting radius ratio (r/r) for both these sites. (*Lucknow 2000*)

5. (*i*) Draw and explain the elements of symmetry present in a face centered cubic crystal.

(*ii*) Draw (100), (110) and (111) planes and explain interplaner distance. (8) (*Bharathiar 2000*)

6. What is lattice energy ? How it is determined? (*Lucknow 2001*)

7. Based on the hard sphere model for the packing of ionic solids, relate the limiting radius ratio of cations and anions to the coordination number of cations. How a defect is created in ionic solids? (*Lucknow 2001*)

8. (*a*) Define unit cell.

(*b*) Explain the law of constancy of interfacial angles. (*Meerut 2005*)

9. (*a*) What is coordination number of Na^+ and Cl^- ions in a NaCl crystal ?

(*b*) Write Bragg's equation. (*Meerut 2006*)

10. What are liquid crystals ? Classify them. (*Meerut 2008*)

11. Explain Born-Hiber Cycle. How can it be used for calculating lattice energy of a compound.

CHAPTER 10

Oxidation-Reduction Reactions

Covalency, Oxidation Number and Oxidation State

Covalency *of an element represents (i) the number of H-atoms with which an atom of that element can combine (ii) the number of single bonds which an atom of that element can form (iii) the number of electrons its atom is able to share with other element. Thus in every case covalency of an element is a pure number and has no plus or minus sign associated with it.* For example covalency of nitrogen in NH_3 is 3.

Oxidation number *of an element is defined as the formal charge which an atom of that element appears to have when electrons are counted. Oxidation number of an atom may be positive or negative.* For example (*i*) oxidation number of K and Br in KBr is +1 and – 1 respectively. (*ii*) Oxidation number of N in NH_3 is – 3.

There may be a difference between the magnitude of covalency and oxidation number of the same element in various compounds. For example, in each of the following compounds, since C-atom shares four pairs of electrons with other atoms, the covalency of carbon is 4 while the oxidation number of carbon in these compounds is – 4, – 2, 0, + 2 and + 4 respectively.

Compound :	**CH_4 Methane**	**CH_3Cl Methyl chloride**	**CH_2Cl_2 Methylene chloride**	**$CHCl_3$ Chloroform**	**CCl_4 Carbon tetra-chloride**
Covalency of carbon :	4	4	4	4	4
Oxidation number of carbon :	– 4	– 2	0	+ 2	+ 4

In ionic compounds the **oxidation state** *of an element is the same as the charge on the ion formed from the atom of the element.* For example in KBr, K is said to be in + 1 oxidation state while Br is said to be in – 1 oxidation state.

Oxidation state of an element is its oxidation number of per atom. For example the oxidation *state of Al in Al_3O_3 = + 6/2 = + 3.*

Differences between Oxidation Number (O.N.) and Valency

The differences between oxidation number and valency has been shown in a tabular form as follows :

Oxidation Number (O.N.)	Valency
(1) O.N. of an element is the number of excess charges that an atom of the element has or appears to have in its combined state, assuming that each electron belongs either to one or another atom.	(1) Valency of an element is its combining capacity and is expressed as the number of H-atoms or Cl atoms or double the number of oxygen atoms that combine with one atom of the element.
(2) With O.N. plus (+) or minus (–) sign is attached as it is the charge left on the atom of the element. For example O.N. of C in CO = +2 while O.N. of C in CH_3Cl = –2.	(2) Since valency is the combining capacity of the element, no plus or minus sign is attached to it. For example valency of C is 4.
(3) O.N. of an element varies from compound to compound. *i.e.* O.N. of an element depends on the nature of the compound in which the element is present. For example O.N. of C in CH_4, C_2H_6, C_2H_4, C_2H_2 and CH_2Cl_2 is –4, –3, –2, –1 and 0 respectively.	(3) Valency of an element is usually fixed. For example the valency of carbon in its compounds is 4, *i.e.* carbon is tetravalent in its compounds.
(4) O.N. of an element may be whole number or it may be fractional. For example O.N. of Mn in Mn_3O_4 is $+\frac{8}{3}$.	(4) Valency of an element is always whole number and is never fractional. For example valency of Mn is 2.
(5) O.N. of an element can be zero. For example O.N. of carbon in CH_2Cl_2 is 0.	(5) Valency of an element is never zero (exception is noble gases), since it is the combining capacity of the element.

Rules for Calculating Oxidation Number

In order to calculate the oxidation number (O.N.) of an element in a given molecule/ion, the following rules are followed. These are arbitrary rules :

(*i*) The oxidation number of an element in the free or uncombined state is zero.

(*ii*) Fluorine, the most electronegative element, has an oxidation number of –1 in all its compounds.

(*iii*) The oxidation number of hydrogen is +1 in all its compound except in the ionic metal hydrides where its O.N. is –1. For example the O.N. of H in H_2O is +1 while its O.N. in NaH is equal to –1.

(*iv*) The O.N. of oxygen is generally equal to –2 except in F_2O in which the O.N. of oxygen is equal to +2. In H_2O_2 molecule whose Lewis structure is

:Ö··Ö: *or* :Ö—Ö:, the electron pair shared between H and O atoms is counted
 ·× ·× | |
 H H H H

with O-atom, since it is more electronegative than H-atom. Therefore the number of electrons counted with each O-atom is equal to seven instead of its outermost shell six electrons. Therefore O-atom in H_2O_2, molecule appears to have –1 charge on it, *i.e.* O.N. of O-atom in H_2O_2 molecule (or its derivatives like BaO_2 etc.) is equal to –1.

(*v*) The net charge on a given ion is equal to the sum of the oxidation numbers of all the atoms present in the ion.

(*vi*) The oxidation number of a neutral molecule is always zero and is equal to the sum of oxidation numbers of the individual atoms, each multiplied by the number of atoms of the element in the molecule.

Solved Examples

The rules given above can be illustrated by studying the following solved examples.

Example 1. *Calculate the oxidation number of (i) Cr in $Na_2Cr_2O_7$ (ii) Mn in $KMnO_4$ (iii) S in H_2SO_4.*

Solution. (*i*) **Cr in $Na_2Cr_2O_7$**

Let the oxidation number of Cr in $Na_2Cr_2O_7$ be x

O.N. of Na = + 1; O.N. of each O = –2

Putting the sum oxidation numbers equal to zero, we have :

$$2 \times 1 + 2x + [7 \times (-2)] = 0$$

or $$x = +6$$

∴ Oxidation number of Cr in $Na_2Cr_2O_7$ is +6.

(*ii*) **Mn in $KMnO_4$.**

Let the oxidation number of Mn in $KMnO_4$ be x

O.N. of K = + 1; O.N. of each O atom = –2

Putting the sum of the oxidation numbers of the atoms in the compound equal to zero, we have :

$$1 + x + 4 \times (-2) = 0$$

or $$x = +7$$

∴ Oxidation number of Mn in $KMnO_4$ is + 7

(*iii*) **S in H_2SO_4.**

Let the O.N. of S in H_2SO_4 be x

O.N. of each hydrogen = + 1; O.N. of each O atom = –2

∴ $$2 + x + 4(-2) = 0$$

or $$x = +6$$

∴ Oxidation number of S in H_2SO_4 is + 6.

Example 2. *Calculate the oxidation number of (i) Mn in MnO_4^- and MnO_4^{2-} ions (ii) Cr in $Cr_2O_7^{2-}$ and CrO_2Cl_2 (iii) S in $S_2O_7^{2-}$.*

(Kanpur B.Sc. 1980; Maduarai 1975)

Solution. (*i*) Let the O.N. of Mn be x.

Now O.N. of each oxygen = –2.

∴ *In MnO_4^- ion :*

$$1 \times x + 4 \times (-2) = -1$$

or $$x = +7$$

$\therefore$ O.N. of Mn in MnO_4^- ion = +7

Similarly, *in MnO_4^{2-} ion* :

$$1 \times x + 4 \times (-2) = -2$$

or $$x = +6$$

$\therefore$ O.N. of Mn ion MnO_4^{2-} ion = + 6.

(*ii*) Let the O.N. of Cr be x

Now O.N. of each O-atom = – 2

and O.N. of each Cl-atom = – 1

$\therefore$ *In $Cr_2O_7^{2-}$ ion* :

$$2 \times x + 7 \times (-2) = -2$$

or $$x = +6$$

$\therefore$ ON. of Cr in $Cr_2O_7^{2-}$ ion = + 6

Similarly, *in CrO_2Cl_2 molecule* :

$$1 \times x + 2 \times (-2) + 2(-1) = 0$$

or $$x = +6$$

$\therefore$ O.N. of Cr in CrO_2Cl_2 = + 6.

(*iii*) Let the O.N. of S be x

$\therefore$ $$2 \times x + 7 \times (-2) = -2$$

or $$x = +6$$

$\therefore$ O.N. of S in $S_2O_7^{2-}$ ion = + 6.

Oxidation and Reduction

Oxidation is a reaction in which an atom or an ion loses one or more electrons and thus increases its valency, i.e. in oxidation the atomic or ionic system loses one or more electrons and is changed into more electropositive or less electronegative state. Due to the loss of electrons, oxidation is also called *de-electronation.*

Reduction is a reaction in which an atom or an ion gains one or more electrons and thus decreases its valency, i.e. in reduction the atomic or ionic system gains one or more electrons and is changed into less electropositive or more electronegative state. Due to gain of electrons, reduction is also called *electronation.*

Illustration. The concept of oxidation and reduction has been illustrated in Fig. 10.1 in which n indicates the number of electrons present in atom/cation/anion. When we proceed upwards from zero valency (Å), the number of electrons (n) goes on decreasing and hence oxidation takes place. Similarly on proceeding downwards from zero valency (Å), the number of electrons (n) goes on increasing and hence reduction takes place.

Redox Reactions and Half Reactions

Since loss or gain of electrons are relative terms, the gain or loss of electrons takes place simultaneously in a chemical reaction. Thus the oxidation (*i.e.* loss of electrons) and reduction (*i.e.* gain of electrons) reactions go hand in hand and

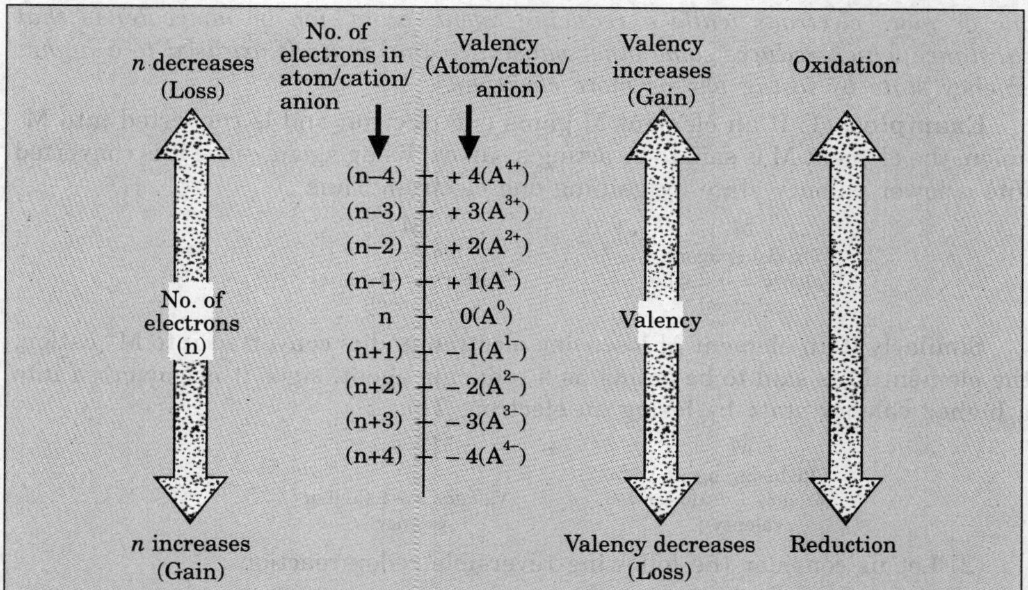

Fig. 10.1. Illustration of the concept of oxidation and reduction. n indicates the number of electrons present in atom/cation/anion.

such reactions in which oxidation and reduction take place simultaneously are known as **oxidation-reduction reactions** or **redox reactions.** For example the reaction between Zn and $CuSO_4$ solution shown as :

$$\left.\begin{array}{ll} & Zn + CuSO_4 \rightarrow ZnSO_4 + Cu \\ \text{or} & Zn + Cu^{2+} \rightarrow Zn^{2+} + Cu \end{array}\right\} \quad ...(A)$$

is a redox reaction, since Zn atom (valency = 0) by losing two electrons is oxidised to Zn^{2+} ion (valency = +2) while Cu^{2+} ion (valency = +2) by gaining the same number of electrons lost by Zn atom gets reduced to Cu atom (valency = 0).

Thus we see that a redox reaction consists of two reactions—one involves oxidation (*e.g.* $Zn \rightarrow Zn^{2+} + 2e^-$) and the other involves reduction (*e.g.* $Cu^{2+} + 2e^- \rightarrow Cu$). Each of these reactions is called **half reaction.** The reaction showing oxidation is called **oxidation half-reaction** while that representing reduction is called **reduction half-reaction.** Thus the two half-reactions of which the redox reaction (A) is composed of are :

$$Zn \rightarrow Zn^{2+} + 2e^- \text{ ...(called } \textit{oxidation half-reaction}) \quad ... (i)$$

and, $$Cu^{2+} + 2e^- \rightarrow Cu \text{ ...(called } \textit{reduction half-reaction}) \quad ...(ii)$$

It may be noted that *the number of electrons lost or gained in two opposite half-reactions of a redox reaction is equal and the reaction mixture of a redox reaction is electrically neutral.*

Oxidising Agent (Oxidant) and Reducing Agent (Reductant)

An oxidising agent (atom, ion or molecule) is that substance which oxidises some other substance, and is itself reduced to a lower valency state by gaining one or more electrons while a reducing agent (atom, ion or molecule) is that substance which reduces some other substance, and is itself oxidised to a higher valency state by losing one or more electrons.

Examples. (1) If an element M gains one electron and is converted into M^- anion, the element M is said to be acting as an oxidising agent, since it is converted into a lower valency state by gaining one electron. Thus :

$$\underset{\substack{\text{Oxidising agent} \\ \text{[Valency = 0 (higher valency)]}}}{M} + e^- \rightarrow \underset{\text{Valency = -1 (lower valency)]}}{M^-}$$

Similarly if an element M loses one electron and is converted into M^+ cation, the element M is said to be acting as a reducing agent, since it is converted into a higher valency state by losing an electron. Thus :

$$\underset{\substack{\text{Reducing agent} \\ \text{[Valency = 0 (lower valency)]}}}{M} \rightarrow \underset{\text{Valency = +1 (higher valency)]}}{M^+ + e^-}$$

(2) Let us consider the following reversible redox reaction :

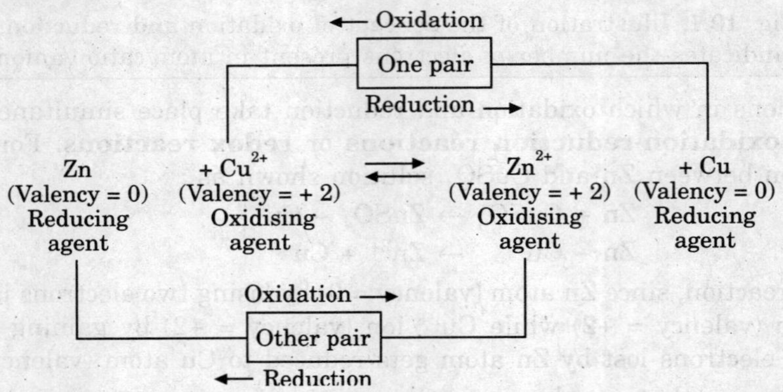

When we consider the left hand side of this reaction, we find that, since Zn metal reduces Cu^+ ions to Cu metal and is itself oxidised to Zn^{2+} ion, Zn metal is said to be acting as a reducing agent. Now we can also say that Cu^+ ion oxidises Zn metal to Zn^{2+} ion and is itself reduced to Cu metal and hence Cu^{2+} ion acts as an oxidising agent. Now when we consider the right hand side of the above reaction, we find that, since Zu^{2+} ion oxidises Cu metal to Cu^{2+} ion and is itself reduced to Zn metal, Zn^{2+} ion acts as an oxidising agent. Similarly since Cu metal reduces Zn^{2+} ion to Zn metal and is itself oxidised to Cu^{2+} ion, Cu metal acts as a reducing agent. On considering both the sides simultaneously we conclude that each side of a redox reaction consists of one oxidising and one reducing agent as written below each of the reactants and products.

The following points may be noted :

(*i*) The oxidising agent has higher oxidation state than its counter part reducing agent lying on the other side of the redox reaction. For example in the above redox reaction Cu^{2+} ion (oxidising agent) is in +2 oxidation state while Cu (counter part reducing agent) is in zero oxidation state. Similarly Zn^{2+} ion (oxidising agent) is in +2 oxidation state while Zn (counter part reducing agent) is in zero oxidation state.

(*ii*) The oxidising agent is obtained when its counter part reducing agent loses electrons. Similarly a reducing agent is obtained when its counter part oxidising agent gains electrons. For example Zn^{2+} ion (oxidising agent) is obtained when Zn (counter part reducing agent) loses two electrons and Cu (reducing agent) is obtained when Cu^{2+} ion (counter part oxidising agent) gains two electrons.

Equivalent Weights of Oxidising Agents

The equivalent weight of an oxidising agent (molecule or ion) is its that weight which can take up one electron, i.e. the equivalent weight of an oxidising agent is equal to its molecular weight or ion weight divided by the number of electrons gained by its per molecule or per ion.

Examples. (*i*) In the reduction of Fe^{3+} ion to Fe^{2+} ion as shown below :

$$Fe^{3+} + e^- \rightarrow Fe^{2+},$$

Fe^{3+} ion gains one electron and hence the equivalent weight of Fe^{3+} ion (oxidising agent) is equal to its ion-weight, *i.e.* equal to 56 gm. Accordingly the equivalent weight of $FeCl_3$ in its reduction to $FeCl_2$ is the same as its molecular weight.

(*ii*) Sn^{4+} ion gains two electrons for its reduction to Sn^{2+} ion.

$$Sn^{4+} + 2e^- \rightarrow Sn^{2+}$$

Hence the equivalent weight of Sn^{4+} ion is half the ion-weight of Sn^{4+} ion. Accordingly the equivalent weight of $SnCl_4$ is equal to half of its molecular weight.

(*iii*) MnO_4^- ion in its reduction to Mn^{2+} ion in acid solution gains five electrons

$$MnO_4^- + 8H^+ + 5e^- \rightarrow Mn^{2+} + 4H_2O$$

Thus the equivalent weight of MnO_4^- ion is *one-fifth of its ion-weight, i.e.* (54.9 + 64)/5 = 23.8 gms. Accordingly the equivalent weight of $KMnO_4$ = (39.1 + 54.9 + 64)/5 = 158.0/5 = 31.6 gms.

(*iv*) The oxidation reaction of $Cr_2O_7^{2-}$ in an acid solution proceeds with the gain of six electrons as shown :

$$Cr_2O_7^{2-} + 14H^+ 6e^- \rightarrow 2Cr^{3+} - 7H_2O$$

Consequently the equivalent weight of $Cr_2O_7^{2-}$ ion is equal to (51.9 × 2 + 16 × 7)/6 = 35.9 gms. Accordingly the equivalent weight of $K_2Cr_2O_7$ is equal to (39 × 2 + 51.9 × 2 + 16 × 7)/6 = 48.9 gms.

Equivalent Weights of Reducing Agents

On the basis of the similar arguments as used for oxidising agents, *the equivalent weight of a reducing agent (molecule or ion) is its that weight which can loss one electron. i.e. the equivalent weight of a reducing agent is equal to its molecular weight or ion weight divided by the number of electrons lost by its per molecule or per ion.*

Examples. (*i*) ***Oxidation of Fe^{2+} ion to Fe^{3+} ion.*** The conversion of Fe^{2+} ion into Fe^{3+} ion is represented by :

$$\underset{\text{Reducing agent}}{Fe^{2+}} \rightarrow \underset{\text{Oxidising agent}}{Fe^{3+}} + e^-$$

Thus the equivalent weight of Fe^{2+} ion (which is a reducing agent in this reaction) is equal to its ion-weight.

(*ii*) ***Oxidation of Sn^{2+} ion to Sn^{4+} ion.*** The oxidation of Sn^{2+} ion into Sn^{4+} ion involves the loss of two electrons as shown below :

$$\underset{\text{Reducing agent}}{Sn^{2+}} - 2e^- \rightarrow \underset{\text{Oxidising agent}}{Sn^{4+}} + 2e^-$$

Thus the equivalent weight of Sn^{2+} ion is equal to its ion-weight/2 because it loses two electrons to convert into Sn^{4+} ion.

(*iii*) ***Oxidation of $S_2O_3^{2-}$ ion to $S_4O_6^{3-}$ ion.*** The equivalent weight of a $S_2O_3^{2-}$ ion whose oxidation proceeds as follows :

$$2S_2O_3^{2-} \rightarrow S_4O_6^{2-} + 2e^-$$

is equal to (2 × ion-weigh)/2, *i.e.* equal to its ion-weight. Hence equivalent weight of $Na_2S_2O_3.5H_2O$ is equal to its molecular weight, *i.e.* equal to 248.

It should be remembered that the equivalent weight of a particular oxidising agent is not a constant quantity. It may change depending on the nature of the oxidation-reduction reaction. For example, the oxidation reactions of $KMnO_4$ solution in acidic and alkaline medium are shown as :

$MnO_4^- + 8H^+ + 5e^- \rightarrow Mn^{2+} \downarrow + 4H_2O$ (*Acidic medium*)

$MnO_4^- + 2H_2O + 3e^- \rightarrow MnO_2 \downarrow + 4OH^-$ (*Alkaline medium*)

The above reactions show that the *equivalent weight of $KMnO_4$ as an oxidising agent in acidic medium is equal to one-fifth of its molecular weight while its equivalent weight in alkaline medium is equal to one-third of its molecular weight.*

Auto-oxidation

There are substances like turpentine, olefinic compounds, phosphorus and some metals (*e.g.* Zn and Pb) which have a tendency to absorb O_2 from the air and then become active. These active substances can oxidise other substances which are not normally oxidised by them. For example :

(*i*) H_2O in presence of lead is oxidised by air to H_2O_2.

$$Pb + O_2 + H_2O \rightarrow PbO + H_2O_2$$

(*ii*) When a dilute solution of KI mixed with a small amount of turpentine is allowed to stand in an open vessel, a small amount of H_2O_2 is formed which is indicated by the liberation of I_2. If starch solution is added, it turns blue.

In both the examples given above the formation of H_2O_2 by the oxidation of H_2O is called **auto-oxidation.**

Explanation of auto-oxidation. In order to explain the phenomenon of auto-oxidation, *Bach* suggested that the other substance is supposed to combine with O_2 to form an *unstable addition compound*, called *moloxide* which then gives up oxygen. This oxygen is used up by H_2O to form H_2O_2. Bach called lead an **activator,** since it activated O_2 to combine with H_2O to form H_2O_2 and water was called an **acceptor.** Thus the formation of H_2O_2 can be represented by the following mechanism :

$$\underset{\text{Activator}}{Pb} + O_2 \rightarrow \underset{\substack{\text{Moloxide}\\ \text{(unstable addition}\\ \text{compound)}}}{PbO_2}$$

$$PbO_2 \rightarrow PbO + O$$

$$\underset{\text{Acceptor}}{H_2O} + O \rightarrow H_2O_2$$

The turpentine or other unsaturated compounds which act as activators are supposed to take up O_2 molecule in the form of —O—O— at the double bond position to form the unstable peroxide (*moloxide*).

$$RHC{-}CHR + O_2 \rightarrow \begin{array}{c} RHC{—}CHR \\ | \quad\;\; | \\ O{—}O \end{array} \text{ (Moloxide)}$$

This moloxide gives up oxygen which is used up by H_2O molecule or any other acceptor. The bleaching and disinfecting action of terpentine are due to the formation of H_2O_2.

Induced Oxidation

It has been observed that if a solution of Na_2SO_3 is exposed to air, it *gets oxidised to* Na_2SO_4. But, if a solution of Na_2AsO_3 is exposed to air, it *does not get oxidised* to Na_2AsO_4. However, if a solution of Na_2SO_3 is mixed with that of Na_2AsO_3 and then the mixture be exposed to air, *both undergo oxidation.*

$$Na_2SO_3 + Na_2AsO_3 + O_2 \rightarrow Na_2SO_4 + Na_2AsO_4$$

Thus we see that Na_2SO_3, while undergoing oxidation on exposure to air, also brings about oxidation of Na_2AsO_3 which itself cannot oxidise when exposed to air. Thus :

The phenomenon in which one substance while undergoing oxidation itself on exposure to air brings about oxidation of another substance also which of itself cannot take up oxygen is known as induced oxidation.

Explanation of induced oxidation. Induced oxidation can be explained by Brodie's theory and by Bach's theory.

(*a*) *Brodie's theory.* According to this theory H_2O_2 is formed which brings about the oxidation of Na_2AsO_3 to Na_2AsO_4.

$$Na_2SO_3 + C_2 + H_2O \rightarrow Na_2SO_4 + H_2O_2$$

$$Na_2AsO_3 + H_2O_2 \rightarrow Na_2AsO_4 + H_2O$$

$$\overline{Na_2SO_3 + Na_2AsO_3 + O_2 \rightarrow Na_2SO_4 + Na_2AsO_4}$$

(*b*) *Bach's theory.* The theory is based on the formation of moloxide (*See auto-oxidation*) by the reaction between Na_2SO_3 and O_2.

$$Na_2SO_3 + O_2 \rightarrow \underset{\text{Moloxide}}{Na_2SO_5}$$

$$Na_2SO_5 + Na_2AsO_3 \rightarrow Na_2SO_4 + Na_2AsO_4$$

$$Na_2SO_3 + Na_2AsO_3 + O_2 \rightarrow Na_2SO_4 + Na_2AsO_4$$

Balancing Redox Equations by Oxidation Number Method.

This method is based on the principle that any increase in oxidation number must be compensated by a decrease. This method consists of the following steps :

(*i*) Note the elements which undergo change in oxidation numbers.

(*ii*) Select the suitable coefficients for the oxidising and reducing agents so that the total decrease in oxidation number of the oxidising agent becomes equal to the total increase in the oxidation number of the reducing agent.

Solved Examples

The following examples illustrate the above method.

Example 1. *Balance the following equation by oxidation number method* :

$$CuO + NH_3 \rightarrow Cu + N_2 + H_2O$$

Solution. The given equation shows that the oxidation number (O.N.) of Cu decreases from +2 (in CuO) to 0 (in Cu) while that of N increases from –3 (in NH_3) to 0 (in N_2) and hence :

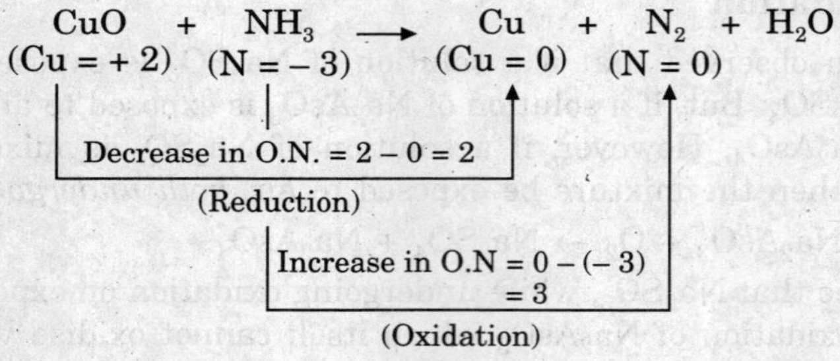

In order to equalise the total increase in O.N. (= 3) to the total decrease in O.N. (= 2), we should have three atoms of Cu for every two atoms of N and hence the equation should be written as :

$$3CuO + 2NH_3 \rightarrow 3Cu + N_2 + H_2O$$

Now in order to balance O-atoms we should add $3H_2O$ molecules to the right hand side. Thus :

$$3CuO + 2NH_3 \rightarrow 3Cu + N_2 + 3H_2O$$

This is the balanced equation.

Example 2. *Balance the following equation by oxidation number method.*

$$K_2Cr_2O_7 + H_2SO_4 + FeSO_4 \rightarrow K_2SO_4 + Cr_2(SO_4)_3 + Fe_2(SO_4)_3$$

(*Lucknow B.Sc. 1981*)

Solution. The given equation shows that O.N. of Cr decreases from +6(in $K_2Cr_2O_7$) to +3 [in $Cr_2(SO_4)_3$] while that of Fe increases from +2 (in $FeSO_4$) to +3 [in $Fe_2(SO_4)_3$] and hence :

$$K_2Cr_2O_7 + H_2SO_4 + FeSO_4 \longrightarrow K_2SO_4 + Cr_2(SO_4)_3 + Fe_2(SO_4)_3$$

(Cr + + 6) (Fe = + 2) (Cr = + 3) (Fe + 3)

Decrease in O.N. = 6 – 3 = 3
(Reduction)

Increase in O.N = 3 – 2 = 1
(Oxidation)

In order to equalise the total increase in O.N. (= 1) to the total decrease in O.N. (= 3) we should have one atom of Cr for every three atoms of Fe or two atoms of Cr for every six atoms of Fe and hence above equation should be written as :

$$K_2Cr_2O_7 + H_2SO_4 + 6FeSO_4 \rightarrow K_2SO_4 + Cr_2(SO_4)_3 + Fe_2(SO_4)_3$$

In order to balance Fe atoms we should write the above equation as :

$$K_2Cr_2O_7 + H_2SO_4 + 6FeSO_4 \rightarrow K_2SO_4 + Cr_2(SO_4)_3 + 3Fe_2(SO_4)_3$$

Now in order to balance O-atoms we add $7H_2O$ molecules to the right hand side and to balance H-atoms we write $7H_2SO_4$ in place of H_2SO_4 on the left hand side. Thus the equation in its balanced form becomes :

$$K_2Cr_2O_7 + 7H_2SO_4 + 6FeSO_4 \rightarrow K_2SO_4 + Cr_2(SO_4)_3 + 3Fe_2(SO_4)_3 + 7H_2O$$

Balancing Redox Equations by Ion-electron Method—By the Use of Half-reactions

This method was developed in 1927 by Jatte and Lamer and various steps involved in the method are as follows :

(*i*) Break up the complete equation in two half reactions, one for the change undergone by the reducing agent and the other for the change undergone by the oxidising agent.

(*ii*) Balance each half reaction as to the number of atoms of each element. For this purpose :

(*a*) Balance the atoms other than H and O for each half-reaction by using simple multiples.

(*b*) In *neutral* and *acid* solutions, H_2O and H^+ are added for balancing oxygen and hydrogen atoms. First balance the oxygen atoms. For each excess oxygen atom on one side of the equation, add one H_2O to the other side. Now use H^+ to balance hydrogen atoms.

In *alkaline* solutions, OH^- may be used. For each excess on one side, balance is secured by adding one H_2O to the *same side* and $2H^-$ to the *other side*. If hydrogen is still unbalanced, balance is secured by adding one OH^- for each excess hydrogen *on the same side as the excess* and one H_2O to the *other side*.

(*iii*) Equalise the charges on both sides by adding electrons to the side deficient in negative charges.

(*iv*) Multiply one or both half-reactions by a suitable number so that on adding the two equations, the electrons are balanced.

(*v*) Add the two balanced half reactions and cancel any terms common to both sides. Also see that all electrons cancel.

Solved Examples

Various steps of the ion-electron method can be illustrated by considering the following examples :

Example 1. *Balance the following redox reaction by ion-electron method :*

$$Fe^{2+} + MnO_4^- + H^+ \rightarrow Mn^{2+} + Fe^{3+} + H_2O$$

(*Udaipur B.Sc. 1978; Kanpur 1980*)

Solution. Obviously the given redox reaction takes place in acidic medium and can be broken into the following two half-reactions :

$$\underset{(Mn\ =\ +7)}{MnO_4^-} \rightarrow \underset{(Mn\ +2)}{Mn^{2+}} \quad ......\textit{Reduction half reaction}$$

and $$\underset{(Fe\ =\ +2)}{Fe^{2+}} \rightarrow \underset{(Fe\ =\ +3)}{Fe^{3+}} \quad ......\textit{Oxidation half-reaction}$$

For reduction half-reaction. (*i*) For balancing O-atoms add $4H_2O$ to right hand side to get :

$$MnO_4^- \rightarrow Mn^{2+} + 4H_2O$$

(*ii*) For balancing H-atoms add $8H^+$ to the left hand side to get :

$$MnO_4^- + 8H^+ \rightarrow Mn^{2+} + 4H_2O$$

(*iii*) For balancing the charges add $5e^-$ to the left hand side to get :

$$MnO_4^- + 8H^+ + 5e^- \rightarrow Mn^{2+} + 4H_2O \quad ...(i)$$

For oxidation half-reaction. Balance the charges on both sides by adding $1e^-$ to the left hand side to get :

$$Fe^{2+} \rightarrow Fe^{3+} + e^-$$

or $$5Fe^{2+} \rightarrow 5Fe^{3+} + 5e^- \quad ...(ii)$$

On adding equations (*i*) and (*ii*), we get

$$MnO_4^- + 8H^+ + 5Fe^{2+} \rightarrow Mn^{2+} + 4H_2O + 5Fe^{3+}$$

which is the balanced equation.

Example 2. *Balance the following equation by using ion-electron method :*

$$Cr_2O_7^{2-} + Fe^{2+} + H^+ \rightarrow Cr^{3+} + Fe^{3+} + H_2O$$

(*Mysore B.Sc. 1980; Kanpur 1980; Raj. 1975; Udaipur 1981*)

Solution. Obviously the given redox reaction takes place in acidic medium and can be broken into the following two half-reactions :

$$\underset{(Cr\ =\ +6)}{Cr_2O_7^{2-}} \rightarrow \underset{(Cr\ =\ +3)}{Cr^{3+}} \quad ......\textit{Reduction half reaction}$$

$$\underset{(Fe\ =\ +2)}{Fe^{2+}} \rightarrow \underset{(Fe\ =\ +3)}{Fe^{3+}} \quad ......\textit{Oxidation half-reaction}$$

For reduction half-reaction. (*i*) Balance the atoms other than oxygen and hydrogen

$$Cr_2O_7^{2-} \rightarrow 2Cr^{3+}$$

(*ii*) For balancing O-atoms add $7H_2O$ to the right hand side to get :

$$Cr_2O_7^{2-} \rightarrow 2Cr^{3+} + 7H_2O$$

(*iii*) For balancing H-atoms add $14H^+$ to the left hand side to get :

$$Cr_2O_7^{2-} + 14H^+ \rightarrow 2Cr^{3+} + 7H_2O$$

(*iv*) For balancing the charges add $6e^-$ to the left hand side to get :

$$Cr_2O_7^{2-} + 14H^+ + 6e^- \rightarrow 2Cr^{3+} + 7H_2O \qquad ...(i)$$

For oxidation half-reaction. Balance the charges on both the sides by adding $1e^-$ to right hand side to get :

$$Fe^{2+} \rightarrow Fe^{3+} + e^-$$

or

$$6Fe^{2+} \rightarrow 6Fe^{3+} + 6e^- \qquad ...(ii)$$

On adding equations (*i*) and (*ii*), we get :

$$Cr_2O_7^{2-} + 6Fe^{2+} + 14H^+ \rightarrow 2Cr^{3+} + 6Fe^{3+} + 7H_2O$$

This is the equation in its balanced form.

Questions with Answers

Q.1 Explain the following reactions :

(*i*) Disproprotionation reactions (ii) Comproprotionation reactions, giving examples.

Ans : (i) Disproprotionation reactions. These reactions are also called *self oxidation-reducation or self redox reactions. A reaction in which the same speciec is simultaneously oxidised as well as reduced is called a diproprotionation reaction.* For such redox reactions to occur, the reacting species must contain an element which has at least three oxidation states. The element in the reacting species is present in the intermediate oxidation state while the higher and lower oxidation states are available for reduction and oxidation to occur.

Examples. Examples of disproprotionation reactions are given below.

Reduced (+4 → +3); Oxidised (+4 → +5)

(a) $\overset{+4}{2NO_2}(g) + 2OH^-(aq) \rightarrow \overset{+3}{NO_2^-}(aq) + \overset{+5}{NO_3^-}(aq) + H_2O(l)$

This is a disproprotionation reaction since here the oxidation state of nitrogen decreases from + 4 in NO_2 to + 3 in NO_2^- ion, as well as increases from + 4 in NO_2 to 5 in NO_3^- ion.

(*b*) The decomposition of hydrogen peroxide (H_2O_2) is a disproprotionation reaction where oxygen atom undergoes disproprotionation.

Reduced (−1 → −2); Oxidised (−1 → 0)

$$\overset{-1}{2H_2O_2}(aq) \rightarrow \overset{-2}{H_2O}(l) + \overset{0}{O_2}(g)$$

In the above reaction, the O.N. of oxygen which is –1 in H_2O_2 increases to zero in O_2 and decreases to –2 in H_2O.

(*c*) Phosphorus (P_4) undergoes disproprotional in alkaline medium as shown below.

Reduced

$$\overset{0}{P_4}\,(s) + 3\,OH^-\,(aq) + 3\,H_2O\,(l) \longrightarrow \overset{-3}{PH_3} + 3\,\overset{+1}{H_2PO_2^-}\,(aq)$$

Oxidised

(*d*) Sulphur (S_8) undergoes disproprotionation reaction in alkaline medium.

Reduced

$$\overset{0}{S_8}\,(s) + 12\,OH^-\,(aq) \longrightarrow 4\,S^{2-}\,(aq) + 2\,\overset{+2}{S_2O_3^{2-}}\,(aq) + 6H_2O\,(l)$$

Oxidised

(*e*) Cl_2 undergoes disproprotionation reaction in presence of dil. cold NaOH and hot NaOH as shown by the following chemical equations.

Reduced

$$\overset{0}{Cl_2}\,(g) + 2\,OH^-\,(aq) \xrightarrow{\text{Cold}} \overset{+1}{ClO^-}\,(aq) + Cl^-\,(aq) + H_2O\,(l)$$

Oxidised

Reduced

$$\overset{0}{Cl_2}\,(g) + 6\,OH^-\,(aq) \xrightarrow{\Delta} 5\,Cl^-\,(aq) + \overset{+5}{ClO_3^-}\,(aq) + 3\,H_2O\,(l)$$

Oxidised

(*f*) $4KCl^{+5}O_3 \rightarrow 3KCl^{+7}O_4 + KCl^{-1}$

(*g*) $2Cu^+ \rightarrow Cu^0 + Cu^{2+}$

(ii) Comproprotionation reactions. Comproprotionation reaction is the reverse of disproprotionation reaction, *i.e.* in comproprotionation reaction two same species in diffirent oxidation states combine together to form a species which has intermediate oxidation state. For example :

$$Ag^{2+}\,(aq) + Ag^{\circ}\,(s) \rightarrow 2Ag^+\,(aq)$$

This reaction can be written as :

Reduction

$$\underset{(Ag\,=\,2)}{Ag^{2+}\,(aq)} + \underset{(Ag\,=\,0)}{Ag^0\,(s)} \longrightarrow \underset{(Ag\,=\,+1)}{Ag^+\,(aq)} + \underset{(Ag\,=\,+1)}{Ag^+\,(aq)}$$

Oxidation

Q.2 Explain why F_2 does not undergo disproprotionation reaction.

Ans : Since F is the most electronegative element, it shows only –1 O.S. in its compounds. It does not show any O.S. less than –1 and greater than –1 and hence it does not undergo disproprotionation reaction.

Q.3 Determine oxidation state of :

(*i*) C-atom in carbon suboxide (C_3O_2).

(*ii*) Br-atom in tribromo octaoxide (Br_3O_8).

(*iii*) S-atom in tetrathionate ion ($S_4O_6^{2-}$).

(*iv*) Fe-atom in magenetic oxide (Fe_3O_4).

(*v*) Mn-atom in Mn_3O_4.

(*vi*) Pb atom in red lead (Pb_3O_4).

(*vii*) O-atom in KO_3.

(*viii*) Fe-atom in ferriferrocynide molecule,$Fe_4[Fe(CN)_6]_3$.

(*ix*) F-atom in $Fe_{0.4}O$.

Ans : (*i*) **O.S. of C-atom in carbon suboxide, C_3O_2.** The structure of C_3O_2 (carbon suboxide) is :

$$\overset{-2}{O} = \overset{+2}{C} = \overset{0}{C} = \overset{+2}{C} = \overset{-2}{O}$$

Since a double bond is formed by mutual sharing of two electrons and since oxygen is more electronegative than carbon, each of the two terminal oxygen atoms have an oxidation state of – 2 and the carbons to which they are attached have an oxidation state of + 2.Furthur, since whenever a covalent bond is formed between similar atoms, each of the atoms is given an oxidation state of zero. Therefore O.S. of the central carbon atom in C_3O_2 is zero.

Thus, the two carbon atoms are present in + 2 oxidation state each,whereas the third one is present is zero oxidation state. Therefore, the average oxidation state of carbon in $C_3O_2 = (2 \times 2 + 1 \times 0) / 3 = 4/3$. **(Ans.)**

However, *in reality , the terminal carbon atoms have an oxidation state of + 2 while the middle carbon has an oxidation state of zero.*

(*ii*) **O.S. of Br-atom in tribromoocta oxide, Br_3O_8.** The structure of tribromooctaoxide (Br_3O_8) is :

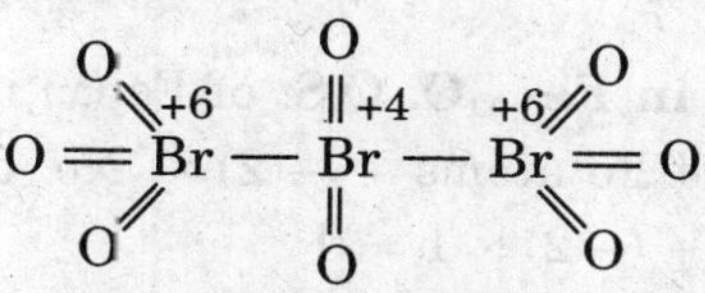

Following the arguments discussed under C_3O_2, we can easily find out that terminal bromine atoms are present in + 6 oxidation state while the middle one has an oxidation state of + 4. Thus the average oxidation number of Br in Br_3O_8 turns out to be [2 (+ 6) + (+ 4)/3 = **16/3. (Ans)**

(*iii*) **O.S. of S-atom in tetrathionate ion, $S_4O_6^{-2}$.** The structure of $S_4O_6^{-2}$ ion is :

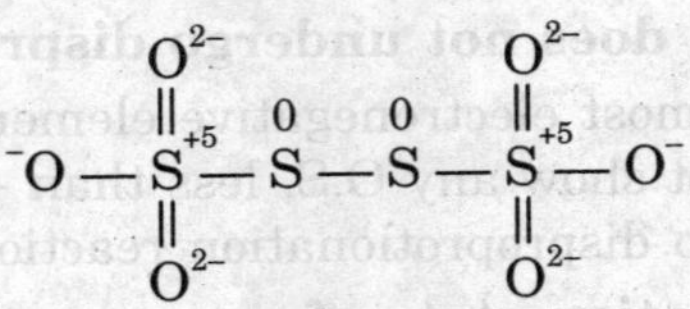

Since each of the two terminal sulphur atoms is connected to two oxygen atoms by a double bond and to one oxygen atom by a single bond, the oxidation state of each of these terminal sulphur atoms is + 5. Since two central sulphur atoms are linked to each other by a single bond and each sulphur is furthur attached to similar species on either side, the electron pair forming the S – S bond remains in the centre and hence each of the two central sulphur atoms has an oxidation state of zero. Thus, the average oxidation state of the four sulphur atoms is $(2 \times 5 + 2 \times 0)/4$ = **2.5 (Ans)**

(*iv*) **O.S. of Fe in magnetic oxide, Fe_3O_4.** Fe_3O_4 is known to be a mixture of FeO and Fe_2O_3 having the composition FeO. Fe_2O_3. Therefore, oxidation number of Fe in FeO is + 2 while in Fe_2O_3 it is + 3. The average value,however, is $[1 (+ 2) + 2 (+ 3)]/3$ = **8/3 (Ans)**

(*v*) **O.S. of Mn in Mn_3O_4.** Mn_3O_4 is regarded as a mixture of MnO and MnO_2 having the composition, $2MnO.MnO_2$. Therefore, the oxidation state of Mn in MnO is + 2 while that is MnO_2 it is + 4. The average is $[2 (+ 2) + (+ 4)]/3$ = **8/3 (Ans)**

(*vi*) **O.S. of Pb in red lead, Pb_3O_4.** Pb_3O_4 is considered to be a mixture with the composition 2PbO. PbO_2. Therefore, the oxidation state of Pb in PbO is + 2 while in PbO_2, it is + 4. Here, again the average value $[2 (+ 2) + 1 (+ 4)]/3$ = **8/3. (Ans)**

(*vii*) **O.S. of O-atom in KO_3.** Let the oxidation state of O-atom in KO_3 be x. Hence :

$$(+1) \times 1 + 3 \times x = 0$$

or $\quad x = -1/3$ **(Ans)**

(*viii*) **O.S. of Fe atom in $Fe_4[Fe(CN)_6]_3$.** In $Fe_4[Fe(CN)_6]_3$ four Fe atoms are in + 3 O.S. and three Fe atoms are in + 2 O.S. Thus the overage O.S. of Fe atom.

$$= \frac{4 \times (+3) + 3 \times (+2)}{7} = \mathbf{\frac{18}{7}} \textbf{ (Ans.)}$$

(*ix*) **O.S. of Fe-atom in $Fe_{0.94}O$.** O.S. of Fe (x) in $Fe_{0.94}O$ is given by :

$$x \times \text{No. of Fe atoms} + (-2) \times \text{No. of O-atoms} = 0$$

or $\quad x \times 0.94 + (-2) \times 1 = 0$

or $\quad 0.94x = 2$

or $\quad x = 2/0.94$ = **2.12 (Ans)**

Q.4 Calculate oxidation state of : (*i*) S-atom in Caro's acid or peroxomono sulphuric acid molecule (H_2SO_5). (*ii*) N-atom in NH_4NO_3. (*iii*) Cr-atom in CrO_5. (*iv*) N-atom in NO^-_3 ion. (*v*) Cl-atoms in bleaching powder molecule ($CaOCl_2$). (*vi*) Fe-atom in $K_4[Fe(CN)_6]$. (*vii*) C-atom in HCN. (*viii*) C-atom in HNC. (*ix*) S-atom in oleum molecule ($H_2S_2O_7$)

(*x*) C-atom in HCOOH molecule. (*xi*) S-atom in sodium thiosulphate molecule ($Na_2S_2O_3$). (*xii*) S-atom in sodium tetrathionate ($Na_2S_4O_6$). (*xiii*) Na atom in NaHg amalgam. (*xiv*) S-atom in Marshall's acid or peroxdisulphuric acid molecule ($H_2S_2O_8$). (*xv*) C-atom in sucrose molecule ($C_{12}H_{22}O_{11}$).

Ans. (*i*) By conventional method, the O.N. of S (x) in H_2SO_5 is given by :

$$2(+1) + x + 5(-2) = 0 \text{ or } x = \mathbf{+8}$$

This is impossible because the maximum O.N. of S cannot be more than six since it has only six electrons in the valence shell. This fallacy is overcome if we calculate the O.N. of S by chemical bonding method.

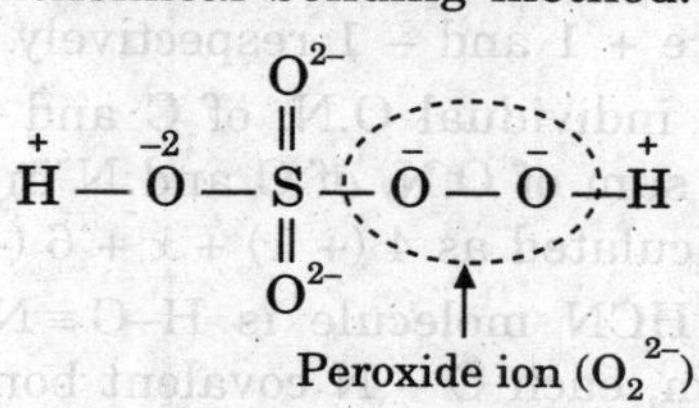

$$\underset{\text{(for H)}}{2 \times (+1)} + \underset{\text{(for S)}}{x} + \underset{\text{for } (O-O)}{2(-1)} + \underset{\text{(for other } O \text{ atoms)}}{3 \times (-2)} = 0 \qquad \text{or } \boldsymbol{x = +6}$$

(*ii*) Oxidation numbers of two N atoms in NH_4NO_3 are different. NH_4NO_3 is made of NH_4^+ and NO_3^-

In NH_4^+ ion, $x + 4(+1) = +1$ ∴ O.N. of N is − 3

In NO_3^- ion $x + 3(-2) = -1$ ∴ O.N. of N is + 5

Average O.S of N-atom $= \dfrac{+5 + (-3)}{2}$

$= \mathbf{+1}$ **(Ans.)**

(*iii*) According to conventional method, O.N. of Cr atom (x) in CrO_5 molecule is given by.

$x + 5x(-2) = 0, \quad \therefore \quad x = +10$

This is impossible because the maximum O.N. of Cr cannot be more than six since it has $3d^5\,4s^1$ outer orbital configuration.

The correct value of O.N. of Cr atom is calculated with the help of the structure of CrO_5 given below :

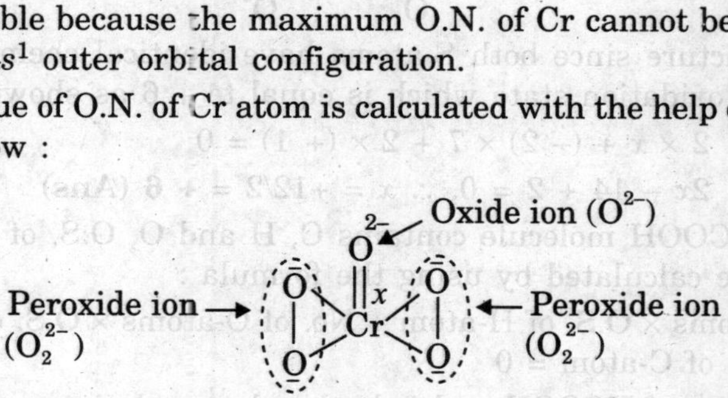

This structure shows that the value of x is given by :

$$x + \underset{\substack{\text{Due to two}\\\text{peroxide ions}\\(O_2^{2-})}}{2 \times (-2)} + \underset{\substack{\text{Due to one}\\\text{oxide ion}\\(O^{2-})}}{1 \times (-2)} = 0$$

$x - 4 - 2 = 0$ or $x = \mathbf{+6}$ **(Ans.)**

(*iv*) According to conventional method, O.N. of N is $NO_3^- = x + 3(-2) = -1$ or $x = +5$

According to chemical bonding method, 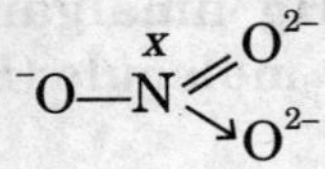

$$x + 1(-1) \quad + \quad 1(-2) \quad + \quad 1(-2) = 0 \qquad \text{or } \boldsymbol{x + 5}$$

$$(\text{for } O^-) \qquad (\text{for } = O) \qquad \text{for} \rightarrow O$$

Thus, there is no fallacy about the O.N. of N in NO_3^- whether one calculates by conventional method or by chemical bonding method.

(*v*) $CaOCl_2$ is made of Ca^{2+}, OCl^- (Cl = +1) and Cl^- (Cl = – 1). Hence O.N. of two Cl atoms in $CaOCl_2$ are + 1 and – 1 respectively.

(*vi*) In $K_4[Fe(CN)_6]$ the individual O.N. of C and N are not known but for calculating O.N. of Fe, the sum of O.N. of C and N in $(CN)^-$ unit is taken to be – 1. Thus, O.N. of Fe is calculated as $4(+1) + x + 6(-1) = 0$ or $\boldsymbol{x = +2}$. **(Ans)**

(*vii*) The structure of HCN molecule is $H{-}C\equiv N$. Since N-atom in more electronegative than C-atom, each C – N covalent bond gives in O.S. of – 1 to N-atom, Now since there are three C – N covalent bonds in HCN molecule, O.S. of N-atom = – 3. Hence HCN can be represented as $\overset{+1}{H}{-}\overset{x}{C}\equiv\overset{-3}{N}$ · Hence $(+1)\times 1 + x \times 1 + (-3)\times 1 = 0$ or $\boldsymbol{x = +2}$ **(Ans)**

(*viii*) The structure of HNC molecule is $H{-}N \overset{\rightarrow}{=} C$. Since N-atom (donor atom) is more electronegative than C-atom (accepter atom), O.S. of N-atom = – 3. Thus HNC can be represented as $\overset{+1}{H}{-}\overset{-3}{N}\overset{\rightarrow}{=}\overset{x}{C}$. Hence $(+1)\times 1 + (-3)\times 1 + x \times 1 = 0$ or $\boldsymbol{x = +2}$ **(Ans)**

(*ix*) The structure of $H_2S_2O_7$ molecule is given below :

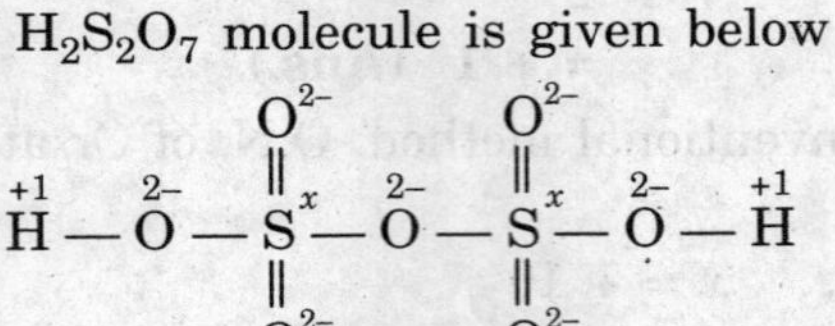

In this structure since both S-atoms have identical chemical bonding, they have the same oxidation state which is equal to + 6 as shown below :

$$2 \times x + (-2) \times 7 + 2 \times (+1) = 0$$

or $$2x - 14 + 2 = 0, \therefore x = +12/2 = \boldsymbol{+6} \text{ (Ans)}$$

(*x*) Since HCOOH molecule contains C, H and O, O.S. of C-atom (*x*) in this molecule can be calculated by using the formula :

No. of H-atoms × O.S. of H-atom + No. of O-atoms × O.S. of O-atom + No. of C-atoms × O.S. of C-atom = 0

Hence in case of HCOOH molecule the above relation can be written as :

$$2 \times (+1) + 2 \times (-2) + 1 \times x = 0$$

$$\therefore \quad \boldsymbol{x = +2} \text{ (Ans)}$$

(*xi*) Conventional method gives O.S. to S-atom in $Na_2S_2O_3$ molecule = + 2 as shown below :

$$2 \times (+1) + 2x + 3 \times (-2) = 0$$

(Due to Na) (Due to O)

$\therefore$ $x = +2$

This oxidation state (= + 2) is wrong since the two S-atom in $Na_2S_2O_3$ molecule are in different oxidation states because of the fact that when $Na_2S_2O_3$ is heated with dil. H_2SO_4, one S-atom gets precipated and the other S-atom is converted into SO_2. The correct O.S. of S-atom can be calculated from the structure of $Na_2S_2O_3$ which is given below.

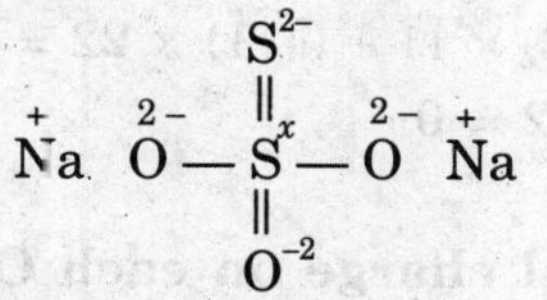

In the structure given above each of the two Na atoms has an O.S. = + 1, each of three O-atoms has O.S. = – 2. Terminal S-atom has O.S. = – 2. Let the O.S. of the central S-atom be x. Thus :

$$(+1) \times 2 \quad + \quad (-2) \times 3 \quad + \quad (-2) \times 1 \quad + \quad x = 0$$

(For Na) (For O-atom) (For terminal S-atom)

$$+2 - 6 - 2 + x = 0, \quad \therefore \mathbf{x = +6}$$

Thus :

O.S. of terminal S-atom = **– 2**
O.S. of central S-atom = **+ 6** } **(Ans)**

$$\text{Average O.S. of S-atom} = \frac{1\times(-2)+1\times(+6)}{2}$$

$$= \frac{-2+6}{2} = \mathbf{+2} \text{ (Ans)}$$

(*xii*) The structure of $Na_2S_4O_6$ molecule is given below :

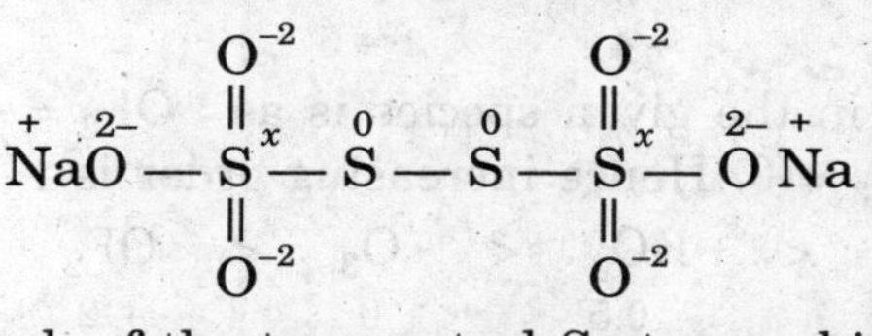

In this structure each of the two central S-atoms which are linked together has O.S. = 0. Let the O.S. of each of the two remaining S-atoms be x. Hence :

$$2 \times (+1) \quad + \quad 6 \times (-2) \quad + \quad 2 \times 0 \quad + \quad 2 \times x = 0$$

(For Na atoms) (For O-atoms) (For central S-atoms)

$$+2 - 12 + 0 + 2x = 0$$

$\therefore$ $x = +10/2 = +5$

Thus :

O.S. of each central S-atom = **0**

In this structure each of the two H-atoms has O.S. = – 1, each of the six terminal O-atoms has O.S. = – 2. This structure has one peroxide ion (O_2^{-2}) whose O.S. = – 2. In this ion each O-atom has O.S. = – 1. Thus :

$$2 \times (+1) \quad + \quad 6 \times (-2) \quad -2 + 2x = 0$$

(Due to H-atoms) (Due to six O-atoms) (Due to peroxide ion, O_2^{-2})

$\therefore$ $x = +12/2 = \mathbf{+6}$ **(Ans)**

(*xv*) Since the given molecule ($C_{12}H_{22}O_{11}$) contains H, C and O, O.S. of C-atom (x) is given by :

$$x \times n_C + (-2) \times n_O + (+1) \times n_H = 0$$

or $x \times 12 + (-2) \times 11 + (+1) \times 22 = 0$

or $12x - 22 + 22 = 0$

$\therefore$ $\mathbf{x = 0}$ **(Ans)**

Q.5 Find out the formal charge on each O-atom in phosphate ion.

Ans : We know that phosphate ion is PO_4^{3-}.

Formal charge on each O-atom in PO_4^{3-} ion

$$= \frac{\text{Total number of charges on } PO_4^{3-} \text{ ion}}{\text{Total number of O-atoms in } PO_4^{3-} \text{ ion}}$$

$$= \frac{-3}{4} = \mathbf{-0.75}$$ **(Ans.)**

Q.6 Arrange the following species in the increasung order of oxidation state of the atom shown in parenthesis.

(*a*) SO_3^{2-}, $S_2O_4^{2-}$, $S_2O_6^{2-}$ (S-atom)

(*b*) OF_2, KO_2, BaO_2, O_3 (O-atom)

(*c*) KO_3, Na_2O_2. (O-atom)

Ans : (*a*) O.S. of S-atom in the given species is as : SO_3^{2-} = + 4, $S_2O_4^{2-}$ = + 3, $S_2O_6^{2-}$ = + 5. Hence the increasing order is :

$$\underset{+3}{S_2O_4^{2-}} < \underset{+4}{SO_3^{2-}} < \underset{+5}{S_2O_6^{2-}}$$ **(Ans)**

(*b*) O.S. of O-atom in the given species is as : OF_2 = + 2, KO_2 = –1/2 = 0.5, BaO_2 = – 2/2 = – 1, O_3 = 0. Hence increasing order is :

$$\underset{-1}{BaO_2} < \underset{-0.5}{KO_3} < \underset{0}{O_3} < \underset{+2}{OF_2}$$

(*c*) O.S. of O-atom in the given species is as : KO_3 = –1/3 = – 0.33, Na_2O_2 = – 2/2 = – 1. Hence increasing order is :

$$\underset{-1}{Na_2O_2} < \underset{-0.33}{KO_3}$$ **(Ans)**

Q.7 Calculate the number of electrons transferred when one mole of $KMnO_4$ is reduced to MnO_4^{2-}, MnO_2, Mn_2O_3 and Mn^{2+}.

Ans : The reduction of one mole of $KMnO_4$ or MnO_4^- to MnO_4^{2-}, MnO_2 , Mn_2O_3 and Mn^{2+} can be represented by the following reduction half-reactions.

(*i*) $MnO_4^- + e^- \rightarrow MnO_4^{2-}$
One mole

(*ii*) $MnO_4^- + 4H^+ + 3e^- \rightarrow MnO_2 + 2H_2O$
One mole

(*iii*) $2MnO_4^- + 10H^+ + 8e^- \rightarrow Mn_2O_3 + 5H_2O$
Two moles

or $MnO_4^- + 5H^+ + 4e^- \rightarrow \frac{1}{2} Mn_2O_3 + \frac{5}{2} H_2O$
one mole

(*iv*) $MnO_4^- + 8H^+ + 5e^- \rightarrow Mn^{2+} + 4H_2O$
one mole

Obviously the number of electrons transferred in the reduction of one mole of MnO_4^- to MnO_4^{2-}, MnO_2, Mn_2O_3 and Mn^{2+} is 1, 3, 4 and 5 respectively.

Q.8 Find out the values of *a*, *b*, *c* and *d* in the following equation.

$$IO_3^- + a\, I^- + b\, H^+ \rightarrow c\, H_2O + d\, I_2$$

Ans : The given equation shows that IO_3^- (I = + 5) is reduced to I_2 (I = 0) and I^- (I = – 1) is oxidised to I_2 (I = 0). Thus the given equatain can be broken into the following two balanced half-reactions :

$$2I^- \xrightarrow{\text{Oxidation}} I_2 + 2e^-] \times 5$$

$$2IO_3^- + 12H^+ + 10e^- \xrightarrow{\text{Reduction}} I_2 + 6H_2O$$

$$10I^- + 2IO_3^- + 12H^+ \longrightarrow 6I_2 + 6H_2O$$

or $$5I^- + IO_3^- + 6H^+ \rightarrow 3I_2 + 3H_2O$$

Thus the values of *a*, *b*, *c* and *d* are as : *a* = 5, *b* = 6, *c* = 3 and *d* = 3 **(Ans)**

University Questions

1. Balance the following equations :

(*i*) $MnO_4^- + H^+ + H_2O \rightarrow Mn^{2+} + O_2 + H_2O$ (*Calcutta 1997*)

(*ii*) $Cr_2O_7^{2-} + H_2SO_3 \rightarrow Cr^{3+} + HSO_4^-$ (*Acidic medium*) (*G. Nanakdev 2000*)

(*iii*) $HCl + KMnO_4 \rightarrow Cl_2 + MnCl_2 + KCl + H_2O$ (*M.D. 2000*)

(*iv*) $P_4 + HNO_3 \rightarrow H_3PO_4 + NO_2 + H_2O$ (*G. Nanakdev 2000*)

2. Determine oxidation number of :

(*i*) Cr in $Cr_2O_7^{2-}$ and CrO_4^{2-} (*Garhwal 2000*)

(*ii*) S in $Na_2S_2O_3$ and H_2SO_4 (*G. Nanak Dev 2000*)

3. Give points of distinction between valency and oxidation number. (*M.D. 2000, Nagpur 2000*)

4. Define oxidation and reduction in terms of oxidation number. (*G. Nanakdev 2000*)

5. Explain why H_2S is both an oxidant and reductant. (*Himachal Pradesh 2000*)

6. What do you understand by the term "Oxidation Number"? Balance the following equation by oxidation number method.

$K_2Cr_2O_7 + FeSO_4 + H_2SO_4 \rightarrow Cr_2(SO_4)_3 + H_2O + Fe(SO_4)_3 + K_2SO_4$

(*Kumaon 2000*)

7. Name the oxy acids of chlorine with their formula. What is the oxidation number of chlorine in these acids? [*Gauhati (General) 2000*]

8. Give electronic interpretation of oxidation and reduction with suitable examples. (*Avadh 2000*)

9. What is oxidation number? How is it determined? (*M.D. Rohtak 2002*)

10. Describe ion-electron method for balancing chemical equations. (*Lucknow 2002*)

11. O.N. of O-atom in O_2 molecule is :

(*a*) 0 (*b*) – 2 (*c*) + 2 (*d*) –1/2 (*Agra 2007*)

12. Oxidation state of N in phosphazene is ... (*H.N. Bahuguna 2007*)

13. In $C + H_2O \rightarrow CO + H_2$, H_2O acts as :

(*a*) Oxidising agent

(*b*) Reducing agent

(*c*) Oxidising and reducing agent both

(*d*) None of these. (*H.N. Bahuguna 2007*)

14. (*a*) Give the O.S. of Cr in CrO_5.

(*b*) Balance the equations :

(*i*) $Fe^{2+} + Cr_2O_7^{2-} + H^+ \rightarrow ...$

(*ii*) $Mn^{2+} + PbO_2 + HSO_4^- \rightarrow ...$

(*iii*) $MnO_4^- + Cr^{3+} + H_2O \rightarrow ...$ (*Agra 2007*)

15. The O.S. of Ni in $Ni(CO)_4$ is :

(*a*) 0 (*b*) 2 (*c*) 3 (*d*) 4 (*Agra 2007*)

CHAPTER 11

Standard Electrode Potentials

Electrode Potential

When a metal piece is placed in water or in water solution containing one of its salts (*i.e.* the solution contains the metal ions. *e.g.* Zn in $ZnSO_4$ solution), the metal atoms tend to lose electrons and hence become positively charged ions. Thus the solution, due to the presence of positively charged metal ions, becomes positively charged.

The conversion of metal atoms into positively-charged ions is due to the attractive force of polar water molecules which remove the metal ions from the metal surface and send them into the solution. The electrons removed from the metal atoms are left on the metal itself and hence the metal becomes negatively charged. Thus :

Metal – Electrons ⇄ *Positively charged metal ions*

↓ *Are left on the metal to make the metal negatively-charged.*

↓ *Go into the solution to make the solution positively-charged.*

The positively-charged metal ions which have passed from the metal into the solution cluster around the surface of the negatively- charged metal and form a *double electric layer* (Fig. 11.1). This double electric layer can be compared with an electric capacitor in which one plate is the charged metal surface and the other is layer of ions near it. The double electric layer thus formed develops a definite potential difference between the metal and the solution. This potential difference is called electrode potential of the metal or electrode. Thus :

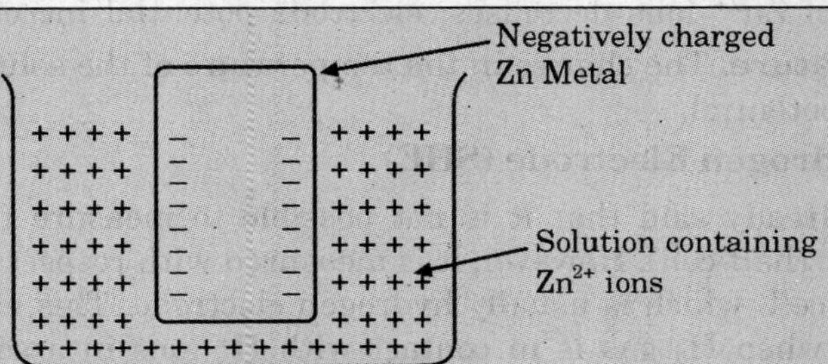

Fig. 11.1. Formation of double electric layer gives rise to potential difference between the metal (Zn metal) and solution containing its own ions (Zn^{2+} ions) which is called electrode potential.

The potential difference between a metal and a solution containing its own ions (in equilibrium) is called **electrode potential** *of that metal or electrode.*

The magnitude of the electrode potential of a metal is a measure of its relative tendency to lose or gain electrons, *i.e.* it is a measure of the relative tendency of the metal to undergo oxidation (loss of electrons) or reduction (gain of electrons).

Oxidation Electrode Potential and Reduction Electrode Potential

Depending on the tendency of a metal to lose or gain electrons, the electrode potential may be of two types namely *oxidation potential and reduction potential. Oxidation potential of a metal indicates its electron losing tendency while reduction potential represents the electron gaining capacity of the metal.*

The unit of electrode potential is *volt which is the potential required to flow one coulomb of electric current per second through a uniform thread of mercury, 106.3 cm. long containing 14.4521 gm. mercury at 0°C.*

The potential of a single electrode cannot be measured directly. Actually what we do is that, first of all, we assign an arbitrary electrode potential to a reference or standard electrode and then the potential of the given electrode is measured with respect to this standard electrode. A universally adopted standard electrode is the hydrogen electrode which will be described a little later.

Factors Affecting the Magnitude of Electrode Potential

The magnitude of electrode potential depends on the following factors :

1. Nature of metal or electrode. The metals have different tendency to lose or accept the electrons and hence have different electrode potential. Na, being extremely active, has a strong tendency to lose electrons and hence has high electrode potential. On the other hand, Cu, being less active, has lower tendency to lose electrons and hence has lower electrode potential.

2. Concentration of metal ions in solution. Consider Zn rod dipped in $ZnSO_4$ solution (*i.e.* in Zn^{2+} ions)

$$Zn \rightleftarrows Zn^{2+} + 2e^-$$

As the concentration of Zn^{2+} ions increases, the equilibrium will shift to the left side and the hence the electrode potential will decrease. Similarly as the concentration of Zn^{2+} ions decreases, electrode potential increases.

3. Temperature. The change in the temperature of the solution also changes the electrode potential.

Standard Hydrogen Electrode (SHE)

We have already said that it is not possible to measure the potential of a singal electrode (half-cell). However, it is measured with respect to some standard electrode (half-cell) which is usually hydrogen electrode. This electrode is a half-cell developed when H_2 gas is in contact with H^+ ions in a solution. *Standard Hydrogen Electrode (SHE) is an electrode when H_2 gas at one atmosphere pressure is in contact with H^+ ions of 1 M concentration at 25°C.* This electrode is also called *Normal Hydrogen Electrode (NHE). It is assigned a potential of zero.* This electrode is obtained and works as follows (Also see Fig. 11.2).

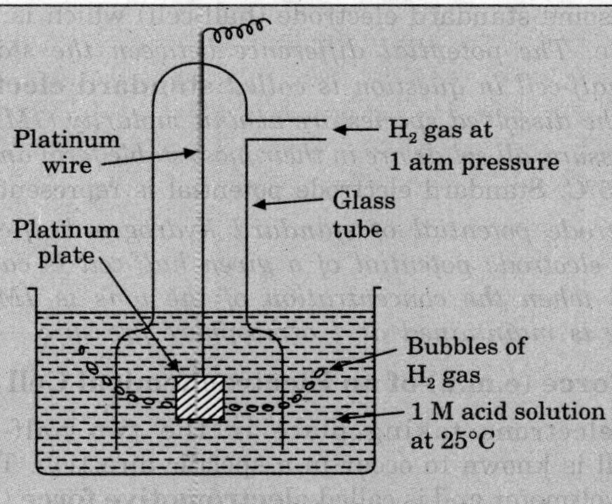

Fig. 11.2. Standard hydrogen electrode.

(*i*) 1M HCl or H_2SO_4 solution is taken in a glass vessel and maintained at 25°C.

(*ii*) A platinum wire fused in a glass tube and having a small platinum plate coated with finely divided platinum black is used as the electrode. The platinum plate provides a large surface for hydrogen to get adsorbed on it.

(*iii*) The electrode (*i.e.* platinum wire and platinum plate) is surrounded by a glass tube which has an inlet at its top for bubbling pure H_2 gas (at 1 atmosphere pressure) into the acid solution contained in the glass vessel and has a number of holes at its base for the escape of excess of H_2 gas. A part of H_2 gas is absorbed on the surface of the platinised Pt electrode while the excess of H_2 gas escapes through the holes.

Hydrogen electrode thus obtained forms one of the two half-cells of a given voltaic cell. When this half-cell is connected with any other half-cell, it begins to work. For example when it is connected with a Zn–Zn^{2+} half-cell, the electrons given up by Zn – Zn^{2+} half-cell are used up at the hydrogen electrode bringing about the reduction of H^+ (*aq*) ions to H_2 gas.

$$Zn\ (s) \rightarrow Zn^{2+}\ (aq) + 2e^-$$

$$2H^+\ (aq) + 2e^- \rightarrow H_2\ (g)$$

Thus we see that here hydrogen electrode acts as a cathode and hence is represented as H_2 (*g*)/$2H^+$(*aq*) and is placed at the right hand of the complete cell.

—— Flow of electrons ⟶

Zn (s) ∣ Zn^{2+}(aq) ∣	*∣ $2H^+$ (aq) ∣ H_2(g)*
Anode (–)	*Cathode* (+)
(Oxidation half-cell)	*(Reduction half-cell)*

Standard Electrode Potential (E°)

We have already pointed out that since it is not possible to determine the electrode potential of a given single electrode (half-cell), it is measured with respect to that of some standard electrode (half-cell) which is usually standard hydrogen electrode. *The potential difference between the standard hydrogen half-cell and the half-cell in question is called* **standard electrode potential,** *provided that all the dissolved species are at unit molarity (1M); all gases are at one atmosphere pressure; all solids are in their most stable form and the temperature is maintained at 25°C.* Standard electrode potential is represented as E°.

Since the electrode potential of standard hydrogen half-cell is arbitrarily taken as zero, the electrode potential of a given half-cell is called its standard electrode potential when the concentration of the ions is 1M, temperature is 25°C and pressure is maintained at 1 atmosphere pressure.

Electromotive Force (e.m.f.) of an Electrochemical Cell

The flow of electrons taking place in the two half-reactions of an electrochemical cell is known to occur in a specific direction. The voltage of the cell is read by the voltmeter and is called **electromotive force (e.m.f.)** of the cell. *The standard e.m.f. of a given cell ($E°_{cell}$) is equal to the difference of the standard reduction potential of the cathode (i.e. right hand electrode), $E°_{cathode}$ and the standard reduction potential of the anode (i.e. left hand electrode), Thus* :

$$\mathbf{E°_{cell} = E°_{cathode} - E°_{anode}}$$
$$\mathbf{= E°_{right} - E°_{left}}$$

Measurement of Standard Electrode Potential

For measuring the standard electrode potential of a given electrode, a voltaic cell is set up with two half-cells one of which has the electrode whose standard electrode potential is to be determined and the other has a standard hydrogen electrode. 1M solutions are used in both the half-cells and the temperature is maintained at 25°C. The electrodes are connected with a voltmeter and the e.m.f. of the cell is noted from the deflection of the needle of the voltmeter. The reading noted in the voltmeter gives the standard electrode potential of the electrode in question with respect to hydrogen electrode. This procedure can be understood by considering the following examples :

1. Determination of standard electrode potential of Zn/Zn^{2+} electrode. In order to determine the electrode potential of Zn electrode dipped in 1M $ZnSO_4$ solution, it is combined with the standard hydrogen electrode to form the complete cell.

The deflection of the needle of the voltmeter towards Zn electrode indicates that the conventional current is flowing from H electrode to Zn electrode and hence *the electrons are flowing from Zn electrode through the connecting wire towards H electrode* as shown in Fig. 11.3. Thus here Zn electrode acts as an *anode* and H electrode acts as a *cathode* and the cell can, therefore, be represented as :

—— ***Flow of electrons*** **⟶**

⟵ ***Flow of current*** **——**

$Zn(s)/Zn^{2+}(aq)$ | | $2H^+(aq)/H_2(g)$

***Anode* (–)** ***Cathode* (+)**

(*Oxidation half-cell*) (*Reduction half-cell*)

The two half-cell reactions are :

Oxidation half-reaction (Anode reaction) :

$$Zn(s) \rightarrow Zn^{2+}(aq) + 2e^-$$

Reduction half-reaction (Cathode reaction) :

$$2H^+(aq) + 2e^- \rightarrow H_2(g)$$

The complete cell-reaction is obtained by adding the above two half-cell reactions.

$$Zn(s) + 2H^+(aq) \rightarrow H_2(g) + Zn^{2+}(aq)$$

The voltmeter reads 0.76 volts which is the observed e.m.f. of the cell, *i.e.*

$$E^\circ_{cell} = 0.76 \text{ volts}$$

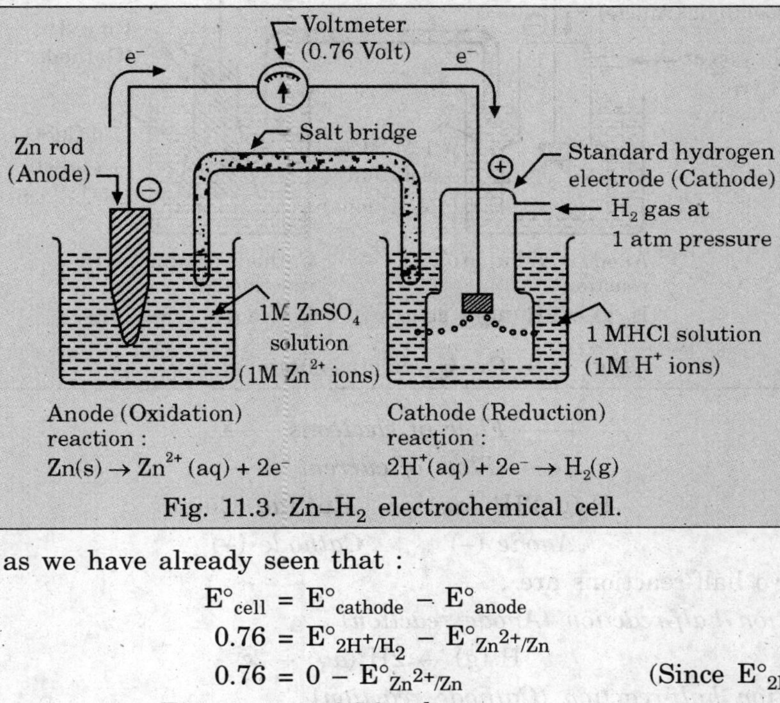

Fig. 11.3. Zn–H_2 electrochemical cell.

Now, as we have already seen that :

$$E^\circ_{cell} = E^\circ_{cathode} - E^\circ_{anode}$$

or $$0.76 = E^\circ_{2H^+/H_2} - E^\circ_{Zn^{2+}/Zn}$$

or $$0.76 = 0 - E^\circ_{Zn^{2+}/Zn} \qquad (\text{Since } E^\circ_{2H^+/H_2} = 0)$$

$\therefore$ $$E^\circ_{Zn^{2+}/Zn} = -0.76 \text{ volt.}$$

Here negative sign indicates that 0.76 volt is the standard reduction potential of Zn^{2+}/Zn electrode, *i.e.*, 0.76 volt is the standard potential corresponding to the reduction reaction :

$$Zn^{2+}(aq) + 2e^- \rightarrow Zn(s);\ E^\circ_{Zn^{2+}/Zn} = -0.76 \text{ volt}$$

In order to get the standard oxidation potential of Zn/Zn^{2+} electrode, (*i.e.* standard electrode potential corresponding to the oxidation reaction, $Zn(s) \rightarrow Zn^{2+}(aq) + 2e^-$) we have to reverse the sign of standard reduction potential obtained as above. Thus :

$$Zn(s) \rightarrow Zn^{2+}(aq) + 2e^-;\ E^\circ_{Zn/Zn^{2+}} = +0.76 \text{ volt.}$$

2. Determination of standard electrode potential of Cu^{2+}/Cu electrode. In order to determine the electrode potential of Cu electrode dipped in 1M $CuSO_4$

solution, it is combined with the standard H-electrode to form the complete cell. Here the deflection of the needle of the voltmeter is towards H-electrode, indicating that the conventional current is flowing from Cu electrode to H-electrode and hence the *electrons are flowing from H-electrode through the connecting wire towards Cu electrode* as shown in Fig. 11.4. Thus here H-electrode acts as anode and Cu-electrode acts as cathode and the cell can be represented as :

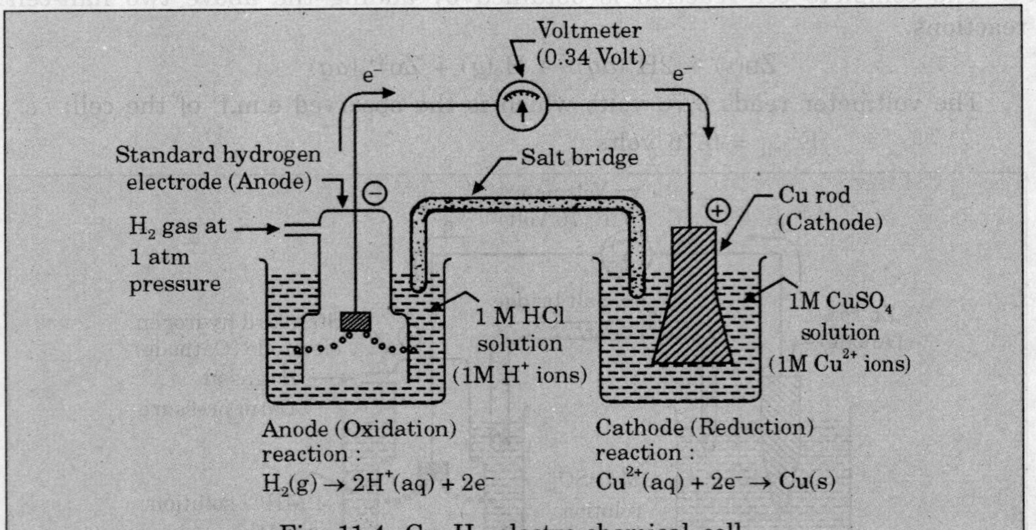

Fig. 11.4. Cu–H_2 electro-chemical cell.

—— *Flow of electrons* ⟶

⟵ *Flow of current* ——

$$H_2(g)/2H^+\ (aq)\,|\quad|\ Cu^{2+}(aq)/Cu(s)$$

Anode (–) *Cathode* (+)

The two half-reactions are :

Oxidation half-reaction (Anode reaction)

$$H_2(g) \rightarrow 2H^+(aq) + 2e^-$$

Reduction half-reaction (Cathode reaction)

$$Cu^{2+}(aq) + 2e^- \rightarrow Cu(s)$$

The *complete cell reaction* is obtained by adding the above two half-cell reactions. Thus :

$$H_2(g) + Cu^{2+}(aq) \rightarrow 2H^+(aq) + Cu(s)$$

Voltmeter reads 3.34 which is the observed e.m.f. of the cell, *i.e.*

$$E^\circ_{cell} = 0.34 \text{ volts}$$

Now, $E^\circ_{cell} = E^\circ_{cathode} - E^\circ_{anode}$

or $0.34 = E^\circ_{Cu^{2+}/Cu} - E^\circ_{2H^+/H_2}$

or $0.34 = E^\circ_{Cu^{2+}/Cu} - 0$

or $E^\circ_{Cu^{2+}/Cu} = 0.34 \text{ volt.}$

0.34 volt is the standard reduction potential of Cu^{2+}/Cu electrode, *i.e.* 0.34 volt is the standard potential corresponding to the reduction reaction :

$$Cu^{2+}(aq) + 2e^- \rightarrow Cu(s)\ ;\ E^\circ_{Cu^{2+}/Cu} = +\ 0.34 \text{ volt.}$$

In order to get the standard oxidation potential of Cu/Cu^{2+} electrode (*i.e.*, standard electrode potential corresponding to the oxidation reaction $Cu(s) \rightarrow Cu^{2+}(aq) + 2e^-$) we have to reverse the sign of standard reduction potential obtained as above. Thus :

$$Cu(s) \rightarrow Cu^{2+}(aq) + 2e^-\ ;\ E^\circ_{Cu/Cu^{2+}} = -\ 0.34 \text{ volt.}$$

In this manner the standard electrode potential (oxidation or reduction) of other electrodes can be obtained by combining the electrode in question with a standard hydrogen electrode and then measuring the e.m.f of the cell formed.

According to I.U.P.A.C. (International Union of Pure and Applied Chemistry) convension *the given value of standard electrode potential is regarded as the standard reduction potential unless it is specifically mentioned that it is standard oxidation potential.*

A list of standard reduction potentials of a number of electrodes (half-cells) measured at 25°C is given in Table 11.1.

Table 11.1. Standard reduction potentials of some electrodes (half-cells)—Electrochemical series

Half-cell (Electrode or couple)	Half-cell reaction (Reduction reaction)	Standard reduction potential (volts) (E° values)
	Oxidising agent → Reducing agent (↓ Oxidising power decreasing; ↓ Reducing power increasing)	(↓ E° values decreasing)
$F_2/2F^-$	$F_2 + 2e^- \rightarrow 2F^-$ (Strongest oxidising agent → Weakest reducing agent)	+ 2.85
O_3/O_2	$O_3 + 2H^+ + 2e^- \rightarrow O_2 + H_2O$	+ 2.07
$S_2O_3^{2-}/SO_4^{2-}$	$S_2O_3^{2-} + 2e^- \rightarrow 2SO_4^{2-}$	+ 2.01
H_2O_2/H_2O	$H_2O_2 + 2H^+ + 2e^- \rightarrow 2H_2O$	+ 1.77
MNO_4^-/MnO_2	$MnO_4^- + 4H^+ + 3e^- \rightarrow MnO_2 + 2H_2O$	+ 1.69
Au^+/Au	$Au^+ + e^- + \rightarrow Au$	+ 1.68
$HClO_3/HClO$	$HClO_3 + 2H^+ + 2e^- \rightarrow HClO + H_2O$	+ 1.64
$HClO/Cl_2$	$2HClO + 2H^+ + 2e^- \rightarrow Cl_2 + 2H_2O$	+ 1.64
Ce^{4+}/Ce^{3+}	$Ce^{4+} + e^- \rightarrow Ce^{3+}$	+ 1.61
$2BrO_3^-/Br_2$	$2BrO_3^- + 12H^+ + 10e^- \rightarrow Br_2 + 6H_2O$	+ 1.52
MnO_4^-/Mn^{2+}	$MnO_4^- + 8H^+ + 5e^- \rightarrow Mn^{2+} + 4H_2O$	+ 1.51
Au^{3+}/Au	$Au^{3+} + 3e^- \rightarrow Au$	+ 1.42
$Cl_2/2Cl^-$	$Cl_2 + 2e^- \rightarrow 2Cl^-$	+ 1.36
$Cr_2O_7^{2-}/2Cr^{3+}$	$Cr_2O_7^{2-} + 14H^+ + 6e^- \rightarrow 2Cr^{3+} + 7H_2O$	+ 1.33
$2IO_3^-/I_2$	$2IO_3^- + 12H^+ + 10e^- \rightarrow I_2 + 6H_2O$	+ 1.20
Pt^{2+}/Pt	$Pt^{2+} + 2e^- \rightarrow Pt$	+ 1.20

Table 11.1. (Contd.)

	Oxidising power decreasing ↓ / Reducing power increasing ↓	E° values decreasing ↓
IO_3^-/I^-	$IO_3^- + 6H^+ + 6e^- \rightarrow I^- + 3H_2O$	+ 1.09
$Br_2/2Br^-$	$Br_2 + 2e^- \rightarrow 2Br^-$	+ 1.06
Hg^{2+}/Hg	$Hg^{2+} + 2e^- \rightarrow Hg$	+ 0.85
Ag^+/Ag	$Ag^+ + e^- \rightarrow Ag$	+ 0.80
Fe^{3+}/Fe^{2+}	$Fe^{3+} + e^- \rightarrow Fe^{2+}$	+ 0.77
O_2/H_2O_2	$O_2 + 2H^+ + 2e^- \rightarrow H_2O_2$	+ 0.68
H_3AsO_4/H_3AsO_3	$H_3AsO_4 + 2H^+ + 2e^- \rightarrow H_3AsO_3 + H_2O$	+ 0.56
$I_2/2I^-$	$I_2 + 2e^- \rightarrow 2I^-$	+ 0.53
Cu^+/Cu	$Cu^+ + e^- \rightarrow Cu$	+ 0.52
O_2/H_2O	$O_2 + 4H^+ + e^- \rightarrow 2H_2O$	+ 0.40
Cu^{2+}/Cu	$Cu^{2+} + 2e^- \rightarrow Cu$	+ 0.34
Bi^{3+}/Bi	$Bi^{3+} + 3e^- \rightarrow Bi$	+ 0.32
Cu^{2+}/Cu^+	$Cu^{2+} + e^- \rightarrow Cu^+$	+ 0.15
Sn^{4+}/Sn^{2+}	$Sn^{4+} + 2e^- \rightarrow Sn^{2+}$	+ 0.15
$S_4O_6^{2-}/S_2O_3^{2-}$	$S_4O_6^{2-} + 2e^- \rightarrow 2S_2O_3^{2-}$	+ 0.09 = 0.00
$2H^+/H_2$	$2H^+ + 2e^- \rightarrow H_2$	(By definition)
Pb^{2+}/Pb	$Pb^{2+} + 2e^- \rightarrow Pb$	– 0.13
Sn^{2+}/Sn	$Sn^{2+} + 2e^- \rightarrow Sn$	– 0.14
Ni^{2+}/Ni	$Ni^{2+} + 2e^- \rightarrow Ni$	– 0.24
Co^{2+}/Co	$Co^{2+} + 2e^- \rightarrow Co$	– 0.28
Cd^{2+}/Cd	$Cd^{2+} + 2e^- \rightarrow Cd$	– 0.40
Cr^{3+}/Cr^{2+}	$Cr^{3+} + e^- \rightarrow Cr^{2+}$	– 0.41
Fe^{2+}/Fe	$Fe^{2+} + 2e^- \rightarrow Fe$	– 0.44
S/S^{2-}	$S + 2e^- \rightarrow S^{2-}$	– 0.51
Cr^{3+}/Cr	$Cr^{3+} + 3e^- \rightarrow Cr$	– 0.71
Zn^{2+}/Zn	$Zn^{2+} + 2e^- \rightarrow Zn$	– 0.76
Mn^{2+}/Mn	$Mn^{2+} + 2e^- \rightarrow Mn$	– 1.05
Al^{3+}/Al	$Al^{3+} + 3e^- \rightarrow Al$	– 1.67
Be^{2+}/Be	$Be^{2+} + 2e^- \rightarrow Be$	– 1.70
$H_2/2H^-$	$H_2 + 2e^- \rightarrow 2H^-$	– 2.24
Mg^{2+}/Mg	$Mg^{2+} + 2e^- \rightarrow Mg$	– 2.38
Na^+/Na	$Na^+ + e^- \rightarrow Na$	– 2.71
Ca^{2+}/Ca	$Ca^{2+} + 2e^- \rightarrow Ca$	– 2.76
Sr^{2+}/Sr	$Sr^{2+} + 2e^- \rightarrow Sr$	– 2.89
Ba^{2+}/Ba	$Ba^{2+} + 2e^- \rightarrow Ba$	– 2.90
M^+/M (M = K, Rb or Cs)	$M^+ + e^- \rightarrow M$	– 2.92
Li^+/Li	$Li^+ + e^- \rightarrow Li$	– 3.04
	Weakest oxidising agent / Strongest reducing agent	

Here it should be noted that the standard reduction potentials given in the table are so called because these are for the reduction reactions :

$$M + e \rightarrow M^-$$

If these values are to be converted into standard oxidation potentials, their magnitude will remain the same but their sign will have to be reversed and the values so obtained are called standard oxidation potentials because they are for the oxidation reactions :

$$M \rightarrow M^+ \times e^-$$

For example for the reduction reaction :

$$Zn^{2+} + 2e^- \rightarrow Zn,$$

standard reduction potential (from Table 11.1), $E^\circ_{Zn^{2+}/Zn} = -0.76$ V.

Now if the E° value is to be written for the oxidation reaction :

$$Zn \rightarrow Zn^{2+} + 2e^-,$$

then the magnitude of E° will remain the same but the sign will be reversed, *i.e.*, *standard oxidation potential*, $E^\circ_{Zn/Zn^{2+}} = +0.76$ V.

Moreover, any reaction equation may be suitably multiplied on both sides in order to balance the electrons gained and lost, in a redox reaction but *values of reduction potential are not multiplied by any coefficient.* For example, Table 11.1. shows that for the reduction reaction :

$$Ag^+(aq) + e^- \rightarrow Ag(s);\ E^\circ_{Ag^+/Ag} = +0.80\text{ V}$$

For two electrons the equation becomes :

$$2Ag^+(aq) + 2e^- \rightarrow 2Ag(s);\ E^\circ_{2Ag^+/2Ag} = +0.80\text{ V}$$

Thus :

$$E^\circ_{Ag^+/Ag} = E^\circ_{2Ag^+/2Ag} = E^\circ_{3Ag^+,3Ag} = \ldots\ldots = +0.80\text{ volts.}$$

Electrochemical Series or e.m.f. Series

The list of elements or ions arranged in the decreasing order of their standard reduction potential values, as in Table 11.1, is called electro-chemical or e.m.f. series or activity series.

From this series the following points should be noted and should be borne in mind while studying the uses of this series.

1. The forward reaction shown in the central column of the series is a *reduction reaction* and hence the standard electrode potential values are *standard reduction potential values*.

2. The species given at the left hand side of the reduction reaction accept electrons and are reduced into the species shown at the right hand side. Thus the species shown at the left hand side act as oxidising agents while those given at the right hand side of the reduction reaction act as reducing agents.

3. The values of standard reduction potential (E° values) are decreasing as we proceed from $F_2/2F^-$ (E° = + 2.85 V) to Li^+/Li (E° = – 3.04 V) through $2H^+/H_2$ (E° = ± 0.00 V). Thus E° values are decreasing from top to bottom in the series, *i.e.* positive E° values (*e.g.* $F_2/2F^-$, E° = + 2.85 V) → zero E° value (*e.g.*

$2H^+/H_2$, E° = ± 0.00 V) → negative E° values (*e.g.* Li^+/Li, E° = – 3.04 V). Thus the species lying above hydrogen have positive E° values while those lying below hydrogen have negative E° values.

4. With the decrease of E° values from + 2.85 V ($F_2/2F^-$) to – 3.04 V (Li^+/Li), the oxidising power of the oxidising agents is decreasing from top to bottom (*i.e.* from F_2 to Li^+ ion) while the reducing power of the reducing agents is increasing in the same direction (*i.e.* from F^- ion to Li metal). Thus F_2 is the strongest oxidising agent while Li^+ ion is the weakest oxidising agent. Similarly F^- ion is the weakest reducing agent while Li metal is the strongest reducing agent.

Uses of Electrochemical Series

1. To compare the oxidising and reducing power of metals or non-metals.

The magnitude of E° value of a metal or a non-metal gives a measure of its tendency to gain electrons to get reduced and hence to act as an oxidising agent or to lose electrons to get oxidised and hence to act as a reducing agent.

Gain of electrons (Reduction)

$$\underset{\text{Oxidising agent}}{M} + e^- \rightarrow \underset{\text{Reducing agent.}}{M^-}$$

Loss of electrons (Oxidation)

$$\underset{\text{Reducing agent}}{M} \rightarrow \underset{\text{Oxidising agent}}{M^+} + e^-$$

Thus the oxidising or reducing property of a given element in solution is measured by the magnitude of its E° value.

Oxidising agents. The species having positive E° values *i.e.*, the species lying above hydrogen in the electrochemical series show a strong tendency to gain electrons to undergo reduction and hence they are strong oxidising agents. As already said, with the decrease of E° value of a given species, the tendency of that species to gain electrons to undergo reduction decreases and hence the oxidising power of that species also decreases.

Examples. (*i*) The metals like Au, Pt, Hg, Ag, Cu etc. which have positive E° values are strong oxidising agents. The oxidising power of these metals is in the order: Au > Pt > Hg > Ag > Cu, since Au has the maximum E° value while Cu has the minimum E° value as shown below :

$$Au^{3+} + 3e^- \rightarrow Au,\ E^\circ = +1.42\ V$$
$$Pt^{2+} + 2e^- \rightarrow Pt,\ E^\circ = +1.20\ V$$
$$Hg^{2+} + 2e^- \rightarrow Hg,\ E^\circ = +0.85\ V$$
$$Ag^{+} + e^- \rightarrow Ag,\ E^\circ = +0.80\ V$$
$$Cu^{2+} + 2e^- \rightarrow Cu,\ E^\circ = +0.34\ V$$

E° values decreasing ↓ Oxidising power decreasing ↓

(*ii*) The non-metals like F_2, Cl_2, Br_2, I_2 and O_2 which have positive E° values are all strong oxidising agents. F_2 which has the maximum E° value is the strongest oxidising agent while O_2 which has the lowest E° value is the weakest oxidising agent. Thus the oxidising power of these non-metals is in the order :

$$F_2 > Cl_2 > Br_2 > I_2 > O_2,$$

since E° values are also decreasing in the same fashion as shown below :

$F_2 + 2e^- \rightarrow 2F^-$, $E° = + 2.85$ V	E° values ↓ decreasing	Oxidising power ↓ decreasing
$Cl_2 + 2e^- \rightarrow 2Cl^-$, $E° = + 1.36$ V		
$Br_2 + 2e^- \rightarrow 2Br^-$, $E° = + 1.06$ V		
$I_2 + 2e^- \rightarrow 2I^-$, $E° = + 0.53$ V		
$O_2 + 2H^+ + 2e^- \rightarrow 2OH^-$, $E° = + 0.40$ V		

(*iii*) Among the oxidising agents shown in Table 11.1, with the decrease of E° values from F_2 to Li^+ ion, the tendency of these species to gain electrons to undergo reduction decreases and hence their oxidising power decreases from F_2 to Li^+ ion. Thus among the oxidising agents shown in Table 11.1 F_2 is the strongest oxidising agent while Li^+ ion is the weakest oxidising agent.

Reducing agents. The species having negative values of standard reduction potentials (E° values), *i.e.* the species lying below hydrogen in the electrochemical series show a strong tendency to lose electrons to undergo oxidation and hence they are strong reducing agents. As already said, with the decreases of E° value of a given species (*i.e.*, as E° value becomes more negative), the tendency of that species to lose electrons to undergo oxidation increases and hence the reducing power of that species also increases.

Examples. (*i*) To illustrate the above rule we can consider the reducing property of alkali metals (group I A) and alkaline earth metals (group II A). E° values of these metals are given below :

Alkali metals (Group I A)	Alkaline earth metals (Group II A)
$Li^+ + e^- \rightarrow Li$, $E° = - 3.04$ V	$Be^{2+} + 2e^- \rightarrow Be$, $E° = - 1.70$ V
$Na^+ + e^- \rightarrow Na$, $E° = - 2.71$ V	$Mg^{2+} + 2e^- \rightarrow Mg$, $E° = - 2.38$ V
$K^+ + e^- \rightarrow K$, $E° = - 2.92$ V	$Ca^{2+} + 2e^- \rightarrow Ca$, $E° = - 2.76$ V
$Rb^+ + e^- \rightarrow Rb$, $E° = - 2.92$ V	$Sr^{2+} + 2e^- \rightarrow Sr$, $E° = - 2.89$ V
$Cs^+ + e^- \rightarrow Cs$, $E° = - 2.92$ V	$Ba^{2+} + 2e^- \rightarrow Ba$, $E° = - 2.90$ V

E° values of alkali metals show that these metals have high negative values and hence are strong reducing agents. Since Li has the minimum E° value (*i.e.* Li has maximum negative E° value), it is the strongest reducing agent of all the alkali metals. Similarly all the alkaline earth metals have high negative E° values and hence, like alkali metals, behave as strong reducing agents. Since E° values are decreasing (*i.e.* E° values are becoming more and more negative) from Be to Ba, the reducing power of these metals is also increasing in the same direction, *i.e.* Be with maximum E° value (*i.e.* with minimum negative E° value) is the weakest reducing agent while Ba which has minimum E° value (*i.e.* maximum negative E° value) is the strongest reducing agent.

When we compare the E° values of alkali metals with those of alkaline earth metals, we find that E° values increase on moving from group I A to group II A (*i.e.* E° values become less negative). Due to this increase in E° values, alkaline earth metals are weaker reducing agents than alkali metals.

(*ii*) Among the reducing agents shown in Table 11.1, with the decrease of E° values from F^- ion to Li metal, the tendency of these species to lose the electrons to undergo oxidation increases and hence their reducing power increases from F^- ion to Li metal. Thus among the reducing agents shown in Table 11.1 F^- ion is the weakest reducing agent while Li metal is the strongest reducing agent.

2. Electropositive Character of Metals.

Metals like K, Ca, Na etc. which lie at the bottom of the series readily lose their outer-most shell electrons to form metal cations and hence are *strongly electropositive* (*i.e. weakly electronegative*) while those lying at the top are not able to lose the outer-most shell electrons to form the cations and hence are *weakly electropositive* (*strongly electronegative*).

3. To Predict the Relative Chemical Reactivity of Metals.

Generally speaking the reactivity of a metal depends on its tendency to lose electrons to change into metal cation (M^+) which in turn, as we have already said, depends on the magnitude of its standard reduction potential. *Greater the value of standard reduction potential of a given metal (M), smaller its tendency to loss electrons to change into metal cation (M^+) and hence smaller its chemical reactivity. Thus with the decrease of standard reduction potential values on descending the electro-chemical series, the chemical reactivity of the metals increases in the same direction.* For example :

(*a*) Although alkali metals (Li, Na, K, Rb) and coinage metals (Cu, Ag and Au) both have one electron in their respective valence-shell, yet alkali metals are *highly reactive* while coinage metals are *least reactive*. For example alkali metals react with cold H_2O to give metallic hydroxides and H_2 gas. Coinage metals, on the other hand, do not react at all with water and steam and hence are called noble metals. These metals are attacked by strong oxidising agents like HNO_3, H_2SO_4 etc. The difference in reactivity is because of the fact that alkali metals have the lowest values (*i.e.* high negative values) of standard reduction potentials while the coinage metals have high positive values of standard reduction potentials.

(*b*) Metals like Pb, Sn, Ni, Co and Cd which lie a little down the series *react very slowly with very hot steam* to liberate H_2 gas.

(c) Metals like, Fe, Cr, Zn, Mn, Al and Mg which lie still down the series *react* with steam to produce the metallic oxides and H_2 gas. For example :

$$3Fe + 4H_2O\ (steam) \rightarrow Fe_3O_4 + 4H_2 \uparrow$$

Example. *Four metals A, B, C and D have their standard reduction potential values (in volt) equal to + 0.40, – 0.54, + 0.14 and – 1.36 respectively. Arrange these metals in the decreasing order of their reactivity.*

Solution : Since the standard reduction potential values of these metals are in the order : A(= + 0.40 V) > C(= + 0.14 V) > B(= – 0.54 V) > D(= – 1.36 V), the reactivity of these metals will be in the reverse order as shown below :

$$A < C < B < D$$

or

$$D > B > C > A$$

4. To predict whether a given metal will displace another metal from aqueous solution of its salt.

A metal with lower standard reduction potential will displace another metal with higher standard reduction potential from the aqueous solution of its salt and the metal with higher standard reduction potential gets precipitated. This means that a *metal will displace another metal from the aqueous solution of its salt that lies above it in the electrochemical series.*

Examples. (*i*) If a piece of Fe is placed in a solution of $CuSO_4$ some of Fe goes into solution as Fe^{2+} (*aq*) ions and Cu metal gets precipitated, *i.e.* Fe displaces Cu from $CuSO_4$ solution as shown below :

$$Fe + CuSO_4 \rightarrow FeSO_4 + Cu$$

or
$$Fe(s) + Cu^{2+}(aq) \rightarrow Fe^{2+}(aq) + \underset{\text{ppt.}}{Cu(s)}$$

The displacement of Cu from $CuSO_4$ solution by Fe is because of the fact that $E^\circ_{Fe^{2+}/Fe}$ (= – 0.44 V) is less than $E^\circ_{Cu^{2+}/Cu}$ (= + 0.34 V).

Why the displacement of Cu from $CuSO_4$ solution by Fe is possible or why the reaction :

$$\underset{\substack{\text{Reducing}\\\text{agent}}}{Fe(s)} + \underset{\substack{\text{Oxidising}\\\text{agent}}}{Cu^{2+}(aq)} \rightarrow \underset{\substack{\text{Oxidising}\\\text{agent}}}{Fe^{2+}(aq)} + \underset{\substack{\text{Reducing}\\\text{agent}}}{Cu(s)}$$

occurs spontaneously can also be explained by saying that since the reducing agent *viz.* Fe lies below the oxidising agent namely Cu^{2+} ions in the electrochemical series, the reaction shown above is possible to occur.

(*ii*) Zn does not displace Mg from a solution of $MgSO_4$, *i.e.* the reaction shown below is not possible to occur.

$$Zn + MgSO_4 \rightarrow ZnSO_4 + Mg$$

or
$$Zn(s) + Mg^{2+}(aq) \rightarrow Zn^{2+}(aq) + \underset{\text{ppt}}{Mg(s)}$$

For this reaction to occur it is essential that $E^\circ_{Zn^{2+}/Zn}$ should be less than $E^\circ_{Mg^{2+}/Mg}$, but actually the former is greater than the latter ($E^\circ_{Zn^{2+}/Zn}$ = – 0.76 V, $E^\circ_{Mg^{2+}/Mg}$ = –2.34 V). This means that the reverse of the above reaction will proceed spontaneously, *i.e.* Mg will displace Zn from $ZnSO_4$ solution as shown below :

$$Mg + ZnSO_4 \rightarrow MgSO_4 + Zn$$

or
$$Mg(s) + Zn^{2+}(aq) \rightarrow Mg^{2+}(aq) + \underset{\text{ppt}}{Zn(s)}$$

In the reaction :

$$\underset{\substack{\text{Reducing}\\\text{agent}}}{Zn(s)} + \underset{\substack{\text{Oxidising}\\\text{agent}}}{Mg^{2+}(aq)} \rightarrow \underset{\substack{\text{Oxidising}\\\text{agent}}}{Zn^{2+}(aq)} + \underset{\substack{\text{Reducing}\\\text{agent}}}{Mg(s)}$$

since the reducing agent *viz.* Zn lies above the oxidising agent namely Mg^{2+} in the electrochemical series, the reaction is not possible to occur in the direction shown. However, the reverse reaction :

$$\underset{\substack{\text{Oxidising}\\\text{agent}}}{Zn^{2+}(aq)} + \underset{\substack{\text{Reducing}\\\text{agent}}}{Mg(s)} \rightarrow \underset{\substack{\text{Reducing}\\\text{agent}}}{Zn(s)} + \underset{\substack{\text{Oxidising}\\\text{agent}}}{Mg^{2+}(aq)}$$

takes place spontaneously because the reducing agent *viz.* Mg lies below the oxidising agent namely Zn^{2+} ions in the electrochemical series.

(*iii*) The reaction :

$$Zn(s) + Fe^{2+}(aq) \rightarrow Zn^{2+}(aq) + Fe(s)$$

is predictable, since $E°_{Zn^{2+}/Zn}$ (= – 0.76 V) is less than $E°_{Fe^{2+}/Fe}$ (= – 0.44 V).

This reaction is used for galvanising Fe with Zn to prevent rusting of Fe. Zn coating on Fe prevents the oxidation of Fe to Fe^{2+} by air. If the galvanised Fe is scratched and some iron is oxidised to Fe^{2+} by air, Fe^{2+} thus produced is immediately reduced by Zn to Fe and thus the rusting of Fe is prevented.

(*iv*) Since $E°_{Cu^{2+}/Cu}$ (= + 0.34 V) > $E°_{Sn^{2+}/Sn}$ (= – 0.14 V) > $E°_{Zn^{2+}/Zn}$ (= – 0.76 V), Zn can displace both Sn and Cu from aqueous solutions of their respective salts; Sn, on the other hand, can displace Cu from aqueous solution of its salt, but it cannot displace Zn from an aqueous solution containing Zn^{2+} ions, *i.e.*, the reactions (*a*), (*b*) and (*c*) are possible but reaction (*d*) is not feasible.

	Reducing agent		Oxidising agent		Oxidising agent		Reducing agent
(*a*)	Zn (*s*)	+	Sn^{2+} (*aq*)	→	Zn^{2+} (*aq*)	+	Sn (*s*) ppt
(*b*)	Zn (*s*)	+	Cu^{2+} (*aq*)	→	Zn^{2+} (*aq*)	+	Cu (*s*) ppt
(*c*)	Sn (*s*)	+	Cu^{2+} (*aq*)	→	Sn^{2+} (*aq*)	+	Cu (*s*) ppt

In reactions (*a*) and (*b*) since the reducing agent *viz.* Zn lies below the oxidising agents namely Sn^{2+} and Cu^{2+} ions and in reaction (*c*) the reducing agent *viz.* Sn lies below the oxidising agent namely Cu^{2+} ion, these reactions are possible to occur. In reaction (*d*) since the reducing agent Sn lies above the oxidising agent *viz.* Zn^{2+}, this reaction is not predictable.

(*v*) Since $E°_{Ag^{+}/Ag}$ (= + 0.80 V) > $E°_{Cu^{2+}/Cu}$ (= + 0.34 V), Cu can displace Ag from an aqueous solution containing Ag^+ ions but Ag cannot displace Cu from an aqueous solution containing Cu^{2+} ions. This means that the reaction :

$$Cu\ (s) + 2Ag^{+}\ (aq) \rightarrow Cu^{2+}\ (aq) + \underset{\text{ppt.}}{2Ag\ (s)}$$

is possible while the reaction :

$$2Ag\ (s) + Cu^{2+}\ (aq) \rightarrow 2Ag^{+}\ (aq) + \underset{\text{ppt}}{Cu\ (s)}$$

is not possible.

5. To Predict Whether a Give Non-metal will Displace Another Non-metal from Aqueous Solution of its Salt.

A non-metal with higher standard reduction potential will displace another non-metal with lower standard reduction potential. This means that *a non-metal will displace another non-metal that lies below it in the electrochemical series.* For example Cl_2 can displace both Br_2 and I_2 from their salt solutions, since Br_2 and I_2 both lie below Cl_2 in the electrochemical series. Thus in the preparation of Br_2 gas when Cl_2 gas is passed through sea-water (containing $MgBr_2$), Br_2 of $MgBr_2$ is displaced by Cl_2 and Br_2 gas is thus obtained.

$$Cl_2 + MgBr_2 \rightarrow MgCl_2 + Br_2$$

or $$Cl_2(g) + 2Br^-(aq) \rightarrow 2Cl^-(aq) + Br_2(g)$$

Note that $E^\circ_{Cl_2/2Cl^-}$ (= + 1.36 V) is greater than $E^\circ_{Br_2/2Br^-}$ (= + 1.06V)

The reaction :

$$\underset{\text{Oxidising agent}}{Cl_2(g)} + \underset{\text{Reducing agent}}{2Br^-(aq)} \rightarrow \underset{\text{Reducing agent}}{2Cl^-(aq)} + \underset{\text{Oxidising agent}}{Br_2(g)}$$

takes place spontaneously in the direction shown, since the reducing agent *viz.* Br^- lies below the oxidising agent namely Cl_2 in the electrochemical series.

Similarly Cl_2 can displace I_2 from KI solution.

$$Cl_2 + KI \rightarrow 2KCl + I_2$$

or $$Cl_2 + 2I^- \rightarrow 2Cl^- + I_2$$

6. To Predict Whether a Metal will Liberate Hydrogen Gas from dil. Acids or not.

The metal (M) which has a strong tendency to lose electrons to form metal ions (M^+) will displace H^+ ions from acid solution and thus will liberate H_2 gas as shown below :

$$M(s) + 2H^+ (aq) \text{ (From acid)} \rightarrow M^{2+} (aq) + H_2 (g) \uparrow$$

We have already pointed out that greater the value of standard reduction potential of a metal, smaller its tendency to lose electrons to form metal ions and hence smaller its tendency to displace H^+ ions from acid solution to liberate H_2 gas. *Thus with the decrease of standard reduction potential values on descending the electrochemical series, the tendency of the metals to liberate H_2 gas from dil. acids increases in the same direction.*

In general it has been observed that the metals like Au, Pt, Ag, Cu etc. which are lying *above* hydrogen in the electrochemical series are not able to liberate H_2 gas from dil. HCl at all while the metals like Zn, Mg, Ca etc. which are lying *below* hydrogen in the electrochemical series can liberate H_2 gas from dil. HCl.

7. To Predict Whether a Given Metal can Liberate H_2 Gas from Water.

Fe and all other metals which lie below it in the electrochemical series can displace H_2 gas from H_2O. Alkali metals and alkaline earth metals liberate H_2 from H_2O at ordinary temperature while the metals from Fe to Mg in the electrochemical series can do so with boiling water or steam.

$$2Na + H_2O \rightarrow NaOH + H_2$$

$$3Fe \text{ (red hot)} + 4H_2O \text{ (steam)} \rightarrow Fe_3O_4 + 4H_2O$$

$$Mg + H_2 \text{ (boiling)} \rightarrow MgO + H_2$$

8. Decomposition of Oxides of Metals.

The oxides of Hg and of the metals lying *above* Hg in the electrochemical series get decomposed on heating to give the free metal and O_2 while the oxides of the metals lying *below* Hg in the electrochemical series do not undergo decomposition. For example HgO gets decomposed on heating to give Hg and O_2

while the oxides of the metals Ag, Mg, Ca, Ba, etc. which are lying below Hg in the electrochemical series do not undergo the decomposition. Thus :

$$2HgO \xrightarrow{\Delta} 2Hg + O_2$$

$$Ag_2O, MgO, CaO \text{ or } BaO \xrightarrow{\Delta} \text{Do not decompose.}$$

9. To Calculate the e.m.f. of a Given Galvanic Cell.

The values of standard reduction potentials given in electrochemical series can also be used to calculate the *e.m.f.* of a given galvanic cell. *The standard e.m.f. of a cell ($E°_{cell}$) is equal to the difference of the standard reduction potential of the cathode (i.e. right hand electrode), $E°_{cathode}$ and the standard reduction potential of the anode (i.e. left hand electrode), $E°_{anode}$.* Thus :

$$\boxed{\begin{aligned} E°_{cell} &= E°_{cathode} - E°_{anode} \\ &= E°_{right} - E°_{left} \end{aligned}}$$

This application can be illustrated by the following solved examples.

Example 1. *Write the half-cell reactions for the following cell and calculate its standard e.m.f.*

$$\underset{\text{Anode}}{Mg \mid Mg^{2+}} \mid\mid \underset{\text{Cathode}}{Cu^{2+} \mid Cu}$$

Solution. Obviously the two half-cell reactions occuring in the cell are :

Oxidation half-reaction (Anode reaction) :

$Mg \rightarrow Mg^{2+} + 2e^- ; E°_{Mg/Mg^{2+}} = +2.38$ V (From Table 11.1)

or $E°_{Mg^{2+}/Mg} = -2.38$ V

Reduction half-reaction (Cathode reaction) :

$Cu^{2+} + 2e^- \rightarrow Cu; E°_{Cu^{2+}/Cu} = +0.34$ V (From Table 11.1)

$$\begin{aligned} \therefore \quad E°_{cell} &= E°_{cathode} - E°_{anode} \\ &= E°_{Cu^{2+}/Cu} - E°_{Mg^{2+}/Mg} = [0.34 - (-2.38)] \text{ volt} \\ &= +2.72 \text{ volt.} \end{aligned}$$

Example 2. *Two half-cell reactions and their standard electrode potentials are given below :*

$$Zn \rightarrow Zn^{2+} + 2e^- ; E° = +0.76 \text{ volt}$$

$$Cu \rightarrow Cu^{2+} + 2e^- ; E° = -0.34 \text{ volt.}$$

What overall chemical reaction will take place in the electrochemical cell formed by the combination of these two half-cells? Calculate the e.m.f. of this cell under standard conditions. *(Punjab B.Sc. 1975; Punjabi 1975)*

Solution. It is obvious from the half-cell reactions given in the question that the E° values are oxidation potential values. Reduction potential values, therefore, are as :

$E°_{Zn^{2+}/Zn} = -0.76$ volt and $E°_{Cu^{2+}/Cu} = +0.34$ volt

Again we know that the electrode having negative value of reduction potential acts as an anode while that having positive value acts as a cathode. Thus Zn/Zn^{2+}

electrode which has negative reduction potential value acts as anode while Cu^{2+}/Cu electrode with positive value acts as a cathode and consequently :

Oxidation or anode reaction :

$Zn \rightarrow Zn^{2+} + 2e^-$; $E^\circ_{Zn/Zn^{2+}} = +\ 0.76$ volt (given)

or $E^\circ_{Zn^{2+}/Zn} = -\ 0.76$ volts.

Reduction or cathode reaction :

$Cu^{2+} + 2e^- \rightarrow Cu$; $E^\circ_{Cu^{2+}/Cu} = +\ 0.34$ volt

∴ Overall cell reaction is represented as :

$$Zn + Cu^{2+} \rightarrow Zn^{2+} + Cu$$

and its e.m.f., E°_{cell} is given by :

$$E^\circ_{cell} = E^\circ_{cathode} - E^\circ_{anode}$$
$$= E^\circ_{Cu^{2+}/Cu} - E^\circ_{Zn^{2+}/Zn} = [0.34 - (-\ 0.76)] \text{ volt}$$
$$= 1.10 \text{ volt}$$

Example 3. *Calculate the e.m.f. of Cu–Ag cell.*

Solution. First of all we have to decide which of the electrodes namely $Cu(s)/Cu^{2+}$ (aq) and Ag $(s)/Ag^+(aq)$ will act as an anode and which will act as a cathode.

The electrode having lower value of reduction potential acts as an anode while that having higher value of reduction potential acts as a cathode. Thus, since $E^\circ_{Cu^{2+}/Cu}$ (= + 0.34 V) is lower than $E^\circ_{Ag^+/Ag}$ (= + 0.80 V), $Cu(s)/Cu^{2+}$ (aq) electrode acts as anode and Ag^+ (aq)/Ag (s) acts as cathode. Thus the cell can be represented as :

⟶ Flow of electrons ⟶

$$Cu(s)\,|\,Cu^{2+}(aq)\ \ ||\ \ Ag^+(aq)\,|\,Ag(s)$$

Anode (–)	Cathode (–)
(Oxidation half-cell)	(Reduction half-cell)

Obviously two half-cell reactions taking place in the above cell are :

(*i*) Oxidation or anode reaction :

$Cu(s) \rightarrow Cu^{2+}$ (aq) + $2e^-$; $E^\circ_{Cu/Cu^{2+}} = -\ 0.34$ volt (From Table 11.1)

or $E^\circ_{cu^{2+}/cu} = +\ 0.34$ volt

(*ii*) Reduction or cathode reaction :

$2Ag^+(aq) + 2e^- \rightarrow 2Ag(s)$; $E^\circ_{2Ag^+/2Ag} = +\ 0.80$ volt (From Table 11.1)

∴ Cell reaction can be represented as :

$$Cu\ (s) + 2Ag^+\ (aq) \rightarrow Cu^{2+}(aq) + 2Ag(s),$$

and
$$E^\circ_{cell} = E^\circ_{cathode} - E^\circ_{anode}$$
$$= E^\circ_{2Ag^+/2Ag} - E^\circ_{cu^{2+}/cu} = 0.80 - 0.34$$
$$= 0.46 \text{ volt}$$

Example 4. *Write the cell reaction and calculate the cell e.m.f. from the following half-reactions :*

$$Ag^+ + e^- \rightarrow Ag;\ E^\circ = +\ 0.799\ V$$
$$Ni^{2+} + 2e^- \rightarrow Ni;\ E^\circ = -\ 0.250\ V$$

Solution. We know that the electrode having negative value of reduction potential acts as anode while that having positive value acts as cathode. Thus the cell can be represented as :

$$\longrightarrow \text{Flow of electrons} \longrightarrow$$

$$\underset{\text{Anode}}{Ni \mid Ni^{2+}} \;\|\; \underset{\text{Cathode}}{Ag^+ \mid Ag}$$

and hence the two half-reactions are :

Oxidation half-reaction (Anode reaction) :

$$Ni \rightarrow Ni^{2+} + 2e^- \;;\; E^\circ_{Ni/Ni^{2+}} = +\,0.250\ V$$

or $$E^\circ_{Ni^{2+}/Ni} = -\,0.250\ V$$

Reduction half-reaction (Cathode reaction) :

$$2 \times [Ag^+ + e^- \rightarrow Ag] \;;\; E^\circ_{2Ag^+/2Ag} = +\,0.799\ V$$

The complete cell reaction is obtained by adding the above two reactions :

$$Ni + 2Ag^+ \rightarrow Ni^{2+} + 2Ag$$

and its e.m.f. is given by :

$$\begin{aligned} E^\circ_{cell} &= E^\circ_{cathode} - E^\circ_{anode} \\ &= E^\circ_{2Ag^+/2Ag} - E^\circ_{Ni^{2+}/Ni} = 0.799 - (-\,0.250) \\ &= 1.049\ \text{volt.} \end{aligned}$$

Example 5. *Represent the cell with the reaction :*

$$Cd + 2Ag^+ \rightarrow Cd^{2+} + 2Ag$$

and calculate the standard e.m.f. of the cell.

Solution. The above reaction is composed of the following two half-reactions;

(*i*) Oxidation or anode reaction :

$$Cd \rightarrow Cd^{2+} + 2e^- \;;\; E^\circ_{Cd/Cd^{2+}} = +\,0.40\ V \qquad \text{(From Table 11.1)}$$

or $$E^\circ_{Cd^{2+}/Cd} = -\,0.40\ V$$

(*ii*) Reduction or cathode reaction :

$$2Ag^+ + 2e^- \rightarrow 2Ag;\; E^\circ_{2Ag^+/2Ag} = +\,0.80\ V \qquad \text{(From Table 11.1)}$$

By convention Cd/Cd^{2+} electrode will act as anode and $2Ag^+/2Ag$ electrode will act as cathode, since oxidation occurs at anode and reduction takes place at cathode. Thus the cell can be represented as :

$$\longrightarrow \text{Flow of electrons} \longrightarrow$$

$$\underset{\substack{\text{Anode }(-)\\ \text{(Oxidation}\\ \text{half-cell)}}}{Cd \mid Cd^{2+}} \;\|\; \underset{\substack{\text{Cathode }(-)\\ \text{(Reduction}\\ \text{half-cell)}}}{Ag^+ \mid Ag}$$

Obviously,
$$\begin{aligned} E^\circ_{cell} &= E^\circ_{cathode} - E^\circ_{anode} \\ &= E^\circ_{2Ag^+/Ag} - E^\circ_{Cd^{2+}/Cd} \\ &= +\,0.80 - (-\,0.40) \\ &= 1.20\ \text{volt.} \end{aligned}$$

10. To Predict Whether a Given Galvanic Cell will Work or Not.

The cell will work only when its standard electrode potential, $E°_{cell}$ is a positive quantity. For example we can show whether the cell represented as :

$$\underset{\text{Anode } (-)}{Pb \mid Pb^{2+}} \mid \mid \underset{\text{Cathode } (+)}{Mg^{2+} \mid Mg}$$

will work or not, *i.e.* whether the cell representation is correct or not.

Now
$$\begin{aligned} E°_{cell} &= E°_{cathode} - E°_{anode} \\ &= E°_{Mg^{2+}/Mg} - E°_{Pb^{2+}/Pb} \\ &= +\ 2.38 - (-\ 0.13) \qquad \text{(From Table 11.1)} \\ &= -\ 2.25 \text{ volt.} \end{aligned}$$

Since $E°_{cell}$ is negative, the representation is not correct. The correct cell representation can be obtained by interchanging the position of cathode and anode. Thus the correct representation will be :

$$\underset{\text{Anode}}{Mg \mid Mg^{2+}} \mid \mid \underset{\text{Cathode}}{Pb^{2+} \mid Pb}$$

The correctness of this representation is evident from the fact that the value of $E°_{cell}$ of this cell comes out to be a positive quantity as shown below :

$$\begin{aligned} E°_{cell} &= E°_{cathode} - E°_{anode} \\ &= E°_{Pb^{2+}/Pb} - E°_{Mg^{2+}/Mg} \\ &= -\ 0.13 - (-\ 2.38) \qquad \text{(From Table 11.1)} \\ &= +\ 2.25 \text{ volt.} \end{aligned}$$

Example. *Referring E° values for the electrodes Sn^{2+}/Sn and Pb^{2+}/Pb in Table 11.1, calculate the e.m.f. of the cell :*

$$Sn \mid Sn^{2+}\ (1.0M) \mid \mid Pb^{2+}(0.001M) \mid Pb$$

Predict whether the cell will function or not.

Solution. E° values for the electrodes are as : $E°_{Sn/Sn^{2+}} = +\ 0.14$ V and $E°_{Pb^{2+}/Pb} = -\ 0.13$ V and $E°_{cell}$ is given by :

$$\begin{aligned} E°_{cell} &= E°_{right} - E°_{left} \\ &= E°_{Pb^{2+}/Pb} - E°_{Sn^{2+}/Sn} \\ &= -\ 0.13 - (-\ 0.14) = +\ 0.01 \text{ volt} \end{aligned}$$

Since $E°_{cell}$ is a positive quantity, the cell will function in the manner shown here.

11. To Predict Whether a Given Redox Reaction is Likely to Occur or Not.

A given redox reaction will take place spontaneously in the direction as represented by a given net cell equation, if $E°_{cell}$ is positive, because in that case the standard free energy change of the reaction (ΔG°) would be negative ($\Delta G° = nFE°_{cell}$). *If E_{cell} is a negative quantity, the given redox reaction will not take place in the direction from left to right but it will take place in the direction from right to left (i.e. in the reverse direction).* In order to illustrate this application let us see whether the reaction :

$$Cu(s) + Zn^{2+}\ (aq) \rightarrow Cu^{2+}(aq) + Zn(s)$$

will take place spontaneously or not. In order to answer this problem we proceed as follows :

The half-cell reactions of which the above reaction is composed are :

Oxidation half-reaction (Anode reaction) :

$Cu(s) \rightarrow Cu^{2+}(aq) + 2e^-$; $E^\circ_{Cu/Cu^{2+}} = -0.34$ volt (From Table 11.1)

Reduction half-reaction (Cathode reaction) :

$Zn^{2+}(aq) + 2e^- \rightarrow Zn(s)$; $E^\circ_{Zn^{2+}/Zn} = -0.76$ V (From Table 11.1)

Thus the cell can be represented as :

$$\underset{\text{Anode}}{Cu(s)\,|\,Cu^{2+}(aq)}\;\|\;\underset{\text{Cathode}}{Zn^{2+}(aq)\,|\,Zn(s)}$$

and
$$\begin{aligned} E^\circ_{cell} &= E^\circ_{cathode} - E^\circ_{anode} \\ &= -0.76 - (+0.34) \\ &= -1.10 \text{ volt.} \end{aligned}$$

Since E°_{cell} is a negative quantity, the cell reaction shown above does not take place spontaneously in the direction as written. However the reaction takes place spontaneously in the reverse direction as shown below :

$$Cu^{2+}(aq) + Zn(s) \rightarrow Zn^{2+}(aq) + Cu(s).$$

An Important Rule.

We know that each reaction consists of one oxidising agent and one reducing agent on the left hand and similarly one oxidising agent and one reducing agent on the right hand side as shown below for the reaction between Cu(s) and $Zn^{2+}(aq)$ discussed above.

$$\underset{\substack{\text{Oxidising}\\\text{agent}}}{Cu^{2+}(aq)} + \underset{\substack{\text{Reducing}\\\text{agent}}}{Zn(s)} \rightarrow \underset{\substack{\text{Oxidising}\\\text{agent}}}{Zn^{2+}(aq)} + \underset{\substack{\text{Reducing}\\\text{agent}}}{Cu(s)} \quad ...(i)$$

We have already seen that the reaction (*i*) can take place spontaneously in the direction as shown. Why this reaction takes place spontaneously in the direction from left to right can also be explained with the help of a rule which says that :

A given oxidising agent (given at the left hand side of the reduction reactions shown in the electrochemical series) will react with any reducing agent (given at the right hand side of the reduction reactions shown in the electrochemical series) whose standard reduction potential is less than that of the oxidising agent. In other words *a given oxidising agent will react (i.e. will oxidise or will be reduced by) with that reducing agent which lies below the oxidising agent in the electrochemical series. Similarly a given reducing agent will react with (i.e. will reduce or will be oxidised by) that oxidising agent which lies above the reducing agent in the electrochemical series.*

Thus, according to the above rule, the reaction (*i*) in which Cu^{2+} ions act as oxidising agent and Zn metal acts as reducing agent, takes place spontaneously in a direction from left to right, since $E^\circ_{Zn^{2+}/Zn}$ (= – 0.76 volt) < $E^\circ_{Cu^{2+}/Cu}$ (= + 0.34 volt)

Now let us use the above rule to explain why the reaction :

$$\underset{\text{Reducing agent}}{Cu(s)} + \underset{\text{Oxidising agent}}{Zn^{2+}(aq)} \rightarrow \underset{\text{Oxidising agent}}{Cu^{2+}(aq)} + \underset{\text{Reducing agent}}{Zn(s)}$$

cannot take place spontaneously in the direction from left to right. The values of $E°_{Cu^{2+}/Cu}$ (= + 0.34 volt) and $E°_{Zn^{2+}/Zn}$ (= – 0.76 volt) show that $E°_{Cu^{2+}/Cu}$ is not less than $E°_{Zn^{2+}/Zn}$ and hence the reaction shown above does not take place in the direction from left to right

Examples to Illustrate the Rule.

The above rule can also be explained with the help of the following examples.

1. *Oxidation of Cu metal to Cu^{2+} ions by Ag^+ ions.* In the oxidation of Cu metal to Cu^{2+} ions by Ag^+ ions, Ag^+ ions are reduced to Ag metal.

$$\underset{\text{Oxidising agent}}{2Ag^+(aq)} + \underset{\text{Reducing agent}}{Cu(s)} \rightarrow \underset{\text{Oxidising agent}}{Cu^{2+}(aq)} + \underset{\text{Reducing agent}}{2Ag(s)}$$

The oxidation of Cu metal by Ag^+ ions is feasible because

$$E°_{Cu^{2+}/Cu} (= + 0.34 \text{ V}) < E°_{2Ag^+/2Ag} = (+ 0.80 \text{ V}).$$

2. *Oxidation of I^- ions to I_2 molecule by Fe^{3+} ions.* In this redox reaction Fe^{3+} ions are reduced to Fe^{2+} ions.

$$\underset{\text{Oxidising agent}}{2Fe^{3+}} + \underset{\text{Reducing agent}}{2I^-} \rightarrow \underset{\text{Reducing agent}}{2Fe^{2+}} + \underset{\text{Oxidising agent}}{I_2}$$

The above reaction takes place in the direction shown because

$$E°_{I_2/2I^-} (= + 0.53 \text{ V}) < E°_{Fe^{3+}/Fe^{2+}} = (+ 0.77 \text{ V}).$$

3. *Oxidation of Cu metal to Cu^{2+} ions by I_2.* In this redox reaction I_2 molecule is itself reduced to I^- ions.

$$\underset{\text{Oxidising agent}}{I_2} + \underset{\text{Reducing agent}}{Cu} \rightarrow \underset{\text{Oxidising agent}}{Cu^{2+}} + \underset{\text{Reducing agent}}{2I^-}$$

The oxidation of Cu metal to Cu^{2+} ions is feasible because $E°_{Cu^{2+}/Cu}$ (= + 0.34 V) < $E°_{I_2/2I^-}$ (= + 0.53 V)

On the same lines of argument it can be explained why the oxidation of Bi metal to Bi^{3+} ions and of H_2 molecule to H^+ ions by I_2 molecule is feasible.

Oxidising agent *Reducing agent* *Oxidising agent* *Reducing agent*

$$3I_2 + 2Bi \rightarrow 2Bi^{3+} + 6I^-;\ E°_{I_2/2I^-} = + 0.53 \text{ V},\ E°_{Bi^{3+}/Bi} = + 0.32 \text{ V}$$

$$I_2 + H_2 \rightarrow 2H^+ + 2I^-;\ E°_{I_2/2I^-} = + 0.53 \text{ V},\ E°_{2H^+/H_2} = + 0.00 \text{ V}$$

On similar grounds with the help of E° values of $I_2/2I^-$, $2Ag^+/2Ag$, $Br_2/2Br^-$, $Cl_2/2Cl^-$ and $F_2/2F^-$ electrodes given as : $E°_{I_2/2I^-}$ = + 0.53 V, $E°_{2Ag^+/2Ag}$ = + 0.80 V, $E°_{Br_2/2Br^-}$ = + 1.06 V, $E°_{Cl_2/2Cl^-}$ = + 1.36 V, $E°_{F_2/2F^-}$ = + 2.85 V, it can be explained why oxidation of Ag to Ag^+ ions, Br^- to Br_2, Cl^- to Cl_2 and F^- to F_2 by I_2 is not possible, *i.e.* the following redox reactions can not occur in the direction as shown below :

Oxidising agent		Reducing agent		Oxidising agent		Reducing agent	
I_2	+	2Ag	→	$2Ag^+$	+	$2I^-$ ...	*Not possible*
I_2	+	$2Br^-$	→	Br_2	+	$2I^-$ ...	*Not possible*
I_2	+	$2Cl^-$	→	Cl_2	+	$2I^-$ ...	*Not possible*
I_2	+	$2F^-$	→	F_2	+	$2I^-$ ...	*Not possible*

4. *Reduction of Fe^{2+} ions to Fe metal by Zn metal.* In this redox reaction Zn metal itself is oxidised to Zn^{2+} ions.

$$\underset{\text{Reducing agent}}{Zn} + \underset{\text{Oxidising agent}}{Fe^{2+}} \rightarrow \underset{\text{Oxidising agent}}{Zn^{2+}} + \underset{\text{Reducing agent}}{Fe}$$

The reduction of Fe^{2+} ions to Fe metal by Zn metal is possible because $E°_{Z^{2+}/Zn}$ (= – 0.76 V) < $E°_{Fe^{2+}/Fe}$ (= – 0.44 V).

On the same lines of argument it can be explained why the reduction of H^+ ions to H_2 and Cu^{2+} ions to metallic Cu by Zn metal is feasible.

Reducing agent		Oxidising agent		Reducing agent		Oxidising agent	
Zn	+	2H+	→	H_2	+	Zn^{2+} ;	$E°_{Zn^{2+}/Zn} = -0.76$ V, $E°_{2H^+/H_2} = +0.00$ V
Zn	+	Cu^{2-}	→	Cu	+	Zn^{2+} ;	$E°_{Zn^{2+}/Zn} = +0.76$ V, $E°_{Cu^{2+}/Cu} = +0.34$ V

On similar grounds with the help of E° values of Al^{3+}/Al, Mg^{2+}/Mg and K^+/K electrodes given as $E°_{Al^{3+}/Al}$ = 1.67 V, $E°_{Mg^{2+}/Mg}$ = – 2.38 V and $E°_{K^+/K}$ = – 2.92 V, it can be explained why the reduction of Al^{3+} to Al, of Mg^{2+} to Mg and of K^+ to K by Zn metal is not possible, *i.e.* the redox reactions given below can not occur in the direction shown.

Oxidising agent		Reducing agent		Oxidising agent		Reducing agent	
Al^{3+}	+	Zn	→	Zn^{2+}	+	Al ...	*Not possible*
Mg^{2+}	+	Zn	→	Zn^{2+}	+	Mg ...	*Not possible*
K^+	+	Zn	→	Zn^{2+}	+	K ...	*Not possible*

Effect of Metal Ion Concentration on the Magnitude of Electrode Potential—Nerst Equation

We know that the standard oxidation potential of Zn/Zn^{2+} electrode in which the Zn rod is dipped in IM solution of $ZnSO_4$ is + 0.76 volt. If the concentration of the solution is not 1M, then the oxidation potential corresponding to the reaction :

$$Zn \rightarrow Zn^{2+} + 2e^-$$

is not equal to + 0.76 volt but is given by :

$$E = E°_{Zn/Zn^{2+}} - \frac{0.0592}{n} \log \frac{[Zn^{2+}]}{[Zn]}$$

where n = No. of electrons in the electrode reaction = 2.

Now since the concentration of the elements and solids = 1, [Zn] = 1 and hence :

$$E = E^\circ_{Zn/Zn^{2+}} - \frac{0.0591}{2} \log [Zn^{2+}]$$

In general :

$$E = E^\circ - \frac{0.0591}{n} \log [\text{Ion}]$$

This is called **Nerst equation.**

This is the equation for a half-cell oxidation reaction and hence :

$$E_{ox} = E^\circ - \frac{0.0591}{n} \log [\text{Ion}]$$

The value of E_{red} for the half-cell reduction reaction is given by reversing the sign of E_{ox}, *i.e.*

$$E_{red} = -E_{ox} = -\left[E^\circ - \frac{0.0591}{n} \log [\text{Ion}]\right]$$

or

$$E_{red} = -E^\circ + \frac{0.0591}{n} \log [\text{Ion}]$$

Effect of Metal Ion Concentration on the Magnitude of e.m.f of a Given Cell

We have seen that the standard e.m.f. of Daniell cell represented as :

$$Zn \mid Zn^{2+}(1M) \parallel Cu^{2+}(1M) \mid Cu$$

in which the concentration of each of the metal ions is unity is given by :

$$\text{Standard e.m.f.} = E^\circ_{cell} = E^\circ_{cathode} - E^\circ_{anode}$$
$$= E^\circ_{Cu^{2+}/Cu} - E^\circ_{Zn^{2+}/Zn}$$
$$= 0.34 - (-0.76)$$
$$= +1.10 \text{ volt}$$

If the concentration of each of the metal ions is different as in the cell :

$$Zn \mid Zn^{2+}(0.1M) \parallel Cu^{2+}(0.01M) \mid Cu,$$

the e.m.f. is given by :

$$\mathbf{E_{cell} = E^\circ_{cell} - \frac{0.0591}{n} \log \frac{[Products]}{[Reactants]}},$$

where n = No. of electrons involved in the redox reaction = 2 (see below)

The value of [Products]/[Reactants] is obtained from the cell reaction, remembering that oxidation occurs at the anode and the reduction occurs at the cathode. This is illustrated for the above cell as follows :

Oxidation half-reaction (Anode reaction)

$$Zn \rightarrow Zn^{2+}(0.1M) + 2e^-$$

Reduction half-reaction (Cathode reaction)

$$Cu^{2+}(0.01M) + 2e^- \rightarrow Cu$$

$\therefore$ *Cell reaction is represented as :*

$$Zn + Cu^{2+}(0.01M) \rightarrow Zn^{2+}(0.1M) + Cu \ (n = 2)$$

$$\therefore \quad \frac{[\text{Products}]}{[\text{Reactants}]} = \frac{[Zn^{2+}][Cu]}{[Zn^{2+}][Cu^{2+}]}$$

Now since concentration of the elements and solids = 1, [Cu] or [Zn) = 1 and hence :

$$\frac{[\text{Products}]}{[\text{Reactants}]} = \frac{[Zn^{2+}]}{[Cu^{2+}]} = \frac{0.1}{0.01}$$

$$\therefore \quad E_{cell} = E^\circ_{cell} - \frac{0.0591}{2} \log \frac{0.1}{0.01}$$

$$= 1.10 - 0.0295 \times 1$$

$$= 1.0705 \text{ volt.}$$

This example clearly shows how the metal ion concentration changes the magnitude of the e.m.f. of the cell.

Questions with Answers

Q.1 Explain the following :

(*i*) Co^{3+} is not stable in aqueous solution. Given that $E^\circ_{Co^{3+}/Co^{2+}} = +1.82$ V and $E^\circ_{H_2O/O_2} = -1.23$ V.

Ans : In aqueous solution, Co^{3+} is reduced to Co^{2+} according to the following equation.

$$\underset{\substack{(Co\,=\,+3)\\ OA}}{4Co^{3+}} + \underset{\substack{(O\,=\,-2)\\ RA}}{2H_2O} \rightarrow \underset{(O\,=\,+2)}{4Co^{2+}} + 4H^+ + \underset{(O\,=\,0)}{O_2}$$

This reaction takes place (*i.e.*, Co^{3+} gets reduced to Co^{2+} ion and hence is unstable), since E° reaction is positive as shown below :

$$E^\circ_{reaction} = (E^\circ_{red})_{OA} - (E^\circ_{red})_{RA}$$

$$= E^\circ_{Co^{3+}/Co^{2+}} - E^\circ_{O_2/H_2O}$$

$$= +1.82 - (+1.23)$$

$$= +1.82 - 1.23 = \mathbf{+0.59\ V}$$

(*ii*) When I_2 and F_2 are added to a soultion containing 1M each of I^- and F^- ions, only I^- ions are oxidised to I_2 but F^- ions are not oxidised to F_2. (Given that : $E^o_{I_2/2I^-} = +0.54$ V and $E^\circ_{F_2/2F^-} = +2.87$ V).

Ans : The oxidation of F^- and I^- ions to F_2 and I_2 respectively can be represented as :

(*a*) $\underset{RA}{2F^-} + \underset{OA}{I_2} \rightarrow F_2 + 2I^-$

(*b*) $2I^- + F_2 \rightarrow I_2 + 2F^-$
RA OA

In reaction (*a*), since E°_{OA} (= $E^\circ_{I_2/2I^-}$ = 0.54 V) < E°_{RA} (= $E^\circ_{F_2/2F^-}$ = +2.87 V), this reaction is not feasible, *i.e.* F^- ions cannot be oxidised to F_2 by I_2. In reaction (*b*), since E°_{OA} (= $E^\circ_{F_2/2F^-}$ = +2.87 V) > E°_{RA} (= $E^\circ_{I_2/2I^-}$ = +0.54 V), this reaction is feasible, *i.e.* I^- ions can be oxidised to I_2 by F_2.

(*iii*) When aqueous solution of NaCl is electrolysed, only H_2 gas and not Na metal is liberated at cathode. (Given that $E^\circ_{Na^+/Na}$ = –2.71 V and $E^\circ_{H_2O/H_2}$ = – 0.83 V).

Ans : On electrolysis aqueons solution of NaCl gives $Na^+(aq)$ and $Cl^-(aq)$ ions.

$$NaCl\ (aq) \xrightarrow{\text{Ionisation}} Na^+\ (aq) + Cl^-\ (aq)$$

On cathode $Na^-\ (aq)$ and $H_2O\ (l)$ both can undergo reduction to give Na metal and $H_2\ (g)$ respectively.

$$Na^+\ (aq) + e^- \xrightarrow{\text{Reduction}} Na\ (s),\ E^\circ_{Na^+/Na} = -2.71\ V$$

(Na = +1) (Na = 0)

$$H_2O\ (l) + 2e^- \xrightarrow{\text{Reduction}} H_2\ (g) + 2OH^-\ (aq),\ E^\circ_{H_2O/H_2} = -0.83\ V$$

(H = +1) (H = 0)

Now since $E^\circ_{H_2O/H_2}$ (= – 0.83 V) > $E^\circ_{Na^+/Na}$ (= –2.71 V), H_2O is more easily reduced to H_2 gas than Na^+ ion is reduced to Na metal.

(*iv*) The reducing power of halide ions (X^-) is in the order : $I^- > Br^- > Cl^- > F^-$.

Ans : When the halide ions (X^-) act as reducing agents, they are oxidised to their respective halogen molecules (X_2) as shown below :

$$2F^- \xrightarrow{\text{Oxidation}} F_2 + 2e^-,\ E^\circ_{2F^-/F_2} = -2.87\ V$$

$$2Cl^- \xrightarrow{\text{Oxidation}} Cl_2 + 2e^-,\ E^\circ_{2Cl^-/Cl_2} = -1.36\ V$$

$$2Br^- \xrightarrow{\text{Oxidation}} Br_2 + 2e^-,\ E^\circ_{2Br^-/Br_2} = -1.08\ V$$

$$2I^- \xrightarrow{\text{Oxidation}} I_2 + 2e^-,\ E^\circ_{2I^-/I_2} = -0.54\ V$$

E°_{ox} values given above show that these values are in the order :

$$E^\circ_{2I^-/I_2} > E^\circ_{2Br^-/Br_2} > E^\circ_{2Cl^-/Cl_2} > E^\circ_{2F^-/F_2}$$

–0.54 V –1.08 V – 1.36 V –2.87 V

Since $E^\circ_{2I^-/I_2}$ is the largest value, I^- ions are most easily oxidised to I_2 molecules, *i.e.* I^- ions are the strongest reducing agents. On the other hand, since $E^\circ_{2F^-/F_2}$ is the smallest value, F^- ions are least readily oxidised to F_2, *i.e.* F^- ions are the weakest reducing agents.

(*v*) The oxidising power of halogen molecules (X_2) is the order : $F_2 > Cl_2 > Br_2 > I_2$.

Ans : When halogen molecules (X_2) act as oxidising agents, they are reduced to their respective halide ions (X^-) as shown below :

$$F_2 + 2e^- \xrightarrow{\text{Reduction}} 2F^-, E^\circ_{F_2/2F^-} = +2.87 \text{ V}$$

$$Cl_2 + 2e^- \xrightarrow{\text{Reduction}} 2Cl^-, E^\circ_{Cl_2/Cl^-} = +1.36 \text{ V}$$

$$Br_2 + 2e^- \xrightarrow{\text{Reduction}} 2Br^-, E^\circ_{Br_2/2Br^-} = +1.08 \text{ V}$$

$$I_2 + 2e^- \xrightarrow{\text{Reduction}} 2I^-, E^\circ_{I_2/2I^-} = +0.54 \text{ V}$$

E°_{red} values given above show that these values are in the order :

$E^\circ_{F_2/2F^-}$	>	$E^\circ_{Cl_2/2Cl^-}$	>	$E^\circ_{Br_2/2Br^-}$	>	$E^\circ_{I_2/2I^-}$
+2.87 V		+1.36 V		+1.08		+0.54V

Since $E^\circ_{F_2/2F^-}$ is the largest value, F_2 is most readily reduced to F^- ions, *i.e.* F_2 is the strongest OA. On the other hand, since $E^\circ_{I_2/2I^-}$ is the smallest value, I_2 is least readily reduced to I^- ions, *i.e.* I_2 is the weakest OA.

Q.2 Arrange the following metals in the order in which they can displace each other. Fe, Cu, Zn, Ag, Mg and Al. Given that $Fe^{2+}/Fe = -0.44$ V, $Cu^{2+}/Cu = +0.34$ V, $Zn^{2+}/Zn = -0.76$ V, $Ag^+/Ag = +0.80$ V, $Mg^{2+}/Mg = -2.37$ V and $Al^{3+}/Al = -1.66$ V.

Ans : The given metals displace each other in the following order :

$$\xrightarrow{\text{Mg, Al, Zn, Fe, Cu, Ag}}$$

This order shows that a given metal can displace all those metals given on its RHS.

Explanation : Arrange E° values in the decreasing order :

$E^\circ_{Ag^+/Ag}$	$E^\circ_{Cu^{2+}/Cu}$	$E^\circ_{Fe^{2+}/Fe}$	$E^\circ_{Zn^{2+}/Zn}$	$E^\circ_{Al^{3+}/AL}$	$E^\circ_{Mg^{2+}/Mg}$
+0.80 V	+0.34 V	–0.44 V	–0.76 V	–1.66 V	–2.37 V

E° values decreasing ⟶

We know that a metal having lower value of E° can displace the metal/metals from its/their salt solution with higher E° value. Now :

(*a*) Since $E^\circ_{Mg^{2+}/Mg}$ (= –2.37 V) value is the lowest (most negative), Mg can displace all other metals from their salt solutions.

(*b*) Since $E^\circ_{Al^{3+}/Al}$ (= –1.66 V) value is lower than E° values for all other metals lying to its left, Al can displace Zn, Fe, Cu and Ag from their salt solutions. Note that Al can not displace Mg from Mg salt solution.

(*c*) Since $E^\circ_{Zn^{2+}/Zn}$ (= –0.76 V) value is lower than E° values for all other metlas lying to its left, Zn can displace Fe, Cu and Ag from their salt solutions. Note that Zn cannot displace Al and Mg from their salt solutions.

(*d*) Since $E^\circ_{Fe^{2+}/Fe}$ (= –0.44 V) is lower than E° values for both the metals lying to its left, Fe can displace Cu and Ag from their salt solutions. Note that Fe cannot displace Zn, Al and Mg from their salt solutions.

(*e*) Since $E^\circ_{Cu^{2+}/Cu}$ (= +0.34 V) is lower than $E^\circ_{Ag^+/Ag}$ (= +0.80 V), Cu can displace Ag from its salt solution. Note that Cu cannot displace Fe, Zn, Al and Mg from the salt solutions of these metals.

(*f*) Since $E^\circ_{Ag+/Ag}$ (= +0.80 V) is the highest, Ag cannot displace any of the other metals from the solutions of their salts.

Q.3 On the basis of the following redox reactions arrange Mg, Zn, Cu and Ag in the decreasing order of their E° values.

$$Cu + 2Ag^+ \rightarrow Cu^{2+} + 2Ag$$
$$Mg + Zn^{2+} \rightarrow Mg^{2+} + Zn$$
$$Zn + Cu^{2+} \rightarrow Cu + Zn^{2+}$$

Ans : Indicate oxidising agents (OA) and reducing agents (RO) in the given reactions.

(*i*) $Cu + 2Ag^+ \rightarrow Cu^{2+} + 2Ag$
RA OA

(*ii*) $Mg + Zn^{2+} \rightarrow Mg^{2+} + Zn$
RA OA

(*iii*) $Zn + Cu^{2+} \rightarrow Cu + Zn^{2+}$
RA OA

We know that in a redox reaction $E°_{OA} > E°_{RA}$. Thus :

In reaction (*i*) $E°_{Ag^+/Ag} > E°_{Cu^{2+}/Cu}$,

In reaction (*ii*) $E°_{Zn^{2+}/Zn} > E°_{Mg^{2+}/Mg}$ and

In reaction (*iii*) $E°_{Cu^{2+}/Cu} > E°_{Zn^{2+}/Zn}$

On combining the above relations we get :

$$E°_{Ag^+/Ag} > E°_{Cu^{2+}/Cu} > E°_{Zn^{2+}/Zn} > E°_{Mg^{2+}/Mg}$$

Q. 4 For M^{2+}/M and M^{3+}/M^{2+} systems, E° values for some metals are as : $Cr^{2+}/Cr = -0.9$ V, $Cr^{3+}/Cr^{2+} = -0.4$ V, $Mn^{2+}/Mn = -1.2$ V, $Mn^{3+}/Mn^{2+} = +1.5$ V, $Fe^{2+}/Fe = -0.4$ V and $Fe^{3+}/Fe^{2+} = +0.8$ V.

(*a*) Compare the stabilty of Fe^{3+} in acid solution with that of Cr^{3+} and Mn^{3+}.

(*b*) Arrange Cr, Mn and Fe in the decreasing order of the ease with which these metals are oxidised.

Ans : (*a*) Write E° values for Fe^{3+}/Fe^{2+}, Cr^{3+}/Cr^{2+} and Mn^{3+}/Mn^{2+} systems as shown below :

$$Mn^{3+} + e^- \xrightarrow{\text{Reduction}} n\ Mn^{2+},\ E° = +1.5\ V$$

$$Fe^{3+} + e^- \xrightarrow{\text{Reduction}} Fe^{2+},\ E° = +0.80\ V$$

$$Cr^{3+} + e^- \xrightarrow{\text{Reduction}} Cr^{2+},\ E° = -0.40\ V$$

Since E° value for Mn^{3+}/Mn^{2+} is higher (= +1.5 V) than that for Fe^{3+}/Fe^{2+} (= +0.80 V), Mn^{3+} ion can more easily be reduced to Mn^{2+} than Fe^{3+} ion can be reduced to Fe^{2+}. Thus Mn^{3+} ion is less stable than Fe^{3+} ion ($Mn^{3+} < Fe^{3+}$). In other words Fe^{3+} ion is more stable than Mn^{3+} ion ($Fe^{3+} > Mn^{3+}$).

Since E° value for Fe^{3+}/Fe^{2+} is positive and that for Cr^{3+}/Cr^{2+} is negative, Fe^{3+} ion can easily be reduced to Fe^{2+} while Cr^{3+} ion cannot be reduced to Cr^{2+} ion. Thus Fe^{3+} ion is less stable than Cr^{3+} ion. ($Fe^{3+} < Cr^{3+}$) In short we say that Fe^{3+} is more stable then Mn^{3-} but less stable then Cr^{3+}.

(*b*) Write $E°_{ox}$ values for Cr/Cr^{2+}, Mn/Mn^{2+} and Fe/Fe^{2+} systems in the decreasing order as shown below :

$$Mn \xrightarrow{\text{Oxidation}} Mn^{2+}, E^\circ_{ox} = +1.2\ V$$

$$Cr \xrightarrow{\text{Oxidation}} Cr^{2+}, E^\circ_{ox} = +\ 0.9\ V$$

$$Fe \xrightarrow{\text{Oxidation}} Fe^{2+}, E^\circ_{ox} = +0.4\ V$$

We know that with the decrease in the values of E°_{ox}, the ease with which the metals are oxidised also decreases. Thus the ease with which the given metals get oxidised is in the order : Mn > Cr > Fe.

Q. 5 Calculate the value of standerd emf of reaction :

$$\mathbf{Fe + 2Fe^{3+} \rightarrow 3Fe^{2+}}$$

Given : $E^\circ_{Fe^{2+}/Fe} = -0.441$ V and $E^\circ_{Fe^{3+}/Fe^{2+}} = 0.771$ V.

Ans : The given reaction can be broken into the following half-reactions.

$$Fe \xrightarrow{\text{Oxidation}} Fe^{2+} + 2e^-, E^\circ_{ox} = E^\circ_{Fe/Fe^{2+}} = +0.441\ V$$

$$Fe^{3+} + e^- \xrightarrow{\text{Reduction}} Fe^{2+}\ ] \times 2, E^\circ_{red} = E^\circ_{Fe^{3+}/Fe^{2+}} = 0.771\ V$$

$$Fe + 2Fe^{3+} \rightarrow Fe^{2+} + 2Fe^{2+}, E^\circ_{cell} = E^\circ_{ox} + E^\circ_{red}$$

$$= 0.441 + 0.771$$

$$= \mathbf{1.212\ V\ Ans.}$$

Q. 6 Two half-reactions with their E° values are given below :

$$\mathbf{2I^- \rightarrow I_2 + 2e^-, E^\circ = -0.54\ V}$$

$$\mathbf{Cr_2O_7^{2-} + 14H^+ + 6e^- \rightarrow 2Cr^{3+} + 7H_2O, E^\circ = +1.33\ V.}$$

(*a*) Write the cell reaction which is composed of these half-reactions.

(*b*) Calculate the value of the cell reaction.

Ans : (*a*) In order to find out the cell reaction proced as follows :

(*i*) Write both the given half-reactions as reduction half-reactions with their E_o values. Both the reduction half-reactions should have the same number of electrons.

$$I_2 + 2e^- \xrightarrow{\text{Reduction}} 2I^-$$

or $$3I_2 + 6e^- \xrightarrow{\text{Reduction}} 6I^-, E^\circ = +\ 0.54\ V$$

$$Cr_2O_7^{2-} + 14H^+ + 6e^- \xrightarrow{\text{Reduction}} 2Cr^{3+} + 7H_2O, E^\circ = +1.33\ V.$$

(*ii*) Now subtract the reduction half-reaction with lower value of E° from that having higher value of E°

$$Cr_2O_7^{2-} + 14H^+ + 6e^- \rightarrow 2Cr^{3+} + 7H_2O, E^\circ = +1.33\ V\ (\textit{Higher value})$$

$$+\ 3I_2 + 6e^- \rightarrow +6I^-, E^\circ = +0.54\ V\ (\textit{Lower value})$$

On subtracting : $Cr_2O_7^{2-} + 14H^+ - 3I_2 \rightarrow 2Cr^{3+} + 7H_2O - 6I^-$

or $Cr_2O_7^{2-} + 14H^+ + 6I^- \rightarrow 2Cr^{3+} + 7H_2O + 3I_2$.

This is the required cell reaction.

(*b*) $E^\circ_{cell\ reaction} = E^\circ_{higher} - E^\circ_{lower}$

$$= +\ 1.33 - 0.54$$

$$= \mathbf{0.79\ V\ Ans.}$$

Q.7 The emf (E°_{cell}) of the cell reaction

$$3Sn^{4+} + 2Cr \rightarrow 3Sn^{2+} + 2Cr^{3+}$$

is 0.89 V. Calculate ΔG° for the reaction

(F = (96, 500 C mol^{-1} and VC = J.)

Ans : The value of ΔG° for a reaction is given by :

$$\Delta G^\circ = -\,n\,F\,E^\circ_{cell}$$

Calculation of the value of *n*

$$Cr \xrightarrow{\text{Oxidation}} Cr^{3+} + 3e^-] \times 2$$

$$Sn^{4+} + 2e^- \xrightarrow{\text{Reduction}} Sn^{2+}] \times 3$$

$$2Cr + 3Sn^{4+} \rightarrow 2Cr^{3+} + 3Sn^{2+} \textbf{ (Cell reaction)}$$

For this reaction n = 6

$\therefore \Delta G^\circ = -\,nFE^\circ_{cell} = -\,6 \times (96500 \text{ C mol}^{-1}) \times (0.89 \text{ V})$

$= -515310 \text{ CV mol}^{-1}$

$= \mathbf{-515310 J\ mol^{-1}}$ **Ans.** (∵ CV = J)

Q.8 The half-cell reactions for rusting of iron are :

$$2H^+ + \frac{1}{2}O_2 + 2e^- \rightarrow H_2O,\ E^\circ = +1.23 \text{ V}$$

$$Fe^{2+} + 2e^- \rightarrow Fe(s),\ E^\circ = -\,0.44 \text{ V}$$

Calculate the value of ΔG° (in KJmol^{-1}) for the overall reaction.

Ans : We know that the value of ΔG° for a reaction is given by :

$$\Delta G^\circ = -n\,FE^\circ_{cell}$$

E^o_{cell} is given by :

$E^\circ_{cell} = E^\circ_{Higher} - E^\circ_{Lower} = 1.23 - (-\,0.44)$

$= 1.23 + 0.44 = 1.67$ V

The given half-reactions show that n = 2. F = 96500 C mol^{-1}.

$\therefore \Delta G^\circ = -n\,FE^\circ_{cell}$

$= -2 \times (96500 \text{ Cmol}^{-1}) \times (1.67 \text{ V})$

$= -2 \times 96500 \times 1.67 \text{ CV mol}^{-1}$

$= -2 \times 96500 \times 1.67 \text{ Jmol}^{-1}$ (CV = J)

$= -322310 \text{ J mol}^{-1}$

$= \dfrac{-322310}{1000} \text{ KJ mol}^{-1}$

$= \mathbf{-322.31\ KJ\ mol^{-1}}$ **Ans.**

Q.10 (*i*) Calculate the cell e.m.f. at 25°C for the following cell :

Mg (*s*) | Mg^{2+} (0.01 M) || Sn^{2+} (0.1 M) Sn (*s*)

Given E° Mg^{2+}/Mg = –2.34 V, $E^\circ_{Sn^{2+}/Sn}$ = –0.136 V, 1F = 96,500 C mol^{-1}

(*ii*) Calculate the maximum work that can be accompalished by the operation of this cell.

Ans : (*i*) Since concentrations of Mg^{2+} and Sn^{2+} ions are different from 1M,e.m.f of the cell will be calculated by Nernst equation.

Calculation of the value of $E°_{cell}$. The value of $E°_{cell}$ is given by :

$$E_{cell} = E_{cathode} - E_{anode} = E_{Sn2+/Sn} - E_{Mg2+/Mg}$$
$$= -0.136 - (-2.340) = -0.136 + 2.340$$
$$= 2.204 \text{ V}$$

Calculation of the value of *n*. Cell reaction is :

$Mg\ (s) + Sn^{2+}\ (0.1\ M) \rightarrow Mg^{2+}\ (0.01\ M) + Sn\ (s)$.

For this reaction $n = 2$

Applying Nernst equation, we get :

$$E_{cell} = E°_{cell} - \frac{0.0591}{n} \log \frac{[Mg^{2+}]}{[Sn^{2+}]}$$
$$= 2.204 - \frac{0.0591}{2} \log \frac{0.01}{0.1}$$
$$= 2.204 - 0.02955 \log \frac{1 \times 10^{-2}}{1 \times 10^{-1}}$$
$$= 2.204 - 0.02955 \times \log 10^{-1}$$
$$= 2.204 - 0.02955 \times (-1)$$
$$= 2.204 + 0.02955 = \mathbf{2.23355\ V\ Ans.}$$

(*ii*) $W_{max} = nF\ E°_{cell}$

$$= 2 \times (96500 \text{ C mol}^{-1}) \times (2.204 \text{ V})$$
$$= 2 \times 96500 \times 2.204 \text{ CV mol}^{-1}$$
$$= 2 \times 96500 \times 2.204 \text{ J mol}^{-1} \qquad (CV = J)$$
$$= \frac{2 \times 96500 \times 2.204}{1000} \text{ KJ mol}^{-1}$$
$$= \mathbf{425.372\ KJ\ mol^{-1}\ Ans.}$$

Q.10 The standard electrode potentials (E°) for OCl^-/Cl^- and $Cl^-/\frac{1}{2}Cl_2$ respectively are 0.94 V and –13.6 V. Calculate E° value for $OCl^-/\frac{1}{2}Cl_2$ electrode.

Ans : The reduction half-reactions for the given electrodes can be written as :

(*i*) $OCl^- + 2H^+ + 2e^- \rightarrow Cl^- + H_2O$ ($n_1 = 2$), $E°_{OCl^-/Cl^-} = +0.94$ V
(Cl = +1) (Cl = –1)

(*ii*) $\frac{1}{2}Cl_2 + e^- \rightarrow Cl^-$ ($n_2 = 1$), $E°_{\frac{1}{2}Cl_2/Cl^-} = +1.36$ V
(Cl = 0) (Cl = –1)

(*iii*) $OCl^- + 2H^+ + e^- \rightarrow \frac{1}{2}Cl_2 + H_2O$ ($n_3 = 1$), $E°_{OCl^-/\frac{1}{2}Cl_2} = ?$
(Cl = +1) (Cl = 0)

EFM diagram showing the reduction process $OCl^- \rightarrow \frac{1}{2}Cl_2 \rightarrow Cl^-$ can be shown in Fig.11.5.

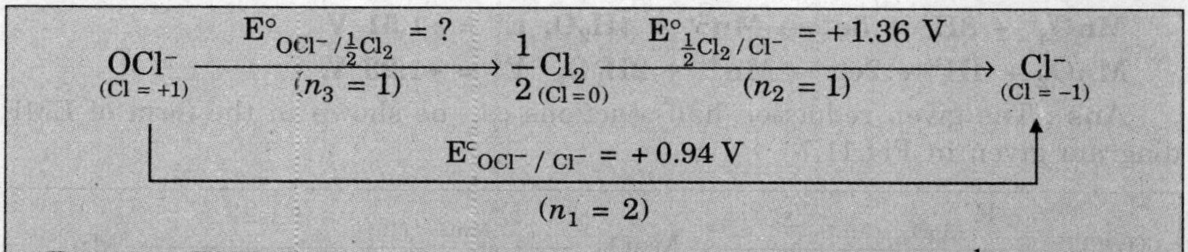

Fig. 11.5. EMF diagram showing the reduction process $OCl^- \rightarrow \frac{1}{2}Cl_2 \rightarrow Cl^-$.

The EMF diagram shows that :

$$E^\circ_{OCl?/\frac{1}{2}Cl_2} \times n_3 + E^\circ_{\frac{1}{2}Cl_2/Cl^-} \times n_1 = E^\circ_{OCl^-/Cl^-} \times n_2$$

or $$E^\circ_{OCl^-/\frac{1}{2}Cl_2} \times 1 + 1.36 \times 1 = 0.94 \times 2$$

or $$E^\circ_{OCl^-/\frac{1}{2}Cl_2} = 0.94 \times 2 - 1 \times 1.36$$

$$= \mathbf{0.52\ V\ (Ans.)}$$

Q.11 Calculate the electrode potential of Cu^+/Cu half cell. Given that the standard reduction protentials of Cu^{2+}/Cu and Cu^{2+}/Cu^+ are o.337 V and 0.153 V respectively.

Ans : Half reduction reactions for the given electrodes can be written as :

(*i*) $Cu^+ + e^- \rightarrow Cu$ $(n_1 = 1)$, $E^\circ_{Cu^+/Cu} = ?$

(*ii*) $Cu^{2+} + 2e^- \rightarrow Cu$ $(n_2 = 2)$, $E^\circ_{Cu^{2+}/Cu} = +0.337$ V

(*iii*) $Cu^{2+} + e^- \rightarrow Cu^+$ $(n_3 = 1)$, $E^\circ_{Cu^{2+}/Cu} = +0.153$ V

Note that the number of Cu-atoms in all three equations is the same. EMF diagram showing the reduction processes $Cu^{2+} \rightarrow Cu^+ \rightarrow Cu$ can be shown in Fig.11.6.

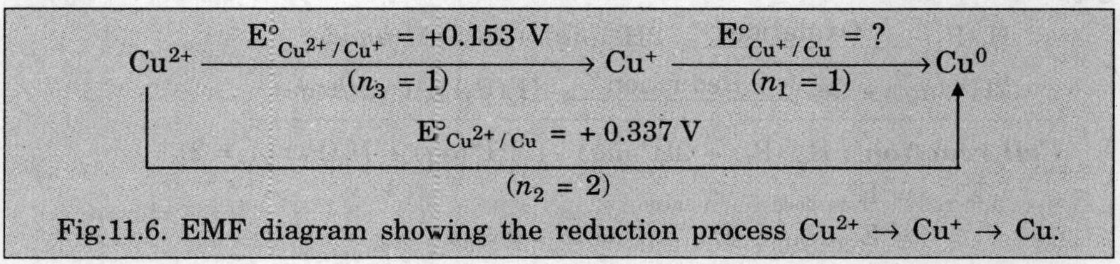

Fig.11.6. EMF diagram showing the reduction process $Cu^{2+} \rightarrow Cu^+ \rightarrow Cu$.

EMF diagram shows that :

$$E^\circ_{Cu^{2+}/Cu^+} \times n_3 + E^\circ_{Cu^+/Cu} \times n_1 = E^\circ_{Cu^{2+}/Cu} \times n_2$$

or $$0.153 \times 1 + E^\circ_{Cu^+/Cu} \times 1 = +0.337 \times 2$$

$$E^\circ_{Cu^+/Cu} = 2 \times 0.337 - 1 \times 0.153$$

$$= 0.674 - 0.153$$

$$= \mathbf{0.521\ V\ Ans.}$$

Q.12 Calculate E° value for the half-cell reaction : $MnO_4^- + 4H^+ + 3e^- \rightarrow MnO_2 + 2H_2O$.

Given that :

$MnO_4^- + 8H^+ + 5e^- \rightarrow Mn^{2+} + 4H_2O$, $E^\circ = +1.51$ V

$MnO_2 + 4H^+ + 2e^- \rightarrow Mn^{2+} + 2H_2O$, $E^\circ = +1.23$ V.

Ans : The given reduction half-reactions can be shown in the form of EMF diagram given in Fig.11.7.

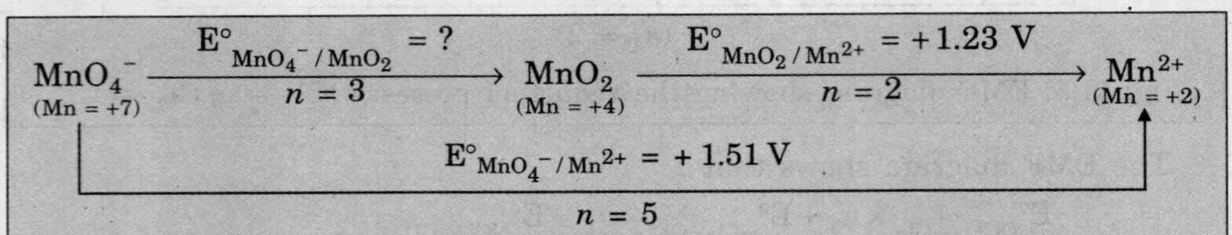

Fig.11.7. EMF diagram for reduction reactions involving MnO_4^-, MnO_2 and Mn^{2+}.

The diagram clearly shows that :

$$3 \times E^\circ_{MnO_4^-/MnO_2} + 2 \times E^\circ_{MnO_2/Mn^{2+}} = 5 \times E^\circ_{MnO_4^-/Mn^{2+}}$$

or $$3 \times E^\circ_{MnO_4^-/MnO_2} + 2 \times 1.23 = 5 \times 1.51$$

or $$E^\circ_{MnO_4^-/MnO_2} = \frac{5 \times 1.51 - 2 \times 1.23}{3}$$

$$= \frac{7.55 - 2.46}{3} = \frac{5.09}{3}$$

$$= \mathbf{1.69\ V\ Ans.}$$

Q.13 Calculate EMF of the following cell.

Pt | $H_2(P_1)$ | H^+ (*aq*) | H_2 (P_2) | Pt.

Ans : Cell reaction can be obtained by adding the oxidation and reduction half-reactions taking place on anode and cathode respectively. Cell reaction also gives the value of *n*.

$$H_2(P_1) \xrightarrow{\text{Oxidation}} 2H^+ (aq) + 2e^- \text{ (At anode)}$$

$$2H^+ (aq) + 2e^- \xrightarrow{\text{Reduction}} H_2(P_2) \text{ (At cathode)}$$

Cell reaction : $H_2 (P_1) + 2H^+ (aq) \rightarrow 2H^+ (aq) + H_2(P_2)$ $(n = 2)$

$$E^\circ_{cell} = E^\circ_{cathode} - E^\circ_{anode}$$

$$= E^\circ_{2H^+/H_2} - E^\circ_{2H^+/H_2} = 0 - 0 = 0$$

$$\therefore \quad E_{cell} = E^\circ_{cell} - \frac{RT}{nF} \log \frac{[H^+]^2 [H_2]}{[H_2] [H^+]^2}$$

$$= 0 - \frac{RT}{2F} \log \frac{(1)^2 \times P_2}{P_1 \times (1)^2} = -\frac{RT}{2F} \log \frac{P_2}{P_1}$$

or $$E_{cell} = \frac{RT}{2F} \log \frac{P_1}{P_2}$$

Q. 14 The standard reduction potential for the half-cell
NO_3^- (*aq*) + $2H^+$ (*aq*) + e^- → NO_2 + H_2O is 0.78 V.
(*i*) Calculate the reduction potential in 8 M H^+.
(*ii*) What will be the reduction potential of the half-cell in a neutral soluiton ? Assume all the other species to be at unit concentration.

Ans : (*i*) Reduction electrode reaction can be written as :

$NO_3^- + 2H^+ (8M) + e^- \rightarrow NO_2 + H_2O$ (n = 1), E°_{red} = 0.78 V
(N = +5) (N = +4)

$$\therefore (E_{red})_{elec} = (E^\circ_{red})_{elec} - \frac{0.0591}{n} \log \frac{[NO_2][H_2O]}{[NO_3^-][H^+]^2}$$

$$= 0.78 - \frac{0.0591}{1} \log \frac{1\times1}{1\times(8)^2}$$
$$= 0.78 - 0.0591 \times [\log 1 - \log 8^2]$$
$$= 0.78 - 0.0591\ [0 - 2 \log 8]$$
$$= 0.78 + 0.0591 \times 2 \times \log 8$$
$$= 0.78 + 0.0591 \times 2 \times 0.903$$
$$= 0.78 + 0.106$$
$$= \mathbf{0.886\ V\ Ans.}$$

(*ii*) Neutral solution means pH = 7
or $[H^+] = 10^{-7}$ M.

Electrode reduction reaction can be written as :
$NO_3^- + 2H^+ (10^{-7} M) + e^- \rightarrow NO_2 + H_2O$ (n = 1)

$$\therefore (E_{red})_{elec} = (E^\circ_{red})_{elec} - \frac{0.0591}{1} \log \frac{[NO_2][H_2O]}{[NO_3^-][H^+]^2}$$

$$= 0.78 - 0.0591 \times \log \frac{1\times1}{1\times(10^{-7})^2}$$
$$= 0.78 - 0.0591 \times [\log 1 - \log 10^{-14}]$$
$$= 0.78 - 0.0591 \times [-(-14) \log 10]$$
$$= 0.78 - 0.0591 \times 14 \times 1$$
$$= -0.78 - 0.8274 = \mathbf{-0.0474\ V\ Ans.}$$

Q. 15 The standard reduction potential of Cu^{++}/Cu and Ag^+/Ag electrodes are 0.337 and 0.799 volt respectively. Construct the galvanic cell using these electrodes so that its standard e.m.f. is positive. For what concentration of Ag^+ will the e.m.f. of the cell at 25°C be zero if the concentration of Cu^{++} is 0.01 M ?

Ans : (*i*) $E^\circ_{Cu^{2+}/Cu}$ = 0.337 V, $E^\circ_{Ag^+/Ag}$ = 0.799 U and E_{cell} = 0. Now since $E^\circ_{Ag^+Ag}$ (= 0.799 V) > $E^\circ_{Cu^{2+}/Cu}$ (= 0.337 V), Ag/Ag^+ electrode will act as cathode and Cu/Cu^{2+} electrode will work as anode. Hence the cell can be represented as :

Cu | Cu^{2+} | | Ag^+ | Ag
Anode Cathode

(*ii*) Cell reaction is :

$Cu + 2Ag^+ (c) \rightarrow Cu^{2+}$ (0.01 M *or* 10^{-2} M) + 2Ag (n = 2)

$$\therefore \quad E_{cell} = E°_{cell} - \frac{0.0591}{n} \log \frac{[Cu^{2+}][Ag]^2}{[Cu][Ag^+]^2}$$

$$\text{or} \quad 0 = (E°_{cathode} - E°_{anode}) - \frac{0.0591}{n} \log \frac{[Cu^{2+}][Ag]^2}{[Cu][Ag^+]^2}$$

$$= (E°_{Ag+/Ag} - E°_{Cu^{2+}/Cu}) - \frac{0.0591}{n} \log \frac{[Cu^{2+}][Ag]^2}{[Cu][Ag^+]^2}$$

$$= (0.799 - 0.337) - \frac{0.0591}{2} \log \frac{10^{-2} \times (1)^2}{1 \times c^2}$$

$$= 0.462 - 0.02955 [\log 10^{-2} - \log c^2]$$

$$= 0.462 + 0.02955 \times [-2 - 2 \log c]$$

$$= 0.462 + 2 \times 00.2955 + 2 \times 0.02955 \log c$$

$$0 = 0.462 + 0.0591 + 0.0591 \log c$$

$$\therefore \quad \log c = \frac{-0.462 - 0.0591}{0.0591} = \frac{-0.5211}{0.0591} = -8.8712$$

$$\therefore \quad c = \text{Antilog} - 8.8172 = \mathbf{1.523 \times 10^{-9}\ M\ Ans.}$$

Q. 16 The EMF of the following cell is found to be 0.20 V at 298 K

Cd | Cd^{2+} (?) || Ni^{2+} (2.0 M) Ni

What is the molar concentraction of Cd^{2+} ions in the solution ?

($E°_{Cd^{2+}/Cd}$ = – 0.40 V, $E°_{Ni^{2+}/Ni}$ = – 0.25 V).

Ans : To calculate $E°_{cell}$.

$E°_{cell} = E°_{higher} - E°_{lower} = -0.25 - (-0.40)$

$= -0.25 + 0.40 = \mathbf{+\ 0.15\ V}$

To write the cell reaction. The representation of the cell given in the question shows that the cell reaction is the sum of the oxidation and reduction half-reactions given below.

$$Cd \xrightarrow{\text{Oxidation}} Cd^{2+} + 2e^-$$

$$Ni^{2+} + 2e^- \xrightarrow{\text{Reduction}} Ni$$

Cell reaction : $Cd + Ni^{2+} \rightarrow Cd^{2+} + Ni$ (n = 2)

or $Cd + Ni^{2+}$ (2.0 M) $\rightarrow Cd^{2+}$ (?) + Ni

Suppose concentration of Cd^{2+} ions is c M. Then from Nernst equation, we have :

$$E_{cell} = E°_{cell} - \frac{0.0591}{2} \log \frac{[Cd^{2+}][Ni]}{[Cd][Ni^{2+}]}$$

$$\text{or} \quad 0.20 = 0.15 - 0.02955 \log \frac{c \times 1}{1 \times 2}$$

or $\log \frac{c}{2} = \frac{0.15 - 0.20}{0.02955} = -\frac{0.05}{0.02955}$

or $\log \frac{c}{2} = -1.69$

or $\log c - \log 2 = -1.69$

or $\log c = -1.69 + \log 2 = -1.69 + 0.3021$

$= -1.3879$

$\therefore$ $c = \text{Antilog } -1.3879$

$= \mathbf{0.0409\ M}$ **Ans.**

Q. 17 Calculate the potential of a hydrogen electrode in contact with a solution whose pH is 10.

Ans : pH = 10

or $-\log [H^+] = 10$

or $\log [H^+] = -10$

or $[H^+] = \text{Antilog } (-10) = 10^{-10}$ M

Reduction reaction taking place on hydrogen electrode can be written as :

$$2H^+ + 2e^- \rightarrow H_2 \ (n = 2)$$

$$\therefore \quad E_{2H^+/H_2} = E^0_{2H^+/H_2} - \frac{0.0591}{n} \log \frac{[H_2]}{[H^+]^2}$$

$$= 0 - \frac{0.0591}{2} \log \frac{1}{(10^{-10})^2}$$

$$= \frac{0.0591}{2} [\log 1 - \log 10^{-20}]$$

$$= + \frac{0.0591}{2} \times (-20) \log 10$$

$$= -0.0591 \times 10 = \mathbf{-0.591\ V}$$ **Ans.**

Q. 18 At what pH of HCl solution will hydrogen gas electrode show electrode potential of – 0.118 V ? H_2 gas is bubbled at 298 K and 1 atm pressure.

Ans : Hydrogen gas electrode is represented as :

$$H^+ / \tfrac{1}{2} H_2 \text{ (1 atom)}, \quad E_{H^+/\frac{1}{2}H_2} = -0.118 \text{ V}$$

Reduction reaction taking place at hydrogen gas electrode is :

$$H^+ + e^- \rightarrow \frac{1}{2} H_2 \text{ (1 atm) } (n = 1)$$

Applying Nernst equation to the above reduction half-reaction, we get :

$$E_{H^+/\frac{1}{2}H_2} = E^\circ_{H^+/\frac{1}{2}H_2} - \frac{0.0591}{n} \log \frac{[H_2]^{1/2}}{[H^+]} \quad ...(i)$$

Now since H_2 gas is at 1 atm. pressure, $[H_2] = 1$. Hence equation (*i*) becomes

$$E_{H^+/\frac{1}{2}H_2} = E^\circ_{H^+/\frac{1}{2}H_2} - \frac{0.0591}{1}\log\frac{1}{[H^+]}$$

or $\quad -0.118 = 0 - 0.0591\ \{-\log [H^+]\}$

or $\quad \dfrac{-0.118}{-0.059} = -\log [H^+]$

or $\quad 1.99 = -\log [H^+]$

or $\quad 2 = pH \qquad (\because pH = -\log [H^+])$

Thus pH of the solution = **2 Ans.**

University Questions

1. Differentiate between ionisation potential and electrode potential of a metal. *(VBS Purvanchal 2002)*

2. Give reason fro the fact that E° of Cu is + 0.34 V while that of Zn is – 0.76 V. *(Delhi 2002)*

Answer. We know that a less reactive metal has higher value of E° than a more reactive metal. Thus Cu has higher value of E° (= +0.34 V) than Zn (= – 0.76 V), since Cu is less reactive than Zn.

3. (*a*) What is electrode potential ? How does it differ from ionisation energy ?

(*b*) Predict whether the following reaction will occur spontaneously or not ?

$Ni\ (s) + Cu^{2+}\ (aq) \rightarrow Ni^{2+}\ (ag) + Cu\ (s)$

Given that : $E^o_{Ni/Ni^{2+}} = +0.25$ V and $E^o_{Cu/Cu^{2+}} = -0.337$ V.

(*c*) What is electrochemical series ? Describe its uses in explaning chemical reactions. *(VBS Purvanchal 2003)*

4. Consider the following EMF diagram for broming and answer the following :

$$E^\circ_A \quad BrO_3^- \xrightarrow{1.49} HOBr \xrightarrow{1.59} \frac{1}{2}Br_2(l) \xrightarrow{1.07} Br^- \qquad (BrO_3^- \to Br^- : 1.44;\ HOBr \to Br^- : E)$$

$$E^\circ_A \quad BrO_3^- \xrightarrow{0.54} BrO^- \xrightarrow{0.45} \frac{1}{2}Br_2(l) \xrightarrow{1.07} Br^- \qquad (BrO_3^- \to Br^- : 0.61)$$

(*a*) Calcualte the value of E.

(*b*) State whether the following statements are true or false. Give reasons to your answer.

(*i*) Bromine disproportionates into bromide and hypobromate ions in basic medium but not in acidic medium.

(*ii*) BrO_3^- is better oxidizing agent in acid solution than in basic solution. *(Rohilkhand 2003)*

5. (*a*) Calculate the equilibrium constant for the reaction

$Zn + Cu^{+2} \rightarrow Cu + Zn^{+2}$ at 25°C

$E^\circ_{Zn/Zn^{+2}} = 0.763$ Volts, $E^\circ_{Cu/Cu^{+2}} = -0.337$ Volts (*b*) The e.m.f. of following cell

Pt, H_2 | H^+ a_{H+} = 1 | Solution | 0.1 M KCl | Hg_2Cl_2 | Hg is 0.93 volts. [$E_{calomel}$ = 0.34 volts]. Calculate the pH of the solution. (*Gujrat 2003*)

6. Calculate e.m.f. of the concentration cell given below at 25°C

Zn | Zn^{2+} (M) || Cu^{2+}(M) | Cu

$E°_{Cu^{2+}/Cu}$ = 0.34 V, $E°_{Zn^{2+}/Zn}$ = 0.76 V (*Meerut 2004*)

7. Calculate free energy change of the following cell at 25°C.

Sn | Sn^{2+}(a = 0.6) || Pb^{2+}(a = 0.3) | Pb

Given that $E°_{cell}$ = 0.014 V (*Meerut 2004*)

8. Calculate the e.m.f. of the following cell at 25°C.

Ag | $AgNO_3$(0.01 M) || $AgNO_3$(0.1 M) | Ag (*C.S.J.M. Kanpur 2005*)

9. Write a note on "Nernst Equation." (*Meerut 2006*)

10. Define electrode potential. Derive Nernst equation for a single electrode potential. How do you decide the sign of electrodes according to IUPAC system of and galvanic cell. (*Meerut 2007*)

11. Write a note on "Glass Electrode". (*C.S.J.M. Kanpur 2008*)

12. Write a note on "Single Electrode Potential". (*Meerut 2008*)

13. Predict whether Zn and Ag react with 1N H_2SO_4 to give out H_2 gas or not. Given that $E°_{Zn^{2+}/Zn}$ = – 0.76 V and $E°_{Ag^+/Ag}$ = 0.80 V. (*C.S.J.M. Kanpur 2009*)

CHAPTER 12

Modern Concepts of Acids and Bases

Following are the important modern concepts of acids and bases :

Arrhenius Concept : Water Ion System

According to this concept, an acid is any hydrogen containing compound which gives H^+ ions in aqueous solution and a base which gives OH^- ions in aqueous solution. Thus HCl is an acid and NaOH is a base and the neutralisation process can be represented by a reaction involving the combination of H^+ and OH^- ions to form H_2O.

$$HCl \underset{}{\overset{H_2O}{\rightleftharpoons}} H^+ + Cl^-$$

$$NaOH \overset{H_2O}{\rightleftharpoons} Na^+ + OH^-$$

$$H^+ + OH^- \longrightarrow H_2O$$

Utility. (*i*) Since the reaction representing neutralisation process involves the combination H^+ and OH^- ions, the approximately constant molar heat of neutralisation would be expected. Thus the constant heat of neutralisation of a strong acid by a strong base is readily understandable in terms of this concept.

(*ii*) This concept has offered a means of correlating catalytic behaviour with the concentration of the H^+ ion.

Limitations. (*i*) According to this concept, HCl is regarded as an acid only when dissolved in H_2O and not in some other solvent such as C_6H_6. or when it exists in the gaseous form.

(*ii*) It cannot account for the acidic and basic character of the materials in non-aqueous solvents, *e.g.*, NH_4NO_3 in liq. NH_3 acts as an acid, though it does not give H^+ ions. Similarly many organic materials and NH_3, which do not have OH^- ions at all, are actually known to show basic character.

(*iii*) The neutralisation process is limited to those reactions which can occur in aqueous solutions only, although reactions involving salt formation do occur in many other solvents and even in the absence of solvents.

(*iv*) It cannot explain the acidic character of certain salts such as $AlCl_3$ in aqueous solution.

Bronsted-Lowry Concept : Proton-Donor Acceptor System

Bronsted and Lowry in 1923 independently proposed a more general definition of acids and bases. According to them, *an acid is defined as any hydrogen containing material* (a molecule or a cation or an anion) *that can release a proton* (H^+) *to any other substance, whereas a base is any substance* (a molecule or a cation or an anion) *that can accept a proton from any other substance.* In short, *an acid is a proton-donor and a base is a proton-acceptor.* As an example consider the reaction between HCl and H_2O as given below :

one proton (H^+) going to H_2O one proton (H^+) going to Cl^-

$$\underset{\text{Acid}}{HCl} + \underset{\text{Base}}{H_2O} \rightleftharpoons \underset{\text{Acid}}{H_3O^+} + \underset{\text{Base}}{Cl^-}$$

In this reaction HCl donates a proton to H_2O and is, therefore, an acid. Water, on the other hand, accepts a proton from HCl, and is, therefore, a base. In the reverse reaction which at equilibrium proceeds at the same rate as the forward reaction, H_3O^+ ion donates a proton to Cl^- ion and hence H_3O^+ ion is an acid. Cl^- ion, because it accepts a proton from H_3O^+ ion, is a base.

Examples of Bronsted-Lowry Acids and Bases

As already stated, Bronsted-Lowry acids and bases may be molecular, cationic and anionic species. Examples of such species are given below :

Type of species	Acids	Bases
Molecular	HCl, HBr, $HClO_4$, HNO_3, H_2SO_4, H_3PO_4, H_2O	NH_3, N_2H_4, amines, H_2O, NH_2OH
Cationic	NH_4^+, $[Fe(H_2O)_6]^{3+}$, $[Al(H_2O)^{3+}$, H_3O^+,	$[Fe(H_2O)_5(OH)]^{2+}$, $[Al(H_2O)_5(OH)]^{2+}$
Anionic	HCO_3^-, HSO_4^-	Cl^-, Pr^-, I^-, OH^-, CO_3^{2-}, SO_4^{2-}, H^-, NH_2^- OH^-, HS^-

Conjugate Acid-Base Pairs

Let us again consider the reaction between HCl and H_2O as given below :

$$\underset{\text{Acid}}{HCl} + \underset{\text{Base}}{H_2O} \rightleftharpoons \underset{\text{Acid}}{H_3O^+} + \underset{\text{Base}}{Cl^-}$$

In this reaction Cl^- ion (base) formed by the loss of one proton from HCl (acid) is called the conjugate base of the acid, HCl. Thus HCl and Cl^- is a conjugate acid-base pair. Similarly if we consider the reverse reaction, H_2O (base) formed by the loss of one proton from H_3O^+ (acid) is called the conjugate base of the acid, H_3O^+. Thus H_3O^+ and H_2O is another conjugate acid-base pair. Now if the acid HCl is labelled as $Acid_1$ and its conjugate base *viz.* Cl^- as $Base_1$ and further, if H_2O is designated as $Base_2$ and its conjugate acid *viz.* H_3O^+ as $Acid_2$, then the two conjugate acid-base pairs can be shown as :

Conjugate acid-base pair (HCl and Cl^-)

$$\underset{Acid_1}{HCl} + \underset{Base_2}{H_2O} \rightleftharpoons \underset{Acid_2}{H_3O^+} + \underset{Base_1}{Cl^-}$$

Conjugate acid-base pair (H_2O and H_3O^+)

It may be noted that an acid and a base which constitute the conjugate acid-base pair differ from each other by a proton and hence can be formed from each other by the loss or gain of a proton as shown below :

$$\text{Acid} \underset{\text{Gain of proton } (+H^+)}{\overset{\text{Loss of proton } (-H^+)}{\rightleftarrows}} \text{Base}$$

e.g. $\underset{Acid}{HCl} \underset{+H^+}{\overset{-H^+}{\rightleftarrows}} \underset{Base}{Cl^-}$ and $\underset{Acid}{H_3O^+} \underset{+H^+}{\overset{-H^+}{\rightleftarrows}} \underset{Base}{H_2O}$

Some common conjugate acid-base pairs are given below :

Conjugate acid-base pair (outer: first Acid and last Base); *Conjugate acid base pair* (inner: Base and Acid)

Acid		*Base*		*Acid*		*Base*
$HClO_4$	+	H_2O	$\rightleftharpoons$	H_3O^+	+	ClO_4
H_2SO_4	+	H_2O	$\rightleftharpoons$	H_3O^+	+	HSO_4^-
HCN	+	H_2O	$\rightleftharpoons$	H_3O^+	+	CN^-
HCl	+	H_2O	$\rightleftharpoons$	H_3O^+	+	Cl^-
HNO_3	+	H_2O	$\rightleftharpoons$	H_3O^+	+	NO_3^-
$[Al(H_2O)_6]^{3+}$	+	H_2O	$\rightleftharpoons$	H_3O^+	+	$[Al(H_2O)_5(OH)]^{2+}$
$[Fe(H_2O)_6]^{3+}$	+	H_2O	$\rightleftharpoons$	H_3O^+	+	$[Fe(H_2O)_5(OH)]^{2+}$
NH_4^+	+	H_2O	$\rightleftharpoons$	H_3O^+	+	NH_3
H_2O	+	CN^-	$\rightleftharpoons$	HCN	+	OH^-
H_2O	+	CO_3^{2-}	$\rightleftharpoons$	HCO_3	+	OH^-
H_2O	+	NH_3	$\rightleftharpoons$	NH_4^+	+	OH^-
HCl	+	NH_3	$\rightleftharpoons$	NH_4^+	+	OH^-
CH_3COOH	+	NH_3	$\rightleftharpoons$	NH_4^+	+	CH_3COO^-
HSO_4^-	+	NH_3	$\rightleftharpoons$	NH_4^+	+	SO_4^{2-}

An Important Rule

It has been observed that acid-base reactions proceed in a direction to form the weak acid and weak base and hence in these reactions a strong acid reacts with a strong base to form a weak acid and a weak base. Consequently in these reactions the strong acid and weak base form one conjugate acid-base pair and similarly a strong base and a weak acid form another conjugate acid-base pair. Thus in the reaction :

$$\underset{\textit{Strong acid}}{HCl} + \underset{\textit{Strong base}}{H_2O} \rightleftharpoons \underset{\textit{Weak acid}}{H_3O^+} + \underset{\textit{Weak base}}{Cl^-},$$

since HCl is a strong acid, its conjugate base *viz.* Cl^- ion is a weak base, *i.e.* HCl (strong acid) Cl^- (weak base) is one conjugate acid-base pair and similarly H_3O^+ (weak acid) —H_2O (strong base) is another conjugate acid-base pair. In general :

Conjugate acid-base pair

$$\underset{\text{Strong acid}}{HA} + \underset{\text{Strong base}}{B} \rightleftharpoons \underset{\text{Weak acid}}{HB^+} + \underset{\text{Weak base}}{A^-}$$

Conjugate acid-base pair

Thus conjugate base of a strong acid is always a weak base and the conjugate acid of a strong base is always a weak acid.

Dual Behaviour of Water

Water can act both as an acid (*i.e.* proton donor) and as a base (*i.e.* proton acceptor). For example H_2O acts as an acid towards NH_3 and acts as a base towards HCl as shown below :

$$\underset{\text{Acid}}{H_2O} + \underset{\text{Base}}{NH_3} \rightleftharpoons \underset{\text{Acid}}{NH_4^+} + \underset{\text{Base}}{OH^-}$$

$$\underset{\text{Base}}{H_2O} + \underset{\text{Acid}}{HCl} \rightleftharpoons \underset{\text{Acid}}{H_3O^+} + \underset{\text{Base}}{Cl^-}$$

The dual behaviour of H_2O has been used to explain the alkaline, acidic and neutral behaviour of the aqueous solution of salts. For example aqueous solution of Na_2CO_3 is alkaline, that of $FeCl_3$, $Al_2(SO_4)_3$, $NiSO_4$, $CuNO_3$ etc. is acidic while that of NaCl, K_2SO_4 etc. is neutral.

When Na_2CO_3 is dissolved in water, CO_3^{2-}ion, being a strong proton-acceptor, acts as a base and H_2O molecule which loses the proton behaves as an acid. Thus OH^- ions are produced and these ions make the solution alkaline.

$$\underset{\text{Acid}}{H_2O} + \underset{\text{Base}}{CO_3^{2-}} \rightleftharpoons \underset{\text{Acid}}{HCO_3^-} + \underset{\text{Base}}{OH^-}$$

In case of the aqueous solution of the salts of Fe^{3+}, Al^{3+} etc. cations, these strongly hydrated and highly charged small cations are strong proton-donors and hence readily give up a proton to H_2O which becomes H_3O^+. Due to the presence of H_3O^+ cations the solution is acidic.

$$\underset{\text{Acid}}{[Fe(H_2O)_6]^{3+}} + \underset{\text{Base}}{H_2O} \rightleftharpoons \underset{\text{Acid}}{H_3O^+} + \underset{\text{Base}}{[Fe(H_2O)_5(OH)]^{2+}}$$

$$\underset{\text{Acid}}{[Al(H_2O)_6]^{3+}} + \underset{\text{Base}}{H_2O} \rightleftarrows \underset{\text{Acid}}{H_3O^+} + \underset{\text{Base}}{[Al(H_2O)_5(OH)]^{2+}}$$

In case of the aqueous solutions of NaCl, K_2SO_4 etc. the hydrated cations are poor proton donors and hence the aqueous solutions are neutral.

Basicity or Protocity of an Acid

The number of H^+ (or H_3O^+) ions liberated by one molecule of an acid in aqueous solution is known as its basicity. On the basis of the basicity of a given acid, it may be a *monobasic or monoprotic* (*e.g.* HCl, HNO_3), *dibasic or diprotic* (*e.g.* H_2SO_4, H_2CO_3), and *tribasic* or *triprotic* (*e.g.* H_3PO_4). These acids liberate one, two and three or H^+ or H_3O^+ ions in aqueous solution respectively.

Acidity or Hydroxity of a Base

The number of OH^- ions liberated by one molecule of a base in aqueous solution is known as its acidity or hydroxity. On the basis of the acidity of a given base, it may be *monoacidic* or *monohydroxic* (*e.g.* NaOH, NH_4OH), *diacidic* or *dihydroxic* [*e.g.* $Ca(OH)_2$, Mg $(OH)_2$] and *triacidic* or *trihydroxic* [*e.g.* Al $(OH)_3$, Fe $(OH)_3$]. These bases liberate one, two and three OH^- ions in aqueous solution respectively.

Relative Strength of Acids and Bases

We know that acids and bases are also electrolytes and hence ionise in water. Acids and bases dissociate to give H^+ or H_3O^+ ions and bases ionise to give OH^- ions in aqueous solutions.

The strength of an acid or a base is defined as its capacity to give H^+ and OH^- ions in equimolar solutions in water respectively. Thus relative strength of two acids or bases can be compared by comparing their H^+ ion or OH^- ion concentrations in their equimolar solutions in water respectively.

Those acids and bases which ionise almost completely in solution are called strong acids and strong bases while those which are ionised only to a limited extent in solution are termed as weak acids and weak bases. Examples of strong acids are HNO_3, H_2SO_4, $HClO_4$, HCl, HBr, HI and those of strong bases are NaOH, KOH, $(CH_3)_4NOH$.

For an aqueous solution of an acid, HA, the equilibrium is shown as :

$$HA + H_2O \rightleftharpoons H_3O^+ + A^-$$

For the above equilibrium the value of equilibrium constant, K is given by :

$$K = \frac{[H_3O^+][A^-]}{[HA][H_2O]}$$

Now since water which is used as a solvent is present in large quantity, $[H_2O]$ = constant. Thus :

$$\frac{[H_3O^+][A^-]}{[HA]} = K \times [H_2O] = K_a \quad ...(i)$$

Here K_a is called *acid dissociation constant.*

Similarly for a weak base, B the corresponding equilibrium is :

$$B + H_2O \rightleftharpoons BH^+ + OH^-$$

and the value of *base dissociation constant*, K_b is given by :

$$K_b = \frac{[BH^+][OH^-]}{[B]} \quad ...(ii)$$

Equation (*i*) shows that *larger the value of K_a, higher the concentration of H_3O^+ ions and consequently stronger the acid. Thus higher the value of acid dissociation constant, K_a of an acid, stronger is the acid.* For example in case of CH_3COOH and HCN acids the values of acid dissociation constants are as follows.

$$CH_3COOH + H_2O \rightleftharpoons CH_3COO^- + H_3O^+$$

$$K_{CH_3COOH} = \frac{[CH_3COO^-][H_3O^+]}{[CH_3COOH]} = 1.8\times10^{-5}$$

$$HCN + H_2O \rightleftharpoons CN^- + H_3O^+$$

$$K_{HCN} = \frac{[CN^-][H_3O^+]}{[HCN]} = 7.1\times10^{-10}$$

Since K_{CH_3COOH} is higher than K_{HCN}, CH_3COOH is much stronger than HCN.

Similarly equation (*ii*) shows that *larger the value of K_b higher the concentration of OH^- ions and consequently stronger the base. Thus higher the value of base dissociation constant, K_b of a base, stronger is the base.* For example since K_b for NH_4OH (= 1.81×10^{-5}) is smaller than that for $C_2H_5NH_2 = 5.60 \times 10^{-4}$), $C_2H_5NH_2$ is a stronger base.

Values of dissociation constants (at 25°C) and relative strengths of some acids and bases are given in Table 12.1.

Table 12.1. Dissociation of some constants of some acids and bases at 25°C and their relative strength

Acid or base	Dissociation	K_a or K_b	Relative strength
Acids			
HCl	$HCl \rightleftharpoons H^+ + Cl^-$	Very large	Very strong
HNO_3	$HNO_3 \rightleftharpoons H^+ + NO_3^-$	Very large	Very strong
H_2SO_4	$H_2SO_4 \rightleftharpoons H^+ + HSO_4^-$	Large	Very strong
HSO_4^-	$HSO_4^- \rightleftharpoons H^+ + SO_4^{2-}$	1.3×10^{-2}	Strong
HF	$HF \rightleftharpoons H^+ + F^-$	6.7×10^{-4}	Weak
CH_3COOH	$CH_3COOH \rightleftharpoons CH_3COO^- + H^+$	1.8×10^{-5}	Weak
H_2CO_3 ($H_2O + CO_2$)	$H_2CO_3 \rightleftharpoons H^+ + HCO_3^-$	4.4×10^{-7}	Weak
H_2S	$H_2S \rightleftharpoons H^+ + HS^-$	1.0×10^{-7}	Weak
NH_4^+	$NH_4^+ \rightleftharpoons H^+ + NH_3$	5.7×10^{-10}	Weak
HCO_3^-	$HCO_3^- \rightleftharpoons H^+ + CO_3^{2-}$	4.7×10^{-11}	Weak
H_2O	$H_2O \rightleftharpoons H^+ + OH^-$	1.8×10^{-16}	Very weak

Bases			
NaOH	$NaOH \rightleftharpoons Na^+ + OH^-$	Very high	Very strong
KOH	$KOH \rightleftharpoons K^+ + OH^-$	Very high	Very strong
$Ca(OH)_2$	$Ca(OH)_2 \rightleftharpoons Ca^{2+} + 2OH^-$	High	Strong
NH_4OH	$NH_4OH \rightleftharpoons NH_4^+ + OH^-$	1.81×10^{-5}	Weak
CH_3NH_2	$CH_3NH_2 + H_2O \rightleftharpoons CH_3NH_3^+ + OH^-$	4.38×10^{-4}	Weak
$C_6H_5NH_2$	$C_6H_5NH_2 + H_2O \rightleftharpoons C_6H_5NH_3^+ + OH^-$	4.7×10^{-10}	Very weak

Acidity and Basicity of Molecules

We have seen that according to Bronsted-Lowry concept, if a molecule is to act as an acid, it should contain H-atom which it should donate as a proton (H^+). The degree of acidity of a given molecule is determined by the case with which the molecule can lose (*i.e.* donate) a proton.

There are many factors which influence the acidity (*i.e.* acidic strength) and basicity of a given molecule. Here we shall consider only one factor namely *electron delocalisation*. It has been observed that with the increase of volume available to the electron, the electron delocalisation also increases. The increase in electron delocalisation decreases the electron density and hence the attraction for a proton also decreases. The decrease in the attraction for a proton means that the basicity also decreases. The effect of the increase of electron delocalisation in decreasing the basicity is evident from the following examples.

1. Relative order of the acidity of CH_4, NH_3, H_2O and HF molecules.

In order to find the order of acidity of these molecules, let us find out the relative order of the basicity of their conjugate bases *viz.* CH_3^- (methide ion), NH_2^- (amide ion), OH^- (hydroxide ion) and F^- (fluoride ion) respectively. The structures of these ions as shown in Fig. 12.1 make it clear that (*i*) in CH_4^- ion 3/4 of the volume of carbon atom carrying the negative charge is occupied by three H-atoms. Thus the negative charge on C-atom exists only on 1/4 volume of the carbon atom (*ii*) in NH_4^+ ion 2/3 of the volume of N-atom carrying negative charge is occupied by two H-atoms. Thus the negative charge on N-atom exists only on 1/3 volume of N-atom (*iii*) in OH^- ion the negative charge on O-atom exists on more volume of O-atom, and (*iv*) in F^- ion the negative charge on F-atom exists on the whole volume of F-atom. Thus the volume available to electron increases as we pass from CH_3^- to F ion. The increase in this volume increases the electron delocalisation from CH_3^- to F^- ion. As the electron delocalisation increases, the electron density and hence the attraction for a proton decreases. The decrease in the attraction for a proton means that the basicity of these conjugate bases decreases from CH_3^- to F^- as shown below :

$$CH_3^- \text{ (Strongest base)} > NH_2^- > OH^- > F^- \text{ (Weakest base)}$$

We have already learnt that a stronger base has a weaker acid as its conjugate acid. Thus the conjugate acid of CH_3^- base (namely CH_4) should be the weakest acid and the conjugate acid of F^- base (namely HF) should be the strongest acid, *i.e.* the acidity of CH_4, NH_3, H_2O and HF molecules should be in the order :

$$CH_4 \text{ (Weakest acid)} < NH_3 < H_2O < HF \text{ (Strongest acid)}$$

—Volume available to electron increasing→
—Electron delocalisation increasing→
—Electron density decreasing→
—Attraction for proton decreasing→
—Basicity decreasing→

Fig. 12.1. The basicity of CH_3^-, NH_2^-, OH^- and F^- ions decreases as we pass from CH_3^- to F^- ion through NH_2^- and OH^- ions.

2. Relative order of the acidity of HF, HCl, HBr and HI acids.

To find the order of the acidity of the halogen acids, let us find out the relative order of the basicity of their conjugate bases, *viz.* F^-, Cl^-, Br^- and I^-.

When we move from F^- to I^-, the volume available to electron increases from F^- to I^-. This results in an increase in electron delocalisation and decrease in electron density from F^- to I^-. Consequently, the attraction for proton and basicity decrease in the order :

F^- (Strongest base) > Cl^- > Br^- > I^- (Weakest base)

The above conclusions are depicted in Fig. 12.2.

—Volume available to electron increasing→
—Electron delocalisation increasing→
—Electron density decreasing→
—Attraction for proton decreasing→
—Basicity decreasing→

Fig. 12.2. The basicity of F^-, Cl^-, Br^- and I^- ions decreases as we pass from F^- to I^- ion through Cl^- and Br^- ions.

It may now be inferred that F^- is the strongest base and its conjugate acid HF is the weakest acid. Similarly, I^- is the weakest base and its conjugate acid HI is the strongest acid among the halogen acids. The order of acidity of halogen acid is, therefore, as shown below :

HF (Weakest acid) < HCl < HBr < HI (Strongest acid)

In general, the acidity of binary acids increases on moving down a periodic group and also across a period from left to right.

Acidity increasing →

CH_4	NH_3	H_2O	HF	Acidity increasing ↓
		H_2S	HCl	
		H_2Se	HBr	
			HI	

3. Relative order of acidity of oxyacids.

The acidic character of the oxy-acids having the same central atom in different oxidation number increases with the increase in the oxidation number of the central atom. For example :

(*a*) $HCl^{+}O < HCl^{3+}O_2 < HCl^{5+}O_3 < HCl^{7+}O_4$

(*b*) $H_2S^{4+}O_3 < H_2S^{6+}O_4$

(*c*) $HN^{+}O < HN^{3+}O_2 < HN^{5+}O_3$.

Let us explain the relative order of acidity of the oxy-acids of chlorine. In order to do so let us find out the relative order of the basicity of the conjugate bases *viz* ClO^-, ClO_2^-, ClO_3^- and ClO_4^- of these acids. With the increase in the number of oxygen atoms in the conjugate bases, the delocalisation of the π bond becomes more and more extended. This results in decrease in electron density. Consequently, proton attraction and basicity also decrease (Fig. 12.3) in the order :

ClO^- (Strongest base) > ClO_2^- > ClO_3^- > ClO_4^- (Weakest base)

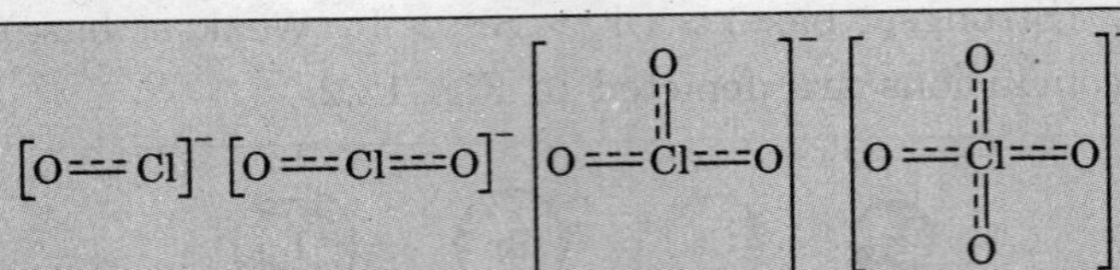

ClO^-	ClO_2^-	ClO_3^-	ClO_4^-
(Cl = + 1)	(Cl = + 3)	(Cl = + 5)	(Cl = + 7)

—Oxidation number of chlorine increasing→
—Number of oxygen atoms increasing→
—Delocalized π bond becomes more extended→
—Electron delocalization increasing→
—Electron density decreasing→
—Attraction for proton decreasing→
—Basicity decreasing→

Fig. 12.3. The basicity of ClO^-, ClO_2^-, ClO_3^- and ClO_4^- ions decreases as we pass from ClO^- ion to ClO_4^- ion through ClO_2^- and ClO_3^-, ions.

It may now be inferred that ClO^- is the strongest base and its conjugate acid HClO is the weakest acid. Similarly ClO_4^- is the weakest base and its conjugate acid $HClO_4$ is the strongest acid among the oxyacids of chlorine. The order of acidity of oxyacids of chlorine is, therefore, as shown below :

HClO (Weakest acid) < $HClO_2$ < $HClO_3$ < $HClO_4$ (Strongest acid)

Cady-Esley Concept : General Theory of Solvent System

The protonic definition of acid and bases given by Bronsted can be extended to the reactions occurring in non-aqueous solvents containing hydrogen such as NH_3, N_2H_4, HF, H_2SO_4, CH_3COOH, HCN, and alcohols.

In an attempt to have a more general definition of acids and bases applicable to protonic and non-protonic solvents, several definitions have been proposed. One of these is due to Cady and Elsey (1928) according to whom *an acid is a solute that, either by direct association or by reaction with the solvent, gives the anion characteristic of the solvent and a base is a solute that either by direct association or by reaction with the solvent, gives the anion characteristic of the solvent.* If, for example, we consider the solvent H_2O, its characteristic cation and anion are H_3O^+ and OH^- respectively as shown below :

$$\underset{\text{Acid}}{H_2O} + \underset{\text{Base}}{H_2O} \rightleftarrows \underset{\text{Acid}}{H_3O^+} + \underset{\text{Base}}{OH^-}$$

Thus all those compounds which can give H_3O^+ ions in H_2O will act as acids and all the compounds which can give OH^- ions in H_2O will behave as bases.

Similarly in N_2O_4 as solvent, substances such as NOCl which yield NO^+ ions are acids and the substances such as $NaNO_3$ which yield NO_3^- ions are bases.

$$\underset{\text{Acid}}{N_2O_4} + \underset{\text{Base}}{N_2O_4} \rightarrow \underset{\text{Acid}}{2NO^+} + \underset{\text{Base}}{2NO_3^-}$$

Evidently this definition of acids applies equally well to protonic and non-protonic solvents.

The auto-ionisation of some protonic and non-protonic solvents is shown below.

Protonic solvents :

$$\underset{\text{Acid}}{H_2O} + \underset{\text{Base}}{H_2O} \rightleftharpoons \underset{\text{Acid}}{H_2O^+} + \underset{\text{Base}}{OH^-}$$

$$NH_3 + NH_3 \rightleftharpoons NH_4^+ + NH_2^-$$

$$CH_3COOH + CH_3COOH \rightleftharpoons CH_3COOH_2^+ + CH_3COO^-$$

Non-protonic solvents :

$$SO_2 + SO_2 \rightleftharpoons SO^{2+} + SO_3^-$$

$$COCl_2 + COCl_2 \rightleftharpoons (COClCOCl_2)^+ + Cl^-$$

$$BrF_3 + BrF_3 \rightleftharpoons BrF_2^+ + BrF_4^-$$

$$N_2C_4 + N_2O_4 \rightleftharpoons 2NO^+ + 2NO_3^-$$

Just as with the Arrhenius definition, neutralisation is a reaction between an acid and a base to produce a salt and the solvent. Neutralisation reactions in some non-aqueous solvents are given below :

	Acid		Base		Salt		Solvent
In liq. NH_3 :	NH_4Cl	+	$NaNH_2$	$\rightleftharpoons$	NaCl	+	$2NH_3$
In N_2O_3 :	NOCl	+	$NaNO_3$	$\rightleftharpoons$	NaCl	+	N_2O_4
In CH_3COOH :	HCl	+	CH_3COONa	$\rightleftharpoons$	NaCl	+	CH_3OOH
In liq. SO_2 :	$SOCl_2$	+	$[N(CH_3)_4]_2SO_3$	$\rightleftharpoons$	$2[N(CH_3)_4]Cl$	+	$2SO_2$

It may be seen from the following reactions that there is a complete analogy between solvolytic and amphoteric behaviours in aqueous solvents.

Solvolytic behaviour :

In liq. NH_3 : $AlCl_3 + NH_3 \rightarrow [Al(NH_3)]^{2+} + H^+ + 3Cl^-$

In H_2O : $AlCl_3 + H_2O \rightarrow [Al(OH)]^{2+} + H^+ + 3Cl^-$

Amphoteric behaviour :

In liq NH_3 : $Zn(NH_2)_2 + 2NH_2^- \rightarrow [Zn(NH_2)_4]^{2-}$

In H_2O : $Zn(OH)_2 + 2OH^- \rightarrow [Zn(OH)_4]^{2-}$

Utility of the concept. Evidently, this concept of solvent system can be used to explain the acid-base reactions occurring in aqueous and non-aqueous solvents (protonic and non-protonic both).

Limitations. (*i*) This theory does not consider a number of acid-base reactions included in the protonic definition.

(*ii*) It limits acid-base phenomena to solvent systems only. Thus it does not explain the acid-base reactions which may occur in the absence of solvent.

(*iii*) It cannot explain the neutralisation reactions occurring without the presence of ions.

Thus this theory can simply be said to be an extension of the Arrhenius water-ion system.

Lux-Flood Concept

This concept was proposed by Lux (1939) and extended by Flood (1947). This concept considers the acid-base reactions in terms of oxide ion. Accordingly, *an acid (like* SiO_2, CO_2, P_4O_{10}, *etc.) is an oxide ion-acceptor while a base (like CaO, BaO, PbO, etc.) is an oxide ion-donor*. Some examples of acid-base reactions, in terms of the Lux-Flood concept, are given below :

Base		*Acid*		*Salt*
CaO	+	SiO_2	$\rightarrow$	$CaSiO_3$
BaO	+	CO_2	$\rightarrow$	$BaCO_3$
$6\ Na_2O$	+	P_4O_{10}	$\rightarrow$	$4Na_3PO_4$
PbO	+	SO_3	$\rightarrow$	$PbSO_4$
TiO_2	+	$Na_2S_2O_7$	$\rightarrow$	$TiOSO_4 + Na_2SO_4$

Amphoteric substances, in terms of this concept, show a tendency to lose and gain an oxide ion. Thus :

$$\underset{\text{(Amphoteric)}}{ZnO} \overset{+O^{2-}}{\rightleftarrows} ZnO_2^{2-}$$

$$ZnO \overset{-O^{2-}}{\rightleftarrows} Zn^{2+} + O^{2-}$$

$$\underset{\text{(Amphoteric)}}{Al_2O_3} \overset{+O^{2-}}{\rightleftarrows} 2AlO_2^-$$

$$Al_2O_3 \overset{-3O^{2-}}{\rightleftarrows} 2Al^{3+} + 3O^{2-}$$

This concept is particularly useful in explaining reactions which take place at high temperatures such as in metallurgical operations and during the manufacture of glass and ceramics. It can be extended to include transfer of any anion like halide, sulphide, carbanion, etc.

Lewis Concept: Electron-Donor Acceptor System

Each of the concepts discussed above had its own weaknesses and hence a need was felt for a new concept which could explain all the hitherto known properties of acids and bases. Lewis who is one of the early exponents of covalent bonding suggested that :

An acid is any species (molecule, radical or ion) which can accept an electron pair and a base is any species that can donate an electron pair. Thus an acid is an electron pair acceptor and a base is an electron-pair donor.

Consequently, a base should have a lone pair of electrons, which is used in making a coordinate covalent bond from the base to the acid, when an acid combines with a base.

Electron pair being donated by base to acid

Acid	+	:Base	$\rightarrow$	[Acid $\leftarrow$ Base]
Lewis acid		Lewis base		Adduct

A Lewis acid is often described as an *electrophilic reagent* (or simply an *electrophile*), since an electrophilic reagent, as its name implies, is an *electron-lover* or *electron-seeker* species (*electro* = electron, *philic* = lover), *i.e.*, it loves to accept the electrons (lone pairs). Thus Lewis acids are deficient of electrons. These may be positively-charged ions (*e.g.* H^+, RCO^+, etc.), metal cations (*e.g.*, Co^{3+}, Fe^{3+} etc.), molecules (*e.g.* CO_2, BF_3, SO_2 etc.) or metals in zero oxidation state.

In keeping with the definition of Lewis acid given above, a Lewis base is often described as a *nucleophic reagent* (or *nucleophile*), since a nucleophilic reagent, as its name suggests, is a *nucleus-lover* or *nucleus-seeker species* (*nucleo* = nucleus, *philic* = lover), *i.e.* it loves to donate electrons to a nucleus—in fact—to a positively-charged ion. Lewis bases have unshared pair of electrons (lone pairs). Since the nucleus is electrically positive, Lewis bases (or nucleophiles) may be negative ions (*e.g.* OH^-, F^-, NH_2^-, etc.) or electron rich molecules (*e.g.* NH_3, H_2O etc.).

Neutralisation Reactions in Terms of Lewis Concept

According to Lewis concept the process of neutralisation is simply the formation of a coordinate covalent bond between an acid and the donor atom of the base. The product formed is called a neutralisation product or an adduct. This adduct may be either non-ionisable or may undergo dissociation or condensation reaction depending on its stability.

The neutralisation reaction between a proton (H^+) and : NH_3 molecule takes place as shown below :

An electron pair from :NH_3 being transferred to H^+

$$\underset{\text{Lewis acid}}{H^+ \text{ (proton)}} + \underset{\text{Lewis base}}{:NH_3} \rightarrow \underset{\text{Adduct}}{[H^+ \leftarrow NH_3]^+ \text{ or } NH_4^+}$$

Evidently in the above reaction proton (H^+) accepts one electron pair from :NH_3 molecule and is, therefore, an acid, whereas :NH_3 molecule which donates an electron pair, is a base. The adduct is NH_4^+ ion.

Arrhenius Acids *Vs* Lewis Acids

Arrhenius considered those substances as acids which could donate proton (H^+). Since proton can accept an electron pair from a base, *all Arrhenius acids are also Lewis acids.*

Bronsted Bases *Vs* Lewis Bases

According to Bronsted a base is a substance which accepts proton while according to Lewis, a base is any species which can donate an electron pair. Bronsted base can accept proton only when it has an unshared pair of electrons in it. *Thus all Bronsted bases are also Lewis bases.* This fact can be made evident from the following neutralisation reaction in which NH_3 behaves both as Bronsted base as well as Lewis base.

NH_3 accepts proton (H^+) and, therefore, is a Bronsted base.

$$H^+ + :NH_3 \rightarrow \underset{\text{Adduct}}{(H^+ \leftarrow NH_3)^+ \text{ or } NH_4^+}$$

NH_3 donates its lone pair of electrons to H^+ and, therefore is a Lewis base.

Solvent System *Vs* Lewis System

Solvent system considers NH_4^+ ion as an acid and NH_2^- ion as a base in liquid NH_3 as shown below :

$$\underset{\textit{Acid}}{NH_4^+} + \underset{\textit{Base}}{NH_2^-} \rightleftarrows \underset{\textit{Acid}}{NH_3} + \underset{\textit{Base}}{NH_3}$$

In the above reaction, NH_2^- ion donates a lone pair of electrons to NH_4^+ ion to produce two molecules of NH_3. Since NH_4^+ ion is nothing but an ammoniated proton and the proton (H^+) can accept a lone pair of electrons from a base (*i.e.* NH_2^- ion), NH_4^+ and NH_2^- ions which are acid and base respectively in liquid NH_3 are also Lewis acid and Lewis base.

Bronsted Acids *Vs* Lewis Acids

There are many Lewis acids (*e.g.* SO_3, halides of B, Al, Fe (III) and Zn) which cannot be treated as Bronsted acids. The central atom of these acids is electron-deficient and hence can accept a pair of electrons from a base to complete its octet.

Similarly HCl which is a Bronsted acid cannot be called a Lewis acid, since it has no vacant orbital to accept an electron pair from a base.

Lewis Acids and Lewis Bases in Coordination Chemistry

In coordination chemistry, the central metal ion, M^{n+} is regarded as a Lewis acid (*i.e.* electrophile), since it accepts an electron pair while the ligand, L is regarded as Lewis base (*i.e.* nucleophile), since it donates an electron pair. The compound formed by the combination of metal cation (M^{n+}) and ligand (L) is called a *complex* or an *adduct*.

Electron pair being donated by L to M^{n+}

M^{n+}	+	:L	$\rightarrow$	$[M^{n+} \leftarrow L]^{n+}$
Lewis acid (Electrophile or Electron pair acceptor)		Lewis base (Nucleophile or Electron pair donor)		Complex or adduct

M^{n+} and L are linked together by a coordinate covalent bond from L to M^{n+} ($M^{n+} \leftarrow$ L bond). Thus the reaction between a metallic cation, M^{n+} and a ligand, L shown above can be considered as an acid-base reaction according to Lewis concept.

The metallic cation (Lewis acid) must have at least one vacant orbital in its valence-shell in which it can accommodate the electron pair donated by the ligand (Lewis base) and the ligand should have at least one lone pair of electrons so that it may be donated to the metallic cation.

Classification of Lewis Acids

Any Lewis acid must contain at least one empty orbital in the valence shell of one of its atoms to accept an electron pair from a Lewis base. Lewis acids may be classified as :

1. Molecules containing a central atom with incomplete octet. Typical examples of this class of acids are electron-deficient molecules such as alkyls and halides of Be, B and Al. Some reactions of this type of Lewis acids with Lewis bases are shown below :

Lewis acid		Lewis base		Adducts
F_3B	+	$O(C_2H_5)_2$	$\rightarrow$	$F_3B \leftarrow O(C_2H_5)_2$
Cl_3Al	+	NC_5H_5	$\rightarrow$	$Cl_3Al \leftarrow NC_5H_5$
Me_3B	+	N_2H_4	$\rightarrow$	$Me_3B \leftarrow N_2H_4$

2. Molecules containing a central atom with vacant d-orbitals. The central atom of the halides such as SiX_4, GeX_4, $TiCl_4$, SnX_4, PX_3, PF_5, SF_4, $TeCl_4$, etc.

have vacant d-orbitals. These substances can, therefore, accept an electron pair from the Lewis base to accommodate in their vacant d-orbital and can thus form adducts with a number of halide ions and organic bases. These substances are, therefore, Lewis acids. These halides are vigorously hydrolysed by H_2O to form an oxy acid or oxide of the central atom and the appropriate HX. The hydrolytic reactions take place presumably through the intermediate formation of unstable adducts with H_2O.

3. Simple cations. Theoretically all simple cations are potential Lewis acids. Reactions of some cations as Lewis acids with Lewis bases are shown below. It will be seen that these reactions are identical with those which produce Werner complexes.

	Lewis acid	Lewis base	Adduct or addition compounds
Ammonation :	Ag^+ +	$2(:NH_3)$ $\longrightarrow$	$[NH_3) \rightarrow Ag \leftarrow NH_3]^+$

$$Cu^{2+} + 4(:NH_3) \longrightarrow \left[\begin{array}{ccccc} & & NH_3 & & \\ & & \downarrow & & \\ H_3N & \rightarrow & Cu & \leftarrow & NH \\ & & \uparrow & & \\ & & NH_3 & & \end{array}\right]^{2+}$$

Hydration :

$$Co^{3+} + 6(:OH_2) \longrightarrow \left[\begin{array}{ccccc} H_2O & \searrow & & \swarrow & OH_2 \\ H_2O & \rightarrow & Co & \leftarrow & OH_2 \\ H_2O & \nearrow & & \nwarrow & OH_2 \end{array}\right]^{3+}$$

Alcoholation :

$$Li^+ + :O\begin{array}{l} \diagup H \\ \diagdown CH_3 \end{array} \longrightarrow \left[Li \leftarrow O\begin{array}{l} \diagup H \\ \diagdown CH_3 \end{array}\right]^+$$

The Lewis acid strength or coordinating ability of the simple cations which, according to Lewis, are Lewis acids, increases with (a) an increase in the positive charge carried by the cation (b) an increase in the nuclear charge for atoms in any period of the periodic table (c) a decrease in ionic radius (d) a decrease in the number of shielding electron shells.

Evidently the acid strength of simple cations increases for the elements on moving from left to right in a period and from bottom to top in a group of periodic table. Thus:

Fe^{2+} < Fe^{3+} (positive charge increases from +2 to +3)

K^+ < Na^+ (on moving from bottom to top in a group)

Li^+ < Be^{2+} (on moving from left to right in a period)

——Strength of Lewis acids increasing——→

4. Molecules having a multiple bond between atoms of dissimilar electronegativity. Typical examples of molecules falling in this class of Lewis acids are CO_2, SO_2 and SO_3. In these compounds the oxygen atoms are more electronegative than S or C-atom. As a result, the electron density of π-electrons is displaced away from carbon or sulphur atoms which are less electronegative than oxygen, towards the O-atom. The C or S-atom thus becomes electron deficient and is, therefore, able to accept an electron pair from a Lewis base such as OH^- ions to form dative bond.

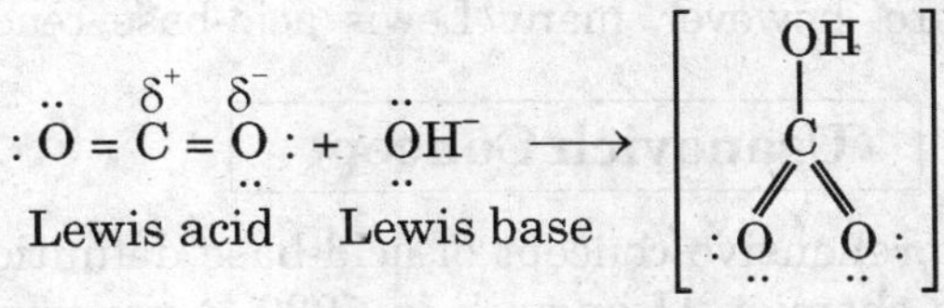

Bicarbonate ion

SO_2 also reacts in the same manner with OH^- ion

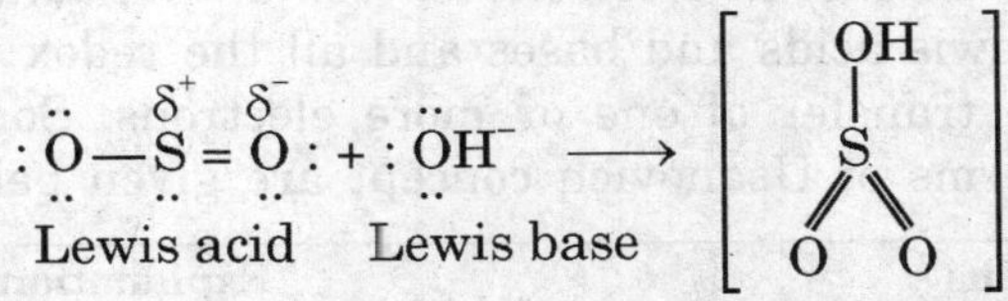

Bisulphate ion

5. Elements with an electron sextent. Oxygen and sulphur atoms contain six electrons in their valence shell and can, therefore, be regarded as Lewis acids. The oxidation of SO_3^{2-} to SO_4^{2-} ion by oxygen and to $S_2O_3^{2-}$ ion by sulphur are the acid-base reactions.

Lewis base		**Lewis acid**		**Adduct**
SO_3^{2-}	+	$:\ddot{\underset{..}{O}}$	$\rightarrow$	$[O \leftarrow SO_3]^{2-}$
SO_3^{2-}	+	$:\ddot{\underset{..}{S}}$	$\rightarrow$	$[S \leftarrow SO_3]^{2-}$

Utility of Lewis Concept

(*i*) This concept also includes those reactions in which no protons are involved.

(*ii*) Lewis concept is more general than the Bronsted-Lowry concept (*i.e.* protonic concept) in that acid-base behaviour is not dependent on the presence of one particular element or on the presence or absence of a solvent.

(*iii*) It explains the long accepted basic properties of metallic oxides and acidic properties of non-metallic oxides.

(*iv*) This theory also includes many reactions such as gas-phase, high temperature and non-solvent reaction as neutralisation processes.

(*v*) The Lewis approach is, however, of great value in cases where the protonic concept is inapplicable, for example, in reaction between acidic and basic oxides in the fused state.

Limitations of Lewis Concept.

(*i*) Since the strength of Lewis acids and bases is found to depend on the type of reaction, it is not possible to arrange them in any order of their relative strength. Thus, for example, experiments show that fluoride complex of Be^{2+} ion is more stable than that of Cu^{2+} ion, indicating that Be^{2+} ion is more acidic than Cu^{2+} ion. On the other hand amine complex of Cu^{2+} is more stable than that of Be^{2+} ion indicating that Cu^{2+} is more acidic than Be^{2+} ion.

(*ii*) According to the phenomenological criteria, an acid-base reaction should be a rapid reaction. There are, however, many Lewis acid-base reactions which are slow.

Usanovich Concept

This is the most comprehensive concept of acid-base definition. This concept was proposed by Russian chemist, Usanovich in 1939. According to this concept *an acid is any chemical species which is capable of combining with anions or electrons or giving up cations. Conversely a base is any chemical species which is capable of giving up anions or electrons or combining with cations.* This definition includes all Lewis acids and bases and all the redox reactions which may consist of complete transfer of one or more electrons. Some examples of acid-base reactions in terms of Usanovich concept are given below :

Acid-base reactions	Explanation
Acid **Base** **Salt**	
$SO_3 + Na_2O \rightarrow Na_2SO_4$	Base, Na_2O gives up O^{2-} ion; acid, SO_3 combines with O^{2-} ion. $Na_2O \rightarrow 2Na^+ + O^{2-}$ $SO_3 + O^{2-} \rightarrow SO_4^{2-}$ $2Na^+ + SO_4^{2-} \rightarrow Na_2SO_4$ (salt)
$Fe(CN)_2 + 4KCN \rightarrow K_4[Fe(CN)_6]$	Base, KCN gives up CN^- ion, acid $Fe(CN)_2$ combines with CN^- ion. $4KCN \rightarrow 4K^+ + 4CN^-$ $Fe(CN)_2 + 4CN^- \rightarrow [Fe(CN)_6]^{4-}$ $4K^+ + [Fe(CN)_6]^{4-} \rightarrow K_4[Fe(CN)_6]$
$Cl_2 + 2Na \rightarrow 2NaCl$	Base, Na atom loses an electron; acid, Cl combines with this electron. $2Na \rightarrow 2Na^+ + 2e^-$ $2Cl + 2e^- \rightarrow 2Cl^-$ $2Na^+ + 2Cl^- \rightarrow 2NaCl$

Utility of the concept.

This concept is particularly advantageous in classifying together all examples of acids and bases. It is thus the most general of all acid-base concepts.

Limitations of the concept.

(*i*) This concept is like a synthesis of all the previous acid-base concepts and hence in this respect this concept becomes extremely general.

(*ii*) It includes many reactions such as oxidation-reduction reactions which can perhaps be better considered from some other point of view.

(*iii*) This concept virtually considers all the chemical reactions as acid-base reactions.

Conclusion

The study of all the acid-base concepts as made above makes it evident that it is difficult to make an appropriate choice for the best concept. As a matter of fact, each concept has its own advantages and disadvantages and the knowledge of all the concepts is essential.

Questions with Answers

Q.1 Indicate the two conjugate acid-base pairs in the following reactions :

(*i*) $HSO_4^- + OH^- \rightleftharpoons SO_4^{2-} + H_2O$

(*ii*) $HCl + CH_3COOH \rightleftharpoons Cl^- + CH_3COOH_2^+$

Ans : (*i*) $HSO_4^- - SO_4^{2-}$ and $H_2O - OH^-$ are two conjugate acid-base pairs as shown below :

+ H⁺ (one pair)

HSO_4^-	+	OH^-	$\rightleftharpoons$	SO_4^{2-}	+	H_2O
Conjugent acid		Conjugate base		Conjugate base		Conjugate acid

+ H⁺ (other pair)

(*ii*) $HCl - Cl^-$ and $CH_3OOH_2^+ - CH_3COOH$ are two conjugate acid-base pairs as shown below

+ H⁺ (one pair)

HCl	+	CH_3COOH	$\rightleftharpoons$	Cl^-	+	$CH_3CHOOH_2^+$
Conjugent acid		Conjugate base		Conjugate base		Conjugate acid

+ H⁺ (other pair)

Q.2 Which of the species given in the following sets act as Bronsted acids and Bronsted bases (amphoteric character) (*i*) H_2O, HCO_3^-, NH_3, H_2S (*ii*) HSO_4^-, NH_3, OH^- (*iii*) HNO_3, HCO_3^-, H_2O, NH_3 (*iv*) HCl, H_3PO_4, HCO_3^-

Ans. (*i*) H_2S can donate a proton (H^+) to form HS^- but cannot accept a proton. Hence H_2S acts only as as Bronsted acid. H_2O, HCO_3^- and NH_3 can donate a proton to form OH^-, CO_3^{2-} and NH_2^- respectively. Thus these species can act as acids. These species can also accept H^+ to give H_3O^+, H_2CO_3 and NH_4^+

respectively. Thus these can also act as bases. In other words HCO_3^-, NH_3 and H_2S show amphoteric character.

(*ii*) All the three species can donate a proton (H^+) to form SO_4^{2-}, NH_2^- and O_2^- respectively. Thus they can act as acids. These species can also accept a proton to give H_2SO_4, NH_4^+ and H_2O respectively. Hence these species can also act as bases. Thus all the given three species show amphoteric character.

(*iii*) HNO_3 can only donate a proton (H^+) to form NO_3^- but cannot accept a proton to give $H_2NO_3^+$. Thus HNO_3 is an acid. HCO_3^-, H_2O and NH_3 can donate as well as accept H^+ to form (CO_3^{2-}, OH^-, NH_2^-) and (H_2CO_3, H_3O^+, NH_4^+) respectively. Thus HCO_3^-, H_2O and NH_3 show amphoteric character.

(*iv*) HCl and H_3PO_4 cannot accept a proton. Hence these species donot act as bases. These species can donate a proton to form Cl^- and $H_2PO_4^-$. Hence these species act acids. HCO_3^- can donate a proton to form CO_3^{2-}. Thus it can act as an acid. This ion can also accept a proton to form H_2CO_3. Hence HCO_3^- can also act as a base. Consequently HCO_3^- can act both as an acid and as a base *i.e.* HCO_3^- shows amphoteric character.

Q.3 Arrange the following bases in the decreasing order of their basic strength : (*i*) $CH \equiv C^-$, CH_3O^- and OH^- (*ii*) CH_3COO^-, OH^- and Cl^- (*iii*) ClO^-, ClO^-_2, ClO_3^- and ClO_4^- (*iv*) Cl^-, HCO_3^- and CH_3COO^- (*v*) OH^-, NH_2^-, $H - C \equiv C^-$ and $CH_3 - CH_2^-$.

Ans : (*i*) The conjugate acids of the given bases are obtained by odding a proton (H^+) to the bases.

$$\left.\begin{array}{l} CH \equiv C^- + H^+ \rightarrow CH \equiv CH \\ CH_3O^- + H^+ \rightarrow CH_3OH \\ OH^- + H^+ \rightarrow H_2O \end{array}\right\} \textit{Conjugate acids}$$

Now since the strength of conjugate acids ($CH \equiv CH$, CH_3OH, H_2O) increases as : $HC \equiv CH < CH_3OH < H_2O$, the strength of their conjugate bases ($CH \equiv C^-$, CH_3O^-, OH^-) decreases as :

$$HC \equiv C^- > CH_3O^- > OH^- \textbf{ Ans.}$$

(*ii*) The conjugate acids of the given bases are obtained by adding a proton (H^+) to the bases.

$$\left.\begin{array}{l} CH_3COO^- + H^+ \rightarrow CH_3COOH \\ OH^- + H^+ \rightarrow H_2O \\ Cl^- + H^+ \rightarrow HCl \end{array}\right\} \textit{Conjugate acids}$$

Now Since the strength of conjugate acids (CH_3COOH, H_2O and HCl) increases as : $H_2O < CH_3COOH < HCl$, the strength of their conjugate bases (CH_3COO^-, OH^-, Cl^-) decreases as :

$$OH^- > CH_3COO^- > Cl^- \textbf{ Ans.}$$

(*iii*) The conjugate acids of the given bases are obtained by adding a proton (H^+) to the bases.

$ClO^- + H^+ \rightarrow HClO$, $\quad ClO_2^- + H^+ \rightarrow HClO_2$

$ClO_3^- + H^+ \rightarrow HClO_3$, $\quad ClO_4^- + H^+ \rightarrow HClO_4$

Now since the strength of the conjugate acids ($HClO$, $HClO_2$, $HClO_3$ and $HClO_4$) increases as : $HClO < HClO_2 < HClO_3 < HClO_4$, the strength of their conjugate bases (ClO^-, ClO_2^-, ClO_3^- and ClO_4^-) decreases as : $ClO^- > ClO_2^- > ClO_3^- > ClO_4^-$ **Ans.**

(*iv*) The conjugate acids of the given bases are HCl, H_2CO_3 and CH_3COOH respectively ($Cl^- + H^+ \rightarrow HCl$, $HCO_3^- + H^+ \rightarrow H_2CO_3$, $CH_3COO^- + H^+ \rightarrow CH_3COOH$). Since the strength of the conjugate acids increases as : $CH_3COOH < H_2CO_3 < HCl$, the strength of their conjugate bases decreases as :

$$CH_3COO^- > HCO_3^- > Cl^-$$

(*v*) The conjugate acids of the given bases are H_2O, NH_3, $H-C \equiv C-H$ and CH_3-CH_3 respectively ($OH^- + H^+ \rightarrow H_2O$, $NH_2^- + H^+ \rightarrow NH_3$, $H-C \equiv C^- + H^+ \rightarrow H-C \equiv C-H$, $CH_3-CH_2^- + H^+ \rightarrow CH_3-CH_3$). Since the strength of conjugate acids increases as : $CH_3-CH_3 < NH_3 < H-C \equiv C-H < H_2O$, the strength of their conjugate bases decreases as :

$$CH_3-CH_2^- > NH_2^- > H-C \equiv C^- > OH^- \textbf{ Ans.}$$

Q.4 Which of the following species is the weakest base (*i*) H^-, CH_3^-, CH_3O^-, Cl^- (*ii*) $C_2H_5O^-$, CN^-, CH_3COO^-, NO_3^- (*iii*) $C_2H_5O^-$, NO_3^-, F^-, CH_3COO^-.

Ans : (*i*) The conjugate acids of the given species (bases) are obtained as shown below : $H^- + H^+ \rightarrow \times$, $CH_3^- + H^+ \rightarrow CH_4$, $CH_3O^- + H^+ \rightarrow CH_3OH$, $Cl^- + H^+ \rightarrow HCl$. Out of CH_4, CH_3OH and HCl, since HCl is the strongest conjugate acid, its conjugate base *viz* Cl^- is the weakest base.

(*ii*) The given acids of the given bases are obtained as shown below :

$C_2H_5O^- + H^+ \rightarrow C_2H_5OH$, $CN^- + H^+ \rightarrow HCN$

$CH_3COO^- + H^+ \rightarrow CH_3COOH$, $NO_3^- + H^+ \rightarrow HNO_3$

Out of C_2H_5OH, HCN, CH_3COOH and HNO_3, HNO_3 is the strongest acid. Hence its conjugate base (NO_3^-) is the weakest base.

(*iii*) The conjugate acids of the given species are C_2H_5OH, HNO_3, HF and CH_3COOH. Out of these acids since HNO_3 is the strongest acid, its conjugate base (*i.e.* NO_3^-) is the weakest base.

Q.5 Which species in the following sets has the highest proton affinity.

(*i*) I^-, HS^-, NH_2^-, F^-

(*ii*) SbH_3, AsH_3, PH_3, NH_3

Ans : (*i*) The species which is the strongest base has the higest proton affinity. Thus we should select the strongest base. The conjugate acids of the given species are HI, H_2S, NH_3 and HF respectively. Among these acids, since NH_3 is the weakest acid, its conjugate base (*i.e.* NH_2^-) is the strongest base. Consequently NH_2^- has the highest proton affinity.

(*ii*) If a substence has proton affinity it can accept the proton and will act as a base. Larger is the size of the central atom (M) in MH_3 molecules, lesser is the tendency of MH_3 molecule to accept a proton and hence lesser is the basic character of MH_3 molecule. Thus since the size of M-atom (central atom) increases from N to Sb, the proton affinity (*i.e.* basic character) of the given molecules decreases from NH_3 to SbH_3 ($NH_3 > PH_3 > AsH_3 > SbH_3$). Thus NH_3 has the highest proton affinity.

Another explanation : The central atom in all the given MH_3 molecules has one *lp* of electrons. Since N atom is the smallest in size, the lone pair electron density on N-atom in NH_3 molecule is the highest and hence the proton affinity of NH_3 is the highest.

Q.6 Which of the following molecules acts as a Lewis acid : $(CH_3)_2O$, $(CH_3)_3P$, $(CH_3)_3N$, $(CH_3)_3B$.

Ans : The structures of $O(CH_3)_2$, $P(CH_3)_3$ and $N(CH_3)_3$ show that O, P and N-atoms (central atom) of these molecules have an octet of electrons (complete octet) round them. Thus these molecules cannot accept electrons. Hence these molecules cannot act as Lewis acids. B-atom (central atom) in $B(CH_3)_3$ has incomplete of octet of electrons (it has six electrons round it), it can accept electrons and hence $B(CH_3)_3$ molecule is Lewis acid.

Q.7 Which of the species given in the follwoing sets is not a Lewis acid ?

(*i*) $SnCl_2$, $AlCl_3$, CCl_4, $SiCl_4$

(*ii*) CO, $SiCl_4$, SO_3, Zn^{2+}

Ans : (*i*) Central atom of $SnCl_2$ and $SiCl_4$ molecules have empty *d*-orbitals. Hence these molecules are Lewis acids. Al atom (central atom) in $AlCl_3$ has incomplete octet of electrons. Hence $AlCl_3$ is also Lewis acis. Since C-atom in CCl_4 has complete octet of electrons, it is not a Lewis acid.

(*ii*) We know that the molecules having empty *d*-orbitals (*e.g.* $SiCl_4$), molecules having multiple bonds between different atoms [e.g. SO_3 or $O{=}S({=}O){=}O$] and simple cations (*e.g.* Zn^{2+}) are Lewis acids. Since CO does not have any of the above characteristics, it is not a Lewis acid.

Q.8 Which of the species given in the following sets is not a Lewis base.

(*i*) CN^-, ROH, NH_3, $AlCl_3$

(*ii*) Ag^+, H_2O, CN^-, C_2H_5OH

Ans : (*i*) CN^-, ROH and NH_3 all have unshared electron pairs which can be donated. Thus these species are Lewis bases. Since Al atom in $AlCl_3$ has empty *d*-orbitals, $AlCl_3$ is a Lewis acid, *i.e.* $AlCl_3$ is not a Lewis base.

(*ii*) H_2O, CN^- and C_2H_5OH all have unshared electron pairs which can be donated. Thus these species act as Lewis bases. Ag^+ is a simple cation. Hence it as a Lewis acid *i.e.* Ag^+ is not Lewis base.

Q.9 Identify Lewis acids and Lewis bases in the following reactions.

(*i*) $BrF_3 + F^- \rightarrow [BrF_4]^-$

(*ii*) $I^- + I_2 \rightarrow I_3^-$

(*iii*) $[SnCl_3]^- + (CO)_5\,MnCl \rightarrow (CO)_5\,Mn\,(SnCl_3) + Cl^-$

Ans : (*i*) Reaction (*i*) is a Lewis acid – Lewis base reaction. In this reaction F^- ion has four *lps* of electrons which are donated to BrF_3 molecule to form $F_3Br \leftarrow F^-$ or BrF_4^-. Thus F^- ion acts a Lewis base (donor). In BrF_3 molecule Br atom is sp^3d hybridised. Br atom has four emply 4*d* orbitals (unhybridised orbitals) which accept electron pair donated by F^- ion (Lewis base). Thus BrF_3 molecule acts a Lewis acid (acceptor).

(*ii*) Reaction (*ii*) is also a Lewis acid-Lewis base reaction. In this reaction I^- ion has four *lps* of electrons and hence acts as Lewis base. In I_2 molecule  each I-atom is sp^3 hybridised. 5*d* orbitals in I-atom are lying empty. These orbitals accept electrons donated by one I^- ion to form 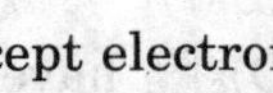 or I_3^- ion. Thus I_2 molecule acts as Lewis acid.

(*iii*) This reaction is not a Lewis acid-Lewise base reaction. Sn atom in $SnCl_3^-$ ion has one *lp* of electrons 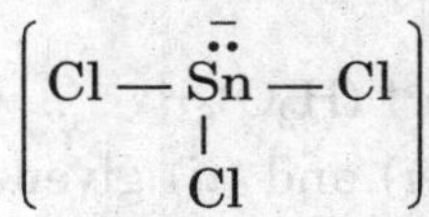. Hence this ion acts as a Lewis base. In $(CO)_5MnCl$, Cl^- ion is also a Lewis base. In the given reaction Lewis base, $SnCl_3^-$ replaces Cl^- ion (Lewis base) present in $(CO)_5MnCl$ to from $(CO)_5$ $Mn(SnCl_3)$.

Q.10 Four acids A, B, C and D have their *pKa* values as A = 1.0, B = 3.0, C = 2.0 and D = 4.5. Which of these acids is the strongest ?

Ans : We Know that lower is the value of *pKa* of an acid, stronger is the acid. Thus since A has the lowest value of *pKa* (= 1.0), it is the strongest acid.

Q.11 Which of the following acids is the strongest acid : ClO_3 (OH), ClO_2 (OH), SO $(OH)_2$, SO_2 $(OH)_2$?

Ans : Given oxy acids can be written as $HClO_4$, $HClO_3$, H_2SO_3, H_2SO_4. Since central atom (Cl-atom) in $HClO_4$ has the highest O.S. (= +7), this acid is the strongest acid.

Q.12 Explain the following :

(*i*) Tricovalent phosphorus compounds like PCl_3 can act both as Lewis acid and Lewis base.

Ans : In PCl_3 molecule, P-atom has a *lp* of electrons. This *lp* of electrons can be donated to a metal ion. Thus PCl_3 can act as a Lewis base. Since P-atom (central atom) in PCl_3 molecule has empty 3*d* orbitals, these orbitals can receive electrons donated by an electron-rich metal in low oxidation state. Thus PCl_3 can also act as Lewis acid. In $[Ni\ (PCl_3)_4]^0$, PCl_3 acts as a Lewis acid.

(*ii*) NH_2^- is better base than PH_2^- towards proton.

Ans : NH_2^- and PH_2^- both act as bases since they can accept a proton (H^+) to form NH_3 and PH_3 respectively. Both the given ions have two *lps* of electrons on their central atom. Since N-atom is smaller in size than P-atom, lone pair electron density on N-atom in NH_2^- is higher than that on P-atom in PH_2^-. Due to higher electron density, NH_2^- is better base towards proton.

(*iii*) The acid strength of $[Fe\ (H_2O)_6]^{2+}$, $[Al\ (H_2O)_6]^{3+}$ and $[Fe\ (H_2O)_6]^{3+}$ ions in aqueons solution decreases as $[Al(H_2O)_6]^{3+} > [Fe(H_2O)_6]^{3+} > [Fe\ (H_2O)_6]^{2+}$

Ans : The given aqua ions have Fe, Al and Fe in +2, +3 and +3 oxidation state respectively. Since Fe in $[Fe\ (H_2O)_6]^{3+}$ has higher charge (= +3) than in $[Fe\ (H_2O)_6]^{2+}$ (Fe = +2), $[Fe\ (H_2O)_6]^{3+}$ has higher acid strength than $[Fe\ (H_2O)_6]^{2+}$.

Thus :

$$[Fe^{3+}\ (H_2O)_6]^{3+} > [Fe^{2+}\ (H_2O)_6]^{2+} \quad ...\ (i)$$

Since Fe and Al in $[Fe\ (H_2O)_6]^{3+}$ and $[Al\ (H_2O)_6]^{3+}$ have the same charge (= +3), the acid strength of these ions dependes on the size of Fe^{3+} and Al^{3+} ions. Since Al^{3+} ion is smaller in size than Fe^{3+}, $[Al\ (H_2O)_6]^{3+}$ has higher acid strength than $[Fe\ (H_2O)_6]^{3+}$

$$[Al^{+3}\ (H_2O)_6]^{3+} > [Fe^{3+}\ (H_2O)_6]^{3+} \quad ...\ (ii)$$

On combining the two relations (*i*) and (*ii*) given above we get :

$$[Al^{3+}(H_2O)_6]^{3+} > [Fe^{3+}(H_2O)_6]^{3+} > [Fe^{2+}(H_2O)_6]$$

(*iv*) When aqueous solution of $CuSO_4$ is added to the aqueous solution of $(NH_4)_2SO_4$, the acidity of $(NH_4)_2SO_4$ solution increases.

Ans : When $CuSO_4$ solution is added to $(NH_4)_2SO_4$, H_2SO_4 or protons (H^+) are released.

$$CuSO_4 + 2(NH_4)_2\ SO_4 \rightarrow [Cu\ (NH_3)_4]\ SO_4 + 2H_2SO_4$$

or $\quad Cu^{2+} + 4NH_4^+ \rightarrow [Cu\ (NH_3)_4]^{2+} + 4H^+$

Due to the production of H^+ ions, the acidity of aqueous soultion of $(NH_4)_2SO_4$ increases.

(*v*) PCl_5 is a better Lewis acid than PCl_3.

Ans : P-atom (central atom) in both the given molecules has vacant *d*-orbitals. Hence both the molecules can act as Lewis acids (electron pair acceptors). Since O.N. of P-atom in PCl_5 (P = +5) is higher than that in PCl_3 (P = +3), EN of P-atom in PCl_5 would be higher than that of the same element (*i.e.* P-atom) in PCl_3. Due to higher EN of P-atom in PCl_5, PCl_5 is a better electron pair acceptor (Lewis acid).

(*vi*) The acidity (acidic character) of H_3PO_2, H_3PO_3 and H_3PO_4 acids decreases as : $H_3PO_2 > H_3PO_3 > H_3PO_4$.

Ans : The decreasing order of acidity of the given acids can not be explained on the basis of oxidation state of P-atom. Their acidity is explained on the basis of inductive effect of the pronated O-atoms in these acids. The structres of these acids are given below in Fig.12.4.

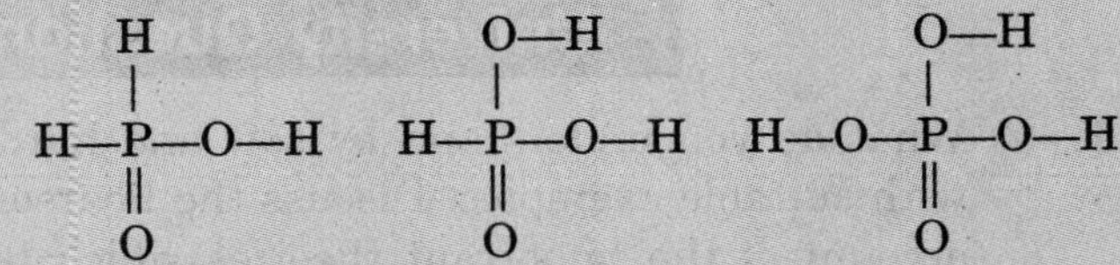

H_3PO_2 (Mono-basic)	H_3PO_3 (Dibasic)	H_3PO_4 (Tri-basic)

		H_3PO_2	H_3PO_3	H_3PO_4
No. of protonated O-atoms	=	1	2	3
No. of unprotonated O-atoms	=	1	1	1

Fig. Structures of H_3PO_2, H_3PO_3 and H_3PO_4 molecules

It may be seen from the structers that the number of unprotonated O-atoms in all three acids is the same (=1) but the number of protonated O-atoms increases as $1 \rightarrow 2 \rightarrow 3$ in H_3PO_2, H_3PO_3 and H_3PO_4 respectively. With the increases in the number of protonated O-atoms, the inductive effect of these O-atoms decreases from H_3PO_2 to H_3PO_4. Due to the decrease in inductive effect of protonated O-atoms, the acidity of acids decreases from H_3PO_2 to H_3PO_4.

Q.13 Classify the following as hard/soft acids or bases. H^-, Ni^{4+}, I^+, H^+.

Ans : Characteristics of the given species are given below. On the basis of these characteristics the nature of the species can be predicated.

Characteristics

Species	Charge	Size	EN	Polaribility	Conclusion
(*i*) H^-	Negative charge	Larger in size than H-atom	Low	Due to large size it has high polarisability.	Soft base
(*ii*) Ni^{4+}	High positive charge	It has smaller size than Ni^{2+}	High	Due to smaller size it has low polarisability.	Hard acid
(*iii*) I^+	Low positive charge	It has larger size	Low	Due to larger size it has high polarisability.	Soft acid
(*iv*) H^+	High positive charge	Smallest size	High	Due to the smallest size it has very low polarisability.	Hard acid

University Questions

1. (*i*) What do you understand by soft and hard acids and bases ? Explain with suitable examples. Discuss the Pearson's principle and give a brief account of the important theories which favour this principle.
   (*ii*) Explain why do Ca and Mg occur as oxides and carbonates but not as sulphides in nature. (*Lucknow 2002*)
2. Lewis strength of BBr_3 is greater than that of BCl_3 and BF_3. Explain. (*Lucknow 2002*)
3. Explain why all Bronsted bases are Lewis bases but all Bronsted acids are not Lewis acids. (*Lucknow 2002*)
4. (*a*) What are hard and soft acids and bases ? Explain with examples.
   (*b*) On the basis of SHAB principle, explain why HgS is insoluble and $Hg(OH)_2$ is soluble in dil. HCl. (*Nagpur 2002*)
5. Discuss the Pearson's hard and soft acid and base principle for the classification of acids and bases. Explain theoretical behaviour of hardness and softness. Predict the relative strength of HF, HBr and HI in aqueous solution in the light of the above concept. (*VBS Purvanchal 2003*)
6. Explain Bronsted-Lowry concept of acids and bases with suitable examples. (*VBS Purvanchal 2003*)
7. Explain Lewis concept of acids and bases. (*VBS Puravanchal 2003*)
8. Discuss the concept of hard and soft acids and bases.
9. (*a*) What are Lewis acids and bases ? How are they classified ?
   (*b*) Discuss the concept of the following, giving suitable examples : (*i*) Lux flood concept of acids and bases (*ii*) Hard and soft acids and bases.
10. (*a*) BCl_3 is stronger acid than BF_3. Explain.
    (*b*) Which is a Lewis acid in the following ? (*i*) NaCl (*ii*) $BaCl_2$ (*iii*) $AlCl_3$. (*VBS Purvanchal 2003*)
11. (*a*) Enumerate the factors which govern the hardness and softness in acids and bases.
    (*b*) Illustrate HSAB principle. Give reasons for the following on the basis of HSAB principle : (*i*) AgI_2^- is stable but AgF_2^- is not ; (*ii*) CsF reacts with Lil but the reverse is not true.
    (*c*) Identity Lewis acid or base in the following : Ag^+, BF_3, CO, PCl_3. (*Rohikhand 2003*)
12. (*a*) Explain the terms conjugate acids and conjugate bases with the help of siutable examples.
    (*b*) Discuss the effect of substituents on the strength of carboxylic acid.
    (*c*) Which is the stronger acid of each pair. H_3O^+ or H_2O, NH_4^+ or H_2S or HS^- and H_2O or OH ? (*Delhi 2003*)

13. (*a*) Apply SHAB principle in : (*i*) Occurrence of metal minerals and (*ii*) predicting the stability of halogen acids.

(*b*) Classify the following as hard and soft acids : Ca^{2+} and Hg^{2+}.

(*Nagpur 2003*)

14. (*a*) Discuss the Bronsted-Lowery concept of acid and bases. What are its applications and limitations ?

(*b*) Explain the following :

(*i*) Arrhenius hypothesis (*ii*) Lux-Flood acid-base hypothesis.

(*c*) Explain the following :

(*i*) Hard acid and soft base (*ii*) Hard base and soft base

(*iii*) Lewis acid and Lewis base. (*Bhopal 2004*)

15. (*a*) Out of BCl_3 and BF_3 which is more acidic ?

(*b*) BF_3 is a Lewis acid where as CCl_4 is not a Lewis acid. Explain.

(*GND Amritsar, 2004*)

16. What are hard and soft acids and bases ? (*Bhopal 2004*)

17. Boron trihalides are called Lewis acids. Explain why. (*Meerut 2004*)

18. Give Lewis definition of acid and base giving suitable examples.

(*Banglore 2005, Nagpur 2005*)

19. What are different kinds of Lewis acids ? (*Calcutta 2005*)

20. What is HSAB principle ? (*Calcutta 2005*)

21. Explain Bronsted-Lowery theory of acids and bases. Give its limitations.

(*Meerut 2006*)

22. What the limitations of HSAB concept ? (*H.N. Bahunguna 2006*)

23. (*a*) What do you understand by conjugate acid-base pair ? Show that a strong acid has a weak conjugate base and a weak acid has a strong conjugate base ?

(*b*) All Arrhenius acids are also Bronsted acids but all Arrhenius bases are not Bronsted bases. Discuss. (*C.S.J.M Kanpur 2007*)

24. Discuss Bronsted-Lowery and Lewis concept of acids and bases with suitable examples. (*Agra 2007*)

25. (*a*) Discuss in brief about conjugate acids and conjugate bases.

(*b*) How does pKa value charge with strength of acids.

(*C.S.J.M. Kanpur 2008*)

26. Describe Lewis acid-base concept with suitable examples. (*Meerut 2008*)

(*a*) Which one is hard base : F^- or I^-.

(*b*) What are characteristics of hard acids ? (*H.N. Bahuguna 2008*)

27. (*a*) Explain Lux-Flood acids and bases.

(*b*) Explain amphoteric substences. (*C.S.J.M Kanpur 2009*)

CHAPTER

13 Non-aqueous Solvents

Classification of Solvents

Solvents have been classified in a number of ways depending on their physical and chemical properties.

1st classification :

This is a common classification which is based on *proton-donor* and *proton-acceptor property* of the solvents. On the basis of this classification the solvents may be classified into three types :

1. Protic or protonic solvents. These solvents have hydrogen atom in their formulae and are of two types. (*a*) *Acidic or protogenic solvents*. These have *strong tendency to donate protons* (*e.g.* H_2SO_4. HF. CH_3COOH, HCN etc.) (*b*) *Basic or protonic solvents.* These have *strong tendency to accept protons* (*e.g.*, NH_3, C_5H_5, amines, N_2H_4).

2. Aprotic or non-protonic solvents. These may or may not have hydrogen in their formula and *neither donate nor accept protons* (*e.g.*, C_6H_6, $CHCl_3$, SO_2, CCl_4, BrF_3).

3. Amphi-protic or amphoteric solvents. These have hydrogen in their formula and donate or accept protons depending on the nature of reacting species, *i.e.*; *these can act both as acids and bases and consequently are amphoteric in nature.* These solvents dissociate feebly into protons and anions. Examples are CH_3COOH, H_2O.

2nd classificaton :

This classification is based on the fact that *"like dissolve like"*. Thus this classification gives (1) *Ionising (polar or ionic) solvents* and *(2) Non-ionising (non-polar or non-ionic) solvents.*

1. Ionising solvents. Ionising solvents are polar or ionic in nature and hence dissolve ionic compounds and initiate ionic reactions. These exist as ions in their pure state and thus are weak conductors of electricity. These have high values of dielectric constants. Because of their polar nature they have strong tendency to form associated structures. Examples are H_2O, NH_3, HF, SO_2 etc. These solvents undergo self-ionisation as shown below.

$$H_2O + H_2O \rightleftarrows H_3O^+ + OH^- : NH_3 + NH_3 \rightleftarrows NH_4^+ + NH_2^- ;$$
$$SO_2 + SO_2 \rightleftarrows SO^{2+} + SO_3^{2-}$$

2. Non-ionising solvents. These are non-polar or non-ionic in nature and hence dissolve only non-polar or neutral compounds and do not initiate ionic reactions. These have low dielectric constants. They have little associating and solvating tendency between the solute and solvents. Examples are C_6H_6, CCl_4 etc. These solvents do not undergo self-ionisation.

3rd Classification : Aqueous and Non-aqueous Solvents.

The solvents other than water are called non-aqueous solvents. The groups obtained as a result of self-ionisatjon of H_2O and non-aqueous solvents are analogous to each other, *e.g.*, H_3O^+ and OH^- groups obtained from the self ionisation of H_2O are analogous to NH_4^+ and NH_2^- groups respectively resulted from the self ionisation of liq. NH_3. This is called *parent solvent concept* and a large number of chemical reactions have been correlated in terms of this concept.

Some of the parent solvents and analogous groups are given below :

$$\underbrace{\begin{array}{c} H_2O + H_2O \\ NH_3 + NH_3 \\ NH_2OH + NH_2OH \\ H_2S + H_2S \end{array}}_{\text{Solvents}} \begin{array}{c} \rightleftarrows \\ \rightleftarrows \\ \rightleftarrows \\ \rightleftarrows \end{array} \underbrace{\begin{array}{c} H_3O^+ + OH^- \\ NH_4^+ + NH_2^- \\ NH_3OH^+ + NHOH^- \\ H_3S^+ + SH^- \end{array}}_{\text{Analogous groups}}$$

In this chapter we shall confine our attention to the study of ionising solvents only in which inorganic substances ionise and give ionic reactions similar to water. Non-ionising solvents are of interest particularly in organic chemistry.

Liquid Ammonia as Solvent

Liquid NH_3 is one of the most comprehensively studied water-like solvent. It is a protonic solvent and is able to dissolve a wide variety of substances. The physical data for liquid NH_3 are as : melting point = 175.3K; boiling point = 239.62K; dielectric constant = 23 at 239.62K; specific conductance = 5×10^{-11} per ohm per cm.

Solubility of Substances in liq. NH_3

1. Inorganic compounds. Since dielectric constant of NH_3 is much less than that of H_2O, liq. NH_3 is a poorer solvent for ionic substances. Molecules having many electrons such as iodine compounds and non-polar substances such as hydrocarbons are, however, soluble in liq. NH_3. Nitrates, thiocyanates, perchlorates and most cyanides are soluble. Oxides, hydroxides, sulphates, carbonates, phosphates, sulphites, and most sulphides, are insoluble. The general order of the solubility of halides in liq. NH_3 is $I^- > Br^- > F^-$. Most iodides are soluble, bromides are less soluble, only NH_4^+, Be^{2+} and Na^+ chlorides are soluble and most fluorides are insoluble.

Most of the ammonium salts (except those of multivalent anions) such as NH_4NO_3, NH_4SCN, $NH_4(CH_3COO)$ are soluble in liq. ammonia. All the metal amides, except those of alkali metals, are insoluble. LiNH is also insoluble.

2. Organic compounds. Alcohols, halogen compounds (*e.g.* chloroform etc.), ketones, esters,, simple ethers and phenol and its derivatives are soluble. The aromatic hydrocarbons are in general sparingly soluble.

3. Non-metals. The non-metals (*e.g.* P, S and I) dissolve in liq. NH_3 reacting with it.

$$\tfrac{5}{4}S_8 + 16NH_3 \rightarrow N_4S_4 + 6(NH_4)_2S$$

Reactions Taking Place in Liquid Ammonia

The various reactions carried out in liquid ammonia are as follows :

1. Auto-ionization of Liquid Ammonia. The extent of auto-ionization of liquid ammonia is less than that of water.

$$2NH_3 \rightleftharpoons NH_4^+ + NH_2^-$$
$$2H_2O \rightleftharpoons H_3O^+ + OH^-$$

It is indicated from its low -specific conductance and also from its ion product constant, *i.e.*,

$$H_2O = [H^+][OH^-] = 1 \times 10^{-14} \text{ at } 293 \text{ K}$$
$$2NH_3 = [NH_4^+][NH_2^-] = 1.9 \times 10^{-33} \text{ at } 218 \text{ K.}$$

2. Acid-base Reactions. On the basis of Bronsted-Lowry concept ammonium ions would behave as acids while amide ions as bases in liquid ammonia.

$$NH_3 + NH_3 \rightleftharpoons NH_4^+ + NH_2^-$$
$$\underset{\text{Acid (I)}}{H_2O} + \underset{\text{Base (I)}}{H_2O} \rightleftharpoons \underset{\text{Acid (II)}}{H_3O^+} + \underset{\text{Base (II)}}{OH^-}$$

The strongest acid available in liquid ammonia is the ammonium ion which is analogue of hydronium ions in aqueous system. A typical neutralization is the reaction of ammonium bromide with potassium amide.

$$\underset{\text{Acid}}{NH_4Br} + \underset{\text{Base}}{KNH_2} \rightarrow \underset{\text{Salt}}{KBr} + \underset{\text{Solvent}}{2NH_3}$$

Reactions of solutions of metal ions in liquid ammonia result in the precipitation of metal amides, imides or nitrides which are analogues to the precipitation of metal hydroxides and oxides in aqueous system.

$$AgNO_3 + KNH_2 \xrightarrow{\text{Liq. } NH_3} AgNH_2 + KNO_3$$
$$PbI_2 + KNH_2 \longrightarrow PbNH + KI + HI$$
$$3HgI_2 + 2KNH_2 \longrightarrow Hg_3N_2 + 2KI + 4HI$$

Many metal amides, imides and nitrides dissolve in solution of potassium amide in liquid ammonia to from complexes.

$$AgNH_2 + KNH_2 \xrightarrow{\text{Liq. } NH_3} K[Ag(NH_2)_2]$$
$$NaNH_2 + 2KNH_2 \longrightarrow K_2[Na(NH_2)_3]$$

3. Precipitation Reactions. Because of the differences in the solubilities of various substances in liquid ammonia and water, a large number of reactions which do not occur in water have been reported to be carried out in liquid ammonia. Following examples are illustrative.

(*i*) When solutions of silver chloride and barium nitrate in liquid ammonia are brought together, a white ppt. of $BaCl_2$ is obtained.

$$2AgCl + Ba(NO_3)_2 \overset{\text{Liq. } NH_3}{\rightleftarrows} \underset{\text{white ppt,}}{BaCl_2} + 2AgNO_3$$

(*ii*) Similarly, a white ppt. of KCl is obtained by mixing solution of KI with NH_4Cl solution in liquid ammonia.

$$KI + NH_4Cl \xrightleftharpoons{\text{Liq. } NH_3} \underset{\text{white ppt.}}{KCl} + NH_4I$$

4. Complex Formation Reactions. Many complex formation reactions have been reported which are similar to reactions occurring in water system. When KCN is added to silver nitrate solution in liquid ammonia or in water, a ppt. of silver cyanide is first formed which dissolves in excess of KCN yielding the complex compound, $[KAg(CN)_2]$.

$$AgNO_3 + KCN \longrightarrow \underset{\text{ppt.}}{AgCN} + KNO_3$$

$$AgCN + KCN \longrightarrow KAg(CN)_2$$

5. Ammonolysis Reactions. These reactions are also termed as solvolysis reactions. Liquid ammonia is regarded as an ideal solvent medium for many ammonolysis reactions. Some of these are as follows :

(*i*) Alkali metal and alkaline earth metal hydrides are ammonolysed to the corresponding metal amides and hydrogen.

$$M^+H^- + NH_3 \longrightarrow MNH_2 + H_2$$

(*ii*) Many covalent halides undergo ammonolysis in liquid ammonia.

$$HgCl_2 \xrightarrow{\text{Liq. } NH_3} HgNH_2Cl$$

$$SiCl_4 \longrightarrow Si(NH_2)_4$$

$$GeI_4 \longrightarrow Ge(NH_2)_4$$

6. Solvation Reactions. In solvation reactions, solvent gets bonded to a solute species (molecule or ion). Such reactions taking place in water are called ammonation reactions. Some examples of solvation reactions are :

$$BF_3 + NH_3 \longrightarrow BF_3.NH_3$$

$$SiF_4 + 2NH_3 \longrightarrow SiH_4.2NH_3$$

$$SO_3 + 2NH_3 \longrightarrow SO_3.2NH_3$$

7. Metal-ammonia Solutions and Their Reactions. (*i*) *Solutions of metals in liquid ammonia.* Metals that have low ionisation energies, low energies of sublimation and high energies of solvation, alkali metals, alkaline earth metals, Al, europium and ytterbium readily dissolve in liq. NH_3 and give metal-ammonia solutions. These are blue in colour in dilute and bronze-coloured in concentrated solutions. The absorption spectra of the blue solutions are alike, regardless of the metal dissolved in the solution. The blue solutions are excellent electrolytes. The bronze-coloured solutions (*i.e.* concentrated solutions) have high thermal coefficient of electrical conductance approaching that of the free metal whereas the electrolytic conductance of the blue-coloured solutions (*i.e.* dilute solutions) is of the magnitude of an electrolytic solution.

By simply evaporating the alkali metal-ammonia solutions, alkali metal can be recovered. Alkaline earth metal-ammonia solutions on evaporation give hexammonate, $M(NH_3)_6$, where M = alkaline earth metal, indicating that the solvent action is chemical.

All metal ammonia solutions, upon long standing or in presence of a suitable catalyst, undergo decomposition and give H_2 gas and the metal amide.

$$M + nNH_3 \longrightarrow M(NH_2)_n + \frac{n}{2}H_2 \uparrow$$

This decomposition reaction proceeds rapidly in presence of small amounts of certain catalysts such as Fe_2O_3 or Pt-black.

The blue metal ammonia dilute solutions are paramagnetic in character (*i.e.* if these solutions are placed in magnetic field, they will be attracted into the field), indicating the presence of unpaired electrons. As the concentration of the solutions increases, the magnetic susceptibility per mole of the solute decreases.

It is now believed that in concentrated solutions the electrons are largely free, *i.e.*

$$M \rightleftarrows M^+ + e^- \qquad ...(1)$$

On the other hand in dilute solutions the electrons and the metal ions, M^+ are solvated (*i.e.* ammoniated) reversibly as :

$$M^+ + xNH_3 \rightleftarrows \underset{\text{Ammoniated cation}}{[M(NH_3)_x]^+} \qquad ...(2)$$

$$e^- + yNH_3 \rightleftarrows \underset{\text{Ammoniated electron}}{[e(NH_3)_y]^-} \qquad ...(3)$$

Thus the properties of metal-ammonia solutions are best accounted for in terms of an equilibrium represented by (4) obtained by adding (1), (2) and (3),

$$e^- + M + (x + y)\ NH_3 \rightleftarrows [M(NH_3)]^+ + [e(NH_3)_y]^-$$

or

$$M \overset{NH_3(l)}{\rightleftarrows} M^+_{(am)} + e^-_{(am)} \qquad ...(4)$$

where the ammoniated cation and ammoniated electron have been represented as $M^+_{(am)}$ and $e^-_{(am)}$ respectively.

The electrons are believed to occupy the cavities in the solvent surrounded by NH_3 molecules whose protons are oriented towards the free electrons. As the concentration increase, the $M^+_{(am)}$ cations tend to bond together by $e^-_{(am)}$ into aggregates such as M_2, M_3, etc. In the concentrated bronze-coloured solutions, the $M^+_{(am)}$ cations and $e^-_{(am)}$ are bonded together in a manner which is similar to that of ordinary metal ions of the same size in a molten metal.

(ii) Reactions of metal-ammonia solutions. Since the blue metal-ammonia solutions contain highly mobile, ammoniated electrons, such solutions can act as strong reducing agents, particularly for reducing those substances which are soluble in liq. NH_3.

Now since the reducing agents which are stronger than hydrogen, react with water to liberate hydrogen, these cannot generally be used in aqueous solutions. It is for this reason that many metal ammonia solutions which are strong reducing agents are widely used to bring about a number of reductions reaction. Some typical reactions are mentioned below :

(*a*) If metal-ammonia solution is treated with non-metallic element, reduction products are obtained by direct combination.

(*b*) Metal-ammonia solutions are quickly decolorised by ammonium salts or by other acids which are reduced.

$$NH_4^+ + e^-_{(am)} \longrightarrow NH_3 + \tfrac{1}{2} H_2$$

This further supports the view that the blue colour is due to ammoniated electrons.

$$SO_2\begin{cases} NH_2 \\ NH_2 \end{cases} + 2e^-_{(am)} \longrightarrow SO_2\begin{cases} NH^- \\ NH^- \end{cases} + H_2$$

(*c*) Many organic compounds undergo reduction in metal ammonia solutions.

$$\underset{\text{Acetylenic hydrocarbon}}{RC \equiv CH} + e^-_{(am)} \xrightarrow[\text{NH}_3(l)\text{ not in excess}]{\text{Metal dissolved in}} \underset{\text{Acetylide ion}}{RC \equiv C^- + H}$$

$$RC \equiv CH + \tfrac{1}{2}H_2 \xrightarrow[\text{in NH}_3(l)\text{ in excess}]{\text{Metal dissolved}} \underset{\text{Ethylenic derivatives}}{RC = CH_2}$$

(*d*) A large number of organometallic compounds of various types can react with metal-ammonia solutions and form a series of compounds containing metal bonds.

Liquid Sulphur Dioxide as Solvent

Liquid sulphur dioxide is a water-like solvent and as such is used extensively for carrying out a number of chemical reactions. Its boiling point is –10°C and freezing point –75.5°C. Since the liquid range is sufficiently high, it can be used as a solvent.

The auto-ionisation occurring in liq. SO_2 takes place as follows :

$$SO_2 + SO_2 \rightleftarrows SO^{2+} + SO_3^{2-}$$

The thionyl ion (SO^{2+}) is analogous to the hydronium ion (H_3O^+) and NH_4^+ ion and SO_3^{2-} ion corresponds to OH^- and NH_3^- ion of the aqueous and liq. NH_3 system.

$$H_2O + H_2O \rightleftarrows H_3O^+ + OH^-$$

$$NH_3 + NH_3 \rightleftarrows NH_4^+ + NH_2^-$$

Dipole moment of SO_2 is 1.61 D. Its dielectric constant is 17.27 (at –16.5°C) and it can dissolve many salts.

Solubility of Inorganic Materials in liq. SO_2

In general, iodides are most soluble followed by thiocyanates. The solubilities of alkali metal halides decrease in the order MI > MBr > MCl > MF. Alkali metal sulphites and acetates are soluble. Metal sulphates, alkali metal thiocyanates and cyanides are soluble. Metal sulphates, sulphides, oxides and hydroxides are virtually insoluble. Many of the ammonium, thallium and mercuric salts are soluble. The substances such as IBr, PBr_3, CCl_4, $SiCl_4$, $GeCl_4$ and $SnCl_4$ are quite soluble in liq. SO_2.

Many substances such as Br_2, ICl, BCl_3, PCl_3, $AsCl_3$, CS_2, $POCl_3$, $SOCl_2$, $SO(CH_3COO)_2$ and many sulphuryl compounds, SO_2X_2, are completely miscible with liq. SO_2. The miscible nature of $SOCl_2$ has been utilised in the preparation of a large number of thionyl derivatives by the metathetical reactions of liq. SO_2.

Solubility of Organic Compounds in liq. SO_2

As might be expected from the value of dielectric constant of SO_2, covalent organic compounds are more soluble in liq. SO_2 than the ionic compounds. Aromatic hydrocarbons and alkenes are more soluble than the alkanes. Liq. SO_2 has also been used as a medium for conducting a number of organic reactions such as Friedel-Crafts' sulphonation and bromination reactions.

Conductivity of Salt Solutions in liq. SO_2

The solutions of alkali metals, ammonium and teraalkyl ammonium chlorides, bromides, iodides and thiocyanates in liq. SO_2 are moderately good conductors. On the other hand the mono-, di- and tri-alkyl ammonium salts are very poor conductors.

The conductivity of electrolyte solutions in liq. SO_2 roughly increases with the size of the cation. Thus the order of conductivity may be shown as: $Na^+ < NH_4^+ < K^+ < Rb^+ < (CH_3)_3S^+ < (CH_3)_4N^+$ etc. The conductivity of the anions increases in the order : $SCN^- < ClO_4^- < Cl^- < I^-$ etc.

Many covalent compounds such as Br_2, I_2, PBr_5, $AsBr_3$, $SbCl_5$, S_2Br_2, IBr, ICl, ICl_3, $SOBr_2$, $(C_6H_5)_3CCl$ etc. behave as electrolytes in liq. SO_2.

The solutions of many tri-substituted hydronium salts such as $[(C_2H_5)_3O]BF_4$, $[(CH_3)_3O]BF_4$ and $[(C_2H_5)_3O]_5SbCl_6$ in liq. SO_2 are moderately good conductors.

Chemical Reactions That Can Occur in Liquid Sulphur Dioxide

Chemical reactions that can occur in liq. SO_2 can be classified under the following headings :

1. Acid-base Reactions. Auto-ionisation of liquid SO_2 is in accordance with following equation.

$$2\ SO_3 \rightleftarrows SO^{2+} + SO_3^{2-}$$

The compounds which contain or make available sulphite ions (SO_3^{2-}) on dissolving in liquid SO_2 would act as typical bases. Therefore, the alkali metal sulphites should behave as typical bases in liquid sulphur dioxide.

Similarly, such compounds which contain or make available thionyl ion (SO^{2+}) would act as typical acids in liquid SO_2. Therefore, thionyl compounds ($SOCl_2$, $SOBr_2$) would act as acids.

Some examples of acid-base reactions are as follows :

$$SOCl_2 + Cs_2SO_3 \xrightarrow{\text{Liq. } SO_2} 2CsCl + 2SO_2$$

$$SOBr_2 + [(CH_3)_4N]_2SO_3 \longrightarrow 2[(CH_3)_4N]Br + 2SO_2$$

$$\underset{\text{Acid}}{SO(SCN)_2} + \underset{\text{Base}}{K_2SO_3} \longrightarrow \underset{\text{Salt}}{2KSCN} + \underset{\text{Solvent}}{2SO_2}$$

2. Solvolytic Reactions. Solvolytic reactions of salts in liquid SO_2 are more complex than the true solvolytic processes in many protonic solvents. For example :

$$Zn(C_2H_5)_2 + 2SO_2 \longrightarrow ZnSO_3 + SO(C_2H_5)_2$$

$$PCl_5 + SO_2 \longrightarrow POCl_3 + SOCl_2$$

$$PBr_5 + SO_2 \longrightarrow POBr_3 + SOBr_2$$

$$NbCl_5 + SO_2 \longrightarrow NbOCl_3 + SOCl_2$$

$$UCl_6 + 2SO_2 \longrightarrow UO_2Cl_2 + 2SOCl_2$$

3. Metathetical Reactions. There are a large number of neutralizations or acid-base reactions which may better be described as metathetical reactions. A few examples are :

$$5CH_3COOAg + SOCl_2 \xrightarrow{\text{Liq. } SO_2} 2AgCl(s) + (CH_3COO)_2SO$$

$$2CH_3COONH_4 + SOCl_2 \xrightarrow{\text{Liq. } SO_2} 2NH_4Cl(s) + (CH_3COO)_2SO$$

$$3KI + SOCl_2 \xrightarrow{\text{Liq. } SO_2} 2KCl + SOI_2$$

4. Reactions with Organic Compounds. Because many organic compounds are soluble in liquid SO_2 and because it is itself inert to many of them, a large number of organic reactions have been carried out in this solvent. In such reactions there is a sulphonation of aomatic hydrocarbons by sulphur trioxide or chlorosulphonic acid.

$$C_6H_6 + SO_3 \xrightarrow{\text{Liq. } SO_2} C_6H_5SO_2OH$$

$$C_6H_6 + ClSO_3H \xrightarrow{\text{Liq. } SO_2} C_6H_5SO_2OH + HCl$$

5. Complex Formation Reactions. A large number of complex ion formation in liquid SO_2 have been reported. For example, the solubility of iodine in liquid SO_2 is greatly increased by the addition of potassium or rubidium iodide. This is due to the formation of complexes like RbI_3 or KI_3.

$$KI + I_2 \longrightarrow KI_3$$

$$RbI + I_2 \longrightarrow RbI_3$$

Similarly, the solubilities of potassium iodide and mercuric iodide in liquid SO_2 are greatly increased owing to the formation of complexes.

$$HgI_2 + 2KI \longrightarrow K_2HgI_4$$

6. Amphoterism in Liquid Sulphur Dioxide. Several reactions have been reported in liquid SO_2 which may be compared to the behaviour of amphoteric hydroxides in aqueous system. When a solution of tetramethylammonium sulphate is added to a solution of aluminium chloride in liquid SO_2, a gelatinous ppt. of aluminium sulphite is formed.

$$2AlCl_3 + 3(Me_4N)_2SO_4 \longrightarrow \underset{\text{ppt.}}{Al_2(SO_3)_3} + 6Me_4NCl$$

Similar reaction has been reported with gallium trichloride. It gives a ppt. with tetramethyl ammonium sulphite.

$$2GaCl_3 + 3\,(Me_4N)_2SO_3 \longrightarrow \underset{\text{ppt.}}{Ga_2(SO_3)_3} + 6Me_4NCl$$

7. Oxidation Reduction Reactions. In many oxidation- reduction reactions studied in SO_2, the solvent acts merely as an inert carrier. A solution of $FeCl_2$ quantitatively oxidizes KI to I_2.

$$2FeCl_3 + 2KI \longrightarrow 2FeCl_2 + 2KCl + I_2$$

Similarly a solution of $SbCl_5$ oxidises KI to I_2

$$6KI + 3SbCl_5 \longrightarrow 2K_3SbCl_6 + SbCl_3 + 3I_2.$$

Questions with Answers

Q.1 (*a*) Classify the following solvents as ionising and non-ionising solvents: HF, liq. SO_2, liq. N_2O_4, H_2SO_4, $POCl_3$, pyridine and acetic acid.

(*b*) Classify the following solvents as protonic (protic) or non-protonic (aprotic) solvents: HF, liq. SO_2, C_6H_6, H_2SO_4 and liq. NH_3.

Ans : (*a*) *Ionising solvents* : HF, H_2SO_4, acetic acid

Non-ionising solvents : liq. SO_2, liq. N_2O_4, $POCl_3$, pyridine.

(*b*) *Protic solvents* : HF, H_2SO_4, liq. NH_3

Aprotic solvents : liq. SO_2, C_6H_6

Q.2 What will be the effect of adding (*a*) KNH_2 (*b*) Li_3N (*c*) CH_3COOK to liq. NH_3 on the acidity/bascity of the solvent.

Ans : (*a*) Since KNH_2 gives NH_2^- ions (amide ions) in liq. NH_3, it behaves as a base in this solvent. Thus the addition of KNH_2 to liq. NH_3 increases the basic character of the solvent

$$KNH_2 \xrightarrow{\text{Liq.}NH_3} K^+ + NH_2^-$$

(*b*) Since Li_3N produces nitride ions (N^{3-} ions) in liq. NH_3, it behaves as a base in this solvent. Thus the addition of Li_3N to liq. NH_3 increases the basic nature of the solvent.

$$Li_3N \xrightarrow{\text{Liq.}NH_3} 3Li^+ + N^{3-}$$

(*c*) Since CH_3COOK does not have any of the ions *viz* NH_4^+, NH_2^-, NH^{2-}, N^{3-}, the addition of this compound to liq. NH_3 has no effect on the acidity/bascity of liq. NH_3.

Q.3 Guess the reaction between $Zn(NH_2)_2$ and CH_3COONH_4.

Ans : Auto-ionisation of liq. NH_3 gives NH_4^+ (ammono acid ion) and NH_2^- (ammono base ion) ions. Thus Zn $(NH_2)_2$ will act as a base and CH_3COONH_4 will act as an acid in liq. NH_3 and the reaction between these substances is a neutralisation reaction.

Ammono base		*Ammono acid*		*Salt*		*Solvent*
$Zn(NH_2)_2$	+	$2CH_3COONH_4$	$\longrightarrow$	$Zn(CH_3COO)_2$	+	$4NH_3$

Zn^{2+} ions obtained by the dissociation of $Zn(CH_3COO)_2$ will react with $4NH_3$ to give tetramminezinc (II) ion, $[Zn(NH_3)_4]^{2+}$

$$Zn^{2+} + 4NH_3 \longrightarrow [Zn(NH_3)_4]^{2+}$$

Q.4 How do the following solutes behave in the solvents given in parentheses ? (*a*) KNH_2 (NH_3) (*b*) $(NH_4)_2SO_4$ (NH_3) (*c*) $SiCl_4$ (NH_3) (*d*) SO_2Cl_2 (NH_3) (*e*) PCl_5 (SO_2) (*f*) AgCl (NH_3) (*g*) $Zn(C_2H_5)_2$ (SO_2) (*h*) Cs_2SO_3 (SO_2) (*i*) $NbCl_5$(SO_2).

Ans : (*a*) KNH_2 behaves as a base in liq. NH_3, since it gives NH_2^- (base anion) in this solvent.

$$KNH_2 \xrightarrow{\text{Liq.}NH_3} K^+ + NH_2^-$$

(*b*) $(NH_4)_2SO_4$ behaves as an acid in liq. NH_3, since NH_4^+ ions are produced.

$$(NH_4)_2SO_4 \xrightarrow{\text{Liq.}NH_3} 2NH_4^+ + SO_4^{2-}$$

(*c*) $SiCl_4$ is covalent and hence undergoes ammonolysis reaction in liq. NH_3. In this reaction the concentration of NH_2^- ions is increased due to the prodution of $Si(NH_2)_4$.

$$SiCl_4 + 8NH_3 \longrightarrow Si(NH_2)_4 + 4NH_4Cl$$

$$\Big\downarrow 0°C$$

$$Si(NH)(NH_2)_2 \xrightarrow{1200°} Si_3N_4$$

(*d*) SO_2Cl_2 undergoes ammonolysis reaction in liq. NH_3, since $SO_2(NH_2)_2$ is produced. This compound increases the concentration of NH_2^- ions.

$$SO_2Cl_2 + 4NH_3 \longrightarrow SO_2(NH_2)_2 + 2NH_4Cl$$

(*e*) In liq. SO_2, PCl_5 undergoes solvolysis reaction and $POCl_3$ is formed.

$$PCl_5 + SO_2 \longrightarrow POCl_3 + SOCl_2$$

(*f*) AgCl dissolves in liq. NH_3 due to the formation of a complex compound, $[Ag(NH_3)_2]Cl$.

$$AgCl + 2NH_3 \longrightarrow [Ag(NH_3)_2]Cl$$

(*g*) The reaction between $Zn(C_2H_5)_2$ and liq. SO_2 is an example of solvolysis reaction

$$Zn(C_2H_5)_2 + 2SO_2\,(l) \longrightarrow ZnSO_3 + (C_2H_5)_2SO$$

(*h*) Since Cs_2SO_3 gives SO_3^{2-} ions in liq. SO_2, this compound acts as a base in this solvent

$$Cs_2SO_3 \xrightarrow{\text{Liq.}SO_2} 2Cs^+ + SO_3^{2-}\ (\textit{Base ion})$$

(*i*) $NbCl_5$ shows solvolysis reaction in liq. SO_2.

$$NbCl_5 + SO_2\,(l) \longrightarrow NbOCl_3 + SOCl_2$$

Q.5 How will the following solutes react with the solvents given in the parenthesis. State whether the solution formed in each case will be more acidic or more basic than the solvent.

$H_2SO_4(H_2O)$, $H_2SO_4(NH_3)$, $CH_3COOH(H_2SO_4)$, $SOCl_2(SO_2)$ and $CH_3CONH_2(NH_3)$.

Ans : The reactions between the solutes and the solvents are given below. The nature of the solution is given in parenthesis.

Solute		*Solvent*					
H_2SO_4	+	H_2O	$\longrightarrow$	HSO_4^-	+	$\boxed{H_3O^+}$	(*Acidic*)
H_2SO_4	+	NH_3	$\longrightarrow$	HSO_4^-	+	$\boxed{NH_4^+}$	(*Acidic*)
CH_3COOH	+	H_2SO_4	$\longrightarrow$	$CH_3COOH_2^+$	+	$\boxed{HSO_4^-}$	(*Basic*)
$SOCl_2$	+	SO_2	$\longrightarrow$	$2Cl^-$	+	$\boxed{SO^{2+}}$	(*Acidic*)
CH_3CONH_2	+	NH_3	$\longrightarrow$	CH_3COONH^-	+	$\boxed{NH_4^+}$	(*Acidic*)

Q.6 Complete the following equations :

(*i*) $C_2H_5OH + HF \longrightarrow$ .... **(Guru Nanak Dev. 96)**

(*ii*) $PbI_2 + KNH_2 \xrightarrow{\text{Liq. NH}_3}$ .... **(P.U. 1977)**

(*iii*) $Zn(NO_3)_2 + KNH_2 \xrightarrow{\text{Liq. NH}_3}$ .... **(P.U. 1997)**

(*iv*) $Na + NH_4Cl \xrightarrow{\text{Liq. NH}_3}$ .... **(Kurukshetra 1977)**

Ans : (*i*) In HF (solvent), C_2H_5OH (solute) behaves as a base, since HF_2^- ions (solvent anions) are produced.

C_2H_5OH (*solute*) + 2HF (*solvent*) $\longrightarrow C_2H_5OH_2^+ + HF_2^-$ (*Solvent anion - base anion*)

(*ii*) PbI_2 is a metal salt. It reacts with the base *viz* KNH_2 and gives the ppt. of lead imide (PbNH).

$$PbI_2 + KNH_2 \text{ (Ammono base)} \xrightarrow{\text{Liq. NH}_3} PbNH\downarrow + KI + HI$$

(*iii*) $Zn(NO_3)_2$ is a metal salt. It reacts with the base, KNH_2 and gives the ppt. of $ZnNH_2$.

$$Zn(NO_3)_2 + 2KNH_2 \text{ (Ammono base)} \xrightarrow{\text{Liq. NH}_3} Zn(NH_2)_2 + 2KNO_3$$

(*iv*) In liq. NH_3, Na metal reduces NH_4Cl to H_2.

$$\underset{(Na\,=\,0)}{2Na} + \underset{(H\,=\,+1)}{2NH_4Cl} \xrightarrow{\text{Liq. NH}_3} \underset{(Na\,=\,+1)}{2NaCl} + 2NH_3 + \underset{(H\,=\,0)}{H_2}$$

Q.7 Complete the following reactions in liq. SO_2.

(*i*) $HgI_2 + KI \longrightarrow$ ....

(*ii*) $SbCl_5 + NOCl_2 \longrightarrow$ ....

(*iii*) $Cr_2(SO_3)_3 + SOCl \longrightarrow$ ....

(*iv*) $C_6H_6 + SO_3 \longrightarrow$ .... **(P.U. April 1996, Kurukshetra 1994)**

Ans : (*i*) When HgI_2 is dissolved in liq. SO_2 in presence of KI, the solubility of HgI_2 in liq. SO_2 is increased due to the formation of $K_2[HgI_4]$ (a complex compound).

$$HgI_2 + 2KI \xrightarrow{\text{Liq. NH}_3} K_2[HgI_4]$$

(*ii*) The reaction between $SbCl_5$ (covalent halide) and NOCl in liq. SO_2 leads to the formation of hexachloro complex, $NO^+[SbCl_6]^-$

$$SbCl_5 + NOCl \xrightarrow{\text{Liq. } NH_3} NO^+[SbCl_6]^-$$

(*iii*) Since $Cr_2(SO_3)_3$ contains SO_3^{2-} ions, this compound acts as a base in liq. SO_2. Similarly, since $SOCl_2$ gives SO^{2+} ions in liq. SO_2, this salt behaves as an acid in liq. SO_2. Thus the reaction between $Cr_2(SO_3)_3$ and $SOCl_2$ in liq. SO_2 is an acid-base (neutralisation) reaction and gives salt and solvent (*i.e.* SO_2)

$$SOCl_2(\textit{acid}) + CsSO_3\ (\textit{base}) \xrightarrow{\text{Liq. } SO_2} 2CsCl\ (\text{salt}) + 2SO_2\ (\text{solvent})$$

(*iv*) In presence of liq. SO_2, C_6H_6 undergoes sulphonation.

$$C_6H_6 + SO_3 \xrightarrow{\text{Liq. } SO_2} C_6H_5SO_3H.$$

Q.8 Show whether each of the following compounds acts as an acid or a base in liq. HF.

BF_3, SbF_5, H_2O, CH_3COOH, C_6H_6.

Ans : We know that HF undergoes self-ionisation as follows:

$$3HF \rightleftharpoons \underset{\substack{\text{Solvent cations}\\ \text{(Acid ions)}}}{H_2F^+} + \underset{\substack{\text{Solvent anions}\\ \text{(Base ions)}}}{HF_2^-}$$

This equation shows that the compounds producing H_2F^+ ions in liq. HF act as acids in this solvent while the compounds giving HF_2^- ions behave as bases in this solvent. Thus:

$$\underset{\text{(Acid)}}{BF_3} + 3HF \rightleftharpoons H_2F^+ + BF_4^-$$

$$\underset{\text{(Acid)}}{SbF_5} + 2HF \rightleftharpoons H_2F^+ + SbF_6^-$$

$$\underset{\text{(Base)}}{H_2O} + 2HF \rightleftharpoons H_3O^+ + HF_2^-$$

$$\underset{\text{(Base)}}{CH_3COOH} + 2HF \rightleftharpoons CH_3COOH_2^+ + HF_2^-$$

$$\underset{\text{(Base)}}{C_6H_6} + 2HF \rightleftharpoons C_6H_7^+ + HF_2^-$$

Q.9 Account for the following, giving appropriate reasons.

(*i*) K_2SO_3 is soluble in liq. SO_2. **[*M.D. Rohtak 1997*]**

(*ii*) SO_2 acts both as a Lewis acid and as a Lewis base. **[*M.D. Rohtak 1998*]**

(*iii*) The electrical conductivity of liq. NH_3 is increased when NH_4Cl is dissolved in it. **[*M.D. Rohtak 1997, Agra 1993*]**

(*iv*) A liq. NH_3 solution of phenolphthalein is colourless but becomes red coloured on addition of KNH_2. **[*M.D. Rohtak 1994*]**

(*v*) Liq. NH_3 exhibits association whereas liq. SO_2 does not. **[*M.D. Rohtak 1992, Punjab 1996*]**

(*vi*) Acetamide behaves as a base in aqueous solution but shows acidic properties in liq. NH_3.

(*vii*) HNO_3 acts as an acid in aqueous solution but in HF it behaves as a base. **[*H.P.W. 1997*]**

(*viii*) CH_3COOH behaves as a base in HF but SbF_6 behaves as an acid in this solvent **[*Guru Nanak pev 1997*]**

(*ix*) Liq. NH_3 behaves as a differentiating solvent for NaH, $NaNH_2$ and $NaOC_2H_5$.

(*x*) Solution of sodium metal in liq. NH_3 is blue and has strong reducing property and high electrical conductance. **[*Punjab 1996*]**

Ans : (*i*) K_2SO_3 is soluble in liq. SO_2, since this compound forms a solvate, $K_2SO_3.SO_2$ (an addition compound) with liq. SO_2. This solvate contains one molecule of SO_2 as sulphur dioxide of crystallisation. In this solvate, K_2SO_3 acts as a solute.

(*ii*) In the formation of solvates like $CaI_2.4SO_2$, $LiI_2.2SO_2$ etc, SO_2 donates electrons to the solutes and hence acts as Lewis base. In the formation of solvates like $C_5H_5N.SO_2$, $C_6H_7N.SO_2$. SO_2 accepts electrons from the solutes and hence acts as Lewis acid. Thus SO_2 acts both as a Lewis acid and as a Lewis base.

(*iii*) Since NH_4Cl gives NH_4^+ ions in liq. NH_3, this compound acts as an acid in this solvent. Due to the production of NH_4^+ ions, the concentration of these ions in liq. NH_3 increases. The increase of NH_4^+ ions increases the electrical conductivity of liq. NH_3.

(*iv*) Addition of KNH_2 to liq. NH_3 increases the concentration of NH_2^- ions and hence this solution behaves as an ammono base. As soon as the concentration of NH_2^- ions increases in solution, the solution behaves as a base and hence imparts red colour to phenolphathlein just as an aqueous basic solution imparts red (pink) colour.

(*v*) Due to small size and high electronegativity of N-atom, many molecules of liq. NH_3 get associated together through H-bonding. This type of association is not possible in case of liq. SO_2, since sulphur atom has bigger size. (N = 0.75 Å, S = 1.02Å) and lower electronegativity (N = 3.0, S = 2.5) than N-atom.

(*vi*) Since acetamide, CH_3CONH_2 gives OH^- ions in aqueous solution, this compound behaves as a base in water. On the other hand, since CH_3CONH_2 gives NH_4^+ ions in liq. NH_3, it behaves as an acid in this solvent.

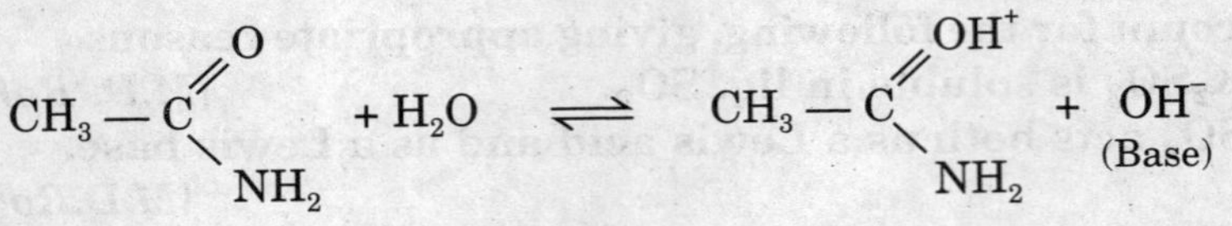

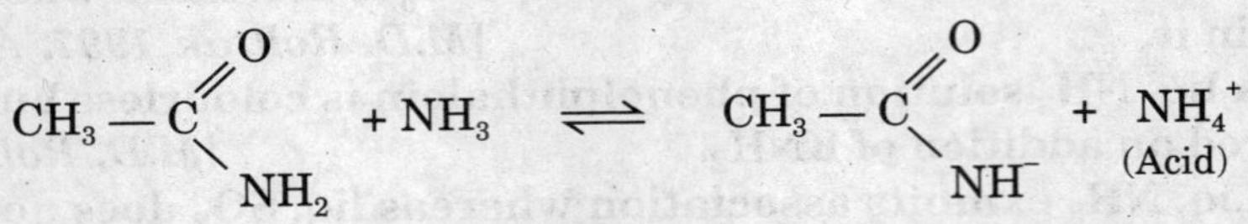

(*vii*) Since, in aqueous solution, HNO_3 gives H^+ or H_3O^+ ions, this compound behaves as an acid in this solution. On the other hand HNO_3 acts as a base in HF, since, in this solvent, this compound gives HF_2^- ions.

$$HNO_3 + H_2O \longrightarrow NO_3^- + H_3O^+ \text{ (Acid)}$$

$$HNO_3 + 2HF \longrightarrow H_2NO^+_3 + HF_2^- \text{ (Base)}$$

(*viii*) CH_3COOH gives HF_2^- ions in HF by accepting H^+ ion.

$$CH_3COOH + 2HF \longrightarrow CH_3COOH_2^+ + HF_2^- \text{ (Base)}$$

SbF_5 ionises as an acid in HF by accepting F^- ion. Thus SbF_5 gives H_2F^+ ions

$$SbF_5 + 2HF \longrightarrow SbF_6^- + H_2F^+ \text{ (Acid)}$$

Here the self ionisation of HF can be noted as given below:

$$3HF \longrightarrow \underset{\substack{\text{Solvent cations} \\ \text{(Acid ions)}}}{H_2F^+} + \underset{\substack{\text{Solvent anions} \\ \text{(Base ions)}}}{HF_2^-}$$

(*ix*) When strong bases like NaH, $NaNH_2$ and $NaOC_2H_5$ react with H_2O (solvent), a proton is donated by H_2O molecule to these bases and NaOH is produced.

(a) $$H_2O + NaH \longrightarrow H_2 + NaOH$$

or $$H_2O + H^- \longrightarrow H_2 + OH^- \quad (H^+ \text{ transferred from } H_2O \text{ to } H^-)$$

(b) $$H_2O + NaNH_2 \longrightarrow NH_3 + NaOH$$

or $$H_2O + NH_2^- \longrightarrow NH_3 + OH^- \quad (H^+ \text{ transferred from } H_2O \text{ to } NH_2^-)$$

(c) $$H_2O + NaOC_2H_5 \longrightarrow C_2H_5OH + NaOH$$

or $$H_2O + OC_2H_5^- \longrightarrow C_2H_5OH + OH^- \quad (H^+ \text{ transferred from } H_2O \text{ to } OC_2H_5^-)$$

Thus we see that these bases are levelled to the base strength of OH^- ions, and hence these bases cannot be differentiated as bases in aqueous solution. On the other hand, since liq. NH_3 is a poor proton donor, these bases can be differentiated as bases in liq. NH_3. Thus liq. NH_3 behaves as a differentiating solvent for NaH, $NaNH_2$ and $NaOC_2H_5$.

(*x*) Blue colour is due to an absorption band having its maximum at 5000Å. Strong reducing property and high electrical conductance shown by the solution of sodium in liq. NH_3 is due to the production of ammoniated cations, $[Na(NH_3)_x]^+$ and ammoniated electron, $[e\ (NH_3)_y]^-$.

$$Na \xrightarrow{\text{Liq. } NH_3} Na^+ + e^-$$

$$Na^+ + x\ NH_3 \longrightarrow [Na\ (NH_3)_x]^+$$

$$e^- + y\ NH_3 \longrightarrow [e\ (NH_3)_y]^-$$

Q.10 What are the principal species formed in the solutions of NaOH, $(NH_4)_2SO_4$ and H_2O in water, anhydrous H_2SO_4 and liq. NH_3?

Ans : NaOH (*a*) *In water* : In water, NaOH dissociates to give Na^+ and OH^- ions.

$$NaOH \xrightarrow{\text{Aqueous solution}} Na^+ + OH^-$$

(*b*) *In anhydrous* H_2SO_4. OH^- ions formed as above further react with H_2SO_4 to produce HSO_4^- and H_3O^+ ions.

$$OH^- + 2H_2SO_4 \longrightarrow 2HSO_4^- + H_3O^+$$

(*c*) *In liq.*NH_3 : In liq. NH_3, no species are formed.

$(NH_4)_2SO_4$ (*a*) *In water* : NH_4^+ and SO_4^{2-} ions are formed.

$$(NH_4)_2SO_4 \xrightarrow{\text{Aqueous solution}} 2NH_4^+ + SO_4^{2-}$$

NH_4^+ ions formed as above react with H_2O to give H_3O^+ ions.

$$NH_4^+ + H_2O \rightleftharpoons NH_4OH + H_3O^+$$

(*b*) *In anhydrous* H_2SO_4: NH_4^+ and SO_4^{2-} ions are obtained.

$$(NH_4)_2SO_4 \longrightarrow 2NH_4^+ + SO_4^{2-}$$

(*c*) *In liq.*NH_3: NH_4^+ and SO_4^{2-} ions are formed. Due to the production of NH_4^+ ions, the solution becomes acidic.

$$(NH_4)_2SO_4 \xrightarrow{\text{Liq. } NH_3} 2NH_4^+ + SO_4^{2-}$$

H_2O. (*a*) *In water* there will be no reaction

(*b*) *In anhydrous* H_2SO_4. Following reaction will occur.

$$H_2O + H_2SO_4 \longrightarrow HSO_4^- + H_3O^+$$

(*c*) *In liq.* NH_3. NH_4^+ ions will be obtained.

$$H_2O + NH_3 \longrightarrow NH_4^+ + OH^-$$

Q.11 Predict the behaviour of urea in water, liq. NH_3 and anhydrous H_2SO_4.

Ans : Urea is NH_2CONH_2.

In water : NH_2CONH_2 is expected to behave as a non-electrolyte.

In liq. NH_3 : NH_2CONH_2 behaves as a weak acid in liq. NH_3, since this compound produces NH_4^+ ion in this solvent.

$$NH_2CONH_2 + NH_3\,(l) \rightleftharpoons NH_4^+ + NH_2CONH^-$$

In anhydrous H_2SO_4 : When NH_2CONH_2 reacts with H_2SO_4, HSO_4^- ion is produced. Thus NH_2CONH_2 behaves as a base in H_2SO_4.

$$NH_2CONH_2 + H_2SO_4 \longrightarrow NH_2CONH_3^+ + HSO_4^-$$

University Questions

1. (*a*) Giving two examples in each, discuss the following reactions in liq. SO_2 (*i*) Acid - base reaction (*ii*) Precipitation reaction (*iii*) Complex formation reaction

 (*b*) Liq. SO_2 is a better solvent for organic liquids. Explain.

 (*c*) Write self - ionisation of liq. NH_3 by giving two examples.

 (*d*) In what respect is liq. NH_3 a better solvent than H_2O ?

 (*e*) Metal - liquid NH_3 solutions are good reducing agents. Explain.

 (*f*) Self ionisation reactions in H_2SO_4. (*Himachal Pradesh 2000*)

2. What are the advantages of non - aqueous solvents ? Discuss the following reactions occurring in liq. SO_2 medium. (*i*) Acid - base reactions (*ii*) Precipitation reactions (*iii*) Complex formation reactions. (*Kumaon 2000*)

3. How are solvents classified ? Taking liq. NH_3 as solvent, explain the following reactions and point out the differences from water as solvent (*i*) Precipitation reactions (*ii*) Solvolysis reactions (*iii*) Neutralisation reactions (*iv*) Solubility of alkali metals. (*Lucknow 2000*)

4. (*a*) Discuss in detail the classification of solvents.

(*b*) Discuss the following in liquid NH_3 and liquid SO_2.

(*i*) Oxidation-reduction reaction.

(*ii*) Precipitation reaction. (*Nagpur 2002*)

5. Compare the self-ionisation of water; liq. NH_3 and anhydrons acetic acid.

(*Delhi 2002*)

6. (*a*) Predict the behaviour of the following solutes in liquid SO_2 :

(*i*) $SOCl_2$ (*ii*) CS_2SO_3 (*iii*) $(CH_3)_4\ N^+\ Br^-$.

(*b*) Predict the solubility of the following soultes in liquid SO_2.

(*i*) NH_4SCN (*ii*) Na_2S.

(*c*) Cmplete the following equations :

(*i*) $PbI_2 + KNH_2 \xrightarrow{NH_3}$ ....

(*ii*) $SO_2\ Cl_2 + NH_3 \xrightarrow{NH_3}$ .... (*Delhi 2002*)

7. (*a*) Explain chemical reactions in non-aqueous solvents with reference to those orruring in liq. NH_3.

(*b*) Discuss the role of non-protonic solvents in chemical reactions with special refrence to the reactions occuring in liquid SO_2. (*VBS, Purvanchal 2003*)

8. (*a*) What is meant by protic and aprotic solvents ? Give examples.

(*b*) Give only one example for each of the following in liquid sulpher dioxide:

(*i*) Acid-base reaction (*ii*) Redox reaction (*iii*) Precipitation reaction (*iv*) Complex formation reaction (*v*) Reaction with organic compounds. (*Rohidkhand 2003*)

9. Explain the following reaction which take place in liq. SO_2 as medium :

(*i*) Acid base reactions (*ii*) Precipitation reactions. (*Delhi 2003*)

10. (*a*) Classify the solvents on the basis of their proton-donor and proton-acceptor property. Give the following reaction in liquid NH_3.

(*i*) Redox reaction (*ii*) Precipitation reaction.

(*b*) Give the following reactions in liquid SO_2.

(*i*) Solvolysis (*ii*) Complex formation. (*Nagpur 2003*)

11. (*a*) Write notes on (in non-aqueous solvent)

(*i*) Ammonolysis (*ii*) Acid-base reactions (*iii*) Redox reactions

(*b*) What are non-aqueous solvents ? Explain the following reactions in liquid HF.

(*i*) Auto ionisation reaction (*ii*) Precipitation reaction (*iii*) Solvation reactions (*iv*) Reaction with C_2H_5OH. (*Bhopal 2004*)

12. What do you understand by ionising and non-ionising solvents ? Give examples. (*GND Amritsar 2004*)

13. Discuss acid-base reaction in liq. SO_2 and metathetical reactions in liq. HF. (*GND Amritsar 2004*)

14. How do the following reactions proceed ?

(*i*) $HgI_2 + KNH_2 \xrightarrow{NH_2}$ ...... (*ii*) $Cs_2SO_3 + SOCl_2 \xrightarrow{SO_2}$ ....

(*GND Amritsar, 2004*)

15. (*a*) Write down the nature of products when AgCl is treated with Ba $(NO_3)_2$ in liq. NH_3.

(*b*) Complete the reaction :

$Zn\ (C_2H_5)_2 + 2SO_2\ (l) \rightarrow$ ...... (*C.S.J.M Kanpur, 2005*)

16. (*a*) Which of the following are protonic solvents : Liq. SO_2, C_6H_6, H_2SO_4, Liq. NH_3, HF ?

(*b*) What the advantages and disadvantages of liq. NH_3 as a solvent ? Using chemical equations illustrate the following types of reactions in liq. NH_3 : Complex formation and precipitation.

(*H.N. Bahuguna 2005*)

17. Explain amphoteric substances in a non-aqueous solvent. (*C.S.J.M. Kanpur 2006*)

18. What are the advantages of non-aqueous solvents ? Discuss the following reactions occuring in liq. SO_2 medium. Acid-base reactions, precipitation reactions and complex formation reactions. (*H.N. Bahuguna 2006*)

19. Discuss the chemistry of liq. NH_3 as a solvent. What are the advantages and disadvantages of liq. NH_3 as a solvent. (*C.S.J.M. 2007*)

20. Explain the classification of non-aqueous solvents and describe their important characteristics. (*Agra 2007*)

21. (*a*) Which of the following is an amphoteric solvent : H_2SO_4, HCl, H_2O, $CHCl_3$?

(*b*) What are non-aqueous solvents ? Explain different types of chemical reactions in liq. NH_3 or in liq. SO_2. (*Agra 2008*)

22. (*a*) Which of the following is a non-aqueous solvent ? Liq. NH_3, Liq. SO_2, Liq HF.

(*b*) Write the following reactions with examples in liq. SO_2. Pricipitation, acid base neutralisation and solvolysis. (*H.N Bahuguna 2008*)

23. What do you mean by solvation and solvolysis in non-aqueous solvents. Explain in liq. NH_3. (*Gurukul Kangri 2008*)

CHAPTER 14

Principles and Processes of Metallurgy

What are Minerals and Ores ?

Metals occur in nature sometimes *free* (*i.e., in uncombined state*) but mostly in the *combined state* (*i.e., as compounds with other elements*). When a metal is found free or in uncombined state it is said to occur as **native.** Metals found in the native state are copper, silver, gold, mercury and platinum. Others are found in the combined state.

The natural materials found in the earth in which the metals occur in the combined state are called **minerals.** *Those minerals from which the metals may be extracted conveniently and economically are called* **ores.** *Thus all the ores are minerals but all the minerals are not ores.* For example, all clays contain aluminium but these are not ores of aluminium, since the extraction of aluminium metal from these clays is too much coumbersome and not economically worthwhile.

What is Mining ?

The process of taking out the ores from mines is called **mining.**

Occurrence of Metals in Nature

Most metals occur in the combined state. Only those metals which have little or no affinity for oxygen, moisture and other chemical reagents occur in the *free* (or *metallic* or *native*) *state*. Thus : (1) *In native state*. Examples are Ag, Au, Pt (2) *As oxide*. Examples are : bauxite, Al_2O_3; tin stone, SnO_2 ; rutile TiO_2 (3) *As carbonate*. Examples are : magnesite, $MgCO_3$; lime stone, $CaCO_3$ calamine, $ZnCO_3$ (4) *As sulphate*. Examples are : gypsum, $CaSO_4$. $2H_2O$; calestine, $SrSO_4$; barytes, $BaSO_4$ (5) *As silicate*. Examples are : beryl, 3 BeO. Al_2O_3. 6 SiO_2; spodumene, LiAl $(SiO_3)_2$ (6) *As phosphate*. Examples are : amblygonite, LiF. $AlPO_4$; triphylite, (Li, Na) PO_4. $(Fe, Mn)_3$ $(PO_4)_2$ (7) *As sulphide*. Examples are : galena, PbS; zinc blende, ZnS; cinnabar, HgS (8) *As halide*. Examples are : rock salt, NaCl; carnallite, KCl.$MgCl_2$. $6H_2O$; horn silver, AgCl.

Occurrence of Metals in India

India is fortunate to have rich mineral resources. This statement is true, since our country is gifted by nature with rich mineral resources. Some common minerals and the places of their occurrence are given in Table 14.1.

What is Metallurgy?

The process of extracting the metals from their ores and refining them is called **metallurgy.**

Table 14.1. Some common minerals and the places of their occurrence in India

Metal	Minerals found in India	Places of occurrence in India
Lithium	Lepidolite or lithium mica, $(Li, Na, K)_2, Al_2(SiO_3)_3.(F.OH)_2$	Bihar, Rajasthan, Mysore and Kashmir
Sodium	Tincal or native borax, $Na_2B_4O_7.10H_2O$	Laddakh and Kashmir
Potassium	Nitre or Indian salt petre, KNO_3	Arid regions of Haryana and Bengal
Beryllium	Beryl, $3BeO.Al_2O_3.6SiO_2$	Bihar, Orissa, Nellore (Tamil Nadu) and Ajmer-Marwara (Rajasthan)
Magnesium	(*i*) Dolomite, $MgCO_3.CaCO_3$ (*ii*) Magnesite, $MgCO_3$	Both the ores occur in Salem (Madras)
Calcium	(*i*) Gypsum, $CaSO_4.2H_2O$	Jodhpur, Bikaner, Nagpur and Jaisalmer districts of Rajasthan, Trichirapalli and Coimbatore districts of Tamil Nadu and Dode and Baramulla districts of Jammu and Kashmir
	(*ii*) Lime stone (chalk or marble), $CaCO_3$	Jaipur and in some places of Madhya Pradesh .
	(*iii*) Fluorspar, CaF_2	Near Jabbalpur
Barium	Barytes or heavy spar, $BaSO_4$	Alwar and Kurnool
Boron	Borax, $Na_2B_4O_7.10H_2O$	In Ladakh and Puga Valley of Kashmir
Aluminium	Bauxite, $Al_2O_3.2H_2O$	In Katni, Kolahpur, Rawa, Orissa, Jabbulpure and Balurghat district, Madras and Jammu
Tin	Tinstone, SnO_2	Hazaribagh (Bihar) and Orissa
Lead		India has no important deposits of lead ores. Recently lead ores have been located at Zawar mines near Udaipur (Rajasthan) and at Hazaribagh (Bihar)
Antimony	(*i*) Stibnite, Sb_2S_3	In small quantities in Punjab and Mysore
	(*ii*) Cu-Sb sulphide	In Madhya Pradesh
Copper	(*i*) Chalcopyrites or copper pyrites, $CuFeS_2$	Bihar, Orissa, Madhya Pradesh
	(*ii*) Malachite, $Cu(OH)_2.CuCO_3$	Singbhum, Hazaribagh, Jabalpur, Salimabad.
Silver		Gold from Kolar fields and Huthi mines (Karnataka) and lead and zinc ores of Zawar mines (Rajasthan) contain some silver
Gold	Native	Found in the free state in Karnataka, Andhra Pradesh in the alluvial sands of the Ganga, the Brahamputra and the Irrawaddy
Zinc	Zinc blende, ZnS	Zewar zinc mines near Udaipur (Rajasthan) containing 8—10% of Zn and 5% of Pb

Cadmium		Found in the zinc ore of Anji Valley of Kashmir to the extent of 0.175%
Chromium	Chromite or chrome iron ore, $FeO.Cr_2O_3$	Occurs in Karnataka, Bihar and Orissa, near Salem in Tamil Nadu and in the Andaman Islands
Manganese	Pyrolusite, MnO_2	Found in Panch Mahal (Bombay), Karnataka and Singbhum (Bihar)
Iron	Haematite, Fe_2O_3	In Mayurbhanj, Singbhum and Mysore
Cobalt	Cobaltite or cobalt glance, (Co, Fe) AsS	Found at Khetri
Nickel	Smaltite, (Fe, Co, Ni) As_2	Mixed with copper ores at Khetri and with gold bearing rocks of Kolar and Travancore

The process used for the metallurgy (*i.e.* extraction) of a metal depends on the nature of the ore from which the metal is being extracted and also on the properties of the metal. *Thus it is not possible to chalk out a universal method for the extraction of all the metals.*

Various Steps of Metallurgy

Various steps used in the metallurgy of metals from their respective ores are discussed below :

1. Grinding or Crushing the Ore

The rock ores after mining are generally obtained as big lumps. These lumps of the ore are crushed to smaller pieces by *hammering* or by means of *jaw-crusher* and *grinders*.

The jaw-crusher consists of two plates forming a sort of jaw. One of the plates is fixed and the other is movable. The lumps of the ore are put in between the plates and movable plate moves with the help of a mechanical device. The crushed ore is collected below. (See Fig. 14.1).

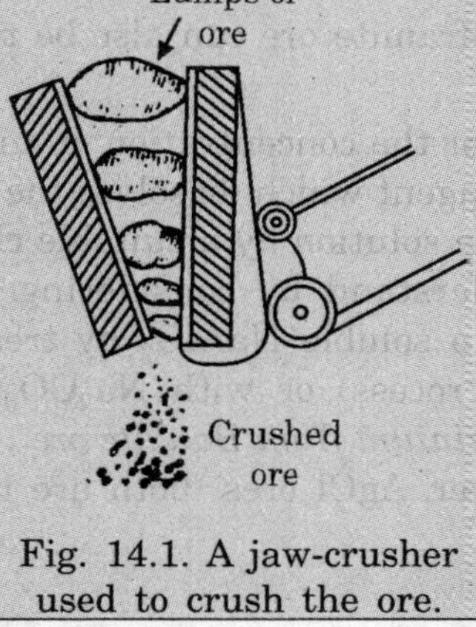

Fig. 14.1. A jaw-crusher used to crush the ore.

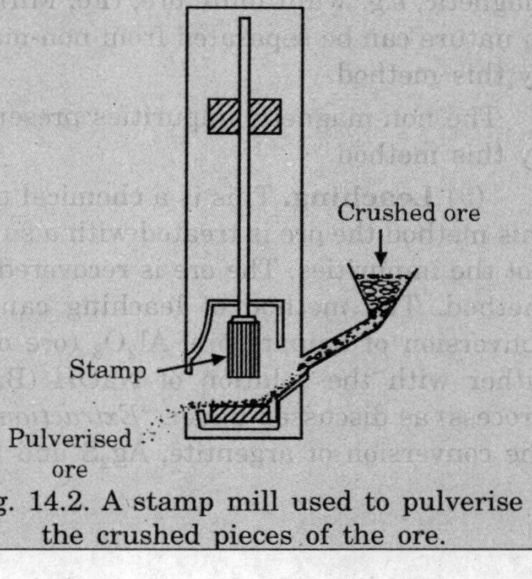

Fig. 14.2. A stamp mill used to pulverise the crushed pieces of the ore.

2. Pulverising the Ore

The crushed pieces of the ore are then taken to *stamp mills* or *pulverisers* in which the crushed pieces of the ore are finely powdered. The coarse ore is fed into the stamp mill from one side (See Fig. 14.2) and the fine powder is taken out by means of a stream of water. Sometimes ball mills are also employed for pulverisation.

3. Concentration of the Ore : Ore-dressing

Generally the ores are contaminated with impurities of earthy matter like sand, clay, lime stone etc. These impurities present in the ore are called **gangue** *or* **martix.** The *removal of these foreign impurities from the ore is called* **concentration** or **ore-dressing** and *the purified are is called* **concentrate.** The concentration is generally carried out by any of the following processes :

(*i*) **Hand picking.** When the impurities present are quite distinct from the ore so as to be easily differentiated with eye, picking by hands is done. This is done usually in the initial stage of concentration.

(*ii*) **Wilfley table and hydraulic classifier methods.** Both these methods are collectly called *gravity separation method*. Both are used for the concentration of *dense ores* like tin stone, SnO_2 (an ore of tin), haematite, Fe_2O_3 (an ore of iron) and are based on the difference in the densities of the gangue particles (*i.e.* impurity particles) and the ore particles.

(*iii*) **Froth flotation process.** This method is commonly used for the concentration of the low grade sulphide ores like *gelana*, PbS (ore of Pb), *copper pyrites*, CuS.FeS (ore of Cu) and *zinc blende*, ZnS (ore of Zn) and is based on the fact that *gangue and ore particles have different degree of wettability with water and oil (e.g. pine oil)—the gangue particles are preferentially wetted by water while the ore particles are wetted by oil.*

(*iv*) **Electromagnetic separation method.** This method is used for the separation of two minerals one of which is magnetic while the other is non-magnetic, *e.g.* wulframite ore, (Fe, Mn)WO_4 (ore of tungsten) which is magnetic in nature can be separated from non-magnetic tin stone ore, SnO_2 (an ore of tin) by this method.

The non-magnetic impurities present in wolframite ore can also be removed by this method.

(*v*) **Leaching.** This is a chemical method for the concentration of an ore. In this method the ore is treated with a suitable reagent which dissolves the ore but not the impurities. The ore is recovered from the solution by a suitable chemical method. The method of leaching can be understand by considering (*a*) the conversion of bauxite ore, Al_2O_3 (ore of Al) into soluble $NaAlO_2$ by treating it either with the solution of NaOH (Baeyer's process) or with Na_2CO_3 (Hall's process) as discussed under. "*Extraction of Aluminium from Bauxite ore*", and (*b*) the conversion of argentite, Ag_2S and born silver, AgCl ores (both are the ores of Ag) into soluble $Na[Ag(CN)_2]$ by treating them with an aqueous solution of NaCN as discussed under *"Extraction of silver"*.

4. Conversion of the Concentrated Ore into Metallic Oxide : Calcination and Roasting.

The conversion of the concentrated ore into the metallic oxide is done by the following methods :

(*i*) **Calcination.** *Calcination is a process in which the ore is heated strongly in the absence of air.* The process does the following things :

(*a*) It removes the volatile impurities like CO_2, SO_2, organic matter, moisture from the ore.

(*b*) It removes water from the hydrated oxide ore, *e.g.*

$$\underset{\text{Bauxite}}{Al_2O_3.2H_2O} \xrightarrow{\text{Calcination}} \underset{\substack{\text{Anhydrous}\\\text{aluminium oxide}}}{Al_2O_3} + 2H_2O$$

$$\underset{\text{Limonite}}{2Fe_2O_3.3H_2O} \xrightarrow{\text{Calcination}} \underset{\substack{\text{Anhydrous}\\\text{iron oxide}}}{2Fe_2O_3} + 3H_2O$$

(*c*) It removes CO_2 from a carbonate ore, *e.g.*

$$\underset{\text{Lime stone}}{CaCO_3} \xrightarrow{\text{Calcination}} \underset{\text{Calcined oxide}}{CaO + CO_2}$$

$$\underset{\text{Malachite}}{Cu(CO_3).Cu(OH)_2} \xrightarrow{\text{Calcination}} \underset{\text{Copper oxide}}{2CuO} + CO_2 + H_2O$$

$$\underset{\text{Dolomite}}{CaCO_3.MgCO_3} \xrightarrow{\text{Calcination}} CaO + MgO + 2CO_2$$

Calcination is generally done in a *reverberatory furnace.*

(*ii*) **Roasting.** *Roasting is a process in which the ore (usually the sulphide ore) is heated strongly in the presence of excess of air. The heating should be done at a temperature below the melting point of the ore.* This process does the following things :

(*a*) It dries the ore.

(*b*) The volatile impurities like CO_2, SO_2 organic matter, moisture are all driven off.

(*c*) The ore is converted into metallic oxide which can easily be reduced to the metal. For example :

$$\underset{\text{Gelana}}{2PbS} + 3O_2 \xrightarrow{\text{Heat}} 2PbO + 2SO_2$$

$$\underset{\text{Cinnabar}}{2HgS} + 3O_2 \xrightarrow{\text{Heat}} 2HgO + 2SO_2$$

$$\underset{\text{Zinc blende}}{2ZnS} + 3O_2 \xrightarrow{\text{Heat}} 2ZnO + 2SO_2$$

$$\underset{\text{Copper glance}}{2Cu_2S} + 3O_2 \xrightarrow{\text{Heat}} 2Cu_2O + 2SO_2$$

Sometimes sulphide ores which are *insoluble* are converted into the corresponding sulphates which are *water soluble*. For example :

$$\underset{\text{(Insoluble)}}{CuS} + 2O_2 \xrightarrow{\text{Heat}} \underset{\text{(Soluble)}}{CuSO_4}$$

$$\underset{\text{(Insoluble)}}{ZnS} + 2O_2 \xrightarrow{\text{Heat}} \underset{\text{(Soluble)}}{ZnSO_4}$$

$$\underset{\text{(Insoluble)}}{PbS} + 2O_2 \xrightarrow{\text{Heat}} \underset{\text{(Soluble)}}{PbSO_4}$$

Roasting is generally done in a *reverberatory furnace* or in a *sinterer* (*or suction reaster*) or in a *blast furnace*.

5. Extraction of the Metal from Roasted or Calcined Ore.

The metal can be obtained from the roasted or calcined ore by any of the following methods :

(*i*) **Carbon reduction method : Smelting.** This method is used for the extraction of Pb, Zn, Fe, Cu and Sn. *Smelting is a process in which the oxide ore in the fused state is reduced by carbon to the free metal. In other words the extraction of a metal from its oxide ore by a process involving melting is called smelting.*

In this method the roasted oxide ore is mixed with carbon (charcoal, coal or coke) and a *flux* (*see below*) and then heated to a high temperature in a suitable furnace. Carbon reduces the oxide ore to the free metal. For reducing cassiterite ore, SnO_2 (an oxide ore of tin) powdered anthracite is used as a reducing agent. The metal obtained may be in the form of vapours (*e.g.* Zn) which are then condensed or in the form of molten metal (*e.g.* Fe, Sn). Examples illustrating the process of smelting are given below :

$$ZnO + C \xrightarrow{\Delta} Zn + \underset{\text{Vapours}}{CO}$$

$$\underset{\text{Cassiterite}}{SnO_2} + \underset{\text{Powered anthracite}}{2C} \longrightarrow \underset{\text{Molten metal}}{Sn} + 2CO$$

$$PbO + C \xrightarrow{\Delta} Pb + CO$$

$$\underset{\text{Heamatite}}{Fe_2O_3} + 3C \xrightarrow{\Delta} 2Fe + 3CO$$

$$CuO + C \xrightarrow{\Delta} Cu + CO$$

Usually the process of smelting is carried out in reverberatory furnace or blast furnace in a controlled supply of air.

In case of metals like Zn which are volatile, smelting cannot be carried out in an open furnace. It is rather done in a fire-clay vertical retorts in which heating is done by producer gas.

A flux *is a substance which is added to the furnace charge (roasted or calcined oxide ore and coke) during the process of smelting to remove the non-fusible impurities of earthy matter (e.g. metallic oxides, silica, silicates etc.) present in the roasted or calcined oxide ore.*

During smelting flux combines with the non-fusible impurity to convert it into a fusible material known as **slag.** Being light, the slag floats over the molten metal from where it can be removed. The slag is a material with low melting point. At high temperature the slag is a liquid which is insoluble in the molten metal.

Thus we see that *the function of flux is to remove the non-fusible impurity as fusible slag.* Fluxes are of two types :

(*a*) *Acidic fluxes* like *silica, borax etc.* which are added to remove *basic impurities* (*e.g.* metallic oxides like CaO, FeO) as fusible slag.

Basic impurity		*Acidic flux*		*Fusible slag*
CaO	+	SiO_2	$\longrightarrow$	$CaSiO_3$
FeO	+	SiO_2	$\longrightarrow$	$FeSiO_3$

(*b*) Basic *fluxes* like *lime stone* ($CaCO_3$), *magnesite* ($MgCO_3$), *haematite* (Fe_2O_3) etc. which are added to the furnace charge to remove the acidic impurities (*e.g.* silica, silicates etc.) as fusible slag.

$$\underset{\text{Acidic impurity}}{SiO_2} + \underset{\text{Basic flux}}{MgCO_3} \longrightarrow \underset{\text{Fusible slag}}{MgSiO_3} + CO_2$$

(*ii*) **Carbon monoxide reduction method.** In some cases CO gas produced in the furnace itself is used as a reducing agent, *e.g.*

$$Fe_3O_4 + 4CO \xrightarrow{\Delta} 3Fe + 4CO_2$$

$$Fe_2O_3 + 3CO \xrightarrow{\Delta} \underset{\text{Molten metal}}{2Fe} + 3CO_2$$

$$PbO + CO \xrightarrow{\Delta} Pb + CO_2$$

$$CuO + CO \xrightarrow{\Delta} Cu + CO_2$$

(*iii*) **Hydrogen reduction method.** H_2 gas has also been used as a reducing agent. For example :

$$WO_3 + 3H_2 \xrightarrow{\Delta} W + 3H_2O$$

$$MoO_3 + 3H_2 \xrightarrow{\Delta} Mo + 3H_2O$$

$$GeO_2 + 2H_2 \xrightarrow{\Delta} Ge + 2H_2O$$

(*iv*) **Magnesium reduction method.** Oxides of some metals are reduced by Mg, *e.g.*

$$Rb_2O_3 + 3Mg \longrightarrow 3MgO + 2Rb$$

Ti is obtained by the reduction of $TiCl_4$ by Mg in an inert atmosphere of argon.

$$TiCl_4 + 2Mg \xrightarrow{\Delta} Ti + 2MgCl_2$$

(*v*) **Aluminium reduction method : Alumino-thermic method.** Certain metallic oxides cannot be reduced by carbon because the affinity of oxygen for the metal is greater than its affinity for carbon. Such metallic oxides can be reduced with the help of aluminum powder. The reduction of a metallic oxide into metal by Al powder is called *alumino-thermic process* and this process has been used to reduce TiO_2, Cr_2O_3 and Mn_3O_4 oxides into Ti, Cr and Mn metals respectively. The details of this process can be studied from "*Extraction of Cr metal from chromite ore, $FeO.Cr_2O_3$*" discussed in chapter 46.

Sr and Ba are obtained by the reduction of their oxides by Al in vacuo.

(*vi*) **Self-reduction method.** This method is also called *auto-reduction method* or *air-reduction method.* If the sulphide ores of some of the less electropositive metals like Hg, Cu, Pb, Sb etc. are heated in air, a part of these sulphide ores is changed into the oxide or sulphate which then reacts with the remaining part of the sulphide ore to give the metal and SO_2. The reactions illustrating the extraction of Hg, Cu and Pb from their sulphide ores are given below :

(*a*) $\underset{\text{Cinnabar}}{2HgS} + 3O_2 \longrightarrow 2HgO + 2SO_2$

$2HgO + HgS \longrightarrow 3Hg + SO_2$

(*b*) $Cu_2S + 3O_2 \longrightarrow 3Cu_2O + 2SO_2$

$2Cu_2O + Cu_2S \longrightarrow 6Cu + SO_2$

(*c*) $\underset{\text{Galena}}{2PbS} + 3O_2 \longrightarrow PbO + 2SO_2$

$2PbO + PbS \longrightarrow 3Pb + SO_2$

The extraction of Pb by heating its sulphide ore (PbS) in air can also be represented as :

$$PbS + 2O_2 \longrightarrow PbSO_4$$

$$PbSO_4 + PbS \longrightarrow 2Pb + 2SO_2$$

(*vii*) **Electrolytic reduction method.** The oxides of the active metals like alkali metals, alkaline earth metals, Al, Zn etc. are very stable and hence in order to reduce their oxides to the free metal, the oxides have to be heated at high temperature with carbon. But at high temperature these metals combine with carbon to form the metallic carbides. Thus such metals cannot be extracted by reducing their oxides with carbon. These are however, extracted by the electrolysis of their oxides, hydroxides or chlorides in the fused state. The metal is liberated at the cathode.

Some other salt may have to be added to lower the melting point of the compound taken.

As an example we may consider the manufacture of sodium metal by the electrolysis of a fused mixture of NaCl and $CaCl_2$ (*Down's process*). Various reactions taking place in the electrolytic cell are :

Fusion : $NaCl \rightleftarrows Na^+ + Cl^-$

These ions are mobile and hence go to the electrodes.

Electrolysis : (*a*) *At cathode (Negative electrode)*

$$Na^+ + e^- \longrightarrow Na \ (\textit{Reduction})$$

(*b*) *At anode (Positive electrode)*

$$Cl^- \longrightarrow Cl + e^- \ (\textit{Oxidation})$$

$$Cl + Cl \longrightarrow Cl_2 \uparrow$$

(*viii*) **Amalgamation process.** This process is used for the extraction of noble metals like silver and gold from their native ores. This process is based on the fact that noble metals dissolve in Hg and form amalgams. Hence the name *amalgamation process*. In this process the mixture of finely powdered ore and water, known as **slurry,** is allowed to flow over copper or brass plates coated with mercury and arranged in a slanting position. The metal particles form an amalgam with Hg and are retained on these plates. The amalgam is scrapped off from the plates and distilled in iron retorts, when mercury distills over leaving the free metal behind.

(*ix*) **Hydro-metallurgy process : Wet process.** In this method concentrated ore is leached with aqueous solution of some suitable chemical reagent whereby the metal is obtained in the form of its soluble salt. The metal is then recovered from this salt solution either by its electrolysis or by adding some electropositive metal to it when the metal to be extracted is obtained as a precipitate. This process is illustrated by the following examples :

(*a*) **Extraction of Cu.** A large heap of *copper glance ore*, Cu_2S is exposed to air and water. In about a year the ore is oxidised to $CuSO_4$. The solution of $CuSO_4$ is either electrolysed or scrap iron is added to it to get Cu as a precipitate.

$$\underset{\text{Copper glance}}{2Cu_2S} + 5O_2 \longrightarrow \underset{\text{Solution}}{2CuSO_4} + 2CuO$$

$$CuSO_4 + \underset{\text{Scrap iron}}{Fe} \longrightarrow FeSO_4 + \underset{\text{ppt.}}{Cu}$$

(*b*) **Extraction of Ag and Au.** For this see the conversion of argentite, Ag_2S or horn silver, AgCl (both are the ores of Ag) and gold into soluble Na $[Ag(CN)_2]$ and $Na[Au(CN)_2]$ complex compounds respectively by the use of an aqueous solution of NaCN as discussed in chapter 46.

6. Purification or Refining of Metals

The metals obtained by any of the methods described above need further purification as they may contain other metals, dissolved oxides, carbon, phosphorus, etc. as impurities.

The refining of metals is done in many ways depending upon the nature of the metal concerned. The common refining methods are :

(*i*) **Liquation.** This method is used to concentrate as well as to purify the metals like Bi, Sn, Pb, Hg, etc. whose melting points are lower than those of the impurities. Impure metal like tin or lead is placed on the sloping hearth of a reverberatory furnace. On being heated to a temperature slightly above its melting point, the metal flows down the hearth while the impurities are left behind on the hearth.

(*ii*) **Distillation.** Metals with low boiling points such as zinc, cadmium, mercury and bismuth can be separated from most impurities by distilling the impure metal in a retort. The vapours of the metal so produced are condensed in a suitable receiver while the impurities are left behind in the retort.

(*iii*) **Oxidation.** The method is used when the impurities present in the metal have a greater affinity for O_2 and are more readily oxidised than the metal. This method is usually employed for refining the metals like Pb, Ag, Cu, etc. In this method the molten impure metal is subjected to oxidation by various ways as given below :

(*a*) **Cupellation process.** In this process the molten impure metal is heated in a cupel (boat shaped dish made of bone ash or cement) and a blast of air is passed over the molted metal. The impurities are oxidised and the volatile oxides thus produced escape with the blast of air. The pure metal remains behind in the cupel. For example, the impurity of lead present in silver is removed by cupellation process.

(*b*) **Bessemerisation.** In this oxidation method the impure molten metal is heated in specially designed furnaces and a blast of compressed air is blown through the molten metal. Molten pig iron is purified by this method. The molten pig iron from the blast furnace is taken in a Bessemer converter and the impurities present in it are oxidised by a blast of compressed air (See "*Manufacture of steel by Bessemer's process*" discussed in chapter 46)

(*iv*) **Poling.** The impure metal containing oxide impurities can be purified by this method. It is melted and stirred with green poles of wood. Hydrocarbon gases escaping from the poles act as reducing agents and reduce the oxide impurities.

This method has been used for the purification of copper metal as discussed under "*Extraction of copper from copper pyrites.*"

(*v*) **Electrolytic method.** Large number of metals like copper, silver, gold, nickel and zinc are purified by this method. The impure metal is made the anode while the thin sheet of pure metal acts as the cathode. The electrolyte is generally an aqueous solution of a salt or a complex of the metal. On passing the electric current, the pure metal starts depositing on the cathode through the solution. The soluble impurities pass in the solution while the insoluble low electropositive impurities collect below the anode as **anode mud** or **anode sludge.** The valuable metals can be recovered from the anode mud.

The method can be understood by studying the purification of copper as discussed under "*Extraction of copper from copper pyrites.*"

(*vi*) **Vapour phase method.** Impure metal is converted to a stable volatile compound taking care that the impurities remain unaffected. The compound produced is then decomposed to obtain the pure metal. Titanium, zirconium and nickel have been purified by this method. The method based on the vapour phase technique used for the purification of Ti and Zr metals is called *Van Arkel method* while that used for the purification of Ni metal is called *Mond's process.* These processes have been discussed under the extraction of these metals at their proper places.

(*vii*) **Zone refining method.** The method of zone refining (also called **fractional crystallization** method) is used to obtain metals of very high purity. It is based on the principle that when impure molten metal begins to solidify, the pure substance crystallises while impurities remain in the liquid. In actual practice the process is carried in an apparatus shown in Fig. 14.3. An impure metal rod is surrounded by a circular heater over a small portion. Here the substance melts. The heater is slowly moved along the rod. As the heater moves along the rod, the pure metal crystallizes out of the melt while the impurities pass into the adjacent molten zone. The process is repeated several times when the impurities are completely swept into one end of the rod which is finally discarded. Semi-conductors, silicon, germanium and gallium are purified by this method. This process is carried in an inert atmosphere to prevent oxidation of the metal.

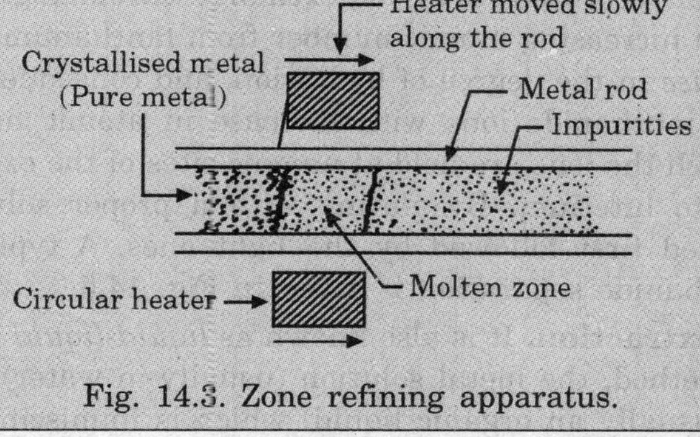

Fig. 14.3. Zone refining apparatus.

(*viii*) **Chromatography method.** The principle of separation or purification by chromatography is that the various components in a mixture are differentially adsorbed on a suitable adsorbent. In column chromatography, an adsorbent like alumina (Al_2O_3) is packed in a glass column. The mixture to be separated or purified (taken in a solvent) is poured on the top of the column. The components are then eluted out stepwise by a suitable eluant (solvent). The weakly adsorbed component is eluted out first followed by more strongly adsorbed components (Fig. 14.4).

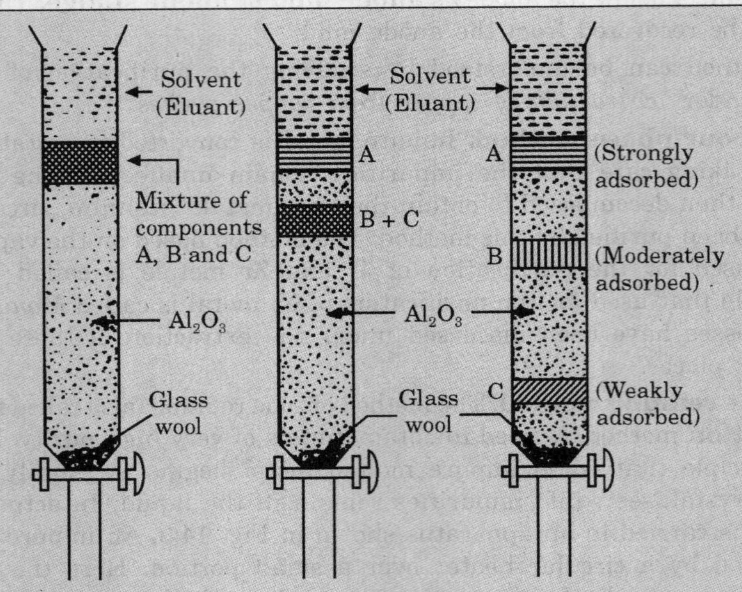

Fig. 14.4. Separation of the components A, B and C of a mixture by column chromatography.

Similarly, in ion-exchange chromatography an ion-exchanger (insoluble organic or inorganic solid containing cationic or anionic groups) capable of undergoing exchange reactions with other ions present in solution is used.

Lanthanides can be resolved by ion-exchange chromatography. Their ionic radii *decrease* with increasing atomic number from lanthanum to lutetium. This gives rise to *increase* in the degree of hydration, and consequent *increase* in the hydrated radii of lanthanide ions with increase in atomic number. Thus, the tendency with which the ions are held at anionic sites of the exchanger *decreases* from lanthanum to lutetium. By eluting with a proper solvent, the heavier members are eluted first followed by the light ones. A typical elution curve depicting the lanthanide separation is given in Fig. 14.5.

(*ix*) **Solvent extraction.** It is also known as *liquid-liquid counter extraction* method. In this method, the metal solution (usually in water) is treated with a suitable solvent (usually an organic liquid) which is immiscible with the metal solution but acts as a better solvent for the metal. In the continuous process of counter extraction, the metal solution and the solvent are made to flow in opposite

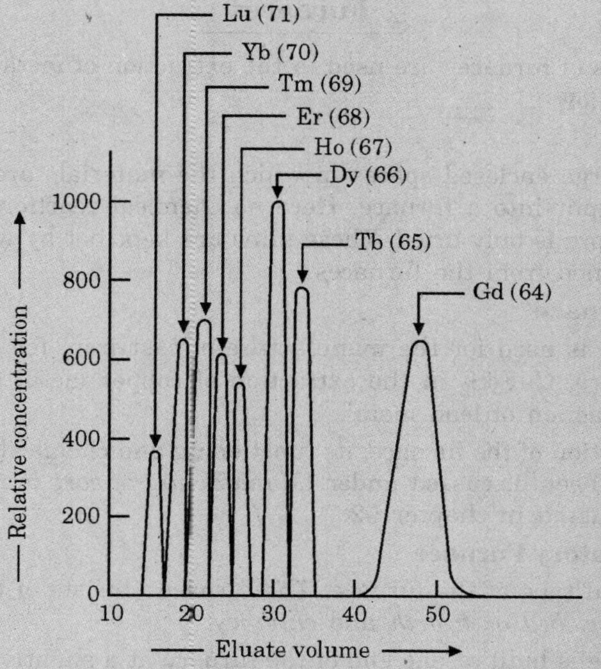

Fig. 14.5. Elution curve of the lanthanides from an ion exchange resin column (separation of lanthanides).

directions through columns packed with suitable inert material (Fig. 14.6). The organic solvent is stripped of its metal content and recycled.

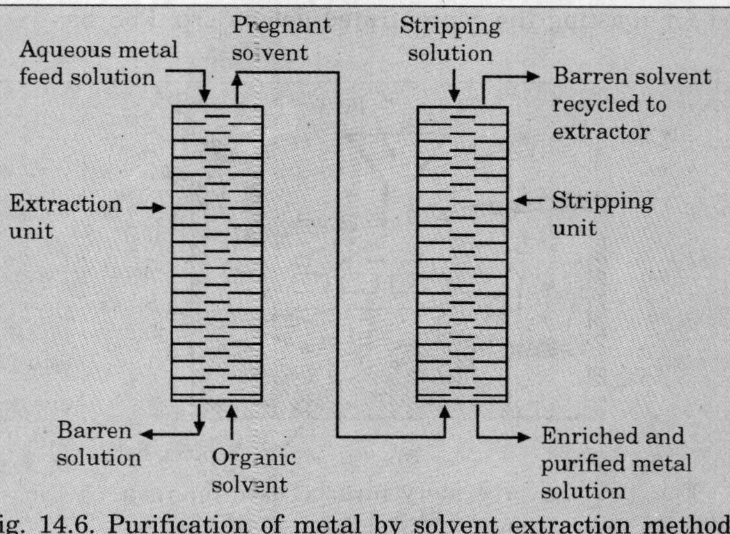

Fig. 14.6. Purification of metal by solvent extraction method.

Uranium and plutonium are purified and separated from each other and from other fission products by this method.

Furnaces

Several types of furnaces are used in the extraction of metals. Some of them are discussed below :

1. Kilns

These are large enclosed spaces in which the materials are mixed together before they are put into a furnace. Here no chemical reaction or fusion takes place. Here the ore is only dried. These kilns are kept hot by waste gases or by waste fuel obtained from the furnaces.

2. Blast Furnace

This furnace is used for the manufacture of cast iron, for smelting roasted copper pyrites ore, $CuFeS_2$ in the extraction of copper metal and for smelting PbO in the extraction of lead metal.

The construction of the furnace, its functioning and chemical reactions taking place in it have been discussed under "*Manufacture of cast iron from haematite ore, Fe_2O_3*" discussed in chapter 52.

3. Reverberatory Furnace

It is a special type of the furnace. The furnace consists of three main parts namely *fire place, bed or hearth* and *chimney.*

The fire place is built at one end of the furnace at a slightly lower level than that of the hearth and is closed at the bottom by fire-bars on which the fuel is placed and burnt. It is covered by a low roof which is made slanting and connects towards the other end with the chimney (flue). The hot gases (waste gases) from the fire place are reflected by the concave ceiling over the hearth. The furnace is surrounded on all sides by walls of fire-bricks. The sketch of a reverberatory furnace used for roasting the concentrated gelana ore, PbS has been shown in Fig. 14.7.

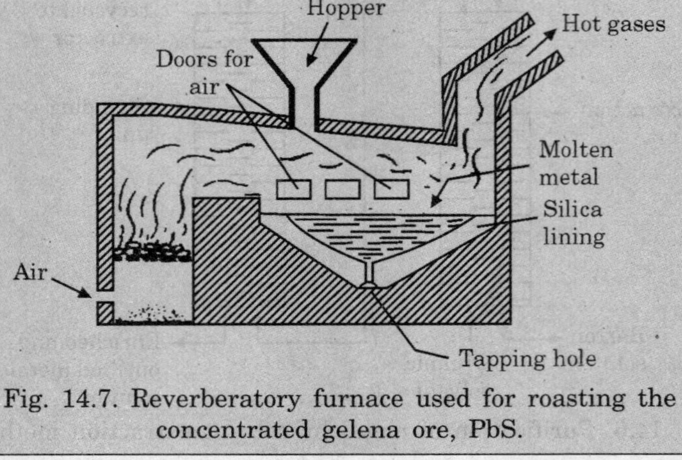

Fig. 14.7. Reverberatory furnace used for roasting the concentrated gelena ore, PbS.

Functioning of the furnace. The charge is placed on the hearth and the fuel is placed on the fire-place. Air enters from the fire-place through the iron grates and waste gases (hot gases) escape out of chimney. There are doors on one side of the furnace for introducing the charge and tap holes on the other side to remove the charge.

In this furnace the ore is not heated directly by the flames of the furnace, but by hot gases which are produced in the fire place. This furnace has been used for smelting (reducing) and roasting the ore. The reduction of the ore is done by the use of some reducing agent like anthracite coal (as in case of tin) while the oxidation is done by means of air. This furnace has been used (*i*) for *reducing* (*smelting*) the roasted tin stone, SnO_2 (an ore of tin) to molten tin metal by the use of anthracite coal (*ii*) for *roasting* the concentrated galena ore, PbS (an ore of Pb) to convert it into PbO and $PbSO_4$ by means of air (*iii*) for *roasting* the concentrated copper pyrites ore, $CuFeS_2$ (an ore of Cu) to convert it into Cu_2O and FeO by means of air.

4. Open-hearth Furnace or Regenerative Furnace

This furnace is like a reverberatory furnace with the dish in the shape of a large oval dish. This furnace works on the regenerative principle of heat economy and hence the name regenerative furnace. It is used for the manufacture of steel from cast iron and the process used is called *open-hearth process* or *Siemen-Martin's process*. Its different parts, functioning and chemical reactions that take place in the furnace during the manufacture of steel have been discussed under "*Manufacture of Steel*" in chapter 46.

5. Electric Furnaces

Electric furnaces are very useful where very high temperatures are required. Further, it is an important advantage of using electric furnaces that the temperatures can be easily controlled by simple regulation of power supply to the furnace or the position of the electrodes etc. These furnaces are particularly useful for the electrolytic reductions. The applicability of electric furnaces is dependent on the availability of the cheap electric power.

Different types of electric furnaces are in use. Of these Heroult's furnace is commonly used. This furnace is used for the manufacture of steel. Different parts of this furnace and the chemical reactions that take place during the manufacture of steel in this furnace have been discussed under "*Manufacture of Steel*" in chapter 46. The process used for the manufacture of steel by this furnace is called *electrical process*.

6. Bessemer's Convertor

Bessemer's convertor was discovered by Sir Henery Bessemer of England in 1856. It is a pear-shaped vessel and is used for the manufacture of steel from cast iron. The construction of the convertor and the reactions that take place in it during the manufacture of steel have been discussed under "*Extraction of Steel*" in chapter 46.

Question with Answers

Q.1 What are Ellingham diagrams ? Discuss their uses and limitations in pyrometallurgy.

Ans : What are Ellingham diagrams ?

We know that the standard free energy change of the formation ($\Delta_f G^\circ$ or ΔG_f°) of metal oxides (solid) is negative. The plots of ($\Delta_f G^\circ$) per mole of O_2 against absolute temperature (T) are called Ellingham diagrams for the formation of metal oxides. These diagrams for the formation of some metal oxides are given in Fig 14.5.

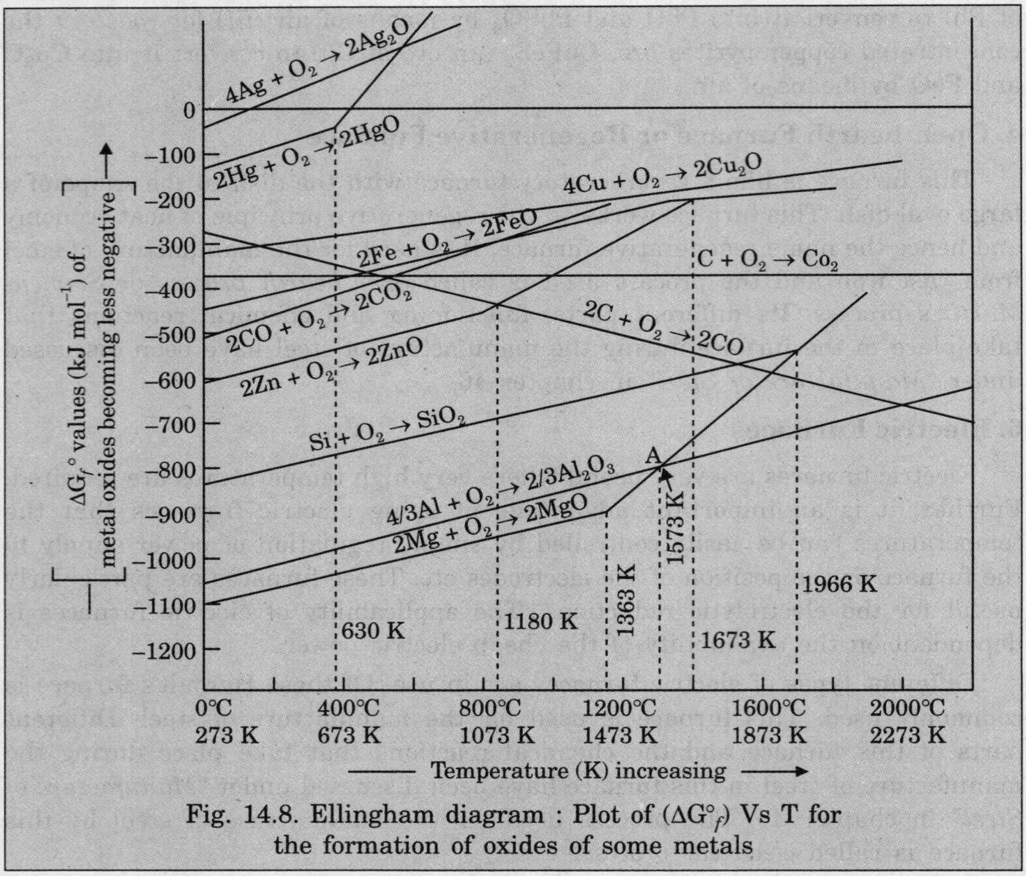

Fig. 14.8. Ellingham diagram : Plot of (ΔG°_f) Vs T for the formation of oxides of some metals

Ellingham diagrams can also be drawn for the conversion of metals to metal sulphides or metal halides using one mole of sulphur or halogen.

Characteristics of Ellingham Diagrams

Following points may be noted from Ellingham diagrams given in Fig 14.8.

(*i*) Each plot in a straight line.

(*ii*) Due to the decrease in entropy (ΔS is negative) during oxide formation reaction, the value of $\Delta_f G^\circ$ increases or becomes less negative with the increase of temperature. In other words curves have positive slopes. However, when some change in phase (*e.g.* solid $\rightarrow$ liquid, liquid $\rightarrow$ gas) takes place, abrupt change in the positive slope of the curve also takes place. For example :

(*a*) In Zn $\rightarrow$ ZnO curve, at the b.pt. of Zn (= 1180K) there is an abrupt increase in the positive slope of the curve.

(*b*) An abrupt increase in the positive slope takes place in Mg $\rightarrow$ MgO curve at the b.pt. of Mg (= 1363K)

(*iii*) For the oxidation reaction like $C(s) + O_2(g) \rightarrow CO_2(g)$ for which $\Delta S = 0$, the slope of C $\rightarrow$ CO_2 curve is zero and hence this curve is parallel to the temperature axis.

(*iv*) All the curves slope upwords since $\Delta_f G°$ values becomes less and less negative with increase in temperature.

(*v*) There is a point on each curve below which the metal oxide is stable and above which the oxide is unstable.

Uses (Applications) of Ellingham diagrams

Simply by looking at Ellingham diagram for the conversion of metals into their oxides we can predict which metal can reduce the oxide of other metal (*s*). At a given temperature, a metal (A) can reduce the oxide of another metal (B), if (A $\rightarrow$ oxide of A) curve lies below (B $\rightarrow$ oxide of B) curve in Ellingham diagram.

Examples : (*i*) Mg can reduce Al_2O_3, ZnO, FeO and Cu_2O to their respective metal, since Mg $\rightarrow$ MgO curve lies below each of the curves *viz* Al $\rightarrow$ Al_2O_3, Zn $\rightarrow$ ZnO, Fe $\rightarrow$ FeO and Cu $\rightarrow$ Cu_2O as is evident from Ellingham diagram given in Fig. 14.8.

(*ii*) Al can reduce ZnO, FeO and Cu_2O to their respective metals, since Al $\rightarrow$ Al_2O_3 curve lies below each of the curves namely Zn $\rightarrow$ ZnO, Fe $\rightarrow$ FeO and Cu $\rightarrow$ Cu_2O curves as is evident from Fig.14.8.

(*iii*) Zn can reduce FeO and Cu_2O to Fe and Cu respectively, since Zn $\rightarrow$ ZnO curve lies below Fe $\rightarrow$ FeO and Cu $\rightarrow$ Cu_2O curves both as is evident from Fig.14.8.

Examples given above show that the tendency of Mg, Al, Zn and Fe to reduce the oxides of other metals is in the order : Mg > Al > Zn > Fe > Cu.

Limitations of Ellingham diagrams

(*i*) Ellingham diagrams simply suggest whether the reduction of metal oxide by a given RA (*e.g.* other metal, C or CO) is posiible or not. These diagrams donot tell anything about the rate at which the reduction reactions take place.

(*ii*) These diagrams are based on the assumption that the reactants and products are in equilibirium. But this is not always true because the reactants/ products may be solid.

Q.2 Give the principle involved in choosing a reducing agent for reducing metal oxide to metal.

Ans : The chemical principle involved in the reduction of a metal oxide to metal by another metal can be understood by considering the following redox reaction.

$$\underset{\text{Metal A (RA)}}{A} + \underset{\text{Oxide of metal B}}{BO} \longrightarrow \underset{\text{Metal B}}{B} + \underset{\text{Oxide of metal A}}{AO}$$

This is a redox reaction in which metal A reduces the oxide of metal B (BO) to B and is itself oxidised to its oxide (AO). Thus metal A acts as a RA. Metal A (RA) reduces BO to B since the value of $\Delta G_f°$ for (A → AO) is more negative than the value of $\Delta G_f°$ for (B → BO). Hence the value of $\Delta G_r°$ of the above reaction becomes negative. [*i.e.* $\Delta G_r° = \Delta G_f° (AO) - \Delta G_f° (BO)$].

From the above discussion we can conclude that any metal having higher negative value of $\Delta G_f°$ for its oxide will reduce the oxide of another metal with lower negative value of $\Delta G_f°$ for its oxide. Without making the calculation of $\Delta G_r°$ for the redox reaction, we can predict the feasibility of the reaction simply by looking at Ellingham diagram. If we look at Ellingham diagram, we find that a metal (A) can reduce the oxide (BO) of another metal B, if A → AO curve lies below B → BO curve.

Above principle can be understood in a better way by considering the following examples :

(*i*) **Reduction of FeO and Fe_2O_3 to Fe by Al.** The reduction of FeO and Fe_2O_3 to Fe by Al can be represented as :

$$\underset{\text{RA}}{2Al} + 3\,FeO \xrightarrow{\Delta} Al_2O_3 + 3Fe$$

$$\underset{\text{RA}}{2Al} + Fe_2O_3 \xrightarrow{\Delta} 2Al_2O_3 + 2Fe$$

In Ellingham diagram (Fig. 14.8), since Al → Al_2O_3 curve lies below Fe → FeO and Fe → Fe_2O_3 curves (not shown in diagram), Al can reduce FeO and Fe_2O_3 both to Fe and Al is oxidised to Al_2O_3.

(*ii*) **Reduction of Al_2O_3 to Al by Fe**. The reduction Al_2O_3 to Al by Fe can be shown as :

$$2\,Fe + Al_2O_3 \rightarrow Fe_2O_3 + Al$$

In Ellingham diagram (Fig. 14.8.), since Fe → FeO curve lies above Al → Al_2O_3 curve, Fe cannot reduce Al_2O_3 to Al, *i.e.* the above redox reaction is not feasable.

(*iii*) **Reduction of FeO to Fe by cabon or CO.** The reduction of FeO to Fe by C and CO be shown as :

$$FeO \rightarrow \underset{\text{RA}}{C} \rightarrow Fe + CO$$

$$FeO + \underset{\text{RA}}{CO} \rightarrow Fe + CO_2$$

Ellingham diagram shows that :

(*a*) Above 1073K, since C → CO curve lies below Fe → FeO curve, (See Fig. 14.9) C is able to reduce FeO to Fe.

$$FeO + C \xrightarrow{< 1073K} Fe + CO$$

(*b*) Below 1073K, since $CO \rightarrow CO_2$ curve lies below $Fe \rightarrow FeO$ curve (Fig. 14.9), CO is able to reduce FeO to Fe. Thus :

$$FeO + CO \xrightarrow{< 1073K} Fe + CO_2.$$

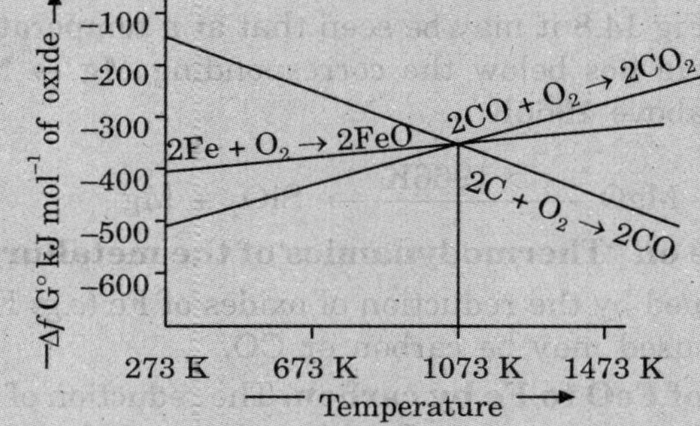

Fig. 14.9 Ellingham diagram for formation of FeO from Fe, CO from C and CO_2 from CO

Q.3 With the help of Ellingham diagram predict whether.

(*i*) Fe can reduce Al_2O_3 to Al and ZnO to Zn.

(*ii*) CO can reduce ZnO to Zn.

Ans : (*i*) $3Fe + Al_2O_3 \rightarrow 3FeO + 2Al$

RA

$Fe + ZnO \rightarrow FeO + Zn$

RA

Ellingham diagram given in Fig. 14.8 shows that since $Fe \rightarrow FeO$ curve lies above $Al \rightarrow Al_2O_3$ curve, Fe cannot reduce Al_2O_3 to Al. Similarly $Fe \rightarrow FeO$ curve lies above $Zn \rightarrow ZnO$ curve, Fe cannot reduce ZnO to Zn. Thus Fe cannot reduce any of the given oxides.

(*ii*) $CO + ZnO \rightarrow Zn + CO_2$

RA

Since $CO \rightarrow CO_2$ curve lies above $Zn \rightarrow ZnO$ curve, CO is not able to reduce ZnO to Zn.

Q.4 Metion the temperature at which :

(*i*) Mg can reduce SiO_2 to Si.

(*ii*) Si can reduce MgO to Mg.

Ans : (*i*) The reduction of SiO_2 to Si by Mg can be shown as :

$$2Mg + SiO_2 \rightarrow 2MgO + Si$$

RA

In this reaction Mg acts as a RA, since it reduces SiO_2 to Si and itself oxidised to MgO. Ellingham diagram given in Fig. 14.8 shows that at a temerature below 1966K, since $Mg \rightarrow MgO$ curve lies below the corresponding $Si \rightarrow SiO_2$ curve, Mg can reduce SiO_2 to Si below 1966 K.

$$2Mg + SiO_2 \xrightarrow{<1966K} 2MgO + Si$$

(*ii*) The reduction of MgO to Mg by Si can be shown as :

$$\underset{RA}{Si} + 2MgO \rightarrow SiO_2 + 2Mg$$

In this reaction Si acts as a RA, since it reduces MgO to Mg and is itself oxidised to SiO_2. In Fig 14.8 it may be seen that at a temperatrure above 1966K, since Si → SiO_2 curve lies below the corresponding Mg → MgO curve, Si can reduce MgO to Mg above 1966K.

$$Si + MgO \xrightarrow{>1966K} SiO_2 + Mg$$

Q.5 Write a note on "Thermodynamics of the metallurgy of iron".

Ans : Fe is obtained by the reduction of oxides of Fe (*e.g.* FeO, Fe_2O_3, Fe_3O_4). The reducing agent used may be carbon or CO.

(*i*) **Reduction of FeO to Fe by carbon.** The reduction of FeO to Fe by C can be shown by the following reaction. In this reaction C is oxidised to CO.

$$FeO + \underset{RA}{C} \longrightarrow Fe\ CO$$

From Fig. 14.9, it may be seen that C → CO and Fe → FeO curves cross each other at a temperature of 1073 K. Thus above this temperature, since C → CO curve lies below Fe → FeO curve, C is able to reduce FeO to Fe at a temperature higher than 1073K.

$$FeO + C \xrightarrow{>1073K} Fe + CO$$

(*ii*) **Reduction of Fe_2O_3 (heamatite) and Fe_3O_4 to Fe by CO.** The reduction of Fe_2O_3 and Fe_3O_4 by CO can be shown by the following reactions. In these reactions CO is oxidised to CO_2 and both iron oxides are converted into FeO.

$$Fe_2O_3 + CO \rightarrow 2FeO + CO_2$$
$$Fe_3O_4 + CO \rightarrow 3FeO + CO_2$$

FeO obtained as above gets reduced to Fe by CO.

$$FeO + CO \rightarrow Fe + CO_2$$

It may be seen from Fig. 14.9 that at a temperature below 1073K, since CO → CO_2 curve lies below Fe → FeO curve, oxides of Fe (FeO, Fe_2O_3, Fe_3O_4) can be reduced to Fe by CO. Thus :

$$FeO + CO \xrightarrow{<1073K} Fe + CO_2$$

Q.6 The values of free energy of formation of MgO(*s*) and CO(*g*) at 1273K and 2273K are given as : (*a*) At 1273K : MgO(*s*) = –914 KJmol^{-1}, CO(*g*) = – 439 KJmol^{-1} (*b*) At 2273K : MgO(*s*) = –314 KJmol^{-1}, CO(*g*) = – 628 KJmol^{-1}. On the basis of these data predict at which temperature carbon can reduce MgO(*s*) to Mg(*s*).

Ans : (*a*) **At 1273K**. Reduction of MgO(*s*) to Mg(*s*) by carbon(*s*) can be shown as :

$$MgO(s) + C(s) \longrightarrow Mg(s) + CO(g).$$

The value of free energy of the above reaction (ΔG_r) is given by :

$$\begin{aligned}\Delta G_r &= \Sigma\Delta G_f \text{ (Products)} - \Sigma\Delta G_f \text{ (Reactants)}\\ &= [\Delta G_f \text{ (Mg)} + \Delta G_f\text{(CO)}] - [\Delta G_f\text{(MgO)} + \Delta G_f\text{(C)}]\\ &= [0 + (-439)] - [-914 + 0]\\ &= -439 + 914 = +502 \text{ KJmol}^{-1}\end{aligned}$$

Since the value of ΔG_r is positive, the reduction of MgO(*s*) to Mg(*s*) by carbon is not feasible at 1273K.

(*b*) **At 2273K.** Reduction of MgO(*s*) to Mg(*s*) by C(*s*) can be shown by the same equation as given above. In this case ΔG_r is given by :

$$\begin{aligned}\Delta G_r &= \Sigma\Delta G_f \text{ (Products)} - \Sigma\Delta G_f \text{ (Reactants)}\\ &= [\Delta G_f \text{ (Mg)} + \Delta G_f\text{(CO)}] - [\Delta G_f\text{(MgO)} + \Delta G_f\text{(C)}]\\ &= [0 + (-628)] - [-314 + 0]\\ &= -628 + 314 = -314 \text{ KJmol}^{-1}\end{aligned}$$

Since the value of ΔG_r is negative, the reduction of MgO(*s*) to Mg(*s*) by carbon is feasible at 2273K.

Q.7 The value of $\Delta_f G°$ for the formation of Cr_2O_3 is –540 KJmol^{-1} and that of Al_2O_3 is –827 KJmol^{-1}. Is the reduction of Cr_2O_3 possible with Al ?

Ans : The reduction of Cr_2O_3 with Al can be represented by the equation :

$$Cr_2O_3(s) + 2Al(s) \xrightarrow{\Delta} 2Cr\,(s) + Al_2O_3(s)$$

The above reaction is feasible, if the value of $\Delta_r G°$ for this reaction is negative. Now let us calculate the value of $\Delta_r G°$ for the above reaction.

$$\begin{aligned}\Delta_r G° &= \Sigma\Delta_f G° \text{ (Products)} - \Sigma\Delta_f G° \text{ (Reactants)}\\ &= [\Delta_f G° (Al_2O_3) + 2 \times \Delta_f G° (Cr)] - [\Delta_f G° (Cr_2O_3) + 2 \times \Delta_f G° (Al)]\\ &= [-827 + 2 \times 0] - [-540 + 2 \times 0]\\ &= -827 + 540 = -287 \text{ KJmol}^{-1}\end{aligned}$$

Since the value of $\Delta_r G°$ is negative (= –287 KJmol^{-1}), the above reaction is feasible, *i.e.* the reduction of Cr_2O_3 by Al is possible.

Q.8 With the help of the following oxidation reactions predict whether :

(*i*) Zn can be oxidised by CO.

(*ii*) ZnO can be reduced by graphite

$$\mathbf{Zn(s) + \frac{1}{2}O_2(g) \rightarrow ZnO(s), \Delta G° = -360 \text{ KJmol}^{-1}}$$

$$\mathbf{C(gr) + \frac{1}{2}O_2(g) \rightarrow CO(g), \quad \Delta G° = -460 \text{ KJmol}^{-1}}$$

Ans : (*i*) Oxidation of Zn to ZnO by CO can be shown as :

$$Zn(s) + CO(g) \rightarrow ZnO(s) + C(s)$$

The value of $\Delta G_r°$ for this reaction is given by :

$$\begin{aligned}\Delta G_r° &= \Delta G_f° \text{ (ZnO)} - \Delta G_f° \text{ (CO)}\\ &= -360 - (-460) = +100 \text{ KJmol}^{-1}\end{aligned}$$

Since the value of $\Delta G_r°$ is positive, Zn cannot be oxidised to ZnO by CO.

(*ii*) The reduction of ZnO by graphite, C(*gr*) can be represented by the equation

$$ZnO\ (s) + C(gr) \rightarrow Zn(s) + CO(g)$$

$$\Delta G_r° = \Delta G_f°\ (CO) - \Delta G_f°\ (ZnO)$$

$$= -460 - (-360) = -100\ KJmol^{-1}$$

Since the value of $\Delta G_r°$ is negative, the reduction of ZnO by carbon is feasible.

Q.9 A mineral consists of PbS and ZnS. NaCN is used as a depressant in preventing ZnS from forming the froth. How does NaCN act as a depressant ?

Ans : When a mineral consisting of PbS and ZnS is concentrated by froth flotatation method, NaCN is used as a depressant. NaCN reacts with ZnS to form a complex, $Na_2\ [Zn(CN)_4]$

$$\underset{\text{Depressant}}{4NaCN} + ZnS\ (\text{Present in mineral}) \rightarrow Na_2\ [Zn(CN)_4] + Na_2S$$

Due to the conversion of ZnS into $Na_2\ [Zn\ (CN)_4]$, this sulphide ore is prevented from forming froth under these conditions, only *PbS* forms froth with air bubbles and can be separated from ZnS ore.

Q.10 Justify the statement : "The extraction of gold by leaching with NaCN in presence of O_2 involves both oxidation and reduction".

Ans : During the leaching process, Au is first oxidised to Au^+ by O_2 *of the air* which then combines with CN^- ions to form the soluble complex, sodium dicyanoaurate (I).

$$\underset{\substack{\text{Gold}\\(\textit{Impure})}}{4Au(s)} + 8NaCN\ (aq) + 2H_2O\ (l) + O_2 \longrightarrow \underset{\substack{\text{Sod. dicyanoaurate (I)}\\(\textit{Soluble complex})}}{4Na\ [Au(CN)_2]\ (aq)} + 4NaOH\ (aq)$$

Gold is then extracted from this complex by displacement method using a more electropositive zinc metal. In this reaction, Zn acts as a reducing agent. It reduces Au^+ to Au while it itself gets oxidized to Zn^{2+} which combines with CN^- ions to form soluble complex, sodium tetracyanozincate (II).

$$2Na\ [Au(CN)_2]\ (aq) + Zn(s) \rightarrow \underset{\substack{\text{Gold}\\(\textit{Pure})}}{2Au(s)} + \underset{\text{Sod. tetracyanozincate (II)}}{Na_2\ [Zn\ (CN)_4]\ (aq)}$$

Q.11 Out of the following metals select the metals which are extracted by carbon-reduction process and which are extracted by electrolytic reduction process. Na, K, Mg, Ca, Al, Pb, Fe.

Ans : Since Na, K, Mg, Ca and Al are highly electropositive metals, these metals are obtained by the electrolytic reduction of their fused oxide (*e.g* MgO, Al_2O_3), hydroxide (e.g., NaOH, KOH) and chloride (e.g. NaCl, KCl, $MgCl_2$, $CaCl_2$). In this process the oxide, hydroxide and chiloride is reduced to the metal which is collected at cathode.

Since Pb and Fe are less electroposive metals, these are obtained by reducing their oxides with carbon (carbon-reduction process) or CO.

$$PbO + C \rightarrow Pb + CO, \quad PbO + CO \rightarrow Pb + CO_2$$

$$Fe_2O_3 + 3C \rightarrow 2Fe + 3CO, \quad Fe_2O_3 + 3CO \rightarrow 2Fe + 3CO_2$$

Q.12 Name the following : (*i*) Two metals that never occur in native state (*ii*) Two metals that always occur both in native and in combined states (*iii*) Two metals which always occur naturally in an uncombined state (*iv*) Two metals which are manufactured by the electrolysis of their fused salts (*v*) Two metals that are used for the reduction in metallurgical process (*vi*) Two metals that can be obtained by the electrolytic reduction of aqueous solution of their salts.
(I.I.T. 1976, Roorkee 1988)

Ans : (*i*) Since Na and K are highly reactive metals, they are never found in nature in the native (free or uncombined) state. These metals are found in nature in the form of their ores (combined state).

(*ii*) Cu and Ag are the metals which are found in nature both in the native (metallic condition) and combined state. In the combined state, Cu is found in the form of its sulphide ores (*e.g.* $CuFeS_2$, Cu_2S, Cu_3FeS_3), oxide ores (Cu_2O) and basic carbonate ores [*e.g.* $CuCO_3$. Cu $(OH)_2$, $2CuCO_3$, $Cu(OH)_2$] while Ag is found in the form of its sulphide ore (*e.g.* Ag_2S etc) and halide ores (*e.g.* AgCl).

(*iii*) Au and Pt are the metals which are found in nature only in the free or native or uncombined state. These are usualy found associated with rock or alluvial materials like clay, sand etc.

(*iv*) Na and Ca are the metals that are obtained by the electrolysis of their fused chlorides viz NaCl and $CaCl_2$. In the electrolysis of NaCl, some amount of $CaCl_2$ and KF is also added to fused NaCl. The addition of $CaCl_2$ and KF reduces the fusion temperature from 800°C to 600°C and the conductivity of the electrolyte (NaCl) is increased. In the electrolysis of $CaCl_2$ some amount of CaF_2 is also added to it. In both the cases the metal (i.e. Na and Ca) is deposited on cathode.

(*v*) Al and Mg are the two metals which are used for the reduction of metallic salts to the metals. For example :

$$TiCl_4 + 2Mg \longrightarrow Ti + 2MgCl_2$$

$$Cr_2O_3 + 2Al \longrightarrow 2Cr + Al_2O_3$$

Al and Mg act as reducing agents.

(*vi*) Cu and Ag are the metals that can be obtained by the electrolytic reduction of aqueous solution of their salts.

Cu is obtained by electrolysing $CuSO_4$ solution using Cu sheet as cathode and lead sheet as anode. $CuSO_4$ is reduced to Cu which is collected on cathode.

Ag can be obtained by electrolysing $AgNO_3$ solution containing 10% HNO_3. Impure Ag is made anode and pure Ag is made cathode. On electrolysing the solution, Ag gets deposited on cathode.

$$AgNO_3 \rightleftharpoons Ag^+ + NO_3^-$$

$$NO_3^- + Ag \rightarrow AgNO_3 + e^- \text{ (At anode)}$$

$$Ag^+ + e^- \rightarrow Ag \text{ (At cathode)}$$

In this way Ag is transferred from anode to cathode. The impurity of Cu present in $AgNO_3$ goes into solution as $Cu(NO_3)_2$ and that of Au gets collected below the anode as anodic mud.

Q.13 Explain the following :

(*i*) Below 673K, CO is a better RA for reducing metal oxides but above this temerature carbon is a better RA.

Ans : When carbon (C) reduces a metal oxide (MO), it is oxidised to CO or CO_2

$$C + MO \rightarrow M + CO$$

$$C + 2MO \rightarrow 2M + CO_2$$

When CO reduces a metal oxide (MO), it gets oxidised to CO_2

$$CO + MO \rightarrow M + CO_2$$

Above redox reaction show that C is oxidised to CO or CO_2 and CO is oxidised to CO_2.

(*i*) $2C(s) + O_2(g) \longrightarrow 2CO(g)$
1 volume → 2 volumes

(*ii*) $C(s) + O_2(g) \longrightarrow CO_2(g)$
1 volume → 1 volume

(*iii*) $\underbrace{2CO(g) + O_2(g)}_{3\text{ volumes}} \longrightarrow 2CO_2(g)$
3 volumes → 2 volumes

In reaction (*i*), since 2 volumes of CO(*g*) are produced by 1 volume of $O_2(g)$, (increase in volume), ΔS for this reaction is positive and hence the value of $\Delta G°_f$ (C → CO) becomes more negative as temperature increases ($\because \Delta H°_f = \Delta H - T\Delta S$). Consequently, C → CO line in Ellingham diagram given in Fig.14.10. slopes downwards.

In reaction (*ii*), since the value of $CO_2(g)$ produced is the same as that of $O_2(g)$ used, ΔS for this reaction is very small and hence the value of $\Delta G_f°$ for (C → CO_2) does not change with tempreature. Consequantly, C → CO_2 line is almost horizontal (*i.e.* parallel to the temperature axis) (See Fig. 14.10)

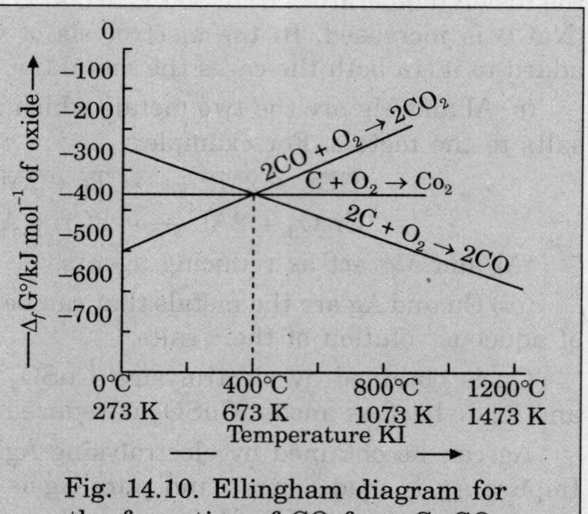

Fig. 14.10. Ellingham diagram for the formation of CO from C, CO_2 from C and CO_2 from CO

In reaction (*iii*), Since 2 volumes of CO_2 are produced by 3 volumes of the reactants, ΔS for this reaction is negative and hence the value of (CO → CO_2) is more positive as the temperature increases. Consequently CO → CO_2 line slopes upwards as shown in Fig. 14.10. C → CO, C → CO_2 and CO → CO_2 lines cross at 673K. Below this tempreature, C and CO both can reduce metal oxides (MO) to metals (M) and C/CO are oxidised to CO_2. Now since CO is more easily oxidised to CO_2 than C is oxidised to CO_2, below 673K CO reduces metal oxides more effectively than carbon, *i.e.* below 673K CO is a better RA than C for reducing metal oxides (MO) to metals (M).

$$CO + MO \xrightarrow{<673K} M + CO_2$$

Above 673K, since CO is more stable than C, its oxidation to CO_2 is more difficult than the oxidation of C to CO_2. Thus above 673K, C is more easily oxidised to CO_2 than CO is oxidised to CO_2. As a result, carbon (C) is a better RA than CO for reducing MO to M at a tempreature higher than 673K.

$$C + 2MO \xrightarrow{>673K} 2M + CO_2$$

(*ii*) Below 1573K Mg can reduce Al_2O_3 to Al but above this temperature Al can reduce MgO to Mg.

Ans : If we look at Ellingham diagram given in Fig. 14.8, we find that $Mg \rightarrow MgO$ and $Al \rightarrow Al_2O_3$ curves intersect at a temperature of 1573K (Point A). Since below this temperature, $Mg \rightarrow MgO$ curve lies below $Al \rightarrow Al_2O_3$ curve, Mg can reduce Al_2O_3 to Al below 1573K *i.e.* the following reaction is feasible at a temperature below 1573K

$$Mg + Al_2O_3 \xrightarrow{<1573K} Al + MgO$$

Above 1573K, since $Al \rightarrow Al_2O_3$ curve lies below $Mg \rightarrow MgO$ curve, Al can reduce MgO to Mg, *i.e.* the following reaction is feasible at a tempreatuer higher than 1573K.

$$2Al + 3MgO \xrightarrow{>1573K} Al_2O_3 + 3Mg$$

(*iii*) The reduction of Cr_2O_3 with Al does not occur at room temperature but proceeds rapidly at high temperature.

Ans : The reduction of Cr_2O_3 with Al can be shown as :

$$Cr_2O_3(s) + Al(s) \rightarrow Al_2O_3(s) + 2Cr(s)$$

This is a redox reaction. In this reaction, since all the reactants and products are in the solid state at room tempreature, there does not exits any equilibiriam between the reactants and products. As a result, this reaction does not occur at room temperature. If the reaction is allowed to take place at high temperature, Cr formed in the reaction melts and the value of $T\Delta S$ increases. Due to the increases in the value of $T\Delta S$, the value of $\Delta G°_r$ becomes more negative ($\Delta G_r° = \Delta H - T\Delta S$) and hence the reaction proceeds rapidly.

(*iv*) In froth floatation process pine oil and cresols/aniline are added to the suspension of ore in water.

Ans : The addition of pine oil enhances the non-wetability of the ore particles in water, *i.e.* ore particles are preferentially wetted by pine oil. As a result, ore particles become lighter and hence rise to the surface along with the froth. The addition of cresols/aniline stabilise the froth. Thus pine oil is used as a collector and aniline is used as froth stabiliser.

(*v*) Out of C and CO, C is a better reducing agent for reducing Zn to ZnO.

Ans : The reduction of ZnO to Zn by C and CO can be shown as :

$$\underset{\text{RA}}{ZnO + C} \longrightarrow Zn + CO$$

$$\underset{\text{RA}}{ZnO + CO} \longrightarrow Zn + CO_2$$

If we look at Ellingham diagram given in Fig. 14.8 , we find that since $CO \rightarrow CO_2$ curve lies above $Zn \rightarrow ZnO$ curve, CO cannot reduce ZnO to Zn. On the other hand, at a temperature of 1673K, since $C \rightarrow CO$ curve lies below $Zn \rightarrow ZnO$ curve, C can reduce ZnO to Zn. Thus out of C and CO, C is a better RA than CO for reducing ZnO.

***(vi)* The reduction of a metal oxide to metal is easier, if the metal formed is in the liquid state.**

Ans : We know that entropy of a metal in liquid state is higher than that of the same metal in solid state ($S_{liquid} > S_{solid}$). Thus when the metal formed is in the liquid state, and the metal oxide being reduced is in the solid state, the value of entropy change (ΔS) for the reduction reaction is more on the positive side.

Now since the value of $T\Delta S$ increases and that of ΔH remains the same, the value of $\Delta G_r°$ for the reduction reaction becomes on the negative side ($\Delta G_r° = \Delta H - T\Delta S$) and hence reduction becomes easier.

***(vii)* In the extraction of metals from their sulphide ores, the sulphide ores are first converted into metal oxides and then these metal oxides are reduced to metal by suitable RA.**

Ans : If a metal sulphide ore (*e.g.* Cu_2S) is directly reduced to metal (Cu) by carbon or H_2, C and H_2 are oxidised to CS_2 and H_2S respectively.

$$2Cu_2S + \underset{\text{RA}}{C} \longrightarrow 4Cu + CS_2$$

$$Cu_2S + \underset{\text{RA}}{H_2} \longrightarrow 2Cu + H_2S$$

These reactions would be feasible, if $\Delta G_f°$ of CS_2 and H_2S both are greater than that of Cu_2S. In fact, the value $\Delta G_f°$ of CS_2 and H_2S both are smaller than that of Cu_2S. Hence above reactions are not feasible, *i.e.* metal sulphides cannot be reduced by C or H_2

When oxide ore (Cu_2O) is reduced by C, C is oxidised to CO

$$Cu_2O(s) + \underset{\text{RA}}{C\ (s)} \longrightarrow 2Cu(s) + CO(g)$$

Now since, in Ellingham diagram, (Fig. 14.8) $C \rightarrow CO$ curve lies below $C \rightarrow Cu_2O$ curve, C can reduce Cu_2O to Cu.

***(viii)* Copper can be extracted by hydrometallyrgy but Zn cannot.**

Ans : Hydrometallurgy is a process in which the ore of the metal is dissolved in a suitable solvent (leaching process) to get soluble compound of the metal. The solution of the compound contains the metal as ions. Metal is obtained from this solution by displacing it by a more reactive or more electropositive metal, *i.e.* the metal acting as displacing metal should have lower E° value than the metal being displaced. Cu is extracted from low grade ores. The ore is leached with an acid in presence of air when Cu metal goes into soultion as Cu^{2+} ions.

$$2Cu\ (s) + 2H_2SO_4\ (aq) + O_2(g) \longrightarrow 2CuSO_4\ (aq) + 2H_2O\ (l)$$

or $$Cu\ (s) + 2H^+\ (aq) + \frac{1}{2}O_2\ (g) \longrightarrow Cu^{2+}\ (aq) + H_2O\ (l)$$

Cu^{2+} ions present in solution are displaced by a metal having lower E^o value than $E^o_{Cu^{2+}/Cu}$ (= + 0.34V). Since Fe has lower value of E^o ($E^o_{Fe^{2+}/Fe}$ = – 0.44V) than $E^o_{Cu^{2+}/Cu}$ (= + 0.34V), this metal is used for displacing Cu metal from the solution of Cu^{2+} ions.

Fe (Displacing metal) + Cu^{2+} (Metal being displaced) $\longrightarrow$ Cu + Fe^{2+}

We know that $E^o_{Zn^{2+}/Zn} = -0.76V$. Since Al, Mg, Ca, K etc have lower E^o values ($E^o_{Al^{3+}/Al} = -1.66V$, $E^o_{Mg^{2+}/Mg} = -2.37V$, $E^o_{Ca^{2+}/Ca} = -2.87V$, $E^o_{K^+/K} = -2.93V$), these metals should be able to displace Zn from the solution of Zn^{2+} ions. But these metals react with H_2O to give their hydroxides and H_2 (*e.g.* Mg + $2H_2O \longrightarrow Mg(OH)_2 + H_2$), these are not able to displace Zn from the solution having Zn^{2+} ions.

Above discussion shows that Zn cannot be extracted from its ore by hydrometallurgy process but Cu can be.

***(ix)* In nature, Cu, Ag, Zn, Hg and Pb occur as sulphides and not as oxides.**

Ans : The cations of Cu, Ag, Zn and Hg *viz* Cu^+, Ag^+, Zn^{2+} and Hg^{2+} have pseudo inert gas configuration, $(n-1)\ s^2\ p^6\ d^{10}$ and the cation of Pb (Pb^{2+}) has inert pair configuration, $4f^{14}\ 5d^{10}\ 6s^2$.

Due to these configurations these cations can easily polarise the bigger S^{2-} ion than the smaller O^{2-} ion. As a result, sulphides of these metals are more stable than their oxides. Consequently, these metals prefer to occur as sulphides and not as oxides in nature.

***(x)* For the recovery of Ag from $[Ag(CN)_2]^-$ ion this ion is reduced with Zn and not with Cu.**

Ans : E^o values of Zn and Cu are $E^o_{Zn^{2+}/Zn} = -0.76V$ and $E^o_{Cu^{2+}/Cu} = +0.34V$. Since $E^o_{Zn^{2+}/Zn} < E^o_{Cu^{2+}/Cu}$, Zn is stronger RA than Cu. Hence $[Ag(CN)_2]^-$ ion is reduced with Zn and not with Cu.

$$\underset{(Ag\ =\ +1)}{[2Ag(CN)_2]^-} + \underset{\underset{(RA)}{(Zn\ =\ 0)}}{Zn} \longrightarrow \underset{(Ag\ =\ 0)}{2Ag} + \underset{(Zn\ =\ +2)}{[Zn(CN)_4]^{2-}}$$

Another reasons is that Zn is cheaper than Cu.

***(xi)* In the metallurgy of Cu, partial roasting of the sulphide ore of the metal is done.**

Ans : In the metallurgy of Cu, copper pyrites ore ($CuFeS_2$) is roasted (oxidation) in a reverberatory furnace when the ore is converted into a mixture of Cu_2S and FeS.

$$2CuFeS_2 + O_2 \xrightarrow{\text{Roasting}} Cu_2S + 2FeS + SO_2$$

Now Cu_2S formed as above is oxidised (roasting) to Cu_2O which combines with the unroasted Cu_2S and forms Cu metal by the self reduction of Cu_2O and Cu_2S.

$$2Cu_2S + 3O_2 \xrightarrow{\text{Partial roasting}} 2Cu_2O + 2SO_2$$

$$\underset{\substack{(Cu\,=\,+1,\\ O\,=\,-2)}}{2\,Cu_2O} + \underset{\substack{(Cu\,=\,+1,\\ S\,=\,-2)}}{Cu_2S\ (\text{unroasted})} \longrightarrow \underset{Cu\,=\,0}{6Cu} + \underset{\substack{(S\,=\,+4,\\ O\,=\,-2)}}{SO_2}$$

Thus it is due to partial roasting (oxidation) of Cu_2S that self reduction of Cu_2O and Cu_2S takes place.

(*xii*) In the metallurgy of Cu, chalcocite ore (Cu_2S) is roasted and not calcined.

Ans : Calcination is a process in which an ore (generally carbonate or hydroxide ore) is heated strongly to convert it into oxide while roasting is a process in which an ore (generally sulphide ore) is heated strongly in the excess of air to convert it into oxide. Since chalcocite is a sulphide ore, it is roasted and not calcined.

(*xiii*) Al cannot be obtained by smelting Al_2O_3.

Ans : We know that smelting is a process in which the oxide of the metal is reduced to metal by heating the oxide with carbon. Since Al has great affinity for O_2, Al_2O_3 is a very stable compound. Al cannot be obtained by heating Al_2O_3 with carbon (smelting), since when Al_2O_3 is heated with carbon, Al_4C_3 (not Al) is obtained.

$$2\,Al_2O_3 + 6C \rightarrow Al_4C_3 + 6\,CO$$

(*xiv*) Graphite is used as an anode but not diamond.

Ans : Graphite is a good conductor of electricity due to the presence of free electrons within its layers which help to conduct electricity. Diamond, on the other hand, is a bad conductor of electricity since no free electrons are present in its structure. Consequently diamond is not used as an anode.

(*xv*) Metals usually do not occur in nature as nitrates.

Ans : Nitrates of all metals are soluble in water. Therefore, if metal nitrates were present in the crust of earth, these would be slowely and gradually washed by rain water into the sea. It is because of this reason that metals usually do not occur in nature as nitrates.

University Questions

1. (*a*) Explain briefly the principle involved in the following : (*i*) Electrolytic refining of metals (*ii*) Van Arkel process (*iii*) Concentration of sulphide ores.

   (*b*) How does calcination differ from roasting? Illustrate using suitable examples. (*Delhi 99*)

2. Describe in short the methods for the extraction and isolation of metals from sulphide and oxide ores. (*Kanpur 2000*)
3. Write notes on : Roasting, Calcination, Smelting and Refining of metals. (*Kumaon 2000, Delhi 2002, Lucknow 2002*)
4. Discuss briefly cyanide process used for the extraction of metals from their ores. [*Gauhati (General) 2000*]
5. The extraction of a metal from its ore is essentially a reduction process represented by :

$$M^n + ne^- \longrightarrow M$$

6. Comment on the above statement and discuss various types of reducing agents employed for the purpose. (*Lucknow 2001*)
7. (*i*) Mention the advantages of powder metallurgy.
   (*ii*) Briefly discuss any two methods used in the refining of metals.
   (*iii*) What is the role of flux in metallurgy ?
   (*iv*) What is the reducing agent employed in the production of tungsten ? Give reason. (*Bangalore 2001*)
8. What different techniques are employed for the purification or refining of a metal ? (*Delhi 2003*)
9. Describe Van Arkel process (Vapour phase process). (*Banglore 2004*)
10. Write in brief the general principles involved in the extraction of metals from their ores. (*Purvanchal 2007*)
11. Explain "Roastring" with examples. (*Meerut 2009*)

CHAPTER 15

Various Reactive Forms and Isotopes of Hydrogen

Various Reactive Forms of Hydrogen

Following are the reactive forms of hydrogen :

1. Nascent Hydrogen

If hydrogen gas is bubbled through acidified potassium permanganate, colour is not discharged. To the same if we add some zinc pieces, tiny bubbles of hydrogen are found to rise and potassium permanganate is decolorised. Hydrogen produced *in situ* in the second case is more active and is called **nascent hydrogen** (*newly born*).

$$Zn + H_2SO_4 \longrightarrow ZnSO_4 + \underset{\text{Nascent hydrogen}}{2H}$$

$$2KMnO_4 + 3H_2SO_4 + 10H \longrightarrow K_2SO_4 + 2MnSO_4 + 8H_2O$$

Ferric chloride solution (*yellow*) is decolorised by adding zinc pieces and hydrochloric acid. This is also due to reduction of ferric chloride to ferrous chloride (*faint green solution*) by nascent hydrogen.

$$FeCl_3 + H \longrightarrow FeCl_2 + HCl$$

Nascent hydrogen produced from Zn and dil. H_2SO_4 can reduce $KClO_3$ to KCl and $AsCl_3$ to AsH_3.

$$KClO_3 + 6H \longrightarrow KCl + 3H_2O$$

$$AsCl_3 + 6H \longrightarrow AsH_3 + 3HCl$$

Preparation of Nascent Hydrogen. Nascent hydrogen can be prepared by the following methods :

(*i*) *By the action of Zn on H_2SO_4*

$$Zn + H_2SO_4 \longrightarrow ZnSO_4 + 2H$$

(*ii*) *By the action of water on sodium amalgam.*

$$Na + H_2O \longrightarrow NaOH + H$$

(*iii*) *A mixture of Sn and conc. HCl or Fe and water acidified with a little of HCl also gives nascent hydrogen which is used for reducing aromatic compounds.*

$$Sn + 2HCl \longrightarrow SnCl_2 + 2H$$

(*iv*) *Metallic sodium and abolute alcohol also gives nascent hydrogen.*

$$Na + C_2H_5OH \longrightarrow C_2H_5ONa + H$$

(*v*) *By the action of NaOH on Zn or Al*

$$Zn + 2NaOH \longrightarrow \underset{\text{Sod. zincate}}{Na_2ZnO_2} + 2H$$

Theories to Explain the Reactivity of Nascent Hydrogen. ***(i) Atomic Hydrogen Theory.*** In one of the explanations it was suggested that nascent hydrogen at the moment of its liberation exists as single atoms while in ordinary hydrogen these atoms have combined to form molecule, H_2. Its activity is due to its atomic state. Single atoms of nascent hydrogen are ready to react. On the other hand, some work is required to be done on molecules of ordinary hydrogen to split them into single atoms before they can react.

This theory fails to account for a number of facts such as the different degrees of reactivity of nascent hydrogen from different reagents. For example, nascent hydrogen from zinc and dilute sulphuric acid can reduce chlorates to chlorides but one from sodium amalgam and water cannot.

(ii) Energy theory. In another explanation, it is believed that difference in the degrees of activity of nascent hydrogen obtained from different sources is due to different amounts of energy associated with it. Some of the energy liberated as a result of chemical reaction appears as heat energy while the rest becomes associated with hydrogen.

(iii) Internal pressure Theory. In another suggested explanation, hydrogen at the moment of its production is believed to be in the form of very minute bubbles with great internal pressure on account of their small size. It has been shown experimentally that activity of ordinary hydrogen can be increased by applying great pressure.

(iv) Theory of association of energy. Recently, an attempt has been made to explain its activity on the basis of association of energy when electrons are transferred from metals of low electronegativity to H^+ ions in order to change its valency from +1 to zero.

$$Zn \longrightarrow Zn^{2+} + 2e^-$$

$$Na \longrightarrow Na^+ + e^-$$

$$Al \longrightarrow Al^{3+} + 3e^-$$

The hydrogen in solution accepts these electrons :

$$H^+ + e^- \longrightarrow H$$

As the energy liberated in the transfer of electrons is different for different reactions, the reducing capacity of various reactions producing nascent hydrogen would be different.

2. Adsorbed or Occluded Hydrogen

Some metals (*e.g.*, Pd, Pt, Au, Fe, Ni) can adsorb under certain conditions relatively large volumes of hydrogen. For example, Graham (1867) found that one volume of palladium adsorbs 935 volumes of hydrogen while cooling from red heat. The gas thus adsorbed by metals is given off when the metal is heated especially under reduced pressure.

The phenomenon was termed **occlusion** or **adsorption** by Grahm. The gas so adsorbed by the metal is termed **occluded** or **adsorbed hydrogen.**

Occluded hydrogen is more active and a stronger reducing agent than ordinary hydrogen. For example :

(*i*) It reduces ferrous salts,

(*ii*) It combines with chlorine and iodine even in the dark.

(*iii*) It precipitates mercury from mercuric chloride.

Theories to Explain the Nature of Occluded Hydrogen. Different views have been put forward from time-to-time to explain the nature of occluded hydrogen. For example :

(*i*) Formerly it was believed that occluded hydrogen forms a chemical compound with the metal.

(*ii*) Holt, Edgar and Firth (1913) suggested that a part of occluded hydrogen is present as a condensed layer on the surface (adsorption) while the rest of it dissolves in the interior of the metal.

(*iii*) According to Gillespie and Hall (1926) there was evidence for the existence of two immiscible solid solutions. They thought that one of these might be a compound, Pd_2H.

(*iv*) At present it is believed that occluded hydrogen gives an *interstitial compound* in which hydrogen occupies the holes in the crystal lattice of the metal.

3. Active Hydrogen

When ordinary hydrogen at room temperature is subjected to the action of silent electric discharge at an electrical pressure more than 30,000 volts, it changes into an active variety of hydrogen. This is called **active hydrogen.**

Active hydrogen is very reactive as shown by the reactions given below :

(*i*) It combines directly with lead or sulphur to form hydrides. The combination occurs at the room temperature.

(*ii*) It reduces cupric and ferric oxides to corresponding metals even in the cold.

Theories to Explain the Nature of Active Hydrogen. Nature of active hydrogen is not clearly understood. Two views put forward are :

(*i*) It consists of molecules H_3 (*cf.* formula of ozone, O_3 which is produced by passing silent electric discharge through ordinary oxygen).

(*ii*) It is atomic hydrogen.

Neither of these explanations has been universally accepted. Active hydrogen is at present regarded merely as an active form of ordinary hydrogen, H_2 (*cf.* active nitrogen).

4. Atomic Hydrogen

In 1922, Langmuir showed that when molecular hydrogen in contact with a tungsten or platinum wire was heated by an electric current at low pressure, it is dissociated into atomic hydrogen. The dissociation of molecular hydrogen into atomic hydrogen is an endothermic reaction.

$$\underset{\text{Molecular hydrogen}}{H_2} \longrightarrow \underbrace{H + H}_{\text{Atomic hydrogen}} - 103.7 \text{ K.cals}$$

The atomic hydrogen so produced was found to be more reactive than ordinary, nascent and adsorbed hydrogen.

Preparation of Atomic Hydrogen. (*i*) *By High temperature dissociation method.* This method is based on the observation made by *Langmuir* who showed that wires of W, Pt and Pd heated to temperature of 1000–2000°C in molecular hydrogen, at pressure of 0.01 mm or less, produce atomic hydrogen, by absorbing a large amount of energy. This method consists in blowing a current of hydrogen across an arc (20 amp. at 300—800 volts) struck between tungsten electrodes in an atmosphere of hydrogen. The degree of dissociation of molecular hydrogen into atomic hydrogen at different temperatures is as shown below :

Temperature (°K)	*Dissociation (%)*
1000	0.00000371
2000	0.122
3000	9.03
4000	62.50
5000	94.69

(ii) By discharge Tube method. *Wood* in 1922 showed that active hydrogen is formed when a glow discharge is passed through hydrogen at a pressure of 0.1 to 1.0 mm.

(*a*) *Bonhoeffer's Method*. Bonhoeffer designed an apparatus for the production of atomic hydrogen by Wood's method and for studying its effect on various substances. The apparatus contains a S-shaped discharge tube, 2 metres long and about 2 centimetres in diameter fitted with two cylindrical electrodes of sheet aluminium. It is connected with a U-tube on one side and to a trap bulb through an experimental tube on the other side. A very slow current of hydrogen is introduced into the U-tube which is surrounded by liquid air to remove water vapour from it. A pressure (0.1 to 1 mm) is maintained in the discharge tube where the gas is subjected to an electric discharge of 3000 to 500 volts when atomic hydrogen is formed. The action of atomic hydrogen on various substances is studied by introducing them in the current of atomic hydrogen coming out of the discharge tube (See Fig. 15.4).

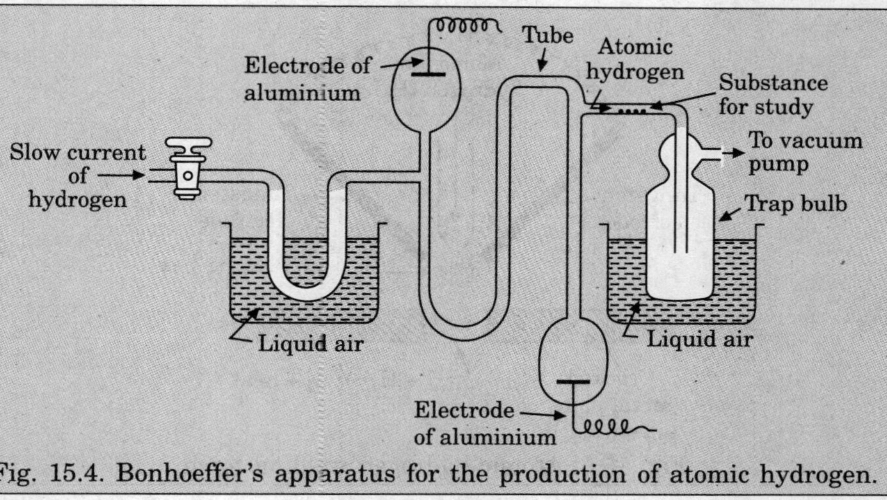

Fig. 15.4. Bonhoeffer's apparatus for the production of atomic hydrogen.

(*b*) *Harteck's Method*. It is a modification of Bonhoeffer's method. In this method a mixture of hydrogen and neon is fed into the discharge tube at pressures upto 20 mm. The presence of neon facilitates the dissociation of hydrogen into atoms.

(iii) By Photochemical Method (Cario and Frank–1922). Ordinary hydrogen mixed with mercury vapour is subjected to the resonance radiation of a mercury arc. The method gives larger yields of atomic hydrogen and at comparatively higher pressure (upto 0.5 atmosphere).

Properties of Atomic Hydrogen. (*i*) *Stability*. The life period of atomic hydrogen is only 1/3rd of a second but can be extended under special circumstances to 10 seconds.

(ii) Recombination of atomic hydrogen to give molecular hydrogen. Atomic hydrogen is extremely unstable (Life period = $\frac{1}{3}$ second) and quickly recombines to give molecular hydrogen. Metals like Pt, Pd etc. accelerate this recombination.

$$H + H \rightleftharpoons H_2 + 103.7 \text{ K.cals}$$

This exothermic reaction is used in *Atomic Hydrogen Welding Torch* which is employed for welding the cracked metals like Mn, Cr, W etc. In atomic hydrogen welding torch when molecular hydrogen is passed through an electric arc struck between two tungsten electrodes, the hydrogen molecules absorb heat and are split into atomic hydrogen.

$$\underset{\text{Molecular hydrogen}}{H_2} + \text{Heat} \longrightarrow \underbrace{H + H}_{\text{Atomic hydrogen}}$$

Just beyond the arc these atoms recombine to form H_2 molecules and evolve the heat which they previously absorbed.

$$\underbrace{H + H}_{\text{Atomic hydrogen}} \longrightarrow \underset{\text{Molecular hydrogen}}{H_2} + \text{Heat}$$

This heat plus the heat of burning hydrogen creates a high temperature (= 5000°C). This heat is used for welding the cracked metals. (See Fig. 16.5) and for melting some of the most refractory materials such as tungsten, tantalum and thorium oxide.

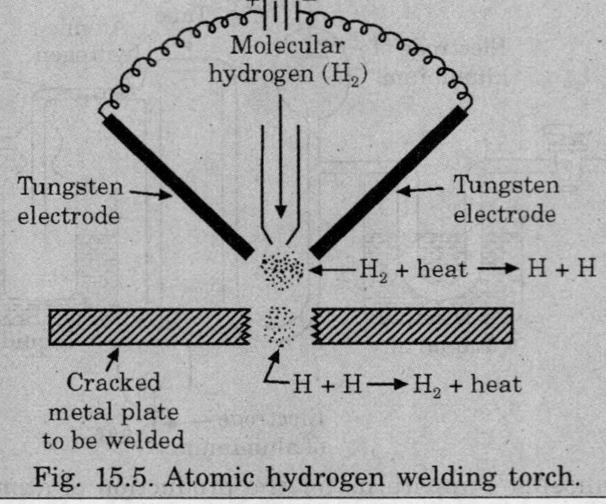

Fig. 15.5. Atomic hydrogen welding torch.

(iii) Formation of hydrides. It combines with metals (*e.g.*, Na, K, Li) and non-metals (*e.g.*, S, P, As, Sb) to give hydrides.

$$S + 2H \longrightarrow H_2S$$
$$Li + H \longrightarrow LiH$$
$$P + 3H \longrightarrow PH_3$$
$$As + 3H \longrightarrow AsH_3$$
$$Sb + 3H \longrightarrow SbH_3$$

(iv) Reducing action. It is an extremely powerful reducing agent and reduces oxides, chlorides and sulphides of Cu, Ag, Hg etc. to metal.

$$HgO + 2H \longrightarrow Hg + H_2O$$

(v) Combination with oxygen. It combines with oxygen to give hydrogen peroxide.

$$2H + O_2 \longrightarrow H_2O_2$$

(vi) Luminiscent effect. A mixture of atomic hydrogen and vapour of alkali metals or mercury gives characteristic glows.

(vii) Combination with organic compounds. It reduces unsaturated organic compounds to saturated compounds, *e.g.* C_2H_4 is converted into C_2H_6 and oleic acid changes into stearic acid.

$$\underset{\text{Ethylene}}{C_2H_4} + 2H \longrightarrow \underset{\text{Ethane}}{C_2H_6}$$

$$\underset{\text{Oleic acid}}{C_{17}H_{33}COOH} + 2H \longrightarrow \underset{\text{Stearic acid}}{C_{17}H_{35}COOH}$$

Uses of Atomic Hydrogen. Atomic hydrogen is used for the production of atomic hydrogen welding torch which is employed for welding and cutting metals as already described under *Properties of atomic hydrogen*.

5. Ortho and Para Hydrogen Molecules

Definition. We know that a hydrogen molecule is composed of two hydrogen atoms. Each of these two H-atoms of H_2 molecule has one proton in its nucleus and one electron revolving round the nucleus. Thus H_2 molecule which is formed from two H-atoms has two protons and two electrons in it. The two electrons in these two H-atoms always spin in *opposite direction* in accordance with Paulie's principle.

$$H(\uparrow) + H(\downarrow) \longrightarrow H \uparrow\downarrow H \text{ or } H—H \text{ or } H_2$$

However, the two protons present in the nucleus of both the H-atoms may spin either in the *same direction* (*i.e.*, *parallel direction*) or in *opposite direction* (*i.e.*, *anti-parallel direction*).

Thus, depending on the direction of the two protons in the nucleus, we have the following two types of hydrogen molecule :

(*i*) **Ortho hydrogen molecule.** *Hydrogen molecule in which both the protons in the nuclei of both H-atoms spin in the same direction (i.e., parallel direction) is called ortho hydrogen* (Fig. 16.6). In this form the resultant nuclear spin is *one*.

(*ii*) **Para hydrogen molecule.** *Hydrogen molecule in which both the protons in the nuclei of both H-atoms spin in opposite direction (i.e., antiparallel direction) is called para hydrogen* (Fig. 16.6). In this form the resultant nuclear spin is zero.

Note that the two electrons in ortho as well as in para hydrogen molecule *always spin in the opposite direction*, otherwise it will not be possible to get stable hydrogen molecule.

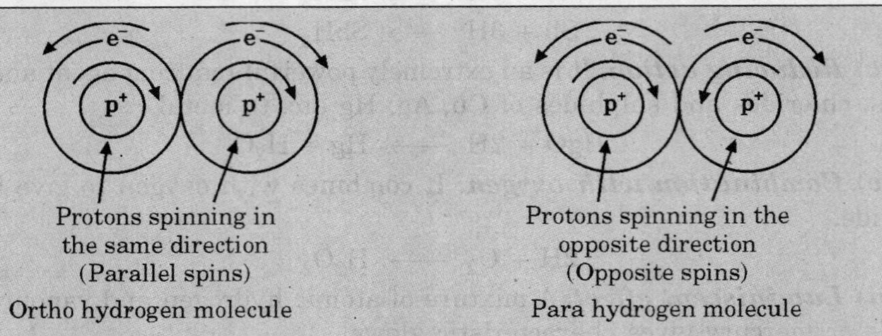

Fig. 15.6. Ortho and para hydrogen molecules. Here P^+ indicates proton.

In 1929, Bonhoeffer and Harteck were also able to show the existence of ortho and para hydrogens from spectral studies.

Differences Between Ortho and Para Hydrogen Molecules.

(i) Internal molecular energy. Mecke discovered in 1934 that the two forms differ in their internal molecular energy. This difference is due to the fact that in case of ortho hydrogen molecule, the spins of the two protons, being in the same direction, increase the molecular energy of ortho form. The resultant spin of ortho hydrogen molecule is *one*.

On the other hand, in case of para hydrogen molecule, the spins of the two protons, being in the opposite direction, neutralise (*i.e.* cancel) each other and thus make the resultant spin equal to *zero*. This neutralisation of spins decreases the molecular energy of para hydrogen molecule.

Thus we see that *para form has lower internal energy than the ortho form.*

(ii) Stability. The ortho form is more stable than the para form and hence the latter (*i.e.*, para form) has a tendency to change to the ortho form.

(iii) Effect of temperature on the relative proportions of the two forms in ordinary hydrogen. It has been shown that the relative proportion of ortho and para hydrogens (*i.e.*, ortho-para ratio) present in ordinary molecular hydrogen is dependent on temperature. *At absolute zero* molecular hydrogen consists *only of para form* (*i.e.*, ratio of para and ortho forms = 1 : 0) which has lower internal energy. *As the temperature increases, the proportion of ortho form increases and that of para form decreases. At the temperature of liquefaction of air, the ratio of para and ortho forms is 1 : 1, while at room temp. and even at elevated temperatures this ratio is 1 : 3. Thus :*

Temperature	*% of para H_2*	*% of ortho H_2*	*Ratio (para : ortho)*
Absolute zero	100%	Nil	1 : 0
Temperature of liquefaction of air	50%	50%	1 : 1
Room temp. and even at elevated temperatures	25%	75%	1 : 3

It is quite evident that at room temperature the ordinary molecular hydrogen is a mixture of para and ortho hydrogen which are present in 1 : 3 ratio.

(iv) Both the forms have different band spectrum. This fact was used to demonstrate the existence of ortho and para hydrogens.

(v) Physical properties. Physical properties like melting point, boiling point, thermal conductivity and specific heat of both the forms are different from each other.

Some physical constants of ortho and para forms are given below. Ortho form is difficult to obtain in a pure state and as such its constants are left blank.

	Ortho form	*Para form*
Boiling point at 760 mm	—	20.26°C
Vapour pressure at 20–39° A :	751 mm (Extrapolated)	737 mm
Triple point	—	13.83°C

Due to the neutralisation of nuclear spins, the magnetic moment of para hydrogen is zero while that of ortho form, as extrapolated from the data for known ortho-para mixture, is approximately twice the momentum of a proton.

The appreciable difference in the thermal conductivities of the two forms has been used to estimate the percentage of ortho and para forms in a given sample of hydrogen.

Although the physical properties of the two forms are different, their chemical properties are similar, because both the forms have same covalent bonding.

Separation of the two Forms from Ordinary Hydrogen : Preparation of Pure Para Hydrogen.

(*i*) We know that ordinary molecular hydrogen at room temperature consists of para and ortho hydrogens in 1 : 3 ratio. In order to separate para form from the ordinary hydrogen, the ordinary hydrogen is passed through a tube packed with activated charcoal and cooled in liquid air or liquid hydrogen for about four hours. The para form thus obtained is pumped off and is about 99.5% pure. It can be stored for a week in glass vessels at room temperature without appreciable change into ortho form.

(*ii*) The two forms can be separated from each other by gas chromatography.

Conversion of Para Hydrogen into Ortho Hydrogen.

Ortho hydrogen is more stable than the para form and the latter has a tendency to change to the ortho form. This process is accelerated under certain conditions and it is possible to convert the para form into ortho.

(*i*) *By heating it to 800°C or above.*

(*ii*) *By passing an electric discharge through it.*

(*iii*) *By treatment with catalysts like plantinum.*

(*iv*) *By mixing it with paramagnetic molecules like oxygen or nitrogen peroxide.*

(*v*) *By mixing with atomic hydrogen.* It was shown by Geib and Herteck that when atomic hydrogen collides with para hydrogen molecule, the para hydrogen molecule is converted into ortho molecule as depicted below in Fig. 15.7.

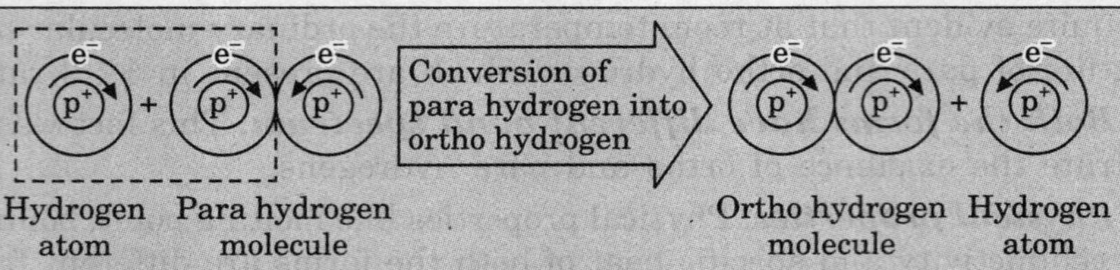

Fig. 15.7. Conversion of para hydrogen into ortho hydrogen

Isotopes of Hydrogen

It has been established by means of mass spectrograph that hydrogen has *three* isotopes with mass number 1, 2 and 3. These isotopes are discussed below :

(*i*) Hydrogen or Protium ($_1{}^1H$ or H). It is represented as $_1H^1$ in which the numbers 1 and 1 indicate the atomic number and mass number (*i.e.* atomic mass or atomic weight) of this isotope. Its exact mass number is, however, 1.0081. Its nucleus consists of one proton and no neutron. It has one electron in 1*s* energy level.

(*ii*) Heavy Hydrogen or Deuterium ($_1{}^2H$ or $_1{}^2D$). Its actual mass number is 2.0147. Its nucleus consists of one proton and one neutron. It has also one electron in 1*s* energy level.

(*iii*) Tritium ($_1{}^3H$ or $_1{}^3T$). Its mass number is 3.0170. Its nucleus consists of one proton and two neutrons. It has one electron in 1*s* energy level.

Some of the properties of these isotopes are given Table 16.1.

Table 15.1. Some properties of the three isotopes of hydrogen

Property	Ordinary hydrogen or protium	Heavy hydrogen or deuterium	Tritium
Symbol	$_1{}^1H$ or H	$_1{}^2H$ or $_1{}^2D$	$_1{}^3H$ or $_1{}^3T$
Atomic number	1	1	1
Mass number (*i.e.* atomic mass or atomic weight)	1.0081	2.0147	3.0170
Atomicity	2	2	2
Molecular formula	H_2	D_2	T_2
No. of protons in the nucleus	1	1	1
No. of neutrons in the nucleus	Nil	1	2
No. of electrons	1	1	1
Relative abundance of their diatomic molecules in ordinary hydrogen	99.984%	0.0156%	10^{-15}%
Stability	Stable	Stable	Radioactive and hence unstable
Electronic configuration	$1s^1$	$1s^1$	$1s^1$

Structure of Isotopes

The structure of three isotopes of hydrogen has already been discussed.

It has been estimated that a sample of ordinary hydrogen contains 99.994% of H_2 and 0.0156% of D_2. 10^{17} molecules of ordinary hydrogen contain only one molecule of tritium. The cause of the presence of tritium in ordinary hydrogen in traces only is that the nucleus of tritium is radioactive and hence unstable.

Since the quantities of D_2 and T_2 present in natural hydrogen are exceedingly small, the properties of natural hydrogen are substantially those of protium.

Isotope Effects

Since the electronic configuration of all the three isotopes is the same *viz.* $1s^1$, their chemical properties are identical. However, these isotopes show a quantitative different in their equilibrium constants and rates of reaction, *e.g.* the dissociation constant (K) for heavy water is significantly smaller than that for ordinary water.

$$H_2O \rightleftharpoons H^+ + OH^- \; ; \; K = 1.0 \times 10^{-10}$$

$$D_2O \rightleftharpoons D^+ + OD^- \; ; \; K = 0.2 \times 10^{-14}$$

The difference in the masses of isotopes of hydrogen have marked effects on the rates of chemical reactions *e.g.*

(*i*) A bond to a protium atom can be broken as much as 18 times faster than that to a deuterium atom. Thus protium reacts with Cl_2 13.4 times faster than deuterium does.

(*ii*) The addition of H_2 to ethylenic compounds takes place two times faster than with deuterium at the same temperature.

(*iii*) Protium is more rapidly adsorbed on the surfaces than deuterium.

The above examples show that the reactions with deuterium are slower than those with protium. These differences in the rates of chemical reaction are due to the difference in the masses of deuterium and protium.

Such differences in properties which are due to the difference in masses of the isotopes are termed **isotope effects.**

Now let us discuss these isotopes in detail.

1. Ordinary Hydrogen or Protium ($_1^1H$ or H)

It is ordinary hydrogen whose nucleus consists of one proton and one electron is revolving in 1*s* orbital round it.

2. Heavy Hydrogen or Deuterium ($_1^2$(H or $_1^2$)D)

Discovery. The credit of the discovery of heavy hydrogen goes to *Urey* who was awarded a Nobel Prize in 1934 for this discovery.

In 1931 *Birge* and *Manzel* reported that the atomic weight of hydrogen as determined by mass spectrograph was 1.0076 while that determined by chemical methods such as displacement of hydrogen from H_2SO_4 by Zn metal, was 1.0078. In order to explain this discrepency in atomic weights, they suggested that ordinary hydrogen contains a very small proportion of an isotope of mass 2.

About the same time, *Allison* and *his coworkers* also reported that ordinary hydrogen is a mixture of two isotopes of hydrogen. The heavier isotope has mass 2 and is present in the ratio of 1 part in about 5000 parts.

Encouraged by the clues and the calculations shown above *Urey, Brickwedde* and *Murphy* evaporated 4 litres of liquid hydrogen under reduced pressure and at a temperature of freezing hydrogen. On concentrating the residue thus obtained, they examined the spectrum and found that it contained very faint lines. They assumed that the faint lines were due the heavy isotopes of hydrogen with atomic weight 2. Afterwards the masses of light and heavy isotopes of hydrogen were more accurately determined as 1.0081 and 2.0147 respectively. They named this heavier isotope as **deuterium** and gave the symbol **D** to it.

Preparation. Deuterium is prepared by the following methods :

***(i)* By separating deuterium from the mixture of protium and deuterium.** Water is decomposed by metals to give a gas which consists of varying amounts of H_2, HD and D_2. Since hydrogen, being lighter than deuterium, is liberated faster than deuterium from water, the first few fractions of the gas are richer in hydrogen while the last fractions contain greater proportion of deuterium. The final fraction of the gas is collected and as electric discharge is passed through it, whereby the atoms of hydrogen and deuterium combine to give a mixture containing H_2 and D_2 molecules. On electrolysis HD also gives H_2 and D_2 molecules.

Now from the mixture of H_2 and D_2, we can separate D_2. The methods used for the separation of D_2 from the mixture are given below :

(*a*) *Diffusion process.* The separation of deuterium from the mixture of deuterium and protium (*i.e.* hydrogen) can be effected by the diffusion process. This *process makes the use of difference in the rates of diffusion of the two isotopes, viz.* protium ($_1H^1$) and deuterium ($_1D^2$). Since protium is lighter, it diffuses more readily than deuterium. Although this method is slow and tedious, yet it gives spectroscopically pure deuterium, *i.e.*, the spectrum of deuterium obtained by this method shows no trace of ordinary hydrogen.

The process is carried out under reduced pressure since lower the pressure, the greater is the efficiency of the process.

The apparatus used in this process consists of a number of porous diffusion units which are called *Hertz diffusion units* (This method is due to Hertz) and are placed in series as shown in Fig. 15.8. The apparatus is called *Hertz apparatus*. Each unit contains a porous membrane shown by dotted portion in the figure and is connected with mercury diffusion pump (P_1, P_2, P_3......). With the help of these diffusion pumps the mixture of the two isotopes is allowed to circulate into the diffusion units at the reduced pressure.

As the mixture of the two isotopes is allowed to pass into the diffusion units under reduced pressure, say, from left to right, with the help of mercury diffusion pumps, the lighter isotope *viz.* protium diffuses faster and its proportion increases on the right side, *i.e.* it is collected in the right reservoir (placed on the right). On the other hand, the heavier isotope *viz.* deuterium is left behind and is collected into the left reservoir (placed on the left). This process is repeated a number of times to get pure deuterium.

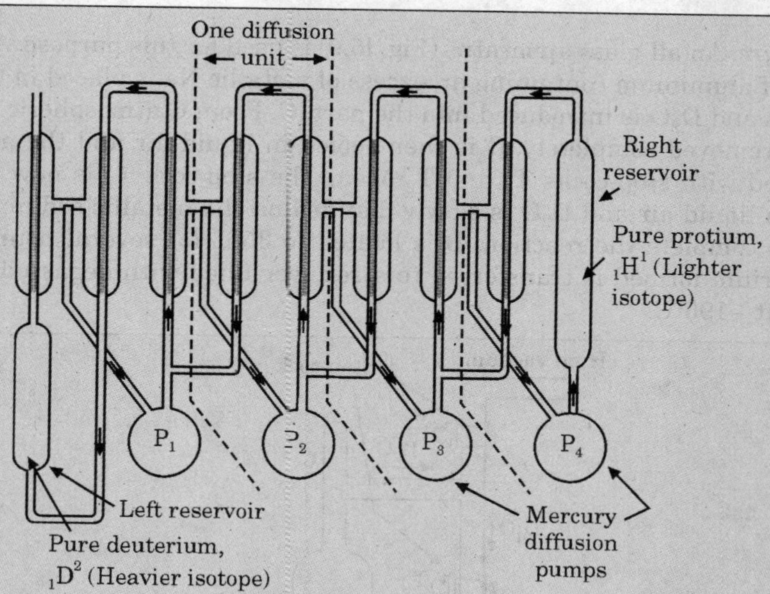

Fig. 15.8. Hertz apparatus for the separation of deuterium from the mixture of protium and deuterium.

(*b*) *Fractional distillation of liquid hydrogen isotopes process.* The mixture of the two isotopes is first of all liquefied by cooling and compression. The liquid so obtained is then slowly distilled under reduced pressure and at the temperature of freezing hydrogen. On concentrating the residue, enriched deuterium is obtained from which pure deuterium can be obtained by the fractional diffusion process as discussed in (*a*).

(*c*) *Adsorption on charcoal process*. This method is based on the fact that protium gets adsorbed on the surface of coconut charcoal more rapidly than deuterium. Thus when a mixture of protium and deuterium is kept in contact with coconut charcoal at the temperature of liquid air for a few hours, most of protium gets adsorbed on the surface of the charcoal while deuterium is collected.

***(ii)* From heavy water.** Ordinary water contains about 1 part in 6000 parts of what is known as *heavy water*, the molecule of which consists of two heavy hydrogen atoms and one oxygen atom (D_2O). Heavy water is obtained by the electrolysis of an aqueous alkali solution, between nicked electrodes, 20 litres of the solution yielding about 0.5 c.c. of pure heavy water. Deuterium can be obtained from this heavy water, D_2O by using any of the following methods :

(*a*) *By the action of sodium on D_2O.* Fairly pure sample of heavy hydrogen may be obtained by the action of sodium on heavy water

$$2D_2O + 2Na \longrightarrow \underset{\text{Sod. deutroxide}}{2NaOD} + D_2$$

The drawback of this method is that a part of deuterium is lost as NaOD. This method is used when a comparatively small quantity (upto 1 lit) of deuterium is required.

Procedure. An all-glass apparatus (Fig. 15.9) is used for this purpose. A crucible (U) made of aluminium containing an excess of metallic Na is placed in the lower part, B of A and D_2O is introduced into the part, C. From C atmospheric moisture should be removed completely. C is then cooled in liquid air and the apparatus is evacuated with stop-cocks T_1 and T_2 open. The stop-cock T_2 is now closed, B is cooled in liquid air and D_2O is slowly distilled on the metallic sodium. Finally in order to complete the reaction, B is heated to 35°C for several hours. At the end, deuterium formed is transferred to a receiver filled with degassed charcoal and kept at –196°C.

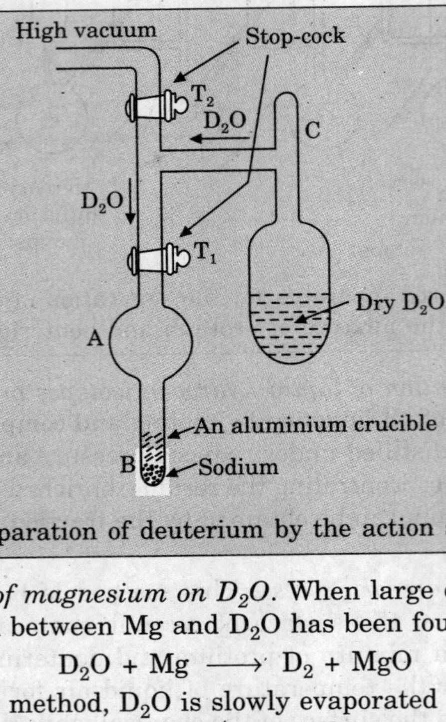

Fig. 15.9. Preparation of deuterium by the action sodium on D_2O.

(*b*) *By the action of magnesium on D_2O*. When large quantities of deuterium are required, reaction between Mg and D_2O has been found to be more suitable.

$$D_2O + Mg \longrightarrow D_2 + MgO$$

Procedure. In this method, D_2O is slowly evaporated from a Pyrex flask that has been previously evacuated to 10^{-4} mm. The vapour passes up a reaction column, set up vertically on the top of the flask. The column is charged with magnesium shavings and is heated to about 500° by means of a furnace. The gas of deuterium produced is very pure and the yield is quantitative.

(*c*) *By the action of uranium on D_2O*. This method not only gives deuterium but also gives us a method for its storage as *uranium deuteride*, UD_3. UD_3 when heated gives D_2 of high purity. The reactions are :

(*i*) Preparation of D_2 : $2D_2O + U \longrightarrow UO_2 + 2D_2$.

(*ii*) Storage of UD_3 : $3D_2 + 2U \longrightarrow 2UD_3$.

(*iii*) Decomposition of UD_3 : $2UD_3 \longrightarrow 3D_2 + 2U$.

With the help of reaction (*iii*) we can get D_2 in any desired quantity.

(*d*) *By the decomposition of D_2O by heated tungsten filament.* This method has been given by *Forkas* and *Farkas* and is useful when small volumes of D_2 are required. The apparatus consists of a decomposition chamber provided with a tungsten filament (*See* Fig. 15.10).

One side of the chamber is connected with a small receiver containing heavy water and the other to a pump. Before beginning the experiment the apparatus is heated to about 1000°C and the tungsten filament to 2000°C as to remove the whole of the absorbed water. The whole apparatus is connected to a high vacuum pump during this preliminary treatment. The receiver containing heavy water is heated and vapours of D_2O so obtained are led to the decomposition chamber where they are decomposed to D_2 by the electrically heated tungsten filament.

D_2O (*Vapours*) + 2W (*Heated*) $\longrightarrow$ W_2O + D_2

W_2O is volatile and is subsequently condensed in the U-tube following the decomposition chamber and D_2 is carried further for collection. It is stated that a wire of tungsten 0.2 mm in thickness and 20 cm long ordinarily gives 80 ml of D_2O.

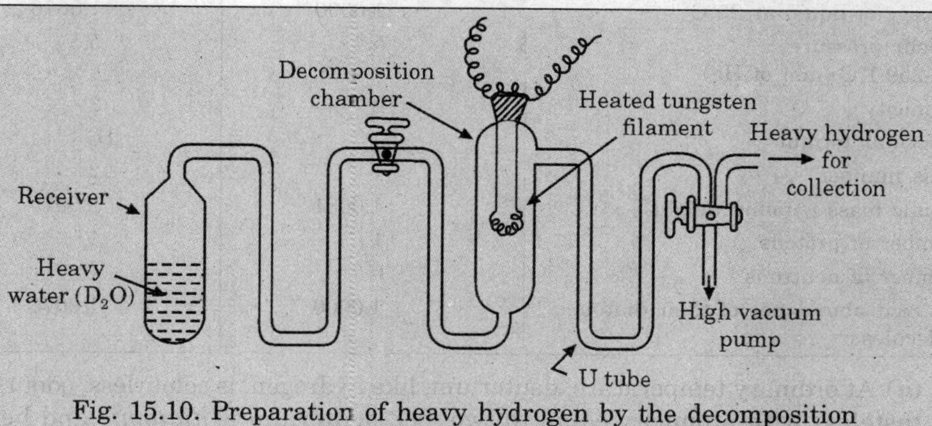

Fig. 15.10. Preparation of heavy hydrogen by the decomposition of heavy water (D_2O) by heated tungsten filament.

(*e*) *By the decomposition of D_2O by heated Zn turnings.* When D_2 is required in large quantities, D_2O is decomposed by heated Zn turnings to get D_2.

$$\text{Zn (Heated)} + D_2O \text{ (Vapours)} \longrightarrow \text{ZnO} + D_2$$

(*f*) *By the electrolysis of D_2O.* D_2 can also be prepared by electrolysing pure D_2O acidified with 35% deuteriumsulphuric acid (D_2SO_4) in a specially designed all-glass apparatus. The apparatus is evacuated before electrolysis and a low current of 5 amperes is passed between platinum electrodes when D_2 is set free at the cathode and O_2 at the anode.

$$2D_2O \longrightarrow \underset{\text{(At cathode)}}{2D_2} + \underset{\text{(At anode)}}{O_2}$$

Deuterium obtained by this method may contain small quantities of O_2 and D_2O vapour. It is purified by heating it over platinised asbestos, followed by drying with liquid air. The yield at 5 amp. is reported to be 2 litres per hour.

Physical Properties. (*i*) *Physical constants.* Physical properties of deuterium are some what different from those of ordinary hydrogen. Some of the physical constants of the two isotopes are given in Table 16.2.

Table 15.2. Physical constant of ordinary (protium) and heavy (deuterium) hydrogen.

Property	Ordinary hydrogen (Protium, H_2)	Heavy hydrogen (Deuterium, D_2)
Symbol	H	D
Boiling point (°C)	–252.6	–249.5
Melting point (°C) (Freezing point)	–259.1	–254.5
Triple point (°C)	–259.1	–254.3
Latent heat of fusion (cal/mole)	28.0	52.3
Heat of dissociation (cal/mole)	104,500	14,900
Heat of vaporisation at triple point (cal/g. mole)	219.7	308
Molecular volume of solid (c.c.)	26.15	23.17
Density of liquid at 25°C	1.0000	1.0076
Vapour pressure at –259.1°C (mm of Hg)	5.4	5.8
Atomicity	2	2
Molecular formula	H_2	D_2
Mass number	1	2
Atomic mass (Atomic weight)	1.0081	2.0015
Number of protons	1	1
Number of neutrons	0	1
Per cent abundance of their diatomic molecules	1.0000	1.0076

(*ii*) At ordinary temperature, deuterium, like hydrogen, is colourless, odourless and tasteless with atomicity equal to two, *i.e.*, deuterium is *diatomic* and hence its molecular formula is written as D_2. Its density is twice that of hydrogen.

(*iii*) It is insoluble in water and bad conductor of heat and electricity.

(*iv*) *Ortho and para forms of deuterium.* The deuterium molecule also, like hydrogen is a mixture of ortho and para forms. At ordinary temperature. It consists of 2 parts of ortho and 1 part of para and this composition is retained even at higher temperatures. At low temperature, it is only ortho form which predominates while in normal hydrogen the para form predominates at ordinary temperature.

Chemical Properties. The chemical properties of deuterium are similar to those of hydrogen, the only difference being that deuterium reacts more slowly and less completely as compared to hydrogen. Some of the chemical properties of deuterium are given below :

(*i*) **Occlusion.** Deuterium is occluded by metallic surfaces more slowly than hydrogen.

(*ii*) **Action of oxygen.** It is combustible and burns in oxygen at 560°C to form heavy water

$$2D_2 + O_2 \longrightarrow 2D_2O \text{ (\textit{Heavy water})}$$

(*iii*) **Action of water.** It reacts with water vapour at 200–400°C in presence of zinc oxide or alumina to form a mixture of deuterium oxide and hydrogen.

$$H_2O + D_2 \rightleftharpoons D_2O + H_2$$

(*iv*) **Action of halogens.** It combines with halogens to form deuterium halides.

$$D_2 + Cl_2 \xrightarrow{\text{In light}} 2DCl \text{ (\textit{Deuterium chloride})}$$

$$D_2 + Br_2 \xrightarrow{\text{On heating}} 2DBr \text{ (\textit{Deuterium bromide})}$$

$$D_2 + F_2 \xrightarrow[\text{dark}]{\text{In cold and}} 2DF \text{ (\textit{Deuterium fluoride})}$$

(*v*) **Combination with nitrogen.** Deuterium combines with nitrogen in the presence of a catalyst to form deutero-ammonia or heavy ammonia.

$$3D_2 + N_2 \longrightarrow 2ND_3$$

(*vi*) **Combination with metals.** It combines with alkali metals at 300°C to form deutrides which closely resemble metallic hydrides in their chemical properties.

$$2Na + D_2 \longrightarrow 2NaD \text{ (Sodium \textit{deuteride})}$$

$$NaD + H_2O \longrightarrow NaOH + HD$$

(*vii*) **Reaction with unsaturated hydrocarbons.** Deuterium gives addition reactions with unsaturated hydrocarbons. The reaction, however, is slower as compared with addition reaction with hydrogen.

$$\underset{\text{Ethylene}}{\begin{matrix} CH_2 \\ \| \\ CH_2 \end{matrix}} + D_2 \xrightarrow[260^\circ]{Ni} \underset{\text{Dideutero ethane}}{\begin{matrix} CH_2D \\ | \\ CH_2D \end{matrix}}$$

(*viii*) **Exchange reactions.** Under suitable conditions, it is possible for deuterium to take the place of hydrogen in some compounds, *e.g.*, at high temperature deuterium reacts with H_2, NH_3, H_2O and CH_4 and slowly exchanges hydrogen atoms of these molecules. The exchange of hydrogen by deuterium may take place partially or completely.

$$2NH_3 + 3D_2 \rightleftharpoons \underset{\substack{\text{Deuterium} \\ \text{ammonia}}}{2ND_3} + 3H_2$$

$$H_2O + D_2 \rightleftharpoons \underset{\text{Deuterium oxide}}{D_2O} + H_2$$

$$CH_4 + 2D_2 \longrightarrow \underset{\text{Deuterium methane}}{CD_4} + 2H_2$$

Some exchange reactions take place in presence of hydrogenation catalyst such as finely divided nickel, palladium or platinum black.

$$C_6H_6 + 3D_2 \underset{\text{black}}{\overset{\text{Platinum}}{\rightleftharpoons}} \underset{\text{Deutero benzene}}{C_6D_6} + 3H_2$$

Exchange reactions occur with great readiness when they involve ions in solutions, *e.g.*, in alkaline solution both acetone and acetyl acetone exchange all of their hydrogen atoms for deuterium because of their ionisation.

$$\underset{\text{Acetone}}{CH_3{-}\overset{\overset{O}{\|}}{C}{-}CH_3} \xrightarrow{\text{Alkali}} CH_3{-}\overset{\overset{OH}{|}}{C} = CH_2 \underset{-3H_2}{\overset{+3D_2}{\rightleftharpoons}} CD_3{-}\overset{\overset{OD}{|}}{C} = CD_2$$

$$\rightleftharpoons \underset{\text{Deutro acetone}}{CD_3{-}\overset{\overset{O}{\|}}{C}{-}CD_3}$$

Similarly, acetylene undergoes deuterium exchange in alkaline solution because of the weakly acidic character of C—H bond in acetylene.

$$C_2H_2 \overset{\text{Alkali}}{\rightleftharpoons} \underset{\text{Deutero acetylene}}{C_2D_2} + H_2$$

(*ix*) **Reducing property.** At 700°C, AgCl is reduced to metallic silver, when deuterium is passed through its solution.

$$D_2 + 2AgCl \longrightarrow 2Ag + 2DCl$$

Uses. (*i*) It has been found as a source of *deutrons* which are capable of initiating several nuclear transformations through the formation of compound nucleus at relatively low energies. Some of nuclear transformations produced by deutrons are given below :

(*a*) (*D*, α) *reactions* :

$$_3Li^6 + {_1D^2} \longrightarrow [_4Be^8] \longrightarrow {_2He^4} + {_2He^4}$$

$$_8O^{16} + {_1D^2} \longrightarrow [_9F^{18}] \longrightarrow {_7N^{14}} + {_2He^4}$$

(*b*) (*D, p*) *reactions* :

$$_{15}P^{31} + {_1}D^2 \longrightarrow [_{16}S^{33}] \longrightarrow {_{15}}P^{32} + {_1}H^1$$

$$_{11}Na^{22} + {_1}D^2 \longrightarrow [_{12}Mg^{32}] \longrightarrow {_{11}}Na^{24} + {_1}H^1$$

(*c*) (*D, n*) *reactions* :

$$_6C^{12} + {_1}D^2 \longrightarrow [_7N^{14}] \longrightarrow {_7}N^{13} + {_0}n^1$$

$$_1T^3 \text{ (or } {_1}H^3) + {_1}D^2 \longrightarrow [_2He^5] \longrightarrow {_2}He^4 + {_0}n^1 + \text{energy}$$

The energy liberated is used in hydrogen bomb.

(*d*) (*D, T*) *reactions* :

$$_3Li^7 + {_1}D^2 \longrightarrow [_4Be^9] \longrightarrow {_3}Li^6 + {_1}T^3$$

$$_{15}P^{31} + {_1}D^2 \longrightarrow [_{16}S^{33}] \longrightarrow {_{15}}P^{30} + {_1}T^3$$

(*ii*) It is used in the preparation of hydrogen bomb.

(*iii*) It has been used in the preparation of deuterium compounds which are very useful, *e.g.*, D_2O is used as a *moderator* in nuclear reactors to slow down the speed of fast moving neutrons.

(*iv*) Deuterium and its compounds are used as a *tracer* in the study of mechanism of various chemical and metabolic reactions taking place in the body, *e.g.*, with the help of deuterium it has been shown that butyric acid (a component of butter) is never stored in the body and is an immediate source of energy which at once gets consumed while stearic acid (a component of lard) is not used up at once and hence is useful in building up reverse stock in the body.

3. Tritium ($_1^3$H or $^3_1$T)

Occurrence. It is an extremely rare isotope of hydrogen. It has been established that 10^{17} molecules of ordinary hydrogen contain only traces of tritium, *i.e.*, only 10^{-15}% of ordinary hydrogen is tritium. The cause of the presence of tritium in ordinary hydrogen only in traces is that the nucleus of tritium is radioactive and hence unstable. Evidently it is very difficult to separate tritium from ordinary hydrogen. 10^{17} parts of ordinary water contain about 7 parts of tritium.

In nature, the traces of tritium are present. These are formed continuously in the upper atmosphere by the action of neutrons (produced by the action of cosmic radiations on the atmospheric gases) on atmospheric nitrogen.

$$_7N^{14} + {_0}n^1 \longrightarrow {_6}C^{12} + {_1}T^3$$

Preparation. (*i*) *Oliphant, Harteek* and *Rutherford* (1934) obtained tritium by bombarding compounds such as D_3PO_4, ND_4Cl, $(ND)_2SO_4$ etc., with high energy deuterons.

$$\underset{\text{Deuterium}}{_1D^2} + \underset{\text{Deuteron}}{_1D^2} \longrightarrow \underset{\text{Tritium}}{_1T^3} + {_1}H^1$$

This is the first method of its preparation which led to its discovery.

(*ii*) *By the nuclear reactions produced by deutrons.*

$$_4Be^9 + {_1D^2} \longrightarrow {_4Be^8} + {_1T^3}$$

$$_6Li^7 + {_1D^2} \longrightarrow [_4Be^9] \longrightarrow {_3Li^6} + {_1T^3}$$

$$_{47}Ag^{109} + {_1D^2} \longrightarrow [_{48}Cd^{111}] \longrightarrow {_{47}Ag^{108}} + {_1T^3}$$

$$_{15}P^{31} + {_1D^2} \longrightarrow [_{16}S^{33}] \longrightarrow {_{15}P^{30}} + {_1T^3}$$

(*iii*) *By neutron bombardment of boron and nitrogen* (*i.e.*, neutron bombarding against saturated aqueous solution of boric acid and ammonium nitrate) :

$$_5B^{11} + {_0n^1} \longrightarrow [_5B^{12}] \longrightarrow {_4Be^9} + {_1T^3}$$

$$_7N^{14} + {_0n^1} \longrightarrow [_7N^{15}] \longrightarrow {_6C^{12}} + {_1T^3}$$

(*iv*) *By slow neutron bombardment of lithium* :

$$_3Li^6 + {_0n^1} \longrightarrow {_1T^3} + {_2He^4}$$

This is probably the best method of obtaining tritium.

Properties. (*i*) Because of the unfavourable neutron-to-proton ratio, tritium is a radioactive isotope with its half-life period of 12.4 years. It emits low energy β-particles and is changed into a lighter isotope of helium, $_2He^3$.

$$_1T^3 \longrightarrow {_2He^3} + \underset{\beta\text{-particle}}{_{-1}e^0}$$

The radioactive decay of tritium is the source of the traces of $_2He^3$ in the atmosphere.

(*ii*) Some of its physical constants are given below :

Boiling point = –248.1°C; Melting point (Tripole point) = –252.4°C; Heat of sublimation at triple point = 393.0 cals/mole; Heat of vaporisation at normal boiling point = 333 cals/mole; Vapour pressure at the triple point = 162.00 mm.

(*iii*) Its atomicity is 2 and hence its molecular formula is T_2.

(*iv*) Its chemistry is very nearly identical with that of hydrogen. However, reactions with tritium are still slower than those with deuterium.

(*v*) *Action of H_2*. It combines with H_2 on heating (300–400°C) to form tritium hydride, HT.

$$H_2 + T_2 \longrightarrow 2HT$$

HT reacts with HCl to give tritium chloride, TCl.

$$HT + HCl \longrightarrow TCl + H_2$$

(*vi*) *Action of O_2*. When burnt in O_2, it forms tritium oxide, T_2O.

$$2T_2 + O_2 \longrightarrow 2T_2O$$

Uses. (*i*) It has the possibility of use as an artificial radioactive tracer in chemistry, medicine and biology.

(*ii*) It is used in nuclear fusion reactions such as :

$$_1D^2 + {_1H^3} \longrightarrow {_2He^4} + {_0n^1} + \text{Energy}$$

In this reaction huge amount of energy is liberated which is used in hydrogen bomb.

(*iii*) It is a source of tritons which are used in several nuclear transformations. For example :

(*a*) (T, p) *reactions* :

$$_{27}Co^{59} + {}_1T^3 \longrightarrow {}_{27}Cu^{64} + {}_1H^1 \text{ (or } {}_1p^1)$$

(*b*) (T, D) *reactions* :

$$_3Li^6 + {}_1T^3 \longrightarrow {}_3Li^7 + {}_1D^2$$

$$_{29}Cu^{63} + {}_1T^3 \longrightarrow {}_{29}Cu^{64} + {}_1D^2$$

(*c*) (T, n) *reactions* :

$$_{16}S^{32} + {}_1T^3 \longrightarrow {}_{17}Cl^{34} + {}_0n^1$$

(*d*) (T, α) *reactions* :

$$_3Li^7 + {}_1T^3 \longrightarrow {}_2He^6 + {}_2He^4$$

(*iv*) It is also of potential use in thermo-nuclear process.

(*v*) Tritium hydride, TH, is a very important compound of tritium. It is used in the preparation of a number of organic compounds containing this radioactive isotope.

Questions with Answers

Q.1 Explain the following :

(*a*) Hydrogen is not prepared by action of concentrated sulphuric acid on zinc.

Ans : Conc. H_2SO_4 reacts with zinc to form SO_2 rather than H_2

$$Zn + 2\,H_2SO_4\,(\text{Conc.}) \xrightarrow{\text{Heat}} ZnSO_4 + SO_2 + 2H_2O$$

(*b*) A solution of ferric chloride is unaffected when hydrogen is bubbled through it, but gets reduced when zinc is added to the same acidified solution.

Ans : Ordinary H_2 is less reactive and hence it does not reduce acidified $FeCl_3$ solution. However, when zinc is added to acidified $FeCl_3$ solution, nascent hydrogen thus produced is associated with more energy. Consequently, it is more reactive and hence reduces acidified $FeCl_3$ solution.

$$FeCl_3 + H_2 \longrightarrow \text{No reaction}$$

$$Zn + H_2SO_4 \longrightarrow ZnSO_4 + H_2^* \text{ (nascent hydrogen)}$$

$$\underset{(Yellow)}{2FeCl_3} + H_2^* \longrightarrow \underset{(Green)}{2FeCl_2} + 2HCl$$

(*c*) Dihydrogen is not preferred in balloons these days ?

Ans : Dihydrogen is highly combustible and hence is likely to catch fire in presence of excess of air.

(*d*) Hydrogen is unreactive at room temperature.

Ans : Due to high enthlapy of H–H bond, hydrogen is quite unreactive at room tempreature. However, at high tempreatures or in presence of catalysts, it combines with many metals and non-metals to form hydrides.

(*e*) Hydrogen occurs in diatomic form rather than in monoatomic form under normal conditions ?

Ans : Hydrogen atom has only one electron and thus has one electron less than the stable inert gas configuration of helium. Therefore, to achieve stable inert gas configuration of helium, it shares its single electron with the electron of other hydrogen atom to form a stable diatomic molecule.

(*f*) Conc. H_2SO_4 cannot be used for drying H_2.

Ans : Conc. H_2SO_4, on absorbing H_2O from anoist H_2, produces so much heat that H_2 catches fire.

Q.2 Name one example of a reaction in which dihydrogen acts (*i*) as an oxidising agent and (*ii*) as a reducing agent.

Ans : (*i*) **As oxidising agent.** In the reaction of dihydrogen with metals to form metal hydrides, it acts as an oxidising agent.

$$\underset{(Na\,=\,0)}{2\,Na(s)} + \underset{(H\,=\,0)}{H_2(g)} \xrightarrow{\text{Heat}} \underset{(Na\,=\,+1,\ H\,=\,-1)}{2\,Na^+\,H^-(s)}$$

Here, Na has been oxidised to Na^+ while dihydrogen has been reduced to hydride (H^-) ion.

(*ii*) **As reducing agent.** In the reaction of heated cupric oxide with dihydrogen to form H_2O and copper metal, dihydrogen acts as a reducing agent.

$$\underset{(Cu\,=\,+2,\ O\,=\,-2)}{CuO\ (s)} + \underset{(H\,=\,0)}{H_2\ (s)} \xrightarrow{\text{Heat}} \underset{(Cu\,=\,0)}{Cu\ (s)} + \underset{(H\,=\,+1,\ O\,=\,-2)}{H_2O\ (g)}$$

Q.3 Write the names of isotopes of hydrogen. What is the mass ratio of these isotopes ?

Ans : Protium ^1_1H, Deuterium ^2_1H or D, Tritium ^3_1H or T. The mass ratio of protium : deuterium : tritium = 1 : 2 : 3.

Q.4 Which istope of hydrogen (*i*) does not contain neturons (*ii*) contains equal number of protons and neutrons (*iii*) is radioactive.

Ans : (*i*) protium (*ii*) deuterium (*iii*) tritium.

Q.5 Name one compound each in which hydrogen exists in (*i*) + 1 (*ii*) –1 (*iii*) zero oxidation state.

Ans : (*i*) HCl or H_2O or NH_3 (*ii*) NaH or CaH_2 (*iii*) H_2.

Q.6 Give two advantages of using hydrogen over gasoline as a fuel.

Ans : High heat of combustion and no pollutants like SO_2, NO_2, CO_2, etc.

Q.7 Which isotope of hydrogen is used as a tracer in organic reactions.

Ans : Hydrogen has three isotopes, *viz*, H, D and T. Due to difference in masses, the rate constants of these isotopes with the same subtrate are different.

In other words, both D and T show isotope effect. But since T is not only radioactive but is also least abundant hydrogen isotope, D is used as a tracer to study the mechanism of organic reactions.

Q.8 How can the production of dihydrogen obtained from coal gasification be increased ?

Ans : The process of producing syngas or synthesis gas from coal is called coal gasification.

$$\underset{\text{Coal}}{C(s)} + \underset{\text{Steam}}{H_2O(g)} \xrightarrow[\text{Ni}]{1270K} \underbrace{CO(g) + H_2(g)}_{\text{Syngas}}$$

The production of hydrogen can be increased by reacting carbon monoxide of syngas with steam in presence of iron chromate as catalyst at 673K.

$$CO(g) + H_2O(g) \xrightarrow{FeCrO_4,\ 673K} CO_2(g) + H_2(g)$$

CO_2 thus produced is removed either by scrubbing with a solution of sodium arsenite or by passing the mixture through water under 30 atm. pressure when CO_2 dissolves leaving behind H_2 which is collected.

Q.9 How does H_2 react with Cl_2 ?

Ans : Dihydrogen reduces chlorine to Cl^- ion and itself gets oxidised to H^+ ions. An electron pair is then shared between these two species to form a cvalent molecule of hydrogen chloride.

$$\underset{(H = 0)}{H_2(g)} + \underset{(Cl = 0)}{Cl_2(g)} \longrightarrow \underset{(H = +1,\ Cl = -1)}{2\,HCl(g)}$$

Q.10 Complete the following reactions :

(*i*) $H_2(g) + M_mO_o(s) \xrightarrow{\Delta}$ ...

(*ii*) $CO(g) + H_2(g) \xrightarrow{\Delta}$ ...

Ans : (*i*) $oH_2(g) + M_mO_o(s) \xrightarrow{\Delta} mM(s) + oH_2O(l)$

(*ii*) $CO(g) + 2H_2(g) \xrightarrow[\text{Catalyst}]{\Delta} \underset{\text{Methanol}}{CH_3OH(l)}$

Q.11 What do you understand by the terms : (*i*) hydrogen economy (*ii*) hydrogenation (*iii*) syngas (*iv*) water-gas shift reaction and (*v*) fuel cell ?

Ans : (*i*) Hydrogen economy. The proposal to use hydrogen as a fuel in industry, power plants and possibly also in homes and motor vehicles is called hydrogen economy. The basic principle of hydrogen economy is the storage and transportation of energy in the form of liquid or gaseous dihydrogen.

(*ii*) Hydrogentation means addition of hydrogen across double and triple bonds to form saturated compounds. The vegetable oils such as soyabean oil, cotton seed oil, groundnut oil, are called polyunsaturated oils since they contain many C = C bonds. When these oils are exposed to air for prolonged periods, the double bonds present in them undergo oxidation, *i.e.* develop unpleasant taste. To avoid this, double bonds are hydrogenated. For this purpose, dihydrogen is bubbled through edible oils in presence of finely divided nickel at 473K when the oils are converted with solid fats.

$$\text{Vegetable oil} + H_2 \xrightarrow{\text{Ni, 473K}} \text{Vegetable Ghee}$$

This process is called hydrogenation or hardening of oils and is used in the manufacture of vegatable ghee like Dalda, Gagan, Rath, etc. from vegetable oils. It may, however, be noted that *hydrogenation reduces the number of double bonds but does not completely eliminate them.*

***(iii)* Syngas.** The mixtures of CO and H_2 are called synthesis gas or syngas. It is so named because it is used for the synthesis of methanol and a number of hydrocarbons. It can be produced by the reaction of steam on hydrocarbon or coke at high temperature in the presence of nickel as catalyst.

$$C_nH_{2n+2} + nH_2O \xrightarrow[\text{Ni}]{\text{1270K}} nCO + (2n + 1)\ H_2$$

e.g. $$CH_4\ (g) + H_2O\ (g) \xrightarrow[\text{Ni}]{\text{1270K}} CO\ (g) + 3\ H_2\ (g)$$

These days syngas is produced from sewage saw dust, scarp wood, newspapers, etc. The process of producing syngas form coal is called '*coal gasification*'

$$C\ (s) + H_2O\ (g) \xrightarrow[\text{Ni}]{\text{1270K}} \underbrace{CO\ (g) + H_2\ (g)}_{\text{Syngas}}$$

***(iv)* Water-gas shift reaction.** The amount of hydrogen in the syngas can be increased by reacting CO of syngas with steam in the presence of iron chromate as catalyst

$$CO\ (g) + H_2O\ (g) \xrightarrow[FeCrO_4]{\text{673 K}} CO_2\ (g) + H_2\ (g)$$

This is called water-gas reaction.

***(v)* Fuel-cell.** Fuel cell is a device which converts the energy produced during the combustion of a fuel directly into electrical energy. Dihydrogen is used in hydrogen-oxygen fuel cells for generating electrical energy. It has many advantages over the conventional fossil fuels. It does not cause any pollution and releases more amount of energy per unit mass of fuel as compared to gasoline and other fuels.

Q.12 What is water gas ? How is it prepared ?

Ans : An equimolar mixture of CO and H_2 is called water gas. It is prepared by passing steam over red hot coke.

$$C\ (s) + \underset{\text{Steam}}{H_2O\ (g)} \xrightarrow{\text{1270 K}} \underbrace{CO\ (g) + H_2\ (g)}_{\text{Water gas}}$$

Q.13 Arrange H_2, D_2 and F_2 molecules in the increasing order of their bond dissociation enthalpy.

Ans : Due to greater nuclear mass, the bond pair in D–D bond is attracted more strongly than in H–H bond. Therefore, bond dissociation energy of D–D bond is higher than that of H–H bond.

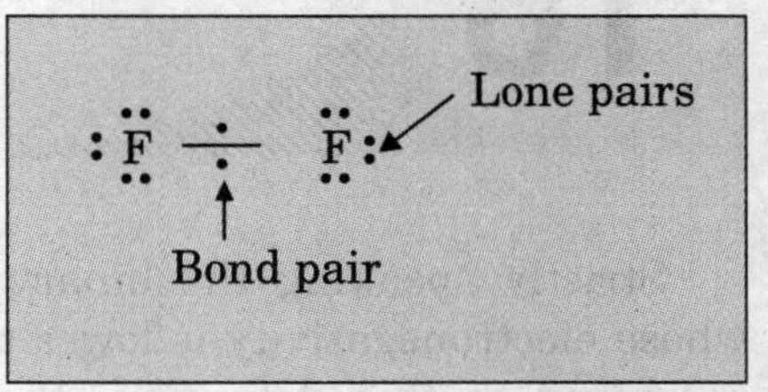

However, due to repulsions between lone pairs of F and the bond pair (See Figure given in the margin), F–F bond dissociation enthaply is the minimum. Thus the bond dissociation enthalpy increases in the order :

$$F-F < H-H < D-D.$$

Q.14 In what respect do ortho and para hydrogen differ from each other ?

Ans : In H_2 molecule, the two electrons of two H-atoms spin in opposite direction (Paulie's exclusion principle) but the spin of two protons in the nucleus of two H-atoms spin in the same direction (parallel spins) in ortho hydrogen and in opposite direction (anti parallel spins) in para hydrogen. Thus ortho and para hydrogen differ in the direction in which the two protons present in the nucleus of two H-atoms spin.

University Questions

1. Justify the position of hydrogen in the periodic table. *(Meerut 89, Garhwal 88)*

2. Outline the manufacture of hydrogen by Lane's process. *(Kerala 80)*

3. What are the different forms of hydrogen? Give two properties and two uses of each. *(Raj. 83; GND 87; MD Rohtak 84)*

4. Write a short note on "ortho and para hydrogen". *(Meerut 80 S, 81, 81 S, 82, 83, 84, 88; Lucknow 80, 81; Kanpur 81; Bundelkhand 88; Rohilkhand 83, 84; Allahabad 87)*

5. (*a*) How is heavy hydrogen prepared by the electrolysis of heavy water and by chemical methods?

(*b*) Write the uses of deuterium.

(*c*) What is difference between ortho and para hydrogen?

(*d*) What is nascent hydrogen? *(Himachal Pradesh 99)*

6. (*a*) How is atomic hydrogen prepared?

(*b*) What are important uses of atomic hydrogen?

(*c*) What happens when : (*i*) Deuterium combines with N_2 (*ii*) Tritium is burnt in air (*iii*) Deuterium is passed over heated alkali metal (*iv*) Atomic hydrogen is kept with molecular oxygen at 20° K.

(*d*) What is the difference between ortho and para hydrogen? *(Himachal Pradesh 2000)*

7. Write a note on isotopes of hydrogen. *(Kanpur 2000)*

CHAPTER

16 General Study of Hydrides

Strictly speaking the binary compounds of hydrogen with other elements whose electronegativity is lower than that of hydrogen are called hydrides. Thus according to this definition, the term hydride is not generally used for the compounds like H_2O, H_2S, NH_3, PH_3, H_2S, HF, HCl, HBr etc. Thus this term is not usually used for the binary compounds of hydrogen with non-metals. On the basis of the type of bonding in the hydrides, these are classified into the following classes by Gibb (1941). (A) *Ionic or salt-like hydrides.* (B) *Covalent* or *molecular hydrides.* (C) *Metallic or interstitial or alloy type hydrides.* (D) *Complex hydrides.* (E) *Borderline hydrides.*

A. Ionic or Salt-like Hydrides

These hydrides are given by the elements of groups IA (Fr is an exception), IIA (Be and Mg are exceptions) and some highly electropositive lanthanides and actinides. Examples are : LiH, CaH_2 etc.

Preperation

(*i*) By heating metal in a current of H_2. The alkali and alkaline earth metals give hydrides when heated at different temperatures in a current of H_2.

$$2M + H_2 \xrightarrow{400°C} 2MH$$

(M = Na or K)

$$M + H_2 \xrightarrow{150\text{–}300°C} MH_2$$

(M = Ca, Sr, Ba)

Li placed in an iron boat kept in a hard glass tube combines with H_2 at 600°C to give LiH.

(*ii*) By heating carbonates in a current of H_2. RbH and CsH are prepared by heating their carbonates in a current of H_2 at 620–650°C. In this reaction the presence of Mg is essential.

$$Rb_2CO_3 + 2H_2 \xrightarrow{\text{In presence of Mg}} 2RbH + CO_2 + H_2O$$

(*iii*) By heating metal nitride in a current of H_2. By this method LiH can be prepared.

$$2LiN_3 + 3H_2 \longrightarrow 6LiH + N_2$$

***(iv)* Alexander's method.** CaH_2 is prepared by heating CaO with metallic Mg in a current of H_2 at 250°C and 50 cm. pressure of hydrogen.

$$CaO + Mg + H_2 \longrightarrow CaH_2 + MgO$$

A yield of 99.2% of CaH_2 is obtained in 2 hours.

Properties

***(i)* Physical Properties.** The saline hydrides are crystalline compounds with ionic lattices. The crystals are somewhat denser than the metal from which they are made because of strong polar bonds existing in the ionic lattices. As ionic compounds, these possess high melting and boiling points and are electrically conducting in the fused state. Excepting LiH they decompose on fusion. They are stoichiometric compounds.

***(ii)* Thermal stability and order of reactivity.** Hydrides of Li, Ca and Sr are most stable while others decompose above 400°C. Their thermal stability decreases with increasing size of the cation in a group. Thus the order of stability of the hydrides of Li to Cs and of Ca to Ba is as follows :

$$LiH > NaH > KH > RbH > CsH$$

$$CaH_2 > SrH_2 > BaH_2$$

The values of heats of formation (in Kcal/mole) of the above hydrides given below support this order.

	I A	II A
Group ⟶		
Hydrides ⟶	LiH, NaH, KH, RbH, CsH	CaH_2, SrH_2, BaH_2
Heat of formation (K.cal/mole) ⟶	21.6, 13.9, 14.4, 13.5, 13.0	45.1, 42.3, 40.9
	⟶ decreasing ⟶	⟶ decreasing ⟶

The relative order of stability given above shows that the order of reactivity is as :

$$LiH < NaH < KH < RbH < CsH$$

***(iii)* Presence of H^- ion.** Electrolysis of these hydrides in the fused state liberates hydrogen at the anode, showing the presence of H^- ion in them.

***(iv)* Action of protonic solvents.** They all vigorously react with protonic solvents (*i.e.* solvents giving H^+ ions) such as H_2O, C_2H_5OH and NH_3 forming H_2 and a base.

$$MH \rightleftarrows M^+ + H^-$$

Protonic solvents

$$H^- + H_2O \longrightarrow OH^- + H_2$$

$$H^- + NH_3 \longrightarrow NH_2^- + H$$

$$H^- + C_2H_5OH \longrightarrow OC_2H_5^- + H_2$$

***(v)* Reducing property.** All these hydrides show reducing properties at high temperatures which is probably due to the formation of atomic hydrogen. Thus they reduce the refractory metal oxides (*e.g.* Fe_3O_4) to free metal, CO_2 to formate, sulphates to sulphides and chlorides to hydrides.

$$NaH + Fe_3O_4 \longrightarrow 3Fe + 4NaOH$$

$$\underset{(M = Li, Na, K)}{MH + CO_2} \xrightarrow{\Delta} \underset{\text{formate}}{HCOOM}$$

$$CaH_2 + 2CO_2 \xrightarrow{\Delta} \underset{\text{metallic formate}}{(HCOO)_2Ca}$$

$$2CaH_2 + PbSO_4 \longrightarrow PbS + 2Ca(OH)_2$$

$$4NaH + SiCl_4 \longrightarrow SiH_4 + 4NaCl$$

(*vi*) Since these hydrides are ionic in nature, their bonds are not directed in space and hence the molecules of thee hydrides have no geometry. Typical type of gattering of these molecules in crystal gives specific type of lattice, *e.g.*, alkali metal hydrides possess face-centred cubic lattice while alkaline earth metal hydrides possess orthorhombic lattice. The radius of H^- ion in these hydrides decreases with the decrease in electropositive character of the metal.

***(vii)* Decomposition by SO_2.** These are decomposed by SO_2 forming hyposulphites.

$$2KH + 2SO_2 \longrightarrow K_2S_2O_4 + H_2\uparrow$$

$$CaH_2 + 2SO_2 \longrightarrow CaS_2O_4 + H_2\uparrow$$

***(viii)* Reaction with N_2.** (*a*) Alkaline earth metal hydrides react with N_2 to form nitrides.

$$3CaH_2 + N_2 \longrightarrow Ca_3N_2 + 3H_2$$

(*b*) LiH is oxidised by N_2, forming $LiNH_2$, Li_2NH and finally Li_3N.

$$LiH \xrightarrow{+N_2} \underset{\text{Li-amide}}{LiNH_2} \xrightarrow{+N_2} \underset{\text{Li-imide}}{LiNH} \xrightarrow{+N_2} \underset{\text{L-nitride}}{Li_3N}$$

CsH, the most reactive hydride, gives $CsNH_2$ when heated with N_2.

$$CsH \xrightarrow[\text{heat}]{+N_2} CsNH_2$$

***(ix)* Reaction with $AlCl_3$.** Only LiH reacts with $AlCl_3$ in ether forming $LiAlH_4$.

$$4LiH + AlCl_3 \text{ (in ether)} \longrightarrow LiAlH_4 + 3LiCl$$

Uses

Ionic hydrides are used as condensing agents in Claisen and Aldol condensations. These hydrides, as shown above, are hydrolysed by water and give large quantities of H_2. Thus these hydrides are used as reducing agents. Other uses of LiH, NaH and CaH_2 which are ionic hydrides can be found in subsequent pages.

Some Ionic Hydrides

Let us discuss the methods of preparation, properties, uses and structure of some ionic hydrides like LiH, NaH, KH, (alkali metal hydrides), CaH_2, (alkaline earth metal hydride). These are discussed as follows :

1. Lithium hydride, LiH

Preparation. (*i*) It is prepared by passing H_2 gas over heated lithium at 700–800°C.

$$2Li + H_2 \xrightarrow{700\text{–}800°C} 2LiH$$

(*ii*) It can also be prepared by heating Li_3N in a current of H_2.

$$Li_3N + 3H_2 \longrightarrow 6LiH + N_2.$$

Properties. (*i*) It is a hard white solid with m.pt. = 680°C.

(*ii*) It reacts with H_2O evolving H_2

$$LiH + H_2O \longrightarrow LiOH + H_2 \uparrow$$

(*iii*) When an etheral solution of LiH is treated with anhydrous $AlCl_3$, lithium aluminium hydride ($LiAlH_4$) is obtained

$$4LiH + AlCl_3 \longrightarrow LiAlH_4 + 3LiCl.$$

(*iv*) Fused LiH, on electrolysis, gives Li at the cathode and H_2 at the anode like other ionic compounds.

(*v*) With B_2H_6 in ether medium it gives lithium borohydride, $LiBH_4$

$$2LiH + B_2H_6 \xrightarrow{\text{Ether}} 2LiBH_4$$

(*vi*) It turns blue on exposure to light without any change in composition.

(*vii*) LiH is the most stable of all the alkali metal hydrides and also chemically active. It is remarkably stable at ordinary temperature and is not affected by dry gases like Cl_2, HCl and O_2.

(*viii*) It is a powerful reducing agent and hence reduces oxides, chlorides, sulphides and also CO_2 to carbon.

Uses. It is used (*i*) as a strong reducing agent in synthetic organic chemistry. (*ii*) in the preparation of $LiAlH_4$ which is a valuable reducing agent. (*iii*) as a convenient source of H_2 for air-sea rescue balloons. (*iv*) for making borohydrides like $LiBH_4$ which are used in jet propulsion.

Structure. Its crystal has face-centred cubic structure (*i.e.* NaCl). The metal-hydrogen distance has been determined accurately by *X*-ray diffraction technique.

2. Sodium hydride, NaH

Preparation. It is prepared by passing pure and dry H_2 over metallic sodium kept in a nickel boat and heated in a glass tube at 365°C. After the reaction, colourless crystals of NaH are deposited on the upper portion of the tube a little away form the nickel boat.

Properties. (*i*) It is a colourles crystalline substance.

(*ii*) On beating above 330°C it is decomposed into its elements.

(*iii*) It is decomposed by H_2O evolving H_2

$$NaH + H_2O \longrightarrow NaOH + H_2$$

(*iv*) It absorbs CO_2 forming sodium formate, HCOONa

$$NaH + CO_2 \longrightarrow HCOONa$$

(*v*) On electrolysis, fused NaH gives H_2 at the anode and Na goes to cathode.

(*vi*) NaH reacts with $B(OCH_3)_3$ to form sodium borohydride, $NaBH_4$ which is used as a reducing agent.

$$4NaH + B(OCH_3)_3 \longrightarrow NaBH_4 + 3NaOCH_3$$

(*vii*) NaH reacts with B_2H_6 in ether medium to give $NaBH_4$.

$$2NaH + B_2H_6 \xrightarrow{\text{Ether}} 2NaBH_4$$

Uses. (*i*) NaH is used to prepare $NaBH_4$ (as shown above) which is a valuable reducing agent.

(*ii*) NaH itself is also used as a reducing agent in organic chemistry.

Structure. NaH has a face-centred cubic lattice.

3. Potassium hydride, KH

Preparation. It is prepared by heating potassium in the atmosphere of H_2 at 360°C.

Properties. (*i*) It forms white needle-like crystalline solid.

(*ii*) It ignites spontaneously in Cl_2 or O_2

(*iii*) It is decomposed by heat

(*iv*) It also gets decomposed by H_2O, liberating H_2 gas

$$KH + H_2O \longrightarrow KOH + H_2$$

(*v*) In presence of moist CO_2 it is converted into HCOOK

$$KH + CO_2 \longrightarrow HCOOK.$$

Structure. It has NaCl like structure.

4. Calcium hydride or hydrolith, CaH_2

Preparation. It is prepared by passing H_2 over fused calcium metal at 400–500°C.

Properties. (*i*) It is a colourless crystalline compound of density 1.90.

(*ii*) On electrolysis of the fused salt, H_2 is evolved at the anode.

(*iii*) It is stable in air but is violently decomposed by cold water, giving slaked lime, $Ca(OH)_2$ and H_2

$$CaH_2 + 2H_2O \longrightarrow Ca(OH)_2 + 2H_2$$

(*iv*) It combines with CO_2 to form calcium formate.

$$CaH_2 + CO_2 \longrightarrow (HCOO)_2Ca$$

Uses. (*i*) It is used for millitary purposes as a source of H_2 for balloons and hence the name of *hydrolith*.

(*ii*) It is used as a reducing agent in the metallurgical processes in organic chemistry and to determine the traces of water in organic solvents.

Structure. X-ray differaction has shown that Ca^{2+} ions in CaH_2 are in a slightly distorted hexagonally close-packed arrangement. The unit cell of this hydride is orthorhombic.

B. Covalent or Molecular Hydrides

These hydrides are given by most of *p*-block elements. Some elements (*e.g.* Sn, Pb, Sb, Bi, Te, Po and halogens) give only mononuclear hydrides while the elements like, B, Al, Ga, In, C, Si, N, O, S give mononuclear as well as polynuclear (or polymeric) hydrides as shown below :

$$B \longrightarrow B_2H_6,\ B_4H_{10},\ B_5H_9 \text{ etc.}$$
$$Al \longrightarrow AlH_3,\ (AlH_3)_n$$
$$Ga \longrightarrow Ga_2H_6,\ Ga_3N_8 \text{ etc.}$$
$$In \longrightarrow (InH_3)_n$$
$$C \longrightarrow CH_4 \text{ and other higher hydrocarbons}$$
$$Si \longrightarrow SiH_4,\ Si_2H_6,\ Si_3H_8 \text{ etc.}$$
$$N \longrightarrow NH_3,\ N_2H_4$$
$$S \longrightarrow H_2S,\ H_2S_3,\ H_2S_5$$

Excepting the hydrides of B, Al and Ga, the simplest hydrides of other elements can be formulated as AH_{8-n} where A is the element whose hydrides are being considered and n is the number of valence electrons in atom, A.

Preparation

(*i*) By the direct union of the free element with H_2 at elevated temperature.

$$N_2 + 3H_2 \longrightarrow 2NH_3$$
$$2H_2 + O_2 \longrightarrow 2H_2O$$
$$H_2 + X_2 \longrightarrow 2HX$$

(*ii*) By the hydrolysis of compounds like borides, silicides, and similar compounds. The hydrolysis may be affected either by H_2O only or by dil. acid solutions.

$$Mg_3B_2 + 6HCl \longrightarrow 3MgCl_2 + B_2H_6$$
$$CaC_2 + 2H_2O \longrightarrow Ca(OH)_2 + C_2H_2$$
$$Al_4C_3 + 6H_2O \longrightarrow 2Al_2O_3 + 3CH_4$$
$$Mg_2Si + 4HCl \longrightarrow 2MgCl_2 + SiH_4$$
$$Mg_3N_2 + 6H_2O \longrightarrow 3Mg(OH)_2 + NH_3$$
$$Ca_3P_2 + 6H_2O \longrightarrow 3Ca(OH)_2 + 2PH_3$$
$$Zn_3As_2 + 6HCl \longrightarrow 3ZnCl_2 + 2AsH_3$$
$$FeS + 2HCl \longrightarrow FeCl_2 + H_2S$$

(*iii*) By the reduction of appropriate anhydrous chlorides in ether solution by $Li[AlH_4]$.

$$4BCl_3 + 3Li[AlH_4] \longrightarrow 2B_2H_6 + 3LiCl + 3AlCl_3$$
$$MCl_4 + Li[AlH_4] \longrightarrow LiCl + AlCl_3 + MH_4$$

(M = Si, Ge or Sn)

(*iv*) By the decomposition of a carbide with alkalies.

$$Be_2C + 4KOH \longrightarrow 2Be(OK)_2 + CH_4$$

(*v*) Halogen hydrides, HX, may be obtained by heating double compounds such as HF. KF

$$HF.KF \xrightarrow{\text{heat}} HF + KF$$

(*vi*) Paneth and Rabino-Witsch's method. These persons prepared SnH_4 by electrolysing a sulphuric acid solution of $SnSO_4$ containing glucose. The cell used had Pb-electrodes and the temperature kept was below 50–60°C.

Properties

(*i*) The hydrogen atoms are covalently bonded to the central atom in these hydrides. The hydrides consist of discrete covalent molecules which are held together by weak van der Waals forces and in some cases by hydrogen bonds. As a result, hydrides are gases, liquids and (in few cases) solids of low melting and boiling points. The volatality is decreased in those hydrides in which hydrogen bonding occurs (*e.g.* NH_3, H_2O, HF) or where the hydrides are either dimeric [*e.g.* $(BH_3)_2$, $(GaH_3)_2$] or polymeric [*e.g.* $(AlH_3)_x$, $(InH_3)_x$].

(*ii*) Thermal stability. The thermal stability of these hydrides within a group decreases with the increasing atomic number (or atomic radius) of the element forming the hydride. Thus as the metallic character of the central atom increases, the thermal stability decreases. For example :

$$NH_3 > PH_3 > AsH_3 > SbH_3 > BiH_3$$

Note that N in NH_3 is a non-metal and Bi in BiH_3 is a metal. Thus the hydrides of the predominantly metallic elements (*e.g.* PbH_4, BiH_3, PoH_2) are particularly unstable.

The stability of the hydrides of the elements of second period increases with the increase in electronegativity of the central atom, *e.g.*,

$$CH_4 < NH_3 < H_2O < HF$$

Note that the atomic radii of the elements decrease on proceeding from left to right in a period.

(*iii*) Hydridic and protonic character. The covalent hydrides of more electropositive elements (*i.e.* of least electronegativity) such as those of Be, B and Al behave as if they possess "*hydridic*" character. Thus these hydrides react with H_2O to form H_2.

$$B_2H_6 + 6H_2O \longrightarrow 2H_3BO_3 + 6H_2$$

The hydrides of more electronegative elements such as those of oxygen family and halogens react as if the H-atoms have protonic character. Thus such hydrides react with bases to form salts.

$$\underset{\text{Hydride}}{HCl} + \underset{\text{Base}}{NH_3} \longrightarrow \underset{\text{Salt}}{NH_4Cl}$$

The formation of the salt, NH_4Cl probably occurs as follows :

$$\underset{\text{Hydride}}{HCl} \longrightarrow H^+ + Cl^-$$

$$\underset{\text{Base}}{NH_3} + H^+ \longrightarrow NH_4^+$$

$$NH_4^+ + Cl^- \longrightarrow \underset{\text{Salt}}{NH_4Cl}$$

The hydrides of the elements of intermediate electronegativity such as those of group IV A show no marked hydridic or protonic character. These hydrides may, however, exhibit either hydridic or protonic character depending on the experimental conditions, *e.g.*, mono-germane, GeH_4, reacts with HBr to form H_2 and with sodium amide, $NaNH_2$ (which is a base) to give sodium germyl, $NaGeH_3$.

$$GeH_4 + HBr \longrightarrow GeH_3Br + H_2$$

$$GeH_4 + NaNH_3 \longrightarrow NaGeH_3 + NH_3$$

***(iv)* Property of behaving as Lewis base.** Covalent hydrides that have one or more lone pairs of electrons behave as Lewis bases. This property of behaving as Lewis base decreases on proceeding down a group and from left to right across a period. Thus this property for the hydrides from NH_3 to SbH_3 and from NH_3 to HF is in the following order :

$NH_3 > PH_3 > AsH_3 > SbH_3$ (*hydrides of V A group elements*)

$NH_3 > H_2O > HF$ (*hydrides of second period elements*)

***(v)* Formation of addition compounds.** The known hydrides of III A group elements [*e.g.* B_2H_6, $(AlH_3)_x$] are electron-deficient molecules and so behave as strong Lewis acids and hence form addition compound with a variety of donor molecules such as CO, PF_3, NR_3 etc.

Lewis acids		**Donor molecules**		**Addition compounds**
B_2H_6	+	$2 : NR_3$	$\longrightarrow$	$2\,[BH_3 : NR_3]$
B_2H_6	+	$2 : PF_3$	$\longrightarrow$	$2\,[BH_3 : PF_3]$
B_2H_6	+	$2 : CO$	$\longrightarrow$	$2\,[CH_3 : OC]$
$(AlH_3)_x$	+	$x : NR_3$	$\longrightarrow$	$x\,[AlH_3 : NR_3]$

(vi) Formation of complex hydrides. The covalent hydrides react with ionic hydrides to form complex hydrides, *e.g.*,

Covalent hydrides		**Ionic hydrides**		**Complex hydrides**
AlH_3	+	LiH	$\longrightarrow$	Li $[AlH_4]$ Li-Al-hydride (complex hydride of Al)
B_2H_6	+	2MH (M = Li, Na)	$\longrightarrow$	2M $[BH_4]$ Li- or Na-borohydride (complex hydride of B)

(vii) In polymeric covalent hydrides like $(AlH_3)_n$, $(InH_3)_n$ etc. the metal atoms are linked together by hydrogen bridges.

Uses

(*i*) Boron hydrides (called boranes) which are covalent hydrides are used as fuels and fuel additives, *e.g.* diborane (B_2H_6) is used as a high energy fuel and propallent. It is also used in welding.

(*ii*) Silane (SiH_4) which is a covalent hydride is used as a source for the preparation of silicon of high purity which is used in transistors. It is also used as a reducing agent.

Hydrides of Boron : Boranes

Hydrides of boron are called *boranes*. A complete description of these hydrides is given in chapter 24. These hydrides are covalent hydrides.

Hydrides of Silicon : Silanes

General Discussion

Hydrides of silicon are called *silanes*. These are also covalent hydrides and are represented by the general formula Si_n (H_{2n+2}). The silanes namely *monosilane* or *silane* (SiH_4), *disilane* (Si_2H_6), *trisilane* (Si_3H_8), *tetrasilane* (Si_4H_{10}), *pentasilane* (Si_5H_{12}), *hexasilane* (Si_6H_{14}) and *heptasilane* (Si_7H_{14}) are known to exist. In addition to these silanes, *silicoacetylene*, $(Si_2H_2)_n$ and *silico-ethylene*, $(Si_2H_4)_n$ are also known. These compounds were studied by Stock and his students in 1926.

When 20% HCl acts on Mg_2Si (prepared by heating a mixture of Mg and SiO_2 in the absence of air) in an atmosphere of H_2, a mixture of silanes is obtained. This mixture which contains the silanes in the proportion, SiH_4 = 40%, Si_2H_6 = 30%, Si_3H_8 = 15%, Si_4H_{10} = 10% and higher silanes = 5% is cooled in liq. N_2 and the resulting liquid is subjected to fractional distillation or vapour phase chromatographic technique in order to separate the individual silanes.

$$Mg_2Si + 4HCl \longrightarrow MgCl_2 + SiH_4 + \text{other silanes in small proportion}$$

$$2Mg_2Si + 8HCl \longrightarrow 4MgCl_2 + Si_2H_6 + H_2$$

These compounds are analogous to alkanes. These are colourless and volatile covalent hydrides. The melting and boiling points of the first four members as given below show that these increase with the increase of their molecular weights.

	SiH_4	Si_2H_6	Si_3H_8	Si_4H_{10}
M.pts.(°*C*)	–185	–132.5	–117	–92.5
B.pts.(°*C*)	–112	–15	53	109

Thermal stability of these hydrides decreases with the increase of their molecular weights. For example SiH_4 is stable and decomposes only at red heat into H_2 and Si; Si_2H_6 decomposes at 400–500°C while Si_6H_{14} decomposes at room temperature. Here it is worth noting that the decomposition of higher silanes is not complete and gives a mixture of simple gaseous silanes and solid unsaturated hydrides. For example :

$$Si_5H_{12} \xrightarrow{\text{Heat}} 2\ SiH + Si_2H_6 + SiH_4$$

SiH_4 and Si_2H_6 are colourless gases at room temperature while higher silanes are colourless liquids. These compounds are very poisnous and produce headache and sickness when inhaled in small quantities. In low concentration they have stupefying and chocolate like odour, but in high concentration their odour resembles that of H_2S.

Chemically silanes are very reactive and burn spontaneously in air or oxygen. For example :

$$SiH_4 + 2O_2 \longrightarrow SiO_2 + 2H_2O$$
$$Si_4H_{10} + 13/2\ O_2 \longrightarrow 4\ SiO_2 + 5H_2O$$

Silanes are stable towards neutral or acidic water, but in the presence of even a trace of a base in water, they form silicates with the liberation of H_2. For example :

$$SiH_4 + 2NaOH + H_2O \longrightarrow Na_2SiO_3 + 4H_2 \uparrow$$
$$Si_2H_6 + 4NaOH + 2H_2O \longrightarrow 2Na_2SiO_3 + 7H_2 \uparrow$$

Now let us discuss some silanes in detail.

1. Monosilane, SiH_4

This compound is also called *silico-methane* or *silicane*.

Preparation. (*i*) SiH_4 can be prepared by passing dry H_2 over hot silicon in an electric arc.

$$Si\ (hot) + 2H_2\ (dry) \longrightarrow SiH_4$$

(*ii*) *Johnson's method* (*1935*). According to this method pure SiH_4 is formed by decomposing Mg_2Si with NH_4Br in liq. NH_3.

(*iii*) Silico-chloroform, $SiHCl_3$, on treatment with C_2H_2OH gives ethyl silico orthoformate, $SiH(OC_2H_5)_3$ which on reduction with Na gives pure SiH_4.

$$SiHCl_3 + 3C_2H_5OH \longrightarrow SiH(OC_2H_5)_3 + 3HCl$$

$$4SiH(OC_2H_5)_3 \xrightarrow[\text{with Na}]{\text{Reduction}} 3Si(OC_2H_5)_4 + SiH_4 \uparrow$$

(*iv*) Pure SiH_4 can also be formed by reducing $SiCl_4$ with $LiAlH_4$ dissolved in ether at 0°C.

$$SiCl_4 + LiAlH_4 \longrightarrow SiH_4 + LiCl + AlCl_3$$

Properties. (*i*) It is a colourless gas which can be liquefied at –112°C and solidified at –185°C. It has faint odour.

(*ii*) *Action of air.* When pure, SiH_4 is not inflammable in air but in the presence of H_2 and other hydrides of silicon, it spontaneously burns in air with a bright flame to form SiO_2 and H_2O.

$$SiH_4 + 2O_2 \longrightarrow SiO_2 + 2H_2O$$

(*iii*) *Action of heat.* (*Decomposition*). When heated to 400°C it breaks down into its elements.

$$SiH_4 \xrightarrow{400°C} Si + 2H_2$$

(*iv*) *Precipitation reactions.* SiH_4 gives a precipitate of copper silicide, Cu_2Si when passed in solution of copper salt like $CuSO_4$. Similarly silver metal is obtained from $AgNO_3$ solution.

$$2CuSO_4 + SiH_4 \longrightarrow \underset{\text{ppt.}}{Cu_2Si} + 2H_2SO_4$$

$$4AgNO_3 + SiH_4 \longrightarrow \underset{\text{ppt.}}{4Ag} + 4HNO_3 + Si$$

(*v*) *Action of alkalies.* SiH_4 reacts with alkalies to form silicate with the liberation of H_2

$$SiH_4 + 2\ NaOH + H_2O \longrightarrow Na_2SiO_3 + 4\ H_2 \uparrow$$

(*vi*) *Action of dry HCl* (*Replacement reaction*). Hydrogen of SiH_4 can be replaced by treating SiH_4 with dry HCl in presence of $AlCl_3$ (catalyst) at 100–125°C to get halogenosilanes and $SiCl_4$ in the end.

$$SiH_4 \xrightarrow[(-H_2)]{+\ HCl} SiH_3Cl \xrightarrow[(-H_2)]{+\ HCl} SiH_2Cl_2 \xrightarrow[(-H_2)]{+\ HCl} SiHCl \xrightarrow[(-H_2)]{+\ HCl} SiCl_4$$

(*vii*) *Action of alkali metals*. When SiH_4 is treated with potassium in glyme (1, 2-dimethoxy ethane) at –78°C, potassium derivative, $KSiH_3$ is obtained.

$$2\ SiH_4 + 2\ K \longrightarrow 2\ KSiH_3 + H_2$$

(*viii*) *Reducing properties*. SiH_4 is a strong reducing agent, since it reduces acidified $KMnO_4$ to $MnSO_4$, Hg (II) to Hg (I or 0), Fe (III) to Fe(II), Ag (I) to metallic Ag etc. Thus when SiH_4 is passed through silver salt solution like $AgNO_3$, precipitate of Ag is obtained.

$$4KMnO_4 + 6H_2SO_4 + 5\ SiH_4 \longrightarrow 2K_2SO_4 + 4MnSO_4 + 16H_2O + 5\ Si$$

$$SiH_4 + 4Ag\ NO_3 \longrightarrow \underset{\text{ppt}}{4Ag \downarrow} + Si \downarrow + 4HNO_3$$

(*ix*) When a mixture of SiH_4 and N_2 is sparked, NH_3 is produced.

$$2SiH_4 + N_2 \longrightarrow 2Si + 2NH_3 + H_2$$

(*x*) *Reaction with X_2, $COCl_2$, $SnCl_4$ etc.* Si is precipitated.

$$SiH_4 + 2\ X_2 \longrightarrow 4HX + \underset{\text{ppt.}}{Si}$$

$$SiH_4 + 2COCl_2 \longrightarrow 4HCl + Si + 2CO$$

$$SiH_4 + SnCl_4 \longrightarrow 4HCl + Si + Sn$$

Excess of Cl_2 gives $SiCl_4$ directly

$$SiH_4 + 4Cl_2 \longrightarrow SiCl_4 + 4HCl$$

In some cases substitution reactions occur with X_2 molecules. For example :

$$SiH_4 + Cl_2 \longrightarrow SiH_3Cl + HCl$$

$$SiH_3Cl + Cl_2 \longrightarrow SiH_2Cl_2 + HCl$$

$$SiH_2Cl_2 + Cl_2 \longrightarrow SiHCl_3 + HCl$$

$$SiHCl_3 + Cl_2 \longrightarrow SiCl_4 + HCl$$

(*xi*) *Reaction with H_2O.* If the gas is bubbled through H_2O, then as it comes in contact with air, each bubble ignites giving vertex ring of finely divided SiO_2. SiO_2 may further form silicic acid, H_2SiO_3

$$SiH_4 + 2H_2O \longrightarrow SiO_2 + 4H_2$$
$$SiO_2 + H_2O \longrightarrow H_2SiO_3$$

Uses. (*i*) SiH_4 is used as a source for the preparation of silicon of high purity which is produced by the pyrolysis and is used in transistors.

(*ii*) It is also used as a reducing agent.

2. Disilane, Si_2H_6

This compound is also called *silicoethane*.

Preparation. (*i*) When Mg_2Si is treated with dil HCl, a mixture of silicon hydrides is obtained. The fraction, obtained at –100°C contains Si_2H_6 and SiH_4. Si_2H_6 is removed by refraction at –100°C. 400 mls. of Si_2H_6 are obtained from 100 gms of Mg_2Si.

(*ii*) It is best prepared by the reaction of Si_2Cl_6 with $Li[AlH_4]$ in ether solution

$$2Si_2Cl_6 + 3Li\ [AlH_4] \longrightarrow Si_2H_6 + 3LiCl + 3AlCl_3$$

(*iii*) It is also prepared by the action of conc. HCl acid on lithium silicide (Li_6Si_2)

$$Li_6Si_6 + 6HCl \longrightarrow 6LiCl + Si_2H_6$$

Properties. (*i*) It is a colourless liquid with more repulsive smell than SiH_4, b.pt = –15°C, m.pt = –132°C.

(*ii*) *Decomposition.* It is stable at ordinary temperature but decomposes into its elements at 200°C in absence of air.

$$Si_2H_6 \xrightarrow{\text{Heat (200°C)}} 2Si + 3H_2$$

(*iii*) *Reactions with alkalies.* It reacts with alkalies to form silicates with the liberation of H_2.

$$Si_2H_6 + 4KOH + 2H_2O \longrightarrow 2K_2SiO_3 + 7H_2$$

(*iv*) *Reaction with $CHCl_3$ and CCl_4.* $SiCl_4$ is formed in each case.

$$Si_2H_6 + 3CHCl_3 \longrightarrow 2\ SiCl_4 + 4H_2 + 3C + HCl$$
$$Si_2H_6 + 2CCl_4 \longrightarrow 2\ SiCl_4 + 3H_2 + 2C.$$

It explodes violently when mixed with CCl_4.

(*v*) *Solubility.* Si_2H_6 is soluble in CS_2, C_6H_6 and alcohol but sparingly soluble in H_2O with which it gives H_2SiO_3

$$Si_2H_6 + 6H_2O \longrightarrow 2H_2SiO_3 + 7H_2$$

(*vi*) *Action of air.* It inflames spontaneously in air and may even explode.

$$Si_2H_6 + 2O_2 \longrightarrow 2\ SiO_2 + 3H_2$$
$$2\ Si_2H_6 + 7O_2 \longrightarrow 4\ SiO_2 + 6H_2O$$

(*vii*) *Hydrolysis.* It is hydrolysed by H_2O containing even traces of alkali, giving H_2 and hydrated silica, $SiO_2.nH_2O$.

$$Si_2H_6 + (4 + 2n)H_2O \longrightarrow 2[SiO_2.nH_2O] + 7H_2 \uparrow$$

(*viii*) *Reducing properties*. It reduces $KMnO_4$, $K_2Cr_2O_7$, $FeCl_3$, $AgNO_3$, $HgCl_2$ etc.

$$6FeCl_3 + Si_2H_6 \longrightarrow 6FeCl_2 + 2Si + 6HCl$$

(*ix*) *Action of halogens*. It reacts explosively with halogens.

$$Si_2H_6 + 3X_2 \longrightarrow 2Si + 6HX$$

3. Trisilane, Si_3H_8

This compound is also called *silico-propane*.

Preparation. When Mg_2Si is treated with dil. HCl acid, a mixture of silanes is obtained. After removal of Si_2H_6 the residue is fractionated at –60°C and refractionated at –70°C to obtain Si_3H_8. 180 mls. of Si_3H_8 are obtained from 100 gms of Mg_2Si.

Properties. (*i*) It is a bright mobile colourless liquid with m.pt. = –117°C and b.pt. = 53°C.

(*ii*) It inflames instantaneously in air with a loud detonation.

$$Si_3H_8 + 3O_2 \longrightarrow 3SiO_2 + 4H_2$$

(*iii*) *Decomposition*. Alkalies decompose it to Ni_2SiO_3 with the liberation of H_2.

$$Si_3H_8 + 6NaOH + 3H_2O \longrightarrow 3Na_2SiO_3 + 10H_2 \uparrow$$

It is also decomposed by H_2O

$$Si_3H_8 + 6H_2O \longrightarrow 3SiO_2 + 10H_2 \uparrow$$

(*iv*) *Reaction with $CHCl_3$*. It also reacts with $CHCl_3$

$$Si_3H_8 + 4CHCl_3 \longrightarrow Si_3H_4Cl_4 + 4CH_2Cl_2$$

4. Tetrasilane, Si_4H_{10}

This compound is also called *silicobutane*.

Preparation. After removing the lower silanes from the mixture obtained by the action of dil. HCl on Mg_2Si, the residue is distilled at a temperature of –25°C which is redistilled at –35°C to get Si_4H_{10}. 100 mls. of Si_4H_{10} are obtained from 100 gms. of Mg_2Si.

Properties. (*i*) m .pt. = 93.5°C, b.pt. = 85°C

(*ii*) It is a colourless liquid.

(*iii*) It resembles Si_3H_8 in chemical properties.

5. Penta and hexa silanes (Si_5H_{12} and Si_6H_{14})

Si_5H_{12} and Si_6H_{14} can also be obtained from the reaction products obtained by treating Mg_2Si with dil. HCl. The liquid obtained after the removal of the series of silicon hydrides by fractional distillation, contains small quantities of Si_5H_{12} and Si_6H_{14}. 50 mls. and 35 mls. of Si_5H_{12} and Si_6H_{14} respectively are obtained from 100 gms. of Mg_2Si.

6. Silico ethylene, $(Si_2H_4)_n$

It is prepared by the action of HCl on calcium monosilicide, Ca_2Si_2

$$nCa_2Si_2 + 4nHCl \longrightarrow 2nCaCl_2 + (Si_2H_4)_n$$

7. Silico acetylene, $(Si_2H_2)_n$

It is prepared by the action of cal. silicide on HCl.

$$nCaSi_2 + 2nHCl \longrightarrow nCaCl_2 + (Si_2H_2)_n$$

Structure of Silanes

These compounds have the structures which are similar to those of the corresponding saturated hydrocarbons. The structure of these compounds are given below :

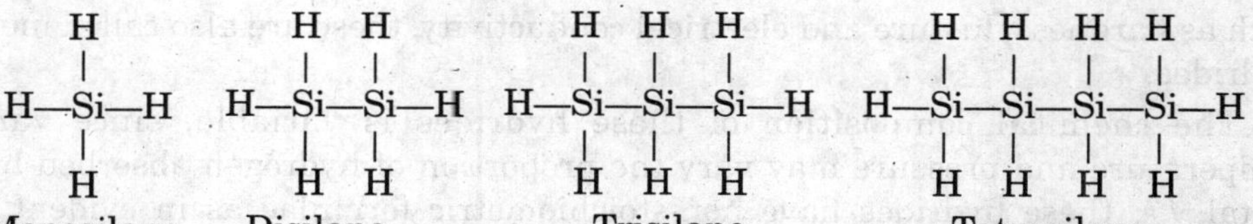

Monosilane (SiH_4) Disilane (Si_2H_6) Trisilane (Si_3H_8) Tetrasilane (Si_4H_{10})

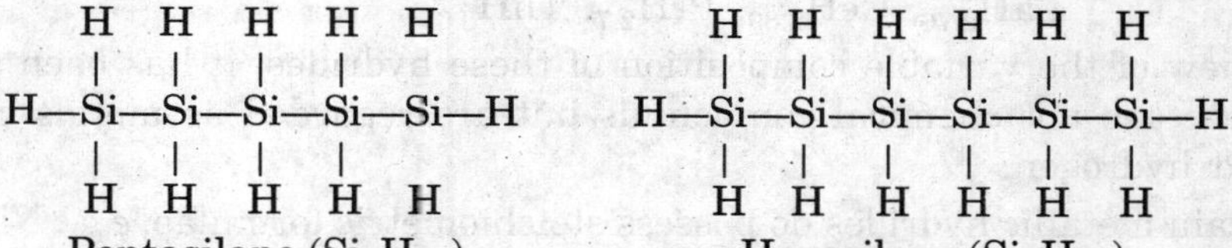

Pentasilane (Si_5H_{12}) Hexasilane (Si_6H_{14})

Differences between Alkanes and Silanes

1. In the hydrocarbon series, the number of C-atoms in a chain can apparently have any value but in case of silanes the maximum number of Si-atoms is seven which is found in hepta silane (Si_7H_{14}). This is because of the fact that relatively weak Si—Si bond makes unstable silanes linked with large number of Si-atoms.

2. SiH_4 gets hydrolysed while CH_4 does not. This is because maximum covalency of carbon is four which is attained in CH_4. Thus CH_4 is resistant to the attack of electron-donating molecules. The maximum covalency of silicon atom is six. In SiH_4 molecule only four covalent bonds are present. It is, therefore, possible for SiH_4 to accept lone pair of electrons from the donor group such as water.

C. Metallic (or Interstitial or Alloy-type) Hydrides

Many *d*-block elements (*e.g.* Ti, Pd, Zr, V, Nb etc.), lanthanides and actinides (*e.g.* La, Ce, Pr etc.) at elevated temperature absorb hydrogen into the holes or interstices existing between the atoms comprising the metallic lattice without changing the original crystal structure of the metal and thus give hydrides which are called metallic hydrides. Examples of such hydrides are : $TiH_{1.73}$, $VH_{0.6}$, $ZrH_{1.92}$, $PdH_{0.6}$, $TaB_{0.76}$, $CeH_{2.69}$, $PrH_{2.7}$ etc.

Thus these hydrides are interstitial combinations of hydrogen and the metals and it is for this reason that these are called interstitial hydrides.

The densities of these hydrides are usually lower than those of the parent metal. This fact suggests that these hydrides are interstitial in nature and the metal lattices expand. On the other hand, their electronegativity values are fairly close to those of IA (alkali metals) and IIA (alkaline earth metals) groups are they have fairly high heats of formation. These facts indicate that these hydrides may be regarded as a transition between the salt-like and interstitial, *e.g.* although CuH, NiH_2, CoH_2, FeH_2, CrH_3 have stoichiometric formulae, these do not resemble salt-like hydrides. Since these hydrides possess metallic properties such as hardness, lusture and electrical conductivity, these are also called metallic hydrides.

The chemical composition of these hydrides is variable, since varying temperature and pressure may vary the proportion of hydrogen absorbed by the metal, *i.e.* these hydrides have non-stoichiometric formulae as in evident from those given below :

$$TiH_{1.73},\ PdH_{0.6},\ ZrH_{1.92},\ VH_{0.6},\ NbH_{0.86},\ TaH_{0.76},$$
$$LaH_{2.76},\ CeH_{2.69},\ PrH_{2.7},\ ThH_{3.07}.$$

In view of the variable composition of these hydrides, it has been suggested that these are not chemical compounds but are regarded as metals containing absorbed hydrogen.

Certain metallic hydrides do possess stoichiometric formulae, *e.g.*. NiH_2, CoH_2, FeH_2 and CrH_3. When phenol magnesium bromide, C_6H_5 Mg Br solution is mixed with metal (II) chloride and H_2 is passed through, black precipitates of NiH_2, CoH_2 and FeH_2 are formed.

$$MCl_2 + 2C_6H_5MgBr \longrightarrow M(C_6H_6)_2 + MgBr_2 + MgCl_2$$
(M = Ni, Co, Fe)
$$M(C_6H_5)_2 + 2H_2 \longrightarrow 2C_6H_6 + MH_2$$

A number of metals possess the remarkable property of combining relatively larger quantities of hydrogen either at room temperature or at high temperatures, *e.g.* Fe, Cu, Ni, Pt etc. absorb increasing quantities of hydrogen at increasing temperature. Certain metals such as Rh, Ta etc. absorb lesser quantities at high temperature.

The absorption of hydrogen by certain metals such as Cu, Ag, Cr, Mo, U, Fe, Co, Ni and Pt is endothermic, while that by other metals such as Ti, Th, Zr, V, Nb, Ta and Pd is exothermic. Pd, after absorbing hydrogen, becomes very reactive and has strong reducing properties. This suggests that hydrogen is held in metal at its surface in the atomic state. The difference in temperatures at which various metals absorb hydrogen is presumably due to the difference in the energy of activation of the reaction $H_2 \rightleftarrows 2H$, occurring at the metallic surface.

The absorbed hydrogen can be expelled from the interstices simply by heating the hydrides and by increasing the gas pressure. Thus these hydrides are good reducing agents.

D. Complex Hydrides

Some of the hydrides contain borohydride (BH_4^-) and aluminium hydride (AlH_4^-) ions. These hydrides are called complex hydrides. Lithium borohydride ($LiBH_4$), sodium borohydride ($NaBH_4$), beryllium borohydride [$Be(BH_4)_2$], aluminium borohydride [$Al(BH_4)_3$] and lithium aluminium hydride ($LiAlH_4$) are the most important examples of complex hydrides. In these hydrides hydrogen atoms are covalently bonded to the trivalent metal atom.

Such hydrides can be obtained by the action of an excess of alkali metal hydride on the appropriate trihalide. For example :

$$4NaH - BF_3 \xrightarrow{\text{Ether}} NaBH_4 + 3NaF$$

$$4LiH + AlCl_3 \xrightarrow{\text{Ether}} LiAlH_4 + 3LiCl$$

Although these hydrides have no free H^- ions, they show many reactions of ionic hydrides, but the reactions are smoother.

Some Complex Hydrides

1. Lithium Borohydride, $LiBH_4$. ***Preparation.*** It is prepared :

(*i*) By the action of $NaBH_4$ on LiCl.

$$LiCl + NaBH_4 \longrightarrow LiBH_4 + NaCl.$$

(*ii*) By the action of B_2H_4 on lithium ethyl (LiC_2H_5) or LiH suspended in ether.

$$2B_2H_6 + 2LiC_2B_5 \longrightarrow 3LiBH_4 + B(C_2H_5)_3$$

$$B_2H_6 + 2LiH \longrightarrow 2LiBH_4$$

(*iii*) By the action of $Al(CH_4)_3$ on LiC_2H_5 in benzene.

$$Al(BH_4)_3 + 3LiC_2H_5 \longrightarrow 3LiBH_4 + Al(C_2H_5)_3.$$

Properties. (*i*) It is a white solid. In the absence of air it melts (decomposes) at 275°C, evolving H_2. It is soluble in polar solvents like H_2O.

(*ii*) It reacts with H_2O even in cold, with CH_2OH down to 100°C and with HCl down to –80°C. In all these reactions H_2 is evolved.

$$LiBH_4 + 2H_2O \longrightarrow LiBO_2 + 4H_2 \uparrow$$

$$LiBH_4 + 4CH_3OH \longrightarrow LiOCH_3 + B(OCH_3)_3 + 4H_2 \uparrow$$

$$2LiBH_4 + 2HCl \longrightarrow 2LiCl + B_2H_6 + 2H_2 \uparrow$$

(*iii*) It reacts with I_2 at 125°C to form BI_3.

(*iv*) It can reduce $FeCl_3$ into free Fe metal and non-metallic halides (*e.g.* $SiCl_4$) to their hydrides (*e.g.* SiH_4)

$$2FeCl_3 + 6LiBH_4 \longrightarrow 6LiCl + 3H_2 + 2B_2H_6 + 2Fe.$$

Thus $LiBH_4$ is a powerful reducing agent.

Uses. Due to the liberation of H_2 gas in some reactions mentioned above, it is used as a reducing agent in organic as well as in inorganic chemistry.

2. Sodium Borohydride, $NaBH_4$

Preparation (*i*) It is obtained by treating NaH with methyl borate, $B(OCH_3)_3$ in tetrahydrofuran (THF) solvent.

$$4NaH + 4B(OCH_3)_3 \xrightarrow{\text{THF solvent}} NaBH_4 + 3NaB(OCH_3)_4$$

(*ii*) It can also be obtained by the action of B_2H_6 on NaH suspended in ether.

$$2NaH + B_2H_6 \longrightarrow 2NaBH_4$$

Properties. (*i*) It is stable upto 397°C and does not decompose water in the cold. Like $LiBH_4$ it is soluble in polar solvents like H_2O.

(*ii*) It reacts with HCl and BF_3 to produce B_2H_6

$$2NaBH_4 + 2HCl \longrightarrow 2NaCl + 2H_2 + B_2H_6$$

$$3NaBH_4 + BF_3 \xrightarrow{\text{Diglyme}} 2NaF + 2B_2H_6.$$

3. Aluminium borohydride, $Al(BH_4)_3$

It is prepared by the following reactions :

$$Al_2(CH_3)_6 + 4B_2H_6 \longrightarrow 2B(CH_3)_3 + 2Al(BH_4)_3$$

$$6LiBH_4 + Al_2Cl_6 \xrightarrow{\text{warm}} 6LiCl + 2Al(BH_4)_3$$

It is a colourless volatile liquid and extra-ordinarily reactive. It dissolves in non-polar solvents. It reacts with H_2O and HCl.

$$Al(BH_4)_3 + 12H_2O \longrightarrow 3H_3BO_3 + Al(OH)_3 + 12H_2$$

$$2Al(BH_4)_3 + 6HCl \xrightarrow{-80°C} Al_2Cl_6 + 2B_2H_6 + 3H_2$$

$Al(BH_4)_3$ also reacts with metallic halides and forms metallic borohydrides, if these are stable or otherwise H-atoms of $Al(BH_4)_3$ are substituted by halogens.

$$nAl(BH_4)_3 + 3M^{n+}X_n \longrightarrow 3M^{n+}(BH_4)_n + nAlX_3$$

$$Al(BH_4)_3 + M^{n+}X_n \longrightarrow Al(BX_4)_3 + M^{n+}H_n$$

If is a good reducing agent. If has covalent character which makes it soluble in non-polar solvents. If can also form addition compounds with amines.

Like B_2H_6 it also adjusts with three-centred bond in the Al—H—B group.

Structure of $[BH_4]^-$ ion. The borohydride ion ($BH^-{}_4$) has tetrahedral shape which results from sp^3 sybridisation of $\overline{B}$ ($\overline{B} = 2s^2sp^2$). For details see chapter 8.

4. Lithium Aluminium Hydride, $Li[AlH_4]$

This compound was discovered by Finholt, Bond and Schlesinger 1947.

Preparation. It is best prepared by the action of LiH on an ether solution of $AlCl_3$

$$4LiH + AlCl_3 \xrightarrow{\text{Ether}} \underset{\text{(soluble)}}{LiAlH_4} + \underset{\text{(ppt)}}{3LiCl \downarrow}$$

$LiAlH_4$ remains in solution in ether while LiCl is insoluble in it. The precipitated LiCl is filtered off the ether is evaporated and $LiAlH_4$ remains in solution. If this solution is allowed to stand before filtration and then evaporated, $LiAlH_4$ of 99% purity is obtained.

Properties. (*i*) It is a non-volatile colourless crystalline solid which is stable to air in the cold. It is soluble in ether and other organic solvents.

(*ii*) If is decomposed by H_2O, liberating H_2 gas

$$LiAlH_4 + 4H_2O \longrightarrow LiOH + Al(OH)_3 + 4H_2 \uparrow$$

(*iii*) When heated, it gets decomposed according to the equation :

$$2LiAlH_4 \xrightarrow{\Delta} 2LiH + 2Al + 3H_2 \uparrow$$

(*iv*) Its etheral solution reduces (*a*) $SiCl_4$ to SiH_4

$$LiAlH_4 + SiCl_4 \longrightarrow LiCl + AlCl_3 + SiH_4$$

(*b*) $Zn(CH_3)_2$ to ZnH_2

$$LiAlH_4 + Zn(CH_3)_2 \longrightarrow ZnH_2 + Li[AlH_2(CH_3)_2]$$

(*c*) aldehydes to alcohols

$$4RCHO + LiAlH_4 \longrightarrow (R—CH_2O)_4LiAl \xrightarrow{+\,4H_2O} 4R—CH_2OH + LiOH + Al(OH)_3$$

(*v*) NH_3, primary amines and secondary amines react with it to from H_2.

$$LiAlH_4 + 4NH_3 \longrightarrow LiAl(NH_2)_4 + 4H_3$$

$$LiAlH_4 + 4RNH_2 \longrightarrow LiAl(RNH)_4 + 4H_2$$

$$LiAlH_4 + 4R_2NH \longrightarrow LiAl(R_2N)_4 + 4H_2$$

(*vi*) It also reacts with diborane (B_2H_6) to form lithium borohydride, $Li[BH_4]$ and aluminum borobydride, $Al[BH_4]_3$.

$$LiAlH_4 + 2B_2H_6 \longrightarrow Li[BH_4] + Al[BH_4]_3$$

(*vii*) BCl_3 reacts to give B_2H_6.

$$3LiAlH_4 + 4BCl_3 \longrightarrow 3LiCl + 3AlCl_3 + 2B_2H_6$$

(*viii*) It converts the chlorides of Si, Sn, B, Al etc. into their hydrides.

$$LiAlH_4 + SiCl_4 \xrightarrow{\text{Ether}} SiH_4 + LiCl + AlCl_3$$

$$LiAlH_4 + SnCl_4 \xrightarrow{\text{Ether}} SnH_4 + LiCl + AlCl_3$$

$$LiAlH_4 + 4BCl_3 \xrightarrow{\text{Ether}} 2B_2H_6 + 3LiCl + 3AlCl_3$$

$$LiAlH_4 + AlCl_3 \xrightarrow{\text{Ether}} 4AlH_3 + 3LiCl.$$

Uses. (*i*) It is used for reducing almost all the compounds containing a carbonyl group, $> C = O$ like ketones, carboxylic acids and their derivatives to alcohols (*See* above).

(*ii*) It is also used for the preparation of the hydrides of Si, B, Al etc from their chlorides (*See* above).

Structure. It is an ionic complex hydride containing the complex anion *viz* alumino hydride anion, $[AlH_4]^-$ ion which is ionically attached with Li^+ cation. $[AlH_4]^-$ ion is formed by the coordination of the hydride ion (H^-) (Lewis base) obtained from LiH molecule to AlH_3 molecule (Lewis acid).

$$LiH \longrightarrow Li^+ + H^-$$

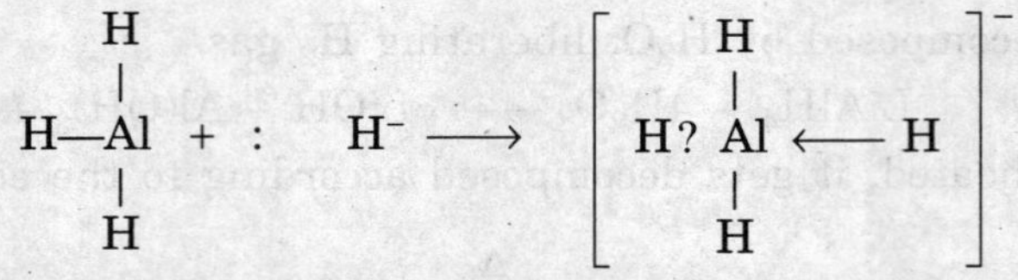

Lewis acid (acceptor) Lewis base (donor) Complex anion

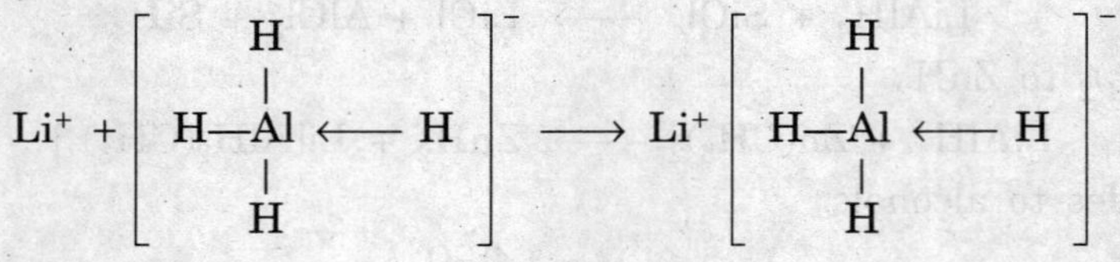

Thus in $[AlH_4]^-$ ion, one of the four hydrogen atoms is attached with Al-atom by a coordinate bond while the remaining three H-atoms are covalently attached.

Shape of $[AlH_4]^-$ ion. Due to the presence of one unit of negative charge on $[AlH_4]^-$ ion, Al atom (central atom) can be regarded as having 4 electrons instead of 3 in its valence-shell $[Al(3s^2p^1) + e^- \longrightarrow Al(3s^2p^2)]$. The Lewis structure of this ion,

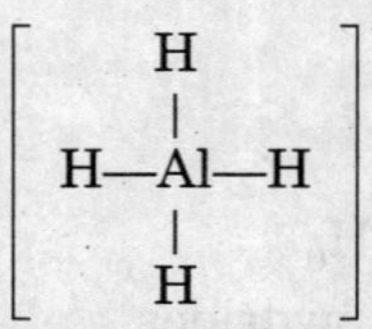

shows that Al atom uses all its 4 electrons in forming four σ-bonds with four H atoms. Thus in this ion σ-*bps* = 4 and *lps* = 0 and hence σ-*bps* + *lps* = 4. Consequently Al atom is supposed to be sp^3 hybridised in the given complex ion as shown in Fig. 16.1.

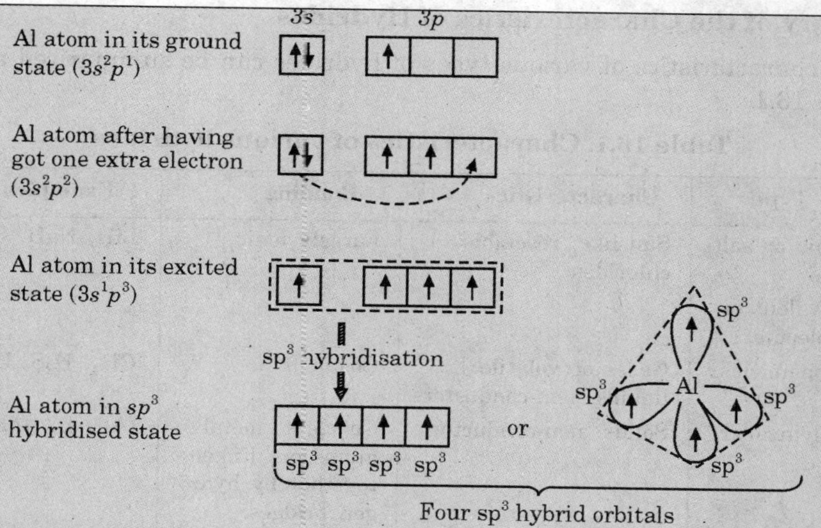

Fig. 16.1. sp^3 hybridisation undergone by Al atom in its excited state.

Each of the four sp^3 hybrid orbitals on Al atom is singly filled. Each of these hybrid orbitals overlaps along the axis with the half-filled 1*s* orbital on H-atom and forms four Al–H σ-bonds and consequently $[AlH_4]^-$ ion has a tetrahedral shape (See Fig. 16.2.)

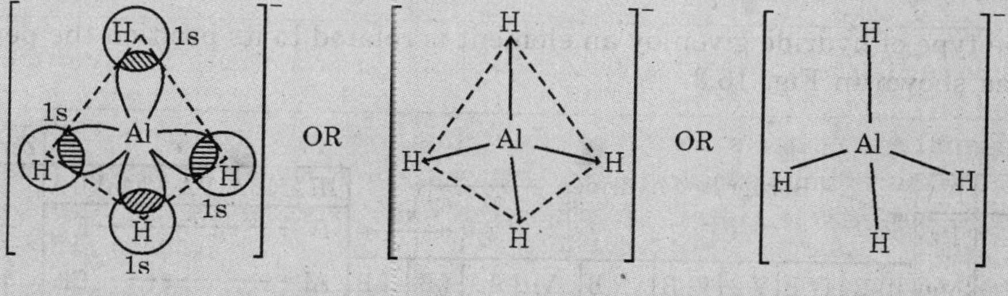

Fig. 16.2. Pictorial representation of the formation of four σ-bonds in $[AlH_4]^-$ ion which has tetrahedral shape.

E. Borderline Hydrides

There are a number of hydrides which do not fall into any of the above four types. Examples of such hydrides are BeH_2, MgH_2, CuH, ZnH_2, CdH_2, HgH_2 etc. Such hydrides are not easily formed from the elements but are generally prepared by special methods. Not all of these hydrides have been obtained in a crystalline state suitable for X-ray examination. These hydrides are volatile but not ionic and relatively much less stable than the ionic hydrides. BeH_2 and MgH_2 have properties of both ionic and covalent hydrides while CuH, ZnH_2, CdH_2, HgH_2 etc. have the properties which are intermediate between those of covalent and metallic hydrides. In none of these hydrides the type of bonding has been established.

Summary of the Characteristics of Hydrides

The characteristics of various types of hydrides can be summarised as given in Table 16.1.

Table 16.1. Characteristics of various hydrides

Type	Characteristics	Bonding	Examples
1. Ionic or salt-like	Salt-like, resemble chlorides	Largely ionic	LiH, NaH, CaH_2
2. Covalent or molecular			
(*i*) Mononuclear	Gases or volatile liquids, non-conductors	Covalent	CH_4, H_2S, HF, H_2O
(*ii*) Polynuclear	Solids, non-conductors	Covalent, metal atoms ore linked together by hydrogen bridges.	$(AlH_3)_n$, $(InH_3)_n$
3. Metallic or interstitial	Possess parental properties such as brittleness, lustrue, electrical conductivity etc.	Partly ionic with residual metallic bonds.	$TiH_{1.73}$, $VH_{0.6}$, $LaH_{2.76}$

Types of Hydrides and Periodic Table

The type of hydride given by an element is related to its position the periodic table as shown in Fig. 16.3.

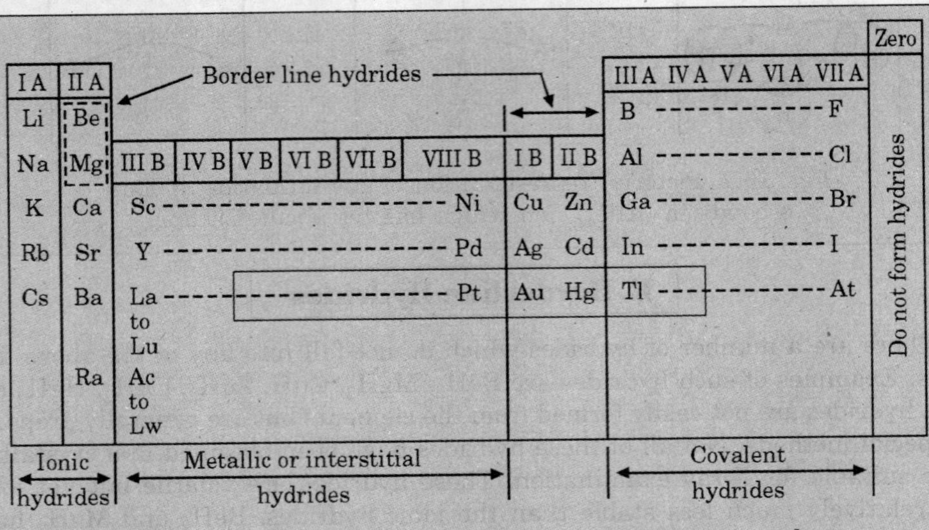

Fig.16.3. Relation between the type of hydrides given by an element and its position in the periodic table.

Question with Answers

Q.1 What characteristics do you expect from an electron-deficient hydride with respect to its structure and chemical reactions ?

Ans : Electron- deficient hydrides do not have sufficient number of electrons to form normal covalent bonds. Therefore, to make up this deficiency, they generally exist in polymeric froms such as B_2H_6, B_4H_{10}, $(AlH_3)_n$, etc.

Further to make up this deficiency of electrons, they react with many metals, non-metals and their compounds. Hence, electron-deficient hydrides are very reactive as shown below :

$$B_2H_6\ (g) + 3O_2\ (g) \longrightarrow B_2O_3\ (g) + 3H_2O\ (g)$$

$$2NaH\ (s) + B_2H_6\ (g) \xrightarrow{\text{Diethyl ether}} \underset{\text{Sod. borohydride}}{2Na^+\ [BH_4]^-\ (s)}$$

Being electron-deficient compounds, they act as Lewis acids and thus form complexes with Lewis bases. For Example.

$$\underset{\text{Diborane}}{B_2H_6} + \underset{\text{Trimethylamine}}{2\ddot{N}Me_3} \longrightarrow \underset{\text{Complex}}{[2H_3B \leftarrow NMe_3]}$$

$$\underset{\ }{B_2H_6} + \underset{\text{Carbon monoxide}}{2 : CO} \longrightarrow \underset{\text{Complex}}{[2H_3B \leftarrow CO]}$$

Q.2 Explain the following :

***(i)* Boiling points of hydrides of N, O and F are much higher than those of their subsequent groups elements.**

Ans : The hydrides of N, O and F are NH_3, H_2O and HF respectively. On the basis of masses of NH_3, H_2O and HF, their boiling points are expected to be lower than the hydrides of the subsequent group members (*i.e.* PH_3, H_2S and HCl). However, due to high electronegativity of N, O and F, their hydrides undergo extensive intermolecular H-bonding. As a result, the boiling points of NH_3, H_2O and HF are much higher than the hydrides of their subseqent group members (*i.e.* PH_3, H_2S and HCl)

***(ii)* Phosphorus atom can form PH_3 but not PH_5.**

Ans : Although P shows oxidation state of +3 and +5, it cannot form PH_5. Due to high Δ_{H-H} (435.88KJ mol^{-1}) and slightly negative $\Delta_{eg}H$ (–73KJ mol^{-1}) dihydrogen acts only as a weak oxidising agent. Therefore, it can oxidise P to + 3 oxidation state but not to its highlest oxidation state of +5. Thus, P forms PH_3 but not PH_5.

***(iii)* H_2 produced by the reaction between saline hydride and H_2O cannot be extingnished by CO_2.**

Ans : Saline hydrides (such as NaH, CaH_2, etc.) react with water voilently to form the corresponding metal hydroxides with the evolution of dihydrogen.

$$NaH\ (s) + H_2O\ (l) \longrightarrow NaOH\ (aq) + H_2\ (g)$$

$$CaH_2\ (s) + 2\ H_2O\ (l) \longrightarrow Ca\ (OH)_2\ (aq) + 2\ H_2\ (g)$$

These reactions are so exothermic that the evoled H_2 catches fire. The fire so produced cannot be exitingished by CO_2 because it gets reduced by the hot metal hydride to form sodium formate.

$$NaH + CO_2 \longrightarrow HCOONa$$

However, sand is useful since it is a highly stable solid.

(*iv*) Beryllium forms a covalent hydride while calcium forms an ionic hybride.

Ans : Because of higher electronegativity (E.N. = 1.5), Be forms covalent hydride while due to lower electronegativity (E.N. = 1.0) Ca forms ionic hydride.

(*v*) The process 1/2 H_2 (*g*) + $e^- \rightarrow$ H^- (*g*) is endothermic (ΔH = +151KJ mol^{-1}), yet salt-like hydrides are known.

Ans : It is true that formation of hydride (H^-) ion is an endothermic process, yet alkali and alkaline earth metals form salt-like hydrides. This is due to the reason that high lattice energy raleased (energy releasd during the formation of solid metal hydrides from their corresponding gaseous ions, *i.e.*, M^+ and H^-) more than compensates the energy needed for the formation of H^- ions from H_2 gas.

Q.3 What is hydride gap ?

Ans : The region of the periodic table from groups 7 to 9 which donot form hydrides is called hydride gap.

Q.4 Arrange the following hydrides in the increasing order of their property mentioned in braket.

(*i*) LiH, NaH, CsH (Inoic character).

(*ii*) NaH, MgH_2, H_2O (reducing property).

(*iii*) CaH_2, BeH_2, TiH_2 (Electrical conductance).

Ans : (*i*) Electronegative decreases down the group from Li to Cs, therefore, the ionic character of their hydrides increases in the same order, *i.e.*, *LiH < NaH < CsH.*

(*ii*) Ionic hydrides are powerful reducing agents. Both MgH_2 and H_2O are covalent hydrides but the bond dissociation energy of H_2O is much higher than that of MgH_2. Therefore, the reducing character increases in the order : H_2O < MgH_2 < NaH.

(*iii*) BeH_2 is a covalent hydride, therefore it does not conduct electricity at all. CaH_2 conducts electricity in the fused state while TiH_2 conducts electricity at room temperature. Thus, the order of increasing electrical conductance is : $BeH_2 < CaH_2 < TiH_2$.

Q.5 (*i*) What is the nature of hydrides formed by the elements with atomic numbers : 15, 19, 23 and 44. (*ii*) Give the chemical equations to show their reaction with H_2O.

Ans : (i) Nature of hydrides. (*a*) Element with Z = 15 is a non-metal (*i.e.* P) and hence forms covalent hydride (*i.e.* PH_3).

(*b*) Element with Z = 19 is an alkali metal (*i.e.* K) nad hence forms saline or ionic hydride (*i.e.* K^+ H^-).

(*c*) Element with atomic number Z = 23 is a transition metal (*i.e* V) of group 3 and hence forms a metallic or interstitial hydride (*i.e.* $VH_{0.56}$).

(*d*) Element with Z = 44 is a transition metal (*i.e.* Ru) of group 8 and hence does not form any hydride (hydride gap).

(*ii*) Reaction with water. Only ionic hydrides react with water evoling H_2 gas. Thus :

$$2\ KH\ (s) + 2\ H_2O\ (s) \longrightarrow 2\ KOH\ (aq) + 2\ H_2\ (q)$$

Q.6 Classify the following hybrides : BeH_2, AsH_3, B_2H_6, LaH_3 and $LiAlH_4$.

Ans : BeH_2 - polymeric, AsH_3 - covalent, B_2H_6 - electron deficient or Lewis acid, LaH_3 - metallic or interstitial, $LiAlH_4$ - complex.

Q.7 How will you explain that metallic hydrides can be used to store hydrogen.

Ans : In metallic hydrides, hydrogen is adsorbed as H-atoms. Due to the inclusion of H-atoms, the metal lattic expands and thus becomes less stable. Therefore, when the metallic hydride is heated, it decomposes to form hydrogen and very finely divided metal.

The hydrogen thus evolved can be used as a fuel. Thus, transition metals or their alloys can be used to store and transport hydrogen to be used as fuel. This is called hydrogen economy.

Q.8 How can saline hydrides remove traces of water from organic compounds ?

Ans : Saline hydrides (*i.e*, NaH, CaH_2, etc.) react with water forming their corresponding metal hydroxides with the liberation of H_2 gas. Thus, traces of water present in organic solvents can be easily removed by distiliing them over saline hydrides when H_2 escapes into the atmosphere, metal hydroxide is left in the flask while dry organic solvent distils over.

Alternatively, organic compounds containing traces of water can be dried by placing them in a desiccator containing saline hydrides at the bottom for a few hours or preferably overnight.

Q.9 Do you expect the carbon hydrides of the type (C_nH_{2n+2}) to act as Lewis acid or base ? Justify your answer.

Ans : Carbon hydrides of the type (C_nH_{2n+2}) are electron-precise hydrides. In other words, they have exact number of electrons required to from covalent bonds. Therefore, they do not have any tendency to either gain or lose electrons and hence they neither act as Lewis acids nor Lewis bases.

Q.10 Give an example each of an ionic hydride and a covalent hydride.

Ans : Ionic : NaH or CaH_2.

Covalent : H_2O, B_2H_6, CH_4, etc.

Q.11 What do you understand by (*i*) electron-deficient hydrides (*ii*) electron-precise hydrides (*iii*) electron-rich hydrides. Give suitable examples.

Ans : (*i*) Hydrides of group 13 (*i.e.*, BH_3, AlH_3, etc.) do not have sufficient number of electrons to form normal covalent bonds and hence are called electron-deficient hydrides. To make up this deficiency, they generally exist in polymeric forms such as B_2H_6, B_4H_{10}, $(AlH_3)_n$, etc.

(*ii*) Hydrides of group 14 (*i.e.*, CH_4, SiH_4, GeH_4, SnH_4, PbH_4) have exact number of electrons to form covalent bonds and hence are called electron-precise hydrides. All these hydrides have tetrahedral shapes.

(*iii*)Hydrides of group 15, 16 and 17 (*i.e.*, NH_3, PH_3, H_2O, H_2S, HF, HCl, etc.) have more electrons than required to form normal covalent bonds and hence are called electron-rich hydrides. The excess electrons in these hydrides are present as lone pairs of electrons.

Q.12 What do you understand by the term "non-stoichiometric hydrides"? Do you expect this type of the hydrides to be formed by alkali metals ? Justify your answer.

Ans : Hydrides which are deficient in hydogen and in which the ratio of the metal to hydrogen is fractional are called non-stoichimetric hydrides. Furthermore, even this fractional ratio is not fixed but varies with the temperature and the pressure. This type of hydrides are formed by *d*-and *f*-block elements. In these hydrides, the hydrogen atoms occupy holes in the metal lattice. Usually some holes always remain unoccupied and hence these metals form non-stoichiometric hydrides.

Since alkali metals are highly reducing, they transfer their lone electron to the H-atom, thereby forming H^- ions. In other words, alkali metal hydrides are ionic in which H^- ions occupy holes in the lattice. Since a hydride ion is formed by complete transfer of an electron, the ratio of metal to hydrogen is always fixed and hence alkali metals form only stoichimetric hydrides. In other words, alkali metals do not form non-stoichometric hydrides.

Q.13 (*a*) What is hydrolith ? How is it prepared ?

(*b*) Write the structures of two complex metal hydrides which are used as reducing agents.

(*c*) What type of elements form interstitial hydrides.

(*d*) Write two uses of interstitial hydrides ?

Ans : (*a*) Hydrolith is calcium hydride, CaH_2. It is prepared as :

$$Ca + H_2 \xrightarrow{\Delta} CaH_2 \text{ (hydrolith)}$$

(*b*) $LiAlH_4$ and $NaBH_4$.

(*c*) *d*-and *f*-block elements.

(*d*) Interestitial hydrides are used : (*i*) in storing H_2 and (*ii*) as catalysts for hydrogenation reactions.

University Questions

1. What are hydrides? How are they classified? Compare the properties of hydrides of B, Si and Be. *(Delhi 86)*
2. What are hydrides ? Give their preparation, properties and uses. *(Delhi 84; Delhi Hons. 86; Raj. 83)*
3. What are covalent hydrides ? How do their stability, basicity and acid strength vary in a group? *(Raj. 85)*
4. What are silicon hydrides? How are they prepared? Describe their properties and applications. *(Meerut B.Sc. 84)*
5. Discuss the structure of borohydride ion. *(Delhi Hons. 80)*
6. (*a*) What are polymeric hydrides ? Discuss the preparation and structure of beryllium hydride.

   (*b*) What are interstial hydrides ? How do these differ from other hydrides? *(Himachal Pradesh 2000)*
7. What are hydrides ? How are they classified ? Give important methods to obtain boron and silicon hydrides. *(Lucknow 2000)*
8. Write a short note on $LiAlH_4$. *(Lucknow 2001)*
9. Hydrides of alkali metals are basic in nature but those of halogens are acidic. Explain why ? *(C.S.J.M. Kanpur 2007)*
10. Give preparation, properties and uses of $LiAlH_4$. *(Meerut 2008)*
11. Complete and balance the equation :

$$2Li + AlH_4 \xrightarrow{\Delta} \ldots + 2Al + \ldots$$ *(Meerut 2009)*

CHAPTER

17

Heavy Water (D_2O)

Occurence of D_2O

(*i*) Washburn and Urey showed that the residual water obtained at the end of prolonged electrolysis of a large volume of ordinary water containing some alkali contains heavy water which is deuterium oxide.

(*ii*) It is also found in minute quantities on the leaves of banyan trees and in rain-water.

(*iii*) It is also found in the last remains obtained by the melting of snow by sun on the hills as in Himalayans.

Preparation of D_2O

Heavy water is prepared by the following methods :

1. By multi-stage electrolysis of ordinary water. (*Taylor, Eyring and Frost process*) This method was described by Taylor, Erying and Frost in 1933. Heavy water is largely prepared by the multi-stage (*i.e.* continuous and prolonged) electrolysis of ordinary water containing calculated quantity of NaOH.

Electrolytic cell and procedure. The cell used in this process is shown in Fig. 17.1 and has been designed by Brown, Degget and Urey. It consists of a steel cell (45 cm long and 10 cm in diameter). The cell itself acts as a cathode while

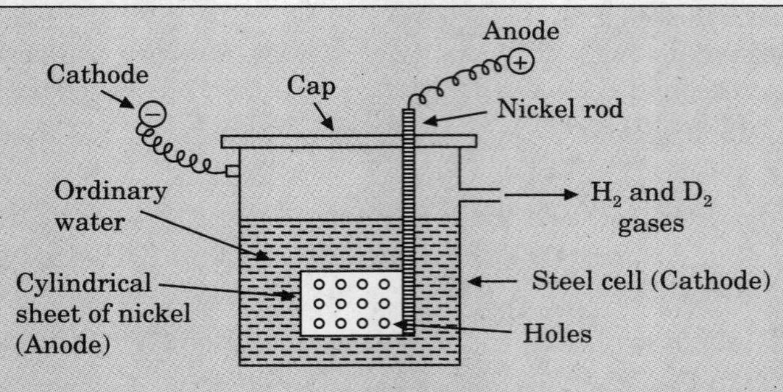

Fig. 17.1. Electrolytic cell used for the preparation of heavy water.

a cylindrical sheet of nickel having a number of holes punched into it (*i.e.* a perforated cylindrical sheet of nickel) serves as the anode.

In actual practice a large number of such cells is used so that the electrolysis of ordinary water may be done in several stages *i.e. the electrolysis of water is continuous and prolonged*. Thus by the continuous and prolonged electrolysis, the concentration of heavy water goes on increasing gradually in the residual water left behind at the end of each stage of electrolysis. Usually the process of electrolysis is complete in seven stages as discussed below :

First stage : Thirty electrolytic cells, joined together in series, are used in this stage. Each cell is charged with 77 litres of a 8% solution of NaOH in ordinary water, *i.e.* 2310 litres of ordinary water are electrolysed in this stage. A current of 110 volts is used and the electrolysis is carried out for about 72 hours. The gases liberated during the electrolysis at the end of this stage contain protium (H_2) and O_2 which are discarded. At the end of 72 hours the total volume of the charge reduces to about $\frac{1}{6}$th of its original volume. The alkali present is neutralised by passing CO_2 in it.

Second stage : The residual water obtained from the first stage is again electrolysed using six cells. The gases evolved during electrolysis in this stage are separately burnt in a burner and the mixture of H_2O and D_2O formed by the burning of gases is sent back to the cells being used in the first stage. The process of electrolysis in this stage is continued till the residual water contains about 0.5% of D_2O.

Third stage : The residual water obtained from the second stage is further electrolysed, till the residual water contains about 2.5% of heavy water. The gases evolved in this stage are separately burnt, as usual, in a burner and the mixture of H_2O and D_2O formed is sent back to the cell being used in the second stage. The residual water obtained at the end of this stage consists of 2.5% of D_2O.

Fourth stage : The residual water obtained from the third stage is further electrolysed so that the residual water contains 8.0% of D_2O. The gases evolved are burnt as usual and the mixture of H_2O and D_2O formed is sent back to the cells being used in the third stage.

Fifth stage : The residual water obtained from the fourth stage is further electrolysed so that the residual water contains about 30.0% of D_2O. The gases produced are burnt as usual and the mixture of H_2O and D_2O formed is sent back to the cells being used in the fourth stage.

Sixth stage : The residual water obtained from the fifth stage is further electrolysed till the residual water obtained contains 93.0% of D_2O. The gases produced are burnt as usual and the mixture of H_2O and D_2O obtained is sent back to the cells being used in fifth stage.

Seventh stage : The residual water obtained from the sixth stage is further electrolysed so that the residual water obtained at the end of this stage contains 99% of D_2O. The gases produced here are burnt as usual and the mixture of H_2O and D_2O obtained is sent back into the cells being used in the sixth stage. 99% pure D_2O obtained in this stage also contains some impurities like alkali metals, ordinary water etc. Therefore, this heavy water is distilled and the distillate obtained is then electrolysed to get pure deuterium. The deuterium is then separately burnt in O_2 to get 100% pure D_2O.

A flow sheet diagram showing the preparation of D_2O by the multistage electrolysis of ordinary water is given in Fig. 17.2.

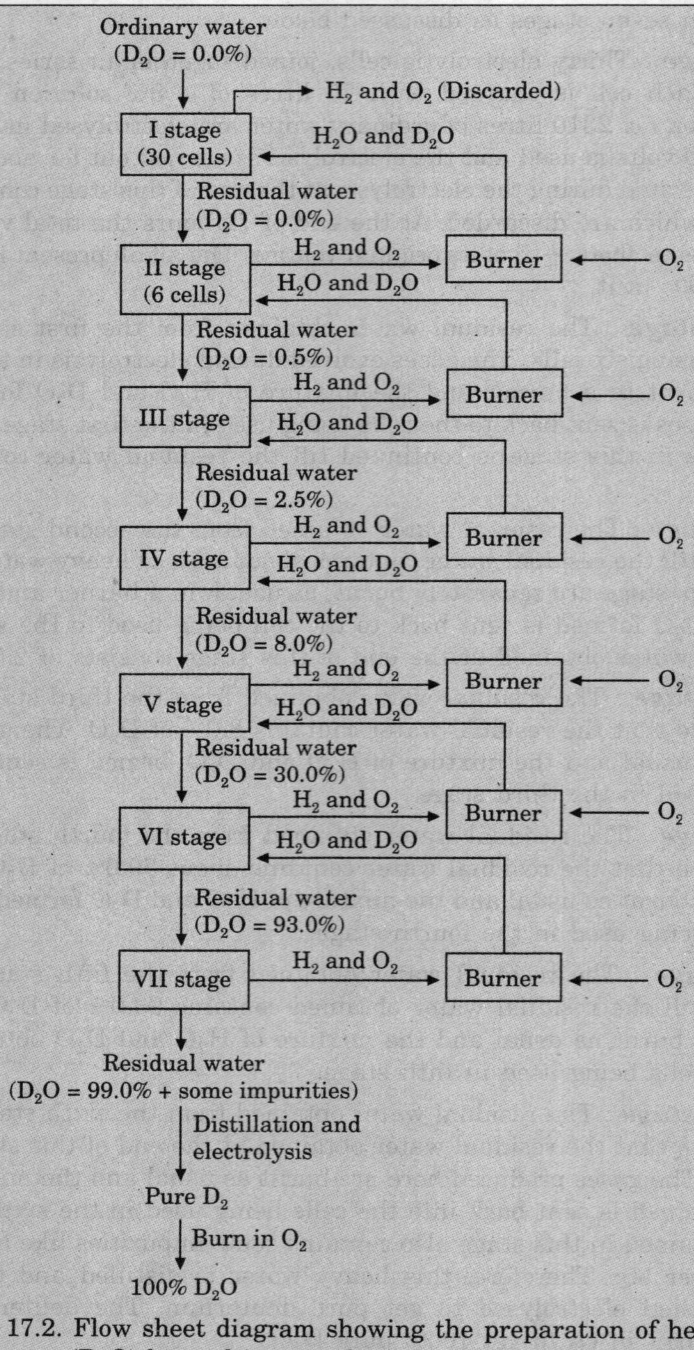

Fig. 17.2. Flow sheet diagram showing the preparation of heavy water (D_2O) by multi-stage electrolysis of ordinary water.

Taylor, Eyring and Frost started the electrolysis taking 2310 litres of ordinary water and obtained about 82% ml of D_2O. Data of their experiment are given in the following table :

Stage of electrolysis	Volume of ordinary water electrolysed (litres)	Density of the end product	Percentage of D_2O at the end of the stages mentioned
I	2310	0.998	—
II	340	0.999	0.5
III	52	1.001	2.5
IV	10.5	1.007	8.0
V	2	1.031	30.0
VI	0.42	1.098	93.0
VII	0.08	1.104	99.0

2. By fractional distillation of ordinary water. Heavy water has been separated from ordinary water by the method which takes the advantage of the difference in the boiling points of ordinary water (= 100°C) and heavy water (= 101.42° C). Since the difference between the boiling points of these two liquids is very small, the fractional distillation of ordinary water is done in a very long fractionating column (13 metre long) and the process is repeated several times. The lighter fraction (*i.e.* H_2O) which has lower boiling point is distilled first while the heavier fraction (*i.e.* D_2O) which has higher boiling point is left behind. The heavier fraction is richer in D_2O.

Distillation plants used for the manufacture of D_2O by this method have been recently established in New Zealand and U.S.A. and it is expected that D_2O on cheaper cost will be obtained.

3. By fractional freezing. Since the freezing point of ordinary water is 0°C and that of D_2O is 3.82°C, the two materials can be separated by fractional freezing. It is completed in several cycles.

4. By preferential adsorption. Based on the fact that the vapours of D_2O are adsorbed more rapidly on charcoal than those of H_2O, Taylor, Gould and Bleakney were able to separate D_2 from H_2O by passing their vapours on charcoal.

5. By exchange reactions. D_2O can also be prepared by the exchange reactions, *e.g.*

$$H_2O + D_2 \rightleftarrows D_2O + H_2$$

One specific exchange reaction used for the preparation of D_2O is discussed below :

When H_2S gas is passed through hot water, the hydrogen atoms exchange their places with deuterium atoms from D_2O present in water and H_2S is enriched with D_2S (Fig. 17.3). On passing this H_2S enriched with D_2S through cold water, the deuterium from D_2S and hydrogen from water (H_2O) again exchange their places and cold water becomes enriched with D_2O. This process can be repeated and cold water goes on becoming enriched with D_2O.

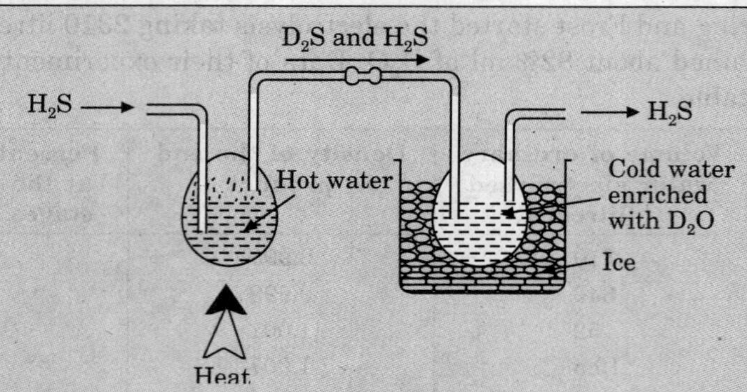

Fig. 17.3. Preparation of heavy water (D_2O) by exchange reaction.

6. Preparation of pure D_2O. Pure D_2O can be prepared by the action of D_2 gas on O_2 gas.

$$2D_2 + O_2 \longrightarrow 2D_2O$$

Manufacture of D_2O in India

In India *The Fertilizer Corporation of India* set up a heavy water plant at Nangal (Punjab) in 1962. Since here the electric supply is cheap, the method being used is of continuous and prolonged electrolysis of ordinary water. D_2O manufactured here is being supplied to the Atomic Energy Commission. Plants at Rourkel, Trombay, Namrup, Neyveli and Naharkatiya have also been set up.

Physical Properties of D_2O

(A) Similarities with H_2O. Like ordinary water, heavy water is a colourless, odourless and tasteless mobile liquid.

(B) Dissimilarities with H_2O. However, there are many physical properties of D_2O which are different from those of H_2O.

This fact is evident from the values of physical constants of H_2O and D_2O given in Table 17.1.

Chemical Properties of D_2O

D_2O behaves like H_2O in most of its chemical properties. However, D_2O reacts more slowly than H_2O in chemical reactions. Some of its chemical reactions are given below :

1. Electrolysis. A solution of D_2O containing P_2O_5 or Na_2CO_3. (The addition of P_2O_5 or Na_2CO_3 makes D_2O conductor), on electrolysis, decomposes into D_2 and O_2 which are evolved at cathode and anode respectively.

$$2D_2O \xrightarrow{\text{Electrolysis}} \underset{\text{(At cathode)}}{2D_2} + \underset{\text{(At anode)}}{O_2}$$

2. Action of metals. D_2O reacts with alkali and alkaline earth metals to liberate D_2 and heavy alkalies are formed.

$$2Na + 2D_2O \longrightarrow 2NaOD + D_2$$

$$Ca + 2D_2O \longrightarrow \underset{\text{Heavy alkalies}}{Ca(OD)_2} + D_2$$

3. Action of metallic oxides. When D_2O reacts slowly with metallic oxides (*i.e.* basic oxides) like Na_2O, CaO etc., heavy alkalies are formed.

$$Na_2O - D_2O \longrightarrow 2NaOD$$

$$CaO - D_2O \longrightarrow Ca(OD)_2$$

4. Action of non-metallic oxides. When D_2O reacts with non-metallic oxides (*i.e.* acidic oxides), deutero-acids are formed.

$$N_2O_5 + D_2O \longrightarrow \underset{\text{Deutero nitric acid}}{2DNO_3}$$

$$SO_3 - D_2O \longrightarrow \underset{\text{Deutero sulphuric acid}}{D_2SO_4}$$

$$P_2O_5 + 3D_2O \longrightarrow \underset{\text{Deutero phosphoric acid}}{2D_3PO_4}$$

Table 17.1. Physical constants of ordinary water (H_2O) and heavy water (D_2O)

Property	Ordinary water (H_2O)	Heavy water (D_2O)
Molecular weight	18.016	20.03
Density at 20°C	0.998	1.017
Boiling point (°C)	100.00	101.42
Melting point (*i.e.*, freezing point) (°C)	0.00	3.82
Temp. of maximum density	4.0	11.6
Max. density	1.0000 (at 4°C)	1.1073 (at 11.6 °C)
Specific heat at 20°C	1.0	1.018
Latent heat of fusion (cals/gm)	79.7	75.5
Latent heat of evaporation (*i.e.* vaporisation) (cals/gm)	539.0	557.0
Surface tension at 20°C (dynes/cm)	72.75	67.80
Refractive index at 20°C	1.3330	1.3284
Viscosity at 20°C (milli poises)	10.09	12.6
Dielectric constant	82.0	80.5
Dissociation constant (*i.e.*, ionic product) at 25°C	1.0×10^{-14}	0.3×10^{-14}
Ionic mobilities at 18°C		
K^+	64.4	54.4
Cl^-	65.5	55.3
Solubility of NaCl (gms/lit)	359	305
Critical temperature (°C)	374.2	374.1
Critical pressure (atm)	218.5	218.6
Molar freezing point depression (°C)	1.85	0.52

5. Action of metallic nitrides, phosphides etc. D_2O decomposes metallic nitrides, phosphides, arsenides and carbides to form the corresponding deutro compounds, *e.g.*

$$2AlN + 6D_2O \longrightarrow 2Al(OD)_3 + \underset{\text{Dutero ammonia}}{2ND_3}$$

$$Mg_3N_2 + 6D_2O \longrightarrow 3Mg(OD)_2 + 2ND_3$$

$$Ca_3P_2 + 6D_2O \longrightarrow 3Ca(OD)_3 + \underset{\text{Deutero phosphine}}{2PD_3}$$

$$Na_3As + 6D_2O \longrightarrow 3NaOD + \underset{\text{Deutero arsine}}{AsD_3}$$

$$Al_4C_3 + 12D_2O \longrightarrow 4Al(OD)_3 + \underset{\text{Deutero methane}}{2CD_4}$$

$$CaC_2 + 2D_2O \longrightarrow Ca(OD)_2 + \underset{\text{Deutero acetylene}}{C_2D_2}$$

The melting points and boiling points of deutero compounds are generally higher than the analogous hydrogen compounds, *e.g.*, the melting and boiling points of ND_3 (m.pt. = 199.5°K, b.pt. = 242.0°K) are higher than those of NH_3 (m.pt. = 195.2°K, b.pt. 239.8°K.)

6. Formation of deuterates. Water (H_2O) gets associated with salts to form the crystalline salts called *hydrates*. Likewise heavy water (D_2O) also gets associated with salts to form crystalline salts called *deuterates or deutero-hydrates*. Examples of some deuterates are : $BeCl_2.4D_2O$; $CuSO_4.5D_2O$; $CoCl_3.6D_2O$; $NiCl_2.6D_2O$; $MgSO_4.7D_2O$.

In these deuterates D_2O acts as heavy water of crystallisation. These deuterates are obtained when solution of salts in D_2O are crystallised. For example :

$$CuSO_4.5H_2O + 5D_2O \xrightarrow[-5H_2O]{\text{Crystalisation}} CuSO_4.5D_2O$$

Deuterates are often analogous in composition with hydrates. The colour of deuterates of transition metals are lighter than those of hydrates, *e.g.* $CuSO_4.5D_2O$ has light blue colour while $CuSO_4.5H_2O$ has deep blue colour.

7. Action of compounds containing hydrogen : Exchange reactions. D_2O reacts with a number of compounds containing labile hydrogen atoms. In these reactions H atoms are partially or completely replaced by deuterium. Such reactions are called *exchange reactions.*

If the compounds contain ionic (*i.e.*, polar) hydrogen atoms, the exchange reactions occur more readily, *e.g.*

$$NaOH + D_2O \rightleftharpoons NaOD + HOD$$

$$HCl\ (aq) + D_2O \rightleftharpoons DCl + HOD$$

$$NH_4Cl + 4D_2O \rightleftharpoons ND_4Cl + 4HOD$$

The exchange reaction of NH_4Cl with D_2O takes place in steps as shown below :

$$NH_4Cl + D_2O \rightleftharpoons NH_3DCl + HOD$$
$$NH_3DCl + D_2O \rightleftharpoons NH_2D_2Cl + HOD$$
$$NH_2D_2Cl + D_2O \rightleftharpoons NHD_3Cl + HOD$$
$$NHD_3Cl + D_2O \rightleftharpoons ND_4Cl + HOD$$

On adding : $NH_4Cl + 4D_2O \rightleftharpoons ND_4Cl + 4HOD$

If the compounds contain non-ionic hydrogen atoms, the exchange reactions take place slowly and require the presence of a catalyst, *e.g.* C_6H_6 when treated with D_2O using nickel dispersed on kieselguhr at a temperature of 200°C leads to the formation of pure *hexa deutero-benzene.*

$$C_6H_6 + 3D_2O \rightleftharpoons \underset{\text{Hexa deutero-benzene}}{C_6D_6} + 3H_2O$$

Hexammine cobaltic chloride, $[Co(NH_3)_6]Cl_3$ shows a slow e change reaction.

$$[Co(NH_3)_6]Cl_3 + D_2O \longrightarrow [Co(NH_3)_5(NH_2D)]Cl_3 + HOD$$

Exchange reactions have been used in determining the number of ionic and non-ionic hydrogen atoms present in a given compound, *e.g.*, treatment of hypophosphorous acid (H_3PO_2) with D_2O brings about the exchange of only one H atom of H_3PO_2 with deuterium.

$$H_3PO_2 + D_2O \longrightarrow H_2DPO_2 + HDO$$

This reaction shows the presence of only one ionic hydrogen atom in H_3PO_2, *i.e.* H_3PO_2 is a *mono basic acid.*

8. Physiological effects. Heavy water of high concentration retards the growth of plants and animals. Lewis has shown that tobacco seeds do not grow in D_2O, though under the same conditions they develop well in ordinary water. Pure D_2O kills small fish, tadpoles and mice, when fed on them. Mice and rats, on drinking D_2O, feel more and more thirsty. According to Taylor, it is a germicide and a bactericide. While water containing higher concentration of D_2O is poisonous, water containing small quantity of D_2O acts as a tonic and stimulates vegetable growth. Although D_2O is injurious to animal life, certain moulds have been found to flourish better in solutions richer in D_2O than in ordinary aqueous solutions.

9. Deuterolysis. Water brings about hydrolysis of certain inorganic chlorides and some other salts. D_2O also brings about a similar reaction which is called *deuterolysis*.

$$AlCl_3 + 3D_2O \longrightarrow Al(OD)_3 + 3DCl$$
$$BaS + 2D_2O \longrightarrow Ba(OD)_2 + D_2S$$
$$BiCl_3 + D_2O \longrightarrow BiOCl + 2DCl$$
$$SiCl_4 + 4D_2O \longrightarrow D_4SiO_4 + 4DCl$$
$$SOCl_2 + 2D_2O \longrightarrow D_4SO_3 + 2DCl$$

Uses of D_2O

Important uses of D_2O are :

(*i*) *As a moderator of neutrons in nuclear reactors*. Neuterons are used for bringing about the fission of uranium atoms. For this purpose the neutrons of *slow speed* are required. The speed of neutrons is slowed down by passing them through D_2O. Such substances which are used for slowing down the speed of neutrons are called *moderators*. Obviously D_2O here acts as a moderator.

(*ii*) *For the preperation of D_2*. D_2 can be prepared either by the electrolysis of DO or by the action of sodium metal on D_2O

$$2D_2O \xrightarrow{\text{Electrolysis}} 2D_2 + O_2$$

$$2D_2O + 2Na \longrightarrow 2NaOD + D_2$$

(*iii*) *As a tracer compound*. D_2O is commonly used as a tracer compound for studying reaction mechanism in organic chemistry *e.g.* D_2O has helped to elucidate the mechanism of electrophilic aromatic substitution reactions.

D_2O is also used as tracer in metabolic processes, *e.g.* when mice fed with labelled butyric acid, within hours all the deuterium was eliminated from the body indicating that it is a ready source of energy.

(*iv*) Although D_2O injurious to animal life, yet certain moulds have been found to flourish better in solutions rich in D_2O than in ordinary aqueous solution.

(*v*) The ability of D_2O is to undergo exchange reactions has been used for studying the structure of some oxy-acids of phosphorus such as H_3PO_2 and H_3PO_3 by finding out the number of ionic hydrogen atoms in these compounds.

Question with Answers

Q.1 Give reactions to show that H_2O_2 can act as an OA as well as a RA.

Ans : H_2O_2 can act as an oxidising as well as a reducing agent both in acidic and basic media as illustrated below :

(*i*) *Oxidising agent (OA) in acidic medium*

$$2\,Fe^{2+}(aq) + 2\,H^+(aq) + H_2O_2(aq) \rightarrow 2Fe^{3+}(aq) + 2\,H_2O\,(l)$$

(*ii*) *Oxidising agent in basic medium*

$$2\,Mn^{2+}(aq) + H_2O_2(aq) + 2\,OH^-(aq) \rightarrow MnO_2(s) + 2\,H_2O\,(l)$$

$$2\,Cr^{3+}(aq) + 3\,H_2O_2(aq) + 10OH^-(aq) \rightarrow 2\,CrO_4^{2-}(aq) + 8\,H_2O\,(l)$$

(*iii*) *Reducing agent (RA) in acidic medium*

$$2\,MnO_4^-(aq) + 6H^+(aq) + 5H_2O_2(aq) \rightarrow 2Mn^{2+}(aq) + 8H_2O\,(l) + SO_2$$

(*iv*) *Reducing agent in basic medium*

$$I_2(s) + H_2O_2(aq) + 2\,OH^-(aq) \rightarrow 2\,I^-(aq) + 2\,H_2O\,(l) + O_2(g)$$

$$2[Fe(CN)_6]^{3-}(aq) + 2\,OH^-(aq) + H_2O_2(aq) \rightarrow 2\,[Fe(CN)_6]^{4-}(aq) + 2H_2O(l) + O_2(g).$$

Q.2 10 ml of a given solution of H_2O_2 contains 0.91 g of H_2O_2. Express its strength in volume.

Ans : 68 g of H_2O_2 produce O_2 = 22400 ml at NTP.

$\therefore$ 0.91 g of H_2O_2 will produce O_2 = $\frac{22400 \times 0.91}{68}$ = 300 ml at NTP

$\therefore$ Volume strength = $\frac{300}{10}$ = **30**

Q.3 Calculate the volume of 10 volume H_2O_2 solution that will react with 200 ml of 2 N $KMnO_4$ in acidic medium :

Ans : Normality of 10 volume H_2O_2 = $\frac{\text{Volume strength}}{5.6} = \frac{10}{5.6}$N

Applying normality equation,

$$N_1V_1 = N_2V_2$$
$$(H_2O_2) \quad (KMnO_4)$$
$$\frac{10}{5.6} \times V_1 = 2 \times 200$$

or $V_1 = \frac{2 \times 200 \times 5.6}{10}$ = **224 cm³**.

Q.4 Calculate the molarity strength of H_2O_2 solution marked '30 volume'.

Ans : Morality = Volume strength/11.2 = $\frac{30}{11.2}$ = **2.68 M.**

Q.5 Name two compounds which retard decomposition of H_2O_2 soultion.

Ans : Acetanilide, glycerol.

Q.6 How does H_2O_2 behave as a bleaching agent ?

Ans : The bleaching action of H_2O_2 is due to the *nascent oxygen* which it liberates on decomposition.

$$H_2O_2 \longrightarrow H_2O + O$$

The nascent oxygen combines with colouring matter which, in turn, gets oxidised. Thus, the bleaching action of H_2O_2 is due to the *oxidation of colouring matter by nascent oxygen*. It is used for the bleaching of delicate materials like ivory, feather, silk, wool, etc.

Q.7 What do you mean by 15 volume H_2O_2 solution ?

Ans : 1 cm^3 of a 15 volume H_2O_2 solution gives 15 ml of O_2 at NTP.

Q.8 Calculate the percentage strength and strength in grams per liter of 10 volume hydrogen peroxide solution.

Ans : Hydrogen peroxide decomposes on heating according to the equation :

$$2H_2O_2 \longrightarrow 2H_2O + O_2$$

2 × 34 g — 22.4 liters at N.T.P.

= 68 g — or 22400 cm^3 at N.T.P.

From the equation, 22.4 liters of O_2 at N.T.P. are obtained from 2 × 34 or 68 g of H_2O_2.

$\therefore$ 10 ml of O_2 at N.T.P. will be obtained from $\frac{68}{22400} \times 10$ g of H_2O_2

But 10 ml of O_2 at N.T.P. are produced from 1 ml of *10 volume* H_2O_2 solution

Thus, 1 ml of *10 volume* H_2O_2 solution contains $\frac{68}{22400} \times 10$ g of H_2O_2

$\therefore$ 100 ml of *10 volume* H_2O_2 solution will contain $\frac{68}{22400} \times \frac{10}{1} \times 100 = 3.036$g

Thus, a *10 volume* H_2O_2 solution is approx. 3%

Alternatively, 1000 ml of 10 volume H_2O_2 will contain $H_2O_2 = \frac{68}{22400} \times 10 \times 1000 = 30.36$ g

Therefore, strength of H_2O_2 in 10 volume H_2O_2 = **30.36 g L^{-1}**

Q.9 Calculate the amount of H_2O_2 present in 10 ml of 25 volume H_2O_2 solution.

Ans : 10 ml of 25 volume H_2O_2 liberate O_2 = 10 × 25 = 250 ml at NTP

$$\underset{68\text{ g}}{2\,H_2O_2} \longrightarrow 2\,H_2O + \underset{22400\text{ml at NTP}}{O_2}$$

Amount of H_2O_2 that will liberate 250 ml of O_2 at NTP = $\frac{68 \times 250}{22400}$ = **0.759 g**.

Q.10 Complete the following equations :

(*i*) PbS (*s*) + H_2O_2 (*aq*) → ....

(*ii*) MnO_4^- (*aq*) + H_2O_2 (*aq*) → ....

Ans : (*i*) PbS (*s*) + 4 H_2O_2 (*aq*) → $PbSO_4$ (*s*) + 4 H_2O (*l*)

(*ii*) MnO_4^- (*aq*) + 5 H_2O_2 (*l*) + 6 H^+ (*aq*) → 2 Mn^{2+} (*aq*) + 8 H_2O (*l*) + 5O_2(*g*)

Q.11 What is perhydrol ?

Ans : Perhydrol is the trade name for H_2O_2 which is used as an antiseptive for washing wounds, teeth and ears.

Q.12 How many moles of H_2O_2 will be required for the neutralistaion of 1 mole of $KMnO_4$?

Ans : The reaction between H_2O_2 and acidified solution of $KMnO_4$ takes place as :

$$\underset{2\text{ moles}}{2\,KMnO_4} + 5\,H_2SO_4 + \underset{5\text{ moles}}{5\,H_2O_2} \rightarrow K_2SO_4 + 2\,MnSO_4 + 8\,H_2O + 5\,O_2$$

This equation shows that :

$\therefore$2 moles of $KMnO_4$ are decolourised by 5 moles of H_2O_2

$\therefore$1 mole of $KMnO_4$ is decolourised by $\frac{5}{2}$ moles of H_2O_2

$\therefore$ Answer is $\frac{5}{2}$ moles of H_2O_2

Q.13 How is heavy water produced from ordinary water ?

Ans : It is obtained by repeated electrolysis of ordinary water containing 3% NaOH.

Q.14 What is the importance of heavy water with regard to nuclear power generation ?

Ans : It is used as a moderator to slow down the neutrons produced as a result of fission and thus helps to control the nuclear reactions.

Q.15 How is D_2O_2 prepared ?

Ans : D_2O_2 is prepared by distillation of potassium persulphate ($K_2S_2O_8$) with D_2O.

Q.16 What happens when chloroform is treated with heavy water in presence of an alkali ?

Ans : Deuterochloform ($CDCl_3$) is formed.

$$CHCl_3 + D_2O \xrightarrow{OH^-} CDCl_3 + DOH.$$

Q.17 To a 25 ml H_2O_2 solution,excess of acidified solution of potassium iodide was added. The iodine liberated required 20 ml of 0.3 N sodium thiosulphate solution. Calculate the volume strength of H_2O_2 solution.

Ans : Step 1. *To determine the normality of H_2O_2 solution.*

Let the normalitiy of the H_2O_2 solution be N_1. According to the question,

25 ml of $N_1H_2O_2 \equiv$ 20 ml of 0.3 N $Na_2S_2O_3$ solution

or $25 \times N_1 = 20 \times 0.3$

or $N_1 = \dfrac{20 \times 0.3}{25} = \mathbf{0.24\ N}$

Thus, the normality of the given H_2O_2 solution = **0.24 N**

Step 2. *To determine the amount of H_2O_2 in 25 ml solution*

1000 ml of $N_1H_2O_2$ solution contain H_2O_2 = 17 g (Eq.wt. of H_2O_2 = 17)

$\therefore$ 25 ml of 0.24 N H_2O_2 solution will contain $H_2O_2 = \dfrac{17 \times 25 \times 24}{1000 \times 100} = \mathbf{0.102\ g}$

Step 3. *To determine the volume strength of H_2O_2 solution.*

Consider the chemical equation,

$$2\,H_2O_2 \rightarrow 2\,H_2O + O_2$$

$2 \times 34 = 68$ g 22.4 liters of O_2 at N.T.P.

68 g of H_2O_2 give O_2 = 22.4 liters of O_2 at N.T.P.

$\therefore$ 0.102 g of H_2O_2 will give $O_2 = \dfrac{22.4 \times 1000 \times 0.102}{68}$ = 33.6 ml at N.T.P.

Now, 25 ml of H_2O_2 solution give O_2 = 33.6 ml at N.T.P.

$\therefore$ 1 ml of H_2O_2 solution will give $O_2 = \dfrac{33.6}{25} = \mathbf{1.344\ ml}$

Thus, the volume strength of the given H_2O_2 solution = **1.344**

Q.18 Explain the following :

(*i*) Electrolysis of ordinary water occurs faster than that of heavy water ?

Ans : Due to lower bond dissociation energy of protium bonds in H–O–H than deuterium bonds in D–O–D, electrolysis of H_2O occurs much faster than that of D_2O.

***(ii)* Although D_2O resembles H_2O chemically, yet it is a toxic subtance.**

Ans : D_2O is toxic, since D^+ reacts at a much slower rate as compared to H^+ in ezyme-catalyzed reactions.

***(iii)* When H_2O_2 is added to blood, rapid evolution of a gas occurs.**

Ans : The enzyme present in the blood catalyses the oxidation of H_2O_2 and hence rapid evolution of O_2 takes place.

$$2H_2O_2 \xrightarrow[\text{in blood}]{\text{Enzyme present}} 2\ H_2O + O_2\uparrow$$

***(iv)* Ionic compounds (e.g. NaCl) are less soluble in D_2O than in H_2O.**

Ans : Since D_2O has lower value of dielectric constant than H_2O, ionic compounds are less soluble in D_2O than in H_2O.

***(v)* Heavy water cannot be used for drinking purpose.**

Ans : Heavy water is injurious to human beings, plants and animals since it slows down the rates of reactions occurring in them. Thus, heavy water does not support life so well as does ordinary water.

***(vi)* Hydrated barium peroxide is used in the preparation of hydrogen peroxide instead of anhydrous peroxide ?**

Ans : If anhydrous barium peroxide is used in the preparation of H_2O_2, the $BaSO_4$ formed during the reaction forms an insoluble protective coating on the surface of solid barium peroxide.

$$BaO_2\ (s) + H_2SO_4\ (aq) \longrightarrow BaSO_4 + H_2O_2\ (aq)$$

$BaSO_4$ thus formed prevents further action of the acid and ultimately the reaction stops. If, however, hydrated barium peroxide (in the form of thin paste) is used, the water of crystallization does not allow $BaSO_4$ to deposite on the surface of BaO_2 and the reaction goes to completion.

***(vii)* Phosphoric acid is preferred to sulphuric acid in the preparation of H_2O_2 from barium peroxide.**

Ans : The aqueous solution of H_2O_2 prepared by the action of dil. H_2SO_4 on hydrated BaO_2 has impurities of heavy metal ions like Ba^{2+}, Pb^{2+}, etc. These catalyse the decomposition of H_2O_2. Therefore, H_2O_2 perpared by the action of dil. H_2SO_4 on hydrated BaO_2 does not have good keeping qualities. If, however, phosphoric acid is used, the impurities of heavy metal ions are precipitated as insoluble phosphates. As a result, the resulting solution of H_2O_2 has good keeping qualities.

***(viii)* Red litmus paper turns white in the solution of Na_2O_2 but the same paper turns blue in the solution of Na_2O.**

Ans : Na_2O_2 reacts with H_2O to produce H_2O_2.

$$Na_2O_2 + H_2O \longrightarrow 2NaOH + H_2O_2$$

It is due to the bleaching property of H_2O_2 that the red litmus paper turns white. Na_2O gives NaOH with H_2O.

$$Na_2O + H_2O \rightarrow 2\ NaOH$$

NaOH thus produced turns red litmus paper blue.

(ix) A mixture of hydrazine ($NH_2 - NH_2$) and H_2O_2 with Cu (II) catalyst is used as a rocket propellant.

Ans : The reaction between hydrazine and H_2O_2 is highly exthermic and is accompaned by a large increase in the volume of the products and hence this mixture is used as a rocket propellant.

$$NH_2–NH_2\ (l) + 2H_2O_2\ (l) \xrightarrow{Cu\ (II)} N_2\ (g) \uparrow + 4\ H_2O\ (g) \uparrow$$

(x) Statues coated with white lead on long exposure to atmosphere turn black and the original colour can be restored on treatment with H_2O_2.

Ans : On long exposure to atmosphere, white lead (PbO_2) is converted into black PbS due to the action of H_2S present in the atmosphere. As a result, statues turn black.

$$PbO_2 + 2\ H_2S \rightarrow PbS + 2\ H_2O$$

On treatment of these blackned statues with H_2O_2, the black PbS gets oxidised to white $PbSO_4$ and the colour is restored.

$$PbS + 4\ H_2O_2 \rightarrow PbSO_4 + 4\ H_2O$$

Q.19 A 5.0 cm³ solution of H_2O_2 liberates 0.508 g of iodine from an acidified KI solution. Calculate the strength of H_2O_2 solution in terms of volume strength at STP.

Ans : (*a*) $2\ KI + H_2SO_4 + H_2O_2 \rightarrow K_2SO_4 + 2\ H_2O + I_2$

From the above equation,

$$H_2O_2 \equiv I_2$$

$$34 \text{ g of } H_2O_2 \equiv 254 \text{ g of } I_2$$

$$\therefore 0.508 \text{ g of } I_2 \text{ will be liberated from } H_2O_2 = \frac{34}{254} \times 0.508 = \mathbf{0.068\ g}$$

(*b*) The decomposition of H_2O_2 occurs as :

$$\begin{array}{ccccc} 2\ H_2O_2 & \longrightarrow & 2\ H_2O & + & O_2 \\ 2 \times 34 = 68 \text{ g} & & 22400 \text{ cm}^3 \text{ at NTP} & & \end{array}$$

$$\therefore 0.068 \text{ of } H_2O_2 \text{ upon decomposition will give } O_2 = \frac{22400}{68} \times 0.068$$

$$= \mathbf{22.4\ ml}$$

(*c*) Now 5.0 cm³ of H_2O_2 solution gives O_2 = 22.4 cm³ at STP

$$\therefore 1.0 \text{ cm}^3 \text{ of } H_2O_2 \text{ solution will give } O_2 = \frac{22.4}{5} = 4.48 \text{ cm}^3 \text{ at STP}$$

Thus, volume strength of given H_2O_2 solution = **4.48.**

University Questions

1. How is anhydrous H_2O_2 obtained from barium peroxide? *(Madras 85)*

2. How does H_2O_2 manufactured? *(Madurai 85)*

3. Discuss the constitution of H_2O_2. *(Meerut 89; Allahabad 86, 87; Bundelkhand 88)*

4. Explain the following :

(i) H_2O_2 acts both as an oxidising as well as a reducing agent.

(ii) The mixture of hydrazine (N_2H_4) and H_2O_2 with a Cu (II) catalyst is used as a rocket propellant ? *(I.I.T. 87)*

(iii) The solution of H_2O_2 cannot be concentrated simply by heating.

(iv) H_2O_2 is used for restoring old paintings in the museum. *(REE 87)*

5. What happens when :

(i) an alkaline solution of potassium ferricyanide is treated with H_2O. *(I.I.T. 82)*

(ii) H_2O_2 reacts with H_2S.

(iii) H_2O_2 reacts with acidified solution of $KMnO_4$ solution. *(I.I.T. 73)*

(iv) H_2O_2 reacts with Ag_2O.

(v) H_2O_2 reacts with PbS

(vi) H_2O_2 and ether is added to an acidic solution of $K_2Cr_2O_7$.

(vii) H_2O_2 is treated with O_3.

(viii) H_2O_2 is added to an acidic solution of $FeSO_4$. *(REE 85)*

(ix) Complete and balance the following equations :

(a) $H_2O_2 + N_2H_4 \longrightarrow .....$ *(Meerut 85)*

(b) $H_2O_2 + C_6H_6 \longrightarrow .....$ *(Madurai 85)*

(c) $H_2O_2 + PbO_2 \longrightarrow .....$ *(Madurai 85)*

(d) $H_2O_2 + H^+ + MnO_4^- \longrightarrow .....$ *(Raj. 85)*

(e) $H_2O_2 + H_2SO_4 + KMnO_4 \longrightarrow MnSO_4 + K_2SO_4 + H_2O + \longrightarrow .....$ *(I.I.T. 75; MLNR 85)*

(f) $H_2O_2 + HO.SO_2.Cl \longrightarrow .....$ *(Delhi 85)*

(x) H_2O_2 reacts with Caro's acid. *(Delhi 87; Calcutta 86)*

(xi) One mole of H_2O_2 reacts with two moles of chlorosulphonic acid. *(Meerut 84 S)*

(xii) A solution of K_2CrO_4 containing H_2O_2 is acidified with H_2SO_4 in presence of ether or pyridine. *(Delhi 87)*

6. Describe the preparation, properties and uses of heavy water. *Allahabad 89; Kanpur 80, 81, 82; Agra 80, 81; Lucknow 81, 84;*

7. Starting from deuterium how would you prepare D_2O? *(MD Rohtak 84)*

8. How does heavy water react with CaC_2, C_2H_2, H_2SO_4, P_2O_5, Mg_3N_2.

9. Discuss the following properties of H_2O_2 *(i)* Oxidising properties *(ii)* Reducing properties *(iii)* Acidic properties. *(Madras 95, Maduria 85, Delhi 83)*

10. Give methods of preparation, properties and uses of heavy water. *(Kanpur 2000, Avadh 2000)*

CHAPTER 18 General Characteristics of Group 1 (I A) Elements : Alkali Metals

Group IA of the long form of periodic table contains six elements (excluding hydrogen) which are *Lithium* (Li), *Sodium* (Na), *Potassium* (K), *Rubedium* (Rb), *Caesium* (C) and *Francium* (Fr). These elements are collectly called alkali metals, since they form strongly alkaline oxides and hydroxides. Fr is a radioactive element.

Position of Alkali Metals in the Periodic Table

The inclusion of the alkali metals in the same sub-group of the periodic table is justified by the following general properties of these elements.

I. Electronic Configurations

The electronic configurations of the atoms of alkali metals are given in the following table :

Element	At. No.	Complete electronic configuration	Valence-shell electronic configuration
Li	3	2, 1	$2s^1$
Na	11	2, 8, 1	$3s^1$
K	19	2, 8, 8, 1	$4s^1$
Rb	37	2, 8, 18, 8, 1	$5s^1$
Cs	55	2, 8, 18, 18, 8, 1	$6s^1$
Fr	87	2, 8, 18, 32, 18, 8, 1	$7s^1$

These configurations show that the ultimate shells of these elements have *one s* electron (s^1 electron) while the penultimate shell contains *eight* (s^2p^6) electrons (Li is an exception). When this valence-shell electron is lost, the alkali metals are converted into M^+ ions which have inert gas configuration. Since the additional electron enters *ns*-orbital, these are *s*-block elements.

It is due the similarity in the electronic configirations of the atoms of alkali metals as shown above in the table that the alkali metals have similar physical and chemical properties as discussed below :

II. Similarities in Physical Properties and Gradation in Them

Some of the physical properties of alkali metals are given in Table 18.1, and are discussed below :

1. Physical state. With the exception of Li, the alkali metals are *extremely soft* and *readily fused*. They are highly malleable (*i.e.*, can be pressed out into sheets) and ductile (*i.e.*, can be drawn into wires). When freshly cut, they have a bright lustre which is quickly tarnished as soon as metal comes in contact with atmosphere.

Table 18.1. Some physical properties of alkali metals

Property	Li	Na	K	Rb	Cs
Atomic weight	6.94	22.99	39.10	85.47	132.91
Abundance (% of earth's crust)	0.0065	2.74	2.47	0.028	3.2×10^{-4}
Atomic volume (c.c)	12.97	23.68	45.36	55.80	69.95
Atomic (*i.e.*, metallic) radius for coordination number 12 (A°)	1.55	1.90	2.35	2.48	2.67
Covalent radius (A°)	1.23	1.54	2.03	2.16	2.35
Ionic (crystal) radius of M^+ ions for coordination number 6 (A°)	0.60	0.95	1.33	1.48	1.69
Specific heat at °C	0.941	0.293	0.17	0.08	0.049
Ionic potential of M^+ ions (*i.e.*, charge/radius value)	1.67	1.05	0.75	0.68	0.59
Melting point (°C)	180.5	97.8	63.7	38.9	28.7
Boiling point (°C)	1330	892	760	688	670
Ionisation energies (KJ/mole)					
I_1	520.3	495.8	418.9	403.0	375.7
I_2	7298.1	4562.4	3051.4	2633.0	2230.0
Standard oxidation potential of $M (s) \rightarrow M^+ (aq) + e^-$ reaction (volt)	+ 3.04	2.71	+ 2.99	+ 2.99	+ 2.99
Sublimation energy (eV/atom)	1.7472	1.2432	1.0320	0.9840	0.9024
Hydration energy (eV/ion)	5.904	3.792	3.696	3.360	0.624
Electronegativity	1.0	0.9	0.8	0.8	0.7
Colour of the flame	Crimson red	Golden yellow	Violet	Violet	Violet
Oxidation states	+ 1	+ 1	+ 1	+ 1	+ 1
Heat of atomisation at 25°C (eV/atom)	1.7472	1.2432	1.0320	0.9840	0.9024
Ionic conduction of M^+ ion (Ω^{-1})	33.5	43.5	64.6	67.5	68.0

2. Atomic volume, atomic and ionic radii. As we move from lithium to caesium, an extra shell of electrons is added to each element. The addition of an extra shell increases the atomic volume. We find, therefore, that *there is an increase of atomic and ionic radii* (of M^+ ions) *as we move from lithium to caesium.*

Elements	*Li*	*Na*	*K*	*Rb*	*Cs*
Atomic volume (c.c.)	*12.97*	*23.68*	*45.36*	*55.80*	*69.95*
Atomic (i.e., metallic) radii for coordination number 12 (A°)	*1.55*	*1.90*	*2.35*	*2.48*	*2.67*
Covalent radii (A°)	*1.23*	*1.54*	*2.03*	*2.16*	*2.35*
Ionic (crystal) radii of M^+ ion for coordination number 6 (A°)	*0.60*	*0.95*	*1.33*	*1.48*	*1.69*

3. Density. The densities of alkali metals are *quite low* due to the large atomic volumes. Li, Na and K are *lighter than water. The densities increase with the increase in atomic number from Li to Cs* indicating that greater atomic weight more than compensates for the bigger size of the atoms. K is, however, lighter than Na which is due to an unusual increase in atomic size of K.

Elements	*Li*	*Na*	*K*	*Rb*	*Cs*
Densities at 0°C (g/c.c)	*0.534*	*0.972*	*0.859*	*1.525*	*1.903*

4. Melting and boiling points. The melting and boiling points are *very low* because of the presence of weak interatomic bonds in the solid state of the alkali metals. These bonds are due to their atomic radii and mainly due to their electronic configuration having a *single valence electron* as compared to large *number of available vacant orbitals*. As the size of the metal atoms increases, the repulsion of the non-bonding electron also increases. This increase in the repulsion of the non-bonding electron *decreases the melting and boiling points of alkali metals, when we move from Li to Cs.*

Elements		*Li*	*Na*	*K*	*Rb*	*Cs*
Melting point (°C)	:	*18.5*	*97.8*	*63.7*	*38.9*	*28.7*
Boiling point (°C)	:	*1330*	*892*	*760*	*688*	*670*

5. Heats of atomisation. Alkali metals have low heats of atomisation which are indicative of the fact that here are weak interatomic bonds in alkali metals.

Elements	*Li*	*Na*	*K*	*Rb*	*Cs*
Heats of atomisation at 25°C (eV/atom)	*1.7472*	*1.2432*	*1.0320*	*0.9840*	*0.9024*

6. Ionisation energies. We know that alkali metals have only one electron in their outermost shell (ns^1 electron). This ns^1 electron is so *weakly held* with the nucleus that it can be removed very easily. Alkali metals, therefore, have *low ionisation energies.*

As the distance of ns^1 electron from the nucleus increases on moving from Li to Cs, its removal becomes more and more easy as we proceed from Li to Cs,. *i.e.*, the amount of energy (ionisation energy) used in the removal of ns^1 electron is maximum in case of Li while in case of Cs it is the minimum. *Consequently the ionisatson energies of alkali metals go on decreasing from Li to Cs* as shown below :

Elements :	*Li*	*Na*	*K*	*Rb*	*Cs*
Ionisation energies (KJ/mole) :	*520.3*	*495.8*	*418.9*	*403.0*	*375.7*

————Values decreasing————→

The second ionisation energies are fairly high, since the loss of the second electron from M^+ cation which has a noble gas configuration is quite difficult.

7. Electropositive (*i.e.*, metallic) character. The elements which have a strong tendency to lose their outermost electrons to form the cations are said to be electropositive or metallic elements. In case of alkali metals we have seen that ns^1 electron in these metals is weakly bound with the nucleus and hence can be easily removed from the atom to form M^+ cation.

$$M \longrightarrow M^+ + e^-$$

Thus these metals have a strong tendency to lose their ns^1 electron to form M^+ cation and hence show *strong electropositive or metallic character.* Because of the successive increase in the distance of ns^1 electron from the nucleus on proceeding from Li to Cs, the tendency of the atoms to lose ns^1 electron goes on increasing from Li to Cs, *i.e.*, Li atom loses its ns^1 electron with great difficulty and hence shows *minimum electropositive character*, while Cs atom loses the same quite readily and hence shows *maximum electropositive character.* Thus the electropositive character of alkali metals *increases* from Li to Cs. Note that ionisation energies of alkali metals *decrease* from Li to Cs.

8. Photo electric effect. (Effect of light). The ns^1 electron in the atoms of alkali metals is so loosely held with the nucleus that even the low energy photons (*i.e.* light) can eject this ns electron from the surface of these metals. This property of emitting the electron even on the exposure of light permits these (especially K and Cs) to be used in photo electric cells which are sensitive to blue light.

9. Reducing properties. It is well known that an oxidising agent is a substance which can accept electrons, while a reducing agent is a substance which can lose electrons. Since, as we have seen above, alkali metals have a strong tendency to lose their ns electron, they act as strong reducing agents.

The tendency of alkali metals to act as strong reducing agents is evident from the fact that these metals can liberate H_2 from H_2O and acids.

$$2M + 2H_2O \longrightarrow 2MOH + H_2 \uparrow$$
$$2M + 2HCl \longrightarrow 2MCl + H_2 \uparrow$$

We have already seen that Li atom loses, its ns^1 electron with great difficulty (*i.e.* its ionisation energy is maximum) while Cs loses the same most readily (*i.e.* its ionisation energy is minimum). Consequently Li atom should have minimum reducing power while Cs atom should have maximum reducing power among the alkali metals. In other words, we can also say that Li atom, because of its maximum ionisation energy, should have minimum reducing power and Cs atom, because of its minimum ionisation energy, should have maximum reducing power.

Now let us examine the values of standard oxidation potentials (E° values) of alkali metals as given below and correlate these values with the reducing power of alkali metals.

$$Li(s) \longrightarrow Li^+ (aq) + e^- ; \qquad E^\circ = +\ 3.04 \text{ volts}$$
$$Na(s) \longrightarrow Na^+ (aq) + e^- ; \qquad E^\circ = +\ 2.71$$
$$K(s) \longrightarrow K^+ (aq) + e^- ; \qquad E = +\ 2.92$$
$$Rb(s) \longrightarrow Rb^+ (aq) + e^- ; \qquad E^\circ = +\ 2.92$$
$$Cs(s) \longrightarrow Cs^+ (aq) + e^- ; \qquad E^\circ = +\ 2.92$$

The high values of oxidation potentials show that alkali metals can lose their ns^1 electron quite readily and hence have a strong tendency to act as reducing agents. Since the value of oxidation potential for Li atom is the highest, it should lose its ns^1 electron most readily and hence should have the maximum reducing power.

Thus we see that on the basis of ionisation energy values, Li atom should have minimum reducing power while on the basis of oxidation potential values it should have maximum reducing power. As a matter of fact Li atom has the maximum reducing power in aqueous condition. Then how to explain this discrepancy regarding the reducing power of Li atom? This can be explained as follows :

As a matter of fact the conversion of an alkali metal in the solid state, M(*s*) in contact with water into M^+ (*aq*) cation represented by :

$$M(s) \longrightarrow M^+(aq) + e^-,$$

which represents the reducing power of M in aqueous medium takes place through the following three steps :

1st step. In this step the metal in the solid state, M(*s*) is evaporated to the gaseous state, M(*g*). In this process *energy is required* which is called *sublimation energy.* For alkali metals sublimation energies are almost the same (Li = 1.7472 eV, Na = 1.2432, K = 1.0320, Rb = 0.9849, Cs = 0.9024)

$$M(s) + \textit{Sublimation energy} \longrightarrow M(g)$$

2nd step. In this step the outer electron of M(*g*) obtained in 1st step is pulled off to form $M^+(g)$ cation. In this process as well *energy is required* which is called *ionisation energy.*

$$M(g) + \textit{Ionisation energy} \longrightarrow M(g) + e^-$$

As already said the value of ionisation energy of Li atom is the maximum

3rd step. In this step the gaseous cation, $M^+(g)$ formed in 2nd step gets hydrated to form $M^+(aq)$. In this process energy is *liberated.* This energy is called *hydration energy.*

$$M^+(g) + H_2O \longrightarrow M^+(aq) + \textit{Hydration energy}$$

Li^+ ion, being the smallest in size, gets hydrated to the maximum extent (*i.e.* maximum water molecules can attach to Li^+ ion) and hence Li^+ ion has the maximum value of hydration energy (Li = 5.904 eV, Na^+ = 3.792, K^+ = 3.696, Rb^+ = 3.360, Cs^+ = 0.624).

The four energy terms *viz.*, standard oxidition potential (E°), sublimation energy (S), ionisation energy (I) and hydration energy (H) are related as :

E	=	H	—	I	—	S
Oxidation potential		Hydration energy		Ionisation energy		Sublimation energy

When we compare with each other the values of sublimation energy, ionisation energy and hydration energy of Li atom, we find that its hydration energy value is maximum. Hence, although ionisation energy of Li atom required in 2nd step is also very high, this high value is more than compensated by the large value of hydration energy (of Li^+ ion) released in 3rd step. This maximum value of hydration energy is responsible for the maximum value of oxidation potential of Li atom which makes this atom have the maximum reducing power in contact with water (*i.e.*, in hydrous condition). In other words *the maximum reducing power of Li atom in contact with water is due to its maximum hydration energy.*

Here it should be noted that the reducing power of alkali metals in the dry state (*i.e.*, in anhydrous condition) depends on their ionisation energy values. Li atom which has the maximum ionisation energy has *maximum* reducing power.

10. Electronegativity. We have seen that alkali metals have a strong tendency to lose their ns^1 electron and hence have electropositive character. Because of this electropositive character, these metals have very little electronegative character. In other words alkali metals have *low values* of electronegativity, *i.e.*, they have little tendency to attract electrons towards themselves.

Now since their electropositive character increases from Li to Cs, their electronegative character (*i.e.* electronegativity values) decreases in the same order.

Elements :	*Li*	*Na*	*K*	*Rb*	*Cs*
Electronegativity :	*1.0*	*0.9*	*0.8*	*0.8*	*0.7*

11. Coloration to the flame. We have seen that the outer electron (*i.e.* ns^1 electron) of the atom of alkali metals is *loosely held with the nucleus* and hence it can be easily *excited* to the higher energy levels even by a small amount of heat energy (*e.g.* by heating the metals or their salts into bunsen burner). During the excitation process the electron absorbs some energy and when this excited electron comes back to its original position, it gives out absorbed energy in the form of light in visible region of the electromagnetic spectrum and hence the colour is imparted by the atom to the flame. Since the amount of energy absorbed during the excitation process is different in different atoms, different colours are imparted by the atoms to the flame, *e.g.* Li—*Crimson red*, Na—*Golden yellow*, K, Rb and Cs—*Violet*. The property of alkali metals to give colouration in the bunsen flame has been used to detect their presence in salts by a test, known as *flame test*.

12. Electrical conductivity. The individual atoms in the crystal structure of the metals (*i.e.* metal structure) are held together by the valence-electrons. In alkali metals, this binding is relatively weak, since the valence-electrons are loosely held. These loosely held electrons are, therefore, free to move throughout the metal structure with the result that the alkali metals have high conductivity, *i.e.* these are good conductors of electricity and heat.

13. Oxidation states. We have already seen that alkali metals can lose their ns^1 electron quite easily to form the unipositive ion, M^+. The loss of second electron from M^+ cation to form M^{2+} cation is very difficult, since M^+ cation has the configuration of its nearest noble gas which is very stable.

Atom (M)	*M⁺ cation*	
Li—2, 1	Li^+— 2	or $[He]_2$
Na—2, 8, 1	Na^+— 2, 8	*or* $[Ne]_{10}$
K—2, 8, 8, 1	K^+— 2, 8, 8	*or* $[Ar]_{18}$
Rb—2, 8, 18, 8, 1	Rb^+— 2, 8, 18, 8	*or* $[Kr]_{36}$
Cs—2, 8, 18, 18, 8, 1	Cs^+— 2, 8, 18, 18, 8	*or* $[Xe]_{54}$

Thus we see that the alkali metals can lose only one electron (*viz.* ns^1 electron) and hence show only + 1 oxidation state. It is evident from the configurations of

M^+ cations given above that these cations have *no unpaired electrons* and hence are *diamagnetic* and *colourless*. All the compounds of alkali metals are *colourless*, with the exception of those in which the anions are coloured, *e.g.* permanganates and dichromates.

14. Formation of ionic compounds. We have seen that ns^1 electron in alkali metals can be removed quite readily to form M^+ cations and hence these metals show highly electropositive character. Because of this character alkali metals react readily with highly electronegative elements by the transfer of their ns^1 electron to form the ionic compounds.

However alkali metals also show covalent bonding in certain compounds, *e.g.*, the vapours of alkali metals contain *diatomic molecules* like Na_2, Cs_2 in which the atoms are covalently bonded with each other. The strength of covalent bond in these diatomic molecules *decreases from Li to Cs.* Organometallic compounds like CH_3Li, C_2H_5Li, $C_6H_5CH_2Na$ are also covalent compounds.

15. Hydration of ions, hydrated radii and hydration energy. The alkali metals are extensively hydrated :

$$M^+(g) + aq \longrightarrow \underset{\text{Hydrated cation}}{[M(aq)]^+}$$

The smaller the cation, the greater is the degree of its hydration. Thus the *degree of hydration of M^+ ions decreases on moving from Li^+ to Cs^+.* As a result of this decrease in hydration degree, the ionic radii of M^+ ions in water (known as **hydrated radii**) decrease from Li^+ to Cs^+.

Hydrated ions	:	$Li^+(aq)$	$Na^+(aq)$	$K^+(aq)$	$Rb^+(aq)$	$Cs^+(aq)$
Hydrated radii (A°)	:	3.40 >	2.76 >	2.32 >	2.28 >	2.27

With the decrease of hydrated radii from Li to Cs^+, the ionic conductance of these hydrated cations increases in the order :

$$Li^+(aq) < Na^+(aq) < K^+(aq) < Rb^+(aq) < Cs^+(aq)$$

Hydration of ions is an *exothermic process.* The energy released when one gm mole of an ion in the gaseous state is dissolved in water to get it hydrated is called **hydration energy,** *i.e.* it is the energy released in the hydration of $M^+(g)$ ions :

$$M^+(g) + aq \longrightarrow \underset{\text{Aqua or hydrated cation}}{[M(aq)]^+} + \text{Energy released (Hydration energy)}$$

Since the degree of hydration of M ions decreases from Li^+ to Cs^+, the hydration energy of these ions also decreases in the same order :

M^+ ions :		L^+	Na^+	K^+	Rb^+	Cs^+
Hydration energy of M^+ ions (Kcal/mole)	:	–121 >	–95 >	–76 >	–69 >	–62

Negative sign associated with these values indicates the release of energy. It is obvious that Li^+ ion has the highest hydration energy and is, therefore, most extensively hydrated. Due to the extensive hydration of Li^+ ion, several lithium salts are hydrated to form *solvates* (*e.g.* $LiClO_4.3H_2O$) while the salts of other alkali metals are not hydrated.

16. Polarising power of alkali metal cations. When alkali metal cations approach near an anion, the cation attracts the outer-most electrons of the anion and repels the nucleus. Thus the distortion or polarisation of the anion takes place. This distortion results in the sharing of electrons between two oppositely charged ions, *i.e.* the bond between the cation and anion becomes partly covalent in character. In general the smaller cations polarise the anions more effectively than bigger ones. Therefore, the lithium salts are slightly covalent while other alkali metal Baits are ionic.

III. Similarities in Chemical Properties and Gradation in Them

1. Reactivity. We have seen that the first ionisation energies and electronegativities of alkali metals are the lowest of all the elements and their heats of atomisation are relatively low. These properties make the alkali metals highly reactive especially towards highly electronegative elements like Cl_2 and O_2 with which they give *ionic compounds. The reactivity of these metals increases with the increase of their atomic number,* since there is a steady decrease in the ionisation energies, *e.g.*, Li reacts *slowly* with H_2O; Na reacts *vigrously*; K, Rb and Cs reat with *increasing violence.*

2. Formation of oxides (Reaction with O_2 or air). Alkali metals react with O_2 or air rapidly and thus get tarnished due to the formation of their oxide on the surface of the metals. It is for this reason that alkali metals are stored in kerosene or paraffin oil.

Preparation. Li when burnt in O_2 gives mainly *lithium monoxide*, (*normal oxide*) Li_2O.

$$4\,Li + O_2 \longrightarrow 2\,Li_2O$$

Na when burnt in O_2 forms *sodium peroxide*, Na_2O_2

$$2Na + O_2 \longrightarrow Na_2O_2$$

Other alkali metals react with O_2 to form *super oxide* of MO_2 type.

$$\underset{(M\,=\,K,\,Rb,\,Cs)}{M} \;+\; O_2 \longrightarrow MO_2$$

Since the normal oxides of alkali metals other than that of Li (Li_2O) are not formed by the direct reaction between the metal and O_2, they are formed by indirect methods, *e.g.*, *by reducing peroxides, nitrites and nitrates with the metal itself.*

$$2Na_2O_2 + 4Na \longrightarrow 4Na_2O$$

$$2NaNO_2 + 6Na \longrightarrow 4Na_2O + N_2$$

$$2NaNO_3 + 10Na \longrightarrow 6Na_2O + N_2$$

Properties. Normal oxides (O^{2-}) react with H_2O to form hydroxides by proton exchange.

Gains proton

$$O^{2-} + H_2O \rightleftarrows OH^- + OH^-$$

Loses proton

The peroxides (O_2^{2-}) and superoxides (O_2^-) are *strong oxidising agents* and react with H_2O to give H_2O_2 and O_2.

$$\underset{\text{Peroxide}}{O_2^{2-}} + 2H_2O \longrightarrow 2OH^- + H_2O_2$$

$$\underset{\text{Super oxide}}{2[O{-}O]^-} + 2H_2O \longrightarrow 2OH^- + H_2O_2 + O_2$$

Normal oxides have *anti-fluorite structure* and are ionic in nature since they contain *monoxide ion,* O^{2-}. Peroxides contain *peroxide ion,* O_2^{2-} or $[—O—O—]^{2-}$.

The super oxide ion has a *three-electron bond* as shown below :

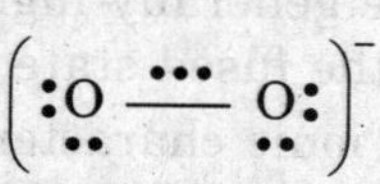

The presence of one unpaired electron in it makes this ion *paramagnetic* and *coloured.*

3. Formation of hydroxides. *Preparation.* Hydroxides of alkali metals can be prepared by treating the normal oxides or the metals with H_2O.

$$Na_2O + H_2O \longrightarrow 2NaOH$$

$$2\ Li + 2H_2O \longrightarrow 2\ LiOH + H_2$$

Properties. (*i*) Alkali metal hydroxides are crystalline solids. These are ionic compounds and are highly soluble in water and alcohol.

(*ii*) Since the alkali metals are highly electropositive, their hydroxides are the strongest bases known. Since the electropositive character of the metals increases from Li to *Cs, the basic character of their hydroxides increases from LiOH to CsOH.*

(*iii*) All the hydroxides are stable towards heat and *their thermal stability increases from* LiOH *to* CsOH. *e.g.*, LiOH decomposes at red heat, forming Li_2O while the other hydroxides do not decompose even at red heat. They melt, without decomposing and are good conductors of electricity in the fused state. At 400°C, the hydroxides consist of dimers, $(MOH)_2$.

(*iv*) With the increase in the size of the cation, the internuclear distance between the oxygen of OH^- ion and the metal cation also increases and hence the ionisation of the hydroxides increases from LiOH to CsOH.

4. Formation of hydrides : Actioia of H_2. *Preparation.* The alkali metals, on heating, react with hydrogen to give white crystalline *salt-like* hydrides. Since the electropositive character of alkali metals increases from Li to Cs, the ease with which these metals react with H_2 to form the hydrides decreases from

Li to Cs, *i.e.*, Li has the maximum tendency to form hydride while Cs has the minimum tendency to do so.

Properties. (*i*) These are salt-like compounds which are ionic in character and hence are represented as M^+H^- indicating that H is present as H^- in these compounds. The ionic character of M —H bond increases from Li—H to Cs—H.

(*ii*) Hydrides react with H_2O to give H_2.

$$MH + H_2O \longrightarrow MOH + H_2$$

Due to this property these are used as reducing agents.

(*iii*) On electrolysis of fused hydrides, H_2 is liberated at the anode

(*iv*) They all dissociate at a very high temperature. The stability of these hydrides decreases from LiH to CsH.

5. Formation of halides : Action of halogens. *Preparation.* Alkali metals combine directly with halogens to form halides of MX type. *As the electropositive character increases from Li to Cs, the ease with which the alkali metals form halides increases from Li to Cs.*

Properties. The halides have generally high melting and boiling points, are good conductors of electricity in the fused state and are readily soluble in water.

The halides essentially have ionic character and are, therefore, represented as M^+X^-. *The ionic character of MX halides increases with the increase of atomic radius on moving down the group.* Thus LiX has the least ionic character (*i.e.*, maximum covalent character) while CsX has the maximum ionic character (*i.e.* minimum covalent character).

Structure. The structure of MX halides is determined by *radius ratio and lattice energy effects.* The radius ratio values for M^+Cl^- chlorides and coordination numbers (C.N's) of M^+ cations corresponding to these values are given below (observed C.N's are also given in the last column).

M^+Cl^- *chlorides*	M^+/Cl^- *radius ratio values*	*Expected C.N's*	*Observed C.N's*
LiCl	*0.328*	*4*	*6*
NaCl	*0.524*	*6*	*6*
KCl	*0.732*	*6*	*6*
RbCl	*0.822*	*8*	*6 or 8*
CsCl	*0.930*	*8*	*8*

An examination of the table shows that the expected C.N's for Na, K, Rb and Cs are the same as their observed C.N's while for Li the observed C.N. is higher. This anomalous behaviour of Li is due to higher lattice energy of LiCl.

6. Action of liquid ammonia. The alkali metals dissolve in liquid ammonia without the evolution of H_2 gas.

Some of the properties of the solution obtained are given below :

(i) *Colour.* *The dilute solutions* of all the alkali metals are blue in colour. The absorption spectra of the solutions of all the alkali metals are the same. The blue colour of the solutions results from the excitation of free electrons to high energy levels ($1s \rightarrow 2p$ *transition*). This type of transition gives absorption band near 1500 A° in red the region of the spectrum and, therefore, the solution looks blue in the transmitted light. The *concentrated solutions* of alkali metals have *deep bronze colour.*

(ii) Electrical conductivity. The concentrated solutions are remarkably good conductors of electricity. As solutions they conduct electric current better than any other known solutions. The high electrical conductivity of these solutions in due to the presence of *ammoniated electrons*, $[e(NH_3)_y]^-$

$$M + (x + y)\, NH_3 \longrightarrow [M(NH_3)_x]^+ + [e(NH_3)_y]^-$$

The presence of *unpaired ammoniated electrons* in the solution makes the solution *paramagnetic.*

(iii) Stability. In the absence of any impurity, these solutions are quite stable, but if kept for a long time or in presence of a catalyst like platinum black, iron oxide etc., or ultraviolet light, these solutions decompose to form metal amide (MNH_2) and H_2.

$$2M + 2NH_3 \xrightarrow{\text{Catalyst}} 2MNH_2 + H_2 \uparrow$$

With the increase in the electropositive character of the metal, the stability of these solutions decreases down the group.

(iv) Reducing power. These solutions are good reducing agents. Their reducing property is due to the presence of free ammoniated electrons. Thus these solutions

(*a*) reduce metal halides to free metals.

(*b*) reduce $NaNO_2$ to Na_2NO_2 (sodium hydronitrite) which is an orange coloured extremely explosive solid.

(*c*) remove hydrogen atoms from acetylenic hydrocarbons.

$$CH_3 \equiv CH + e^- \longrightarrow CH_3 \equiv C^- + H$$

(*d*) remove halogen atoms from organic molecules.

$$C_2H_5Cl + 2e^- \longrightarrow C_2H_5^- + Cl^-$$

(v) Reversible nature. By careful evaporation of these solutions, the alkali metals can be recovered.

7. Formation of amalgams : Action of mercury. Alkali metals dissolve readily in mercurry and form the *amalgams*. This process is very *exothemic*. The formation of amalgams is an example of the formation of an alloy of alkali metals with Hg.

8. Nature of carbonates, bicarbonates and nitrates. The carbonates (M_2CO_3) and bicarbonates ($MHCO_3$) are highly stable to heat. *With increase of electropositive character from Li to Cs, the stability of these salts increases.*

Their nitrates decompose on strong heating to the corresponding nitrite and O_2 (Exception is $LiNO_3$).

$$2NaNO_3 \longrightarrow 2NaNO_2 + O_2 \uparrow$$

Anomalous Properties of Lithium : Dissimilarities with Other Alkali Metals

From the general properties of alkali metals discussed above it may be seen that Li differs in many properties from the other alkali metals.

The main reasons due to which it shows anomalous properties are :

(*a*) The size of Li atom and Li^+ ion is the smallest of all the alkali metal ions.

(*b*) The polarising power of Li ion is the greatest of all the alkali metal ions.

(*c*) Li has the highest ionisation energy and electronegativity as compared to other alkali metals.

Points of Differences. Following are the main points of differences between Li and other alkali metals.

1. The greatest polarising power of Li^+ ion results in its tendency towards *solvation* and *covalent bond formation*. Thus the salts of Li are covalent and are, therefore, more soluble in non-polar solvents (*e.g.* organic solvents). However, the salts of other alkali metals are insoluble in organic solvents.

2. Li is the only alkali metal which reacts with nitrogen and silicon to form ionic lithium nitride (Li_3N) and lithium silicide (Li_6Si_2). Li_6Si_2 is a dark violet and hygroscopic compound and is formed when Li and Si are heated together.

3. Li is much harder and lighter than the other alkali metals.

4. Its melting and boiling points are rather high and it can be melted in dry air without losing its brilliancy.

5. Due to its high ionisation energy, Li does not react with O_2 below 0°C; reacts *slowly* with H_2O to liberate H_2 or with liquid Br_2.

6. Li is the *least reactive* metal of all the alkali metals.

7. Li^+ ion shows a tendency to form complexes with NH_3, H_2O (to form hydrates) and forms a large number of organo lithium compounds.

8. When burnt in air, Li gives only the *monoxide*, Li_2O (O^{2-}) while other alkali metals form *peroxide* (O_2^{2-}) and *super oxides* (O_2^{-}).

9. When acetylene gas is passed over heated Li metal, it does not form lithium acetylide, while other alkali metals do form metal acetylides.

10. Li when heated in NH_3 forms imide, Li_2NH while other alkali metals form amides, MNH_2.

11. Li is the only alkali metal whose salts (*e.g.* LiCl) may undergo hydrolysis.

12. Due to the *low electropositive character* of Li, its various salts (*e.g.* Li_2CO_3, $LiNO_3$) are less stable and, therefore, decompose to give oxide, *e.g.*

$$2\,LiOH \xrightarrow{\text{Red heat}} Li_2O + H_2O$$

$$4\,LiNO_3 \longrightarrow 2\,Li_2O + 4\,NO_2 + O_2$$

$$Li_2CO_3 \longrightarrow Li_2O + CO_2$$

13. Some of the salts of Li like Li_2CO_3, Li_3PO_4 and LiF are *insoluble* in H_2O while the corresponding salts of other alkali metals are *soluble*. The solubility of the salts of other alkali metals is due to the weak electrostatic force of attraction.

14. Li_2O dissolves in H_2O quietly while other oxides do go more energetically.

15. LiOH is considerably *less soluble* in H_2O and is a *much weaker base* than NaOH or KOH.

LiOH decomposes at red heat forming Li_2O, whereas other alkali metal hydroxides sublime unchanged as $(MOH)_2$.

16. LiCl is deliquescent and is soluble in alcohol as well as in pyridine. It forms the hydrate, $LiCl.2H_2O$. Chlorides of other alkali metals do not show this behaviour.

17. $LiCO_3$ decomposes on heating to from Li_2O and CO_2 while other carbonates do not decompose. The small size of Li^+ ion makes Li_2O lattice *more stable* than Li_2CO_3 lattice. However, the large size of other alkali metal ions make M_2O lattices *less stable* than M_2CO_3 lattice and hence they decompose.

18. $LiNO_3$, on heating, forms Li_2O, NO_2 and O_2 while other alkali metal nitrates form nitrites and O_2.

$$4LiNO_3 \longrightarrow 2Li_2O + 4NO_2 + O_2$$
$$NaNO_3 \longrightarrow 2NaNO_2 + O_2$$

19. Lithium hydro suiphide, LiSH is thermally unstable while the hydro sulphides of other alkali metals are so stable that they can be fused without extensive decomposition.

20. Lithium per chlorate, $LiClO_4$ is remarkably soluble in alcohol, acetone and ethyl acetone, while the per chlorates of other alkali metals are almost insoluble in these solvents. The solubility of $LiClO_4$ is due to strong solvation of Li^+ ion.

21. Li_2SO_4 is the only alkali metal sulplate which does not form double salts such as alums.

Diagonal Relationship of Li with Mg : Similarities between Li and Mg

It has been observed that Li resembles more in its many properties with the element, Mg, lying at its right in 3rd period (*i.e.* diagonally opposite element) than with other members of its own sub-group I A. This type of similarity between Li and Mg is called diagonal relationship.

Following are the points which show that Li resembles Mg in many respects, *i.e.* Li and Mg have diagonal relationship.

1. Atomic and ionic radii. The atomic radii of Li and Mg are comparable (Li = 1.225 Å, Mg = 1.365 Å). The ionic radious of Li^+ (= 0.60 Å) is closer to that of Mg^{2+} (= 0.65 Å) than to that of Na^+ (= 0.95 Å).

2. Atomic volume. The atomic volumes of solid Li and Mg are quite similar (Li = 12.97 c.c., Mg = 13.97 c.c.)

3. Electronegativity. Electronegativities of Li and Mg are quite comparable (Li = 1.00, Mg = 1.20).

4. High polarising power (*i.e.* ionic potential) and its effects. Li^+ and Mg^{2+} both the ions have high polarising power, *i.e.*, charge/radius ratio of both the ions is high.

It is due to this high polarising power that both the ions have great tendency to form hydrates with a definite number of water molecules, *i.e.* a number of salts of these ions crystallise from water with a definite number of water molecules of crystallisation to form the hydrates. Some of the hydrates are given below :

$$LiCl.2H_2O;\ LiClO_4.3H_2O;\ LiNO_3.3H_2O$$

$$MgCl_2.6H_2O;\ Mg(ClO_4)_2.6H_2O;\ Mg(NO_3)_2.6H_2O$$

It is also because of the strong polarising power of Li^+ and Mg^{2+} ions on the charge cloud of halide ions that their halides have appreciable covalent character.

Because of the strong polarising power of Li^+ and Mg^{2+} ions on the charge cloud of the ions like NO_3^- and CO_3^{2-}, their nitrates and carbonates show *low thermal stability*, *i.e.* they decompose easily on heating to give out NO_2 and CO_2 respectively.

$$4LiNO_3 \xrightarrow{\Delta} 2Li_2O + 4NO_2 + O_2$$

$$2Mg(NO_3)_2 \xrightarrow{\Delta} 2MgO + 4NO_2 + O_2$$

$$Li_2CO_3 \xrightarrow{\Delta} Li_2O + CO_2$$

$$MgCO_3 \xrightarrow{\Delta} MgO + CO_2$$

5. Enthalpies of vaporisation. Enthalpies of both the elements are comparable.

6. Melting points. Li and Mg have higher melting points than the other members of their respective groups. Their melting points are comparable with each other (Li = 1336°C, Mg = 1100°C).

7. Hardness. Li and Mg are harder than the other members of their respective groups. Hardness of Li is comparable with that of Mg.

8. Action of H_2O. Like Mg, Li decomposes water *slowly* to liberate H_2.

$$2Li + 2H_2O \longrightarrow 2LiOH + H_2$$

$$Mg + 2H_2O \longrightarrow Mg(OH)_2 + H_2$$

9. Action of N_2. Both the elements combine with N_2 on heating to form their *ionic nitrides*, Li_3N and Mg_3N_2.

Both the nitrides readily decompose H_2O to liberate NH_3.

$$Li_3N + 3H_2O \longrightarrow 3LiOH + NH_3$$

$$Mg_3N_2 + 6H_2O \longrightarrow 3Mg(OH)_2 + 2NH_3$$

10. Action of Cl_2. Both the elements form their chlorides, LiCl and $MgCl_2$, on heating with dry Cl_2,

$$2Li + Cl_2 \longrightarrow 2LiC1$$

$$Mg + Cl_2 \longrightarrow MgCl_2$$

Both the chlorides have many similar properties, *e.g.* (*i*) Both are deliquescent solids (*ii*) Both are soluble in water and alcohol (*iii*) Both form hydrates *viz.* LiCl. $2H_2O$ and $MgCl_2$. $6H_2O$ (*iv*) Both undergo hydrolysis in hot water.

11. Combination with carbon. On heating, both the elements combine with carbon to form the carbides, Li_2C_2 and MgC_2.

12. Action of air or O_2. Li, when burnt in air or O_2, forms *lithium oxide*, Li_2O (which is a *normal oxide*). Mg ribon, under similar conditions, forms *magnesium oxide*, MgO (which is also a *normal oxide*). Like Li_2O, MgO does not combine with further quantity of O_2 to form the *peroxide* (M_2O_2) or *superoxide* (MO_2).

13. Nature of hydroxides. Of all the alkali metal hydro-oxides, LiOH is *less soluble in water, less basic and decomposes on heating* to form Li_2O. Likewise $Mg(OH)_2$ is also *sparingly soluble, less basic* and *decomposes on heating* to form MgO.

$$2LiOH \xrightarrow{\Delta} Li_2O + H_2O$$

$$Mg(OH)_2 \xrightarrow{\Delta} MgO + H_2O$$

14. Solubility of salts. The solubility of lithium salts resembles that of the corresponding magnesium salts, *e.g. fluorides, carbonates* and *phosphates* of the both the metals are *insoluble in water* while the *chlorides, bromides, iodides and perchlorates are soluble in oxygenated organic solvents.*

15. Unstability of hydrosulphides. Lithium hydrosulphide, LiSH is unstable under ordinary conditions. Magnesium hydrosulphide, $Mg(SH)_2$, is also unstable in aqueous solution and decomposes to form $Mg(OH)_2$, liberating H_2S

$$Mg(SH)_2 + 2H_2O \longrightarrow Mg(OH)_2 + 2H_2S$$

16. Like $MgSO_4$, Li_2SO_4 does not form alums.

17. Due to their covalent nature, the alkyls of Li and Mg are soluble in organic solvents, *e.g.* $LiAlH_4$, RMgX are soluble in either. Alkyls of Li undergo reactions similar to Grignard reagents (RMgX).

18. Li^+ and Mg^{2+} ions are strongly hydrated.

Cause of Diagonal Relationship

Why Li resembles in its many properties with its diagonally opposite element *viz.*, Mg can be explained as follows on the basis of the *concept of electronegativity.*

We know that electronegativity increases from left to right in a period and decreases on descending a group. Thus, Be is more electronegative than Li (Be = 1.5; Li = 1.0; Be > Li) and Mg is slightly less electronegative than Be. (Be = 1.5, Mg = 1.2; Be > Mg). Li and Mg being slightly less electronegative than Be have almost the same value of electronegative and hence have many similar properties, *i.e.* they show diagonal relationship with each other.

Position of Lithium in the Periodic Table

Li is present in group IA with Na, K, Rb, Cs and Fr, Li resembles not only with the elements of its own sub-group but also resembles with its diagonally related element namely Mg. The position of Li in the periodic table is, therefore, justified because of the following facts.

(*i*) It resembles with the elements of its own sub-group.

(*ii*) It shows a regular gradation with the elements of its own sub-group.

(*iii*) It resembles with the diagonally related element namely Mg.

(*i*) and (*ii*) have been discussed under *"Position of Alkali Metals in the Period Table"* while (*iii*) has been discussed under *"Diagonal Relationship of Li with Mg"*.

Questions with Answers

Q.1 Explain the following :

(*i*) Alkali metals are strong reducing agents in the gaseous state and in the aqueous solution ?

Ans : (*i*) **In the gaseous state.** Reducing power of alkali metals (M) in the gaseous state depends on their IE values. Since these metals have low IE values, they can lose their ns^1 electrons quite readily to get oxidised to M^+ (*g*) ion and hence act as strong reducing agents.

(*ii*) In aqueous solution. Reducing power of alkali metals in aqueous solution depends on their E^o_{ox} values for the oxidation half reaction, M (*s*) $\rightarrow$ M^+ (*aq*) + e^-. Since E^o_{ox} values for alkali metals are fairly high, these metals can readily be oxidised to M^+ (*aq*) in aqueous solution and hence act as strong reducing agents.

(*iii*) In the gaseous state, the reducing power (R.P.) of alkali metals is in the order : Li < Na < K < Pb < Cs.

Ans : We know that R.P. of alkali metals in the gaseous state depends on the value of their IE values. Lower is the value of IE of an alkali metal [M (*s*)], more quickly it will be oxidised to M^+ (*g*) ion and hence higher will be its R.P. Thus with the decrease of IE values, the R.P. of alkali metals increases. Now since IE values decrease on moving down the group from Li to Cs (Li = 520.3, Na = 495.3, K = 418.9, Rb = 403.0, Cs = 375.7 KJ mol^{-1}), the R.P. of these metals increases in the same direction. Thus :

$$\text{Li} < \text{Na} < \text{K} < \text{Rb} < \text{Cs}$$

R.P. increase $\longrightarrow$

This order shows that Li metal is the weakest R.A. in the gaseous state.

(*iv*) In the aqueous solution, the reducing power (R.P.) of alkali metals is in the order : Li > Na < K < Rb.

Ans : We know that R.P. of alkali metals in aqueous solution depends on E^o_{ox} value for the oxidation half reaction, M (*s*) M^+ (*aq*) + e^-. Higher is the value of E^o_{ox}, more quickly the metal, M (*s*) is oxidised M^+ (*aq*) ion and hence higher is the R.P. of the metal. E^o_{ox} values for M/M^+ couples are as : Li/Li^+ = +3.04 V, Na/Na^+ = +2.71V, K/K^+ = +2.92V and Rb/Rb^+ = +2.99V. Now since E^o_{ox} for Li/Li^+ couple is the maximum, Li metal is oxidised to Li^+ (*aq*) ion most readily and hence is the strongest reducing agent (R.A.). Since E^o_{ox} values for other couples increase from Na/Na^+ to Rb/Rb^+, the R.P. of Na, K and Rb also increases from Na to Rb (Na < K < Rb). Thus the overoll order of R.P. of alkali metals in aqueous solution is as :

Li	> Na < K < Rb
Maximum R.P.	— R.P. increases →
(Strongest R.A.)	

Note : The oxidation half-reaction, $M\ (s) \rightarrow M^+\ (aq) + e^-$ is actually the sum of the following reactions :

$M\ (s) \longrightarrow M\ (g)$; ΔH = Energy absorbed
= Sublimition energy of M (s) atom.

$M\ (g) \longrightarrow M^+\ (g) + e^-$; ΔH = Energy absorbed
= IE of M (g) atom

$M^+\ (g) + aq \longrightarrow M^+\ (aq)$; ΔH = Energy released
= Hydration energy of M^+ (g) ion

On adding : $M\ (s) \longrightarrow M^+\ (aq) + e^-$

Since Li^+ ion is the smallest in size, the hydration energy of this ion is maximum and hence E^o_{ox} value for the oxidation half-reaction, $Li\ (s) \rightarrow Li^+\ (aq) + e^-$ or for Li/Li^+ couple is also maximum.

(*iv*) Alkali metals are paramagnetic but their salt are diamagnetic.

Ans : Alkali metals contain one upaired electron (ns^1) and hence are paramagnetic. However, during salt formation, this unpaired electron is transferred to the non-metalic atom forming its anion. As a result, salt has paired electrons both in the cation as well as in the anion and hence alkali metals salts are diamagnetic in nature.

(*v*) Alkali metals impart colour to the flame.

Ans : Alkali metals have low ionization energies. Their valence electrons easily absorb energy from the flame and are excited to higher energy levels. When they return to the ground state, the energy is emitted back.

(*vi*) Sodium has higher melting point than potassium.

Ans : On going from Na to K, the size of the atom increases and hence the metallic bonding weakens. In other words sodium has higher melting point than potassium because of stronger metallic bonding.

(*vii*) Sodium fire in the laboratory should not be extinguished by pouring water.

Ans : Sodium reacts violently with water producing H_2 gas ($2\ Na + 2\ H_2O \rightarrow 2\ NaOH + H_2$) which also cathes fire. As a result, the fire spreads rather than being extinguished. Therefore, H_2O should not be used for extinguishing sodium fire. Instead pyrene (CCl_4) should be used.

(*viii*) Table salt gets wet in rainy season.

Ans : Pure NaCl is not hygroscopic but table salt is impure NaCl containing impurities of Na_2SO_4, $CaSO_4$, $MgCl_2$ and $CaCl_2$. All of these, being hygroscopic, absorb moisture from air in rainy season. As a result, table salt gets wet.

(*ix*) Alkali metals donot occur free in nature ? **(*Roorkee 1980*)**

Ans : Since alkali metals are placed at the extrem left of the periodic table, they have the lowest IE values. Due to lowest IE values, these metals are extremely reactive. For example they readily combine with O_2, moisture and CO_2 of the atmosphere and many other substances. Due to their high reactivity, these matals donot occur free in nature.

(*x*) Sodium salts in aqueous solution are either neutral or alkaline in nature.

Ans : The anions present in sodium salts are either from strong acids (*e.g.* HCl) or from weak acids (*e.g.* H_2CO_3). When the anions are from strong acids, the salt does not undrego hydrolysis and hence the aqueous solution is neutral. NaCl is an example of such salts. On the other hand, when the anions are from the weak acids, then the anions of these salts undergo hydrolysis and give alkaline solution due to the formation of strong akali. Na_2CO_3 is an example of this type of salts.

$$Na_2CO_3 + 2H_2O \rightleftharpoons \underset{\text{Strong alkali}}{2NaOH} + \underset{\text{Weak acid}}{H_2CO_3}$$

or

$$CO_3^{2-} + 2H_2O \rightleftharpoons \underset{\text{Strong alkali}}{2OH^-} + \underset{\text{Weak acid}}{H_2CO_3}$$

(*xi*) Alkali metals are obtained by the electrolysis of their moten salt and not by the electrolysis of their aqueous solution.

Ans : The aqueous solution of alkali metal salt contains alkali metal cation, salt anions, H^+ and OH^- ions. When this solution is electrolysed, H^+ ions present in the solution are discharged at cathode as H_2 in preference to alkali metals cations, since discharge potential of H^+ ions is lower than that of alkali metal cations. Thus instead of getting alkali metal on cathode, we get H_2 gas. On the other hand, when a molten salt of alkali metal is electrolysed, only alkali metal cation gets discharged on cathode, since the salt does not contain H^+ ions.

(*xii*) When K, Rb and Cs are heated in the excess of air, they prefer to form superoxides (O_2^-) and not oxides (O^{2-}) and peroxides (O_2^{2-}).

Ans : K^+, Rb^+ and Cs^+ ions are large cations. Hence these cations stabilise large anions. Now since superoxide ion (O_2^-) is larger than oxide (O^{2-}) and peroxide (O_2^{2-}) ions, K^+, Rb^+ and Cs^+ ions stabilise the larger anoin *viz* superoxide ion (O_2^-).

(*xiii*) Superoxides of alkali metals are paramagnetic but normal oxides are diamagnetic.

***or* KO_2 is paramagnetic.**

Ans : Superoxides of alakli metals contain superoxide ion (O_2^-) whose structure is 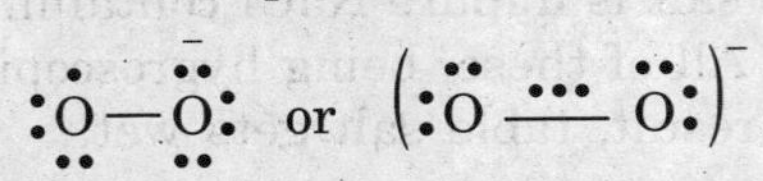 . Since this structure has one unpaired electron, alkali metal superoxides are paramagnetic. Normal oxides of alkali matals contain oxide ion (O^{2-}) whose structure is $[:\ddot{\underset{..}{O}}:]^{2-}$. Since this structure has all the electrons in the paired state, alkali metal oxides are diamagnetic.

Q.2 Give the reasons for the following :

(*i*) Alkali metals are used in photoelectric cells. (*Kurukshetra 1995*)

***or* Cerium can be used in photoelectric cells but lithium cannot.**

Ans : Since, in each period of the periodic table, the alkali metals have the lowest values of IE, these metals can release electrons readily when exposed to light even of low energy. Since Li has the highest value of IE amongst alkali metals, this metal cannot release the electron readily and hence cannot be used in photoelectric cells. On the other hand, since Cs has the lowest value of IE, this metal can readily release the electron and hence is used in photoelectric cells.

(*ii*) Lithium is most extensively hydrated among alkali metals.

(*M.D. Rohtak 1994*)

***or* Several salts of lithium are hydrated while the salts of other alkali metals are not hydrated.**

Ans : We know that alkali metal ions are hydrated in aqueous solution. The degree of hydration of these ions decreases with the increase in the size of these ions. Thus since Li^+ ion is the smallest in size, it is hydrated in aqueous solution to the maximum extent. It is due to the maximum hydration of Li^+ ion that its several salts are hydrated (*e.g.* $LiClO_4 . 3 H_2O$). Since other alkali metals ions are hydrated to lesser extent, their salts are not hydrated.

(*iii*) In aqueous solution, Li^+ ion has the lowest mobility.

Ans : Among alkali metal cations (M^+), since Li^+ ion is the smallest ion, it is hydrated in aqueous solution to the maximum extent. Due to its maximum hydration, the hydrated Li^+ ion, Li^+ (*aq*) becomes heaviest and hence ionic mobility (conductance) of Li^+ ion in aqueous solution is the lowest.

(*iv*) In aqueous solution, Li^+ ion moves through the solution less rapidly than other alkali metal ions.

Ans : Since Li^+ ion is the smallest in size than other alkali metals ions, it gets hydrated by H_2O molecules to the maximum extent and hence its hydrated ion, [Li (*aq*)]$^+$ is the heaviest than other hydrated ions. Due to its heaviest nature, [Li (*aq*)]$^+$ ion moves through the solution less rapidly than each of the other hydrated ions.

(*v*) Conductivity of Li^+ ion in aqueous solution is smaller than that of Cs^+ ion, though the former ion is much lighter than the latter ion.

(*Kurukshetra 1994, Madras 1995*).

Ans : We know that alkali metal ions are hydrated in aqueous solution. The degree of hydration of these ions decreases with the increase in their size. Thus, since Li^+ ion is smaller in size than Cs^+ ion, Li^+ ion is hydrated to greater extent than Cs^+ ion. Due to greater hydration of Li^+ ion, Li^+ (*aq*) ion becomes heavier than Cs^+ (*aq*) ion [Li^+ (*aq*) > Cs^+ (*aq*)]. Being heavier, Li^+ (*aq*) ion has smaller conductivity than Cs^+ (*aq*) ion. Thus the conductivity of Li^+ ion in aqueous solution is smaller than that of Cs^+ ion.

(*vi*) Na is less reactive than K.

Ans : We know that electropositive character and reactivity of alkali metals depend on the ease with which the metal loses its ns^1 electron. Since the ionisation energies of alkali metals decrease down the group, the ease with which these metals lose their ns^1 electron increases in the same direction and hence electropositive character and reactivity of these metals also increase in the same direction. Thus, since Na has higher IE value than K, this metal is less reactive than K.

(*vii*) The softness of alkali metals increases down the group with increasing atomic number. **(*I.I.T. 1986*)**

Ans : Softness of metals depends on the strength of metallic bond holding the atoms of the metal together. Since the size of alkali metal atoms increases down the group, the metallic bond holding the atoms together becomes weaker and weaker. With the weakening of metallic bond, the softness of akali metals increases.

(*viii*) Na metal can be used for drying directly ether but not ethanol ? **(*I.I.T. 1982*)**

Ans : Na-metal can be usd for drying diethyl ether, since the metal does not react with this compound. On the other hand, since Na-metal reacts with ethanol (C_2H_5OH) as shown below, this metal cannot be used for drying ethanol.

$$Na + 2\ C_2H_5OH \longrightarrow 2\ C_2H_5ONa + H_2$$

(*ix*) Alkali metals are normally kept in kerosene oil.

Ans : This is because in the air they are easily oxidized to oxides which may dissolve in the moisture of the air to from hydroxides. These hydroxides combine with CO_2 present in the atmosphere and form M_2CO_3.

$$4M + O_2 \xrightarrow{\text{Oxidation}} 2\ M_2O$$

$$M_2O + H_2O \longrightarrow 2\ MOH$$

$$2\ MOH + CO_2 \longrightarrow M_2CO_3 + H_2O$$

(*x*) Alkali metals do not form bivalent cation (M^{2+}). **(*Banglore 1995, Punjab 1996, Magadh 1996*).**

Ans : General electronic configuration of alkali metal atoms is..., $(n-1)s^2$ $(n-1)p^6$, ns^1. For the formation of bivalent cation (M^{2+}) two electrons have to be removed from the alkali metal atoms as shown below :

$$M\ (g) \xrightarrow[(-e^-)]{+IE_1} M^+\ (g) + e^-$$

$$M^+\ (g) \xrightarrow[(-e^-)]{+IE_2} M^{2+}\ (g) + e^-$$

For the formation of M^+ (*g*) ion the electron is removed from *ns* orbital. Due to low IE values ns^1 electron is easily lost and hence M^+ (*g*) ion is quite easily formed.

On the other hand, for the formation of M^{2+} (*g*) ion the electron is removed from $(n-1)s^2$ $(n-1)p^6$ configuration (noble gas configuration). Since this configuration is extremely stable, IE_2 values are fairly high and hence M^{2+} (*g*) ion is not formed.

(*xi*) Solution of alkali metals in liq. NH_3 are blue coloured, conducting, paramagnetic and has reducing natrue. ***(G.N. Dev 1994)***

Ans : All alkali metals dissolve in anhydrous liq. NH_3 and give alkali metal – NH_3 solutions. In these solutions ammoniated cations, $[M\ (NH_3)_x]^+$ and ammoniated electrons, $[e\ (NH_3)_y]^-$ are fromd.

$$M \longrightarrow M^+ + e^-$$

$$M^+ + xNH_3 \longrightarrow [M\ (NH)_x]^+$$

$$e^- + yNH_3 \longrightarrow [e\ (NH_3)_y]^-$$

On adding : $M + (x + y)\ NH_3 \longrightarrow [M\ (NH_3)_x]^+ + [e\ (NH_3)_y]^-$

These solutions have the follwoing characteristics which can be explained on the basis of the presence of $[M\ (NH_3)_x]^+$ and $[e\ (NH_3)_y]^-$ in the solution.

(*i*) *Blue Colour.* (*a*) These solutions have blue colour. This colour is due to the excitation of free $[e\ (NH_3)_y]^-$. When ordinary light falls on these solutions, the free ammoniated electrons, $[e\ (NH_3)_y]^-$ get exited to higher energy levels by absorbing energy. Since the absorption of energy takes place in red portion of the visible region, the colour of the transmitted light is blue due to which the solution appears blue. In short we can say that the blue colour of the solution is due to the presence of ammoniated electrons.

(*b*) With the increase in the concentration of alkali metal in liq. NH_3, the blue colour starts changing to that of metallic copper after which no more metal dissoves in liq. NH_3.

(*ii*) *Conducting nature.* The blue solution is conducting due to the presence of $[M\ (NH_3)_x]^+$ and $[e\ (NH_3)_y]^-$ in solution.

(*iii*) *Paramagnetic nature.* The blue solution is paramagnetic due to the presence of unpaired electrons in the cavities of ammonical solution.

(*iv*) *Reducing property.* The blue solution slows strong reducing properties due to the presence of free ammoniated electrons, $[e\ (NH_3)_y]^-$ in the solution.

(*xii*) LiF is most soluble among the fluorides of alkali metals.

Ans : Due to lesser degree of polarization, LiF will be most ionic hence most soluble in water.

(*xiii*) $NaHCO_3$ is known in solid state but $Ca(HCO_3)_2$ is not isolated in solid state.

Ans : Calcium bicarbonate is fromed only when CO_2 is passed through aqueous solution of $Ca(OH)_2$. On evaportion, calcium bicarbonate decomposes to calcium carbonate.

(*xiv*) Li is the only alkali metal that forms nutride directly.

Ans : Li and Mg have diagonal relationship with each other. Now since Mg forms nitride (Mg_3N_2), Li also forms nitide (Li_3N)

$$6Li\ (s) + N_2\ (g) \xrightarrow{\Delta} 2Li_3N\ (s)$$

Q.3 Atomic radius of Li atom is 1.23A° and ionic radius of Li^+ ion is 0.76 A°. Calculate the volume occupied by valence electron.

Ans : The electronic configurations of Li atom and Li^+ ion are as : Li = $1s^2$, $2s^1$ and Li^+ = $1s^2$. Obviously valence electron is $2s^1$ electron. Hence :

Volume occupied by valence electron = Volume of Li atom – Volume of Li^+ ion

$$= \frac{4}{3}\pi (r_{Li})^3 - \frac{4}{3}\pi (r_{Li^+})^3$$

$$= \frac{4}{3}\pi (1.23 \times 10^{-8})^3 - \frac{4}{3}\pi (0.76 \times 10^{-8})^3$$

$$= \frac{4}{3}\pi \times 10^{-24} [(1.23)^3 - (0.76)^3] \text{ (Ans).}$$

Q.4 Arrange the following as indicated :

(*a*) LiOH, NaOH, KOH (Increasing order of solubility in water)

(*b*) $LiHCO_3$, $NaHCO_3$, $KHCO_3$ (Increasing order of solubility in water)

(*c*) Li_2CO_3, Na_2CO_3, K_2CO_3 (Increasing order of solubility in water)

(*d*) LiCl, NaCl, KCl, RbCl, CsCl (Increasing order of lattice energy)

(*e*) LiF, LiCl, LiBr, LiI (Increasing order of lattice energy)

(*f*) NaF, NaCl, NaBr, NaI (Increasing order of lattice energy)

(*g*) Li^+, Na^+, K^+, Rb^+, Cs^+ (Increasing order of size of hydrated ion).

Ans : (*a*) LiOH < NaOH < KOH

(*b*) $LiHCO_3$ < $NaHCO_3$, $KHCO_3$

(*c*) Li_2CO_3 < Na_2CO_3 < K_2CO_3

(*d*) CsCl < RbCl < KCl < NaCl < LiCl

(*e*) LiI < LiBr < LiCl < LiF

(*f*) NaI < NaBr < NaCl < NaF

(*g*) Cs^+ < Rb^+ < K^+ < Na^+ < Li^+.

Q.5 Arrange the follwoing as specified :

(*i*) MgO, SrO, K_2O, NiO and Cs_2O (increasing order of basic character) [I.I.T 1991]

(*ii*) LiCl, LiBr, LiI (decreasing order of order of covalent character) [Roorkee 1987]

(*iii*) KCl, KBr, KI (decreasing order of solubility in water)

(*iv*) $NaHCO_3$, $KHCO_3$, Mg $(HCO_3)_2$, Ca $(HCO_3)_2$ (Increasing order of solubility) [M.L.N.R. 1984]

(*v*) LiF, NaF, KF, RbF and CsF (increasing order of lattic energy)

(*vi*) Li, Na, K (decreasing order of reducing nature in solution)

Ans : (*i*) NiO < MgO < SrO < K_2O > Cs_2O

(*ii*) LiI > LiBr > LiCl

(*iii*) KI > KBr > KCl

(*iv*) $NaHCO_3 < KHCO_3 < Mg(HCO_3)_2 < Ca(HCO_3)_2$

(*v*) CsF < RbF < KF < NaF < LiF

(*vi*) Li > K > Na

Q.6 Amongst LiF and LiI which has more covalent character and why ? [Roorkee 1980; I.I.T. 1986]

Ans : LiI is more covalent . According to Fajan's rule, Li^+ will be able to polarise the bigger ion, I^-, more than the smaller ion, F^-. More polarisation brings more covalent character.

Q.7 Calculate heat of formation of NaCl from the following data. Hydration energy of Na^+ = –389.4 KJ mol^{-1}, hydration energy of Cl^- = –382.3 KJ mol^{-1} and lattice energy of NaCl = –776 KJ mol^{-1}.

Ans : We know that heat of solution of NaCl, $[\Delta H_{solu}]_{NaCl}$ is given by :

$$[\Delta H_{solu}]_{NaCl} = \text{Hydration energy of NaCl} - \text{Lattic energy of NaCl}$$

$$= [\Delta H_{hyd}]_{NaCl} - [\Delta H_{lat}]_{NaCl}$$

$$= [(\Delta H_{hyd})_{Na^+} + (\Delta H_{hyd})_{Cl^-}] - [\Delta H_{lat}]_{NaCl}$$

$$= [-389.4 + (-382.3)] - (-776.0)$$

$$= -7771.7 - 776.0 = + 4.3 \text{ KJ mol}^{-1} \text{ (Ans)}.$$

University Questions

1. Explain the following :

(*i*) Alkali metals are obtained by the electrolysis of their molten salts and not by the electrolysis of their aqueous solution. (*REE 84*)

(*ii*) Alkali metals show on oxidation station + 1 only.

OR

Alkali metals are univalent. (*Agra 89*)

(*iii*) Lattice energy of the fluorides of alkali metals decreases from LiF (= 1032 kJ $mole^{-1}$) to CsF (726 kJ mol^{-1}).

(*iv*) Degree of hydration of alkali metal ions decreases from Li^+ to Cs^+.

(*v*) Alkali metals are good reducing agents. (*Agra 89, REE 81, 83*)

(*vi*) Alkali metals have low m.pts.

(*vii*) Alkali metals give characteristic colour to the bunsen flame.

(*viii*) Alkali metals have body-centered cubic structures.

(*ix*) Na is stored under kerosene. (*IIT 70*)

(*x*) Alkali metals do not occur free in nature. (*REE 80*)

(*xi*) Alkali metals form M^+ cations instead of M^{2+} cations.

(*xii*) Alkali metals do not form dipositive ions. (*Raj. 82S, 86; Jodhpur 84*)

(*xiii*) Solutions of alkali metals in liquid NH_3 are blue in colour.

(*xiv*) Group IA elements are poor complexing agents. (*Jodhpur 86*)

(*xv*) Alkali metals produce different colours in Bunsen flame. (*Delhi 83*)

(*xvi*) Alkali metals are active in metallic character. (*MD Rohtak, 83*)

(*xvii*) Alkali metals lose electrons even when exposed to light. (*MD Rohtak 84*) (*xviii*) Li^+ ions is most extensively hydrated amongst alkali metals. (*Delhi 85*)

(*xix*) Of all the alkali metals, Li has the highest oxidation potential. (*Meerut 83*)

(*xx*) Li^+ ion in aqueous solution are the poorest conductor of electricity as compared to other alkali aqua ions. (*Delhi 88*)

(*xxi*) Alkali metals have low ionisation potentials. (*Agra 89*)

2. Explain the following :

(*i*) Salts of lithium are covalent while those of other alkali metals are ionic.

(*ii*) Li-salts are insoluble in H_2O while those of other alkali metals are soluble.

(*iii*) In general alkali metals form ionic compounds. (*Agra 89*)

3. Explain the following :

(*i*) Li, on burning in O_2, gives normal oxide (mono oxide, Li_2O), Na forms peroxide (Na_2O_2) and Rb and Cs form Super oxides (MO_2). (*Punjab 85*)

(*ii*) Super oxides of alkali metals are paramagnetic in character. (*Delhi Hons. 85*)

4. How does Li_2CO_3 differ from K_2CO_3? (*Madras 86*)

5. Explain the following:

(*i*) Li_2CO_2, on heating, gives CO_2 but the carbonates of other alkali metals do not.

(*ii*) L_2CO_2 decomposes on heating to give CO_2 while Na_2CO_3 does not decompose. (*GND 82*)

(*iii*) $NaHCO_3$ is quite stable and exists as solid while $LiHCO_3$ exists only in aqueous solution. (*Delhi Hons. 82*)

6. Explain the following :

(*i*) In spite of the fact that Li is an alkali metal, lithium halide is covalent in nature. (*Punjab 85, Jodhpur 86*)

(*ii*) LiCl is soluble in organic solvents while the chlorides of other alkali metals are not. (*Meerut 83*)

(*iii*) Li is soluble in water while LiF is not. (*Jodhpur 87*)

7. Explain the following giving appropriate reason for your answer : Super oxides of alkali metals are coloured. (*Delhi 99*)

8. Which is a stronger base: KOH or $Ba(OH)_2$? (*Delhi 99*)

9. Hydroxides of 1st group are strong bases. Explain why. (*Himachal Pradesh 99*)

10. Explain the following :

(*a*) Lithium forms normal oxide, sodium forms peroxide and potassium, rubedium and caesium for superoxides.

(*b*) Li_2CO_3 is unstable while other alkali metal carbonates are relatively more stable.

(*c*) Alkali metals give coloured solutions in liq. NH_3. (*Himachal Pradesh 2000*)

11. Explain why Li resembles Mg in certain properties. (*Lucknow 2001*)

12. Alkali metals are univalent strong reducing agents and have low ionisation potential. Explain. (*C.S.J.M. Kanpur 2007*)

13. Write a short note on "Diagonal relationship between Li and Mg". (*C.S.J.M. Kanpur 2005, 2006, 2009*)

14. Discuss the position of Li in the period table. (*Meerut 2009*)

CHAPTER 19

Chemistry of Lithium and Its Compounds

Lithium

History

While analysing the minerals *petalite* and *spodumene* in the laboratory of *Berzelius, Arfvedson* in 1817 resported the discovery of a new metal which was called *lithium*. Later on *Bunsen* and *Mathiessen* isolated lithium by the electrolysis of fused lithium chloride in 1855. The word lithium which means *stony* was given to this metal because it was then believed that this metal was confined to the mineral kingdom only.

Occurrence

Lithium is rare but widely distributed in the earth crust. In traces it occurs in soils, milk, blood, muscular tissues, lungs, seawater, meteorites and ashes of plants like sugar, beat, tobacco, coffee and in many mineral springs etc. Important minerals of lithium are :

1. Aluminosilicates. These include the following minerals.

(*a*) *Lepidolite or lithia mica*, $(Li, Na, K)_2 Al_2 (SiO_3)_3 (F, OH)_2$. It contains 2–6% of Li.

(*b*) *Spodumene*, $LiAl(SiO_3)_2$. It contains 3.8–5.6% of Li.

(*c*) *Petalite*, $Li Al(Si_2O_5)_2$. It contains 2–3% of Li.

2. Phosphates. These are mainly of the following minerals.

(*a*) *Triphylite*, $(Li, Na)_3PO_4. (Fe, Mn)_3(PO_4)_2$. It is a double phosphate of Li Na, Fe and Mn and contains 1.6–3.7% of Li.

(*b*) *Amblygonite*, $Li(AlF) PO_4$. It contains 7–10% of Li.

Occurence in India

Lithium mainly occurs in India as *lepidolite* in Bihar, Kashmir Karnataka and Rajasthan.

Extraction of Lithium

The extraction of lithium involves the following two steps :

1. Preparation of LiCl. LiCl can be prepared either from *spodumene ore*, $LiAl (SiO_3)_2$ or from triphylite ore as shown below :

(*i*) *Preparation of LiCl from spodumene ore, $LiAl(SiO_3)_2$*. LiCl can be prepared from spodumene ore by using any of the following two methods :

(*a*) *Acid treatment method*. The finely powdered spodumene is digested (boiled) with conc. H_2SO_4.

$$\underset{\text{Spodumene}}{2LiAl(SiO_3)_2} + 4H_2SO_4 \longrightarrow Li_2SO_4 + Al_2(SO_4)_3 + \underset{\text{Silicic acid}}{4H_2SiO_3}$$

The digested mass is evaporated to dryness to convert H_2SiO_3 formed during the process into insoluble *hydrated silica*, $SiO_2.H_2O$.

$$HSiO_3 \longrightarrow SiO_2.H_2O$$

The filtrate is heated with Na_2CO_3 to precipitate the impurities of Mg, Al and Fe. At this stage, the solution is, however, too dilute for the precipitation of Li_2CO_3. The filtrate after concentration is treated with excess of Na_2CO_3 to precipitate Li_2CO_3. The precipitate of Li_2CO_3 is filtered and converted into LiCl by the treatment of HCl.

$$Li_2SO_4 + Na_2CO_3 \longrightarrow Li_2CO_3 + Na_2SO_4$$

$$Li_2CO_3 + 2HCl \longrightarrow 2LiCl + H_2O + CO_2$$

(*b*) *Fusion method*. The finely powdered spodumene is fused with a mixture of $BaCO_3$ and $BaSO_4$.

$$\underset{\text{Spodumene}}{2LiAl(SiO_3)_2} + BaSO_4 \longrightarrow Li_2SO_4 + Al_2(SiO_3)_3 + BaSiO_3$$

$$2LiAl(SiO_3)_2 + BaCO_3 \longrightarrow Li_2CO_3 + Al_2(SiO_3)_3 + BaSiO_3$$

From the fused mass two layers are separated : *upper layer* which consists of Li_2SO_4 and *lower layer* which contains the silicates of Al and Ba. The lower layer is rejected whereas the upper layer is separated, dissolved in water and treated with $BaCl_2$ when insoluble $BaSO_4$ and soluble LiCl are obtained.

$$Li_2SO_4 + BaCl_2 \longrightarrow \underset{\text{(Soluble)}}{2LiCl} + \underset{\text{(Insoluble)}}{BaSO_4}$$

The filtrate of LiCl is evaporated to dryness and then is purified by the extracting it with alcohol.

(*ii*) *Preparation of LiCl from triphylite ore*. The finely powdered triphylite mineral is digested with conc. HCl. H_3PO_4 so liberated is removed by precipitation with $FeCl_3$ in presence of CH_3COONH_4 and CH_3COOH as in qualitative analysis. The filtrate which consists mostly of LiCl, NaCl and $MnCl_2$ is evaporated to dryness, extracted with hot water and Mn is removed as MnS by the treatment of *excess* of BaS solution.

$$MnCl_2 + BaS \longrightarrow \underset{\text{ppt.}}{MnS} + BaCl_2$$

The filtrate is then treated with dil. H_2SO_4 to remove the excess of barium as insoluble $BaSO_4$ which is again removed by filtration. Now the filtrate containing Li_2SO_4 and Na_2SO_4 is evaporated along with oxalic acid. Lithium and sodium oxalates are formed which on ignition yield Li_2CO_3, Na_2CO_3 and CO. When these carbonates are treated with HCl, a mixture of LiCl and NaCl is obtained. From the mixture, LiCl is dissolved out in pyridine while NaCl remains insoluble. From this solution LiCl is obtained by distilling away pyridine.

2. To get Li metal by the electrolysis of LiCl. LiCl obtained as above is electrolysed to get Li metal. If an aqueous solution of LiCl is electrolysed, no lithium metal is obtained but there is only liberation of hydrogen at the cathode. In electrolytic process, therefore, aqueous solution of LiCl is not used. A fused mixture of LiCl and KCl is electrolysed at 400°C. The addition of KCl lowers the melting point (fusion temperature) of LiCl from 610°C to about 400°C so that the formation of lithium metal fog is prevented and increases the electrical conductivity of LiCl.

The electrolysis is carried out in an electrolytic cell shown in Fig. 19.1. The cell is made of steel with a refractory lining and consists of a steel cathode and graphite anode. The anode and cathode are separated from each other by a cast iron enclosure having iron gauze in its lower portion. The cell is operated at 400–420°C and the voltage is maintained between 8 to 9 volts. On passing the electric current, Li^+ ions are discharged in preference to K^+ ions at the cathode and Cl^- ions at the anode. The molten lithium containing about 1.3% of potassium, being lighter, rises to the surface of the fused mixture of electrolyte in the cast iron enclosure surrounding the cathode and is collected under a bell to protect it from contact with air. The metal is removed intermittently and stored either in well-stoppered bottles or in liquids like water-free ether, naphtha or petroleum. Metal obtained by this method is 99% pure.

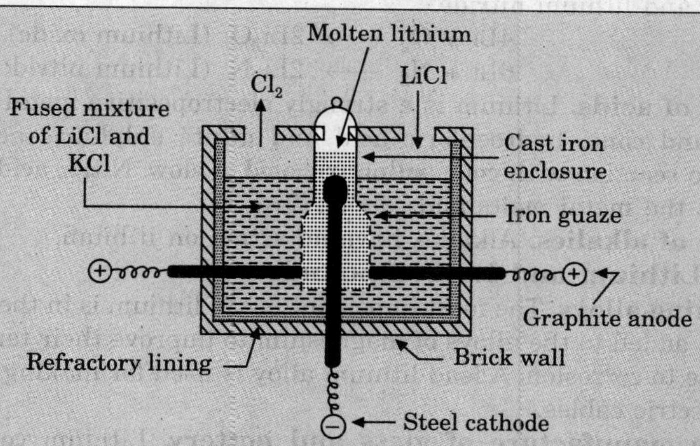

Fig. 19.1. Prepartion of lithium by the electrolysis of fused mixture of LiCl and KCl.

The following reactions take place :

Ionisation : $LiCl \rightleftarrows Li^+ + Cl^-$

Cathode process : $Li^+ + e^- \longrightarrow Li$

Anode process : $Cl^- \longrightarrow Cl + e^-$

$Cl + Cl \longrightarrow Cl_2$

Cl_2 gas which is a valuable by-product is removed from the outlet of the anodic compartment.

Lithium may also be obtained by the electrolysis of conc. solution of LiCl in pyridine, ethanol or acetone.

Physical Properties

Lithium is a silvery white metal which melts at 186°C. It is harder than sodium or potassium but is extremely light. Its density which is 0.534 at 20°C *is the lowest of all the metals*. When volatilized, it imparts a red colour to the flame. It is a good conductor of heat and electricity. It forms an amalgam and alloys with a number of other metals. It has the highest specific heat of any element (0.9408).

Chemical Properties

1. Formation of compounds. When heated above its fusion point, it combines with hydrogen, nitrogen, sulphur, halogen and carbon dioxide.

$$2Li + H_2 \longrightarrow 2LiH \text{ (Lithium hydride)}$$
$$6Li + N_2 \longrightarrow 2Li_3N \text{ (Lithium nitride)}$$
$$2Li + Cl_2 \longrightarrow 2LiCl \text{ (Lithium chloride)}$$

2. Action of water. It reacts with water forming the hydroxide and liberating hydrogen but unlike sodium or potassium it does not fuse on water as its m.pt. is higher

$$2Li + 2H_2O \longrightarrow 2LiOH + H_2$$

3. Action of Air. It is not affected in dry air but is oxidised when moisture is present. Heated in air above 180°C, it burns with a brilliant light forming lithium oxide and lithium nitride.

$$4Li + O_2 \longrightarrow 2Li_2O \text{ (Lithium oxide)}$$
$$6Li + N_2 \longrightarrow 2Li_3N \text{ (Lithium nitride)}$$

4. Action of acids. Lithium is a strongly electropositive metal and it reacts with dilute and conc. hydrochloric acid and dilute sulphuric acid liberating hydrogen. The reaction with conc. sulphuric acid is slow. Nitric acid attacks it so violently that the metal melts and catches fire.

5. Action of alkalies. Alkalies have no action on lithium.

Uses of Lithium and Its Compounds

1. In making alloys. The most important use of lithium is in the manufacture of alloys. It is added to the alloys of magnesium to improve their tensile strength and resistance to corrosion. A lead-lithium alloy is used for making bearings and sheaths of electric cables.

2. In the manufacture of glass and pottery. Lithium compounds are employed to increase the fluidity of glass. Its carbonate is employed to increase the strength and resistance of glass. Its salts are used in ceramics to prevent surface cracking of pottery.

3. As deoxidiser. It is employed as a deoxidiser in the purification of nickel and copper.

4. In medicine. Its citrate and salicylate are employed in medicine for relieving gout since lithium urate is fairly soluble in water. Its compounds have also been used in the synthesis of vitamin A and some pharmaceuticals.

5. In lubricants. Its hydroxide is used in the manufacture of high quality lubricating greases which can withstand extreme variations of temperature.

6. In thermonuclear energy. It is finding increasing use as an ingredient of high energy fuels for propulsion of intercontinental rockets.

Compounds of Lithium

1. Lithium Chloride, LiCl

Preparation. It is prepared by burning the metal in Cl_2 or by adding HCl to its hydroxide or carbonate

$$2Li + Cl \longrightarrow 2LiCl$$
$$LiOH + HCl \longrightarrow LiCl + H_2O$$
$$Li_2CO_3 + 2HCl \longrightarrow 2LiCl + H_2O + CO_2 \uparrow$$

The resulting solution is completely evaporated at white heat in a current of hydrogen chloride when anhydrous salt is obtained.

Manufacture. It can be manufactured from spodumene and triphylite minerals as described under "*Extraction of Lithium*".

Properties. It is a white crystalline solid with m.pt. = 606°C and b.pt. = 1350°C. Unlike the other alkali metal chlorides, it is the most *deliquescent* substance and is extremely solube in water. The fact that LiCl is also soluble in organic solvents shows that it has partial covalent character. It is due to its covalent character that it is hydrolysed by water. It crystallises as anhydrous salt from its aqueous solution above 98°C and as $LiCl.H_2O$, $LiCl.2H_2O$ and $LiCl.3H_2O$ at lower temperatures. It absorbs NH_3 both in solution and in the dry state to give the *ammoniates* like $LiCl.NH_3$, $LiCl.2NH_3$, $LiCl.3NH_3$ and $LiCl.4NH_3$. It also forms a number of double salts with chlorides of Fe, Co, Ni, Cu, Mn etc. It is the most volatile of all the alkali metal chlorides and, on being heated above the fusion point, it loses its chlorine.

Uses. It is used (*i*) in the preparation of metal by electrolysis (*ii*) in air-conditioning plants in the form of aqueous solution (35-40%) for humidity control (*iii*) in pyrotechny (*iv*) as a constituent for cleaning and soldering fluxes for Al and Mg (*v*) In Ni-Fe accumulators.

Structure. It is primarily an ionic compound. The formation of ionic bond by the transfer of one electron from Li atom to Cl atom can be explained as we do in NaCl.

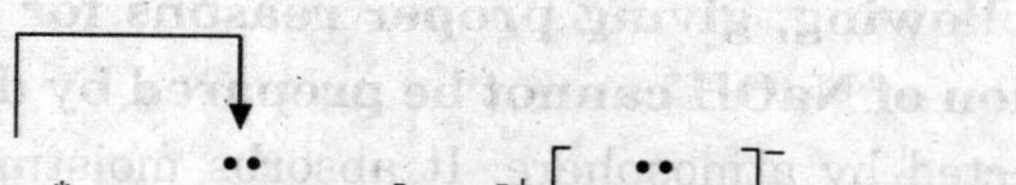

2. Lithium Carbonate, Li_2CO_3

Preparation. When $(NH_4)_2CO_3$ solution is added to a solution of lithium salt like LiCl in presence of NH_3, white preciptate of Li_2CO_3 is produced.

$$2\,LiCl + (NH_4)_2CO_3 \longrightarrow \underset{ppt.}{Li_2CO_3} + 2NH_4Cl$$

The preciptate is filtered and dried.

Manufacture. (a) From amblygonite, Li(AlF) PO$_4$. The mineral is ground to fine powder which is heated with $CaSO_4$ solution so that Li_3PO_4 and $AlPO_4$ present in the mineral are converted into soluble Li_2SO_4 and $Al_2(SO_4)_3$ respectively

$$2Li_3PO_4 + 3CaSO_4 \longrightarrow \underset{\text{soluble}}{2Li_2SO_4} + \underset{\text{ppt.}}{Ca_3(PO_4)_2}$$

$$2AlPO_4 + 3CaSO_4 \longrightarrow \underset{\text{soluble}}{Al_2(SO_4)_3} + \underset{\text{ppt.}}{Ca_3(PO_4)_2}$$

The filtrate containing Li_2SO_4 and $Al_2(SO_4)_3$ is treated with Na_2CO_3 whereby Li_2CO_3 is precipated.

$$Li_2SO_4 + Na_2CO_3 \longrightarrow \underset{\text{ppt.}}{Li_2CO_3} + Na_2SO_4$$

(b) From spodumene, LiAl $(SiO_3)_2$. The preparation of Li_2CO_3 from spodumene can be obtained by the acid treatment method as described under the extraction of lithium.

Properties. *(i)* It is a white crystalline substance, very sparingly soluble in water and insoluble in alcohol. It is more soluble in cold water than in hot. Its m.pt. is 618°C and density is 2.11.

(ii) Action of heat. When heated above 600°C, it gets decomposed into its oxide (Li_2O) and CO_2.

(iii) Action of CO_2. When CO_2 is passed through a suspension of Li_2CO_3 in water, the latter dissolves forming $LiHCO_3$.

$$Li_2CO_3 + H_2O + CO_2 \longrightarrow 2LiHCO_3$$

The solution of $LiHCO_3$ is used in medicine as a remedy for gout and rheumatism under the name of *lithia water*.

(iv) Action H_3PO_4. When Li_2CO_3 is dissolved in H_3PO_4 and the solution so obtained is evaporated, Li_3PO_4 is obtained.

$$3Li_2CO_3 + 2H_3PO_4 \longrightarrow 2Li_3PO_4 + 3Ha_2O + 3CO_2 \uparrow$$

Uses. Li_2CO_3 is used *(i)* as a starting material for the production of other lithium compounds *(ii)* in glass industry for preparing glass transparent to ultra-violet light.

Questions with Answers

Q.1 Explain the following, giving proper reasons for your answer :

***(i)* Standard solution of NaOH cannot be prepared by direct weighing.**

Ans : NaOH is affected by atmosphere. It absorbs moisture and CO_2 from atmosphere. Thus, accurate weighing of NaOH is not possible.

***(ii)* The hydoxides of aluminium and iron are insoluble in water but NaOH is used to separate one from the other.** **[I.I.T 1991]**

Ans : $Al(OH)_3$ dissolves in excess of NaOH forming a soluble compound, sodium meta aluminate, $NaAlO_2$

$$Al(OH)_3 + NaOH \rightarrow NaAlO_2 + 2\ H_2O$$

***(iii)* The inside surface of a glass bottle containing caustic soda becomes dull.**

Ans : Silica is present in glass. It dissolves in NaOH slowly and forms soluble sodium silicate (Na_2SiO_3) and thus inside surface becomes dull.

$$2\ NaOH + SiO_2\ (\text{silica}) \rightarrow Na_2SiO_3 + H_2O$$

Q.2 A certain compound (*X*) is used in the laboratory for analysis. Its aqueous solution gives th following reactions :

(*i*) On addition to copper sulphate, a brown precipitate is obtained which turns white on addition of excess of $Na_2S_2O_3$ solution.

(*ii*) On addition of Ag^+ ion solution, a yellow curdy precipitate is obtained which is insoluble in ammonium hydroxide.

Identify (*X*) and give equations for the reactions at steps (*i*) and (*ii*).

(Roorkee 1991)

Ans : *X* is KI.

(*i*) $2\ CuSO_4 + 4KI \longrightarrow 2\ CuI_2 + 2\ K_2SO_4$

$$CuI_2 \longrightarrow \underset{\text{(White)}}{Cu_2I_2} + I_2$$

$$I_2 + 2\ Na_2S_2O_3 \longrightarrow 2\ NaI + Na_2S_4O_6$$

(*ii*) $Ag^+ + KI \longrightarrow AgI + K^+$

Yellow ppt.
(Insoluble in NH_4OH)

Q.3 When 16.78 g of a white solid *X* were heated, 4.4 g of acid gas *A* that turned lime water milky was driven off together with 1.8 g of a gas *B* which condensed to a colourless liquid. The solid that remained, *Y*, dissolved in water to give an alkaline solution, which with excess of barium chloride solution gave a white precipitate, *Z*. The precipitate effervesced with acid giving off carbon dioxide. Identity *A*, *B* and *Y* and write down the equation for the thermal decomposition of *X*.

[I.I.T. 1984]

Ans : X is $NaHCO_3$

$$\underset{(X)}{2\ NaHCO_3} \longrightarrow \underset{(Y)}{Na_2CO_3} + \underset{(A)}{CO_2} + \underset{(B)}{H_2O}$$

Q.4 Solvay process used for the manufacture of Na_2CO_3 can be represented by the follwoing scheme.

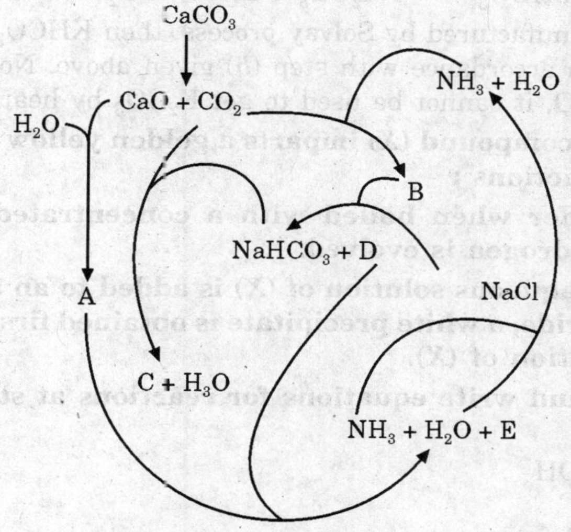

Identity A, B, C, D and E. (*I.I.T 1999*)

Ans : The reactions taking place in the scheme are :

(*i*) $NH_3 + H_2O + CO_2 \longrightarrow \underset{(B)}{NH_4HCO_3}$

(*ii*) $\underset{(B)}{NH_4HCO_3} + NaCl \longrightarrow NaHCO_3 + \underset{(D)}{NH_4Cl}$

(*iii*) $2NaHCO_3 \longrightarrow \underset{(C)}{Na_2CO_3} + H_2O + CO_2$

(*iv*) $CaO + H_2O \longrightarrow \underset{(A)}{Ca(OH)_2}$

(*v*) $\underset{(A)}{Ca(OH)_2} + \underset{(D)}{2NaCl} \longrightarrow 2\,NH_3 + 2\,H_2O + \underset{(E)}{CaCl_2}$

Thus : A = $Ca(OH)_2$, B = NH_4HCO_3, C = Na_2CO_3, D = NH_4Cl and E = $CaCl_2$.

Q.5 Na_2CO_3 is made by Solvay process but the same process cannot be extended to the manufacture of K_2CO_3. Explain. (*I.I.T. 1981*)

Ans : We know than Solvay process used for the manufacture of Na_2CO_3 consists of the following steps :

(*i*) CO_2 gas is passed through commercial brine solution (NaCl) when $NaHCO_3$ (an insoluble compound) is produced as an intermediate compound.

$$NaCl + NH_3 + H_2O + CO_2 \longrightarrow NaHCO_3 + NH_4Cl.$$

The formation of $NaHCO_3$ actually takes place through the following two steps :

(*a*) $NH_3 + H_2O + CO_2 \longrightarrow NH_4HCO_3$

(*b*) $NH_4HCO_3 + NaCl \longrightarrow \underset{\substack{\text{Intremediate} \\ \text{(Insoluble)}}}{NaHCO_3} + NH_4Cl$

(*ii*) $NaHCO_3$ obtained as above gets decomposed, on heating, to give Na_2CO_3

$$2\,NaHCO_3 \xrightarrow{\Delta} Na_2CO_3 + H_2O + CO_2$$

If K_2CO_3 is manufactured by Solvay process, then $KHCO_3$ will be formed as an intermediate in accordance with step (*b*) given above. Now since $KHCO_3$ is very soluble in H_2O, it cannot be used to get K_2CO_3 by heating it.

Q.6 A certain compound (*X*) imparts a golden yellow flame and shows the following reactions :

(*i*) Zinc powder when boiled with a concentrated solution of (*X*) dissolves and hydrogen is evolved.

(*ii*) When an aqueous solution of (*X*) is added to an aqueous solution of stannous chloride, a white precipitate is obtained first which dissolves in excess of solution of (*X*).

Identify (*X*) and write equations for reactions at steps (*i*) and (*ii*). (***Roorkee 1991***)

Ans : *X* is NaOH

(*i*) $Zn + 2\ NaOH \longrightarrow Na_2ZnO_2 + H_2$

(*ii*) $SnCl_2 + 2\ NaOH \longrightarrow Sn\ (OH)_2 + 2\ NaCl$

$$Sn\ (OH)_2 + 2\ NaOH \longrightarrow \underset{\text{Soluble}}{Na_2SnO_2} + 2\ H_2O$$

Q.7 Gradual addition of potassium iodide solution to Bi $(N_3)_3$ solution initially produces a dark brown precipitate which dissolves in excess of KI to give a clear yellow solution. Write the chemical equation for the above reactions. [I.I.T. 1996]

Ans : At first, Bi $(NO_3)_3$ undergoes hydrolysis. Nitric acid is formed. Which oxidises KI to iodine. The liberated iodine dissolves in KI to form yellow solution of KI_3.

$$Bi\ (NO_3)_3 + H_2O \longrightarrow [Bi\ (OH)\ (NO_3)_2] + HNO_3$$

$$2\ HNO_3 \longrightarrow H_2O + 2\ NO_2 + O$$

$$2\ KI + H_2O + O \longrightarrow 2\ KOH + \underset{\text{Dark brown}}{I_2}$$

$$I_2 + KI \longrightarrow \underset{\text{Yellow solution}}{KI_3}$$

Q.8 Write balanced equations for the reaction between caustic soda and (*a*) Zn (*b*) $AgNO_3$ (*c*) Phosphorus (*d*) Arsenion oxide (As_2O_3) (*e*) I_2 (*f*) $ZnSO_4$ solution (*g*) $AlCl_3$ solution. *(MLNR 1979)*

Ans : (*a*) NaOH reacts with Zn to form sodium zincate (Na_2ZnO_2) and H_2 gas is evolved.

$$Zn + 2\ NaOH \longrightarrow Na_2ZnO_2 + H_2$$

(*b*) $AgNO_3$ reacts with NaOH in two stages. In the first stage unstable hydroxide of Ag (AgOH) is formed. In the second stage this hydroxide gets decomposed into insoluble oxide of Ag (Ag_2O).

$$2\ AgNO_3 + 2\ NaOH \longrightarrow \underset{\text{(Unstable)}}{2\ AgOH} + NaNO_3$$

$$2\ AgOH \longrightarrow \underset{\text{Stable (Brown)}}{Ag_2O} + H_2O$$

(*c*) White (yellow) phosphorus, when heated with NaOH solution, evolves phosphine (PH_3) and sodium hypophosphite (NaH_2PO_2) is also formed.

$$P_4 + 3\ NaOH + 3\ H_2O \longrightarrow 3\ NaH_2PO_2 + PH_3$$

(*d*) Since As_2O_3 is an acidic oxide, salt *viz.* sodium arsenite ($NaAsO_2$) is formed.

$$2\ NaOH + As_2O_3 \longrightarrow 2\ NaAsO_2 + H_2O$$

(*e*) With cold and dil. NaOH solution, sodium iodide (NaI) and sodium hypoiodite (NaIO) are formed. But with hot concentrated NaOH solution, sodium iodide (NaI) and sodium iodate $(NaIO)_3$ are obtained.

$$2\ NaOH + I_2 \longrightarrow NaI + NaIO + H_2O$$

Dilute and cold.

$$6\ NaOH + 3\ I_2 \longrightarrow 5NaI + NaIO_3 + 3\ H_2O$$

Hot and concentrated

(*f*) NaOH reacts with $ZnSO_4$ in two stages. In the first stage Zn $(OH)_2$ (white ppt.) is formed. In the second stage Zn $(OH)_2$ formed as above gets dissolved in the excess of NaOH and gives a soluble oxy salt namely sodium zincate, Na_2ZnO_2.

$$2\,NaOH + ZnSO_4 \longrightarrow \underset{\text{White ppt.}}{Zn(OH)_2} + Na_2SO_4$$

$$Zn(OH)_2 + 2\,NaOH \longrightarrow \underset{\text{(Soluble)}}{Na_2ZnO_2} + 2\,H_2O$$

$$4\,NaOH + ZnSO_4 \longrightarrow \underset{\text{(Soluble)}}{Na_2ZnO_2} + 2\,H_2O + Na_2SO_4$$

(*g*) NaOH reacts with $AlCl_3$ in two stages. In the first stage Al $(OH)_3$ (white ppt.) is formed and in the second stage white ppt. of Al $(OH)_3$ formed gets dissolved in the excess of NaOH and gives a soluble oxy salt namely sodium aluminate $(NaAlO_2)$.

$$3\,NaOH + AlCl_3 \longrightarrow \underset{\text{White ppt.}}{Al(OH)_3} + 3\,NaCl$$

$$Al(OH)_3 + NaOH \longrightarrow \underset{\text{(Soluble)}}{NaAlO_2} + 2\,H_2O$$

$$4\,NaOH + AlCl_3 \longrightarrow \underset{\text{(Soluble)}}{NaAlO_2} + 2\,H_2O + 3\,NaCl$$

Q.9 What are the compounds *U, V, W, X, Y, Z, A, B, C* in the following road map problems ?

$$NaCl + H_2SO_4\ (\text{conc.}) \longrightarrow U + \text{Gas } V$$

$$V + MnO_2 \xrightarrow{\Delta} \text{Gas } W + MnCl_2 + H_2O$$

$$U + NaCl \xrightarrow[\Delta]{} X + V$$

$$X + \text{Coke} + CaCO_3 \longrightarrow Y + Z$$

$$CaCO_3 \xrightarrow[\Delta]{} A + B \text{ (colourless and odourless gas)}$$

$$A + H_2O \longrightarrow C$$

$$C + Y \rightarrow NaOH + CaCO_3$$

Ans :

$$NaCl + H_2SO_4\ (\text{conc.}) \longrightarrow \underset{(U)}{NaHSO_4} + \underset{(V)}{HCl\ (g)}$$

$$MnO_2 + \underset{(V)}{4\,HCl\ (g)} \longrightarrow MnCl_2 + \underset{(W)}{Cl_2} + 2\,H_2O$$

$$\underset{(U)}{NaHSO_4} + NaCl \xrightarrow{\Delta} \underset{(X)}{Na_2SO_4} + \underset{(V)}{HCl}$$

$$\underset{(X)}{Na_2SO_4} + CaCO_3 \longrightarrow \underset{(Y)}{Na_2CO_3} + \underset{(Z)}{CaSO_4}$$

$$CaCO_3 \xrightarrow{\Delta} \underset{(A)}{CaO} + \underset{(B)}{CO_2\ (g)}$$

$$\underset{(A)}{CaO} + H_2O \longrightarrow \underset{(C)}{Ca(OH)_2}$$

$$\underset{(C)}{Ca(OH)_2} + \underset{(Y)}{Na_2CO_3} \longrightarrow 2\,NaOH + CaCO_3$$

Q.10 From the following reactions identity A, B, C and D and write their formulae.

$$A\ (aq) + Zn \longrightarrow B\ (g)$$

$$A\ (aq) + C \xrightarrow{\Delta} PH_3$$

$$A\ (aq) + NH_4Cl \longrightarrow D\ (g)$$ **(Roorkee 1987)**

Ans : A = NaOH, sodium hydroxide; B = H_2, hydrogen; C = P_4, phosphorus; D = NH_3, ammonia.

Q.11 State the reasons :

(*i*) The addition of NH_4OH to $ZnSO_4$ solution produces white precipitate but no precipitate is formed if it contain NH_4Cl. **(*Roorkee 1980*)**

(*ii*) The addition of NaOH to a solution of $ZnCl_2$ produces a white precipitate which dissolves on further addition of NaOH. **(*Roorkee 1980*)**

Ans : (*i*) NH_4OH is a weak hydroxide. It ionises slightly furnishing OH^- ions. However, OH^- ions produced are sufficient to cause precipitation of $Zn(OH)_2$ as its solubility product is exceeded.

$$ZnSO_4 + 2NH_4OH \longrightarrow \underset{\text{White ppt.}}{Zn(OH)_2} + (NH_4)_2SO_4$$

In presence of NH_4Cl, the ionisation of NH_4OH is further suppressed and sufficient OH^- ions are not available to cause precipitation as the solubility product is not exceeded.

(*ii*) $$2NaOH + ZnCl_2 \longrightarrow Zn(OH)_2 + 2NaCl$$

$$Zn(OH)_2 + 2NaOH \longrightarrow Na_2Zn(OH)_4 \longrightarrow Na_2ZnO_2 + 2H_2O$$

NaOH pricipitates insoluble hydroxides of metals from their soluble salts since zinc metal is amphoteric in nature. Since $Zn(OH)_2$ precipitate so formed is amphoteric in nature, on adding excess of NaOH a soluble complex $Na_2Zn(OH)_4$ is formed.

Q.12 What happens when :

(*i*) Excess of NaOH solution is added slowly to $ZnCl_2$ solution.

(*ii*) Excess of NH_4OH solution is added to $ZnCl_2$ solution.

(*iii*) ZnO is treated with excess of NaOH solution.

Ans : (*i*) First a white precipitate of $Zn(OH)_2$ is formed which dissolves in excess of NaOH forming $Na_2Zn\ (OH)_4$.

(*ii*) A white Zn $(OH)_2$ precipitated is first formed of which dissolves in excess of NH_4OH forming $[Zn\ (NH_3)_4]\ (OH)_2$, which is a soluble complex.

(*iii*) ZnO dissolves in NaOH forming a soluble complex, $Na_2Zn\ (OH)_4$,

$$\underset{\text{insoluble}}{ZnO} + 2\ NaOH + H_2O \longrightarrow \underset{\text{soluble}}{Na_2Zn\ (OH)_4}$$

Q.13 Identity A, B, C and D and give their chemical formulae :

$$A + NaOH \xrightarrow{\Delta} NaCl + NH_3 + H_2O$$

$$NH_3 + CO_2 + H_2O \longrightarrow B$$

$$B + NaCl \longrightarrow C + NH_4Cl$$

$$C \longrightarrow Na_2CO_3 + H_2O + D.$$

Ans : A = NH_4Cl, ammonium chloride ; B = NH_4HCO_3, ammonium bicarbonate; C = $NaHCO_3$, sodium bicarbonate ; D = CO_2, carbon dioxide.

University Questions

1. Give the ores, extraction and isolation of pure lithium. (*Meerut 89; Garhwal 88*)

2. Clarify the following points of the extraction of lithium from its ores :

(*i*) How is LiCl obtained from NaCl and KCl.

(*ii*) How is metallic Li obtained from LiCl ? (*Gorakhpur 84*)

3. Describe the properties of lithium. (*Meerut 88*)

4. What are the uses of lithium metal and its compounds. (*Delhi 86; Madras 85; Agra 84, 85; Rohilkhand 84; Gorakhpur 80, 84; Lucknow 81; Bhopal 80; Kanpur 82; Meerut 84, 86, 88*)

5. Give an account of the preparation, properties, uses and structure of LiCl. (*Meerut 82, 82 S, 89*)

6. What happens when Li metal is heated with H_2, N_2 and Cl.

7. How is Li_2CO_3 manufactured from amblygonite ore.

8. What happens when Li_2CO_3 is treated with H_3PO_4.

9. What is the action of heat on Li_2CO_3.

10. Give the method of preparation, properties and uses of KI. (*MLNR 78*)

11. (*a*) Discuss the role of alkali metals as a source of energy.

(*b*) Give the chemical principles involved in the metallurgical extraction of lithium from Lepidolite.

(*c*) Describe the preparation, important properties, and structure of lithium tetrahydroaluminate. (*Allahabad 99*)

12. Explain why alkali metals are good reducing agents. (*Kanpur 2000*)

13. How will you get LiCl from lepidolyte? (*Kumaon 2000*)

14. How is lithium obtained from its ores ? Discuss its anomolous behaviour. (*Lucknow 2001*)

CHAPTER 20

General Characteristics of Group 2 (IIA) Elements : Alkaline Earth Metals

Group IIA of the long form of the periodic table consists of *six* elements which are : *Breyllium* (Be), *Magnesium* (Mg), *Calcium* (Ca), *Strontium* (Sr), *Barium* (Ba), and *Radium* (Ra).

The oxides of the three metals *viz.*, Ca, Sr and Ba were known much earlier than the metals themselves and were called *alkaline earths*, since they were alkaline in character and occured in nature as earths [*lime* (CaO), *strontia* (SrO) and *baryta* (BaO)]. Later, when Ca, Sr and Ba were discovered, they were named *alkaline earth metals*. Now this term is used to include all the elements of Group II A. Although Ra has similar properties as alkaline earth metals, but being radioactive, it is studied separately along with other radioactive elements.

Position of Alkaline Earth Metals in the Periodic Table

The inclusion of the alkaline earth metals in the same subgroup is justified by the following points :

I. Electronic Configurations.

The electronic configurations of the alkaline earth metals (Group II A) are given below :

Element	At No.	Complete electronic configuration	Valence-shell electronic configuration
Be	4	2, 2	$2s^2$
Mg	12	2, 8, 2	$3s^2$
Ca	20	2, 8, 8, 2	$4s^2$
Sr	38	2, 8, 18, 8, 2	$5s^2$
Ba	56	2, 8, 18, 18, 8, 2	$6s^2$
Ra	88	2, 8, 18, 32, 18, 8, 2	$7s^2$

These configurations show that the ultimate shell of these elements has *two* *s*-electrons (s^2 electrons) while the penultimate shell contains *eight* (s^2p^6 electrons.) When the two *s*-electrons of the valence-shell are lost, alkaline earth metals are converted into M^{2+} ions which have inert gas configuration. Since the additional electron enters the *ns*-orbital, these elements are called *s*-block elements.

II. Similarities in Physical Properties and Gradation in Them

Some of the physical properties of alkaline earth metals are given in Table 20.1 and are discussed as follows :

Table 20.1. Some physical properties of alkaline earth metals

Property	Be	Mg	Ca	Sr	Ba	Ra
Atomic weight	9.01	24.31	40.08	87.62	137.34	226
Abundance (% of earth's crust)	6.0×10^{-4}	2.0	3.45	0.915	0.040	1.3×10^{-10}
Density (gm/c.c.)	1.84	1.74	1.55	2.54	3.75	6.00
Melting point (°C)	1277	650	838	763	714	700
Boiling point (°C)	2770	1107	1440	1380	1640	—
Atomic volume (c.c.)	4.90	13.97	25.9	34.54	36.7	38.0
Atomic (*i.e.*, metallic) radius for coordination number 12(Å)	1.12	1.60	1.97	2.15	2.22	—
Covalent radius (Å)	0.90	1.36	1.74	1.91	1.98	—
Ionic (crystal) radius of M^{2+} ion for coordination number 6(Å)	0.31	0.65	0.99	1.13	1.35	140
Ionisation energies (KJ/mole)						
I_1	899.5	737.7	829.8	547.5	502.9	509.4
I_2	1757.1	1450.7	1145.4	1064.3	965.2	979.06
$I_1 + I_2$	2656.6	2188.4	1735.2	1613.8	1468.1	1488.46
Oxidation state	+ 2	+ 2	+ 2	+ 2	+ 2	+ 2
Electronegativity	1.5	1.2	1.0	1.0	0.9	0.9
Flame colouration	None	None	Brick red	Crim-son	Apple green	Red
Oxidation potentials (volts) for $M(s) \longrightarrow M^{2+}(aq) + 2e^-$	1.70	2.37	2.87	2.89	2.90	2.92
Heat of automisation at 25°C and 1 atm. pressure (KJ/mole)	327.26	146.89	181.21	163.21	175.77	—
Heat of hydration (KJ/mole)	2385.45	1925.1	1653.07	1485.67	1276.42	—
Ionic potential of M^{2+} ion (*i.e.*, charge/radius ratio)	6.66	3.08	2.12	1.82	1.55	1.33

1. Physical state. The elements of this group have a greyish white lustre when freshly cut and are malleable and ductile.

2. Density. Density decreases slightly upto Ca after which it increases. Irregular variation of the density is due to the difference in the crystal structure of these elements. By comparing the densities of these elements with those of alkali metals, it may be seen that these elements are *denser* (*i.e.*, have higher density) than the alkali metals in the same period. This is due to the fact that the elements of Group IIA can be packed more tightly due to their *greater nuclear charge* and *smaller size*.

3. Melting and boiling points. Melting and boiling points of these elements do not show any regular trend. *These are, however, higher than those of alkali* *metals.* This is because the alkaline earth metals have *two valence electrons* and are *much strongly bonded* in the solid state than the alkali metals.

4. Atomic volume, atomic and ionic radii. Because of the addition of an extra shell of electrons to each element from Be to Ra, the atomic volume *increases* from Be to Ra. With the increases of atomic volume *the atomic and ionic radii* (of M^{2+} ions) *also increase from Be to Ra.* The atomic radii of these elements are, however, smaller than those of alkali metals in the same period. This is due to the fact that the alkaline earth metals have higher nuclear charge which tends to draw the orbit electrons towards the nucleus. The smaller values of atomic radii result in that the alkaline earth metals are *harder, have higher densities* and *higher melting points* than the alkali metals.

Elements	:	Be	Mg	Ca	Sr	Ba	Ra
Atomic volume (*c.c.*)	:	4.90	13.97	25.9	34.54	36.7	38.09
Atomic radii (Å)	:	1.12	1.60	1.97	2.15	2.22	—
Ionic radii of M^{2+} ions (Å)	:	0.31	0.65	0.99	1.13	1.35	1.40

5. Colour and magnetic properties of M^{2+} ions. Since the outer-shell electronic configuration of M^{2+} ions are similar to those of inert gases (s^2 or s^2p^6 configuration), these ions have *no unpaired electrons* and consequently are *diamagnetic and colourless.*

6. Ionisation energies. *The first and second ionisation energies of these elements decrease with the increase of atomic radii from* Be to Ba. However, both these values for Ra are slightly higher than those of Ba.

Elements		:	Be	Mg	Ca	Sr	Ba	Ra
Ionisation energies								
(*KJ/mole*)	I_1	:	899.5	737.7	589.8	549.5	502.9	509.4
	I_2	:	1757.1	1450.7	1145.4	1064.3	965.2	979.06

7. Electropositive character. Although the ionisation energies of these elements are higher than those of alkali metals, yet these are sufficiently low to make these metals to *lose readily their ns^2 electrons to form M^{2+}* ions :

$$M_1 \longrightarrow M^{2+} + 2e^-$$

These metals, therefore, show strong electropositive character which increases from Be to Ba. Since these metals have higher ionisation energies than the alkali metals, *these are not strongly electropositive as the alkali metals.*

8. Formation of divalent ions and oxidation states. In view of the fact that first ionisation potenials of these metals are lower than the second ionisation potentials, it appears that these metals should form univalent ions rather than divalent ions. But actually these elements give divalent ions. This anomaly is because of the fact that M^{2+} ions are extensively hydrated to form the hydrated ions, $[M(H_2O)_x]^{2+}$ and in this process a large amount of energy, known as *energy of hydration* is released.

$$M^{2+} + xH_2O \longrightarrow \underset{\text{Hydrated cation}}{[M(H_2O)_x]^{2+}} + \textit{Energy of hydration}$$

This energy of hydration more than counterbalances the higher value of second ionisation energy.

Because of the fact that Be has the highest ionisation energy, it has the least tendency to form Be^{2+} ion. Consequently, its compounds with nitrogen, oxygen, sulphur, halogens etc., are *covalent* while the corresponding compounds of Mg, Ca, Sr, and Be are *ionic*.

9. Hydration of alkaline earth metals. The values of the heats of hydration of M^{2+} ions (*i.e.* hydration energy) show that *these decrease with the increase in ionic radii of M^{2+} ions.*

Alkaline earth metal ions	:	Be^{2+}	Mg^{2+}	Ca^{2+}	Sr^{2+}	Ba^{2+}
Radii for M^{2+} ions (Å)	:	0.31	0.65	0.92	1.13	1.35
Heats of hydration of M^{2+} ions (KJ/mole)	:	2385.45	1925.1	1653.07	1485.67	1276.42

Since the hydration energies for alkaline earth metal ions are larger than those of alkali metals ions, the alkaline earth metal compounds are more extensively hydrated than those of alkali metals, *e.g.* magnesium chloride and calcium chloride exist as $MgCl_2.6H_2O$ and $CaCl_2.6H_2O$ respectively while sodium chloride and potassium chloride do not from hydrates.

With the decrease of hydration energies from Be^{2+} to Ba^{2+}, the ionic mobilities (*i.e.* ionic conductance) of these ions increase in the same order.

10. Electronegativities. These are *small* and *decrease from Be to Ba*.

11. Flame colouration. In the flame tests their electrons absorb energy and are excited to higher energy levels. When reverting to lower levels, they give out extra energy which appears as visible light of characteristic colours as shown below :

Ca—*Brick red*; Sr—*Crimson*;

Ba—*Apple green*; Ra—*Red*.

Be and Mg do not give any colour to the flame. This is because the atoms are smaller and hence their electrons are more strongly attracted by the nucleus. Hence these electrons are not excited to the higher energy levels by the energy of the flame.

12. Polarising power of M^{2+} cations. In this group the ionic radius increases from Be to Ra, whereas the ionic charge remains the same. Therefore, polarising power of M^{2+} cations decreases from Be^{2+} to Ra^{2+}. In general, smaller ions polarise the anions more effectively than the bigger ones. Therefore, Be^{2+} has a tendency to form covalent compounds.

13. Conductivity. Since two valence electrons are free to move throughout the metal structure, the alkaline earth metals are good conductors of heat and electricity.

14. Reducing properties : Oxidation potentials. Alkaline earth metals have two electrons in their valence shell. Due to large size they are comparatively easily oxidised to M^{2+} ions. They are, therefore, *strong reducing agents*. On moving down, the oxidation potentials increase due to increase in electropositive character, since the strength as a reducing agent *increases* downwards from Be to Ba.

The oxidation potentials of alkaline earth metals are *lower* than those of alkali metals because of their *higher sublimation* and *ionisation energies*. Hence alkaline earth metals are *weaker* reducing agents than the alkali metals.

III. Similarities in Chemical Properties and Gradation in Them

1. Formation of oxides (*a*) Normal oxides of MO type. *Preparation.* The alkaline earth metals form the normal oxides of MO type which are obtained by heating the metal in O_2 or by heating their carbonates at high temperature, *e.g.*

$$Ca + O_2 \xrightarrow{\Delta} CaO$$

$$CaCO_3 \xrightarrow{\Delta} CaO + CO_2 \uparrow$$

Properties. (*i*) These oxides are extremely stable white crystalline solids due to their *high crystal lattice energy* obtained by packing *doubly-charged ions* in a sodium chloride type of lattice.

(*ii*) Excepting BeO which has a *wurtzite* (4 : 4) structure, all other oxides have the sodium chloride (6 : 6) structure.

(*iii*) BeO and MgO are quite insoluble in H_2O while CaO, SrO and BaO react with H_2O to give soluble hydroxides, $M(OH)_2$ which are *strong bases.*

(*iv*) BeO is *not* at all *basic in nature*; in fact it is *amphoteric* since it reacts with acids to form salts and with alkalies to give beryllates.

$$BeO + 2HCl \longrightarrow BeCl_2 + H_2O$$

$$BeO + 2NaOH \longrightarrow \underset{\text{Sod. beryllate}}{Na_2BeO_2} + H_2O$$

The oxides of other metals are basic in character. Their basic character increases on moving down the group.

(*v*) Due to *small size* of Be^{2+} ion, BeO is *covalent* while other oxides are ionic. Although BeO is covalent, yet is has a higher melting point and is harder than the oxides of other metals, as it is *polymeric*. Each Be atom is *tetrahedrally* coordinated by four oxygen atoms.

(*b*) **Peroxides. *Preparation.*** The peroxides of heavier metals (Ca, Sr, Ba etc.) can be obtained on heating the normal oxides with O_2 at high temperature.

$$2BaO + O_2 \longrightarrow 2BaO_2$$

Properties. The peroxides are white, ionic solids having peroxide anion, $[O—O]^{2-}$. They react with acids to produce H_2O_2.

$$BaO_2 + H_2SO_4 \longrightarrow BaSO_4 + H_2O_2$$

2. Formation of hydroxides. *Preparation.* The hydroxides, $M(OH)_2$ are obtained :

(*a*) by the reaction of the metals with H_2O, *e.g.*

$$Ca + 2H_2O \longrightarrow Ca(OH)_2 + O_2$$

Be does not react with H_2O even at elevated temperature, Mg reacts only with steam while other metals react with cold water as well.

(*b*) by the reaction of the oxides (excepting MgO) with H_2O, *e.g.*

$$MO + H_2O \longrightarrow M(OH)_2$$

Properties. (*i*) The heats of formation of hydroxides increase from $Be(OH)_2$ to $Ba(OH)_2$.

(*ii*) $Be(OH)_2$ is *not at all basic*; in fact it is *amphoteric*, since it reacts with acids to form salts and with alkalies to give beryllates.

$$Be(OH)_2 + 2HCl \longrightarrow BaCl_2 + 2H_2O$$

$$Be(OH)_2 + 2NaOH \longrightarrow \underset{\text{Sod. beryllate}}{Na_2BeO_2} + 2H_2O$$

The hydroxides of other metals are basic in character. Their basic character increases on moving down the group. Thus $Mg(OH)_2$ is *weakly basic* while $Ba(OH)_2$ is the *strongest base*. The increase in basic character of the hydroxides on moving down the group is due to the fact that with the increase in size of M^{2+} cation both the *polarity* of M—OH bond and the *internuclear distance* between oxygen of OH^- ion and the metal atom *increase.* As a result of this, there is greater ionisation of $M(OH)_2$ and hence basic character increases.

(*iii*) Due to *high polarising power of small Be^{2+} ion*, $Be(OH)_2$ is *covalent* while other hydroxides are *ionic.*

(*iv*) $Be(OH)_2$ and $Mg(OH)_2$ are almost insoluble in H_2O while the hydroxides of other metals are slightly soluble. Their solubility increases on moving down the group as shown by the increasing value of the solubility products of these hydroxides.

$Be(OH)_2 = 1.6 \times 10^{-26}$; $Mg(OH)_2 = 8.9 \times 10^{-12}$; $Ca(OH)_2 = 1.3 \times 10^{-4}$;

$Sr(OH)_2 = 3.2 \times 10^{-4}$; $Ba(OH)_2 = 5.4 \times 10^{-2}$

3. Formation of halides. ***Preparation.*** The halides, MX_2 of these metals can be obtained :

(*a*) by heating the metals with halogens at appropriate temperatures,

(*b*) by reacting the metal carbonates with dil. halogen acids.

Properties. Halides of Be (*e.g.* $BeCl_2$, BeF_2) are covalent and, therefore, do not conduct electricity in fused state. They are hygroscopic and fume in air due to hydrolysis. They have low melting points. The covalent character of the halides of Be is due to *high polarising power of small Be^{2+} cation.*

The chlorides and fluorides of other alkaline earth metals are *ionic solids* and hence good conductors of electricity in the fused as well as in the dissolved state. The melting points and conductivity increases on moving down the group due to the increase in the ionic character of the halides. They are hygroscopic and readily form hydrates, *e.g.* $MgCl_2.6H_2O$, $CaCl_2.6H_2O$. The solubility of halides in water decreases on moving down the group due to the decrease in their hydration energy with the increasing size of the metal ion. The fluorides, MF_2 are almost insoluble in water.

Most of the anhydrous halides of alkaline earth metals crystallise in the fluorite type structure.

4. Formation of hydrides. All the elements of this group form the hydrides of MH_2 type.

Preparation. BeH_2 cannot be prepared by the direct combination of Be and H_2. It is, however, prepared by reducing $BeCl_2$ by LiH or $LiAlH_4$.

$$BeCl_2 + 2LiH \xrightarrow{\text{Ether}} BeH_2 + 2LiCl$$

$$2BeCl_2 + LiAlH_4 \longrightarrow 2BeH_2 + LiCl + AlCl_3$$

Mg, Ca, Sr and Ba form their hydrides when heated with H_2.

$$M(M = Mg, Ca, Sr, Ba) + H_2 \longrightarrow MH_2$$

Properties. (*i*) BeH_2 is a *covalent compound*, MgH_2 is *partly ionic* (*i.e. almost covalent*) in nature while the hydrides of other metals are ionic solids due to their greater electropositive character.

(*ii*) They react with H_2O, liberating H_2. Thus they are used as *reducing agents*.

$$MH_2 + 2H_2O \longrightarrow M(OH)_2 + H_2$$

5. Formation of carbonates. (*i*) All the carbonates decompose on heating at appropriate temperature, evolving CO_2.

$$MCO_3 \xrightarrow{\Delta} CO_2 + MO$$

The stability of the carbonates of these metals *increases on moving down the group.* This is illustrated by the values of the decomposition temperatures of these carbonates as given below :

$$BeCO_3 = 25°C,\ MgCO_3 = 540°C,\ CaCO_3 = 900°C,$$
$$SrCO_3 = 1290°C,\ BaCO_3 = 1360°C.$$

The marked instability of $BeCO_3$ is due to *strong polarising effect of small* Be^{2+} *cation* on the large polarisable CO_3^{2-} ion.

(*ii*) Carbonates are insoluble in water and, therefore, occur as solid rock minerals in nature. However, they dissolve in H_2O containing CO_2 due to the formation of bicarbonates.

$$\underset{\text{(Insoluble)}}{CaCO_3} + CO_2 + H_2 \longrightarrow \underset{\text{(Soluble)}}{Ca(HCO_3)_2}$$

6. Formation of nitrides. All the elements burn in nitrogen to form nitrides, M_3N_2, *e.g.*

$$3Ca + N_2 \longrightarrow Ca_3N_2$$

These react with H_2O to liberate NH_3 *e.g.*

$$Ca_3N_2 + 6H_2O \longrightarrow 3Ca(OH)_2 + 2NH_3$$

Be_3N_2 is *volatile* while other nitrides are *not so.*

7. Formation of carbides. When the elements from Mg to Ba or their oxides are heated with carbon, carbides (MC_2) are obtained, *e.g.*

$$CaO + 3C \longrightarrow CaC_2 + CO$$

All these carbides are *ionic in nature* and have NaCl type structure with M^{2+} replacing Na^+ and $[C \equiv C]^{2-}$ replacing Cl^-. On treatment with H_2O, they liberate *acetylene*.

$$CaC_2 + 2H_2O \longrightarrow Ca(OH)_2 + CH \equiv CH$$

When Be is heated with carbon, it gives the carbide, Be_2C and not BeC_2. This carbide with H_2O gives *methane*.

$$Be_2C + 2H_2O \longrightarrow 2BeO + CH_4$$

8. Formation of complexes. The tendency of M^{2+} ions to form the complexes *decreases with the increases of the size of M^{2+} ions*. Thus Be^{2+} ion, on account of its small size, forms many stable complexes such as $[BeF_3]^-$, $[BeF_4]^{2-}$, while Ba^{2+} forms *very few complexes*.

9. Solubility of sulphates and chromates. Calcium sulphate is sufficiently soluble in water, whilst strontium and barium sulphates are almost insoluble but calcium and strontium chromates are soluble in acetic acid medium while barium chromate is insoluble.

Anomalous Behaviour of Beryllium : Dissimilarities with Other Alkaline Earth Metals

Beryllium, the first element of the group differs from rest of alkaline earth metals due to its small atomic size (0.90 Å) and comparitively high electronegativity (1.5). The main points of difference are :

1. Hardness. Beryllium is the hardest of all the elements of its group.

2. Melting and boiling points. The melting and boiling points of beryllium are the highest.

3. Formation of covalent compounds. Beryllium has a tendency to form covalent compounds. Thus, when it reacts with another element, the electronegativity difference is not so large and the bond is, therefore, covalent.

4. Reaction with water. Beryllium does not react with water even at high temperatures. Other alkaline earth metals decompose water liberating H_2 gas.

$$Mg + H_2O \longrightarrow MgO + H_2$$

5. Reaction with hydrogen. Beryllium does not react with hydrogen directly to form its hydride. Its hydride, however, has been prepared indirectly. The rest of the alkaline earth metals combine with hydrogen to form hydrides. The hydrides of Be and Mg are covalent, whereas the hydrides of other metals are ionic.

6. Reaction with alkalies. Beryllium reacts with alkalies to form hydrogen.

$$Be + 2NaOH \longrightarrow \underset{\text{Sodium beryllate}}{Na_2BeO_2} + H_2 \uparrow$$

Other alkaline earth metals do not react with alkalies.

7. Behaviour of oxides and hydroxides. The oxides and hydroxides of beryllium are amphoteric, *i.e.*, dissolve in both acids and alkalies to form salts.

$$BeO + H_2SO_4 \longrightarrow BeSO_4 + H_2O$$

The oxides and hydroxides of other alkaline earth metals are basic.

8. Behaviour of carbides. Beryllium carbide is decomposed by water to form methane (CH_4).

$$Be_2C + 4H_2O \longrightarrow 2Be(OH)_2 + CH_4$$

The carbides of other alkaline earth metals are decomposed by water to form acetylene (C_2H_2). For example :

$$CaC_2 + 2H_2O \longrightarrow Ca(OH)_2 + C_2H_2$$

9. Behaviour of nitrides. Be_3N_2 is volatile while the nitrides of other alkaline earth metals are non-volatile.

10. Number of molecules of water of crystallisation. The salts of Be^{2+} ion cannot have more than four molecules of water of crystallisation while other alkaline earth metals have more than four molecules of water of crystallisation. This is explained as follows. In case of Be^{2+} ion there are only four orbitals (namely one 2*s* and three 2*p* orbitals) available in its outermost shell. These four orbitals can accept lone pairs of electrons denoted by *O*-atom on each of the water molecules as shown below :

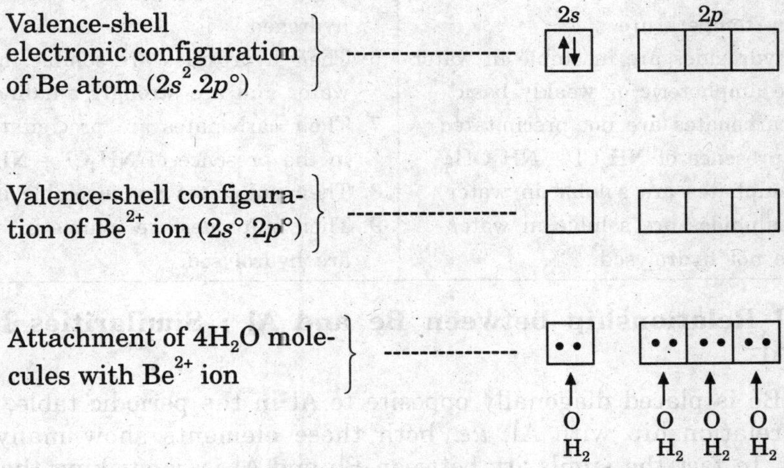

On the other hand, other alkaline earth metals like Mg can extend their co-ordination number to six by using one 3*s*, three 3*p* and two 3*d* orbitals belonging to their outermost shell.

11. Formation of complex compounds. Be^{2+} ion, on account of its small size, forms many stable complex compounds like $[BeF_3]^-$, $[BeF_4]^{2-}$ while M^{2+} ions derived from other alkaline earth metals form very few complex compounds.

Comparison of Be and Mg with Other Alkaline Earth Metals (Ca, Sr, and Ba)

Beryllium and magnesium, although placed in the same group, differ from rest of the alkaline earth metals in a number of properties. Their similarities and dissimilarities with each other are given below.

Similarities

(*i*) Both Be, Mg and Ca, Sr, Ba have two electrons in their outermost shells.

(*ii*) Both have valency equal to two.

(*iii*) Both react with halogens to form halides of the general formula, MX_2.

(*iv*) Both combine directly with nitrogen to form nitrides.

(*v*) Both react with dilute acids to form hydrogen.

Dissimilarities

Be, Mg	Ca, Sr, Ba
1. They are hard metals, Be is the hardest.	1. They are soft metals, Ba is the softest.
2. They do not give any colour to the flame.	2. They give characteristic colours to the flame.
3. They do not react with hydrogen.	3. They react with hydrogen to form hydrides.
4. They are not affected by dry air.	4. They are affected by dry air to form oxides.
5. They do not react with water at ordinary temperature.	5. They react with water to form hydrogen.
6. Their hydroxides are insoluble in water and are amphoteric or weakly basic.	6. Their hydroxides are soluble in water and are strongly alkaline.
7. Their carbonates are not precipitated in the presence of $NH_4Cl + NH_4OH$.	7. Their carbonates are precipitated in the presence of $NH_4Cl + NH_4OH$.
8. Their sulphates are soluble in water.	8. Their sulphates are insoluble in water.
9. Their sulphides are soluble in water and are not hydrolysed.	9. Their sulphides are soluble in water and are hydrolysed.

Diagonal Relationship between Be and Al : Similarities between Be and Al

Since Be is placed diagonally opposite to Al in the periodic table, it shows diagonal relationship with Al, *i.e.* both these elements show many similar properties. In fact the similarity between Be and Al is so striking that Be—Al pair is considered to be an ideal diagonal pair in the periodic table and hence the similarity between Be and Al is the best example of diagonal relationship. Their diagonal relationship is attributed because of the following reasons :

(*i*) Polarising powers (*i.e.* ionic potentials), which are equal to charge/radius ratio, of Be^{2+} (= 6.45) and Al^{3+} (= 6.00) are of similar magnitude as shown below :

Ion :	Be^{2+}	Al^{3+}
Charge on ion :	+ 2	+ 3
Ionic radius (*A*°) :	0.31	0.50
Polarising power (*i.e. charge/radius ratio*) :	2/0.31 = 6.45	3/0.50 = 6.0

(*ii*) The standard oxidation potential of Be(= + 1.70 volts) for the electrode reaction :

$$Be \longrightarrow Be^{2+} + 2e^-, \; E^\circ_{ox} = +\,1.70 \text{ volts}$$

is very close to that of Al (= + 1.67 volts) for the electrode reaction :

$$Al \longrightarrow Al^{3+} + 3e^-, E^\circ_{ox} = + 1.67 \text{ volts}.$$

(*iii*) The electronegativity values of both the elements are the same (Be = 1.5, Al = 1.5)

(*iv*) Both the metals are weakly electropositive.

The following points show that Be and Al have many similar properties, *i.e.* they show diagonal relationship.

1. Ionic radii. The ionic radius of Be^{2+} (= 0.31 A°) is comparable with that of Al^{3+} (= 0.50 A°).

2. Extraction. Both the metals are obtained by the electrolysis of their fused salts in the presence of fluoride.

3. Heat of vaporisation. Heats of vaporisation of both the elements at their boiling points are comparable (Be = 73.9 k.cal/g.atom, Al = 67.9 k.cal/g.atom).

4. Occurrence. In nature both the elements occur together in beryl, $3BeO.Al_2O_3.6SiO_2$.

5. Formation of polymerised covalent hydrides. Both the elements form the polymerised hydrides which are covalent in nature.

6. Formation of covalent compounds. Both Be and Al have small size and higher charge density and as such they have a strong tendency to form *covalent compounds.* Thus the compounds of both the elements have lower melting points, are soluble in organic solvents and get hydrolysed by water.

7. Action of HNO_3 (Passivity). Both Be and Al are rendered passive by HNO_3 due to the formation of an inert and impervious film of their oxide on the surface of the metals.

8. Action of alkalies. Both the metals dissolve in boiling alkalies liberating H_2 and forming *beryllates* and *aluminates* respectively, *e.g.*

$$Be + 2NaOH \longrightarrow \underset{\text{Sod. beryllate}}{Na_2BeO_2} + H_2 \uparrow$$

$$Al + 2NaOH + 2H_2O \longrightarrow \underset{\text{Sod. aluminate}}{2NaAlO_2} + 3H_2 \uparrow$$

9. Properties of oxides and hydroxides. Both the metals burn in O_2 and form oxides. The oxides and hydroxides of both the metals are *amphoteric in nature*, *i.e.* they are soluble in acids and alkalies as shown below :

$$BeO + 2HCl \longrightarrow BeCl_2 + H_2O$$

$$Al_2O_3 + 6HCl \longrightarrow AlCl_3 + 3H_2O$$

$$Be(OH)_2 + 2HCl \longrightarrow BeCl_3 + 2H_2O$$

$$Al(OH)_3 + 2HCl \longrightarrow AlCl_3 + 3H_2O$$

$$BeO + 2NaOH \longrightarrow Na_2BeO_2 + 2H_2O$$

$$Al_2O_3 + 2NaOH \longrightarrow 2NaAlO_2 + H_2O$$

$$Be(OH)_2 + 2NaOH \longrightarrow Na_2BeO_2 + 2H_2O$$

$$Al(OH)_3 + NaOH \longrightarrow NaAlO_2 + 2H_2O$$

Oxides of both the metals are extremely hard, non-volatile and have *very high melting point.*

10. Halildes. Both the metals react with halogens and form halides.

$$Be + Cl_2 \longrightarrow BeCl_2 \; ; \; 2Al + 3Cl_2 \text{ —} AlCl_3$$

These halides have low melting points and are hydrolysed by water. The anhydrous halides (excepting fluorides) are covalent in nature and as such they dissolve in organic solvents and fume in moist air. Their halides also behave as Lewis acids.

11. Formation of carbides. Both the metals form carbides (Be_2C and Al_4C_3) which evolve methane on hydrolysis.

$$Be_2C + 4H_2O \longrightarrow CH_4 + 2Be(OH)_2$$

$$Al_4C_3 + 12H_2O \longrightarrow 3CH_4 + 4Al(OH)_3$$

12. Formation of complexes. Be^{2+} and Al^{3+} ions have strong tendency to form chelate type complexes. For example Be^{2+} gives tetrahedral complexes like BeF_4^{2-} and $[Be(C_2O_4)_2]^{2-}$ while Al^{3+} forms octahedral complexes like AlF_6^{3-} and $[Al(C_2O_4)_3]^{3-}$

13. Properties of salts. Salts of both the metals like nitrates, sulphates and chlorides crystallise from aqueous solutions as hydrates. They get easily hydrolysed in solution. The extent of hydrolysis of the salts of both the metals is almost comparable.

The carbonates of both the metals are unstable, except in an atmosphere of CO_2.

14. Formation of hydrated ions. Both the ions give hydrated ions in aqueous solution like $[Be(H_2O)_4]^{2+}$ and $[Al(H_2O)_6]^{3+}$.

15. Formation of bridged chlorides. Both the metals form bridged chlorides in vapour form. For example :

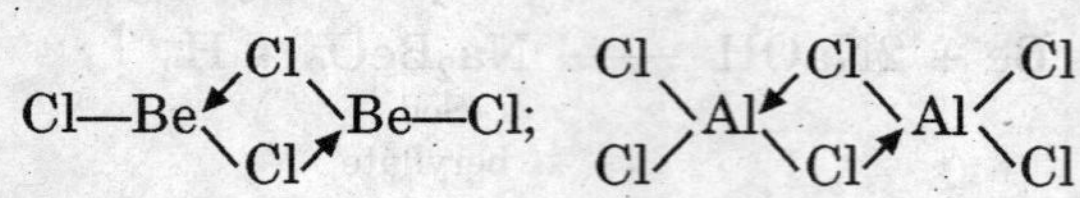

16. Formation of nitrides. Both the metals combine with N_2 and form nitrides. These nitrides are decomposed by water to form NH_3.

$$3Be + N_2 \longrightarrow Be_3N_2$$

$$2Al + N_2 \longrightarrow 2AlN$$

$$Be_3N_2 + 6H_2O \longrightarrow 3Be(OH)_2 + 2NH_3$$

$$AlN + 3H_2O \longrightarrow Al(OH)_3 + NH_3$$

Comparison of Alkaline Earth Metals with Alkali Metals

(1) The metals calcium, barium and strontium have been isolated from their stable salts by similar electrolytic methods which are applied to the isolation of alkali metals. Beryllium and magnesium are also prepared by the electrolysis of their chlorides.

(2) These metals are, likewise, soft with silvery lustre and very light.

(3) They also *react with water* or *oxygen* with readiness and show much activity towards other substances.

(4) Magnesium and beryllium are comparatively *harder* and *more stable* metals, but in absorbing nitrogen, magnesium shows an analogy to lithium. They possess higher density and higher melting and boiling points.

(5) The *electropositive nature* continuously increases from beryllium to radium in the second group, as in the first group from lithium to caesium.

(6) The *oxides* of alkaline earth metals have less affinity towards water when compared to the oxides of alkali metals but they also dissolve in water to a fair extent (though much less when compared to Na_2O) with the evolution of a considerable amount of heat, and the solution obtained is alkaline. Their *hydroxides*, like alkali hydroxides, are good absorbents for carbon dioxide forming carbonates. But there is a difference between alkali *carbonates* and alkaline earth carbonates. With the exception of lithium carbonate, other alkali carbonates dissolve fairly in water and produce mild alkali solutions. But the carbonates of magnesium, calcium, strontium and barium are only sparingly soluble and come down as precipitates whenever they are formed in solution.

(7) Another contrast is in the behaviour of *bicarbonates*. Sodium and potassium bicarbonates are comparatively less soluble than their carbonates, but, on the contrary, magnesium and calcium bicarbonates are more soluble than the carbonates, and therefore, a fresh precipitate of magnesium or calcium carbonate can be redissolved by passing an excess of carbon dioxide.

(8) Alkali *sulphates* and *phosphates* are also soluble in water but calcium, strontium and barium sulphates and phosphates are much less soluble, and so is the case with their oxalates.

(9) In other salts as chlorides, acetates, nitrates, cyanides etc., the two series of salts are similar. But all these salts are not so important and not so extensively used as the alkali salts.

Magnesium Acts as a Bridge Element Between Groups II A and II B Elements

The sub-group A of the second group of periodic classification consists of the elements Be, Mg, Ca, Sr, Ba and Ra, whereas sub-group B consists of Zn, Cd and Hg. The branching of the two sub-groups occurs at magnesium as shown below :

$$\text{Be—Mg}\begin{cases}\text{Ca—Sr—Ba—Ra}\ldots\text{Group IIA}\\ \text{—Zn—Cd—Hg}\ \ldots\text{Group IIB}\end{cases}$$

As we have said in the case of the first group, the metals of the sub-group B are the transitional ones of a long series, and the first metal in the second group of this transition is zinc and not magnesium, and necessary therefore, magnesium does not belong to sub-group B. It is a member of the main group, but being at the position where the branches of two sub-groups just begin, it shows similarity with calcium as well as with zinc. In fact it *resembles zinc more than it resembles calcium* (*See below*). It is also seen that the electropositive character of the elements increases from magnesium to radium whereas decreases gradually from magnesium to mercury. In other words, it gradually decreases from radium to mercury via magnesium. Magnesium thus links the two sub-groups and is therefore, regarded as a bridge element between them.

A. Magnesium resembles calcium. Both the metals resemble in the following respects :

(1) Both of them have similar electronic configuration, *i.e.* they have two electrons in their ultimate shell and eight electrons in their penultimate shell.

$$Mg_{12} \longrightarrow 1s^2, 2s^2p^4, 3s^2$$
$$Ca_{20} \longrightarrow 1s^2, 2s^2p^4, 3s^2p^6, 4s^2$$

The metals are, therefore, highly electropositive and bivalent.

(2) Both are light silvery and white light metals which are mallable and ductile.

(3) Both of them occur as carbonates and sulphates in nature.

(4) Both form a number of deliquescent salts.

(5) When these metals are heated in a current of nitrogen, both form nitrides (Mg_3N_2 and Ca_3N_2).

(6) Both liberate H_2 when treated with very dil. HNO_3.

(7) Sulphides of both the metals are soluble in water and are not precipitated by passing H_2S in their salt solutions.

(8) Nitrates of both give oxide when heated.

(9) The oxides of both the metals (MgO and CaO) are basic in nature and stable towards heat.

B. Magnesium resembles zinc. Both of these metals resemble in the following respects :

(1) Both are malleable and ductile metals.

(2) Both Mg and Zn *slowly* tranish in air while Ca *quickly* tranishes in air.

(3) Both have the same electronic configuration, *i.e.* they have two electrons in their outermost shell.

$$Mg_{12} \longrightarrow 1s^2, 2s^2p^6, 3s^2$$
$$Zn_{20} \longrightarrow 1s^2, 2s^2p^6, 3s^2p^6d^{10}, 4s^2$$

These metals are, therefore, highly electropositive and bivalent.

(4) Both have *no action on cold water*. They, however, react with steam to give metal oxides and H_2. Calcium, however, *decomposes cold water*.

(5) Salts of both the metals are *readily hydrolysed*.

$$MgCl_2 + H_2O \longrightarrow \underset{\text{Mag. oxychloride}}{Mg(OH)Cl} + HCl$$

$$ZnCl_2 + H_2O \longrightarrow \underset{\text{Zinc oxychloride}}{Zn(OH)Cl} + HCl$$

Calcium salts are *quite stable towards water*.

(6) Their sulphates are isomorphous (*e.g.* $MgSO_4.7H_2O$ and $ZnSO_4.7H_2O$). These sulphates are *readily soluble in water and decompose on strong heating.*

$$2MgSO_4 \longrightarrow 2MgO + 2SO_2 + O_2$$
$$2ZnSO_4 \longrightarrow 2ZnO + 2SO_2 + O_2$$

The sulphates also form double salts of the type

$$K_2SO_4.MgSO_4.6H_2O \text{ and } K_2SO_4ZnSO_2.6H_2O.$$

The sulphate of calcium. $C_2SO_4.2H_2O$ is *insoluble in water* and gives the anhydrous sulphate, C_2SO_4 on heating at 400°. Moreover, it does not form double salts as $MgSO_4$ and $ZnSO_4$ do.

(7) Both form double phosphates, *e.g.* $Mg(NH_4)PO_4$ and $Zn(NH_4)PO_4$. These are crystalline compounds, insoluble in water and give pyrophosphates on heating.

$$2Mg(NH_4)PO_4 \longrightarrow Mg_2P_2O_7 + H_2O + 2NH_3$$
$$2Zn(NH_4)PO_4 \longrightarrow Zn_2P_2O_7 + H_2O + 2NH_3$$

Calcium does not form such double phosphates.

(8) Their salts do not impart any characteristic colour to the flame while calcium salts give brick red colour to the flame.

(9) They do net form hydrides with hydrogen while calcium unites directly with hydrogen to form a salt-like hydride, CaH_2.

Questions with Answers

Q.1 Explain the following giving appropriate reasons for your answer.

***(i)* The solubility of M_2^+ CO_3 carbonates (M = alkali metals) increases while that of $M^{2+}CO_3$ carbonates (M = alkaline earth metals) decreases on moving down the group.**

Ans : The variation in solubility of M_2CO_3 and MCO_3 carbonates on moving down the group can be explained on the basis of lattice energies of carbonates and hydration energies of M^+/M^{2+} ions.

In case of $M^+(CO_3)_2$ carbonates since the lattice energies of carbonates decrease down the group, the solubility of carbonates increases down the group. The decrease in lattic energies is due to the increase in the size of M^+ ions from Li^+ to Cs^+ down the group.

In case of $M^{2+}CO_3$ carbonates, since the size of CO_3^{2-} ion is much larger than that of M^{2+} ion, the lattice energies of $M^{2+}CO_3$ carbonates remain almost the same on moving down the group. Hence the decrease in solubility of $M^{2+}CO_3$ carbonates cannot be explained on the basis of lattice energies of $M^{2+}CO_3$ carbonates. The decrease in solubility of these carbonates down the group is due to the decrease of hydration energies of M^{2+} ions from Be^{2+} to Ba^{2+} ions (Be^{2+} = 2455, Mg^{2+} = 1920, Ca^{2+} = 1620, Sr^{2+} = 1480 and Ba^{2+} = 1360 KJ mol^{-1}). Hydration energies of M^{2+} ions decrease down the group because of the gradual increase in the size of M^{2+} ions on moving down the group.

***(ii)* $Be(OH)_2$ is insoluble but $Ba(OH)_2$ is fairly soluble in water.**

(Kurukshetra 1995)

Ans : We know that, due to the increase in the size of M^{2+} ions on moving down the group from Be^{2+} to Ba^{2+}, lattice energies of hydroxides decrease from $Be(OH)_2$ to Ba $(OH)_2$. Thus Be $(OH)_2$ has maximum value of lattic energy and $Ba(OH)_2$ has the minimum value. Since Be $(OH)_2$ has maximum value of lattice energy, this hydroxide is insoluble. On the other hand, since Ba $(OH)_2$ has minimum value of lattice energy, this hydroxide is fairly soluble.

The answer of this question can also be given on the basis of covalent/ionic character of the hydroxides. Since Be^{2+} is smallest in size, $Be(OH)_2$ is primarily covalent and hence is insoluble. On the other hand, since Ba^{2+} is the largest in size, $Ba(OH)_2$ has least covalent character and hence is fairly soluble in water.

***(iii)* Ba $(OH)_2$ is highly soluble in water $BaSO_4$ is almost insoluble.**

Ans : The size of the SO_4^{2-} ion in $BaSO_4$ is quite big as compared to that of OH^- ion in $Ba(OH)_2$. The SO_4^{2-} ion has masked the Ba^{2+} ion in $BaSO_4$ to a large extent with the result that the cation has a very little tendcy to get hydrated. On the other hand, the OH^- ions due to smaller size masks the Ba^{2+} ions to lesser extent which means that hydration energy relased when $Ba(OH)_2$ dissolves in water is quite large. Therefore Ba $(OH)_2$ readily dissolves in water while $BaSO_4$ is almost insoluble.

***(iv)* BeO is insolube in water but $BeSO_4$ is soluble in water.**

Ans : The lattice energy BeO is higher as compared to $BeSO_4$ because the size of O^{2-} ion is very small while SO_4^{2-} ion has bigger size. Since high lattice energy opposes the solubility of a substance in water, BeO is almost insoluble while $BeSO_4$ is isoluble in water.

***(v)* Na_2SO_4 is water soluble while $BaSO_4$ is not.**

Ans : Lattice energy of Na_2SO_4 is less compared with that of $BaSO_4$ because of smaller charge density on Na^+ ion as compared to Ba^{2+} ions. Therefore, hydration energy released easily overcomes the lattice energy when Na_2SO_4 dissolves in water. But this is not possible in case of $BaSO_4$ when tried to be dissolved in water. Therefore, Na_2SO_4 is readily soluble in water while $BaSO_4$ in almost insoluble.

***(vi)* The hydroxides and carbonates of sodium and potassium are easily soluble in water while the corresponding salts of magnesium and calcium are sparingly soluble.**

Ans : The solubility of a particular salt in water is inversely proportional to the lattice energy. Now, the sizes of both Mg^{2+} and Ca^{2+} ions are smaller as compared to Na^+ and K^+ ions. This means that there is a stronger metallic bonding. As a result, the lattice energies of the compounds containing these ions are more and they are expected to be less soluble in water compared with the corresponding compounds of sodium and potassium.

***(vii)* $BaSO_4$ is insoluble where as $BeSO_4$ is soluble in water**

Ans : The lattice energy of $BaSO_4$ is much more than its hydration energy and hence it is insoluble in water. But hydration energy of $BeSO_4$ is much higher than its lattice energy because of small size of Be^{2+} ions. Therefore, it is highly soluble in water

(*viii*) BaO is soluble but $BaSO_4$ in insoluble in water.

Ans : Both BaO and $BaSO_4$ are ionic compounds. However, the size of O^{2-} ion is much smaller than that of the SO_4^{2-} ion. *Since a bigger anion stabilizes a bigger cation more than a smaller anion stabilizes a bigger cation*, the lattice energy of BaO is much smaller than that of $BaSO_4$ and hence BaO is soluble while $BaSO_4$ is insoluble in water.

(*ix*) LiI is more soluble than KI in ethanol.

Ans : Li^+ is much smaller than K^+ ion. Therefore, according to Fajans rule, Li^+ ion can polarize bigger I^- ion to a greater extent than K^+ ion. As a result, LiI is more covalent than KI and hence is more soluble in organic solvents like enthanol.

Q.2 Be and Mg donot give characteristic colour to the flame while other alkaline earth metals do give. Explain why.

(Punjab 1996 ; M.D. Rohtak 1997 ; Delhi 1997)

Ans : When alkali earth metals or their salts are heated in a flame, they give a characteristic colour to the flame as shown below :

Ca	Sr	Ba	Ra
Brick red	*Crison*	*Apple green*	*Crimson*

Reason. When alkaline earth metals or their salts are heated in a flame, the electrons present in the outer most shell absorb energy of the flame. Since ionisation energies of Ca, Sr, Ba and Ra are comparatively low, the energy absorbd by the electrons is suficient to excite the electrons to higher energy levels. When these excited electros come back to the ground state levels, they emit the absorbed energy in the form of radiations having particular frequency or wavelength. It is due to the emission of energy that Ca, Sr, Ba and Ra give colour in the flame.

Since Be and Mg atoms are smaller in size, they have high values of ionisation energies and hence their electrons are bound to the nucleus very strongly. They, therefore, need greater amount of energy to get excited to higher enrgy levels. Now since the energy absorbed by the electrons from the flame is not sufficient to excite the electrons to go to higher energy levels, Be and Mg donot impart any colour to the flame.

Q.3 IE_1 values of alkali metals are lower than those of the corresponding alkaline earth metals but IE_2 values of alkali metals are higher than those of the corresponding alkaline earth metals. Explain why ?

Ans : This can be explained by considering IE_1 and IE_2 values of Na (alkali metal) and Mg (alkaline earth metal) as given below :

Since Na atom is larger in size than Mg atom (Na = 186 pm, Mg = 160 pm), Na atom has lower value of IE_1 than Mg atom.

In case of alkali metals (*i.e.* Na) the second electron is to be removed from the monovalent cation (Na^+) which has already acquired a noble gas configuration *viz.* $1s^2$, $2s^2$ $2p^6$ which is a stable configuration. On the other hand, in alkaline earth metals (*i.e.* Mg), the second electron is to be removed from a monovalent cation (Mg^+) which still has one electron in the outer most shell (Mg^+ = $1s^2$, $2s^2$,

$2p^6$, $3s^1$). Thus the second electron in Mg can be removed easily and hence IE_2 of Mg is lower than that of Na.

$$\underset{1s^2,\ 2s^2 2p^6,\ 3s^1}{Na} \xrightarrow[IE_1 = 496\ KJ/mol]{-3s^1} \underset{1s^2,\ 2s^2 2p^6}{Na^+}$$

(Noble gas configuration : stable configuration)

$$\xrightarrow[IE_2 = 4562\ KJ/mol]{-2p^1} Na^{2+}$$

$$\underset{1s^2,\ 2s^2 2p^6,\ 3s^2}{Mg} \xrightarrow[IE_1 = 737\ KJ/mol]{-3s^1} \underset{1s^2,\ 2s^2 2p^6,\ 3s^1}{Mg^+}$$

(Unstable configuration)

$$\xrightarrow[IE_2 = 1450\ KJ/mol]{-3s^1} \underset{1s^2,\ 2s^2 2p^6}{Mg^{2+}}$$

IE_3 of Mg will be very high because the electron is to be removed from the stable noble gas configurtion of Mg^{2+}.

Q.4 Why do alkaline earth metals form bivalent ions (M^{2+}) through their second ionisation energy (IE_2) is much higher than their first ionisation energy (IE_1). (M.D. Rohtak 1997)

Ans : Since the values of IE_1 for alkaline earth metal atoms are much lower than those of IE_2, these metals should prefer to form monovalent ions (M^+) and not bivalent ions (M^{2+}). However, in practice, these metals form bivalent ions like Mg^{2+}, Ca^{2+} etc. Greater stability of M^{2+} ions as compared to that of M^+ ions can be explained as follows :

(*i*) Greater stability of M^{2+} ions as compared to M^+ ions can be explained on the basis of electronic configuration of these ions. We know that the E.C. of the atoms of alkaline earth metals (excepting Be) is 2, 8,... 8, 2. Thus E.C. of M^+ and M^{2+} ions is 2, 8, ...8, 1 and 2, 8, ...8 respectively. These configurations show that M^+ ions have one electron in their outer most shell while M^{2+} ions have 8 electrons (noble gas configuration) in their outer most shell. Since 8-electron configuration (noble gas configuration) of M^{2+} ions is more stable than 1-electron configuration of M^+ ions, M^{2+} ions are more stable than M^+ ions. Consequently alkaline earth metals prefer to form M^{2+} ions rather than M^+ ions.

***(ii)* In the solid state,** the lattices (ionic solids) formed by M^{2+} ions are stronger (more stable) than those formed by M^+ ions. This is due to the fact that lattice energy released in the formation of lattices of M^{2+} ions is higher than that which is released in the formation of lattices of M^+ ions. Higher value of lattice energy released in the formation of lattices of M^{2+} ions is because of the fact that M^{2+} ions are smaller in size than M^+ ions (Smaller is the size of the cation, higher is the value of lattice energy of its lattice). This higher value of lattice energy of lattices of M^{2+} ions conterbalances the high value of IE_2. Thus it is *due to the higher values of lattice energy of the lattices of M^{2+} ions that M^{2+} ions have higher stability than M^+ ions.* Thus the alkaline earth metal atoms (M) prefer to form M^{2+} ions and not M^+ ions.

***(iii)* In aqueous solution,** it has been observed that the amount of energy releasd when M^{2+} ion is dissolved in H_2O (hydration energy) is much more than that which is released when M^+ ion is dissolved in H_2O, since M^{2+} ion is smaller in size than M^+ ion. This large value of hydration energy of M^{2+} ions counterbalances the high value of IE_2 required for removing two electrons from M atom to form M^{2+} ions. Thus we see that *it is due to higher value of hydration energy of M^{2+} ions that these ions (i.e. M^{2+} ions) have higher satbility than M^+ ions in aqueous solution.*. Consequently alkaline earth metal atoms (M) prefer to form M^{2+} ions and not M^+ ions in aqueous solution.

Q.5 Calcium forms CaF_2, but neither CaF nor CaF_3. Explain why ?

(M.D. Rohtak 1994)

Ans : Electronic configuration of Ca (Z = 20), is 2, 8, $3s^2$, $3p^6$, $4s^2$. In CaF, CaF_2 and CaF_3, Ca atom is present as Ca^+, Ca^{2+} and Ca^{3+} ion respectively. These ions are obtained by removing one electron from 4*s* orbital, both electrons from 4*s* orbital and two electrons from 4*s* orbital and one electron from 3*p* orbitals. The energies required for the removal of these electrons are equal to IE_1 IE_2 and IE_3 of Ca atom respectively. Value of IE_1 is much smaller than that of IE_2 and the value of IE_3 is very large ($IE_3 > IE_2 >> IE_1$).

Now since the value of IE_1 is much smaller than that of IE_2, Ca should prefer to form CaF (Ca is present as Ca^+ in CaF) and not CaF_2 (Ca is present as Ca^{2+} in CaF_2). But Ca prefers to form CaF_2 instead of CaF. The preference for the formation of CaF_2 is due to the fact that CaF_2 has much more lattice energy than CaF has. Thus CaF_2 is more stable than CaF.

Formation of CaF_3 (Ca is present as Ca^{3+} in CaF_3) is not possible because the removal of three electrons from Ca atom to form Ca^{3+} ion requires a large amount of energy which is not available in chemical reaction.

Q.6 Explain why in aqueous solution $MgCl_2$ is more readily formed than MgCl.

***or* Explain why in aqueous solution $MgCl_2$ is more stable than MgCl.**

***or* Explain why in aqueous solution $MgCl_2$ exists but MgCl does not.**

Ans : The formation of one mole of MgCl and $MgCl_2$ in aqueous solution can be represented by the following equations :

$$Mg\ (s) + \frac{1}{2}\ Cl_2\ (g) + aq \rightarrow MgCl\ (aq)$$

$$Mg\ (s) + Cl_2\ (g) + aq \rightarrow MgCl_2\ (aq)$$

It has been observed that the values of $\Delta H°_f$ for MgCl (*aq*) and $MgCl_2$ (*aq*) both are negative but this value for $MgCl_2$ (*aq*) is more negative than that for MgCl (*aq*). Hence $MgCl_2$ (*aq*) is more readily fromed in aqueous solution than

MgCl (*aq*). In other words we can say that in aqueous solution $MgCl_2$ is more stable than MgCl or in aqueous solution $MgCl_2$ exists but MgCl dose not exist.

Q.7 Explain the following, giving reasons for your answer.

(*i*) Compounds of alkaline earth metals are more hydrated than those of alkali metals.

or **The crystalline salts of alkaline earth metals contain more water of crystallisation than the corresponding alkali metal salts. (I.I.T. 1997)**

Ans : Since alkaline earth metal ions (M^{2+}) are smaller in size than alkali metal ions (M^+), M^{2+} ions are more extentively hydrated than M^+ ions and hence compounds of M^{2+} ions are more hydrated than those of M^+ ions. For example magenisum chloride and calcium chloride exist as $MgCl_2 . 6H_2O$ and $CaCl_2 . 6H_2O$ while sodium chloride and potassium chloride exist as NaCl and KCl. Among alkali metals, only Li, because of its small size, forms hydrated salts like $LiCl. 2H_2O$.

(*ii*) Mg^{2+} ion is more highly hydrated than Na^+ ion.

Ans : The positive charge density around Mg^{2+} ion is more than that around Na^+ ion due to its smaller size and more number of positive charges. Therefore, Mg^{2+} ion has a greater tendency to attract surrounding H_2O molecules than Na^+ ion. In other words, Mg^{2+} is more highly hydrated than Na^+ ion.

Q.8 How would you explain the following ?

(*i*) Group 2 metals (Mg and Ca) are harder and denser than group 1 metals (Na and K).

Ans : In metals belonging to the group 2, the atomic sizes are less as compared to the metals of the group 1 present in the same period. For example, size of Mg is smaller as compared to Na. This means that the atoms in group 2 metals are more closely packed as compared to those belonging to group 1 and the metallic bonds in them are also stronger. Therefore, the metals of group 2 are denser as compared to metals of group 1.

(*ii*) Halides of beryllium fume in moist air but other alkaline earth metal halides do not.

Ans : $BeCl_2$ being a salt of a weak base, $Be(OH)_2$ and a strong acid, HCl undergoes hydrolysis by water to form HCl which fumes in air. $BaCl_2$, on the other hand, being a salt of a strong base, Ba $(OH)_2$ and strong acid, HCl does not undergo hydrolysis by water to form HCl and hence does not fume in air.

$$BeCl_2 + 2\,H_2O \longrightarrow Be(OH)_2 + 2HCl\ ;$$
$$BaCl_2 + H_2O \not\longrightarrow Ba(OH)_2 + 2HCl$$

(*iii*) Halides of Be dissolve in organic solvents while those of Ba do not.

Ans : Halides of Be are covalent because of high $\Delta_i H$ of Be while those of Ba are ionic due to low $\Delta_i H$ of Ba.

(*iv*) Beryllium halides are polymeric.

Ans : The halides of Be are electron deficient covalent compounds since they have only four electrons in the valence shell. Therefore, to complete their octets, they undergo polymerization to form polymerisd molecule. For example in the solid state, $BeCl_2$ molecule exists as a polymerised molecule whose structure is given at (*a*) of Fig. 20.1. In the vapour state $BeCl_2$ molecule exits as a chloro-bridged dimer, $(BeCl_2)_2$ as shown at (*b*) of Fig. 20.1.

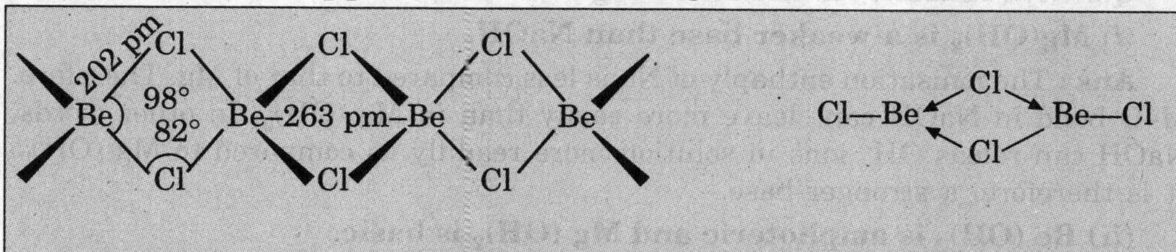

(*a*) Polymerised structure of $BeCl_2$ molecule in the solid state

(*b*) Dimeric structure of $BeCl_2$ molecule in the vapour state.

Fig. 20.1. Structure of $BeCl_2$ molecule in the solid and vapour state.

(*v*) MgO is used as a refractory material.

Ans : MgO has high lattice energy due to greater charge and smaller ionic size of Mg^{2+} and O^{2-} ions. As a result, it has a very high melting point and hence is used as a refractory material.

(*vi*) $MgCl_2$ is more covalent than NaCl.

Ans : Since Mg^{2+} ion has higher charge than Na^+ ion, this ion palarises Cl^- ion more strongly in $MgCl_2$ than Na^+ ion polarises Cl^- ion in NaCl. Hence $MgCl_2$ is more covalent than NaCl.

(*vii*) The chemistry of beryllium is not essentially ionic. Justify the statement by making a reference to the nature of oxide, chloride and fluoride of beryllium.

Ans : The compounds of alkaline earth metals of group 2 are generally of ionic nature but those of beryllium are mostly covalent. Actually Be^{2+} ion due to very small size has a high positive charge density around it. It can easily polarise the O^{2-} ion or F^- or Cl^- ion in BeO, BeF_2 and $BeCl_2$ respectively. The oppositely charged ions mutually netralise their charges to large extent. Therefore, these compounds are almost covalent in nature.

(*viii*) Mg_3N_2 when reacted with H_2O gives NH_3 but $MgCl_2$ does not give HCl with H_2O.

Ans : Since Mg_3N_2 is a salt of strong base [Mg $(OH)_2$] and weak acid (NH_3), it gets hydrolysed by H_2O to give $Mg(OH)_2$ and NH_3

$$Mg_3N_2 + 6\ H_2O \xrightarrow{\text{Hydrolysis}} \underset{\text{Strong base}}{Mg\ (OH)_2} + \underset{\text{Weak acid}}{2\ NH_3}$$

Since $MgCl_2$ is a salt of strong base [$Mg(OH)_2$] and strong acid (HCl), it is not hydrolysed by H_2O to give HCl

$$MgCl_2 + 2\ H_2O \not\longrightarrow \underset{\text{Strong base}}{Mg\ (OH)_2} + \underset{\text{Strong acid}}{2\ HCl}$$

(*ix*) Alkaline earth metals have greater tendency to form complexes than alkali metals.

Ans : Since alkaline earth metal cations (M^{2+}) have smaller size and higher charge than the alkali metal cations (M^+), M^{2+} ions have greater tendency to form complexes.

Q.9 Give reasons for the following :

(*i*) $Mg(OH)_2$ is a weaker base than NaOH.

Ans : The ionisation enthaply of Na is less compared to that of Mg. Therefore, M–O bond in NaOH can cleave more easily than in Mg $(OH)_2$. In other words, NaOH can releas OH^- ions in solution more readilly as compared to Mg $(OH)_2$. It is therefore, a stronger base.

(*ii*) Be $(OH)_2$ is amphoteric and Mg $(OH)_2$ is basic.

Ans : Since IE of Mg < IE of Be, Mg–OH bond in Mg $(OH)_2$ is broken more easily than Be–OH bond in Be $(OH)_2$ to give OH^- ions. Thus Mg $(OH)_2$ gives OH^- ions more easily and hence shows basic character. Be–OH bond in Be $(OH)_2$ is not broken and hence OH^- ions are not obtained, *i.e.* Be $(OH)_2$ shows amphoteric character.

(*iii*) Be $(OH)_2$ dissolves in NaOH but $Mg(OH)_2$ does not.

Ans : Be $(OH)_2$ is amphoteric and hence it dissolves in NaOH forming sodium beryllate (Na_2BeO_2).

$$Be(OH)_2 + 2\,NaOH \rightarrow Na_2BeO_2 + 2\,H_2O$$

$Mg(OH)_2$, on the other hand, being basic does not dissolves in NaOH.

Q.10 The enthalpy of formation of hypothetical CaCl (*s*) is found to be –180 $KJmol^{-1}$ and that of $CaCl_2$ (*s*) is – 800 $KJmol^{-1}$. Calculate ΔH°_f for the disproportation reaction :

$$\mathbf{2\,CaCl\,(s) \rightarrow CaCl_2\,(s) + Ca\,(s)}$$

$$(Ca = H) \qquad (Ca = +2) \qquad (Ca = 0)$$

Ans : ΔH°_f of the given reaction

$= \Delta H^\circ_f$ (Products) $- \Delta H^\circ_f$ (Reactants)

$= [\Delta H^\circ_f (CaCl_2) + \Delta H^\circ f (Ca)] - 2 \times \Delta H^\circ_f (CaCl)$

$= [-800 + 0] -2 \times (-180)$

$= -800 + 360 =$ **-440 $KJmol^{-1}$ (Ans)**

Q.11 Answer the following with appropriate reason.

(*i*) Which are more electropositive ? Alkali metals or alkaline earth metals.

(*ii*) Which one is a covalent hydride ? BeH_2, NaH, CaH_2 ?

(*iii*) Whose aqueous solution is acidic ? NaCl, $BaCl_2$, $BeCl_2$?

(*iv*) Which has the least hardness ? Na, Be, Li, Ca.

(*v*) Which are stronger reducing agents ? Alkali metals or alkaline earth metals.

Ans : (*i*) *Alkali metals are more electropositive than alkaline earth metals.*

Reason. Since alkali metals have lower IE values than alkaline earth metals, these metals can lose their ns^1 electron more easily and hence show more electropositive character than alkaline earth metals.

(*ii*) *BeH_2 is a covalent hydride.*

Reason. Since Be is the first element in group II A, it has the smallest size and hence Be^{2+} ion has high polarising power. Consequently BeH_2 has high covalent character.

(*iii*) *Aqueous solution of* $BeCl_2$ *is acidic.*

Reason. Since $BeCl_2$ is a salt of weak base [$Be(OH)_2$ – amphoteric hydroxide] and strong acid (HCl), the hydrolysis of aqueous solution of this salt gives weak base and strong acid. Due to the formation of strong acid, the aqueous solution of $BeCl_2$ shows acidic property.

$$BeCl_2 + 2\,H_2O \longrightarrow \underset{\text{(Amphotoeric hydroxide)}}{\underset{\text{Weak base}}{Be(OH)_2}} + \underset{\text{Strong acid}}{2\,HCl}$$

NaCl is a salt of strong acid (HCl) and strong base (NaOH). Simmilary $BaCl_2$ is also a salt of strong acid (HCl) and strong base *viz* Ba $(OH)_2$.

(*iv*) *Na metal has the least hardness.*

Reason. We know that, since with the increase in the size of atoms of a metal, the strength of the metallic bond holding the atoms of the metal together decreases, the hardness of the metal also decreases. On this basis, the relation between hardness of the given metals can be written as (*i*) Li > Na (*ii*) Be > Ca and (*iii*) Li < Be or Be > Li. These relations show that Na metal has the least hardness.

(*v*) *Alkali metals are stronger reducing agents than alkaline earth metals.*

Reason. Since E^o_{ox} values for alkali metals are higher than those for alkaline earth metals, alkali metals are stronger reducing agents than alkaline earth metals.

Q.12 Two metals (A) and (B) belong to the same group of the periodic table. Metal (A) froms an insoluble oxide (AO) but soluble sulphate (ASO_4) while metal (B) forms soluble oxide (BO) and insoluble sulphate (BSO_4). The hydroxide of (A) is soluble in NaOH while that of (B) is insoluble in NaOH. Identity the metals (A) and (B).

Ans : The formulae of oxides (AO, BO) and sulphates (ASO_4 and BSO_4) given in the question indicate that (A) and (B) metals belong to group II A. Now we know that among the oxides of alkaline earth metals (group II A), BeO is insoluble and BaO is soluble. Similarly, among the sulphates of alkaline earth metals, $BeSO_4$ is soluble and $BaSO_4$ is insoluble. Being amphoteric in nature,BeO reacts with NaOH to form Na_2BeO_2 while since BaO is strongly basic, it does not react with NaOH.

$BeO + 2\,NaOH \rightarrow Na_2BeO_2 + 2\,H_2O$

$BaO + NaOH \rightarrow$ No reaction

Thus the metals are A = Be and B = Ba.

University Questions

1. Explain the following :

(*i*) Alkaline earth metals form M^{2+} ions but not M^+ ions. (*Raj. 87; Agra 83*)

(*ii*) Although the value of 2nd ionisation potential of alkaline earth metal is higher than that of 1st ionisation potential, yet these metals form M^{2+} ions instead of M^+ ion. (*Gorakhpur 82; Allahabad 89*)

(*iii*) Be^{2+} ion has a much stronger tendency to form complexes than Mg^{2+} ion. (*Meerut 80S*)

(*iv*) Beryllium salts have never more than four molecules of water of crystallisation. (*Meerut 80*)

(*v*) Be and Mg do not import any colour to the flame. (*Mysore 80*)

(*vi*) The solubility of the sulphates of alkaline earth metals decreases whereas the solubility of their hydroxides increases on moving down the group. (*Gorakhpur 82*)

(*vii*) The thermal stability of carbonates of alkaline earth metals increases in atomic numbers. (*Allahabad 87*)

(*viii*) Alkaline earth metal salts import colour to the bunsen flame. (*Delhi 89*)

(*ix*) $MgSO_4$ is more soluble in water than $BaSO_4$. (*Raj. 87*)

(*x*) $BaSO_4$ is insoluble in water whereas $MgSO_4$ is soluble. (*Raj. 79*)

(*xi*) Be^{2+} ion has a much stronger tendency to form complexes than Mg^{2+} ion. (*Meerut 80S*)

(*xii*) CaO is a basic oxide whereas BeO is amphoteric. (*Allahabad 89*)

(*xiii*) The first ionisation energy of Ca is much smaller than the second ionisation energy; but calcium forms only Ca^{2+} ions in solution. (*Allahabad 86*)

(*xiv*) Alkaline earth metals form M^{2+} ions but not M^{3+} ions. (*Meerut 88*)

(*xv*) Basic character of the hydroxides of alkaline earth metals increases on descending the group. (*Delhi 89*)

2. Compare a typical alkali metal with a typical alkaline earth metal with respect to ionisation potential, ionic size, melting points and hardness. Account for the similarities and differences. (*Jodhpur 85*)

3. Compare alkali metals with alkaline earth metals with respect to their electronic configuration, atomic and ionic radii, ionisation potential, oxidation state, polarising power and hydration energies. (*Meerut 84; GND 87*)

4. Explain the following :

(*i*) The ionisation potential values of alkaline earth metals are higher than those of neighbouring alkali metals.

(*ii*) Alkaline earth metals have greater tendency to form complexes than those of alkali metals. (*Delhi B.Sc. Hons. 81*) (*iii*) The metals of group IIA are harder and have higher m. pts than those of group IA elements. (*Raj. 84; Delhi 88, 89*)

(*iv*) The alkaline earth metals are harder than alkali metals.

(*v*) The hydroxide of group IIA metals are weaker than those of group IA metals. (*Raj. 84*)

(*vi*) The carbonates of group IIA metals are less stable to heat than those of group IA metals. (*Raj. 84*)

(*vii*) KOH is stronger base than $Ba(OH)_2$. (*Punjab 85*)

(*viii*) Group IIA elements are smaller in size than group IA elements. (*Delhi 88*)

5. (*a*) Explain why the solubility of sulphates decreases while that of hydroxides increases as the size of cations increases from beryllium to barium in group 2 of the periodic table.

(*b*) Compare and contrast the alkali metals with alkaline earth metals with respect to : (*i*) Electronic configuration and hardness (*ii*) Reducing properties and oxidation potentials (*iii*) Their carbonates and bicarbonates. (*Allahabad 99*)

6. Discuss the anomalous behaviour of Be. [*Gauhauti* (*General*) *2000*]

7. How will you obtain ? (*i*) Beryllium from beryl (*ii*) Radium bromide from pitchblende. (*Lucknow 2001*)

8. Explain why solubilities of alkaline earth metal sulphates decrease down the group. (*Delhi 2002*)

9. Explain the hybridisation and structure of $BeCl_2$ molecule. (*C.S.J.M. Kanpur 2006*)

10. (*a*) Explain why $BeCl_2$ is more covalentthan $MgCl_2$.

(*b*) Explain physical and chemical properties of II A group elements. (*Meerut 2006*)

11. Justify the position of alkaline earth metals in the periodic table. (*Meerut 2008*)

12. Which of the following is the most covalent? $BeCl_2$, $MgCl_2$, $CeCl_2$ and $SrCl_2$. (*Agra 2008*)

CHAPTER

21

Chemistry of Beryllium, Radium and Their compounds and Portland Cement

Beryllium

Occurrence

Beryllium does not occur free in nature. In the combined state this metal is largely found as silicates. Its important minerals are : *beryl*, 3 $BeO.Al_2O_2$. $6SiO_2$ (containing 11–15% of BeO); *chryso beryl*, $BeO.Al_2O_3$ (containing 19% of BeO); and *phenacite*, $2BeO.SiO_2$ (containing 45% of BeO). The chief and the important mineral from which beryllium is extracted is beryl. In India it is found in Bihar, Orissa, Madras and Rajsthan.

Extraction.

Beryllium is obtained from beryl mineral by any of the following two methods:

1. Electrolysis Method. This method consists of the following two steps:

(a) Conversion of beryl into sodium beryllium fluoride, Na_2BeF_4. The finely powdered beryl is mixed with sodium fluosilicate, Na_2SiF_6 and a little sodium fluoride and then fused at a temperature of 800–900°C. As a result, beryllium and aluminium present in the mineral are converted into soluble sodium beryllium fluoride and sodium aluminium fluoride respectively.

$$3BeO.Al_2O_3.6SiO_2 + 6Na_2SiF_6 \xrightarrow[800-900^\circ C]{NaF} \underset{\substack{\text{Sodium}\\\text{beryllium fluoride}\\\text{(soluble)}}}{3Na_2BeF_4} + \underset{\substack{\text{Sodium aluminium}\\\text{fluoride}\\\text{(insoluble)}}}{2Na_2AlF_6} + 3SiF_4 + \underset{\text{Silica}}{9SiO_2} \downarrow$$

The added sodium fluoride reacts with silicon tetrafluoride so formed to give sodium fluosilicate again.

$$2NaF + SiF_4 \longrightarrow \underset{\substack{\text{Sodium}\\\text{fluosilicate}}}{Na_2SiF_6}$$

The fused mass is boiled with water and filtered. The filterate contains sodium beryllium fluoride while the residue contains mostly sodium fluoride and silica.

(b) Electrolysis of sodium beryllium fluoride, Na_2BeF_4. The filtrate is evaporated to dryness and the residue obtained is mixed with sodium fluoride and barium fluoride. The fluorides of sodium and barium are added to increase the conductivity and to decrease the volatility of the fused mass at high temperatures. The mixture is fused at 1400°C and electrolysed in a graphite pot which itself acts as anode and a water cooled iron pipe dipping in the melt acts as cathode. The beryllium is deposited on the cathode. The fluorides of sodium and barium are added at intervals to maintain the concentration of beryllium constant in the electrolyte.

2. Electrolytic Reduction Method. This method also consists of the following two steps:

(*a*) ***Conversion of beryl into beryllium oxide (beryllia), BeO.*** The finely powdered beryl is fused with potassium carbonate at about 1500°C in an electric furnace and then evaporated with conc. H_2SO_4. The resulting mass is extracted with boiled water and filtered. The filtrate contains the sulphates of beryllium, potassium and aluminium while the residue contains mostly silica.

$$3BeO.Al_2O_3.6SiO_2 + 6K_2CO_3 \longrightarrow 3BeO + Al_2O_3 + 6K_2SiO_3 + 6CO_2$$

$$BeO + H_2SO_4 \longrightarrow \underset{\text{(Soluble)}}{BeSO_4} + H_2O$$

$$Al_2O_3 + 3H_2SO_4 \longrightarrow \underset{\text{(Soluble)}}{Al_2(SO_4)_3} + 3H_2O$$

$$K_2SiO_3 + H_2SO_4 \longrightarrow \underset{\text{(Soluble)}}{K_2SO_4} + H_2O + \underset{\text{Residue}}{SiO_2 \downarrow}$$

The filtrate is concentrated when most of the potassium and aluminium sulphates crystallize out as potash alum.

$$K_2SO_4 + Al_2(SO_4)_3 + 24H_2O \longrightarrow \underset{\text{Potash alum}}{K_2SO_4.Al_2(SO_4)_3.24H_2O}$$

The concentrated solution now contains beryllium sulphate along with some potassium and aluminium sulphates. This is added to a hot and concentrated solution of $(NH_4)_2CO_3$ when remaining aluminium is precipitated as its hydroxide.

$$Al_2(SO_4) + 3(NH_4)_2CO_3 + 3H_2O \longrightarrow \underset{\text{ppt.}}{2Al(OH)_3} + 3(NH_4)_2SO_4 + 3CO_2 \uparrow$$

The filterate on concentration deposits basic beryllium carbonate, $Be(OH)_2.BeCO_3$ which on ignition gives beryllium oxide.

$$Be(OH)_2.BeCO_3 \xrightarrow{\Delta} 2BeO + H_2O + CO_2 \uparrow$$

(*b*) ***Reduction of beryllium oxide, BeO to Be metal.*** The beryllium oxide is mixed with powdered coke and introduced into an alternating current arc furnace (Fig. 21.1). The furnace consists of a big carbon crucible provided with a tapping hole and is enclosed in the fire bricks. The crucible acts as one of the electrodes. The other carbon electrode dips in the charge and is held vertically. On striking the electric arc between the electrodes, beryllium oxide is reduced to beryllium which in the molten state is taken away through the tapping hole. CO escapes from the hole.

$$BeO + C \longrightarrow Be + CO \uparrow$$

Physical Properties.

Beryllium is silvery white in colour, malleable, easily forged and takes a high polish. It is a light metal (S.G. = 1.842) and possesses a high melting point (1280°). Its specific heat is probably the highest of all the known metals and this accounts for the non-metallic nature of the metal (*cf. Boron*).

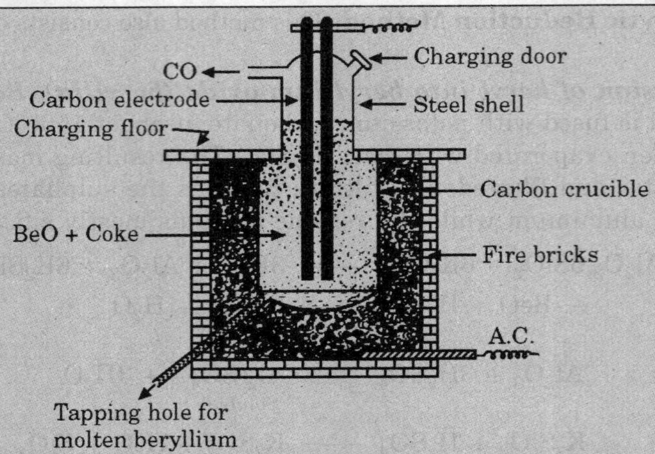

Fig. 21.1. Preparation of beryllium by the electrolytic reduction of beryllium oxide, BeO.

Chemical Properties.

(i) Action of air. The metal is more permanent in the air than magnesium, but is oxidised in the moist air. When heated to redness in oxygen, it forms a protective layer of the oxide round the metal. The metal powder, however, burns in air brilliantly forming the oxide, BeO.

(ii) Action of water. Water or steam is not decomposed even at red heat (compare with magnesium).

(iii) Combination with elements. It does not directly combine with sulphur but unites with halogens (difficultly with iodine, but with incandescence with chlorine and bromine). It unites directly with carbon and boron and gives alloys with metals.

(iv) Action of acids. It is easily acted upon by hydrochloric acid and dilute sulphuric acid, evolving hydrogen. It liberates sulphur dioxide with strong sulphuric acid.

$$Be + 2H_2SO_4 \longrightarrow BeSO_4 + SO_2 + 2H_2O$$

Cold concentrated nitric acid makes it passive while hot acid produces nitrogen peroxide.

$$Be + 4HNO_3 \longrightarrow Be(NO_3)_2 + 2H_2O + 2NO_2$$

(v) Action of alkalies. It is attacked readily by alkalies with the formation of *beryllates* and liberation of hydrogen.

$$Be + 2NaOH \longrightarrow Na_2BeO_2 + H_2$$

Uses

(i) In making alloys. *Beryllium bronzes*, which contain 2.5 to 2.25% of beryllium are hard and can be tempered just like steel. These are being used for making springs and slip-rings for high speed motors. *(ii)* Because of its high transparency to X-rays, it is used for making "windows" of X-ray tubes. *(iii)* It is also used for making electrodes of neon signs. *(iv) It is used as a moderator in nuclear reactors.* This use depends on the fact that beryllium absorbs fewer neutrons than any other known structural material. *(v)* Beryllium also finds special applications in

the fields of developing rockets, missiles and high speed air crafts. (*vi*) Recently beryllium has been used as an ingredient of a new textile fibre of rayon type.

Compounds of Beryllium

1. Basic beryllium acetate, $3Be(CH_3COO)_2.BeO$ *or* $Be_4O(CH_3COO)_6$.

It is a complex compound of beyllium with acetic acid, CH_3COOH. In this complex compound beryllium is *tetra-covalent*.

Preparation. It is prepared by dissolving beryllium hydroxide or carbonate in glacial acetic acid. The resulting solution is diluted with water and then evaporated to dryness. The residue is extracted with chloroform which dissolves basic beryllium acetate which can be obtained from the solution by crystallization.

$$4Be(OH)_2 + 6CH_3COOH \longrightarrow 3Be(CH_3COO)_2.BeO + 7H_2O$$

$$4BeCO_3 + 6CH_3COOH \longrightarrow 3Be(CH_3COO)_2.BeO + 3H_2O + 4CO_2 \uparrow$$

Properties. It is a colourless volatile crystalline solid with m.pt. = 283°C and b.pt. = 330°C, It is insoluble in water but soluble in the organic solvents such as glacial acetic acid, chloroform and ether. It is a covalent compound and is slowly hydrolysed by boiling water. It reacts with mineral acids to form the corresponding beryllium salt and acetic acid.

$$3Be(CH_3COO)_2.BeO + 8HCl \longrightarrow 4BeCl_2 + 6CH_3COOH + H_2O$$

$$3Be(CH_3COO)_2.BeO + 4H_2SO_4 \longrightarrow 4BeSO_4 + 6CH_3COOH + H_2O$$

Uses. (*i*) It is used as a starting source for the preparation of other beryllium compounds required for research work. (*ii*) It is also used as a means for purifying beryllium by solvent extraction.

Structure. X-ray diffraction studies have shown that the compound has a tetrahedral structure. The four beryllium atoms occupy the four corners of a regular tetrahedron and the oxygen atom lies at the centre. The six acetate groups are arranged along the six edges of the tetrahedron as shown in Figure 21.2.

Thus each Be-atom is attached to the central oxygen atom and three acetate-group-oxygen atoms. It may be noted that oxygen is the first short period element and hence its maximum covalency is 4 which is shown in this compound.

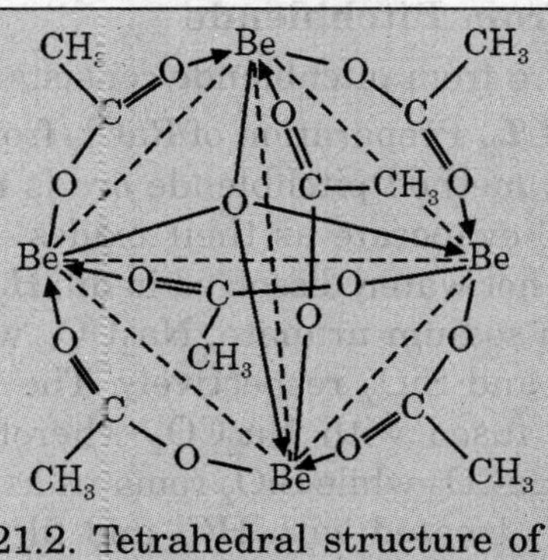

Fig. 21.2. Tetrahedral structure of basic beryllium acetate, $Be_4O\ (CH_3COO)_6$

2. Beryllium chloride, $BeCl_2$.

The *anhydrous* chloride is prepared: (*i*) by heating metallic beryllium in dry Cl_2 or HCl (*ii*) by passing CCl_4 or Cl_2 and SCl_2 vapour over heated BeO (*iii*) by the action of Cl_2 or HCl an heated Be_4C (*iv*) by heating a mixture of BeO and C in a current of Cl_2.

$$BeO + C + Cl_2 \xrightarrow{\Delta} BeCl_2 + CO$$

Properties. It forms snow-white fusible mass which sublimes readily. Melting and boiling points are 405° and 488°C respectively. It fumes strongly in moist air.

$$BeCl_2 + 2H_2O \rightleftarrows Be(OH)_2 + 2HCl$$

The fused salt is a poor conductor of electricity. It dissolves in water, evolving heat and crystallises out on drying in vacuum as tetrahydrate, $BeCl_2.4H_2O$. Anhydrous salt dissolves in alcohol and ether. With ether it forms an addition compound, $BeCl_2.2(C_2H_5)_2O]$. The sublimed chloride forms a tetraammoniate, $BeCl_2.4NH_3$ with NH_3.

Uses. It may be used in organic synthesis in place of anhydrous aluminium chloride though it is less efficient.

Uses. It is used (*i*) in surgery for plastering (*i.e.* setting) the fractured bones. (*ii*) in making casts for stalues, in dentistry, surgical instruments, toys etc. (*iii*) in making black board chalks.

Radium

Occurrence

Radium in minute quantities is found widely distributed throughout the earth's crust. It occurs in traces in almost all rocks and water of seas and oceans. All uranium minerals contain radium as it is a disintegration product in the uranium series. The two chief minerals of radium are: (*i*) *Pitchblende* or *uraninite*, U_3O_8. It contains barium, radium, arsenic, silicon, sulphur etc. This mineral occurs in Congo and Canada. (*ii*) *Carnotite*, $K(UO_2). (VO_4).4H_2O$. It is a potassium vanadate containing high percentage of radium and occurs in Katanga mines of Congo, U.S.A. and Russia.

Extraction of Radium from Pitchblende

The extraction of radium from pitchblende consists of the following two steps:

1. *Preparation of $RaCl_2$*. Preparation of $RaCl_2$ from pitchblende can be done by Curies' method. In this method pitchblende ore is roasted in a reverberatory furnace whereby As and S evaporate as their oxides. The residue is then fused with Na_2CO_3, washed with hot water, dissolved in dil. H_2SO_4 and filtered. Uranium passes into the solution as sodium uranate, Na_2UO_4 while Ra Ba and Si go into residue as $RaSO_4$, $BaSO_4$ and SiO_2 respectively. The residue containing $RaSO_4$ $BaSO_4$ and SiO_2 is again fused with Na_2CO_2 whereby $RaSO_4$ and $BaSO_4$ are converted into $RaCO_3$ and $BaCO_3$ while SiO_2 remains as such. The mass containing the carbonates and SiO_2 is treated with HCl and filtered. The filtrate contains $RaCl_2$ and $BaCl_2$ while SiO_2 is collected as residue. The solution containing $RaCl_2$ and $BaCl_2$ is concentrated and cooled when less soluble $RaCl_2$ crystallises out first, while more soluble $BaCl_2$ remains dissolved in mother liquor. This process of fractional crystallisation is repeated a number of times to get pure $RaCl_2$.

2. *Electrolysis of fused $RaCl_2$.* Radium metal is prepared by the electrolysis of fused $RaCl_2$, using a Pt–Ir anode and Hg cathode. The metal collects as an amalgam which is heated in an atmosphere of H_2 at 400–700°C whereby Hg distils over while radium is left behind as a white metal. The electrolytic cell used is shown in Fig. 21.3.

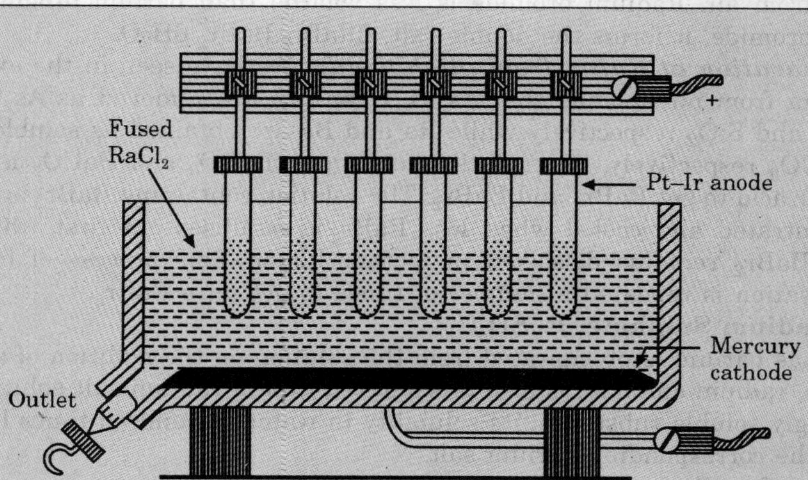

Fig. 21.3. Electrolytic cell used for the extraction of radium by the electrolysis of fused radium chloride.

Properties of Radium

Radium is a white metal which melts at 700°C, volatilises at 960°C and boils at 1140°C. When exposed to air, it blackens due to the formation of a nitride. It is luminescent and its blue luminescence becomes more pronounced when barium is present as impurity in it. It is acted upon by water, when its hydroxide is obtained. It is soluble in hydrochloric acid, forming the chloride. It gives out alpha, beta, and gamma rays. and this property of radio-activity remains unaffected between temperatures 2000°C and –250°C, and under other physical conditions. The radiations affect the photographic plate and discharge the electroscope. Radium salts give a caramine colour to the flame.

Compounds of Radium

Radium closely resembles barium in all its salts. Radium furnishes oxalate, phosphate, ferrocyanide, platinocyanide and other salts very much resembling the corresponding salts of barium. Its important compounds are discussed below:

1. Radium chloride, $RaCl_2.2H_2O$.

It is isomorphous with barium chloride, B $Cl_2.2H_2O$; both of them contain two molecules of water. The salt is prepared by dissolving radium carbonate in hydrochloric acid. When heated to 100° for many hours, it furnishes the anhydrous salt. It is less soluble in water than barium chloride. From the solution it can be precipitated by addition of alcohol, or by strong hydrochloric acid.

2. Radium Bromide, $RaBr_2.2H_2O$.

Like the chloride it also contains two molecules of water and is prepared by dissolving radium carbonate in hydrobromic acid. It is isomorphous with barium bromide, $BaBr_2$. $2H_2O$. When exposed to air, it loses bromine and becomes alkaline after absorbing moisture, and then finally it becomes carbonate after taking carbon dioxide from air. Radium bromide is less volatile than barium bromide. With barium bromide, it forms the double salt, $2RaBr_2.BaBr_2.6H_2O$.

Preparation of $RaBr_2$ from pitchblende. We have seen in the extraction of radium from pitchblende that As, S, U and Si are removed as As_2O_3. SO_2, Na_2UO_4 and SiO_2 respectively while Ra and Ba are obtained as soluble $RaCO_3$ and $BaCO_3$ respectively. The solution containing $RaCO_3$ and $BaCO_3$ is treated with HBr acid to get $RaBr_2$ and $BaBr_2$. The solution containing $RaBr_2$ and $BaBr_2$ is concentrated and cooled when less $RaBr_2$ crystallises out first while more soluble $BaBr_2$ remains dissolved in mother liquor. This process of fractional crystallisation is repeated a number of times to get pure $RaBr_2$.

3. Radium Sulphate, $RaSO_4$.

Just as barium is precipitated from the solution on the addition of a soluble sulphate, radium sulphate is also precipitated from a radium salt solution. It is a sparingly soluble substance; its solubility in water is hundred times less than that of the corresponding barium salt.

Uses of radium salts.

(*i*) In the manufacture of *luminous prints* used for painting clock and watch dials, push buttons etc. The paint consists of a mixture of sulphides of Zn, Cu, Mn etc. containing a trace of radium bromide. (*ii*) In medicine in the treatment of cancer.

Portland Cement

What is Portland Cement ?

Cement is a very important building material which was first of all introduced in 1824 by Joseph Aspdin, a mason of Leeds, England. Be found that when a strongly heated mixture of limestone and clay was mixed with water and allowed to stand, it hardened to a stone like mass which resembled Portland rock which is a famous building stone of England. It is since then that the name *Portland cement* is given to a mixture of lime (obtained from lime stone), silica, iron oxide and alumina. Thus:

Cement is a dirty greenish heavy powder which contains lime in the form of aluminates and silicates. The various silicates and aluminates which constitute more than 90% of the cement are: *Tri-calcium silicate, $3CaO.SiO_2$; Dicalcium silicate, $2CaO.SiO_2$; Tricalcium aluminate, $3CaO.Al_2O_3$; Tetracalcium alumino ferrite, $4CaO.Al_2O_3$. Fe_2O_3.* Out of these, tricalcium silicate is the most important constituent of cement.

What is the Composition of Portland Cement ?

An average composition of a good simple of cement is: lime (CaO)—62%, silica (SiO_2)—22%, alumina (Al_2O_3)—7.5%, magnesia (MgO)—2.5%, iron oxide (Fe_2O_3)—

2.5%, sulphur trioxide (SO_3)—1.5%, sodium oxide (Na_2O)—1.0%, potassium oxide (K_2O)—1.0%. The essential constituents are lime (obtained from *lime stone*), silica and alumina (present in clay). For a good quality of cement the following ratios should be maintained;

$$\frac{\%SiO_2}{\%Al_2O_3} = 2.5 - 4.0, \quad \frac{\%CaO}{\%SiO_2 + \%Al_2O_3 + \%Fe_2O_3} = 1.9 - 2.1.$$

Cement containing no iron is white but hard to burn. If less lime is present than given by the above ratio, the cement is low in strength and sets very soon. If more lime is there, the cement cracks. Excess of silica produces a slow-hardening cement while excess of alumina gives a quick-setting product.

Raw Materials Used in the Manufacture of Cement.

The important raw materials used in the manufacture of cement are.

1. *Lime stone, $CaCO_3$.* It provides lime, CaO. Chalk, marble and alkali waste can also be used, since these also supply CaO.

2. *Clay.* The composition of clay is $Al_2O_3.SiO_2.Fe_2O_3.2H_2O$. Clay supplies silica ($SiO_2$) and alumina ($Al_2O_3$) to the cement. Some of the clays do not contain Fe_2O_3 and the cement obtained in this case is white and hard to burn.

3. *Gypsum, $CaSO_4.2H_2O$.* The addition of gypsum decreases the setting time of cement.

Manufacture of Cement

The manufacture of cement consists of the following steps:

1. *Preparation of raw meal or slurry.* The raw materials *viz.* lime stone and clay are mixed in proper proportion by any of the following methods:

(*a*) *Dry process.* This process is employed when the raw materials *viz.* lime stone and clay are hard. In this process the lime stone is first broken into small pieces. It is then mixed with clay in the proper proportions and finally pulverised to such a fineness that 90–95% passes through a 100 mesh sieve. The mixture is made homogenous to produce what is known as **raw meal.**

(*b*) *Wet process.* This process is used when the raw materials *viz.* lime stone and clay are soft, the climate is fairly damp and the fuel is cheap. In this process, the lime stone is crushed to suitable size and the clay is washed with water in wash mills to remove foreign materials like flint. The powdered limestone is then mixed with the clay paste in the proper proportions (lime stone 75%, clay 25%) and the mixture is finely ground and made homogenous by means of a compressed air mixing arrangement. The resulting paste is known as **slurry.** The slurry contains about 40% water.

2. *Burning (calcination) of raw meal or slurry in rotary kiln: Preparation of cement clinker.* Raw meal or slurry prepared as above is introduced into the rotary kiln with the help of a screw conveyer. The rotary kiln consists of a long cylinder 6 to 8 feet in diameter and 100 to 250 feet in length. It is made of steel and is lined inside with fire bricks (Fig. 21.4). The kiln rotates on its axis at the rate of $\frac{1}{2}$ to 1 revolution per minute. As the kiln rotates, the charge slowly moves downwards due to the rotary motion of the kiln. Now the charge is heated by burning coal dust which enters at the lower end with the help of a blower. As the charge moves forward, it meets higher temperatures. When it reaches the lowest part of the kiln, the temperature is about 1500°C which is the maximum temperature in the kiln. The zone having this temperature is called *maximum temperature zone*. The charge takes 2-3 hours to cover the journey in the kiln.

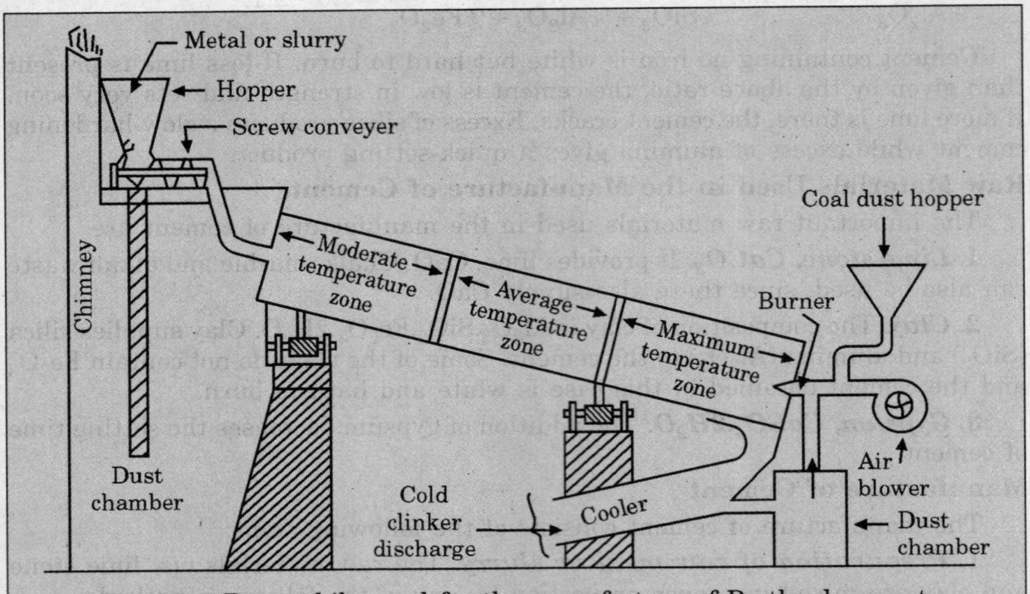

Fig. 21.4. Rotary kiln used for the manufacture of Portland cement.

Reactions Reactions taking place in the rotary kiln can be divided into the following three parts:

(i) Reactions taking place in moderate temperature zone. In this zone the temperature is upto 800°C. In this zone free moisture is removed and clay is broken into Al_2O_3, SiO_2 and Fe_2O_3.

$$\underbrace{Al_2O_3.2SiO_2.Fe_2O_3.2H_2O}_{\text{Clay}} \longrightarrow Al_2O_3 + 2SiO_2 + Fe_2O_3 + 2H_2O$$

(ii) Reactions taking place in average temperature zone. Here the temperature is 800–1000°C. In this zone lime stone ($CaCO_3$) decomposes into lime (CaO) and CO_2

$$\underset{\text{Lime stone}}{CaCO_3} \longrightarrow \underset{\text{Lime}}{CaO} + CO_2 \uparrow$$

(iii) Reactions taking place in maximum temperature zone. In this zone where the temperature is 1000–1500°C, the oxides *viz.* CaO, SiO_2, Al_2O_3 and Fe_2O_3 combine together and form calcium silicates *viz* $2CaO.SiO_2$, $3CaO.SiO_2$; calcium aluminates *viz.* $2CaO.Al_2O_3$; $3CaO.Al_2O_3$ and calcium ferrite *viz.* $4CaO.Al_2O_2.\ Fe_2O_3$.

$$2CaO + SiO_2 \longrightarrow \underset{\text{Dicalcium silicate}}{2CaO.SiO_2}$$

$$3CaO + SiO_2 \longrightarrow \underset{\text{Tricalcium silicate}}{3CaO.SiO_2}$$

$$2CaO + Al_2O_3 \longrightarrow \underset{\text{Dicalcium aluminate}}{2CaO.Al_2O_3}$$

$$2CaO + Al_2O_3 \longrightarrow \underset{\text{Tricalcium aluminate}}{3CaO.Al_2O_3}$$

$$4CaO + Al_2O_3 + Fe_2O_3 \longrightarrow \underset{\text{Tetracalcium alumino ferrite}}{4CaO.Al_2O_3.Fe_2O_3}$$

The resulting product is known as **cement clinker** and as it comes out into the cooler, it has the appearance of small greenish black or grey-coloured balls varying in size from small nuts to peas.

Quite obviously the composition of cement clinker is: $2CaO.SiO_2$, $3CaO.SiO_2$, $2CaO.Al_2O_3$, $3CaO.Al_2O_3$, $4CaO.Al_2O_3.Fe_2O_3$.

3. *Mixing the cement clinker with gypsum.* The cement clinker is cooled in the coolers by cold air. The cooled clinker is mixed with 2–3% of its weight of gypsum ($CaSO_4.2H_2O$).$3CaO.Al_2O_3$ which is a fast setting constituent of the clinker reacts with gypsum to form the crystals of calcium sulpho-aluminate, $3CaO.Al_2O_2.3CaSO_4.2H_2O$.

$$3CaO.Al_2O_3.\underset{\text{Gypsum}}{3(CaSO_4.2H_2O)} + 2H_2O \longrightarrow \underbrace{3CaO.Al_2O_3.3CaSO_4.2H_2O}_{\text{Calcium sulpho-aluminate.}} + 6H_2O$$

Thus we see that the addition of gypsum removes the fast setting $3CaO.Al_2O_3$ as $3CaO.Al_2O_3.3CaSO_4.2H_2O$ and hence the process of setting cement gets retarded and this results in better strength of the mass which sets.

Setting of Cement

The use of cement in the construction of buildings is based on its property of setting to a hard mass when its paste with water is allowed to stand for some time. When the cement is mixed with water, it absorbs water and the mass becomes hard and very resistant to pressure. This is known as the *setting of cement.* The reactions involved in the setting of cement are:

(*i*) *Reactions taking place during first 24 hours.* A short time after the cement is mixed with H_2O, the following reactions take place:

(*a*) $3CaO.Al_2O_3$ absorbs H_2O (hydration) and forms a colloidal gel of the composition, $3CaO.Al_2O_3.6H_2O$

$$3CaO.Al_2O_3 + 6H_2O \xrightarrow{\text{Hydration}} \underset{\text{Hydrated colloidal gel of tricalcium aluminate}}{3CaO.Al_2O_3.6H_2O}$$

The gel of $3CaO.Al_2O_3.6H_2O$ so formed starts crystallising slowly.

(*b*) $3CaO.Al_2O_3$ which is a fast-setting material reacts with gypsum ($CaSO_4.2H_2O$) to form the crystals of calcium sulphoaluminate, $3CaO.Al_2O_3.3CaSO_4.2H_2O$.

$$3CaO.Al_2O_3 + \underset{\text{Gypsum}}{3(CaSO_4.2H_2O)} + 2H_2O \longrightarrow \underbrace{3CaO.Al_2O_3.3CaSO_4.2H_2O}_{\text{Calcium sulpho-aluminate}} + 6H_2O$$

Thus we see that the addition of gypsum removes the fast setting $3CaO.\ Al_2O_3$ as $3CaO.\ Al_2O_3.\ 3CaSO_4.\ 2H_2O$ and hence the process of setting cement gets ratarded and this result in better strength of the mass which sets.

(ii) Reactions taking place between 1 and 7 days. $3CaO.\ SiO_2$ and $3CaO.\ Al_2O_3$ get hydrolysed (hydrolysis) according to the equations:

$$3CaO.SiO_2 + H_2O \xrightarrow{\text{Hydrolysis}} \underset{\text{ppt.}}{Ca(OH)_2} + \underset{\text{(Amorphous)}}{2CaO.SiO_2}$$

$Ca(OH)_2$ formed starts changing into needle-shaped crystals which get studded in the colliodal gel, $2CaO.SiO_2$. formed as above and thus impart strength to it.

$$3CaO.Al_2O_3 + 6H_2O \xrightarrow{\text{Hydrolysis}} \underset{\text{ppt.}}{3Ca(OH)_2} + \underset{\text{ppt.}}{2Al(OH)_3}$$

$Al(OH)_3$ so formed fills the interstices resulting in hardening the mass.

(iii) Reactions taking place between 7 and 28 days. $2CaO.SiO_2$ begins to hydrate (a very slow reaction) and forms the hydrated colloidal gel of the composition, $2CaO.SiO_2.xH_2O$.

$$2CaO.SiO_2 + xH_2O \xrightarrow{\text{Hydration}} \underset{\substack{\text{Hydrated colloidal gel} \\ \text{of dicalcium silicate}}}{2CaO.SiO_2.xH_2O}$$

Needles of $Ca(OH)_2$ formed in (*ii*) above get studded in the hydrated colloidal gel of $2CaO.SiO_2.xH_2O$ and thus impart strength to it.

$4CaO.Al_2O_3.Fe_2O_3$ also gets hydrated to form a colloidal gel of the composition, $3CaO.Al_2O_3.6H_2O$

$$4CaO.Al_2O_3.Fe_2O_3 + 6H_2O \xrightarrow{\text{Hydration}} \underset{\text{Hydrated colloidal gel}}{3CaO.Al_2O_3.6H_2O} + CaO.Fe_2O_3$$

The gel formed as above starts losing water partly by evaporation and partly by forming hydrates with the hydrated constituents. Thus cement sets to a hard mass.

Manufacture of Cement in India

In our country at present there are 56 factories in private as well as in public sectors which are manufacturing cement. Following are the important factories running in the country.

1. Factories in public sector. The important factories of this type are: (*i*) *Cement Corporation of India.* This Corporation has set up the plants at Rajpur

(M.P.), Kurkunta (Mysore) and Bokajan (Assam). (*ii*) *Assam Cement Corporation of India Ltd., Cherpaunji (Assam).* (*iii*) *U.P. State Cement Corporation Ltd., Dalla (U.P.)* (*iv*) *T.N. industrial Development Corporation Ltd., Alangulam (Tamil Nadu).*

2. Factories in private sector. Important factories of this type are: (*i*) *Associated Cement Co. Ltd., Bombay.* It is the biggest manufacturing unit in India having 24 factories located in Madhya Pradesh (6), Gujarat (4), Bihar (3), Andhra Pradesh (3), Rajasthan (3), Mysore (2), Madras (1), Maharashtra (1) and Haryana (1). (*ii*) *Andhra Cement Co. Ltd., Vijayawada (Andhra Pradesh);* (*iii*) *Ashoka Cement Co. Ltd., Dalmianagar (Bihar);* (*iv*) *Bagalkot Udyog Co. Ltd., Bagalkot (Mysore);* (*v*) *Century Cement, Tilda (Madhya Pradesh);* (*vi*) *Dalmia Cement Ltd., Dalmiapuram (Tamil Nadu);* (*vii*) *J.K. Synthetics Ltd., Nimbahera (Rajsthan);* (*viii*) *Jaipur Udyog Ltd., Sawai Madhopur and Beawar (Rajsthan);* (*ix*) *Madras Cement Ltd., (Madras);* (*x*) *Mysore Cement Ltd., Bangalore;* (*xi*) *Orissa Cement Ltd., Sundergarh (Orissa);* (*xii*) *Punyam Cement and Minerals Industries Ltd., Cement Nagar (Andhra Pradesh);* (*xiii*) *Saurashtra Cement and Chemical Industries Ltd., Porbandar (Gujrat);* (*xiv*) *Shri Digvijay Cement Ltd., Digvijaygram, (Gujrat);* (*xv*) *Sone Valley Portland Cement Ltd., Japla (Bihar);* (*xvi*) *Travancore Cement Ltd., Kothayam (Kerala).*

Questions with Answers

Q.1 Explain the following, giving appropriate reason for your answer.

(*i*) Hydrated magnesium chloride cannot be dehydrated by heating.

or Dehydration of hydrated magnesium chloride is done in presence of HCl gas.

Ans : Formula of hydrated magnesium chloride is $MgCl_2 . 6H_2O$. When heated, it gets hydrolysed by its own water of crystallisation to give MgO.

$$MgCl_2 . 6H_2O \xrightarrow{\Delta} MgO + 2HCl + 5H_2O$$

Therefore, the salt can be dehydrated by passing HCl gas through the hydrated salt at about 650 K.

$$\underset{\text{Hydrated}}{MgCl_2 . 6H_2O} \xrightarrow[650K]{HCl\ (g)} \underset{\text{Anhydrous}}{MgCl_2} + 6H_2O$$

(*ii*) It is necessary to add gypsum in the final stages in the preparation of cement.

Ans : Gypsum ($CaSO_4 . 2H_2O$) is added in the final stages of the preparation of cement in order to regulate the process of setting of cement. When water is added to cement, it sets into a hard mass. In the absence of gypsum, the setting will occur immediatly. The process is highly exothermic and as a result, cracks will appear after sometime. It is , therefore, necessary to control or regulate the time of setting.

(*iii*) Calcium is preferred over sodium to remove last traces of moisture from alcohol ?

Ans : Both Na and Ca react with water to form their respective hydroxides. However, Na is normally not preferred since it reacts with alcohol (C_2H_5OH) and evolves H_2

$$2\ C_2H_5OH + 2\ Na \rightarrow 2\ C1_2H_5OH\ \text{(Sod.ethoxide)} + H_2$$

On the other hand, calcium has no action with alcohol and is, therefore, preferred over sodium to remove water or moisture from alcohol.

***(iv)* Halides of Be dissolve in organic solvents while those of Ba do not.**

Ans : Since Be^{2+} ion is the smallest in size and Ba^{2+} ion is the largest in size, among alkaline metal cations (M^{2+}), halides of Be^{2+} are covalent and those of Ba^{2+} are ionic. Due to covalent character, halides of Be^{2+} dissolve in organic solvents (non polar solvents) and due to ionic character, halides of Ba^{2+} donot dissolve.

***(v)* The solubility of calcium acetate in water decreases with rise in temperature while that of lead nitrate increases. (I.I.T. 2000)**

Ans : The dissolution of calcium acetate in water is of exothermic nature while that of lead nitrate is of endothermic nature. Therefore, the solubility of lead acetate in water decreases with rise in temperature while that of lead nitrate increases.

***(vi)* $BeCl_2$ can be easily hydrolysed. (I.I.T. 1999)**

Ans : Since Be^{2+} ions are small, they have high value of charge/size ratio and hence are easily hydrolysed. The hydrolysis occurs because Be–O bond is very strong.

***(vii)* $BeCl_2$ fumes in moist air but the chlorides of other alkaline earth metals donot.**

***or* Solution of $BeCl_2$ is acidic in nature.**

Ans : $BeCl_2$ reacts with the moisture present in air and produces the fumes of HCl gas. Thus $BeCl_2$ fumes in air.

$$BeCl_2 + 2\ H_2O\ \text{(moisture)} \longrightarrow \underset{\text{Fumes}}{2\ HCl} + Be\ (OH)_2$$

OR

Since $BeCl_2$ is a salt of weak base *viz* Be $(OH)_2$ and strong acid (HCl), it gives weak base and strong acid when dissolved in water.

$$BeCl_2 + 2H_2O \longrightarrow \underset{\text{Strong acid}}{2HCl} + \underset{\text{Weak base}}{Be\ (OH)_2}$$

Due to the production of strong acid, aqueous solution of $BeCl_2$ shows acidic character.

***(viii)* The reaction between marble and dil. H_2SO_4 is not used to prepare CO_2.**

Ans : When marble ($CaCO_3$) reacts with dil. H_2SO_4, ppt. of $CaSO_4$ is obtained and CO_2 gas is evolved.

$$CaCO_3 + H_2SO_4 \longrightarrow \underset{\text{ppt.}}{CaSO_4} + H_2O + CO_2$$

The ppt. of $CaSO_4$ formed as above gets deposited on the surface of $CaCO_3$ and hence further reaction between $CaCO_3$ and dil. H_2SO_4 is prevented and the formaton of CO_2 ceases after some time.

(*ix*) Although Ba^{2+} ion is posinuous, yet it is safer to use $BaSO_4$ internally in patients prior to taking their stomach X-ray.

Ans : Since $BaSO_4$ is extremely insoluble, it does not pass from digestive system to circulatory system.

(*x*) Sodium and potassium can be obtained by the electrolysis of fused NaOH and KOH respectively but calcium cannot be obtained by electrolysing $Ca(OH)_2$.

Ans : This is because of the fact that $Ca(OH)_2$ decomposes into CaO (quick lime) which is diffcult to fuse.

Q.2 Give chemical equations involved in the isolation of $MgCl_2$ from sea water. (I.I.T. 1993)

Ans : Different steps involved in the isolation of Mg from sea water are :

(*i*) Sea water contains $MgCl_2$ and $MgSO_4$. Sea water is treated with Ca $(OH)_2$ so that ppt. of Mg $(OH)_2$ is abtained.

$$\text{Sea water containing } MgCl_2 \text{ and } MgSO_4 + Ca(OH)_2 \rightarrow \underset{\text{Percipitate}}{Mg(OH)_2} + CaCl_2 + MgSO_4$$

(*ii*) $Mg(OH)_2 + 4H_2O + 2HCl \longrightarrow MgCl_2.\ 6H_2O$

(*iii*) $MgCl_2.\ 6H_2O \xrightarrow{\Delta} MgCl_2 + 6H_2O$

Q.3 What happens when the following are heated ? (*i*) Hydrated magnesiam chloride (*ii*) Hydrated magnessium chloride in presence of NH_4Cl.

Ans : (*i*) MgO is obtained.

$$MgCl_2\ 6H_2O \xrightarrow[(-4H_2O)]{\Delta} MgCl_2.\ 2H_2O \xrightarrow{\Delta} \underset{\text{Mag. oxy chloride}}{Mg(OH).\ Cl} + HCl + H_2O$$

$$Mg(OH)\ Cl \xrightarrow{\Delta} MgO + HCl$$

(*ii*) Anhydrous magnesium chloride ($MgCl_2$) is obtained.

$$MgCl_2.\ 6H_2O + NH_4Cl \xrightarrow{\Delta} \underset{\text{Double salt}}{MgCl_2.\ NH_4Cl.\ 6H_2O}$$

$$\downarrow \Delta\ (-6H_2O)$$

$$MgCl_2 + NH_4Cl \xleftarrow{\Delta} MgCl_2.\ NH_4Cl$$

Q.4 Complete the following :

(*i*) $Be_2C + H_2O \rightarrow$ ....

(*ii*) $BeCl_2 + LiAlH_4 \rightarrow$ ....

(*iii*) Be $(OH)_2$ + 2NaOH $\rightarrow$ ....

(*iv*) $CaSO_4.\ 2H_2O \xrightarrow{\Delta}$ ....

(*v*) Ca + $H_2O \rightarrow$ ....

(*vi*) BeO + NaOH $\rightarrow$ ....

(vii) $CaC_2 + H_2O \rightarrow ....$

(viii) $CaC_2 + N_2 \rightarrow ....$

(ix) $Ca\ (OH)_2 + Cl_2 \rightarrow ....$

(x) $BaO_2 + H_2SO_4 \rightarrow ....$

Ans : (*i*) $Be_2C + 2H_2O \rightarrow 2Be + CH_4$

(*ii*) $2BeCl_2 + LiAlH_4 \rightarrow 2BeH_2 + LiCl + AlCl_3$

(*iii*) $Be\ (OH)_2 + 2NaOH \rightarrow \underset{\text{Sod. beryllate}}{Na_2BeO_2} + 2H_2O$

(*iv*) $CaSO_4.\ 2H_2O \xrightarrow{\Delta} \underset{\text{Plaster of Paris}}{CaSO_4.\ \frac{1}{2}\ H_2O} + \frac{3}{2}\ H_2O$

(*v*) $Ca + 2\ H_2O \rightarrow Ca\ (OH)_2 + H_2$

(*vi*) $BeO + 2NaOH \rightarrow \underset{\text{sod. beryllate}}{Na_2BeO_2} + H_2O$

(*vii*) $CaC_2 + 2H_2O \rightarrow Ca\ (OH)_2 + HC \equiv CH$

(*viii*) $CaC_2 + N_2 \rightarrow \underset{\text{Cal. cyanamide}}{CaCN_2} + C$

(*ix*) $Ca\ (OH)_2 + Cl_2 \rightarrow \underset{\text{Bleaching powder}}{CaOCl_2} + H_2O$

(*x*) $BaO_2 + H_2SO_4 \rightarrow BaSO_4 + H_2O_2$

Q.5 What is the diffrence between milk of lime and lime water.

Ans : A suspension of slaked lime, *i.e.* $Ca\ (OH)_2$ in water is called milk of lime but a clear decanted solution of slaked lime in water is called lime water.

Q.6 On treatment with cold H_2O, an element (A) reacted with H_2O, liberating a colourless gas (B) and a solution (C). A reacted with (B) yielding a solid product (D) which reated with H_2O to give a basic solution due to the formation of (E). When CO_2 was bubbled through solution (C), initially a white ppt. (F) was formed but this ppt. redissolved forming solution (G) when more CO_2 was added. Precipitate (F) effervesced when moitered with conc. HCl. When (F) was heated with carbon at 1000 °C, a white compound (H) which when heated with carbon at 1000 °C gave a solid (I) of some commoercial importance. Identify the substances (A) to (I).

Ans : (A) is calcium metal which reacts with H_2O and evolves H_2 (B) and $Ca(OH)_2$ solution (C). Thus all the reactions can be written as :

$\underset{(A)}{Ca} + 2\ H_2O \longrightarrow \underset{\text{Solution (C)}}{Ca\ (OH)_2} + \underset{(B)}{H_2}$ (*Colourless and odourless gas*)

$\underset{(A)}{Ca} + \underset{(B)}{H_2} \longrightarrow \underset{(D)}{CaH_2}$ (*Solid product*)

$\underset{(D)}{CaH_2} + 2H_2O \longrightarrow \underset{\text{Basis solution (E)}}{Ca\ (OH)_2}$ (White ppt) + $2H_2$

$$\underset{\text{Solution (C)}}{Ca(OH)_2} + CO_2 \longrightarrow \underset{(F)}{CaCO_3} + H_2O$$

$$\underset{(F)}{CaCO_3} + CO_2 + H_2O \longrightarrow \underset{\text{Soluble (G)}}{Ca(HCO_3)_2}$$

$$\underset{(F)}{CaCO_3} + 2HCl \longrightarrow CaCl_2 + H_2O + CO_2$$

$$\underset{(F)}{CaCO_3} + C \xrightarrow{1000°C} \underset{(H)}{CaO} \text{ (\textit{White product})} + 2CO$$

$$\underset{(H)}{CaO} + 2C \xrightarrow{1000°C} \underset{(I)}{CaC_2} \text{ (\textit{Solid})} + \frac{1}{2} O_2.$$

Thus the substances (A) to (I) are :

$A = Ca$, $B = H_2$, $C = Ca(OH)_2$, $D = CaH_2$,
$E = Ca(OH)_2$, $F = CaCO_3$, $G = Ca(HCO_3)_2$, $H = CaO$, $I = CaC_2$

Q.7 Element (A) burns in N_2 to give an ionic compound (B). Compound (B) with water gives (C) and (D). Solution of (C) becomes milky on bubbling CO_2 through it. Identity (A), (B), (C) and (D) and write the chemical equations for the reactions involved. (I.I.T. 1997)

Ans : $\underset{(A)}{\text{Element}} \xrightarrow{+N_2} \underset{(B)}{\text{Nitride of the element, A}} \xrightarrow{+H_2O} (C) + (D)$

$$(C) \xrightarrow{+CO_2} \text{Milkiness.}$$

Since (C) becomes milky on bubbling CO_2 through it, (C) should be Ca $(OH)_2$. Now we know that when a nitride reacts with H_2O, NH_3 is evolved. Thus (D) should be NH_3, nitride (B) should be Ca_3N_2 (an ionic compound) and hence the element (A) should be Ca.

Thus all the reactions involved can be written as :

$$\underset{(A)}{3Ca} + N_2 \longrightarrow \underset{(B)}{Ca_3N_2}$$

$$\underset{(B)}{Ca_3N_2} + 6H_2O \longrightarrow \underset{(C)}{3Ca(OH)_2} + \underset{(D)}{2NH_3}$$

$$\underset{(C)}{Ca(OH)_2} + CO_2 \longrightarrow \underset{\text{Milkiness}}{CaCO_3} + H_2O$$

Q.8 Calcium burns in air to produce a white powder which dissolves in sufficient water to produce a gas (A) and an alkaline solution. The solution on exposure to air produces a thin solid layer of (B) on the surface. Identity the compounds A and B. (I.I.T. 1996)

Ans : Ca burns in air to form CaO and Ca_3N_2.

$$2\,Ca + O_2 \longrightarrow 2\,CaO$$
$$3\,Ca + N_2 \longrightarrow Ca_3N_2 \text{ (White powder)}$$

Calcium nitride on hydrolysis with H_2O gives ammonia, NH_3 (A)

$$Ca_3N_2 + 6\,H_2O \longrightarrow 3\,Ca(OH)_2 \text{ (Gas)} + \underset{A}{2\,NH_3} \text{ (Alkaline solution)}$$

The alkaline solution of Ca $(OH)_2$ thus formed reacts with CO_2 present in the air to form $CaCO_3$ (B).

$$Ca(OH)_2 + CO_2 \longrightarrow \underset{B}{CaCO_3} + H_2O$$

Thus: A = NH_3 and B = $CaCO_3$

Q.9 Metal nitrate (A) on heating, decomposes to give a solid residue (B) which goes into solution with dil. HCl. The solution thus obtained gives a white ppt. (C) with $(NH_4)_2CO_3$. The ppt. (C) is dissolved in dil. HCl and treated with K_2CrO_4 solution to get yellow ppt. (D). Identity (A), (B), (C) and (D). Also give chemical equations.

Ans : The reactions can be shown as :

$$\underset{(A)}{\text{Metal nitrate}} \xrightarrow{\Delta} \underset{\text{Residue (B)}}{\text{Metal oxide}} \xrightarrow{+HCl} \underset{\text{Solution}}{\text{Metal chloride}}$$

$$\xrightarrow{+(NH_4)_2CO_3} \underset{\text{White ppt (C)}}{\text{Metal carbonate}} \xrightarrow{+HCl} \text{Metal chloride}$$

$$\xrightarrow{+K_2CrO_4} \underset{\text{Yellow ppt. (D)}}{\text{Metal chromate}}$$

Metal chromate may be either $PbCrO_4$ or $BaCrO_4$ and hence the metal nitrate may be either Pb $(NO_3)_2$ or Ba $(NO_3)_2$. Since $(NH_4)_2$ CO_3 has been used in the above reactions, the metal nitrate is Ba $(NO_3)_2$. The reactions taking place at different steps can be written as :

$$\underset{(A)}{Ba(NO_3)_2} \xrightarrow{\Delta} \underset{\substack{\text{Solid residue}\\(B)}}{BaO} + 2\,NO_2 + O_2$$

$$\underset{(B)}{BaO} + 2\,HCl\text{ (Dil)} \longrightarrow \underset{\text{Solution}}{BaCl_2} + H_2O$$

$$\underset{\text{Solution}}{BaCl_2} + (NH_4)_2CO_3 \longrightarrow \underset{\substack{\text{White ppt.}\\(C)}}{BaCO_3} + 2\,NH_4Cl$$

$$\underset{(C)}{BaCO_3} + 2\,HCl \longrightarrow BaCl_2 + H_2O + CO_2$$

$$BaCl_2 + K_2CrO_4 \longrightarrow \underset{\substack{\text{Yellow ppt.}\\(D)}}{BaCrO_4} + 2\,KCl$$

Thus : $A = Ba(NO_3)_2$, $B = BaO$,
$C = BaCO_3$ and $D = BaCrO_4$

Q.10 Lime water becomes turbid on passing CO_2 through it, but becomes clear when more CO_2 is passed. Explain.

Ans : When CO_2 is passed through lime water [Ca $(OH)_2$], turbiditly due to the formation of insoluble $CaCO_3$ is obtained. This ppt. of $CaCO_3$ gets dissolved in excess of CO_2 due to the formation of soluble Ca $(HCO_3)_2$

$$\underset{\text{Lime water}}{Ca(OH)_2} + CO_2 \longrightarrow \underset{\text{ppt.}}{CaCO_3} + H_2O$$

$$CaCO_3 + H_2O \xrightarrow{+CO_2} \underset{\text{Soluble}}{Ca(HCO_3)_2}$$

University Questions

1. Give the preparation, properties, uses and structure of beryllium chloride. (*Agra 82, 83; Kanpur 81*)
2. Describe the preparation, properties, uses and structure of basic berryllium acetate. (*Meerut 80, 82, 84, 87, 88, 89; Agra 80, 81, 83; Kanpur 81*)
3. Describe the discovery, extraction and uses of radium. (*Madras 85*)
4. What is the role of gypsum in cement? (*Himachal Pradesh 99*)
5. How does beryllium occur in nature? Using flow sheet diagram give one method for its extraction. (*Gorakhpur 99*)
6. How will you obtain radium chloride from carnotite? (*Kumaon 2000*)
7. (*a*) Justify the position of radium in second group of periodic table. How it is obtained from pitchblende and how it is separated from Hg cathode?
   (*b*) How is beryllium obtained from beryl? Give anomalous behaviour and its diagonal relationship with aluminium. (*Lucknow 2000*)
8. Give line diagram of the extraction of beryllium from its important ore. (*Avadh 2000*)
9. Discuss the manufacture of Portland Cement (*Bangalore 2009*)

CHAPTER 22 General Characteristics of Group 13 (III A) Elements: Boron Group Elements

Group III A of the long form of the periodic table consists of *five* elements which are: *Boron* (B), *Aluminium* (Al), *Gallium* (Ga), *Indium* (In) and *Thallium* (Tl). These elements are collectively known as *boron group elements*.

Position of Boron Group Elements in the Periodic Table

The inclusion of boron group elements namely B, Al, Ga, In and Tl in the same sub-group of the periodic table is justified by the following general characteristics of these elements.

I. Electronic Configurations.

The electronic configurations of the atoms of boron group elements are: B (Z = 5) → 2, 3 ($2s^2p^1$); Al (Z = 13) → 2, 8, 3 ($3s^2p^1$); Ga(Z = 3) → 2, 8, 18, 3 ($3d^{10}4s^2p^1$); In (Z = 49) → 2, 8, 18, 18, 3 ($4d^{10}5s^2p^1$); Tl (Z = 81) → 2, 8, 18, 32, 18, 3 ($4f^{14}5d^{10}6s^2p^1$). The configurations given in bracket are the valence-shell configurations. On the basis of the electronic configuration of the penultimate shell, these elements can be divided into two groups namely (*i*) 1*st* group of elements consists of B and Al elements whose penultimate shells have *inert gas type configuration* (s^2 or s^2p^6 configuration) (B has 2 electrons and Al has 8 electrons in their penultimate shell) (*ii*) 2*nd* group consists of Ga, In and Tl whose penultimate shells have *pseudo inert gas type configuration* ($s^2p^6d^{10}$ configuration). This difference in electronic configuration is responsible for the difference in properties of the elements of both the groups and also of boron and aluminium.

II. Similarities in Physical Properties and Gradation in Them.

Some of the physical properties of boron group elements are discussed below:

1. Atomic volume, atomic and ionic radii and density. Atomic volume, atomic and ionic radii (of M^{3+} ions) and density increase on moving from B to Tl.

2. Melting points, boiling points and heats of sublimation. Melting points *decrease* from B to Ga and then increase upto Tl. The low value of melting point of Ga is explained on the basis that its structure consists of Ga_2 molecules. Boiling points and heats of sublimation show a continuous *decrease*. The successive decrease in heat of sublimation indicates that the atoms of these elements are held less and less closely as we move from B to Tl.

3. Ionisation energies. First ionisation energies of these elements correspond to the removal of np^1-electron from ns^2p^1 configuration while the second and third ionisation energies correspond to the removal of ns^2-electrons in succession. Because of the increase in size of Al, the ionisation energy decreases from B to

Al. But in going from Al to Ga the value of ionisation energy increases. Thus increase in case of Ga is duo to the fact that the intervening *d*-electrons in Ga are not able to shield the nuclear charge effectively and hence the valence electrons in Ga are more firmly held to the nucleus and its ionisation energy increases.

Although the nuclear charge of these elements is *larger* and their size is *smaller* than that of the elements of Group II A, the first ionisation energies (*i.e.*, I_1 values) of these elements are lower than those of the elements of Group II A. This is due to the fact that the *p*-electrons (Group III A) are *less penetrating* and more shielded than *s*-electrons (Group II A), *i.e.*, *p*-electrons, being at a greater distance from the nucleus than the *s*-electrons, are held *less tightly* with the nucleus and hence can be removed more easily than the *s*-electrons.

The second and third ionisation energies (*i.e.*, I_2 and I_3 values) are considerably higher because the successive electron are to be removed from *s*-orbital.

In this group since the sum of I_1, I_2 and I_3 is very high for boron, boron is not able to exist as B^{3+} ion in its compounds. In fact the sum, $(I_1 + I_2 + I_3)$ for boron is the highest because of its extremely small size and because of the fact that its valence-electrons are not effectively shielded from the nuclear charge by the intervening electrons. However, as we move down the group, the total ionisation energies decrease but not in a regular order.

4. Oxidation states. Since boron atom has *small size* and *high ionisation energy*, it does not lose all its three valence-shell electrons (ns^2p^1 electrons) and hence does not give B^{3+} ion. It is for this reason that boron atom combines with other elements through covalent bonds, *i.e. boron atom gives mainly covalent compounds.* For convenience it is still preferred to treat boron atom in its compounds as B^{3+} ion. The small size and high charge of Al^{3+} ion give it high charge density. This charge density is responsible for the facts namely: (*a*) The compounds of Al in the gaseous state are *covalent*. (*b*) Due to *high hydration energy of Al^{3+}*, its compounds are *stabilised in solution*. (*c*) Its compounds in the solid state have *high lattice energy*.

Other elements namely Ga, In and Tl show +1 as well as +3 oxidation states, *i.e.* these elements form M^+ as well as M^{3+} cations. M^{3+} cations are obtained when all the three ns^2p^1 electrons are lost while M^+ cations are formed when only one np^1 electron is lost. In the formation of M^+ cations the electron pair in *ns* orbital remains inert. This is called **inert pair effect**. B and Al atoms do not show inert pair effect and hence do not form M^+ cations. Inert pair effect increases when we move from Ga to In to Tl. Thus Tl shows maximum inert pair effect and hence forms only Tl^+ cation which is the most stable ion in case of Tl element.

Relative stability of M^+ and M^{3+} cations. When we move down the group from B to Tl, the stability of +3 state *decreases* while that of +1 state *increases*, *i.e.* the stability of M^{3+} cations decreases from B^{3+} to Tl^{3+} while that of M^+ cations increases from B^+ to Tl^+ as shown below :

$$\mathbf{B^{3+} > Al^{3+} > Ga^{3+} > In^{3+} > Tl^{3+}}$$
$$\mathbf{B^{+} < Al^{+} < Ga^{+} < In^{+} < Tl^{+}}$$

Thus when we compare the stability of the compounds formed by Ga^+, In^+ and Tl^+ ions, the compounds of Ga^+ are unstable, those of In^+ are moderately stable while those of Tl^+ are the most stable.

Compounds of Tl^+ are more stable than those of Tl^{3+} ($Tl^+ > Tl^{3+}$) and hence the compounds of Tl^{3+} get immediately changed (reduced) into those of Tl^+. In other words compounds of Tl^{3+} act as oxidising agents and those of Tl^+ act as reducing agents.

$$\underset{\text{Unstable (oxidising agent)}}{Tl^{3+} \text{ compounds}} \xrightarrow[(+\,2e^-)]{\text{Reduction}} \underset{\text{Stable (reducing agent)}}{Tl^{+} \text{ compounds}}$$

5. Nature of M^{3+} compounds. From the electronic configurations of these elements it is clear that all of them would be expected to form M^{3+} ions. The nature of the compounds of M^{3+} ions is decided by Fajan's rule according to which *the smaller the cation, the greater is its tendency to form covalent compounds.* Thus with the *increase* of the size of M^{3+} ions from B^{3+} to Tl^{3+}, the tendency of these ions to form covalent compounds *decreases*. Consequently the compounds of B^{3+} are predominantly covalent while Al^{3+}, Ga^{3+}, In^{3+} and Tl^{3+} give ionic compounds as well.

6. Electropositive (metallic) character. Electropositive (metallic) character of the elements *increases* as we move from B to Tl. This is indicated from the ionisation energies of these elements. Thus B is a *semi-metal i.e.*, it is closer to non-metals than to metals in its properties (its crystalline form is hard, inert and a poor conductor of electricity), while other elements are metallic and good conductors of electricity. Al, Ga, In and Ti have almost equal electropositive character as is evident from the fact that the ionisation energies of these elements do not differ much from each other.

7. Oxidation potentials. With the exception of boron which does not form B^{3+} ion in aqueous solution due to large quantity of hydration energy associated with B^{3+} ion, the standard oxidation potentials of other elements corresponding to the reaction:

$$M\,(s) \longrightarrow M^{3+}\,(aq) + 3e^-$$

are quite high in spite of their large values of total ionisation energies. These high values of standard oxidation potential are due to *large heats of hydration* which result from the *high charges* and *small radii* of M^{3+} ions.

III. Similarities in Chemical Properties and Gradation in Them.

Some chemical properties of these elements and gradation in them are discussed below:

1. Formation of oxides and hydroxides. All the elements of this group form oxides and hydroxides. B^{3+} ion, being very small in size, has high (positive) charge density. Therefore, when this ion is placed in H_2O, it tends to pull off electrons from H_2O molecule towards itself sufficiently strongly by rupturing the O—H bond present in H_2O molecule $\left(\begin{matrix}H\\H\end{matrix}\!\!>\!O\right)$. This results in a rapid hydrolysis of B^{3+} ion by H_2O to release H^+ ions. Hence $B(OH)_3$ and B_2O_3 are *acidic in character*. Al^{3+} and Ga^{3+} ions being relatively large than B^{3+} ion, have smaller ability to rupture the O—H bond in H_2O molecule by pulling off electrons. Thus Al^{3+} and Ga^{3+} ions are hydrolysed to a smaller extent than B^{3+} ion. This means that the oxides and hydroxides of Al^{3+} and Ga^{3+} are *amphoteric*. In^{3+} and Tl^{3+} ions are even larger than Al^{3+} and Ga^{3+} ions and hence their hydrolysis by H_2O is very low, *i.e.*, the O—H bond in water is hardly affected by In^{3+} and Tl^{3+} ions and this makes the oxides and hydroxides of In^{3+} and Tl^{3+} ions *distinctly basic*. Thus we see that *on moving down the group from boron to thallium, the oxides and hydroxides of these elements change from acidic through amphoteric to basic in character*. This trend is inconsistent with the change from *non-metallic to metallic character of these elements*. This trend shows that the oxides and hydroxides of B are *acidic*, those of Al and Ga are *amphoteric* while those of In and Ti are *basic*.

2. Formation of trihalides. All the elements of this group react with halogens to form the trihalides, MX_3.

Properties. (*i*) *Covalent and ionic character*. Boron trihalides are *covalent in character* due to the *small size* and *high charge density* on B^{3+} ion.

The fluorides of Al, Ga, In and Tl are *ionic* and have high melting points while the other halides of these elements are *covalent* when *anhydrous*. For example anhydrous $AlCl_3$ is covalent but in water it gets hydrolysed to give Al^{3+} (*aq*) ions. This change from a covalent compound in an aqueous solution is mainly due to the high hydration energy of Al^{3+} ions. However, BCl_3, on hydrolysis, does not give B^{3+} ions, because the ionisation energy of B^{3+} ion is very large.

(*ii*) *Geometry* (*i.e.*, *shape*) *of trihalides*. The normal trihalides are *planar triangular* in shape which results from sp^2 hybridisation of the central atom. The central atom in these trihalides has a *vacant* *p*-orbital which remains *unhydridised*. As an example BF_3 can be considered. The shape of BF_3 molecule has already been discussed.

(*iii*) *Formation of $BX_3.L$ complexes*. The boron trihalides (BX_3) form complexes of $BX_3.L$ type where L is a *neutral* or *anionic donor* (*i.e.*, ligand). These complexes are *tetrahedral in shape* and in their formation the unhybridised vacant *p*-orbital accepts the lone pair of electrons denoted by the donor, L. Thus:

$$\underset{\substack{\text{Planar triangular} \\ (sp^2)}}{BX_3} + L \longrightarrow \underset{\substack{\text{Tetrahedral} \\ (sp^3)}}{BX_3.L}$$

(*iv*) *Formation of dimeric molecules*. Boron trihalides exist as *discrete species* (*i.e.*, *monomeric moiecules*, BX_3) while the trihalides of other elements (*e.g.* $AlCl_3$, $AlBr_3$, $GaCl_3$, $InCl_3$, $InBr_3$ and InI_3) exist as *dimers* both in vapour sate and in non-polar solvents. In the formation of dimers the trihalides make use of the vacant *p*-orbital which remained unhybridised in the formation of MX_3 trihalides. In these dimeric molecules the halogen atoms are arranged *tetrahedrally* around the central atom (*i.e.*, each central atom is surrounded by four halogen atoms) and form halogen bridges between MX_2 units (M = central atom *viz.*, Al, Ga, In). Dimeric structure of aluminium trihalide, $(AlCl_3)_2$ is shown below:

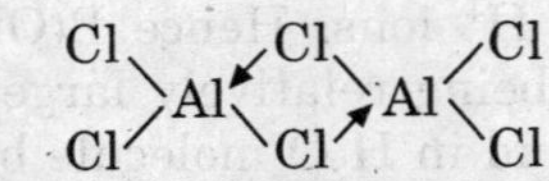

In this structure the four terminal Cl atoms and two Al atoms lie in the same plane while the bridging Cl atoms are above and below this plane. Thus each Al atom is surrounded by halogen atoms. The arrows indicate the donation of an electron pair by the chloride ion to Al atom. This electron pair goes into the vacant *p*-orbital of Al atom. In the solid state $AlCl_3$ has a layer structure in which each Al atom is surrounded by *six* Cl atoms.

Here a question arises: *Why the trihalides of aluminium exist as dimers while the trihalides of boron exist as monomers*? This is explained on the basis of the *smaller size of boron atom relative to that of aluminium.* Owing to its small size, boron atom cannot coordinate to four larger halide ions and hence cannot form the dimeric molecules.

The dimeric structures are retained when the halides are dissolved in non-polar solvents like C_6H_6. But when the halides are dissolved in H_2O, due to heat of hydration, the dimeric structure, $(MX_3)_2$ splits into $[M(H_2O)_6]^{3+}$ and X^- ions which yield a conducting solution.

$$(MX_3)_2 + 12H_2O \longrightarrow 2\ [M(H_2O)_6]^{3+} + 6X^-$$

The covalent dimers may also be splited by reactions with donor molecules giving complexes of $R_3N.AlCl_3$ type.

(*v*) *Formation of complex halides.* Boron trihalides (*e.g.*, BF_3) also form complex halides of the type, $[BF_4]^-$ in which boron atom (central atom) has a co-ordination number equal to 4. It can not extend its co-ordination number beyond 4 due to the non-availability of *d*-orbitals in its configuration. However, the trihalides of other elements of this group form the complex halides of $[AlF_6]^{3-}$ $(GaCl_6]^{3-}$, $[InCl_6]^{3-}$ etc., type in which the central atom has co-ordination number of 6 which is because of the presence of *d*-orbitals in these central atoms.

(*vi*) *Electron-acceptor property: Formation of addition compounds.* The trihalides of all the elements of this group act as Lewis acids, *i.e.*, they react with donor molecules and form addition compounds. The Lewis acidity (*i.e.*, electron-acceptor ability) of the trihalides is in the order:

$$BX_3 > AlX_3 > GaX_3 > InX_3$$

The position of the halides of thallium is not clear.

3. Formation of hydrides. None of the elements of this group reacts directly with H_2 to form hydrides. However, hydrides of these elements have been prepared by indirect methods. Boron forms a large number of polymeric hydrides which are called *boranes*. Some examples are B_2H_6, B_4H_{10}, B_5H_9, B_5H_{11} etc. These hydrides are covalent in nature. Al forms only one high molecular weight polymeric hydride, $(AlH_3)_x$ which is an insoluble substance and, on heating, decomposes into its elements. X-ray and neutron diffraction studies have shown that this compound contains Al atoms surrounded octahedrally by six H-atoms. The structure consists of a network of AlH_2Al bridges. Ga gives GaH_4 which is called *gallane*. It is an unstable and volatile liquid. Indium gives polymerised solid hydride, $(InH_3)_x$. The hydrides given by Tl are extremely unstable. The elements of this group also form the *complex anionic hydrides* like $Li^+[BH_4]^-$, $Li^+[AlH_4]^-$,

$Li^+[GaH_4]^-$ etc. In the formation of these hydrides H^- ion acts as an electron pair donor while MH_3 hydride acts as an electron pair acceptor.

Electron pair acceptor		***Electron pair donor***		***Complex hydride***
BH_3	+	H^-	$\longrightarrow$	$[BH_4]^{4-}$
AlH_3	+	H^-	$\longrightarrow$	$[AlH_4]^-$
GaH_3	+	H^-	$\longrightarrow$	$[GaH_4]^-$

The electron pair donated by H^- ion goes into the vacant *p*-orbital on the central atom. These complex hydrides are prepared as follows:

$$4LiH + AlCl_3 \xrightarrow{\text{Ether}} Li[AlH_4] + 3LiCl$$

The complex hydrides, particularly $Li[AlH_4]$ and $Na[AlH_4]$ are useful reducing agents. These are used to reduce the aldehydes and ketones to alcohols and nitrates to amines.

Anomalous Properties of Boron: Dissimilarities of Boron with Other Elements of Group III A.

In many properties boron differs from the other elements of its own sub-group. The main points of difference are:

(1) B has a very small atomic radius. The hypothetical B^{3+} ion has very small size, high charge and hence high charge density. This value is so high that B^{3+} ion does not exist. All the compounds of boron are, therefore, covalent.

(2) Boron has less than four valence-electrons and this gives it a great electron accepting power, *i.e.*, its compounds behave as strong Lewis acids and hence forms a large number of complex compounds exhibiting its tendency to acquire a stable octet.

(3) Boron shows a maximum covalency of four while all other elements show a covalency of six or more.

(4) It does not exihibit inert pair effect.

Comparison between Boron and Aluminium.

(A) Similarities. (*i*) *Electronic configuration.* Both have similar electronic configurations having three electrons in their outermost shells.

$$B(Z = 5) = 1s^2 2s^2 2p^1 \ (2, 3)$$
$$Al(Z = 13) = 1s^2 2s^2 2p^6 3s^2 3p^1 \ (2, 8, 3)$$

(*ii*) *Valency.* Both the elements are trivalent. They do not show variable valency because their penultimate shells are complete and stable. Both have generally no tendency to form B^{3+} and Al^{3+} ions and, therefore, form tri-covalent compounds. Aluminium, however, may also form electrovalent compounds with strong electron accepting atoms or groups.

(*iii*) *Oxidation state.* The common oxidation state of these elements is +3. Boron, however, also exhibits an oxidation state of –3 in the metal borides.

(*iv*) *Action of conc. H_2SO_4.* Both react with conc. H_2SO_4 to form SO_2.

$$2B + 3H_2SO_4 \longrightarrow 2H_3BO_3 + 3SO_2 \uparrow$$
$$2Al + 6H_2SO_4 \longrightarrow Al_2(SO_4)_3 + 6H_2O + 3SO_2 \uparrow$$

(*v*) *Action of alkalies.* They react with alkalies to form H_2.

$$2B + 6NaOH \longrightarrow 2Na_3BO_3 + 3H_2 \uparrow$$
$$2Al + 2NaOH + 2H_2O \longrightarrow 2NaAlO_2 + 3H_2 \uparrow$$

(*vi*) *Formation of oxides.* Both form oxides when heated with oxygen at high temperature.

$$4B + 3O_2 \xrightarrow{700°C} 2B_2O_3$$
$$4Al + 4O_2 \xrightarrow{800°C} 2Al_2O_3$$

These oxides are amphoteric and dissolve in acids as well as in alkalies to form salts.

$$B_2O_3 + 6HCl \longrightarrow 2BCl_3 + 3H_2O$$
$$B_2O_3 + 6NaOH \longrightarrow \underset{\text{Sodium borate}}{2Na_3BO_3} + 3H_2O$$
$$Al_2O_3 + 6HCl \longrightarrow 2AlCl_3 + 3H_2O$$
$$Al_2O_3 + 2NaOH \longrightarrow \underset{\text{Sodium meta-aluminate}}{2NaAlO_2} + H_2O$$

(*vii*) *Formation of nitrides.* Both, when heated with nitrogen or ammonia, form nitrides.

$$2B + N_2 \longrightarrow 2BN$$
$$2Al + N_2 \longrightarrow 2AlN$$
$$2B + 2NH_3 \longrightarrow 2BN + 3H_2$$
$$2Al + 2NH_3 \longrightarrow 2AlN + 3H_2$$

These nitrides are decomposed by steam to form ammonia.

$$BN + 3H_2O \longrightarrow H_3BO_3 + NH_3$$
$$AlN + 3H_2O \longrightarrow Al(OH)_3 + NH_3$$

(*viii*) *Formation of sulphides.* Both from sulphides on heating with sulphur at high temperatures. These are hydrolysed by water.

$$B_2S_3 + 6H_2O \longrightarrow 2H_3BO_3 + 3H_2S$$
$$Al_2S_3 + 6H_2O \longrightarrow 2Al(OH)_3 + 3H_2S$$

(*ix*) *Formation of chlorides*. They form trichlorides by direct combination with Cl_2 or by passing Cl_2 over heated mixture of their oxides and charcoal.

$$2B + 3Cl_2 \longrightarrow 2BCl_3$$
$$B_2O_3 + 3C + 3Cl_2 \longrightarrow 2BCl_3 + 3CO$$
$$2Al + 3Cl_2 \longrightarrow 2AlCl_3$$
$$Al_2O_3 + 3C + 3Cl_2 \longrightarrow 2AlCl_3 + 3CO$$

These trichlorides are covalent and are hydrolysed by water.

$$BCl_3 + 3H_2O \longrightarrow H_3BO_3 + 3HCl$$
$$AlCl_3 + 3H_2O \longrightarrow Al(OH)_3 + 3HCl$$

(*x*) *Formation of alkyl compounds.* Both form similar types of alkyl compounds.

$$2BCl_3 + 3Zn(CH_3)_2 \longrightarrow 2B(CH_3)_3 + 3ZnCl_2$$
$$2Al + 3Hg(CH_3)_2 \longrightarrow 2Al(CH_3)_3 + 3Hg$$

(B) Dissimilarities. Although B and Al both belong to the same group of the periodic table, yet they show many dissimilarities. The difference in their properties is due to the big difference in their size and ionisation potential. The points showing dissimilarities between boron and aluminium are given in the following table.

Boron	Aluminium
1. There are two electrons in the penultimate shell (2, 3).	There are eight electrons in the penultimate shell (2, 8, 3).
2. It is a non-metal and is a bad conductor of heat and electricity.	It is a metal and is a good conductor of heat and electricity.
3. It has high melting point (2300°C).	It has low melting point (660°C).
4. It exhibits allotropy and is found in two forms *viz.* crystalline and amorphous.	It does not exhibit allotropy.
5. Its crystalline form is very hard.	It is sufficiently soft.
6. It forms covalent compounds only.	It forms both covalent and electrovalent compounds.
7. It reacts with hydrogen to form a number of hydrides like B_2H_6, B_5H_9, etc.	It does not form any stable hydride.
8. It does not decompose water or steam.	It decomposes steam to form hydrogen. $2Al + 6H_2O \rightarrow 2Al(OH)_3 + 3H_2$
9. It does not react with dil. acids.	It reacts with dil. acids to form hydrogen. $2Al + 3H_2SO_4 \rightarrow Al_2(SO_4)_3 + 3H_2 \uparrow$
10. It is oxidised by conc. HNO_3 to form boric acid. $B + 3HNO_3 \rightarrow H_3BO_3 + 2NO_2$	It becomes passive with conc. HNO_3
11. It dissolves in fused alkalies to form hydrogen. $2B + 6NaOH \rightarrow 2Na_3BO_3 + 3H_2$	It dissolves in hot alkalies to form hydrogen. $2Al + 2NaOH + 2H_2O \rightarrow 2NaAlO_2 + 3H_2$
12. Borates are very stable.	Aluminates are lese stable.
13. $B(OH)_3$, *i.e.*, H_3BO_3, is weakly acidic.	$Al(OH)_3$ is amphoteric.
14. BCl_3 is a fuming liquid.	$AlCl_3$ is a solid.
15. It combines with metals to form borides. $3Mg + 2B \rightarrow Mg_3B_2$	It combines with metals to form alloys.
16. Its maximum covalency is 4.	Its maximum covalency is 6.

Diagonal Relationship between B and Si: Similarities between B and Si.

Boron resembles silicon in many of its properties with which it is diagonally related. Some common properties showing the similarities between B and Si are given below:

1. Preparation. Both the elements can be prepared by reducing their oxides with Mg.

$$B_2O_3 + 3Mg \longrightarrow 2B + 3MgO$$
$$SiO_2 + 2Mg \longrightarrow Si + 2MgO$$

2. Action of alkalies. Both the elements dissolve in alkalies (boron only in fused while silicon in fused as well as aqueous), forming *metaborate* (BO_2^-) and *metasilicate* (SiO_3^{2-}) end liberate H_2.

$$2B + 2NaOH + 2H_2O \longrightarrow 2NaBO_2 + 3H_2 \uparrow$$
$$Si + 2NaOH + H_2O \longrightarrow Na_2SiO_2 + 2H_2 \uparrow$$

3. Action of water and non-oxidising acids. None of the elements is attacked by water and non-oxidising acids.

4. Occurrence. None of these elements occurs in the free state. Both are found as oxy-compounds, *e.g. borates* and *silicates*.

5. Non-metallic character. Chemically both the elements are typical non-metals and are bad conductors of heat and electricity.

6. Allotropic forms. Both the elements exist in two allotropic forms *viz. amorphous* and *crystalline*. The crystalline form of both the elements is harder and less reactive than the amorphous form.

7. Density, electronegativity, ionisation potential and boiling points. Density, electronegativity, ionisation potential and boiling point of both the elements are almost the same.

8. Melting points. Both the elements have high melting points (B = 2300°C, Si = 1420°C).

9. Formation of covalent compounds. Both the elements form mainly covalent compounds like hydrides (*e.g.* B_2H_6, SiH_4), halides (*e.g.* BX_3, SiX_4), oxides (*e.g.* B_2O_3, SiO_2).

10. Reactivity. Both the elements are normally inert at ordinary temperature. For example they are not oxidised in air and do not react with other elements or with dil. acids. However, the reactivity can be increased by increasing the temperature. At high temperatures these elements react with a number of metals and non-metals.

11. Combination with metals. Both the elements combine with metals like Mg and form borides and silicides respectively.

$$2B + 3Mg \longrightarrow Mg_3B_2 \quad (\textit{Boride})$$
$$Si + 2Mg \longrightarrow Mg_2Si \quad (\textit{Silicide})$$

12. Oxides. *Preparation.* Boron and silicon combine with O_2 at elevated temperature to form the stable oxides *viz.* B_2O_3 and SiO_2.

$$4B + 3O_2 \longrightarrow 2B_2O_3$$
$$Si + O_2 \longrightarrow SiO_2$$

Properties. (*i*) These oxides are solids of high melting points and are acidic in nature, since they give salts (metaborates, BO_2^- and metasilicates, SiO_3^{2-} respectively) with strong alkalies.

$$B_2O_3 + 2NaOH \longrightarrow \underset{\text{Sod. meta-borate}}{2NaBO_2} + H_2O$$

$$SiO_2 + 2NaOH \longrightarrow \underset{\text{Sod. meta-silicate}}{Na_2SiO_3} + H_2O$$

(*ii*) These oxides also react with fused metal oxides forming metaborate (BO_2^-) and meta silicate (SiO_3^{2-}).

$$B_2O_3 + MO \longrightarrow M(BO_2)_2$$
$$SiO_2 + MO \longrightarrow MSiO_3$$

(*iii*) Oxides also appear to have some *feeble basic character*, since they also react with HF.

$$B_2O_3 + 6HF \longrightarrow 2BF_3 + 3H_2O$$
$$SiO_2 + 4HF \longrightarrow SiF_4 + 2H_2O$$

These reactions have been used to prepare the fluorides of these elements.

(*iv*) Both the oxides have macro-molecular structure.

(*v*) When these oxides are reduced by Mg metal, elements are obtained in the free state.

$$B_2O_3 + 3Mg \longrightarrow 2B + 3MgO$$
$$SiO_2 + 2Mg \longrightarrow Si + 2MgO$$

(*vi*) Oxides, in association with water, give the corresponding oxy acids *viz. boric acids* and *silicic acids.*

(*vii*) When heated with carbon in an electric furnace, carbides are obtained.

$$2B_2O_3 + 7C \longrightarrow B_4C + 6CO$$
$$SiO_2 + 3C \longrightarrow SiC + 2CO$$

13. Halides. *Preparation.* (*i*) BCl_3 and $SiCl_4$ can be prepared by passing dry Cl_2 over a heated mixture of the oxide (B_2O_3 or SiO_2) and carbon.

$$B_2O_3 + 3C + 3Cl_2 \longrightarrow 2BCl_3 + 3CO$$
$$SiO_2 + 2C + 2Cl_2 \longrightarrow SiCl_4 + 2CO$$

(*ii*) BF_3 and SiF_4 can be prepared as follows:

$$B_2O_3 + 6HF \longrightarrow 2BF_3 + 3H_2O$$
$$SiO_2 + 4HF \longrightarrow SiF_4 + 2H_2O$$

(*iii*) Both the elements burn in the vapours of halogens and form their respective halides.

Properties. (*i*) Halides of both the elements are covalent compounds.

(*ii*) Chlorides (BCl_2 and $SiCl_4$) are liquids, fume in moist air end are readily hydrolysed by water to give *orthoboric acid*, H_3BO_3 and *orthosilicic acid*, H_4SiO_4.

$$BCl_3 + 3H_2O \longrightarrow \underset{\text{Orthoboric acid}}{H_3BO_2} + 3HCl$$

$$SiCl_4 + 4H_2O \longrightarrow \underset{\text{Orthosilicic acid}}{H_4SiO_4} + 4HCl$$

(*iii*) Fluorides (BF_3 and SiF_4) are colourless fuming gases and are readily hydrolysed by water.

$$BF_3 + 3H_2O \longrightarrow H_3BO_3 + 3HF$$
$$[BF_3 + HF] \longrightarrow HBF_4] \times 3$$

On adding:
$$4BF_4 + 3H_2O \longrightarrow \underset{\text{Orthoboric acid}}{H_3BO_3} + \underset{\text{Hydro fluoboric acid}}{3HBF_4}$$

$$SiF_4 + 4H_2O \longrightarrow H_4SiO_4 + 4HF$$
$$[SiF_4 + 2HF] \longrightarrow H_2SiF_6] \times 2$$

On adding:
$$3SiF_4 + 4H_2O \longrightarrow \underset{\text{Orthosilicic acid}}{H_4SiO_4} + \underset{\text{Fluosilicic acid}}{2H_2SiF_6}$$

Both the acids *viz*. H_3BO_3 and H_4SiO_4 are weak acids.

(*iv*) Halides can act as *Lewis acids* towards suitable donor molecules to form *adducts*, *e.g.*

Halides		***Donor molecules***		***Adducts***
BX_3	+	py	$\longrightarrow$	BX_3.py
SiX_4	+	2py	$\longrightarrow$	SiX_4.2py

Here py = pyridine which is a donor molecule.

(*v*) When benzene solution of anhydrous chlorides is refluxed with alcohol, alkoxides (*i.e.*, esters) are obtained.

$$BCl_3 + 3C_2H_5OH \longrightarrow (C_2H_5)_3\ BO_3 + 3HCl$$
$$SiCl_4 + 4C_2H_5OH \longrightarrow (C_2H_5)_4SiO_4 + 4HCl$$

(*vi*) Halides can be reduced by *lithium aluminium hydride*, $LiAlH_4$ to get the hydrides of B and Si *viz*. B_2H_6 and SiH_4.

$$4BCl_3 + 3LiAlH_4 \longrightarrow 3LiCl + 3AlCl_3 + 2B_2H_6$$
$$SiCl_4 + LiAlH_4 \longrightarrow LiCl + AlCl_3 + SiH_4$$

14. Non-existence of oxy salts. No oxy salts of boron and silicon containing cationic boron and silicon are known.

15. Hydrides. B and Si form similar hydrides known as *boranes* and *silanes* respectively.

Preparation. These can be prepared by their magnesium salts *e.g.*

$$Mg_3B_2 + 6HCl \longrightarrow 3MgCl_2 + B_2H_6 + \text{Higher boranes}$$
$$Mg_2Si + 4HCl \longrightarrow 2MgCl_2 + SiH_4 + \text{Higher silanes}$$

(ii) B_2H_6 and SiH_4 can also be prepared by the reduction of their chlorides by *lithium aluminium hydride*, $LiAlH_4$

$$4BCl_3 + 3LiAlH_4 \longrightarrow 3LiCl + 3AlCl_3 + 2B_2H_6$$
$$SiCl_4 + LiAlH_4 \longrightarrow LiCl + AlCl_3 + SiH_4$$

Properties *(i)* All the hydrides are covalent compounds which are either volatile liquids or gases and possess a disagreeable smell.

(ii) They instantly catch fire on exposure to air, forming the oxides of the elements and H_2O.

$$B_2H_6 + 3O_2 \longrightarrow B_2O_3 + 2H_2O$$
$$SiH_4 + 2O_2 \longrightarrow SiO_2 + 2H_2O$$

(iii) They can be hydrolysed by water

$$B_2H_6 + 6H_2O \longrightarrow 2H_3BO_3 + 6H_2$$
$$SiH_4 + 4H_2O \longrightarrow H_4SiO_4 + 4H_2$$

(iv) They are strong reducing agents *e.g.*

$$B_2H_6 + Cl_2 \longrightarrow B_2H_5Cl + HCl$$
$$SiH_4 + Cl_2 \longrightarrow SiH_3Cl + HCl$$

16. Sulphides. Both the elements combine with sulphur at elevated temperatures to form sulphides (B_2S_3 and SiS_2). These sulphides are attacked by water with the liberation of H_2S.

$$B_2S_3 + 6H_2O \longrightarrow 2H_3BO_3 + 3H_2S$$
$$SiS_2 + 4H_2O \longrightarrow H_4SiO_4 + 2H_2S$$

17. Nitrides. Nitrides (BN and Si_3N_4) can be obtained either by the direct combination of the elements with N_2 or by passing N_2 over heated mixture of the oxide and carbon. These nitrides decompose steam, liberating NH_3.

$$BN + 3H_2O\ (steam) \longrightarrow H_3BO_3 + NH_3$$
$$Si_3N_4 + 12H_2O\ (steam) \longrightarrow 3H_4SiO_4 + 4NH_3$$

18. Carbides. *Preparation.* The carbides *viz.* B_4C and SiC are obtained by heating the elements with carbon in an electric furnace. On a commercial scale these are obtained by heating their oxides with carbon in an electric furnace.

$$2B_2O_3 + 7C \longrightarrow B_4C + 6CO$$
$$SiO_2 + 3C \longrightarrow SiC + 2CO$$

Properties. These are covalent network solids. These are inert towards water. Both the carbides are hard substances and are used as abrasive.

19. Formation of complexes. Both the elements form complexes, *e.g. hydrofluoboric acid*, HBF_4 and *fluosilicic acid*, H_4SiF_6.

Comparison between Boron and Silicon

(A) Similarities. For this see "*Diagonal relationship between boron and silicon: Similarities between boron and silicon*".

(B) Dissimilarities. The following points show the dissimilarities between boron and silicon.

Boron	Silicon
1. Boron has three electrons in its valence-shell (2, 3)	Silicon has four electrons in its valence-shell (2, 8, 4)
2. Boron in the form of minerals constitutes only 0.001% of the earth's crust	Silicon is the second most abundant element and constitutes 26% of the earth's crust
3. Boron is tri-covalent in its compounds	Silicon is tetra-covalent in its compounds
4. Boron on heating with N_2 gives BN $2B + N_2 \longrightarrow 2BN$	Silicon does not combine with N_2
5. Boron is oxidiged by hot conc. H_2SO_4 or conc. HNO_3 to form H_3BO_3 $2B + 3H_2SO_4 \longrightarrow 2H_3BO_3 + 3SO_2$ $B + 3HNO_3 \longrightarrow H_3BO_3 + 3NO_2$	Silicon is attacked only by HF, forming hydrofluosilicic acid, H_2SiF_6 $Si + 6HF \longrightarrow H_2SiF_6 + 2H_2$
6. B_3O_3 is amphoteric in nature	SiO_2 is acidic in nature.

Position of Boron in the Periodic Table

We know hat boron is present in Group III A of the periodic table. The element lying below it is aluminium and the element which is diagonally related to boron is silicon which is present in Group IVA (period number = 3) as shown below :

III A	IV A
B	C
Al	Si
Ga	Ge

(B → Si diagonal relationship)

Position of boron in the periodic table

Thus the position of boron in the periodic table is justified by discussing (1) *Similarities between B and Al,* and (2) *Similarities between B and Si.* These points have already been discussed in this chapter.

Questions with Answers

Q.1 BCl_3 is trigonal planar while $AlCl_3$ is tetrahedral in dimeric state. Explain.

Ans : Both BCl_3 and $AlCl_3$ are electron deficient molecules having six electrons in the valence shell of their respective central atoms. To complete their octets, the central atom in each case can accept a pair of electrons from the chlorine atom of another molecule froming dimeric structures. However, because of small size of B, it cannot accommodate four big sized Cl atoms around it. Therefore, *BCl_3 prefers to exist as monomeric planar molecule in which B atom is sp^2 hybridized.*

On the other hand, Al because of its bigger size can easily accommodate four Cl atoms around it. As a result, $AlCl_3$ exsts as a dimer. In this dimer, since the covalency of Al has increased to 4, *Al is sp^3-hybridized and the four Cl atoms are held tetrahedrally around it.*

Q.2 Aqueous solution of hydrous $AlCl_3$ is acidic in natuer. Explain.

Ans : Aqueous solution of $AlCl_3$ is acidic due to its hydrolysis.

$$\underset{}{AlCl_3} + 3H_2O \longrightarrow \underset{\text{weak base}}{Al(OH)_3} + \underset{\text{strong acid}}{3\,HCl}$$

or $$3Cl^- + 3H_2O \longrightarrow 3OH^- + 3HCl$$

Since the acid produced in the above reaction is a strong acid, the solution of $AlCl_3$ in water is acidic in nature.

Q.3 In non polar solvents $AlCl_3$ is covalent but in polar solvents this compound is electrovalent. Explain.

Ans : In non-polar solvents or organic solvents like benzene, $AlCl_3$ exists as a dimeric molecule, $(AlCl_3)_3$ or Al_2Cl_6 which has covalent bonds and hence is covalent in nature. On the other hand in polar solvents like water, $AlCl_3$ gets hydrated to form hydrated salt, $AlCl_3.\ 6H_2O$. In the formation of $AlCl_3.\ 6H_2O$, $AlCl_3$ gets hydrated by H_2O molecules and a large amount of hydration energy is also liberated. This hydration energy helps in the ionisation of hydrated salt, $AlCl_3.\ 6H_2O$ to produce $[Al(H_2O)_6]^{3+}$ (hydrated ion) and Cl^- ions. Due to the production of ions, $AlCl_3$ in polar solvents behaves as an electrovalent compound.

$$AlCl_3 + 6H_2O \longrightarrow AlCl_3.\ 6H_2O \longrightarrow \underset{\text{Hydrated ion}}{[Al(H_2O)_6]^{3+}} + 3Cl^- + \text{Hydration energy}$$

Q.4 When BF_3 is added to a solution of AlF_3 in KF, AlF_3 is precipitated. Explain.

Ans : Since KF is an ionic fluorid, it gives F^- ions which combine with AlF_3 to form AlF_6^- ion (complex ion)

$$3KF \longrightarrow 3K^+ + 3F^-$$

$$AlF_3 + 3F^- \longrightarrow [AlF_6]^{3-}$$

$$\overline{3KF + AlF_3 \longrightarrow 3K^+ [AlF_6]^{3-} \textit{ or } K_3[AlF_6]}$$

Now since B-atom has smaller size and higher electronegativity than Al atom, B-atom has higher tendency to form complexes.

Thus BF_3 will react with $K_3[AlF_3]$ formed as above and forms the complex, $K[BF_4]$. AlF_3 is obtained as precipitate.

$$K_3[AlF_6] + 3BF_3 \longrightarrow \underset{\text{Pot. tetra-fluoro borate}}{3K[BF_4]} + AlF_3(s)\downarrow$$

Q.5 AlF_3 dissolves in KF but not in anhydrous HF. Explain.

Ans : Anhydrous HF is a covalent compound and is strongly H-bonded. Therefore, it does give F^- ions and hence AlF_3 does not dissolve in HF. In contrast, KF contains F^- ions and hence combines with AlF_3 to form the soluble complex.

$$3KF \longrightarrow 3K^+ + 3F^-$$
$$AlF_3 + 3F^- \longrightarrow [AlF_6]^-$$

On adding : $3KF + AlF_3 \longrightarrow 3K^+ + [AlF_6]^{3-}$ or $K_3[AlF_6]$
Pot. hexafluoro aluminate (soluble complex)

Q.6 BF_3 and BCl_3 do not form dimeric molecules but BH_3 forms a dimer, B_2H_6. Explain.

Ans : Since F and Cl atoms are big in size, these atoms cannot be accomdated around small B-atom in the bridged position and hence BF_3 and BCl_3 donot form dimeric molecules. On the other hand, since H-atoms are small in size, these atoms can be accomdated round small B-atom in the bridged position and hence dimeric molecule, B_2H_6 is easily fromed.

Q.7 Boron trihaldes (excepting BF_3) and $AlCl_3$ fume in moist air. Explain. (Roorkee 1986)

Ans : All the given halides are hydrolysed by H_2O present in moist air and fumes of HCl are produced. (Also see the action of moist air on $BeCl_2$)

$$BX_3\ (X = Cl, Br, I) + 3H_2O \rightarrow B(OH)_3 + 3HX$$
$$AlCl_3 + 3H_2O \rightarrow Al(OH)_3 + 3HCl$$

BF_3 reacts with H_2O in the following way :

$$F_3B + :\ddot{O}H_2 \rightarrow F_3B \leftarrow OH_2 \text{ (Adduct)}$$

Q.8 Explain why B–F bond energy in BF_3 is 646 KJ/mole but that of N–F bond energy in NF_3 is only 280 KJ/mole.

***or* Explain why B–F bond energy in BF_3 molecule is higher than N–F bond energy in NF_3 molecule.**

Ans : We know that BF_3 molecule (sp^2 hydridisation of B-atom) has three B–F bonds each of which has double bond character due to the formation of an additional $[2p\ (F) \rightarrow 2p\ (B)]p$ bond which is a $(p\pi - p\pi)$ bond. This bond is formed by the donation of an electron pair from filled $2p$ orbital of F-atom to the vacant $2p$ orbital of B-atom. In case of NF_3 molecule (sp^3 hybridisation of N-atom), since there is no vacant $2p$ orbital on N-atom, there is no scope of the formation of a $[2p(F) \rightarrow 2p\ (N)]\pi$ bond in NF_3 molecule. Thus none of the three N–F bonds in NF_3 molecule has double bond character. Now since bond energy increases with the increase in double bond character, B–F bond energy in BF_3 molecule is higher than N–F bond energy in NF_3 molecule.

Q.9 Write all the possible octet structures for BF_3 molecule and say which of these structures are the most stable ?

Ans : For BF_3 molecule, the following octet structures can be written.

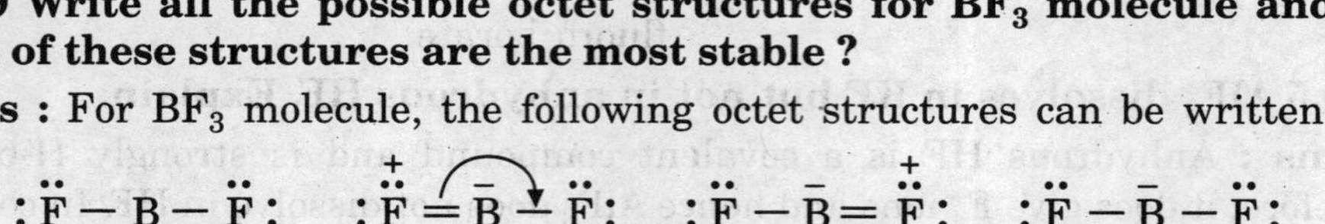

(*a*) (*b*) (*c*) (*d*)

Structure (*a*) is a covalent structure while structures (*b*), (*c*) and (*d*) are ionic structures. We know that greater is the number of B–F covalent bonds present in a contributing structure, greater is the stability of that structure. Now since each of the ionic structures has four B–F covalent bonds while covalent structure has only three B–F covalent bonds, ionic structures are more stable than covalent staructure. BF_3 molecule has 54% π-character.

Q.10 Anhydrous $AlCl_3$ is more soluble in diethyl ether than hydrous (hydrated) $AlCl_3$. Explain.

Ans : Higher solubility of anhydrous $AlCl_3$ in diethyl ether, $O(C_2H_5)_2$ as compared to that of hydrous $AlCl_3$ can be explained as follows on the basis of bonding. Anhydrous $AlCl_3$ is an electron-deficient compound while hydrated $AlCl_3$ is not. Therefore, Anhyd. $AlCl_3$ is more soluble in diethyl ether because the oxygen atom of ether donates a pair of electrons to the vacant *p*-orbital on the Al atom in $AlCl_3$ froming a coordinate bond.

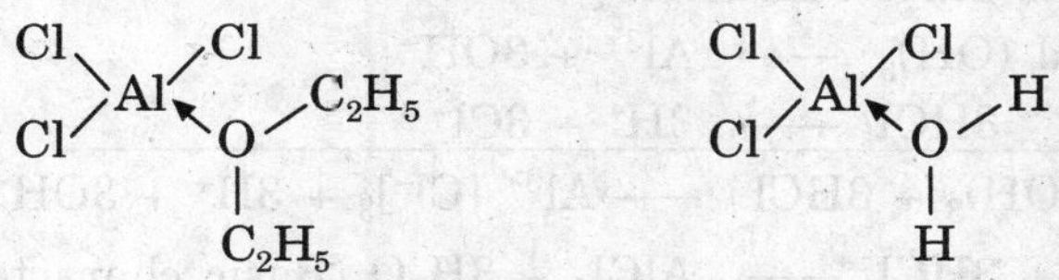

Hydrated $AlCl_3$

($AlCl_3.H_2O$)

In case of hydrated $AlCl_3$ Al is not electron deficient since H_2O has already donated a pair of electrons to it.

Q.11 Aluminium forms AlF_6^{3-} ion but boron does not form BF_6^{3-} ion. Explain why? (Punjab 1995 April)

Ans : (*i*) Explanation based on steric hinderance.

Since B-atom is much smaller in size than Al-atom, this atom cannot accomdate six F^- ions around it and hence cannot form BF_6^{3-} ion. On the other hand, since Al is a bigger atom, it can easily accomdate six F^- ions and can from AlF_6^{3-} ion.

(*ii*) Explanation based on the presence/absence of *d*-orbitals. Lewis structures of the given ions are given in Fig. 22.1.

BF_6^{3-} ion (wrong structure)

AlF_6^{3-} ion (correct structure)

Fig. 22.1. Lewis structures of BF_6^{3-} and AlF_6^{3-} ions.

These structures show that the central atom (B or Al) is sp^3d^2 hybridised. Now since B-atom does not have *d*-orbitals (B = $2s^1\ 2p^1\ 2p^1\ 2p^\circ$), it cannot undergo sp^3d^2 hybridisation and BF_6^{3-} ion cannot be fromed. On the other hand, since Al has $3d$ orbitals (Al = $3s^1\ 3p^1\ 3p^1\ 3p^\circ\ 3d^\circ\ 3d^\circ$), it can undergo sp^3d^2 hybridisation and can accept three electron pairs donated by three F^- ions to form AlF_3^{3-} ion.

***(iii)* Explanation based on the rule of maximum covalency.**

In both Lewis structures, B and Al atoms show a covalency of six. Now since B is an element of *2nd* period, it cannot extend its covalency beyond four, *i.e.* B can form BF_4^- but not BF_6^{3-}. On the other hand, since Al is an element of *3rd* period (higher period), it can achieve a covalency of six, *i.e.* it can from AlF_6^{3-}.

Q.12 Explain why Al $(OH)_3$ is amphoteric. (MLNR 1995)

Ans : Al $(OH)_3$ is an amphoteric compound since it reacts both with acids as well as with alkalies to form salt and water (solvent). With acids it behaves as a weak base, since it ionises to a slight extent to give OH^- ions which combine with H^+ ions obtained by the ionisation of the acid to form H_2O. Thus its reaction with HCl shows its basic character.

$$\text{Al }(OH)_3 \longrightarrow Al^{3+} + 3OH^-$$
$$3HCl \longrightarrow 3H^+ + 3Cl^-$$

On adding : $\text{Al }(OH)_3 + 3HCl \longrightarrow Al^{3+}[Cl^-]_3 + 3H^+ + 3OH^-$

or $\text{Al }(OH)_3 + 3HCl \longrightarrow \underset{\text{Salt}}{AlCl_3} + \underset{\text{Solvent}}{3H_2O}$ (Basic character)

In presence of an alkali, Al $(OH)_3$ acts as an acid, since it gives metaborate ion (AlO_2^-) and H^+ ions. These H^+ ions combine with OH^- ions obtained by the ionisation of the alkali to form H_2O (solvent). Thus the reaction between Al $(OH)_3$ and NaOH shows the acidic character of Al $(OH)_3$.

$$\text{Al }(OH)_3 \longrightarrow AlO_2^- + H_2O + H^+$$
$$NaOH \longrightarrow Na^+ + OH^-$$

$\text{Al }(OH)_3 + NaOH \longrightarrow Na^+[AlO_2]^- + H_2O + H^+ + OH^-$

or $\text{Al }(OH)_3 + NaOH \longrightarrow \underset{\text{Salt}}{Na^+AlO_2^-} + \underset{\text{Solvent}}{2H_2O}$ (Acidic character)

The formation of AlO_2^- ion takes place through the following steps :

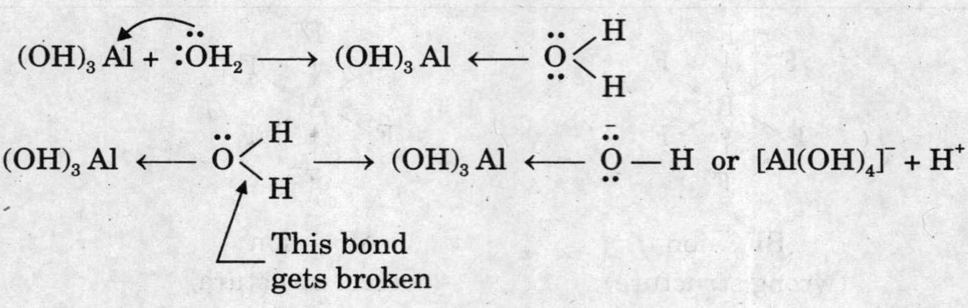

$$[\text{Al }(OH)_4]^- \xrightarrow{\text{Dissociation}} AlO_2^- + 2H_2O$$

On adding : $\text{Al }(OH)_3 + H_2O \longrightarrow AlO_2^- + 2H_2O + H^+$

or $\text{Al }(OH)_3 \longrightarrow AlO_2^- + H_2O + H^+$

Q.13 Boron atom does not from B^{3+} ion Explain. **(Punjab 1996 Oct)**

Ans : In order to get B^{3+} ion, we will have to remove three electrons from the valence-shell of B-atom ($B = 2s^2\ 2p^1$). Obviously, energy required to remove three electrons is equal to the sum of IE_1, IE_2 and IE_3 of B-atom. Now since B-atom has very small size and the sum of IE_1, IE_2 and IE_3 is so high that lattice energy liberated in the interaction of B^{3+} ion with anions is not enough for the fromation of B^{3+} ion.

University Questions

1. Justify the inclusion of B, Al, Ga, In and Tl in the same group of the periodic table on the basis of the following properties : (*i*) ionisation energies (*ii*) oxidation states (*iii*) covalent character (*iv*) hydrated ions (*v*) size of atoms (*vi*) reactivity (*vii*) electropositive or metallic character. (*Jodhpur 83*)
2. Arrange BF_3, BCl_3 and BBr_3 in the increasing order of their Lewis acid strength. (*Delhi Hons. 86, 87*)
3. Explain the following: (*i*) Acceptor behaviour of boron halides (*ii*) B—X bond shortening in boron halides (*iii*) Lewis acid character of boron halides. (*Raj. 87, 87 S*)
4. Discuss the position of boron in the period table. (*Meerut 80 S, 82 S, 83 S, 84; Agra 83; Raj. 81; Utkal 86; Madras 86*)
5. Discuss the similarities and dissimilarities between boron and aluminium. (*Meerut 88, 89; Agra 88*)
6. Discuss the similarities and dissimilarities between boron and silicon. (*Meerut 80, 83; Jiwaji 80; Agra 83; Garhwal 81; Indore 81; Lucknow 81; Delhi 86*)

CHAPTER 23

Chemistry of Group 13 (III A) Elements, Their Compounds and Ceramics Industry

Boron

Occurrence.

Boron does not occur in the free state. In the combined state it occurs as salts of boric acid, the important of which are : *borax or tincal*, $Na_2B_4O_7.10H_2O$; *colemanite*, $Ca_2B_5O_{11}.5H_2O$; *kernite or razorite*, $Na_2B_4O_7.4H_2O$; *boracite*, $2Mg_3B_8O_{15}$, $MgCl_2$; *barocalcite*, $CaB_4O_7.4H_2O$. The important minerals from which boron is extracted are borax and colemanite. Borax occurs in Ceylon, Tibet, California and Puga valley in Kashmir. Colemanite occurs in California and South Africa. Boron is also present to some extent in some plants and sometimes in coal ash.

Extraction of Boron From Minerals.

Boron is generally extracted from borax and colemanite minerals. The extraction from these minerals involves the following two steps :

(i) Preparation of boric anhydride; B_2O_3. (a) From borax. The finely powdered mineral is boiled with conc. HCl or conc. H_2SO_4 when sparingly soluble boric acid separates out.

$$\underset{\text{Borax}}{Na_2B_4O_7} + 2HCl + 5H_2O \longrightarrow 2NaCl + 4H_3BO_3$$

$$Na_2B_4O_7 + H_2SO_4 + 5H_2O \longrightarrow Na_2SO_4 + 4H_3BO_3$$

Boric acid on strong heating forms boric anhydride.

$$2H_3BO_3 \longrightarrow B_2O_3 + 3H_2O$$

(b) From colemanite. The finely powdered mineral is dissolved in the boiling water and sulphur dioxide gas is passed through the solution. Calcium bisulphite being highly soluble remains in solution and the sparingly soluble boric acid separates on cooling.

$$\underset{\text{Colemanite}}{Ca_2B_6O_{11}} + 11H_2O + 4SO_2 \longrightarrow \underset{\substack{\text{Calcium bisulphite}\\\text{(highly Soluble)}}}{2Ca(HSO_3)_2} + \underset{\text{(Insoluble)}}{6H_3BO_3}$$

Boric acid on strong heating forms boric anhydride.

$$2H_3BO_3 \longrightarrow B_2O_3 + 3H_2O$$

Alternatively, the finely powdered colemanite mineral is boiled with calculated quantity of sodium carbonate solution.

$$2Ca_2B_6O_{11} + 2Na_2CO_2 + H_2O \longrightarrow 3Na_2B_4O_7 + 3CaCO_3 \downarrow + Ca(OH)_2$$

The precipitate of calcium carbonate is removed by filteration and borax is crystallized from the filtrate: From borax, boric anhydride may be obtained as above.

(*ii*) *Reduction of boric anhydride.* The boric anhydride so obtained is mixed with sodium, potassium or magnesium powder and heated in a covered crucible when it is reduced to boron.

$$B_2O_2 + 6K \longrightarrow 2B + 3K_2O$$
$$B_2O_3 - 3Mg \longrightarrow 2B + 3MgO$$

The heated mass is stirred with an iron rod so as to oxidise the excess of sodium or potassium. The mass is now boiled with water, then with hydrochloric acid and finally with hydrofluoric acid, when dark brown powder of amorphous boron only remains as residue and rest everything is dissolved. This residue is now washed with water.

Crystalline form of boron is similarly obtained by the reduction of boric anhydride with aluminium powder.

$$B_2O_3 + 2Al \longrightarrow 2B + Al_2O_3$$

The crystals are treated with NaOH solution to remove the impurity of aluminium.

Extraction of Boron by Electrolysis.

Now-a-days boron is obtained by the electrolysis of a fused mixture of boric anhydride, magnesium oxide and magnesium fluoride at 1100°C. The electrolysis is carried out in a carbon crucible which itself acts as anode and cathode is of iron which remains in the centre of the crucible. The magnesium obtained by electrolysis of its oxide reduces boric anhydride to boron which is deposited on the cathode.

$$2MgO \longrightarrow 2Mg + O_2$$
$$B_2O_3 + 3Mg \longrightarrow 2B + 3MgO$$

Boron so obtained still contains a number of impurities. It is purified by heating it electrically in vacuum at 1100°C. At this temperature, the impurities are volatized off leaving behind pure boron.

Physical Properties

Boron exists in two allotropic forms *viz. crystalline or adamantine* and *amorphous boron.* Crystalline form can be obtained by the reduction of B_2O_3 or KBF_4 by Al.

$$B_2O_3 + 2Al \longrightarrow Al_2O_3 + 2B \text{ (crystalline boron)}$$
$$KBF_4 + Al \longrightarrow AlF_3 + KF + B \text{ (crystalline boron)}$$

Crystalline boron can also be prepared by dropping amorphous boron in molten Al at 1500° when it appears to dissolve. On cooling yellow coloured crystals deposit on this surface of Al. The mass is broken up and heated with HCl to dissolve away Al when very hard crystals are left behind. Boron obtained by this method is not pure since it contains some aluminium as impurity.

Crystalline boron is black and possesses metallic lustre. It is extremely inert. It is scarcely attacked by conc. HNO_3 or H_2SO_4 and not at all by boiling conc.

HCl or HF. It is, however, vigorously attacked by fused sodium peroxide and by fused oxidizing fusion mixture. Crystalline boron is very hard and is second to diamond in hardness. Its specific gravity is 3.3. Both needle-like and plate-like hexagonal crystals of crystalline boron are known. It is a bad conductor of electricity. Amorphous boron is brown powder of density 2.45. It is very hard to fuse. The m.pt. and b.pt. are 2300°C and 2550°C respectively.

Chemical Properties.

1. Action of air and O_2. Amorphous boron is very reactive. When it is heated at 700°C in air or O_2, it burns with a reddish flame forming a mixture of oxide and nitrate.

$$4B + 3O_2 \longrightarrow 2B_2O_3$$
$$2B + N_2 \longrightarrow 2BN$$

2. Action of acids. Halogen acids have no action on boron but it is attacked by oxidising acids like HNO_3 and H_2SO_4,

$$B + 3HNO_3 \longrightarrow H_3BO_3 + 3NO_2$$
$$2B + 3H_2SO_4 \longrightarrow 2H_3BO_3 + 3SO_2$$

3. As reducing agent. Boron is powerful reducing agent, since it can replace carbon from CO_2, silicon from SiO_2 and N_2 from NO.

$$4B + 3CO_2 \longrightarrow 2B_2O_3 + 3C$$
$$4B + 3SiO_2 \longrightarrow 2B_2O_3 + 3Si$$
$$4B + 6NO \longrightarrow 2B_2O_3 + 3N_2$$

4. Action of alkalies. It is vigorously attacked by fused alkalies or by fusion with oxidising fusion mixture.

$$2B + 6NaOH \longrightarrow 2Na_3BO_3 + 3H_2 \uparrow$$

5. Action of metals. Excepting with Cu, Ag and Au, boron combides with other metals at high temperatures to form borides. These borides are extremely hard substances. Mg_3B_2 and Be_3B_2 are attacked by acids with the formation of boron hydrides.

6. Action of non-metals. Boron also reacts with non-metals. For example : (*a*) Boron combines with N_2 at a temperature greater than 900° to form boron nitride, BN. (*b*) Boron forms boron sulphide B_2S_3 when heated with S. (*c*) On heating with carbon in an electric furnace an extremely hard substance namely boron carbide, B_4C is formed.

7. Action of steam. Boron reacts with steam at red heat liberating H_2.

$$2B + 3H_2O \longrightarrow B_2O_3 + 3H_2 \uparrow$$

Uses of Boron

Boron is used (*i*) as a deoxidiser in the casting of copper, (*ii*) for making boron steels which are very hard and are used as control rods in atomic reactors. (*iii*) the traces of boron compounds are essential for plant metabolism and for the fertility of soil, (*iv*) as a catalytic agent, (*v*) in the manufacture of thermometers and thermoregulators, (*vi*) as scavanger to remove gaseous impurities from molten copper and as a semi-conducting material, (*vii*) the hydrides of boron are used as fuels for rockets because of their very high energy/weight ratio, (*viii*) metallic borides and boron carbide are used as abrasives.

Compounds of Boron

1. Boric Acid, H_3BO_3.

Several boric acids like *orthoboric acid* (H_3BO_3 or $\frac{1}{2}B_2O_3.\frac{3}{2}H_2O$ or $B(OH)_3$], *metaboric acid* (HBO_2 or $\frac{1}{2}B_2O_3.\frac{1}{2}H_2O$), *pyroboric acid* ($H_6B_4O_9$ or $2B_2O_3.3H_2O$) and *tetraboric acid* ($H_2B_4O_7$ or $2B_2O_3.H_2O$) are known. Out of these orthoboric acid which is generally referred to as **boric acid** is the most important and is stable both in the solid state and in solution. The remaining acids are stable in the solid state and gradually change into orthobaric acid in solution

$$\underset{\text{Metaboric acid}}{HBO_2} + H_2O \longrightarrow \underset{\text{Boric acid}}{H_3BO_3}$$

$$\underset{\text{Tetraboric acid}}{H_2B_4O_7} + 5H_2O \longrightarrow 4H_3BO_3$$

Occurrence : Small quantities of boric acid are present in the jets of steam called *soffioni* which is mixed with other gases in the volcanic district of Tuscany. Besides this, it is also found in many mineral waters. It is also present in traces of hops, in berries and fruits and often in wines.

Extraction : Boric acid can be extracted from borax and colemanite minerals as discussed under the extraction of boron.

Properties : (*i*) Boric acid forms colourless nacreous laminae which are soapy to the touch. It is slightly soluble in cold water, more so in hot and is volatile in steam. 100 grams of water dissolve 4.01 gms. at 12°C and 39.95 gms. at 100°C. The solution in water is a very weak acid.

(*ii*) *Acidic nature* : *Action of alkalies.* It is a weak acid and gives only a wine red colour with litmus. It reacts with alkalies to form salts

$$NaOH + H_3BO_3 \longrightarrow \underset{\text{Sod. meta-borate}}{NaBO_2} + 2H_2O$$

$$2NaOH + 4H_3BO_3 \longrightarrow \underset{\text{Sod. tetra-borate}}{Na_2B_4O_7} + 7H_2O$$

The end point during the titration of $B(OH)_3$ against NaOH is not sharp, since excessive hydrolysis of $Na^+BO_2^-$ occurs. However, if the titration is carried out in presence of polyhydroxy compounds like glycerol or sugar, $B(OH)_3$ behaves as a strong monobasic acid and the end point can be detected as easily as it can be done during the titration of strong acid with a strong alkali.

(*iii*) *Action of acetic anhydride.* When treated with acetic anhydride, boron acetate is formed.

$$2B(OH)_3 + \underset{\text{Acetic anhydride}}{5(CH_3CO)_2O} \longrightarrow (CH_3COO)_2BOB(OOCCH_3)_2.6CH_3COOH$$

(*iv*) *Action of Na_2O_2.* With Na_2O_2 it gives sod. metaborate peroxohydrate, $NaBO_2.H_2O_2.3H_2O$.

(*v*) It is a very weak acid and ionises mainly as monobasic acid and that too, to a very limited extent. It is not a proton donor but behaves as a *Lewis acid, i.e.* it can accept a lone pair of electrons from an OH^- ion.

$$H_2O \longrightarrow OH^- + H^+ \quad ...(i)$$

$$B(OH)_3 + OH^- \longrightarrow [B(OH)_4]^- \quad ...(ii)$$

On adding (*i*) and (*ii*), we get :

$$B(OH)_3 \text{ (or } H_3BO_3) + H_2O \rightleftarrows [B(HO)_4]^- + H^+$$

The above equilibrium lies to the left, showing that it is a weak acid.

(*vi*) *Action of heat.* It decomposes on heating to form metaboric acid at 100°C, tetraboric acid at 140°C and a glassy mass of boric anhydride at red heat.

$$4 \times [\underset{}{H_3BO_3} \xrightarrow{100°C} \underset{\text{Metaboric acid}}{HBO_2} + H_2O] \quad ...(i)$$

$$4HBO_2 \xrightarrow{140°C} \underset{\text{Tetraboric acid}}{H_2B_4O_7} + H_2O \quad ...(ii)$$

$$H_2B_4O_7 \xrightarrow{\text{Read heat}} \underset{\text{Boric anhydride (Glassy mass)}}{2B_2O_3} + H_2O \quad ...(iii)$$

On adding (*i*), (*ii*) and (*iii*), we get

$$4H_3BO_3 \longrightarrow 2B_2O_3 + 6H_2O$$

$$\text{or } 2H_3BO_3 \longrightarrow B_2O_3 + 3H_2O$$

This equation represents that on strong heating H_3BO_3 gives B_2O_3

(*v*) *Action of C_2H_5OH.* When H_3BO_3 is heated with alcohol (C_2H_5OH) in presence of conc. H_2SO_4, volatile ester namely triethyl borate, $(C_2H_5)_3\ BO_3$ is formed. This ester burns with a green-edged flame.

$$H_3BO_3 + 3C_2H_5OH \xrightarrow{\text{Conc. } H_2SO_4} (C_2H_5)_3\ BO_3 + 3H_2O$$

(*vi*) *Action of CaF_2 and conc. H_2SO_4.* When H_3BO_3 is heated with CaF_2 and conc. H_2SO_4, BF_3 is obtained.

$$3CaF_2 + 3H_2SO_4 + 2H_3BO_3 \xrightarrow{\text{Heat}} 3CaSO_4 + 2BF_3 + 6H_2O$$

(*vii*) *Action of soda-ash, Na_2CO_3.* When H_3BO_3 is neutralised by Na_2CO_3, borax is obtained.

$$4H_3BO_3 + Na_2CO_3 \longrightarrow \underset{\text{Borax}}{Na_2B_4O_7} + 6H_2O + CO_2$$

(*viii*) *Action of alkali acid fluorides.* When an alkali acid fluoride (MHF_2) reacts with H_3BO_3, alkali fluoborate (MBF_4) is obtained. For example :

$$2NaHF_2 + H_3BO_3 \longrightarrow \underset{\text{Sod. fluo-borate}}{NaBF_4} + NaOH + 2H_2O$$

Uses. Boric acid is used (*i*) in the manufacture of pottery glazes, glass and enamels, (*ii*) in medicine as an antiseptic and as an eye lotion (*iii*) as preservative in food industry, (*iv*) in tanning industry, (*v*) in candle industry to stiffen wicks, (*vi*) in the manufacture of borax and pigments.

Structure of BO_3^{3-} ion. This ion has *triangular planar shape* which results from sp^2 hybridisation of B-atom in the excited state (configuration of B-atom in the excited state is $2s^1$. $2p^1$. $2p^1$. $2p^0$). How triangular planar shape is obtained from sp^2 hybridisation has already been explained.

Boric acid is covalent compound and contains planar triangular BO_3^{3-} units which are bonded together through *hydrogen bonds* into *two-dimensional* sheets as shown in Fig. 23.1.

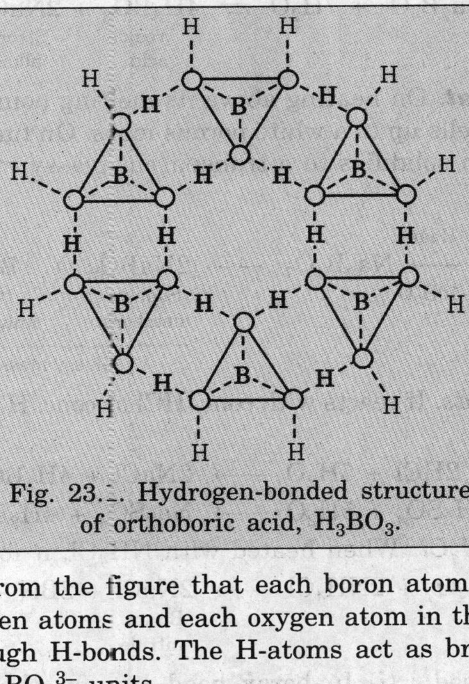

Fig. 23.1. Hydrogen-bonded structure of orthoboric acid, H_3BO_3.

It may be seen from the figure that each boron atom of each BO_3^{3-} unit is bonded to three oxygen atoms and each oxygen atom in this unit is bonded to a hydrogen atom through H-bonds. The H-atoms act as bridge between the two O-atoms of different BO_3^{3-} units.

2. Sodium Tetraborate or Borax, $Na_2B_4O_7.10H_2O$.

It is the sodium salt of tetraboric acid, $H_2B_4O_7$. It occurs naturally as *tincal* or *suhaga* in the dried-up lakes of Tibet, Ceylon, California (U.S.A.) and India. In India it is found in *Ladakh* and *Puga Valley of Kashmir.* Tincal contains about 55% of borax, $Na_2B_4O_7.10H_2O$. Borax can be obtained from colemanite (see the extraction of boron) and by the action of Na_2CO_3 on H_3BO_3. Some of the properties of borax are discussed below :

(i) Different forms of borax. Borax exists in three forms (*a*) *Ordinary* or *prismatic borax* which is *decahydrate* or *monoclinic*, $Na_2B_4O_7.10H_2O$. This is the common form of borax and is obtained when a solution of the salt is crystallised at room temperature (*i.e.* below 60°) (*b*) *Octahedral or Jeweller's borax* which is *pentahydrate*, $Na_2B_4O_7.5H_2O$ and is obtained when the solution is crystallised above 60° (*c*) *Borax glass* which is the *anhydrous* form, $Na_2B_4O_7$ and is obtained by heating the ordinary borax above its melting point until all the water of crystallisation is given off. It is a colourless glassy mass (density = 2.37), absorbs moisture from air and is gradually converted into decahydrate form.

(ii) Basic nature. It is sparingly soluble in cold water and fairly soluble in hot water. The solution is alkaline in reaction because borax is hydrolysed and it forms a mixture of sodium hydroxide (a strong base) and boric acid (a weak acid).

$$Na_2B_4O_7 + 3H_2O \rightleftharpoons 2NaBO_2 + 2H_3BO_3$$
$$NaBO_2 + 2H_2O \rightleftharpoons NaOH + H_3BO_3] \times 2$$
$$\overline{Na_2B_4O_7 + 7H_2O \rightleftharpoons \underset{\text{Weak acid}}{4H_3BO_3} + \underset{\text{Strong alkali}}{2NaOH}}$$

(iii) Action of heat. On heating above its melting point, it loses its water of crystallization and swells up to a white porous mass. On further heating, it melts to a liquid which then solidifies to a transparent glassy mass which consists of $NaBO_2$ and B_2O_3.

$$Na_2B_4O_7.10H_2O \xrightarrow[-10H_2O]{\text{Heat}} Na_2B_4O_7 \longrightarrow \underbrace{\underset{\text{Sodium metaborate}}{2NaBO_2} + \underset{\text{Boric anhydride}}{B_2O_3}}_{\text{Glassy mass}}$$

(iv) Action of acids. It reacts with conc. HCl or conc. H_2SO_4 to form sparingly soluble boric acid.

$$Na_2B_4O_7 + 2HCl + 5H_2O \longrightarrow 2NaCl + 4H_3BO_3$$
$$Na_2B_4O_7 + H_2SO_4 + 5H_2O \longrightarrow Na_2SO_4 + 4H_3BO_3$$

(v) Action of NH_4Cl. When heated with NH_4Cl, it forms boron nitride.

$$Na_2B_4O_7 + 2NH_4Cl \longrightarrow 2NaCl + \underset{\text{Boron nitride}}{2BN} + B_2O_3 + 4H_2O$$

Uses. Borax is used : (*i*) In borax bead test for the detection of basic radicals. (*ii*) In the manufacture of enamels and glazes for pottery. (*iii*) In making optical glass and borosilicate glassware. (*iv*) As a flux in metallurgy. (*v*) As a preservative for food. (*vi*) In the manufacture of washing powders and soaps. (*vii*) In leather and match industries.

3. Borazine or Borazole, $B_3N_3H_6$.

This compound is iso-electronic with benzene and hence has been called *Inorganic Benzene* by Weberg. In our discussion we shall represent this compound by the structure given in the margin.

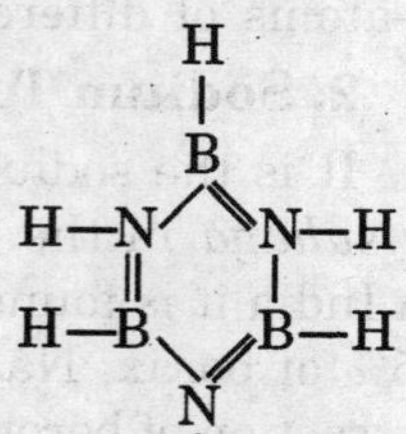

Structure of borazine, $B_3N_3H_6$

The suffix *ole* added with the name of this compound indicates the presence of OH group in the compound, but the fact is that the compound does not contain any OH group. It is, therefore, wrong to call this compound borazole. The word borazone has been derived from the older name *benzol* that was used for benzene with which borazine is iso-electronic.

Preparation. Borazine can be prepared by the following methods :

(i) By Stocks' method. In this method diborane (B_2H_6) and NH_3 are heated in 1 : 2 molar ratio at low temperature (= – 120°C) to get diammoniate of diborane ($B_2H_6 . 2NH_3$) which is an addition compound. This compound, on being heated at 200°C, gives borazine.

$$\underset{\text{(Ratio = 1 : 2)}}{B_2H_6 + 2NH_3} \xrightarrow{-120°C} \underset{\substack{\text{Diammoniate} \\ \text{of diborane}}}{B_2H_6.2NH_3}$$

$$2(B_2H_6.2NH_3) \xrightarrow{200°C} \underset{\text{Borazine}}{2B_3N_3H_6} + 12H_2$$

This method gives 40% yield of $B_3N_3H_6$.

(ii) By heating BCl_3 with NH_4Cl. When BCl_3 is heated with NH_4Cl in chlorobenzene (C_6H_5Cl) in the presence of Fe, Ni or Co (used as catalyst) at about 140°C, B, B, B-trichloroborazine (chloro groups attached with B-atoms) is formed. This derivative of borazine, on being reduced by lithium borohydride ($LiBH_4$) in polyether gives borazine, $B_3N_3H_6$

$$3NH_4Cl + 3BCl_3 \xrightarrow[C_6H_5Cl \text{ (Fe, 140°C)}]{\text{Heat in presence of}} \underset{\substack{\text{B, B, B-trichloro-} \\ \text{borazine}}}{Cl_3B_3N_3H_3} + 9HCl$$

$$3Cl_3B_3N_3H_3 + 6LiBH_4 \xrightarrow{\text{Reduction}} \underset{\text{Borazine}}{2B_3N_3H_6} + 6LiCl + 3B_2H_6$$

The yield of borazine obtained by this method is about 50–60%.

(iii) By heating a mixture of $LiBH_4$ and NH_4Cl.

$$3LiBH_4 + 3NH_4Cl \xrightarrow{275°C} \underset{\text{Borazine}}{B_3N_3H_6} + 3LiCl + 9H_2$$

Properties (*i*) Physical properties. Borazine is a colourless, mobile volatile liquid. It freezes at –58°C. Its boiling and melting points are 63°C and –58°C respectively.

(ii) Addition reactions. (*a*) One molecule of $B_3N_3H_6$ adds three molecules of H_2O, CH_3OH, CH_3I, HCl or HBr in the cold without a catalyst. The more negative group of these molecules is generally attached with boron, since boron atom is less electronegative than nitrogen atom in B—N bond. When hydrogen chloride derivative is heated at 50–100°C, it loses three H_2 molecules to give B-*trichloroborazine*.

$$\underset{\text{Borazine}}{B_3N_3H_6} \xrightarrow{+3HCl} \underset{\substack{\text{Hydrogen}\\\text{chloride}\\\text{derivative}}}{B_3N_3H_9Cl_3} \xrightarrow[50-100^\circ]{-3H_2} \underset{\substack{\text{B-trichloro-}\\\text{borazine}}}{B_3N_3H_3Cl_3}$$

Addition reaction shown above for borazine is not shown by benzene, *i.e.* C_6H_6 does not add HCl.

(*b*) One molecule of borazine also adds three molecules of Br_2 at 0°C and gives *B-tribromo-N-tribromoborazine* which, on being heated at 60°C, loses three molecules of HBr and forms B-tribromo-borazine.

$$\underset{\text{Borazine}}{B_3N_3H_6} \xrightarrow[0^\circ]{+3Br_2} \underset{\substack{\text{B-tribromo-}\\\text{N-tribromo-}\\\text{borazine}}}{B_3N_3H_6Br_6} \xrightarrow[60^\circ]{-3HBr} \underset{\substack{\text{B-tribromo}\\\text{borazine}}}{B_3N_3H_3Br_3}$$

This reaction can be compared with that shown by benzene where substitution takes place

$$C_6H_6 \xrightarrow[\substack{\text{In presence of}\\FeBr_3}]{+Br_2} \underset{\substack{\text{Bromo}\\\text{benzene}}}{C_6N_5Br} + HBr$$

(iii) Hydrolysis. (*a*) Borazine gets slowly hydrolysed by water to produce H_2, boric acid [$B(OH)_3$ or H_3BO_3] and NH_3. Hydrolysis is favoured by the increase in temperature.

$$B_3N_3H_6 + 9H_2O \longrightarrow 3B(OH)_3 \text{ or } 3H_3BO_3 + 3NH_3 + 3H_2$$

(*b*) It is also reported that under proper conditions, borazine reacts with three molecules of water to produce *B-trihydroxyl borazine*, $(OH)_3B_3N_3H_3$ in which OH groups are attached with boron atoms.

$$\underset{\text{Borazine}}{B_3N_3H_6} + 3H_2O \longrightarrow \underset{\substack{\text{B-trihydroxyl}\\\text{borazine}}}{B_3N_3H_3(OH)_3} + 3H_2$$

(iv) Hydrogenation. Benzene can be hydrogenated to produce saturated *cyclohexane*, C_6H_{12}. Borazine can also be converted into *cyclotriborazane*, $B_3N_3H_{12}$ by special techniques as shown below :

$$\underset{\text{Borazine}}{B_3N_3H_6} \xrightarrow{+3HCl} \underset{\substack{\text{Hydrogen}\\\text{derivative}}}{B_3N_3H_9Cl_3} \xrightarrow[\text{chloride}]{+3NaBH_4} \underset{\substack{\text{Cyclotri-}\\\text{borazane}}}{B_3N_3H_{12}} + \tfrac{3}{2}B_2H_6 + 3NaCl$$

Structure. Various chemical reactions and electron diffraction studies of borazine have shown that this compound is iso-electronic with benzene and hence its structure is similar to that of benzene. Thus, like benzene, borazine also has a *planar hexagonal structure* containing six-membered ring in which B and N atoms alternate.

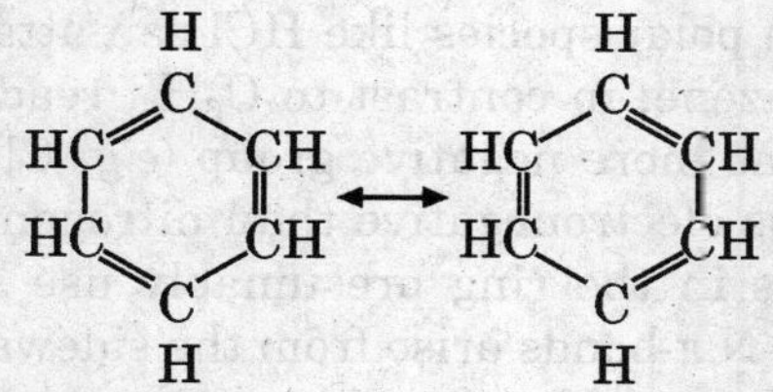

Benzene (C_6H_6)

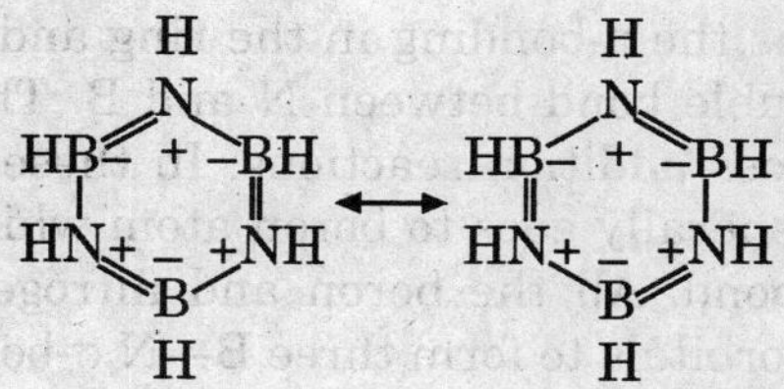

Borazine ($B_3N_3H_6$)

In the structure of borazine each of the B—N distances is equal to 1.44A° which is between the calculated single B—N (= 1.54 A°) and double B = N (= 1.36 A°) bond distances. The angles are equal to 120°. The ring structure of borazine is related to the layer lattice structure of boron nitride in the same way in which the ring structure of benzene is related to the layer lattice structure of graphite (*see* below)

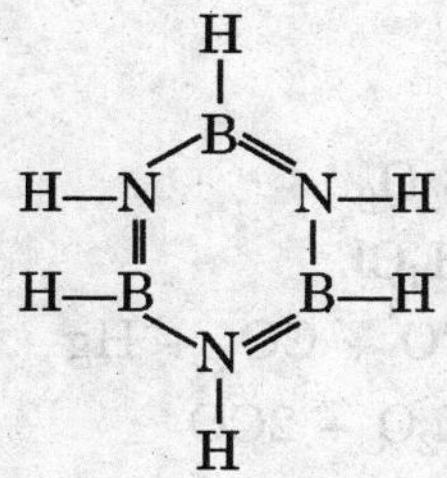

Borazine
($B_3N_3H_6$)

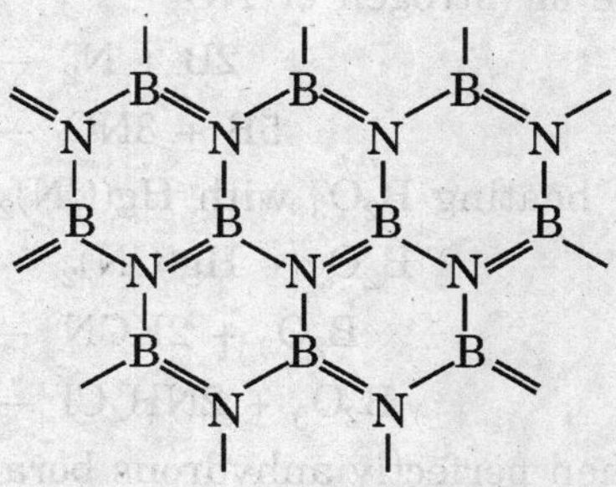

Boron nitride
$[(BN)_n]$

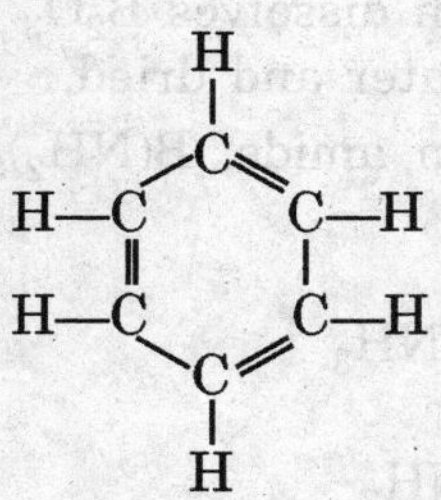

Benzene
(C_6H_6)

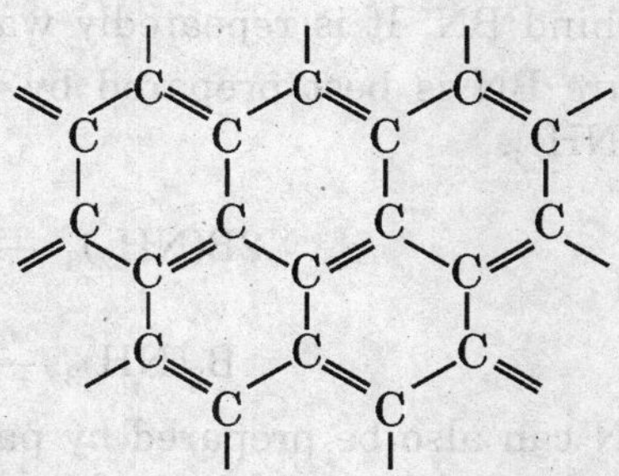

Graphite
(C_n)

Since borazine is isoelectronic with benzene, both the compounds have aromatic π-clouds of electron density which is delocalised over the atoms of the ring. In C_6H_6, C = C bonds are non-polar while in case of $B_3N_3H_6$, due to the difference in electronegativity between B and N atoms, B—N bond is polar ($B^{\delta+}$ – $N^{\delta-}$) and hence π cloud in $B_3N_3H_6$ is more lumpy with more electron density localised on N atom. This partial localisation of electron density on N atom weakens the π-bonding in the ring and as such polar species like HCl can attack this double bond between N and B. Thus borazene, in contrast to C_6H_6, readily undergoes addition reactions. In these reactions more negative group (*e.g.* Cl in HCl) generally goes to boron atom which is less electronegative than nitrogen in B—N bond. All the boron and nitrogen atoms in the ring presumably use sp^2 hybrid orbitals to form three B—N σ-bonds. B—N π-bonds arise from the sideways overlapping of the unhybridised *p*-orbitals of B- and N-atoms which are normal to the plane of the ring. Unlike in benzene, the π-electrons are not derived from all the six atoms of the ring but from the three N-atoms. Molecular orbital calculations indicate that the π-electrons are partially delocalised. These calculations also have shown that N—B π-electron drift is largely counterbalanced by the B—N σ drift.

4. Boron Nitride, BN.

Preparation. It is formed (*i*) by heating boron to a white heat in an atmosphere of nitrogen or NO

$$2B + N_2 \longrightarrow 2BN$$

$$5B + 3NO \longrightarrow 3BN + B_2O_3$$

(*ii*) by heating B_2O_3 with $Hg(CN)_2$, KCN or NH_4Cl.

$$B_2O_3 + Hg(CN)_2 \longrightarrow 2BN + CO + CO_2 + Hg$$

$$B_2O_3 + 2KCN \longrightarrow 2BN + K_2O + 2CO$$

$$B_2O_3 + 2NH_4Cl \longrightarrow 2BN + 3H_2O + 2HCl$$

(*iii*) when perfectly anhydrous borax ($Na_2B_4O_7$) is heated with dry NH_4Cl to a red heat in a platinum crucible.

$$Na_2B_4O_7 + 2NH_4Cl \longrightarrow 2BN + 2NaCl + B_2O_3 + 4H_2O$$

The resulted mass is treated with dil. HCl which dissolves B_2O_3 and NaCl, leaving behind BN. It is repeatedly washed with water and dried.

(*iv*) *Pure* BN is best prepared by heating boron amide, $B(NH_2)_3$ or boron imide, $B_2(NH)_3$.

$$2B(NH_2)_3 \xrightarrow{\Delta} 2BN + 4NH_3$$

$$B_2(NH)_3 \xrightarrow{\Delta} 2BN + NH_3$$

(*v*) BN can also be prepared by passing N_2 gas through a mixture of B_2O_3 and carbon heated in an electric furnace.

$$B_2O_3 + 3C + N_2 \longrightarrow 2BN + 3CO \uparrow$$

Properties. (*i*) Boron nitride is a white powder of density 2.34. It melts under pressure at 3000°C. It is very stable and unreactive substance. It remains unaffected by mineral acids, solutions of alkalies and Cl_2 at red heat.

(*ii*) *Decomposition.* (*a*) It gets decomposed when heated in steam, evolving NH_3.

$$BN + 3H_2O \longrightarrow B(OH)_3 \text{ or } H_3BO_3 + NH_3$$

(*b*) It is decomposed, but slowly, by HF, forming ammonium borofluoride, NH_4BF_4

$$BN - 4HF \longrightarrow NH_4BF_4$$

(*c*) It is also decomposed when fused with KOH.

$$BN + 3KOH \longrightarrow K_3BO_3 + NH_3 \uparrow$$

(*iii*) *Reaction with K_2CO_3*. When BN is fused with K_2CO_3, potassium cyanate (KCNO) and potassium metaborate (KBO_2) are obtained.

$$BN + K_2CO_3 \longrightarrow KCNO + KBO_2$$

Uses. BN possesses same hardness as diamond and can withstand temperature of more than 300°C. Due to this property it is used for coating crucible linings.

Structure. Boron nitride has a layered lattice structure. Each layer consists of a hexagonal arrangement of B and N atoms (Fig. 23.2). This structure of BN resembles that of graphite. The layer gets superimposed on one another so that B and N-atoms are alternately above one another. The inter-atomic distances are 1.45 A° and the distance between the layers is 8.33 A°. BN, like graphite, is able to act as host; it is able to occlude oxides, sulphides, oxyhalides and chlorides in the same way as graphite does. BN, when heated to 3000°C at 70,000 atmospheric pressure yields a cubic form with a diamond structure.

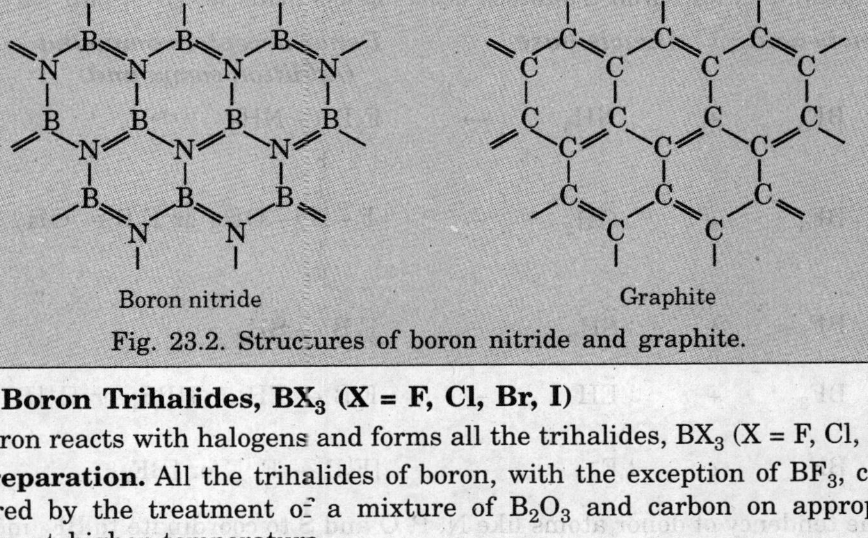

Fig. 23.2. Structures of boron nitride and graphite.

5. Boron Trihalides, BX_3 (X = F, Cl, Br, I)

Boron reacts with halogens and forms all the trihalides, BX_3 (X = F, Cl, Br, I).

Preparation. All the trihalides of boron, with the exception of BF_3, can be prepared by the treatment of a mixture of B_2O_3 and carbon on appropriate halogen at higher temperature.

$$B_2O_3 + 3C + 3X_2 \longrightarrow 2BX_3 + 3CO$$
$$(X = Cl, Br, I)$$

Properties. ***(i) Physical state.*** BF_3 and BCl_3 are gases, BBr_3 is a liquid and BI_3 is a solid at room temperature.

(ii) Covalent character. Boron trihalides are covalent in character due to the *small size* and *high charge density* on B^{3+} ion. Some other points in favour of the covalent character of boron trihalides are : (*a*) As liquids they do not conduct electricity *i.e.*, they are *non-electrolytes*. (*b*) Their boiling points are very low as compared to the halides of the elements of group IA and IIA. (*c*) They exist as discrete molecular species (*i.e.*, monomeric molecules) in all the states. (*d*) The boiling points of boron halides *increase as the atomic number of the halogen atoms increases.* This behaviour is very similar to that shown by covalent halogen compounds of carbon.

(*iii*) *Geometry.* Boron trihalides have planar triangular geometry which results from sp^2 hybridisation of B-atom (central atom). B-atom has a vacant 2*p*-orbital which remains unhybridised. As an example the geometry of BF_3 can be considered.

(*iv*) *Formation of complex halides.* Boron trihalides (*e.g.* BF_3) also form complex halides of the type $[BF_4]^-$ in which boron atom (central atom) has a coordination number equal to 4. Boron cannot extend its coordination number beyond 4 due to the non-availability of *d*-orbitals in its configuration.

(*v*) *Electron-acceptor property : Formation of addition compounds.* The structure of BX_3 molecule consisting of three B—X bonds indicates that boron atom is surrounded by a sextent (*i.e.* six electrons), *i.e. there are two electrons short of an octet.* Hence B-atom (central atom) in BX_3 molecule can accept an electron pair from the donor atoms like N, P, O, S, F etc. in NH_3, PH_3, H_2O, H_2S, HF, F^- etc. respectively to form *donor-acceptor compounds* (*addition compounds or adducts*), *i.e.*, all boron trihalides act as *Lewis acids* (*electron-pair acceptors*).

Lewis acid		***Lewis base***		***Donor-acceptor compound (Addition compound)***
BF_3	+	$:NH_3$	→	$F_3B \leftarrow NH_3$
BF_3	+	$:\ddot{O}H_2$	→	$F\text{—}B(F)_2 \leftarrow \ddot{O}H_2$ or $F_3B \leftarrow \ddot{O}H_2$
BF_3	+	$:\ddot{S}H_2$	→	$F_3B \leftarrow \ddot{S}H_2$
BF_3	+	$:\ddot{F}H$	→	$F_3B \leftarrow \ddot{F}H$ *or* HBF_4 *or* $H^+[BF_4]^-$
BF_3	+	$:\ddot{F}^-$	→	$[F_3B \leftarrow \ddot{F}:]^-$ or $[BF_4]^-$

The tendency of donor atoms like N, P, O and S to coordinate to BF_3 molecule makes this compound an extremely useful catalyst for Friedel-Crafts reactions like alkylations and acylations, esterifications and polymerisation of olefins.

The addition compounds like $F_3B \leftarrow NH_3$, $F_3B \leftarrow OH_2$ have tetrahedral shape and in their formation the unhybridised *p*-orbital accepts the lone pair of electrons donated by the donor atom.

Relative order of electron-acceptor ability of boron trihalides. Since F atom has the highest electron-attracting power (*i.e.* highest electronegativity), it should create the greatest positive charge on B-atom which would enhance its capacity for accepting the electrons from a donor atom. Thus the order of electron-acceptor ability of boron trihalides should be as follows :

$$\boxed{BF_3 > BCl_3 > BBr_3 > BI_3}$$

But the measurement of the molar heats of formation of solutions of boron trihalides in nitrobenzene and dipole moments of their adducts with pyridine has shown that the electron-acceptor power of boron trihalides is opposite to that shown above, *i.e.* the observed order is :

$$\boxed{BF_3 < BCl_3 < BBr_3 < BI_3.}$$

This anomaly has been explained as follows :

We have seen that in BF_3 molecule boron atom (configuration of boron atom in the excited state is $2s^1.2p^1.2p^1.2p^0$) is sp^2 hybridised. All the three sp^2 hybrid orbitals on boron atom overlap with three singly filled 2*p*-orbitals of three F atoms ($F \rightarrow 2s^2, 2p^2\,2p^2\,2p^1$) to form three sp^2 (B)—2*p* (F) σ-bonds. Since filled 2*p*-orbital on F-atom and vacant 2*p*-orbital on B-atom are of *the same energy*, they overlap *laterally* with each other to form a *p*(B)—*p*(F) π bond (or more correctly *p*π (B)—*p*π (F) π-bond) and as a result of this overlap 2*p*-electrons of F atom are transferred into the vacant 2*p*-orbital of B-atom (see Fig 23.3) and the B—F bond acquires some double bond character.

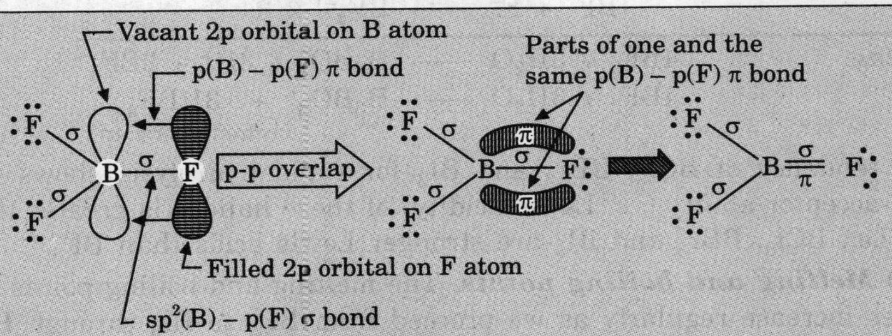

Fig. 23.3. Formation sp^2 (B)—*p* (F) σ bond and *p* (B)—*p* (F) π bonds in BF_3 molecule.

This type of bond formation is known as *dative* or *back bonding.* The effect of back bonding from F atom to B-atom is that positive charge on B-atom is reduced, *i.e.* the electron deficiency of B-atom is compensated and as such electron-acceptor property (*i.e.* Lewis acidity) of BF_3 is *decreased.* The overlapping of 2*p*-orbital of B-atom with filled 2*p*-orbital of F-atom is maximum in case of BF_3, because both the 2*p*-orbitals are at a nearer distance from each other and of equivalent energy. Thus the tendency of *p*π—*p*π back bonding in BF_3 is maximum and *decreases* as we move to BCl_3 and BBr_3, because of the increasing distance and energy difference between 2*p*-orbital of B-atom and 3*p* and 4*p*-orbitals of Cl and Br atoms respectively. Hence BBr_3 would give more stable adducts than BCl_3 and BCl_3 more stable than BF_3. In other words the electron-acceptor ability of boron trihalides is in the order :

$$\boxed{BF_3 < BCl_3 < BBr_3 < BI_3}$$

In order to explain the equality of all the three observed B—X bond lengths in BX_3, resonance between the three equivalent structures has been suggested.

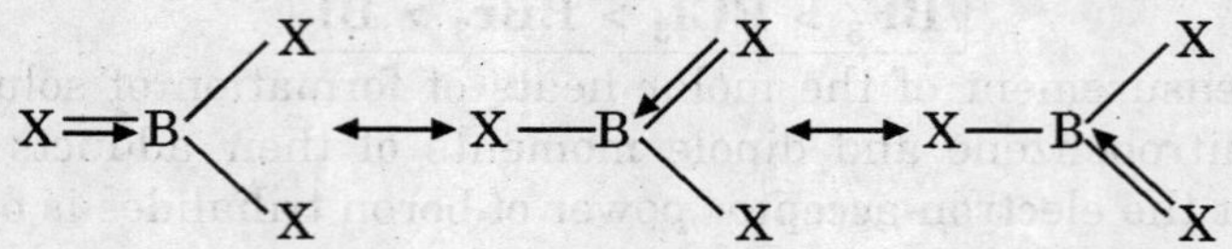

(vi) Hydrolysis. We have seen that B-atom BX_3 molecule has a hunger for two electrons to fill its vacant 2*p*-orbital. This hunger initiates the hydrolysis of BX_3 molecule by H_2O whose oxygen atom can easily donate its electron pair to B-atom. Thus BX_3 halides (X = Cl, Br and I) are easily hydrolysed to boric acid (H_3BO_3) and corresponding halogen acids (HX).

$$BX_3 + 3H_2O \longrightarrow H_3BO_3 + 3HX$$
(X = Cl, Br, I)

BF_3, however, gives hydrofluoboric acid, HBF_4 (also called fluoboric acid) which is produced by the combination of H^+ and BF_4^- ions.

$$BF_3 + 3H_2O \longrightarrow H_3BO_3 + 3HF$$
$$[HF \longrightarrow H^+ + F^-] \times 3$$
$$[BF_3 + F^- \longrightarrow BF_4^-] \times 3$$

On adding $$4BF_3 + 3H_2O \longrightarrow H_3BO_3 + 3H^+ + 3BF_4^-$$
or $$4BF_3 + 3H_2O \longrightarrow H_3BO_3 + \underset{\text{Hydrofluoboric acid}}{3HBF_4}$$

The tendency of BCl_3, BBr_3 and BI_3 for rapid hydrolysis shows that the electron-acceptor ability (*i.e.* Lewis acidity) of these halides is greater than that of BF_3, *i.e.*, BCl_3, BBr_3 and BI_3 are stronger Lewis acids than BF_3.

(vii) Melting and boiling points. The melting and boiling points of boron trihalides increase regularly as we proceed from BF_3 to BI_3 through BCl_3 and BBr_3.

(viii) Solvolytic reactions. While BF_3 forms addition compounds with certain donor molecules containing labile hydrogen, the other trihalides of boron undergo solvolytic reactions with such donor molecules. The extent of solvolysis depends on the *concentration of the donor molecules and temperature.* Some solvolytic reactions of BCl_3 are shown below :

$$BCl_3 + 3C_2H_5OH \longrightarrow B(OC_2H_5)_3 + 3HCl$$
$$BCl_3 + 4NH_3 \longrightarrow BN + 3NH_4Cl$$
$$BCl_3 + 3N_2H_4 \longrightarrow B(N_2H) + 3N_2H_5Cl$$
$$BCl_3 + 6C_6H_5NH_2 \text{ (in benzene)} \longrightarrow B(NHC_6H_5) + 3C_6H_5NH_2.HCl$$

6. Boron Trifluoride, BF_3

Among the boron trihalides **boron trifluoride** is the most important. It is prepared by heating a mixture of CaF_2, B_2O_3 and conc. H_2SO_4.

$$3CaF_2 + B_2O_3 + 3H_2SO_4 \longrightarrow 2BF_3 + 3CaSO_4 + 3H_2O$$

The gas is collected over mercury since it is exceedingly soluble in H_2O. *Pure* BF_3 is best prepared by the action of potassium fluoborate, KBF_4 on B_2O_3

$$KBF_4 + 2B_2O_3 \longrightarrow BF_3 + K[B_4O_6F]$$

Boron trifluoride is a colourless gas (density 2.37) with a suffocating odour, m.pt. = – 127°C, *b.pt.* = – 101°C. It is not combustible. It fumes strongly in moist air and is readily decomposed by water into boric acid and fluoboric acid (HBF_4)

$$4BF_3 + 3H_2O \longrightarrow H_3BO_3 + 3HBF_4$$

It forms addition compounds with NH_3, PH_3, alcohols, ether etc. which act as electron pair donors and thus behave as a *Lewis acids* (electron pair acceptors)

$$H_3N : + BF_3 \longrightarrow H_3N \longrightarrow BF_3$$

$$R_2O : + BF_3 \longrightarrow R_2O \longrightarrow BF_3$$

It is used as a catalyst in organic chemistry.

7. Fluoboric Acid (HBF_4) and its Salts.

Preparation. Fluoboric acid is prepared by dissolving BF_3 in water or by dissolving H_3BO_3 in conc. HF.

$$4BF_3 + 3H_2O \longrightarrow H_3BO_3 + 3HBF_4$$

$$H_3BO_3 + 4HF \longrightarrow HBF_4 + 3H_2O$$

Properties. HBF_4 is a strong acid, stronger than HF. It is decomposed when heated. The salts given by this acid are *fluoborates* or *borofluorides.* These salts are prepared by dissolving the corresponding metallic oxides, hydroxides or corbonates in aqueous hydrofluoboric acid. Fluoborates are also obtained in solution by treating alkali acid fluorides with H_3BO_3.

$$2NaHF_2 + H_3BO_3 \longrightarrow NaBF_4 + NaOH + 2H_2O$$

Excepting KBF_4, the fluoborates are crystalline solids and fairly soluble in water. They decompose on heating, evolving BF_3.

$$KBF_4 \xrightarrow{\Delta} KF + BF_3$$

Here it may be noted that in acid as well as in its salts, boron is *tetracovalent.*

Uses. (*i*) HBF_4 is used as a catalyst in organic chemistry (*ii*) HBF_4 is used in the preparation of fluoborates. (*iii*) The acid and its salts are used in electroplating. (*iv*) KBF_4 is now used for the extraction of boron from its minerals.

Aluminium

Occurrence of Aluminium.

Aluminium does not occur free in nature. It is a constituent of clay, slate and many types of silicate rocks and forms 7.28% of the earth crust. It occurs : (*i*) *As oxides, e.g.* (*a*) *Corundum, Ruby Sapphire, Emerald,* Al_2O_3 (*b*) *Bauxite* which is the hydrated oxide, $Al_2O_3.2H_2O$ and is the chief ore of aluminium. (*c*) *Gibbsite,* $Al_2O_3.3H_2O$, and (*d*) *Diaspore,* $Al_2O_3.H_2O$. Usually all hydrated oxide ores containing 50 to 70% of Al_2O_3, about 25–30% of water with varying amounts of ferric oxide, titanic oxide and silica are called bauxite ores. Bauxite is the chief mineral used for the extraction of Al and is found in Kashmir, Bihar, Maharashtra, Madhya Pradesh and Tamil Nadu. (*ii*) *As fluoride, e.g. Cryolite,* Na_3AlF_5. (*iii*) *As silicates e.g.* (*a*) *Felspar,* $K_2O.Al_2O_3.\ 6SiO_2$ or $KAlSi_3O_8$. (*b*) *Koalin, porcelain, mica, china clay, slate* etc., $Al_2O_3.2SiO_2.2H_2O$. (*iv*) *As basic phosphate e.g. Turquoise* which is a hydrated basic phosphate, $AlPO_4.Al(OH)_3.H_2O$, coloured blue with copper phosphate, (*v*) *As basic sulphate e.g. alumite* or *alum stone*, $K_2SO_4.Al_2(SO_4)_3.4Al(OH)_3$.

Extraction of Aluminium from Bauxite.

The extraction of aluminium from bauxite consists of the following *three* steps:

1. Purification of bauxite and preparation of $Al(OH)_3$. Bauxite is associated with silica, SiO_2 and ferric oxide, Fe_2O_3 (*red bauxite*) and these impurities must be completely removed in order to get good quality of aluminium. A number of processess have been suggested for this purpose.

(a) Baeyer's process. It is especially used for red bauxite containing ferric oxide as chief impurity. It is first roasted at a low temperature to convert any ferrous oxide to ferric oxide and then digested at 150°C with a concentrated solution of caustic soda in an autoclave. The aluminium oxide present in bauxite dissolves in caustic soda forming soluble sodium meta-aluminate, $NaAlO_2$, while the impurities are left behind.

$$Al_2O_3.2H_2O + 2NaOH \longrightarrow 2NaAlO_2 + 3H_2O$$

The solution is filtered and agitated with some freshly precipitated $Al(OH)_3$ when the whole of aluminium present in solution in the form of sodium meta-aluminate precipitates in the form of $Al(OH)_3$.

$$NaAlO_2 + 2H_2O \longrightarrow NaOH + \underset{\text{ppt.}}{Al(OH)_3}$$

(b) Hall's process. (*Dry* or *fusion process*). In this process, ore is fused with sodium carbonate when the aluminium oxide present is converted into soluble sodium meta-aluminate. The silica and ferric oxide are left as insoluble residue when the fused mass is extracted with water.

$$Al_2O_3.2H_2O + Na_2CO_3 \longrightarrow 2NaAlO_2\ (\textit{soluble}) + CO_2 + 2H_2O$$

The filtrate is warmed at a temperature of 50–60°C and a current of carbon dioxide is passed through it when aluminium hydroxide is precipitated.

$$2NaAlO_2 + 3H_2O + CO_2 \longrightarrow \underset{\text{ppt.}}{2Al(OH)_3} + Na_2CO_3$$

(c) Serpek's process. This is used chiefly for ores containing silica as the chief impurity. The ore is mixed with carbon and nitrogen gas passed over the mixture heated to 1800°C when aluminium nitride is formed. It is hydrolysed to get aluminium hydroxide. Silica (SiO_2) is reduced to silicon

$$Al_2O_3 + 3C + N_2 \longrightarrow 2AlN + 3CO$$

$$AlN + 3H_2O \longrightarrow Al(OH)_3 + NH_3$$

$$SiO_2 + 2C \longrightarrow Si + 2CO$$

2. Conversion of $Al(OH)_3$ into Al_2O_3. The precipitate of $Al(OH)_3$ obtained by any of the above methods is washed, dried and ignited to get Al_2O_3.

$$2Al(OH)_3 \xrightarrow{\text{Ignite}} Al_2O_3 + 3H_2O$$

3. Recovery of Al from Al_2O_3 by Hall's electrolytic process. Al is obtained from Al_2O_3 by electrolysing it. This method was discovered by American chemist, Charless Martin Hall in 1886. Hence the name *Hall's electrolytic process.*

In this method a solution of Al_2O_3 in fused cryolite, Na_3AlF_6 is taken in an electrolytic cell in which it is electrolysed. The cell consists of a cast iron box, measuring 6' × 3' × 3' (Fig. 23.4). This box is lined with a refractory material which acts as a thermal insulation. Within this lining there is a layer of carbon (carbon lining) which acts as cathode. A number of carbon rods are suspended from a copper bus bar and can be raised and lowered, as desired. These carbon rods act as anodes (carbon anodes). These rods are immersed in the electrolyte (solution of Al_2O_3 in fused crysolite, Na_3AlF_6). The bath containing the electrolyte is maintained at a temperature of 875–950°C. A glow lamp (control lamp) is connected in parallels with the electrolyte cell. This lamp begins to glow when the quantity of Al_2O_3 gets exhausted. Now more quantity of Al_2O_3 as added. Thus it is a continuous process. When electric current is passed, AlF_3 [from cryolite, Na_3AlF_6 or $3NaCl.AlF_3$)] ionises as :

$$AlF_3 \rightleftarrows Al^{3+} + 3F^-$$

Al is liberated at the cathode and F_2 goes to anode where it reacts with Al_2O_3 to from AlF_3 and O_2 is liberated. The oxygen liberated at anode reacts with carbon anodes to form CO and CO_2.

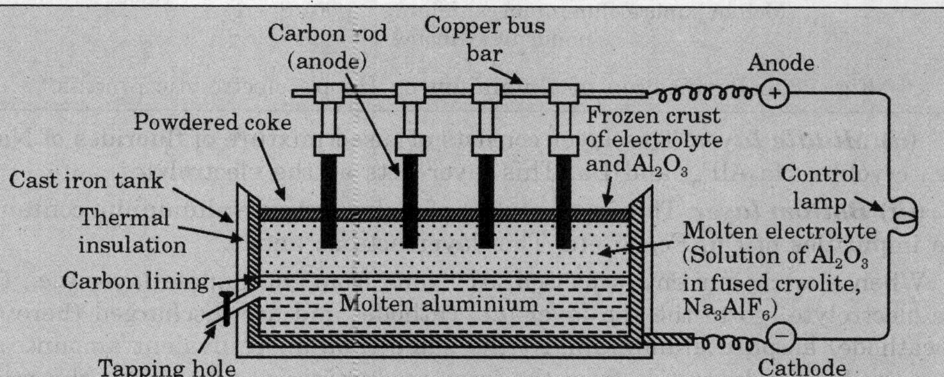

Fig. 23.4. Recovery of aluminium by the electrolysis of molten Al_2O_3 (Hall's electrolytic process).

At cathode : $Al^{3+} + 3e^- \longrightarrow Al$

At anode : $3F^- \longrightarrow 3F + 3e^-$

$$3F + 3F \longrightarrow 3F_2$$

$$2Al_2O_3 + 6F_2 \longrightarrow 4AlF_3 + 3O_2$$

$$2C + O_2 \longrightarrow 2CO$$

$$C + O_2 \longrightarrow CO_2$$

Molten aluminium collects at the bottom of the cell from where it is withdrawn periodically through the tapping hole. The formation of CO and CO_2 by the reaction between O_2 and carbon anodes causes a serious consumption of the anodes and hence these have to be replaced frequently. The electrolyte is covered with a layer of powdered coke to avoid the loss of heat by radiation.

4. Purification of aluminium by Hoop's electrolytic process. Aluminium obtained as above is not perfectly pure, since it contains the impurities of Cu, Si, Fe etc. Aluminium obtained as above is, therefore, purified. The purification is done by *Hoop's electrolytic process.* The cell used consists of a carbon-lined iron tank (Fig. 23.5). This tank contains the following three liquid layers one over the other of different specific gravity.

(i) Top layer. This layer consists of pure molten aluminium. This layer acts as cathode.

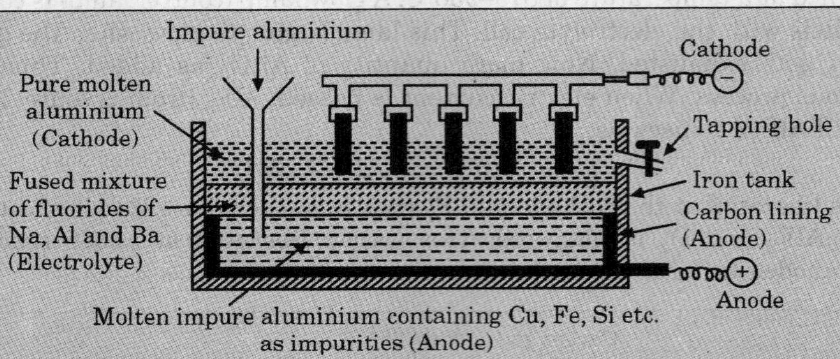

Fig. 23.5. Purification of aluminium by Hoop's electrolytic process.

(ii) Middle layer. This layer consists of fused mixture of fluorides of Na, Al (*i.e.*, cryolite, Na_3AlF_6) and Ba. This layer acts as the electrolyte.

(iii) Bottom layer. This layer consist of molten impure aluminium containing the impurities of Cu, Fe, Si etc. This layer acts as anode.

When electric current is passed, Al^{3+} ions from the middle layer (*i.e.*, from the electrolyte) go to the top layer (*i.e.* cathode) and are discharged there (*i.e.*, at cathode) as pure aluminium. At the same time an equivalent amount of Al from the bottom layer (*i.e.* from the impure aluminium) passes into the middle layer (*i.e.*, into the electrolyte), while the impurities of Cu etc. present in the bottom layer do not pass, but remain below *i.e.* they do not dissolve into the electrolyte. Al obtained in this way is 99.99% pure. It is periodically removed from the tapping hole.

Impure Al containing the impurities of Cu, Fe, Si etc., are introduced into the bottom layer as shown in the figure.

Physical Properties of Aluminium.

Aluminium shows a bluish white lustre which is destroyed on long exposure to air due to the formation of a layer of oxide on it. It is a light metal (density 2.7), malleable and ductile at 100—150°C but brittle below its melting point which is 651°C. It is a good conductor of heat and electricity.

Chemical Properties of Aluminium :

1. Action of air. Aluminium is not affected in dry air but in moist air a thin film of oxide is formed over its surface. It burns in oxygen producing brilliant light.

$$4Al + 3O_2 \longrightarrow Al_2O_3 + 772{,}000 \text{ cals.}$$

2. Action of water. Aluminium, if pure, is not affected by pure water but is corroded by water containing salts if some impurities are present in the metal. It decomposes boiling water. Aluminium-mercury couple is used as a reducing agent because it decomposes water evolving hydrogen.

$$2Al + 6H_2O \longrightarrow 2Al(OH)_3 + 3H_2$$

3. Action of alkalies. Aluminium dissolves in alkalies with the evolution of hydrogen and formation of soluble aluminates viz. $NaAlO_2$ and Na_3AlO_3

$$2Al + 2NaOH + 2H_2O \longrightarrow 2NaAlO_2 + 3H_2$$

$$2Al + 6NaOH \longrightarrow 2Na_3AlO_3 + 3H_2$$

4. Action of acids. Aluminium dissolves in hydrochloric and dil sulphuric acid producing hydrogen

$$2Al + 6HCl \longrightarrow 2AlCl_3 + 3H_2$$

$$2Al + 3H_2SO_4 \longrightarrow Al_2(SO_4)_3 + 3H_2 \text{ (\textit{Reaction very slow})}.$$

Hot concentrated sulphuric acid produces sulphur dioxide

$$2Al + 6H_2SO_4 \longrightarrow Al_2(SO_4)_3 + 3SO_2 + 6H_2O$$

Concentrated nitric acid has no action on aluminium. Dilute HNO_3 reacts only slowly on impure Al to produce $Al(NO_3)_3$

$$8Al + 30HNO_3 \longrightarrow Al(NO_3)_3 + 3NH_4NO_3 + 9H_2O$$

5. Action of halogens. Finely powered heated aluminium combines with halogens, if the latter are passed over it.

$$2Al + 3Cl_2 \longrightarrow 2AlCl_3$$

6. Action of nitrogen. Aluminium combines with nitrogen forming aluminium nitride.

$$2Al + N_2 \longrightarrow 2AlN$$

7. Action of carbon monoxide and carbon dioxide. Red hot aluminium is attacked by both carbon monoxide and carbon dioxide.

$$6Al + 3CO \longrightarrow Al_2O_3 + \underset{\text{Aluminium carbide}}{Al_4C_3}$$

$$4Al + 3CO_2 \longrightarrow 2Al_2O_3 + 3C$$

8. Reducing properties. Aluminium metal is a good reducing agent and reduces oxides of iron, chromium and manganese etc. to corresponding metals.

$$2Al + Cr_2O_3 \longrightarrow 2Cr + Al_2O_3 + \text{heat}$$

$$2Al + Fe_2O_3 \longrightarrow 2Fe + Al_2O_3 + \text{heat}$$

$$3Mn_3O_4 + 8Al \longrightarrow 4Al_2O_3 + 9Mn + \text{heat}$$

This property is made use of in the extraction of chromium, manganese etc. by *Gold Schmidt's aluminothermic process.*

Uses of Aluminium.

(*i*) Because of its lightness, good conductivity and resistance to corrosion, aluminium is used for making *alloys* which find extensive application in industries and arts (*ii*) For making utensils, frames, coils for motors and wire for electrical transmission. (*iii*) *Aluminium foil* is used for wrapping confectionary, cigarettes etc. (*iv*) Aluminium powder is used in making silvery paints, in fireworks, flashlight powders and in thermit welding. (*v*) In the extraction of chromium and manganese. (*vi*) As a deoxidiser and for preventing blow holes in metallurgy. (*vii*) It is used for the manufacture of electric cables. (*viii*) Aluminium amalgam is used as a reducing agent. (*ix*) It is also used for making alloys. Some of its important alloys are given in the table below :

Alloy	Approx. composition	Uses
Magnalium	Al = 94.5%, Mg = 5.5%	Scientific apparatus, machined articles.
Duralumin	Al = 95%, Cu = 4%, Mg = 0.5%, Mn = 0.5%	Aircraft parts, railroad cars, boat machinery.
Nickeloy	Al = 95%, Cu = 4%, Ni = 1%	Aircraft part.
Y-alloy	Al = 93%, Cu = 4%, Ni = 2%, Mg = 1%	Pistons and machinery parts.
Alnico	Steel = 50%, Ni = 2%, Al = 20%, Co = 1%	For making permanent magnets.

Compounds of Aluminium

1. Aluminium Chloride, $AlCl_3$ or Al_2Cl_6.

Preparation. When aluminium metal or $Al(OH)_3$ is treated with HCl, solution of *hydrated* aluminium chloride, $AlCl_3.6H_2O$ is obtained.

$$2Al + 6HCl \longrightarrow 2AlCl_3 + 3H_2$$

$$Al(OH)_3 + 3HCl \longrightarrow AlCl_3 + 3H_2O$$

On evaporating the solution, the crystals of $AlCl_3.6H_2O$ are obtained. On heating these crystals, anhydrous aluminium chloride cannot be obtained because of the formation of Al_2O_3.

$$2(AlCl_3.6H_2O) \xrightarrow{\text{Heat}} Al_2O_3 + 6HCl + 9H_2O$$

The anhydrous aluminium chloride is, however, prepared :

(*i*) By passing dry chlorine or hydrochloric acid gas over heated aluminium powder. The vapour of aluminium chloride is condensed in the receiver as shown in Fig. 23.6.

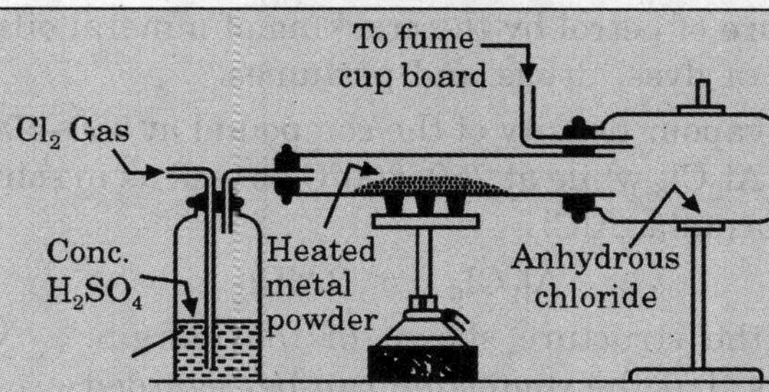

Fig. 23.6. Preparation of anhydrous aluminium chloride.

$$2Al + 3Cl_2 \longrightarrow 2AlCl_3$$
$$2Al + 6HCl \longrightarrow 2AlCl_3 + 3H_2$$

(*ii*) By heating a mixture of alumina and coke in a current of chlorine (*Mac Affe process*),

$$Al_2O_3 + 3C + 3Cl_2 \longrightarrow 2AlCl_3 + 3CO$$

(*iii*) By heating alumina (Al_2O_3) in a current of carbonyl chloride ($COCl_2$).

$$Al_2O_3 + 3COCl_2 \longrightarrow 2AlCl_3 + 3CO_2$$

(*iv*) By heating alumina in a current of S_2Cl_2 vapour and Cl_2. (*Commercial preparation*)

$$4Al_2O_3 + 3S_2Cl_2 + 9Cl_2 \longrightarrow 8AlCl_3 + 6SO_2$$

Properties. (*i*) Anhydrous salt is a white crystalline solid. It is hygrocopic and fumes in moist air. It sublimes at 180°C. (*m.pt.* = 193°C at 2 atom, pressure).

(*ii*) It is a typical covalent compound as is shown by its volability, solubility in organic solvents like C_2H_6, CS_2 etc. and poor conductivity of the fused state.

(*iii*) An aqueous solution of the salt is acidic in nature because in water it undergoes hydrolysis and forms HCl.

$$AlCl_3 + 3H_2O \rightleftarrows Al(OH)_3 + 3HCl$$

With moist air, the salt gives the fumes of HCl. The equation representing the reaction is the same as given above.

(*iv*) When NH_4OH is added to $AlCl_3$ solution, white precipitate of $Al(OH)_3$ is obtained.

$$AlCl_3 + 3NH_4OH \longrightarrow \underset{\text{White ppt.}}{Al(OH)_3} + 3NH_4Cl$$

(*v*) With excess of NaOH, the salt solution gives sodium meta-aluminate, $NaAlO_2$ (soluble)

$$AlCl_3 + 4NaOH\ (\textit{excess}) \longrightarrow \underset{\text{(soluble)}}{NaAlO_2} + 2H_2O + 3NaCl$$

(*vi*) It forms addition compounds with a number of donor molecules like NH_3, PH_3, $COCl_2$ etc.

$$AlCl_3 + 6NH_3 \longrightarrow AlCl_3.6NH_3 \quad (\textit{Addition compound}).$$

Uses. It is used : (*i*) in organic chemistry for introducing alkyl radicals originally linked with halogen into benzene nucleus (*Friedel Craft's reaction*).

(*ii*) in the manufacture of petrol by the cracking of mineral oils, (*iii*) as a catalyst in the manufacture of dyes, drugs and perfumes.

Structure. The vapour density of the compound at 350–400°C corresponds to the dimeric formula, Al_2Cl_6 while at 750–800°C as well as in solution it corresponds to the monomeric formula, $AlCl_3$.

$$Al_2Cl_6 \rightleftarrows 2AlCl_3.$$

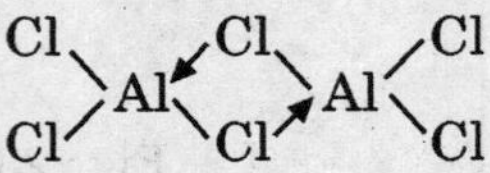

The dimer has the structure shown in the margin. This structure shows that the compound can be regarded as an auto-complex, $Al(AlCl_6)$. In this structure, the halogen atoms are tetra-hedrally arranged about each aluminium atom. The dimeric formula is retained when it is dissolved in non-polar solvents such as benzene, but because of higher heat of hydration, the covalent dimer is broken into $[Al.6H_2O)^{3+}$ or $[Al\ (H_2O)_6]^{3+}$ and $3Cl^-$ ions on dissolution in water.

2. Aluminium Nitride, AlN.

Preparation. (*i*) By heating strongly aluminium powder in an atmosphere of nitrogen or ammonia.

$$2Al + N_2 \xrightarrow{800°} 2AlN$$

(*ii*) By heating alumina mixed with coke in a current of nitrogen.

$$Al_2O_3 + 3C + N_2 \xrightarrow{1800°} 2AlN + 3CO$$

Properties. It is a colourless crystalline solid which is decomposed by water or alkalies.

$$AlN + 3H_2O \longrightarrow Al(OH)_3 + NH_3$$

It is thermally stable upto 800°C. When heated with HCl at 900°C and with Cl_2 at 300°, it gives $AlCl_3$.

$$AlN + 3HCl \xrightarrow{900°C} AlCl_3 + NH_3$$

$$AlN + 3Cl_2 \xrightarrow{300°C} 2AlCl_3 + N_2$$

Uses. Its formation during purification of bauxite by Serpeck's process is utilised for the manufacture of ammonia.

Structure. Crystalline AlN has wurtzite structure.

3. Aluminium Sulphate, $Al_2(SO_4)_3.18H_2O$.

This compound occurs as *hair salt* or *feather alum.*

Preparation. Commercially it is prepared from bauxite ore, $Al_2O_3.2H_2O$. Bauxite is treated with conc. H_2SO_4 and the resulting solution is cooled and the product is recrystallised from water containing alcohol when lustrous crystals of $Al_2(SO_4)_3.18H_2O$ are obtained.

$$\underset{\text{Bauxite}}{Al_2O_3.2H_2O} + 3H_2SO_4 \longrightarrow Al_2(SO_4)_3 + 5H_2O$$

Properties. Anhydrous salt is a white powder (density 2.71) while hydrated salt forms colourless crystals (density 1.62) having a sour astringent taste. It is soluble in H_2O and the solution is acidic in nature due to hydrolysis.

$$Al_2(SO_4)_3 + 6H_2O \longrightarrow 2Al(OH)_3 + 3H_2SO_4$$

On heating, crystals of the salt lose water of crystallisation leaving anhydrous salt which decomposes at 100°C into Al_2O_3 and SO_3.

$$\underset{\text{Hydrated}}{Al_2(SO_4)_3.18H_2O} \xrightarrow[(-18H_2O)]{\text{Heat}} \underset{\text{Anhydrous}}{Al_2(SO_4)_3} \xrightarrow{800°C} Al_2O_3 + 3SO_3$$

With sulphates of alkali metals it forms double sulphates called *alums* having the composition $R_2SO_4.M_2(SO_4)_3.24H_2O$ where R stands for univalent basic radical like Na, Rb, Cs, NH_4^+ etc., and M is a trivalent metal like Al, Cr, or Fe.

Uses. It is used : (*i*) in the purification of sewage and water, since it possesses remarkable power of precipitating collides (*ii*) as a mordant in dyeing (*iii*) in sizing of paper (*iv*) in foam fire-extinguishers (*v*) in the tanning of kid-leather (*vi*) as a starting material in the preparation of other aluminium salts (*vii*) in the preparation of alums.

4. Alums.

Originally the name alum was given to the double salt, potassium aluminium sulphate, $K_2SO_4.Al_2(SO_4)_3.24H_2O$. But now this name is used for all the double salts having the composition : $M_2^I\ SO_4\ M_2^{III}(SO_4)_3.24H_2O$, where M^I stands for univalent basic radical like Na^+, K^+, Rb^+, Cs^+, NH_4^+, Ag^+, Tl^+ and M^{III} is a trivalent metal Al^{3+}, Cr^{3+}, Fe^{3-}, Co^{3+} etc. The crystals of alums are made up of $[M^I(H_2O)_6]^+$, $[M^{III}(H_2O)_6]^{3+}$ and two SO_4^{2-} ions. Some common examples of alums are (*i*) *Potash alum,* (or *common alum*)—$K_2SO_4.Al_2(SO_4)_3.24H_2O$; (*ii*) *Sodium alum*—$Na_2SO_4.Al_2(SO_4)_3.24H_2O$; (*iii*) *Ammonium alum* $(NH_4)_2SO_4.Al_2(SO_4)_3.24H_2O$; (*iv*) *Chrome alum*, $K_2SO_4.Cr_2(SO_4)_3.24H_2O$ etc. The crystals of alums are generally obtained when hot solutions of equimolecular quantities of their constituent sulphates are mixed and the resulting solution is cooled. All the alums are isomorphous, form mixed crystals in all proportions and also layer-crystals. They are usually much more soluble in hot water than in cold water. In solution they give the ions of the component salts.

Alums are often used for the removal of colloidal impurities from water (as coagulants).

5. Potash Alum or Common Alum, $K_2SO_4.Al_2(SO_4)_3.24H_2O$ *or* $KAl(SO_4)_2.12H_2O$.

Preparation. Commercially it is prepared as follows from alunite or alumstone mineral, $K_2SO_4.Al_2(SO_4)_3.4Al(OH)_3$.

The mineral alunite is calcined and treated with H_2SO_4 when $Al(OH)_3$ component is converted into $Al_2(SO_4)_3$. Now calculated quantity of K_2SO_4 is added to the solution which is then crystallised for alum. Finally it is purified by recrystallisation.

$$\underset{\text{Alunite mineral}}{K_2SO_4.Al_2(SO_4)_3.4Al(OH)_3} + 6H_2SO_4 \longrightarrow K_2SO_4 + 3Al_2(SO_4)_3 + 12H_2O$$

$$K_2SO_4 + Al_2(SO_4)_3 + 24H_2O \longrightarrow \underset{\text{Potash alum}}{K_2SO_4.Al_2(SO_4)_3.24H_2O}$$

Properties. Potash alum is a white crystalline solid, soluble in water but insoluble in alcohol. The aqueous solution is acidic due to the hydrolysis of aluminium sulphate. When heated, it melts at 92°C and on further heating loses the whole of its water of crystallisation and swells up. The swollen mass so produced is called *burnt alum* (*phul phatkiri*).

Uses. It is used in purification of water, water proofing of textiles and in dyeing and paper industry. It is also employed as a styptic to arrest bleeding.

Thallium

Occurrence.

Thallium is one of the rare elements which constitutes only 10^{-5} per cent of the earth's crust. It is present in traces in Cu, Zn and Fe blendes and pyrites. It is also found in mica, carnallite, alluminite etc. Flue dust obtained from the roasting of thalliferrous pyrites contains about 0.25% or more of thallium and it is probably the best source of thallium.

Extraction from Flue Dust.

Flue dust contains Tl_2O and the oxides of other elements like Cu, Pb, As, Bi, Fe etc. Flue dust is dissolved in aqua regia and the resulted solution contains TlCl, $CuCl_2$, $PbCl_4$, $AsCl_3$, $BiCl_3$ and $FeCl_3$. This solution is evaporated to drive away HNO_3 and then heated with H_2S. By this operation $CuCl_2$, $PbCl_4$, $AsCl_3$ and $BiCl_3$ are precipitated as sulphides while TlCl and $FeCl_3$ go into the solution. The solution which contains TlCl and $FeCl_3$ is heated with NH_4OH to get the precipitate of $Fe(OH)_3$ and the filtrate which contains TlOH is treated with KI to get the precipitate of TlI. The precipitate of TlI is treated with hot conc. H_2SO_4 to get Tl_2SO_4. Tl_2SO_4, on electrolysis, gives Tl metal which is collected at the cathode. Alternatively Tl_2SO_4, on being treated with Zn metal gives the Tl metal. The essential reactions are shown below :

$$Fe_2O_3 \text{ (From flue dust)} + 6HCl \text{ (From aqua regia)} \rightarrow 2FeCl_3 + 3H_2O$$

$$Tl_2O \text{ (From flue dust)} + 2HCl \text{ (From aqua regia)} \rightarrow 2TlCl + H_2O$$

$$FeCl_3 + 3NH_4OH \longrightarrow \underset{\text{ppt.}}{Fe(OH)_3} + 3NH_4Cl$$

$$TlCl + NH_4OH \longrightarrow \underset{\text{(In solution)}}{TlOH + NH_4Cl}$$

$$TlOH + KI \longrightarrow \underset{\text{ppt.}}{TlI} + KOH$$

$$2TlI + H_2SO_4 \text{ (hot and concentrated)} \longrightarrow Tl_2SO_4 + 2HCl$$

$$Tl_2SO_4 \xrightarrow{\text{Electrolysis}} Tl \text{ is collected at cathode}$$

$$Tl_2SO_4 + Zn \longrightarrow 2Tl + ZnSO_4$$

Physical Properties.

Thallium is a soft grey metal. Its melting and boiling points are 303° and 1457°C. It is so soft that it can be scratched even by finger nails. It is malleable but not ductile. It exists in two allotropic forms, namely, α, and β forms. It alloys with a number of metals like Cu, Ag, Au, Zn, Cd, etc.

Chemical Properties.

Tl is more reactive than In and Ga due to its tendency to form unipositive ion, Tl^+. It oxidises in moist air and decomposes steam at red heat. It dissolves readily in dil. mineral acids except HCl, giving Tl (I) compounds. It does not dissolve readily in HCl because TlCl is insoluble. It combines with halogens at ordinary temperature and with S, Se and Te on heating. It does not combine with B, C, N_2 and H_2 even on heating.

Similarities between Thallium and Other Elements Outside its Group.

(*a*) Similarities with alkali metals. Some of Tl (I) compounds are similar to the corresponding compounds of alkali metals. This similarity is due to the fact that Tl^+ ion has the same charge and approximately the same size as the ions derived from the heavier alkali metals. Some points showing the similarity are : (*i*) Tl (+ 1) cyanide, perchlorate, sulphate, phosphate and carbonate are stable salts and are isomorphous with alkali salts. For example, Tl_2SO_4, $TlClO_4$ and Tl_2PtCl_6 are isomorphous with K_2SO_4, $KClO_4$ and K_2PtCl_6 respectively. (*ii*) TlOH is soluble in water giving strong alkaline solution which is very similar in behaviour to NaOH. (*iii*) The aqueous solution of Tl_2CO_3 is strongly alkaline like that of Na_2CO_3. (*iv*) TlF has distorted NaCl type structure while other Tl (+ 1) halides crystallise with CsCl crystal.

(*b*) Similarities with lead. The cause of similarity between Tl (I) and Pb (II) is that Tl^+ and Pb^{2+} ions are isoelectronic, *i.e.* these ions have identical electronic configuration. Some points showing the similarity are: (*i*) Thallium and lead both are white metals with a bluish type. (*ii*) Both the metals are so soft that they can be cut with a knife and scratched with a finger's nail. (*iii*) Both the metals mark the paper and have low melting points. (*iv*) The halides, chromates and sulphides of both the elements are formed by the same method. These salts have the same appearance and are sparingly soluble in H_2O. TlCl, like $PbCl_2$, is soluble in hot water.

(*c*) Similarities with silver. (*i*) Chlorides, bromides and iodides of Tl(I) and Ag(I) have similar colour and are sparingly soluble in H_2O. (*iii*) Like AgCl, TlCl turns violet on exposure to light.

Uses of Thallium.

Thallium does not have any worthwhile application as a metal. However, its salts are very useful. Some uses of its salts are : (*i*) Thallium salts are toxic and hence are used as fungicide, germicide, and for killing rats and destroying vermin. (*ii*) Thallium salts have been used in ointments for the treatment of ring worms. (*iii*) Thallium compounds have unusually high refractive index and are, therefore, used for the preparation of optical glasses like Crooke's glasses. (*iv*) Tl_2CO_3 has been recommended as a standard in acidimetry. (*v*) Thallium oxysulphide is used in thallosulphide photoelectric cells. (*vi*) Certain thallium salts have been used as depilatories. (*vii*) Alloys of Tl with Ag are lustrous and remain unattacked in air. An alloy containing Tl = 10%, Al = 10% and Ag = 80% is resistant to the action of H_2S.

Ceramics Industry

The art of making the articles from china clay or kaoline (*e.g.* chinaware, earthware stoneware etc.) comes under ceramics industry.

Raw Materials Required for the Manufacture of Whiteware.

Following are the raw materials which are required for the manufacture of whiteware.

1. Clay. It is *hydrated aluminium silicate.* Some common clays are : (*i*) *Kaolin*, $Al_2O_3.2SiO_2.2H_2O$ (*ii*) *Beidellite*, $Al_2O_3.2SiO_2$. H_2O (*iii*) *Halloysite*, $Al_2O_3.2SiO_2.3H_2O$. Clay is used for the manufacture of whiteware due to its following properties : (*i*) It is plastic and can be moulded when it is wet. (*ii*) It retains its shape when dried. (*iii*) When mould mass is heated to a high temperature, it sinters together forming a hard, coherent mass without losing its original colour.

2. Felspar or spar. It is double silicate of K or Ca with Al. Some common felspars are : (*i*) *Potash felspar*, $K_2O.Al_2O_3.6SiO_2$ (*ii*) *Soda felspar*, $Na_2O.Al_2O_3.6SiO_2$ (*iii*) *Lime felspar*, $CaO.Al_2O_3.6SiO_2$. Felspar is fusible and serves as a flux as well as a binder for the clay particles. It is not present in the naturally occurring clays and hence is added from outside.

3. Sand or flint. Sand (SiO_2) or flint is used as an opener for reducing the shrinkage of the clay.

4. Fluxes. Fluxes are easily fusible materials. Borax, boric acid, soda ash, pearl ash (K_2CO_3), fluourspar (CaF_2), cryolite ($Na_3 AlF_6$), iron oxide, lead oxides etc. are used as fluxes.

5. Refractories. These are substances which can withstand high temperature (at least 600°) in the furnaces without melting. Al_2O_3, FeO, Cr_2O_3, $MgCO_3$, TiO_2, SiC, ZrO_2 etc. are the important examples of refractories.

Manufacture of Whiteware.

The manufacture of whiteware consists of the following steps :

1. Preparation of slip. Clay, felspar, sand etc. are mixed with H_2O and stirred in a revolving *blunger* to get a cream-like paste which is known as *slip*. The slip so prepared is filtered through vibrating screen to remove the coarse materials (*e.g.* bits of wood, metals, coal etc.) present in it.

2. Filtering and aging. The slip obtained as above is passed through the filter paper so that almost all water is removed and a moist cake is obtained. This moist cake contains only 10–30% water. The moist cake is stored in dampt cellars for a few weeks. This is called *aging of the slip*. This process increases the plasticity of the clay.

3. Shaping of the articles. The shaping of the articles is done by hand moulding, by potter wheel or by die pressing. When cast is required, Na_2CO_3 or K_2CO_3 or water glass is added to the clay.

4. Drying the articles. The articles are now dried in air or chambers through which warm air circulates.

5. Firing. The dried articles are placed in clay boxes called *saggers* which are placed in the furnace called *hovel oven* so that the articles may not coins in direct contact of the flames. During the process of firing the following changes occur : (*i*) At 150°–600°C water is completely removed. (*ii*) At 800–900°C lime stone ($CaCO_3$), if any, gets decomposed. ($CaCO_3 \rightarrow CaO + CO_2 \uparrow$). (*iii*) At 350–900°C, oxidation of ferrous salts and organic matter present in the raw materials takes place. (*iv*) At 1200–1300°C, silicates of different metals may be formed by the combination of metallic oxides with silica (SiO_2).

The article obtained after it has been fired is porous like a biscuit and hence is called *biscuit*.

6. Glazing of the biscuit. In order to make the earthware impervious to liquids, glazes are applied on it. *Dipping* or *immersion, pouring, spraying, dusting, volatilization* etc. are the various methods of applying the glazes on the articles. The biscuit is again fired in a furnace at a temperature 700–990°C. During the firing, the article becomes smooth and shining in appearance. SiO_2, Al_2O_3, lead oxide, ceric oxide etc. are the examples of the glazes. The glazes contain at least one alkaline earth metal.

7. Colouring of the articles. The colouring agent is sometimes mixed in clay, felspar, sand etc. or in glaze components. The materials used as colouring agents are cobalt oxide (blue), copper oxide (green or red), chromium (green or red), iron oxide (yellow, orange or red), gold chloride (pink), MnO_2 (violet), TiO_2 (faint yellow) etc. The colours imparted by them are given in brackets.

Questions with Answers

Q.1 Explain the following with reasons :

(*i*) Al metal cannot be obtained by carbon reduction method.

Ans : In carbon reduction method pure alumina (Al_2O_3) obtained from bauxite ore (Al_2O_3. $2H_2O$) is heated with carbon. In this reaction Al_2O_3 is not reduced to Al by carbon. Instead Al_4C_3 is formed.

$Al_2O_3 + 3C \xrightarrow{\Delta} 2Al + 3CO$ (*This reaction does not take place*)

$2Al_2O_3 + 9C \xrightarrow{\Delta} Al_4C_3 + 6CO$

(*ii*) Aqueous solution of anhydrous $AlCl_3$ is a good conductor of electricity.

Ans : We know that anhydrous $AlCl_3$ exists as a dimer, Al_2Cl_6. When it is dissolved in water, dimeric structure gets broken into $[Al(H_2O)_6]^{3+}$ and Cl^- ions. Due to the production of these ions, the aqueous solution of dimer acts as a good conductor of electricity.

$Al_2Cl_6 + 12H_2O \rightarrow 2[Al(H_2O)_6]^{3+} + 6Cl^-$

(*iii*) Aluminium forms AlF_6^{3-} ion but boron does not form BF_6^{3-} ion.

(Punjab 1995 April)

Ans : Since AlF_6^{3-} ion has octahedral geometry, in the formation of this ion Al-atom in its excited state (Al = $3s^{10}\ 3p^1\ 3p^1\ 3p^\circ\ 3d^\circ\ 3d^\circ$) undergoes sp^3d^2 hybridisation. Further since Al-atom has *d*-orbitals in its valence shell, the occurrence of sp^3d^2 hybridisation of Al-atom in AlF_6^{3-} ion is possible. On the other hand, since B-atom does not have *d*-orbitals, the occurrence of sp^3d^2 hybridisation of B-atom in BF_6^{3-} ion is not possible and hence BF_6^{3-} ion is not formed.

(*iv*) Like Na, Al cannot be obtained by the electrolysis of fused $AlCl_3$.

Ans : We know that Na can be obtained by the electrolysis of fused NaCl containing $CaCl_2$ and KF, but Al cannot be obtained by electrolysing fused $AlCl_3$, since $AlCl_3$ is a covalent solid and hence sublimes at 453K.

Q.2 What happens when :

(*i*) Borax is heated strongly. **(Roorkee 1986)**

(*ii*) A mixture of borax and CaO is heated in a falme. **(Roorkee 1984)**

Ans : (*i*) A glass transparent mass of boric anhydride (B_2O_3) is formed.

$$\underset{\text{Borax}}{Na_2B_4O_7.10H_2O} \xrightarrow[(-10H_2O)]{\Delta} \underset{\substack{\text{Anhydrous}\\\text{borax (white}\\\text{porous mass)}}}{Na_2B_4O_7} \xrightarrow{\Delta} 2NaBO_2 + \underset{\substack{\text{Boric}\\\text{anhydride}}}{B_2O_3}$$

(*ii*) Blue bead of cobalt metaborate $Co(BO_2)_2$ is formed

$$CaO + Na_2B_4O_7.10H_2O \rightarrow 2NaBO_2 + 10H_2O + \underset{\substack{\text{Cobalt}\\\text{metaborate}\\\text{(blue bead)}}}{Co(BO_2)_2}$$

Actually the above reaction takes place through the following steps :

$$Na_2B_4O_7.10H_2O \xrightarrow{\Delta} 2NaBO_2 + 10H_2O + \underset{\text{Glassy bead}}{B_2O_3}$$

$$CoO + B_2O_3 \rightarrow CoO.B_2O_3 \ \textit{or} \ Co(BO_2)_2$$

$$\overline{Na_2B_4O_7.10H_2O + CoO \rightarrow 2NaBO_2 + 10H_2O + \underset{\text{Blue bead}}{Co(BO_2)_2}}$$

Q.3 Starting from borax, how would you obtain the following (*i*) boric acid (*ii*) boron trioxide (*iii*) boron (*iv*) boron intride (*v*) boron trichloride.

Ans : The preparation of the given compounds from borax can be shown in the flow sheet given in Fig. 23.7.

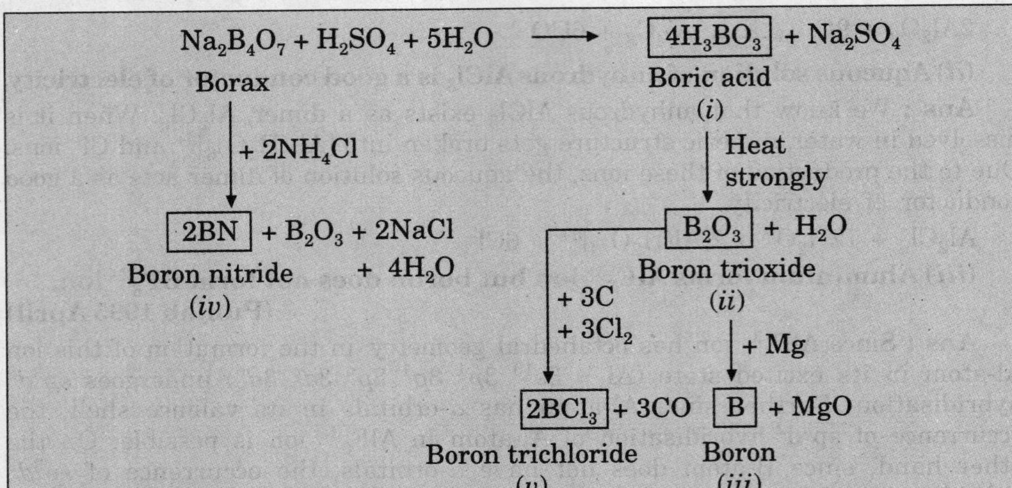

Fig. 23.7. Preparation of compounds from borax.

Q.4 A certain salt, X gives the following tests :

(*i*) Its aqueous solution is alkaline to litmus.

(*ii*) On strongly heating, it swells to give a glassy bead.

(*iii*) When conc. H_2SO_4 is added to a hot concentrated solution of X, white crystals of a weak acid separates out.

Identity X and write down the chemical equation for reactions at (*i*) (*ii*) and (*iii*). (Roorkee 1992)

Ans : (*i*) X is borax, $Na_2B_4O_7 \cdot 10H_2O$

$$Na_2B_4O_7 + 7H_2O \rightleftharpoons \underset{\text{(strong alkali)}}{2NaOH} + \underset{\text{(weak acid)}}{4H_3BO_3}$$

Due to the presence of strong alkali *viz* NaOH, aqueous solution of $Na_2B_4O_7$ is alkaline to litmus.

(*ii*) $Na_2B_4O_7 \cdot 10H_2O \xrightarrow[(-10H_2O)]{\Delta} Na_2B_4O_7 \xrightarrow{\Delta} 2NaBO_2 + \underset{\text{Glassy bead}}{B_2O_3}$

(*iii*) $Na_2B_4O_7 + H_2SO_4 + 5H_2O \longrightarrow \underset{\text{White crystals (Weak acid)}}{4H_3BO_3} + Na_2SO_4$

Q.5 (*i*) A white ppt. (B) is formed when a mineral (A) is bolied with Na_2CO_3 sloution.

(*ii*) The ppt. is filltered and the filtrate contains two compounds (C) and (D). The compound (C) is removed by crystalisation and when CO_2 is passed through the mother liquor left, (D) changes to (C).

(*iii*) The compound (C), on strong heating, gives two compounds (D) and (E).

(*iv*) (E), on heating with CoO, producess blue coloured substance (F).

Identity (A) to (F) and give chemical equation for the reactions at steps (*i*) to (*iv*) (*Roorkee 1995*)

Ans : (*i*) (A) is colemanite ore, $Ca_2B_6O_{11} \cdot 5H_2O$

$$\underset{\text{(A)}}{Ca_2B_6O_{11}} + 2Na_2CO_3(aq) \xrightarrow{\text{Boil}} \underset{\text{(B) Insoluble}}{2CaCO_3} + \underset{\text{(C)}}{Na_2B_4O_7} + \underset{\text{(D)}}{2Na_2BO_2}$$

(C) + (D): Soluble (Solution)

(*ii*) (C) which is borax is obtained by crystallising the solution containing (C) and (D). Mother liquor contains (D) (sodium metaborate). (D) is treated with CO_2.

$$\underset{\text{(D)}}{4NaBO_2} + CO_2 \rightarrow \underset{\text{(C)}}{Na_2B_4O_7} + Na_2CO_3$$

(*iii*) $\underset{\text{(C)}}{Na_2B_4O_7} \xrightarrow{\Delta} \underset{\text{(D)}}{2NaBO_2} + \underset{\text{(E)}}{B_2O_3}$

(*iv*) $B_2O_3 + CoO \xrightarrow{\Delta} Co(BO_2)_2$
(E) (F)
(Blue coloured cobalt metaborate)

Above discussion shows that : A = $Ca_2B_6O_{11}.5H_2O$, B = $CaCO_3$; C = $Na_2B_4O_7$, D = $NaBO_2$, E = B_2O_3, F = $Co(BO_2)_2$.

Q.6 Identity A and B in the following reactions :

Colemanite ore + A $\longrightarrow$ $Na_2B_4O_7$ + ...

$Na_2B_4O_7$ + B $\rightarrow$ H_3BO_3 + ... **(Roorkee 1989)**

Ans : A is sodium carbonate, Na_2CO_3

$$\underset{\text{Colemanite}}{Ca_2B_6O_{11}} + 2Na_2CO_3 \longrightarrow Na_2B_4O_7 + 2NaBO_2 + 2CaCO_3$$

B is either conc. H_2SO_4 or HCl

$$Na_2B_4O_7 + \underset{(B)}{H_2SO_4} + 5H_2O \longrightarrow Na_2BO_4 + 4H_3BO_3$$

$$Na_2B_4O_7 + \underset{(B)}{2HCl} + 5H_2O \longrightarrow 2NaCl + 4H_3BO_3$$

Q.7 How will you convert $AlCl_3$ into $AlBr_3$.

Ans : Reaction are given below :

(*i*) $AlCl_3 + 3NH_4OH \longrightarrow \underset{\text{ppt}}{\boxed{Al(OH)_3}} + 3NH_4Cl$

$Al(OH)_3 + 3HBr \longrightarrow \boxed{AlBr_3} + 3H_2O$

(*ii*) $AlCl_3 + 3NH_4OH \longrightarrow \boxed{Al(OH)_3} + 3NH_4Cl$

$Al(OH)_3 \xrightarrow{\Delta} \boxed{Al_2O_3} + 3H_2O$

$Al_2O_3 + 3C + 3Br_2 \longrightarrow \boxed{2AlBr_3} + 3CO$

University Questions

1. How does boron occur in nature? How is boron extracted from its minerals? Describe the properties and uses of boron. (*Meerut 82 S, 83 S, 84, 85 S*)
2. Give the method for the preparation and uses of borax. (*R.E.E. 81*)
3. How will you get the following from borax : boron, BN, boric acids, BCl_3, BF_3, B_2H_6, B_2O_3, B_2H_6 and inorganic benzene? (*Meerut 1982, 82 S, 84, 85, 87; Agra 81, 82*)
4. Give the preparation, properties and uses of boric acid. (*Meerut 83, 89; Rohilkhand 84; Kanpur 81; Allahabad 87; R.E.E. 74*)
5. Explain the following giving appropriate reasons :
   (*i*) Although H_3BO_3 has negligible tendency to give protons, even then it behaves as a weak monobasic acid in solution. (*Raj. 86; Gorakhpur*)

(*ii*) On addition of glycerol or ethyl glycol, the acidity of boric acid increases. (*Allahabad 87*)

(*iii*) The pH of orthoboric acid solution decreases in presence of ethylene glycol.

6. Write a note on "Borazine". (*Delhi 89*)

7. Explain the following, giving appropriate reasons :

(*i*) Borazine is more reactive than benzene. (*Delhi 80; Delhi Hons. 86*)

(*ii*) Borazine is called inorganic benzene. (*Madras 86; Delhi Hons. 87; Madurai 86*)

8. Name the four important ores of Al. Describe the extraction of Al from bauxite. (*Raj. 86, 87; Meerut 85*)

9. Explain why : (*i*) in the electrolytic manufacture of Al-metal, carbon anodes are consumed. (*ii*) pure alumina cannot be electrolysed for the extraction of Al-metal. (*R.E.E. 80*)

10. Draw the structure of borazine. (*Delhi 99*)

11. How is boric acid produced? Give its two chemical properties, uses and chemical equations. (*Delhi 99*)

12. (*a*) How will you prepare boric acid ? Discuss its (*i*) acidic nature (*ii*) action of heat.

(*b*) Discuss the structure of borazine. Why it undergoes addition reactions readily in contrast to benzene ? Explain giving two reactions.

(*c*) Explain in brief "back bonding" and its effect on reactivity of the substance. (*Himachal Pradesh 99*)

13. (*a*) Explain why orthoboric acid is strongly acidic in presence of glycerol.

(*b*) Describe the method of preparation and uses of borazole. (*Allahabad 99*)

14. (*a*) What is back bonding? Explain why does it occur in boron halides and not in aluminium halides.

(*b*) Discuss the structure of diborane.

(*c*) Draw the structure of orthoboric acid. Why is it a weak acid but in the presence of ethylene glycol, it acts as a strong acid? (*Himachal Pradesh 2000*)

15. Compare the properties of Ga and Tl pair. (*Kanpur 2000*)

16. Give preparation, properties and applications of borax. (*Lucknow 2000*)

CHAPTER 24

Hydrides of Boron : Boranes

Boron does not combine directly with hydrogen to form its hydrides. Nevertheless a number of interesting hydrides of boron are known. The hydrides of boron are also called **boranes** to establish an analogy with alkanes.

Although the trivalency of boron suggests that a hydride of the molecular formula, BH_3 should also exist. This hydride, due to its unstable nature at room temperature, does not exist in the free condition, although the boric radical (BH_3) is found in several compounds like $OC.BH_3$ (borine carbonyl), $(CH_3)_3N.BH_3$. A systematic investigation of boranes was made by Stock in 1914.

The boranes are named as *di-, tri-, tetra-, penta-, hexa-boranes* and so on according to the number of boron atoms followed by the numeral which indicates the number of hydrogen atoms. For example B_5H_9 is called penta borane-9. Here penta indicates the number of boron atoms and 9 indicates the number of H-atoms.

The boranes have been grouped into two classes namely B_nH_{n+4} (n = 2, 5, 6, 8, 10 and 18) and B_nH_{n+6} (n = 4, 5, 6, 9 and 10). The boranes of the former type are quite stable and have high m. pts. while those of the latter type are unstable and have low m. pts. Different members of both the types are : (*i*) B_nH_{n+4} *type*, *e.g.* diborane, B_2H_6; pentaborane–9, B_5H_9; hexaborane–10, B_6H_{10}; octaborane–12, B_8H_{12}; decaborane–14, $B_{10}H_{14}$ etc. (*ii*) B_nH_{n+6} *type e.g.*, tetraborane, B_4H_{12}; pentaborane–11, B_5H_{11}; hexaborane–12, B_6H_{12}; enneaborane–15, B_9H_{15}; decaborane–16, $B_{10}H_{16}$ etc.

Boranes are the examples of electron-deficient compounds. Electron-deficient compounds are those in which the total number of atomic orbitals available on all the atoms of the compound is more than the total number of valence-electrons.

Diborane, B_2H_6

Preparation of Diborane.

B_2H_6 is prepared :

1. *By the action of ionic hydrides like NaH, CaH_2 etc. on BCl_3*

$$6NaH + 2BCl_3 \longrightarrow 6NaCl + B_2H_6$$
$$3CaH_2 + 2BCl_3 \longrightarrow 3CaCl_2 + B_2H_6$$

2. *By the action of boron trifluoride ether complex, $(C_2H_5)_2O.BF_3$ on metallic hydrides like LiH, $Li[BH_4]$, $Na[BH_4]$ etc.*

(*i*) $6LiH + 8(C_2H_5)_2O.BF_3 \longrightarrow B_2H_6 + 6LiBF_4 + 8(C_2H_5)_2O$

(*ii*) $3Li[BH_4] + (C_2H_5)_2O.BF_3 \longrightarrow 2B_2H_6 + 3LiF + (C_2H_5)_2O$

(*iii*) $3Na[BH_4] + 4(C_2H_5)_2O.BF_3 \longrightarrow 2B_2H_6 + 3NaBF_4 + 4(C_2H)_2O$

Reaction (*i*) is used for the manufacture of B_2H_6 in U.S.A.

3. *By reducing BF_3 with LiH, Li[BH_4] or Na[BH_4].*

(*a*) $8BF_3 + 6LiH \longrightarrow 6LiBF_4 + B_2H_6$

(*b*) $BF_3 + 3Li[BH_4] \xrightarrow{\text{Ether}} 3LiF + 2B_2H_6$

(*c*) $BF_3 + 3Na[BF_4] \longrightarrow 3NaF + 2B_2H_6$

B_2H_6 is generally produced *in situ* in reactions (*b*) and (*c*).

4. *By reducing BCl_3 with Li[AlH_4].*

$$4BCl_3 + Li[AlH_4] \longrightarrow 2B_2H_6 + 3LiCl + 3AlCl_3$$

5. *By reducing vapours of BCl_3 by molecular hydrogen.*

$$2BCl_3\ (vapours) + 6H_2 \longrightarrow B_2H_6 + 6HCl$$

6. *By treating an alkali metal borohydride with HCl.*

$$2Li[BH_4] + 2HCl \longrightarrow B_2H_6 + 2LiCl + 2H_2\uparrow$$

$$2Na[BH_4] + 2HCl \longrightarrow B_2H_6 + 2NaCl + 2H_2\uparrow$$

Properties of Diborane.

1. It is an inflammable colourless gas with a sticky sweet odour and is extremely toxic. It is an extremely reactive gas and hence should be handled in a special apparatus so that it comes in contact only with glass and mercury.

2. Decomposition. (*a*) *By heat.* It is stable at low temperatures only in the absence of moisture. At higher temperatures B_2H_6 gets decomposed into a number of higher boranes. For example :

$$5B_2H_6 \xrightarrow{115°C\ \text{(U-tube)}} 2B_5H_{11} + 4H_2$$

$$2B_2H_6 \xrightarrow{-78\ \text{to}\ 100°C} B_4H_{10} + H_2$$

$$5B_2H_6 \xrightarrow{100°C\ \text{(sealed bulb)}} B_{10}H_{14} + 8H_2$$

$$5B_2H_3 \xrightarrow[\text{(tube)}]{250°C,\ 120\ \text{mm.}} B_5H_9 + B_5H_{11} + 5H_2$$

$$5B_2H_6 \xrightarrow{200\text{–}250°C} 2B_5H_9 + 6H_2$$

(*b*) *By water* (*Hydrolysis*). It is readily decomposed by water into boric acid and H_2.

$$B_2H_6 + 6H_2O \longrightarrow 2H_3BO_3 + 6H_2$$

(*c*) *By aqueous alkalies* (*Hydrolysis*). When B_2H_6 is passed into a *concentrated* aqueous solution of KOH at 0°C, reaction occurs and the solution obtained on evaporation in vacuo deposits potassium hypoborate, $K(BH_2OH.BH_2OH)$ or $K_2(B_2H_6O_2)$.

$$B_2H_6 + 2KOH \xrightarrow{0°C} K_2(B_2H_6O_2)$$
Potassium hypoborate

If KOH solution is *diluted*, then potassium hypoborate, $K_2(B_2H_6O_2)$ is converted into potassium metaborate, KBO_2.

$$K_2(B_2H_6O_2) + 2H_2O \longrightarrow 2KBO_2 + 5H_2$$

The evolution of H_2 by the action of aqueous alkalies makes B_2H_6 as a useful reducing agent.

3. Action of lithium ethyl (LiC_2H_5) and alkali metal hydrides. LiC_2H_5 and alkali metal hydrides suspended in ether react with B_2H_6 and produce alkali metal borohydrides, MBH_4

$$2B_2H_6 + 2LiC_2H_5 \longrightarrow 3LiBH_4 + B(C_2H_5)_3$$
$$B_2H_6 + 2LiH \longrightarrow 2LiBH_4$$
$$B_2H_6 + 2NaH \longrightarrow 3NaBH_4$$

4. Reaction with trimethyl aluminium, $Al(CH_3)_3$. Aluminium borohydride, $Al(BH_4)_3$ is produced

$$Al(CH_3)_3 + 2B_2H_6 \longrightarrow B(CH_3)_3 + Al(BH_4)_3$$

5. Action of O_2 (Ignition or combustion). Pure B_2H_6 undergoes no change when mixed with dry air or O_2 at room temperature, but it may ignite (*i.e.* burn), if impure. In this reaction a large amount of energy is produced.

$$B_2H_6 + 3O_2 \longrightarrow B_2O_3 + 3H_2O + 480 \text{ Kcal}$$

The production of large amount of heat in the above reaction makes B_2H_6 a useful rocket fuel.

6. Action of halogens. It reacts with halogens (not iodine) under different conditions to give different products.

$$B_2H_6 + 6Cl_2 \xrightarrow{25°C} 2BCl_3 + 6HCl \quad \textit{(Vigorous reaction)}$$
$$B_2H_6 + Br_2 \xrightarrow{100°C} B_2H_5Br + HBr \quad \textit{(Slow reaction)}$$

7. Action of halogen acids. B_2H_6 reacts readily with HCl acid in presence of $AlCl_3$ as catalyst to form chlorodiborane (B_2H_5Cl). With HBr it reacts at 90°C in presence of $AlBr_3$ to give B_2H_5Br while with HI at 50°C in the absence of a catalyst to give B_2H_5I.

$$B_2H_6 + HX\ (X = Cl, Br, I) \longrightarrow B_2H_5X + H_2$$

8. Action of sodium or potassium amalgam. When B_2H_6 is treated with sodium or potassium amalgam (*reducing agents*) in cold, crystalline product of the composition, $B_2H_6.2Na$ or $B_2H_6.2K$ is formed.

$$2Na\ \textit{(in the form of amalgam)} + B_2H_6 \rightarrow B_2H_6.2Na$$
$$2K\ \textit{(in the form of amalgam)} + B_2H_6 \rightarrow B_2H_6.2K$$

9. Reaction with NH_3. Diborane reacts with NH_3 giving various products depending on the experimental conditions. For example :

(*a*) When B_2H_6 reacts with *excess of NH_3 at low temperature* (–120°C), it forms an addition compound called *diammoniate of diborane*, $B_2H_6.2NH_3$ which forms a conducting solution in liquid ammonia.

$$B_2H_6 + 2NH_3\ (Excess) \xrightarrow{\text{Low tempt. } (-120°C)} \underset{\text{Diammoniate of diborane}}{B_2H_6.2NH_3}$$

$B_2H_6.2NH_3$ is a white non-volatile solid and is soluble in water. It is an ionic compound and hence is represented, as $[BH_2(NH_3)_2]^+.\ [BH_4]^-$ or as follows :

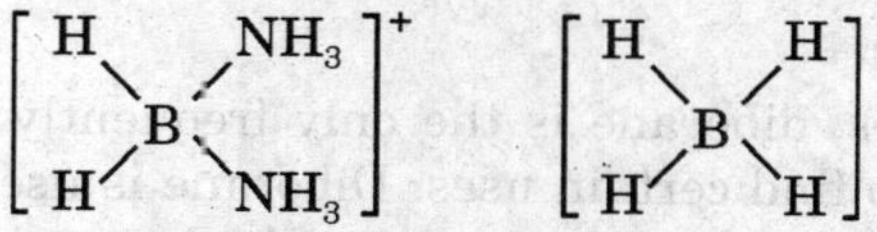

The formation of $B_2H_6.2NH_3$ represents the only case of non-symmetrical cleavage of B_2H_6 by consecutive substitution.

(*b*) With *excess* of NH_3 and at *high temperature*, boron nitride, $(BN)_x$ is formed.

$$xB_2H_6 + 2xNH_3\ (Excess) \xrightarrow{\text{High temp.}} \underset{\text{Boron nitride}}{2(BN)_x} + 6xH_2$$

(*c*) When B_2H_6 and NH_3 are treated at high temperature in 1 : 2 ratio, *borazole* (also called *borazine*), $B_3N_3H_6$ is formed.

$$3B_2H_6 + 6NH_3 \xrightarrow{\text{High temp.}} \underset{\text{Borazole}}{2B_3N_3H_6} + 12H_2$$

Borazole has a flat hexagonal ring structure containing —BH— and —NH groups alternately. Thus it has an analogous electronic structure to that of C_6H_6 and is, therefore called *inorganic benzene*.

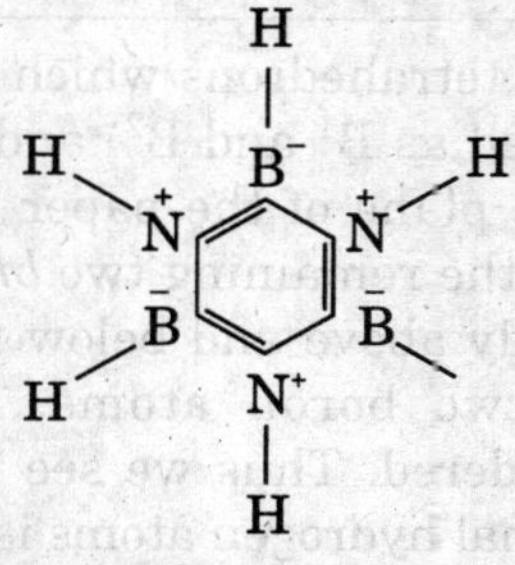

Borazine
($B_3N_3H_6$)

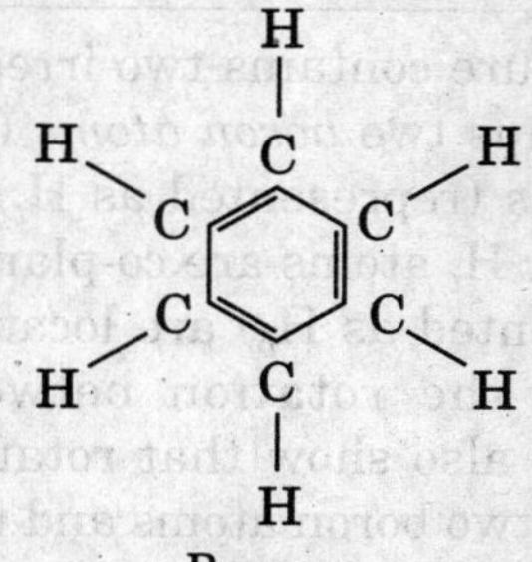

Benzene
(C_6H_6)

10. Formation of adducts. Because of its electron-deficient nature, B_2H_6 reacts with a number of molecules, having lone pairs of electrons [*e.g.*, $(CH_3)_2O$, $(CH_3)_2S$, CO, PF_3, $N(CH_3)_3$ etc.] (Lewis bases) to form adducts or addition compounds. Thus in these reactions B_2H_6 acts as an acceptor (Lewis acid) and the molecules like $(CH_3)_2O$, $(CH_3)_2S$ etc. act as donors (Lewis bases).

Acceptor (Lewis acid)		***Donor (Lewis base)***		***Adducts or addition compounds***
$\frac{1}{2}$ B_2H_6	+	$(CH_3)_2O$ Ether	$\rightarrow$	$(CH_3)_2O \rightarrow BH_3$ Dimethyl ether borane
$\frac{1}{2}$ B_2H_6	+	$(CH_3)_2S$ Thioether	$\rightarrow$	$(CH_3)_2S \rightarrow BH_3$
$\frac{1}{2}$ B_2H_6	+	CO	$\xrightarrow[\text{20 atom.}]{200°C}$	$CO \rightarrow BH_3$ Carbonyl borine

Uses of Diborane.

Among the boranes, diborane is the only frequently used borane, although other boranes may also find certain uses. Diborane is used : (*i*) as rocket fuel for supersonic bombers, (*ii*) for the preparation of high energy fuels and propellents, (*iii*) for the preparation of higher boranes, organic boron compounds and metal boron hydrides like $LiBH_4$, $NaBH_4$ etc. (*iv*) as a reducing agent in organic reactions, (*v*) as a catalyst in polymerisation reactions (*vi*) for welding torches.

Hydrogen-bridge Structure of Diborane.

Electron diffraction and infra-red spectroscopic studies of this molecule have shown that B_2H_6 has *hydrogen-bridge structure* shown in Fig. 24.2.

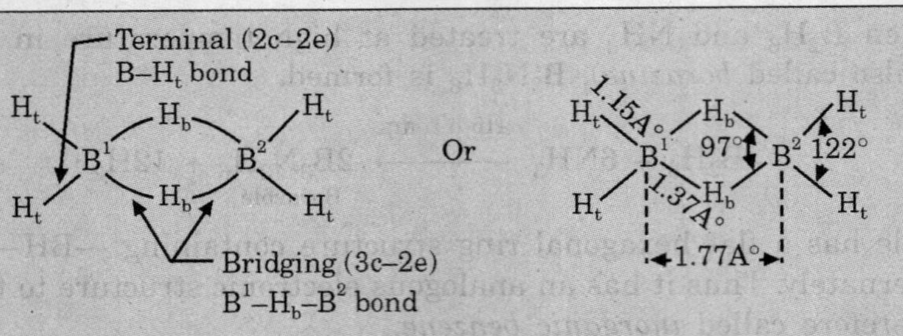

Fig. 24.1. Hydrogen-bridge structure of diborane molecule, B_2H_6

This structure contains two irregular BH_4 tetrahedrons which have one edge in common. Thus two *boron atoms* (represented as B^1 and B^2) and *four terminal hydrogen atoms* (represented as H_t) lie in the plane of the paper (*i.e.* two boron atoms and four H_t atoms are co-planar) while the remaining two *bridge hydrogen atoms* (represented as H_b) are located centrally above and below this plane and thus prevent the rotation between the two boron atoms. Specific heat measurements also show that rotation is hindered. Thus we see that the plane containing the two boron atoms and four terminal hydrogen atoms is perpendicular to that containing the two bridging hydrogen atoms. The following points may be noted from the hydrogen-bridge structure :

(*i*) B^1—H_t bond length = B^2—H_t bond length = 1.19A°. This bond length is almost equal to that expected for a single bond given by the sum of the single-bond normal covalent radii of boron and hydrogen. *i.e.* $r_B + r_H = 0.85 + 0.30 = 1.15A° \simeq 1.19A°$.

(*ii*) B^1—H_b bond length = B^2—H_b bond length = 1.37A° which is greater than $r_B + r_H$ (= 1.15A°).

(*iii*) B^1—B^2 bond length = 1.77A°.

(*iv*) $H_tB^1H_t$ bond angle = $H_tB^2H_t$ bond angle = 122° while $H_bB^1H_b$ bond angle = $H_bB^2H_b$ bond angle = 97°.

(*v*) B_2H_6 molecule has two types of bonds discussed below :

(*a*) *Four terminal (2c–2e)B—H_t bonds.* Each of these bonds is formed by the sharing of two electrons between two atoms (or nuclei) namely B and terminal H-atom (*i.e.* H_t atoms). This bond as called *terminal two-centre-two-electron* B—H_t bond and is denoted as (2*c*—2*e*) B—H_t bond. It is a normal σ-covalent bond.

(*b*) *Two bridging* (3*c*–2*e*) B^1—H_b—B^2 *bonds.* Each of these bonds is formed by the sharing of two electrons between three atoms (or nuclei) namely B^1, B^2 and bridging H (H_b) atoms. This bond is called bridging *three-centre-two-electron* B^1—H_b—B^2 *bond* and is denoted as (3*c*—2*e*) B—H_b—B bond. This bond makes a bridge between two B-atoms and is also called *bannana bond.*

The above discussion makes it evident that the terminal (2*c*—2*e*) B—H_t bond holds two atoms together while the bridging (3*c*—2*e*) B^1—H_b—B^2 bond holds three atoms (B^1, B^2 and H_b) together.

Explanation of the Formation of Bridging (3*c*–2*e*) B^1—H_b—B^2 and Terminal (2*c*–2*e*) B—H_t Bonds in B_2H_6 on the Basis of Molecular Orbital Theory.

On the basis of molecular orbital theory the formation of two bridging (3*c*–2*e*) B^1—H_b—B^2 and four terminal (2*c*–2*e*) B—H_t bonds can be explained as follows :

Each of two boron atoms is sp^3 hybridised and gives four sp^3 hybrid orbitals one of which is vacant and the other three are singly filled.

(*a*) Formation of two bridging (3*c*—2*e*) B^1—H_b—B^2 bonds. One singly-filled sp^3 hybrid orbital on B^1 and one vacant sp^3 hybrid orbital on B^2 atom overlap with the singly-filled 1*s* orbital on H_b atom to form two (3*c*–2*e*) B^1—H_b—B^2 bonds. Each of these bonds contains a pair of electrons and forms a bridge between two B^1 and B^2 atoms. Obviously each of the bonds results from sp^3 (B^1)—1*s* (H_b)—sp^3 (B^2) overlap (Fig. 24.2). This bond is generally represented as :

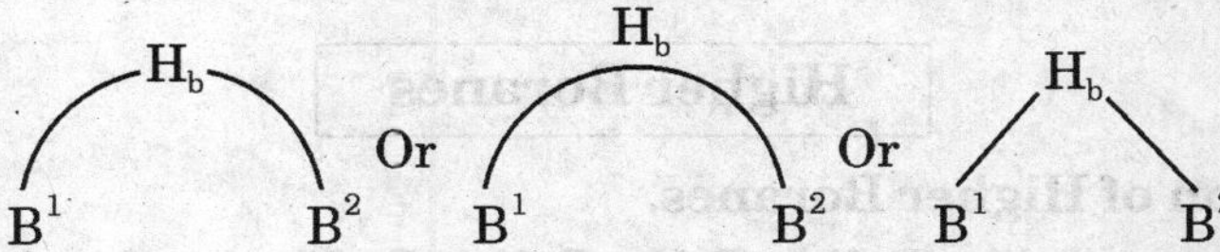

(*b*) Formation of four terminal (2*c*—2*e*) B—H_t bonds. We have seen that two of the four sp^3 hybrid orbitals on each boron atom are consumed in the formation of one bridging (3*c*–2*e*) B^1—H_b—B^2 bond while the remaining two sp^3 hybrid orbitals on each boron atom are left with us. Each of these hybrid orbitals is singly-filled. Each of these two singly-filled sp^3 hybrid orbitals of each boron atom overlaps with singly-filled 1*s* orbital of H_t atom to form terminal (2*c*–2*e*) B—H_t bond (Fig. 24.2). This bond is a normal σ-bond and results from sp^3 (B)—1*s* (H_t) overlap. Thus we see that B_2H_6 has four terminal (2*c*–2*e*) B—H_t bonds.

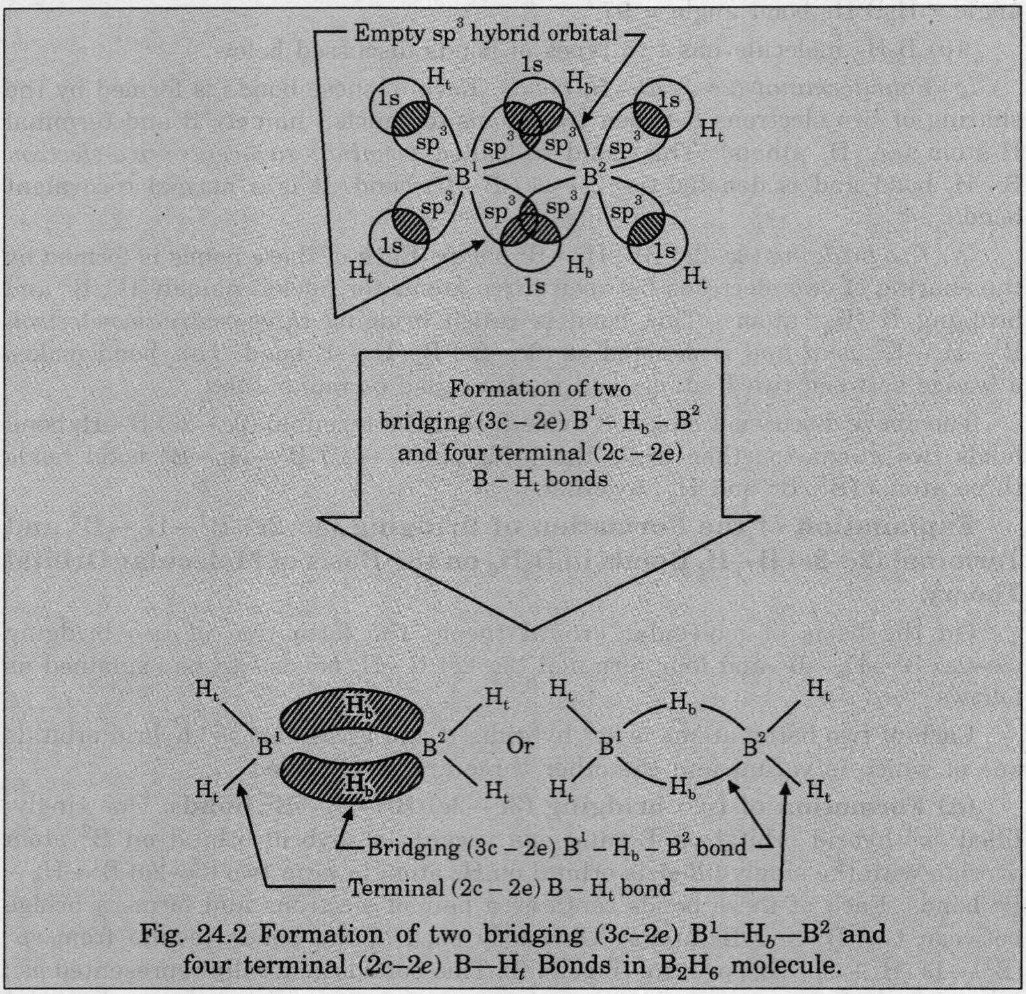

Fig. 24.2 Formation of two bridging (3*c*–2*e*) B^1—H_b—B^2 and four terminal (2*c*–2*e*) B—H_t Bonds in B_2H_6 molecule.

Higher Boranes

Preparation of Higher Boranes.

(*i*) Higher boranes like B_4H_{10}, B_5H_9, B_5H_{11}, $B_{10}H_{14}$ can be obtained : (*a*) by the thermal decompositian of B_2H_6 at different temperatures and (*b*) by the conversion of one higher borane into the other by heating it. Both these methods have been shown in Fig. 24.3.

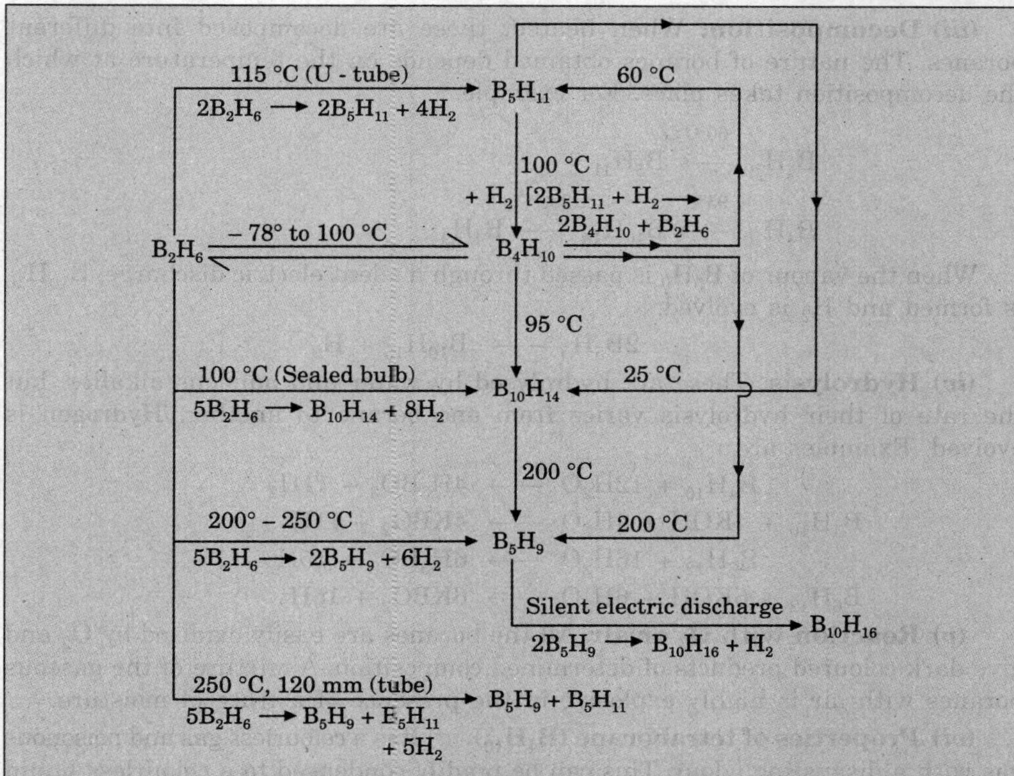

Fig. 24.3. Preparation of various boranes by the thermal decomposition of B_2H_6.

(*ii*) The higher boranes can also be obtained by the action of of dil. HCl on Mg_3N_2. The product obtained by this reaction is a volatile gaseous mixture which is then condensed to a liquid by cooling it in liq. air. The liquid condensates so formed contains the higher boranes such as B_4H_{10}, H_5H_9, H_5H_{11} etc. These hydrides can be separated from each other by the fractional distillation under reduced pressure.

(*iii*) **Wurtz-type synthesis.** This method has been used to prepare B_4H_{10}.

$$2B_2H_5I + 2Na(Hg) \longrightarrow B_4H_{10} + 2NaI$$

Properties of Higher Boranes.

***(i)* Physical state.** All the higher boranes are volatile gases, liquids or solids.

***(ii)* Melting and boiling points.** These are given as follows :

Borane	*m.pt.* (°C)	*b.pt.* (°C)	*Borane*	*m.pt.* (°C)	*b.pt.* (°C)
B_4H_{10}	–120	18	B_6H_{10}	–62.3	108
B_5H_9	–46.6	48	B_6H_{12}	–82.3	80–90
B_5H_{11}	–123	63	$B_{10}H_{14}$	–92.7	213

***(iii)* Decomposition.** When heated, these are decomposed into different boranes. The nature of boranes obtained depends on the temperature at which the decomposition takes place. For example :

$$B_4H_{10} \xrightarrow{60°C} B_5H_{11}$$

$$B_4H_{10} \xrightarrow{95°C} B_{10}H_{14} \xrightarrow{200°C} B_4H_9$$

When the vapour of B_5H_9 is passed through a silent electric discharge, $B_{10}H_{16}$ is formed and H_2 is evolved.

$$2B_5H_9 \longrightarrow B_{10}H_{16} + H_2 \uparrow$$

***(iv)* Hydrolysis.** These are hydrolysed by water and aqueous alkalies, but the rate of their hydrolysis varies from one borane to another. Hydrogen is evolved. Examples are :

$$B_4H_{10} + 12H_2O \longrightarrow 4H_3BO_3 + 11H_2$$

$$B_4H_{10} + 4KOH + 4H_2O \longrightarrow 4KBO_2 + 11H_2$$

$$B_6H_{12} + 16H_2O \longrightarrow 6H_3BO_3 + 15H_2$$

$$B_6H_{12} + 6KOH + 6H_2O \longrightarrow 6KBO_2 + 15H_2$$

***(v)* Reaction with O_2 or air.** All the boranes are easily oxidised by O_2 and give dark coloured preducts of determined composition. A mixture of the gaseous boranes with air is highly explosive in the presence of a trace of moisture.

***(vi)* Properties of tetraborane (B_4H_{10}).** (*a*) It is a colourless gas and poisonous gas with a disgusting odour. This can be readily condensed to a colourless liquid with m.pt. = –120°C and b.pt. = 18°C.

(*b*) It is soluble in dry benzene but slightly so in cold water.

(*c*) *Decomposition.* It is stable in the absence of air and moisture. It decomposes at ordinary temperature, giving H_2 and lower boranes.

(*d*) *Action of HCl.* With HCl, H_2 is liberated

$$B_4H_{10} + HCl \longrightarrow B_4H_9Cl + H_2$$

(*e*) *Reaction with CO.* B_4H_{10} reacts with CO at 120°C in a hot-cold flow reactor and gives the carbonyl adduct, B_4H_8CO :

$$B_4H_{10} + CO \longrightarrow B_4H_8CO + H_2$$

***(vii)* Properties of pentaborane—9 (B_5H_9).** Some of the reactions shown by B_5B_9 are :

$$B_5H_9 + C_2H_4 \xrightarrow{AlCl_3} B_5H_8C_2H_5$$

$$B_5H_9 + 2LiH \xrightarrow{\text{Diethyl ether}} Li_2B_5H_{11}$$

$$B_5H_9 + 2Na(Hg) \xrightarrow{\text{Ether}} Na_2B_5H_9 \xrightarrow{\text{Vacuum}} B_5H_9$$

$$B_5H_9 + 2Li \xrightarrow{\text{Liq. } NH_3} Li_2B_5H_9$$

***(viii)* Properties of pentaborane—11 (B_5H_{11}).** In presence of weak Lewis bases like ethers, amines etc., B_5H_{11} gives B_6H_{10} and B_2H_6

$$2B_5H_{11} \xrightarrow[\text{a Lewis base}]{\text{In presence of}} B_6H_{10} + 2B_2B_6$$

B_5H_{11} reacts with CO in a hot-cold flow reactor and gives the carbonyl adduct, B_4H_8CO.

$$B_5H_{11} + 2CO \longrightarrow B_4H_8CO + BH_3CO$$

When heated with H_2 at 100°C, it gives B_4H_{10}.

***(ix)* Properties of decaborane—l4 ($B_{10}H_{14}$).** Some reactions shown by this hydride are :

$$B_{10}H_{14} + NaH \xrightarrow{(C_2H_5)_2O} NaB_{10}H_{13}.(C_2H_5)_2O + H_2$$

$$B_{10}H_{14} \xrightarrow[\text{liq. } NH_3]{2Na^+ \text{ in}} (Na^+)_2\ [B_{10}H_{14}]^{2-}$$

$$B_{10}H_{14} + 2L\ (L = CH_3CN, R_3N, R_3P) \xrightarrow{\Delta} B_{10}H_{12}L_2 + H_2$$

L is the weakly basic liquid.

$$B_{10}H_{14} + CH_3MgI \longrightarrow B_{10}H_{13}MgI + CH_4$$

Question with Answers

Q.1 Explain the following, giving appropriate reasons for your answer.

***(i)* BF_3 exists but BH_3 does not.**

***(ii)* B–F bond length in BF_3 is shorter than that in BF_4^-.**

***(iii)* Borazine or borazole ($B_3N_3H_6$) is called inorganic benzene.**

Ans : *(i)* Due to pπ – pπ back bonding, the lone pair of electrons of F is denoted to the B-atom. This delocalization reduces the deficiency of electrons on B, thereby increasing the stability of BF_3 molecule.

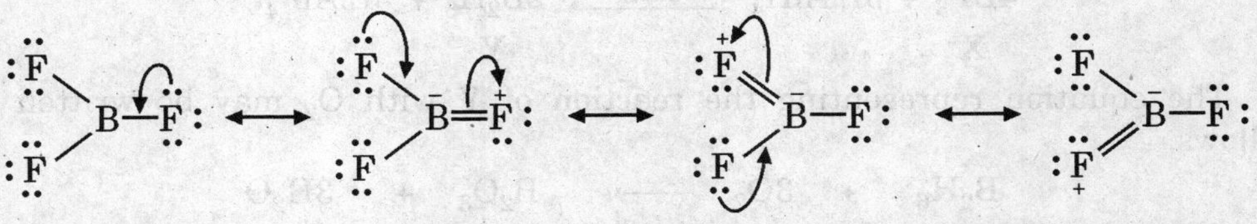

Due to absence of lone pair of electrons on H-atom, this compensation does not occur in BH_3. In other words, electron deficiency of B stays and hence to reduce its electron deficiency, BH_3 dimerises to form B_2H_6.

(ii) Because of double bond character of B–F bonds in BF_3, it has a shorter B–F bond length than in BF_4^-.

(iii) Borazine is called inorganic benzene because the structure of borazine is similar to that of benzene. It is also isoelctronic and isosteric with benzene. Like C-atom in C_6H_6, both N^+ and B^- in $B_3N_3H_6$ are sp^2 hybridised.

Q.2 How is B_2H_6 converted into borazine ($B_3N_3H_6$) ?

Ans : At low temperature, diborane combines with ammonia to form an addition compound.

$$B_2H_6 + 2NH_3 \xrightarrow{\text{Low temp.}} B_2H_6.2NH_3 \text{ (Addition Compound)}$$

When this addition compound is heated to 473K, it decomposes to form a volatile compound called borazine.

$$3[B_2H_6.\ 2NH_3] \xrightarrow{473K} \underset{\text{Borazine}}{2B_3N_3H_6} + 12H_2$$

Q.3 Compound X on reduction with $LiAlH_4$ gives a hydride Y containing 21.72% hydrogen along with other products. The compound Y reacts with air explosively resulting in boron trioxide. Identity X and Y. Give balanced equations involved in the formation of Y and its reaction with air. Draw the structure of Y.

Step 1. *To determine the molecular formula and structure of compound* **Y**.

(*i*) Since the hydride **Y** reacts with air forming boron trioxide, **Y** *must be hydride of boron.*

(*ii*) %H = 21.72% (*Given*)

$\therefore$ %B = 100 – 21.72 = 78.28%

$$\text{Now B : H} = \frac{78.28}{11} : \frac{21.72}{1} = 7.12 : 21.72 = 1 : 3$$

$\therefore$ E.F of **Y** = BH_3

Since boron froms two types of hydrides, *i.e.* B_nH_{n+4} (nidoboranes) and B_nH_{n+6} (arachnoboranes), **Y** must be a nidoborane with n = 2. Thus, M.F. of **Y** = B_2H_6. If **Y** is B_2H_6 (diborane), then its structure must be as follows :

Bridges B.....H = 134 pm Terminal B – H = 119 pm

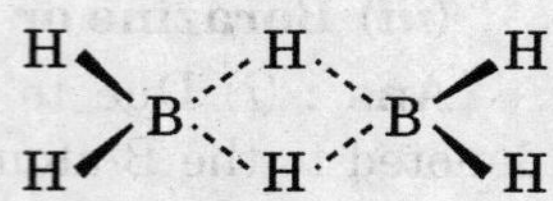

Step 2. *To detrmine the structure of the compound* **X**.

Since compound **Y** *i.e.*, B_2H_6 is formed by reduction of compound **X** with $LiAlH_4$, **X** must either BCl_3 or BF_3.

$$\underset{\textbf{X}}{4BF_3} + 3LiAlH_4 \xrightarrow{\text{ether}} \underset{\textbf{Y}}{2B_2H_6} + 3LiAlF_4$$

The equation representing the reaction of **Y** with O_2 may be written as follows.

$$\underset{\text{Diborane (Y)}}{B_2H_6} + 3O_2 \longrightarrow \underset{\text{Boron trioxide}}{B_2O_3} + 3H_2O$$

Thus **X** = BF_3 and **Y** = B_2H_6.

University Questions

1. Describe the hydrides of boron. *(Meerut B.Sc. 86; Punjab 82)*
2. Give the preparation, properties and uses of boron hydrides. *(Lucknow 80)*
3. Discuss the preparation of B_2H_6. *(Delhi Hons 87)* *Allahabad 87, 88; Delhi 88; Agra 88)*
4. How does diborane react with water, ammonia and NaH. *(Delhi 82)*
5. Discuss the structure and bonding of B_2H_6. *(Raj 87; Meerut 83, 84, 85, 88; Delhi Hons. 87; Utkal 85; Madras 86; Gorakhpur 82; Allahabad 87, 88; Delhi 88; Bundelkhand 88)*
6. Complete and balance the equation :
   $B_2H_6 + LiAlH_4 \rightarrow$ ...... *(Meerut 83)*
7. Explain why : BF_3 is monomeric while BH_3 is dimeric. *(Delhi 99)*
8. Discuss the structure and constitution of diborane. *(Allahabad 99)*
9. Write note on "Hydrides of boron." *(Kanpur 2000)*
10. Describe the preparation, properties and constitution of diborane. *(Avadh 2000)*
11. (*i*) Explain why diborane is electron deficient compound.
    (*ii*) Discuss the structure of B_2H_6 molecule. *(Lucknow 2001)*
12. Write the formula and structure of diborane. *(Meerut 2009)*

CHAPTER

25

General Characteristics of Group 14 (IV A) Elements : Carbon Group Elements

Group IV A of the long form periodic table consists of *five element viz.* Carbon (C), *Silicon* (Si), *Germanium* (Ge), *Tin* (Sn) and *Lead* (Pb). Carbon is an essential constituent of all organic matter while silicon is the main constituent of inorganic matter.

Position of Carbon Group Elements in the Periodic Table

The inclusion of carbon group elements in the same sub-group of the periodic table is justified by the following general characteristics of these elements.

I. Electronic Configurations.

The electronic configurations of these elements are : C(Z = 6) $\rightarrow$ 2, 4 ($2s^2p^2$); Si (Z = 14) $\rightarrow$ 2, 8 4($3s^2p^2$); Ge(Z = 32) $\rightarrow$ 2, 8, 18, 4 ($4s^2p^2$); Sn (Z = 50) $\rightarrow$ 2, 8, 18, 18, 4($5s^2p^2$); Pb(Z = 82) $\rightarrow$ 2, 8, 18, 32, 18, 4 ($6s^2p^2$). These configurations show that these elements have four electrons in their ultimate (valence) shell, two of which are in *s*-orbital while the remaining two are in *p*-orbitals. Hence these elements possess s^2p^2 configuration in the valence shell. The penultimate shell of C contains s^2 electrons (saturated), of Si contains s^2p^6 electrons (saturated), of Ge contains $s^2p^6d^{10}$ electrons (saturated) while of Sn and Pb contain $s^2p^6d^{10}$ electrons (unsaturated). This shows why C differs from Si and these two (*i.e.*, C and Si) from the other members of this sub-group.

II. Physical Properties and Gradation in Them.

Some of the physical properties of these elements are discussed below :

1. Occurrence. Carbon is the only element of this group which occurs in the *free state* as *diamond, graphite* and *coal.* In the combined state it occurs as *hydrocarbons, carbohydrates* and *carbon dioxide* in the atmosphere. Silicon is found as a major constituent in rocks either in the form of *silica* or *silicates.* Germanium is a *rare element.* Tin is mainly found in the form of *tinstone*, SnO_2 while lead occurs as mineral *galena*, PbS.

2. Allotropy. Excepting Pb, all other elements of this group show *allotrophy*, *i.e.* these elements exist in different allotropic forms. For example : (*i*) Carbon exists into the following two allotropic forms. (*a*) Crystalline forms: diamond and graphite. (*b*) Amorphous forms which are microcrystalline forms, *e.g.*, coal, coke, charcoal etc. (*ii*) silicon exists into two forms. (*a*) Crystalline form. (*b*) Amorphous form. It is the common variety of silicon. (*iii*) Germanium exists into two crystalline forms. (*iv*) Tin exists into three allotropic forms. (*a*) White tin (*b*) Grey tin (*c*) Rhombic tin.

3. Atomic volumes. These increase on moving down the group from C to Pb.

4. Density. Density of these elements shows a regular increase from C to Pb.

5. Atomic and ionic radii. Like atomic volumes, these values also increase from C to Pb.

6. Metallic and non-metallic character. The change from non-metallic to metallic character with the increase in atomic number is best illustrated by the elements of this group. The first two elements viz., C and Si are distinctly *non-metals,* although electrical properties of Si are like those of a semi-metal (*i.e.*, a *metalloid*); the third element *viz.*, Ge is partly a metal and partly a non-metal (*i.e.*, it is a *metalloid*) while the last two elements *viz.*, Sn and Pb are distinctly *metals.* Thus :

$$\underbrace{C,\ Si}_{\text{Non-metals}} \qquad \underbrace{Ge}_{\text{Metalloid}} \qquad \underbrace{Sn,\ Pb}_{\text{Metals}}$$

——— Metallic character increases ⟶

The change from non-metallic to metallic character with the increase in atomic number is due to the *reduction* in *effective nuclear charge* and *increase in the number of available orbitals* with the increase in the size of the atoms.

7. Melting and boiling points. As we move down the group from C to Pb, the melting points as well as boiling points *generally decrease*, although the decrease is not in a regular order. This decrease in melting points as well as in boiling points indicates that interatomic forces also decrease in the same direction. The melting and boiling points of C and Si are notably *high* because of the tendencies of these elements to form *giant molecules*.

8. Ionisation energies. The ionisation energy values decrease on moving down the group from C to Pb, although the decrease does not occur in a regular order. The irregularity in the decrease of these values is due to the filling of intervening *d*-orbitals in case of Ge and Sn end *f*-orbitals in case of Pb which are not able to screen the valence-electrons *effectively* in the elements following them.

9. Electronegativity. Carbon is the most electronegative element of this sub-group and the electronegativities decrease with the rise of atomic number but not in a regular manner. This is probably due to the filling of the *d*-orbitals in case of Ge and Sn and *f*-orbitals in case of Pb.

10. Valency. The elements are all tetracovalent. Their ionization potentials being very high, the existence of simple M^{4+} ions is unlikely. On account of their low electronegativity values M^{4-} ions also do not normally exist. Carbon does form C_2^{2-} ions in salt-like carbides, *e.g.*, calcium carbide, $Ca^{2+}\ C_2^{2-}$. C and Si do not form M^{4+} or M^{2+} ions. The rest of the elements, on the other hand, apart from showing *tetracovalency*, exhibit *bielectrovalency* also. Thus Ge, Sn and Pb give stable bivalent ions *viz.*, Ge^{2+}, Sn^{2+} and Pb^{2+}.

11. Oxidation states. (*a*) positive oxidation states and inert pair effect. Carbon and silicon show + 4 oxidation states while Ge, Sn and Pb show + 2 as well as + 4 oxidation states. The occurrence of + 2 and + 4 oxidation states in case of Ge, Sn and Pb is explained as follows : when only two *np* electrons from the ns^2p^2 configuration are lost, we get the elements in + 2 oxidation state *i.e.* M^{2+} cations are obtained. In this case the two ms^2 electrons remain inert and hence are not lost in the formation of M^{2+} cations. This pair of ns^2 electrons is called inert pair of electrons. Since the stability of inert pair of *ns* electrons increases on moving down the group, the stability of + 2 oxidation state also increases in the same direction, *i.e.* the stability of M^{2+} cations inreases from Ge^{2+} to Pb^{2+}, *i.e.* $Ge^{2+} < Sn^{2+} < Pb^{2+}$.

When all the four ns^2p^2 electrons are lost, we get the elements in + 4 oxidation state, *i.e.* M^{4+} cations are formed. On descending the group stability of + 4 oxidation state decreases, *i.e.* the stability of M^{4+} cations decreases from Ge^{4+} to Pb^{4+}, *i.e.* $Ge^{4+} > Sn^{4+} > Pb^{4+}$.

Compounds of Ge^{2+} are less stable than those of Ge^{4+} ($Ge^{2+} < Ge^{4+}$) and hence the compounds of Ge^{2+} are readily changed (oxidised) into those of Ge^{4+}. In other words compounds of Ge^{2+} act as strong reducing agents while those of Ge^{4+} act as oxidising agents.

$$\underset{\text{(Reducing agent)}}{\underset{\text{Less stable}}{Ge^{2+}\text{ compounds}}} \xrightarrow[(-2e^-)]{\text{Oxidation}} \underset{\text{(Oxidising agent)}}{\underset{\text{More stable}}{Ge^{4+}\text{ compounds}}}$$

On similar grounds it can be shown that the compounds of Sn^{2+} (*e.g.* $SnCl_2$) are *less stable* than those of Sn^{4+} (*e.g.* $SnCl_4$) ($Sn^{2+} < Sn^{4+}$) and hence the compounds of Sn^{2+} are readily changed (oxidised) into those of Sn^{4+}. In other words, compounds of Sn^{4+} act as strong reducing agents while those of Sn^{4+} act as oxidising agents.

$$\underset{\text{(Reducing agent)}}{\underset{\text{Less stable}}{Sn^{2+}\text{ compounds (e.g. }SnCl_2)}} \xrightarrow[(-2e^-)]{\text{Oxidation}} \underset{\text{(Oxidising agent)}}{\underset{\text{More stable}}{Sn^{4+}\text{ compounds (e.g. }SnCl_4)}}$$

When we compare the stability of the compounds of Pb^{2+} and Pb^{4+} ions, we find that Pb^{2+} compounds (*e.g.* $PbCl_2$) are *more stable* than those of Pb^{4+} (*e.g.* $PbCl_4$) ($Pb^{2+} > Pb^{4+}$) and hence the compounds of Pb^{4+} are readily changed (reduced) into those of Pb^{2+}. In other words compounds of Pb^{4+} act as strong oxidising agents while those of Pb^{2+} act as reducing agent.

$$\underset{\text{(Oxidising agent)}}{\underset{\text{Less stable}}{Pb^{4+}\text{ compounds (e.g. }PbCl_4)}} \xrightarrow[(+2e^-)]{\text{Reduction}} \underset{\text{(Reducing agent)}}{\underset{\text{More stable}}{Pb^{2+}\text{ compounds (e.g. }PbCl_2)}}$$

Thus when we compare the stability of M^{2+} and M^{4+} cations of Ge, Sn and Pb, we find that their stability is in the order $Ge^{2+} < Ge^{4+}$; $Sn^{2+} < Sn^{4+}$; $Pb^{2+} > Pb^{4+}$.

***(b)* Negative oxidation states.** Since the electronegativities of these elements are low, they do not have much tendency to form the negative ions. However, carbon forms C^{4+} and C_2^{2-} ions in certain compounds, *e.g.* $Be_4^{2+}\ C_2^{4-}$ or Be_2C(Be^{2+} and C^{-4} ions), $Al_4^{3+}\ C_3^{4-}$ (Al^{3+} and C^{4-} ions), $Na^+\ CH_3^+$ (Na^+, C^{-4} and H^+ ions), $Na_2^+\ C_2^{2-}$ (Na^+ and C_2^{2-} ions), $Ca^{2+}\ C_2^{2-}$ (Ca^{2+} and C_2^{2-} ions).

12. Nature of M^{2+} and M^{4+} compounds. The nature of the compounds of M^{2+} and M^{4+} cations can be predicted by *Fajan's rule* which states that *the smaller the cation, the greater is the amount of covalent character in its compounds.* Since Sn^{4+} ion is smaller than Sn^{2+}, the compounds of Sn^{4+} are *covalent in nature* while those of Sn^{2+} are *ionic.* Similarly Pb^{4+} compounds are covalent and those of Pb^{2+} are ionic. In general the compounds of M^{4+} are covalent and those of M^{2+} are ionic. As we move down the group, the tendency of given cations to form ionic compounds increases.

13. *sp*³ hybridisation and cavalency of 4. The valence shell electronic configuration of these elements *viz.* $ns^2\ np_x^1\ np_y^1\ np_z^0$ shows that these elements have two unpaired electrons and, therefore, should be bi-covalent. But, in fact, all these elements show a covalency of 4. This is explained as due to the excitation of the atom which causes the promotion of one of the paired *ns-electrons* to the vacant *np* level in the same shell. In this way four half-filled orbitals become available in the excited state of the atom which mix together to give four sp^3 hybrid orbitals (sp^3 hybridisation) resulting in a tetrahedral structure and each having one unpaired electron.

14. Tendency of the elements to form homo chains (M—M chains) : catenation or self-linkage. All the elements of this group have a tendency to link with each other and thus form long chains of identical atoms. Such long chains of identical atoms are called *homo chains.* This type of linking of identical atoms with each other is called *catenation* or *self-linkage.* This property decreases on moving down the group from C to Pb. Thus carbon has the *maximum* property of catenation, silicon has much *lesser* tendency, germanium has still *lesser tendency* whereas tin and lead *hardly* show this property. The maximum tendency of catenation associated with carbon forms the basis of the multiplicity of carbon compounds in the field of organic chemistry.

It has been observed that the tendency of catenation of these elements *decreases* with the *decrease* of energy of M—M bond. Thus the M—M bond energy values (in KJ/mole) are : C—C, 348; Si—Si, 222; Ge—Ge, 167; Sn—Sn, 155. These values show that these tendency of these elements towards catenation is in the order : C > Si > Ge > Sn > Pb.

Thus it is easy to prepare compounds containing C—C chains upto twenty carbon atoms, while for Si and Ge it is not possible to extend Si—Si or Ge—Ge chains beyond six units. For Sn and Pb it is not possible to extend the Sn—Sn or Pb—Pb chain even beyond two units.

III. Chemical Properties and Gradation in Them.

Some of the properties shown by these elements and gradation in them are discussed below :

1. Formation of hydrides. All the elements of group IV A form the hydrides which are covalent in nature. Hydrides of MH_4 type are in the gaseous state. The volatility of these hydrides decreases with the increasing atomic number of the central atom. Thermal stability of these hydrides decreases steadily from CH_4 to PbH_4 as is evident from their decomposition temperatures given as CH_4 = 800°C,

SH_4 = 450°C, GeH_4 = 285°C, SnH_4 = 150°C, PbH_4 = 0°C. The decrease in stability from CH_4 to PbH_4 is due to the decrease in M—H bond (M = C, Si, Ge, Sn, Pb) energy in going from C to Pb. This order also shows that the hydrides of the heavier elements of this group are thermally unstable.

2. Formation of oxides. Elements of group IV A. form the oxides which are of three types *viz.* (*i*) Monoxides (MO type oxides) (*ii*) Dioxides (MO_2 type oxides), and (*iii*) other oxides like C_3O_2, Pb_3O_4, Pb_2O, Pb_2O_3 etc.

3. Formation of halides. All the elements of this group form halides like MX_4, MHX_3 (*e.g.* $CHCl_3$, $SiHF_3$ etc), MX_2 (*e.g.* $SiCl_2$, Si_2I_2 etc.) and catenated halides (*e.g.* $Si_{14}F_3$, $Si_{10}Cl_{22}$ etc.). The tetrahalides may be prepared either by heating the element with the appropriate halogen or by passing dry halogen over a heat-mixture of the dioxide and carbon.

$$M + 2X_2 \longrightarrow MX_4$$

$$MO_2 + 2C + 2X_2 \longrightarrow MX_4 + 2CO$$

(M = C, Si etc.) (X = Halogen)

The thermal stability of tetrahalides *decreases* from CX_4 to PbX_4. Thus : $CX_4 > SiX_4 > GeX_4 > SnX_4 > PbX_4$. Thermal stability and volatality of the tetrahalides of the same element *decreases* from F^- to I^-. Thus $MF_4 > MCl_4 > MBr_4 > MI_4$.

The fluorides of carbon (6uoro.carbons) are, therefore, the *most, stable* and *inert*.

The tetrahalides of C are not hydrolysed by water while those of Si, Ge and Sn *get readily hydrolysed.* Thus :

$$CX_4 + H_2O \longrightarrow \text{No action}$$

$$MX_4 + 4H_2O \longrightarrow M(OH)_4 + 4HX$$

(M = Si, Ge, Sn;
X = Halogen atom)

Actually the hydrolysis of tetrahalides takes place through the following two steps :

1st step. In this step oxygen atom of H_2O which acts as a *donor* attacks the central atom of the halide to form a coordinate bond with it and thus produces an *unstable intermediate compound*, $MX_4.4H_2O$.

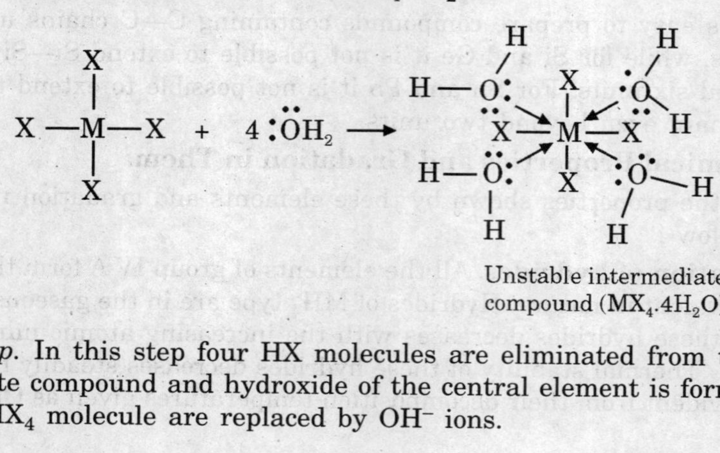

Unstable intermediate compound ($MX_4.4H_2O$)

2nd step. In this step four HX molecules are eliminated from the unstable intermediate compound and hydroxide of the central element is formed. Thus X atoms of MX_4 molecule are replaced by OH^- ions.

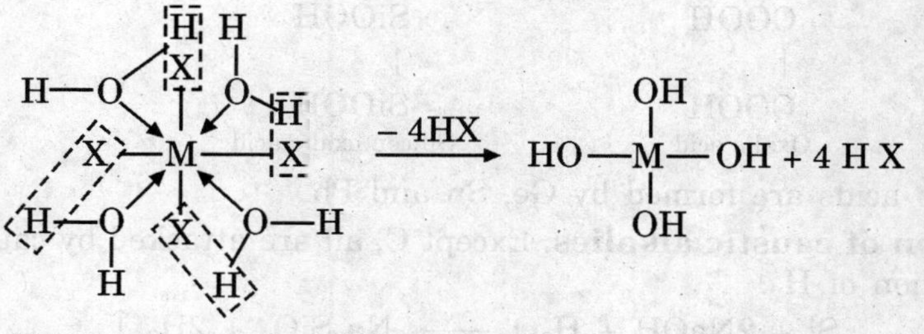

Unstable intermediate compound
($MX_4.4H_2O$)

Why the tetrahalides of C are not hydrolysed while those of Si, Ge and Sn get readily hydrolysed can be explained as follows :

We know that C atom, being a member of 2nd period of the periodic table, has *no d*-orbitals in its valence shell and hence is unable to accommodate the lone pair donated by the donor oxygen atom of H_2O molecule to form an unstable intermediate compound. Thus the tetrahalides of C are not hydrolysed. On the other hand Si, Ge and Sn have vacant *d*-orbitals which can accept the lone pair and thus their tetrahalides get readily hydrolysed.

The ease with which the tetrahalides are hydrolysed by H_2O decreases from Si to Sn as the metallic character of the central atom increases in this order. Thus GeX_4 and SnX_4 tetrahalides are *less readily* hydrolysed than SiX_4 tetrahalides.

It may be mentioned here that empty orbitals are always available with any atom and they can be utilised, if sufficient energy is provided for the reaction to occur, *e.g.*, CCl_4 undergoes hydrolysis when superheated steam is used.

$$CCl_4 + H_2O\ (\textit{superheated steam}) \longrightarrow \underset{\text{Phosgene}}{COCl_2} + 2HCl$$

Hydrolysis of tetrahalides of Pb follows essentially the same pattern, but due to the instability of tetravalent compounds of Pb, some decomposition of $PbCl_4$ to $PbCl_2$ also takes place.

$$PbCl_4 \longrightarrow PbCl_2 + Cl_2$$

$PbCl_4$ is hydrolysed by H_2O as follows :

$$PbCl_4 + 2H_2O \longrightarrow PbO_2 + 4HCl$$

Excepting the tetrahalides of C, those of Si, Ge, Sn and Pb react with halide ions and form the hexahalo complex ions like $[SiF_6]^{2-}$, $[GeX_6]^{2-}$ (X = F, Cl), $[SnX_6]^{2-}$ (X = F, Cl, Br I) and $[PbCl_6]^{2-}$ For example :

$$SiF_4 + 2F^- \longrightarrow [SiF_6]^{2-}$$

4. Formation of acids. (*a*) All the elements give acids of H_2MO_3 type, *e.g.*, H_2CO_3 (carbonic acid) and H_2SiO_3 (silicic acid), H_2GeO_3, (germanic acid), H_2SnO_3 (meta stannic acid) and H_2PbO_3 (meta plumbic acid). Sodium salts of these acids are stable.

(*b*) Carbon and silicon also form acids of the $(MOOH)_2$ type as shown below :

$$\begin{array}{ccc} \text{COOH} & & \text{SiOOH} \\ | & & | \\ \text{COOH} & & \text{SiOOH} \\ \text{Oxalic acid} & & \text{Silicon oxalic acid} \end{array}$$

No such acids are formed by Ge, Sn and Pb.

5. Action of caustic alkalies. Except C, all are attacked by caustic alkalies with evolution of H_2.

$$Si + 2NaOH + H_2O \longrightarrow Na_2SiO_3 + 2H_2 \uparrow$$
$$Ge + 2NaOH + H_2O \longrightarrow K_2GeO_3 + 2H_2 \uparrow$$
$$Sn + 2NaOH + H_2O \longrightarrow Na_2SnO_3 + 2H_2 \uparrow$$
$$Pb + 2NaOH \longrightarrow Na_2PbO_2 + H_2 \uparrow$$

Differences of Carbon and Silicon From Other Elements of Group IVA.

Carbon and silicon differ from the remaining members of group IVA in the following respects :

1. Electronic configuration. Both carbon and silicon have respectively 2 and 8 electrons (configuration of inert gas) in their penultimate shell while the other elements of this group possess 18 electrons in their penultimate shell.

2. Metallic and non-metallic character. Both carbon and silicon are non-metals while Ge is a metalloid and Sn and Pb are metals.

3. Valency. Both C and Si show tetra-valency and do not give tetravalent or bivalent ions M^{4+} or M^{2+} ions). On the other hand Ge, Sn and Pb show tetra-covalency as well as form bivalent ions *viz.* Ge^{2+}, Sn^{2+} and Pb^{2+}.

4. Oxidation states. Both C and Si show an oxidation state of + 4 only while Ge, Sn and Pb show two types of oxidation states which are + 2 and + 4.

5. Catenation. Carbon and, to some extent, silicon have a tendency for catenation while other elements do not have this property to a great extent.

6. Nature of dioxide. CO_2 and SiO_2 are acidic while other dioxides (GeO_2, SnO_2 and PbO_2) are amphoteric.

7. Formation of acids. Carbon and silicon form the acids of the type.

$$\begin{array}{ccc} \text{COOH} & & \text{SiOOH} \\ | & & | \\ \text{COOH} & & \text{SiOOH} \\ \text{Oxalic acid} & & \text{Silicon oxalic acid} \end{array}$$

while other elements do not give the acids of this type.

Comparison between Carbon and Silicon.

(A) Similarities. Both these elements belong to the same sub-group (group IVA) of the periodic table and have the same number of electrons (= 4) in their valence shell. It is for this reason that these elements have many similarities as discussed below :

(i) Electronic configuration. Both the elements have 4 electrons in their valence-shell as shown :

$$C\ (Z = 6) \longrightarrow 2,\ 4;\ Si\ (Z = 14) \longrightarrow 2,\ 8,\ 4$$

(ii) Occurrence. Both are widely distributed in nature. Carbon is an essential constituent of vegetable and animal kingdom, *i.e.*, of the organic compounds. Silicon is an essential constituent of mineral kingdom, *i.e.*, of the inorganic compounds.

(iii) Valency. Both are tetravalent. These elements show a covalency of four and form compounds by sharing of electrons.

(iv) Allotropy. Both show allotropy, *i.e.*, they exist in more than one form. Thus, carbon. occurs in two crystalline forms namely diamond and graphite and a number of amorphous forms such as coal, coke, charcoal etc. Silicon exists in two forms *viz.*, crystalline and amorphous.

(v) Formation of oxides. Both burn in oxygen to form acidic oxides.

$$CO_2 + 2NaOH \longrightarrow Na_2CO_3 + H_2O$$
$$SiO_2 + 2NaOH \longrightarrow Na_2SiO_3 + H_2O$$

(vi) Formation of hydrides. Both form a number of covalent hydrides.

CH_4	C_2H_6	C_3H_8	C_4H_{10}
SiH_4	Si_2H_6	Si_3H_8	Si_4H_{10}

These are all colourless gases and burn in air to form oxides.

$$CH_4 + 2O_2 \longrightarrow CO_2 + 2H_2O$$
$$SiH_4 + 2O_2 \longrightarrow SiO_2 + 2H_2O$$

Both CH_4 and SiH_4 can be prepared by similar methods, *i.e.*, by the action of water or acids on carbides and silicides, respectively.

$$Be_2C + 4HCl \longrightarrow 2BCl_2 + CH_4$$
$$Mg_2Si + 4HCl \longrightarrow 2BeCl_2 + SiH_4$$

(vii) Formation of halides. Both form similar types of covalent halides.

CF_4	CCl_4	CBr_4	$CHCl_3$
SiF_4	$SiCl_4$	$SiBr_4$	$SiHCl_3$

(viii) Formation of acids. Both form similar types of acids, *e.g.*,

COOH	SiOOH
|	|
COOH	SiOOH
Oxalic acid	Silicon oxalic acid

(B) **Dissimilarities.** In some properties these elements differ from each other. The dissimilarity is due to the difference in the number of electrons in the penultimate shell (carbon has 2 electrons while silicon has 8 electrons). The points showing dissimilarity between these elements are given in the following table.

Carbon	Silicon
1. It has two electrons in the penultimate shell.	It has eight electrons in the penultimimate shell.
2. It is a black solid, melting point 3500°C.	It is a reddish brown solid, melting point 1410°C.
3. Some varieties of carbon, *i.e.*, graphite are good conductors of heat and electricity.	Bad conductor of heat and electricity.
4. It has a tendency to form long chains of identical atoms	It has a lesser tendency to form such chains.
5. No action with alkalies.	It reacts with fused alkalies to form H_2 $Si + 2NaOH + H_2O \rightarrow Na_2SiO_3 + 2H_2$
6. CO_2 is a gas. It has a linear structure. The carbon atom is attached to two oxygen atoms by double covalent bond. $O = C = O$	SiO_2 is a solid. It has a giant structure. The silicon atom is attached with oxygen atoms by single covalent bonds. 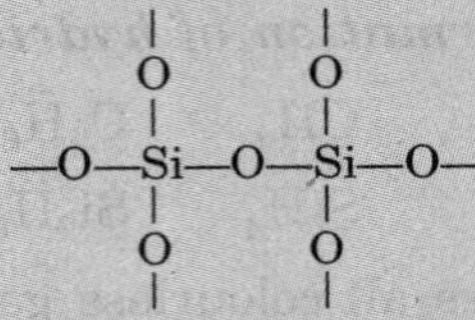
7. The lower oxide (CO) is known.	The lower oxide SiO is unknown.
8. It forms a large number of stable hydrates known as hydrocarbons.	It forms lesser number of hydrides which are unstable.
9. It does not react with Cl_2 or Br_2 directly.	It reacts with Cl_2 or Br_2 at 500°C. $Si + 2Cl_2 \xrightarrow{500°C} SiCl_4$
10. CCl_4 is quite stable and is not hydrolysed by water.	$SiCl_4$ is less stable and hence is hydrolysed by water. $SiCl_4 + 3H_2O \longrightarrow \underset{\text{Silicic acid}}{H_2SiO_3} + 4HCl$
11. The oxy-acid H_2CO_3 is unstable.	The oxy-acid H_2SiO_3 is stable.
12. It does not form complex compounds.	It forms a number of complex compounds because of availability of *d*-orbital in the valency shell.

Comparison between Silicon and Tin.

(A) Similarities. The following similarities are important.

(i) Allotropy. Both tin and silicon show allotropy.

(ii) Oxides. Oxides of both (SiO_2 and SnO_2) are acidic.

(iii) Hydrides. Both form hydrides which are gases. Tin gives a hydride SnH_4 by the action of HCl on an alloy of tin and magnesium. The hydride of silicon, SiH_4 is, however, more stable.

(iv) Halides. Both give similar halides. Their chlorides are colourless, volatile liquids which are easily hydrolysed. The fluorides of both, SiF_4 and SnF_4 are known to form similar silicofluorides, K_2SiF_6 and stannifluorides, K_2SnF_6.

(v) Action of alkalies. Both of them react with hot alkalies solution and liberate hydrogen.

$$Si + 2NaOH + H_2O \longrightarrow Na_2SiO_3 + 2H_2$$

$$Sn + 2NaOH + H_2O \longrightarrow Na_2SnO_3 + 2H_2$$

(*vi*) A similarity also exists between the silicates and meta-stannates.

B. Dissimilarities. The following dissimilarities may be noted :

(i) Metallic character. Tin is a metalloid whereas silicon is a non-metal.

(ii) Melting point. Tin metals at 232°C whereas the melting point of silicon is very high.

(iii) Action of dilute acids. Tin reacts with dilute acids to liberate hydrogen whereas silicon has no such action.

(iv) Salts. Nitrates, sulphates and other salts of tin are known but such compounds are rare in the case of silicon. Many varieties of silicon are found in nature (olivine Mg_2SiO_4) and can also be prepared but orthostannates are almost unknown (with the exception of cobalt stannate Co_2SnO_4). Most of the stannates are meta as Na_2SnO_3.

Similarities between Tin and Lead.

The following points show the similarities between the two elements.

(*i*) Although both the elements have electro-positive character, yet, being in the middle of the periodic table, they also show feeble electronegative character in some of its compounds like halides, metastannates, metaplumbates etc.

(*ii*) Both the elements form highly unstable hydrides.

(*iii*) They form quite stable halides. The formation of such halides indicates that these elements have feeble electronegative character.

(*iv*) Both the elements form similar complexes namely stannates and plumbates respectively.

(*v*) When caustic alkali is added to a Sn + 2) salt solution in the absence of air, $Sn(OH)_2$ is first precipitated which dissolves in the excess of alkali, forming stannite (SnO_2^{2-}). Similarly a plumbite (PbO_2^{2-}) is formed by the addition of an excess of an alkali to a Pb(+2) salt solution.

$$Sn^{2+} + 2NaOH \longrightarrow \underset{\text{ppt.}}{Sn(OH)_2} + 2Na^+$$

$$Sn(OH)_2 + 2NaOH \longrightarrow \underset{\text{Sod. stannite}}{Na_2SiO_2} + 2H_2O$$

$$Pb^{2+} + 2NaOH \longrightarrow \underset{\text{ppt.}}{Pb(OH)_2} + 2Na^+$$

$$Pb(OH)_2 + 2NaOH \longrightarrow \underset{\text{Sod. plumbite}}{Na_2PbO_2} + 2H_2O$$

(*vi*) Metastannates (*e.g.* Na_2SiO_3) and metaplumbates (*e.g.* Na_2PbO_3) are stable and well-defined compounds.

Question with Answers

Q.1 Elemental silicon does not form graphite-like structure as carbon does. Explain.

Ans : The crystal structure of graphite consists of hexagonal layers (sheets) of C-atoms. These layers are linked together by weak vander Waals forces. In each hexagonal layer each C-atom is linked with other C-atoms by C–C covalent bond (sp^2 hybridisation of C-atom) and one electron of C-atom is thus left unsused in $2p$ orbital of each C-atom. The singly-filled $2p$ orbital on one C-atom overlaps with the singly-fulled $2p$ orbital on the other C-atom and forms $p\pi - p\pi$ bond. Thus many $p\pi - p\pi$ bonds are formed. This results in that an extended delocalised π – bonding system encompasing the entire layer is produced. Thus graphite has a two-dimentional sheet-like structure. Si-atom, on the other hand, is not able to form graphite-like structure due to the following reasons :

(*a*) Since Si-atom is bigger in size than C-atom (Si = 117 pm, C = 77 pm), it has no tendency to undergo sp^2 hybridisation (It only undergoes sp^3 hybridisation) and does not form Si–Si bonds. Due to the bigger size of $3p$ orbitals of Si-atom than $2p$ orbitals of C-atom, $p\pi$ – pπ bond cannot be formed by the overlap of singly filled $3p$ orbitals on two Si-atoms.

(*b*) Silicon has much lesses tendency for catenation (self-linking of atoms) than carbon because Si–Si bonds are much weaker (bond energy of Si–Si bond = 200 $KJmol^{-1}$) than C–C bonds (bond energy of C–C bond = 355 $KJmol^{-1}$).

Above discussion shows that Si does not form an analogue of carbon since neither it has a tendency to undergo sp^2 hybridisation nor to form Si–Si bonds.

Q.2 Diamond is extremely hard while graphite is very soft. Explain.

Ans : In diamond each C-atom is covalenty bonded to four additional C-atoms in a rigid three-dimentional network solid. This makes diamond very hard. On the other hand, graphite is a two-dimentional covalent network solid consisting of hexagonal carbon atoms sheets (layers) which are linked together by weak vander Waals forces and hence can slide over each other. This sliding property of graphite makes it soft.

Q.3 Explain the following : (*i*) $PbCl_2$ is more stable than $PbCl_4$ (*ii*) $PbCl_4$ is less stable than $SnCl_4$ but $PbCl_2$ is more stable than $SnCl_2$.

Ans : (*i*) Due to inert pair effect, + 2 oxidation state of Pb is more stable than its + 4 oxidation state., *i.e.* Pb^{2+} ion is more stable than Pb^{4+} ion. Thus since $PbCl_2$ has Pb^{2+} ion and $PbCl_4$ has Pb^{4+} ion, $PbCl_2$ is more stable than $PbCl_4$.

(*ii*) Since the inert pair effect increases on moving down the group 4A elements, the tendency of ns^2 electron pair to participate in bond formation decreases and hence the stability of + 4 oxidation state decreases and that of + 2 oxidation state increases. Consequently $PbCl_4$ is less stable than $SnCl_4$ and $PbCl_2$ is more stable than $SnCl_2$.

Q.4 CO_2 is a gas while SiO_2 is a soild of high melting point. Explain why ?
(*Punjab 1994 ; Kurukshetra 1994 ; Delhi 1995*)

Ans : We know that since the forces of attraction existing between CO_2 molecules are weak vander Walls forces, CO_2 exists as a monomeric (discrete) linear molecule (O = C = O). Due to weak forces, CO_2 exists as a gas at ordinary temperature.

On the other hand, since silicon dioxide has three-dimentional polymeric structure containing many Si–O bonds, there are no discrete SiO_2 molecules. SiO_2 is extremely stable, since considerable amount of energy is required to break Si–O bonds present in the structure of SiO_2. Due to extreme stability associated with the polymeric structure, silica exsist as a solid with high melting point.

Q.5 Structure of SiO_2 is similar to that of diamond but different from that of graphite. Explain.

Ans : In the crystal structuer of SiO_2 each Si-atom is linked tetrahedrally with four O-atoms and hence all the valence shell four electrons of Si-atom are consumed in forming four Si–O bonds. Thus Si-atom in SiO_2 has no free electrons. In the structure of diamond, each C-atom is linked tetrahedrally with four C-atoms placed at the four corners of a tetrahedron. Thus C-atom in diamond has no free (mobile) electrons. Thus SiO_2 and diamond both have similar structure.

In graphite, hexagonal layers are attached with each other by weak vander Waal's forces. Each C-atom in each hexagonal layer is linked with three C-atoms by C–C bond. Thus each C-atom is left with a free (mobile) electron and hence graphite has different structure from SiO_2.

Q.6 Explain the following, giving appropriate reasons for your answer.

(*i*) $SnCl_2$ has higher melting point (is a solid) while $SnCl_4$ has lower melting point (is a volatile liquid).

(*ii*) PbF_4 is an ionic solid but $PbCl_4$ is a covalent liquid.

Ans : (*i*) In $SnCl_2$, Sn has a charge = + 2 and in $SnCl_4$ this atom has charge = + 4. Due to higher charge on Sn in $SnCl_4$, $SnCl_4$ is covalent and $SnCl_2$ is ionic (Fajan's rules). Being ionic, $SnCl_2$ exists as a solid and has high melting point. On the other hand, since $SnCl_4$ is covalent it exists as a liquid and has low melting point.

(*ii*) Since F^- ion is smaller than Cl^- ion, F^- ion gets polarised by Pb^{4+} ion in PbF_4 molecule to smaller extent than Cl^- ion is polarised by Pb^{4+} ion in $PbCl_4$. Due to smallar polarisation of F^- ion than Cl^- ion, PbF_4 is ionic and $PbCl_4$ is covalent. Being ionic, PbF_4 exists as a solid. On the other hand since $PbCl_4$ is covalent,it exists as a liquid at room temperature.

Q.7 Explain the following, giving appropriate reasons for your answer.

(*i*) Lead (II) chloride does not react with Cl_2 to give $PbCl_4$

(*ii*) Lead (IV) chloride is highly unstable towards heat.

Ans : (*i*) Due to inert pair effect, Pb is more stable in + 2 oxidation state than in + 4 oxidation state. Therefore, lead (II) chloride is more stable than lead (IV) chloride and hence lead (II) chloride does not react with Cl_2 to form lead (IV) chloride

$$\underset{\text{More stable}}{\underset{(Pb\ =\ +2)}{PbCl_2(s)}} + Cl_2(g) \not\longrightarrow \underset{\text{Less stable}}{\underset{(Pb\ =\ +4)}{PbCl_4(l)}}$$

(*ii*) Due to inert pair effect, Pb is less stable in + 4 oxidation state than in +2 oxidation state. Thus $PbCl_4$ is less stable than $PbCl_2$ and hence decomposes, on heating, to give more stable $PbCl_2$.

$$\underset{\substack{(Pb = +4) \\ \text{(Less stable)}}}{PbCl_4(l)} \xrightarrow{\Delta} \underset{\substack{(Pb = +2) \\ \text{(More satble)}}}{PbCl_2(s)} + Cl_2(g)$$

Q.8 Explain which of the following tetraiodides is the least stable and has doubtful existence ? CI_4, GeI_4, SnI_4 and PbI_4.

Ans : *PbI_4 is the least stable and hence has doubtful existance.*

Explanation. All the given tetraiodides have C^{4+}, Ge^{4+}, Sn^{4+} and Pb^{4+} cations. Since C-atom does not show inert pair effect, this element is the most stable in its group oxidation state (= + 4). Hence CI_4 is the most stable tetraiodide. Ge, Sn and Pb (heavier elements) exhibit inert pair effect. As we move down the group from Ge to Pb, inert pair effect becomes more and more prominant and hence stability of + 2 oxidation state of these elements increases as $Ge^{2+} < Sn^{2+} < Pb^{2+}$ and that of + 4 oxiadtion state decreases as $Ge^{4+} > Sn^{4+} > Pb^{4+}$. The decreasing order of stability of + 4 O.S. shows that Pb^{4+} ion is the least stable. Consequently PbI_4 is the least stable and hence has doubtful existence.

University Questions

1. Compare the properties of carbon with those of silicon. *(M.D Rohtak 82, 85, 87)*
2. Compare the properties of silicon and tin. *(Delhi 86)*
3. Compare between tin and lead. *(Delhi Hons. 86)*
4. Explain the following giving appropriate reasons :

(1) $SnCl_2$ is a solid while $SnCl_4$ is a liquid.

(2) Carbon and silicon are always tetravalent but Ge, Sn and Pb show divalency.

(3) CCl_4 is resistant to hydrolysis but $SiCl_4$ is readily hydrolysed. *(Raj. 87; Allahabad 87; Delhi Hons. 87; Jodhpur 86; Meerut 88)*

(4) Si—Cl bond is stronger than C—Cl bond, still $SiCl_4$ is hydrolysed easily bit CCl_4 is not. *(Jodhpur 83; Delhi Hons. 82, 86)*

(5) CCl_4 does not act as a Lewis acid while $SiCl_4$ and $SnCl_4$ do so.

(6) $[CF_6]^{2-}$ ion does not exist while $[SiF_6]^{2-}$ ion exists.

(7) Saturated carbon compounds are not reactive whereas the corresponding silicon compounds are reactive. *(Agra 85; Raj. 85)*

(8) All the tetrahalides of lead are known, except PbI_4. *(Allahabad 88)*

(9) $PbCl_4$ exists while $PbBr_4$ and PbI_4 do not.

(10) The boiling point of CCl_4 is greater than that of $SiCl_4$.

(11) $[SiF_6]^{2-}$ ion is known but $[SiCl_6]^{2-}$ ion is not known.

OR

$SiCl_4$ does not react with Cl^- ions to form $[SiCl_6]^{2-}$ while SiF_4 reacts with F^- ions to give $[SiF_6]^{-2}$. *(Punjab Hons. 86)*

(12) Silicon has much less tendency to catenation than carbon. *(Delhi 88)*

(13) The elements *viz.* Si, Ge, Sn and Ph do not show the property of catenation to that extent to which carbon shows.

(14) Carbon possesses greater catenation property than other members of its group. *(Meerut 88)*

(15) The tendency for catenation decreases with the increasing atomic number in the carbon family. *(Agra 89)*

(16) Carbon frequently forms double bond with another carbon atom in its compounds while silicon atom does not do so. *(Gorakhpur 84)*

(17) Tin forms a large number of salts like sulphates, nitrates etc. While the corresponding compounds of silicon are unknown.

(18) Silicon does not form multiple bonds like carbon.

(19) Si—O bond is stronger than C —O bond. *(Jodhpur 86)*

(20) Properties of CO_2 are different from those of SiO_2. *(Delhi Hons. 86)*

(21) CO_2 is a gas while SiO_2 is a solid of high m.pt. *(Meerut 84, 88; Allahabad 88; Raj. 87S; Delhi 86; Punjab 83)*

(22) Solid CO_2 is known as dry ice.

5. Explain the following giving appropriate reasons for your answer : Tetrahalides of carbon are not hydrolysed whereas other members of the group are readily hydrolysed. *(Delhi 1999)*

6. Explain the following :

(*i*) Carbon shows catenation. *(Himachal Pradesh 1999)*

(*ii*) SiO_2 is solid while CO_2 is a gas at room temperature. *(Allahabad 99, Himachal Pradesh 2000)*

7. Give a comparative account of the chemistry of the following elements : C, Si, Ge, Sn, Pb. *(Allahabad 99)*

8. Compare the properties of Ge and Sn pair. *(Kanpur 2000)*

9. $SiCl_4$ is hydrolysed but CCl_4 is not. Explain. *(Lucknow 2000)*

10. Explain the following :

(*i*) CO_2 is a gas while SiO_2 is a solid.

(*ii*) Graphite conducts electricity whereas diamond is a bad conductor of electricity. *(Avadh 2000)*

CHAPTER

26 Compounds of Carbon

Following are the important compounds of carbon which are worth studying.

1. Percarbonic Acids and Percarbonates.

Carbon forms two percarbonic acids which are : (*i*) *Permonocarbonic acid*, H_2CO_4, and (*ii*) *Perdicarbonic acid*, $H_2C_2O_6$. These acids can be regarded as being derived from H_2O_2 by replacing its H-atom by COOH group. Thus :

O—H | O—H	O—H | O—COOH	O—COOH | O—COOH
Hydrogen peroxide (H_2O_2)	Permonocarbonic acid (H_2CO_4)	Perdicarbonic acid ($H_2C_2O_6$)

Permonocarbonic acid, H_2CO_4 is not known in the free state but perdicarbonic acid, $H_2C_2O_6$, is formed as an unstable solution by the action of phosphoric acid on an etheral solution of potassium salts of $H_2C_2O_6$ ($K_2C_2O_6$). Salts derived from these acids are called permonocarbonates and perdicorbonates and in general these are called percarbonates. These salts are of three different types given below :

(A) Permonocarbonates. These are the salts of H_2CO_4. Na_2CO_4, K_2CO_4 and $BaCO_4$ are the important examples of such type of salts.

Preparation. Na_2CO_4 and $BaCO_4$ are formed when CO_2 is absorbed by Na_2O_2 or BaO_2 (*Wolffenstein and Peltner, 1908*).

$$CO_2 + Na_2O_2 \longrightarrow Na_2CO_4$$

or $$O=C=O + Na—O—ONa \longrightarrow O=C\begin{matrix}\diagup ONa \\ \diagdown OONa\end{matrix}$$

$$BaO_2 + CO_2 \longrightarrow BaCO_4$$

Na_2CO_4 is also obtained by the reaction of Na_2O_2 with phosgene, $COCl_2$.

$$2Na_2O_2 + COCl_2 \longrightarrow Na_2CO_4 + 2NaCl + \tfrac{1}{2}O_2$$

A disodium per monocarbonate having the formula, $Na_2CO_4.\ \tfrac{1}{2}H_2O$ has been obtained by passing CO_2 into Na_2O_2 mixed with ice. Sodium permonocarbonate of the formula, $NaHCO_4.H_2O$ and having the structure shown

Structure of $NaHCO_4$

in the margin has been obtained by the action of CO_2 on sodium hydroperoxide perhydrate, $NaOOH.\frac{1}{2}H_2O_2$.

Properties. Freshly prepared permonocarbonates contain water of crystallisation. These are soluble in H_2O and on hydrolysis they give H_2O_2. They liberate I_2 from a solution of KI.

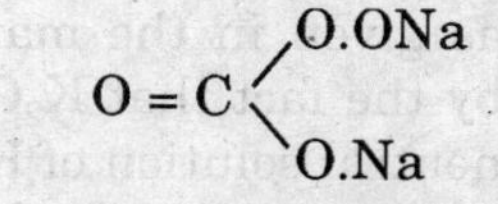

Structure of Na_2CO_4

Structure. Na_2CO_4 has the structure shown in the margin.

(B) Perdicarbonates. These are derived from $H_2C_2O_6$. These are best represented by potassium and sodium salts whose formulae are $K_2C_2O_6$ and $Na_2C_2O_6$ respectively.

Preparation. (*i*) *By anodic oxidation.* $K_2C_2O_6$ is obtained by the electrolysis of a cold saturated solution of K_2CO_3 or $KHCO_3$ at –12°C using a platinum anode enclosed in a porous cell, when a bluish-white amorphous ppt. of $K_2C_2O_6$ is deposited on the anode.

$$\underset{\text{Cathode}}{K^+} \longleftarrow K_2CO_3 \longrightarrow \underset{\text{Anode}}{KCO_3^-}$$

$$2K^+ + 2e^- + 2H_2O \quad 2KOH + H_2 \quad KCO_3^- + KCO_3^- \longrightarrow K_2C_2O_6 + 2e^-$$

$Na_2C_2O_6$ is also obtained in a similar way.

(*ii*) Wolffenstein and Peltner in 1908 obtained $Na_2C_2O_6$ by the action of CO_2 on Na_2O_2 and ice at 0°C.

$$Na_2O_2 + 2CO_2 \longrightarrow Na_2C_2O_6$$

(*iii*) It is also obtained by passing F_2 in a conc. solution of Na_2CO_3 at –13°C to –16°C.

$$2Na_2CO_4 + F_2 \longrightarrow Na_2C_2O_6 + 2NaF$$

Properties. (*i*) $K_2C_2O_6$ is a hygroscopic, skyblue powder which, on drying, loses its colour and becomes white.

(*ii*) $K_2C_2O_6$ is unstable and decomposes slowly at room temperature and radily on heating.

$$2K_2C_2O_6 \longrightarrow 2K_2CO_3 + 2CO_2 + O_2$$

(*iii*) With dil. H_2SO_4 we obtain a quantitative yield of H_2O_2.

$$K_2C_2O_6 + H_2SO_4 \longrightarrow K_2SO_4 + H_2O_2 + 2CO_2$$

(*iv*) It bleaches indigo and oxidises PbS to $PbSO_4$

$$PbS + 2K_2C_2O_6 \longrightarrow 2K_2CO_3 + PbSO_4 + O_2$$

(*v*) It reduces $KMnO_4$, PbO_2 and MnO_2

$$2K_2C_2O_6 + 8H_2SO_4 + 2KMnO_4 \longrightarrow 2MnSO_4 + 6K_2SO_4 + 8H_2O + 10CO_2 + 5O_2$$

$$PbO_2 + K_2C_2O_6 \longrightarrow K_2CO_3 + PbCO_3 + O_2$$

(*vi*) It liberates I_2 quantitatively in cold from a solution of KI.

$$K_2C_2O_6 + 2KI \longrightarrow 2K_2CO_3 + I_2$$

(*vii*) Its solution in ice-cold H_2O is also fairly stable, but on long standing or on raising the temperature, it is hydrolysed to $KHCO_3$ and H_2O_2

$$K_2C_2O_6 + 2H_2O \longrightarrow 2KHCO_3 + H_2O_2$$

Structure. The structure of $K_2C_2O_6$ or $Na_2O_2O_6$ is as given in the margin. This structure is confirmed by the fact that $K_2C_2O_6$ or $Na_2O_2O_6$ liberates I_2 from neutral solution of KI which is equivalent to its active oxygen.

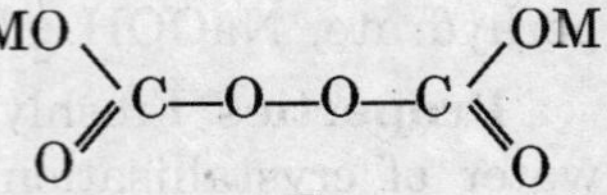

(I)

Symmetrical structure of $M_2C_2O_6$ (M = Na or K)

In 1908 Wolffenstein and Peltner passed CO_2 into an ice cold solution of K_2O_2 in absolute alcohol and obtained a salt. Partington and Fathallah (1934) have shown that the salt formed in the above reaction is isomeric with the bluish potassium perdicarbonate discussed above and represented by structure (I). The new compound liberates only 50% of the theoretical amount of iodine from the neutral solution of KI and loses only one-half of its active oxygen as oxygen gas. Thus it has an unsymmetrical structure given in the margin. Thus we see that perdicarbonates are of two types whose molecular formulae are the same (*i.e.* these are isomeric with each other) but their structures are different. One is represented by structure (I) while the other, by structure (II).

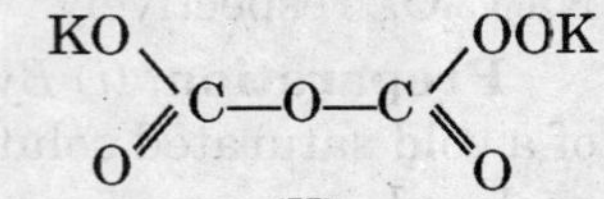

(II)

Unsymmetrical structure of $K_2C_2O_6$

(C) Perhydrates of simple carbonates. These compounds contain H_2O_2 of crystallization. Some permonocarbonates when freshly prepared contain water of crystallisation. They are soluble in H_2O and the solution obtained behaves as if it is a mixture of simple carbonate (*i.e.* CO_3^{2-}) containing water of crystallisation and H_2O_2. Thus these salts may be regarded as simple perhydrates of simple carbonates containing H_2O_2 as a molecule of crystallisation. The most important compound of this class of compounds is sodium carbonate perhydrate having the composition, $Na_2CO_3.H_2O_2.\frac{1}{2}H_2O$. This compound is obtained as a ppt. by the action of H_2O_2 on Na_2CO_3 and subsequent addition of alcohol. It is stabilised by the addition of sodium silicate (water glass) or magnesium silicate. This compound is used in washing powders, etc.

The formula of this compound, $Na_2CO_3.H_2O_2.\frac{1}{2}H_2O$, is supported by the fact that it slowly liberates iodine from an aqueous solution of KI and that CO_2 can be extracted from it by shaking its aqueous solution with ether.

2. Carbonyl Chloride or Phosgene, $COCl_2$.

This compound was discovered by John Davy in 1811. Since it is produced by the direct combination of CO with Cl_2 in presence of bright sunlight, the name *phosgene* (from Greek *phos*, light and *gennao*, I produce) has been given to this compound.

Preparation. $COCl_2$ is prepared (*i*) by the direct combination of equal volumes of CO and Cl_2 in the presence of bright sunlight

$$CO + Cl_2 \longrightarrow COCl_2$$

(*ii*) *in the laboratory* by dropping 80% fuming H_2SO_4 into boiling CCl_4 contained in a flask provided with a reflex condenser and dropping funnel. The resulting gas is passed through conc H_2SO_4 and condensed in a U-tube kept in freezing mixture.

$$CCl_4 + 2SO_3 \longrightarrow COCl_2 + \underset{\text{Disulphuryl chloride}}{S_2Cl_2O_5}$$

(*iii*) by the oxidation of chloroform ($CHCl_3$) with $K_2Cr_2O_7$ and conc. H_2SO_4.

$$K_2Cr_2O_7 + 4H_2SO_4 \longrightarrow K_2SO_4 + Cr_2(SO_4)_3 + 4H_2O + 3O$$
$$3CHCl_3 + 3O \longrightarrow 2COCl_2 + H_2O + Cl_2$$

Properties. (*i*) It is a colourless gas with a penetrating and suffocating odour and is highly poisonous.

(*ii*) It is very soluble in benzene and toluene.

(*iii*) It can be condensed to a liquid which boils at 8.2°C.

(*iv*) *Action of heat.* It is decomposed by heat and the decomposition is complete at 800°C.

$$COCl_2 \longrightarrow CO + Cl_2$$

(*v*) *Hydrolysis.* It does not fume in moist air but is readily hydrolysed by water, giving CO_2 and HCl.

$$COCl_2 + H_2O \longrightarrow CO_2 + 2HCl$$

(*vi*) *Action of NH_3 and C_2H_5OH.* When it is passed into a solution of NH_3 in toluene, urea, $NH_2.CO.NH_2$ is formed.

$$COCl_2 + 4NH_3 \longrightarrow \underset{\text{urea}}{NH_2.CO.NH_2} + 2NH_4Cl$$

With alcohol it forms ethylchloroformate or diethyl carbonate.

$$COCl_2 + 2C_2H_5OH \longrightarrow CO(OC_2H_5)_2 + 2HCl$$

Uses. It is used : (*i*) in dyestuff industry (*ii*) for synthetic work in laboratories. (*iii*) in warfare.

3. Carbon Disulphide, CS_2.

This compound was discovered by W. A. Lampadins in 1796.

Manufacture. CS_2 is manufactured by passing sulphur vapours over red hot charcoal or coke.

$$C + 2S \text{ (vapours)} \longrightarrow CS_2 - 19.0 \text{ K.Cals.}$$

Following two processes are used for the manufacture of CS_2.

(i) Zahn's process. In this process a vertical cast iron retort set in a furnace is filled with charcoal and sulphur kept in the fused state by the waste heat of furnace is slowly fed in through a tube. The temperature is kept between 800–900°C. The sulphur volatalises and the vapour passes over red hot charcoal forming CS_2. CS_2 so formed escapes at the top of the furnace and is condensed in long condensing coils.

(ii) Taylor's electrical process. In this process a cylindrical furnace 40 feet high and 16 feet in diameter is packed with coke or charcoal from the top. Separate hoppers are provided for adding fresh coke and sulphur. Heating is done with the help of an electric are struck between two carbon electrodes.

Sulphur is vapourised on the floor of the furnace; the vapour of sulphur passes through red hot charcoal forming CS_2. CS_2 so formed escapes at the top and is condensed in long condensing coils.

Purification. The crude CS_2 obtained by either of the above processes is purified by distillation over $(CH_3COO)_2Pb$ or by agitation with Hg until it no longer blackens it, after which it is redistilled over P_2O_5.

Properties. (*i*) Pure CS_2 is a colourless mobile highly refractory poisonous liquid of specific odour, density = 1.2923 at 0°C.

(*ii*) Boiling and freezing points are 46.25°C and –116°C respectively.

(*iii*) The commercial product possesses yellow colour and usually disagreeable and rancid odour due to the presence of sulphur and H_2S.

(*iv*) It is almost insoluble in H_2O but is completely miscible with absolute alcohol, ether, beuzene and essential oils.

(*v*) *Action of air.* The vapour of CS_2 is highly inflammable and catches fire when healed in air producing CO_2 and SO_2.

$$CS_2 + 3O_2 \longrightarrow CO_2 + 2SO_2$$

(*vi*) *Action of NO.* With NO, CS_2 forms an explosive mixture

$$2CS_2 + 10NO \longrightarrow 2CO + 4SO_2 + 5N_2$$

(*vii*) *Action of K.* When heated with K, CS_2 vapour gets decomposed

$$CS_2 + 4K \longrightarrow 2K_2S + C$$

(*viii*) *Action of slaked lime, $Ca(OH)_2$ and NaOH.* When CS_2 vapour is passed over slaked lime, a mixture of $CaCO_3$ and calcium thiocarbonate ($CaCS_3$) is produced.

$$3Ca(OH)_2 + 3CS_2 \longrightarrow CaCO_3 + 2CaCS_3 + 3H_2O$$

Similarly, with NaOH solution. CS_2 vapour gives a mixture of Na_2CO_3 and $NaCS_3$.

$$6NaOH + 3CS_2 \longrightarrow Na_2CO_3 + 2Na_2CS_3 + 3H_2O$$

(*ix*) *Action of basic sulphides.* CS_2 combines with basic sulphides to form thiocarbonates. For example :

$$Na_2S + CS_2 \longrightarrow Na_2CS_3$$

(*x*) *Action of Cl_2.* When Cl_2 is passed into boiling CS_2 containing a little of I_2, CCl_4 and S_2Cl_2 are formed.

$$CS_2 + 3Cl_2 \longrightarrow CCl_4 + S_2Cl_2$$

(*xi*) *Action of PCl_5.* When CS_2 is heated with PCl_5, thiocarbonyl chloride, $CSCl_2$ (analogous to $COCl_2$) is formed.

$$CS_2 + PCl_5 \longrightarrow CSCl_2 + PSCl_3$$

(*xii*) *Action of SO_3.* CS_2 reacts with SO_3 to form carbonyl sulphide, COS.

$$CS_2 + 3SO_3 \longrightarrow COS + 4SO_2$$

(*xiii*) *Action of H_2.* When a mixture of CS_2 and H_2 is passed over heated Ni at 450°C, H_2S is evolved.

$$CS_2 + 2H_2 \xrightarrow{\text{Ni (450°C)}} C + 2H_2S$$

(*xiv*) *Action of Cu.* When CS_2 vapour is passed over red hot copper, CuS is formed.

$$CS_2 + 4Cu\ (\textit{red hot}) \longrightarrow 2Cu_2S + C$$

(*xv*) *Action of steam and H_2S.* When a mixture of CS_2 and steam (H_2O) or H_2S is passed over red hot copper, CH_4 is formed.

$$CS_2 + 2H_2O\ (\textit{steam}) + 6Cu\ (\textit{red hot}) \longrightarrow 2Cu_2S + 2CuO + CH_4$$

$$CS_2 + 2H_2S + 8Cu\ (\textit{red hot}) \longrightarrow 4Cu_2S + CH_4$$

Uses. CS_2 is used (*i*) as a solvent for sulphur, white phosphorus, iodine, bromine, camphor, gums, resins, waxes, fats, oils etc. and also for S_2Cl_2 in cold vulcanization of rubbur. (*ii*) in the manufacture of artificial silk and CCl_2. (*iii*) to kill moths in furs and rats and mice in grain elevators. (*iv*) as an insecticide for curing the insect infected seeds and for killing ants.

University Questions

1. Give the preparation, properties and uses of carbon disulphide. (*Agra 86; Madras 86*)
2. Discuss the preparation, properties and structure of peroxy compounds of carbon. (*Meerut M.Sc. 82*)
3. Give methods of preparation, properties and uses of carbonyl chloride.

CHAPTER 27

Carbides

What are Carbides ?

The binary compounds of carbon with the elements which are more electropositive than carbon are called carbides. Thus this definition excludes the binary compounds of carbon with N, P, O, S and halogens.

General Methods of Preparation.

1. By heating metal with carbon. Many metals like Be, Ag, Mn, Li, Ta, V etc. combine with carbon at high temperature to form carbides. For example :

$$Be + 2C \longrightarrow BeC_2 \; ; \quad 2Ag + 2C \longrightarrow Ag_2C_2$$
$$3Mn + C \longrightarrow Mn_3C \; ; \quad 2Li + C \longrightarrow Li_2C_2$$
$$3Fe + C \longrightarrow Fe_3C \; ; \quad 2Mn + C \longrightarrow Mn_3C$$

2. By heating oxides, hydroxides, nitrides, phosphates, sulphates etc. with carbon. For example :

$$MgO + 3C \longrightarrow MgC_2 + CO\uparrow \; ; \quad CeO_2 + 4C \longrightarrow CeC_2 + 2CO\uparrow$$
$$2Al_2O_3 + 9C \longrightarrow Al_4C_3 + 6CO\uparrow \; ; \quad SiO_2 + 3C \longrightarrow SiC + 2CO\uparrow$$
$$U_3O_8 + 14C \longrightarrow 3UC_2 + 8CO\uparrow \; ; \quad 2Mg(OH)_2 + 4C \longrightarrow 2MgC_2 + 2H_2O + O_2$$
$$Ca_3H_2 + 6C \longrightarrow 3CaC_2 + N_2 \; ; \quad Ca_3(PO_4)_2 + 14C \longrightarrow 3CaC_2 + 2P + 8CO$$
$$4AlPO_4 + 19C \longrightarrow Al_4C_3 + 4P + 16CO \; ; \quad BaSO_4 + 2C \longrightarrow BaC_2 + SO_2 + O_2$$

3. By passing acetylene (C_2H_2) or ethylene (C_2H_4) on heated element or its oxide. For example :

$$4Li + C_2H_2 \longrightarrow Li_2C_2 + 2LiH \; ; \quad 2Si + C_2H_2 \longrightarrow 2SiC + H_2$$
$$2Ag_2O + C_2H_2 \longrightarrow Ag_2C_2 + H_2O \; ; \quad 6Li + C_2H_4 \longrightarrow Li_2C_2 + 4LiH$$

4. By heating the metallic oxide or halide with CaC_2. For example :

$$MgO + CaC_2 \longrightarrow MgC_2 + CaO \; ; \; CuCl_2 + CaC_2 \longrightarrow CuC_2 + CaCl_2$$
$$MgF_2 + CaC_2 \longrightarrow MgC_2 + CaF_2$$

5. By passing CO over metal at high temperature.

$$2Li + 2CO \longrightarrow Li_2C_2 + O_2 \; ; \quad 6Al + 3CO \longrightarrow Al_4C_3 + Al_2O_3$$

6. By passing acetylene (C_2H_2) through ammoniacal solution of the metal salt.

$$2CuCl + 2NH_4OH + C_2H_2 \longrightarrow Cu_2C_2 + 2NH_4Cl + 2H_2O$$
$$2AgNO_3 + 2NH_4OH + C_2H_2 \longrightarrow Ag_2C_2 + 2NH_4NO_3 + 2H_2O$$

7. By heating metallic carbonate with carbon in presence of an oxidisable metal.

$$BaCO_3 + C + 3Mg \longrightarrow 3MgO + BaC_2$$

General Properties.

1. Physical state. Generally carbides are transparent crystalline solids. In the solid state they are non-conductors of electricity.

2. Colour. Carbides of alkali metals and of Ca, Sr and Ba are only colourless while most of the remaining carbides are coloured.

3. Softness and hardness. Alkali metal carbides are soft while others are usually hard. For example Be_2C and UC_2 are so hard that they can scratch glass and quartz.

4. Explosive nature. Carbides of U, Cu, A, An, Hg (*ic*) etc. are explosive substances, *e.g.* Hg (*ic*) carbide explodes on rapid heating. Uranium carbide emits sparks when struck and takes fire even when powdered quickly.

5. Reducing property. The carbides of alkali metals and of Ca, Sr and Ba are strong reducing agents, *e.g.* MgO and $MgCl_2$ are reduced to the metals on heating with CaC_2.

$$3MgO + CaC_2 \xrightarrow{\Delta} 3Mg + CaO + 2CO \uparrow$$

$$MgCl_2 + CaC_2 \xrightarrow{\Delta} Mg + CaCl_2 + 2C$$

6. Hydrolysis. Ionic carbides can easily be hydrolysed by water or dil. acids with the formation of different types of hydrocarbons.

Classification of Carbides.

Carbides are classified into four types depending on the nature of bonding between carbon atoms and other elements. These four types are : (1) *Ionic or salt-like carbides* (2) *Interstitial or metallic carbides* (3) *Iron type or borderline carbides* (4) *Covalent carbides.*

Ionic Carbides

These carbides are given by the metals of groups IA, IIA and III A (except boron), coinage metals, Zn, Cd and some lanthanides.

Preparation.

These are obtained by heating the metal or its oxide with carbon, CO or a hydrocarbon like acetylene

$$2Be + C \longrightarrow Be_2C \ ; \qquad CaO + 3C \longrightarrow CaC_2 + CO$$

$$4Al + 3C \longrightarrow Al_4C_3 \ ; \qquad 2Al_2O_3 + 9C \longrightarrow Al_4C_3 + 6CO$$

$$2Li + 2CO \longrightarrow Li_2C_2 + O_2 \ ; \qquad 6Al + 3CO \longrightarrow Al_4C_3 + Al_2O_3$$

$$4Li + C_2H_2 \longrightarrow Li_2C_2 + 2LiH \ ; \ 2Ag_2O + C_2H_2 \longrightarrow Ag_2C_2 + H_2O$$

Properties.

(i) Physical state. These carbides form transparent colourless crystals with ionic lattices containing metal cations in the interstices between carbon anions.

(ii) Conductance of electricity. These are non-conductors of electricity.

(iii) Hydrolysis. These are easily hydrolysed by water or dil acids to give different hydrocarbons. Depending on the nature of the hydrocarbon formed during the hydrolysis of ionic carbides by water, these carbides can be classified into the following types :

(a) Acetylides. These are the ionic carbides which give acetylene ($CH \equiv CH$) on hydrolysis. These carbides are, therefore, regarded as derivatives of acetylene and are called *acetylides.* These carbides are said to contain *carbide ion,* $(C \equiv C)^{2-}$

$$CaC_2 + 2H_2O \longrightarrow Ca(OH)_2 + CH \equiv CH$$

Some of the carbides like Cu_2C_2, Ag_2C_2 do not hydrolyse to form acetylene and are, therefore, not regarded as true acetylides.

Acetylides have a NaCl type crystal structure. However, in CaC_2 the shape of $(C \equiv C)^{2-}$ ion elongates the unit cell in one direction.

(b) Methanides. These are the ionic carbides which give methane (CH_4) on hydrolysis. These carbides are regarded as derivatives of methane and hence are said to contain C^{4-} ion. These carbides are called *methanides.*

$$Al_4C_3 + 12H_2O \longrightarrow 4Al(OH)_3 + 3CH_4 \uparrow$$

$$Be_2C + 4H_2O \longrightarrow 2Be(OH)_2 + CH_4 \uparrow$$

Mn_3C probably has a related character, since its hydrolysis also gives CH_4.

$$Mn_3C + 6H_2O \longrightarrow 3Mn(OH)_2 + CH_4 + H_2$$

(c) Allylides. These are the ionic carbide, which give allylene ($H_2C = C = CH_2$) on hydrolysis. These carbides are regarded as derivatives of allylene and hence are said to contain $C_3{}^{4-}$ group.

$$Mg_2C_3 + 4H_2O \longrightarrow 2Mg(OH)_2 + C_3H_4$$

(d) Mixed carbides. A few carbides like those of Th and U of the formula MC_2 give, on hydrolysis, a mixture of several hydrocarbons including acetylene, olefins and H_2. These carbides are related in structure to intermetallic compounds.

Structure. The ionic carbides posses ionic lattice. in which the metallic cations are packed into the cavities between carbon anions *viz.* C^{-4}, $C_2{}^{2-}$ and $C_3{}^{4-}$

Interstitial Carbides

These carbides are also called *refractory carbides.* These carbides are formed by those transition metals whose atomic radii (metallic radii for coordination number 12) are generally greater than 1.3A° ($R_M > 1.3$A°), since the radius ratio namely R_C/R_M must be either 1 : 0.41 or 1 : 0.59 to permit C-atoms to enter into the octahedral holes (also called cavities, interstices or interstitial positions) of the cubic close packing of the metallic lattice without distorting it. The radius of the metals Ti, Zr, Hf (Group IV B); V, Nb. Ta (Group V B) : Mo and W (Group VI B) is greater than 1.3 A° and hence these metals give the interstitial carbides which are of two types *viz.* MC type (M = Ti, Zr, Hf, V, Nb, Ta, Mo, W) and MC_2 type (M = V, Mo, W).

Preparation.

Interstitial carbides are prepared :

(i) by direct combination of the metal with carbon. In this method powdered metal is heated with powdered carbon at high temperature. Carbides obtained by this method are purified and made compact by sintering them in vacuum or in an atmosphere of inert gases.

(ii) by reducing the metallic oxide with carbon in an electric furnace at about 2000°C. This method has been used for the preparation of TiC, ZrC, HfC, VC, NbC, TaC, WC.

$$TiO_2 + 3C \longrightarrow TiC + 2CO$$
$$WO_3 + 4C \longrightarrow WC + 3CO$$
$$MoO_3 + 4C \longrightarrow MoC + 3CO$$

(iii) by heating the metal in an atmosphere of CH_4 at different temperatures. HfC, MoC, TaC. WC etc. carbides have been prepared by this method, *e.g.*

$$Hf + CH_4 \longrightarrow HfC + 2H_2$$
$$Mo + CH_4 \longrightarrow MoC + 2H_2$$

Properties.

(*i*) The melting points of these carbides are very high. For example the melting points of TaC and ZrC are 3900° and 3800°C respectively.

(*ii*) These are chemically inert, and extremely hard like diamond. For example WC is used for cutting tools on account of its hardness. The hardness on Moh's scale is between 9 and 10.

(*iii*) The crystal lattice of these carbides is a close-packed type with C-atoms occupying the octahedral holes in the metal lattice. Presence of C-atoms, therefore, does not affect the electrical conductivity of the metal. Thus these carbides possess metallic lustre and high electrical conductivity which increases at lower temperature and may become infinite at absolute zero.

(*iv*) These are weakly paramagnetic which is characteristic of the presence of metal lattice in their crystal structure.

(*v*) They are attacked by strong oxidising agents at red heat.

(*vi*) Carbides of Mo and W are somewhat more reactive than the metals from which they are derived but other carbides are less reactive than the parent metals.

Borderline Carbides

Fe, Mn, Co, Ni and Cr have atomic radii less than 1.3 A° (Mn is an exception) and hence octahedral holes in the metallic lattices of these metals are too small to accommodate C-atoms in them without producing a great deal of distortion in the metallic lattice (or metal structure). Here the metallic lattice is distorted and these carbides are intermediate in properties between the ionic and interstitial carbides but their radius ratio namely R_C/R_M has been found to be equal to 0.60–0.61 which places these carbides closer to the interstitial carbides.

These carbides are relatively less stable and liberate a mixture of hydrocarbons and H_2 on hydrolysis with H_2O or acids.

In these carbides C-atoms are located inside the trigonal prisms formed by metal atoms. The metal lattice is distorted and long chains of C-atoms are running through distorted metal structures. C–C distance is almost equal to 1.65A°

Covalent Carbides

The only covalent carbides are those of silicon and boron (SiC and B_4C).

Preparation.

SiC and B_4C are prepared .by reducing their oxides with carbon in an electric furnace.

$$SiO_2 + 3C \longrightarrow SiC + 2CO \uparrow$$

$$2B_2O_3 + 7C \longrightarrow B_4C + 6CO \uparrow$$

Properties.

These are not attacked by H_2O, dil. and conc. acids. These are extremely hard and decompose at high temperatures. Because of their hardness they are used for cutting and as abrasives.

Silicon carbide or Carborundum , SiC

Preparation.

It is prepared by heating a mixture of coke and SiO_2 in an electric furnace at 2000°C by passing acetylene on heated silicon.

$$SiO_2 + 3C \longrightarrow SiC + 2CO \uparrow$$

$$2Si + C_2H_2 \longrightarrow 2SiC + H_2 \uparrow$$

Manufacture.

SiC is manufactured by Acheson's process. In this process a mixture of sand (54%), coke (34%), sawdust (10%) and salt (2%) is heated in an electric furnace made of fire bricks to 1550–2200°C. The bed of the furnace and the end walls of it are permanent while the side walls are built up with the charge and are pulled down after the completion of the process to take out the product. It is provided at each end with carbon electrodes consisting of sixty rods of carbon. A heavy current is passed for 36 hours, whereby a high temperature is rapidly reached. At this high temperature, the following reaction occurs resulting in the formation of SiC.

$$SiO_2 + 3C \longrightarrow SiC + 2CO \uparrow$$

For about two hours in the beginning, the emf gradually decreases from 165 volts to 125 volts and the current increases from 1700 amperes to 6000 amperes due to the gradual decrease in resistance. These conditions persist for the remaining period of time. At the end of the operation, the side walls are pulled down and the dark coloured mass of black crystals of SiC is crushed and washed successively with H_2SO_4 and NaOH solution to remove the impurities. It is finally dried in kilns and graded into various portions according to the size of the particles.

The salt acts as a flux while saw dust increases the porousity of the charge which enables a continuous escape of CO that burns at the top of the charge.

Properties.

It is colourless when pure. The commercial sample is yellow, green or blue. It is nearly as hard as diamond and does not decompose below 2200°C. Chemically it is extremely inert and even at high temperatures it is not attacked even by HF, HCl, O_2 or S. Even a mixture of fuming HNO_3 and HF has no action on it. It is decomposed by fused NaOH in presence of O_2.

$$SiC + 4NaOH + 2O_2 \longrightarrow Na_2CO_3 + Na_2SiO_3 + 2H_2O$$

Uses.

(*i*) It is used as an abrasive for cutting and grinding glasses. (*ii*) It has a very high m.pt. and is, therefore, used in furnace-lining. (*iii*) On account of its refractory nature and high heat conductivity, it is used for making crucibles used for melting metals. (*iv*) It is also used as carbon rods in resistant heaters. (*v*) It is also used for making wheels, hones, wheatstones by mixing SiC with moistened china clay and felspar, moulding under pressure and firing the article in a kiln. (*vi*) It is also used as a de-oxidant in metallurgy and as resistor for electrical furnaces.

Structure.

SiC exists in three forms which are related to another as diamond, zinc blende and wurtzite. These forms are different combinations of "layers" corresponding to zinc blende and wurtzite structures. These three forms are : (*i*) *Carborundum I* represented as *aaa bb* (*ii*) *Carborundum II* represented as *a a a b b b* (*iii*) *Carborundum III* represented as *aa bb*. Here *a* a zinc blende layer and *b* is a wurtzite layer.

The lattice of SiC consists of C-atoms at points corresponding to those occupied by atoms in a close-packed face-centred cubic or hexagonal structure, with Si-atoms at half the points corresponding to the positions of the tetrahedral holes. This type of structure of SiC has been confirmed by X-ray studies.

Uses of Carbides

(*i*) Many metal carbides have been used as reducing agents in metallurgical and electro-thermal process.

(*ii*) In general, carbides are very hard and used as efficient abrasives, drilling tools and cutting tools.

(*iii*) Carbides can withstand high temperature and hence are used widely in furnace lining.

(*iv*) Calcium carbide, CaC_2 is used : (*a*) in the manufacture of calcium cyanamide, $CaCN_2$ which is employed as a nitrogenous fertiliser and as a source of NH_3. (*b*) in the preparation of acetylene which is employed in oxy-acetylene blow pipes for welding and illuminating purposes and for the manufacture of a large number of organic compounds like acetaldehyde, acetic acid etc. (*c*) as a reducing agent.

(*v*) Boron carbide, B_4O is used (*a*) for cutting diamonds, since it is the hardest abrasive ever made artificially. It is a better abrasive than carborundum, SiC, (*b*) for drilling holes in rockets, (*c*) for making electrodes for electric furnaces, (*d*) for making lamp filaments, (*e*) as a shield against radiation.

(*vi*) Aluminium carbide, Al_4C_2 is used for the preparation of methane.

(*vii*) Tungsten carbide, WC is used for the manufacture of high speed tools.

(*viii*) ZrC is used (*a*) as a component of cemented hard metal cutting tools, (*b*) as an electrode material, for refractory crucibles and as incandascent filament, (*c*) as a coating (1–2 mm thick) to delay the release of fission products from uranium carbide fuel elements.

Questions with Answers

Q.1 Classify the following ionic carbides as methanides, acetylides and allylides : Be_2C, Al_4C_3, BeC_2, MgC_2, CaC_2, $Al_2(C_2)_3$ and Mg_2C_3.

Ans : Methanides : Be_2C and Al_4C_3 are called methanides, since they contain C^{4-} ion and give mathane (CH_4) on hydrolysis.

$$Be_2C + 4H_2O \longrightarrow 2Be(OH)_2 + CH_4$$

$$Al_4C_3 + 12H_2O \longrightarrow 4Al(OH)_3 + 3CH_4$$

Acetylides : BeC_2, MgC_2, CaC_2 and $Al_2(C_2)_3$ are called acetylides, since these contain acetylide ion (C_2^{2-}) and give acetylence ($H-C\equiv C-H$ or C_2H_2) on hydrolysis.

$$MC_2 + 2H_2O \longrightarrow M(OH)_2 + C_2H_2$$

(M = Be, Mg, Ca)

$$Al_2(C_2)_3 + 6H_2O \longrightarrow 2Al(OH)_3 + 3C_2H_2$$

Allylides : Mg_2C_3 is called magnesium allylide, since it contains allylide ion (C_3^{4-}) and gives alllyene ($CH_3-C\equiv CH$ or C_3H_4) on hydrolysis.

$$Mg_2C_3 + 4H_2O \longrightarrow 2Mg(OH)_2 + C_3H_4$$

Q.2 Give the name of the anion present in Be_2C, Al_4C_3, MC_2 (M = Be, Mg, Ca), $Al_2(C_3)_3$ and Mg_2C_3 and give their structure.

Ans : Be_2C and **Al_4C_3** contain **C^{4-} ion** which is derived from methane molecule (CH_4)

$$CH_4 \xrightarrow{-4H} [:\ddot{\underset{..}{C}}:]^{4-} \text{ or } C^{4-}$$

MC_2 (M = Be, Mg, Ca) and **$Al_2(C_2)_3$** contain **acetylide ion (C_2^{2-})** which is derived from acetylene molecule ($H-C\equiv C-H$ or C_2H_2).

$$H-C\equiv C-H \xrightarrow{-2H} \overset{-}{\ddot{C}}\equiv\overset{-}{\ddot{C}} \text{ or } C_2^{2-}$$

Mg_2C_3 contains **allylide ion (C_3^{4-})** which is derived from allylene molecule ($CH_3-C\equiv CH$ or C_3H_4)

$$CH_3-C\equiv C-H \xrightarrow{-4H} \overset{3-}{\underset{..}{\ddot{C}}}\equiv\overset{-}{\ddot{C}} \text{ or } C_3^{4-}$$

Q.3 Discuss how different bonds are formed in acetylide ion, C_2^{2-}.

Ans : The structure of acetylide ion (C_2^{2-}) is $\ddot{\bar{C}} \equiv \ddot{\bar{C}}$. This structure shows that each C-atom is surrounded by one σ-*bp* and one *lp*. Hence each $\bar{C}$ -atom is *sp* hydridised [$\bar{C}$ = $(sp)^2\ (sp)^1\ 2p^1\ 2p^1$]. (C – C) σ bond is obtained by the overlap between two singly filled *sp* hybrid orbitals on two $\bar{C}$ atoms and two (C – C) π bonds are obtained by the parallel overlap of two unhybridised 2*p* orbitals on two $\bar{C}$ atoms. Thus both π bonds are (*p*π – *p*π)π bonds. (See Fig. 27.1).

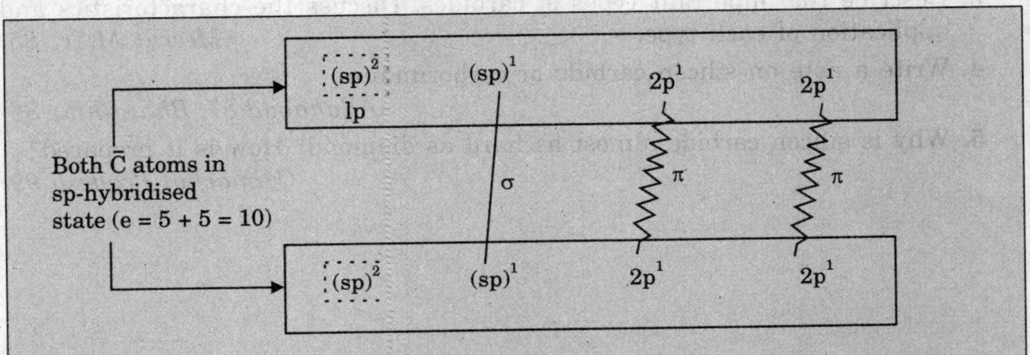

Fig. 27.1. Formation of different bunds in acetylide in, or $:\bar{C}\equiv\bar{C}:$ or C_2^{2-}.

Q.4 Find out the number and type of bonds and type of hydridisation of C-atom in allylide ion.

Ans : Allylide ion is derived by removing all the four H-atoms from allylene molecule (also called propyne or methyl acetylene), $CH_3 - C \equiv CH$.

$$\underset{\text{Allylene}}{CH_3 - C \equiv CH} \xrightarrow{-4H} \underset{\text{Allyide ion}}{:\overset{3-}{\ddot{\underset{\cdot\cdot}{C}}} - C \equiv \bar{\ddot{C}}:} \quad \text{or} \quad C_3^{4-}$$

The structure of C_3^{4-} given above shows that it has **two σ-bonds and two π bonds**. $\overset{3-}{C}$, C and $\bar{C}$ atoms are ***sp*³, *sp*** and ***sp* hybridised** respectively. (See Fig. 27.2).

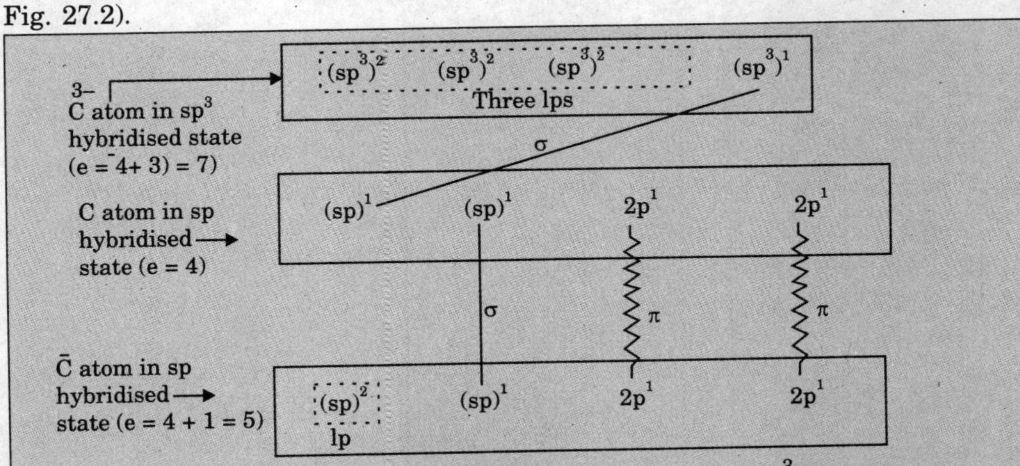

Fig. 27.2. Formation of different bonds in allylide ion, $:\overset{3-}{\ddot{\underset{\cdot\cdot}{C}}}-C\equiv\bar{C}:$ or C_3^{4-}.

University Questions

1. What are carbides ? Give their general methods of preparation. Discuss about three different types of carbides. *(Raj. 86 S)*

2. Discuss the classification of carbides. Give suitable examples. *(Andhra 82; Madurai 86)*

3. Describe the important types of carbides. Discuss the characteristics and application of each type. *(Meerut M.Sc. 85)*

4. Write a note on silicon carbide or carborundum. *(Allahabad 87; Bharathiar 86)*

5. Why is silicon carbide almost as hard as diamond? How is it prepared? *(Himachal Pradesh 99)*

CHAPTER

28 Metallic Carbonyls

What are Metallic Carbonyls ?

The electronic configuration of CO molecule shows that it has a lone pair of electrons on carbon and. oxygen atom each. Carbon atom can donate its electron pair to a transition metal atom (M), forming OC $\rightarrow$ M coordinate bond. The compounds formed by the combination of CO molecules with transition metals are known as metallic carbonyls. Since the electrons forming OC $\rightarrow$ M bond are supplied solely by CO molecules, metal atom in carbonyls is said to be in zero oxidation state. In metal carbcnyls CO molecules act as neutral ligands.

Types of Carbonyls

Depending on the number of metal atoms in a given carbonyl, carbonyls have been classified into the following two types :

1. *Mononuclear* (or *monomeric*) *carbonyls.* *These contain only one metallic atom per molecule* and are of the type $M(CO)_y$. Examples are : $V(CO)_6$. $Cr(CO)_6$ etc.

2. *Polynuclear carbonyls.* *These contain two or more metallic atoms per molecule* and are of the type $M_x(CO)_y$. However some authors call the carbonyls containing two metal atoms as *bridged carbonyls* and those containing more than two metal atoms as *polynuclear carbonyls.* Polynuclear carbonyls may be *homonuclear* [*e.g.* $Fe_3(CO)_{12}$] or *heteronuclear* [*e.g.* $MnCo(CO)_9$, $MnRe(CO)_{10}$].

General Methods of Preparation

1. By direct synthesis. Of the simple carbonyls, nickel tetracarbonyl, $Ni(CO)_4$ and iron pentacarbonyl, Fe $(CO)_5$ can be prepared by the direct combination of CO with finely divided metals at suitable temperature and pressure :

$$Ni + 4CO \xrightarrow[\text{one atm. pressure}]{\text{Room temp. and}} Ni\ (CO)_4$$

$$Fe + 5CO \xrightarrow[\text{pressure}]{200° \text{ and } 100 \text{ atm.}} Fe\ (CO)_5$$

2. By the reduction of metal compounds in presence of CO. Some carbonyls may be prepared by treating the transition metal compounds (usually metal halides) in suspension in organic solvents like tetrahydrofuran (THF), ether, diethylene glycol dimethyl ether (diglyme) etc with CO it a pressure of 200–300 atm. and temperature upto 300°C in presence of some suitable reducing agent. The substances used as reducing agents are H_2, metals (*e.g.* Na, Al, Mg, Cu etc.) or compounds such as trialkyl aluminium or sodium benzophenone ketyl, $Ph_2CO^-Na^+$

$$VCl_3 + CO \xrightarrow[\text{100° at 15 atm.}]{\text{Na (excess) in diglyme}} \underset{\text{Unstable organo-metallic carbonyl intermediate}}{[Na\ (diglyme)_2]^+\ [V(CO)_6]^-}$$

$$\xrightarrow[\text{(ii) Sublimation at 50°}]{\text{(i) Acidification by } H_3PO_4} V(CO)_6$$

$$CrCl_3 + 6CO \xrightarrow{(C_2H_5)_3Al \text{ in ether}} Cr(CO)_6$$

$$MoCl_5 + 6CO \xrightarrow{\text{5 Na in diglyme}} Mo(CO)_6 + 5NaCl$$

3. From $Fe(CO)_5$. Since CO ligands in $Fe(CO)_5$ are labile, $Fe(CO)_5$ may be used to form certain carbonyls, *e.g.*

$$\underset{(M = W, Mo)}{MCl_6} + 3Fe(CO)_5 \xrightarrow[\text{ether}]{100°} M(CO)_6 + 3FeCl_2 + 9CO$$

4. By the thermal or photochemical decomposition of simple carbonyls. Polynuclear carbonyls of Fe, Rh, Os, Cu, Rh and Ir are obtained by the thermal or photochemical decomposition of simple carbonyls, *e.g.*

$$\underset{(M = Fe, Os)}{2M(CO)_5} \xrightarrow{\text{ultraviolet light}} M_2(CO)_9 + CO$$

Physical Properties.

With the exception of $Ni(CO)_4$, $Fe(CO)_5$, $Ru(CO)_5$ and $Os(CO)_5$ which are liquids at ordinary temperatures, all other carbonyls are crystalline solids. They melt or decompose at low temperatures. All are typical covalent compounds and for this reason they are soluble in non-polar solvents. With the exception of $Fe_2(CO)_9$, all are more or less soluble in organic solvents. Excepting $V(CO)_6$, all the carbonyls are *diamagnetic.* $V(CO)_6$ is *paramagnetic* and its paramagnetic property corresponds to the presence of unpaired electron. The metals in carbonyls are in zero oxidation state.

Chemical Properties.

1. Substitution reactions. The most important general reactions shown by carbonyls are substitution reactions in which CO molecules are replaced by ligands like PX_3, PR_3, $P(OR)_3$, SR_2, NR_3, OR_2, RNC etc. or by unsaturated organic molecules such as C_6H_6 or cycloheptatriene. These reactions may proceed through photochemical activation according to which the carbonyl group absorbs a photon (*hv*) to eliminate a CO molecule followed by the entry of the substituent into the co-ordination sphere, *e.g.*

$$Cr(CO)_6 \xrightarrow[(-\ CO)]{hv} C(CO)_5 \xrightarrow{+\ L} Cr(CO)_5L$$

$$Fe(CO)_5 \xrightarrow[(-\ 2CO)]{hv} Fe(CO)_3 \xrightarrow{+\ 2PR_3} Fe(CO)_3(PR_3)_2$$

$$Fe_2(CO)_9 \xrightarrow[(-5CO)]{hv} Fe_2(CO)_4 \xrightarrow{+4NO} 2[Fe(CO)_2(NO)_2]$$

Bidentate ligands like *diars* replace two CO groups at a time to give the substituted carbonyls like $Mo(CO)_4$(*diars*), $Fe(CO)_3$ (*diars*), $Ni(CO)_2$ (*diars*) etc. Substitution reaction in case of $Cr(CO)_6$ gives $Cr(CO)_2$ (*diars*)$_2$.

2. Formation of anionic carbonyl complexes (also called carbonylate anions). Many of the carbonyls form carbonylate anions, *e.g.* $[Fe(CO)_4]^{2-}$, $[Mo(CO)_5Cl]^-$, $[HFe(CO)_4]^-$, $[Fe_2(CO)_8]^{2-}$, $[Co(CO)_4]^-$, $[Mn(CO)_5]^-$ etc. These complex anions can be prepared by any of the following methods :

(*a*) *By the action of NaOH or nitrogenous bases* (*e.g.* NH_3, amines, pyridine) *on metal carbonyls, e.g.*

$$Fe(CO)_5 \xrightarrow{NaOH} Na_2^+[Fe(CO)_4]^{2-}$$

$$Fe(CO)_5 \xrightarrow[Et_4NI]{NH_3/H_2O} [Et_4N]^+[HFe(CO)_4]^-$$

$$Fe_2(CO)_9 \xrightarrow[Et_4NI]{KOH/MeOH} [Et_4N]_2^+\ [Fe_2(CO)_8]^{2-}$$

$$Fe_3(CO)_{12} \xrightarrow{(en)/H_2O} [Fe(en)_2]^{3+}\ [Fe_3(CO)_{11}]^{3-}$$

$$3Co_2(CO)_8 + 12py \longrightarrow 2[Co(py)_6]^{2+}\ [Co(CO)_4]_2^{2-} + 8CO$$

(*b*) *By the reduction of carbonyls with alkali metals, alkali metal amalgams or borohydrides, e.g.*

$$Co_2(CO)_8 \xrightarrow[THF]{Na/Hg} Na^+[Co(CO)_4]^-$$

$$Mn_2(CO)_{10} \xrightarrow[THF]{Na/Hg} Na^+[Mn(CO)_5]^-$$

$$Re_2(CO)_{10} \xrightarrow[THF]{NaBH_4} Na_2[Re_4(CO)_{16}]$$

$$Cr(CO)_6 \xrightarrow{Na/NH_3} Na_2[Cr(CO)_5]$$

$$Cr(CO)_6 \xrightarrow[THF]{NaBH_4} Na[HCr_2(CO)_{10}]$$

$$Cr(CO)_6 \xrightarrow[NH_3]{NaBH_4} Na_2[Cr_2(CO)_{10}]$$

3. Formation of cationic carbonyl complexes (also called carbonylate cations). Carbonylate cations may be prepared from the carbonyls by the following methods :

(*a*) By the protonation of carbonyls in strong acids, *e.g.*

$$Fe(CO)_5 \xrightarrow{HCl/BCl_3} [FeH(CO)_5]^+ \; [BCl_4]^-$$

n.m.r. spectroscopy has shown that the proton becomes bounded to the metal atom in $[FeH(CO)_5]^+$.

(*b*) By the action of CO and a Lewis acid like $AlCl_3$ or BF_3 on carbonyl halide, *e.g.*

$$Mn(CO)_5Cl + CO + AlCl_3 \longrightarrow [Mn(CO)_6]^+[AlCl_4]^-$$

4. Formation of carbonyl halides. The action of Cl_2, Br_2, I_2 or F_2 on some carbonyls gives carbonyl halides, *e.g.*,

$$2Mo(CO)_6 + 2Cl_2 \longrightarrow [Mo(CO)_4Cl_2] + 4CO$$

$$Mn_2(CO)_{10} + Br_2 \overset{\text{Pet. ether, } 120°}{\rightleftarrows} [Mn(CO)_4Br]_2 + 2CO$$

$$Fe(CO)_5 + I_2 \xrightarrow{(-\,2CO)} [Fe(CO)_4I_2] + CO$$

$$4Ru(CO)_5 + 4F_2 \longrightarrow [Ru(CO)_3F_2]_4 + 8CO$$

5. Formation of carbonyl hydrides. Some of the metallic carbonyls [*e.g.* $Fe(CO)_5$, $Mn_2(CO)_{10}$,$Os_3(CO_{12}$ etc] dissolve in an alkali and give an unidentified colourless solution. On acidification with a mineral acid, this solution gives the carbonyl hydrides. $H_2Fe(CO)_4$ [from $Fe(CO)_5$], $H_3Mn_3(CO)_{12}$ [from $Mn_2(CO)_{10}$]. $H_4Os_4(CO)_{12}$, $H_2Os_4(CO)_{13}$ [from $Os_3(CO)_{12}$] and $Ni(CO)_3H_2$ [from $Ni(CO)_4$] have been obtained by this method.

$$Fe(CO)_5 + 2OH^- \longrightarrow Fe(CO)_4H_2 + CO_3^{2-}$$

$$Ni(CO)_4 + 2OH^- \longrightarrow Ni(CO)_3H_2 + CO_3^{2-}$$

Carbonyl hydrides can also be prepared by treating the carbonyls with H_2, *e.g.*

$$Mn_2(CO)_{10} + H_2 \xrightarrow{200°/200 \text{ atm.}} 2HMn(CO)_5$$

Structure and Bonding in Mononuclear Carbonyls

The structure of mononuclear carbonyls *viz.* $V(CO)_6$, $Cr(CO)_6$, $Fe(CO)_5$ and $Ni(CO)_4$ have been studied by X-ray diffraction, infrared spectroscopy and [except for $V(CO)_6$] electron-diffraction. The last method has also been used for $Mo(CO)_6$ and $W(CO)_6$. All the mono-nuclear carbonyls have *linear* M—CO bonds in which CO group is linked to the metal atom through the carbon atom.

Now let us discuss the structure of $Cr(CO)_6$, $Fe(CO)_5$ and $Ni(CO)_4$ carbonyls.

1. Chromium Hexacarbonyl, $Cr(CO)_6$. In this carbonyl Cr-atom which is in zero oxidation state is d^2sp^3 hybridised as shown in Fig. 28.1. Since the energy of 3*d* and 4*s* orbitals is not very different from each other, both the electrons from 4*s* orbital go to 3*d* orbital to make $3d_{x^2-y^2}$, $3d_{s^2}$ and 4*s* orbitals empty. Now $3d_{x^2-y^2}$, $3d_{z^2}$, 4*s* and all the three 4*p* orbitals hybridise together to form six d^2sp^3 hybrid orbitals. These vacant six hybrid orbitals overlap with lone-pair containing σ-orbitals on CO ligards to form six Cr ← CO σ bonds as shown in Fig. 28.1.

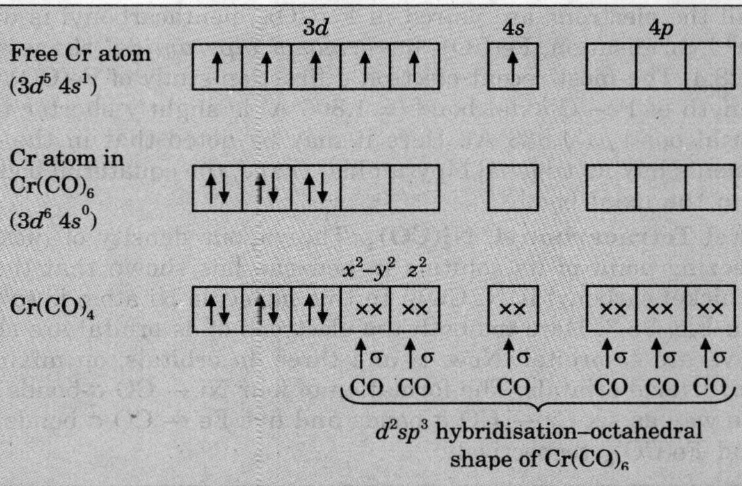

Fig. 28.1.d^2sp^3 hybridisation of Cr atom in octahedral $Cr(CO)_6$ carbonyl. Crosses (xx) represent elecgron pair donated by CO groups. These electrons are in opposite spin

Due to the presence of all paired electrons, $Cr(CO)_6$ shows diamagnetic character. Due to d^2sp^3 hybridisation. $Cr(CO)_6$ has the expected *octahedral* shape as shown at (*c*) of Fig. 28.4. Cr—C and C—O bond lengths have been found to be 1.92A° and 1.16A°.

2. Iron Pentacarbonyl, $Fe(CO)_5$. In this carbonyl Fe atom which is in zero oxidation state is dsp^3 hybridised as shown in Fig. 28.2. Here too, both the electrons from 4*s* orbital go to 3*d* orbitals. Thus one 3*d*, 4*s* and all the three 4*p* orbitals, on mixing together, give five dsp^3 hybrid orbitals. The formation of five Fe ← CO σ bonds takes place in the same manner as six Cr ← Co σ bonds are formed in $Cr(CO)_6$ carbonyl.

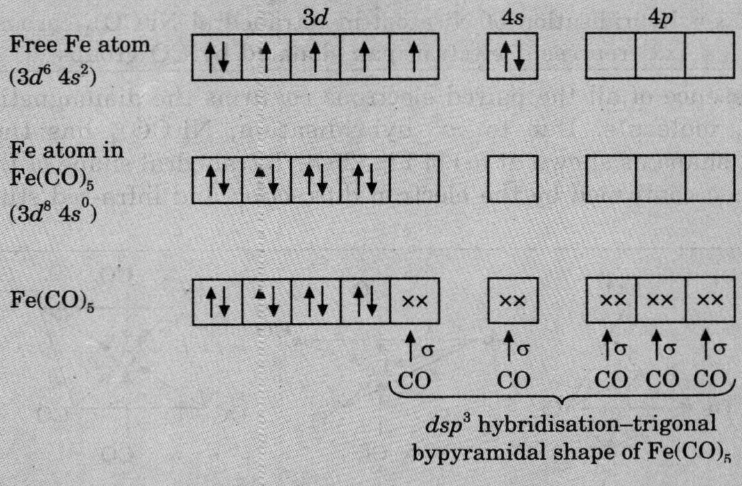

Fig. 28.2. dsp^3 hybridisation of Fe atom in trigonal bipyramidal $Fe(CO)_5$ carbonyl. Crosses (xx) represent electron pair donated by CO group

Since all the electrons are paired in $Fe(CO)_5$, pentacarbonyl is diamagnetic. Due to dsp^3 hybridisation, $Fe(CO)_5$ has *trigonal bipyramidal* shape as shown at (*b*) of Fig. 28.4. The most recent electron diffraction study of $Fe(CO)_5$ has shown that the length of Fe—C axial bond (= 1.806 Å) is slightly shorter than that of the equatorial bond (= 1.833 Å). Here it may be noted that in the molecule of typical elements having trigonal bipyramidal shape, the equatorial bond is usually shorter than the axial bond.

3. Nickel Tetracarbonyl, $Ni(CO)_4$. The vapour density of nickel carbonyl and the freezing point of its solution in benzene has shown that the molecular formula of nickel carbonyl is $Ni(CO)_4$. In this molecule Ni atom is sp^3 hybridised as shown in Fig. 28.3. Here too both the electrons of 4*s* orbital are shifted to 3*d* orbitals to vacate 4*s* orbital. Now 4*s* and three 4*p* orbitals, on mixing together, give four sp^3 hybrid orbitals. The formation of four Ni ← CO σ-bonds takes place in the same way as six Cr ← CO σ bonds and five Fe ← CO σ bonds are formed $Cr(CO)_6$ and $Fe(CO)_5$ respectively.

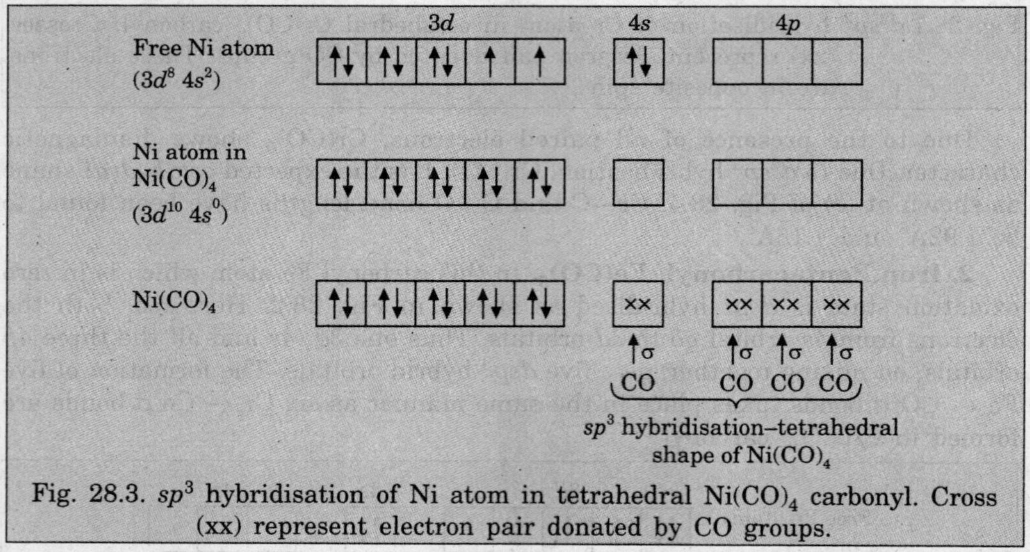

Fig. 28.3. sp^3 hybridisation of Ni atom in tetrahedral $Ni(CO)_4$ carbonyl. Cross (xx) represent electron pair donated by CO groups.

The presence of all the paired electrons confirms the diamagnetic character of $Ni(CO)_4$ molecule. Due to sp^3 hybridisation, $Ni(CO)_4$ has the expected *tetrahedral* shape as shown at (*a*) of Fig. 28.4. Tetrahedral shape of the molecule has also been confirmed by the electron diffraction and infra-red studies of this compound.

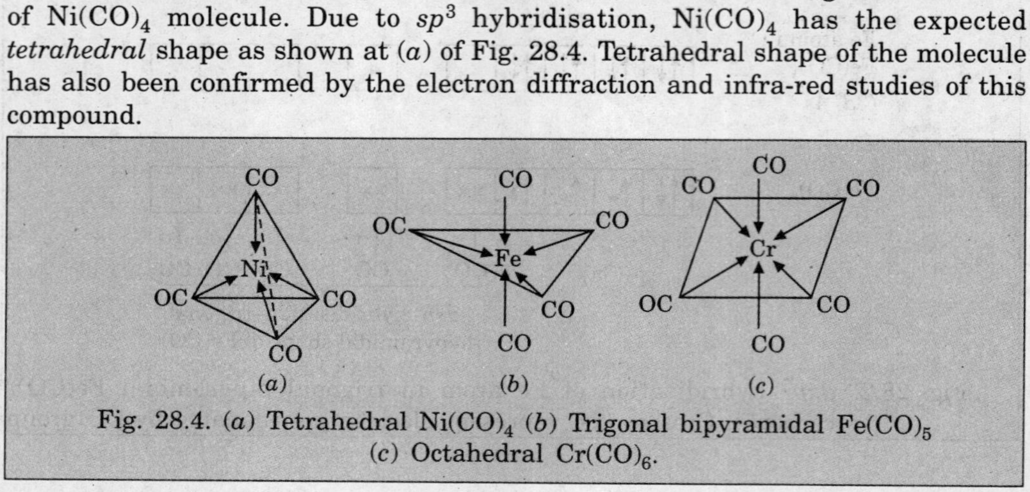

Fig. 28.4. (*a*) Tetrahedral $Ni(CO)_4$ (*b*) Trigonal bipyramidal $Fe(CO)_5$ (*c*) Octahedral $Cr(CO)_6$.

The Ni—C bond length in this molecule has been found to be equal to 1.50 A° which is shorter by 0.32 A° in comparison to Ni—C single bond length (= 1.82 A°) found in carbonyls. The C — O bond length in this carbonyl has been found to be equal to 1.15 A° which is larger than the C — O bond length in CO molecule (= 1.128 A°).

It is to be pointed out here that, due to the formation of four OC → M bonds, a large negative charge is accumulated on the central Ni atom. This is most unlikely. In such a situation, Pauling suggested that the double bonding takes place with the back sonation of *d*-electron from Ni atom to CO ligands to such an extent that electroneutrality principle is obeyed. According to this principle the electron pair is not shared equally between Ni and C-atoms of CO ligands but is attracted more strongly by C-atom which prevents the accumulation of negative charge on Ni-atom and is in keeping with the greater electronegativity of C-atom compared to Ni atom (C = 2.5, Ni = 1.8).

Structure and Bonding in Binuclear Carbonyls

All the binuclear carbonyls contain metal—metal bonds. Here we shall discuss the following binuclear carbonyls :

1. Dimanganese Decacarbonyl, $Mn_2(CO)_{10}$. The molecular weights of this compound indicates the dimeric formula, $Mn_2(CO)_{10}$. The structure of this carbonyl in the solid state contains *octahedrally coordinated Mn atoms.* Each Mn atom is linked to five CO groups and is also directly bonded to the other Mn atom. Thus this carbonyl contains ten CO groups which are called terminal CO groups ($-C \equiv O$) and one Mn—Mn bond See Fig. 28.5.

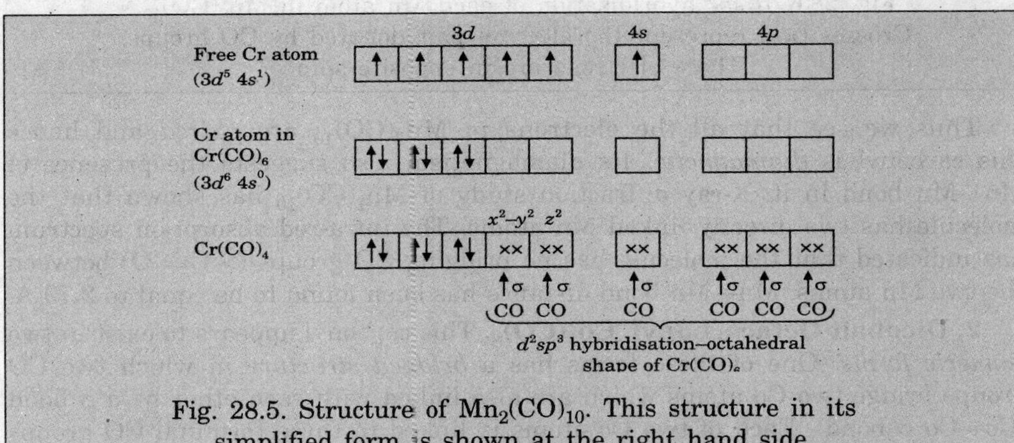

Fig. 28.5. Structure of $Mn_2(CO)_{10}$. This structure in its simplified form is shown at the right hand side.

The OC—Mn—Mn—CO chain is linear and the two square planes containing four CO groups and one Mn atom are *staggered*, *i.e.* the two linked octahedral units are staggered from the eclipsed position by an angle of 45°. Thus the structure of $Mn_2(CO)_{10}$ may be supposed to contain two staggered octahedra sharing one corner and in each of the octahedra the four equatorial Mn—CO bonds are bent inwards slightly. Sixth position of the octahedron is occupied by

Mn—Mn bond so that the steric repulsion between the CO groups is reduced. Structure of $Mn_2(CO)_{10}$ in its simplified form can be depicted as shown at the right hand side of Fig. 28.5.

In $Mn_2(CO)_{10}$ both Mn atoms are d^2sp^3 hybridised. Five (which are empty) out of six d^2sp^3 hybrid orbitals on each Mn-atom accommodate a lone pair of electrons donated by each of the five CO groups and the sixth hybrid orbital on one Mn atom which contains one unpaired electron overlaps with a similar d^2sp^3 hybrid orbital of the other Mn atom to form Mn —Mn σ bond (See Fig. 28.6).

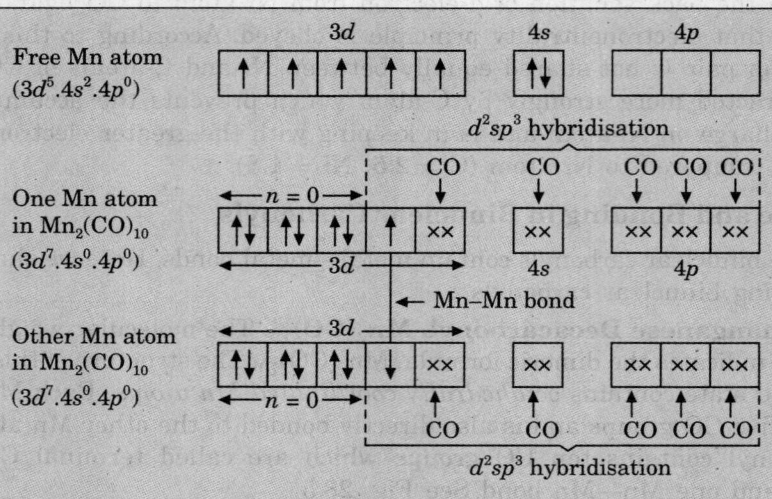

Fig. 28.6. d^2sp^3 hybridisation of each Mn atom in $Mn_2(CO)_{10}$. Crosses (××) represent the electron pair donated by CO groups. These electrons are in opposite spin.

Thus we see that all the electrons in $Mn_2(CO)_{10}$ are *paired* and hence this carbonyl is *diamagnetic.* Its diamagnetism also suggests the presence of Mn—Mn bond in it. X-ray diffraction study of $Mn_2(CO)_{10}$ has shown that the molecule has two directly linked Mn atoms. The infra-red absorption spectrum has indicated that the molecule has no bridging CO groups (> C = O) between the two Mn atoms. Mn—Mn bond distance has been found to be equal to 2.79 Å.

2. Dicobalt Octacarbonyl, $Co_2(CO)_8$. This carbonyl appears to exist in two *isomeric forms.* One of these forms has a *bridged structure* in which two CO groups bridge two Co atoms which are also linked with each other by a σ bond (Co—Co σ bond). Each of two Co atoms is linked to three terminal CO groups (See Fig. 28.7 (*a*)]. The other form has a *non-bridged structure* with a Co—Co σ bond linking two $Co(CO)_4$ groups (See Fig. 28.7 (*b*)]. Thus both the structures have Co—Co σ bonding and the diamagnetic nature of $Co_2(CO)_8$ also confirms the spin-pairing between two Co atoms, *i.e.* diamagnetic character of $Co_2(CO)_8$ is confirmed by the presence of Co—Co σ bond in it. There is very little difference in energy between the two structures.

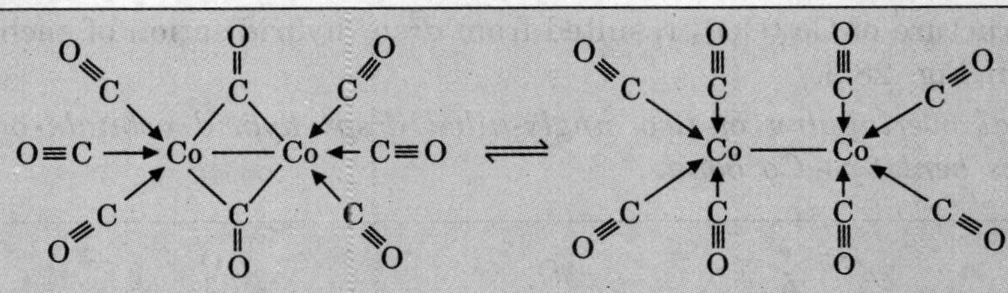

Fig. 28.7. Two isomeric forms of $CO_2(CO)_8$. Both the forms contain Co—Co bond.

The infra-red study of the bridged structure has shown that it has two types of CO groups [*i.e. terminal* CO groups (– CO ≡ O) and *bridging* CO groups (> C = O)]. Co—Co bond distance in this structure is 2.52 Å. There is infra-red evidence, showing that in solution both the structures exist in equilibrium with each other. At very low temperature the bridged structure predominates and as the temperature is raised, the non-bridged structure appears. If the temperature is increased to over 100°C, a third form appears. This third form is of uncertain structure. Bridged structure shows that coordination number of each Co-atom in the-molecule is *six*.

Each of the Co atoms is d^2sp^3 hybridised as shown in Fig. 28.8. The Co—Co bond is *bent* which arises due to the unusual overlapping of two singly-filled d^2sp^3 hybrid orbitals. One of these two singly-filled d^2sp^3 hybrid orbitals is from one Co atom while the other similar hybrid orbital is from the other Co atom.

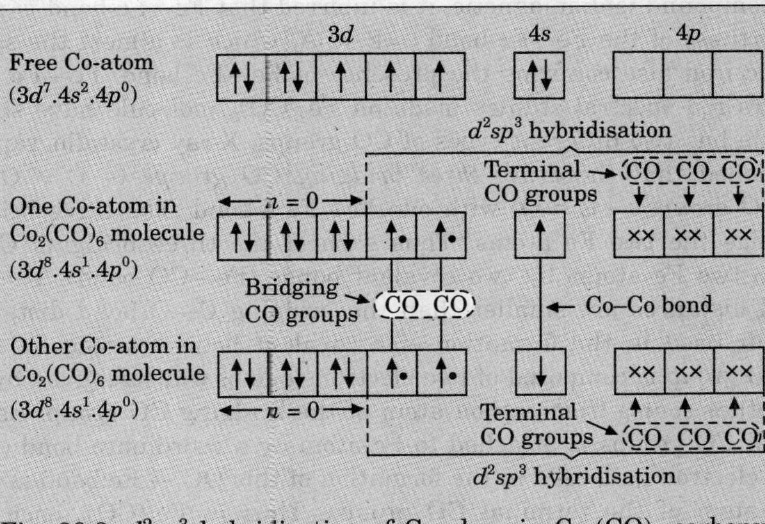

Fig. 28.8. d^2sp^3 hybridisation of Co alum in $Co_2(CO)_8$ carbonyl Crosses (××) represent the electron pair denoted by terminal CO groups, while the dot (•) represents an electron on bridging CO groups. ↑↓ represent electrons on Co atom.

The structure of $Co_2(CO)_8$ resulted from d^2sp^3 hybridisation of each Co atom is shown in Fig. 28.9.

Unusual overlapping of two singly-filled d^2sp^3 hybrid orbitals on two Co atoms gives bent Co—Co bond.

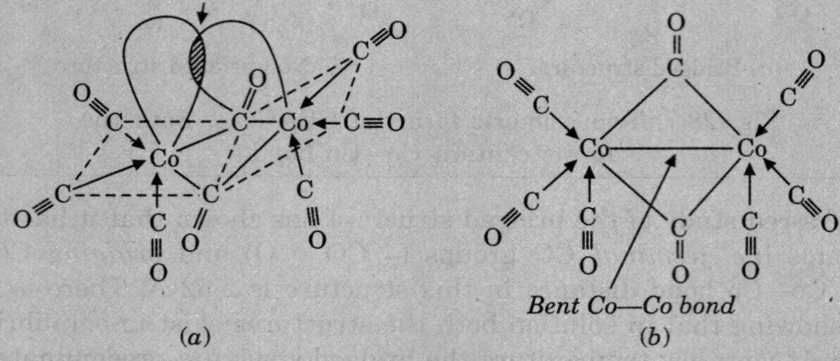

Fig. 28.9. Structure of $Co_2(CO)_8$. (*a*) shows the unusual overlapping of the singly filled d^2sp^3 hybrid orbitals on two Co atoms to form bent Co—Co bond. (*b*) shows the bond structure of $Co_2(CO)_8$.

3. Diiron Enneacarbonyl, $Fe_2(CO)_9$. This carbonyl base structure which is related to the of $Co_2(CO)_8$ with ninth CO group occupying a bridging position so that there is three-fold symmetry around the Fe—Fe axis. Its structure consists of two octahedra around each Fe atom and share their faces with each other. Since the compound is diamagnetic, it is inferred that Fe—Fe bond is present in it. The shortness of the Fe—Fe bond (= 2.46 Å) which is almost the same as in the metallic iron also confirms the presence of Fe—Fe bond. Fe—Fe bond is a δ bond. Infra-red spectral studies made on $Fe_2(CO)_9$ molecule have shown that the molecule has two different types of CO groups. X-ray crystallographic study has established that there are *three bridging CO groups* (– C = O) and *six terminal CO groups* (– C ≡ O) with one Fe—Fe δ-bond. The three bridging CO groups bridge the two Fe atoms. Thus each of the three bridging CO groups is linked to two Fe atoms by two covalent bonds (Fe—CO bond). The terminal C—O bond distances are smaller than the bridging C—O bond distances. The electron pair used in the formation of a covalent bond between Fe atom and bridging CO group is composed of two electrons one of which is given by Fe atom while the other comes from carbon atom of the bridging CO group. Each of the six terminal CO groups is attached to Fe atom by a coordinate bond (OC → Fe bond). The electron pair used in the formation of this OC → Fe bond is furnished by carbon atom of the terminal CO groups. Thus in $Fe_2(CO)_9$ each Fe atom accepts three electron pairs from the three carbon atoms of three terminal CO groups. Based on the above description the structure of $Fe_2(CO)_9$ can be shown as given in Fig. 28.10.

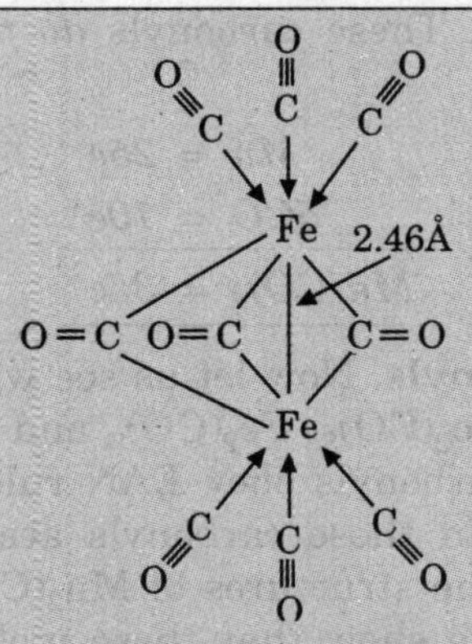

Fig. 28.10. Structure of $Fe_2(CO)_9$.

Effective Atomic Number (EAN) Rule As Applied to Metallic Carbonyls

(A) Mononuclear carbonyls having the metallic atom with even atomic number. $Cr(CO)_6$, $Mo(CO)_6$, $W(CO)_6$, $Fe(CO)_5$, $Ni(CO)_4$ etc. are the examples of such carbonyls. All such carbonyls obey the EAN rule according to which *after CO groups have donated a certain number of electron pairs to the zero valent metal atom through OC → M σ-bonding, the total number of electrons on the metal atom including those gained from CO molecules becomes equal to the atomic number of the next inert gas.* This is shown in Table 28.1.

Table 28.1. EAN rule as applied to mononuclear carbonyls having metallic atoms with even atomic number. Note that in all the carbonyls metal atom is in zero oxidation state.

Metal carbonyl	No. of electrons on the central metal atom = At. No. of the metal atom = Z	No. of electrons donated by CO molecules = x	EAN of the metal atom in carbonyl = Z + x (symbol of the next inert gas is given in parentheses)
$Cr(CO)_6$	24	6 × 2 = 12	24 + 12 = 36 (Kr)
$Mo(CO)_6$	42	6 × 2 = 12	42 + 12 = 54 (Xe)
$W(CO)_6$	74	6 × 2 = 12	74 + 12 = 86 (Rn)
$Fe(CO)_5$	26	5 × 2 = 10	26 + 10 = 36 (Kr)
$Ru(CO)_5$	44	5 × 2 = 10	44 + 10 = 54 (Xe)
$Os(CO)_5$	70	5 × 2 = 10	76 + 10 = 86 (Rn)
$Ni(CO)_4$	28	4 × 2 = 8	28 + 8 = 36 (Kr)

On the basis of EAN rule it can be explained why Ni atom does not form a hexacarbonyl, $Ni(CO)_6$. Non-formation of $Ni(CO)_6$ is because of the fact that EAN of Ni atom in $Ni(CO)_6$ would be equal to 28 + 2 × 6 = 40 which is not the atomic number of any of the noble gases.

(B) Mononuclear carbonyls having the metallic atom with odd atomic number. $V(CO)_6$ and hypothetical carbonyls *viz.* $Mn(CO)_5$ and $Co(CO)_4$ are the examples of such carbonyls. These carbonyls *do not obey* EAN rule as shown below :

$V = 23e^-$	$Mn = 25e^-$	$Co = 27e^-$
$6CO = 12e^-$	$5CO = 10e^-$	$4CO = 8e^-$
$V(CO)_6 = 35e^-$	$Mn(CO)_5 = 35e^-$	$Co(CO)_4 = 35e^-$

(C) Polynuclear carbonyls. Now let us see whether or not the polynuclear carbonyls like $Mn_2(CO)_{10}$, $Co_2(CO)_8$, $Fe_2(CO)_9$ and $Fe_5(CO)_{12}$ obey EAN rule. It has been seen that these carbonyls obey EAN rule, if two electrons from each metal–metal bond present in these carbonyls are included in calculating the electrons per metal atom. The structures of $Mn_2(CO)_{10}$, $Co_2(CO)_8$, $Fe_2(CO)_9$ and $Fe_3(CO)_{12}$ as discussed above show that these molecules contain *one* Mn—Mn, *one* Co—Co, *one* Fe—Fe, *three* Fe—Fe bonds respectively and hence the EAN per metal atom in these molecules is calculated as follows :

(i) $\boldsymbol{Mn_2(CO)_{10}}$: Electrons from 2Mn atoms = 25 × 2 = 50

Electrons from 1000 molecules = 10 × 2 = 20

Electrons from one Mn—Mn bond = 1 × 2 = 2

$Mn_2(CO)_{10}$ = 72

∴ EAN per Mn atom = 72/2

= 36(Kr)

(ii) $\boldsymbol{Co_2(CO)_8}$: Electrons from 2Co atom = 2 × 27 = 54

Electrons from 8CO molecules = 8 × 2 = 16

Electrons from one Co—Co bond = 1 × 2 = 2

$Co_2(CO)_8$ = 72

∴ EAN per Co atom = 72/2

= 36(Kr)

(iii) $\boldsymbol{Fe_2(CO)_9}$: Electrons from 2Fe atoms = 2 × 26 = 52

Electrons from 9CO molecules = 9 × 2 = 18

Electrons from one Fe—Fe bond = 1 × 2 = 2

$Fe_2(CO)_9$ = 72

∴ EAN per Fe atom = 72/2

= 36(Kr)

(iv) $\boldsymbol{Fe_3(CO)_{12}}$: Electrons from 3Fe atoms = 3 × 26 = 78

Electrons from 12CO molecules = 12 × 2 = 24

Electrons from three Fe—Fe bonds = 3 × 2 = 6

$Fe_3(CO)_{12}$ = 108

∴ EAN per Fe atom = 108/3

= 36(Kr)

Some Metallic Carbonyls

1. Chromium Hexacarbonyl, $Cr(CO)_6$

Preparation. In can be obtained by the following methods :

(*i*) CO at 50 atm pressure and at room temperature is passed into a suspension of $CrCl_3$ in ether, which has been treated with C_6H_5MgBr at –70° (*Job's reaction*).

(*ii*) A solution of Cr(+3) salt dissolved in ether is treated with CO at a high temperature and pressure in the presence of a reducing agent such as $Al(C_5H_5)_3$, $LiAlH_4$ etc.

(*iii*) $CrCl_3$ in presence of $AlCl_3$, powdered Al and benzene is treated with CO. The yield is 88%.

Properties. (*i*) Chromium hexacarbonyl is a white solid.

(*ii*) Crystals of $Cr(CO)_6$ are soluble in ether, chloroform, CCl_4 and C_6H_6.

(*iii*) It is resistant to chemical attack. For example it does not react with air, cold aqueous alkalies and acids (except conc. HNO_3).

(*iv*) *Decomposition.* It is decomposed by F_2 at –75°C to form CrF_6 by Cl_2 or by conc. HNO_3.

(*v*) *Reduction.* It is reduced with alkali metals in liq. NH_3 and alkali metals borohydrides. For example.

$$Cr(CO)_6 \xrightarrow{Na/NH_3} Na_2[Cr(CO)_5]$$

$$Cr(CO)_6 \xrightarrow[(THF)]{NaBH_4} Na[HCr_2(CO)_{10}]$$

$$Cr(CO)_6 \xrightarrow[(NH_3)]{NaBH_4} Na_2[Cr_2(CO)_{10}]$$

(*vi*) *Reaction with amines.* $Cr(CO)_6$ reacts with amines like *py, phen, en* to form the products in which CO groups in $Cr(CO)_6$ are replaced by molecules of the amine. For example above 140°, *py* reacts with $Cr(CO)_6$ to give the following products in succession :

$$Cr(CO)_6 \longrightarrow \underset{\text{yellow brown solid}}{Cr(CO)_4(py)_2} \longrightarrow \underset{\text{yellow red solid}}{Cr_2(CO)_7(py)_5} \longrightarrow \underset{\text{bright red solid}}{Cr(CO)_3(py)_3}$$

2. Dimanganese Decacarbonyl, $Mn_2(CO)_{10}$

Preparation. (*i*) Mn_2 (CO) is obtained when MnI_2 (prepared by special method) is reduced at a high pressure of CO by Mg in diethyl ether. The yield of $Mn_2(CO)_{10}$ is only 1%.

$$Mn + 2CuI \xrightarrow{500°C} MnI_2 + 2Cu$$

$$(MnI_2 + 2Cu) + Mg \xrightarrow[25°C/16\ hrs]{CO,\ 210\ atm} Mn_2(CO)_{10} + MgI_2 + CuI_2$$

(*ii*) $Mn_2(CO)_{10}$ has also been prepared in better yield by the reaction of Mn $(CH_3COO)_2$ with triethyl aluminium, $(C_2H_5)_3Al$ under CO at a pressure of 20 atmospheres at 100°C for 5 hours.

$$Mn(CH_3COO)_2 + (C_2H_5)_3\ Al \xrightarrow[100°C/5\ hrs]{CO,\ 20\ atm} Mn_2(CO)_{10}$$

The yield of carbonyl by this method is 53–60%.

(*iii*) Cabonylation of $MnCl_2$ in presence of benzophenone ketyl like $(C_6H_5)_2CONa$ at 165°C gives 32% yield of $Mn_2(CO)_{10}$.

$$MnCl_2 + 2(C_6H_5)_2CONa \longrightarrow NaCl + Mn[CO(C_6H_5)_2]_2$$

$$Mn[OC(C_6H_5)_2] \xrightarrow[165°C]{CO,\ 140\ atm} Mn_2(CO)_{10} + (C_6H_5)_2CO$$

(*iv*) More recently 48% yield of $Mn_2(CO)_{10}$ has been obtained by the carbonylation of methyl cyclopentadienyl manganese dicarbonyl, $(CH_3.C_5H_4)$ Mn $(CO)_3$ in presence of Na at 50 atmosphere pressure of CO and 125°C for 8 hours.

$$(CH_3.C_5H_4)Mn(CO)_3 + Na \xrightarrow[125°C/8\ hrs]{CO,\ 50\ atm} Mn_2(CO)_{10}$$

Properties. (*i*) It is a golden yellow crystalline substance. It has m.pt. = 155°C and sublimes in vacuo.

(*ii*) It is soluble in organic solvents.

(*iii*) It is slowly oxidised in air, especially in solution.

(*iv*) *Action of halogens.* Halogens form carbonyl halides, Mn $(CO)_5$ X as shown below :

$$Mn_2(CO)_{10} + Br_2\ (l) \xrightarrow{40°C} 2Mn(CO)_5\ Br$$

$$Mn_2(CO)_{10} + I_2 \longrightarrow 2Mn(CO)_5\ I$$

The order of activity of halogens is as : I > Br > Cl.

(*v*) *Action of Li and Na.* Mn—Mn bond is broken by lithium in the presence of tetrahydrofuran (THF) and by Na in presence of liq. NH_3 and $Li^+ [Mn(CO)_5]^-$ is formed.

$$Mn_2(CO)_{10} + 2Li \xrightarrow{THF} 2Li^+\ [Mn(CO)_5]^-$$

$$Mn_2(CO)_{10} + 2Na \xrightarrow{liq.\ NH_3} 2Na^+\ [Mn(CO)_5]^-$$

(*vi*) *Reduction.* When $Mn_2(CO)_{10}$ is reduced by hydrogen under 200 atmospheric pressure at a temperature of 200°C, carbonyl hydride in formed.

$$Mn_2(CO)_{10} + H_2 \xrightarrow[200°C]{200\ atm.} 2H\ Mn\ (CO)_5$$

(vii) Action of phosphines, arsines and stibines. With these materials $Mn_2(CO)_{10}$ gives monomeric paramagnetic compounds of the type, Mn $(CO)_4$ (PR_3).

(viii) Diamagnetic nature. $Mn_2(CO)_{10}$ is a diamagnetic substance. Diamagnetic character confirms the facts that all the electrons in $Mn_2(CO)_{10}$ are paired and Mn—Mn bond is also present in it.

3. Carbonyls of Iron

Iron forms these carbonyls which are : *(A) Iron pentacarbonyl,* Fe $(CO)_5$. *(B) Diiron enneacarbonyl,* $Fe_2(CO)_9$ and *(C) Tri-iron dodecacarbonyl,* $Fe_3(CO)_{12}$.

A. Iron pentacarbonyl, $Fe(CO)_5$. Preparation. It is prepared :

(i) by the action of CO at 200 atm. pressure on iron heated to 200°.

(ii) by the action of CO at 200 atm. pressure and 200° on FeI_2 in the presence of Cu which acts as an acceptor for I_2. It is thought that iron carbonyl iodide, $Fe(CO)_4I_2$ is first. formed which reacts a with Cu in presence of CO to form $Fe(CO)_5$.

$$FeI_2 + 4CO \longrightarrow Fe(CO)_4I_2$$

$$CO + Fe(CO)_4I_2 + 2Cu \longrightarrow Fe(CO)_5 + Cu_2I_2$$

In place of FeI_2, FeS can also be used.

Properties. *(i)* Fe $(CO)_5$ is a yellow liquid which is soluble in methyl alcohol, ether, acetone and C_6H_6. It is insoluble in H_2O.

(ii) Decomposition. On thermal decomposition at 250°C, it gives pure Fe.

$$2Fe(CO)_5 \longrightarrow Fe + 5CO$$

(iii) Action of ultra-violet light. When it is irradiated with near ultra violet light, $Fe_2(CO)_9$ is formed.

$$2Fe(CO)_5 \longrightarrow Fe_2(CO)_9 + CO$$

The above reaction is reversed in darkness.

(iv) Hydrolysis. $Fe(CO)_5$ is hydrolysed by H_2O and acids.

$$Fe(CO)_5 + H_2SO_4 \longrightarrow FeSO_4 + 5CO + H_2$$

(v) Action of alkalies. When heated with aqueous NaOH and the solution is acidified, carbonylate anion, $[Fe^{2-}(CO)_4H^+]^-$ is formed. The hydrogen atom in this anion is acidic which means that Fe is in –2 oxidation state.

$$\underset{(Fe\ =\ 0)}{Fe(CO)_5} + 3NaOH \longrightarrow \underset{(Fe\ =\ -2)}{Na^+[Fe^{2-}(CO)_4H^+]^-} + Na_2CO_3 + H_2O$$

(vi) Action of NH_3. With NH_3 it gives $Fe(CO)_4H_2$ and carbonic acid, NH_2COOH

$$Fe(CO)_5 + H_2O + NH_3 \longrightarrow Fe(CO)_4H_2 + NH_2COOH$$

(vii) Reaction with halogens. It reacts with halogens in non- aqueous solvents to form the stable tetra-carbonyl halides, $Fe(CO)_4X_2$.

$$Fe(CO)_5 + X_2 \longrightarrow Fe(CO)_4X_2 + CO$$

The velocities of these reactions follow the order Cl < Br < I.

(*viii*) *Reaction with amines.* When $Fe(CO)_5$ reacts with ethylene diamine (*en*), an addition product, $Fe(CO)_5$ (*en*) is formed. Pyridine (*py*) reacts with it to give $[Fe(py)_6]$ $[Fe_4(CO)_{13}]$.

(*ix*) *Reaction with NO.* With NO, under pressure below 45°, it gives $Fe(NO)_4$ as the final product which is a black crystalline material.

$$Fe(CO)_5 + 2NO \longrightarrow Fe(CO)_2(NO)_2 + 3CO$$
$$Fe(CO)_2(NO)_2 + 2NO \longrightarrow Fe(NO)_4 + 2CO$$

(*x*) *Reactions with metallic and non-metallic halides.* In these reactions, $Fe(CO)_4$ acts as a dehalogenating agent and a carbonylating agent; Fe becomes bi-valent and CO groups are either liberated or transferred to some other atom.

$$Fe(CO)_5 + 2CCl_4 \longrightarrow C_2Cl_6 + FeCl_2 + 5CO$$
$$Fe(CO)_5 + SO_2Cl_2 \longrightarrow SO_2 + FeCl_2 + 5CO$$

$SnCl_4$ and $SbCl_5$ are reduced to $SnCl_2$ and $SbCl_3$.

$$Fe(CO)_5 + SnCl_4 \longrightarrow SnCl_2 + Fe(CO)_4Cl_2 + CO$$
$$Fe(CO)_5 + SbCl_5 \longrightarrow SbCl_3 + Fe(CO)_4Cl_2 + CO$$

$Fe(CO)_4Cl_2$ then forms addition compounds with $SnCl_2$ and $SbCl_3$.

$$SnCl_2 + Fe(CO)_4Cl_2 \longrightarrow \underset{\text{(I)}}{SnCl_2.Fe(CO)_4Cl_2}$$
$$SbCl_3 + Fe(CO)_4Cl_2 \longrightarrow \underset{\text{(II)}}{SbCl_3.Fe(CO)_4Cl_2}$$

Addition compounds

The addition compounds (I) and (II) are halogen bridged compounds as shown below :

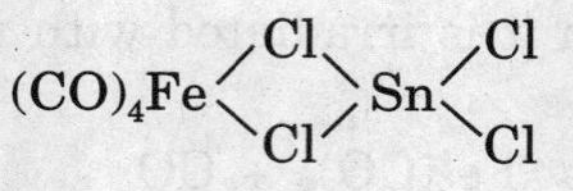

(I)

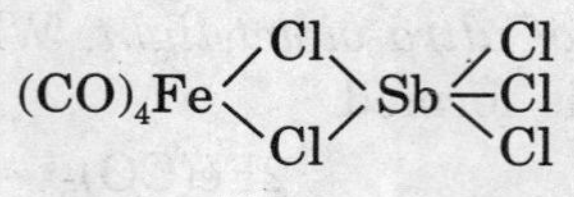

(II)

With WCl_6 and $MoCl_6$, $Fe(CO)_5$ gives $W(CO)_6$ and $Mo(CO)_6$ respectively.

$$\underset{(M = W,\ Mo)}{MCl_6} + 2Fe(CO)_5 \longrightarrow M(CO)_6 + 4CO + 2FeCl_3$$

(*xi*) When it is burnt in air, finely divided ferric oxide is obtained which is used as a pigment and a polishing agent.

B. Diiron enneacarbonyl, $Fe_2(CO)_9$. Preparation. It is made by the irradiation of $Fe(CO)_5$ with near ultraviolet light. $Fe(CO)_5$ is dissolved in its own volume of glacial acetic acid and the solution thus obtained is cooled and then irradiated for six hours. Golden crystals of $Fe_2(CO)_9$ are precipitated and are obtained by filtration.

Properties. (*i*) $Fe_2(CO)_9$ forms triclinic crystals which are diamagnetic.

(*ii*) It is slightly soluble in alcohol and acetone and more so in toluene and pyridine and almost insoluble in water, ether, C_6H_6 and many other organic solvents.

(*iii*) It is non-volatile.

(*iv*) *Action of heat.* When warmed to 50°, it gives $Fe_3(CO)_{12}$. At 100° it decomposes to give Fe, CO, and some $Fe(CO)_5$ and $Fe_3(CO)_{12}$.

(*v*) *Action of NO.* With NO it gives $Fe(CO)_2(NO)_2$ together with $Fe(CO)_5$ and $Fe_3(CO)_{12}$.

(*vi*) *Action of phen.* In acetone or benzene solution at 80° $Fe_2(CO)_9$ reacts with *phen* to form $[Fe(phen)_3]$ $[Fe_2(CO)_8]$. In pyridine, with *phen* it gives $[Fe(phen)_3]$ $[Fe_4(CO)_{13}]$.

(*vii*) *Action of alkali metals.* Carbonylate anion is obtained in its reaction with Na in NH_3 solution

$$Fe_2(CO)_9 + 4Na \longrightarrow 2Na_2[Fe(CO)_4] + CO$$

C. Tri-iron dodecacarbonyl, $Fe_3(CO)_{12}$. Preparation. It is prepared : (*i*) by the disproportionation of $Fe_2(CO)_9$. A toluene solution of $Fe_2(CO)_9$ is heated at 70°. On cooling, green crystals of $Fe_2(CO)_{12}$ separate from the solution.

$$3Fe_2(CO)_9 \xrightarrow{\Delta} 3Fe(CO)_5 + Fe_3(CO)_{12}$$

(*ii*) by the oxidation of $Fe(CO)_4H_2$ with H_2O_2 or MnO_2.

$$3Fe(CO)_4H_2 + 3H_2O_2 \longrightarrow Fe_3(CO)_{12} + 6H_2O$$

$$3Fe(CO)_4H_2 + 3MnO_2 + 3H_2SO_4 \longrightarrow Fe_3(CO)_{12} + 3MnSO_4 + 6H_2O$$

(*iii*) by treating $Fe(CO)_5$ with aqueous tri-ethylamine to form the dark red $[Et_3\ NH]$ $[Fe_3(CO)_{11}H]$ which is then acidified and $Fe_3(CO)_{12}$ is extracted with petroleum ether.

Properties. (*i*) $Fe_3(CO)_{12}$ forms deep green crystals which are soluble in organic solvents like toluene, alcohol, ether and pyridine.

(*ii*) *Action of heat.* When heated alone above 100°C, $Fe_3(CO)_{12}$ decomposes to give metallic iron and CO.

$$Fe_3(CO)_{12} \xrightarrow{> 100°C} 3Fe + 12CO$$

(*iii*) *Disproportionation reaction.* It undergoes disproportionation reactions. For example it reacts with CH_3OH at room temperature and $Fe(CO)_5$ is thereby produced.

$$3Fe(CO)_{12} + nCH_3OH \longrightarrow [Fe(CH_3OH)_n]\ [Fe_3(CO)_{11}] + 5Fe(CO)_5$$

(*iv*) *Reaction with Na.* Carbonylate anion is obtained when $Fe_3(CO)_{12}$ reacts with Na metal in liq. NH_3.

$$Fe_3(CO)_{12} + 6Na \longrightarrow 3Na_2\ [Fe(CO)_4]$$

(*v*) *Substitution reactions.* These reactions occur with pyridine (*py*) and methyl alcohol as shown below :

$$2Fe_3(CO)_{12} + 3py \longrightarrow Fe_3(CO)_9\ (py)_3 + 3Fe(CO)_5$$

$$2Fe_3(CO)_{12} + 3CH_3OH \longrightarrow Fe_3(CO)_9\ (CH_3OH)_3 + 3Fe(CO)_5$$

(*vi*) *Action of NO.* At 85° it reacts with NO to form iron dicarbonyl dinitrosyl, $Fe(CO)_2\ (NO)_2$.

$$Fe_3(CO)_{12} + 6NO \longrightarrow 3Fe(CO)_2\ (NO)_2 + 6CO$$

4. Dicobalt Octacarbonyl, $Co_2(CO)_8$.

Preparation. It can be prepared : (*i*) by the reaction between CO and reduced metallic cobalt at 220° and 250 atm.

$$2Co + 8CO \xrightarrow[220°C]{250\ atm.} Co_2(CO)_8$$

(*ii*) by the reaction of dry CO at 200° and 200 atm in presence of metallic copper on certain binary compounds of cobalt such as CoS and CoX_2. Here Cu forms Cu_2S or CuX.

$$2CoS + 8CO + 4Cu \longrightarrow Co_2(CO)_8 + 2Cu_2S$$

$$2CoX_2 + 8CO + 4Cu \longrightarrow Co_2(CO)_8 + 4CuX$$

(*iii*) by the action of CO and H_2 on $CoCO_3$ under high pressure and high temperature.

$$2CoCO_3 + 2H_2 + 8CO \xrightarrow[120\text{–}200°C]{250\text{–}300\ atm.} Co_2(CO)_8 + 2CO_2 + 2H_2O$$

(*iv*) by the action of an acid on a solution of $Co(CO)_4H$. Hydrogen is evolved and $Co_2(CO)_8$ is left behind.

$$2Co(CO)_4H \longrightarrow Co_2(CO)_8 + H_2$$

(*v*) by the thermal decomposition of $Co(CO)_4H$.

Properties. (*i*) $Co_2(CO)_8$ is an orange brown crystalline substance having m.pt. = 51°C. It is soluble in alcohol, ether and CCl_4. The carbonyl is air sensitive both in the solid and solution states.

(*ii*) *Action of heat.* It is thermally decomposed at 50°C in an inert atmosphere to give tetra cobalt dodecacarbonyl, $Co_4(CO)_{12}$.

$$2Co_2(CO)_8 \xrightarrow{50°C} Co_4(CO)_{12} + 4CO$$

(*iii*) *Action of air.* On exposure to air, dicobalt octacarbonyl is converted into deep violet basic carbonate of cobalt.

(*iv*) *Reduction reactions.* It is reduced to cobalt carbonyl hydride, $H^+[Co(CO)_4]$ by H_2 at 165°C and 120 atmospheric pressure.

$$Co_2(CO)_8 + H_2 \xrightarrow[165°C]{120\ atm.} 2H^+[Co(CO)_4]^-$$

$Co_2(CO)_8$ is also reduced by Na metal in liq. NH_3 below −75°C or tetrahydrofuran (THF).

$$\underset{(Co\ =\ 0)}{Co_2(CO)_8} + 2Na \xrightarrow[\text{or liq. } NH_3]{THF} \underset{(Co\ =\ -1)}{2Na^+[Co(CO)_4]^-}$$

(*v*) *Disproportionation reactions.* There are two types of such reactions in case of $Co_2(CO)_8$.

(*a*) Strong bases having nitrogen or oxygen donor atoms cause disproportionation into Co(+2) and Co(–1). For example :

$$\underset{(Co\,=\,0)}{2Co_2(CO)_8} + 12NH_3 \longrightarrow \underset{(Co\,=\,+2)}{2[Co(NH_3)_6]}\ \underset{(Co\,=\,-1)}{[Co(CO)_4]_2} + 8CO$$

(*b*) With isocyanides, phosphines, arsines and stibines, however, the disproportionation reaction gives penta-coordinate cobalt (I) cation.

$$\underset{(Co\,=\,0)}{Co_2(CO)_8} + 5CNR \longrightarrow \underset{(Co\,=\,+1)}{[Co(CNR)_5]}\ \underset{(Co\,=\,-1)}{[Co(CO)_4]} + 4CO$$

(*vi*) *Reaction with NO.* $Co_2(CO)_8$ also reacts with NO at 40° and forms $Co(CO)_3(NO)$

$$\underset{(Co\,=\,0)}{Co_2(CO)_8} + 2NO \longrightarrow \underset{(Co\,=\,-1)}{Co(CO)_3(NO)} + 2CO$$

In $Co(CO)_3(NO)$, NO group appears to be more firmly held with Co metal atom than CO groups. This is confirmed by the fact that CO groups can be replaced by the amines very easily, *e.g.* the action of *phen* on $Co(CO)_3(NO)$ gives [Co(CO)(*phen*)(NO)].

$$Co(CO)_3(NO) + phen \xrightarrow{-2CO} [Co(CO)(phen)(NO)]$$

(*vii*) *Reaction with halogens.* The halogens decompose $Co_2(CO)_8$ according to the following reaction

$$Co_2(CO)_8 + 2X_2 \longrightarrow 2CoX_2 + 8CO$$

5. Nickel Tetracarbonyl, $Ni(CO)_4$

It was the first of the metallic carbonyls to be discovered and has found application in the metallurgy of nickel by *Mond's process.* It has zero valent nickel.

Preparation. (*i*) It is made by passing carbon monoxide over finely-divided metallic nickel at a temperature below 100°C.

$$Ni + 4CO \longrightarrow Ni(CO)_4$$

(*ii*) When CO at atmospheric pressure is passed over reduced nickel at 30–50°C, we get $Ni(CO)_4$.

(*iii*) $Ni(CO)_4$ is also obtained when CO is passed through an alkaline suspension of $Ni(CN)_2$ or NiS.

$$Ni(CN)_2 + 4CO \longrightarrow Ni(CO)_4 + C_2N_2$$

$$NiS + 4CO \longrightarrow Ni(CO)_4 + S$$

(*iv*) *By the disproportionation of univalent nickel.* (*a*) On reducing potassium tetracyano nickelate (II), $K_2[Ni(CN)_4]$ with potassium amalgam, a red solution is obtained. From this solution potassium tricyano nickelate (I), $K_2[Ni(CN)_3]$ may be isolated. This compound absorbs CO at 90°C, probably forming $K_2[Ni(CN)_3(CO)]$. On treating the solution of this compound with HCl, $Ni(CO)_4$ is obtained.

$$\underset{(Ni\,=\,+1)}{4K_2[Ni(CN)_3(CO)]} + 2HCl \longrightarrow \underset{(Ni\,=\,0)}{Ni(CO)_4} + \underset{(Ni\,=\,+2)}{3K_2[Ni(CN)_4]} + 2KCl + H_2$$

(*b*) It is most conveniently prepared in the laboratory by the action of CO on nickel(II) phenyl dithiocarbamate, $Ni(S.SC.NH.C_6H_5)_2$.

$$\underset{\text{Nickel (II) phenyl dithiocarbamate (Ni = +2)}}{2Ni(S.SC.NH.C_6H_5)_2} + 4CO \longrightarrow \underset{(Ni = 0)}{Ni(CO)_4} + \underset{\text{Nickel (IV) phenyl dithiocarbamate (Ni = +4)}}{Ni(S.SC.NH.C_6H_5)_4}$$

Properties. (*i*) It is a colourless liquid boiling at 43° and solidifying at –25°C. It is miscible with benzene and is almost insoluble in water. The benzene solution of the nickel carbonyl does the same on boiling, but in the cold, the compound is not acted on by dilute acids or alkalies. Nickel carbonyl is highly poisonous.

(*ii*) *Action of H_2SO_4.* H_2SO_4 reacts with it as follows :

$$Ni(CO)_4 + H_2SO_4 \longrightarrow NiSO_4 + H_2 + 4CO$$

(*iii*) *Action of NO.* Slightly moist nitric oxide reacts with $Ni(CO)_4$ in the gaseous state or in solution in chloroform to give a compound of intense blue colour, Ni(NO)(OH).

$$2Ni(CO)_4 + 2NO + 2H_2O \longrightarrow 2Ni(NO)(OH) + 8CO + H_2$$

When $Ni(CO)_4$ reacts with NO in the absence of H_2O, the blue solution of the composition, $Ni(NO)(NO_2)$ is obtained.

$$Ni(CO)_4 + 4NO \longrightarrow Ni(NO)(NO_2) + 4CO + N_2O$$

(*iv*) *Substitution reactions.* CO groups can be replaced by NO_2, PCl_3 and CNR molecules. o-phenanthroline (o.phen) replaces only two CO molecules

$$Ni(CO)_4 + 2NO_2 \longrightarrow Ni(NO_2)_2 + 4CO \uparrow$$

$$Ni(CO)_4 + 4PCl_3 \longrightarrow \underset{\text{Yellow non-volatile}}{Ni(PCl_3)_4} + 4CO$$

$$Ni(CO)_4 + \underset{\text{Aryl isocyanide}}{4CNR} \longrightarrow \underset{\text{Yellow crystalline substance}}{Ni(CNR)_4} + 4CO \uparrow$$

$$Ni(CO)_4 + \textit{o-phen} \longrightarrow \underset{\text{Ruby-red needles}}{Ni(CO_2)\ (\textit{o-phen})} + 2CO \uparrow$$

(*v*) *Action of heat.* $Ni(CO)_4$, when heated to 180°, gives nickel.

$$Ni(CO)_4 \xrightarrow{180°} Ni + 4CO \uparrow$$

In the pure state nickel carbonyl explodes to form nickel, carbon dioxide and carbon.

$$Ni(CO)_4 \xrightarrow{\Delta} Ni + 2CO_2 + 2C$$

Under high atmospheric pressures (say, 100 atm) it is, however, not decomposed even up to 250°C.

(*vi*) *Oxidation reactions.* Solution of this carbonyl in organic solvent is readily oxidised by air developing a pale green gelatinous precipitate. Such solutions are also readily oxidised to Ni (II) salts by the halogens, cyanogen and sulphur.

$$\underset{(Ni = 0)}{Ni(CO)_4} + Br_2 \longrightarrow \underset{(Ni = +2)}{NiBr_2} + 4CO \uparrow$$

Gaseous HCl decomposes the solution of $Ni(CO)_4$ evolving H_2 and CO

$$\underset{(Ni\,=\,0)}{Ni(CO)_4} + 2HCl(g) \longrightarrow \underset{(Ni\,=\,+2)}{NiCl_2} + H_2 + 4CO\uparrow$$

(*vii*) *Reduction reactions.* $Ni(CO)_4$ gets reduced by several reducing agents. Examples are :

(*a*) $N(CO)_4$ can be reduced to $[Ni^-H^+(CO)_3]$ by the reaction of $Ni(CO)_4$ with Na in liq. NH_3.

$$\underset{(Ni\,=\,0)}{2Ni(CO)_4} + 2Na + 2NH_3 \longrightarrow \underset{(Ni\,=\,-1)}{2[Ni^-H^+(CO)_3]_2} + 2CO + 2NaNH_3$$

$[Ni^-H^+(CO)_3]$ is isolated as a red tetraammoniate, $[NiH(CO)_3]_2\ 4NH_3$.

(*b*) When $Ni(CO)_4$ solution in THF is reduced by Na, K or Mg amalgams, the enneacarbonyl tetranickelate is formed.

$$4Ni(CO)_4 + 2Na \longrightarrow Na_2[Ni_4(CO)_9] + 7CO$$

With Li, $Ni(CO)_4$ gives $Li_2\ [Ni_3(CO)_8]$

$$3Ni(CO)_4 + 2Li \longrightarrow Li_2[Ni_3(CO)_8] + 4CO$$

Uses. (*i*) Since $Ni(CO)_4$ is decomposed to metallic nickel, it is used in the production of nickel by Mond's process. (*ii*) It is also used in gas plating and also as a catalyst.

Questions with Answers

Q.1 What are the different types of bonds found in metal carbonyls.

Ans : Following are the important types of bonds found in metal carbonyls.

1. $:O\equiv C\rightarrow M$ σ bond. This bond can also be represented as $OC\rightarrow M\sigma$ or OC–Mσ bond. In this bond CO acts as a terminal carbonyl group. This bond is found in all types of carbonyls.

2. 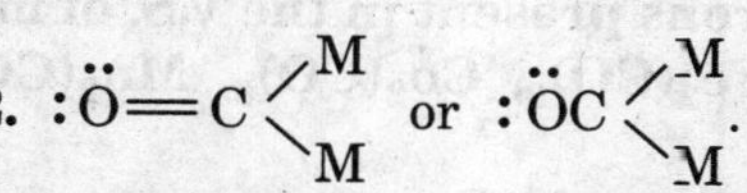 . This system is found in bridged polynuclear carbonyls like $Co_2(CO)_8$ (in the solid state), $M_2(CO)_9$ (M = Fe, Os), $Fe_3(CO)_{10}$ etc. C-atom in the system is sp^2 hybridised.

3. M – M bond. This bond is found in $M_2\ (CO)_9$ (M = Fe, Os), $Co_2\ (CO)_8$ (in solution and solid), $M_3\ (CO)_{12}$ (M = Fe, Os) etc. The length of this bond is larger than that of normal M–M bond. The presence of this bond in metal carbonyls makes all electrons paired and hence carbonyls containing M–M bonds are diamagnetic.

Q.2 Determine the number of M–M bonds present in $Fe_2(CO)_9$, $Fe_3(CO)_{12}$, $Co_4(CO)_{12}$ and $[(\eta^5 - Cp)\ Mo\ (CO)_3]_2$.

Ans : (*i*) **$Fe_2(CO)_9$.** Electrons from two Fe-atoms = 2 × 26 = 52

Electrons from 9 CO groups = 2 × 9 = 18

EAN for two Fe atoms in $Fe_2(CO)_9$ = 52 + 18 = 70

∴ EAN for one Fe atom = 70/2 = 35

Since this EAN is less than 36 (Kr_{36}) by one electron, there is one Fe–Fe bond in $Fe_2(CO)_9$ molecule.

***(ii)* $Fe_3(CO)_{12}$.** Electrons from 3Fe atoms = 3 × 26 = 78

Electrons from 12 CO groups = 12 × 2 = 24

EAN for three Fe-atoms = 78 + 24 = 102

∴ EAN for one Fe atom = 102/3 = 34

Since this EAN is less than 36 by two electrons, each Fe-atom in $Fe_3(CO)_{12}$ molecule is linked with other two Fe-atoms. Thus $Fe_3(CO)_{12}$ has three Fe – Fe bond.

***(iii)* $Co_4(CO)_{12}$.** Electrons from 4CO atoms = 4 × 27 = 108

Electrons from 12 CO groups = 12 × 2 = 24

EAN of four Co atoms = 108 + 24 = 132

∴ EAN for one Co-atom = 132/4 = 33

Since this EAN is less than 36 by three electrons, each Co-atom in $Co_4(CO)_{12}$ molecule is linked with other three Co-atom. Thus $Co_4(CO)_{12}$ has five Co–Co bonds.

(iv) **$[(\eta^5 - Cp)\ Mo(CO)_3]_2$.** Electrons from two Mo-atoms = 2 × 42 = 84

Electrons from two η^5 – Cp = 2 × 5 = 10

Electrons from 6CO = 6 × 2 = 12

EAN for two Mo-atoms in $[(\eta^5 - Cp)\ Mo\ (CO)_3]_2$ = 84 + 10 + 12 = 106

∴ EAN for one Mo-atom = 106/2 = 53

Since this EAN is less than 54 (Xe_{54}) by one electron, there is one Mo–Mo bond in the given mixed carbonyl.

Q.3 Show that the total number of electrons present in the V.S. of metal atom in the following carbonyls is 18. $Fe_2(CO)_9$, $Co_2(CO)_8$, $Mn_2(CO)_{10}$, $Fe_3(CO)_{12}$ and $Co_4(CO)_{12}$.

Ans : (i) $Fe_2(CO)_9$. 2Fe = 2 × 8 = 16 (Fe = $3d^6 4s^2$)

9CO = 9 × 2 = 18

One Fe–Fe bond = 1 × 2 = 2

∴ Total electrons in two Fe atoms = 16 + 18 + 2 = 36

∴ Electrons in the V.S. of one Fe atom = $\frac{36}{2}$ = 18

(ii) **$Co_2(CO)_8$.** 2Co = 2 × 9 = 18 (Co = $3d^7 4s^2$)

8CO = 8 × 2 = 16

One Co–Co bond = 1 × 2 = 2

Total electrons in two Co atoms = 18 + 16 + 2 = 36

Electrons in the V.S. of one Co-atom = $\frac{36}{2}$ = 18

(iii) **$Mn_2(CO)_{10}$.** 2Mn = 2 × 7 = 14 (Mn = $3d^5 4s^7$)

$$10CO = 10 \times 2 = 20$$
$$\text{One Mn–Mn bond} = 1 \times 2 = 2$$

Total electorns on two Mn-atoms = 14 + 20 + 2 = 36

∴ Electrons on one Mn-atom = 36/2 = 18

(*iv*) **$Fe_3(CO)_{12}$.** $3Fe = 3 \times 8 = 24$ ($Fe = 3d^6 4s^2$)

$$12CO = 12 \times 2 = 24$$
$$\text{Three Fe–Fe bonds} = 3 \times 2 = 6$$

Total electrons on three Fe-atoms = 24 + 24 + 6 = 54

$$\text{Electrons on one Fe-atom} = \frac{54}{3} = 18$$

(*v*) **$Co_4(CO)_{12}$.** $4Co = 4 \times 9 = 36$ ($Co = 3d^7 4s^2$)

$$12CO = 12 \times 2 = 24$$
$$\text{Six Co–Co bonds} = 6 \times 2 = 12$$

Total electrons in four Co-atoms = 36 + 24 + 12 = 72

$$\therefore \text{Electrons on one Co-atom} = \frac{72}{4} = 18$$

University Questions

1. Give one method for the preparation of each of the following : $Ni(CO)_4$, $Fe_2(CO)_9$, $Co_2(CO)_8$, $Fe(CO)_5$. (*Himachal Pradesh 99*)

2. What are metal carbonyls ? Give few examples and discuss the nature of M—CO bonding in carbonyls. (*Gorakhpur 99*)

3. Discuss the structure and constitution of $Ni(CO)_4$. (*Allahabad 99*)

4. (*a*) Why does CO form stable compounds with transition metals though it is a poor donor?

(*b*) Define π acidity and give an example.

(*c*) Discuss the structure of mononuclear carbonyls.

(*d*) Discuss the preparation and structure of $Ni(CO)_4$ and $Fe_2(CO)_9$.

(*e*) What is meant by back bonding in metal carbonyls?

(*Himachal Pradesh 2000*)

5. Explain the following :

(*i*) CO is a Lewis acid.

(*ii*) $Fe(CO)_5$ has trigonal bipyramidal structure. (*Lucknow 2000*)

6. Discuss the general methods for the preparation of metal carbonyls. Describe the nature of bonding in binary and polynuclear metal carbonyls with special reference to the structures of $Fe(CO)_5$, $Mn_2(CO)_{10}$ and $Fe_2(CO)_9$. (*Lucknow 2000*)

7. What are metal carbonyls ? Classify them. Give one method of preparation of each class of metal carbonyls. Discuss the structure and bonding in $Fe(CO)_5$, $Cr(CO)_6$ and $Mn_2(CO)_8$. *(Lucknow 2001)*

8. Describe the preparation, properties and structure of $Ni(CO)_4$. *(C.S.J.M. Kanpur, 2005)*

9. (*a*) Write a short note on "18-electron rule in metal carbonyls".

   (*b*) Describe methods of preparation, properties and structure of $Fe(CO)_5$. *(C.S.J.M. Kanpur 2006)*

10. (*a*) Describe the preparation and constitution of $Cr(CO)_6$.

    (*b*) What is EAN rule ? How is it applicable in metal carbonyls ? *(C.S.J.M. Kanpur 2007)*

11. Discuss bonding in linear metal carbonyls (M–C–O). How do vibrational spectra help to understand bonding in carbonyls ? *(M.J.P. Rohilkhand, 2008)*

12. Find out EAN in $Cr(CO)_6$ molecule. *(M.J.P. Rohilkhand 2009)*

13. What are metal carbonyls ? *(Meerut 2009)*

CHAPTER

29

Compounds of Silicon and Glass Industry

Compounds of Silicon

1. Silicic Acids.

Two silicic acids of silicon are known. These acids are *orthosilicic acid*, H_4SiO_4 or $Si(OH)_4$ and *meta silicic acid*, H_2SiO_3 or H_2SiO_3 or $SiO(OH)_2$. The existence of these acids is not definite but many salts of these acids (called silicates) of definite composition are known. Orthosilicic acid is obtained as a gelatinous ppt. when SiF_4 or $SiCl_4$ is passed into water.

$$3SiF_4 + 4H_2O \longrightarrow \underset{\text{Gelatinous ppt.}}{H_4SiO_4} + 2H_2SiF_6$$

The gelatinous ppt. is filtered, washed, first with water and then with ether and is finally dried between folds of filter paper when a white amorphous powder of H_4SiO_4 is obtained.

Metasilicic acid is obtained as a gelatinous ppt. by adding HCl to a concentrated solution of an alkali silicate like Na_2SiO_3.

$$Na_2SiO_3 + 2HCl \longrightarrow 2NaCl + \underset{\text{Gelatinous ppt.}}{H_2SiO_3}$$

The gelationous ppt. of H_2SiO_3 so formed is filtered, dehydrated with 90% alcohol and finally dried when a white amorphous powder is obtained.

The colloided solution of metasilicic acid is known as *silicic acid sol.* It was discovered by Graham in 1861 and is formed by slowly pouring a cold dilute solution of Na_2SiO_3 (called *water glass*) into an excess of dil HCl.

Na_2SiO_3 (*solution*) + HCl (*dil. and in excess*) $\longrightarrow$ 2NaCl + H_2SiO_3

The resulting mixture of liquids is dialysed in a parchment bag placed in distilled water when NaCl and HCl diffuse out of the bag while a pure colloidal solution of silicic acid is left in the dialyser.

2. Sodium Silicate or Water Glass, Na_2SiO_3.

This is a salt of *metasilicic* acid, H_2SiO_3 and is also called *water glass or soluble glass.* It consists of sodium metasilicate containing an excess of silica; the composition may vary from $Na_2SiO_3.SiO_2$ to $Na_2SiO_3.3SiO_2$.

Preparation. (*i*) It is prepared by fusing together sodium carbonate with pure sand in a reverberatory furnace.

$$Na_2CO_3 + SiO_2 \longrightarrow Na_2SiO_3 + CO_2 \uparrow$$

(*ii*) It is also obtained by fusing together sodium sulphate, charcoal and sand.

$$Na_2SO_4 + 2C \longrightarrow Na_2S + 2CO_2 \uparrow$$
$$Na_2S + 3Na_2SO_4 + 4SiO_2 \longrightarrow 4Na_2SiO_3 + 4SO_2 \uparrow$$

The resulting mass is extracted with water and the solution evaporated to a syrup known as water glass having the Composition $Na_2SiO_3.SiO_2$.

Properties. Water glass is soluble in water and the solution is strongly alkaline due to hydrolysis. Crystals of various soluble coloured salts like $CuSO_4$, $FeSO_4$, $Co(NO_3)_2$, $NiCl_2$, $MnCl_2$ etc. when placed in a solution of water glass having a specific gravity of 1.1 produçe beautiful growth which is called *silica garden* or *chemical garden.* The preparation of silica garden is done as follows :

Prepare a concentrated solution of sodium silicate water of density 1.1. Place this liquid in a beaker, sprinkle some crystals of coloured salts like $FeSO_4$, cobalt nitrate, nickel sulphide etc. on the surface of the liquid. Coloured shoots will rise up and grow on long standing like a plant. This is called silica garden or chemical garden.

Uses. Sodium silicate is used : (*i*) in calicoprinting, (*ii*) in fire- proofing wood and textiles, (*iii*) as a. preservative for egg, (*iv*) as a furniture polish, and (*v*) in the soap industry as a filler for a cheaper variety of soap.

3. Silicon Tetrachloride, $SiCl_4$.

Preparation. This compound is prepared:

(*i*) by passing dry Cl_2 over heated amorphous silicon or magnesium silicide (Mg_2Si) or an intimate mixture of SiO_2 and coke.

$$Si + 2Cl_2 \longrightarrow SiCl_4$$
$$Mg_2Si + 4Cl_2 \longrightarrow SiCl_4 + 2MgCl_2$$
$$2SiO_2 + 2C + 2Cl_2 \longrightarrow SiCl_4 + 2CO$$

(*ii*) by passing a mixture of dry Cl_2 and S_2Cl_2 vapour over heated SiO_2.

$$2SiO_2 + 3Cl_2 + S_2Cl_2 \longrightarrow 2SiCl_4 + 2SO_2$$

(*iii*) by passing CCl_4 vapour over heated SiO_2

$$SiO_2 + CCl_4 \longrightarrow SiCl_4 + CO_2$$

Properties (*i*) Physical properties. It is a colourless fuming liquid of density 1.50 (at 0°C). Its b.pt. and m.pt. are 56.8°C and 70°C respectively.

(ii) Hydrolysis. $SiCl_4$ is hydrolysed by H_2O with the formation of orthosilicic acid (H_4SiO_4) and HCl

$$SiCl_4 + 4H_2O \longrightarrow H_4SiO_4 + 4HCl \quad ...(A)$$

The hydrolysis of SiC_4 takes place through the following steps :

(*a*) $SiCl_4$ coordinates with two H_2O molecules through their O-atoms to form the unstable intermediate compound, $SiCl_4.2H_2O$. In this compound Si atom attains its maximum covalency of 6.

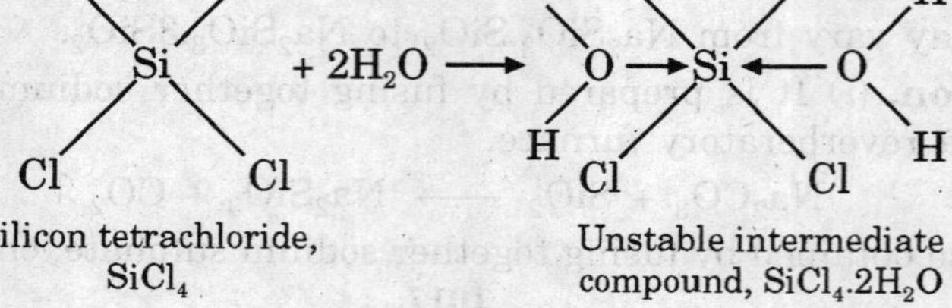

Silicon tetrachloride, $SiCl_4$ — Unstable intermediate compound, $SiCl_4.2H_2O$

(*b*) The intermediate compound, $SiCl_4.2H_2O$ then eliminates two HCl molecules to give $SiCl_2(OH)_2$

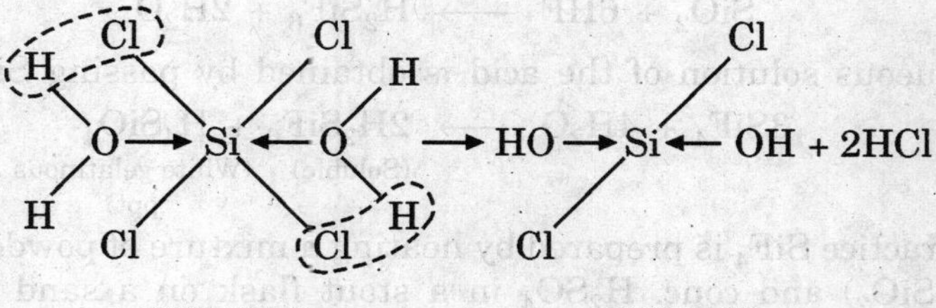

$SiCl_4.2H_2O$ $SiCl_2(OH)_2$

(*c*) The compound, $SiCl_2(OH)_2$ then combines with two H_2O molecules and forms $SiCl_2(OH)_2.2H_2O$.

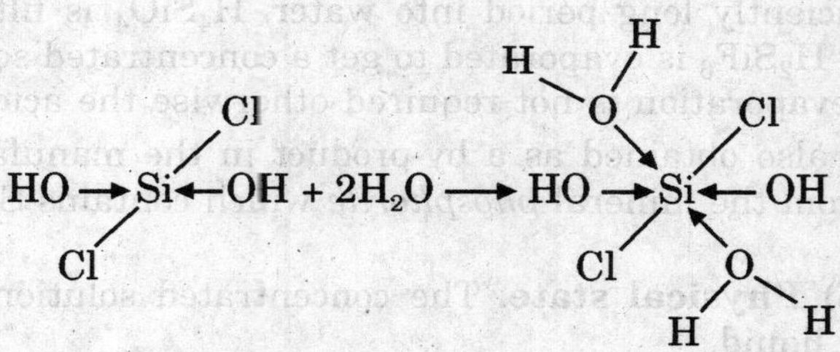

$SiCl_2(OH)_2$ $SiCl_2(OH)_2.2H_2O$

(*d*) The compound, $SiCl_2(OH)_2.2H_2O$ eliminates two HCl molecules and gives rise to the formation of orthosilicic acid, $Si(OH)_4$ or H_4SiO_4.

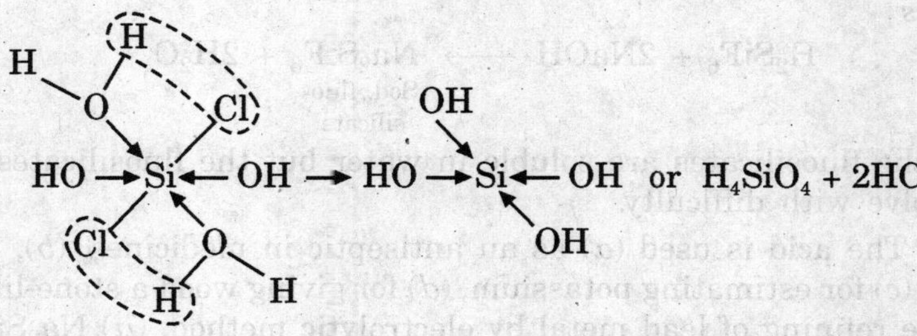

$SiCl_2(OH)_2.2H_2O$ Orthosilicic acid

On combining the above equations, we get the equation (A) given above.

Since carbon atom has already attained its maximum covalency of 4 in CCl_4, it cannot coordinate with water molecules to form the compound, $CCl_4.2H_2O$. Hence CCl_4 cannot be hydrolysed and in extremely stable.

Uses. $SiCl_4$ is used in warfare in producing smoke screen. It is more effective when used along with NH_3, as in that case white clouds of NH_4Cl are also formed.

4. Hydrofluosilicic Acid, H_2SiF_6.

Preparation (*i*) The acid is made in solution by dissolving SiO_2 in aqueous HF.

$$SiO_2 + 6HF \longrightarrow H_2SiF_6 + 2H_2O$$

(*ii*) An aqueous solution of the acid is obtained by passing SiF_4 in water.

$$3SiF_4 + 4H_2O \longrightarrow \underset{\text{(Soluble)}}{2H_2SiF_6} + \underset{\text{(White gelatinous ppt)}}{H_4SiO_4}$$

In actual practice SiF_4 is prepared by heating a mixture of powdered fluorspar (CaF_2), sand (SiO_2) and conc. H_2SO_4 in a stout flask on a sand bath. SiF_4 so produced is then passed into H_2O contained in a cylinder by means of a delivery tube, the end of the dry delivery tube dipping under mercury placed at the bottom of the cylinder. This arrangement prevents the chocking of the delivery tube by gelatinous ppt. of H_2SiO_4 formed in this reaction. After passing the gas of SiF_4 for a sufficiently long period into water, H_4SiO_4 is filtered off and the filtrate containing H_2SiF_6 is evaporated to get a concentrated solution. Too much concentration by evaporation is not required otherwise the acid will decompose.

(*iii*) H_2SiF_6 is also obtained as a by-product in the manufacture of calcium superphosphate from the mineral *phosphorite* which contains SiO_2 and fluorides as impurities.

Properties (*i*) Physical state. The concentrated solution of the acid ii a colourless fuming liquid.

(ii) Decomposition. It decomposes on evaporation into HF and SiF_4.

$$H_2SiF_6 \longrightarrow 2HF + SiF_4$$

(iii) Acidic character. It is a strong acid and it is as highly ionised in solution as H_2SO_4. With alkalies it forms salts which are called *fluosilicates* or *silicofluorides*

$$H_2SiF_6 + 2NaOH \longrightarrow \underset{\text{Sod. fluo-silicate}}{Na_2SiF_6} + 2H_2O$$

Most of the fluosilicates are soluble in water but the fluosilicates of Na, K, Ba etc. dissolve with difficulty.

Uses. (*i*) The acid is used (*a*) as an antiseptic in medicines, (*b*), for testing barium salts (*c*) for estimating potassium, (*d*) for giving wood a stone-like surface, and (*e*) in the refining of lead metal by electrolytic method. (*ii*) Na_2SiF_4 is used as an insecticide for chickens. (*iii*) $MgSiF_6$ is used for hardening and polishing the surface of concrete floors.

Silicones

What are silicones ?

Silicones are synthetic *polymerised organo-silicon polymers* containing —Si—O—Si—O—Si— linkages along with —C—C—C—C linkages present as the side chains.

Types of silicones on the basis of their structure.

The silicones may be of the following types : (1) *Linear* (*straight chain*) *silicones*, (2) *Cyclic silicones*, (3) *Cross-linked* (*three dimensional*) *silicones*.

Structures of the three types are given below.

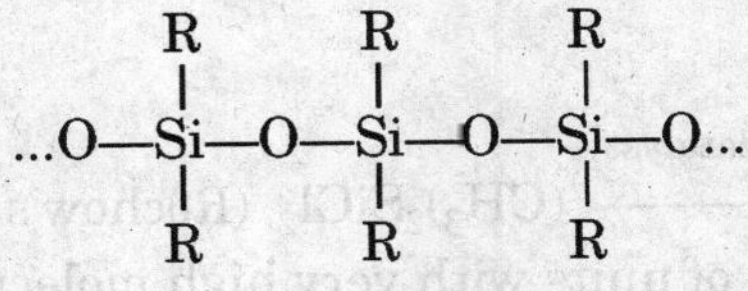

Linear silicone

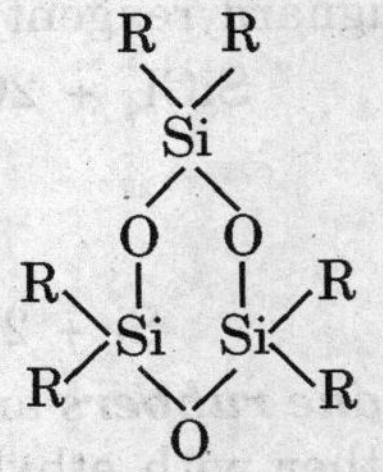

Cyclic silicone (Trimer)

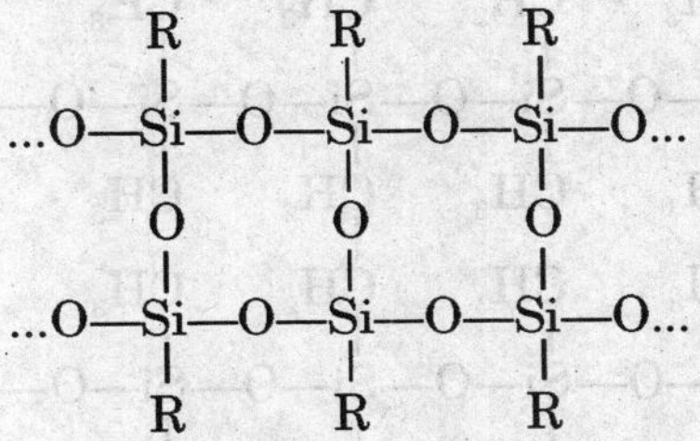

Cross-linked silicone

Here R stands for CH_3, C_2H_5 or C_6H_5 groups.

1. Linear Silicones.

These silicones are obtained by the hydrolysis and subsequent condensation of dialkyl or diaryl substituted silicon chloride, R_2SiCl_2 (B = alkyl or aryl group). For example when $(CH_3)_2$ $SiCl_2$ is hydrolysed, an *unstable silicol*, $(CH_3)_2Si(OH)_2$ is obtained.

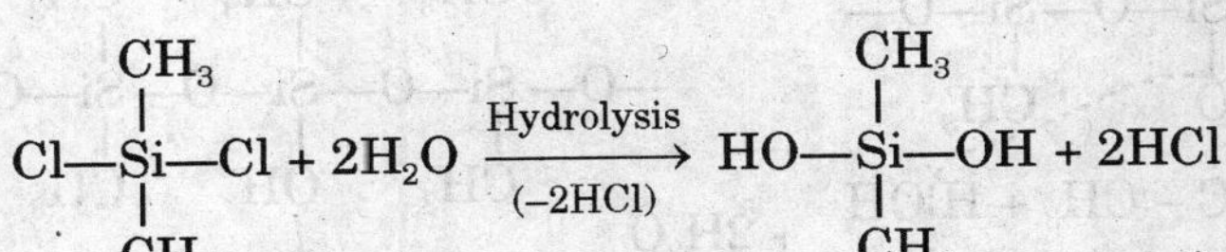

Unstable silicol

Now a number of molecules of this silicol undergo condensation by intermolecular elimination of some H_2O molecules to form linear silicones.

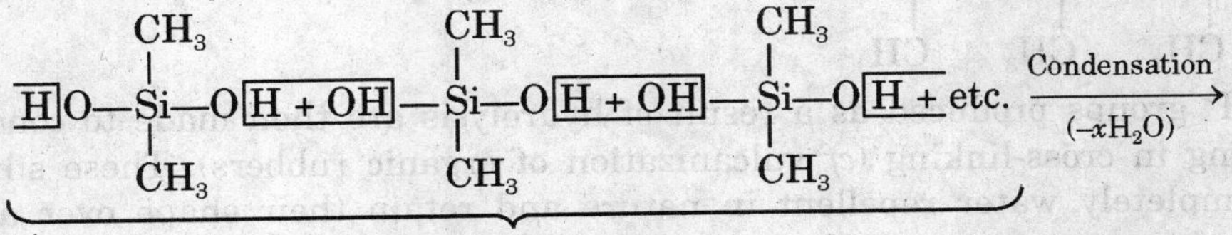

Many molecules of silicol

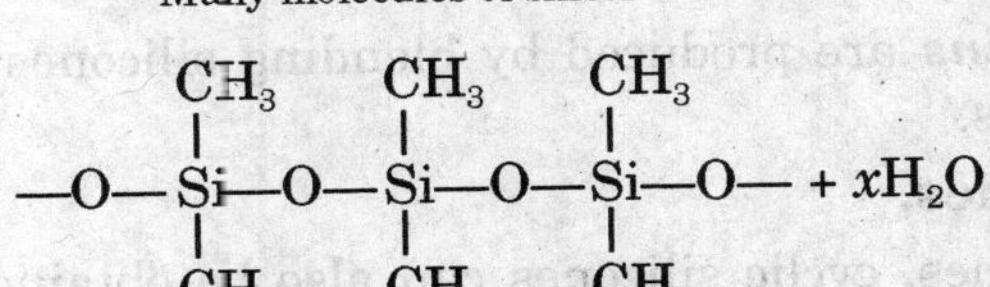

Linear silicone

$(CH_3)_2SiCl_2$ required in the preparation of linear silicones is obtained by the action of Grignard reagent on $SiCl_4$ or by Rochow's method (1945)

$$SiCl_4 + \underset{\text{Grignard reagent}}{2CH_3MgCl} \longrightarrow (CH_3)_2SiCl_2 + 2MgCl_2$$

$$Si + 2CH_3Cl \xrightarrow{\text{Cu catalyst}} (CH_3)_2SiCl_2 \text{ (Rochow's method)}$$

(a) Silicone rubbers are composed of units with very high molecular weights bridged together with ethylene or similar groups as shown below :

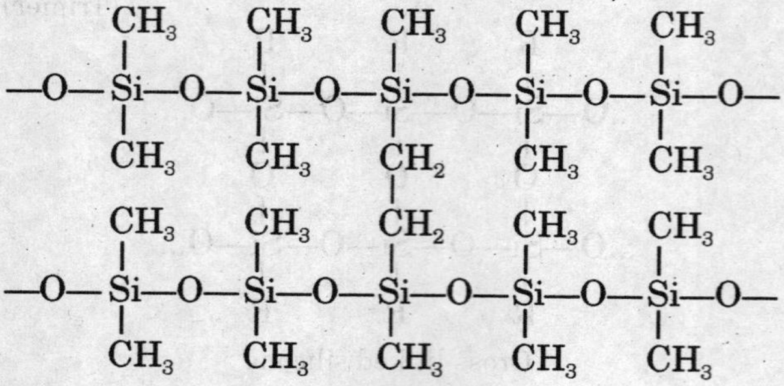

Silicone rubber

(b) Room-temperature-vulcanizing (RTV) silicone rubbers are also available. These contain groups like CH_3COO^- which can easily be hydrolysed and cross link in the presence of atmospheric moisture.

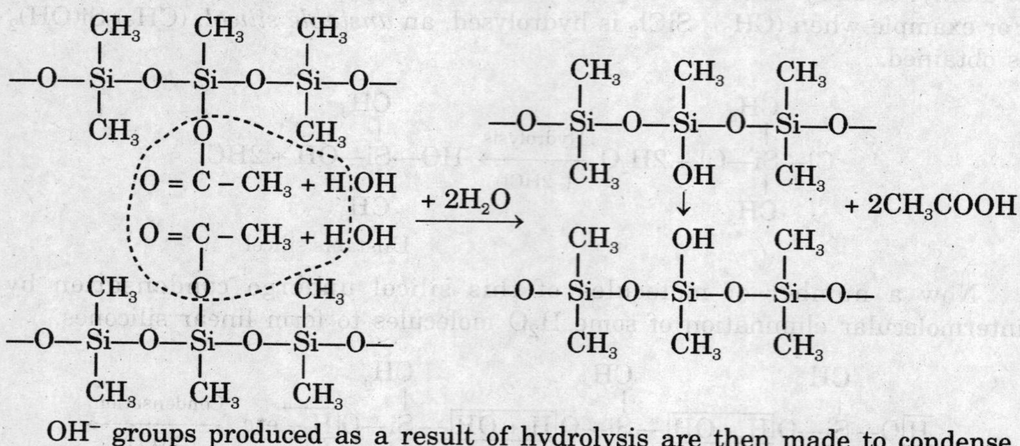

OH^- groups produced as a result of hydrolysis are then made to condense, resulting in cross-linking (*cf* vulcanization of organic rubbers). These silicones are completely water repellent in nature and retain their shape over a wide range of temperatures.

(c) Silicone resins are produced by blending silicones with organic resins such as acrylic esters.

2. Cyclic Silicones.

Like linear silicones, cyclic silicones can also be obtained by the hydrolysis and subsequent condensation of R_2SiCl_2. For example :

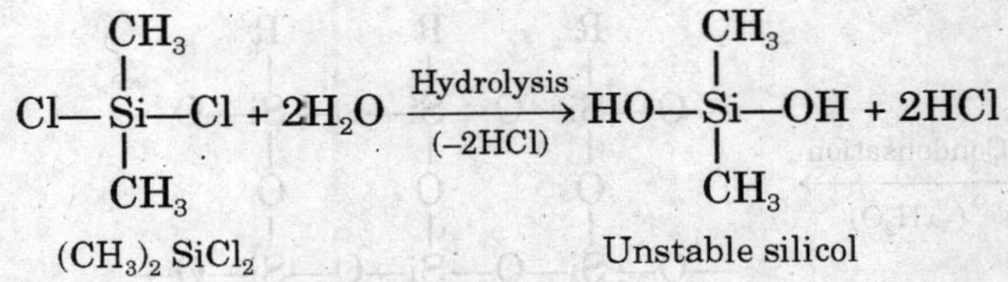

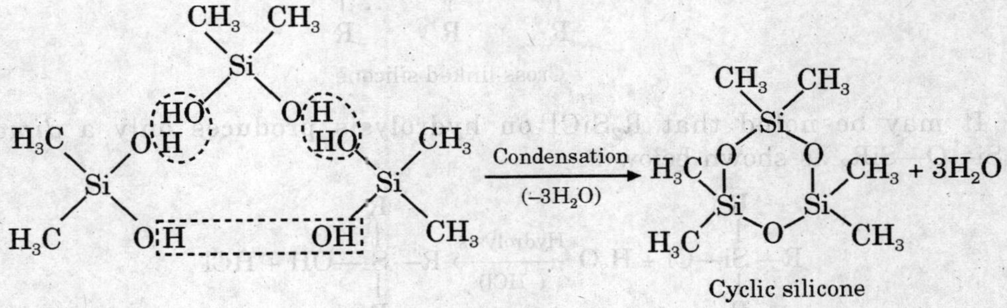

3. Cross-linked Silicones.

Such silicones are obtained by the hydrolysis and subsequent condensation of mono alkyl or aryl substituted silicon chloride, $RSiCl_3$. Its formation takes place through the following steps :

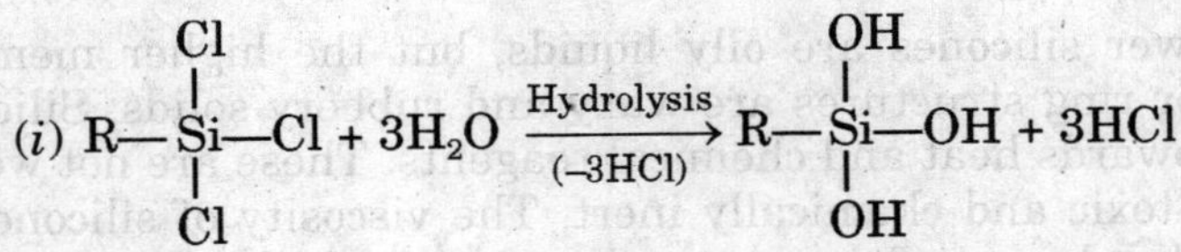

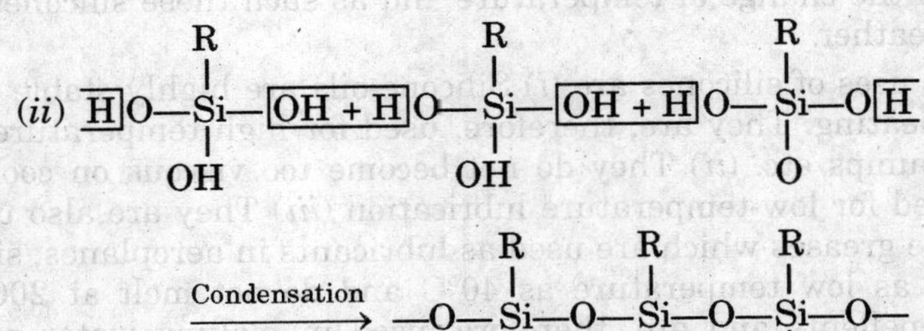

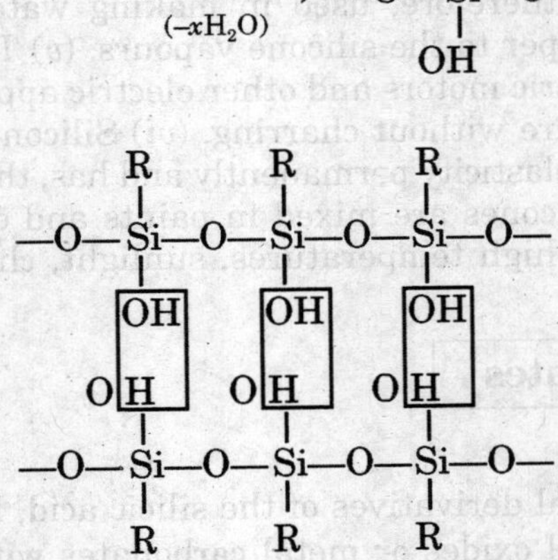

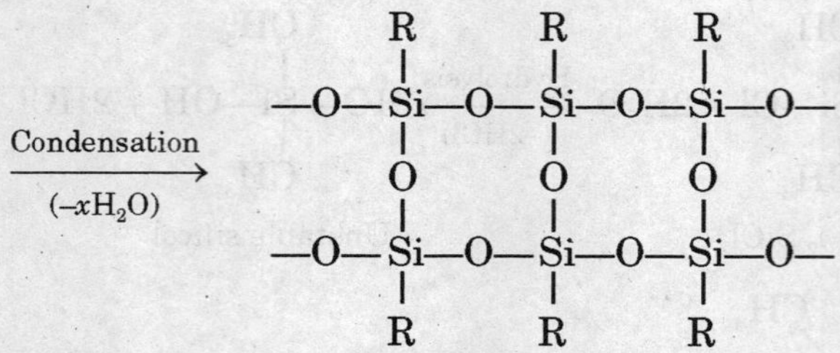

Cross-linked silicone

It may be noted that R_2SiCl on hydrolysis produces only a *dimer*, $R_3Si—O—SiR_3$ as shown below :

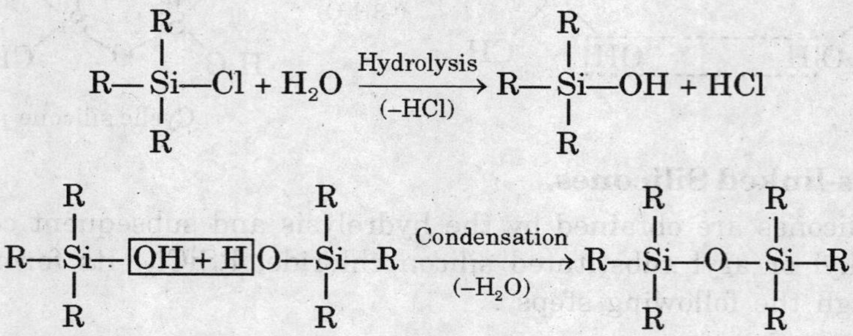

Properties. The lower silicones are oily liquids, but the higher members containing long chains or ring structures are waxy and rubbery solids. Silicones are remarkably stable towards heat and chemical reagents. These are not wetted by water. They are non-toxic and chemically inert. The viscosity of silicone oils remains constant with the change of temperature and as such these silicones do not thicken in cold weather.

Uses. Some of the uses of silicones are (*i*) Silicone-oils are highly stable and non-volatile even on heating. They are, therefore, used for high temperature oil baths, high vacuum pumps etc. (*ii*) They do not become too viscous on cooling and are, therefore, used for low-temperature lubrication (*iii*) They are also used in making vaseline like greases which are used as lubricants in aeroplanes, since they do not freeze at as low temperature as 40°C and do not melt at 200°C. (*iv*) They are water-repellents and are, therefore, used in making water proof cloth and paper by exposing cloth and paper to the silicone vapours. (*v*) They are also used as insulating materials for electric motors and other electric appliances, since they can withstand high temperature without charring. (*vi*) Silicon-rubber, after vulcanization, retains its shape and elasticity permanently and has, therefore, been used in a number of ways. (*vii*) Silicones are mixed in paints and enamels to make them resistant to the effects of high temperatures, sunlight, chemicals and damp.

Silicates

What are Silicates ?

Silicates may be regarded as the metal derivatives of the silicic acid, H_4SiO_4. Silicates can be prepared by fusing metal oxides or metal carbonates with sand (SiO_2). For example :

$$Na_2CO_3 \xrightarrow[\text{sand } (SiO_2)]{\text{Fuse with}} Na_2SiO_4, (Na_2SiO_3)_n \text{ etc.}$$

Only some silicates of alkali metals are soluble in water while others are insoluble. The insolubility is due to the great strength of Si—O bond. When silicates are treated with HF, Si—O bond is broken.

Classification and Structure of Silicates.

X-ray diffraction studies of various types of silicates have shown that all the silicates have SiO_4^{4-} anions which are formed by sp^3 hybridisation of Si atom. Si atom in its excited state contains 4 unpaired electrons and hence forms 4 covalent bonds with four negatively-charged O-atoms, resulting in the formation of SiO_4^{4-} anion. Each O-atom becomes negatively charged by picking up an electron from some metal.

Since Si atom undergoes sp^3 hybridisation, SiO_4^{4-} anion has tetrahedral shape as shown in Fig. 29.1.

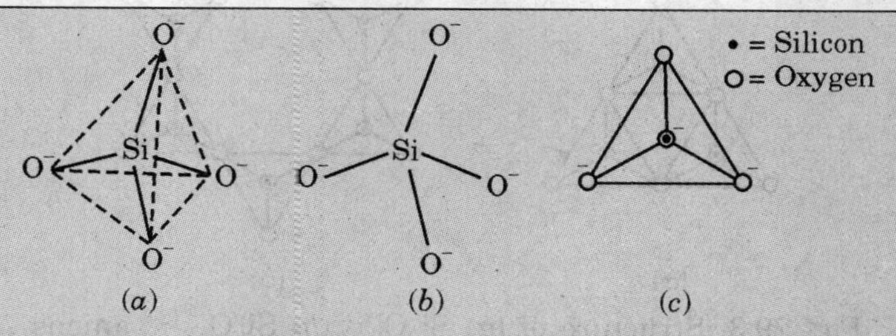

Fig. 29.1. Various ways of showing the tetrahedral shape of SiO_4^{4-} anion.

It is evident from the above discussion that all silicates contain the tetrahedral SiO_4^{4-} anions. These anions differ from one another in the manner in which these anions are linked together.

Silicates are classified according to the nature of linking between the tetrahedral SiO_4^{4-} anions. This basis of classification gives the following types of silicates.

1. Orthosilicates. These silicates contain discrete SiO_4^{4-} tetrahedral anions. The O-atoms of each SiO_4^{4-} anion is also coordinated to the metal ion to impart electrical neutrality to the structure. Examples of such silicates are *phenacite*, Be_2SiO_4; *willemite*, Zn_2SiO_4; *olivine*, Mg_2SiO_4; *zircon*, $ZrSiO_4$; *garnets*, M_3^{2+} M_2^{3+} $(SiO_4)_3$ (where M^{2+} = Mg^{2+}, Ca^{2+}, Fe^{2+} etc. and M^{3+} = Al^{3+}, Cr^{3+}, Fe^{3+} etc.). In garnets SiO_4 tetrahedra are arranged about M^{2+} and M^{3+} so they become 8-and 6-coordinated respectively. Garnets are very hard and find extensive uses as industrial abrasives.

2. Pyrosilicates. These silicates contain the discrete $Si_2O_7^{6-}$ anion which is formed by joining two SiO_4 tetrahedral units through one oxygen atom (See Fig. 29.2).

Examples of silicates containing $Si_2O_7^{6-}$ anion are *thorteveitite*, $Sc_2(Si_2O_7)$ and *hemimorphite*, $Zn_3(Si_2O_7).Zn(OH)_2.H_2O$.

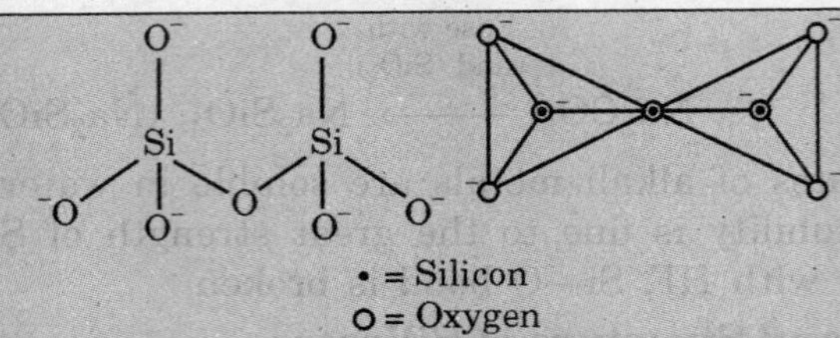

Fig. 29.2. Structure of $Si_2O_7^{-6}$ ion found in pyrosilicates.

3. Cyclic or ring silicates. Such silicates contain the cyclic or ring anions like $Si_3O_9^{6-}$ or $Si_6O_{18}^{12-}$. The structure of these anions is given in Fig. 29.3. These anions are obtained when each SiO_4 tetrahedral unit shares an O-atom with each of the two neighbouring SiO_4 units.

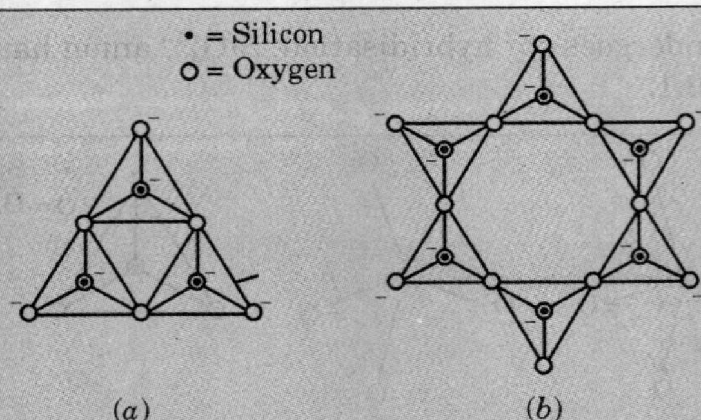

Fig. 29.3. Structure of (*a*) $Si_3O_9^{6-}$ (*b*) $Si_6O_{18}^{12-}$ anions found in cyclic silicates.

Examples of cyclic silicates containing $Si_3O_9^{6-}$ anion are *benitoite*, $BaTiSi_3O_9$; *wollastonite*, $Ca_3Si_3O_9$; *catapleite*, $Na_2ZrSi_3O_9.2H_2O$ while the example containing $Si_6O_{18}^{12-}$ anion is beryl, $Be_3Al_2Si_6O_{18}$.

4. Chain silicates. Such silicates contain the anions which are formed by the sharing of two oxygen atom by each tetrahedron The anions may be one of the types (*a*) $(SiO_3)_n^{2n-}$ (*b*) $(Si_4O_{11})_n^{6n-}$ whose structures are shown in Fig. 29.4.

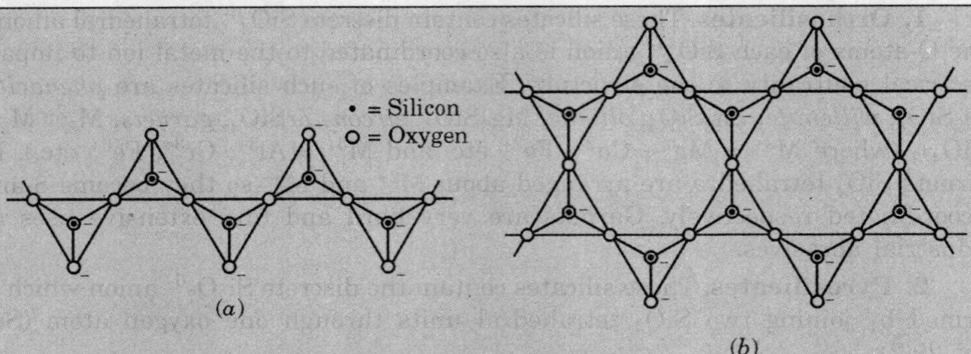

Fig. 29.4. Structure of (*a*) $(SiO_3)_n^{2n-}$ (*b*) $(Si_4O_{11})_n^{6n-}$ anions found in chain silicates.

The chains in the silicates containing $(SiO_3)_n^{2n-}$ anions lie parallel to each other and the cations lie between the chains and bind them together. Such silicates are presented by the *pyroxene* minerals and several synthetic silicates. Examples are (*a*) *synthetic silicates e.g.* Li_2SiO_3, Na_2SiO_3 (*b*) *pyroxene minerals e.g.* spodimene, $LiAl(SiO_3)_2$; jadeite, $NaAl(SiO_3)_2$; enstatite, $MgSiO_3$; diopside, $CaMg(SiO_3)_2$.

The silicates containing $(Si_4O_{11})^{6n-}$ anions have double chains in which the simple cations are held together by shared oxygens. Such silicates are represented by the *amphibole minerals* which include most asbestos minerals. Structurally amphiboles are similar to pyroxenes though they contain some OH groups which are attached to the cations. *Tremolite*, $Ca_2Mg_5(Si_4O_{11})_2(OH)_2$ is an example of amphibole.

5. Two-dimensional sheet silicates. When three 0.atoms of each SiO_4 tetrahedron are shared with adjacent SiO_4 tetrahedra, an infinite two-dimensional sheet structure of the composition $(Si_2O_5)_n^{2n-}$ results. (See Fig. 29.5). The metal ions present hold the layers together by the weak electrostatic forces. As a result, the minerals containing $(Si_2O_5)_n^{2n-}$ are soft and cleave easily. Such sheet-like anions are found in micas and different types of clay minerals. *Talc*, $Mg_2(Si_2O_5)_2.Mg(OH)_2$ and *kaolin*, $Al_2(OH)_4(Si_2O_5)$ are the common examples of clay minerals.

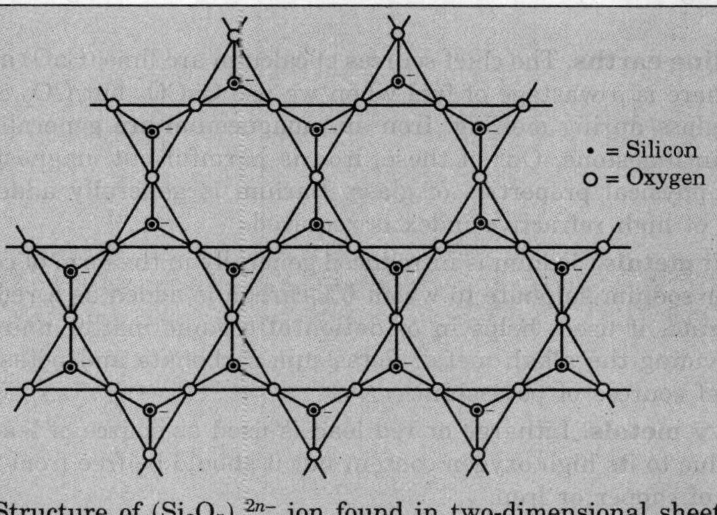

Fig. 29.5 Structure of $(Si_2O_5)_n^{2n-}$ ion found in two-dimensional sheet silicates.

6. Three dimensional silicates. When all the four oxygens of a SiO_4 tetrahedron are shared with adjacent tetrahedra and the process is repeated, an infinite three-dimensional structure results. Since all the oxygens are the bridge atoms, the silicate is neutral. In case Si is not replaced by other metal atoms, the silicate is neutral and will have the neutral formula SiO_2. Such structure is found in *quartz, tridymite* and *crystobalite*. This explains the structure of silica, SiO_2.

However, if some Si^{4+} ions are replaced by Al^{3+} ions in the tetrahedral positions in the SiO_2 structure, in order to maintain electrical neutrality, some other monovalent cations (*e.g.* K^+, Na^+ etc.) must be introduced. Such a replacement of Si^{4+} cations by Al^{3+} and monovalent cations gives the minerals like *feldspars, zeolites* and *ultra marines*.

Glass Industry

Any substance which has solidified from the liquid state without crystallization is known as glass. Glass is therefore, classified as a 'super cooled liquid'. Commonly the word glass means the transparent substance obtained by the fusion of certain metal oxides with white sand. Glass does not have a sharp melting point. When such a substance is heated, it begins to soften at a temperature far below the one at which it flows like a liquid.

Raw Materials Used in the Manufacture of Glass.

Following are the main raw materials used in the manufacture of glass :

1. Silica. It is added in the form of sand which must be of uniform size (too coarse variety prevents reaction while too fine one makes it violent). The chief impurities with which natural sand is frequently associated are alumina, iron and organic matter. Small quantities of alumina are not harmful but iron acts as a colouring matter, and, therefore, must be removed along with the organic matter.

2. Alkaline earths. The chief sources of calcium are lime (CaO) and limestone ($CaCO_3$). There is a wastage of fuel when we use $CaCO_3$ but CO_2 evolved helps in stirring glass during melting. Iron and magnesium are generally present as impurities in limestone. Out of these, iron is harmful but magnesium helps to control the physical properties of glass. Barium is generally added as $BaCO_3$ when glass of high refractive index is required.

3. Alkali metals. Sodium is introduced generally in the form of coarse sodium carbonate or sodium sulphate to which 5% carbon is added as a reducing agent. Sodium nitrate, if used, helps in oxidation of organic matter and ferrous iron besides providing the alkali metal. Potassium carbonate and potassium nitrate are the chief sources of potassium.

4. Heavy metals. Litharge or red lead is used as source of lead. The latter preferable due to its high oxygen content but it should be free from metallic lead and oxides of copper or iron.

In the modern heat resisting glass, calcium or lead is replaced by zinc which is added in the form of zinc oxide along with boric acid or borax which decreases the coefficient of expansion.

Phosphates when required are added as $Ca_3(PO_4)_2$ while arsenic and antimony are added as oxides. Addition of these helps to produce glass which is opalescent.

5. Colouring materials. Certain metallic oxides if added to the molten mass produce different shades of colours, and are used in the manufacture of coloured glasses *e.g.*, the following colours are produced by substances noted against each (*i*) Red glass contains Cu_2O and selenium metal, (*ii*) *Green glass* contains CuO and Cr_2O_3, (*iii*) *Yellow glass* contains CdS, carbon, (*iv*) *Blue glass* contains CuO, CoO. (*v*) *Violet glass* contains MnO_2, (*vi*) *Milky glass* contains SnO, $Ca_3(PO_4)_2$, (*vii*) *Black glass* contains oxides of cobalt, manganese, nickel etc.

6. Borax. It is used as a flux. Borax also lowers the coefficient of expansion of the glass and increases its durability.

7. Alumina. Alumina makes glass resistant to sudden changes of temperature. Alumina is often used in the form of feldspar. It is a naturally occurring mineral with a formula $Na_2O\ Al_2O_3.6SiO_2$ (soda feldspar) or $K_2O.Al_2O_3.6SiO_2$ (potash feldspar).

8. Cutlets. These are pieces of waste glass. These serve two purposes—these facilitate fusion and secure utilization of waste.

Manufacture of Glass.

The manufacture of glass consists of the following steps :

1. Fusion of raw materials. The raw materials to be used depend upon the variety of glass to be obtained. For ordinary soda-lime glass the raw materials are sand, washing soda and limestone. These are mixed in the proper proportion and ground together. The resulting mass known as "*batch*" is mixed with broken glass called "*cullet*" and introduced either into the pots of a pot furnace or into the tank of a *tank furnace*.

(*a*) *Pot furnace.* In this furnace the pots are placed in a circle around a central opening in its bottom. Hot gases and the flames come out of this opening. The roof over the pots is of refractory material which reflects the heat on to the pots. Pots are of two types *viz.* open and closed. Clay rings are placed in pots which facilitate withdrawal of the molten material. (See Fig. 29.6.)

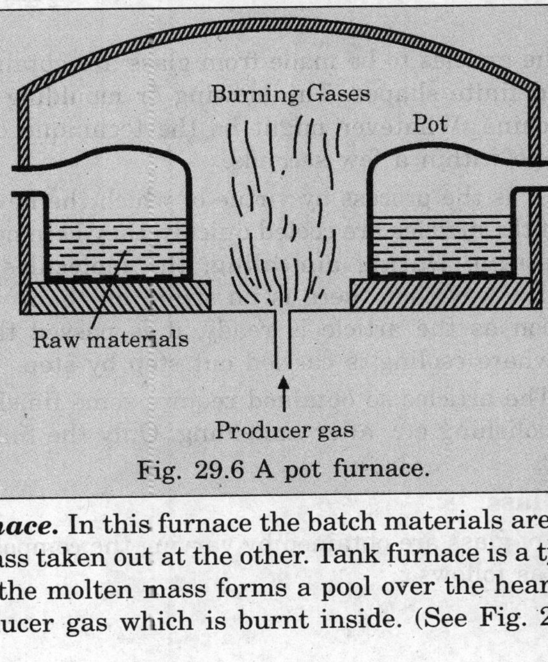

Fig. 29.6 A pot furnace.

(*b*) *Tank furnace.* In this furnace the batch materials are charged at one end and the molten glass taken out at the other. Tank furnace is a type of reverberatory furnace in which the molten mass forms a pool over the hearth. Heating is done by means of producer gas which is burnt inside. (See Fig. 29.7).

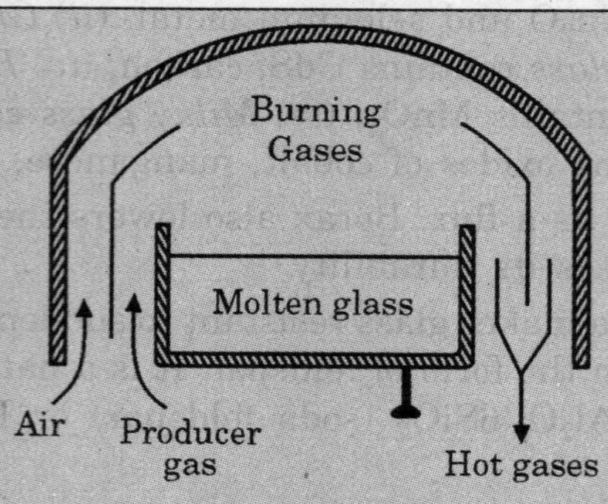

Fig. 29.7. A tank furnace

During melting a good deal of frothing is caused owing to the evolution of gases like CO_2, SO_2, O_2 etc. When froth subsides, the temperature is raised and the molten glass is allowed to stand for some time. In this way a homogeneous mass free from gas bubbles is obtained.

In tank furnace the products obtained may not be of a superior quality because the molten glass cannot be stirred and the glass is not of a uniform composition.

The chemical reactions taking place during the process are :

$$Na_2CO_3 + n_1SiO_2 \longrightarrow Na_2O.n_1SiO_2 + CO_2$$

$$Na_2SO_4 + C \longrightarrow Na_2O + CO + SO_2$$

$$Na_2O + n_2SiO_2 \longrightarrow Na_2O.n_2SiO_2$$

$$CaCO_3 + n_3SiO_2 \longrightarrow Na_2O.n_3SiO_2 + CO_2$$

Commercial glass is the mixture of all the silicates *viz.* $Na_2O.n_1SiO_2$. $Na_2O.n_2SiO_2$ and $Na_2O.n_3SiO_2$.

2. Shaping. The articles to be made from glass are obtained by blowing the molten glass into definite shapes. The blowing or moulding is done by skilled workers or by machine. Whatever might be the technique of work, the whole process is to be done within a few seconds.

3. Annealing. It is the process by virtue of which the newly shaped articles are cooled slowly. If the articles are cooled quickly, they become brittle on account of high internal strain. During annealing, the molecules of glass arrange themselves in such a way that there is no strain on them when the article is cooled. Thus as soon as the article is ready, it is passed through a series of cooling chambers where cooling is carried out step by step.

4. Finishing. The articles so obtained require some finishing touch such as cutting, grinding, polishing etc. after annealing. Only the finished products are sent to the market.

Varieties of Glass.

Large varieties of glass are obtained by varying the composition of the batch. Some of these are as follows :

1. Soft glass. It is a mixture of sodium and calcium silicates (raw materials: SiO_2, Na_2CO_3 + CaO_3) and is the common variety employed for making ordinary glassware and glass plates.

2. Hard glass. It is a mixture of potassium and calcium silicates (raw materials : SiO_2, K_2CO_3 + $CaCO_3$) and is used for making hard glass apparatus.

3. Jena glass. It is a mixture of zinc and barium borosilicates and is very resistant to heat, chemicals and shock. It has a high alumina content.

4. Pyrex glass. It possesses properties similar to Jena glass but it is a mixture of sodium aluminium borosilicates. It is rich in silica (about 80%) but poor in alumina.

5. Optical glass. It is generally made by fusing red lead, potassium carbonate and sand. Special varieties contain boron trioxide in place of silica and barium and zinc oxide in place of red lead. *Crookes glass* contains cerium oxide which cuts off ultraviolet light harmful to eyes.

6. Safety glass. It is obtained by placing a layer of butyral plastic with a suitable adhesive between two layers of glass and cementing them by heat and pressure. It does not shatter and is used in making automobile wind shields. Fine layer laminated glass prepared as above is used in building construction.

7. Glass wool. It is glass fibre obtained by forcing molten glass through tiny holes in presence of a jet of high pressure system.

Glass being a thermal insulator is wrapped round various articles in the form of glass wool. Glass fibres are used for making fire proof clothing. These would incidentally be stain resistant as well s bad conductors of electricity. Hollow glass bricks are used in constructing walls where light would come in but cold could be kept out. Glass fibres have also been used as the base of several useful plastic moulded articles.

8. Coloured glass. To produce glass of different colours, various mineral substances are mixed with it during the process of manufacture. Examples have already been given on page 969.

Questions with Answers

Q.1 Give different groups of silicones which are commercially important.

Ans : Silicones which are commercially important can be classified into four groups namely *silicone fluids* or *oils, silicone greases, silicone rubbers and silicon resins.*

1. Silicone Fluids or Oils.

These silicone fluids are mixtures of straight chain polymers differing in viscosities. Most of these are made from dimethyldichlorosilane, hence have methly groups attached to the silicon atoms of the polymers chain. Other organic groups may be introduced to alter the properties of the fluids.

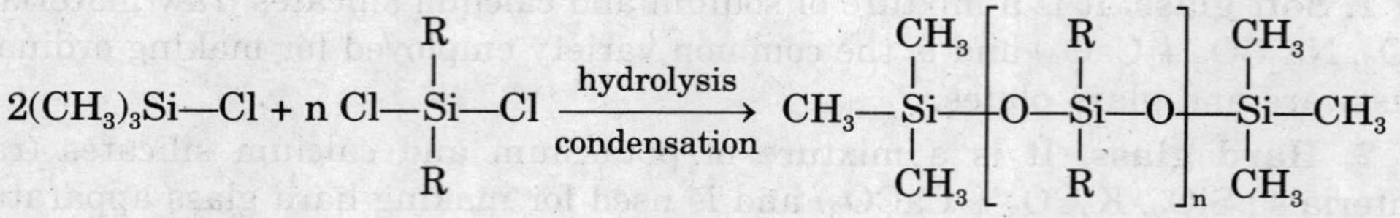

silicon oil or fluid

Properties and uses of silicone fluids. (*i*) Silicon fluids are *good lubricants under vacuum* and at high temperature. They are used in vacuum pumps, jet turbines and guided missiles as lubricants.

(*ii*) They are used in heating baths for industrial *heating and as damping fluid* in instruments and electrical switchgears.

(*iii*) They are used as *antifoaming agents* in several industries such as dyes, textiles, brewing, paint, glue, paper, food processing and sewage disposal.

(*iv*) The *water repellent properties* of silicones have become a boon in the manufacture of water proofing textiles, leather,paper, glass and other matericals.

(*v*) They are also used for coating glass and ceramics insulators *to prevent electrical leakage during rain*.

(*vi*) Silicones added to *wax polishes give a high gloss finish*.

(*vii*) Silicones resist adhesion to several surfaces and are used *as release agents*. They are coated in moulds for easy removal of plastic articles and rubber tyers form moulds.

(*viii*) *Greases are made from silicone* fluids by mixing with meatllic soap, graphite or silica.

2. Silicone Greases :

Silicon grease is the modified form of silicon fluid. It can be prepared by adding some fillers like silica, carbon black, etc. to silicone fluids.

Properties and Uses : (*i*) Silicon greases are stable to a wide range of temperature from –100°C to 450°C. So they are used in industries as very effective lubricants.

(*ii*) When treated with silica they are used as electrical greases in car ignition systems.

(*iii*) Greases with lithium soap filler are used for ball-bearings, etc.

(*iv*) Greases with carbon black filler are used in high temperature conveyers and oven doors, etc.

3. Silicone Rubbers :

Silicone elastomers are made form high molecular weight linear polymers containing about 9000 monomers by cross-linking. General purpose silicone elastomers are made from polydimethylsioxanes. Other special ealstomers contain a small proportion of groups other than methyl. The linear polymer silicon oil is mixed with filler such as silica, nodifying additives and organic peroxide catalyst and cured by heating.

The high moleculear weight polymers are made by dissoling dimethydichlorosilane in ether and mixing with water. Cyclic polymers are fomed which are recovered by distillation. The tetramer *b.p.* 175°C is polymerised at 150-200°C in presence of traces of sodium hydroxide and a very small amount of monofunctional silane to yield highly viscous dimethyl silicone gum.

Properties : *The silicone elastomers combine the properties of rubber and silicones.* They retain flexibility over –90°C to 250°C temperature range. They can withstand exposure to ozone, ultraviolent light and weathering. They have excellent dielectric properties and are not affected by oils. They get more easily torn than rubber and dissolve in aromatic solvents.

Silicone rubbers are masticated and cured with 2, 4-dichloro-benzoyl peroxide at higher temperature. Prolonged curing makes the rubber less elastic but hard and more resistant to oil and chemicals.

Vinyl silicone rubbuers are made by incorporating a small percentage of vinylsubstituted silanes. These are more elastic at higher temperature. Curing is done with ditertiarybutyl and dicumyl peroxides.

Curing of silicone rubbers can be done with high-energy radiation such as gammasays.

Uses : (*i*) Silicone rubbers are extensively used *in aircraft industry* in diaphargms, gaskets, insulation, hot air ducting, tyers etc.

(*ii*) They are used for insulation for electrical wiring in naval ships.

(*iii*) They are used *for coating metal roller used in several industries,* for coating, pipes and ducting carring chemicals.

(*iv*) Silicone rubber is *used in moulds* for making chocolate or jelly, sterilizable baby bottle nipples, insulation for domestic machines etc.

(*v*) It has become important in *medical appliactions e.g.* transfusion tubing, heart valves, plastic surgey, nerves repairs etc.

4. Silicone Resins :

These are compact cross-linked silicones which are hard. They are made by hydrolysis of a mixture of dimethyldichlorosilan and silicon tetrachloride or methylsilicon trichloride. The resins are polymersied only upto gel points, then they are dissolved in a solvent and applied to a surface or shared and polymerished further to a hard infusible solid. Phenyl methyl silicon resins can withsatnd 300°C for long time.

Uses : (*i*) They can be reinforced with glass fibre to provide structural material resistant it heat.

(*ii*) Silicone resin varnishes are used for coating armatures and transformer windings.

(*iii*) Silicone resin based paints are used where resistant to weathering, light, water and oxidation are required *e.g.* factory chimneys, ventillators, exhaust pipes, open plants in chemicals industries.

(*iv*) Silicone resins give a non-stick surface to frying pans, baking vessels etc.

(v) Silicone laminate with gass fiber or asbestos is fireproof, damp proof with good electrical insulation properties. It is used for making fire barries in aircrafts etc.

(vi) Silicone resin emulsions are sprayed on bricks, concrete mortan or stone to give water repellent finish.

Q.2 Starting from $SiCl_4$, prepare in not more than four steps, a linear silicone containing methyl groups only. ***(I.I.T. 2001)***

Ans : Various steps involved in the preparation of linear silicone by $SiCl_4$ consists of the following steps :

(*i*) First of all $SiCl_4$ is treated with CH_3MgCl (Grignard reagent) so that it is converted into dimethyl dichloro silane, $(CH_3)_2SiCl_2$.

$$SiCl_4 + 2CH_3MgCl \longrightarrow \underset{\text{Dimethyl dichloro silane}}{(CH_3)_2SiCl_2} + 2MgCl_2$$

(*ii*) $(CH_3)_2SiCl_2$ obtained as above is hydrolysed to get dimethyl silandiol, $(CH_3)_2\ Si(OH)_2$.

$$(CH_3)_2\ SiCl_2 + 2H.OH \xrightarrow{\text{Hydrolysis}} \underset{\text{Dimethyl silandiol}}{(CH_3)_2Si(OH)_2} + 2HCl$$

(*iii*) Many molecules of $(CH_3)_2Si(OH)_2$ obtained as above in step (*ii*) get polymerised to form linear silicone.

$$n\left[HO-\overset{\displaystyle CH_3}{\underset{\displaystyle CH_3}{Si}}-OH\right] \xrightarrow[(-nH_2O)]{\text{Polymerisation}} \underset{\text{Linear silicon}}{\left[-O-\overset{\displaystyle CH_3}{\underset{\displaystyle CH_3}{Si}}-O-\right]_n} + nH_2O$$

Q.3 Classify the following silicates : Zircon, hemimorphite, wallastonite, beryl, spodumene, asbestos, talc and silica.

Ans : Zircon ($ZrSiO_4$) – *orthosilicate*, hemimorphite [$Zn_3(Si_2O_7)$. $Zn(OH)_2$ or $Zn_4(Si_2O_7)\ (OH)_2$] – *pyrosilicate*, wallastonite ($Ca_3Si_3O_9$ or $3CaO.\ 3SiO_2$) – *cyclic (ring) silicate*, beryl ($Be_3\ Al_2\ Si_6\ O_{18}$ or $3BeO.\ Al_2O_3.\ 6SiO_2$) – *cyclic or ring silicate*, spodumene [$LiAl\ (SiO_3)$ or $\frac{1}{2}$ ($Li_2O.\ Al_2O_3.\ 4SiO_2$)] – *single chain silicate*, asbestos [$CaMg_3\ O(Si_4O_{11})$ or $CaO.3MgO.\ 4SiO_2$] – *double chain silicate*, talc [$2Mg\ (Si_2O_5)$. $Mg(OH)_2$ or $3MgO.\ 4SiO_2.\ H_2O$] – *two dimensional sheet silicate*, silica (SiO_2) – *three dimensional silicate*.

Q.4 Beryl and pyroxenes both share two O-atoms with other SiO_4^{4-} tetrahedral ions and have the same anion *viz* $[(SiO_3)_n]^{2n-}$. How do they differ ?

Ans : Beryl ($Be_3\ Al_2\ Si_6\ O_{18}$) contains $Si_6O_{18}^{12-}$ ion which is $[(SiO_3)_n]^{2n-}$ ion having $n = 6$. $Si_6O_{18}^{12-}$ ion is obtained when two O-atoms of each SiO_4^{4-} ion are shared with other SiO_4^{4-} ions in such a way that cyclic ring is fromed. Thus $Si_6O_{18}^{12-}$ ion is a cyclic ion.

Pyroxenes contain $[(SiO_3)_2]^{4-}$ ion which is a $[(SiO_3)_n]^{2n-}$ ion having $n = 2$. $[(SiO_3)_2]^{4-}$ ion is obtained when two O-atoms of each SiO_4^{4-} ion are shared with other SiO_4^{4-} ions in such away that single chain is formed. Thus $[(SiO_3)_2]^{4-}$ ion found in pyroxenes is a single chain ion.

Above discussion shows that beryl is a cyclic silicate whereas pyroxenes are chain silicates.

University Questions

1. How will you prepare the following : Silicic acid, Silicon tetrachloride and Hydrofluosilicic acid ? *(Utkal 86; Bharathidasan 83; Gorakhpur 85)*
2. Write a note on "Sodium silicate". *(Madras 86)*
3. What is silica gel ? *(Delhi 87)*
4. Describe the preparation, properties and uses of silicones. *(Allahabad 87, 89)*
5. How are silicates classified? Discuss in brief the structure of chain silicates. *(Himachal Pradesh 99)*
6. Write a short note on "Glass". *(R.E.E. 82)*
7. Write short notes on "Silicates". *(Delhi 99)*
8. How are silicates classified? Discuss in brief the structure of chain silicates. *(Himachal Pradesh 99)*
9. Describe the method of preparation and uses of silicone. *(Allahabad 99)*
   (*a*) What are silicones ? Give methods for the formation of different types of silicones. Give their applications in technology.
   (*b*) Discuss the physico-chemical changes which occur during the setting of cement. *(Himachal Pradesh 2000)*
10. Write a note on "Glass". *(Kanpur 2000)*
11. Write the composition of borosilicate. *(Banglore 2001)*
12. Describe different varieties of glass. *(Banglore 2002)*
13. Discuss the preparation, properties and uses of silicones. *(Meerut 2009)*

CHAPTER

30

General Characteristics of Group 15 (VA) Elements : Nitrogen Group Elements

Group VA of the periodic table consists of five elements *viz. Nitrogen* (N), *Phosphorous* (P), *Arsenic* (As), *Antimony* (Sb) and *Bismuth* (Bi).

Position of Nitrogen Group Elements in the Periodic Table

The inclusion of nitrogen group elements namely N, P, As, Sb and Bi in the same sub-group of the periodic table is justified by the following general characteristics of these elements.

I. Electronic Configurations.

The electronic configurations of these elements are N(Z = 7) → 2, 5 ($2s^2p^3$); (P = 15) → 2, 8, 5($3s^2p^3$); As (Z = 33) → 2,8, 18, 5 ($4s^2p^3$); Sb(Z = 51) → 2, 8,18,18, 5($5s^2p^3$); Bi(Z = 83) → 2, 8, 18, 32, 18, 5 ($6s^2p^3$). It is evident from these configurations that all of these elements have five electrons in the outermost orbit which are distributed as ns^2, np^3 electrons. The three P electrons are equally distributed in accordance with *Hunds rule* amongst the three orbitals np_x^1, np_y^1 and np_z^1 and since the orbitals are exactly half-filled, the elements are fairly stable and not so reactive, *e.g.* nitrogen behaves almost as an inert element. The shell previous to outermost shell contains in nitrogen the group s^2 (*saturated*), in phosphorus s^2p^6 (*saturated*), in arsenic $s^2p^6d^{10}$ (*saturated*) whereas in antimony and bismuth $s^2p^6d^{10}$ (*unsaturated*). This shows why nitrogen differs from phosphorus and these two from the remaining elements of this sub-group.

II. Similarities in Physical Properties and Gradation in Them.

Some of the physical properties of these elements are discussed below :

1. Occurrence. Nitrogen occurs in the free state in nature while all the remaining elements occur in the combined state. Nitrogen occurs as the diatomic gas (N_2) and constitutes approximately four-fifth part of the earth's atmosphere and in chemical combination it occurs mainly as a nitrate. Phosphorus is very reactive and hence does not occur in the native state. Rather it occurs as phosphates in its ores. As, Sb and Bi are less abundant and occur chiefly as sulphide ores such as arsenical pyrites (FeAsS), stibnite (Sb_2S_3) and bismuth glance (Bi_2S_3j)

2. Allotropy. All the elements except bismuth show allotropy. Nitrogen exists in two solid allotropic forms *viz.* α*-nitrogen* and β*-nitrogen* Phosphorus exists in the allotropic forms such as *white, red, scarlet,* α*-black,* β*-black* and *violet.* Arsenic exists in three allotropic forms namely grey, yellow and black whilst antimony exists as *yellow, black,* β and *explosive antimony.*

3. Physical State and Elemental Structure. Nitrogen is a gas, phosphorus is a soft, waxy and lustreless solid, As is a hard lustreless solid and Sb and Bi are hard solids and have characteristic metallic lustre. Nitrogen exists as diatomic molecule, N_2 while phosphorus, arsenic and antimony exist as tetratomic molecules P_4, As_4 and Sb_4. Here a very important question arises : *Why do phosphorus and arsenic atoms not form P_2 and As_2 species while nitrogen atoms form N_2 species?* The answer to this question is that it is not possible for P and As atoms to form $p\pi$–$p\pi$ bonds because these atoms cannot come closer to each other to form $p\pi$–$p\pi$ bonds due to the increased repulsions from the non-bonding electrons of penultimate shell. However, N-atoms do not encounter any repulsion, since they contain $1s^2$ electrons in their penultimate shell. Thus N-atoms can form N_2 species. Moreover all the elements form giant lattices in which the bonding changes form covalent to metallic, *e.g.* violet and red phosphorus have layered covalent structure with a coordination of three; As and Sb have more closely packed layers which have intermediate bonding while Bi has truly metallic bonding.

4. Atomic Volume and Atomic Radius. Both the atomic volume and atomic radius increase as we move down the group. However, As has an exceptionally low value of atomic volume.

5. Ionisation Potentials. With the increase of size from N to Bi, the values of first ionisation energies decrease on descending the group.

6. Metallic and Non-metallic Character. Due to the decrease in the values of ionisation energies from N to Bi, the metallic character of these elements increases in the same direction. For example, the first two elements *viz.* N and P are exclusively non-metals, the next two elements *viz.* As and Sb are metalloids while the last element namely Bi is definitely a metal.

7. Electronegativity and Heat of Atomisation. From N to Bi electronegativity and heat of atomisation decrease. This decrease confirms the gradual increase in the metallic character of the elements from N to Bi.

8. Melting and Boiling Points. The melting points (except for Sb and Bi) and the boiling points both increase as we move down the group. The melting point of Bi is usually low. Low melting point of Bi suggests that there is little possibility of the availability of the pair of electrons in *s*-orbital. The elements of this group are more volatile than their intermediate neighbours. This is because of the fact that these elements have 5 electrons in their valence shell.

9. Hardness and Density. There is a pronounced increase in hardness and density in going from P to Bi.

10. Oxidation States. (*a*) *Formation of M^{3+} and M^{5+} cations.* M^{3+} cations are formed only when all the three electrons of *np* orbital from ns^2p^3 configuration are lost and ns^2 electrons remain inert (*inert pair effect*). Since the inert pair effect increases down the group, the heavier elements namely Sb and Bi lose their three electrons from *np* orbital and form M^{3+} cations. For example $Sb_2(SO_4)_3$ contains Sb^{3+} ion and similarly Bi F_3 and $Bi(ClO_4)_3.5H_2O$ both contain Bi^{3+} ions. Here it should be noted that Bi^{3+} and Sb^{3+} ions in $Bi(ClO_4)_3.4H_2O$ and $Sb_2(SO_4)_3$

respectively are not stable in presence of water because of hydrolysis to form bismuthyl (BiO^+) and antimonyl (SbO^+) ions.

In order to get noble gas configuration it is not possible for N and P to lose all its five ns^2p^3 electrons to form M^{5+} cations, since the energy requirements for the formation of M^{5+} cations are prohibited. Thus in case of nitrogen and phosphorus M^{5+} cations do not exist.

(b) Formation of M^{3-} anions. We have seen that the valence shell configuration of these elements is ns^2p^3. Thus, in order to achieve noble gas configuration, these elements accept three electrons from strongly electropositive elements and form M^{3-} anions. As the size of the atoms increases on moving down the group, the attraction of the nucleus, for the newly added electrons decreases. Thus, on moving down the group, the tendency of the elements to give M^{3-} ions decreases Thus, N-atom which is of the smallest size has the strongest tendency to give M^{3-} ion, P forms P^{3-} ion less readily while other members of the group show little or no tendency to form M^{3-} ions.

(c) Covalency of 3 and 5. As is evident from the valence-shell configuration, each of these elements has two paired *s*-electrons and three unpaired *p*-electrons. The elements form *three covalent bonds* and thereby attain the noble gas electronic configuration (ns^2p^6) when they use all their *p*-electrons. Except nitrogen all other elements have *d* orbitals as well. As a result, one of the two *ns*-electrons may be promoted to the vacant *nd*-orbitals to have five unpaired electrons in the valence-shell. These five unpaired electrons can be used in forming *five covalent bonds.* Thus P, As, and Sb show 5-covalent state. This is not possible for N-atom, since it has no *d*-orbital. This explains why nitrogen forms only trihalide and not pentahalide.

Due to the presence of lone pair of electrons on N and P atoms in NH_3 and PH_3 molecules, these molecules can act as Lewis bases towards H^+ and BF_3 to form NH_4^+, PH_4^+, $H_3N \rightarrow BF_3$ and $H_3P \rightarrow BF_3$ species in which N and P atoms show a covalency of 4. P, As and Sb show a covalency of 6 in $[PCl_6]^-$, $[AsF_6]^-$ and $[SbF_6]^-$ complexions.

11. Tendency of the Elements for Catenation. Nitrogen and phosphorus have a tendency to form M—M links. For example N-atoms are linked together in hydrazine ($H_2N—NH_2$) and azide [(N—N—N) skeleton]. In case of phosphorus P—P links are limited only to two atoms (*e.g.* P_2H_4) while As has no tendency for catenation. The decrease in the tendency from N to As to form M—M links is due to the decrease in their M—M bond energies as shown : N—N = 83.7 kJ/mole, P—P = 79.09 kJ/mole, As—As = 63.19 kJ/mole.

12. Electrical and Thermal Conductivity. Both these properties increase as we move down the group because the delocalisation of electrons increases on moving from nitrogen (which is represented as: $N \equiv N$: with a coordination number of one) to bismuth (which has metallic crystal with a coordination number of six). Thus N and P are non-conductors, As is a. poor conductor, Sb is a good conductor and Bi is an excellent conductor.

III. Similarities in Chemical Properties and Gradation in Them.

Some chemical properties are discussed below :

1. Formation of Hydrides. All the elements of this group form the hydrides like MH_3 (M = N, P, As, Sb, Bi), M_2H_4 (M = N, P). It has been seen that all the elements form MH_3 type hydrides. Their methods of preparation, properties etc. are discussed below :

Preparation. MH_3 type hydrides are prepared by the action of water or dil. acids on the binary metallic compounds such as Mg_3N_2, Ca_3P_2, Zn_3As_2, Mg_3Sb_2 and Mg_3Bi_2.

Properties. ***(i) Physical state.*** All the hydrides are colourless and volatile gases.

(ii) Basic character. The basic character, *i.e.* the ability to act as electron pair donor, decreases on proceeding from NH_3 to BiH_3 Thus NH_3 readily coordinates with H^+ to give the stable cation, NH_4^+ whereas PH_4^+ ion is much less stable and MH_4^+ ion for As, Sb and Bi elements do not exist. The electron donor ability of the hydrides is increased by the replacement of the H-atom by alkyl groups. For example $P(CH_3)_3$ shows stronger basic character than PH_3.

NH_3 is the strongest electron pair donor, because N-atom has the smallest size and hence the electron density of the Lone pair of electrons present in one of the four sp^3 hybrids on N-atom is concentrated over a small region. As we move downwards in the group, due to the increase in the size of the central atom, the electron density of the lone pair gets diffused over a large region and hence the ability to donate electron pair (*i.e.* basic character) decreases.

(iii) Thermal stability. Thermal stability of the hydrides decreases gradually from NH_3 to BiH_3 as is evident from a decrease in the temperature of their decomposition, *i.e.* NH_3(1300°C) > PH_3 (440°C) > AsH_3 (280°C) > SbH_3 (room temperature) > BiH_3 (very unstable). The temperatures given in brackets are the decomposition temperatures of the hydrides. The least stable nature of BiH_3 confirms the fact that only trace quantities of this hydride have been identified. The decrease in thermal stability from NH_3 to BiH_3 means that M—H bond goes on weakening.

(iv) Covalent character. Since the electronegativity difference between nitrogen and hydrogen (in NH_3) is more than that between phosphorus and hydrogen (in PH_3), NH_3 molecule has more ionic character than PH_3 ($NH_3 > PH_3$). Similarly it can be shown that the ionic character of MH_3 molecules decreases on moving from NH_3 to BiH_3, *i.e.* the covalent character of these hydrides increases from NH_3 to BiH_3.

(v) Ruducing property. We have seen that the thermal stability of MH_3 molecules goes on decreasing as we move from NH_3 to BiH_3. This means that M—H bond goes on weakening or the reducing property of MH_3 hydrides goes on increasing. Thus NH_3 is a mild reducing agent while SbH_3 and BiH_3 are powerful reducing agents.

(vi) Shape of hydrides and HMH bond angles. The shape of MH_3 type hydrides is *trigonal pyramidal.* This shape may be regarded as tetrahedral shape having one lone pair of electrons occupying one of the four tetrahedral positions. Trigonal pyramidal shape results from sp^3 hybridisation of the central atom (M atom). The HMH bond angle in all the four MH_3 hydride (M = N, P, As, Sb) is never equal to the ideal tetrahedral angle (= 109.5°) but less than this value. HMH bond angles are as : NH_3 = 107.5°, PH_3 = 93.3°, AsH_3 = 91.8°, SbH_3 = 91.3°. The decrease in HMH bond angle from NH_3 to SbH_3 can be explained on the basis of electronegativity values of the central atom (M atom) as discussed on pages 326–327.

2. Formation of Halides. These elements form two types of halides namely trihalides (MX_3) and pentahalides (MX_5). With the exception of BiF_3, all other trihalides have covalent character which decreases as we descend the group. Except NF_3 and BiF_3, all other trihalides are readily hydrolysed by water. NX_2 trihalides (X = Cl, Br, I) are completely hydrolysed by water, giving NH_3 and hypohalous acid (HOX). For PX_3, AsX_3 and SbX_3 trihalides the ease and extent of hydrolysis generally increase with the decreasing electronegativity of halogen atom with a given central atom and decrease with the increasing metallic character of the central atom.

The structure of MX_3 trihalides is similar to that of NH_3. In these molecules the central atom (M) is sp^3 hybridised. Out of four sp^3 hybrid orbitals, one contains a lone pair of electrons. Although the spatial arrangement of four electron pairs (*bps* = 3, *lp* = 1) round the central atom is tetrahedral, the shape of MX_3 molecule gets distorted and becomes trigonal pyramidal. The distortion in shape is due to the presence of a lone pair of electrons in one of the sp^3 hybrid orbitals. Due to the presence of a lone pair of electrons, the trihalides (except NF_3) can act as electron donors (Lewis bases). NF_3 is not able to act as electron donor, since the highly electronegative F-atom draws electron pair away from N-atom and thus makes NF_3 molecule a poor electron donor.

Nitrogen and bismuth cannot form pentahalides. Nitrogen cannot form pentahalides because it cannot expand its octet due to the non-availabilty of *d*-orbitals. Bismuth cannot form pentahalides because the electron pair present in 6*s* orbital is inert (inert pair effect) and hence does not participate in bond formation. PX_5 halides are rapidly hydrolysed, forming H_3PO_4 and HX. SbX_5 halides are hydrolysed with excess of H_2O to hydrated $Sb_2O_5.2H_2O$.

$$PX_5 + 4H_2O \longrightarrow H_3PO_4 + 5HX$$

$$2SbX_5 + 7H_2O \longrightarrow Sb_2O_5.2H_2O + 10HX$$

X-ray studies of MX_5 type halides (*e.g.* PCl_5) in the liquid and gaseous state have shown that those molecules have *trigonal bipyramidal* shape. This shape results from sp^3d^2 hybridisation of the central atom (M atom). In the solid state PCl_5 is ionic and exists as $[PCl_4]^+$ $[PCl_6]^-$ in which $[PCl_4]^+$ is tetrahedral and $[PCl_6]^+$ is octahedral. Similarly PBr_5 is also ionic and exists as $[PBr_4]^+$ $[Br]^-$.

3. Formation of Oxides. All the elements of this group form oxides. More important oxides are *trioxides* (M_2O_3), *tetraoxides* (M_2O_4) and *pentoxides* (M_2O_5). Besides these three types of oxides nitrogen also gives *nitrous oxide* (N_2O) and *nitric oxide* (NO). Both these oxides are neutral. Bismuth also forms bismuth monoxides (BiO).

Thermal stability and acidic character of M_2O_3 type oxides decreases from N_2O_3 to Bi_2O_3. Thus N_2O_3 and P_2O_3 are completely acidic, since with water they give nitrous and phosphorus acid respectively. As_2O_3 and Sb_2O_3 are amphoteric and hence form salts with acids as well as with alkalies.

$$As_2O_3 + 6HCl \longrightarrow 2AsCl_3 - 3H_2O; \; As_2O_3 + 6NaOH \longrightarrow \underset{\text{Sod. arsenite}}{2Na_3AsO_3} + 3H_2O$$

$$Sb_2O_3 + 6HCl \longrightarrow 2SbCl_3 + 3H_2O; \; Sb_2O_3 + 6NaOH \longrightarrow \underset{\text{Sod. antimonite}}{2Na_3SbO_3} + 3H_2O$$

Bi_2O_3 is predominantly basic and forms salts with acids. Bi_2O_3 also shows feeble acidic character, since it slightly dissolves in conc, alkalies due to the formation of bismuthites (BiO_2^-)

$$Bi_2O_3 + 6HCl \longrightarrow 2BiCl_3 + 3H_2O$$

$$Bi_2O_3 + 2NaOH \longrightarrow \underset{\text{Sod. bismuthite}}{2NaBiO_2} + H_2O$$

The decrease in acidic character from N_2O_3 to Bi_2O_3 is accounted for by saying that N^{3+} ion, being smaller in size than Bi^{3+} ion, interacts with H_2O more strongly and thus gives an acidic N_2O_3 oxide while Bi^{3+} ion forms basic Bi_2O_3 oxide. This also explains the fact that an oxide in lower oxidation state is less acidic than that in higher oxidation state. For example N_2O (oxidation state of N = + 1) is neutral while N_2O_4 (oxidation state of N = + 4) is acidic. The decrease in acidic character from N_2O_3 to Bi_2O_3 is also consistent with the increasing metallic character of the elements in passing from nitrogen to bismuth.

P_2O_3 is readily oxidised to P_2O_5 while the trioxides of other elements are rather stable towards oxidation to pentaoxides. Bi_2O_3 shows maximum stability towards oxidation. The acidic character of pentaoxides decreases from N_2O_5 to Bi_2O_5. Thus N_2O_5, P_2O_5 and As_2O_5 are exclusively acidic and react with H_2O or alkalies to form oxyacids or their salts. Sb_2O_5 is acidic and forms salts called antimonates. Bi_2O_5 is feebly acidic and forms salts when fused with alkalies. These salts are called bismuthates. The decrease in the acidic character from N_2O_5 to Bi_2O_5 is due to the decrease in the size of M^{5+} ion from N^{5+} to Bi^{5+}. The decrease in acidic character is also in accordance with the increase of metallic character of the elements from N to Bi.

N_2O_5, P_2O_5 and As_2O_5 are readily soluble in water and form acids while Sb_2O_5 and Bi_2O_5 are relatively insoluble. Thus the solubility of these oxides decreases from N_2O_5 to Bi_2O_5.

N_2O_5 is the least stable and hance gets dissociated into nitrogen dioxide (NO_2) and O_2 even at room temperature. P_2O_5 is thermally stable : it sublimes on heating. As_2O_5 decomposes into trioxide (As_2O_3) and O_2 at red heat. Sb_2O_5 is less stable and decomposes into trioxide (Sb_2O_3) and O_2 above 450°C. Excepting P_2O_5, all other pentoxides show oxidizing property. N_2O_5 is the strongest oxidizing agent.

4. Formation of Oxy-acids. Excepting bismuth, all other elements form oxy-acids in various oxidation states. Oxy-acids of nitrogen and phosphorus are quite common and useful. These will be discussed separately in chapters 34 and 37 respectively. The strength and solubility of oxy-acids having the element in the same oxidation state (*e.g.* HNO_3, H_3PO_4, H_3AsO_4, H_3SbO_4) decreases gradually.

5. Combination with Metals. All these elements form compounds with metals, *e.g.* N forms *nitrides* : Li_3N, Mg_3N_2 etc. ; P gives *phosphides* : Ca_3P_2 etc.; As forms *arsenides* : Na_3As etc. ; Sb gives *antimonides* : Mg_3Sb_2 etc. and Bi forms *bismuthides*, Mg_3Bl_2 etc. Of these compounds the nitrides are most stable. Antimonides and bismuthides are rare. In fact Sb and Bi, on combining with metals, form alloys.

Difference Between Nitrogen and Other Elements of Group VA.

Nitrogen which is the first element of VA group differs considerably, like carbon (first element of IVA group), from the other elements of this group. This is because of the following inherent properties of nitrogen : (*a*) *small size* (*b*) *higher electronegativity* (*c*) *non-availability of d-orbitals.* The main points of difference are : (*i*) Nitrogen is a gas while other elements are solids (*ii*) Nitrogen molecule is diatomic (N_2) while other elements are tetra-atomic and are, therefore, represented as P_4, As_4 and Sb_4. (*iii*) Nitrogen does not show allotropy while other elements (excepting Bi) show allotropy. (*iv*) Nitrogen occurs in the free state while none of the other elements occurs in the free state. (*v*) Nitrogen shows a large number of oxidation states such as + 1 (in N_2O), + 2 (in NO), + 3 (in N_2O_3), + 4 (in N_2O_4), + 5 (in N_2O_5), –1 (in NH_2OH), –2 (in N_2H_4) and –3 (in NH_3). Other elements do not show such a variety of oxidation states. (*vi*) NH_3 which is a hydride of N is highly stable and non-poisonous while the hydrides of the other elements are progressively less stable and poisonous. Moreover, NH_3 forms an associated molecule $(NH_3)_x$ due to the occurrence of hydrogen bonding in this molecule, while other hydrides do not show such behaviour due to the absence of H-bonding. (*vii*) NCl_3 is highly reactive and unstable while the trichlorides of the remaining elements are progressively more stable and less reactive. N does not form pentachloride while P and Sb form such chlorides. (*viii*) The phenomenon of catenation is more common in phosphorus compounds than in nitrogen compounds. This is due to the relative weakness of the N—N single bond as compared to P—P single bond (*ix*) The oxides of N are gaseous and contain multiple bonds between N and O (*e.g.* NO) whereas the oxides of phosphorus are solids containing single P—O bonds. This is due to the relatively high N = O and N ≡ O bond strength and relatively low N—O bond strength but reverse is true for P—O, P = O and P ≡ O bond strengths. (*x*) Due to the absence of *d*-orbitals in nitrogen atom, it can not expand its octet while other elements of this group can do so. Therefore N cannot have coordination number greater than 4 whereas other elements can have coordination number greater than 4. Thus other elements can form the compounds like PF_5, PCl_5 etc. and complexes like PF_6^- while N does not form compounds like NF_5 and complexes like NF_6^-. (*xi*) Nitrogen is chemically inert in spite of its being the third most electronegative element. This is due to the high dissociation energy of N ≡ N bond. However, other elements of the group are quite reactive as they have low M—M bond dissociation energy.

Transition From Non-metallic Character to Metallic Character of Group VA Elements.

We have already said that, on proceeding from N to Bi in group VA, there is a transition from non-metallic to metallic character of the elements. The variation (*i.e.* increase or decrease) of the properties like (*i*) physical state of the elements (*ii*) electrical and thermal conductivity of the elements (*iii*) hardness and density (*iv*) stability of hydrides of these elements (*v*) acid character and solubility of oxides (*vi*) strength and solubility of oxy-acids (*vii*) hydrolysis of halides as studied on the previous pages confirms the fact that there is a transition from non-metallic to metallic character from N to Bi in group VA.

Questions with Answers

Q.1 Elemental nitrogen exists as a diatomic molecule (N_2) while white phoshorus exists as a tertratomic molecule (P_4). Explain the difference

or Nitrogen exists as a gas but phosphoras exsits as a solid. Explain the difference.

Ans : Since N-atom is small in size, it forms two $(p\pi - p\pi)\pi$ bonds with other N-atom to give $N \equiv N$ or N_2 molecule and thus nitrogen exists as a triply bonded diatomic gaseous molecule at room temperature. On the other hand, since P-atom is larger in size, it does not form $p\pi - p\pi$ bonds with other P-atoms to give $P \equiv P$ or P_2 molecule. Instead, it prefers to form single P–P bonds with other three P-atoms and gives discrete tetratomic tetrahedral P_4 molecule in which each P-atom is linked with other three P-atoms by three P–P single covalent bonds and four P-atoms of P_4 molecule lie on the corners of a regular tetrehedron with PPP bond angle = 60° and P–P bond length = 221 pm as shown in Fig. 32.1 Each P-atom in P_4 molecule completes its octet. Each P-atom has a lone pair of electrons on it.

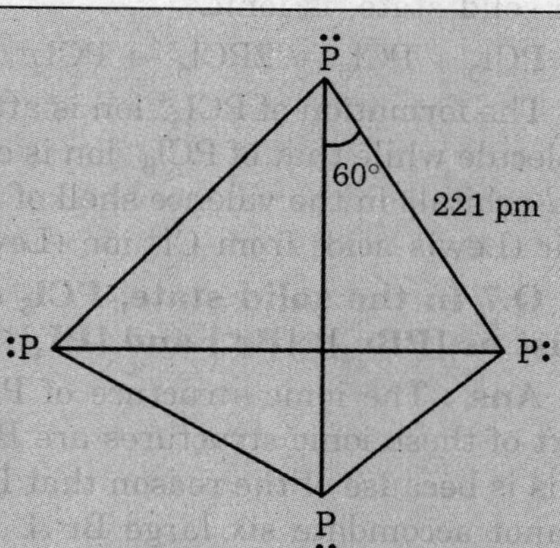

Fig.30.1 Tetrahedral structure of P_4 molecule (white phosphorus)

In white phosphorus individual P_4 tetrahedrons are linked together by weak vender Waals forces. Structure of P_4 molecule given in Fig.30.1 shows that this structure contains six P–P sigma bonds, four lone pairs of electrons and PPP angle equal to 60°.

Since P_4 molecule has high molecular mass, phosphorus exists as a solid.

Q.2 White phophorus catches fire in air but red phosphorus does not catch fire easily. Explain the difference.

Ans : White phophorus catches fire in air due to its low ignition temperature which is equal to 303K. On the other hand, since ignition temperature of red phophorus (= 543K) is much higher that that of white phosphorus, this phosphorus does not catch fire easily.

Q.3 NF_3 is an exothermic compound but NCl_3 is an endothermic compound. Explain.

Ans : NF_3 is an exothermic compound, since in the formation of one mole of NF_3 heat is released (negative value of heat of formation). Heat is released, since F_2 is stronger oxidising agent.

$$\frac{1}{2}N_2 + \frac{3}{2}Fe \longrightarrow NH_3 \text{ (one mole); } \Delta H = \text{Heat released} = -\Delta H_f \text{ of } NF_3$$

On the other hand, NCl_3 is an endothermic compound, because in the formation of one mole of NCl_3 heat is obsorbed (positive value of heat of formation). Heat is absorbed, since Cl_2 is comparatively weaker oxidising agent.

$$\frac{1}{2}N_2 + \frac{3}{2}Cl_2 \rightarrow NCl_3 \text{ (one mole); } \Delta H = \text{Heat absorbed} = +\Delta H_f \text{ of } NCl_3$$

Q.5 PCl_5 is known but PI_5 is unknown. Why ?

Ans : Since Cl-atoms are smaller in size, these atoms can easily be accomdated round P-atom to form PCl_5. On the other hand, since I-atoms are larger in size, these atoms cannot be accomdated round P-atom and hence PI_5 cannot be formed.

Q.6 Expalin why PCl_5 is ionic in the solid state.

Ans : X-ray study of the structure of PCl_5 molecule in the solid state has shown that two molecules of this compound unite together and then get splitted into $[PCl_4]^+$ and $[PCl_6]^-$ ions. Due to the presence of these ions, PCl_5 molecule, in the solid state, is ionic.

$$PCl_5 + PCl_5 \rightarrow 2PCl_5 \rightarrow PCl_4^+ + PCl_6^- \rightarrow [PCl_4]^+ [PCl_6]^- \text{ (Ionic solid)}$$

The formation of PCl_4^+ ion is attributed to the cleavage of a P–Cl bond in PCl_5 molecule while that of PCl_6^- ion is on accound of the fact that, due to the presence of *d*-orbitals in the valence shell of P-atom, this atom in PCl_5 accepts one electron pair (Lewis acid) from Cl^- ion (Lewis base) and forms PCl_6^- ion.

Q.7 In the solid state, PCl_5 exists as $[PCl_4]^+ [PCl_6]^-$ but PBr_5 and PI_5 exist as $[PBr_4]^+ [Br^-]$ and $[PI_4]^+ [I^-]$ respectively. Explain this difference.

Ans : The ionic structure of PBr_4 and PI_4 molecules show that the anionic part of these ionic structures are Br^- and I^- ions and not $[PBr_6]^-$ and $[PI_6]^-$ ions. This is because of the reason that Br^- and I^- ions are so large in size that P-atom cannot accomdate six large Br^-/I^- ions round it to form $[PBr_6]^-$ and $[PI_6]^-$ ions respectively.

Q.8 Black phosphorus has graphite-like structure. Why ?

Ans : Black phophorus has a layered structure in which each phosphorus atom is covalently bonded to three neighbouring phophorus atoms as shown in Fig.30.2. The P – P – P angles are of 99° and P – P distance is 218 pm.

The adjacent layers are held 368 pm apart. The atoms within a layer are more strongly bound than the atoms in adjacent layers. This gives black phosphorus graphite-like structure.

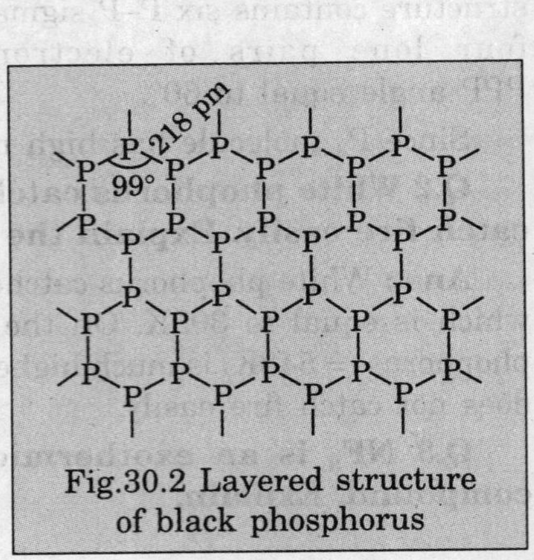

Fig.30.2 Layered structure of black phosphorus

The black phosphorus is a highly polymerised form of phosphorus.

It is due to its graphite-like structure that black phoshorus is a fairly good conductor of electricity. Black phosphorus is the only allotropic form of phosphorus which is a good conductor of electricity.

Q.9 Bond angle in PH_4^+ is larger than that in PH_3. Why ?

Ans : P in PH_3 is sp^3 hybridized. PH_3 has three bond pairs and one lone pair around P. Due to stronger lone pair–bond pair repulsions than bond pair–bond pair replulsions, the tetrahedral angle decreases from 109.5° to 93.3°. As a result, PH_3 is pyramidal. However, when it reacts with a proton, it forms PH_4^+ ion which has four bond pairs and no lone pair. Due to the absence of lone pair–bond pair repulsions and presence of four identical bond pair–bond pair interactions, PH_4^+ assumes tetrahedral geometry with bond angle of 109.5°. This explains why the bond angle in PH_4^+ is higher than in PH_3. (See Fig. 32.3)

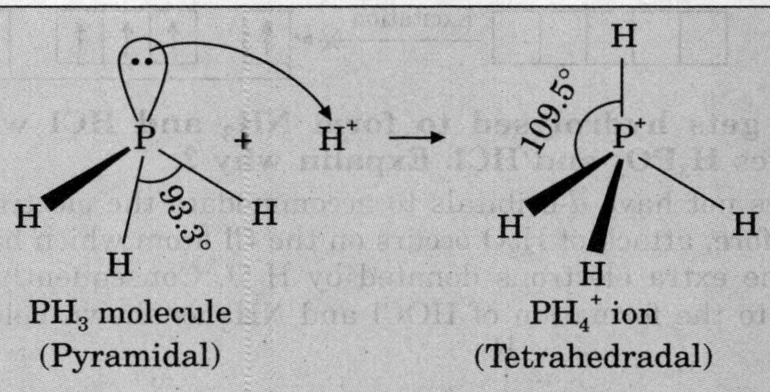

Fig 32.3 Bond angle in PH_4^+ is larger than that in PH_3.

Q.10 Explain why Bi_2O_3 is not acidic in any of its reactions.

Ans : The size of Bi^{3+} is very large and thus has a very weak +ve field around it. Therefore, it does not interact with H_2O to release H^+ ions.

$$Bi^{3+} + 3\ H{-}O{-}H \not\longrightarrow Bi(OH)_3 + 3H^+$$

In contrast, the O^{2-} is small and has a strong –ve electric field around it. It, therefore, interacts with water releasing OH^- ions.

$$O^{2-} - H - OH \longrightarrow 2OH^-$$

Thus, *Bi_2O_3 is not acidic in any of its reactions but acts as a base.*

Q.11 Expalin why both N and Bi do not form pentahalides while phophorus does.

Ans : N does not form a pentahalide because it does not have *d*-orbitals to which 2*s* electron may be excited to make available five half-filled orbitals needed for the formation of five N-halogen bonds.

In contrast, P forms a pentahalide because it has vacant 3*d*-orbitals to which 3*s* electron can be excited to make available five half-filled orbitals needed for fromation of five P-halogen bonds.

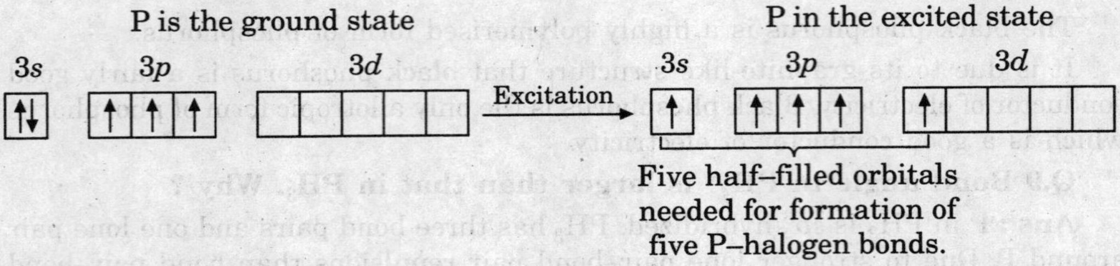

Like P, Bi also has vacant 6*d*-orbitals in the valence shell. But due to inert pair effect, the energy needed for excitation of 6*s*-electron to 6*d*-orbital is much more than the energy released when two additional Bi-hlaogen bonds are formed. Thus, bismuth does not form a pentahalide (except BiF_5) but forms only a trihalide.

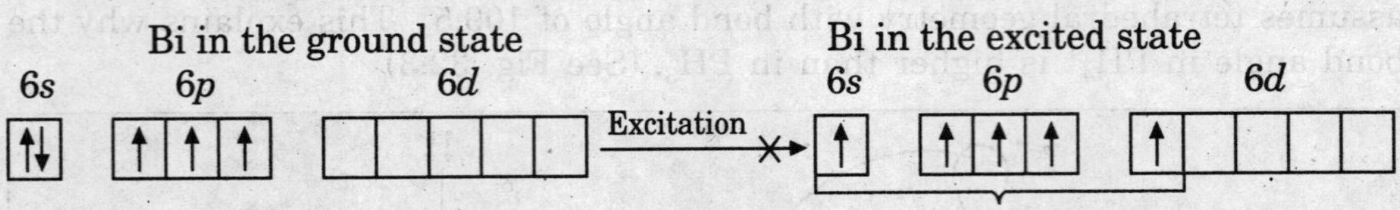

Q.12 NCl_3 gets hydrolysed to form NH_3 and HCl while PCl_3 on hydrolysis gives H_3PO_3 and HCl. Expalin why ?

Ans : N does not have *d*-oribitals to accommodate the electrons donated by O of H_2O, therefore, attack of H_2O occurs on the Cl atom which has *d*-orbitals to accommodate the extra electrons donated by H_2O. Consequently Cl–O bond is formed leading to the formation of HOCl and NH_3 as shown below :

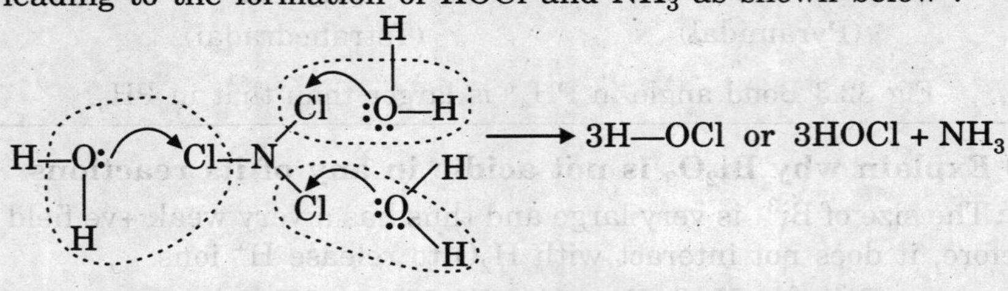

In contrast, both P and Cl have *d*-orbitals to accommodate electrons donated by H_2O. But P–O bond is much stronger than Cl–O bond. As a result, attack of H_2O molecules occurs preferentially on P of PCl_3 to form H_3PO_3 and HCl.

$$PCl_3 + 3H_2O \rightarrow 3HCl + P(OH)_3 \text{ or } H_3PO_3 \text{ . . . . . (A)}$$

As a matter of fact the above reaction takes place through the following steps:

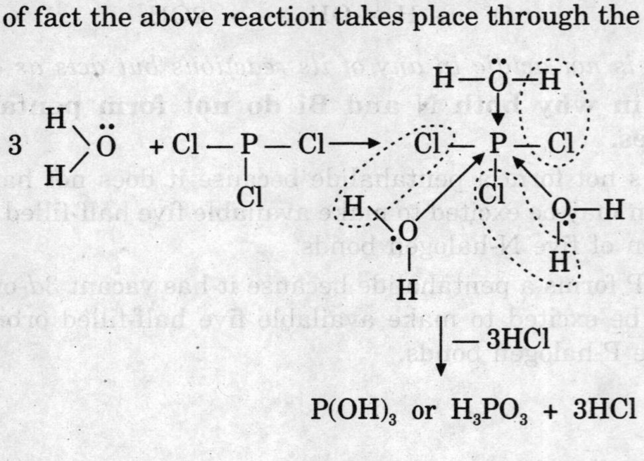

Q.13 A translucent white waxy solid (A) on heating in an inert atmosphere is converted into its allotropic form (B). Allotrope (A) on reaction with very dilute aqueous KOH liberates a highly poisonous gas (C) having rotten fish smell. With excess of chlorine (C) forms (D) which hydrolyses to compound (E). Identity (A) to (E).

Ans : (*i*) Since a translucent white waxy solid (A), on heating in an inert atmosphere, is converted into its allotrope (B), *(A) is white or yellow phophorus while (B) is red phosphorus.*

$$\underset{\text{White phosphorus}}{\underset{(A)}{P_4(s)}} \xrightarrow{\text{Heat, inert gas}} \underset{\text{Red phosphorus}}{\underset{(B)}{P_4(s)}}$$

(*ii*) Since allotrope (A) on boiling with dilute aqueous KOH gives a highly poisonous gas (C) having rotten fish smell, it is confirmed that *allotrope (A) is white phosphorus and the poisonous gas (C) is phosphine (PH_3)*

$$\underset{\text{White phosphorus (A)}}{P_4(s)} + 3KOH(aq) + 3H_2O \xrightarrow{\text{Heat}} \underset{\text{Pot. hypophosphite}}{3KH_2PO_2(aq)} + \underset{\text{Phosphine (C)}}{PH_3(g)}$$

(*iii*) Since phosphine (C) reacts with excess of chlorine to form a compound (D) which upon hydrolysis gives compound (E), *(D) must be phosphorus pentachloride (PCl_5) and (E) must be phosphoric acid (H_3PO_4).*

$$\underset{\text{Phosphine (C)}}{PH_3(g)} + 4Cl_2(g) \longrightarrow \underset{\text{Phosphorus pentachloride (D)}}{PCl_5(s)} + 3HCl(g)$$

$$\underset{(D)}{PCl_5(g)} + 4H_2O \longrightarrow \underset{\text{Phosphoric acid (E)}}{H_3PO_4(aq)} + 5HCl(aq)$$

Thus, (A) is white phosphorus, (B) is red phophorus, (C) is phosphine (PH_3) (D) is phosphorus pentachloride (PCl_5) and (E) is phosphoric acid (H_3PO_4).

Q.14 Bismuth is a strong oxidising agent in the pentavalent state. Expalin.

Ans : Since the inert pair effect is very prominent in Bi, its + 5 oxidation state is much less stable than its + 3 oxidation state. In other words, bismuth in the poentavalent state can easily accept two electrons and thus gets reduced to trivalent bismuth.

$$\underset{\text{Less stable (OA)}}{Bi^{5+}} + 2e^- \xrightarrow{\text{Reduction}} \underset{\text{More stable}}{Bi^{3+}}$$

Thus Bi^{5+} acts as a strong oxidising agent (OA).

Q.15 Give the chemical reaction to support that + 5 oxidation state of Bi is less stable than + 3 state.

Ans : Due to inert effect, Bi can show + 3 and + 5 oxidation states. Since inert pair effect is maximum in case of Bi, its + 5 oxidation state is less stable than its + 3 oxidation state. This is evident from the observation that $BiCl_3$ even on prolonged heating with Cl_2 does not form $BiCl_5$

$BiCl_3 + Cl_2 \rightarrow BiCl_5$ (*This reaction does not take place*).

Q.16 Of Bi(V) and Sb(V) which may be a stronger oxidising agent and why ?

Ans : On moving down the group, the stability of + 5 oxidation state decreases while that of + 3 oxidation state increases due to inert pair effect. Thus, + 5 oxidation state of Bi is less stable than + 5 oxidation of Sb. Therfore, Bi(V) is a stronger oxidising agent than Sb(V).

University Questions

1. Justify the inclusion of N, P, As, Sb and Bi in the same group of the periodic table. (*Raj. 84 S, 86, 87, 87 S; Pune 86; Kashmir 85*)

2. In which oxidation state (+ 3 or + 5) the elements of group VA are generally more acidic? Illustrate your answer from examples of HNO_4 and HNO_5. (*Punjab 86*)

3. Explain the following, giving appropriate reasons.

(*i*) Nitrogen forms only diatomic molecule unlike phosphorus and other members of group VA. (*Allahabad 87*)

(*ii*) Nitrogen is quite inert, particularly at temperature near 300 K. (*Allahabad 89*)

(*iii*) Nitrogen is a gas while phosphorus is a solid at room temperature. (*Delhi 89*)

(*iv*) PCl_5 exists but PH_5 does not. (*Delhi Hons. 87*)

(*v*) Pentahalides of P, As and Sb are known but their pentahydrides are not known. (*Allahabad 88*)

(*vi*) Black phosphorus and nitrogen have low chemical activity. (*Delhi Hons. 87*)

(*vii*) White phosphorus is more reactive than red phosphorus. (*Delhi 87*)

(*viii*) White phosphorus is soluble in CS_2. (*Delhi Hons. 87*)

(*ix*) Nitrogen molecule is diatomic (N_2) while phosphorus molecule is tetratomic (P_4).

(*x*) Reactions of P, As, Sb and Bi with oxidising agents reflect metallic character. (*Delhi Hons. 82*)

(*xi*) The stability of the trihalides of MX_3 type (M = N, P, As etc.) increases down the group.

(*xii*) $BiCl_3$ is more stable than $BiCl_5$.

(*xiii*) NF_3 is extremely stable while NCl_3 is readily hydrolysed.

(*xiv*) The trihalides of group VA elements give different products on hydrolysis. (*Delhi Hons. 86*)

(*xv*) NF_3 has little tendency to act as electron donor but PF_3 is a good donor. (*Allahabad 88*) (*xvi*) NCl_3 is hydrolysed but NH_3 is not.

(*xvii*) All P—F bonds in PF_5 molecule are not equivalent. (*Punjab 86*)

(*xviii*) PI_5 is unknown.

(*xix*) (*a*) PCl_5 is stable while NCl_5 cannot be prepared. (*Delhi 82*)

(*b*) PCl_5 exists while NCl_5 does not exist. (*Delhi 80*)

(*c*) Nitrogen exhibits 3-covalent state while P, As and Sb exhibit 5-covalent state. (*Himachal 84*)

(*d*) Nitrogen does not form pentahalides while P and Sb give PCl_5 and $SbCl_5$ respectively. (*Delhi 86; Raj. 87*)

(*e*) NF_3 is known but NF_5 is not known. (*Delhi 84*)

(*f*) N-atom forms only one chloride *viz.* NCl_3 while P-atom forms both PCl_3 and PCl_5. (*Gorakhpur 84*)

(*g*) PCl_5 is more stable than NCl_5. (*Poona 86*)

(*xx*) $[NF_6]^-$ ion does not exist but $[PF_5]^-$ ion exists.

(*xxi*) N-atom shows a maximum coordination number of 4 but phosphorus how a coordination number of 6. (*Allahabad 88*)

4. Compare the elements of nitrogen group elements with respect to : (*i*) Electronic configuration (*ii*) Electronegative character (*iii*) Oxidation states (*iv*) Stability of MH_3 hydrides. (*v*) Acidic properties of oxides. (*Gorakhpur 99*)

5. Explain why NF_3 is stable while NCl_3 and NI_3 are explosive. (*Allahabad 99*)

6. Explain why white phosphorus is more reactive than red phosphorus. (*Delhi 99*)

7. Compare the properties of Sb and Bi pair. (*Kanpur 2000*)

CHAPTER

31 Fertilisers

Fertilizers

What are Fertilizers ?

Fertilizers are the substances which are added to the soil in order to make up the deficiency of essential elements like nitrogen, phosphorus and potassium required for the growth of the plants.

Essential Qualities of a Good Fertilizer.

Every substance containing plant nutrients cannot be used as a fertilizer. However, the essential requisites of a good fertilizer are *(i)* The nutrient elements present in it must be readily available to the plants. *(ii)* It must be fairly soluble in water so as to enable it to dissolve in soil. *(iii)* By rain or water, the fertilizer must be converted into a form which the plant can assimilate easily. *(iv)* It must be stable so that it is available for a long time to the growing plant. *(v)* It should not be injurious to plants. *(vi)* It should be able to correct the acidity of the soil. *(vii)* It should be cheap.

Classification of Fertilizers.

Fertilizers are classified according to the nature of the element/elements like *N*, *P* or *K* they provide to the soil. This classification gives the following types of fertilizers

1. Nitrogenous fertilizers. These fertilizers mainly supply nitrogen to the plants or soil. $(NH_4)_2\ SO_4$, calcium ammonium nitrate (CAN), basic calcium nitrate, calcium cyanamide, urea etc. are the examples of nitrogenous fertilizers.

2. Phosphatic fertilizers. These fertilizers provide phosphorus to the soil. Superphosphate of lime, triple superphosphate and phosphate slag are the examples of phosphatic fertilizers.

3. Potash fertilizers. These fertilizers supply potassium to the plant. These are useful to the plants especially to mendow, glass, tobacco, cotton, coffee, potatoes and corn. All these fertilizers occur in nature. KCl, KNO_3, K_2SO_4 etc. are the important examples.

4. NP fertilizers. These fertilizers contain two elements namely nitrogen and phosphorus. These are obtained by mixing together nitrogenous and phosphatic fertilizers in suitable proportions. Examples of NP fertilizers are : *dihydrogen ammoniated phosphate*, $NH_4H_2PO_4$; *calcium superphosphate nitrate* $Ca(H_2PO_4)_2$, $2Ca(NO_3)_2$ and *ammoniated phosphate sulphate*, $(NH_4)H_2PO_4$, $(NH_4)_2SO_4$.

5. NPK or complete fertilizers. These fertilizers supply all the three essential elements namely nitrogen, phosphorus and potassium to the soil and are produced by mixing nitrogeneous, phosphatic and potash fertilizers in suitable proportions. It is observed that these fertilizers produce much better results. NPK fertilizers are mainly produced by *Zuari Agrochemicals Ltd; Goa.*

Important Fertilizers

1. Ammonium Sulphate or Sindri Fertilizer, $(NH_4)_2SO_4$.

This fertilizer is being manufactured at Sindri Fertilizer Factory, Sindri (Bihar) and hence is called *Sindri fertilizer.*

Manufacture (*i*) *From air and steam : Sindri process.* Air and steam are passed over red hot coke when a mixture of CO, N_2 and H_2 gases is formed.

$$C + Air \longrightarrow CO + N_2$$
$$C + H_2O \longrightarrow CO + H_2$$

The gaseous mixture of CO, N_2 and H_2 is then passed over hot Fe_3O_4 and Cr_2O_3 when CO gets oxidised to CO_2. The mixture of CO_2, N_2 and H_2 is then compressed to about 20 atmospheric pressure and passed into water in which CO_2 gets dissolved and is thus removed from the gaseous mixture. The remaining gases namely N_2 and H_2 are mixed together in the ratio of 1 : 3 and then converted into NH_3 by Haber's process. In Haber's process the mixture of H_2 and N_2 is heated to 450°–550°C, compressed to 150–200 atmospheric pressure and passed over hot Fe_2O_3, Al_2O_3 and K_2C mixture. Here Fe_2O_3 acts as a catalyst and the mixture of Al_2O_3 and K_2O acts as a promotor

$$N_2 + 3H_2 \underset{\substack{150\text{–}200 \text{ at. pressure} \\ (Fe_2O_3 + Al_2O_3 + K_2O)}}{\overset{\text{temperature } = 450\text{–}500°C}{\rightleftarrows}} 2NH_3 + \text{Heat}$$

Besides from Haber's process, NH_3 can also be obtained from coal distillation as is done at Sindri Fertilizer Factory. Now NH_3 gas is passed into a suspension of finely powdered $CaSO_4$ (calcined gypsum) in water through which a stream of CO_2 is also passed. CO_2 is obtained by heating $CaCO_3$.

$$CaCO_3 \longrightarrow CaO + CO_2$$
$$2NH_3 + H_2O + CO_2 + CaSO_4 \longrightarrow \underset{\text{soluble}}{(NH_4)_2SO_4} + \underset{\text{ppt.}}{CaCO_3} \downarrow$$

The precipitate of $CaCO_3$ is filtered off and the filtrate containing $(NH_4)_2SO_4$ is concentrated by evaporation under vacuum and cooled when crystals of $(NH_4)_2SO_4$ are obtained. These crystals are packed in jute bags and sold in the market under the name of *Sindri ferlilizer.*

$CaCO_3$ obtained as a by-product in this process is used as a raw material for the manufacture of cement and another fertilizer called *nitrochalk* which is a mixture of chalk and NH_4NO_3.

$CaSO_4$ used in the above reaction is obtained from Rajasthan where large deposits of gypsum occur.

The process described above is being carried out successfully at Sindri Fertilizer Factory, Sindri (Bihar). Hence the name *Sindri process.* Similar types of plants for the manufacture of $(NH_4)_2SO_4$ are also working in England, Germany, Austria, Italy and U.S.S.R.

***(ii)* From NH_3.** $(NH_4)_2SO_4$ is also made on a large scale by absorbing synthetic NH_3 gas in H_2SO_4.

$$2NH_3 + H_2SO_4 \longrightarrow (NH_4)_2SO_4$$

$(NH_4)_2SO_4$ thus obtained is purified by crystallisation.

(iii) From ammonical liquir *In steel factories*, ammonium sulphate is obtained from ammoniacal liquor obtained as a by-product in the manufacture of coke by the dry distillation of coal. Then, it is heated with milk of lime whereby ammonia gas is passed into 20% sulphuric acid to get ammonium sulphate.

$$2NH_4Cl + Ca(OH)_2 \longrightarrow CaCl_2 + 2NH_3 + 2H_2O$$

$$2NH_3 + H_2SO_4 \longrightarrow (NH_4)_2SO_4$$

Besides organic compounds, ammoniacal liquor containts free NH_3 and ammonium salts which are of two types : *volatile* [*e.g.* $(NH_4)_2CO_3$, NH_4HS, NH_4CN etc.] and *fixed* [*e.g.* $(NH_4)_2SO_4$, NH_4Cl etc.]. Volatile salts are decomposed by boiling alone while fixed salts are decomposed by lime water, $Ca(OH)_2$. The total NH_3 in combination is about 17 g/litre. The ammoniacal liquor is heated in ammonia still (Fig. 31.1) first with steam to remove free and volatile NH_3 and then the residue is treated with lime water in order to decompose the fixed salts.

$$\begin{array}{ll} \textit{Volatile} & \left\{\begin{array}{l} (NH_4)_2CO_3 \longrightarrow 2NH_3 + CO_2 + H_2O \\ (NH_4)_2HS \longrightarrow NH_3 + H_2S \end{array}\right. \\ \textit{salts} & \end{array}$$

$$\begin{array}{ll} \textit{Fixed} & \left\{\begin{array}{l} (NH_4)_2SO_4 + Ca(OH)_2 \longrightarrow 2NH_3 + CaSO_4 + 2H_2O \\ 2NH_4Cl + Ca(OH)_2 \longrightarrow 2NH_3 + CaCl_2 + 2H_2O \end{array}\right. \\ \textit{salts} & \end{array}$$

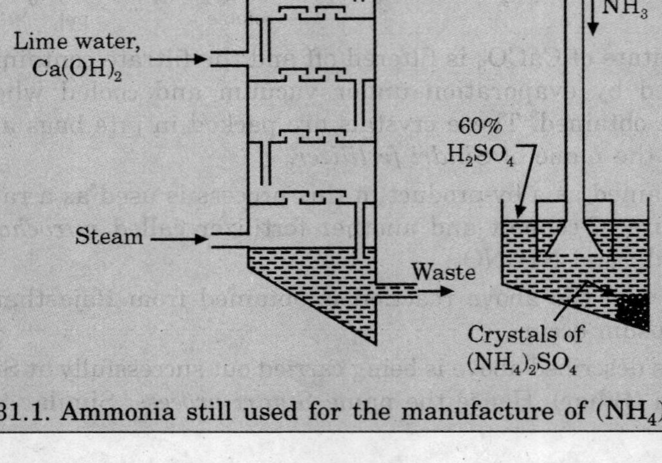

Fig. 31.1. Ammonia still used for the manufacture of $(NH_4)_2SO_4$

The mixture of NH_3 and steam so produced is passed through a lead-lined tank containing 60% H_2SO_4, when crystals of $(NH_4)_2SO_4$ separate on cooling.

$$2NH_3 + H_2SO_4 \longrightarrow (NH_4)_2SO_4$$

This method is being used in all the major steel plants in India, since ammoniacal liquor is obtained as a by-product in the manufacture of coke needed for the production of steel.

Manufacture in India. The fertilizer factories at Sindri, Durgapore, Bernpore, Jamshedpur, Bhilai, Alwaye, Bansjora, Digboi and Hanumangarh are manufacturing $(NH_4)_2SO_4$. Many other factories are being installed at various places for the production of $(NH_4)_2SO_4$.

Use of $(NH_4)_2SO_4$ as a Fertilizer. This fertilizer contains 24–25% NH_3 which is converted by nitrifying bacteria present in the basic soil into nitrates. These nitrates can be taken up by the plants easily from the soil.

$$(NH_4)_2SO_4 \xrightarrow[\text{in the soil}]{\text{Basic matters}} NH_3 \xrightarrow[\text{bacteria}]{\text{Nitrifying}} NO_3^-$$

This fertilizer is appropriate especially for rice and potato crops and basic soils. In acidic soils it should be used in combination with ground lime stone or chalk which neutralises the acidity the soil. Addition of NaCl has been found to increase the fertilising power of $(NH_4)_2SO_4$. This fertilizer may be applied prior to sowing or at the sowing time or during the growing period of the crop. After it is applied to the soil, it is absorbed and retained by soil particles and hence there is no danger of its being lost from the soil in drainage.

Methods of its application. The following precautions should be observed when $(NH_4)_2SO_4$ is being used as a fertilizer : (*i*) It should not be applied along with seeds as it is harmful to the germinating seeds. (*ii*) It should not be applied to wet clay soils, since in these soils the want of good aeration makes the nitrifying process slow and imperfect. (*iii*) It should not be used to soil soon after liming, since the fertilizer gets decomposed by lime. (*iv*) The fertilizer should not be used too frequently on soils which are poor in lime since loss of NH_3 makes the soil acidic. In such soils the fertilizer should be used in combination with ground lime stone or chalk which neutralises the acidity of the soil.

$(NH_4)_2SO_4$ is less hygroscopic than urea and NH_4NO_3 and, therefore, has better storage properties. It can be applied prior to sowing or at the sowing time or during the growing period of the crop.

2. Calcium Ammonium Nitrate (CAN) or Nitro Lime Stone or Nangal Fertilizer, $Ca(NO_3)_2.NH_4NO_3$.

This fertilizer is being manufactured at Nangal in Punjab by the process which consists of the following steps :

(*i*) Production of NH_3. N_2 (obtained from liquefaction of air) and H_2 (obtained by the electrolysis of H_2O) are mixed together in the ratio of 1 : 3 and are converted into NH_3 by Haber's process.

$$N_2 + 3H_2 \underset{\substack{150\text{–}200 \text{ at. pressure} \\ (Fe_2O_3 + Al_2O_3 + K_2O)}}{\overset{\text{temp.} = 450\text{–}500°C}{\rightleftarrows}} 2NH_3$$

***(ii)* Production of HNO_3.** A part of NH_3 produced as above is used in the manufacture of HNO_3 by the Ostwald's process. In this process NH_3 and air are mixed together in the ratio of 1 : 10 by volume. The mixture is introduced into a converter having a cage of fine platinum gauge (80 mesh) which is kept electrically heated between 820–920°C. By this treatment nitric oxide, NO is produced.

$$4NH_3 + 5O_2 \text{ (From air)} \underset{820\text{–}920°C}{\overset{\text{Pt guage}}{\rightleftarrows}} \underset{\text{Nitric oxide}}{4NO} + 6H_2O$$

NO is cooled and mixed with air so that NO is oxidised to nitrogen dioxide, NO_2. NO_2 is dissolved in H_2O to get the solution of HNO_3.

$$2NO + O_2 \text{ (From air)} \longrightarrow 2NO_3 \text{ (Nitrogen dioxide)}$$

$$3NO_2 + H_2O \longrightarrow 2HNO_3 + NO$$

NO produced in the above reaction is reoxidised to NO_2 which can be used to prepare HNO_3 by dissolving it into H_2O as shown above.

***(iii)* Formation of NH_4NO_3.** Solution of HNO_3 obtained as above is concentrated and heated to about 75°C and is then allowed to fall in the form of a fine spray from the top of a tower called neutraliser while NH_3 gas pre-heated to about 70°C is introduced from the bottom of the neutraliser. This operation results in the neutralisation of HNO_3 by NH_3 and formation of NH_4NO_3.

$$NH_3 + HNO_3 \longrightarrow NH_4NO_3$$

NH_4NO_3 is obtained in the form of a liquor at the bottom of the neutraliser. This liquor contains about 84% NH_4NO_3. This liquor is taken to vacuum concentrator where this liquor gets concentrated upto 92–94%.

***(iv)* Formation of calcium ammonium nitrate (CAN) pellets.** 92% concentrated solution of NH_4NO_3 is stirred with finely powdered lime stone ($CaCO_3$) in a granulator. The solution of NH_4NO_3 also contains some HNO_3 which reacts with lime stone to produce $Ca(NO_3)_2$ which, in turn, combines with NH_4NO_3 to form the pellets of calcium ammonium nitrate, $Ca(NO_3)_2.NH_4NO_3$.

$$CaCO_3 + 2HNO_3 \longrightarrow Ca(NO_3)_2 + H_2O + CO_2$$

$$Ca(NO_3)_2 + NH_4NO_3 \longrightarrow \underbrace{Ca(NO_3)_2.NH_4NO_3}_{\text{Calcium ammonium nitrate (CAN)}}$$

Lime stone is added as a filler. Its function is to render NH_4NO_3 safe to handle, since NH_4NO_3 is an explosive substance.

***(v)* Coating of CAN pellets with soap stone.** Since CAN is very hygroscopic, it has to be protected from the atmospheric moisture during its transportation and storage. For this the pellets of CAN are stirred with concentrated solution of finely powdered soap stone (sodium silicate). This treatment produces a thin layer of soap stone on the surface of pellets. The pellets are dried and packed in polythene lined bags.

Manufacture in India. The fertilizer factories at Nangal (Punjab) and Rourkela are manufacturing calcium ammonium nitrate. The production of Nangal factory is 3.2 lac tons per year while the target production of Rourkela factory is 6 lac tons per year.

Use of CAN as a fertilizer. (*i*) CAN contains about 20% nitrogen.

(*ii*) It is directly assimilated by plants and does not undergo any changes in the soil.

(*iii*) It is highly soluble in water and as such finds its way into the soil very easily.

3. Basic Calcium Nitrate or Norwegian Salt Petre or Nitrate of Lime, $Ca(NO_3)_2.CaO$.

It is manufactured by a process which consists of the following two steps.

(*i*) Production of $CaNO_3$. When lime stone ($CaCO_3$) is neutralised by HNO_3 and the solution thus obtained is evaporated, monoclinic crystals of $Ca(NO_3)_2.4H_2O$ are obtained at ordinary temperature.

$$CaCO_3 + 2HNO_3 \longrightarrow Ca(NO_3)_2 + CO_2 + H_2O$$

(*ii*) Production of $Ca(NO_3)_2.CaO$. When $Ca(NO_3)_2$ is mixed with excess of quick lime (CaO), basic calcium nitrate, $Ca(NO_3)_2.CaO$ is obtained.

$$Ca(NO_3)_2 + CaO\ (\textit{Excess}) \longrightarrow \underbrace{Ca(NO_3)_2.CaO}_{\text{Basic calcium nitrate}}$$

Use of $Ca(NO_3)_2.CaO$ as a Fertiliser. Basic calcium nitrate is hygroscopic and is very soluble in water. It is basic in nature. It is directly assimilated by the plants and does not undergo any changes in the soil. It is less deliquescent than ordinary calcium nitrate, $Ca(NO_3)_2.4H_2O$ and hence is used in preference to $Ca(NO_3)_2.4H_2O$ as a fertiliser.

4. Calcium Cyanamide or Nitrolim, $CaCN_2$.

This compound is a derivative of cynamide, $H_2N—C = N$ or H_2CN_2 in which two hydrogen atoms have been replaced by one bivalent calcium atom.

Manufacture. Lime stone ($CaCO_3$) coal and air are the raw materials used in the manufacture of $CaCN_2$. The limestone is burnt in kilns to obtain lime (CaO). A mixture of lime and coal is heated in an electric furnace at 2500°C to produce CaC_2. CaC_2 so produced is finely powdered and then is heated in a cylindrical electric furnace at 1000°C in the atmosphere of N_2 under pressure obtained from air. By this operation $CaCN_2$ is formed.

$$\underset{\text{Lime stone}}{CaCO_3} \xrightarrow{\text{Heat}} \underset{\text{Lime}}{CaO} + CO_2$$

$$\underset{\text{Lime}}{CaO} + \underset{\text{Coal}}{3C} \xrightarrow[\text{2500°C}]{\text{Heat}} CaC_2 + CO$$

$$CaC_2 + N_2 \xrightarrow[1000°C]{Heat} \underbrace{CaCN_2 + C}_{\substack{\text{Nitrolim} \\ \text{(Grey)}}} + Heat$$

The grey mixture of $CaCN_2$ and C thus obtained is sold under the name of *nitrolim.*

Process. Since the reaction representing the formation of nitrolim is highly exothermic, the heating is done only at one time and the reaction further proceeds by itself. The heating is done in a cylindrical electric furnace (Fig. 31.2). The charge consisting of finely powdered CaC_2 is placed in the furnace and N_2 gas is passed from below. The electric current is passed through a thin carbon rod which is surrounded by a tube made of carbon board. On passing the electric current through the carbon rod, all the card board partitions get carbonised due to the production of heat, leaving openings through which N_2 can be freely circulated. The heat produced in the reaction is enough to maintain a temperature of 1000°C which is necessary for the reaction to proceed. After 36 hours, the reaction is complete.

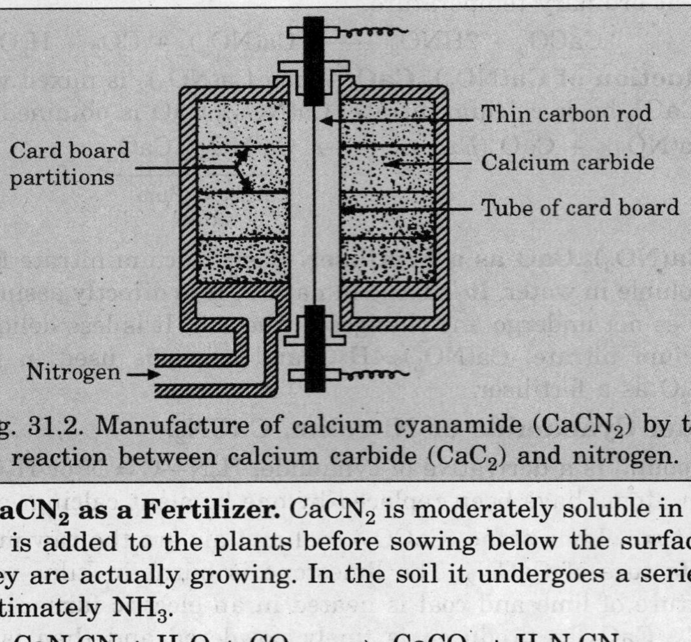

Fig. 31.2. Manufacture of calcium cyanamide ($CaCN_2$) by the reaction between calcium carbide (CaC_2) and nitrogen.

Use of $CaCN_2$ as a Fertilizer. $CaCN_2$ is moderately soluble in water and as a fertiliser it is added to the plants before sowing below the surface of soil and not when they are actually growing. In the soil it undergoes a series of changes producing ultimately NH_3.

$$CaN.CN + H_2O + CO_2 \longrightarrow CaCO_3 + \underset{\text{Cyanamide}}{H_2N.CN}$$

$$H_2N.CN + H_2O \longrightarrow \underset{\text{Urea}}{CO(NH_2)_2}$$

$$\underset{\text{Urea}}{CO(NH_2)_2} + H_2O \longrightarrow 2NH_3 + CO_2$$

NH_3 thus produced is converted into nitrates by the nitrifying bacteria and these nitrates are finally assimilated by plants. Since the conversion of $CaCN_2$ into NH_3 and that of NH_3 into nitrates by nitrifying bacteria takes place slowly,

$CaCN_2$ is a slow acting fertilizer and is preferred to soluble compounds like $(NH_4)_2SO_4$ and $NaNO_3$ as it imparts fertility to the soil of a more permanent nature.

The formation of $CaCO_3$ during the reaction between $CaCN_2$ and moist CO_2 shown above is also useful from the agricultural point of view.

5. Urea or Carbamide, NH_2CONH_2.

Manufacture of urea can be studied from some standard book on Organic Chemistry.

Use of urea as a fertilizer. (*a*) *Advantages of its Use.* (*i*) Urea has the highest nitrogen content equal to 46.6%. This percentage of N_2 is much higher than that for $(NH_4)_2SO_4$ and CAN. Higher nitrogen content in urea results in economy in freight and packing. (*ii*) The cost of production of urea is relatively low since CO_2 required for its manufacture is obtained from crude naphtha. (*iii*) Urea is not subject to fire or explosion hazards as is NH_4NO_3 and hence there is no risk in the storage of urea. (*iv*) It does not change the *p*H of the soil. (*v*) Urea can be used to all types of crops and soils, since after its assimilation by plants through the interaction of nitrifying bacteria, it leaves behind only CO_2 in the soil. This CO_2 does no harm to the soil.

$$O = C\begin{matrix} NH_2 \\ NH_2 \end{matrix} + H_2O \longrightarrow 2NH_3 + CO_2$$

On the other hand, $(NH_4)_2SO_4$ and CAN produce SO_4^{2-} and NO_3^- ions which make the soil acidic.

(*b*) *Disadvantages of its use.* (*i*) Urea is very soluble in H_2O and hygroscopic and hence requires better quality package than $(NH_4)_2SO_4$. (*ii*) It is not so stable as other solid nitrogenous fertilisers are, since urea decomposes even at lower temperatures, particularly in humid atmosphere, releasing NH_3 and CO_2. The formation of NH_3 and CO_2 results in serious loss. (*iii*) If urea contains impurities more than 2%, it cannot be used as a fertiliser since the impurities are toxic to certain crops, particularly to citrus crops.

(*c*) *Methods of its application.* (*i*) Urea should be applied at the sowing time. It should not come in contact with the seeds. (*ii*) It can also be applied as a top dressing. (*iii*) Since urea is highly concentrated, it should be used in combination with earth or sand before its application. (*iv*) It should not be applied when the soil contains free water or is likely to remain wet for three or four days after urea application.

6. Super-phosphate of Lime, or Calcium Super-phosphate, $[Ca(H_2PO_4)_2 + 2\ (CaSO_4.2H_2O)]$.

It is a mixture of *calcium dihydrogen phosphate* $Ca(H_2PO_4)_2$ and *dihydrate calcium sulphate* (*gypsum*), $CaSO_4.2H_2O$.

Manufacture. It is manufactured by treating calcium phosphate (obtained from phosphate rock or bone ash) with calculated quantity of commercial (*i.e.*, chamber H_2SO_4.

$$Ca_3(PO_4)_2 + 2H_2SO_4 + 4H_2O \longrightarrow Ca(H_2PO_4)_2 + 2(CaSO_4.2H_2O) + \text{Heat}$$

(Insoluble) — Calcium dihydrogen phosphate — Gypsum

Super-phosphate of lime

Plant and procedure. The plant used for the manufacture is shown in Fig. 31.3.

The phosphate rock is well powdered and is placed in a cast iron mixer. Now a calculated quantity of chamber H_2SO_4 is added to it. The whole mass is stirred by means of paddles for 2 to 5 minutes and then dumped mechanically into one of the brick-lined pits or dens D_1 or D_2 through the valves V_1 and V_2. Here the mixer is left for about 24 hours. The reaction which actually started in the mixer continues in the pot or den and the temperature gradually rises to 100–110°C because of the exothermic nature of the reaction. Since the phosphate rook invariably contains $CaCO_3$ and CaF_2 as impurities, they react with H_2SO_4 to evolve CO_2 and HF.

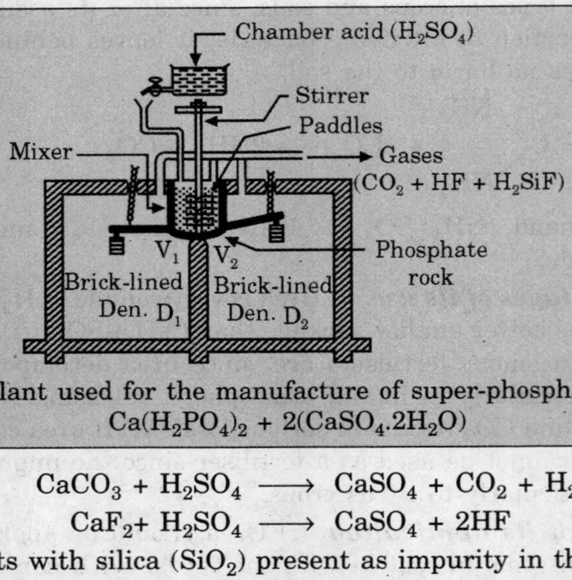

Fig. 31.3 Plant used for the manufacture of super-phosphate of lime, $Ca(H_2PO_4)_2 + 2(CaSO_4.2H_2O)$

$$CaCO_3 + H_2SO_4 \longrightarrow CaSO_4 + CO_2 + H_2O$$
$$CaF_2 + H_2SO_4 \longrightarrow CaSO_4 + 2HF$$

Free HF reacts with silica (SiO_2) present as impurity in the phosphate rock to produce SiF_4 which at once gets decomposed in presence of H_2O, forming hydrofluosilic acid, H_2SiF_6.

$$4HF + SiO_2 \longrightarrow SiF_4 + 2H_2O$$
$$3SiF_4 + 2H_2O \longrightarrow SiO_2 + 2H_2SiF_6$$

The production of poisonous HF gas in the manufacture indicates that CaF_2 should not be present in large quantities in the phosphate rock. The presence of CaF_2 in large quantities also consumes huge quantities of extra H_2SO_4. The gases like CO_2, HF and H_2SiF_6 are led into a suitable absorption tower. As a result of the reaction, the mass becomes hard and is finally solidified to a porous dry solid due to the formation of gypsum. The mass is allowed to stand for a few days in the den until it becomes perfectly dry. The dried mass is removed from the den and is kept in the open for 4–6 weeks. During this period the excess of H_2SO_4 is completely consumed and the mass is completely dried. Finally the dry mass is crushed to a fine powder and is sold in the market under the name of *super-phosphate of lime.*

Use of super-phosphate of lime as a fertilizer. It contains about 16–20% of P_2O_5, depending on the purity of the phosphate rock used in the manufacture of the fertilizer and its active ingredient is calcium dihydrogen phosphate which is soluble in water. Thus it provides phosphate in a water-soluble form which is readily assimilated by the plants for their growth. In this respect it has an advantage over the normal calcium phosphate which is insoluble in water and cannot be assimilated by the plants.

Super-phosphate of lime has some disadvantages of its being used as a fertilizer. These disadvantages are : (*i*) It is relatively dilute fertiliser containing only 16–20% of P_2O_5 depending on the purity of the phosphate rock used in its manufacture. (*ii*) Being a dilute product, it requires relatively heavy freight and handling charges and a large number of expensive bags. (*iii*) It has a setting tendency and is converted into heavy lumps. (*iv*) The acidity in it causes the package to rot and contents to spill.

Manufacture of Fertilizers in India

In our country at present there are more than 60 factories in private as well as in public sectors which are manufacturing fertilizers. Following are the important factories running in the country.

1. Factories in Public Sector. The important factories of this type are: (*i*) Fertilizer Corporation of India. This corporation is a government undertaking and is running nine units under its management. These units are situated at Sindri (Bihar), Nangal (Punjab), Namrup (Assam), Gorakhpur (U.P.), Trombay (Maharashtra), Korba (M.P.), Barauni (Bihar), Talohar (Orissa) and Durgapur (West Bengal). (*ii*) The fertilizer and Chemical Travancore Ltd., Alwaye. (*iii*) Madras Fertilizer, Madras.

2. Factories in Private Sector. The important units being run by private sector are: (*i*) D.C.M. Chemical Works, Delhi and Kota. (*ii*) Indian Explosive Ltd., Kanpur. (*iii*) Zuari Agrochemicals Ltd., Goa. (*iv*) Coromandal Fertilizer Ltd., Visakhapatnam. (*v*) E.I.D. Parry's Ltd., Madras. (*vi*) Phosphate Company Ltd., Calcutta.

In 1982–83 the production of nitrogenous, phosphatic and potash fertilizer was 5.2 (in terms of nitrogen) 1.40 (in terms of P_2O_5) and 100 (in terms of K_2O) million tons. Although the number of factories manufacturing fertilizers is quite large, yet the production is too low to meet the total requirements of the country. In order to meet the increasing demand of fertilizers, 12 new factories being set up. Three of these have been set up at Bhatinda (Punjab), Panipat (Haryana) and Mathura (U.P.)

University Questions

1. Discuss the various methods for the fixation of nitrogen.
(*Indore 85; Jodhpur 85; Delhi 84*)

2. Give a brief account of the fertilizer industry in India.
(*Gorakhpur 81; Meerut 89*)

3. What are manures ? Name some natural and artificial manures.
(*Punjab 84*)

4. What is a fertilizer ? What is a mixed fertilizer ? (*Madras 85, 86; Jodhpur 84*)

5. Give an account of different types of fertilizer. (*Madras 85, 86; Jodhpur 84*)

6. Write a note on "Fertilizers and their maufacture in India". (*Bundelkhand 88*)

7. Explain the following :

(*i*) Calcium superphosphate is a hotter fertilizer than triple superphosphate. (*Delhi 87; M.D. Rohtak 84*)

(*ii*) A fertilizer that contains phosphorus has only limited utility. (*Bharathidasan 85*)

8. Give the methods for the industrial preparation, properties and uses of :

(*i*) Calcium ammonium nitrate (CAN) (*Delhi 82, 89*)

(*ii*) Calcium cyanamide. (*Jiwaji 84*)

(*iii*) Urea. (*Delhi 88, 89*)

(*iv*) Calcium superphosphate. (*Delhi 87; M.D. Rohtak 84; Bundelkhand 88; Raj. 85; Meerut 89 S*)

(*v*) Triple superphosphate. (*Delhi 87; M.D. Rohtak 84*)

(*vi*) NPK fertilizers. (*Madras 85, 86; Jodhpur 84*)

(*vii*) $(NH_4)_2\ SO_4$. (*Delhi 82; Punjab 86; Poona 87*)

9. Write short note on CAN as a fertilizer. (*Himachal Pradesh 99, 2000*)

CHAPTER 32

Hydrides, Oxides and Oxyacids of Nitrogen

In this chapter we shall discuss the following compounds of nitrogen.

Hydrides

The well known hydrides of nitrogen are : (1) *Ammonia*, NH_3 (2) *Hydrazine*, $NH_2.NH_2$ or N_2H_4, and (3) *Hydrazoic acid*, HN_3. Out of these the first two hydrides are basic while the last one is acid'. Here we shall discuss only N_2H_4 and HN_3.

1. Hydrazine or Diammine, $NH_2.NH_2$ or N_2H_4.

This compound can be regarded as being formed by replacing one H-atom of NH_3 molecule by NH_2 group. This was first prepared by Curtius in 1889.

Preparation (*i*) *By Raschig's method* (1907). In this method a strong aqueous solution of NH_3 is boiled with sodium hypochlorite (NaOCl) in presence of a little glue or gelatin. The overall reaction can be represented as :

$$2NH_3 + \underset{\text{Sod. hypochlorite}}{NaOCl} \longrightarrow N_2H_4 + NaCl + H_2O$$

(*ii*) *Anhydrous* N_2H_4 is more easily prepared by adding a solution of sodium methoxide, CH_3ONa in methyl alcohol to a solution of hydrazine hydrochloride, $NH_2.NH_2.HCl$ in methyl alcohol, filtering off NaCl and distilling under reduced pressure when CH_3OH first distilles and then $NH_2.NH_2$.

$$NH_2.NH_2.HCl + CH_3ONa \longrightarrow NH_2.NH_2 + CH_3OH + NaCl$$

Properties. (*a*) **Physical.** Free hydrazine is colourless liquid, very hygroscopic and soluble in alcohol and water (bpt. 113.5°; m.pt. 1.4°). It is not very stable. On exposure to air, it readily absorbs carbon dioxide and moisture. It dissolves sulphur, arsenic, selenium and phosphorus.

(*b*) **Chemical.** (*i*) *Decomposition.* Hydrazine decomposes on heating into ammonia and nitrogen.

$$3NH_2.NH_2 \longrightarrow N_2 + 4NH_3$$

(*ii*) *Oxidation.* When heated in air, it gives nitrogen.

$$NH_2.NH_2 + O_2 \longrightarrow N_2 + 2H_2O$$

Ozone and hydrogen peroxide oxidise it to hydrazoic acid, N_3H,

$$3N_2H_4 + 5O_3 \longrightarrow 2N_3H + 5H_2O + 5O_2$$

$$3N_2H_4 + 5H_2O_2 \longrightarrow 2N_3H + 10H_2O.$$

(*iii*) *Reaction with sodium.* In the absence of oxygen, it reacts with sodium to form sodium hydrazine.

$$2NH_2NH_2 + 2Na \longrightarrow 2NH_2.NH.Na + H_2$$

(*iv*) It inflames in dry O_2 and reacts readily with halogens.

$$N_2H_4 + 2I_2 \longrightarrow 4HI + N_2$$

(*v*) *Basic nature.* It is a very weak base (much weaker than NH_3) and liberates, NH_3 from NH_4Cl. It forms two series of salts *viz. hydrazine monochloride*, $NH_2.NH_2.HCl$ and *hydrazine dichloride*, $ClH.NH_2.NH_2.HCl$

(*vi*) *Action of HNO_2.* It reacts with HNO_2, giving hydrazoic acid, HN_3

$$N_2H_4 + HNO_2 \longrightarrow HN_3 + 2H_2O.$$

(*vii*) *Formation of complexes.* Since N-atoms of N_2H_4 have a lone pair of electrons, N_2H_4 can act as a ligand and hence can form donor-acceptor complexes with metal ions like Ni^{2+}, Co^{2+} etc. The complexes given by this ligand with Ni^{2+} and Co^{2+} are $[Ni^{2+}(NH_2.NH_2)_4]^{2+}$ and $[Co^{2+}(NH_2.NH_2)]^{2+}$.

(*viii*) *Reducing properties.* Hydrazine and its salts are powerful reducing agents. For example :

(*a*) It precipitate Ag, Au and Pt from their salts.

$$4AgNO_3 + N_2H_4 \longrightarrow 4Ag + 4HNO_3 + N_2$$

$$4AuCl_3 + 3N_2H_4 \longrightarrow 4Au + 12HCl + 3N_2$$

$$PtCl_4 + N_2H_4 \longrightarrow Pt + 4HCl + N_2$$

(*b*) It reduces an alkaline solution of CuO (*i.e.* Fehling's solution) to Cu_2O.

$$4CuO + N_2H_4 \longrightarrow 2Cu_2O + 2H_2O + N_2$$

(*c*) It reduces I_2 to HI quantitatively in presence of $NaHCO_3$.

$$N_2H_4 + 2I_2 \rightleftarrows N_2 + 4HI$$

(*d*) It also reduces $KMnO_4$ in presence of dil. H_2SO_4.

$$4KMnO_4 + 6H_2SO_4 + 5N_2H_4 \longrightarrow 2K_2SO_4 + 4MnSO_4 + 16H_2O + 5N_2$$

The above reaction can be used for the titration of hydrazine and its salts.

(*e*) It reduces ferric salts to ferrous salts and iodates (IO_3^-) to iodides (I^-).

$$4Fe^{3+} + N_2H_4 \longrightarrow 4Fe^{2+} + 4H^+ + N_2$$

$$2IO_3^- + 3N_2H_4 \longrightarrow 2I^- + 6H_2O + 3N_2$$

(*f*) It reduces halogens to the corresponding halogen acids.

$$NH_2NH_2 + 2Cl_2 \longrightarrow N_2 + 4HCl$$

$$NH_2NH_2 + 2Br_2 \longrightarrow N_2 + 4HBr$$

$$NH_2NH_2 + 2I_2 \longrightarrow N_2 + 4HI$$

Thus hydrazine is on the one hand a base, and on the other hand it is a good reducing agent just in contrast to nitric acid which is a strong acid and powerful oxidising agent.

Uses. (*i*) An organic derivative of hydrazine known as *phenyl hydrazine*, $C_6H_5NH.NH_2$. has been largely used in the preparation of hydrazones, osazones and nitrogen ring compounds in organic chemistry. (*ii*) Hydrazine has recently been employed as a rocket fuel and as an anti-T.B. drug. (*iii*) It is also used as a reducing agent in analytical chemistry.

Structure. In structure hydrazine resembles H_2O_2 and is its *ammino analogue.*

HO—OH	$H_2N—NH_2$
Hydrogen peroxide	Hydrazine
(H_2O_2)	(N_2H_4)

If it is assumed that each nitrogen atom is sp^3 hybridised, it follows that there are two lone pairs of electrons in N_2H_4 molecule, one on each of the two nitrogen atoms. Raman spectra and high value of dipole moment (about 1.85 debye units) show that N_2H_4 has an *unsymmetrical structure* as shown in Fig. 32.1. In this structure NH_2 groups do not rotate freely. This is due to the tendency of the lone pairs of electrons to get as far apart as possible. In this molecule.

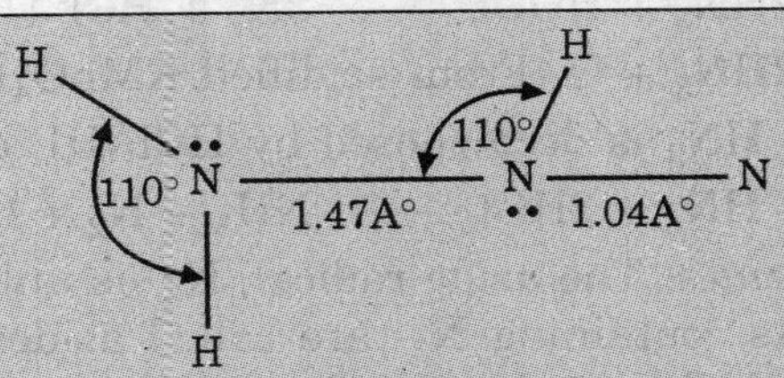

Fig. 32.1. Unsymmetrical structure of N_2H_4 molecule.

The N—N distance = 1.47 ± 0.02 Å, N—H distance = 1.04 Å and the NNH and HNH angles are about 110° each. The formation of two types of salts *viz.* : $NH_2NH_2.HCl$ and $ClH.NH_2NH_2.HCl$ can easily be explained on the basis of the above structure.

2. Hydrazoic Acid or Azoimide, NH_3.

Preparation. (*i*) *From sodamide.* When Na metal is heated to about 200°C and a current of dry NH_3 is passed over it, sodamide, $NaNH_2$ is obtained.

$$2Na + 2NH_3 \xrightarrow{200°C} 2NaNH_2 + H_2$$

When no more H_2 is evolved, in the same test tube is passed nitrous oxide (N_2O). This gives sodium azide, NaN_3 which on being distilled with dil. H_2SO_4 gives a solution of hydrazoic acid.

$$NH_2Na + N_2O \longrightarrow NaN_3 + H_2O$$

$$2NaN_3 + H_2SO_4 \longrightarrow Na_2SO_4 + 2HN_3$$

The HN_3 solution obtained as above is fractionated and finally distilled over fused calcium chloride, when anhydrous acid is obtained.

(*ii*) *From hydrazine.* HN_3 can also be prepared by the careful oxidation of hydrazine $(NH_2.NH_2)$ with NO, H_2O_2 or O_3.

$$3NH_2.NH_2 + 5O \longrightarrow 2HN_3 + 5H_2O$$

Properties. (*i*) HN_3 is a colourless volatile liquid with b.pt. = 37°. It is extremely poisonous and explosive and dissociates into N_2 and H_2 with the evolution of a large amount of heat.

$$2HN_3 \longrightarrow 3N_2 + H_2 + \text{Heat } (= 123.2 \text{ kcal})$$

(*ii*) *Reduction.* It is reduced to NH_3 or $NH_2.NH_2$ by sodium amalgam or finely divided platinum.

$$HN_3 + 4H_2 \longrightarrow 2NH_3$$
$$2HN_3 + 5H_2 \longrightarrow 2NH_2.NH_2$$

In acid solution, Ti (III), Cr (II) and Sn (II) readily reduce the acid.

$$Sn^{2+} + 3H^+ + HN_3 \longrightarrow Sn^{4+} + NH_4^+ + N_2$$

(*iii*) *Oxidation.* HNO_2, HOCl, Ce^{4+} ion and acidified $KMnO_4$ oxidise the acid to N_2 and these reactions can be used for the volumetric estimation.

$$HN_3 + HNO_2 \longrightarrow N_2 + N_2O + H_2O$$
$$2HN_3 + HOCl \longrightarrow 3N_2 + HCl + H_2O$$
$$2HN_3 + O \text{ (From acidified } KMnO_4) \longrightarrow 3N_2 + H_2O$$

(*iv*) *Decomposition.* HN_3 is decomposed by HCl, HI, etc., giving N_2

$$HN_3 + 2HCl \longrightarrow NH_3 + N_2 + Cl_2$$

(*v*) *Formation of azides.* The azide radical, N_3^- is unknown in the free state or as a dimer. The salts containing N_3^- are called azides which are formed by various ways some of which are shown below :

$$3HN_3(aq) + M(M = Zn, Cu \text{ etc.}) \longrightarrow M(N_3)_2 + N_2 + NH_3$$
$$2HN_3(aq) + Mg \longrightarrow Mg(N_3)_2 + H_2$$
$$2HN_3 \text{ (in ether)} + AlH_3 \longrightarrow Al(N_3)_3 + 3H_2$$
$$HN_3 \text{ (vapour)} + AgNO_3\text{(solution)} \longrightarrow AgN_3 \downarrow + HNO_3$$
$$HN_3 + NH_3 \longrightarrow [NH_4]^+N_3^- \text{ (Ammonium azide)}$$
$$HN_3 + N_2H_4 \longrightarrow [N_2H_5]^+N_3^- \text{ (Hydrazonium azide)}$$

Structure of HN_3 molecule and N_3^- ion. HN_3 molecule has an unsymmetrical (bent) structure with linear N—N—N skeleton while the X-ray and electron diffraction studies of the ionic metal azides have shown that N_3^- ion in these azides has symmetrical structure with linear N—N—N skeleton, *i.e.* all the N-atoms in N_3^- ion lie in a straight line with both bond lengths equal to 1.16 A° (See Fig. 32.2).

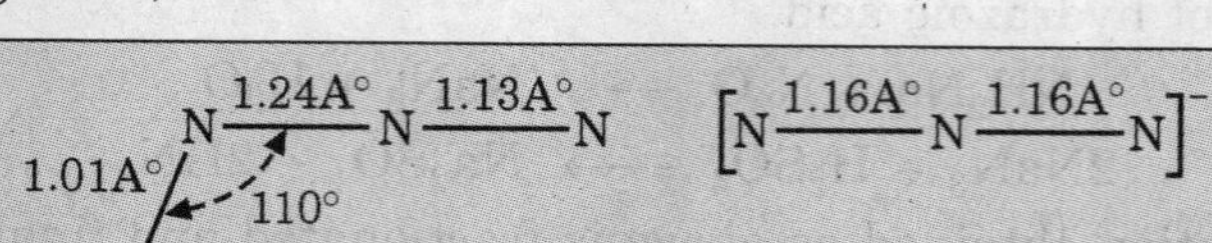

(*a*) (*b*)

Fig. 32.2. (*a*) Unsymmetrical (bent) structure of HN_3 molecule (*b*) Symmetrical structure of azide ion, N_3^-.

Electron structure of N_3^- ion is a resonance hybrid of the following three structures :

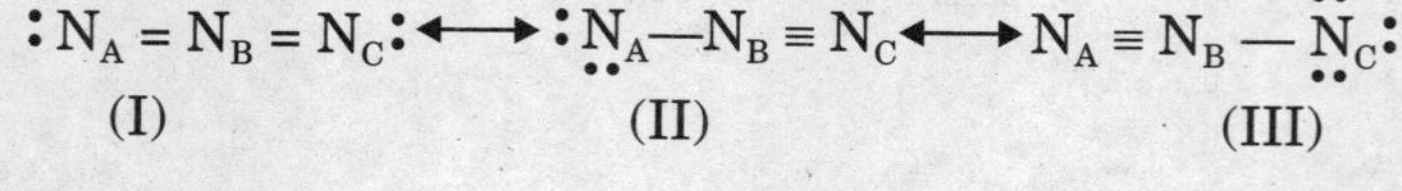

(I) (II) (III)

All the three N-atoms in structure (I) use *sp* hybrid orbitals to form σ-bonds. Each of the terminal N-atoms (N_A and N_C atoms) has a lone pair of electrons in one of the *sp* hybrid orbitals. The formation of two σ- and two π-bonds in N_3^- ion has been shown in Fig. 32.3. σ-bonds are *sp—sp* σ-bonds while π-bonds are (*p*π—*p*π) π-bonds.

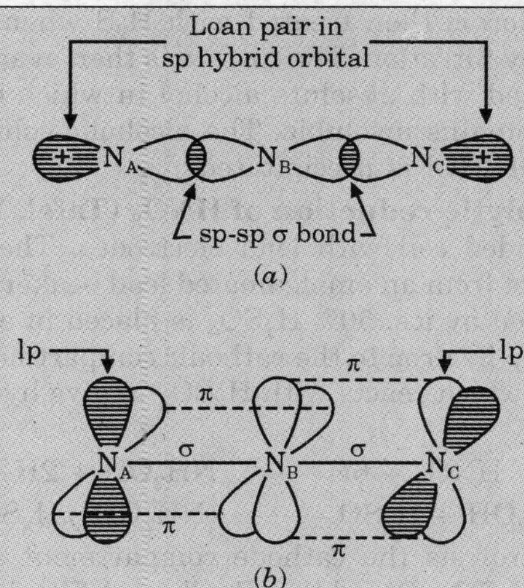

Fig. 32.3. Formation of (*a*) two σ-bonds (*b*) to π-bonds in Ni_3^- ion, σ-bonds are *sp—sp* σ-bonds while π-bonds are *p–p* π-bonds.

For the acid, HN_3 itself which is largely covalent the following three resonance structures *viz.* (A), (B) and (C) corresponding to the above three structures *viz.* (I), (II) and (III) may be written, but one of these structures *viz.* structure (C) is excluded, since in this structures two adjacent N-atoms have the same charge which is against the Pauling's adjacent charge rule. Thus HN_3 is a resonance hybrid of structures (A) and (B) only and structure (C) is a wrong structure.

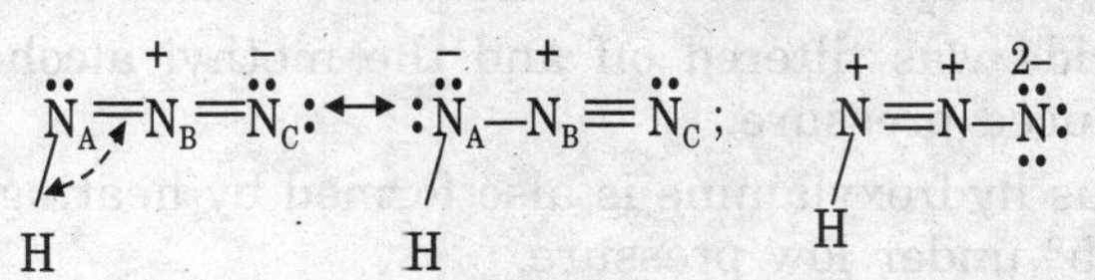

(A) (B) (C)(Wrong structure)

H–N_A–N_B bond angle equal to 110° in HN_3 molecule suggests that both the end N-atoms in structure (A) are sp^2 hybridised. One of the three sp^2 hybrids on N-atom linked with H-atom has a lone pair of electrons.

Hydroxylamine, NH_2OH

It is a hydroxy derivative of NH_3 and was discovered by Lossen in 1865.

Preparation.

As a rule only the salts of hydroxylamine are prepared because the preparation of the free hydroxylamine is difficult and not without hazard under some conditions. Salts of hydroxylamine are prepared by the following methods :

***(i)* By the reduction of NO with nascent hydrogen.** Nitric oxide (NO) is passed through mixture of granulated tin and conc. HCl when hydroxylamine hydrochloride, $NH_2OH.HCl$ or $[NH_3OH]Cl$ and a little of NH_4Cl are formed.

$$NO + 3H \longrightarrow NH_2OH$$
$$NH_2OH + HCl \longrightarrow NH_2OH.HCl$$

The resulting solution is then treated with H_2S when tin is precipitated as SnS which is removed by filtration. The filtrate is then evaporated to dryness and the dry mass is extracted with absolute alcohol in which only $NH_2OH.HCl$ gets dissolved and NH_4Cl remains insoluble. The alcoholic solution is finally treated with ether when $NH_2OH.HCl$ is precipitated.

***(ii)* By the electrolytic reduction of HNO_3 (Tafel, 1902).** The electrolysis is carried out in a divided cell with lead electrodes. The cooled lead anode is separated by a porous pot from an amalgamated lead beaker which acts as cathode. The whole thing is cooled by ice. 50% H_2SO_4 is placed in each compartment and 50% HNO_3 is added drop by drop to the cathodic compartment Cathodic reduction of HNO_3 gives HN_2OH which reacts with H_2SO_4 to give hydroxylamine sulphate, $(NH_2OH)_2.H_2SO_4$.

$$HNO_3 + 6H \longrightarrow NH_2OH + 2H_2O$$
$$2NH_2OH + H_2SO_4 \longrightarrow (NH_2OH)_2.H_2SO_4$$

At the end of electrolysis the cathode compartment contains a solution of $(NH_2OH)_2.H_2SO_4$ and H_2SO_4. By adding $BaCl_2$ and filtering off the precipitated $BaSO_4$, a solution of $NH_2OH.HCl$ is obtained.

$$(NH_2OH)_2.H_2SO_4 + BaCl_2 \longrightarrow \underset{\text{solution}}{2(NH_2OH.HCl)} + \underset{\text{ppt.}}{BaSO_4}$$

Preparation of Anhydrous Hydroxylamine.

(i) It was prepared by Lobry de.Bruyn in 1891 by treating a solution of hydroxylamine hydrochloride in methyl alcohol with a solution of sodium methoxide in methyl alcohol.

$$CH_3ONa + NH_2OH.HCl \longrightarrow CH_3OH + NaCl + NH_2OH$$

The sodium chloride was filtered off and the methyl alcohol separated by distillation under reduced pressure.

(ii) The anhydrous hydroxylamine is also formed by heating hydroxylamine orthophosphate to 125° under low pressure.

$$(NH_2OH)_3.H_3PO_4 \longrightarrow 3NH_2OH + H_3PO_4$$

Physical Properties.

It is hard needle-like white substance, melting at 33° and boiling at 58° at 22 mm pressure. It decomposes at the steam temperature. It is readily soluble in water but not much in alcohol. It can be crystallised out from its ethereal solution. Its solution in water is a weak base.

$$NH_2OH + H_2O \longrightarrow \underset{\text{Hydroxylamonium ion}}{NH_3OH^+} + OH^-$$

Chemical Properties.

(i) **Explosive nature** *(a)* It explodes with halogens.

$$2NH_2OH + 2Cl_2 \longrightarrow N_2O + H_2O + 4HCl$$

(*b*) It also explodes with permanganate and dichromate.

(*ii*) **Stability.** It is very unstable and the solid decomposes slowly above 15°C.

$$3NH_2OH \longrightarrow N_2 + NH_3 + 3H_2O$$

$$4NH_2OH \longrightarrow N_2O + 2NH_3 + 3H_2O$$

(*iii*) **Action of HNO_3 and HNO_2.** NH_2OH when heated with HNO_3 gives nitric oxide, NO

$$NH_2OH + HNO_3 \longrightarrow 2NO + 2H_2O$$

But with HNO_2 it gives nitrous oxide, N_2O

$$NH_2OH + HNO_2 \longrightarrow \underset{\text{Hyponitrous acid}}{H_2N_2O_2} + H_2O$$

$$H_2N_2O_2 \longrightarrow N_2O + H_2O$$

(*iv*) **Basic nature.** It is soluble in water and the resulting solution is a weak monoacid base (even weaker than NH_3).

$$NH_2OH + H_2O \rightleftarrows \underset{\text{Hydroxyl-amonium cation}}{[NH_3OH]^+} + OH^-$$

$[NH_3\ OH]^+$ cation is comparable to NH_4^+ ion. NH_2OH forms salts with acids which are formulated as the addition compounds of NH_2OH and the acid. For example hydroxylamine hydrochloride and hydroxylamine sulphate are formulated as $NH_2OH.HCl$ and $(NH_2OH)_2.H_2SO_4$ respectively.

(*v*) **Reducing properties.** In acidic medium, NH_2OH is powerful reducing agent. Potential equation is :

$$2NH_2OH \longrightarrow N_2O + H_2O + 4H$$

Thus it reduces :

(*a*) Fehling's solution (CuO) to a red precipitate of Cu_2O.

$$2CuO + 2H \longrightarrow Cu_2O + H_2O$$

(*b*) $AuCl_3$ to metallic Au

$$AuCl_3 + 2H \longrightarrow Au + 3HCl$$

(*c*) Fe (III) salts to Fe (II) salts

$$Fe^{3+} + H \longrightarrow Fe^{2+} + H^+$$

(*e.g.*, $FeCl_3 + H \longrightarrow FeCl_2 + HCl$)

(*d*) Halogens (X_2) to the corresponding halogen acids (HX)

$$X_2 + 2H \longrightarrow 2HX$$

Generally explosion occurs in the above reduction reactions.

(*vi*) **Oxidising properties.** In *alkaline medium*, NH_3OH acts as an oxidising agent. The potential equation is :

$$NH_2OH \longrightarrow NH_3 + O$$

Thus it oxidises :

(*a*) $Fe(OH)_2$ to $Fe(OH)_3$

$$2Fe(OH)_2 + H_2O + O \longrightarrow 2Fe(OH)_3$$

(*b*) Sodium arsenite, Na_3AsO_3 to sodium arsenate, Na_3AsO_4

$$Na_3\ AsO_3 + O \longrightarrow Na_3AsO_4$$

(*vii*) **Reaction with aldehydes and ketones.** NH_2OH reacts with aldehydes and ketones to give oximes.

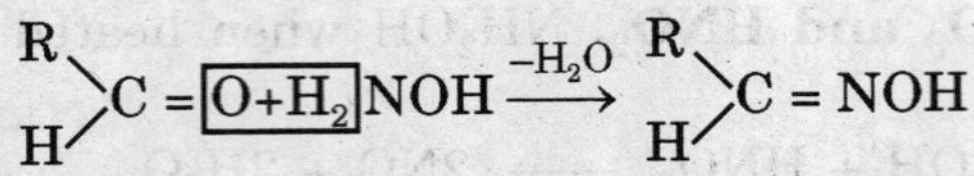

Aldehyde Aldoxime

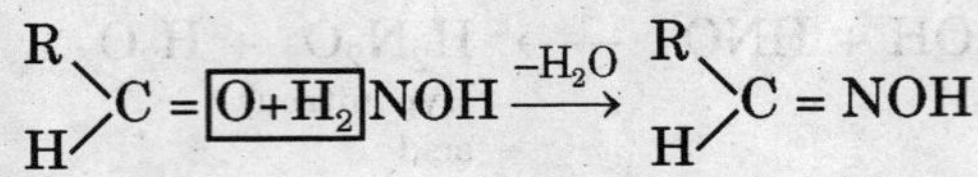

Ketone Ketoxime

Uses.

It is used as a reagent in organic chemistry.

Structure.

It has been given the following structure (I) which is similar to that of NH_3 (II)

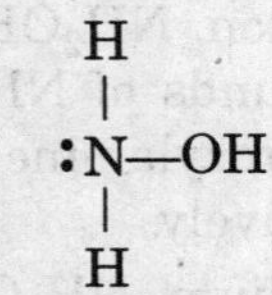

(I)

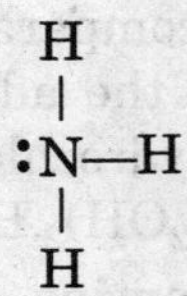

(II)

It is quite obvious that N-atom has a lone pair. In the reaction with an acid (say HCl) the lone pair gets attached with the H^+ ion obtained from the acid. Thus :

$$HCl \longrightarrow H^+ + Cl^-$$

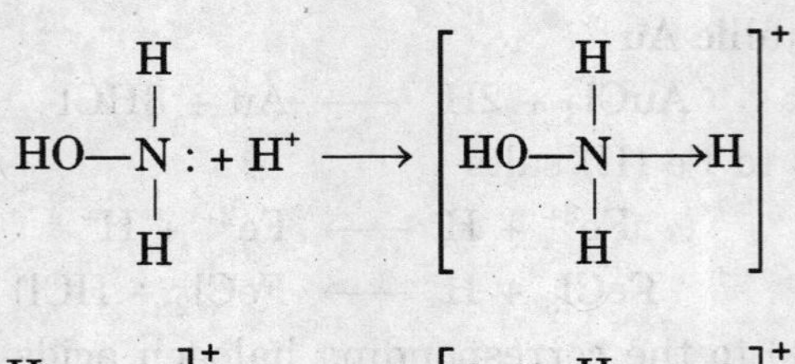

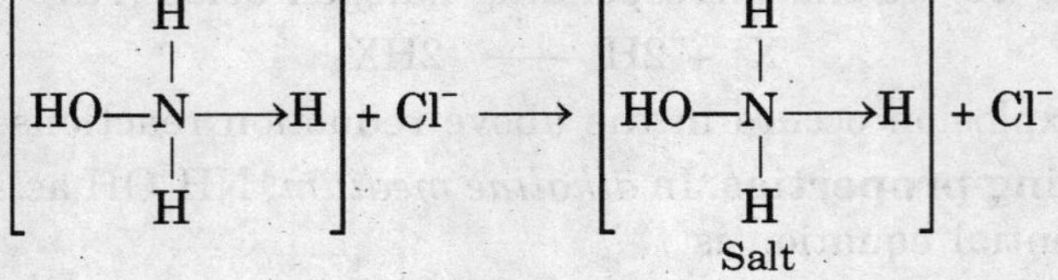

Salt

Structure (I) which is that of NH_2OH can also explain the behaviour of the molecule to act as a base, as an acid and as an oxidising agent as shown below :

(*a*) *As a base* : 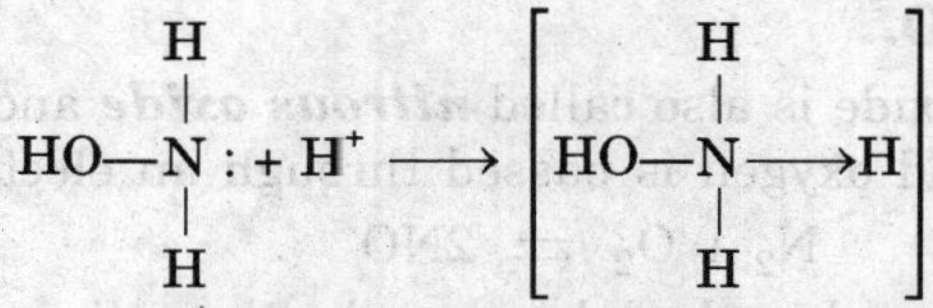

(*b*) *As an acid* : 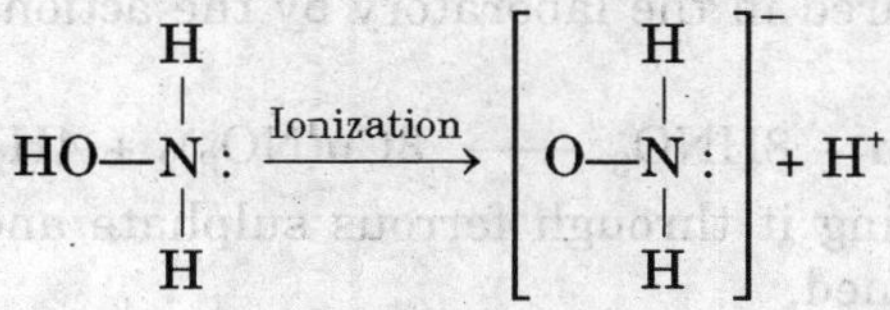

(*c*) *As an oxidising agent* : 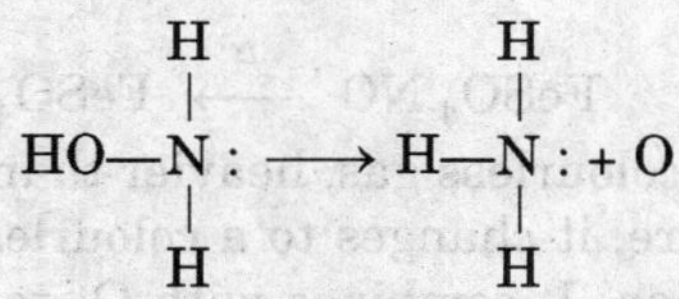

Ammonia

Oxides

1. Nitrous Oxide, N_2O.

When this oxide is inhaled in small quantities, it produces hysterical laughter and for this reason this oxide is also called ***laughing gas***.

Preparation. N_2O is prepared by heating a mixture of $NaNO_3$ and $(NH_4)_2SO_4$.

$$2NaNO_3 + (NH_4)_2SO_4 \xrightarrow{\Delta} Na_2SO_4 + 2N_2O + 4H_2O$$

Properties. Nitrous oxide is a colourless gas, having a faint sweetish smell and a sweet taste. When inhaled for long, it produces insensibility. Large quantities may prove fatal. It is soluble in water (63 volumes at 20°C) and the solution has a sweetish taste. That is why it is collected over hot water. It is heavier than air (vapour density 22). It can be easily liquefied at 0°C and a pressure of 30 atmospheres. The liquid boils at –89.5°C. The gas is neutral to litmus. It supports the combustion of substances like, sulphur, phosphorus, sodium, magnesium, candle and a splinter just as well as oxygen. This is because of its ready decomposition into its elements. Itself it is incombustible.

$$S + 2N_2O \longrightarrow SO_2 + 2N_2; \quad C + 2N_2O \longrightarrow CO_2 + 2N_2$$

$$2P + 5N_2O \longrightarrow P_2O_5 + 5N_2; \quad 2Na + N_2O \longrightarrow Na_2O + N_2$$

$$Mg + N_2O \longrightarrow MgO + N_2; \quad Cu\ (\textit{red hot}) + N_2O \longrightarrow CuO + N_2$$

A mixture of hydrogen and nitrous oxide (equal volumes) explodes with violence.

$$N_2O + H_2 \longrightarrow N_2 + H_2O$$

Ammonia and nitrous oxide explode violently.

$$3N_2O + 2NH_3 \longrightarrow 4N_2 + 3H_2O$$

Uses. Nitrous oxide is used (*i*) as the propellant gas in *whipped cream bombs*. (*ii*) as a mild anaesthetic in dental and other minor surgical operations when mixed with 10% oxygen.

2. Nitric Oxide, NO.

Preparation. This oxide is also called ***nitrous oxide*** and is produced when a mixture of nitrogen and oxygen is passed through an electric arc.

$$N_2 + O_2 \rightleftarrows 2NO$$

The gas may be prepared in the laboratory by the action of cold dilute nitric acid on copper.

$$3Cu + 8HNO_3 \longrightarrow 3Cu(NO_3)_2 + 4H_2O + 2NO$$

It is purified by passing it through ferrous sulphate and heating the black compound $FeSO_4.NO$ formed.

$$FeSO_4.NO \xrightarrow{\Delta} FeSO_4 + NO$$

Properties. It is a colourless gas, heavier than air and very sparingly soluble in water. Under pressure, it changes to a colourless liquid at –151°C. The gas is neutral to litmus solution. It combines with O_2 to form brown fumes of nitrogen peroxide (NO_2). It reacts with burning charcoal, S, P to give oxides.

$$2NO + O_2 \longrightarrow 2NO_2;\ 2NO + C \longrightarrow N_2 + CO_2$$

$$2NO + S \longrightarrow N_2 + SO_2;\ 10NO + 4P \longrightarrow 2P_2O_5 + 5N_2$$

NO reacts with ferrous sulphate, potassium sulphate etc. to form addition products.

$$FeSO_4 + NO \longrightarrow \underset{\text{Nitroso ferrous sulphate}}{[Fe(NO)]SO_4} \text{ or } FeSO_4.NO$$

$$K_2SO_4 + 2NO \longrightarrow \underset{\text{Nutroso potaesium sulphate}}{K_2SO_4.2NO}$$

$$HCl + NO \longrightarrow \underset{\text{Nitrosonium chloride}}{NO.HCl} \text{ or } [NOH]Cl$$

It reduces acidified solution of $KMnO_4$ to $MnSO_4$ and HNO_3 to NO_2.

$$6KMnO_4 + 9H_2SO_4 + 10NO \longrightarrow 3K_2SO_4 + 6MnSO_4 + 10HNO_3 + 4H_2O$$

$$2HNO_3 + NO \longrightarrow 3NO_2 + H_2O$$

It oxidises SO_2 to H_2SO_4 and H_2S to S.

$$2NO + SO_2 + H_2O \longrightarrow H_2SO_4 + N_2O;$$

$$H_2S + 2NO \longrightarrow H_2O + S + N_2O$$

It combines with Cl_2 and Br_2 to form nitrosyl chloride (NOCl) and nitrosyl bromide (NOBr) respectively.

Uses. The gas is used (*i*) for the manufacture of nitric acid. (*ii*) as a catalyst in the lead chamber process for the manufacture of sulphuric acid. (*iii*) to distinguish oxygen from, nitrous oxide.

3. Dinitrogen Trioxide, N_2O_3.

This oxide is also called ***nitrogen sesquioxide.*** It is also called ***nitrous anhydride***, *i.e.* anhydride of HNO_2, since it gives HNO_2 when treated with H_2O.

$$N_2O_3 + H_2O \longrightarrow 2HNO_2$$

Preparation. Nitrogen trioxide is prepared by the reduction of nitric acid with arsenious oxide

$$As_2O_3 + 2HNO_3 + 2H_2O \longrightarrow N_2O_3 + 2H_3AsO_4$$

Properties. It is a red coloured gas. On condensing, it gives a dark blue liquid. It is acidic in nature and reacts with sodium hydroxide to form sodium nitrite.

$$2NaOH + N_2O_3 \longrightarrow 2NaNO_2 + H_2O$$

On heating it is dissociated to nitric oxide (NO) and nitrogen peroxide (NO_2).

$$N_2O_3 \xrightarrow{\Delta} NO + NO_2$$

4. Nitrogen Dioxide $\rightleftharpoons$ Dinitrogen Tetroxide, $2NO_2 \rightleftharpoons N_2O_4$.

This oxide is also called ***nitrogen peroxide***. Since it gives a mixture of HNO_3 and HNO_2, it is also called ***mixed acid anhydride*** *i.e. anhydride of HNO_2 and HNO_3.*

$$2NO_2 + H_2O \longrightarrow HNO_2 + HNO_3$$

Preparation. NO_2 is prepared by the action of heat on $Pb(NO_3)_2$ (laboratory preparation) or by the action of Cu on conc. HNO_3.

$$2Pb(NO_3)_2 \xrightarrow{\Delta} 2PbO + 4NO_2 + O_2$$
$$Cu + 4HNO_3 \longrightarrow Cu\,(NO_3)_2 + 2NO_2 + 2H_2O$$

Properties. Nitrogen dioxide is a reddish-brown gas with a pungent smell. At 14°C it has a *deep brown* colour and a density which corresponds to the formula, NO_2. As the temperature falls, its colour goes on fading. On further cooling, it condeneses to a *pale-yellow* liquid (b.p. = 22°C) which freezes to *colourless* crystals (m.p. = –9°C). The density of the solid corresponds to the formula. N_2O_4. Quite evidently, *progressive association of the molecules occurs with the decrease in temperature.*

$$2NO_2 \rightleftarrows N_2O_4$$

The colour of nitrogen dioxide begins to fade above 140°C also. This is due to its dissociation into nitric oxide (NO) and oxygen (O_2).

$$2NO_2 \rightleftharpoons 2NO + O_2$$

The dissociation is complete at 620°C when the gas becomes completely colourless. These changes are reversed on cooling.

$$\underset{\text{Solid (Colourless)}}{N_2O_4} \overset{-90^\circ}{\rightleftharpoons} \underset{\text{Liquid (Pale-yellow)}}{N_2O_4} \overset{22^\circ}{\rightleftharpoons} \underset{\text{Vapour (Pale-yellow)}}{N_2O_4} \overset{140^\circ}{\rightleftharpoons} \underset{\text{Gas (Reddish-brown)}}{2NO_2} \overset{620^\circ}{\rightleftharpoons} \underbrace{2NO + O_2}_{\text{Gaseous mixture (Colourless)}}$$

N_2O_4	N_2O_4	N_2O_4	$2NO_2$	$2NO + O_2$
Solid	Liquid	Vapour	Gas	Gaseous mixture
(Colourless)	(Pale-yellow)	(Pale-yellow)	(Reddish-brown)	(Colourless)

At any temperature below 22° and 140°, the gas exists as an equilibrium mixture of *coloured* NO_2 molecules and *colourless*, N_2O_4 molecules.

$$2NO_2 \rightleftharpoons N_2O_4$$

It is soluble in water. This can be shown by *fountain experiment.* It is much denser than air and can be *poured* downwards like CO_2. It is poisonous and sets up septic pneumonia, if inhaled. It also corrodes the skin.

It is acidic towards litmus. When dissolved in water, it forms a mixture of nitrous and nitric acids. It is, therefore, a *mixed anhydride*.

$$N_2O_4 + H_2O \longrightarrow HNO_2 + HNO_3$$

It neutralizes alkali solution with the formation of mixture of alkali nitrate and nitrite, thus proving its acidic nature.

$$2NaOH + 2NO_2 \longrightarrow NaNO_2 + NaNO_3 + H_2O$$

It supports the combustion of brightly being phosphorus, but does not support the combustion of candle, a splinter or sulphur. Metals like iron and tin burn in it when heated to 500°C.

It acts as a strong *oxidising agent*. Thus :

(*a*) It liberates iodine from potassium iodide, and turns starch iodide paper blue.

$$2KI + 2NO_2 \longrightarrow 2KNO_2 + I_2.$$

(*b*) Suphuretted hydrogen is oxidised to sulphur and carbon monoxide to carbon dioxide.

$$NO_2 + H_2S \longrightarrow NO + H_2O + S$$
$$CO + NO_2 \longrightarrow CO_2 + NO$$

(*c*) Ferrous sulphate is oxidised to ferric sulphate.

$$2FeSO_4 + H_2SO_4 + NO_2 \longrightarrow Fe_2(SO_4)_3 + H_2O + NO$$

Due to the combination of the nitric oxide formed with the unreacted ferrous sulphate, the whole mass turns black.

(*d*) It oxidises metals like potassium, sodium, mercury and tin. Finely divided copper gives cuprous oxide which then absorbs more of it forming "*nitro copper*".

$$NO_2 + 2Na \longrightarrow Na_2O + NO$$
$$NO_2 + 2K \longrightarrow K_2O + NO$$
$$2Cu + NO_2 \longrightarrow Cu_2O + NO$$

(*e*) Phosphorus, sulphur and charcoal burn in it to form their oxides.

$$5NO_2 + 2P \xrightarrow{\text{heat}} P_2O_5 + 5NO$$

$$2NO_2 + S \xrightarrow{\text{heat}} SO_2 + 2NO$$

$$2NO_2 + C \xrightarrow{\text{heat}} 2O_2 + 2NO$$

(*f*) A mixture of carbon disulphide and nitrogen dioxide explodes violently.

$$2CS_2 + 6NO_2 \longrightarrow 2CO_2 + 4SO_2 + 3N_2$$

(*g*) Hydrogen reacts with NO_2 in the presence of catalysts like platinum or finely divided nickel to give ammonia and water.

$$7H_2 + 2NO_2 \xrightarrow{\text{heat}} 2NH_3 + 4H_2O$$

It reduces ozone to oxygen and is itself oxidized to nitrogen pentoxide.

$$N_2O_4 + O_3 \longrightarrow N_2O_5 + O_2$$

Uses. It is used (*i*) as a catalyst in the manufacture of H_2SO_4 by lead chamber process. (*ii*) in the production of HNO_2. (*iii*) as a rocket fuel.

5. Nitrogen Pentoxide, N_2O_5.

This oxide is also called ***nitric anhydride***, *i.e.* anhydride of HNO_3, since it makes HNO_3 with H_2O

$$N_2O_5 + H_2O \longrightarrow 2HNO_3$$

Preparation. Nitrogen pentoxide is prepared by distiling concentrated nitric acid over phosphorus pentoxide

$$2HNO_3 + P_2O_5 \longrightarrow 2HPO_3.$$

The vapours of N_2O_5 so obtained are cooled in a U-tube which is cooled by solid carbon. dioxide and ether.

It is also obtained by passing dry chlorine over solid silver nitrate at 95°C and as soon as the reaction starts, cooling the mixture to 60°C.

$$4AgNO_3 + 2Cl_2 \longrightarrow 4AgCl + 2N_2O_5 + O_2$$

Properties. N_2O_5 forms colourless hard crystals which are deliquescent and which melt at 30°C. N_2O_5 is very unstable and may explode even without heating. It decomposes to give nitrogen dioxide and oxygen. It dissolves in water with a hissing giving nitric acid and is, therefore, called *nitric anhydride*.

$$2N_2O_5 \longrightarrow 4NO_2 + O_2$$

$$N_2O_5 + H_2O \longrightarrow 2HNO_3$$

Structure of Oxides

1. Nitrous Oxide, N_2O.

Nitrous oxide is a linear and unsymmetrical molecule with a very small value of dipole moment (= 0.116D). Thus N_2O molecule is a resonance hybrid of the two structures shown in the margin.

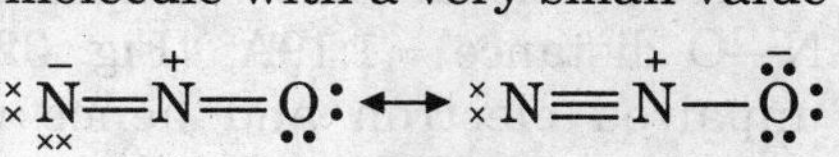

2. Nitric oxide, NO.

According to valence bond theory, NO molecule is an *odd electron* molecule, since it contains an odd number of electrons (= 11 electrons = 5 from N-atom + 6 from O-atom). NO molecule in the *gaseous state* has one unpaired electron which makes this molecule paramagnetic. Thus its structure can be represented as shown in the margin. Since the observed nitrogen–oxygen bond length (= 1.14 A°) is in between that of a N = O bond (= 1.15 A°) and N ≡ O (= 1.05 A°), the true structure of NO molecule is supposed to be a resonance hybrid of the structures shown in the margin. According to magnetic evidence the odd electron in NO molecule spends half of its time on N-atom and half of its time on O-atom. This situation can be represented by assuming that NO molecule contains one *three-electron bond* [represented as (3c—2e) bond] and two *two-electron bonds* [represented as (2c—2e) bond] as shown in the margin.

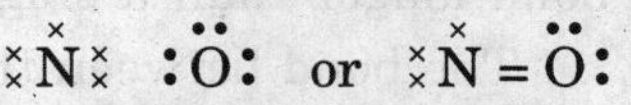

Structure of nitric oxide molecule, NO

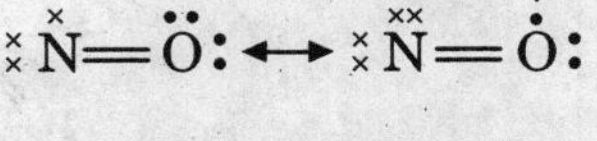

(3c—2e) bond is about half as strong as a normal (2c—2e) bond. This structure is in agreement with small value of dipole moment (= 0.16 D) of NO molecule.

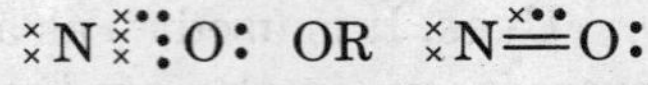

NO molecule contains one (3c–2c) and two (2c–2e) bonds.

X-ray analysis has shown that in the *solid state* NO molecule contains *rectangular dimers* whose dimensions are given in the margin.

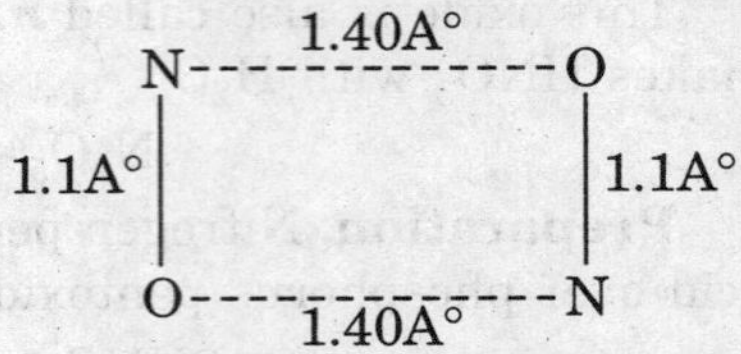

NO molecule contains rectangular dimers in the solid state.

The structure of NO molecule has also been explained on the basis of molecular orbital theory as discussed on page 602 in Chapter 8.

5. Dinitrogen Trioxide, N_2O_3.

Since in liquid and gaseous states N_2O_3 dissociates to give NO and NO_2 as shown below :

$$N_2O_3 \rightleftharpoons NO + NO_2,$$

it may be assumed that N_2O_3 has the structure :

$$:\ddot{O}=\overset{\times\times}{N}-\ddot{\underset{\cdot\cdot}{O}}-\overset{\times\times}{N}=\ddot{O}: \quad or \quad ON—NO_2$$

This structure is confirmed by the tracer studies which have shown that when nitrogen dioxide labelled with N^{15} ($N^{15}O_2$) is mixed with $N^{14}O$, a rapid exchange between N^{14} and N^{15} atoms takes place as shown below :

$$N^{14}O + N^{15}O_2 \rightleftarrows N^{15}O + N^{14}O_2$$

N_2O_3 molecule is *diamagnetic* in spite of its intense blue colour and hence it is not an *odd electron molecule.*

4. Nitrogen Dioxide (NO_2) and Dinitrogen Tetroxide (N_2O_4).

NO_2 molecule has *V-shaped* structure with O—N—O bond angle = 132° and N—O distance = 1.19A° [Fig. 32.4(*a*)]. This structure shows that NO_2 has one unpaired electron and hence it is *an odd electron molecule*. Thus NO_2 is *paramagnetic*. NO_2 molecule has a tendency to dimerise to colourless dimer N_2O_4 which is *planar* with long N—N bond = 1.75A° and O—N—O bond angle = 132° as shown in Fig. 32.4(*b*). The planar structure of N_2O_4 admits some δ+ charge on each nitrogen atom. The repulsion between the adjacent δ+ charges make N—N bond longer than a single bond.

The bond between the two nitrogen atoms in N_2O_4 is established by the two unpaired electrons residing on nitrogen atom in NO_2. The structure of N_2O_4 as shown below shows that it has no unpaired electron and hence is *diamagnetic.* Note that NO_2 is paramagnetic.

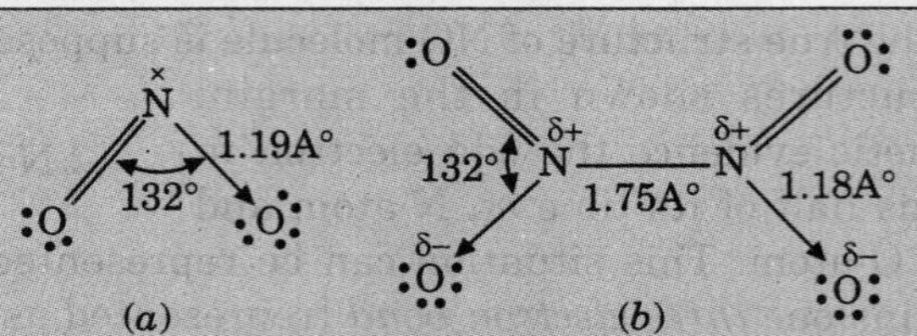

Fig. 32.4 (*a*) V-shaped structure of NO_2 molecule
(*b*) Planar structure of N_2O_4 molecule.

5. Nitrogen Pentoxide, N_2O_5.

X-ray studies of N_2O_5 suggest that it is an ionic solid and hence is *nitronium nitrate*, $NO_2^+NO_3^-$. But in its vapour state it exists as a *symmetrical molecule* having the structure $O_2N—O—NO_2$ as shown below :

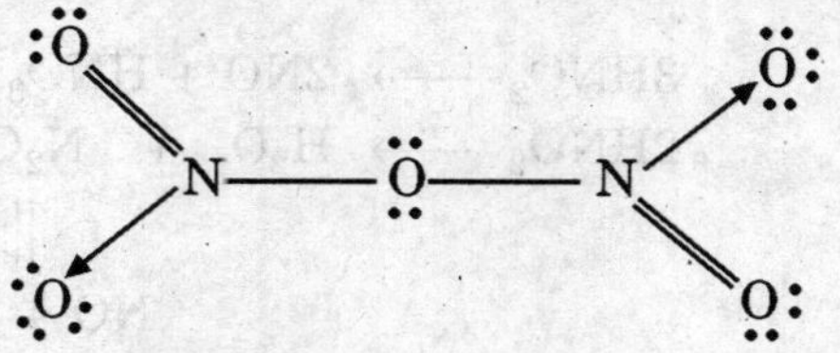

N—O—N bond is almost linear *i.e.* N—O—N bond angle = 80°.

Oxyacids

1. Hyponitrous Acid, HNO or $H_2N_2O_2$.

Preparation. Free acid is prepared by treating silver hyponitrite. $Ag_2N_2O_2$ with a solution of HCl in dry ether or with H_2S.

$$Ag_2N_2O_2 + 2HCl \longrightarrow \underset{\text{solution}}{H_2N_2O_2} + \underset{\text{ppt.}}{2AgCl}$$

$$Ag_2N_2O_2 + H_2S \longrightarrow \underset{\text{solution}}{H_2N_2O_2} + \underset{\text{ppt.}}{Ag_2S}$$

The solution containing $H_2N_2O_2$ is evaporated to dryness in vacuum when hygroscopic crystals (white leaflets when pure) of the acid are obtained.

Properties. Pure acid forms white leaflets which explode when rubbed. When exposed to air, the acid decomposes slowly giving HNO_2 and HNO_3. The aqueous solution of the acid decomposes on heating evolving N_2O.

$$2H_2N_2O_2 + 3O_2 \longrightarrow 2HNO_2 + 2HNO_3$$

$$H_2N_2O_2 \longrightarrow N_2O + H_2O$$

The acid is soluble in water, alcohol, chloroform, ether and benzene. It is a weak dibasic acid and hence forms two types of salts *viz.*, *acid salts* like $Ba(HN_2O_2)_2$ and *normal salts* (called *hyponitrites* like BaN_2O_2. The acid is a reducing agent and is oxidised to nitrate by acidified $KMnO_4$.

$$5H_2N_2O_2 + 8KMnO_4 + 12H_2SO_4 \longrightarrow 10HNO_3 + 4K_2SO_4 + 8MnSO_4 + 12H_2O$$

2. Nitrous Acid, HNO_2.

Preparation. This is an unstable acid and is known only in solution. A solution of nitrous acid may be prepared :

(*i*) by the action of water on its anhydride, N_2O_3.

$$N_2O_3 + H_2O \longrightarrow 2HNO_2$$

(*ii*) by the action of barium nitrite with calculated quantity of dilute sulphuric acid. The precipitate of barium sulphate so formed is filtred off and the filtrate contains nitrous acid. Since the acid is very unstable, the reaction is carried out at the temperature of freezing mixtures.

$$Ba(NO_2)_2 + H_2SO_4 \longrightarrow \underset{\text{ppt.}}{BaSO_4} + 2HNO_2$$

Properties. (*i*) Nitrous acid is a weak acid. Aqueous solution of nitrous acid is pale blue due to the presence of nitrogen trioxide. This colour fades on standing for some time. The acid is unstable and even in cold solution undergoes auto-oxidation. At higher temperatures it decomposes into nitric oxide and nitrogen peroxide.

$$3HNO_2 \longrightarrow 2NO + HNO_3 + H_2O$$

$$2HNO_2 \longrightarrow H_2O + N_2O_3 \rightleftharpoons NO + NO_2$$

(*ii*) It acts as a good oxidising agent on account of the ease with which it can part with an atom of oxygen.

$$2HNO_2 \longrightarrow 2NO + H_2O + O$$

For example : (*a*) *It oxidised hydrogen sulphide to sulphur*

$$2HNO_2 + H_2S \longrightarrow 2H_2O + 2NO + S$$

(*b*) *Iodine is liberated from potassium iodide.*

$$2KI + 2HNO_2 \longrightarrow 2KOH + I_2 + 2NO$$

(*c*) *Stannous chloride is oxidised to stannic chloride.*

$$SnCl_2 + 2HCl + 2HNO_2 \longrightarrow SnCl_4 + 2NO + 2H_2O$$

(*d*) *It can oxidise sulpur dioxide to sulphuric acid.*

$$2HNO_2 + SO_2 \longrightarrow 2NO + H_2SO_4$$

(*e*) It can oxidise $FeSO_4$ to $Fe_2 (SO_4)_3$

$$2FeSO_4 + H_2SO_4 + 4HNO_2 \longrightarrow Fe_2(SO_4)_3 + 2NO + 2H_2O$$

(*f*) *Hydriodic acid is oxidised to iodine.*

$$2HI + 3HNO_2 \longrightarrow I_2 + 2NO + 2H_2O$$

(*g*) *Sodium arsenite is oxidised to sodium arsenate.*

$$Na_3AsO_3 + 2HNO_2 \longrightarrow H_2O + 2NO + Na_3AsO_4$$

(*iii*) It can be easily oxidised to nitric acid and as such reduces strong oxidising agents.

$$HNO_2 + O \longrightarrow HNO_3$$

For example :

(*a*) *It reduces bromine and iodine to hydrobromic acid and hydriodic acid respectively.*

$$HNO_2 + I_2 + H_2O \longrightarrow 2HI + HNO_3$$

$$HNO_2 + Br_2 + H_2O \longrightarrow 2HBr + HNO_3$$

(*b*) *Acidified potassium permanganate is reduced to manganese sulphate*

$$KMnO_4 + 3H_2SO_4 + 5HNO_2 \longrightarrow K_2SO_4 + 2MnSO_4 + 5HNO_3 + 3H_2O$$

(*c*) *Acidified potassium dichromate is reduced to chromium sulphate (green colour).*

$$K_2Cr_2O_7 + 4H_2SO_4 + 5HNO_2 \longrightarrow K_2SO_4 + Cr_2(SO_4)_3 + 3HNO_3 + 4H_2O$$

(*d*) *It also reduces hydrogen peroxide*

$$H_2O_2 + HNO_2 \longrightarrow HNO_3 + H_2O$$

(*iv*) Nitrous acid reacts with alkalies to form nitrites. Nitrites are much more stable than nitrous acid.

$$NaOH + HNO_2 \longrightarrow NaNO_2 + H_2O$$

Uses. Nitrous acid is used in organic chemistry especially in the preparation of diazo compounds which are the basic substances in the manufacture of aniline dyes. It is also used both as an oxidising and reducing agent in analytical chemistry.

3. Nitric Acid, HNO_3.

In the laboratory it is prepared by heating conc. H_2SO_4 and an alkali nitrate in a glass retort.

Manufacture. (*i*) *From nitre ($NaNO_3$).* Nitric acid is prepared on a large scale by heating sodium nitrate with concentrated sulphuric acid. The mixture is placed in a cast iron retort (Fig. 32.5) and subjected to distillation at a temperature of about 200°C.

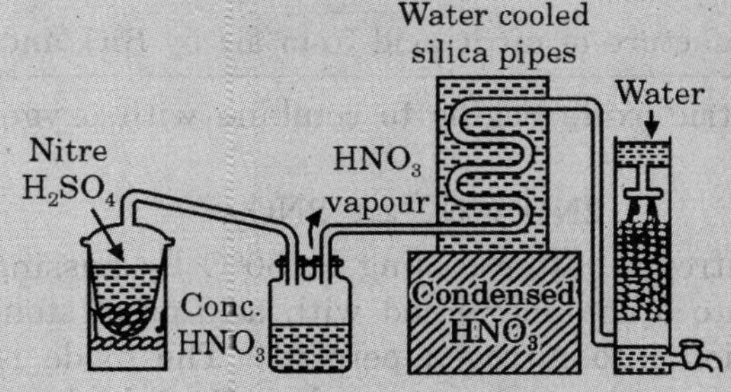

Fig. 32.5. Manufacture of nitric acid from nitre, $NaNO_3$.

The vapours of nitric acid from the pot are conducted to a trap and then passed through water-cooled silica pipes. The acid condensed in the pipes is sent back to the trap. The uncondensed vapours of nitric acid are then finally passed up a tower packed with quartz from the top of which a steam of water is trickling down. The vapours are dissolved in water and the dilute nitric acid thus produced collects at the bottom.

The residue left in the pot is known as *nitre cake* and is used in bleaching and dyeing industries. It is also used for the preparation of hydrochloric acid.

The acid so obtained is *concentrated* by distilling the dilute acid with concentrated sulphuric acids. Commercial nitric acid contains hydrochloric acid, sulphuric acid, iodine, iron salts and nitrogen pentoxide as impurities. A current of dry air passed through the liquid removes N_2O_5. Further *purification* is carried out by distilling the acid under reduced pressure.

(ii) From air by Birkland-Eyde process. Nitrogen and oxygen of the air combine to form nitric oxide when air is passed through an electric arc.

$$N_2 + O_2 \rightleftarrows 2NO - 43{,}200 \text{ cals.}$$

The reaction is endothermic and is favoured by a high temperature.

In the Birkland-Eyde process, air is blown into an electric arc struck between two water cooled copper electrodes and spread into a large disc with the help of a strong magnetic field applied at right angles (Fig. 32.6)

The issuing gases contain about 1.25 per cent of nitric oxide. The gases escaping from the furnace have to be rapidly cooled down to a temperature of about 1000°C. Their heat is used for steam raising.

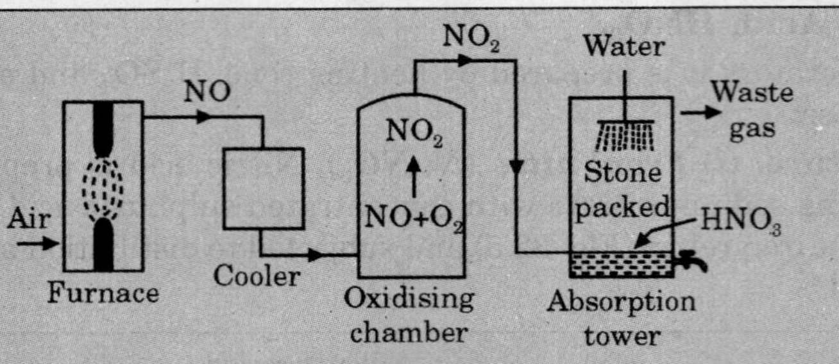

Fig. 32.6. Manufacture of nitric acid from air by Birkland-Eyde process.

At 600°C the nitric oxide begins to combine with oxygen to from nitrogen peroxide.

$$2NO + O_2 \rightleftarrows 2NO_2.$$

The oxides of nitrogen, after cooling to 50°C by passing the pipe through water boiler pass into a chamber lined with acid-proof stone in order to carry further the conversion into nitrogen peroxide. The oxide is then absorbed in water trickling over quartz stone in towers about 70 ft. high and 20 ft. in diameter.

The nitric acid in the tower reaches a strength of 30 to 40 per cent while in the succeeding towers, the strength remains at 20, 10 and 5 per cent respectively. In case of cold dilute solutions, the nitric acid is formed according to the equation

$$2NO_2 + H_2O \longrightarrow HNO_2 + HNO_3$$

In the hot state, the nitrous acid under goes autoxidation to nitric acid, and the nitric oxide formed again takes up oxygen and forms nitrogen peroxide.

$$2HNO_2 \longrightarrow H_2O + 2NO + O$$
$$HNO_2 + O \longrightarrow HNO_3$$
$$\overline{3HNO_2 \longrightarrow HNO_3 + 2NO + H_2O}$$

(iii) By catalytic oxidation of ammonia (Ostwald's process). Owing to the great wear and tear of the plant because of high temperature employed in the Birkland-Eyde process, Ostwald's process is finding preference.

In this method a mixture of ammonia gas with air in the ratio 1 : 8 is passed through a catalyst chamber packed with platinum gauze maintained at a temperature of 803°C. The ammonia is oxidised under these conditions to nitric oxide.

$$4NH_3 + 5O_2 \longrightarrow 4NO + 6H_2O$$

The nitric oxide so formed is led to an oxidising tower (Fig. 32.7) where it combines with oxygen of the air to form nitrogen peroxide, NO_2

$$\underset{\text{Nitric oxide}}{2NO} + O_2 \longrightarrow \underset{\text{Nitrogen peroxide}}{2NO_2}$$

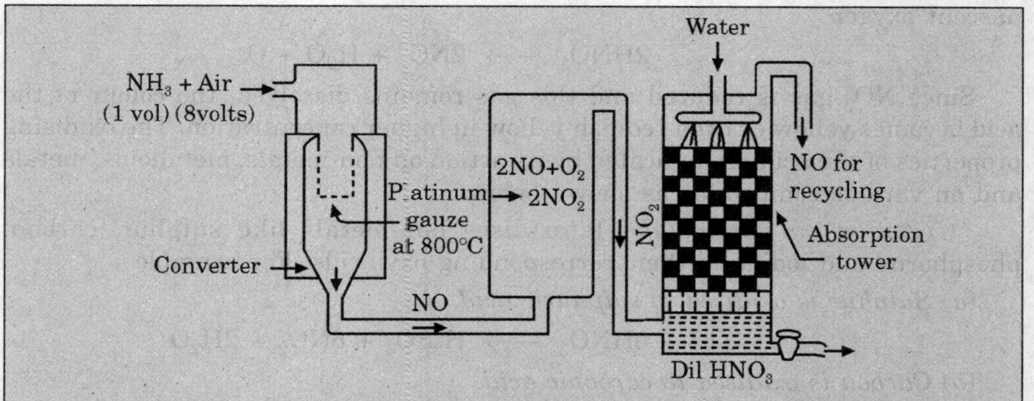

Fig. 32.7. Ostwald's process for the manufacture of nitric acid.

The resulting gas passes up a tower (called *absorption tower*) packed with earthen balls down which a spray of water trickles. Nitrogen peroxide reacts with water in the presence of oxygen forming nitric acid.

$$4NO_2 + 2H_2O + O_2 \longrightarrow 4HNO_3$$

The acid so obtained is dilute and concentrated by distillation under reduced pressure.

Preparation of concentrated nitric acid. The dilute HNO_3 can be concentrated to 68% by distillation, when constant boiling mixture is formed. This constant boiling mixture of HNO_3 has a specific gravity of 1.42 and is the commercial concentrated HNO_3. More concentrated acid needed for certain uses such as nitration is produced by distillation with conc. H_2SO_4 which holds back the water. The distillate is 98% HNO_3. The cent percent HNO_3 is obtained by strongly cooling 98% acid when the pure acid is deposited as colourless crystals at –42°C. These crystals are separated and melted to 100% HNO_3.

A recently developed method for the direct production of conc. HNO_3 involves the reaction between the liquid NO_2 and water (or dil. HNO_3) in the presence of O_2 at 50 atm. pressure and a temperature of 75°C. The product is 88% HNO_3.

$$4NO_2 + O_2 + 2H_2O \longrightarrow 4HNO_3$$

Preparation of fuming nitric acid. It is made by distilling conc. HNO_3 with a little starch. The starch reduces some acid to NO_2 which dissolves in the remaining acid to form fuming nitric acid. This acid is yellow in colour due to the presence in it of dissolved oxides of nitrogen highly corrosive liquid consisting of HNO_3. Fuming nitric acid is a much more powerful oxidising agent and nitrating agent than conc. HNO_3.

Physical properties of nitric acid. Pure HNO_3 is a colourless liquid with a characteristic choking smell. It boils at 86°C and freezes at –42°C into a transparent crystalline mass. It is soluble in water in all proportions. In the moist air, it fumes as its vapours form a mist of minute droplets of aqueous HNO_3.

Chemical properties of HNO_3 (1) **Oxidising property.** Nitric acid is a strong oxidising agent because of the ease with which it decomposes to give nascent oxygen.

$$2HNO_3 \longrightarrow 2NO_2 + H_2O + O$$

Since NO_2 gas is reduced and this gas remains dissolved, the colour of the acid becomes yellow or oven reddish yellow in higher concentration. The oxidising properties of the acid are indicated by its action on non-metals, metalloids, metals and on various compounds as shown below :

(*i*) *Action on non-metals.* It oxidises non-metals like sulphur, carbon, phosphorus and iodine to their corresponding oxy-acids. For example :

(*a*) *Sulphur is oxidised to sulphuric acid.*

$$S + 6HNO_3 \longrightarrow H_2SO_4 + 6NO_2 + 2H_2O$$

(*b*) *Carbon is oxidised to carbonic acid.*

$$C + 4HNO_3 \longrightarrow H_2CO_3 + 4NO_2 + H_2O$$

(*c*) *Iodine is oxidised to iodic acid (HIO_3).*

$$I_2 + 10HNO_3 \longrightarrow 2HIO_3 + 10NO_2 + 4H_2O$$

(*d*) *Phosphorus is oxidised to ortho-phosphoric acid (H_3PO_4).*

$$P + 5HNO_3 \longrightarrow H_3PO_4 + 5NO_2 + H_2O$$

(*ii*) *Action on metalloids.* Metalloids like arsenic and antimony are oxidised to the corresponding oxy-acids. *e.g.*, arsenic is oxidised to ortho-arsenic acid (H_3AsO_4).

$$As + 5HNO_3 \longrightarrow H_3AsO_4 + 5NO_2 + H_2O$$

(*iii*) *Action on metals.* Most of the metals, with the exception of noble metals like gold and platinum, are attacked by nitric acid. Some like tin and antimony give *oxides* whilst the others form nitrates. During the reaction, a part of the acid is reduced to give products like NO_2, NO, N_2, N_2O, NH_2OH or NH_3 and the nature of products depends upon the concentration of the acid, temperature and the nature of the metal.

It is believed that the *primary reaction* in each case is the liberation of nascent hydrogen which then reduces the acid to different reaction products mentioned above.

$$M + HNO_3 \longrightarrow MNO_3 + H$$

According to another view, this reduction occurs only in case of metals which are above hydrogen in the electrochemical series. Metals like Cu, Ag, Hg which cannot displace hydrogen are oxidised by nitric acid and the oxide formed, being more basic, reacts with more of nitric acid to form the nitrate. Examples are :

(*a*) *Copper.* With dilute nitric acid, it gives nitric oxide.

$$3Cu + 8HNO_3 \longrightarrow 3Cu(NO_3)_2 + 2NO + 4H_2O$$

With concentrated nitric acid, it forms nitrogen peroxide.

$$Cu + 4HNO_3 \longrightarrow Cu(NO_3)_2 + 2NO_2 + 2H_2O$$

Silver, mercury and lead give similar reactions.

(*b*) *Tin.* With dilute nitric acid, it forms stannous nitrate and ammonium nitrate.

$$4Sn + 10HNO_3 \longrightarrow 4Sn(NO_3)_2 + 3H_2O + NH_4NO_3$$

With concentrated nitric acid it gives meta-stannic acid and nitrogen peroxide (*cf.* Action on non-metals).

$$Sn + 4HNO_3 \longrightarrow \underset{\text{Meta stannic acid}}{H_3SnO_3} + H_2O + 4NO_2$$

(*c*) *Zinc.* With very dilute nitric acid it gives ammonium nitrate.

$$4Zn + 10HNO_3 \longrightarrow 4Zn(NO_3)_2 + 3H_2O + NH_4NO_3$$

With dilute acid, it produces N_2O while concentrated acid gives NO_2.

(*d*) *Magnesium and manganese* are the only metals that liberate hydrogen from dilute acid.

$$\underset{\text{dilute}}{2HNO_3} + Mg \longrightarrow Mg(NO_3)_2 + H_2$$

(*e*) *Gold* and *platinum* dissolve in a mixture of 1 part HNO_3 and 3 parts HCl known as *aqua regia.* The dissolution occurs as a result of the action of nascent chlorine on the metals.

$$HNO_3 + 3HCl \longrightarrow NOCl + 2H_2O + 2Cl$$

(*iv*) *Action on compounds.* A number of compounds are oxidised. For example it oxidises :

(*a*) *sulphur dioxide to the sulphuric acid.*

$$SO_2 + 2HNO_3 \longrightarrow H_2SO_4 + 2NO_2$$

(*b*) *sulphuretted hydrogen to sulphur.*

$$H_2S + 2HNO_3 \longrightarrow 2H_2O + 2NO_2 + S$$

(*c*) *ferrous sulphate to ferric sulphate* in the presence of sulphuric acid.

$$6FeSO_4 + 2HNO_3 + 3H_2SO_4 \longrightarrow 3Fe_2(SO_4)_3 + 4H_2O + 2NO$$

The nitric oxide formed gives a dark brown ring (*Ring test for nitrates*).

(*d*) *cane sugar to oxalic acid.*

$$C_{12}H_{22}O_{11} + 18O \longrightarrow 6(COOH)_2 + 5H_2O$$

(2) **Nitrating property.** It reacts with organic compounds and forms their nitro compounds. The reaction takes place in the presence of stronger acids like conc. H_2SO_4.

In the presence of an acid which is stronger than nitric acid, NO_2^+ ions (*nitronium ions*) are produced.

$$HNO_3 + 2H_2SO_4 \longrightarrow NO_2^+ + 2HSO_4^- + H_3O^+$$

When NO_2^+ ions replace the hydrogen ions in organic compounds, nitration is said to take place. Thus :

$$\underset{\text{benzene}}{C_6H_6} + HNO_3 \xrightarrow{H_2SO_4} \underset{\text{nitrobenzene}}{C_6H_5NO_2} + H_2O$$

The yellow strain produced on wood or skin is due to this reaction.

(3) **Action on HCl.** A mixture of conc. HCl and conc. HNO_3 (3 : 1) is called aqua regia. It reacts considerably more vigorously than either of these acids taken separately.

$$HNO_3 + 3HCl \longrightarrow 2H_2O + \underset{\text{Nitrosyl chloride}}{NOCl} + Cl_2$$

Aqua regia dissolves even noble metals like gold and platinum.

$$Au + HNO_3 + 3HCl \longrightarrow AuCl_3 + NO + 2H_2O$$
$$3Pt + 4HNO_3 + 12HCl \longrightarrow 3PtCl_4 + 4NO + 8H_2O$$

The cause of greater activity of aqua regia is the formation of nascent chlorine at the moment of formation.

(4) **Acidic property.** Nitric acid is a strong acid. It ionises in aqueous solution as :

$$HNO_3 \rightleftharpoons H^+ + NO_3^-$$

It decomposes metallic oxides, carbonates, bicarbonates, sulphides, sulphites, bisulphites and hydroxides forming metal nitrates. H_2S liberated in reaction between HNO_3 and a sulphide further reacts with HNO_3 to give sulphur.

$$CuO + 2HNO_3 \longrightarrow Cu(NO_3)_2 + H_2O$$
$$Na_2CO_3 + HNO_3 \longrightarrow 2NaNO_3 + H_2O + CO_2$$
$$NaHCO_2 + HNO_3 \longrightarrow NaNO_3 + H_2O + CO_2$$
$$Na_2S + 2HNO_3 \longrightarrow 2NaNO_3 + H_2S$$
$$H_2S + 2HNO_3 \longrightarrow 2H_2O + 2NO_2 + S$$
$$NaSO_3 + 2HNO_3 \longrightarrow 2NaNO_3 + H_2O + SO_2$$
$$NaHSO_3 + HNO_3 \longrightarrow NaNO_3 + H_2O + SO_2$$
$$KOH + HNO_3 \longrightarrow KNO_3 + H_2O$$

Uses. It is used : (*i*) in the manufacture of *nitrates* which are important chemicals of commerce. Basic calcium nitrate is used as a fertilizer. Silver nitrate is used in photography and sodium nitrate in the manufacture of gun powder and fireworks. (*ii*) in the manufacture of *explosives* like nitroglycerine, dynamite, trinitrotoluene (T.N.T.), picric acid etc. (*iii*) in the manufacture of artificial silk, dyes, medicines and perfumes. (*iv*) in the purification of gold and silver. (*v*) as an important reagent in the laboratory. (*vi*) in the manufacture of H_2SO_4. (*vii*) in etching designs on wares of brass, bronze, etc. (*viii*) in the preparation of aqua regia.

4. Pernitric Acid, HNO_4.

It is prepared by dissolving N_2O_5 in anhydrous H_2O_2.

$$N_2O_5 + H_2O_2 \longrightarrow HNO_4 + HNO_3$$

It is unstable and explosive. It smells like bleaching powder. It is a powerful oxidising agent and oxidises aniline to nitrobenzene and liberates Br_2 from KBr solution.

Structure of Oxyacids

1. Hyponitrous Acid ($H_2N_2O_2$) and Hyponitrite Ion ($N_2O_2^{2-}$).

The molecular formula of the acid has been determined by the freezing point method. Because of the decomposition of its aqueous solution on heating into nitrous oxide, N_2O, and decomposition of its salts and esters on heating, giving N_2 as shown below :

$$H_2N_2O_2 \longrightarrow N_2O + H_2O$$
$$3Na_2N_2O_2 \longrightarrow 2NaNO_2 + 2Na_2O + 2N_2$$
$$(C_2H_5)_2N_2O_2 \longrightarrow C_2H_5OH + CH_3CH + N_2,$$

the acid is said to contain an azo group, —N = N— and hence is assigned the structure, HO—N = N—OH. Since this structure involves a double bond, *cis-trans* isomerism may be expected for this molecule. Thus the acid may have either of the following two structures :

cis-structure *trans*-structure

The value of dipole moment of the diethyl ester, $(C_2H_5)_2N_2O_2$ has been found to be equal to zero, showing that the acid has *trans* structure. Thus the *trans* structure of hyponitrite ion ($N_2O_2^{2-}$) can be shown as given in the margin. The *trans* structure of $N_2O_2^{2-}$ ion as given has also been confirmed by the infra-red spectra of sodium, mercury and silver salts of the acid.

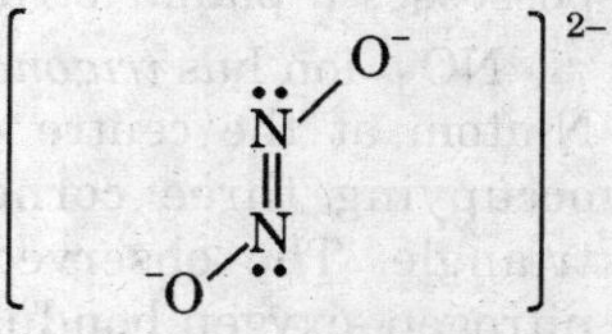

trans structure of $N_2O_2^{2-}$ ion

2. Nitrous Acid (HNO_2) and Nitrite Ion (NO_2^-).

When ethyl bromide, C_2H_5Br is distilled with $AgNO_2$, a mixture consisting of ethyl nitrite, C_2H_5—ONO and nitro ethane, $C_2H_5NO_2$ is obtained. The formation of these two compounds shows that nitrous acid forms two types of organic derivatives *viz.*, nitrites (*e.g.* C_2H_5—ONO) and nitro, compounds (*e.g.* C_2H_5—NO_2). On the basis of these two types of organic derivatives, nitrous acid is supposed to be an equilibrium mixture of the following two tautomeric forms :

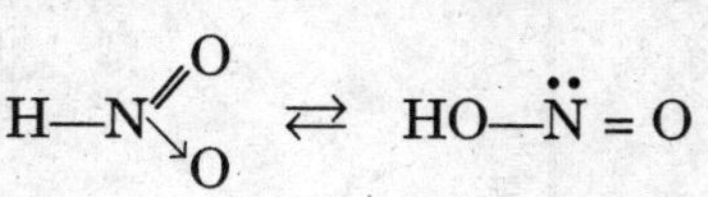

There is, however, no physical evidence supporting the above view. Rather it is against it.

L.H. Jones, from an infra-red study of nitrous acid vapour (1951), came to the conclusion that the acid has a *trans* configuration with the dimensions shown in the margin.

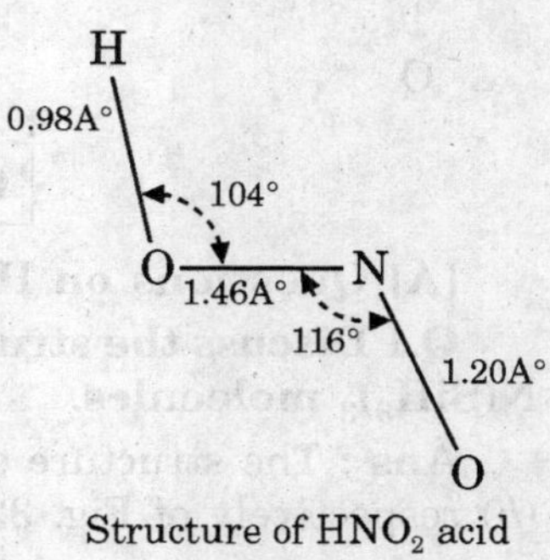

Structure of HNO_2 acid

X-ray and Raman studies of crystalline $AgNO_2$ and $NaNO_2$ have shown that NO_2^- ion has an angular structure [Fig. 32.8(*a*)] with ONO bond angle and N—O bond length equal to 116° or 182° and 1.13 Å or 1.23 Å respectively. These two sets of results have been reported by different workers. The discrepancy has not yet been explained. Since nitrogen–oxygen bonds in NO_2^- ion are equivalent, the ion is said to be a resonance hybrid of the two structures shown at (*b*) and (*c*) of Fig. 32.8.

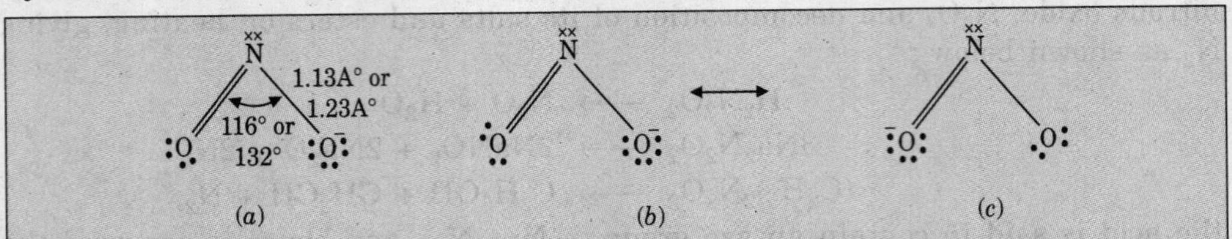

Fig. 32.8. (*a*) Angular structure of NO_2^- ion. (*b*) and (c) NO_2^- ion is said to be a resonance hybrid of these two structures.

3. Nitric Acid (HNO_3) and Nitrate Ion (NO_3^-).

HNO_3 is a monobasic acid and can be represented as HO—N = O (with ↓ O on N). As indicated

by electron diffraction studies, the molecule of nitric acid, *in the vapour state*, possesses a planar structure shown in Fig. 32.9.

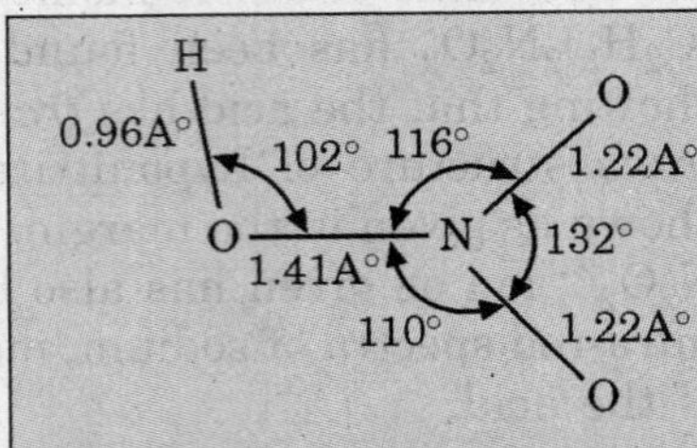

Fig. 34.9. Structure of HNO_3 molecule in the vapour state.

NO_3^- ion has *trigonal planar geometry* with N-atom at the centre and three O-atoms are occupying three corners of the equilateral triangle. The observed bond length of each nitrogen–oxygen bond is 1.21 Å and each ONO bond angle is equal to 120°. The triangular planar geometry of NO_3^- ion results from sp^2 hybridisation of N-atom. Since the observed nitrogen–oxygen bond length (= 1.21Å) is intermediate between the calculated N—O (= 1.36 Å) and N = O (= 1.15Å) bond lengths, the actual structure of NO_3^- ion is supposed to be a resonance hybrid of various resonating structures.

4. Pernitric Acid, HNO_4.

Like HNO_3. HNO_4 also has a simple structure which is represented as O = N—O—OH (with ↓ O on N).

Questions with Answers

[A] Questions on Hydrides of Nitrogen.

Q.1 Discuss the structers of trimethylamine, $N(CH_3)_3$ and trisilylamine, $N(SiH_3)_3$ molecules.

Ans : The structure of $N(CH_3)_3$ and $N(SiH_3)_3$ molecules are given at (*a*) and (*b*) respectively of Fig. 32.10.

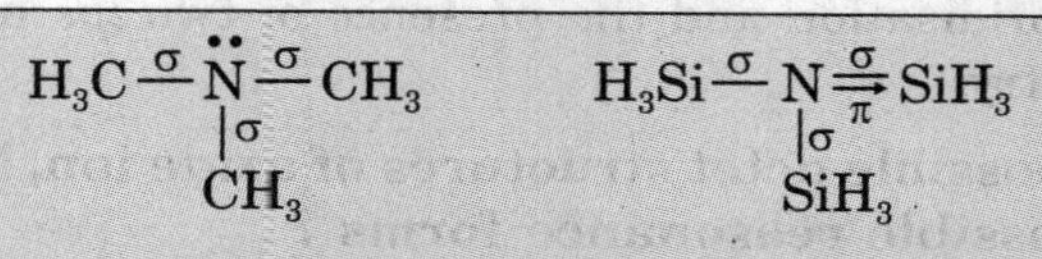

(*a*) $N(CH_3)_3$ (*b*) $N(SiH_3)_3$

Fig. 32.10 Structure of (*a*) trimethylamine molecule, $N(CH_3)_3$ and (*b*) trisilylamine molecule, $N(SiH_3)_3$.

Structure of $N(CH_3)_3$ molecule shown at (*a*) shows that since N-atom (central atom) is surrounding by three σ-*bps* and one *lp* of electrons, this atom is sp^3 hybridised [N = $(sp^3)^1$ $(sp^3)^1$ $(sp^3)^1$ $(sp^3)^2$]. Due to sp^3 hybridistion and the presence of one *lp* of electrons in one of the four sp^3 hybrid orbitals, $N(CH_3)_3$ molecule has **trigonal pyramidal shape**.

The structure of $N(SiH_3)_3$ molecule shown at (*b*) shows that since N-atom (central atom) is surrounded by *three* σ-*bps* and *no lp of electrons*, this atom is sp^2 hybridised [N → $(sp^2)^1$ $(sp^2)^1$ $(sp^2)^1$ $2p^2$].

The presence of π bond [(N → Si)π bond] has no effect on sp^2 hybridsation. In sp^2 hybridsation one 2*p* orbital which contains one electron pair remains unhybridised. The electron pair present in 2*p* orbital of N-atom is donated to the vacant 3*d* orbiatl of Si-atom (Si = $3s^2$ $3p^2$ $3d°$ = $3s^1$ $3p^1$ $3p^1$ $3p^1$ $3d°$) to form (*p*π – *d*π)π bond. This π bond is represented as [$2p^2$ (N) → $3d°$ (Si)]π bond. The donation of an electron pair from the filled 2*p* orbital to the vacant 3*d* orbital is called **(*p*π – *d*π) back bounding.** Due to sp^2 hybridisation of N-atom and absence of *lp* of electrons on N-atom, $N(SiH_3)_3$ molecule has **trigonal planar geometry**.

Q.2 Discuss the structure/geometry of amide ion (NH_2^-) on the basis of hybridisation.

Ans : Since N-atom has higher EN than H-atom, negative charge on NH_2^- ion can be assumed to be present on N-atom. Consequently structure of NH_2^- ion can be written as H — $\ddot{\underset{\cdot\cdot}{N}}$ — H. Since N-atom (central atom) is surrounded by σ-*bps* + *lps* = 2 + 2 = 4, N-atom is sp^3 hybridised as shown in Fig. 32.11.

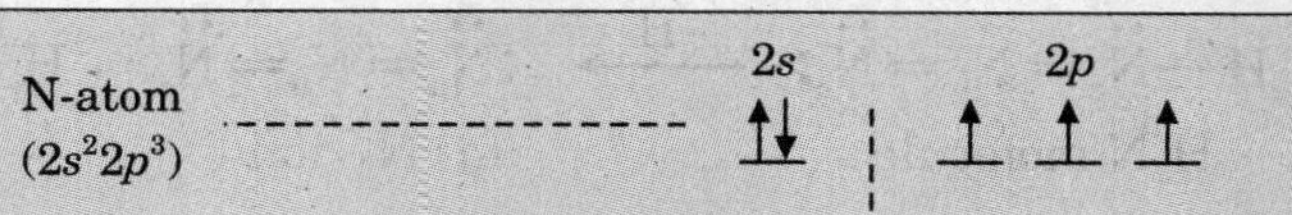

Fig 32.11 sp^3 hybridisation of N^- in NH_2^- ion.

Due to sp^3 hybridisation and the presence of two *lps* of electrons NH_2^- ion has angular geometry.

Q.3 Draw all possible octet structures of azide ion, N_3^-. Which of these structures are possible reasonance forms ?

Ans : Possible structures of N_3^- ion satisfying octet rule (*i.e.* octet structures) can be written as :

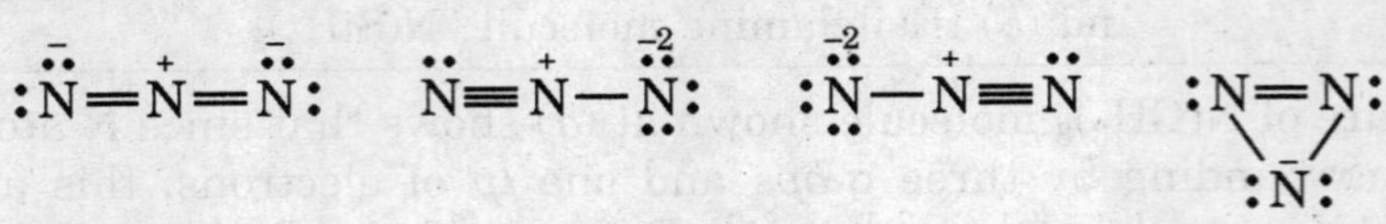

(*a*) (*b*) (*c*) (*d*)

Structures (*a*), (*b*) and (*c*) have been obtained by removing H-atom from the structures of hydrazoic acid molecule (HN_3). Structure (*d*) is a ring structure. Ring structure (*d*) is ruled out because N_3^- ion has linear geometry. Besides the bond angle of 60° or less demanded by the ring structure requires considerable strain from the normal angles of bonding. This discussuin shows that structures (*a*), (*b*) and (*c*) are the only structures which can be called resonance froms of N_3^- ion, *i.e.* the true structure of N_3^- ion is a reasonance hybrid of structures (*a*), (*b*) and (*c*).

Q.4 (*i*) Write the three resonating octet structures of azide ion (N_3^-).

(*ii*) Explain the linear structure of N_3^- ion and formation of different bonds in it.

(*iii*) Determine the formal charge of each N-atom in each of the three structures of N_3^- ion.

(*iv*) The observed nitrogen-nitrogen bond length in N_3^- ion is 1.16A° which is shorter than the predicted value of 1.26A°. Explain.

Ans : (*i*) The resonating structures of N_3^- ion in which all the three N-atoms have an octet of electrons can be obtained by removing H-atom from the resonating structures of hydrazoic acid molecule, H_3N. In order to differentiate the three N-atoms they have been designated as N_a, N_b and N_c.

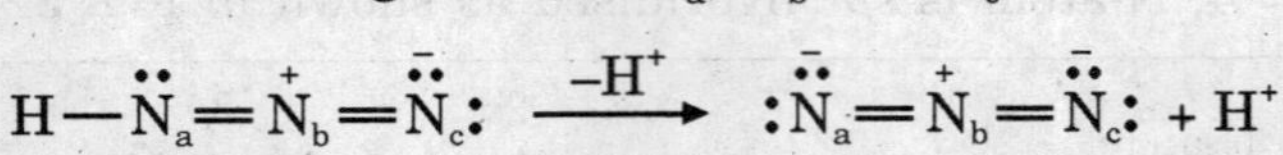

H_3N molecule (*a*) N_3^- ion

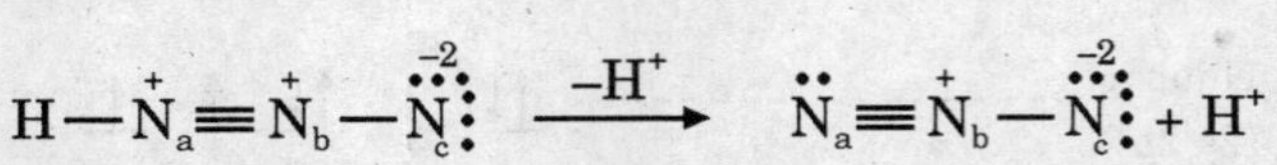

H_3N molecule (*b*) N_3^- ion

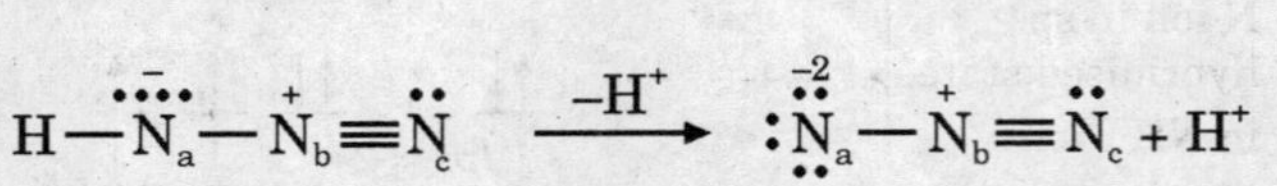

H_3N molecule (*b*) N_3^- ion

(*ii*) In each of the three resonating structures, since the central N-atom (N_b^+ atom) is surrounded by σ-*bps* + *lps* = 2 + 0 = 2, this atom is *sp* hybridised. Due to *sp* hybridisation and absence of any *lp* of electrons on N_b^+ atom, N_3^- ion has linear geometry. The formation of differnt bonds in structure (*a*) has been shown in Fig. 32.12.

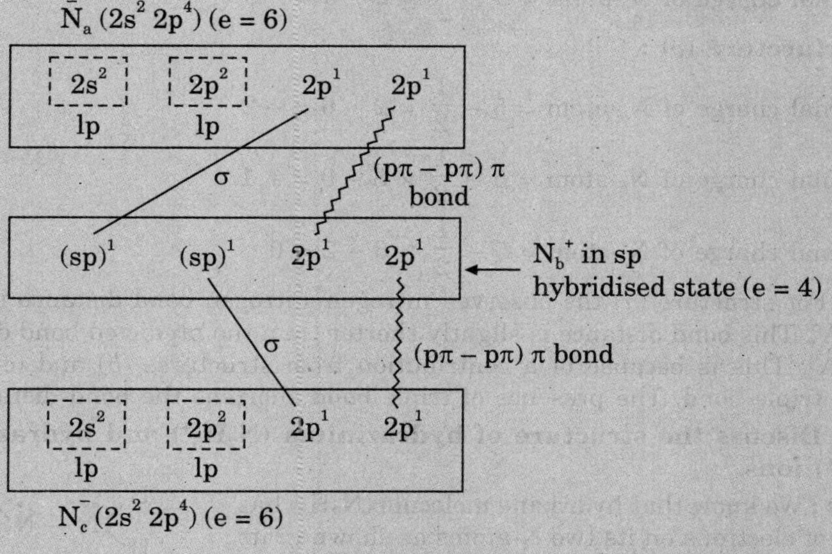

Fig. 32.12 Formation of different bonds in N_3^- ion having the structure $:\ddot{\bar{N}}_a = \overset{+}{N}_b = \ddot{\bar{N}}_c:$

(*iii*) We know that :

Formal charge of an atom in a given molecule

= No. of electrons present in the valence-shell of the atom – $\frac{1}{2}$ × No. of electrons shared by that atom with the neighbouring atom – No. of electrons from the lone pairs of electrons on the atom.

Thus :

In structure (*a*) :

Fromal charge of N_a-atom $= 5 - \frac{1}{2} \times 4 - 4$

$= 5 - 2 - 4 = -1$

Fromal charge of N_b-atom $= 5 - \frac{1}{2} \times 8 - 0 = +1$

Formal charge of N_c-atom $= 5 - \frac{1}{2} \times 4 - 4 = 5 - 6 = -1$

In structure (*b*) :

Formal charge of N_a-atom $= 5 - \frac{1}{2} \times 6 - 2 = 0$

Formal charge of N_b-atom $= 5 - \frac{1}{2} \times 8 - 0 = +1$

Formal charge of N_c-atom $= 5 - \frac{1}{2} \times 2 - 6 = -2$

In sturcture (*c*) :

Fromal charge of N_a-atom $= 5 - \frac{1}{2} \times 2 - 6 = -2$

Formal charge of N_b-atom $= 5 - \frac{1}{2} \times 8 - 0 = +1$

Formal charge of N_c-atom $= 5 - \frac{1}{2} \times 6 - 2 = 0$

(*iv*) For structure (*a*) the observed nitrogen–nitrogen bond distance is equal to 1.16A°. This bond distance is slightly shorter than the predicted bond distance of 1.26A°. This is because of a contribution from structures (*b*) and (*c*) which contain triple bond. The presence of triple bond shortens the bond distance.

Q.5 Discuss the structure of hydrazinium ($N_2H_5^+$) and hydrazoinum ($N_2H_6^{2+}$) ions.

Ans : We know that hydrazine molecule (N_2H_4) has two *lps* of electrons on its two N-atoms as shown in its structure given in the margin. When one N-atom of N_2H_4 donates its *lp* of electrons to H^+ ion ($H^+ = 1s^\circ$), $N_2H_5^+$ ion is obtained. Similary when both N-atoms of N_2H_4 donate their *lp* of electrons to two H^+ ions, $N_2H_6^{2-}$ ion is obtained.

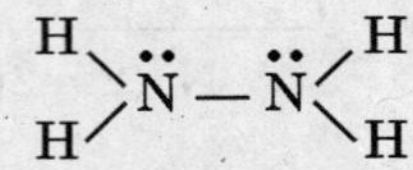

Structure of hydrazine molecule (N_2H_4)

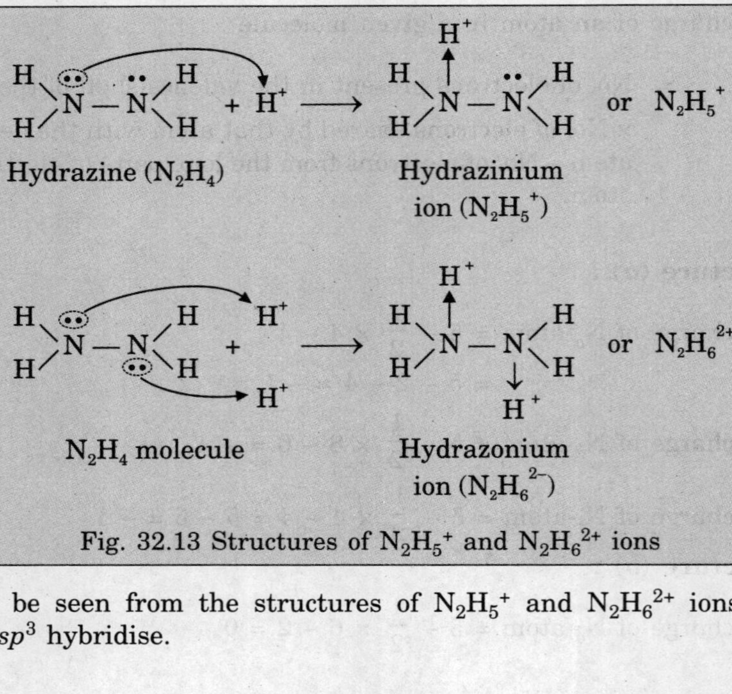

Fig. 32.13 Structures of $N_2H_5^+$ and $N_2H_6^{2+}$ ions

It may be seen from the structures of $N_2H_5^+$ and $N_2H_6^{2+}$ ions that each N-atom is sp^3 hybridise.

Q.6 Substance (A) is a colourless pungent smelling gas of vapour density 8.5. It turned Nessler's reagent brown (A) on oxidation at high temperature with platium catalyst gave a colurless gas (B), which readily turned brown in air, forming a gas (C). (B) and (C) were condensend together to give substance (D), which reacted with water to form an acid (E). When (E) was heated with a solution of NH_4Cl, a stable gas (F) was evolved. (F) did not support combustion, but magnesium continued to burn in it. Identity substences (A) to (F) and explain all the reactions involved.

Ans : (*i*) $2K_2[HgI_4] + 2NH_3(g) \longrightarrow 4KI + NH_4I + \left[\begin{matrix} I{-}Hg \\ I{-}Hg \end{matrix} \rangle NH_2\right]^+ I^-$

Nessler's reagent; Ammonia (A); Iodide of Millon's base (*Brown ppt*)

(*ii*) $4NH_3 + 5O_2 \xrightarrow{\text{Pt. Heat}} 4NO + 6H_2O$

Ammonia (A); Nitric oxide (B)

(*iii*) $2NO + O_2 \longrightarrow 2NO_2$

(B); Nitrogen dioxide (C) (*Brown*)

(*iv*) $NO + NO_2 \xrightarrow{\text{Condensed}} N_2O_3$

(B) (C); Dinitrogen trioxide (D)

(*v*) $N_2O_3 + H_2O \longrightarrow 2HNO_2$

(D); Nitrous acid (E)

(*vi*) $HNO_2 + NH_4Cl \longrightarrow N_2 + 2H_2O + HCl$

(E); Nitrogen (F)

(*vii*) $Mg + N_2 \xrightarrow{\text{Heat}} Mg_3N_2$

Magnesium (F); Magnesium nitride

Thus A = NH_3, B = NO, C = NO_2, D = N_2O_3, E = HNO_2 and F = N_2.

[B] Questions on Oxides of Nitrogen.

Q.1 Expalin why the following structures of N_2O molecule are not possible (according to the rules of resonance)

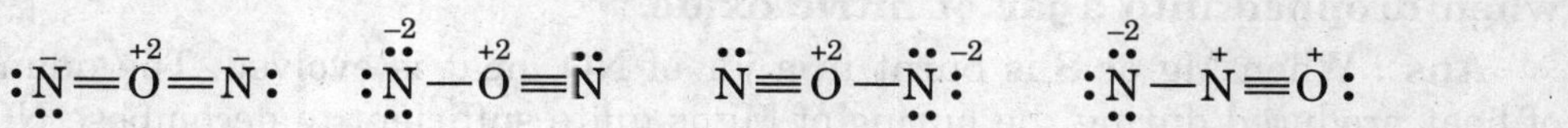

(a) (b) (c) (d)

Ans : Accoridng to the rules of resonance, the structures of N_2O molecule given above are not possible. since in structure (*a*), (*b*) and (*c*) positive charge is present on the more electronegative O-atom and in structure (*d*), + + chrages (which are like chrages) are present on the adjacent N and O-atoms.

Q.2 Nitrous oxide (N_2O) supports combustion more vigrously than air. Explain.

Ans : Nitrous oxide (N_2O) decomposes to give N_2 and O_2.

$$2N_2O \longrightarrow 2N_2 + O_2$$

Ratio = 2 : 1

Total volume = 3

We know that O_2 present in air is 1/5 of the total volume of air while, as is evident from the above equation, O_2 produced by the decomposition of N_2O is 1/3 of the total volume of the gases mixture, (N_2 + O_2). It is due to the greater content of O_2, N_2O supports combustion more vigorously than air.

Q.3 Write two resonating structures of N_2O that satisfy octet rule. ***(I.I.T. 1985, 90; MLNR 1993)***

Ans : Theoretically, the true structure of N_2O molecule is supposed to be a resonance hybrid of four structures. All these structures obey octet rule. Out of these structures, two are covalent and two are ionic. Covalent structures donot contribute much to the actual structure of N_2O molecule. Thus we are left with only two ionic structures which contribute to real structure of N_2O. These two ionic structures are given below :

(I) (II)

Each atom in both the structures sutisfies octet rule.

Q.4 Nitric oxide is a netral oxide. Explain why ?

Ans : Nitric oxide (NO) is a netural oxide due to the following properties shown by NO.

(*i*) NO reacts neither with an acid nor with an alklali.

(*ii*) NO does not give any acid with water.

Q.5 Nitric oxide turns brown in air. Why ? ***(B.I.T. 1987)***

Ans : Nitric oxide (NO) (colourless) directly reacts with O_2 of the air and brown fumes of NO_2 are formed.

$$\underset{\text{(Colourless)}}{2NO} + O_2 \text{ (From air)} \longrightarrow \underset{\text{(Brown fumes)}}{2NO_2}$$

Q.6 Burning Mg continues to burn while burning S gets extingunished when dropped into a jar of nitric oxide.

Ans : When Mg or S is burnt in a jar of NO, heat is evolved. The quantity of heat produced during the buring of Mg is quite sufficient to decompose NO to N_2 and O_2 ($2NO \xrightarrow{\Delta} N_2 + O_2$). O_2 thus produced keeps Mg burning. On the other hand, the quantity of heat produced during the burning of S is not quite sufficient and NO does not get decompose to produce O_2. Due to the absence of O_2, S stops burning.

Q.7 Nitric oxide (NO) is paramagetic in the gaseous state but diamagnetic in the liquid and solid state. Expalin.

or Nitric oxide (NO), though an odd-electron molecule, is diamagnetic in the liquid state. Explain (I.I.T. 1991)

Ans : We know that in the gaseous state, nitric oxide exists as a monomeric molecule, NO. This monomeric molecule has one unpaired electron on N-atom (:N═O:). It is due to the presence of one unpaired electron that NO molecule in the gaseous state shows paramagnetic character.

In the solid and liquid state, the monomeric NO molecule gets dimerised to form a symmetrical or asymmetrical dimer, :O═N—N═O:. This dimer is formed through the pairing of two unpaired electrons present on N-atoms of two NO molecules.

:O═N· + ·N═O: —Dimerisation→ :O═N—N═O:

Two NO molecules — Dimer

Since this dimer has all the electrons in the paired state, NO molecule in the solid and liquid state is diamagnetic.

Q.8 Reddish brown NO_2, on cooling undergoes dimerisation to give colourless dinitrogen tetraoxide, N_2O_4. Explain the change.

Ans : NO_2 molecule is an odd-electron molecule ($e = 5 + 6 \times 2 = 17$) and hence has one unpaired electron. Due to the presence of unpaired electron, NO_2 has colour (reddish brown). On cooling, NO_2 molecule gets dimerised and forms dinitrogen tetraoxide (N_2O_4) which contains one N–N bond. Due to the presence of N–N bond in N_2O_4, all the electrons in this molecule become paired and hence N_2O_4 is colourless.

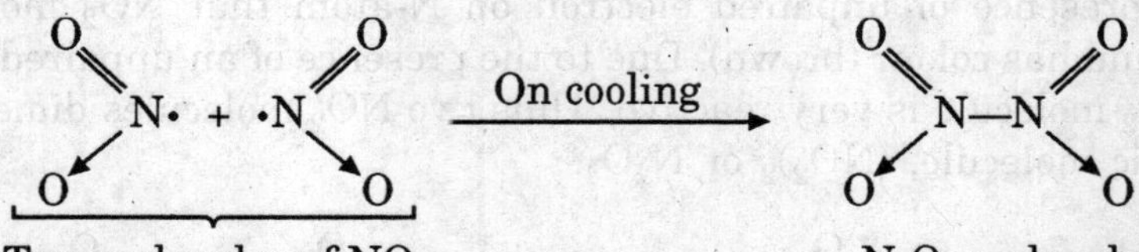

Two molecules of NO_2 (Reddish - Less stable) — N_2O_4 molecule (Colourless - More stable)

Q.9 NO_2 cannot be dried by means of NaOH solution. Expalin why ?

Ans : At 300°C, NO_2 is an equimolar mixture of NO and NO_2. When NO_2 is treated with NaOH solution, NO and NO_2 gases react with NaOH and produce $NaNO_2$

$NO_2 + NO + 2NaOH \rightarrow 2NaNO_2 + H_2O$

Since NO_2 reacts with NaOH as shown above, this gas cannot be dried by NaOH.

Q.10 Expalin why the following structure of NO_2 molecule is wrong ? :O═N═O:.

Ans : Since nitrogen is an element of *2nd* period of the periodic table, it should have an octet of electrons in its valence shell. However, in the given structure N-atom has 4 + 4 + 1 = 9 electrons in its valence-shell which is beyond an octet.

Q.11 Althrough NO and NO_2 both contain one unpaired electron, yet NO does not dimerise while NO_2 is readily dimerised.

Ans : The structures of NO and NO_2 molecules show that both the molecules have one unpaired electron. In case of NO molecule the unpaired electron is not localised on a particular atom but is spread over the whole molecule while in NO_2 molecule the unpaired electron is localised on N-atom in its sp^2 hybrid orbital. Since in NO molecule the unpaired electron is spread over the whole molecule, there is a possibilty of the dimersation of two NO molecules to form an asymmetrical loose dimer in which N...O bond is a weaker bond.

In case of NO_2 molecule, since the unpaired electron is localised on N-atom, there is a possibility for the dimerisation of two monomeric NO_2 molecules to form the dimeric molecule, $(NO_2)_2$ or N_2O_4. The dimerisation (*i.e.* formation of N_2O_4 molecule) takes palce through the pairing of two unpaired electrons on two N-atoms of two NO_2 molecules.

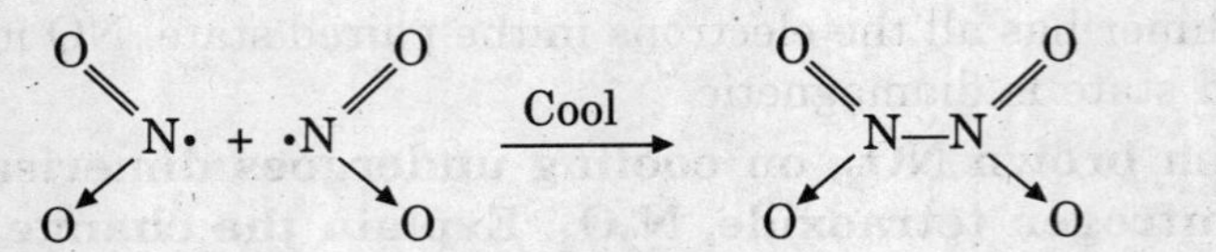

Two molecules of NO_2 (*Monomers, less stable*) Dinitrogen tetroxide (N_2O_4) (*Dimer, more stable*)

Q.12 NO_2 is paramagnetic and brown gas but N_2O_4 is diamagnetic and colourless. Explain.

Ans : We know that NO_2 molecule has one unpaired electron on N-atom. It is due to the presence of unpaired electron on N-atom that NO_2 molecule is paramagnetic and has colour (brown). Due to the presence of an unpaired electron on N-atom, NO_2 molecule is very reactive. Thus two NO_2 molecules dimerise and give the dimeric molecule, $(NO_2)_2$ or N_2O_4.

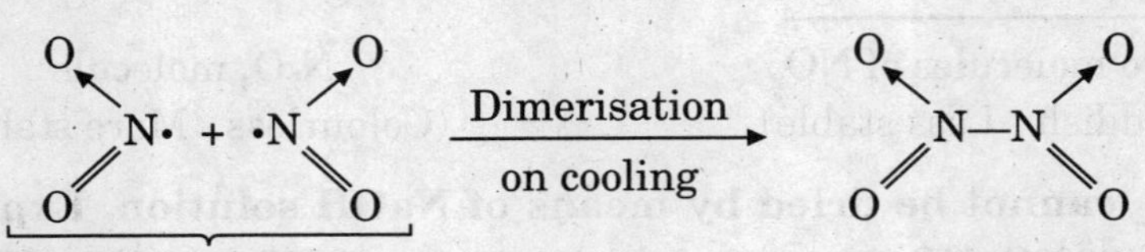

Two molecules of NO_2 N_2O_4 molecule

Since in the formation of N–N bond present in N_2O_4 molecule, the two unpaired electrons residing on two N-atoms of two NO_2 molecules get paired, N_2O_4 molecule has all the electrons in the paired state and hence N_2O_4 molecule is diamagnetic and colourless.

Q.13 P_2O_5 is used in the preparation of N_2O_5 from HNO_3. Explain why ?

Ans : N_2O_5 is prepared by heating HNO_3 in presence of P_2O_5

$$2HNO_3 \xrightarrow[\text{of } P_2O_5]{\text{Heat in presence}} N_2O_5 + H_2O$$

In the above reaction, P_2O_5 is used since this oxide, being a strong dehydating agent, removes H_2O by reacting with it to form HPO_3 and H_3PO_4

$$H_2O + P_2O_5 \longrightarrow 2HPO_3, \quad 3H_2O + P_2O_5 \longrightarrow 2H_3PO_4$$

Q.14 Arrange the oxides of nitrogen in the increasing order of their acidic nature.

Ans : We know that oxides formed by N are : N_2O, NO, N_2O_3, NO_2 and N_2O_5. The acidic nature of these oxides increases with the increase of oxidation state (O.S.) of N-atom. Thus :

Oxides :	N_2O	NO	N_2O_3	NO_2	N_2O_5
O.S. of N :	+ 1	+ 2	+ 3	+ 4	+ 5

——— Increases ———→

Acidic nature : ——— Increases ———→

Q.15 How would you distingnish among NO, N_2O_5 and NO_2.

Ans : NO is colourless gas which turns brown in air due to the oxidation of NO to NO_2

$$\underset{\text{Colourless gas}}{2NO} + O_2 \text{ (Form Air)} \longrightarrow \underset{\text{Brown gas}}{2NO_2}$$

N_2O_5 is a volatile solid which, due to its acidic nature, dissolves in H_2O with a hissing noise and gives HNO_3

NO_2 is a brown gas which condenses to a yellow liquid (N_2O_4) below – 20°C.

$$\underset{\text{Brown gas}}{2NO_2} \xrightarrow[\text{below } -20°C]{\text{Condensation}} \underset{\text{Yellow liqiud}}{N_2O_4}$$

Q.16 A colourless inorganic salt (A) decomposes at about 250°C to give only two products, (B) and (C), leaving no residue. The oxide (C) is a liquid at room temperature and neutral to moist litmus paper while the gas (B) is a neutral oxide. White phosphorus burns in the excess of (B) to produce a strong white dehydrating agent. Write balanced equations for the reactions involved in the above process. ***(I.I.T. 1996)***

Ans : Balanced equations for the reactions involved are :

$$\underset{\text{(A)}}{NH_4NO_3} \xrightarrow{250°C} \underset{\text{Nitrons oxide-gas-neutral oxide (B)}}{N_2O} + \underset{\text{Water-liquid-neutral oxide (C)}}{2H_2O}$$

$$\underset{\text{White phosphorus}}{P_4} + \underset{\text{(B)}}{N_2O} \longrightarrow \underset{\text{Phosphorus pentoxide (white dehydrating agent)}}{P_4O_{10}} + 10N_2$$

[C] Questions on Oxy acids of Nitrogen.

Q.1 Which has more acidic strength : HNO_2 or HNO_3 ?

Ans : HNO_3 has greater acidic strength (more acidic) than HNO_2. (HNO_3 > HNO_2)

Explanation. Higher acidic character of HNO_3 is due to higher O.S. of N-atom in HNO_3. ($H^{+5}NO_3 > H^{+3}NO_2$).

Q.2 HNO_3 acts only as an oxidising agent (O.A.) while HNO_2 can act both as an oxidising agent and reducing agent (R.A.). Explain why ? ***(Roorkee 1987)***

Ans : In HNO_3, N-atom is in + 5 oxidation state. + 5 oxidation state is the highest oxidation state exhibited by N-atom in its compounds. If HNO_3 acts as an O.A., this oxidation state of N-atom (= + 5) will be decreased. The decrease in O.S. is possible, since N-atom also shows oxidation states lower than + 5. If HNO_3 acts as a R.A., O.S. of + 5 of N-atom will be increased. The increase in O.S. of N-atom is not possible, since N-atom is not able to show an O.S. higher than +5. Thus HNO_3 can act only as an O.A.

In HNO_2, N-atom is in + 3 oxidation state. If HNO_2 acts as an O.A., the O.S. of N-atom (= + 3) will be decreased. The decrease in O.S. is possible , since N-atom also shows oxidation states lower than + 3. If HNO_2 acts as a R.A., the O.S. of N-atom (= + 3) will be increased. The increase in O.S. is also possible, since N-atom also exhibits oxidation states higher than + 3. Thus HNO_2 can act both as O.A. and as R.A.

Q.3 What happens when P_2O_5 reacts with conc. HNO_3 ?

Ans : Since P_2O_5 is a powerful dehydarting agent, it extracts water from HNO_3 and HNO_3 gets dehydarted resulting in the formation of N_2O_5.

$$2HNO_3 \xrightarrow[\text{of } P_2O_5]{\text{In presence}} H_2O + N_2O_5$$

Now since P_2O_5 has strong affinity for H_2O, it combines with H_2O lost by HNO_3 and forms metaphosphoric acid (HPO_3)

$$H_2O + P_2O_5 \longrightarrow 2HPO_3$$

On adding the above equation we get the final equation as :

$$2HNO_3 + P_2O_5 \longrightarrow N_2O_5 + 2HPO_3$$

Q.4 Fresh conc. HNO_3 is colourless but it turnes yellow on standing in sun light.

or Fresh conc. HNO_3 is colourless but old (*i.e.* commercial) sample of HNO_3 is yellow.

Ans : When conc. HNO_3 is allowed to stand in sunlight, it gets decomposed to give NO_2. NO_2 thus formed gets dissolved in HNO_3 and gives yellow colour to HNO_3

$$4HNO_3(l) \xrightarrow[\text{in sunlight}]{\text{Decomposition}} 4NO_2(g) + O_2(g) + 2H_2O(l)$$

HNO_3 containing dissolved NO_2 is called **fuming nitric acid**.

Q.5 Aluminium containers can be used for storing conc. HNO_3. Expalin why ?

Ans : In presence of conc. HNO_3, Al becomes passive due to the formation of a thin protective layer of its oxide (Al_2O_3) on its surface. This layer of Al_2O_3

prevents the further action of the acid on Al metal. Therefore, Al containers can be used for storing conc. HNO_3.

Q.6 Complete and balance the following equations in basic medium.

(i) $Al + NO_3^- \rightarrow Al(OH)_4^- + NH_3$

(ii) $Zn + NO_3^- \rightarrow ZnO_2^{2-} + NH_3$

Ans : (*i*) The given equation shows that Al (Al = 0) is being oxidised to $Al(OH)_4^-$ (Al = + 3) and NO_3^- ion (N = + 5) is being reduced to NH_3 (N = – 3). Thus reaction (*i*) in its complete/balanced form can be obtained by adding half-reactions (*a*) and (*b*) given below :

(*a*) $[Al + 4OH^- \xrightarrow{\text{Oxidation}} Al(OH)_4^- + 3e^-] \times 8$
(Al = 0) (Al = + 3)

(*b*) $[NO_3^- + 6H_2O + 8e^- \xrightarrow{\text{Reduction}} NH_3 + 9OH^-] \times 3$
(N = + 5) (N = – 3)

$$8Al + 3NO_3^- + 18H_2O + 5OH^- \longrightarrow 8Al(OH)_4^- + 3NH_3$$

(*ii*) The given equation shows that Zn (Zn = 0) is oxidised to ZnO_2^{2-} ion (Zn = + 2) and NO_3^- ion (N = + 5) is reduced to NH_3 (N = – 3). Thus reaction (*ii*) in its complete/balanced form can be obtained by adding half-reactions (*a*) and (*b*) given below :

(*a*) $[Zn + 4OH^- \xrightarrow{\text{Oxidation}} ZnO_2^{2-} + 2H_2O + 2e^-] \times 4$
(Zn = 0) (Zn = + 2)

(*b*) $NO_3^- + 6H_2O + 8e^- \xrightarrow{\text{Reduction}} NH_3 + 9OH^-$
(N = + 5) (N = – 3)

$$4Zn + NO_3^- + 7OH^- \longrightarrow 4ZnO_2^{2-} + NH_3 + 2H_2O$$

Q.7 Write the chemical equation for the reaction in which concentrated HNO_3 is mixed with conc. H_2SO_4.

Ans : According to one view the reaction between conc. HNO_3 and conc. H_2SO_4 produces nitronium ion (NO_2^+)

$$HNO_3 \text{ (conc.)} + H_2SO_4 \text{ (conc.)} \longrightarrow NO_2^+ + H_3O^+ + HSO_4^-$$

According to another view, when conc. HNO_3 is added to conc. H_2SO_4, HNO_3 gets dehydrated by H_2SO_4 and N_2O_5 is formed.

$$2HNO_3 \xrightarrow[\text{Dehydratetion } (-H_2O)]{\text{Conc. } H_2SO_4} N_2O_5 + H_2O$$

In polar solvent, N_2O_5 formed as above gives nitronium ion (NO_2^+) and nitrate ion (NO_3^-)

$$N_2O_5 \xrightarrow[\text{solvent}]{\text{In polar}} NO_2^+ + NO_3^-$$

Q.8 Write the chemical equation showing the reaction between conc. HNO_3 and conc. HCl.

Ans : When 1 part of conc. HNO_3 is mixed with 3 parts of conc. HCl, aqua regia is obtained. Aqua regia produces nescent chlorine (Cl) along with nitrosyl chloride (NOCl) or nitric oxide (NO).

$$HNO_3 + 3HCl \longrightarrow 2H_2O + NOCl + 2Cl$$
$$HNO_3 + 3HCl \longrightarrow 2H_2O + NO + 3Cl$$

Q.9 Arrange the following species in the decreasing order of ONO bond angles : NO_2, NO_2^+ and NO_2^-. Give appropriate reason for answer.

Ans : NO_2 molecule $(:\underset{\cdot\cdot}{\ddot{O}} \leftarrow \dot{N} = \ddot{O}:)$. In this molecule N-atom is sp^2 hybridised but ONO bond angle is 132° which is greater that the expected value of 120° for sp^2 hybridisation. The increase in bond angle is due to the fact that $(bp - bp)$ replusions are stronger than $(lp - lp)$ repulsions.

NO_2^+ ion. $(:\ddot{O} = \overset{+}{N} = \ddot{O}:)$. In this ion N^+ is sp hybridised and hence ONO bond angle = 180°.

NO_2^- ion $(:\ddot{O} = \ddot{N} - \underset{\cdot\cdot}{\overset{-}{\ddot{O}}}:)$. In this ion N-atom is sp^2 hybridised but ONO bond angle is less than (= 115°) the expected bond angle of 120° for sp^2 hybridisation. The decrease in bond angle is due to the fact that $(lp - bp)$ repulsions are stronger than $(bp - bp)$ repulsions.

Thus the decreasing order of ONO bond angles is :
NO_2^+ (= 180°) > NO_2 (= 132°) > NO_2^- (= 115°)

Q.10 (*i*) Write the chemical equations involved in ring test for NO_3^- ion.

(*ii*) What is oxidation state of Fe and NO in $[Fe(H_2O)_5(NO)]^{2+}$.

Ans : (*i*) The presence of NO_3^- ion is detected by ring test. In this test a freslly prepared solution of $FeSO_4$ is added to the aqueous solution of nitrate taken in a test tube. Pure conc. H_2SO_4 is then added carefully dropwise along the walls of test tube. The appeareance of a dark brown ring at the junction of the two layers indicates the presence of NO_3^- ion. The brown ring is due to the formation of a complex ion, $[Fe(H_2O)_5(NO)]^{2+}$. The formation of this complex ion takes place through the following steps :

(*a*) NO_3^- oxidises Fe^{2+} ion (from $FeSO_4$ solution) in presence of H_2SO_4 to Fe^{3+} ion and is itself reduced to NO.

$$\underset{\text{(From } FeSO_4 \text{ solution)}}{3Fe^{2+}} + NO_3^- + \underset{\text{(From } H_2SO_4)}{4H^+} \longrightarrow 3Fe^{3+} + 2H_2O + NO$$

$$[6FeSO_4 + 2KNO_3 + 4H_2SO_4 \longrightarrow 3Fe_2(SO_4)_3 + 4H_2O + K_2SO_4 + 2NO$$

(*Molecular equation*)

(*b*) In this step NO evolved in reaction (*a*) reacts with Fe^{2+} ions (from unreacted $FeSO_4$ solution) and forms brown coloured complex ion, $[Fe(H_2O)_5(NO)]^{2+}$

$$\underset{\text{(From unreacted } FeSO_4 \text{ solution)}}{Fe^{2+}} + 5H_2O + NO \longrightarrow \underset{\text{Brown complex ion.}}{[Fe(H_2O)_5(NO)]^{2+}}$$

$[FeSO_4 + 5H_2O + NO \longrightarrow [Fe(H_2O)_5(NO)]SO_4$ (*Molecular equation*)]

(*ii*) O.S. of Fe and NO both is + 1 in $[Fe(H_2O)_5(NO)]^{2+}$

The formation of $[Fe(H_2O)_5(NO)]^{2+}$ as shown by chemical equation given at (*b*) above suggests that since $[Fe(H_2O)_5(NO)]^{2+}$ results from Fe^{2+} ion (O.S. = + 2) and NO (neutral molecules, O.S. = 0), O.S. of Fe and NO in $[Fe(H_2O)_5(NO)]^{2+}$ should be + 2 and zero respectively. Experiments have, however, shown that O.S. of Fe and NO both is + 1. This can be explained as follows.

Explanation. As a matter of fact, the formation of $[Fe(H_2O)_5 (NO)]^{2+}$ takes through the following steps :

(*a*) We know that NO molecule is an odd-electron molecule and hence has one unpaired electron on N-atom. This molecule loses its unpaired electron (oxidation) and is converted into nitrosonium ion, NO^+.

$$:\dot{N}=\ddot{O}: \xrightarrow{\text{Oxidation } (-e^-)} :\overset{+}{N}=\ddot{O}: + e^- \ (\textit{Nitrosonium ion})$$

(*b*) In this step the electron obtained in step (*a*) is accepted by Fe^{2+} (present in unreacted $FeSO_4$) which is reduced to Fe^+ ion.

$$Fe^{2+} \text{ (From } FeSO_4) + e^- \xrightarrow{\text{Reduction } (+e^-)} Fe^+$$

(*c*) Fe^+ ion formed in step (*b*) combines with NO^+ ion formed in step (*a*) and gives brown complex ion, $[Fe^+(H_2O)_5 (NO^+)]^{2+}$,

$$Fe^+ + NO^+ + 5H_2O \longrightarrow \underset{\text{Brown complex ion}}{[Fe^+(H_2O)_5 (NO^+)]^{2+}}$$

Above discussion shows that the brown ring complex ion, $[Fe(H_2O)_5(NO)]^{2+}$ is formed due to charge-transfer, since an electron is transfered from NO to Fe^{2+}. In $[Fe(H_2O)_5(NO)]^{2+}$ ion, since Fe and NO both are in + 1 oxidation state, this complex ion is called *pentaquanitrosonium iron (I) ion*.

Alternative explanation based on magnetic moment value. The presence of Fe and NO as Fe^+ and NO^+(NO^+ ion is a two-electron donor ligandions in $[Fe(H_2O)_5(NO)]^{2+}$ ion is also confirmed by the magnetic property of $[Fe(H_2O)_5 (NO)]^{2+}$ ion. It has been found that the complex ion is paramenetic and its magnetic moment value is equal to 3.9 B.M.[$\mu = \sqrt{n(n+2)}$) B.M. = $\sqrt{3\,(3+2)}$ B.M. = 3.87 BM] (n = 3)]. Now the valence-shell configurations of Fe^+, Fe^{2+} and Fe^{3+} ions (= 1, + 2 and + 3 are common oxidation states of Fe) as given below show that it is only Fe^+ ion which has three unpaired electrons.

Fe ($3d^6 4s^2$) = $3d^2\ 3d^1\ 3d^1\ 3d^1\ 3d^1\ 4s^2$ (n = 4)

Fe^+ ($3d^6\ 4s^1$) = $3d^7\ 4s^\circ$ = $3d^2\ 3d^2\ 3d^1\ 3d^1\ 3d^1\ 4s^\circ$ (n = 3)

Fe^{2+} ($3d^6 4s^0$) $3d^2\ 3d^1$ = $3d^1\ 3d^1\ 3d^1\ 4s^0$ (n = 4)

Fe^{3+} ($3d^5\ 4s^0$) = $3d^1\ 3d^1\ 3d^1\ 3d^1\ 3d^1\ 4s^\circ$ (n = 5)

$[Fe(H_2O)_5(NO)]^{2+}$ ion is an octahedral complex ion, since C.N. of Fe^+ ion (central metal ion) is 6. Since H_2O and NO^+ are weak ligands, $[Fe(H_2O)_5(NO)]^{2+}$ ion results from sp^3d^2 hybridisation of Fe^+ ion. This hybridisation gives n = 3. Due to sp^3d^2 hybridisation $[Fe(H_2O)_5(NO)]^{2+}$ ion is a HS octahedral complex ion.

Q.11 When conc. H_2SO_4 was added into an unknown salt present in a test tube, a brown gas (A) was evolved. This gas intensified when copper turnings were also added into this test tube. On cooling, the gas

(A) changed into a colourless gas (B). Identity the gases (A) and (B). Write the equations for the reactions involved.

Ans : When a nitrate salt is heated with conc. H_2SO_4, vapours of HNO_3 are first produced which undergo decomposition to produce nitrogen dioxide (NO_2) which is a brown gas (A). When Cu turnings are added, the gas (A) intersified due to the reduction of HNO_3 by Cu.

$$\underset{\text{Sod. nitrate}}{2NaNO_3} + H_2SO_4 \xrightarrow{\text{Heat}} \underset{\substack{\text{Nitric acid}\\ (\textit{Colourless})}}{2HNO_3} + Na_2SO_4$$

$$4HNO_3 \xrightarrow{\text{Heat}} \underset{\substack{\text{Nitrogen dioxide}\\ (\textit{Brown gas A})}}{4NO_2} + O_2 + 2H_2O$$

$$\underset{\text{Copper turnings}}{Cu} + 4HNO_3 \xrightarrow{\text{Heat}} Cu(NO_3)_2 + \underset{\substack{\text{Nitrogen}\\ \text{dioxide (A)}}}{2NO_2} + 2H_2O$$

Since NO_2 gas formed in the above reaction is an odd electron molecule ($e = 7 + 2 \times 8 = 23$), this molecule, on cooling, dimeries to form a colourless gas, N_2O_4. N_2O_4 is more stable than NO_2.

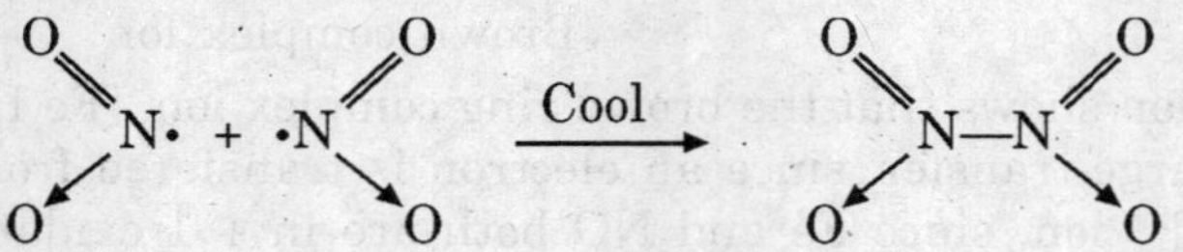

Nitrogen dioxide (NO_2) Dinitrogen tetroxide (N_2O_4)
(*Brown gas, A*) (*Colourless gas, B*)

Thus, (A) is nitrogen dioxide (NO_2) and (B) is dinitrogen tetroxide (N_2O_4).

University Questions

1. Write a note on "Hydrides of nitrogen". (*Utkal 86*)
2. Explain why a mixture of hydrazine and H_2O_2 in presence of Cu (II) catalyst is used as a rocket propellant. (*I.I.T. 87*)
3. Draw the resonating structures of N_3^- ion. (*Punjab 83*)
4. Write a detailed note on "Hydroxylamine". (*Delhi 88*)
5. Discuss the structure of various oxides of nitrogen. (*G.N.D. 82*)
6. Draw the electron dot structures of the oxy acids of nitrogen. (*Meerut 85*)
7. Write a note on "Pernitric acid". (*Madras 86*)
8. Write short note on "Hydrazoic acid". (*Delhi 99*)
9. Describe the action of heat on hydrazine. Give balanced chemical equation. (*Delhi 99*)
10. Discuss the structure and constitution of hyponitrous acid. (*Allahabad 99*)
    (*a*) How does hydrazine react with (*i*) $AgNO_3$ (*ii*) HONO.
    (*b*) Give two methods to prepare hydroxylamine.(*Himachal Pradesh 2000*).
11. (*i*) Describe the method of preparation, structure and uses of hydrazoic acid or hyponitrous acid.
    (*ii*) Explain why PH_3 is more volatile than NH_3 ?[*Gauhati (General) 2000*]

CHAPTER

33

Nitrides

What are Nitrides ?

The binary compounds of nitrogen with electropositive elements or with the elements which are less electronegative than nitrogen are called nitrides.

This definition does not include the salts of hydrazoic acid, HN_3 (*e.g.* NaN_3). Rather the salts of hydrazoic acid are called *azides.* Thus, NaN_3, AgN_3 etc. are the examples of azides while Na_3N, Ag_3N etc. are the examples of nitrides.

General Methods of Preparation

1. By direct combination of elements. Combination is brought about generally by heating the element in a current of nitrogen to a suitable temperature.

In the case of alkaline earth metals, amalgam may be used in place of the pure metal.

$$6Li + N_2 \longrightarrow 2Li_3N$$
$$3Mg + N_2 \longrightarrow Mg_3N_2$$
$$2Al + N_2 \longrightarrow 2AlN$$

In some cases, as for instance aluminium and chromium, the combination is effected by striking an electric arc between metal electrodes in an atmosphere of nitrogen. In the case of alkali metals, *active nitrogen* is used in place of ordinary nitrogen.

2. By heating the element, its oxide or chloride with NH_3. The elements or their oxides are heated with NH_3 at 300–400°C to get their nitrides. The nitrides of Mg, Zn, Ag, B, Al, Fe, Co, Ni etc. can be obtained by this method.

$$3Mg + 2NH_3 \longrightarrow Mg_3N_2 + 3H_2$$
$$6Cu + 2NH_3 \longrightarrow 2Cu_3N + 3H_2$$
$$3Fe + 2NH_3 \longrightarrow Fe_3N_2 + 3H_2$$
$$3ZnO + 2NH_3 \longrightarrow Zn_3N_2 + 3H_2O$$
$$3Ag_2O + 2NH_3 \longrightarrow Ag_3N + 3H_2O$$
$$3CuO + 2NH_3 \longrightarrow Cu_3N_2 + 3H_2O$$
$$B_2O_3 + 2NH_3 \longrightarrow 2BN + 3H_2O$$

The nitrides of the elements like Ti, Zr, Th, Cr, Mo, W, U, B etc. have been prepared by passing their chloride vapour with NH_3 through a tube heated to 800°C.

$$BCl_3 + NH_3 \xrightarrow{800°C} BN + 3HCl$$

3. By the action of nitrogen on a mixture of metallic oxide and carbon. The reaction is carried out at a high temperature in an electric furnace. At the high temperature carbon reduces the oxide to the metal which reacts with nitrogen to form the nitride. Nitrides of aluminium, zirconium, scandium, niobium and boron are prepared in this way.

$$Al_2O_3 + 3C + N_2 \longrightarrow 2AlN + 3CO \uparrow$$
$$B_2O_3 + 3C + N_2 \longrightarrow 2BN + 3CO \uparrow$$
$$ZrO + C + N_2 \longrightarrow ZrN_2 + 3CO \uparrow$$

4. By heating amides and imides . Amides and imides, on heating, decompose to give the nitrides. For example :

$$3Ba(NH_2)_2 \xrightarrow{\Delta} 4Ba_3N_2 + NH_3 \uparrow$$
$$3Ca(NH_2)_2 \xrightarrow{\Delta} Ca_3N_2 + 4NH_3 \uparrow$$
$$3Zn(NH_2)_2 \xrightarrow{\Delta} Zn_3N_2 + 4NH_3 \uparrow$$
$$B_2(NH)_3 \xrightarrow{\Delta} 2BN + NH_3 \uparrow$$
$$As_2(NH)_3 \xrightarrow{\Delta} 2AsN + NH_3 \uparrow$$

5. By heating the element with NO. Under suitable conditions, certain elements combine with NO to form nitrides.

$$5B + 3NO \longrightarrow 3BN + B_2O_3$$

6. By the action of N_2 on hydrides. The nitrides of certain metals can be prepared by heating their hydrides in a current of N_2.

$$3MgH_2 + 2N_2 \longrightarrow Mg_3N_2 + 2NH_3 \uparrow$$
$$3BaH_2 + 2N_2 \longrightarrow Ba_3N_2 + 2NH_3 \uparrow$$

7. By the action of a halide on amide. When halides and amides of certain metals are mixed together, nitride of the metal (whose halide has been taken) is formed.

$$BiCl_3 + 3KNH_2 \longrightarrow BiN + 3KCl + 2NH_3$$

8. By the action of cyanides on nitrates. Some nitrides have been obtained by the action of cyanides on nitrates.

$$3Cu(CN)_2 + 12NH_4NO_3 \longrightarrow Cu_3N_2 + 24H_2O + 6CO_2 + 14N_2$$

General Properties

1. Action of water. Nitrides react with water to produce ammonia gas. The rate of hydrolysis varies with the nature of the nitride and some of them (*e.g.*, alkaline earth metals, Mg and Zn nitrides) are decomposed even by the moisture present in air and the reaction is quite vigorous with water. Where as Mn_4N_3 and W_2N_3 are only slowly hydrolysed by water, CrN is not decomposed even by steam at 220°C.

$$Li_3N + 3H_2O \longrightarrow 3LiOH + 2NH_3$$

$$Mg_3N_2 + 6H_2O \longrightarrow 3Mg(OH)_2 + 2NH_3$$
$$AlN + 3H_2O \longrightarrow Al(OH)_3 + NH_3$$
$$BN + 3H_2O \longrightarrow B(OH)_3 \text{ or } H_3BO_3 + NH_3$$

2. Action of hydrogen. Nitrides are reduced by hydrogen with the formation of hydrides and evolution of ammonia.

$$Mg_3N_2 + 6H_2 \longrightarrow 3MgH_2 + 2NH_3$$
$$CeN + 3H_2 \longrightarrow CeH_3 + NH_3$$

3. Action of acids. Nitrides are decomposed by dilute acids producing the salt of the metal atom present in the nitride.

$$2Cu_3N + 8HCl \longrightarrow 2NH_4Cl + 6CuCl$$
$$2CeN + 4H_2SO_4 \longrightarrow Ce_2(SO_4)_3 + (NH_4)_2SO_4$$

4. Action of alkalies. Caustic alkalies decompose nitrides liberating ammonia gas.

$$AlN + 3KOH \longrightarrow K_3AlO_3 + NH_3 \uparrow$$
$$BN + 3KOH \longrightarrow K_3BO_3 + NH_3 \uparrow$$

5. Action of halogens. Nitrides react with halogens at 700°C giving N_2 gas.

$$2AlN + 3Cl_2 \longrightarrow 2AlCl_3 + N_2 \uparrow$$

6. Action of CO and CO_2. CO and CO_2 decompose nitrides at about 200°C to form metal oxides. For example :

$$Mg_3N_2 + 3CO \longrightarrow 3MgO + 3C + N_2$$
$$Mg_3N_2 + 3CO_2 \longrightarrow 3NgO + 3CO + N_2$$

7. Sublimation. There are many nitrides which, like oxides, sublime. For example gallium nitride sublimes without decomposition above 800°C.

Classification of Nitrides

The nitrides have been classified into three types depending on the nature of bonding between nitrogen and the other element. These three types are: (1) *Ionic or salt-like nitrides* (2) *Covalent nitrides*, (3) *Interstitial or metallic or alloy nitrides.*

Ionic Nitrides

These are ionic in nature and contain the nitride ion, N^{3-}. These are given by the elements of groups IA (alkali metals), IIA (alkaline earth metals), Zn, Cd, Cu(*ous*), Ag.

Preparation.

Ionic nitrides are prepared (*i*) by direct union of the elements.

$$6Li + N_2 \longrightarrow 2Li_3N$$
$$3Mg + N_2 \longrightarrow Mg_3N$$

(*ii*) by passing NH_3 over heated metal or its oxide.

$$3Mg + 2NH_3 \longrightarrow Mg_3N_2 + 3H_2 \uparrow$$

$$3Ag_2O + 2NH_3 \longrightarrow 2Ag_3N + 3H_2O$$

(*iii*) by decomposing the amide at elevated temperature

$$3Ba(NH_2)_2 \xrightarrow{\Delta} Ba_3N_2 + 4NH_3$$

Properties.

These are crystalline colourless compounds having high melting points. These are hydrolysed by H_2O to form the hydroxides and NH_3.

$$Li_3N + 3H_2O \longrightarrow 3LiOH + NH_3$$

$$Ca_3N_2 + 6H_2O \longrightarrow 3Ca(OH)_2 + 2NH_3$$

When heated in a current of CO_2 at 200°C, N_2 gas is evolved.

$$Mg_3N + 3CO_2 \longrightarrow 3MgO + 3CO + N_2$$

When heated in a current of H_2S, sulphides are obtained.

$$Mg_3N_2 + H_2S \longrightarrow 3MgS + (NH_4)_2S$$

Structure.

Ionic nitrides contain metal ions in combination with nitride ion, N^{3-}. The structure of Li_3N is written as $Li^+[Li^+ - N^{3-} - Li^+]^-$. This nitride has a hexagonal structure.

The nitrides of M_3N_2 type are *anti-isomorphous* with oxides of M_2O_3 type, metal atoms occupying oxygen positions and N-atoms occupying metal positions in the oxide structure.

Covalent Nitrides

These nitrides are formed by the sharing of electrons between nitrogen and the metal or non-metal. This type of nitrides some volatile nitrides of non-metals such as H, C, Si. P, Cl, F etc. and some non-volatile nitrides, of the elements of group IIIA like B, Al, Ga and In. Thus the compounds like NH_3, cyanogen and oxides and halides of nitrogen are the examples of volatile covalent nitrides while BN, AlN, Si_3N_4, P_3N_5 are non-volatile covalent nitrides. The non-volatile nature of the nitrides is attributed to their macro-molecular structures. For example BN has a layer-like structure which is analogous to that of graphite and crystal structure of AlN is very similar to that of diamond. Non-volatile covalent nitrides undergo hydrolysis to give NH_3.

$$BN + 3H_2O \longrightarrow B(OH)_3 \text{ or } H_3BO_3 + NH_3$$

$$AlN + 3H_2O \longrightarrow Al(OH)_3 + NH_3$$

Interstitial Nitrides

These nitrides are given by transition metals. On heating, transition metals of groups IIIB. IVB and VB absorb N_2 to give the nitrides of MN type. Mo_2N, W_2N, Fe_4N and Mn_4N nitrides are also known. In these nitrides N-atoms occupy the holes in the metallic lattice. The atomic radius of transition metals is sufficiently greater to allow the inclusion of N-atoms in the holes in the metallic lattice.

Preparation.

These nitrides are obtained (*i*) by heating powdered metal in N_2 or NH_2 at 1100–1300°C (*ii*) by heating metal filaments in N_2 (*iii*) by heating a carrier filament (*e.g.* Pt, W or C) in a mixture of N_2 and metal halide vapours and subsequently removing the carrier by high temperature volatilisation.

Properties.

Being nitrogen deficient, these nitrides are often not exactly stoichiometric. They have some metallic properties, *e.g.* they are extremely hard and electrical conductors, since electrical structure of metal persists in these nitrides. They are stable and very inert. They do not hydrolyse. These have very high melting points. For example UN has its melting point equal to 2570°C. Some nitrides are isomorphous to oxides and carbides. For example TiN is isomorphous to TiO and TiC.

Structure.

The structure of nitrides is usually cubic like rock-salt. Nitrogen atoms in the nitrides of MN type occupy all the octahedral holes while in Mo_2N and W_2N nitrides only one-half of the octahedral holes are occupied by N-atoms and in Mn_4N and Fe_4N nitrides only one-quarter of octahedral holes are occupied by N-atoms. Thus we see that the nitrides of MN type have NaCl type structure.

Uses of Nitrides

1. BN possesses same hardness as diamond and can withstand temperature of more than 300°C. Due to this property it is used for coating crucible lining.

2. The formation of AlN during the purification of bauxite by Serpeck's process is utilised for the manufacture of NH_3.

Questions with Answers

Q.1 Element A burns in N_2 to give an ionic compound, B. Compound B reacts with H_2O to give C and D. A soution of C becomes milky on bubbling CO_2. Identity A, B, C and D. (*I.I.T. 1997*)

Ans : A = Ca, B = Ca_3N_2, C = $Ca(OH)_2$ and D = NH_3

Explanation (*i*) Since A burns in N_2 to give an ionic compound B, B must be the nitride of a metal, *i.e.* A must be a metal.

$$\text{A (Metal)} + N_2 \longrightarrow \underset{\text{(B)}}{\text{Metallic nitride of A}}$$

(*ii*) Since compound B is a metallic nitride, it should give hydroxide of A and NH_3 with H_2O. Thus D is NH_3

$$\text{B} + H_2O \longrightarrow \underset{\text{(C)}}{\text{Hydroxide of A}} + \underset{\text{(D)}}{NH_3}$$

(*iii*) Since solution of C (*i.e.* hydroxide of A) becomes milky on bubbling CO_2, milkiness is due to the formation of $CaCO_3$. Thus A metal is Ca and C must be $Ca(OH)_2$. Obviously B is Ca_3N_2.

Reactions can now be shown as :

$$\underset{(A)}{2Ca} + N_2 \xrightarrow{\Delta} \underset{(B)}{Ca_3N_2} \quad (\textit{Ionic compound})$$

$$\underset{(B)}{Ca_3N_2} + 6H_2O \longrightarrow \underset{(C)}{Ca(OH)_2} + \underset{(D)}{2NH_3}$$

$$\underset{(C)}{Ca(OH)_2} + CO_2 \longrightarrow \underset{\substack{(D)\\(\text{Milkiness})}}{CaCO_3} + H_2O$$

Q.2 Calcium burns in nitrogen to produce a white powder which dissolves in suffecient water to produce a gas (A) and an alkaline solution. The solution on exposure to air produces a thin solid layer (B) on the surface. Identity the compounds (A) and (B). ***(I.I.T. 1996)***

Ans : A = NH_3, B = $CaCO_3$

Explanation : Ca burns in N_2 to give white powder of Ca_3N_2

$$3Ca + N_2 \xrightarrow{\Delta} Ca_3N_2 \ (\textit{White powder})$$

This white powder of Ca_3N_2 dissolves in H_2O to produce NH_3 (which is a gas) and alkaline solution of $Ca(OH)_2$. Thus A is NH_3.

$$Ca_3N_2 + 6H_2O \longrightarrow 3Ca(OH)_2 + \underset{(A)}{2NH_3} \ (\textit{Gas})$$

Alkaline solution of $Ca(OH)_2$ reacts with CO_2 present in the atmosphere (air) and produces $CaCO_3$ which gets deposited as a layer on the surface. Thus B is $CaCO_3$.

$$Ca(OH)_2 + CO_2 \ (\textit{From air}) \longrightarrow \underset{(B)}{CaCO_3} + H_2O$$

University Questions

1. How are nitrides classified ? Give a brief account of interstitial nitrides. *(Madras 86; Bharathiar 86)*
2. Describe the preparation, properties, uses and structure of boron nitride. *(Bharathiar 86; Jodhpur 85; Agra 80, 82, 84)*
3. How does BN resemble with and differ from graphite ? *(Bharathiar 86)*
4. Give the preparation, properties and uses of magnesium nitride. *(Gorakhpur 85)*
5. Write balanced equation for :

   $Mg_3 N_2 + H_2O \longrightarrow . . . .$ *(Bharathiar 86)*
6. Describe the action of heat of Li_3N. *(Delhi 99)*
7. Write the action of H_2O, H_2, acids, alkalies, halogens, CO and CO_2 on nitrides.

CHAPTER

34 Oxides, Oxy acids, Halides and Oxy halides of Phosphorus

Oxides

1. Phosphorus Trioxide, P_2O_3 or P_4O_6.

Preparation. This oxide is also called *phosphorus sesquioxide* and is the *anhydride of phosphorus acid*, H_3PO_3. It is prepared when white phosphorus is treated with a restricted supply of O_2. It is better prepared in the low pressure reaction of phosphorus vapour with N_2O at 600°C.

$$P_4 \text{ (vapour)} + 6N_2O \xrightarrow{600°C} 2P_2O_3 + 6N_2$$

Properties. It is a snow white wax-like solid and has garlic smell. When impure it is waxy. It is poisonous. P_2O_3 melts at 23.6° and boils at 17.4°C.

When heated to about 440°C, it dissociates to give phosphorus dioxide and red phosphorus.

$$2P_2O_3 \longrightarrow 3PO_2 + P$$

When P_2O_3 is shaken with ice water, it gives phosphorous acid. (H_3PO_3). Hence this oxide is also called anhydride of H_3PO_3

$$P_2O_3 + 2H_2O \longrightarrow 2H_3PO_3$$

With hot water, the reaction is violent and phosphoric acid with some phosphine is produced.

$$2P_2O_3 + 6H_2O \longrightarrow 3H_3PO_4 + PH_3$$

P_2O_3 oxidises in air or O_2 to peroxide (P_2O_6), inflames at 70°C in air and at 40°C in O_2

$$2P_2O_3 + 3O_2 \longrightarrow 2P_2O_6$$

It inflames spontaneously in Cl_2 and forms phosphoryl chloride, $POCl_3$ and perhaps meta phosphoryl chloride, PO_2Cl.

It combines violently with sulphur forming crystalline $P_4O_6S_4$.

Structure. The vapour density of the oxide and the lowering of the freezing point of its solution in benzene correspond to the formula P_4O_6. Thus phosphorus trioxide exists as the dimer, P_4O_6. The structure of P_4O_6 molecule in the vapour state can be derived from that of P_4 molecule which has tetrahedral structure. If in the tetrahedral structure of P_4 molecule six oxygen atoms are placed just outside the mid points of the six edges of the tetrahedron to form six 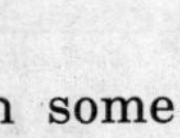 bridges, we get the structure of P_4O_6 molecule. Thus in the structure of P_4O_6 molecule each P-atom is covalently bonded to three O-atoms and each O-atom is covalently bonded to two P-atoms. In this way all the four P-atoms and six O-atoms have eight electrons in their valence shells and we have *six non-linear* P—O—P arrangements. Here it may be noted that P—P bonds are absent in this molecule, although in the structure of P_4 molecule there are six P—P bonds. The P—O bond length is 1.65A° which is about the P—O single bond distance and the bond angles POP and OPO are 127.6 and 99° respectively. Since each P-atom is tricovalent in this molecule, an unused electron pair (called lone pair) is left on each P-atom (Fig. 34.1).

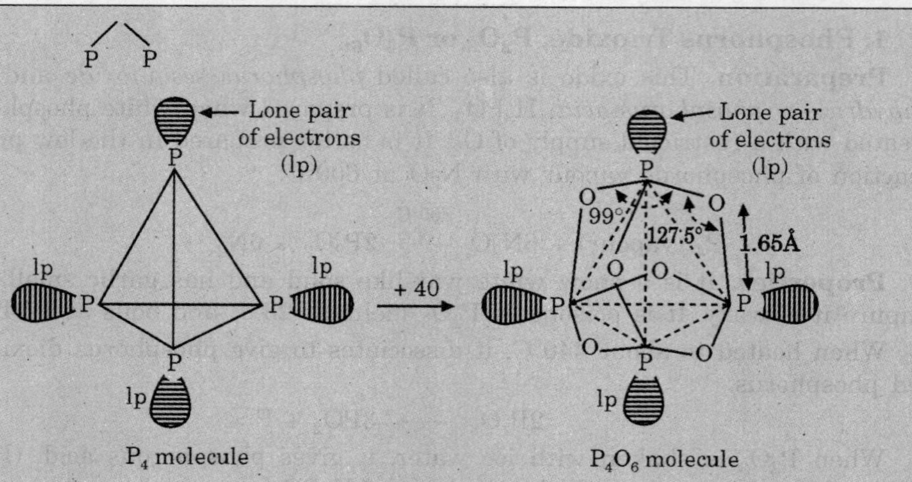

Fig. 34.1. Structure of P_4 ant P_4O_6 molecules.

2. Phosphorus Tetroxide, P_2O_4.

This oxide was discovered by Thrope and Tutton in 1886. It is prepared by heating P_2O_3 in a sealed tube at 440°C or by heating a mixture of P_2O_3 and P_2O_5 in a sealed tube to 290°C.

(*a*) $4P_2O_3 \longrightarrow 3P_2O_4 + 2P$

(*b*) $P_2O_3 + P_2O_5 \longrightarrow 2P_2O_4$

In reaction (*a*) P_2O_4 appears as a crystalline sublimate while red phosphorus remains as a residue. It is a colourless crystalline solid. The vapour density at 1800°C corresponds with the formula P_5O_{16}. With H_2O it gives a mixture of H_3PO_3 and H_3PO_4.

$$P_2O_4 + 3H_2O \longrightarrow H_3PO_3 + H_3PO_4$$

Thus P_2O_4, like N_2O_4, may be regarded as a *mixed anhydride* of phosphorus acid (H_3PO_3) and phosphoric acid (H_3PO_4).

3. Phosphorus Pentoxide, P_2O_5 or P_4O_{10}.

Preparation. It is obtained when white phosphorus burns in excess of air in iron drums or when white phosphorus is burnt in CO_2 at 100°C.

$$4P + 5O_2 \longrightarrow 2P_2O_5$$

$$2P + 5CO_2 \longrightarrow P_2O_5 + 5CO$$

It can be purified by sublimation in oxygen at about 360°C.

Properties. Phosphorus pentoxide is a white powdery solid, having a garlic smell which is due to presence of traces of the trioxide. Pure pentoxide has no smell. Its vapour density corresponds to the formula P_4O_{10} but it is generally represented by the formula P_2O_5. It exhibits a strong phosphorescence after illumination especially at low temperatures.

It gives phosphoric acid with water.

$$P_2O_5 + 3H_2O \longrightarrow 2H_3PO_4$$

The main product is definitely ortho phosphoric acid.

It, however, gives different acids with different quantities of water *e.g.*,

$$P_2O_5 + H_2O \longrightarrow 2HPO_3 \text{ (Metaphosphoric acid)}$$

$$P_2O_5 + 2H_2O \longrightarrow H_4P_2O_7 \text{ (Pyrophosphoric acid)}$$

$$P_2O_5 + 3H_2O \longrightarrow 2H_3PO_4 \text{ (Orthophosphoric acid)}$$

P_2O_5 is hygroscopic. It absorbs water from various compounds like H_2SO_4, HNO_3 etc. and thus dehydrates them

$$H_2SO_4 + P_2O_5 \longrightarrow 2HPO_3 + SO_3$$

$$2HNO_3 + P_2O_5 \longrightarrow 2HPO_3 + N_2O_5$$

It also dehydrates many organic compounds. For example amides are dehydrated to nitrides.

$$RCONH_2 + P_2O_5 \longrightarrow RCN + 2HPO_3$$

For the same reason, P_2O_5 chars wood, paper, sugar etc.

It gives meta phosphates with alkalies

$$P_2O_5 + 2NaOH \longrightarrow \underset{\text{sodium meta phosphate}}{2NaPO_3} + H_2O$$

It is reduced to red phosphorus when heated with carbon.

$$P_2O_5 + 5C \longrightarrow \underset{\text{Red phosphorus}}{2P} + 5CO$$

When fused with basic oxides it gives solid phosphates.

$$6CaO + 2P_2O_5 \longrightarrow 2Ca_3(PO_4)_2$$

It is extensively used as a *drying* and a *dehydrating* agent in the laboratory.

Structure. The vapour density of this oxide corresponds to the formula P_4O_{10}. The structure of this molecule can be derived from that of P_4O_6 molecule. If four additional oxygen atoms are bonded to four phosphorus atoms by a P→O co-ordinate bond, we get the structure of P_4O_{10} (Fig. 34.2). P→O co-ordinate bond is formed by the donation of a lone pair on phosphorus atom. The length of this P—O bond is 1.39A° which is much shorter than the length of other P—O bonds (= 1.65A°) in the molecule. This much shorter value is explained by $p\pi(O)$—$d\pi(P)$ *back bonding via* lateral overlap of a filled *p*-orbital of oxygen atom with a vacant *d*-orbital of phosphorus atom.

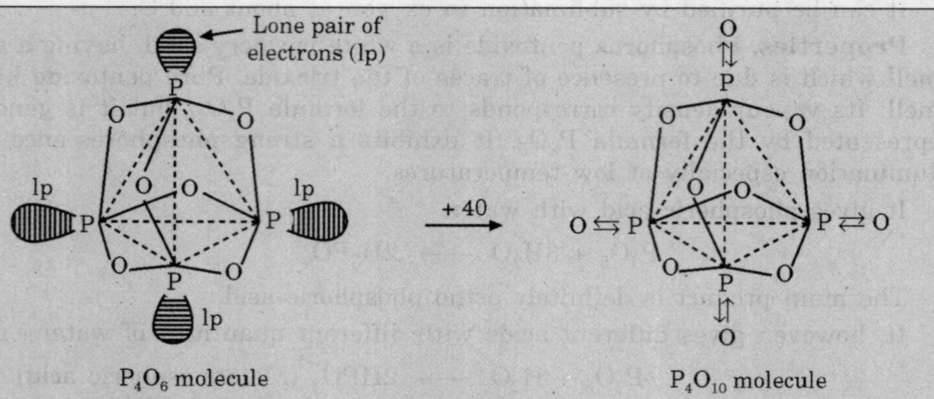

Fig. 34.2. Structure of P_4O_{10} molecule.

4. Phosphorus Peroxide, PO_3 or P_2O_6.

It is formed when an electric discharge is passed through a mixture of P_2O_5 vapour and oxygen at about 1 mm. pressure. It is a violet solid which when heated to 130°C gives P_2O_5 and O_2. Its aqueous solution liberates iodine from K solution.

Oxy-acids

1. Hypophosphorus Acid, H_3PO_2.

Preparation. This acid is manufactured from borium hypophosphite, $Ba(H_2PO_2)_2$. When phosphorus is boiled with baryta solution (barium hydroxide solution), $Ba(H_2PO_2)_2$ is formed, which is filtered out.

$$3Ba(OH)_2 + 8P + 6H_2O \longrightarrow \underset{\text{ppt.}}{3Ba(H_2PO_2)_2} + 2PH_3$$

If to this barium salt, sulphuric acid be added in calculated amounts, and the solution filtered, free hypophosphorus acid comes down 'in the filtrate.

$$Ba(H_2PO_2)_2 + H_2SO_4 \longrightarrow \underset{\text{ppt.}}{BaSO_4} + 2H_3PO_2$$

The filtrate, when evaporated leaves the syrupy acid, and when carefully cooled below 0°, can be crystallised out also.

Properties. (*i*) It is a colourless crystalline solid having specific gravity 1.492 and m.pt = 26.5°. It is readily soluble in water.

(*ii*) *Action of heat.* When strongly heated, the acid is decomposed to phosphine.

$$2H_3PO_2 \longrightarrow PH_3 + H_3PO_4$$

The salts of the acid also decompose on heating and evolve PH_3.

(*iii*) *Action of nascent hydrogen.* H_3PO_2 is reduced to PH_3 by nascent hydrogen (Zn + dil. H_2SO_4).

$$H_3PO_2 + 4H \longrightarrow PH_3 + 2H_2O$$

(*iv*) *Reducing properties.* The acid and its salts are powerful reducing agents. The potential equation for these reductions is :

$$H_3PO_2 + 2H_2O \longrightarrow H_3PO_4 + 4H$$

Thus :

(*a*) It precipitates silver and gold from their solutions.

$$4AgNO_3 + H_2O + H_3PO_2 \longrightarrow \underset{\text{ppt.}}{4Ag} + 4HNO_3 + H_3PO_3$$

(*b*) Mercuric chloride is reduced to calomel (mercurous chloride) and finally to metallic mercury (grey ppt.)

$$H_3PO_2 + 2H_2O \longrightarrow H_3PO_4 + 4H$$
$$4HgCl_2 + 4H \longrightarrow 2Hg_2Cl_2 + 4HCl$$
$$4Hg_2Cl_2 + H_3PO_2 + 2H_2O \longrightarrow \underset{\text{Calomel}}{2Hg_2Cl_2} + 4HCl + H_3PO_4$$
$$2Hg_2Cl_2 + H_3PO_2 + 2H_2O \longrightarrow \underset{\substack{\text{Metallic mercury}\\ \text{(grey ppt.)}}}{4Hg} + H_3PO_4 + 4HCl$$

(*c*) I_2 reduces Cl_2 and I_2 to HCl and HI respectively.

$$H_3PO_2 + 2H_2O \longrightarrow H_3PO_3 + 4H$$
$$2X_2 + 4H \longrightarrow 4HX$$
$$2X_2 + H_3PO_2 + 2H_2O \longrightarrow 4HX + H_3PO_4$$
$$(X = Cl, I)$$

(*d*) Copper salts on warming with the acid give a brown precipitate of cuprous hydride, CuH

$$4CuSO_4 + 3H_3PO_2 + 6H_2O \longrightarrow 4CuH + 3H_3PO_4 + 4H_2SO_4$$

(*v*) *Action of active metals.* Active metals like Zn, Mg etc. dissolve in the acid and H_2 is evolved.

$$Zn + 2H_3PO_2 \longrightarrow Zn(H_2PO_2)_2 + H_2$$

(*vi*) *Acidic nature.* It is a strong *monobasic* acid and ionises as follows :

$$H_3PO_2 \rightleftarrows H^+ + \underset{\text{Hypophosphite ion}}{[H_2PO_2^-]};\ \text{Ionisation constant} = 1 \times 10^{-2}$$

Hypophosphites. The salts of the acid are called *hypophosphites.* These salts are mostly soluble in water. They possess reducing properties as the acid itself has.

Uses. Some of its salts *e.g.*, those of sodium, potassium and calcium are used in medicine as nerve tonics.

2. Orthophosphorus Acid or Phosphorus Acid, H_3PO_3.

Preparation. It is prepared : (*i*) by adding phosphorus trichloride to cold water. Since the reaction is very violent, the addition should be done slowly. (*ii*) by adding PCl_3 to anhydrous oxalic acid. In this case the reaction is not violent and by-products are gases.

$$PCl_3 + 3H.OH \longrightarrow H_3PO_3 + 3HCl$$
$$PCl_3 + 3(COOH)_2 \longrightarrow H_3PO_3 + 3CO + 3CO_2 + 3HCl$$

The solution in both cases is concentrated to get orthophosphorus acid as white crystals.

Properties. (*i*) It forms white deliquescent crystals (m.pt. = 73.6°C) and is extremely soluble in water.

(*ii*) *Acidic nature.* It is a *dibasic acid* and ionizes as :

$$H_3PO_3 \rightleftarrows H^+ + H_2PO_3^- \; ; \; K_1 = 1 \times 10^{-2}$$
$$H_2PO_3^- \rightleftarrows H^+ + HPO_3^{2-} \; ; \; K_2 = 2 \times 10^{-7}$$

From the mode of its ionisation shown above it is evident that the acid gives two types of salts *viz. primary phosphite*, (*e.g.* MH_2PO_3) and *secondary phosphites* (*e.g.* M_2HPO_3).

(*iii*) *Decomposition.* When heated it gives ortho phosphoric acid and phosphine.

$$4H_3PO_3 \xrightarrow{200°} 3H_3PO_4 + PH_3$$

(*iv*) *Reducing properties.* Ortho phosphorus acid is a strong reducing agent. The potential equation is :

$$H_3PO_3 + H_2O \longrightarrow H_3PO_4 + 2H$$

Thus : (*a*) *It reduces mercuric chloride to mercurous chloride,*

$$2HgCl_2 + H_3PO_3 + H_2O \longrightarrow HgCl_2 + 2HCl + H_3PO_4$$

(*b*) *Copper sulphate is reduced to metallic copper.*

$$CuSO_4 + H_3PO_3 + H_2O \longrightarrow H_3PO_4 + Cu + H_2SO_4$$

(*c*) *Silver nitrate is reduced to metallic silver*

$$2AgNO_2 + H_3PO_3 + H_2 \longrightarrow 2Ag + 2HNO_3 + H_3PO_4$$

(*d*) *It decolorises acidified potassium permanganate solution.*

$$2KMnO_4 + 3H_2SO_4 + 5H_3PO_3 \longrightarrow K_2SO_4 + 2MnSO_4 + 5H_3PO_4 + 3H_2O$$

(*e*) *It reduces iodine to hydriodic acid*

$$I_2 + H_3PO_3 + H_2O \longrightarrow H_3PO_4 + 3HI$$

(*f*) *Sulphur dioxide is reduced to sulphur.*

$$2H_3PO_3 + SO_2 \longrightarrow 2H_3PO_4 + S$$

(*v*) *Action of nascent hydrogen.* It is reduced to PH_3 by ascent hydrogen (Zn + dil. H_2SO_4)

$$H_3PO_3 + 6H \longrightarrow PH_3 + 3H_2O$$

(*vi*) *Action of PCl_5.* It reacts with PCl_5 and forms PCl_3 and $POCl_3$.

$$H_3PO_3 + 3PCl_5 \longrightarrow PCl_3 + 2POCl_3 + 3HCl$$

3. Hypophosphoric Acid, $H_4P_2O_6$.

Preparation. When phosphorus is exposed to a limited supply of moist air, the acid, along with phosphorus and phosphoric acid, is produced. The resulting solution is partly neutralised with Na_2CO_3 when sparingly soluble sodium salt, $Na_2H_2P_2O_6$. $6H_2O$ separates slowly. Now $Pb(NO_3)_2$ solution is added to the solution of the salt when lead hypophosphate, $Pb_2P_2O_6$ is precipitated. The precipitate of $Pb_2P_2O_6$ is suspended in water and a current of H_2S a passed through the solution when PbS is precipitated and a solution of free $H_4P_2O_6$ is obtained. The precipitated PbS is filtered off and the filtrate is evaporated in a vacuum desiccator over H_2SO_4 when rhombic crystals of the dihydrate. $H_4P_2O_6$. $2H_2O$ separate. On further desiccation of the crystals over H_2SO_4 in vacuum, anhydrous $H_4P_2O_6$ acid is produced.

Properties. (*i*) The dihydrated acid, $H_4P_2O_6.2H_2O$ forms rhombic crystals which melt at 62°C. The anhydrous acid melts at 70°C.

(*ii*) *Acidic nature.* It is a tetrabasic acid and ionises as follows :

$$H_4P_2O_6 \rightleftarrows H^+ + H_3P_2O_6^- \ ; K_1 = 6.4 \times 10^{-3}$$
$$H_3P_2O_6^- \rightleftarrows H^+ + H_2P_2O_6^{2-} \ ; K_2 = 1.5 \times 10^{-3}$$
$$H_2P_2O_6^{2-} \rightleftarrows H^+ + HP_2O_6^{3-} \ ; K_3 = 5.4 \times 10^{-8}$$
$$HP_2O_6^{3-} \rightleftarrows H^+ + P_2O_6^{4-} \ ; K_4 = 9.4 \times 10^{-11}$$

Its salts like $Na_2H_2P_2O_6$, $Pb_2P_2O_6$ etc. and the ester $(C_2H_5)_2P_2O_6$ have been prepared.

(*iii*) *Action of heat.* When heated above its melting point, it decomposes into H_3PO_3 and HPO_3

$$H_4P_2O_6 \longrightarrow H_3PO_3 + HPO_3$$

On heating at 180°, PH_3 is evolved

$$4H_4P_2O_6 \longrightarrow 3H_3PO_4 + 4HPO_3 + PH_3$$

(*iv*) *Hydrolysis.* The aquecus solution of the acid undergoes hydrolysis on being heated with mineral acid

$$H_4P_2O_6 + H_2O \longrightarrow H_3PO_3 + H_3PO_4$$

(*v*) The acid has no reducing properties.

4. Orthophosphoric Acid or Phosphoric Acid, H_3PO_4.

Manufacture. H_3PO_4 is manufactured from bone ash. The process consists of the following steps :

(*a*) When bone ash obtained by the destructive distillation of bones is treated with concentrated sulphuric acid and heated, orthophosphoric acid and calcium sulphate are produced. Calcium sulphate being insoluble is filtered off. The filtrate which contains phosphoric acid is concentrated.

$$\underset{\text{Bone ash}}{Ca_3(PO_4)_2} + 3H_2SO_4 \longrightarrow \underset{\text{ppt.}}{3CaSO_4} + 2H_3PO_4$$

(*b*) When bone ash is oxidised with concentrated nitric acid, all phosphorus present in the bone ash is converted into phosphoric acid. The resulting liquid is treated with lead acetate which precipitates insoluble lead phosphate. Phosphoric acid is then recovered from lead phosphate by treating with H_2S gas.

$$P + 5HNO_3 \longrightarrow H_3PO_4 + 5NO_2 + H_2O$$

$$2H_3PO_4 \ (\textit{impure}) + 3(CH_3COO)_2Pb \longrightarrow Pb_3(PO_4)_2 + 6CH_3COOH$$

$$Pb_3(PO_4)_2 + 3H_2S \longrightarrow 3PbS + 2H_3PO_4 \ (\textit{pure})$$

The acid so obtained is separated from PbS by filtration. It is concentrated by heating. Nitric acid employed in this process should not be more than 30 to 35% since higher concentration makes the reaction too violent.

Properties (*a*) **Physical.** Orthophosphoric acid forms bard, colourless (glassy) rhombic prisms. The crystals deliquesce in air and form a syrupy liquid. Commercial acid is 82.98%. It is soluble in all proportions. Its density is 1.87 and melts at 42.3°C. (anhydrous). The concentrated phosphoric acid (usually 85%) is a colourless syrupy liquid which, on being cooled in vacuum, yields crystals of anhydrous acid, H_3PO_4. The acid also forms hemihydrate, H_3PO_4. H_2O.

(*b*) **Chemical.** (*i*) *Reaction with alkalies.* It is tribasic acid and forms three series of salts with alkalies.

$$H_3PO_4 + NaOH \longrightarrow \underset{\text{Primary sodium phosphate}}{NaH_2PO_4} + H_2O$$

$$H_3PO_4 + 2NaOH \longrightarrow \underset{\text{Secondary sodium phosphate}}{Na_2HPO_4} + 2H_2O$$

$$H_3PO_4 + 3NaOH \longrightarrow \underset{\text{Tertiary sodium phosphate}}{Na_3PO_4} + 3H_2O$$

(*ii*) *Action of heat.* When heated to 250°C, it gives pyrophosphoric acid $H_4P_2O_7$

$$2H_3PO_4 \longrightarrow H_4P_2O_7 + H_2O$$

When heated to 320°C, it further loses one molecule of water giving metaphosphoric acid, HPO_3

$$H_4P_2O_7 \xrightarrow{320°} 2HPO_3 + H_2O$$

When strongly heated (to red heat), it gives phosphorus pentoxide

$$3HPO_3 \xrightarrow{600°-700°} P_2O_5 + H_2O$$

(*iii*) *Action of bromides and iodides.* Phosphoric acid liberates hydrobromic and hydriodic acids from bromides and indides respectively.

$$3NaBr + H_3PO_4 \longrightarrow N_3PO_4 + 3HBr$$

$$3NaI + H_3PO_4 \longrightarrow Na_3PO_4 + 3HI$$

These reactions afford a very good method for the laboratory preparation of these halogen acids.

(*iv*) *Action on silver nitrate.* Silver nitrate gives a yellow precipitate of silver ortho-phosphate with ortho-phosphoric acid.

$$3AgNO_3 + H_3PO_4 \longrightarrow \underset{\text{Yellow ppt.}}{Ag_3PO_4} + 3HNO_3$$

(*v*) *Action on barium chloride.* Barium chloride gives a white precipitate of barium ortho-phosphate in neutral or slightly alkaline solution.

$$2BaCl_2 + 2H_3PO_4 \longrightarrow \underset{\text{White ppt.}}{Ba_3(PO_4)_2} + 6HCl$$

(*vi*) *Action on magnesia mixture.* Ortho-phosphoric acid reacts with magnesia mixture (*i.e.* $MgCl_2 + NH_4OH$) to give a white crystalline precipitate of ammonium magnesium phosphate, NH_4MgPO_4.

$$H_3PO_4 + MgCl_2 + NH_4OH \longrightarrow \underset{\text{White crystalline ppt.}}{NH_4MgPO_4.} + 2HCl + H_2O$$

(*vii*) *Oxidising properties.* It has no oxidising properties below 350–400°C but at higher temperature it is fairly reactive towards metals and is reduced.

(*viii*) *Acidic nature.* The acid is a tribasic acid and ionises as follows :

$$H_3PO_4 \rightleftharpoons H^+ + [H_2PO_4]^- \; ; K_1 = 7.5 \times 10^{-3}$$
$$[H_2PO_4]^- \rightleftharpoons H^+ + [HPO_4]^{2-} \; ; K_2 = 6.2 \times 10^{-8}$$
$$[HPO_4]^{2-} \rightleftharpoons H^+ + [PO_4]^{3-} \; ; K_3 = 1.1 \times 10^{-12}$$

The acid, therefore, forms three types of salts namely *primary phosphates* (*e.g.* NaH_2PO_4), *secondary phosphates* (*e.g.* Na_2HPO_4) and tertiary phosphates (*e.g.* Na_3PO_4).

(*ix*) When phosphoric acid and phosphates are treated with ammonium molybdate, $(NH_4)_2MoO_4$ and excess of conc. HNO_3, a canary yellow precipitate of ammonium phosphomolybdate, $(NH_4)_3PO_4.12MoO_3$ is obtained, on warming to about 60°C

$$H_3PO_4 + 12(NH_4)_2MoO_4 + 21HNO_3 \longrightarrow \underset{\text{Canary yellow ppt.}}{(NH_4)_3PO_4.12MoO_3} + 21NH_4NO_3 + 12H_2O$$

Uses. (*i*) Pure ortho-phosphoric acid has pharmaceutical applications and is used in the preparation of nervous tonics. (*ii*) It serves as a stabilizer for hydrogen peroxide. (*iii*) It is used for the manufacture of fertilizers, dyes, enamels and porcelain cement. (*iv*) It is used in the dental fillings.

5. Pyrophosphoric Acid, $H_4P_2O_7$.

Preparation. It is prepared by prolonged heating of orthophosphoric acid between 200–300°C.

$$2H_3PO_4 \longrightarrow H_4P_2O_7 + H_2O$$

Properties. (*a*) **Physical.** It is a white crystalline mass with melting point 61°C. It is readily soluble in water.

(*b*) **Chemical.** (*i*) *Action on alkalies.* It is a tetra-basic acid and forms two series of salts with alkalies.

$$2NaOH + H_4P_2O_7 \longrightarrow \underset{\text{Disodium acid pyrophosphate}}{Na_2H_2P_2O_7} + 2H_2O$$

$$4NaOH + H_4P_2O_7 \longrightarrow \underset{\text{Tetra sodium neutral pyro-phosphate}}{Na_4P_2O_7} + 4H_2O$$

(*ii*) *Action on silver nitrate.* It gives a white precipitate of silver pyrophosphate.

$$2AgNO_3 + H_4P_2O_7 \longrightarrow \underset{\text{white ppt.}}{Ag_4P_2O_7} + 4HNO_3$$

(*iii*) *Action on water.* Pyrophosphoric acid dissolves in cold water and is slowly converted to ortho-phosphoric acid but in hot water and especially in the presence of nitric acid, it is easily converted into ortho-phosphoric acid.

$$H_4P_2O_7 + H_2O \longrightarrow 2H_3PO_4$$

(*iv*) *Acidic nature.* The acid is tetrabasic and is stronger than ortho-phosphoric acid. It ionises as :

$$H_4P_2O_7 \rightleftarrows H^+ + H_3P_2O_7^- \; ; K_1 = 1.4 \times 10^{-1}$$
$$H_3P_2O_7^- \rightleftarrows H^+ + H_2P_2O_7^{2-} \; ; K_2 = 1.1 \times 10^{-2}$$
$$H_2P_2O_7^{2-} \rightleftarrows H^+ + HP_2O_7^{3-} \; ; K_3 = 2.7 \times 10^{-7}$$
$$HP_2O_7^{3-} \rightleftarrows H^+ + P_2O_7^{4-} \; ; K_4 = 2.4 \times 10^{-10}$$

Although the acid is tetrabasic, it forms two types of salts *viz. normal salts* (*e.g.* $M_4P_2O_7$) and *diacid salts* (*e.g.* $M_2H_2P_2O_7$).

(*v*) *Action of heat.* On heating at 316°C, it gives metaphosphoric acid, HPO_3

$$H_4P_2O_7 \xrightarrow{316°C} 2HPO_3 + H_2O$$

6. Metaphosphoric Acid or Glacial Phosphoric Acid, HPO_3.

Preparation. When $(NH_4)_3$ PO_4 or H_3PO_4 or $H_4P_2O_7$ is heated to about 32°C, meta phosphoric acid is formed.

$$(NH_4)_3PO_4 \longrightarrow HPO_3 + 3NH_3 + H_2O$$
$$H_3PO_4 \longrightarrow HPO_3 + H_2O$$
$$H_4P_2O_7 \longrightarrow 2HPO_3 + H_2O$$

Properties. (*a*) **Physical.** Meta-phosphoric acid is a hard, sticky mass. The acid is, therefore, called *glacial phosphoric acid.* It dissolves in water with a cracking noise. It volatilises on heating. It melts at 384°C.

(*b*) **Chemical** (*i*) *Action on silver nitrate.* Meta phosphoric acid gives a white precipitate with silver nitrate solution forming silver meta-phosphate, $AgPO_3$.

$$AgNO_3 + HPO_3 \longrightarrow AgPO_3 + HNO_3$$

(*ii*) *Action on barium chloride.* Barium chloride gives a white precipitate in neutral or alkaline solutions.

$$BaCl_2 + 2HPO_3 \longrightarrow Ba(PO_3)_2 + 2HCl$$

The precipitate of barium phosphate is insoluble in acetic acid.

(*iii*) *Action on water.* Aqueous solution of meta-phosphoric acid is gradually converted to ortho-phosphoric acid on prolonged standing. This conversion is accelerated in the presence of acids and on heating.

$$HPO_3 + H_2O \longrightarrow H_3PO_4$$

(*iv*) The alkali meta-phosphate, $NaPO_3$, when heated with metal oxides, forms mixed phosphate. The mixed phosphates of heavy metals have characteristic colours.

$$NaPO_3 + CuO \longrightarrow NaCuPO_4$$

This reaction finds an application in the basic radical analysis as an alternative of the borax bead test. The mixed phosphates so formed are often referred to as microcosmic beads.

(*v*) Meta-phosphoric acid or its salts are not present in the form of simple molecules but are often polymerised. In the vapour phase, meta-phosphoric acid has been reported to have the formula $(HPO_3)_2$ but in solutions or in the melt, it is $(HPO_3)_n$ where n = 3 or 4.

(vi) Acidic nature. It is a monobasic acid and its salts are called metaphosphates.

(vii) Like other phosphoric acids, it also gives ammonium molybdate test.

(viii) Unlike other phosphoric acids, it at once coagulates albumin (white part of the egg).

7. Peroxy Phosphoric Acids.

Phosphorus forms two peroxy phosphoric acids which are analogous to peroxy sulphuric acids and are discussed below :

(A) Peroxy Monophosphoric Acid, H_3PO_5. This acid is known in solution only. It is formed by the action of 30% H_2O_2 on P_2O_5 in acetonitrile solution at a low temperature.

$$P_2O_5 + 2H_2O_2 - H_2O \longrightarrow 2H_3PO_5$$

This acid is a powerful oxidising agent and oxidises manganous salts to permanganates.

(B) Peroxy Diphosphoric Acid, $H_4P_2O_8$. This acid is prepared by the action of H_2O_2 on the excess of pyrophosphoric acid, $H_4P_2O_7$.

$$H_4P_2O_7 + H_2O_2 \longrightarrow H_4P_2O_8 + H_2O$$

Potassium salt of the acid is stable and is prepared by the electrolysis of potassium hydrogen phosphate in presence of KF. The solution of the acid oxidises aniline nitrobenzene and nitrosobenzene. In acidic solution the acid is first hydrolysed to H_3PO_5 and finally to H_2O_2.

Distinction Between Hypophosphorus Acid and Ortho, Pyro and Meta phosphoric Acids.

The various reactions showing the distinction between these acids are given below :

Reaction	Hypo-phosphorus acid (H_3PO_2)	Ortho-phosphoric acid (H_3PO_4)	Pyro-phosphoric acid ($H_4P_2O_7$)	Meta-phosphoric acid (HPO_3)
(i) Magnesia mixture	No ppt.	White ppt.	White ppt. soluble in excess reagent	No ppt.
(ii) Silver nitrate	Reduced to metallic silver (ppt.)	Yelllow ppt. of Ag_3PO_4	White ppt. of $Ag_4P_2O_7$	White gela. tinous ppt. of $AgPO_3$
(iii) Barium chloride	No action	White ppt. soluble in acetic acid	White ppt. insoluble in acetic acid	White ppt. insoluble in acetic acid
(iv) Copper sulphate (solution in acid)	Reduced to metallic copper (ppt.)	No action	No action	No action
(v) Egg albumin with dil. acetic acid	No action	No action	No action	Coagulation takes place
(vi) Acidified $KMnO_4$	Decolorises	No action	No action	No action
(vii) Cadmium chloride acidified with dil. acetic acid	No action	No action	White ppt.	No action

Salts of the Acids

1. Sodium Ammonium Hydrogen Phosphate or Microcosmic Salt, $Na(NH_4)HPO_4.4H_2O$.

This phosphate is also called microcosmic salt since it is present in the urine of the "*microcosm*" *i.e.* man. It is prepared by dissolving molecular proportion of Na_2HPO_4 and NH_4Cl in a little hot water, filtering off NaCl precipitated and crystallising.

$$\underset{}{Na_2HPO_4} + NH_4Cl \longrightarrow \underset{\text{In solution}}{Na(NH_4)HPO_4} + \underset{\text{ppt.}}{NaCl}$$

It forms colourless needle-like crystals. When heated strongly it loses NH_3 and water and forms a transparent glassy bead of sodium meta-phosphate, $NaPO_3$.

$$Na(NH_4)HPO_4 \xrightarrow{\Delta} NH_3 + H_2O + NaPO_3$$

This bead of $NaPO_3$ in the fused state combines with metallic oxides (*e.g.* CuO) to form coloured ortho-phosphates.

$$CuO + NaPO_3 \longrightarrow \underset{\text{Blue bead}}{CuNaPO_4}$$

The salt is used for testing SiO_2 with which a cloudy bead containing floating particles of SiO_2 is obtained.

2. Sodium Hexametaphosphate or Graham's Salt, $(NaPO_3)_n$.

It is a polymeric compound with high molecular weight (12000–18000). It is miscalled sodium hexametaphosphate. At present it is called *sodium (1 : 1) phosphate glass* and its preferred formula is Na_2O (1 : 1) P_2O_5. Graham's salt is prepared by heating NaH_2PO_4, until a molten material, $(NaPO_3)_n$ is formed. This material, on chilling, gives a glassy solid which is called Graham's salt. The changes taking place can be shown as follows :

$$NaH_2PHO_4 \xrightarrow{230°} \underset{\substack{\text{Sod. cyclo-}\\ \text{triphosphate}}}{Na_2(P_3O_9)} \xrightarrow{638°} \text{Molten } (NaPO_3)_n \xrightarrow{\text{Chilling}} \text{Graham's salt (Glassy solid)}$$

Graham's salt is a transparent glassy solid. It is hygroscopic and readily soluble in water without limit. On divitrification, the material loses water. This water is regarded as being present in Graham's salt as OH groups attached to terminal phosphorus atom of a chain of PO_4 tetrahedra. Graham's salt is used for softening water. It acts by sequestering the Ca^{2+} and Mg^{2+} ions *i.e.* it forms complexes with Ca^{2+} and Mg^{2+} ions present in hard water. A hard water, after the treatment with Graham's salt, becomes soft since Ca^{2+} and Mg^{2+} ions present in it have been eliminated; hence Graham's salt is also given a trade name "*calgon*" (derived from "*calcium gone*") Graham's salt is also used with other alkalies for cleansing sinks, drain's floors etc.

Structure of the Acids

While discussing the structure of the oxy-acids of phosphorus, the following points should be kept in mind. (*i*) The structure of these acids is such that each phosphorus atom is tetrahedrally surrounded by neighbouring atoms. (*ii*) The number of H-atoms directly linked to oxygen atoms are responsible for the basicity of oxyacids. In other words we can say that the basicity of oxy-acids is determined by the number of —OH groups present in the acid. (*iii*) The hydrogen atoms directly linked to phosphorus atom are responsible for the reducing properties of the oxy-acids. The larger the number of H-atoms, the greater is the reducing property of the acid.

1. Hypophosphorus Acid, H_3PO_2.

The measurement of heat of neutralisation, conductivity and hydrogen concentration have shown that the acid is a monobasic acid, *i.e.* the acid has only one —OH group attached directly to P-atom. The strong reducing character of the acid confirms the presence of P—H bonds in the molecule. In the light of these properties and the evidences obtained from Raman spectra and X-ray analysis of its salts, the structure shown in Fig. 34.3 has been assigned to this acid. Obviously it has a tetrahedral structure.

2. Phosphorus Acid, H_3PO_3.

Nuclear magnetic resonance spectroscopy of the acid, its salts and esters shows that there is one H-atom directly linked to P-atom. This H-atom is not ionisable and confers reducing properties to the acid while other two H-atoms which are ionisable exist as —OH groups. The dibasic nature of the acid has been confirmed from heat of neutralisation data. X-ray study of the crystalline acid has shown that three oxygen atoms and one of the three H-atoms are arranged tetrahedrally round P-atom. Keeping in view the above points, the acid has been assigned the tetrahedral structure shown in Fig. 34.3. This structure can be regarded as being obtained from the structure of H_3PO_2 by the replacement of one H-atom by one —OH group.

3. Orthophosphoric Acid or Phosphoric Acid, H_3PO_4.

It is a *tribasic acid* and forms the ester, trietheyl phosphate $PO(OC_2H_5)_2$. On this basis, the acid has been assigned the tetrahedral structure shown in Fig. 34.3. In this structure tetrahedral PO_4 groups are connected together by hydrogen bonds which have been shown by dotted lines (......)

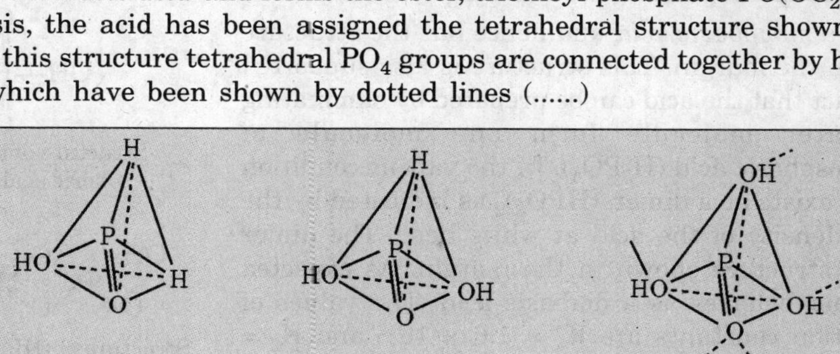

Hypophosphorus acid (H_3PO_2) | Phosphorus acid (H_3PO_3) | Orthophosphoric acid (H_3PO_4)

Fig. 34.3. Tetrahedral structures of hypophosphorus acid (H_3PO_2), phosphorus acid (H_3PO_3) and orthophosphoric acid (H_3PO_4). In the structure of H_3PO_4 acid the dotted lines ...... show the hydrogen bonds.

Structures of various orthophosphate anions can be derived easily from the structure of orthophosphoric acid by the step-wise removal of H-atom from OH groups. The tetrahedral structures of H_3PO_2, H_3PO_3 and H_3PO_4 acids can be explained on the basis of sp^3 hybridisation of P-atom.

4. Hypophosphoric Acid, $H_4P_2O_6$.

The double formula of the acid, $H_4P_2O_6$ has been indicated by the molecular weight determination of the acid and its salts, and by the diamagnetic nature of its salts. The acid has been given the structure shown in the margin. This structure is supported by the Raman spectrum, X-ray studies of diammonium hypophosphate, the absence of reducing properties and tetrabasic nature of the acid. Due to its tetrabasic nature four H-atoms are bonded to O-atoms.

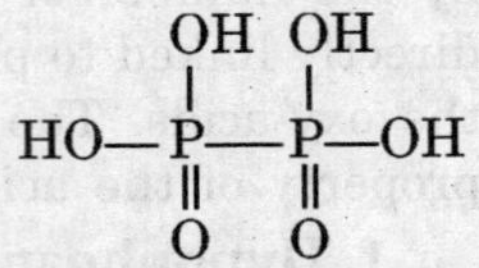

Structure of hypophosphoric acid, $H_4P_2O_6$

5. Pyrophosphoric Acid, $H_4P_2O_7$.

It is a *tetrabasic acid* and hence has the structure shown in Fig. 34.4. This structure follows directly from the fact that this acid can be prepared by removing one H_2O molecule from two molecules of orthophosphoric acid (H_3PO_4) which has tetrahedral shape.

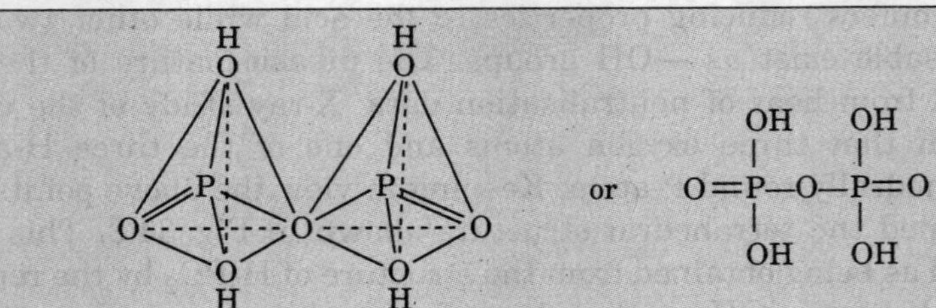

Fig. 34.4. Structure of pyrophosphoric acid, $H_4P_2O_7$.

6. Metaphosphoric Acid, HPO_3.

It is a *monobasic acid* and hence has the structure shown in the margin. This structure is also supported by the fact that the acid can be prepared by eliminating one water molecule form one molecule of orthophosphoric acid (H_3PO_4). In the vapour condition the acid exists as a dimer, $(HPO_3)_2$ as indicated by the vapour density of the acid at white heat. The dimer has the structure shown in the margin. As expected this dimer behaves as a diabasic acid. The values of dissociation constants are $K_1 = 1.6 \times 10^{-2}$ and $K_2 = 7.0 \times 10^{-7}$. Metaphosphoric acid also exists as a polymer which is represented as $(HPO_3)_n$. The higher polymers have cyclic structures.

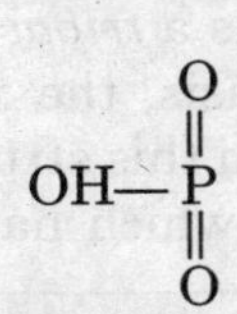

Structure of metaphosphoric acid (HPO_3)

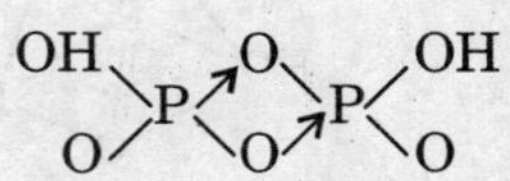

Structure of $(HPO_3)_2$ dimer

7. Peroxy phosphoric Acids.

Both the peroxy-phosphoric acids namely H_3PO_5 and $H_4P_2O_8$ have the following structures.

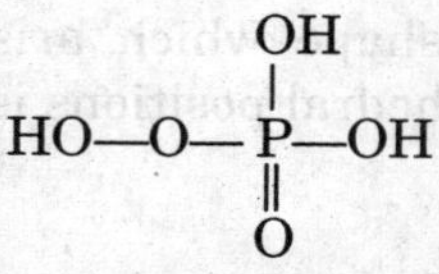

Peroxy-monophosphoric acid, H_3PO_5

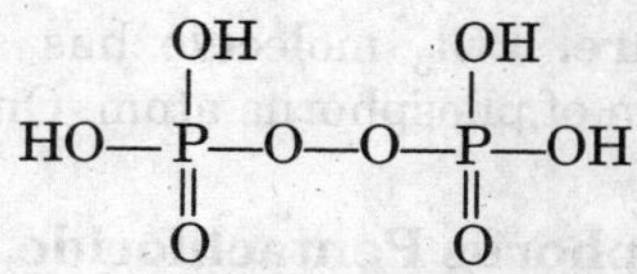

Peroxy-diphosphoric acid, $H_4P_2O_8$

Halides

With halogens phosphorus forms two types of halides *viz. trihalides*, PX_3 and *pentahalides*, PX_5. Diphosphorus tetrachloride, P_2Cl_4 and diphosphorus tetraiodide, P_2I_4 are also known. Here we shall discuss only PCl_3 and PCl_5.

1. Phosphorus Trichloride. PCl_3

Preparation. The trichloride of phosphorus is obtained by passing dry chlorine over red phosphorus gently heated in a retort over a water bath. The trichloride formed distils over and is collected in a receiver cooled in a freezing mixture.

$$P_4 + 6Cl_2 \longrightarrow 4PCl_3$$

Properties. (*a*) **Physical.** PCl_3 is a colourless liquid which fumes strongly in air. Its density is 1.57. It boils at 76°C.

(*b*) **Chemical.** (*i*) *Action of water.* PCl_3 reacts violently with water forming phosphorus acid.

$$PCl_3 + 3HOH \longrightarrow H_3PO_3 + 3HCl$$

(*ii*) *Action of conc. H_2SO_4.* Conc. sulphuric acid reacts with PCl_3 forming chlorosulphonic metaphosphoric acid.

$$4H_2SO_4 + 2PCl_3 \longrightarrow \underset{\text{Chlorosulphonic acid}}{2HSO_3.Cl} + P_2O_5 + 2SO_2 + 4HCl + H_2O$$

(*iii*) *Action of Cl_2.* PCl_3 reacts with chlorine forming phosphorus pentachloride.

$$PCl_3 - Cl_3 \longrightarrow PCl_5$$

(*iv*) *Action of sulphur trioxide.* It reacts with SO_3 forming phosphoryl chloride and sulphur dioxide.

$$SO_3 + PCl_3 \longrightarrow POCl_3 + SO_2$$

(*v*) *Action of thionyl chloride ($SOCl_2$).* It reacts with $SOCl_2$ forming phosphoryl chloride, thio-phosphoryl chloride and phosphorus pentachloride.

$$SOCl_2 + 3PCl_3 \longrightarrow POCl_3 + PSCl_3 + PCl_5$$

(*vi*) *Action of organic compounds.* It reacts with all such organic compounds which contain an alcoholic group forming corresponding chloro derivatives. Thus —OH group is replaced by —Cl. For example :

$$\underset{\text{Ethyl alcohol}}{3C_2H_5OH} + PCl_3 \longrightarrow \underset{\text{Ethyl chloride}}{3C_2H_5Cl} + H_3PO_3$$

$$\underset{\text{Benzene sulphuric acid}}{3C_6H_5SO_2.OH} + PCl_3 \longrightarrow \underset{\text{Benzene sulphonly chloride}}{3C_6H_5SO_2.Cl} + H_3PO_3$$

Uses. PCl_3 is used in the preparation of organic compounds.

Structure. PCl_3 molecule has a pyramidal shape which arises from sp^3 hybridisation of phosphorus atom. One of the tetrahedral positions is occupied by a lone pair.

2. Phosphorus Pentachloride, PCl_5.

Preparation. Phosphorus pentachloride is prepared by the action of dry chlorine (*excess*) on phosphorus trichloride.

$$PCl_3 + Cl_2 \longrightarrow PCl_5$$

Properties. (*i*) It is an almost colourless crystalline solid with pungent odour when pure.

(*ii*) On heating, it sublimes below 100°C and can be melted (m.pt. = 148°) only by heating under pressure.

(*iii*) *Dissociation.* It is inferred from its vapour density data that phosphorus pentachloride dissociates on heating into phosphorus trichloride and free chlorine.

$$PCl_5 \longrightarrow PCl_3 + Cl_2$$

(*iv*) *Action of water.* It combines with H_2O violently forming phosphorus oxychloride, $POCl_3$ (with insufficient water) and phosphoric acid, H_3PO_4 (with excess of water).

$$PCl_5 + H_2O \longrightarrow POCl_3 + 2HCl$$

$$PCl_5 + 4H_2O \longrightarrow H_3PO_4 + 5HCl$$

(*v*) *Action of hydroxy compounds.* It reacts with hydroxy compounds and replaces the hydroxyl groups by chlorine atoms. For example :

$$\underset{\text{Ethyl alcohol}}{C_2H_5OH} + PCl_5 \longrightarrow \underset{\text{Ethyl chloride}}{C_2H_5Cl} + POCl_3 + HCl$$

$$\underset{\text{Acetic acid}}{CH_3COOH} + PCl_5 \longrightarrow \underset{\text{Acetyl chloride}}{CH_3COCl} + POCl_3 + HCl$$

$$\underset{\text{Sulphuric acid}}{SO_2(OH)_2} + 2PCl_5 \longrightarrow \underset{\text{Sulphuryl chloride}}{SO_2Cl_2} + 2POCl_3 + 2HCl$$

(*vi*) *Action of concentrated sulphuric acid.* Sulphuric acid forms chlorosulphonic acid.

$$H_2SO_4 + PCl_5 \longrightarrow POCl_3 + HCl + \underset{\text{Chlorosulphonic acid}}{Cl.SO_3H}$$

(*vii*) *Action of boric acid (H_3BO_3).* Boric acid forms its anhydride and boron trioxide with PCl_5.

$$2H_3BO_3 + 3PCl_5 \longrightarrow 3POCl_3 + 6HCl + B_2O_3$$

(*viii*) *Action of metals.* Many metals like Zn, Cd, Au, Pt etc. are converted into chlorides when heated with PCl_5.

$$PCl_5 + Zn \longrightarrow PCl_3 + ZnCl_2$$

Uses. It is extensively used in organic chemistry for replacement of hydroxy groups for chlorine atoms, *i.e.* it is used as a chlorinating agent in organic chemistry.

Structure. X-ray studies have shown that solid PCl_5 consists of ionic lattice of tetrahedral $[PCl_4]^+$ cations and octahedral $[PCl_6]^-$ anions. (*See* Fig. 34.5). In the *vapour state* PCl_5 consists of descrete PCl_5 molecules which have a trigonal bipyramidal shape. (*See* Fig. 34.5). The axial P—Cl distance = 2.19A° and equatorial P—Cl = 2.04A°. *The axial bonds are relatively weak and have considerable ionic character.*

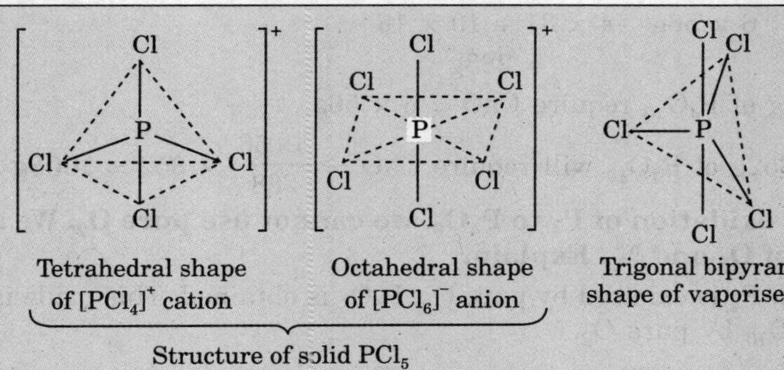

Fig. 34.5. Shape of PCl_5 in the solid and vaporised state.

Oxy-halides

All the oxy-halides of the POX_3(X = F, Cl and Br) are known. These compounds are also called *phosphoryl halides*. Here we shall discuss only $POCl_3$.

Preparation. It is prepared by the action of water on phosphorus pentachloride or by heating PCl_5 with oxalic acid or boric acid.

$$PCl_5 + H_2O \longrightarrow POCl_3 + 2HCl$$
$$PCl_5 + H_2C_2O_4 \longrightarrow POCl_8 + HCl + CO_2 + CO$$
$$3PCl_5 + 2H_3BO_3 \longrightarrow 3POCl_3 + B_2O_3 + 6HCl$$

Properties. It is a colourless liquid (b. pt. = 107.2°, m.pt. = 1.37°). With water, it forms orthophosphoric acid.

$$POCl_2 + 3H_2O \longrightarrow H_3PO_4 + 3HCl$$

It reacts with organic compounds containing hydroxyl group.

$$3ROH + POCl_3 \longrightarrow 3RCl + H_3PO_4$$

Structure. $POCl_3$ molecule has a *tetrahedral shape* with oxygen atom occupying one of the tetrahedral positions (See Fig. 34.6).

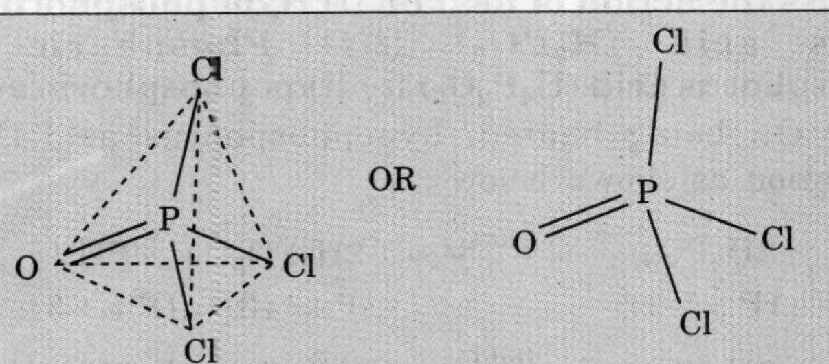

Fig. 34.6. Tetrahedral shape of phosphorus oxychloride, $POCl_3$.

Questions with Answers

[A] Questions on Oxides

Q.1 How many grams of CaO are required to nutralize 852g of P_4O_{10}?

Ans : The chemical equation for the reaction is :

$$6CaO + P_4O_{10} \longrightarrow 2Ca_3(PO_4)_2$$
$$6 \times 56g \quad 4 \times 31 + 10 \times 16 = 288g$$

288g of P_4O_{10} require CaO = 6×56g

$$\therefore 852g \text{ of } P_4O_{10} \text{ will require CaO} = \frac{6 \times 56}{288} \times 852 = \mathbf{1008g}$$

Q.2 In the oxidation of P_4 to P_4O_6, we cannot use pure O_2. We actually use mixture of O_2 and N_2. Explain.

Ans : When P_4 is oxidised by pure O_2, P_4O_6 is obtained. This oxide is further oxidised to P_4O_{10} by pure O_2.

$$P_4 \xrightarrow{+3O_2} P_4O_6 \xrightarrow{+2O_2} P_4O_{10}$$

In order to prevent the oxidation of P_4O_6 to P_4O_{10}, we use mixture of N_2 and O_2 in place of pure O_2.

$$P_4 + 3O_2 \ (\textit{Containing } N_2) \longrightarrow P_4O_6$$

Q.3 Oxides of nitrogen have open chain structures while those of phoshphorus have closed chain or cage structures. Explain your answer with one example of each type of oxides.

or The oxides of phosphorus have cage structres and not open ones. Explain.

Ans : Since N-atom has smaller size and higher electronegativity than P-atom, this atom forms $p\pi - p\pi$ multiple bonds easily with O-atom and hence oxides of N have open chain structure. On the other hand, since P-atom has larger size and lower electronnegativity than N-atom, this atom does not form $p\pi - p\pi$ multiple bonds with O-atoms. Thus oxides of P-atom donot have open chain structure. P-atom prefers to form single bonds with O-atoms. This tendency of phosphorus gives closed chain or cage sturcture to its oxides. For example N_2O_5 which is an oxide of N has open chain structure but P_4O_6 and P_4O_8 (oxides of P) have closed chain structures.

[B] Questions on Oxy acids

Q.1 Discuss the action of heat on (*i*) Hypophosphorus acid (H_3PO_2) (*ii*) Phosphorus acid (H_3PO_3) (*iii*) Phosphoric acid (H_3PO_4). (*iv*) Pyrophosphorus acid ($H_4P_2O_5$) (*v*) Hypophosphoric acid ($H_4P_2O_6$).

Ans : (*i*) On being heated, hypophosphorus acid (H_3PO_2) undergoes disproportionation as shown below :

$$\underset{(P = +1)}{3H_3PO_2} \xrightarrow{140°C} \underset{(P = +3)}{2H_3PO_3} + \underset{(P = -3)}{PH_3}$$

$$\underset{(P = +1)}{2H_3PO_2} \xrightarrow{160°C} \underset{(P = +5)}{H_3PO_4} + \underset{(P = -3)}{PH_3}$$

(*ii*) On heating, phosphorus acid (H_3PO_3) undergoes disproportionation (*self oxidation-reduction reaction*)

$$\underset{(P = +3)}{4H_3PO_3} \xrightarrow{160°C} \underset{(P = +5)}{3H_3PO_4} + \underset{(P = -3)}{PH_3}$$

(*iii*) When H_3PO_4 is heated, different products are obtained. The nature of products obtained depends on the temperature.

$$2H_3PO_4 \xrightarrow[(-H_2O)]{250°C} H_4P_2O_7 + H_2O$$

$$H_3PO_4 \xrightarrow[(-H_2O)]{600°C} HPO_3 + H_2O$$

$$2H_3PO_4 \xrightarrow[(-H_2O)]{\text{Red heat}} P_2O_5 + 3H_2O$$

(*iv*) On heating strongly, $H_2P_2O_5$ gives metaphosphorus acid (HPO_2).

$$H_4P_2O_5 \xrightarrow[(-H_2O)]{\text{Strong heating}} 2HPO_2 + H_2O$$

(*v*) (*a*) On being heated above its m.pt; $H_4P_2O_6$ undregoes disproportionation to phosphorus acid (H_3PO_3) and metaphosphoric acid (HPO_3)

$$H_4P_2O_6\ (P = +4) \xrightarrow[m.pt.]{\text{Heat above}} H_3PO_3\ (P = +3) + HPO_3\ (P = +5)$$

(*b*) On heating at about 180°C, the acid gives orthophosphoric acid (H_3PO_4), metaphosphoric acid (HPO_3) and phosphene (PH_3)

$$4H_4P_2O_6 \xrightarrow{\text{Above } 180°} 3H_3PO_4 + 4HPO_3 + PH_3$$

Q.2 What happens when :

(*i*) Phosphoric acid is heated with HNO_3 and ammonium molybdate.

(*ii*) Phosphoric acid reacts with $Ca_5(PO_4)_3F$. ***(I.I.T. 1998)***

(*iii*) Hypophorus acid reacts with NaOH solution.

Ans : (*i*) On heating H_3PO_4, in presence of HNO_3, with ammonium molybdate [$(NH_4)_2MoO_4$], a canary yellow ppt. of ammonium phosphomolybdate $(NH_4)_3PO_4.12MoO_3$ is formed.

$H_3PO_4 + 21HNO_3 + 12(NH_4)_2MoO_4 \rightarrow 21NH_4NO_3 + 12H_2O +$

$(NH_4)_3PO_4.12\ MoO_3$ (*yellow ppt. of ammonium phosphomolybdate*)

(*ii*) HF is formed.

$Ca_5(PO_4)_3F + 7H_3PO_4 + 5H_2O \rightarrow 5Ca(H_2PO_4)_2.\ H_2O + HF$

(*iii*) Hypophorus acid is H_3PO_2. Since this acid is a monobasic acid, only one H-atom will be replaced by Na-atom. Thus NaH_2PO_2 will be formed.

$H_3PO_2 \rightarrow H^+ - H_2PO_2^-$

$NaOH \rightarrow Na^+ + OH^-$

$H_3PO_2 + NaOH \rightarrow Na^+H_2PO_2^- + H_2O$

Q.3 What is the difference between the structure of NO_3^- and PO_3^- ions.

Ans : NO_3^- ion (nitrate ion) exists in the free state. Since N-atom in this ion is sp^2 hybridised, NO_3^- ion has trigonal planar shape with ONO bond angle equal to 120° as shown in Fig. 34.7

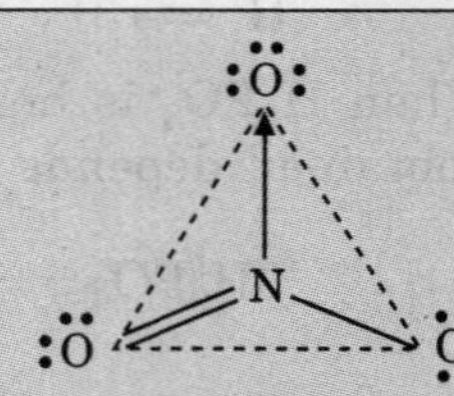

Fig. 34.7 Trigonal planar structure of NO_3^- ion

On the other hand PO_3^- ion (metaphosphate ion) does not exist in the free state, since experiments have shown that this ion gets polymerised and forms polymerised ion, $[(PO_3^-)_n]^{n-}$. Thus PO_3^- ion exists as a polymerised ion, $[(PO_3^-)_n]^{n-}$ which is called *polymeric metaphosphate ion*. $[(PO_3^-)_n]^{n-}$ ion has polymeric structure which may be either linear structure or cyclic structure as shown below in Fig. 34.8 and Fig. 34.9.

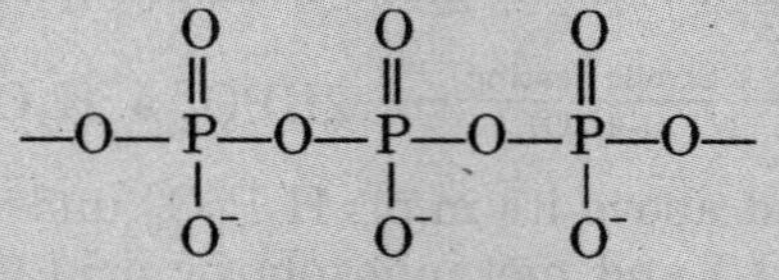

Fig. 34.8 Linear structure of polymeric metaphosphate ion, $[(PO_3^-)_n]^{n-}$

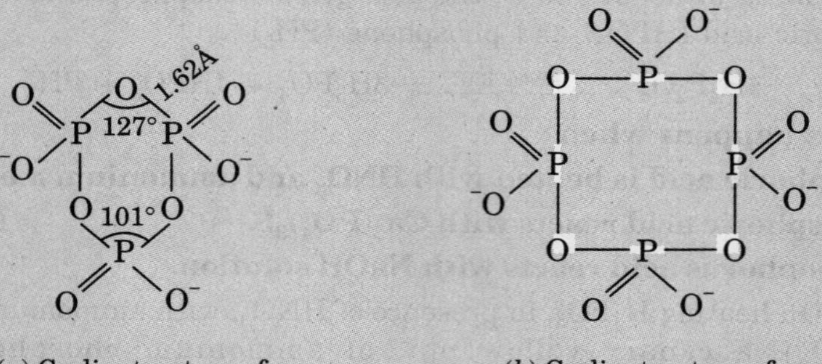

(*a*) Cyclic structure of trimetaphosphate ion, $[(PO_3)_3]^{3-}$

(*b*) Cyclic structure of tetrametaphosphate ion, $[(PO_3)_4]^{4-}$

Fig. 34.9 Cyclic structres of trimetaphosphate and tetrametaphosphate ions.

In these polymetaphosphate ions, various PO_3^- units are linked together through P – O – P bonds. In these PO_3^- units, P-atom is sp^3 hybridised and thus each PO_3^- unit has tetrahedral shape. It may also be seen that for getting cycilc metaphosphate the minimum value of n = 3. Thus $(PO_3^-)_3$ is the smallest cyclic metaphosphate ion This cyclic metaphosphate ion [*i.e.* $(PO_3^-)_3$] has three P–O–P bonds.

Q.4 Explain the following, giving appropriate reason for your answer.

(*i*) Orthophosphorus acid is not tribasic though it contains three H-atoms. ***(I.I.T.1987)***

(*ii*) Orthophoric acid is tribasic but phosphorus acid is dibasic. ***(I.I.T. 1982)***

(*iii*) Concentrated solution of orthophosphoric acid is syrupy liquid.

(*iv*) H_3PO_2 and H_3PO_3 act as reducing agents but H_3PO_4 does not.

(*v*) H_3PO_3 (Phosphorus acid) is diprotic (or dibasic).

Ans : (*i*) Orthophoshorus acid (H_3PO_3) is not tribasic acid, since it contains only two OH groups attached directly with central P-atom or since it contains only two H-atoms bound to O-atoms or since it contians only two ionisable H-atoms as is evident from its structure. Due to the presence of two OH groups H_3PO_3 dissociates in two stages given below :

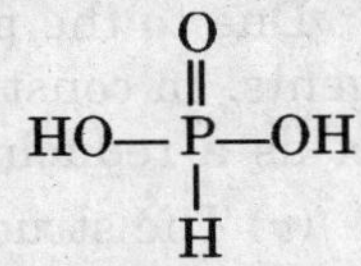

Structure of orthophosphorus acid molecule, H_3PO_3

$$H_3PO_3 \rightleftharpoons H^+ + H_2PO_3^-$$

$$H_2PO_3^- \rightleftharpoons H^+ + HPO_3^{2-}$$

It is due to the presence of two OH groups in H_3PO_3 molecule that it behaves as a dibasic acid and not as a tribasic acid.

(*ii*) The aqueous solution of orthophosphoric acid (H_3PO_4) ionises in three stages and hence gives three types of anions.

$H_3PO_4 \rightleftharpoons H^+ + H_2PO_4^-$ (*Dihydrogen phosphate ion*)

$H_2PO_4^- \rightleftharpoons H^+ + HPO_4^{2-}$ (*Hydrogen phosphate ion*)

$HPO^{2-} \rightleftharpoons H^+ + PO_4^{3-}$ (*Phospahte ion*)

Due to the production of three types of anions, H_3PO_4 gives three types of salts. Due to the formation of three types of salts, H_3PO_4 is tribasic. Its tribasic nature is also confirmed by the fact that H_3PO_4 has three OH groups attached directly with P-atom

Although phosphorus acid (H_3PO_3) has three H-atoms, its aqueous solution gives only two H^+ ions on ionisation

$H_3PO_3 \rightleftharpoons H^+ + H_2PO_3^-$ (*Dihydrogen phosphite ion*)

$H_2PO_3^- \rightleftharpoons H^+ + HPO_3^{2-}$ (*Hydrogen phosphite ion*)

Due to the production of two H^+ ions, H_3PO_3 behaves as a dibasic acid. Its dibasic character is also confirmed by the fact that H_3PO_3 has only two OH groups attached directly with P-atom.

(*iii*) The solid and concentrated solution of orthophosphoric acid has various H_3PO_4 tetrahedral molecules connetced together by H-bonds (Fig. 34.10). The existence of H-bonds between different tetrahedral H_3PO_4 molecules increases the viscosity of the concentrated solution of the acid and thus accounts for the syrupy nature of the acid. H-bonds have been shown by dotted lines.

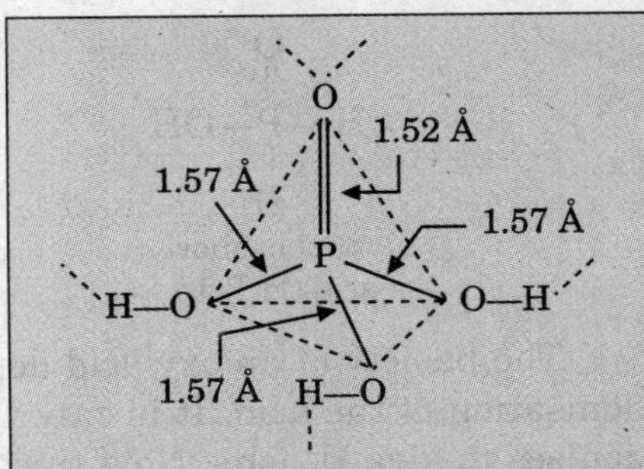

Fig. 34.10 Structure of H_3PO_4 in solid and concentarted solution form.

(*iv*) The structures of H_3PO_2, H_3PO_3 and H_3PO_4 are :

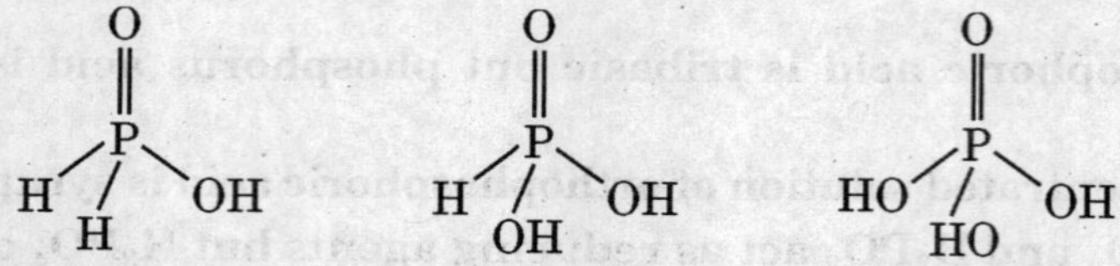

Hypophosphorus acid (H_3PO_2) Phosphorus acid (H_3PO_3) Orthophosphoric acid (H_3PO_4)

Due to the presence of P – H bonds, both H_3PO_2 and H_3PO_3 act as reducing agents. In constrast, H_3PO_4 does not have any P – H bond and hence it does not act as a reducing agent.

(*v*) The structure of H_3PO_3 molecule shows that it contains two – OH groups attached directly with the central P-atom. Thus H_3PO_3 ionises to give two H^+ ions and hence is dibasic.

$$H_3PO_3 \rightleftharpoons H^+ + H_2PO_3^-$$
$$H_2PO_3^- \rightleftharpoons H^+ + HPO_3^{2-}$$
$$H_3PO_3 \rightleftharpoons 2H^+ + HPO_3^{2-}$$

Q.5 How will you get the following from orthophoric acid : (*i*) yellow phosphorus (*ii*) metaphophoric acid.

Ans : (*i*) $H_3PO_4 \xrightarrow[(-H_2O)]{600°C} HPO_3 + H_2O$

HPO_3 obtained as above is mixed with powered coke and distilled in fireclay retorts at a bright red heat. Coke reduces HPO_3 to phosphorus and is itself oxidised to CO.

$$4HPO_3 + 10C \rightarrow P_4 + 10CO + 2H_2O$$

(*ii*) $H_3PO_4 \xrightarrow[(-H_2O)]{600°C} HPO_3$ (*Meta phosphoric acid*) $+ H_2O$

Q.6 Predict whether the basicity of H_3PO_2, H_3PO_3 and H_3PO_4 is the same or different. Give reasons for your answer.

Ans : The structure of the given molecules are given below :

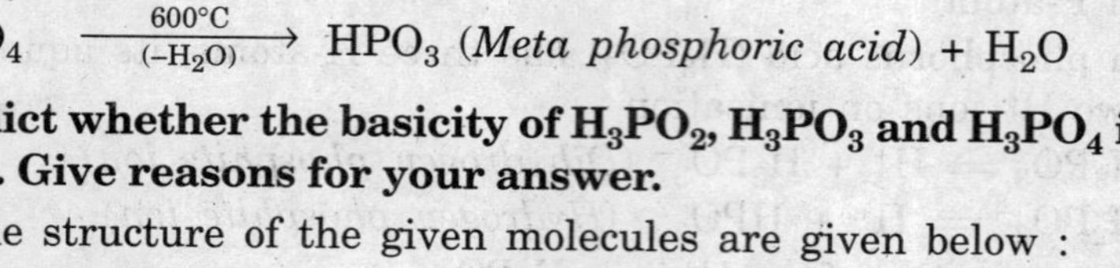

Hypophosphorus acid (H_3PO_2) Orthophosphorus acid (H_3PO_3) Orthophosphoric acid (H_3PO_4)

The basicty of an oxy acid depends on the number of H^+ ions obtained by the ionisation of the acid. It is only OH group (*s*) attached to the central atom which ionises to give H^+ ions. Now since H_3PO_2, H_3PO_3 and H_3PO_4 molecules have one, two and three OH groups attached with central P-atom, these molecules, on ionisation, give one, two and three H^+ ions. Thus these acids have differnt bascity.

Q.7 How will you obtain pyrophosporic acid from calcium phosphate.

Ans : This can be obtained as follows :

$$\underset{\text{Cal. phosphate}}{Ca(PO_4)_2} + 3H_2SO_4(conc.) \longrightarrow CaSO_4 + 2H_3PO_4$$

$$2H_3PO_4 \xrightarrow[(-H_2O)]{250°C} \underset{\text{Pyrophosphoric acid}}{H_4P_2O_7} + H_2O$$

[C] Questions on Halides.

Q.1 What is the structure of PCl_5 molecule in the solid state.

Ans : In the solid state PCl_5 exists as an ionic compound which can be represented as $[PCl_4]^+ [PCl_6]^-$ in which the cation, PCl_4^+ is tetrahedral and the anion, PCl_6^- is octahedral.

Q.2 Give the structure of PBr_5 and PI_5 molecules in the solid state.

Ans : In the solid state PBr_5 exists as an ionic solid whose structure is $[PBr_4]^+ Br^-$. The cation, $[PBr_4]^+$ has tetrahedral geometry. Similary PI_5 molecule also exists as an ionic solid, $[PI_4]^+ I^-$ in which the cation, $[PI_4]^+$ has tetrahedral geometry.

Q.3 P–X bonds in PX_5 molecules (X = F, Cl) are not of the same length. Explain.

Ans : In PX_5 molecules, P-atom is sp^3d hybridised. In this molecule, each of the two axial P – X bonds.is slightly larger than each of the three equatorial P – X bonds. The difference in bond lengths is because of the following reasons :

(*i*) sp^3d hybridisation gives trigonal bipyramidal geometry to PX_5 molecule. This geometry is not a symmetrical geometry, since two bond angles are of 90°, three bond angles are of 120° and one bond angle is of 180°. Due to unsymmetrical trigonal bipyramidal geometry of PX_5 molecule, all P – X bonds are not of the same length.

(*ii*) All the P – X bonds are not of the same length, since axial X-atoms experience greater repulsion as compared to that experienced by equatorial X-atoms.

Q.4 PCl_5 acts a chlorinating agent in organic reactions. Why ?

Ans : In the gaseous and liquid states, PCl_5 molecule has a trigonal bipyramidal shape in which the two axial P – Cl bonds are slightly larger in length than the three equatonal P – Cl bonds. Due to the difference in bond lengths, PCl_5 becomes unstable and hence dissociates to give two Cl-atoms making axial bonds.

$$PC_5 \rightleftharpoons PCl_3 + Cl_2$$

Due to the production of Cl_2, PCl_5 acts as a chlorinating agent.

Q.5 Can PCl_5 act as an oxidising agent (OA) or as a reducing agent (RA). Give appropriate reasons for your answer.

Ans : Maximum oxidation state (O.S.) shown by P-atom is +5 which is shown by this elements in PCl_5. Thus :

(*i*) If PCl_5 (P = +5) acts as an OA, then it would be reduced to some compound in which P-atom would show an O.S. lower than +5. This is possible as is evident from the following reactions :

$$\underset{(Ag = 0)}{2Ag} + \underset{(P = +5)}{PCl_5} \longrightarrow \underset{(Ag = +1)}{2AgCl} + \underset{(P = +3)}{PCl_3}$$

$$\underset{(Sn = 0)}{Sn} + \underset{(P = +5)}{2PCl_5} \longrightarrow \underset{(Sn = +4)}{SnCl_4} + \underset{(P = +3)}{2PCl_3}$$

$$\underset{(H = 0)}{H_2} + \underset{(P = +5)}{PCl_5} \longrightarrow \underset{(H = +1)}{HCl} + \underset{(P = +3)}{PCl_3}$$

(*ii*) If PCl_5 (P = +5) acts as a RA, then it would be oxidised to some other compound in which P-atom would show an O.S. higher than +5. This is not possible because maximum O.S. shown by P-atom is only +5.

Q.7 Which one of PCl_3 and PCl_5 is a better Lewis acid ?

Ans : The central P-atom in PCl_3 molecule has a *lp* of electrons and hence PCl_3 molecule acts as a Lewis base and not as a Lewis acid. Lewis base character of PCl_3 molecule is evident from the following reactions.

$$\underset{\text{Lewis base}}{Cl_3P\!:} + \underset{\substack{\text{Lewis acid} \\ (Cl^+ = 3s^2 3p^2 3p^2 3p^0)}}{Cl^+} \longrightarrow \underset{\substack{\text{Complex} \\ \text{ion}}}{[PCl_4]^+}$$

$$\underset{\text{Lewis base}}{Cl_3P\!:} + \underset{\text{Lewis acid}}{BF_3} \longrightarrow \underset{\text{Adduct}}{Cl_3P \rightarrow BF_3}$$

$[PCl_4]^+$ ion is found in the solid state of PCl_5 (solid $PCl_5 \equiv [PCl_4]^+ [PCl_6]^-$)

In PCl_5 molecule, since central P-atom does not have *lp* of electrons, this molecule does not show basic character. However, since P-atom has vacant *d*-orbitals, PCl_5 molecule shows acidic character. The acidic character is evident from the following reaction.

$$\underset{\text{Lewis acid}}{PCl_5} + \underset{\substack{\text{Lewis base} \\ (Cl^- = 2s^2 2p^2 2p^2 2p^2)}}{:\ddot{Cl}^-} \longrightarrow Cl_5P \leftarrow Cl^- \text{ or } [PCl_6]^-$$

$[PCl_6]^-$ ion is found in PCl_5 when PCl_5 is in the solid state. (Solid $PCl_5 \equiv [PCl_4]^+ [PCl_6]^-$)

Above discussion shows that PCl_5 is a Lewis acid but PCl_3 is a Lewis base.

Q.6 PCl_5 is known but PI_5 is unkown. Why ?

Ans : Since Cl-atoms are smaller in size, these atoms can easily be accomdated round P-atom to from PCl_5. On the other hand, since I-atoms are larger in size, these atoms cannot be accomdated round P-atom and hence PI_5 cannot be formed.

Q.7 Which is more stable PCl_5 or PCl_3 ?

Ans : PCl_3 is more stable than PCl_5 because on heating PCl_5 decomposes to form PCl_3 and Cl_2.

$$PCl_5 \xrightarrow{\Delta} PCl_3 + Cl_2$$

Q.7. When PCl_5 is heated, it is broken to form PCl_3 and Cl_2. Expalin this reaction.

Ans : We know that in PCl_5 molecule, due to greater (*bp* – *bp*) repulsions, the two axial P – Cl bonds are less stable than the three equatorial P – Cl bonds. It is for this reason that when PCl_5 is heated, less stable axial bonds are removed from PCl_5 molecules as Cl_2.

$$PCl_5 \xrightarrow{\Delta} PCl_3 + Cl_2$$

University Questions

1. Explain the formulae and structures of various oxides of phosphorus. (*Delhi 87; Punjab 84*)

2. Write a note on "oxy-acids of phosphorus". (*Jiwaji 83, 84*)

3. What are polymetaphosphates? (*Madras 86*)

4. Give brief account of the chlorides of phosphorus. (*Madras 86*)

5. Describe the action of heat on orthophosphorus acid. Give balanced chemical equation. (*Delhi 99*)

6. H_3PO_3 is dibasic acid whereas H_3PO_4 is a tribasic acid. Explain. (*Gorakhpur 99*)

7. Describe the method of preparation and uses of hypophosphorus acid. (*Allahabad 99*)

8. (*a*) Name the oxyacids of phosphorus and write down their structural formulae.

(*b*) Discuss the basicity and reducing character of these acids. (*Lucknow 2000*)

CHAPTER 35

General Characteristics of Group 16 (VI A) Elements : Oxygen Group Elements

Group VI A of the periodic table consists of five elements *viz. Oxygen* (O), *Sulphur* (S), *Selenium* (Se), *Tellurium* (Te) *and Polonium* (Po). Oxygen and sulphur are the *typical elements* of this group. The elements O, S, Se and Te are often collectively called *chalcogens, i.e.* the ore forming elements.

Position of Oxygen Group Elements in the Periodic Table

The inclusion of the elements of Group VI A in the same sub-group of the periodic table is justified by the following general characteristics of these elements.

I. Electronic Configurations.

The electronic configurations of the atoms of these elements are : O(Z = 8) $\rightarrow$ 2, 6 ($2s^2p^4$); S(Z = 16) $\rightarrow$ 2, 8, 6 ($3s^2p^4$); Se(Z = 34) $\rightarrow$ 2, 8, 18, 6 ($4s^2p^4$); Te(Z = 52) $\rightarrow$ 2, 8, 18, 18, 6 ($5s^2p^4$); Po(Z = 84) $\rightarrow$ 2, 8, 18, 32, 18, 6($6s^2p^4$). It is clear that all the elements have ns^2p^4 configuration in the outermost shell. The four electrons in *p*-orbitals are distributed as $p_x^{\,2}p_y^{\,1}p_z^{\,1}$ in accordance with *Hund's rule.* The penultimate shell in oxygen atom contains s^2 group (*saturated*), in sulphur s^2p^6 (*saturated*), in selenium $s^2p^6d^{10}$ (*saturated*), whereas in tellurium and polonium $s^2p^6d^{10}$ (*unsaturated*). This shows that oxygen differs from sulphur and these two elements from the remaining elements of this group. We thus conclude that the elements of this group are *p*-block elements.

II. Similarities in Physical Properties and Gradation in Them.

Some of the physical properties of these elements are discussed below :

1. Density, Atomic Volume, Melting and Boiling Point. With the rise of atomic number the *density*, the *atomic volume, the melting point* and the *boiling point* all show a steady increase. Polonium shows exception when we consider its melting and boiling points. Due to diminished availability of *s* electron pair, the melting and boiling point of Po are less than those of Te. The large difference in melting and boiling points between O and S is attributed, in part, to the fact that whereas oxygen exists in the form of O_2 molecules, sulphur exists as S_8 molecules.

2. Ionisatian Energies. Due to the increased nuclear charge, the first and subsequent ionisation energies of these elements are higher than the corresponding elements of group VA. These elements have high values of ionisation energies. These high values indicate that the elements of this group show reluctance to lose the electrons to form positive ions. As the size of the atom increases, the values of ionisation energies decrease down the group.

3. Metallic and Non-metallic Character. Due to the decrease of ionisation energy from oxygen to polonium, the metallic character of these elements increases down the group. For example oxygen and sulphur are distinctly non-metallic and some of the metallic properties begin to develop in selenium and tellurium Polonium which is radioactive is definitely a metal.

4. Atomicity. (*Molecular structure*). All the elements are polymeric. Oxygen molecule is diatomic molecule (O_2) while the molecules of other elements are more complex, *e.g.* sulphur, selenium and polonium are *octatomic molecules viz.* S_8, Se_8 and Po_8 with *puckered ring structures.* These structures contain 8-atom rings or zig-zag chains. In these staggered rings 8-atoms are joined together by covalent bonds. The tendency of these elements to exist in 8-atom rings is greatest with sulphur and decreases as we go down the group. Such type of structure accounts for oxygen being as gas and the other members being solids. Puckered ring structure of sulphur molecule, S_8 is shown in Fig. 35.1. Such rings give molecular crystal in which the rings are held together by relatively weak polar forces. Consequently both the forms of sulphur (rhombic or mono-clinic forms) have low melting points.

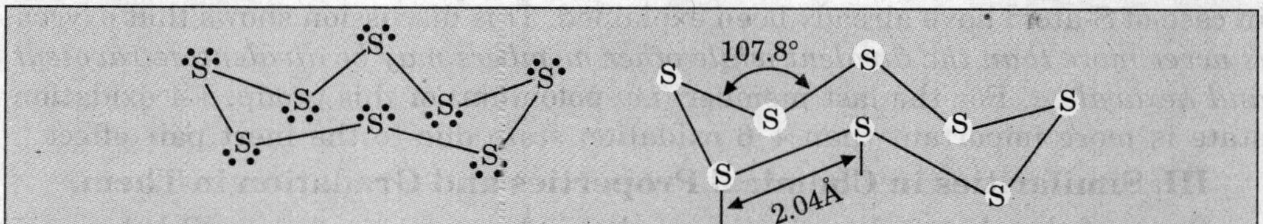

Fig. 35.1. Puckered ring structure of sulphur S_8 molecule.

5. Polymorphism. (*Allotropy*). All the elements are polymorphic, *i.e.* they show allotropy, *e.g.* oxygen exists in two nonmetallic forms *viz.* O_2 and O_3. Sulphur has several allotropic forms all of which are non-metallic, *e.g. rhombic*, *monoclinic* and *plastic sulphur*. Se has two forms: *red* (*non-metallic*) and *grey* (*metallic*). Te exists in two forms *non-metallic and metallic* (more stable), Po also exists in two forms α and β (*both metallic*).

6. Catenation. Oxygen and to a greater extent sulphur shows the property of catenation *i.e.* they have a tendency to form chains of identical atoms *e.g.* chains of sulphur atoms are present in persulphides, polythionic acids and in S_8 molecule. Oxygen gives polyoxides similar to poly sulphides given by sulphur. Examples are : H_2O_2, H—O—O—H; H_2S_2, H—S—S—H; H_2S_3, H—S—S—S—H; H_2S_4, H—S—S—S—S—H.

7. Oxidation States. Since all the elements of this group have ns^2p^4 configuration in their outermost orbit, they attain noble gas configuration *viz.* ns^2p^6 either by gaining or by sharing two electrons. Thus these elements show two types of oxidation states.

***(a)* Negative oxidation states.** The high electronegativity of oxygen indicates that it will tend to complete its octet preferably by gaining electrons. These shared electrons, for the purpose of determining the oxidation state, will be counted towards oxygen because it forms the electronegative atom in all its compounds (excepting, of course, OF_2 molecule in which oxygen is as O^{2+} ion).

Hence most of the metal oxides have O^{2-} ions and are ionic, giving oxygen an oxidation state equal to –2.

The electronegativities of S, Se and Te are low and hence their compounds even with most electropositive elements will not be more than 50% ionic. As a result the formation of S^{2-}, Se^{2-} and Te^{2-} ions is less probable. The least electronegative element Po does not form Po^{2-} but it forms Po^{2+}.

(*b*) Positive oxidation states. Oxygen, being highly electronegative, shows only –2 oxidation state in its compounds, *i.e.* does not show positive oxidation state, except in OF_2. As we move down the group, electronegativity decreases and the tendency of the elements to show the positive oxidation state increases. S, Se, Te and Po can also show, in addition to + 2 oxidation state, + 4 and + 6 oxidation states. This is due to the availability of *d*-orbitals in these elements. Oxygen atom has no *d*-orbital and hence cannot show + 4 and + 6 oxidation states. In case of sulphur unpairing of electrons can take place, if the promotion of 3*p* and 3*s* electrons is made to 3*d* orbitals. + 2, + 4 and + 6 oxidation states in case of S-atom have already been explained. This discussion shows that *oxygen is never more than the divalent while other members may be divalent, tetravalent and hexavalent.* For the last member (*i.e.* polonium) of this group, + 4 oxidation state is more important than + 6 oxidation state due to the inert pair effect.

III. Similarities in Chemical Properties and Gradation in Them.

Some of the chemical properties of these elements are discussed below :

1. Formation of Hydrides. All the elements of this group form the hydrides of H_2M type, where M = O, S, Se, Te and Po.

Preparation. H_2O is prepared by burning H_2 in the atmosphere of O_2 while H_2S, H_2Se and H_2Te are prepared by the action of acids on metal sulphides, selenides and tellurides, *e.g.*

$$FeS + H_2SO_4 \longrightarrow FeSO_4 + H_2S$$
$$K_2Se + H_2SO_4 \longrightarrow K_2SO_4 + H_2Se$$

H_2Po has been prepared only in traces by dissolving Mg foil plated with Po in 0.2 N HCl.

Properties. (*i*) **Physical state.** Water is a colourless, odourless liquid while the hydrides of other elements are colourless, poisonous gases with unpleasant odours. These are inflammable gases which burn with a blue flame.

(*ii*) **Volatility.** The volatility of these hydrides increases markedly from H_2O to H_2S and then decreases. Abnormal low volatility of H_2O (*i.e.*, high boiling point) is due to its association as a result of hydrogen bonding both in solid and liquid states. As a consequence H_2O has the highest boiling point and melting point among the hydrides as shown here : H_2O = 100°, 0.0° ; H_2S = –60.3°, –85.5°; H_2Se = –41.3°, –65.7°; H_2Te = –23°, –51.0°. The former numbers indicate the boiling points (in °C) while the later number indicate the melting points (in °C). Due to lower electronegativity of S-atom, H-bonding in H_2S is rare and hence the association of H_2S molecules does not occur, but H_2S gas remains as unassociated molecule. This makes H_2S to have more volatility than H_2O. In going down the group, the volatility of H_2M hydrides decreases as $H_2S \rightarrow H_2Se \rightarrow H_2Te$ due to the increase in molecular weights of the hydrides.

***(iii)* Thermal stability and reducing power.** The thermal stability of the hydrides decreases from H_2O to H_2Te. Thus water dissociates at about 2000°C, H_2S at 400–600°C, H_2Se at about 150°C while H_2Te decomposes even at ordinary temperature. The decrease in thermal stability from H_2O to H_2Te is because the size of M^{2-} ion in H_2M hydrides increases from O^{2-} to Te^{2-} ion and hence the charge is distributed over a large volume and this makes the H_2M hydride thermally unstable. The decrease in thermal stability from H_2O to H_2Te is also because of an increase in M—H bond length as we move from H_2O to H_2Te as shown here : H_2O, 0.96A°; H_2S, 1.32A°; H_2Se, 1.47A°, H_2Te, 1.73A° The increase of M—H bond length from O—H (in H_2O) to Te—H (in H_2Te) weakens the M—H bond and hence the hydride becomes thermally unstable. With the decrease of thermal stability of these hydrides from H_2O to H_2Te, the reducing power of these hydrides gradually increases in the same direction.

***(iv)* Covalent character.** Since the electronegativity of O, S, Se and Te decreases from O to Te, the electronegativity difference between M (M = O, S, Se, Te) and H decreases and hence the ionic character of M—H bond also decreases from O—H to Te—H bonds in H_2M hydrides. In other words *the covalent character of these hydrides increases from H_2O to H_2Te.*

The increase in covalent character from H_2O to H_2Te can also be explained by Fajan's rules. According to one of these rules as the size of M^{2-} ion in H_2M hydrides increases from O^{2-} to Te^{2-}, the tendency of these ions to form covalent hydrides increases.

***(v)* Acidic character.** Aqueous solutions of the hydrides of these elements behave as weak, diprotic acids and hence ionise as :

$$H_2M \rightleftharpoons H^+ + HM^-$$
$$HM^- \rightleftharpoons H^+ + M^{2-}$$
$$\overline{H_2M \rightleftharpoons 2H^+ + M^{2-}}$$

The acidic character of these acids increases in the sequence from H_2O to H_2Te. The increase in acidic character from H_2O to H_2Te is also confirmed by the fact that the value of dissociation constant of these hydrides also increases in the same direction as shown here : $H_2O = 1.0 \times 10^{-11}$, $H_2S = 1.1 \times 10^{-7}$, $H_2Se = 2.0 \times 10^{-4}$, $H_2Te = 2.3 \times 10^{-3}$. As the charge density on the conjugate bases *viz.* OH^-, SH^-, SeH^- and TeH^- decreases from OH^- to TeH^-, the proton is less tightly held with higher members and hence the acidic character increases from H_2O to H_2Te.

***(vi)* Shape of H_2M hydrides and HMH bond angles.** All these hydrides have *V-shape* (*i.e.* angular or bent shape). In these molecules the central atom (M atom) is sp^3 hybridised. Out of the four sp^3 hybrid orbitals two have lone pairs of electrons while the remaining two have bonding electron pairs. Although the spatial arrangement of the electron pairs is tetrahedral round the central M atom, due to the presence of two lone pairs in the two hybrid orbitals, the shape of MH_2 hydrides gets distorted and becomes *V-shape*.

Due to the distortion in shape, HMH bond angle in all the hydrides is never equal to the ideal tetrahedral angle (= 109.5°), but less than this value. HMH bond angles are as : H_2O = 104.5°, H_2S = 92.5°, H_2Se = 91.0°, H_2Te = 90.0°. The decrease in HMH bond angle from H_2O to H_2Te has been explained on the basis of the concept of electronegativity values of the central atom (M atom).

2. Formation of Halides. All the elements of this group form binary halides. Since fluorine is more electronegative than oxygen, fluorine compounds with oxygen are called *fluorides of oxygen* rather than *oxides of fluorine*, *e.g.* OF_2 is called *oxygen difluoride* and not difluorine oxide. Similarly O_2F_2 is called *dioxygen difluoride* and not *difluorine dioxide*. On the other hand, since oxygen is more electronegative than Cl, Br and I, oxygen compounds with Cl, Br and I are called oxides of these halogens rather than halides of oxygen, *e.g.*, ClO_2 is called *chlorine dioxide* rather than *dioxygen chloride*. Important halides are discussed below :

(A) Hexafluorides (MF_6). Except oxygen, all other elements form hexa fluorides. In these fluorides the element (S, Se or Te) shows its maximum covalency of six by using its *d*-orbitals. These can be prepared by the direct combination of the elements. Hexafluorides are relatively stable and their stability decreases from SF_6 to TeF_6 due to the increase in size from S to Te atoms. Thus SF_6 is extremely stable and inert both thermally and chemically. SeF_6 is quite reactive (*i.e.* less stable) while TeF_6 is still more reactive as is evident from the fact that TeF_6 is completely hydrolysed by H_2O and adds F^- ion and other Lewis bases like $(CH_3)_3N$ to form 8-coordinated complex.

$$TeF_6 + 6H_2O \longrightarrow 6HF + H_2TeO_6$$

$$TeF_6 \xrightarrow{+F^-} [TeF_7]^- \xrightarrow{+F^-} [TeF_8]^{2-}$$

$$TeF_6 + 2(CH_3)_3N \longrightarrow \underset{\text{Adduct}}{TeF_6 \cdot 2(CH_3)_3N}$$

All the hexafluorides are colourless gases. All have low boiling points and hence show a degree of covalency. All the hexafluorides have *regular octahedral geometry* which results from sp^3d^2 hybridisation of the central atom, M(M = S, Se or Te). All the M—F bonds are σ-bonds. All the FMF bond angles and M—F bond lengths are equal. The octahedral shape of SF_6 molecule on the basis of sp^3d^2 hybridisation has already been explained. Of all the hexafluorides, SF_6, due to its chemical inertness and high value of dielectric constant, is industrially important, since it is used as a gaseous insulator in high voltage generators.

(B) Tetrahalides (MX_4). Preparation. PTetrafluorides cannot be prepared by the direct reaction between the element and F_2, since this reaction gives hexafluorides. Rather these are obtained by indirect methods like those given below :

$$S + 4CoF_3 \longrightarrow SF_4\ (g) + 4CoF_2$$

$$SeCl_4 + 4AgF \longrightarrow SeF_4\ (l) + 4AgCl$$

$$TeO_2 + 2SeF_2 \longrightarrow TeF_4\ (s) + 2SeOF_2$$

Tetrachlorides, tetrabromides and tetraiodides are prepared by the direct reaction between the element and X_2 molecule.

$$M + 2X_2 \longrightarrow MX_4$$
$$(M = S, Se, Te)$$

Properties (*i*) *Physical state.* Among the tetrafluorides SF_4 is a gas, SeF_4 is a liquid while TeF_4 is solid. Among the tetrachlorides SCl_4 is an unstable liquid while $SeCl_4$ and $TeCl_4$ are solids. TeI_4 is a dark grey powder and PoI_4 is a black solid.

(*ii*) *Thermal stability.* Tetrafluorides are more stable to thermal decomposition than tetrahalides. $TeCl_4$ is more stable than SCl_4 and $SeCl_4$. $SeBr_4$ is unstable and dissociates into Se_2Br_2 and Br_2.

$$2SeBr_4 \longrightarrow Se_2Br_2 + 3Br_2$$

$SeBr_4$ is also readily hydrolysed as :

$$SeBr_4 + 4H_2O \longrightarrow H_2SeO_3 + H_2O + 4HBr$$

(*iii*) *Fluorinating property.* All the tetrafluorides are strong fluorinating agents.

(*iv*) *Lawis acid.* Tetrafluorides can act as strong Lewis acid, since they can accept electrons and are converted into hexafluorides. For example :

(*a*) SF_4 reacts with F_2 and is converted into SF_6. SeF_4 and $TeCl_4$ combine with F^- and Cl^- ions respectively to form complex ions.

$$SF_4 + F_2 \longrightarrow SF_6; \; TeCl_4 \xrightarrow{+\,2Cl^-} [TeCl_6]^{2-}$$

$$SeF_4 \xrightarrow{+\,F^-} [SeF_5]^- \xrightarrow{+\,F^-} [SeF_6]^{2-}$$

(*b*) Both SF_4 and SeF_4 form solid 1 : 1 addition compounds with BF_3, AsF_5 and SbF_5 which act as Lewis bases in these reactions.

(*v*) *Structure.* All the tetrahalides (MX_4) possess see-saw geometry which can be regarded as a distorted trigonal bipyramidal structure having one lone pair (*lp*) of electrons in the basal (*i.e.* equatorial) position of the trigonal bipyramid. The structure of SF_4 molecule (in the gaseous state) has already been explained. See-saw geometry of MX_4 molecules arises due to the sp^3d hybridisation of the central atom (M atom). The distortion in shape is due to the presence of one lone pair of electrons. Lone pair occupies the equatorial position of the trigonal bipyramid, since in this case the *lp—lp* repulsion in minimum. In MX_4 molecule the two axial M—X bonds are slightly bent towards the equatorial M—X bonds, showing the repulsion effect of lone pair. The axial M—X bonds are longer than the equatorial M—X bonds.

(*C*) Dihalides (MX_2). Only oxygen forms difluoride (OF_2). Excepting oxygen, all the remaining elements form dichlorides (SCl_2, $SeCl_2$, $TeCl_2$, $PoCl_2$). Excepting oxygen and sulphur, all other elements form dibromides ($SeBr_2$, $TeBr_2$, $PoBr_2$). None of the elements forms diodides. SCl_2 is prepared by saturating S_2Cl_2 with Cl_2 at room temperature and $PoCl_2$ is prepared by reducing $PoCl_4$ with SO_2. $TeBr_2$ is prepared by reacting tellurium with Br_2. $SeCl_2$ and $SeBr_2$ are unstable and decompose in the vapour state. The stability of dihalides decreases from SX_2 to SeX_2 and that of the dihalides of the same element decreases from MCl_2 to MBr_2. Thus : $SCl_2 > SeCl_2 > TeCl_2 > PoCl_2$; $SeCl_2 > SeBr_2$; $TeCl_2 > TeBr_2$; $PoCl_2 > PoBr_2$. All the dihalides (MX_2) possess *angular* (*V-shaped*) structure which is similar to that of H_2O molecule. The angular structure can be regarded as a distorted tetrahedral structure having two lone pairs of electrons. The angular structure of MX_2 molecules arises due to sp^3 hybridisation of the central atom (M atom). The distortion in shape is due to the presence of two lone pairs of electrons. As an example, the structure of SCl_2 molecule is explained as follows :

The Lewis dot structure of SCl_2 molecule is $Cl{-}\ddot{\underset{\bullet\bullet}{S}}{-}Cl$ which shows that the central atom (*i.e.* S-atom) uses two of its six valence-shell electrons ($S \rightarrow 3s^2p^4$) in forming two S—Cl σ-bonds while the remaining four electrons (*i.e.* two electron pairs) remain as non- bonding electrons on S-atom. Thus σ-*bps* = 2 and *lps* = 2 and hence σ-*bps* + *lps* = 2 + 2 = 4. Now since the sum of σ-*bps* and *lps* is 4, S-atom in SCl_2 molecule is supposed to be sp^3 hybridised. Out of the four sp^3 hybrid orbitals, two contain *lps* while each of the remaining two has one electron. These two singly-filled sp^3 hybrid orbitals make a head-to-head (linear) overlap with the singly-filled $3p_x$ orbitals on each of the two Cl-atoms ($Cl \rightarrow 3s^2, 3p_x^1\ 2p_y^2\ 3p_z^2$) and form two sp^3 (S)—2*p* (Cl) σ-bonds as shown in Fig. 35.2.

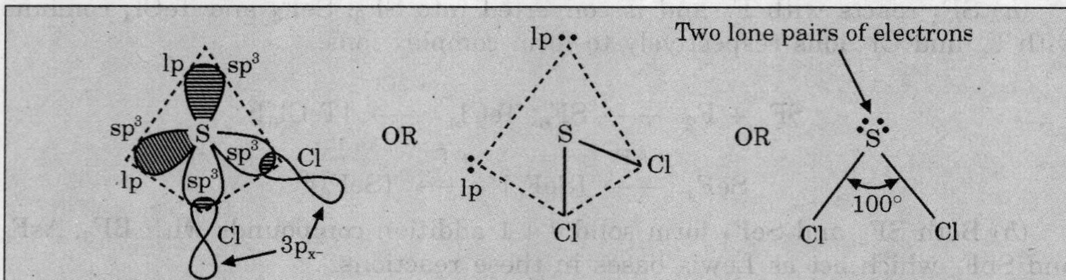

Fig. 35.2. Pictorial representation of the formation of two σ-bonds in SCl_2 molecule (σ-*bps* = 2, *lps* = 2) which has angular shape (V-shape).

Although the spatial arrangement of four electron pairs (*bps* = 2 and *lps* = 2) round the central S-atom is tetrahedral, duo to the presence of two lone pairs of electrons in two hybrid orbitals, the shape of SCl_2 molecule gets distorted and becomes V-shape and Cl—S—Cl bond angle decreases from the expected tetrahedral angle (= 109.5°) to 103°. The distortion in shape and the decrease in bond angle can be explained on the basis of same reasoning as used in case of H_2O molecule. Bond angles in case of OF_2 and $TeBr_2$ dihalides are 103° and 98° respectively.

(D) Dimeric monohalides (M_2X_2). The well known dimeric monohalides are O_2F_2, S_2F_2, S_2Cl_2, S_2Br_2, Se_2F_2, Se_2Cl_2, Se_2Br_2. S_2Cl_2 is prepared by the action of dry Cl_2 on molten sulphur

$$2S + Cl_2 \longrightarrow S_2Cl_2$$

The dimeric monohalides are slowly hydrolysed and undergo disproportionation. For example :

$$2S_2F_2 + 2H_2O \xrightarrow{\text{Hydrolysis}} 4HF + SO_2 + 2S$$

$$\underset{(Se\,=\,+1)}{2Se_2Cl_2} \xrightarrow{\text{Disproportionation}} \underset{(Se\,=\,+4)}{SeCl_4} + \underset{(Se\,=\,0)}{3Se}$$

O_2F_2 and S_2Cl_2 molecules, like H_2O_2 molecule, have *non-linear* and *non-planar* structures (Fig. 35.3) in which the two O—F and S—Cl bonds are asymmetrically distributed, *i.e.* all the four atoms in both the molecules are lying in different planes.

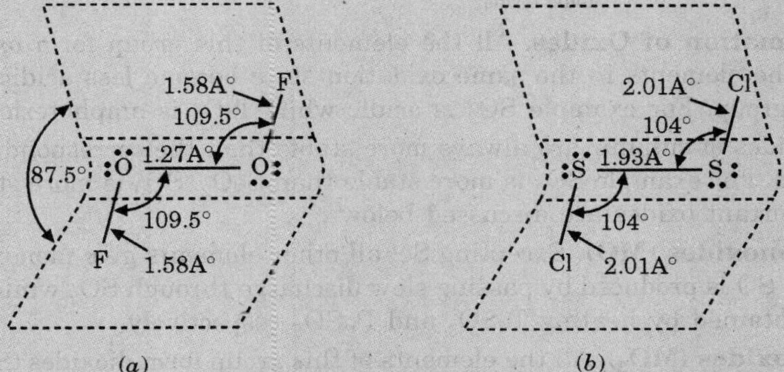

Fig. 35.3. Non-linear and non-planar structures of (*a*) O_2F_2 and (*b*) S_2Cl_2 molecules.

Each of the two O-atoms in O_2F_2 molecule and each of the two S-atoms in S_2Cl_2 molecule has two lone pairs of electrons on it. Thus each O-atom in O_2F_2 and each S-atom in S_2Cl_2 molecule is sp^2 hybridised. How O—O and O—F bonds in O_2F_2, and S—S and S—Cl bonds in S_2Cl_2 are formed has been shown in Fig. 35.4.

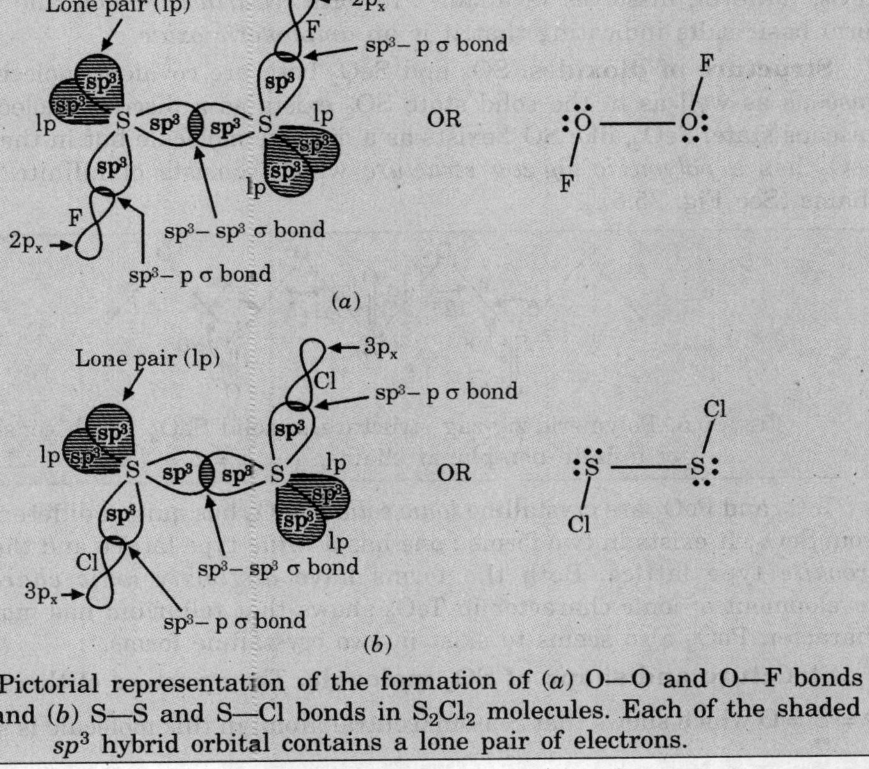

Fig. 35.4 Pictorial representation of the formation of (*a*) O—O and O—F bonds in O_2F_2, and (*b*) S—S and S—Cl bonds in S_2Cl_2 molecules. Each of the shaded sp^3 hybrid orbital contains a lone pair of electrons.

In O_2F_2 molecule it appears that the ionic forms also contribute with the covalent form.

$$\underbrace{F{-}O^+ = OF^-,\ F^-O = O^+ - F}_{\text{Ionic forms}} \qquad \underset{\text{Covalent form}}{F{-}O{-}O{-}F}$$

3. Formation of Oxides. All the elements of this group form oxides. The oxides of the elements in the same oxidation state become less acidic as we go down the group. For example SO_2 is acidic while TeO_2 is amphoteric.

The oxides of sulphur are always more stable than the corresponding oxides of selenium. For example SO_2 is more stable than SeO_2. SO_3 is more stable than SeO_3. Important oxides are discussed below :

(A) Monoxides (MO). Excepting Se, all other elements give monoxides (SO, TeO, PoO). SO is produced by passing slow discharge through SO_2 while TeO and PoO are obtained by heating $TeSO_3$ and $PoSO_3$ respectively.

(B) Dioxides (MO_2). All the elements of this group form dioxides (SO_2, SeO_2, TeO_2, PoO_2). These oxides can be prepared by burning the elements in air.

These oxides differ from one another in their reaction towards H_2O. Gaseous SO_2 dissolves in H_2O giving *sulphurous acid* (H_2SO_3) which is known in solution only. SeO_2 gives an acidic aqueous solution which on evaporation in vacuum gives *selenous acid* (H_2SeO_3) in colourless hexagonal prisms. The formation of H_2SO_3 and H_2SeO_3 indicate that SO_2 and SeO_2 both are *acidic* in nature. TeO_2 is almost insoluble in H_2O giving very dilute solution of *tellurous acid* (H_2TeO_3), TeO_2, however, dissolves in alkalies to form *tellurites*, $MTeO_3$ and in acids to form basic salts indicating that it is an *amphoteric oxide.*

Structure of dioxides. SO_2 and SeO_2 both are covalent molecules. In the gaseous as well as in the solid state SO_2 exists as a discrete molecule. In the gaseous state, SeO_2, like SO_2, exists as a discrete molecule but in the solid state SeO_2 has a *polymeric zig-zag structure* which consists of infinite non-planar chains (See Fig. 35.5).

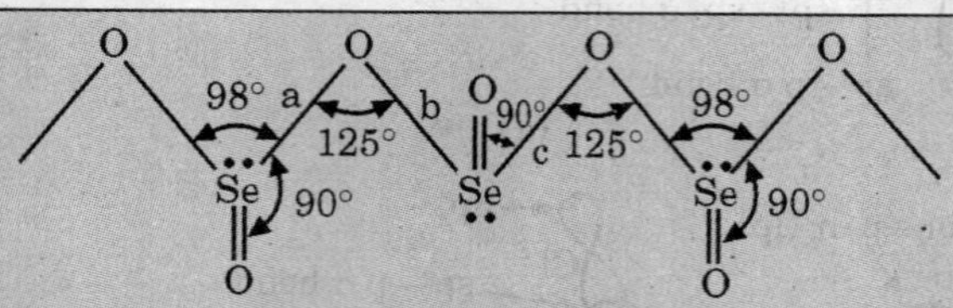

Fig. 35.5. Polymeric zig-zag structure of solid SeO_2 which consists of infinite non-planar chains, $a = b = c = 1.78$A°.

TeO_2 and PoO_2 are crystalline *ionic solids.* TeO_2 has quite a different structure from SeO_2. It exists in two forms : one has a *rutile* type lattice and the other has *brookite* type lattice. Both the forms have *distinctly ionic character.* The development of ionic character in TeO_2 shows that tellurium has more metallic character. PoO_2 also seems to exist in two crystalline forms.

Structure and shape of SO_2 molecule. The structure of this molecule is $O = \underset{\bullet\bullet}{S} = O$ which shows that S-atom (central atom) in this molecule is surrounded by two σ-*bps* and one *lp*. Thus σ-*bps* + *lp* = 2 + 1 = 3 and hence S-atom is sp^2 hybridised.

One of the three sp^2 hybrid orbitals is occupied by a lone pair of electrons while the remaining two have one electron each. One 3*p* orbital and one 3*d* orbital on S-atom do not participate in the hybridisation process. Each of the two singly-filled sp^2 hybrid orbitals overlaps with the singly-filled 2*p* orbitals on both O-atoms and form two σ-bonds while 3*p* and 3*d* orbitals on S-atom make a sidewise overlap with 2*p* orbitals on both O-atoms and form *p*—*p* and *d*—*p* π-bonds respectively. The sidewise overlap of 3*p* orbital on S-atom with 2*p* orbital on O-atom to form *p*—*p* π bond and that between 3*d* orbital on S-atom with 2*p* orbital on O-atom to form *d*—*p* π bond is shown in Fig. 35.6.

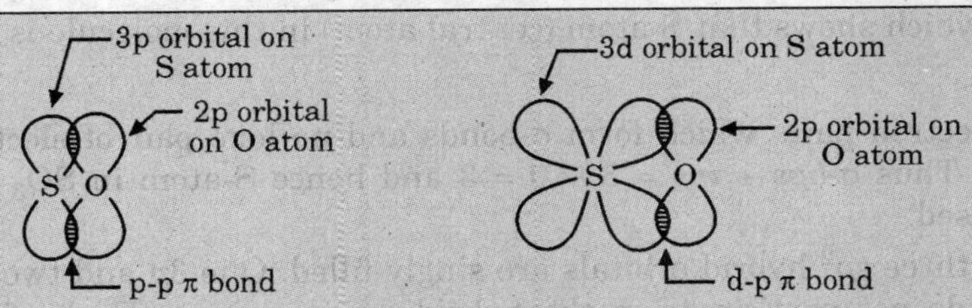

(*a*) (*b*)

Fig. 35.6. Formation of (*a*) *p*—*p* and (*b*) *d*—*p* π bonds in SO_2 molecule.

Although the arrangement of three electron pairs (two electron pairs used in σ-bonds and one lone pair) round S-atom is triangular, due to the presence of one lone pair of electrons on S-atom, the shape of SO_2 molecule gets distorted and becomes *angular* (V-shape) (Fig. 35.7) OSO bond angle has been found to be equal to 119.5° which is nearly equal to 120°. This value of bond angle and the polar nature of SO_2 molecule confirms its angular geometry. The formation of *p*—*p* and *d*—*p* π bonds in SO_2 molecule results in bringing S-and O-atoms closer to each other and hence the S—O bond length becomes shorter than the single S—O bond length. The shortening of S—O bond lengths suggests multiple bonding between S and O-atoms. The difference in the nature of π-bonds between S- and two O-atoms must mean that both S—O bond lengths must be different from each other. Actually both the bonds are of the same length. Their same length is due to the resonance between the two structures shown in the margin.

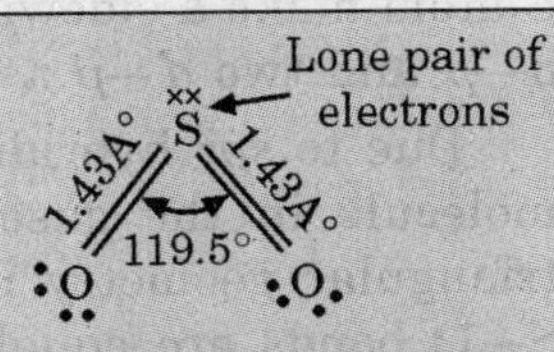

Fig. 35.7. Angular geometry of SO_2 molecule.

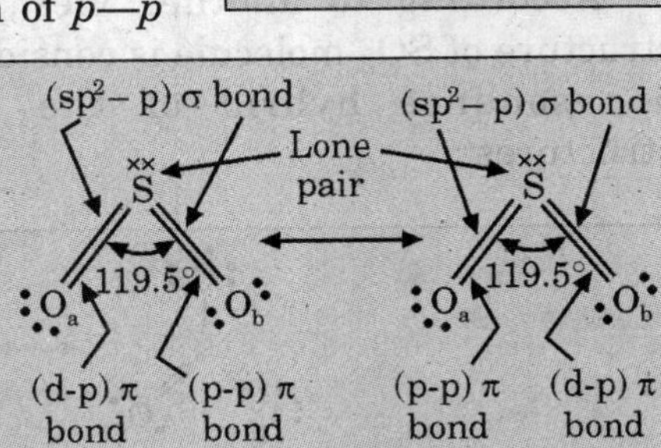

Resonance structures of SO_2 molecule

Since one of the sp^2 hybrid orbitals on S-atom is occupied by a lone pair of electrons and also since the (*lp*—*bp*) repulsion is more than (*bp*—*bp*) repulsion, OSO bond angle gets reduced from 120° to 119.5°.

(C) Trioxides (MO_3). Preparation. S, Se and Te also form the trioxides of MO_3 type (SO_3, SeO_3, TeO_3). SO_3 is prepared by catalytic oxidation of SO_2 with air, SeO_3 is formed by passing an electric discharge through selenium vapours and O_2 at 4 mm pressure while TeO_3 is made by heating telluric acid, H_6TeO_6 strongly.

Properties. SO_3 and SeO_3 are soluble in water and give sulphuric (H_2SO_4) and selenic (H_2SeO_4) acids. Thus these are acidic oxides. TeO_3 is practically insoluble in water due to the diminished tendency of Te to form multiple bonds with oxygen and its lower electronegativity resulting in greater polarity in Te—O bond. It is for this reason that TeO_3 is polymeric and insoluble in water.

Structure and Shape of SO_3 molecule. The structure of SO_3 molecule is 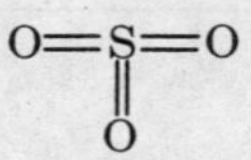 which shows that S-atom (central atom) in this molecule is surrounded by three electron pairs which form σ-bonds and no lone pair of electrons is left on S-atom. Thus σ-*bps* + *ips* = 3 + 0 = 3 and hence S-atom in SO_3 molecule is sp^2 hybridised.

All the three sp^2 hybrid orbitals are singly-filled. One 3*p* and two 3*d* orbitals on S-atom do not participate in the hybridisation process. Each of the singly-filled sp^2 hybrid orbitals overlaps with the singly-filled 2*p*-orbitals on all the three O-atoms and form three sp^2 (S)—*p* (O) σ-bonds while 3*p* and 3*d* orbitals on S-atom make a sidewise overlap with 2*p* orbitals on three O-atoms and form one *p*—*p* and two *d*—*p* π bonds.

Due to sp^2 hybridisation of S-atom, SO_3 molecule in the gaseous state has planar triangular geometry in which all the three S—O bonds are equal to 1.43Å and all the OSO bond angles are equal to 120° (See Fig. 35.8).

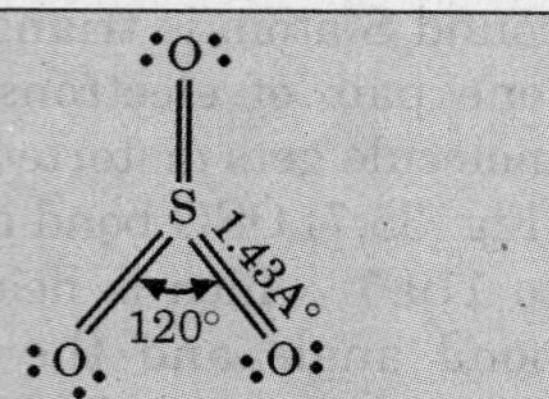

Fig. 35.8. Triangular planar geometry of SO_3 molecule in the gaseous state.

According to another view, the true structure of SO_3 molecule is considered to be a resonance hybrid of the following structures :

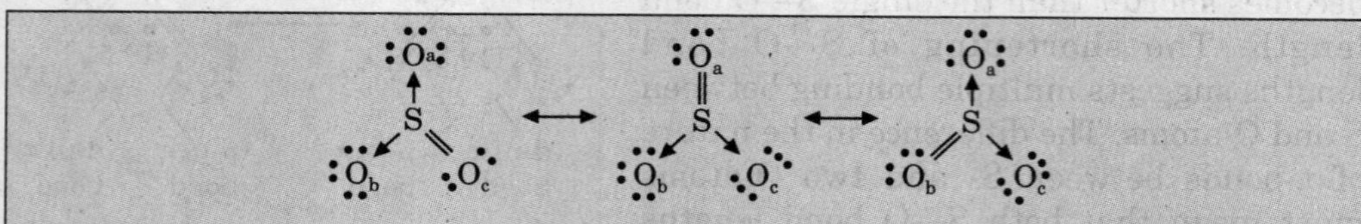

Solid SO_3 has either a puckered cyclic (or ring) trimeric structure (α-form) or a chain (or layer) structure (β- and γ-forms) as shown below in Fig. 35.9. SO_3 chain is made up of linked SO_4 tetrahedra. Each tetrahedron shares two oxygen atoms.

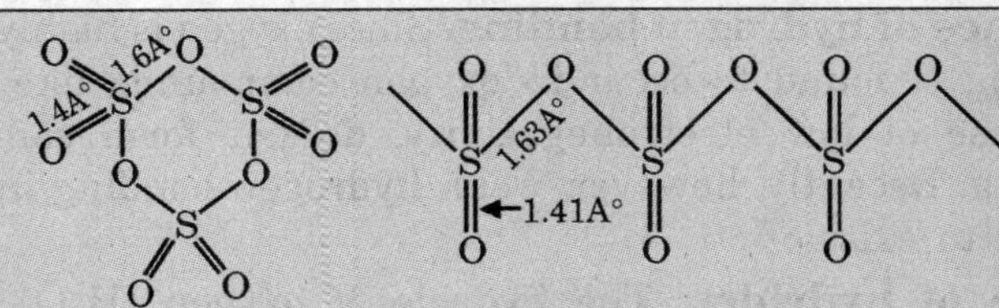

(a) (b)

Fig. 35.9. Structure of solid SO_3 (a) Puckered ring trimeric structure (α-form) (b) Chain (or layer) structure (β- and γ-forms).

(D) Heptoxide (S_2O_7). It is only sulphur which gives heptoxide, S_2O_7. It is formed by passing a silent electric discharge in a mixture of SO_2 and O_2.

$$4SO_2 + 3O_2 \longrightarrow 2S_2O_7$$

It liberates O_2 from H_2O

$$2S_2O_7 + 4H_2O \longrightarrow 4H_2SO_4 + O_2$$

4. Formation of Oxy-acids. Sulphur, selenium and tellurium form a number of oxy acids. Out of these oxy-acids, those of sulphur are the most important and have been discussed in chapter 36.

Anomalous Properties of Oxygen : Dissimilarities with Other Elements of Group VI A.

Oxygen, the first element of group VI A, like carbon (group IV A) and nitrogen (group V A) differs considerably from the rest of the elements of group VI A because of the following inherent characteristics of oxygen : (*i*) *Small size* (*ii*) *Higher electrontgativity* (*iii*) *Non-availability of d-orbitats in the electronic configuration of oxygen atom.* The main points of difference between oxygen and the remaining members of group VI A are given below :

***(i)* Physical state.** Oxygen is a *gas* while the other members are *solids.*

***(ii)* Abundance.** Oxygen is the *most abundant* of all the elements and constitutes 46-6% of the earth's crust.

***(iii)* Atomicity.** Oxygen molecule is *diatomic* (O_2) while the molecules of the other elements are more complex, *e.g.* sulphur and selenium have *octatomic* molecules (*i.e.* S_8 and Se_8) with puckered ring structures.

***(iv)* Oxidation states.** Oxygen does not show any positive oxidation state except in oxygen difluoride, OF_2 in which it shows + 2 oxidation state, while the other members of this group show both positive and negative oxidation states. Positive oxidation states are more frequent.

***(v)* Formation of M^{2-} ions.** On account of higher electronegativity of oxygen atom, *dinegative anion* O^{2-} (*oxide anion*) is more common while the S^{2-}, Se^{2-} and Te^{4-} ions are less probable.

***(vi)* Maximum covalency.** The maximum covalency of oxygen is two because of the absence of *d*-orbitals in its valence-shell. Other elements, on the other hand, exhibit a maximum covalency of *six* due to the presence of *d*-orbitals in which *s*- and *p*-electrons can be promoted and thus can be made unpaired.

***(vii)* Existence of hydrogen bonding.** Since oxygen is highly electronegative, hydrogen bonding is found to occur in oxygen compounds like H_2O etc. Other elements, because of low electronegativity, do not form compounds having hydrogen bonding. Recently, however, weak hydrogen bonding involving sulphur has been proved to exist.

***(viii)* Nature of hydrides.** The hydride of oxygen (H_2O) is a colourless, odourless and non-poisonous liquid while the hydrides of other elements are inflammable gases with an unpleasant smell and are poisonous.

***(ix)* Magnetic behaviour.** Oxygen shows paramagnetic behaviour.

***(x)* Reactivity.** Metals like Cu, Ag, Hg etc. show lesser tendency to react with O_2 to form oxides than with sulphur to form sulphides. This is because of the fact that a stronger bond is formed when larger S^{2-} ion is polarised by the small Cu^{2+}, Ag^+, Hg^{2+} ions. It is because of this reason that in nature weaker metals tend to occur as sulphides rather than as oxides.

Comparison of Oxygen and Sulphur.

These are the typical elements of group VI A and show many similar and dissimilar properties.

(A) Similarities. *(i)* Both are non-metals *(ii)* They have six electrons in their outermost shell ($O_8 \rightarrow 2, 6$; $S_{16} \rightarrow 2, 8, 6$) *(iii)* It shows analogy with sulphur in compounds like *(a)* C_2H_5OH and C_2H_5SH *(b)* CO_2 and CS_2 *(b)* $(C_2H_5)_2O$ and $(C_2H_2)_2S$ *(d)* H_2O and H_2S *(iv)* O and S combine with H_2 to form the hydrides *viz.* H_2O and H_2S *(v)* They both form oxides of RO_2 type. *e.g.* O_3 ($O\ O_2$) and SO_2. *(vi)* Both the elements are obtained by heating their higher oxides and sulphides.

$$2BaO_2 \longrightarrow 2BaO + O_2 \uparrow \; ; \; FeS_2 \longrightarrow FeS + S \downarrow$$

(vii) They combine with carbon on heating to give CO_2 and CS_2.

(B) Dissimilarities. *(i)* O_2 is colourless and odourless gas while sulphur is a pale yellow solid with a faint odour, *(ii)* Oxygen is more widely distributed in nature than sulphur, *(iii)* Maximum covalency of oxygen is 2 while that of sulphur is 6 *(iv)* Oxygen does not react with conc. HNO_3 while sulphur is oxidised by conc. HNO_3. *(v)* Oxygen, due to non-availability of *d*-orbitals, does not form co-ordination compounds while S gives a large number of co-ordination compounds.

Conceptual Questions with Answers

Q.1 Arrange the following elements as directed :

***(i)* Po, Se, Te, S and O in the increasing order of their metallic character.**

Ans : O < S < Se < Te < Po

***(ii)* Po, Se, Te, S and O in the decreasing order of their ionisation energy.**

Ans : O > S > Se > Te > Po

Q.2 Explain the following :

***(i)* Liq. O_2 sticks to magnet pole but liq. N_2 does not.**

Ans : O_2 is paramagnetic but N_2 is diamagnetic. Hence O_2 is attracted towards magnet but N_2 is not.

(*ii*) Sulphur exhibits greater tendency for catenation than selenium.

Ans : As we move from S to Se, the atomic size increases and hence the strength of M–M bond decreases. As a result, S–S bond is much stronger than Se–Se bond and consequently S shows greater tendency for catenation than selenium.

(*iii*) Oxygen forms diatomic molecules (O_2) while sulphur forms octaatomic molecules (S_8).

Ans : Due to small size of oxygen, the non-bonding electrons on the two oxygen atoms in O–O bond strongly repel each other while such repulsions are much less in S S bond. As a result O — O bond dissociation energy is much less than S — S bond dissociation energy. In other words, sulphur has much higher tendency for catenation than oxygen. Further, oxygen because of its small size has a high tendency to form $p\pi - p\pi$ multiple bonds but sulphur because of its bigger size normally does not form $p\pi - p\pi$ S = S multiple bonds.

Thus, oxygen because of its less tendency for catenation and higher tendency for $p\pi - p\pi$ multiple bonds form stable diatomic (O_2) molecules whereas sulphur because of its higher tendency for catenation and lesser tendency for $p\pi - p\pi$ multiple bonds forms octaatomic (S_8) molecules having eight membered puckered ring structure.

(*iv*) Oxygen is diatomic (O_2) and gaseous in nature while sulphur is a solid.

Ans : The oxygen molecule is diatomic involving a double bond. Its small size produces weak intermolecular forces amongst its molecules and is, thus, a gas at room temperature. Sulphur is incapable of forming π-bonds. Its molecule has a puckered ring structure of eight sulphur atoms joined by single covalent bonds. The intermolecular forces acting between S_8 molecules are stronger and hence, sulphur is a solid.

(*v*) The tendency of the elements from S to Po to show –2 oxidation state diminishes on moving down the group from S to Po.

Ans : Since the electronegativity of the elements decreases on moving down the group from S to Po, the tendency of these elements to accept two electrons to form M^{2-} ions which have stable noble gas configuration ($ns^2\ np^6$ configuration) also decreases. Due to the decrease in this tendency, the tendency of the elements to show –2 oxidation state also decreases. Thus since Po is the least electronegative, this element does not show –2 oxidation state.

(*vi*) Oxygen generally exhibts an oxidation state of –2 only while other members of its family show oxidation states of +2, +4 and +6 as well.

Ans : The valence-shell of O-atom has six electrons (O = $2s^2\ 2p^2\ 2p^1\ 2p^1$). Since O-atom is highly electronegative, it can accept two electrons to form oxide ion (O^{2-}) which has noble gas stable configuration. Thus O-atom generally shows –2 oxidation state. On the other hand, since other elements like sulphur (S = $3s^2\ 3p^2\ 3p^1\ 3p^1\ 3d^\circ\ 3d^\circ$.....) have empty $3d$-orbitals in their valence-shell, an electron from the completely-filled $3p$ orbital and one electron from the completely-filled $3s$ orbital to the empty $3d$ orbitals can be promoted to get 4 and 6 unpaired electrons. The presence of 2 [S = $3s^2$ $3p^2$ $3p^1$ $3p^1$ $3d^\circ$ $3d^\circ$ (n = 2)], 4 [S* = $3s^2$ $3p^1$ $3p^1$ $3p^1$ $3d^1$ $3d^\circ$ (n = 4)] and 6 [S** = $3s^1$ $3p^1$ $3p^1$ $3p^1$ $3d^1$ $3d^1$(n = 6)] unpaired electrons in S-atom explains its +2, +4 and +6 oxidation states.

(*vii*) Oxygen cannot show oxidation states of +4 and +6 but sulphur atom can do so.

Ans : Since O-atom (O = $2s^2$ $2p^2$ $2p^1$ $2p^1$) does not have *d*-orbitals in its valence shell, the promotion of an electron from the completely-fillled 2*p* or 2*s* orbitals is not possible and hence 4 or 6 unpaired electrons cannot be obtained. Thus O-atom cannot show oxdation states of +4 and +6. On the other hand, in S-atom (S = $3s^2$ $3p^2$ $3p^1$ $3p^1$ $3d^\circ$ $3d^\circ$...), one electron from 3*p* orbital can be promoted to the empty 3*d* orbital to get 4 unpaired electrons. Similarly the promotion of an electron from 3*s* orbital to the empty 3*d* orbital can be done to get 6 unpaired electrons. Thus S-atom can show +4 and +6 oxidation states due to the presence of 4 and 6 unpaired electrons respectively.

(*viii*) Negative electron gain enthalpy of O-atom is lower than that of S-atom (O = –141 KJmol^{-1}, S = – 200KJmol^{-1}).

Ans : Since 2*p* orbitals in O-atom are relatively smaller in szie than 3*p* orbitals in S-atom, electron-electron repulsions are comparatively larger in O-atom. Hence O-atom cannot accept the incoming electron so easily as S-atom can do. Consequently energy released in the formation of $O^-(g)$ from O(*g*) is less than that released in the formation of $S^-(g)$ from S(*g*). In other words *negative electron gain enthalpy of O-atom is less than that of S-atom.*

$$O(g) + e^- \rightarrow O^-(g),\ \Delta H = -\Delta eg = -141\ \text{KJmol}^{-1}$$
$$S(g) + e^- \rightarrow S^-(g),\ \Delta H = -\Delta eg = -200\text{KJmol}^{-1}$$

(*ix*) The compounds of fluorine with oxygen are called fluorides of oxygen and not oxides of fluorine.

Ans : Since F-atom has higher electronegativity than O-atom, compounds of F with O are called fluorides of oxygen and not oxides of fluorine. For example F_2O is called *oxygen difluoride* and not *difluorine oxide*. This compound is, therefore, represented as OF_2 and not F_2O.

(*x*) Elements of group 16 generally show lower value of first ionization enthaply compared to the corresponding elements of group 15.

Ans : This is because of the following two reasons :

(*i*) Due to extra stability of the exactly half-fillled *p*-orbitals electronic configuration of group 15 elements, they require larger amount of energy to remove an electron as compared to those of group 16 elements.

(*ii*) After removal of an electron, the group 16 elements yield species which have more stable exactly half-filled *p*-orbitals electronic configuration while this is not so in case of group 15 elements.

Q.3 Name the element of group 16 which has : (*i*) highest electronegativity (*ii*) highest metallic character (*iii*) highest melting point. (*iv*) Maximum ionisation energy (*v*) Maximum catenation property.

Ans : (*i*) *Oxygen has the highest electronegativity.*

Explanation. Since the atomic size of the elements increases down the group, the electronegativity values of the elements decrease down the group. Thus oxygen has the highest electronegativity.

(*ii*) *Polonium has the highest metallic character.*

Explanation. Due to the decrease in IE values down the group, metallic character of the elements increases. Thus Po has the highest metallic character.

(*iii*) *Tellurium has the highest melting point.*

Explanation. Since the van der Waal's forces of attraction between the atoms of elements increases down the group, the melting points of the elements also increase down the group from O to Te. Po has lower m.pt. than Te. Thus the actual order of m.pts. is O < S < Se < Te > Po.

(*iv*) *Oxygen has the highest ionisation energy.*

Explanation. Due to the increase in the size of the atoms of the elements on moving down the group from O to Po, ionisation energies of the elements decrease down the group. Thus oxygen has the highest ionisation energy.

(*v*) *Sulphur has the maximum catenation property.*

Explanation. Catenation property of an element (M) increases with the increase of the strength of M–M bond. Since among the elements of group 16, S–S bonds are the strongest, sulpher has the maximum catenation property.

Q.4 Explain the following :

(*i*) The two sulphur-oxygen bonds in SO_2 molecule are equal (= 143 pm).

Ans : In SO_2, S is sp^2 hybridized. Two of the three sp^2 orbitals form two bonds while the third contains the lone pair of electrons. S is now left with one half-fillled *p*-orbital and one half-filled *d*-orbital. These form one $p\pi - p\pi$ and $d\pi - p\pi$ double bond with oxygen atom. Thus, SO_2 has bent structure with OSO bond angle of 119.5. Due to resonance, the two π-bonds are equal (143 pm).

(*ii*) SO_2 is a more powerful reducing agent (RA) in alkaline medium than in acidic medium. (I.I.T.1992)

Ans : In presence of moisture, SO_2 can produce nescent hydrogen and hence acts as a reducing agent. When SO_2 reduces a certain susbtance, it gets oxidised to SO_3 which is further oxidised to H_2SO_4.

$$SO_2 + H_2O \rightleftharpoons SO_3 + 2H$$

$$SO_3 + H_2O \rightleftharpoons H_2SO_4$$

$$\overline{SO_2 + 2H_2O \rightleftharpoons H_2SO_4 + 2H}$$

This equation shows that when SO_2 acts as a reducing agent, it gets oxidised to H_2SO_4. In presence of an alkali, H_2SO_4 formed as above reacts with the alkali and gets neutralised by it. Thus the equlibrium of the above reaction is shifed towards right and hence more quantity of nescent hydrogen is produced. Thus in presence of alkali, SO_2 acts as a stronger R.A. In presence of an acid, the equilibrium gets shifted towards left and hence lesser quantity of nescent hydrogen is formed. This makes SO_2 to act as a weaker R.A. Thus in presence of an acid SO_2 acts as a weaker R.A.

Oxidation of SO_2 to H_2SO_4 in acidic and basic medium can be shown as :

$$\left.\begin{aligned} & SO_2 + 2H_2O \longrightarrow H_2SO_4 + 2H \\ \text{or} \quad & SO_2 + 2H_2O \longrightarrow SO_4^{2-} + 4H^+ + 2e^- \end{aligned}\right\} \textit{In acidic medium}$$

$$\left.\begin{aligned} & SO_2 + 2NaOH + \frac{1}{2}O_2 \longrightarrow Na_2SO_4 + H_2O \\ \text{or} \quad & SO_2 + 4OH^- + \longrightarrow SO_4^{2-} + 2H_2O + 2e^- \end{aligned}\right\} \textit{In basic medium}$$

***(iii)* All the three sulphur-oxygen bonds in SO_3 molecule are identical.**

Ans : The equivalence of all the three sulphur-oxygen bonds in SO_3 molecule suggests that the actual structure of SO_3 molecule is a resonance hybrid of the three structures.

Q.5 Find out the oxidation state of S-atom in SF_6 and SH_6 molecules.

Ans : Electronegativity values of S, F and H are S = 2.44, F = 4.00 and H = 2.20. Since S-atom has lower electronegativity than F-atom, in SF_6 molecule S-atom shows positive oxidation state (= +6) and F-atom shows negative oxidation state (= –1). On the other hand, since S-atom has higher electronegativity than H-atom, in SH_6 molecule S-atom shows negative oxidation state (= –6) and H-atom has positive oxidation state (= +1). Thus oxidation state of S-atom in SF_6 = +6 and oxidation state of S-atom in SH_6 = –6.

Q.6 Although the central O-atom in water (H_2O) and diethylether, $(C_2H_5)_2O$ has the same hybridisation, get these molecules have different bond angles. (H_2O = 104.5°, $(C_2H_5)_2O$ = 110°). Expalin.

Ans : The central O-atom in H_2O and $(C_2H_5)_2O$ both is sp^3 hybridised. Hence both molecules are expected to have the same bond angles. In $(C_2H_5)_2O$ molecule, the repulsion between two bonding electron pairs (*i.e.* electron pairs involved in two O – C_2H_5 bonds) is greater than that between two bonding electron pairs in H_2O molecule (*i.e.* electron pairs involved in two O – H bonds). Hence C_2H_5 – O – C_2H_5 bond angle in $(C_2H_5)_2O$ is greater (= 110°) than H – O – H bond angle in H_2O (= 104.5°).

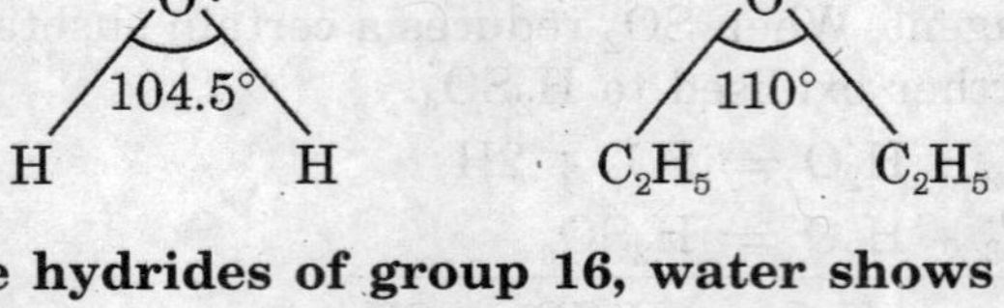

Q.7 Among the hydrides of group 16, water shows unusual physical properties. Why ? ***(D.B.S 2003)***

Ans : Due to high eiectronegativity of O, the O–H bond in H_2O froms strong intermolecular H-bonds. As a result, water exists as an associated molecule while other hydrides of group 16 do not form H-bonds and hence exist as discrete molecules. As a result, water shows unusual physical properties, *i.e.* high b.p., high thermal stability and weaker acidic character as compared to other hydrides of group 16.

Q.8 Why H_2O is a liquid while H_2S is a gas ? ***(H.S.B 2001)***

Ans : There is H-bonding in H_2O due to high electronegativity and small size of O-atom but there is no H-bonding in H_2S.

Q.9 H_2S is acidic while H_2O is neutral. Explain.

Ans : The S–H bond is weaker than O–H bond because size of S-atom is bigger than that of O-atom. Hence, H_2S can dissociate to give H^+ ions in aqueous solution.

Q.10 Among the hydrides of group 16 elements :

***(i)* Which has the lowest boling point.**

Ans : *H_2S has the lowest boiling point. ($H_2O > H_2Te > H_2Se > H_2S$)*

Explanation. Due to the increase in van der Waals forces of attraction from H_2S to H_2Te, the boiling points of these hydrides also increase from H_2S to H_2Te ($H_2S < H_2Se < H_2Te$). However, the boiling point of H_2O is the highest. Thus the overall order of boiling points is $H_2S < H_2Se < H_2Te < H_2O$. This order shows that H_2S has the lowest boiling point.

***(ii)* Which has the maximum thermal stability.**

Ans : *H_2O has the maximum thermal stability ($H_2O > H_2S > H_2Se > H_2Te$)*

Expalnation. Due to the increase in the size of M-atom from O to Te in H_2M molecules (M = O, S, Se, Te), M – H bond becomes weaker and weaker on moving down the group from H_2O to H_2Te and hence thermal stability of H_2M molecules decreases from H_2O to H_2Te ($H_2O > H_2S > H_2Se > H_2Te$). Obiviously H_2O has the maximum thermal stability.

***(iii)* Which has the weakest acidic character.**

Ans : *H_2O has the weakest acidic character. In fact H_2O is neutral [H_2O (neutral) $< H_2S < H_2Se < H_2Te$)].*

Explanation. The acidic character of H_2M molecules (M = O, S, Se, Te) increases from H_2O to H_2Te. Due to the increase in the size of M-atom from O to Te, M – H bond distance in H_2M molecules also increases from O – H to Te – H. Because of the increase in M–H bond distance, this bond becomes weaker and weaker and hence the cleavage of this bond to produce H as proton (H^+) becomes easier and easier. Thus the tendency of H_2M molecules to give H^+ (or the acidic character or acidic strength of H_2M molecules) increases from H_2O to H_2Te. Thus H_2O has the minimum acidic character. In fact H_2O does not show acidic character. It is a neutral hydride.

***(iv)* Which is liquid at room temperature.**

Ans : *Water (H_2O) is liquid at room temperature. Other hydrides are gases*

Explanation. Due to the highest EN of O-atom, H_2O molecule undergoes intermolecular H-bonding and forms associated molecule, $(H_2O)_n$. Since in order to break H-bonds present in $(H_2O)_n$ a large amount of heat energy is required, H_2O exists as a liquid at room temperature.

***(v)* Which has the highest bond angle.**

Ans : *H_2O has the highest bond angle.*

Explanation. The highest value of bond angle in H_2O is due to the fact that the distance between the two adjacent bps in H_2O molecule is minimum. Hence repulsion between these bps is maximum (strongest) and HOH bond angle is also maximum.

Q.11 Arrange the oxoacids of S, Se and Te given in the following sets in the decreasing order of their acid strength: (*a*) H_2SO_3, H_2SeO_3 and H_2TeO_3 (*b*) H_2SO_4, H_2SeO_4 and H_2TeO_4.

Ans : The acid strength of oxoacids of S, Se and Te with the same oxidation state decreases in moving down the group from S to Te. Thus :

(*a*) $\overset{+4}{H_2SO_3} > \overset{+4}{H_2SeO_3} > \overset{+4}{H_2TeO_3}$ (O.S. of element = +4)

(*b*) $\overset{+6}{H_2SO_4} > \overset{+6}{H_2SeO_4} > \overset{+6}{H_2TeO_4}$ (O.S. of element = +6)

Q.12 Arrange the oxyacids given the following sets in the increasing order of their acid strength : (*a*) H_2SO_3 and H_2SO_4 (*b*) H_2SeO_3 and H_2SeO_4 (*c*) H_2TeO_3 and H_2TeO_4.

Ans : Acid strength of oxoacids of an element of group VI A in different oxidation states increases with the increase in oxidation state of the element. Thus :

(*a*) $\overset{+4}{H_2SO_3} < \overset{+6}{H_2SO_4}$

(*b*) $\overset{+4}{H_2SeO_3} < \overset{+6}{H_2SeO_4}$

(*c*) $\overset{+4}{H_2TeO_3} < \overset{+6}{H_2TeO_4}$

University Questions

1. Discuss the position of sixth group elements namely O, S, Se and Te in the periodic table. (*Jiwaji 84*)
2. Discuss the anomalous behaviour of oxygen in it group. (*Agra 84*)
3. Explain the following, giving appropriate reasons :
   (1) The bond lengths of covalent bonds in SO_2 and SeO_2 molecules are much shorter than expected. (*Delhi 87*)
   (2) Sulphur does not form S_2 molecule like O_2 molecule given by oxygen. (*Raj. 86 S*)
   (3) OF_6 does not exist while SF_6 exists. (*Punjab; Raj. 86*)
   (4) SF_6 is chemically inert. (*Poona 86*)
   (5) The formula of telluric acid is H_6TeO_6 or $Te(OH)_6$. The analogous sulphuric acid H_6SO_6 or $S(OH)_6$ is not known.
   or Tellurium forms H_6TeO_6 but sulphur does not form H_6SO_6. (*Delhi Hons. 87*)

(6) SF_6 is chemically inert whereas SF_4 is hydrolysed even though the average bond energy is SF_4 is greater than that in SF_6. *(Delhi Hons. 87)*

(7) Oxygen forms only a difluoride, while sulphur forms a hexafluoride. *(Allahabad 89)*

(8) Although the central atom in the hexafluorides SF_6, SeF_6 and TeF_6 belongs to the same group, it is only TeF_6 which can be hydrolysed by H_2O. SF_6 and SeF_6 both remain unreactive. *(Allahabad 87)*

(9) Oxygen molecule is O_2 but sulphur molecule is S_8. *(Madurai 86; M.D. Rohtak 83)*

(10) Unlike SO_2, SeO_2 is a solid that has a chain structure. *(Delhi Hons. 87)*

(11) Oxygen is a gas while sulphur and selenium are solids. *(Madras 86)*

(12) The strength of hydroxides of elements of oxygen family increases with increase in atomic number of element. *(Poona 86)*

(13) SF_6 ii known but SCl_6 is not known. *(Delhi 85)*

(14) Hydride of oxygen is less acidic than the hydride of tellurium. *(GND 82)*

(15) The hydride of oxygen is a liquid of high b. pts. While that of sulphur is a gas at room temperature. *(Punjab 85; Madras 86; Raj. 86 S; Delhi 81, 87, 88; Bundelkhand 88; Madurai 84; MLNR 82; Allahabad 87, 89; Calcutta 81; Meerut 84 S)*

(16) H_2O has higher b.pt than H_2S. *(Madras 85; Delhi 81; Raj. 83)*

(17) The boiling point of H_2O is much higher than the values of the outer hydrides of VI group. *(Bharathidasan 85; Andhra 86)*

(18) H_2S more polar than H_2O. *(M.D. Rohtak 84)*

(19) H_2S is a better reducing agent than H_2O.

(20) H_2S less, stable than H_2O.

(21) The bond angles of the hydrides of oxygen family decrease with the increase in atomic number of element.

or Bond angles decrease from H_2O, H_2S, H_2Se, to H_2Te. *(GND 85, 86)*

or Bond angles in H_2O, H_2S, and H_2Se are 104.5°, 92.2° and 91.0° respectively. *(Punjab 84)*

CHAPTER

36 Oxides, Oxyacids, Halides and Oxyhalides of Sulphur

Oxides

I. Sulphur Dioxide, SO_2.

It may be prepared by the reactions given below :

$$S + O_2 \longrightarrow SO_2$$

$$2ZnS + 3O_2 \longrightarrow 2ZnO + 2SO_2$$

$$4FeS_2 + 11O_2 \longrightarrow 2Fe_2O_3 + 8SO_2$$

It is a colourless gas having a pungent suffocating odour. It is highly soluble in H_2O and the solution so obtained is called *susphurous acid.* Thus SO_2 is an acidic oxide and is called *sulphurous acid anhydride.* Aqueous solution of SO_2 is a strong reducing agent.

$$2H_2O + SO_2 \longrightarrow H_2SO_4 + 2H$$

Thus this gas reduces halogens to their respective halogen acids, $Fe_2(SO_4)_3$ to $FeSO_4$, iodates to I_2 etc. SO_2 also acts as an oxidising agent and hence oxidises H_2S to S, Fe to FeO, Mg to MgO, H_2 to H_2O, CO to CO_2, $SnCl_2$ to $SnCl_4$, Hg_2Cl_2 to $HgCl_2$ etc. In the presence of moisture, SO_2 bleaches vegetable colouring matter and hence is used as a bleaching agent. SO_2 is used : (*i*) for bleaching delicate articles like wool, straw silk and hair which are likely to be damaged by chlorine. (*ii*) for household fumigation. (*iii*) in the manufacture of sulphuric acid and sulphites, (*iv*) for the refining and bleaching for sugar, (*v*) in the liquid form for purifying petroleum and as a solvent for iodine, sulphur and phosphorus, and (*vi*) as a refrigerant. Cooling is caused when liquid sulphur dioxide is allowed to evaporate.

SO_2 molecule is a covalent molecule. In the gaseous as well as in the solid state it exists as a discrete molecule. The structure of this molecule is $O{=}\overset{\times\times}{S}{=}O$ and it has *angular shape* which results from sp^2 hybridisation of S-atom (central atom). The structure and shape of SO_2 molecule on the basis of sp^2 hybridisation of S-atom has already been explained.

II. Sulphur Trioxide SO_3.

This oxide is prepared by the reactions shown below :

$$2SO_2 + O_2 \xrightarrow[450°C]{\text{Platinised asbestos}} 2SO_3$$

$$Fe_2(SO_4)_3 \xrightarrow[1150]{\Delta} Fe_2O_3 + 3SO_3$$

At room temperature it is liquid which strongly fumes in air. SO_3 exists in three varieties which are α-SO_3, β-SO_3, and γ-SO_3. It combines with H_2O with a hissing sound, forming H_2SO_4. It oxidizes S to SO_2, P to P_2O_5 and KI to I_2. SO_3 is used as a solvent, for the manufacture of sulphuric acid and as a drying agent for gases.

SO_3 molecule is a covalent molecule. In the gaseous state it exists as a discrete molecule. The structure of this molecule is O=S=O (with a further =O on S, shown as ‖ O below it) and it has *trigonal planar geometry* which results from sp^2 hybridisation of S-atom (central atom). The structure and shape of SO_3 molecule on the basis of sp^2 hybridisation of S-atom has already been explained.

Oxy Acids

Sulphur forms a large number of oxy-acids. To emphasize their structural similarities, these acids have been classified into five groups (1) *Sulphoxylic acid, H_2SO_2* which is not known in the free state (2) *Sulphurous acids.* Examples are sulphurous acid, H_2SO_3 (not known in the free state); hyposulphurous acid or hydrosulphurous acid or dithionous acid, $H_2S_2O_4$ (not known in the free state); thiosulphurous acid, $H_2S_2O_2$; disulphurous or pyrosulphurous acid, $H_2S_2O_5$ (not known in the free state) (3) *Sulphuric acids.* Examples are sulphuric acid, H_2SO_4; thiosulphuric acid, $H_2S_2O_3$; pyrosulphuric acid or disulphuric acid, $H_2S_2O_7$ (4) *Thionic acids.* Examples are dithionic acid, $H_2S_2O_6$; polythionic acids. $H_2S_nO_6$ (n = 3, 4, 5, 6) (5) *Peroxy (or peroxo or per) sulphuric acids.* Examples are peroxomonosulphuric acid or permono sulphuric acid (also called Caro's acid), H_2SO_5 and peroxodisulpuric acid or perdisuiphuric acid (also called Marshall's acid), $H_2S_2O_8$. Here we shall discuss only a few acids.

I. Sulphurous Acid, H_2SO_3.

This acid is known only in solution and is formed by dissolving SO_2 in H_2O. This acid acts as an oxiding agent towards strong reducing agents. Thus H_2S is oxidised to S, HI to I_2 etc. It is a dibasic acid and ionises as :

$$H_2SO_3 \rightleftarrows H^+ + HSO_3^- \; ; \; HSO_3^- \rightleftarrows H^+ + SO_3^{2-}$$

Following two structures have been suggested for the acid. One of these structures is symmetrical and has tetrahedral shape while the other is unsymmetrical and has pyramidal shape.

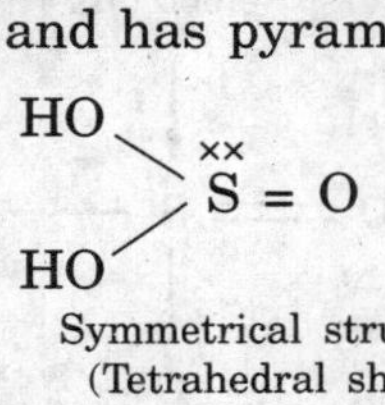

Symmetrical structure
(Tetrahedral shape)

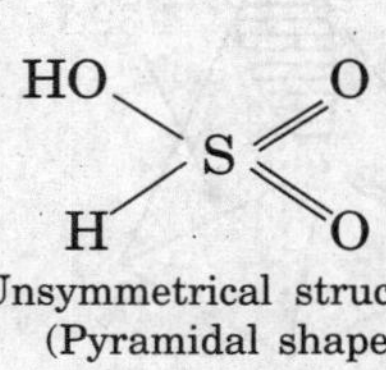

Unsymmetrical structure
(Pyramidal shape)

The following points are in favour of symmetrical structure of the acid :

(*i*) The preparation of sulphurous acid by the action of water on thionyl chloride suggests the symmetrical formula.

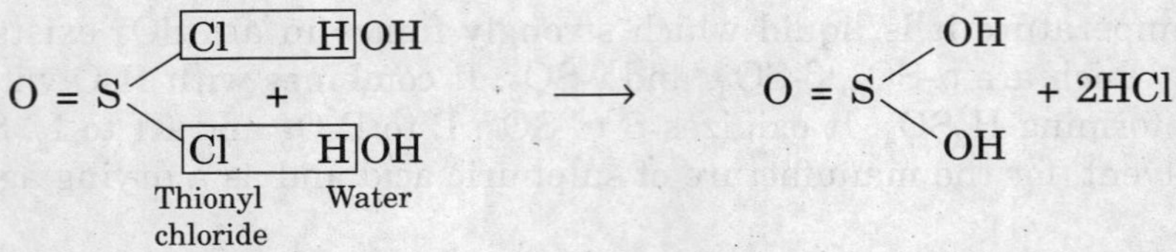

(*ii*) H_2SO_3 readily takes up one oxygen or sulphur atom which gets attached with the sulphur atom of H_2SO_3 to give sulphuric acid (H_2SO_4) or thiosulphuric acid ($H_2S_2O_5$). This property of H_2SO_3 can be explained if we assume that H_2SO_3 has the symmetrical formula shown above. The lone pair of electrons on sulphur atom in H_2SO_3 can easily be shared by another oxygen or sulphur atom to produce H_2SO_4 or $H_2S_2O_3$.

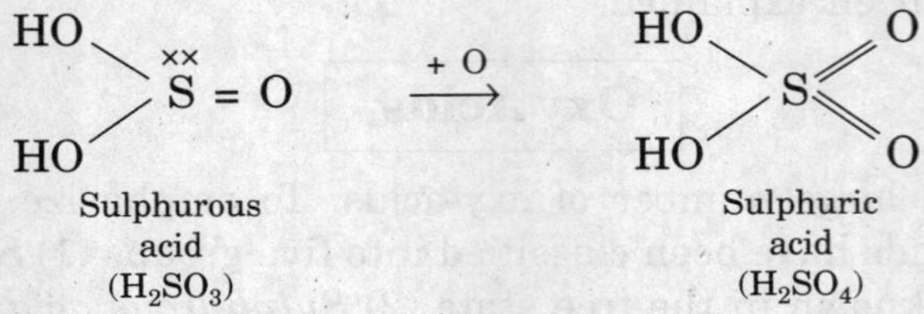

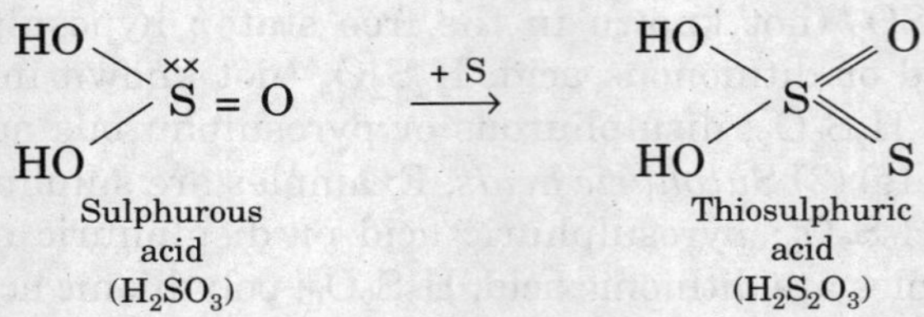

The unsymmetrical structure which contains S—H bond explains the reducing property of H_2SO_3. Now it is believed that the two forms of the acid shown above exist in equilibrium with each other. X-ray analysis of the ionic crystals containing SO_3^{2-} ion has shown that SO_3^{2-} ion has *pyramidal shape* (Fig. 36.1). This shape has three O-atoms in a triangle and results form sp^3 hybridisation of S-atom (central atom). In this structure all the S—O bond lengths are 1.39 A°. S—O π bond is d(S)—p(O) π bond.

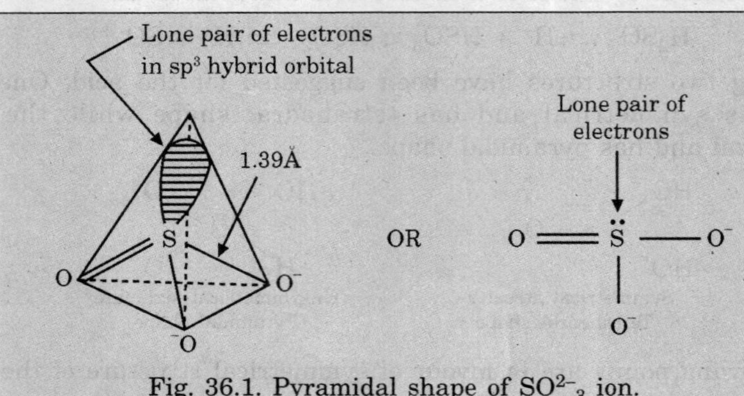

Fig. 36.1. Pyramidal shape of $SO^{2-}{}_3$ ion.

III. Sulphuric Acid, H_2SO_4.

This acid is manufactured by the following processes.

1. Lead Chamber Process. In this process sulphuric acid is obtained by the oxidation of SO_2 by nitrogen dioxide, NO_2. The simplest explanation of the changes taking place is due to Berzelius. According to him sulphur dioxide is first oxidised by nitrogen dioxide to sulphur trioxide which reacts with steam forming sulphuric acid. Nitrogen dioxide itself is reduced to nitric oxide, NO which is subsequently oxidised back to nitrogen dioxide by atmospheric oxygen. The various reactions are represented as follows :

$$SO_2 + NO_2 \longrightarrow SO_3 + NO$$
$$SO_3 + H_2O\ (\textit{Steam}) \longrightarrow H_2SO_4$$
$$2NO + O_2\ (\textit{From air}) \longrightarrow 2NO_2$$

Thus NO and NO_2 act as oxygen carriers. *i.e.* they transfer oxygen from the air to sulphur dioxide.

It is, however, observed that if the chambers are *insufficiently supplied with steam, lead chamber crystals are formed* which consist of nitroso-sulphuric acid, $HSO_4.NO$. The simple theory given by Berzelius as outlined above does not explain the formation of $HSO_4.NO$. Hence it has been suggested by Davy and Lunge that the nitroso acid is formed as an intermediate product during the formation of sulphuric acid as follows :

$$NO + NO_2 \longrightarrow N_2O_3$$
$$2SO_2 + N_2O_3 + O_2 + H_2O \longrightarrow 2HSO_4.NO$$
$$2HSO_4.NO + H_2O \longrightarrow 2H_2SO_4 + NO + NO_2$$

Here it must be noted that the fact that nitroso-sulphuric acid is formed if insufficient steam is present, does not prove that it is formed as an intermediate product under normal working conditions.

Plant Used and Procedure. The various parts of the plant along with their functions are described below. The plant is shown diagrammatically in Fig. 36.2.

(*i*) *Pyrites burners or sulphur burners.* These are brick furnaces in which sulphur or iron pyrites in lumps is burnt, the supply of air being regulated by sliding doors below and above the pyrites. Spent oxide or pyrites powder is burnt in vertical iron cylinders lined with firebricks and provided with series of shelves. The powder is raked from shelf to shelf by rotating scrapers until the burnt ore is discharged at the bottom. The issuing gas contains about 7% SO_2, 83% N_2, and 10% O_2.

SO_2 is also obtained by burning sulphur in special burners of large inclined rotating cylinders.

$$4FeS_2 + 11O_2 \longrightarrow 2Fe_2O_3 + 8SO_2$$
$$S + O_2 \longrightarrow SO_2$$

(*ii*) *Dust catcher.* The burner gas is passed, if pyrites are used, through a dust catcher, which is provided with baffle walls where dust particles separate due to gravity. This may also be done by passing through Cottrell electrostatic dust precipitator.

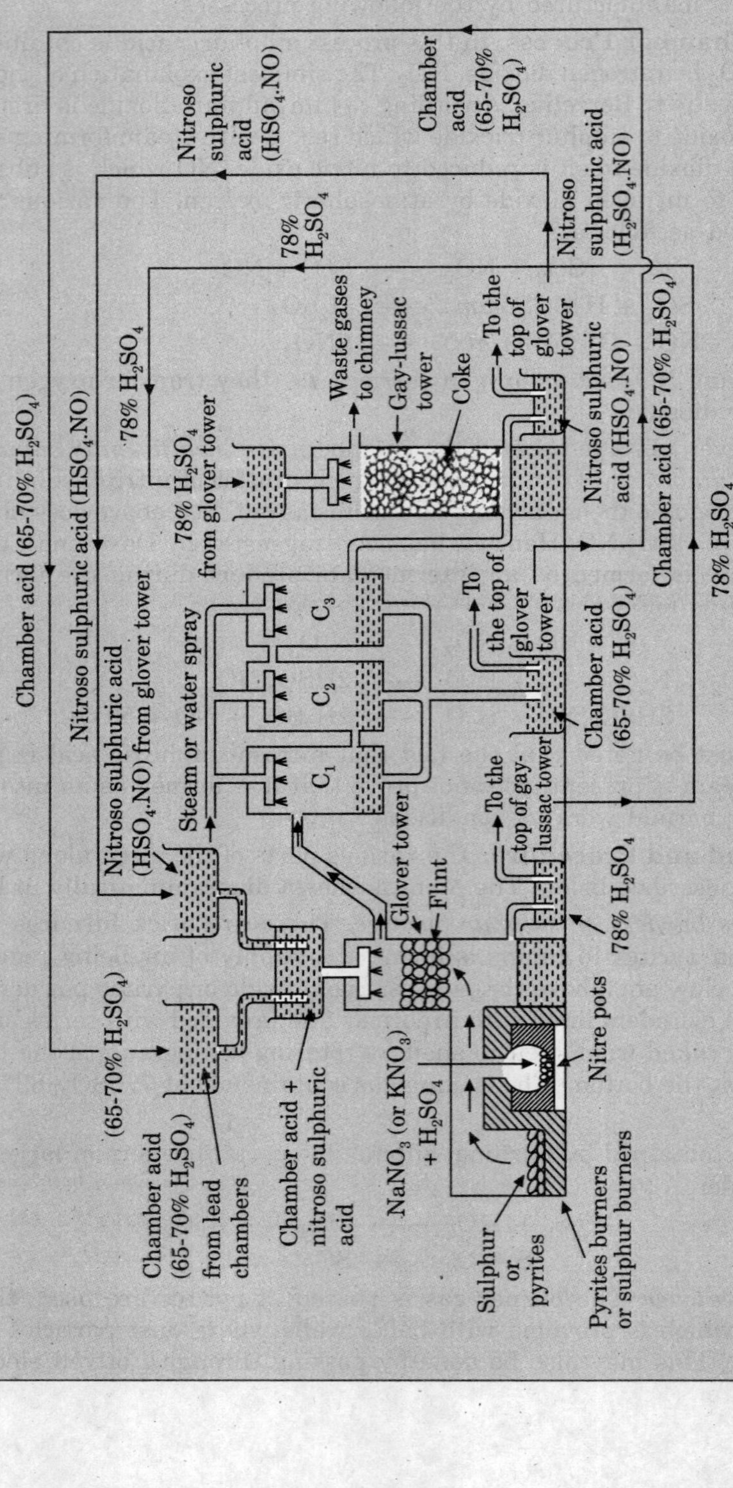

Fig. 36.2 Parts of the plant used for the manufacture of sulphuric acid by lead chamber process.

(*iii*) *Nitre pots.* Here $NaNO_3$ or KNO_3 is heated with conc. H_2SO_4 to produce nitric acid vapour.

$$NaNO_3 + H_2SO_4 \longrightarrow NaHSO_4 + HNO_3$$

This nitric acid vapour mixes with the hot gases coming from the sulphur burners and going to glover tower and is reduced by SO_2 to give a mixture of NO and NO_2

$$2HNO_3 + SO_2 \longrightarrow H_2SO_4 + 2NO_2$$

In modern plants, the oxides of nitrogen are produced by the catalytic oxidation of NH_3.

$$4NH_3 + 5O_2 \xrightarrow[750\text{–}900°C]{\text{Pt guaze}} 4NO + 6H_2O + 216 \text{ K.cals.}$$

$$2NO + O_2 \longrightarrow 2NO_2$$

(*iv*) *Glover tower* (or *Denitrating tower*). The gases now enter the glower tower near the base. The tower is of lead, lined inside with acid resisting material and packed with flint. There are two tanks at the top. One contains the dilute chamber acid and the second the nitrated acid (*i.e.* nitroso-sulphuric acid, $NO.HSO_4$) from the base of Gay-Lussac tower. Both these acids flow down slowly over flint pieces and meet an ascending stream of hot gases. There are four functions of this tower :

(*a*) The burner gases at 330—440°C are cooled to 50–80°C.

(*b*) The nitrated acid loses the dissolved oxides of nitrogen because these are taken by the burner gases.

$$2NO.HSO_4 + H_2O \longrightarrow 2H_2SO_4 + NO + NO_2$$

(*c*) The chamber acid is concentrated to about 80%.

(*d*) Sulphur dioxide is partly oxidised to sulphuric acid.

Sulphuric acid is always drawn from the bottom of the glover tower and is about 80%.

(*v*) *Lead chambers.* The gases (SO_2, O_2, oxides of nitrogen and moisture) are then led to a set of three or more large lead chambers (shown as C_1, C_2 and C_3 in Figure 36.2) which are made of lead and supported in wooden frames. The capacity of each lead chamber is 2503–7503 cu. feet and they are connected by means of wide lead pipes. Steam from a low-pressure boiler or more usually a very fine spray of water is admitted into these chambers from the top as shown in the figure when dil. H_2SO_4 is formed by the interaction of SO_2, oxides of nitrogen and H_2O.

$$SO_2 + H_2O \longrightarrow H_2SO_3$$
$$H_2SO_3 + NO_2 \longrightarrow H_2SO_4 + NO$$
$$2NO + O_2 \longrightarrow 2NO_2$$

Dilute H_2SO_4 so formed flows down in a common receiver placed at the floor of the chambers. It is called *chamber acid* and contains 65–70% H_2SO_4. It is pumped to the glover tower for its further concentration. In these chambers about 98% of SO_2 is converted into H_2SO_4.

(*vi*) *Gay-Lussac tower.* The residual gases coming from the last lead chamber contain N_2, a little O_2, oxides of nitrogen and traces of SO_2 and are led to the Gay-Lussac tower. This tower is lined with lead, and is 40 to 60 feet high and 8 to 15 feet in diameter. It is packed with coke down which flows a stream of cold 78% H_2SO_4 coming from the base of glover tower. This acid absorbs oxides of nitrogen in the exit gases of the lead chambers forming nitroso.sulphuric acid, $HSO_4.NO$.

$$2H_2SO_4 + NO + NO_2 \longrightarrow 2HSO_4.NO + H_2O$$

The nitroso-sulphuric acid so produced collects at the base of Gay-Lussac tower and is pumped to the top of the glover tower for the recovery of the oxides.

(*vii*) *Chimney.* The waste gas from the Gay-Lussac tower is left off into the atmosphere by passing through a chimney which is placed on the extreme right of the plant (not shown in the figure), thus creating a draught (partial vacuum) which helps in drawing the gases through the plant.

Impurities Present in the Chamber Acid and Their Removal. The chief impurities present in the chamber acid are arsenious oxide, lead sulphate and oxides of nitrogen. These are removed as follows :

(*i*) *Lead sulphate* comes in as an impurity sometimes by the action of acid on the lead of the chambers. It is removed by diluting the acid when the insoluble lead sulphate is separated.

(*ii*) *Arsenious oxide* (from the pyrites) is removed by passing sulphuretted hydrogen through the dilute acid, when it is precipitated in the form of sulphide.

$$As_2O_3 + 3H_2S \longrightarrow As_2S_3 + 3H_2O$$

It is removed by filtration under reduced pressure through unglazed earthenware plates or by flotation with paraffin which comes to the surface gathering all the precipitate with it.

(*iii*) *The oxides of nitrogen* are expelled by distilling the diluted acid with ammomum sulphate. The oxides of nitrogen are reduced to elementary nitrogen which escapes.

$$(NH_4)_2SO_4 + NO + NO_2 \longrightarrow 2N_2 + H_2SO_4 + 3H_2O$$

It is not always necessary to purify the chamber acid. For most purposes it is used as such. Whenever pure acid is needed it is prepared by the *contact process* rather than by the purification of chamber acid.

Concentration of the Chamber Acid. The acid obtained from glover tower contains about 78% H_2SO_4. Acid of this strength is known as *Brown Oil of Vitriol* (*B.O.V.*) on account of its colour which is due to the presence of impurities in it. It is further concentrated to form the acid containing 93–95% H_2SO_4. This concentrated acid is called *Rectified Oil of Vitriol* (*R.O.V.*)

The chamber acid is concentrated by : (*i*) *Cascade process* and (*ii*) *Gaillard process.* The principle involved in both the processes is to evaporate water present in the acid by bringing the dilute acid in contact with hot gases. The difference lies in the exposure of the dilute acid to the hot gases. The dilute acid and the hot gases come in contact with each other on the counter-current principle.

(*i*) *Cascade process.* (Fig. 36.3). The concentration of the acid is done in this process by heating the acid in fused silica basins arranged in such a way that the acid passes from the upper basin to the lower one as it goes on being concentrated. The heating is done by means of hot gases playing below the basins and as the acid passes downwards it is subjected to stronger heat, so that by the time it reaches the lowermost basin it has been concentrated to the maximum. The acid vapour along with the steam is taken to a recovery *plant* while the *concentrated* acid is taken to storage tanks.

(*ii*) *Gaillard process.* (Fig. 36.4). The acid is introduced in the form of a fine spray from the top of a tower (called *Gaillard tower*) built of acid-resisting stone, up which hot gases are passed. During the passage downwards the droplets of the dilute acid come in contact with the hot gases and lose their water so that by the time the acid reaches the bottom, it has been concentrated to about 95% pure acid.

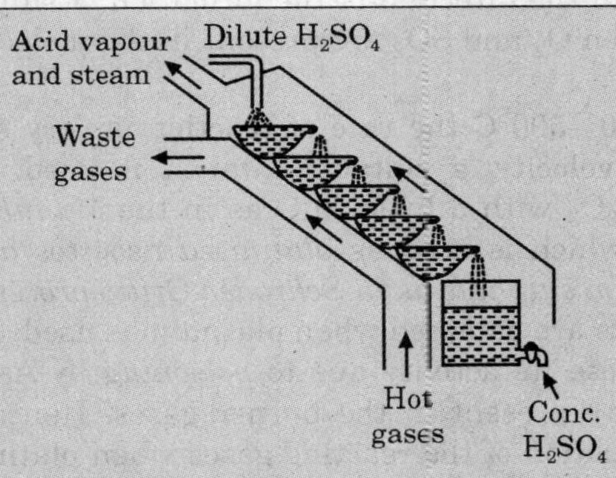

Fig. 36.3. Concentration of H_2SO_4 by Cascade process.

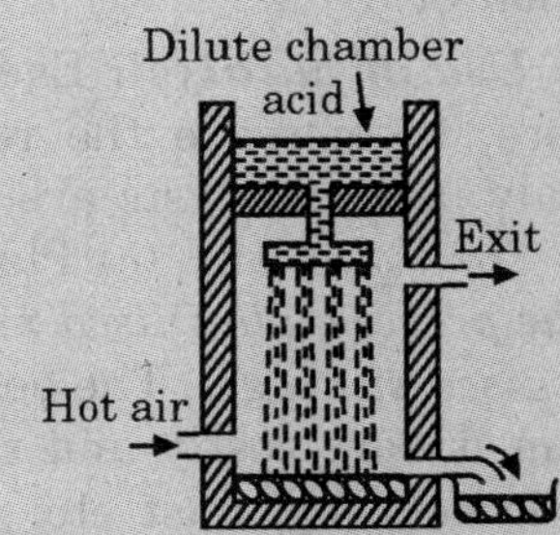

Fig. 36.4. Concentration of H_2SO_4 by Gaillard process.

The last traces of water are removed by adding sulphur trioxide to the acid which combines with the excess of water to form sulphuric acid.

2. Contact Process. Sulphur dioxide obtained by the burning of sulphur or pyrites (FeS_2) is made to combine with oxygen under suitable conditions in order to get sulphur trioxide.

$$S + O_2 \longrightarrow SO_2$$

$$4FeS_2 + 11O_2 \longrightarrow 2Fe_2O_3 + 8SO_2$$

$$2SO_2 + O_2 \rightleftarrows 2SO_3 + 45{,}000 \text{ cals.}$$

The sulphur trioxide so formed is dissolved in sulphuric acid and the *oleum* or *pyrosulphuric acid* ($H_2S_2O_7$) which results is diluted with water to get sulphuric acid of required concentration

$$H_2SO_4 + SO_3 \longrightarrow H_2S_2O_7$$

$$H_2S_2O_7 + H_2O \longrightarrow 2H_2SO_4$$

Conditions for Maximum Yield of SO_3. We have seen that in this process we make use of the reaction,

$$2SO_2 + O_2 \rightleftarrows 2SO_3 + 45{,}000 \text{ cals.}$$

which is *reversible, exothermic* and proceeds with a *decrease* in *volume* (*i.e.* 3 volumes change to 2 volumes). According to Le-Chatelier's principle the formation of SO_3 is favoured by :

(i) Low temperature. This reaction is exothermic, naturally the formation of SO_2 does not require high temperature. At low temperatures no reaction takes place between SO_2 and O_2. Thus a temperature between 400°—600°C is maintained and the reaction occurs with adequate speed.

(ii) High pressure. Since the reaction proceeds in the forward direction with decrease in volume, *i.e.* 3 volumes change to 2 volumes, a high pressure will favour the formation of SO_3. The gases are given a pressure of two atmospheres. High pressures may bring about greater corrosion of the chambers.

(iii) Excess of oxygen. A slight excess of oxygen in the reacting mixture of $SO_2 + O_2$ helps to carry the reaction in the direction of the formation of sulphur trioxide. Best results are obtained when O_2 and SO_2 are present in the molecular proportion of 3 : 2.

(iv) Use of catalyst. Even at 400°–500°C the rate of reaction is very slow. Therefore, to increase the reaction velocity, a suitable catalyst is used. The commonly used catalysts are : (*a*) Fe_2O_3 with a little CuO as in the *Mannheim process.* (*b*) V_2O_5, and (*c*) Platinum which is used as *platinised asbestos as in Badische proress, platinised magnesium sulphate as in Schroder-Grillo process or as platinised silica gel.* The best results are obtained when platinum is used. But, platinum is very costly and rapidly lose its activity due to *poisoning* by As_2O_3, sulphuric acid fog, dust etc. which are present in the burner gases. Hence an utmost care has to be taken for purification of the reacting gases when platinum is used as a catalyst. Platinised silica gel is, however, immune to arsenic poisoning. Now-a-days most of the sulphuric acid plants use V_2O_5 as a catalyst in place of platinum because it is cheaper and is not easily poisoned, How ever, it cannot be recovered after use.

Plant Used and Procedure. The plant used is shown diagrammatically in Fig. 36.5. It consists of four essential parts :

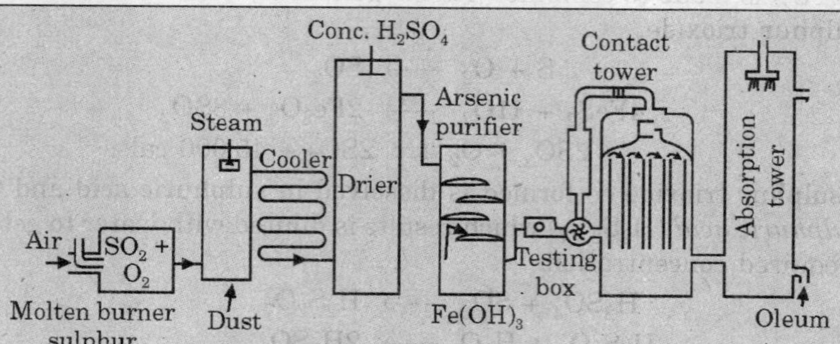

Fig. 36.5. Parts of the plant used for the manufacture of sulphuric acid by contact process.

(i) Sulphur burners. In these sulphur or pyrites (FeS_2) is burnt to get sulphur dioxide.

$$S + O_2 \longrightarrow SO_2$$
$$4FeS_2 + 11O_2 \longrightarrow 2Fe_2O_3 + 8SO_2$$
$$2S + 3O_2 \longrightarrow 2SO_3$$

Traces of SO_3 are also produced which give sulphuric acid fog with water vapours present in air.

(ii) Purification unit. The gases from sulphur burners containing about 8% SO_2, 1% O_2 and the rest N_2 are not pure. They contain impurities of As_2O_3, sulphur dust and sulphuric acid fog which act as poisons on the catalytic surface. Hence they are *purified* thoroughly by passing in succession through : (*a*) *a dust remover* in which the heavy dust particles are removed by mechanical precipitation aided by steam or electrically by using the *Cottrell process.* (*b*) *a series of cooling pipes* (*i.e., cooler*) to reduce the temperature to about 100°C. (*c*) *a scrubber* (*washing tower—not shown in Fig. 36.5*) where the gases meet a descending spray of water which removes the *acid fog.* Most of the dust which could not be removed in the dust remover gets washed off here with a spray of water. The gases, however, *moist.* (*d*) *a drying tower* (*drier*) which is packed with coke or flint stone and down which a stream of conc. H_2SO_4 flows. Here the gases are dried. (*e*) *an arsenic purifier* which contains precipitated $Fe(OH)_3$ placed on horizontal shelves. $Fe(OH)_3$ absorbs As_2O_3, if present as impurity in the gases. (*f*) *a testing box* in which a strong beam of light is thrown against the gases in order to be sure that the gases are absolutely free from dust and As_2O_3 particles. If any particles are present, they are illuminated due to the Tyndall effect.

(iii) Preheater and contact converter (or tower). The purified gases are then led into the *preheater* (P) where the gases are preheated to 400°–450°C (*optimum temperature*) by the hot gases (mainly SO_3 and N_2) from the converter which are themselves thereby cooled. The hot gases now enter into an iron cylinder which is called a *contact converter* or *tower* and is fitted inside with vertical iron tubes which are packed with platinised asbestos catalyst in the *Badische process.* The Badische contact converter is separately shown in Fig. 36.6.

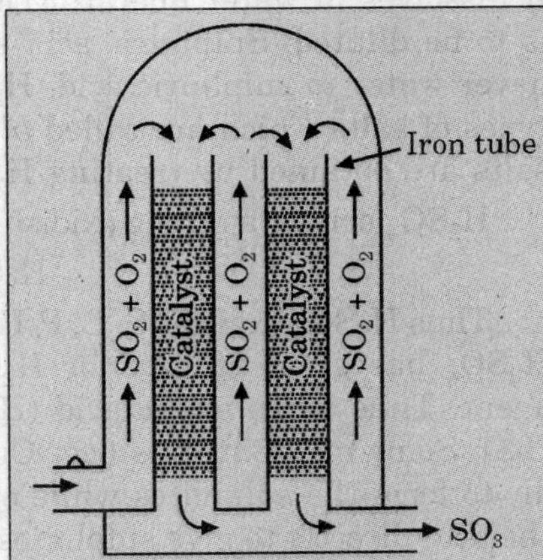

Fig. 36.6. Badische contact converter

The incoming gases circulate round the hot tubes and then pass down them when SO_2 is oxidised to SO_3. Since the reaction is exothermic, the heat produced raises the temperature to about 580°C. The rate of flow of gases is so adjusted that the temperature in the converter is kept to about 400°–450°C.

(iv) Absorption tower. The sulphur trioxide formed in the converter is passed up an absorption tower where it meets a descending stream of concentrated sulphuric acid. The acid absorbs the sulphur trioxide to form *fuming sulphuric acid* or *oleum* which is then diluted with the calculated quantity of water to get sulphuric acid of the required concentration.

$$H_2SO_4 + SO_3 \longrightarrow H_2S_2O_7$$
$$H_2S_2O_7 + H_2O \longrightarrow 2H_2SO_4$$

Water cannot be used for absorbing SO_3 because in doing so a large amount of heat is evolved resulting in the formation of a dense fog of minute particles of sulphuric acid which does not easily condense down.

Comparison Between the Contact Process and Chamber Process.

(*i*) Sulphuric acid obtained by the contact process is of higher purity than that obtained by the chamber process.

(*ii*) Sulphuric acid obtained by the contact process needs no further concentration. It is highly concentrated. The acid obtained by the chamber process is maximum 78% concentrated and needs further concentration.

(*iii*) The cost of production of sulphuric acid by chamber process is considered to be lower than that of the contact process because of the high cost of platinum metal but the use of vanadium pentoxide in place of platinum has reduced the cost of production in the contact process as well.

Properties of H_2SO_4.

Sulphuric acid is a colourless syrupy liquid of specific gravity 1.84 at 15°C. It fumes strongly in moist air and is highly corrosive. It is highly soluble in water. It dissolves in water liberating heat. Great care should be taken when the acid is to be diluted. Sulphuric acid should be added in a thin stream to water and never water to sulphuric acid. H_2SO_4 is strong dibasic acid and hence gives two types of salts which are called *bisulphates* (HSO_4^-) and *sulphates* (SO_4^{2-}). These salts are obtained by treating H_2SO_4 with alkalies.

H_2SO_4 acts as a strong oxidising agent, because it can supply an atom of oxygen.

$$H_2SO_4 \longrightarrow H_2O + SO_2 + O$$

Thus H_2SO_4 oxidises S, C, P, HI and HBr to SO_2, H_3PO_3, I_2 and Br_2 respectively. H_2SO_4 has great affinity for H_2O and hence is used as a strong dehydrating agent. Thus when oxalic acid ($H_2C_2O_4$) is heated with H_2SO_4, the former loses H_2O. Some of the metals (*e.g.* Cu, Ag, Hg etc.) react with H_2SO_4 in presence of air to form the sulphates while other metals (*e.g.* Zn, Al, Zn, Mg etc.) react with the acid directly to give sulphates. Noble metals like Au and Pt. do not react with the acid. The acid dissolves sulphur trioxide forming fuming sulphuric acid, also called *oleum* or *pyrosulpluric acid*, $H_2S_2O_7$ which is much more active than sulphuric acid and forms the salts pyrosulphates which are extensively used in the manufacture of aniline dyes and explosives.

Uses of H_2SO_4.

H_2SO_4 is used (*i*) in the manufacture of chemical manures like *ammonium sulphate* and *calcium superphosphate.* (*ii*) in the manufacture of other acids like nitric acid, hydrochloric acid, phosphoric acid etc. (*iii*) in refining petroleum. (*iv*) in the manufacture of coal-tar dyes and a number of pharmaceuticals. (*v*) in the manufacture of explosives. (*vi*) in the textile, paper and dyeing industries. (*vii*) for cleansing metals before enamelling, electroplating, galvanising etc. (*pickling*). (*viii*) in storage batteries. (*ix*) in the manufacture of sulphates, bisulphates and sodium carbonate. (*x*) in the laboratory as an important reagent and as a drying and dehydrating agent.

Structure of H_2SO_4 Molecule and SO_4^{2-} Ion.

The structure of H_2SO_4 molecule is as shown in the margin.

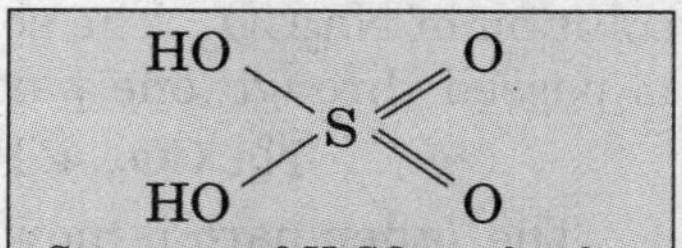

Structure of H_2SO_4 molecule.

This structure is confirmed by the following facts :

(*i*) Sulphuric acid (H_2SO_4) contains two hydroxyl (OH) groups because it reacts with PCl_5 to give sulphuryl chloride (SO_2Cl_2) which in turn is hydrolysed by water to sulphuric acid. The structure of sulphuryl chloride is Cl_2SO_2 which is confirmed by the fact that SO_2Cl_2 can be obtained by the direct combination of SO_2 and Cl_2.

$$\underset{\text{Sulphuric acid}}{(HO)_2SO_2} + 2PCl_5 \longrightarrow \underset{\text{Sulphuryl chloride}}{Cl_2SO_2} + 2POCl_3 + 2HCl$$

$$Cl_2SO_2 + 2H.OH \longrightarrow (OH)_2SO_2 + 2HCl$$

$$SO_2 + Cl_2 \longrightarrow Cl_2SO_2$$

(*ii*) The two hydroxyl groups are symmetrically arranged in H_2SO_4 molecule because we get only one type of organic derivative like diethyl sulphone, $C_2H_5—SO_2—C_2H_5$. Both OH groups are directly linked to S-atom as is evident from the formation of diethyl sulphone, $(C_2H_5)_2SO_2$ by different routes :

$$(C_2H_5)_2S \xrightarrow{\text{Oxidation}} \underset{\text{Diethyl sulphone}}{(C_2H_5)_2SO_2} \xleftarrow{C_2H_6} Cl_2SO_2 \xleftarrow{PCl_5} H_2SO_4$$

(*iii*) When we consider the S—O bond lengths in H_2SO_4 molecule, we are suggested to believe that O-atoms are attached with S-atom by double bond.

X-ray analysis of the ionic crystals containing SO_4^{2-} ions has shown that SO_4^{2-} ion has *tetrahedral shape*. This shape results from sp^3 hydridisation of S-atom (central atom). Both S—O π bonds are d(S)—p(O) π bonds.

III. Thioulphuric Acid, $H_2S_2O_3$.

The free acid is quite unstable at ordinary temperature but it has been isolated free from the solvent by reaction (*a*). It can also be obtained by treating $Na_2S_2O_3$ with very dil HCl as shown in reaction (*b*).

(*a*) $\underset{\text{Chlorosulphuric acid}}{OHSO_2Cl} + H_2S \longrightarrow H_2S_2O_3 + HCl;$

(*b*) $Na_2S_2O_3 + 2HCl \longrightarrow H_2S_2O_3 + 2NaCl$

The acid is very unstable and at once decomposes into SO_2 and S. The salts of $M_2S_2O_3$ type are called thiosulphates. Among thiosulphates, the most important is sodium thiosulphate (also called hypo), $Na_2S_2O_3$.

Sodium Thiosulphate, $Na_2S_2O_3.5H_2O$.

It was formerly (wrongly) termed as *sodium hypo sulphate* or simply ***hypo*** and is still so called by photographers.

Preparation. (*i*) *By heating a solution of sodium sulphite with sulphur.* Sodium sulphite for this purpose may be obtained by passing sulphur dioxide through a solution of Na_2CO_3. The solution is divided into two equal parts. Sulphur dioxide is passed through one part till it is saturated with it.

$$Na_2CO_3 + H_2O + SO_2 \longrightarrow 2NaHSO_3 + CO_2$$

The other part is now added to convert the bisulphite into normal sodium sulphite.

$$2NaHSO_3 + Na_2CO_3 \longrightarrow 2Na_2SO_3 + H_2O + CO_2$$

This solution is boiled with sulphur until the alkaline reaction has disappeared.

$$Na_2SO_3 + S \longrightarrow Na_2S_2O_3$$

The solution is filtered and the filtrate on concentration gives the crystals of sodium thiosulphate.

(*ii*) *By Spring reaction.* It is also prepared when requisite iodine is added to a mixture of sodium sulphite and sulphide solutions

$$Na_2S + Na_2SO_3 + I_2 \longrightarrow 2NaI + Na_2S_2O_3$$

In fact, any method which leads to the addition of sulphur to sodium sulphite would give thiosulphate.

(*iii*) It is obtained by the oxidation of soluble sulphides or hydrosulphides in air.

$$2NaHS + 2O_2 \longrightarrow Na_2S_2O_3 + H_2O$$

(*iv*) *Pure* sodium thiosulphate is formed by the interaction of sodium hydrosulphide (NaHS) and $NaHSO_3$ in solution and crystalling,

$$2NaHS + 4NaHSO_3 + I_2 \longrightarrow 2Na_2S_2O_3 + 3H_2O$$

Manufacture. It is manufactured from the waste liquor obtained in the manufacture of Na_2S. The waste liquor, still contains some Na_2S, Na_2CO_3, Na_2SO_3 and Na_2SO_4. The liquor is concentrated by evaporation and SO_2 is passed through it when $Na_2S_2O_3$ is formed.

$$Na_2SO_3 + 2Na_2S + 3SO_2 \longrightarrow 2Na_2S_2O_3$$

$$Na_2CO_3 + Na_2S + 4SO_2 \longrightarrow 3Na_2S_2O_3 + CO_2$$

The solution is filtered, concentrated and allowed to cool when colourless monoclinic crystals of $Na_2S_2O_2.5H_2O$ separate out.

Properties (*i*) *State.* It crystallises with five molecules of water of crystallisation. The crystals are colourless, dissolve in water and melt at 48°C. The water in combination is lost completely at 215°C.

(ii) Action of dilute acids. When treated with hydrochloric acid, thiosulphuric acid is temporarily formed but it decomposes almost immediately into sulphur dioxide and sulphur.

$$Na_2S_2O_3 + 2HCl \longrightarrow 2NaCl + H_2S_2O_3$$
$$H_2S_2O_3 \longrightarrow H_2O + SO_2 + S$$
$$\overline{Na_2S_2O_3 + 2HCl \longrightarrow 2NaCl + H_2O + SO_2 + S}$$

(iii) Action of heat. On heating it melts at 48°C, loses its water of crystallisation at 215°C and decomposes above 223°C into Na_2SO_4 and sodium pentasulphide (Na_2S_5)

$$4Na_2S_2O_3 \xrightarrow[> 223°]{\Delta} 3Na_2SO_4 + Na_2S_5$$

At higher temperature, Na_2S_5 loses some sulphur

$$Na_2S_5 \longrightarrow Na_2S_4 + S$$

(iv) Action of silver nitrate. With silver nitrate solution dilute solution of hypo forms an unstable white precipitate of $Ag_2S_2O_3$ which changes colour and eventually becomes black due to the formation of Ag_2S.

$$2AgNO_3 + Na_2S_2O_3 \longrightarrow 2NaNO_3 + Ag_2S_2O_3$$
$$\underset{\text{White}}{Ag_2S_2O_3} + H_2O \longrightarrow \underset{\text{Black}}{Ag_2S} + H_2SO_4$$

Excess of $Na_2S_2O_3$, however, dissolves the precipitate of $Ag_2S_2O_3$ forming the complex sodium argento-thiosulphate, $Na[AgS_2O_3]$

$$Ag_2S_2O_3 + Na_2S_2O_3 \longrightarrow 2Na[AgS_2O_3]$$

(v) Action of silver halides. A solution of sodium thiosulphate dissolves silver halides (AgCl, AgBr, AgI) forming colourless complex—*sodium argento thiosulphate*, $Na_3[Ag(S_2O_3)_2]$. This property is utilized in the '*fixing*' of negatives and prints in photography as the thiosulphate solution removes the unaffected silver bromide.

$$AgBr + 2Na_2S_2O_3 \longrightarrow Na_3[Ag(S_2O_3)_2] + NaBr$$

(vi) Action of halogens. $Na_2S_2O_3$ is oxidised by chlorine or bromine water and sulphur is precipitated.

$$Na_2S_2O_3 + Cl_2 + H_2O \longrightarrow Na_2SO_4 + 2HCl + S$$

But with excess of halogens, S is oxidised to H_2SO_4

$$Na_2S_2O_3 + 4Cl_2 + 5H_2O \longrightarrow 2NaCl + 6HCl + 2H_2SO_4$$

It decolourises a solution of iodine giving a quantitative yield of *sodium tetrathionate*, $Na_2S_4O_6$

$$2Na_2S_2O_3 + I_2 \longrightarrow \underset{\text{Sod. tetra-thionate}}{Na_2S_4O_6} + 3NaI$$

Hence this reaction is used in the volumetric estimation of iodine.

(vii) Action of KMnO₄. In *neutral* solution, $KMnO_4$ oxidises $Na_2S_2O_3$ to Na_2SO_4

$$Na_2S_2O_3 + 2KMnO_4 \longrightarrow Na_2SO_4 + K_2SO_4 + Mn_2O_3$$

In *acidic* solution, some sodium dithionate, $Na_2S_2O_6$ is also formed.

(viii) Action of ferric chloride solution. It gives a violet coloration with ferric chloride solution. The violet coloration is due to the formation of *ferric thiosulphate*, $Fe_2(S_2O_3)_3$.

$$3Na_2S_2O_3 + 2FeCl_3 \rightleftharpoons Fe_2(S_2O_3)_3 + 6NaCl$$

The colour rapidly disappears due to the reduction of ferric chloride to ferrous chloride.

$$[FeCl_3 \longrightarrow FeCl_2 + Cl] \times 2$$
$$2Na_2S_2O_3 + 2Cl \longrightarrow Na_2S_4O_6 + 2NaCl$$
$$\overline{2Na_2S_2O_3 + 2FeCl_3 \longrightarrow 2NaCl + 2FeCl_2 + Na_2S_4O_6}$$

(ix) Reaction with cupric salts. $Na_2S_2O_5$ first reduces cupric salts to cuprous salts which then dissolve in the excess of hypo forming a complex salt, *e.g.*

$$\underset{(ic)}{2CuCl_2} + Na_2S_2O_3 \longrightarrow \underset{(ous)}{2CuCl} + Na_2S_4O_6 + 2NaCl$$
$$2CuCl + Na_2S_2O_3 \longrightarrow Cu_2S_2O_3 + 2NaCl$$
$$3Cu_2S_2O_3 + 2Na_2S_2O_3 \longrightarrow \underset{\text{Complex salt}}{Na_4[Cu_6(S_2O_3)_5]}$$

(x) Reaction with auric salts. The action of $Na_2S_2O_3$ on auric salts is similar to that on cupric salts. Thus $Na_2S_2O_3$ first reduces auric salts to aurous salts which then dissolve in the excess of hypo forming a complex salt, *e.g.*

$$\underset{(ic)}{AuCl_3} + 2Na_2S_2O_3 \longrightarrow \underset{(ous)}{AuCl} + Na_2S_4O_6 + 2NaCl$$
$$AuCl + 2Na_2S_2O_3 \longrightarrow \underset{\text{Sod. aurothio sulphate}}{Na_3[Au(S_2O_3)_2]} + NaCl$$

(xi) Reaction with $HgCl_2$. $Na_2S_2O_3$ reacts with the excess of $HgCl_2$ liberating H_2SO_4 and forming a white precipitate of $HgCl_2.2HgS$

$$[Na_2S_2O_3 + HgCl_2 \longrightarrow HgS_2O_3 + 2NaCl] \times 2$$
$$[HgS_2O_3 + H_2O \longrightarrow HgS + H_2SO_4] \times 2$$
$$HgCl_2 + 2HgS \longrightarrow HgCl_2.2HgS$$
$$\overline{2Na_2S_2O_3 + 3HgCl_2 + 2H_2O \longrightarrow 4NaCl + 2H_2SO_4 + \underset{\text{White ppt.}}{HgCl_2.2HgS}}$$

(xii) Reaction with bismuth salts. $Na_2S_2O_3$ with a solution of bismuth salt gives a clear solution of *sodium bismuth thiosulphate*, $Na_3[Bi(S_2O_3)_3]$

$$3Na_2S_2O_3 + Bi^{3+} \longrightarrow Na_3[Bi(S_2O_3)_3] + 3Na^+$$

The solution quickly decomposes and deposits black bismuth sulphide.

(xiii) Reaction with $BaCl_2$ and $CaCl_2$ solutions. A moderately concentrated solution of $Na_2S_2O_3$ gives a white precipitate of BaS_2O_3 with $BaCl_2$ solution.

$$Na_2S_2O_3 + BaCl_2 \longrightarrow \underset{\text{White ppt.}}{BaS_2O_3} + 2NaCl$$

$CaCl_2$ gives no precipitate, since CaS_2O_3 is fairly soluble in H_2O

$$Na_2S_2O_3 + CaCl_2 \longrightarrow \underset{\text{Soluble}}{CaS_2O_3} + 2NaCl$$

Uses. It is used (*i*) *in photography.* Due to its property of dissolving Ag-halides, it is used as fixer in photography under the name Sodium thiosulfate or hypo. It removes the excess of silver halides. (*ii*) *in textile industry.* During bleaching it is used as *antichlor* for removing excess of chlorine from bleaching fabrics. (*iii*) *in metallurgy.* It is used in the extraction of Au and Ag from their ores. (*iv*) *in the laboratory.* It is used for the estimation of iodine and as a reagent. (*v*) *in medicine.* It is also used in medicine.

Structure of $H_2S_2O_3$ molecule and $S_2O_3^{2-}$ ion.

The structures of $H_2S_2O_3$ molecule and $S_2O_3^{2-}$ ion are as shown below :

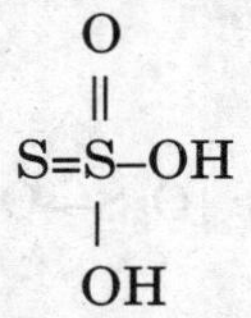

Structure of $H_2S_2O_3$ molecule

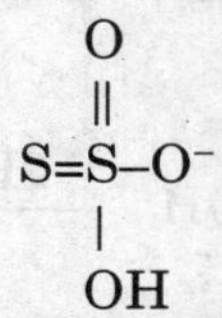

Structure of $S_2O_3^{2-}$ ion (Tetrahedral shape)

X-ray analysis of the ionic crystals containing $S_2O_3^{2-}$ ion has shown that $S_2O_3^{2-}$ has tetrahedral shape. This shape results from sp^3 hybridisation of S-atom (central atom).

The structure of $S_2O_3^{2-}$ ion as shown above indicates that the two S-atoms in $S_2O_3^{2-}$ ion are not equivalent in bonding and hence are not interchangeable. The presence of two different types of S-atoms in $S_2O_3^{2-}$ ion has been verified by the fact that a thiosulphate prepared from sulphite and radio-active sulphur (*e.g.* S^{35})

$$SO_3^{2-} + S^{35} \longrightarrow [S^{35} - SO_3]^{2-}$$

gets decomposed by acid to give exclusively radio active S^{35} and a non-radio active aqueous solution of SO_3^{2-} ions arising from SO_2 evolved.

$$[S^{35} - SO_3]^{2-} + 2H^+ \longrightarrow S^{35} + H_2O + SO_2$$

Distinguishing Tests for Sulphides, Sulphites, Sulphates and Thiosulphates.

The distinguishing tests for these salts have been summarised in Table given 36.1.

IV. Pyrosulphuric acid, $H_2S_2O_7$.

This acid is also called disulphuric acid. This acid is commonly known as *oleum* or *fuming sulphuric acid* as, in presence of moist air, it releases white fumes. The acid can be regarded as the dehydrated product of H_2SO_4 as shown below :

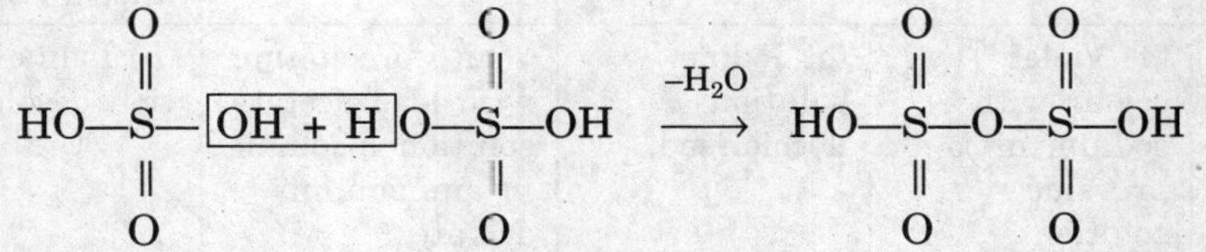

Two molecules of H_2SO_4 — Pyrosulphuric acid or disulphuric acid ($H_2S_2O_7$)

This acid is isolated as a colourless crystaline solid (m. pt. = 36°C) by mixing equimolecular quantities of anhydrous H_2SO_4 and SO_3. The acid is very unstable and when dissolved in H_2O it gives H_2SO_4 molecule.

$$H_2S_2O_7 + H_2O \longrightarrow 2H_2SO_4$$

The fact that $H_2S_2O_7$ molecule can be regarded as being formed by the removal of one H_2O molecule from two H_2SO_4 molecules suggests that $H_2S_2O_7$ has the structure as shown below. X-ray diffraction studies of the ionic salts containing $S_2O_7^{2-}$ ion have shown that $S_2O_7^{2-}$ ion has $[O_3S—O—SO_3]^{2-}$ structure which has been shown at the right of that of $H_2S_2O_7$ molecule.

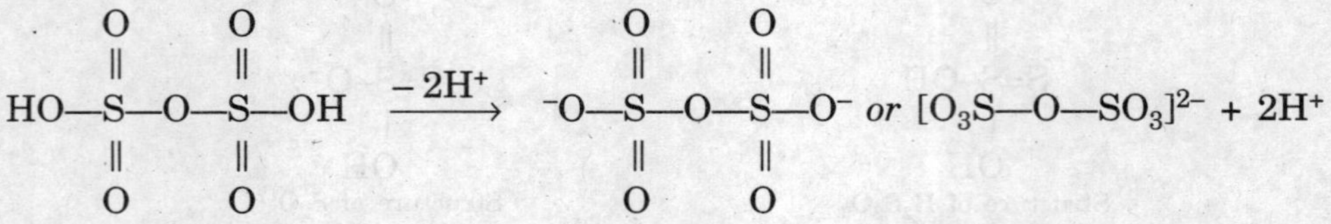

Pyrosulphuric acid ($H_2S_2O_7$) Pyrosulphate ion ($S_2O_7^{2-}$)

Table 36.1. Comparison between sulphides, sulphites, sulphates and thiosulphates.

Test	Sulphide (S^{2-})	Sulphite (SO_3^{2-})	Sulphate (SO_4^{2-})	Thiosulphate ($S_2O_3^{2-}$)
1. *Action of dil. HCl*	H_2S is given off which turns lead acetate paper black.	SO_2 is given off which turns dichromate paper green.	No action	SO_2 and S are produced.
2. *Action of $BaCl_2$ solution*	No ppt.	Precipitate of $BaSO_3$ soluble in HNO_3 evolving SO_2.	White ppt. of $BaSO_4$ insoluble in HNO_3.	White ppt. of BaS_2O_3 from conc. solutions soluble in HNO_3 with ppt. of sulphur.
3. *Action of $AgNO_3$*	Black ppt. of Ag_2S insoluble in cold but soluble HNO_3.	White precipitate of Ag_2SO_3 soluble in HNO_3.	No ppt.	White ppt. but quickly changing to yelldw brown
4. *Action of $FeCl_3$ solution*	Black ppt.	Dark red coloration.	No action	Violet coloration. Colour disappears at once.
5. *Confirmatory Test*	(*a*) Violet colour with sodium nitroprusside solution. (*b*) Black ppt. with lead acetate solution.	(*a*) Iodine solution decolorised. (*b*) $KMnO_4$ solution decolorised.	White precipitate with lead acetate solution insoluble in ammonium acetate.	(*a*) Iodine solution decolorised. (*b*) $KMnO_4$ solution decolorised.

Thionic Acids (H_2SnO_6)

The term thionic acids is used for the compounds in which two sulphonic groups, —SO_2OH *or* —SO_3H are linked together either directly or through one or more sulphur atoms. All the members of this class of acids can be represented by a general formula $H_2S_nO_6$ where $n = 2, 3, 4, 5, 6$. Thionic acids can be classified into two groups namely (*i*) *Dithionic acid*, $H_2S_2O_6$ ($n = 2$) in which two —SO_2OH groups are linked together directly. (*ii*) *Polythionic acids* in which two sulphonic groups, —SO_2OH or —SO_3H are linked together through one or more S-atoms. Thus all the members of polythionic acids can be represented by the general formula, $H_2S_nO_6$ or $HOO_2S—S_{n-2}—SO_2OH$ ($n = 3, 4, 5, 6$). Different members of polythionic acids are : *trithionic acid* ($H_2S_3O_6$ or $HOO_2S—S—SO_2OH$), *tetrathionic acid* ($H_2S_4O_6$ or $HOO_2S—S_2—SO_2OH$), *pentathionic acid* ($H_2S_5O_5$ *or* $HOO_2S—S_3—SO_2OH$) and *hexathionic acid* ($H_2S_6O_6$ *or* $HOO_2S—S_4—SO_2OH$).

Dithionic Acid, $H_2S_2O_6$.

This acid can be prepared by the slow oxidation of H_2SO_3 by MnO_2.

$$2H_2SO_3 + O \longrightarrow H_2S_2O_6 + H_2O$$

The aqueous solution of $H_2S_2O_6$ can be concentrated on a water bath and then in vacuum over H_2SO_4 until its specific gravity is 1.347 Excess of concentration of the acid should be avoided, since the acid decomposes under this condition.

$$H_2S_2O_6 \xrightarrow{\text{Decomposition}} H_2SO_4 + SO_2$$

Dithionic acid exists in solution only. The acid decomposes on concentrating its aqueous solution above a specific gravity of 1.347 into H_2SO_4 and SO_2. The acid is remarkably stable towards oxidising agents. On hydrolysis $H_2S_2O_6$ gives H_2SO_4 and H_2SO_3.

$$H_2S_2O_6 + H_2O \longrightarrow H_2SO_4 + H_2SO_3$$

Dithionates ($S_2O_6^{2-}$).

Although $H_2S_2O_6$ has been shown to be *dibasic* by conductivity measurement of its sodium salt, it gives only one type of salts, called *dithionates* ($S_2O_6^{2-}$). Acid thionates (*i.e.* $NaHS_2O_6$) are *unknown.*

Dithionates can be prepared by decomposing BaS_2O_6 solution with a metal sulphate or metal carbonate, filtering off $BaSO_4$ or $BaCO_3$ and crystallising. Dithionates are also prepared by the electrolytic oxidation of neutral or alkaline solutions of sulphites.

$$2SO_3^{2-} \longrightarrow S_4O_6^{2-} + 2e^-$$

All dithionates are soluble in H_2O and mostly crystallise with H_2O. The dithionates of alkaline earth metals are very stable and their solutions may be boiled without decomposition. Dithionates of heavy metals are not so stable. Dithionates also decompose on heating, *e.g.*

$$K_2S_2O_6 \xrightarrow{\Delta} K_2SO_4 + SO_2$$

Sodium amalgam reduces the solution of a dithionate to the corresponding sulphites, *e.g.*

$$Na_2S_2O_6 + 2Na \longrightarrow 2Na_2SO_3$$

Structure of $H_2S_2O_6$ Molecule and $S_2O_6^{2-}$ Ion.

In $H_2S_2O_6$ two sulphonic acid groups, $—SO_2OH$ are directly linked with each other. Since the acid is dibasic, it has two OH groups linked directly with S-atom. Thus the $H_2S_2O_6$ and $S_2O_6^{2-}$ have the following structures :

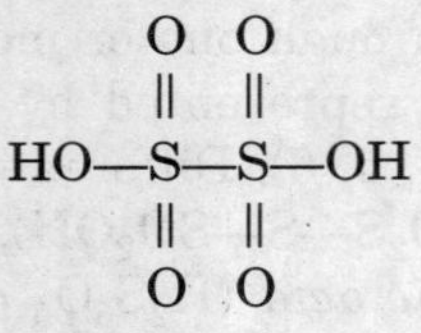

Dithionic acid
($H_2S_2O_6$)

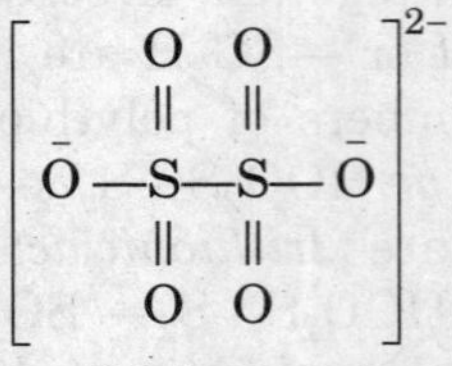

Dithionate ion
($S_2O_6^{2-}$)

The structure of $S_2O_6^{2-}$ ion has approximately *tetrahedral bond angles* about each sulphur atom. The S—O bond distance is 1.43 Å as compared to 1.44 Å in SO_4^{2-} ion. The short bond length suggests considerable double bond character in $S_2O_6^{2-}$ ion.

Polythionic Acids ($H_2S_nO_6$)

Preparation.

Polythionic acids are prepared by various methods. The following methods are, however, important.

1. By the action of sulphur chlorides on sulphites (SO_3^{2-}) and thiosulphates ($S_2O_3^{2-}$) in etheral solution.

$$SCl_2 + 2SO_3^{2-} \longrightarrow \underset{\text{(tri)}}{S_3O_6^{2-}} + 2Cl^-$$

$$SCl_2 + 2S_2O_3^{2-} \longrightarrow \underset{\text{(penta)}}{S_5O_6^{2-}} + 2Cl^-$$

$$S_2Cl_2 + 2SO_3^{2-} \longrightarrow \underset{\text{(tetra)}}{S_4O_6^{2-}} + 2Cl^-$$

$$S_2Cl_2 + 2S_2O_3^{2-} \longrightarrow \underset{\text{(tetra)}}{S_4O_6^{2-}} + 2Cl^-$$

2. By passing H_2S through a saturated aqueous solution of SO_2. When H_2S gas is passed very slowly through a solution of SO_2 kept cooled by *ice, Wakenroder solution* is obtained. This solution contains a mixture of polythionic acids, suspended and colloidal sulphur, $H_2S_2O_3$ etc. and hence can be used to prepare polythionic acids. The following mechanism for the formation of polythionic acids has been suggested by Stawrn (1944).

$$H_2S + H_2SO_3 \longrightarrow H_2S_2O_2 + H_2O$$

$$H_2S_2O_2 + 2H_2SO_3 \longrightarrow H_2S_4O_6 + 2H_2O$$

$$2H_2S + H_2SO_3 \longrightarrow 3H_2O + 3S \downarrow$$

$$H_2S_4O_6 + H_2SO_3 \longrightarrow H_2S_2O_3 + H_2S_3O_6$$
$$H_2S_2O_3 + 2H_2S_2O_3 \longrightarrow H_2S_6O_6 + 2H_2O$$

In less acidic solutions, $H_2S_6O_6$ undergoes decomposition as follows :

$$H_2S_6O_6 \longrightarrow H_2S_5O_6 + S\downarrow$$
$$H_2S_5O_6 + H_2SO_3 \longrightarrow H_2S_4O_6 + H_2S_2O_3$$

3. Specific methods for the preparation of individual polythionic acids. Preparation of individual polythionic acids by specific methods is done by the following reactions :

(*a*) $H_2S_3O_6$: $\underset{\text{(solution)}}{K_2S_3O_6} + H_2SiF_6 \rightleftharpoons \underset{\text{(soluble)}}{H_2S_3O_6} + \underset{\text{(insoluble)}}{K_2SiF_6}$

(*b*) $H_2S_4O_6$: $\underset{\text{(solution)}}{BaS_4O_6} + H_2SO_4 \longrightarrow \underset{\text{(insoluble)}}{BaSO_4} + \underset{\text{(soluble)}}{H_2S_4O_6}$

(c) $H_2S_5O_6$: $5S_2Cl_2 + 6H_2O \xrightarrow{\text{Hydrolysis}} H_2S_5O_6 + 10HCl + 5S$

(*d*) $H_2S_6O_6$: $2K_2S_2O_3 \text{ (in acetone)} + S_2Cl_2 \xrightarrow{-78°C} H_2S_6O_6 + 2KCl$

Properties

(*i*) Decomposition. The free polythionic acids exist only in solution. The solutions of $H_2S_4O_6$ and $H_2S_5O_6$ are sufficiently stable to be concentrated, without decomposition, upto a density of 1.5. On keeping, the solutions of all polythionic acids decompose to give H_2SO_4, SO_2 and sulphur as the final products.

For example :

$$H_2S_3O_6 \longrightarrow H_2SO_4 + SO_2 + S$$
$$H_2S_4O_6 \longrightarrow H_2SO_4 + SO_2 + 2S$$
$$H_2S_5O_6 \longrightarrow H_2SO_4 + SO_2 + 3S$$

In general the decomposition may be shown as :

$$H_2S_nO_6 \longrightarrow H_2SO_4 + SO_2 + (n-2)S.$$

(*ii*) Change in stability. The stability of solutions of polythionic acids is increased by the addition of a Little mineral acid while the same is diminished in the presence of free alkali.

(*iii*) Dibasic nature. All the polythionic acids are dibasic but the acid salts of these acids have not been prepared so far.

(*iv*) Stability. The alkali polythionates are quite stable in aqueous solution; those of alkaline earth metals are less so. The heavy metal polythionates, as a rule, are unstable.

(*v*) Solubility. Polythionates are all soluble in water. However, polythionates of Ag and Hg are insoluble. Of the potassium salts which are the best known, the trithionate is the most soluble, the hexathionate comes next; while tetra- and pentathionates are the least soluble, almost insoluble. The difference in solubilities of potassium salts may be utilised in the separation of a mixture of polythionic acids.

***(vi)* Action of oxidising agent.** Aqueous solution of all polythionates are converted into sulphates when treated with oxidising agents. For example :

$$S_3O_6^{2-} + 2H_2O + 4O \longrightarrow SO_4^{2-} + 2H_2SO_4$$
$$S_4O_6^{2-} + 3H_2O + 7O \longrightarrow SO_4^{2-} + 3H_2SO_4$$

In general :

$$S_nO_6^{2-} + (n-1)\ H_2O + (3n-5)\ O \longrightarrow SO_4^{2-} + (n-1)\ H_2SO_4.$$

***(vii)* Reduction.** These are reduced to thiosulphates on treatment with alkali sulphides. For example :

$$S_3O_6^{2-} + S^{2-} \longrightarrow 2S_2O_3^{2-}$$
$$S_4O_6^{2-} + S^{2-} \longrightarrow 2S_2O_3^{2-} + S$$
$$S_5O_6^{2-} + S^{2-} \longrightarrow 2S_2O_3^{2-} + 2S$$
$$S_6O_6^{2-} + S^{2-} \longrightarrow 2S_2O_3^{2-} + 3S$$

Polythionates, when treated with alkali sulphite, get gradually reduced to give ultimately the trithionate and thiosulphate.

$$S_6O_6^{2-} + SO_3^{2} \longrightarrow S_5O_6^{2-} + S_2O_3^{2-}$$
$$S_5O_6^{2-} + SO_3^{2-} \longrightarrow S_4O_6^{2-} + S_2O_3^{2-}$$
$$S_4O_6^{2-} + SO_3^{2-} \longrightarrow S_3O_6^{2-} + S_2O_3^{2-}$$

***(viii)* Action of aqueous KCN solution.** When treated with KCN in aqueous solution, a trithionate is converted into thiosulphate ($S_2O_3^{2-}$), sulphate (SO_4^{2-}) and free hydrocyanic acid (HCN).

$$S_3O_6^{2-} + 2CN^- + H_2O \longrightarrow S_2O_3^{2-} + SO_4^{2-} + 2HCN$$

Higher polythionates react similarly. The excess of sulphur combines with a further quantity of KCN to form sulphocyanide (CNS^-).

$$S_4O_6^{2-} + 3CN^- + H_2O \longrightarrow S_2O_3^{2-} + SO_4^{2-} + 2HCN + CNS^-$$
$$S_5O_6^{2-} + 4CN^- + H_2O \longrightarrow S_2O_3^{2-} + SO_4^{2-} + HCN + 2CNS^-$$

***(ix)* Action of $Hg_2(NO_3)_2$ and $AgNO_3$.** Polythionates give precipitates with $Hg_2(NO_3)_2$ or $AgNO_3$.

$$S_3O_6^{2-} + Hg_2(NO_3)_2 \longrightarrow \text{Black ppt.}$$
$$S_4O_6^{2-} \text{ or } S_5O_6^{2-} + Hg_2(NO_3)_2 \longrightarrow \text{Yellow ppt.}$$

$S_4O_6^{2-}$ does give any precipitate with $Hg_2(NO_3)_2$ or $AgNO_3$.

***(x)* Action of sodium or potassium amalgam.** With sodium or potassium amalgam, their solutions are reduced to thiosulphate ($S_2O_3^{2-}$)

$$S_3O_6^{2-} + 2Na/Hg \longrightarrow S_2O_3^{2-} + 2Na^+ + SO_3^{2-}$$
$$S_4O_6^{2-} + 2Na/Hg \longrightarrow 2S_2O_3^{2-} + 2Na^+$$

***(xi)* Action of alkali solution.** $S_3O_6^{2-}$ and $S_4O_6^{2-}$, when treated with an aqueous solution of an alkali, decompose into $S_2O_3^{2-}$ and SO_3^{2-}. $S_6O_6^{2-}$, however, gives only $S_2O_3^{2-}$.

$$2S_3O_6^{2-} + 6OH^- \longrightarrow S_2O_3^{2-} + 4SO_3^{2-} + 3H_2O$$
$$2S_4O_6^{2-} + 6OH^- \longrightarrow 2S_2O_3^{2-} + 2SO_3^{2-} + 3H_2O$$
$$2S_5O_6^{2-} + 6OH^- \longrightarrow 5S_2O_3^{2-} + 3H_2O$$

***(xii)* Action of Cu^{2+} ions.** It is only $S_3O_6^{2-}$ which reacts with Cu zone at 70°C to form CuS etc.

$$S_3O_6^{2-} + 2H_2O + Cu^{2+} \xrightarrow{70°C} CuS + 2SO_4^{2-} + 4H^+$$

Structure of Polythionic Acids and Polythionate Ions.

The structure of polythionic acids is obtained by attaching two —SO_2OH groups together through one or more S-atoms. Thus the structure of these acids and the anions derived from them are as given below :

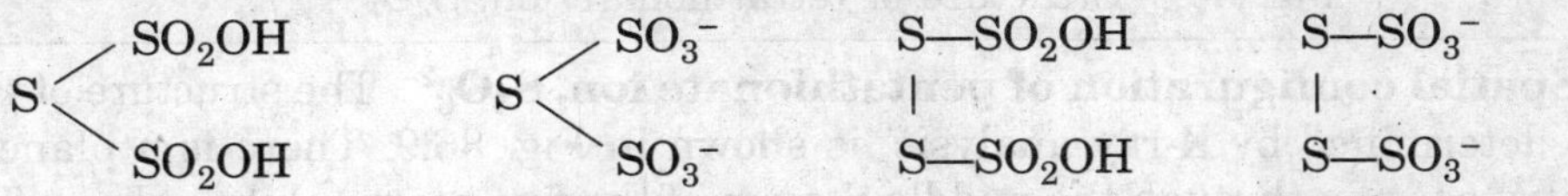

Trithionic acid ($H_2S_3O_6$) | Trthionate ion ($S_3O_6^{2-}$) | Tetrathionic acid ($H_2S_4O_6$) | Tetrathionate ion ($S_4O_6^{2-}$)

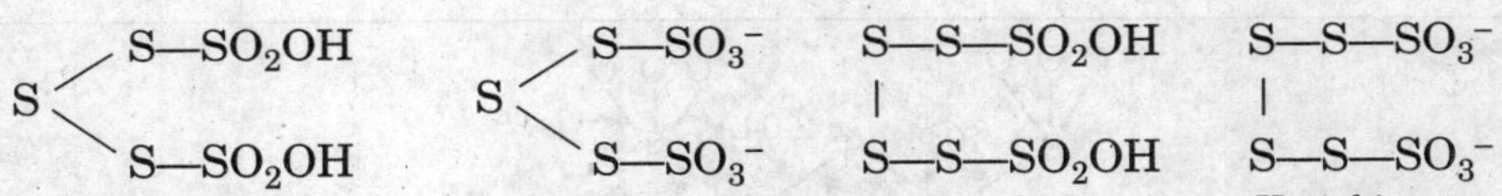

Pentathionic acid ($H_2S_5O_6$) | Pentathionate ion ($S_5O_6^{2-}$) | Hexathionic acid ($H_2S_6O_6$) | Hexathionate ion ($S_6O_6^{2-}$)

1. Spatial configuration of trithionate ion, $S_3O_6^{2-}$. X-ray analysis of trithionates (*e.g.* $Na_2S_3O_6$, $K_2S_3O_6$ etc.) has shown that the structure of $S_3O_6^{2-}$ ion is as shown in Fig. 36.7. In this structure the two end S-atoms lie in different planes which are inclined at an angle of 99° to each other while the middle S-atom lies on the common axis of the two planes. The SSS bond angle is 108° and both S—S bond lengths are equal to 2.0A°. All the six S—O bond distances are equal to 1.4A°. The disposition of the three O-atoms round each of the end S-atoms is approximately *tetrahedral*.

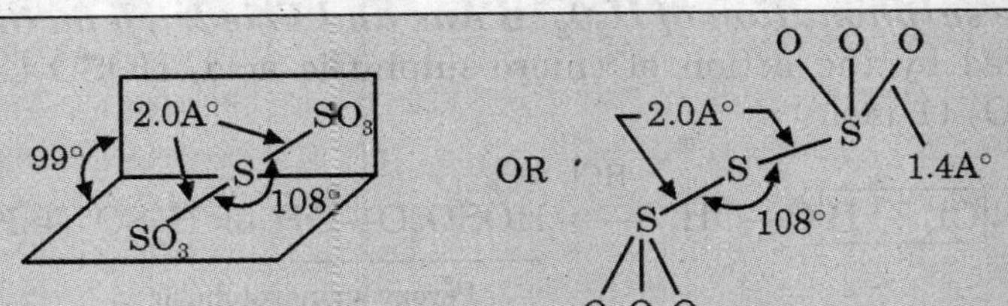

Fig. 36.7. Structure of trithionate ion, $S_3O_6^{2-}$ (Diagrammatic).

2. Spatial configuration of tetrathionate ion, $S_4O_6^{2-}$. X-ray analysis has shown that the structure of this ion is as shown in Fig. 36.8. In this structure the two end S-atoms lie in different planes which are inclined at an angle of 90° to each other while the two middle S-atoms lie on the common axis of the two planes. The bond lengths and bond angles are as shown in the figure. The disposition of the three O-atoms about each of the two end S-atoms is approximately tetrahedral.

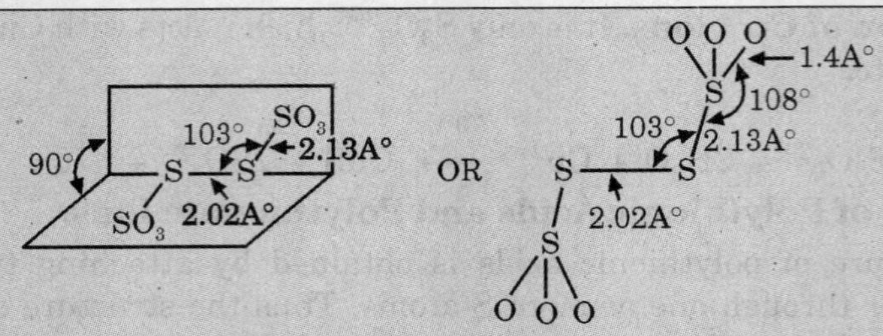

Fig. 36.8. Structure of tetrathionate ion, $S_4O_6^{2-}$.

3. Spatial configuration of pentathionate ion, $S_5O_6^{2-}$. The structure of this ion, as determined by X-ray analysis, is shown in Fig. 36.9. There is a plane of symmetry passing through the middle S-atom of the five-atom sulphur chain. The disposition of the three O-atoms on each of the end S-atom is, as usual, approximately, *tetrahedral.*

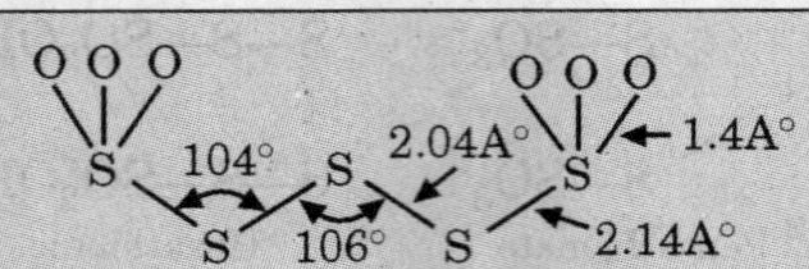

Fig. 36.9. Structure of pentathionate ion, $S_5O_6^{2-}$.

4. Spatial configuration of hexathionate ion, $S_6O_6^{2-}$. The structure of this ion is the same as that of $S_4O_6^{2-}$ and $S_5O_6^{2-}$ ions. The disposition of the three O-atoms about each of the two end S-atoms is approximately tetrahedral.

Peroxy Sulphuric Acids

1. Peroxy Monosulphuric Acid, H_2SO_5.

Preparation. This acid is also called ***Cero's acid.*** It is prepared :

(i) By the sulphonation of H_20_2 : d'Ans and Friedrich method. Pure acid can be prepared by the action of chioro-sulphuric acid, $HOSO_2Cl$ (1 mole) on anhydrous H_2O_2 (1 mole)

$$HOSO_2\boxed{Cl + H}O\text{—}OH \xrightarrow{-HCl} \underbrace{HOSO_2O\text{—}OH \text{ or } H_2SO_5}_{\text{Peroxy monosulphuric acid}} + HCl$$

(ii) By the anodic oxidation of H_2SO_4. When conc. H_2SO_4 is submitted to electrolysis in a diaphragm cell at a low temperature, using a small smooth platinum anode and a high current density, a solution of H_2SO_5 is obtained.

$$H_2SO_4 + O \longrightarrow H_2SO_5$$

Properties. *(i) Physical state.* Caro's acid in the anhydrous state is a white crystalline solid (m. pt. = 45°C) and hygroscopic. It is stable for some days but gradually loses ozonised oxygen.

(ii) Action of H_2O_2. On treatment with H_2O_2, it evolves O_2,

$$H_2SO_5 + H_2O_2 \longrightarrow H_2SO_4 + H_2O + O_2$$

(iii) Hydrolysis. When diluted with dil. H_2SO_4, it gives H_2O_2

$$H_2SO_5 + H_2O \longrightarrow H_2SO_4 + H_2O_2$$

(iv) Oxidising property. It is a powerful oxidising agent stronger than $H_2S_2O_8$. Thus it oxidises Mn(II), Ag(I), Co(II), Ni(II) and Pb(II) salts to peroxides and liberates Cl_2, Br_2 and I_2 from the respective halides

$$H_2SO_5 + 2AgNO_3 \longrightarrow 2H_2SO_4 + 2HNO_3 + \underset{\text{ppt.}}{Ag_2O_2}$$

$$2KI + H_2SO_5 \longrightarrow K_2SO_4 + H_2O + I_2$$

It also oxidises SO_2 to SO_3, Fe (II) salts to Fe (III) salts, HCl to Cl_2. aniline to nitrosobenzene and nitrobenzene.

(*v*) It carbonises wool and cellulose.

(vi) Monobasic nature. No solid salts of H_2SO_5 of M_2SO_5 type are known. However, a solution of $KHSO_5$ is said to be formed by the action of H_2O_2 on $KClSO_3$

$$KClSO_3 + H_2O_2 \longrightarrow KHSO_5 + HCl$$

Thus H_2SO_5 is a *monobasic acid.*

2. Peroxydisulphuric Acid, $H_2S_2O_8$.

Preparation. This acid is also called **Marshall's acid.** It is prepared :

(i) By the sulphonation of H_2O_2 : d'Ans and Friedrich's method. Anhydrous $H_2S_2O_8$ can be prepared by the action of H_2O_2 (1 mole) on chlorosulphuric acid, $ClSO_2.OH$ (2 moles)

$$2ClSO_2OH + 2H_2O_2 \xrightarrow{-2HCl} \underbrace{HOSO_2—O—O—SO_2OH \text{ or } H_2S_2O_8}_{\text{Per oxydisulphuric acid}} + 2HCl$$

(ii) By the anodic oxidation of H_2SO_4. $H_2S_2O_8$ is obtained, by the electrolysis of H_2SO_4 in cold. About 50% H_2SO_4 is taken in a large glass test tube which is placed in a glass cylinder open at both ends. A spiral of copper wire is wound round this cylinder which acts as a cathode. A fine platinum wire fused in a capillary tube is placed inside the inner tube which acts as an anode. The large glass test tube containing H_2SO_4 is cooled by putting in an ice bath. On passing electric current of high density, $H_2S_2O_8$ is formed in solution at the anode. The reactions taking place at cathode and anode are :

Anode *Cathode*

$$HSO_4^- \longleftarrow H_2SO_4 \longrightarrow H^+$$

$$HSO_4^- + HSO_4^- \rightarrow H_2S_2O_8 + 2e^- \qquad H^+ + e^- \rightarrow H$$

$$H + H \rightarrow H_2$$

Another explanation for the formation of the acid at the anode is that HSO_4^- ions liberated at the anode are oxidised by nascent oxygen that appears at the anode as a result of electrolysis.

$$2HSO_4^- + H_2O + O \longrightarrow H_2S_2O_8 + 2OH^-$$

Properties. ***(i) Physical state.*** It is a colourless, hygroscopic, crystalline compound (m.pt. = 65°C) which is fairly stable at ordinary temperature :

(ii) Decomposition. On heating, $H_2S_2O_8$ decomposes into SO_3 and O_2

$$2H_2S_2O_8 \xrightarrow{\text{heat}} 2H_2SO_4 + 2SO_3 + O_2$$

It is also decomposed by light.

(iii) Hydrolysis. In aqueous solution, $H_2S_2O_8$ is slowly hydrolysed into H_2SO_4 and H_2SO_5.

$$H_2S_2O_8 + H_2O \longrightarrow H_2SO_4 + H_2SO_5$$

When distilled with dil. H_2SO_4, it is also hydrolysed but the product is H_2O_2.

$$H_2S_2O_8 + 2H_2O \longrightarrow 2H_2SO_4 + H_2O_2$$

(iv) The anhydrous acid vigorously reacts with organic compounds.

(v) Oxidising property. The acid and its salts (which are called *perdisulphates*) are strong oxidising agents. The ion-electron equation representing their oxidising property is :

$$M^{2+} + 2H_2O + S_2O_8^{2-} \longrightarrow MO_2 + 4H^+ 2SO_4^{2-}$$

or

$$S_2O_8^{2-} + 2e^- \rightleftharpoons 2SO_4^{2-}$$

Following examples show the oxidising property of perdisulphates :

(a) They oxidise Mn (II), Ag (I), Pb (II), Co (II) and Ni (II) salts to the corresponding metal peroxides (in presence of an alkali).

$$Mn^{2+} + 2H_2O + S_2O_8^{2-} \longrightarrow MnO_2 + 4H^+ + 2SO_4^{2-}$$

$$Pb^{2+} + 2H_2O + S_2O_8^{2-} \longrightarrow PbO_2 + 4H^+ + 2SO_4^{2-}$$

$$\underset{\text{salts}}{2Ag^+} + 2H_2O + S_2O_8^{2-} \longrightarrow \underset{\text{peroxides}}{Ag_2O_2} + 2H^+ + 2HSO_4^-$$

(b) They oxidise Fe (II) salts to Fe (III) salts and Cr (III) salts to chromates (CrO_4^{2-}) even in alkaline medium. Metallic Zn and Cu are oxidised to their sulphates.

$$2Fe^{2+} + S_2O_8^{2-} \longrightarrow 2Fe^{3+} + 2SO_4^{2-}$$

$$Zn^0 + S_2O_8^{2-} \longrightarrow Zn^{2+} + 2SO_4^{2-}$$

(c) They liberate Cl_2, Br_2 and I_2 from the solution of the corresponding halide and even oxidises free I_2 to iodic acid and bleaches indigo. The bleaching of indigo and liberation of I_2 from an iodide proceeds slowly.

$$2Cl^- + S_2O_8^{2-} \longrightarrow Cl_2 + 2SO_4^{2-}$$

$$2I^- + S_2O_8^{2-} \longrightarrow I_2 + 2SO_4^{2-}$$

$$I_2 + 6H_2O + 5S_2O_8^{2-} \longrightarrow 2IO_3^- + 10SO_4^{2-} + 12H^+$$

Structure of Peroxysulphuric Acids and Peroxysulphate Ions.

These acids can be regarded as being obtained from H_2O_2 by the replacement of one or both H-atoms by sulphonic acid group, $-SO_2OH$. Thus, these acids can be represented as :

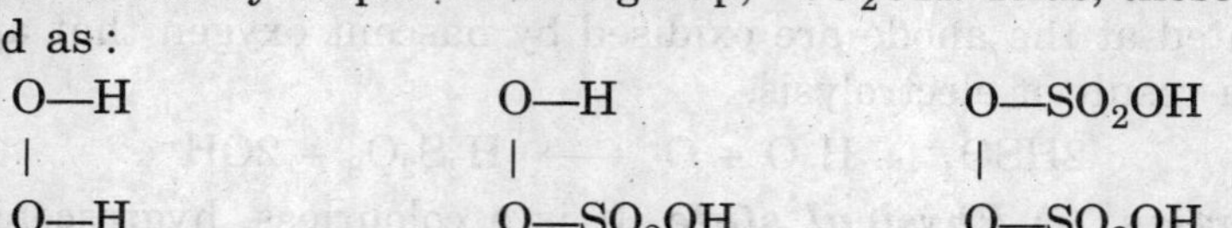

Hydrogen peroxide (H_2O_2)   Permonosulphuric acid or Caro's acid (H_2SO_5)   Perdisulphuric acid or Marshall's acid ($H_2S_2O_8$)

Permonosulphate (SO_5^{2-}) and perdisulphate ($S_2O_8^{2-}$) ions have the following structures :

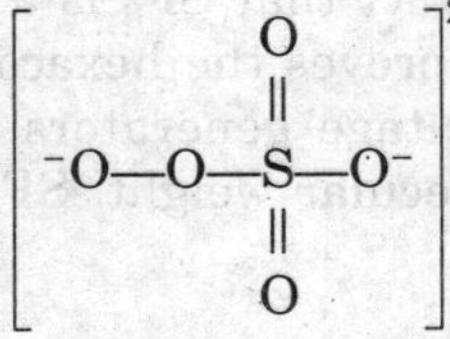

Permonosulphate ion, SO_5^{2-}

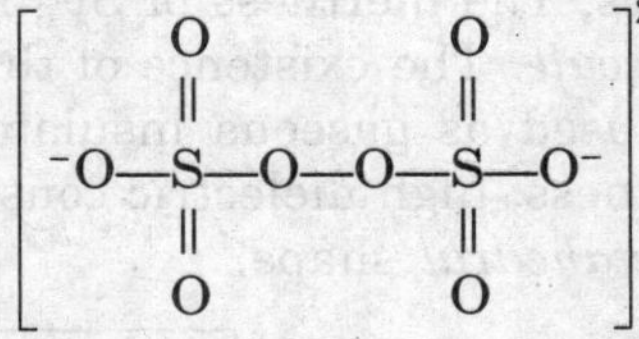

Perdisulphate ion, $S_2O_8^{2-}$

X-ray anlysis of $(NH_4)_2S_2O_8$ and $Cs_2S_2O_8$ has shown that $S_2O_8^{2-}$ ion consists of two SO_4 tetrahedra joined together by a covalent bond. The S—O bond distance (= 1.50 A°) is short. The short S—O bond distance indicates a high degree of double bond character.

Distinction Between H_2O_2 and Peroxysulphuric Acids.

The distinction between H_2O_2 and peroxy sulphuric acids can be made by the following tests :

1. $KMnO_4$ test. When H_2O_2 is added to a solution of $KMnO_4$ acidified with H_2SO_4, $KMnO_4$ decolorises and O_2 is evolved. Persuiphuric acids do not react.

2. $K_2Cr_2O_7$ test. When H_2O_2 is added to a solution of $K_2Cr_2O_7$, acidified with dil. H_2SO_4 and to which a little diethyl ether has been added, bright blue *chromium peroxide,* CrO_5 is formed that passes into etheral layer on gentle shaking. Persulphuric acids do not show this test.

3. $Ti(SO_4)_2$ test. H_2O_2 reacts with an acidified solution of $Ti(SO_4)_2$ to give a yellow or orange-red colouration, due to the formation of *pertitanic acid*, $TiO_2.xH_2O$. Persulphuric acids do not show this reaction.

4. Aniline test. Coro's acid (H_2SO_5) forms nitroso benzene and nitrobenzene with aniline while Marshall's acid ($H_2S_2O_8$) gives aniline black. H_2O_2 does not show this reaction.

Halides

Sulphur forms a number of halides. Out of these, sulphur hexafluoride, SF_6 is the most important. It is prepared by allowing sulphur to burn spontaneously in F_2. The gaseous SF_6 so formed is usually mixed with low fluorides of sulphur and hence the resulting gaseous mixture of fluorides is liquified at —80°C and then fractionally distilled, separating the lower fluorides by caustic potash solution. SF_6 is a colourless and odourless gas (m.pt. = 50.8°C and b. pt. = + 68.8°C). It is chemically inert like N_2 and hence remains unaffected by heat. It also remains undecomposed by water, acids or alkalies. Even fused caustic potash and ignited lead chromate or copper have nc effect on it. It is decomposed by molten (boiling) Na at 250°C and also by H_2S.

$$SF_6 + 8Na \longrightarrow Na_2S + 6NaF$$

$$SF_6 + 3H_2S \longrightarrow 6HF + 4S$$

It is also decomposed by H_2 at 400°C and by sparking with H_2 or O_2. In stability and non-reactivity, SF_6 resembles CCl_4 and differs from all other non-metallic halides. The inertness of SF_6 is due to the fact that SF_6 is a *covalently saturated molecule.* The existence of this compound proves the hexacovalency of S-atom. It is used as gaseous insulator in high voltage generators due to its chemical inertness, high dielectric constant and molecular weight. SF_6 molecule has *regular octahedral* shape.

Oxyhalides

Sulphur forms a number of oxyhalides. Here we shall discuss only thionyl chloride ($SOCl_2$) and sulphonyl chloride (SO_2Cl_2).

1. Thionyl Chloride, $SOCl_2$.

Preparation. It is prepared by adding sulphur into Cl_2O at –12°C or by the action of SO_3 on S_2Cl_2 at 75°–80° (*Commercial method*).

$$S + Cl_2O \xrightarrow{-12^\circ C} SOCl_2$$

$$SO_3 + C_2Cl_2 \xrightarrow{72^\circ - 80^\circ} SOCl_2 + SO_2 + S$$

Properties. It is a colourless, pungent smelling liquid of density 1677 (at 0°C) b.pt. = 78°C and m.pt. = – 99.5°C. It fumes in moist air and gets decomposed by H_2O forming H_2SO_3 and HCl. Organic compounds containing OH groups react with $SOCl_2$ and OH group is replaced by Cl atom.

$$SOCl_2 + 2H_2O \longrightarrow H_2SO_3 + 2HCl$$

$$\underset{\text{Organic compound}}{ROH} + SOCl_2 \longrightarrow RCl + SO_2 + HCl$$

It reacts with KBr, forming thionyl bromide, $SOBr_2$

$$SOCl_2 + 2KKr \longrightarrow SOBr_2 + 2KCl$$

$SOCl_2$ reacts with SbF_3 in presence of $SbCl_5$ (Swart's reagent) to form thionyl fluoride, SOF_2

$$3SOCl_2 + 2SbF_3 \xrightarrow{\text{In presence of } SbCl_5} 3SOF_2 + 2SbCl_3$$

Uses. It is used a reagent in organic chemistry to replace OH groups present in organic compounds by Cl atom.

Structure. In $SOCl_2$ molecule S-atom is sp^3 hbridised. Due to the presence of one lone pair of electrons on S-atom, the molecule has distorted tetrahedral shape, *i.e., trigonal pyramidal shape*. This shape contains two Cl^- and one O-atoms in a traingle (Fig. 36.10). S—O π bond is *d—p* π bond.

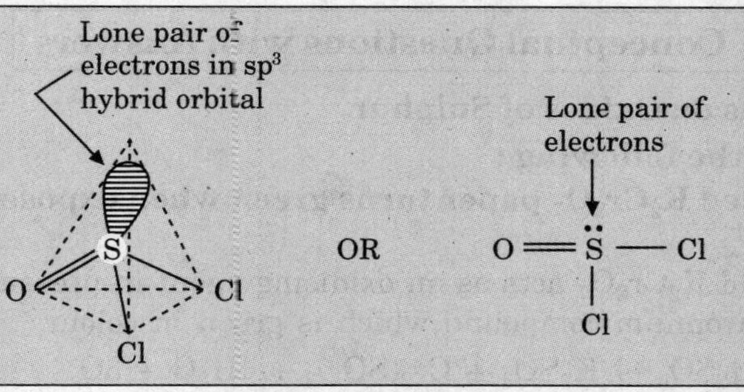

Fig. 36.10. Trigonal pyramidal shape of thionyl chloride, $SOCl_2$.

2. Sulphonyl Chloride, SO_2Cl_2.

Preporation. This compound is also called ***sulphuryl chloride*** and is formed by the direct combination of SO_2 and Cl_2 in sunlight or in the presence of catalyst like camphor, charcoal or acetic anhydride.

$$SO_2 + Cl_2 \longrightarrow SO_2Cl_2$$

Properties. It is a white fuming liquid of specific gravity of 1.667 at 20°C. It boils at 69.1°C without decomposition; the vapour is much dissociated at 330°C.

$$SO_2Cl_2 \rightleftharpoons SO_2 + Cl_2$$

With ice-cold water it gives a crystalline hydrate, $SO_2Cl_2.15H_2O$.

When warmed with water, it is first hydrolysed to chlorosulphuric acid, $(OH)SO_2Cl$ and then to H_2SO_4.

$$SO_2Cl_2 + H_2O \longrightarrow (OH)SO_2Cl + HCl$$
$$(OH)SO_2Cl + H_2O \longrightarrow SO_2(OH)_2 \text{ or } H_2SO_4 + 3HCl$$
$$\overline{SO_2Cl_2 + 2H_2O \longrightarrow SO_2(OH)_2 \text{ or } H_2SO_4 + 3HCl}$$

In benzene, SO_2Cl_2 reacts with NH_3 to form sulphamide, $SO_2(NH_2)_2$. With CH_3COONa, it forms acetyl chloride, CH_3COCl.

Uses. Its solution in petrol is used in making the wool unshrikable. It is used as a chlorinating agent in organic chemistry, *i.e.*, it is used for introducing sulphonic acid group.

Structure. In SO_2Cl_2 molecule S-atom is sp^3 hybridised, and hence the molecule has *tetrahedral shape* (See Fig. 36.11). It may be noted that one S—O π bond is *p–p* π bond while the other S—O bond is *d—p* π bond.

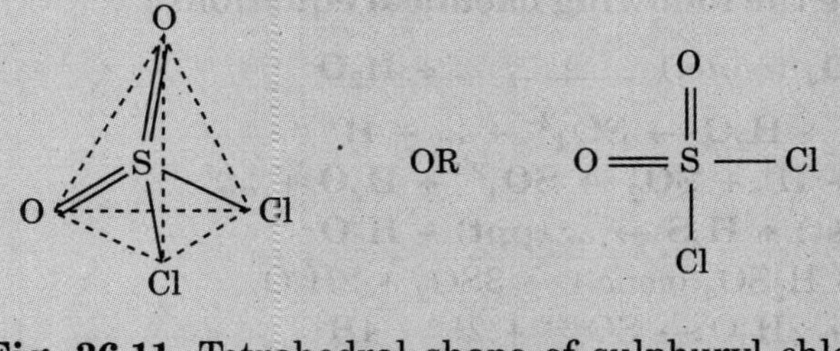

Fig. 36.11. Tetrahedral shape of sulphuryl chloride or sulphonyl chloride, SO_2Cl_2.

Conceptual Questions with Answers

[A] Questions on Oxides of Sulphur.

Q.1 Explain the following :

(*i*) An acidified $K_2Cr_2O_7$ paper turns green when exposed to SO_2.

(*Roorkee 1986*)

Ans : Acidified $K_2Cr_2O_7$ acts as an oxidising agent in presence of SO_2 and is converted into chromium compound which is green in colour.

$$K_2Cr_2O_7 + 4H_2SO_4 \rightarrow K_2SO_4 + Cr_2(SO_4)_3 + 4H_2O + 3O$$

$$3SO_2 + 3O + 3H_2O \rightarrow 3H_2SO_4$$

$$K_2Cr_2O_7 + H_2SO_4 + 3SO_2 \rightarrow K_2SO_4 + \underset{\text{Green}}{Cr_2(SO_4)_3} + H_2O$$

(*ii*) SO_2 acts as a bleaching agent.

Ans : SO_2 in presence of moisture is oxidised to H_2SO_4 liberating nascent hydrogen. The nascent hydrogen bleaches the coloured substance.

$$SO_2 + 2H_2O \rightarrow H_2SO_4 + 2H$$

Coloured substance + 2H $\rightarrow$ Colourless

(*iii*) SO_2 does not bleach dry flowers.

Ans : Only in presence of moisture, the nascent hydrogen responsible for bleaching (redcution) is produced.

$$SO_2 + 2H_2O \rightarrow H_2SO_4 + 2H$$

Coloured flower + H $\rightarrow$ Colourless flower

If H_2O or moisture is not present, nescent hydrogen will not be prodcued and hence flowers will not be bleached.

(*iv*) SO_2 is neither conbustible nor helps in burning, yet burning Mg continues to burn in it.

Ans : The heat of reaction is suffecient to dissociate SO_2 into S and free O_2. O_2 thus obtained burns Mg.

Q.2 How is SO_2 obtained from H_2SO_4 ?

Ans : SO_2 can be obtained by heating Cu-turnings with conc. H_2SO_4 or by boiling sulphur with conc. H_2SO_4.

$$Cu + 2H_2SO_4 \rightarrow CuSO_4 + 2H_2O + SO_2 \;; \qquad S + 2H_2SO_4 \rightarrow 3SO_2 + 2H_2O$$

Q.3 Complete the following chemical equations :

(*a*) $S + H_2SO_4$ (*conc.*) $\xrightarrow{\Delta}$ ... + H_2O

(*b*) $I_2 + SO_2 + H_2O \rightarrow SO_4^{2-}$ + ... + H^+

(*c*) $Cr_2O_7^{2-} + H^+ + SO_2 \rightarrow SO_4^{2-} + H_2O$ + ...

(*d*) SO_2(Moist) + $H_2S \rightarrow$ ... (ppt) + H_2O

Ans : (*a*) $S + H_2SO_4$ (conc.) $\rightarrow 3SO_2 + 2H_2O$

(*b*) $I_2 + SO_2 + 2H_2O \rightarrow SO_4^{2-} + 2I^- + 4H^+$

(*c*) $Cr_2O_7^{2-} + 2H^+ + 3SO_2 \rightarrow 3SO_4^{2-} + H_2O + 2Cr^{3+}$

(*d*) SO_2(Moist) + $2H_2S \rightarrow 3S$(ppt) + $2H_2O$

Q.4 Give exampels to show that :

(*i*) SO_2 acts as a reducing agent (RA).

Ans : $Cl_2 + 2H_2O + SO_2 \rightarrow H_2SO_4 + 2HCl$
(Cl = 0) (S = +4) (S = +6) (Cl = –1)

$2FeCl_3 + 2H_2O + SO_2 \rightarrow H_2SO_4 + 2HCl + 2FeCl_2$
(Fe = +3) (S = +4) (S = +6) (Fe = +2)

In both the reactions, since SO_2 is oxidised to H_2SO_4, SO_2 acts as a RA.

(*ii*) SO_3 is an anhydride of H_2SO_4.

Ans : Since the combination of SO_3 with H_2O gives H_2SO_4, this oxide is the anhydride of H_2SO_4

$$SO_3 + H_2O \rightarrow H_2SO_4$$

Q.5 H_2S acts only as a reducing agent (RA) but SO_2 acts both as a reducing agent as well as an oxidising agent (O.A). Expalin why ? (***Roorkee 1986***)

Ans : Minimum O.N of S-atom in – 2 and maximum O.N. of this atom is +6.

In H_2S, O.N. of S-atom is equal to – 2. When H_2S acts as a reducing agent, it gets oxidised and hence O.N. of S-atom increases from – 2 to some higher O.N. Thus H_2S can easily act as a reducing agent. If H_2S acts as an O.A., O.N. of S-atom should decrease from – 2 to some lower O.N. which is not possible. Thus H_2S cannot act as an O.A.

In SO_2, O.N. of S-atom is equal to + 4. When SO_2 acts as a R.A., it gets oxidised and hence O.N. of S-atom increases from + 4 to higher O.N. (= + 6) which is possible. Thus SO_2 can act as a RA. When SO_2 acts as an OA, it gets reduced and hence O.N. of S-atom decreases from + 4 to lower O.N. (= + 2). This is also possible. Thus SO_2 can also act as an OA. The above discussion clearly indicates SO_2 can act both as a RA and as an OA but H_2S can act only as a RA.

[B] Questions an Oxy acids of Sulphur.

1. Explain the following :

(*i*) The value of K_{a2} is less than that of K_{a1} for H_2SO_4.

Ans : H_2SO_4 is a dibasic acid. Hence it ionizes in two stages and has two dissociation constants.

(*i*) $H_2SO_4(aq) + H_2O(l) \rightarrow H_3O^+(aq) + HSO_4^-(aq)$; $K_{a1} > 10$

(*ii*) $HSO_4^-(aq) + H_2O(l) \rightarrow H_3O^+(aq) + SO_4^{2-}(aq)$; $K_{a2} = 1.2 \times 10^{-2}$

The value of K_{a2} is less than K_{a1} because the negatively charged HSO_4^- ion has much less tendency to donate a proton to H_2O as compared to neutral H_2SO_4.

(*ii*) (*a*) In the contact process for the manufacture of H_2SO_4, some amount of sulphuric acid is used as a starting material. Explain briefy.

(*b*) What is the catalyst used for oxidation of SO_2 to SO_3 ?

Ans : (*a*) In contact process, H_2SO_4 is used initially to absorb SO_3 to form oleum.

$$H_2SO_4 + SO_3 \rightarrow H_2S_2O_7 \ (\textit{Oleum})$$

Oleum is then diluted with calculated amount of water to get H_2SO_4 of desired concentration. It may be noted that SO_3 is not directly absorbed in water to form H_2SO_4 because it forms a dense fog of H_2SO_4 which does not condens easily.

(*b*) V_2O_5 is used as catylst for oxidation of SO_2 to SO_3.

(*iii*) H_2S can not be dried by passing it through conc. H_2SO_4.

Ans : H_2S can not be dried by passing it through conc. H_2SO_4, since it is oxidised by H_2SO_4 to S.

$$H_2SO_4 \rightarrow H_2O + SO_2 + O$$

$$H_2S + O \rightarrow H_2O + S$$

$$\underset{(OA)}{H_2SO_4} + \underset{(RA)}{H_2S} \rightarrow 2H_2O + SO_2 + S$$

(*iv*) In the manufacture of H_2SO_4 by contact process, SO_3 is not directly dissolved in H_2O.

Ans : SO_3 reacts with H_2O to form H_2SO_4 which is not easily condensed but escapes in the atmoshphere. SO_3 is, therefore, dissolved in H_2SO_4 (not in H_2O) to form oleum ($H_2S_2O_7$) which, on dilution, gives H_2SO_4

$$SO_3 + H_2SO_4 \rightarrow H_2S_2O_7 \ (\textit{oleum})$$

$$H_2S_2O_7 + H_2O \rightarrow 2H_2SO_4.$$

(*v*) For the dilution of conc. H_2SO_4, water should not be added to conc. H_2SO_4.

Ans : Conc. H_2SO_4 has great affinity for H_2O molecules. During addition of water to conc. H_2SO_4, large amount of heat released rises the temperature suddenly and hence bumping takes place. To avoid any accident due to bumping, water should not be added to the acid. Instead, we should add acid to water with constant stirring.

(*vi*) Conc. H_2SO_4 cannot be used for drying H_2.

Ans : H_2 is oxisised by H_2SO_4 to H_2O ($H_2SO_4 + H_2 \rightarrow 2H_2O + SO_2$). Sometimes H_2O formed as above is absorbed by H_2SO_4 and heat is released. This heat ignites H_2.

(*vii*) $KMnO_4$ should not be dissolved in conc. H_2SO_4.

Ans : Mn_2O_7 is formed in the reaction between $KMnO_4$ and H_2SO_4. Mn_2O_7 is unstable and explosive in nature.

$$2KMnO_4 + 2H_2SO_4 \rightarrow K_2SO_4 + (MnO_3)_2SO_4 + 2H_2O$$

$$(MnO_3)_2SO_4 + H_2O \rightarrow Mn_2O_7 + H_2SO_4$$

(*viii*) H_2SO_4 has high boiling point and viscosity.

Ans : High b.pt. and viscosity of H_2SO_4 is due to the fact that many H_2SO_4 molecules unit together by intermolecular H-bonding and a cluster is formed. It is due to the formation of a cluster that H_2SO_4 has high b.pt. and high viscosity.

(*ix*) A fresh strain of I_2 is generally washed away with hypo solution.

Ans : Hypo solution ($Na_2S_2O_3$) reduces I_2 to colourless NaI and , therefore, fresh strain of I_2 is washed away.

$$2Na_2S_2O_3 + I_2 \rightarrow Na_2S_4O_6 + 2NaI$$

or $$2S_2O_3^{2-} + I_2 \rightarrow S_4O_6^{2-} + 2I^-$$

Q.2 State with balanced equation what happens when :

(*i*) Pot. ferrocyanide is heated with conc. H_2SO_4. (*I.I.T. 1988*)

Ans : CO is evolved.

$$K_4Fe(CN)_6 + 6H_2SO_4 + 6H_2O \rightarrow 2K_2SO_4 + FeSO_4 + 3(NH_4)_2SO_4 + 6CO$$

(*ii*) A mixture of pot. chlorate, oxalic acid and H_2SO_4 is heated. (*I.I.T.1985*)

Ans : Potassium chlorate is reduced while oxalic acid is oxidised.

$$KClO_3 + H_2SO_4 \rightarrow KHSO_4 + HCl + 3O$$

$$3H_2C_2O_4 + 3O \rightarrow 6CO_2 + 3H_2O$$

$$KClO_3 + 3H_2C_2O_4 + H_2SO_4 \rightarrow KHSO_4 + HCl + 6CO_2 + 3H_2O$$

(*iii*) H_2SO_4 is heated with PCl_5.

Ans : $H_2SO_4 + PCl_5 \xrightarrow{\Delta} SO_2Cl_2 + POCl_3 + {}_2HCl$

(*vi*) KI is hetaed with conc. H_2SO_4.

Ans : $2KI + H_2SO_4 \rightarrow K_2SO_4 + 2HI$

$$2HI + H_2SO_4 \rightarrow I_2 + SO_2 + 2H_2O$$

$$\underset{(RA)}{2KI} + \underset{(OA)}{2H_2SO_4} \rightarrow K_2SO_4 + I_2 + SO_2 + 2H_2O$$

In this reaction KI is oxidised to I_2 and H_2SO_4 is reduced to SO_2.

(*v*) Concentrated sulphuric acid is added to a mixture of sodium chloride and manganese dioxide.

Ans : $2NaCl(s) + 2H_2SO_4(aq) + 2MnO_2(s) \rightarrow Na_2SO_4(aq) + 2H_2O(l) + Cl_2(g)$

(*vi*) Sugar turns black on addition of conc. H_2SO_4.

Ans : Conc. H_2SO_4 has great affinity for water molecules, *i.e.*, it acts as a dehydrating agent. When in contact with sugar, all water molecules are taken up and carbon is left behind.

$$C_{12}H_{22}O_{11} \xrightarrow{\text{Conc. } H_2SO_4} 12C + 11H_2O$$

Q.3 Determine the bond order (B.O.) of SO_4^{2-} ion.

Ans : We know that SO_4^{2-} ion is a reasonance hybrid of the six resonating structures shown in Fig. 36.12. In these structures four O-atoms have been designated as O_a, O_b, O_c and O_d. Obviously :

B.O. of S–O_a bond SO_4^{2-} ion.

$$= \frac{1+1+1+2+2+2}{6} = \frac{9}{6} = \mathbf{1.5\ (Ans)}$$

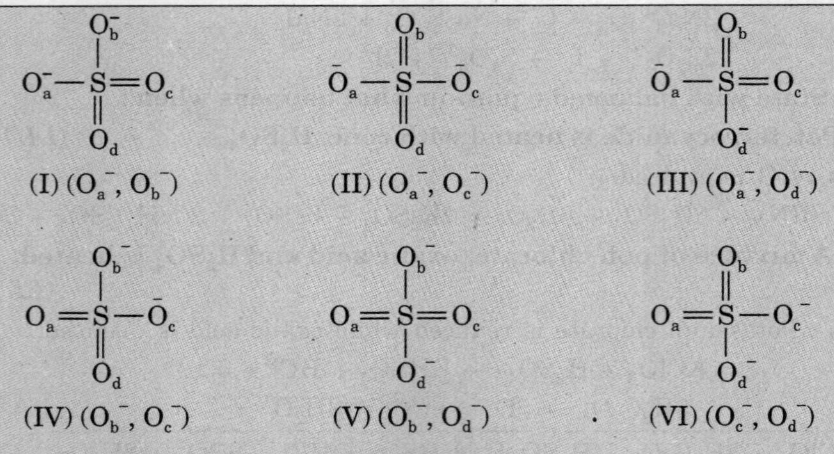

Fig. 36.12. Various resonating structures of SO_4^{2-} ion.

Q.4 H_2SO_4 acts as an acid, as an oxidising agent and as a dehydrating agent. In which way does it behave when it acts on :

(*a*) Glucose (*b*) Sulphur (*c*) $CaCO_3$ (*d*) Cu
(*e*) HCOOH (*f*) HI (*g*) NaOH (*h*) C.

Ans : (*a*) Dehydrating agent (*b*) Oxidising agent (*c*) An acid (*d*) Oxidising agent (*e*) Dehydrating agent (*f*) Oxidising agent (*g*) An acid (*h*) Oxidising agent.

Q.5 Why is sulphurous acid a reducing agent (RA)?

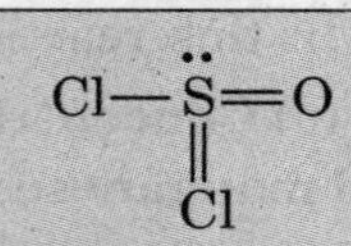

Structure of $SOCl_2$ molecule

Ans : The structure of sulphurous acid molecule (H_2SO_3) is given in the margin in a box. This structure shows that the central S-atom has a *lp* of electrons. It is due to the presence of this *lp* of electrons that H_2SO_3 can easily be oxidised to H_2SO_4. Consequently H_2SO_3 can act as a RA.

$$\underset{(S = +4)}{H_2SO_3} + H_2O + \underset{(Br = 0)}{Br_2} \longrightarrow \underset{(Br = -1)}{2HBr} + \underset{(S = +6)}{H_2SO_4}$$

Q.6 How you will obtain the following from sulphuric acid ?

(*i*) SO_2 (*ii*) SO_3 (*iii*) SO_2Cl_2 (*iv*) H_2S (*v*) S.

Ans : (*i*) (*a*) SO_2 is obtained by heating copper with conc. H_2SO_4.

$$Cu + 2H_2SO_4 \longrightarrow CuSO_4 + SO_2 + 2H_2O$$

(*b*) It can also the obtained by boiling sulphur with conc. H_2SO_4.

$$S + 2H_2SO_4 \longrightarrow 3SO_2 + 2H_2O$$

(*ii*) H_2SO_4 when treated with P_2O_5 loses water and forms SO_3.

$$H_2SO_4 \xrightarrow{P_2O_5} SO_3 + H_2O$$

$$P_2O_5 + H_2O \longrightarrow 2HPO_3$$

$$H_2SO_4 + P_2O_5 \longrightarrow 2HPO_3 + SO_3$$

(*iii*) SO_2Cl_2 is formed when conc. H_2SO_4 is treated with excess of PCl_5.

$$H_2SO_4 + 2PCl_5 \rightarrow SO_2Cl_2 + 2POCl_3 + 2HCl$$

(*iv*) and (*v*) H_2S and S both can be obtained by oxidising HI with H_2SO_4 as shown below :

(*a*) $H_2SO_4 + 8HI \longrightarrow H_2S + 4H_2O + 4I_2$

or $SO_4^{2-} + 10H^+ + 8I^- \longrightarrow H_2S + 4H_2O + 4I_2$

(*b*) $H_2SO_4 + 6HI \longrightarrow S + 4H_2O + 3I_2$

or $SO_4^{2-} + 8H^+ + 6I^- \longrightarrow S + 4H_2O + 3I_2$

Q.7 Starting from elemental suphur how will you prepare H_2SO_4 ?

Ans :

$$S + O_2 \longrightarrow SO_2$$

$$SO_2 + O_2 \xrightarrow[\text{2 bar pressure}]{V_2O_5,\ 720\ K} 2SO_3$$

$$SO_3 + H_2SO_4 \longrightarrow H_2S_2O_7$$

$$H_2S_2O_7 + H_2O \longrightarrow 2H_2SO_4$$

Q.8 From the reactions given below, identity (A), (B), (C) and (D) and write their formulae.

$$\textbf{(A) + dil. } H_2SO_4 + K_2Cr_2O_7 \longrightarrow \textbf{(B) (Green coloured)}$$

$$\textbf{(A) + dil. } H_2SO_4 + \textbf{(C)} \longrightarrow MnSO_4$$

$$\textbf{(A)} + O_2 \xrightarrow{H_2O} \textbf{(D)}$$

$$\textbf{(D)} + BaCl_2 \longrightarrow \textbf{White ppt.}$$

(Roorkee 1987)

Ans : (A) = Na_2SO_3; (B) = $Cr_2(SO_4)_3$; (C) = $KMnO_4$; (D) = Na_2SO_4

Q.9 Concentrated sulphuric acid is added to each of the following test tubes labelled (*i*) to (*v*).

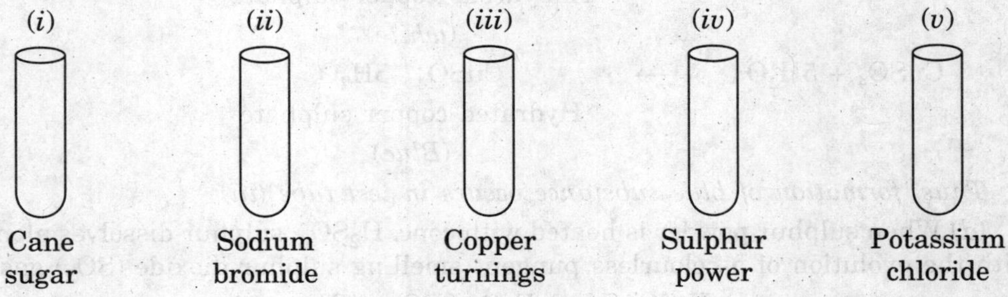

Identify in which of the above test tubes, the following change will be observed. Support your answer with the help of a chemical change.

(*a*) formation of black substance

(*b*) evolution of brown gas

(*c*) formation of brown substance which on dilution becomes blue

(*d*) disappearance of yellow powder along with the evolution of a colourless gas

(*e*) evolution of colourless gas.

Ans : (*a*) When conc. H_2SO_4 reacts with cane sugar, water is removed from cane sugar and sugar charcoal (black substance) is obtained. In this reaction conc. H_2SO_4 acts as a dehydrating a gent.

$$\underset{\text{Cane sugar}}{C_{12}H_{22}O_{11}} \xrightarrow{\text{Conc.}H_2SO_4} \underset{\substack{\text{Sugar charcoal} \\ (\textit{black substance})}}{12C} + \underset{\text{Water}}{11H_2O}$$

Thus we see that in the reaction occuring in test tube (i) the formation of a black substance (i.e. sugar charcoal) takes place.

(*b*) When sodium bromide is heated with conc. H_2SO_4, a brown gas (Br_2) is obtained. In fact, first HBr is produced. Since HBr is a moderate reducing agent, it reduces H_2SO_4 to SO_2 and is itself oxidised to bromine gas (Br_2). Conversely, H_2SO_4 is a moderate oxidising agent, it oxidises HBr to Br_2

$$2NaBr + H_2SO_4 \longrightarrow Na_2SO_4 + 2HBr$$

$$2HBr + H_2SO_4 \longrightarrow Br_2 + SO_2 + 2H_2O$$

$$\underset{\text{Sod. bromide}}{2NaBr} + 2H_2SO_4\,(\textit{conc.}) \xrightarrow{\text{Heat}} Na_2SO_4 + \underset{\substack{\text{Bromine} \\ (\textit{Brown gas})}}{Br_2} + \underset{\substack{\text{Sulphur} \\ \text{dioxide}}}{SO_2} + 2H_2O$$

Thus, evolution of brown gas occurs in test tube (ii)

(*c*) When copper turnings are heated with conc. H_2SO_4, first a brownish substance (presumably a mixture of unreacted copper turnings and colourless anhydrous copper sulphate) is formed along with evolution of pungent smelling sulphur dioxide (SO_2) gas. The initially formed brown substance on dilution turns blue due to the formation of hydrated copper sulphate while copper turnings remain undissolved.

$$Cu + 2H_2SO_4 \xrightarrow{\text{Heat}} \underset{\substack{\text{Anhydrous copper sulphate} \\ (\textit{white})}}{CuSO_4} + SO_2 + 2H_2O$$

$$CuSO_4 + 5H_2O \longrightarrow \underset{\substack{\text{Hydrated copper sulphate} \\ (\textit{Blue})}}{CuSO_4 \cdot 5H_2O}$$

Thus, formation of blue substance occurs in test tube (iii)

(*d*) When sulphur powder is heated with conc. H_2SO_4, sulphur dissolves along with the evolution of a colourless pungent smelling sulphur dioxide (SO_2) gas.

$$H_2SO_4 \longrightarrow H_2O + SO_2 + O] \times 16$$

$$S_8 + 16O \longrightarrow 8SO_2$$

$$\underset{\substack{\text{Sulphur} \\ (\textit{Yellow powder})}}{S_8} + 16\,H_2SO_4 \longrightarrow \underset{\substack{\text{Sulphur dioxide} \\ (\textit{Colourless pungent} \\ \textit{smellling gas})}}{24SO_2} + 16\,H_2O$$

Thus, colourless pungent smelling sulphur dioxide gas is evolved is test tube (iv)

(*e*) When potassium chloride is heated with conc. H_2SO_4, a colourless gas hydrogen chloride (HCl) is produced.

$$\underset{\text{Pot. chloride}}{2KCl} + H_2SO_4\,(conc.) \xrightarrow{\text{Heat}} \underset{\substack{\text{Hydrogen chloride} \\ (\textit{Colourless gas})}}{2HCl} + K_2SO_4$$

It may be noted here that unlike HBr, HCl is not a reducing agent. Therefore, it does not reduce H_2SO_4 to SO_2. Conversely, H_2SO_4 is a moderate oxidising agent. Therefore it does not oxidise HCl to Cl_2.

Thus, evolution of colourless gas occurs in test tube (v)

[C] Questions on Halides and Oxy halides of Sulphur.

Q.1 Explain the following, giving appropriate reasons for your answer.

(*i*) SF_6 is known but SCl_6 is not known.

Ans : Fluorine is a much stronger oxidising agent than chlorine, therefore, it can easily oxidise sulphur to its maximum oxidation state of + 6 and hence forms SF_6. Chlorine, on the other hand, being a weaker oxidising agent can oxidise sulphur at the maximum to its + 4 oxidation state and hence can form SCl_4 but not SCl_6.

$$\underset{(S = 0)}{S} + \underset{(F = 0)}{3F_2} \longrightarrow \underset{(S = +6,\ F = -1)}{SF_6} \quad (\textit{Possible reaction})$$

$$S + 3Cl_2 \longrightarrow SCl_6 \quad (\textit{Not possible})$$

$$\underset{(S = 0)}{S} + \underset{(Cl = 0)}{2Cl_2} \longrightarrow \underset{(S = +4,\ Cl = -1)}{SCl_4} \quad (\textit{Possible reaction})$$

Since Cl^- ion is larger in size than F^- ion, S-atom cannot accomdate large-sized Cl^- ions round it and hence cannot form SCl_6 molecule.

(*ii*) SF_6 is known but SH_6 is not known.

Ans : This can be explained in any of the following two ways :

(*i*) We know that the valence-shell of S-atom ($S = 3s^2\ 3p^2\ 3p^1\ 3p^1$) has only two unpaired electrons ($n = 2$). Now for the formation of SF_6 and SH_6 molecules paired electrons present in 3*s* and one of the three 3*p* orbitals must be made unpaired by promoting them in the vacant 3*d* orbitals so that we get six unpaired electrons. Since F-atom is highly electronegative (F = 4.0), it can easily cause the promotion of electrons of S-atom from the filled 3*s* and 3*p* orbitals to the vacant 3*d* orbitals and thus six unpaired electrons ($n = 6$) are obtained. These six unpaired electrons are used in forming SF_6 molecule.

However, H-atom is less electronegative, even lesser than S-atom (H = 2.20, S = 2.44), it cannot cause the promotion of paired electrons to occupy 3*d* orbitals. Thus S-atom has only two unpaired electrons ($n = 2$) and hence can form only SH_2 and not SH_6.

(*ii*) The formation of SF_6 and SH_6 molecules can be represented by the following chemical equations.

(*a*) $$\underset{(S = 0)}{S} + \underset{(F = 0)}{3F_2} \xrightarrow{\text{Oxidation}} \underset{(S = +6,\ F = -1)}{SF_6} \ (\textit{Possible reaction})$$

(*b*) $$\underset{(S = 0)}{S} + \underset{(H = 0)}{3H_2} \xrightarrow{\text{Reduction}} \underset{(S = -6,\ H = +1)}{SH_6} \ (\textit{Not possible})$$

In reaction (*a*), S (S = 0) is oxidised to SF_6(S = + 6) by F_2 which acts as an oxidising agent. Now since F_2 is the strongest oxidising agent, it can oxidise S to SF_6 quite readily, *i.e.*, SF_6 is fromed quite easily.

Now in reaction (*b*), S is being reduced to SH_6 by H_2 which acts as a reducing agent. Since H_2 is a poor reducing agent, it cannot reduce S to SH_6, *i.e.* SH_6 will not be formed.

(*iii*) SF_6 is known but OF_6 is unknown.

Ans : The formation of OF_6 molecule requires the presence of six unpaired electrons in the valence-shell of O-atom. Due to the absence of *d*-orbitals in the valence-shell of O-atom, electrons from the completely-filled 2*s* and $2p_x$ orbitals ($O = 2s^2\ 2p^2{}_x\ 2p^1{}_y\ 2p^1{}_z$) cannot be promoted to get six unpaired electrons. Thus OF_6 molecule cannot be formed.

The formation of SF_6 molecule also requires six unpaired electrons. In this case, since S-atom has vacant 3*d* orbitals in its valence-shell, paired electrons present in 3*s* and $3p_x$ orbitals ($S = 3s^2\ 3p^2{}_x\ 3p^1{}_y\ 3p^1{}_z\ 3d^\circ$) can be promoted to the vacant 3*d* orbitals to get six unpaired electrons in the valence-shell. These six unpaired electrons can be used to form SF_6 molecule.

The whole discussion can be summarised by saying that OF_6 is not known due to the absence of *d*-orbitals in the valence-shell of O-atom while SF_6 is known because of the presence of *d*-orbitals (3*d*-orbitals) in the valence-shell of S-atom.

(*iv*) SF_6 is used as a gaseous electrical insulator.

Ans : SF_6 is a colourless, odourless and non-toxic gas at room temperature. It is thermally stable and chemically inert. Because of its interness and high tendency internal discharges, it is used as a gaseous electrical insulator in high voltage generators and switch gears.

(*v*) $SOCl_2$ can act as a weak acid as well as weak base.

Ans : The structure of $SOCl_2$ molecule is given in the margin. This structure shows that S-atom (central atom) has a *lp* of electrons on it. It is due to the presence of this *lp* of electrons that $SOCl_2$ acts as a base. Further, since S-atom has empty *d*-orbitals, it can accept electrons donoted by Lewis base, $SOCl_2$ also act as a Lewis acid.

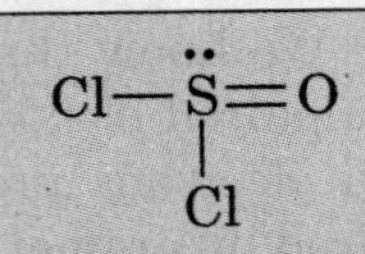

Structure of $SOCl_2$ molecule

(*vi*) SF_6 is not easily hydrolysed though thermodynamically it should be but SF_4 can undergo hydrolysis.

Ans : In SF_6, S is sterically protected by six F atoms and hence does not allow H_2O molecules to attack the S-atom. Furthermore, F does not have *d*-orbitals to accept the electrons donated by H_2O molecules. As a result of these two reasons, SF_6 does not undergo hydrolysis. In contrast, in SF_4 S is not sterically protected, since it is surrounded by only four F atoms. As a result, attack of H_2O molecules can take place easily and hence hydrolysis occurs.

Q.2 Starting from elemental sulphur how would you prepare (*a*) SCl_2 and (*b*) SF_6 ?

Ans : Heating S and Cl_2 gives S_2Cl_2 which upon saturation with Cl_2 gives SCl_2. SF_6 is obtained by the direct combination of $S_8(g)$ and $F_2(g)$.

(*a*) $S_8(l) + 4Cl_2(g) \xrightarrow{\text{Heat}} 4S_2Cl_2(l)$

$S_2Cl_2(l) + Cl_2(g) \longrightarrow 2SCl_2(l)$

(*b*) $S_8(g) + 24F_2(g) \longrightarrow 8SF_6(g)$.

3. Among the hexafluorides SF_6, SeF_6 and TeF_6 which fluoride is not hydrolysed.

Ans : SF_6 is not hydrolysed.

University Questions

1. Write a note on the structure of oxides of sulphur. (*Delhi 86; GND 84; Meerut 86; Calcutta 86*)

2. Write a note on the structure of sulphurous acid molecule. (*Calcutta 86*)

3. Explain the structure of $H_2S_2O_4$ molecule.

4. How ii sod. hyposulphite ($Na_2S_2O_4$) prepared? What are its uses? (*Bharthiar 85, Madras 86*)

5. (*a*) Name the basic raw materials required for H_2SO_4 manufacture.

(*b*) How is dilute and concentrated H_2SO_4 stored?

(*c*) Why is 97% H_2SO_4 used for the absorption of SO_2 in contact process.

(*d*) How does SO_2 act as a food preservative? (*Delhi (I.C.) 87*)

6. Compare the structures of SO_3^{2-} and SO_4^{2-} ions. (*Delhi B.Sc. (Hons.) 85, 87*)

7. Write a note on "sodium thiosulphate". (*Jiwaji 84*)

8. Write a note on "Thionic acids". Give their structures. (*Utkal 86; Calicut 85; Andhra 85*)

9. Give an account of the preparation, properties, uses and structure of peroxyacids of sulphur. (*Allahabad 87, 89; Delhi 89; GND 84; Calcutta 86*)

10. Give methods of preparation, properties and uses of Marshall's perdisulphuric acid. (*Kanpur 2000*)

11. Give preparation, properties and applications of sodium thiosulphate. (*Lucknow 2000*)

12. Describe the preparation, properties and constitution of Care's acid and sodium thiosulphate. (*Avadh 2000*)

13. How does $Na_2S_2O_3.5H_2O$ react with I_2 solution and $AgNO_3$ solution? (*Bharahias 2000*)

14. Give the preparation, properties and uses of sodium thiosulphate. Give its structure. Why is it called hypo ? (*Meerut 2009*)

CHAPTER

37

General Characteristics of Group 17 (VII A) Elements : Halogens

Group VII A of the periodic table consists of five elements *viz. Fluorine* (F), *Chlorine* (Cl), *Bromine* (Br), *Iodine* (I) and *Astatine* (At). The elements F, Cl, Br and I are collectively called *Halogens.* The term *halogen* is derived from the Greek *halos* (= salt) and *genes* (= born) meaning salt producers because most of these elements exist in sea water, notably in the form of their sodium compounds. Astatine was discovered in 1940 and is an unstable element of radioactive origin. The other four elements are stable and resemble each other in physical and chemical properties.

Position of Halogens in the Periodic Table

The inclusion of halogens in the same sub-group VII A is justified by the following points :

I. Electronic Configurations.

The electronic configurations of the atoms of halogens are given as : F(Z = 9) → 2, 7 ($2s^2p^5$) ; Cl (Z = 17) → 2, 8, 7 ($3s^2p^5$); Br (Z = 35) → 2, 8, 18, 7 ($4s^2p^5$); I (Z = 53) → 2, 8, 18, 18, 7 ($5s^2p^5$); At (Z = 85) → 2, 8, 18, 32, 18, 7 ($6s^2p^5$). These configurations show that all the halogens have *seven* electrons (ns^2p^5) in their outer most shell. Two electrons are in *ns*-orbital and the remaining five electrons are in *np*-orbitals and are distributed as $np^2{}_x$, $np^2{}_y$, $np^1{}_z$ in accordance with Hunde's rule. The shell previous to the outermost shell contains in F the group s^2 (*saturated*), in Cl s^2p^6 (*saturated*), in Br $s^2p^6d^{10}$ (*saturated*) while in I and At $s^2p^6d^{10}$ (*unsaturated*). This shows why F differs from Cl and these two from the remaining elements. Since the additional electron in these elements enters the *p*-orbital of the outermost shell, these are called *p-block elements* of the long form of the periodic table.

II. Similarities in Physical Properties and Gradation in Them.

Some physical properties are discussed below :

1. Occurrence. These elements do not occur in the free state. But in the combined state they occur fairly abundantly in nature.

2. Diatomic Nature. Heat capacity and combining volume data have shown that all the halogens exist as covalent diatomic molecules (X_2 molecule) in the gaseous, liquid and crystalline states. These discrete X_2 molecules are held together by weak van der. Waals forces which explain the volatile nature of these elements. At high temperatures, X_2 molecules dissociate into X atoms. The stablility of diatomic molecules towards heat decreases from F_2 to I_2. No dissociation occurs in case of F_2 gas. Cl_2 is slightly dissociated above 1100°C, Br_2 about 6% while I_2 quite extensively above 1100°C.

3. Heat of Dissociation of X_2 Molecules. Heat of dissociation (ΔH_d) of X_2 molecule is the amount of heat required to dissociate X_2 molecule in the gaseous state into X gaseous atoms or to break X—X bond in X_2 molecule. Thus ΔH_d can be defined by the equation :

$$X_2\ (g) + \Delta H_d \longrightarrow 2\ X\ (g)$$

Obviously ΔH_d is equal to X—X bond energy. Excepting for F_2, ΔH_d values decrease from Cl_2 to I_2. The lower value of heat of dissociation of F_2 molecule as compared to Cl_2 molecule is due to the reasons namely : (*i*) Due to very small F—F bond length in F_2 molecule, repulsion between the non-bonding electrons is very large as compared to other X—X bond lengths (X = C1, Br, I). This makes the F atoms in F_2 molecule repel each other and helps the dissociation of F_2 molecule into F atoms. (*ii*) X—X bond in Cl_2, Br_2 and I_2 molecules is stronger than F—F bond in F_2 molecule. This is due to the possibility of the existence of multiple bonding in X—X bond involving *d*-orbitals.

4. Oxidising Property. A substance that has a tendency to accept one or more electrons is said to show oxidising property. The halogens, due to high electron affinity values, have a great tendency to accept an electron and hence act as strong oxidising agent. The oxidising property of a halogen molecule, X_2 is represented by :

$$\tfrac{1}{2}X_2\ (s, l \text{ or } g) + e^- + aq \longrightarrow X^-\ (aq) + \text{Energy released (E)}$$

The above reaction is accompanied by the release of energy (E) which is made up of many energy terms like heat of fusion, vaporisation etc. It has been seen that since the values of E are decreasing from F_2 to I_2, the oxidising power of halogens is also decreasing in the same direction, *i.e.* the oxidising power of halogens is in the order : F_2 (strongest oxidising agent) > Cl_2 > Br_2 > I_2 (weakest oxidising agent). Since F_2 is the strongest oxidising agent in the series, it will *oxidise* other halide ions to halogens in solution or when dry, *i.e.* F_2 displaces other halogens from their corresponding halides. For example :

$$F_2 + 2X^- \longrightarrow 2F^- + X_2$$
$$(X^- = Cl^-, Br^-, I^-)$$

Similarly Cl_2 will displace Br^- and I^- ions from their solutions and Br_2 will displace I^- ions from their solutions.

$$Cl_2 + 2X^- \longrightarrow 2Cl^- + X_2$$
$$(X^- = Br^-, I^-)$$

$$Br_2 + 2I^- \longrightarrow 2Br^- + I_2$$

5. Heat of Hydration (ΔH_{hyd}) of X^- Ions. Heat of hydration of X^- ion (ΔH_{hyd}) is the amount of energy liberated in the process of converting $X^-(g)$ ion into hydrated $X^-(aq)$ ion.

$$X^-(g) + aq \longrightarrow X^-(aq) + \Delta H_{hyd}$$

Since ΔH_{hyd} represents the heat liberated, its value is shown by putting negative sign before it. It has been found that the value of ΔH_{hyd} becomes less negative as the size of X^- ion increases from F^- ion to I^- ion. It is because of the fact that the electrostatic interaction of a large X^- ion with water dipole becomes weaker than a smaller X^- ion and hence the magnitude of ΔH_{hyd} goes on decreasing.

6. Formation of Clathrates. Water solutions of the halogens on cooling give the crystals of their clathrates of $X_2 . xH_2O$ type. Examples are : $Cl_2 . 7.3\ H_2O$, $Br_2 .10H_2O$, $I_2 .xH_2O$ etc.

7. Eectronegativity. Halogens have large values of electronegativity. These values decrease as we proceed from F to I in the group. Large electronegativity values of halogen atoms indicate that X atoms have a strong tendency to form X^- ions.

8. Physical State and Density. There is a change of state from gas to solid and hence density increases as we move from F to I.

9. Atomic and Ionic Radii. These parameters increase as we descend the group form F to I.

10. Melting and Boiling Points. The melting and boiling points of the halogens regularly increase form F to I. This indicates that the attractive forces (van der Waals forces) between molecules become progressively more prominent as the molecules increase in size and mass. This effect is also reflected in the physical state of halogens. F and Cl are gases at ordinary temperatures, Br is a heavy liquid while I is a solid.

11. Colour. All the halogens are coloured. Their colour is due to the fact that their molecules absorb light in the visible region by means of which the outer electrons are excited to higher energy levels. The amount of energy required for the excitation (*i.e. excitation energy*) decreases progressively form fluorine to iodine as the size of the atom increases. Thus there is a progressive depeening of colour of the halogens from fluorine to iodine. Fluorine absorbs violet light (higher excitation energy) and hence appears *yellow*, while iodine vapour absorbs yellow light (lower energy) and appears *deep violet.* Similarly we can account for *greenish yellow* colour of chlorine and *reddish brown* colour of bromine.

12. Odour. All the halogens have very pungent and unpleasant odours. If inhaled even in very small quantities, they cause inflammation of the mucous membranes of the throat and nose and serious irritation of respiratory duets. Large quantities may have fatal effect.

13. Solubility. The halogens are slightly soluble in water and their solubility decreases from Cl to I. F cannot be dissolved in H_2O, since it decomposes H_2O liberating O_2 and some O_3. The halogens dissolve much better than H_2O in many organic solvents like carbon disulphide, chloroform, ether etc. Their higher solubility in organic solvents has been utilised to extract them from aqueous solutions. In organic solvents, iodine gives violet colour, bromine yields brown colour and chlorine produces yellow colour. It is assumed that in organic solvents halogens exist as free molecule as in the gas phase.

In nucleophilic (electron donating) polar solvent like alcohols, ketones, liquid SO_2, the halogens give coloured solutions whose colours are due to the formation of solvent $\rightarrow$ halogen type complexes. These complexes are charge-transfer compounds. In such compounds an electron located mainly in an orbital of the solvent molecule jumps into orbital located mainly on the halogen and absorbs energy from the light, causing the compound to look coloured. Such compounds have been isolated at low temperatures.

14. Ionisation Potentials. The ionisation potentials of halogens are very high which indicate that the halogens have a little tendency to lose an electron to form X^+ cation. These values progressively decrease from F to I, as the atomic size of these elements increases in the same order. With the decrease of ionisation potential from F to I, the tendency of these elements to lose an electron to form X^+ cation increase, *i.e.* F is not able to give F^+ cation while I forms I^+ cation easily. In other words, due to minimum value of ionisation potential of iodine, this elements shows *basic (non-metallic) character*.

15. Electron Affinity. Electron affinity values decrease from Cl to I. Why the electron affinity value of F is less than that of Cl has already been explained.

16. Oxidation States. Since these elements have ns^2p^5 configuration in their outermost shell, each halogen atom in its ground state can complete its octet either by accepting an electron from an element which is less electronegtive than it (*e.g.* alkali metals) to form a halide ion X^- or by sharing its unpaired electron in *p*-orbital with another element to form a covalent bond. When a halogen atom *in its ground state combines* with an element of lesser electronegativity than it, it shows –1 oxidation state. On the other hand when it combines with an element having greater electronegativity it exhibits +1 oxidation state. Since F is the most electronegative element, it always shows –1 oxidation state only. It does not exhibit any positive oxidation state. Cl, Br and I have vacant *d*-orbitals in their valence-shell and, therefore, can have excited states (*i.e.* CI, Br and I can promote their *s*-and *p*-electrons to the vacant *d*-orbitals), which make it possible for these elements to exhibit higher positive oxidation states of + 3, + 5 and + 7. Higher oxidation states of I equal to + 3, + 5 and + 7 have already been explained.

Since F atom has no *d*-orbital in its valence-shell, it cannot have any excited state and consequently cannot exhibit any of the higher oxidation states. In case of bromine the penultimate shell *viz.* $3s^2p^6d^{10}$ is weakly screened and hence the energy required to promote an electron from 4*s*-orbital of the atom in the second excited state to 4*d*-orbital is markedly higher than that required in chlorine. Thus Br does not exhibit + 7 oxidation state.

Since all the halogens, excepting fluorine, are less electronegative than oxygen, they show positive oxidition states in their oxygen compounds *viz.* oxides and oxy acids, *e.g.*, Cl in ClO_2 and Br in BrO_3 show + 4 and + 6 oxidation states and similary Cl in HClO, $HClO_2$, $HClO_3$ and $HClO_4$ show + 1, + 3, + 5 and + 7 oxidation states respectively. In interhalogen compounds the halogen with lower electronegativity (having higher atomic number or atomic radius) shows positive oxidation state, *e.g.* I in ICl and Cl in ClF show + 1 oxidation state each. Note that Cl in ICl shows –1 oxidation, state while in ClF it shows + 1 oxidation state.

III. Similarities in Chemical Properties and Gradation in Them.

Some chemical properties are discussed below :

1. Reactivity. All the halogens are the most reactive elements as a family. The main reasons for their extreme reactivity are : (*a*) Dissociation energies in (KJ/mole) of X_2 molecules and other diatomic molecules like N_2, O_2 and H_2 are : N_2 = 941.62, O_2 = 495.08, H_2 = 431.89, F_2 = 157.77, Cl_2 = 237.38, Br_2 = 189.99, I_2 = 148.56. These values show that X_2 molecules have *low values* and hence it is much easier for these molecules to dissociate into X atoms and react with other substances more readily as compared to other diatomic molecules like N_2, O_2 and H_2. (*b*) Because of *high electron affinity values* of halogen atoms, halogens have a strong tendency to add on electron to form X^- ion.

The reactivity of halogen atoms decreases as we move down the group from F to I. For example fluorine decomposes H_2O vigorously even in dark, chlorine decomposes H_2O in sunlight, bromine also decomposes H_2O in sunlight but very slowly while iodine does not decompose water at all.

$$2F_2 + 2H_2O \longrightarrow 4HF + O_2; \quad Cl_2 + H_2O \longrightarrow HClO + HCl$$

$$Br_2 + H_2O \longrightarrow HBrO + HBr; \quad I_2 + H_2O \longrightarrow \text{No reaction}$$

2. Formation of Hydracids. All the elements of this group combine with hydrogen and form hydrogen halides (HX) which are called hydracids or halogen acids. The affinity of these elements for hydrogen decreases from fluorine to iodine. Thus F_2 combines with H_2 explosively and even in the dark, Cl_2 reacts with H_2 in diffused sunlight, Br_2 reacts with H_2 only on heating above 500°C while I_2 reacts with H_2 only on heating and in presence of Pt as a catalyst. This clearly shows that the tendency of the halogens to react with H_2 decreases from F_2 to I_2. The reaction of I_2 with H_2 is reversible and endothermic. Hydracids can be prepared by the direct combination of the elements, by heating the metal halides with conc. H_2SO_4, by the hydrolysis of non-metallic halides etc. Some important properties of hydracids are given below :

(i) Acidic properties. All the halogen acids in the gaseous state are essentially covalent but in the aqueous solution they ionise to give the solvated proton (H_3O^+) and hence act as acids.

$$H_2O + HX \longrightarrow H_3O^+ + X^-$$

HF ionises only slightly while HCl, HBr and HI ionise completely. Hence HF is the weakest acid and the strength of these acids increases from HF to HI, *i.e.* HF (weakest acid) < HCl < HBr < HI (strongest acid). The weakest acidic nature of HF is due to the fact that the dissociation energy of H—F bond in HF molecule is the highest (= 564.97 $KJ.mol^{-1}$) and hence this molecule has least tendency to split up into H^+ and F^- ions in aqueous solution. Another explanation of the above order of the acidic strength of HX acids can be given by finding out the relative order of the basicity of the conjugate bases *viz.* F^-, Cl^-, Br^- and I^- of these acids. This explanation has already been given. The hydrides show no acidic character when perfectly dry.

(ii) Covalent character. The bond between hydrogen and the halogen in these hydrides is *largely covalent* as is indicated by their low melting and boiling points and also by the fact that the pure liquid hydrogen halides are poor conductors of electricity. The bonds do, however, have small amount of polar character which decreases with the decrease of electronegativity of the halogen atom. Thus the polar character of H—X bond in HX molecules decreases in the order : H—F > H—Cl > H—Br > H—I.

(iii) Physical properties. Anhydrous HF is a liquid at ordinary temperature whereas other HX (X = Cl, Br, I) are colourless gases. These have sharp choking odour. Excepting HF, the melting and boiling points of these hydrides increase with the increase of the atomic weight of the halogen atom. The abnormally high melting point of HF is explained by saying that, due to H-bonding, HF molecule gets associated to form $(HF)_6$ which is an aggrevated molecule.

The values of heat of formation increase with the increase of atomic weight of halogen atom. Thus the quantity of energy released during the formation of HX molecules decreases from HF to HI.

(iv) Stability and reducing property. As we go from HF to HI, the stability of these acids decreases from HF to HI. (HF > HCl > HBr > HI). Thus HF and HCl are stable upto 1200°C while HBr dissociates slightly (1%) and HI dissociates at 440°C to 20%.

The decrease in the stability of HX molecules from HF to HI implies that the reducing properties of these acids increase from HF to HI, *i.e.* HF < HCl < HBr < HI.

The above order shows that HF has no appreciable reducing properties while HI is the strongest reducing agent. Even dilute solution of HI, on exposure to the air, slowly acquires a brown colour due to the liberation of I_2.

$$4HI + O_2 \text{ (From atmosphere)} \longrightarrow 2H_2O + 2I_2$$

3. Formation of Binary Halogen-oxygen Compounds. Halogens do not combine directly with O_2. However, binary halogen-oxygen compounds can be prepared by indirect methods. Examples of such compounds are : OF_2, O_2F_2, Cl_2O, ClO_2, Br_2O, BrO_2, I_2O_4, I_2O_5 etc. F being more electronegative than oxygen, OF_2 and O_2F_2 are called fluorides of oxygen and not oxides of fluorine.

4. Formation of Oxy-acids. Excepting F, all other halogens form oxy-acids. These are mainly of four types *viz.* HXO (*e.g.* HClO, HBrO, HIO); HXO_2 (*e.g.* $HClO_2$); HXO_3 (*e.g.* $HClO_3$, $HBrO_3$, HIO_3) and HXO_4 (*e.g.* $HClO_4$, $HBrO_4$, HIO_4). Some of their properties are given below :

(i) Acidic character. (*a*) In the series of oxy-acids of the same halogen in different oxidation number the acidic character of such acids increases with the increase of the oxidation number of the halogen. Thus the acidic character of HClO, $HClO_2$, $HClO_3$, $HClO_4$ acids is in the order. HClO (+ 1) < $HClO_2$(+ 3) $HClO_3$(+ 5) < $HClO_4$(+ 7). Oxidation number of Cl atom in each acid is given in parentheses.

Explanation. The above order of acidic character of the acids is explained by finding out the relative order of the basicity of the conjugate bases (*viz.* ClO^-, ClO_2^-, ClO_3^- and ClO_4^-) of these acids. This has already been explained.

(*b*) The acidic character of the oxy-acids of different halogens in the same oxidation state decreases, in general, with decrease of the electronegativity of the halogen atom. Thus the acidic character of HClO, HBrO and HIO acids is in the order : HClO(Cl = 3.0) > HBrO (Br = 2.8) > HIO(I = 2.5). The electronegativity of halogen atoms is given in bracket.

Explanation. As the atomic number of the halogen atom increases, its electronegativity decreases and size increases; As a result of this, the tendency of these acids to lose a proton to water decreases. This makes the acid a weaker acid.

(ii) Oxidising power. As the oxidation number of the halogen atom increases, X—O bond becomes more covalent, resulting in a decrease of the oxidising power of the acid. For example under usual laboratory conditions, the oxidising action of HClO, $HClO_2$, $HClO_3$, and $HClO_4$ acids is in the order HClO(+ 1) > $HClO_2$(+ 3) > $HClO_3$(+ 5) > $HClO_4$(+ 7). In brackets are given the oxidation number of chlorine atom in the respective acid.

(iii) Thermal stabitity. With the increase of oxidation number of the halogen atom, X—O bond in the oxy-acids becomes more covalent and hence thermal stability of the acids increases. Thus the thermal stability of HClO, $HClO_2$, $HClO_3$ and $HClO_4$ acids is in the order : HClO(+ 1) < $HClO_2$(+ 3) < $HClO_3$(+ 5) < $HClO_4$(+ 7).

In the series of oxy-acids of different halogens in the same oxidation state, thermal stability of these acids decreases, in general, with the decrease of the electronegativity of the halogen atom. For example thermal stability of HXO acids (X = Cl, Br, I) is in the order : HClO(Cl = 3.0) > HBrO(Br = 2.8) > HIO(I = 2.5)

5. Formation of Interhalogen Compounds. Two different halogen atoms, due to the difference in their electronegativity values, combine with one another to form a number of covalent compounds which are called interhalogen compounds. These are of four types namely AB (*e.g.* ClF, BrF etc); AB_3 (*e.g.* ClF_3, BrF_3, ICl_3); AB_5(*e.g.* BrF_5, IF_5) and AB_7 (*e.g.* IF_7) where A and B are two different halogen atoms.

6. Formation of Polyhalides. In some cases halogens combine with simple metallic halides (*e.g.* KI, KBr, CsCl) and form polyhalides like $K^+[I_3]^-$, $K^+[Br_3]^-$, $Cs^+[ClF_4]^-$ etc.

Anomalous Behaviour of Fluorine : Dissimilarities of Fluorine with Other Halogens.

Like other elements such as C, N and O which are the first members of their respective groups, fluorine also differs from the other members of its own sub-group due to (*a*) *its small size* (*b*) *its high electronegativity and* (*c*) *non-availability of d-orbitals in its valence shell. The main differences* are : (*i*) Due to its maximum electronegativity, *it shows only a negative oxidation state of –1. It does not show any positive oxidation state.* The other members show negative as well as positive oxidation states of + 1, + 3, + 5 and + 7. (*ii*) Fluorine is the most reactive element among halogens. This is due to the minimum value of F—F bond dissociation energy. (*iii*) On account of high electronegativity of fluorine atom hydrogen bonding occurs in HF molecule while it is absent in HCl, HBr and HI molecules. The presence of hydrogen bonding in HF explains why (*a*) HCl, HBr and HI are gaseous while HF is a liquid with an abnormally high boiling point (*b*) HF forms a number of compounds containing HF_2^- ion (*e.g.* $K^+HF_2^-$) while such compounds (*e.g.* $K^+HCl_2^-$, $K^+HBr_2^-$ etc.) are not given by other HX molecules. (*iv*) HF ionises only slightly in aqueous solution and is, therefore a weak acid. HCl, HBr and HI, on the other hand, ionise almost completely and are thus strong acids. (*v*) Fluoride of silver is soluble while other halides of silver are insoluble. Similarly fluoride of calcium is insoluble and other halides of calcium are soluble. (*vi*) Fluorides are ionic while other halides have frequently *molecular lattices.* (*vii*) F ion has a great tendency to give complex ions like $[AlF_6]^{3-}$, $[FeF_6]^{3-}$ etc. The remaining halide ions show this tendency to a much lesser extent. (*viii*) Fluorine, being the most electronegative, gives SF_6 while other members do not form hexahalides with sulphur. (*ix*) Because of the absence of *d*-orbitals in its valence shell, fluorine does not combine with F ions to give polyhalide ion, F_3^- while other members of the group give such ions (*e.g.* Cl_3^-, Br_3^-, I_3^-, I_5^- etc.) because they contain *d*-orbitals.

Comparison of Fluorine and Oxygen.

Fluorine resembles oxygen which is the first member of the neighbouring group *viz.* VIA in some respects on account of the following factors : (*a*) *The atomic and ionic radii of both the elements are very similar (b) Both are highly electronegative.*

(A) Similarities. (*i*) Both of them exist in gaseous state at ordinary temperature (*ii*) Both react with metals (*e.g.* Na) and non- metals (*e.g.*, H, C, Si, S etc.) to give fluorides and oxides which resemble each other in solubilty, volatility, fusibility etc., *e.g.* HF and H_2O resemble each other in the following respects (*a*) Both possess associated molecules, $(H_2O)_x$ and $(HF)_x$ respectively due to hydrogen bonding (*b*) The boiling points of both are abnormally high (*c*) HF like H_2O is a weak acid and both are ionising solvents. (*d*) Heats of formation of HF and H_2O are close to each other. (*iii*) Certain elements like sulphur exhibit higher valencies in both fluorides and oxides, *e.g.* the valency of sulphur in SF_6 and SO_3 is 6.

(B) Dissimilarities. However these elements differ from each other in the following respects : (*i*) F is always monovalent while oxygen is bivalent (*ii*) F is more reactive than oxygen.

Basic Properties of Halogens

In general, the metallic (*i.e.* basic) character of the elements increases on descending a group while the same decreases on proceeding from left to right in a period. Thus the halogens should be expected to show the minimum basic character and among halogens themselves astatine (the last member of the halogen family) should be the most basic in character. Unfortunately no evidence is available which may prove the basic character of astatine. Iodine which is the next heaviest halogen shows maximum basic properties in halogen family elements. Note that *iodine is not a metal.* A progressive increase in the basic character from F to I is due to the factors *viz.* (*i*) Electronegativity *decreases* from F to I (*ii*) Electron affinity *decreases* from Cl to I (*iii*) Ionisation potential *decreases* from F to I.

Flourine is the most electronegative element and as such it does not show any basic character at all, *i.e.* it has no tendency to form positive ions in any of its known compounds. The nonexistence of the oxy-acids of F further confirms our belief that F is unable to bear a positive charge. There is, however, evidence for the existence of basic properties in Cl, Br and I. Chlorine has a slight tendency to form positive ions, *e.g.* Cl shows + 1 oxidation state in OCl^-, Cl_2O, ClF etc. species and exhibits + 3 oxidation state in ClO^-_2, ClF_3 etc. species, since O and F are more electronegative than Cl (O = 3.5, F = 4.0 and Cl = 3.0). Bromine exists as Br^+ in Br_2O and BrF (Br = 2.8) and as Br^{3+} in complexes containing $[BrF_2]^+$ ion such as $[BrF_2]^+ [Sb^{5+}F_6]^-$, $[BrF_2]^+ [Sn^{4+}F_6]^{2-}$. The compounds in which iodine functions as cationic (*i.e.* positive) constituent are numerous and well established. Some of the compounds containing *unipositive iodine*, I^+ are ICl, IBr, INO_3, I_2SO_4, $[Ipy_2]^+X^-$ ($X = NO_3^-$, ClO_4^-). $I(ClO_4)_3$, $I(CH_3COO)_3$, IPO_4, $I(NO_3)_3$, ICl_3 etc. are some compounds presumably containing *tripositive* iodine, I^{3+}. I_2O_4 or $[I^{3+}O]^+[I^{5+}O_3]^-$ which is called iodyl iodate contains both I^{3+} and I^{5+} cations.

Compounds Containing Cationic Iodine

Following are the important compounds containing unipositive iodine (I^+) or tripositive iodine (I^{3+}).

1. ICl and IBr interhalogen compounds. For the preparation and properties of these compounds see chapter 41.

2. Iodine nitrate (INO_3) and iodine sulphate (I_2SO_4). Solutions of these compounds are obtained by passing a solution of H_2SO_4 and of HNO_3 respectively in absolute alcohol through exchanger containing sorbed I^+ ion.

$$2I^+R^- + H_2SO_4 \longrightarrow I_2SO_4 + 2HR$$
$$I^+R^- + HNO_3 \longrightarrow INO_3 + HR$$

When the aqueous solutions of these compounds are electrolysed, iodine is liberated at cathode, proving that both of these salts contain unipositive iodine, I^+.

3. Complex compounds of $[Ipy_2]^+X^-$ type ($X = NO_3^-$, ClO_4^-). In these compounds I^+ ion gets stabilised by coordination. These compounds are obtained by treating Ag(I) salt with I_2 in chloroform in presence of excess of pyridine (py). For example :

$$AgNO_3 + I_2 + 2py \longrightarrow \underset{\text{ppt.}}{AgI} + [Ipy_2]NO_3$$

Complexes of this type have the following properties. We shall consider $[Ipy_2]NO_3$ as an example.

(*i*) *Reaction with acidified KI solution.* When $[Ipy_2]NO_3$ reacts with acidified KI solution, I_2 is liberated.

I^+ (From complex compound) + I^- (From KI) $\longrightarrow I_2$

I_2 liberated in the above reaction can be quantitatively estimated with sod. thiosulphate.

$$2Na_2S_2O_3 + I_2 \longrightarrow Na_2S_4O_6 + 2NaI$$

(*ii*) *Electrolysis.* Electrolysis of $[Ipy_2]NO_3$ in chloroform liberates iodine at the cathode.

$2I^-$ (From the complex compound) + $2e^- \longrightarrow I_2$ (*At cathode*)

These reactions show the presence of I^+ ion in the complex compound.

(*iii*) *Hydrolysis.* The complex compound is hydrolysed by water as follow :

$$5[Ipy_2]NO_3 + 3H_2O \longrightarrow 5HNO_3 + 10py + HIO_3 + 2I_2$$

(*iv*) *Reaction with NaOH.* Treatment of the complex compound with NaOH gives the base, $[Ipy_2]OH$ which is converted into its anhydride.

$$[Ipy_2]NO_3 + NaOH \longrightarrow \underset{\text{Base}}{[Ipy_2]^+(OH)^-} + NaNO_3$$

$$\underset{\text{Base}}{2[Ipy_2]OH} \longrightarrow \underset{\text{Anhydride of the base}}{Ipy{-}O{-}Ipy} + H_2O + 2py$$

(*v*) *Reaction with phenol (C_6H_5OH) gives tri-iodophenol.*

$$C_6H_5OH + 3[Ipy_2]NO_3 \longrightarrow \underset{\text{Tri-iodophenol}}{C_6H_2I_3OH} + 3HNO_3 + 6py$$

(*vi*) *Noble metals like Au get dissolved in chloroform solution of the complex compound.*

$$3[Ipy_2]NO_3 + Au \longrightarrow Au^{3+} + 3NO_3^- + \tfrac{1}{2}I_2 + 6py$$

In $[Ipy_2]\,NO_3$ and other similar compounds I^+ appears to be two-coordinate and hence the complex, $[Ipy_2]^+NO_3^-$ should be formulated as $[pyIpy]^+NO_3^-$.

4. Iodine perchlorate, $I(ClO_4)_3.2H_2O$. It is formed as yellowish-green needles by the action of O_3 on a cooled solution of I_2 in anhydrous $HClO_4$.

$$I_2 + 6HClO_4 + O_3 \longrightarrow 2I(ClO_4)_3 + 3H_2O$$

When the aqeous solution of this compound is electrolysed, iodine is liberated at cathode.

$$I(ClO_4)_3 \longrightarrow I^{3+} + 3ClO_4^-$$

$$2I^{3+} + 6e^- \longrightarrow I_2 \text{ (At cathode)}$$

5. Iodine triacetate, $I(CH_3COO)_3$. It is obtained by oxidising I_2 in acetic anhydride, $(CH_3.CO)_2O$ with fuming HNO_3.

It is a crystalline colourless substance, stable in cold but decomposes at 100°C. The saturated solution of the compound in acetic anhydride conducts electricity and during electrolysis I is liberated at the cathode (of silver), forming AgI.

$$(CH_3COO)_3\,I \rightleftharpoons I^{3+} + 3CH_3COO^-$$

$$I^{3+} + 3e^- + Ag \longrightarrow AgI \text{ (At silver cathode)}$$

6. Iodine trichloride, ICl_3. For the preparation, properties and uses of this compound see chapter 47.

Conceptual Questions with Answers

Q.1 Fluorine does not undergo disproportionation reactions but other halogens do. Explain.

Ans : F-atom shows only – 1 oxidation state. When this atom undergoes disproportionation reaction, its O.S. should be decreased from – 1 to some lower O.S. (reduction) and also should be increased from – 1 to some higher O.S. (oxidation). None of these processes is possible, since F-atom does not show any O.S. other than – 1. On the other hand, since other halogens show – 1, + 1,

+ 3, + 5 and + 7 oxidation states, the oxidation and reduction of these halogens both are possible. Hence other halogens can undergo disproportionation reactions.

Q.2 Explain the following, giving appropriate reasons for your answer.

***(i)* HF and HCl can be prepared by the action of conc. H_2SO_4 on CaF_2 and NaCl respectively but HBr and HI cannot be prepared by the action of conc. H_2SO_4 on NaBr and NaI (or KI) respectively.**

Ans : When CaF_2 and NaCl are heated separately with conc. H_2SO_4, HF and HCl are produced.

$$CaF_2 + H_2SO_4 \xrightarrow{\Delta} CaSO_4 + \boxed{2HF}$$

$$2NaCl + H_2SO_4 \xrightarrow{\Delta} Na_2SO_4 + \boxed{2HCl}$$

When NaBr and NaI are heated with conc. H_2SO_4, HBr and HI are first formed which, being moderate reducing agents (RA), reduce H_2SO_4 to SO_2 and are themselves oxidised to Br_2 and I_2 respectively.

$$2NaBr + H_2SO_4 \xrightarrow{\Delta} Na_2SO_4 + 2HBr$$

$$\underset{\substack{(Br = -1)\\(RA)}}{2HBr} + \underset{\substack{(S = +6)\\(OA)}}{H_2SO_4} \longrightarrow \underset{(S = +4)}{SO_2} + 2H_2O + \underset{(Br = 0)}{Br_2}$$

$$2NaBr + 2H_2SO_4 \longrightarrow Na_2SO_4 + SO_2 + 2H_2O + \boxed{Br_2}$$

or $$\underset{\substack{(Br = -1)\\(RA)}}{2Br^-} + \underset{(S = +6)}{SO_4^{2-}} + 4H^+ \longrightarrow \underset{(S = +4)}{SO_2} + 2H_2O + \underset{(Br = 0)}{Br_2}$$

Similary,

$$2NaI\ (or\ 2KI) + H_2SO_4 \xrightarrow{\Delta} Na_2SO_4 + 2HI$$

$$\underset{\substack{(I = -1)\\(RA)}}{2HI} + \underset{\substack{(S = +6)\\(OA)}}{H_2SO_4} \longrightarrow \underset{(S = +4)}{SO_2} + 2H_2O + \underset{(I = 0)}{I_2}$$

$$2NaI + 2H_2SO_4 \longrightarrow Na_2SO_4 + SO_2 + 2H_2O + \boxed{I_2}$$

or $$\underset{\substack{(I = -1)\\(RA)}}{2I^-} + \underset{(S = +6)}{SO_4^{2-}} + 4H^+ \longrightarrow \underset{(S = +4)}{SO_2} + 2H_2O + \underset{(I = 0)}{I_2}$$

***(ii)* In the preparation of HBr and HI from NaBr and KI respectively, phosphoric acid (H_3PO_4) is preferred to H_2SO_4.**

Ans : When NaBr and NaI are heated with conc. H_2SO_4, HBr and HI are produced.

$$2NaBr + H_2SO_4 \xrightarrow{\Delta} Na_2SO_4 + 2HBr$$

$$2NaI + H_2SO_4 \xrightarrow{\Delta} K_2SO_4 + 2HI$$

Now since H_2SO_4 is an OA and HBr/HI is RA, H_2SO_4 oxidises HBr and HI to Br_2 and I_2 respectively.

$$\underset{\text{OA}}{H_2SO_4} + \underset{\text{RA}}{2HBr} \longrightarrow SO_2 + 2H_2O + Br_2$$

$$\underset{\text{OA}}{H_2SO_4} + \underset{\text{RA}}{2HI} \longrightarrow SO_2 + 2H_2O + I_2$$

If we use H_3PO_4 in place of H_2SO_4, this acid also gives HBr and HI but, since this acid is non-oxidising, this acid does not oxidise HBr and HI as H_2SO_4 does.

$$3NaBr + H_3PO_4 \longrightarrow Na_3PO_4 + 3HBr$$

$$3KI + H_3PO_4 \longrightarrow K_3PO_4 + 3HI$$

$$HBr + H_3PO_4 \longrightarrow \text{No reaction}$$

$$HI + H_3PO_4 \longrightarrow \text{No reaction}$$

***(iii)* Among the tetrahalides of lead, only PbF_4 and $PbCl_4$ are quite stable while $PbBr_4$ and PbI_4 are unstable (unknown or donot exist).**

Ans : We know that the halide ions are reducing agents and their reducing power is in the order : $F^- < Cl^- < Br^- < I^-$.

On ionisation, PbI_4 gives Pb^{4+} and I^- ions. Now since I^- ion is a very strong RA, this ion reduces Pb^{4+} to Pb^{2+} and is itself oxidised to I_2.

$$PbI_4 \longrightarrow Pb^{4+} + 4I^-$$

$$Pb^{4+} + \underset{\text{(RA)}}{2I^-} \longrightarrow Pb^{2+} + I_2$$

$$PbI_4 \longrightarrow Pb^{2+} + 2I^- + I_2$$

or

$$\underset{\text{(unstable)}}{PbI_4} \longrightarrow PbI_2 + I_2$$

Thus we see that PbI_4 is reduced to PbI_2 and hence is unstable.

Similary, since Br^- ion is also a strong RA, Pb^{4+} ion gets reduced to Pb^{2+} and Br^- ion oxidised to Br_2.

$$PbBr_4 \longrightarrow Pb^{4+} + 4Br^-$$

$$Pb^{4+} + \underset{\text{(RA)}}{2Br^-} \longrightarrow Pb^{2+} + Br_2$$

$$PbBr_4 \longrightarrow Pb^{2+} + 2Br^- + Br_2$$

or

$$\underset{\text{(unstable)}}{PbBr_4} \longrightarrow PbBr_2 + Br_2$$

Thus $PbBr_4$ is also unstable.

On ionisation, PbF_4 and $FbCl_4$ both give Pb^{4+} and X^- ions ($X^- = F^-, Cl^-$). Now since these halide ions are very weak reducing agents, these ions are not able to reduce Pb^{4+} ions to Pb^{2+} ions and hence PbX_2 is not formed.

$$PbX_4 \longrightarrow Pb^{4+} + 4X^- \ (X^- = F^-, Cl^-)$$

$Pb^{4+} + 2X^- \longrightarrow Pb^{2+} + X_2$ (*This reaction does not take place*)

Thus PbF_4 and $PbCl_4$ are stable compounds.

Q.3 KHF_2 is well known while $KHCl_2$, $KHBr_2$ and KHI_2 donot exist. Explain with reasons.

Ans : The answer of this question can be given in the following two ways.

(*a*) KHF_2 can be regarded as having been obtained by the combination of KF and HF.

$$KF + HF \longrightarrow K^+[HF_2]^-$$

As a matter of fact the formation of $K^+[HF_2]^-$ takes place through the following two steps :

$$KF \longrightarrow K^+ + F^- \quad ...(i)$$
$$F^- + H-F \longrightarrow [F...H\text{–}F]^- \text{ or } [HF_2]^- \quad ...(ii)$$

On adding : $KF + HF \longrightarrow K^+[F...H\text{–}F]^- \text{ or } K^+[HF_2]^-$

Because of high electronegativity of F^- ion, the combination of this ion with HF molecule to form $[HF_2]^-$ ion as shown at step (*ii*) takes place through H-bonding. Since the electronegativity values of Cl^-, Br^- and I^- are comparatively smaller, these ions cannot combine with HCl, HBr and HI molecules through H-bonding and hence $[HCl_2]^-$, $[HBr_2]^-$ and $[HI_2]^-$ ions are not formed. In other words we can say that following reactions are not feasable.

$$\left.\begin{array}{l} KCl + HCl \longrightarrow K^+[HCl_2]^- \\ KBr + HBr \longrightarrow K^+[HBr_2]^- \\ KI + HI \longrightarrow K^+[HI_2]^- \end{array}\right\} \text{These reactions donot take place.}$$

(*b*) We know that hydrofluoric acid exists as a dimeric molecule (H_2F_2) due to H-bonding. Thus this acid is a dibasic acid and hence ionises in the following two ways :

$$H_2F_2 \longrightarrow H^+ + HF_2^-$$
$$HF_2^- \longrightarrow H^+ + 2F^-$$

Obviously H_2F_2 can form two types of salts like $K^+[HF_2]^-$ or $K^+[F...H\text{–}F]^-$ and K^+F^-. Because of the absence of H-bonding in HCl, HBr and HI molecules, these acids exists as monomeric molecules. Consequently these acids give only one type of salts like KCl, KBr and KI.

University Questions

1. Discuss the position of the elements of group VIIA in the periodic table. (*Agra 84*)
2. Compare the methods of preparation and properties of various halogen acids. (*Meerut 86*)
3. Arrange the hydrides of halogens with reasons, in the increasing order of heat of vaporisation, bond energy, ionic character dipole moment, reducing power, strength, heat of formation, acidic character, thermal stability and bond length. (*R.E.E. 84*)
4. How does fluorine differ from other halogens. (*Avadh 2000*)
5. Discuss the abnormal behaviour of fluorine. (*Meerut 2009*)

CHAPTER

38 Fluorine and Fluorocarbons

Fluorine

History of its Isolation

It was known to *Schwankhard* (1670) that glass is attacked when exposed to the fumes obtained by heating fluorspar with conc. H_2SO_4. In 1771 *Scheele* heated fluorspar with conc. H_2SO_4 in a tin retort and obtained an acid solution. By his experiment he concluded that fluorspar is the calcium salt of this acid solution which he named *fittoric acid. Gay Lasac and Thenard* in 1809 prepared this acid and showed that the acid resembles hydrochloric acid in many properties and on analogy they named it *hydrofluoric acid.* In 1813 *Day* showed that hydrofluoric acid was a compound of H_2 and an unknown element which he named *fluorine.* Davy tried to prepare fluorine by the electrolysis of aqueous solution of hydrofluoric acid just as Cl_2 is obtained of the electrolysis of aqueous solution of HCl. To his great surpr1se he obtained H_2 and ozone mixed with O_2, instead of fluorine, at the anode. For being most resistant to oxidation, H_2O gets oxidised at the anode liberating O_3 mixed with O_2.

$$2H_2O \longrightarrow 4H^+ + O_2 + 4e^-$$
$$3H_2O \longrightarrow 6H^+ + O_3 + 6e^-$$

Gore then attempted to electrolyse the anhydrous hydrofluoric acid to isolate fluorine from it, but he found this acid to be non-conductor and hence failed to isolate fluorine from it. In 1860 he showed that the anhydrous acid dissolves in KHF_2 and gives a conducting solution. But another problem was that of great chemical reactivity of fluorine. It attacked the apparatus whether made of glass, carbon and platinum. When platinum vessel was used, it gave a chocolate powder of PtF_4 while with carbon it gave the gaseous CF_4. It was in 1886 that the long standing problem was finally solved by *Moisson.* He prepared fluorine by electrolysing a solution of KHF_2 (1 part) in anhydrous hydrofluoric acid (5 parts) in a U-tube made of Pt—Ir alloy using electrodes of the same alloy. He reduced the great chemical reactivity of fluorine by keeping the bath and collecting the gas at very low temperature.

What Were the Difficulties Encountered in the Discovery and Isolation of Fluorine?

For a long time a number of difficulties were experienced in the discovery and isolation of fluorine. The main difficulties (or causes) for the late discovery and isolation of fluorine are (*i*) *Extreme reactivity of fluorine.* Fluorine is so reactive that it reacts with the material of all the vessels, *e.g.* F_2 attacks carbon and silicon present in the glass vessels with the formation of CF_4 and SiF_4

respectively. . Similarly F_2 also attacks the electrodes and vessels made of Pt with the formation of chocolate coloured powder, PtF_4. Pb, Fe and Sn vessels were also tried and found to be unsuitable for the preparation of fluorine. Thus no metal could be employed for the preparation of F_2 as vessel or as electrode. (*ii*) *Greater stability and non-conducting nature of hydrofluoric acid.* Hydrofluoric acid is very stable, highly poisonous and corrosive in nature. It is so stable that all attemps to isolate F_2 from hydrofluoric acid by oxidising agents failed. Moreover, electrolysis of aqueous solution of hydrofluoric acid gives H_2 and ozone mixed O_2 while anhydrous hydrofluoric acid is non-conductor.

Occurrence

Fluorine is very reactive and hence does not occur free in nature but is widely found in the combined state as fluorides. Its chief minerals are : *fluorspar*, CaF_2; *cryolite*, Na_3AlF_6 and *fluoraptite*, $CaF_2.3Ca_3(PO_4)_2$. Small quantities of fluorine are present as fluoride in plant-ash, soil, sea-water, river water plants, bones and teeth of animals. One litre of sea-water contains about 2 mg. of fluorine. Fluorine is also found in the phosphate *rocks. Fluorspar*, CaF_2 is the principal source of fluorine and its compounds. A six mile long belt of fluorspar has been recently discovered in Durgapur (Rajasthan) and it is considered to be sufficient to meet the need of the country.

Isolation

Different methods employed for the isolation of F_2 are given below :

1. Moissan's Method. (*Historical method—1886*). This is a historical method, since this method was used for the first time by Moissan to isolate fluorine. Essential parts of the electrolytic cell used by Moissan are : (*a*) *Electrolytic cell.* Electrolytic cell consists of a U-shaped tube made of Pt—Ir alloy cooled at –23°C by placing it in a bath of CH_3Cl (b.pt. = –23°C) which was constantly renewed. (*b*) *Electrodes.* These are made of Pt—Ir alloy and are insulated with fluorspar stoppers covered with shellac. (*c*) *Electrolyte.* Electrolyte consists of a solution of *potassium hydrogen fluoride*, KHF_2 (1 part) in anhydrous hydrofluoric acid, H_2F_2 (5 parts).

Process. Moissan took the solution consisting of KHF_2 and anhydrous H_2F_2 (in 1 : 5 ratio) in a U-shaped electrolytic cell at a temperature of –23°C by placing the cell in a bath of CH_3Cl and electrolysed this solution (Fig. 38.1). On electrolysis H_2 was liberated at the cathode and F_2 at the anode which is passed through platinum spiral tube similarly placed in the bath of methyl chloride (CH_3Cl). The vapours of hydrofluoric acid got condensed in the spiral tube. F_2 gas was next passed through a tube containing fused sodium fluoride (NaF) which absorbed even the last traces of hydrofluoric acid forming $NaHF_2$. F_2 obtained by this method is pure.

$$2KHF_2 \text{ (In anhydrous } H_2F_2) \longrightarrow 2KF + H_2F_2$$

$$H_2F_2 \rightleftarrows 2H^+ + 2F^-$$

At anode : $F^- - e^- \longrightarrow F;\ F + F \longrightarrow F_2$

At cathode : $H^+ + e^- \longrightarrow H;\ H + H \longrightarrow H_2$

$$2NaF + H_2F_2 \longrightarrow 2NaHF_2$$

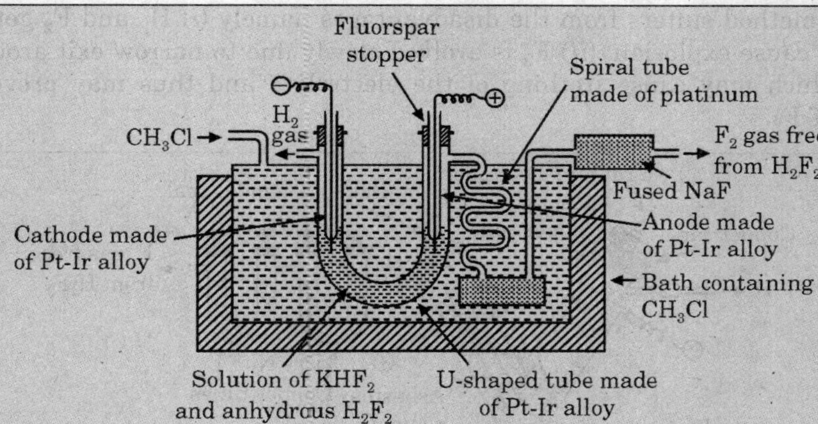

Fig. 38.1. Moissan's method for the isolation of fluorine.

In spite of low temperature *viz.* –23°C employed in this method, there was severe corrosion and wastage of about 5 grams of platinum for the preparation of one gram of F_2. In 1899 he observed that a Cu-tube could be used in place of Pt—Ir tube, since Cu-tube gets coated with a projective layer of CuF_2 which prevents further action of F_2 on Cu, but he continued the use of Pt—Ir electrodes.

2. Deniss Method. This method was devised by *Deniss, Veeder* and *Rockow* in 1931. Essential parts of the electrolytic cell used in this method are : (*a*) *Electrolytic cell.* Electrolytic cell consists of V-shaped tube made of Cu about 2" wide and fitted with copper caps into which graphite electrodes are fixed with bakelite cement The V-shaped copper tube is thickly covered with an insulting layer of asbestos cement. This layer is wound by a resistance wire for heating the V-shaped tube electrically. The whole apparatus is then thickly lagged to prevent the loss of heat. (*b*) *Electrolyte.* Pure and dry KHF_2 (m.pt. = 227°C) is used as an electrolyte. KHF_2 should be perfectly dry, since the presence of traces of moisture results in the formation of O_2 or OF_2. KHF_2 is usually dried in a current of air at 140°–150°C. (*c*) *Electrodes.* Graphite electrodes fixed into the V-shaped copper tube by bakelite cement are used.

Process. When a current of 5 amperes and 12 volts is passed through perfectly dry electrolyte of KHF_2, H_2 is liberated at the cathode and F_2 at the anode and passes out through an exit near the anode (Fig. 38.2). F_2 so produced attacks the copper walls forming the layer of CuF_2 which protects the deeper layers of copper from further attack. The reactions taking place are :

$$2KHF_2 \longrightarrow 2KF + H_2F_2$$

$$H_2F_2 \longrightarrow 2H^+ + 2F^-$$

At cathode : $2H^+ - 2e^- \longrightarrow H_2$

At anode : $2F^- + 2e^- \longrightarrow F_2$

F_2 so obtained is passed through copper tubes containing dry NaF which absorbs HF forming $NaHF_2$

$$2NaF + H_2F_2 \longrightarrow 2NaHF_2$$

This method suffers from the disadvantages namely (*i*) H_2 and F_2 get mixed and may cause explosion. (*ii*) F_2 is evolved slowly due to narrow exit around the anode which may cause frothing of the electrolyte and thus may prevent the escape of F_2.

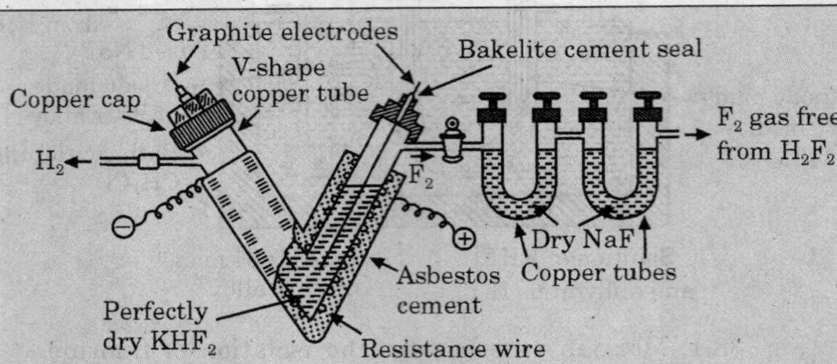

Fig. 38.2. Deniss method for the isolation of fluorine.

3. Whytlaw-Gray's Method. The essential part of the apparatus (Fig. 38.3) used in this method is a copper cell wound with resistance wire for electrical heating. It also acts as the cathode resistance and has an outlet for hydrogen near the top.

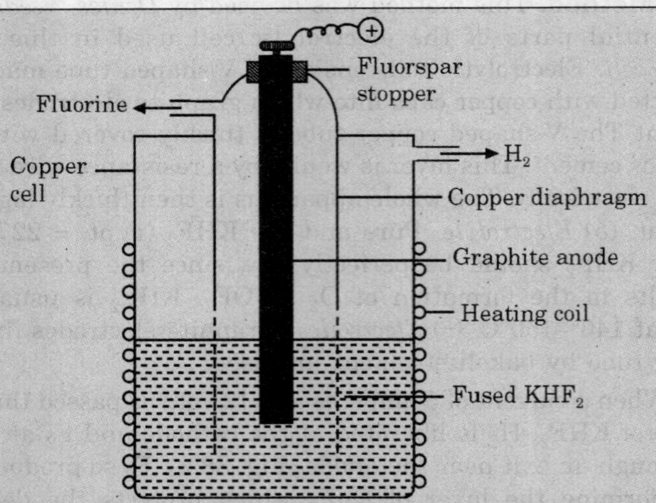

Fig. 38.3. Whytlaw-Gray's apparatus for the isolation of fluorine.

The anode is a rod of graphite enclosed in a copper cylinder. This cylinder has an outlet near the top for the escape of fluorine gas and is perforated near the bottom.

The electrolyte is fused potassium hydrogen fluoride (KHF_2). It undergoes electrolysis to give hydrogen and fluorine.

$$2KHF_2 \longrightarrow 2KF + F_2 + H_2$$

F_2 and H_2 gases escape out of the cell as shown without any possibility of mixing with each other as the anode and the cathode are separated by the cylindrical diaphragm.

Precautions. *(i)* The electrolyte must be absolutely dry, because the presence of moisture hinders electrolysis and the fluorine evolved may react with it to produce oxygen and ozone. All parts of the apparatus which come into contact with fluorine must also be free from oil and grease. *(ii)* The gas must be freed from HF by using NaF. If HF is present, it will attack the glass vessels in which the fluorine is kept. *(iii)* The vessels in which fluorine is collected must also be absolutely dry. *(iv)* As it attacks organic matter of all types, it should be very carefully handled.

Advantages of this method. This method is superior to the above methods in the following respects : *(i)* Copper cylinder (diaphragm) separates the anode from the cathode with the result that the liberated H_2 and F_2 are prevented from mixing with each other and thus the possibility of explosions is avoided. *(ii)* There is no chance of explosion due to the frothing of the electrolyte by the accumulation of gases as the exits for gases are quite wide to allow the gases to escape out quite promptly.

4. Modern Method. The manufacture of F_2 by modern method consists of electrolysing fused KHF_2 mixed with 2 to 3 moles of H_2F_2 at 70-80°C in a rectangle steel vessel using a current of 1000–2000 amperes at 8.5 to 11 volts. The steel vessel acts as a cathode and the anode was earlier made of special graphite form of carbon. Now anodes of petroleum coke impregnated with copper are preferred. A steel cylinder attached to the lid acts as a diaphragm (Fig. 38.4) which does not allow F_2 and H_2 to come in contact with each other and thus the two gases are led out through separate exits. Consequently the possibility of explosion is avoided. The valves used are of monel metal or nickel with packing of teflon, $(C_2F_4)_x$ which is inert to chemical reagents like boiling aqua regia. F_2 obtained at the anode is freed from H_2F_2 by passing it through dry NaF and is stored in steel cylinders in compressed conditions.

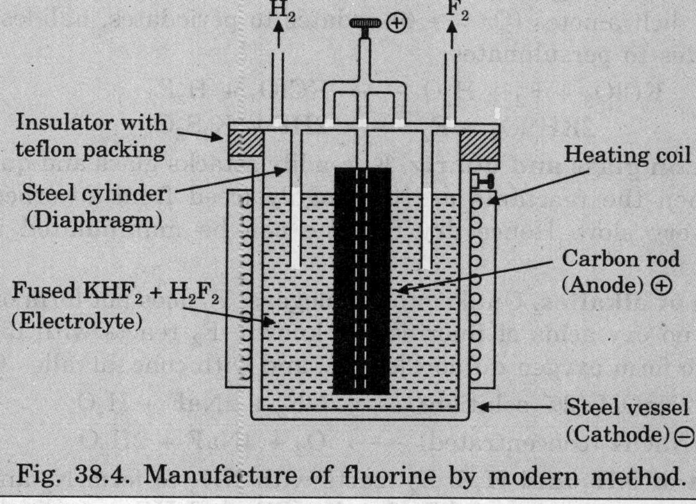

Fig. 38.4. Manufacture of fluorine by modern method.

Physical Properties

Fluorine is a pale yellow gas which condenses to a clear yellow liquid having a density of 1.108 and b.pt. = –187.7°C, and freezes to a pale yellow solid (m.pt. = 223°C). At –252°C the solid becomes colourless. It is a pungent smelling and poisonous gas but less poisonous than HF gas. It attacks organic matter and hence should be handled carefully. Its vapour density is 18.91 which corresponds to the molecular formula F_2.

Chemical Properties

1. Reactivity. Fluorine is the most reactive element as is evident from the fact that it ignites in contact with most of the oxidising substances.

2. Action on halides. When we compare the values of standard reduction potential values (E° values) of halogens corresponding to the reaction :

$$X_2 + 2e^- \longrightarrow 2X^-,$$

we find that E° value for fluorine is the maximum and hence F_2 can displace all the halogens from their halides, *i.e.* F_2 liberates Cl_2 from chlorides, Br_2 from bromides and I_2 from iodides.

3. Combination with elements. Fluorine combines with most metals (*e.g.* Na, Ca, Mg, Al, Ni, Au, Pt etc), non-metals (*e.g.* H, C, 8, P, Cl, Br, I etc.) and metalloids (*e.g.* As, Sb) to form the fluorides.

4. Combination with xenon. Fluorine combines with xenon to form three definite fluorides, XeF_2, XeF_4 and XeF_6. The final product depends on the temperature, pressure and mixing ratio of the constituents.

5. Action of H_2O. F_2 reacts vigorously with H_2O giving O_2 and O_3.

$$2F_2 + 2H_2O \longrightarrow 4HF + O_2 \; ; \; 3F_2 + 3H_2O \longrightarrow 6HF + O_3$$

At the same time, some oxygen fluoride (OF_2) is also formed. A jet of steam burns in an atmosphere of F_2 like an illuminating gas.

6. Oxidising action. Since F_2 readily attacks water, giving O_2 and O_3, it acts as a strong oxidising agent. It oxidises chlorates to perchlorates, chromic salts (Cr = + 4) to dichramotes (Cr = + 6), iodates to periodates, halides to halogens and bisulphates to persulphates.

$$KClO_3 + F_2 + H_2O \longrightarrow KClO_4 + H_2F_2$$
$$2KHSO_4 + F_2 \longrightarrow 2HF + K_2S_2O_8$$

7. Action on glass and quartz. F_2 readily attacks glass and quartz forming SiF_4. But when the reactants are thoroughly freed from adsorbed water, the reaction is very slow. Hence dry fluorine can be manipulated in dry glass apparatus.

8. Action of alkalies. Unlike other halogens, F_2 does not form oxy salts with alkalies and no oxy acids of fluorine are known. F_2 reacts with dilute alkalies (2% NaOH) to form oxygen difluoride, OF_2 and with conc. alkalies O_2 is formed.

$$2F_2 + 2NaOH \text{ (2\% solution)} \longrightarrow OF_2 + 2NaF + H_2O$$
$$2F_2 + 4NaOH \text{ (concentrated)} \longrightarrow O_2 + 4NaF + 2H_2O$$

9. Action of NH_3 and H_2S. F_2 reacts with NH_3 to form N_2 and some NF_3. H_2S burns in an atmosphere of F_2 forming SF_6 and HF.

$$2NH_3 + 3F_2 \longrightarrow N_2 + 6HF$$
$$NH_3 + 3F_2 \longrightarrow NF_3 + 3HF$$
$$H_2S + 4F_2 \longrightarrow SF_6 + 2HF$$

Uses of F_2. (*i*) F_2 is used as an insecticide. (*ii*) It gives a number of compounds which are very useful. Some of the more important compounds and their uses are : (*a*) *Fluorocarbons.* Fluorocarbons like freon-12 and teflon are very useful compounds. For their uses see the properties of fluorine (*Reaction with hydrocarbons and formation of fluoro carbons* as discussed above. (*b*) UF_6 is used in the separation of U^{235} from natural uranium which is used in atom and hydrogen bomb. (*c*) SF_6 has a remarkable property of being a high voltage insulation material and hence is used as an insulating gas in nuclear physics and electrical and X-ray works. (*d*) *Freon*, CCl_2F_2 (*dichlorofluoromethane*) is used in refrigerators and cold storage plants. (*e*) H_2F_2 is used for etching glass, in brewing industry and for removing SiO_2 from iron castings. (*f*) *Cryolite* (Na_3AlF_6) and *fluorspar* (CaF_2) are used in the extraction of Al. (*g*) *Fuorspar* is used in the preparation of H_2F_2. (*h*) *NaF* and Na_3AlF_6 in powdered form are used as insecticides. (*i*) *Presence of small quantities of fluorides*, less than one part per million, in drinking water prevents dental decay. (*j*) *Antimony fluoride* is used in the manufacture of pottery and porcelain and *sodium antimony fluoride* which is a double fluoride of Na and Sb is used as a mordant in the textile industry (*k*) *Sodium fluoro acetate* finds use as rat poison. (*l*) CuF_2 is used in the ceramic industry and as a flux in soldering, welding and blazing. (*m*) MgF_2 is used in optical industry. (*n*) *KF* is widely used in the preparation of fluorocarbons. *Many inorganic fluorine compounds* are used in industry and technology. (*o*) D.D.F.T. which is similar to D.D.T. is an extremely active fungicide and fumigant

Fluorocarbons

F_2 also reacts with hydrocarbons explosively and giver rise to the formation of fluorinated hydrocarbons like CHF_3, CH_2F_2, CHF_3, CF_4, C_2F_8, C_3F_3 etc which are called **fluorocarbons.** The formation of fluorocarbons has been successfully carried out by diluting fluorine with N_2.

Synthesis.

Fluorocarbons can be synthesised :

(*i*) *By the replacement of halogen atom (other than F) in alkyl halides by P-atom in a metal fluoride.* For example :

$$RCl + AgF \longrightarrow AgCl + RF$$

$$C_6H_5PCl_3 + AsF_3 \xrightarrow{25°C} C_6H_5PF_3 + AsCl_3$$

(*ii*) *By the replacement of halogen atom by F-atom in anhydrous HF*

$$CCl_4 \xrightarrow{+HF} \underbrace{CCl_3F + CCl_2F_2 + CClF_3 + CF_4}_{\text{Freons}} + HCl$$

$$CHCl_3 + 2HF \longrightarrow CHCl_2F + 2HCl$$

Chloro-fluorocarbons obtained from CCl_4 are called **freons.** The reactions shown above can be activated by using $SbCl_5$ as catalyst at 50–150°C and pressure of 50–500 lb. inch^{-2}.

(*iii*) *By electrolytic replacement of H_2 by F_2.* In this method organic compounds are subjected to electrolysis in liquid HF at voltage (4–6 volts) below that required for the liberation of F_2. Steel cells are used with Ni anode and steel cathode. Fluorination takes place at anode. This method has been used in laboratory and also as an industrial method.

$$(C_2H_5)_2O \longrightarrow (C_2F_5)_2O$$

$$CH_3COOH \longrightarrow CF_3COOF \xrightarrow{H_2O} CF_3COOH$$

(*iv*) *By the direct replacement of H_2 by F_2.* The reacting compound and F_2 are subjected to electrolytic fluorination in a nitrogen atmosphere. The catalysts used are Cu-gauze, silver-coated copper gauze or caesium fluoride.

$$C_6H_6 + 9F_2 \xrightarrow[265°C]{\text{Cu-gauze}} C_6F_{12} + 6HF$$

Inorganic fluorides such as CoF_3 have also been used for vapour phase fluorination of organic compounds.

$$(CH_3)_3N \xrightarrow{CoF_3} (CH_3)_2FN + (CH_3)F_2N + NF_3$$

Properties and Uses of Fluorocarbons.

(*i*) Chloro-fluorocarbons obtained from CCl_4 are called **freons.** These are used as non-toxic and non-corrosive inert refrigerants, aerosol bomb propellants and heat transfer and fire extinguishing agents. Freons differ from the hydrocarbons in that they are chemically very inert. This property makes them useful solvents, lubricants and insulators.

Difluorodichioromethane, CF_2Cl_2 which is called **freon-12** is used in refrigeration and air conditioning in place of NH_3 and SO_2. It is non-toxic and non-flammable. It is also used as a solvent for D.D.T. and other insecticides.

(*ii*) Just as ethylene (C_2H_4), tetrafluoroethylene (C_2F_4) can also thermally be polymerised to form poly tetrafluoroethylene, $(C_2F_4)_n$. This polymerised compound is called **Teflon.** This is a plastic-like inert material. It is insoluble in any solvent and is not attacked by strong acids, alkalies and oxidizing agents, Due to these properties, this material is used in the construction of chemical plants. It has a very high electrical resistance and is used as an insulating material in cables.

(*iii*) CF_3OF is prepared by the fluorination of CH_3OH or CO by F_2 in presence of AgF_2. It is a stable compound upto 50°C and has been used as a strong oxidising agent.

$$CF_3FO + SO_2 \longrightarrow CF_3OSO_2F$$

$$CF_3FO + SO_3 \longrightarrow CF_3OOSO_2F$$

$$CF_3OF + SF_4 \longrightarrow CF_3OSF_5.$$

Conceptual Questions with Answers

Q.1 Explain why it is not possible to obtain F_2 by the electrolysis of aqueous NaF

or Explain why the electrolysis of KBr(aq) gives Br_2 at anode but the electrolysis of KF(aq) does not give F_2.

Ans : On electrolysis, the aqueous solution of NaF gives Na metal on cathode and F_2 on anode.

$$NaF \rightleftharpoons Na^+ + F^-$$

$$F^- \longrightarrow F + e^-, \; F + F \longrightarrow F_2 \; (\textit{At anode})$$

$$Na^+ + e^- \longrightarrow Na \; (\textit{At cathode})$$

Now, since F_2 is a strong C.A., F_2 obtained at anode is consumed in oxidising H_2O (to O_2) present in the aqueous solution of NaF.

$$\underset{\substack{(F = 0)\\(OA)}}{2F_2} + \underset{(O = -2)}{2H_2O} \longrightarrow \underset{(O = 0)}{O_2} + \underset{(F = -1)}{4HF}$$

F_2 also produces O_3 with H_2O.

$$3F_2 + 3H_2O \longrightarrow O_3 + 6HF$$

Thus at anode, we get ozonised O_2 instead of F_2.

Q.2 Write the balanced chemical equation for the reaction between Cl_2 and hot and conc. NaOH. Justify that this reaction is a disproportionation reaction.

Ans : $\underset{(Cl = 0)}{3Cl_2} + 6NaOH\,(\textit{conc}) \xrightarrow{\text{Hot}} \underset{(Cl = -1)}{5NaCl} + \underset{(Cl = +5)}{NaClO_3} + 3H_2O$

It is obvious from the reaction that the O.S. of Cl decreases from zero to –1 in NaCl and increases from zero to +5 in $NaClO_3$. Hence this reaction is a disproportionation reaction.

Q.3 Addition of Cl_2 to KI solution gives a brown colour but excess of Cl_2 turns it colourless. Explain with reasons.

Ans : Since Cl_2 is stronger O.A. than I_2, it oxidises I^- ion to I_2 when Cl_2 is a added to KI solution.

$$\underset{(Cl = 0)}{Cl_2} + \underset{(I = -1)}{KI\,(aq)} \longrightarrow \underset{(Cl = -1)}{2KCl}\,(aq) + \underset{(I = 0)}{I_2} \quad (\textit{Brown colour})$$

If Cl_2 gas is passed in excess, then Cl_2 oxidises I_2 evolved in the above reaction to iodic acid (HIO_3) which is colourless.

$$\underset{(Cl = 0)}{5Cl_2} + \underset{(I = 0)}{I_2} + 6H_2O \longrightarrow \underset{(Cl = -1)}{10HCl} + \underset{(I = +5)}{2HIO_3} \quad (\textit{Colourless})$$

or $$5Cl_2 + I_2 + 6H_2O \longrightarrow 10Cl^- + 2IO_3^- + 12H^+$$

Q.4 How can you prepare Cl_2 from HCl and HCl from Cl_2. Write reactions only.

Ans : HCl can be oxidised to Cl_2 by a number of oxidising agents such as MnO_2, $KMnO_4$, $K_2Cr_2O_7$, etc.

$$MnO_2 + 4HCl \longrightarrow MnCl_2 + Cl_2 + 2H_2O$$

Cl_2 can be reduced to HCl by reaction of H_2 in presence of diffused sunlight.

$$H_2 + Cl_2 \xrightarrow{\text{Diffused sunlight}} 2HCl$$

Q.5 Sea is the greatest source of halogens. Comment.

Ans : Sea water contains chlorides, bromides and iodides of sodium, potassium, magnesium and calcium, but mainly sodium chloride (2.5% by mass). Dried up sea beds contain sodium chloride and carnallite, $KCl.MgCl_2.6H_2O$. Certain seaweeds contain upto 0.5% of iodine as sodium iodide and chile saltpetre ($NaNO_3$) contains upto 0.2% of sodium iodate. Thus, sea is the greatest source of halogens.

Q.6 Write the reactions of F_2 and Cl_2 with water.

Ans : F_2 being a strong oxidising agent oxidises H_2O to O_2 or O_3.

$$2F_2(g) + 2H_2O(l) \longrightarrow 4H^+ (aq) + 4F^- (aq) + O_2(g)$$

$$3F_2(g) + 3H_2O(l) \longrightarrow 6H^+ (aq) + 6F^- (aq) + O_3(g)$$

Cl_2 on the other band, reacts with H_2O to form hydrochloric acid and hypochorous acid

$$\underset{}{Cl_2(g) + H_2O(l)} \longrightarrow \underset{\text{Hydrochloric acid}}{HCl\ (aq)} + \underset{\text{Hpochlorous acid}}{HOCl\ (aq)}$$

Q.7 Name two poisonous gases which can be prepared from chlorine gas.

Ans : Two poisonous gases which can be prepared from Cl_2 are : (*i*) phosgene (*ii*) mustard gas. These are prepared as follows :

$$(i)\ CO + Cl_2 \xrightarrow{hv,\ \text{charcoal}} \underset{\text{Phosgene}}{COCl_2}$$

(*ii*) Cl_2 is passed through boiling S, when S_2Cl_2 is formed. This when reacted with ethene gives mustard gas.

$$\underset{\text{(Boiling)}}{S_8} + 4Cl_2 \longrightarrow \underset{\text{Sulphur monochloride}}{4S_2Cl_2}$$

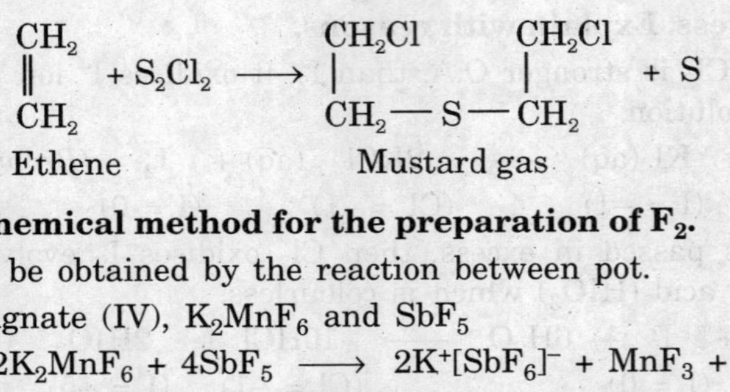

Ethene Mustard gas

Q.8 Give a chemical method for the preparation of F_2.

Ans : F_2 can be obtained by the reaction between pot.
hexafluoromagnate (IV), K_2MnF_6 and SbF_5

$$2K_2MnF_6 + 4SbF_5 \longrightarrow 2K^+[SbF_6]^- + MnF_3 + F_2$$

Above reaction takes place through the following steps :

$$2KMnF_6 \longrightarrow 2K^+ + 2F^- + MnF_4\ ] \times 2$$

$$SbF_5 + F^- \longrightarrow SbF_6^-\ ] \times 4$$

$$2MnF_4 \longrightarrow 2MnF_3 + F_2$$

$$2K_2MnF_6 + 4SbF_5 \longrightarrow 4K^+ + 4SbF_6^- + 2MnF_3 + F_2$$

or $2K_2MnF_6 + 4SbF_5 \longrightarrow 4K^+[SbF_6]^- + 2MnF_3 + F_2$

Q.9 Chlorine water has both oxidising as well as bleaching properties. Explain.

Ans : In presence of moisture or water, Cl_2 gives nascent oxygen which is responsible for its oxidising and bleaching properties as shown below :

$$Cl_2 + H_2O \longrightarrow [HCl + HClO] \longrightarrow 2HCl + \underset{\text{Nascent oxygen}}{[O]}$$

(*i*) It oxidises acidified ferrous sulphate to ferric sulphate

$$Cl_2 + H_2O \longrightarrow 2HCl + O$$
$$2FeSO_4 + H_2SO_4 + O \longrightarrow Fe_2(SO_4)_3 + H_2O$$
$$\underset{(Green)}{2FeSO_4} + H_2SO_4 - Cl_2 \longrightarrow \underset{(Brown)}{Fe_2(SO_4)_3} + 2HCl$$

(*ii*) It bleaches vegetable and organic colouring matter to colourless substances by oxidation.

$$Cl_2 + H_2O \longrightarrow 2HCl + [O]$$

Vegetable colouring matter + O $\longrightarrow$ Oxidised colourless substances.

Q.10 (*a*) Why does chlorine water lose its yellow colour on standing? (*b*) What happens when Cl_2 reacts with cold dilute solution of sodium hydroxide? Write equations only.

Ans : (*a*) Chlorine water is yellow due to the presence of hypochlorous acid (HClO) in it. On standing, HClO, being unstable, decomposes to form HCl. As a result, yellow colour disappears.

$$Cl_2 + H_2O \longrightarrow HCl + \underset{(Yellow)}{HClO} \longrightarrow \underset{(Colourless)}{2HCl} + [O]$$

(*b*) With cold dilute NaOH solution, Cl_2 gives a mixture of sodium chloride (NaCl) and sodium hypochlorite (NaClO)

$$2NaOH\ (dil.) + Cl_2 \xrightarrow{\text{Cold}} NaCl + NaClO + H_2O$$

Q.11 What is superhalogen ? What happens when this halogen reacts with a cold solution of NaOH ?

Ans : F_2 is called superhalogen. It is so called because it differs from other halogens in many properties. For example it has highest electronegativity, smallest size, highest E° value, no vacant *d*-orbitals in its valence shell etc.

When F_2 reacts with NaOH, OF_2 is obtained.

$$2F_2 + 2NaOH \longrightarrow OF_2 + 2NaF + H_2O$$

Q.12 How will you prepare : (*a*) Cl_2 gas in the laboratory (*b*) Br_2 from sea water.

Ans : (*a*) Chlorine gas is prepared in the laboratory by heating a mixture of manganese dioxide (MnO_2) and conc. HCl.

$$MnO_2\ (s) + 4HCl\ (aq) \xrightarrow{\Delta} MnCl_2\ (aq) + Cl_2\ (g) + H_2O(l)$$

(*b*) Br_2 is manufactured from sea water by passing Cl_2 through it.

$$2Br^-\ (aq) + Cl_2\ (g) \longrightarrow Br_2\ (l) + 2Cl^-\ (aq)$$

University Questions

1. What were the difficulties encountered in the isolation of fluorine? How were these overcome? *(Agra 89)*
2. Write the names of chief ores of fluorine. Describe the modern methods for the isolation of this element. *(Meerut 85, 86; Sambalpur 8; Utkal 85; Jiwaji 84; Agra 89)*
3. Write a note on "Fluorocarbons". *(Meerut 85)*
4. Describe how Cl_2 is prepared on a large scale. Mention its important uses. *(Delhi Hons. 78 S)*
5. What are the important sources of Br_2? How is it obtained on a large scale? *(Punjab 86)*
6. How is iodine manufactured?*(Punjab 86; Kurukshetra 85; Hyderabad 86)*
7. Write a short note on "Basic iodine". *(Raj. 84, 86, 87; Madras 85, 86)*
8. Explain the following :
   (*i*) F is a non-metal while I shows metallic character.
   (*ii*) A we go down the column of halogens, there is an increase in the metallic character of the halogens.
9. Describe the action of heat on HBr and conc. H_2SO_4. *(Delhi 99)*
10. What is main source of fluorine ? Give modern method for the isolation of fluorine. *(Gorakhpur 99, Allahabad 99)*
11. Write a note on "Anomalous behaviour of fluorine". *(Lucknow 2000)*
12. Discuss the difficulties met in the isolation of fluorine. Describe the modern method for its isolation. *(Avadh 2000)*

CHAPTER

39 Halogen Acids : Hydracids or Hydrogen Halides

All the halogens combine with hydrogen and form the volatile covalent hydrides of HX type (X = F, Cl, Br and I) in which the halogen (X) shows an oxidation state of –1. These hydrides are called by various names like **hydrogen acids, hydracids, hydrohalic acids and hydrogen halides.** The salts of these acids are called **halides.** General methods of the preparation and properties of these acids have already been discussed in chapter 43. The four halogen acids given by halogens are *hydrofluoric acid*, H_2F_2; *hydrochloric acid*, HCl; *hydrobromic acid*, HBr and *hydroiodic acid*, HI. Now these acids are discussed separately as below.

Hydrofluoric Acid, H_2F_2

This acid is an associated molecule and hence is represented as H_2F_2.

Preparation

Pure anhydrous acid is prepared by distilling dry KHF_2 in a platinum, copper or steel vessel cooled in freezing mixture (laboratory preparation) or by passing dry H_2 over dry AgF.

$$2KHF_2 \longrightarrow H_2F_2 + 2KF; 2AgF + H_2 \longrightarrow H_2F_2 + 2Ag$$

An aqueous solution of hydrofluoric acid is mannfatured by heating calcium fluoride, CaF_2 (*fluorspar*) with 90% H_2SO_4

$$CaF_2 + H_2SO_4 \longrightarrow H_2F_2 + CaSO_4$$

The vapours of hydrofluoric acid are often dissolved in water contained in a lead receiver and coated on the inside by wax to get an aqueous solution of hydrofluoric acid.

Storage of the Acid

The anhydrous acid is stored in well-cooled platinum, gold or silver vessels or steel cylinders. The commercial aqueous acid is kept in gutta-percha bottles when dilute and in ceresin wax bottles when concentrated.

Physical Properties

Anhydrous acid is a colourless, strongly fuming liquid which boils at 194°C and freezes at –83°C. It is extremely poisonous and dangerous to handle, as in contact with skin it forms ulcerated sores. It is highly soluble in water and forms a constant boiling mixture containing 36% of the acid which boils at 120°C. The anhydrous acid is a bad conductor of electricity. The liquid acid and its vapour at low temperatures consist of polymeric form, $(HF)_x$ whereas at 90°C the vapour density has the normal value.

Chemical Properties

(i) Combustibility. H_2F_2 is neither combustible nor supporter of combustion.

(ii) Acidic nature. It is a weak dibasic acid (even weaker than H_3PO_4 but stronger than HNO_2) and hence forms two types of salts namely *normal salts* (*e.g.* NaF) and *acid salts* (*e.g.* $NaHF_2$) The formation of acid salts is due to the fact that aqueous solution of the acid contains hydrogen-bonded species like

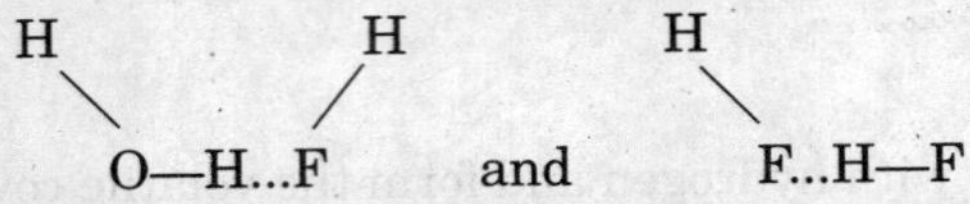

Note that acid salts are not given by other halogen acids.

(iii) Action on metals. Anhydrous acid does not attack any metal under ordinary conditions except potassium while aqueous concentrated solution of the acid attacks most of metals (*e.g.* Na, Ag, Al, Ca, Sc, Zn, Mg) forming their fluorides with the liberation of H_2 gas.

(iv) Action on oxides, hydroxides, carbonates etc. Concentrated aqueous solution of the acid reacts with these species to form the fluorides. For example :

$$B_2O_3 + 3H_2F_2 \longrightarrow 2BF_3 + 3H_2O$$

$$2NaOH + H_2F_2 \longrightarrow 2NaF + H_2O$$

$$CaCO_3 + H_2F_2 \longrightarrow CaF_2 + H_2O + CO_2$$

Reaction of the concentrated aqueous solution of the acid with silica (SiO_2) is of importance. In this reaction H_2SiF_6 is formed.

$$SiO_2 + 3H_2F_2 \longrightarrow H_2SiF_6 + 2H_2O$$

Since glass contains SiO_2, H_2F_2 cannot be kept in glass containers.

(v) Action on glass. Pure anhydrous acid does not attack glass but a concentrated aqueous solution of the acid attacks glass. In this reaction H_2F_2 reacts with silica (SiO_2), sodium silicate (Na_2SiO_3) and calcium silicate ($CaSiO_3$) which are the constituents of glass, forming SiF_4 and H_2SiF_6 and its salts.

(*a*) *With* SiO_2 : $SiO_2 + 3H_2F_2 \longrightarrow H_2SiF_6 + 3H_2O$

(*b*) *With* Na_2SiO_3 : $Na_2SiO_3 + 3H_2F_2 \longrightarrow Na_2SiF_6 + 3H_2O$

(*c*) *With* $CaSiO_3$: $CaSiO_3 + 3H_2F_2 \longrightarrow CaSiF_6 + 3H_2O$

The etching of glass is based on the above reactions. For this purpose commercial H_2F_2(40–60%) is used.

(vi) Action of $AgNO_3$ and $BaCl_2$ solutions. With $AgNO_3$ solution, it gives no precipitate since AgF formed is soluble in water (*cf.* other halogen acids). With $BaCl_2$ solution, it forms a white precipitate of BaF_2 which is insoluble in dil. HNO_3 but soluble in conc. HCl (difference from other halogen acids)

$$H_2F_2 + 2AgNO_3 \longrightarrow \underset{\text{(Soluble in water)}}{2AgF} + 2HNO_3$$

$$H_2F_2 + BaCl_2 \longrightarrow \underset{\text{ppt.}}{BaF_2} + 2HCl$$

(vii)*Stability.* H_2F_2 is extremely stable and is not oxidised by strongly oxidising agents or decomposed by heat. Thermal stability is due to the strength of H—F bond in H_2F_2 molecule.

Etching of Glass by H_2F_2

In etching, glass surface is covered with a thin film of paraffin wax and the design to be produced is scratched through the film with a sharp needle. The surface is then exposed to the action of the gas or of an aqueous solution (40–60%) of the acid. The paraffin coating protects the unexposed glass. Burettes, thermometers and other glass wares are engraved in this manner. Etching with solution is clear but it is opaque with the gas.

Uses

H_2F_2 is used : *(i)* In the etching of glass and making scales on the glass instruments. *(ii)* As an antiseptic in dyeing and brewing industries, since its effect on yeast is very little as compared with its action on other organisms. *(iii)* In the manufacture of fluorine. *(iv)* In the manufacture of a class of fluorine-containing dyes that have proved very fast and remarkably resistant to light. *(v)* In the pickling of alloy steels. *(vi)* In the manufacture of fluorine-containing plastics like Teflon etc. which are characterised by extreme chemical inertness, high thermal stability and very good dielectric properties. *(vii)* Because of its action on silica H_2F_2 is extensively used for removing silica from artificial graphite and Swedish filter paper. *(viii)* By the action of H_2F_2 on sodium aluminate, *synthetic (artificial) cryolite* is prepared. H_2F_2 is also used for the preparation of AlF_3. Both these substances are used in the metallurgy of aluminium. *(ix)* H_2F_2 is also used in the production of uranium hexafluoride (UF_6) which, because of its thermal stability and sufficiently high vapour pressure at the room temperature, is employed for the separation of uranium isotopes by the gaseous diffusion process. UF_6 is prepared by the action of H_2F_2 on U_2O_3 to get UF_3 which is then treated with F_2 to get UF_6 *(x)* Fluorides of alkali metals and antimony are used as mordant in dyeing industry. *(xi)* H_2F_2 is also used as a refrigerant in many of the cooling processes.

Structure

The vapour density of hydrofluoric acid measured near about its boiling point *viz.* 30°C indicates the formula H_2F_2. Measurement of the lowering of freezing point of water by hydrogen fluoride compounds gives a molecular formula H_2F_2 in aqueous solutions. The formula of the salts like KHF_2 also confirms this view. It may, therefore, be concluded that the hydrofluoric acid vapour just above the boiling point may be a mixture of H_2F_3, H_2F_2 and HF molecules but as the temperature rises, more and more dissociation takes place but at 100°C it is only monomeric. In concentrated aqueous solutions the molecule is represented as H_2F_2. Dilute solutions and the vapour above 88°C are mostly composed of HF molecules.

Comparison of H_2F_2 with Other Halogen Acids

(1) Acid strength. All the halogen acids ionise in aqueous solution and hence behave as acids. H_2F_2 ionises only slightly while other acids ionise completely. Hence H_2F_2 is the weakest acid of all the acids.

(2) Ionic character of H—X bond. H—F bond in H_2F_2 acid has the highest degree of polarity, since F atom has the highest value of electronegativity of all the halogen atoms.

(3) Physical state. H_2F_2 is a liquid at ordinary temperature while other halogen acids are colourless gases.

(4). Reducing property. H_2F_2 has no reducing property while other acids are good reducing agents.

(5) Hydrogen bonding and formation of acid salts. Aqueous solution of HF contains hydrogen bonded species. On account of the presence of H-bonding in HF molecule, this molecule forms acid salts of KHF_2 type which contains the anion $[F—H...F]^-$. No such acid salts are given by other HX acids, since other acids have no hydrogen bonding. The occurrence of H-bonding between HF molecules is due to the high electronegativity and small size of F-atom. No hydrogen bonds are formed between HCl, HBr and HI molecules, since Cl, Br and I atoms in these molecules are much less electronegative and larger in size than F-atom.

(6) Action on silica or glass. Strong solution of H_2F_2 attacks silica or glass readily forming SiF_4 which gives complex fluosilicic acid, H_2SiF_6 with excess of H_2F_2. Other halogen acids do not act upon silica or glass.

$$SiO_2 + 2H_2F_2 \longrightarrow SiF_4 + 2H_2O$$

$$SiF_4 + H_2F_2 \longrightarrow H_2SiF_6.$$

(7) Action on metals. All the halogen acids react with metals like Zn and Mg to form H_2. The action of H_2F_2 on metals is, however, less violent than with other halogen acids.

(8) Action of mixture of MnO_2 and H_2SO_4. When H_2F_2 is heated with a mixture of MnO_2 and H_2SO_4, no gas is evolved. On the other hand, in case of HCl, HBr and HI acids, Cl_2 (greenish-yellow), Br_2 (reddish brown vapour) and I_2 (violet vapour) gases are evolved.

(9) Action of $AgNO_3$ solution. With $AgNO_3$ solution, H_2F_2 gives no precipitate, since AgF formed is soluble in water. On the other hand, HCl, HBr and HI give the precipitates of AgCl (white), AgBr (pale yellow) and AgI (yellow).

$$H_2F_2 + 2AgNO_3 \longrightarrow 2AgF \text{ (soluble)} + 2HNO_3$$

$$HX(X = Cl, Br, I) + AgNO_3 \longrightarrow AgX \text{ (coloured ppt.)} + HNO_3$$

(10) Action of $BaCl_2$ solution. With $BaCl_2$ solution, H_2F_2 gives a white precipitate of BaF_2 while other halogen acids do not give any precipitate, since BaX_2(X = Br and I) formed are soluble.

$$H_2F_2 + BaCl_2 \longrightarrow BaF_2 \text{ (white ppt)} + 2HCl$$

$$2HX(X = Br, I) + BaCl_2 \longrightarrow BaX_2 \text{ (soluble)} + 2HCl$$

(11) Action of $Pb(CH_3COO)_2$ solution. H_2F_2 gives no precipitate with $Pb(CH_3COO)_2$ solution while other halogen acids give the precipitates of $PbCl_2$ (white), $PbBr_2$ (white) and PbI_2 (yellow) respectively.

(12) Dissocation. H_2F_2 does not dissociate on heating while other halogen acids dissociate at definite temperatures.

Similarities between HF and H_2O

(*i*) Both the compounds resemble each other in solubility, fusibility etc.

(*ii*) Both form associated molecules, $(HF)_x$ and $(H_2O)_x$. The formation of such molecules is due to hydrogen bonding in HF and H_2O molecules.

(*iii*) The boiling point of HF is exceptionally high in comparison to those of other hydrogen halides. (H_2F_2 = 19.4°C, HCl = –85°C, HBr = –67°C, HI = –36°C). Similarly the boiling point of H_2O is also exceptionally high in comparison to those of H_2S and H_2Se (H_2O = 100°C, H_2S = –61.8°C, H_2Se = –42°C). Just as the exceptionally high boiling point of HF is ascribed due to strong H-bonds between HF molecules, the high boiling point of H_2O is also attributed to the H-bonds between H_2O molecules.

(*iv*) Like H_2O, H_2F_2 is a weak acid and both are ionising solvents.

(*v*) Heats of formation of H_2F_2 and H_2O are close to each other.

Hydrochloric Acid, HCl

Manufacture

Large quantities of commercial HCl today are being obtained by the synthesis method which is based on the direct combination of hydrogen with chlorine in accordance with the equation :

$$H_2 + Cl_2 \rightleftharpoons 2HCl + 44{,}000 \text{ K.cals.}$$

This reaction is very slow under ordinary conditions but proceeds very smooth in the presence of high temperatures and sunlight. Both the gases, *viz.*, H_2 and Cl_2 are obtained as by-products during the manufacture of NaOH by the electrolysis of a solution of NaCl. Hence the process is quite economical. In this process pure electrolytic chlorine is burnt in an atmosphere of electrolytic hydrogen in large vertical silica tubes. The acid obtained by this process is quite pure.

Physical Properties

It is a colourless gas, heavier than air with pungent smell and acidic taste. It is extremely soluble in water and fumes strongly in moist air. 450 volumes of the gas dissolve in one volume of water. The gas can be easily liquefied and the liquid so produced boils at –83°C and freezes to a white crystalline solid at –11.3°C. It forms a constant boiling mixture with water when it contains 22.2 per cent of the acid. The mixture boils at 110°C. Dilute solution of the acid cannot, therefore, be concentrated by boiling beyond 22.2 per cent.

Chemical Properties

(*i*) Dissociation. HCl is quite stable but dissociates to give H_2 and Cl_2 at about 1500°C.

$$2HCl \longrightarrow H_2 + Cl_2$$

(*ii*) Action on litmus paper. Hydrogen chloride does not affect litmus in the absence of moisture but in the moist state or in solution it turns blue litmus red. This is due to the fact that it dissolves in water to give hydrochloric acid.

(*iii*) Combustibility. It is neither combustible nor a supporter of combustion.

(*iv*) Action on NH_3. It forms dense white fumes of ammonium chloride with ammonia.

$$HCl + NH_3 \longrightarrow NH_4Cl$$

(v) Acidic nature. The solution of the gas is highly acidic and one of the strongest acid. It reacts with metals like iron, zinc, with sodium, potassium and calcium, forming the corresponding *chlorides* with the evolution of hydrogen. HCl gas also reacts with metals like zinc, aluminium, tin etc., when heated, forming *anhydrous chlorides.*

(vi) Action on oxides, hydroxides, carbonates and bicarbonates. HCl neutralises oxides and hydroxides while carbonates and bicarbonates get decomposed with the evolution of CO_2. For example :

$$MgO + 2HCl \longrightarrow MgCl_2 + H_2O; NaOH + HCl \longrightarrow NaCl + H_2O$$

$$2HCl + No_2CO_3 \longrightarrow 2NaCl + H_2O + CO_2; HCl + NaHCO_3 \longrightarrow NaCl + H_2O + CO_2$$

(vii) Action of oxidising agents. HCl gets oxidised to Cl_2 by oxidising agents like MnO_2, $KMnO_4$, $K_2Cr_2O_7$ and PbO_2.

(viii) Precipitation reactions. When HCl is added to the solution of a soluble salt of Ag (I), Pb (II) and Hg (I), AgCl, $PbCl_2$ and Hg_2Cl_2 are precipitated.

(ix) Formation of aqua regia. HCl forms aqua regia with HNO_3 when mixed in the ratio of 3 : 1.

$$3HCl + HNO_3 \longrightarrow NOCl + 2Cl + 2H_2O$$

Uses

(*i*) Hydrochloric acid is used in the manufacture of chlorides and chlorine. (*ii*) It is used in the textile and dyeing industries, as also in tanning. (*iii*) In the form of *aqua regia* it is used for dissolving metals like gold and platinum. (*iv*) It is used as an important reagent in the laboratory. (*v*) It is also used in medicine. (*vi*) It is also used for cleaning iron sheets during tin plating, galvanisation etc. (*vii*) It is also used for the extraction of glue from animal tissues and bones.

Hydrobromic Acid, HBr

Preparation

HBr can be prepared by passing a mixture of H_2 and Br_2 over a platinum spiral heated to redness by an electric current or its aqueous solution can be obtained by passing SO_2 or H_2O through Br_2 water.

$$H_2 + Br_2 \longrightarrow 2HBr; Br_2 + H_2S \longrightarrow 2HBr + S$$

$$Br_2 + SO_2 + 2H_2O \longrightarrow H_2SO_4 + 2HBr$$

Physical Properties

Hydrogen bromide is a colourless pungent smelling gas heavier than air. It is very soluble in water and fumes strongly in moist air. At low temperature and high pressure it is changed to a liquid which boils at – 67°C and solidifies at –87°C. It forms a constant boiling mixture at 126°C when it contains 48% of the gas in solution.

Chemical Properties

***(i)* Action on litmus.** Dry HBr has no action on litmus but moist gas turns it red.

***(ii)* Combustibility.** It is neither combustible nor a supporter of combustion.

***(iii)* Acidic nature.** It reacts with metals like zinc, iron and tin to give hydrogen gas. It is a monobasic acid and reacts with alkalies, carbonates and oxides to give *bromides.*

$$Zn + 2HBr \longrightarrow ZnBr_2 + H_2$$
$$Na_2CO_3 + 2HBr \longrightarrow 2NaBr + H_2O + CO_2$$
$$NaOH + HBr \longrightarrow NaBr + H_2O$$

***(iv)* Action with ammonia.** With concentrated ammonia solution, white fumes of ammonium bromide are produced.

$$NH_3 - HBr \longrightarrow NH_4Br$$

***(v)* Reducing action.** It is more easily decomposed than hydrochloric acid and as such acts as a *reducing agent.* Sulphuric acid, hydrogen peroxide and air oxidise it to bromine.

$$H_2SO_4 + 2HBr \longrightarrow Br_2 + SO_2 + 2H_2O$$
$$2HBr + H_2O_2 \longrightarrow Br_2 + 2H_2O$$
$$4HBr + O_2 \text{ (From air)} \longrightarrow 2Br_2 + 3H_2O.$$

***(vi)* Action with $AgNO_3$ and $Pb(CH_3COO)_2$.** With silver nitrate it gives a pale yellow precipitate insoluble in nitric acid and sparingly soluble in ammonium hydroxide. With lead acetate it gives a white precipitate soluble in hot water.

$$HBr + AgNO_3 \longrightarrow \underset{\text{Pale yellow ppt.}}{AgBr} + HNO_3$$
$$2HBr + (CH_3COO)_2Pb \longrightarrow \underset{\text{White ppt.}}{PbBr_2} + 2CH_3COOH$$

***(vii)* Action of halogens.** Bromine is liberated from hydrobromic acid by fluorine and chlorine but iodine has no action.

$$2HBr + X_2(X = F, Cl) \longrightarrow 2HX + Br_2$$

Uses

HBr is an important laboratory regent and is used in preparing bromo derivatives of unsaturated organic compounds and some inorganic bromides. AgBr and KBr are used in photography and in medicine.

Hydroiodic Acid, HI

Preparation

HI can be prepared by the reactions shown below :

$$H_2 + I_2 \rightleftharpoons 2HI$$
$$2P + 3I_2 + 3H_2O \longrightarrow H_3PO_3 + HI \text{ (\textit{Laboratory preparation})}$$
$$H_2S + I_2 \longrightarrow 2HI + S$$
$$SO_2 + I_2 + 2H_2O \longrightarrow 2HI + H_2SO_4$$

Physical Properties

Hydrogen iodide is a colourless, pungent smelling gas, heavier than air and highly soluble in water. Like other halogen hydracids, it fumes in moist air and can be liquefied (b. pt. –35.5°C) and solidified (m. pt. –50.8°C). It forms a constant boiling mixture at 127°C when the solution contains 57% of hydrogen iodide.

Chemical Properties

***(i)* Combustibility.** It is neither combustible nor a supporter of combustion.

***(ii)* Solubility.** HI gas is highly soluble in water like HCl or HBr. The solution saturated at 10°C contains 70% of HI by weight. The solution fumes strongly in air and is acidic in nature. Thus the solution reacts with alkalies and bicarbonates to form salts called *iodides.*

$$NaOH + HI \longrightarrow NaI + H_2O$$
$$NaHCO_3 + HI \longrightarrow NaI + H_2O + CO_2$$

The solution is colourless when prepared but turns brown on keeping owing to the formation of free I_2 by the oxidation of HI.

$$4HI + O_2 \longrightarrow 2H_2O + 2I_2$$

***(iii)* Action on NH_3.** White fumes of NH_4I are obtained.

$$NH_3 + HI \longrightarrow NH_4I$$

***(iv)* Action on halogens.** F_2, Cl_2 or Br_2 liberate I_2 when passed through an aqueous solution of HI.

$$2HI + X_2(X_2 = F_2, Cl_2 \text{ or } Br_2) \longrightarrow 2HX + I_2$$

***(v)* Stability.** HI is the least stable of all the halogen acids and decomposes into H_2 and I_2 in sunlight or on heating.

$$2HI \rightleftharpoons H_2 + I_2$$

***(vi)* Reducing properties.** HI is the most powerful reducing agent among all the halogen acids. During this reaction HI is itself oxidised to free I_2. Thus HI reduces O_2 to H_2O, H_2O_2 to H_2O, Fe(III) salts to Fe(II) salts, HNO_2 to NO, $SnCl_4$ to $SnCl_2$, H_2SO_4 to SO_2,S or H_2S, HNO_3 to NO_2, HIO_3 to I_2, acidified $KMnO_4$ solution to $MnSO_4$, acidified $K_2Cr_2O_7$ to $Cr_2(SO_4)_3$ and Cl_2 to HCl.

***(vii)* Reaction with salts.** HI gives a yellow precipitate of AgI with $AgNO_3$ solution. With the solutions of $Pb(CH_3COO)_2$, $CuSO_4$ and $HgCl_2$, HI gives the pricipitates of PbI_2 (yellow), CuI and HgI_2 (scarlet) respectively. Here it may be noted that HF, HCl and HBr do not react with $CuSO_4$ and $HgCl_2$ solutions.

Conceptual Questions with Answers

Q.1 How will you prepare : *(a)* HF from CaF_2 *(b)* HBr from NaBr.

Ans : (*a*) HF is prepared by heating a mixture of calcium fluoride and conc. H_2SO_4.

$$CaF_2\ (s) + H_2SO_4\ (l) \xrightarrow{\Delta} CaSO_4\ (s) + 2HF\ (g)$$

(*b*) NaBr on heating with a non-oxidising acid such as H_3PO_4 gives HBr.

$$3NaBr\ (s) + H_3PO_4\ (l) \xrightarrow{\Delta} 3HBr\ (g) + Na_3PO_4\ (s)$$

Q.2 Name the products formed by the action of conc. H_2SO_4 on (*i*) CaF_2 (*ii*) NaCl (*iii*) NaBr and (*iv*) NaI.

Ans : (*i*) $CaF_2 + H_2SO_4 \xrightarrow{\Delta} CaSO_4 + \boxed{2HF}$

(*ii*) $2NaCl + H_2SO_4 \xrightarrow{\Delta} Na_2SO_4 + \boxed{2HCl}$

(*iii*) $2NaBr + H_2SO_4 \xrightarrow{\Delta} Na_2SO_4 + 2HBr$

$2HBr + H_2SO_4 \longrightarrow SO_2 + 2H_2O + Br_2$

$2NaBr + 2H_2SO_4 \longrightarrow Na_2SO_4 + SO_2 + 2H_2O + \boxed{Br_2}$

(*iv*) $2NaI + H_2SO_4 \xrightarrow{\Delta} Na_2SO_4 + 2HI$

$2HI + H_2SO_4 \longrightarrow SO_2 + 2H_2O + I_2$

$2NaI + 2H_2SO_4 \longrightarrow Na_2SO_4 + SO_2 + 2H_2O + \boxed{I_2}$

University Questions

1. How is hydrofluoric acid prepared? (*Raj. 87; Meerut 88; Utkal 86*)
2. Discuss the properties cf HF. (*Raj. 87; Meerut 88*)
3. Discuss the uses and structure of HF. (*Meerut 88*)
4. Give reasons for the following :
   (*i*) HF cannot be stored in glass bottles. (*Burdwan 86; Poona 85; Meerut 84*)
   (*ii*) Aqueous solution of HF_2 on electrolysis, does not give fluorine at anode. (*Delhi 87*)
   (*iii*) The solutions of organic compounds in HF become conductive. (*Raj. 86 S*)
5. Describe with the help of a neat diagram the manufacture of HCl. How is it purified? (*Punjab 85; GND 84*)
6. How is pure HBr prepared? How does it differ from HF, HCl and HI? (*Calicut 86*)
7. Explain the following :
   (*i*) HF is a liquid (b.pt. = 19°C) while HCl is a gas.
   (*ii*) HBr cannot be prepared by the action of conc. H_2SO_4 on a bromide salt. (*Kanpur 2000*)
8. Explain why, in aqueous solution, HF behaves as a weak acid in comparison to other hydrogen acids. (*Avadh 2000*)
9. Give IUPAC names of HClO, $HClO_2$, $HClO_3$ and $HClO_4$. (*Lucknow 2001*)

CHAPTER

40 Binary Halogen-Oxygen Compounds and Oxyacids of Halogens

Binary Halogen-Oxygen Compounds

Halogens do not combine directly with oxygen. However, binary halogen-oxygen compounds have been prepared by indirect methods. The known compounds of this type are : (oxidation number of halogen atom is given in bracket) $OF_2(-1)$, $O_2F_2(-1)$; $Cl_2O(+ 1)$, $ClO_2(+ 4)$, ClO_3 or $Cl_2O_6(+ 6)$, $Cl_2O_7(+ 7)$; $Br_2O(+ 1)$, BrO_2 (+ 4), $BrO_3(+ 6)$; I_2O_4 (+ 4), I_2O_5 (+ 5), I_4O_9 or $I^{3+}[I^{5+}O_3]_3$. Since F is more electronegative than oxygen, fluorine-oxygen compounds are called *fluorides of oxygen* rather than the oxides of fluorine, *e.g.* OF_2 is called *oxygen difluoride* and not *difluorine oxide.* Similarly O_2F_2 is called *dioxygen difluoride* and not difluorine dioxide. On the other hand, since oxygen is more electronegative than Cl, Br and I, oxygen compounds with these halogens are called the oxides of these halogens rather than the halides of oxygen, *e.g.* ClO_2 is called *chlorine dioxide* rather than dioxygen chlorine.

Fluorides of Oxygen

1. Oxygen Difluoride, OF_2.

It is prepared by passing F_2 through 2% NaOH solution.

$$2NaOH + 2F_2 \longrightarrow OF_2 + 2NaF + H_2O$$

The gas obtained by this method is collected over water. It is a pale yellow poisonous gas (b.pt. = –145°C, m.pt. = –223.8°C). It is very slightly soluble in water. It is a stable gas. It is relatively unreactive and can be mixed with H_2, CH_4 or CO without reaction although explosion will take place on sparking. It forms an explosive mixture with Cl_2, Br_2 and I_2. It dissociates into its elements on heating. It reacts slowly with H_2O but explodes with steam.

$$OF_2 + H_2O \longrightarrow O_2 + 2HF$$

It is readily hydrolysed by base. It is powerful oxidising agent. For example it liberates I_2 from KI solution

$$OF_2 + 2OH^- \longrightarrow O_2 + 2F^- + H_2O$$

$$OF_2 + 4KI + H_2O \longrightarrow 2KF + 2KOH + I_2$$

Metals and non-metals are oxidised and/or fluorinated by it. Xenon reacts with it in an electric discharge to form a mixture of fluoride and oxy-fluoride.

2. Dioxygen Difluoride, O_2F_2.

It is prepared by passing an electric discharge through a mixture of F_2 and O_2 cooled to about –185°C. It is a solid (m.pt. = –163.5°). On increasing the temperature O_2F_2 is converted into a pale-brown gas. Above –100°C, O_2F_2 is decomposed into F_2 and O_2. It is extremely potent fluorinating and oxidising agent. Many substances explode on exposure at low temperatures and even C_2F_4 is changed into COF_2, CF_4 etc. With Cl_2, O_2F_2 gives a purple and fairly stable intermediate compound, $(O_2ClF_2)_n$. O_2F_2 is used for oxidising primary aliphatic amines to the corresponding nitroso compounds.

Oxides of Chlorine

1. Chlorine Monoxide, Cl_2O

It is prepared by passing dry Cl_2 over freshly precipitated yellow mercuric oxide (HgO) previously heated to 300–400°C and contained in a cooled tube.

$$HgO + 2Cl_2 \text{ (dry)} \longrightarrow HgCl_2 + Cl_2O$$

It is a brownish-yellow gas which condenses in a freezing mixture to an orange coloured liquid (b.pt. = 2°C and m.pt. = –120.6°C). It has characteristic penetrating odour. Liquid Cl_2O explodes readily on heating or sparking giving Cl_2 and O_2. It also explodes when the sides of the test-tube containing it are scratched with a file or a glass rod.

$$2Cl_2O \longrightarrow 2Cl_2 + O_2$$

It dissolves in H_2O giving a golden yellow solution of hypochlorous acid, HClO.

$$Cl_2O + H_2O \rightleftharpoons 2HClO$$

It is, therefore, known as *anhydride of hypochlorous acid.* On cooling the aqueous solution of Cl_2O strongly, crystalline hydrate, $Cl_2O.H_2O$ (m.pt. = –36°C) is obtained. It is a strong oxidising agent and oxidises HCl to Cl_2.

$$Cl_2O + 2HCl \longrightarrow 2Cl_2 + H_2O$$

It attacks mercury but only slowly.

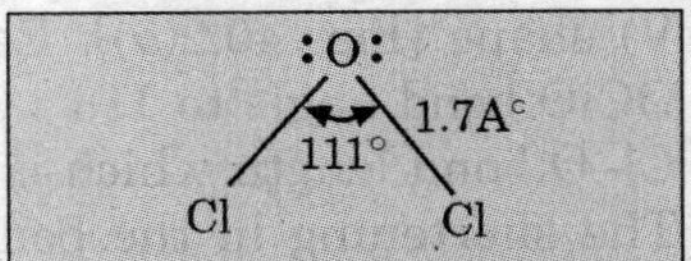

Fig. 40.1. V-shape of Cl_2O molecule. The dots shown at O-atom indicate the lone pairs of electrons.

The structure of Cl_2O molecule is the same as that of H_2O molecule. Thus Cl_2O molecule has V-shape (or angular or bent shape) as shown in Fig. 40.1 given in the margin. This shape results from sp^3 hybridisation of Cl atom. Since the electronegativity value of Cl atom is less than that of O-atom, the electron pairs used in the formation of O—Cl bonds in the molecule are nearer to O-atom and also at the same time (*lp*—*lp*) repulsion becomes less than (*bp*—*bp*) repulsion. Hence ClOCl bond angle increases from the expected tetrahedral angle to 111°.

2. Chlorine Dioxide, ClO_2

Preparation. Pure ClO_2 is obtained by passing dry Cl_2 over $AgClO_3$ heated to 90°C.

$$2AgClO_3 + Cl_2 \text{ (dry)} \longrightarrow 2AgCl + 2ClO_2 + O_2$$

The resulting gaseous mixture is passed through a freezing mixture when Cl_2 condenses while ClO_2 passes on condensation.

Properties. It is a dark yellow pungent smelling gas which can be condensed to a dark red liquid (b.pt. = 11°C). This liquid freezes at –9°C to an orange red crystalline solid. The liquid and solid both are fairly stable in dark but are slowly decomposed when exposed to light. The gas explodes and is decomposed into Cl_2 and O_2 by an electric spark, on heating with a glass rod or hot wire and in contact with turpentine, alcohol or ether

$$2ClO_2 \longrightarrow Cl_2 + 2O_2$$

It dissolves in H_2O giving a mixture of chlorous ($HClO_2$) and chloric acids ($HClO_3$).

$$2ClO_2 + H_2O \longrightarrow HClO_2 + HClO_3$$

With alkalies it gives a mixture of chlorite and chlorate (*e.g.* $KClO_2$ and $KClO_3$ with KOH)

$$2ClO_2 + 2KOH \longrightarrow KClO_2 + KClO_3 + H_2O$$

It is a powerful oxidising and bleaching agent. It liberates I_2 from acidified KI solution. Phosphorus, sulphur and many organic substances spontaneously catch fire in the gas. Like Cl_2O, ClO_2 also reacts with Hg but slowly. With O_3 at 0°C, it gives Cl_2O_6.

$$2ClO_2 + 2O_3 \longrightarrow Cl_2O_6 + 2O_2$$

Uses. It is used (*i*) for bleaching the highest grade paper pulp (*ii*) in water purification and for odour control. (*iii*) for improving the quality of low-grade fats and oils. (*iv*) in making the wool unshrinkable.

Structure. ClO_2 molecule has one coordinate, one three- electron and one two-electron (normal covalent) bond. Due to the presence of three-electron bond, ClO_2 is an odd-electron molecule. The presence of one unpaired electron in three-electron bond explains the paramagnetic character, colour and reactivity of the molecule.

Like NO_2 molecule, ClO_2 molecule also has a symmetrical angular (or bent or V) shape [Fig. 40.2(*a*)]. Each of the Cl–O bond distane is equal to 1.49A° and OClO bond angle to 118°. Cl–O bond length (= 1.49A°) is shorter than the single Cl–O bond length which is equal to the sum, $r(Cl) + r(O) = 0.99 + 0.73 = 1.72$A°. The shortening in the bond length is due to the presence of the three-electron bond in the molecule.

Here it should be noted that, unlike NO_2, ClO_2 has no tendency to dimerise. This is perhaps due to the fact the unpaired electron present in the three-electron bond is less localised on the central chlorine atom than in other odd-electron molecules. According to another view, ClO_2 molecule has the structure shown in Fig. 40.2 (*b*). This structure arises from sp^3 hybridisation of Cl-atom.

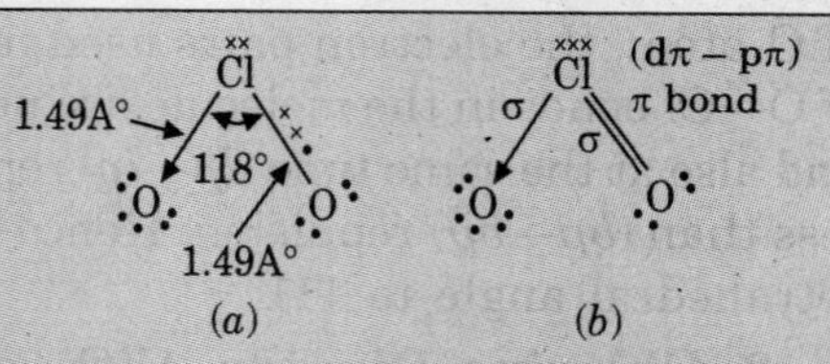

Fig. 40.2. Structure of ClO_2 molecule (*a*) Angular shape of ClO_2 molecule. (*b*) Structure of ClO_2 molecule based on sp^3 hybridisation of Cl-atom.

3. Dichlorine Hexoxide, $Cl_2O_6 \rightleftharpoons 2ClO_3$

It is best prepared by the action of O_3 on ClO_2 at 0°C atom.

$$2ClO_2 + 2O_3 \longrightarrow Cl_2O_6 + 2O_2$$

It is a dark red liquid (density = 1.65) which melts at 35°C. It is an explosive liquid, but less so than Cl_2. It is unstable and decomposes into ClO_2 and O_2 even at its melting point. It reacts violently with cooled water vapours and produces perchloric acid monohydrate ($HClO_4.H_2O$) or oxomium perchlorate, (H_3O^+) (ClO_4^-) and chloric acid, $HClO_3$.

$$Cl_2O_6 + 2H_2O \longrightarrow HClO_4.H_2O \text{ or } (H_3O^+)(ClO_4^-) + HClO_3$$

With an alkali, it gives perchlorate and chlorate.

$$Cl_2O_6 + 2NaOH \longrightarrow NaClO_4 + NaClO_3 + H_2O$$

With HF, it gives $HClO_4$ and chloryl fluoride, ClO_2F.

$$Cl_2O_6 - HF \longrightarrow HClO_4 + ClO_2F$$

In the vapour (gaseous) state it exists principally as ClO_3 molecule which has odd number of electrons (= 7 + 6 × 3 = 25) and hence paramagnetic while in the liquid state it gets dimerised and exists as Cl_2O_6 which has even number of electrons (= 7 × 2 + 6 × 6 = 50) and hence is diamagnetic.

$$\underset{\text{liquid}}{Cl_2O_6} \underset{\text{Cool}}{\overset{\text{Heat}}{\rightleftarrows}} \underset{\text{vapour}}{2ClO_3}$$

The structure of Cl_2O_6 is unknown.

4. Chlorine Heptoxide, Cl_2O_7. Preparation. It is prepared by the dehydration of anhydrous $HClO_4$ with P_2O_5 at –10°C followed by vacuum distillation with precautions against explosions.

$$2HClO_4 + P_2O_5 \longrightarrow Cl_2O_7 + 2HPO_3.$$

Properties. It is a colourless oily liquid (b.pt. = 82°C) which is extremely explosive in nature. It is more stable than Cl_2O or ClO_2 and may be poured on phosphorus, sulphur, wood or paper without explosion. It explodes on heating or on percussion in a few days. It slowly reacts with H_2O forming $HClO_4$. Thus Cl_2O_7 is the anhydride of perchloric.

$$Cl_2O_7 + H_2O \longrightarrow 2HClO_4$$

Structure. Infra red spectrum of Cl_2O_7 in vapour state has shown that this molecule has two ClO_3 groups which are linked together by an O-atom, *i.e.* Cl_2O_7 molecule has the structure, $O_3Cl—O—ClO_3$ in which two tetrahedra are sharing one O-atom (See Fig. 40.3). The molecule is polar (μ = 0.72).

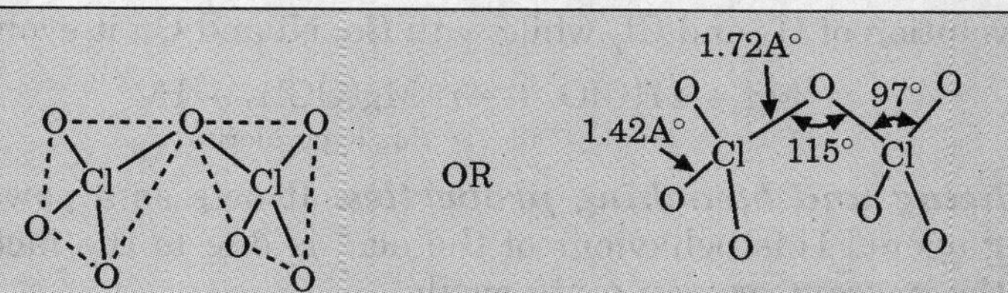

Fig. 40.3. Structure of Cl_2O_7 molecule.

Oxy-acids of Halogens

Except fluorine, all the remaining halogens form oxy-acids, which are grouped into four classes namely (1) *Hypo halous acids*, HXO(+ 1), *e.g.* hypochlorous acid (HClO), hypobromous acid (HBrO), hypoiodous acid (HIO) (2) *Halous acids*, HXO_2(+ 3), *e.g.* chlorous acid ($HClO_2$) (3) *Halic acids*, HXO_3(+ 5), *e.g.* chloric acid ($HClO_3$), bromic acid ($HBrO_3$), iodic acid (HIO_3) (4) *Perhalic acids*, HXO_4 (+ 7) *e.g.* perchloric acid ($HClO_4$), perbromic acid ($HBrO_4$), periodic acid (HIO_4). The numbers given in brackets indicate the oxidation state of the halogen atom in the acid. Here we shall discuss the oxy-acids of chloride and their important salts only.

Oxy Acids of Chlorine

I. Hypochlorous Acid, HClO

Preparation. This acid is known only in solution. It is best prepared by shaking chlorine water with freshly precipitated yellow mercuric oxide, HgO

$$2Cl_2 + 2HgO + H_2O \longrightarrow \underset{\substack{\text{Oxy chloride}\\ \text{of mercury}\\ \text{(insoluble)}}}{Hg_2OCl_2} + 2HClO$$

Commercially it is prepared by passing CO_2 into a suspension of bleaching powder, $CaOCl_2$ in water and then distilling.

$$2CaOCl_2 + CO_2 + H_2O \longrightarrow CaCl_2 + CaCO_3 + 2HOCl$$

Properties. (*i*) **Physical state and colour :** The concentrated solution of the acid is yellow in colour while the dilute solution is colourless. It is a weak acid, even weaker than H_2CO_3. Its dissociation constant value is equal to 3×10^{-8} at 20°C.

(ii) Decomposition. The dilute solution of the acid is fairly stable in the dark while concentrated solution is unstable and hence decomposes into HCl and O_2.

$$2HClO \longrightarrow 2HCl + O_2$$

The decomposition is accelerated by platinum black, manganese oxide and cobalt oxide.

On distillation, HClO decomposes into H_2O and Cl_2O.

$$2HClO \xrightarrow{\text{Distillation}} Cl_2O + H_2O$$

(iii) Action on metals. HClO dissolves Mg with the evolution of H_2; Fe and Al with the evolution of H_2 and Cl_2 while with Co, Ni and Cu it evolves Cl_2 and O_2.

$$Mg + 2HClO \longrightarrow \underset{\text{Mg.hypochlorite}}{Mg(OCl)_2} + H_2$$

(iv) Oxidising and bleaching properties. It acts as a powerful oxidising and bleaching agent. This behaviour of the acid is due to the fact that the acid can liberate the nascent oxygen quite easily.

$$HClO \longrightarrow HCl + O$$

Thus the acid liberates I_2 from the acidified KI solution, oxidises $Cr(OH)_3$ to chromate (CrO_4^{2-}) in alkaline medium and oxidises white $Mn(OH)_2$ to brown hydrated maganic oxide, $MnO_2.2H_2O$ or $Mn(OH)_4$.

$$2KI + HClO + HCl \longrightarrow 2KCl + H_2O + I_2$$

$$2Cr(OH)_3 + 3NaClO + 4NaOH \longrightarrow 2Na_2CrO_4 + 3NaCl + 5H_2O$$

$$Mn(OH)_2 + NaOCl + H_2O \longrightarrow MnO_2.2H_2O + \text{ or } Mn(OH)_4 + NaCl.$$

(v) Action of H_2O_2. The acid reacts with H_2O_2 to give O_2.

$$HClO + H_2O_2 \longrightarrow HCl + H_2O + O_2$$

(vi) Action on metallic mercury. When the acid is shaken with metallic mercury, a light brown precipitate of basic mercuric chloride, HgCl(OH) which is soluble in HCl is obtained.

$$2Hg + 2HClO \longrightarrow \underset{\text{Basic mercuric chloride}}{2HgCl(OH)}$$

(vii)Action of alkalies. With alkalies it forms salts which are called *hypochlorites*.

Uses. The acid is used for bleaching paper pulp etc.

Structure of hypochlorite ion, ClO^-. ClO^- ion has linear shape (Fig. 40.4) with Cl—O bond distance = 1.70A° and Cl—O bond energy is equal to 209 KJ/mole. Linear shape results from sp^3 hybridisation of Cl atom (central atom).

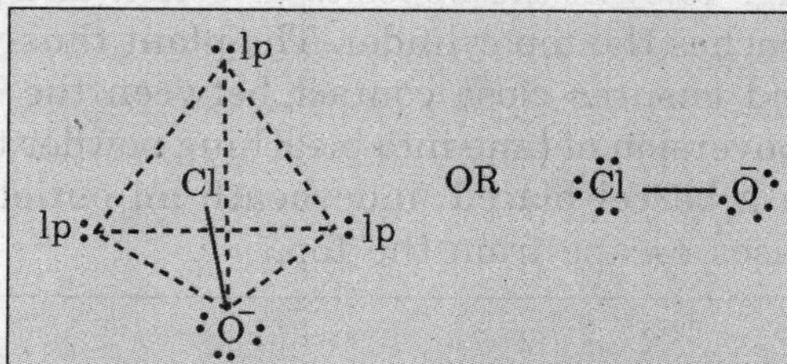

Fig. 40.4. Linear shape of hypochlorite ion, ClO^-.

Bleaching Powder, $CaOCl_2$ or Ca(OCl)Cl (Mixture of salts of HCl and HClO)

Bleaching powder is probably not a single compound but is a mixture of salts of hydrochloric acid (HCl) and hypochlorous acid (HClO) and hence has the formula $Ca^{2+}Cl^-(ClO)^-$ or $CaOCl_2$. The formula, CaCl(ClO)shows that bleaching powder is calcium *chloro-hypochlorite.* The formula *viz.* $CaOCl_2$ explains all the properties of bleaching powder.

Manufacture. Bleaching powder is manufactured by the action of Cl_2 on dry slaked lime, $Ca(OH)_2$. Following are the two main views regarding the formation of bleaching powder by this method :

(i) Odling's view (1861). According to this view, bleaching powder is calcium chloro-hypochlorite, $Ca^{2+}Cl^-(ClO)^-$.

$$Ca(OH)_2 + Cl_2 \longrightarrow \underset{\text{Calcium chloro-hypochlorite (Bleaching powder)}}{CaCl(ClO)} + H_2O$$

(ii) Bunn, Clark and Clifford's view (1935). According to this view, bleaching powder is usually supposed to be a mixture of calcium hypochlorite, $Ca(OCl)_2$ and basic calcium chloride, $CaCl_2.Ca(OH)_2.H_2O$.

Thus the formation of bleaching powder can be shown as :

$$2Ca(OH)_2 + 2Cl_2 \longrightarrow \underset{\text{Cal . hypo-chlorite}}{Ca(OCl)_2} + CaCl_2 + 2H_2O$$

$$CaCl_2 + Ca(OH)_2 + H_2O \longrightarrow \underset{\text{Basic calcium chloride}}{CaCl_2 \, . \, Ca(OH)_2 \, . \, H_2O}$$

On adding : $$3Ca(OH)_2 + 2Cl_2 \longrightarrow \underbrace{Ca(OCl)_2 + CaCl_2 \, . \, Ca(OH)_2 \, . \, H_2O}_{\text{Bleaching powder}} + H_2O$$

The chlorine used in the manufacture of bleaching powder should be dilute and the temperature should be maintained below 40°C.

The manufacture of bleaching powder is carried in any of the following two plants :

1. Hasenclever's plant (*Old method*). The Hasenclever's plant used for the manufacture, consists of a number of horizontal cast iron cylinders each containing a rotating shaft, fitted with blades. Slaked lime is fed in at the hopper and is pushed from one end of the cylinder to the other with the help of the blades when the shaft rotates. A stream of chlorine is introduced from near the bottom (*See* Fig. 40.5) in the opposite direction and it is completely absorbed by the time it reaches the top cylinder. The plant thus works on the *principle of counter-current* and ensures close contact between the slaked lime and Cl_2 and thus complete conversion of lime into bleaching powder takes place. Bleaching powder is collected in a barrel placed underneath an outlet in lowermost cylinder while the waste gases escape from the top.

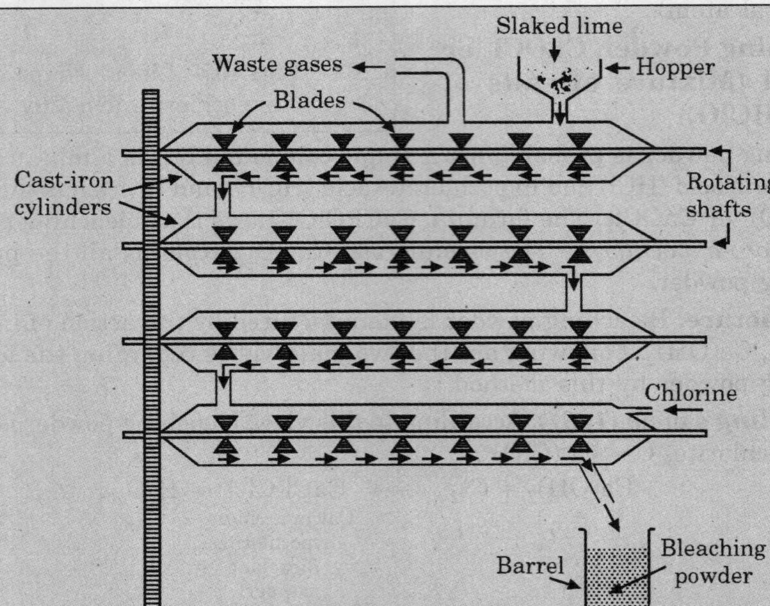

Fig. 40.5. Hasenclever's plant for the manufacture of bleaching powder (Old method).

2. Bachmann's plant (*Modern method*). Now-a-days bleaching powder is prepared in Bachmann's plant (*See* Fig. 40.6). It consists of a vertical cast iron tower fitted with eight shelves at different heights each equipped with rotating rakes.

Slaked lime is conveyed to the top of the tower with the help of a suitable air pumping arrangement and fed into the tower through a hopper at the top. A current of chlorine is introduced at the seventh shelf. The slaked lime thus added moves downward with the help of rotating rakes and meets an upcoming current of Cl_2. Slaked lime and Cl_2 react together and form bleaching powder which is collected in a barrel placed at the base. A current of hot air is blown into the eighth shelf to remove the uncombined free chlorine. This uncombined free chlorine goes out of the exit at the top.

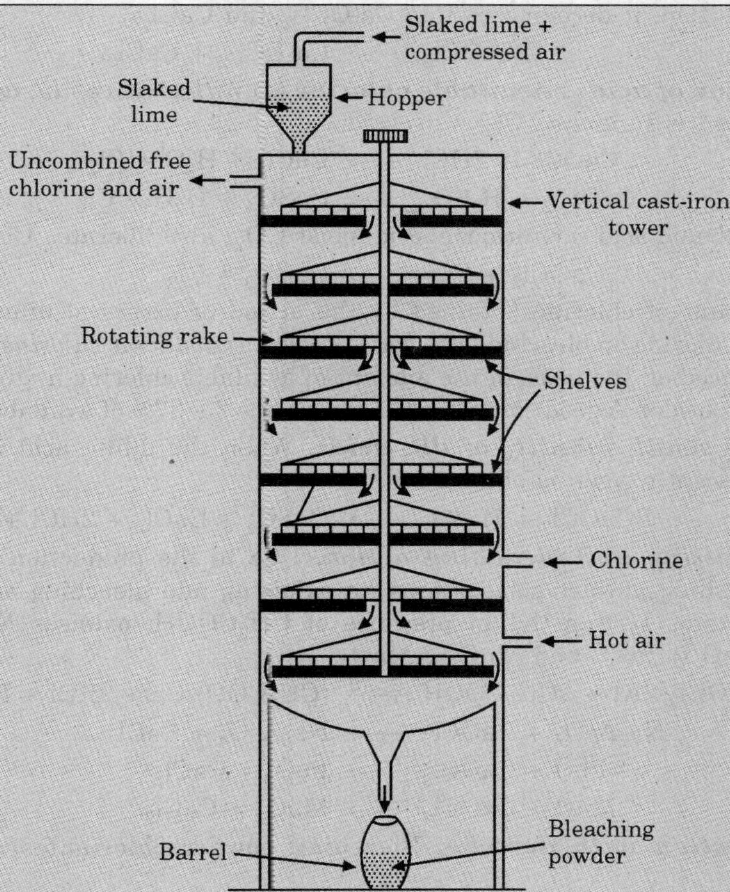

Fig. 40.6. Bachmann's plant for the manufacture of bleaching powder (Modern method).

Properties (*i*) *Physical state and colour*. It is a yellowish, white powder which strongly smells of chlorine.

(ii) Solubility in water. It is soluble in H_2O but a clear solution is never obtained due to the presence of impurities of lime in it.

With H_2O, it gives $Ca(OCl)_2$ which, on decomposition, gives O_2.

$$2CaOCl_2 + H_2O \longrightarrow CaCl_2 + Ca(OCl)_2 + H_2O$$
$$Ca(OCl)_2 \longrightarrow CaCl_2 + O_2$$

(iii) Decomposition. (*a*) On long standing, it slowly decomposes into a mixture of $Ca(ClO_3)_2$ and $CaCl_2$ (*Auto-oxidation*).

$$6CaOCl_2 \longrightarrow Ca(ClO_3)_2 + 5CaCl_2$$

(*b*) In presence of $CoCl_2$ (catalyst), it loses its oxygen.

$$6CaOCl_2 \xrightarrow{CoCl_2} 2CaCl_2 + O_2$$

(*c*) In solution, it decomposes into $Ca(OCl)_2$ and $CaCl_2$

$$2CaOCl_2 \longrightarrow Ca(OCl)_2 + CaCl_2$$

(iv) Action of acids : Available chlorine (*a*) *With excess of dil. acids.* When the dilute acid is in *excess*, Cl_2 is evolved.

$$CaOCl_2 + 2HCl \longrightarrow CaCl_2 + H_2O + Cl_2$$
$$CaOCl_2 + H_2SO_4 \longrightarrow CaSO_4 + H_2O + Cl_2$$

Even carbonic acid (*i.e.* atmospheric moist CO_2) also liberates Cl_2

$$CaOCl_2 + CO_2 \longrightarrow CaCO_3 + Cl_2$$

The amount of chlorine obtained by the action of excess of dilute acids or moist carbon dioxide on bleaching powder is known as *available chlorine.* Bleaching powder is priced on the basis of the amount of available chlorine in given sample of bleaching powder. A good sample usually contains 35–37% of available chlorine.

(b) With small quantity of dil. acids. When the dilute acid is in small quantity, nascent oxygen is obtained.

$$2CaOCl_2 + H_2SO_4 \longrightarrow CaSO_4 + CaCl_2 + 2HCl + 2O$$

(v) Oxidising and bleaching action. Due to the production of nescent oxygen, bleaching powder acts both as an oxidising and bleaching agent. Thus $CaOCl_2$ liberates I_2 from KI in presence of CH_3COOH, oxidises Na_3AsO_3 to Na_3AsO_4, PbO to PbO_2 and MnO to MnO_2.

$$CaOCl_2\ 2KI + 2CH_3COOH \longrightarrow (CH_3COO)_2Ca + 2KCl + H_2O + I_2$$
$$Na_3AsO_3 + CaOCl_2 \longrightarrow Na_3AsO_4 + CaCl_2$$
$$PbO + CaOCl_2 \longrightarrow PbO_2 + CaCl_2$$
$$MnO + CaOCl_2 \longrightarrow MnO_2 + CaCl_2$$

(vi) Reaction with acetone. Bleaching powder chlorinates acetone to chloroform.

$$3CaOCl_2 + \underset{\text{Acetone}}{2CH_3.CO.CH_3} \longrightarrow \underset{\text{Chloro-form}}{2CHCl_3} + (CH_3COO)_2Ca + 2Ca(OH)_2$$

(vii) Reaction with alcohol. Alcohol is chlorinated and oxidised by bleaching powder

$$4CaOCl_2 + 2C_2H_5OH \longrightarrow 2CHCl_3 + (HCOO)_2Ca + Ca(OH)_2 + 2H_2O + CaCl_2$$

(viii) Action of NH_4OH. When bleaching powder is heated with NH_4OH, N_2 gas is liberated

$$3CaOCl_2 + 2NH_4OH \longrightarrow CaCl_2 + 5H_2O + N_2$$

Uses. It is used (*i*) For bleaching cotton, linen and wood pulp in textile and paper factories. Delicate articles like straw or silk are not bleached by bleaching powder as these are injured by it. (*ii*) As a disinfectant and germicide for sewers and drains and for sterilization of water. (*iii*) For making wool unshrinkable. (*iv*) In the manufacture of chloroform. (*v*) As an oxidising agent in industry. (*vi*) For removing poisonous gases from the atmosphere during the war time.

Constitution of bleaching powder

Following different views have been put forward to explain the constitution of bleaching powder.

1. Dalton's view. Dalton in 1813 suggested that bleaching powder is a loose combination of quick lime (CaO) and chlorine (Cl_2). But this view was rejected because of the fact that quick lime has no tendency to absorb Cl_2.

2. Balard's view. In 1835 Balard suggested that bleaching powder is an equimolecular mixture of calcium chloride ($CaCl_2$) and calcium hypochlorite [$Ca(OCl)_2$]. Thus according to this view bleaching powder has the mixed formula, $CaCl_2.Ca(OCl)_2$. If this formula is supposed to be true, bleaching powder should exhibit the properties of both $CaCl_2$ and $Ca(OCl)_2$. But, in fact, bleaching powder does not possess the properties of both the salts as indicated by the facts : (*a*) $CaCl_2$ is extremely daliquescent while bleaching powder is not so. (*b*) $CaCl_2$ is soluble in alcohol while bleaching powder is not soluble. (*c*) Moist CO_2 or dil. acid has no action on $CaCl_2$ while they liberate the whole of the available chlorine from bleaching powder.

$$CaCl_2 + CO_2 \longrightarrow \text{No reaction}$$

$$\text{Bleaching powder} + CO_2 \longrightarrow CaCO_3 + Cl_2$$

(*d*) Bleaching powder cannot be obtained by mixing $CaCl_2$ and $Ca(OCl)_2$ in equimolecular proportions. These points show that bleaching powder cannot be regarded as a mixture of $CaCl_2$ and $Ca(OCl)_2$.

3. Odling's view. In 1861, Odling suggested that bleaching powder is a mixed salt of HCl and HClO and hence gave the formula, CaCl(OCl) or $Ca^{+2}\begin{matrix} / Cl^- \\ \backslash OCl^- \end{matrix}$.

Thus according to this view bleaching powder is *calcium chloro-hypochlorite*. This formula explains all the properties of bleaching powder. For example :

(*a*) In solution bleaching powder decomposes into $Ca(OCl)_2$ and $CaCl_2$.

$$2Ca\begin{matrix} / Cl \\ \backslash OCl \end{matrix} \longrightarrow Ca(OCl)_2 + CaCl_2$$

(*b*) On long standing, bleaching powder undergoes auto- oxidation into $Ca(ClO_3)_2$ and $CaCl_2$.

$$6Ca\begin{matrix} \diagup Cl \\ \diagdown OCl \end{matrix} \longrightarrow Ca(ClO_3)_2 + 5CaCl_2$$

(*c*) The action of dil. acids or moist CO_2 on bleaching powder gives all the available chlorine.

$$Ca\begin{matrix} \diagup Cl \\ \diagdown OCl \end{matrix} + CO_2 \longrightarrow CaCO_3 + Cl_2$$

4. Bunn, Clark and Clifford's view. In 1935, these chemists suggested that when Cl_2 gas is passed over dry slaked lime [$Ca(OH)_2$], basic calcium hypochiorite, $Ca(ClO)_2.2Ca(OH)_2$ and non-deliquescent basic calcium chloride, $CaCl_2.Ca(OH)_2.H_2O$ are obtained.

$$\underset{\text{Slaked lime}}{5Ca(OH)_2} + 2Cl_2 \longrightarrow \underset{\text{Basic calcium hypochlorite}}{Ca(ClO)_2.2Ca(OH)_2} + \underset{\text{Basic cal. chloride}}{CaCl_2.Ca(OH)_2.H_2O} + H_2O$$

Basic calcium hypochlorite, $Ca(ClO)_2$. $2Ca(OH)_2$ on further chlorination gives calcium hypochlorite, $Ca(ClO)_2.4H_2O$. Hence according to this view bleaching powder is a *mixture of calcium hypochlorite*, $Ca(ClO)_2.4H_2O$ and *basic calcium choride*, $CaCl_2.Ca(OH)_2$. H_2O. This view is in full agreement with the physical and chemical properties of bleaching powder and is confirmed by phase rule study, and microscopic and X-ray analysis.

Available Chlorine in Bleaching Powder

The amount of chlorine obtained by the action of excess of dilute acids or carbon dioxide on bleaching powder is known as available chlorine. Calculated on the basis of Odling formula, the maximum percentage of available chlorine from bleaching powder is about 49%.

$$\underset{\substack{40+16+71+18 \\ =145}}{CaOCl_2.H_2O} + H_2SO_4 \longrightarrow CaSO_4 + 2H_2O + \underset{\substack{35.5 \times 2 \\ =71}}{Cl_2}$$

145 gm. of $CaOCl_2.H_2O$ contains chlorine = 71 gm.

1 ,, ,, ,, ,, $= \frac{71}{145}$ gm.

100 ,, ,, ,, ,, $= \frac{71 \times 100}{145} = 49$

i.e., Percentage of available chlorine = 49%.

The actual percentage of available chlorine in the commercial sample of bleaching powder has, however, been found to be much lower. It is nearly 35–40%. The low percentage is due to :

(*i*) Loss of chlorine on exposure of bleaching powder to air.

(*ii*) Incomplete reaction between slaked lime and chlorine during it formation.

(*iii*) Impurities present in the slaked lime used for the manufacture.

Estimation of Available Chlorine in Bleaching Powder

The amount of available chlorine in bleaching powder can be estimated by two methods :

(*i*) **Arsenite method** (*Pento's method*). Chlorine can oxidise sodium arsenite to sodium arsenate. Thus the amount of sodium arsenite oxidised to sodium arsenate by given weight of bleaching powder sample is a measure of the available chlorine present in it.

$$Na_3AsO_3 + CaOCl_2 \longrightarrow Na_3AsO_4 + CaCl_2$$

A standard solution of sodium arsenite is prepared. A weighed quantity of bleaching powder is suspended in distilled water. A known volume of standard arsenite solution is then added to this suspension. A part of sodium arsenite is oxidised to sodium arsenate by the available chlorine. Tile unused sodium arsenite solution is estimated volumetrically by titration against a standard iodine solution in the presence of sodium bicarbonate and using starch solution as an indicator. The percentage of available chlorine can be calcualated from the volume of sodium arsenite used. Finally the amount of available chlorine in the sample of bleaching powder is calculated with the help of the following relationship :

1 ml. of $Na_3AsO_3 \equiv 0.0354$ g. of available chlorine.

(*ii*) **Iodometric method** (*Bunsen and Wagner's method*). It consists in treating a suspension of bleaching powder in distilled water with an excess of potassium iodide and acetic acid. Available chlorine in bleaching powder libarates an equivalent quantity of iodine from potassium iodide.

$$CaOCl_2 + 2CH_3COOH \longrightarrow Ca(CH_3COO)_2 + Cl_2 + H_2O$$

$$2KI + Cl_2 \longrightarrow 2KCl + I_2$$

Iodine thus liberated is estimated by titrating against a standard (N/10) sodium thiosulphate solution using starch solution as an indicator.

$$\underset{\text{Sod. thiosulphate}}{2Na_2S_2O_3} + I_2 \longrightarrow \underset{\text{Sod. tetrathionate}}{Na_2S_4O_6} + 2NaI$$

1 ml. of N/10 $Na_2S_2O_3 \equiv 0.00355$ g. of available chlorine

II. Chlorous Acid, $HClO_2$

It is the only definitely known halous acid. Being a weak acid, it cannot be isolated in the free state. In aqueous solution it is obtained by treating a suspension of barium chlorite, $Ba(ClO_2)_2$ with dil. H_2SO_4 and filtering off the precipitate of $BaSO_4$.

$$Ba(ClO_2)_2 + H_2SO_4 \longrightarrow 2HClO_2 + \underset{\text{ppt.}}{BaSO_4}$$

The freshly prepared solution of the acid is colourless but it soon decomposes to ClO_2 which colours the solution yellow. The acid gives a violet colour with ferrous sulphate. The acid undergoes auto-oxidation at ordinary temperature and changes into a mixture of hypochlorous acid (HClO) and chloric acid ($HClO_3$).

$$2HClO_2 \longrightarrow HClO + HClO_3$$

The acid liberates I_2 from an iodide.

$$4KI + HClO_2 + H_2O \longrightarrow 4KOH + HCl + 2I_2$$

Chlorite ion, ClO_2^- has *angular shape* (Fig 40.7) which results from sp^3 hybridisation of Cl atom (central atom). Each Cl—O bond distance is equal to 1.64A° and each OClO bond angle is equal to 111°.Cl—O bond energy is equal to 245 KJ/mole.

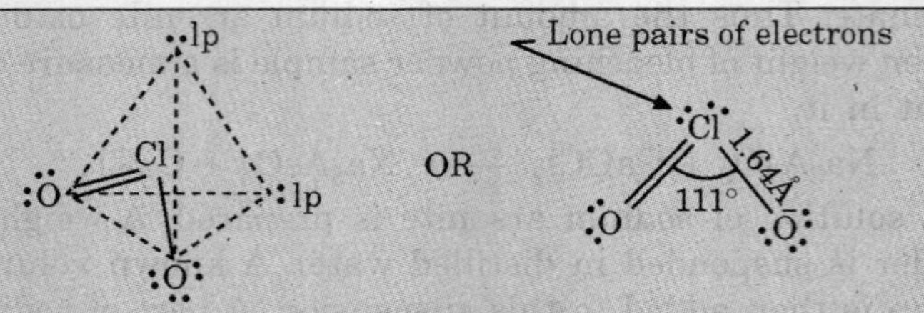

Fig. 40.7 Angular shape of chlorite ion, ClO_2^-

III. Chloric Acid, $HClO_3$

This acid is known only in solution. The acid is conveniently prepared by the action of dil. H_2SO_4 on barium chlorate, $Ba(ClO_3)_2$.

$$Ba(ClO_3)_2 + H_2SO_4 \longrightarrow \underset{\text{Insoluble}}{BaSO_4} + 2HClO_3$$

Insoluble $BaSO_4$ is filtered off and the excess (*i.e.* unused) of H_2SO_4 is precipitated with baryta water. The filtrate is evaporated in a vacuum desicater over conc. H_2SO_4 until it contains 40% $HClO_3$. Note that on further concentration, the acid decomposes and hence further concentration is not required.

$$3HClO_3 \longrightarrow HClO_4 + Cl_2 + 2O_2 + H_2O$$

Concentrated acid is colourless and pungent smelling liquid. It is fairly stable in the dark. When exposed to light, it decomposes and gives O_2. It is for this reason that in the light, it acts as a powerful oxidising and bleaching agent. Organic substances like paper, cotton, wool etc. catch fire in contact with the acid. On evaporation or distillation, the acid undergoes disproportionation to give perchloric acid ($HClO_4$) and ClO_2.

$$\underset{\text{(O.N. of Cl = + 5)}}{3HClO_3} \longrightarrow \underset{\text{(O.N. of Cl = + 7)}}{HClO_4} + \underset{\text{(O.N. of Cl = + 4)}}{2ClO_2} + H_2O$$

Chlorate ion, ClO_3^- has trigonal pyramidal shape (Fig. 40.8) which results from sp^3 hybridisation of Cl atom. Two neutral O-atoms are linked with Cl-atom by double bonds while the negatively-charged O-atom is attached by a single bond. Each Cl–O bond distance is equal to 1.57 Å and each OClO bond angle has been found to be equal to 106°. Cl—O bond energy is equal to 244 KJ/mole.

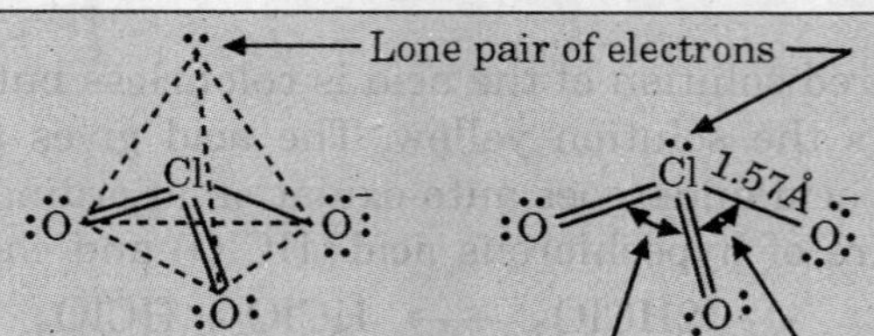

Fig. 40.8. Trigonal pyramidal shape of chlorate ion, ClO_3^-.

Potassium Chlorate, $KClO_3$ (Salt of $HClO_3$)

Preparation : It is an important salt of chloric acid, $HClO_2$ and *in the laboratory* it is prepared as :

$$6KOH \text{ (hot concentration solution)} + 3Cl_2 \longrightarrow KClO_3 + 5KCl + 3H_2O$$

Commercially it is prepared by electrolytic process which consists of electrolysing a hot concentrated solution of NaCl in a partitionless cell consisting of a steel tank which also acts as a cathode. The anodes are graphite rods which are arranged very close to each other so that the products obtained as a result of electrolysis mix and react with each other. The temperature is maintained at 40°C by circulating H_2O through cooling pipe.

Reactions. The Cl_2 set free at the anode reacts with NaOH formed at the cathode to produce $NaClO_3$

Ionisation : $NaCl \longrightarrow Na^+ + Cl^-$; $H_2O \longrightarrow H^+ + OH^-$

At cathode : $Na^+ + OH^- \longrightarrow NaOH$

$H^+ + e^- \longrightarrow H$; $H + H \longrightarrow H_2$

At anode : $Cl^- - e^- \longrightarrow Cl$; $Cl + Cl \longrightarrow Cl_2$

$$6NaOH - 3Cl_2 \longrightarrow NaClO_3 + 5NaCl + 3H_2O$$

$NaClO_3$ obtained as above is treated with an equivalent quantity of KCl to convert $NaClO_3$ into $KClO_3$ which is least soluble and hence crystallises out on cooling.

$$NaClO_3 + KCl \longrightarrow KClO_3 + NaCl$$

Electrolysis of KCl solution gives $KClO_3$ directly but, because it is sparingly soluble, it crystallises during the course of electrolysis and creates trouble. Hence it is best to use NaCl for electrolysis instead of KCl.

Properties. ***(i) Physical state and solubility.*** $KClO_3$ is a white crystalline solid with m.pt. = 370°C. It is sparingly soluble in cold water but more so in hot water.

(ii) Decomposition. On heating above 400°C, $KClO_8$ decomposes as :

$$2KClO_3 \longrightarrow 2KCl + 3O_2$$

(iii) Action of conc. *sulphuric acid.* When it is heated with concentrated sulphuric acid, long explosions occur due to the breaking up of the molecules of chlorine dioxide obtained during the reaction.

$$3KClO_3 + 3H_2SO_4 \longrightarrow 3KHSO_4 + HClO_4 + 2ClO_2 + H_2O$$

(iv) Action of conc. *hydrochloric acid.* With concentrated hydrochloric acid, a yellow gas which is called *euchlorine* and is a mixture of chlorine and chlorine dioxide is obtained.

$$2KClO_3 + 4HCl \longrightarrow 2KCl + 2H_2O + \underbrace{Cl_2 + 2ClO_2}_{\text{Euchlorine gas}}$$

(v) Oxidising properties. $KClO_3$ is a strong oxidising agent owing to the ease with which it parts with its oxygen. In these reactions, $KClO_3$ is reduced to KCl. Thus $KClO_3$ oxidises HCl to Cl_2.

$$2KClO_3 + 4HCl \longrightarrow 2KCl + Cl_2 + 2ClO_2 + 2H_2O$$

(vi) Reaction with S, P or charcoal. With sulphur, phosphorus or charcoal, $KClO_3$ forms an explosive mixture.

(vii)Action of iodine. Iodine in presence of nitric acid converts it into potassium iodate (KIO_3).

$$2KClO_3 + I_2 \longrightarrow 2KIO_3 + Cl_2$$

Uses. *(i)* Potassium chlorate is used in the manufacture of gunpowder, matches, fire works and photographic flash powders and is preferred over $NaClO_3$, since the latter is hydroscopic. *(ii)* It is a strong insecticide, and is used as an antiseptic, especially in sore throat. A very dilute solution of the salt is used as a gargle for sore throat. *(iii)* In the laboratory, it is employed in the preparation of oxygen and potassium perchlorate and as an oxidising agent.

IV. Perchloric Acid, $HClO_4$

It is the most stable oxy-acid of chlorine. *Anhydrous* perchloric acid is prepared by distillating potassium perchlorate, $KClO_4$ with 96–97.5% H_2SO_4 under 10–20 mm. pressure at 90–160°C.

$$KClO_4 + H_2SO_4 \longrightarrow KHSO_4 + HClO_4$$

It is purified by distilling at 40–60°C under 60 mm. pressure.

An aqueous solution of the acid is prepared by treating $Ba(ClO_4)_2$ with calculated quantity of dil. H_2SO_4 and removing the insoluble $BaSO_4$ by filtration.

$$Ba(ClO_4)_2 + H_2SO_4 \longrightarrow 2HClO_4 + \underset{\text{Insoluble}}{BaSO_4}$$

Anhydrous $HClO_4$ is a colourless mobile, hygroscopic and oily liquid. It fumes strongly in moist air and dissolves in water with a hissing sound due to the liberation of much heat. It forms hydrates with 1, 2, 2.5, 3, 3.5 molecules of water of crystallisation. It is unstable and decomposes with explosion on heating and sometimes merely on standing for a few days even in the dark. Aqueous solution of the acid is quite stable and does not decompose and hence can be kept indefinitely. It is highly dangerous acid and produces severe wounds on the skin. It is a powerful oxidising agent and inflames paper and wood. On dehydration with P_2O_5, it gives Cl_2O_7.

$$2HClO_4 + P_2O_5 \longrightarrow Cl_2O_7 + 2HPO_3$$

It is not reduced by nascent hydrogen but is reduced to chloride by strong reducing agents like $SnCl_2$, $CrCl_2$ etc. The aqueous acid is used in analysis for estimating potassium gravimetrically.

Perchlorate ion, ClO_4^- has tetrahedral shape (Fig. 40.9) which results from sp^3 hybridisation of Cl atom. Each Cl—O bond distance is equal to 1.45 Å and each of the OClO bond angles is equal to the tetrahedral angle (= 109.5°). Cl—O bond energy is equal to 364 KJ/mole.

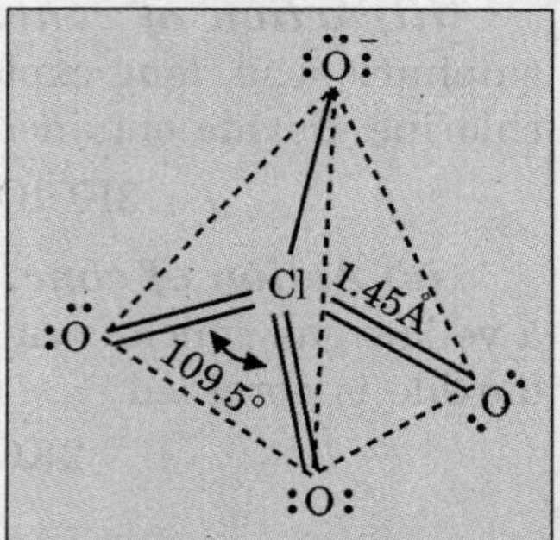

Fig. 46.9. Tetrahedral shape of perchlorate ion, ClO_4^-.

Potassium Perchlorate, $KClO_4$ (Salt of $HClO_4$)

It is the most important salt of perchloric acid. On *commercial scale*, it is prepared by electrolytic (anodic) oxidation of a conc. aqueous solution of $NaClO_3$ at room temperature. Sodium perchlorate, $NaClO_4$ thus formed is treated with KCl when sparingly soluble $KClO_4$ is precipitated. It is a colourless crystalline solid, sparingly soluble in water. On heating it decomposes as follows :

$$KClO_4 \xrightarrow{\Delta} KCl + 2O_2$$

With hot conc. H_2SO_4, it gives fumes of perchloric acid.

$$KClO_4 + H_2SO_4 \longrightarrow KHSO_4 + HClO_4$$

It is used in fireworks and eplosives.

Conceptual Questions with Answers

Q.1 Write two uses of ClO_2.

Ans : (*i*) *ClO_2 is a powerful oxidising agent and chlorinating agent.* Large quantities of ClO_2 are used for bleaching wood pulp and cellulose and for purifying drinking water.

(*ii*) *It is an excellet bleaching agent.* Its bleaching powder is about 30 times higher than that of Cl_2 and is used for bleaching fluor or make write bread.

Q.2 Balance and complete the following redox reaction in basic reaction.

$$\mathbf{ClO_2 + SbO_2^- \longrightarrow ClO_2^- + Sb(OH)_6^-}$$

Ans : The skeleton half-reactions can be written as :

$$\underset{(Sb\,=\,+3)}{SbO_2^-} \xrightarrow{\text{Oxidation}} \underset{(Sb\,=\,+5)}{Sb(OH)_6^-}$$

$$\underset{(Cl\,=\,+4)}{ClO_2} \xrightarrow{\text{Reduction}} \underset{(Cl\,=\,+3)}{ClO_2^-}$$

In the balanced form the above reactions can be written as :

$$SbO_2^- + 2H_2O + 2OH^- \xrightarrow{\text{Oxidation}} Sb(OH)_6^- + 2e^-$$

$$ClO_2 + e^- \xrightarrow{\text{Reduction}} ClO_2^-\ ] \times 2$$

On adding : $SbO_2^- + 2H_2O + 2OH^- + 2ClO_2 \longrightarrow Sb(OH)_6^- + 2ClO_2^-$

Q.3 Discuss the structure of the following oxides (*i*) I_2O_4 (*ii*) I_4O_9 (*iii*) I_2O_5 (*iv*) Cl_2O_6

Ans : (*i*) **Structure of I_2O_4 molecule.** This molecule is believed to be the true oxide of iodine but is an ionic compound which is represented as $[IO]^+ [IO_3]^-$ and is called **iodosyl iodate.** In IO^+ ion I is in +3 oxidation state while in IO_3^- ion, I is in +5 oxidation state.

(*ii*) **Structure of I_4O_9 molecule.** This molecule is not believed to be a true oxide of iodine but is regarded as **iodine iodate,** $I^{+3}[IO_3]_3^-$ in which iodine is in +3 and +5 oxidation state.

(*iii*) **Structure of I_2O_5 molecule.** The infrared spectrum of I_2O_5 molecule indicates that this molecule has two IO_2 units which are joined together by O-atom. Thus I_2O_5 can be represented as $O_2I—O—IO_2$, *i.e.* I_2O_5 molecule has oxygen-bridged structure shown in Fig. 40.10. Each I-atom has *lp* of electrons. I-atom is sp^3 hybridised.

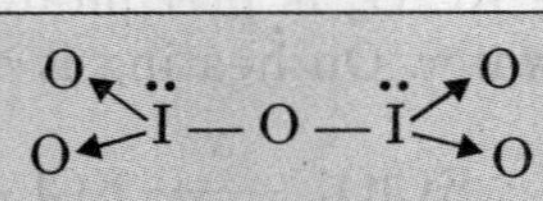

Fig. 40.10. Structure of I_2O_5 molecule

(*iv*) **Structure of Cl_2O_6 molecule.** The structure of Cl_2O_6 molecule is not known with certainty. However, its dissociation into two ClO_3 molecules shows that Cl_2O_6 is obtained by combining two ClO_3 molecules. Thus Cl_2O_6 molecule can be represented as $O_3Cl—ClO_3$ as shown in Fig. 40.11 Cl_2O_6 has one Cl—Cl bond which is formed by the sharing of two unpaired electrons on two Cl-atoms of two ClO_3 molecules. Thus in Cl_2O_6 molecule all the electrons are paired and hence this molecule is diamagnetic. Each Cl-atom is sp^3 hybridised. In this structure Cl-atom shows a covalency of seven.

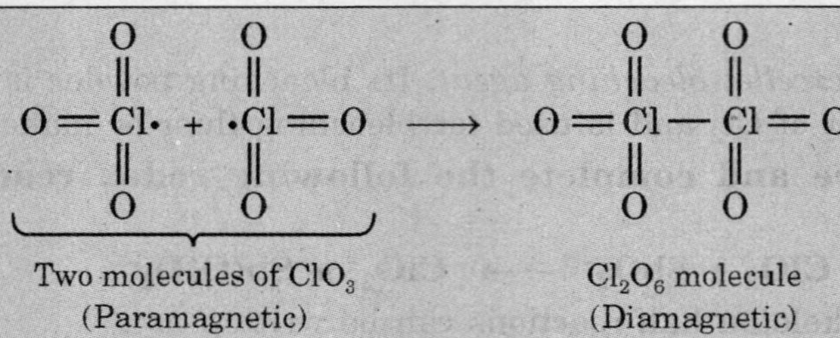

Fig. 40.11. Structure of Cl_2O_6 molecule

Q.4 Make a comparative study of NO_2 and ClO_2 molecules.

Ans. (*i*) The structures of the given molecules can be written as :

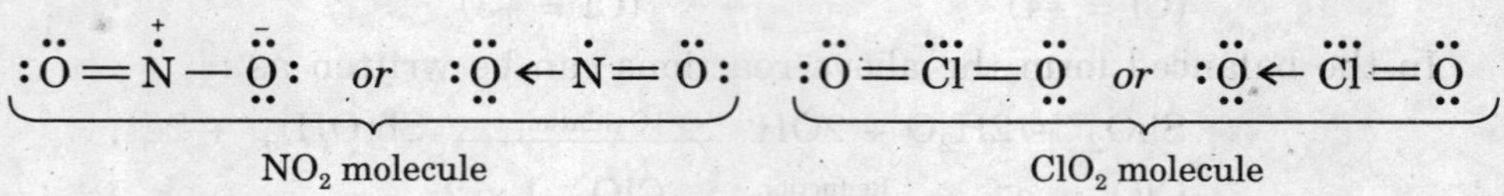

(*ii*) Both the molecules are odd electron molecules (NO_2 = 17, ClO_2 = 19)

(*iii*) N-atom in NO_2 molecule is sp^2 hybridised (σ-*bps* + *lps* = $2 + \frac{1}{2} = 2\frac{1}{2}$), but in ClO_2 molecule Cl-atom is sp^3 hybridised (σ-*bps* + *lps* = $2 + 1\frac{1}{2} = 3\frac{1}{2}$).

(*iv*) In both the molecules the central atom (*i.e.* N and Cl atoms) has one unpaired electron. Hence both the molecules are paramagnetic.

(*v*) On cooling, NO_2 dimerises to form N_2O_4 but ClO_2 does not dimerise.

(*vi*) NO_2 in the gaseous state has brown colour. ClO_2 also has marked colour.

(*vii*) NO_2 molecule has one ($p\pi$–$p\pi$)π bond while ClO_2 molecule has one ($p\pi$–$d\pi$)π bond.

(*viii*) NO_2 and ClO_2 both have angular geometry.

(*ix*) ONO bond angle = 132° and OClO bond angle = 118°.

(*x*) Both bond lengths in NO_2 = 120 pm and both bond lengths in ClO_2 = 149 pm.

(*xi*) B.O. in $NO_2 = \frac{2+1}{2}$ = 1.5 but B.O. of ClO_2 = 3.5

Q.5 Name a compound of fluorine which shows +1 oxidation state. How is this compound prepared ? Is this a disproportionation reaction ?

Ans. The compound of F which shows an oxidation state of +1 is HOF. It is prepared by passing F_2 over ice at 233 K.

$$\overset{0}{F_2} + H_2O\ (ice) \xrightleftharpoons{233\ K} HO\overset{+1}{F} + H\overset{-1}{F}$$

This is a disproportionation reaction since the oxidation state of F decreases from zero in F_2 to –1 in HF and increases to +1 in HOF.

Q.6 Perchloric acid is stronger acid than sulphuric acid. Explain.

Ans : The structures of perchloric acid ($HClO_4$) and sulphuric acid (H_2SO_4) are given below :

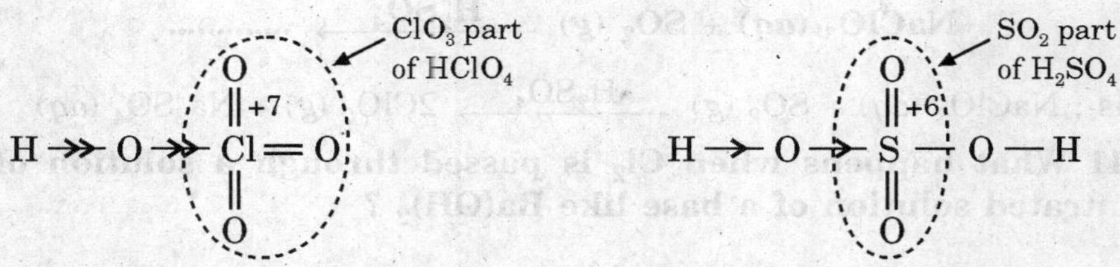

$HClO_4$ molecule H_2SO_4 molecule

O.S. of Cl-atom in $HClO_4$ is +7 and that of S-atom in H_2SO_4 is +6. Since O.S. of Cl is higher than that of S, ClO_3 part of $HClO_4$ molecule pulls the electrons of O—H bond more strongly and hence O—H bond can break more easily to liberate H as proton (H^+) than SO_2 part in H_2SO_4. Thus $HClO_4$ is a stronger acid than H_2SO_4.

Q.7 Balance the following reaction taking place in alkaline medium.

$$\mathbf{Cl_2 + KIO_3 \longrightarrow Cl^- + KIO_4}$$

Ans : The skeleton oxidation and reduction half-reactions for the given reaction are :

$$\underset{(I\ =\ +5)}{KIO_3} \xrightarrow{\text{Oxidation}} \underset{(I\ =\ +7)}{KIO_4}$$

$$\underset{(Cl\ =\ 0)}{Cl_2} \xrightarrow{\text{Reduction}} \underset{(Cl\ =\ -1)}{Cl^-}$$

Above half-reactions, in their balanced form, can be written as :

$$KIO_3 + 2OH^- \xrightarrow{\text{Oxidation}} KIO_4 + H_2O + 2e^-$$

$$Cl_2 + 2e^- \xrightarrow{\text{Reduction}} 2Cl^-$$

On adding : $KIO_3 + Cl_2 + 2OH^- \longrightarrow KIO_4 + 2Cl^- + H_2O$

Q.8 Write chemical equation for the reaction of alkaline perborate with Zn, giving tetrahydroxzincate anion.

Ans : In this reaction perborate ion, BrO_4^- (Br = +7) oxidises Zn(Zn = 0) to tetrahydroxo-zincate anion, $[Zn(OH)_4]^{2-}$ (Zn = +2) in alkaline medium and is itself gets reduced to bromate ion, BrO_3^- (Br = +5). Thus the two half-reactions can be written as :

$$Zn + 4OH^- \xrightarrow{\text{Oxidation}} [Zn(OH)_4]^{2-} + 2e^-$$

$$BrO_4^- + H_2O + 2e^- \xrightarrow{\text{Reduction}} BrO_3^- + 2OH^-$$

On adding : $Zn + 2OH^- + BrO_4^- + H_2O \longrightarrow [Zn(OH)_4]^{2-} + BrO_3^-$

Q.9 How is ammonium perchlorate (NH_4ClO_4) prepared ? What is the use of this compound ?

Ans : NH_4ClO_4 is obtained as a precipitate by treating in an aqueous solution of $NaClO_4$ with an aqueous solution of NH_4Cl.

$$NaClO_4 + NH_4Cl \longrightarrow NH_4ClO_4 + NaCl$$

NH_4ClO_4 is used as an oxidiser in solid rocket propellants in missiles.

Q.10 Complete the following equation

$$\mathbf{NaClO_3\ (aq) + SO_2\ (g) \xrightarrow{H_2SO_4} ....}$$

Ans : $NaClO_3\ (aq) + SO_2\ (g) \xrightarrow{H_2SO_4} 2ClO_2\ (g) + Na_2SO_4\ (aq)$

Q.11 What happens when Cl_2 is passed through a solution of hot concentrated solution of a base like $Ba(OH)_2$?

Ans : $6Cl_2 + 6Ba(OH)_2\ (\text{conc}) \xrightarrow{\Delta} 5BaCl_2 + Ba(ClO_3)_2 + 6H_2O$

Q.12 Write the balanced equations for the following reactions :

(i) Sodium iodate is added to a solution of sodium bisulphite.

(ii) Iodate ion reacts with bisulphite ion to liberate I_2.

Ans : (*i*) The reaction between sodium iodate ($NaIO_3$) and sodium bisulphite ($NaHSO_3$) is a redox reaction in which $NaHSO_3$ (S = +4) is oxidised to $NaHSO_4$ (S = +6) and Na_2SO_4 (S = +6) while $NaIO_3$ (I = +5) gets reduced to I_2 (I = 0).

$$\underset{(I = +5)}{2NaIO_3} + \underset{(S = +4)}{5NaHSO_3} \longrightarrow \underset{(S = +6)}{3NaHSO_4} + \underset{(S = +6)}{2Na_2SO_4} + H_2O + \underset{(I = 0)}{I_2}$$

As a matter of fact the above reaction takes place through the following steps :

(*a*) $NaIO_3$ and $NaHSO_3$ combine together to produce HIO_3

$$NaIO_3 + NaHSO_3 \longrightarrow Na_2SO_3 + HIO_3\ ] \times 2 \quad ...(i)$$

(*b*) HIO_3 produced as above gives nescent oxygen and is itself reduced to I_2.

$$2HIO_3 \longrightarrow I_2 + H_2O + 5O \quad ...(ii)$$

On adding equations (*i*) and (*ii*), we get :

$$2NaIO_3 + 2NaHSO_3 \longrightarrow 2Na_2SO_3 + I_2 + H_2O + 5O \quad ...(iii)$$

(*c*) Nescent oxygen as obtained in equation (*iii*) oxidises $NaHSO_3$ to $NaHSO_4$ and Na_2SO_3 to Na_2SO_4

$$NaHSO_3 + O \longrightarrow NaHSO_4] \times 3 \quad ...(iv)$$
$$(S = +4) \qquad (S = +6)$$
$$Na_2SO_3 + O \longrightarrow Na_2SO_4] \times 2 \quad ...(v)$$
$$(S = +4) \qquad (O = +6)$$

On adding equations (*iii*), (*iv*) and (*v*), we get :

$$2NaIO_3 + 5NaHSO_3 \longrightarrow I_2 + H_2O + 3NaHSO_4 + 2Na_2SO_4$$

(*Molecular equation*)

This molecular equation in its ionic form can be obtained as follows :

$$2IO_3^- + 12H^+ + 10e^- \xrightarrow{\text{Reduction}} I_2 + 6H_2O$$
$$(I = +5) \qquad (I = 0)$$
$$HSO_3^- + H_2O \xrightarrow{\text{Oxidation}} HSO_4^- + 2H^+ + 2e^-] \times 3$$
$$(S = +4) \qquad (S = +6)$$
$$HSO_3^- + H_2O \xrightarrow{\text{Oxidation}} SO_4^{2-} + 3H^+ + 2e^-] \times 2$$
$$(S = +4) \qquad (S = +6)$$

On adding : $5HSO_3^- + 2IO_3^- \longrightarrow 3HSO_4^- + 2SO_4^{2-} + I_2 + H_2O$

(*ii*) The reaction between iodate ion (IO_3^-) and bisulphite ion (HSO_3^-) leading to the liberation of I_2 can be represented by the following ionic equation.

$$2IO_3^- + HSO_3^- \longrightarrow 3HSO_4^- + 2SO_4^{2-} + H_2O + I_2$$
$$(I = +5) \quad (S = +4) \qquad (S = +6) \quad (S = +6) \qquad (I = 0)$$

University Questions

1. Write a note on "Oxides of iodine and fluorine". (*Madras 87*)
2. Describe the preparation and properties of the oxides of bromine. (*Bharathidasan 85*)
3. Write a short note on "Oxy-acids of chlorine". (*Raj. 86*)
4. Write a short note on bleaching powder. (*Jiwaji 85*)
5. What is available chlorine? Describe the chemistry involved in the method for the determination of percentage of available chlorine in a given sample of bleaching powder. (*Delhi 87; Agra 84*)

CHAPTER 41

Interhalogen Compounds, Polyhalides and Pseudohalogens

Interhalogen Compounds

We know that halogen atoms have different electronegativity. Due to this difference in electronegativity, the halogen atoms combine with each other and give rise to the formation of binary covalent compounds which are called *interhalogen compounds.* These compounds are of four types : (*i*) *AB type* (+ 1), *e.g.* ClF, BrF, IF (detected spectroscopically), BrCl, ICl, IBr (*ii*) AB_3 *type* (+ 3), *e.g.* ClF_3, BrF_3, IF_3, ICl_3 (*iii*) AB_5 *type* (+ 5), *e.g.* ClF_5, BrF_5, IF_5 (*iv*) AB_7 *type* (+ 7), *e.g.* IF_7. The number given in the bracket indicates the oxidation number of A atom. A halogen atom is larger than B halogen atom, *i.e.* A atom is less electronegative than B atom. Interhalogen compounds are named as the halides of less electronegative halogen atom. Thus ClF is called *chlorine monofluoride* and not *fluorine monochloride*, since the electronegativity of Cl atom is less than that of F.

General Methods of Preparation.

Following are the important general methods by which interhalogen compounds can be prepared.

1. By the direct combination of halogens. All the interhalogens, except IF_7, can be prepared by the direct combination of the halogens under appropriate conditions. The following examples are noteworthy. For the preparation of AB_3 and AB_5 interhalogen compounds excess of B atom must be used in the reaction mixture.

$$\underbrace{Cl_2 + F_2}_{\text{(equal volumes)}} \xrightarrow{250^\circ} 2ClF$$

$$Cl_2 + 3F_2 \text{ (excess)} \xrightarrow{250^\circ} 2ClF_3$$

$$\underbrace{I_2 + Cl_2 \text{ (liquid)}}_{\text{in equimolecular amount}} \longrightarrow 2ICl$$

$$I_2 + 3Cl_2 \text{ (liquid and in excess)} \longrightarrow 2ICl_3$$

$$Br_2 \text{ (vapour)} + 3F_2 \text{ (diluted with } N_2) \longrightarrow 2BrF_3$$

$$I_2 + 3F_2 \text{ (in freon)} \xrightarrow{-78^\circ C} 2IF_3$$

$$Br_2 + 5F_2 \text{ (excess)} \longrightarrow 2BrF_5$$

2. From lower interhalogens. Higher interhalogens can be prepared by the interaction of lower interhalogens with halogens. This method is particularly used for the preparation of *halogen fluorides.* The following examples illustrate the method.

$$ClF + F_2 \longrightarrow ClF_3$$

$$ClF_3 + F_2 \text{ (excess)} \xrightarrow{350°} ClF_5$$

$$BrF_2 + F_2 \xrightarrow{200°} BrF_5$$

$$IF_5 + F_2 \xrightarrow{270°} IF_7$$

3. By miscellaneous methods. Some miscellaneous methods have been illustrated by the following examples :

$$6HCl + KIO_3 + 2KI \longrightarrow 2KCl + 3H_2O + 3ICl$$

$$Cl_2 + ClF_3 \xrightarrow{250\text{–}350°} 3ClF$$

$$KCl + 3F_2 \xrightarrow{200°} KF + ClF_5$$

$$3I_2 + 5AgF \longrightarrow 5AgI + IF_5$$

$$8F_2 + PbI_2 \longrightarrow PbF_2 + 3IF_7$$

General Properties.

1. Physical state. The interhalogen compounds may be covalent gases (*e.g.* ClF, BrF, ClF_3, IF_7), liquids (*e.g.* BrF_3, BrF_5, IF_5) or solids (*e.g.* ICl, IBr, IF_3, ICl_3).

2. Colour. Although many of the interhalogen compounds containing fluorine are *colourless*, yet those made up of heavier halogens are *coloured.* The colour becomes deeper with the increase of the molecular weight of the compound. In this behaviour these compounds resemble the halogens themselves.

3. Diamagnetic nature. Since all the valence electrons in interhalogen compounds are present as bonding or non-bonding (*i.e.* lone pairs) electron pairs, these compounds are diamagnetic in nature.

4. Heat of formation. All these compounds have low heats of formation. For example these values for IBr and BrCl are –25 and –0.34 kcal/mole respectively. These are extremely volatile compounds. In general, compounds containing fluorine are more volatile than those containing chlorine, bromine or iodine.

5. Boiling points. For each type of interhalogen compounds, the boiling points increase with the increase in the electronegativity difference between A and B atoms.

6. Thermal stabitity. Thermal stability of AB type interhalogen compounds decreases with the decrease in electronegativity difference between A and B atoms. Thus the order of stability of some AB compounds is as :

IF (= 1.5) > BrF(1.2) > ClF(1.0) > ICl (0.5) > IBr (0.3) > BrCl(0.2)

In parentheses are given the electronegativity difference between A and B atoms. The above order is also explained by saying that greater is the difference between A and B, the more polar is the A—B bond and hence greater is the thermal stability of AB compound.

7. Reactivity. AB type compounds are more reactive than A_2 and B_2 molecules, since A—B bond in AB compounds is weaker than A—A and B—B bonds in A_2 and B_2 molecules respectively. Thus AB type compounds convert the metals into a mixture of two halides. For example :

$$ICl + 2Na \longrightarrow NaI + NaCl$$

The order of reactivity of some interhalogen compounds has been found as :

$$ClF_3 > BrF_3 > IF_7 > BrF_5 > IF_5 > BrF.$$

8. Hydrolysis. Hydrolysis gives halogen acid and oxy-halogen acid. The oxy-halogen acid is of larger (*i.e.* central) halogen atom. Examples are :

				Halogen acid		*Oxy-halogen acid*
BrCl	+	H_2O	$\longrightarrow$	HCl	+	HOBr
ICl	+	H_2O	$\longrightarrow$	HCl	+	HIO
ICl_3	+	$2H_2O$	$\longrightarrow$	3HCl	+	HIO2
IF_5	+	$3H_2O$	$\longrightarrow$	5HF	+	HIO_3
BrF_5	+	$3H_2O$	$\longrightarrow$	5HF	+	$HBrO_3$
IF_7	+	$6H_2O$	$\longrightarrow$	7HF	+	H_5IO_6

It may be noted that the oxidation state of A atom does not change during hydrolysis.

9. Reaction with non-metallic and metallic oxides. Non-metallic and metallic oxides are fluorinated by halogen fluorides to give the corresponding fluorides.

$$4BrF_3 + 3SiO_2 \longrightarrow 3SiF_4 + 2Br_2 + 3O_2$$
$$4BrF_3 + 2WO_3 \longrightarrow 2WF_6 + 2Br_2 + 3O_2$$

10. Addition reactions. The diatomic interhalogens (AB type compounds) add at olefinic double bond sites.

$$\text{—CH}=\text{CH— } + ICl \longrightarrow \text{—CH(I)—CH(Cl)—}$$

$$CH_2=CH_2 + BrCl \longrightarrow CH_2(Br)\text{—}CH_2(Cl)$$

11. Reaction with alkali metal halides. Reaction with alkali metal halides gives polyhalides.

$$\left.\begin{array}{l} IBr + NaBr \longrightarrow NaIBr_2 \\ ICl + KCl \longrightarrow KICl_2 \\ BrCl + CsI \longrightarrow CsIBrCl \\ ICl_3 + KCl \longrightarrow KICl_4 \\ BrF_3 + KF \longrightarrow KBrF_4 \end{array}\right\} \textit{Polyhalides}$$

Uses of Interhalogen Compounds.

Uses of some interhalogen compounds are given below :

(*i*) (*a*) A solution of ICl in glacial acetic acid is used for the estimation of iodine number which is a measure of the unsaturation of oils and fats.

$$\text{—CH}=\text{CH— + ICl} \longrightarrow \text{—}\underset{\text{I}}{\underset{|}{\text{CH}}}\text{—}\underset{\text{Cl}}{\underset{|}{\text{CH}}}\text{—}$$

(*b*) ICl is also used to iodonate and chlorinate organic compounds. For example :

Salicylic acid + ICl (vapours) ⟶ chlorinated product of salicylic acid

Salicylic acid + ICl (solution in nitrobenzene) ⟶ Iodonated product of salicylic acid

(*b*) The solution of ICl is cormmonly used as a catalyst in oxidising As_2O_3 by $Ce(SO_4)_2$. ICl is also used to prepare polyhalides.

(*ii*) (*a*) BrF_3 is used as a fluorinating agent since it converts many metals, metallic oxides and chlorides into their fluorides. For example :

$$6M + 2BrF_3 \longrightarrow 6MF + Br_2$$
$$3MCl + BrF_3 \longrightarrow 3MF + BrCl_3$$

(*b*) Many metal fluorides have been converted into complex fluorides. For example :

$$\left.\begin{array}{l} KF + BrF_3 \longrightarrow K^+[BrF_4]^- \\ SbF_3 + BrF_3 \longrightarrow Sb^{3+}[BrF_6]^{3-} \end{array}\right\} \text{polyhalides}$$

Thus BrF_3 is used to prepare polyhalides.

(*iii*) ClF_3 has been used as a fluorinating agent, as an incendiary and in cutting oil well tubes. It is also used as an oxidiser for propellants. It is also used to prepare polyhalides.

(*iv*) ICl_3 is used in medicine. It is also used to prepare polyhalides.

(*v*) BrF_5 is used in organic synthesis. It is also used as an oxidiser for propellants.

Some Important Interhalogen Compounds

1. Iodine Monochloride, ICl.

Preparation. It is prepared as a dark red liquid by passing Cl_2 over I_2 or by heating I_2 with $KClO_3$.

$$I_2 + Cl_2 \longrightarrow 2ICl$$
$$KClO_3 + I_2 \longrightarrow KIO_3 + ICl$$

Properties. (*i*) It solidifies on standing even at room temperature (15–25°). From this solid two forms of the compound may be obtained. These are (*a*) α-form which is ruby-red. It is a stable form (m.pt. = 27.2°) (*b*) β-form which is red rhombic. It is a meta stable form (m.pt. = 14.0°, b.pt. = 97.4°). These two forms are identical in the gaseous state. These can be separated by cooling the liquid ICl below 14° and by adding to the liquid a crystal of either of the two forms.

(*ii*) ICl is soluble in water and gets hydrolysed by it.

$$5ICl + 3H_2O \longrightarrow 2I_2 + HIO_3 + 5HCl$$

(*iii*) It also dissolves in alkalies to form I_2, iodate and chloride

$$5ICl + 6NaOH \longrightarrow 2I_2 + NaIO_3 + 5NaCl + 3H_2O$$

(*iv*) ICl is an electrical conductor when in the molten state. Thus when ICl is in the molten state, it ionises as :

$$2ICl \rightleftarrows I^+ + ICl_2^-$$

$$2I^+ + 2e^- \longrightarrow I_2 \text{ (At cathode)}$$

$$2ICl_2^- \longrightarrow I_2 + 2Cl_2 + 2e^- \text{ (At anode)}$$

The liberation of I_2 at cathode and both I_2 and Cl_2 at anode shows that ICl is an ionic compound and in it iodone is present as I^+ ion (*unipositive iodine*).

Uses. (*i*) A solution of ICl in glacial acetic acid is used for the dertermination of iodine number which measures the unsaturation of oils and fats. (*ii*) RICl is also used to iodate and chlorinate organic compounds. (*iii*) The solution of ICl is used as a catalyst in oxidising As_2O_3 by $Ce(SO_4)_2$. (*iv*) It is also used to prepare polyhalides (*v*) Because of the formation of I^+ cation, ICl is also used as an electrophilic iodinating agent. For example, it converts acetanitride into 4-iodoacetanitride and salicylic acid into 3, 5-diodosalicylic acid in nitrobenzene.

2. Iodine Monobromide, IBr.

Preparation. It is obtained by the direct combination of the elements.

$$I_2 + Br_2 \longrightarrow 2IBr$$

Properties. (*i*) It is a hard, crystalline solid of grey-violet colour.

(*ii*) Its vapour dissociates to a small extent on heating.

$$2IBr \rightleftarrows I_2 + Br_2$$

(*iii*) IBr is an electrical conductor when in the molten state. It ionises as :

$$IBr \longrightarrow I^+ + Br^-$$

Thus in IBr, iodine is present as I^+.

3. Iodine Trichloride, ICl_3.

Preparation. (*i*) It is obtained by the action of excess of Cl_2 on I_2 or on ICl at 100°C.

$$3Cl_2 \text{ (Excess)} + I_2 \longrightarrow ICl_3$$

$$Cl_2 \text{ (Excess)} + ICl \longrightarrow ICl_3$$

(*ii*) It is also formed when dry HCl gas reacts with heated I_2O_5.

$$I_2O_5 \text{ (heated)} + 10HCl \text{ (dry gas)} \longrightarrow 2ICl_3 + 5H_2O + 2Cl_2$$

(*iii*) It is also formed along with ICl by the action of HI on Cl_2

$$4HI + 5Cl_2 \longrightarrow 4HCl + I_2 + ICl + ICl_3$$

Properties. (*i*) It is a lemmon-yellow crystalline solid which fumes readily. It is soluble in organic solvents, liq.NH_3 and liq.SO_2.

(*ii*) It forms crystalline additive compounds with metallic chlorides. These compounds dissociate on exposure to air, giving ICl_3 and MCl.

$$MCl + ICl_3 \rightleftarrows \underset{\text{Additive compound}}{MICl_4}$$

(*iii*) With fluorides of K, Rb, Cs, it gives polyhalides of $MICl_3F$ type.

(*iv*) ICl_3 is completely hydrolysed by H_2O.

$$2ICl_3 + 3H_2O \longrightarrow 5HCl + HIO_3 + ICl$$

(*v*) ICl_3 dissociates on heating at 68° giving ICl and Cl_2.

$$ICl_3 \longrightarrow ICl + Cl_2$$

(*vi*) It reacts with CS_2 as :

$$4ICl_3 + 3CS_2 \longrightarrow CCl_4 + 2CSCl_2 + 2S_2Cl_2 + 2I_2$$

(*vii*) Like ICl, ICl_3 is also decomposed by KOH solution

$$3ICl_3 + 12KOH \longrightarrow 9KCl + KI + 2KIO_3 + 6H_2O$$

(*viii*) Molten ICl_3 has a high conductivity (= 8.4×10^{-3} ohm^{-1} cm^{-1}). The liberation of I_2 and Cl_2 at both electrodes indicates that ICl_3 is ionised into $[I^{3+}Cl_2]^+$ and $[I^{3+}Cl_4]^-$ ions.

$$2ICl_3 \rightleftarrows [ICl_2]^+ + [ICl_4]^-$$

$$2[ICl_2]^+ + 2e^- \longrightarrow I_2 + 2Cl_2 \text{ (At cathode)}$$

$$2[ICl_4]^- - 2e^- \longrightarrow I_2 + 4Cl_2 \text{ (At anode)}$$

Uses. ICl_3 is used in medicines. It is also used in the preparation of polyhalides.

4. Chlorine Trifluoride, ClF_3.

Preparation. It is prepared when Cl_2 is heated with excess of F_2 in a copper vessel at 250°C.

$$Cl_2 + 3F_2 \longrightarrow 2ClF_3$$

Properties. It is a colourless gas which condenses to give a pale-green liquid (b.pt. = – 12°C). It is exceedingly reactive and attacks a number of metals. It also attacks glass vigorously. It reacts with water to give a red liquid which crystallises at –70°C. The compound formed is ClFO.

Uses. (*i*) ClF_3 has been used as a fluorinating agent, as an incendiary and in cutting oil well tubes. (*ii*) It is also used as an oxidiser for propellants (*iii*) It is also used for the preparation of polyhalides.

5. Iodine Pentafluoride, IF_5.

Preparation. It is formed by the direct combination of the elements or by the action of F_2 on heated I_2O_5.

$$2I_2O_5 \text{ (heated)} + 10F_2 \longrightarrow 4IF_5 + 5O_2$$

It may also be prepared by heating together I_2 and AgF.

$$5AgF + 3I_2 \xrightarrow{\text{Red heat}} IF_5 + 5AgI$$

or

$$10AgF + I_2 \longrightarrow 2IF_5 + 10Ag$$

Properties. It is a colourless liquid (b.pt. = 98°). Its vapours attack the respiratory organs. It is a very poor conductor. It decomposes on heating to I_2 and IF_7

$$7IF_5 \longrightarrow I_2 + 5IF_7$$

When heated with H_2O, it is hydrolysed giving successively IOF_3, IO_2F and finally I_2O_5.

$$2IF_5 + 5H_2O \longrightarrow I_2O_5 + 10HF$$

Sulphur is converted to SF_6. Phosphorus in an atmosphere of IF_5 forms I_2 and PF_3. I_2O_5 when boiled with IF_5 gives white crystals of IOF_3.

$$3IF_5 + I_2O_5 \longrightarrow \underset{\text{iodine oxy-fluoride}}{5IOF_3}$$

These crystals decompose at 100° to give back IF_5 and IO_2F which is a white stable solid.

$$2IOF_3 \xrightarrow{100°} IF_5 + IO_2F$$

I_2O_4 dissolves in IF_5 to give IOF_3 and some I_2.

The oxides P_2O_5, V_2O_5, Sb_2O_5, Cr_2O_3 and WO_3 dissolve in hot IF_5 to form POF_3, VOF_3, $SbF_5.IO_2$, CrO_2F_2 and $WO_3.2IF_5$ respectively. Nitrogen (IV) oxide, NO_2, dissolves in IF_5 to give a cream coloured crystalline compound, $IF_5.NO_2$.

It is a good conductor of electricity and ionises as :

$$2IF_5 \rightleftarrows IF_4^+ + IF_6^-$$

6. Iodine Heptafluoride, IF_7.

Preparation. It may be prepared by refluxing IF_5 in a current of excess of F_2 through a tube of Pt heated to 270°

$$IF_5 + F_2 \xrightarrow{270°} IF_7$$

When iodine reacts with excess of F_2 at 300°C, IF_7 is formed.

Properties. It is a colourless gas having mustly odour. It can be condensed to a colourless liquid or to a colourless crystalline solid. It reacts with most metals (except the platinum group), non- metals including Cl_2 and I_2, water to give periodate and fluoride ions, many inorganic compounds and a variety of organic materials. It explodes when heated with H_2. It is slowly decomposed by water. Two of fluorine atoms present in IF_7 can easily be removed.

Structure of Interhalogen Molecules

1. Structure and Shape of AB Type Interhalogen Molecules.

The structure and shape of AB type molecules can be illustrated by considering ICl molecule. In this molecule the central atom *viz.* I atom is sp^3 hybridised and the molecule has *linear shape* with I—Cl bond length = 2.32A°. For the details see chapter 8.

2. Structure and Shape of AB_3 Type Interhalogen Molecules.

The structure and shape of this type of intorhalogen compounds can be illustrated by considering ClF_3 molecule. In this molecule the central atom *i.e.* Cl atom is sp^3d hybridised and ClF_3 molecule has *slightly bent T-shape*. For details see chapter 8. Like ClF_3, BrF_3 molecule also has *T-shaped structure.*

4. Structure and Shape of AB_5 Type Molecules.

In IF_5 molecule the central atom *viz.* I atom is sp^3d^2 hybridised and the molecule has *square pyramidal* shape. For details see chapter 8. Like IF_5, ClF_5 and BrF_5 molecules also have square pyramidal shape which results from sp^3d^2 hybridisation of Cl and Br atoms respectively.

4. Structure and Shape of IF_7 Molecule.

This molecule has *pentagonal bipyramidal shape* which results from sp^3d^3 hybridisation of I atom (central atom). For details see chapter 8.

Polyhalide Ions and Polyhalides

The ions like I_3^-, IBr_2^-, $IBrF^-$, ICl_4^-, I_7^-, I_9^- (anions), ICl_2^+, BrF_2^+, ClF_2^+ (cations) etc. which contain more than two same or different halogen atoms are called *polyhalide ions.* The ionic compounds like $Cs^+[IBr_2]^-$, $K^+[ICl_4]^-$, $[N(CH_3)_4]^+[ICl_4]^-$, $[ICl_2]^+[SbCl_6]^-$, $[BrF_2]^+[SbF_6]^-$ etc. which contain polyhalide cations or anions are called *polyhalides.* As in case of interhalogens, here too the heavier halogen atom plays the role of central atom.

Preparation of Polyhalides

(*i*) Simple polyhalides like $K^+Cl_3^-$, $NH_4^+I_5^-$ etc. can be prepared by the action of halogen on metallic halides, *e.g.*

$$Cl_2 + KCl \longrightarrow K^+Cl_3^- \;;\; 2I_2 + NH_4I \longrightarrow NH_4^+I_5^-$$

(*ii*) Mixed polyhalides like $Cs^+[ClF_4]^-$, $K^-[ICl_2]^-$ etc. can be prepared by the action of an appropriate gaseous halogen or interhalogen on a solid metallic halide in the absence of a solvent, *e.g.*

$$2F_2 + CsCl \longrightarrow Cs^+[ClF_4]^- \;;\; ICl + KCl \longrightarrow K^+[ICl_2]^-.$$

(*iii*) Polyhalides of one type can be converted into the other type by the action of halogen or interhalogen on them. *e.g.*

$$K[ICl_2] + Cl_2 \longrightarrow K[ICl_4]$$
$$K[ClIBr] + ICl \longrightarrow K[Cl_2] + IBr$$

Properties

(*i*) **Colour.** All the polyhalides are coloured compounds. The depth of the colour increases with the increase of the atomic number of the halogen atoms present. Thus the bromoiodides are red and chlorobromides are yellow.

(*ii*) **Dissociation in aqueous solution.** The polyhalides are usually very soluble in water from which they can often be crystallised. Some of them get dissociated in water to free halogen and halide ion (metallic monohalide). For example :

$$I_3^- \longrightarrow I^- + I_2$$

The stability of metallic trihalides of MX_3 type having the same cation in the same oxidation state and different trihalide anions in aqueous solution towards dissociation to free halogen and metallic monohalide (MX halide) is in the order :

$$MI_3 > MBr_3 > MCl_3.$$

For the metallic trihalides having the same trihalide anion and different cations in the same oxidation state, the stability increases with the increase of the size of the cation. For example : $NaI_3 < KI_3 < RbI_3 < CsI_3$. Note that the size of Na^+, Rb^+ and Cs^+ cations increases from Na^+ to Cs^+.

The polyhalides containing ICl_4^- anion dissociate in aqueous solution to metal chloride and ICl_3; the latter then hydrolyses to iodate, IO_3^-.

$$M^+[ICl_4]^- \longrightarrow MCl + ICl_3] \times 5$$
$$5ICl_3 + 9H_2O \longrightarrow 3HIO_3 + 15HCl + I_2$$
$$\overline{5M^+[ICl_4]^- + 9H_2O \longrightarrow 5MCl + 3HIO_3 + 15HCl + I_2}$$

or
$$5ICl_4^- + 9H_2O \longrightarrow 20Cl^- + 18H^+ + 3IO_3^- + I_2$$

***(iii)* Formation of complexes.** Some of the polyhalide ions form complexes with organic donor molecules. Some examples of such complexes are : $LiI_3 \cdot 4C_6H_5CN$, $MI_3 \cdot 2C_6H_5CN$ (M = Na or K).

***(iv)* Effect of heat : Thermal decomposition.** When the solid polyhalides are heated, they dissociate more or less readily. The ease of dissociation of a given polyhalide decreases with the increase of the size of the cation with which the polyhalide anion is associated. This means that *the stability of the solid polyhalides towards dissociation is maximum when the cation with which the polyhalide anion is associated is large.*

On thermal dissociation the polyhalides give the metal monohalide and halogen molecule or interhalogen molecule. For example :

$$CsI_3 \xrightarrow{\Delta} CsI + I_2$$
$$RbICl_2 \longrightarrow RbCl + ICl$$

When a polyhalide containing more than one type of halogen atoms undergoes thermal dissociation, the metal monohalide formed is always that which *has highest lattice energy.* Thus when $Cs^+[ICl_2]^-$ undergoes thermal dissociation, the products formed are CsCl + ICl and not CsI + Cl_2, since CsCl has higher lattice energy than CsI.

For trihalide ions formed by the same metal, the order of thermal stability is : $I_3^- > IBr_2^- > ICl_2^- > I_2Br^- > Br_3^- > BrCl_2^- > Br_2Cl^-$

***(v)* Reaction with halogens.** Sometimes when a polyhalide is allowed to react with a halogen molecule, one or two atoms of halogen present in the polyhalide are substituted by one or two atoms of the reacting halogen, as the case may be.

$$KIBr_2 + Cl_2 \longrightarrow KICl_2 + Br_2$$
$$CsBr_3 + I_2 \longrightarrow CsIBr_2 + IBr$$

Sometimes a higher polyhalide is formed.

$$KICl_2 + Cl_2 \longrightarrow KICl_4$$

***(vi)* Dissolution in liq. halogen or interhalogen.** Polyhalides may dissolve in liquid halogen or interhalogen to give solution from which the original polyhalide may be crystallised on evaporation. The polyhalides like $K^+[IBr_2]^-$, $NH_4^+[IBr_2]^-$ and $Rb^+[IBr_2]^-$ form solution in bromine.

Structure of Polyhalide Ions

1. Structure and Shape of Trihalide Anions.

Trihalide anions like ICl_2^-, I_3^- etc. have *symmetrical linear shape* (see chapter 8). This shape results from sp^3d hybridisation of I atom.

2. Structure and Shape of Trihalide Cations.

The trihalide cations like ICl_2^+, BrF_2^+, ClF_2^+ etc. have *angular shape* which results from sp^3 hybridisation of the central atom. For the details of the structure of ICl_2^+ ion see chapter 8.

Pseudohalides and Pseudohalogens

There are certain uninegative ions (*e.g.* CN^-, OCN^-, N_3^- etc.) which are made up of two or more electronegative atoms and resemble halide ions in some respects. These uninegative ions are called *pseudohalide ions.* Examples of some important pseudohalide ions are : cyanide ion (CN^-), isocyanide ion (NC^-), fulminate ion (ONC^-), cyanate or oxycyanide ion (OCN^-), isocyanate ion (NCO^-), thiocyanate ion (SCN^-), isothiocyanate ion (NCS^-), tellurocyanate ion ($TeCN^-$), selenocyanate ion ($SeCN^-$), isoselenocyanate ion ($NCSe^-$), azide ion (N_3^-), azidothiocarbonate ion or azido carbondisulphide ion ($SCSN_3^-$). The salts (ionic compounds) given by the pseudohalide ions [*e.g.* AgCN, $Pb(CNS)_2$ etc.] are called *pseudohalides.*

As the dimers of halide ions are called halogens, the covalent dimers of the pseudohalide ions are called *pseudohalogens* or *halogenoids.* So far only a limited number of pseudohalogens have been isolated and characterised. Examples of some pseudohalogens or halogenoids are : cyanogen, $(CN)_2$; oxy cyanogen $(OCN)_2$; thiocyanogen, $(SCN)_2$; tellurocyanogen, $(TeCN)_2$; selcnocyanogen, $(SeCN)_2$; azidocarbun disulphide $(SCSN_3)_2$.

Similarities between Halide Ions and Pseudohalide Ions.

Following are the important similarities between halide ions and pseudohalide ions.

1. Formation of ionic as well as covalent compounds. Like halide ions, pseudohalide ions also form ionic as well as covalent compounds. Ionic compounds are given by these ions with metals while covalent compounds ore given by the combination of these ions with non-metals or metalloids. Examples are :

Ionic compounds : AgCl(AgCN), $PbCl_2[Pb(CNS)_2]$, AgCl(AgCNS)

Analogous covalent compounds : ICl(ICN), $SiCl_4[Si(NCS)_4]$, $COCl_2[CO(N_3)_2]$, $SO_2Cl_2[SO_2(N_3)_2]$

The compounds given in parentheses are the analogous compounds given by pseudohalide ions.

2. Oxidation of hydracids. HX acids ($X = Cl^-, Br^-, I^-$) can he oxidised to free halogen (X_2). Similarly HY acids ($Y = SCN^-, CN^-$ etc.) can also be oxidised to free pseudohalogen (Y_2). For example :

$$4HCl + MnO_2 \longrightarrow MnCl_2 + 2H_2O + Cl_2$$
$$4HSCN + MnO_2 \longrightarrow Mn(SCN)_2 + 2H_2O + (SCN)_2$$

3. Thermal decomposition of Pb (IV) salts. On heating some of the Pb(IV) salts of halide ions (*e.g.* $PbCl_4$) decompose to give Pb (II) salts and free halogen molecule. Similarly some Pb(IV) salts of pseudohalide ions [*e.g.* $Pb(SCN)_4$, $Pb(SeCN)_4$] decompose to give Pb(II) salts and free pseudohalogens. For example :

$$PbCl_4 \xrightarrow{\Delta} PbCl_2 + Cl_2$$
$$Pb(SCN)_4 \xrightarrow{\Delta} Pb(SCN)_2 + (SCN)_2$$

4. Formation of complex ions. Like halide ions, pseudohalide ions also form complex ions with transition metal ions. Examples are given below :

Complex ions given by halide ions : $[FeF_6]^{3-}$, $[CoCl_4]^{2-}$, $[Hg1_4]^{2-}$

Analogous complex ions given by pseudohalide ions : $[Fe(CN)_6]^{3-}$, $[Co(SCN)_4]^{2-}$, $[Hg(CN)_4]^{2-}$

5. Formation of inter-pseudohalogen and inter-halogen compounds. Halide ions can combine together to form inter-halogen compounds like ClF, ICl, IBr etc. Similarly pseudohalides can also combine together to form inter-pseudohalogen compounds like $CN.N_3$, CN.SCN etc.

6. Formation of monobasic hydracids. Like halide ions, pseudohalide ions also combine with H_2 to form monobasic hydracids.

$$H_2 + 2Cl^- \longrightarrow 2HCl + 2e^-$$
$$H_2 + 2CN^- \longrightarrow 2HCN + 2e^-$$

7. Formation of insoluble salts. Like halide ions, pseudohalide ions give insoluble salts with Ag^+, Pb^{2+} and Hg^+ cations.

(*a*) $Ag^+ + Cl^- \longrightarrow AgCl \downarrow$
$Ag^+ + CN^- \longrightarrow AgCN \downarrow$

(*b*) $Hg^+ + Cl^- \longrightarrow HgCl \downarrow$
$Hg^+ + CNS^- \longrightarrow HgCNS \downarrow$

(*c*) $Pb^{2+} + 2Cl^- \longrightarrow PbCl_2 \downarrow$
$Pb^{2+} + 2CNS^- \longrightarrow Pb(CNS)_2 \downarrow$

8. Acting as bridging ligands. Like halide ions, pseudohalide ions also have more than one pair of electrons and hence can coordinate with two metal ions simultaneously. Consequently like halide ions, pseudohalide ions also can act as

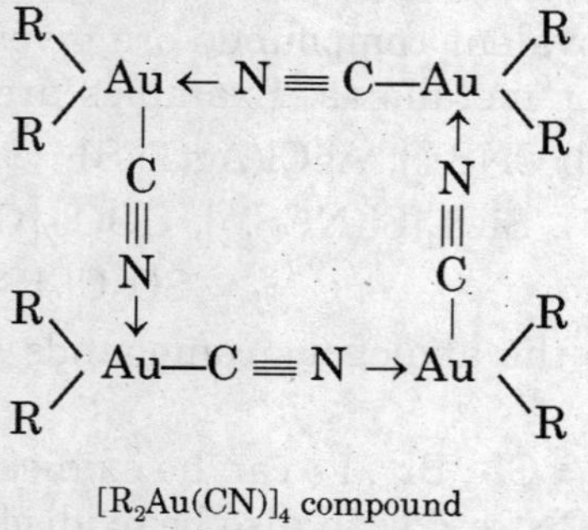

$[R_2Au(CN)]_4$ compound
(Tetrameric)

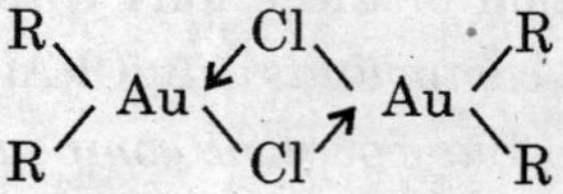

$[R_2AuCl)]_2$ compound
(Dimeric)

bridging ligands. For example in $[R_2Au(CN)]_4$ which is a tetrameric compound, four CN^- ions act as bridging ligands. Similarly in $[R_2AuCl]_2$ which is a dimeric compound, two Cl^- ions at as bridging ligands.

Note that in each of the above compounds the geometry of Au atoms is square planar.

Dissimilarities between Halide Ions and Pseudohalide Ions.

The differences between the two classes of ions are given as follows :

(*i*) The hydracids formed by the pseudohalide ions on combination with hydrogen are relatively weaker than the analogous hydracids formed by the halide ions. For example HCN is a very weak acid while HCl is a very strong acid in aqueous medium. The weak nature of hydracids given by pseudohalide ions may be due to the poor electronegativity of the pseudohalogens.

(*ii*) Pseudohalide ions are stronger coordinating ligands than the halide ions and hence the complexes formed by pseudohalide ions are *low spin* (*i.e.* inner-orbital involving d^2sp^3 hybridisation) while those formed by halide ions are *high spin* (outer-orbital involving sp^3d^2 hybridisation). For example $[Fe(CN)_6]^{3-}$ is a low spin complex while $[FeF_6]^{3-}$ is a high spin complex.

That the pseudohalide ions are stronger ligands than the halide ions is due to the ability of the electron deficient pseudo halides to form a bond (from the pseudohalide to the metal) as well as π-bond (from the metal to the pseudohalide).

(*iii*) Pseudohalide ions, being made up of two hetero atoms, can function as *ambidentate* ligands. For example class *a* metals form isothiocyanate complexes like $[Co(NCS)_4]^{2-}$ in which N atom is coordinated to Co^{2+} ion while class *b* metals form thiocyanate complexes like $Pd(SCN)_4]^{2-}$ in which S atom is bonded to Pd^{2+} ion. Halide ions have no tendency to act as ambidentate ligands.

Similarities between Halogens and Pseudohalogens.

Following are the points showing the similarities between the two classes of compounds.

1. Volatile and dimeric nature. Like halogens, pseudohalogens are also dimeric and fairly volatile (with the exception of polymeric thiocynogen) in the free state.

2. Isomorphous nature. Pseudohalogens are isomorphous to halogens when in the free or solid state. For example Cl_2 is isomorphous to $(CN)_2$ and similarly Br_2 is isomorphous with $(SCN)_2$.

3. Addition to ethylenic double bond. Like halogens, pseudohalogens also add to ethylenic double bond linkage. For example :

$$H_2C=CH_2 + Cl_2 \longrightarrow \underset{\displaystyle Cl\quad Cl}{H_2C{-}CH_2}$$

$$H_2C=CH_2 + (SCN)_2 \longrightarrow \underset{\displaystyle SCN\quad SCN}{H_2C - CH_2}$$

4. Reaction with alkalies. Like halogens. pseudohalogens also react with alkalies. For example :

(*i*) $(SCN)_2 + 2KOH$ (*cold and dilute*) $\longrightarrow KSCN + KOSCN + H_2O$

$Cl_2 + 2KOH$ (*cold and dilute*) $\longrightarrow KCl + KOCl + H_2O$

(*ii*) $3(SCN)_2 + 6KOH$ (*hot and concentrated*) $\longrightarrow 5KSCN + KSCNO_3 + 3H_2O$

$3Cl_2 + 6KOH$ (*hot and concentrated*) $\longrightarrow 5KCl + KClO_3 + 3H_2O$

5. Formation of monobasic hydracids. Like halogens, pseudohalogens also combine with H_2 to form monobasic hydracids.

$$Cl_2 + H_2 \longrightarrow 2HCl$$

$$(CN)_2 + H_2 \longrightarrow 2HCN$$

Dissimilarity between Halogens and Pseudohalogens.

Pseudohalogens undergo polymerisation and form polymerised species. For example :

$$n(CN)_2 \xrightarrow{500°C} 2(CN)_n$$

$$n(SCN)_2 \xrightarrow{\text{Room temp.}} 2(SCN)_n$$

$$n(SeCN)_2 \xrightarrow{\text{On standing}} 2(SeCN)_n$$

Halogens have no tendency to undergo polymerisation.

Conceptual Questions with Answers

Q.1 Give reasons for the following :

(*i*) ClF_3 exists but FCl_3 does not.

Ans. Let us write down Lewis structure of the given molecules

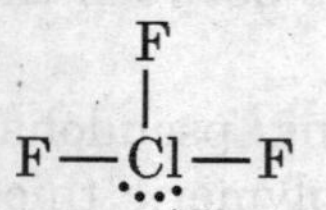

σ-*bps* = 3, *lps* = 2,
Sum = 3 + 2 = 5 (sp^3d hybridisation of Cl-atom)
(correct structure)

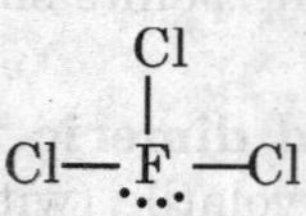

σ-*bps* = 3, *lps* = 2,
Sum = 3 + 2 = 5 (sp^3d hybridisation of F-atom)
(wrong structure)

These structures show that central halogen atom should be sp^3d hybridised. sp^3d hybrisation of F-atom for the formation of FCl_3 is not possible since F-atom does not have d-orbitals. Thus due to the absence of d-orbitals in F-atom, F-atom is not able to undergo sp^3d hybridisation and hence cannot form FCl_3 molecule.

Since Cl-atom is bigger in size than F-atom, this atom can accomdate three small F-atoms round it but F-atom (smaller atom) cannot accomdate three larger Cl-atoms round it.

(*ii*) Interhalogens are more reactive than halogens.

Ans. We know that in the formation of X–Y bond in interhalogen molecule (XY), the overlapping between the two orbitals of dissimilar halogen atoms takes place while in the formation of X–X bond in halogen molecule (X_2), the overlapping between the two orbitals of similar halogen atoms takes place. Now, since the overlapping between the orbitals of similar atoms in X_2 molecule is more effective than that between the orbitals of dissimilar atoms in XY molecule, X–X bond in X_2 molecule is strong than X–Y bond in XY molecule. Due to weaker X–Y bond, interhalogen (XY) are more reactive than halogen (X_2).

Q.2 Discuss the structure of I_3^+ ion on the basis of hydrisation.

Ans. Positive charge on I_3^+ ion may be regarded as being present on central I-atom. Hence the structure of I_3^+ can be written as $:\ddot{\underset{..}{I}}—\overset{+}{\ddot{\underset{..}{I}}}—\ddot{\underset{..}{I}}:$. Since the central I^+ atom is surrounded by σ-*bps* + *lps* = 2 + 2 = 4, I^+ atom is sp^3 hybridised. Due to sp^3 hybridisation and the presence of two *lps* of electrons, I_3^+ ion has ***angular*** geometry.

Q.3 Write the structures of the following pseudohalide ions (*i*) Cyanide ion (*ii*) Isocyanide ion (*iii*) Cyanate ion (*iv*) Isocyanate ion (*v*) Thiocyanate ion (*vi*) Isothiocyanate ion (*vii*) Fulminate ion.

Ans. The structures of the given pseudohalide ions can be obtained by removing H^+ from the parental acids of these ions. Thus :

(*i*) $H—C\equiv N: \xrightarrow{-H^+} :\bar{C}\equiv N: + H^+$

Hydrocyanic acid or hydrogen cyanide → Cyanide ion

(*ii*) $H—N\rightleftarrows C: \xrightarrow{-H^+} :\bar{N}\rightleftarrows C: + H^+$

Isohydrocyanic acid or isohydrogen cyanide → Isocyanide ion

(*iii*) $H—\ddot{\underset{..}{O}}—C\equiv N: \xrightarrow{-H^+} :\overset{-}{\ddot{\underset{..}{O}}}—C\equiv N: + H^+$

Cyanic acid → Cyanate ion or oxycyanide ion

(*iv*) $H—\ddot{N}=C=\ddot{O}: \xrightarrow{-H^+} :\overset{-}{\ddot{N}}=C=\ddot{O}: + H^+$

Isocyanic acid → Isocyanate ion

(*v*) $H—\ddot{\underset{..}{S}}—C\equiv N: \xrightarrow{-H^+} :\overset{-}{\ddot{\underset{..}{S}}}—C\equiv N: + H^+$

Thiocyanic acid → Thiocyanate ion

(*vi*) $H—\ddot{N}=C=\ddot{S}: \xrightarrow{-H^+} :\overset{-}{\ddot{N}}=C=\ddot{S}: + H^+$

Isothio cyanic acid → Isothiocyanate ion

(*vii*) $\overset{-}{\ddot{C}}\equiv\overset{+}{N}—\overset{-}{\ddot{\underset{..}{O}}}:$

Fulminate ion

University Questions

1. Write a note on "Interhalogen Compounds". *(Bundelkhand 88)*

2. Give the reasons for the following :

(*i*) All known interhalogen compounds are diamagnetic and contain an even number of atoms. *(Delhi Hons. 87)*

(*ii*) Interhalogen compounds have covalent character.

(*iii*) ClF_3 can be prepared but FCl_3 cannot be prepared. *(Punjab 85)*

(*iv*) In halogen fluorides the number of fluorine atoms is always odd.

(*v*) Interhalogen compounds are diamagnetic. *(Delhi 84, 85)*

(*vi*) Interhalogen compounds are more reactive than halogens.

3. Write a concise account of polyhalides. *(Calcutta 86; Meerut M.Sc. 83, 84, 85)*

4. Discuss the similarities between halogens and pseudohalogens. *(Delhi 99)*

5. Give a brief account of interhalogen compounds. *(Kanpur 2000)*

6. What are interhalogen compounds? Discuss their preparation, important properties and structure. *(Lucknow 2000)*

7. How are IF_5 and BrF_3 prepared? *(Bharathiar 2000)*

CHAPTER 42

Elements of Group 18 (Zero) : Noble Gases

Zero group of the periodic table consists of six elements namely *helium* (Z = 2); *neon*, Ne (Z = 10); *argon*, Ar (Z = 18); *krypton*, Kr (Z = 36); *xenon*, Xe(Z = 54) and *radon*, Rn (Z = 86). These elements are also called by other names like *inactive gases, inert gases, rare gases* and *noble gases.*

Position of Inert Gases in the Periodic Table

The inert gases were not known at the time of Mendeleef when he gave his original classification of elements. He did not leave gaps for these gases although he left gaps for several other unknown elements. After their discovery a new group was introduced into the periodic table. In the modern periodic table, these may be placed in the zero group as well as in the VIII group because of the following reasons.

1. Reasons for placing in zero group. (*i*) *Chemical inertness.* These elements are chemically inert. The existence of such a group is expected from the fact that there must be a separating group between strongly electronegative halogens of VII A group and strongly electropositive alkali metals of IA group. Therefore, these elements should be placed in between VII A group and I A group, *i.e.*, in the zero group.

(*ii*) *Valency.* Valenoy is defined as the combining capacity of the element. The electronic configurations of these elements given as : He $\rightarrow$ 2(1*s*), Ne $\rightarrow$ 2, 8 ($2s^2p^6$), Ar $\rightarrow$ 2, 8, 8 ($3s^2p^6$), Kr $\rightarrow$ 2, 8, 18, 8 ($4s^2p^6$), Xe $\rightarrow$ 2, 8, 18, 18, 8 ($5s^2p^6$), Rn $\rightarrow$ 2, 8, 18, 32, 18, 8 ($6s^2p^6$) indicate that the atoms of these elements have saturated shells. They have, therefore, no tendency either to lose, gain or hare electrons with the atoms of other elements. In other words, their combining capacity or valency is zero. Therefore, these elements should be placed in the zero group.

2. Reasons for placing in VIII group. (*i*) *Electronic cofiguration.* Electronic configurations of inert gases as given above, indicate that, except helium which has $1s^2$ configuration, all the elements have eight electrons in the outermost shell. Therefore, on this basis they should be placed in the VIII group.

(*ii*) *Analogy with other groups.* In the modern periodic table, all the groups, except VIII, are divided into two sub-groups A and B each. The elements of A sub-groups are normal elements. whereas those of B sub-groups are transitional. If the VIII group is also considered parallel to others, it should also be divided into two sub-groups *viz.* VIII A and VIII B. The VIII A group should contain inert elements and VIII B group should contain transitional elements. In such a case, the periodic table will be more symmetrical. Thus, from the above discussion it is clear that the inert gases may be placed in both the groups, *i.e.*, zero and VIII. However, for the general reference they are still considered as the members of the zero group. The position of these gases is further justified because of the facts namely : (*a*) They are all colourless, odourless and tasteless gases, under ordinary conditions. (*b*) The specific heat ratio, *i.e.*, C_p/C_v is the same (1.66) for all gases, which indicates their mono-atomic character. (*c*) They show a regular gradation in their physical and chemical properties. (*d*) They do not have any resemblance with the elements of their neighbouring groups. (*e*) They are all chemically inert.

Occurrence of Noble Gases.

Excepting radon, all other noble gases occur throughout the universe in atomic state. Since radon is a radioactive element and hence decays rapidly, it does not occur in the free state. Other rare gases occur in the following sources :

(*i*) In the stars and interstellar space. The masses of the major planets Jupiter and Saturn are estimated to be due to 23% and 27% of helium respectively. The atmosphere of Mars consists of mostly N_2 and argon.

(*ii*) In the earth's atmosphere. The non-radioactive rare gases are present in the atmosphere to the extent of 1% by volume. The volume of each rare gas in 100 litres of this is as : He = 0.524 c.c., Ne = 1.82 c.c., Ar = 9.34 c.c., Kr = 0.114 c.c., Xe = 0.0087 c.c.

(*iii*) In natural gas. Helium and traces of neon are found in natural gas which is found in the wells of certain oil fields in U.S.A. At present most of helium is isolated from natural gas as the concentration of helium in it varies from 2 to 7%.

(*iv*) In minerals and spring waters. Noble gases are also found in certain minerals. Except helium, the concentration of other noble gases in minerals is quite low. For example, the concentration of helium in thorianite is 10.5 c.c. gm. Certain spring waters give off gases which contain helium and argon. The percentage of these gases in spring water is so low that these cannot be isolated.

Why Are the Rare Gases Chemically Inert?

The chemical inertness of these elements is due to the following reasons :

(*i*) We have already seen that the atoms of rare gases have saturated shells. They all have eight electrons in their outermost shells, except helium, which has two. They have, therefore, no tendency either to lose, gain or share electrons with atoms of other elements. These elements, thus, do not enter into chemical combination.

(*ii*) The atoms of these elements have all the electrons paired. Thus, the electrons are not available for sharing with the electrons of other atoms. Hence, they can not form covalent compounds.

(*iii*) The electronegativity and electron affinity of these elements are zero. Thus) these elements have no tendency to attract or accept electrons and hence, are unable to form compounds.

(*iv*) The ionization potentials of these gases are very high. Hence, the removal of the electron from the outermost shell is very difficult.

Facts Indicating the Mono-atomic Character of Inert Gases.

The mono-atomic character of inert gases can be established by the following facts :

***(i)* Specific heat ratio.** The specific heat ratio, *i.e.*, Cp/Cu for all inert gases has been found to be 1.66, which is the theoretical ratio for all mono-atomic gases.

***(ii)* Value of physical constants.** The values of other physical constants such as dielectric constant, refractive index etc., also confirm their .mono-atomic character.

***(iii)* Electronic configuration.** The atoms of these elements have saturated shells. They have, therefore, no tendency to combine together to form polyatomic molecules and hence exist as single atoms.

History of Discovery of Rare Gases.

The credit for the discovery of rare gases goes to Lord Rayleigh and William Ramsay.

1. Discovery of argon—Cavendish in 1785, repeatedly sparked a mixture of purified air and excess of oxygen and absorbed the resulting oxides of nitrogen and unused oxygen in caustic potash and potassium pentasulphide solutions, respectively. He found that a small volume of the gas which was about 1/120 part of the original air, was left. It was neither nitrogen nor oxygen. Cavendish, however, could not draw any conclusion regarding the nature of the residual gas.

The experimental observation of Cavendish was not given any importance for a long time. Rayleigh in 1892, observed that the density of nitrogen isolated from air was 0.5% more than the density of nitrogen obtained by chemical means, *i.e.*, by the decomposition of urea or ammonium nitrite. Ramsay suggested that this difference of density was due to the presence of some heavier gas of the atmosphere. Rayleigh and Ramsay repeated Cavendish's experiment in a modified apparatus and found that the residual gas was chemically inert. They also removed various constituents of air. The residual gas thus obtained was the same as obtained by Cavendish's experiment. This gas was completely unreactive and different from all the known gases. On sparking in a discharge tube, it gave red light and a spectral line never observed before. This indicated the presence of a new element. It was named argon derived from the Greek word argon, meaning lazy.

2. Discovery of helium—Lockeyer and Frankland in 1868, observed a new yellow line in the chromosphere of the sun, during total solar eclipse. This line did not coincide with the D_1 and D_2 lines of sodium and Lockeyer called it D_3 line. This indicated the presence of a new element. It was named helium derived from the Greek word helios, meaning sun.

Hillebrand in 1888, observed that uranium minerals such as clevite, on treatment with dil. H_2SO_4 gave out a gas which he thought to be nitrogen. Ramsay who was constantly in the search of some other sources of argon. suspected the same gas to be argon. He examined the spectrum of the gas which indicated the presence of helium, the element already discovered by Lockeyer and Frankland in the chromosphere of the sun.

3. Discovery of neon, krypton and xenon. After the discovery of argon and helium, Ramsay and Travers collected large amounts of argon by the fractional evaporation of liquid air. This argon was then subjected to systematic fractional evaporation under different reduced pressures. From the first fraction, they isolated a new element which gave a number of red lines in the spectrum. It was named neon (meaning new). From the final fraction they isolated two more inert gases—krypton (meaning hidden) which gave green spectral lines and xenon (meaning strange) which gave a blue glow.

4. Discovery of radon—Dorn in 1900, discovered radon as one of the disintegration products of radium.

$$^{226}_{88}Ra \xrightarrow{-\,^{4}_{2}He} \,^{222}_{86}Rn + \,^{4}_{2}He$$

Isolation of Rare Gases From Dry Air (*Chemical Method*)

The method used for the isolation of rare gases from dry air is the chemical method. This method consists of the following two steps :

1. Isolation of Inert Gases Mixture from the Dry Air.

Different constituents of the air *viz.* CO_2, water vapour, O_2 and N_2 are removed from the air by any of the following methods and thus inert gases mixture which is free from CO_2, water vapour, O_2 and N_2 is obtained.

(i) Rayleigh and Ramsay first method. In this method, the different constituents of air, *i.e.*, carbon dioxide, water vapour, oxygen and nitrogen are removed, one by one, as shown below :

(*a*) ***Carbon dioxide.*** By passing over soda lime.

$$2NaOH + CO_2 \longrightarrow Na_2CO_3 + H_2O$$

(*b*) ***Water vapour.*** By passing over P_2O_5.

$$P_2O_5 + 3H_2O \longrightarrow 2H_3PO_4$$

(*c*) ***Oxygen.*** By passing over red hot Cu.

$$2Cu + O_2 \longrightarrow 2CuO$$

(*d*) ***Nitrogen.*** By passing over heated Mg.

$$3Mg + N_2 \longrightarrow Mg_3N_2$$

The process of passing air successively over red hot copper and heated magnesium is repeated till it is completely free from nitrogen and oxygen. The residual gas is a mixture of inert gases.

(ii) Rayleigh and Ramsay second method. In this method, 9 volumes of air and 11 volumes of oxygen are introduced into a glass globe of 50 litres capacity. This glass globe is fitted with two heavy platinum electrodes through which a transformer discharge of 600–800 volts is passed (Fig. 42.1). The nitrogen of the air combines with excess of oxygen to form oxides of nitrogen. Now a continuous stream of caustic soda solution is introduced into the glass globe through a tube. The oxides of nitrogen are absorbed by the alkali and used up caustic soda solution is taken away by another tube.

$$N_2 + O_2 \xrightarrow[\text{Discharge}]{\text{Electric}} 2NO$$

$$2NO + O_2 \longrightarrow 2NO_2$$

$$2NaOH + 2NO_2 \longrightarrow NaNO_2 + NaNO_3 + H_2O$$

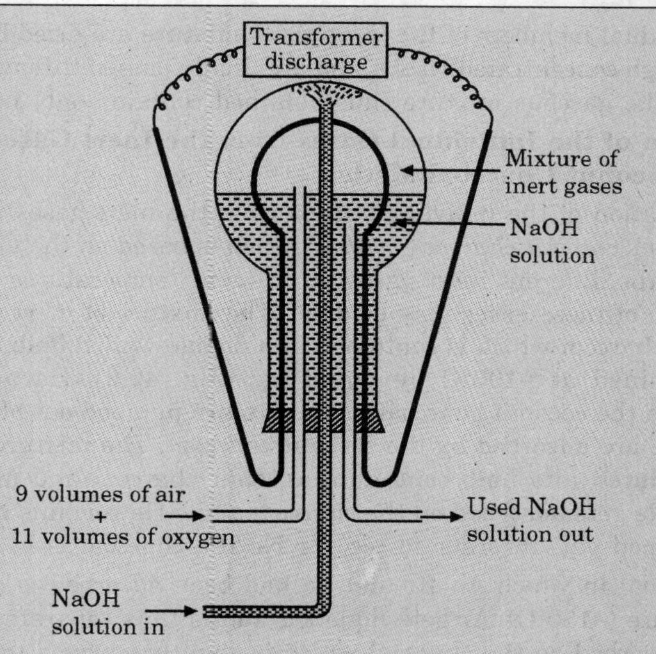

Fig. 42.1. Rayleigh and Ramsey second method for the isolation of inert gases mixture from dry air.

As a result of the above method, most of O_2 and N_2 is removed from the air in the glass globe. The mixture of residual gases contains noble gases with traces of O_2. The mixture of residual gases is passed through tubes containing alkaline pyrogallol in which O_2 gas gets absorbed and the mixture of rare gases which is free from O_2 gas comes out.

(iii) Fisher Ringe's method. In this method, the air free from moisture and carbon dioxide is passed over a heated (800°C) mixture containing 90% calcium carbide and 10% calcium chloride in an iron tube, when the following reactions take place :

$$CaC_2 + N_2 \xrightarrow{800°C} \underset{\text{calcium cyanamide}}{CaCN_2} + C$$

$$C + \tfrac{1}{2}O_2 \longrightarrow CO$$

$$C + O_2 \longrightarrow CO_2$$

$$2CaC_2 + 3CO_2 \longrightarrow 2CaCO_3 + 5C$$

The mixture of gases containing CO and CO_2 so obtained is passed over heated copper oxide when CO is oxidized to CO_2 and CO_2 thus formed is absorbed by caustic potash solution.

$$CuO + CO \longrightarrow Cu + CO_2$$
$$2KOH + CO_2 \longrightarrow K_2CO_3 + H_2O$$

The individual members of the inert gases mixture are dried by bubbling the mixture through concentrated H_2SO_4 and are finally passed through tubes packed with P_4O_{10}. The gaseous mixture thus obtained contains only noble gases.

2. Separation of the Individual Gases from the Inert Gases Mixture by Dewar's Coconut Charcoal Method.

The separation of the individual gases from the inert gases mixture can be done by *Dewar's coconut charcoal method* which is based on the fact that coconut charcoal adsorbs different inert gases at different temperatures and, therefore, the separation of these gases may be done. The mixture of inert gases is passed over coconut charcoal which is contained in a double walled bulb (called Dewar's flask) maintained at –100°C by liquid air (Fig. 42.2). He and Ne remain unadsorbed on the coconut charcoal and hence are pumped out of the bulb while Ar, Kr and Xe are adsorbed by the coconut charcoal. The mixture of He and Ne is then introduced into bulb containing coconut charcoal and maintained at –180°C when Ne gets adsorbed on the charcoal while He remains unadsorbed and hence is pumped out. In order to recover Ne the charcoal is heated.

The charcoal in which Ar, Kr and Xe had been adsorbed is cooled to liquid air temperature (–180°C). Ar gets liquefied and is thus separated while Kr and Xe remain adsorbed on the charcoal which is slightly warmed to a temperature equal to –90°C. At this temperature Kr ramains unadsorbed and hence comes out of the charcoal. Xe remains adsorbed on the charcoal which, on being heated at 0°C, gives Xe.

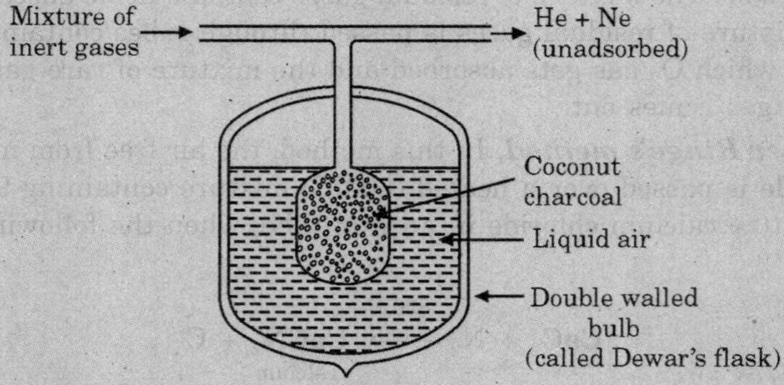

Fig. 42.2. Dewar's coconut charcoal method for the separation of inert gases from the inert gases mixture.

Flow sheet given below illustrates the various steps of this method.

Flow sheet for the separation of inert gases by Dewar's coconut charcoal method

Inert gases mixture
(He, Ne, Ar, Kr, Xe)

- Brought in contact with coconut charcoal at –100°C
  - **He, Ne** — Remain unadsorbed and hence are pumped out
    - Brought in contact with another sample of coconut chacoal at –180°C
      - **He** — Remains unadsorbed and hence is pumped out
      - **Ne** — Gets adsorbed on the charcoal
        - Heat the charcoal → **Ne**
  - **Ar, Kr, Xe** — Adsorbed on the charcoal
    - The coconut charcoal is cooled to –180°C
      - **Ar** — Gets liquefied
      - **Kr, Xe** — Remain unadsorbed
        - Heat at –90°C
          - **Kr** — Remains unadsorbed and hence comes out of the charcoal
          - **Xe** — Remains adsorbed on the charcoal
            - Heat the charcoal at 0°C → **Xe**

Isolation of Inert Gases From Liquid Air (*Physical Method*).

Liquid air contains He, Ne, Ar, Kr, Xe, N_2 and O_2. Inert gases can be obtained from this liquid air by its fractional evaporation. The method used is based on the fact that the boiling points of the various constituents of liquid air widely differ from each other as shown : He = –269°C, Ne = –246°C, N_2 = –195.7°C, Ar = –186°C, O_2 = –183°C, Kr = –152°C, Xe = –107°C. Important conclusions drawn from these values can be summarised as : (*i*) He and Ne have very low b.pts. and hence during the process of liquefying the air, He and Ne will remain as gases while other constituents will change into liquids. (*ii*) There is little difference between the b.pts of Ar and O_2 and hence these gases will be collected more or less as a single fraction, (*iii*) Kr and Xe have fairly high boiling points as compared to those of N_2, Ar and O_2. Therefore, Kr and Xe are always found in the less volatile portion of the liquid air and are always easily separated from other gases.

The apparatus used in this method consists of two fractionating columns *viz. lower fractionating column* and *upper fractionating column.* These columns are provided with baffle plates on the inner side. There is a reservoir between these columns. This reservoir contains liquid O_2. (Fig. 42.3). Air free from CO_2 and moisture (as these substances, on cooling, change into solid and hence clog the pipes) is compressed and cooled and then is introduced into the fractionating column. The cooling results in the liquefaction of less volatile gases namely O_2, Ar, Kr, Xe and a little of N_2 (*Fraction A*). These gases get collected at the bottom of the lower fractionating column. The remaining gases (which are more volatile) namely He, Ne and N_2 rise upwards in the column where they are cooled and are liquefied (*Fraction B*).

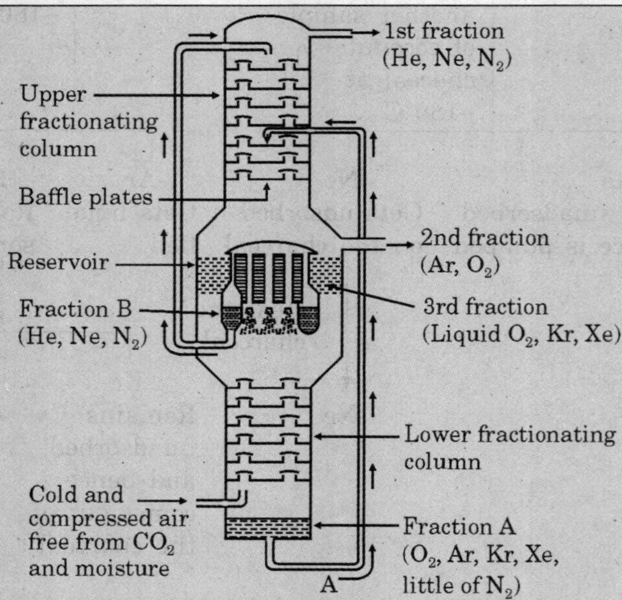

Fig. 42.3. Isolation of inert gases from liquid air.

Now the fraction A (O_2, Ar, Kr, Xe and a little of N_2) is pumped up from the bottom and is dropped from the middle of the upper fractionating column. The fraction B (He, Ne and N_2) is also pumped and is dropped from the top of the upper fractionating column. By this operation the liquid air gets separated into three fractions *viz.* (*a*) *1st fraction.* This fraction contains most volatile gases namely He, Ne and N_2. The mixture of these gases escapes from the top of the upper fractionating column. (*b*) *2nd fraction.* This fraction contains Ar and O_2. The mixture of these gases escapes from the top of central reservoir. (*c*) *3rd fraction.* This fraction contains liquid O_2, Kr and Xe. The mixture of these gases gets collected at the bottom of the reservoir.

Now from these fractions the individual inert gases are obtained as follows :

(i) Separation of helium and neon from the 1st fraction. The first fraction containing He, Ne and N_2 is passed through a spiral tube cooled by liquid nitrogen. Here most of the nitrogen is condensed and a mixture of He, Ne and some N_2 passes out. The remaining N_2 is removed by passing the mixture over heated calcium carbide.

$$CaC_2 + N_2 \xrightarrow{800°C} CaCN_2 + C$$

The gaseous mixture containing He and Ne is then passed through a spiral tube cooled by liquid H_2 at –253°C when Ne solidifies (b. pt. –246°C) and is collected at the bottom, while helium in the gaseous state escapes from the top (Fig. 42.4).

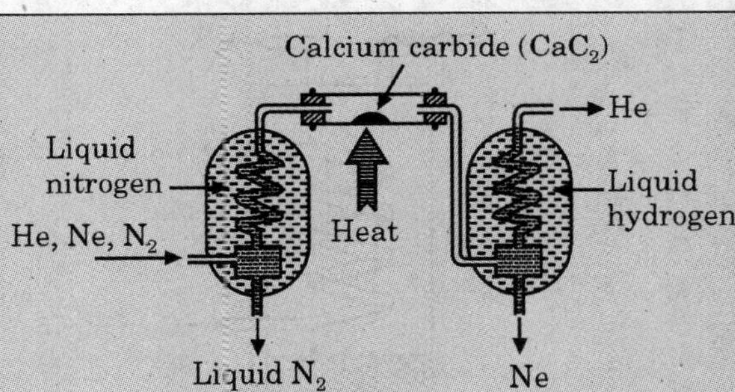

Fig. 42.4. Separation of He and Ne from *1st* fraction.

(ii) Separation of argon from the 2nd fraction. The second fraction containing argon and oxygen is passed through a column which is provided with a coil in which liquid N_2 is being circulated. Here most of the oxygen is condensed while more of argon escapes at the top (Fig. 42.5). The argon, so obtained, still contains some oxygen which is removed by passing over heated copper.

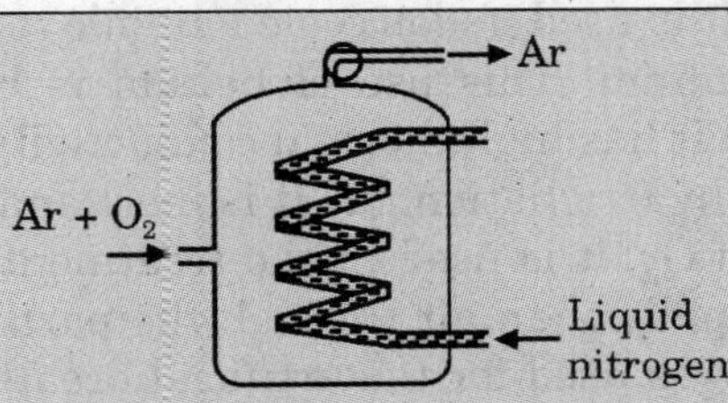

Fig. 42.5. Separation of Ar from the *2nd* fraction.

(iii) Separation of krypton and xenon from the 3rd fraction. From the third fraction containing liquid O_2, Kr and Xe, oxygen is removed by evaporation. The remaining liquid becomes richer in krypton and xenon. These two may be separated easily by fractional evaporation as there is a wide difference in their boiling points (Kr = –152°C, Xe = –107°C) (Fig. 42.6).

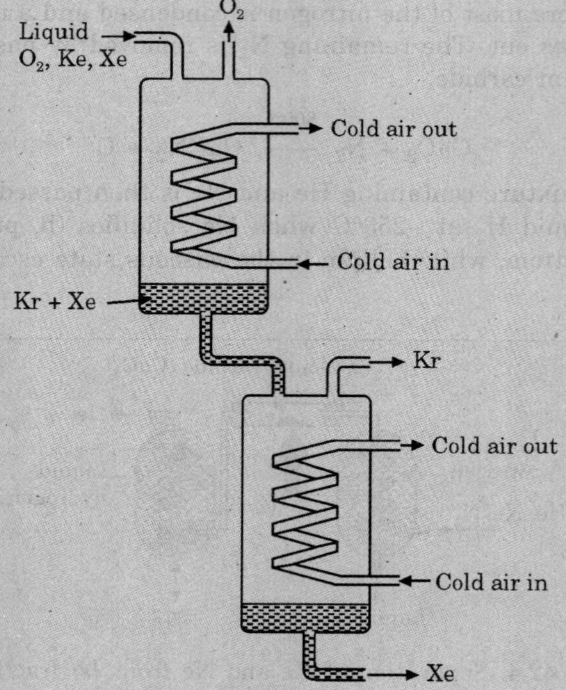

Fig. 42.6. Separation of Kr and Xe from in nature *3rd* fraction

Uses of Noble Gases.

Of all the noble gases, only He and Ar which are available easily and in plenty have got many uses. The uses are due to their chemical inertness and low boiling points. Important uses of noble gases are given below :

Uses of Helium. (*i*) It is used in gas cooled atomic reactors as a heat transfer gas, *i.e.* as a cooling medium. This use of helium is because of the following properties of this gas : (*a*) It has high thermal conductivity. (*b*) It has low viscosity. (*c*) It remains unaffected by irradiation. (*d*) It is inert and hence does not undergo corrosion. (*ii*) Mixed with O_2, it is used in the treatment of asthma. Being light, this gas diffuses more rapidly than air through the partly choked lung passages. The mixture of He and O_2 is used for the artificial breathing of asthma patients. This mixture keeps breathing steady. (*iii*) It is used for filling in balloons and airships as it is a light and non-combustible gas. For this purpose, a mixture of 85% helium and 15% hydrogen is generally used. (*iv*) The mixture of helium and oxygen is used for respiration by sea divers. The air proves harmful because nitrogen of the air dissolves in the blood at high pressure when the diver is under deep sea. When he comes to the surface, the pressure is suddenly decreased.

The bubbles of nitrogen gas now escape from the blood and are collected at joints which cause pains. Helium taken in the place of air is much less soluble in blood at high pressure. (*v*) Liquid helium is used for producing low temperatures required for research work. (*vi*) For producing inert atmosphere in the welding and metallurgy of certain metals (*e.g.* Mg, Al, Ti etc.) which are easily oxidised. (*vii*) In the preservation of food (*viii*) It is used for filling electrical transformers. (*ix*) It is used as a pressuring agent in rockets to expel liquid O_2 and liquid H_2. (*x*) It is used as a shielding gas for arc welding as it protects both the arc and the metal from atmosphere. (*xi*) It is suitable for low temperature gas thermometry because this gas has low boiling point and has nearly ideal gas behaviour. (*xii*) Helium nucleus (α-particles) is used as a bombarding particle for the artificial disintegration of atoms.

Uses of Neon. (*i*) It is used in neon discharge lamps and signs which are used for advertising purposes, since the neon lights are visible from long distances and even visible through fog and mist. Since the light of neon signs has a better penetrating power through fog and mist, the neon signs are also used in beacon lights for the safety of air navigation. (*ii*) Mixture of He and Ne does not conduct electric current until a certain limiting voltage is applied to it and thereafter it starts conducting electric current. Based on this property, it is used in certain protective electrical devices like voltage stabilisers, current rectifiers etc.

Uses of Argon. (*i*) A mixture of argon with 15% nitrogen is filled in electric bulbs to increase the life of the filament. (*ii*) Fr producing inert atmosphere in welding and metallurgy of certain metals which are easily oxidised. (*iii*) In filling fluorescent tubes and radio-valves. (*iv*) It is mixed with neon in neon signs to obtain the lights of varying colours.

Uses of Krypton. (*i*) It is used in high efficiency miner's cap lamps. (*ii*) It is used in filling luminous tubes. (*iii*) Krypton-85, one of the isotopes of Kr, is used in electronic tubes for regulating the voltage and for testing the leaks. (*iv*) Krypton-85 is also used for the measurement of the thickness of sheets of metals and plastics.

Uses of Xenon. (*i*) It is used in discharge tubes for producing high speed flash of bluish light used in quick photography. (*ii*) Liquid xenon is used in bubble chamber for the detection of gamma photon and neutral mesons.

Uses of Radon. (*i*) It is used in the preparation of ointment for the treatment of cancer and other diseases. (*ii*) It is used for photographing the interior of opaque materials for locating the defects in steel castings.

Importance of Inert Gases in Theoretical Chemistry.

The inert gases have an important role in the development of theoretical chemistry as follows :

(*i*) **In the periodic table.** The introduction of these gases between strongly electronegative halogens and strongly electropositive alkali metals, has removed one of the anomalies of Mendeleef's periodic table. The zero group of inert gases acts as a separating group between these two classes of elements.

(*ii*) **In the elucidation of distribution of electrons in atoms.** The outermost shell of the atoms of inert gases has eight electrons, except helium which has two. This configuration accounts for the inertness of these elements. On this basis Bohr and Bury postulated that outermost shells of the atoms cannot have more than eight electrons.

(*iii*) **In the development of electronic theory of valency.** The modern concept of valency is based upon the structure of inert gases. According to the concepts of Kossel and Lewis, the atoms tend to acquire inert gas type of electronic configuration by transference of electrons from one atom to another (electrovalency), by sharing of electrons between two atoms (covalency) and by the combination of both the processes, *i.e.*, transference and sharing (co-ordinate).

(*iv*) **In the discovery of isotopes.** Neon was the first non-radioactive substance whose isotopes were first of all separated. On this basis, it was concluded that other elements may also consist of a mixture of two or more isotopes and thereafter, a number of isotopes of other elements have been identified and separated.

(*v*) **In radioactivity.** Helium is one of the disintegration products of some radioactive elements. This fact has been useful in the development of the theory of the radioactive disintegration and group displacement law.

Types of Compounds Formed by Inert Gases.

The atoms of inert gases have saturated shells and, therefore, they are chemically inert. Recent researches have shown that under certain specified conditions, they enter into chemical combinations and form some rare chemical compounds. The specified conditions and types of compounds formed by these gases are described below :

1. Compounds formed under excited conditions : Compounds of helium. These compounds known as helides are formed by sparking helium at low pressure in presence of mercury, tungsten etc., to form $HgHe_2$, $HgHe_{10}$, WHe_2. In addition, metal electrodes in the discharged tubes also form compounds, with helium like $BiHe_2$, FeHe, Pt_3He, PdHe. In these compounds, it is supposed that helium is adsorbed on the surface of the metal and, therefore, these compounds are not considered as true chemical compounds.

2. Compounds formed through co-ordination : Compounds of argon. Booth and Wilson in 1935 reported that argon forms a number of unstable compounds with varying number of BF_3 molecules, *e.g.*, $Ar.BF_3$, $Ar.2BF_3$, $Ar.3BF_3$, $Ar.6BF_3$, $Ar.8BF_3$ and $Ar.16BF_3$. In these compounds, argon atom donates a pair of electrons to boron atom. In the case of higher compounds, the fluorine atoms also donate pair of electrons. The structure of $Ar.3BF_3$ and $Ar.8BF_3$ are given below.

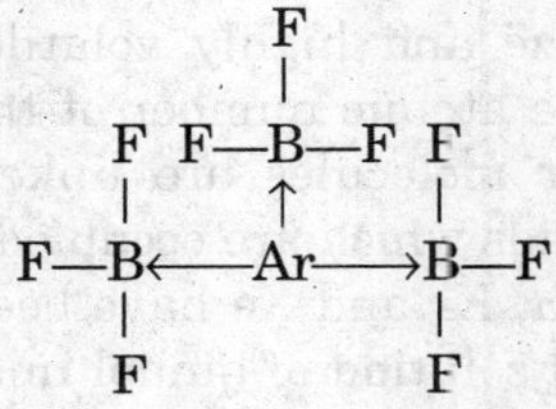

Structure of $Ar.3BF_3$

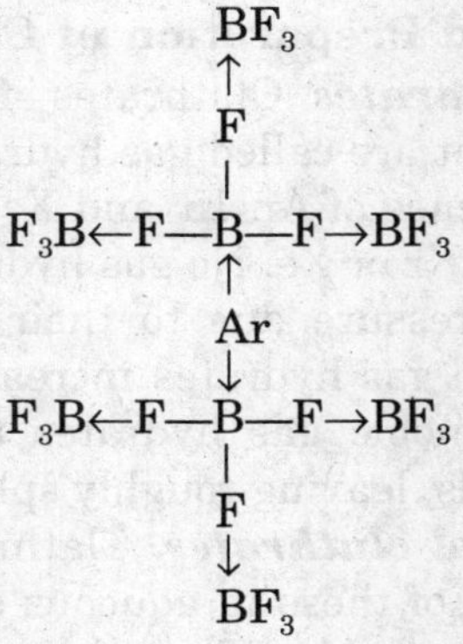

Structure of $Ar.8BF_3$

3. Compounds of inert gases with water : Formation of hydrates. The inert gases Ar, Kr and Xe form compounds with water at low temperature and high pressure. These compounds are known as hydrates. The compounds like $Ar.6H_2O$, $Kr.6H_2O$ and $Xe.6H_2O$ have been obtained and studied.

4. Compounds formed by physical trapping : Clathrate compounds. The inert gases argon, krypton and xenon form solid compounds with certain organic molecules such as phenol, hydroquinone, $C_6H_4(OH)_2$ under pressure. In these compounds the inert gases are enclosed into the crystal lattices of the organic molecules. These compounds are known as clathrates. (Latin : *clathrate* = enclosed or protected by cross bars or grating). These are also referred to as *cage compounds.*

In clathrates *the noble gases are trapped into the cavities of crystal lattices of certain organic and inorganic compounds.*

The organic or inorganic substance having the cavities of crystal lattices is called the *'host'* while the atom of noble gases entrapped in it is called the *'guest'*.

Organic or inorganic molecule (Host component) + Inert gas (Guest component) $\longrightarrow$ *Clathrate.*

Thus a simple clathrate can be represented by the formula $mH.n$G where H and G are the host und guest components . m and n represent the number of their molecules.

In these compounds the guest components are held by Van der Waals forces in crystalline cages formed by the host substances and consequently there is no chemical bonding involved in these compounds. It is not necessary that all the cavities in the host structure are filled. As a result the clathrates are normally *nonstoichiometric compounds.* The 'guest' can escape from the 'host' by heating the crystals or by dissolving them in a suitable solvent such as alcohol.

Conditions for the formation of clathrates.

(*i*) The crystal lattice of the host component should contain cavities of appropriate size. (*ii*) Size of the guest component should be such that it should fit in the cavities of the host component without bringing any atoms closer together than which corresponds to the van der Waals radii of the atoms.

Types and Preparation of Clathrates.

(i) Gas hydrates. Clathrates of Ar, Kr and Xe in which water molecule is the host component are called gas hydrates. These have been prepared by solidifying water in presence of Ar, Kr and Xe. The general formula of these is $5.75\ H_2O.G$ where G = Ar. Kr or Xe. No gas hydrate of He and Ne has yet been prepared even under high pressure due to their small size and highly volatile nature. The stability of the gas hydrates increases as the atomic number of the *'guest'* atom increases. In noble gas hydrates the water molecules are linked together by hydrogen bonds, leaving roughly spherical voids which are occupied by gas atoms.

(ii) Quinol clathrates. Clathrates of Ar, Kr and Xe have been obtained by dissolving any of these in aqueous saturated solution of quinol under a pressure of 10–40 atmospheres followed by slow cooling.

These compounds are crystalline and can be stored for several years. However, when heated or dissolved in water, the gas atoms escape and quinol remains behind as such. Quinol molecules which form the cavities are held together by H-bonds.

(iii) Phenol and β-hydroquinone clathrates. Clathrates given by phenol and β-hydroquinone with Ar, Kr and Xe have been prepared by dissolving any of these gases in phenol or β-hydroquinone followed by slow crystallisation. These clathrates are also quite stable. Clathrate of β-hydroquinone can be represented by the formula, 3.β-hydroquinone. G or $3.\beta\text{-}C_6H_4(OH)_2.G$ where G = Ar, Kr or Xe.

Reasons for the Stability of Clathrates.

The stability of clathrates is due to the reasons namely : (*i*) The guest molecules are tightly held in the cavities of the host molecules. (*ii*) The guest molecules within the cages are at minimum potential energy. The guest molecules can be made to leave the cavities in host lattices only when the forces holding the molecular cages together are overcome. The forces can be overcome either by heating the crystals, or by dissolving the crystals in a suitable solvent such as alcohol.

Uses of clathrates.

(*i*) In the separation of noble gases, *e.g.*, Ne can be separated from Ar, Kr and Xe, since Ne is the only gas that does not form a clathrate with quinol. (*ii*) Kr-85 clathrate is used as a safe and useful source of beta radiations which are used for measuring thickness of guages. (*iii*) Xe-133 clathrate provides a compact source for gamma radiations. (*iv*) Clathrates play an important role in some physiological actions.

Compounds of Noble Gases

1. Xenon Difluoride, XeF_2.

Preparation. It has been prepared :

(*i*) by irradiating a mixture of xenon and fluorine with light from a high pressure mercury arc.

$$Xe + F_2 \xrightarrow[\text{mercury lamp}]{\text{Light from a}} XeF_2$$

(*ii*) by heating xenon and fluorine in the ratio of 1 : 3 in a nickel tube at 400°C. On cooling the vapours, it is obtained as colourless solid.

$$Xe + F_2 \xrightarrow[400°C]{\text{Ni tube}} XeF_2$$

(*iii*) by the action of dioxygen difluoride on xenon at –118°C

$$Xe + C_2F_2 \xrightarrow{118°C} XeF_2 + O_2$$

Properties. *(i) Physical state.* XeF_2 is a white crystalline solid at room temperature and can be sublimed. Its melting point is 140°C. It can be stored for an indefinite period in a nickel or Monel metal container.

(ii) Fluorinating properties. XeF_2 is a fluorinating agent, since it can fluorinate ethylene (C_2H_4) to $FH_2C—CH_2F$ and $F_2HC—CH_3$, C_6H_6 to C_6H_5F, propylene (C_3H_6) to $F_2HC—CH_2—CH_3$, NO to NOF, NO_2 to NO_2F and SO_3 to $S_2O_6F_2$.

$$2XeF_2 + 2H_2C = CH_2 \longrightarrow FH_2C—CH_2F + F_2HC—CH_3 + 2Xe$$
$$C_6H_6 + XeF_2 \longrightarrow C_6H_5F + Xe + HF$$
$$C_3H_6 + XeF_2 \longrightarrow F_2HC—CH_2—CH_3 + Xe$$
$$2NO + XeF_2 \longrightarrow 2NOF + Xe$$
$$2NO_2 + XeF_2 \longrightarrow 2NO_2F + Xe$$
$$2SO_3 + XeF_2 \longrightarrow S_2O_6F_2 + Xe$$

(iii) Substitution reactions. XeF_2 undergoes substitution reactions when reacted with strong protonic acids like HSO_3F, $HClO_4$, CF_3COOH etc.

$$XeF_2 + 2HX \longrightarrow XeX_2 + 2HF$$

Here X is SO_3F^-, ClO_4^-, CF_3COO^- etc.

(iv) Oxidising properties. XeF_2 can oxidise a number of compounds and is itself reduced to Xe. For example XeF_2 oxidises H_2 to HF, I_2 to IF (in presence of BF_3), Cl^- to Cl_2, Ce^{3+} to Ce^{4+}, NH_3 to N_2, HCl to Cl_2 etc.

$$H_2 + XeF_2 \longrightarrow 2HF + Xe$$
$$I_2 + XeF_2 \xrightarrow{BF_3} 2IF + Xe$$
$$2Cl^- + XeF_2 \longrightarrow Cl_2 + 2F^- + Xe$$
$$2Ce^{3+} + XeF_2 \longrightarrow 4Ce^{4+} + 2F^- + Xe$$
$$8NH_3 + 3XeF_2 \longrightarrow N_2 + 6NH_4F + 3Xe$$
$$2HCl + XeF_2 \longrightarrow Cl_2 + 2HF + Xe$$

(v) Hydrolysis reactions. (*a*) XeF_2 dissolves unchanged in water or in acidic solutions, but, on standing hydrolysis occurs slowly.

$$2XeF_2 + 2H_2O \longrightarrow 2Xe + 4HF + O_2$$

(*b*) It is rapidly hydrolysed in an aqueous solution of a base.

$$2XeF_2 + 4OH^- \longrightarrow 2Xe + 4F^- + 2H_2O + O_2$$

(vi) Formation of addition compounds. XeF_2 can react with covalent pentafluorides like PF_5, AsF_5, SbF_5 and transition metal fluorides like NbF_5,

TaF_5 etc. In these reactions XeF_2 acts as a fluoride ion-donor (*Lewis base*) and the pentafluoride acts as a fluoride ion-acceptor (*Lewis acid*). The products are called adducts. These *adducts* are *ionic compounds.*

Lewis base		***Lewis acid***		***Adduct***
XeF_2	+	MF_5 (M = Sb, As)	$\longrightarrow$	$XeF_2.MF_5$ or $[XeF]^+[MF_6]^-$
XeF_2	+	$2SbF_5$	$\longrightarrow$	$XeF_2.2SbF_5$ or $[XeF]^+[Sb_2F_{11}]^-$
$2XeF_2$	+	MF_5	$\longrightarrow$	$2XeF_2.MF_5$ or $[Xe_2F_3]^+[MF_6]^-$
XeF_2	+	IF_5	$\longrightarrow$	$XeF_2.IF_5$

$XeF_2.IF_5$ has molecular lattice while other adducts are ionic compounds.

(vii) Action of HF. XeF_2 dissolves in HF without any reaction. The solution does not show any change in conductivity.

Structure. See chapter 8.

2. Xenon Tetrafluoride, XeF_4.

Preparation. It is prepared by heating a mixture of fluorine and xenon in the ratio of 5 : 1 in a nickel vessel at 400°C and 6 atmospheric pressure for a few hours. After heating, the reaction is quenched by cooling the nickel vessel rapidly in cold water. The excess of fluorine is removed by pumping at –78°C.

$$Xe + 2F_2 \xrightarrow[\text{400°C, 6 atoms}]{\text{Nickel vessel}} XeF_4$$

The tetrafluoride prepared under the above mentioned conditions contains less than 8% XeF_3 and XeF_6 as impurities. These impurities can be removed from XeF_4 by sublimation.

Properties. ***(i) Physical State.*** XeF_4 is a white crystalline solid at room temperature and can be sublimed readily. Its melting point is 117.1°C. It is less volatile than XeF_6. It dissolves in CF_3COOH.

(ii) Storing. It can he stored for an idefinite period in a nickel or Monel metal container, but since XeF_4 is very susceptible to hydrolysis, even traces of water must be excluded.

(iii) Fluorinating properties. XeF_4 is a strong fluorinating agent, since it can fluorinate a number of compounds. For example XeF_4 fluorinates SF_4 to SF_6, Pt to PtF_4, C_6H_6 to C_6H_5F, ethylene (C_2H_4) to $FH_2C—CH_2F$ and $F_2HC—CH_3$, NO to NOF and NO_2 to NO_2F.

$$2SF_4 + XeF_4 \longrightarrow 2SF_6 + Xe$$
$$Pt + XeF_4 \longrightarrow PtF_4 + Xe$$
$$2C_6H_6 + XeF_4 \longrightarrow 2C_6H_5F + 2HF + Xe$$
$$2C_2H_4 + XeF_4 \longrightarrow FH_2C—CH_2F + F_2HC—CH_3 + Xe$$
$$4NO + XeF_4 \longrightarrow 4NOF + Xe$$
$$4NO_2 + XeF_4 \longrightarrow 4NO_2F + Xe$$

(iv) Reaction with Xe. Whan XeF_4 is warmed with excess of Xe, XeF_2 is formed.

$$XeF_4 + Xe\ (excess) \xrightarrow{\text{Warm}} 2XeF_2$$

(v) Oxidising properties. XeF_4 can oxidise a number of compounds and itself is reduced to Xe. For example XeF_4 oxidises H_2 to HF at 130°C, Hg to HgF_2, I^- to I_2, BCl_3, to Cl_2, Cl^- to Cl_2, NH_3 to N_2 etc.

$$XeF_4 + 2H_2 \xrightarrow{130°C} 4HF + Xe$$
$$XeF_4 + 2Hg \longrightarrow 2HgF_2 + Xe$$
$$XeF_4 + 4I^- \longrightarrow 2I_2 + 4F^- + Xe$$
$$3XeF_4 + 4BCl_3 \longrightarrow 4BF_3 + 6Cl_2 + 3Xe$$
$$XeF_4 + 4Cl^- \longrightarrow 2Cl_2 + 4F^- + Xe$$
$$XeF_4 + 6NH_3(g) \longrightarrow N_2 + 4NH_4F + H_2 + Xe$$

(vi) Hydrolysis reactions. (*a*) XeF_4 undergoes *disproportionation* in H_2O to form a highly explosive solid compound, XeO_3.

$$2XeF_4 + 3H_2O \longrightarrow Xe + XeO_3 + 6HF + F_2$$
$$\underset{(Xe\,=\,+4)}{6XeF_4} + 12H_2O \longrightarrow \underset{(Xe\,=\,0)}{4Xe} + \underset{(Xe\,=\,+6)}{2XeO_3} + 24HF + 3O_2$$

(*b*) Hydrolysis of XeF_4 at –80°C gives xenon oxydifluoride, $XeOF_2$

$$XeF_4 + H_2O \xrightarrow{-80°C} XeOF_2 + 2HF$$

(vii) Formation of addition compounds. XeF_4 can react with covalent pentafluorides like PF_5, AsF_5, SbF_5 and transition metal fluorides like NbF_5, TaF_5, RuF_5, OsF_5, RhF_5, IrF_5 and PtF_5. In these reactions XeF_4 acts as a fluoride ion-donor (*Lewis base*) and the covalent pentafluoride acts as a fluoride ion-acceptor (*Lewis acid*). The products are called *adducts*. These adducts are *ionic compounds*.

Lewis base		***Lewis acid***		***Adduct***
XeF_4	+	SbF_5	$\longrightarrow$	$XeF_4.SbF_5$ or $[XeF_3]^+[SbF_6]^-$
XeF_4	+	$2MF_5$	$\longrightarrow$	$XeF_4.2MF_5$ or $[XeF_3]^+[M_2F_{11}]^-$

Structure. See chapter 8.

3. Xenon Hexafluoride, XeF_6.

Preparation. (*i*) It is very difficult to prepare. However, it is prepared by heating a mixture of xenon and fluorine in the ratio of 1 : 20 in a nickel vessel and at 200–250°C and 50 atmospheric pressure.

$$Xe + 3F_2 \xrightarrow[50\ \text{atm}]{200\text{–}250°C} XeF_6$$

The yield of this method is 95%. It becomes 100% when the pressure is 2 atmospheres and the temperature is 702°C.

(*ii*) XeF_6 is also formed by the reaction of XeF_4 with O_2F_2 at –130°C.

$$XeF_4 + O_2F_2 \longrightarrow O_2 + XeF_6$$

Properties (*i*) *Physical state.* XeF_6 is white solid and can be sublimed at room temperature. It can be stored for an indefinite period in nickel or Monel metal container, but, since XeF_6 is more susceptible to hydrolysis, even traces of water must be excluded. The melting point of XeF_6 is equal to 495°C at which XeF_6 melts to give a yellow liquid,. It is best stored in nickel container, since it attacks glass and silica (SiO_2)

$$2XeF_6 + SiO_2 \longrightarrow 2XeOF_4 + SiF_4$$
$$[XeOF_4 + SiO_2 \longrightarrow SiF_4 + XeO_3] \times 2$$
$$\overline{2XeF_6 + 3SiO_2 \longrightarrow 3SiF_4 + 2XeO_3}$$

(ii) Fluorinating properties. Out of the three xenon fluorides, XeF_6 is the best fluorinating agent.

(iii) Oxidising properties. XeF_6 can oxidise a number of compounds and is itself reduced to Xe. For example :

$$XeF_6 + 3H_2 \longrightarrow 6HF + Xe$$
$$XeF_6 + 6HCl \longrightarrow 3Cl_2 + 6HF + Xe$$
$$XeF_6 + 8NH_3 \longrightarrow N_2 + 6NH_4F + Xe$$

(iv) Hydrolysis reactions. (*a*) Slow hydrolysis of XeF_6 by atmospheric moisture gives the solid XeO_3

$$XeF_6 + 3H_2O \longrightarrow XeO_3 + 6HF$$

(*b*) Small quantities of water bring about partial hydrolysis of XeF_6 to give a colourless liquid containing xenon oxyfluoride, $XeOF_4$

$$XeF_6 + H_2O \longrightarrow XeOF_4 + 2HF$$

(*c*) In *strongly basic solution* (2.5 M NaOH solution), XeF_6 is converted into xenate ion, $HXeO_4^-$ which slowly disproportionates into perxenate ion. XeO_6^{4-} and Xe.

$$[XeF_6 + 7OH^- \longrightarrow \underset{\text{Xenate ion}}{HXeO_4^-} + 3H_2O + 6F^-] \times 2 \quad ...(i)$$

$$2HXeO_4^- + 2OH^- \longrightarrow \underset{\substack{\text{Perxenate ion} \\ (Xe = +8)}}{XeO_6^{4-}} + \underset{(Xe = 0)}{Xe} + 2H_2O + O_2 \quad ...(ii)$$

On adding (*i*) and (*ii*) we get the final equation as :

$$2XeF_6 + 16OH^- \longrightarrow 8H_2O + 12F^- + XeO_6^{4-} + Xe + O_2$$

(v) Formation of addition compounds. XeF_6 reacts with covalent fluorides like PF_5, AsF_5, SnF_4, BF_3 etc. and transition metal fluorides like NbF_5, TaF_5 etc. In these reactions XeF_6 acts as a fluoride ion-donor (*Lewis base*) and the covalent pentafluoride acts as fluoride ion-acceptor (*Lewis acid*). The products are called adducts which are ionic compounds.

$$XeF_6 + AsF_5 \longrightarrow XeF_6.AsF_5 \text{ or } [XeF_5]^+[AsF_6]^-$$
$$4XeF_6 + SnF_4 \longrightarrow 4XeF_4.SnF_4 \text{ or } [Xe_4F_{15}]^+[SnF_5]^-$$
$$XeF_6 + BF_3 \longrightarrow XeF_6.BF_3 \text{ or } [XeF_5]^+[BF_4]^-$$
$$2XeF_6 + MF_5 \longrightarrow 2XeF_6.MF_5 \text{ or } [Xe_2F_{11}]^+[MF_6]^-$$

The formation of addition compounds shows that XeF_6 acts as a F^- ion-donor. However, there are many reactions in which XeF_6 acts as a F^- ion-acceptor as is evident from the fact that XeF_6 can accept F^- ion from alkali metal fluorides (except LiF) like RbF, CsF etc. to form heptafluoro xenates (VI), $M^+[XeF_7]^-$

$$XeF_6 + \underset{(M = Rb, Cs)}{MF} \longrightarrow [M]^+[XeF_7]^-$$

On heating $M^+[XeF_7]^-$ salts decompose to form octafluoro xenates (VI), $M_2{}^+[XeF_8]^{2-}$

$$2M^+[XeF_7]^- \xrightarrow{50°C} XeF_6 + M_2{}^+[XeF_8]^{2-}$$

XeF_6 also reacts with NOF as follows :

$$2NOF + XeF_6 \longrightarrow [NO]_2{}^+[XeF_8]^-$$

(vi) Reaction with HF. XeF_6 reacts with HF and forms $[XeF_5]^+[HF_2]^-$ which is appreciably ionised.

$$HF + XeF_6 \longrightarrow [XeF_5]^+[HF_2]^- \longrightarrow [XeF_5]^+ + [HF_2]^-$$

(vii) Action of heat. On strong heating, it decomposes to give XeF_2 and XeF_4.

$$2XeF_6 \xrightarrow{\Delta} XeF_2 + XeF_4$$

Structure. See chapter 8.

4. Xenon Trioxide, XeO_3.

Preparation. (*i*) It can be obtained in small quantity by the hydrolysis of XeF_4 or XeF_6.

(*ii*) It can also be prepared by allowing the reaction between XeF_6 and quarts (SiO_2) to run for a prolonged period.

(*iii*) It can also be prepared by hydrolysing $XeOF_4$

$$XeOF_4 + 2H_2O \longrightarrow 4HF + XeO_3$$

Great care must be taken in the preparation of XeO_3, since it is highly explosive.

Properties (*i*). XeO_3 is a white non-volatile deliquescent solid and soluble in water. It explodes violently, when dry, on gentle heating, rubbing or pressing. It produces the same effect as T.N.T.

(*ii*) Its aqueous solution is weakly acidic, since it gives xenic acid, $H(HXeO_4)$ in water.

$$XeO_3 + H_2O \longrightarrow \underset{\text{Xenic acid}}{H_2XeO_4} \longrightarrow H^+ + \underset{\text{Xenate ion}}{HXeO_4{}^-}$$

With more quantity of H_2O, it gives H_6XeO_6 acid

$$XeO_3 + 3H_2O \longrightarrow H_6XeO_6 \longrightarrow H^+ + H_5XeO_6{}^-$$

(*iii*) In basic medium XeO_3 slowly disproportionates.

$$XeO_3 + OH^- \longrightarrow HXeO_4{}^-] \times 2 \quad ...(i)$$

$$2HXeO_4{}^- + 2OH^- \longrightarrow XeO_6{}^{4-} + Xe + O_2 + 2H_2O \quad ...(ii)$$

On adding (*i*) and (*ii*), we get the final equation as :

$$\underset{(Xe = +6)}{2XeO_3} + 4OH^- \longrightarrow \underset{\substack{(Xe = +8)\\ \text{Perxenate}\\ \text{ion}}}{XeO_6{}^{4-}} + \underset{(Xe = 0)}{Xe} + O_2 + 2H_2O$$

(*iv*). XeO_3 is a strong oxidising agent. For example it oxidises Pu^{3+} to Pu^{4+} in presence of H^+ ions

$$6Pu^{3+} + XeO_3 + 6H^+ \longrightarrow 6Pu^{4+} + Xe + 3H_2O$$

XeO_3 can quantitatively be reduced by I^- ion

$$XeO_3 + 6H^+ + 9I^- \longrightarrow 3I_3^- + Xe + 3H_2O$$

(*v*) When a solution of XeO_3 is treated with KF or CsF, an oxo-fluoro salt, $M^+[XeO_3F]^-$ is obtained

(*vi*) With $XeOF_4$, XeO_3 gives XeO_2F_2

$$XeO_3 + XeOF_4 \longrightarrow 2XeO_2F_2$$

Structure. XeO_3 molecule has pyramidal shape as shown in Fig. 42.7. This shape results from sp^3 hybridisation of Xe atom. Each of Xe—O bond distances is equal to 1.76 A° and each OXeO bond angle is equal to 103°.

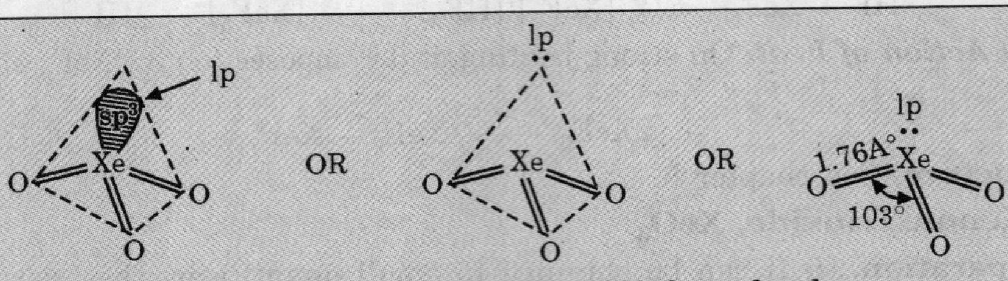

Fig. 42.7. Pyramidal shape of XeO_3 molecule.

5. Xenon Oxytetrafluoride, $XeOF_4$.

Preparation. (*i*) It is prepared by the controlled hydrolysis of XeF_6 with equimolar quantity of water (See the hydrolysis reactions of XeF_6).

(*ii*) $XeOF_4$ may also be prepared by heating XeF_6 in a silica tube at 50°C. When the yellow colour of XeF_6 disappears, the bulb is immediately cooled with solid CO_2. The vapours of SiF_4 are pumped out and $XeOF_4$ is collected.

$$2XeF_6 + SiO_2 \xrightarrow{50°C} 2XeOF_4 + SiF_4$$

Properties. (*i*) It is a clear, colourless mobile liquid which can be stored unchanged in nickel containers for long period.

(*ii*) It reacts with hydrogen to form xenon gas at 300°C,

$$XeOF_4 + 3H_2 \xrightarrow{300°C} Xe + H_2O + 4HF$$

(*iii*) With silica, it results in the formation of explosive trioxide, XeO_3.

$$XeOF_4 + SiO_2 \longrightarrow SiF_4 + XeO_3$$

(*iv*) With water it undergoes hydrolysis forming the explosive oxide, XeO_3 (See the preparation of XeO_3).

(*v*) (*a*) $XeOF_4$ also forms a few complexes with fluoride-ion acceptors (*e.g.* $XeOF_4.2SbF_5$). These complexes are probably ionic compounds and can, therefore, be represented as $[XeOF_3]^+[Sb_2F_{11}]^-$.

(*b*) $XeOF_4$ also forms complexes with fluoride-ion donors (*e.g.* $CsF.XeOF_4$; $3KF.XeOF_4$). In these complexes $[XeOF_5]$ ions or more complex ions containing fluorine-bridges may be present.

(*vi*) The interaction of $XeOF_4$ and XeO_3 gives XeO_2F_2

$$XeOF_4 + XeO_3 \longrightarrow 2XeO_2F_2$$

Structure. $XeOF_4$ molecule has square pyramidal shape as shown in Fig. 42.8. This shape results from sp^3d^2 hybridisation of Xe atom.

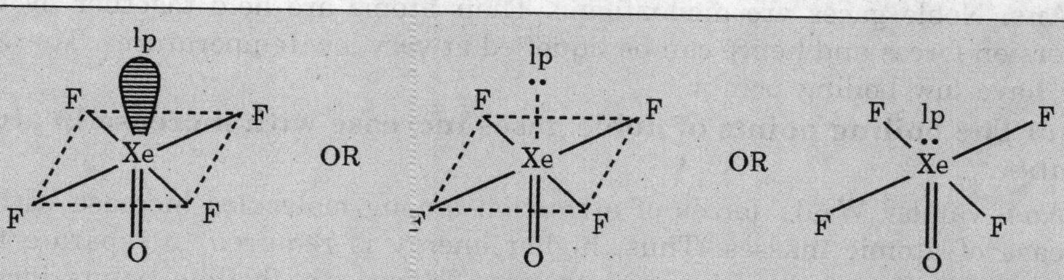

Fig. 42.8. Square pyramidal shape of $XeOF_4$ molecule.

6. Fluorides of Krypton.

Recently, it is observed that krypton also combines with F_2 above 400°C to from krypton difluride (KrF_2) and krypton tetrafluoride (KrF_4). These fluorides are lees stable as compared to xenon fluorides. These are hydrolysed by water according to the equations :

$$2KrF_2 + 2H_2O \longrightarrow 2Kr + 4HF + O_2$$
$$KrF_4 + 2H_2O \longrightarrow Kr + 4HF + O_2$$

KrF_2 is a colourless solid. It is a powerful oxidising and fluorinating agent. KrF_2 also forms addition compound with SbF_5

$$KrF_2 + 2SbF_5 \longrightarrow KrF_2.2SbF_5 \longrightarrow [KrF]^+[Sb_2F_{11}]^-$$

Just like XeF_2, KrF_2 has linear shape which results from sp^3d hybridisatinn of Kr atom.

Conceptual Questions with Answers

Q.1 Explain the following :

(*i*) Noble gases are mostly inert.

Ans. Noble gases are chemically inert because of the following reasons.

(*i*) They have completely filled ns^2np^6 electronic configurations in thier valence shells.

(*ii*) They have high ionization enthalpies.

(*iii*) Electron affinities of noble gases are almost zero.

(*ii*) Noble gases have comparatively large atomic sizes.

Ans. The atomic radii of noble gases are by far the largest in their respective periods. This is due to the reason that noble gases (because they do not form molecules) have only *vander Waals radii* while other elements have covalent radii. vander Waals radii, by definition, are larger than covalent radii. That is why noble gases have comparatively large atomic sizes.

(*iii*) Elements of group 18 are known as noble gases.

Ans. The elements of group 18 have their valence shell orbitals completely filled. As a result, they react with only a few elements (oxygen and fluorine) only under certain conditions. Therefore, they are called noble gases.

(*iv*) Noble gases have very low boiling points.

Ans. Noble gases are monoatomic. Their atoms are held together by weak dispersion forces and hence can be liquefied at very low temperatures. Therefore, they have low boiling points.

(*v*) The boiling points of noble gases increase with increase in atomic number.

Ans. vander Waals' forces of attraction among molecules increase with the increase of atomic masses. Thus, higher energy is required to separate these molecules as to get them in gaseous state. Hence, the boiling points increase with increase of atomic masses.

(*vi*) Argon does not form diatomic molecules like oxygen and nitrogen?

Ans. All the orbitals which are occupied by the electrons are completely filled in argon and hence it has no tendency to lose, gain or share electrons with other atom of Ar to form diatomic molecules. O and N, on the other hand, have unpaired electrons and hence share with other O and N atom respectively to form O_2 and N_2 molecules.

(*vii*) Helium is used in diving apparatus?

Ans. Since helium has low solubility (as compared to N_2) in blood, a mixture of oxygen and helium is used in diving apparatus used by deep sea divers.

(*viii*) Neon is generally used for warning signals.

Ans. Neon lights are visible from long distances even in fog and mist and hence neon is generally used for warning signals.

(*ix*) The group in which inert gases are acommodated was called zero group.

Ans. Inert gases do not show chemical activity under ordinary conditions due to possession of completed outermost shell *i.e.*, they show zero valency. On account of zero valency, the group in which inert gases were placed was called zero group.

(*x*) Zero group elements do not form compounds under ordinary conditions.

Ans. Zero group elements have complete electronic configuration of their outermost shells, *i.e*, s^2p^6 (except helium which has $1s^2$). Hence they do not have any tendency either to lose or accept electron/electrons or share electrons with other atoms.

Q.2 Explain the following :

(*i*) Helium and neon do not form compounds with fluorine.

Ans. He and Ne do not contain *d*-orbitals in thier respective valence shells and hence their electrons cannot be promoted to higher energy levels like that in Xe to form bonds. Therefore, *He and Ne do not form compounds with fluorine.*

(*ii*) Noble gases form compounds with fluorine and oxygen only.

(A.I.S.B. 1998)

Ans. The electronegativity of both fluorine and oxygen is very high. Moreover, fluorine is a highly reactive element.

***(iii) (a)* Xenon does not form fluorides such as XeF, XeF_3 and XeF_5.**

***(b)* Out of noble gases, only xenon is known to form established chemical compounds. (D.S.B. 2003, A.I.S.B. 2004 S, D.S.B. 2005)**

Ans. *(a)* All the filled orbitals of Xe have paired electrons. The promotion of one, two or three electrons from the *p*-filled orbitals to the 5*d*-vacant orbitals will give rise to two, four and six half-filled orbitals. So Xe can combine with even but not with odd number of F-atoms. Hence, it cannot form XeF, XeF_3 and XeF_5.

(b) Except radon which is radioactive, Xe has least ionization energy among noble gases and hence it readily forms chemical compounds particularly with O_2 and F_2.

***(iv)* Xenon forms compounds with fluorine and oxygen even though xenon has a closed-shell electronic configuration.**

Ans. Since Xe has low ionisation energy (= 1170KJ/mol) and flurine and oxygen both have high electronegativity, these elements can excite paired electrons present in 5*s* and 5*p* orbitals to vacant 5*d* orbitals so that unpaired electrons become available. These unpaired electrons are used in forming bonds with fluorine and oxygen atoms.

***(v)* He, Ne and Ar donot form chemical compound while O and F do.**

Ans. Since the ionization enthalpies of He, Ne and Ar are sufficiently higher than those of oxygen and fluorine.

***(vi)* He and Ne donot form clathrate compounds with quinol.**

Ans. The size of the cavities formed during crystallisation of quinolol is more than the size of He and Ne atoms.

Q.3 Complete the following equations :

***(i)* $XeF_2 + H_2O \longrightarrow$ ....**

***(ii)* $XeF_4 + H_2O \longrightarrow$ ....**

***(iii)* $XeF_6 + H_2O \longrightarrow$ ....**

***(iv)* $XeF_6 + NaF \longrightarrow$ ....**

***(v)* $XeF_2 + PF_5 \longrightarrow$ ....**

***(vi)* $XeF_6 + SiO_2 \longrightarrow$ ....**

***(vii)* $XeF_6 + NH_3 \longrightarrow$ ......**

Ans. *(i)* XeF_2 undergoes slow hydrolysis with H_2O evolving O_2.

$$2XeF_2 + 2H_2O \longrightarrow 2Xe + 3HF + O_2$$

(ii) XeF_4 reacts with H_2O and undergoes disproportionation reaction giving Xe and XeO_3.

$$\underset{(Xe\ =\ +4)}{6XeF_4} + 12H_2O \longrightarrow \underset{(Xe\ =\ 0)}{4Xe} + \underset{(Xe\ =\ +6)}{2XeO_3} + 24HF + 3O_2$$

(iii) XeF_6 undergoes hydrolysis with H_2O in two ways :

(a) Complete hydrolysis of XeF_6 gives XeO_3

$$XeF_6 + 3H_2O \xrightarrow[\text{hydrolysis}]{\text{Complete}} XeO_3 + 6HF$$

(*b*) Partial hydrolysis of XeF_6 gives $XeOF_4$ and XeO_2F_2

$$XeF_6 + H_2O \longrightarrow XeOF_4 + HF$$
$$XeF_6 + 2H_2O \longrightarrow XeO_2F_2 + 2HF$$

(*iv*) In this reaction NaF acts as fluoride ion donor and XeF_6 behaves as fluoride ion acceptor. Hence the reaction between XeF_6 and NaF gives fluoroanion, $[XeF_7]^-$.

$$NaF \longrightarrow Na^+ + F^-$$
$$XeF_6 + F^- \longrightarrow [XeF_7]^-$$

$$\underset{\substack{F^-\text{ ion}\\ \text{donor}}}{NaF} + \underset{\substack{F^-\text{ ion}\\ \text{acceptor}}}{XeF_6} \longrightarrow Na^+ + [XeF_7]^- \text{ or } Na^+[XeF_7]^-$$

(*v*) In this reaction XeF_2 acts as F^- ion donor and PF_5 behaves as F^- ion acceptor. Hence the reaction between XeF_2 and PF_5 gives caionic species, $[XeF]^+$

$$XeF_2 \longrightarrow XeF^+ + F^-$$
$$PF_5 + F^- \longrightarrow PF_6^-$$

$$\underset{\substack{F^-\text{ ion}\\ \text{donor}}}{XeF_2} + \underset{\substack{F^-\text{ ion}\\ \text{acceptor}}}{PF_5} \longrightarrow [XeF]^+[PF_6]^-$$

(*vi*) $2XeF_6 + SiO_2 \longrightarrow 2XeOF_4 + SiF_4$

$$2XeOF_4 + SiO_2 \longrightarrow 2XeO_2F_2 + SiF_4$$
$$2XeO_2F_2 + SiO_2 \longrightarrow \underset{\text{Explosive}}{2XeO_3} + SiF_4$$

(*vii*) $XeF_6 + 8NH_3 \longrightarrow Xe + N_2 + 6NH_4F$

Q.4 Discuss the reaction between fluorides of Xe and pentafluorides like AsF_5, SbF_5 etc. and alkali metal fluorides.

Ans. (*i*) **XeF_2.** In BrF_3 solution, XeF_2 reacts with AsF_5, SbF_5 etc. In this reaction XeF_2 molecule acts as F^- donor (Lewis base) and MF_5 molecule (M = As, Sb) acts as F^- ion acceptor [Lewis acid]

F^- ion donor (Lewis base)		F^- ion acceptor (Lewis acid)		Adduct
XeF_2	+	SbF_5	$\longrightarrow$	$[XeF]^+[SbF_6]^-$
$2XeF_2$	+	AsF_5	$\longrightarrow$	$[Xe_2F_3]^+[AsF_6]^-$

The formation of $[XeF]^+[SbF_6]^-$ takes place through the following steps :

$$XeF_2 \longrightarrow [XeF]^+ + :\ddot{\underset{\cdot\cdot}{F}}:^-$$

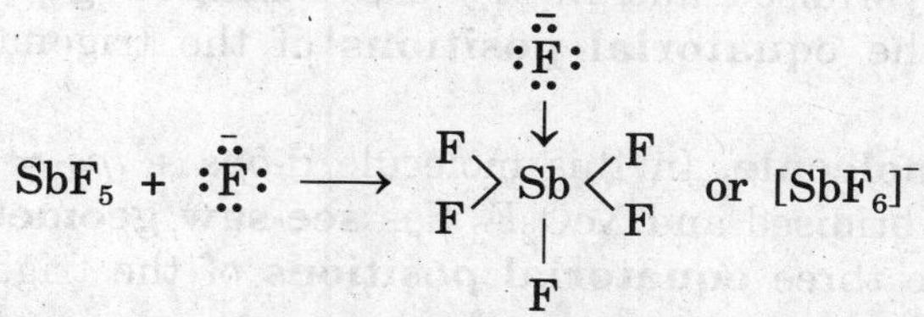

On adding :	XeF_2	+	SbF_5	$\longrightarrow$	$[XeF]^+ [SbF_6]^-$
	F^- ion donor (Lewis base)		F^- ion acceptor (Lewis acid)		Adduct

(ii) **XeF_4.** This molecule reacts with SbF_5. In this reaction XeF_4 acts as F^- ion donor and SbF_5 molecule behaves as F^- ion acceptor.

XeF_4 (*F^- ion donor*) + SbF_5 (*F^- ion acceptor*) $\longrightarrow$ $[XeF_3]^+ [SbF_6]^-$ (*Adduct*)

(iii) **XeF_6** *(a)* XeF_6 reacts with AsF_5, SbF_5 etc. In this reaction XeF_6 acts as F^- ion donor and MF_5 molecule (M = As, Sb) behaves as F^- ion acceptor.

XeF_6(*F^- ion donor*) + MF_5 (*F^- ion acceptor*) $\longrightarrow$ $[XeF_5]^+ [MF_6]^-$ (*Adduct*)

(b) XeF_6 molecule also reacts with alkali metal fluorides. In this reaction XeF_6 acts as F^- ion acceptor and MF molecule (M = Na, K, Rb, Cs) acts as F^- ion donor.

MF (*F^- ion donor*) + XeF_6 (*F^- ion acceptor*) $\longrightarrow$ $M^+ [XeF_7]^-$ (*Adduct*)

Q.5 In the following molecules *(i)* Write the hybridisation of Xe atom *(ii)* Draw the structure/geometry of the molecule *(iii)* Locate the position of *lps* of electrons, if any. XeO_4, $XeOF_2$, XeO_2F_2, XeO_3F_2 and XeO_2F_4.

Ans. The structure of the given molecules are given in Fig. 42.9. Form this figure the following points may be noted.

(i) **XeO_4 molecule.** In this molecule, σ-*bps* + *lps* = 4 + 0 = 4. Hence Xe atom XeO_4 molecule is sp^3 hybridised. XeO_4 has **tetrahedral geometry** shown in Fig. 42.10. This molecule has no *lp* of electrons.

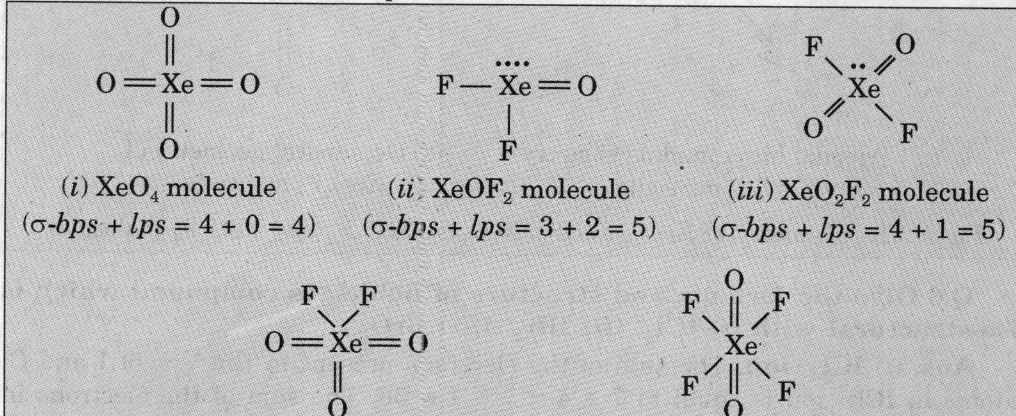

(i) XeO_4 molecule (σ-*bps* + *lps* = 4 + 0 = 4)

(ii) $XeOF_2$ molecule (σ-*bps* + *lps* = 3 + 2 = 5)

(iii) XeO_2F_2 molecule (σ-*bps* + *lps* = 4 + 1 = 5)

(iv) XeO_3F_2 molecule (σ-*bps* + *lps* = 5 + 0 = 5)

(v) XeO_2F_4 molecule (σ-*bps* + *lps* = 6 + 0 = 6)

Fig.42.9 Structure of XeO_4, $XeOF_2$, XeO_2F_2, XeO_3F_2 and XeO_2F_4 molecule.

(*ii*) $XeOF_2$ molecule. In this molecule, σ-*bps* + *lps* = 3 + 2 = 5. Hence Xe-atom is sp^3d hybridised and $XeOF_2$ has **T-shaped geometry**. Two *lps* of electrons occupy the **equatorial positions** of the trigonal bipyramid. (See Fig. 42.10)

(*iii*) XeO_2F_2 molecule. In this molecule σ-*bps* + *lps* = 4 + 1 = 5. Hence Xe-atom is sp^3d hybridised and XeO_2F_2 has **see-saw geometry**. *lp* of electrons occupies one of the three **equatorial positions** of the trigonal pyramid. (See Fig. 42.10)

(*iv*) XeO_3F_2 molecule. In this molecule, σ-*bps* + *lps* = 5 + 0 = 5. Hence Xe atom is sp^3d hybridised and XeO_3F_2 has **trigonal bipyromidal geometry**. This molecule has no *lp* of electrons (See Fig. 42.10)

(*v*) XeO_2F_4 molecule. In this molecule, σ-*bps* + *lps* = 6 + 0 = 6. Hence Xe atom is sp^3d^2 hybridised and XeO_2F_4 has **octahedral geometry**. This molecule has no *lp* of electrons. (See Fig 42.10)

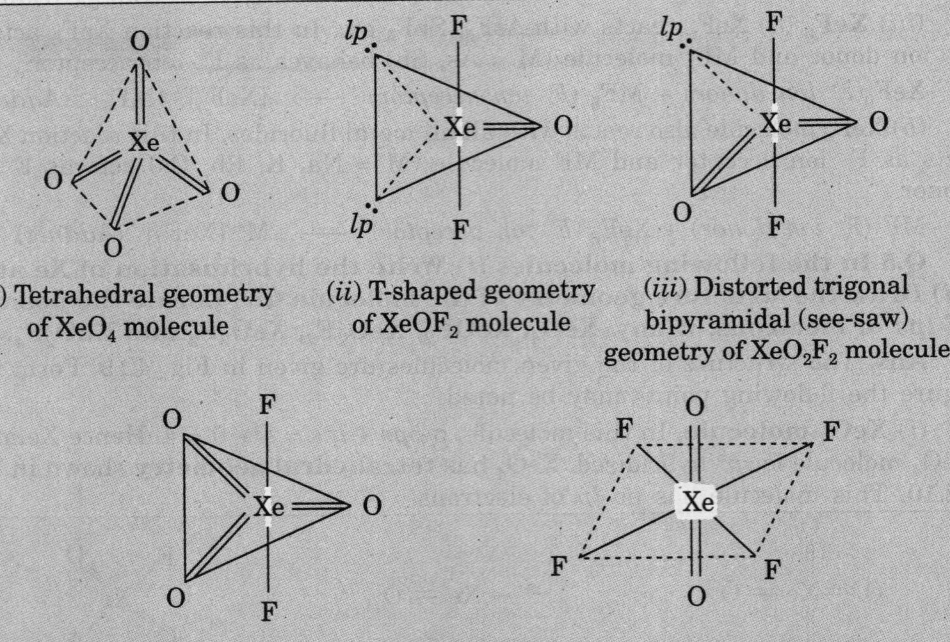

(*i*) Tetrahedral geometry of XeO_4 molecule

(*ii*) T-shaped geometry of $XeOF_2$ molecule

(*iii*) Distorted trigonal bipyramidal (see-saw) geometry of XeO_2F_2 molecule

(*iv*) Trigonal bipyramidal geometry of XeO_3F_2 molecule

(*v*) Octahedral geometry of XeO_2F_4 molecule

Fig. 42.10 Geometry of XeO_4, $XeOF_2$, XeO_2F_2, XeO_3F_2 and XeO_2F_4 molecules.

Q.6 Give the formula and structure of noble gas compound which is iso-structural with (*i*) ICl_4^- (*ii*) IBr_2^- (*iii*) BrO_3^-.

Ans. (*i*) **ICl_4^- ion.** The sum of the electrons present in the V.S. of I and Cl atoms in ICl_4^- ion is equal to 7 + 4 × 7 + 1 = 36. The sum of the electrons in the V.S. of Xe and F atoms in XeF_4 molecule is also 36 electrons (8 + 4 × 7 = 36). Since ICl_4^- and XeF_4 have the same number of electrons in thier V.S. (iso-electronic species), these species have the same geometry (iso-sturctural species). Both these species have **square planar geometry.**

(*ii*) **IBr_2^- ion.** The sum of the electrons present in the V.S. of I and Br atoms in IBr_2^- ion is equal to 7 + 7 × 2 + 1 = 22. The V.S. of Xe and F atoms in XeF_2 also has 22 electrons (8 + 2 × 7 = 22). Since IBr_2^- and XeF_2 have the same number of electrons in their V.S. (iso-electronic species), these species have the same geometry (iso-structural species). Both these species have **linear geometry.**

(*iii*) **BrO_3^- ion.** The sum of the electrons present in the valence shell of Br and O-atom in BrO_3^- ion is equal to 7 + 3 × 6 + 1 = 26. The V.S. of Xe and F atom in XeO_3 also has 26 electrons (8 + 6 × 3 = 26). Since BrO_3^- and XeO_3 have the same number of electrons in their V.S. (iso-electronic species), these species have the same geometry. Both these species have **pyramidal geometry.**

University Questions

1. Discuss the position of inert gases in the periodic table. (*Meerut 82, 83; Raj. 84, 86; Jiwaji 84; Panjab 84; Andhra 86; Bharathiar 86*)
2. State the important industrial uses of noble gases. (*Indore 86; Sagar 85; Punjab 86; Madurai 85; Raj. 86; M.D. Rohtak 84*)
3. Write a note on the history of the discovery of noble gages. (*Raj. 87, 87 S*)
4. Give an account of different methods for the isolation of inert gases from the atmosphere. (*Meerut 82, 85; Indore 86; Raj. 87 S; Jiwaji 84; Sagar 85*)
5. Write a note on "Compounds of inert gases". (*Meerut 83, 86, 87, 88, 89 S; Allahabad 86, 89; Agra 89; Utkal 85; Andhra 86; Gorakhpur 83; Raj. 87*)
6. How is XeO_3 prepared ? Describe its properties and structure. (*Madurai 85, 86*)
7. Write a note on xenon fluorides, mentioning their preparation, properties uses and structure. (*Meerut 84, 85, 88; Madurai 85; Jodhpur 84; Utkal 86; Bundelkhend 88*)
8. (*a*) Give one method of preparation and structure of the following : XeF_6, XeO_2F_6 and KrF_2.

   (*b*) What are clathrate compounds of noble gases? Give conditions to form stable clathrates. Which inert gases do not form clathrates and why? (*Himachal Pradesh 2000*)
9. Give a brief description of the compounds of xenon. Write the structures of XeF_2 and XeF_4. Why He and Ne do not form such compounds. (*Kanpur 2000*)

10. Discuss the position of inert gases in the periodic table, their isolation and structure of any two important compounds of xenon. *(Kumaon 2000)*

11. Write a note on "Compounds of noble gases." *[Gauhati (General) 2000]*

12. What are possible oxidation states of xenon ? Give preparation and structure of one compound in each of its oxidation states. *(Lucknow 2000)*

13. What are noble gases? Write their electronic configuration. *(Avadh 2000)*

14. Write notes on the following : (*a*) Clatherate compounds (*b*) Hybridization and geometrical shape of : XeF_2, XeF_4, XeF_6. *(Avadh 2000)*

15. (*a*) Explain the isolation of noble gases in detail.

(*b*) Write the structure and shape of XeF_4 molecule *(Meerut 2009)*

CHAPTER

43 Co-ordination Compounds

Definitions of Some Terms

A. Addition or Molecular Compounds ?

When solutions of two or more salts in simple molecular ratio are mixed together and allowed to evaporate, crystals of a new compound are obtained. This new compound is called *addition or molecular compound.* Molecular compounds are of two types as discussed below :

1. Double Salts. There are the addition compounds which have the properties *viz.* (*i*) These compounds are stable in the solid state only but are broken into their individual compounds when dissolved in water or any other ionic solvent. Their solutions, therefore, have the same properties as a mixture of the individual compounds. For example, a solution of Mohr's salt, $FeSO_4.(NH_4)_2SO_4.6H_2O$ which is a double salt, shows the same physical and chemical properties as a mixture of the solutions of $FeSO_4$ and $(NH_4)_2SO_4$. (*ii*) The physical and chemical properties of a double salt are the same as those of their individual compounds. Examples of double salts are carnallite, $KCl.MgCl_2.6H_2O$; Mohr's salt, $FeSO_4.(NH_4)_2SO_4.6H_2O$; potash alum, $K_2SO_4.Al_2(SO_4)_3.24H_2O$.

2. Coordination or complex compounds. These are also the addition compounds which have the following properties : (*i*) These compounds retain their identities (*i.e.* do not lose their properties) in the solid as well as when dissolved in water or any other ionic solvent. (*ii*) Their properties are completely different from those of their individual constituents. For example when aqueous NH_3 is added to a green solution of $NiCl_2$ and the solution thus obtained is evaporated, purple crystals of a new compound having the composition, $[Ni(NH_3)_6]Cl_2$ are obtained. This new compound, when dissolved in water, ionizes as :

$$[Ni(NH_3)_6]Cl_2 \rightleftharpoons [Ni(NH_3)_6]^{2+} + 2Cl^-,$$

showing that the solution of the complex compound has neither Ni^{2+} ions nor NH_3 molecules. The properties of the complex compound, $[Ni(NH_3)_6]Cl_2$ are entirely different from those of Ni^{2+} ions and NH_3 molecules. The ion, $[Ni(NH_3)_6]^{2+}$ is called complex ion. A complex ion can be defined as *an electrically charged radical which is obtained by the combination of a metal cation with one or more neutral molecules or simple anions.* For example $[Ni(NH_3)_6]^{2+}$ which is a complex ion is obtained by combining one Ni^{2+} cation with six NH_3 molecules Similarly $[Fe(CN)_6]^{4-}$ which is a complex ion is obtained by combining one Fe^{2+} cation and six CN^- ions.

B. Ligands or Coordinating Groups.

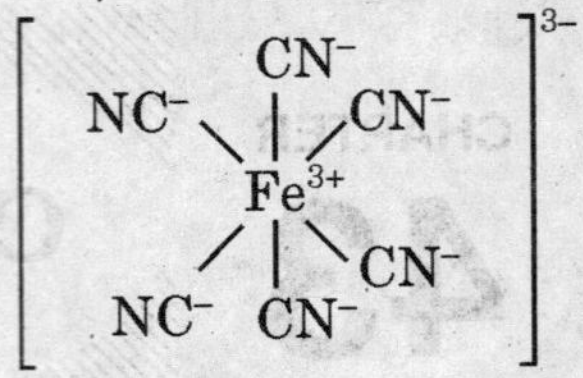

The neutral molecules or ions (usually anions) which are attached with the central ion in complex compounds are called *ligands* or *co-ordinating grounds*. For example in the complex ion, $[Fe(CN)_6]^{3-}$ the six CN^- ions which are attached with the central Fe^{3+} ion as shown in the margin act as ligands. In Lewis sense, in most of the complex compounds the ligands act as Lewis bases (electron pair donors) and the central metal ion acts as a Lewis acid (electron pair acceptor), *i.e.* in most of the complex compounds the ligands donate one or more electron pairs to the central metal ion.

M^{n+}	$+\ xL$	$\longrightarrow$	$[ML_x]^{n+}$
Central metal ion (Lewis acid or electron pair acceptor)	Ligand (Lewis base or electron pair donor)		Complex compound

In some complex compounds the ligand acts both as donor and acceptor. For example in metallic carbonyls CO molecules which act as ligands act both as donor and acceptor ($M \rightleftarrows CO$). In a ligand the atom which actually donates the electron pair to the central metal ion is called *donor atom*. The ligands are attached with the central metal ion through their donor atom (or atoms).

The ligands are arranged round the central metal ion inside the first sphere of attraction in preferred geometries which are : *linear, equilateral triangular, tetrahedral, square planar, trigonal bipyramidal, square pyramidal and octahedral.*

C. Monodentate or Unidentate Ligands.

The ligands which have only one donor atom or are co-ordinated through one electron pair are called *monodentate* or *unidentate ligands*. Such ligands are coordinated to the central metal ion at one site or by one metal—ligand bond only. These ligands may be neutral molecules or in anionic form. Examples of such ligands are given below. Abbreviations, if any, for the ligands are given in parenthesis and the donor atoms are indicated by asterisk. (*A*) *Neutral monodentate ligands*. Examples are : thiourea (abbreviated as *tu*), $\overset{*}{S}=C\begin{smallmatrix} NH_2 \\ NH_2 \end{smallmatrix}$; pyridine (*py*) $C_5H_5\overset{*}{N}$; ammine, $\overset{*}{N}H_3$; aqua or aquo, $H_2\overset{*}{O}$; $\overset{*}{S}PMe_3$; $\overset{*}{P}H_3$; $\overset{*}{P}R_3$; (*e.g.* $PEtPh_2$); $\overset{*}{C}O$; $R_2\overset{*}{S}$; $R_3\overset{*}{As}$; $\overset{*}{N}O$ etc. (*B*) *Anionic monodentate ligands*. Examples are F^-, Cl^-, Br^-, I^-, CN^-, NCS^-, NO_2^-, $\overset{*}{N}H_2^-$: $\overset{*}{N}H^{2-}$, N^{3-}, $\overset{*}{O}H^-$, O^{2-}, $CH_3CO\underset{*}{O}^-$, OCN^-, H^-, S^{2-} etc.

D. Polydentate Ligands.

The ligands having two or more donor atoms are called polydentate ligands. These ligands may be of the following types, depending on the number of donor atoms present in them.

1. Bidentate ligands. The ligands having two donor atoms are called bidentate ligands. Examples of such ligands are :

(*A*) *Anionic bidentate ligands*. Examples are : (*i*) *Oxalato*, $C_2O_4^{-2}$ (abbreviated as ox^{2-}). This is a bidentate dibasic ligand and its co-ordination to the metal ion occurs through its both negatively-charged O-atoms (acidic donor atoms), forming a 5-membered chelate as is evident from the structure of dioxalato platinate (II) ion, $[Pt^{2+}(C_2O_4)_2]^{2-}$ or $[Pt^{2+}(ox)_2]^{2-}$ shown below. (*ii*) *Glycinato*, $\overset{*}{N}H_2.CH_2—\overset{O}{\overset{\|}{C}}—\overset{*}{O^-}$ (gly^-). It is a didentate monobasic ligand and its co-ordination to the metal ion occurs through neutral N-atom (neutral donor atom) and negatively-charged O-atom (acidic donor atom), forming a 5-memberod ring as is avident from the structure of bis-(glycinato) copper (II) ion, $[Cu^{2+}(NH_2.CH_2.COO)_2]^0$ or $[Cu(gly)_2]^0$ as shown below.

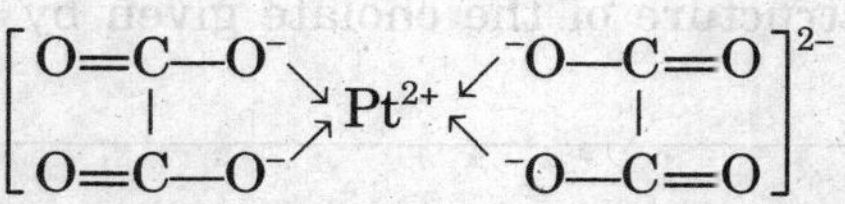

Dioxalato platinate (II) ion,
$[Pt^{2+}(C_2O_4)_2]^{2-}$ or $[Pt^{2+}(ox)_2]^{2-}$

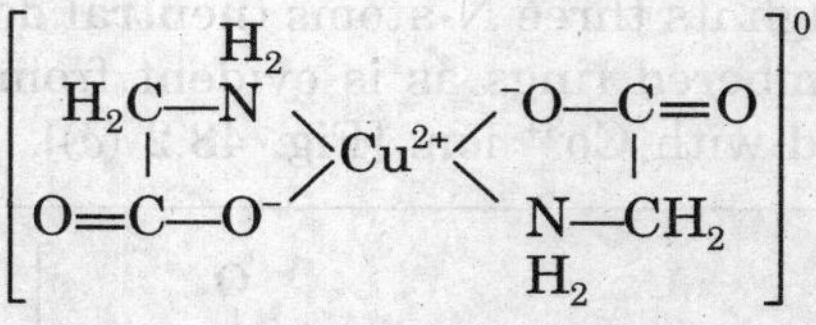

Bis (glycinato) copper (II) ion,
$[Cu^{2+}(NH_2.CH_2.COO)_2]^0$ or $[Cu^{2+}(gly)_2]^0$

(*B*) *Neutral bidentate ligands. o-phenanthroline* (*phen or phenan*) is an important example of this type of bidentate ligand. It has two N-atoms as its donor atoms (neutral donor atoms). Its co-ordination to the metal ion takes place through these N-atoms as is evident from the structure of ferrous orthophenanthroline ion, $[Fe^{2+}(phen)_3]^{2+}$ shown in Fig. 43.1.

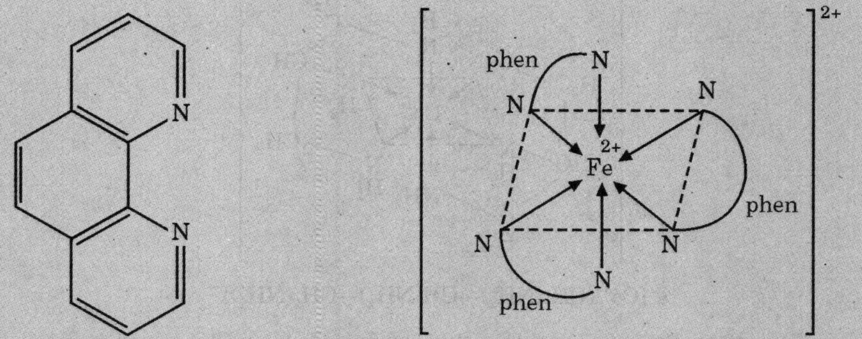

Ortho phenanthroline (phen or phenan)

Ferrous orthophenanthroline ion, $[Fe^{2+}(phen)_3]^{2+}$

Fig. 43.1. Structure of $[Fe^{2+}(phen)_3]^{2+}$ ion.

2. Tridentate ligands. The ligands having three donor atoms are called tridentate ligands. Examples are:

(A) *Anionic tridentate ligands*. Example of such ligands are (*i*) *Anion of aspartic acid* whose structure is $^{-}\overset{*}{O}OC—CH_2—CH(\overset{*}{N}H_2)— CO\overset{*}{O}^{-}$ and is abbreviated as Asp^{2-}. It is a tridentate dibasic ligand and its coordination to the metal ion occurs through its N-atom (neutral donor atom) and two negatively-charged O-atoms (acidic donor atoms) as is evident from the structure of the chelate formed by this ligand with Co^{3+} ion [Fig. 43.2 (*a*)]. (*ii*) *Anion of diaminopropionic acid*, $H_2C(\overset{*}{N}H_2)—CH(\overset{*}{N}H_2)—CO\overset{*}{O}^{-}$. Obviously it is a tridentate mono-basic ligand and its coordination to the metal ion takes place through its two N-atoms (neutral donor atoms) and one negtively-charged O-atom (acidic donor atom) as is evident from the structure of the chelate formed by this ligand with Co^{3+} ion [Fig. 43.2 (*b*)]

(*B*) *Neutral tridentate ligands.* Such ligands can be exemplified by triamino propane, $CH_2(\overset{*}{N}H_2)—CH(\overset{*}{N}H_2)—CH_2(\overset{*}{N}H_2)$ which coordinates to the metal ion through its three N-atoms (neutral donor atoms), leading to the formation of two 5-membered rings as is evident from the structure of the cholate given by this ligand with Co^{3+} ion. [Fig. 43.2 (*c*)].

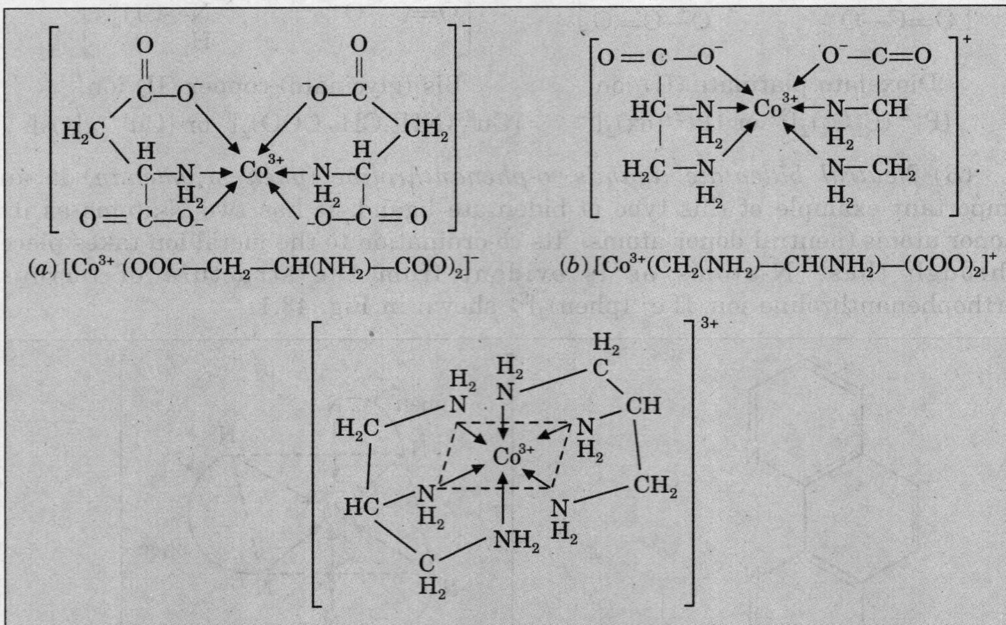

(*a*) $[Co^{3+}(OOC—CH_2—CH(NH_2)—COO)_2]^{-}$ (*b*) $[Co^{3+}(CH_2(NH_2)—CH(NH_2)—(COO)_2]^{+}$

© $[Co^{3+}(CH_2(NH_2)—CH(NH_2)—CH_2(NH_2)]^{3+}$

Fig. 43.2. Structure of the chelates given by tridentate ligands

3. Tetradentate ligands. The ligands having four donor atoms are called tetradentate ligands. These may be anionic or neutral ligands.

(*A*) *Anionic tetradentate ligands.* Nitrilotriacetate (nta^{3-}) is an important example of the anionic tetradentate ligand. Its structure is

$$\overset{*}{N}\begin{cases} CH_2—CO\overset{*}{O}^{-} \\ CH_2—CO\overset{*}{O} \\ CH_2—CO\overset{*}{O}^{-} \end{cases}$$

which

shows that it is a tribasic tetradentate ligand. The coordination of this ligand to the metal ion takes place through N-atom (neutral donor atom) and three negatively-charged O-atoms (acidic donor atoms) of COO^- groups. This type of coordination gives rise to the formation of a chelate having three 5-membered rings [Fig. 43.3 (*a*)].

(B) Neutral tetradentate ligands. Triethylentetramine (trien), $\underset{*}{N}H_2—CH_2—CH_2—\underset{*}{N}H—CH_2—CH_2—\underset{*}{N}H—CH_2—CH_2—\underset{*}{N}H_2$ is an important example of this type of tetradentate ligands. This ligand has four N-atoms as its donor atoms (neural donor atoms). Coordination of this ligand to the metal ion takes place through these N-atoms, forming a chelated complex containing 5-membered rings as shown in Fig. 43.3 (*b*).

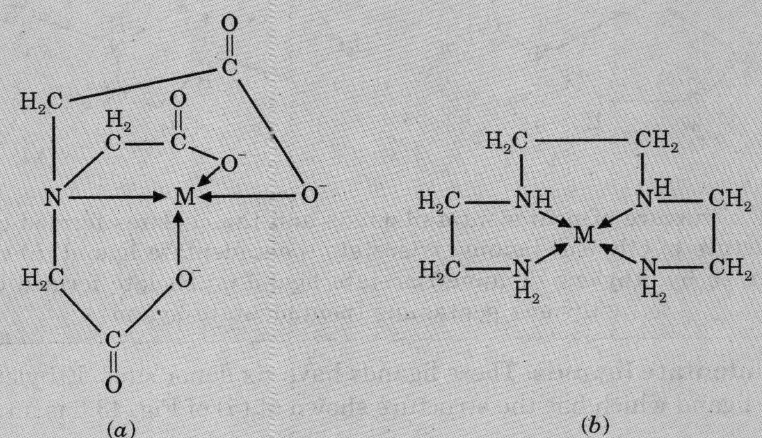

Fig. 43.3. Structure of the chalates formed by tetradentate ligands. (*a*) Chelate formed by the coordination of nitrilotriacetato ion with the metal atom, M (*b*) Chelate given by triethylene tetramine with metal atom, M.

4. Pentadentate ligands. The ligands having five donor atoms are called pentadentate ligands. These may be anionic or neutral ligands.

(*A*) *Anionic pentadentate ligands.* Ethylenediaminetriacetato whose structure is shown at (*a*) of Fig. 43.4 is an important example of this type of ligands. Its coordination to the metal atom occurs through two N-atoms (neutral donor atoms) and three negatively-charged O-atoms (acidic donor atoms) of COO^- groups, forming four 5-membered rings as shown at (*b*) of Fig. 43.4.

(*B*) *Neutral pentadentate ligands.* Tetraethylenepentamine (tetraen) is an important example of this type of pentadentate ligands. Its structure is $\underset{*}{N}H_2—(CH_2)_2—\underset{*}{N}H—(CH_2)_2—\underset{*}{N}H—(CH_2)_2—\underset{*}{N}H—(CH_2)_2—\underset{*}{N}H_2$, showing that it has five N-atoms as its donor atoms (neutral donor atoms). Thus this ligand coordinates to the metal atom through these five N-atoms and forms four 5-membered rings as shown at (*c*) of Fig. 43.4.

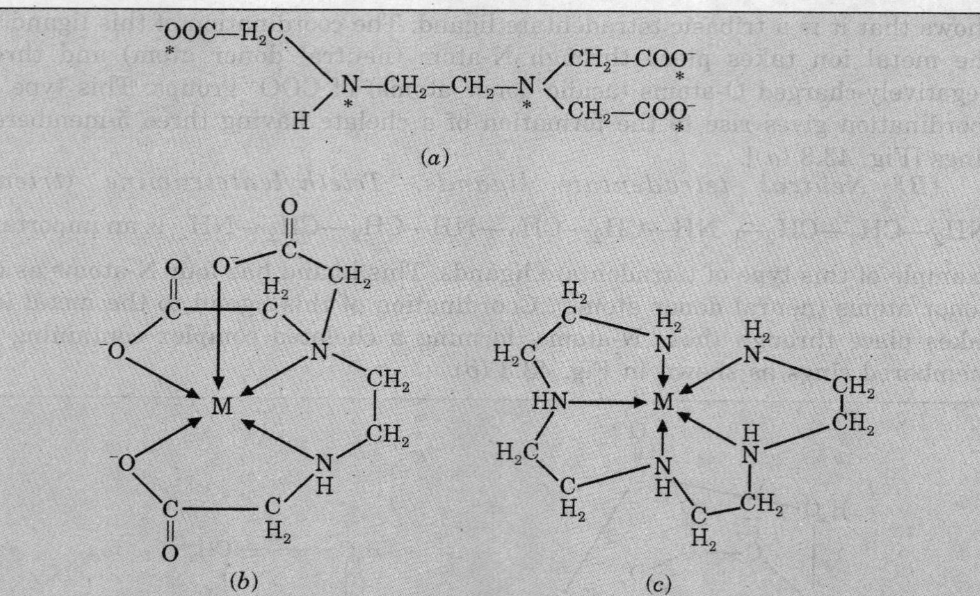

Fig. 43.4. Structure of pentadentate ligands and the chelates formed by them. (*a*) structure of ethylenediamine triacetato (pentadentate ligand (*b*) chelate formed by ethylene diaminetriacetato ligand (*c*) chelate formed by tetraethylene pentamine (pentadentate ligand).

5. Hexadentate ligands. These ligands have six donor sites. Ethylenediamine tetracetato ligand which has the structure shown at (*a*) of Fig. 43.5 is an important

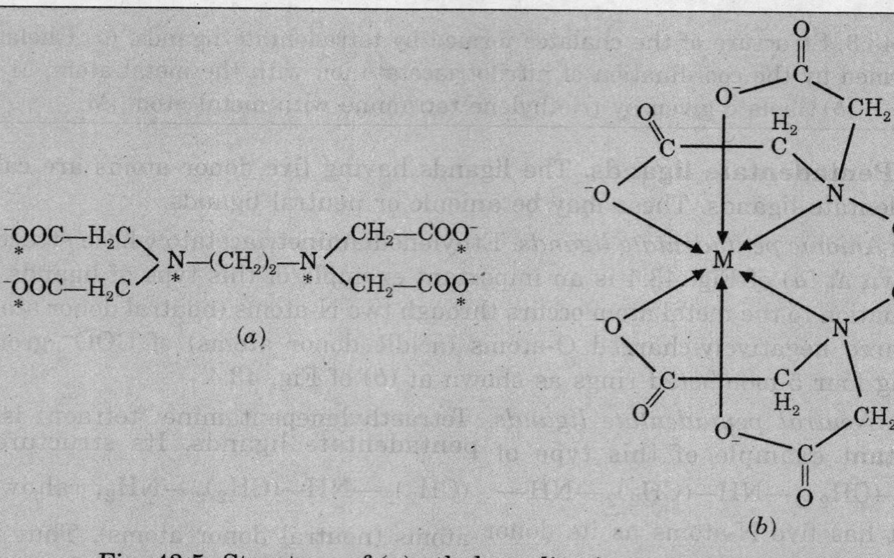

Fig. 43.5. Structure of (*a*) ethylene diamine tetraacetato ion (hexa dentate ligand) (*b*) chelate complex given by ethylene diamine tetraacetato ligand with metal atom, M.

example of hexadentate ligands. The structure of the ligand clearly shows that it has two neutral N-atoms and four negatively-charged O-atoms as its donor sites. Thus it is a tetrabasic hexadentate ligand. The coordination of this ligand to the metal atom takes place through two N-atoms and four O-atoms, forming five 5-membered rings as shown at (*b*) of Fig. 43.5.

E. Coordination Number.

The number of ligands which are directly attached with the central metal atom/ion is known as the coordination number of that metal atom/ion. For example the co-ordination number of Ag^+ ion in $[Ag(NH_3)]^{2+}$ is two. 2, 3, 4 and 6 coordination numbers are very common. Light transition metal atoms prefer to attain 4 and 6 coordination numbers while the heavier transition metal atoms exhibit coordination numbers greater than 6. A metal atom may exhibit more than one coordination number in different complexes. For example in the complexes *viz.* $[FeCl_4]^-$ and $[FeCl_6]^{3-}$, Fe^{3+} ion shows coordination number equal to 4 and 6 respectively. Normally the coordination number of a metal ion is twice the oxidation state of the metal ion. For example the coordination numbers of Ag^+, Cu^{2+} and Cr^{3+} ions are 2, 4 and 6 respectively.

F. Coordination Sphere.

While writing the structural formula of a given complex ion, the ligands which are directly linked with the metal atom are enclosed in square bracket which is called coordination sphere. Thus coordination sphere consists of the central metal atom/ion and the ligands attached with it. Thus in the complex compound, $[Co(NH_3)_6]Cl_3$, the coordination sphere consists of one Co^{3+} ion and six NH_3 molecules which are acting as ligands.

G. Tendency of Polydentate Ligands to Form Chelates : Chelation.

When all the donor atoms of a polydentate ligand get coordinated with the same metal ion, we get a complex compound which contains one or more rings in its structure and hence is called a *chelated* (or *cyclic*) *complex* or *chelate.* The process of the formation of a chelated complex is called *chelation* or *cyclisation.* For example when oxalato ligand, $C_2O_4^{2-}$ (bidentate ligand) gets coordinated with Pt^{2+} ion, we get dioxalato platinate (II) ion, $[Pt^{2+}(C_2O_4)_2]^{2-}$ which contains two 5-membered rings and hence is a chelated complex. In simple words the complexes given by the polydentate ligands with the central metal ion are called chelates or chelated complexes. It is for this reason that the polydentate ligands are also called chelating ligands. The complexes given by a monodentate ligand do not contain any ring in their structure and hence are called *non-chelated complexes.* The word chelating ligand does not apply to a bidentate ligand which forms a bridge system with its two donor atoms bonded to separate metal ions. A chelating ligand must have two or more lone pairs which should be at such a distance from each other that the ligand may form suitable strain-free rings at the metal ion. Polydentate ligands may further be subdivided according to the nature of the donor atoms present in them. A polydentate ligand may contain all neutral donor atoms or all acidic (anionic) donor atoms or some of the donor atoms may be neutral and the remaining may be acidic. Neutral donor atoms are generally N, O, S etc atoms which are from the groups like $-NH_2$ (primary, secondary and tertiary amines), = NOH (oxime), –OH (alcoholic hydroxyl), = CO (carbonyl) and –S–(thioether). The acidic donor atoms are from the anions obtained by removing one or more H-atoms from the acidic groups like –COOH (carboxyl), $-SO_3H$ (sulphonic), –OH (enolia hydroxyl) and = NOH (oxime). Diehal (1937) has classified the chelates on the basis of the fact whether the chelating ligand has all neutral donor atoms or all acidic (anionic) donor atoms or some of the donor atoms are neutral and the remaining arc acidic.

H. Applications of Chelate Formation.

Applications of chelate formation can be studied under the following heads.

1. Formation of chelates in analytical chemistry. Examples are :

(i) *Estimation and identification of Ni^{2+} ions by dimethyl glyoxime.* Dimethyl glyoxime solution reacts with Ni^{2+} ions in ammoniacal medium and forms a red coloured precipitate of *bis-* (dimethyl glyoximato) nickel (II). Thus the formation of this precipitate can be used for the identification, estimation of Ni^{2+} and also for the separation of Ni^{2+} ions from Co^{2+} ions

(ii) *Estimation of Mg^{2+} and Ca^{2+} ions by EDTA.* Disodium salt of ethylene diamine tetraacetic acid (abbreviated as $EDTAH_2Na_2$) has been used for the estimation of M^{2+} metal ions like Mg^{2+}, Ca^{2+} etc. by complexometric titrations. In these titrations an organic dye is used as an indicator. A good indicator should meet the following requirements (*a*) The indicator must form a coloured complex with the metal ion at the same *p*H at which the $EDTA^{4-}$ ion forms complex with the same metal ion. (*b*) The colour of the metal- indicator complex should be different from that of the free indicator ion. (*c*) The stability of the metal-indicator complex should be low compared to that of metal-EDTA complex.

Suppose in the complexometric titration of Mg^{2+} or Ca^{2+} ions with $EDTAH_2Na_2$, we use Erio-chrome Black T (abbreviated as H_3D) as an indicator. In this titration the solution containing Mg^{2+} or Ca^{2+} ions is buffered to *p*H = 10 (This *p*H is obtained by using $NH_4Cl—NH_4OH$ buffer), a few drops of the indicator are added and the solution is titrated with a standard solution of $EDTAH_2Na_2$. In this titration, indicator ions (D^{3-} ions) form red metal-indicator complex, $[MD]^-$ with M^{2+} (M^{2+} = Mg^{2+} or Ca^{2+}) ions.

$$\underset{\substack{\text{Metal ion}\\(M^{2+} = Mg^{2+} \text{ or } Ca^{2+})}}{M^{2+}} + \underset{\substack{\text{Indicator}\\\text{ion (blue)}}}{D^{3-}} \xrightleftharpoons{pH = 10} \underset{\substack{\text{Metal-indicator}\\\text{complex (red)}}}{[M^{2+}D]^-}$$

Metal-indicator complex reacts with $EDTA^{4-}$ ion and forms metal-EDTA complex.

$$\underset{\substack{\text{Metal-indicator}\\\text{complex}\\\text{(red)}}}{[M^{2+}D]^-} + \underset{\substack{\text{EDTA}\\\text{ion}}}{EDTA^{4-}} \rightleftharpoons \underset{\substack{\text{Metal-EDTA}\\\text{complex}}}{[M^{2+}(EDTA)]^{2-}} + \underset{\substack{\text{Indicator}\\\text{ion}\\\text{(blue)}}}{D^{3-}}$$

$EDTA^{4-}$ ion also forms metal-EDTA complex with the free Ca^{2+} or Mg^{2+} ions (present. in the titration solution) that have not reacted with D^{3-} ions to form $[MD]^-$ complex.

Thus we see that free indicator ion is obtained at the end point. The release of the free indicator is marked by a change form red (colour of $[M^{2+}D]^{2-}$ complex) to blue (colour of D^{3-} ion) colour.

2. Formation of chelates in softening of water. Ethylenediamine tetracetic acid ($EDTAH_4$) has been used for softening water which does a lot of harm in industrial boilers and other equipments by forming a scale in them. Ethylenediamine tetracetic acid is added to the boilers in which $EDTA^{4-}$ forms water-soluble chelated complex with Ca^{2+} and Mg^{2+} ions present In hard water and thus these ions are removed.

3. Formation of chelates in medicines. Some applications are :

(*i*) *In the removal of poisonous metals from the body.* In this case injection of $CaNa_2EDTA$ is given to the patient. $CaNa_2EDTA$ reacts with Pb in the body and forms Pb-EDTA complex which is rapidly excreted in urine. Since EDTA is not readily metabolised, most of Ca-EDTA complex also goes out of the body in urine.

(*ii*) *Disorder of lever.* The desease regarding the disorder of the lever leads to the malfunctioning of the central nervous system. This disease is due to the fact that free Cu^{2+} ions accumulate in the lever. The excess of Cu^{2+} ions is removed by giving penicillamine (I) to the patient. (The structure of (I) is shown in the margin). Cu^{2+} ions form a soluble complex with (I) by coordinating with S, N and O donor atoms of (I).

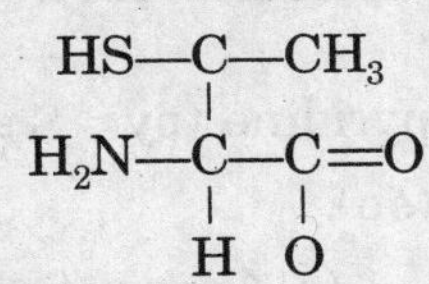

Penicillamine (I)

4. Formation of chelates in living systems. Metal chelates play vital role in living systems. For example :

(*i*) The green colouring matter present in plants which is called chlorophyll is a square planar complex of Mg^{2+}. This substance converts CO_2 and water to starches and sugars in the plants. Solar system provides the energy for this conversion.

(*ii*) Hemoglobin imparts red colour to the human blood and is porphyrin complex of Fe^{2+} ion. In human livings blood is saturated with air and the hemoglobin is completely changed to the oxyhemoglobin form.

$$\textit{Hemoglobin} + O_2 \rightleftarrows \textit{Oxyhemoglobin} + H_2O$$

As the blood runs through the arteries to the tissues, the oxygen pressure dwindles and the bound oxygen gets liberated. This liberated O_2 is used in the combustion of food. By this reversible process hemoglobin serves as an oxygen carrier to all the cells of the organism. Oxygen carriers in the blood of various invertebrates are the chelates of Cu, Mn and V.

(*iii*) At present it is believed that all the essential trace metals in human body probably function through processes involving chelate formation.

(*iv*) Within the animal cell another Fe(II)–Fe(III) protein, coordination compound of hemoglobin family, called *cytochrome-C* acts as a catalyst for the utilisation of oxygen in the basic cell metabolism of the fundamental life process.

Nomenclature of Coordination Compounds

Coordination compounds are named according to the rules suggested by the *International Union of Pure and Applied Chemistry*, IUPAC (1976). For naming all types of complex compounds the ligands are named first and the central metal atom is named in the end.

1. Naming the Ligands. The ligands are named as :

(i) Naming the anionic ligands. The name of anionic ligands ends in *o*. Examples are CH_3COO^-, acetato; CN^-, cyno; NO_2^- (ONO^-), nitrito; NO_3^-, nitrato; $NH_2CH_2.COO^-$, glycinato; Cl^-, chloro; $C_2O_4^{2-}$, oxalato etc.

(ii) Naming the neutral inorganic ligands. For neutral inorganic ligands the names are not systematic. Examples are N_2, dinitrogen; H_2O, aquo or aqua; NH_3, ammine; CO, carbonyl etc.

(iii) Naming the cationic ligands. Positively-charged ligands have the suffix *ium* in the end. Examples are $NH_2.NH_3^+$, hydrazinium; NO^+, nitrosylium; NH_4^+, ammonium etc.

(iv) Naming the organic ligands. Organic free radicals are given their own names. Examples are CH_3, methyl; C_2H_5, ethyl; C_6H_5, phenyl. For organic neutral molecules their names are used, *e.g.*, $P(C_6H_5)_3$, triphenylphosphine; C_5H_5N, pyridine (py); $S = C\begin{smallmatrix} NH_2 \\ NH_2 \end{smallmatrix}$, thiourea (*tu*); $NH_2.CH_2.CH_2.NH_2$, ethylene diamine (en).

(v) Order of naming ligands. If there are two or more different kinds of ligands, the order of naming such ligands is *negative (anionic) ligands, neutral ligands* and *positive (cationic) ligands* without separation of hyphens (*i.e.* as single words). Within the ligands of the same category, the ligands are named in the order of their increasing complexity. The anionic ligands are named in the order : H^-, O^{2-}, OH^-, other simple monoatomic anionic ligands, polyatomic anionic ligands and finally organic anions in alphabetical order. Where the two ligands have the same number of atoms, the order is that of decreasing atomic number of the central metal species in the ligand, *e.g.* CrO_4^{2-} first and then SO_4^-. Neutral ligands are named in the following order : *first* : H_2O (*aquo*), then NH_3 (*ammine*), *second* : other neutral coordinated ligands in the order in which the coordinated atom falls in the series : B, Si, C, Sb, As, P. N, H, Te, Se, S, At, I, Br, Cl, O, F, *third* : neutral organic ligands in alphabetical order.

(vi) Indication of the number of ligands. The number of simple ligands such as Cl^-, CH_3COO^-, $C_2O_4^{2-}$ etc. is indicated by using before them the Greek prefixes : *di-*, *tri-*, *tetra-*, *penta-*, *hexa-* etc. In case of chelating ligands like ethylenediamine, trialkyl phosphine which contain the prefixes *di-*, *tri-*, etc. in their ligand names, the prefixes *bis-* (for two), *tris-*, (for three), *tetrakis-* (for four), *pentakis*. (for five), *hexakis-* (for six) etc. are used before their names.

(vii) Naming the bridging ligands. There are many monodentate ligands like OH^-, NH^{2-}, NH_2^-, NO_2, Cl^-, O_2^{2-}, CO etc. which can coordinate simultaneously with more than one metal ion and thus form a bridge between the metal ions. Such ligands are called *bridging ligands* and the complexes containing such ligands are called *bridged* or *poly-nuclear complexes*. For naming a bridging ligand in such complexes the Greek prefix μ is used before its name and for naming two or more bridging ligands of the same kind the prefixes *di*-μ. *tri*-μ, etc. are used before their name. The bridging ligand (or ligands) is written in the name of the complex by separating it from the remaining name of the complex by a hyphen.

(*viii*) *Naming the ambidentate ligands.* Some of the ligands possess two different donor atoms. For example CN^- ion contains C and N as donor atoms and SCN^- ion contains S and N as donor atoms. When such ligands get coordinated to the metal ion as monodentate ligands, they use either of their two donor atoms, *i.e.* these ligands bond to the metal ion through one type of donor atom in one complex and through another type of donor atom in another complex. Such ligands are called *ambidentate* (or *ambident*) *ligands.* Ambidentate ligands are named by adding the italicised symbol (or symbols) of the element (or elements) coordinated to the metal ion at the end of the name of the ligand. However, some such ligands are also given special names.

2. Naming the Central Metal Atom. Different rules are used for naming the central metal atom in different complexes. For example :

(*i*) *In cationic and neutral complexes.* In cationic and neutral complexes the name of the central metal atom remains the same. Its name is followed by its oxidation number or state in Roman numeral in parentheses. For example nickel (0), iron (II), copper (III) etc.

(*ii*) *In anionic complexes.* In such complexes the name of the central metal atom is modified to end in *ate.* For example chromate (for Cr), ferrate (for Fe), cobaltate (for Co), cuperate (for Cu), argentate (for Ag), palladate (for Pd), rhenate (for Re), stannate (for Sn), iridate (for Ir), ruthenate (for Ru), osinate (for Os), platinate (for Pt), cadmate (for Cd), zincate (for Zn), silicate (for Si). *Marcurate* and *tungstate* are the exceptions. The modified name of the central metal atom is followed by its oxidation state (or number) in Roman numerals in parantheses.

3. Naming the Complex Compounds. In naming the complex compounds the ligands are named first and then the central metal atom is named. The rules for naming the ligands have already been discussed.

(*i*) *Naming the cationic and neutral complexes.* Cationic complexes such as $[CrCl_2(H_2O)_4]^+$, $[Co(H_2O)_6]^{3+}$ etc. and neutral complexes like $[Co\,(NH_3)_2\,(NO_2)_3\,Cl]$, $[Co\,(NH_3)_3\,(SCN)_3]$ etc. are named by giving the number and the name of the ligands followed by the name of the central metal atom with its oxidation number (or state) in Roman numerals in parentheses. *e.g.* $[CrCl_2\,(H_2O)_4]^+$, tetra-aquadichloro chromium (III) ; $[Co(NH_3)_4\,(NO_2)Br]$, tetraamrninenitro cobalt (III);

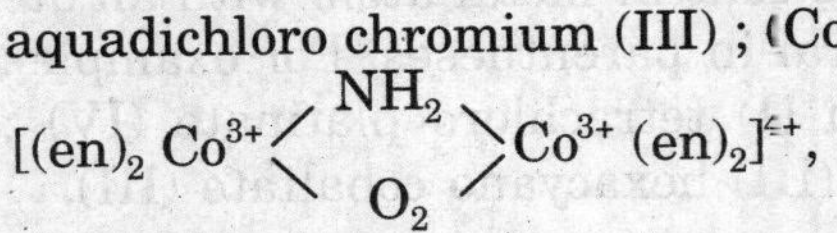, μ-amido μ-superoxo *tetrakis* (ethylenediamine) cobalt (III) (Cationic complexes); *cis*-$[Pt\,(NH_3)_2Cl_2]$, *cis*-dichloro diammine platinum (II); $[Co\,(en)_2\,Cl_2]$, dichloro *bis* (ethylenediamine) cobalt (II); $[Co\,(NH_3)_3\,(SCN)_3]$, tristhiocyanate (*S*) triammine cobalt (III) (Neutral complexes).

(*ii*) *Naming the anionic complexes.* The anionic complexes like $[Ag\,(CN)_2]^-$, $[Fe\,(CN)_6]^{3-}$, $[Si\,F_6]^{2-}$ etc. are named by giving the number and the name of the ligands followed by the name of the central metal atom with an *ate* ending and its oxidation state in Roman numeral in parentheses. For example $[Ag\,(CN)_2]^-$, dicyano argentate (I); $[Fe\,(CN)_6]^{3-}$, hexacyanoferrate (III);

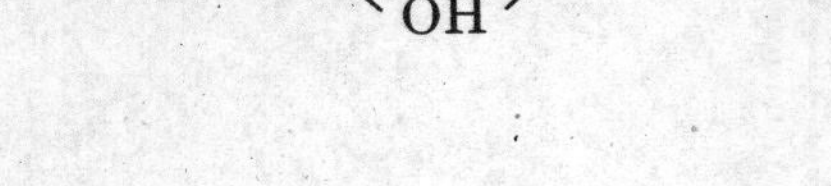, μ-dihydroxo *tatrakis* oxalato chromate (III).

(*iii*) *Naming the complex compounds having complex cation and simple anion.* $[Co(NH_3)_4Cl_2]Cl$, $[Cr(H_2O)_4Cl_2]NO_3$, $[Cu(NH_3)_4]SO_4$ etc. are the examples of such complex compounds. In naming such complex compounds the complex cation is named first and then the simple anion is named as we do for an ionic simple salt like NaCl (sodium chloride). The complex cation of such complex compounds is named according to the rule used for naming the cationic complexes, *i.e.* this complex cation is named by giving the number and the name of the ligands followed by the name of the central metal atom with its oxidation state in Roman numerals in parentheses. Simple anion present in the complex compound is named as usual. For example : $[Co (NH_3)_4 Cl_2] Cl$, dichlorotetraaminine cobalt (III) chloride; $[(NH_3)_5\ Cr^{3+} —OH—Cr^{3+}\ (NH_3)_5]\ Cl_5$, μ-hydroxo decaammine chromium (III) chloride; $[Cu (NH_3)_4] [SO_4$, tetraammine copper (II) sulphate.

(*iv*) *Naming the complex compounds having simple cation and complex anion.* $K [Ag (CN)_2]$, $Fe [Fe (CN)_6]$, $Hg [Co (CNS)_4]$. $K_2 [HgI_4]$ etc. are the examples of such complex compounds. In naming such complex compounds the simple cation is named first and then the complex anion is named as we do for a simple ionic salt like NaCl (sodium chloride). The complex anion of such complex compounds is named according to the rule used for naming anionic complexes, *i.e.* this complex anion is named by giving the number and the name of the ligands followed by the name of the central metal atom with an *ate* ending and its oxidation state in Roman numeral in parentheses. Simple cation is named as usual. For example $K [Ag (CN)_2]$, potassium dicyano argentate (I); $K_4 [Fe (CN)_6]$, potassium hexacyano ferrate (II); $Na_2 [Fe (NO) (CN)_5]$, sodium pentacyano nitrosyl ferrate (II); $Hg[Co(CNS)_4]$, mercury tetrathiocyanato (N) cobaltate (III).

(*v*) *Naming the complex compounds having complex cation and complex anion.* $[Pt (py)_4] [Pt Cl_4]$, $[Co (NH_3)_6] [Cd Cl_5]$ etc are the examples of such complex compounds. In naming such complex compounds the complex cation is named first and then the complex anion is named. The complex cation of such complex compounds is named by giving the number and the name of the ligands followed by the name of the central metal atom with its oxidation state in Roman numerals in parentheses while the complex anion is named by giving the number and the name of the ligands followed by the name of the central metal atom with an *ate* ending and its oxidation state in Roman numeral in parentheses. For example : $[Pt (py)_4] [Pt Cl_4]$, tetrakisphyridine platinum (I) tetrachloro platinate (IV) ; $[Cr (NH_3)_6] [Co(CN)_6]$, hexaammine chromium (III) hexacyano cobaltate (III).

Isomerism

The compounds having the same molecular formula but having different structural arrangement of their atoms are called isomers and the phenomenon that gives rise to isomers is called isomerism. Coordination compounds show two types of isomerism namely *structural isomerism* and *stereo-isomerism or space isomerism*.

Structural Isomerism

This isomerism is of the following types :

1. Conformation Isomerism. In this two isomers have different geometries but are otherwise identical. For example $[Ni^{2+}(P.Et.Ph_2)_2Br_2]$ complex (four coordinated complex) gives two conformation isomers. One of these is green and paramagnetic while the other is brown and diamagnetic. The green form is tetrahedral while the brown form is square planar. Thus these have different geometries but the coordination number of Ni^{2+} ion in both the isomers is the same (= 4)

2. Ionisation Isomerism. Complexes which have the same empirical formula and are produced by the interchange of the position of the ligands inside the complex zone (coordination sphere) and the anions outside the complex zone are called isonisation isomers. They give different ions in solution on ionisation. Thus $[Co^{3+}(NH_3)_5Br]SO_4$ and $[Co^{3+}(NH_3)_5(SO_4)]Br$ are the two ionisation isomers, since they give different ions in solution on ionisation as shown below :

$$[Co^{3+}(NH_3)_5Br]SO_4 \rightleftharpoons [Co^{3+}(NH_3)_5Br]^{2+} + SO_4^{2-}$$

$$[Co^{3+}(NH_3)_5(SO_4)]Br \rightleftharpoons [Co^{3+}(NH_3)_5(SO_4)]^{+} + Br^{-}$$

3. Hydrate Isomerism. This type of isomerism arises out of different disposition of water molecules inside and outside the coordination sphere. For example, the compound $CrCl_3.6H_2O$ exists in three hydrate isomers which are : (*a*) $[Cr(H_2O)_6]Cl_3$—violet, does not lose water over H_2SO_4 and all Cl^- ions are immediately precipitated by Ag^+ ions. (*b*) $[Cr(H_2O)_5Cl]Cl_2.H_2O$—green, loses H_2O over H_2SO_4 and two Cl^- ions are precipitated by Ag^+ ion. (*c*) $[Cr(H_2O)_4 Cl_2]Cl.2H_2O$—green, loses two water molecules over H_2SO_4 and only one Cl^- ion is precipitated by Ag^+ ions.

4. Ligand Isomerism. Some ligands themselves are capable of existing as isomers, *e.g.*, diamino propane can exist both as 1, 2-diamino propane (pn) and 1, 3-diamino propane, also called trimethylene diamino (tn).

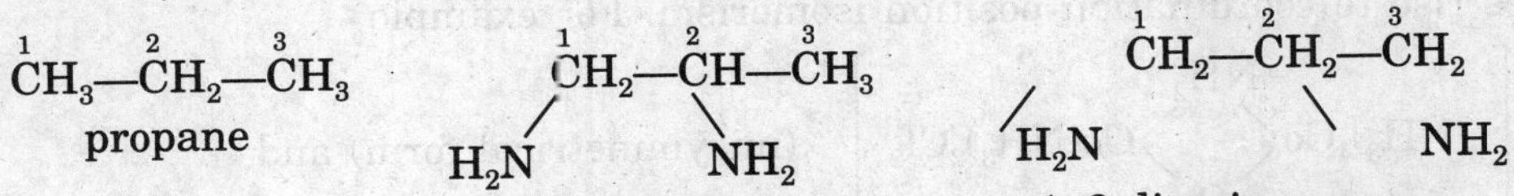

propane — 1, 2-diamino propane (pn) — 1, 3-diamino propane (tn)

When these ligands (*i.e.*, pn and tn) form complexes, the complexes are isomers of each other. One example of isomeric complexes having this ligand is : $[Co(pn)_2Cl_2]^+$ and $[Co(tn)_2 Cl_2]^+$ ions. The structural formulas of these complex ions are shown below in Fig. 43.6.

5. Linkage Isomerism. The tendency of ambidentate ligands due to which they can be linked to the metal ion in two different ways gives rise to linkage isomerism in the complexes given by these ambidentate ligands. As an example we can consider the linkage isomers given by NO_2^- ion. NO_2^- ion is an ambidentate ligand and possesses two different donor atoms *viz.* N and O. This ion can, therefore, coordinate to the metal ion either through N-atom or through O-atom. Thus NO_2^- ion gives two linkage isomers which are called *nitro-N* or *nitro isomer* and *nitro-O* or *nitrito isomer.*

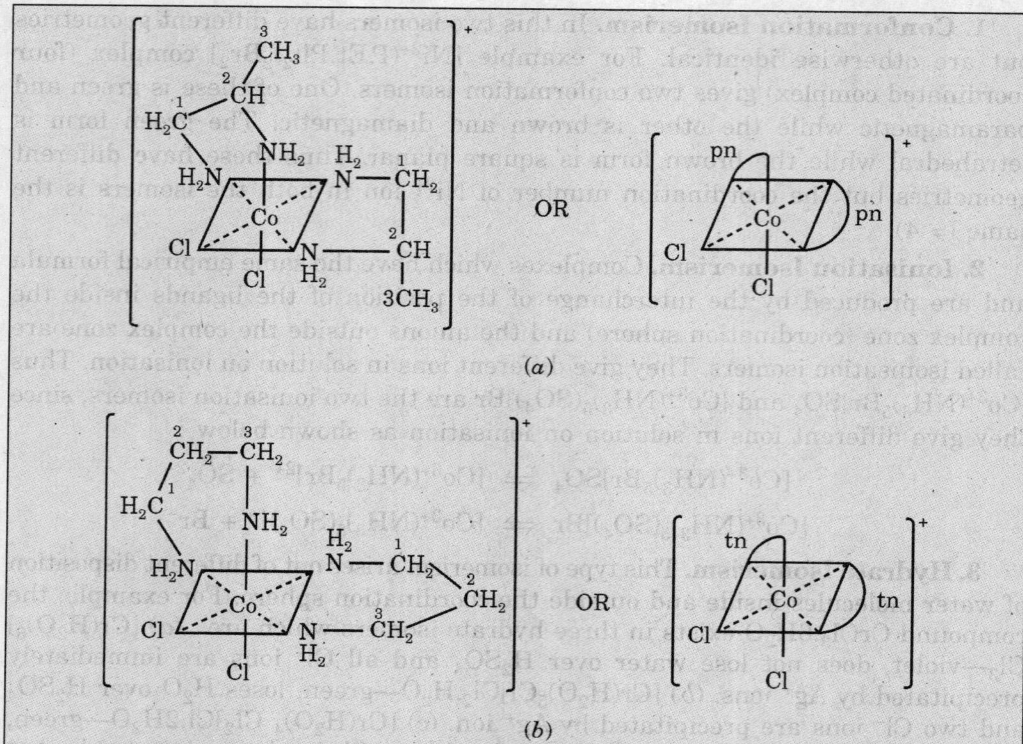

Fig. 43.6. Structural formula of (*a*) $[Co(pn)_2Cl_2]^+$ and (*b*) $[Co(tn)_2Cl_2]^+$ ions.

6. Coordination Position Isomerism. In some poly-nuclear complexes, to interchange the ligands between the metal atoms which are present as part of the complex is possible. This type of interchange of ligands between the metal atoms gives rise to coordination position isomerism. For example

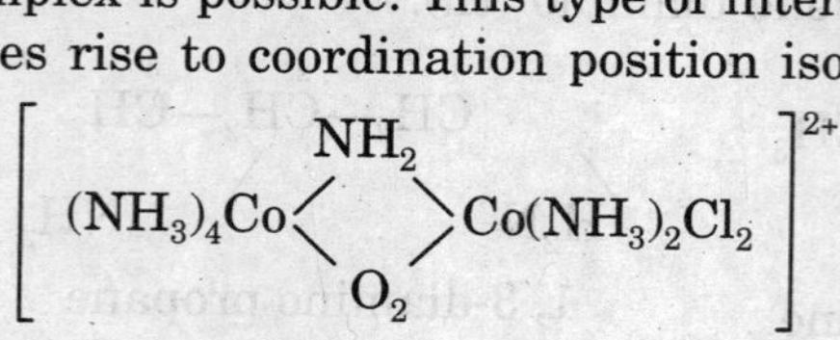

(unsymmetrical form) and

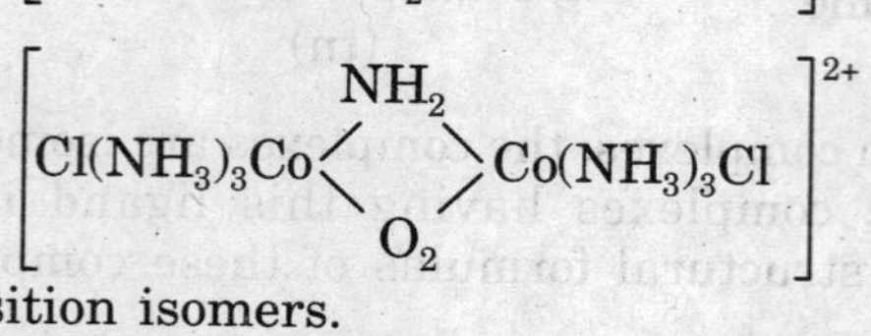

(symmetrical form) are coordination position isomers.

7. Coordination Isomerism. If both cation and anion of a complex compound are complex, there may be an exchange of ligands between the two coordination spheres, giving rise to isomers known as coordination isomers and the isomerism thus obtained is called coordination isomerism. For example $[Cr(NH_3)_6]^{3+}$ $[Cr(SCN)_6]^{3-}$ and $[Cr(NH_3)_4(SCN)_2]^+[Cr(NH_3)_2(SCN)_4]^-$ are coordination isomers to each other.

8. Polymerisation Isomerism. This type of isomerism is found in those complex compounds whose formulae appear to be polymers of some simple complex compound. All these complex compounds have the same ratio of different metal atoms and ligands in them. For example the following complex compounds are polymerisation isomers to each other, since (*ii*) and (*iii*) complexes appear to be dimer of (*i*) complex and (*iv*) complex appears to be the pentamer of (*i*) complex. The ratio Co^{3+} : NH_3 : NO_2^- in all the complexes is 1 : 3 : 3.

Complex compound	***Number of***		
	Co^{3+}	NH_3	NO_2^-
(*i*) $[Co(NH_3)_3(NO_2)_3]$	1	3	3
(*ii*) $[Co(NH_3)_6][Co(NO_2)_6]$	2	6	6
(*iii*) $[Co(NH_3)_4(NO_2)_2][Co(NH_3)_2(NO_2)_4]$	2	6	6
(*iv*) $[Co(NH_3)_5(NO_2)]_3[Co(NO_2)_6]_2$	5	15	15

Stereo (or Space) Isomerism

When two compounds contain the same ligands coordinated to the same central ion, but the arrangement of ligands in space is different, the two compounds are said to be stereo-isomers and the type of isomerism is called *stereo-isomerism.* It is also called *space isomerism.* Stereisomers are quite similar. *e.g.*, optical isomers (which are stereisomers) differ only in their optical activity. On the other hand structural isomers differ from each other in various ways. Stereo isomerism is of two types : (*i*) *Geometrical or cis-trans isomerism*, and (*ii*) *Optical or mirror-image isomerism.*

Geometrical Isomerism

In this isomerism, the ligands occupy different positions round the central metal ion. If the two identical ligands occupy adjacent positions. the isomer is called a *cis*-isomer (*cis* means *same* in Latin), *i.e.* in *cis*-isomer the ligands in question are next to one another. On the other hand, when the two identical ligands are placed opposite to each other, the isomer is called a *trans*-isomer (*trans*-means *across* in Latin). *i.e.* in trans-isomer the line connecting the two identical ligands passes through the central point at which the metal ion is placed.

Geometrical Isomerism in Tetrahedral Complexs (C.N. = 4)

Geometrical isomerism cannot arise in tetrahedral complexes (C.N. = 4) because this geometry contains all the ligands in cis (*i.e.*, adjacent) position with respect to each other, *i.e.* each ligand is equidistant from the other three ligands and all the bond angles are the same (= 109.5°). Thus we see that in tetrahedral complexes all the four positions of the ligands are adjacent to each other and hence geometrical isomerism can not be expected.

Geometrical Isomerism in Square Planar Complexes (C.N. = 4)

Square planar complexes of $[Ma_4]$, $[Ma_3b]$ and $[Mab_3]$ type (*a* and *b* are monodentote ligands) do not show geometrical isomerism, since every conceivable spatial arrangement of ligands round the metal ion is exactly equivalent. However, the square planar complexes of $[Ma_2b_2]$, [Mabcd] etc. type do show geometrical isomerism as discussed below :

1. $[Ma_2b_2]$ type complexes. Square planar complexes of this type (M is the central metal ion and *a* and *b* are monodentate ligands) exist as *cis*-and *trans* isomers. For example $[Pt^{2+}(NH_3)_2Cl_2]$ exists in *cis*- and *trans*- isomers as shown below in Fig. 43.7. $[Pt^{2+}(NH_3)_2Br_2]$ and $[Pd^{2+}(NH_3)_2(NO_2)_2]$ are other examples of square planar complexes which, like $[Pt(NH_3)_2Cl_2]$, also exist in *cis*- and *trans*-isomers.

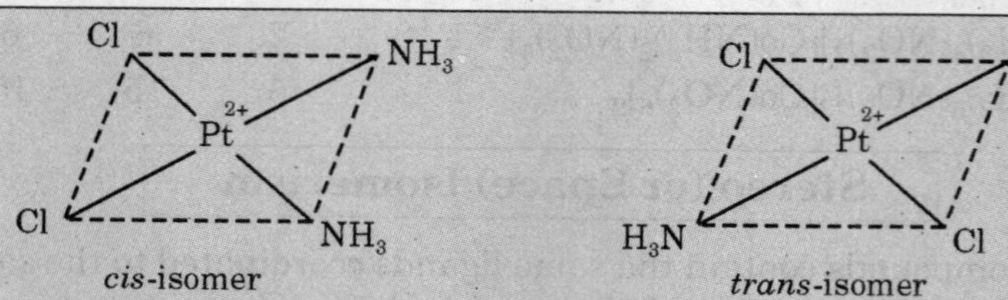

Fig. 43.7. Cis- and trans-isomers of $[Pt^{2+}(NH_3)_2Cl_2]$.

2. [Mabcd] type complexes. Square planar complexes of this type exist in three isomeric forms. For example $[Pt^{2+}(NH_3)$ (py)(Cl) (Br)] exists in three isomeric forms shown below in Fig. 43.8. These isomeric forms can be obtained by selecting one ligand, say NH_3, and then placing the remaining three ligands, one by one, *trans* to NH_3.

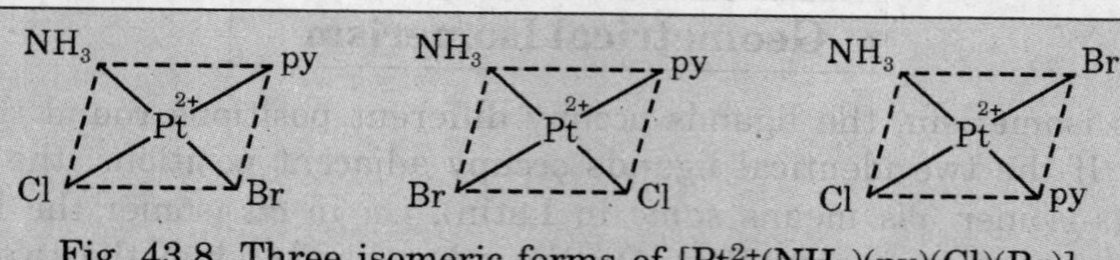

Fig. 43.8. Three isomeric forms of $[Pt^{2+}(NH_3)(py)(Cl)(Br)]$

$[Pt^{2+}(NO_2)(py)(NH_3)(NH_2OH)]^+$ and $[Pt^{2+}(C_2H_4)(NH_3)(Cl)(Br)]$ are other examples of square planar complexes which exist in three isomeric forms.

The existence of three isomeric forms in case of the complexes mentioned above indicates that these complexes have square planar geometry.

3. $[Ma_2bc]$ type complexes. Square planar complexes of this type also show *cis-trans* isomerism. For example $[Pd^{2+}Cl_2BrI]^{2-}$ ion exists in *cis* and *trans*-isomers as shown below in Fig. 43.9.

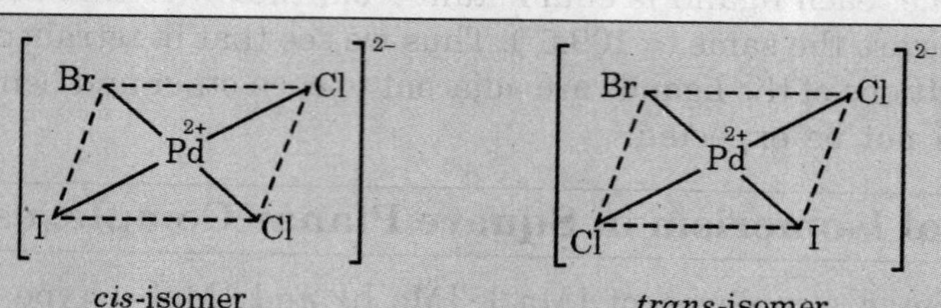

Fig. 43.9. Cis-and trans-isomers of $[Pd^{2+}Cl_2BrI]^{2-}$ ion.

4. $[M(AB)_2]$ type complexes. Here M is the central metal ion and AB represents an unsymmetrical bidentate ligand in which *A* and *B* are two different coordinating (donor) atoms. Square planar complexes of this type also exist in *cis*- and *trans*-isomers. For example $[Pt^{2+}(gly)_2]$ exists in *cis*- and *trans*-isomers as shown below in Fig. 43.10. Here gly represents the glycinato ligand, $NH_2.CH_2COO^-$ which has N and O atoms as its donor atoms (*i.e.* A = N and B = O).

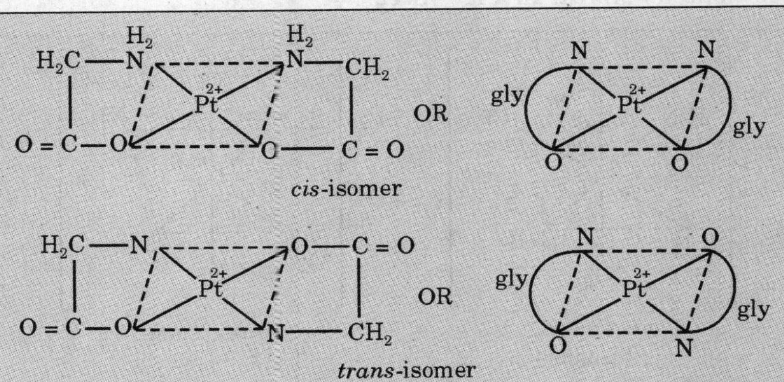

Fig. 43.10. Cis- and trans-isomer of $[Pt^{2+}(gly)_2]$.

Geometrical Isomerism in Octahedral Complexes (C.N. = 6)

We know that a complex compound having the central metal ion with coordination number equal to 6 has octahedral shape. The system used for numbering different positions of the ligands in an octahedral geometry has been shown below in Fig. 43.11.

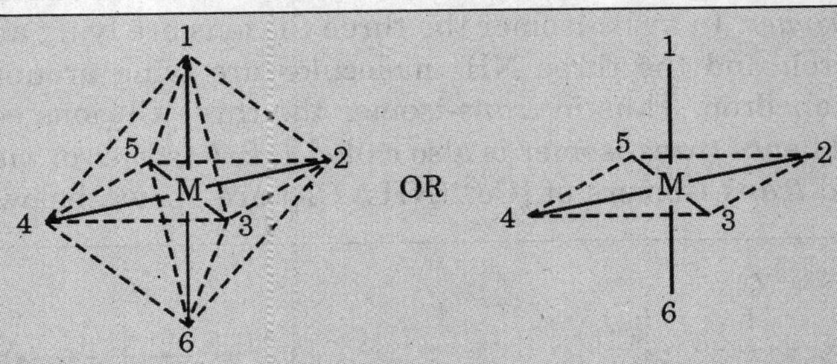

Fig. 43.11 Numbering of six ligands in a regular octahedral complex round the central metal ion, M.

It has been shown by Werner that octahedral complexes of $[Ma_6]$, $[M(AA)_3]$ and $[Ma_5b]$ type give only single geometrical isomer, *i.e.* the complexes of these types do not show geometrical isomerism. However, the following octahedral complexes give two or more geometrical isomers.

1. $[Ma_4b_2]$ type complexes. Octahedral complexes of this type exist in *cis* and *trans* isomers. In *cis*-isomer the two monodentate *b* ligands occupy adjacent positions (*i.e.* 1 and 2 positions) of the octahedron while in *trans*-isomer these ligands occupy the opposite positions (*i.e.* 1 and 6 positions) of the octahedron. For example $[Co^{3+}(NH_3)_4Cl_2]^+$ ion exists in *cis*- and *trans*-isomers. In *cis*-isomer the two Cl^- ions occupy 1 and 2 positions while in *trans*-isomer these ions occupy 1 and 6 positions as shown in Fig. 43.12.

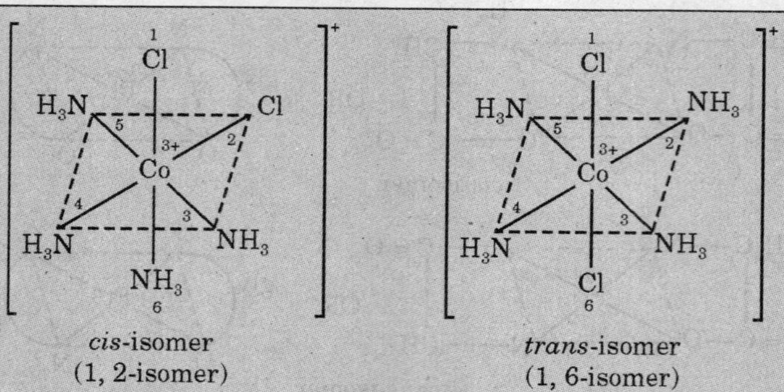

Fig. 43.12. Cis-and trans-isomers of $[Co^{3+}(NH_3)_4Cl_2]^+$ ion.

$[Co^{3+}(NH_3)_4(NO_2)_2]^+$ is another example of $[Ma_4b_2]$ type octahedral complex which exists in *cis*- and *trans*- isomers.

2. $[Ma_3b_3]$ type complexes. Octahedral complexes of this type also exist in *cis*- and *trans*-isomers. For example $[Co^{3+}(NH_3)_3Cl_3]$ exists in *cis*- and *trans*-isomers. In *cis*-isomer the three Cl^- ions lie on the corners of one of the triangular face of the octahedron and the three NH_3 molecules are lying on the corners of the opposite triangular face of the octahedron. Thus in *cis*-isomer the three Cl^- ions occupy 1, 2 and 3 positions and hence this isomer is also called *1, 2, 3 isomer* or *facial (fac) isomer.* In *trans*-isomer the three Cl^- ions are lying around one edge of the octahedron and the three NH_3 molecules are lying around the opposite edge of the octahedron. Thus in *trans*-isomer the three Cl^- ions occupy 1, 2 and 6 positions and hence *trans*-isomer is also called *1, 2, 6 isomer* or *meridional (mer) isomer. Cis* and *trans*-isomers of $[Co^{3+}(NH_3)_3Cl_3]$ are shown below in Fig. 43.13.

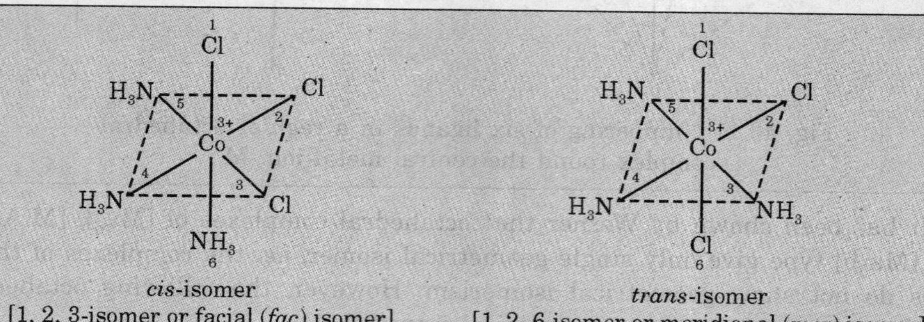

Fig. 43.13. Cis- and trans- isomers of $[Co^{3+}(NH_3)_3Cl_3]$

$[Cr^{3+}(H_2O)_3F_3]$, $[Rh^{3+}(py)_3Cl_3]$, $[Cr^{3+}(NH_3)_3Cl_3]$, $[Ru^{3+}(H_2O)_3Cl_3]$, $[Pt^{4+}(NH_3)_3Br_3]^+$, $[Pt^{4+}(NH_3)_3I_3]^+$, $[Ir^{3+}(H_2O)_3Cl_3]$, $[Ru^{3+}(MeCN)_3Cl_3]$ and $[Co^{3+}(NH_3)_3(NO_2)_3]$ are other examples of $[Ma_3b_3]$ type octahedral complexes which exist in *cis*- and *trans*-isomers.

3. $[Ma_2b_2c_2]$ type complexes. Octahedral complexes of this type can exist theoretically in five geometrical isomers. Out of these only three have been prepared. All the five geometrical isomers for $[Pt^{4+}(NH_3)_2(py)_2Cl_2]^{2+}$ ion have been shown in Fig. 43.14.

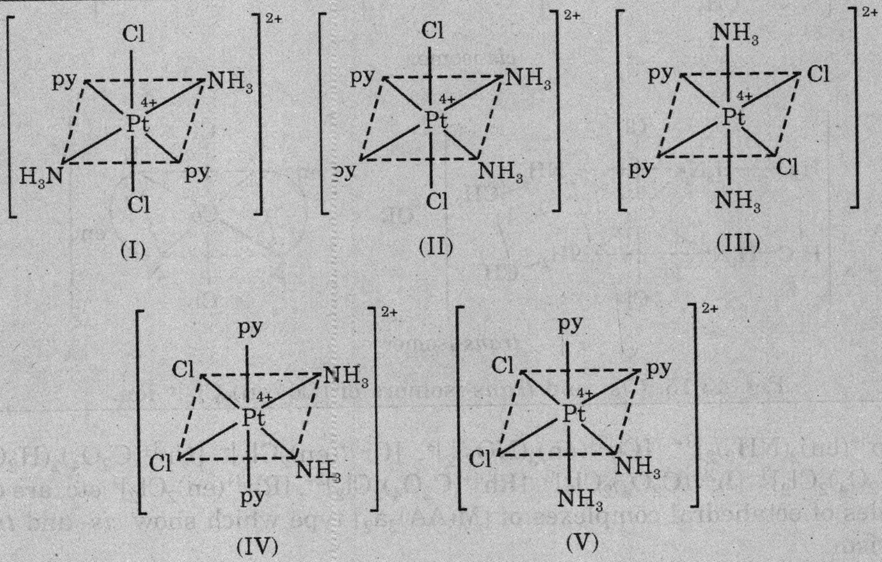

Fig. 43.14. All possible five geometrical isomers of $[Pt^{4+}(NH_3)_2(py)_2Cl_2]^{2+}$ ion. Note that the two like groups in isomer (*V*) are *cis* to each other.

4. [Mabcdef] type complexes. Octahedral complexes which have different monodentate ligands have been prepared for Pt(IV). $[Pt^{4+}(py)(NH_3)(NO_2)(Cl)(Br)(I)]$ is the only complex of this type of complexes. Theoretically 15 geometrical isomers are possible for this complex compound. In fact only three geometrical isomers have been isolated.

5. $[M(AA)_2a_2]$ type complexes. Here (AA) represents a symmetrical bidentate ligand in which A and A are two identical coordinating (donor) atoms. Octahedral complexes of $[M(AA)_2a_2]$ type exist in *cis*-and *trans*-isomers. In *cis*-isomer the two monodenate *a* ligands occupy adjacent positions of the octahedron while in *trans*-isomer these ligands occupy the opposite positions. Thus the *cis*- and *trans*-isomers of $[Co^{3+}(en)_2Cl_2]^+$ ion can be shown as follows in Fig. 43.15.

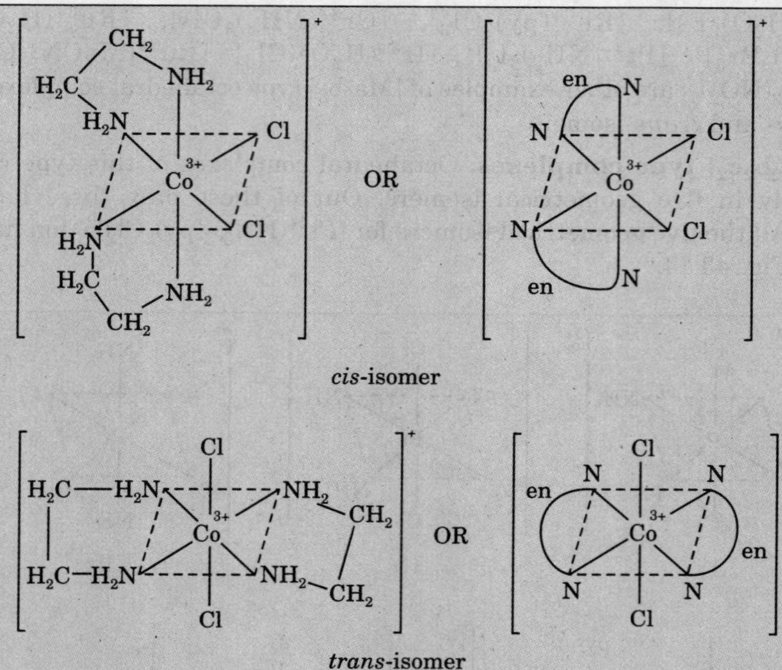

cis-isomer

trans-isomer

Fig. 43.15. *Cis*- and *trans*-isomers of $[Co^3(en)_2Cl_2]^+$ ion.

$[Co^{3+}(en)_2(NH_3)_2]^{3+}$, $[Co^{3+}(en)_3(NO_3)_2]^+$, $[Cr^{3+}(en)_2Cl_2]^+$ $[Cr^{3+}(C_2O_4)_2(H_2O)_2]^-$, $[Ir^{3+}(C_2O_4)_2Cl_2]^{3-}$ $[Ir^{4+}(C_2O_4)_2Cl_2]^{2-}$ $[Rh^{4+}(C_2O_4)_2Cl_2]^{2-}$. $[Rh^{3+}(en)_2Cl_2]^+$ etc. are other examples of octahedral complexes of $[M(AA)_2a_2]$ type which show *cis*- and *trans*-isomerism.

6. $[M(AA)_2ab]$ type conplexes. Octahedral complexes of this type exist in *cis*- and *trans*-isomers. In *cis*-isoiner the two monodentate *a* and *b* ligands occupy the adjascent (*i.e. cis*) positions of the octahedron while in *trans*-isomer these ligands occupy the opposite (*i.e. trans*) positions. Thus *cis*- and *trans*-isomers of $[Co^{3+}(en)_2(NH_3)(Cl)]^{2+}$ ion can be shown as follows in Fig. 43.16.

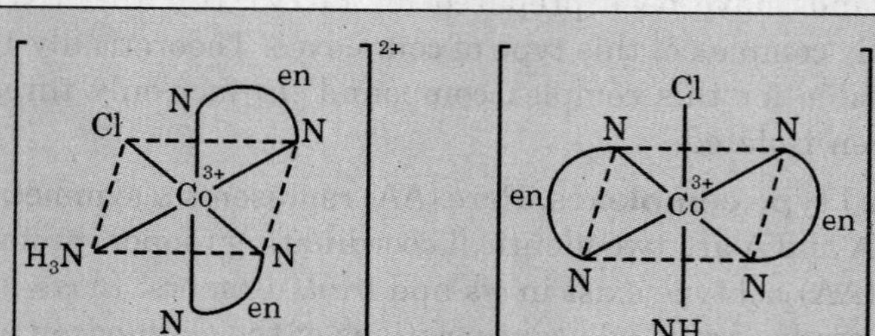

cis-isomer *trans*-isomer

Fig. 43.16. *Cis*- and *trans*-isomers of $[Co^{3+}(en)_2(NH_3)(Cl)]^{2+}$ ion.

$[Ru^{3+}(C_2O_4)_2(py)(NO)]^-$ is another example of octahedral complex of $[M(AA)_2ab]$ type which shows *cis*- and *trans*-isomerism.

7. [M(AA)a_2b_2] type complexes. Octahedral complexes of this type exist in *cis-* and *trans*-isomers. In *cis*-isomer both monodentate *a* and *b* ligands occupy *cis* positions while in *trans*-isomer these ligands occupy *trans* positions. For example in *cis-* isomer of $[Co^{3+}(en)(NH_3)_2Cl_2]^+$ ion both NH_3 and Cl^- ion are *cis* to each other while in *trans*-isomer both NH_3 and Cl^- ion are *trans* to each other (See Fig. 43.17).

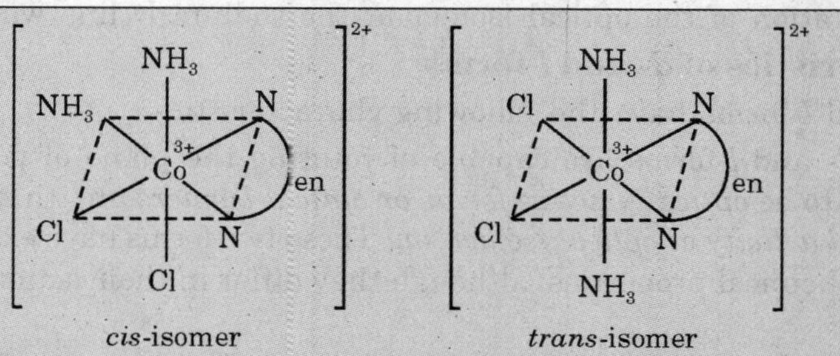

Fig. 43.17. *Cis-* and *trans*-isomers of $[Co^{3+}(en)(NH_3)_2Cl_2]^+$ ion.

8. [M(AB)$_3$] type complexes. Here (AB) represents an unsymmetrical bidentate ligand in which A and B are two different coordinating (donor) atoms. Octahedral complexes of this type exist in *cis-* and *trans*-isomers. As an example the *cis-* and *trans*-isomers of $[Cr^{3+}(gly)_3]$ has been shown in Fig 43.18. Each of these forms is optically active and hence each has a pair of optical isomers.

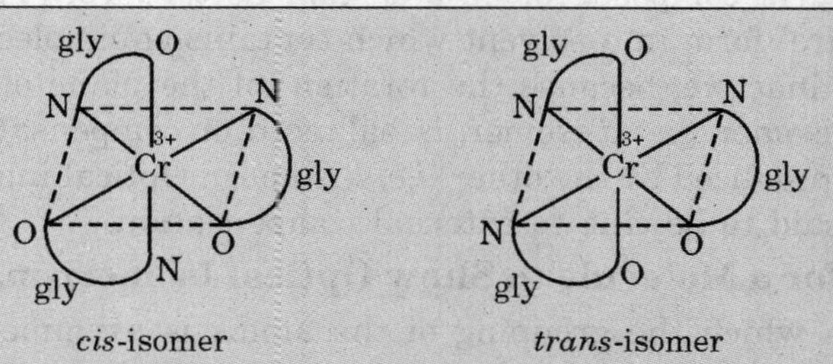

Fig 43.18. *Cis-* and *trans-* isomers of $[Cr^{3+}(gly)_3]$.

Optical Isomerism

Definition of Some Terms.

Before discussing the optical isomerism shown by various types of complexes we will define some important terms related to this isomerism.

When the solutions of certain complex compounds are placed in the path of a *plane-polarised light* (the waves of the plane-polarised light vibrate only in one direction; vibrations in other directions are cut off), they rotate its plane through a certain angle which may be either to the left or to the right. *This property of a complex of rotating the plane of polarised light is called its optical activity and the complex possessing this property is said to be optically active.* Optically active complexes are said to exist in the following forms (*a*) One which rotates the plane of polarised light towards right (*i.e.* in clockwise direction) is said to be *dextro-rotatory* or *d*-form. *d-form* is also represented by placing (+) sign before its name or formula. (*b*) One which rotates the plane of polarised light towards left (*i.e.* in anti-clockwise direction) is called *levo-rotatory* or *l*-form. *l*-form is also represented by putting (–) sign before its name or formula. (+), dextro, and (–), levo, refer to the sign of rotation of the optical isomer at the sodium D line wavelength.

Characteristics of *d*- and *l*-forms

The *d*- and *l*-forms have the following characteristics :

(*i*) Since *d*- and *l*-forms are capable of rotating the plane of polarised light, these are said to be *optically active forms or optical isomers* and this phenomenon is called *optical activity* or *optical isomerism.* These two forms have exactly identical physical and chemical properties, although they differ in their action on polarised light.

(*ii*) *d*- and *l*-forms are mirror-images to each other just as left hand is the mirror image of the right hand. Thus *d*- and *l*-forms can be superimposed on each other. Since these forms are related to each other as mirror-images, they are commonly called *enantiomorphs* (Latin : *enantio* = opposite, *morphs* = forms) or *enantiomers.* Thus optical isomerism is now often referred to as *enantiomerism.*

(*iii*) The isomer which is not capable of rotating the plane of polarised light is called *optically inactive.* Such an isomer is called *racemic, dl-or* (±) *form.* A racemic substance is composed of 50% *d*- and 50% *l*-form. The solution of a racemic (*dl* mixture) form in a solvent which contains equimolecular amounts of *d*- and *l*-forms is inactive, because the rotation (of the plane of polarised light) produced by one isomer, say *d*-isomer, is balanced or compensated by equal but opposite rotation produced by the other (*i.e.* *l*-isomer). Optical inactivity produced in *dl*-mixture is said to be clue to external compensation.

Conditions for a Molecule to Show Optical Isomerism.

A molecule in which the grouping of the atoms is asymmetric is called an *asymmetric* or *dissymmetric molecule.* An asymmetric molecule has the following features :

(*i*) An asymmetric molecule never has a plane of symmetry (also called mirror-image plane) which is defined as an imaginary plane dividing the molecule in such a way that the part of it on one side of the plane is the mirror image of that on the other side of the plane. The molecules possessing such a plane of symmetry are always *inactive while those having no plane of symmetry are optically active and hence show optical isomerism.*

(*ii*) An asymmetric molecule cannot be superimposed on its mirror image. Thus the most necessary and sufficient condition for a molecule to show optical isomerism (*i.e.* to exist in *d*- and *l*-forms) is that *the molecule should be asymmetric,* (*i.e., it should have no plane of symmetry*) *and should not be superimposable on its mirror image.*

Optical Isomerism in Tetrahedral Complexes (C.N. = 4)

Tetrahedral complexes of [Ma_4] and [Ma_3b] type are not able to show optical isomerism because all the possible arrangements of the ligands round the central metal ion, M are exactly equivalent. By analogy with organic tetrahedral molecules in which carbon atom is surrounded by four different atoms or groups, it is expected that a tetrahedral metal complex of [*Mabcd*] type should also exhibit optical isomerism, *i.e.* it should be possible for us to isolate optically active *dextro* and *levo* forms of tetrahedral metal complex of [*Mabcd*] type. However it has not been possible to isolate dextro and levo forms of this complex due to its *labile* nature. Here it may be noted that 4 different groups round the central metal ion are not the only requirement to make the complex to show mirror-image isomerism. All that is required is that the molecule should be asymmetric (*i.e.*, unsymmetrical), *i.e.* it should have no plane of symmetry so that it can exist in two mirror-image forms. Tetrahedral complexes of Be(II), B(III) and Zn(II) with symmetrical bidentate ligands have been made and resolved into optical isomers. Some examples of tetrahedral complexes of this type are *bis* (salicylaldehydato) boron (III) ion, [$B^{3+}(C_6H_4.O.CHO)_2]^+$ and *bis* (benzoylacetonato) beryllium (II), [Be^{2+} $(C_6H_5CO.CH.CH_3CO)_2]^\circ$ whose mirror-image isomers are given in Fig. 43.19 and 43.20 respectively.

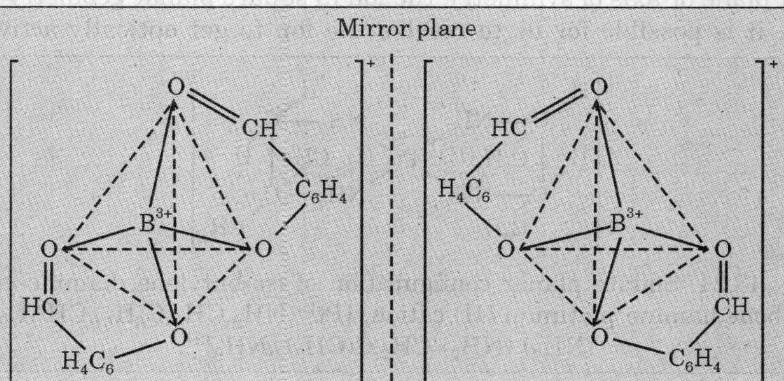

Fig. 43.19. Mirror-image isomers of tetrahedral *bis* (salicyl aldehydato) boron (III) ion, [B^{3+} $(C_6H_4.O.CHO)_2]^+$.

Optical Isomerism in Squar Planar Complexes (C.N. = 4)

Optical isomerism rarely occurs in square planar complexes, since they have all the four ligands and the central metal ion in the same plane and hence contain a plane or axis of symmetry. Thus these complexes are optically inactive, *i.e.* they cannot show optical isomerism even if all the four ligands are different. Although square planar complexes seldom show optical isomerism, yet the square planar complex *viz. iso*-butylene diamine-*meso*-stilbenediamine platinum (II) cation,

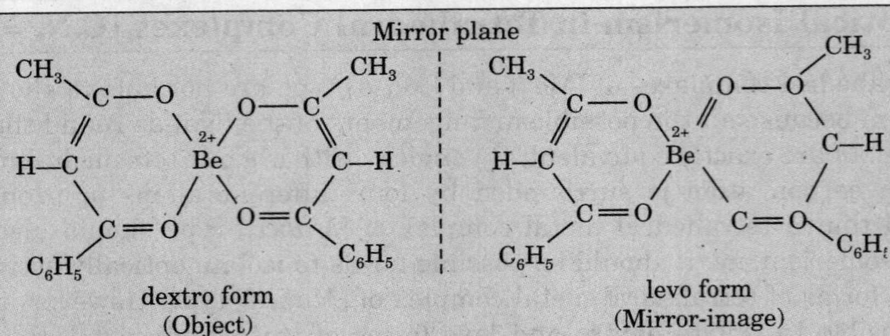

Fig. 48.20. Mirror-image isomers of tetrahedral *bis* (benzoylacetonato) beryllium (II), $[Be^{2+} (C_6H_5CO.CH.CH_3CO)_2]^{\circ}$.

$[Pt^{2+} (NH_2.CH (C_6H_5).CH(C_6H_5).NH_2) (NH_2.CH_2.C(CH_3)_2.NH_2)^{2+}$ has no plane or axis of symmetry (Fig. 43.21) and hence is unsymmetric and optically active, *i.e.* this square planar complex can be resolved into optically active isomers. Why this square planar complex can be resolved into optically active isomers can be explained as follows : In the square planar configuration of this ion as shown in Fig. 43.21 with two rings (I) and (II), since the C_6H_5 groups are below and H-atoms are above the plane of the ring (II), this structure has no plane or axis of symmetry. In the absence of plane or axis of symmetry, the ion in square planar geometry is optically active, *i.e.* it is possible for us to resolve the ion to get optically active isomers.

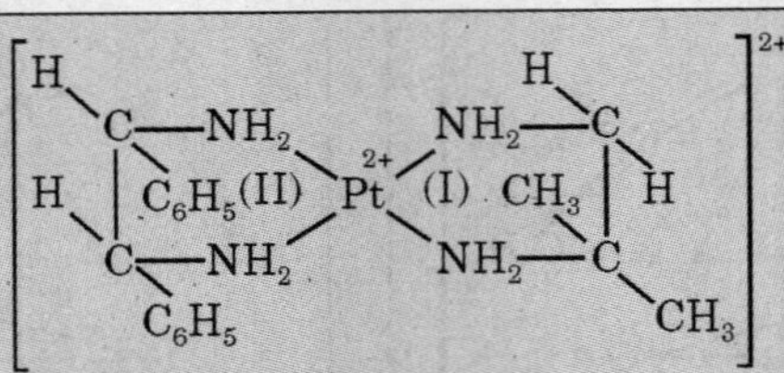

Fig. 43.21. Square planar configuration of *iso*-butylene diamine-*meso*-stilbenediamine platinum (II) cation, $[Pt^{2+}(NH_2.CH (C_6H_5).CH (C_6H_5).(NH_2) (NH_2) CH_2.C(CH_3)_2.NH_2]^{2+}$.

Optical Isomerism in Octahedral Complexes (C.N. = 6)

Both *cis*- and *trans*-isomers of $[Ma_4b_2]$ type octahedral complexes are optically inactive due to the presence of symmetry and hence do not show optical isomerism. The facial and meridional isomers of $[Ma_3b_3]$ type octahedral complexes are optically inactive and hence do not exist in optical isomers. The following octahedral complexes show optical isomerism.

1. $[Ma_2b_2c_2]$ type complexes. We have seen that the octahedral complexes of this type can exist theoratically in five geometrical isomers which have already been shown in Fig. 43.14. Out of these only three have been prepared. The geometrical isomer with all like groups *cis* to one another occurs in two optical isomers. For example the geometrical isomer of $[Pt^{4+} (NH_3)_2 (py)_2 Cl_2]^{2+}$ which has all the like groups *cis* to one another occurs in mirror-image isomers as shown in Fig. 43.22.

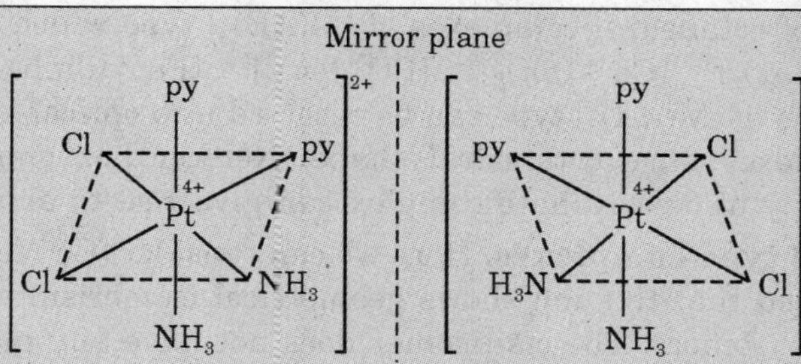

cis-d-isomer (Object) *cis-l*-isomer (Mirror-image).

Fig. 43.22. Mirror-image isomers of $(Pt^{4+}(NH_3)_2(py)_2Cl_2]^{2+}$ ion.

2. [Mabcdef] type complexes. Octahedral complexes of this type for Pt (IV) have only 15 geometrical isomers each of which exists in *d*-and *l*-forms to give a total of 30 isomers. For example *d*- and *l*-forms obtained for one geometrical isomer of $[Pt^{4+}(py)(NH_3)(NO_2)(Cl)(Br)(I)]^{\circ}$ are shown in Fig. 43.23.

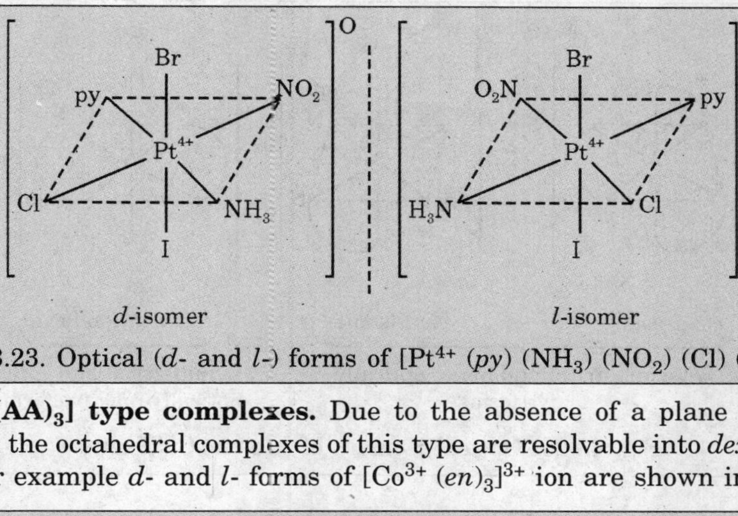

d-isomer *l*-isomer

Fig. 43.23. Optical (*d*- and *l*-) forms of $[Pt^{4+}(py)(NH_3)(NO_2)(Cl)(Br)(I)]^0$.

3. $[M(AA)_3]$ type complexes. Due to the absence of a plane or centre of symmetry, the octahedral complexes of this type are resolvable into *dextro* and *levo* forms. For example *d*- and *l*- forms of $[Co^{3+}(en)_3]^{3+}$ ion are shown in Fig. 43.24.

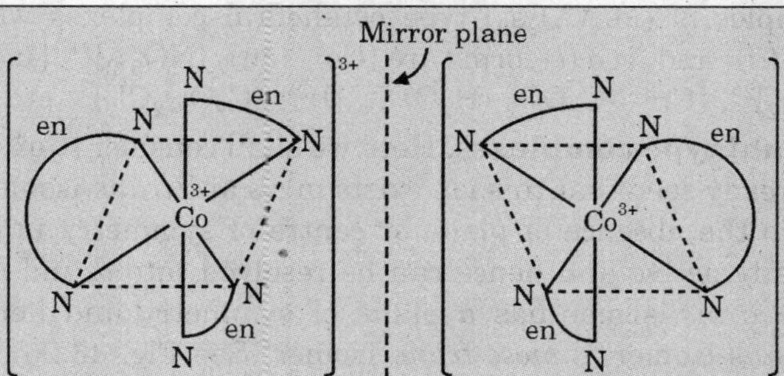

d-form *l*-form

Fig. 43.24. Optical isomers of $[Co^{3+}(en)_3]^{3+}$ ion. These cannot be superimposed on each other.

Other examples of octahedral complexes of $[M(AB)_3]$ type which exist in *d* and *l* forms are : $[Cr^{3+} (ox)_3]^{3-}$, $[Co^{3+} (pn)_3]^{3+}$, $[Pt^{4+} (en)_3]^{4+}$, $[Fe^{2+} (diph)_3]^{2+}$ etc. The fact that the complexes of $[M(AA)_3]$ type can be resolved into optical isomers confirms that these complexes have octahedral shape. Neither hexagonal nor trigonal prismatic geometry of 6-coordinate complex can give rise to activity.

4. $[M(AA)_2a_2]$ type complexes. Here we can consider $[Co^{3+}(en)_2Cl_2]^+$ ion. We have already stated that this ion shows geometrical isomerism and hence exists as *cis-* and *trans-* isomers. The *cis-* isomer does not have any plane or centre of symmetry and hence is *optically active*, *i.e. cis-* isomer can be resolved into two optically active isomers (*d-* and *l-* isomers). On the other hand the *trans-* isomer has a plane of symmetry and hence is *optically inactive*, *i.e.*, it cannot be resolved into optically active isomers. Thus trans isomer is optically inactive (meso form). Consequently $[Co(en)_2Cl_2]^+$ ion has three optical isomers *viz.* : (Also see Fig. 43.25) (*a*) Two optically active isomers corresponding to cis-configuration (*i.e. d-* and *l-* forms). One form is the mirror image of the other. (*b*) One optically inactive form corresponding to trans-form (meso-form).

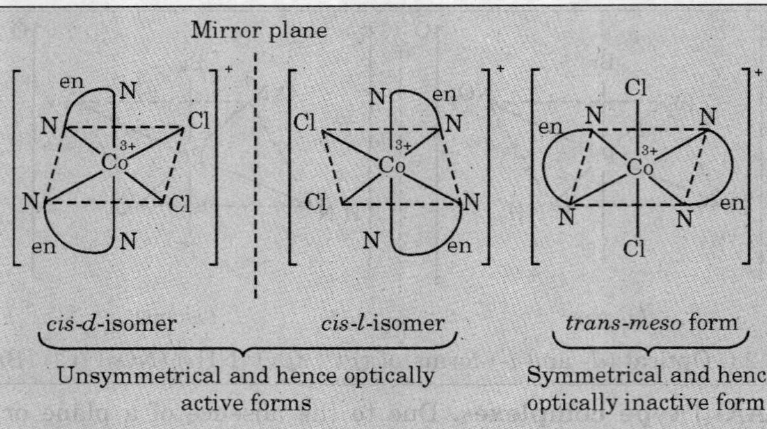

Fig. 43.25. Optical isomers of $[Co(en)_2Cl_2]^+$ ion.

Other examples of $[M(AA)_2a_2]$ type octahedral complexes whose *cis*-isomer exists in *dextro* (*d*)- and *levo* (*l*) forms are $[Co^{3+} (en)_2 (NO_3)_2]^{3+}$, $[Ir^{4+} (C_2O_4)_2Cl_2]^{2-}$, $[Rh^{4+} (C_2O_4)_2Cl_2]^{2-}$, $[Cr^{3+} (C_2O_4)_2 (H_2O)_2]^-$, $[Ir^{3+} (C_2O_4)_2Cl_2]^{3-}$ etc.

5. $[M(AA)_2$ ab] type complexes. Here we shall consider $[Co^{3+}(en)_2(NH_3)Cl]^{2+}$ ion. We have already seen that this ion exists in *cis* and *trans*-isomers (geometrical isomers). Due to the absence of plane or centre of symmetry in *cis*-isomer, this isomer is optically active and hence can be resolved into *d* and *l*-forms. On the other hand the *trans*-isomer has a plane of symmetry and hence is optically inactive, *i.e.*, *trans*-isomer is *meso-trans* isomer (See Fig. 43.26).

Another example of $[M(AA)_2$ *ab*] type octahedral complexes whose *cis*-isomer exists in *d-* and *l*-forms is $[Ru^{3+}(C_2O_4)_2(py)(NO)]^-$.

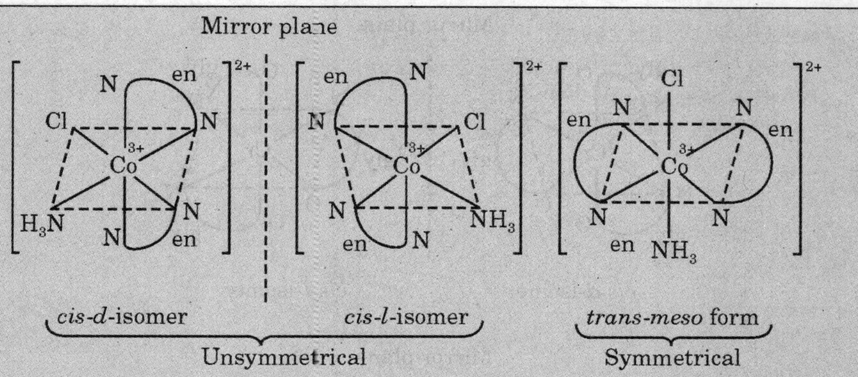

Fig. 43.26. Optical isomers of $[Co^{3+}(en)_2(NH_3)Cl]^{2+}$ ion.

6. $[M(AA)a_2b_2]$ type complexes. Here we shall consider $[Co^{3+}(en)(NH_3)_2Cl_2]^+$ ion. We have seen that this ion exists in *cis-* and *trans*-isomers (geometrical isomers). *Cis*-isomer is optically active and hence gives optically active *d-* and *l*-forms. On the other hand *trans*-isomer is optically inactive and hence does not give any optical active isomer. (See Fig 43.27).

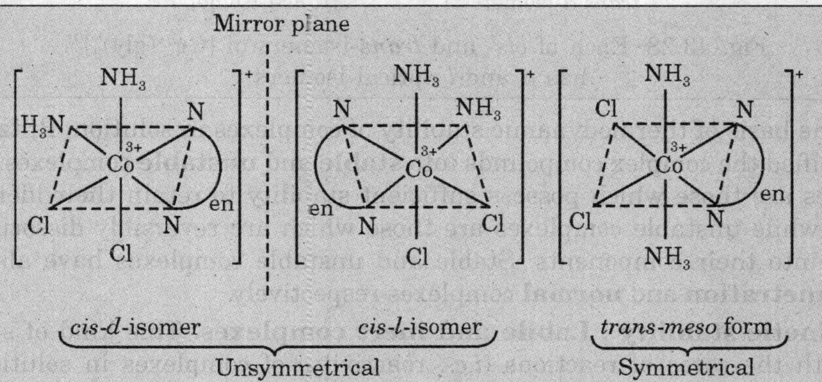

Fig. 43.27. Optical isomers of $[Co^{3+}(en)(NH_3)_2Cl_2]^+$ ion

$[Co^{3+}(C_2O_4)(NH_3)_2(NO)_2]^-$ ion is another important example of $[M(AA)\ a_2b_2]$ type octahedral complexes whose *cis*-isomer gives *d-* and *l*-isomers.

7. $[M(AB)_3]$ type complexes. We have seen that an octahedral complex of this type exists in *cis-* and *trans*-isomers (geometrical isomers). Each of these forms is optically active and hence each gives a pair of optical isomers (*d-* and *l*-forms) as shown below in Fig. 43.28 for $[Cr^{3+}(gly)_3]$ complex compound.

Stability of Complex Compounds in Aqueous Solution

In studying the formation of complex compounds in solution two kinds of stability have to be distinguished. These are :

1. Thermodynamic stability : Stable and unstable complexes. This kind of stability deals with the properties like bond energies, stability constants and redox potentials that affect the equilibrium conditions.

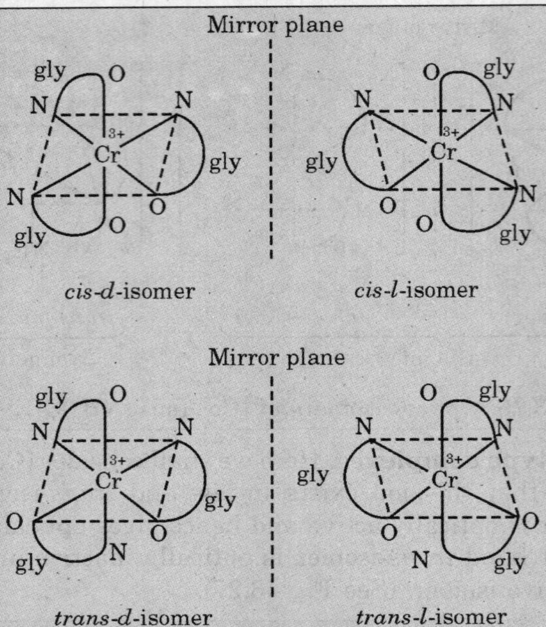

Fig. 43.28. Each of *cis*- and *trans*-isomers of $[Cr^{3+}(gly)_3]^0$ has *d* and *l* optical isomers

On the basis of thermodynamic stability of complexes in solution, Biltz (1927) has classified the complex compounds into **stable** and **unstable** complexes. Stable complexes are those which possess sufficient stability to retain their identity in solution while unstable complexes are those which are reversibly dissociated in solution into their components. Stable and unstable complexes have also been called **penetration** and **normal** complexes respectively.

2. Kinetic stability : Labile and inert complexes. This kind of stability deals with the rates of reactions (*i.e.*, reactivity) of complexes in solution, the mechanisms of chemical reactions, formation of intermediate complexes, activation energies for the process etc. On the basis of the rate of reactions (*i.e.* kinetic stability) of the complexes in solution, Taube (1950) has classified the complexes into **labile** and **inert** complexes. Labile complexes are those whose one or more ligands in the co-ordination sphere can be rapidly replaced by other ligands and the ability of a complex to replace its one or more ligands by other ligands is called its **lability**. Inert complexes are those whose one or more ligands can either not be replaced or can be replaced with difficulty by other ligands.

Factors Affecting the Stability of Metal Complexes.

The stability (or stability constants) of metal complexes depends on the following factors :

1. Properties of central metal ion. The following properties of central metal ion affect the stability of the metal complexes :

***(i)* Size of central metal ion.** For a given ligand the stability (or stability constant) of the complexes of the metallic ions having the same charge on them decreases with the increase of the size of the central metal ion. Thus the stability of complexes given by the cations belonging to the same group and having the same charge decreases as we proceed from top to bottom in the group, since the size of the metallic cations increases in the same order. For example the stability of hydroxide complexes given by alkali metal ions (Li^+, Na^+ etc.), alkaline earth metal ions (Be^{2+}, Mg^{2+} etc.) and III B group ions (Sc^{3+}, Y^{3+} and La^{3+}) is in the order : (*a*) $Li^+(r = 0.60$ Å) > Na^+ ($r = 0.95$ Å) > $K^+(r = 1.33$ Å) > $Rb^+(r = 1.48$ Å) > $Cs^+(r = 1.69$Å). (*b*) Be^{2+} ($r = 0.31$ Å) > Mg^{2+} ($r = 0.65$ Å) > Ca^{2+} ($r = 0.99$ Å) > Sr^{2+} ($r = 1.13$ Å) >$Ba^{2+}(r = 1.35$ Å) > $Ra^{2+}(r = 1.40$ Å). (*c*) $Sc^{3+}(r = 0.81$ Å) > Y^{3+} ($r = 0.93$ Å) > $La^{3+}(r = 1.15$ Å). There are also some exceptions to the above order, *e.g.* EDTA complex of $Mg^{2+}(r = 0.65$ Å) is less stable than that of Ca^{2+} ($r = 0.99$ Å).

***(ii)* Charge on the central metal ion.** For a given ligand, the stability of the complexes of the metallic ions having almost the same size but different charges on them decreases with the decrease of the charge on them. Thus the stability of complexes given by : (*a*) La^{3+}, Sr^{2+} and K^+ ions. (*b*) Co^{3+} and Co^{2+} ions. (*c*) Fe^{3+} and Fe^{2+} ions, and (*d*) Th^{4+}, Y^{3+} Ca^{2+} and Na^+ ions with the same ligand is in the order : (*a*) $La^{3+}(r = 1.15$ Å) > Sr^{2+}(= 1.13 Å) > K^+(= 1.33 Å) (*b*) $Co^{3+}(r = 0.63$ Å) > Co^{2+}(= 0.74 Å) (*c*) $Fe^{3+}(r = 0.64$ Å) > $Fe^{2+}(r = 0.76$ Å) (*d*) $Th^{4+}(r = 0.95$ Å) > Y^{3+}(0.93 Å) > Ca^{2+}(= 0.99 Å) > Na^+(= 0.95 Å).

If the factors (*i*) and (*ii*) mentioned above are combined, then we can say that with the increase of ionic potential of the central metal ion (Ionic potential of the metal ion = charge on the metal ion/ size of the ion), the stability of the complexes with the same ligand also increases. For example the stability of hydroxide complexes of Li^+, Ca^{2+}, Ni^{2+}, ...Be^{2+} ions whose ionic potential increase from Li^+ to Be^{2+} ions also increases in the same direction as shown below :

Central metal ion (charge on the ion is shown in parentheses)	Ionic radius (A°)	Ionic potential	Order of stability of complexes
Li^+ (+ 1)	0.60	1/0.60 = 1.6	Increasing ↓
Ca^{2+} (+ 2)	0.99	2/0.99 = 2.0	
Ni^{2+} (+ 2)	0.72	2/0.72 = 2.97	
Y^{3+} (+ 3)	0.93	3/0.93 = 3.22	
Th^{4+} (+ 4)	0.95	4/0.95 = 4.20	
Al^{3+} (+ 3)	0.50	3/0.50 = 6.0	
Be^{2+} (+ 2)	0.31	2/0.31 = 6.45 (Increasing ↓)	

***(iii)* Electronic configuration of the central metal ion : Irving-William order of stability.** In case of metallic ions which have the same radius and same charge (*e.g.* Na^+ and Cu^+; Ca^{2+} and Cd^{2+} or Sn^{2+} ; Sc^{3+} and Ga^{3+} or Sb^{3+} ions), the outer electronic configuration of the ions plays an important role in determing the stability of the complexes. The complexes given by metal ions possessing inert gas (8-electron ions), pseudo inert gas (18-electron ions) and pseudo inert gas plus two (18 + 2 electron ions) electronic configuration are less stable than those given by transition metal ions which have $(n-1)d$ orbitals available for accepting the electron pairs donated by the ligands. The stability of transition metal complexes depends on, besides ionic potential, other factors like crystal field stabilisation energy, electron pairing energy, stereochemistry, availability of empty $d_z{}^2$, and $d_{x^2-y^2}$ orbitals for π bonding or back donation. The use of inner *d*-orbitals gives stronger metal-ligand bonding than the use of *s*-and *p*-orbitals or outer *d*-orbitals. The various factors influencing the stability of complexes given by transition metal ions ultimately depends on the number of electrons present in $(n-1)$ *d* orbitals. The stability constants of the complexes given by the transition metal d^n ions for the equilibirium : $[M(H_2O)_n] + nL \longrightarrow ML_n + nH_2O$ follow the order : $d^0 < d^1 < d^2 < d^3 \gtreqless d^4 > d^5 < d^6 < d^7 < d^8 \gtreqless d^9 > d^{10}$. The sequence $d^3 > d^4$ and $d^8 > d^9$, is expected when the John-Teller effect is low and the coordination number is 6. The sequence $d^3 < d^4$ and $d^8 < d^9$, however, is expected when the John-Teller effect is large and the coordination number is 4. The stability of complexes of the divalent ions of the first transition series from Mn^{2+} to Zn^{2+} with a given ligand increases from Mn^{2+} to Cu^{2+} and then decreases at Zn^{2+}. Thus : $Mn^{2+} < Fe^{2+} < Co^{2+} < Ni^{2+} < Cu^{2+} > Zn^{2+}$. This sequence of stability is commonly known as **Irving-William order of stability** of complexes of M^{2+} ions (M^{2+} = Mn^{2+}, Fe^{2+}, Co^{2+}, Ni^{2+}. Cu^{2+}, Zn^{2+}) with a given ligand.

***(iv)* Electronegativity and polorising power of the central metal ion.** The bonding between a central metal ion and a ligand is, to some extent, due to the donation of electrons by the ligand to the central ion. Thus, it might be concluded that a central ion with higher electronegativity (*i.e.* a central ion with a strong electron-attracting tendency) would give more stable complexes. Similarly the central metal ion with high polarising power gives more stable complexes. For example complexes given Cu^+ and Cd^{2+} ions are more stable than those given by Na^+ and Ca^{2+} ions respectively.

***(v)* Ionisation energy of the central metal ion.** Generally the stability of complexes given by M^{2+} ions increases with the increase of first plus second ionisation potential for M^{2+} ions.

***(vi)* Class *a* and class *b* metals.** Chatt and Ahrland have classified the metals into three categories : *a*, *b* and *borderline*, on the basis of their electron-acceptor properties. This classification is shown below (normal valence states are assumed). (*a*) *Class a metals* : H, the alkali and alkaline earth metals, the elements Sc ⟶ Cr, Al ⟶ Cl, Zn ⟶ Br, In, Sn, Sb and I, the lanthamides and actinides (*b*) *Class b metals* : Rh, Pd, Ag, Ir, Pt, Au, Hg. (*c*) *Borderline metals* : The elements Mn ⟶ Cu, Tl ⟶ Po, Mo, Te, Ru, W, Re, Os, Cd.

Class *a* metals form *more stable* complexes with ligands having the coordinating atoms from the second period elements (*e.g.* N, O, F) than those of an analogous ligand in which the donor atom is from third or later period (*e.g.* P, S, Cl). Class *b* metals have the relative stabilities reversed. If the ligand contains the heavier donor atoms, class *a* and *b* metals are characterised by the stability order given below :

$$\left.\begin{array}{l} F^- > Cl^- > Br > I^- \\ O >> S > Se > Te \\ N >> P > As > Sb > Bi \end{array}\right\} \text{Class } a \text{ metals}$$

$$\left.\begin{array}{l} F^- < Cl^- < Br^- < I^- \\ O << S \approx Se \approx Te \\ N << P < As < Sb < Bi \end{array}\right\} \text{Class } b \text{ metals}$$

Class *b* metals are characterised by the presence of a number of *d*-electrons beyond an inert-gas core. These *d*-electrons are used to form π-bond with ligand atoms. It is believed that the stability of the complexes of class *b* metals results from covalent contribution to metal-ligand bonds and from the transfer of electron density from the metal to the ligand *via* π-bonding. The most stable complexes of class *b* metals are formed with ligands like PMe_3, S^{2-} and I^- which have vacant *d*-orbitals or like CO, CN^- which have vacant molecular orbitals of low energy.

For *borderline metals* the stability constants do not display either class *a* or class *b* behaviour uniquely.

2. Properties of the ligand. The following properties of the ligand affect the stability of the metal complexes.

***(i)* Size and charge of ligand.** If a ligand is smaller, it can approach the metal ion more closely forming a stable bond. Similarly, a highly charged ligand would also form a strong bond with the metal. Thus the high charge and small size of a ligand leads to the formation of stable complexes. For example the stability of the complexes of a given metal ion with halide ions used as ligands is in the order : $F^- > Cl^- > Br^- > I^-$. This order is applicable for class *a* metals. When class *b* metals (*e.g.* Pd, Ag, Pt, Hg etc) are used, the above order of stability is reversed, *i.e.* for class *b* metals the order is : $F^- < Cl^- < Br^- < I^-$.

***(ii)* Dipole moment of ligands.** For neutral ligands, the larger the magnitude of permanent dipole moment, the greater is the stability of the complexes. For example the order of stability of complexes formed by some neutral ligands is as : ammonia > ethylamine > diethylamine > triethylamine.

***(iii)* Basic character of ligands.** The more basic is the ligand, more easily it can denote electron pairs to the central ion and hence more easily it can form complexes of greater stability. The ligands that bind H^+ firmly form stable complexes with metal ions. Thus F^- should form more stable complexes than Cl^-, Br^- or I^-, and NH_3 should be better ligand than H_2O which in turn should be better than HF. ($NH_3 > H_2O > HF$). This behaviour is observed for alkali, alkaline earth and other electropositive metals like first-row transition elements, lanthanides and actinides.

***(iv)* π-bonding capacity of ligands.** The ligands like CN^-, CO, PR_3, AsR_3, SR_2, alkenes, alkynes which are capable of forming π-bonds with transition metal ions give more stable complexes

***(v)* Steric hindrance of bulky groups.** When a bulky group is either attached to or present near a donor atom of a ligand, repulsion between the donor atom of the ligand and the bulky group is caused and this mutual repulsion weakens the inetal-ligand bonding and hence makes the complex less stable. For example the complex of Ni^{2+} ion with 2-methyl-8-hydroxy quinoline ($\log_{10} \beta = 17.8$) is less stable than that with 8-hydroxy quinoline ($\log_{10} \beta = 21.4$). The effect of the presence of bulky group on the stability of a complex is commonly called *steric hindrance.*

3. Stability of chelates : Chelate effect. Chelates are more stable than similar non-chelated complexes of the same metal ion. Greater stability of chelated complex means that there will be smaller dissociation of the chelated complex into its components in a particular solvent. The enhanced stability conferred on a chelate is called *chelate effect.* The greater stability of a chelated complex as compared to that of a non-chelated complex is due to the increase in entropy. Thus when we consider the formation of $[Co(NH_3)_6]^{2+}$ (non-chelated complex) and $[Co(en)_3]^{2+}$ (chelated complex) from $[Co(H_2O)_6]^{2+}$ as represented by reactions (*a*) and (*b*) given below :

$$(a)\quad \underbrace{[Co(H_2O)_6]^{2+} + 6NH_3}_{\text{No. of particles} = 1 + 6 = 7} \rightleftharpoons \underbrace{\underset{\text{Non-chelated complex}}{[Co(NH_3)_6]^{2+} + 6H_2O}}_{\text{No. of particles} = 1 + 6 = 7}$$

$$(b)\quad \underbrace{[Co(H_2O)_6]^{2+} + 3\text{ en}}_{\text{No. of particles} = 1 + 3 = 4} \rightleftharpoons \underbrace{\underset{\text{Chelated complex}}{[Co(en)_3]^{2+} + 6H_2O,}}_{\text{No. of particles} = 1 + 6 = 7}$$

we find that number of particles on the right hand side of reaction (*b*) which represents the formation of chelated complex is greater than that on the left hand side while the number of particles on both sides of reaction (*a*) representing the formation of non-chelated complex is the same. Consequently the ehelatel complex, $[Co(en)_3]^{2+}$ is more stable than the non-chelated complex, $[Co(NH_3)_6]^{2+}$.

The stability of chelated complexes increases with the increase of the number of donor atoms present in the chelating ligand. The stability of chelated complexes also depends on the size of the ring formed in the complex. The chelated complexes containing 4- memberod ring are not stable while those containing 5- and 6- membered ring are quite stable. A chelated complex given by a polydentate ligand containing double bonds is more stable than that given by a polydentate ligand containing only single bonds. For example an octahedral complex of acetylacetonato ligand, $\left[M^{2+}(CH_3-\overset{\overset{O}{\|}}{C}-CH=\overset{\overset{O}{|}}{C}-CH_3)_3\right]$ is more stable than the octahedral complex of ethylenediamine with the same metal ion, $[M^{2+}(NH_2-CH_2-CH_2-NH_2)_3]^{2+}$. Note that both the complexes are chelated complexes. With the increase of the number of the rings present in the chelated complex, the complex becomes more stable.

Nature of Metal—Ligand Bonding in Complexes

In order to explain the nature of metal-ligand bonding in complexes the following theories have been suggested : (1) *Werner's Coordination Theory and Sidgwick's Electronic Concept of Coordination Compounds* (2) *Valence Bond Theory* (*VBT*), also called *Atomic Orbital Theory* (*AOT*) (3) *Crystal Field Theory* (*CFT*), and (4) *Ligand Field Theory* (*LFT*), also called *Molecular Orbital Theory* (*MOT*).

Werner's Coordination Theory

In 1893 Werner put forward a theory, called Werner's coordination theory, to explain the formation of complex compounds like the ammines of Pt (IV) (*e.g.* $PtCl_4.6NH_3$, $PtCl_4.5NH_3$, $PtCl_4.4NH_3$, $PtCl_4.3NH_3$, $PtCl_4.2NH_3$) and Co (III) (*e.g.* $CoCl_3.6NH_3$, $CoCl_3.5NH_3$, $CoCl_3.4NH_3$, $CoCl_3.3NH_3$).

Salient Features of Werner's Coordination Theory.

The salient features of Werner's coordination theory are summarized below :

1. Every metal atom or ion exhibits two types of valencies, *viz.*, *primary valency* and *secondary valency*. The primary valency of the central metal ion is satisfied by negative ions. Its attachment to the metal ion is shown by dotted lines. It corresponds to the *oxidation state* of the metal atom or ion and is ionizable. The secondary valency is satisfied by either negative ions or neutral molecules. Its attachment to the metal is shown by thick lines. It corresponds to the *coordination number* of the metal atom or ion and is non-ionizable.

2. Every metal atom or ion wants to satisfy both its primary and secondary valencies. In order to meet this requirement, a negative ion may perform a dual function of satisfying both types of valencies.

3. Every metal has a fixed number of secondary valencies.

4. The ligands which satisfy the secondary valencies are always directed towards fixed positions in space. The geometry of a complex is determined by the number and arrangement of such ligands in space. If a metal ion has a secondary valency of four, the geometry of the complex can be tetrahedral or square palanar. On the other hand, if its secondary valenoy is six, the complex should be octahedral. The primary valencies are non-directional.

Designation and Formulation of Co(III) Ammines on the Basis of Werner's Coordination Theory.

The ammines given by Co(III) are $CoCl_3.6NH_3$, $CoCl_3.5NH_3$, $CoCl_3.4NH_3$ and $CoCl_3.3NH_3$. On the basis of Werner's coordination theory, these ammines can be designated as shown in Fig. 43.29.

1. $CoCl_3.6NH_3$. This ammine has six NH_3 molecules that satisfy the secondary valency of the metal ion, (*viz.*, Co^{3+} ion) and their attachment with the central metal ion is shown by thick lines. The primary valency (*i.e.*, oxidation state of +3) is satisfied by three Cl^- ions whose attachment to Co^{3+} ion has been shown by dotted lines. As all the three Cl^- ions are loosely bound, they are immediately precipitated as AgCl on the addition of $AgNO_3$ solution. Thus the solution of this compound should conduct current to give four ions in all *viz.* $[Co(NH_3)_6]^{3-}$ and $3Cl^-$ which has been confirmed by conductivity measurements. Thus this ammine is formulated as $[Co(NH_3)_6]^{3+}$ $(Cl^-)_3$.

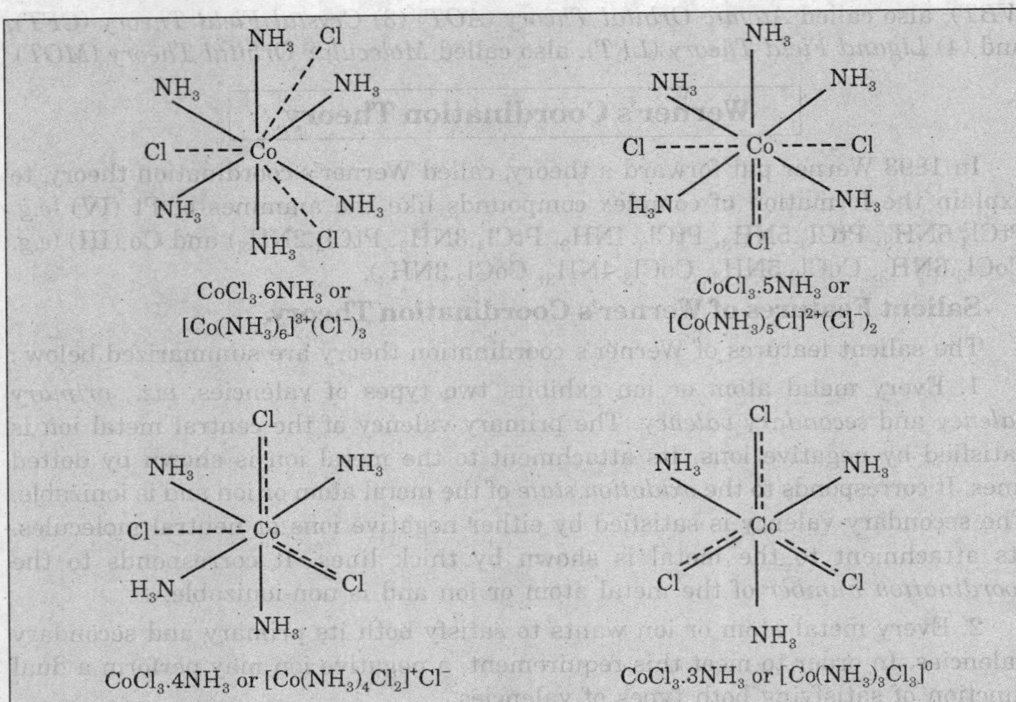

Fig. 43.29 Designation of Co (III) ammines on the basis of Werner's coordination theory.

2. $CoCl_3.5NH_3$. This ammine has five NH_3 molecules. On the basis of Werner's theory one Cl^- ion does the dual function, since it satisfies both primary and secondary valency. Werner, therefore, showed its attachment with the central metal ion by a combined dashed-solid line (----). This Cl^- ion, being non-ionic, is not precipitated as AgCl by Ag^+ ions and hence it is different from the other two Cl^- ions which, being ionic, are precipitated as AgCl by Ag^+ ions and the total number of ions obtained is three one complex ion, $[Co(NH_3)_5Cl]^{2+}$ and two simple ions, $2Cl^-$. Thus $CoCl_3.5NH_3$ is formulated as $[Co(NH_3)_5Cl]^{2+}(Cl^-)_2$ which satisfies both primary (= +3) and secondary valencies (= C.N. = 6) of Co^{3+}.

3. $CoCl_3.4NH_3$. This ammine has four NH_3 molecules. One ionizable chlorine (primary valency) is shown to be attached by dotted lines. Two non-ionizable chlorine ligands perform dual function of satisfying both types of valencies. Thus this ammine has only one ionic Cl^- ion which gets precipitated is AgCl by $AgNO_3$ solution. Consequently this ammine is formulated as $[Co(NH_3)_4Cl_2]^+$ Cl^-.

4. $CoCl_3.3NH_3$. All the three chlorine ligands are non-ionizable and perform dual function of satisfying both the primary and secondary valencies. Thus this ammine has no ionic Cl^- ions and hence behaves as a non-electrolyte (neutral molecule). Consequently it is formulated as $[Co(NH_3)_3Cl_3]^0$.

The formulation of all these ammines has been shown as follows :

Werner's formulation	**Modern formulation**	**Cation**	**Anion**	**Total number of ions**
$CoCl_3.6NH_3$	$[Co(NH_3)_6]^{3+}(Cl^-)_3$	$[Co(NH_3)_6]^{3+}$	$3Cl^-$	4
$CoCl_3.5NH_3$	$[Co(NH_3)_5Cl]^{2+}(Cl^-)_2$	$[Co(NH_3)_5Cl]^{2+}$	$2Cl^-$	3
$CoCl_3.4NH_3$	$[Co(NH_3)_4Cl_2]^+Cl^-$	$[Co(NH_3)_4Cl_2]^+$	Cl^-	2
$CoCl_3.3NH_3$	$[Co(NH_3)_3Cl_3]^0$	— (Non-isonisable)	—	—

Experimental Verification of Werner's Theory.

Six coordination number is the most common coordination number of metal complexes. Six ligands in a 6-coordination compound may be arranged round the central metal ion, M, either at the corners of *hexagonal plane* or at the apices of a *trigonal prism* or at the apices of a *regular octahedron.* These arrangements together with mumbers designating substitution positions may be depicted as shown in Fig. 43.30.

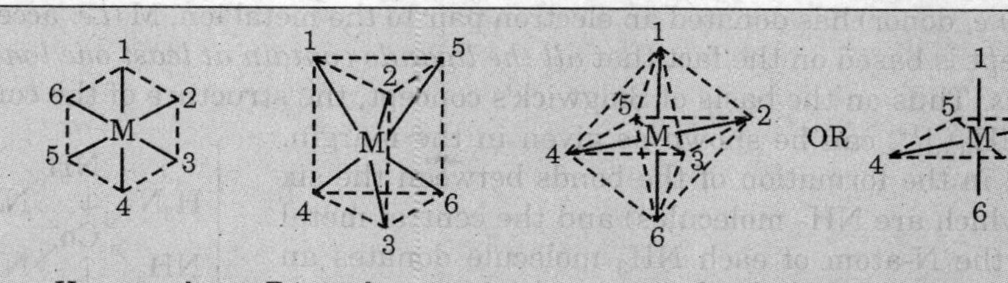

Fig. 43.30. Three possible arrangements of six ligands in a six coordinated complex round the central metal atom, M.

An extensive study of the geometrical and optical isomers of complexes with coordination number six has, however, shown that arrangement of six ligands in a 6-coordination compound is always octahedral and that the arguments concerning other possible geometries (*i.e.* hexagonal planar and trigonal prismatic geometries) are of historical interest only. In order to show that other geometries *viz.* hexagonal planar and trigonal prismatic are not preferred geometries for 6-coordination complex, Werner worked out the theoretical number of geometrical isomers that each of these geometries could give for the complexes Ma_6, Ma_5b, Ma_4b_2 and Ma_3b_3 (a and b are monodentate ligands). He found that the number of geometrical isomers predicted theoretically for these complexes in different geometries are as : *Hexagonal geometry* ⟶ Ma_6 = 1, Ma_5b = 1, Ma_4b_2 = 3 [(1, 2), (1, 3), (1, 4)], Ma_3b_3 = 3 [(1, 2, 3), (1, 2, 4), (1, 3, .5)]; *Trigonal prismatic geometry* ⟶

Ma_6 = 1, Ma_5b = l, Ma_4b_2 = 3 [(1, 2), (1, 3), (1, 4)], Ma_3b_3 = 3 [(1, 2, 3), (1, 2, 5), (1, 2, 6)], *Octahedral geometry* $\longrightarrow$ Ma_6 = 1, Ma_5b = 1, Ma_4b_2 = 2 [(1, 2), (1, 6)], Ma_3b_3 = 2 [(1, 2, 3), (1, 2 6)]. Werner then took some six-coordinated Co(III) ammines and isolated experimentally their geometrical isomers whose numbers are given as : Ma_6 = 1, Ma_5b = l, Ma_4b = 2 and Ma_3b_3 = 2. When he compared these numbers with those predicted theoretically for these geometries, he was convinced that the shape of a six-coordinated compound is octahedral, although it is not an absolute proof of this shape, since the correct experimental conditions for preparing all the isomers may not have been found, bot the shape is now known to be octahedral.

Sidgwick's Electronic Concept of Coordinate Bond in Coordination Compounds

In 1916 Lewis introduced a concept of two-electron covalent bond between two atoms in a molecule. According to this concept a covalent bond is formed between two atoms of a molecule by the sharing of an electron pair. Sidgwick (1927) extended this Lewis concept of covalent bond and introduced a new concept of *coordinate bond* which is also some times called *polar* or *dative bond.* According to this Sidgwick's concept the *ligands donate the electron pair to the central metal ion and thus form a coordinate bond,* M $\longleftarrow$ L which indicates that the ligand, L (*i.e.* donor) has donated an electron pair to the metal ion, M (*i.e.* acceptor). This concept is based on the fact that *all the ligands contain at least one lone pair of electrons.* Thus on the basis of Sidgwick's concept, the structure of the complex ion, $[Co(NH_3)_6]^{3+}$, can be shown as given in the margin. Evidently, in the formation of the bonds between the six ligands (which are NH_3 molecules) and the central metal Co^{3+} ion, the N-atom of each NH_3 molecule donates an electron pair to Co^{3+} ion. Thus in all twelve electrons (*i.e.* six electron pairs) are donated by six NH_3 ligands to Co^{3+} ion. The donation of an electron pair by NH_3 molecule has been represented by an arrow. Coordinate bond formed is not very different from a covalent bond. The only difference lies in the mode of its formation.

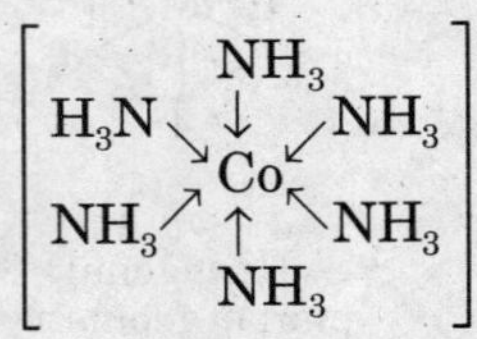

Limitations of Sidgwick's Concept of Coordinate Bond.

The donation of one electron pair by each of the ligands to the central metal ion to form L $\longrightarrow$ M coordinate bond in complexes accumulates negative charge on the central metal ion which is most unlikely. For example in $[Co^{3+}(NH_3)_6]^{3+}$, six NH_3 molecules donate half share of 12 electrons (*i.e.* 6 electrons) to Co^{3+} ion and thus +3 charge on cobalt atom is reduced to –3 (+ 3 – 6 = –3). Such an accumulation of negative charge on the central atom is unlikely.

Pauling suggested that the accumulation of undue negative charge on the central metal ion can be prevented, if it is assumed that *the metal-ligand bond has some ionic character (Electro-neutrality principle) and in addition to σ-bond, the central atom also forms a π-bond with the ligand* (*π-bonding in complexes*)

Sidgwick's Effective Atomic Number Rule (EAN Rule).

On the basis of his concept of coordinate bond, Sidgwick suggested that after the ligands have donated a certain number of electrons to the central metal ion through L $\longrightarrow$ M bonding, the total number of electrons on the central atom, including those gained from ligands in the bonding is called the *effective atomic number* (EAN) of the central metal ion and in many cases this total number of electrons (*i.e.* EAN) surrounding the coordinated metal ion is equal to the atomic number of the inert gas which follows the central metal atom in the periodic table. This is called *Effective Atomic Number Rule* or *Noble Gas Rule.* When the EAN is 36 (Kr), 54 (Xe) or 86 (Rn), the rule is said to be followed. EAN of a central metal atom or ion in a given complex is given by the formula *viz.* EAN $= (Z-x) + n \times y$ where Z = atomic number of the central metal atom, x = positive oxidation state of the central metal ion, n = number of ligands and y = number of electrons donated by one ligand. For example EAN of Fe^{2+} ion in $[Fe^{2+}(CN)_6]^{4-}$ is equal to $(26 - 2) + 6 \times 2 = 36$. On the basis of this formula, the EAN's of the central metal atom/ion of some complexes have been calculated and are given as $[Mn^{+}(CO)_5\ (C_2H_4)]^{+} = 36$. $[Fe^{2+}(CN)_6]^{4-} = 36$, $[Co^{3+}(NH_3)_6]^{3+} = 36$, $[Ag^{+}(NH_3)_4]^{+} = 54$, $[Pd^{4+}(NH_3)_6]^{4+} = 54$. $[Pt^{4+}(NH_3)_6]^{4+} = 86$, $[Pt^{4+}Cl_6]^{2-} = 86$.

The EAN of the central metal atom/ion of some complexes has been found to be some units more or less than the atomic number of the next inert gas. Examples of such complexes are : $[Fe^{3-}Cl_4]^{-} = 31$, $[Cu^{+}Cl_2]^{-} = 32$, $[Cr^{3+}(NH_3)_6]^{3+} = 33$, $[Ni(CN)_4]^{2-} = 34$, $[Mn^{2+}(H_2O)_6]^{2+} = 35$, $[Co(CN)_6]^{4-} = 37$, $[Ni^{2+}(NH_3)_6]^{2+} = 38$, $[Ag^{+}X_2]^{-} = 50$, $[Pd^{2+}Cl_4]^{2-} = 52$, $[Pt^{2+}(NH_3)_4]^{2+} = 84$.

Valence Bond Theory (VBT)

This theory is mainly due to Pauling. It deals with the electronic structure of the central metal ion in its ground state, kind of bonding, geometry and magnetic properties of the complexes.

Assumptions of Valence Bond Theory.

This theory is based on the following assumptions : (1) *The central metal atom or ion (as the case may be) makes available a number of empty s, p and d atomic orbitals equal to its coordination number. These vacant orbitals hybridise together to form hybrid orbitals which are the same in number as the atomic orbitals hybridising together. These hybrid orbitals are vacant, equivalent in energy and have definite geometry.* Important types of hybridisation occuring in the first row transition metal (*3d* metals) complexes and the geometry of the complexes are given in Table 43.1. (2) The ligands have at least one σ-orbital containing *a lone pair of electrons.* (3) Vacant hybrid orbitals of the metal atom or ion *overlap with the filled* (containing lone pair of electrons) σ-orbitals of the ligands to form ligand → metal σ-bond (represented as $M \overset{\sigma}{\leftarrow} L$). This bond which is generally known as coordinate bond is a special type of covalent bond and shows the characteristics of both the overlapping orbitals. However, it also possesses a considerable amount of polarity because of the mode of its formation.

(4) The non-bonding electrons of the metal atom or ion are then rearranged in the metal orbitals (*viz.*, pure *d*, *s* or *p* orbitals as the case may be) which do not participate in forming the hybrid orbitals. The rearrangement of non-bonding electrons takes place according to Hund's rule.

VBT As Applied to Octahedral Complexes (C.N. = 6, d^2sp^3 or sp^3d^2 Hybridisation)

On the basis of VBT, octahedral complexes are of two types : (1) *Inner-orbital octahedral complexes* which result form d^2sp^3 hybridisation of the central metal atom/ion (2) *Outer- orbital octahedral complexes* which result from sp^3d^2 hybridisation.

Table 43.1. Important types of hybridisation in the first row transition metal (3d-metal) complexes and the geometry of the complexes. C.N. indicates the coordination number of the central metal atom/ion, M (indicated by black circles) and L (white circles) represents ligands which may be the same or different.

C.N. of the central metal ion/ atom	Type of hybridisation and the atomic orbitals used in hybridisation	Geometry of the complex	Illustration of the geometry and examples
2	*sp* ($4s$, $4p_x$)	Linear or diagonal	 *Examples* : $[Cu^+Cl_2]^-$, $[Cu^+(NH_3)_2]^+$, $[Ag(NH_3)_2]^+$, $[Ag(CN)_2]^-$ etc.
3	sp^2 ($4s$, $4p_x$, $4p_y$)	Trigonal planar or equilateral triangular	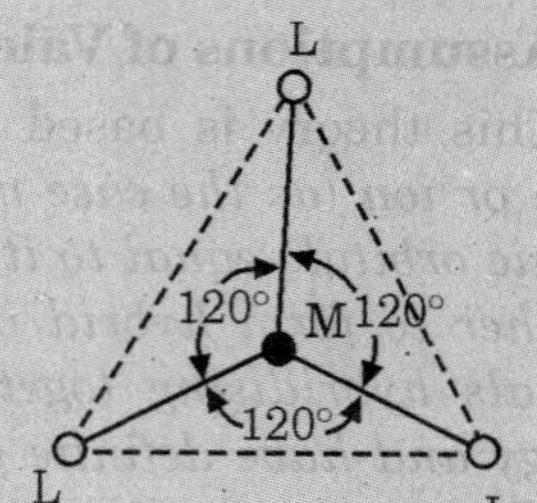 *Examples* : $[HgI_3]^-$, $[Cu^+(S{=}C(NH{-}CH_2)(NH{-}CH_2))_3]^+$, $[Cu^+Cl(tu)_2]^0$ (distorted trigonal planar geometry) etc.

(*Continued*)

4	sp^3 ($4s$, $4p_x$, $4p_y$, $4p_z$)	Tetrahedral	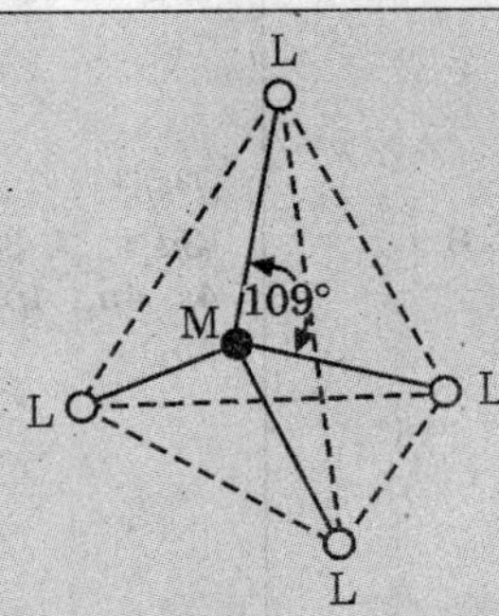 *Examples* : $[NiCl_4]^{2-}$, $[Ni(CO)_4]^0$, $[Zn(NH_3)_4]^{2+}$, $[Cu(CN)_4]^{3-}$ etc.
4	dsp^2 $3d_{x^2-y^2}$, $4s$, $4p_x$, $4p_y$)	Square planar	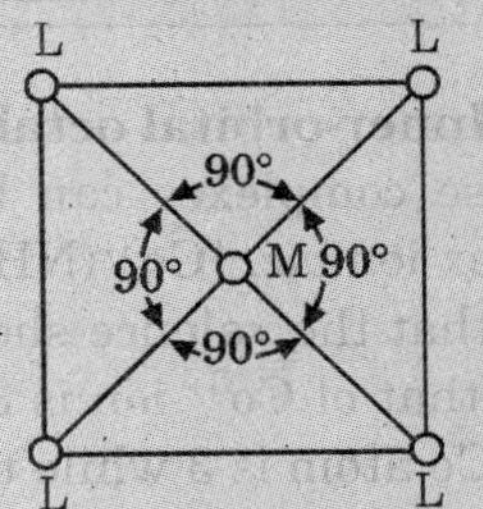 *Examples* : $[Cu(CN)_4]^{2-}$, $[Pt(CN)_4]^{2-}$, $[Ni(CN)_4]^{2-}$ etc.
5	dsp^3 ($3d_{z^2}$, $4s$, $4p_x$, $4p_y$, $4p_z$)	Trigonal bipyramidal	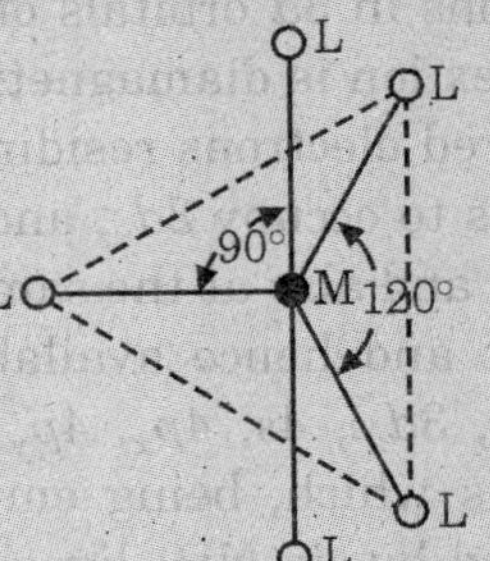 *Examples* : $[Fe^0(CO)_5]^0$, $[Cu^{2+}Cl_5]^{3-}$, $[Ni^{2+}($*triars*$)Br_2]^0$ etc.
5	sp^3d ($3d_{x^2-y^2}$, $4s$, $4p_x$, $4p_y$, $4p_z$)	Square pyramidal	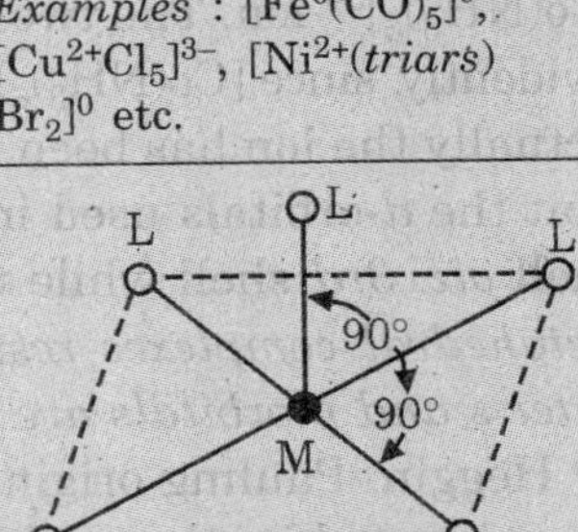 *Examples* : $[Co^{2+}($*triars*$)I_2]^0$, $[Ni^{2+}(CN)_5]^{3-}$, $[SbF_5]^{2-}$etc.

(Continued)

6	d^2sp^3 ($3d_{x^2-y^2}$, $3d_{z^2}$, $4s$, $4p_x$, $4p_y$, $4p_z$)	Inner-orbital octahedral	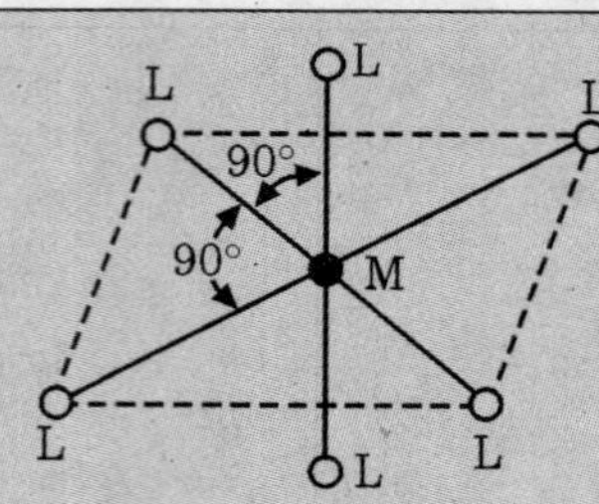 *Examples* : $[Ti(H_2O)_6]^{3+}$, $[Co(NH_3)_6]^{2+}$, $[Fe(CN)_6]^{4-}$ etc.
	sp^3d^2 ($4s$, $4p_x$, $4p_y$, $4p_z$, $4d_{x^2-y^2}$, $4d_{z^2}$)	Outer-orbital octahedral	*Examples* : $[Cr(H_2O)_6]^{2+}$, $[CoF_6]^{3-}$, $[Zn(NH_3)_6]^{2+}$ etc.

1. Inner-orbital octahedral complexes (d^2sp^3 hybridisation). The formation of these complexes can be explained on the basis of VBT by considering the complex ion *viz.* $[Co^{3+}(NH_3)_6]^{3+}$. In this complex ion Co is present as Co^{3+} ion. We know that the valence-shell electronic configuration of Co atom is $3d^7.4s^2.4p^0$ and hence that of Co^{3+} ion is $3d^6.4s^0.4p^0$. Note that the number of unpaired electrons (n) in Co atom is 3 while that in Co^{3+} ion is 4 as shown at (a) and (b) of Fig. 43.31. The measurement of magnetic moment value (μ value) for $[Co(NH_3)_6]^{3+}$ ion has shown that μ = 0, *i.e.*, $[Co(NH_3)_6]^{3+}$ ion has no unpaired electron or all the electrons in $3d$ orbitals of Co^{3+} ion in this complex ion are paired and hence the complex ion is diamagnetic. In order to make all the $3d$ electrons paired, the two unpaired electrons residing in $3d_{z^2}$ and $3d_{x^2-y^2}$ orbitals are forced by the six NH_3 ligands to occupy $3d_{yz}$ and $3d_{zx}$ orbitals. By doing so, all the $3d$ electrons become paired and also at the same time two $3d$ orbitals namely $3d_{z^2}$ and $3d_{x^2-y^2}$ are made vacant and hence available for d^2sp^3 hybridisation [See(c) of Fig. 43.31]. Now $3d_{x^2-y^2}$, $3d_{z^2}$, $4s$, $4p_x$, $4p_y$ and $4p_z$ orbitals hybridise together to give six d^2sp^3 hybrids which, being empty, accept the six electron pairs (represented as ••) donated by six NH_3 ligand molecules to form six $H_3N \rightarrow Co$ coordinate bonds in $[Co(NH_3)_6]^{3+}$ ion. The two electrons of each electron pair (••) have opposite spins. Evidently, since $[Co(NH_3)_6]^{3+}$ ion has no unpaired electron, it should be diamagnetic. Actually the ion has been found to be diamagnetic. It may be seen from Fig. 43.31 that the d-orbitals used in the formation of six d^2sp^3 hybrids belong to the inner shell *viz. 3rd* shell while the s and p orbitals belong to outer shell *viz. 4th* shell. *Octahedral complexes resulted form d^2sp^3 hybridisation, using the inner d and outer s and p orbitals are called inner-orbital octahedral complexes,* as suggested by Heggin. Pauling originally called such complexes as *covalent complexes*, since according to him the metal-ligand bonds in such complexes are primarily covalent.

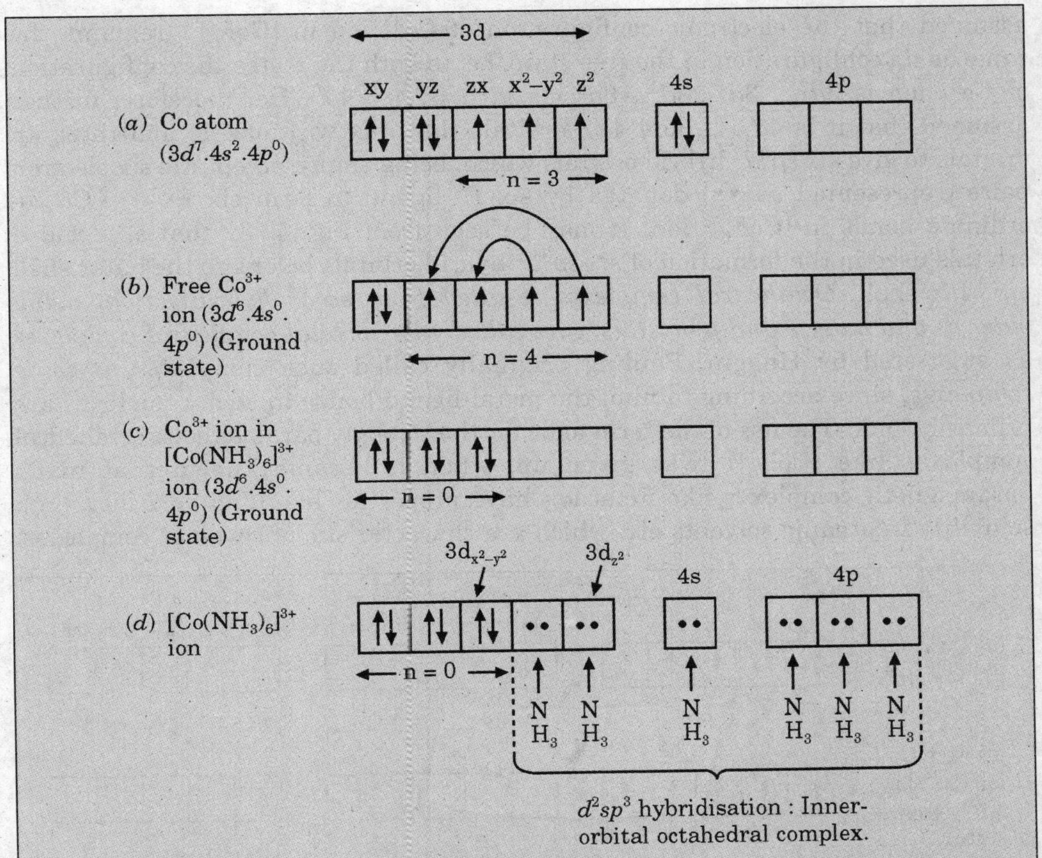

Fig. 43.31. Formation of inner-orbital octahedral complex ion *viz.* $Co(NH_3)_6]^{3+}$ on the basis of valence-bond theory. *n* represents the number of unpaired electrons. Dots (••) represent an electron pair donated by each of the six NH_3 ligand moleoules. The two electrons of this electron pair have opposite spins.

Examples of inner-orbital octahedral *paramagnetic* complexes are : $[Ti(H_2O)_6]^{3+}$(n = 1), $[Cr(NH_3)_6]^{3+}$ (n = 3), $[Mn(CN)_6]^{4-}$ (n = 1), $[Mn(CN)_6]^{3-}$ (n = 2), $[Fe(CN)_6]^{3-}$ (n = 1), $[Co(CN)_6]^{4-}$ (n = 1 in 5*s* orbital) etc. while the examples of inner-orbital octahedral *diamagnetic* complexes are : $[Fe(CN)_6]^{4-}$ (n = 0), $[Co(CN)_6]^{3+}$ (n = 0), $[Co(H_2O)_6]^{3+}$ (n = 0), $[Co(NH_3)_6]^{3+}$ (n = 0), $[Co(C_2O_4)_3]^{3-}$ (n = 0), $[Co(NO_2)_6]^{3-}$ (n = 0), $[Pt(NH_3)_6]^{4+}$ (n = 0), $[PtCl_6]^{2-}$ (n = 0) etc. In all these complexes *n* represents the number of unpaired electrons in the complex ion. All these complexes result from d^2sp^3 hybridisation of the central metal ion.

2. Outer-orbital octahedral complexes (sp^3d^2 hybridisation). The formation of these complexes can be explained on the basis of VBT by considering the complex ion *viz.* $[Co^{3+}F_6]$. In this complex ion Co is present as Co^{3+} whose valence-shell configuration is $3d^6.4s^0.4p^0$. $[CoF_6]^{3-}$ is a paramagnetic ion and the value of magnetic moment corresponds to the presence of four unpaired electrons. Now in order to account for the presence of four unpaired electrons in this ion Huggin assumed that the electronic configuration of Co^{3+} ion in $[CoF_6]^{3-}$ remains the same as its configuration in the free state, *i.e.* in both the states the configuration of Co^{3+} ion is $3d^2_{xy}.\ 3d^1_{yz}.3d^1_{zx}\ 3d^1_{x^2-y^2}.3d^1_{z^2}.4s^0.4p^0\ 4d^0$. He, therefore, further assumed that it is $4d_{x^2-y^2}$, and $4d_{z^2}$ orbitals that mix with one 4*s* and three 4*p* orbitals to give six sp^3d^2 hybrid orbitals which, being empty, accept the six electron pairs (represented as ••) donated by six F^- ligand to form six $F^- \longrightarrow Co$ co-ordinate bonds in $[CoF_6]^3$ ion, It may be seen from Fig. 43.32 that *s*, *p* and *d* orbitals used in the formation of six sp^3d^2 hybrid orbitals belong to the outer shell *viz.* 4*th* shell. *Octahedral complexes resulted from sp^3d^2 hybridisation, using outer d- and outer s and p orbitals, care called outer-orbital octahedral complexes* as suggested by Huggin. Pauling originally called such complexes as *ionic complexes*, since according to him, the metal-ligand bonds in such complexes are primarily ionic. The use of the term ionic for the strongly paramagnetic octahedral complexes like $[CoF_6]^{3-}$ was given up, when it became known that many paramagnetic complexes like $Fe(acac)_3$ have properties like high volatility, high solubility in organic solvents etc. which are characteristic of covalent complexes.

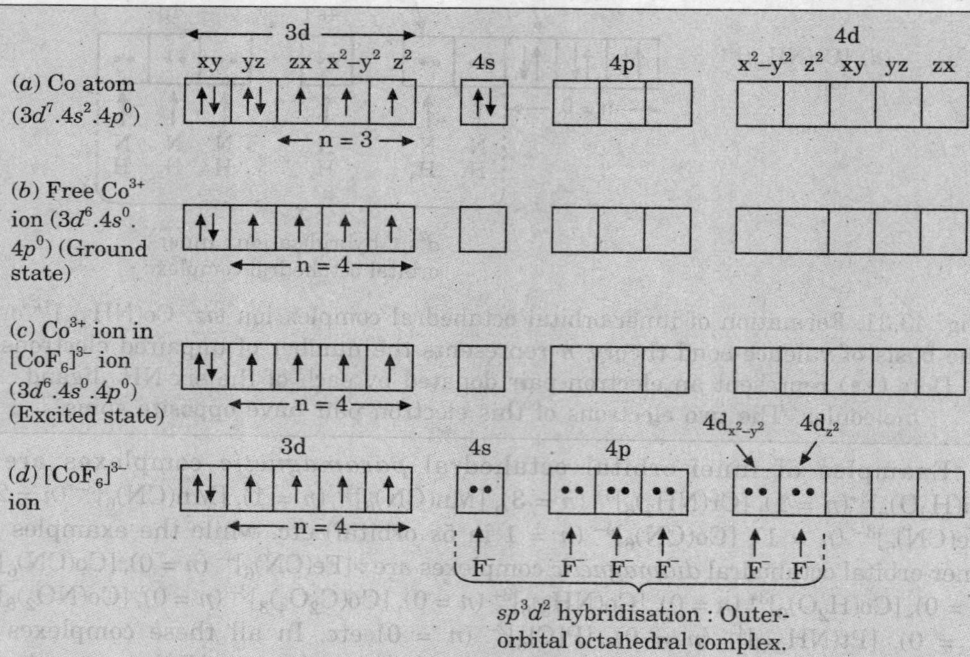

Fig. 43.32. Formation of outer-orbital octahedral complex ion, $[CoF_6]^{3-}$ on the basis of valence bond theory. *n* represents the number of unpaired electrons. Dots (••) represent an electron pair donated by each of the six F^- ligand ions. The two electrons of this electron pair have opposite spins.

Examples of outer-orbital octahedral *paramagnetic* complexes are : $[Cr(H_2O)_6]^{2+}$ ($n = 4$), $[Mn(H_2O)_6]^{2+}$ ($n = 5$), $[Fe(H_2O)_6]^{2+}$ ($n = 4$), $[Fe(NH_3)_6]^{2+}$ ($n = 4$), $[Fe(NH_3)_6]^{3+}$

(n = 4), $[Fe(H_2O)_6]^{3-}$ (n = 5), $[FeF_6]^{3-}$ (n = 5), $[CoF_6]^{3-}$ (n = 4), $[Co(NH_3)_6]^{2+}$ (n = 3), $[Co(H_2O)_6]^{2+}$ (n = 3), $[Ni(NH_3)_6]^{2+}$ (n = 2), $[Ni(H_2O)_6]^{2+}$ (n = 2), $[Ni(NCS)_6]^{4-}$ (n = 2), $[Ni(NO_2)_6]^{4-}$ (n = 2), $[CuF_3]^{3-}$ (n = 2). $[Zn(NH_3)_6]^{2+}$ (n = 0), is an example of outer-orbital octahedral diamagnetic complex. All these complexes result from sp^3d^2 hybridisation.

VBT As Applied to Tetrahedral Complexes (C.N. = 4, sp³ hybridisation)

Tetrahedral complexes result from sp^3 hybridisation. In sp^3 hybridisation s- and three p-orbitals should belong to the same shell. The formation of tetrahedral complexes by VBT can be explained by considering the complex ion like $[MnCl_4]^2$. This complex ion is parmagnetic corresponding to the presence of five unpaired electrons and hence the configuration of Mn^{2+} ion in the free state and in the complex ion remains the same as shown in Fig. 43.33.

Examples of some *paramagnetic* tetrahedral complexes are : $[NiCl_4]^{2-}$ (n = 2) $[Ni(NH_3)_4]^{2+}$ (n = 2), $[MnBr_4]^{2-}$ (n = 5), $[FeCl_4]^{2-}$ (n = 5), $[CoCl_4]^{2-}$ (n = 3), $[CuCl_4]^{2-}$ derived from $Cs_2[CuCl_4]$ (n = 1) etc. while the examples of some *diamagnetic tetrahedral* complexes are : $[Ni(CO)_4]^0$ (n = 0), $[Cu(CN)_4]^{3-}$ (n = 0), $[Zn(NH)_4]^{2+}$ (n = 0), $[ZnCl_4]^{2-}$ (n = 0), $[Cd(CN)_4]^{2-}$ (n = 0) etc. All these complexes result from sp^3 hybridisation of the central metal atom/ion.

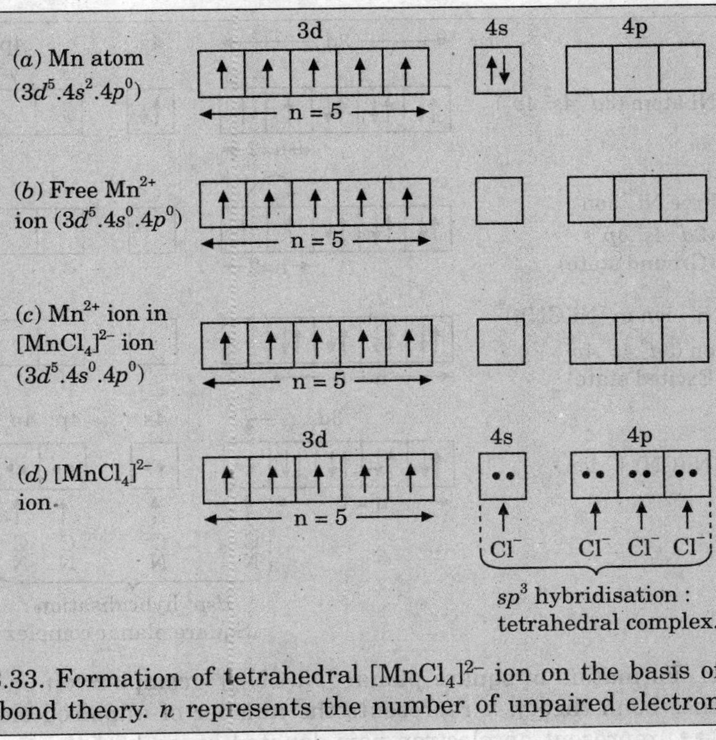

Fig. 43.33. Formation of tetrahedral $[MnCl_4]^{2-}$ ion on the basis of valence bond theory. n represents the number of unpaired electrons.

VBT As Applied to Square Planar Complexes (C.N. = 4, dsp^2 hybridisation)

Square planar complexes result from dsp^2 hybridisation. In dsp^2 hybridisation d-orbital should be $d_{x^2-y^2}$ orbital belonging to the lower shell while s and p orbitals should be from the higher shell. The two p-orbitals should be p_x and p_y orbitals. The selection of $d_{x^2-y^2}$, p_x and p_y orbitals is based on the fact that all these orbitals lie in xy plane. The formation of square planar complexes by VBT can be explained by considering the complex ion like $[Ni(CN)_4]^{2-}$ in which Ni is present as Ni^{2+} whose valence-shell configuration is $3d^8.4s^0.4p^0$ having two unpaired electrons. The measurement of magnetic moment value for $[Ni(CN)_4]^{2-}$ ion has shown that $\mu = 0$, *i.e.* the complex ion has no unpaired electron or all the electrons in $3d$ orbitals of Ni^{2+} ion in $[Ni(CN)_4]^{2-}$ ion are paired and hence the complex ion is diamagnetic. In order to make all the $3d$-electrons paired, one unpaired electron residing in $3d_{x^2-y^2}$ orbital is forced by the four CN^- ligands to occupy $3d_{z^2}$ orbital. By doing so, all the $3d$-electrons become paired and also at the same time $3d_{x^2-y^2}$ orbital becomes vacant and hence available for dsp^2 hybridisation. Now $3d_{x^2-y^2}$, $4s$, $4p_x$ and $4p_y$ orbitals mix together to form four dsp^2 hybrid orbitals which, being empty, accept the four electron pairs (represented as ••) donated by four CN^- ligand ions to form four $NC^- \longrightarrow Ni$ co-ordinate bonds in $[Ni(CN)_4]^{2-}$ ion (See Fig. 43.34).

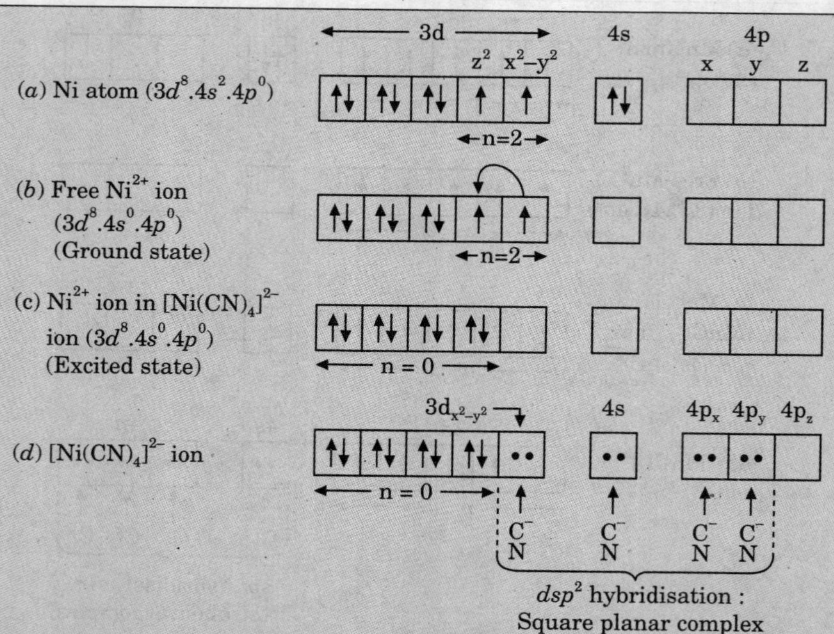

Fig. 43.34. Formation of square planar $[Ni(CN)_4]^{2-}$ complex ion on the basis of valence bond theory. n represents the number of unpaired electrons. Dots (••) represent an electron pair donated by each of the four CN^- ligand ions. The two electrons of this electron pair have opposite spins.

Examples of *paramagnetic* square-planar complexes are : $[Cu(CN)_4]^{2-}$ (n = 1), $(Cu(NH_3)_4]^{2+}$(n = 1), $[CuCl_4]^{2-}$ derived from $(NH_4)_2[CuCl_4]$(n = 1) etc. while the examples of *diamagnetic* square planar complexes are : $[Ni(CN)_4]^{2-}$ (n = 0), $[CuO_2]^-$ (n = 0), $[PtCl_4]^{2-}$ (n = 0), $[Pt(NH_3)_4]^{2+}$ (n = 0) etc.

Limitations of VBT.

(1) VBT cannot account for the relative stabilities for different shapes and different coordination numbers in metal complexes, *e.g.* it cannot explain satsifactorily as to why Co(+ 2)(d^7 system) forms both octahedral and tetrahedral complexes while Ni(+2) (d^8 system) rarely forms tetrahedral complexes. Similarly Fe(+3)(d^5 system) forms both tetrahedral and octahedral complexes while Cr(+3) (d^3 system) gives only octahedral complexes. **(2)** VBT cannot explain as to why Cu(+2) forms only distorted octahedral complexes even when all the six ligands are identical. **(3)** This theory cannot account for the relative rates of reactions of analogous metal complexes. *e.g.* $[Mn(phen)_3]^{2+}$ dissociates *instantaneusly* in acidic aqueous solution while $[Fe(phen)_3]^{2+}$ dissociates at a stow rate. **(4)** The classification of metal complexes on the basis of their magnetic behaviour into *covalent* (*inner-orbital*) and *ionic* (*outer-orbital*) complexes is not satisfactory and is often misleading, *e.g.* many of the so-called ionic complexes like $[Fe^{3+}(acac)_3]^0$ have properties like high volatility, high solubility in organic solvents etc. which are characteristic of covalent compounds, Also it is known that the aquo ion, $[Cr(H_2O)_6]^{3+}$ (pK_a = 3.8 at 25°C) is a much weaker acid than $[Fe(H_2O)_6]^{3+}$(pK_a = 2.2) and $[Co(H_2O)_6]^{3+}$(pK_a = 0.7) ions, despite the fact that only $[Fe(H_2O)_6]^{3+}$ is of outer-orbital type with ionic bonding while the other two aquo ions are of outer-orbital type with covalent bonding. Both $[Fe(H_2O)_6]^{3+}$ and $[Co(H_2O)_6]^{3+}$ ions exchange with H_2O^{18} in solvent instantaneously while $[Cr(H_2O)_6]^{3+}$ ion exchanges H_2O^{18} very slowly. The acidity of an aqueo metal ion is expected to increase with the increase of covalent character of the metal–oxygen bond. But neither the sequence of acidities nor the exchange rates of the above mentioned three aqueo metal ions are in the order expected from their classification into ionic and covalent types on the basis of Pauling's VBT. **(5)** VBT fails to explain the finer details of magnetic properties including the magnitude of the orbital contribution to the magnetic moments, *e.g.* although both tetrahedral (sp^3 hybridisation) and outer-orbital octahedral (sp^3d^2 hybridisation) complexes of Co(+2)(d^7 system) have three unpaired electrons and are, therefore, expected to have μ value equal to 3.87 B.M.; the tetrahedral complexes generally have μ value in the range of 4.4–4.8 B.M. while the octahedral complexes have still higher value of μ in the range of 4.7–5.2 B.M. . The increase in the value of μ is due to the orbital contribution, Similar is the case with tetrahedral and octahedral complexes of Ni(+2) (d^8 system). VBT cannot explain the increase in the value of μ. **(6)** VBT cannot interpret the spectra (colour) of the complexes. **(7)** This theory does not predict or explain the magnetic behaviour of complexes. This theory only predicts the number of unpaired electrons. Its prediction even for the number of unpaired electrons and their correlation with stereochemistry is misleading. For example it was assumed for long that all square planar complexes of Ni(+2) (d^8 system) formed by dsp^2 hybridisation were diamagnetic (n = 0), while all tetrahedral complexes of Ni(+2) formed by sp^3 hybridisation were paramagnetic due to the presence of two unpaired electrons (n = 2). X-ray study of 4-coordinate Ni (+ 2) complexes has shown that one and the same complex can he obtained in both a *paramagnetic form* (having blue colour) and a *diamagnetic form* (having yellow colour). VBT has no explanation for it. VBT cannot explain the temperature dependent paramagnetism of the complexes. **(8)** VBT cannot give any explanation for the order of reactivities of the inner-orbital inert complexes of d^3, d^4 d^5 and d^6 ions and of the observed differences in the energies of activation in a series of similar complexes. **(9)** The magnetic moment values of the complexes of certain ions (*e.g.* Co^{2+}, Ni^{2+}, etc.) are much higher than those expected by spin only formula. VBT can not explain the enhanced values of magnetic moment.

Crystal Field Theory (CFT)

This theory determines the effect of different electrostatic fields caused by the different geometries of the complexes and different basic strengths of the ligands, on the energies of *d*-orbitals of the central metal ion.

Salient Features of Crystal Field Theory.

(1) The central metal cation is surrounded by ligands which contain one or more lone pairs of electrons (2) The ionic ligands (*e.g.* F^-, Cl^-, CN^- etc.) are regarded as negative point charges (also called point charges) and the neutral ligands (*e.g.* H_2O, NH_3 etc.) are regarded as point dipoles or simply dipoles, *i.e.* according to this theory neutral ligand are dipolar. If the ligand is neutral, the negative end of this ligand dipole is oriented towards the metal cation. (3) The interaction between the metal cation and the ligands is regarded as purely electrostatic, *i.e.* the metal–ligand bond is considered to be 100% ionic.

CFT As Applied to Octahedral Complexes

Splitting of d-orbitals in Octahedral Complexes.

In case of free metal ion all the five *d*-orbitals are degenerate, *i.e.* these have the same energy. Now let us consider an octahedral complex, $[ML_6]^{n+}$ in which the central metal cation, M^{n+} is placed at the centre of the octahedron and is surrounded by six ligands which reside at the six corners of the octahedron as shown in Fig. 43.35. The three axes *viz. x-*, *y-*, and *z-* axes which point along the corners have also been shown. Now suppose both the ligands on each of the three axes are allowed to approach towards the metal cation, M^{n+} from both the ends of the axes. In this process the electrons in *d*-orbitals of the metal cation are repelled by negative point charge or by the negative end of the dipole of the ligands. (Remember CFT ragards the ionic ligands as negative point charges and neutral ligands as dipoles). This repulsion will raise the energy of all the five *d*-orbitals. If all the ligands approaching towards the central cation are at an equal distance from each of the

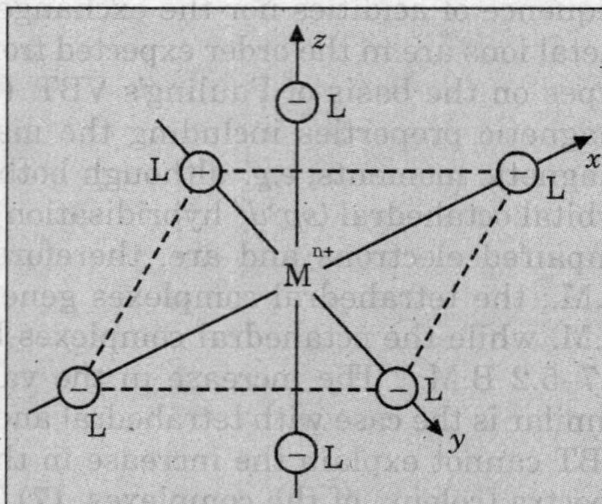

Fig. 49.35. Position of the central metal cation, M^{n+} and six ligands, L's in an octahedral complex, $[ML_6]^{n+}$.

d-orbitals (*i.e.* the ligand field is spherically symmetrical), the energy of each of five *d*-orbitals will raise by the same amount, *i.e.* all the *d*-orbitals will still remain degenerate, although they will have now higher energy than before. This is only a *hypothetical situation. Since the lobes of the two e_g orbitals* (*i.e.* d_{z^2} and $d_{x^2-y^2}$ orbitals) *lie directly in the path of the approaching ligands, the electrons in these orbitals experience greater force of repulsion than those in three t_{2g} orbitals* (*i.e.* d_{xy}, d_{yz} and d_{zx} orbitals) *whose lobes are directed in space between the path of the approaching ligands i.e.* the energy of e_g orbitals is increased while that of t_{2g} is decreased (Remember : *greater is the repulsion, greater the increase in energy*). Thus we find that under the influence of approaching ligands the five *d*-orbitals which were originally degenerate in the free metallic cation are now split (or resolved) into two levels *viz.* t_{2g} level which is triply degenerate and is of lower energy, and e_g level which is doubly degenerate and is of higher energy (*See* Fig. 43.36). In other words the degeneracy of the five *d*-orbitals is removed under the influence of the ligands. *The separation of five d-orbitals of the metal ion into two sets having different energies is called crystal field splitting or energy level splitting.* This concept of crystal field splitting makes the basis of CFT.

The energy gap between t_{2g} and e_g sets is denoted by Δ_0 or 10Dq (Δ_0 = 10Dq) where Δ_0 indicates an octahedral arrangement of the ligands round the central metal cation. This energy difference arises because of the difference in electrostatic field exerted by the ligands on t_{2g} and e_g sets of orbitals of the central metal cation. Δ_0 or 10Dq is called *crystal field splitting energy.* With the help of simple geometry it can be shown that the energy of t_{2g} orbitals is 0.4 Δ_0 (= 4Dq) *less* than that of hypothetical degenerate *d*- orbitals (No splitting state shown by dotted line in Fig. 43.36) and, hence, that of e_g orbitals is 0.6Δ_0 (= 6 Dq) *above* that of the hypothetical degenerate *d*-orbitals. Thus, we find that t_{2g} set *loses* an energy equal to 0.4 Δ_0 (= 4Dq) while e_g set gains an energy equal to 0.6Δ_0 (= 6Dq). In Fig. 43.36 the loss and gain in energies of t_{2g} and e_g orbitals is shown by negative and positive signs respectively.

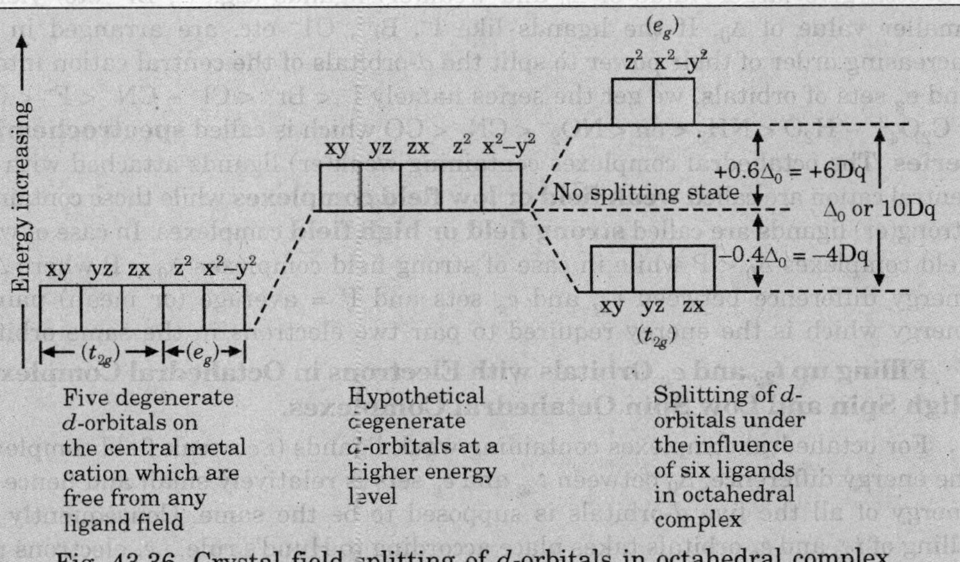

Fig. 43.36. Crystal field splitting of *d*-orbitals in octahedral complex.

Factors Affecting the Magnitude of Δ_0.

Following are important factors which influence the magnitude of Δ_0.

1. Nature of the central cation. Here we shall consider the complex compounds having the same geometry and same ligands. (*a*) The complex compound having the central cation with higher oxidation state (*i.e.* higher charge) has higher value of Δ_0 than that having the central cation with lower oxidation state. For example Δ_0 for $[Fe^{3+}(H_2O)_6]^{3+}$ (= 13,700 cm^{-1}) > Δ_0 for $[Fe^{2+}(H_2O)_6]^{2+}$ (= 10,400 cm^{-1}). Similarly $[Cr^{3+}(H_2O)_6]^{3+}$ (= 17,400 cm^{-1}) > $[V^{2+}(H_2O)_6]^{2+}$ (= 12,400 cm^{-1}). The increase in the value of Δ_0 with the increase of oxidation state is explained by saying that the metal cation with higher oxidation state polarises the ligands more effectively and hence the ligands approach this cation more closely than they approach the cation with lower oxidation state, resulting in a larger splitting of *d*-orbitals. (*b*) In case of complexes having the central cations with the same oxidation state but with different number of *d*-electrons in the central cation, the magnitude of Δ_0 decreases with the increase of the number of *d*-electrons. For example Δ_0 for $[Co^{2+}(H_2O)_6]^{2+}$ (= 9,300 cm^{-1}) in which Co^{2+} ion has seven d-electrons is higher than that for $[Ni(H_2O)_6]^{2+}$ (= 8,500 cm^{-1}) in which Ni^{2+} ion has eight d-electrons. (*c*) The value of Δ_0 for the complexes having the central cations with the same oxidation state but with different principal quantum number (*n*) of their d-orbitals increases with the increase of the value of *n*. For example $[Co^{2+}(NH_3)_6]^{3+}$ (= 23,000 cm^{-1}, $Co^{3+} \rightarrow 3d^6$ with $n = 3$) < $[Rh^{3+}(NH_3)_6]^{3+}$ (= 34,000 cm^{-1}, $Rh^{3+} \rightarrow 4d^6$ with $n = 4$) < $[Ir^{3+}(NH_3)_6]^{3+}$ (= 41,000 cm^{-1}, $Ir^{3+} \rightarrow 5d^6$ with $n = 5$).

2. Nature of the ligand : Spectrochernical series. The ligands which split the *d*-orbitals of the central metal cation into t_{2g} and e_g sets of orbitals to a greater extent are called *strong(er) ligands* while those which split the *d*-orbitals to a lesser extent are called *weak(er) ligands.* Thus strong (er) ligands (*e.g.* CN^-, CO etc.) give larger value of Δ_0 and weak(er) ligands (*e.g.* I^-, Br^- etc) yield a smaller value of Δ_0. If the ligands like I^-, Br^-, $C1^-$ etc. are arranged in the increasing order of their power to split the *d*-orbitals of the central cation into t_{2g} and e_g sets of orbitals, we get the series namely $I^- < Br^- < Cl^- \sim CN^- < F^- < OH^- < C_2O_4^{2-} \sim H_2O < NH_3 <$ en $< NO_2^- < CN^- <$ CO which is called **spectrochemical series**. The octahedral complexes containing weak(er) ligands attached with the central cation are called **weak field or low field complexes** while those containing strong(er) ligands are called **strong field or high field** complexes. In case of weak field complexes $\Delta_0 < P$ while in case of strong field complexes $\Delta_0 > P$ where Δ_0 = energy difference between t_{2g} and e_g sets and P = average (or mean) pairing energy which is the energy required to pair two electrons in the same orbital.

Filling up t_{2g} and e_g Orbitals with Electrons in Octahedral Complexes : High Spin and Low Spin Octahedral Complexes.

For octahedral complexes containing weak ligands (*i.e.* weak field complexes) the energy difference, Δ_0 between t_{2g} and e_g sets is relatively small and hence the energy of all the five *d*-orbitals is supposed to be the same. Consequently the filling of t_{2g} and e_g orbitals takes place according to Hund's rule, *i.e.* electrons pair up only when each of the five d-orbitals is at least singly filled. Thus: (a) in case of weak field complexes the distribution of d^1 to d^{10} electrons t_{2g} and e_g orbitals takes place as : $d^1 = t_{2g}{}^1e_g{}^0$ (n = 1, $S = n/2$ = 1/2); $d^2 = t_{2g}{}^2e_g{}^0$ (n = 2, S = 1); $d^3 = t_{2g}{}^3e_g{}^0$ (n = 3, S = 3/2); $d^4 = t_{2g}{}^3e_g{}^1$ (n = 4, S = 2); $d^5 = t_{2g}{}^3e_g{}^2$ (n = 5, S = 5/2); $d^6 = t_{2g}{}^4e_g{}^2$ (n = 4, S = 2); $d^7 = t_{2g}{}^5e_g{}^2$ (n = 3, S = 3/2); $d^8 = t_{2g}{}^6e_g{}^2$ (n = 2, S = 1); $d^9 = t_{2g}{}^6e_g{}^3$ (n = 1, S = 1/2); $d^{10} = t_{2g}{}^6e_g{}^4$ (n = 0, S = 0). Here n = number of unpaired electrons and S = resultant spin which is equal to n/2.

(b) For octahedral complexes containing strong ligands ($i.e.$ strong field complexes) all the three d-orbitals of low energy t_{2g} set are filled up first and then the two d-orbitals of high energy e_g set are filled. Thus in this ease the distribution of d^1 to d^{10} electrons in t_{2g} and e_g orbitals takes places as : $d^1 = t_{2g}{}^1e_g{}^0$ (n = 1, S = 1/2); $d^2 = t_{2g}{}^2e_g{}^0$ (n = 2, S = 1/2); $d^3 = t_{2g}{}^3e_g{}^0$ (n = 3, S = 3/2); $d^4 = t_{2g}{}^4e_g{}^0$ (n = 2, S = 1); $d^5 = t_{2g}{}^5e_g{}^0$ (n = 1, S = 1/2); $d^6 = t_{2g}{}^6e_g{}^0$ (n = 0, S = 0); $d^7 = t_{2g}{}^6e_g{}^1$ (n = 1, S = 1/2); $d^8 = t_{2g}{}^6e_g{}^2$ (n = 2, S = 1); $d^9 = t_{2g}{}^6e_g{}^3$ (n = 1, S = 1/2); $d^{10} = t_{2g}{}^6e_g{}^4$ (n = 0, S = 0).

It may be seen that the weak field octahedral complexes of d^4, d^5, d^6 and d^7 ions have greater number of unpaired electrons (n) than those of (same ions) strong field complexes and hence have higher value of S. It is for this reason that weak field complexes of d^4, d^5, d^6 and d^7 ions are also called **high spin** or **spin free** complexes. Similarly strong field complexes are also called **low spin** or **spin paired** complexes. Recall that VBT has called high spin complexes as ionic (Pauling) or outer orbital (Huggin) complexes while the same theory has called low spin complexes as covalent or inner orbital complexes.

Crystal Field Stabilisation Energies (CFSEs) of Octahedral Complexes.

We have seen that, according to CFT, under the influence of the six ligands approaching towards the central metal ion during the formation of an octahedral complex, the d-orbitals of the central metal ion are split into two sets of orbitals *viz., lower energy triplet* (or *trio*) t_{2g} and *higher energy doublet* (or *pair*) e_g sets. t_{2g} set of orbitals is triply degenerate while e_g set is doubly degenerate The energy gap between these two sets is equal to Δ_0 (or 10Dq). The energy of t_{2g} set is lowered by 2/5 Δ_0 (= 0.4Δ_0) or 4Dq while that of e_g set is raised by 3/5 Δ_0 (= 0.6Δ_0) or 6Dq relative to the energy of hypothetical degenerate d-orbitals. Thus each electron occupying t_{2g} orbitals decreases the energy of d-orbitals by –0.4 Δ_0 (= –4Dq) while that going into e_g orbitals increases its energy by + 0.6 Δ_0 (= + 6Dq). – and + signs indicate respectively the decrease and increase in the energy of d-orbitals caused by their splitting under the influence of six ligands. Now let us consider a d^x ion containing $t_{2g}{}^pe_g{}^q$ configuration in which p is the number of electrons in t_{2g} set, q is the number of electrons in e_g set and $x = p + q$. Quite obviously :

Change in energy (in terms of Δ_0) for $t_{2g}{}^pe_g{}^q$ configuration

= Loss in energy due to p electrons in t_{2g} set

+ gain in energy due to q electrons in e_g set.

= –0.4 $\Delta_0 \times p$ + 0.6 $\Delta_0 \times q$

= [–0.4 p + 0.6q] Δ_0 ...(i)

Now since Δ_0 = 10Dq, the above expression can also be written as :

Change in energy (in terms of Dq) for $t_{2g}^p e_g^q$ configuration

$$= [-0.4p + 0.6q] \times 10Dq$$

$$= [-4p + 6q]Dq \qquad ...(ii)$$

Thus equations (*i*) and (*ii*) give the energies of d^x ion containing $t_{2g}^p e_g^q$ configuration. The change in energy for d^x ion containing $t_{2g}^p e_g^q$ configuration calculated as above is called *crystal field stabilisation energy* (CFSE) of d^x ion, since it stabilises *d*-orbitals by lowering their energy which results from their splitting into t_{2g} and e_g orbitals.

In the derivation of equations (*i*) and (*ii*) we have not considered the pairing energy, P of d^x ion which is the energy required to pair two electrons against electron-electron repulsion in the same orbital. If the pairing energy of the ion is also involved in any $t_{2g}^p e_g^q$ configuration of a given d^x ion, then CFSE of the ion is given by the expression. :

$$= [-0.4p + 0.6q]\,\Delta_0 \times m\,p \qquad ...(iii)$$

$$= [-4p + 6q]\,Dq + mp \;(\because \Delta_0 = 10\,Dq) \qquad ...(iv)$$

Here *m* is the total number of paired electrons in t_{2g} and e_g sets of orbitals. Equations (*iii*) and (*iv*) have been used to calculate the CESE values (in terms of Δ_0 and Dq respectively) for d^0 to d^{10} ions of high spin and low spin octahedral complexes. These values, as calculated from the above equations, are given below :

(*a*) For high spin (or spin free or weak field) octahedral complexes : d^0 ($t^0{}_{2g}$ $e^0{}_g$, $p = 0$, $q = 0$) = 0.0 Δ_0 or 0.0 Dq; d^1 ($t_2{}^1{}_g$ $e^0{}_g$, $p = 1$, $q = 0$) = – 0.4 Δ_0 or – 4D*q*; d^2 ($t^2{}_{2g}$ $e^0{}_g$, $p = 2$, $q = 0$) = – 0.8 Δ_0 or –8 Dq; d^3 ($t^3{}_{2g}$ $e^0{}_g$, $p = 3$, $q = 0$) = – 1.2 Δ_0 or – 12 Dq; d^4 ($t^3{}_{2g}$ $e_g{}^1$, $p = 3$, $q = 1$) = – 0.6 Δ_0 or – 6 Dq; d^5 ($t^3{}_{2g}$ $e_g{}^2$, $p = 3$, $q = 2$) = 0.0 Δ_0 or 0.0 Dq; d^6 ($t_{2g}{}^4$ $e_g{}^2$, $p = 4$, $q = 2$, $m = 1$) = – 0.4 Δ_0 + P or – 4 D*q* + P; d^7 ($t^5{}_{2g}$ $e_g{}^2$, $p = 5$, $q = 2$, $m = 2$) = – 0.8 Δ_0 + 2P or – 8 Dq + 2P; d^8 ($t^6{}_{2g}$ $e_g{}^2$, $p = 6$, $q = 2$, $m = 3$) = – 1.2 Δ_0 + 3P or – 12 D*q* + 3P; d^9 ($t^6{}_{2g}$ $e_g{}^3$, $p = 6$, $q = 3$, $m = 4$) = – 0.6 Δ_0 + 4P or – 6 D*q* + 4P; d^{10} ($t^6{}_{2g}$ $e_g{}^4$, $p = 6$, $q = 4$, $m = 5$) = 0.0 Δ_0 + 5P or 0.0 D + 5P.

(*b*) For low spin (or spin paired or strong field) octahedral complexes : d^0 ($t^0{}_{2g}$ $e_g{}^0$, $p = 0$, $q = 0$) = 0.0 Δ_0 or 0.0 D*q*; d^1 ($t^1{}_{2g}$ $e_g{}^0$, $p = 1$, $q = 0$) = – 0.4 Δ_0 or – 4D*q*; d^2 ($t^2{}_{2g}$ $e_g{}^0$, $p = 2$, $q = 0$) = – 0.8 Δ_0 or – 8 Dq; d^3 ($t^3{}_{2g}$ $e_g{}^0$, $p = 3$, $q = 0$) = – 1.20 Δ_0 or – 12 Dq; d^4 ($t^4{}_{2g}$ $e_g{}^0$, $p = 4$, $q = 0$, $m = 1$) = – 1.6 Δ_0 + P or – 16 Dq + P; d^5 ($t^5{}_{2g}$ $e_g{}^0$, $p = 5$, $q = 0$, $m = 2$) = 2.0 Δ_0 + 2P or – 20 D*q* + 2P; d^6 ($t^6{}_{2g}$ $e_g{}^0$, $p = 6$, $q = 0$, $m = 3$) = – 2.4 Δ_0 + 3P or – 24 Dq + 3P; d^7 ($t^6{}_{2g}$ $e_g{}^1$, $p = 6$, $q = 1$, $m = 3$) = – 1.8 Δ_0 + 3P or – 18 D*q* + 3P; d^8 ($t^6{}_{2g}$ $e_g{}^2$, $p = 6$, $q = 2$, $m = 3$) = – 1.2 Δ_0 + 3P or – 12 D*q* + 3P; d^9 ($t^6{}_{2g}$ $e_g{}^3$, $p = 6$, $q = 3$, $m = 4$) = – 0.6 Δ_0 + 4P or – 6 D*q* + 4P; d^{10} ($t^6{}_{2g}$ $e_g{}^4$, $p = 6$, $q = 4$, $m = 5$) = 0.0 Δ_0 + 5P or 3.0 D*q* + 5P.

CFSE values as given above show that, (*i*) When Δ_0 = P, each of d^4, d^5, d^6 and d^7 ions gives the same value of CFSE for both spin states. (*ii*) When Δ_0 > P, the

CFSE of the low spin state becomes smaller than that of the high spin state and hence the low spin state becomes the preferred state of the ion. In this condition (*i.e.* when $\Delta_0 > P$), the electrons tend to pair and hence low spin complexes are obtained (*iii*) When $\Delta_0 < P$, the CFSE of the high spin becomes smaller than the low spin state and hence high spin state becomes the preferred state of the ion. Under this condition (*i.e.* when $\Delta_0 < P$) the electrons tend to remain unpaired and hence high spin complexes are obtained.

Thus we see that the spin state of an ion in a complex depends on whether the crystal field splitting (Δ_0) is bigger or smaller than the pairing energy (P).

CFT As Applied to Tetrahedral Complexes

Splitting of d-orbitals in Tetrahedral Complexes.

In order to understand the splitting of *d*-orbitals in tetrahedral complexes, a tetrahedron should be pictured as being placed in a cube as represented in Fig 43.37. It may be seen from the figure that the four negative ligands (represented by circles with negative signs) placed at the four corners of the tetrahedron are located at the four corners of the cube. Thus these four ligands are lying between the three axes *viz.* *x*, *y* and *z* axes which pass through the centres of the six faces of the cube and thus go through the centre of the cube. Now since the lobes of t_2 orbitals (d_{xy}, d_{yz} and d_{zx}) are lying between the axes, *i.e.* are lying directly in the path of the ligands, these orbitals will experience greater force of repulsion from the ligands than those of *e* orbitals (d_{z^2} and $d_{x^2-y^2}$) whose lobes are lying along the axes, *i.e.* are lying in space between the ligands. Thus the energy of t_2

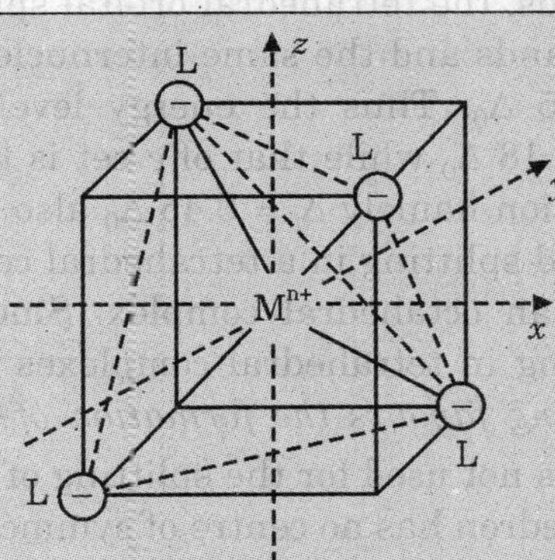

Fig. 43.37. A tetrahedron placed in a cube. The centre of the cube is the centre of the tetrahedron at which is placed central metallic cation, M^{n+}. Four corners of the cube are the four corners of the tetrahedron at which are placed the four negative ligands which have been shown by circles with negative signs.

orbitals will be *increased* while that of *e* orbitals will be *decreased*. Consequently the *d* orbitals are again split into two sets as shown in Fig. 43.38 from which it may be seen that the order of energy of t_2 and *e* sets is the reverse of that seen for t_{2g} and e_g sets in octahedral complexes.

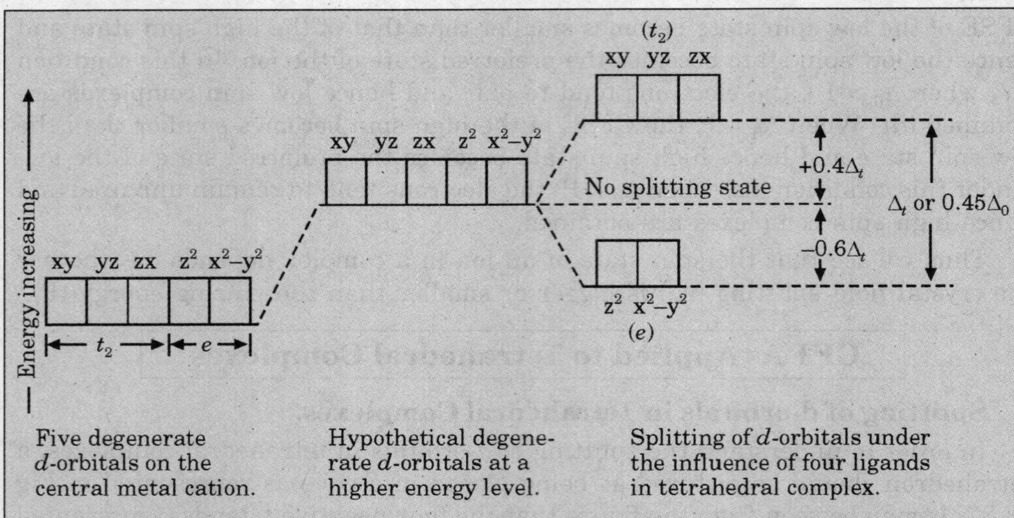

Fig. 43.38. Crystal field splitting of *d*-orbitals in tetrahedral complex.

The energy difference between t_2 and e sets for tertrahedral complex is represented as Δ_t. In has been shown that $\Delta_t < \Delta_0$, the cause of which is that t_2-orbitals, although now closest to the ligands, do not point directly at the ligands, *i.e.* in an octahedral complex there is a ligand along each axis and in a trahedral complex no ligand lies directly along any axis. For this reason and also because there are only four ligands in the tetrahedral complex, while in an octahedral complex there are six ligands, the tetrahedral orbital splitting, Δ_t is less than Δ_0, for the same metal and ligands and the same internuclear distances. It has also been shown that $\Delta_t = 0.45\ \Delta_0$. Thus the energy level of the t_2 set is raised by $0.4\ \Delta_t = 0.4 \times 0.45\Delta_0 = 0.18\ \Delta_0$ while that of e set is lowered by $0.6\ \Delta_t = 0.6 \times 0.45\ \Delta_0 = 0.27\ \Delta_0$. The relation namely $\Delta_t = 0.45\ \Delta_0$ also shows that, other things being equal, the crystal field splitting in a tetrahedral complex will be about half the magnitude of that in an octahedral complex. Since Δ_t is relatively small orbital splitting, spin-pairing in tetrahedral complexes is rather unlikely. Since $\Delta_t < \Delta_0$, *crystal field splitting favours the formation of octahedral complexes.*

Note : The subscript g is not used for the splitting of *d*-orbitals in tetrahedral complexes because a tetrahedron has no centre of symmetry. The symbol *g* is used for the ligand fields which have centre of symmetry.

Splitting of *d*-orbitals in Tetragonally Distorted Octahedral and Square Planar Complexes.

Before considering the splitting of *d*-orbitals of the central metal cation in these complexes, we should undrstand how tetragonally distorted octahedral and square planar geometries are obtained from regular octahedral geometry. Consider a regular (symmetrical) octahedral complex, $[M(L_b)_4\ (L_a)_2]$ in which L_a are two *trans*-ligands (*i.e.* L_a are the ligands lying along the *z*-axis) and L_b are the basal ligands lying in *xy* plane. In this complex all the six bond distances (four $M–L_b$

and two M–L_a distances) are equal [See Fig. 43.39(*a*)]. Now if two L_a ligands are moved slightly away from the central metal cation, M so that each of the two M–L_a distances becomes slightly greater than each of the four co-planar M–L_b distances, the symmetrical shape of octahedral complex becomes tetragonally distorted octahedral shape [See Fig 43.39 (*b*)]. Now if the two L_a ligands are completely removed away from the *z*-axis, the tetragonally distorted octahedral shape becomes square planar (Fig. 43.39 (*c*)].

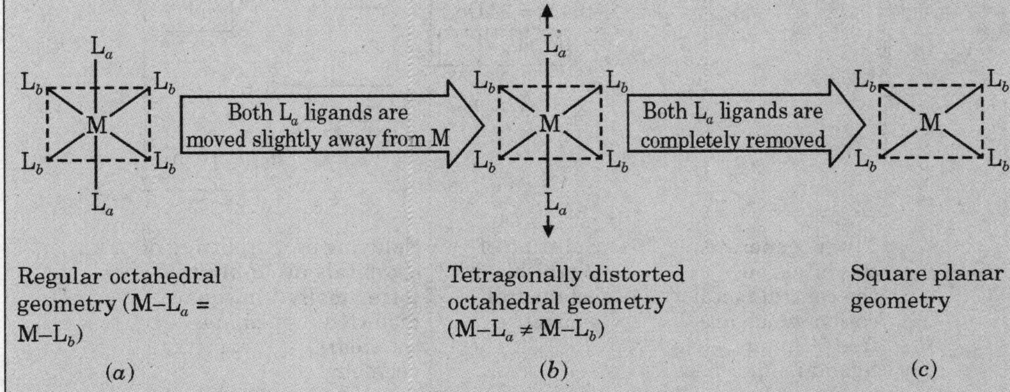

Fig. 43.39. Distortion in regular octahedral geometry leading to the formation of tetragonally distorted octahedral geometry and square planar geometry.

Now in order to consider the splitting of *d*-orbitals in tetragonally distorted octahedral and square planar complexes, we start with the splitting of *d*-orbitals in octahedral complexes. We have already seen that in octahedral complexes, the energy of d_{xy}, d_{yz} and d_{zx} orbitals (t_{2g} orbitals) is decreased while that of d_{z^2} and $d_{x^2-y^2}$ orbitals (e_g orbitals) is increased [Fig. 43.40(*a*)]. Now in tetragonally distorted octahedral complex, since the distance of the *trans*-ligands (L_a ligands) is increased from the central metal ion by removing them away along the *z*-axis, *d*-orbitals along the *z*-axis (*i.e.* d_{z^2} orbital), in *xz* (*i.e.* d_{xz}) and *yz* (*i.e.* d_{yz}) planes experience less repulsion from the ligands than they do in the octahedral complex while the *d*-orbitals in *xy* plane (*i.e.* d_{xy} and $d_{x^2-y^2}$ orbitals) experience more repulsion than they do in the octahedral complex. Consequently the energy of d_{zy}, d_{yz} and d_{zx} orbitals decreases while that of $d_{x^2-y^2}$ and d_{z^2} orbitals increases [Fig.43.40(*b*)]. Note that d_{yz} and d_{zx} orbitals still remain degenerate as they are in the octahedral complex. In square planar geometry the energies of d_{z^2}, d_{yz} and d_{zx} orbitals again fall down while those of $d_{x^2-y^2}$ and d_{xy} orbitals rise up [see Fig. 43.40 (*d*)].

The crystal field splitting energy in square planar complexes is denoted by Δ_{sp} and is given by :

$$\Delta_{sp} = E(d_{x^2-y^2}) - E(d_{xz}, d_{yz}) \qquad \text{[See Fig. 43.40(d)]}$$
$$= \Delta_1 + \Delta_2 + \Delta_3$$

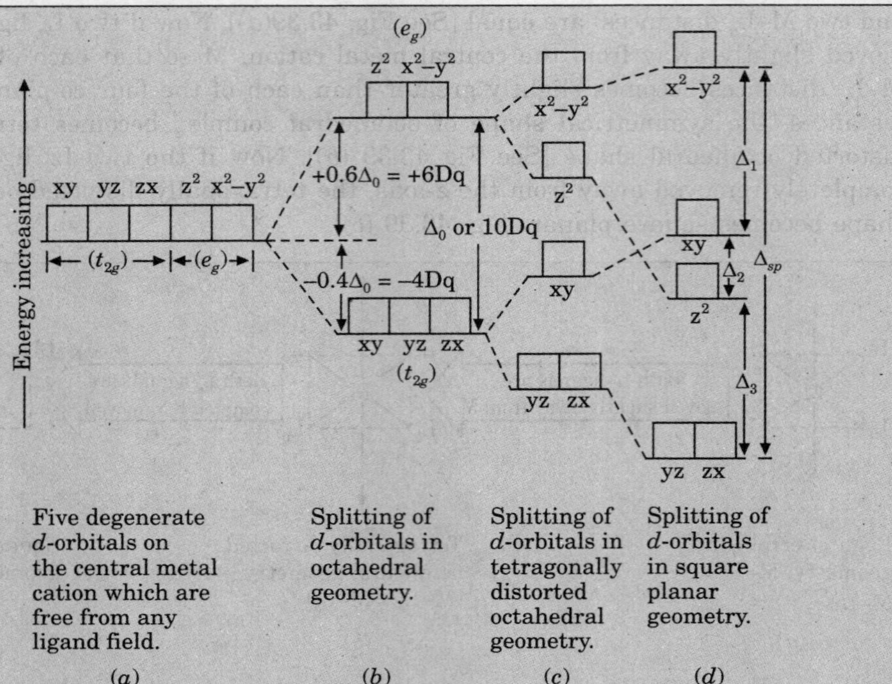

Fig. 43.40. Crystal field splitting of *d*-orbitals in octahedral, tetragonally distorted octahedral and square planar geometries.

The value of Δ_{sp} has been found larger than Δ_0 (Δ_{sp} = 1.3 Δ_0) because of the reason that d_{xz} and d_{yz} orbitals interact with only two ligands in the square planar complexes, while in octahedral complexes the interaction takes place only with four ligands.

Applications of Crystal Field Theory

The following properties of transition metal complexes can well be explained on the basis of CFT.

1. Colour of transition metal complexes.

It has been seen that the solution of transition metal complexes formed by metal cations having no unpaired electrons in *d*-orbitals (*i.e.* having completely filled or vacant *d*-orbitals) are colourless while the solution of transition metal complexes given by transition metal cations having one or more unpaired electrons in *d*-orbitals are coloured. The colour of transition metal complexes having unpaired electrons in *d*-orbitals can well be explained by considering the example of $[Ti(H_2O)_6]^{3+}$ ion which is an octahedral complex and has purple colour. Now when white light is allowed to fall on this complex ion, the ion absorbs radiation from the white light (*i.e.* visible spectrum). The maximum absorption takes place at a wavelength of 5000Å. The absorbed light of this wavelength is green and the transmitted light corresponding to this absorbed light it purple which is the colour of $[Ti(H_2O)_6]^{3+}$ solution.

The cause of colour can also be explained on the basis of CFT as follows : We know that Ti^{3+} ion has the configuration $t_{2g}{}^1\, e_g{}^0$. Now when white light is allowed to fall on $[Ti(H_2O)_6]^{3+}$ ion, the unpaired electron present in t_{2g} set of orbitals moves from lower energy t_{2g} set to higher energy e_g set, since the energy of the radiation absorbed by unpaired t_{2g} electron is the same as the energy difference between t_{2g} and e_g sets (Fig. 43.41). This type of electronic transition from t_{2g} to e_g level is called *d—d* or *ligand—field transition* which is the cause of the colour of $[Ti(H_2O)_6]^{3+}$ ion.

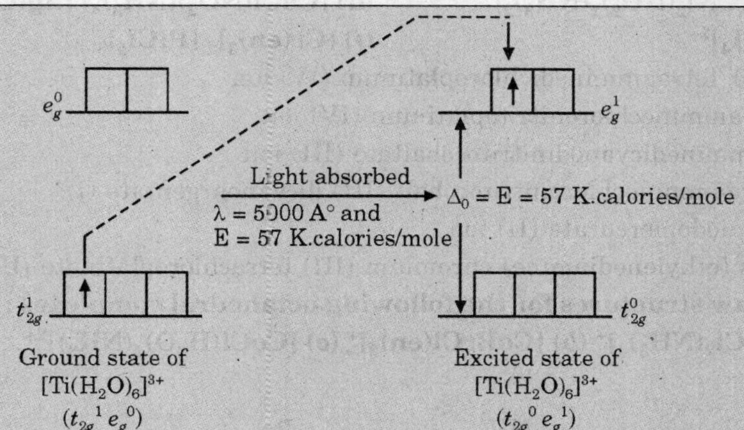

Fig. 43.41. Absorption of light by $[Ti(H_2O)_6]^{3+}$ ion involving a shift of an electron from t_{2g} level to e_g level. This transition $[t_{2g}{}^1\, e_g{}^0 \rightarrow t_{2g}{}^0\, e_g{}^1)$ gives the $[Ti(H_2O)]^{3+}$ ion its purple colour.

2. Magnetic moment values of complex compounds.

CFT can be used to find out the number of unpaired electrons (n) in a given octahedral complex. The substitution of n value in the spin-only formula, $\mu = \sqrt{n(n+2)}$ gives the value of magnetic moment of the given complex compound. From the knowledge of μ-value the valence state of the central metal cation and the nature of bonding in the complex (*i.e.* whether the complex is high spin or low spin) can be known.

Comparison between VBT and CFT.

The points showing the comparison between the two theories are given below.

1. Inner orbital octahedral complexes of VBT are the same as the spin-paired or low-spin octahedral complexes of CFT. Simalarly outer-orbital complexes of VBT are the same as the spin-free or high-spin octahedral complexes of CFT.

2. In the formation of some inner-orbital octahedral complexes of VBT, the promotion of an electron from *d*-orbital to *s*-orbital is required, while in the formation of spin-paired octahedral complexes of CFT no such promotion is required.

3. According to VBT, the metal-ligand bonding in complexes is only covalent, since VBT assumes that ligand electrons are donated to the vacant *d*-orbitals on the central cation. On the other hand, CFT considers the bonding to be entirely electrostatic. Thus CFT does not allow the ligand electrons to enter the metal *d*-orbitals.

Questions with Answers

Q.1 Name the following species :

(*a*) $[PtCl_2(NH_3)_4]^{2+}$ **(*b*) $[PtCl(NO_2)(NH_3)_4]^{2+}$**

(*c*) $[Co(CN)_2(NO_2)_2(NH_3)_2]^-$ **(*d*) $[CoCl(NO_2)(NH_3)_4]\ [Ag(CN)_2]$**

(*e*) $[HgI_4]^{2-}$ **(*f*) $[Cr(en)_3]_2\ [PtCl_4]_3$**

Ans. (*a*) Tetraamminedichloroplatinum (IV) ion

(*b*) Tetramminechloronitritoplatinum (IV) ion

(*c*) Diamminedicyanodinitritocobaltate (III) ion

(*d*) Tetraamminechloronitritocobalt (III) dicyanoargentate (I)

(e) Tetraiodomercurate (II) ion

(*f*) Tris (ethylenediamine) chromium (III) tetrachloroplatinate (II)

Q.2 Draw structures for the following octahedral complexes :

(*a*) $[CoCl_2(NH_3)_4]^+$ (*b*) $[CoBrCl(en)_2]^+$ (*c*) $[CoCl(H_2O)_4(NH_3)]^{2+}$

Ans.

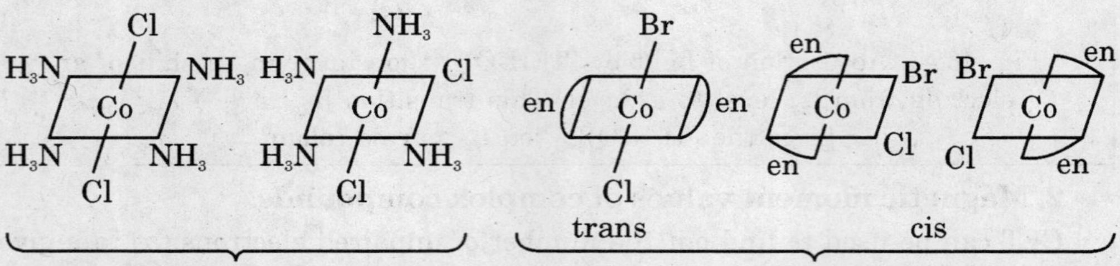

(*a*) $[Co^{3+}Cl_2(NH_3)_4]^+$ (*b*) $[Co^{3+}BrCl(en)_2]^+$ (optical isomers)

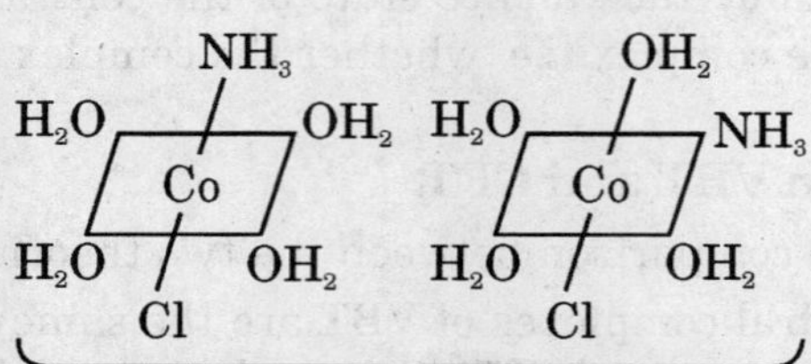

(*c*) $[Co^{3+}Cl(H_2O)_4(NH_3)]^{2+}$

Q.3 How do you expect $(C_6H_5)_3P$, $(C_6H_5)_3PO$ and $(C_6H_5)_2SO$ to behave as coordinating ligands ?

Ans. The structures of the given molecules are given below :

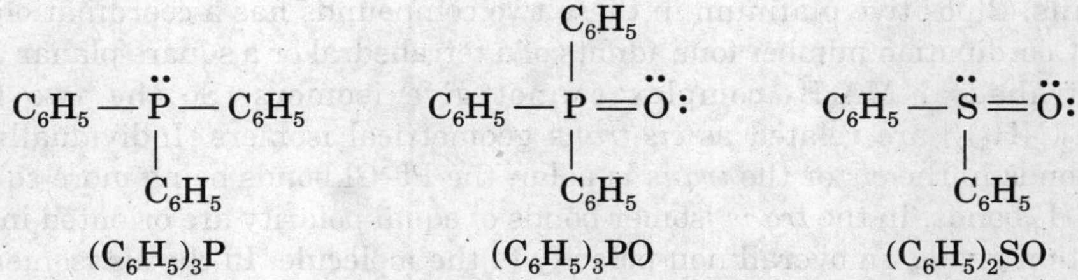

In the structure of $(C_6H_5)_3P$, since P-atom has a *lp* of electrons, this molecule can act as a monodentate ligand. P-atom is the donor atom.

In the structure of $(C_6H_5)_3PO$, since O-atom has two *lps* of electrons, this molecule can act as a ligand, coordinating through O-atom. Now since two *lps* reside on the same atom (*i.e.* O-atom), this molecule can act only as a monodentate ligand and not as a bidentate ligand.

In the structure of $(C_6H_5)_2SO$, since S and O both have *lps* of electrons, this molecule can act as a ligand. Now since *lps* reside on different atoms, this molecule can act as an ambidentate ligand, coordinating through S or through O-atom. This molecule coordinates with Pt(II), Pd(II) through S-atom and with Co(II), Ni(II), Zn(II)etc. through O-atom.

Q.4 Molecular N_2 is a weaker sigma donor as well as weaker pi acceptor than CO molecule, although both molecules are isoelectronic and have the same bonding patter. Explain.

Ans. It is true that CO and N_2 are isoelectronic and have the same bonding pattern. Both have a triple bond and a lone pair on each of the two atoms. Oxygen has more protons in the nucleus than carbon and hence oxygen orbitals are more stable than those of carbon. Therefore the bonding orbitals of CO have more oxygen character while the anti-bonding orbitals have more carbon character. Bonding orbitals of N_2 are little more stable than those of CO while the antibonding orbitals are less stable than those of CO. (Remember that in N_2 the two atomic orbitals are the same). Thus molecular nitrogen is a weaker sigma donor as well as a weaker pi acceptor than carbon monoxide.

Q.5 (*a*) How would you distinguish between *cis* and *trans* isomer of $[CrCl_2(NH_3)_4]Cl$?

(*b*) How would you decide whether MA_2B_2 complex is square planar or tetrahedral?

Ans. (*a*) The *cis* isomer has the two coordinated chloro groups as near neighbours. Hence it will react readilly with oxalate ion ($C_2O_4^{2-}$) giving tetrammine (oxalato) chromium (III) ion, $[Cr^{3+}(NH_3)_4C_2O_4]^+$ .The *trans* isomer will not react as readily as the *cis* isomer. A bidenate oxalato group can span only two adjacent positions as in the *cis* but not in the *trans*.

(*b*) A square planar MA_2B_2 complex is capable of giving *cis* and *trans* isomers while a tetrahedral MA_2B_2 complex cannot do so.

Q.6 $[PtCl_2(NH_3)_2]$ has been isolated in two isomer. One isomer has a good dipole moment while the other has nearly zero dipole moment. Elucidate their structures.

Ans. Bipositive platinum in these two compounds has a coordination number four. Coordination number four admits of a tetrahedral or a square planar geometry. A tetrahedral MA_2B_2 complex cannot give isomers. So the two forms of $[PtCl_2(NH_3)_2]$ are related as *cis-trans* geometrical isomers. Individually each of the bonds in the *cis* or the *trans* is polar, the Pt–Cl bonds being more so than the Pt–NH_3 bonds. In the *trans* isomer bonds of equal polarity are oriented in opposite direction giving an overall non-polarity to the molecule. In the *cis* isomer the two chlorine ends will have more negative charge than the ammonia end. Effectively then there will be separation of charge : the positive centre being on the platinum and the negative centre being in between the two chlorine atoms. Therefore that isomer which has nearly zero dipole moment is the *trans* while the other with a good value of dipole moment is the *cis* variety.

Q.7 10 *Dq* of $[Mn(H_2O)_6]^{3+}$ is known from electronic spectrum to be 21,000 cm^{-1}. The pairing enrgy of Mn(III) is 28,800 cm^{-1}. Predict whether the complex is high-spin or low-spin.

Ans. $[Mn(H_2O)_6]^{3+}$ has Mn^{3+} ion which is a d^4 ion.

CFSE for d^4 ion in H.S. state ($t_{2g}{}^3\, e_g{}^1$ having m = 0),

$$(CFSE)_{H.S.} = -6Dq + mP = -6 \times \frac{21000}{10} + 0 \times P$$
$$= -12,600\ cm^{-1}$$

CFSE for d^4 ion in L.S. state ($t_{2g}{}^4 e_g{}^\circ$ having m = 1),

$$(CFSE)_{L.S.} = -16Dq + mP = -16 \times \frac{21000}{10} + 1 \times 28,800$$
$$= -33600 + 28800$$
$$= -4800\ Cm^{-1}$$

Now since $(CFSE)_{H.S.}$ is more negative than $(CFSE)_{L.S.}$, the given complex ion is more stable in H.S. state.

Alternatively, since 10*Dq* (= 21,000 cm^{-1}) < P (= 28,000 cm^{-1}), the given complex is a high spin complex.

Q.8 10*Dq* values of $[Ru(H_2O)_6]^{2+}$ and $[RuCl_6]^{3-}$ are nearly the same but according to spectrochemical series, the value for 10*Dq* for the chloride complex should be lower than H_2O complex. How do you explain the similarity of the 10*Dq* values of the two complexes ?

Ans : $[Ru(H_2O)_6]^{2+}$ ion has Ru in +2 O.S. while $[Ru(H_2O)_6]^{3-}$ ion has Ru in +3 O.S. It is true that chloride ion is a far weaker ligand than H_2O. In one of the two complexes we have ruthenium (II) associated with H_2O and in the other ruthenium (III) is associated with the weaker ligand, chloride. 10*Dq* depends on several factors of which a very important one is the oxidation state. It is empirical fact that with an increase in the oxidation state by one unit (from +2 to +3), 10*Dq* increases by about 100%. This explains why even a weaker lignad (chloride) produces with ruthenium(III) nearly the same 10*Dq* as does hexaaquaruthenium(II).

Q.9 Chromium(III) under forcing conditions forms hexachloro complex ion, $[CrCl_6]^{3-}$ but the corresponding complexes are never found in Mn(III) and Fe(III). Why ?

Ans. Cr(III), Mn(III) and Fe(III) are d^3, d^4 and d^5 systems respectively. Since chloride ion is a weak field ligand, all the complexes, if found at all, must be high-spin. The *CFSE's* are $-12Dq$, $-6Dq$ and $0Dq$ (in respective *Dq's*) respectively. These values show that since Cr(III) (d^3 ion) has the most negative value of CFSE, Cr(III) has the maximum possibilty of forming hexachloro complex ion, $[CrCl_6]^{3-}$.

Q.10 Arrange the following species in the order of increasing Jhan Teller distortion in an octahedral enviroment : high-spin Co(III), low-spin Co(II), copper(II) and Ti(III).

Ans. Jhan Teller effect becomes more severe when the e_g level is unsymmetrically occupied than when the t_{2g} level is unsymmetrically occupied. $t_{2g}{}^p\ e_g{}^2$ configurations of the given ions are given below :

H.S. Co(III) = $3d^6 = t_{2g}{}^4\ e_g{}^2$,

L.S. Co(II) = $3d^7 = t_{2g}{}^6\ e_g{}^1$

Cu(II) = $3d^9 = t_{2g}{}^6\ e_g{}^3$,

Ti(III) = $3d^1 = t_{2g}{}^1\ e_g{}^\circ$

These configurations show that Jhan Teller distortion of the given ions is in the order :

Ti(III) $\simeq$ H.S. Co(III) < Cu(II) $\simeq$ L.S. Co(II)

Q.11 $(NH_4)_2\ [VF_6]$ shows one electronic transition around 20.1 *KK* while $[VO(acac)_2]$ in chloroform gives three bands around 14.8, 16.8 and 25.0 *KK*. Explain.

Ans. Both the complexes have $3d^1$ vanadium(IV). $(NH_4)_2\ [VF_6]$ may be viewed as octahedral with some slight Jhan-Teller distrotion while $[VO(acac)_2]$ is under heavy tetragonal compression. The vanadly oxygen bond is much too short compared to the other equatorial vanadium-oxygen bonds. Hence the spiliting of the d orbitals is unlike what obtains in octahedral complexes. So in the former compound one transition ($t_{2g} \longrightarrow e_g$) is expected while in the latter as many as three ($d_{xy} \rightarrow d_{xz}$, d_{yz}; $d_{xy} \rightarrow d_{x2-y2}$; $d_{xy} \rightarrow d_z{}^2$) are expected. The three bands around 14.8, 16.8 and 25.0 *KK* are due to these electronic transitions.

Q.12 Magnetic moment measurement of potassium hexa-cyanomanganate(III), $K_3[Mn^{3-}(CN)_6]$ indicates the presence of two unpaired electrons. Predict whether the complex compound is outer orbital octahedral or inner orbital octahedral.

Ans. Mn^{3+} ion is a $3d^4$ ion. The formation of outer orbital octahedral complex ion, $[Mn^{3+}(CN)_6]^{3-}$ and that of inner orbital octahedral complex ion, $[Mn^{3+}(CN)_6]^{3-}$ has been at (*a*) and (*b*) respectively of Fig. 43.42. It may be seen from the figure that $[Mn(CN)_6]^{3-}$ ion is an inner orbital octahedral complex ion.

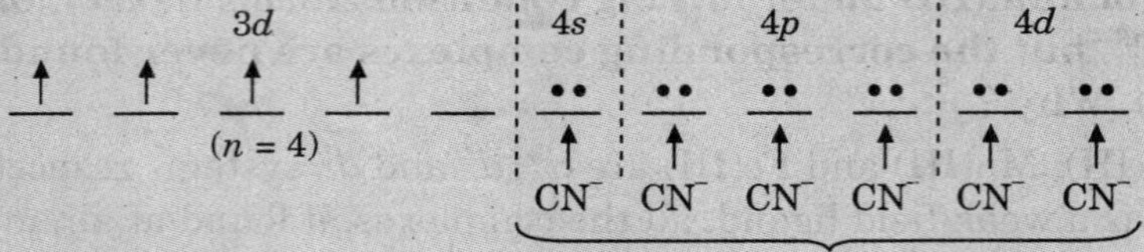

(a) sp^3d^2 hybridisation : Outer orbital octahedral complex ion, $[Mn(CN)_6]^{3-}$

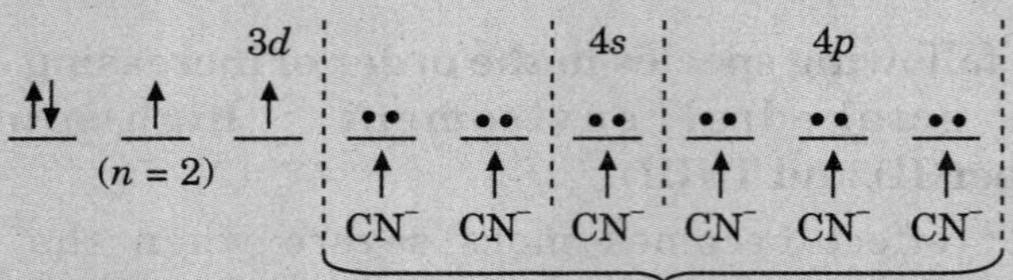

(b) d^2sp^3 hybridisation : Inner orbital octahedral complex ion, $[Mn(CN)_6]^{3-}$

Fig.43.42. Formation of (*a*) outer orbital octahedral complex ion, $[Mn(CN)_6]^{3-}$ and (*b*) inner orbitals octahedral complex ion, $[Mn(CN)_6]^{3-}$.

Q.13 $Fe(CO)_5$ is a diamagnetic complex. Predict the geometry of the complex.

Ans. In $Fe(CO)_5$, Fe is present in zero oxidation state. Hence Fe is present as Fe° whose V.S.E.C. is $3d^64s^2$ or $3d^8$. Since C.N. of Fe is 5 in $Fe(CO)_5$, $Fe(CO)_5$ can have trigonal bipyramidal geometry (dsp^3 hybridisation of Fe atom) or square pyramidal geometry (d^2sp^2 hybridisations of Fe atom). Both these hybridisations shown in Fig.43.43. indicate that $Fe(CO)_5$ molecule has $n = 0$ and hence is diamagnetic. For d^2sp^2 hybridisation two $3d$ electrons have to be promoted to higher energy $4p$ orbitals. This process is unfavourable since the promotion of electrons to higher energy orbitals will make these electrons valnerable to oxidation. Hence trigonal bipyramidal geometry is preferred.

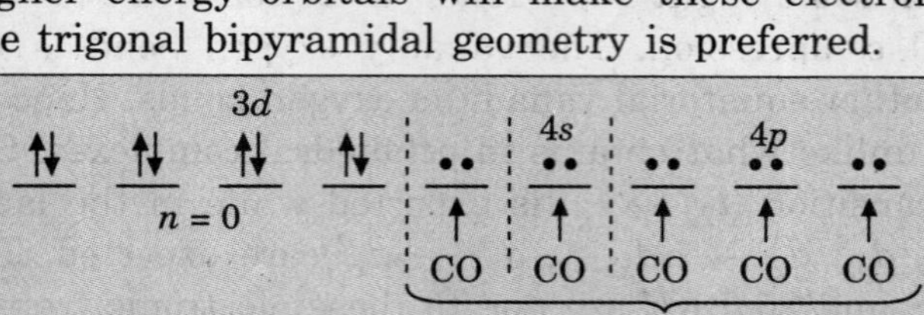

dsp^3 hybridisation : Trigonal bipyramidal geometry of $Fe(CO)_5$

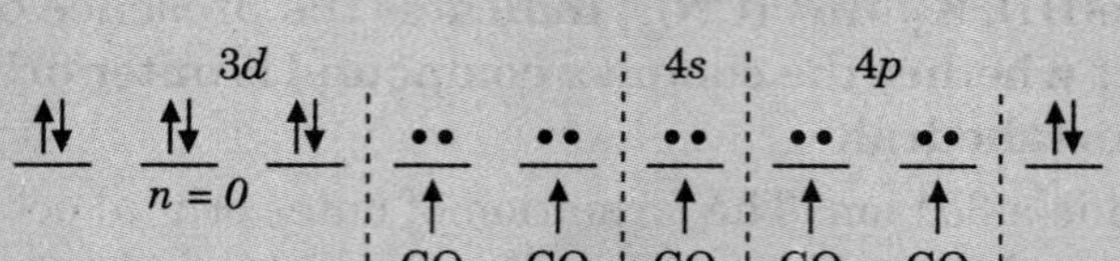

d^2sp^2 hybridisation : Square pyramidal geometry of $Fe(CO)_5$

Fig. 43.43. dsp^3 and d^2sp^2 hybridisation of Fe atom in $Fe(CO)_5$.

Q.14 $[NiCl_4]^{2-}$ is paramagnetic tetrahedral while $[PdCl_4]^{2-}$ and $[PtCl_4]^{2-}$ are diamagnetic square palnar. Explain.

Ans. The given complex ions are the complex ions of Ni(II), Pd(II) and Pt(II). As we move from Ni(II) ($3d^8$) to Pd(II) ($4d^8$) to Pt(II) ($5d^8$), the distance between d-orbitals increases as Ni(II) < Pd(II) < Pt(II) but the parring energy decreases as Ni(II) > Pd(II) > Pt(II). The increasing trend of $10Dq$ and decreasing trend of pairing energy both favour L.S. complexes as we move as Ni(II) → Pd(II) → Pt(II).

Q.15 Why $K_3[CuF_6]$ is paramagnetic while $K[AgF_4]$ is diamagnetic ?

Ans. In $K_3[CuF_6]$ and $K[AgF_4]$ both the metal ions are trivalent ($3d^8$ and $4d^8$ respectively). In general Δ increases as we move down from the first to the second and then to the third transition series. Also square palnar geometry has a higher Δ than an octahedral one having the same ligand.

Higher value of Δ in Ag(III) favours a square planar diamagnetic $K[AgF_4]$ compared to an octahedral paramagnetic $K_3[CuF_6]$.

$K_3[CuF_6]$: $t_{2g}{}^6\ e_g{}^2$: 2 unpaired electrons.

$K[AgF_4]$: $d^2{}_{z2}\ d^2{}_{xz}\ d^2{}_{yz}\ d^2{}_{xy}\ d^0{}_{x2-y2}$: no unpaired electron.

Q.16 The divalent metal ions in the oxides MO are octahedraly surrounded by weak field oxide ions in the crystal lattice. Explain the trend seen in the following lattice enthalpies :

CaO	TiO	VO	MnO	
3460	3878	3913	3810	KJ/mole

Ans. Calcium(II), titanium(II), vanadium(II) and managanese(II) are d^0, d^2, d^3 and d^5 ions respectively. Calcium (d^0) and high-spin managnese(II) (d^5) have $CFSE = 0$. $CFSE$ of titanium(II) and vanadium(II) are $-8Dq$ and $-12Dq$. So the MO oxides of Ti(II) and V(II) will be stabilished compared to Ca(II). MnO will be having no extra stablisation due to the crystal field of the oxide ions and lattice enthalpy will drop in MnO.

University Questions

1. Explain why $K_3[Fe(CN)_6]$ is considered to be a complex compound but $K_2SO_4 . Al_2(SO_4)_3.24H_2O$ is not? *(Delhi Hons. 82)*
2. Define with examples, the following terms : (*i*) Ligand (*ii*) A pentadentate ligand. (*iii*) Coordination number. *(Allahabad 86; Delhi 89; Raj. 84)*
3. Explain, giving examples, the following terms : (*i*) Chelation (*ii*) A chelate (*iii*) Chelate effect in coordination chemistry. *(Allahabad 88)*
4. Write a note on "stepwise stability constants". *(Allahabad 88)*
5. (*a*) Write IUPAC name of the following : (*i*) $[Pt\ Cl\ (NO_2)\ (NH_3)_4]\ SO_4$.

(ii) $\left[(en)_2 Co \begin{matrix} NH_2 \\ NH_2 \end{matrix} Co\,(en)_2\,Br_4\right]$. (iii) $NH_4[Cr(NCS)_2(NH_3)_3]$ Also calculate the EAN of the central metal atom (Cr = 24, Co = 27, Pt = 78).

(b) Write the structural formula for dichloro*bis* (ethyleneamine) platinum (IV) tetrachloroplatinate (II). *(Allahabad 99)*

6. (a) Write the formulae of the following compounds : (i) Hexa aquiron (II) sulphate (ii) Potassium tetracyanonickelate (III) (iii) Trans-Chloronitro-diammine platinum (II).

(b) Write the IUPAC names of $Fe(CO)_5$ and $[(CO)_3\,Co\,(CO)_2Co\,(CO)_3]$ *(Himachal Pradesh 2000)*

7. What are various types of isomerism exhibited by coordination compounds? Write in detail giving examples. *(Lucknow 2001)*

8. (i) $[Co\,F_6]^{3-}$ is paramagnetic while $[CO(NH_3)_6]^{3+}$ is diamagnetic.

(ii) $[Fe\,(CN)_6]^{3-}$ is less paramagnetic than $[Fe\,(H_2O)_6]^{3+}$ *(Avadh 2000)*

9. (i) Explain why $[FeF_6]^{3-}$ and $[Fe(CN)_6]^{3-}$ ions are paramagnetic but their magnetic moments are quite different.

(ii) Write a note on : High spin and low spin complexes.

(iii) What are inner and outer complexes? Give two examples of each.

(iv) Write a note on "Crystal field theory" *(Lucknow 2001)*

10. (a) What is CFSE ? Explain.

(b) Discuss in brief the limitations of VBT.

(c) Give an elementry idea of CFT and explain the spliting of d-orbitals in octahedral and tetrahedral complexes. *(Agra 2008)*

11. Which of the following has larger value of CFSE ? $[Co(NH_3)_6]^{2+}$ and $[Co(NH_3)_6]^{3+}$. *(H.N. Bahugura 2008)*

12. $[FeF_6]^{3-}$ in has less magnetic moment than $[Fe(CN)_6]^{3-}$, though both are octahedral complex ions. Explain. *(C.S.J.M. Kanpur 2009)*

13. (a) Discuss VBT of complex compounds. Explain different geometric and magnetic character of complex compounds on the basis of this theory with suitable examples.

(b) Write oxidation number and coordination number of Cr-atom in $[Cr)NH_3)_6]Cl_3$.

14. (c) Explain :

(i) Zn^{2+} salts are white but Cu^{2+} salts the blue

(ii) Cu+ is diamagnetic but Cu^{2+} is paramagnetic *(Meerut 2009)*

15. (a) What are the important features of CFT. Discuss its applications and limitations.

(b) Write a note on "Jahn – Teller Effect".

(c) Write a note on "Spectrochemical Series". *(Meerut 2009)*

CHAPTER 44

Characteristics of d- and f- Block Elements (Transition and Inner-transition Elements)

d-block Elements : Transition Elements

What are d-Block Elements ?

The definition of *d*-block elements has already been given in chapter 3.

Why are d-Block Elements Called Transition Elements ?

The *d*-block elements are called transition elements because they exhibit transitional behaviour between highly reactive ionic-compound forming *s*-block elements (electropositive elements) on one side and mainly covalent-compound forming *p*-block elements (electronegative elements) on the other side.

Position of d-Block Elements in the Periodic Table.

d-block elements are located in between *s*-(ionic-compound forming elements) and *p*-(covalent-compound forming elements) block elements as shown in Fig. 44.1. Thus these elements are located in the middle of the periodic table and occur in fourth ($n = 4$) and subsequent periods of the periodic table. It is because of their inclusion in between *s*- and *p*-block elements that the properties of *d*-block elements are intermediate between those of *s*- and *p*-block elements. Thus the elements of groups 3 (IIIB), 4 (IVB), 5 (VB), 6 (VIB), 7 (VIIB), 8, 9, 10 (VIII), 11 (IB) and 12 (IIB) belong to this block.

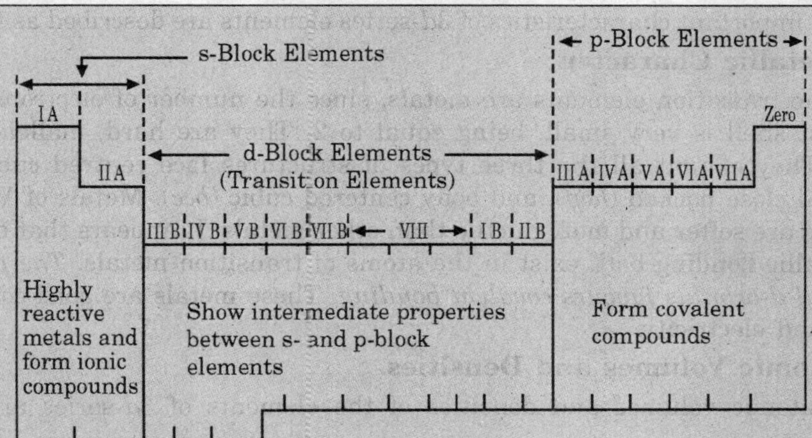

Fig. 44.1. Position of *d*-block elements in the periodic table.

Electronic Configurations of d-Block Elements.

Complete and valence-shell configurations of *d*-block elements namely (*a*) Sc (Z = 21) to Zn (Z = 30); (*b*) Y (Z = 39) to Cd (Z = 48); (*c*) La(Z = 57), Hf (Z = 72) to Hg (Z = 80); and (*d*) Ac (Z = 89), Ku(Z = 104), Ha (Z = 105) have already been given in chapter 3. From these configurations these points may be noted (*i*) In atoms of *d*-block elements two outermost shells [*i.e. nth* and (*n* – *1*)*th* shells] are partially filled while the remaining inner-shells [*i.e. 1st, 2nd*......(*n* – *2*)*th* shells] are completely filled. (*ii*) With the exception of Cr, Cu, Nb, Mo, Ru, Rh, Pd, Ag, Pt and Au, in the atoms of these elements the *ns*-orbital is completely filled (ns^2 configuration) and the differentiating electron enters (*n* – 1)d orbital Thus the valence- shell configuration of most of these elements varies from $(n-1)d^1.\ ns^2$ (Group IIIB) to $(n-1)d^{10}.ns^2$ (Group IIB) configuration. The elements of the groups IIIB, IVB, VB, VIB, VIIB, VIII, IB and IIB belong this block as is evident from their configurations. (*iii*) It may be noted from the valence-shell configurations of the atoms of these elements that in the elements like Cr, Cu, Nb, Mo, Ru, Rh, Pd, Ag, Pt and Au one or both the electrons from *ns* orbital get shifted to (*n*–1)*d* orbital and hence such elements have one or no electron in *ns* orbital (instead of two electrons in *ns* orbital). This type of shifting makes the (*n*–l) *d* orbitals half-filled or completely filled and such (*n*–1) *d* orbitals are more stable than other (*n*–1) *d* orbitals. Thus the shifting of one or both the electrons from *ns* orbital to (*n*–1)*d* orbital gives more stable configuration to the atom.

Classification of d-Block Elements.

Based on whether the differentiating electron goes to *3d, 4d, 5d* or *6d* orbitals, *d*-block elements have been classified into *four series viz. 3d, 4d, 5d* and *6d* series as discussed in chapter 3.

General Characteristics of d-Block Elements

Some important characteristics of 3*d*-series elements are described as follows :

1. Metallic Character.

All the transition elements are metals, since the number of electrons in the outermost shell is very small, being equal to 2. They are hard, malleable and ductile. They exhibit all the three types of structures face centred cubic (*fcc*), hexagonal close packed (*hcp*), and body centered cubic (*bcc*). Metals of VIII and IB groups are softer and more ductile than other metals. It appears that covalent and metallic bonding both exist in the atoms of transition metals. *The presence of unfilled d-orbitals favours covalent bonding.* These metals are good conductor of heat and electricity.

2. Atomic Volumes and Densities.

The atomic volumes and densities of the elements of 3*d*-series are given below :

Elements :	Sc	Ti	V	Cr	Mn	Fe	Co	Ni	Cu	Zn
Atomic volume : *(cm³)*	15.0	1.66	8.3	7.2	7.3	7.1	6.7	6.6	7.1	9.2
Density (g/cm³) :	3.0	4.5	6.1	7.2	7.4	7.8	8.9	8.9	8.9	7.1

In d-block elements $(n-1)d$ sub-shell is filled and the increased nuclear charge pulls the electrons in. This leads to the comparatively low atomic volumes of these elements compared to those of the neighbouring s- and p-block elements. The densities of d-block elements are very high in comparison to those of group IA and IIA elements. This is understandable because of their decreasing volumes.

3. Melting and Boiling Points.

The melting and boiling points of the elements of 1*st* transition series are given below :

Elements :	Sc	Ti	V	Cr	Mn	Fe	Co	Ni	Cu	Zn
Boiling point (°C) :	2730	3260	3450	2665	2150	3000	2900	2730	2595	906
Melting point (°C) :	1539	1668	1900	1876	1245	1536	1495	1453	1083	419.5

It is evident from the values that the d-block elements have high melting and boiling points. It shows that they are held by strong forces. It can be seen from the table that the melting and boiling points have highest values in the middle of the series. In this region the elements have maximum number of unpaired d-electrons available for bonding. The elements also exhibit the highest oxidation states in this region. The trend in melting and boiling points shown by $3d$-series elements is also followed by $4d$ and $5d$-series elements. However Zn, Cd and Hg have relatively low values (melting and boiling points for these metals are : Zn = 419.5°C, 906°C; Cd = 320.9°C, 765°C and Hg = –38.4°C, 357°C respectively). The reason of these low values is that these metals have completely filled d-orbitals with no unpaired electron (Zn = $3d^{10}.4s^2$ with $n = 0$; Cd = $4d^{10}.4s^2$ with $n = 0$ and Hg = $5d^{10}.6s^2$ with $n = 0$) that may be available for *covalent bonding* amongst the atoms of these metals and is responsible for high melting and boiling points.

Generally the melting and boiling points of d-block elements increase on descending a group. Melting and boiling points of Ag are lower than those of Cu and Au. This can be explained on the basis that both Cu and Au have greater tendency to acquire higher oxidation state than +1, suggesting that in these metals the involvement of d-orbitals for metallic bonding is to a greater extent than in Ag.

4. Atomic (Covalent) Radii.

The atomic (covalent) radii of d-block elements are given below :

Sc	Ti	V	Cr	Mn	Fe	Co	Ni	Cu	Zn
1.44 A°	1.32	1.22	1.18	1.17	1.17	1.16	1.15	1.17	1.25
Y	**Zr**	**Nb**	**Mo**	**Tc**	**Ru**	**Rh**	**Pd**	**Ag**	**Cd**
1.62 A°	1.45	1.34	1.30	1.27	1.25	1.25	1.28	1.34	1.48
La	**Hf**	**Ta**	**W**	**Re**	**Os**	**Ir**	**Pt**	**Au**	**Hg**
1.68 A°	1.44	1.34	1.30	1.28	1.26	1.27	1.30	1.34	1.49

These points may be noted : (*i*) As is evident, these values generally decrease on moving from left to right in the series (*i.e.* period). This is due to the fact that as the nuclear charge increases with increasing atomic number across any series, the attraction of the electrons towards the nucleus increases and this increase in attraction leads to a decrease in the radii values across each period. There are, however, a few exceptions. The atomic radii of the elements form Cr to Cu are very close to one another. This is explained as follows. As we move across the period, there is an increase in nuclear charge which tends to pull the 4*s*-electron inward near the nucleus, *i.e. it tends to reduce the size.* At the same time the addition of extra electrons to the 3*d*-orbitals also provides a screening effect which shields the 4*s*-electrons from the inward pull of the positive nucleus on these electrons. This screening effect increases as the number of extra electrons being added to 3*d*-orbitals increases. This effect *opposes the increase in size.* Thus there are operating two effects namely *screening effect* and the *nuclear charge effect* which *oppose each other*, resulting in that the effect of increase in the screening effect *almost* compensates the corresponding effect in increase in nuclear charge and hence very slight variations in the atomic radii from Cr to Cu are observed. (*ii*) In a particular series the atomic radius attains a minimum value for the elements of VIII group and increases again towards the end of the series. This is explained in terms of the increased force of repulsion among the added electrons which exceeds the attractive force due to the increased nuclear charge and results in the expansion of the electron cloud. (*iii*) On descending a group *n* increases. The atomic size is, therefore, expected to increase from top to bottom. But this increase is not the same for all the members. The difference in radius of the second and third series elements is very small (~ 0.02 A°) as compared to the first and second series members. This is due to lanthanide contraction. Due to the inclusion of 14 lanthanides between La (Z = 57, Group IIIB) and Hf (Z = 72, Group IV B), Hf is not as large as it would have been, if it had followed directly after La. The nuclear charge in the second series members are 18 units greater than the first series members, but this difference between the 2*nd* and 3*rd* series elements is 32. Thus, the increase in *n* [from 2*nd* series ($n = 5$) to 3*rd* series ($n = 6$)], though tends to increase the size, the excessive increase in nuclear charge annuls this effect. Consequently *the elements of 2nd and 3rd series are very similar in size.*

5. Ionic Radii.

The ionic radii follow the same trend as the atomic radii. It may be seen that the radii of the ions having the same charge (valency) decrease with the increase in atomic number.

Sc^{2+}	Ti^{2+}	V^{2+}	Cr^{2+}	Mn^{2+}	Fe^{2+}	Co^{2+}	Ni^{2+}	Cu^{2+}	Zn^{2+}
0.95 A°	0.90	0.88	0.74	0.80	0.76	0.74	0.72	0.69	0.74
Sc^{3+}	Ti^{3+}	V^{3+}	Cr^{3+}	Mn^{3+}	Fe^{3+}	Co^{3+}	Ni^{3+}	—	—
0.81 A°	0.76	0.74	0.69	0.66	0.64	0.63	0.62	—	—

——— *Decreasing* ———→

6. Ionisation Potentials.

The ionisation potential values of most of the *d*-block elements lie in between those of *s*- and *p*-block elements. This indicates that the *d*-block elements are *less electropositive than s*-block elements *and more so than p-block elements.* Thus *d*-block elements do not form ionic compounds so readily as the alkali and alkaline earth metals do. Unlike *s*-block elements, *d*-block elements form covalent compounds as well. *1st*, *2nd* and *3rd* ionisation potentials (in KJ/ mole) of the elements of 3*d* series are given below :

	Sc	Ti	V	Cr	Mn	Fe	Co	Ni	Cu	Zn
1st	631	658	650	652.8	717.4	759.4	758	736.7	745.5	906.4
2nd	1235	1310	1414	1635	1509.1	1561	1646	1753	1957.9	1733.3
3rd	2389	2852	2828	2987	2248.4	2957.4	3232	3393	3554	3832.7

From these values the following points may be noted : (*i*) The ionisation potentials of 3*d* series elements increase as we move across the series from left to right, although the increase is not quite regular, *e.g.*, the values for Sc, Ti, V and Cr differ very slightly. Similarly the values for Fe, Co, Ni and Cu are fairly close to one another. The value for Zn is appreciably higher due to the additional stability associated with completely-filled 3*d*-level in Zn (Zn $\longrightarrow$ $3d^{10}\ 4s^2$). Slight variations that occur in ionisation potentials across the series are mainly due to the slight changes in atomic radii which are on account of the fact that the screening effect caused by the addition of extra electrons to 3*d*-level almost compensates the effect in increase in nuclear charge, as we move from left to right in the series. (*ii*) As can be seen, second ionization energies for Cr and Cu are sufficiently higher than those of the neighouring elements on the left as well as on the right. This is because the outer electronic configuration of Cr is $3d^5$ (half-filled) and that of Cu is $3d^{10}$ (completely-filled). Thus, the second ionization of chromium means the removal of one electron from the *highly stable exactly half-filled* sub-shell. Similarly, the second ionization of copper involves the removal of one electron from the *highly stable completely-filled 3d* sub-shell. This explains why the second ionization energies of Cr and Cu are higher than those of the neighbouring elements.

7. Different Oxidation States : Variable Valency.

The term oxidation state is preferred to valency. Most of the transition metals show several oxidation states or variable valency in their compounds. Oxidation states shown by the elements of 3*d*-series are : Sc (Z = 21) → (+2), +3; Ti (Z = 22) = +2, +3, +4; V (Z = 23) → +2, +3, +4, +5; Cr (Z = 24) → +2, +3, (+4), +6; Mn (Z = 25) → +2, +3, +4, (+6), +7; Fe (Z = 26) → +2, +3, (+4), (+6); Co (Z = 27) → +2, +3, (+4); Ni (Z = 28) → +2, (+3); Cu(Z = 29) → +1, +2; Zn (Z = 30) → +2. Oxidation states shown in brackets are less common oxidation states, *i.e.* these oxidation states are exhibited in a few compounds.

The cause of showing different oxidation states is that these elements have several (*n*–1)*d* and *ns* electrons and that the energies of (*n*–1)*d* and *ns* orbitals are fairly close to each other. Hence (*n*–1)*d* electrons are as easily lost as *ns* electrons. The following points may be noted : (*i*) Among the elements of 3*d* series, Mn shows the maximum number of oxidation states because it is at Mn that the number of unpaired electrons in $3d$ orbitals is maximum, On proceeding towards right or left from Mn, the number of unpaired electrons in $3d$ orbitals decreases and hence the number of oxidation states shown by the elements lying at the right or left of Mn decreases. (*ii*) With the exception of Cr ($3d^5.4s^1$), the minimum oxidation state shown by the elements of $3d$-series is equal to the number of electrons in $4s$ orbital. (*iii*) Excepting the elements of group IB, other elements show maximum oxidation state which is never more than their group number. (*iv*) In going down a group, the higher oxidation states become more stable. For example +2, and +3 oxidation states shown by Fe are stable oxidation states while +4 and +6 are unstable ones. Similarly +2, +3, +4, and +6 are stable while +5, +7 and +8 are unstable oxidation states shown by Ru. (*v*) The relative stability of various oxidation states of a given element can be explained on the basis of the stability of d^0, d^5 and d^{10} configurations. For example : (*a*) Ti^{4+} ($3d^{10}.4s^0$) is more stable than Ti^{3+} ($3d^1.4s^0$), because of the presence of $3d^0$ orbital in Ti^{4+} ion. (*b*) Mn^{2+} ($3d^5.4s^0$) is more stable than Mn^{3+} ($3d^4.4s^0$), since Mn^{2+} ion has $3d^5$ orbital. (*c*) Ag^+ ($4d^{10}.5s^0$) is more stable than Ag^{2+} ($4d^9.5s^0$) because of the presence of $4d^{10}$ orbital in Ag^+ ion. (*vi*) Transition elements can form ionic bonds in their lower oxidation states and the ionic character of the bond decreases (or covalent character of the bond increases) with the increase of oxidation state. With the decrease of ionic character, the acidic character of oxides or chlorides increases. For example :

(*a*) With the decrease of the ionic character of M–O bond in the oxides, the acidic character of the oxides increases as shown below :

⟵ *Oxidation states* ⟶

+2	+3	+4	+6
TiO	Ti_2O_3	TiO_2	—
VO	V_2O_3	—	—
CrO	Cr_2O_3	—	CrO_3
MnO	MnO_3	MnO_2	MnO_3 (unstable)
Basic	*Amphoteric*		*Acidic*

— *Oxidation state increasing* ⟶

— *Ionic character of M—O bond decreasing* ⟶

— *Acidic character of oxides increasing* ⟶

(*b*) With the decrease of the ionic character of M—Cl bond in the chlorides, the acidic character of the chlorides increases as shown below :

⟵ *Oxidation states* ⟶

+2	+3	+4	+6
$TiCl_2$	$TiCl_3$	$TiCl_4$	—
VCl_2	VCl_3	VCl_4	$VOCl_3$

— Oxidation state increasing ⟶

— Ionic character of M—Cl bond decreasing ⟶

— Acidic character of chlorides increasing ⟶

8. Standard Reduction Potentials and Reducing Properties.

Standard reduction potentials (in volts) of some of the *d*-block elements are given below. These values are given in the decreasing order.

	E° (*Volts*)		E° (*Volts*)
$Hg_2^{2+} + 2e^- \longrightarrow 2Hg$;	+ 0.79	$Cr^{3+} + 3e^- \longrightarrow Cr$;	– 0.71
$Cu^{2+} + 2e^- \longrightarrow Cu$;	+ 0.34	$Zn^{2+} + 2e^- \longrightarrow Zn$;	– 0.76
$Ni^{2+} + 2e^- \longrightarrow Ni$;	– 0.24	$Mn^{2+} + 2e^- \longrightarrow Mn$;	– 1.05
$Co^{2+} + 2e^- \longrightarrow Co$;	– 0.28	$V^{2+} + 2e^- \longrightarrow V$;	– 1.20
$Cd^{2+} + 2e^- \longrightarrow Cd$;	– 0.40	$Ti^{2+} + 2e^- \longrightarrow Ti$;	– 1.60
$Fe^{2+} + 2e^- \longrightarrow Fe$;	– 0.44	$Sc^{3+} + 3e^- \longrightarrow Sc$;	– 2.08
		$La^{3+} + 3e^- \longrightarrow La$;	– 2.37

These values make the following points evident; (*i*) Excepting for Hg and Cu, these values for other elements are negative, *i.e.* less than that of hydrogen whose value is taken as zero. Their values being less than that of hydrogen, make these metals able to liberate hydrogen from dilute acids.

$$M + 2H^+ \text{ (from acid)} \longrightarrow M^{2+} + H_2\ (g) \uparrow$$

In actual practice, the rate with which most of these metals react with acid and liberate H_2 gas is very slow. Some of the metals get protected from further attack of the acids due to the formation of a thin protective layer of their oxide which prevents the acid to come in further contact with the metal, *e.g.*, although the electrode potential of Cr is much less than that of hydrogen ($E°_{Cr^{3+}/Cr} = -0.71$ volt), it gets coated with its non-reactive oxide, Cr_2O_3 which makes it so unreactive that can be used as a protective non-oxidising metal. (*ii*) In spite of the fact that all the metals of first transition series, excepting Cu, have high negative values of standard electrode potentials, these are not as good reducing agents as the metal of groups IA, IIA and IIIA. The poor reducing capacity of the transition metals is due to high heats of vaporisation, high ionisation potentials and low heats of hydration of their ions.

9. Formation of Coloured Compounds.

Normally the compounds of *s*- and *p*-block elements are colourless while those of transition metals in the solid or in solution state are usually coloured. The colours of hydrated cations of transition metals of the first transition series are : Sc^{3+}, Ti^{4+} ($3d^\circ$, $n = 0$) → colourless; Ti^{3+} ($3d^1$, $n = 1$) → purple; V^{3+} ($3d^2$, $n = 2$) → green; V^{2+} ($3d^3$, $n = 3$) → violet; Cr^{3+} ($3d^3$, $n = 3$) → green; Cr^{2+} ($3d^4$, $n = 4$) → blue ; Mn^{3+} ($3d^4$, $n = 4$) → violet; Mn^{2+} ($3d^5$, $n = 5$) → faintly pink or faintly green; Fe^{3+} ($3d^5$, $n = 5$) → yellow; Fe^{2+} ($3d^6$, $n = 4$) → pale green or faintly green; Co ($3d^6$, $n = 4$) → blue; Co^{2+} ($3d^7$, $n = 3$) → pink; Ni^{2+} ($3d^8$, $n = 2$) → green; Cu^{2+} ($3d^9$, $n = 1$) → blue; Cu^+ ($3d^{10}$, $n = 0$) → colourless; Zn^{2+} ($3d^{10}$, $n = 0$) → colourless. Here n denotes the number of unpaired electrons in the respective cations. For simplicity water molecules associated with each cation have not been shown.

It may be seen from the colours given above that the cations (*e.g.* Sc^{3+}, Ti^{4+}, Cu^+, Zn^{2+}) having vacant or completely-filled *d*-orbitals (d^0 and d^{10} orbitals) are colourless while those (*e.g.* Ti^{3+}, V^{3+}, Cr^{3+} etc.) having partially-filled *d*-orbitals (d^1, d^2, d^3 etc. orbitals) are coloured. In other words, the cations having all the electrons paired in *d*-orbitals (*i.e.* completely-filled *d*-orbitals) or no electrons in *d*-orbitals (*i.e.* empty *d*-orbitals) are colourless while those having some or all the electrons unpaired in *d*-orbitals are coloured. If n is the number of unpaired electrons in *d*-orbitals, then the ions having $n = 0$ are colourless while those having $n = 1, 2$ etc. are coloured.

10. Magnetic Properties.

There are, in general, three types of magnetism : (*i*) *Paramagnetism* (*ii*) *Diamagnetism* (*iii*) *Ferromagnetism.*

(*i*) **Paramagnetism.** This property is shown by the substances which, when, placed between the magnetic poles allow the magnetic lines of force to pass through them rather than through vacuum, *i.e.* a paramagnetic bar will tend to set itself with its length parallel to the field, when it is placed between the magnetic poles. In other words a *paramagnetic substance is attracted into a magnetic field.* Paramagnetism is common among the transition metal complexes, *i.e.* most of the transition metals are paramagnetic.

Origin of paramagnetism. The paramagnetism is caused by the presence of unpaired electrons in the substance. How do the unpaired electrons render the substance paramagnetic ? We know that an electron has two types of motion : *spin motion* (*i.e.*, spinning of the electron on its axis) and *orbital motion* (*i.e.*, moving of the electron in an orbital round the nucleus). It is also known from elementary Physics that the movement of an electric charge produces a magnetic field. Consequently an electron spinning on its axis which may be regarded as a small sphere of negative charge also creates a magnetic moment which is called *spin magnetic moment* or simply *spin moment* of the electron. The orbital motion produces a magnetic moment just like an electric current travelling in a loop of wire. The magnetic moment produced in this way is called *orbital moment* of the electron. The observed magnetic moments are combination of these two moments. Both these moments arise due to the presence of unpaired electrons.

In the substances containing *one or more unpaired electrons*, the magnetic moment which is a combination of spin moment and orbital moment is sufficiently of high magnitude to overcome the small magnetic moments induced by an approaching magnetic field. Such a substance, therefore, instead of experiencing repulsion, experiences *attraction* in a magnetic field, *i.e.*, it shows *paramagnetic behaviour. The compounds having one or more unpaired electrons are paramagnetic.*

If two electrons with opposite spins are paired in the same orbital, the magnetic field caused by one electron is cancelled by that caused by the other electron, because each of the two electrons will have equal but opposite magnetic momenta, thus giving zero resultant magnetic moment. Consequently the substances having *paired electrons* will not show paramagnetism but, in fact, they will show *diamagnetism.* The magnetic moment is expressed in *Bohr magneton* (BM) which is given by :

$$1 \text{ BM} = \frac{eh}{4\pi mc}$$

where h = Planck'a constant, e = electronic charge, c = velocity of light and m = mass of the electron.

(*ii*) **Diamagnetism.** This property is exhibited by the substances which allow the magnetic lines of force to pass through the vacuum rather than through themselves. Thus a diamagnetic substance is *repelled* by a magnetic field and hence tends to set itself at right angles to the field. *The compounds having all the electrons paired are diamagnetic, i.e. diamagnetic compounds have no unpaired electron, and hence their magnetic moment is zero.* The property of diamagnetism is produced by the orbital motion of the electrons of the atoms, ions or molecules of the sample. It is characteristic of all substances that contain paired electrons. Diamagnetism is exhibited by all substances. Even paramagnetic substances have some diamagnetic character and the net magnetic moments for paramagnetic substances are obtained after making due corrections for the diamagnetism. Diamagnetic behaviour is due to small magnetic moments which are induced by the applied magnetic field but does not exist in the absence of the field.

(*iii*) **Ferromagnetism.** In case of some transition metals (*e.g.* Fe, Co and Ni) the unpaired electron spins are more pronounced and hence such metals are much more paramagnetic than the other transition metals. Such metals are called ferromagnetic metals and the magnetic behaviour shown by these metals is known as ferromagnetism. Such metals can be magnetised.

11. Tendency to Form Complexes.

The cations of *d*-block elements have a strong tendency to form complexes with certain molecules (*e.g.* CO, NO, NH_3 etc.) or ions (*e.g.* F^-, Cl^-, CN^- etc.). The molecules or ions with which the cations attach themselves to form the complexes are called *ligands.* The molecules or ions acting as ligands have one or more lone pairs of electrons on their central atom. In the process of formation of complexes the central atom of the ligands donates readily lone pairs of electrons to the cation to form M ← L coordinate bond (M = metal cation and L = ligand) The tendency of the cations of *d*-block elements to form complexes with ligands is due to two reasons: (*i*) The cations are relatively very small in size and hence have high positive charge density which makes it easy for the cations to accept the lone pairs of electrons from the ligands. (*ii*) The cations have vacant $(n-1)d$ orbitals which are of appropriate energy to accept lone pair of electrons from the ligands for bonding with them.

12. Formation of Non-stoichiometric Compounds.

Many of the compounds given by the transition metals are non-stoichiotnetric compounds, *i.e.* these compounds have indefinite or variable composition. Transition metals form non-stoichiotnetric compounds due to their variable valency and due to the defects existing in the solid structures. Some examples of non-stoichiometric compounds given by transition metals are given below: (*i*) The formula of Fe(II) oxide is FeO which shows that it has Fe and O atoms in 1 : 1 ratio. But the analysis of this oxide has revealed that the actual formula varies between $Fe_{0.94}O$ and $Fe_{0.84}O$ and hence the oxide is a non-stoichiometric compound. In order to show the non-stoichiometric nature of the compound, it is written with a bar over its formula, *i.e.* this oxide is written as $\overline{FeO}$, (*ii*) Vanadium selenide is another non-stoichiometric compound whose composition varies between $VSe_{0.98}$ and VSe_2. (*iii*) Small nonmetallic atoms (*e.g.*, H, B, C, N etc.) are able to fit into the interstitial sites of transition metal lattices to form the compound which are called interstitial compounds. These have variable composition and cannot, therefore, be expressed by simple formulae, *i.e.*, these are also *non-stoichiometric compounds*. Interstitial compounds of transition metals, *viz.*, hydrides, borides, carbides and nitrides are usually prepared by heating the metal with hydrogen, boron, carbon (or hydrocarbon), and nitrogen (or ammonia), respectively. Interstitial compounds have the properties of alloys. They are extremely hard, possess good electrical conductivity and are with high melting point (titanium carbide melts at 3410°C, titanium nitride melts at 3200°C). Except towards strong oxidizing agents, the compounds are inert chemically.

13. Catalytic Properties.

Most of the transition metals and their compounds are used as catalysts. Transition metals show a variety of oxidation states and thereby can form intermediate products with various reactants. They are also capable of forming interstitial compounds which can absorb and activate the reacting species. Thus, they act as catalysts. Some examples of catalysts are : Ni acts as a catalyst in hydrogenation reactions; MnO_2 catalyses the decomposition of H_2O_2; $TiCl_4$ is employed as a catalyst in the manufacture of polythene ; V_2O_5 is used for catalytic oxidation of SO_2, to SO_3, etc.

14. Formation of Alloys.

Transition metals form alloys with each other. This is because they have almost similar sizes and the atoms of one metal can easily take up positions in the crystal lattice of the other. Thus, we have alloys of Cr—Ni, Cr—Ni—Fe, Cr—V—Fe, Mn—Fe, etc. Alloys are usually harder and have higher melting point than the parent metals. They are more resistant to corrosion than their constituents.

Comparison of Transition Elements with Non-transition Elements.

A comparison of important properties of transition elements (*d*-block elements) with those of non-transition elements (*s*- and *p*-block elements) is given Table 44.1. There are, however, two generalizations that will apply equally to both the transition and non-transition elements : (*i*) High oxidation states are usually stabilized in alkaline media. (*ii*) The capacity of the halogens to bring out high oxidation states varies in the order : F > Cl > Br > I.

Table 44.1. Comparison of the important properties of transition elements (*d*-block elements) with those of non-transition elements (*s*- and *p*-block elements).

Transition elements (*d*-block elements)	**Non-transition elements (*s*- and *p*-block elements)**
(*i*) All the transition elements are metals.	(*i*) Non-transition elements may be metals, metalloids or non-metals. The metallic character in a group increases with the increase in atomic weight.
(*ii*) The atomic radii show irregular decrease in a series of transition elements. The decrease is not pronounced.	(*ii*) The atomic radii show regular and pronounced decrease along a period of non-transition elements.
(*iii*) The ionization energies show irregular increase in a series of transition elements. The increase is not pronounced.	(*iii*) The ionization energies show regular and pronounced increase on moving across a period of non-transition elements.
(*iv*) They exhibit a variety of oxidation states in their compounds. The oxidation states differ from each other by one unit. For example : Cu^+, Cu^{2+} Fe^{2+}, Fe^{3+} Ti^{3+}, Ti^{4+} V^{4+}, V^{5+}	(*iv*) They usually exhibit the group valency or another valency which is less by two units than the group valency (inert pair effect). For example : Ti^+, Ti^{3+} Sn^{2+}, Sn^{4+} Pb^{2+}, Pb^{4+}
(*v*) Most of the transition metal ions are coloured in aqueous solution. For example : Cu^{2+}, V^{4+} — Blue Ni^{2+}, Cr^{3+} — Green Co^{2+} — Pink Fe^{3+} — Yellow	(*v*) Aqueous solutions of the ions of non-transition elements are colourless. For example : Na^+ — Colourless Ba^{2+} — Colourless Al^{3+} — Colourless Cl^- — Colourless
(*vi*) They form a large number of complexes. For example : $[Ag(NH_3)_2]^+$, $[Cu(NH_3)_4]^{2+}$ $K_4[Fe(CN)_6]$	(*vi*) They do not form many complexes.
(*vii*) Transition elements or their compounds show catalytic activity. For example nickel catalyses hydrogenation reactions and V_2O_5 is used as a catalyst for the oxidation of SO_2 to SO_3.	(*vii*) They lack catalytic activity.
(*viii*) They form interstitial compounds with the non-metals like H, C and N. These compounds may have non-stoichiometric ratios, For example : $TiH_{1.7}$, $VH_{0.56}$, etc.	(*viii*) They do not form non-stoichiometric compounds.
(*ix*) Their compounds are mostly *paramagnetic*.	(*ix*) Their compounds are mostly *diamagnetic*.

4f Block Elements : Lanthanides

What are f-block Elements ?

The elements in which the extra electron (also called differentiating electron) enters $(n-2)f$ orbitals are called *f*-block elements. Since the extra electron goes to $(n-2)$ *f*-orbitals, these elements have been given the name *f*-block elements. Since $(n-2)$ *f*-orbitals lie comparatively deep within the kernel (being inner to the penultimate shell), these elements are also called ***inner-transition elements.***

Classification of f-Block Elements.

We have already said that in case of *f*-block elements, the extra electron enters $(n-2)$ *f*-orbitals. These $(n-2)$ *f*-orbitals may be either 4*f*-orbitals or 5*f*-orbitals, Depending on whether the extra electron enters 4*f* or 5*f*-orbitals, *f*-block elements have been classified into two blocks namely (*i*) 4*f*-block elements. These elements are also called by other names like *1st inner-transition series elements, lanthanides* and *lanthanones*, (*ii*) *5f-block elements.* These elements are also called by other names like *2nd inner-transition series elements, actinides* and *actinones.*

What are 4f-Block Elements ?

In the atoms of the 4*f*-block elements the extra electron enters 4*f*-orbital, *i.e.* in the atoms of the elements of this block, 4*f*-orbitals are being progressively filled. A group of 14 elements from cerium, Ce(Z = 58) to lutetium, Lu (Z = 71) constitute this series. The elements of this series are also called *lanthanides* or *lanthanones*, since they follow lanthanum, La (Z = 57) in the periodic table and show close similarities with La. Although, strictly speaking the fourteen elements from Ce(Z = 58) to Lu(Z = 71) are called lanthanides, yet, due to the similarities in properties among all the fifteen elements from La(Z = 57) to Lu (Z = 71), all these fifteen elements are considered to be the members of lanthanide series, Thus La is also studied along with lanthanides. The lanthanide elements constitute the first *inner-transition series.* Lanthanide elements lie in *7th* period (incomplete period with $n = 7$) and group III B of the periodic table.

Position of Lanthanides in the Periodic Table.

All the fifteen lanthanides have atomic weights between those of barium (atomic number, Z = 56) and hafnium (Z = 72) and, therefore, must be placed between these two elements as was also proved by Moseley.

Barium has exactly the same outer electronic configuration as calcium and strontium and resembles them very closely. In a similar way hafnium (Z = 72) is similar to zirconium (Z = 40). Therefore Ba must be placed below Sr (Group IIA) and Hf below Zr (Group IV B), thus leaving only one place between them, which lies exactly below Y (Z = 39, Group III B).

Since all the fifteen lanthanides resemble one another in many respects, these must be placed in the same group. These elements also resemble Y, because of the following additional similarities : (*a*) Owing to lanthanide contraction (to be discussed subsequently), the ionic radius of Y^{3+} ion is almost similar to that of Er^{3+}(r_{Y3+} = 0.93Å, r_{Er3+} = 0.96Å). (*b*) Y generally occurs in nature associated with the ores of heavier lanthanides and resembles Tb(III) and Dy(III) in its compounds. It, therefore, became necessary to accommodate all the fifteen lanthanides together at one place. This has been done by placing the first element *viz.* La_{57} below Y_{39} and remaining fourteen elements (*viz.* Ce_{58}, to Lu_{71}) have been placed separately in the lower part of the periodic table as shown below in the margin.

General Characteristics of Lanthanides

1. Occurrence.

II A	III B	IV B
Be_4		
Mg_{12}		
Ca_{20}	Sc_{21}	Ti_{22}
Sr_{36}	Y_{39}	Zr_{40}
Ba_{56}	La_{57}	Hf_{72}
	⋮ ....→	Ce_{58}......Lu_{71} Lanthanides

Position of lanthanides in the periodic table.

Lanthanides are not so scare in the earth's crust as the name rare earth implies. Rather, their sources are unlimited. Each lanthanide mineral contains all the lanthanides (some in traces) excepting promethium, Pm (Z = 61) which is an unstable radioactive element Lanthanide minerals can be divided into two sub-groups : (*i*) *Cerium sub-group minerals.* These minerals contain largely the lighter lanthanides with atomic numbers 57 to 62 (La to Sm). Lanthanides with atomic number 57 to 62 are also called *cerium earths. Monazite sand* and *bastanesite* are the minerals which contain cerium earths. Monazite mineral is a mixture of phosphates mainly the phosphates of rare earths (RPO_4 where R is any lanthanide). It contains 49–74% of cerium earths as oxides and 1–4% of yttrium earths (yttrium earths are lanthanides with atomic number 63 to 71 and yttrium) as oxides. It also contains 5–9% of ThO_2, 1–2% of SiO_2 and traces of uranium, Monazite sand occurs mainly in the sand beaches of Travancore (India), Brazil, Union of South Africa, U.S.A. Bastanesite mineral contains mixed fluorides and carbonates ($RFCO_3$) and contains 65–70% of cerium earths, < 1% of yttrium earths and traces of thorium. This mineral is found in California, New Mexico and Sweden. (*ii*) *Yttrium sub-group minerals.* These minerals contain largely yttrium and heavier lanthanides with atomic number 63 to 71, *i.e.* these minerals contain yttrium earths. *Euxenite* and *gadolinite* are minerals which contain yttrium earths. Euxenite mineral is represented as R(Nb, Ra) $TiO_6.xH_2O$. It may contain 13–35% of yttrium earths as oxides, 2–3% cerium earths as oxides, 20–23% of ThO_2 and 25–35% of $(Nb,Ta)O_5$. This mineral is found in Australia and Idaho. Gadolinite is represented as $R_2\ M_3^{II}\ S_2O_{19}$, where R = lanthanide atom and M = divalent atom like Fe or Be. This mineral contains 35–48% of yttrium earths as oxides, 2–17% of cerium earths, 11.6% of BeO and traces of Te. This mineral is found in Sweden, Norway and U.S.A.

2. Electronic Configuration.

Complete and valence-shell configurations of the atoms of lanthanides in their ground state are : La (Z = 57) → 2, 8, 18, 18, 9, 2 ($4f^0.5d^1.6s^2$), Ce (Z = 58) → 2, 8, 18, 19, 9, 2 ($4f^1.5d^1.6s^2$) *or* 2, 8, 18, 20, 8, 2 ($4f^2.5d^0.6s^2$), Pr (Z = 59) → 2, 8, 18, 21, 8, 2 ($4f^3.5d^0.6s^2$), Nd (Z = 60) → 2, 8, 18, 22, 8, 2 ($4f^2.5d^0.6s^2$), Pm (Z = 61) → 2, 8, 18, 23, 8, 2 ($4f^5.5d^0.6s^2$), Sm (Z = 62) → 2, 8, 18, 24, 8, 2 ($4f^6.5d^0.6s^2$), Eu (Z = 63) → 2, 8, 18, 25, 8, 2 ($4f^7.5d^0.6s^2$), Gd (Z = 64) → 2, 8, 18, 25, 9, 2 ($4f^7.5d^1.6s^2$), Tb (Z = 65) → 2, 8, 18, 27, 8, 2 ($4f^9.5d^0.6s^2$), Dy (Z = 66) → 2, 8, 18, 28, 8, 2 ($4f^{10}.5d^0.6s^2$), Ho (Z = 67) → 2, 8, 18, 29, 8, 2 ($4f^{11}.5d^0.6s^2$), Er (Z = 68) → 2, 8, 18, 30, 8, 2 ($4f^{12}.5d^0.6s^2$), Tm (Z = 69) → 2, 8, 18, 31, 8, 2 ($4f^{13}.5d^0.6s^2$), Yb (Z = 70) → 2, 8, 18. 32, 8, 2 ($4f^{14}.5d^0.6s^2$), Lu (Z = 71) → 2, 8, 18, 32, 9, 2 ($4f^{14}.5d^1.6s^2$). These configurations show that the atoms of lanthanides have a total of *six* shells. Out of these six shells, first three shells namely 1*st*, 2*nd* and 3*rd* are completely filled while the remaining three shells *viz.* 4*th*, 5*th* and 6*th* are partially filled as shown below :

$$\mathbf{2, 8, 18, 4s^2p^6d^{10}\ f^{0...7},\ ^{9...14}, 5s^2p^6d^{0,1}, 6s^2}\ \textit{(six shells)}$$

The configuration given above clearly shows that 4*f* and 5*d* orbitals in the atoms of lanthanides are partially filled while 6*s* orbital is completely filled ($6s^2$ configuration). Thus the number of electrons in 4*th* shell [*i.e.* (*n*–2)*th* shell] is increasing from 18 ($4s^2 + 4p^6 + 4d^{10}$) to 32 ($4s^2 + 4p^6 + 4d^{10} + 4f^{14}$) by adding an electron, one by one, to 4*f* orbital. Consequently the atoms of lanthanides have partly filled 4*f* orbitals, since the differentiating electron enters these orbitals. Due to the fact that the extra electron goes to 4*f* orbital, the lanthanides are also called 4*f*-block *elements.* Complete electronic configurations as given above also indicate that the valence shell configuration of these elements can be represented as :

$$\mathbf{4f^{0...7},\ ^{9...14}, 5d^{0,1}, 6s^2}$$

3. Oxidation States.

Important oxidation states shown by lanthanides and their valence-shell configurations (given in brackets) are given as : La (Z = 57) → +3 ($4f^0.5d^0.6s^0$); Ce (Z = 58) → +3 ($4f^1.5d^0.6s^0$), + 2 ($4f^2.5d^0.6s^0$), + 4 ($4f^0.5d^0.6s^0$); Pr (Z = 59) → +3 ($4f^2.5d^0.6s^0$), + 4 ($4f^1.5d^0.6s^0$); Nd (Z = 60) → +3 ($4f^3.5d^0.6s^0$), +2 ($4f^4.5d^0.6s^0$), + 4 ($4f^2.5d^0.6s^0$); Pm (Z = 61) → +3 ($4f^4.5d^0.6s^0$); Sm (Z = 62) → +3 ($4f^5.5d^0.6s^0$), +2 ($4f^6.5d^0.6s^0$); Eu (Z = 63) → +3 ($4f^6.5d^0.6s^0$), +2 ($4f^7.5d^0.6s^0$), Gd (Z = 64) → +3 ($4f^7.5d^0.6s^0$); Tb (Z = 65) → +3 ($4f^8.5d^0.6s^0$), +4 ($4f^7.5d^0.6s^0$); Dy (Z = 66) → +3 ($4f^9.5d^0.6s^0$), +4 ($4f^8.5d^0.6s^0$); Ho (Z = 67) → +3 ($4f^{10}.5d^0.6s^0$); Er (Z = 68) → +3 ($4f^{11}.5d^0.6s^0$); Tm (Z = 69) → +3 ($4f^{12}.5d^0.6s^0$), +2 ($4f^{13}.5d^0.6s^0$); Yb (Z = 70) → +3 ($4f^{13}.5d^0.6s^0$), +2 ($4f^{14}.5d^0.6s^0$); Lu (Z = 71) → +3 ($4f^{14}.5d^0.6s^0$). It may be seen that +3 oxidation state is common to all the lanthanides. In addition to +3 oxidation state, some lanthanides also exhibit +2 and +4 states. In case of some lanthanides, the formation of M^{2+}, M^{3+} and M^{4+} cations can be explained on the basis of stability associated with $4f^0$, $4f^7$ and $4f^{14}$ configurations. For example : (*a*) La (Z = 57), Gd (Z = 64) and Lu (Z = 71) give La^{3+}, Gd^{3+} and Lu^{3+} cations by attaining $4f^0$, $4f^7$ and $4f^{14}$ configurations. Consequently in case of these elements any oxidation state other than +3 is not possible. (*b*) Ce (Z = 58) and Tb (Z = 65) form Ce^{4+} and Tb^{4+} cations by attaining $4f^0$ and $4f^7$ configuration. (*c*) Eu (Z = 63) and Yb (Z = 70) form Eu^{2+} and Yb^{2+} cations by attaining $4f^7$ and $4f^{14}$ configurations. Although the formation of La^{3-}, Gd^{3+}, Lu^{3+}, Ce^{4+}, Tb^{4+}, Eu^{2+} and Yb^{2+} cation can be explained on the basis of the special stability associated with $4f^0$, $4f^7$ and $4f^{14}$ configurations, yet the formation of many cations like Ce^{3+}, Ce^{2+}, Pr^{3+}, Pr^{4+} etc. can not be explained on this basis. In explaining the formation of these cations thermodynamic and kinetic factors are to be taken into consideration.

4. Radii of Tripositive Lanthanide Cations : Lanthanide Contraction.

A look at the values of the radii of tripositive lanthanide cations (M^{3+}) reveals that these values decrease as we move from La^{3+} to Lu^{3+} in the lanthanide series. *This steady decrease in the values of the radii of M^{3+} cations in the tanthanide series is called lanthanide contraction.*

$$\xrightarrow{\textit{Decreasing}}$$

La^{3+} (r = 1.06 A°) > Ce^{3+} (1.03 A°) > Pr^{3+} (1.01 A°) > Nd^{3+} (0.99 A°) > Pm^{3+} (0.98 A°) > Sm^{3+} (0.96 A°) > Eu^{3+} (0.95 A°) > Gd^{3+} (0.94 A°) > Tb^{3+} (0.92 A°) > Dy^{3+} (0.91 A°) > Ho^{3+} (0.89 A°) > Er^{3+} (0.88 A°) > Tm^{3+} (0.87 A°) > Yb^{3+} (0.86 A°) > Lu^{3+} (0.85 A°).

Cause of lanthanide contraction.

We know that as we proceed from one element to the next one in the lanthanide series, the nuclear charge (*i.e.* atomic number) increases by +1 at each next element and the addition of extra electron takes place in $4f$ orbital. Again it is also known that the shielding of one electron in $4f$ orbital by the other electron residing in the same orbital is very poor. Due to the negligible amount of mutual shielding effect between the electrons residing in $4f$ orbital, the increase in nuclear charge by + 1 at each next element in the lanthanide series brings the valence shell nearer to the nucleus and hence the size of M^{3+} cations goes on decreasing as we move from one element to the next one in the series.

Consequences of lanthanide contraction.

Lanthanide contraction plays an important role in determining the chemistry of lanthanides and heavier transition elements. Some important consequences of lanthanide contraction are discussed below :

(*i*) Basic character of lanthanide hydroxides, $M(OH)_3$. There is a *decrease in basic strength of the hydroxides of lanthanides with increase in atomic number.* Thus $La(OH)_3$ is the most basic while $Lu(OH)_3$ is the least basic. Due to lanthanide contraction the decrease in size of M^{3+} cations increases the covalent character (*i.e.* decreases the ionic character) between the M^{3+} ion and OH^- ion (according to Fajan's rules), thereby reducing the basic character of the lanthanide hydroxides.

***(ii)* Atomic radii of post-lanthanide elements.** Normally in the same sub-group the atomic (covalent) radii increase as the atomic number increases. This fact is evident when are compare these values for the elements of first and second transition series as given in Table 44.2. On the basis of this rule the atomic radii of the elements of 3*rd* transition series should be greater than those of the elements of 2*nd* transition series, but, as the table shows, although the atomic radius of La is greater than that of Y (La = 1.69 A°, Y = 1.62 A°), the atomic radii of Hf, Ta, W, Re etc. (elements of 3*rd* transition series) are not greater than those of Zr, Nb, Mo, Tc etc. respectively (elements of 2*nd* transition series). The expected increase in the atomic radii on proceeding from the elements of 2*nd* transition series to those of 3*rd* transition series is cancelled by the presence of 14 lanthanides (Ce with Z = 58 to Lu with Z = 71) between La(Z = 57) and Hf (Z = 72). Consequently the atomic radii of the elements of 3*rd* transition series become almost equal to those of the elements of 2*nd* transition series. Due to the similarity in size of the elements of the two series, the elements of 2*nd* and 3*rd* series resemble each other more closely than do the elements of 1*st* and 2*nd* transition series.

Table 44.2. Atomic (covalent) radii (in A°) of the elements preceding and following lanthanides.

III B		IVB	VB	VIB	VIIB		VIII		IB	IIB
Sc_{21}		Ti_{22}	V	Cr	Mn	Fe	Co	Ni	Cu	Zn_{30}
1.44		1.32	1.22	1.18	1.17	1.17	1.16	1.15	1.17	1.25
Y_{39}		Zr_{40}	Nb	Mo	Tc	Ru	Rh	Pd	Ag	Cd_{48}
1.62		1.45	1.34	1.30	1.27	1.25	1.25	12.8	1.34	1.48
La_{57}	Ce_{55}–Lu_{71}	Hf_{72}	Ta	W	Re	OsI	r	Pt	Au	Hg_{80}
1.69	1.65 1.56	1.44	1.34	1.30	1.28	12.6	1.27	1.30	1.34	1.49
	↑ 14 Lanthanides									

***(iii)* High density of post-lanthanide elements.** Because of lanthanide contraction, the atomic sizes of the post-lanthanide elements become very small. Consequently the packing of atoms in their metallic crystal becomes so much compact that their densities are very high. Thus, while the densities of the elements of second transition series are only slightly higher than those of the elements of first series, the densities of the elements of third transition series are almost double to those of the elements of second transition series.

***(iv)* Occurrence of Y with heavy lanthanides.** The crystal radii of Y^{3+} and Er^{3+} are equal (Y^{3+} = 0.88 A° and Er^{3+} = 0.88 A°). This similarity in ionic size of these two cations coupled with the equality in ionic charge (= +3 in both the ions) accounts for the invariable occurrence of Y with heavier lanthanides. The marked similarities in the crystal structure, chemical properties and solubility between yttrium compounds and the corresponding compounds of the heavier lanthanides make it difficult to separate yttrium from the heavier lanthanides. It is because of these similarities that yttrium is regarded for all practical purposes as a member of lanthanide series.

5. Colours of M^{3+} Cations.

Some of the tripositive lanthanide cations (M^{3+} cations) are coloured. Their colours are : La^{3+} ($4f^0.5d^0.6s^0$, n = 0) → colourless; Ce^{3+} ($4f^1.5d^0.6s^0$, n = 1) → colourless; Pr^{3+}($4f^2.5d^0.6s^0$, n = 2) → yellow green; Nd^{3+}($4f^3.5d^0.6s^0$, n =3) → bright pink; Pm^{3+}($4f^4.5d^0.6s^0$, n = 4) → pink yellow; Sm^{3+}($4f^4.5d^0.6s^0$, n = 5) → yellow; Eu^{3+}($4f^6.5d^0.6s^0$, n = 6) → very pale pink; Gd^{3+}($4f^7.5d^0.6s^0$, n = 7) → colourless; Tb^{3+}($4f^8.5d^0.6s^0$, n = 6) → very pale pink; Dy^{3+}($4f^9.5d^0.6s^0$, n = 5) → yellow; Ho^{3+}($4f^{10}.5d^0.6s^0$, n = 4) → pink yellow; Er^{3+}($4f^{11}.5d^0.6s^0$, n = 3) → pink; Tm^{3+}($4f^{12}.5d^0.6s^0$, n = 2) → pale green; Yb^{3+}($4f^{13}.5d^0.6s^0$, n = 1) → colourless Lu^{3+}($4f^{14}.5d^0.6s^0$, n = 0) → colourless. Here n is the number of unpaired electrons in 4f orbitals. Following points may be noted from these colours. (*i*) M^{3+} cations having *empty, half-filled or completely filled 4f-orbttal* (*i.e.*, $4f^0$, $4f^7$ and $4f^{14}$ configuration) are colourless, *e.g.*, La^{3+}($4f^0$), Gd^{3+}($4f^7$) and Lu^{3+}($4f^{14}$) ions are colourless. The fact that the cations namely Ce^{3+}($4f^1$) and Yb^{3+}($4f^{13}$) are also colourless can not be explained on the basis of the above concept. (*ii*) Pairs of M^{3+} cations having the same number of unpaired electrons in 4f orbitals have the same colour. The colours of following five pairs of M^{3+} cations confirm this rule. (*a*) Pr^{3+} (yellow green)—Tm^{3+} (pale green), n = 2. (*b*) Nd^{3+} (bright pink)—Er^{3+}(pink), n = 3. (*c*) Pm^{3+} (pink yellow)—Ho^{3+} (pink yellow), n = 4. (*d*) Sm^{3+} (yellow)—Dy^{3+} (yellow), n = 5. (*e*) Eu^{3+} (very pale pink)—Tb^{3+} (very pale pink), n = 6. Here n indicates the number of unpaired electrons in 4f orbitals.

6. Magnetic Properties of M^{3+} Cations.

We know that ions having unpaired electrons are paramagnetic while those having all the electrons paired are diamagnetic. Thus, since La^{3+}($4f^0.5d^0$. $6s^0$) and Lu^{3+}($4f^{14}.5d^0.6s^0$) ions have all the electrons paired, these ions are diamagnetic. On the other hand, the remaining tripositive lanthanide ions (M^{3+} ions) are paramagnetic, since they contain 1, 2,...6 unpaired electrons.

Cause of paramagnetism. In case of M^{3+} cations, since 4f orbitals are quite deeply seated, these are well shielded from the ligands surrounding the M^{3+} ions by the overlapping 5s and 5p orbitals and hence the electrons residing in 4f orbitals are not able to interact with the surrounding ligands. Thus the orbital motion of electrons in 4f orbitals is not quenched as in case of d-block elements, and hence the observed paramagnetism in 4f-blook elements is due to both factors namely orbital motion of the electron and its spinning round its axis.

7. Formation of Complexes.

Although tripositive lanthanide cations have a high charge equal to +3 on them, yet their size is so large that their charge-to-radius ratio becomes so small that these ions have very poor tendency to form complexes. Common ligands with which M^{3+} ions form stable complexes are : (*i*) *chelating oxygen containing ligands* like EDTA, citric acid, oxalic acid, acetylacetone (*ii*) *nitrogen containing ligands* like ethylenediamine, NCS^- etc.

8. Properties Dependent on Standard Reduction Potential Values.

The values of standard reduction potential (E° values) for the reduction reaction :

$$M^{3+}(aq) + 3e^- \rightleftarrows M(s)$$

increase from La to Lu, *i.e.* E° values become less negative as we move from La to Lu as shown below (values are given in volts) :

$$\xrightarrow{\hspace{3em} E^\circ \text{ values become less negative} \hspace{3em}}$$

La = –2.52, Ce = –2.48, Pr = –2.46, Nd = –2.43, Pm = –2.42, Sm = –2.41, Eu = –2.40, Gd = –2.39, Tb = –2.39, Dy = –2.35, Ho = –2.32, Er = –2.30, Tm = –2.28, Yb = –2.27, Lu = –2.25.

These E° values explain the following properties of lanthanides :

(*a*) Reducing property. Since lanthanides have high negative E° values, these elements have a strong tendency to lose their three electrons to undergo oxidation and hence are strong reducing agents. With the increase of E° values from La to Lu, *i.e.* as E° values become less negative from La to Lu, the reducing power of these elements decreases from La to Lu.

(*b*) Electropositive character. Due to negative E° values, the lanthanides readily lose their electrons and hence show strong electropositive (or metallic) character. With the increase of E° values from La to Lu, the electropositive character of these elements decreases from La to Lu.

(*c*) Liberation of H_2 from H_2O. Due to highly electropositive character of lanthanides, these elements slowly react with cold water and rapidly with hot water to liberate H_2 from it.

Extraction of Lanthanides from Monazite Sand

The method used for the extraction of lanthanides from monazite sand consists of the steps which have been shown in flow sheet 1 given on next page. The solution obtained after removing cerium as basic ceric nitrate contains the nitrates of La, Pr, Nd and Sm. The individual lanthanides are separated from this solution by suitable methods.

Separation of Lanthanide Elements

All the lanthanide elements form tripositive ions (M^{3+} ions) which are almost identical in size. The chemical properties of lanthanides which depend on the size and charge of M^{3+} ions are almost the same. Due to the similarity in chemical

Flow sheet 1. Extraction of lanthanides from monazite sand.

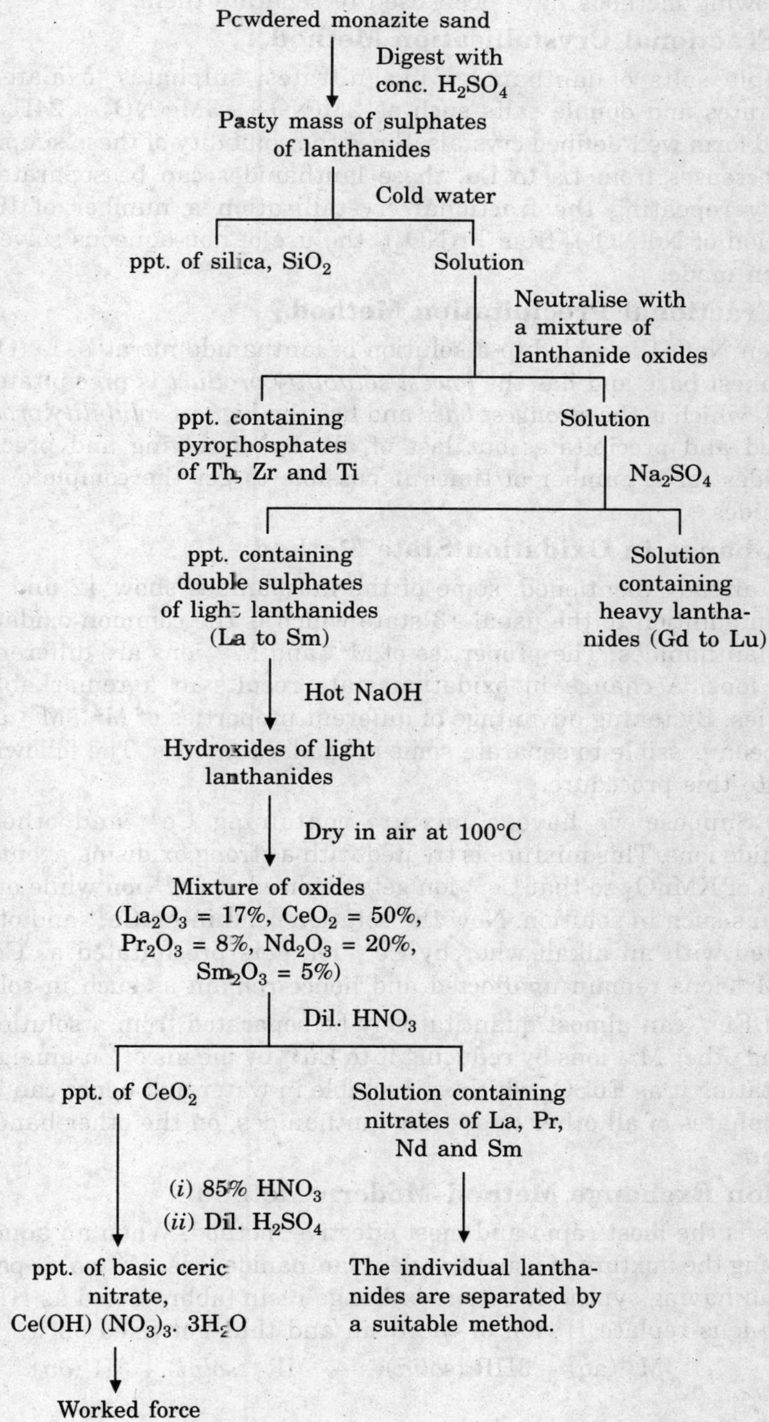

properties, it is difficult to separate the lanthanides from each other. However, the following methods have been used to separate them.

1. Fractional Crystallisation Method.

Simple salts of lanthanides like nitrates, sulphates, oxalates, bromates, perchlorates and double salts such as $2M(NO_3)_3 . 3Mg(NO_3)_2 . 24H_2O$ crystallise well and form well-defined crystals. Since the solubility of these simple and double salts decreases from La to Lu, these lanthanides can be separated from each other by repeating the fractional crystallisation a number of times. In the separation of $Nd(NO_3)_3$ from $Pr(NO_3)_3$ the use of non-aqueous solvent like ether has been made.

2. Fractional Precipitation Method.

When NaOH is added to a solution of lanthanide nitrates, $Lu(OH)_3$ which is the weakest base and has the *lowest solubility product* is precipitated first while $La(OH)_3$ which is the *strongest base* and has the *highest solubility product* remains dissolved and precipitates out last of all. By dissolving and precipitating the hydroxides for a number of times it possible to get the complete separation of lanthanides.

3. Change in Oxidation State Method.

As already mentioned, some of the lanthanides show +2 and +4 oxidation states in addition to the usual +3 state which is the common oxidation state for all the lanthanides. The properties of M^{2+} and M^{4+} ions are different from those of M^{3+} ions. A change in oxidation state results in a remarkable change in properties. By taking advantage of different properties of M^{2+}, M^{3+} and M^{4+} ions, it has been possible to separate some of the lanthanides. The following examples illustrate this procedure.

(*i*) Suppose we have a mixture containing Ce^{3+} and other tripositive lanthanide ions. This mixture is treated with a strong oxidising agent like alkaline solution of $KMnO_4$ so that Ce^{3+} ion gets oxidised to Ce^{4+} ion while other M^{3+} ions remain as such in solution. Now the solution containing Ce^{4+} and other M^{3+} ions is treated with an alkali whereby Ce^{4+} ion gets precipitated as $Ce(OH)_4$ while other M^{3+} ions remain unaffected and hence remain as such in solution.

(*ii*) Eu^{3+} can almost quantitatively be separated from a solution containing Eu^{3+} and other M^{3+} ions by reducing it to Eu^{2+} by means of Zn-amalgam and then precipitating it as $EuSO_4$ which is insoluble in water and hence can be separated. The sulphates of all other tripositive lanthanides, on the other hand, are soluble in water.

4. Ion Exchange Method–Modern Method.

This is the most rapid and most effective method. When an aqueous solution containing the mixture of usual trivalent lanthanide ions, $M^{3+}(aq)$ is passed through a column having synthetic cation-exchange resin [abbreviated as HR (solid)], the $M^{3+}(aq)$ ions replace H^+ ion of the resin and thus get fixed on it.

$$M^{3+}(aq) + 3HR\ (solid) \longrightarrow MR_3\ (solid) + 3H^+(aq)$$

Since $Lu^{3+}(aq)$ is largest in size and $La^{3+}(aq)$ is smallest, $La^{3+}(aq)$ is attached to the column with maximum and $Lu^{3+}(aq)$ with minimum firmness.

In order to recover the M^{3+} ions fixed on the resin, the column is *eluted* (*i.e.*, leached) with a citric acid–ammonium citrate solution (called **eluant** or **eluante**). During elution process NH_4^+ ions replace M^{3+} ions and M-citrate complexes are formed.

$$MR_3 + 3NH_4^+ \longrightarrow 3NH_4R + M^{3+}$$
$$M^{3+} + \text{citrate ions} \longrightarrow \text{M-citrate complex}$$

We have seen that since $La^{3+}(aq)$ is attached to the resin with maximum and $Lu^{3+}(aq)$ with minimum firmness, *Lu-citrate complex comes out of the column first and La-citrate complex comes out last.* In actual practice the process has to be repeated several times by careful control of concentration of citric acid-ammonium citrate solution.

5. Solvent Extraction Method.

This method makes the use of the difference in the value of partition coefficient of lanthanides between two solvents. $La(NO_3)_3$ and $Gd(NO_3)_3$ have been separated form each other by this method. The partition coefficient of $Gd(NO_3)_3$ between water and tributyl phosphate in kerosene is different from that of $La(NO_3)_3$ between the same solvents. This means that $Gd(NO_3)_3$ can be separated from $La(NO_3)_3$ by continuous extraction with water from a solution of these salts in tributyl phosphate in kerosene.

Production of Lanthanide Metals

The following methods may be used for this purpose :

1. Electrolysis of Fused Chlorides.

This method is similar to that used in the metallurgy of Ca by the electrolysis of $CaCl_2$.

2. Reduction of Anhydrous Chlorides with Na.

Lighter lanthanides such as La, Ce and Gd can easily be prepared by the reduction of their anhydrous chlorides with Na at 100°C.

$$LaCl_3 + 3Na \xrightarrow{100°C} La + 3NaCl$$

3. Reduction of Anhydrous Fluorides and Chlorides with Mg or Ca.

Heavier lanthanides such as Lu are prepared when anhydrous fluorides and chlorides are reduced by Ca or Mg metals at a temperature above 1000°C, since the fluorides are less volatile than the chlorides and consequently the loss caused by evaporation in case of fluorides is small.

Uses of Lanthanides and Their Compounds

(*i*) Certain alloys of the lanthanide elements, known as ***mishmetals*** containing predominantly 30–35% of Ce together with smaller quantities of other light lanthanides are used as reducing agents in metallurgical operations. Mg-alloys containing about 30% misch metal and 1% Zr are used for making parts of jet engine. (*ii*) La, Ce, Pr, Nd mixed with steel are used in cigarette lighters, toys, flame throwing tanks. (*iii*) Nd_2O_3 and PrO_3 are used as colouring agents for glass and in the production of standard filters. Lanthanum oxides are used for polishing glass. Ceria (CeO_2) is used in gas mantles. (*iv*) Cerium phosphate is used as a catalyst in petroleum cracking. (*v*) Cerium salts are used in analysis, dyeing cotton, lead accumulators, medicines etc. (*vi*) Lanthanide compounds are used as good catalysts in a number of reactions like hydrogenation, dehydrogenation, oxidation and cracking of petroleum. (*vii*) Lanthanide compounds are filled into arc carbon electrodes to give more brilliant light.

Comparison Between d- and f-Block Elements (*particularly between 3d and 4f-block elements*).

Some of the differences are given below :

1. Entry of extra electron. In case of *d*-block elements the extra electron enters (*n*–1)*d*-orbitals while in case of *f*-block elements the extra electron occupies (*n*–2) *f*-orbitals.

2. Energy of orbitals. (*n*–1) *d*-orbitals have higher energy than (*n*–2) *f*-orbitals.

3. Cause of paramagnetism. The maximum value of magnetic moment (μ value) for *d*-block elements is due to the presence of 5 unpaired electrons while the maximum value of μ for *f*-block elements is due to the presence of 7 unpaired electrons. Further, in case of *d*-block elements, since *d*-orbitals project out, these are not well shielded from the ligands surrounding the transition metal atoms or ions, *i.e.* the electrons residing in *d*-orbitals are able to interact with the surrounding ligands. Thus in case of transition metals, the orbital motion of the electron is quenched and hence the observed paramagnetism in these elements is only due to the spin motion of the electron. Due to the quenching of orbital motion, the value of μ is calculated from the spin-only formula, $\mu = \sqrt{n\ (n + 2)}$.

Now consider the case of tripositive lanthanide ions. In case of these ions since 4 *f*-orbitals are quite deeply seated, these are well shielded from the ligands surrounding these ions by the overlapping 5*s* and 5*p* orbitals, *i.e.* the electrons residing in 4*f* orbitals are not able to interact with the surrounding ligands. Thus in case of lanthanides. the orbital motion of the electron is not quenched as in case of *d*-block elements and hence the observed paramagnetism in these elements is due to both the motions of the electron *viz.* orbital motion and spin motion. In other words, the value of μ for M^{3+} ion, cannot be calculated by using spin-only formula. Rather μ is calculated from : $\mu = g\sqrt{J\ (J + 1)}$.

4. Nature of absorption bands. *d*-block elements give *broad bands* in the absorption spectra in the visible region while, excepting La^{3+}, and Lu^{3+}, all the remaining tripositive lanthanide ions give *sharp almost line-like bands* in the absorption spectra.

5. Oxidation states. *d*-block elements show much greater number of oxidation states while *f*-block elements, particularly lanthanides, show a limited number of oxidation states.

6. Formation of complex compounds. *d*-block elements form complexes more easily while *f*-block elements, particularly lanthanides, do so with great difficulty.

7. Coordination number. Because of their larger size, lanthanide ions generally have higher coordination numbers than transitional metal ions. Coordination numbers greater than 6 for transition metal ions are only formed with difficulty because of the strong ligand repulsion in the coordination spheres.

5f-block Elements : Actinides

What are 5*f*-Block Elements ?

In the atoms of 5*f*-block elements the extra electron enter 5f-orbitals, *i.e.* in the atoms of the elements of this block 5*f* orbitals are being progressively filled. A group of 14 elements from thorium, Th (Z = 90) to lawrencium, Lw (Z = 103) constitute this series. The elements of this series are also called *actinides* or *actinones*, since these elements follow actinium, Ac (Z = 89) in the periodic table and show close similarities with Ac. Although, strictly speaking, the fourteen elements from Th (Z = 90) to Lw (Z = 103) are called actinides, yet, due to the similarities in properties among all the fifteen elements from Ac (Z = 89) to Lw (Z = 103), all these fifteen elements are considered to be the members of actinide series. Thus Ac (Z = 89) is also studied along with actinide elements. The actinide elements constitute the *second inner transition series.* Actinide elements lie in 7*th* period (incomplete period with *n* = 7) and IIIB group of the periodic table.

Position of Actinides in the Periodic Table.

Prior to the discovery of trans-uranium elements (trans-uranium elements are the elements lying beyond uranium in the periodic table), the naturally occurring heaviest known elements namely Th_{90}, Pa_{91}, and U_{92}, were placed below Hf_{72}, Ta_{73} and W_{74} in IV B, V B and VI B groups of the periodic table, because these elements showed +4, +5 and +6 oxidation states and resembled Hf, Ta and W respectively in many of their properties. The then undiscovered trans-uranium elements with atomic numbers 93 to 100 were thus expected to occupy the positions in the periodic table below Re_{75}, Os_{76}, Ir_{77}, Pt_{78}, Au_{79}, Hg_{80}, Tl_{81}, and Pb_{82} respectively as shown below :

IV B	V B	VI B	VII B	VIII			I B	II B	III A	IVA
—	—	—	—	—	—	—	—	—	—	—
Hf_{72}	Ta_{73}	W_{74}	Re_{75}	Os_{76}	Ir_{77}	Pt_{78}	Au_{79}	Hg_{80}	Tl_{81}	Pb_{82}
Th_{90}	Pa_{91}	U_{92}	93	94	95	96	97	98	99	100

The discovery of the element neptunium (Np_{93}) came in 1940 and this discovery was followed shortly by the discovery of plutonium (Pu_{94}) in 1941. The tracer chemical experiments with Np_{93} and Pu_{94} showed that the chemical properties of these two elements very much resemble those of U_{92}, and not at all those of Re_{75} and Os_{76}. On this basis in 1914. all the three elements namely U_{92}, Np_{93} and Pu_{94} were placed under W_{74} in group VIB as shown below. It was thought that the undiscovered elements with atomic numbers 95 and 96 should be very much like U_{92}, Np_{93}, and Pu_{94} in their chemical properties, and hence the elements with atomic numbers 95 and 96 were also placed along with U_{92}, Np_{93}, and Pu_{94}, below W_{74} in group VI B as shown below :

IVB	VB	VIB
Hf_{72}	Ta_{73}	W_{74}
Th_{90}	Pa_{91}	U_{92}, Np_{93}, Pu_{94}, 95, 96

This assumption, however, proved to be wrong, since the experiments directed towards the discovery of elements with atomic numbers 95 and 96 on the pattern of discovery of Np_{93} and Pu_{94} failed. Later on, in the same year (1944) Seaborg suggested that all the elements having atomic number greater than 89 (Ac_{89}) constitute a second series of inner-transition elements. This series is similar to lanthanide series. The elements of this new series which have atomic number greater than 89 were called *actinides*. The elements of this series were placed below lanthanides out of the main body of the periodic table as shown below :

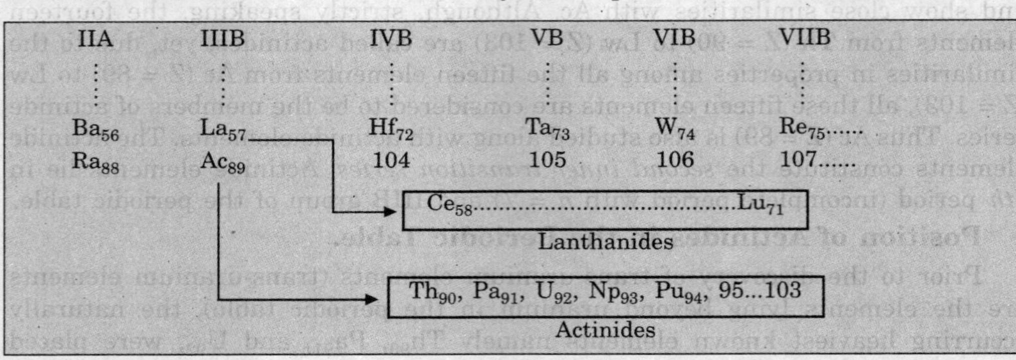

The position of actinides shown above was confirmed by the discovery of the elements from atomic numbers 95 to 103 in 1961 and by the electronic configuration of these elements.

General Characteristics of Actinides

1. Occurrence.

Excepting Ac, Tb, Pa and U which occur in nature in uranium minerals, all the remaining actinides are unstable and made artificially by nuclear transmutations, The actinide elements lying beyond U (Z = 92), *i.e.* the actinide elements with atomic numbers 93 to 103 are called *trans-uranium* or *trans-uranic elements*. All the actinides are radioactive.

2. Electronic Configurations

Complete and valence-shell configurations of the atoms of actinides in their ground state are : Ac(Z = 89) $\rightarrow$ 2, 8, 18, 32, 18, 9, $2(5f^0.6d^1.7s^2)$; Th(Z = 90) $\rightarrow$ 2, 8, 18, 32, 19, 9, $2(5f^1.6d^1.7s^2)$ *or* 2, 8, 18, 32, 18, 10, $2(5f^0.6d^2.7s^2)$; Pa(Z = 91) $\rightarrow$ 2, 8, 18, 32, 20, 9, $2(5f^2.6d^1.7s^2)$ *or* 2, 8, 18, 32, 19, 10, $2(5f^1.6d^2.7s^2)$; U(Z = 92) $\rightarrow$ 2, 8, 18, 32, 21, 9, $2(5f^3.6d^1.7s^2)$; Np(Z = 93) $\rightarrow$ 2, 8, 18, 32, 23, 8, $2(5f^5.6d^0.7s^0)$ *or* 2, 8, 18, 32, 22, 9, $2(5f^4.6d^1.7s^2)$; Pu(Z = 94) $\rightarrow$ 2, 8, 18, 32, 24, 8, $2(5f^6.6d^0.7s^2)$ *or* 2, 8, 18, 32, 23, 9, $2(5f^5.6d^1.7s^2)$; Am(Z = 95) $\rightarrow$ 2, 8, 18, 32, 25, 8, $2(5f^7.6d^0.7s^2)$;

Cm(Z = 96) → 2, 8, 18, 32, 25, 9, 2($5f^7.6d^1.7s^2$); Bk(Z = 97) → 2, 8, 18, 32, 27, 8, 2($5f^9.6d^0.7s^2$) *or* 2, 8, 18, 32, 25, 10, 2($5f^7.6d^2.7s^2$); Cf(Z = 98) → 2, 8, 18, 32, 28, 8, 2($5f^{10}.6d^0.7s^2$) *or* 2, 8, 18, 32, 27, 9, 2($5f^9.6d^1.7s^2$); Es(Z = 99) → 2, 8, 18, 32, 29, 8, 2($5f^{11}.6d^0.7s^2$); Fm(Z = 100) → 2, 8, 18, 32, 30, 8, 2($5f^{12}.6d^0.7s^2$); Md(Z = 101) → 2, 8, 18, 32, 31, 8, 2($5f^{13}.6d^0.7s^2$); No(Z = 102) → 2, 8, 18, 32, 32, 8, 2($5f^{14}.6d^0.7s^2$); Lw(Z = 103) → 2, 8, 18, 32, 32, 9, 2($5f^{14}.6d^1.7s^2$). These configurations show that the atoms of actinides have a total of *seven* shells. Out of these seven shells, first four shells namely 1*st*, 2*nd*, 3*rd* and 4*th* are completely-filled while the remaining three shells *viz.* 5*th*, 6*th* and 7*th* shells are partially-filled as shown below :

2, 8, 18, 32, $5s^2p^6d^{10}f^{0,\ 1......7,\ 9......14}$, $6s^2p^6d^{0,\ 1,\ 2}$, $7s^2$ (seven shells)

The configuration given above clearly shows that 5*f* and 6*d* orbitals in the atoms of actinides are partially-filled while 7*s* orbital is completely-filled ($7s^2$ configuration). Thus the number of electrons in 5*th* shell [*i.e.* (*n*–2)*th* shell) is increasing from 18 ($5s^2 + 5p^6 + 5d^{10}$) to 32 ($5s^2 + 5p^6 + 5d^{10} + 5f^{14}$) by adding an electron, one by one, to 5*f* orbital. Consequently the atoms of actinides have partly-filled 5*f* orbital, since the differentiating electron enters in this orbital. Due to the fact that the extra electron goes to 5*f* orbital, the actinides are also called 5*f*-block elements. Complete electronic configurations as given above also indicate that the valence-shell configuration of these elements can be represented as :

$5f^{0,\ 1......7,\ 9......14}$, $6d^{0,\ 1,\ 2}$, $7s^2$

3. Oxidation States.

Important oxidation states shown by actinides are (Most stable oxidation states are given in circles while the unstable ones are given in brackets) given below :

Ac(Z = 89) → + 2, (+ 3), + 4; Th(Z = 90) → + 2, (+ 3), (+ 4), + 5; Pa(Z = 91) → (+ 3), (+ 4), (+ 5) ; U(Z = 92) → (+ 3), (+ 4), + 5, (+ 6); Np(Z = 93) → + 3, (+ 4), (+ 5), (+ 6), (+ 7); Pu(Z = 94) → (+ 3), (+ 4), + 5, + 6 + 7; Am(Z = 95) → + 2, (+ 3), (+ 4), + 5, + 6; Cm(Z = 96) → (+ 3), (+ 4); Bk(Z = 97) → (+ 3); Cf(Z = 98) → (+ 3); Es(Z = 99) → (+ 3); Fm(Z = 100) → (+ 3); Md(Z = 101) → (+ 3); No(Z = 102) → (+ 3); Lw(Z = 103) → (+ 3). It may be seen from these oxidation states that +2 state is shown by two elements namely Am and Th in its few compounds like $ThBr_2$, ThI_2, ThS etc. +3 oxidation state is shown by all the actinides. +3 state becomes more and more stable as the atomic number increases. +4 oxidation state is shown by Th, Pa, U, Np, Pu, Am and Cm while +5 state is shown by Th, Pu, U, Np, Pu and Am. +6 oxidation state is exhibited by U, Np, Pu and Am while +7 oxidation state is shown by Np and Pu.

The principal cations given by actinide elements are M^{3+}, M^{4+} and oxo-cations like MO_2^+ (oxidation state of M = +5) and MO_2^{2+} (oxidation state of M = + 6). Examples of MO_2^+ ion are UO_2^+ and PuO_2^+ while those of MO_2^{2+} are UO_2^{2+} and PuO_2^{2+}. These oxo-cations are stable in acid and aqueous solutions.

U^{3+} ion in aqueous solution liberates H_2 on standing. Np^{3+} and Pu^{3+} are stable in water but are readily oxidized by air to Np^{4+} and Pu^{4+} respectively. All the remaining M^{3+} ions upto Md^{3+} are stable in aqueous solution. U^{4+} and Np^{4+} ions are stable in aqueous solution but are slowly oxidized by air to UO_2^{2+}(U = + 6) and PuO_2^{2+} (Pu = + 6) ions respectively.

4. Radii of Tripositive and Tetrapositive Actinide Cations : Actinide Contraction.

A look at the values of the radii of tripositive and tetrapositive actinide cations (M^{3+} and M^{4+} cation.) reveals that these values for both the cations decrease as we move from Ac to Cm. This steady decrease in the size of M^{3+} and M^{4+} cations in the actinide series is called *adinide contraction* which is analogous to lanthanide contraction found in lanthanides.

	——— Decreasing ——→
Radii of M^{3+} cations (A°) :	Ac^{3+}(1.11) > Th^{3+}(1.08) > Pa^{3+}(1.05) > U^{3+}(1.03) > Np^{3+}(1.01) > Pu^{3+}(1.00) > Am^{3+}(0.99) > Cm^{3+}(0.98)
	——— Decreasing ——→
Radii of M^{4+} cations (A°) :	Ac^{4+}(0.99) > Th^{4+}(0.96) > Pa^{4+}(0.93) > U^{3+}(0.92) > Np^{4+}(0.91) > Pu^{4+}(0.90) > Am^{4+}(0.89) > Cm^{4+}(0.88)

Cause of Actinide Contraction. We know that as we proceed from one element to the next one in the actinide series, the nuclear charge (*i.e.* atomic number) increases by +1 at each next element and the addition of extra electron takes place in 5*f* orbital. Again it is also known that the shielding of one electron in 5*f* orbital by the other electron residing in the same orbital is very poor. Due to the negligible amount of mutual shielding effect between the electrons residing in 5*f* orbital, the increase in nuclear charge by + 1 at each next element in the actinide series brings the valence-shell nearer to the nucleus and hence the size of M^{3+} and M^{4+} cations goes on decreasing as we move from one element to the next one in the series.

5. Colours of M^{3+} and M^{4+} Cations.

Most of the tripositive and tetrapositive actinide cations (M^{3+} and M^{4+} cation.) are coloured. The colours are given below :

M^{3+} cations : Ac^{3+} ($5f^0.6d^0.7s^0$, n = 0) → colourless; U^{3+} ($5f^3.6d^0.7s^0$, n = 3) → red; Np^{3+} ($5f^4.6d^0.7s^0$, n = 4) → purple; Pu^{3+} ($5f^5.6d^0.7s^0$, n = 5) → violet; Am^{3+} ($5f^6.6d^0.7s^0$, n = 6) → pink; Cm^{3+} ($5f^7.6d^0.7s^0$, n = 7) → colourless.

M^{4+} cations : Th^{4+} ($5f^0.6d^0.7s^0$, n = 0) → colourless; Pa^{4+} ($5f^1.6d^0.7s^0$, n = 1) → colourless; U^{4+} ($5f^2.6d^0.7s^0$, n = 2) → green; Np^{4+} ($5f^3.6d^0.7s^0$, n = 3) → yellow-green; Pu^{4+} ($5f^4.6d^0.7s^0$, n = 4) → orange; Am^{4+} ($5f^5.6d^0.7s^0$, n = 5) → red.

Here n is the number of unpaired electrons in 5*f* orbitals. These colours show that M^{3+} and M^{4+} cations having $5f^0$, $5f^1$ and $5f^7$ configuration are colourless while those containing $5f^2$, $5f^3$, $5f^4$, $5f^5$ and $5f^6$ configuration are coloured. These colours are produced when an electron jumps from one energy level to the other within 5*f* orbitals.

6. Formation of Complexes.

Most of the actinide halides form complex compounds with alkali metal halides. For example $ThCl_4$ forms complexes such as $K[ThCl_5]$, $K_2[ThCl_6]$ etc. with KCl. $ThCl_4$ and $ThBr_4$ also form complexes with pyridine. Actinides also form chelates with organic compounds like EDTA and oxine.

The degree of complex formation for the ions M^{4+}, MO_2^{2+}, M^{3+} and MO_2^+ decreases in the order :

$$M^{4+} > MO_2^{2+} > M^{3+} > MO_2^+$$

The complexing power of different anions with the above cations is in the order :

Singly-charged anions : $F^- > NO_2^- > Cl^-$

Doubly-charged anions : $CO_3^{2-} > C_2O_4^{2-} > SO_4^{2-}$

7. Properties of Actinides.

The actinides are silvery white metals which are highly reactive. All the actinides are highly electropositive like lanthanides. Tripositive (*e.g.* U^{3+}, Np^{3+} etc.) and tetrapositive (*e.g.* Pa^{4+}, U^{4+} etc.) actinide cations are paramagnetic like the tripositive lanthanide cations like Ce^{3+}, Pr^{3+}, Nd^{3+} etc. When actinides combine with O_2, oxides are formed. Examples are ThO_2, PaO_2, UO_2, NpO_2 AmO_2, Cm_2O_3 etc. When H_2 acts on actinides, hydrides are formed. Examples are ThH_4, PaH_3, UH_3, PuH_2, AmH_2 etc. Chlorides, bromides and iodides are obtained when the metal is acted on by Cl_2, Br_2 and I_2 respectively.

Comparison between Lanthanides and Actinides.

(A) Similarities. (*i*) In the atoms of the elements of both the series, the extra electron enters (*n*–2)*f* orbitals. (*ii*) The elements of both the series show +3 oxidation state. (*iii*) Like lanthanide contraction seen in case of lanthanides, we also have actinide contraction in actinides. Both the contractions are due to the poor shielding effect between the electrons residing in (*n*–2) *f* orbitals. (*iv*) The absorption bands of the elements of both the series are so sharp that they appear to be almost line-like bands. Both the bands are produced due to the jump of an electron from one energy level to the other within (*n*–2)*f* orbitals. (*v*) Elements of both the series have low electronegativities and are very reactive. (*vi*) The nitrates, perchlorates and sulphates of trivalent actinides as well as lanthanides are soluble while the hydroxides, fluorides and carbonates of both cations are insoluble. (*vii*) Most of the lanthanide and actinide cations are paramagnetic. (*viii*) In the atoms of the elements of both the series, three outermost shells are partly-filled while the remaining inner-shells are completely-filled.

(B) Dissimilarities. The elements of both the series also show some dissimilarities which are due to the difference in nature of 4*f* and 5*f* orbitals. The points showing dissimilarity are given below in a tabular form :

Lanthanides	Actinides
(*i*) Due to large energy difference between 4*f* and 5*d* orbitats, the properties of lanthanides are fairly similar.	(*i*) Due to small energy difference between 5*f* and 6*d* orbitals, the properties of actinides are considerably different from each other.
(*ii*) Binding energies of 4*f* orbitals are higher.	(*ii*) Binding energies of 5*f* orbitals are lower.
(*iii*) The additional electron enters 4*f* orbitals.	(*iii*) The addition electron enters 5*f* orbitals.
(*iv*) These elements show only +2, +3 and +4 oxidation states.	(*iv*) These elements show a variety of oxidation states like +2, +3, +, +5, +6 and +7.
(*v*) The mutual shielding effect between two electrons residing in 4*f* orbitals is greater.	(*v*) The mutual shielding effect between two electrons residing in 5*f* orbitals is poor.
(*vi*) Most of the tripositive lanthanide cations are colourless.	(*vi*) Most of the tripositive and tetrapositive actinide cations are coloured.
(*vii*) The paramagnetic character of lanthanides cart easily be explained.	(*vii*) The paramagnetic character of actinides can not be explained easily.
(*viii*) They do not form complex compounds easily. They do not form complexes with p-bonding ligands like alkyl phosphines, thioethers etc.	(*viii*) They form complex compounds comparatively more easily. They do form complexes with p-bonding ligands.
(*ix*) Except Pm, all the remaining lanthanides are non-radioactive.	(*ix*) All the actinides are radioactive.
(*x*) The compounds of lanthanides are less basic.	(*x*) The compounds of actinides are more basic.
(*xi*) Lanthanides do not form oxo-cations.	(*xi*) Some actinide form oxo-cations of MO_2^+ (*e.g.* UO_2^+, PuO_2^+) and MO_2^{2+} (*e.g.* UO_2^{2+}, PuO_2^{2+}) type.
(*xii*) The atoms of lanthanides have a total of six shells. Out of these, first three shells *viz.* 1*st*, 2*nd* and 3*rd* are completely filled while the remaining three shells namely 4*th*, 5*th* and 6*th* are partially-filled.	(*xii*) The atoms of actinides have a total of seven shells. Out of these, first tour shells *viz.* 1*st*, 2*nd*, 3*rd* and 4*th* are completely-filled while the remaining three shell. *viz.* 5*th*, 6*th* and 7*th* are partially-filled.

Conceptual Questions with Answers

Q.1 Write down the electronic configuration of (*a*) Cr^{3+} (*b*) Cu^+ (*c*) Co^{2+} (*d*) Mn^{2+}

Ans. (*a*) $Cr^{3+} = [Ar]_{18}\ 3d^3$ (*b*) $Cu^+ = [Ar]_{18}\ 3d^{10}$ (*c*) $Co^{2+} = [Ar]_{18}\ 3d^7$ (*d*) $Mn^{2+} = [Ar]_{18}\ 3d^5$.

Q.2 Atomic radius of Cu is greater than that of Cr but ionic radius of Cr^{2+} is greater than that of Cu^{2+}. Give suitable explanation.

Ans. In Cu, all the *d*-electrons are paired ($3d^{10}\ 4s^1$). In Cr, all the *d*-electrons are unpaired ($3d^5\ 4s^1$). Hence, *d–d* electron repulsions in Cu are much greater than those in Cr. Hence, Cu atom is larger in size than Cr. In Cu^{2+} ($3d^9$), *d–d* electron repulsions decrease due to presence of one unpaired *d*-electron. Moreover in Cu^{2+} ion the electrons are attracted by 29 protons of the nucleus whereas in Cr^{2+} ($3d^4$), four unpaired electrons are still present but they are attracted by only 24 protons of the nucleus. Thus Cu^{2+} is smaller in size than Cr^{2+}.

Q.3 Why are IEs of 5*d* elements greater than 3*d* elements ?

Ans. In 5*d* elements, after La_{57} there is lanthanide contraction. In each group the size of 5*d* elements is small while nuclear charge is greater than 3*d* elements. Hence IEs of 5*d* elements are greater than 3*d* elements.

Q.4 The valence shell E.C. of Cr atom is $3d^5\ 4s^1$. Write (*i*) the most stable oxidation state of Cr (*ii*) most oxidising state of Cr.

Ans. (*i*) The most stable oxidation state of Cr is +2 which is obtained when Cr atom loses one electron from 4*s* orbital and one electron from 3*d* orbitals.

(*ii*) Most oxidising state of an element is its highest oxidation state which it can exhibit. Obviously, since the highest oxidation state of Cr is +6, the most oxidising state of Cr is also +6. This state is obtained when all the six electrons present in the valence shell of Cr are lost. This oxidation state is shown by Cr in $K_2Cr_2O_7$ which is an oxidising sgent. During its oxidising action, $K_2Cr_2O_7$ (Cr = +6) gets reduced to Cr^{3+} (Cr = +3)

$$Cr_2O_7^{2-} + 14H^+ + 6e^- \xrightarrow{\text{Reduction}} 2\,Cr^{3+} + 7H_2O$$

Q.5 Predict which of the following will be coloured in aqueous solution ? Ti^{3+}, V^{3+}, Sc^{3+}, Mn^{2+}, Fe^{3+}, Co^{2+} and MnO_4^-.

Ans. Only those ions will be coloured which have incompletely filled d-orbitals. Those with fully-filled or empty d-orbitals are colourless. Thus, Ti^{3+}, V^{3+}, Mn^{2+}, Fe^{3+} and Co^{2+} are all coloured. ($Ti^{3+} = 3d^1$, $V^{3+} = 3d^2$, $Mn^{2+} = 3d^5$, $Fe^{3+} = 3d^5$, $Co^{2+} = 3d^7$). MnO_4^- is also coloured but due to charge transfer (from O to Mn) which reduces the oxidation state of Mn from +7 to +6 momentarily.

Q.6 Out of Co^{2+} and Zn^{2+} salts which is attracted in a magnetic field. Explain with reasons.

Ans. Co^{2+} and Zn^{2+} ions are $3d^7$ (*n* = 3) and $3d^{10}$ (*n* = 0) ions. Since Co^{2+} ion has unpaired electrons (*n* = 3) and Zn^{2+} ion has no unpaired electrons (*n* = 0), Co^{2+} ion is paramagnetic and Zn^{2+} ion is diamagnetic. Being paramagnetic Co^{2+}

ion is attracted in magnetic field. On the other hand, Zn^{2+} ion, being diamagnetic, is not attracted by the magnetic field.

Q.7 Why hydrated copper sulphate is blue while anhydrous copper sulphate is white ?

Ans. In hydrated copper sulphate, four water molecules are present as ligands. In the presence of these ligands *d*-orbitals are no longer degenerate in energy. Hence *d–d* transition takes palce absorbing red wavelength. The complementary colour, *viz*., blue is reflected. In anhydrous $CuSO_4$, *d*-orbitals remain degenerate. Hence, no *d–d* transition can occur. The white light is completely reflected back. Hence,it looks white.

Q.8 Write the formulae of different oxides of manganese. What is the oxidation state of manganese in each of them ? Arrange them in order of their decreasing acidic charater.

	+2	+8/3	+3	+4	+7
Ans.	MnO	Mn_3O_4	Mn_2O_3	MnO_2	Mn_2O_7
	Basic	Amphotetic			Acidic

Thus, their acidic character decreases as follows :

$Mn_2O_7 > MnO_2 > Mn_2O_3 > Mn_3O_4 > MnO$.

Q.9 Why Zr and Hf or Nb and Ta exhibit similar properties ?

Or Zirconium (atomic number 40) and hafnium (atomic number 72) occur together in minerals and they exhibits similar properties. Give reasons.

Ans. Due to lanthanoid contraction, Hf (Z = 72) has size similar to that of Zr(Z = 40). Hence, their properties are similar and therefore, occur together. For the same reason, Nb and Ta have similar size and hence similar properties.

Q.10 The E° values in respect of the electrodes of chromium (Z = 24), manganese (Z = 25) and iron (Z = 26) are :

$$Cr^{3+} / Cr^{2+} = -0.4V,\ Mn^{3+} / Mn^{2+} = +1.5V,\ Fe^{3+} / Fe^{2+} = +0.8V.$$

On the basis of the above information compare the feasibilities of further oxidation of their +2 oxidation states.

Ans. –ve value of E° for Cr^{3+} / Cr^{2+} shows that Cr^{2+} is least stable. Greater +ve value for Mn^{3+} / Mn^{2+} than that for Fe^{3+} / Fe^{2+} shows that Mn^{2+} is more stable than Fe^{2+}. Hence, stability of +2 oxidation state is in the order :

$$Cr^{2+} < Fe^{2+} < Mn^{2+}$$

or the feasibility of oxidation of thier +2 state to +3 state is in the order :

$$Cr^{2+} > Fe^{2+} > Mn^{2+}$$

Q.11 Why Sm^{2+}, Eu^{2+} and Yb^{2+} ions in solutions are good reducing agents but an aqueous solution of Ce^{4+} is a good oxidizing agent ?

Ans. The most stable oxidation state of lanthandies is +3. Hence, ions in +2 state tend to change to +3 state by loss of electron and those in +4 state tend to change to +3 state by gain of electron.

Q.12 Expalin why lanthnoids are paramagnetic in nature ?

Ans. All lanthanoids except La^{3+} and Lu^{3+} contain unpaired electrons and hence are paramagnetic.

Q.13 The +3 oxidation state of lanthanum (Z = 57), gadolinium (Z = 64) and lutetium (Z = 71) are especially stable. Why ?

Ans. This is because they have empty, half-filled and completely filled $4f$ sub-shell respectively.

Q.14 Among the ionic species, Sc^{3+}, Ce^{4+} and Eu^{2+}, which one is a good oxidizing agent ? Give a suitable reason for your answer. (Atomic nos : Sc = 21, Ce = 58, Eu = 63)

Ans. The typical (stable) oxidation state of lanthanoids is +3. Ions with oxidation state +2 or +4 tend to change to +3 by loss or gain of electrons. Ce^{4+} gains electron to change to Ce^{3+}. Hence, it is a good oxidizing agent.

Q.15 Among the lanthanides, Ce (III) can be easily oxidized to Ce (IV) (At. No. of Ce = 58). Explain why ?

or Of the lanthanides, cerium (At. No. 58) forms a tetrapositive ion, Ce^{4+} in aqueous solutions. Why ?

Ans. Ce (III) having the configuration $4f^1\ 5d^0\ 6s^0$ can easily lose an electron to acquire the configuration $4f^0$ and form Ce(IV). In fact, this is the only (+ IV) lanthanide which exists in solution.

Q.16 The outer electronic configuration of two members of the lanthanoid series are as follows :

$$4f^1\ 5d^1\ 6s^2 \text{ and } 4f^7\ 5d^0\ 6s^2$$

What are their atomic number ? Predict the oxidation states exhibited by these elements in their compounds.

Ans. Complete E.C. of 1st lanthanoid = $[Xe]^{54}\ 4f^1\ 5d^1\ 6s^2$. Atomic No. = 58 (Ce)

Complete E.C. of 2nd lanthanoid = $[Xe]^{54}\ 4f^7\ 5d^0\ 6s^2$. Atomic No. = 63 (Eu)

Oxidation states of 1st lanthanoid = + 2 ($4f^2$), + 3 ($4f^1$), +4 ($4f^0$)

Oxidation states of 2nd lanthanoid = + 2 ($4f^7$) and +3 ($4f^6$)

Q.17 Which out of the two, $La(OH)_3$ and $Lu(OH)_3$, is more basic and why ?

Ans. $La(OH)_3$ is more basic than $Lu(OH)_3$. As the size of the lanthanoid ions decreases from La^{3+} to Lu^{3+}, the covalent character of the hydroxides increases (Fajan's rules). Hence, the basic strength decreases from $La(OH)_3$ to $Lu(OH)_3$.

University Questions

1. Explain, with examples, what you understand by transition (or *d*-block elements. *(Meerut 82, 84; Raj. 88; Utkal 86; Agra 84; Andhra 85; MD Rohtak 84; Punjab 86; Saugar 85)*

2. Discuss the general characteristics of *d*-block elements. (*Meerut 82, 89; Raj. 82, 83, 86, 87, 87S*)
3. What do you understand by the shielding effect of electrons? How does this effect affect the properties of transition elements. (*Raj. 78*)
4. Discuss the radii of *d*-block elements. (*Raj. 86*)
5. Discuss the trends and variation in oxidation states (valeney) of transition metals. (*Raj. 82, 86; Meerut 82; Bundelkhand 88; Utkal 86; Agra 84; Andhra 85; R.D. Rohtak 84*)
6. Discuss the trends and variations in ionisation potential of transition metals. (*Meerut 81*)
7. Discuss the tendency of transition metals to form complex compounds. Explain with examples. (*Utkal 86; Agra 84; Andhra 86; M.D. Rohtak 84*)
8. Discuss the magnetic properties of *d*-block elements. (*Raj. 86*)
9. Distinguish between transition elements and main group elements in term of their electronic configuration. (*Andhra 85; Madras 85*)
10. How do *d*-block elements differ from *f*-block elements ? (*Meerut 82*)
11. (*a*) Discuss the properties of actinide elements in respect of the following: (*i*) Electronic configuration (*ii*) Oxidation states (*iii*) Ionic radius (*iv*) Magnetic properties.
    (*b*) Give an account of physico-chemical principles involved in the separation of lanthanides by the following methods : (*i*) Fractional precipitation method (*ii*) Fractional crystallisation method.
    (*c*) Give an account of separation of La^{3+}, Pr^{3+}, Nd^{3+} and Sm^{3+} ions by ion exchange chromatographic method. (*Kanpur 99*)
12. What are lanthanides ? Describe ion exchange method for the separation of lanthanides. (*Himachal Pradesh 99*)
13. (*a*) Why do some lanthanides exhibit unusual oxidation states ? Discuss this on the basis of their electronic configuration.
    (*b*) How do transition elements differ from inner transition elements ? Explain. (*Gorakhpur 99*)
14. What are actinides and how do they differ from lanthanides with reference to the following : (*a*) Electronic configuration (*b*) Tonic radii (*c*) Oxidation states. (*Kumaon 2000*)
15. (*a*) What are lanthanide Elements?
    (*b*) Write brief notes on the oxidation states by these elements and also the changes in the magnetic properties of their compounds.
    (*c*) Explain why trivalency oxidation state of the lanthanides elements is common and most stable. (*Avadh 2000*)
16. Discuss the principle and procedure of determination of magnetic moments by Gouy method. How are these helpful in deciding the structure of molecules? (*Avadh 2000*)

17. (*a*) Giving electronic configuration justify the position of *f*-block elements in the periodic table.

(*b*) Write note on "Lanthanide Contraction". (*Lucknow 2001*)

18. Arrange the following actinide ions in the increasing order of their size : Th^{3+}, Am^{3+}, Np^{3+} and U^{3+}. (*H.N. Bahuguna 2005*)

19. What are lanthanides ? Explain in detail their oxidation states and lanthanide contraction. (*Gurukul Kangri 2008*)

20. Explain lanthanide contraction. (*Agra 2008*)

21. (*a*) What are inner transition elements ? Discuss their electronic configurations and oxidation states.

(*b*) Why do actinides show higher oxidation states than lanthanides ?

(*c*) List important uses of actinides. (*M.J.P. Rohilkhand 2008*)

22. Describe general characteristics of lanthanides. (*M.J.P. Rohilkhand 2009*)

23. What are actinides ? Explain resemblances and differences of these from lanthanides. (*Meerut 2009*)

24. Discuss d-block elements in the following respects. (ii) Electronic configuration (ii) Coloured ion formation (iii) oxidation states (iv) Magnetic properties (v) Complex formation. (*Meerut 2009*)

25. Write a note on "Lanthanide Contraction". (*Meerut 2009*)

CHAPTER

45

Group Discussion of d-Block Element

Group 3 (III B) : Sc_{21}, Y_{39}, La_{57} and Ac_{89}

Group III B of the periodic table consists of four elements which are *Scandium* (Sc), *Yttrium* (Y), *Lanthanum* (La) and *Actinium* (Ac). These elements are generally called *Scandium group elements.*

Position of Scandium Group Elements in the Periodic Table

The inclusion of scandium group elements in the same group of the period table is justified by the similarity and gradation in their properties as discussed below :

The electronic configurations of these elements in the ground state given as : Sc (Z = 21) $\rightarrow$ 2, 8, 9, 2 ($3d^1.4s^2$); Y (Z = 39) $\rightarrow$ 2, 8, 18, 9, 2 ($4d^1.5s^2$); La (Z = 57) $\rightarrow$ 2, 8, 18, 18, 9, 2 ($5d^1.6s^2$); Ac (Z = 89) $\rightarrow$ 2, 8, 18, 32, 18, 9, 2, ($6d^1.7s^2$) show that the valence-shell configurations of these elements can be represented by $(n-1)d^1.ns^2$, indicating that the extra electron goes in $(n-1)d$ orbital. Thus these are *d*-block (transition) elements. All the elements have 2 electrons in the outermost shell. All the elements show +3 oxidation state in solution, *i.e.* in solution all the elements exist as M^{3+} ions. These ions are obtained by removing both *ns* and one $(n-1)d$ electrons. The salts of M^{3+} ions have *paired electrons* and hence are *colourless* and *diamagnetic.* The metals have high melting points. Due to their high electropositive character, these metals are highly reactive. For example they react with H_2O to form basic oxides, MO.OH and liberate H_2 gas. These metals react slowly with cold water and more rapidly with hot water. The oxides of these metals can be prepared by heating hydroxides or salts such as carbonate, nitrate or sulphate, while hydroxides can be prepared by treating these metals with H_2O. Oxides and hydroxides have basic character. *The basic character of the hydroxides increases with the increase in size of M^{+3} ion.* [$Sc(OH)_3 < Y(OH)_3 < La(OH)_3$]. Thus $La(OH)_3$ is the strongest base. Because of their basic character the oxides and hydroxides react with acids to form salts. Their oxides are very stable and hence cannot be reduced by thermit reaction. The size of the atoms (atomic radii) and M^{3+} ions increase as we proceed from Sc to Ac. *The tendency of M^{+3} ions to form complexes decreases with the increase of their size.* The Sc^{3+} ion which is the smallest shows some tendency to form the complexes like $[ScF_4]^-$, $[ScF_5]^{2-}$, $[ScF_6]^{3-}$ etc, while the other members do not have a strong tendency to form complexes.

Group 4 (IV B) : Ti_{22}, Zr_{40}, Hf_{72} and Ku_{104}

Group IVB of the periodic table consists of four elements which are *Titanium* (Ti), *Zirconium* (Zr), *Hafnium* (Hf) and *Kurchatovium* (Ku). These elements are generally called *titanium group elements.*

Position of Titanium Group Elements in the Periodic Table : Position of Titanium in the Periodic Table

The inclusion of titanium group elements in the same group of the periodic table is justified by the similarity and gradation in their properties as discussed below :

The electronic configurations of these elements in the ground state given as Ti (Z = 22) → 2, 8, 10, 2 ($3d^2.4s^2$); Zr (Z = 40) → 2, 8, 18, 10, 2 ($4d^2.5s^2$); Hf (Z = 72) → 2, 8, 18, 32, 10, 2 ($5d^2.6s^2$) indicate that the valence-shell configurations can be represented by $(n–1)d^2.ns^2$ showing that the extra electron goes to $(n–1)d$ orbitals. Thus these are *d*-block (transition) elements. All the elements have 2 electrons in the outer most shell. More common oxidation states are +2, +3 and +4. +2 oxidation state is very unstable and so strongly reducing that it reduces water and itself is oxidised to +3 state. +3 oxidation state is comparatively more stable and hence the compounds in this state are reducing agents. Since M^{3+} ions (+3 oxidation state) have one unpaired electron [$M^{3+} \rightarrow (n–1)d^1$], the compounds in this oxidation state are paramagnetic. Due to *d–d* transition of one $(n–1)d$ electron, the compounds of M^{3+} ions show one band in the visible spectrum and hence are purple coloured. +4 oxidation state is most common oxidation state for these elements. Compounds of M^{4+} ion have the following properties : (*i*) Since the amount of energy to remove all the four $(n–1)d^2.ns^2$ electrons to get M^{4+} ions is large, the anhydrous compounds having M^{4+} ion (*e.g.* $TiCl_4$) are covalent. (*ii*) The covalent compounds are tetrahedral. (*iii*) Since the element in +4 oxidation state has no electron (d° configuration), the compounds in this state are diamagnetic and colourless or white. Lower oxidation states undergo disproportionation into higher and more lower oxidation states. Zero, –1 and –2 oxidation states are shown in $[Ti(dipy)_3]^\circ$, $Li[Ti(dipy)_3]$. 3.5 THF and $Li_2[Ti(dipy)]$, 5 TIIF respectively. (THF tetrahydrofuran). The covalent radii increase from Ti to Zr and then decrease to Hf. Ionic radii of M^{4+} ions increase from Ti^{4+} to Hf^{4+}. No definite hydroxides, $M(OH)_4$ are known. The hydroxides are really hydrated oxides. These metals, on heating with O_2 above 800°C, give dioxides (MO_2) which are stable, non-volatile and insoluble and become refractory on heating. TiO_2 is amphoteric (though more acidic than basic). It dissolves in alkalies to form titanates and in acids to form titanyl compounds. ZrO_2 and HfO_2 are more basic and less acidic. The metals also form peroxides and proxy compounds when reacted with H_2O_2. Ti and Zr give complete series of tetrahalides. Hf gives only HfF_4 and $HfCl_4$. MX_4 tetrahalides are formed by the interaction of the metal with halogen. $TiCl_4$ is formed by heating TiO_2 with carbon and chlorine. Other tetrahalides of Ti can be prepared from $TiCl_4$ by the reaction with appropriate HX. Tetrafluorides are largely ionic while other tetrahalides are predominantly covalent volatile substances. Except tetrafluorides all other tetrahalides are readily hydrolysed by H_2O. The extent of hydrolysis decreases from Ti to Hf. TiX_4 compounds are hydrolysed to TiO_2 while ZrX_4 and HfX_4 give $ZrOX_2$ and $HfOX_2$. Tetrahalides act as electron pair acceptor (Lewis acid) donated by the donor atoms like F^-, Cl^-, phosphines (R_3P), arsines (R_3As), R_2O, NH_3, pyridine and $(CH_3)_3$ N and form complexes or addition compounds. These elements are very hard metals, good conductors and possess high melting and boiling points. These are unreactive at ordinary temperature but combine on heating with O_2 to give dioxides (MO_2) and with halogens to give tetrahalides (MX_4). Hydrogen, nitrogen, carbon and boron give hydrides ($MH_{1.7, 2.0}$), nitrides (MN), carbides (MC) and borides (MB and MB_2) respectively. These are interstitial compounds, very hard and refractory (can withstand high temperatures). These metals are not attacked by dil. acids or alkalies. This is due to the formation of a protective oxide film. Ti dissolves slowly in hot and conc. acids and explodes in fuming HNO_3. Zr dissolves in acqua regia and to a smaller extent in hot conc. H_2SO_4. The best solvent for Ti is HF which forms hexafluoro complex with it. As a result of *lanthanide contraction* Hf has almost the same atomic (covalent) ionic radius as Zr. The similarity in size of the atoms and ions of these two elements is responsible for similarity in their properties. The chemical properties of the two elements are so similar that till recently their separation from each other was considered difficult. Now these are separated by ion exchange technique. Ti^{4+} does not form any normal salt with oxy-acids. However the normal salts of Zr^{4+} ion like $Zr(SO_4)_2$ have been obtained.

Group 5 (V B) : V_{23}, Nb_{41} and Ta_{73}

Group VB of the periodic table consists of three elements which are *Vanadium* (V), *Niobium* (Nb) and *Tentalum* (Ta).

Position of Vanadium Group Elements in the Periodic Table : Position of Vanadium in the Periodic Table

The inclusion of vanadium group elements in the same group is justified by the similarity and gradation in their properties. Some of these properties are discussed below :

The electronic configurations of these elements in the ground state given as V (Z = 23) $\rightarrow$ 2, 8, 11, 2 ($3d^2.4s^2$); Nb (Z = 41) $\rightarrow$ 2, 8, 18, 12, 1 ($4d^4.5s^1$); Ta (Z = 73) $\rightarrow$ 2, 8, 18, 32, 11, 2 ($5d^3.6s^2$) show that, excepting Nb, the valence-shell configuration of the remaining two elements can be represented as $(n\text{–}1)d^3.ns^2$. In case of Nb one electron from 5*s* orbital gets shifted to 4*d* orbital and hence the valence-shell configuration of Nb becomes $4d^4.5s^1$ instead of the expected configuration namely $4d^3.5s^2$. The valence-shell configurations also show that the extra electron goes to $(n\text{–}1)d$ orbital and hence these are *d*-block (transition) elements. It is also evident that these elements have 5 electrons in their valence-shell. They show a variety of oxidation states. The stability of +5 state increases from V to Ta. Thus V(+ 5) compounds are reduced all the way to V(+2) compounds and Nb (+5) and Ta (+5) compounds remain stable towards reduction. Consequently V(+5) compounds are good oxidising agents while those of Nb (+5) and Ta (+5), being stable, are not, Lower oxidation states become less stable on moving down from V to Ta. The ionic character of compounds of each element decreases with the decrease of cationic size (*Fajans's rule*). Thus, since V (+2) > V (+3) > V (+4), V(+2) compounds are more ionic than those of V (+3) and (V +4). Consequently V (+4) compounds (*e.g.* VCl_4) are covalent. These metals are unreactive at ordinary temperatures. But, on heating, they show their reactivity. On heating with O_2 these elements give M_2O_5 type pentoxides. Vanadium also gives V_2O_4 in addition to V_2O_5, The basic character of M_2O_5 type oxides increases from V_2O_5 to Ta_2O_5. Thus V_2O_5 is amphoteric (but more acidic than basic) while Nb_2O_5 and Ta_2O_5 are less acidic than basic. This type of variation in the basic character of M_2O_5 oxides shows that the metallic character of the elements increases from V to Ta on descending the group. The ionic character of their compounds decreases as the oxidation state of the element increases. For example V (+2) compounds are ionic. V (+3) compounds are less ionic while V (+4) compounds (*e.g.* VCl_4) are definitely covalent. Since the compounds with lower oxidation states have incomplete *d*-orbitals, these are *coloured*, while those with higher oxidation states which have *completely-filled d-orbitals* are *colourless*. For example the pentoxides namely Nb_2O_5 and Ta_2O_5 have completely filled *d*-orbitals and are, therefore, colourless. V_2O_5 is however, red-orange. Its orange colour is considered to be due to some defects in the crystal structure of the oxide. The colour of NbX_5 is considered to be due to charge transfer. Vanadium, on heating with halogens, gives VF_5, VCl_5, VCl_4, VCl_3, VI_3 halides while Nb and Ta give only MX_5 pentahalides (M = Nb, Ta). All these halides are covalent and are, therefore, volatile and get readily hydrolysed On heating with N_2, H_2 and carbon, these elements form nitrides (MN), hydrides and carbides (MC and MC_2). Nitrides are interstitial compounds. MC type carbides are interstitial, refractory and hard while MC_2 type carbides are ionic and react with water to liberate acetylene. Hydrides are non-stoichiometric and the amount of hydrogen absorbed depends on the temperature and pressure. $VH_{0.71}$, $NbH_{0.86}$ and $TaH_{0.76}$ are the examples of non-stoichiometric hydrides. The tendency of these elements to forms the complexes decreases with the increase of atomic size. Thus vanadium which has the smallest size forms the largest number of complexes while Ta gives the smallest number. Tetraperoxy compounds of $[M(O_2)_4]^{3-}$ type (M = V, Nb and Ta) are known. In these compounds V, Nb and Ta are in +5 oxidation state. $K_3[V(O_2)_4]$ is obtained by adding alcohol to the blue solution obtained by treating V_2O_5 with KOH and H_2O_2. $K_3[V(O_2)_4]$ is isomorphous with $K_3[Cr(O_2)_4]$. $K_3[Nb(O_2)_4]$ and $K_3[Ta(O_2)_4]$ can also be prepared by similar methods. In addition to $[V(O_2)_4]^{3-}$ vanadium also gives $[HV(O)(O_2)_3]^{2-}$, $[HV(O)_2(O)_2]^{2-}$ and $[V(O)(O)_2]^+$ peroxy compounds.

Group 6 (VI B) : Cr_{24}, Mo_{42} and W_{74}

Group VI B of the periodic table consists of three elements which are *Chromium* (Cr). *Molybdenium* (Mo) and *Tungsten* (W). These are called *Chromium group elements.*

Position of Chromium Group Elements in the Periodic Table

The inclusion of chromium group elements in the same group is justified by the similarity and gradation in their properties. Some of these properties are discussed below :

The electronic configurations of these elements in the ground state given as : Cr (Z = 24) → 2, 8, 13, 1 ($3d^5.4s^1$); Mo (Z = 42) → 2, 8, 18, 13, 1 ($4d^5.5s^1$); W (Z = 74) → 2, 8, 1, 2, 12, 2 ($5d^4.6s^1$) show that, excepting W, the valence-shell configuration of the remaining two elements can be represented as $(n-1)d^5.ns^1$. In case of these two elements one electron from ns orbital gets shifted to $(n-1)d$ orbital and hence the observed configuration becomes $(n-1)\ d^5.ns^1$ instead of the expected configuration, $(n-1)\ d^4.ns^2$. The observed valence-shell configuration of W is the same as the expected configuration. The valence-shell configuration of all the three elements shows that the extra electron goes to $(n-1)\ d$ orbital and hence these are *d*-block (transition) elements. It is also evident that these elements have six electrons in their valence- shell. These elements show variable oxidation states. This is because of the fact that the energy of $(n-1)d$ orbitals is almost the same as that of *ns* orbitals and hence the electrons from both the orbitals can be used for bonding. Most stable oxidation state for Cr is +3 while that for Mo and W is +5. This shows that with increase of atomic number the higher oxidation state becomes more stable while the lower one becomes less stable. *The stability of the hexavalent compounds increases with the atomic number.* Chromic acid is one of the well known powerful oxidising agents; it gives out oxygen and reverts to a chromic salt in which it is positively trivalent. Molybdic acid is only a weak oxidising substance but tungstic and uranic acid are stable acids, devoid of oxidising character. In metal carbonyls like $Cr(CO)_6$, $Mo(CO)_6$, and $W(CO)_6$ these metals have *zero oxidation state.* The most stable chromium salts have the valency 2 (*chromous*) and 3 (*chromic*). These are hard silvery metals. These are harder than glass and are refractory. Ionisation energy values of these elements gradually increases from Cr to W. The density of these elements increases gradually from Cr to W. All the three metals have high melting and bonding points and these values increase with the increase in atomic number. All the three elements have almost the same electronegativity. The size of M atoms and M^{6+} ions increase from Cr to W. The size of Mo and W are almost the same because of lanthanide contraction. All the three metals form excellent alloys. For example Cr is used in making stainless steel, Mo and W, when alloyed with steel, give very hard alloys. All the three metals combine with S and N_2 and form sulphides and nitrides respectively. All the three metals combine with O_2 and form trioxides of RO_3 type in which R(R = Cr, Mo or W) is in +6 oxidation state. All the three metals form oxy-acids of H_2RO_4 type. The strength of these acids decreases from H_2CrO_4 (very strong acid) to H_2WO_4 (very weak acid). Their oxidising properties decrease from H_2CrO_4 to H_2WO_4. All the three metals form oxychlorides of RO_2Cl_2 type. These are obtained by dissolving RO_3 oxides in conc. HCl in presence of H_2SO_4 or by direct addition of Cl_2 to RO_3 oxide.

Position of Cr in the Periodic Table

A look at the position of Cr in the periodic table shows that (1) Cr lies with Mo and W in the same sub-group (2) Mn and Fe lie to its right hand side (3) S is acting as a bridge element between the elements of sub-groups VIA and VI B. Thus the position of Cr in the periodic table is justified by discussing the similarities of Cr with the above said elements. These similarities are discussed as follows :

1. Similarities of Cr with Mo and W. The similarities of Cr with Mo and W have already been discussed under "*Position of Chromium Group Elements in the Periodic Table*".

2. Similarities of Cr with Mn and Fe. The points showing the similarity between the three metals are (*i*) None of them occurs in nature in the free state. (*ii*) These are transition elements and show all the properties of transition elements. (*iii*) Their hydroxides *viz.* $Cr(OH)_3$, $Mn(OH)_2$ and $Fe(OH)_3$ are insoluble in H_2O. (*iv*) These are hard and shining metals. The melting and boiling points of all the three metals are high. (*v*) They form the oxides of MO (*e.g.* CrO, MnO, FeO) and M_2O_3 (*e.g.* Cr_2O_3 Mn_2O_3, Fe_2O_3) type. (*vi*) Oxides of Cr and Fe (Cr_2O_3 and Fe_2O_3) are isomorphons. (*vii*) CrO_3 and Mn_2O_3 are acidic in nature, since they dissolve in water, giving chromic (H_2CrO_4) and manganic (H_2MnO_4) acids. The salts of H_2CrO_4 (called chromates) and H_2MnO_4 (called manganates) are isomorphous *e.g.* K_2CrO_4 and K_2MnO_4. (*viii*) They give excellent alloys. (*ix*) Cr and Fe form isomorphous alums, *e.g.* K_2SO_4. $Fe_2(SO_4)_3$. $24H_2O$ (called *ferric alum*) and K_2SO_4. $Cr_2(SO_4)_3$ $24H_2O$ (*chrome alum*) are isomorphous. (*x*) Both Cr and Fe become passive on treatment with conc. HNO_3.

3. Similarities of Cr with S. The points showing similarities between Cr and S are : (*i*) Both show hexavalency. (*ii*) Trioxides of both (*i.e.*, CrO_3 and SO_3) are acidic and react with water to give chromic and sulphuric acid. The salts of these acids, chromates and sulphates are isomorphous. (*iii*) $BaSO_4$ and $BaCrO_4$, and $PbSO_4$ and $PbCrO_4$ are insoluble in water. (*iv*) CrO_2Cl_2 and SO_2Cl_2 both react with water to give H_2CrO_4 and H_2SO_4 respectively.

Comparison between Cr and S.

The comparison between Cr and S consists of making a discussion of the similarities and dissimilarities between Cr and S.

(*a*) **Similarities.** These have already been discussed above.

(*b*) **Dissimilarities.** (*i*) Sulphur is a *typical element* with *six* electrons in its outermost orbit ($3s^2p^4$) while chromium is a *transition element* with *one* electron in its outermost orbit ($3d^54s^1$). (*ii*) Sulphur is a typical non-metal while chromium shows *metallic behaviour.* It is hard, lustrous and has high density, melting and boiling point. (*iii*) Sulphur readily burns in air and forms SO_2 while chromium is not acted upon by air. (*iv*) Sulphur dissolves in alkalies to form thiosuiphate while chromium remains unaffected by alkalies. (*v*) Dilute acids have on action on sulphur but chromium dissolves in them to form salts. Concentrated HNO_3 oxidises sulphur to form H_2SO_4 but chromium remains passive. (*iv*) Sulphur combines with metals and non-metals to give sulphides. Chromium forms alloys with metals.

Group 7 (VII B) : Mn_{25}, Tc_{43} and Re_{75}

Group VII B of the periodic tae consists of three elements which are *Manganese* (Mn), *Technitium* (To) and *Rhenium* (Re). Of these the last two are rare metals. The elements of this group are called manganese group elements. Out of these metals Tc does not occur in nature and was the first man-made element. It is radioactive and its chemistry is not fully known.

Position of Manganese Group Elements in the Periodic Table

The inclusion of manganese group elements in the same group of the periodic table is justified by the similarity and gradation in their properties. Some of these properties re discussed below:

The electronic configurations of these elements in the ground state given as : Mn (Z = 25) → 2, 8, 13, 2 ($3d^5.4s^2$); Tc (Z = 43) → 2, 8, 18, 13, 2 ($4d^5.5s^2$) : Re (Z = 75) → 2, 8. 18, 32, 13, 2 ($5d^5.6s^2$), show that the valence-shell configurations of these elements can be represented by $(n-1)d^5.ns^2$ showing that the extra electron goes to $(n-1)d$ orbital and hence these are *d*-block (transition) elements. In accordance with their electronic configurations these metals show a maximum oxidation state of +7 in compounds like $KMnO_4$. Mn_2O_7, Re_2O_7 etc. In addition to this state, they also show lower oxidation states. The stability of higher oxidition states increases gradually from Mn to Re, *e.g.* the permanganates, MnO_4^-, decompose readily in acidic, alkaline and neutral solutions evolving O_2 and thus act as strong oxidising agents, while pertechnetates, TcO_4^- and perrhenates, ReO_4^- are more stable and hence have less oxidising power. The stability of lower oxidation states decreases from Mn to Re. All these elements show zero oxidation states in carbonyls like $Mn_2(CO)_{10}$, $Tc_2(CO)_{10}$ and $Re_2(CO)_{10}$.

Both Mn and Re form a number of oxides while the most important oxide of Tc is Tc_2O_7. The basic character of the oxides decreases gradually with the increase of oxidation state of the element. For example, the oxides *viz.* MnO and Mn_2O_3 (in which Mn is in +2 and +3 state respectively) are *basic and ionic*, MnO_2 (in which Mn is in +4 state) is *amphoteric* and *does not exist as Mn^{4+} ion*, MnO_3 (in which Mn is in +6 state) is an *unstable acidic oxide* and Mn_2O_7 (in which Mn is in +7 state) is *strongly acidic.* The permanganic acid, $HMnO_4$ which corresponds to Mn_2O_7 is the *strongest acid known.* This acid exists only in solution. Thus we see that *with the increase of oxidation state of Mn atom in its oxides, the character of these oxides changes from basic to amphoteric and finally to acidic.* All the three elements form the oxy-acids of HMO_4 type (M = Mn, Tc and Re), *e.g.* $HMnO_4$ (*permanganic acid*), $HTcO_4$ (*pertechnetic acid*) and $HReO_4$ (*perrhenic acid*). The salts of these acids are called permanganates (MnO_4^-), *pertechnetates* (TcO_4^-) and *perrhenats* (ReO_4^-) respectively. These salts are analogous to each other.

Position of Mn in the periodic Table

A look at the position of Mn in the periodic table shows that (1) Mn lies with Tc and Re in the same sub-group (2) Cr lies to its left hand side (3) Fe lies to its right hand side (4) Cl is acting as a bridge element between the elements of sub-groups VII A and VII B. Thus the position of Mn in the periodic table is justified by discussing the similarities of Mn with the above said elements. These similarities are discussed below :

1. Similarities of Mn with Tc and Re (*Vertical relationship*). This has already been discussed above under "*Position of manganese group elements in the periodic table*".

2. Similarities of Mn with Cr (*Horizontal relationship*). For this see "*Position of Cr in the periodic table* (*Group VI B*)".

3. Similarities of Mn with Fe (*Horizontal relatsonship*). For this see "*Position of Cr in the periodic table (Group VI B)*".

4. Similarities of Mn with Cl. Mn also shows some similarities with chlorine which is the bridge element. These similarities are very slight and are shown in their highest oxidation states, *e.g.* (*i*) The heptoxides of both the elements (Mn_2O_7 and Cl_2O_7) are similar to each other, *e.g.* both these oxides are acidic, volatile and explosive. (*ii*) Permanganates (MnO_4^-) and perchlorates (ClO_4^-) are isomorphous salts and have similar solubilities, *e.g.* both $AgMnO_4$ and $AgClO_4$ are sparingly soluble in water.

Groups 8, 9 and 10 (VIII) : Ferrous Metals (Fe_{26}, Co_{27}, Ni_{28}) and Platinum Metals (Ru_{44}, Rh_{45}, Pd_{46}; Os_{76}, Ir_{77}, Pt_{78})

Group VIII of the periodic table consists of nine elements which are grouped into three triads namely : (*i*) First triad which consists of iron (Fe), cobalt (Co) and nickel (Ni) (*ii*) Second triad, consisting of ruthenium (Rh), rhodium (Rh) and palladium (Pd) (*iii*) Third triad, consisting of osmium (Os), iridium (Ir) and platinum (Pt). The elements of the first triad belong to the first transition series and are called *ferrous metals* while those of second and third triads are the members of the second and third transition series respectively and are collectively known as the *platinum metals.* Ru, Rh and Pd are called *light platinum metals* while Os, Ir and Pt are known as *heavy platinum metals* on the basis of their density as compared to platinum.

Position of Ferrous Metals (Fe, Co, Ni) in the Periodic Table

We know that the elements lying to the left of Fe are Mn (Group VII B) and Cr (Group VI B), the element lying to the right of Ni is Cu (Group I B) while the elements lying below Fe, Co and Ni are Ru, Os (below Fe), Rh, Ir (below Co) and Pd, Pt (below Ni). Thus the inclusion of ferrous metals in the same group is justified by discussing the following similarities

1. Similarities between Fe, Co and Ni. The electronic configurations of these element given as : Fe (Z = 26) $\rightarrow$ 2, 8, 14, 2 ($3d^6.4s^2$); Co (Z = 27) $\rightarrow$ 2, 8, 15, 2 ($3d^7.4s^2$); Ni (Z = 28) $\rightarrow$ 2, 8, 16, 2 ($3d^8.4s^2$) show that Fe, Co and Ni have two electrons in their outermost shell (*i.e. 4th shell*) but their kernels are incomplete. As the atomic number increases from Fe to Ni, the number of electrons in the *d*-orbital of the penultimate shell (*i.e. 3rd shell*) increases by one in each succeeding element. All the ferrous metals particularly Fe and Ni occur in nature in large quantities. Their natural abundance is attributed to the high stability of their atomic nuclei. These are all greyish white hard metals of high densities. The hardness and density increases from Fe to Ni. These have fairly close melting and boiling points which are uniformly high. Very similar melting points seem to indicate that the average number of electrons per atom contributing to the metallic bonding is nearly the same per each one of them. Their oxidation potentials, particularly of Co and Ni, are very close and uniformly positive indicating that they can liberate hydrogen gas from acids very slowly. Fe reacts readily while Ni and Co react only slowly. This behaviour is in accordance with similar values of electrode potentials of these metals. All the three metals are *ferromagnetic, i.e.*, these are strongly attracted by a magnet and remain magnetised even after the removal of the magnetic field. The ferromagnetic property *decreases* from Fe to Ni. At elevated temperatures called their *curie points*, the metals lose ferromagnetic property and become paramagnetic. All the metals show allotropy and exist in *two solid phases.*

Fe shows both +2 (in *ferrous salts*) and +3 (in *ferric salts*) oxidation states, cobalt mostly exhibits +2 oxidation state but also forms alums and complexes [*e.g.*, $CoCl_6^{3-}$, $Co(CN)_6^{3-}$] in which it shows +3 state while nickel usually shows +2 state. In certain complexes nickel also shows +1 state, *e.g.*, $K_2[Ni(CN)_3]$, NiCN etc. Nickel does not form Ni^{3+} ion. The removal of a 3*d*-electron from M^{2+} cations to oxidise them into M^{3+} cations becomes more and more difficult as we proceed from Fe^{2+} to Ni^{2+}, *i.e.*, the stability of M^{2+} cations is in the order : $Fe^{2+} < Co^{2+} < Ni^{2+}$ and hence the reducing properties of these ions is in the order : $Fe^{2+} > Co^{2+} > Ni^{2+}$. The increase in stability from Fe^{2+} to Ni^{2+} ions is evident from the behaviour of their M(II) oxide on heating. *e.g.*, FeO is oxidised to Fe_2O_3 simply on heating, CoO changes to Co_3O_4 but does not undergo further oxidation to Co_2O_3 while NiO is stable. *i.e.* it remains unaffected. The tendency of these elements to show + 3 state decreases sharply from Fe to Ni, *i.e.*, the stability of M cations is in the order : $Fe^{3+} > Cb^{3+} > Ni^{3+}$, indicating that the oxidising properties of Fe^{3+}, Co^{3+} and Ni^{3+} cations are in the order : $Fe^{3+} < Co^{3+} < Ni^{3+}$. The reason for this difference in the +3 state is that, while Fe ion has a half-filled 3*d*-orbital ($Fe^{3+} \rightarrow 3d^5 4s^0$) and therefore stable, Co^{3+} and Ni^{3+} have relatively unstable $3d^6$ and $3d^7$ configurations respectively ($Co^{3+} \rightarrow 3d^7 4s^0$, $Ni^{3+} \rightarrow 3d^7 4s^0$). The +4 state is relatively uncommon for these elements. While +6 oxidation state is reported for iron in ferrate ion, $[FeO_4]^{2-}$ which is unstable and oxidising agent, this state has not been reported for Co and Ni. Lower oxidation states such as zero are found in complexes having *pi*-bonding ligands such as CO. Thus in the carbonyls, $Fe(CO)_5$, $Co_2(CO)_8$ and $Ni(CO)_4$ these metals show zero oxidation state.

They are weak metals and, therefore, their salts react acidic in solution. All of them occlude hydrogen, Ni being superior to the other two in this respect. As a result of this property they are used as catalyst in certain reactions involving the addition of hydrogen. They give rise to similar *complex cyanides*. All the three elements from isomorphous vitriol type sulphates $MSO_4.7H_2O$, where M = Fe(II), Co(II) or Ni(II). For example $FeSO_4.7H_2O$ is isomorphous with $CoSO_4.7H_2O$. All these sulphates form schonite type double sulphates, $M^+_2\ SO_4.M_2^{2+}\ SO_4.\ 6H_2O$. Of these Mohr's salt, $(NH_4)_2SO_4.\ FeSO_4.6H_2O$ is very important. These schonites are also isomorphous with each other, *e.g.*, Mohr's salt is isomorphous with nickel ammonium sulphate, $(NH_4)_2SO_4.\ NiSO_4.6H_2O$. Since Fe(III) and Co(III) [but not Ni(III)] are known as aquo ions, these form alum type sulphates : $M_2^+SO_4.\ M_2^{3+}(SO_4)_3.\ 24H_2O$ M^+ = Na, K, NH_4^+ etc. and M^{3+} = Fe(III) or Co(III). Under the influence of conc. HNO_3 all the three metals become passive. This tendency increases in the order : Fe < Co < Ni. All the three metals are reactive and reactivity decreases from Fe to Ni.

2. Similarities of Fe with Cr and Mn. These similarities have already been discussed under "*Position of Cr in the periodic table*".

3. Similarities between Ni and Cu. Both the metals give coloured salts. Copper salts are bluish and green, while the salts of nickel are green. The stable salts of nickel and copper (one-*ous* and the other *-ic*) are both bivalent. Both give two series of salts : *cuprous* and *cupric* ; *nickelous* and *nickelic*. Both form *isomorphous* sulphates, *e.g.* $CuSO_4.5H_2O$; $NiSO_4.7H_2O$. Both give black sulphides. Both give complex compounds, *e.g.* $[Cu(NH_3)_4]SO_4$, $K_2[Ni(CN)_4]$. The oxides of both the elements can be obtained by heating their carbonates or nitrates. Both the elements form schonite type double sulphates, $M^+_2\ SO_4\ M^{2+}SO_4.\ 6H_2O$ where $M = NH_4^+$ etc., $M' = Cu(II)$ or $Ni(II)$. Both the elements show a co-ordination number of 4. Their M^{2+} ions are almost similar in size (Cu^{2+} = 0.69A, Ni^{2+} = 0.72Å). Their hydrated salts in solution react with acid due to hydrolysis. Basic carbonates of the general formula $MCO_3.M(OH)_3$ are precipitated on treating their salt solutions with Na_2CO_3. Both of them show catalytic properties. Salts of both the metals give precipitates of their hydroxides when treated with NH_4OH. The hydroxides of both dissolve in excess of NH_4OH to form complex salts.

4. Similarities between Fe, Ru and Os. All the three metals form isomorphous complex cyanides. Ferrites. (FeO_3^{2-}) and ruthenites (RuO_3^{2-}) are similar. Ferrates, rutherates and osmates (MO_4^{2-}) are also similar.

5. Similarities between Co, Rh and Ir. These elements form similar complex cyanides, ammines, nitrites etc. The sulphates of all the elements form true alums.

6. Similarities between Ni, Pd and Pt. These elements form similar complex ammines, cyanides and nitrites. These elements are relatively inert in their respective triads. All show variable valency. All can occlude H_2 and hence act as catalysts. They give similar oxides and chlorides.

Similarities between Pt and Au

Both the metals occur native. They have high melting points and show metallic lustre. They are malleable, ductile and good conductors of heat and electricity. Both are noble metals and are not acted upon by air, water, alkalies and acids. Both have similar electronic configurations. Both form similar complex halogen acids like $HAuCl_4$, $H_2Pt\ Cl_6$. Colloidal solutions of both the metals can be prepared by Bredig's method.

Position of Cobalt in the Periodic Table

Cobalt is present in group VIII of the periodic table. To its left and right are lying Fe arid Ni respectively which belong to the same group and period. Below it are lying Rh and Ir. Thus the position of Co in the periodic table is justified by discussing the following similarities.

1. Similarities between Fe, Co and Ni. These have already been discussed under "*Position of ferrous metals in the periodic table*".

2. Similarities between Co, Rh and Ir. This has been discussed above.

Position of Nickel in the Periodic Table

Nickel is present in group VIII of the periodic table. To its left are lying Fe and Co (same group) while to its right is lying Cu (Group IB). Below it are Pd and Pt which are in the same sub-group. Thus the position of nickel in the periodic table is justified by discussing the following similarities :

1. Similarities between Fe, Co and Ni. This has been discussed above.

2. Similarities between Ni, Pd and Pt. These have been discussed above.

3. Similarities between Ni and Cu. See *Position of ferrous metals*.

Position of Platinum in the Periodic Table

We know that Pt is present in group VIII of the periodic Table. To its right is lying Au (Group IB) and above it are lying Ni and Pd (same group). Thus the position of Pt in the periodic table is justified by discussing the following similarities :

1. Similarities between Pt and Au. These have been discussed above.

2. Similarities between Ni, Pd and Pt. These have been discussed above.

Group 11 (IB) : Cu_{29}, Ag_{47} and Au_{79} (Coinage Metals)

Group IB of the long form of periodic table consists of three elements which are : *Copper* Cu), *Silver* (Ag) and *Gold* (Au). These are collectively known as *coinage metals* or *currency metals* because they have been used in the past for the manufacture of coins for currency. However, now their main uses are quite different. These are also called *copper group metals.*

Position of Coinage Metals in the Periodic Table

We know that Cu, Ag and Au belong to group IB and the elements lying to their left are Ni, Pd and Pt (Group VIII) while those lying to their right are Zn, Cd and Hg (Group II B).

Thus the inclusion of Cu, Ag and Au in the same sub-group is justified by discussing the following similarities :

1. Similarities between Cu, Ag and Au. On account of their great inert character, all these metals are found in the native state in nature. All are hard, malleable and ductile metals. They possess characteristic metallic lustre and take high polish. They have high density, melting points and boiling points and low atomic volume. All are good conductors of heat and electricity. Ag is the best conductor of heat and electricity. The electronic configurations of these elements given as Cu(Z = 29) $\rightarrow$ 2, 8, 18, 1 ($3d^{10}.4s^1$) Ag(Z = 47) $\rightarrow$ 2, 8, 18, 18, 1 ($4d^{10}.5s^1$); Au(Z = 79) $\rightarrow$ 2, 8, 18, 32, 18, 1 ($5d^{10}.6s^1$) show that the ultimate shell of these elements has *one* electron (s^1 electron) while the penultimate shell contains 18 (s^2 p^6 d^{10}) electrons. All the elements are weakly electropositive and hence are almost unreactive. These metals show variable valency. These metals are not able to displace H_2 from the acids, since they are less electropositive than hydrogen. Zn, Fe and Mg can displace these metals from their salt solutions. The *monovalent salts* of these metals are colourless and generally insoluble in water. For example, cuprous chloride (CuCl) and silver chloride (AgCl) are both colourless and insoluble in water. All of them form *complex salts* and in them metal may be a part of the complex cation or of the complex anion.

2. Similarities between Cu and Zn. Both the metals show bivalency. Sulphides, phosphates and carbonates of both the metals are insoluble in H_2O while their chlorides and sulphates are soluble in H_2O. Both the metals form isomorphous salts like $CuSO_4$. $5H_2O$; $ZnSO_4$ $7H_2O$; form double salts like $MSO_4.(NH_4)_2SO_4.6H_2O$ and complex compounds.

3. Similarities between Cu and Ni. See *Position of ferrous metals.*

4. Similarities between Ag and Cd. Both are white metals. Both form complexes like $[Ag(NH_3)_2]^+$, $[Cd(NH_3)_4]^{2+}$. Their oxides are coloured and get reduced when heated with carbon.

5. Similarities between Ag and Pd. Both are metals and are white in colour. Both are noble metals. Both act as catalysts in their finely-divided state. Both the metals form complex compounds.

Group 12 (II B) : Zn_{30}, Cd_{80} and Hg_{48}

Group II B of the long form of periodic table consists of three elements namely *zinc* (Zn), *cadmium* (Cd) and *mercury* (Hg). These elements are collectively known as *zinc group elements.*

Position of Zinc Group Elements in the Periodic Table

The inclusion of Zn, Cd and Hg in the same sub-group II B of the periodic table is justified by the similarities in their properties and gradation in them with the rise of atomic number from Zn to Hg as shown below :

The electronic configurations of these elements given as : Zn(Z = 30) $\rightarrow$ 2, 8, 18, 2 ($3d^{10}.4s^2$); Cd(Z = 48) $\rightarrow$ 2, 8, 18, 18, 2 ($4d^{10}.5s^2$) ; Hg(Z = 80) $\rightarrow$ 2, 8, 18, 32, 18, 2 ($5d^{10}.6s^2$) show that the ultimate shell has two electrons (s^2 electrons) while the penultimate shell contains 18 electrons ($s^2p^6d^{10}$ electrons). Thus the valence-shell configuration can he represented as $(n-1)d^{10}ns^2$. Thus *these are transitional elements*, but they do not show many properties which are shown by transition elements, *e.g.*, (*i*) They do not show variable valency (*ii*) Many of their compounds are white. (*iii*) Their melting and boiling points are very low.

These elements do not occur free in nature. Important ore of the metals is sulphide. We have seen that the valence-shell configuration of these elements can be represented as $(n–1)d^{10}.ns^2$. When s^2 electrons are lost by these elements, M^{2+} ion produced possesses $(n–1)d^{10}$ configuration. This configuration is stable and hence cannot lose any more electrons. Thus all the elements show +2 oxidation state. It may be noted that Hg forms two series of salts–mercurous salts containing Hg_2^{2+} ion and *mercuric* salts containing Hg^{2+} ion. Both these ions are in +2 oxidation state. These are greyish white lustrous metals, malleable end ductile. The malleability and ductility gradually decrease as the atomic number increases from Zn to Gd to Hg. Except Hg which is a liquid, other two elements are moderately hard. The hardness of Cd and Zn is in the order : Cd < Zn. All the three metals are good conductors of electricity. The atomic radii of these elements are fairly large. Atomic and ionic radii both increase with the increase of atomic number from Zn to Cd to Hg. This accounts for the covalent character in them. The atomic volume increases from Zn to Hg due to an increase in the atomic radii in the same direction. All these elements have high values of density which gradually increases from Zn to Hg. These elements have very low melting and boiling points as compared to other elements because electrons from $(n\text{–}1)d$ orbital are not available for bonding. Excepting Hg, both the metals have higher melting and boiling points than the alkali metals present in the same period.

These metals have fairly high values of ionisation energy which indicate that these metals should be less reactive. In fact, they are even more reactive than the coinage metals (Cu, Ag and Au). This is due to their high values of reduction electrode potential for M^{2+}/M couple. It can also be noted that ionisation potential values do not show regular trends. These metals have low electronegativity values and these values decrease from Zn to Cd to Hg. These metals are not very reactive and their chemical reactivity decreases from zinc to mercury with the decrease in standard electrode potential which is negative in the case of mercury. Accordingly, zinc and cadmium dissolve in dilute mineral acids with the evolution of H_2 while Hg does not.

The action of air on these metals increases from Zn to Hg. For example in air, zinc is slowly coated with a bluish white film of its oxide which prevents further action while cadmium is much less affected and mercury is not affected at all. They *do not react with cold water.* Zinc and cadmium slightly decompose boiling water and decompose steam at red heat while mercury does not react with water at any temperature.

Metallic zinc and cadmium dissolve readily in dilute mineral acids forming corresponding salts with the liberation of H_2 (oxides of nitrogen or NH_3 in case of HNO_3). Hg remains unaffected by dilute acids. It dissolves in cold HNO_3 whether dilute or concentrated evolving respectively nitric oxide and nitrogen dioxide. It also remains unaffected by cold and concentrated H_2SO_4 but dissolves in boiling H_2SO_4 forming $HgSO_4$ and evolving SO_2.

Metallic zinc dissolves in caustic alkalies forming zincate (ZnO_2^{2-}) with the evolution of H_2 while Cd and Hg remain unaffected. When a caustic alkali is added to solutions of zinc, cadmium and mercuric salts, Zn^{2+} and Cd^{2+} ions are preciptated as their hydroxides while Hg^{2+} is precipitated as HgO. (The hydroxide of Hg is unknown)

All the three elements form stable complexes readily. The important *ligands* (*i.e.* lone pair donors) are cyanide ions, ammonia amines and chloride ions. The complexes of Hg^{2+} are much more stable than those of Zn^{2+} and Cd^{2+}. The co-ordination number in case of zinc and cadmium ions is 4 whereas in the case of Hg_2^{2+} ion, it is 4 as well as 2.

Their halides may be prepared by the direct combination of elements and have very low melting points and are easily volatile. The chlorides are freely soluble in organic solvents. The nature of the bond in the halides depends on the polarising power of M^{2+} cation (M^{2+} = Zn^{2+}, Cd^{2+} or Hg^{2+}) and the polarisability of the halide anion. It may be noted that the polarising power of M^{2+} cation increases as $Zn^{2+} < Cd^{2+} < Hg^{2+}$ while the polarisability of halide ions increases as $F^- < Cl^- < Br^- < I^-$. Therefore CdI_2 is much less ionic than CdF_2. Similarly HgF_2 is ionic while $HgCl_2$ is covalent.

University Questions

1. Give reasons for the following :

(*i*) The salts of scandium group elements are colourless and diamagnetic.

(*ii*) Scandium group elements undergo hydrolysis readily.

(*iii*) Scandium forms complexes easily.

(*iv*) Sc^{3+} ion has greater polarising power.

2. Taking La and Ac as examples, compare and contrast the properties of *d*-block elements with *f*-block elements. (*Delhi Hons. 85*)

3. Lanthanium belongs to *f*-block although it has no electron in 5*f* orbital. Comment. (*Bombay 84*)

4. Discuss briefly the general characteristics of the elements of titanium group under the following heads : (*i*) oxidation states (*ii*) reactivity and passive behaviour (*iii*) nature of oxides (*iv*) formation of complexes. (*Bharathiar 86*)

5. Discuss the position of V, Nb and Ta in the periodic table. (*Delhi 85*)

6. How do the following change when we move from V to Ta ma Nb?

(*i*) Stability of various oxidation states shown by V, Nb and Ta.

(*ii*) Ionic character of the compounds of V, Nb and Ta with change in their oxidation states.

(*iii*) Basic character of the oxides *viz.* V_2O_5, Nb_2O_5 and Ta_2O_5. (*Delhi Hons. 86*)

7. Give the reason for the following :

(*i*) Vanadium has a greater tendency to form complexes than do niobium and tantalum.

(*ii*) The compounds of V B group elements in their lower oxidation states are coloured while those in higher oxidation states are colourless. (*Banaras 84; Lucknow 86*)

8. Give a comparative account of the general chemistry of Cr, Mo and W. Also discuss the position of these elements in the periodic table. (*Allahabad 88*)

9. Discuss the position of Mn in the periodic table. (*Madras 86; Bharathiar 86; M.D. Rohtak 84*)

10. Indicate the resemblance in properties between (*i*) Mn and Cr (*ii*) Mn and Fe. (*Madras 86*)

11. Give reason for the following :

(*i*) The aqueous solution of Mn^{2+} has only a pale-pink colour. (*Madras 85*)

(*ii*) Mn with five unpaired electrons is less ferromagnetic than Fe with four unpaired electrons. (*Delhi Hons. 87*)

(*iii*) Mn is an example of an element exhibiting a high degree of paramagnetic character. (*Andhra 85*)

12. Make a comparative study of Fe, Co and Ni and their compounds. (*Madras 86; Bharathiar 86; Madurai 86*)

13. Give the reasons for the following :

(*i*) Coinage metals show variable valency. (*Madras 85*)

(*ii*) The first iomisation potential of copper is higher than those of alkali metals but second and third ionisation potentials are lower. (*Punjab 85*)

14. Discuss the position of Cu, Ag and Au in the periodic table. (*Delhi 85*)

15. Why do you write murcurous ion as Hg_2^{2+} and cuprous ion as Cu^+ ? (*Delhi 85*)

16. Discuss the position of Zn in the periodic table with reference to electronic structure and characteristics of its oxide and chloride. (*Madras 86*)

17. Discuss the position of chromium or copper family elements in the periodic table. (*Kanpur 2000*)

CHAPTER

46 Extraction, Properties and Uses of Some d- and f-Block Elements

Titanium (Ti_{22})

Occurrence

Titanium is the ninth most abundant element in the earth's crust. Its main ores are : (*i*) *Rutile*, TiO_2 which is one of the three forms of TiO_2, and (*ii*) *Ilmenite*, $FeTiO_3$.

Extraction of Titanium Metal From Rutile Ore, TiO_2

Since titanium has high melting point and it reacts quite readily with air, O_2, N_2, H_2 and carbon at elevated temperatures, the extraction of the metal is a difficult task. The extraction of the metal from the rutile mineral consists of the following steps :

1. Conversion of rutile ore into $TiCl_4$. Rutile (TiO_2) mineral is heated with carbon in a current of Cl_2.

$$\underset{\text{Rutile ore}}{TiO_2} + C + 2Cl_2 \longrightarrow TiCl_4 + CO_2$$

2. Reduction of $TiCl_4$ to Ti metal. $TiCl_4$ obtained as above is reduced to Ti metal by any of the following methods :

(*a*) Kroll process. In this process vapours of $TiCl_4$ are passed over molten magnesium, when $TiCl_4$ is reduced to spongy Ti metal.

$$TiCl_4 + 2Mg \xrightarrow{830°C} \underset{\text{Spongy}}{Ti} + 2MgCl_2$$

The spongy metal is heated n an inert atmosphere of Ar to get a compact mass.

(*b*) I.C.I. process. In this process $TiCl_4$ is reduced by heating it with Na in an atmosphere of Ar.

$$TiCl_4 + 4Na \xrightarrow{\text{Heat}} Ti + 4NaCl$$

Extraction of Titanium Metal From Ilmenite Ore, $FeTiO_3$

The extraction of titanium metal from ilmenite ore consists of the following steps :

1. Conversion of ilmenite ore, $FeTiO_3$ into TiO_2. TiO_2 can be obtained from the ilmenite ore by the following processes :

(*a*) When ilmenite ore is heated with HCl in a current of Cl_2, TiO_2 is obtained.

$$\underset{\text{Ilmenite}}{2FeTiO_3} + 4HCl + Cl_2 \xrightarrow{\Delta} 2FeCl_3 + 2TiO_2 + 2H_2O$$

(*b*) The powdered ilmenite ore is fused with Na_2S. The product obtained consists of FeS, TiO_2 and Na_2O.

$$\underset{\text{Ilmenite}}{FeTiO_3} + Na_2S \xrightarrow{\text{Fuse}} \underbrace{FeS + TiO_2}_{\text{Insoluble}} + \underset{\text{Soluble}}{2Na_2O}$$

Now when the product consisting of FeS, TiO_2 and Na_2O is extracted with H_2O, FeS and TiO_2 remain insoluble and Na_2O goes into solution. The insoluble portion containing FeS and TiO_2 is treated with H_2SO_3 in presence of atmospheric O_2 when FeS is converted into soluble FeS_3O_3 but TiO_2 remains unaffected.

$$FeS + H_2SO_3 + \tfrac{1}{2}O_2 \text{ (From air)} \longrightarrow \underset{\text{Soluble}}{FeS_2O_3} + H_2O$$

$$TiO_2 + H_2SO_3 + \tfrac{1}{2}O_2 \text{ (From air)} \longrightarrow \text{No reaction}$$

(*c*) When ilmenite ore is mixed with carbon and heated in a current of Cl_2 at 360° a mass is obtained which consists of $FeCl_3$ and $TiCl_4$. Now when this mass is heated at 550°C, volatile $TiCl_4$ distils over. $TiCl_4$ thus obtained is hydrolysed by H_2O to get TiO_2.

$$\underset{\text{Ilmenite}}{2FeTiO_3} + 2C + 7Cl_2 \xrightarrow{350^\circ} \underbrace{2FeCl_3 + 2TiCl_4}_{} + 3CO_2\uparrow$$

$$\downarrow 550^\circ C$$

$$TiCl_4 \text{ distils over}$$

$$TiCl_4 + 2H_2O \xrightarrow{\text{Hydrolysis}} TiO_2 + 4HCl$$

2. Reduction of TiO_2 to Ti metal. TiO_2 obtained as above is reduced to titanium metal by any of the following methods :

(*a*) By alumino-thermic process. In this method TiO_2 is mixed with Al and the mixture is ignited with a suitable fuse.

$$3TiO_2 + 4Al \longrightarrow 2Al_2O_3 + 3Ti$$

(*b*) By heating with carbon. TiO_2 can also be reduced to free Ti metal by heating it with carbon.

$$TiO_2 + C \longrightarrow Ti + CO_2\uparrow$$

Purification of Titanium Metal

Titanium metal obtained either from rutile or from ilmenite contains many impurities. Impure metal can be purified by ***Van Arkel process*** in which impure titanium is heated with a limited amount of I_2 in an evacuated glass apparatus. TiI_4 vapours formed are allowed to diffuse on to a tungsten filament at a very high temperature. The vapours of TiI_4 decompose and the pure Ti metal gets deposited on the W-filarnent.

$$\text{Impure Ti metal} + 2I_2 \xrightarrow{150\text{–}250°} TiI_4(g) \xrightarrow{1800°,\ \text{Tungsten filament}} \underset{\text{(gets deposited on W-filament)}}{\text{Pure Ti metal}} + 2I_2(g)\uparrow$$

Properties

Metallic titanium has a silvery lusture and when polished resembles steel appearance. Amorphous form is a dark grey powder resembling iron (reduced) in appearance. It is brittle at ordinary temperature but at low red heat it is readily malleable. The pure metal is very ductile. Its m.pt. and b.pt. are 1850°C and 3000°C. The fused metal is very hard and crystalline and scratches quartz. When rubbed against steel, it gives bright sparks. Titanium metal is stable in air at room temperature, but, when heated. it reacts with O_2 and N_2 present in air to form TiO_2 and TiN. Hot titanium metal decomposes steam to form TiO_2 with the evolution of H_2. With acids like H_2SO_4, HF, HNO_3, CH_3COOH etc. Ti (*III*) salts are formed with the evolution of H_2. With halogens, the metal gives tetrahalides. The metal combines with non-metals like H_2, S, C, Si etc. to form TiH_2, TiS, TiC, TiSi etc.

Uses of Titanium Metal and its Alloys

(*i*) Due to the strength, lightness, corrosion, resistance and low thermal expansion of Ti metal, it is used in high speed air- craft, naval and military projects and also in chemical industry. (*ii*) It is used to remove the last traces of gases like O_2, H_2 and N_2 from evacuated vessels. (*iii*) A titanium cathode is used in special photoelectric cell. (*iv*) Pure Ti metal is used in making incandescent filament. (*v*) A number of titanium alloys have been found to be very useful in industry. For example : (*a*) Ferrotitanium alloy is used in steel industry as a scavanger to remove adsorbed O_2, H_2 and N_2, and as a cleansing and deoxidising agent in the metallurgical operations. Ferrotitanium is obtained in the electric furnace by reducing ilmenite ore, $FeTiO_3$ with carbon or Al metal.

$$\underset{\text{Ilmenite}}{FeTiO_3} + 3C\ (\text{or } 2Al) \longrightarrow \underbrace{Fe + Ti}_{\text{Ferrotitanium alloy}} + 3CO\ (\text{or } Al_2O_3)$$

(*b*) The alloys of Ti with Cr, Fe and Mo are used for making rails, gears, propellers, car wheels etc. (*c*) The alloys of Ti with Co and Ni are used as filaments in vacuum tubes in place of Pt—Ir alloys which are very costly.

Vanadium (V_{23})

Occurence

Compounds of vanadium are abundantly distributed in the earth's crust. Some of the prominent minerals are : (*i*) *Vanadinite*, $3Pb_3(VO_4)_2 . PbCl_2$ or $2Pb_5(VO_4)_3Cl$

(*Lead chloro orthotvanadate*) (*ii*) *Carnotite*, $K_2O.2UO_3.V_2O_5.3H_2O$ or $K_2(UO_2)_2(VO_4)_2.3H_2O$ (*Potassium uranyl orthovanadate*) (*iii*) *Patronite*, $V_2S_5.3CuS_2$. Vanadium is also found in small amounts in clays, rocks, coal and crude oil.

Extraction

The extraction of vanadium metal from its ores consists of the following steps :

1. Preparation of ammonium metavanadate, NH_4VO_3. Ammonium metavanadate can be obtained either from carnotite or vanadinite ore as shown below :

(*a*) From carnotite ore, $K_2(UO_2)_2(VO_4)_2.3H_2O$. Carnotite ore is boiled with Na_2CO_3 solution and concentrated. On cooling, the crystals of sodium uranyl carbonate, $Na_4(UO_2)(CO_3)_3$ are obtained and sodium orthovanadate, Na_3VO_4 is left in solution. The solution containing Na_3VO_4 is treated with NH_4Cl when ammonium metavanadate, NH_4VO_3 is precipitated.

$$\underset{\text{Carnotite ore}}{K_2(UO_2)_2(VO_4)_2} + 7Na_2CO_3 \longrightarrow \underset{\substack{\text{Sod. uranyl car-}\\\text{bonate (crystals)}\\\text{(Solution)}}}{2Na_4(UO_2)(CO_3)_3} + \underset{\substack{\text{Sod. ortho}\\\text{vanadate}\\\text{(Insoluble)}}}{2Na_3VO_4} + K_2CO_3$$

$$\underset{\text{(In solution)}}{Na_3VO_4} + 3NH_4Cl \longrightarrow \underset{\substack{\text{Ammonium}\\\text{metavana-}\\\text{date (ppt)}}}{NH_4VO_3} + 3NaCl + H_2O + 2NH_3\uparrow$$

(*b*) From vanodinite ore, $Pb_5(VO_4)_3Cl$. Powdered vanadinite ore is treated with conc. HCl when $PbCl_2$ is obtained as precipitate and a complex salt, $[V(OH)_4]Cl$ goes into solution. When the solution containing $[V(OH)_4]Cl$ is boiled with NH_4Cl, ammonium metavanadate, NH_4VO_3 is obtained as precipitate.

$$\underset{\text{Vanadinite ore}}{Pb_5(VO_4)_3Cl} + 12HCl \longrightarrow \underset{\text{ppt.}}{5PbCl_2} + \underset{\text{In solution}}{3[V(OH)_4]Cl}$$

$$\underset{\text{(In solution)}}{[V(OH_4)]Cl} + NH_4Cl \longrightarrow \underset{\substack{\text{Ammonium}\\\text{motavenadale}\\\text{(ppt)}}}{NH_4VO_3} + 2HCl + H_2O$$

2. Conversion of NH_4VO_3 to V_2O_3. When NH_4VO_3 obtained from carnotite or vanadinite as shown above is ignited, vanadium pentoxide (V_2O_5) is obtained.

$$\underset{\substack{\text{Ammonium}\\\text{megavanadate}}}{2NH_4VO_4} \xrightarrow{\text{Ignite}} V_2O_5 + 2NH_3 + H_2O$$

3. Reduction of V_2O_5 to vanadium metal. Vanadium metal can be obtained by reducing V_2O_5

(*a*) with Ca and $CaCl_2$ (used as a flux) in a steel bomb at 900–950°C (*Marden and Rich process*)

$$V_2O_5 + 5Ca + 5CaCl_2 \longrightarrow 2V + 5(CaO.CaCl_2)$$

(*b*) with Al (*Alumino-thermic process*)

$$3V_2O_5 + 10Al \longrightarrow 5Al_2O_3 + 6V$$

(*c*) with mishmetal (*Thermite process*). Mischmetal is an alloy of certain rare earth metals like Ce, La, Nd, Pr etc obtained from the residues left after the extraction of Th.

4. Purification of vanadium metal. Vanadium metal obtained as above is impure, since it contains some impurities associated with it. The impure metal is purified by *Van Arkel method* in which the impure vanadium metal is heated with a limited amount of I_2 in a evacuated glass apparatus. VI_4 vapours so formed are allowed to diffuse on a tungsten filament at a very high temperature. The vapours of VI_4 decompose and the pure vanadium metal gets deposited on the filament.

$$\text{Impure V metal} + 2I_2 \longrightarrow VI_4(g) \xrightarrow{\text{W-filament}} \underset{\text{Pure metal}}{V(s)} + 2I_2(g)\uparrow$$

Properties of Vanadium Metal

It is a steel-grey hard metal which melts above 1700°C and boils at 3400°C. When pure, it is malleable as well as ductile. When heated in air, the metal is converted into V_2O_5. When heated upto 1400°C, it forms VH_2. When treated with excess of dry Cl_2, VCl_4 is formed. Bromine water has no action on vanadium metal. HCl and H_2SO_4 both have no action on the metal. However, with hot and conc. H_2SO_4 it gives SO_2. With HF it gives H_2. It is easily attacked by HNO_3 and forms $V(NO_3)_3$. Aquea regia attacks the metal very vigorously. Fused alkalies give orthovanadates (Na_3VO_4), with the liberation of H_2.

$$2V + 6NaOH + 2H_2O \longrightarrow 2Na_3VO_4 + 5H_2$$

Oxidising agents like chloric acid, perchloric acid, bromic acid, $KClO_3$, bromate, iodate and perchiorate convert vanadium into V_2O_5. Vanadium metal reduces the solution of $AgNO_3$, $AuCl_3$, $PtCl_4$ and $IrCl_4$. It combines with non-metals like N_2, C, S anti Si to form nitride, carbide, sulphide and silide respectively. Vanadium combines with a number of metals like Fe, Cu, Co, Ni and Al and forms alloys. For example :

(*a*) *Ferro-vanadium alloy* is obtained by heating V_2O_5 with iron ore (Fe_2O_3) and coke in presence of a flux (lime or fluorspar) in an electric furnace.

$$V_2O_5 + 5C \longrightarrow 2V + 5CO\uparrow$$

$$Fe_2O_3 + 3C \longrightarrow 2Fe + 3CO\uparrow$$

This alloy contains 40–80% vanadium.

(*b*) *Nickel-vanadium alloy* is obtained by heating a mixture of V_2O_5 and NiO.

(*c*) *Alumino-vanadium alloy* is prepared by adding Al and then thermite mixture of V_2O_5 and aluminium soot in the electric furnace having melting fluorspar.

Uses of Vanadium.

The most important use of vanadium is in the form of its alloys. For example, ferro-vanadium alloy is used as a scavanger by removing efficiently N_2 and O_2 in the manufacture of special steels. The presence of a small amount of vanadium (0.1 to 0.3%) increases the toughness, tensite strength and resistance to shock which are valuable characteristics for the manufacture of machinery parts subjected to vibration. Thus ferrovanadium is used in the construction of motor parts. Aluminovanadium alloy is used in the manufacture of aircraft parts. The addition of 0.2% vanadium in glass manufacture is recommended for cutting of all the wave lengths below 3580°A which are harmful for the eyes and fade fabrics.

Chromium (Cr_{24})

Occurrence

Metallic chromium does not occur in the native state. The ores of chromium are : (*i*) *chromite* or *chrome ore*, FeO. Cr_2O_3 (*ii*) *chrome-ochre*, Cr_2O_3 (*iii*) *crocosite*, $PbCrO_4$. Chromite ore is the most common and is the industrial source of the metal. In India the chromite ore occurs in Singhblum district of Bihar, Mysore, Orissa, Madras and Bombay.

Extraction of Chromium Metal from Chromite Ore, FeO.Cr_2O_3

The extraction of Cr metal from chromite ore consists of the following steps :

1. Concentration of the chromite ore. The ore is first crushed and then concentrated by *Wilfley table method* (*Gravity separation method*). In this method the crushed ore is collected in the form of a heap on a sloping floor. A stream of water is allowed to flow through the heap. By this operation the lighter impurities of earthy matter go with water flowing down the slope while the heavier impurities remain with the ore in the heap.

2. Conversion of the concentrated chromite ore into Na_2CrO_4 : Roasting of the concentrated ore. The concentrated ore is mixed with excess of Na_2CO_3 and roasted in a reverberatory furnace at 900–100°C in the presence of free supply of air. By this operation the chromite ore, FeO.Cr_2O_3 is converted into soluble sodium chromate, Na_2CrO_4.

$$\underset{\text{Chromite ore}}{4(FeO.Cr_2O_3)} + 8Na_2CO_3 + 7O_2 \text{ (From air)} \xrightarrow{900-100°C} \underbrace{\underset{\text{Soluble}}{8Na_2CrO_4} + \underset{\text{Insoluble}}{2Fe_2O_3}}_{\text{Roasted mass}} + 8CO_2\uparrow$$

The roasted mass containing Na_2CrO_4 and Fe_2O_3 is extracted with water when Na_2CrO_4 goes into solution while Fe_2O_3 is left behind as a residue.

In the process of roasting, a small amount of lime (CaO) is also added with the mixture of the ore and Na_2CO_3. The addition of CaO makes the mass porous so that the process of roasting is facilitated.

3. Conversion of Na_2CrO_4 into $Na_2Cr_2O_7$. The solution containing Na_2CrO_4 is treated with a calculated quantity of H_2SO_4 whereby Na_2CrO_4 is converted into sodium dichromate, $Na_2Cr_2O_7$ and less soluble Na_2SO_4 is also obtained.

$$2Na_2CrO_4 + H_2SO_4 \longrightarrow \underset{\text{More soluble}}{Na_2Cr_2O_7} + \underset{\text{Less soluble}}{Na_2SO_4} + H_2O$$

The solution containing $Na_2Cr_2O_7$ (more soluble) and Na_2SO_4 (less soluble) is concentrated when the less soluble Na_2SO_4 crystalises out leaving behind more soluble $Na_2Cr_2O_7$ in solution. The solution containing $Na_2Cr_2O_7$ is further concentrated when crystals of $Na_2Cr_2O_7$ are obtained.

4. Conversion of $Na_2Cr_2O_7$ into Cr_2O_3. $Na_2Cr_2O_7$ obtained as above can be converted into Cr_2O_3 by the following methods :

(*a*) $Na_2Cr_2O_7$ is heated with carbon to get sodium chromite, $Na_2Cr_2O_4$ which, on being treated with H_2O, gives the precipitate of chromic oxide, Cr_2O_3.

$$Na_2Cr_2O_7 + 3C \xrightarrow{\Delta} \underset{\text{Sod. chromite}}{Na_2Cr_2O_4} + 3CO \uparrow$$

$$\underset{\text{Sod. chromite}}{Na_2Cr_2O_4} + H_2O \longrightarrow \underset{\text{ppt.}}{Cr_2O_3} + 2NaOH$$

(*b*) $Na_2Cr_2O_7$ is heated with NH_4Cl when the precipitate of Cr_2O_3 is obtained.

$$Na_2Cr_2O_7 + 2NH_4Cl \xrightarrow{\Delta} \underset{\text{ppt.}}{Cr_2O_3} + 2NaCl + 4H_2O + N_2 \uparrow$$

5. Reduction of Cr_2O_3 to chromium metal. Cr_2O_3 obtained as above is reduced to Cr metal

(*i*) by heating Cr_2O_3 with Al powder (Alumino-thermite process)

(*ii*) by heating Cr_2O_3 with a calculated quantity of silicon in presence of CaO which forms a slag of $CaSiO_3$ with SiO_2

$$2Cr_2O_3 + 3Si \longrightarrow 4Cr + 3SiO_2; \qquad SiO_2 + CaO \longrightarrow \underset{\text{Slag}}{CaSiO_3}$$

(*iii*) by passing H_2 over Cr_2O_3 (industrial method) (*iv*) by heating Cr_2O_3 with sugar charcoal (Daville's process).

Properties

The metal is silver white and crystalline. It is malleable and can take a very high polish producing a brilliant surface which is not tarnished even after long exposures to the atmosphere. Its sp. gr. is 7.14, m.pt. = 1830°C and b.pt. = 2200°C. It is extremely hard. Air and water have no action on chromium at the ordinary temperature. Heated in air, it is superficially oxidised to Cr_2O_3 but it burns in the oxy-hydrogen flame forming Cr_2O_3. It decomposes steam. Chromium dissolves in dilute hydrochloric acid forming chromous and ohromic chloride and liberating hydrogen. With dilute sulphuric acid, chromous sulphate and hydrogen are produced. Hot concentrated sulphuric acid produces sulphur dioxide forming chromic sulphate. It is insoluble in dilute nitric acid. Concentrated HNO_3 makes it *passive i.e.*, after immersion in it, it no longer reacts with hydrochloric or sulphuric acid. When heated with NH_3 or N_2, the metal gives the nitride, CrN. The metal reacts directly with halogens, sulphur and carbon etc.

Alloys of Chromium

Chromium combines with a number of metals and forms alloys which are very useful. Its important alloys are given in the Table 46.1.

Table 46.1. Alloys of chromium.

Name of the alloy	Composition	Properties and Uses
(*i*) Ferrochrome	Cr = 65%, Fe = 35%. It is prepared by heating the chrommite ore, $FeO.Cr_2O_3$ with carbon in an electric furnace. $\xrightarrow{\Delta} \underbrace{Fe + 2Cr}_{\text{Ferrochrome}} + 4CO\uparrow$	It is used in the manufacture of chrome-steel. Chrome-steel is drill proof and is, there-fore, used in the manufacture of burglar-proof safes.
(*ii*) Stainless steel	Cr = 11–13%, C = 0.1–0.4%.	It is used for cutlery and house hold wares.
(*iii*) Chromium-vanadium steel		It is very hard and strong and is used for gears, locomotive wheels, oxide and bearings etc.
(*iv*) Nichrome	Cr = 15%, Ni = 60%, Fe = 25%	It is resistant to atmospheric corrosion and can withstand high temperature. It is used in resistance wires for electrical heating.
(*v*) Stellite	Cr = 20–35%, Co = 40–80%, Ni = 0.25%, C = 0.75–2.5%.	It is extremely hard, bright and very resistant to corrosion and is used in cutlery, surgical instruments, cutting and high speed tools etc.
(*vi*) Chrome-nickel steel		It is used for armour pintes.

Uses of Chromium

(*i*) The metal is used in the preparation of various types of alloys which have been put to many uses as shown in the above table. (*ii*) The metal is also used in *chrome plating*, since even very thin layer imparts remarkable resistance to corrosion and has great protective power. Generally iron articles on which chromium plating is desired are first coated with layer of Cr or Ni before chrome plating is done.

Molybdenum (Mo_{42})

Occurrence

The chief ores of molybdenum are *molybdenite*, MoS_2; *wulf-fenite*, $PbMoO_4$; and *molybdite*, also known as *molybdic ochre*, $Fe_2O_3.3MoO_3\ 7\frac{1}{2}H_2O$.

Extraction of Molybdenum From Melybdenite Ore MoS_2

The extraction of molybdenum from molybdenite ore consists of the following steps :

1. Concentration of the ore. Molybdenite ore is crushed in jaw-crushes and powdered in ball mills. Now the powdered ore is concentrated by froth floatation process.

2. Conversion of molybdenite ore, MoS_2 into MoO_3 : Roasting. Concentrated molybdenite ore is roasted in a reverberatory furnace in a plenty supply of air to get the crude MoO_3.

$$\underset{\text{Molybdenite}}{2MoS_2} + 7O_2 \xrightarrow{\text{Roasting}} 2MoO_3\ (\text{Impure}) + 4SO_2$$

The crude MoO_3 obtained as above is extracted with hot conc. NH_4OH to get the crystals of ammonium molybdate, $(NH_4)_2MoO_4$.

$$MoO_3\ (\text{Impure}) + 2NH_4OH \longrightarrow (NH_4)_2MoO_4 + H_2O$$

Now the crystals of $(NH_4)_2$ MoO_4, on heating, give relatively pure MoO_3.

$$(NH_4)_2MoO_4 \xrightarrow{\Delta} MoO_3 + 2NH_3 + H_2O$$

3. Reduction of MoO_3 into Mo metal. MoO_3 obtained as above is reduced to Mo metal either by H_2, carbon or Al (*Alumino-thermic process*). Mo metal obtained by the second method is contaminated with some carbide of molybdenum.

$$MoO_3 + 3H_2 \longrightarrow Mo + 3H_2O;\ MoO_3 + 3C \longrightarrow Mo + 3CO$$

$$MoO_3 + 2Al \longrightarrow Mo + Al_2O_3.$$

4. Purification of Mo metal. Mo metal obtained as above is contaminated with impurities. In order to remove impurities, crude Mo metal obtained as above is heated with CO at 200°C under pressure of about 200 atmospheres when $Mo(CO)_6$ is formed.

$$Mo\ (\text{Impure}) + 6CO \longrightarrow Mo(CO)_6$$

$Mo(CO)_6$, on being heated at 360–410°C, gets decomposed to form pure Mo metal.

$$Mo(CO)_6 \xrightarrow{\Delta} Mo\ (\text{pure}) + 6CO \uparrow$$

Preparation of Pure Molybdenum Metal

Thermal decomposition of molyodenum pentachloride on an electrically heated molybdenum filament gives pure metal.

$$2MoCl_5 \xrightarrow[\text{filament}]{\text{Molybdenum}} 2Mo\ (\text{pure})\ 5Cl_2$$

Properties

Massive Mo has typical silver white appearance and metallic properties. Mo has density 10.2 and high m.pt. (= 2600°C). The metal is extremely resistant to the action of dil. or conc. acids except moderately conc. HNO_3 and hot conc. H_2SO_4 which oxidises it. Mo remains unaffected by solutions of alkalies but is rapidly attacked by fused alkalies in presence of oxidising agents like $NaNO_3$. The metal does not combine at ordinary temperature with non-metals but at elevated temperature it readily combines with O_2, S, C, N_2 and halogens to form MoO_3, MoS_2, MoC, MoN, MoF_6, $MoCl_5$ etc.

Uses of Molybdenum and its Compounds

(*i*) Molybdenum finds most important use in the production of alloy steels and high speed tools. (*ii*) Molybdenum is used in thermo-couples for high temperature measurement, (*iii*) Both in radio valves and bulbs molybdenum is used as filament support. (*iv*) Mo is used in the manufacture of standard weights. (*v*) In potteries, Mo-compounds are used as blue pigments. (*vi*) Molybdenum compounds are used for dyeing silk and wool and for colouring leather and rubber. (*vii*) $(NH_4)_2\ MoO_4$ is used to detect and estimate phosphate and arsenic in chemical analysis.

Tungsten (W_{74})

Occurrence

Tungsten is a rare element, found combined in minerals which are usually associated with tin ores. The chief minerals are *wolframite* (Fe, Mn) WO_4, *scheelite*, $CaWO_4$ and *hubnerite*, $MnWO_3$, all the three containing 70 to 80 per cent of WO_3.

Extraction of Tungsten From Wolframite Ore, (Fe, Mn) WO_4

The extraction of tungsten metal from wolframite involves the following steps :

1. Concentration of the ore. The powdered wolframite ore is concentrated by electromagnetic separation method to remove the non-magnetic impurities.

2. Conversion of wolframite ore into WO_3. The concentrated wolframite ore is fused with Na_2CO_3 in air. This operation gives a fused mass consisting of sodium tungstate, Na_2WO_4 (soluble) and a mixture of Fe_2O_3 and MnO_2 (insoluble).

$$\underbrace{4[(Fe, Mn)WO_4] \text{ or } (4FeWO_4 + 4MnWO_4)}_{\text{Wolframite ore}} + 8Na_2CO_3$$

$$+ 3O_2 \text{ (from air)} \xrightarrow{\text{Fusion}} \underbrace{\underset{\text{Soluble}}{8Na_2WO_4} + \underbrace{2Fe_2O_3 + 4MnO_2}_{\text{Insoluble}}}_{\text{Fused mass}} + 8CO_2 \uparrow$$

Now the fused mass is leached with water so that Na_2WO_4 goes into solution and the mixture of Fe_2O_3 and MnO_2 is left as residue. The solution containing Na_2WO_4, on being acidified with HCl, gives a precipitate of tungstic acid, H_2WO_4 which on ignition gives WO_3.

$$Na_2WO_4 + 2HCl \longrightarrow \underset{\text{ppt.}}{H_2WO_4 \downarrow} + 2NaCl$$

$$H_2WO_4 \xrightarrow{\text{Ignition}} WO_3 + H_2O$$

3. Reduction of WO_3 into W metal. WO_3 obtained as above is reduced to W metal by either of the following methods.

$$WO_3 + 3H_2 \xrightarrow{1200° \text{ C}} W + 3H_2O;\ WO_3 + 3C \xrightarrow{\text{White heat}} W + 3CO$$

$$WO_3 + 2Al \longrightarrow Al_2O_3 + W$$

Preparation of Pure Tungsten Metal

Pure tungsten metal can be obtained :

(*i*) by heating the tungstic acid, H_2WO_4 in a current of H_2.

(*ii*) by the thermal decomposition of WCl_6 vapour on a tungsten electric arc.

$$H_2WO_4 + 3H_2 \longrightarrow W \text{ (pure)} + 4H_2O \text{ ; } WCl_6 \longrightarrow W \text{ (pure)} + 3Cl_2$$

Properties

Massive tungsten possesses typical white appearance and the metallic properties. Its density and m.pt. are 19.9 and 3400°C respectively. The metal is not acted upon by dil. and conc. acids and solutions of alkalies. It is, however, readily attacked by fused alkalies in presence of oxidising agents like KNO_3. The metal does not combine with elements at ordinary temperature but at high temperatures it interacts with O_2, S, N_2, halogens, C and H_2.

$$2W + 3O_2 \longrightarrow 2WO_3; \quad W + 2S \longrightarrow WS_2; \ W + C \longrightarrow WC$$

$$2W + C \longrightarrow W_2C; \quad 4W + N_2 \longrightarrow 2W_2N; \ 2W + N_2 \longrightarrow 2WN$$

$$W + 3X_2 \longrightarrow WX_6 (X = F, Cl, Br) \quad ; \ W + I_2 \longrightarrow WI_2.$$

Uses to Tungsten and Its Compounds

The metal is used in making high speed tool steels, alloy steels, filaments of electric lamps, pen-points, strings of musical instruments, contact points, telegraphic keys, gramophone needles, voltage regulators etc. Surgical instruments are manufactured from its alloy namely "stellite" which contains W along with Co and Cr.

Manganese (Mn_{25})

Occurrence

Manganese does not occur in the free state in nature. The chief ores of manganese are oxide ores which are : (*i*) *Pyrolusite*, MnO_2. This ore contains 60–63% manganese and is black in colour. It is found mainly in India, Russia, South Africa, U.S.A. and Germany. (*ii*) *Braunite*, Mn_2O_3. This ore contains 60–70% manganese and is brownish-black in colour. (*iii*) *Hausmannite*, Mn_3O_4 which contains 72.1% manganese and is brownish black. (*iv*) *Manganite*, $Mn_2O_3.H_2O$ which contains 62.4% manganese and is grey or black. Other less important ores of manganese are *manganese blende* (MnS), *rhodonite* ($MnSiO_3$) and *rhodothrosite* ($MnCO_3$). Manganese is commonly associated with iron ores. Its estimated abundance is 0.01% of the earth's crust. Traces of Mn occur in water, plants, animals and in many amethyst coloured minerals.

Occurrence in India

India is the biggest producer of manganese ores in the world. Pyrolusite, MnO_2 which is the most important ore is found in India at Panch Mahal (Bombay), Karnatak, Singbhum (Bihar), Madhya Pradesh, Orissa, Mysore and Rajasthan.

Extraction of Mn From Oxide Ore

Manganese may be extracted from its oxide ores like *pyrolusite* (MnO_2) and *hausmannite* (Mn_3O_4) by any of the following methods :

1. Carbon-reduction process. When any oxide ore of Mn is heated with carbon, the reduction of the oxide ore stops at the stage of manganese monoxide, MnO. The reduction of MnO to the metallic manganese requires a very high temperature. The molten metal at such a high temperature of the furnace dissolves much carbon and other impurities. High grade oxide ores (*e.g.* MnO_2 and Mn_3O_4) are, therefore, mixed with charcoal and a little borax and heated to a white heat in an electric furnace when the oxide ores are reduced to metallic manganese.

$$MnO_2 + 2C \xrightarrow{\text{Heat to white heat}} Mn + 2CO$$

$$Mn_3O_4 + 4C \xrightarrow{\text{Heat to white heat}} 3Mn + 4CO$$

Mn obtained by this method is not pure, since it is contaminated with carbon. For obtaing pure metal, less than the theoretical amount of carbon is used in this method.

2. Alumino-thermic process. The reaction between the pyrolusite ore (MnO_2) and aluminium is an explosive reaction. Hence MnO_2 is first converted into Mn_3O_4 by heating it to redness and then reduced to Mn by metallic aluminium.

$$3MnO_2 \xrightarrow{\text{Heat of redness}} Mn_3O_4 + O_2$$

$$3Mn_3O_4 + 8Al \longrightarrow 9Mn + 4Al_2O_3 + \text{Heat evolved}$$

The heat produced in the reaction makes the reaction proceed very rapidly and vigorously and Mn melts and collects as a liquid at the bottom of the crucible and alumina (Al_2O_3) remains on the surface as a slag which solidifies to form artificial corundum. Mn in the molten state is prevented from being oxidised by the layer of Al_2O_3. Molten Mn may be allowed to flow out through the tapping plug. Molten manganese obtained by this method is fairly pure.

3. Elcctrolytic method : Production of pure manganese. In this method, the oxide ores of Mn namely MnO_2 and Mn_3O_4 are converted into $MnSO_4$ and the oxide, MnO_2 is converted into $MnCl_2$. These conversions are made as follows :

(*i*) $MnO_2 + 2CO \xrightarrow{\text{Reduction}} 2MnO + 2CO_2 \uparrow$

MnO thus obtained is treated with H_2SO_4 to get $MnSO_4$.

$$MnO + H_2SO_4 \longrightarrow MnSO_4 + H_2O$$

(*ii*) $Mn_3O_4 + H_2SO_4 \longrightarrow \underset{\text{Solution}}{2MnSO_4} + \underset{\text{Residue}}{MnO_2.H_2O}$

(*iii*) $MnO_2 + 4HCl \longrightarrow MnCl_2 + 2H_2O + Cl_2 \uparrow$

Concentrated solution of $MnSO_4$ or $MnCl_2$ obtained as above is prepared and electrolysed, using a mercury cathode, and carbon as anode. Mn is liberated on the cathode and forms an amalgam with Hg. The amalgam thus obtained is distilled in vacuum at 250° to remove Hg.

Properties

In the solid state Mn exists in three allotropic modifications which are α, β and γ. The α-form is stable upto 742°C and has a complex cubic structure. It is hard and brittle metal which scratches glass. α-form is a silvery white metal in the pure state and grey when impure. It melts at 1260°C, boils at 1900° and its specific gravity is 7.39. The metal becomes very hard if traces of carbon or silicon are present in it as impurity. Mn is not magnetic at any temperature but its compounds like carbides and nitrides are magnetic. β-form is stable between 742° and 1100°C. Like α-form it has also a complex cubic structure and is hard and brittle metal which scratches glass. γ-form is stable above 1100°C and forms face-centred cubic tetragonal crystals. It is ductile and tends to pass into α-form under the ordinary conditions. It is flexible and soft with a density of 720. Mn is superficially oxidised when exposed in air but is readily oxidised on heating in air. The finely divided metal burns in air (*pyrophoric*). Pure Mn decomposes cold water slowly but hot water rapidly, forming $Mn(OH)_2$ and evolving H_2. Mn has a tendency to form Mn^{2+} ions and the oxidation potential is equal to +1.18 volts. Thus Mn can be oxidised to Mn^{2+} ions by H^+ ions from the dilute acids or Mn can displace H_2 from acids. Mn dissolves in conc. H_2SO_4 with the evolution of SO_2 and in HNO_3 evolving NO. It combines with carbon at a temperature of electric arc (1300°C) to form Mn_3C. When heated to 1200°C in an atmosphere of *nitrogen*, it forms Mn_3N_2. When heated with halogens, it forms halides. With *sulphur* it gives sulphide.

It gives Mn_3N_2 with NH_3 and at 330° it gives carbonyl, $Mn_2(CO)_{10}$ with CO. Because of its position in electrochemical series, the metal can displace As, Sb, Bi, Sn, Pb, Cu, Fe, Ni, Co, Cd and Zn from the solutions of the salts of these elements. This property is used for cathodic protection and prevention of corrosion in steel equipment.

Uses

Manganese, in the form of pure metal, is used only for special purposes. It is generally used in the form of its alloys with other metals. Thus, about 80% of Mn produced is used in the manufacture of its alloys which are more useful than Mn itself. Its important alloys are : (*i*) Ferro-manganese and spiegeleisen are the alloys of Mn with Fe and carbon These are obtained by strongly heating a mixture of MnO_2, iron oxide (*e.g.* Fe_2O_3) and a calculated quantity of carbon in a blast furnace. For example ferro-manganese is obtained as follows :

$$3MnO_2 \longrightarrow Mn_3O_4 + O_2 \uparrow$$

$$Mn_3O_4 + 4C \longrightarrow 3Mn + 4CO \uparrow$$

$$Fe_2O_3 + 3C \longrightarrow 2Fe + 3CO \uparrow$$

The approximate composition of these alloys is : *Ferromanganese*— Mn = 70–80%, C = 0.2–0.3%, rest is Fe; *Spiegeleisen*— Mn = 20–30%, C = 5%, rest is Fe. These alloys are used in the manufacture of steel by the Bessemer process. When Mn, in the form of these alloys, is added to steel in small amounts, it acts as *scavenger* (cleaning agent), *i.e.* it removes dissolved O_2, N_2, P and S in the form of oxide, nitride, phosphide and sulphide of Mn which can easily be removed.

The removal of dissolved O_2, N_2, P and S gives us steel of greater strength. (*ii*) When Mn, in the form of ferro-manganese and spiegeleisen, is added to steel in large amounts upto 14%, we get steel of special hardness and toughness. This steel which contains 14% Mn is called *manganese steel* and is used for making rock crushes, steel halmets etc. (*iii*) Manganese is an alloy of Mn with Cu, Ni and Fe. Its approximate composition is : Mn = 10–14%, Cu = 83–85%, Ni = 3–4% and Fe = 0.4%. It has a very low coefficient of resistance (*i.e.* high specific resistance and hence is used in electrical instruments for making resistance wires. (*iv*) Manganese bronze is an alloy of Mn with Cu Zn and Fe. Its approximate composition is : Mn = 0.5%h Cu = 60–65%, Zn = 35–40% and Fe = 0.5–1%. It contains no tin and hence the name bronze is misnomer. It is really manganese brass and not manganese bronze. It is resistant towards water and is, therefore, used in making parts required in shipping industry.

Iron (Fe_{26})

Occurrence of Iron

Iron is the fourth most abundant element and second most abundant metal—the first three elements being oxygen, silicon and aluminium. Iron rarely occurs in the free state, since it is highly reactive.

The most important ores of iron are : (*i*) *Magnetite*, Fe_3O_4 which contains 70% or even more of Fe (*ii*) *Haematite*, Fe_2O_3. (*iii*) *Limonite* or *hydrated oxide*, $Fe_2O_3.3H_2O$. (*iv*) *Copper-iron pyrites*, $CuFeS_2$. India possesses about one fourth of the total world reserve of iron ore, compaired to U.S.A. (14%) and U.S.S.R. (24.7%). Huge deposits of iron are found in Singhbhum and Bonai. The iron ores of these places are of highest quality. Iron ores are widely distributed in Madhya Pradesh (Durg, Bastar, Katni, Bijori). Important deposits have been found in Maharashtra (Chanda, Loharu, Akola, Ratanpur, Ratangiri), Mysore (Hospet, Bellary districts), Orissa (Cuttack, Koraput, Sambalpur, Mayurbhanj, Brrybil-Koira Valley). Small deposits of iron ore have also been located in Kerala, Himachal Pradesh (Mandi district), Assam (Kashi Hills), Punjab (Mahendragarh), Rajasthan (Alwar, Bundi, Udaipur and Jaipur), Uttar Pradesh and West Bengal.

Passivity of Iron

When a piece of iron is brought in contact with dilute sulphuric acid, a vigorous reaction begins and hydrogen gas starts coming out. The same piece of iron will displace copper from a solution of copper sulphate. It is not iced, however, that if the iron piece be kept for sometime in conc. HNO_3, then washed free from nitric acid and allowed to react in the above manner it can neither liberate hydrogen from H_2SO_4 nor precipitate copper from copper sulphate solution. In other words, by treatment with nitric acid iron has so changed that it would not behave in the same manner as any other untreated piece of iron would do. *Iron in this form is said to be passive, i.e. it shows lesser reactivity than it would normally show.* Iron can also be made passive (*i.e.* can be passivated) (*i*) by dipping it in a solution of other oxidising agents like H_2CrO_4, acidified $KMnO_4$, H_2O_2 (*ii*) by treating it with dry NO_2 (*iii*) by subjecting it to anodic oxidation, *i.e.* in electroplating or electrolysis, the anodes sometimes become passive, if a high current density is employed.

Degree of Passivity

Passivity is not an absolute property and a metal exhibits variable degree of passivity depending on the chemical conditions. For example :

(*i*) Oxidising conditions tend to promote passivity while reducing conditions tend to decrease it.

(*ii*) Fe coupled with Pt is more easily passivated than iron alone while Fe coupled with Zn cannot be passivated.

Theories of Passivity

(*i*) Smith and Russet's theory. According to Smith and Russel (1926) the passivity is due to the change in electron configuration of atoms on the surface of the metal.

(*ii*) Oxide film theory. *Faraday* suggested that the passivity of iron is due to the formation of an extremely thin and impervious film on the surface of iron. This film is usually of ferroso-ferric oxide, Fe_3O_4. This theory is confirmed by the fact that if the oxide film on the surface is removed by scratching or by heating it in a reducing atmosphere of H_2 or CO or by dissolving it in iodine solution, iron again becomes active and begins to give its usual reactions. *Evans* strongly supported this theory in 1927 by his researches on the subject of passivity.

Rusting or Corrosion of Iron

When a piece of iron is left exposed to ordinary moist air, it is found covered with a reddish-brown coating which can easily be detached and is called *rust. The process of formation of rust is called rusting.* Partington suggests that rust is chiefly *hydrated ferric oxide*, $2Fe_2O_3.3H_2O$. Some authorities believe that rust is an indefinite mixture which varies in its composition according to its age. Analysis shows that it is probably a mixture of *ferric oxide, hydrated ferrous and ferric oxides and basic ferrous and ferric carbonates.* If the rust is long exposed to the air, the amounts of ferrous compounds becomes very small and ultimately hydrated ferric oxide predominates.

Conditions Required for Rusting

Conditions essentially required for the rusting of iron are : (*i*) *Presence of moisture* (*ii*) *Presence of a weakly acidic atmosphere* (*iii*) *Presence of impurity in the iron.*

Theories of Rusting

A number of theories have been put forward to explain the phenomena of rusting, But the following theories are mentioned here,

(*i*) Acid theory (Carbonate formation theory). According to this theory put forward by *Calvert* and *Crum Brown* in 1888, iron reacts with H_2CO_3 formed by the dissolution of CO_2 in H_2O to form $FeCO_3$ which is then hydrolysed in presence of dissolved oxygen to give a precipitate of $Fe(OH)_3$ and an acidic solution of H_2CO_3 which continues to attack iron.

$$CO_2 + H_2O \longrightarrow H_2CO_3$$

$$H_2CO_3 + Fe \longrightarrow FeCO_3 + H_2$$

$$FeCO_3 + 10H_2O + O_2 \longrightarrow \underset{\text{ppt.}}{4Fe(OH)_3} + 4H_2CO_3$$

This theory fails to explain rusting in the absence of CO_2 and even in alkaline solutions. Alkalies prevent rusting, since they neutralise the CO_2 present.

(*ii*) Electrochemical theory. According to this theory rusting is an electrochemical phenomena and takes place due to local action. Iron contains a number of impurities of lower oxidation potential and hence a kind of tiny voltaic cell (or an *electronic couple* or a *corrosion couple*) is set up between the impurities and iron itself in contact with H_2O containing dissolved O_2 and CO_2 in which iron acts as an anode.

Reactions : The surface of the moist iron is covered with water film which contains dissolved O_2 and CO_2. CO_2 reacts with H_2O to give H_2CO_3. Thus the surface of Fe is covered with a solution of an electrolyte *viz.* H_2CO_3 and a depoliser *viz.*, dissolved O_2. H_2O and H_2CO_3 give H^+, HCO_3^- and OH^- ions as follows :

$$CO_2 + H_2O \longrightarrow H_2CO_3$$

At cathode : $H_2CO_3 \rightleftharpoons H^+ + HCO_3^-$

At anode : $H_2O \rightleftharpoons H^+ + OH^-$

H^+ ions originated from H_2CO_3 and H_2O go to the cathode where these are converted into H_2 molecules. These molecules of hydrogen are oxidized to H_2O by the dissolved O_2 which acts as a depolariser.

$$2H^+ + 2e^- \longrightarrow H_2;\ 2H_2 + O_2 \longrightarrow 2H_2O$$

HCO_3^- and OH^- ions obtained from H_2CO_3 and H_2O respectively migrate to the anode where OH^- ions combine with Fe^{3+} ions (formed at the anode from Fe^{2+} ions–see below) to form $Fe(OH)_3$—a powdery deposit of rust.

$$Fe^{3+} + 3OH^- \longrightarrow \underset{\text{deposit of rust}}{Fe(OH)_3}$$

Fe^{3+} ions are formed at the anode as follows : Fe atoms at the anode are oxidised to Fe^{2+} ions by losing two electrons and these Fe^{2+} ions pass into solution and then are oxidised to Fe^{3+} ions by the dissolved O_2

$Fe \longrightarrow Fe^{2+} + 2e^-$ (*Oxidation of Fe atoms to Fe^{2+} ions*)

$Fe^{2+} + H_2O + O_2$ (*Dissolved oxygen*) $\longrightarrow 4Fe^{3+} + 4OH^-$ (*Oxidation of Fe^{2+} ions to Fe^{3+} ions*).

Protection of Iron From Rusting

The protection of iron and steel from rusting may be brought about :

(*i*) By coating iron with an impervious film of oils, paints (specially aluminium paints, red oxide), coal tar, grease, pitch etc. which does not allow iron to come in contact with air and moisture and thus protects iron from rusting.

(*ii*) By blowing steam over red-hot iron. An adherent film of Fe_3O_4 gets deposited on the surface of iron (Russian iron) and thus iron is protected from rusting (*Barf process*).

***(iii)* By alloying iron water with Ni, Cr, Si etc.** The alloys obtained by alloying iron water with Ni, Cr, Si etc have excellent resistance towards rusting due to homogenous make up and formation of a film of the oxide on the surface.

***(iv)* By metallic coatings.** In this method the surface of iron is covered with a layer of a metal more active than iron, *e.g.* Zn, Sn, Al, Cr etc. The zinc coating is usually obtained by galvanising which is carried out by dipping iron into a bath of molten zinc. The zinc coating of the galvanised iron does not allow iron to come in contact with air and moisture and thus protects it from rusting. Even when the zinc coating is partially removed from the surface of iron, iron does not get rusted. In this case both the metals *viz.* Fe and Zn come in contact with air. But, because Zn has higher oxidation potential, it gets oxidised to Zn^{2+} in preference to Fe, *i.e.*, reaction (*a*) takes place while the reaction (*b*) which is necessary for rusting does not take place.

(*a*) $Zn \longrightarrow Zn^{2+} + 2e^-$ (*Takes place*)

(*b*) $Fe \longrightarrow Fe^{2+} + 2e^-$ (*Does not take place*)

In other words it is the zinc and not iron that goes into solution.

***(v)* By other coatings.** Some other types of coating like phosphate coatings and concrete coatings are also effective in protecting iron from rusting.

***(iv)* By using corrosion inhibitors.** *The substances that bring about chemical alteration of the metal surface so as to make the same rustproof are called corrosion inhibitors.* The most commonly used inhibitor of this class is K_2CrO_4. In a sulution of this salt iron will remain free from corrosion indefinitely.

Commercial Forms of Iron

Iron is worked for the following three varieties of iron which *differ in their carbon content.*

1. Wrought iron or malleable iron. It is the purest form of commercial iron and contains the *lowest percentage* of carbon, *viz.*, 0.12–0.25% and some impurities like S, P, Si and Mn (not more than 0.3%).

2. Steel. It contains 0.25–2.5% of carbon and traces of S, P, Si and Mn.

3. Cast iron or pig iron. This is the *least pure* form of commercial iron and contains the *highest percentage* of carbon *viz.*, 2.5–4.5% and traces of impurities like S, P, Mn and Si.

Thus we see that the *amount of carbon in steel is midway between that in wrought iron and in cast iron.* First the cast iron is manufactured from iron ores and then both wrought iron and steel are obtained from cast iron.

Cast Iron

This is the *least pure* form of commercial iron and contains the highest percentage of carbon *viz.* 2.5 to 4.5% and traces of impurities like S, P, Mn and Si The average composition of cast iron is : iron = 93–95% *phosphorous* = 0.4–1%; *carbon* = 2.5–4.5%; *sulphur* = 01.–0.3%; *silicon* = 0.6–2.8% ; *manganese* = 0.3–1.2%. There are two varieties of cast iron namely (*i*) *White cast iron* which contains carbon mostly as *iron carbide,* Fe_3C and very small amount of graphite. It is obtained by cooling the molten pig iron quickly. (*ii*) *Grey cast iron* which contains carbon mostly as graphite and very small amount of Fe_3C. It is obtained by cooling the molten pig iron slowly.

Manufacture of Cast Iron From Haematite Ore, Fe_2O_3

The manufacture of cast iron from haematite ore involves the following steps :

1. Concentration of the ore. The ore is first crushed to small pieces of 3–5 cm size, screened and sifted. This removes most of the gangue. In a few cases the ore is washed with water to remove silicious impurities. The ore is thus concentrated.

2. Roasting or calcination of the concentrated ore. The concentrated ore is roasted with a little of coal in shallow kilns in the excess of air whereby the following reactions take place.

$$2Fe_2O_3.3H_2O \longrightarrow 2Fe_2O_3 + 3H_2O \uparrow$$
$$FeCO_3 \longrightarrow FeO + CO_2 \uparrow$$
$$4FeO + O_2 \longrightarrow 2Fe_2O_3$$

These reactions result in the following : (*a*) *Moisture and CO_2 are removed.* (*b*) *Sulphur, arsenic and phosphorus ore oxidised to their oxides and are thus removed.* (*c*) FeO is converted into Fe_2O_3 which does not form stag of $FeSiO_3$ with silica (SiO_2) during smelting while FeO forms such slag. *Thus the loss of iron is prevented.* (*d*) In addition to the above chemical reactions, *the entire mass becomes porous and thus the reduction of Fe_2O_3 to spongy iron is facilitated.*

3. Smelting in the blast furnace. The roased ore is mixed with limestone and desulphurised coke in the ratio of 8 : 1: 4 and smelted in a blast furnace.

Construction of the Blast Furnace. (Fig. 46.1). It. is a tall, cylindrical furnace the dimension and construction of which varies from place to place. It is from 80 to 120 feet in diameter. Following are its main parts :

(*a*) *A double cup and cone arrangement.* At the top, the furnace carries a hopper which rests over a funnel-shape double cup and cone arrangement. This arrangement gets lowered by a lever arrangement worked out by chains and wheels (not fully shown in Fig. 46.1). By lowering it, a fresh charge is introduced into the furnace. This arrangement prevents the exit of gases during the charging of the furnace. The charge is fed at the cup and when it is full, the cone is depressed so that the charge automatically goes down into the furnace and subsequently the throat of the furnace is automatically closed.

(*b*) *Body, bosh and tuyers.* The furnace has a *shaft* which is about 100 feet high made of steel plates and is formed by joining two cones: the upper one is called *body* and the lower one is called *bosh*. The upper one is longer than the lower one. The furnace is lined inside with a casting of ordinary bricks and then there is a lining of fire bricks 1.5 feet thick to withstand high temperature of the furnace. At the upper part of the furnace there is a hole through which the hot burnt mixture of waste gases escapes. This mixture of waste gases is called *furnace gas*.

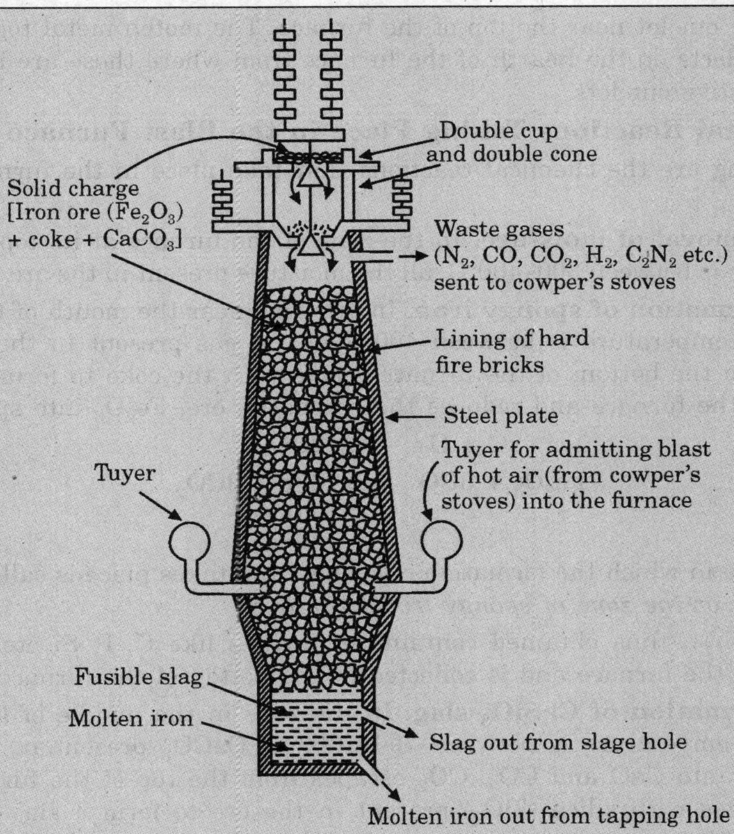

Fig. 46.1. Blast furnace for the manufacture of cast iron.

Between 6 and 8 feet from the base of the furnace there are openings for the insertion of the water-cooled pipes called *tuyers* through which a blast of hot air is sent upwards in the furnace.

(*c*) *Hearth or crucible.* The furnace narrows down below the bosh to an internal width of 8 feet at the base which is called *hearth* or *crucible* of the furnace. The hearth serves as a large crucible to hold the molten iron and is provided with two holes—one is for flowing away the lighter slag (called *slag notch* or *slag hole*) and the other taps out the molten metal (called *metal hole*).

Functioning of the Furnace. The charge consisting of roasted ore (8 parts), limestone (1 part) and desulphurised coke (4 parts) is carried to the top of the furnace. By lowering the cup and cone arrangement the charge in introduced into the furnace. At the same time the furnace is lit and a blast of hot air is sent upwards through the *tuyers.* The blast of hot air for this purpose is obtained by heating air in a series of *cowper stoves* or *preheaters* and is conveyed through a steel pipe lined with fire-bricks to the tuyers from where it is discharged into the furnace. As the charge moves down the furnace, it meets high temperature and various gases are liberated as a result of the chemical reaction between the charge and the air blast rising from the below. The waste gases leave the furnace through the out let near the top of the furnace. The moten metal together with the slag collects on the hearth of the furnace from where these are removed by their respective out lets.

Chemical Reactions Taking Place in the Blast Furnace

Following are the chemical reactions that take place in the furnace during smelting.

(*i*) Removal of moisture. In the zone of the furnace at its top where the temperature is between 300–400°C, all the moisture present in the are is removed.

(*ii*) Formation of spongy iron. In the zone near the mouth of the furnace where the temperature is between 400–700°, O_2 gas present in the hot gases coming from the bottom of the furnace reacts with the coke to form CO which goes up in the furnace and reduces the haematite öre, Fe_2O_3 into spongy iron.

$$2C + O_2 \longrightarrow 2CO \uparrow$$

$$Fe_2O_3 + 3CO \longrightarrow \underset{\text{Spongy iron}}{2Fe} + 3CO_2$$

The zone in which the formation of spongy iron takes place is called *the zone of reduction or the zone of spongy iron formation.*

Spongy iron thus obtained contains impurities like C, P, Si etc. This iron slides down the furnace and is collected at the botton of the furnace.

(*iii*) Formation of $CaSiO_3$ slag. In the zone in the middle of the furnace where the temperature is between 700–1000°C, $CaCO_3$ present in the charge decomposes into CaO and CO_2. CO_2 escapes from the top of the furnace while CaO combines with silica (SiO_2) present in the ore to form a slag of calcium silicate, $CaSiO_3$.

$$CaCO_3 \longrightarrow CaO + CO_2 \uparrow \; ; \; CaO + SiO_2 \longrightarrow \underset{\text{Slag}}{CaSiO_3}$$

This slag sinks down and floats over the molten iron collected at the bottom of the furnace. The slag floating over the molten iron protects iron from oxidation. The molten iron and fusible slag are removed from their respective holes. The molten metal is run into moulds and allowed to solidify. It is called *pig iron.* It is remelted in a vertical furnace called *cupola* and can be cast or poured into moulds and hence is known as *cast iron.*

The zone in which the decomposition of $CaCO_3$ and formation of $CaSiO_3$ slag takes place is called *the zone of slag fonnation.*

(*iv*) Reduction of oxides. In the zone near the tuyers, where the temperature is between 1240–1500°C the more stable oxides like P_4O_{10}, SiO_2 and MnO_2 are reduced by red-hot coke to the free elements.

$$P_4O_{10} + 10C \longrightarrow 4P + 10CO \uparrow$$

$$SiO + 2C \longrightarrow Si + 2CO \uparrow$$

$$MnO_2 + 2C \longrightarrow Mn + 2CO \uparrow$$

Si and Mn go into the molten iron as such while P combines with Fe to form iron sulphide, Fe_3P.

By-Products of Blast Furnace

Following are the by-products obtained from the blast furnace.

(*i*) Pig iron. The molten iron produced in the blast furnace is withdrawn through a hole (*slag hole*) into a large ladle lined with fire-bricks which is packed with hundreds of tons of the metal. The molten metal is cast into small semicylindrical moulds called *pigs* and hence the name *pig iron* for the product.

(*ii*) Blast furnace gas. The mixture of waste gases coming out of the hole near the mouth of the furnace is called *blast furnace gas* which contains N_2 (58%), CO(25%), CO_2(10.5%) and H_2, C_2N_2, hydro-carbons etc. (about 6.5%). The furnace gas is first made free from dust particles by passing it through a dust *chamber* and *water scrubber* and then brought to the bottom of one of the two *cowper stoves* and *iron tower* packed with a checker. Checker is a work of firebricks with a channel on one side to act as combustion chamber. Here the furnace gas is burnt with excess of air and the heat of combustion thus obtained heats up this stove. When this stove has been heated up in this manner, the furnace gas is directed to the second stove to heat it up and at the same time the blast of dry air to be used in the furnace is passed into the first stove which heats up the blast of air which is then sent to the furnace through the tuyers. The stove gets cooled and is reheated by burning the furnace gas with excess of air as before. In the meanwhile the second stove has been heated to raise the temperature of fresh blast of air. Thus the two stoves are used alternatively, one for generating heat by burning the furnace gas with excess of air and the other for heating the blast of air to be used in the furnace. This economises fuel to a very large extent. *Thus the cowper stoves serve as heat-generators and fuel-economisers.*

(*iii*) Slag. The fusible slag which is withdrawn from the furnace through the slag hole is granulated by running it into water. It is then. (*a*) used as ballast for rail roads. (*b*) mixed with tar for road-making. (*c*) used in cement manufacture. For this purpose it is burnt with lime and then finely powdered.

Properties and Uses of Cast Iron

It is the most impure form of commercial iron and contains the highest percentage of carbon and traces of impurities of S P, Mn and Si. The impurities present in cast iron have remarkable influence on its properties. Cast iron melts at about 1250°C whereas pure iron melts at 1530°C. The presence of carbon and silicon increases the hardness and strength of the material. Phophorus increases the fluidity of iron when molten while sulphur makes it brittle when cold. Such an iron is called *cold short.* Sulphur increases the brittleness when red hot. Such a specimen is called red short. Cast iron is very hard and cannot be welded. It is so brittle that it cannot be used for making parts of machinery that are subjected to great stress. It expands on solidification and is used for casting various articles such as stoves, pipes, radiators, railway sleepers, gutter pipes, toys etc. A large quantity of cast iron is used in the manufacture of wrought iron and steel.

Wrought Iron

It is the *purest* from of commercial iron and contains the *lowest percentage* of carbon *viz.* 0.12 to 0.25% and 0.3% of impurities like S, P, Si and Mn.

Manufacture of Wrought Iron From Cast Iron

It is manufactured from cast iron by *pudlling i.e.* by heating cast iron in a special type of reverberatory furnace called *puddling furnace* (Fig. 46.2).

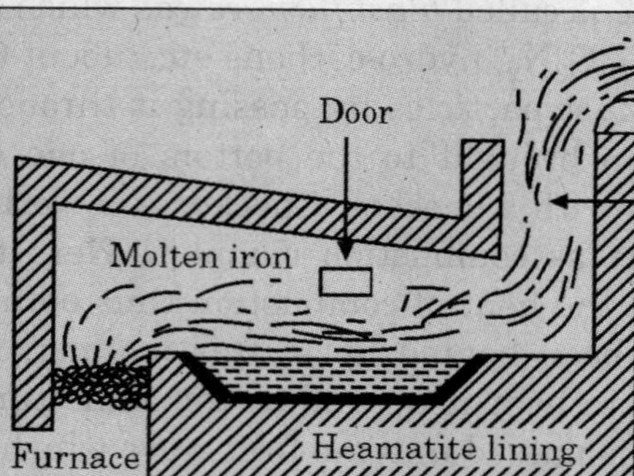

Fig. 46.2. Puddling furnace for the manufacture of wrought iron.

This furnace has a low roof to deflect the hot gases and flames downwards and to melt cast iron. The hearth of furnace is lined with haematite (Fe_2O_3). The cast iron is piled on the hearth, melted by the hot gases and stirred or puddled with long iron rods called *rabbles* through the door to bring it in thorough contact with the lining of the hearth. The heamatite (Fe_2O_3) lining supplies oxygen necessary for the oxidation of carbon, sulphur, silicon, manganese and phosphorus present in the cast iron. Oxides of carbon and sulphur, being volatile, escape while those of manganese, silicon and phoaphorus form slags. Thus :

$$3C + Fe_2O_3 \longrightarrow 2Fe + 3CO \uparrow$$

$$3S + 2Fe_2O_3 \longrightarrow 4Fe + 3SO_2 \uparrow$$

$$3Si + 2Fe_2O_3 \longrightarrow 4Fe + 3SiO_2$$

$$2Mn + O_2 \longrightarrow 2MnO; \; MnO + SiO_2 \longrightarrow \underset{\text{Slag}}{MnSiO_3}$$

$$4P + 5O_2 \longrightarrow 2P_2O_5; \; Fe_2O_3 + P_2O_5 \longrightarrow \underset{\text{Slag}}{2FePO_4}$$

With the removal of impurities, the melting point of the metal rises and it becomes a semi-solid mass. As this stage, it is taken out in the form of balls or blooms on the ends of rabbles. While still hot, these balls are subjected to hammering to squeeze out as much of slag as possible. The products so obtained is known as *wrought iron.*

Properties and Uses of Wrought Iron

It is the purest form of iron containing the lowest percentage of carbon, *viz.*, 0.12–0.25% and other impurities of S, P, Si and Mn (not more than 0.3%). It is grey and extremely tough and hence is used in the manufacture of articles which are to stand severe and sudden stresses such as archers, wires, bolts, chains and agricultural implements. Its tough nature and resistance towards rusting and corrosion is due to the presence of slag in it. Its melting point is 1500°C. The most valuable property of wrought iron is that it is brittle and can be easily welded and worked under the hammer, so as to assume any desired shape. For most purposes, however, the use of mild steel (*i.e.* steel containing a low proportion of carbon) has replaced that of wrought iron.

Steel

What is Steel ?

Steel is an alloy of iron (as ferrite) containing 0.25 to 2.5% of carbon (as cementite, Fe_3C) and traces of S, P, Si and Mn. Thus we see that the percentage of carbon in steel is intermediate between that in wrought iron and in cast iron. S and P are objectionable impurities. The percentage of S and P is usually below 0.05%, depending on the method used for the manufacture of steel. Si is usually between 0.15 to 0.35% and is present as *iron silicide*, FeSi which is soluble in ferrite. High content of Si gives a fibrous structure to steel. Mn is added to steel during its manufacture to serve as a deoxidiser and to neutralise the harmful effect of sulphur by forming MnS which is inert, while FeS makes steel brittle in hot working.

Varieties of Steel

There are many varities of steel depending on the amount of carbon present in it.

1. Mild and quenched steels. Mild steel is a veriety which contains *lower percentage* of carbon. It possesses the properties of wrought iron along with elasticity and hardness of steel. If mild steel is heated to a high temperature (*i.e.* to bright redness) and then suddenly cooled by plunging it in oil or water, it becomes as hard and brittle as glees. This is called *quenched steel* and process is called *quenching* or *hardening*.

2. Hard steel. This variety contains *higher percentage* of carbon. It is hard and brittle like glass.

3. Special steels or alloy steels. Addition of small amounts of nickel, cobalt, chromium, tungsten, molybdenum, manganese and silicon confer special properties on steels considerably altering its hardness, tenacity, resistance to corrosion and coefficient of expansion. Such products are called *special steels or alloy steels* and find extensive use in industry.

Some of the alloy steels along with their composition, properties and uses are listed in Table 46.2 given below :

Table. 46.2 Some important alloy steels

Name of the alloy steel	Metal added	Proprties	Uses
(*i*) Manganese steel	1.2–1.5% Mn	Hard, tough, resistant to wear	Rock crushing machinery, safes, helmets etc.
(*ii*) Nickel steel	2.5–5% Ni	Hard, tough, rustless	Armour plate, cables, propeller shafts automobile parts, heavy guns.
(*iii*) Invar	3.6% Ni	Same coef. of expension as glass	Pendulum, messuring tapes, instruments.
(*iv*) Chrome steel	1.5–2% Cr	Extremely hard	Cutting tools crushing machinery
(*v*) Stainless steel	13–14% Cr 1% Ni	Resists corrosion	Cutery, food processing machenery
(*vi*) Tungsten steel	14–20% W	Hard and strong	High speed tools
(*vii*) Chrome-vanadium steel	1% Cr, 0.15% V	High tensile strength	For making spring shafts, axles etc.

4. Steel containing S and P. Iron or steel containing excess of S is brittle when hot (*hot short*) and that containing excess of P is also brittle when cold (*cold short*).

Manufacture of Steel

We have already seen that the amount of carbon in steel is mid-way between that in cast iron and in wrought iron. So the steel can be obtained either by removing a part of carbon form cast iron or by adding some carbon to wrought iron. The following methods are in use for the manufacture of steel : (*i*) *Cementation process* (*From wrought iron*). (*ii*) *Crucible process* (*From wrought iron*). (*iii*) *Bessemer's process* (*From cast iron*). (*iv*) *Open-hearth process* (*From cast iron and wrought iron or mild steel scrap*). (*v*) *Electrical process* (*From pure ore and carbon*).

Here we shall discuss *Bessemer's process, open hearth process, Duplex process and electrical process.*

1. Bessemer's Process.

This process was discovered by Sir Henry Bessemer of England in 1856 and hence the name of the process. The furnace used in this process is called *Bessemer's convertor* which is a pear-shaped vessed made of riveted steel plates. The convertor is provided at the botton with a number of holes through which a hot blast of air can be admitted. The convertor is mounted on trunnions or pivots so that it can be tilted in any position. A schemetic diagram of Bessemer's convertor used for the manufacture of steel from cast iron is shown in Fig. 46.3.

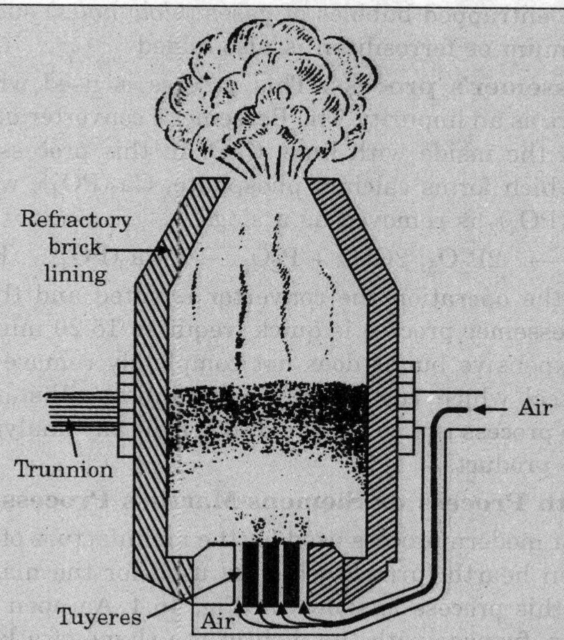

Fig. 46.3. Bessemer's converter used for the manufacture of steel from cast iron.

Depending on the nature of the impurities present in the cast iron being used for the manufacture of steel, Bessemer's process is of two types :

***(i)* Acidic Bessemer's process.** This process is used when the cast iron being used for the manufacture of steel contains Si and Mn as impurities. The Bessemer's converter used in this process should be lined on the inside with silica, SiO_2. To start with, the converter is brought in the horizontal position and then cast iron (which is free from P and S) from the blast furnace is put into it. Now it is brought back into the vertical position and a blast of air is admitted into it. Si and Mn present as impurities in cast iron are oxidised to SiO_2 and MnO respectively. These two oxides combine together to form manganese silicate, $MnSiO_2$ which is removed as slag.

$$Si + O_2 \longrightarrow SiO_2; 2Mn + O_2 \longrightarrow 2MnO$$

$$SiO_2 + MnO \longrightarrow MnSiO_3 \text{ (Removed as slag)}$$

Later on, carbon is oxidized to carbon monoxide which burns with a blue flame at the mouth of the converter. *When the flame dies out, it indicates the complete oxidation of impurities.* At this stage, *spiegeleisen* (an alloy of iron, mananese and carbon) is added and the blast of air continued for a moment to ensure thorough mixing. Manganese and carbon serve, in the first instance, as *deoxidizers* and reduce any FeO that may be present. The excess of these elements then dissolves in the molten metal, the carbon being the necessary component of steel while manganese serves to make it harder and to increase its tensile strength.

In order to remove entrapped bubbles of gases (*blow holes*) such as O_2, N_2, CO_2 etc., a little aluminium or ferrosilicon is also added.

***(ii)* Basic Bessemer's process.** This process is used when the cast iron contains phosphours as an impurity. The Bessemer's converter used in this process should be lined on the inside with lime, CaO. In this process phosphorus gets oxidised to P_2O_5 which forms calcium phosphate, $Ca_3(PO_4)_2$ with CaO lining of the convertor. $Ca_3(PO_4)_2$ is removed as a slag.

$$4P + 5O_2 \longrightarrow 2P_2O_5; \; 3CaO + P_2O_5 \longrightarrow Ca_3(PO_4)_2 \text{ (Removed as slag)}$$

At the end of the operation the converter is tilted and the molten steel is poured out. The Bessemer process is quick (requires 15.20 minutes to complete) and relatively inexpensive but it does not completely remove acidic impurities and produces a steel which rusts easily and cannot withstand too much and sudden strain. The process is also too rapid to permit any analytical control of the composition of the product.

2. Open-hearth Process or Siemens-Martin's Process

This is the most modern process used for the manufacture of steel. A schemetic diagram of an open hearth furnace which is used for the manufacture of steel from cast iron by this process is shown in Fig. 46.4. An open hearth furnace is like a reverberatory furnace with the dish in the shape of a large oval dish. An open hearth is of two types : (*i*) furnace whose hearth is lined with a basic lining like dolomite, CaO.MgO. Such a furnace is used when the impurities present in the cast iron are of P. A furnace with a basic lining is used in the manufacture of steel (*ii*) furnace whose hearth is lined with an *acidic lining* like silica, SiO_2. Such a furnace is used when the impurities are of Mn, Si etc. An open hearth furnace utilises the heat from the waste gases obtained from the blast furnace. Hot waste gases obtained from the blast furnace are passed through a brick-work chamber which gets heated and the gases going out of this chamber are at a much lower temperature. After some time fuel gases are passed into this hot chamber and the fuel gases are made to heat another brick work chamber. The fuel gas preheated in the hot chamber burns and thus supplies more heat. After some time the direction of the gases is reversed so that the incoming gases are always preheated. Thus we see that an open hearth furnace works on the *regenerative principal of heat economy* and hence this furnace is also called *regenerative furnace.* In order to manufacture steel, the charge consisting of a mixture of *pig iron* (*or cast iron*), *scrap iron* and *haematite ore* (Fe_2O_3) is melted on the hearth of the furnace by forcing pre-heated air and burning gaseous fuels. The reactions taking place in the furnace can be shown as follows :

$$2Fe_2O_3 + 3S \longrightarrow 4Fe + 3SO_2 \uparrow \text{ (Sulphur removed)}$$
$$2Fe_2O_3 + 3C \longrightarrow 4Fe + 3CO_2 \uparrow \text{ (Carbon removed)}$$
$$5Fe_2O_3 + 6P \longrightarrow 10Fe + 3P_2O_5$$
$$2Fe_2O_3 + 3Si \longrightarrow 4Fe + 3SiO_2$$
$$Fe_2O_3 + 3Mn \longrightarrow 2Fe + 3MnO$$

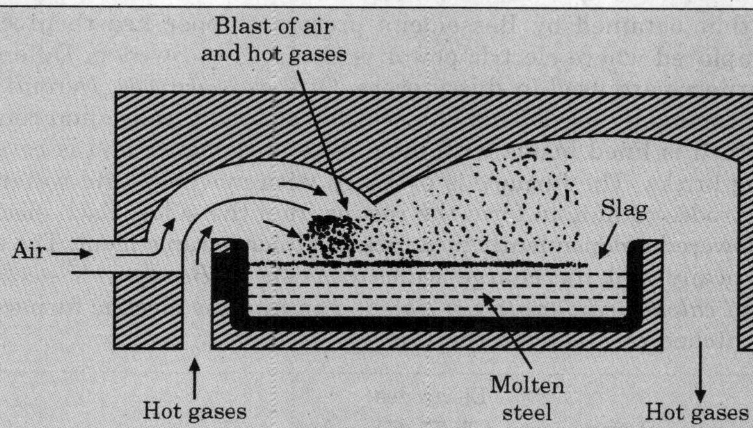

Fig. 46.4. An open-hearth furnace used for the manufacture of steel from cast iron.

SO_2 and CO_2 formed go out of the furnace while the oxides namely P_2O_5 and SiO_2 form $Ca_3(PO_4)_3$ and $CaSiO_3$ slags with CaO lining of the furnace. Similarly MnO also forms $MnSiO_3$ slag with SiO_2 lining of the furnace. Thus the impurities of P, Si and Mn are removed as slags.

$$P_2O_5 + 3CaO \longrightarrow \underset{\text{Slag}}{Ca_3(PO_4)_2} \ \textit{(Phosphorus removed)}$$

$$SiO_2 + CaO \longrightarrow \underset{\text{Slag}}{CaSiO_3} \ \textit{(Silicon removed)}$$

$$MnO + SiO_2 \longrightarrow \underset{\text{Slag}}{MnSiO_3} \ \textit{(Manganese removed)}$$

Open-hearth process has the following advantages over Beasemer's process. (*i*) Better quality of steel of the desired composition is obtained by this process, since the composition of the product can be controlled, *i.e.* the process can be stopped when the analysis of the *sample of the charge* product meets our requirement. (*ii*) Scrap iron, iron ore and pig iron (or cast iron) of low grade can be directly utilised in this process. In other words in this process pig iron of any composition can be used. (*iii*) The temperature can be easily controlled, since heat is supplied from external source. (*iv*) Loss of iron as slag is very little (about 4% against 15% in Bessemer's process), since no blast of air is passed.

3. Duplex Process

Naw-a-days Duplex process is being used for the manufacture of large quantities of steel. This process is a combination of Bessemer's process and open-hearth process. In this process the molten pig iron is first treated in an acid Bessemer's converter to remove Si, Mn and a part of C. The molten pig iron is then transferred to open-hearth furnace with a basic lining to remove P and remaining C and is finished off as usual. Tata Iron and Steel Works are using Duplex process for the manufacture of steel.

4. Electrical Process.

Steel obtained by this process is free from sulphur, phosphorus and gas bubbles and contains fewer solid, non-matallic impurities (known as *sonims*) and hence is superior to that obtained by Bessemer's process or open-hearth process. This process is employed where electric power is cheap as in Sweden. Different types of electric furnaces are used in this process. Of these furnaces, *Heroult's furnace* (Fig. 46.5) is commonly used. This is a crucible shaped furnace and consists of a steel shell which is lined inside with magnesite or dolomite and is covered with heat resisting bricks. The furnace is provided with movable kind water jacketed graphite electrodes coming in form the roof or from the sides. Each electrode can be raised or lowered independently by a rack or pinion arrangement. The electrodes are held vertically and the charge which consists of *Beasemer's steel, iron ore heamatite and calculated quantity of lime* is run into the electric furnace and the current is switched on to strike the arc.

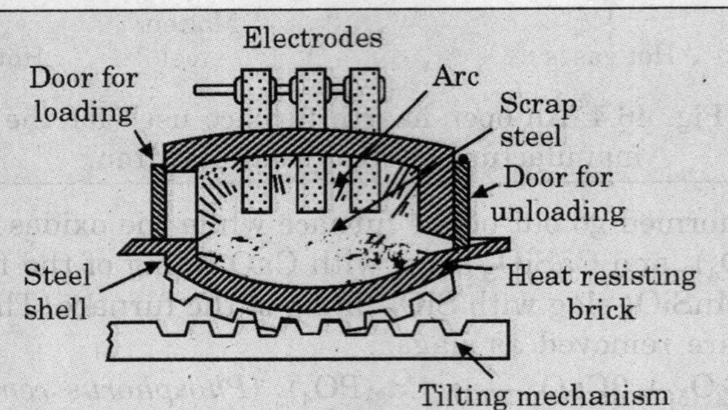

Fig. 46.5. Heroult's furnace for the manufacture of steel.

Since high temperatures are possible in electric furnaces, alloy steels with high melting points are manufactured in electric furnaces.

Reactions. The process involves the following two reactions :

(i) Oxidation and formation of slag. At the temperature of the electric arc, the charge melts and vigorous reaction starts. In this reaction Si, Mn, C and P are oxidised to SiO_2, MnO, CO and P_4O_{10} respectively and $CaCO_3$ gets decomposed.

(*a*) $3Si + 2Fe_2O_3 \longrightarrow 4Fe + 3SiO_2$
(Iron ore)

(*b*) $3Mn + Fe_2O_3 \longrightarrow 2Fe + 3MnO$

(*c*) $C + O_2 \longrightarrow CO_2, CO_2 + O \longrightarrow 2CO$

(*d*) $P_4 + 5O_2 \longrightarrow 2P_4O_{10}$

(*e*) $CaCO_3 \longrightarrow CaO + CO_2$

The oxides *viz.* SiO_2, MnO, CO and P_4O_{10} are removed as slags.

(*a*) $SiO_2 + CaO \longrightarrow \underset{\text{Slag}}{CaSiO_2}$

(*b*) $MnO + SiO_2 \longrightarrow \underset{\text{Slag}}{MnSiO_3}$

(*c*) $Fe_2O_3 + CO \longrightarrow CO_2 + 2FeO;\quad FeO + SiO_2 \longrightarrow \underset{\text{Slag}}{FeSiO_3}$

(*d*) $P_4O_{10} + 6CaO \longrightarrow \underset{\text{Slag}}{2Ca_3(PO_4)_2}$

The slag being lighter floats over the metallic portion. As soon as the oxidation and formation of slag in complete, the furnace is carefully tilted and the slag is poured out of it.

***(ii)* Deoxidation and desulphurisation.** After the slag is poured out of the furnace, charge consisting of *lime, sand, fluorspar* and some *coke* and *ferro silicon* is added to the furnace. Any FeO present in the steel is removed as Fe by reducing it with the added coke.

$$FeO + C \longrightarrow Fe + CO \text{ (\textit{Removal of FeO})}$$

Sulphur present as FeS is removed as CaS by treating it with

$$FeS + CaO \longrightarrow Fe + CaS \text{ (\textit{Removal of S})}$$

FeO thus produced is at once reduced to Fe as before while CaS goes into the slag. When desulphurization (*i.e.* removad of sulphur) is complete, the steel is poured out into moulds.

This method is also used for the manufacture of *alloy steels, tool steels, stainless and many special quality steel.*

Properties of Steel.

1. Composition. Steel is an alloy of iron containing 0.25 to 2.5% of carbon and traces of S, P, Si and Mn. Thus we see that percentage of carbon in steel is intermediate between that in wrought iron and in cast iron.

2. Mechanical properties. Steel combines the useful properties of cast iron and wrought iron, being both *hard* and *elastic* and provided the proportion of carbon is not high, it can be *forged* and *welded.*

3. Action of heat : Annealing, hardening or quenching and tempering of steel. A characteristic behaviour of steel which distinguishes it from other commercial forms of iron is that its hardness and elasticity can be varied by heating the steel at different temperatures and then allowing it to cool in different ways. For example :

(*i*) When steel is heated to bright redness (200–350°C) and then cooled very slowly, it becomes soft, weak and ductile. The process of heating and cooling the steel in this way is called *annealng of the steel* and the steel obtained by this process is called *annealed steel.* Annealed steel is used in fabrication process.

(*ii*) When steel is heated to bright redness and then cooled suddenly by plunging it into water or oil, steel becomes as hard and brittle as glass. The process of heating and cooling the steel in this way is called *quenching or hardening of steel* and the steel obtained by this process is called *hardened or quenched steel.*

(*iii*) When the hardened or quenched steel obtained as above is heated to a temperature much below redness and then cooled slowly, steel becomes neither very hard nor very brittle. The process of heating and cooling the hardened steel in this way is called *tempering of steel* and the steel obtained by this process is called *tempered steel.* In the process of tempering, a thin film of the oxide is formed on the surface of the hardened steel. The colour of the oxide film depends on the temperature at which the hardened steel is heated as shown below :

Temperature range (°C) :	200–230	225–270	300
Colour of the oxide film :	yellow	brown	blue

Steel required for cutting blades and tools is tempered yellow while that required for springs, saws etc. is tempered blue.

4. Action of charcoal : Case hardening of steel. When mild steel is heated with charcoal and then plunged (quenched) into oil, a thin coating of hardened or quenched steel is produced on the surface of the mild steel. The process of producing a thin coating of hardened steel on the surface of a mild steel in this way is called *case hardening* and the steel obtained by this process is called *case-hardened steel.* The case-hardened steel becomes resistant to wear and tear and hence is largely used for the manufacture of armour plates, locomotive axles etc.

5. Action of dry NH_3 : Nitriding of steel. When steel is heated in an atmosphere of dry NH_3 in sealed boxes for about 3–4 days, hard coating of iron nitride is produced on the surface of the steel. The process of producing a hard coating of iron nitride on the surface of steel in this way is called *nitriding of the steel.*

Uses of Steel

(*i*) Steel is employed for numerous purposes. It is used for making girders, armour plates, machinery, knives, razors, swords, tools, springs, dies, etc. (*ii*) It is also used for making bar magnets and magnetic needles.

Comparison of Cast Iron, Wrought Iron and Steel

Comparison of cast iron, wrought iron and steel is given in Table 46.3.

Table 46.3. Comparison, of cast, wrought iron and steel.

Property and uses	Cost iron	Wrought iron	Steel
1. Chemical composition	Iron = 93–95% Carbon = 2.5–4.5% Impurities = 0.1 to 2.8% (P, Si, S, Mn)	Iron = 98.5–98.8% Carbon = 0.12–0.25% Impurities : not more than 0.3%	Iron = 98.5–99.5% Carbon = 0.25–2.5%
2. Melting point	Lowest about 1200°C	Highest, about 1500°C	Between 1300–1400°C
3. Hardness	Very hard	Soft	Medium hardness.
4. Behaviour when struck	Brittle	Malleable	Malleable also brittle
5. Magnetisation	Cannot be permanently magnetised.	Magnetisation not permanent but it is easy	Can be permanently magnetised
6. Tempering and welding	Can neither be tempered nor welded.	Cannot be tempered but can be welded.	Can be tempered and welded; can be brought in any state of hardness and malleability by proper tempering.
7. Structure	Crystalline	Fibrous	Granular
8. Uses	Uses for the manufacture of wrought iron and steel.	Used for making electric wires, pipes, sheets, rods etc.	Used for the manufacture of utensils, magnets, knives, engines, surgical instruments, springs, cranes and other important materials.

Prospects of Iron and Steel Industry in India

Although India has very large iron ore reserves in the world, yet it produces only 1% of the total steel production of the world. The main reasons for the poor development of iron and steel industry in India are the lack of technical knowledge and huge capital needed, in putting up steel plants. Major plants set up in private sector and producing iron and steel are *Tata Iron and Steel Company, Jamshedpur; India Iron and Steel Company, Hirapur* (*Asansol*); and *Mysore Iron and Steel Works, Bhadravati*. Steel plants set up in public sector are *Hindustan Steel Ltd, Bhilai* (in collaboration with U.S.S.R); *Hindustan Steel Ltd, Durgapur* (in collaboration with U.K.) and *Hindustan Steel Ltd, Rourkela* (in collaboration with West Germany). Other plants have been set up at *Bokaro* (*Orissa*) and *Salem* (*Tamil Nadu*). Plants are being set up at *Visakhapatnam* (*Andhra Pradesh*) and *Vijayanagar* (*Karnataka*). We hope that a time will come soon when India would become one of the biggest producers of steel in Asia and would export steel to other countries.

Cobalt (Co_{27})

Occurrence

Cobalt does not occur free in nature. It constitutes about 0.001% of the earth's crust. Cobalt occurs mainly as *arsenide* and *sulphide* ores with which other metals like Fe, Ni and Ag are also associated. Its principal ores are : *smaltite* or *speiss cobalt*, (Ni, Co, Fe)As_2; *cobaltite* or *cobalt glance*, (Co, Fe, As) S; *crythrite* or *cobalt bloom*, $Co_3(AsO_4)2.8H_2O$ and *linnacite*, $(Fe, Co, Ni)_3\ S_4$. All these ores are associated with Fe, Ni, Cu and sometimes with traces of Ag. Many cobalt ores also contain profitable amounts of Ag. Small amounts of Co are also present in the ores of some other metals and are obtained from them as by-product. *e.g.* Co is also obtained as by-product during the refining of Cu, Ni, Fe and Pb. Cobalt is also present in vitamin-12 and its presence in this vitamin is of physiological interest. Cobalt ores are found in Canada, Northern Rhodesia and Belgian Congo. In India cobalt ores are found in Rajasthan.

Extraction

Due to the presence of several other metals in the ores of cobalt, the extraction of cobalt is a very complex process. Cobalt ores are generally worked for the manufacture of cobalt salts and not for the free metal. The extraction of cobalt from cobaltite ore, (Co, Fe, As) S involves the following steps :

1. Concentration. The ore is crushed and finely powdered and then concentrated by the froath-flotation process.

2. Roasting. The concentrated ore contains Fe, Co, Ni, Ag. As and S as the chief constituents. It is roasted in a free supply of air which removes S as SO_2 and As as As_2O_3 and FeS is oxidised to FeO.

$$S + O_2 \longrightarrow SO_2;\ 4As + 3O_2 \longrightarrow 2As_2O_3 \uparrow$$
$$2FeS + O_2 \longrightarrow 2FeO + 2S$$

As_2O_3 is condensed and used as a source of As.

3. Smelting. The roasted mass is smelted in a small furnace with a fuel *e.g.* coke and fluxing material (*e.g.* sand, SiO_2 and lime stone, $CaCO_3$). Here FeO produced during roasting and CaO produced from $CaCO_3$ combine with SiO_2 to form a fusible slag at the top of the molten mass.

$$CaCO_3 \longrightarrow CaO + CO_2$$

$$\left.\begin{array}{l} CaO + SiO_2 \longrightarrow CaSiO_3 \\ FeO + SiO_2 \longrightarrow FeSiO_3 \end{array}\right\} \text{slag}$$

The molten mass obtained at the end of smelting separates into the following three layers which contain different materials : (*i*) *Slag*. This is the top layer and consists of earthy matter present in the ore and slag of $FeSiO_3$ and $CaSiO_3$ formed during smelting. This is rejected. (*ii*) *Speiss*. This is the middle layer and consists of fused arsenides of Co, Ni, Fe and Cu and some As and S. (*iii*) *Bullion*. This is the bottom layer and consists of easily reducible and impure metals, mainly Ag and Cu. It is used for the recovery of 99% pure Ag.

4. Roasting the speiss with NaCl. The speiss is finely ground and roasted is air in a reverberatory furnace to convert As and S, if any, into As_2O_3 and SO_2 which are driven off. It is then mixed with NaCl and then roasted whereby the metals *viz.* Co, Ni, Fe and Cu are converted into their chlorides.

5. Separation of Co, Ni, Fe and Cu. The roasted speiss is extracted with H_2O to get the aqueous solution of the chlorides. This aqueous solution is then treated with lime atone ($CaCO_3$) to precipitate Fe as $FeCO_3$ and the filtrate (after filtration) with Na_2CO_3 to precipitate Cu as $CuCO_3$. By this treatment the chlorides of Co and Ni remain unaffected and hence remain in solution. Ag (in the form of AgCl), if present, is extracted by using a dil. solution of NaCN (*Cyanide process*).

6. Separation of Ni from Co. The solution containing the chlorides of Ni and Co is cautiously treated with lime, $Ca(OH)_2$ and bleaching powder, $CaOCl_2$ whereby $CoCl_2$ is converted into a precipitate of $Co(OH)_3$ and $NiCl_2$ remains in solution.

$$CoCl_2 + \underset{\text{Lime}}{Ca(OH)_2} \longrightarrow Co(OH)_2 + CaCl_2$$

$$2Co(OH)_2 + H_2O + \underset{\substack{\text{Bleaching}\\\text{powder}}}{CaOCl_2} \longrightarrow \underset{\text{ppt.}}{2Co(OH)_3} + CaCl_2$$

$NiCl_2$ that remains in solution can be converted into basic carbonate, if necessary.

7. Formation of Co_3O_4. The precipitate of $Co(OH)_3$ obtained as above is dried and ignited, when cobalto-cobaltic oxide, Co_3O_4 is obtained.

$$12Co(OH_3) \xrightarrow{\text{Ignite}} 4Co_3O_4 + 18H_2O + O_2$$

8. Reduction of Co_3O_4 to Co. Generally Co_3O_4 is used as such for the manufacture of cobalt compounds. In case cobalt metal is needed, it may be obtained by reducing Co_3O_4 by heating it with coke in an oil-fired furnace or with aluminium (*Alumino thermic process*).

$$Co_3O_4 + 4C \longrightarrow 3Co + 4CO$$

$$3Co_3O_4 + 8Al \longrightarrow 9Co + 4Al_2O_3$$

Preparation of Pure Cobalt

Pure cobalt can be prepared by electrolysing the solution of $CoSO_4$ containing $(NH_4)_2SO_4$ and NH_3 using platinum electrodes.

Properties

Cobalt is similar to iron in appearance but is harder, malleable and tough. It is a white metal, and is feebly magnetic. It melts at 1480°C. Cobalt absorbs upto a hundred times its volume of hydrogen, thus showing, analogy to nickel or palladium. Cobalt is not affected by dry or moist air at ordinary temperature. It gets slowly oxidised when red hot. Dilute hydrochloric and sulphuric acids have only a slow action on this metal. The metal is, however, readily soluble in nitric acid forming cobalt nitrate and liberating oxides of nitrogen. Red hot cobalt decomposes steam. Heated in ammonia, it forms the nitride Co_4N_2.

Uses

Cobalt is used in electroplating and in the formation of alloys (*e.g.*, alloys for high speed tools). *Cochrome* is an alloy similar to *nichrome* and contains Co 60%, Cr 15%, and Fe 25%. *Stellite*, an alloy of cobalt, chromium and tungsten is used for making surgical instruments.

Nickel (Ni_{28})

Occurrence

Nickel constitutes about 0.02% of the earth's crust. Metallic nickel is formed only in some of the meteorites. Chief ores of Ni are (*i*) *Pentlandite*, (Ni, Cu, Fe) S. It contains about 3% of Ni and is a mixture of sulphides of Fe, Cu and Ni. It is the main source of nickel. It is found in New Sudbury district of the province of Ontario in Canada. For this reason, this ore s also called *Sudbury ore.* (*ii*) *Nickel glance, NiAsS.* (*iii*) *Garnierite,* (Ni, Mg) $SiO_3.xH_2O$. It contains 5–8% of Ni and is found in New Caledonia (Australia), Cuba, Brazil, Russia etc. (*iv*) *Smaltite*, (Ni, Ce, Fe) As_2. (*v*) *Kupfer nickel* or *niccolite*, *NiAs* (*vi*) *White nickel ore*, Ni As_2. The arsenide ores are found in Norway, Canada, Spain and U.S.A. Nickel ores are not found in India but small amounts of the metal are recovered from the copper ores found in Khetri and other places in Rajsthan and from gold bearing rocks of Kolar and Travancore.

Extraction of Nickel from Pentlandite ore, (Ni, Cu, Fe)S

The extraction of nickel from *pentlandite ore* (Ni, Cu, Fe) S consists of the following operations.

1. Concentration of the ore. The ore is crushed, finely powdered and then concentrated by the *froth flotation process.* The concentrated ore consist of FeS, NiS and CuS, and some impurities like sulphur etc.

2. Roasting. The concentrated ore is roasted in a free supply of air in tall furnaces (about 30 feet high and 22 feet wide) for 8–10 weeks. This operation results in that : (*a*) Excess of free sulphur is oxidised to SO_2. (*b*) Most of FeS is oxidised to FeO; some FeS remains unaffected. (*c*) Moisture is removed. During roasting NiS and CuS remain unaffected and the roasted mass thus obtained contains NiS, CuS and a little of free sulphnr and FeS. The roasting operation once started is maintained by the heat liberated by the burning of sulphur.

3. Smelting. The roasted mass containing NiS, CuS and a little of free sulpur and FeS is mixed with coke and fluxing material (*e.g.* silica, SiO_2 and limestone, $CaCO_3$) and then smelted in a blast furnace so that most of the remaining FeS and free sulphur are oxidised to FeO and SO_2 respectively. FeO produced during roasting and smelting combine with silica to form ferruos silicate, $FeSiO_3$.

$$FeO + SiO_2 \longrightarrow FeSiO_3 \text{ (\textit{Fusible slag})}$$

At the same time lime, CaO obtained from the decomposition of $CaCO_3$ also combines with silica to form calcium silicate, $CaSiO_3$

$$CaO + SiO_2 \longrightarrow CaSiO_3 \text{ (\textit{Fusible slag})}$$

$CaSiO_3$ and $FeSiO_3$ both form a fusible slag which, being lighter, floats on the molten mass which consists of NiS, CuS and a little of free sulphur and FeS. The slag is continuously removed.

4. Besemerisation. The molten mass obtained at the end of smelting operation is mixed with silica and then introduced into a Bessemer type converter with a basic lining and fitted with tuyers to admit hot air under pressure. A blast of hot air is admitted into the converter under pressure through the layers. This hot air oxidises almost all the remaining sulphur and FeS into SO_2 and FeO respectively. FeO combines with SiO_2 to form the slag of $FeSiO_3$ as mentioned above. This slag is continuously removed. The resulting mass obtained at this stage is called *bessemerised NiS–CuS matte* or simply *matte* which consists almost entirely of the sulphides of Ni and Cu. The composition of the matte is as Ni = 55%, Cu = 25–30%, S = 14–17% and Fe = 0.1–0.5%.

5. Recovery of nickel from NiS–CuS bessemerieed matte : Refining of nickel. The process of recovering Ni metal from the bessemerised NiS—CuS matte obtained as above is termed as *refining of matte or refining of nickel.* Refining of nickel is generally done by Mond's process in which the matte is roasted in presence of O_2 so that NiS and CuS are converted into NiO and CuO respectively. The residue consisting of oxide mixture of NiO and CuO is treated with dil. H_2SO_4 at 80°C which converts CuO into $CuSO_4$ (soluble) while NiO remains unaffected (as residue). The residue of NiO is treated with water gas (CO + H_2) when NiO is reduced to crude nickel.

$$2NiO + \underbrace{H_2 + CO}_{\text{Water gas}} \longrightarrow \underset{\substack{\text{Crude}\\\text{nickle}}}{2Ni} + H_2O + CO_2$$

Crude nickel is treated with CO so that Ni is converted into the vapours of nickel tetracerhonyl, $Ni(CO)_4$ which, on being heated to 180°C, decomposes into 99.9% pure nickel.

$$Ni(CO)_4 \xrightarrow{180°} Ni(pure) + 4CO \uparrow$$

Properties

Nickel is a soft silvery white metal, quite malleable and ductile. It is slightly magnetic and possesses a high melting and boiling point (m. pt. = 1420°C, b. pt. = 2900°C). Like iron and cobalt, it absorbs a large volume of hydrogen gas (one volume of the metal absorbing 17 volumes of the gas). The action of air on the metal is slow at the ordinary temperature, but it vigorously burns in oxygen. It is attacked by chlorine, when nickel chloride, $NiCl_2$ is formed. Nickel decomposes steam at a red heat, forming NiO and H_2 Dilute hydrochloric and sulphuric acid solutions have a very slow action on this metal. But it readily dissolves in nitric acid and aqua regia.

$$3Ni + 8HNO_3 \longrightarrow 3Ni(NO_3)_2 + 2NO + H_2O$$

When a nickel sample is kept for some time under the surface of fuming nitric acid, the metal becomes *passive.*

The metal remains unaffected by caustic alkalies, even by fused alkalies. Nickel crucibles are, therefore, commonly used for alkali fusion in qualitative experiments. When CO gas is passed over nickel kept at about 50°C, a volatile colourless liquid, nickel tetracarbonyl, $Ni(CO)_4$ is formed The formation of $Ni(CO)_4$ finds an industrial application in the purification (or extraction) of nickel by Mond's process.

Uses

(*i*) Due to high melting point, great resistance to corrosion, acids and alkalies, nickel is largely used in making crucibles, dishes and other chemical wares and also in electrical industry, (*ii*) It is used for nickel plating in which $NiSO_4.7H_2O$ or $NiSO_4.(NH_4)_2SO_4.6H_2O$ solution is used as an electrolyte. (*iii*) Finely divided nickel is used as a catalyst in the hydrogenation of oils and fats (*iv*) Nickel metal combines with many other metals and forms alloys which find a number of applications. Some important alloys of Ni and their uses are given in Table 46.4.

Table 46.4. Some important alloys of nickel and their uses.

Alloy	Approx. composition	Properties and uses
1. Inwar	Ni = 30%, C = 0.3%, rest iron	Coeffeciant of expansion is almost equal to zero, used in pendulums and scientific instruments.
2. Platinite	Ni = 46%, Fe = 54%	Same coeffecient of expansion as that of glass and platinum, used in electric bulbs and radio valves.
3. Cupro nickel	Ni = 25%, Cu = 75%	Coinage
4. Coustantan	Ni = 40%, Cu = 60%	Electrical resistance
5. Platinoid	Ni = 14%, Cu = 60%, Zn = 24%, W = 2%	Used for thermnocouples and resistance coils.
6. Permalloy	Ni = 21.5%, Fe = 78.5%	Has unusual magnetic properties, used in telephony and telegraphy.

Platinum (Pt_{78})

Occurrence

Platinum usually occurs in the native form along with Cu, Au, Fe and other Pt-metals (Os, Ir, Rh, Pd) in sands and gravels. The main deposits of native Pt are found in Russia (Ural mountains), Columbia and South Africa. In the combined state it is found in a few minerals like *sperrylite*, $PtAs_2$; *cooperite*, $Pt(AsS)_2$ and *braggite*, (Pt, Pd, Ni) S. Sperrylite is found in Sudbury district of Canada and Transvaal (South Africa).

Extraction

Platinum is generally extracted from Sudbery nickel ore. The ore is concentrated to remove the sulphides of Fe, Ni and Cu and then it is digested with dil. aqua regia under pressure. The solution thus obtained contains $PdCl_2$ and $PtCl_4$. This solution is treated with NH_4Cl when $PtCl_4$ is converted into a yellow precipitate of $(NH_4)_2[PtCl_6]$ while $PdCl_2$ remains unaffected in solution. The yellow precipitate of $(NH_4)_2[PtCl_6]$, on heating, gives spongy platinum metal.

$$2NH_4Cl + PtCl_4 \longrightarrow (NH_4)_2[Pt(Cl_6] \xrightarrow{\text{Heat}} Pt + 2NH_4Cl + 2Cl_2$$

Properties

In colour it appears as white as tin. With the exception of gold and silver it is the most malleable of the metals. It joins well with glass because its coefficient of expansion is the same as that of glass. At red heat temperature, it easily allows hydrogen to pass through it. The pure metal is not attacked by hot concentrated sulphuric acid. It, however, dissolves in hot *aqua regia*. The alloy of lead and platinum dissolves in nitric acid forming *platinum nitrate*.

Various Types of Platinum

Platinum may be obtained in various forms to be used in different industries.

1. Spongy platinum. When the double salt ammonium platinochloride $(NH_4)_2PtCl_6$ is very *gently heated*, a sort of grey porous form with a large surface is obtained which is known as spongy platinum. It has great surface activity and is, therefore. used for absorption of gases.

2. Platinum black. (*a*) When an alloy of platinum with metals like copper or zinc is treated with nitric acid, the nascent hydrogen reduces the platinum salt and falls back as a black powder. (*b*) It is best obtained by reducing platinic chloride with formaldehyde, or reducing neutral chloroplatinate with sodium formate. The platinum black absorbs much more hydrogen than the spongy platinum (160 volumes), the maximum absorption is at 0°C; and, therefore, it is commercially used in reduction processes. It also serves as a good catalyst. Alcohol is oxidised by it on account of its occluded oxygen. In its presence, a mixture of hydrogen and oxygen explodes.

3. Platinised asbestos. When asbestos, soaked with platinic chloride, a little hydrochloric and some ammonium chloride, is heated in a crucible, or, if reduced with sodium formate, platinum precipitates out in the fibres of the asbestos. This is now known as platinised asbestos and finds extensive uses as a catalyst.

4. Colloidal platinum. It is obtained either by Bredig's method by passing an electric arc between two platinum terminals dipped in water; or it may be prepared by reducing platinic chloride with hydrasine hydrate in presence of some protective colloid (*e.g.*, sodium lysalbate). The colloidal solution is also a catalyst; it decomposes hydrogen peroxide, catalytically.

Uses of Platinum

1. In scientific instruments and laboratory apparatus. Due to its remarkable properties like high melting point and corrosion resistance, platinum is largely used to manufacture scientific instruments and laboratory apparatus such as crucibles, dishes, basins etc. Pt is also used for contacts in electrical apparatus and for making electrodes.

2. As a catalyst. Pt is also used as a catalyst in a number of inorganic and organic reactions. Some important applications of platinum as catalyst are given below. (*a*) *Manufacture of H_2SO_4.* SO_3 and O_2 combine to form SO_3 in the presence of platinised asbestos in the manufacture of H_2SO_4 by contact process. (*b*) *Manufacture of HNO_3.* NH_3 is oxidised to NO in the presence of platinum gauge in the manufacture of HNO_3 by *Ostwald's process.* (*a*) *Preparation of formaldehyde.* Vapours of CH_3OH and O_2 are passed on platinum black to get HCHO. (*d*) *Synthesis of HBr and HI*, Mixture of H_2 and halogens is passed over red hot finely divided platinum. (*e*) *In hydrogenation and dehydrogenation industries.* Since finely divided platinum has the valuable property of absorbing several hundred times its own volume of hydrogen gas, the metal is widely used for many hydrogenation and dehydrogenation reactions. Large quantities of the metal are also used in the petroleum industry synthetic fibre industry and glass industry.

3. In jewellery. Pt is also used in jewellery as a setting for diamonds. For this purpose the metal is alloyed with metals like Ir, Ru or Cu. Pt—Ir alloy is also used in pen points, standard weights, coins etc.

4. In dentity. The metal is used for making artificial teeth, pins dental filling and support for dentures.

5. Other uses. Pt-salts are used in printing and toning in photography Pt wire is also used in the qualitative analysis in chemistry. In the medical fields many instruments like X-ray tubes, radium applicators, lypodermic needles require platinum.

Copper (Cu_{29})

Occurrence

Copper is found in both the native and combined forms. The native copper is remarkably pure (99.9%) and is only a minor source of copper. Native copper is found in U.S.A. Mexico, Russia and China. In the combined form it is found as (*i*) *Oxide ore.* Example is *Cuprite* or *Ruby Copper*, Cu_2O containing about 88% copper, (*ii*) *Carbonate ores*. Examples are (*a*) *Malachit* (*green*), $CuCO_3.Cu(OH)_2$ or $CuCO_3.CuO.H_2O$ containing about 57% of copper (*b*) *Azurite* (blue), $2CuCO_3.Cu(OH)_2$ containing about 65% copper (*iii*) *Sulphide ores.* Examples are (*a*) *Chalcocite* or *copper glance* Cu_2S containing about 80% copper (*b*) *Copper pyrites* or *Chalcopyrites*, $CuFeS_2$ ($Cu_2S + Fe_2S_3$) containing about 34% copper. Out of the ores mentioned above, copper pyrites is the principle ore of copper which is found in India mainly in Singbhum district (Bihar), Dharwar and Matigara.

Extraction of Cu from Oxide and Carbonate Ores

The ore is crushed, concentrated by gravity process and roasted. The carbonate ore [*e.g.* malachite, $CuCO_3.Cu(OH)_2$] decomposes to form the oxide and the impurities either volatilize or are oxidised.

$$\underset{\text{Malachite}}{CuCO_3.Cu(OH)_2} \longrightarrow 2CuO + CO_2 + H_2O$$

The oxide so formed is reduced to copper with carbon in reverberatory furnace.

$$CuO + C \longrightarrow Cu + CO$$

The oxide ore *viz.* cuprite, Cu_2O is directly reduced by carbon to metallic copper.

$$\underset{\text{Cuprite}}{Cu_2O} + C \longrightarrow 2Cu + CO$$

Extraction of Cu from Sulphide Ore viz., Copper Pyrites, ($CuFeS_2$) by Dry Process

Copper is extracted form copper pyrites by *dry process* which is also called thermal or smelting process. In this process the crushed copper pyrites ($CuFeS_2$) is concentrated by froth flotation process so that the silicious impurities are removed. The concentrated ore is then heated strongly in a current of air on the hearth of a reverberatory furnace. In this operation (*i*) a considerable portion of S, As and Sb present as impurities are oxidized to their oxides which are then removed. (*ii*) $CuFeS_2$ is converted into a mixture of Cu_2S and FeS ($2CuFeS_2 + O_2 \longrightarrow Cu_2S + 2FeS + SO_2$) (*iii*) Cu_2S and FeS are furher *partially* oxidised to Cu_2O and FeO respectively.

$$2Cu_2S + 3O_2 \longrightarrow 2Cu_2O + 2SO_2;\ 2FeS + 3O_2 \longrightarrow 2FeO + 2SO_2$$

The roasted ore is transferred to a small blast furnace and heated there with coke and sand (silica, SiO_2) in the presence of excess of air. Here FeO combines with SiO_2 to form the slag of iron silicate, $FeSiO_3$ which is removed. The unreacted FeS reacts with Cu_2O and form Cu_2S and FeO [$FeS + Cu_2O \longrightarrow Cu_2S + FeO$]. The molten mass obtained at the end of this operation consists of Cu_2S, FeS and SiO_2. This mass is called **matte or course metal.** Now the moten matte is introduced in a Bessemer convertor and heated in a blast of air blown through tuyers. CuS and FeS which have not been completely oxidised in the reverberatory furnace are oxidised here into Cu_2O and FeO.FeO thus formed forms a slag of $FeSiO_3$ with SiO_2 When the whole of iron has been removed, cuprous oxide reacts with cuprous sulphide to form copper [$Cu_2S + 2Cu_2O \longrightarrow 6Cu + SO_2$]. The molten mass is poured off from the converter, and allowed to cool when it slowly gives up the dissolved SO_2 and produces blisters on the surface of solidified copper. The copper so produced is called **Blister copper.** It contains about 2%

impurities consisting of Ag, Au, Au, Ni, Zn, Pb, Sn, As and S. It is purified either by poling method or by electrolytic method. In poling method the blister copper is melted on the hearth of a reverberatory furnace when oxygen present in the metal as Cu_2O oxidises impurities *e.g.* iron to FeO which forms a slag with silica lining of the furnace, while oxides of others either pass out of as gases or form a scum which can be removed. Excess of Cu_2O renders copper brittle and removed by covering the surface of fused metal with powdered anthracite and stirring with poles of green wood when reducing gases bubble up through the metal and the oxide is reduced. The copper is 99.5% pure. In electrolytic method a large piece of impure copper (blister copper) is made anode in the electrolytic cell and a thin piece of very pure copper is made the cathode. An acidified solution of $CuSO_4$ is used as an electrolyte. On passing an electric current of low voltage through the electrolytic solution Cu^{2+} ions obtained from $CuSO_4$ solution go to the cathode where they are reduced to the free metal which gets deposited on the cathode. As the process goes on anode becomes thinner and the cathode becomes thicker. The impurities like Ag, Au, So, Te etc. settle down at the bottom of the electrolytic cell as **anode mud.**

Uses

Copper is one of the most important technical metals. Its chief uses are (*i*) It is used for making electric cables and other electrical appliance. (*ii*) Copper sheets are used for making utensils. (*iii*) Copper forms a number of alloys like brass, bronze etc. Its alloys find various uses in arts and industry. (*iv*) It is used for the preparation of various salts which are used as insecticides, pesticides pigment etc. (*v*) it is used in electroplating, electrotyping and in making coins.

Alloys of Copper

Copper forms a large number of alloys which are very useful. Some of these are given in Table 46.5.

Silver (Ag_{47})

Occurrence

Native silver is found in large quantity in Mexico and Norway usually associated with Cu and Ag. The principal ores of Ag are : (1) *sulphides, e.g. argentite or silver glance*, Ag_2S; *silver-copper glannce*, $(Cu, Ag)_2S$; *ruby silver*, $3Ag_2S.Sb_2S_3$. (2) *halides, e.g., horn silver,* AgCl; *bromyrite* AgBr. Most Cu and Pb ores also contain small amount of Ag and constitutes one of the important sources of Ag. U.S.A., Mexico Peru. Canada, Bolivia, New South Wales, Barma and Japan are the major silver producing countries of the world. In fact, there are no silver ores in India, the metal being obtained entirely as by-product in the mining of gold and lead. A small amount of silver is also obtained from gold ores of Kolar fields in Mysore and Anantapur mines at Madras. The lead and zinc ore of Zewar mines in Rajasthan also contains silver which is being smelted (Since 1954) at Tundoo (Bihar) by the Metal Corporation of India.

Table 46.5. Some alloys of copper

Name	Composition	Important properties	Uses
1. Brass	Cu = 60–80% Zn = 20–40%	Harder, palar and more brittle with more zinc, can be moulded, malleable and ductile. High tensile strength.	Domestic utensils, condensers, tubes, casting cart ridges.
2. Bronze	Cu = 75–90% Sn = 10–75%	Hard, brittle tenacious, durable resistant to corrosion, takes high polish.	Statues, coins, utensils, bells, machine parts, propellers, blades.
3. Aluminium bronze	Cu = 90% Al = 10%	Light, strong, tenacious, resistant to corrosion, golden yellow.	Utensils, coins, frames, statues, jewellery, bearings and castings.
4. Phosphorus bronze	Cu = 85% Sn = 13% P = 2%	Very hard, tough, elastic, resistant to corrosion,	Suspension wire instrument, gears, aerials, ships, propellers.
5. Silica bronze	CU = 97% Sn = 2% Si = 1%	Very hard, tensile, tough, non-corrosive, good conductor.	Telephone and telegraph wires, suspension wires.
6. German silver or nickel silver	Cu = 50% Zn = 30% Ni = 20%	White shining, malleable and ductile, tough, resistance to corrosion.	Utensils, plates, ornament.
7. Gun metal	Cu = 88% Sn = 10% Zn = 2%	Strength increases with the increase of the percentage of Sn, very hard and strong, resistant to corrosion.	For making cannons, guns, machine parts.
8. Monel metal	Cu = 30% Ni = 67% Fe + Mn = 3%	Very resistant to chemical action high tensile strength.	Alkali industry resistance wires, domestic wires, values, automobile engine parts.

Extraction of Silver from Argentite (Ag_2S) and Horn Silver (AgCl) by Cyanide Process

Cyanide process is also called **Mac-Arthur and Forest's cyanide process.** In this press the finely powdered ore is concentrated by froth flotation process and then is treated with a dilute aqueous solution of NaCN and a current of air is passed into the solution so that Ag present in the ore is converted into soluble sodium argento cyanide complex, $Na[Ag(CN)_2]$.

$$\underset{\text{Argentite ore}}{Ag_2S} + 4NaCn\ (\text{solution}) \rightleftharpoons \underset{\text{Sod. argento cyanide (soluble)}}{2Na[Ag(CN)_2]} + Na_2S$$

The current of air passed into the solution oxidises. Na_2S formed in the above reaction into $Na_2S_2O_3$, Na_2SO_4 and S and thus enables the above reversible reaction to proceed into the forward direction.

$$2Na_2S + 2O_2 + H_2O \longrightarrow Na_2S_2O_3 + 2NaOH$$

$$4Na_2S + 5O_2 + 2H_2O \longrightarrow 2Na_2SO_4 + 4NaOH + 2S$$

Horn silver ore, AgCl is also converted into soluble $Na[Ag(CN)_2]$ by NaCN.

$$\underset{\text{Horn silver ore}}{AgCl} + 2NaCN \text{ (solution)} \longrightarrow Na[Ag(CN)_2] + NaCl$$

The solution containing $Na[Ag(CN)_2]$ is treated with Zn dust so that Ag gets precipitated.

$$2Na[Ag(CN)_2] + Zn \text{ (dust)} \longrightarrow Na_2[Zn(CN)_4] + \underset{\text{ppt.}}{2Ag}$$

Ag obtained as above contains a number of impurities and hence is required to be purified. It is purified by electrolytic method. The cathode consists of a pure silver plate and the anode is a block of *impure* Ag. A solution of 6% $AgNO_3$ with 1% HNO_3 is used as the electrolyte. On passing the current, silver along with some impurities dissolves at the anode while only Ag gets deposited on the cathode. Zn and Cu remain in solution while Au, if any, settles down as anode mud.

To Get Silver From Commercial Lead (Argentiferous Lead) by Parke's Process : Desileverisation of Lead

Lead extracted from galena ore, PbS either by air-reduction process or by carbon-reduction process contains the impurities of Cu, Ag, Bi, As, Fe, Zn, Sn etc. and is called *commercial lead* or *argentiferous lead*. This commercial lead contains Ag upto 2%. Ag can be obtained from this commercial lead by *Parke's process*. To get Ag from this commercial lead is called *desilverisation of lead*. In Parke's process commercial lead is mixed with a little quantity of Zn and the mixture is melted in iron pots. In this operation Ag forms Zn—Ag alloy with Zn while Pb is obtained in the molten condition. On cooling, Zn—Ag alloy solidifies while Pb still remains in the molten state. Zn—Ag alloy is strongly heated in fire-clay retorts to distil off Zn as its vapours and Ag is obtained in the free state. Ag obtained by this method contains some quantity of Pb as impurity which can be removed by cupellation process. In cupellation process the molten impure Ag metal is taken in a cupel (boat shaped dish made of bone ash or cement) which is placed on the bead of a reverberatory furnace provided with a number of openings in the sides so that air freely circulates through it. Now the cupel is heated and a blast of air is passed over the molten metal. Lead present in impure Ag metal is oxidised to litharge, PbO which is carried away by the blast of air along with waste gases and pure Ag is left in the cupel.

To Get Silver From Silver Coins or Silver Ornaments

Silver coins or silver ornaments are Ag—Cu alloys. Ag can be obtained from such an alloy by dissolving the alloy in HNO_3 when a solution containing the mixture of $AgNO_3$ and $Cu(NO_3)_2$ is obtained. After boiling off the excess of nitric acid, the solution is treated with hydrochloric acid when a precipitate of silver chloride is obtained. This is separated and converted into silver by any of the following three methods : (*i*) By reduction with nascent hydrogen [$AgCl + H \longrightarrow Ag + HCl$] (*ii*) By fusion with sodium carbonate [$4AgCl + 2Na_2CO_3 \longrightarrow 4Ag +$

$4NaCl + 2CO_2 + O_2$] (*iii*) By boiling with caustic soda and glucose, $C_6H_{12}O_6$ [$2AgCl + 2NaOH \longrightarrow Ag_2O + 2NaCl + H_2O$; $Ag_2O + C_6H_{12}O_6 \longrightarrow 2Ag + C_6H_{12}O_7$]. The silver thus obtained is purified by fusion with borax and nitre followed by electrolytic purification.

Uses

Silver alloyed with copper is used in making coins and jewellery. Silver leaf is used in medicine while silver amalgam is employed in silvering mirrors and hi filling teeth. Salts of silver are used in photography and silverplating.

Gold (Au_{79})

Occurrence

Gold is found mostly in the *free state* in quartz veins mixed or alloyed with silver with traces of platinum, iron or bismuth as impurities. Such quartz is known as *auriferous quartz.* It is also present in the sands of those rivers which pass over auriferous rocks. In the combined state it is found as a telluride in *sylvanite* (Ag, Au) Te_2. In quantities too small to be profitably extracted, it is found in sea water. India produces about 2% of the world's gold production and Mysore is the richest centre, the *Kolar gold fields* there producing 99.97% of our gold output. Alluvial gold washing is carried out in small scale in Bihar, Assam and Orissa.

Extraction of Au From Alluvial Sands

From alluvial sands, gold is recovered by making use of its high specific gravity (= 19.3). The primitive method was known as *pan washing.* In big pans made of iron or zinc the gold-bearing sand is agitated with water by giving a rotatory motion to the pan when the heavier gold particles settle down while the sand is washed away. This has now been replaced by *placer mining* in which the gold-bearing sand is introduced in a sluice way system and powerful streams of water passed over it. *A sluice way system* is an inclined channel provided with a series of sluices *i.e.*, troughs and cross wire strips called *riffles.* The gold particles are retained in the sluices while the sand is carried away by the water. Whenever loose gold-bearing deposits occur near the edge of a river, the process of *hydraulic mining* is employed. Water under high pressure is forced on the deposit and the auriferous sand which flows down with the water stream is washed into sluices as in placer mining.

Extraction of Au From Auriferous Quartz

The original pieces of auriferous quartz are crushed and broken into pieces in jaw or gyratory crushers. These pieces are now fed in a battery of stamp mills working in series. The powdered ore is washed out of stamp mill by a current of water. Only those particles which are fine enough can pass through the screen. Now the crushed ore is treated by any of the three processes namely :

(*a*) Amalgamation process. This process which is in common use in the South Indian gold fields consists in passing the gold paste over amalgamated copper plates placed just in front of mortar boxes. The gold combines with mercury to form amalgam. This is periodically removed from the mill plates, squeezed to remove superfluous mercury and then distilled when the mercury passes over leaving gold behind. This process can remove 70–80% of the gold. The residual ore called "*tailings*" is subjected to the above process again in order to recover more gold from it.

(*b*) **Plattner's Chlorination Process.** The powdered ore is roasted in order to remove sulphur and arsenic and is then placed in wooden vats with false perforated bottoms moistened with water and is treated with chlorine. The gold chloride formed is leached with water and gold recovered from the solution by the addition of ferrous sulphate, [$AuCl_3 + 3FeSO_4 = Au + Fe_2(SO_4)_3 + FeCl_3$]

(*c*) **Mac-Arthur and Forrest's cyanide process.** Tailings from the amalgamation process or powdered ore from stamp mill is placed in large vats made of wood or cement provided with perforated false bottoms covered with coconut mattings. It is then treated with 0.1% solution of NaCN in the presence of excess of air which is passed in from the bottom of the tank when the following reactions occur :

$$4Au + 8NaCN + 2H_2O + O_2 \longrightarrow 4Na[Au(CN)_2] + 4NaOH$$

The gold goes in solution in the form of a double cyanide of sodium and gold. From this solution gold is recovered either by the addition of zinc or by electrolysis.

$$2Na[Au(CN)_2] + Zn = 2NaCN + Zn(CN)_2 + \underset{\text{ppt.}}{2Au}$$

The precipitated gold is treated with dilute sulphuric acid to remove the excess of zinc and changed into compact mass by fusion.

Refining of Gold

The gold obtained by any of the above processes usually contains traces of Fe, Sb, Cu, Pb and other base metals as impurities. These impurities can be removed by any of the processes mentioned as : (*1*) *Fusion process.* The impure metal is fused with borax and nitre in a crucible, the base metals in the process are oxidised and form a scum at the molten surface, and are removed. (*2*) *Parting process.* The refined metal obtained by fusion, may contain silver as an impurity. This can be removed by treatment with concentrated sulphuric acid. It is found, however, that if the percentage of gold in the impure sample is more than 30%, the sample is not attacked by the acid. More silver is, therefore, added to the sample till an alloy containing one quarter of its weight of gold, *i.e.* 25% is obtained (process called *quartation*). This is treated with sulphuric acid when the silver dissolves to form silver sulphate leaving gold behind. Silver is recovered from this solution by introducing coper plates in it. (*3*) *Electrolyc refining.* Gold may be refined still further by electrolytic process using 4% gold chloride solution acidified with 10% hydrochloric acid as the electrolyte. Impure gold is made the anode while the cathode is a plate of pure gold.

Properties.

(*a*) **Physical.** Gold is a lustrous yellow metal which melts at 1063°C to a green liquid and boils at 2610°C. Its specific gravity is 19.3. It is the most malleable and ductile of all metals and is a good conductor of heat and electricity. With mercury, it forms an amalgam.

(*b*) **Chemical.** It is not affected by dry or moist air or oxygen. It is unaffected by water or steam. Alkalis have no action on gold. Single dilute or concentrated acids except selenic acid have no action on it. It dissolves, however, in aqua regia forming chlorauric acid, $HAuCl_4$

$$2Au + 2HNO_3 + HCl \longrightarrow 2HAuCl_4 + 3NOCl + 6H_2O$$

At 200°C, it is readily attacked by chlorine forming auric chloride. It dissolves in alkali cyanide solution in presence of oxygen of air.

$$4Au + 8KCN + 2H_2O + O_2 \longrightarrow 4K[Au(CN)_2] + 4KOH$$

Uses

(*i*) In coinage, jewellery and ornamental vessels. (*ii*) In gold plating. (*iii*) Gold leaf is used for decorations, lettering and in medicine as tonic. (*iv*) "Purple of Cassius" which is a combination of gold with colloidal stannic acid is used for making ruby red glass and high class pottery.

Fineness of Gold

Gold content of an alloy of gold is termed its fineness and is generally expressed in carats. Pure gold is 24 carats. Ornaments available these days are made of 18 carat gold, which is an alloy containing 18 parts by weight of gold alloyed with 6 parts by weight of another metal generally copper.

Colloidal Gold

Colloidal gold is obtained by two methods.

1. Bredig's arc method. It is a dispersion process in which an electric arc is struck between two gold electrodes dipped under cold water. As a result of sparking the gold particles disintegrate forming a colloidal solution. A trace of KOH is added to cold water and helps to stabilise the colloidal solution.

2. Reduction process. Gold salt solutions when reduced with ferrous sulphate or stannous chloride, produce colloidal gold solution by reduction.

$$2AuCl_3 + 3SnCl_2 \longrightarrow 3SnCl_4 + 2Au$$

$$2AuCl_3 + 3FeSO_4 \longrightarrow Fe_2(SO_4)_3 + FeCl_3 + Au$$

Purple of Cassius

It is only form of colloidal gold prepared by mixing very dilute solution of gold chloride with stannous chloride solution.

$$2AuCl_3 + 3SnCl_2 \longrightarrow \underset{\text{ppt.}}{2Au} + 3SnCl_4$$

The gold thus precipitated is absorbed by stannic hydroxide formed by the hydrolysis of $SnCl_4$.

$$SnCl_4 + 4H_2O \longrightarrow Sn(OH)_4 + 4HCl$$

It is purple or red in colour and after the name of Cassius who discovered it, it is called Purple of Cassius. According to Moissan this is a colloidal form of stannic oxide with absorbed colloidal gold.

$$2AuCl_3 + 2SnCl_2 + 6H_2O \longrightarrow 2Au + 3SnO_2 + 12HCl$$

It is used in making ruby-red glass and high class pottery.

Zinc (Zn_{30})

Occurrence

Zinc is not found native. In the form of minerals, it occurs as *zinc blende ZnS, calamine ZnCO₃, zincite ZnO and franklinite* $ZnO.Fe_2O_3$. In India zinc is found in Rajasthan. A belt of zinc ore covers an area of about thirty square miles in the Zawar and Udaipur region of Rajasthan. It has also been located in certain parts of Kashmir.

Extraction of Zn by Reduction Process

If the ore contains iron oxide as an impurity, the latter is removed magnetically. The silicious impurities are removed by gravity separation and in case of zinc blende, the concentration is done by froth floatation process. The concentrated ore is heated in excess of air when zinc sulphide is oxidised to zinc oxide.

$$2ZnS + 3O_2 \longrightarrow 2ZnO + 2SO_2$$

Some of the ore is converted to zinc sulphate as well.

$$ZnS + 2O_2 \longrightarrow ZnSO_4$$

To avoid this, the roasting is done at a high temperature (900°C) in a regulated supply of air. At this temperature the sulphate is decomposed.

$$ZnSO_4 \longrightarrow 2ZnO + 2SO_2 + O_2$$

If *calamine* is the ore, it decomposes into oxide and CO_2.

$$ZnCO_3 \longrightarrow ZnO + CO_2$$

ZnO obtained as above is reduced to Zn metal by any of the following processes.

(*a*) Belgian process. The roasted ore is heated with coke to about 1100°C in small fire clay retorts. The heating is often done by producer gas. Each retort is provided with a clay pipe which serves as a condenser and an earthenware pot in which is collected the condensed metal. Zinc, being fairly volatile, is distilled in the retort and collected in the pot. The chemical reactions involved in this process are :

$$ZnO + C \longrightarrow Zn + CO$$

$$ZnO + CO \longrightarrow Zn + CO_2 \; ; \; CO_2 \rightleftharpoons 2CO$$

There are a number of such retorts which are arranged in three or four rows one above the other. The lowermost is supported on the hearth.

(*b*) New Jersy or Vertical Retort Process. It is a modern process and is much more economical than the Belgian process. It is a continuous process and yields metal of much greater purity (95–99%). Powdered ZnO and coal (called charge) is made into small brickets (small blocks) with the help of molasses which bends the particles together. These brickets are fed into the tall vertical retort from the top. The retort is lined on the inside with fire clay bricks. The retort is heated externally by means of a heating jacket in which producer gas is burnt. The vapours of zinc metal obtained by the reduction of zinc oxide with carbon rise to the top and are condensed in the condenser. The crude metal obtained is called **spelter.**

Zinc metal obtained above contains iron, cadmium, antimony and arsenic as the main impurities. The crude metal is *redistilled* in silica retorts. More volatile cadmium, arsenic and antimony are distilled first. Zinc is distilled as a next fraction and is collected in a separate receiver. The non-volatile impurities remain in the retort. Very pure zinc is obtained by *electro refining* in which $ZnSO_4$ serves as an electrolyte, crude zinc is made the anode and cathode is a plate of pure zinc metal.

Extraction of Zn by Electrolytic Process

The zinc smelter at Debari (Rajasthan) utilises this process. The concentrated ore is roasted under controlled temperature conditions.

$$2ZnS + 3O_2 \longrightarrow 2ZnO + 2SO_2$$

The sulphur dioxide evolved is purified and converted into sulphuric acid. The roasted ore is dissolved out with dilute sulphuric acid as also with spent electrolyte. The zinc goes into solution as zinc sulphate, Other soluble sulphates are also dissolved. [$ZnO + H_2SO_4 \rightarrow ZnSO_4 + H_2O$]. The filtered extract is first treated with milk of lime to precipitate iron, aluminium, silica, arsenic and antimony. The filtrate is warmed to 75°C and treated with zinc dust to remove copper and cadmium. [$CdSO_4 + Zn \rightarrow ZnSO_4 + Cd$]. The filterate is then electrolysed in cells having pure lead inert anodes and aluminium cathodes. The zinc is deposited on the cathode while sulphuric acid is regenerated. The zinc is removed from the cathodes by melting (m. pt. of Zn 420°C; Al 660°C) and made into ingots. The purity of the metal is 99.96%.

Properties

Zinc is a bluish white metal brittle at ordinary temperature but malleable and ductile between 100° and 150°C. At 200°C, it becomes brittle again. It has a sp. gra. = 6.9, m.pt. = 420°C, and b. pt. = 930°C. *Granulated zinc* is obtained by pouring molten zinc in a thin steam in cold water. Zinc is not affected in dry air but in moist air a protective coating of basic carbonate is formed on its surface. When strongly heated in air it burns with a greenish blue flame forming clouds of light zinc oxide (*Philosopher's wool*).

$$2Zn + O_2 \longrightarrow 2ZnO$$

Pure zinc is not dissolved by water. Boiling water is slowly decomposed by impure zinc, *e.g.*, zinc-copper couple. It decomposes steam.

$$Zn + H_2O \longrightarrow ZnO + H_2$$

Pure zinc is attacked very slowly by acids but the presence of impurities accelerates the rate of reaction due to the formation of electro-chemical couples. It dissolves readily in dilute and concentrated hydrochloric acid.

$$Zn + 2HCl \longrightarrow ZnCl_2 + H_2$$

Dilute sulphuric acid produces hydrogen.

$$Zn + H_2SO_4 \longrightarrow ZnSO_4 + H_2$$

Hot and concentrated sulphuric acid produces sulphur dioxide.

$$Zn + 2H_2SO_4 \longrightarrow ZnSO_4 + SO_2 + 2H_2O$$

Dilute nitric acid gives nitrous oxide or ammonium nitrate according to the strength of the acid, very dilute acid giving ammonium nitrate.

$$4Zn + 10HNO_3 \longrightarrow 4Zn(NO_3)_2 + 5H_2O + N_2O$$

$$4Zn + 10HNO_3 \longrightarrow 4Zn(NO_3)_2 + NH_4NO_3 + 3H_2O$$

With concentrated acid, nitrogen peroxide is produced.

$$Zn + 4HNO_3 \longrightarrow Zn(NO_3)_2 + 2H_2O + 2NO_2$$

Zinc dissolves in hot solutions of caustic alkalies forming zincates and liberating hydrogen.

$$Zn + 2NaOH \longrightarrow \underset{\text{Sodium zincate}}{Na_2ZnO_2} + H_2$$

It displaces less electro-positive metals like copper, silver, gold end lead from their salt solutions.

$$CuSO_4 + Zn \longrightarrow ZnSO_4 + Cu$$

$$2Na[Ag(CN)_2] + Zn \longrightarrow \underset{\text{ppt.}}{2Ag} + Na_2[Zn(CN)_4]$$

Zinc decomposes ammonia at red heat.

$$3Zn + 2NH_3 \longrightarrow Zn_3N_2 + 3H_2$$

Uses

The most important use of zinc is for the **galvanising (or galvanisation)** of iron. (*Galvanising is the process of depositing a thin layer of zinc over the surface of iron*). It is used in the desilverisation of lead and the extraction of gold and silver. Granulated zinc is used in the laboratory. It is also employed as cathode plate in dry cells and in photo-engraving and printing plates. Its alloys with copper find extensive application in industry and arts.

Alloys of zinc

As already said, zinc metal combines with copper and forms many useful alloys. Some useful alloys of Zn with Cu are given below :

1. *Brass.* For this see *"Alloys of copper"*

2. *Muntz metal.* It is an alloy containing Zn = 60% and Cu = 40%.

3. *German silver.* For this see *"Alloys of copper"*

4. *Die casting metal.* Zinc alloyed with 4% Al, upto 3.5% Cu and 0.05% Mn is used for the production of die castings.

Mercury (Hg_{80})

Occurrence

The principal ore of mercury is *cinnabar*, HgS. Mercury also occurs as *tremannite*, HgSe; *calomel*, Hg_2Cl_2. Small drops of metallic mercury are also found distributed throughout the rock. The principal sources of world supply of mercury are Spain, Italy, California, Russia, Mexico, China and Japan.

Extraction of Mercury From Cinnabar Ore, HgS

Hg is extracted from cinnabar ore by using the following two processes.

1. Old process (Idrian process). This process is only used in Spain. The ore is crushed and finally powdered and then concentrated by the froth floatation process. The concentrated ore is placed on the perforated arches constructed over one another in the central chamber. The larger lumps of the concentrated ore are placed on the lowest arch and finer ones, on the uppermost. Now the concentrated ore is heated by the flames of the furnace rising from below. HgS is oxidised to HgO and SO_2, the former decomposes at about 300°C to give Hg and O_2.

$$2HgS + 3O_2 \longrightarrow 2HgO + 2SO_2 \; ; \; 2HgO \xrightarrow{300°} 2Hg + O_2$$

Mercury vapours and SO_2 are passed into a series of condensing (cooling) chambers on both sides of the furnace. Hg vapours get condensed at the botton of the chamber while SO_2 etc. escape out through the chimney.

2. Modern process. In this process HgS ore is mixed with charcoal and is charged into a shaft furnace from the top fitted with cup and cone arrangement. The charge (*i.e.* HgS + charcoal) is heated on the sloping bars of iron in a current of air drawn through the furnace. Gases coming out of the furnace are Hg vapours, SO_2 and CO.

$$2HgS + 3O_2 \longrightarrow 2HgO + 2SO_2$$

$$2HgO + 2Hg \text{ (vapours)} + O_2 \; ; \; HgO + C \longrightarrow Hg \text{ (vapours)} + CO\uparrow$$

The gases of Hg vapours, SO_2 and CO are passed in Y-shaped tubes cooled in water when vapours of Hg get condensed. The advantages of this process over the old one are : (*i*) it is a continuous process (*ii*) loss of Hg vapours is negligible, and (*iii*) it can be used for the extraction of Hg from all grades of HgS ore.

Purification of Mercury

Mercury obtained by any of the above methods is usually contaminated with metallic impurities like Zn, Sn, Pb, Bi etc. Some of these impurities get oxidised in air and form a black scum of their oxides on the surface of Hg. The suspended impurities can be removed by filtering impure Hg through chamois leather and filtered matter is obtained. Hg can further be purified by treating the filtered matter with dil. HNO_2. In this method the filtered matter is dropped in a long tube containing dil. HNO_3. The base metal impurities like Pb etc. react with $Hg_2(NO_3)_2$ (which is formed by the action of Hg on HNO_3) to give their nitrates which go into the solution and pure Hg is obtained at the other end. Finally the metal is purified by distillation in vacuum.

$$Pb \text{ (base metal impurity)} + Hg_2(NO_3)_2 \longrightarrow \underset{\text{Soluble}}{Pb(NO_3)_2} + 2Hg$$

Uses of Mercury

(1) Because of its high boiling point and low freezing point it is used as a barometric and thermometric liquid. (2) In the extraction of gold and silver by

amalgamation process. (3) Compounds of mercury are used in medicine. (4) In the manufacture of caustic soda by Castner-Kellner process or Solvay-Kellner process. (5) Mercury vapour lamp is used as a source of ultraviolet rays. (6) Sodium and aluminium amalgams are used as reducing agents. Tin amalgam is used as a dental filling.

Amalgams

Mercury dissolves nearly in all the metals forming alloys which are called *amalgans.* As a rule, amalgams are the solutions in mercury of the compounds of metals and mercury. Sodium amalgam is formed by dissolving clean pieces of sodium in mercury. The metal dissolves with a bright flash and a solid compound Na—Hg is formed. Sodium amalgam is a useful reducing agent, evolving hydrogen in presence of water. When Na—Hg amalgam is placed in a concentrated solution of NH_4Cl, HgS swells up into a buttery mass thirty times its original volume which is known as ***ammonium amalgam*** which at higher temperature decomposes to give Hg, N_2 and NH_3. ***Iron amalgam,*** Fe—Hg is prepared by placing Na—Hg or Zn—Hg amalgam in solution of $FeSO_4$.

$$FeSO_4 + 2Na\text{–}Hg \longrightarrow Na_2SO_4 + \underset{\text{Iron amalgam}}{Fe\text{–}Hg}$$

Copper amalgam, Cu—Hg is formed by dipping the Cu metal in a solution of $HgNO_3$ or $HgCl_2$.

$$HgCl_2 + Cu \longrightarrow Hg + CuCl_2 \; ; \; Cu + Hg \longrightarrow Cu\text{–}Hg \; (\textit{copper amalgam})$$

Cu—Hg amalgam is used in filling teeth.

Thorium (Th_{90})

Occurrence

Thorium occurs widely distributed in nature though in small proportions. Its important minerals are *thorite*, $(SiO_2.\ ThO_2).\ nH_2O$ and *thorianite* which contains 55–80% ThO_2, 11–32% UO_2 and a little ceria earths. Thorium is found in Ceylon and Travancore *monazite* sands of India (about 9% of ThO_2). 80% of the world's thorium is located in India as monazite sand.

Extraction of Th from Monazite Sand

1. Preparation of thoria (ThO_2). Monazite sand is dissolved in conc. H_2SO_4 by heating at 200°C. The residue is leached with H_2O to have a solution containing sulphates of Th and some lanthanides (mainly La and Ce).

$$ThO_2 + 2H_2SO_4 \longrightarrow \underset{\text{(Solution)}}{ThSO_4} + 2H_2O$$

$$2(La, Ce)PO_4 + 3H_2SO_4 \longrightarrow \underset{\text{(Solution)}}{(La,Ce)_2(SO_4)_2} + 3H_3PO_4$$

Due to the formation of H_3PO_4 the solution containing the sulphates is sufficiently acidic. The solution is, therefore, neutralised by adding NaOH or MgO and pH adjusted so that $Th(HPO_4)_2$ gets precipitated, Leaving the phosphates of La and Ce, for the most part, in solution.

$$Th(SO)_4)_2 + 2H_3PO_4 \underset{H^+}{\overset{HO^-}{\rightleftarrows}} Th(HPO_4)_2 + 2H_2SO_4$$

Now the precipitate of $Th(HPO_4)_2$ is heated with Na_2CO_3 to form crude thoria (ThO_2) containing La and Ce as chief impurities.

$$Th(HPO_4)_2 + 3Na_2CO_3 \longrightarrow \underset{\text{(Crude)}}{ThO_2} + 2Na_3PO_4 + 3CO_2 + H_2O$$

The crude thoria is treated with oxalic acid and ammonium oxalate solution. Thorium oxalate so formed remains in solution, while the oxalate of La and Ce are left behind as residue. Solution containing $Th(C_2O_4)_2$ is dried and ignited when ThO_2 is formed. It is dissolved in HNO_2 and $Th(NO_3)_4 . 4H_2O$ formed is crystallised. The nitrate on heating gives pure ThO_2.

2. Preparation of thorium metal. The oxide ThO_2 is heated with carbon in a current of chlorine when $ThCl_4$ is precipitated which on being reduced with Ca metal gives Th.

$$ThO_2 + C + 2Cl_2 \longrightarrow ThCl_4 + CO_2$$
$$ThCl_4 + 2Ca \longrightarrow Th + 2CaCl_2$$

Extraction of Th by Electrolysis

The electrolysis of a mixture of thorium chloride and potassium chloride in a graphite crucible using graphite electrode furnishes thorium of about 88% purity.

Properties

Thorium is available as a dark grey or iron grey podwer; sp, gr. 10.11. The metal is stable in air at 120°, but it takes fire just below redness. Water does not react with the metal, unless the metal contains carbide. When thorium is heated in chlorine, thorium chloride is formed with incandescence. Bromine and iodine have also a similar action on it. Hydrofluoric acid has no action on it; hydrogen, however, evolves, when hydrochloric acid acts on it. Thorium sulphide is formed when thorium is heated with sulphur. Hydrogen sulphide also reacts with red-hot thorium to give thorium sulphide. Sulphuric acid reacts with thorium slowly, and sulphur dioxide, evolves. Thorium nitride is formed when the metal is heated with nitrogen. Solution of alkalies have no marked action on the metal.

Thoria Mantles and Uses of Thorium

Thoria has an immense use in the preparation of incandescent gas mantles. Mixtures of thoria with traces of ceria have been used in the form of pastilles for searchlights and headlights of cars. Magnesium flash light powders also often contain some thorium preparations. Thorium alloys have been used as reducing agents and thoria as a contact catalyst. In the nuclear chemistry, thorium containing substances have a very important future awl hence controls have been imposed on thoria industry which is everywhere the state monopoly.

Uranium (U_{92})

Occurrence

Uranium is not found free in nature. It occurs in the combined state in a few scarce minerals. Its most important minerals are : (*i*) *Pitchblende* or *uraninite* in which uranium is present as U_3O_8 or $2UO_2$. $2UO_3$ or uranium uranate, $U(UO_4)_2$. It contains 75–90% of the oxide and is found in Australia. Bohemia, Hungary Norway, India, East Africa etc : (*ii*) *Carnotite* $K_2O.2UO_3.V_2O_5$. $3H_2O$ or $2K(UO_2)VO_4.3H_2O$ (potassium uranyl ortho vanadate) which contains both uranium and radium. It occurs in *Jaduquda mines* of Bihar (*iii*) *Antunite*, $K(UO_2)PO_4$ $8H_2O$ (*potassium uranyl orthophosphate*). It is found in Congo. In India at present the principal sources of uranium ore are the *Jaduguda mines* of Bihar. In 1964 rich deposits of high grade uranium ore have been found in Bhatin area of Bihar. Another important source of uranium in India is the *monazite sand* of Travancore.

Extraction of Uranium From Pitchblende

1. Acid digestion process. First of all pitchblende is concentraed by gravity process to remove sand, clay etc. and then concentrated ore is roasted in excess of air whereby S, As and Sb are removed as their volatile oxides. Silver, if present in the mineral, is removed, as insoluble AgCl by roasting the mineral at 800°C with NaCl. The roasted ore is digested with dil. H_2SO_4 in presence of MnO_2 (an oxidising agent) for about 24 hours when Ba, Pb, Ba, etc. are precipitated as sulphates and uranium goes into solution as *uranyl sulphate*, UO_2SO_4.

$$\underset{\text{(In mineral)}}{2U_3O_8} + 6H_2SO_4 + \tfrac{1}{2}O_2 \longrightarrow \underset{\substack{\text{Uranyl sulphate}\\\text{(Soluble)}}}{3UO_2SO_4} + 3H_2O$$

UO_2SO_4 present in solution is converted into *sodium uranyl carbonate*, $N_4UO_2(CO_3)_3$ by the addition of Na_2CO_3.

$$UO_2SO_4 + 3Na_2CO_3 \longrightarrow \underset{\substack{\text{Sod. uranyl car-}\\\text{bonate (solution)}}}{Na_4UO_2(CO_3)_3} + Na_2SO_4$$

The solution containing $Na_4UO_2(CO_3)_3$ is now neutralised with HCl in order to decompose it into *uranyl chloride*, UO_2Cl_2 and then saturated with H_2S gas so that Pb, Cu etc. are precipitated as their sulphides. The solution of UO_2Cl_2 is now treated with excess of NH_4OH whereby we get the precipitate of *ammonium diuranate*, $(NH_4)_2U_2O_7$ which, when strongly ignited in air, gives the oxide, U_3O_8.

$$Na_4UO_2(CO_3)_3 + 6HCl \longrightarrow \underset{\substack{\text{Uranyl chloride}\\\text{(Solution)}}}{UO_2Cl_2} + 4NaCl + 3H_2O + 3CO_2$$

$$2UO_2Cl_2 + 6NH_4OH\ \text{(excess)} \xrightarrow{\text{Ignite}} \underset{\substack{\text{Ammonium}\\\text{diuranate}}}{(NH_4)_2U_2O_7} + 4NH_4Cl + 3H_2O$$

$$3(NH_4)_2U_2O_7 \longrightarrow 2U_3O_8 + 6NH_3 + 3H_2O + O_2$$

When U_3O_8 is reduced by carbon by heating in an electric furnace, uranium is obtained.

$$U_3O_8 + 8C \longrightarrow 3U + 8CO$$

98–99% of pure uranium can be obtained by reducing U_3O_8 by Mg, Ca or Al (*Allumino-thermic process*).

$$3U_3O_8 + 16Al \longrightarrow 9U + 8Al_2O_3$$

Burger reduced U_3O_8 by the vapours of calcium in vacuo.

2. Alkali digestion process. The roasted ore is digested with Na_2CO_3—$NaHCO_3$ mixture solution whereby U_3O_8 present in the mineral is converted into soluble $Na_4UO_2(CO_3)_3$. The solution containing $Na_4UO_2(CO_3)_3$ is treated with NaOH which precipitates uranium as *sodium diuranate*, $Na_2U_2O_7$.

$$\underset{\text{(In mineral)}}{U_3O_8} + 3Na_2CO_3 + 6NaHCO_3 + \tfrac{1}{2}O_2 \longrightarrow \underset{\text{(Soluble)}}{3Na_4UO_2(CO_3)_3} + 3H_2O$$

$$Na_4UO_2(CO_3)_3 + 6NaOH \longrightarrow \underset{\text{Sod. diuranate (ppt.)}}{Na_2U_2O_7} + 6Na_2CO_3 + 3H_2O$$

The precipitate of $Na_2U_2O_7$ is dissolved in dil. H_2SO_4 and the solution is treated with NH_4OH. The precipitate of $(NH_4)_2U_2O_7$ thus obtained on strong ignition in air gives U_3O_8.

$$\underset{\text{(ppt)}}{Na_2U_2O_7} + H_2SO_4 \longrightarrow Na_2SO_4 + \underset{\text{(Soluble)}}{H_2U_2O_7}$$

$$H_2U_2O_7 + 2NH_4OH \longrightarrow \underset{\text{Ammon. diuranate (ppt)}}{(NH_2)_2U_2O_7 \downarrow} + 2H_2O$$

$$3(NH_4)_2U_2O_7 \xrightarrow{\text{Ignition}} 2U_3O_8 + 6NH_3 + 3H_2O + O_2$$

The oxide, U_3O_8, is reduced, as in acid digestion process, to uranium.

Extraction of Uranium From Carnotite

In addition to uranium, carnotite also contains Ba, V, Fe, Al, Ra etc. From this ore uranium is obtained as a by-product. The ore is treated with conc. HNO_3 at 100°C. The heating is done by the steam. Most of the ore dissolves in HNO_3 and the solution thus obtained is noutralised by NaOH and then treated with $BaCl_2$ and H_2SO_4 whereby Ba and Ra are precipitated as their sulphate, and U and V are converted into *soluble* UO_2SO_4 and $VOSO_4$ respectively. This filtrate which also contains Fe and Al is treated with excess of boiling solution of Na_2CO_3 which precipitates Fe and Al (*Reject*), and UO_2SO_4 and $VOSO_4$ are converted into *solub1e* $Na_4UO_2(CO_3)_3$ and Na_2VO_4 (*sod. vanadate*) respectively.

$$UO_2SO_4 + 3Na_2CO_3 \longrightarrow \underset{\text{(Soluble)}}{Na_4UO_2(CO_3)_3} + Na_2SO_4$$

$$2VOSO_4 + 5Na_2CO_3 + \tfrac{1}{2}O_2 \longrightarrow \underset{\text{(Soluble)}}{2Na_3VO_4} + 3Na_2SO_4 + 5CO_2$$

The solution containing $Na_4UO_2(CO_3)_3$ and Na_3VO_4 is neutralised by HNO_3 and treated with NaOH. This treatment precipitates uranium as $Na_2U_2O_7$.

$$Na_4UO_2(CO_3)_3 + 6HNO_3 \longrightarrow UO_2(NO_3)_2 + 4NaNO_3 + 3H_2O + 3CO_2$$

$$2UO_2(NO_3)_2 + 6NaOH \longrightarrow \underset{\text{ppt}}{Na_2U_2O_7} + 4NaNO_3 + 3H_2O$$

The precipitate of $Na_2U_2O_7$ is converted into $(NH_4)_2U_2O_7$ as in alkali digestion process.

$$Na_2U_2O_7 + H_2SO_4 \longrightarrow Na_2SO_4 + H_2U_2O_7$$

$$H_2U_2O_7 + 2NH_4OH \longrightarrow \underset{\text{ppt.}}{(NH_4)_2U_2O_7 \downarrow} + 2H_2O$$

$(NH_2)_2U_2O_7$ on strong ignition in air gives U_3O_8 which can be reduced to uranium as already discussed.

Properties

The metal uranium resembles nickel in appearance. The pure sample is white, but on contamination with some nitride, it assumes yellow tint. It is malleable, takes polish, is softer than steel, but becomes hard and brittle when alloyed with carbon and chilled. It melts at 1850°. The powdered metal burns in air when heated to 160°, in chlorine at 150° and in fluorine at the room temperature. It also burns in iodine, sulphur vapour and nitrogen at higher temperature. It also decomposes water though slowly at the room temperature, yet rapidly on boiling. It reacts with ammonia at higher temperatures to give out nitrogen, dissolves in dilute mineral acids, giving out hydrogen, and evolves sulphur dioxide from concentrated sulphuric acid. The caustic alkalies have no action upon the metal. Uranium will displace mercury, silver, copper and tin from the solution of their salts. When a solution of uranium salt is exposed to light, it shows fluorescence. Uranium is a radioactive element and gives a series of disintegration products.

Nitrogen combines directly with uranium at 1000° forming *uranium nitride*, U_3N_4. Carbon also directly combines with uranium giving *uranium carbide*, UC_2.

Uses

In ceramics, uranium compounds are sometimes used to give coloured glasses, orange or yellow. Its salts have also been used as mordants for silk and wool. The metal and its carbide are one of the best catalysts in the Haber's synthesis of ammonia. Uranyl nitrate is used as a reagent in volumetric determination of phosphate and arsenates. Uranium steels are also of technical importance. The metal is used in nuclear reactors.

University Questions

1. How is titanium metal extracted in a pure form from ilimenite and rutile ores. (*Meerut 82, 88 Jodhpur 85; Allahabad 88; Bundelkhand 88*)
2. How does vanadium occur in nature? Discuss the extraction of vanadium from either vanadinite or carnotito. (*Raj. 87; Meerut 82; Indore 86*)
3. What are important ores of chromium? Describe the extraction of the metal from it chief ores. Give the chemical principles involved in the extraction process. (*Meerut 81, 82, 87; Garhwal 88; Allahabad 89; Gorakhpur 86; Sambalpur 86; Agra 85*)
4. Name the important ores of molybdenum. How is the metal extracted? What are its uses? (*Meerut M.Sc. 82, 85; Bharathiar 85*)

5. What are the important ores of tungsten? How is the metal obtained from its ores ? *(Madras 85, 86; Madurai 85)*

6. Outline the method for the extraction of Mn-metal from its ores. Give all the equations of reactions involved. *(Allahabad 87; Punjab 86; Meerut 86; Bundelkhand 88; Madras 85; Lucknow 85)*

7. Give the names of important ores of iron. *(Raj. 86)*

8. How is pig iron produced from iron ore? *(Agra 86; Raj. 86)*

9. How is pig iron converted into wrought iron and steel? *(Agra 86)*

10. What are alloy steels? Mention three elements that are used in preparing them? What special properties are produced in these metals ? *(Madras 86; Bharathiar 86)*

11. Write a note on alloy steels. Name their uses. *(Bharathiar 86; Punjab 84; Delhi 85; Madurai 85)*

12. Mention two important ores of cobalt. Explain with equations, how is the metal is extracted from its ores ? *(Madras 86; Punjab 85)*

13. What are the chief ores of nickel ? Describe the extraction of nickel (*i*) by Mond's process (*ii*) from Sudbery ores. What by-products are obtained in (*ii*) process. Give the utility of the metal in industry. *(Jiwaji 84; Allahabad 88; Meerut 82, 84; Garhwal 88; Gorakhpur 85)*

14. Discuss the method and give the chemical principles involved in the extraction of Pt from native ore. *(Allahabad 85, 88; Jodhpur 84; Agra 82, 84, 87, 88; Bharathiar 86)*

15. How is Pt obtained from Sudbury nickel ores? *(Meerut 83)*

16. Discuss briefly the extraction of Cu from its ore. Give all equations involved. *(Punjab 86; Madurai 86)*

17. Name the important ores of Silver. Describe the extraction of this metal by cyanide process. Give equations. *(Raj. 87)*

18. How is gold extracted from its ores? Discuss the chemical reactions involved in methods employed. *(Allahabad 89)*

19. What are the important ores of Zn? Describe the commercial method of the extraction of Zn. What are the common uses of Zn? *(Madras 85; Sambalpur 86; Agra 86; Poona 85)*

20. Discuss the methods of extraction of chromium from one of its important ores. *(Allahabad 99)*

21. (*i*) What happens when gold is heated with aqua regia.

(*ii*) Give important ores of Ni. Describe Mond's process of extracting Ni from sulphide ore.

(*iii*) Write notes on (*a*) Ni and Pt catalysts (*b*) Alloys of Cu and Zn. *(Kanpur 2000)*

22. Name the two ores of titanium. How is this metal extracted? *(Himachal Pradesh 99)*

23. Name two important ores of cobalt and discuss the chemistry of its extraction. *(Himachal Pradesh 2000)*

24. How is uranium extracted from one of its main ores ? Discuss the chemical principle of the extraction and give method of preparation and use of one of its compounds. *(Kumaon 2000)*

25. How is vanadium extracted from one of its important ores 7 Explain all the steps giving chemical equations. Give method of preparation of V_2O_5. *(Kumaon 2000)*

26. (*a*) Name two chief ores of Ni with their formulae.

(*b*) How would you extract Ni from its important ores? *(Gauhati (General) 2000)*

27. (*a*) What element is prepared from chromite ore? Discuss its different oxidation states. Mention its two important alloys and their uses.

(*b*) Name the members of titanium family. How is titanium prepared from rutile ? Give preparation of one of its important compounds. *(Lucknow 2000)*

28. (*a*) With particular reference to the electronic configuration of 4*f* block elements discuss the isolation of oxides of lanthanones from monazite sand.

(*b*) Discuss ion exchange method for the separation of lanthanones. *(Lucknow 2000)*

29. Name the important ores of Cr. How is this metal obtained from Chromite ore in pure form ? How is chromealum obtained from $K_2Cr_2O_7$? Explain the reactions giving equations. *(Avadh 2000)*

30. (*i*) Explain how the properties of iron are affected by the addition of carbon?

(*ii*) Give an account of 'heat treatment of steel'. *(Bharathiar 2000)*

31. (*i*) Discuss the extraction of chromium from its oxide ore.

(*ii*) Discuss briefly (*a*) Tempering of steel (*b*) Influence of chromium on the properties of steel. *(Bangalore 2001)*

32. What are the important ores of Th ? Write in brief the extraction of Th from monazite ore. Give its important uses. *(M.J.P. Rohilkhand M.Sc. 2008)*

33. How is uranium obtained from its ores ? State the important properties and uses of the metal. *(M.J.P. Rohilkhand 2009)*

34. What are important ores of nickel ? How is nickel extracted from pentandite. Describe its important properties. *(Meerut 2009)*

CHAPTER 47

Some Compounds of d- and f-Block Elements

Compounds of Titanium

Different Oxidation States of Titanium in Its Compounds

Titanium forms compounds in 0, +2, +3, +4 and –1 oxidation states. The electronic configuration of Ti^0 atom and of, Ti^{2+}, Ti^{3+}, Ti^{4+} and

Ti^{1-} ions are as : $Ti^0 \rightarrow 2, 8, 3s^2p^6d^2, 4s^2$; $Ti^{2+} \rightarrow 2, 8, 3s^2p^6d^2, 4s^0$; $Ti^{3+} \rightarrow 2, 8, 3s^2p^6d^1, 4s^0$; $Ti^{4+} \rightarrow 2, 8, 3s^2p^6d^2, 4s^0$; $Ti^{1-} \rightarrow 2, 8, 3s^2p^6d^3, 4s^2$. Compounds having Ti metal in +4 oxidation state are the most stable and hence the compounds in lower oxidation states are readily oxidised to +4 state even by water or air.

Ti (+4) compounds are generally covalent and resemble, in certain respects, the corresponding compounds of group IVA elements. Ti (+3) and Ti (+2) compounds are mainly ionic. They act as active reducing agents.

$$2Ti^{3+} \xrightarrow{\text{Heat}} Ti^{4+} + Ti^{2+}$$

$$2Ti^{2+} \xrightarrow{\text{Water}} Ti^{4+} + Ti^0$$

Ti (0) and Ti (–1) compounds are rare. Examples of compounds having Ti metal in different oxidation state are : *Ti (0) compounds* : $[Ti(dipy)_3]^0$; *Ti (+2) compounds*; TiO, $TiCl_2$; Ti (+3) *compounds* (*called titanous compounds*) : $TiCl_3$, Ti_2O_3, $Ti(OH)_3$, $[Ti(H_2O)_6]^{3+}$, $[TiF_6]^{3-}$; *Ti (+4) compounds* (*called titanic compounds*) : TiO_2, $TiCl_4$, $TiBr_4$, TiI_4, $TiOCl_2$, $Ti(SO_4)_2$, $TiOSO_4$; *Ti (–1) compounds* : $[Ti(dipy)_3]^-$. Now let us discuss some important compounds of titanium.

1. Titanium Trichloride or Titanous Chloride, $TiCl_3$

It is the most important compound of Ti (+3). It is prepared by reducing $TiCl_4$ with H_2 in an electrically heated spiral in a water cooled tube or with Ag at 200°C.

$$2TiCl_4 + H_2 \longrightarrow 2TiCl_3 + 2HCl$$

$$TiCl_4 + Ag \longrightarrow TiCl_3 + AgCl$$

Anhydrous $TiCl_3$ is a violet powder. It disproportionates into $TiCl_4$ and $TiCl_2$.

$$2TiCl_3 \longrightarrow TiCl_3 + TiCl_2$$

It is a powerful reducing agent since it reduces chlorates and perchlorates to chlorides, Fe^{3+} to Fe^{2+}, Cu^{2+} to Cu, and persulphate to $SO_4{}^{2-}$. Due to this property $TiCl_3$ is used in volumetric analysis of the above mentioned compounds. An aqueous solution of $TiCl_3$, when added to a dilute solution of a gold salt, produces an intense violet colour, due to the reduction of the gold salt followed by the absorption of colloidal gold on hydrous TiO_2. This reaction has been used as a very delicate test for gold.

2. Titanium Dioxide, TiO_2

Preparation. This compound is also called ***titanic dioxide*** or ***titania.*** This is prepared by adding an aqueous solution of NH_3 to $TiCl_4$ and then heating the precipitate of $Ti(OH)_4$ thus obtained.

$$TiCl_4 + 4NH_3 + 4H_2O \longrightarrow \underset{\text{ppt.}}{Ti(OH)_4} + 4NH_4Cl$$

$$Ti(OH)_4 \xrightarrow{\Delta} TiO_2 + 2H_2O$$

TiO_2 can also be obtained from rutile ore. This ore is first converted into $TICl_4$ by heating the mixture of rutile ore and carbon with Cl_2 and then $TiCl_4$ is oxidised to pure TiO_2 with O_2.

$$TiO_2 + C + 2Cl_2 \xrightarrow{\Delta} TiCl_4 + CO_2$$

$$TiCl_4 + O_2 \longrightarrow TiO_2 + 2Cl_2$$

TiO_2 can also be obtained from ilmenite ore, $FeTiO_3$ as discussed in the extraction of Ti.

Properties : TiO_2 occurs in nature in three crystalline forms which are *rutile, anatase* and *brookite*. Rutile is the commonest form and forms tetragonal transparent to opaque, red or yellowish crystals. This form is stable above 1040°. The anatase modification also crystallises in the tetragonal system but with the different axial ratio. This form is stable upto 860°. Brookite form crystallises in rhombic form and is stable between 860–1040°. The rutile form of TiO_2 is considered to be the most stable. The rutile or brookite forms are metastable below 860° and thus are converted into anatase by suitable heat treatment.

$$\textit{Anatase} \overset{860°C}{\rightleftarrows} \textit{Brookite} \overset{1040°C}{\rightleftarrows} \textit{Rutile}$$

Natural TiO_2 is impure due to the presence of iron oxide as an impurity contaminated with it. Pure TiO_2 is colourless (amorphous white powder) when cold, and yellowish when hot.

It melts at about 1800° and looses O_2 to form Ti_2O_3 [$4TiO_2 \xrightarrow{>1800°} 2Ti_2O_3 + O_2$]. It is insoluble in H_2O and dil. acids but dissolves in hot conc. H_2SO_4 and HF due to the formation of soluble Ti (+4) salts. It is amphoteric oxide but more acidic than basic. It dissolves in alkalies to form titanates (TiO_3^{2-}). [$TiO_2 + 2NaOH \rightarrow Na_2TiO_3 + H_2O$]. On heating with carbon in a current of Cl_2, it forms $TiCl_4$. With dry Cl_2, it gives $TiCl_4$ [$TiO_2 + 2Cl_2 \text{ (dry)} \rightarrow TiCl_4 + O_2$]. It is reduced to Ti metal by heating it with carbon or Al metal.

$$TiO_2 + C \longrightarrow Ti + CO_2$$

$$3TiO_2 + 4Al \longrightarrow 3Ti + 2Al_2O_3$$

On passing the vapours of CCl_4 over TiO_2 at high temperature. $TiCl_4$ is formed.

$$TiO_2 + CCl_4 \text{ (vapours)} \longrightarrow TiCl_4 + CO_2$$

Uses. (*i*) Because of its high opacity, inertness towards H_2S, high covering power, light weight, non-poisonous nature, ease of wetting and disperssion in oil as well as resistance to high temperature it is used as a white pigment under the name *titan white.* Mixed with $BaSO_4$ it is sold under the trade name *titanox.* (*ii*) It can be also be used for interior protection purpose. When applied externally the film gets chalked out. (*iii*) It is also used (*a*) in making non-greasable paper (*b*) in the manufacture of arc lamp electrodes (*c*) in gas mantles, (*d*) for the preparation of glazes, (*e*) in making components for use in radio and television sets (*f*) as an additive in the manufacture of fibre glass (*g*) in the manufacture of artificial gems (*h*) in enamelling of porcelain goods (*i*) in the surface of white leather (*j*) in rubber industry to soften rubber (*k*) as a whitener in cosmetics (*e.g.* in face powder, cold creams etc.) (*l*) in preparing printing and decorating inks.

3. Titanium Tetrachloride, $TiCl_4$

Preparation : This compound is also called ***titanic chloride*** and is prepared (*i*) by heating the rutile ore with carbon in a current of Cl_2 (see the preparation of TiO_2) (*ii*) by the action of of dry Cl_2 on TiO_2 (see the properties of TiO_2) (*iii*) by passing vapours of CCl_4 at high temperature over TiO_2 (see the properties of TiO_2) (*iv*) from ilmenite ore (see the extraction of Ti metal).

Properties : It is a colourless, strongly fuming liquid. It has m.pt. = –30°C and b.pt. = 136.4°C. It is rapidly hydrolysed by excess of H_2O. [$TiCl_4 + 2H_2O \rightarrow TiO_2 + 4HCl$]. If water is not present in excess, hydrolysis is incomplete and *titanium oxy-chloride*, $TiOCl_2$ is formed [$TiCl_4 + H_2O \rightarrow TiOCl_2 + 2HCl$].

It is a strong *Lewis acid* (*i.e.* electron pair acceptor) and hence can accept electron pairs form donor molecules like NH_3 and PCl_5, forming the addition compounds like $TiCl_4.6NH_3$, $TiCl_4.8NH_3$ and $TiCl_4.PCl_5$ etc.

$TiCl_4$ can be reduced to free metal by heating it with Mg, Na, NaH etc. When $TiCl_4$ is treated with amalgamated Na or with H_2 in the silent electric discharge, it is reduced to $TiCl_2$. $TiCl_4$ can also be reduced to $TiCl_2$ by heating it with titanium shavings.

$$TiCl_4 + H_2 \longrightarrow TiCl_2 + 2HCl$$

$$TiCl_4 \xrightarrow[\text{Titanium shavings}]{900°C} 2TiCl_2$$

When $TiCl_4$ is heated with H_2 by an electrically heated spiral in a water-cooled tube, anhydrous $TiCl_3$ gets deposited on the cold walls of the tube. The reduction of $TiCl_4$ with Ag at 200°C also gives $TICl_3$.

$$2TiCl_4 + H_2 \longrightarrow 2TiCl_3 + 2HCl.$$

$$TiCl_4 + Ag \longrightarrow TiCl_3 + AgCl.$$

With NH_4Cl yellow crystals of *ammonium chlorotitanate*, $(NH_4)_2[TiCl_6].2H_2O$ are obtained.

$$TiCl_4 + 2NH_4Cl + 2H_2O \longrightarrow (NH_4)_2[TiCl_6].2H_2O$$

$TiCl_4$ dissolves in conc. HCl, forming chloro complexes of the formulae, $TiCl_4$. HCl. H_2O or $H^+[TICl_5(H_2O)]$ and $TiCl_4$. 2HCl or $2H^+[TiCl_6]^{2-}$. On heating $TiCl_4$ with anhydrous HF, TiF_4 is obtained.

$$TiCl_4 + 4HF \longrightarrow TiF_4 + 4HCl$$

Uses. It is used : (*i*) to form irritating smoke-clouds and dense fogs in war time. (*ii*) for the production of titanium metal by reducing it with Mg (*Kroll process*). *Na*(*I.C.I. process*), NaH etc. (*iii*) in the manufacture rutile gems and organic titanium compounds.

Uses of titanium compounds. (*i*) The uses of TiO_2 and $TiCl_4$ have already been described. (*ii*) Double salts of Ti like $TiOC_2O_4.K_2C_2O_4.H_2O$, $Na_2SO_4.Ti_2(SO_4)_3.5H_2O$ etc. are used as mordants in the dyeing of textile and leather. (*iii*) $TiCl_2$ is used as a reducing agent in the laboratory for the titration of Fe^{3+}. $TiCl_3$ and $Ti_2(SO_4)_3$ are used for removing certain dyes from the goods so that they may be redyed. They are also used for removing stains. (*iv*) TiC has been used as a catalyst during the preparation of nitrogen- hydrogen and nitrogen-oxygen compounds. TiC is also used for the preparation of hard metals or cutting alloys. (*v*) Titanium boride is one of the hardest substances known and hence is used as an abrasive.

Compounds of Vanadium

Different Oxidation States of Vanadium in Its Compounds

Vanadium forms compounds in 0, +1, +2, +3, +4, +5 and –1 oxidation states. The electronic configuration V^0 atom and of V^+, V^{2+}, V^{3+}, V^{4+}, V^{5+} and V^- ions are : $V^0 \rightarrow 2, 8, 3s^2p^6d^3, 4s^2$; $V^+ \rightarrow 2, 8. 3s^2p^6d^3, 4s^1$; $V^{2+} \rightarrow 2, 8, 3s^2p^6d^3, 4s^0$; $V^{3+} \rightarrow 2, 8, 3s^2p^6d^2, 4s^0$; $V^{4+} \rightarrow 2, 8, 3s^2p^6d^1, 4s^0$; $V^{5+} \rightarrow 2, 8, 3s^2p^6d^0, 4s^0$; $V^- \rightarrow 2, 8, 3s^2p^6d^4, 4s^2$. Examples of compounds having vanadium in different oxidation states are : *V*(0) *compounds* : $V(CO)_6$, V(*dipy*)$_3$; *V* (*+1*) *compounds* : [V(*dipy*)$_3$]$^+$; *V* (*+2*) *compounds* : VO, VS, VX_2 (X = Cl, Br, I), VSO_4, $[V(H_2O)_6]^{2+}$, $[V(CN)_6]^{4-}$; *V* (*+3*) *compounds* : V_2O_3, V_2S_3, $V(OH)_3$, VX_3 (X = F, Cl, Br, I), VOX (X = Cl, Br), $[VCl_4]^-$ $[VF_6]^{3-}$, $[V(H_2O)_6]^{3+}$; *V* (*+4*) *compounds* : VO_2, VX_4 (X = F, Cl), $VOSO_4$, VOX_2 (X = F, Cl, Br), $V(NO_3)_2$, $[VCl_6]^{2-}$; *V*(*+5*) *comounds* : V_2O_5, VOX_3 (X = F, Cl, Br), VF_5, $[VF_6]^-$, NH_4VO_3, Na_3VO_4. $12H_2O$; *V* (*–1*) *compounds* : $[V(CN)_5 (NO^+)]^{5-}$, $[V(CO)_6]^-$. Compounds in lower oxidation states are very unstable and act as powerful reducing agents; sometimes so sufficiently powerful that they can liberate H_2 even from H_2O. The most stable oxidation states are + 4 and. +5. Vanadium forms compounds of three main types : (*i*) Interstitial compounds with C, N_2 or H_2 (*ii*) Giant molecular compounds with electronegative elements like O, S, F etc. Examples of such compounds are : VO, VS, V_2O_3. V_2S_3, VF_3 etc. (*iii*) Salts in which vanadium atom is the central atom in complex anions. Now let us discuss some important compounds of vanadium.

1. Vanadium Pentoxide, V_2O_5

Preparation. It can be prepared (*i*) by heating ammonium metavanadate, NH_4VO_3 or (*ii*) by the hydrolysis of vanadium (V) oxychloride, $VOCl_3$

$$2NH_4VO_3 \xrightarrow{\Delta} V_2O_5 + 2NH_3 + H_2O$$
$$2VOCl_3 + 3H_2O \longrightarrow V_2O_5 + 6HCl$$

Properties. It is a yellowish red powder and undergoes reversible dissociation in the temperature range 700–1125°C

$$2V_2O_5 \rightleftharpoons 2V_2O_4 + O_2$$

When V_2O_5 is reacted with Cl_2 at elevated temperature, $VOCl_3$ is obtained. When Cl_2 gas is passed over a mixture of V_2O_5 and carbon, $VOCl_2$ is formed.

$$2V_2O_5 + 6Cl_2 \longrightarrow 4VOCl_3 + 3O_2$$
$$V_2O_5 + 3C + 3Cl_2 \longrightarrow 2VOCl_3 + 3CO$$

When V_2O_5 is reduced with K, *hypovanadium oxide* (V_2O_2) is obtained.

$$V_2O_5 + 6K \longrightarrow V_2O_2 + 3K_2O$$

When V_2O_5 is reduced by H_2 or sulphur, *vanadium trioxide* (V_2O_3) is obtained.

$$V_2O_5 + 2H_2 \xrightarrow{\Delta} V_2O_3 + 2H_2O$$
$$V_2O_5 + S \longrightarrow V_2O_3 + SO_2$$

When V_2O_5 is reduced with SO_2, H_2S or HX, or by fusion with oxalic acid ($H_2C_2O_4$), *vanadinin tetoxide* (V_2O_4) is obtained.

$$V_2O_5 + SO_2 \xrightarrow{\Delta} V_2O_4 + SO_3$$
$$V_2O_5 + H_2S \longrightarrow V_2O_4 + H_2O + S$$
$$V_2O_5 + 2HX \longrightarrow V_2O_4 + H_2O + X_2$$

When V_2O_5 is heated with Ca metal, V_2O_5 gets reduced to *vanadium metal.*

$$V_2O_5 + 5Ca \longrightarrow 5CaO + 2V$$

Reduction of a solution of V_2O_5 in conc. H_2SO_4 by means of SO_2 gives a blue solution which, on evaporation, gives the bright blue crystals of *acid vanadyl sulphate*, $2VOSO_4 . H_2SO_4 . xH_2O$. When this acid *vanadyl sulphate* is heated with conc. H_2SO_4 at 260°, a grey-green crystalline powder of *anhydrous vanadyl sulphate*, $VOSO_4$ is obtained.

V_2O_5 dissolves in HCl to give salts like vanadyl (IV) oxychloride, $VOCl_2$ or VCl_5

$$V_2O_5 + 6HCl \longrightarrow 2VOCl_2 + 3H_2O + Cl_2$$
$$V_2O_5 + 10HCl \longrightarrow 2VCl_5 + 5H_2O$$

With HNO_3, V_2O_5 gives metavanadic acid, HVO_3

$$V_2O_5 + 2HNO_3 \longrightarrow 2HNO_3 + N_2O_5$$

V_2O_5 dissolves in the excess of acids to give the complex anions like $[H_2V_{10}O_{28}]^{4-}$, $[HVO_{10}O_{28}]^{6-}$ and $[V_{10}O_{28}]^{6-}$ within the pH range = 1.3 to 6.5. V_2O_5 dissolves in alkalies to form a series of vanadates like *orthovanadates* (*e.g.* Na_3VO_4), *metavanadates* (*e.g.* $NaVO_3$) and *pyrovanadates* (*e.g.* $Na_4V_2O_7$). These vanadates are analogous to phosphates. At high pH value, sodium ortho vanadate, Na_3VO_4 is obtained by the action of V_2O_5 on NaOH.

$$V_2O_5 + 6NaOH \longrightarrow 2Na_3VO_4 + 3H_2O$$

As the pH is lowered, protonation of VO_4^{3-} ion and subsequent polymerisation takes place, and different types of anions are obtained as shown below :

$$VO_4^{3-} \longrightarrow [VO_2(OH)_3]^{2-} \longrightarrow [(VO_2)_2(OH)_5]^{3-} \longrightarrow [(VO_2)_3(OH)_6]^{3-}$$

When heated with carbon at high temperature, vanadium carbide (VC) is obtained. When V_2O_5 is melted with Na_2CO_3, Na_3VO_4 is obtained.

$$2V_2O_5 + 4C \longrightarrow 4VC + 5O_2$$
$$V_2O_5 + 3Na_2CO_3 \longrightarrow 2Na_3VO_4 + 3CO_2 \uparrow$$

Uses : V_2O_5 is used : (*i*) as a catalyst in a number of reactions. Some such reactions are : (*a*) oxidation of SO_2 to SO_3 used in the manufacture of H_2SO_4 by contact process. (*b*) oxidation of anthracene to anthraquinone. (*c*) oxidation of phenanthrene to phenanthraquinone (*d*) oxidation of toluene to benzaldehyde to benzoic acid. (*e*) oxidation of ethyl alcohol to acetaldehyde. (*ii*) in the preparation of vanadium glass. This glass is prepared by adding 0.02% vanadium as V_2O_5 to ordinary glass. This glass cuts out completely the harmful actinic rays which injure the eye and brings about fadings of organic compounds. Such glass is used particularly in textile store windows. Such glasses protect the eyes against day-light glare. (*iii*) in the preparation of ferro-vanadium alloy which is used as a scavenger in the manufacture of special steels. (See *uses of vanadium*).

2. Sodium Orthovanadate, $Na_3VO_4.12H_2O$

Preparation : It is prepared : (*i*) by dissolving V_2O_5 in NaOH, followed by crystallisation or by treating sodium pyrovanadate, $Na_4V_2O_7$ with NaOH (*ii*) by melting V_2O_5 with Na_2CO_3

$$V_2O_5 + 6NaOH \longrightarrow 2Na_3VO_4 + 3H_2O$$
$$Na_4V_2O_7 + 2NaOH \longrightarrow 2Na_3VO_4 + H_2O$$
$$V_2O_5 + 3Na_2CO_3 \longrightarrow 2Na_3VO_4 + 3CO_2 \uparrow$$

Properties and uses : It is soluble in H_2O. It is isomorphous with sodium orthophosphate dodecahydrate, $Na_3PO_4.12H_2O$. This compound is used (*i*) as a catalyst in the dyeing of leather to black colour. (*ii*) in making good quality, fire-flowing blue-black ink and marking inks. (*iii*) in volumetric analysis for certain oxidation-reduction reactions. The estimations by means of this reagent include metals such as Fe, Mo, Sn and Cu, and organic compounds such as citric acid, lactic acid, tartaric acid, aldehydes, ketones and sugars.

3. Ammonium Metavanadate, NH_4VO_3

It is obtained : (*i*) by adding an ammonium salt like NH_4Cl to a solution of $NaVO_3$, (*ii*) by treating metavanadic acid, HVO_3 with NH_3.

$$NaVO_3 + NH_4Cl \longrightarrow NH_4VO_3 + NaCl$$
$$HVO_3 + NH_3 \longrightarrow NH_4VO_3$$

Compounds of Chromium

Different Oxidation States of Chromium in Its Compounds

Chromium forms compounds in 0, +2, +3 and + 6 oxidation states. The electronic configuration of Cr^0 atom and of Cr^{2+}, Cr^{3+} and Cr^{6+} ions are as; $Cr^0 \rightarrow 2, 8,$

$3s^2p^6d^5$, $4s^1$; $Cr^2 \rightarrow 2, 8, 3s^2p^6d^4, 4s^0$; $Cr^{3+} \rightarrow 2, 8, 3s^2p^6d^3, 4s^0$; $Cr^{6+} \rightarrow 2, 8, 3s^2p^6d^0, 4s^0$. Compounds having Cr metal in +3 state are the most stable. Cr (+2) compounds are strong reducing agents while those of Cr (+6) are strong oxidising agents.

Examples of compounds of Cr in different oxidation states are : *Cr(0) compounds* : $Cr(CO)_6$; *Cr(+2) compounds* (called *chromous compounds*) : CrX_2 (X = Cl, Br, I), $(CH_3COO)_2Cr$, CrO, $Cr(OH)_2$; *Cr(+3) compounds* (called *chromic compounds*) CrX_3 (X = Cl, Br, I), $Cr_2(SO_4)_3.18H_2O$, $Cr_2(SO_4)_3.6H_2O$, $K_2SO_4.Cr_2(SO_4)_3.24H_2O$, Cr_2O_3, $Cr(OH)_3$; *Cr(+6) compounds* : CrO_3, K_2CrO_4, $K_2Cr_2O_7$, CrO_2Cl_3.

Acidic character of oxides. With the increase in the oxidation state of Cr, the acidic character of the oxides of this metal also increases. For example :

Oxides :	CrO	Cr_2O_3	CrO_3
Oxidation state of Cr :	+2	+3	+6
Nature :	Basic	Amphoteric	Acidic

CrO is basic, since it gives Cr (+2) salt with an acid.

$$\underset{\text{(Basic)}}{CrO} + H_2SO_4 \longrightarrow CrSO_4 + H_2O$$

Cr_2O_3 is amphoteric because it gives salts with both acids and alkalies.

$$\underset{\text{(Amphoteric)}}{Cr_2O_3} + 3H_2SO_4 \longrightarrow Cr_2(SO_4)_3 + 3H_2O$$

$$\underset{\text{(Amphoteric)}}{Cr_2O_3} + 2KOH \longrightarrow \underset{\text{Pot. chromite}}{2KCrO_2} + H_2O$$

CrO_3 is the anhydride of chromic acid (H_2CrO_4) and is distinctly acid. It gives chromates with alkalies.

$$\underset{\text{(Acidic)}}{CrO_3} + 2NaOH \longrightarrow Na_2CrO_4 + H_2O$$

Covalent character of compounds. With the increase in the oxidation state of Cr, the covalent character of the compounds of Cr also increases. Thus Cr(+2) compounds are ionic, Cr(+3) compounds ionise to some extent while those of Cr(+0) are covalent.

Now let us discuss some important compounds of chromium.

1. Chromous Chloride, $CrCl_2$

Preparation : *Anhydrous $CrCl_2$* is prepared : (*a*) by heating chromic chloride, $CrCl_3$ in a current of H_2 or (*b*) by heating Cr metal in a current of dry HCl gas

$$2CrCl_3 + H_2 \longrightarrow 2CrCl_2 + 2HCl$$

A solution of $CrCl_2$ is obtained by reducing $K_2Cr_2O_7$ solution with Zn and HCl in the absence of air. When $K_2Cr_2O_7$ is heated with conc. HCl, Cl_2 is given off and $CrCl_3$ is obtained.

$$K_2Cr_2O_7 + 14HCl \longrightarrow 2KCl + 2CrCl_3 + 7H_2O + 3Cl_2 \uparrow$$

Now when Zn metal is added to the above solution, nascent hydrogen is evolved which reduces $CrCl_3$ to blue chromous chloride, $CrCl_2$

$$CrCl_3 + H \text{ (From Zn/HCl)} \longrightarrow CrCl_2 + HCl$$

Properties. $CrCl_2$ forms white deliquescent silky needles which dissolve in H_2O to give a blue solution. It is a powerful reducing agent and is readily oxidised in air to $CrCl_3$. Its vapour density, even above 1500°C, indicates appreciable association [$Cr_2Cl_4 \rightleftharpoons 2CrCl_2$]. When treated with HF, it gives CrF_2.

2. Chromous Acetate, $(CH_3COO)_2Cr$

Preparation. This compound is prepared as red precipitate by adding a saturated solution of CH_3COONa to the blue solution of $CrCl_2$.

$$CrCl_2 + 2CH_3COONa \longrightarrow \underset{\text{Red ppt.}}{(CH_3COO)_2Cr} \downarrow + 2NaCl$$

$CrCl_2$ used in the above reaction is obtained by reducing $CrCl_3$ by nascent hydrogen obtained by the action of Zn on HCl. $CrCl_3$ is obtained by the action of HCl on $K_2Cr_2O_7$ in the absance of air.

$$K_2Cr_2O_7 + 14HCl \longrightarrow 2KCl + 2CrCl_3 + 7H_2O + 3Cl_2 \uparrow$$

$$CrCl_3 + H \text{ (From Zn/HCl)} \longrightarrow CrCl_2 + HCl$$

Properties and uses : $(CH_3COO)_2Cr$ forms red crystals which are fairly stable in acids and hence is used for the preparation of other Cr(+2) salts.

3. Chromic Oxide or Chromic Sesquioxide or Chrome Green, Cr_2O_3

Preparation : This oxide is used as a green pigment for oil paints and hence is also called *chrome green*. It is prepared by the reactions represented by the following chemical equations :

$$2Cr(OH_3) \xrightarrow{\Delta} Cr_2O_3 + 3H_2O$$

$$(NH_4)_2Cr_2O_7 \xrightarrow{\Delta} Cr_2O_3 + 4H_2O + N_2 \uparrow$$

$$K_2Cr_2O_7 + 2NH_4Cl \xrightarrow{\Delta} Cr_2O_3 + 2KCl + 4H_2O + N_2 \uparrow$$

$$Na_2Cr_2O_7 + S \xrightarrow{\text{Heat in an iron pot}} Cr_2O_3 + Na_2SO_4$$

$$4CrO_2Cl_2 \text{ (vapours)} \xrightarrow{\text{Pass vapour in red hot tube}} Cr_2O_3 + O_2 + 4Cl_2$$

Properties and uses : It is a green powder with very high melting point and is very refractory. It dissolves in fused borax or glass and gives a green colour and hence is used in tining glass and painting procelain. It is not decomposed by heat nor reduced by H. It is reduced to metal when heated with carbon (better with Al powder) at a high temperature. It is an amphoteric oxide because it forms salts with acids and chromites when fused with alkalies

$$Cr_2O_3 + 3H_2SO_4 \longrightarrow Cr_2(SO_4)_3 + 3H_2O$$

$$Cr_2O_3 + 2NaOH \longrightarrow 2NaCrO_2 + H_2O$$

When it is fused with an alkali in presence of air or an oxidising agent, a chromate is formed.

$$2Cr_2O_3 + 8NaOH + 3O_2 \longrightarrow 4Na_2CrO_4 + 4H_2O$$

If it is ignited too strongly, it becomes passive towards both acid and base. When fused with an oxide of MO type, it gives a crystalline compound of the composition, $MO.Cr_2O_3$. It is used (*i*) as a green pigment for oil, paints under the name *chrome-green*. *Guignet's green*, $Cr_2O_3.2H_2O$ is also used as a pigmant and is formed by fusing $K_2Cr_2O_7$ with boric acid and treating the resulting product with water. (*ii*) in tining glass and painting porcelain.

4. Potassium Chromic Sulphate or Chrome Alum

K_2SO_4. $Cr_2(SO_4)_3$. $24H_2O$

Preparation : It crystallises out when $K_2Cr_2O_7$ is reduced by sulphur dioxide in presence of sulphuric acid.

$$K_2Cr_2O_7 + H_2SO_4 + 3SO_2 \longrightarrow K_2SO_4 + Cr_2(SO_4)_3 + H_2O$$

On standing, beautiful dark purple octahedral crystals of chrome alum separate.

$$K_2SO_4 + Cr_2(SO_4)_3 + 24H_2O \longrightarrow \underbrace{K_2SO_4.Cr_2(SO_4)_3.24H_2O}_{\text{Chrome alum}}$$

Properties and uses : The crystals of chrome alum possess a deep violet colour and are isomorphous with other alums. It forms a violet solution in cold water but if heated to 70°C, the colour of the solution becomes green. It is used (*i*) as a mordant in dyeing and calico-printing. (*ii*) for tanning leather. (*iii*) in photography (during fixing) for hardening of the negatives.

5. Chromium Trioxide or Chromic Oxide, CrO_3

Preparation. This oxide is the anhydride of unknown chromic acid, H_2CrO_4 and hence is also called ***chromic acid anhydride***. It is obtained as deep red needle-like crystals when a well-cooled solution of $K_2Cr_2O_7$ is treated with conc. H_2SO_4.

$$K_2Cr_2O_7 + H_2SO_4 \text{ (cold)} \longrightarrow \underset{\substack{\text{Red}\\\text{crystals}}}{2CrO_3} + K_2SO_4 + H_2O$$

The crystals of CrO_3 are filtered by means of a Buchner funnel containing asbestos, washed with pure HNO_3 and dried in a current of warm air to remove the adhering HNO_3.

Properties : This compound forms deep red needlelike crystals which are deliquescent. It is highly poisonous and damages to the kidneys. When heated to 198°C it melts and gives a dark liquid and then, on further heating, O_2 gas is evolved.

$$4CrO_3\ (\textit{solid}) \xrightarrow{198^\circ} 4CrO_3 \text{ (liquid)} \xrightarrow{>198^\circ} 2Cr_2O_3 + 3O_2$$

It is soluble in water, giving a red solution. This red solution is strongly acidic and contains several chromic acids. The simplest of these acids is *chromic acid*. H_2CrO_4

$$CrO_3 + H_2O \rightleftarrows H_2CrO_4$$

It is a powerful oxidising agent and its oxidising action is based on the following equation.]

$$2CrO_3 \longrightarrow Cr_2O_3 + O$$

When CrO_3 oxidises any substance, the oxide itself is reduced to Cr(+3) salt. Thus it oxidises SO_2, acidified H_2S, $FeSO_4$, $SnCl_2$ as shown by the following equations :

$$2CrO_3 + 3SO_2 \longrightarrow Cr_2(SO_4)_3$$

$$2CrO_3 + 3H_2S + 3H_2SO_4 \longrightarrow Cr_2(SO_4)_3 + 6H_2O + S$$

$$2CrO_3 + 6FeSO_4 + 6H_2SO_4 \longrightarrow Cr_2(SO_4)_3 + 3Fe_2(SO_4)_3 + 6H_2O$$

$$2CrO_3 + 3SnCl_2 + 12HCl \longrightarrow 2CrCl_3 + 3SnCl_4 + 6H_2O$$

When CrO_3 is dissolved in calculated amount of NH_3, ammonium dichromate, $(NH_4)_2Cr_2O_7$ is formed.

$$2CrO_3 + 2NH_3 + H_2O \longrightarrow (NH_4)_2Cr_2O_7$$

Uses : CrO_3 in the form of $H_2CrO_4(K_2Cr_2O_7)$ and conc. H_2SO_4 is used as an oxidising agent in volumetric analysis and for cleaning glass apparatus. CrO_3 dissolved in glacial acetic acid is used as an oxidising agent and us a disinfectant to wash foul ulcers and ores.

6. Potassium Chromate, K_2CrO_4

Preparation : On commercial scale is prepared from chromite ore, $FeO.Cr_2O_3$. The ore is first concentrated by Wilfley table method (gravity separation method). The concentrated ore is mixed with excess of K_2CO_3 and roasted in a reverberatory furnace at a high temperature in the presence of free supply of air. By this operation the chromite ore, $FeO.Cr_2O_3$ is converted into soluble potassium chromate, K_2CrO_4

$$\underset{\text{Chromite ore}}{4[FeO.Cr_2O_3]} + 8K_2CO_2 + 7O_2 \text{ (From air)} \xrightarrow{\Delta} \underbrace{\underset{\text{Soluble}}{8K_2CrO_4} + \underset{\text{Insoluble}}{2Fe_2O_3}}_{\text{Roasted mass}} + 8CO_2 \uparrow$$

The roasted mass containing K_2CrO_4 and Fe_2O_3 is extracted with water when K_2CrO_4 goes into solution and Fe_2O_3 is left behind as a residue. In the process of roasting some quantity of CaO is also added so that the mixture of chromite ore and K_2CO_3 may become porous.

K_2CrO_4 can also be prepared from $K_2Cr_2O_7$. To the water solution of $K_2Cr_2O_7$ calculated amount of KOH or K_2CO_3 is added. On evaporating the resulting solution lemon-yellow crystals of anhydrous potassium chromate are obtained.

$$K_2Cr_2O_7 + 2KOH \longrightarrow 2K_2CrO_4 + H_2O$$

$$K_2Cr_2O_7 + K_2CO_3 \longrightarrow 2K_2CrO_4 + CO_2 \uparrow$$

Properties : It forms lemon-yellow crystals (m.p. = 968°C) which is isomorphous with K_2SO_4. It is highly soluble in water and its solution is yellow in colour and on the addition of acids, it becomes red due to the formation of dichromate.

$$2K_2CrO_4 + H_2SO_4 \longrightarrow K_2Cr_2O_7 + K_2SO_4 + H_2O$$

Conversely, a red dichromate solution turns yellow on the addition of alkali on account of the chromate formation.

$$K_2Cr_2O_7 + 2KOH \longrightarrow 2K_2CrO_4 + H_2O$$

On being heated above 670°C it changes to a *red modification* which melts at 970°C. An acidified solution of K_2CrO_4 acts as an oxidising agent.

$$2K_2CrO_4 + 5H_2SO_4 \longrightarrow 2K_2SO_4 + Cr_2(SO_4)_3 + 5H_2O + 3O$$

Thus it liberates I_2 from KI, oxidises H_2S to S, SO_2 to H_2SO_4, HCl to Cl_2 etc.

When potassium chromate solution is added to silver nitrate solution, brick red precipitate of *silver chromate*, Ag_2CrO_4 is formed. Soluble barium and lead salts give with potassium chromate yellow precipitates of barium and lead chromates, $BaCrO_4$ and $PbCrO_4$.

Uses : It is used (*i*) as a reagent in the laboratory (*ii*) as an indicator in Mohr's volumetric method of silver estimation (*iii*) as an inhibitor for corrosion.

7. Lead Chromate or Chrome Yellow, $PbCrO_4$

Preparation : This compound occurs naturally as *crocosite*. $PbCrO_4$ is obtained as a yellow precipitate by mixing an aqueous solution of K_2CrO_4 with that of $PbCH_3COO)_2$.

$$(CH_3COO)_2Pb + K_2CrO_4 \longrightarrow \underset{\text{Yellow ppt.}}{PbCrO_4} + 2CH_3COOK$$

Properties : It is a yellow solid and is least soluble of all the lead salts. It is precipitated in presence of ammonium acetate. It is insoluble in dilute but dissolves in concentrated HNO_3. It is also soluble in concentrated NaOH forming sodium plumbite and chromate.

$$PbCrO_4 + 4NaOH \longrightarrow \underset{\text{Sod. plumbite}}{Na_2PbO_2} + Na_2CrO_4 + 2H_2O$$

When $PbCrO_4$ is boiled with cold dilute solution of NaOH, *basic lead chromate*, $Pb(OH)_2.PbCrO_4$ (*chrome red*) is formed as red precipitate.

$$2PbCrO_4 + 2NaOH\ (\textit{cold and dilute solution}) \longrightarrow \underset{\text{Basic lead chromate (red ppt.)}}{Pb(OH)_2.PbCrO_4} + Na_2CrO_4$$

Uses : Lead chromate is used as a yellow pigment under the name of *chrome yellow, cologne yellow* or *lemon chrome.*

8. Basic Lead Chromate or Chrome Red, $PbCrO_4$ $Pb(OH)_2$.

As already stated, this compound is prepared by boiling $PbCrO_4$ with cold dilute solution of NaOH.

$$2PbCrO_4 + 2NaOH\ (\textit{cold and dilute solution}) \longrightarrow \underset{\text{Basic lead chromate (red ppt.)}}{Pb(OH)_2.PbCrO_4} + Na_2CrO_4$$

It is used as a red pigment under the name of *chrome red.*

9. Potassium Dichromate, $K_2Cr_2O_7$

Manufacture : $K_2Cr_2O_7$ is manufactured from chromite ore, $FeO.Cr_2O_3$ by a method which consists of four steps *viz.* (*i*) *Concentration of the chromite ore.* (*ii*) *Conversion of the concentrated chromite ore into* Na_2CrO_4 *(Roasting of the concentrated ore).* (*iii*) *Conversion of* Na_2CrO_4 into $Na_2Cr_2O_7$. For these three steps see "*Extraction of chromium metal from chromite ore*". (*iv*) *Conversion of* $Na_2Cr_2O_7$ *to* $K_2Cr_2O_7$. Hot solution of $Na_2Cr_2O_7$ obtained as above in step (*iii*) is treated with the calculated quantity of KCl whereby less soluble $K_2Cr_2O_7$ separates out from the solution on crystallisation.

$$\underset{}{Na_2Cr_2O_7} + 2KCl \longrightarrow \underset{\text{More soluble}}{2NaCl} + \underset{\text{Less soluble}}{K_2Cr_2O_7}.$$

Properties : $K_2Cr_2O_7$ forms orange-red prismatic crystals which melt at 396°C. It is moderately soluble in cold water and freely so in hot water. When heated to a white heat, it decomposes to give potassium chromate, K_2CrO_4 and O_2 gas is evolved.

$$4K_2Cr_2O_7 \xrightarrow{\Delta} 4K_2CrO_4 + 2Cr_2O_3 + O_2 \uparrow$$

When an alkali is added to an orange-red solution of $K_2Cr_2O_7$, a yellow solution results. The yellow colour of the solution is due to the formation of K_2CrO_4.

$$\underset{\text{Orange-red}}{K_2Cr_2O_7} + 2KOH \longrightarrow \underset{\text{Yellow}}{2K_2CrO_4} + H_2O$$

$$\underset{\text{Orange-red}}{K_2Cr_2O_7} + 2NaOH \longrightarrow \underset{\text{Yellow}}{K_2CrO_4} + H_2O + Na_2CrO_4$$

On acidifying, the yellow colour of K_2CrO_4 again changes to orange-red due to the reformation of $K_2Cr_2O_7$.

$$\underset{\text{Yellow}}{2K_2CrO_4} + H_2SO_4 \longrightarrow \underset{\text{Orange-red}}{K_2Cr_2O_7} + K_2SO_4 + H_2O$$

This interconversion is explained on the basis of the fact that in $K_2Cr_2O_7$ solution, orange-red $Cr_2O_7^{2-}$ ions are in equilibirium with yellow $Cr_2O_4^{2-}$ ions.

$$\underset{\text{Orange-red}}{Cr_2O_7^{2-}} + H_2O \rightleftarrows \underset{\text{Yellow}}{2CrO_4^{2-}} + 2H^+$$

On adding an alkali, the concentration of H^+ ions decreases and hence the above reaction proceeds in the forward direction producing yellow solution containing CrO_4^{2-} ions. On the other hand when an acid is added, the concentration of H^+ ions increases and hence the above reaction proceeds in the backward direction producing the orange solution containing $Cr_2O_7^{2-}$ ions. The above description makes it evident that since both $Cr_2O_7^{2-}$ and CrO_4^{2-} ions are present in a dichromate solution, $K_2Cr_2O_7$ will precipitate a salt of lower solubility product with Pb^{2+} or Ba^{2+} ions. Thus when $K_2Cr_2O_7$ is treated with $Pb(NO_3)_2$, $PbCrO_4$ and not $PbCr_2O_7$ is precipitated. This is because of the fact that $PbCrO_4$ has less solubility product than $PbCr_2O_7$.

$$K_2Cr_2O_7 + 2Pb(NO_3)_2 + H_2O \longrightarrow \underset{\text{ppt.}}{2PbCrO_4} + 2KNO_3 + 2HNO_3$$

With cold H_2SO_4, red crystals of chromium trioxide, CrO_3 are formed. When $K_2Cr_2O_7$ is heated with H_2SO_4, O_2 gas is evolved.

$$K_2Cr_2O_7 + H_2SO_4 \text{ (cold)} \longrightarrow \underset{\text{Red crystals}}{2CrO_3} + K_2SO_4 + H_2O$$

$$2K_2Cr_2O_7 + 8H_2SO_4 \xrightarrow{\Delta} 2K_2SO_4 + 2Cr_2(SO_4)_3 + 8H_2O + 3O_2$$

When solid $K_2Cr_2O_7$ is heated with conc. H_2SO_4 and a soluble chloride like NaCl, deep red vapours of chromyl chloride, CrO_2Cl_2 are produced.

$$K_2Cr_2O_7 + 4NaCl + 3H_2SO_4 \text{ (concentrated)} \xrightarrow{\Delta} \underset{\text{Red vapours}}{2CrO_2Cl_2} + K_2SO_4 + 2Na_2SO_4 + 3H_2O$$

$K_2Cr_2O_7$ is a powerful oxidising agent. In presence of dil. H_2SO_4, one molecule of this compound gives three atoms of oxygen as shown below :

$$K_2Cr_2O_7 + 4H_2SO_4 \text{ (dil)} \longrightarrow K_2SO_4 + Cr_2(SO_4)_3 + 4H_2O + 3O$$

Thus in presence of dil H_2SO_4, it liberates I_2 from KI solution, oxidises Fe(+2) salts to Fe(+3) salts, oxidises H_2S to S, oxidises SO_2 to H_2SO_4.

$$K_2Cr_2O_7 + 7H_2SO_4 + 6KI \longrightarrow 4K_2SO_4 + Cr_2(SO_4)_3 + 7H_2O + 3I_2$$

or $$Cr_2O_7^{2-} + 14H^+ + 6I^- \longrightarrow 2Cr^{3+} + 7H_2O + 3I_2$$

$$K_2Cr_2O_7 + 7H_2SO_4 + 6FeSO_4 \longrightarrow K_2SO_4 + Cr_2(SO_4)_3 + 7H_2O + 3Fe_2(SO_4)_3$$

or $$Cr_2O_7^{2-} + 14H^+ + 6Fe^{2+} \longrightarrow 2Cr^{3+} + 7H_2O + 6Fe^{3+}$$

$$K_2Cr_2O_7 + 4H_2SO_4 + 3H_2S \longrightarrow K_2SO_4 + Cr_2(SO_4)_3 + 7H_2O + 3S$$

or $$Cr_2O_7^{2-} + 8H^+ + 2H_2S \longrightarrow 2Cr^{3+} + 7H_2O + 3S$$

$$K_2Cr_2O_7 + H_2SO_4 + 3SO_2 \longrightarrow K_2SO_4 + Cr_2(SO_4)_3 + H_2O$$

or $$Cr_2O_7^{2-} + 2H^+ + 3SO_2 \longrightarrow 2Cr^{3+} + 3SO_4^{2-} + H_2O$$

When $K_2Cr_2O_7$ is heated with conc. HCl, Cl_2 is given off and chromic chloride ($CrCl_3$) is obtained.

$$K_2Cr_2O_7 + 14HCl \longrightarrow 2KCl + 2CrCl_3 + 7H_2O + 3Cl_2 \uparrow$$

Now when Zn metal is added to the above solution, nascent hydrogen is evolved which reduces $CrCl_3$ to blue chromous chloride, $CrCl_2$.

$$CrCl_3 + H \text{ (From Zn/HCl)} \longrightarrow CrCl_2 + HCl$$

Now when a saturated solution of CH_3COONa is added to the blue solution of $CrCl_2$ obtained as above, a red precipitate of chromous acetate, $(CH_3COO)_2Cr$ is obtained.

$$CrCl_2 + 2CH_3COONa \longrightarrow \underset{\text{Red ppt.}}{(CH_3COO)_2Cr \downarrow} + 2NaCl$$

Uses. Potassium dichromate finds extensive use in dyeing, calico-printing and chrome-tanning. It is used as an oxidising agent and in the preparation of various chromium compounds like Cr_2O_3, CrO_3, $K_2SO_4.Cr_3(SO_4)_3.24H_2O$, CrO_2Cl_2, K_2CrO_4, $CrCl_3$, $Cr(CH_3COO)_2$ etc. It is also used in photography. A mixture of $K_2Cr_2O_7$ and conc. H_2SO_4, usually known as *chromic acid mixture*, has strong oxidising properties and is used as a cleaning agent for glassware etc. It is also used in leather industry (*chrome tanning*).

10. Chromyl Chloride, CrO_2Cl_2

Preparation : When solid $K_2Cr_2O_7$ is heated with conc. H_2SO_4 and a soluble salt (*e.g.* NaCl), deep red vapours of CrO_2Cl_2 are obtained.

$$K_2Cr_2O_7 + 4NaCl + 3H_2SO_4 \text{ (concentrated)} \xrightarrow{\Delta} \underset{\text{Chromyl chloride (red vapours)}}{2CrO_2Cl_2} + K_2SO_4 + 2Na_2SO_4 + 3H_2O$$

Properties : It is a dark red liquid (b.pt. = 116.7°C and m.pt. = 16.5°C). It is miscible with organic solvents like CCl_4 and CS_2. It is photosensitive but otherwise it is stable. It is hydrolysed by H_2O, giving a yellow solution of H_2CrO_4.

$$CrO_2Cl_2 + 2H_2O \longrightarrow \underset{\text{Yellow solution}}{H_2CrO_4} + 2HCl$$

The solution of CrO_2Cl_2 gives a yellow precipitate of $PbCrO_4$ with $(CH_3COO)_2Pb$.

$$CrO_2Cl_2 + 2H_2O + (CH_3COO)_2Pb \longrightarrow \underset{\text{Yellow ppt.}}{PbCrO_4} + 2HCl + 2CH_3COOH$$

It is a powerful oxidising agent. A solution of CrO_2Cl_2 in CS_2 oxidises toluene to benzaldehyde (*Etard's reaction*). When CrO_2Cl_2 is added to a saturated solution of KCl, red crystals of *potassium chloro-chromate*, $KCrO_3Cl$ (*Petigot's salt*) are formed. With NaOH, it gives Na_2CrO_4.

$$CrO_2Cl_2 + KCl + H_2O \longrightarrow KCrO_3Cl + 2HCl$$

$$CrO_2Cl_2 + 4NaOH \longrightarrow Na_2CrO_4 + 2NaCl + 2H_2O$$

Uses : It is used : (*i*) as an intermediate product in the identification of Cl^- ions. (*ii*) as an excellent oxidising agent in organic chemistry,. *e.g.* its solution in CS_2 is used to oxidise toluene to benzaldehyde.

CrO_2Cl_2 is an acid chloride of chromic acid, $CrO_2(OH)_2$ in which OH groups have been replaced by Cl-atoms.

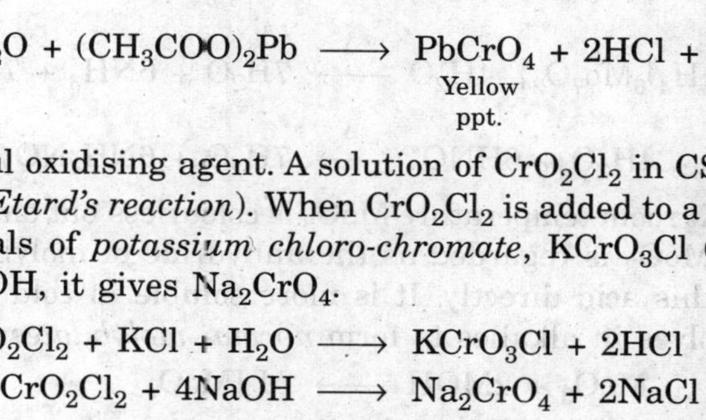

Compounds of Molybdenum

Different Oxidation States of Molybdenum in its Compounds

Molybdenum forms compounds in –2, 0, +2, +3, +4, +5 and +6 oxidation states. The electronic configuration of Mo^0 atom and of Mo^{2+}, Mo^{3+}, Mo^{4+}, Mo^{5+} and Mo^{6+} ions are : $Mo^0 \rightarrow 2, 8, 18, 4s^2p^6d^5, 5s^1$; $Mo^{2+} \rightarrow 2, 8, 18, 4s^2p^6d^4, 5s^0$; $Mo^{3+} \rightarrow 2, 8, 18, 4s^2p^6d^3, 5s^0$; $Mo^{4+} \rightarrow 2, 8, 18, 4s^2p^6d^2, 5s^0$; $Mo^{5+} \rightarrow 2, 8, 18, 4s^2p^6d^1, 5s^0$; $Mo^{6+} \rightarrow 2, 8, 18, 4s^2p^6d^0, 5s^0$. Compounds having Mo metal in +6 oxidation state are the most stable. Examples of compounds of Mo in different oxidation states are : *Mo (0) compounds* : $Mo(CO)_6$, $[Mo(py)_3(CO)_3]$; *Mo(+ 2) compounds* : $[Mo(diars)_2 X_2]$, $[Mo(CN)_6]^{4-}$, *Mo(+3) compounds* : Mo_2O_3, Mo_2S_3, $[Mo(H_2O)_3 Cl_3]^0$, $[MoCl_6]^{3-}$; *Mo(+4) compounds* : MoO_2, $MoCl_4$, $[Mo(NCS)_6]^{2-}$, MoS_2

; *Mo(+5) compounds* : Mo_2S_5, MoF_5. $MoOCl_3$, $MoCl_5(g)$, $(Mo\ (CN)_8]^{3-}$; *Mo(+6) compounds* : MoO_3, MoS_3, MoF_6, MoO_2F_2, $(MoO_4]^{2-}$, $[MoF_8]^{2-}$. Now let us discuss some important compounds of molybdenum.

1. Molybdenum Trioxide, MoO_3

Preparation : It is prepared : (*i*) by heating Mo metal or MoS_2 in air or O_2. (*ii*) by heating MoO_2 with HNO_3, (*iii*) by heating normal ammonium molybdate, $(NH_4)_2MoO_4$ alone or by beating ammonium paramolybdate, $(NH_4)_6Mo_7O_{24}.\ 4H_2O$ alone or with HNO_3.

$$2Mo + 3O_2 \longrightarrow 2MoO_3\ ;\ 2MoS_2 + 7O_2 \longrightarrow 2MoO_3 + 4SO_2$$

$$MoO_2 + 2HNO_3 \longrightarrow MoO_3 + H_2O + 2NO_2$$

$$(NH_4)_2MoO_4 \xrightarrow{\Delta} 2NH_3 + H_2O + MoO_3$$

$$(NH_4)_6Mo_7O_{24}.\ 4H_2O \xrightarrow{\Delta} 7H_2O + 6NH_3 + 7MoO_3$$

$$(NH_4)_6Mo_7O_{24}.\ 4H_2O + 6HNO_3 \xrightarrow{\Delta} 7H_2O + 6NH_4NO_3 + 7MoO_3$$

Properties : At room temperature MoO_3 is colourless but on heating it becomes yellow. Although MoO_3 is regarded as the anhydride of molybdic acid, H_2MoO_4, it does not form this acid directly. It is more soluble in cold water than in hot water. MoO_3 dissolves in alkalies to form *normal molybdates* of $M_2[MoO_4]$ type.

$$MoO_3 + 2MOH \longrightarrow M_2[MoO_4]$$

A solution of MoO_3 in HNO_3, when evaporated, gives a white crystalline powder of molybdic acid, H_2MoO_4. When heated with dry HCl at about 150-200°C, it gives a pale yellow water-soluble sublimate of the composition, $MoO_3.2HCl$ or $MoO(OH)_2Cl$. When MoO_3 is dissolved in an alkali sulphide and then the solution so obtained is acidified, MoS_3 is formed.

$$MoO_3 \xrightarrow{+\ Na_2S} Na_2MoS_4 \xrightarrow{+\ H^+} M_2S_3$$

On being reduced by H_2, it gives MoO_2 and then Mo metal.

$$MoO_3 + H_2 \xrightarrow{300\text{–}470°C} MoO_2 + H_2O;\ MoO_2 + 2H_2 \xrightarrow{600°C} Mo + 2H_2O$$

With hot conc. NH_4OH, normal ammonium molybdate, $(NH_4)_2MoO_4$ is formed.

$$MoO_3 + 2NH_4OH \longrightarrow (NH_4)_2MoO_4 + H_2O$$

Now when the solution containing $(NH_4)_2MoO_4$ is evaporated, crystals of *ammonium paramolybate*, $(NH_4)_6[Mo_7O_{24}]$ or $6NH_3.\ 7MoO_3.3H_2O$ are obtained.

$$7MoO_3 + 6NH_4OH \longrightarrow (NH_4)_6[Mo_7O_{24}] + 3H_2O$$

2. Molybdic Acid, $H_2MoO_4.\ H_2O$

It is obtained by treating ammonium molydate with HNO_3 and allowing the solution to crystallise in *cold*. It is a yellow crystalline powder slightly soluble in water, giving an acidic solution. This acid is much weaker than H_2CrO_4. It dissolves in alkalies, forming molybdates.

3. Ammonium Molybdate, $(NH_4)_2MoO_4$

Preparation : This compound is obtained by the treatment of MoO_3 with NH_4OH and evaporating the solution to crystallisation.

Properties : It is a white crystalline powder, soluble in water and insoluble in alcohol. $(NH_4)_2MoO_4$ easily loses NH_3 while $(NH_4)_6[Mo_7O_{24}]$ is quite stable in air. The solution of ammonium molybdate in dil. HNO_3 slowly deposits yellow α-molybdic acid, $H_2MoO_4 . H_2O$ which, on warming at 70°C with H_2O forms white β-molybdic acid, H_2MoO_4. The two acids have different X-ray spectra and vapour pressure. When a solution of ammonium molybdate and HCl is dialysed, colloidal molybdic acid is obtained. On heating the salt alone or with HNO_3, MoO_3, is obtained (see the preparation of MoO_3). When a phosphate or an arsenate is heated with conc. HNO_3 and $(NH_4)_2MoO_4$ solution, yellow precipitate of ammonium phosphomolybdate, $(NH_4)_3PO_4.12MoO_3$ or ammonium arsenomolybdate, $(NH_4)_3AsO_4. 12MoO_3$ is obtained.

$$Ca_3(PO_4)_2 + 6HNO_3 \longrightarrow 3Ca(NO_3)_2 + 2H_3PO_4$$

$$12(NH_4)_2MoO_4 + 21HNO_3 + H_3PO_4 \longrightarrow \underset{\text{Ammonium phospho-molybdate (yellow ppt.)}}{(NH_4)_3PO_4.12MoO_3} + 12H_2O + 21NH_4NO_3$$

$$12(NH_4)_2MoO_4 + 21HNO_3 + H_3AsO_4 \longrightarrow \underset{\text{Ammonium arseno-molybdate (yellow ppt.)}}{(NH_4)_3AsO_4.12MoO_3} + 12H_2O + 21NH_4NO_3$$

Uses (*i*) On the basis of the above reactions, $(NH_4)_2MoO_4$ is used in the laboratory for the detection and estimation PO_4^{3-} and AsO_4^{3-} ions. When As_2S_3 or As_2S_5 is boiled with conc. HNO_3, arsenic acid, H_3AsO_4 is produced which gives a yellow precipitate of ammonium arsenomolybdate, $(NH_4)_3AsO_4.12MoO_3$ with $(NH_4)_2MoO_4$.

$$As_2S_3 + 10HNO_3 \longrightarrow 2H_3AsO_4 + 10NO_2 + 2H_2O + 3S$$

$$As_2S_5 + 10HNO_3 \longrightarrow 2H_3AsO_4 + 10NO_2 + 2H_2O + 5S$$

$$H_3AsO_4 + 12(NH_4)_2\,MoO_4 + 21HNO_3 \longrightarrow \underset{\text{yellow ppt.}}{(NH_4)_3AsO_4.12MoO_3} + 21NH_4NO_3 + 12H_2O$$

(*ii*) $(NH_4)_3\ MoO_4$ is also used as a catalyst for dehydrogenation in petroleum and coal technology.

4. Molybdenum Blue

Preparation : It is obtained : (*i*) by treating molybdic acid or an acidic solution of a molybdate with a reducing agent like SO_2, H_2S, HI, hydrazine, glucose, zinc or molybdenum metal or (*ii*) by reducing MoO_3 in 25–75% H_2SO_4 with Cu^+ ions or, Sn^{2+} ions in presence of sunlight.

Properties : Molybdenum blue is a deep blue colloidal solution. The actual colour of the solution depends on the concentration of molybdenum, pH of the medium, nature of the reducing agent and temperature of the reaction. The solution resembles indigo blue in appearance. It is very soluble in water and also in some of the organic solvents. The exact structure of molybdenum blue is not known. It is supposed to be a mixture of hydrated oxides of molybdenum with hydroxyl groups. The mean oxidation state (O.5) of Mo atom is between 5 and 6. Some molybdenum blues are crystalline while others are amorphous. Examples are : (*i*) *Crystalline molybdenum blues* : $MoO_{2.0}(OH)$ (Mean O.S. of Mo = 5); $MoO_{2.6}.\ nH_2O$ (O.S. = 5.20); $MoO_{2.5}\ (OH)_{0.5}$ (O.S = 5.50); $MoO_{2.38}.H_2O$(O.S = 5.76) (*ii*) *Amorphous molybdenum blues* : $MoO_{2.75}.\ nH_2O$ (O.S. = 5.50), $MoO_{2.88}.nH_2O$(O.S. = 5.76). On heating molybdenum blue above 350°C. only H_2O molecules are lost while the OH groups remain attached. This may suggest that OH groups are involved in bridging of different groups.

Uses : Formation of molybdenum blue is utilised in the detection and calorimetric determination of Mo. Phosphorus is indirectly estimated by this method. Because of the colloidal nature of molybdenum blue, it is easily absorbed on vegetable and animal fibres and hence is used for dyeing those fibres.

Compounds of Tungsten

Different Oxidation States of Tungsten in Its Compounds

Tungsten forms compounds in 0 *e.g.* [($W(CO)_6$, $[W(CO)_5X]^-$ etc.], + 2 (*e.g.* WX_2, $[W(diars)\ (CO)_4I]^+$ etc.), +3 (*e.g.* $[W(diars)(CO)_3\ Br_2]^+$), +4 (*e.g.* WO_2, WCl_4, $[W(OH_2Cl_5]^{2-}$). +5 (*e.g.* WF_6^-) and +6 (*e.g.* WO_3, WX_6, WO_2Cl_2, WO_4^{2-} etc.) oxidation states. Compounds containing tungsten in + 6 oxidation state are the most stable. Important compounds are discussed below :

1. Tungsten Trioxide or Tungstic Oxide, WO_3

Preparation : It is obtained by the following methods :

$$H_2WO_4 \xrightarrow{\text{Dehydration}} WO_3 + H_2O$$

$$2W + 3O_2 \xrightarrow{\Delta\ (500°C)} 2WO_3$$

$$2WS_2 + 7O_2\ (\text{from air}) \xrightarrow{\text{Roasting}} 2WO_3 + 4SO_2$$

$$\underset{\text{Amm. tungstate}}{(NH_4)_2\ WO_4} \xrightarrow{\Delta} WO_3 + 2NH_3 + H_2O$$

$$Na_2WO_4 + 2HNO_3 \longrightarrow WO_3 + 2NaNO_3 + H_2O$$

WO_3 can also be obtained from wolframite ore, (Fe, Mn)WO_4 as discussed earlier.

Properties : It is a soft lemon-yellow coloured powder, but turns orange on heating. It melts at 1470° to a green liquid. It is insoluble in water and acids (excepting HCl).

$$WO_3 + 4HF \longrightarrow \underset{\text{Hydrofluo-tungstic acid}}{H_2WO_2F_4} + H_2O$$

It dissolves in alkalies to form tungstate.

$$WO_3 + 2NaOH \longrightarrow Na_2WO_4 + H_2O$$

When heated with H_2 it is reduced to W metal *via* W_2O_5 and WO_2.

$$\underset{\text{Yellow}}{WO_3} \xrightarrow{H_2} \underset{\text{Blue}}{W_2O_5} \xrightarrow{H_2} \underset{\text{Brown}}{WO_2} \xrightarrow{H_2} W$$

When heated with C, Mg etc., W metal is obtained directly. When heated with Cl_2 it gives volatile oxychloride, WO_2Cl_2 while with H_2S it gives WS_3

$$WO_3 + 3C \longrightarrow 3CO + W \; ; \; WO_3 + 3Mg \longrightarrow 3\,MgO + W$$

$$2WO_3 + 2Cl_2 \longrightarrow 2WO_2Cl_2 + O_2 \; ; \; WO_3 + 3H_2S \longrightarrow WS_3 + 3H_2O$$

Uses : It is used (*i*) as a mordant in dyeing and ceramic industry (*ii*) as a pigment in oil and water colours.

2. Tungstic Acid, H_2WO_4

Preparation : A precipitate of this acid may be obtained by treating a solution of any soluble tungstate with acid. The precipitate of the acid may be white ($H_2WO_4.H_2O$) or yellow (H_2WO_4) according as the precipitation has been made from cold solution or hot solution.

$$(NH_4)_2WO_4 + H_2SO_4 \longrightarrow \underset{\text{Yellow ppt.}}{H_2WO_4} + (NH_4)_2SO_4$$

Properties and uses : The hydrated acid ($H_2WO_4.H_2O$) is somewhat soluble in water but the anhydrous acid (H_2WO_4) is neither soluble in H_2O nor in any acid, excepting HF. With HF hydrofluotungstic acid, $H_2WO_2F_4$ is formed. With alkalies, H_2WO_4 forms tungstates. On heating WO_3 is obtained.

$$H_2WO_4 + 4HF \longrightarrow H_2WO_2F_4 + 2H_2O$$

$$H_2WO_4 + Na_2OH \longrightarrow Na_2WO_4 + 2H_2O$$

$$H_2WO_4 \xrightarrow{\Delta} WO_3 + H_2$$

The composition of various tungstic acids is as : *meta tungstic acid*—$WO_3.H_2O$ or H_2WO_4 *para tungstic acid*—$WO_3.2H_2O$ or $H_2WO_4.2H_2O$ or H_2WO_5; *ortho tungstic acid*—$WO_3.3H_2O$ or $H_2WO_4.2H_2O$ or H_6WO_6. Tungstic acid is used in ceramic industry and in making pigments.

3. Tungsten Bronzes, M_xWO_3

Preparation. These are the *non-stoichiometirc compounds* having a general formula, M_xWO_3(M = Li, Na, K etc. and $x < 1$). These are also called *metal-tungstate bronzes.* They are obtained by reducing alkali metal or alkaline earth metal tungstates by H_2, metallic Na, metallic W or electrically. These are insoluble in water, chemically inert, have metallic lustre and are conductors of electricity.

Properties and uses : These are highly coloured. Their colour depends on the value of x as shown below :

x =	*0.93*	*0.64*	*0.46*	*0.32*
Colour =	*Gole yellow*	*Orange*	*Red-violet*	*Blue-violet*

These are resistant to the action of acids excepting HF. In the presence of an alkali, these are oxidised by O_2. In this oxidation reaction W^{5+} in M_xWO_3 is oxidised to W^{5+} state. Tungsten bronzes are used as pigments.

4. Tungsten Halides

The known halides given by tungsten are : (*i*) *Fluorides* : WF_6. (*ii*) *Chlorides* : WCl_6, WCl_5, WCl_4, WCl_2. (*iii*) *Bromides* : WBr_6 (very unstable), WBr_5, WBr_2. (*iv*) *Iodides* : WI_4, WI_2. Of these WF_6 and WCl_6 are the most important.

WF_6 can be prepared by any of the following methods :

$$W + 3F_2 \longrightarrow WF_6$$

$$WCl_6 + 6HF \text{ (anhydrous)} \longrightarrow WF_6 + 6HCl$$

$$WCl_6 + 2AsF_3 \longrightarrow WF_6 + 2AsCl_3$$

$$WCl_6 + 3SbF_5 \longrightarrow WF_6 + 2SbF_3Cl_2$$

It is colourless or slightly yellow coloured gas. It is hydrolysed by H_2O, forming WO_3.

$$WF_6 + 3H_2O \longrightarrow WO_3 + 6HF$$

It is extremely volatile diamagnetic compound. It attacks all metals except Pt and Au. in presence of traces of moisture, it attacks glass because of the production of HF (due to hydrolysis). It behaves a Lewis acid and forms addition compounds with NH_3, pyridine (py) and methylamine ($CH_3.NH_2$) (*e.g.* $WF_6.4NH_3$, $WF_6.3py$, $WF_6.3CH_3NH_2$). With NaF it gives crystalline solid $WF_6.2NaF$ in which W is 8-coordinated. With $SbCl_6$ it gives WCl_6.

Tungesten hexachloride WCl_6 can be prepared by the following reactions :

$$W + 3Cl_2 \xrightarrow{300^\circ} WCl_6$$

$$WS_2 + 4Cl_2 \xrightarrow{\Delta} WCl_6 + S_2Cl_2$$

$$2WO_3 + 6Cl_2 \xrightarrow{\text{In } CCl_4 \text{ in a sealed tube at } 280^\circ} 2WCl_6 + 3O_2$$

Pure WCl_6 is steel blue coloured compound, slowly attacked by cold water, but rapidly by hot H_2O.

$$WCl_6 + 4H_2O \longrightarrow H_2WO_4 + 6HCl$$

Other reactions of WCl_6 are given below :

$$WCl_6 + 5HBr \longrightarrow WBr_5 + HCl + \tfrac{1}{2}Cl_2$$

$$WCl_6 + 2NH_3 \longrightarrow WN_2 + 6HCl$$

$$WCl_6 \xrightarrow{H_2} WCl_5 \longrightarrow WCl_4 \longrightarrow WCl_2$$

Compounds of Manganese

Different Oxidation States of Manganese in Its Compounds

Manganese forms compounds in –1, 0, +1, +2, +3, +4, +5, +6 and +7 oxidation states. The electronic configuration of Mn^0 atom and of Mn^+, Mn^{2+}, Mn^{3+}, Mn^{4+},

Mn^{5+}, Mn^{6+} and Mn^{7+} ions are : $Mn^0 \rightarrow 2, 8, 3s^2p^6d^5, 4s^2$; $Mn^+ \rightarrow 2, 8, 3s^2p^6d^5, 4s^1$; $Mn^{2+} \rightarrow 2, 8, 3s^2p^6d^5, 4s^0$; $Mn^{3+} \rightarrow 2, 8, 3s^2p^6d^4, 4s^0$; $Mn^{4+} \rightarrow 2, 8, 3s^2p^6d^3, 4s^0$; $Mn^{5+} \rightarrow 2, 8, 3s^2p^6d^2, 4s^0$; $Mn^{6+} \rightarrow 2, 8, 3s^2p^6d^1, 4s^0$; $Mn^{7+} \rightarrow 2, 8, 3s^2p^6d^0, 4s^0$. Compounds having Mn atom in +2 oxidation state are the most stable. Examples of compounds of Mn in different oxidation states are : *Mn(–1) compounds* : $[Mn(CO)_5]^-$; *Mn(o) compounds* : $Mn_2(CO)_{10}$; *Mn(+1) compounds* : $[Mn(CN)_6]^{5-}$, $[Mn(CO)_5]^+$; *Mn(+2) compounds* (called *manganous compounds*) : MnO, $MnCl_2$, $MnCO_3$, $MnSO_4$, $Mn(NO_3)_2$, $MnCl_4{}^{2-}$, $[Mn(H_2O)_4]SO_4.4H_2O$; *Mn(+3) compounds* (called *manganic compounds*) : Mn_2O_3, MnF_3, $Mn_2(SO_4)_3$ $[Mn(CN)_6]^{3-}$; *Mn(+4) compounds* : MnO_2, $Mn(SO_4)_2$, $[MnCl_6]^{2-}$; *Mn(+5) compounds* : $Na_3MnO_4.10H_2O$; *Mn(+6) compounds* : K_2MnO_4 ; *Mn(+7) compounds* : $KMnO_4$, Mn_2O_7. Out of the above compounds the oxides of Mn (*viz.* MnO, Mn_2O_3, Mn_3O_4. MnO_2 and Mn_2O_7) and potassium permanganate ($KMnO_4$) are the most important. These are discussed below.

1. Manganeous Oxide MnO

Preparation : This oxide is prepared by the following reactions :

$$MnO_2 + H_2 \xrightarrow{\Delta} MnO + H_2O; \quad Mn(COO)_2 \xrightarrow{\Delta} MnO + CO + O_2$$

Properties and uses : This is a grayish-green powder and insoluble in water. It is strongly basic and forms manganous salts [*i.e.* Mn(II) salts] with acids. It is quickly oxidised by air in the cold to Mn_2O_3. It has a NaCl lattice. It is used as an intermediate for the preparation of Mn(II) salts, as a nutrient in agriculture, and in welding.

2. Manganic Oxide, Mn_2O_3

Preparation : This compound is also called ***manganese sesquioxide.*** It can be obtained by the following reactions :

$$4MnO + O_2 \xrightarrow{\Delta} 2Mn_2O_3 \; ; \; 4MnO_2 \xrightarrow{\Delta} 2Mn_2O_3 + O_2$$

The *hydrous oxide*, $Mn_2O_3.H_2O$ is obtained as :

$$2Mn(OH)_2 + \tfrac{1}{2}O_2 \longrightarrow Mn_2O_3.H_2O$$

Properties : It is a black amorphous powder. Above 1000°C it decomposes into Mn_3O_4, liberating O_2. This oxide is weakly basic and dissolves in hot dil. acids to form Mn(II) salts and MnO_2.

$$6Mn_2O_3 \xrightarrow{> 1000°C} 4Mn_3O_4 + O_2$$

$$Mn_2O_3 + H_2SO_4 \longrightarrow MnSO_4 + MnO_2 + H_2O$$

$$Mn_2O_3 + 2HNO_3 \longrightarrow Mn(NO_3)_2 + MnO_2 + H_2O$$

Mn_2O_3 dissolves in conc. HCl, forming $MnCl_2$ with the evolution of Cl_2.

$$Mn_2O_3 + 6HCl \longrightarrow 2MnCl_2 + Cl_2 + 3H_2O$$

Structure : Mn_2O_3 has O = Mn—O—Mn = O structure which is supported by magnetic susceptibility. In this oxide Mn is in +3 oxidation state.

3. Manganese Tetroxide. Mn_3O_4

Preparation : It is also called ***mangano-manganic oxide*** or ***red oxide of manganese.*** It may be prepared by heating any oxide of Mn.

$$6MnO + O_2 \xrightarrow{\Delta} 2Mn_3O_4$$

$$3MnO_2 \xrightarrow{\text{Heat in the absence of air}} Mn_3O_4 + O_2$$

$$6Mn_2O_3 \xrightarrow[\text{(in air)}]{> 95°C} 4Mn_3O_4 + O_2$$

Properties : It is a brownish-red powder. It slowly dissolves in conc. cold H_2SO_4, forming a red solution containing $MnSO_4$ and $Mn_2(SO_4)_3$. With dil. HNO_3 it gives $Mn(NO_3)_2$ and MnO_2 while with HCl, $MnCl_2$ is formed and Cl_2 is evolved.

$$Mn_3O_4 + 4H_2SO_4 \longrightarrow MnSO_4 + Mn_3(SO_4)_3 + 4H_2O$$
$$Mn_3O_4 + 4HNO_3 \longrightarrow 2Mn(NO_3)_2 + MnO_2 + 2H_2O$$
$$Mn_3O_4 + 8HCl \longrightarrow 3MnCl_2 + Cl_2 + 4H_2O$$

Structure : Mn_3O_4 has the structure shown in the margin.

4. Magnanese Dioxide, MnO_2

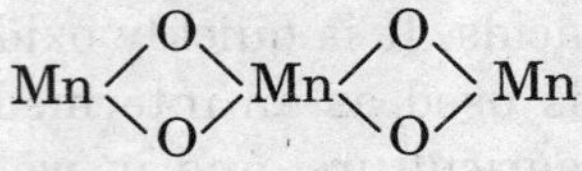

Structure of Mn_3O_4

Preparation. The pure MnO_2 is prepared by heating $Mn(NO_3)_2$ while the hydrated form, $MnO_2.H_2O$ (called maganous acid, H_2MnO_3) is prepared by the alkaline oxidation of Mn(II) salt by sodium hypochlorite or bleaching powder.

$$Mn(NO_3)_2 \xrightarrow{\Delta} MnO_2 + 2NO_2$$
$$Mn^{2+} + 2OH^- + ClO^- \longrightarrow MnO_2.H_2O + Cl^-$$

Properties : It is a black powder, insoluble in water and a good conductor of electricity. When heated above 530°C in air, it loses O_2 to form Mn_3O_4. It acts as a basic oxide towards cold acids like HCl, forming salts containing Mn in +4 oxidation state. With conc. H_2SO_4, $MnSO_4$ is formed and O_2 is evolved. HNO_3 has no action on MnO_2

$$3MnO_2 \xrightarrow{>530°C} Mn_3O_4 + O_2$$

$$MnO_2 + 4HCl \text{ (cold and concentrated)} \longrightarrow MnCl_4 + 2H_2O$$
$$2MnO_2 + 2H_2SO_4 \text{ (concentrated)} \longrightarrow 2MnSO_4 + 2H_2O + O_2$$

In the presence of excess of H^+ ions, MnO_2 acts as a strong oxidising agent. In showing this behaviour Mn^{4+} changes to Mn^{2+} ion.

$$\underset{(Mn\ =\ +4)}{MnO_2} + 4H^+ + 4e^- \longrightarrow \underset{(Mn\ =\ +2)}{Mn^{2+}} + 2H_2O$$

Thus in presence of hot dil. H_2SO_4, MnO_2 oxidises oxalic acid to CO_2, Fe(II) salt to Fe (III) salt and chloride to Cl_2

$$C_2O_4^{2-} + MnO_2 + 4H^+ \longrightarrow Mn^{2+} + 2H_2O + 2CO_2$$

$$2Fe^{2+} + MnO_2 + 4H^+ \longrightarrow 2Fe^{3+} + Mn^{2+} + 2H_2O$$

$$MnO_2 + 4H^+ + 2Cl^- \longrightarrow Mn^{2+} + 2H_2O + 2Cl_2$$

In presence of air or oxidising agents (*e.g.* KNO_3, $KClO_3$) MnO_2 reacts with fused KOH to give potassium manganate, K_2MnO_4.

$$2MnO_2 + 4KOH + O_2 \longrightarrow 2K_2MnO_4 + 2H_2O$$

Structure : Since in MnO_2, Mn is in +4 state, the structure of MnO_2 is O = Mn = O which has also been supported by magnetic susceptibility.

Uses : MnO_2 is used : (*i*) as a depolariser in Lechanehe cell and dry cells. In these cells MnO_2 oxidises H_2 liberated at the cathode and thus counteracts polarisation. (*ii*) as a decolouriser in glass. (*iii*) in forming, with ferric oxide, a dark brown glaze which is used in pottery. (*iv*) as a catalyst in the preparation of O_2 from $KClO_3$. (*v*) as a dies in paints and varnishes. (*vi*) in the preparation of Cl_2 in the laboratory. (*vii*) in dyestuff industry and in varnish manufacture. (*viii*) in the manufacture of metallic manganese, manganates, permanganates and a number of its salts.

5. Manganese Heptoxide, Mn_2O_7

Preparation : It is also called ***permanganic anhydride,*** since in water it gives a solution of *permanganic acid*, $HMnO_4$. When a solution containing concentrated sulphuric acid and potassium permanganate is cooled, and water is carefully added, a dark reddish brown liquid separates out, which is probably Mn_2O_7. It is very unstable.

$$2KMnO_4 + 2H_2SO_4 \longrightarrow (MnO_3)_2SO_4 + K_2SO_4 + 2H_2O$$

$$(MnO_3)_2SO_4 + H_2O \longrightarrow Mn_2O_7 + H_2SO_4$$

Properties : It is a dark-brown only liquid (density = 2.4) which solidifies on cooling to 5.9°C. It gives violet vapours at 40–50°C, but explodes violently on warming or in presence of organic matter giving MnO_2 and O_2.

$$2Mn_2O_7 \longrightarrow 4MnO_2 + 3O_2$$

It is acidic in character, since it dissolves in H_2O to give a violet solution of *permanganic add*, $HMnO_4$. It is, therefore, known as *permanganic anhydride*.

$$Mn_2O_7 + H_2O \longrightarrow 2HMnO_4$$

$$(cf. Cl_2O_7 + H_2O \longrightarrow 3HClO_4)$$

It is a powerful oxidising agent and organic matter like dry wood, paper etc. catches fire when dropped into it.

6. Potassium Permanganate, $KMnO_4$

This compound is the salt of permanganic acid, $HMnO_4$ which is an unstable acid and exists only in solution.

Preparation : It may be prepared by acidifying the solution of potassium manganate, K_2MnO_4 by H_2SO_4.

$$3K_2MnO_4 + 2H_2SO_4 \longrightarrow 2K_2SO_4 + 2KMnO_4 + MnO_2 \downarrow + 2H_2O$$

or

$$3MnO_4^{2-} + 4H^+ \longrightarrow 2MnO_4^- + MnO_2 \downarrow + 2H_2O$$

Manufacture. *On a large scale* it is prepared from the mineral pyrolusite, MnO_2. The finely powdered pyrolusite mineral is fused with KOH in presence of air or an oxidising agent like KNO_3. $KClO_3$ etc. This treatment gives us fused green coloured potassium manganate, K_2MnO_4.

$$2MnO_2 + 4KOH + O_2 \longrightarrow 2K_2MnO_4 + 2H_2O$$

or $$2MnO_2 + 4OH^- + O_2 \longrightarrow 2MnO_4^{2-} + 2H_2O$$

$$MnO_2 + 2KOH + KNO_3 \longrightarrow K_2MnO_4 + KNO_2 + H_2O$$

or $$MnO_2 + 2OH^- + NO_3^- \longrightarrow MnO_4^{2-} + NO_2^- + H_2O$$

$$3MnO_2 + 6KOH + KClO_3 \longrightarrow 3K_2MnO_4 + KCl + 3H_2O$$

or $$3MnO_2 + 6OH^- + ClO_3^- \longrightarrow 3MnO_4^{2-} + Cl^- + 3H_2O$$

The fused K_2MnO_4 obtained as above is extracted with water and the solution, after filtration, is converted into potassium permanganate ($KMnO_4$) by any of the following methods :

(*a*) Stadeler's process. In this method Cl_2 is passed through the green solution of K_2MnO_4 until it becomes purple due to the formation of K_2MnO_4. Here Cl_2 oxidises K_2MnO_4 into $KMnO_4$.

$$2K_2MnO_4 + Cl_2 \longrightarrow 2KCl + 2KMnO_4$$

or $$\underset{\text{Green}}{2MnO_4^{2-}} + Cl_2 \longrightarrow 2Cl^- + \underset{\text{Purple}}{2MnO_4^-}$$

(*b*) By the action of CO_2 or O_3. In this process CO_2 or O_3 is passed though the green solution of K_2MnO_4 until it becomes purple.

$$3K_2MnO_4 + 2H_2O + 4CO_2 \longrightarrow 2KMnO_4 + MnO_2 + 4KHCO_3$$

or $$\underset{\text{Green}}{3MnO_4^{2-}} + 2H_2O + 4CO_2 \longrightarrow \underset{\text{Purple}}{2MnO_4^-} + MnO_2 + 4HCO_3^-$$

$$2K_2MnO_4 + H_2O + O_3 \longrightarrow 2KMnO_4 + 2KOH + O_2$$

or $$\underset{\text{Green}}{2MnO_4^{2-}} + H_2O + O_3 \longrightarrow \underset{\text{Purple}}{2MnO_4^-} + 2OH^- + O_2$$

(*c*) Electrolytic oxidation process. Now-a-days the conversion of K_2MnO_4 into $KMnO_4$ is done by the electrolytic oxidation process in which K_2MnO_4 solution is placed in a cell in which a rotating iron wire-net cylinder acts as the anode and an inlet pipe made of iron acts as the cathode. Anode and cathode are properly insulated from each other. Anode is made to revolve, during electrolysis, mechanically by a revolving pulley. K_2MnO_4 solution is sent inside the cell through the inlet pipe and electric current is passed.

Reactions : On passing electric current H_2O ionises to give hydrogen (*at cathode*) and oxygen (*at anode*).

$$H_2O \longrightarrow 2H^+ \text{ (At cathode) } + O^{2-} \text{ (At anode)}$$

Oxygen liberated at anode oxidises manganate (MnO_4^{2-}) into permanganate (MnO_4^-) while hydrogen gets collected at the cathode.

$$2 \times [MnO_4^{2-} \longrightarrow MnO_4^- + e^-] \quad \text{(At anode) ...(i)}$$

$$2H^+ + 2e^- \longrightarrow H_2 \quad \text{(At cathode) ...(ii)}$$

On adding (*i*) and (*ii*), we get :

$$2MnO_4^{2-} + 2H^+ \longrightarrow 2MnO_4^- \text{ (At anode) } + H_2 \text{ (At cathode)}$$

or $$\underset{\text{Green solution}}{2K_2MnO_4} + 2H_2O \longrightarrow \underset{\text{Purple solution}}{2KMnO_4} + H_2 + 2KOH$$

The purple solution of $KMnO_4$ obtained as above is filtered through asbestos, concentrated and allowed to crystallise when $KMnO_4$ deposits as deep purple-red rhombic prisms.

Properties : Potassium permanganate forms dark purple lustrous crystals giving deep pink colour in solution. It is isomorphous with potassium perchlorate. Its solubility in water at 20° is only about 7% whilst it dissolves more at higher temperatures (25% at 63°C). It gives off O_2 at 200°C.

$$2KMnO_4 \longrightarrow K_2MnO_4 + MnO_2 + O_2$$

At red heat K_2MnO_4 formed as above is decomposed into permanganite, K_2MnO_3 and O_2.

$$2K_2MnO_4 \longrightarrow 2K_2MnO_3 + O_2$$

When an alkaline solution of $KMnO_4$ is heated, O_2 is evolved.

$$4KMnO_4 + 4KOH \longrightarrow 4K_2MnO_4 + 2H_2O + O_2$$

$$\text{or } 4MnO_4^- + 4OH^- \longrightarrow 4MnO_4^- + 2H_2O + O_2$$

With *well cooled* conc. H_2SO_4, it gives Mn_2O_7 which decomposes on warming.

$$2KMnO_4 + 2H_2SO_4 \longrightarrow Mn_2O_7 + 2KHSO_4 + H_2O$$

$$2Mn_2O_7 \longrightarrow 4MnO_2 + 3O_2$$

With *warmed* conc. H_2SO_4 oxygen gas is evolved.

$$4KMnO_4 + 6H_2SO_4 \longrightarrow 2K_2SO_4 + 4MnSO_4 + 6H_2O + 5O_2$$

When heated in a current of H_2, solid $KMnO_4$ gives KOH, MnO and steam.

$$2KMnO_4 + 5H_2 \longrightarrow 2KOH + 2MnO + 4H_2O$$

A mixture of $KMnO_4$ with sulphur and charcoal burns when heated, and explodes like a gun powder when rubbed. A well powdered mixture of $KMnO_4$ and oxalic acid catches five spontaneously in a few seconds. When a drop of glycerine is poured over powdered $KMnO_4$, the same thing happens. $KMnO_4$ is a powerful oxidising agent. Its oxidising property depends on whether the oxidation takes place in alkaline, neutral or acid solution.

(a) In alkaline solution, MnO_4^- ion (Mn = +7) is first reduced to MnO_4^{2-} ion (Mn = +6) and then to insoluble MnO_2 (Mn = +4). The solution first changes from purple (due to MnO_4^-) to green (due to MnO_4^{2-}) and finally it becomes colourless (MnO_2).

$$\underset{\substack{\text{Purple}\\(\text{Mn} = +7)}}{2KMnO_4} + 2KOH \longrightarrow \underset{\substack{\text{Green}\\(\text{Mn} = +6)}}{2K_2MnO_4} + H_2O + O \quad ...(i)$$

$$\underset{\substack{\text{Green}\\(\text{Mn} = +6)}}{2K_2MnO_4} + 2H_2O \longrightarrow \underset{\substack{\text{Colourless}\\(\text{Mn} = +4)}}{2MnO_2} + 4KOH + 2O \quad ...(ii)$$

On adding : $$\underset{\substack{\text{Purple}\\(\text{Mn} = +7)}}{2KMnO_4} + H_2O \xrightarrow{\text{Alkali}} \underset{\substack{\text{Colourless}\\(\text{Mn} = +4)}}{2MnO_2} + 2KOH + 3O \quad ...(iii)$$

This is the complete equation which represents the oxidising action of $KMnO_4$ in alkaline medium.

From equation *(iii)* it is evident that two molecules of $KMnO_4$ in alkaline solution give three oxygen atoms (*i.e. six equivalents*) and, therefore, *the equivalent weight of* $KMnO_4$ *as an oxidising agent in alkaline solution is one-third of its molecular weight.*

Equations *(i)*, *(ii)* and *(iii)* can also be written in the ionic forms as follows :

$$\underset{\substack{\text{Purple}\\(Mn\,=\,+7)}}{2MnO_4^-} + 2OH^- \longrightarrow \underset{\substack{\text{Green}\\(Mn\,=\,+6)}}{2MnO_4^{2-}} + H_2O + O$$

$$\underset{\substack{\text{Green}\\(Mn\,=\,+6)}}{2MnO_4^{2-}} + 2H_2O \longrightarrow \underset{\substack{\text{Colourless}\\(Mn\,=\,+4)}}{2MnO_2} + 4OH^- + 2O$$

On adding :

$$\underset{\substack{\text{Purple}\\(Mn\,=\,+7)}}{2MnO_4^-} + H_2O \xrightarrow{\text{Alkali}} \underset{\substack{\text{Colourless}\\(Mn\,=\,+4)}}{2MnO_2} + 2OH^- + 3O$$

or

$$MnO_4^- + 2H_2O + 3e^- \longrightarrow MnO_2 + 4OH^-$$

Thus in alkaline medium, $KMnO_4$ oxidises iodides (I^-) to iodates (IO^-_3), NH_3 to N_2 and nitrotoluene to nitrobenzoic acid.

$$2MnO^-_4 + H_2O + I^- \xrightarrow{\text{Alkali}} 2MnO_2 + 3OH^- + IO_3^-$$

$$2MnO^-_4 + C_6H_4\begin{matrix}\diagup NO_2\\ \diagdown CH_3\end{matrix} \longrightarrow 2MnO_2 + 2OH^- + C_6H_4\begin{matrix}\diagup NO_2\\ \diagdown COOH\end{matrix}$$

$$2MnO_4^- + 2NH_3 \longrightarrow MnO_2 + 2OH^- + N_2 + 2H_2O$$

(*b*) In neutral solution, $KMnO_4$ is directly reduced to MnO_2.

$$2KMnO_4 + H_2O \longrightarrow 2KOH + 2MnO_2 + 3O$$

or

$$2MnO_4^- + H_2O \longrightarrow 2OH^- + 2MnO_2 + 3O$$

or

$$MnO_4^- + 2H_2O + 3e^- \longrightarrow MnO_2 + OH^-$$

This equation is the same as for alkaline medium. Evidently, *in neutral medium as well, the equivalent weight of* $KMnO_4$ *is one-third of its molecular weight.* It is obvious from the equation representing the oxidising action of $KMnO_4$ in neutral medium that an alkali is generated in the reaction. This alkali makes the medium alkaline even if we start with the neutral solution. It is for this reason that the equations representing the oxidising property of $KMnO_4$ in alkaline and neutral media are the same. Thus in neutral medium, $KMnO_4$ oxidises Mn^{2+} salt to MnO_2, $Na_2S_2O_3$ to Na_2SO_4, H_2S to S.

$$2MnO_4^- + 2Mn^{2+} + 2H_2O \longrightarrow 5MnO_2 + 4H^+$$

$$8MnO_4^- + 3S_2O_3^{2-} + H_2O \longrightarrow 8MnO_2 + 6SO_4^{2-} + 2OH^-$$

$$2MnO_4^- + 4H_2S \longrightarrow 2MnS + S + SO_4^{2-} + 4H_2O$$

(c) In acidic medium (dil. H_2SO_4), MnO_4^- ion is reduced to Mn^{2+} ion and five O-atoms are rendered available from two MnO_4^- ions.

$$2KMnO_4 + 3H_2SO_4 \longrightarrow K_2SO_4 + 2MnSO_4 + 3H_2O + 5O$$

or
$$\underset{\substack{\text{Purple}\\(\text{Mn} = +7)}}{2MnO_4^-} + 6H^+ \longrightarrow \underset{\substack{\text{Colourless}\\(\text{Mn} = +2)}}{2Mn^{2+}} + 3H_2O + 5O$$

The reduction of MnO_4^- ion into Mn^{2+} ion as shown above can also be represented by the following ionic equation :

$$\underset{\substack{\text{Purple}\\(\text{Mn} = +7)}}{MnO_4^-} + 8H^+ + 5e^- \longrightarrow \underset{\substack{\text{Colourless}\\(\text{Mn} = +2)}}{Mn^{2+}} + 4H_2O$$

It is obvious that *in acid solution the equivalent weight of $KMnO_4$ is one-fifth of its molecular weight.*

In acidic medium $KMnO_4$ oxidises Fe(II) salts to Fe(III) salts, oxalic acid to CO_2, H_2S to S, SO_2 to H_2SO_4, sulphites to sulphates, KI to I_2, H_2O_2 to O_2 halogen acids to halogens.

$$MnO_4^- + 8H^+ + 5Fe^{2+} \longrightarrow Mn^{2+} + 5Fe^{3+} + 4H_2O$$

$$2MnO_4^- + 16H^+ + 5C_2O_4^{2-} \longrightarrow 2Mn^{2+} + 10CO_2 + 8H_2O$$

$$2MnO_4^- + 16H^+ + 3S^{2-} \longrightarrow 2Mn^{2+} + 5S + 8H_2O$$

$$2MnO_4^- + 2H_2O + 5SO_2 \longrightarrow 2Mn^{2+} + 5SO_4^{2-} + 4H^+$$

$$2MnO_4^- + 6H^+ + 5SO_3^{2-} \longrightarrow 2Mn^{2+} + 5SO_4^{2-} + 3H_2O$$

$$2MnO_4^- + 16H^+ + 10I^- \longrightarrow 2Mn^{2+} + 5I_2 + 8H_2O$$

$$2MnO_4^- + 6H^+ + 5H_2O_2 \longrightarrow 2Mn^{2+} + 5O_2 + 8H_2O$$

$$2MnO_4^- + 16H^+ + 10X^- \longrightarrow 2Mn^{2+} + 5X_2 + 8H_2O$$

Uses : $KMnO_4$ is used (*i*) as an oxidising agent in the laboratory as well as in industry; (*ii*) in volumetric estimations of ferrous salts, oxalic acid, H_2O_2 etc.; (*iii*) as a disinfectant and a germicide on account of its oxidising property; (*iv*) in qualitative analysis for detecting halides and oxalates. (*v*) in the manufature of saccharin, benzoic acid, acetaldelyde etc.; (*vi*) for bleaching waxes; (*vii*) for the detection of double bond in organic compounds. For this purpose alkaline $KMnO_4$ solution under the name *Bayer's reagent* is used. (*viii*) as a source of reactive form of MnO_2 which is used in place of pyrolusite in the manufacture of dry cells.

Compounds of Iron

Different Oxidation States of Iron in Its Compounds

The most important compounds of iron are those in which Fe is in +2 and +3 oxidation state. Compounds containing Fe in +2 (divalent) and +3 (trivalent) oxidation state are called *ferrous* and *ferric* compounds respectively. Compounds containing Fe in –2, 0, +1, +4 and +6 oxidation state are also known, although their number is very little. Examples are : *compound containing Fe (–2)* : $[Fe(CO)_4]^{2-}$; *compounds containing Fe(0)* : $Fe(CO)_5$. $Fe_2(CO)_9$, $Fe_3(CO)_{12}$; *compounds containing Fe (+1)* : $[Fe^+(NO^+)]^{2+}SO_4^{2-}$; *compounds containing Fe(+4)* :

$[Fe(diars)_2Cl_2]^{2+}$; *compound containing Fe* (+5) : $[FeO_4]^{3-}$; *compound containing Fe* (+6) : $[FeO_4]^{2-}$. Let us discuss some important compounds of iron.

1. Ferrous Sulphate or Green Vitriol, $FeSO_4.7H_2O$

Preparation : This is the most important ferrous salt. It is also called ***green vitriol.*** We also call it as ***copperas*** because it is a by-product of the hydrometallurgy of copper. In the laboratory it is prepared :

(*i*) *By dissolving iron in dil. H_2SO_4, filtering, evaporating and crystallising*

$$Fe + H_2SO_4 \longrightarrow FeSO_4 + H_2$$

(*iii*) *From Kipp's waste.* Kipp's waste (or residue) contains $FeSO_4$ also with free H_2SO_4. The waste is heated with a small quantity of scrap iron which reacts with excess of H_2SO_4 to give $FeSO_4$ and nascent hydrogen. This nascent hydrogen reduces any ferric iron, if present, to ferrous state. On allowing the hot solution to stand, crystals of $FeSO_4.7H_2O$ separate out.

Manufacture. On a large scale it is obtained by the slow oxidation of iron pyrites in the, presence of air and moisture. The pyrites are exposed to air in big heaps and the following reaction takes place :

$$2FeS_2 + 2H_2O + 7O_2 \longrightarrow 2FeSO_4 + 2H_2SO_4$$

The free H_2SO_4 is removed by the addition of scrap iron which also reduces any ferric sulphate formed. . On erystallisation we get green coloured crystals containing seven molecules of water of crystallisation. These crystals are monoclinic and are isomorphous with $MgSO_4.7H_2O$. Ferrous sulphate containing 6, 5. 3, 2, 1 molecules and no water of crystallisation is also known.

Properties (*i*) Ferrous sulphate forms pale green monoclinic crystals of heptahydrate, $FeSO_4.7H_2O$ which is isomorphous with the corresponding heptahydrates of cobalt; nickel, chromium, manganese, magnesium, zinc etc. *i.e.* with $MSO_4.7H_2O$ where M = Co, Ni, Cr, Mn, Mg, Zn etc. Crystalline hydrates having 6, 5, 4, 3 and 2 water mole. cules are also known. $FeSO_4.5H_2O$ is isomorphous with blue vitriol, $CuSO_4.5H_2O$. Anhydrous salt (*i.e.* $FeSO_4$) is white.

(*ii*) *Action of heat.* When heated to redness, $FeSO_4$ and $FeSO_4.7H_2O$ both decompose giving a residue of Fe_2O_3 which is used as a pigment under the name of *rogue* or *venetian red.*

$$2FeSO_4 \xrightarrow{\Delta} Fe_2O_3 + SO_2 + SO_3$$

$$2[FeSO_4.7H_2O] \xrightarrow{\Delta} Fe_2O_3 + H_2SO_4 + SO_2 + 13H_2O$$

(*iii*) *Reducing properties.* it is an active reducing agent, *e.g.*

(*a*) *It decolorises acidified $KMnO_4$ solution*

$$2KMnO_4 + 10FeSO_4 + 8H_2SO_4 \longrightarrow K_2SO_4 + 2MnSO_4 + 5Fe_2(SO_4)_3 + 8H_2O$$

(*b*) *It turns acidified $K_2Cr_2O_7$ green*

$$K_2Cr_2O_7 + 7H_2SO_4 + 6FeSO_4 \longrightarrow K_2SO_4 + Cr_2(SO_4)_3 + 3Fe_2(SO_4)_3 + 7H_2O$$

(*c*) *It reduces NO_2 to NO which then gives nitrosoferrous sulphate, Fe(NO)SO$_4$*

$$2FeSO_4 + H_2SO_4 + NO_2 \longrightarrow Fe_2(SO_4)_3 + H_2O + NO$$

$$FeSO_4 + NO \rightleftarrows \underset{\text{Nitrosoferrous sulphate}}{Fe(NO)SO_4}$$

(*d*) *It reduces silver and gold salts to metal.*

$$Ag^+ + Fe^{2+} \longrightarrow Ag \downarrow + Fe^{3+}$$

$$Au^{3+} + 3Fe^{2+} \longrightarrow Au \downarrow + 3Fe^{3+}$$

(*e*) *It reduces $HgCl_2$ to Hg_2Cl_2*

$$2HgCl_2 + 2Fe^{2+} \longrightarrow Hg_2Cl_2 + 2Fe^{3+} + 2Cl^-$$

(*iv*) *Formation of addition compound with NO.* A cold solution of ferrous sulphate absorbs NO forming dark brown nitroso ferrous sulphate, $Fe(NO)SO_4$ which is destroyed by the action of heat.

$$FeSO_4 + NO \rightleftarrows \underset{\text{Nitroso ferrous sulphate}}{Fe(NO)SO_4}$$

(*v*) *Action of air.* When exposed to air, the green crystals of $FeSO_4$ effloresce and turn brown due to the formation of basic ferric sulphate, $Fe(OH)SO_4$. Its formation is due to the surface oxidation of $FeSO_4$ to $Fe(OH)SO_4$.

$$\underset{\text{Green}}{4FeSO_4} + 2H_2O + O_2 \longrightarrow \underset{\text{Brown}}{4Fe(OH)SO_4}$$

(*vi*) *Formation of double salts.* With sulphates of alkali metals it readily forms double salts with the composition, $R_2SO_4.FeSO_4.\ 6H_2O$ where, R = an alkali metal or NH_4^+ radical. Out of these double salts, *ferrous ammonium sulphate* (called *Mohr's salt*), $(NH_4)_2SO_4.\ FeSO_4.6H_2O$ is of great use and has been discussed below.

(*vii*) *Acid character.* It is readily soluble in H_2O, the solution is acidic in nature duo to its hydrolysis

$$FeSO_4 + 2HOH \rightleftarrows Fe(OH)_2 + H_2SO_4$$

The solution readily passes into ferris state by atmospheric oxidation.

$$4FeSO_4 + 2H_2SO_4 + O_2 \longrightarrow 2Fe_2(SO_4)_3 + 2H_2O$$

(*viii*) *Action of KCN.* When excess of KCN is added to ferrous sulphate solution, potassium ferrocyanide, $K_4[Fe(CN)_6]$ is formed.

$$FeSO_4 + 2KCN \longrightarrow \underset{\text{Brown ppt.}}{Fe(CN)_2} + K_2SO_4$$

$$Fe(CN)_2 + 4KCN \longrightarrow K_4[Fe(CN)_6]$$

$$FeSO_4 + 6KCN \longrightarrow K_4[Fe_2(CN)_6] + K_2SO_4$$

Uses. It is used (*i*) in the manufacture of writing ink, ferrous ammonium sulphate (*Mohr's salt*), iron alum, ferric oxide (*rouge*) etc. (*ii*) as a mordant in dyeing. (*iii*) as a weed-killer in agriculture. (*iv*) as a reducing agent. (*v*) as a reagent in the laboratory. (*vi*) for clarifying water.

2. Ferrous Ammonium Sulphate (Mohr's Salt), $(NH_4)_2SO_4.FeSO_4.6H_2O$

This compound is also called *Mohr's salt* and is a double salt. Ferrous ammonium sulphate is obtained by mixing ferrous sulphate and ammonium sulphate in molecular proportions and crystallising the double salt from the saturated solution of the mixture. A few drops of H_2SO_4 and a little iron wire is added before concentration to prevent oxidation of ferrous sulphate to ferric sulphate. It forms pale green crystals which do not easily effloresce. Its solution is also not as readily oxidised as that of ferrous sulphate. It is, therefore. employed in volumetric analysis in place of ferrous sulphate in preference to ferrous sulphate.

3. Potassium Ferrocyanide, $K_4[Fe(CN)_6]$

Preparation : Its IUPAC name is potassium hexacyanoferrate (II). On a small scale, it is prepared by mixing a solution of ferrous sulphate with that of potassium cyanide until a small amount of the precipitated ferrous cyanide remains undissolved. The solution on evaporation followed by cooling yields crystals of potassium ferrocyanide

$$FeSO_4 + 2KCN \longrightarrow Fe(CN)_2 + K_2SO_4$$
$$Fe(CN)_2 + 4KCN \longrightarrow K_4[Fe(CN)_6]$$

Manufacture *(a)*. On a large scale, it is prepared from coal gas which contains 0.2 percent of hydrogen cyanide as an impurity. The coal gas when allowed to pass through a washer containing ferrous sulphate and potassium hydroxide yields potassium ferrocyanide. The following reactions take place :

$$HCN + KOH \longrightarrow KCN + H_2O$$
$$6KCN + FeSO_4 \longrightarrow K_4Fe(CN)_6 + K_2SO_4$$

(b) On a large scale, it is also obtained from coal gas by another method. This gas, when allowed to pass through hydrated iron oxide [a mixture of $Fe(OH)_2$ and $Fe(OH)_3$], gets purified by absorbing hydrogen cyanide on hydrated iron oxide and forms ferric ferro cyanide, also known as *Prussian's blue*, $Fe_7C_{18}N_{18}$. The following reactions take place :

$$Fe(OH)_2 + 2HCN \longrightarrow Fe(CN)_2 + 2H_2O$$
$$Fe(OH)_3 + 3HCN \longrightarrow Fe(CN)_3 + 3H_2O$$
$$3Fe(CN)_2 + 4Fe(CN)_3 \longrightarrow \underset{\text{Prussian blue}}{Fe_4[Fe(CN)_6]_3} \text{ or } Fe_7C_{18}N_{18}$$

Prussian blue, formed is treated with $Ca(OH)_2$ when calcium ferrocyanide is formed.

$$Fe_7C_{18}N_{18} + 6Ca(OH)_2 \longrightarrow \underset{\text{soluble}}{3Ca_2[Fe(CN)_6]} + \underset{\text{ppt.}}{4Fe(OH)_3}$$

Calcium ferrocyanide is extracted with water and then treated with potassium carbonate solution when potassium ferrocyanide is obtained.

$$\underset{\text{cal. ferro cyanide}}{Ca_2[Fe(CN)_6]} + K_2CO_3 \longrightarrow 2CaCO_3 + K_4[Fe(CN)_6]$$

Calcium carbonate is removed by filtration and the filtrate when evaporated yields the crystals of potassium ferrocyanide.

Properties. *(i)* Potassium ferrocyanide forms lemon yellow crystals which are associated with 3 molecules of water of crystilization. Thus, the molecular formula is $K_4[Fe(CN)_6].3H_2O$.

(ii) Action of heat. When heated to 42°C, trihydrate loses its three molecules of water of crystallization. On further heating, it decomposes to yield iron carbide, potassium cyanide and nitrogen.

$$K_4[Fe(CN)_6] \longrightarrow 4KCN + FeC_2 + N_2$$

(*iii*) *Action of H_2SO_4.* When it is boiled with dil. sulphuric acid, hydrogen cyanide is evolved.

$$K_4[Fe(CN)_6] + 3H_2SO_4 \longrightarrow FeSO_4 + 3K_2SO_4 + 6HCN$$

When treated with conc. H_2SO_4, it yields carbon monoxide.

$$K_4[Fe(CN)_6] + 6H_2SO_4 + 6H_2O \longrightarrow 2K_2SO_4 + 3(NH_4)_2SO_4 + 6CO + FeSO_4$$

(*iv*) *Reducing property.* It is a reducing agent and is oxidized by Cl_2, O_3, H_2O_2 or acidified $KMnO_4$ to potassium ferricyanide.

$$2K_4[Fe(CN)_6] + Cl_2 \longrightarrow 2KCl + 2K_3[Fe(CN)_6]$$

$$2K_4[Fe(CN)_6] + O_3 + H_2O \longrightarrow 2KOH + 2K_3[Fe(CN)_6] + O_2$$

(*v*) *Action of metallic salts.* (*a*) With zinc salts, it yields a white ppt. of zinc ferrocyanide.

$$K_4[Fe(CN)_6] + 2ZnSO_4 \longrightarrow \underset{\text{Zinc ferrocyanide}}{Zn_2[Fe(CN)_6]} + 2K_2SO_4$$

(*b*) With copper salts, it yields a red brown ppt. of cupric ferrocyanide.

$$2CuSO_4 + K_4[Fe(CN)_6] \longrightarrow Cu_2[Fe(CN)_6] + 2K_2SO_4$$

(*c*) With ferric salts, it produces Prussian blue.

$$Fe^{3+} + K_4Fe(CN)_6 \longrightarrow \underset{\text{Soluble Prussian blue}}{KFe[Fe^{2+}(CN)_6]} + 3K^+$$

Uses : (*i*) It is used as a reagent in the laboratory for testing Fe^{3+}, Cu^{2+} and Zn^{2+} ions. (*ii*) It is used for making blue prints. (*iii*) It is need for making Prussian blue pigment. (*iv*) It is used for the preparation of pure CO and for the manufacture f $K_3[Fe(CN)_6]$.

4. Ferric Chloride, $FeCl_3$

Preparation : *Anhydrous ferric chloride* is prepared by passing dry Cl_2 over heated iron filing when the salt sublimes at 315°C in a receiver as iron-black crystals with a green iridescence.

$$2Fe + 3Cl_2 \longrightarrow 2FeCl_3$$

In solution it is obtained by dissolving Fe_2O_3 or $Fe(OH)_3$ in HCl or by saturating a solution of $FeCl_2$ with Cl_2.

$$Fe_2O_3 + 6HCl \longrightarrow 2FeCl_3 + 3H_2O$$

$$Fe(OH)_3 + 3HCl \longrightarrow FeCl_3 + 3H_2O$$

$$2FeCl_2 + Cl_2 \longrightarrow 2FeCl_3$$

The solution on evaporation and cooling deposits yellow crystals of the hydrate, $FeCl_3.6H_2O$ or $Fe_2Cl_6.12H_2O$. Other hydrates that separate from the aqueous solution contain Fe_2Cl_6 with 7, 5 and 4 molecules of water.

Properties. (*i*) $FeCl_3$ is not an ionic compound but is a covalent compound as is indicated by its volatility and solubility in non-polar solvents like ether and alcohol. However, it dissolves in H_2O to give Fe^{3+} and Cl^- ions.

(*ii*) Anhydrous $FeCl_3$ forms iron black crystals with a green iridescence. These crystals sublime at 283°C and its vapour density at 450°C corresponds to the formula Fe_2Cl_6. At high temperatures dissociation of Fe_2Cl_6 occurs into $FeCl_3$

molecules so that its vapour density progressively decreases on heating and corresponds to the formula $FeCl_3$ at 750°C. At higher temperatures the vapour density further decreases, probably due to the dissociation of $FeCl_3$ into $FeCl_2$ and Cl_2.

$$Fe_2Cl_6 \underset{}{\overset{750°}{\rightleftarrows}} 2FeCl_3 \overset{\text{Above } 750°C}{\rightleftarrows} 2FeCl_2 + Cl_2 \uparrow$$

(*iii*) *Formation of double salts.* It forms a number of garnet- red double salts with other chlorides, *e.g.* $FeCl_3.2KCl.H_2O$; $FeCl_3. 2NH_4Cl.H_2O$; $FeCl_3.MgCl_2.H_2O$ etc.

(*iv*) *Oxidising properties.* It is an oxidising agent and oxidises various substances. In these reactions it itself is reduced to $FeCl_2$. Thus it oxidises $SnCl_2$ to $SnCl_4$, SO_2 to H_2SO_4, H_2S to S, nascent hydrogen to HCl and liberates I_2 from KI solution.

$$2FeCl_3 + SnCl_2 \longrightarrow SnCl_4 + 2FeCl_2$$
$$SO_2 + 2FeCl_3 + 2H_2O \longrightarrow H_2SO_4 + 2HCl + 2FeCl_2$$
$$H_2S + 2FeCl_3 \longrightarrow S + 2HCl + 2FeCl_2$$
$$FeCl_3 + H \longrightarrow FeCl_2 + HCl$$
$$2FeCl_3 + 2KI \longrightarrow 2KCl + I_2 + 2FeCl_2$$

(*v*) *Action of $Na_2S_2O_3$.* $FeCl_3$ reacts with sodium thiosulphate, $Na_2S_2O_3$ to form sodium tetrathionate, $Na_2S_4O_6$

$$2FeCl_3 + 2Na_2S_2O_3 \longrightarrow 2NaCl + Na_2S_4O_6 + 2FeCl_2$$

(*vi*) It is hygroscopic and dissolves readily in H_2O, the solution is strongly acidic due to hydrolysis

$$FeCl_3 + 3HOH \rightleftharpoons Fe(OH)_3 + HCl$$

(*vii*) *Action of NH_3 and NOCl.* It absorbs NH_3 and nitrosyl chloride, NOCl to form $FeCl_3.6NH_3$ and $FeCl_3.NOCl$ respectively.

(*viii*) *Formation of various hydrates.* $FeCl_3$ solution deposits crystal of various hydrates at various temperatures $Fe_2Cl_6.12H_2O$ (37°); $Fe_2Cl_6.7H_2O$ (32.5°); $Fe_2Cl_6.5H_2O$ (56°); $Fe_2Cl_6.2H_2O$ (73.50°)

(*ix*) *Action of heat.* On heating $FeCl_3.6H_2O$ is changed into Fe_2O_3, and H_2O and HCl are liberted.

$$2[FeCl_3.6H_2O] \overset{\Delta}{\longrightarrow} Fe_2O_3 + 9H_2O + 6HCl$$

This reaction clearly shows that anhydrous $FeCl_3$ cannot be prepared by simply heating the hydrated ferric chloride, $FeCl_3.6H_2O$.

Structure. Ferric chloride is a covalent compound and a *dimerised molecule*, Fe_2Cl_6 which has the structure shown below.

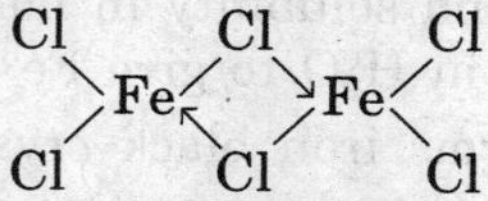

Structure of Fe_2Cl_6

Uses. (*i*) Ferric chloride solution coagulotes blood and hence is used as a styptic, *i.e.*, in stopping bleeding. (*ii*) It is also used as an oxidising agent in organic dye stuff industry. (*iii*) It is also used as a mordant in dyeing.

5. Potassium Ferricyanide, $K_3[Fe(CN)]_6$

Preparation : Its IUPAC name is potassium hexacyanoferrate (III). It is prepared by the oxidation of potassium ferrocyanide solution by chlorine, hydrogen peroxide or nitric acid. The ferricyanide is separated by repeated fractional crystallization.

$$2K_4[Fe(CN)_6] + Cl_2 \longrightarrow 2K_3[Fe(CN)_6] + 2KCl$$

$$2K_4[Fe(CN)_6] + H_2O_2 \longrightarrow 2K_3[Fe(CN)_6] + 2KOH$$

It is also prepared by the electrolytic oxidation of slightly alkaline solution of potassium ferrocyanide.

$$2[Fe(CN)_6]^{4-} + 2H_2O \longrightarrow 2[Fe(CN)_6]^{3-} + H_2 + 2OH^-$$

Properties : It forms dark red crystals. It is comparatively less stable than potassium ferrocyanide. As it undergoes easy reduction to potassium ferrocyanide in alkaline medium. it acts as a useful oxidizing agent in alkaline medium.

$$2K_3Fe(CN)_6 + 2KOH \longrightarrow 2K_4Fe(CN)_6 + H_2O + O$$

Thus, it oxidizes hydrogen sulphide to sulphur, litharge into lead dioxide, chromic oxide into potassium chromate and sulphur dioxide into sulphuric acid.

$$2K_3[Fe(CN)_6] + 2KOH + H_2S \longrightarrow 2K_4[Fe(CN)_6] + 2H_2O + S$$

$$2K_3[Fe(CN)_6] + 2KOH + PbO \longrightarrow 2K_4[Fe(CN)_6] + H_2O + PbO_2$$

$$6K_3[Fe(CN)_6] + 10KOH + Cr_2O_3 \longrightarrow 6K_4[Fe(CN)_6] + 5H_2O + 2K_2CrO_4$$

$$2K_3[Fe(CN)_6] + 2KOH + SO_2 \longrightarrow 2K_4[Fe(CN)_6] + H_2SO_4$$

The alkaline solution, of potassium ferricyanide is readily reduced to potassium ferrocyanide by H_2O_2, sodium amalgam or hydrazine.

$$2K_3[Fe(CN)_6] + H_2O_2 + 2KOH \longrightarrow 2K_4[Fe(CN)_6] + 2H_2O + O_2$$

$$4K_3[Fe(CN)_6] + 4KOH + N_2H_4 \longrightarrow 4K_4[Fe(CN)_6] + 4H_2O + N_2$$

When potassium ferricyanide is treated with ferrous sulphate solution, ferrous ion is first oxidized to ferric icn while ferricyanide ion is reduced to ferrocyanide ion. Then, the ferric ions react with ferrocynanide ions to form patassium ferric ferrocyanide (Turnbull's blue).

$$Fe^{2+} + \underset{\text{Ferricyanide icn}}{[Fe(CN)_6]^{3-}} \longrightarrow Fe^{3+} + \underset{\text{Ferrocyanide ion}}{[Fe(CN)_6]^{4-}}$$

$$K^+ + Fe^{3+} + [Fe(CN)_6]^{4-} \longrightarrow \underset{\substack{\text{Potassium ferric ferrocyanide} \\ \text{(Turnbull's blue)}}}{K^+Fe^{3+}[Fe^{2+}(CN)_6]^-}$$

Uses : $K_3[Fe(CN)_6]$ is used (*i*) in making blue prints. (*ii*) as an external indicator in the titration of Fe^{2+} with $K_2Cr_2O_7$. (*iii*) in qualitative analysis to distinguish ferrous salt from a ferric salt. With ferrous salts, it yields a deep blue ppt. of Turnbull's blue while with ferric salts it yields deep brown ppt.

Compounds of Cobalt

Like iron the important oxidation states exhibited by cobalt in its compounds are +2 and +3. The compounds in +2 and +3 oxidation states are called *cobaltous* and *cobaltic compounds* respectively. Important compounds of Co are discussed below :

1. Cobalt Nitrate, $Co(NO_3)_2.6H_2O$

It is obtained by dissolving the metal, its oxide or carbonate in nitric acid. The anhydrous salt is made by the action of nitrogen pentoxide on the hydrated salt. The hexahydrate forms pink slightly deliquescent crystals, highly soluble in water. On heating it decomposes to give Co_3O_4. It is used as a reagent in blow-pipe analysis 'and in the preparation of cobalt pigments and ceremics.

2. Complex Compounds

Cobalt forms complexes in –1,0 +1, +2, +3 and +4 oxidation states. Complexes in +3 state are important. All known Co(+3) complexes are *octahedral* except the one *viz.* heteropoly tungetate which is *tetrahedral.* Octahedral complexes like $[CoF_3]^{3-}$ which contain weak F^- ligands are *high-spin complexes* (*outer-orbital–sp^3d^2* hybridisation) and, therefore, show *paramagnetic character corresponding to the presence of four unpaired electrons.* Other octahedral complexes are *low-spin* (*inner-orbital–d^2sp^3* hybridisation) and, hence, are *diamagnetic due to the presence of all paired electrons.* Important complex compounds of Co(+3) are discussed below. It may be noted that the coordination number of Co(+3) in these complexes is six.

(*i*) Sodium cobaltinitrite or sodium hexanitro cobaltate (III), $Na_3[Co(NO_2)_6]$. It is prepared by mixing well cooled solutions of $Co(NO_3)_2$ and $NaNO_2$ and acidifying the mixture with glacial CH_3COOH. In this reaction HNO_2 produced by the interaction between $NaNO_2$ and CH_3COOH oxidises $Co(NO_2)_2$ to $Co(NO_2)_3$ which combines with $NaNO_2$ to give $Na_3[Co(NO_2)_6]$. Thus this reaction takes place through the following steps :

$$[NaNO_2 + CH_3COOH \longrightarrow HNO_2 + CH_3COONa] \times 2$$
$$2NaNO_2 + Co(NO_3)_2 \longrightarrow Co(NO_2)_2 + 2NaNO_3$$
$$Co(NO_2)_2 + 2HNO_2 \longrightarrow Co(NO_2)_3 + H_2O + NO$$
$$Co(NO_2)_3 + 3NaNO_2 \longrightarrow Na_3[Co(NO_2)_6]$$

$$Co(NO_3)_2 + 7NaNO_2 + 2CH_3COOH \longrightarrow Na_3[Co(NO_2)_6] + 2CH_3COONa + 2NaNO_3 + H_2O + NO$$

or
$$Co^{2+} + 7NO_2^- + 2H^+ \longrightarrow [Co(NO_2)_6]^{3-} + H_2O + NO$$

On treating the solution with alcohol, sodium cobaltinitrite is separated out as a yellow crystalline powder. The complex is fairly soluble in water. With aqueous solution containing soluble ammonium or potassium salts a yellow precipitate of slightly soluble complex is formed.

$$[Co(NO_2)_6]^{3-} + 3K^+ \longrightarrow \underset{\text{yellow precipitate}}{K_3[Co(NO_2)_6]}$$

This reaction is used for the detection and estimation of K^+ and NH_4^+ ions.

***(ii)* Potassium cobaltinitrite or potassium hexanitro cobaltate (III) or Fisher's salt, $K_3[Co(NO_2)_6]$.** It is prepared when KNO_2 is added to $Co(NO_3)_2$ solution acidified with CH_3COOH.

$$Co(NO_3)_2 + 7KNO_2 + 2CH_3COOH \longrightarrow K_3[Co(NO_2)_6] + 2CH_3COOK + 2KNO_3 + H_2O + NO$$

It is insoluble in water. It is a bright yellow powder which may be washed with potassium acetate solution and alcohol. $[Co(NO_2)_6]^{3-}$ ion is decomposed by alkalies to form hydroxides while strong acids decompose it into oxides of nitrogen. It is used (*i*) as a yellow pigment (*Indian yellow or cobalt yellow*) (*ii*) in painting porcelain blue.

***(iii)* Hexammine cobaltic chloride, $[Co(NH_3)_6]Cl_3$.** When a vigorous stream of air is drawn for several hours through a solution of a cobalt (II) chloride, $CoCl_2$, containing ammonia, ammonium chloride and some activated charcoal, good yield of hexamine cobalt (III) chloride is obtained.

$$4CoCl_2 + 4NH_4Cl + 20NH_3 + O_2 \longrightarrow 4[Co(NH_3)_6]Cl_3 + 2H_2O$$

In the absence of charcoal, replacement usually occurs to give, for example, $[Co(NH_3)_5Cl]^{2+}$ and $[Co(NH_3)_4(CO_3)]^+$. This forms orange-yellow crystals. Conductance measurement indicates the presence of four ions in its solution, *i.e.*, one cobalt and three chloride ions (ammonia molecules are netural). This is in accordance with the dissociation :

$$[Co(NH_3)_6]Cl_3 \rightleftarrows [Co(NH_3)_6]^{3+} + 3Cl^-$$

Compounds of Nickel

Different Oxidation States of Nickel in Its Compounds

Nickel forms compounds in 0, +2, +3 and +4 oxidation states. Examples are *Ni(0) compounds* : $Ni(CO)_4$, $Ni(PF_3)_4$, $[Ni(CN)_4]^{4-}$; *Ni(+2) compounds* (called *nicklous compounds*) : NiO, $NiSO_4$, $[Ni(CN)_4]^{2-}$, $[Ni(dmg)_2]^0$; *Ni(+3) compounds* : Ni_2O_3, $[NiBr_3(PEt_3)_2]^0$; *Ni(+4) compounds* : NiO_2. $K_2[NiF_6]$. Compounds of Ni in +2 oxidation state are the most stable. Some compounds of Ni are discussed below :

1. Nickel Sulphate, $NiSO_4.7H_2O$

This compound forms heptahydrate rhombic crystals, isomorphous with cobalt and ferrous sulphates. Nickel sulphate is prepared by dissolving its hydroxide or carbonate in dilute sulphuric acid or by heating nickel nitrate, $Ni(NO_3)_2.6H_2O$ with concentrated sulphuric acid. If a saturated solution of nickel sulphate is heated to 54°, monoclinic crystals of $NiSO_4.6H_2O$ are obtained. The crystals are green. When heated above 280°C, yellow anhydrous $NiSO_4$ is formed. If anhydrous nickel sulphate is dissolved in a concentrated solution of ammonia, an unstable dark blue salt, tetrammine nickel sulphate, $Ni(NH_3)_4SO_4.2H_2O$ is formed similar to cuprammonium salts. Besides this complex salt, nickel sulphate also forms a double salt with ammonium sulphate, known as nickel ammonium sulphate $NiSO_4.(NH_4)_2SO_4.6H_2O$, which is extensively used in nickel plating.

2. Nickel Ammonium Sulphate, $NiSO_4.(NH_4)_2SO_4.6H_2O$

This compound is obtained by mixing, the solution of $NiSO_4.7H_2O$ with that of $(NH_4)_2SO_4$ and then evaporating the solution to crystallisation. The compound forms bluish green crystals. This compound is used for nickel electroplating.

3. Bis-(dimethyl glyoximato) Nickel (II), $Ni(dmg)_2$

It is an important complex compound of Ni(+2). In this complex dmg^- ion acts as a bidentate ligand. dmg^- is the abbreviation used for dimethylglyoxime ion which is obtained by removing one H-atom from the oxime group (> N—OH) present in dimethylglyoxime (abbreviated as Hdmg).

When dimethylglyoxime is added to an alkaline solution of a Ni^{2+} salt, red precipitate of *bis*-(dimethyl glyoximato) nickel (II), $Ni(dmg)_2$ is obtained.

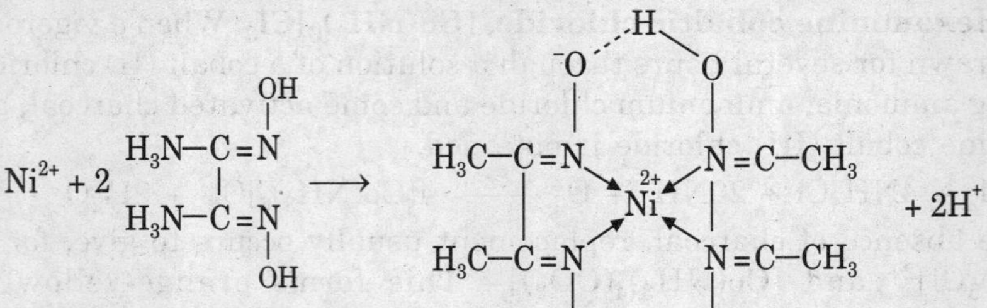

Dimethyl glyoxime, Hdmg

Bis-(dimethylglyoximato) nickel (II), $Ni(dmg)_2$ (red ppt.)

It is a chelate complex. It is electrically neutral and covalent in character. Its covalent character accounts for its low solubility in water. It may be seen from the structure that the co-ordination of dmg^- ion (ligand) to Ni^{2+} ion occurs through both its N-atoms making two 5-membered rings. Further two 6-membered rings are also formed by H-bonding between NOH and ON. H-bonding stabilises the structure of the complex. The complex is diamagnetic and has square planar shape. It is an inner-orbital complex. The formation of $Ni(dmg)_2$ is used in detecting as well as estimating nickel.

Compounds of Platinum

Different Oxidation States of Platinum

Platinum chiefly forms two series of compounds—*platinous* and *platinic* corresponding to +2 and +4 oxidation states of platinum respectively. The compounds in +1 (*e.g.* PtCl), +3 ($PtCl_3$), +5(PtF_5) and +6(PtO_3) oxidation states are also known but these are rare. Here we shall discuss only platinic compounds which are more important than the platinous compounds.

1. Platinic Chloride, $PtCl_4$. It is formed when chloroplatinic acid, H_2PtCl_6 is heated at 370°C in chlorine.

$$H_2PtC_6 \longrightarrow PtCl_4 + 2HCl$$

It forms reddish-brown slightly hygroscopic crystalline mass. It dissolves in water giving yellowish-red solution which is strongly acidic and seems to contain a complex acid, $H_2[PtCl_4(OH)_2]$ or $PtCl_4.2H_2O$, because it forms silver salt, $Ag_2[PtCl_4(OH)_2]$ and on electrolysis the platinum is discharged at the anode as a part of the anion.

$$H_2[PtCl_4(OH)_2] \rightleftarrows 2H^+ + [PtCl_4(OH)_2]^{2-}$$

$PtCl_4$ liberates I_2 from KI and gives H_2PtCl_6 with HCl.

$$PtCl_4 + 4KI \longrightarrow PtI_2 + I_2 + 4KCl$$

$$PtCl_4 + 2HCl \longrightarrow H_2PtCl_6$$

With KOH, $Pt(OH)_4$ is precipitated and with KCN it gives potassium platinocyanide, $K_2[Pt(CN)_4]$.

$$PtCl_4 + 4KOH \longrightarrow Pt(OH)_4 \downarrow + 4KCl$$

$$PtCl_4 + 6KCN \longrightarrow K_2[Pt(CN)_4] + 4KCl + (CN)_2 \downarrow$$

Platinic chloride is used for the preparation of potassium chioroplinate which is used in photography.

2. Complex Compounds

Pt forms complex compounds in +2 and +4 oxidation states. Important examples of such complex are (*i*) *Platinous ammines* like $[Pt(NH_3)_4]Cl_2$. $[Pt(NH_3)_3Cl]$ Cl, $[Pt(NH_3)_2Cl_2]^0$, $K[Pt(NH_3)Cl_3]$ (*ii*) *Tetrachloro complexes* like $H_2[PtCl_4]$ (*iii*) *Tetracyano complexes* like $K_2[Pt(CN)_4].3H_2O$, $Ba[Pt(CN)_4].4H_2O$. The majority of Pt(+2) complexes are four coordinate. They are diamagnetic with dsp^2 hybridisation and have square planar geometry. These are quite inert. Important examples of complex compounds formed by Pt(+4) are : (*i*) *Platinic ammines* like $[Pt(NH_3)_6]Cl_4$, $[Pt(NH_3)_5Cl]Cl_3$, $[Pt(NH_3)4Cl_2]Cl_2$, $[Pt(NH_3)_3Cl_3]Cl$, $[Pt(NH_3)_2Cl_4]^0$, $K[Pt(NH_3)Cl_5]$ (*ii*) *Hexachloro complexes* like H_2 $[PtCl_6]$, $(NH_4)_2[PtCl_6]$. Complexes of Pt(+4) are remarkably stable and kinetically inert. All these are *octahedral* resulting from d^2sp^3 hybridisation (*inner orbital*). Since Pt(+4) is a d^6 system, all the electrons are paired and hence the complexes are *diamagnetic*. Now let us discuss some of the above complex compounds in detail :

(*i*) Chloroplatinous acid, $H_2[PtCl_4]$. In this complex compound Pt is in +2 oxidation state. Here the coordination number of Pt is 4. It is obtained by dissolving PCl_2 in HCl or by the reduction of H_2PtCl_6 with SO_2.

$$PtCl_2 + 2HCl \longrightarrow H_2PtCl_4$$

$$H_2PtCl_6 + 2H_2O + SO_2 \longrightarrow H_2PtCl_4 + H_2SO_4 + 2HCl$$

The acid does not crystallise but forms a dark-brown amorphous solid on evaporation. Its salts are called *chloroplatinites*. Potassium chloro platinite, K_2PtCl_4 may be better prepared by reducing chloroplatinate, K_2PtCl_6 by cuprous chloride.

$$K_2PtCl_6 + Cu_2Cl_2 \longrightarrow 2CuCl_2 + K_2PtCl_4$$

The salt forms dark red crystals and is used in platinum printing in photography. The corresponding ammonium, silver or lead salts are also known.

(*ii*) Potassium platinocyanide, $K_2[Pt(CN)_4].3H_2O$. It is obtained as yellow needles by mixing solutions of $PtCl_2$ and KCN in proper proportions and evaporating the solution to crystallisation.

$$PtCl_2 + 4KCN \longrightarrow K_2[Pt(CN)_4] + 2KCl$$

***(iii)* Barium platinocyanide, $Ba[Pt(CN)_4].4H_2O$.** It is prepared by warming $H_2[PtCl_6]$ solution with baryta water, $Ba(OH)_2$ and HCN and passing in SO_2 until the solution becomes colourless.

$$H_2[PtCl_6] + 5Ba(OH)_2 + 4HCN + SO_2 \longrightarrow Ba[Pt(CN)_4] + BaSO_4 + 3BaCl_2 + 8H_2O$$

$BaSO_4$ is filtered off and the filtrate is crystallised to get Ba $[Pt(CN)_4].4H_2O$. It is a lemon-yellow powder which is soluble in water. It is used for fluorescent screens in X-ray work.

***(iv)* Chloroplatinic acid, $H_2[PtCl_6]$. Preparation :** It is an important complex compound in which platinum metal is in +4 oxidation state and the co-ordination number of Pt is 6. It is best prepared by the action of aqua-regia on platinum.

$$\underbrace{2HNO_3 + 6HCl}_{\text{Aqua-regia}} + Pt \longrightarrow H_2PtCl_6 + 4H_2O + 4NOCl$$

The solution is evaporated to remove excess of acid and on cooling reddish brown needles of $H_2PtCl_2.6H_2O$ are left behind.

Properties. (*a*) It forms reddish-brown deliquescent crystals which dissolve in water to give a bright yellow coloured solution. It is a strong dibasic acid and ionises as :

$$H_2PtCl_6 \rightleftarrows 2H^+ + [PtCl_6]^{2-}$$

(*b*) *Action of heat.* When heated to 370°C in Cl_2 or at 165° in H_2, $PtCl_4$ is formed. At 390°C in Cl_2, $PtCl_3$ is formed while at 580°C in Cl_2 it gives $PtCl_2$.

$$H_2PtCl_6 \xrightarrow[\text{}]{\text{Heat in } Cl_2 \text{ at } 370°C} PtCl_4 + 2HCl$$

(*c*) *Electrolysis.* When its solution is electrolysed, metallic platinum is deposited on the cathode as a result of the reducing action of the liberated hydrogen.

(*d*) *Reduction.* When reduced with Zn or formic acid (HCOOH) finely divided powder of *platinum black* is obtained, and when reduced with SO_2, chloroplatinous acid, H_2PtCl_4 is formed.

$$H_2PtCl_6 + 2HCOOH \longrightarrow \underset{\text{platinum black}}{Pt} + 6HCl + 2CO_2$$

$$H_2PtCl_6 + SO_2 + 2H_2O \longrightarrow H_2SO_4 + 2HCl + H_2PtCl_4$$

(*e*) *Formation of chloroplatinates.* It forms stable salts which are called *chloroplatinates*. Important examples are $Li_2PtCl_6.6H_2O$, $Na_2PtCl_6.6H_2O$, K_2PtCl_2, $(NH_4)_2PtCl_2$, Ag_2PtCl_6 etc. Out of these chloroplatinates ammonium chloroplatinate, $(NH_4)_2PtCl_2$ is the most important and is discussed on the next page.

(*f*) *Action of bases.* With a number of bases (*e.g.* $C_6H_5NH_2$), chloroplatinic acid forms stable crystalline compound and hence it is used in their estimations.

$$2C_6H_5NH_2 + H_2PtCl_6 \longrightarrow [C_6H_5NH_2]_2H_2PtCl_6$$

(*g*) *Action of H_2S.* When H_2S is passed into a hot solution of H_2PtCl_6, platinic sulphide (PtS_2) is obtained as a black precipitate.

$$H_2PtCl_6 + 2H_2S \longrightarrow PtS_2 + 6HCl$$

(*h*) *Action of KI.* When hot solution of H_2PtCl_6 is treated with KI, platinic iodide (PtI_4) is formed.

$$H_2PtCl_6 + 4KI \longrightarrow PtI_4 + 4KCl + 2HCl$$

(*i*) *Conversion of $H_2[PtCl_6]$ into PtO_2.* When the solution of $H_2[PtCl_6]$ is treated with a caustic alkali, alkali metal chloroplatinate, $M_2[PtCl_6]$ is formed. The action of CH_3COOH on this alkali metal chloroplatinate gives $H_2[Pt(OH)_6]$ which on heating gives PtO_2.

$$H_2[Pt(OH)_6] \xrightarrow{\Delta} 4H_2O + PtO_2$$

(*j*) *Action of baryta water, HCN and SO_2.* When H_2PtCl_6 is warmed with baryta water, $Ba(OH)_2$ and HCN and SO_2 is passed in until the solution becomes colourless, barium platinocyanide, $Ba[Pt(CN)_4]$ is formed.

$$H_2PtCl_6 + 5Ba(OH)_2 + 4HCN - SO_2 \longrightarrow \underset{\text{solution}}{Ba[Pt(CN)_4]} + \underset{\text{ppt.}}{BaSO_4} + 3BaCl_2 + 8H_2O$$

$BaSO_4$ is filtered off and the filtrate is crystallised to get the crystals of $Ba[Pt(CN)_4]$.

Uses. H_2PtCl_6 is used : (*i*) for the determination of molecular weight of bases (*ii*) for preparing spongy Pt and many platinum compounds like $PtCl_4$ etc. (*iii*) for the gravimetric estimation of organic bases (*iv*) as a tanning agent in photography.

(*v*) Ammonium chloroplatinate, $(NH_4)_2[PtCl_6]$. Preparation : It is an important salt of chloroplatinic acid, H_2PtCl_6. It is also called *ammonium platinicchloride.* It is prepared by adding NH_4Cl to a solution of H_2PtCl_6 or by treating H_2PtCl_6 with NH_3.

$$H_2[Pt\ Cl_6] + 2NH_4Cl \longrightarrow (NH_4)_2[PtCl_6] + 2HCl$$
$$H_2[Pt\ Cl_6] + 2NH_3 \longrightarrow (NH_4)_2[PtCl_6]$$

Properties. (*a*) it is a yellow substance which is sparingly soluble in H_2O.

(*b*) *Action of heat.* On heating *spongy platinum* is obtained.

$$(NH_4)_2Pt\ Cl_6 \longrightarrow \underset{\substack{\text{Spongy}\\\text{platinum}}}{Pt} + 2NH_3 + 2HCl + 2Cl_2 \uparrow$$

(*c*) *Action of liq. NH_3.* When $(NH_4)_2[PtCl_6]$ is treated with liq. NH_3, hexammino platinic chloride, $[Pt(NH_3)_6Cl_4$ is obtained

$$(NH_4)_2[PtCl_6] + 6NH_3 \longrightarrow [Pt(NH_3)_6]Cl_4 + 2NH_4Cl.$$

(*d*) *Action of hydrazine hydrochloride, $N_2H_4.2HCl$.* When hydrazine hydrochloride is added to an ammoniacal solution of $(NH_4)_2[PtCl_6]$, ammonium chloroplatinate is reduced to metallic platinum some of which gets deposited on the inside of the tube as a *platinum mirror.*

$$(NH_4)_2[Pt\ Cl_6] + N_2H_4.2HCl + 6NH_3 \longrightarrow Pt + N_2 + 8NH_4Cl$$

Uses. The formation of $(NH_4)_2[PtCl_6]$, followed by its decomposition, forms the basis of the *refining of platinum metal.*

Compounds of Copper

Copper forms two types of compounds : (*i*) *Copper (+1) or cuprous compounds* in which copper is *monovalent,* and (*ii*) *Copper (+2) or cupric compounds* in which copper is *divalent.* When cupric or cuprous is not stated in the case of a salt of copper, cupric is implied *e.g.* copper sulphate implies cupric sulphate. Most of the *cuprous compounds are usually colourless and diamagnatic* as 3*d* orbital is completely filled ($Cu^+ \rightarrow 3d^{10}$). Cu_2O, Cu_2S etc. which are red and black are exceptions. These are generally insoluble in water. The soluble compounds of Cu(+1) are *unstable in aqueous solution, since they disproportionate* to Cu^{2+} and Cu.

$$2Cu^+ \longrightarrow Cu + Cu^{2+}$$

Most of the *anhydraus cupric compounds are colourless* while the *hydrated cupric compounds are blue* due to the formation of blue hydrated ion, $[Cu(H_2O)_4]^{2+}$ or $[Cu(H_2O)_6]^{2+}$. Important compounds of Cu are Cu_2O, CuCl, etc. which are discussed below :

1. Cuprous Oxide, Cu_2O

Preparation : It is also called ***red oxide of copper.*** It is obtained by heating copper in air or by very strongly heating cupric oxide alone or with copper. It is also precipitated in red or orange form when alkaline cupric sulphate solution (or Fehling'a solution) is reduced by substances like glucose or aldehydes.

$$CuSO_4 + 2NaOH \longrightarrow Na_2SO_4 + Cu(OH)_2 \longrightarrow CuO + H_2O$$

$$\underset{\text{Cupric oxide}}{2CuO} + \underset{\text{Glucose}}{C_5H_{11}O_5.CHO} \longrightarrow \underset{\text{Cuprous oxide}}{Cu_2O} + \underset{\text{Gluconic acid}}{C_5H_{11}O_6.COOH}$$

$$2Cu(OH)_2 + \underset{\text{Acetaldehyde}}{CH_3CHO} \longrightarrow Cu_2O + \underset{\text{Acetic acid}}{CH_3COOH} + 2H_2O$$

Properties. (*i*) It is a dull red powder. It is practically insoluble in water.

(*ii*) *Action of acids.* Cuprous oxide reacts with dilute acids, forming cupric salts

$$Cu_2O + H_2SO_4 \longrightarrow CuSO_4 + H_2O$$

$$Cu_2O + 4HCl \longrightarrow 2H[CuCl_2] + H_2O$$

(*iii*) *Action of ammonia.* It dissolves in ammonia and gives complex compounds called amines.

$$Cu_2O + 4NH_3 + 4H_2O \longrightarrow 2[Cu(NH_3)_2]OH$$

(*iv*) *Action of alkali cyanide.* It dissolves in alkali cyanide (*e.g.* KCN) solution forming a colourless complex, $K_2[Cu(CN)_3]$

$$Cu_2O + 6KCN + H_2O \longrightarrow 2K_2[Cu(CN)_3] + 3KOH$$

Uses. It is used (*i*) for making ruby glass and enamel. (*ii*) as pigment. (*iii*) in the manufacture of antirust paints. (*iv*) as a pesticide.

2. Cuprous Chloride, CuCl or Cu_2Cl_2

Preparation : It is obtained by the following reactions :

$$CuCl_2 + Cu \xrightarrow[\text{of conc. HCl}]{\text{Heat in presence}} 2CuCl.$$

$$2CuCl_2 + 2H \xrightarrow{\text{Reduction}} 2CuCl + 2HCl$$

$$2CuCl_2 + SO_2 + 2H_2O \xrightarrow{\text{Reduction}} 2CuCl + H_2SO_4 + 2HCl$$

$$2CuCl_2 \xrightarrow{450°} 2CuCl + Cl_2$$

Properties. (*i*) Cuprous chloride is a colourless crystalline substance, insoluble in water but soluble in HCl and NH_4OH. When exposed to air, it is slowly oxidised to green basic cupric chloride, $3CuO. CuCl_2.3H_2O$.

(*ii*) *Action of oxidising agents.* With oxidising agents, it is converted into cupric chloride.

$$2CuCl + 2HCl + O \longrightarrow 2CuCl_2 + H_2O$$

(*iii*) An acidified solution of cuprous chloride absorbs oxygen and carbon monoxide while an ammoniacal solution absorbs acetylene.

$$4CuCl + 4HCl + O_2 \longrightarrow 4CuCl_2 + 2H_2O$$

This reaction is utilised in the manufacture of chlorine by *Deacon's process.*

$$CuCl + 2CO \longrightarrow \underset{\text{Addition compound}}{CuCl.2CO}$$

This addition compound decomposes on heating, regenerating carbon monoxide. This reaction is, therefore, *employed in gas analysis* for separation of carbon monoxide.

$$2CuCl + \underset{\text{Acetylene}}{C_2H_2} \longrightarrow \underset{\text{Copper acetylide}}{C_2Cu_2} + 2HCl$$

Acetylene can be regenerated by treating the acetylide with strong HCl. The reaction is, therefore, *used in the purification and separation of acetylene.*

(*iv*) *Action of alkalies.* Cuprous chloride solution gives with alkalies a yellow ppt. of cuprous oxide which gradually changes to red colour.

$$2CuCl + 2NaOH \longrightarrow \underset{\substack{\text{Yellow ppt.}\\ \text{changing to red}}}{Cu_2O} + 2NaCl + H_2O$$

(*v*) *Action of H_2S.* With H_2S, a black precipitate of cuprous sulphide Cu_2S is produced.

$$CuCl + H_2O \longrightarrow \underset{\text{black ppt.}}{Cu_2S} + 2HCl$$

Uses. Ammoniacal solution of cuprous chloride is used in gas analysis for absorbing acetylene while HCl solution of it is used for absorbing CO and oxygen.

3. Basic Cupric Acetate or Verdigris, $Cu(CH_3COO)_2 . Cu(OH)_2$

Preparation : It is prepared by exposing thin sheets of copper to the action of vinegar, CH_3COOH and air.

$$2Cu + 2CH_3COOH + O_2 \longrightarrow (CH_3COOH)_2Cu.Cu(OH)_2$$

Properties : It is greenish blue powder. It is sparingly soluble in cold water but dissolves in dilute acids and ammonia.

Uses. (*i*) Because of its fairly good covering power, it is used as green pigment, (*ii*) used in dyeing and calico-printing, (*iii*) used in making insecticide and fungicides, (*iv*) when verdigris is mixed with As_2O_3, cupric aceto arsenite, (called *paris green*) $Cu_3(AsO_3)_2 . Cu(CH_3COO)_2$ which is poisonous is obtained.

4. Cupric Sulphate or Blue Vitriol, $CuSO_4.5H_2O$

Preparation : This compound is generally called ***copper sulphate*** and is *Nila Thotha*. In the laboratory, it is prepared by the action of cupric oxide, carbonate or hydroxide on dilute sulphuric acid followed by evaporation and crystallisation.

$$CuO + H_2SO_4 \longrightarrow CuSO_4 + H_2O$$

$$CuCO_3 + H_2SO_4 \longrightarrow CuSO_4 + H_2O + CO_2$$

$$Cu(OH)_2 + H_2SO_4 \longrightarrow CuSO_4 + 2H_2O$$

Manufacture. (*i*) On a large scale copper sulphate is obtained by boiling copper turnings with concentrated sulphuric acid. Blue solution of copper sulphate is formed.

$$Cu + 2H_2SO_4 \longrightarrow CuSO_4 + SO_2 + 2H_2O$$

(*ii*) It is also obtained by treating copper scrappings in hot dilute sulphuric acid in presence of air.

$$2Cu + H_2SO_4 + O_2 \longrightarrow 2CuSO_4 + 2H_2O.$$

Copper scrap or copper turnings are packed in a lead lined vertical tower. Dilute sulphuric acid is sprayed at the top and percolates through copper chips. A current of air and steam is blown from below. Copper dissolves in dilute sulphuric acid in the presence of air to give a solution of copper sulphate which is collected at the base. This solution of copper sulphate contains a lot of dilute acid and is recirculated in the tank by means of a pump to get a concentrated solution which is led into crystallisation tanks. On cooling, crystals of *pentahydrate copper sulphate*. $CuSO_4.5H_2O$ are deposited.

(*iii*) In another process, the mineral copper pyrites, $(Cu_2S.Fe_2S_3)$ is roasted in air. Iron oxide and copper sulphate are formed. The roasted mass is treated with water. Copper sulphate dissolves and is separated by filtration. On crystallisation crystals of $CuSO_4.5H_2O$ are formed.

Anhydrous copper sulphate, which is a white powder, is obtained by heating pure $CuSO_4.5H_2O$ in the powdered form in a U-tube at 230°C in a current of dry air until the weight corresponds with $CuSO_4$.

Properties. (*i*) Copper sulphate forms deep blue crystals of the composition $CuSO_4.5H_2O$. Copper sulphate is fairly soluble in water.

(*ii*) *Action of heat.* On heating copper sulphate loses its water of crystallisation. On heating to 100°C it gives $CuSO_4.H_2O$ and when heated to 230°C it becomes white anhydrous $CuSO_4$.

The white anhydrous copper sulphate decomposes at about 720°C to give cupric oxide and sulphur trioxide.

$$\underset{\text{Blue}}{CuSO_4.5H_2O} \xrightarrow{100°C} \underset{\text{Pale blue}}{CuSO_4.H_2O} \xrightarrow{230°C} \underset{\substack{\text{Colourless}\\\text{(amorphous)}}}{CuSO_4} \xrightarrow{720°C} CuO + SO_3$$

(*iii*) Copper sulphate gives all the general reactions of sulphates.

(*iv*) *Action of potassium iodide.* Copper sulphate gives white cuprous iodide with potassium iodide.

$$2CuSO_4 + 4KI \longrightarrow 2K_2SO_4 + 2CuI + I_2$$

(*v*) *Action of alkalies.* Alkalies produce a pale blue precipitate of copper hydroxide.

$$CuSO_4 + 2NaOH \longrightarrow \underset{\text{Pale blue ppt.}}{Cu(OH)_2} + Na_2SO_4$$

(*vi*) *Action of NH_4OH.* With ammonium hydroxide, precipitate of $Cu(OH)_2$ is first produced but it dissolves in excess of it forming a bright blue solution, the colour being due to the formation of *cuprammonium ion*, $[Cu(NH_3)_4]^{2+}$

$$CuSO_4 + \underset{\text{Small amount}}{2NH_4OH} \longrightarrow \underset{\text{ppt.}}{Cu(OH)_2} + (NH_4)_2SO_4$$

$$CuSO_4 + \underset{\text{Large amount}}{4NH_4OH} \longrightarrow \underset{\substack{\text{Cuprammonium}\\\text{sulphate}\\\text{(Deep blue soln.)}}}{Cu(NH_3)_4SO_4} + 4H_2O$$

(*vii*) *Action of potassium ferrocyanide.* Reddish brown precipitate of cupric ferrocyanide is formed.

$$2CuSO_4 + K_4[Fe(CN)_6] \longrightarrow 2K_2SO_4 + Cu_2[Fe(CN)_6]$$

(*viii*) *Action of potassium sulphocyanide.* Oupric sulphocyanide is formed.

$$CuSO_4 + 2KCNS \longrightarrow K_2SO_4 + \underset{\text{Cupric sulphocyanide}}{Cu(CNS)_2}$$

(*ix*) *Action of KCN.* With KCN it gives a yellow ppt. of cupric cyanide which decomposes to give cuprous cyanide and cyanogen. Cuprous cyanide dissolves in excess of KCN to give potassium cuprocyanide.

$$CuSO_4 + 2KCN \longrightarrow Cu(CN)_2 + K_2SO_4$$

$$\underset{\substack{\text{Cupric}\\\text{cyanide}}}{2Cu(CN)_2} \longrightarrow \underset{\substack{\text{Cuprous}\\\text{cyanide}}}{Cu_2(CN)_2} + \underset{\text{Cyanogen}}{(CN)_2}$$

$$Cu_2(CN)_2 + 6KCN \longrightarrow \underset{\text{Potassium cuprocyanide}}{2K_3Cu(CN)_4}$$

(*x*) *Addition of electropositive metals.* Electropositive metals like zinc or iron precipitates copper from a solution of copper sulphate.

$$CuSO_4 + Fe \longrightarrow FeSO_4 + Cu$$

(*xi*) *Action of H_2S.* With H_2S it gives a black ppt. of copper sulphide.

$$CuSO_4 + H_2S \longrightarrow \underset{\text{Black ppt.}}{CuS} + H_2SO_4$$

Uses. It is used : (*i*) in electroplating, in electrotyping and in the purification of copper. (*ii*) in agriculture as a germ killer and fungicide. *Bordeaux mixture* is copper sulphate mixed with lime and is extensively used for dusting vines and potatoes. (*iii*) in calico-printing, dyeing and the preparation of other compounds of copper. (*iv*) in analysis (as anhydrous salt). (*v*) in medicine as an antiseptic.

5. Cupric Cyanide, $Cu(CN)_2$

Preparation : This compound as such has not been isolated so for. It is formed as a yellow precipitate when a small quantity of KCN is added to a cold solution of $CuSO_4$

$$CuSO_4 + 2KCN \longrightarrow \underset{\text{Yellow ppt.}}{Cu(CN)_2} + K_2SO_4$$

Properties. (*i*) It is extremely unstable and rapidly decomposes even at ordinary temperature to give cuprous cyanide, CuCN and free cyanogen, $(CN)_2$

$$2Cu(CN)_2 \longrightarrow 2CuCN + (CN)_2$$

(*ii*) It dissolves in the excess of KCN, forming potassium cupri cyanide, $K_2[Cu(CN)_4]$.

$$Cu(CN)_2 + 2KCN \text{ (excess)} \longrightarrow K_2[Cu(CO)_4]$$

(*iii*) It forms coordination compound, $[Cu(py)_2](CN)_2$, with pyridine,

6. Complex Compounds of Cuprous Ion

(*i*) Complexes with hydrochloric acid. Cuprous oxide Cu_2O or cuprous chloride, Cu_2Cl_2, when dissolved in hydrochloric acid gives *hydrocuprochloric acid*, $HCuCl_2$, in which copper goes to the anion $[CuCl_2]^-$:

$$Cu_2O + 2HCl \rightleftharpoons Cu_2Cl_2 + H_2O$$

$$Cu_2Cl_2 + 2HCl \rightleftharpoons 2HCuCl_2 \rightarrow 2H^+ + 2[CuCl_2]^-$$

(*ii*) Complexes with ammonium hydroxide. When cuprous oxide is treated with ammonium hydroxide, a complex is formed in which copper goes to the cation, $[Cu(NH_3)_2]^+$

$$Cu_2O + H_2O \rightleftharpoons 2CuOH$$

$$CuOH \rightleftharpoons Cu^+ + OH^-$$

$$Cu^+ + 2NH_3 + OH^- \rightleftharpoons [Cu(NH_3)_2]OH$$

A similar complex is obtained when cuprous chloride dissolves in ammonia. The complex is *cuproamine chloride* which is a salt of the cupromine base

$$Cu^+ + Cl^- + 2NH_3 \rightleftharpoons [Cu(NH_3)_2]Cl \rightleftharpoons [Cu(NH_3)_2]^+ + Cl^-$$

(*iii*) Complexes with alkali cyanides. When potassium cyanide, cuprous oxide as well as its salt gives *potassium cuprocyanide*, $K_3Cu(CN)_4$ or $K_2Cu(CN)_4$:

$$Cu_2O + 6KCN + H_2O \rightleftharpoons 2K_2Cu(CN)_3 + 2KOH$$

When potassium cyanide is added to copper(*ic*) sulphate, cupric cyanide, $Cu(CN)_2$, first precipitated, at once decomposes to cuprous cyanide and cyanogen :

$$CuSO_4 + 2KCN \longrightarrow Cu(CN)_2 + K_2SO_4$$
$$2Cu(CN)_2 \longrightarrow Cu_2(CN)_2 + (CN)_2$$

Cuprous cyanide dissolves in an excess of KCN to give cuprocyanide :

$$Cu_2(CN)_2 + 6KCN \rightleftharpoons 2K_3Cu(CN)_4$$
$$K_3Cu(CN)_4 \rightleftharpoons 3K^+ + [Cu(CN)_4]^{3-}$$

The second dissociation constant of $[Cu(CN)_4]^{3-} \rightleftharpoons Cu^+ + 4CN^-$ is very small; the complex is thus so stable that it gives no precipitate with H_2S (copper and cadmium separation).

***(iv)* Complexes with hypo solution.** When cupric sulphate solution is mixed with hypo solution, the blue colour of the former is discharged forming sodium cuprous thiosulphate (cupric sulphate is reduced to cuprous) :

$$CuSO_4 + Na_2S_2O_3 \longrightarrow Na_2SO_4 + CuS_2O_3$$
$$2CuS_2O_3 + Na_2S_2O_3 \longrightarrow 2NaCuS_2O_3 + Na_2S_4O_6$$

On boiling or on acidification, cuprous sulphide is precipitated :

$$2NaCuS_2O_3 + H_2O \longrightarrow Cu_2S + Na_2S_2O_3 + H_2SO_4$$

7. Complex Compounds of Cupric Ion

Cupric hydroxide, when dissolved in aqueous ammonia, gives a blue soluble complex of cupriamine hydroxide, $[Cu(NH_3)_4]$ (OH).

$$Cu(OH)_2 + 4NH_3 \longrightarrow Cu(NH_3)_4(OH)_2 \rightleftharpoons [Cu(NH_3)_4]^{2+} + 2(OH)^-$$

If ammonium hydroxide be added in excess to a solution of copper sulphate, the precipitate hydroxide again dissolves yielding an intense blue solution. If a layer of alcohol is carefully poured over this solution, and evaporation prevented by corking, deep blue rhombic crystals of *cupriamine sulphate* are obtained :

$$CuSO_4 \rightleftharpoons Cu^{2+} + SO_4^{2-}$$
$$Cu^{2+} + 4NH_3 \rightleftharpoons Cu(NH_3)_4^{2+}$$
$$Cu(NH_3)_2^{2+} + SO_4^{2-} + 4H_2O \longrightarrow Cu(NH_3)_4SO_4.4H_2O$$

If a hot solution of cupric chloride saturated with ammonia gas be cooled, *cupriamine chloride*, $Cu(NH_3)_4Cl_2$ crystallises. The structure of *cupriamine ion* is

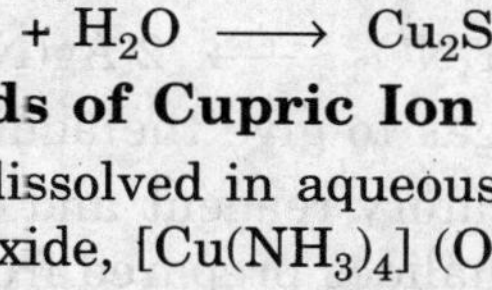

Compounds of Silver

In its familiar compounds, Ag is in +1 oxidation state (*monovalent*) while in a few compounds it is in +2 state (*divalent*). Important compounds of Ag are discussed below :

1. Silver Nitrate or Lunar Caustic, $AgNO_3$

Preparation. It is usually prepared by dissolving Ag in hot dil. HNO_3 in a porcelain or stainless steel vessel

$$3Ag + 4HNO_3 \longrightarrow 3AgNO_3 + 2H_2O + NO$$

Properties. It forms colourless transparent rhombic crystals of density 4.35 and m.pt. = 212°C. In the fused state it can be cast into sticks called *lunar caustic* due to its strong caustic action. It is highly soluble in water and its solubility increases with the rise of temperature. Its aqueous solution slowly darkens when exposed to light. It is also soluble in alcohol, pyridine etc. On heating above its melting point, it decomposes into $AgNO_2$ and O_2. When heated to red, it decomposes to metallic Ag.

$$2AgNO_3 \longrightarrow 2AgNO_2 + O_2;\ 2AgNO_3 \longrightarrow 2Ag + 2NO_2 + O_2$$

It is decomposed by organic matter such as skin, paper, cloth or cork especially under the action of light to finely divided metallic silver (black). With NH_4OH solution a brown precipitate of Ag_2O is obtained which dissolves in the excess of NH_4OH forming a complex, $[Ag(NH_3)_3]NO_3$.

$$2AgNO_3 + 2NH_4OH \longrightarrow \underset{\text{Brown ppt.}}{Ag_2O} + 2NH_4NO_3 + H_2O$$

$$Ag_2O + 2NH_4OH + 2NH_4NO_3 \longrightarrow 2[Ag(NH_3)_2]NO_3 + 3H_2O$$

Solid $AgNO_3$ absorbs NH_3 gas to give the addition compound, $AgNO_3.3NH_3$.

Uses. It is used as a laboratory reagent and in volumetric analysis for the estimation of chlorides. Silver halides prepared from it are used in photography. It is also used as an indelible ink for marking purposes, in silvering mirrors, as a caustic in surgery under the name of *lunar caustic*, as a starting material for the preparation of other silver componnds and as a reagent for arsine in the Gutzeit test.

2. Silver Nitride or Fulminating Silver, Ag_3N

It is obtained by the action of ammonia on freshly precipitated silver oxide. It is an explosive black powder. Above 140°C, it explodes spontaneously. It is decomposed by HNO_3 to Ag and NH_4NO_3. Similarly it reacts with H_2SO_4 and HCl. When added to KCN solution, it decomposes with the evolution of NH_3

$$Ag_3N + 6KCN + 3H_2O \longrightarrow 3K[Ag(CN)_2] + 3KOH + NH_3$$

Compounds of Gold

Gold forms two series of compounds (*i*) *aurous* compounds in which gold shows + 1 oxidation state, and (*ii*) *auric* compounds wherein gold is in +3 oxidation state. Compounds in +3 oxidation state are more common. The following compounds of gold are important.

1. Aurous chloride, AuCl

Preparation. It is prepared by heating auric chloride ($AuCl_3$) to 175°C.

$$AuCl_3 \xrightarrow{175°C} AuCl + Cl_2$$

Properties. It is a light-yellow powder which is insoluble in water. When heated above 175°C, it gets decomposed into metallic gold and free Cl_2.

$$2AuCl \xrightarrow{> 175^\circ C} 2Au + Cl_2$$

When treated with H_2O at ordinary temperature, it is decomposed into metallic Au and $AuCl_3$.

$$3AuCl \longrightarrow AuCl_3 + 2Au$$

AuCl combines with alkali metal chlorides (*e.g.* KCl) to form chloroaurates (I), $M[AuCl_2]$. When treated with Cl_2, $AuCl_3$ is formed.

$$AuCl + KCl \longrightarrow K[AuCl_2] \text{ [Pot. dichloroaurate (I)]}$$

$$AuCl + Cl_2 \longrightarrow AuCl_3$$

2. Auric Chloride, $AuCl_3$

Preparation : When the yellow crystals of hydrated chloroauric acid, $HAuCl_4.4H_2O$ are heated at 200°C, deep red crystals of *anhydrous auric chloride*, $AuCl_3$ are obtained.

$$HAuCl_4.4H_2O \xrightarrow{200^\circ C} AuCl_3 + HCl + 4H_2O$$

The preparation of *anhydrous auric chloride* can also be done as follows : Gold is first reacted with chlorine to give auric chloride which is dissolved in a little water. This is gently heated, in order to convert any contaminated aurous chloride into metallic gold which is thrown down as a precipitate and is filtered out, and then the solution evaporated and heated to 150°C. This gives anhydrous auric chloride. It is soluble in ether.

Properties : Anhydrous $AuCl_3$ is a deep red crystalline substance which is freely soluble in water. When crystallised from water solution, orange crystals of dihydrate, $AuCl_3.2H_2O$ are obtained. Auric chloride is non-polar (covalent) and does not ionise in aqueous solutions. The non-polar character is further shown by its low melting point (290°), volatility and its solubility in organic solvents as alcohol and ether. It is susceptible to photo-reduction and is reduced to metallic gold by H_2O_2 in presence of an alkali metal or by SO_2. Gold is easily displaced from $AuCl_3$ solution by the metals like Fe, Zn, Mg, Al, Ag, Pb etc. When KOH or NaOH is added to a solution of $AuCl_3$, $Au(OH)_3$ is precipitated which dissolves in the excess of the alkali forming potassium or sodium aurate (III), $M[AuO_2]$ (M = K or Na).

$$AuCl_3 + 3NaOH \longrightarrow \underset{\text{ppt.}}{Au(OH)_3} + 3NaCl$$

$$Au(OH)_3 + NaOH \longrightarrow \underset{\text{Sod. aurate (III)}}{Na[AuO_2]} + 2H_2O$$

When $AuCl_3$ is digested with NH_3, fulminating gold. $HN = Au—NH_2$ is obtained.

$$AuCl_3 + 2NH_3 \longrightarrow \underset{\text{Fulminating gold}}{HN = Au—NH_2} + 3HCl$$

Auric chloride with hydrochloric acid gives auri-chloric acid or chloroauric acid. Similarly with calculated amounts of potassium chloride, it gives potassium aurichloride, also known as potassium chloroaurate; and similarly the sodium salt with common salt :

$$AuCl_3 + HCl \longrightarrow \underset{\text{Chloroauric acid}}{H[AuCl_4]}$$

$$AuCl_3 + NaCl \longrightarrow \underset{\text{Sod. chloroaurate}}{Na[AuCl_4]}$$

On heating $AuCl_3$ decomposes to form AuCl. [$AuCl_3 \xrightarrow{\Delta} AuCl + Cl_2$]. Gold chloride solution is not very stable. It decomposes readily in neutral solution to form aurous chloride and hydrogen peroxide. [$AuCl_3 + 2H_2O \rightarrow AuCl + 2HCl + H_2O_2$]. It decomposes on exposure to sunlight or in presence of reducing substances to metallic gold.

The vapour density of auric chloride shows that it is a dimeric molecule, Au_2Cl_6

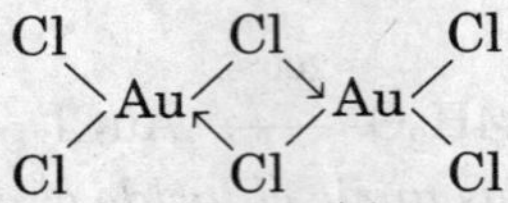

Uses : Hydrochloroauric acid, $H[AuCl_4].4H_2O$ under the name of gold chloride is used in photography as a toning agent.

3. Fulminating Gold, HN = Au—NH$_2$

It is obtained by treating auric chloride solution with ammonia :

$$AuCl_3 + 2NH_3 \longrightarrow HN = Au\text{—}NH_2 + HCl$$

It is also obtained by treating auric hydroxide with ammonia.

$$Au(OH)_3 + 2NH_3 \longrightarrow HN = Au\text{—}NH_2 + 3H_2O$$

Properties : It is an olive green powder of variable composition. The dry mass explodes violently on heating or percussion, forming Au, N_2, NH_3 and steam. It dissolves in HCl to form tetrachloro auric acid, $H[AuCl_4]$ and NH_4Cl. It readily dissolves in alkali cyanide solutions. It is converted to non-explosive compounds by the action of H_2S and by careful heating with S. It is reduced to Au with explosion by $SnCl_2$. Raschig (1980) formulated it as $HN = Au\text{—}NH_2.\frac{3}{2}H_2O$. According to Weitz (1915) the fulminating gold produced by the action of NH_3 on $AuCl_3$ solution contains some chlorine and, therefore, he suggested the following structure to this compound.

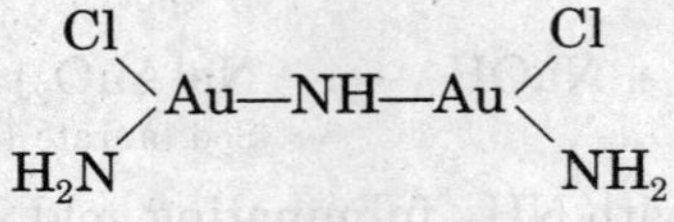

4. Complex Compounds

Gold forms stable complexes in +1 and +3 oxidation states. 2-coordinate complex compounds given by Au(+1) are *linear* involving *sp*-hybridisation whereas 4-

coordinate complexes are probably *tetrahedral* resulting from sp^3 hybridisation. Both the types of complexes are *diamagnetic*. The 4-coordinate complexes given by Au(+3) are *diamagnetic* and *square planar* with dsp^2 hybridisation while the 5-coordination complexes are *trigonal bipyramidal* and 6-coordinate complexes are *octahedral*. The following are important complexes formed by gold.

***(i)* Potassium aurocyanide, $K[Au(CN)_2]$. Preperation.** It is obtained by the following methods :

(*a*) Gold is dissolved in aqua regia and precipitated with ammonia forming fulminating gold. It is washed and dissolved in potassium cyanide and crystallised.

(*b*) Fine gold is treated with potassium cyanide solution in contact with atmospheric oxygen :

$$4Au + 8KCN + O_2 + 2H_2O \longrightarrow 4KAu(CN)_2 + 4KOH$$

Properties. It is a colourless substance and soluble in water. When heated with HCl at 50°C, yellow crystals of AuCN are produced.

$$KAu(CN)_2 + HCl \longrightarrow KCl + HCN + AuCN$$

When Zn is added, gold is precipitated.

$$KAu(CN)_2 + Zn \longrightarrow K_2[Zn(CN)_2] + \underset{\text{ppt.}}{2Au}$$

Uses : It is used in gold plating.

***(ii)* Sodium aura *(i)* thiosulphate, $Na_3[Au(S_2O_3)_2]$. Preparation.** Just as hypo forms a complex with silver chloride, a similar complex is obtained by adding auric chloride to hypo solution.

$$4Na_2S_2O_3 + AuCl_3 \longrightarrow Na_3[Au(S_2O_3)_2] + Na_2S_4O_6 + 3NaCl$$

Hypo is side by side oxidised to sodium tetrathionate as in iodine reaction.

Properties and uses. It crystallises in colourless needles, fairly soluble in water. It is not reduced by $FeSO_4$, oxalic acid or $SnCl_2$. It is used in medicine as *sanochrysin* (Greek *chrusos*).

***(iii)* Tetrachloroauric acid or gold chloride, $H[AuCl_4]$. Preparation.** When metallic gold is dissolved in aqua regia, a bright yellow solution of $HAuCl_4$ is obtained

$$2Au + 3HNO_3 + 11HCl \longrightarrow \underset{\text{Solution}}{2HAuCl_4} + 6H_2O + 3NOCl$$

On evaporating the solution, yellow deliquescent crystals of tetrahydrate chloroauric acid, $HAuCl_4.4H_2O$ deposit. This as called *gold chloride* of commerce. It can also be prepared by the action of HCl on $AuCl_3$,

$$HCl + AuCl_3 \longrightarrow HAuCl_4$$

Properties. (*a*) It soluble in water and in donor organic solvents. It is converted into $AuCl_3$ when heated at 200°C in a current of Cl_2.

(*b*) *Reduction.* It is easily reduced to metallic gold by reducing agents like $FeSO_4$, $SnCl_2$ and even by H_2.

$$HAuCl_4 + 3FeSO_4 \longrightarrow Fe_2(SO_4)_3 + FeCl_3 + HCl + Au$$

$$2HAuCl_4 + 3SnCl_2 \longrightarrow 3SnCl_4 + 3HCl + 2Au$$

$$2HAuCl_4 + 3H_2 \longrightarrow 8HCl + 2Au$$

(c) *Action of KI.* Reaction with KI precipitates AuI.

$$HAuCl_4 + 3KI \longrightarrow \underset{\text{ppt.}}{AuI} + 3KCl + HCl + I_2$$

(d) *Reaction with alkalies.* Alkalies precipitate $Au(OH)_3$ which dissolves in excess of the alkali solution to form soluble complex, $Na[Au(OH)_4]$

$$HAuCl_4 + 4NaOH \longrightarrow \underset{\text{ppt}}{Au(OH)_3} + 4NaCl + H_2O$$

$$Au(OH)_3 + NaOH \longrightarrow \underset{\text{Soluble complex}}{Na[Au(OH)_4]}$$

(e) *Reaction with KCN.* First AuCN is obtained as precipitate which in excess of KCN dissolves due to the formation of soluble complex, $K[Au(CN)_2]$

$$HAuCl_4 + 3KCN \longrightarrow \underset{\text{ppt}}{AuCN} + 3KCl + HCl + (CN)_2$$

$$AuCN + KCN \longrightarrow \underset{\text{Soluble complex}}{K[Au(CN)_2]}$$

(f) *Reaction with H_2S.* With H_2S the acid yields a precipitate which is probably a mixture of Au_2S and Au_2S_3.

(g) *Formation of chloroaurates.* The salts of the acid are called *chloroarates.* When a solution of the acid reacts with an alkali chloride, alkali chloreurate, $M[AuCl_4].nH_2O$ is obtained in solution which on evaporation gives the crystals of the salt.

Uses. The acid and its sodium and potassium salt are used in photography for toning and printing.

Structure : $[AuCl_4]^-$ ion is diamagnetic and has square planar geometry arising from dsp^2 hybridisation of Au atom.

***(iv)* Potassium auricyanide, $K[Au(CN)_4]$.** A hot concentrated solution of auric chloride when mixed with potassium cyanide gives auricyanide which crystallises out.

$$AuCl_3 + 4KCN \longrightarrow KAu(CN)_4 + 3KCl$$

The salt gives a precipitate of $AgAu(CN)_4$ with silver nitrate which on treatment with hydrochloric acid gives silver chloride and auricyanic acid, $HAu(CN)_4$. $KAu(CN)_4$ is used in gold plating.

Compounds of Zinc and Cadmium

Zinc and cadmium both are always in +2 (bivalent) oxidation state in their compounds. Some of their important compounds are discussed below.

1. Zinc Oxide or Philosopher's Wool, ZnO

Preparation : It occurs in nature as the mineral *zincite* or *red zinc ore.* It is obtained by the calcination of zinc carbonate, nitrate or hydroxide,

$$ZnCO_3 \longrightarrow ZnO + CO_2$$

$$2Zn(NO_3)_2 \longrightarrow 2ZnO + 4NO_2 + O_2$$

$$Zn(OH)_2 \longrightarrow ZnO + H_2O$$

Zinc oxide prepared by burning zinc vapour in air is collected as white wooly flocks and hence is also known as *philosopher's wool.*

Properties : It is a white amorphous powder which becomes sulphur-yellow on heating but becomes white on cooling. It is insoluble in water. It is an amphoteric oxide since it dissolves readily in acids forming the zinc salt and in alkalies forming zincates.

$$ZnO + 2HCl \longrightarrow ZnCl_2 + H_2O$$

$$ZnO + NaOH \longrightarrow \underset{\text{Sod. zincate}}{Na_2ZnO_2} + H_2O$$

When heated in dry hydrogen above 450° or with carbon, it is reduced to metallic zinc. When it is heated with a solution of $Ca(NO_3)_2$, a green mass is formed which is due to the formation of *cobalt zincate*, $CoZnO_2$ (double oxide, ZnO.CoO) which is known as *Rinmann's green.*

$$ZnO + H_2 \longrightarrow H_2O + Zn$$

$$ZnO + C \longrightarrow Zn + CO \uparrow$$

$$2Co(NO_3)_2 \longrightarrow 2CoO + 4NO_2 + O_2$$

$$ZnO + CoO \longrightarrow ZnO.CoO \text{ or } CoZnO_2$$

Uses : Zinc oxide is used (*i*) as a white paint under the name of *zinc white* or *chinese white* although it does not possess as good a covering powder as white lead. It is, however, sometimes preferred because it is not blackened by H_2S. (*ii*) as a catalyst when mixed with Cr_2O_3 in the manufacture of methyl alcohol from water gas and H_2. (*iii*) as a filler for soft rubber (*iv*) as an alsorbent in surgical dressing. (*v*) for the preparation of cement and hard resins. (*vi*) for the preparation of other zinc compounds. (*vii*) The green mass obtained by heating ZnO with $Co(NO_3)_2$ is used as a pigment under the name of *Rinmann's green.*

2. Zinc Chloride, $ZnCl_2.2H_2O$

Preparation. It is prepared by dissolving zinc oxide, carbonate or hydroxide in dilute hydrochloric acid

$$ZnO + 2HCl \longrightarrow ZnCl_2 + H_2O$$

The solution so obtained on concentration and cooling gives the crystals of $ZnCl_2.2H_2O$. *Anhydrous zinc chloride* cannot be prepared merely by heating the crystals of $ZnCl_2.2H_2O$ because a basic chloride Zn(OH)Cl is formed during decomposition which on further heating gives ZnO.

$$ZnCl_2.2H_2O \longrightarrow Zn(OH)Cl + HCl + H_2O$$

$$Zn(OH)Cl \longrightarrow ZnO + HCl$$

Hence anhydrous zinc chloride is obtained either by heating zinc in a current of chlorine gas or by distilling a mixture of zinc powder and mercuric chloride.

$$Zn + Cl_2 \xrightarrow{\Delta} ZnCl_2$$

$$Zn + HgCl_2 \longrightarrow ZnCl_2 + Hg \uparrow$$

Properties. It is a white crystalline solid, deliquescent and soluble in water. On beating it forms the oxychloride and finally the oxide. Its concentrated solution sets to a hard mass when mixed with zinc oxide and the product is used as a dental filling.

Uses. (*i*) Zinc chloride finds us as a dehydrating agent and a catalyst. (*ii*) Timber is impregnated with a solution of zinc chloride to stop dry rot. (*iii*) It is employed in making parchment paper, vulcanised fibre, cold water glues and adhesives. (*iv*) It is also used in the manufacture of dry cells.

3. Zinc Sulphate or White Vitriol $ZnSO_4.7H_2O$

Preparation : It is found in nature as an anhydrous salt *viz. zincosite* and as hydrated salt *viz., zinc vitriol.* It is obtained by dissolving metallic zinc, its oxide, or carbonate in dil. H_2SO_4. On concentrating the solution to crystallisation below 39°, colourless efflorescent crystals of *heptahydrate*, $ZnSO_4.7H_2O$ deposit

$$Zn + H_2SO_4 \longrightarrow ZnSO_4 + H_2$$

Above 39°C, *hexahydrate*, $ZnSO_4.6H_2O$ and above 70°C *monohydrate*, $ZnSO_4.H_2O$ is deposited. *On a large scale* it may be obtained by dissolving zinc scrap in dil. H_2SO_4 or by roasting *zinc blende* (ZnS) and extracting the mass with water.

$$ZnS + 2O_2 \xrightarrow{\Delta} ZnSO_4$$

Properties : It is a colourless, crystalline solid and effloreacent substance. It is freely soluble in water.

On heating water of crystallisation is lost according to the following scheme.

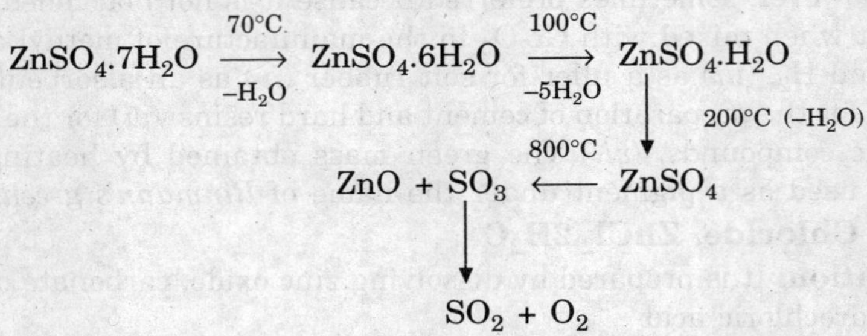

$ZnSO_4.7H_2O$ is isomorphous with $MgSO_4.7H_2O$, $FeSO_4.7H_2O$, $NiSO_4.7H_2O$, $CoSO_4.7H_2O$, $CrSO_4.7H_2O$ and $MnSO_4.7H_2O$. With alkali metal sulphates and $(NH_4)_2SO_4$ it forms double salts of the type, $M_2SO_4.ZnSO_4.6H_2O$.

Uses : Zinc sulphate is used (*i*) as an eye wash (*ii*) as a mordant in dyeing and calico printing (*iii*) for making lithopone–a white paint obtained by mixing barium sulphide with a solution of zinc sulphate (*iv*) in zinc plating baths and in the electrolytic refining of zinc.

4. Zinc Sulphide, ZnS

Preparation : It occurs in nature as *zinc blende* and is the principal source of zinc metal. It is prepared by passing hydrogen sulphide through an alkaline solution of zinc salt in the presence of ammonium chloride. It can also be obtained as a white precipitate when H_2S is passed through a solution of a zincate.

$$ZnCl_2 + H_2S \longrightarrow ZnS + 2HCl$$

$$Na_2ZnO_2 + H_2S \longrightarrow \underset{\text{white ppt.}}{ZnS} + 2NaOH$$

Properties : Zinc sulphide occurs in two forms *viz.* α-ZnS and β-ZnS. It is a white amorphous substance, insoluble in water and alkali. It readily dissolves in all dilute mineral acids but not in acetic acid, liberating sulphuretted hydrogen. Pure ZnS does not show any phosphorescenece property but in presence of traces of sulphides of Cu, Ag, Mn, Bi, etc. it becomes strongly phosphorescent. The phosphorescent suiphide is called *Sidot's blende.*

Uses : A mixture of ZnS and $BaSO_4$ is called ***lithopone*** which is used as a white pigment in paint. It is also used as a pigment in paint, rubber and paper. The phosphorescent ZnS which is called *Sidot's blends* is used in making fluorescent screens for X-ray and radioactivity works. The phosphorescent ZnS is prepared by passing H_2S through a solution obtained by dissolving 5 mgs. of zinc ammonium sulphate, 0.1 gm. of $MnCl_2$ and 1 gm. of NaCl in 100 ml. of H_2O. The white ppt. so obtained is filtered off without washing and then heated to whiteness in a covered crucible in an atmosphere of H_2S. The phosphorescent ZnS is also used for preparing luminous paints of watches. Luminous paint of watches is a mixture of phosphorescent ZnS and trace of radium salt.

5. Lithopone, (ZnS + $BaSO_4$)

It is a mixture of ZnS and $BaSO_4$ and is obtained by the simultaneous precipitation of zinc sulphide and barium sulphate by mixing solutions of barium sulphide and zinc sulphate.

$$BaS + ZnSO_4 \longrightarrow \underbrace{BaSO_4 + ZnS}_{\text{Lithopone}}$$

The precipitate is thoroughly washed and dried. It is then heated to red heat in specially constructed retorts. The hot mass is quenched in cold water and then ground wet. This process of quenching increases the whiteness of the pigment and its covering power. The whole mass is then dried and finally ground to a fine powder. It is a valuable white pigment and is used in wall paints and cheaper grades of enamel paints. It does not get blackened when expossed to H_2S but slowly drakens on exposure to light due to the liberation of metallic zinc from ZnS. In dark the metal is again changed into white ZnO by the atmospheric oxidation and therefore the paint regains its original whiteness. It is much cheaper than white lead $2PbCO_3.Pb(OH)_2$ and is non-poisonous. It has good covering power but is inferior to white lead and zinc oxide. In bright light it darknes to some extent due to the separation of zinc and is, therefore, preferred for interior decoration. It is also finding use in the rubber industry in the manufacture of white goods, also in the manufacture of lacquers, oil cloth and linolenm.

6. Zinc Carbonate, $ZnCO_3$

It ocours in nature as *calamine.* It is obtained as a precipitate when sodium bicarbonate is added to zinc sulphate solution.

$$ZnSO_4 + 2NaHCO_3 \longrightarrow ZnCO_3 + Na_2SO_4 + H_2O + CO_2 \uparrow$$

Sodium carbonate is not used because it precipitates basic zinc carbonate, $2ZnCO_3.3Zn(OH)_2.H_2O$. It is a white powder, insoluble in water which decomposes on heating into ZnO and CO_2 at 300°C

$$ZnCO_3 \xrightarrow{300°} ZnO + CO_2$$

It is used (*i*) in making ointment for curing skin diseases (*ii*) in the preparation of cosmetics (*iii*) as a fire proofing agent (*iv*) as a reinforcing pigment in rubber.

7. Cadmium Sulphide, CdS

It occurs as *greenokite* in nature. It is prepared by passing H_2S gas into a cadmium salt solution in neutral, alkaline or feebly acidic media. It is usually yellow in colour but the colour varies from lemon yellow to orange depending upon the conditions of precipitation. It is used : (*i*) as a yellow pigment under the name of *cadmium yellow* in making oil and water colours (*ii*) in the manufacture of *cadmipone* which is a mixture of CdS and $BaSO_4$ and is used as a red pigment (*iii*) in luminoscent pigments usually with ZnS to produce yellow and fluorescence and phosphorescence.

Compounds of Mercury

Mercury forms compounds in + 1 and + 2 oxidation states. Compounds in these states are called **mercurous** and **mercuric compounds** respectively. Of these, the mercuric salts are more stable and important. A mercuric salt can easily be converted to mercurous either by the action of some reducing agent and in some cases by the action of mercury. The action of an acid on excess of mercury generlly produces mercurous salts.

Experimental Evidences to Show That Mercurous Ion is Hg_2^{2+}

There has been some doubt regarding the structure of mercurous ion in mercurou salts. Experiments of H.B. Baker on the vapour density of very dry mercurous chloride, of Beckman on the freezing point of a solution of mercuous chloride in fused mercuric chloride and of Ogg on the E.M.F. of suitable concentration cells have indicated beyond doubt that the mercurous compounds contain the group—Hg—Hg—. *i.e.* the two atoms in mercurous ion are joined by a covalent linkage and one valency of each atom is free from combination Thus the metal is actually diatomic in the mercurous ion which can be written as Hg_2^{++} and exists in equilibrium :

$$Hg^{2+} + Hg^0 \longrightarrow Hg_2^{2+} \text{ or } (\text{—Hg—Hg})^{2+}$$

Following are some additional evidences which support Hg_2^{2+} structure for the mercurous ion.

(*i*) Raman spectra. Raman spectra of the nitrates of all the metals are the same due to covalent bonding in NO_3^- ion but that of an aqueous solution of mercurous nitrate contains *an extra strong line* which is due to $(Hg—Hg)^{2+}$ or Hg_2^+ ion.

(*ii*) X-ray study. The X-ray study of crystalline mercurous chloride shows that the molecule has a linear structure, Cl—Hg—Hg—Cl in which the mercury atoms are always associated in pairs. This proves the presence of Hg—Hg bond in mercurous compounds.

(*iii*) Electrical conductivity. The value of electrical conductivity of an aqueous solution of mercurous nitrate has been found similar to that of $Pb(NO_3)_2$ and not to that of $AgNO_3$. Hence on the analogy of Pb^{2+}, the mercurous ion should be Hg_2^{2+} and consequently mercurous nitrate should be $Hg_2(NO_3)_2$ and not $Hg(NO_3)$.

Now let us discuss some important compounds of mercury :

1. Mercurous Chloride or Calomel, Hg_2Cl_2

Preparation : This is the most important mercurous compound. This compound occurs in nature as the mineral *horn quick silver.* It is prepared : (*i*) by heating an intimate mixture of $HgCl_2$ and metallic Hg in appropriate proportions in an iron pot. (*ii*) by adding a soluble chloride (*e.g.* NaCl) or dil. HCl to the solution of $Hg_2(NO_3)$. (*iii*) by reducing $HgCl_2$ solution with SO_2.

$$HgCl_2 + Hg \longrightarrow Hg_2Cl_2$$
$$Hg_2(NO_3)_2 + 2NaCl \longrightarrow Hg_2Cl_2 + 2NaNO_3$$
$$Hg_2(NO_3)_2 + 2HCl \longrightarrow Hg_2Cl_2 + 2HNO_3$$
$$2HgCl_2 + 2H_2O + SO_2 \longrightarrow Hg_2Cl_2 + 2HCl + H_2SO_4$$

Properties : It is a white amorphous substance of density 7.16 It is very sparingly soluble in H_2O but dissolves in aqua regia, Cl_2 water, hot conc. HNO_3 etc. With HNO_3 it gives $HgCl_2$.

$$3Hg_2Cl_2 + 8HNO_3 \longrightarrow 3HgCl_2 + 3Hg(NO_3)_2 + 2NO + 4H_2O$$

When gently heated or rubbed, it becomes yellow, and darkens when exposed to light. This is due to the partial decomposition of Hg_2Cl_2 into $HgCl_2$ and Hg. When treated with NH_3, it turns black due to the formation of finely divided metallic Hg and mercuric amino chloride, $Hg(NH_2)Cl$. On this basis Hg_2Cl_2 is also called *calomel* from the Greek word *Kalomelas* meaning beautiful black.

$$Hg_2Cl_2 + 2NH_3 \longrightarrow Hg + Hg(NH_2)Cl + NH_4Cl.$$

Uses : Hg_2Cl_2 is used : (*i*) as a purgative in medicine. (*ii*) in making calomel electrodes which are used in potentiometric titrations.

2. Mercurous Nitrate, $Hg_2(NO_3)_2$. $2H_2O$

It is prepared by the action of cold dil. HNO_3 (density = 1.14) on an excess of metallic mercury which readily crystallises from the solution on standing.

$$6Hg + 8HNO_3 \longrightarrow 3Hg_2(NO_3)_2 + 4H_2O + 2NO$$

Properties : It is a colourless crystalline substance, soluble in water containing a little HNO_3 (to prevent hydrolysis). With excess of H_2O basic nitrate, $Hg_2(OH)NO_3$ is precipitated.

$$Hg_2(NO_3)_2 + H_2O \rightleftarrows Hg_2(OH)NO_3 + HNO_3$$

Dry $Hg_2(NO_3)_2$ decomposes on heating into HgO and NO_2. With HCl or soluble chlorides (*e.g.* NaCl) the solution of $Hg_2(NO_3)_2$ gives a white precipitate of Hg_2Cl_2

$$Hg_2(NO_3)_2 + 2HCl \longrightarrow Hg_2Cl_2 + 2HNO_3$$
$$Hg_2(NO_3)_2 + 2NaCl \longrightarrow Hg_2Cl_2 + 2NaNO_3$$

The solution of $Hg_2(NO_3)_2$ has strong reducing properties.

3. Mercuric Chloride or Corrosive Sublimate, $HgCl_2$

Preparation : This compound is known as *corrosive sublimate*, since it possesses corrosive properties and is prepared in the dry process by *sublimation.* It can be prepared by the action of Cl_2 on metallic Hg. *Commercially* it is prepared by heating a dry mixture $HgSO_4$, NaCl and little MnO_2.

$$HgSO_4 + 2NaCl \longrightarrow HgCl_2 + Na_2SO_4$$

$HgCl_2$ separates on sublimation on the cooler part of the flask. The cake of the sublimate is removed and $HgCl_2$ is purified by recrystallisation from hot water. The addition of MnO_2 prevents the formation of Hg_2Cl_2 from $Hg_2(SO_4)_2$ which is usually present in $HgSO_4$.

Properties : $HgCl_2$ forms colourless needles with m.pt. = 227°C and b.pt. = 302°C. It is sparingly soluble in cold H_2O but readily in hot water. It is a covalent. compound and dissolves in alcohol, ether and benzene. With metal chlorides crystalline compounds are formed, *e.g.* $KHgCl_3$, Na_2HgCl_4 etc. With alkalies HgO is precipitated.

$$HgCl_2 + 2NaOH \longrightarrow HgO + 2NaCl + H_2O$$

With PCl_5 a volatile crystalline compound, $2HgCl_2.2PCl_5$ is formed. When $HgCl_2$ is treated with $SnCl_2$, it is first reduced to Hg_2Cl_2 and then to metallic Hg.

$$2HgCl_2 + SnCl_2 \longrightarrow Hg_2Cl_2 + SnCl_4$$

$$Hg_2Cl_2 + SnCl_2 \longrightarrow 2Hg + SnCl_4$$

When treated with KI solution, a yellow precipitate of HgI_2 is obtained. This precipitate dissolves in the excess of KI to form colourless *potassium tetra-iodo mercurate*, $K_2[HgI_4]$.

$$HgCl_2 + 2KI \longrightarrow \underset{\text{Yellow ppt.}}{HgI_2} + 2KCl$$

$$HgI_2 + 2KI \longrightarrow \underset{\text{Colourless}}{K_2[HgI_4]}$$

When treated with NH_3, a bulky white precipitate of mercuric amino chloride, $[Hg(NH_2)]Cl$ is obtained. This white precipitate is infusible and is known as infusible white precipitate.

$$HgCl_2 + 2NH_3 \longrightarrow [Hg(NH_2)]Cl + NH_4Cl$$

If $HgCl_2$ is treated with gaseous NH_3, a white precipitate of *diammine mercuric chloride,* $[Hg(NH_3)_2]Cl_2$ which is called fusible white precipitate is obtained. When SO_2 is passed through Hg_2Cl_2 solution, a white precipitate of Hg_2Cl_2 is obtained.

$$2HgCl_2 + SO_2 + 2H_2O \longrightarrow Hg_2Cl_2 + 2HCl + H_2SO_4$$

When a solution of $HgCl_2$ is boiled with mercuric oxide (HgO), mercuric oxy-chlorides like $2HgCl_2.HgO$ (red), $HgCl_2.2HgO$ (black), $HgCl_2.3HgO$ (yellow), $HgCl_2.4HgO$ (black) are obtained. When a bright copper foil is immersed in $HgCl_2$ solution, a grey film of Hg gets deposited on Cu foil.

$$HgCl_2 + Cu \longrightarrow CuCl_2 + Hg$$

Uses : $HgCl_2$ is used (*i*) as its 0.1% solution for sterilising the hands and instruments in surgery. (*ii*) for preserving wood, skins and furs against insect pests. (*iii*) in the preparation of dry cells. (*iv*) as a reagent in laboratory. (*v*) in preventing growth of fungi and bacteria in water, *e.g.* in thermostats. (*vi*) as an intensifier for negatives.

4. Mercuric Iodide, HgI_2

Preparation : It is obtained as a precipitate when solution of KI is added to that of $HgCl_2$.

$$HgCl_2 + 2KI \longrightarrow HgI_2 + 2KCl.$$

Properties : HgI_2 exists in two forms red and yellow. The red form is tetragonal and stable below 126°C while the yellow form is rhombic and is stable above 126°C. When the red form is heated to 126°C, it is changed to yellow form. HgI_2 is sparingly soluble in water but more soluble in alcohol. It dissolves in the excess of KI forming a colourless complex *viz.* potassium tetra-iodo mercurate, $K_2[HgI_4]$.

$$HgI_2 + 2KI \longrightarrow K_2[HgI_4]$$

Uses. HgI_2 is used as an ointment in the treatment of skin diseases.

5. Mercuric Sulphide or Vermilion, HgS

Preparation. This compound exists in two forms *red* and *black.* Both these forms occur native. The red form occurs as the mineral *cinnabar* which is the chief source of Hg. Red form is used as a pigment under the name of *vermilion.* Black variety is found native in small amounts with cinnabar as *metacinnabariate.* The black variety is obtained by passing H_2S gas through a solution of $HgCl_2$ acidified with HCl.

$$HgCl_2 + H_2S \longrightarrow \underset{\text{ppt.}}{HgS} + 2HCl$$

In the beginning the precipitate of HgS is white which changes to yellow, brown and finally black. On sublimation this black precipitate changes to red and is called vermilion. Vermilion (red variety) can also be prepared (*a*) by subliming metallic Hg with S (*b*) by heating Hg with potassium pentasulphide (K_2S_5) solution.

$$Hg + K_2S_5 \longrightarrow HgS + K_2S_4$$

Properties. (*i*) It is insoluble in H_2O, dil. and hot HNO_3 and boiling conc. HCl but dissolves in aqua regia forming $HgCl_2$.

$$3HgS + 2HNO_3 + 6HCl \longrightarrow 3HgCl_2 + 4H_2O + 2NO + 3S$$

(*ii*) *Action of heat.* On heating in air it gets decomposed into metallic Hg and SO_2.

$$HgS + O_2 \longrightarrow Hg + SO_2$$

(*iii*) *Action of boiling conc. H_2SO_4.* $HgSO_4$ is obtained.

$$HgS + 2H_2SO_4 \longrightarrow HgSO_4 + 2H_2O + SO_2 + S$$

(*iv*) *Reduction.* It is reduced to metallic Hg when heated with iron filings, Na_2CO_3 etc.

$$HgS + Fe \longrightarrow Hg + FeS$$

$$2HgS + 2Na_2CO_3 \longrightarrow 2Hg + 2Na_2S + 2CO_2 + O_2$$

Uses. Red variety of HgS is used (*i*) as a pigment under the name of *vermilion* (*ii*) in Ayurvedic medicines like *Makardhwaja* and *Rasa-Sindur.*

6. Mercury Fulminate, $2Hg(CNO)_2.H_2O$

It is obtained by adding alcohol to a solution of mercury in an excess of nitric acid. This substance is a violent explosive and is employed in detonators.

7. Mercuric Thiocyanate, $Hg(CNS)_2$

It is obtained as a white precipitate by mixing mercuric chloride solution with ammonium thiocyanate. On igniting it yields voluminous snake like tubes of ash due to the formation of oxides. It is used in making Pharaoh's serpent.

8. Millon's Base, $[Hg_2N]OH.2H_2O$ *or* $(OH)_2Hg_2NH_2(OH)$

It is obtained as a yellow compound by warming HgO with aqueous ammonia. It is a basic compound. With acids it gives a series of salts of the composition,

$[Hg_2N]X$ (X = Cl, Br, NO_3). Dehydration of this compound gives $[Hg_2N]OH.H_2O$ and explosive $[Hg_2N]OH$. The structure shown below has been suggested for Millon's base.

$$\left[\begin{matrix} HO\text{—}Hg \\ HO\text{—}Hg \end{matrix} \rangle NH_2 \right] OH.$$

Structure of Millon's base, $(OH)_2\ Hg_2\ NH_2(OH)$

9. Nesseler's Reagent, $K_2[HgI_4]$

When excess of KI is added to $HgCl_2$ solution, a solution of potassium tetra-iodo mercurate, $K_2[HgI_4]$ is obtained. This solution of $K_2[HgI_4]$, when made strongly alkaline with KOH or NaOH, is known as *Nesseler's reagent.* Nesseler's reagent gives a brown precipitate of *iodide of Millon's base*, $NH_2Hg_2I_3$ with excess of NH_3.

$$\underset{\text{Nesseler's reagent}}{2K_2[HgI_4]} + 2NH_3 \longrightarrow \underset{\substack{\text{Iodide of} \\ \text{Millon's base} \\ \text{(brown ppt.)}}}{NH_2Hg_2I_3} + 4KI + NH_4I$$

The formation of iodide of Millon's base has been utilised for the detection and determination of NH_3 by using Nesseler's reagent.

10. Iodide of Millon's Base, $NH_3Hg_2I_3$ *or* $I_2Hg_2NH_2I$

It is obtained when Nesseler's reagent reacts with excess of NH_3.

$$\underset{\text{Nesseler's reagent}}{2K_2[HgI_4]} + 2NH_3 \longrightarrow \underset{\substack{\text{Iodide of} \\ \text{Millon's base}}}{NH_2Hg_2I_3} + 4KI + NH_4I$$

The formation of iodide of Millon's base represented by the above chemical equation actually takes place through the following steps :

$$[K_2[HgI_4] \longrightarrow 2KI + HgI_2] \times 2$$

$$HgI_2 + 2NH_3 \longrightarrow \underset{\substack{\text{Mercuric amino} \\ \text{iodide}}}{Hg(NH_2)I} + NH_4I$$

$$Hg(NH_2)I + HgI_2 \longrightarrow \underset{\substack{\text{Iodide of} \\ \text{Millon's base}}}{NH_2Hg_2I_3}$$

On adding : $2K_2[HgI_4] + 2NH_3 \longrightarrow NH_2Hg_2I_3 + 4KI + NH_4I$

Corresponding to the structure of Millon's base, the structure of its iodide can be written as follows :

$$\left[\begin{matrix} HO\text{—}Hg \\ HO\text{—}Hg \end{matrix} \rangle NH_2 \right] OH \qquad \left[\begin{matrix} I\text{—}Hg \\ I\text{—}Hg \end{matrix} \rangle NH_2 \right] I$$

Structure of Millon's base, $(OH)_2\ Hg_2.NH_2(OH)$ — Structure of iodide of Millon's base

Compounds of Cerium

In aqueous solution, cerium exists in +3 and +4 oxidation states. Its salts in +3 and +4 salts are called *cerous* and *ceric* salts respectively. Ceric sulphate, $Ce(SO_4)_2$ and ceric ammonium sulphate, $Ce(SO_4)_2.2(NH_4)_2SO_4.2H_2O$ are its very important compounds and are discussed below :

1. Ceric Sulphate, $Ce(SO_4)_2$

The anhydrous salt is prepared by heating CeO_2 with cone. H_2SO_4 and forms an orange-yellow crystalline powder, insoluble in conc. H_2SO_4. It dissolves in H_2O, giving a brown-yellow solution from which the salt separates below 3°C as the dodecahydrate, between 3° and 42.5° as the octahydrate and above the latter temperature as the tetrahydrate. On diluting an aqueous solution of the salt, hydrolysis occurs and ultimately sparingly soluble basic salts get precipitated. Hence for use in volumetric analysis, the salt is dissolved in dil H_2SO_4(0.5 N or stronger), $Ce(SO_4)_2$ is widely used in volumetric analysis as an oxidising agent. The salt is used for the titration of Fe^{2+}, $C_2O_4^{2+}$, NO_2^-, Cu^+, Mo^{5+}, Te^{4+} and U^{4+} ions. It is also used as an oxidiser in preparative organic chemistry.

2. Ceric Ammonium Sulphate, $Ce(SO_4)_2.2(NH_4)_2SO_4.2H_2O$

This compound is also called *ammonium tetrasulphatocerate* (*IV*) and hence is represented as $(NH_4)_4[Ce(SO_4)_4].2H_2O$. It is prepared by adding $(NH_4)_2SO_4$ to $Ce(SO_4)_2$ solution containing excess of dil. H_2SO_4 to prevent the hydrolysis, followed by crystallisation This double salt is more soluble than $Ce(SO_4)_2$ and is, therefore, prepared for use as a volumetric reagent. Ionic equations showing the quantitative estimations of $C_2O_4^{2-}$, NO_2^-, I^-, H_2O_2, etc. made with this salt are given below :

$$C_2O_4^{2-} + 2Ce^{4+} \longrightarrow 2Ce^{3+} + 2CO_2$$
$$NO_2^- + 2Ce^{4+} + H_2O \longrightarrow 2Ce^{3+} + 2H^+ + NO_3^-$$
$$2I^- + 2Ce^{4+} \longrightarrow 2Ce^{3+} + I_2$$
$$H_2O_2 + 2Ce^{4+} \longrightarrow 2Ce^{3+} + 2H^+ + O_2$$

Compounts of Throium

Thorium is *tetravalent* in its compounds. Following are its important compounds.

1. Thorium Dioxide or Thoria, ThO_2

Preparation : It can be obtained by the following reactions :

$$Th + O_2 \xrightarrow{\Delta} ThO_2 \; ; \; Th(OH)_4 \xrightarrow{\Delta} ThO_2 + 2H_2$$

$$Th(NO_3)_2 \xrightarrow{\Delta} ThO_2 + 4NO_2 + O_2$$

It can also be obtained from monazite sand.

Properties and uses : It is white hygroscopie powder. Crystallised or strongly ignited thoria is almost insoluble in acids but on evaporation with conc. H_2SO_4 it is dissolved in it to form $Th(SO_4)_2$.

$$ThO_2 + 2H_2SO_4 \longrightarrow Th(SO_4)_2 + 2H_2O$$

When heated with carbon and boron, it forms carbide and boride of thorium respectively. Due to it resistance to heat and involatility it is used in the preparation of gas mantles. It is also used as a refractory material as well as a catalyst in many organic reactions. It is employed in welding operations and also in medicines. Since ThO_2 increases the life of the tungsten filament, it is used as an additive to tungsten used for metal filament lamps. Since ThO_2 has high absorptive power for X-rays, it is used in X-ray diagnostic work as a contrast medium in the examination of the digestive organs. It is also used in ceramic industry. Mixtures of ThO_2 and ceria have been used for search lights, head lights of motor cars etc.

2. Thorium Nitrate, $Th(NO_3)_2$

Preparation : It is prepared by dissolving $Th(OH)_4$ in HNO_3.

$$Th(OH)_4 + 4HNO_3 \longrightarrow Th(NO_3)_2 + 4H_2O$$

Properties and uses : It is a white deliquescent salt which is soluble in water. On heating it gives ThO_2 in a very fine state of subdivision. Due to this property thorium nitrate is used in mantle industry. Thorium nitrate mixed with Mg powder is used for flash powders.

Compounds of Uranium

Uranium forms compounds in +2, +3, +4, +5 and +6 oxidation states. Following are the important compounds.

1. Uranium Dioxide, UO_2 (U = +4)

It is obtained by reducing higher oxides such as UO_3 or U_3O_8 with H_2, C or CO at elevated temperature [$UO_3 + CO \xrightarrow{500°} UO_2 + CO_2\uparrow$]. It is a brown, strongly basic oxide, insoluble in water and alkalies. When heated in air it burns to give U_3O_8. It dissolves slowly in most of the acids, but HNO_3 oxidises it to $UO_2(NO_3)_2$. It reduces $AgNO_3$ to metallic silver.

$$2AgNO_3 + UO_2 \longrightarrow UO_2(NO_3)_2 + 2Ag$$

2. Triuranium Octoxide or Urano-uranic Oxide, U_3O_8

This oxide is regarded as a double oxide of $U^{6+}O_3$ with either $U^{4+}O_2$ (*i.e.* $U_3O_8 = 2U^{6+}O_3.U^{4+}O_2$) or $U_2^{5+}O_5$ (*i.e.* $U_3O_8 = U^{6+}O_3.U_2^{5+}O_5$). This oxide occurs in pitchblende and is the most stable oxide. It is obtained by heating other oxides in O_2 or by heating uranyl acetate. It is green-coloured oxide. When it is reduced with H_2 or CO, UO_2 is obtained. It is not acted upon by HCl or H_2SO_4 but H_2F, gives UF_4 whereas HNO_3 gives $UO_2(NO_3)_2$. This oxide is used to some extent for colouring ceramic enamels.

3. Uranium Trioxide, UO_3 (U = +6)

It is obtained by heating $UO_2(NO_3)_2$ in O_2. It is yellow when pure but impure sample is orange-red. When strongly heated, it loses O_2 and passes to U_3O_8. It is amphoteric, forms uranates with alkalies and uranyl salts with acids.

$$UO_3 + 2NaOH \longrightarrow \underset{\text{Sodium uranate}}{Na_2UO_4} + H_2O$$

$$UO_3 + 2HNO_3 \longrightarrow \underset{\text{Uranyl nitrate}}{UO_2(NO_3)_2} + H_2O$$

4. Uranium Hexafluoride, UF_6

It is obtained by the direct combination of elements in presence of a little of Cl_2 or by the action of F_2 on UF_4. It is a colourless crystalline volatile substance and melts at 64.1°C. It attacks many metals and a large number of organic compounds. It is readily hydrolysed by water to uranyl fluoride, UO_2F_2 and is reduced to UF_4.

$$UF_6 + 2H_2O \longrightarrow UO_2F_2 + 4HF \; ; \; UF_6 \xrightarrow{\text{Reduction}} UF_4 + F_2$$

It is used in separating the uranium isotopes and thus in producing pure U^{235} nuclear fuel. It is a powerful fluorinating agent.

5. Uranyl Salts

These are the only common uranium salts which are derived from the *bivalent basic uranyl radical,* UO_2^{2+}. In these salts uranium is in +6 oxidation state. These salts are formed by the action of acids on uranium oxides especially UO_3. These are generally yellow crystalline substances, soluble in water. A uranyl salt with potassium ferrocyanide solution gives a red precipitate of *uranyl ferrocyanide.* Following are the important uranyl salts :

(*i*) Uranyl nitrate, $UO_2(NO_3)_2$, $6H_2O$. It is also known as uranium nitrate and is prepared by any of the following methods :

$$UO_3 + 2HNO_3 \longrightarrow UO_2(NO_3)_2 + H_2O$$

$$\underset{\text{Uranyl chloride}}{UO_2Cl_2} + 2AgNO_3 \longrightarrow UO_2(NO_3)_2 + 2AgCl$$

This substance crystallises as a lemon yellow hexahydrate, $UO_2(NO_3)_2.6H_2O$. It is soluble in water, alcohol and ether. In contact with H_2SO_4 it is first converted into $UO_2(NO_3)_2.3H_2O$ and then into $UO_2(NO_3)_2.2H_2O$. When $UO_2(NO_3)_2.3H_2O$ is heated at about 130° in a current of CO_2 having HNO_3 to check hydrolysis, $UO_2(NO_3)_2$ is obtained.

(*ii*) Uranyl acetate. $UO_2(CH_3COO)_2.2H_2O$. It is obtained by dissolving $UO_2(OH)_2$ or UO_3 in CH_3COOH.

$$UO_2(OH)_2 + 2CH_3COOH \longrightarrow UO_2(CH_3COO)_2 + 2H_2O$$

$$UO_3 + 2CH_3COOH \longrightarrow UO_2(CH_3COO)_2 + H_2O$$

When dihydrate is heated to 110°C it is converted into anhydrous salt. It gets decomposed at 275°C to UO_3. It is sparingly soluble in H_2O. This compound is of technical importance. It is used for making *Zn* or *Mg uranyl acetate* which is used as a qualitative reagent for the detection of sodium.

6. Uranates

These are the salts given by uranic acid, H_2UO_4. These are of two types namely *monouranates* (*e.g.* $NaUO_4$) and *diuranates* (*e.g.* $Na_2U_2O_7.6H_2O$, $(NH_4)2U_2O_7$). These are prepared as follows :

$$UO_3 + 2NaOH \longrightarrow Na_2UO_4 + H_2O$$

$$2UO_3 + 2NaOH \longrightarrow Na_2U_2O_7 + H_2O$$

$$UO_3 + Li_2CO_3 \longrightarrow Li_2UO_4 + CO_2 \uparrow$$

$$2UO_3 + Li_2CO_3 \longrightarrow Li_2U_2O_7 + CO_2 \uparrow$$

Uranates are insoluble in water and are orange or yellow in colour. Following are the important uranates :

(*i*) Sodium diuranate, $Na_2U_2O_7.6H_2O$. It is also called *sodium uranate* and is obtained as a yellow precipitate by adding NaOH to a uranyl salt solution. When hexahydrate salt is heated, anhydrous salt is obtained. Anhydrous salt is used in commerce as a colouring matter for glass and porcelain glazes under the name *uranium yellow.*

***(ii)* Ammonium diuranate, $(NH_4)_2U_2O_7$.** It is obtained as a yellow precipitate by adding NH_3 to a uranyl salt solution. The formation of this substance provides a convenient method for the gravimetric estimation of uranium.

University Questions

1. Give the preparation, properties and uses of the following : (*i*) titanox (*ii*) titanium dioxide (*iii*) titanous chloride. (*Meerut M.Sc. 84; Lucknow 82; Madras 86; Meerut 82*)

2. Discuss the preparation, properties and uses of : (*i*) Oxosalts of vanadium (*ii*) Vanadyl chloride (*iii*) Vanadium trichloride (*iv*) Vanadyl sulphate (*v*) V_2O_5 (*vi*) Ammonium vanadato. (*Lucknow 80; Madras 86; Madurai 85; Jiwaji 84; Meerut 82, 89; Bharathiar 86; Poona 86*)

3. Give the reasons for the following :

(*i*) V_2O_5 is an important compound of vanadium. (*Raj. 79*)

(*ii*) In V_2O_5 vanadium is in +5 oxidation state and hence should be colourless, but it is actually orange in colour. (*Delhi Hons. 86*)

4. Give the preparation, properties, uses and electronic structure of $K_2Cr_2O_7$. (*Allahabad 89; Meerut 82; Lucknow 80; Gorakhpur 81; M.D. Rohtak 84; Bharathiar 86; Jodhpur 84; Utkal 85*)

5. How can you get chronic anhydride, chromyl chloride and chrome alum from $K_2Cr_2O_7$? (*Gorakhpur 86; Meerut 89; Delhi 87; Madras 86; Jodhpur 84*)

6. A solution salt (X) when heated with conc. H_2SO_4 and solid $K_2Cr_2O_7$ gave a volatile compound of chromium (Y). This compound Y when passed through e solution NaOH and lead acetate containing CH_3COOH gave a yellow coloured precipitate (Z). Name X, Y and Z. Give chemical equations for the above stated reactions. (*Delhi 89*)

7. What happens when :

(*i*) SO_2 reacts with acidified $K_2Cr_2O_7$ (give ionic equation). (*Jodhpur 86; Raj. 86; Andhra 86; Kerala 87; Agra 85*)

(*ii*) $K_2Cr_2O_7$ is treated with KI in presence of acid. (*Agra 84, 85; Delhi 89*)

(*iii*) $K_2Cr_2O_7$ acts on HCl. (*Punjab 85; GND 84*)

(*iv*) A chloride and $K_2Cr_2O_7$ mixture is heated with conc. H_2SO_4. (*Meerut 88; Indore 86; Agra 85, 89*)

(*v*) Chloroform is oxidised with acidified $K_2Cr_2O_7$. (*Meerut 89*)

(*vi*) $K_2Cr_2O_7$ is treated with Fe^{2+} ions in an acidic medium. (*Delhi 88; Utkal 86*)

(*vii*) An alkali is added to a solution of $K_2Cr_2O_7$. (*Delhi Hons. 86*)

8. Complete and balance the following equations :

(*a*) $Cr_2O_7^{2-} + H^+ + I^- \rightarrow ......$ (*Utkal 85, 86*)

(*b*) $Cr_2O_7^{2-} + H^+ + Br^- \rightarrow ......$ (*Delhi 87*)

(*c*) $Cr_2O_7^{2-} + Fe^{2+} + H^+ \rightarrow ......$ (*Punjab 86*)

(*d*) $Cr_2O_7^{2-} + C_2O_4^{2-}$ (in acidic medium) $\rightarrow$ (*Raj. 81, 83*)

(*e*) $K_2Cr_2O_7 + KI + HCl \rightarrow ......$

(*f*) $FeCl_2 + K_2Cr_2O_7 + HCl \rightarrow ......$ (*Raj. 85*)

9. Give chemical equations to explain what happens when :

(*i*) a solution of K_2CrO_4 containing H_2O_2 is acidified with H_2SO_4 in presence of ether or pyridine. (*Delhi 87; Delhi Hons. 87*)

(*ii*) H_2O_2 is added to an aqueous solution of $K_2Cr_2O_7$ and the solution is acidified with H_2SO_4. (*Sambalpur 86*)

10. Starting from chrome iron ore, how would you obtain a pure specimen of $K_2Cr_2O_7$, $(NH_4)_2Cr_2O_7$, K_2CrO_4, Cr_2O_3, CrO_2Cl_2 and metallic chromium. *(Gorakhpur 84; Meerut 89)*

11. Describe the proprties and uses of ammonium molybdenate. *(Gorakhpur 81; Utkal 86)*

12. Describe the preparation, properties and uses of : (*i*) sodium silicotungstate (*ii*) tungsten hexachloride (*iii*) tungstic acid (*iv*) sodium tungstate. *(Madurai 86; Meerut M.Sc. 84; Madras 86)*

13. Write a note on "Hetercpoly acids of tungsten". *(Meerut 82)*

14. What happens when :

(*i*) $KMnO_4$ reacts with H_2S in neutral medium. *(Meerut 87)*

(*ii*) $KMnO_4$ interacts with MnO_2 in the presence of KOH. (I.I.T. 87)

(*iii*) $KMnO_4$ reacts with HCl. *(Punjab 85; GND 84)*

(*iv*) Acidified $KMnO_4$ reacts with SO_2. *(Jodhpur 86; Raj. 86; Adndhra 86; Kerala 87)*

(*v*) $KMnO_4$ reacts with KI in alkaline medium. *(Meerut 88)*

(*vi*) Acidified $KMnO_4$ reacts with hydrazoic acid. *(Raj. 86)*

(*vii*) $KMnO_4$ is added to $MnSO_4$ solution in neutral medium. *(Utkal 85)*

15. Complete and balance the following equations :

(*i*) $MnO_4^- + H^+ + Cl^- \rightarrow$ ...... *(Utkal 85, 86)*

(*ii*) $MnO_4^- + H^+ + Fe^{2+} \rightarrow$ ...... *(Utkal 85, 86; Calicu 79)*

(*iii*) $MnO_4^- + H^+ + H_2O_2 \rightarrow$ ...... *(Raj. 85)*

(*iv*) $C_2H_4 + KMnO_4 + H_2SO_4 \rightarrow$ ...... *(Punjabi 85, 86)*

(*v*) $HCHO + KMnO_4 + HCl \rightarrow$ ...... *(Punjabi 85, 86)*

(*vi*) $MnO_4^- + SO_3^{2-} + H^+ \rightarrow$ ...... *(Raj. 81, 83)*

(*vii*) $MnO_4^- + H_2S + H^- \rightarrow$ ...... *(Raj. 84)*

(*viii*) $KMnO_4 + KBr + H_2SO_4 \rightarrow$ ...... *(Calcutta 77)*

(*ix*) (*a*) $MnO_4^- + SO_2 + H_2O \rightarrow$ ......

(*b*) $MnO_4^- + Mn^{2-} + H_2O \rightarrow$ ......

(c) $MnO_4^- + NO_2^- + H_2O \rightarrow$ ...... *(Punjab 88; Meerut 86; Delhi 86, 87)*

16. Give the reasons for the following :

(*i*) $KMnO_4$ solutions is kept in dark bottles and freshly prepared where possible. *(Calicut 86)*

(*ii*) HCl cannot be used to acidify a permanganate solution in volumetric estimation of Fe^{2+} or $C_2O_4^{2-}$. *(Madras 86)*

(*iii*) $KMnO_4$ disproportionates in the presence of acids according to the equation : $4H^+ + 3[Mn^{+6}O_4]^{2-} \rightarrow 2[Mn^{+5}O_4]^- + Mn^{+4}O_2 + 2H_2O$. *(Poona 86)*

19. What happens when :

(*i*) Co (II) salt solution is treated with KNO_2 and dil. CH_3COOH. *(Delhi 89)*

(*iii*) CH_3COOH is added to a mixture of $NaHO_2$ and $Co(NO_3)_2$ solutions. *(Delhi 88)*

18. How is dimethylglyoxime complex of nickel prepared? What are its uses ? *(Madras 86; Madurai 85)*

19. What happens when :

(*i*) Chloroplatinic acid reacts with SO_2. *(Meerut 88)*

(*ii*) NH_4Cl is added to chloroplatinic acid. *(Agra 85)*

20. Starting from Pt, how will you get the following :

(*i*) Chlorides of Pt in different oxidation states. *(Madras 84; Madurai 86)* (*ii*) Ammonium chloroplatinate. (*Indore 86; Punjab 82*)

21. What happens when :

(*i*) $CuSO_4$ solution reacts with excess of NH_4OH. (*MLNR 79; I.I.T. 73; Poona 86*)

(*ii*) $CuSO_4$ solution reacts with excess of KI solution. (*Bhararthiar 86; Meerut 89 S*)

(*iii*) $CuSO_4$ reacts with excess of KCN solution. (*Bharathiar 86*)

(*iv*) To $CuSO_4$ solution is added KI and then $Na_2S_2O_3$ solution. (*R.E.E. 83*)

(*v*) $CuSO_4$ solution is treated with hypo solution. (*Agra 85*)

(*vi*) $CuSO_4$ solution is treated with NH_4CNS. (*Madras 85*)

(*vii*) $CuSO_4$ solution in treated with $K_4[Fe(CN)_6]$ solution. (*Bharathiar 86*)

(*viii*) $CuSO_4$ solution reacts with H_2SO_4 and NH_4CNS solution. (*Bharathiar 86*)

(*ix*) Balance the following equations :

(*a*) $CuSO_4 + NH_4CNS \rightarrow$ ...... (*Madras 85*)

(*b*) $CuSO_4 + KI \rightarrow$ ...... (*Utkal 85, 86*)

22. Give the preparation, properties and uses of : (*i*) Basic copper acetate (*ii*) Malachit green; (*iii*) Cuprous oxide (*iv*) Cuprous chloride (*v*) CuCN. (*Raj. 85; Delhi 87, 87 S*

23. What happens when :

(*i*) PH_3 is passed into a solution of $AgNO_3$ (*MLNR 76*)

(*ii*) $AgNO_3$ is heated (*R.E.E. 83, 86*)

(*iii*) Solution of $AgNO_3$ is added to the solution of sodium thiosulphate. (*R.E.E. 82*)

(*iv*) $AgNO_3$ reacts with ammonia. (*Indore 85*)

(*v*) $AgNO_3$ reacts with HCl. (*Punjab 85, GND 84*)

(*vi*) The solution of NH_3 is treated with AgCl solution. (*I.I.T. 74*)

(*vii*) AgCl reacts with $Na_2S_2O_3$ solution. (*R.E.E. 73*)

24. Give the reasons for the following :

(*i*) On adding NaCl to $AgNO_3$ solution, a white precipitate of $AgNO_3$ is obtained but CCl_4 does not produce any precipitate

(*ii*) AgCl precipitate dissolves in NH_4OH. (*Agra 85*)

OR

AgCl precipitate is soluble in ammonia. (*Delhi 85*)

(*iii*) AgCl is soluble in saturated KI, but AgI is soluble in saturated KCl. (*Delhi Hons. 87*)

25. Give the preparation, properties and uses of :

(*i*) chloroauric acid. (*Bharathiar 86*)

(*ii*) gold chloride. (*Bhagalpur 72; Madras 74; Gorakhpur 71*)

(*iii*) gold fulminate. (*Vikram 86*)

26. Write the preparation, properties and uses of :

(*i*) Lithopone (*Bharathiar 86; Jodhpur 85*)

(*ii*) $ZnSO_4$ (*Madurai 85; Andhra 85*)

(*iii*) $ZnCl_2$ (*Jammu 85*)

(*iv*) $CdSO_4$ (*Delhi 83*)

27. Give the reasons for the following :

(*i*) ZnO is yellow in colour when hot but is white when cold. (*Delhi Hons. 87; Punjab 85*) (*ii*) It is usually not possible to prepare ZnO corresponding exactly to steichicometric formula, ZnO. (*Madras 85*)

(*iii*) Anhydrous $ZnCl_2$ cannot be prepared merely by heating the crystals of hydrated $ZnCl_2.2H_2O$.

28. What happens when :

(*i*) Hg_2Cl_2 is heated with NH_4OH. (*Meerut 89 S; MLNR 76*)

(*ii*) Hg_2Cl_2 reacts with aqua regia. (*MLNR 76*)

(*iii*) SO_2 is passed through corrosive sublimate. (*R.E.E. 73*)

(*iv*) NH_3 is added to $HgCl_2$. (*R.E.E. 73, 77*)

(*v*) $HgCl_2$ is heated with Hg in an iron vessel. (*R.E.E. 75*)

(*vi*) $HgCl_2$ is treated with NaOH solution. (*Delhi 83*)

29. Give the preparation, properties and uses of : (*i*) ceric ammonium sulphate (*ii*) ceric sulphate.

30. Give the preparation, properties and uses of : (*i*) Thoria (*ii*) Thorium nitrate. (*Meerut M.Sc. 84*)

Compounds of uranium

31. Discuss the preparation properties and uses of : (*i*) uranyl nitrate (*ii*) ammonium diuranate (*iii*) uranium hexafluorid (*iv*) uranyl acetate. (*Meerut 83; Lucknow 85; Madras 86; Madurai 86*)

32. (*a*) How is sodium nitroprusside prepared? How does it react with (*i*) Alkali metal sulphide (*ii*) $AgNO_3$ solution (*iii*) Alkalies

(*b*) Give the chemistry of chromyl chloride test.

(*c*) How does potasssium ferricyamide react with : (*i*) Ferrous sulphate (*ii*) Alkaline solution of lead oxide (*iii*) Ferric chloride. (*Himachal Pradesh 99*)

33. Give a brief account of preparation and properties of $K_2Cr_2O_7$. (*Gorakhpur 99*)

(*a*) Describe the preparation of chromyl chloride and discuss its structure. How does it react with water, NaOH and toluene?

(*b*) Draw the structure of TiO_2.

(*c*) Discuss the extraction of thorium from monazite sand.

(*d*) How is chromium acetate monohydrate prepared? (*Himachal Pradesh 2000*)

34. (*i*) Write down the methods of preparation and uses of : (*a*) $K_2Cr_2O_7$ (*b*) Sodium nitroprusside (*c*) Milon's base

(*ii*) What happens when :

(*a*) Cobalt nitrate reacts with $NaNO_3$ acidified with CH_3COOH.

(*b*) Cl_2 gas is passed through potassium ferrocyanide solution.

(*c*) A mixture of $Na_2Cr_2O_7$ and NH_4Cl a strongly heated. (*Kanpur 2000*)

35. Give preparation, important reactions, uses and structure of the following : (*i*) $TiCl_4$ (*ii*) $KMnO_4$. (*Lucknow 2000*)

36. Describe the method of preparation, properties and uses of : (*i*) Chlorauric acid (*ii*) $KMnO_4$. (*Avadh 2000*)

37. Calculate equivalent weight of $K_2Cr_2O_7$ in acidic medium, giving required reactions. (Meerut 2009)

38. Discuss the prepration, properties and uses of any two of the following : (i) $KMnO_4$ (ii) Sodium cobaltnitrite (ii) Potassium ferro cyanide. (*Meerut 2009*)

INDEX

A

B

C

D

E

F

G

H

I

K

L

M

N

O

P

R

S